OZONE IN THE ATMOSPHERE

Studies in
Geophysical Optics and Remote Sensing

Series Editor: *Adarsh Deepak*

Published Volumes and Volumes in Preparation

Advances in Remote Sensing Retrieval Methods
Adarsh Deepak, Henry E. Fleming, and Moustafa T. Chahine (Eds.)

Aerosols and Climate
Peter V. Hobbs and M. Patrick McCormick (Eds.)

Aerosols and Their Climatic Effects
Hermann E. Gerber and Adarsh Deepak (Eds.)

Applications of Remote Sensing for Rice Production
Adarsh Deepak and K.R. Rao (Eds.)

Atmospheric Aerosols: Their Formation, Optical Properties, and Effects
Adarsh Deepak (Ed.)

Hygroscopic Aerosols
Lothar H. Ruhnke and Adarsh Deepak (Eds.)

IRS '84: Current Problems in Atmospheric Radiation
Giorgio Fiocco (Ed.)

IRS '88: Current Problems in Atmospheric Radiation
J. Lenoble and J.F. Geleyn (Eds.)

Light Absorption by Aerosol Particles
Hermann E. Gerber and Edward E. Hindman (Eds.)

Microwave Remote Sensing of the Earth System
A. Chedin (Ed.)

Optical Properties of Clouds
Peter V. Hobbs and Adarsh Deepak (Eds.)

Polar and Arctic Lows
Paul F. Twitchell, Erik A. Rasmussen, and Kenneth L. Davidson (Eds.)

Polarization and Intensity of Light in the Atmosphere
Kinsell L. Coulson

Proceedings in Atmospheric Electricity
Lothar H. Ruhnke and John Latham (Eds.)

Ozone in the Atmosphere
R.D. Bojkov and P. Fabian (Eds.)

Radiative Transfer in Scattering and Absorbing Atmospheres: Standard Computational Procedures
Jacqueline Lenoble (Ed.)

RSRM '87: Advances in Remote Sensing Retrieval Methods
Adarsh Deepak, Henry E. Fleming, and John S. Theon (Eds.)

Spectral Line Shapes, Volume 4
R.J. Exton (Ed.)

Tropical Rainfall Measurements
John S. Theon and Nobuyoshi Fugono (Eds.)

This publication contains only peer reviewed papers presented at the Quadrennial Ozone Symposium and the Tropospheric Ozone Workshop held on 4–13 August 1988 in Göttingen, Federal Republic of Germany.

OZONE IN THE ATMOSPHERE

Proceedings of the

QUADRENNIAL OZONE SYMPOSIUM 1988
and TROPOSPHERIC OZONE WORKSHOP
Göttingen, Federal Republic of Germany
4–13 August 1988

Edited by

Rumen D. Bojkov
World Meteorological Organization
Geneva, Switzerland

Peter Fabian
Max-Planck-Institute for Aeronomy, Katlenburg-Lindau
and
University of Munich
Federal Republic of Germany

A. DEEPAK Publishing 1989
A Division of Science and Technology Corporation
Hampton, Virginia USA

A Rapid Manuscript Publication.
To produce this book as quickly as possible, some papers whose presentations might have been refined, were accepted as is, given that the concepts and data were communicated adequately.

A. DEEPAK Publishing
A Division of Science and Technology Corporation
101 Research Drive
Hampton, Virginia 23666–1340 USA

Library of Congress Cataloging-in-Publication Data

Ozone Symposium (1988: Göttingen, Germany)
Ozone in the atmosphere: proceedings of the Quadrennial Ozone Symposium 1988 and Tropospheric Ozone Workshop, Göttingen, Federal Republic of Germany, 4–13 August 1988/edited by Rumen D. Bojkov, Peter Fabian.
p. cm.
Includes bibliographical references.
ISBN 0–937194–15–8
1. Atmospheric ozone—Congresses. I. Bojkov, Rumen D. II. Fabian, Peter. III. Tropospheric Ozone Workshop (1988: Göttingen, Germany). IV. Title.
QC879.7.0947 1988
551.5'11—dc20 89–77829
CIP

Printed in the United States of America

International Ozone Commission
International Association for Meteorology and Atmospheric Physics

QUADRENNIAL OZONE SYMPOSIUM 1988 and TROPOSPHERIC OZONE WORKSHOP

Scientific Program Committee:

J.S. Chang and P. Fabian (Co-Chairmen) • R.D. Bojkov • B. Boville • A. Ghazi • A.Kh. Khrigian • G.J. Mégie • A.J. Miller • J.A. Pyle • R.S. Stolarski

Local Organizing Committee:

Peter Fabian (Chairman) • Reinhard Borchers • Margit Deppe • Harald Duscha • Maike Fabian • Silke Fabian • Günter Flentje • Michaela Hirschberg • Konstantinos Kourtidis • Andrea Macke • Karin Peschke • Kristian Schlegel • Manfred Schmidt • Onkar Singh • Jürgen Wendt

in collaboration with:

Max-Planck-Institut für Aeronomie
Georg-August-Universität Göttingen
Stadt Göttingen—Fremdenverkehrsamt
Studentenwerk Göttingen

Substantial support provided by the following sponsors is gratefully acknowledged:

Bundesministerium für Forschung und Technologie, Bonn
Niedersächsisches Umweltministerium, Hannover
Commission of European Communities, Brussels
World Meteorological Organization
American Meteorological Society
United Nations Environmental Program
Max-Planck-Institut für Aeronomie, Katlenburg-Lindau
Alcan Deutschland GmbH, Werk Göttingen
Commerzbank, Göttingen
B.E.S.T. Ventil + Fitting GmbH, Hamburg
Deutsche Bank, Filiale Göttingen
Fischer & Porter, Göttingen
Lambda-Physik, Göttingen
Messer Griesheim, Kassel
Spindler & Hoyer, Göttingen
Städtische Sparkasse, Göttingen
Adolf Thies GmbH + CoKG, Göttingen

CONTENTS

POLAR OZONE

OZONE OBSERVATIONS FROM SATELLITES

OZONE AND RADIATION, SOLAR EFFECTS AND MESOSPHERE

TROPOSPHERIC OZONE

REACTION KINETICS AND LABORATORY STUDIES

DEVELOPMENTS IN OBSERVATIONAL TECHNIQUES

PREFACE

The Ozone Symposium 1988, held from August 8 to 13 at the University of Göttingen (Federal Republic of Germany), was the sixteenth quadrennial symposium organized by the International Ozone Commission and was by far the largest Ozone Symposium ever. Almost 500 scientists from 34 countries participated in this international event, which was preceded by a Workshop on Tropospheric Ozone held from August 4 to 6, both at the University of Göttingen and at the Max-Planck Institut für Aeronomie. The Workshop, attended by 185 scientists, most of whom stayed over for the Symposium, demonstrated the increasing importance given to the studies of tropospheric ozone and its possible modification by human activities, a field that has been largely neglected in the past.

The papers presented during the meetings covered the following topics: tropospheric ozone, polar ozone, ozone observations (from ground and space), observations of relevant trace constituents, reaction kinetics, mesospheric ozone, chemical-radiative-dynamical models studies, and new observational techniques.

In this volume are published nearly two hundred of the presented papers. All those have passed an initial review process; however, in order to produce this book as quickly as possible, it was left to the authors to accommodate the reviewers' comments without editorial scrutiny of their final submission.

It is the pleasant duty of the Editors to acknowledge the assistance rendered by many colleagues who responded with time and efforts to review the numerous manuscripts, thus undoubtedly improving the appearance of most of the published papers. The excellent collaboration with the publisher A. Deepak and his staff is also acknowledged.

Rumen D. Bojkov
Geneva

Peter Fabian
Munich

INTRODUCTION

Rumen D. Bojkov and Peter Fabian

An excellent local organization was evident from the formal opening of the Ozone Symp 1988 which took place in the University Aula with a Sonata from Händel. Some of the mo portant co-sponsors—the World Meteorological Organization (Geneva), the Federal Minis Research and Technology (Bonn) and the State Minister of the Environment (Hanover)—deli messages emphasizing the importance of the continuous ozone observations and studies. The dent of the Ozone Commission announced that, in order to commemorate the 30th annive of the International Geophysical Year (IGY 1957–58), the Commission extended formal rec tion to 26 scientists who have continuously made major contributions to the study of atmosp ozone during the last three decades. The recognized scientists are: Marcel Nicolet, Ernest Vig Victor H. Regener, Rudolf Penndorf, Hans-Karl Paetzold, Alan W. Brewer, Hans U. Dütsch, A Vassy, Sorren H.H. Larsen, Anna Mani, Alexander Khr. Khrgian, Warren L. Godson, C. mond Walshaw, Julius London, Carlton L. Mateer, Genady Iv. Gustin, Walter D. Komhyr, R D. Bojkov, Wolfgang Warmbt, S. Ichtiaque Rasool, Joseph C. Farman, Karl-Heinz Grasnick sumi Kitaoka, Yashiro Sekiguchi, R.N. Kulkarni, and Edith Farkas.

The Opening Lecture was presented by the 50-year veteran in ozone studies, Baron Dr. Marcel Nicolet, and was followed by an introductory review of the findings by the NASA/\ Ozone Trends Panel presented by Dr. D.L. Albritton.

The impact of anthropogenic emissions on ozone behaviour was one of the major i discussed. The reports presented during both the Workshop and the Symposium clearly sugg and, in certain aspects, demonstrated that the ozone layer has already been modified by h activities and that these changes are bound to grow during the coming years. Ozone inc in large portions of the troposphere of the northern hemisphere and global ozone losses stratosphere are consequences of the release of man-made pollutants, mainly oxides of nit and halogenated hydrocarbons, respectively. Moreover, it was shown that the dramatic sprin ozone decline observed over the Antarctic continent has as its prerequisite the specific ci polar circulation creating extremely low (below $-80\,^{\circ}C$) lower stratospheric temperatures com with the increased level of chlorine in the atmosphere, which is a result of anthropogenic r of halocarbons.

A total of 341 papers were presented, 127 of which were oral (35 Workshop and 92 posium) and the remaining 214 were posters. In addition, a few "Last-Minute-Posters" accommodated. Following the tradition of the Ozone Commission to avoid parallel session large number of poster presentations was necessary. Ample poster space was provided to e that all posters could remain on display for the duration of the Symposium. In order to give a high priority in the programme, five special Happy Hour Poster Sessions were held. These extremely well attended and appreciated.

Whereas four years ago at the Symposium in Halkidiki less than twenty papers disc tropospheric ozone, in Göttingen nearly 70 presentations demonstrated the increased impo given by researchers to tropospheric ozone, mainly due to concerns of its possible modifi by human activities and noticeable potential contribution to the greenhouse effect.

New data on the tropospheric ozone distribution from satellite, ground-based, aircraft and ship measurements were presented. Many area gaps noted in the past, in particular over the oceans, were discussed with data for the first time. Moreover, measurements from the tropical regions of South America, Africa and Asia showed that considerable regional increases of tropospheric ozone are often observed due to photochemical production from precursors released by biomass burning. Solar radiation and tropospheric ozone studies demonstrated that elevated ozone levels measured in Europe, North America and the Far East are due to photochemical production based on precursor substances such as NO_x, carbon monoxide and hydrocarbons. Precursor distributions and their relationships with ozone were discussed in detail, both for urban as well as for rural areas.

Several reports discussed the tropospheric ozone trends. They clearly demonstrated that surface ozone concentrations have increased in many areas of the northern hemisphere over the past two decades by 1% to 3% per year. At the same time, study state of the surface ozone was observed over Antarctica, which may be partially due to substantially lower concentrations of NO_x and other precursors and partially related to the springtime decline of stratospheric ozone. Model studies relating the tropospheric ozone budget to NO_x and hydrocarbon emissions showed that enhanced ozone levels are not confined to the pollution source regions on the continents but also are spread over the oceans and other non-industrial areas of the northern hemisphere.

The earlier-established[1] at least two-fold increase of surface ozone concentrations from measurements taken during the second half of the last century was re-confirmed by both a verification of the measurement techniques[2] and by modelling of surface emissions of NO_x and NHMC[3]. Past and future tropospheric ozone changes due to the release of pollutants and the feedback of future climatic changes were assessed as well.

In more than 40 papers on polar ozone, details mainly of the spectacular Antarctic-spring ozone decline observed since the late 1970's clarified and confirmed the view that the unique features of atmospheric circulation during the Antarctic winter establishes an isolated air mass (stratospheric polar-night vortex) with very low (−80 °C and lower) temperatures. Such conditions are favourable for the formation of polar stratospheric clouds. This is the prerequisite for perturbed atmospheric chemistry during September and early October (when the sun starts to appear over the most southerly parts of Antarctica). Laboratory studies show that polar stratospheric clouds are capable of converting photochemically inactive chlorine species into active forms which are capable of destroying ozone. These studies are entirely consistent with the high levels of active chlorine observed within the polar vortex in 1986 and 1987. The observed strong anti-correlation between active chlorine and ozone from the 1987 aircraft measurements indicates that these largely industrially derived chlorine species are primarily responsible for the observed decrease of ozone within the stratospheric polar vortex. Most of the depletion observed is in the 12–22 km region.

The Antarctic ozone decline could be much smaller if there were to be a significant increase in the temperature of the Antarctic stratosphere and an early stratospheric vortex breakdown, thus precluding the formation of the polar stratospheric clouds. It was shown that the rapid and frequent meridional exchange of air in the Arctic lower stratospheric levels preclude the possibility of a strong closed circumpolar vortex, which is typical over the Antarctic region during the polar night. This could explain why the available observations do not indicate over the Arctic the existence of an extensive ozone reduction similar to the one observed over Antarctica since the late 1970's. It is interesting to mention that for the first time extremely low ozone values observed in Antarctica in 1982 were reported to the previous ozone symposium by Chubachi. However, this paper passed unnoticed at that time.

[1]Bojkov, 1986
[2]Volz and Kley, 1988
[3]Makeen et al.

As usual, the large part of the presentations discussed the information collected by both the ground- and satellite-based components of the WMO Global Ozone Observing System (GO_3OS). For the first time, detailed studies utilizing re-evaluated ozone records demonstrated that there is a statistically significant total ozone decrease of 1.7% to 3.0% from 1969 through 1986 in 30° to 64 °N latitudes. The decreases are largest during the winter-spring months (2.3% to 6.2%) and contrast with smaller losses of less than 1% during the summer. A noticeable decrease of ozone, not only in the upper stratosphere but also in the 16–26 km layer, detected from ozone soundings at Hohenpeissenberg and Payerne may indicate some possible shortcomings in the present modelling simulations. The variations in ozone related to the Quasi Biennial Oscillation (QBO) and El Niño Southern Oscillation (ENSO) related fluctuations in the stratospheric circulation were also discussed. Strong emphasis was given to the urgent need for re-evaluating the published total ozone records, taking into consideration corrections deduced from periodic instrument calibrations or intercomparisons of instruments. A critical assessment of a large number of filter ozonometer measurements indicated that reasonably good ozone distribution and changes similar to those shown by the Dobson instruments could be deduced from the data after 1972.

Thirty papers presented a critical reassessment of the satellite ozone observations which demonstrated that, after appropriate well-documented corrections, the adjusted nine years of global total ozone measurements by the TOMS instrument on Nimbus 7 show a decrease of approximately 3% in concurrence with the results from the Dobson data for the same time period. Numerous interesting details in the global ozone distribution were demonstrated with TOMS and SAGE data; however, it was clear that substantial revisions in the currently existing SBUV data set would be necessary in view of possible degradation of the diffuser plate used by both SBUV and TOMS instruments.

The session on other relevant trace constituents contained various reports on new measurements of NO_x and HO_x radicals and reservoir species including NO_3 and N_2O_5. Results of intercomparison experiments of ground-based, balloon-borne and satellite NO_x sensors, in situ and remote, were presented. Furthermore, an extensive data base on the distribution of long-lived source gases, in particular various halogenated hydrocarbons, was reported.

In 14 papers presented at the session on radiation, solar effects and mesosphere, co-sponsored by IAGA, diurnal variation and responses of the middle atmosphere to solar variability and to gravity waves were discussed. Most reports focussed on the mesosphere where significant ozone variability is observed. A few papers discussed the atmospheric ozone and the role of the solar UV radiation.

In nearly 40 papers on chemical-radiative-dynamical models studies, it was encouraging to see that results of 2-D and 3-D models were dominating. Specific problems of parameterization in the 2-D models and solar heating due to O_3 and other constituents were discussed in great detail. A large fraction of the papers discussed the problems of stratospheric changes as a result of anthropogenic emissions. It is estimated that the ozone depletion would be at least twice stronger in the middle and higher latitudes than over the tropical belt, much stronger again during the winter-spring, and insignificant during summer. These model estimates are in broad agreement with observational analysis. Various scenarios were presented for effects of reduction of CFC emissions, as called for by the Montreal Protocol on Substances that Deplete the Ozone Layer, from which it seems that the foreseen reductions of CFCs in their present form would not be sufficient to recuperate and/or prevent an ozone reduction during the next century. Furthermore, it was estimated that Bromine compounds increasingly released from man-made halons (used for example in fire distinguishers) could contribute to about 15% to 20% of the total depletion.

Estimates of sensitivity of ozone to radiative flux changes, due to, for example, a projected doubling of the CO_2 concentration indicated that an 8% to 16% ozone increase in the middle stratosphere and 1% to 2% in the lower stratosphere could result—mainly related to the calculated temperature decrease in the same region and related dynamic perturbations. Dynamical effects of wave perturbations and the QBO on the composition were discussed. Some interesting model testing by means of ATMOS and SAGE II satellite data was also presented.

Reaction kinetics and related laboratory studies have never been well represented during past Quadrennial Ozone Symposia. The Göttingen Symposium was no exception—only a few papers were presented dealing with various aspects relevant to the O_x, HO_x, NO_x, ClO_x, and BrO_x systems. About twenty papers discussed new observational techniques including application of microwave sensors, lidar using XeCl excimer laser, and application of an automated Dobson and a Brewer spectrophotometer.

At the end of the Ozone Symposium 1988 the International Ozone Commission held its formal meeting: for the next four years Dr. G.J. Mégie (France) and Dr. A.J. Miller (USA) were elected President and Vice-President and Dr. R.D. Bojkov (WMO) was re-elected Secretary. The concluding meeting noted that after many years of doubt and research, there is now clear evidence of mankind having affected the whole global ozone layer. This is a significant point in history as it is the first firm evidence that humans have significantly and harmfully altered the atmosphere on a global scale. It further recognized that it was only as a result of scientific research that we have come to recognize and understand the basic ozone problem. The scientists urged all national and international agencies (such as WMO, CEC, UNEP) to continue to support and develop their programmes for scientific research and monitoring of ozone and associated atmospheric parameters, including studies of impacts and related public policy issues in order to reach a better understanding of the future impacts of ozone change and associated changes to the chemistry and physics of the atmosphere.

POLAR OZONE

ANTARCTIC OZONE DEPLETION MEASURED BY BALLOON SONDES AT HALLEY BAY

Brian G. Gardiner
British Antarctic Survey, Natural Environment Research Council
High Cross, Madingley Road, Cambridge, CB3 OET, UK

ABSTRACT

Profiles from a series of balloon-borne ozonesonde ascents are used to chart the development of the Antarctic ozone depletion over Halley Bay in the austral spring of 1987. Flights were launched in the unaffected winter period and continued until well after the end of the active phase. The vertical structure of the ozone layer is discussed, including the presence of stratification, which occurs at all stages of development. The onset of depletion in late August is described, and features of the development phase are illustrated. The depletion process, though irregular throughout September, eventually culminates in the establishment of a deep minimum in ozone mixing ratio. By mid-October, most of the ozone has been destroyed throughout a deep layer of the lower stratosphere.

1. INTRODUCTION

In order to study the temporal development of the Antarctic spring ozone depletion, a programme of ozonesonde flights was undertaken at Halley Bay (76° S, 27° W) in 1987. The ozonesondes and operational procedures are described in Komhyr (1986). More than sixty ascents were completed in the period April to November, of which over fifty were in the crucial months of August, September and October. The fine vertical resolution achieved by ozonesondes is not readily obtainable by other methods. Information can be obtained not only about the rate of loss of ozone as the polar night comes to an end, and about the altitude range in which ozone destruction takes place, but also about the complex interplay of chemistry and dynamics, which is central to an understanding of the ozone depletion problem.

The magnitude of the phenomenon is made clear in Figure 1. The profile of August 15 can be regarded as showing the stratosphere in its normal winter state, with a maximum ozone partial pressure of about 170 nbar at the 65 mbar level. By contrast, the lower stratosphere on October 13 has lost nearly all the ozone at that level, and is severely depleted throughout a vertical range of many kilometres. From October 7 to 13, the depletion is rather more severe than that observed on 16 October 1986 at McMurdo by Hofmann et al. (1987).

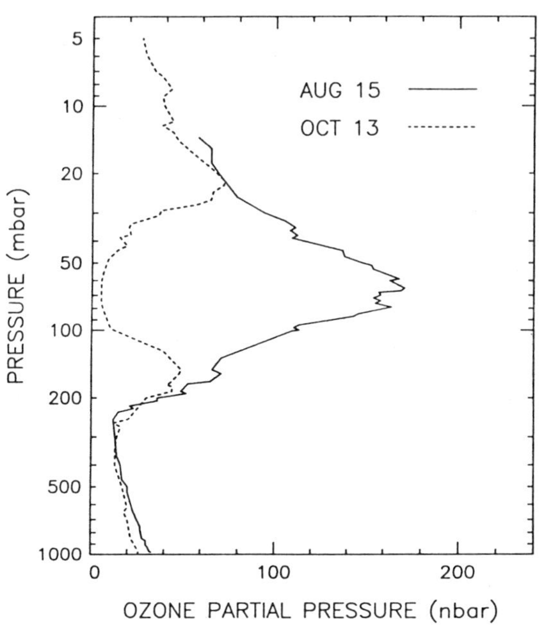

Fig. 1. Ozone partial pressure on August 15 and October 13.

2. UNPERTURBED PROFILES

Before examining the depleted conditions of September and October, it is necessary to have some understanding of the unperturbed profiles of early August. In particular, it should not be thought that the circulation of the polar vortex in the depth of winter is so uniform as to be devoid of structure or temporal change.

Figure 2 shows the ascents of August 4 and August 9. At first sight it appears as if the lower stratosphere is in a relatively steady state: the profiles agree with each other, and the ozone mixing ratio increases more or less monotonically with height. The profile of August 15 in Figure 1, which corresponds almost exactly to those in Figure 2, confirms this picture. This consistency in the flights of August 4, 9 and 15 may result from sampling of air which has completed one or more circuits of the polar vortex. However, the flight of August 6 tells a different story (Figure 3). Here the profile in the lower stratosphere is dominated by structural features with a vertical scale of 1 km or less. Irregular structure is also seen in the profile of July 26 (Figure 3), and on many other occasions during the winter. As the tropopause is found at about 260 mbar on August 6 and July 26, coincident with a sharp increase in ozone mixing ratio with height, these pronounced

ISBN 0-937194-15-8

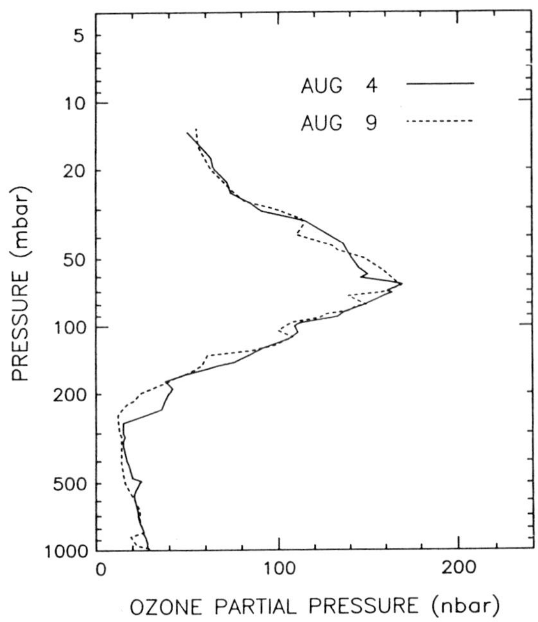

Fig. 2. Ozone partial pressure on August 4 and August 9.

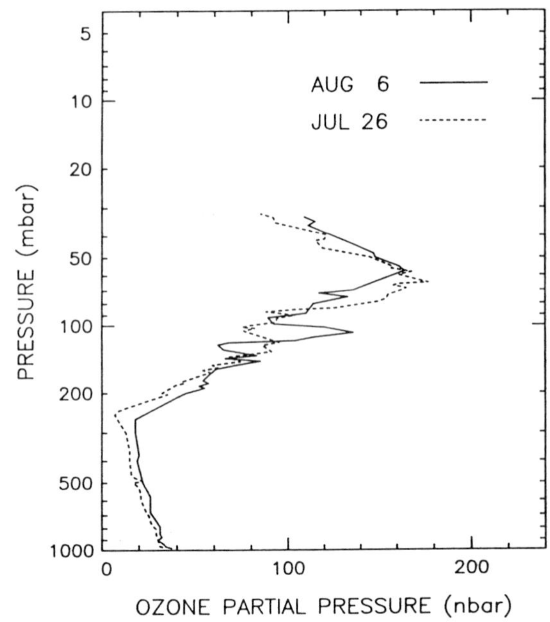

Fig. 3. Ozone partial pressure on August 6 and July 26.

structural features are well inside the stratosphere. The profile of August 6 contains two minima in ozone mixing ratio, at 125 mbar and 95 mbar. As there is insufficient sunlight to drive the chemistry of ozone depletion, the structure can only arise from the effects of wind shear. The ozone profile reflects the recent history of the wind trajectory at each level.

Despite such variations, the winter profiles display a consistent maximum in ozone partial pressure of about 160 to 180 nbar at around 70 mbar, and a mixing ratio of nearly 3 ppmv in the region between 50 mbar and 20 mbar. These conditions persist unaltered until August 17.

3. ONSET OF DEPLETION

The first phase of ozone depletion is observed in the last two weeks of August. The process is swift but erratic. In the layer between 65 and 25 mbar, the ozone mixing ratio falls to 2.5 ppmv on August 22 and remains low in the flights of August 25 and 26. The effect on the next day is not so marked, and is limited to parts of the layer, but the air column sampled on August 28 shows a much deeper depletion, down to 2 ppmv in the layer between 60 and 50 mbar. Beyond 70 mbar this profile is virtually the same as that obtained at McMurdo by Hofmann et al. (1987) on the same date in 1986, confirming their suspicion that depletion was already present at the 20 km (40 mbar) level. At Halley Bay in 1987, the depleted layer extends downwards on the following day, August 29, to remove the whole of the ozone peak (Figure 4). Some 40% of the ozone has been lost between 90 and 50 mbar, and there is no air with 2 ppmv of ozone at any level between the surface and 50 mbar. The depletion of August 29 is confirmed on the next two days. Nevertheless, the first two weeks of September exhibit considerable variation. In the flights of September 2, 3, 4 and 5, an ozone partial pressure maximum of about 130 to 140 nbar reappears around 60 mbar, showing that the vortex is not affected uniformly. On September 5, the top of the depletion layer occurs at 16 mbar (24.7 km), but in general depletion is found in the region from 150 mbar to 20 mbar from August 31 onwards.

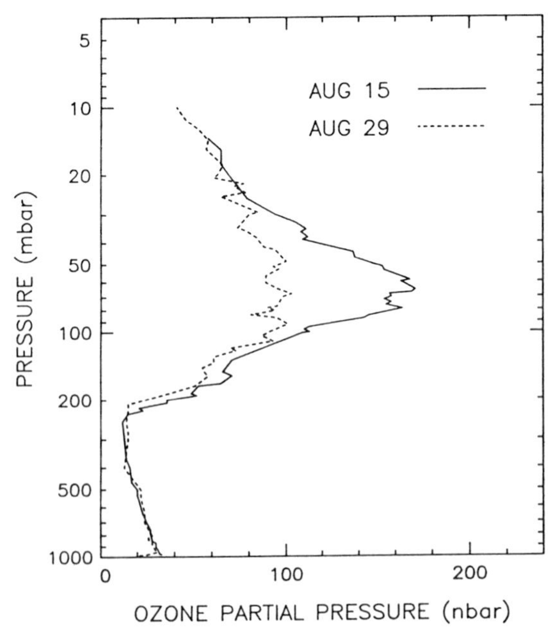

Fig. 4. Ozone partial pressure on August 15 and August 29.

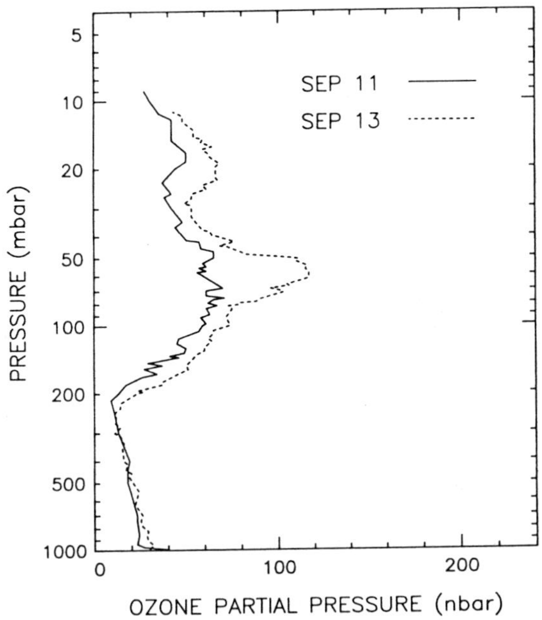

Fig. 5. Ozone partial pressure on September 11 and September 13.

4. DEVELOPMENT

After September 5, ozone is progressively lost throughout the lower stratosphere. By September 11, 60% of the ozone has been destroyed between 90 and 50 mbar, but again not all of the vortex is affected equally. The column observed on September 13 shows milder depletion than that of September 11 at all heights, and much of the original ozone peak has survived (Figure 5). The contrast is quite instructive. The profiles have a remarkably similar form up to 80 mbar and beyond 45 mbar, but twice as much ozone remains at 60 mbar on September 13. There is no evidence of vertical motion, the temperatures on the two days being virtually the same throughout the lower stratosphere. Either the layer between 80 and 45 mbar on September 13 has been advected from a column with less advanced depletion, or the ozone loss process has been temporarily less effective in that height range. It is a characteristic of the ozone profile in September that, while depletion may be delayed in one part of the vortex relative to another, and may vary from layer to layer, the delay is never more than a few days. Progress is irregular but inexorable. For example, ozone partial pressures greater than 100 nbar, as seen on September 13 between 70 and 50 mbar, are not found at any level in any ascent after September 19. This implies that the solar forcing of the depletion mechanism is fairly direct, as it appears that an air parcel cannot resist for more than a few days the depletion appropriate to its latitude and date, whatever its dynamical and chemical history might be.

There is a tendency throughout September for depletion to occur in two distinct layers, centred roughly at the 90 mbar and 30 mbar levels. This effect, which is quite clear on some days, and completely absent on others, becomes more pronounced in the second half of September, and can still be detected as late as October 7.

In the last week of September, ozone destruction in the 120 to 60 mbar layer has advanced to the stage at which a minimum has appeared in ozone mixing ratio as a function of height. This minimum, which deepens progressively to encompass the whole depletion region (Figures 6 and 7), is present in every

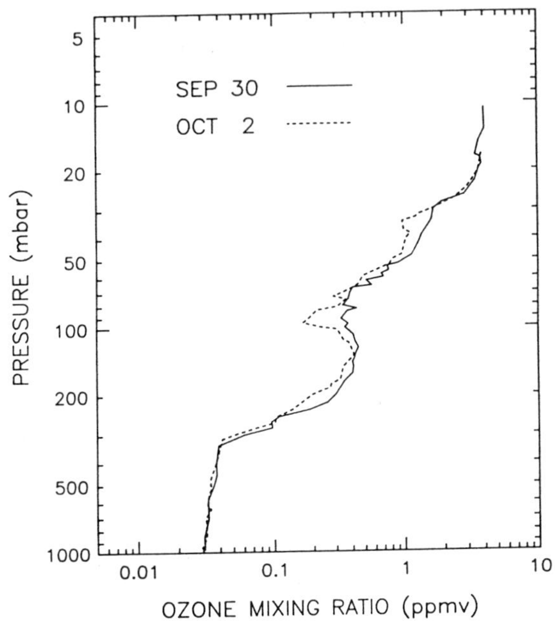

Fig. 6. Ozone mixing ratio on September 30 and October 2.

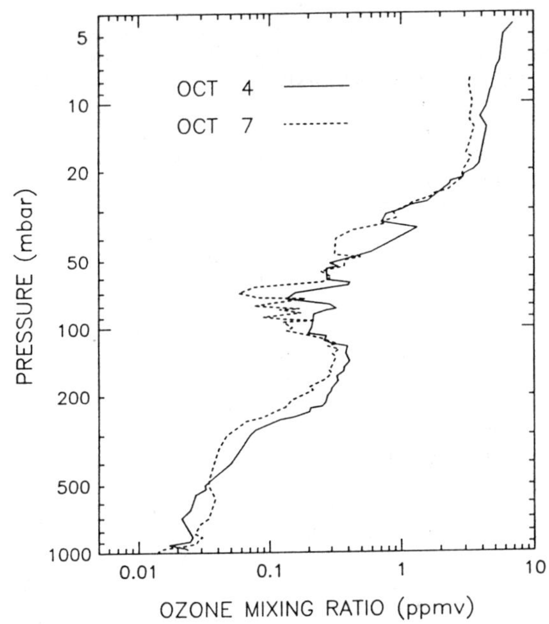

Fig. 7. Ozone mixing ratio on October 4 and October 7.

profile from the equinox until the end of October, and is even seen in late November. Such a persistent minimum is at variance with the requirements of the upwelling theory of ozone depletion suggested by Tung et al. (1986).

It would be easy to infer that the stratification shown in Figure 7 provides evidence of chemical activity in restricted layers, and indeed it may be so, but we saw earlier that such laminations are by no means confined to the spring. They may arise simply by advection of air from locations where the depletion has reached a different stage, owing to its dependence on solar elevation and temperature.

5. FINAL PHASE

When the depletion is fully developed, in the second week of October, the layer between 90 and 50 mbar has lost over 95% of the ozone which was present in mid-August (Figure 1). The laminations which were evident during the advance of the depletion (Figure 7) are no longer present on October 11 and 13. It is as though the depletion were in a sense complete, each level having reached its pre-determined minimum concentration.

In the second half of October, the depleted region recovers rather slowly. Eventually, laminations reappear as less-depleted air intrudes into the column, and the affected region suffers some descent over a period of a few weeks (Figure 8). The depleted layer is still clearly present in the flight of November 23.

6. LIMITS OF DEPLETION

In the fully-developed phase, the depleted layer extends from about 150 mbar to 20 mbar (12 to 24 km). During development, the upper boundary is often rather well defined. As the boundary marks a critical condition for the onset of depletion, it might be profitable to study its characteristics. A preliminary examination of the temperature soundings shows that a sharp edge in the ozone profile is often associated with a recorded temperature of -80° C, depletion occurring always on the colder side of the boundary. Instances can be found for upper and lower edges of a depleted layer, though the relationship is by no means consistent and may be spurious. In any event it does not occur after September 16. Nevertheless, as the region of very cold temperatures becomes progressively shallower, the deepest development continues to occur at temperatures colder than -80° C, even in October, when the extreme boundaries of the depleted layer are rather warmer.

7. CONCLUSIONS

The progression of the Antarctic spring ozone depletion in 1987 has been examined from inception to completion by a series of ozonesondes flown from Halley Bay. Depletion begins in late August and progresses irregularly throughout September. Considerable fine structure is found in the ozone profiles before and during the development phase, and depletion

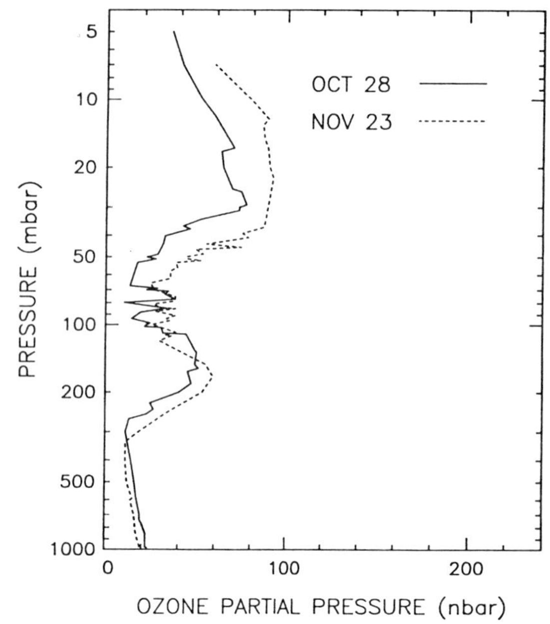

Fig. 8. Ozone partial pressure on October 28 and November 23.

sometimes occurs in two separate layers, but no level in the affected region (150 to 20 mbar) escapes ozone destruction. A minimum in ozone mixing ratio appears at the equinox, and deepens to encompass the whole of the depletion layer by mid-October, when nearly all the ozone has been lost at the 70 mbar level. About 80% of the ozone is destroyed in the height range 13 to 22 km. The total ozone minimum cannot therefore fall much lower in future years unless there is a substantial increase in the vertical extent of the depleted layer. The strongest depletion is accompanied by temperatures colder than -80° C, but its effect can still be seen in late November when the air is warmer than -70° C throughout the stratosphere.

ACKNOWLEDGMENTS

This work was supported by the Fluorocarbon Program Panel of the Chemical Manufacturers Association and the U.K. Department of the Environment.

REFERENCES

Hofmann, D.J., J.W. Harder, S.R. Rolf, and J.M. Rosen, 1987: Balloon-borne observations of the development and vertical structure of the Antarctic ozone hole in 1986. Nature, **326**, 59-62.

Komhyr, W.D., 1986: Operations handbook -- ozone measurements to 40-km altitude with model 4A electrochemical concentration cell (ECC) ozonesondes. NOAA Technical Memorandum ERL ARL-149, 49 pp., NOAA/GMCC, Boulder, Colo.

Tung, K.K., M.K.W. Ko, J.M. Rodriguez, and N.D. Sze, 1986: Are Antarctic ozone variations a manifestation of dynamics or chemistry? Nature, **322**, 811-814.

ANTARCTIC AND ARCTIC TOTAL OZONE EXTREME MINIMA

Arlin J. Krueger
NASA Goddard Space Flight Center
Greenbelt, Maryland 20771, USA

ABSTRACT

Total ozone in the Antarctic ozone hole reached an unprecedented low value of 109 Dobson units, as measured with the Nimbus 7 TOMS instrument on October 5, 1987 over the East Antarctic highlands. This extremely low amount appears to be the result of ozone destruction in the lower stratosphere by chlorine compounds. The continental-scale, month-duration ozone hole is preceded by smaller total ozone minima which live only for a few days. Short-lived total ozone minima as low as 165 DU also appear in the Arctic, primarily during fall and early winter. However, during the decade of TOMS observations, Arctic features the size, duration, and depth of the Antarctic ozone hole were not found.

1. INTRODUCTION

The appearance of the Antarctic ozone hole in recent years has produced a heightened interest in the biological effects of enhanced UVB radiation under reduced total ozone conditions. While the extreme decreases over the Antarctic continent and surrounding oceans are remote from population centers, similar decreases in the Arctic are cause for concern. The flux of biologically-active uv radiation depends on the absolute amount of total ozone.

The Total Ozone Mapping Spectrometer (TOMS) data base was used to determine the location and characteristics of total ozone minima since the launch of Nimbus 7 in October 1978. The contiguous coverage of the instrument assures that all ozone minima of 50 km and larger size can be detected under sunlit conditions. For the purposes of this study of biologically important low ozone episodes, only extremely low values, defined as those lying below 200 DU, are considered. The Version 5.0 archival data base, which is based on the internal calibration of the instrument, was used. However, the TOMS calibration may have drifted relative to the Dobson network by 3 to 4 % after nine years of operation (Stolarski, et al, this publication). The effects are considered in the analysis.

The TOMS total ozone results also depend to a small extent on the difference between an actual ozone vertical distribution and the climatological distribution assumed for purposes of constructing albedo tables for data processing. Very unusual ozone distributions, such as found in the Antarctic ozone hole can produce total ozone underestimates of 10% (Bhartia, et al., 1988).

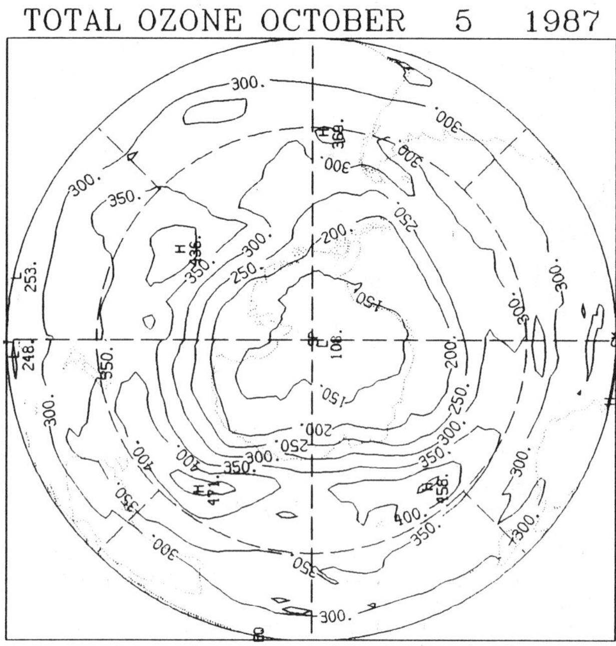

Fig. 1. Total ozone over the southern hemisphere on October 5, 1987 illustrating the ozone hole at its greatest development.

2. ANTARCTIC OZONE MINIMA

Extremely low total ozone values in the higher latitudes of the southern hemisphere have been found in spring months of recent recent years (Farman, et al, 1985; Stolarski, et al, 1986; Krueger, et al, 1987). The location is within the "ozone hole", a roughly circlar, continental-scale minimum which forms in September and reaches its deepest level in October. The characteristic shape and size of the ozone hole are illustrated in Fig. 1, a southern hemisphere plot of total ozone on October 5, 1987. Contours are shown for every 50 Dobson units (DU) in this polar orthographic projection. On this day the 200 DU contour surrounds an area larger than the Antarctic continent (14 million sq km) that is pole-centered and roughly circular. The lowest value, 109 DU, is found near the South Pole. This minimum is lowest ever detected with the TOMS instrument (Krueger, et al, 1988). With the possible errors noted above, the mimimum could be underestimated by as much as 15 DU. For

ISBN 0-937194-15-8

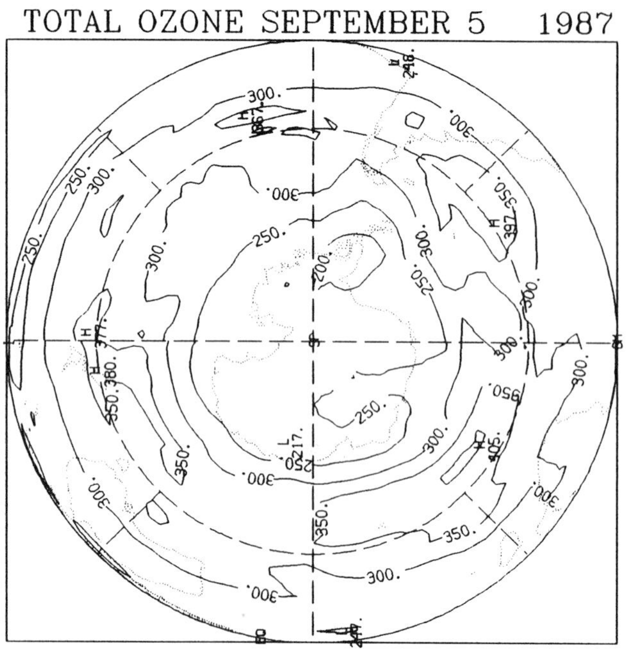

Fig. 2. Total ozone over the southern hemisphere on September 5, 1987. A Type II ozone minimum is present over the Antarctic Peninsula.

purposes of classification, this ozone minimum with a dimension of greater than 5000 km can be designated a Type I minimum.

Extremely low total ozone has been found in two other types of minima which occur over and adjacent to the Antarctic continent. One type of minimum (Type II) has spatial dimensions of about 1000 km, appears to be created at mountain ranges, and propagates downwind with time. An example is presented in Fig. 2, the southern hemisphere total ozone distribution on September 5, 1987 where the Type II feature is the 200 DU contour which is embedded in the 250 DU polar minimum located over the Antarctic Peninsula. On the next two days this minimum drifted across the Weddell Sea to the vicinity of the Greenwich meridian.

The third type of minimum (Type III) is the smallest with dimensions of about 150 km, is stationary, and occurs in mountain waves primarily over the Antarctic Peninsula. The lifetime of the smaller scale minima is about 3 days, compared with about 60 days for the Type I ozone hole. These phenomena have very different origins and life cycles.

3. ARCTIC OZONE MINIMA

In comparison with the extensive areas covered by extremely low total ozone in the Southern hemisphere, cases of total ozone less than 200 DU in the Northern hemisphere are rare, occurring only once or twice per year since 1982 in the TOMS data set. These occurrences are limited to the fall and winter months and, with one set of exceptions, to latitudes greater than 55°N. The exceptions occurred in December 1987 in the tropics on five days over the western Pacific Ocean and on one day over Mexico. Table 1 is a list of the Arctic cases that have been identified including the depth of the minimum, the date, and the location.

TABLE 1. TOTAL OZONE EXTREME MINIMA IN THE NORTHERN POLAR REGIONS

YEAR	DATE	CENTRAL AMOUNT	LATITUDE	LONGITUDE
1987	Jan 7	165	60	17W
1986	Oct 13 to	189	76	102E
	Oct 15			132E
1985	Oct 30	166	58	12E
1984	Feb 10	192	64	12W
	Nov 15	192	64	47E
1983	Sept 27 to	195	82	2W
	Oct 1			60E
1982	Nov 12	183	68	77E
1978–1981	no minima < 200 DU observed			

The normal pattern for polar ozone in the Northern Hemisphere is a pole-centered minimum which is bracketed and sometimes surrounded by midlatitude maxima. This pattern is subject to rapid changes due to meandering of the lower stratospheric planetary waves and is frequently disrupted in midwinter by amplification of the maxima and encroachment (and sometimes bridging) of the polar night region by very high total ozone features. As a result the highest

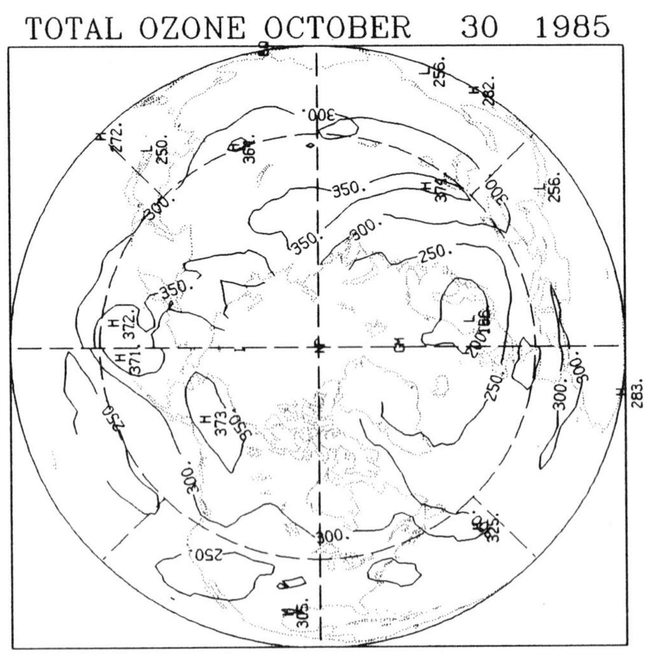

Fig. 3. Total ozone over the northern hemisphere on October 30, 1985. A deep minimum (166 DU), similar to the Antarctic Type II minima, is located over the Scandinavian countries.

Northern Hemisphere total ozone amounts can be found above 60°N at times during late winter and spring. By the end of May the high amplitude total ozone modulations have relaxed and a persistent polar minimum begins forming in July.

All of the deep minima are found between the autumnal equinox and early February. One reason for this is related to the annual cycle which reaches its minimum at Arctic latitudes in the fall months. The minima are either features which emerge from the polar night, affecting the immediate boundary regions, or are circular or elliptical features, similar to the Type II Antarctic minima in scale and appearance. The minima are not locked in position but generally move toward the east. An example of the second type of minimum is presented in Fig. 3, the Northern Hemisphere total ozone on October 30, 1985. In this polar stereographic projection the Greenwich meridian is at the right and the 30° and 60° latitude circles are shown with dashed lines. The minimum on this day is centered over Sweden near 60°N, 20°E and covers longitudes from 60°W to 60°E at 60°N. The feature was first recognizable on October 27 as a minimum centered at 70°N, 10°W. It moved to 70°N, 90°E in early November and disappeared after November 10.

The minimum total ozone on October 30 is 166 DU, located at 58°N, 12°E. On the same day the total ozone measured with the Dobson Spectrophotometer at Oslo, Norway (59.54°N, 10.43°E) was 170 DU. The 2.4% difference between the TOMS and Dobson readings is consistent with the drift between the TOMS and the Dobson network (Stolarski, et al. this issue).

On October 30 the polar vortex was well developed and centered over Novaya Zemlya (75°N,

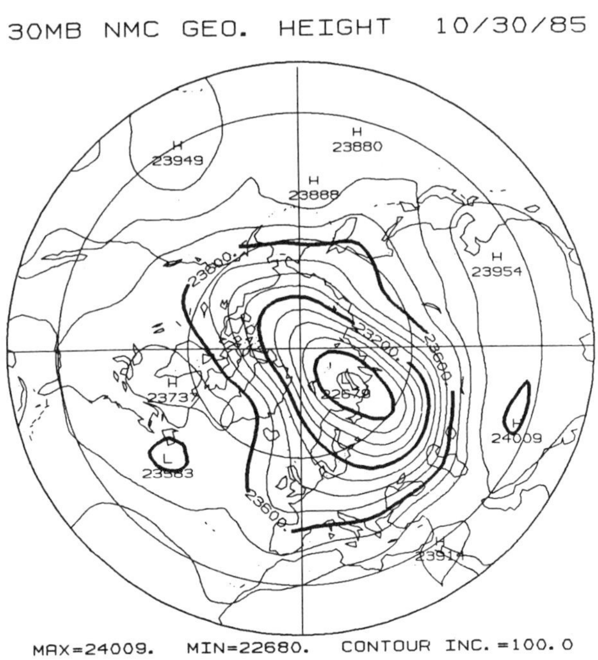

Fig. 4. Geopotential height field for the Northern hemisphere 30 mb surface on October 30, 1985.

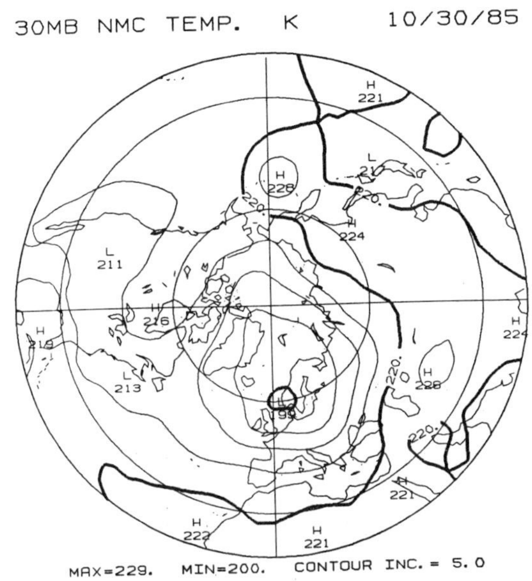

Fig. 5. Temperature field for the Northern hemisphere 30 mb surface on October 30, 1985.

60°E), about 1800 km to the northeast of the total ozone minimum, as shown in Fig. 4, the 30 mb geopotential height field. The depression in geopotential height extends across the lower stratosphere to the tropopause at the 200 mb level. The lack of co-location between the vortex center and the total ozone minimum indicates that the vortex circulations are not directly controlling the total ozone. The air temperature, however, exhibits a relationship at levels above 70 mb. For example, the 30 mb temperature field is shown in Fig. 5. The temperature contours follow the ozone contours remarkably well and the temperature at the location of the ozone minimum is the lowest in the hemisphere (199°K). This association diminishes at lower altitudes, having disappeared at the 100 mb level. These circumstances suggest that the total ozone decrease is produced in secondary circulations on the boundary of the polar vortex and that the upward motions are highest at the highest altitudes.

The closest ozonesonde flight on October 30 was at Hohenpeissenberg, while a second sounding was made at Lindenberg on October 31. Both soundings show ozone density maxima in the 70 to 50 mb region. As shown in Fig. 6, the Lindenberg sounding exhibits lower ozone mixing ratios than the Hohenpeissenberg sounding at all altitudes above the 100 mb surface. The difference in mixing ratio also increases with altitude. Noting that the Lindenberg location is closer to the total ozone minimum, this behavior is further supporting evidence for a disturbance that is maximum in the middle stratosphere.

The October 30 event is similar in spatial dimensions to a Type II ozone minimum in the Antarctic (see Fig. 2). However, a comparison of the ozonesonde profile obtained at Halley Bay

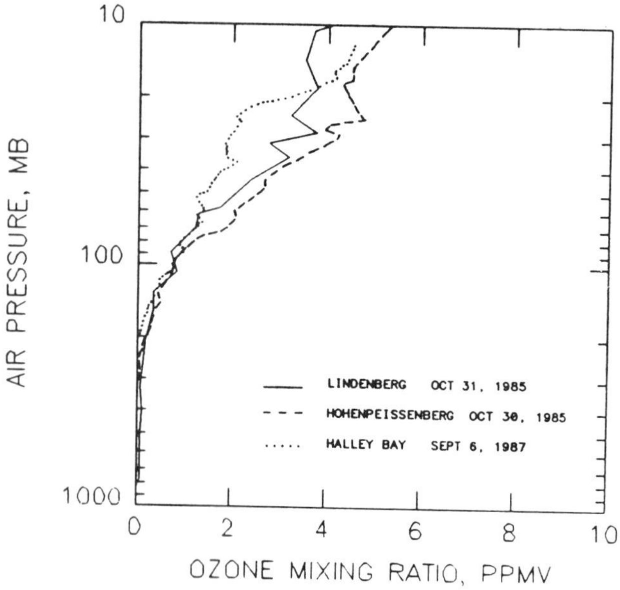

Fig. 6. Ozone mixing ratio profiles at Lindenberg on October 31, 1985 (solid line), Hohenpeissenberg on October 30, 1985 (dashed line), and Halley Bay, Antarctica on September 6, 1987 (dotted line). The Lindenberg sounding is closest to the center of the total ozone minimum shown in Fig. 3, while Hohenpeissenberg is on the periphery. The Halley Bay sounding was taken within the total ozone minimum shown in Fig. 2.

on September 6, 1987, also shown in Fig. 6, with the Lindenberg sounding shows that the change has occurred at different altitudes. The Antarctic case shows a decrease at levels below 20 mb while in the Arctic case the largest decrease is at levels above 30 mb.

The mesoscale Arctic ozone minima described above are not recent phenomena although the absolute total ozone levels may be lower than in prior years. An analysis of Arctic ozone variations based on ground-based total ozone observations (Bojkov, 1988) indicates that transient ozone minima approaching 200 DU were present at Tromso in the 1957/58 winter. Thus, these features are likely to be a normal variation in the northern fall and winter polar ozone.

4. SYNOPTIC-SCALE TOTAL OZONE MINIMA IN THE NORTHERN HEMISPHERE

While "ozone holes" of the Antarctic Type I minima are not found in the Arctic, the Northern hemisphere does exhibit wintertime, synoptic scale fluctuations of total ozone that are transient, moving with the speed of planetary waves, and, at some times very persistent. For example, in February 1984 a region extending from the North Atlantic Ocean, across Scandinavia, into central Russia had a total ozone decrease averaged over the month by as much as 120 DU compared with a reference average. This decrease is comparable to the 140 DU decreases found in the Antarctic spring during the 1987 record year. However, the lowest monthly mean value in the Northern Hemisphere case was 288 DU, far from an alarmingly low value. This same region had above normal total ozone in the previous month (January 1984) and only small decreases (20 - 40 DU) in the succeeding month (March 1984). In February 1985 the North Atlantic region had small total ozone decreases while the Scandinavian and Russian regions had increases. The largest decreases (80 DU) during January and February 1985 were over the Pacific Ocean and northwest United States.

In general, although the general distribution of total ozone over the Northern Hemisphere forms a stationary pattern which during each month persists from year to year, the deviations appear to be non-stationary. The large negative deviations found in 1984 are not unusual, appearing over the North Atlantic in January 1981, February 1983 and in February 1986. Similar minima were found over northern Europe in February 1982, February 1984, and March 1986.

5. SUMMARY

Total ozone extreme minima in the Southern Hemisphere are found in the late winter preceeding the formation of the ozone hole as mesoscale minima embedded in the polar minimum and in springtime in the ozone hole. The embedded minima are of two types: small stationary minima associated with mountain ranges and larger, drifting minima.

The lowest total ozone values found in the nine year TOMS data set over the Northern Hemisphere are near 165 DU. They occur during fall or early winter in ozone depressions near the polar night boundary or in small circular minima at high latitudes. These minima move eastward with time, similar to the drifting minima of the Southern Hemisphere. No minima comparable to the Antarctic ozone hole are found in the Arctic although features approaching the size but not the magnitude or duration of the ozone hole exist in the fall and winter months of some years over the North Atlantic Ocean and over Europe.

ACKNOWLEDGEMENTS

The author appreciates the assistance of Scott D. Doiron and Reginald N. Galimore in the analysis of the TOMS data and the help of Dr. Paul A. Newman in the preparation of NMC upper air charts.

REFERENCES

Bhartia, P. K., A. J. Krueger, S. Taylor, and C. Wellemeyer, 1988: Estimation of errors in the TOMS total ozone measurement during the

Antarctic ozone campaign of August/September 1987, J. Geophys. Res., submitted.

Bojkov, R. D., 1988: Ozone variations in the northern polar region, Meteorol. Atmos. Phys., 38, 117-130.

Farman, J. C., Gardiner, B. C., and J. D. Shanklin, 1985: Large losses of total ozone in Antarctic reveal seasonal ClOx/NOx interaction, Nature, 315, 207-210.

Krueger, A. J., M. R. Schoeberl, and R. S. Stolarski, 1987: TOMS observations of total ozone in the 1986 Antarctic spring, Geophys. Res. Lett., 15, 527-530.

Krueger, A. J., M. R. Schoeberl, and R. S. Stolarski, 1988: The 1987 Antarctic ozone hole: A new record low, Geophys. Res. Lett., submitted.

Stolarski, R. S., A. J. Krueger, M. R. Schoeberl, R. D. McPeters, P. A. Newman and J. C. Alpert, 1986: Nimbus-7 SBUV/TOMS measurements of the springtime Antarctic ozone hole, Nature, 322, 808-811.

Stolarski, R. S, A. J. Krueger, and M. R. Schoeberl, Total ozone trends from TOMS data, this issue.

THE BREAKUP OF THE SOUTHERN HEMISPHERE SPRING POLAR OZONE AND TEMPERATURE MINIMUMS FROM 1979 TO 1987

Paul A. Newman
Applied Research Corporation
Landover, Maryland 20783, USA

Mark R. Schoeberl
NASA Goddard Space Flight Center
Greenbelt, Maryland 20771, USA

ABSTRACT

The date of the southern hemisphere (SH) polar ozone minimum breakup is diagnosed using National Meteorological Center (NMC) heights and temperatures in conjunction with Total Ozone Mapping Spectrometer (TOMS) data. Zonal mean gradients of 100mb NMC temperatures and TOMS total ozone are used to objectively determine the dates of these minimum breakups for the years 1979 to 1987. Typically, the polar total ozone and the lower stratosphere temperature minima disappear during November. The breakup dates indicate that October polar total ozone amounts are clearly related to the November breakup dates. High October polar ozone (e.g. 1979-1982) correlate with an early breakup date (mid-November), while low polar ozone (e.g. 1985 and 1987) correlate with a late breakup (early December).

1. INTRODUCTION

During the southern hemisphere (SH) spring, the total ozone field has a broad scale polar minimum surrounded by a circumpolar maximum located in the 40-60°S latitude band (London et al., 1972). During the late-spring, the polar ozone minimum increases, and is usually displaced off of the pole into the western hemisphere as the circumpolar ozone maximum intrudes onto the pole from the eastern hemisphere. As the polar ozone minimum is moving off of the pole and filling, it breaks up. Because the lower stratospheric temperature is highly correlated with total ozone, the polar ozone field breakup is accompanied by an identical evolution of the lower-stratospheric thermal field (Godson, 1963; Bojkov, 1986). This breakup is conventionally referred to as the final warming.

Recently, it has been suggested that the October Antarctic ozone depletion may be influencing the timing of the polar vortex breakup (Kiehl et al., 1988). The timing change would result from reduced heating of the polar lower stratosphere because of depleted lower stratospheric ozone mixing ratios. Since the vortex would be colder, zonal mean winds would be stronger, and these stronger zonal winds may inhibit vertical propagation of waves into the stratosphere. The net effect is the stabilization of the colder vortex against erosion by Rossby waves, resulting in a delayed polar vortex breakup.

The purposes of this study are: 1) to document the behavior of the lower stratospheric thermal and total ozone fields during the breakup, 2) develop an objective method for determining the breakup, and 3) determine whether low October ozone is influencing the breakup. The data used in this study consist of Total Ozone Mapping Spectrometer (TOMS) data (see Schoeberl et al., 1986), and National Meteorological Center (NMC) analyses (see Newman and Randel, 1988). The final warming will be diagnosed using the difference between zonal means at 80° and 50°S for temperature, ozone, and layer mean temperature. The polar vortex breakup can also be diagnosed by the onset of weak zonal mean zonal winds (i.e. [u], brackets denotes a zonal average) at 60°S.

2. ANALYSIS

The final warming can be diagnosed using a variety of meteorological quantities. The disappearance of the polar total ozone and temperature minimums can be diagnosed using station data, but this technique is limited by the spatial distribution of stations, transient waves, and interannual variability. Figure 1 displays TOMS data co-located to three stations for the SH spring of 1987: Amundsen-Scott (90°S, solid line), Halley Bay (75°30'S, 26°39'W, dotted line), and Syowa (69°S, 39°36'E, dash-dot line). In addition, the thick line displays the minimum values of the TOMS data found in the 60-90°S region for each day, with the long-short dashed line denoting the 100mb temperatures co-located to the minimum TOMS (note the good correlation of the minimum temperature and the TOMS minimum, except for September). Large transient wave events are reflected by the stations as the large ozone excursions. As in other years, Syowa shows the greatest day-to-day variability. In general, the minimum is best represented by Amundsen-Scott since it remains close to the pole over most of the spring season. The breakup period begins in mid-

ISBN 0-937194-15-8

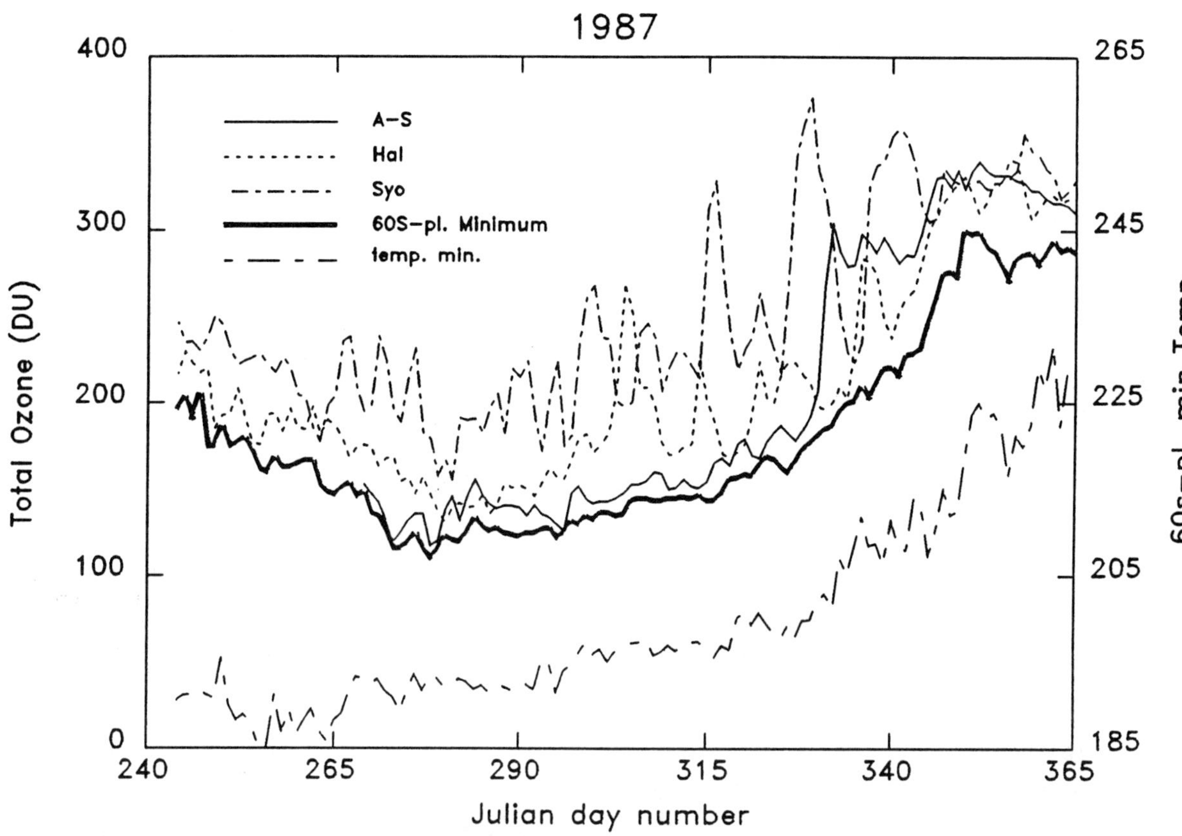

Fig 1. TOMS total ozone data co-located to various station locations plotted against the Julian date. The solid line is for Amundsen-Scott (90°S), the short dash is for Halley Bay (75°S, 27°W), and the dash-dot is for Syowa (69°S, 40°E). The heavy solid line is the minimum value of ozone for each day in the 60°S to pole region, while the long dash-short dash line is for the minimum value of the 100 mb temperature.

November, as the polar minimum rapidly increases (day 316), is displaced from the pole into the western hemisphere by November 27 (day 331), moves eastward into the Weddell Sea/South Atlantic region by December 10 (day 344), and continues to move eastward and evanesce by December 16 (day 350). Although the station data yield important information on the polar minimum disappearance, they do not provide an objective measurement of its disappearance.

Meteorological quantities provide a second source of information on the polar minimum disappearance. The polar vortex begins to erode in September at the high stratospheric altitudes; this is subsequently followed by irreversible erosion at lower levels later in the season as the jet descends. The final 100 mb breakup usually occurs in late-November (Farrara and Mechoso, 1986). The lower stratospheric temperature data is highly correlated spatially with the total ozone field, having a polar minimum with the surrounding circumpolar maximum. The final warming during 1987 is illustrated in Fig. 2a using NMC [u] at 60°S and 100 mb (long dash), [T(80°S)]-[T(50°S)] (hereafter referred to as the temperature difference) at 100 mb (solid), and the 80°-50°S 100-150 mb layer mean temperature difference (short dash). Because of the cold polar temperatures and warmer mid-latitude temperatures, this temperature difference is negative early in the spring. As the warm circumpolar air intrudes on the pole from the eastern hemisphere, this gradient reverses sign on December 9, 1987. Similarly, the 150-100 mb layer mean temperatures reverse sign on December 6, 1987. As the thermal gradient weakens, the vertical wind shear is also reduced, hence weak zonal winds (i.e. winds less than 20m/s) appear on December 12, 1987. At the higher levels, the temperature gradients reverse at an earlier date (Newman, 1986). Figure 2b displays the ozone hole breakup using the zonal mean 80°-50°S total ozone difference (solid) and the the polar map minimum total ozone values (dashed). Again, as the total ozone polar minimum increased, and the circumpolar maximum intruded onto pole from the eastern hemisphere, the 80-50°S gradient rapidly reverses on December 11, 1987. The rapid breakup is accompanied by an increase of total ozone values in the polar minimum. Other years show similar evolution with time. These sign reversals of the thermal and total ozone gradients provides the best indicators of the polar minimum breakup.

The breakup dates of the polar ozone

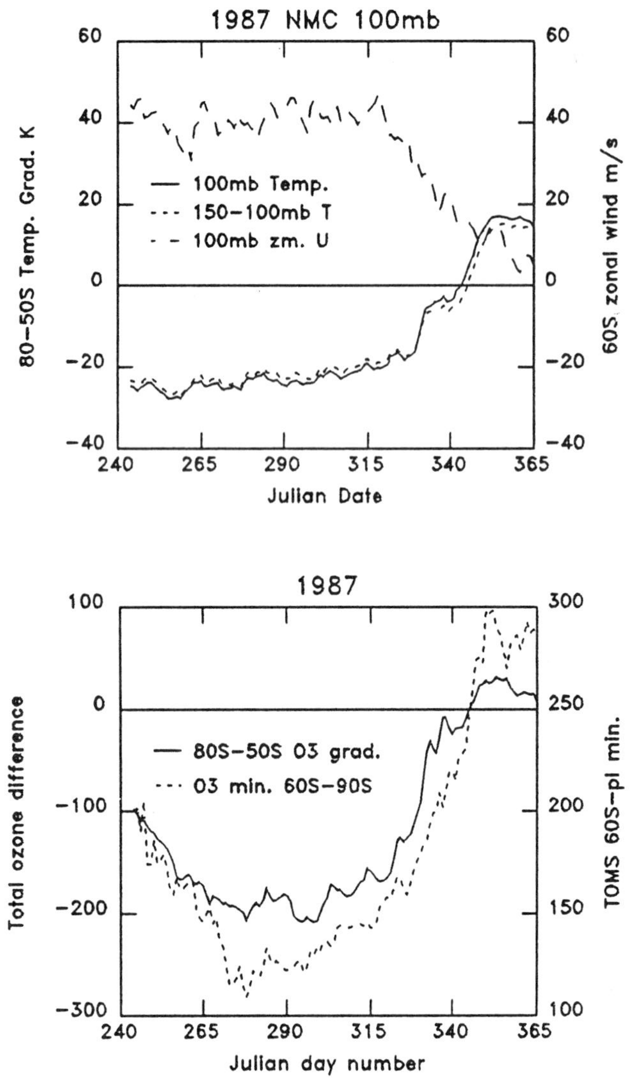

Fig. 2. Spring 1987 line plots of, top: NMC 80°-50°S 100 mb temperature difference (solid), NMC 80°-50°S zonal mean 100-150 mb layer mean temperature difference (short dash), and zonal mean zonal wind at 60°S (long dash); and bottom: TOMS total ozone 80°-50°S difference (solid), and TOMS total ozone map minimums (short dash).

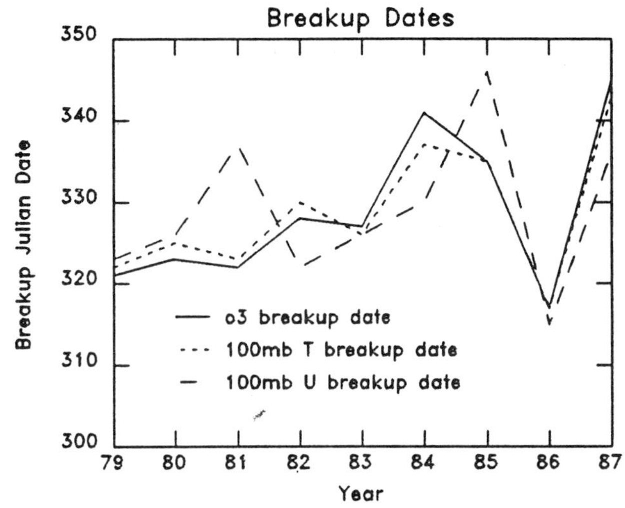

Fig. 3. Polar vortex breakup dates (Julian day) versus year as diagnosed from: a) 100 mb 80°-50° temperature differences (short dash), b) total ozone 80°-50°S zonal mean difference (solid), and c) 100 mb zonal winds (long dash).

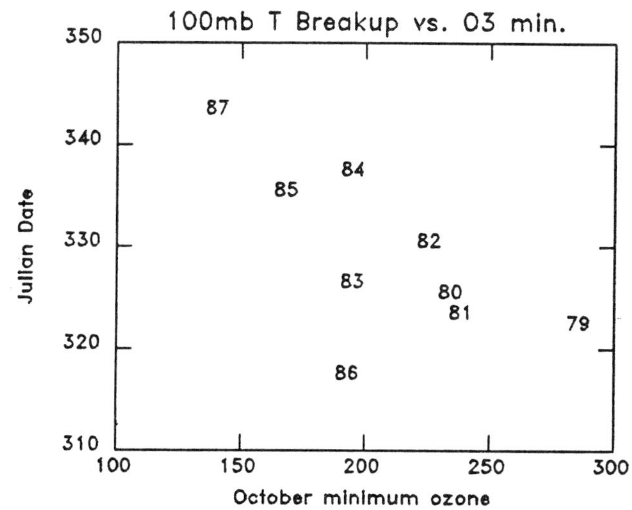

Fig. 4. Breakup date from the 80°-50°S 100mb temperature gradient plotted against the TOMS October average map minimum. The single outlier (86) should be displaced to approximately day 330, since the vortex was highly displaced, and the objectively determined breakup date does not represent the true disappearance of the polar minimum.

minimum for all years 1979-1987 are shown in Fig. 3 using the 80°-50°S zonal mean total ozone differences (solid), 100 mb temperature differences (short dash), and the onset date of zonal mean winds weaker than 20 m/s at 100 mb and 60°S (long dash). The weak winds onset dates do not show a clear general trend toward a later breakup date because of the large interannual variability. The total ozone and temperature gradients show strong coherence, and also show a general trend toward later final warmings. The earliest breakup seems to have taken place in 1986 (a year of relatively low ozone), but further analysis of TOMS data indicated that the polar minimum was highly displaced from the pole and remained intact for an additional 10-15 days. Comparisons at other stratospheric levels are inconclusive because of the large interannual variance of the breakup date and the fact that total ozone reflects lower stratospheric processes.

Finally, Fig. 4 shows a plot of the breakup date (as diagnosed from the 100 mb temperature difference, see the short dashed line in Fig. 3) versus the total ozone October map minimum value. Generally, the figure reveals that reduced ozone in October is associated with a later breakup date. This trend is significant at the 95% confidence interval (CI). When the breakup is diagnosed from temperatures at higher altitudes, the trend is not significant. The single outlier point is 1986, but as mentioned earlier, 1986 was a year in which the minimum persisted longer than is indicated in the graph. If the breakup date for 1986 is displaced 10-15 days, the linear data fit is improved.

Individual plots of TOMS total ozone data for late November (not shown here) indicate that the ozone minimum remains intact for a longer period in later years of the TOMS data set (i.e. 1985, 1986, 1987) than earlier years (i.e. 1979, 1980, 1981, and 1982).

3. SUMMARY AND CONCLUSIONS

Computations of polar minimum breakup dates using NMC meteorological data and TOMS total ozone data indicate that the breakup is occurring later in the spring in the lowest portion of the stratosphere. At altitudes above 100 mb, large interannual variance inhibits the analysis of a clear trend of breakup dates. Individual plots of TOMS total ozone show that the total ozone minimum remains intact for a longer period of time than is observed in earlier years.

REFERENCES

Bojkov, R. D., 1986: Spring-Ozone Change in Antarctica and the Role of the Polar Vortex, Adv. Space Res., 6, 89-98.

Farrara, J. D., C. R. Mechoso, 1986: An Observational Study of the Final Warming and Polar Vortex Disappearance During the Southern Hemisphere Spring, Geophys. Res. Lett., 13, 1232-1235.

Godson, W. L., 1963: A Comparison of middle stratosphere behavior in the Arctic and Antarctica with special reference to the Final Warming, Meteor. Abhandl., 36, 161-206.

Kiehl, J. T., B. A. Boville, B. P. Briegleb, 1988: Response of a General Circulation Model to a Prescribed Antarctic Ozone Hole, submitted to Nature.

London, J., R. D. Bojkov, S. Oltmans, J. I. Kelley, 1976: Atlas of the Global Distribution of Total Ozone, July 1957 - June 1967, NCAR Tn 113, 276 pp.

Newman, P. A., 1986: The Final Warming and Polar Vortex Disappearance During the Southern Hemisphere Spring, Geophys. Res. Lett., 13, 1228-1231.

Newman, P. A., W. J. Randel, 1988: Coherent Ozone-Dynamical Changes During the Southern Hemisphere Spring, 1979-1986, in press J. Geophys. Res.

Schoeberl, M. R., A. J. Krueger, P. A. Newman, 1986: The Morphology of Antarctic total ozone as seen by TOMS, Geophys. Res. Lett., 13, 1217-1220.

OZONE AND STRATOSPHERIC TEMPERATURE CHANGE IN ANTARCTICA CORRELATED TO THE SOLAR ACTIVITY

Yoshiro Sekiguchi
Japan Weather Association
Tokyo, Japan

ABSTRACT

Total ozone data obtained by ground-based measurements (Dobson spectrophotometer) at the Japanese Antarctic Station, Syowa (69° 00′ S, 39° 35′ E), are analyzed together with the lower stratospheric temperature data at the same site. The data period is 22 years(1966-87), corresponding to the two sunspot cycles.

Harmonic components of 22-year and 11-year period, corresponding to the predominant periods of sunspot cycles, are major features of the power spectra of ozone and stratospheric temperature. The contribution of these components is estimated to exceed 30 % of the temperature or ozone variance in the spring.

The phase relation between the solar cycle and corresponding components of temperature or ozone variations are found to be compatible with that in the lag correlation between these quantities, as obtained by the present author. The phase relation between the sunspot cycle and ozone and the stratospheric temperature at Syowa suggests the influence of solar activity not only on ozone but also on the atmospheric circulation.

It is of note that the influence of the quasi-biennial oscillation is found in the ozone and temperature variations. The contribution of components of which periods of 2.4, 2.2 and 2 years in the temperature variance in October are estimated to be 37 %.

1. INTRODUCTION

The present author demonstrated the existence of a relation between the sunspot cycle and the antarctic ozone amount [Sekiguchi, 1987]. He calculated the lag correlation between the relative sunspot number (S.N.) and the total ozone amount in the spring at Syowa (Ω) . The peak negative correlation (r= - 0.70) was found at a lag of three years of Ω to S.N.(the maximum of Ω appears about 2 to 3 years before the maximum of S.N.), and r shows a clear cyclic change with a period of 11 years.

In the past, interrelation between solar activity and atmospheric ozone has been discussed by a number of investigators. Among them, Willett (1962) presented certain statistical evidence of a significant relationship between the global total ozone and the relative sunspot number, for a 27-year period, 1933-1959 inclusive. His study was widely debated because the data of total ozone he used were geophysically and chronologically spotty. Willett and Prohaska (1965) found, however, a similar relationship with little reduction of significance for single stations (Arosa and Tromso), suggesting that the previous result by Willett could well be reliable. It is noteworthy that there is a slight difference between the lag relation of their two consecutive studies. Willett revealed that the peak positive correlation (r = + 0.74) appeared at a lag of about 4 years of ozone relative to the sunspots. This fact may suggest that the global atmospheric ozone is relatively abundant during the quiet sun period. In the second paper it was indicated that the peak positive correlations (r = +0.70) appeared at lags of about 2.5 to 3 years in the case of two single stations, suggesting that ozone relative abundance appears at higher latitudes during the period when the solar activity is decreasing but still moderate.

The present author recently made a similar analysis for Tateno, Japan, where ozone data have been accumulated for 30 years. The result was similar to that for Arosa and Tromso for the period, 1962-1980 inclusive, the peak positive correlation (r = + 0.52) appearing at a lag of 3 years of ozone relative to sunspots, while r = + 0.46 appeared at a lag of only 1 year for the 22-year period, 1966-1987 inclusive. This fact suggests that the linkage between the solar activity and ozone, if it exists, is likely rather complicated. The initial impact of the variation of the solar ultraviolet radiation on the ozone amount might occur in the upper stratosphere, and then atmospheric circulation, which could be influenced by the solar activity through some unknown complicated mechanism, might possibly redistribute ozone in the whole stratosphere. Therefore, the phase relation between the solar activity and ozone need not be invariable. The circulation in the stratosphere is expected to be more active in the higher latitudes, so that the earlier appearance of ozone increase in the higher latitudes might be reasonable. Further. the phase between sunspots and ozone might be variable according to the impact of the atmospheric circulation, whose phase relation to sunspots also could also be variable.

The present study will pursue the relation between ozone at Syowa and sunspots by means of harmonic analysis, taking into consideration the results mentioned above.

2. HARMONIC ANALYSIS-SUNSPOT NUMBER

In this study the relative sunspot number is

ISBN 0-937194-15-8

adopted as the index of solar activity. The 22-year time series of the annual mean relative sunspot number (S.N.) from 1966 to 1987 is analyzed by means of Fourier series . In spite of availability of S.N. data for a longer period. we choose two sunspot cycles (20th and 21st) only, corresponding to the period when the total ozone data are available at Syowa. In order to confirm that this period was in no way unusual, harmonic components of 22- and 11-year periods obtained with 44-year S.N. data (1937-80) were compared with those obtained with the 22-year S.N. data mentioned above and it was found that there is little difference between them. The composite of the two harmonic components of 22- and 11-year periods is shown together with the original time series in Fig.1. The 22- and 11-year period components of S.N. are expressed by the following formulae respectively:

$$(S.N.)_{22} = 15.12 \cos [\ \pi (t-1965)/11+2.17] \qquad (1)$$
$$(S.N.)_{11} = 60.48 \cos [2\pi (t-1965)/11-2.42] \qquad (2)$$

where t are years A.D. The phase angles of 2.17 and 2.42 correspond to 7.6 and 4,2 years, respectively.

The powers of 22- and 11-year components of S.N. are 114 and 1829 , respectively, representing 5 and 85 % of the total power(= 2147). The phase difference between 11- and 22-year components results in the enhancement of the maximum of S.N. towards 1980.

3. HARMONIC ANALYSIS-TOTAL OZONE

The total ozone amount has been measured at Syowa since 1966 except in 1973 . We have a 22-year time series of the monthly mean total ozone (Ω), provided that it is possible to estimate the value of Ω in 1973. As shown by Sekiguchi (1986), the highly positive correlation exists between Ω and the monthly mean stratospheric temperature (T) at Syowa, especially in the spring. The correlation coefficients between Ω and T at 100 hP are 0.92, 0.96 and 0.80 for Oct.,Nov. and Dec., respectively. Ω is estimated to be 299, 328 and 360 D.U. for Oct., Nov. and Dec. 1973, respectively, using the following regression equations:

$$\Omega = 12.4T\ (℃) + 1160\ (D.U.) \quad \text{October} \qquad (3)$$
$$\Omega = 7.2T\ (℃) + 734\ (D.U.) \quad \text{November} \qquad (4)$$
$$\Omega = 7.4T\ (℃) + 688\ (D.U.) \quad \text{December} \qquad (5)$$

Thus we obtain the 22-year consecutive data set of Ω at Syowa for harmonic analysis.

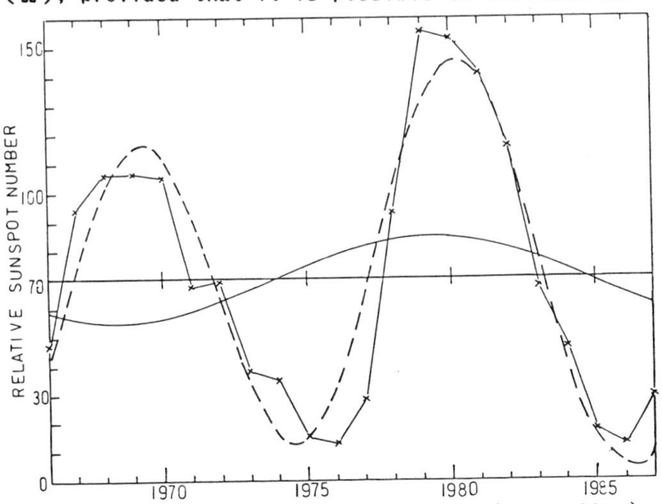

Fig. 1. Year-to-year change of S.N. (solid line). Broken curve: $(S.N.)_{22} + (S.N.)_{11}$; Solid curve: $(S.N.)_{22}$.

As well known, there exists strong decresing tendency in the antarctic ozone, especially in recent years. When harmonic analysis is applied to such a small sample as 22 year data, we need preliminary trend removal from the original data in advance of analysis. The actual trend in the ozone data at Syowa can be seen in Fig. 2. By applying a parabolic trend removal to the original time series of ozone data, we may remove not only a linear trend but also a trend of a half cycle of about 44-year period. A parabolic trend in ozone time series of each month is obtained by means of the least square method (as a quadratic curve fitting with the original data set).

After removing such a parabolic trend from the original data set, we obtained harmonic components of the 22-year time series (1966-1987) at Syowa. In Table 1 are summarized the obtained result, in which we can realize that considerable powers are concentrated in components of 11-year period. To sum up, 22, 9 and 20 % of the total power, i.e. variance, of Ω are contained in the components of 11-year period for October, November and December, respectively. The proportion of the energy in 22-year component is not so large and may not be so significant.

As the main purpose of the present study is to investigate the influence of the solar activity on the ozone depletion during the antarctic spring, our analysis will be limited to the three months of October, November and December. Although a significant ozone decrease in September is known to have occured in recent years,the number of ozone data in this month are very few in every year as a result of the poor observsational conditions. Therefore, our analysis must inevitably be limited to the three months of Oct., Nov. and Dec..

Table 1. POWER PROPORTION OF MONTHLY MEAN TOTAL OZONE $(D.U.)^2$ IN EACH HARMONIC COMPONENT AT ANTARCTIC STATION,SYOWA.

wave	period	October			November			December		
number	(years)	power	proportion		power	proportion		power	proportion	
1	22	15.9	1.4	%	29.4	4.4	%	10.7	4.0	%
2	11	248.8	21.9	24.3	58.3	8.8	13.2	54.6	20.1	24.1
3	7.3	38.2	2.8		53.2	8.0		49.2	18.5	
4	5.5	3.4	0.3		7.3	1.1		11.0	4.1	
5	4.4	166.0	14.7		18.3	2.8		33.8	12.7	
6	3.7	113.0	10.0		192.0	28.9		3.6	1.3	
7	3.3	287.1	25.4		48.6	3.6		31.8	12.0	
8	2.8	32.5	2.9		43.7	6.5		3.6	1.4	
9	2.4	87.5	7.7		8.9	1.3		42.1	15.9	
10	2.2	139.8	12.3		36.1	5.4		19.4	7.3	
11	2.0	0.7	0.0	20.0	167.8	25.3	32.0	6.8	2.6	23.2
total		1132.2			663.7			266.6		

The 22-year components of Ω for these three months are expressed by the following formulae:

$$\Omega_{22} = 5.6 \cos[\pi (t-1965)/11+2.506] \quad \text{Oct.} \qquad (6)$$
$$\Omega_{22} = 7.5 \cos[\pi (t-1965)/11+1.785] \quad \text{Nov.} \qquad (7)$$
$$\Omega_{22} = 4.6 \cos[\pi (t-1965)/11+2.296] \quad \text{Dec.} \qquad (8)$$

where t are years A.D.. The phases of 2.506, 1.785 and 2.296 correspond to 8.8, 6.3 and 8.0 years (7.7 ± 1.4 years) respectively. The 11-year components for the same months are expressed by the following formulae:

$$\Omega_{11} = 22.3 \cos[2\pi(t\text{-}1965)\text{-}0.754] \quad \text{Oct.} \qquad (9)$$
$$\Omega_{11} = 10.8 \cos[2\pi(t\text{-}1965)\text{-}1.101] \quad \text{Nov.} \qquad (10)$$
$$\Omega_{11} = 10.5 \cos[2\pi(t\text{-}1965)\text{-}0.641] \quad \text{Dec.} \qquad (11)$$

where phase angles 0.754, 1.101 and 0.641 correspond to 1.3, 1.9 and 1.1 years (1.4±0.5 years). As seen in Table 1, the shorter-period components of Ω such as τ = 2.4, 2.2 and 2.0 years (nearly biennial) comprise a relatively large proportion of the total variance.

4. HARMONIC ANALYSIS-STRATOSPHERIC TEMPERATURE

Data of the stratospheric temperature (T) at Syowa are available for a period longer than that for ozone. The harmonic analysis of the temperature in the lower stratosphere, however, is confined to the same period as that for Ω, in order to guarantee the compatibility between them. We selected the monthly mean temperature of the 100-hP level (T{100}) as representing the temperature field in the lower stratospher. The parabolic trend removal was applied to each original temperature time series in advance of the harmonic analysis. The analyzed result is summarized in Table 2.

The harmonic components of T{100}, for a period of 22 years, for the three months are expressed by the following formulae, respectively:

$$T\{100\}_{22}=1.47\cos[\pi(t\text{-}1965)/11+1.610] \quad \text{Oct.} \qquad (12)$$
$$T\{100\}_{22}=1.12\cos[\pi(t\text{-}1965)/11+1.154] \quad \text{Nov.} \qquad (13)$$
$$T\{100\}_{22}=0.97\cos[\pi(t\text{-}1965)/11+1.588] \quad \text{Dec.} \qquad (14)$$

where phase angles 1.610, 1.154 and 1.588 correspond to 5.6, 4.0 and 5.6 years (5.1±1.1years), respectively. The corresponding components of T{100}, for a period of 11 years, for the same three months are expressed by the following formulae:

$$T\{100\}_{11}=1.95\cos[2\pi(t\text{-}1965)/11\text{-}1.038] \quad \text{Oct.} \qquad (15)$$
$$T\{100\}_{11}=1.13\cos[2\pi(t\text{-}1965)/11\text{-}1.602] \quad \text{Nov.} \qquad (16)$$
$$T\{100\}_{11}=0.84\cos\{2\pi(t\text{-}1965)/11\text{-}2.978] \quad \text{Dec.} \qquad (17)$$

where phase angles 1.038, 1.602 and 2.978 correspond to 1.8, 2.8 and 1.5 years (2.0 ±0.8 years), respectively. The shorter period components of T{100} such as τ = 2.4, 2.2 and 2.0 years contain a considerable proportion of its total power, i.e. variance, as for Ω, as seen in Table 2.

5. DISCUSSION

The years of the sunspot maximum are determined as 1969-70 and 1980-81 from the relation of (1). The composite of 22- and 11-year components, is also has its maxima at the same years as of the 11-year component, as shown in Fig. 1. The contribution of 22-year component to the actual variation of S.N. is only the enhancement of the maximum toward 1980-81. The actual maxima of S.N. appear at 1968-70 and 1979-80, nearly half a year earlier than those of the composite of two components, resulted from the influence of contribution of shorter period components.

As seen in Figs. 2 and 3, the composite of 11- and 22-year components of both Ω and T{100} can express the general tendency of the year-to-year variation of both quantities, suggesting that the variation of the solar activity seems to activate the variation of ozone amount to some extent. The maxima of Ω_{11} appear in 1966-67 and 1977-78, 1966-67 and 1976-77, and 1965-66 and 1976-77 for Oct., Nov. and Dec. according to relations (9), (10) and (11), respectively. Accordingly, the maxima of Ω_{11} appear two year or one earlier than those of $(S.N.)_{11}$. This relation between $(S.N.)_{11}$ and Ω_{11}

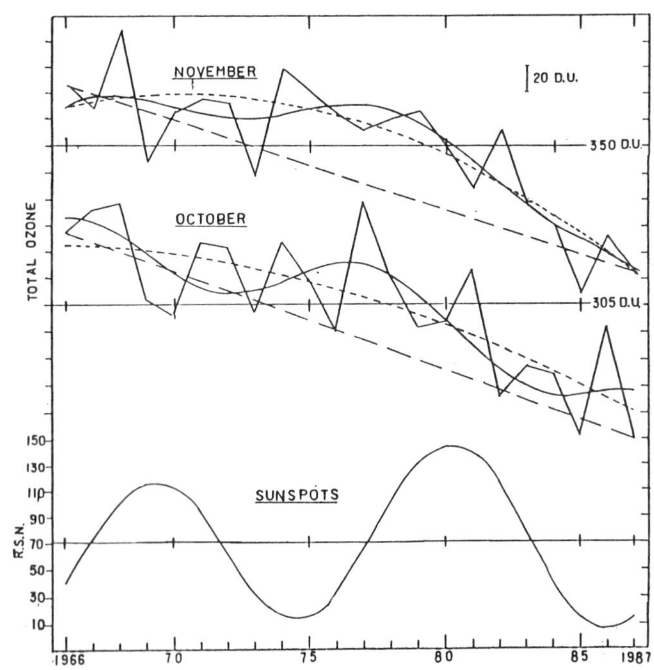

Fig. 2. Year-to-year change of Ω for Oct.and Nov. Solid curve: $\Omega_{11}+\Omega_{22}$. $(S.N.)_{11}+(S.N.)_{22}$ is shown at the bottom for comparison.

Table 2. POWER PROPORTION OF MONTHLY MEAN TEMPERATURE AT 100-HP (DEGREE)^2 IN EACH HARMONIC COMPONENT AT SYOWA.

wave	period	October			November			December		
number	(years)	power	proportion		power	proportion		power	proportion	
1	22	1.08	10.5	%	0.63	4.2	%	0.47	22.4	%
2	11	1.90	18.5	29.0	0.64	4.2	8.4	0.35	16.8	39.2
3	7.3	0.47	4.6		0.52	3.5		0.46	22.0	
4	5.5	1.01	9.9		0.84	5.7		0.38	17.9	
5	4.4	0.25	2.4		0.27	1.8		0.03	1.4	
6	3.7	0.91	8.8		6.06	41.0		0.01	0.3	
7	3.3	0.64	6.2		0.71	4.8		0.08	3.7	
8	2.8	0.21	2.0		1.31	8.9		0.07	3.3	
9	2.4	2.05	20.0		0.01	0.1		0.03	1.3	
10	2.2	1.72	16.8		1.25	8.5		0.23	10.8	
11	2.0	0.02	0.2	37.0	2.64	17.9	26.5	0.00	1.3	13.4
total		10.25			14.79			2.10		

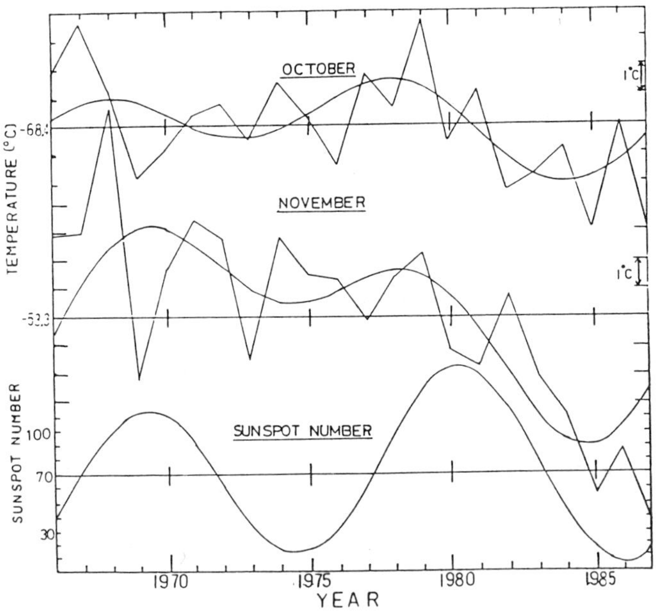

Fig. 3. Year-to-year change of T_{100} for Oct. and Nov. Solid curve: $T\{100\}_{11} + T\{100\}_{22}$. $(S.N.)_{11} + (S.N.)_{22}$ is shown at the bottom for comparison.

is consistent with the lag correlation for which the negative peak correlation appears nearly three years after the sunspot maximum [Sekiguchi, 1987].

The maxima of Ω_{22} appear in 1978-79, 1980-81 and 1978-79 for Oct., Nov. and Dec. according to relations (6),(7) and (8), respectively. Similarly minimum of Ω_{22} appears in 1967-68, 1969-70 and 1969-70 for Oct., Nov. and Dec., respectively. The amplitudes of Ω, however, are so small that the influence of the 22-year cyclic variation of the solar activity must be ignored at least as far as the 22-year ozone or temperature data at Syowa are concerned.

As expected from the high correlation between $T\{100\}$ and Ω, the phase relations both between $(S.N.)_{11}$ and $T\{100\}_{11}$ and between $(S.N.)_{22}$ and $T\{100\}_{22}$ are realized to be similar to those between S.N. and Ω (cf. Figs. 2 and 3).

In Fig. 4 are shown the year-to-year variation of the annual mean total ozone $\langle\Omega\rangle$ at Tateno, Japan, where is no remarkable linear or parabolic trend. Therefore, we applied harmonic analysis to the original time series of $\langle\Omega\rangle$ at Tateno. The formulae expressing $\langle\Omega\rangle_{11}$ and $\langle\Omega\rangle_{22}$ are as follows:

$$\langle\Omega\rangle_{22} = 1.29 \cos[\ \pi(t-1965)/11+1.286] \qquad (19)$$

$$\langle\Omega\rangle_{11} = 4.10 \cos[2\pi(t-1965)/11+2.549] \qquad (20)$$

where phase angles of 1.286 and 2.549 are equally 4.5 years for $\langle\Omega\rangle_{22}$ and $\langle\Omega\rangle_{11}$, respectively. This periodical change of $\langle\Omega\rangle$ is nearly in phase

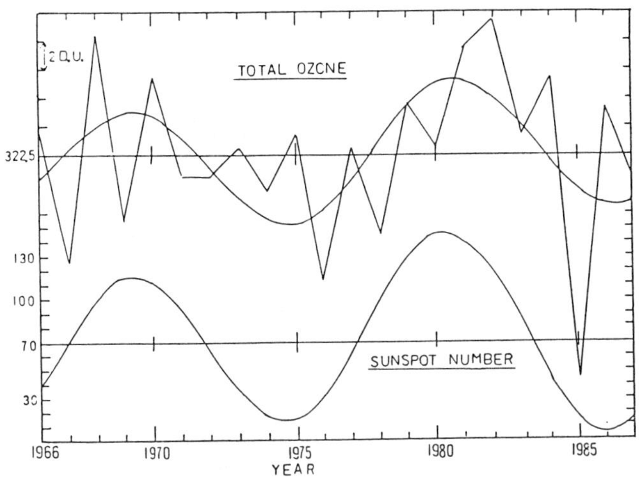

Fig. 4. As in Fig. 2 for annual mean total ozone at Tateno, Japan.

with the sunspot cycle , while phases of $\langle\Omega\rangle$ for Syowa are advanced about two years from those of (S.N.). The difference of the phase relation at different locations between Ω and S.N. must be the problem to be studied in the future.

The present study could not clearly reveal the influence of the solar activity on the decrease of the ozone amount at Syowa. We need accumulate ozone data in a longer period at a number of stations in the Antarctic in order to reveal the mechanism of tha antarctic ozone depletion. In conclusion, we need assess significanc of the evaluated result in order to discuss the interrelation between S.N. and Ω in detail.The present analysis, however, has been dealing with data of small sample , so that further inquiry should be left until sufficient data will be accumulated.

References

Sekiguchi,Y., Antarctic ozone change correlated to the stratospheric temperature field, Geophs. Res. Lett., 13, 1202-1205,1986

Sekiguchi,Y., Antarctic ozone change correlated to the stratospheric temperature field and the solar activity,Proc. NIPR Sympo.,Polar Meteorol.Geociol.,1 18-23;1987

Willett, H.C., The relationship of total ozone to the sunspot cycle, J.Geophys.Res.,67, 661-670, 1962

Willett, H.C.and Prohaska,J., Further evidence of sunspot-ozone relationship, J. Atmos. Sci.,22, 493-497, 1965

INTERANNUAL VARIATIONS OF ZONAL MEAN TEMPERATURE IN THE SOUTHERN HEMISPHERE STRATOSPHERE: A LINK TO THE ANTARCTIC OZONE DEPLETION

Kohji Kawahira
Toyama National College of Technology, 13 Hongo-cho, Toyama 939, Japan

Toshihiko Hirooka
Meteorological College, 7-4-81 Asahi-cho, Kashiwa-city 277, Japan

ABSTRACT

Interannual variations of the daily zonal mean temperature in the Southern Hemisphere stratosphere are analyzed for the past seven years(1980-86) by using NMC data. It is found that the Antarctic lower stratosphere tends to cool nearly all months, more noticeable in winter and spring seasons. From these evidences the possible interrelation between the temperature decline and the ozone depletion has been discussed.

1. INTRODUCTION

It is well known that the year-to-year spring ozone depletion in the Antarctic lower stratosphere, especially in October, is accompanied by a temperature decline(Angell, 1986; Kawahira and Hirooka, 1987). The temperature decline has been a basis of ideas to account for the cause of the ozone hole. Tung et al.(1986) proposed the ozone depletion due to diabatic upward flow in the lower stratosphere; the upward flow becomes more remarkable as the temperature falls. The heterogeneous reactions to reduce ozone necessitates the polar stratospheric clouds which appears in lower temperature below 195 K(Solomon et al, 1986; McElroy et al., 1986). It is important for studies of the cause of the ozone hole to analyze the detailed behavior of temperature changes in the Southern Hemisphere(SH) stratosphere.

The aims of the analysis is to clarify what temperature changes have occured during the past seven years, when ozone depletion was apparent in spring season. Furthermore the daily temperature changes in the lower stratosphere including middle and low latitudes would show how temperature in the region north of the Antarctic did change and in which period these changes are significant.

In the present paper, the geopotential heights at 100 and 200 mb analyzed by the National Meteorological Center(NMC) are used to calculate stratospheric temperatures from the relative thickness of the layer.

2. INTERANNUAL TEMPERATURE CHANGES

Figure 1 shows the interannual changes of the zonal mean temperature in the layer between 100 and 200 mb levels(center is at about 14 km) calculated from the geopotential thickness, for each year 1980 to 1986.

On this figure it is clearly seen that the region with temperature over 230 K was seen in high latitudes in January, February and even December in the year 1980, but in the year 1985 such region disappeared. Furthermore, if one follows the isoline of 215K seen at about 50°S in the winter of 1980, the line gradually moved equatorward as the year progressed; the temperature in the midlatitude of the 1985 winter became lower than 215 K.

As seen from the change of the thick line(195 K) the winter polar region remakably tends to cool. This tendency of the cooling is strongest in April to November. Temperature under 195 K is adequate to the formation of the Polar Stratospheric Clouds(PSC)(McCormick et al., 1986), the figure shows that the conditions favoring formation of the PSC becomes longer and its area also become wider in recent years.

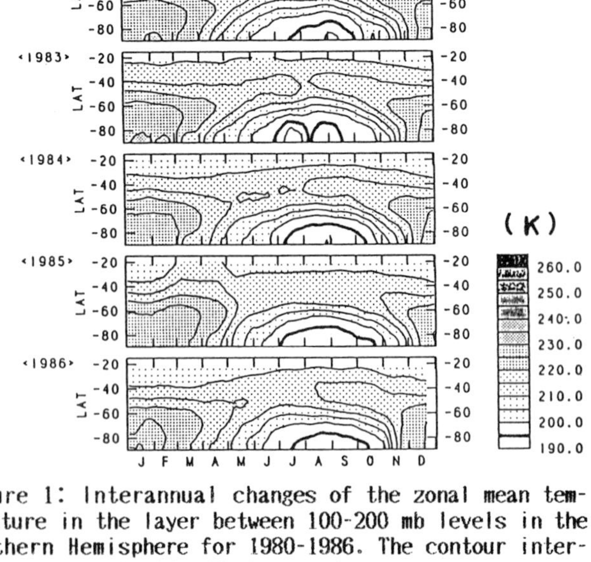

Figure 1: Interannual changes of the zonal mean temperature in the layer between 100-200 mb levels in the Southern Hemisphere for 1980-1986. The contour interval is 5 K smoothed by 20-day running means.

ISBN 0-937194-15-8

During 1980-1986 the strongest decline is seen in 1985. The temperature difference between the 1980 and 1985 are shown in Figure 2 to see time and space feature of the cooling tendency. Shaded region indicates the temperature in 1985 is lower than in 1980. The results show remarkable temperature difference in the winter polar region. It is apparent that in nearly all seasons and nearly all latitudes poleward of 20°S the cooling has occurred from 1980 to 1985 except March and April. Temperature decline more than 4 K are seen in June, August, October, November and December at the polar region.

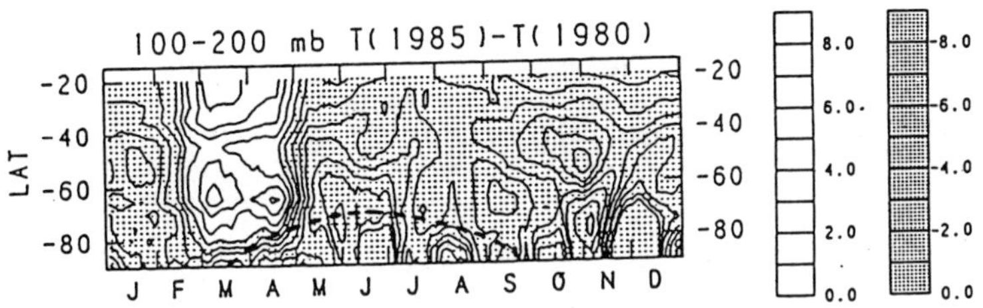

Figure 2: Temperature differences from 1985 to 1980. The contour interval is 1.0 K. Shaded area means the temperature decline.

The thick dotted line is the boundary of the polar night. Then it is clear that the cooling in the polar night is more remarkable compared to other season. This feature may suggest the cooling may be caused not by less chemical heating due to ozone depletion, but by dynamical effects due to less poleward heat transport or by enhanced cooling due to radiative effects.

Remarkable temperature changes in the Antarctic polar region may have an important effect on the spring ozone depletion. The quantitative analysis of the cooling in the polar region is done by determining how long is the period per year when it is cooler than specified temperature. The results are shown in Figure 3a, b and c. For example, the period under 205 K at 80°S(Fig.3a) changed 130 days duration in 1980 to 170 days in 1985. The expanding of the period with below 195 K is seen in Figure 1; it changes from about 20 days in 1980 to about 70 days in 1985. Therefore, the period favoring the formation of PSC at 80°S becomes recently longer.

The thick line shown in Figure 3 is the year-to-year change of October monthly mean total ozone averaged poleward of 40°S (Newman and Schoeberl, 1986). The year-to-year changes of the temperature and ozone present similar tendency, suggesting possible relation between them.

Same results as Fig.3a but for 70 and 60°S are shown in Fig.3b and c. At 70°S, the period under 200 K changed gradually but slight oscillation like as quasi-biennial oscillation is seen. On the othar hand, the period under 210 K changed about 140 days in 1980 to 180 days in 1985. At 60°S, the remarkable chages are seen in the period under 215 K. However, those of 210 K show gradual expansion with some exceptions. When one compare the period changes with the total ozone changes, similar tendency is found between them as seen in Figure 3a.

The cooling tendency shown in Figure 3 occurred in nearly all seasons at 80°S, but at 70 and 60°S only in winter and spring seasons as noticed also in Figure 1. It is noteworthy that the period with temperatures below than 195 K, adequate for the PSC formation, becomes longer at 80 °S; this tendency is coinciding with spring ozone hole deepening. This strengthened the view that the PSC have increased and may play an important role in the ozone depletion in the Antarctic spring.

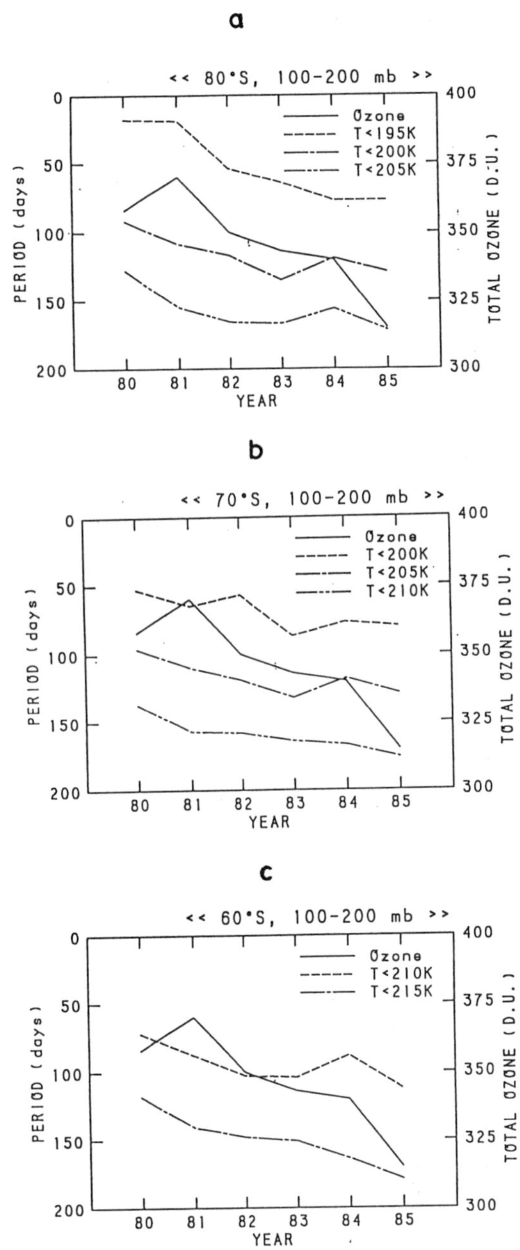

Figure 3: Year-to-year variations of the period during which the temperature is under 195, 200 and 205 K, respectively at 80°S(70 and 60°S). The averaged October monthly mean total ozone south of 40°S(after

Newman and Schoeberl, 1986) is also shown.

3. INTERANNUAL CHANGES OF THE MERIDIONAL GRADIENT OF ZONAL MEAN TEMPERATURES

The analysis of temperature changes has clearly showed that temperature decline becomes stronger at the higher latitude. This suggests that the meridional gradient of the mean temperature also excites year-to-year variation; the gradient may have important bearing in considering the cause of the ozone hole, because vertical shear of geostrophic zonal wind is proportional to the meridional gradient of the mean temperature. Therefore, the change of the gradient may provide an evidence of dynamical effects on the Antarctic spring-ozone depletion.

Figure 4 presents the interannual changes of the meridional temperature gradient at the 100-200 mb layer, where the positive(negative) sign corresponds to the temperature decrease (increase)toward the south pole(the equator). From now on, we define the onset of the winter(summer) pattern is the beginning of the positive(negative) gradient. For example, in 1980 the change at 80°S occurred in late November or early December, but it happened in late December in 1985. This clearly demonstrate a delay of the beginning of final warming during most recent years.

Also the onset of the winter pattern has occurred earlier in recent years. This change means that the westerlies in the Antarctic lower stratosphere tend to persist for longer period per year. The longer persisting strong westerlies in the winter polar region

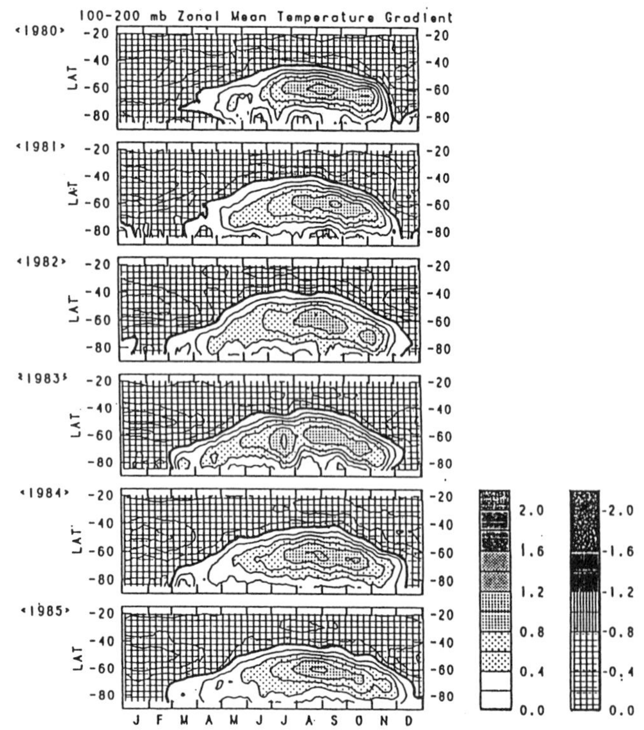

Figure 4: Interannual variation of the meridional gradient of the zonal mean temperature at 100-200 mb levels of the Southern Hemisphere. The contour is K/latitude.

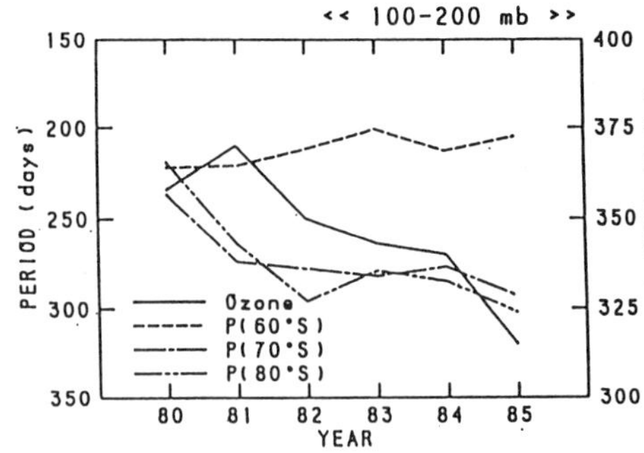

Figure 5: The interannual variations of the period during which the positive meridional gradient of the zonal mean temperature(winter pattern) persists at 80, 70 and 60°Srespectively. Total ozone trend as Fig.1 is also shown.

have a important effect on the poleward propagation of planetary-scale disturbances via inhibition of further penetration into the polar region (e.g., Kawahira, 1982).

Figure 5 shows the length of period per year when the positive gradient, i.e., the winter pattern persists. Although the time at 60°S is nearly constant, the period at 70 and 80°S becomes longer: the period at 80°S in 1985 is about 300 days per year but in 1980 only 240 days, i.e., has expanded by about 2 months. This feature is very interesting and important when we consider possible causes of the ozone hole.

The expansion of the winter pattern may result in inadequate condition for poleward propagation of large scale waves, which also transport ozone toward the pole(e.g., Kawahira, 1982; Bojkov, 1986). These notions are consistent with recent analysis on the dynamical condition at the ozone hole period(Yamazaki, 1987) showing the less poleward erosion of the trajectory of air parcel due to strong circumpolar westerlies.

4. DISCUSSION AND SUMMARY

The observation analysis in the present paper suggest a very intimate relationship between the deepening of the spring-ozone hole and the year-to-year temperature decline in nearly all seasons of the SH lower stratosphere. In the study of possible mechanism causing the ozone hole, it is important to consider what is the links between the ozone depletion and the temperature decline in the polar region and what induces the continuous temperature decline.

The recent temperature fall in the Antarctic lower stratosphere is low enough to produce PSC(McCormick and Larsen, 1986). When PSC are formed, its effect on radiative flux causes the cooling in the polar lower stratosphere as demonstrated by numerical analysis of Pollack and McKay(1985); the PSC causes more cooling in higher latitudes, especially after the polar night. This chain of effects may be a cause of the stratospheric temperature decline in Antarctic spring. This cooling in high latitudes induces longer persistence of strong westerlies through thermal wind relation. This wind can play a role of trapping of meridional propagation of planetary waves from middle to higher latitudes. As a result, poleward transport of ozone and heat by the waves may be reduced (Kawahira, 1982; Bojkov, 1986). This situation can contribute to both partial ozone depletion and temperature decline in the polar region.

On the other hand, Shine(1986) proposed that the temperature decline is due to ozone depletion. This view can not accommodate the fact observed in the present analysis indicating year-to-year cooling in winter high latitudes, i.e., in the period of polar night during which there is not uv ozone heating not contribution to any temperature changes. If the origin of the temperature decline is elucidated, the ozone depletion may be caused by chemical and dynamical effects being proposed: both processes can well contribute to the depletion when the temperature is low enough to form the PSCs.

It is very important question why the temperature decline has occurred remarkably since the year of about 1979. From the results in the present paper, it can be possible that recent increase of greenhouse gases as CO_2,CFC-11, CFC-12, CH_4 and N_2O could induce the temperature decline(Kawahira and Hirooka, 1987). This speculative assumption seems to be an important problem to be clarified in near future, because recent remarkable increase of the greenhouse gases is apparent(a review of Solomon, 1988).

The main results in the present study are:

(1) The temperature in the Antarctic lower stratosphere has been decreasing in nearly all months;

(2) Characteristics of the temperature gradient changes show a recent expanding of winter pattern(temperature decrease toward the south pole for longer time); which suggest that strong westerlies persist longer year after year;

(3) The deepening of the ozone hole is considered as partially due to planetary wave effects; the wave transport of ozone and heat toward the pole can be reduced by enhanced westerlies due to stronger meridional temperature gradient induced recent remarkable cooling in the Antarctic polar region.

ACKNOWLEDGEMENTS

The authors thanks to Dr.R.D.Bojkov, for his invaluable comments.

REFERENCES

Angell, J.K., 1986: The close relation between Antarctic total-ozone depletion and cooling of the Antarctic low stratosphere, Geophys. Res. Lett., 13,1240-1243.

Bojkov, R.D., 1986: Spring-ozone change in Antarctica and the role of polar vortex, Adv. Space Res., 6(10), 89-98.

Bojkov, R.D., 1988: Polar ozone changes and the role of the stratospheric vortex, Proc. Ozone Symposium-88 Gottingen, Bojkov & Fabian Edit., Deepak Publ. Co.

Kawahira, K., 1982: A quasi-one-dimensional model of the ozone transport by planetary waves in the winter stratosphere,J.Meteorol. Soc. Jpn.,60, 831-849.

Kawahira, K. and T. Hirooka, 1987: Interannual variations of zonal mean temperature in the Southern Hemisphere stratosphere: Withreference to the Antarctic ozone depletion, Proc. NIPR Symp. Polar Meteorol. Glaciol.,1, 31-38.

McElroy, M. B., R. J. Salawitch, S. C. Wofsy, and J.A. Logan, 1986: Antarctic ozone: Reductions due to synergetic interactions in chlorine and bromine, Nature, 321, 759-762.

McCormick, M. P. and J. C. Larsen, 1986: Antarctic springtime measurements of ozone,nitrogen dioxide and aerosol extinction by SAM II, SAGE, and SAGE II,Geophys. Res. Lett., 13, 1276-1279.

Newman, P.A. and M. R. Schoeberl, 1986: October Antarctic temperature and total ozone trends from 1979-1985, Geophys. Res. Lett.,13, 1206-1209.

Pollack, J. B. and C. P. McKay, 1985: The impact of polar stratospheric clouds on the heating rates of the winter polar stratosphere, J. Atmos. Sci., 42, 245-262.

Shine, K. P., 1986: On the modelled thermal response of the Antarctic stratosphere to a depletion of ozone, Geophys. Res. Lett.,13, 1331-1334.

Shine, K. P., 1987: The middle atmosphere in the absence of dynamical heat fluxes, Q. J. R. Meteorol. Soc.,113,603-633.

Solomon, S., R. R. Garcia, F. S. Rowland, and D. Wuebbles, 1986: On the depletion of Antarctic ozone, Nature, 321, 755-758.

Solomon, S., 1988: The mystery of the Antarctic ozone "hole", Rev.Geophys.,26,131-148.

Tung, K. K., M. K. W. Ko, J. M. Rodriguez, and N. D. Sze, 1986: Are Antarctic ozone variations a manifestation of dynamics or chemistry, Nature, 322, 8811-8813.

Yamazaki, K., 1987: Observations of the stratospheric final warmingsin the two hemispheres, J. Meteorol. Soc. Jpn., 65, 51-66.

AIRBORNE OBSERVATIONS OF CHEMICAL CONSTITUENTS IN THE ANTARCTIC WINTER STRATOSPHERE

William G. Mankin and M.T. Coffey
National Center for Atmospheric Research
Boulder, Colorado 80307, USA

1. INTRODUCTION

Between 1979 and 1984, the October mean ozone column at Halley Bay, on the coast of Antarctica, showed a progressive decrease amounting to about 50% (Farman *et al.*, 1985); satellite data (Stolarski *et al.*, 1986) subsequently indicated that the depletion occured over the entire Antarctic continent. To investigate the source of this striking and potentially dangerous phenomenon, NASA, NOAA, NSF, and CMA organized and sponsored an airborne expedition to Antarctica (Tuck *et al.*, 1988). Aircraft were based in Punta Arenas, Chile, from 15 August to 29 September 1987. As part of an extensive measurement program an NCAR experiment remotely measured a number of gaseous constituents; O_3, NO, NO_2, HNO_3, $ClONO_2$, HCl and HF are discussed below.

2. INSTRUMENTATION AND OBSERVATIONS

During the Airborne Antarctic Ozone Expedition, we operated the NCAR Fourier transform spectrometer on board the DC-8, obtaining high resolution infrared spectra on ten flights over Antarctica. The instrument has been described by Mankin (1978). The spectrometer operates at wavelengths of 2-14 μm with an apodized resolution of 0.06 cm^{-1}.

3. ANALYSIS

For analysis, interferograms are first transformed to spectra using the fast Fourier transform routines on the Cray computers at NCAR. The relevant regions of a number of successive spectra are averaged to produce an average spectrum, with its assigned average value of latitude, longitude, and solar elevation.

The observed spectra are compared with synthetic spectra calculated using the spectroscopic line parameters from the HITRAN 1986 database (Rothman *et al.*, 1987) in the computational method of Mankin (1979). In this method, the atmosphere is divided into four isothermal layers in which the pressure and mixing ratio of absorbers vary continuously.

To compute atmospheric transmittance, we need a model for the temperature and mixing ratio profiles with altitude or pressure. For the temperature profiles, we have used the middle atmosphere reference model applicable to 80° S. from Barnett and Corney (1985), similar to that derived from Halley Bay sonde data.

Most profiles of stratospheric trace species have been measured in northern midlatitudes; they may be far from representative of the Antarctic stratosphere. For ozone we have averaged ozonesonde profiles from Halley Bay for three periods from 4 Aug. to 23 Sept. 1987, and used the average in the analysis of our ozone data for the corresponding periods. For other constituents, we have used profiles deduced from the spectra on 24 September. Since the line-of-sight amount as a function of solar elevation depends upon the distribution of absorber, we can obtain some information about the distribution, although not a detailed profile, if we measure in the same air mass at different elevations, as we did on 24 September 1987. For HCl, NO and NO_2 we have used profiles which are significantly depleted below 20 km. A profile with the bulk of HNO_3 below 20 km was used for HNO_3 analysis and produced a much more consistent retrieval for the three elevations of 24 September.

The next step in the analysis is the comparison of the observed spectra and the calculated spectra to determine which calculated spectrum is the best match for the observation. We have used two methods for this, one of which essentially compares equivalent width (integrated area of the absorption line) while the second uses nonlinear least squares curve fitting to determine the amount. The equivalent width method is used where we have a few well isolated absorption lines for the analysis, as is the case with HCl, HF, NO, and NO_2.

Least squares curve fitting is appropriate for species for which a large number of lines are available such as ozone and nitric acid.

After the line-of-sight amount is obtained by fitting the spectra, it is divided by a factor, computed from the viewing geometry and the assumed height distribution of absorber, relating the vertical column and line-of-sight amount to produce the vertical column amount reported.

The spectral region chosen for ozone retrievals,1140.0 to 1142.0 cm^{-1}, is in the O_3 ν_1 band system centered near 9.6 μm. We utilize the R_1 and R_2 lines in the fundamental absorption band of $H^{35}Cl$ in the 3 μm region. $ClONO_2$ is measured with a single

ISBN 0-937194-15-8

feature at 780.22 cm^{-1}. HF measurements use the R_1 line of the fundamental band. NO is measured with three lines in the fundamental, while NO_2 measurements are based on three groups of lines in the ν_3 band at 1600 cm^{-1}. Nitric acid is fit in both the 870 and 1720 cm^{-1} regions.

3.1 PRECISION AND ACCURACY

During a few flights we recorded more than 100 consecutive spectra in the O_3 absorption region. Variations from observation to observation give us an indication of the precision of our O_3 retrieval. The precision of the results from this analysis is about 1%. We do not have as extensive a set of contiguous observations for the other species as we do for ozone, but the precision can be estimated similarly. For HCl, the precision is approximately 5% while for NO_2 and HNO_3 it is 3% and 13% respectively.

Absolute accuracy of our retrievals depends on the accuracy of spectral line data used to calculate spectra, the assumed profiles for mixing ratio and temperature and errors in the regression fitting. We believe the O_3 line data in this spectral region to be accurate to about 10% (Pickett *et al.*, 1985).

Numerical experiments with different assumed profiles produced results differing by about 7% with elevation angle. Overall, we believe our O_3 results to be accurate to better than 10 %.

Similar error estimates for the other species lead us to believe that our errors are less than 20% for HCl, 15% for HF, 25% for $ClONO_2$, 14% for NO, 16% for NO_2, and 18% for HNO_3.

4. RESULTS

Ozone column amounts derived by the methods described above have been compared on a number of days with ground based, satellite and other airborne observations (Margitan *et al.*,1988). Good agreement is found between the various observations. Figure 1 shows the retrieved ozone columns for one day along with a contour plot of the TOMS columns for the same day.

Several other species observed show dramatic effects, compared to northern hemisphere high latitudes. HCl outside the polar vortex has values similar to its northern hemisphere values, but the column amount drops sharply inside the vortex. NO_2 and HNO_3 show similar behavior.

To put the data on a common footing, we have attempted to plot the columns as a function of the distance from the vortex boundary. The best two dimensional picture we have is the TOMS total ozone maps (Krueger, *et al.*, 1988). Therefore, we chose the 250 Dobson unit contour in the TOMS maps, near the

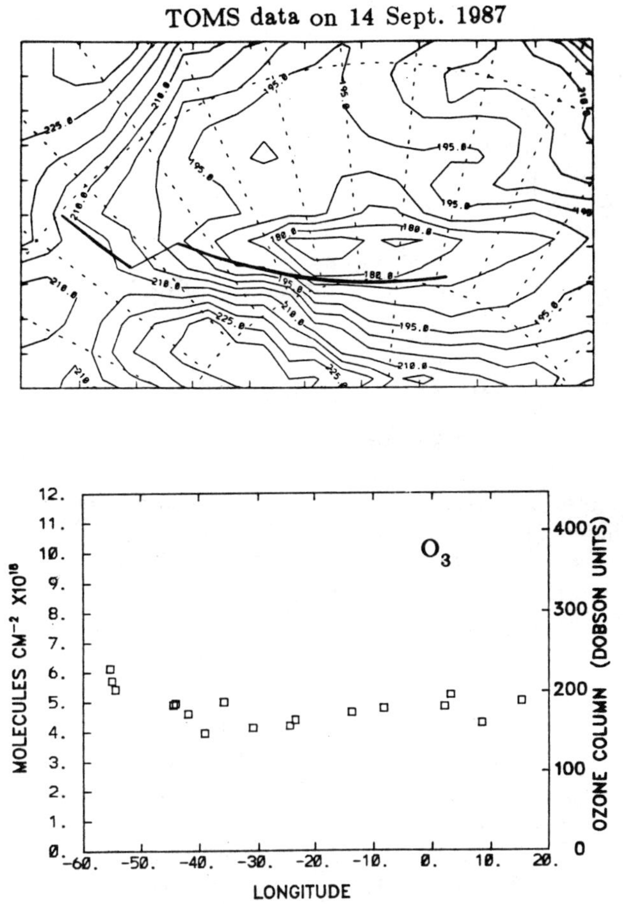

Figure 1. Observations of vertical column abundances of ozone on 14 Sept. 1987; the upper panel shows the TOMS data (Krueger, *et al.*, 1988) with the flight path of the DC-8 superimposed. Heavier line represents the portion of the path during which observations were made. The lower panel presents the results from the infrared spectrometer.

region of steepest gradient in the ozone, as a working definition of the vortex boundary. In Fig. 2, we have plotted all of the HCl column amounts as a function of the distance from the vortex boundary. Different symbols have been used for data recorded before 8 September and for later data. The overall pattern is clear. The point outside the vortex, near -1500 km, was recorded near Cape Horn on 2 September, and is very close to the corresponding values from similar northern hemisphere latitudes when the temporal rate of increase of HCl is considered (Mankin and Coffey, 1983). Near the vortex boundary, the HCl column decreases sharply, although the gradients are not as steep as shown in the TOMS ozone columns late in September. The gradient persists for a distance of ~1000 km inside the boundary and then flattens for observations near the center of the vortex. The values later in the observation period are significantly larger than the earlier values at all distances.

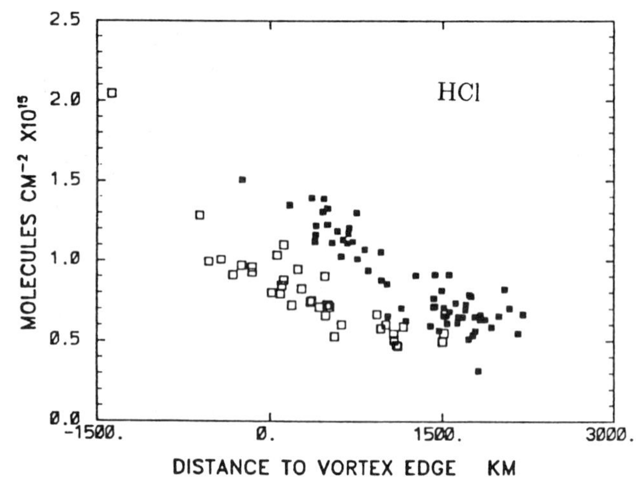

Figure 2. Measured HCl vertical column from all flights plotted as a function of distance from the vortex boundary, defined as the location of the 250 DU contour on the TOMS data. The open symbols represent data before 7 September 1987; the closed symbols, data after 7 September 1987.

Figure 3 shows a similar picture for NO_2 and HNO_3. Values outside the vortex are similar to northern hemisphere columns at the corresponding latitude and season (Coffey *et al.*, 1981); there is a transition region near the vortex edge of approximately 1000 km where NO_2 and HNO_3 show a sharp decrease. Column amounts deep in the vortex are more uniformly low.

After mid-September, the stratospheric temperatures are too high to allow the formation of PSCs and so PSC related chemistry comes to an end. This might be expected to change the HCl columns. Beginning about the middle of September, the HCl column slowly increases. This probably reflects the formation of additional HCl by the attack of enhanced levels of chlorine on methane, without the concomitant destruction mechanism provided by the PSCs earlier. It was seen that the $ClONO_2$ concentration is low deep in the core of the ozone hole, compared to its values near the boundary. Still, the values, $\sim 10^{15}$ molecules-cm^{-2}, are much larger than midlatitude values measured during DC-8 ferry flights. The high levels, $\sim 30 \times 10^{14}$ molecules-cm^{-2}, near the boundary of the vortex may be due to mixing at the edge of the vortex. As is shown in Brune *et al.* (1988), ClO is very high within the vortex, with a very steep gradient at the boundary, while the NO_2 concentration is very small within the vortex and comparatively large outside (Coffey and Mankin, 1988). Mixing at the boundary could produce a chlorine nitrate peak by reaction of these species.

Hydrogen fluoride results, as does hydrogen chloride, from the photolysis of chlorofluorocarbons. It does not, so far as is known, participate further in stratospheric chemistry. The variation of the observed HF column is similar to that observed at northern high latitudes (Mankin and Coffey, 1983), except that the values are around 75% higher, reflecting the increase in atmospheric HF levels in the seven years since the earlier data. There is no evidence of abnormal chemistry affecting antarctic HF.

The HCl:HF ratio was about 5 in 1980 at midlatitudes (Mankin and Coffey, 1983), whereas in the antarctic measurements it is much lower, on the order of unity or sometimes even less. This argues that the low values of gases with stratospheric origins, e.g., HCl, NO_2, and O_3, cannot simply be the result of atmospheric motions, specifically the upwelling of ozone poor air from lower altitudes (Tung *et al.*, 1986), for such a cause would result in a more or less uniform lowering of HF and HCl and little change in

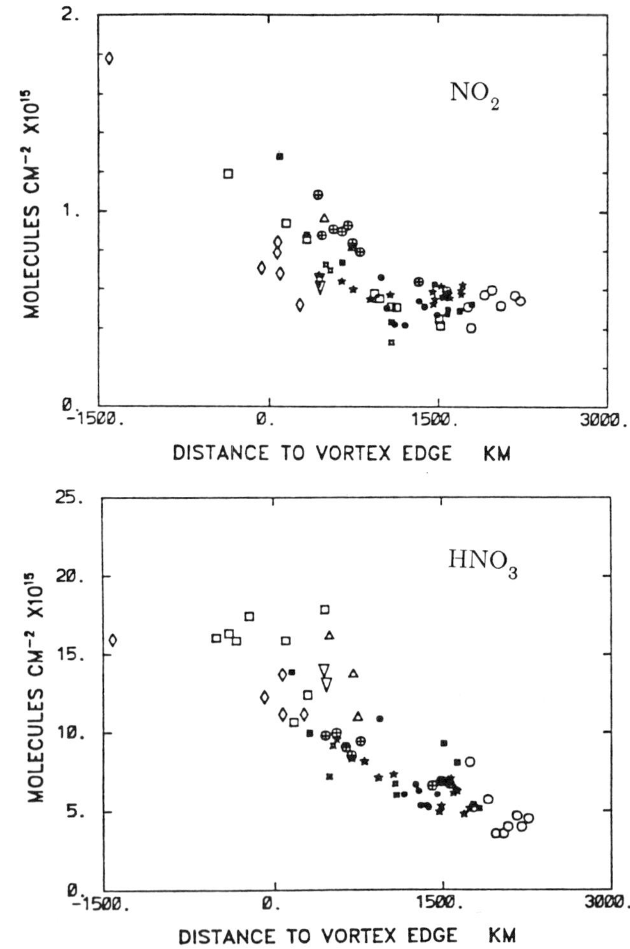

Figure 3. Vertical column above the aircraft for NO_2 and HNO_3 as a function of distance from the vortex edge. Symbols represent typically one minute of observation: 28 August, ▽ ; 30 August, ⌑ ; 2 September, ◇ ; 5 September, □ ; 8 September, △ ; 14 September, ○ ; 19 September, ● ; 21 September, ■ ; 24 September, ★ ; and 26 September, ⊕ .

their ratio. The observed ratio and the agreement of the HF columns with values measured at high northern latitudes implies that there is at least some component of chemical or physical removal of HCl in the antarctic stratosphere. The evidence from fitting the profiles to the 24 September data supports the conclusion that HCl has been removed in the lower layers (~10-20 km) of the stratosphere.

5. CONCLUSIONS

Again in 1987, serious ozone depletion occurred in the south polar vortex. Deep in the vortex, column amounts declined by 40% during our observation period. The columns observed were generally quite close to the TOMS values. The column measurements of HCl and $ClONO_2$ in the Antarctic spring clearly show that the halogen chemistry is quite perturbed. HCl is reduced by a factor of five compared to its value at high northern latitudes, and at times is actually lower than HF. The results are consistent with the idea that HCl is converted to free chlorine by heterogeneous processes in the winter and that the resulting chlorine destroys ozone. The reduction in column amounts of NO_2 and HNO_3 shows that a denitrification process occurs in the stratosphere. This is a requirement for high levels of ClO; otherwise, the ClO would rapidly recombine with NO_2 producing chlorine nitrate. This process has may have occured to produce the $ClONO_2$ levels seen within the vortex.

REFERENCES

Barnett, J.J., and M. Corney, 1985: Middle atmosphere reference model derived from satellite data, Middle Atmosphere Program, Handbook for MAP, v. 16.

Brune, W.H., J.G. Anderson and K.R. Chan, 1988: InSitu observations of ClO in the antarctic: ER-2 aircraft results from 54°S to 72°S latitude, *J. Geophys. Res.,* in press.

Coffey, M.T., W.G. Mankin and A. Goldman, 1981: Simultaneous spectroscopic determination of the latitudinal, seasonal, and diurnal variability of stratospheric N_2O, NO, NO_2, and HNO_3, *J. Geophys. Res., 86,* 7331-7341.

Coffey, M.T., W.G. Mankin, 1988: Column amounts of stratospheric NO, NO_2 and HNO_3 in the antarctic spring 1987, *J. Geophys. Res.,* in press.

Farman, J.C., B.G. Gardiner and J.D. Shanklin, 1985: Large losses of total ozone in Antarctica reveal seasonal ClO_x/NO_x interaction, *Nature, 315,* 207-210.

Krueger, Arlin J., P.E. Ardanuy, F.S. Sechrist, L.M. Penn, D.E. Larko, S.D. Doiron, and R.N. Galimore, 1988: *The 1987 Airborne Antarctic Ozone Experiment: The Nimbus-7 TOMS Data Atlas,* NASA Reference Publication.

Mankin, W.G., 1978: Airborne Fourier transform spectroscopy of the upper atmosphere, *Opt. Eng., 17,* 39-43.

Mankin, W.G., 1979: Fourier transform method for calculating the transmittance of inhomogeneous atmospheres, *Appl. Opt., 18,* 3426-3433.

Mankin, W.G. and M.T. Coffey, 1983: Latitudinal distributions and temporal changes of stratospheric HCl and HF, *J. Geophys. Res., 88,* 10,776-10,784.

Margitan, J.J., G.A. Brothers, E.V. Browell, D. Cariolle, M.T. Coffey, J.C. Farman, C.B. Farmer, G.L. Gregory, J.W. Harder, D.J. Hofmann, W. Hypes, S. Ismail, R.O. Jakoubek, W. Komhyr, S. Kooi, A.J. Krueger, J.C. Larsen, W.G. Mankin, M.P. McCormick, G.H. Mount, M.H. Proffitt, A.R. Ravishankara, A.L. Schmeltekopf, W.L. Starr, G.C. Toon, A. Torres, A.F. Tuck, A. Wahner and I. Watterson, 1988: Intercomparison of ozone measurements over Antarctica, *J. Geophys. Res.,* in press.

Pickett, H.M., E.A. Cohen and J.S. Margolis, 1985: The infrared and microwave spectra of ozone for the (0,0,0), (1,0,0) and (0,0,1) states, *J. Mol. Spectrosc., 110,* 186-214.

Rothman, L.S., R.R. Gamache, A. Goldman, L.R. Brown, R.A. Toth, H.M. Pickett, R.L. Poynter, J.-M. Flaud, C. Camy-Peyret, A. Barbe, N. Husson, C.P. Rinsland, and M.A.H. Smith, 1987: The HITRAN database: 1986 edition, *Appl. Opt., 26,* 4058-4097.

Stolarski, R.S., A.J. Krueger, M.R. Schoeberl, R.D. McPeters, P.A. Newman and J.C. Alpert, 1986: Nimbus 7 SBUV/TOMS measurements of the springtime Antarctic ozone hole, *Nature, 322,* 808-811.

Tuck, A.F., R.J. Watson, E.P. Condon, J.J. Margitan and O.B. Toon, 1988: The planning and execution of ER-2 and DC-8 aircraft flights over Antarctica, August and September 1987, *J. Geophys. Res.,* in press.

Tung, K.K., M.K.W. Ko, J.M. Rodriguez, and N.D. Sze, 1986: Are antarctic ozone variations a manifestation of dynamics or chemistry?, *Nature, 322,* 811-813.

TOTAL OZONE BY LUNAR DOBSON OBSERVATION AT SYOWA, ANTARCTICA

Shigeru Chubachi
Meteorological Research Institute
1-1 Nagamine Tsukuba Ibaraki 305, Japan

Ryoichi Kajiwara and Kouji Kondoh
Aerological Observatory
1-2 Nagamine Tsukuba Ibaraki 305, Japan

ABSTRACT

The accuracy of the Dobson observation by focussed moon image at Syowa is described. The total ozone values obtained by the Dobson moon observation at night are usually about 6% higher than the direct sun measurement of the day before or the day after the night moon measurement. The annual change of total ozone at Syowa including polar night season can be shown during the period from February 1982 to January 1987. The difference in the monthly means of vertical profiles of ozone partial pressure between the period from 1966 to January 1982 and from February 1982 to December 1986 is also shown.

1. INTRODUCTION

The Dobson observation by focussed moon image is almost the only way to get the total ozone in the polar night season at high latitudes where the total ozone by direct sun observation are not possible. Dobson moon observations were carried out at Syowa Station (69°S, 40°E), Antarctica in 1969 (Ishida et al., 1971), and from 1982 (Chubachi, 1984) to 1986, in the months from March to October. In this paper, are described the method, the accuracy and the results of these observations.

2. INSTRUMENT AND METHOD

The Dobson instrument #122 was used for Dobson moon observation at Syowa until January 1986 and #119 thereafter. The intensity of the moonlight is not so strong that the sun director used for moon observation was is equipped with a quartz lens (f=45cm) to condense the moonlight. The position of the lens is set to get the maximum output in the "A" pair wavelengths. No microscope is used, however we are able to see the focused image of the moon on the input slit of the Dobson instrument. Although, most of the total ozone measurements were carried on clear sky at night, it was possible even if thin cloud covers the moon to focus the clear image of the moon on the input slit. The observation should be carried out near the culmination of the moon. Five measurements on "A" and "D" pair wavelengths are used. The value of the extraterrestrial constant for lunar Dobson observation is the same value as that for the sun. The horizontal parallax is taken into consideration. And the calculation is done the usual way according to the WMO manual (Komhyr, 1980).

3. DATA REPRESENTATIVE

Since the accuracy of the Dobson moon observation is not established, the total ozone amounts when the height of the moon is largest is called "night representative" value. This value is assumed to be representative in the night 12 hour before and after the culmination of the moon. The comparison of 15 "night representative" values with the corresponding direct sun measurements (if both daily representative are available, it shows the mean value of the daily representative before and after the night when

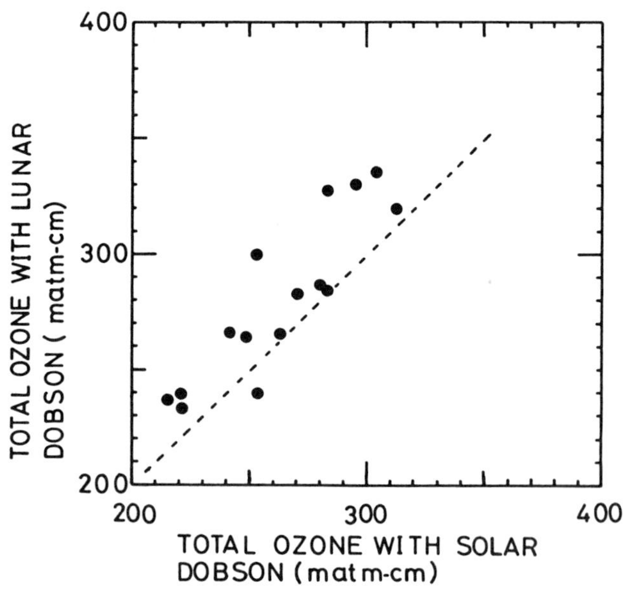

Fig.1 Comparison of the total ozone of night representative with that of the daily representative. The moon values are usually about 6 % high than the direct sun measurement of the day before or the day after the night moon measurement.

This paper has been already published; S. Chubachi and R. Kajiwara, Geophys. Res. Lett. Vol.15, No.8, 905-906, (1988), Copyright by the American Geophysical Union.

ISBN 0-937194-15-8

the lunar Dobson was carried out; otherwise, it shows the only one direct sun daily representative) is carried out. The result of the comparison is shown on Figure 1. This shows that the total ozone at night by Dobson moon observation is about 18 matm-cm greater than that in the daytime by solar Dobson (actually it is about 7% greater at 300 matm-cm and 6.5-5.5% at 200 and 250 matm-cm respectively).

4. POSSIBILITY OF THE DIURNAL VARIATION OF TOTAL OZONE

There are two explanation for the bias in Figure 1: the first is that this bias is observational error, and the second is that this bias is real. In the former case, the bias could be related to the light source (moonlight or sunlight), and/or difficulties with the focused image method without ground quartz plate. If the bias is real, possibility exist for a diurnal variation. Recently solar and moon observations in 1987 at Syowa provided similar results as in Figure 1. So the possibility of existence of a diurnal variation of total ozone should be checked further although on general theoretical grounds such a large difference of 6-7 % is not expected.

5. TOTAL OZONE DURING THE POLAR NIGHT AT SYOWA

Figure 2 shows the 5 year mean of the daily representative total ozone at Syowa between February 1982 to January 1987, superimposed with the 16 year mean of the daily representative from 1961 to January 1982. It is shown that the Antarctic spring time decrease starts in late August or early September and reaches its minimum in middle October. Then the ozone starts to increase and reaches its maximum in early December as shown by Bojkov(1986).

6. VERTICAL PROFILE OF OZONE DECREASE

Figure 3 shows the difference of the monthly means of the vertical profiles between above two periods., i.e. 1966 to January 1982 and February 1982 to December 1986. This figure shows that the maximum decrease occurs between 150 mb and 70 mb in October and in November as demonstrated earlier by Bojkov(1986). It should be noted that an ozone increasing at the altitude above than 50 mb during the austral summer.

7. REFERENCES

Bojkov,Rumen D., (1986): The 1979-85 Ozone Decline in the Antarctic as Reflected in Ground Based Observations, Geophys. Res. lett., 13, 1236-1239.

Chubachi, S., (1984): Preliminary result of ozone observation at Syowa Station from February 1982 to January 1983, Mem. Natl. Inst. Polar Res., Spec. Issue, 34, 13-19.

Ishida, K., Suzuki, T., and Sakai, S., (1971): Syowa Kiti ni okeru 1969-nen no ozon zenryo kansoku Total ozone observation at Syowa Station, Antarctica in 1969), (in Japanese) Nankyoku Shiryo (Antarct Rec.), 39, 32-38.

Komhyr, W. D.(1980): WMO Global Ozone Research and Monitoring Project Report No. 6 (Operation handbook ozone observations with a Dobson Spectrophotometer).

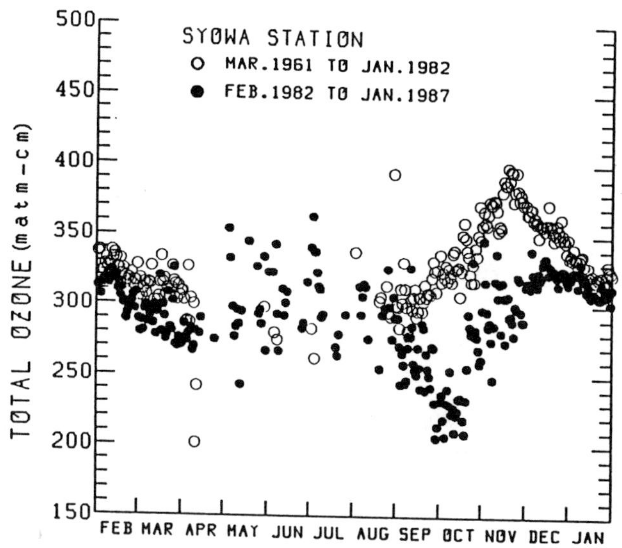

Fig.2 Closed circles shows the 5 year mean of the daily representative total ozone at Syowa Station from February 1982 to January 1987. Open circles shows the 16 year mean of the ones from 1961 to January 1982.

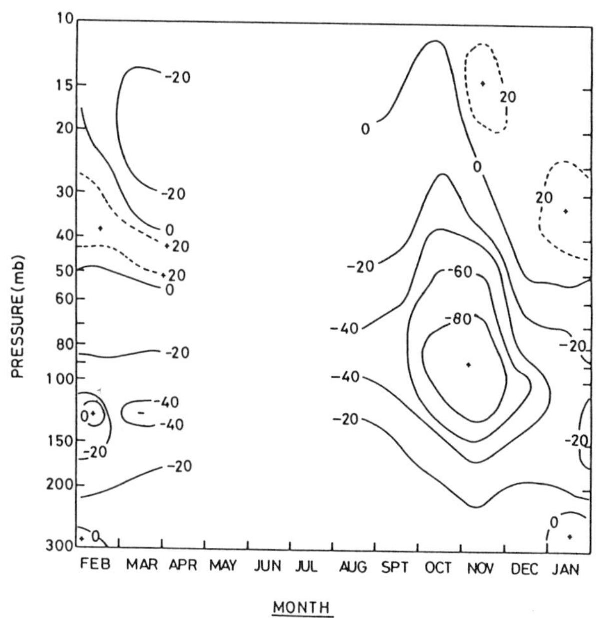

Fig.3 The difference of the monthly means of the vertical profiles of ozone partial pressure between the period 1966 to January 1982 and February 1982 to December 1986. (unit: micro mb).

CHANGES IN TOTAL OZONE AND OZONE VERTICAL DISTRIBUTION AT SOUTH POLE, ANTARCTICA, 1962–1987

W.D. Komhyr, S.J. Oltmans, R.D. Grass
NOAA/Air Resources Laboratory
Boulder, Colorado 80303, USA

P.R. Franchois and R.K. Leonard
Cooperative Institute for Research in Environmental Sciences
University of Colorado
Boulder, Colorado 80309, USA

ABSTRACT

Ozone profile measurements made at South Pole with Regener chemiluminescent ozonesondes in 1962-1965 and with ECC ozonesondes in 1967-1971 and 1986-1987 indicate a decrease in the magnitudes of the primary ozone maxima and an increase in the maxima altitudes, with time, for all seasons. The most marked ozone reduction occurred in recent years during October months. A record low ozone amount of 127 DU was observed at South Pole in early October 1987. The 1986 and 1987 springtime ozone decreases at South Pole differed in magnitude as well as in the times of ozone recovery.

1. INTRODUCTION

Total ozone observations have been made at South Pole since 1962. Since 1964, Dobson ozone spectrophotometers used for the observations have been calibrated periodically relative to World Standard Dobson instrument no. 83 maintained by NOAA/GMCC in Boulder, Colorado (Komhyr et al., 1986). The calibration level of instrument no. 83 has been maintained to within an uncertainty of ±0.5% since 1962 (Komhyr et al., 1988a).

Balloonborne measurements of the vertical distribution of ozone were made at South Pole with Regener chemiluminescent ozonesondes (Regener, 1964) in 1962-1965, and with electrochemical concentration cell (ECC) ozonesondes (Komhyr, 1969) in 1967-1971 and 1986-1987. Table 1 lists the number of ozonesonde launches made during summer (DJF), autumn (MAM), winter (JJA), spring (SON), September (SEP), and October (OCT) months of the three periods of record, as well as the total (ANN) number of soundings made, during each of the observing periods.

The Regener chemiluminescent sonde ozone profiles were normalized to Dobson spectrophotometer total ozone. Where direct comparison data were unavailable for normalization (e.g., when soundings were made during the polar night in the absence of moonlight), total ozone amounts were estimated from South Pole climatological total ozone means. Prior to normalization, an additional correction was applied to Regener instrument soundings which exhibited sensor decay above the primary ozone maximum following sensor sensitization by water vapor. This correction, determined by Hering and Dütsch (1965), involved multiplying ozone partial pressures starting at the 40th minute of flight by an exponential factor $e^{a(t-4)}$ where a is 0.2 and t is minute.

ECC sonde data lacking Dobson spectrophotometer ozone values for normalization were left unnormalized. On average, the normalization factor for such sounding is ~1.0 (Komhyr et al., 1988b).

TABLE 1. PERIODS OF RECORD AND NUMBER OF OZONESONDE SOUNDINGS USED IN DERIVING SOUTH POLE AVERAGE SEASONAL, ANNUAL, AND SEPTEMBER- AND OCTOBER-MONTH OZONE PROFILES

Period of Record	Number of Soundings						
	DJF	MAM	JJA	SON	ANN	SEP	OCT
1962-1965	17	27	24	34	102	13	12
1967-1971	19	24	21	21	85	12	6
1986-1987	22	19	23	58	122	19	22

We present South Pole total ozone monthly means for 1962-1987; compare average seasonal, annual, and September- and October-month ozone vertical distributions at South Pole for the three periods of record; and examine select August-December 1986 South Pole ozonesonde profiles relative to those obtained in 1987--a year of record ozone lows during the months of October and November.

2. TOTAL OZONE DECREASE

Monthly mean total ozone values for South Pole, for 1962-1987, are plotted in Fig. 1. The plots do not include data for 1972 and 1983, or for the twilight periods of March to mid-April and September to mid-October of each year when Dobson spectrophotometer total ozone observations cannot be made. Monthly means obtained from observations on moonlight during April-August months are, furthermore, based on relatively few observations per month that have at times in the past been made sporadically.

The decrease in ozone at South Pole, beginning in 1980, is clearly evident in Fig. 1, especially for October (October 15-31) months. Mean February total ozone values for the period of record, however, show no downward trend.

Total ozone decreased at South Pole to a record low of 127 Dobson units (DU) in early October 1987 (Komhyr et

ISBN 0-937194-15-8

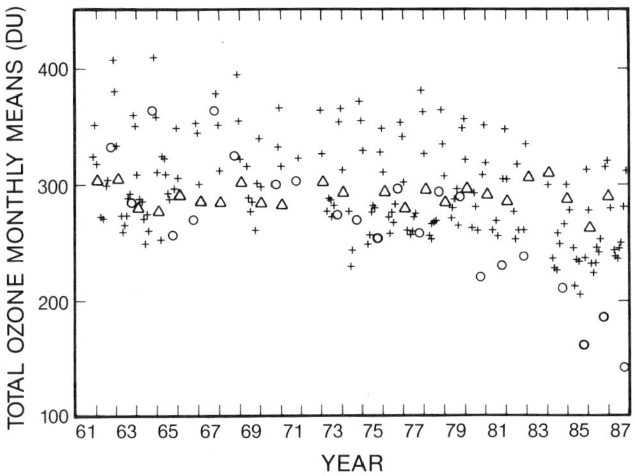

Fig. 1. Mean monthly total ozone amounts at South Pole, from 1962-1987 Dobson spectrophotometer observations. Circles (o) denote October (October 15-31) means; triangles (Δ) denote February means; and pluses (+) denote means for all other months.

al., 1988c). The 1987 October mean total ozone amount was 138 DU. The previous mean low of 161 DU occurred in 1985. The average total ozone amount for October months at South Pole during 1962-1979 was 292 DU.

The 1987 South Pole November total ozone mean, of 184 DU, was also unusually low. The previous November mean low of 238 DU occurred in 1985. The 1962-1979 average total ozone amount for November months at South Pole was 351 DU.

3. LONG-TERM CHANGES IN SOUTH POLE OZONE PROFILES

Average seasonal, annual, and September- and October-month ozone profiles for 1962-1965, 1967-1971, and 1986-1987 are compared in Figs. 2(a)-2(g). The 1967-1971 and 1986-1987 data sets were obtained with the same kind of instruments (ECC sondes), and are directly comparable. Less comparable 1962-1965 data were obtained with Regener chemiluminescent ozonesondes which have faster response to ozone but, as indicated earlier, are subject to sensitization by water vapor during preparation for flight.

Figures 2(d), 2(f), and 2(g) portray the springtime decrease in ozone that has progressed at South Pole with time during 1962-1987 in the 10-30 km stratospheric region; maximum ozone decrease occurred at about 17 km. During summer months (Fig. 2(a)), a reduction in ozone with time occurred in the stratosphere below 20 km, but an increase appears to have occurred above 20 km. Ozone maxima in the profiles for all seasons exhibit a decrease with time in the magnitude of the maxima, but an increase with time in the altitude of maxima. Because of the large, natural variability of stratospheric ozone, the profile standard deviation intervals (not shown in Fig. 2) generally overlap for adjacent period data, except for averaged 1967-1971 and 1986-1987 October data for which mean ozone partial pressures at, e.g., 70 mb were 121.5 ± 24.7 and 24.8 ± 20.2 nb, respectively.

Figures 2(h) and 2(i) compare air temperatures measured with the ECC sondes. Temperatures in the region of the maximum ozone decrease were several degrees colder in 1986-1987 than they were in 1967-1971.

4. COMPARISON OF AUGUST-DECEMBER 1986 AND 1987 SOUTH POLE OZONE PROFILES

Select ECC ozonesonde data obtained at South Pole in 1986 and 1987, during the time of ozone hole formation and ozone recovery, are shown in Fig. 3. The springtime reduction of stratospheric ozone was more pronounced in 1987 than in 1986. Minimum ozone volume mixing ratios in the heart of the ozone depletion region ranged from 0.2 to 0.4 parts per million by volume (ppmv) in 1986, but approached 0.04 ppmv in 1987. The depletion of ozone in a 4-km-thick layer of the atmosphere centered at about 17 km was, therefore, virtually complete in early October 1987.

Two exponential ozone decrease rates, with half-lives of 19.5 and 4.5 days, were observed in the heart of the ozone depletion region prior to and after about September 22, 1987, respectively. Whether the half-life of 4.5 days was photochemical in origin, or resulted from advection to South Pole of air from which ozone had been depleted elsewhere in Antarctica, is unknown. Although ozone began to decrease during both years in late August, the 1987 depletion persisted into early December, 3 weeks longer than it did in 1986.

The November 20, 1986, temperature profile of Fig. 3 clearly portrays the rapid stratospheric warming that accompanied the return of ozone to South Pole. Because of the difference in times of polar vortex breakdown during 1986 and 1987, temperatures in the stratosphere above 11 km were considerably colder in 1987. At 17 km, air temperatures in November 1987 were colder than in November 1986 by 10° to 40°C. By mid-December 1987, the ozone and temperature vertical distributions assumed the basic characteristics of corresponding data obtained in 1986.

ACKNOWLEDGMENTS

Logistics support for the South Pole ozone sounding program, as well as helium gas for the balloon ozone soundings made in 1986 and 1987, were provided by the National Science Foundation.

REFERENCES

Hering, W. S., and H. U. Dütsch, 1965: Comparison of chemiluminescent and electrochemical ozonesonde observations. *J. Geophys. Res., 70,* 5483-5490.

Komhyr, W.D., 1969: Electrochemical concentration cells for gas analysis. *Ann. Geophys., 25* (1), 203-210.

Komhyr, W.D., R.D. Grass, and R.K. Leonard, 1986: Total ozone decrease at South Pole, Antarctica, 1964-1985. *Geophys. Res. Lett., 13,* 1248-1251.

Komhyr, W.D., R.D. Grass, and R.K. Leonard, 1988a: Dobson spectrophotometer No. 83: A standard for total ozone measurements, 1962-1987. Submitted for publication to the *J. Geophys. Res.*

Komhyr, W.D., S.J. Oltmans, P.R. Franchois, W.F.J. Evans, and W.A. Matthews, 1988b: The latitudinal distribution of ozone to 35 km altitude from ECC ozonesonde observations. *Proc. Ozone Symposium-88,* Editors: R. D. Bojkov and P. Fabian, A. Deepak Publishing, Hampton, Virginia.

Regener, V.H., 1964: Measurement of atmospheric ozone with the chemiluminescent method. *J. Geophys. Res., 69,* 3795-3800.

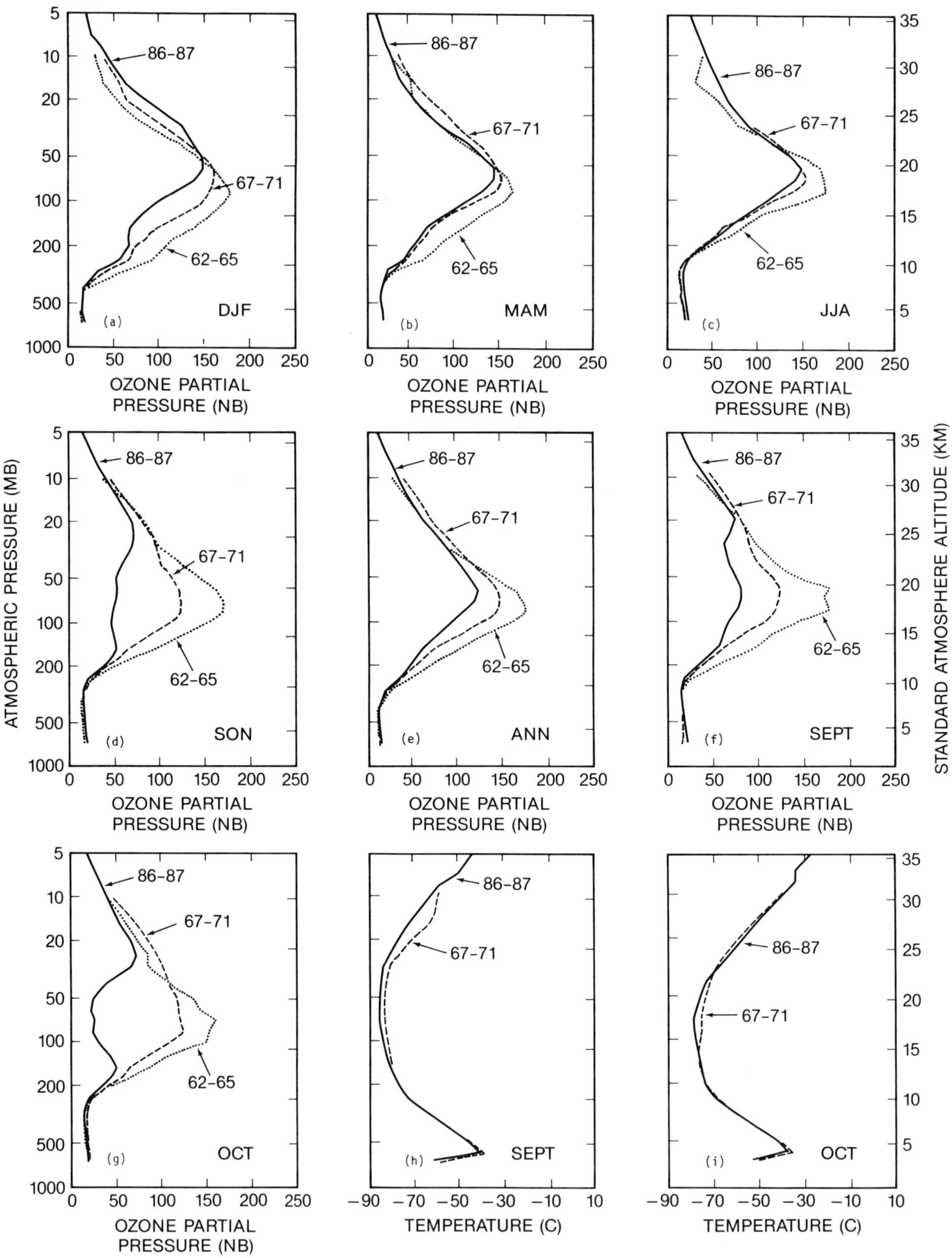

Fig. 2. (a)-(g) Comparison of average seasonal, annual, and September- and October-month ozone profiles for 1962-1965, 1967-1971, and 1986-1987. (h)-(i) Average temperature profiles, from September and October ozone soundings made in 1967-1971 and 1986-1987.

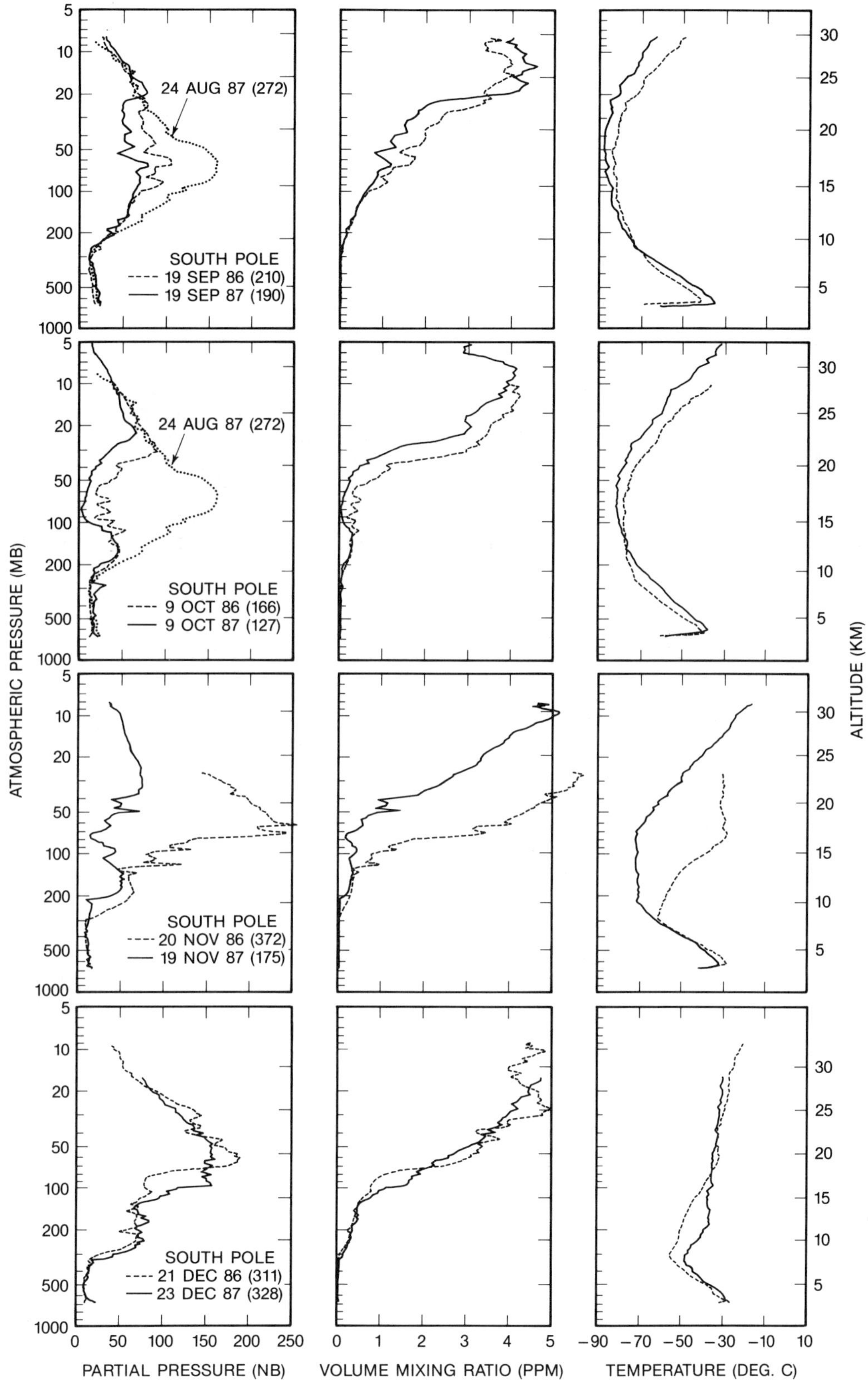

Fig. 3. Select South Pole ozonesonde data showing ozone hole formation and ozone recovery in 1986 and 1987. Values in parentheses in the figure legends are total ozone amounts in Dobson units.

A COMPARISON OF ANTARCTIC OZONE PROFILES DURING SPRING 1986 AND 1987

*D.J. Hofmann and J.W. Harder**
Department of Physics and Astronomy
University of Wyoming
Laramie, Wyoming 82071, USA

ABSTRACT

Ozone and temperature profiles were measured in 50 balloon flights at McMurdo Station (78°S) during the spring of 1987. Compared to similar data obtained in 1986, stratospheric temperatures were lower and the springtime Antarctic ozone reduction was greater in magnitude, extended to higher altitude, and proceeded at a higher rate in 1987. The major cause of the reduction was an ozone sink in the 11-23 km region. Adiabatic vertical motions over 1-2 km intervals between 12 and 20 km with consequent ozone reductions were observed in association with the formation of nacreous clouds, indicating these to be rare events on a local scale probably associated with mountain lee waves.

1. INTRODUCTION

During the Antarctic spring of 1986 the first National Ozone Expedition (NOZE I) was conducted at McMurdo Station. The purpose of that program was to study springtime depletion of stratospheric ozone, first reported by Farman et al. [1985]. Ozone soundings on 33 occasions from August 25 to November 3, in 1986, indicated that the springtime decrease in ozone was limited to the 12-20 km region and that most of the reduction occurred in September. An exponential halflife of about 25 days for the 18 km ozone mixing ratio was measured during September [Hofmann et al., 1987]. In 1986, ozone partial pressures at 18 km reached values as low as 10 nbar in mid-October, down from values of about 150 nbar in late August. Considerable structure in the ozone profiles, especially in October, suggested that either the depletion process was not uniform in space or that ozone was being extruded into the ozone hole from the vortex boundary region. In order to determine if these features were characteristic of the ozone hole phenomenon, the measurements were expanded during NOZE II in 1987 at McMurdo Station with 50 soundings between 29 August and 9 November. This paper reports the results of these measurements.

*now at CIRES/Univ. of Colorado, Boulder, CO 80309-0449, USA

2. OBSERVATIONS

Digital conversions of commercial electrochemical cell ozonesondes (Science Pump Corp.), digital radiosondes (Vaisala Corp.), and small polyethylene balloons [Hofmann et al., 1987] were again employed during the 1987 program. Pressure, temperature, humidity, ozone pump temperature, and ozone partial pressure were transmitted approximately every 5 s during balloon ascent providing about 15 m resolution.

Temperatures in the 12-24 km region were of the order of 5°C lower in 1987 as compared to 1986. Periods with temperatures below -85°C were encountered in late August, early and mid-September and early October. From about 7 October to 24 October the edge of the ozone hole was over McMurdo resulting in higher and variable temperatures. After 24 October the region of coldest stratospheric temperature was once again in the vicinity of McMurdo, although it had warmed to about -75°C, and remained there until measurements terminated on 9 November.

Height-time contours of ozone partial pressure and mixing ratio are shown in Fig. 1. As in 1986, the mixing ratio contours again indicate that the depletion begins at high altitude and proceeds to lower altitudes later. The progression in altitude of the depletion region is shown in Fig. 2 indicating a rapid spread in altitude in mid-September. As will be discussed later, the depletion magnitude also changed rapidly at this time. The altitude interval of depletion in Fig. 2 has a magnitude of about 12 km covering the 12-24 km region until about mid-October and 11 km covering the 11-22 km region after that. In 1986 the general region of depletion was 12-20 km [Hofmann et al., 1987] so it appears that the upper boundary was 2-3 km higher in 1987.

Most of the reduction in ozone occurred during 2 short periods, i.e., 4-6 September and 11-15 September with the second period being more severe with reductions by a factor of 2 in only a few days at 18 km. This episodal phenomenon can be visualized more clearly in the ozone partial pressure and mixing ratio surface plots in Fig. 3. Similar rapid reductions were observed in the TOMS (Total Ozone Mapping Spectrometer) total ozone daily minima on several

ISBN 0-937194-15-8

occasions between 5 and 27 September (A. Krueger, personal communication, 1987).

Data shown in Fig. 1 and 3 appear to negate the idea of mixing of ozone-poor air into the ozone hole region. Mixing ratios decrease with altitude in the hole region suggesting a local sink. Similar

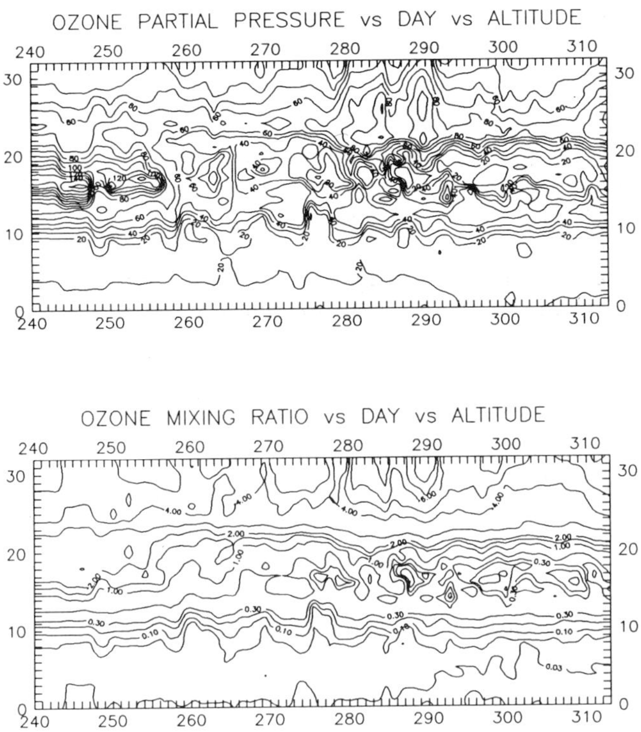

Fig. 1. Ozone partial pressure (nbar) and mixing ratio (ppmv) contours from 50 balloon soundings at McMurdo Station between 29 August and 9 November 1987. The abscissa units are day numbers in 1987 and the ordinate units are kilometers. The partial pressure contour interval is 10 nbar and the mixing ratio contour intervals are variable to cover the large dynamic range.

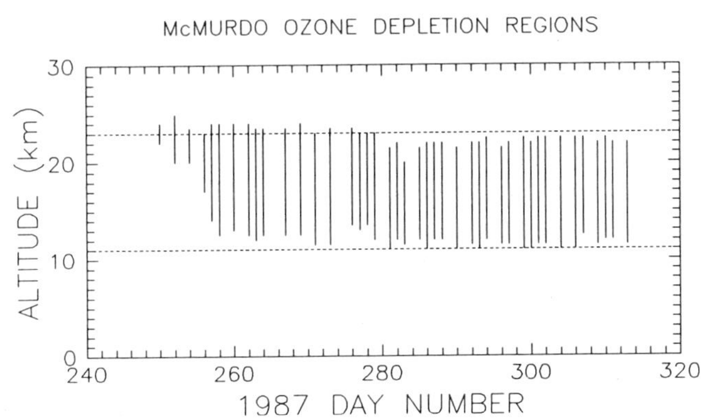

Fig. 2. Altitude intervals of ozone depletion at McMurdo Station in 1987. The dashed lines are reference lines drawn at 11 and 23 km.

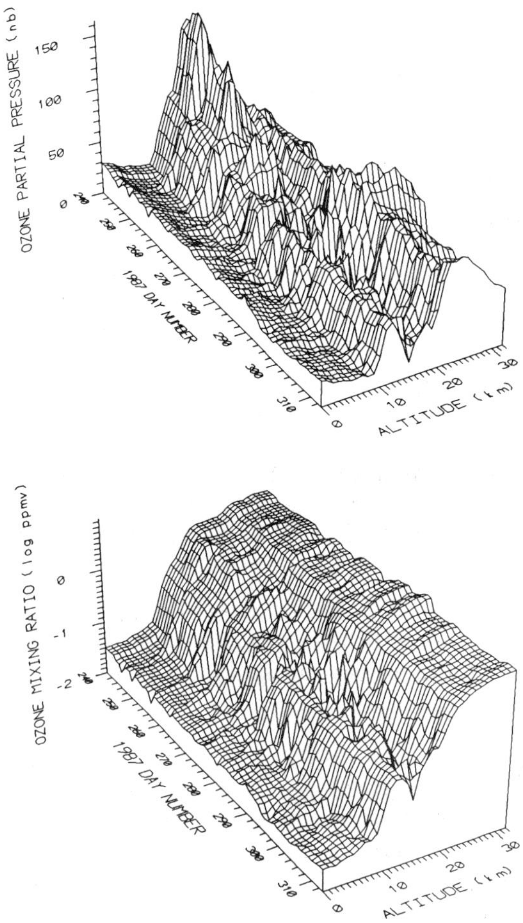

Fig. 3. Three-dimensional ozone partial pressure and mixing ratio surfaces versus time and altitude constructed from 50 balloon soundings at McMurdo Station between 29 August and 9 November 1987.

conclusions were reached in 1986 [Hofmann et al., 1988]. On rare occasions, as will be discussed later, adiabatic mixing in 1-2 km layers, probably associated with mountain generated waves, results in layers of reduced ozone. While such processes may reduce ozone in the lower stratosphere on occasion, they do not appear to occur frequently enough to be a major factor in the overall ozone depletion phenomenon.

Figure 1 shows a good deal of structure in the ozone distribution at 14-18 km from about 7-23 October. Ozone layers within the ozone hole region were present in 12 of 13 soundings during this period. The fact that the period of unusual ozone layering in the ozone hole occurred at about the same time that the edge of the ozone depletion region was situated near McMurdo suggests that this structure, and probably that observed in 1986 [Hofmann et al., 1987], is most likely associated with ozone extrusion from the walls of the ozone cavity. However, as will be shown later, such wall effects do not add substantially to the amount of ozone in the 12-20 km

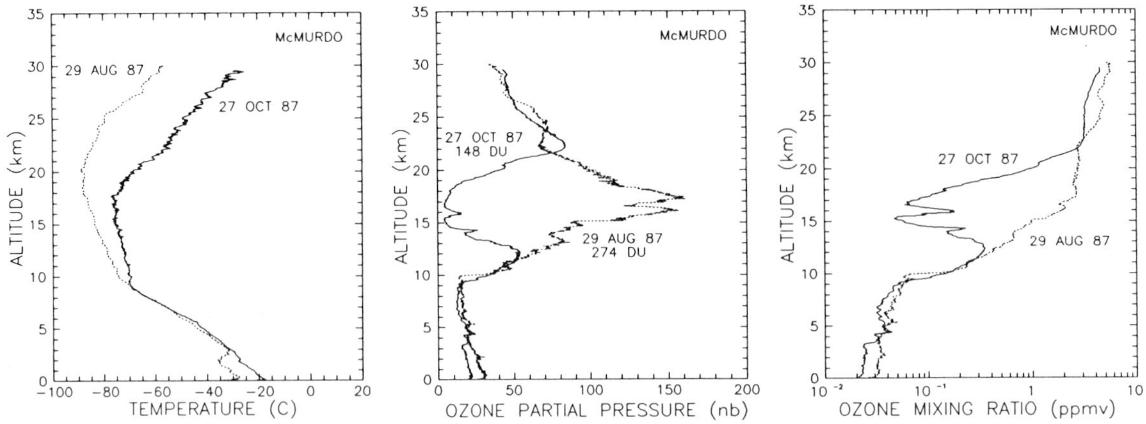

Fig. 4. Comparison of temperature, ozone partial pressure, and ozone mixing ratio profiles before major ozone depletion had begun (29 August) and when the lowest value of total ozone was observed (27 October) at McMurdo Station in 1987.

region but increase total ozone dramatically. The latter is due to the high ozone values above 20 km associated with recovery of the ozone hole at these altitudes in the wall region as indicated clearly in Fig. 1 and 3.

Data from soundings on 29 August and 27 October, corresponding to maximum (274 DU) and minimum (148 DU) total ozone values observed in the 1987 soundings, are shown in Fig. 4 and indicate that removal of ozone in the 15-18 km region was nearly complete. The ozone partial pressure decreased from about 150 nbar in late August to about 3-5 nbar in early October representing a reduction in excess of 95%. This may be compared to a minimum of about 10 nbar at McMurdo in 1986. Since the minimum ozone partial pressures in the ozone hole region in 1987 were near the ozonesonde background level, and since some ozone may have diffused into the minimum region, the depletion may in fact have been total at some earlier time in this air mass.

Figure 5 compares total ozone as measured by TOMS and from integrations of the vertical ozonesonde profiles. In general, agreement is very good except for the period 3-9 September when the soundings indicated more ozone than derived from TOMS data. These differences are larger than can be accounted for by ozonesonde uncertainties. Since TOMS data are daily averages and the ozonesonde data represent, in effect, about 1 hour averages sometime during that day, fluctuations of the order of 5% are to be expected. These fluctuations average to zero, as they should, after day 280 (7 October) but do not before this time. From day 246 to 280, TOMS data are an average of about 5% lower than the ozonesonde data and considerably more during the early September period. The apparent discrepancy with TOMS data may be related to the effect of high clouds on remotely sensed data during the early September period when stratospheric temperatures were below -85°C at McMurdo.

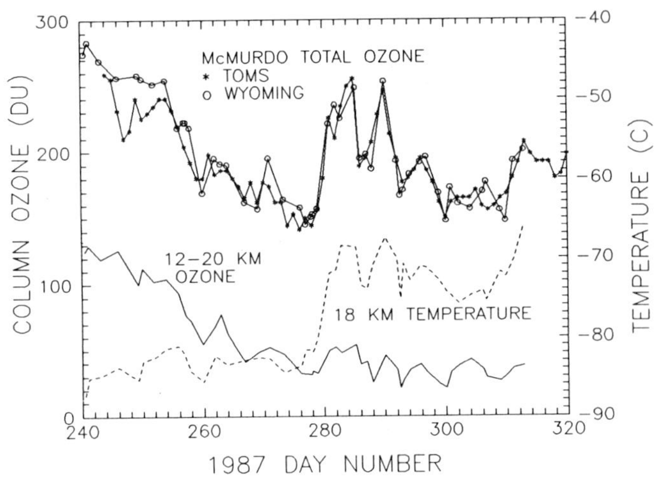

Fig. 5. Comparison of total ozone over McMurdo Station as determined by integrating ozonesonde profiles and TOMS satellite data between 29 August and 9 November 1987. Also shown is the 18±1 km average temperature and the 12-20 km ozone column as determined from the ozone soundings. Note the correlation between total ozone and temperature between days 280 and 300.

The 18±1 km average temperature as measured at McMurdo is also shown in Fig. 5. The large values of total ozone between days 280 and 300 (7 and 27 October) and the high temperatures are related to the boundary of the cold vortex region being near McMurdo during this time. Fig. 5 also shows the 12-20 km integrated column ozone. The period of warmer temperatures at 18 km, accompanied by high ozone above this altitude, did not affect the 12-20 km ozone column as much as total ozone, contributing perhaps 20-30 DU in the 12-20 km region whereas total ozone increased about 100 DU at this time. The 12-20 km column ozone fell at an average rate of 80

DU per mo. in September and remained low throughout October and early November.

Figure 6 shows the rate of decay of the ozone mixing ratio for 2 km averages during September at 12, 18, and 22 km. Halflifes, for an exponential decay are 21.6, 12.4, and 42.5 days, respectively. Figure 7 compares ozone partial pressure profiles on 16 October 1986 and 27 October 1987 when the lowest total ozone was observed at McMurdo, 166 DU in 1986 and 148 DU in 1987. It is clear that ozone depletion in the 15-20 km region was considerably more complete in 1987. Figure 8 compares the rate of decay of the 18±1 km average ozone mixing ratio in September of 1986 and 1987. The halflife, for an exponential decay, was 12.4 days in 1987, less than half the 1986 value.

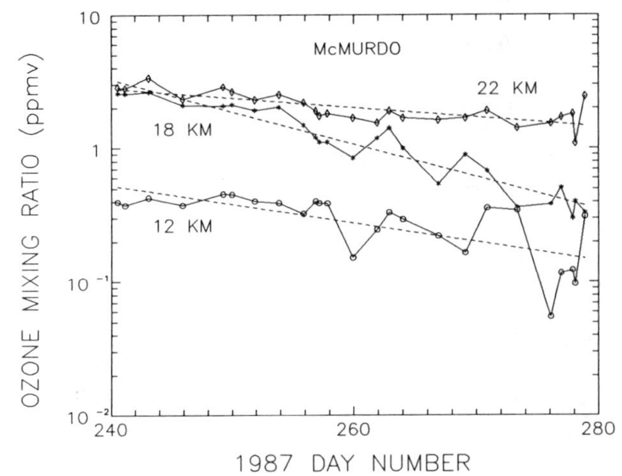

Fig. 6. Ozone mixing ratio (2 km averages) versus time at McMurdo Station during September 1987 at 12, 18, and 22 km. Halflifes for an exponential decay (dashed lines) are 21.6, 12.4, and 42.5 days, respectively.

On 20 September, reductions in ozone in 1-2 km layers in the 12-20 km region were correlated with reductions in temperature. The latter were near the frost point for water and cloud layers, as observed in lidar data at McMurdo at this time (B. Morley, personal communication, 1987), were correlated with minima in the temperature profiles. Figure 9 shows the temperature, ozone partial pressure, potential temperature, and ozone mixing ratio profiles at this time. Except for the ozone mixing ratio decrease at about 20 km, indicating an ozone sink in this region, the associated ozone variations appear to be the result of adiabatic mixing in 1-2 km thick layers and are probably related to mountain generated waves and the formation of nacreous clouds. Major nacreous cloud events were observed at McMurdo on this day and on 5 October in 1987. Thus such events appear to be relatively rare and probably do not play a major role in ozone depletion except possibly in

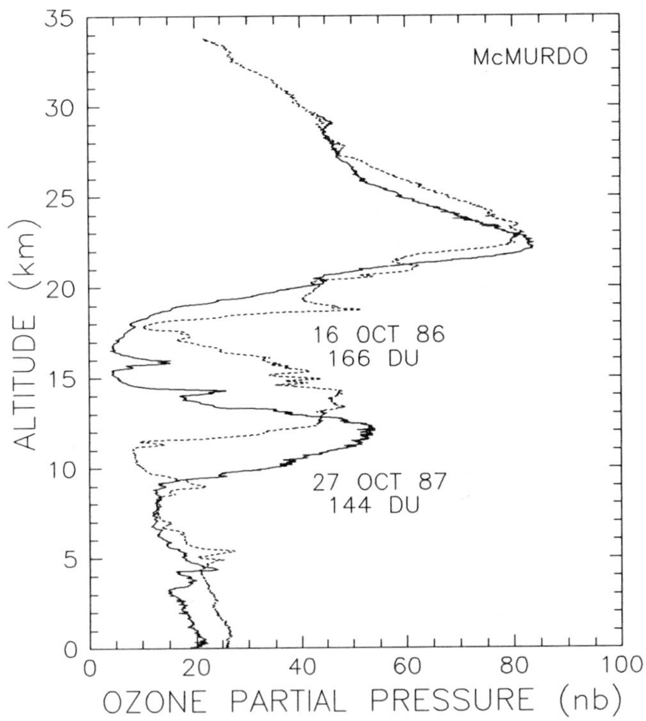

Fig. 7. Comparison of ozone partial pressure profiles during periods of maximum ozone depletion at McMurdo Station in 1986 and 1987.

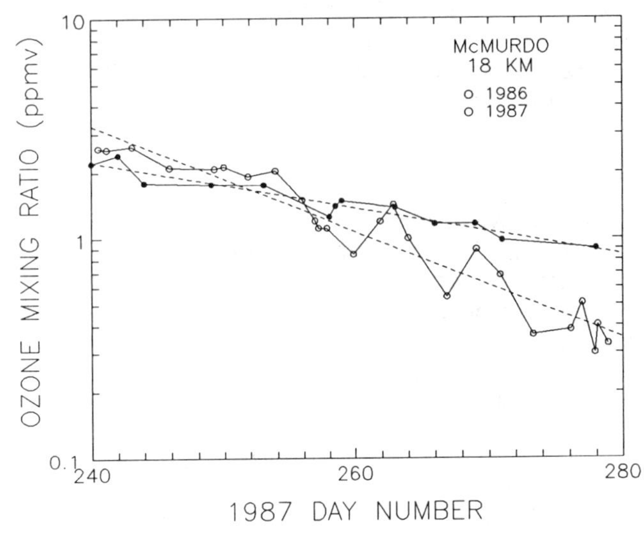

Fig. 8. Comparison of the decay rate of the 18±1 km average ozone mixing ratios at McMurdo Station in September of 1986 and 1987. Halflifes for an exponential decay are 28 and 12.4 days, respectively.

localized regions in the lower stratosphere after mid-September. However, the frequency of such events over the entire continent is unknown and should be determined and included in the overall ozone reduction picture.

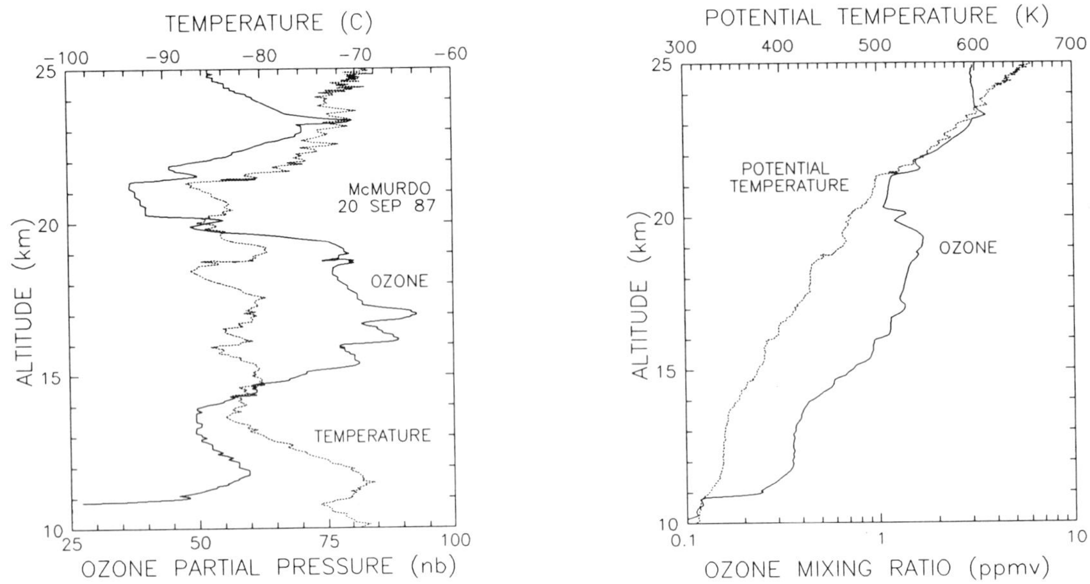

Fig. 9. Temperature-ozone partial pressure, and potential temperature-ozone mixing ratio profiles observed between 10 and 25 km at McMurdo Station on 20 September 1987. Note the correlations between decreasing temperature and decreasing ozone partial pressure (constant potential temperature and ozone mixing ratio) in 1-2 km layers between 12 and 20 km.

3. CONCLUSIONS

Balloonborne ozone measurements at McMurdo Station, Antarctica indicate that the 1987 spring ozone reduction phenomenon was unusual in several ways. Temperatures in the stratosphere were lower than in 1986 and ozone depletion was more severe in rate, absolute magnitude, and vertical extent. TOMS measurements indicate that the reduction was also greater in horizontal extent and in duration with recovery not occurring until late November. While vertical motions may affect ozone distributions on a limited, local scale, the main reduction was caused by an ozone sink in the 11-23 km region.

ACKNOWLEDGEMENTS

This research was supported by the National Science Foundation, Division of Polar Programs and Atmospheric Sciences and the National Aeronautics and Space Administration.

REFERENCES

Farman, J. C., B. G. Gardiner, and J. D. Shanklin, 1985: Large losses of total ozone in Antarctica reveal seasonal ClO_x/NO_x interaction. *Nature*, **315**, 207-210.

Hofmann, D. J., J. W. Harder, S. R. Rolf, and J. M. Rosen, 1987: Balloonborne measurements of the development and vertical structure of the Antarctic ozone hole in 1986. *Nature*, **326**, 59-62.

Hofmann, D. J., J. M. Rosen, and J. W. Harder, 1988: Aerosol measurements in the winter/spring antarctic stratosphere 1. Correlative measurements with ozone. *J. Geophys. Res.*, **93**, 665-676.

OZONE SOUNDINGS AT THE ANTARCTIC STATION GEORG FORSTER (70°46'S, 11°50'E)

U. Feister and P. Plessing
Meteorological Service of the GDR
Main Meteorological Observatory
Telegrafenberg, Potsdam, GDR–1561

H. Gernandt
Meteorological Service of the GDR
Aerological Observatory
Lindenberg, GDR–1231

ABSTRACT

Since May 1985 regular ozone soundings have been carried out at the Antarctic station Georg Forster. About 200 soundings were made between May 1985 and May 1988. An analysis of ozone and temperature profiles of the period September to February 1985/86 and 1986/87, respectively has shown some pecularities in their distribution. The strong ozone decrease at the end of September is characterized by the development of a layer of very small ozone values between 100 and 25 hPa (16-25 km). This ozone minimum layer shifts to lower height levels at the end of November. Therefore, a secondary ozone maximum occurs in the lower stratosphere from the end of September to November. Short-term warmings in September/October are accompanied by increasing total ozone values and a downward shift of the minimum ozone layer. The weaker spring ozone minimum in 1986 as compared to 1985 can be explained by a different pattern of temperature waves in both years.

1. INTRODUCTION

A programme of regular ozone soundings at the GDR research base* near the USSR station Novolazarevskaya, which is located in the Schirmacher oasis 100 km south from the shelf ice barrier, was started in May 1985. A stimulus for such a sounding programme came from the results of total ozone observations that were made by G. Skeib and Ch. Popp at the Antarctic station Mirny in 1960/61. These early total ozone measurements shown in Fig. 1 reveal a strong ozone decrease to about 320 D followed by an increase to more than 470 D at the end of October 1960.

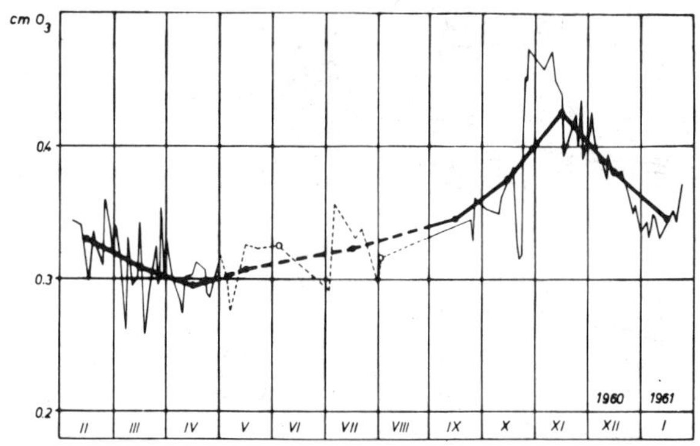

Fig. 1. Total ozone values determined by G. Skeib and Ch. Popp (1961) using Dobson instrument No. 71 at the Antarctic station Mirny (66°33' S, 93°01' E) from February 1960 to January 1961

The height of the ozone minimum in the lower stratosphere with the sudden ozone increase and the upward movement of the ozone maximum was identified by Chubachi (1984) on the basis of ozone soundings made at Syowa station in 1982. Recognizing the strong ozone variability at the time of the circulation change in spring, we decided to have a higher sounding frequency from September to November.

*Since 1 July 1987 this research base at 70°46' S, 11°50' E (111 m asl) is an independent GDR station called Georg Forster. G. Forster (1754-1794) took part in James Cook's second journey round the world, during which they sailed round the Antarctic continent.

ISBN 0–937194–15–8

A further stimulus to continue the soundings on a longer term came from the well-known report of Farman et al. (1985) on the Antarctic ozone decline, which has led to enormous efforts in designing national and international campaigns for trace gas measurements over the Antarctic continent. More than 200 ozone sondes of the type OSK-2 (Rönnebeck and Sonntag 1976; Feister et al. 1985) were launched between May 1985 and April 1988. Total ozone values of Syowa station (69° S, 39°35' E), which is located 1070 km from Georg Forster station, have been used for the correction of ozone profiles.

In this paper we present results of the two periods September to January 1985/86 and 1986/87, respectively.

2. TIME-HEIGHT CROSS-SECTIONS OF OZONE AND TEMPERATURE FROM SEPTEMBER 1985 TO JANUARY 1986

Fig. 2 shows a time-height cross-section of ozone and temperature over Georg Forster station.

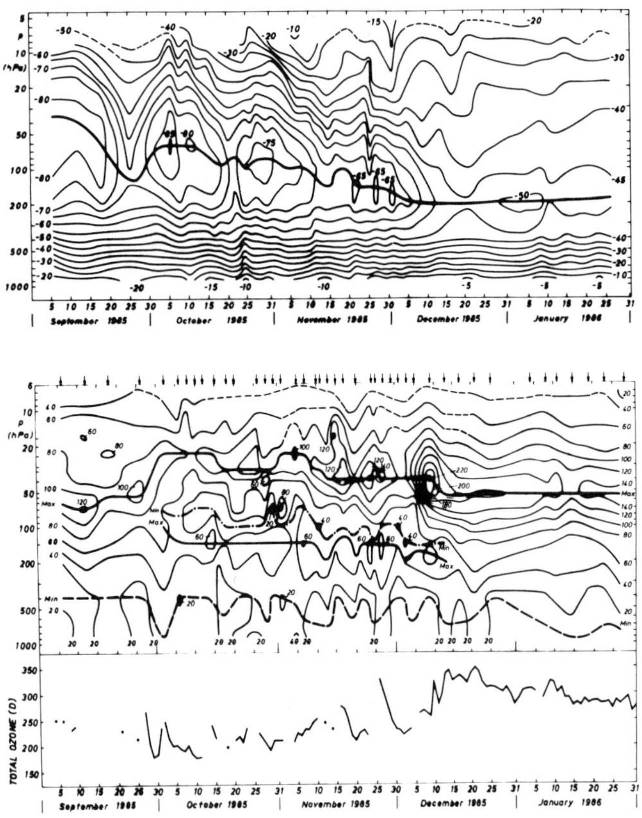

Fig. 2. Time-height cross-section of temperature (°C) and ozone partial pressure (nbar) over Georg Forster station (70°46' S, 11°50' E, 111 m) from September 1985 to January 1986. Total ozone values in Dobson units (bottom) are from Syowa station.

The little arrows on top designate dates of ozone soundings. Minima and maxima of ozone and temperature are drawn as thicker solid or dashed lines. The temperature shows a wave of warmer air at the end of September 1985, when the temperature minimum shifts downwards from about 40 hPa (22 km) to 120 hPa (15 km) and becomes weaker at the same time. This warming period is followed by a strong cooling and upward movement of the temperature minimum to 65 hPa (19 km) in October with temperatures below -85° C. At the end of September the main ozone maximum shifts upwards from 60 - 70 hPa (18.5 - 19.5 km) to 25 hPa (25 km), and at the beginning of October at about 100 hPa (16 km) a secondary ozone maximum appears, which moves downwards to 140 hPa (14 km). The October ozone minimum between both maxima is at about 70 - 80 hPa (17.5 - 18.5 km) just below the level where the main ozone maximum was in September. Minimum ozone values are reached between about 5 and 10 October when the temperature is at its minimum. A second weaker warming period starts in the second half of October with maximum values of -65° C ... - 70° C between 60 and 150 hPa (13.5 - 19.5 km) around 22 October. A second cooling period, which is weaker (-75° C) and lower (around 100 hPa = 16 km) than that in the first half of October, can be seen at the end of October. Both ozone maxima and the ozone minimum reveal a wave-like pattern superimposed on a downward shift. A third warming beginning in November is accompanied by increasing total ozone values with the strengthening and a downward shift of the main ozone maximum to 40 hPa (22 km). It can be seen that also in November short-term ozone drops coincide with temperature decreases. The final warming period starts in December with minimum temperatures of -45° C ... - 50° C at the lowest height of 200 hPa (12 km). At the same time ozone values increase and reach maxima on 10 December. Ozone minimum and secondary ozone maximum disappear, and both ozone and temperature distribution show a smoothed pattern afterwards.

The layer of the strongest ozone reduction can readily be seen in Fig. 3, which shows the percentage portion of ozone - averaged for Umkehr layers - to the total ozone change. The total ozone change is referred to the period September 1985, when ozone values are at their "normal" level. During October and the first half of November layers 4 (125 - 62.5 hPa = 15 - 19 km) and 5 (62.5 - 31.25 hPa = 19 - 23.5 km) contribute about 100% to the total ozone decrease, which is about 15 - 25 % below the total ozone level of September.

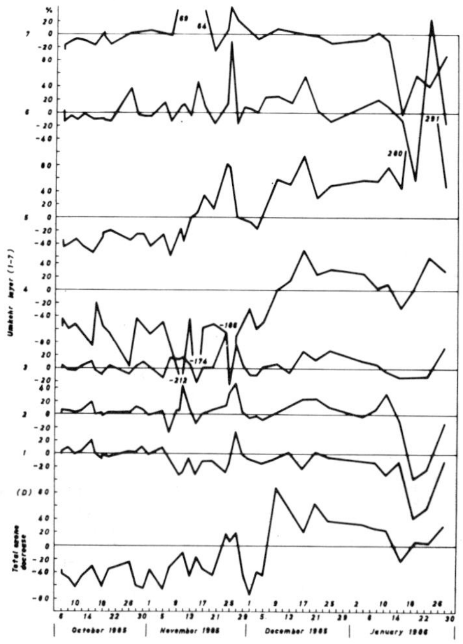

Fig. 3. Portion of ozone in each Umkehr layer to the overall change of total ozone difference as compared to September 1985 at Georg Forster station

On some occasions Umkehr layer 4 contributes more than 100% to the total ozone decrease. This is a result of a compensating ozone increase in the higher layers. The ozone increase above the September values after 7 December is again mainly produced by layers 4 and 5. It should be noted that ozone in layer 5 is above normal a few days earlier than ozone in layer 4. This may be an indication for a recovery proceeding from higher to lower altitudes.

3. TIME-HEIGHT CROSS-SECTIONS OF OZONE AND TEMPERATURE FROM SEPTEMBER 1986 TO JANUARY 1987

Fig. 4 shows that temperature and ozone distribution over Georg Forster station in the period September 1986 to January 1987 differ from the distribution one year before. A warming less than -85° C at about 50 hPa (20.5 km) starts in September, and this warming continues to about 24 October 1986 with minimum temperature values of -62° C at 140 hPa (14 km). October 1986 was thus by 10 K warmer in the lower stratosphere than October 1985. In spite of these higher temperatures reaching minimum values of -75 ... - 78° C between 40 and 150 hPa (13.5 - 22 km) at the end of September, the main ozone maximum moves upward from 70 hPa (18.5 km) to 25 hPa (25 km) in the period 15 - 27 September, and a secondary ozone maximum appears around 27 September at 100 hPa (16 km) with an ozone minimum in between at 40 hPa (22 km).

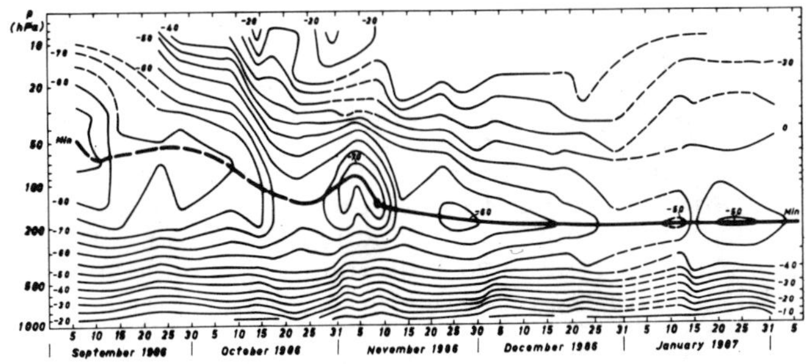

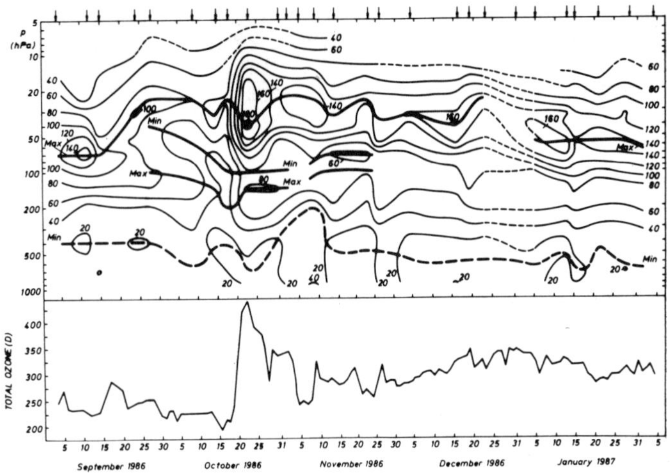

Fig. 4. As in Fig. 2, but for the period September 1986 to January 1987. Total ozone change is referred to the period September 1986

Both the secondary ozone maximum and the ozone minimum shift downward reaching their lowest level at 200 hPa (12 km) and 100 hPa (16 km), respectively on 18 October just before the strong total ozone increase from less than 200 D to more than 425 D between 18 and 22 October at the time of the maximum warming. Cooling of the lower stratosphere starts late in October, thus such a strong cooling as that in the first half of October 1985 does not occur. This first cooling is similar in magnitude to the second cooling in October 1985, though it occurs some days later than in 1985 with a minimum temperature at 140 hPa (14 km) on 9 November. At the same time ozone values in the lower stratosphere and consequently total ozone values are low, but the ozone concentration in the layer of the main ozone maximum between 20 and 40 hPa (22 - 26.5 km) remains fairly high. Secondary ozone maximum and minimum in the lower stratosphere disappear around 24 November - almost 3 weeks earlier than in 1985. It should be noted that the lower stratosphere was by about 5 K warmer at the end of November and the beginning of December in 1986 than in the corresponding months of 1985. Minimum temperature values of 210 K occur between 140 and 220 hPa (11 - 14 km). A gradual ozone increase is observed from the end of November to the end of December. Unfortunately, there is a gap of observations from 24 December 1986 to 5 January 1987, but in January the main ozone maximum is again at its lower level of 60 hPa (19.5 km). This behaviour is in contrast to 1985, when the ozone maximum shifted downwards from about 40 hPa (22 km) to 70 hPa (18.5 km) already on 12 December with the strong sudden warming.

4. DISCUSSION

If we compare the period September to December of 1985/86 with that of 1986/87, we recognize that

- there is a higher variability in ozone and temperature in 1985,
- in October and November the lower stratosphere is by 5 - 10 K colder in 1985,
- the layer of minimum ozone values occurs from the end of September to the beginning of December in 1985, while it lasts only to the second half of November in 1986,
- there is an indication for a general downward propagation of the warming and the episodes of ozone changes between October and January,
- a second ozone maximum with a pronounced ozone minimum between both maxima appears in October/November 1986 in the lower stratosphere, although the minimum temperatures were between -60 ... - 75° C, i.e. above the frost point of water and nitric acid,
- there is a close correspondence between the height of the temperature minimum and the height of the ozone minimum with the ozone minimum being 2 - 3 km higher than the temperature minimum.

It is interesting to note that Polar Stratospheric Clouds (PSC) over the Antarctic continent also show a downward slope in thickness from July to October as is shown by the SAM II weekly averaged extinction ratio at 1 μm (McCormick and Trepte 1986). On the other hand, while in early October 1985 the minimum temperature regimes are below the levels of water frost point (around -85° C) and nitric acid ice (around -80° C), the higher temperature in October and November 1986 of -65 ... -75° C are above the level to form nitric acid/ice aerosols (PSCs). It remains to be seen, whether heterogeneous chemical processes in the liquid phase such as those proposed by Crutzen and Arnold (1986) and Krüger et al. (1987) - under the specific circulation regime of the Antarctic polar vortex - contribute to the causes of the ozone deficiency in the Antarctic spring.

REFERENCES

Crutzen, P.J., Arnold, F., 1986: Nitric acid cloud formation in the cold Antarctic atmosphere: major cause for the springtime "ozone hole". Nature, 324, 1986, 651-655.

Chubachi, S., 1984: Preliminary result of ozone observations at Syowa station from February 1982 to January 1983. Memoirs of National Institute of Polar Research. Special Issue No. 34. Tokyo, December 1984, 13-20.

Feister, U., Grasnick, K.-H., Peters, G., 1985: Performance of the electrochemical ozone sonde OSR. Pure Appl. Geophys., 123, 422-440.

Krüger, B.C., Wang, G., Fabian, P., 1987: The Antarctic ozone depletion caused by heterogeneous photolysis of halogenated hydrocarbons. Geophys. Res. Lett., 14, 523-526

McCormick, M.P., Trepte, C.R., 1986: SAM II measurements of Antarctic PSCs and aerosols. Geophys. Res. Lett., 13, 1276-1279.

Rönnebeck, K., Sonntag, D., 1976: Eine weiterentwickelte elektrochemische Ozonradiosonde. Zeitschr. Meteorol., 26, 15-19.

Skeib, G., Popp, Ch., 1961: Messungen des Gesamtozongehalts der Atmosphäre in Mirny, Antarktika. Zeitschr. Meteorol., 15, 287-291.

ATMOSPHERIC OZONE VARIATIONS IN SPRING-SUMMER PERIOD ACCORDING TO THE DATA OF SOVIET ANTARCTIC STATIONS

A.I. Voskresensky, V.V. Nesterov, A.M. Sveshnikov, and L.N. Yurganov
Arctic and Antarctic Research Institute, 199226, Leningrad, USSR

I.L. Karol and A.M. Shalamyansky
Main Geophysical Observatory, 194018, Leningrad, USSR

ABSTRACT

Atmospheric ozone measurements conducted at 3 Soviet Antarctic stations are presented. The longest series of total ozone content (TOC) is examined for Mirny observatory (67°S, 93°E). The character of seasonal TOC variations is found to be different at stations in the region of stratospheric vortex and at stations in the outlying regions. TOC at Mirny station in the period of 1979-1988 appeares to demonstrate a continuous decrease with the maximum trend in Spring (October trend is 7% per year) and the minimum trend in Summer (January trend is 1% per year). The period of spring anomaly, so cold "ozone hole" period, is characterized by very low values of TOC, e. g. at Novolazarevskaya station TOC drops down to 150 DU. The largest changes of the ozone content were observed in the layers from 16 to 20 km. Possible reasons of the Antarctic spring decrease of TOC have been discussed.

The so-called effect of "ozone hole" attracts enormous attention (Farman et al. 1985, Gardiner and Shanklin, 1986). The effect means that total ozone content (TOC) decreases in Antarctica from September through October, showing its minimum in mid-October, then it recovers. The magnitude of this depletion is increasing beginning from 1980 up to nowadays. This report presents preliminary TOC data, as well as data on the vertical ozone distribution, obtained at Soviet Antarctic stations in 1987. The data show the dynamics of TOC depletion at Mirny station during the period from 1976 to 1987.

TOC measurements have been conducted by filter ozonometers of M-83 type (Gushchin et al., 1976). Before transportation to Antarctica the instruments had been calibrated by Dobson spectrophotometer installed at Voyeikovo Main Geophysical Observatory, near Leningrad. Every individual measurement has an error of about 5-8% (Romashkina, 1985). Mirny station appears to have the longest period of measurements (since 1958)(Karimova, 1975, Karimova, 1986). Unfortunately, up to 1976 the Soviet network of stations, measuring ozone content, was equipped with an old version of M-83 device, having some disadvantages (Gushchin et al.,1976, Romashkina, 1985). Hence, the data, obtained prior to 1976 could be used only after special processing.

The vertical distribution of ozone was measured at Novolazarevskaya station by electrochemical ozonesondes made in GDR (Gernandt, 1987).

Fig.1 gives TOC mean monthly values at Mirny station (67°S, 93°E) in Dobson

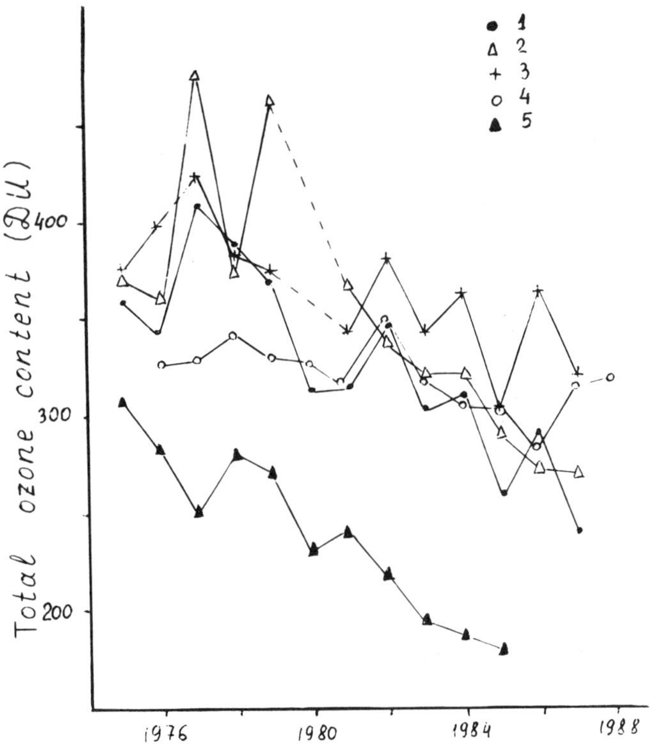

Fig. 1. Variations of TOC mean values in different months for 1976 - 1988 at Mirny. 1 - September, 2 - October, 3 - November, 4 -February. Data for Hally (5, October) are given for comparison.

ISBN 0-937194-15-8

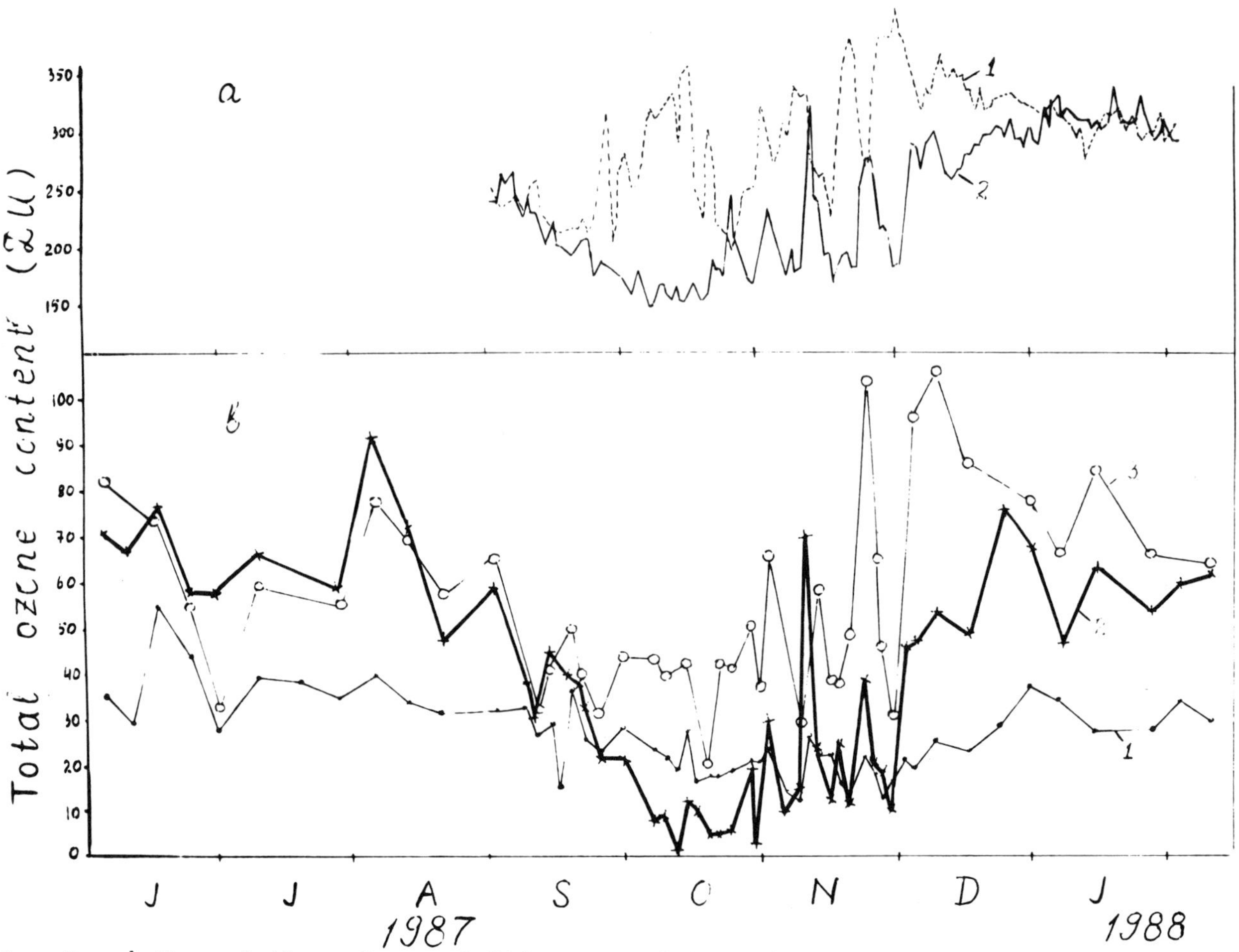

Fig. 2. a) Mean daily values of TOC measured by M-83 in Antarctica in 1987. 1 - Mirny, 2 - Novolazarevskaya.
b) Integrated ozone contents in DU for the following layers: 100-200 mb, 11-16 km (1); 50 - 100 mb, 16-20 km (2); 20-50 mb, 20-26 km (3) at Novolazarevskaya.

Units (1 DU = 10^{-3} atm·cm). A year to year depletion of TOC occurs through a whole sunny period of a year. The Table gives the values of slopes of regression lines. In Spring (September - October) TOC decreases at a rate of 3-4% per year and in Summer (December-February) - at arate of 1 - 1.5 % per year. TOC quasi-biannual oscillations have recently been recorded in September and November; odd years give lower TOC values then even ones. In September and October of 1987 minimal ever observed TOC has been registered at Mirny for these months. In November 1987 TOC values were lower than in 1986, though higher than the values registered in 1985. For a comparison Fig.1 gives monthly mean October TOC values for Hally station (Gardiner and Shanklin, 1986).

TOC variations over Antarctica show some interesting features in the period from September through December. Fig.2 presents TOC daily values for Mirny (thin line) and Novolazarevskaya (71°S, 12°E, thick line). From September 1 to

TABLE. LINEAR TRENDS OF TOTAL OZONE CONTENT ($b/\bar{\Omega} \pm \sigma/\bar{\Omega}$) IN % PER YEAR

Month	1976-1987	1979-1987
January	-0.7±0.9	-1.7±0.5
February	-1.0 ±0.6	-1.3±0.8
March	-1.0± 0.6	-1.7±0.8
April	-0.6 ±0.8	-1.5±0.9
September	-3.2 ±0.6	-3.9±0.5
October	-3.4 ±1.2	-7.2±0.9
November	-1.4 ±0.7	-1.8±1.1
December	-1.4 ±0.4	-3.0±0.5

b - a slope of regression line (DU/year)
$\bar{\Omega}$ - mean TOC for the given period,
σ - standard deviation of b.

23 TOC appeared to decrease at a rate of about 2 DU per day at Mirny and Novolazarevskaya in accordance with Gardiner and Shanklin (1986). By the end of September TOC changes at Mirny by 1.5 fold per 1 or 2 days, that seem to result either

from ozone decrease in "ozone hole" or due to its removal towards West Antarctica with Mirny remaining at its periphery. At Novolazarevskaya TOC continues to decrease until mid-October, giving values of 150 DU.

Fluctuations similar to these observed at Mirny are found to start at Novolazarevskaya in the middle of November. The disappearance of fluctuations is known to concur with a break of the stratospheric vortex (Karimova, 1975) observed on the 27-th of November at Mirny and on the 2-nd of December at Novolazarevskaya. The TOC at Vostok station remained low in November, being equal to 16o - 200 DU, the final stratospheric warming over Vostok station occured at the same time as at Mirny. In January 1988 the TOC at all three stations differed but slightly, the difference not exceeding 2% on an average.

The analysis of Fig.2b allows us to estimate the contributions of various stratospheric layers into the observed depletion of the TOC in spring months. As seen from Fig.2a the TOC display the decrease reaching 100 DU at Novolazarevskaya from the beginning of September to the middle of October. The integrated ozone content in the upper layer (2o-50 mb) decreased by 25 DU, in the middle layer(50--100 mb) this decrease amounted to 60 DU and in the lower layer (100-200 mb) it was 15 DU at the same time, that is the contribution of the middle layer in the TOC is responsible for a half of the total stratospheric ozone depletion. The ozonesonde measurements indicate that the ozone depletion starts in August, then this decrease continues till the middle of September due to its decrease in the upper and middle layers. The upper layer of the ozonesphere becomes stable in the middle of September, while in the middle layer ozone content continues to decrease in the following month till the middle of October when the TOC decreases another 40 DU. The changes in the lower layer are not so pronounced and lag behind those in the upper and middle layers; this is in accord with the concept of the ozone transfer from the upper layers of the ozonosphere to the lower ones. The behaviour of the middle layer in September-October does not agree with the assumption of downward transfer of ozone:after the upper layer stabilization the ozone content in the middle layer continues to decrease indicating the ozone destruction within the layer. It is interesting to note that except the period from the middle of September to the middle of October ozone variations in the middle layer occur simultaneously with those in the upper layer.

Fig.3 shows the changes in the vertical ozone distribution from August to November. The largest changes of the ozone content are observed in the layer from

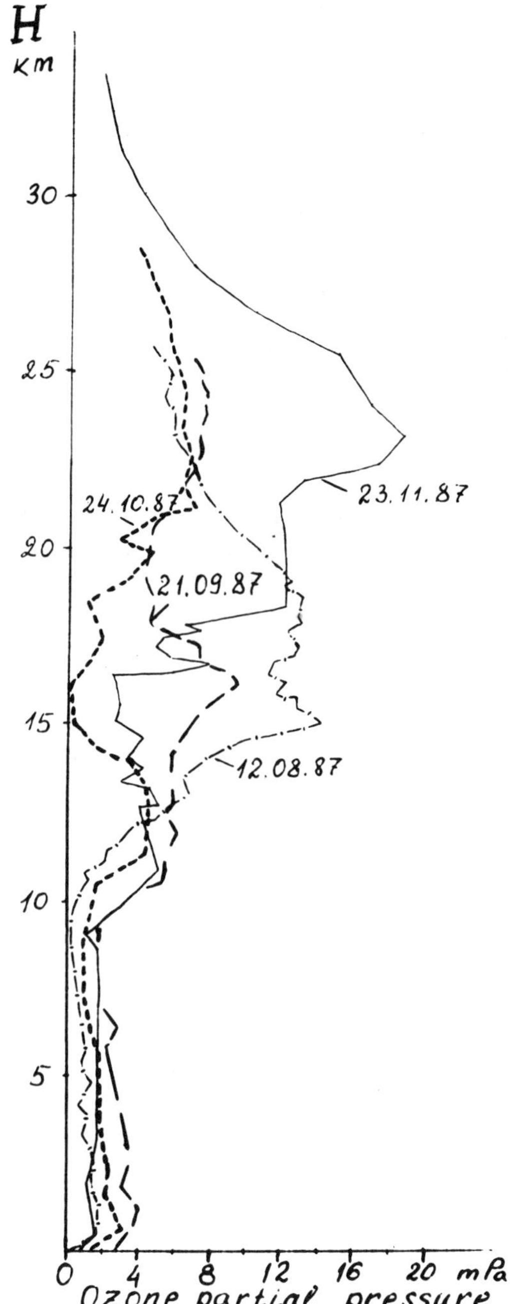

Fig. 3. Vertical distribution of ozone partial pressure, Novolazarevsk.

12 to 22 km, that is in the vicinity of the ozone maximum and below it; moreover at the height of 15-16 km the ozone content in certain days is below the detection limit.

Thus we can state that in the austral Spring of 1987 TOC in Antarctica was found to be lower then in all preceeding years.The pattern of ozone variations in

Antarctica suggests a complex mechanism resulting in the "ozone hole". There is a strong evidence indicating a dynamical nature of the ozone variations throughout the year, including spring months. The decrease of the TOC in September-October mainly at 16-20 km, however, can be attributed to the processes occuring within the ozone layer. The data available are not sufficient to verify existing hypotheses that would explain such a rapid and strong ozone concentration decrease.

ACKNOWLEGMENT.

The authors thank K.I.Romashkina for preparing instruments and R.M.Sayetov, A.V.Gubin for taking part in observation programme.

REFERENCES.

Farman, J.C., B.J.Gardiner, J.D.Shanklin, 1985: Large losses of ozone in Antarctica reveal seasonal ClO_x/NO_x interaction. Nature, 315, 207-210.

Gardiner, B.G., J.D.Shanklin, 1986; Recent measurements of antarctic ozone depletion. Geophys.Res.Lett., 13, 1199-1201.

Gernandt, H., 1987: The vertical ozone distribution above the GDR-base, Antarctica, in 1985. Geophys.Res.Lett., 14, 84 - 86.

Gushchin, G.P., K.I.Romashkina, A.M.Shalamyansky, 1976: Experience of total ozone content measurements using an improved version of ozonometer M-83 in Voyeikovo, 1971-1974. Trudy GGO, 357, 106-120 (in Russian).

Karimova,G.U.,1975: Atmospheric ozone in polar regions. Tr. Arkt. Antarkt. In. Leningrad, Gidromtizdat, 168pp.

Karimova,G.U.,1986: Results and prospects of atmospheric ozone studies in Antarctica. Proceed. of the 2-nd Symp. Antarctic meteorological studies, pt.1, 124-130 (in Russian)

Romashkina, K.I., 1985:Comparison of analysed results of observations conducted by ozonometers M-83 before and after its improvement. Trudy GGO, 487,101-111 (in Russian).

RADIATIVE RELAXATION IN THE POLAR STRATOSPHERE: THE ROLE OF OZONE

Pi-Huan Wang
Science and Technology Corporation, Hampton, Virginia 23666, USA

*A. Ghazi**
Commission of the European Communities, 1049 Brussels, Belgium

M.P. McCormick
Atmospheric Sciences Division, NASA Langley Research Center
Hampton, Virginia 23665, USA

ABSTRACT

Radiative relaxation processes of the thermal disturbances associated with planetary waves in the Antarctic stratospheric polar region have been studied by using satellite ozone measurements of the Stratospheric Aerosol and Gas Experiment (SAGE II) and the associated temperature data from the National Meteorological Center (NMC).

1. INTRODUCTION:

Stratospheric radiative relaxation processes have been theoretically examined by many investigators. In the case of infrared radiation, the thermal emission of CO_2 15 μm band plays an important role in suppressing the temperature perturbations (Dickinson 1973; Fels 1982). Approximately, this relaxation process is proportional to the local temperature perturbation. It is also realized that wave structures in the thermal field enhance the relaxation processes (Fels, 1982). As to the short wave radiation, the stratospheric ozone absorption of the solar radiation is important in damping thermal disturbances in the upper stratosphere (Ghazi et al., 1979, 1985). This radiative relaxation is primarily a consequence of the coupling between temperature and the ozone heating. By using daily data sets from the Stratospheric Aerosol and Gas Experiment (SAGE I), Ghazi et al. (1985) studied these damping processes in great detail. The objective of this paper is to extend their analysis to examine radiative relaxation processes in the Antarctic stratosphere during the spring by using recent ozone measurements from the SAGE II satellite instrument. The period of observational analysis corresponds to that of formation of the "ozone hole" over the South pole.

2. Computation

Following Ghazi et al. (1985), the governing equation of the temperature departure (T′) from the zonal average can be written as

$$\frac{\partial T'}{\partial t} = PW + Q'_{IR} + Q'_S \qquad (1)$$

where PW represents the effect of planetary waves on T′, and Q′ is the perturbation radiative heating (or cooling) with the subscripts S and IR denoting solar and infrared terms, respectively. At a certain time and given latitude, T′ and Q′ are functions of altitude and longitude. By multiplying both sides of Eq. (1) by T′ and then averaging over longitudes, we have:

$$\frac{1}{2}\frac{\partial}{\partial t}\overline{T'^2} = \overline{(PW)T'} + \overline{Q'_{IR}T'} + \overline{Q'_S T'} \qquad (2)$$

The LHS of Eq. 2 can be interpretated as the time rate of change of zonally averaged inten-sity of the thermal disturbances. The first term on the RHS is the effect of planetary waves on $\overline{T'^2}$. The last two terms are the contribu-tions of the infrared and solar radiation of the perturbation field, respectively, to the temporal change of the intensity of the thermal disturbances through the zonal average of their correlation with the thermal disturbances. The main objective of this analysis is to study the Antarctic behavior of temperature and ozone waves, to examine those last two terms in Eq. 2, and to investigate the radiative damping coefficients a and b defined by:

$$a = -\left(\frac{\overline{Q'_{IR}T'}}{\overline{T'^2}}\right); \quad b = -\left(\frac{\overline{Q'_S T'}}{\overline{T'^2}}\right) \qquad (3)$$

To achieve this objective, the following steps have been taken:

1) Calculate the longitudinal distributions of Q_S based on SAGE II ozone profiles;

2) Determine the longitudinal distribution of Q_{IR} using associated temperature distributions;

3) Determine the fluctuations of T′, O'_3, Q'_{IR}, Q'_S based on a harmonic analysis;

4) Compute the zonal average quantities $\overline{T'^2}$, $\overline{O'_3T'}$, $\overline{Q'_{IR}T'}$, and $\overline{Q'_ST'}$;

5) Calculate a and b coefficients

As in the study of Ghazi et al. (1985), we adopted the radiative transfer model of Ramanathan (1976) for heating/cooling calculation.

As to the data sets, the SAGE II ozone density profiles and the temperature data from

*Privatdozent, University of Cologne, FRG.

ISBN 0-937194-15-8

the National Meteorological Center (NMC) during the period from October 1 to 12, 1987 in the south polar region, are employed in this analysis. This SAGE II ozone data are used to determine Q_S at 11 pressure levels between 50 and 0.5 mb. Below 50 mb, we have used the ozone values from the vertical ozone distribution given in the U.S. Standard Atmosphere (1976). The solar absorption by CO_2 and H_2O is also included. Their concentrations are assumed to be uniformly distributed with 320 ppmv and 3 ppmm, respectively. Since we are mainly interested in the effect of ozone solar heating associated with the waves on the damping processes, and since the solar absorption by CO_2 and H_2O in the stratosphere are of secondary nature, we emphasize mainly the contribution of O_3 in determination of Q'. Fig. 1 shows the SAGE II sampling locations on October 6, 1987. The daily mean latitude of the SAGE II measurements is listed in Table 1. Table 1 indicates that all the daily mean SAGE II latitudes during that period are within 25° of the south pole. Since the variation of the latitudes in Table 1 is relatively smaller than the meridional scale of the stratospheric planetary waves, this variation of latitude is ignored in the following discussions. For the reasons that, normally, there are 15 sunrise/sunset measurements per day, only the wave components which are less and equal to 3 are included in this analysis.

3. RESULTS:

(a) Evolution of $\overline{T'^2}$, $\overline{O_3'^2}$, $\overline{Q_{IR}'^2}$, $\overline{Q_S'^2}$

The results of $\overline{T'^2}$ are displayed in Fig. 2. As mentioned in the previous section, $\overline{T'^2}$ represents the intensity of the temperature perturbations associated with stratospheric long wave disturbances. Its temporal changes indicate intensification or decay of the disturbances. From Fig. 2, we can see that the intensity of thermal disturbances intensified on October 4, especially in the region below 4mb (~43 km) pressure level. This feature is also reflected in the calculated infrared cooling waves, $\overline{Q_{IR}'^2}$ (Fig. 3). Fig. 4 shows the corresponding temporal variation of ozone disturbances, $\overline{O_3'^2}$. The intensity of ozone perturbations exhibits generally a similar development as that of the temperature perturbations, with an outstanding amplification in the region below about 4 mb pressure level on October 4, 1987. The calculated disturbances associated with ozone absorption of solar radiation, Q_S', are shown in Fig. 5. The notable features in Fig. 5 are the intensification of

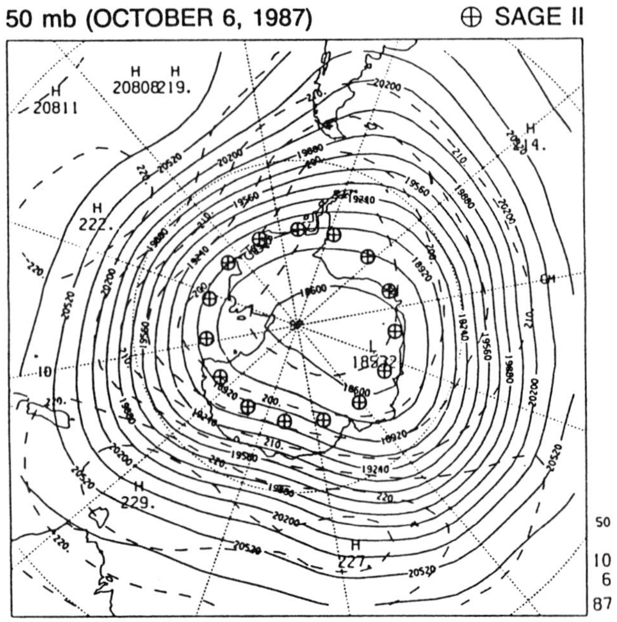

Fig. 1. SAGE II sampling locations on October 6, 1987.

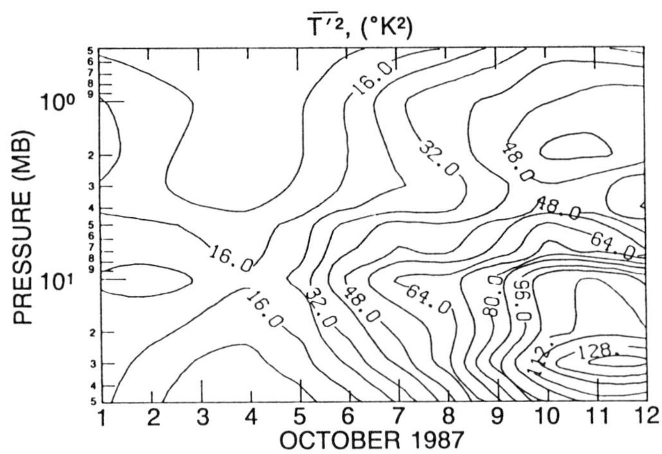

Fig. 2. Evolution of $\overline{T'^2}$ (October 1-12, 1987)

TABLE I

Daily Mean SAGE II sampling latitude*

	October 1987											
	1	2	3	4	5	6	7	8	9	10	11	12
Latitude (Deg)	-65.2	-68.6	-70.1	-71.8	-72.4	-72.4	-72.1	-71.5	-70.7	-69.6	-68.4	-67.0

*Sunrise event

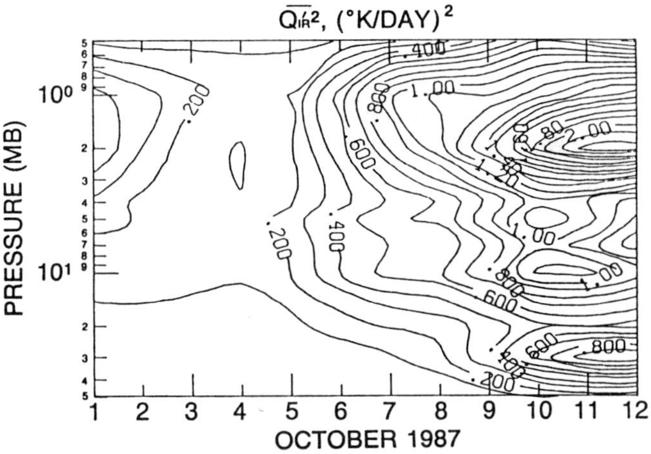

Fig. 3. Evolution of $\overline{Q_{IR}'^{2}}$ (October 1-12, 1987)

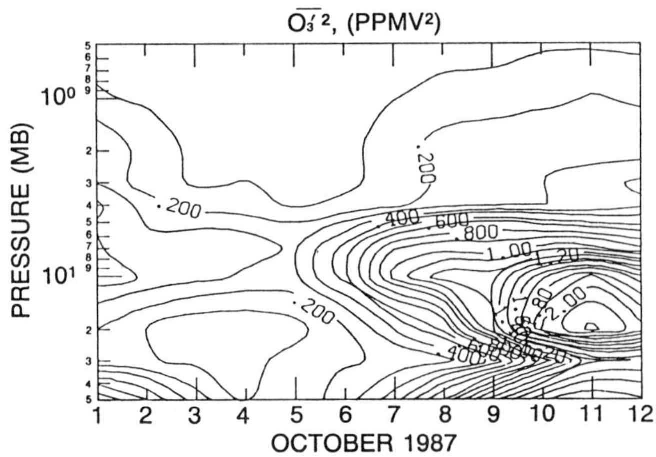

Fig. 4. Evolution of $\overline{Q_3'^{2}}$ (October 1-12, 1987)

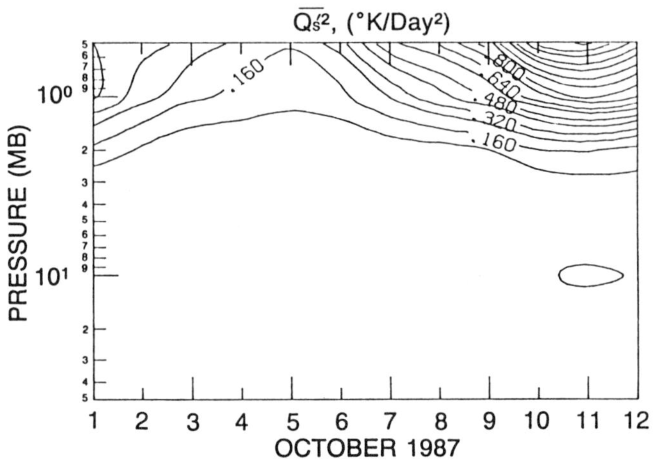

Fig. 5. Evolution of $\overline{Q_S'^{2}}$ (October 1-12, 1987)

the perturbation field of Q_S' above 2 mb pressure level beginning on the 5th of October. From Fig. 5, it is evident that a larger damping effect of about 1°k per day due to ozone solar heating occurred on the 11th of October, 1987, at 0.5 mb. Below 2 mb, the ozone absorption of solar radiation associated with ozone waves is very small.

(b) Evolution of $\overline{O_3'T'}$, $\overline{Q_S'T'}$, $\overline{Q_{IR}'T'}$

The calculated zonal mean covariance between ozone and temperature disturbances, $\overline{O_3'T'}$, is given in Fig. 6. In general, Fig. 6 indicates that the ozone and temperature perturbations are positively correlated below about 3 mb pressure level, and are negatively correlated above. The positive correlation below 3 mb could be mainly due to dynamical processes, and the negative correlation above 3 mb is generally believed to be induced by ozone photochemistry. It should be noted that wave dynamics may also introduce negative correlation between temperature and ozone if their gradient in the transport direction show opposite signs in the unperturbed state. The calculated results of mean covariance between temperature and infrared radiation $\overline{Q_{IR}'T'}$ are given in Fig. 7. Fig. 7 indicates that the thermal disturbances are negatively correlated with the perturbation of infrared radiation. Thus, the infrared radiation acts as a damping mechanism to suppress the thermal disturbances. This association is made on the ground that, to a large extent, the infrared radiation of the perturbation field depends on the local thermal emission associated with the thermal disturbances. The derived results of the mean covariance between temperature and ozone solar heating, $\overline{Q_S'T'}$, are displayed in Fig. 8. It is interesting to note that Figs. 6 and 8 show a similar distribution pattern, with negative correlation in the upper stratosphere and positive in the middle and lower stratosphere. However, the physical meaning of the results shown in Fig. 8 is different from that of the results in Fig. 6. Here, when the correlation between temperature and ozone solar heating, $\overline{Q_S'T'}$, is negative, the effect of solar heating

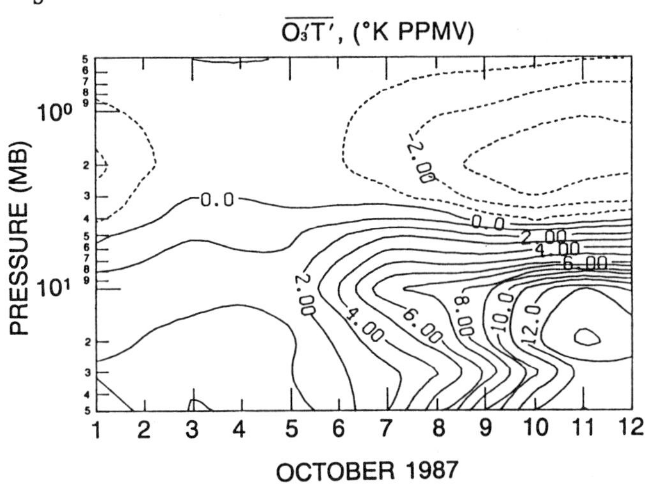

Fig. 6. Evolution of $\overline{Q_3'^{2}}$ (October 1-12, 1987)

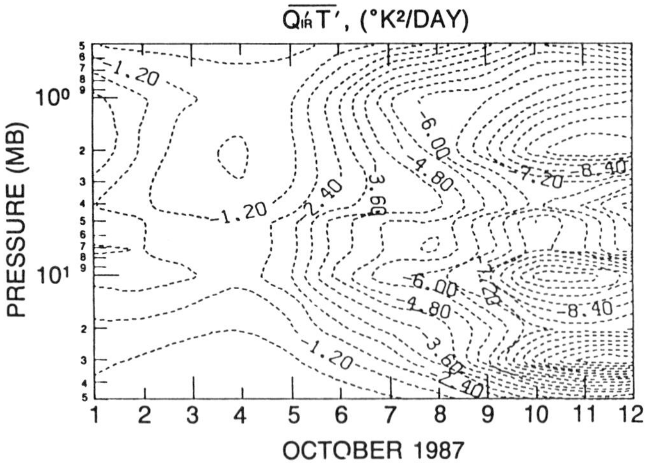

Fig. 7. Evolution of $\overline{Q'_{IR}T'}$ (October 1-12, 1987)

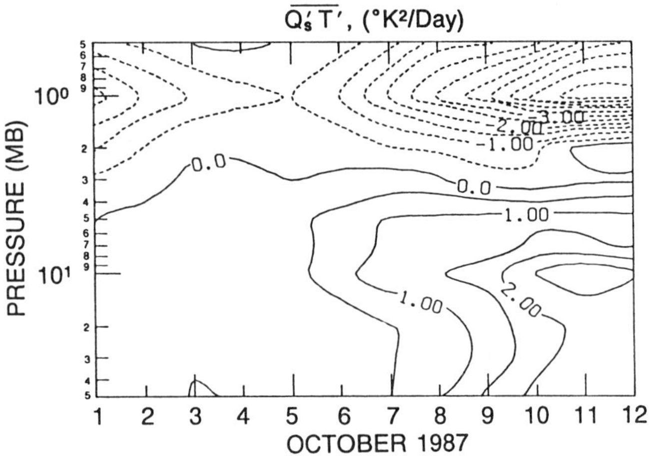

Fig. 8. Evolution of $\overline{Q'_S T'}$ (October 1-12, 1987)

associated with ozone waves is to damp the thermal disturbances, and is to amplify the disturbances when the correlation is positive. The negative correlation in the upper stratosphere is a consequence of two important factors. One is the negative correlation between the temperature and ozone disturbances (Fig. 6). The other one is the fact that, in the upper stratosphere, the solar absorption due to ozone is not saturated, thus the perturbation of ozone solar heating is proportional mainly to the perturbation of local ozone concentration, and is out-of-phase with the thermal disturbances. As to the positive correlation in the middle and lower stratosphere (Fig. 8), it is primarily a consequence of the so-called "opacity effect" of ozone solar heating in that altitude region (Ghazi et al., 1985). However, in this region, the damping effect due to infrared radiation (Fig. 7) is much larger in magnitude than the amplifying effect of ozone solar absorption (Fig. 8).

(c) Radiative relaxation coefficients a and b

Fig. 9 summarizes the radiative relaxation effect due to infrared radiation (a) and ozone solar heating (b). As can be seen from Fig. 9, ozone solar heating contributes about 38, 38, and 10% of the stratospheric radiative relaxation at .5, 1, and 2 mb pressure levels, respectively, over the period from October 1 to 12, 1987. Thus, the inclusion of this radiative relaxation effect of ozone solar heating can be important to the results of model simulation studies of the polar stratospheric circulation in the Antarctic region.

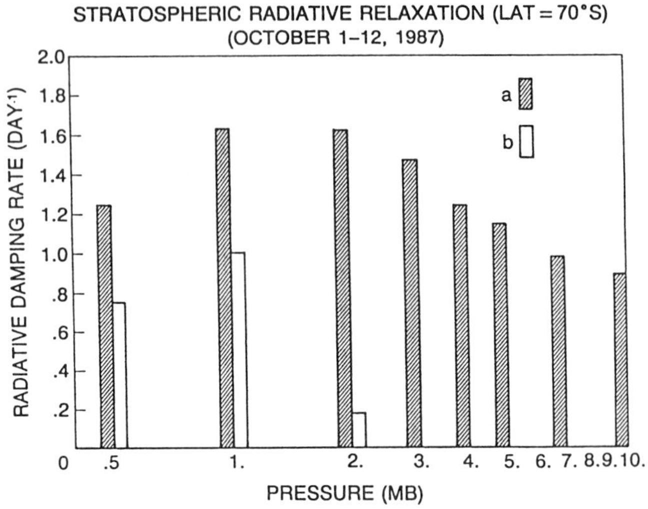

Fig. 9. Mean radiative relaxation coefficients a. and b.

ACKNOWLEDGMENTS

The authors are grateful to C. Trepte for providing us one of the illustrations of this study (Fig. 1). One of us (P.-H. Wang) is supported by NASA contract NAS1-18252.

REFERENCES

Cogley, A.C., and W.J. Borucki, 1976: Exponential approximation for daily average solar heating of photolysis. J. Atmos. Sci., 33, 1347-1356.

Craig, R.A., and G. Ohring, 1958: The temperature dependence of ozone radiational heating in the vicinity of the mesospeak. J. Meteor., 15, 59-62

Dickinson, R., 1973: Method of parameterization for infrared cooling between altitudes of 30 and 70 kilometers. J. Geophys. Res., 78, 4451-4457.

Fels, S. B., 1982: A parameterization of scale-dependent radiative damping rates in the middle atmosphere. J. Atmos. Sci., 39, 1141-1152.

Ghazi, A., V., Ramanathan, and R.E. Dickinson, 1979: Acceleration of upper stratospheric radiative damping: Observational evidence. Geophys. Res. Lett., 6, 437-440.

Ghazi, A., Pi-Huam Wang, and M.P. McCormick, 1985: A study on radiative damping of planetary waves utilizing stratospheric observations, J. Atmos. Sci., 42, 2032-2042.

Ramanathan, V., 1976: Radiative transfer within the earth's troposphere and stratosphere: A simplified radiative-convective model. J. Atmos. Sci., 33, 1330-1346.

MEASUREMENTS OF ANOMALOUS ClO CONCENTRATION IN THE SPRING STRATOSPHERE OVER MCMURDO STATION, 1986 AND 1987, AND CONSEQUENCES FOR OZONE DEPLETION

*R.L. de Zafra, M. Jaramillo, J. Barrett, L. Emmons, P. Solomon, and A. Parrish**
State University of New York
Stony Brook, New York 11794, USA

ABSTRACT

As much as 1.85 ± 0.4 parts per billion by volume of ClO has been found in the Antarctic spring vortex region, at about 20 km altitude, using a ground-based remote sensing mm-wave spectrometer. This is more than two orders of magnitude greater than normally present in this altitude range, and proves that highly anomalous chlorine chemistry, made possible by denitrification of the antarctic stratosphere, is responsible for the seasonal ozone hole. Measurements were made in both 1986 and 1987 and yield the diurnal behavior, long term behavior, and mixing ratio profile for ClO from about 17 to 45 km. Calculations show that the large concentration of ClO is enough to yield efficient closure of the chlorine catalytic cycle solely through the dimer formation mechanism proposed by Molina and Molina. Resulting ozone depletion rates calculated for various altitudes agree well with ozone observations at McMurdo Station in 1987 by the University of Wyoming.

1. INTRODUCTION

Soon after discovery of the antarctic springtime ozone hole was announced by Farman, Gardiner, and Shanklin (1985), a variety of suggestions were offered as the primary cause for its occurance. Among these were changes in the meridional transport dynamics of the south polar vortex, possibly enhanced by the unusually strong El Niño activity of the early 1980s; downward transport of increased mesospheric concentrations of NO_x arising from sunspot activity; upward transport of ozone-deficient air from the lower stratosphere induced by heating from increased dust or aerosol loading; and several chemical theories which had as common elements the supposition that removal of NO_x by mechanisms peculiar to Antarctica was being coupled with an increasing stratospheric halocarbon burden, to yield unprecidented and rapidly increasing destruction of ozone in the antarctic spring season.

The first National Ozone Expedition (NOZE I), conducted during the period from late August to November, 1986, at McMurdo Station, Antarctica (78° S), returned a wealth of data with which all but chemical theories could be eliminated as major factors in the formation of the ozone hole. Moreover, earlier evidence that stratospheric NO_x is strongly depleted within the antarctic spring vortex region (Keys and Johnson, 1986, and references therein) was confirmed in further detail (S.Solomon et al., 1987), and the evidence discussed in the present paper unequivocably showed that chlorine is the active agent in ozone destruction during the formation of the hole (de Zafra et al., 1987, and P. Solomon et al. 1987). The second National Ozone Expedition in 1987 enabled us to refine our measurements of chlorine monoxide (ClO) in the middle and lower stratosphere and to characterize its mixing ratio and column density over the important altitude range ~ 17 to 25 km.

Chemically active chlorine cannot deplete ozone without forming ClO as an intermediate product in any catalytic cycle yet concieved. Therefore the key to proving chlorine chemistry to be the driving force behind antarctic ozone destruction is to look for, and find, ClO at altitudes congruent with those of greatest ozone depletion. This was accomplished during NOZE I with our discovery of a layer of ClO in the vicinity of 20 km altitude with a mixing ratio more than two orders of magnitude greater than that normally found at mid-latitudes (de Zafra et al., 1987), and the demonstration that it was a seasonal event, declining below detection limits by early October when active depletion of ozone had essentially ceased (P. Solomon et al., 1987).

New data was acquired during NOZE II at McMurdo Station, using a spectral bandwidth twice that which was employed in 1986. This allowed a significantly greater fraction of the pressure broadened emission lineshape from low altitude stratospheric ClO to be measured, and thus permits a relatively accurate ClO profile to be determined by deconvolution of the lineshape down to ~ 17 km altitude, well below the peak of the mixing ratio. The ground-based emission technique which we employ, making use of the J=15/2 to 13/2 rotational transition at 278.63 GHz, may be carried out night or day, and therefore also gives a characterization of the diurnal change in ClO distribution, as well as its vertical profile.

We find a strongly bimodal structure to the total mixing ratio profile for ClO in the springtime antarctic stratosphere, with a "normal" high altitude layer of ClO occuring in the vicinity of 35 km (lower in geometric

*Also at Millitech, Inc., South Deerfield, MA, USA.

ISBN 0-937194-15-8

altitude here than at mid-latitudes, but more comparable in pressure altitude) in addition to the above-mentioned anomalous low altitude layer. The high altitude layer shows a different diurnal behavior, indicative of different chemistry, and remains after the low altitude component begins to deplete by early October.

2. OBSERVATIONS

The diurnal behavior and general characterization of the vertical profile, though already clearly exhibited in our NOZE I results, are best seen in data taken during the period Sept. 20-24, 1987, using the doubled spectral bandpass (512 MHz) referred to above. Observing conditions (dependent on very low tropospheric water vapor content) were quite good for these five consecutive days, (an unusual situation during 1987 at McMurdo), and the ozone depletion process was well advanced and the ClO layer well developed by this time.

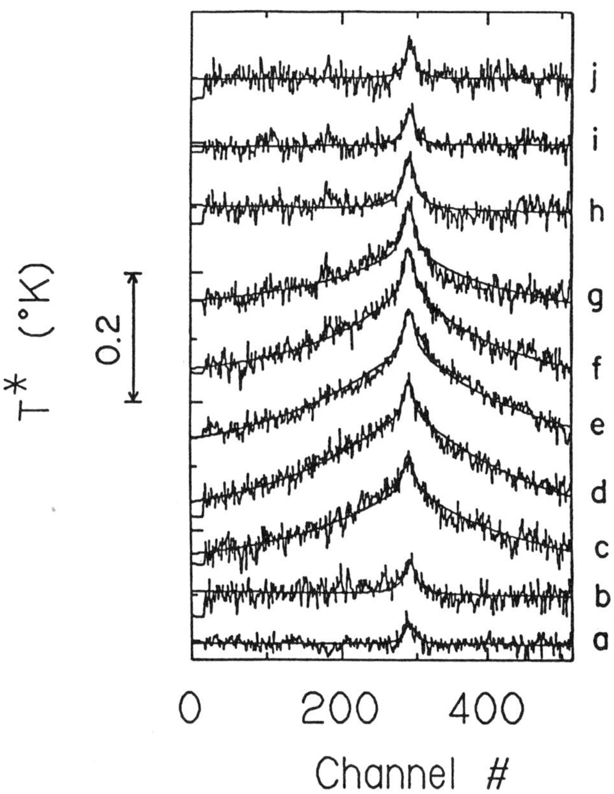

Fig. 1: Diurnal variation of the ClO emission lineshape at 278 GHz. The abscissa measures frequency at 1 MHz per channel number, and intensity is measured in Rayleigh-Jeans equivalent black body emission temperature. The various spectra are averages over two hour time intervals, except for "e". See text for explanation. Broad linewings which form during daylight hours are caused by low altitude ClO, indicating anomalous destruction of ozone.

Figure 1 shows the diurnal development of the total ClO emission lineshape over a series of daily time intervals into which the data has been binned, after processing to remove ozone background emission and low amplitude instrumental artifacts (de Zafra et al., 1988). The daytime growth of very broad line wings, indicative of highly pressure broadened - and thus low altitude - ClO emission, is very prominently displayed here. In the binning process, stratospheric dawn and sunset at 20 km were used as primary time markers, since the length of day changed somewhat during the period. All but "e" are two hours in length. Bin "a" contains the last two pre-dawn hours, "b" covers sunrise to two hours after dawn, etc. Bin "h" ends at sunset. Bin "e" is a midday interval begining 6 hours after dawn and ending 6 hours before sunset, and thus covering on average nearly a four hour interval. (Stratospheric daylight is a little over 15 hours in length on Sept. 20 and a little over 16 hours by the 24th.)

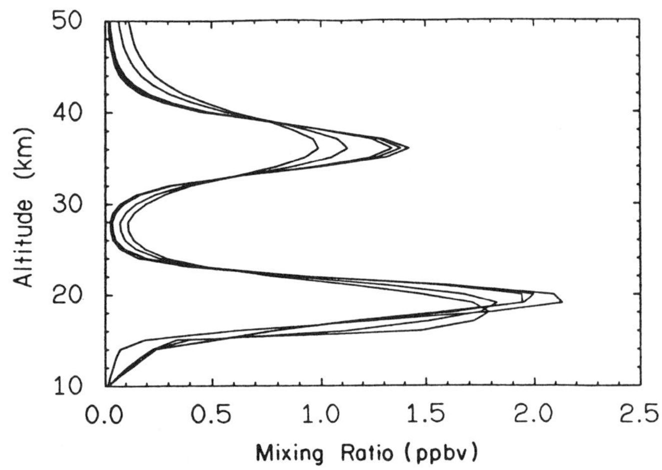

Fig. 2: Envelope of volume mixing ratio profiles obtained by deconvolution of mid-day lineshape "e" of Fig. 1, using various initial inputs to the deconvolution algorithm. See text.

The smooth curves through the data in Fig. 1 are computer-generated lineshapes which follow from the vertical profiles obtained by deconvolving the lineshape data, and are included to indicate the quality of fit and self-consistency of the profile recovery process. In Fig. 2 we display an envelope of recovered mixing ratio profiles for the midday ("e") data bin. These differ somewhat from one-another because the output of the deconvolution algorithms is somewhat sensitive to the original input "guess" for the vertical profile. Here, input profiles varying in mixing ratio by 300 %, and containing no initial bimodality, have all resulted in a series of strongly bimodal profiles which differ about the mean mixing ratio by only ~ 18 % at the peak of the lower component.

After allowing for additional uncertainties in instrumental sensitivity calibration, we arrive at a peak mixing ratio of 1.85 ± 0.4 parts per billion by volume at an altitude of 19.5 ± 1 km, where the uncertainties quoted are 2 standard deviation estimates (95% confidence level). The recovered low altitude profile is ~ 6 km in thickness, measured at its half-maximum, and this is very nearly the vertical resolution of the deconvolution algorithm used for profile recovery, raising the possibility that the layer is actually thinner. It should be noted however that the area under the profile is approximately conserved over a local range in altitude - a layer actually thinner than recovered would require a larger peak mixing ratio to be consistent with the observed lineshape, and this seems unlikely.

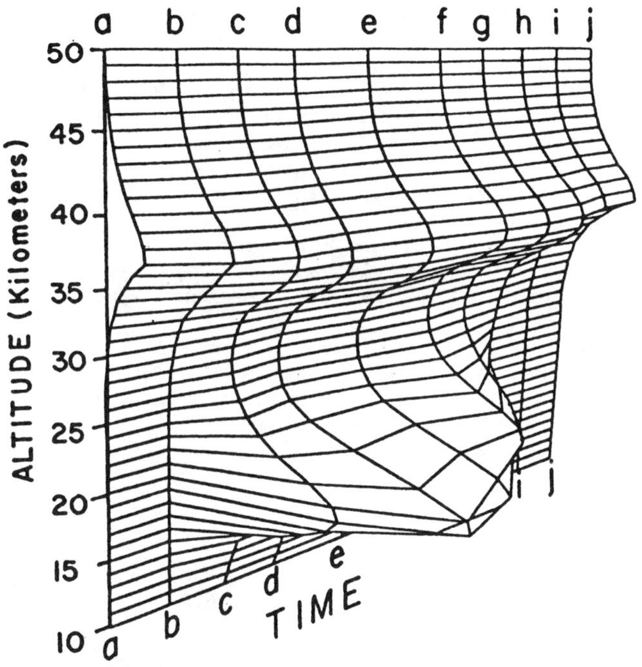

Fig. 3: Mixing ratio profiles recovered from pressure-broadened profiles of Fig. 1, displayed as a function of time. Letters refer to the same two-hour time averages used in Fig. 1. Note the different behavior of upper and lower ClO components.

The diurnal variation in the recovered mixing ratio profiles is shown in a three dimensional plot in Fig. 3, where the same labeling of time bins is employed as in Fig. 1. The considerable difference in behavior of the low and high altitude components is clearly revealed.

The column densities obtained by integrating the recovered profiles over the range 12-20 km are shown in Fig. 4, in comparison with a model prediction by Salawitch et al. (1988) for expected diurnal behavior over the same altitude interval. The arrowed data bars represent upper limits only, determined by the sensitivity limit of the detection system. General agreement is good over the daylight hours, although the modeled ClO column density predictions have been multiplied by 1.9 to obtain a match to the data. (Integration of our recovered profiles over the range 15-25 km, encompassing virtually the entire lower component, increases column densities by only ~ 5%.)

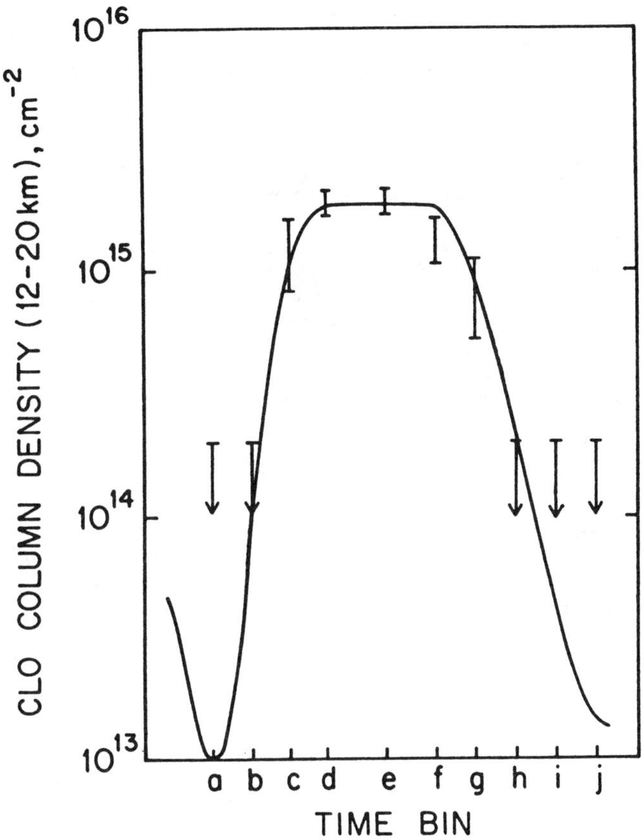

Fig. 4: Diurnal variation of ClO column density integrated from 12 to 20 km. Time intervals are the same as in Figs. 1 and 3. Bars with arrows indicate upper limits only. Solid curve is from an antarctic diurnal model by Salawitch et al., multiplied by a factor of 1.9 to match present data.

3. DISCUSSION

The extraordinary value for the mixing ratio of ClO found here at 19 to 20 km is in good agreement with our estimate of ~ 1.5 ppbv from the data of NOZE I (de Zafra et al., and P. Solomon et al. 1987), and is also in agreement with the maximum of 1.3 ppbv ± 30% found in late September, 1987, at 18.5 km altitude during the *in situ* ER-2 aircraft measurements described by

Brune et al. (1988), especially when it is considered that the aircraft probably did not reach the altitude of local maximum mixing ratio, and that the mixing ratio appeared to be still increasing as a function of latitude when the plane reached its southern limit of 72° S over the Palmer Penninsula.

The present results yield a midday ClO concentration of ~$3x10^9$ cm^{-3} at 19 to 20 km, enough to promote rapid ClO dimer formation at the currently reported rate for this process (Hayman et al., 1986), and to allow the mechanism proposed by Molina and Molina, namely dimer formation followed by release of both Cl atoms via photolysis and thermal decomposition, to provide the entire means for catalytically recycling chlorine. This scheme is particularly attractive at low altitudes in the early antarctic spring, given the absence of the atomic oxygen reaction which closes the chlorine cycle at high altitudes and normal latitudes. The alternative BrO + ClO reaction proposed by McElroy et al. (1986) appears not to be rapid enough in view of low concentrations of BrO found during NOZE II (Carroll et al.,1988). The dimer formation cycle, because of its quadratic dependence on ClO concentration, also explains how a roughly linear increase in the stratospheric chlorine burden can have led to a far from linear onset and progression of antarctic ozone depletion.

Finally, because ClO dimer formation depends on a three body collision to carry away excess energy, it is dependent on the background gas pressure as well as the square of the ClO concentration. This clarifies how a rapid rate of O_3 depletion can be sustained at altitudes well below that at which the ClO mixing ratio peaks. A detailed analysis shows that predictions of ozone removal rates based on our 1987 ClO concentration profile and the dimer formation cycle agree well with the 1987 rates of ozone depletion at various altitudes over McMurdo observed by Hofmann et al. (1988). A brief outline of these computations is given in de Zafra et al. (1988) and in detail in Barrett et al.,(1988).

ACKNOWLEDGEMENTS

We are grateful to the National Science Foundation's division of Polar Programs, and particularly to Drs. P. Wilkness and J. Lynch for the excellent logistical support received, and their help and encouragement during both NOZE I and II. We are equally grateful to the Chemical Manufacturer's Association's support under contracts FC 86-607 and FC 87-658, and to NASA's Upper Atmospheric Research Program under Dr. Robert Watson, for support of this work under grant NAGW-353.

REFERENCES

Barrett, J.W., P.M. Solomon, R.L. de Zafra, M. Jaramillo, L. Emmons, and A. Parrish, 1988: Formation of the Antarctic Ozone Hole by the ClO Dimer Mechanism, (in press, Nature, 334).

Brune, W.H., J.G. Anderson, and K.R. Chan, 1988: In situ observations of ClO in the antarctic: ER-2 aircraft results from 54° S to 72° S latitude, (in press, J. Geophys. Res. 93).

Carroll, M.A., R.W. Sanders, S. Solomon, and A.L. Schmeltekopf, 1988: Visible and near-ultraviolet spectroscopy at McMurdo Station, Antarctica: 5. Observations of BrO, (in press, J. Geophys. Res. 93).

de Zafra, R.L., M.Jaramillo, A.Parrish, P.Solomon, and J.Barrett, 1987: High concentrations of chlorine monoxide at low altitudes in the antarctic spring stratosphere: diurnal variation, Nature, 328, 408-411.

de Zafra, R.L., M. Jaramillo, J. Barrett, L.K. Emmons, P.M. Solomon, and A. Parrish, 1988: New observations of large concentrations of ClO in the springtime lower stratosphere over Antarctica, (in press, J. Geophys. Res. 93).

Farman,J.C., B.G.Gardiner, and J.D. Shanklin, 1985: Large losses of total ozone in Antarctica reveal seasonal ClO_x/NO_x interaction, Nature, 315, 207-210.

Hayman, G.D., J.M. Davies, and R.A. Cox, 1986: Kinetics of the reaction ClO+ClO->products and potential relevance to antarctic ozone, Geophys. Res. Lett.,13, 1347-1350.

Hofmann, D.J., J.W. Harder, J.M. Rosen, J.V. Hereford, and J.R. Carpenter, 1988: Ozone Profile Measurements at McMurdo Station Antarctica During the Spring of 1987. (in press, J.Geophys.Research)

McElroy,M.B., R.J. Salawitch, S.C. Wofsy, and J.A. Logan, 1986: Antarctic ozone: reductions due to synergistic interactions of chlorine and bromine, Nature, 321, 759-762.

Molina, L.T. and M.J. Molina, 1987: Production of Cl_2O_2 from the self reaction of the ClO molecule, J.Phys. Chem., 91, 433-436.

Salawitch,R.J., S.C.Wofsy, and M.B.McElroy, 1988: Chemistry of OClO in the antarctic stratosphere: Implications for bromine. Planetary Space Sci., 36, 213-224.

Solomon,P.M., B.Connor, R.L.de Zafra, A.Parrish, J.Barrett, and M.Jaramillo, 1987: High concentrations of chlorine monoxide at low altitudes in the antarctic spring stratosphere: secular variation, Nature, 328, 411-413.

2-D MODEL SIMULATIONS OF ANOMALOUS SPRING VARIATIONS OF OZONE, TEMPERATURE AND OTHER MINOR TRACE CONSTITUENTS IN THE SOUTHERN POLAR REGION

R.K.R. Vupputuri
Atmospheric Environment Service
Canadian Climate Centre
Downsview, Ontario, Canada

Abstract

A 2-D time dependent Eulerian model which takes into account the major interaction between chemistry and dynamical transports in the stratosphere is used to simulate, in a self consistent manner, the anomalous spring variations of ozone, temperature and other relevant minor constituents in the southern polar region. The normal transport coefficients and the standard chemistry have been altered to represent the anomalous dynamical and chemical conditions in the extremely cold lower stratosphere of Southern Hemisphere (south of 65°S) spring. The results show that it is necessary to invoke both dynamical and chemical mechanisms in order to explain the observed rapid spring decline of total ozone in the southern polar region.

1. Introduction

Since the first report of observations of a large decline in spring total ozone in the Antarctic by Farman et al., (1985) a number of studies have been made to explain the phenomenon, now widely known as the "Antarctic Ozone Hole". Some of the studies are chemically oriented while others are purely dynamical in nature. Possible dynamical mechanisms to account for the decreasing trend of total ozone have been discussed by Tung et al., (1986) and Mahlman and Fels (1986). Tung et al. argue that the increased net heating and subsequent upward motion caused by the enhanced aerosol loading during spring can produce a total ozone minimum similar to that observed by Farman et al., (1985) and Stolarski et al., (1986). Mahlman and Fels suggest that the weak planetary wave activity associated with much colder southern polar vortex in late winter and early spring would produce roughly the same effect by reducing the poleward and downward transport of ozone. The chemical theories, on the other hand, suggest the involvement of active chlorine species which are formed through heterogeneous reactions on ice particle surfaces in the much colder lower stratosphere as the cause of spring time ozone decline in the southern polar region (Solomon et al., 1986; McElroy et al., 1986; Solomon et al., 1987). Since the spring decline of total ozone in the southern polar region is accompanied by pronounced cooling of the lower stratosphere and changes in the other minor trace constituents such as N_2O, NO_x, CFC's and Cl_x, any theory that attempts to explain spring ozone depletion should also consider these associated changes.

Attempts to model the anomalous spring ozone variations in the southern polar region (Tung et al., 1986; Cariolle and Déque, 1986; Isaksen and Stordal, 1986; Evans et al., 1986; Pyle, 1986) have focussed either on chemistry or dynamics for the cause of ozone decline and they do not consider the simultaneous variations of temperature and the relevant minor constituents. In this study a classical type Eulerian 2-D time dependent model which takes into account the major interactions between stratospheric chemistry and dynamics is used to simulate, in a self consistent manner, the anomalous variation of ozone, temperature and the relevant minor trace constituents in the southern polar region during the spring season.

2. Brief Description of the Model

The 2-D model which extends from 10 km to 55 km in the vertical and pole to pole in the meridional direction uses the thermodynamic energy and the minor trace constituents representing odd oxygen, odd hydrogen, N_2O, odd nitrogen, CFC's and odd chlorine as prognostic equations. The zonal wind and the mean meridional circulation are determined diagnostically by the integration of the thermal wind equation and the streamfunction equation. Large scale quasi-horizontal eddy mixing is parameterized using generalized diffusion formulation on a sloping surface (Reed and German, 1965). Radiative heating and cooling and photochemical sources and sinks are incorporated in a form which allows for the major interactions among the minor trace constituents, temperature and mean circulation. As described in Vupputuri (1978/79) the model provides a credible description of seasonal and latitudinal variations of temperature, mean circulation, ozone and other minor trace constituents for the climatological stratosphere.

3. Altered Transports and Chemistry

For the purpose of this study the normal seasonal and latitudinal variation of transport coefficients and ozone photochemical sinks are altered to represent the anomalous dynamical and chemical conditions associated with a much colder southern polar vortex in late winter and early spring. More precisely, transport coefficients south of 65° S are modified in such way that they produce poleward and downward transports that are weak from August to October and strong in November and December following the vortex breakdown

ISBN 0-937194-15-8

It is generally believed that the weak quasi-horizontal eddy mixing from August to October represent the stable vortex conditions which prevent poleward transports of heat, ozone and other minor trace gas substances. In order to represent the anomalous chemical conditions, the photochemical scheme has been modified to include additional ozone loss reactions $ClO + HO_2$ and $ClO + BrO$. These reactions, particularly the later, have been proposed to be one of the potential chemical mechanisms contributing to the spring ozone decline in southern polar region (Hills et al., 1987). It is important to realize that although the above reactions can operate under the conditions of little or no sun, they would still require high concentrations of ClO and the presence of BrO in order to be effective in destroying ozone. Despite the fact that the chemical mechanisms of ClO formation are still speculative, there is strong observational evidence for abnormally high concentrations of ClO over the McMurdo station during the Antarctic spring season (Hills et al., 1987; Solomon et al., 1987). It is suggested that the heterogeneous reaction between HCl and $ClNO_3$ on ice particle surfaces in the extremely cold lower stratosphere in the southern polar region during early spring can release Cl and tie up the odd nitrogen resulting in enhanced ClO levels (Isaksen and Stordal, 1986). Due to the speculative nature of these heterogeneous reactions and the uncertainties in the reaction rate constants the anomalous variation of ClO concentration is not simulated in the present model study. Instead, enhanced levels of ClO, south of 65°S are specified based on observations during the months of September and October (Solomon et al., 1987). The photochemical model also assumes a 5 pptv mixing ratio for BrO in the lower stratosphere while assigning a value of 3.7×10^{-12} cm^3 mol^{-1} sec^{-1} for the rate constantof reaction between BrO and ClO.

4. Model Results

In order to explore the possible roles of dynamical and chemical mechanisms in the spring decline of total ozone over the southern polar region the following model experiments have been carried out by changing the declination of the sun daily and using a 6 hour time step in the model.

1) Experiment 1 uses the standard oxygen, hydrogen, nitrogen and chlorine chemistry with normal background stratospheric values for the seasonal and latitudinal variation of transport coefficients.

2) In experiment 2, the transport coefficients (K_{yy}, K_{yz} and K_{zz}) south of 65°S are modified to allow poleward transports that are weak from August to October and strong in November and December.

3) In experiment 3, the standard chemistry was also altered to include additional ozone loss reactions $ClO + HO_2$ and $ClO + BrO$.

In each of the above experiments the integration was continued until equilibrium is reached. The results are then analysed to produce monthly mean total ozone and latitude-height cross-sections of temperature, mean circulation and the mixing ratios of ozone and other minor trace constituents.

The results in the first experiment show that the calculated monthly mean meridional cross-sections of temperature, flow pattern of mean meridional circulation, ozone mixing ratio and the mixing ratios of other minor trace gases are in satisfactory agreement with observed variations in the background stratosphere. As shown in Fig. 1 the calculated monthly mean total ozone amounts are higher in the polar regions than at lower latitudes and near the pole the ozone column shows a distinctive annual cycle with the maximum in the spring and the minimum in the fall season. This variation in total ozone is typical of observations under normal atmospheric conditions. These results for the first experiment reveal the fundamental importance of large scale quasi-horizontal eddy mixing and mean meridional motions in maintaining the seasonal and latitudinal variations of temperature, ozone and other minor constituents in the stratosphere in the presence radiative-photochemical processes.

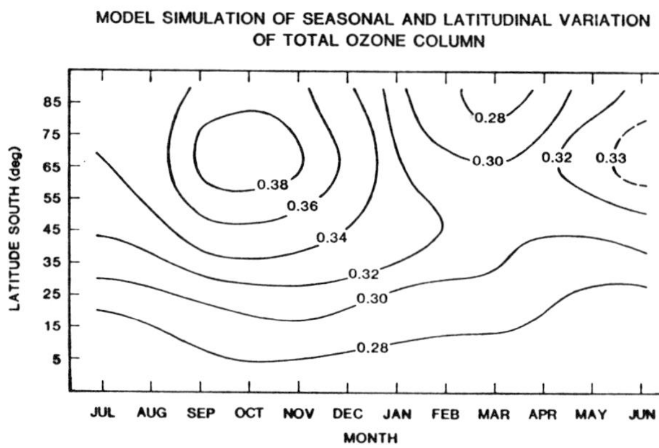

Figure 1: The computed distribution of zonal mean total ozone (cm STP) as a function of latitude and month in Case 1 experiment.

a) Experiment 2. Results with Altered Transports

Figs. 2 and 3 show the latitude-height cross-sections of the predicted changes in ozone mixing ratio and temperature resulting from the modified transport coefficients in Experiment 2 for the month of October while the seasonal and latitudinal variation of total ozone column change is illustrated in Fig. 4. It is evident from Figs. 2 and 3 that the reduced northward and downward transports by the large scale quasi-horizontal eddies south of 65°S during late winter and early spring could lead to substantial reduction in the ozone mixing ratio and temperature in the lower stratosphere. It is important to note that for temperature, ozone and other minor constituents such as NO_x and $C\ell_x$ with positive equilibrium meridional slopes the weaker poleward transports cause a decrease in meridional slopes (they become flatter) resulting in lower values of temperature and mixing ratios near the pole (Mahlman and Fels, 1986). For minor trace gases with negative meridional slopes in the lower stratosphere (N_2O and CFC's) the weak poleward transports cause an enhancement in the negative slopes resulting in higher values of mixing ratio near the pole. As illustrated in Fig. 4, the altered transports in Experiment 2 result in a substantial decline of total ozone in a column during the months of September and October over the southern polar region although the amplitude of total ozone decline is less than that observed on a zonally averaged basis.

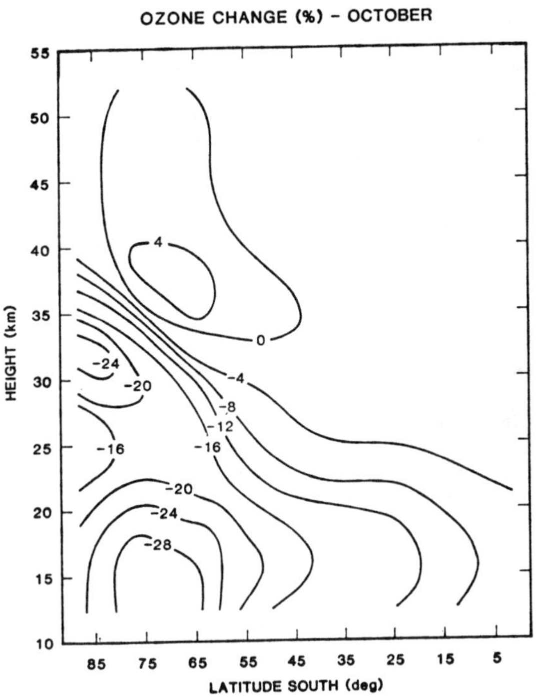

Figure 2: The computed latitude-height cross-section of ozone mixing ratio changes (in %) in Case 2 experiment.

One of the most significant feature of the observations is that the spring time variations of total ozone and temperature in the lower stratosphere are highly correlated over the Antarctic (Angell, 1986; Bozkov, 1986). To test whether the model can reproduce this feature, the computed annual cycle variation of zonal mean total ozone and temperature in the lower stratosphere (21 km) are compared in Fig. 5. It may be seen from Fig. 5 that the observed stable correlation between total ozone and temperature in the lower stratosphere is indeed preserved in the model calculation particularly in southern polar regions during the spring season. The model calculations indicate that each 1K cooling leads to a 3 to 4% change in the ozone mixing ratio in the lower stratopshere (see Figs. 2 and 3). In terms of total ozone decline, each 1 K cooling leads to a decrease of about 5 Dobson units.

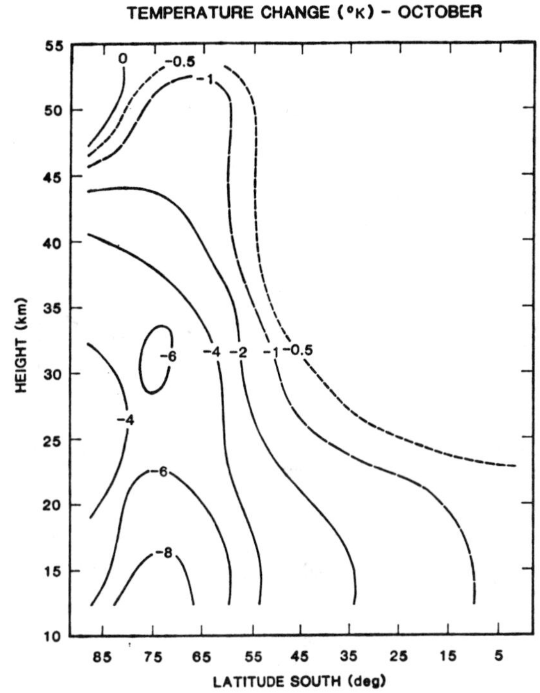

Figure 3: As in Fig. 2 but for the temperature changes (°K).

An analysis of ozone and heat budgets suggests that, in the lower stratosphere, the radiative-photochemical processes and dynamical transports are tightly coupled and that in the absence poleward transports of ozone and heat south of 65°S during late winter and early spring, both ozone mixing ratio and temperature continue to decrease until the net ozone and temperature tendencies are balanced by the ozone photochemical sources and sinks and net radiation respectively.

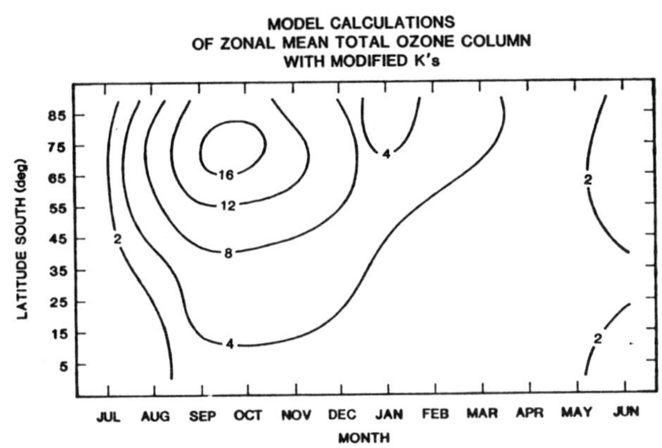

Figure 4: The computed distribution of zonal mean total ozone reduction (in %) as a function of latitude and month in Case 2 experiment.

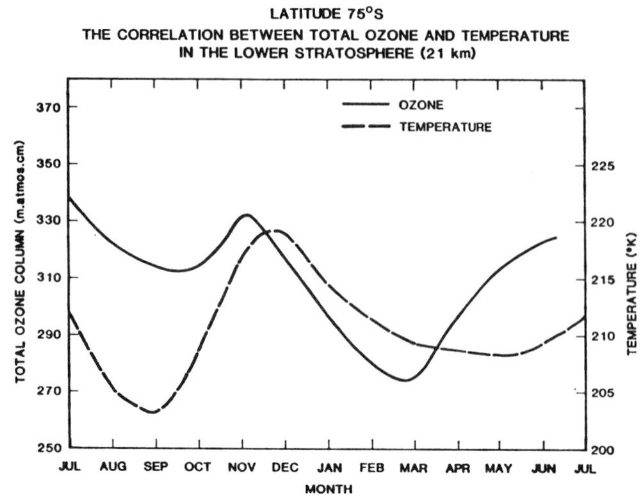

Figure 5: The comparison of computed annual cycle variation of total ozone with temperature in the lower stratosphere.

b) Experiment 3. Results with Altered Transports and Chemistry

The fact that the altered transports alone cannot produce the observed amplitude of the total ozone decline suggests that additional ozone loss mechanisms related to the heterogeneous chemical reactions on ice particle surfaces in the extremely cold Antarctic lower stratosphere could be responsible for further ozone decline. This combined dynamical and chemical hypothesis has been tested in Experiment 3. Fig. 6 gives the calculated total ozone changes as a function of latitude and season due to the incorporation of the combined effects of altered transports and chemistry. It is of considerable significance to note that the incorporation of additional ozone loss processes results in a much more rapid decline of total ozone in the southern polar region during the months of September and October and the computed amplitude of about 30% is in better agreement with observations. This suggests that the combination of both dynamical and chemical effects could be responsible for the rapid spring ozone decline over the Antarctic.

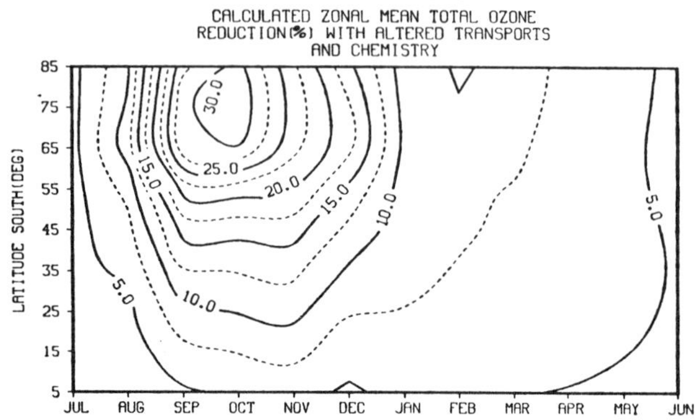

Figure 6: As in Fig. 4 except in Case 3 experiment with altered transports and chemistry.

To determine the extent of chemistry contribution to the ozone column decline in the southern polar region Experiment 3 has been repeated by changing only the chemistry. The results indicate that the altered chemistry alone contributes less than 10% to the ozone column decline and that it is not by itself sufficient to explain the observed spring ozone column decline in the southern polar region. More importantly that in the chemistry alone experiment, the ozone depletion does not lead to the associated temperature cooling of the lower stratosphere. This suggests that it is not possible to explain the observed simultaneous variations of ozone and temperature in southern polar lower stratosphere during the spring season by chemical theories alone without invoking dynamical changes.

5. Concluding Remarks

Despite certain drawbacks of using Reed and German's K-theory approach to parameterize the eddy transport quantities in a Eulerian model and the uncertainties in the ozone chemical loss reactions, the 2-D model results indicate that both transport and chemical mechanisms are necessary in order to explain the observed rapid spring total ozone decline over the south polar region. Additional information on the transport characteristics and conditions in the extremely cold lower stratosphere over south polar region would help to place the combined dynamical-chemical hypothesis on a more quantitatively understandable level.

Acknowledgements

The author would like to thank Mr. J.A.W. McCulloch, Director General of the Canadian Climate Centre for encouragement and support; Dr. G.J. Boer for helpful comments and suggestions on the manuscript; F. Szekely who assisted with computer programming and Lynda Smith for efficient preparation of the manuscript.

References

Angell, J.K. The Close relation between Antarctic total ozone depletion and cooling of the Antarctic lower stratosphere. Geophy. Res. Lett., 13, 1240-1243, 1986.

Bojkov, R.D. The 1979-1985 ozone decline in the Antarctic as reflected in ground based observations. Geophy. Res. Lett., 13, 1236-1239, 1986.

Cariolle, D., M. Déqué and J.J. Morcrette. A GCM simulation of the ozone seasonal variations at high latitudes in the southern hemisphere. Geophy. Res. Lett., 13, 1304-1307.

Evans, W.F.J., B.W. Boville, J.C. McConnell and G.S. Henderson. The simulation of Antarctic ozone with chemical and dynamical effects. Geophy. Res. Lett., 13, 1323-1326, 1986.

Farman, J.C., B.G. Gardiner and J.D. Shanklin. Large losses of total ozone in Antarctica reveal seasonal ClO_x/NO_x interaction. Nature, 315, 207-210, 1985.

Hills, A.J., R.J. Cicerone, J.G. Calvert, and J.W. Birks. Kinetics of the Bro + CℓO reaction and implications for stratospheric ozone. Nature, 30, 405-408, 1987.

Isaksen, Ivar S.A. and F. Stordal. Antarctic ozone depletion: 2-D model studies. Geophy. Res. Lett., 13, 1327-1330, 1986.

Mahlman, J.D. and S.B. Fels. Antarctic ozone decrease: A dynamical cause? Geophy. Res. Lett., 13, 1316-1319, 1986.

McElroy, M.B., R.J. Salawitch, S.C. Wofsy and J.A. Logan. Reductions of Antarctic ozone due to synergistic interactions of chlorine and bromine. Nature, 321, 759-762, 1986.

Pyle, J.A. Large ozone losses in Antarctica: The role of heating perturbations. Geophy. Res. Lett., 13, 1320-1322, 1986.

Reed, R. and K.E. German. A contribution to the problem of stratospheric diffusion by large scale mixing. Mon. Weather Rev., 93, 313, 1965.

Solomon, S., R.R. Garcia, F.S. Rowland and D.J. Wuebbles. On the depletion of Antarctic ozone. Nature, 321, 755-758, 1986.

Solomon, P.M., B. Connor, R. de Zafra, A. Parrish, J. Barrett and M. Jaramille. Observations of high concentrations of chlorine monoxide in the lower stratosphere during the Antarctic spring, II, Secular variation. Nature, 328, 411-413, 1987.

Stolarski, R.S., A.J. Krueger, M.R. Schoeberl, R.D. McPeters, P.A. Newman and J.C. Alpert. Nimbus 7 satellite measurements of the spring time Antarctic ozone decrease. Nature, 322, 808-811, 1986.

Tung, K.K., M.K.W. Ko, J.M. Rodriguez and N.D. Sze. Antarctic ozone variations: A manifestation of dynamical and chemical processes. Nature, 322, 811-814, 1986.

Vupputuri, R.K.R. The structure of the natural stratosphere and the impact of chlorofluoromethanes on the ozone layer investigated with a 2-D time dependent model. Pure Appl. Geophys., 117, 448-485, 1978/79.

STRATOSPHERIC HYDROGEN CYANIDE MEASUREMENTS AT MCMURDO STATION DURING THE ANTARCTIC SPRING

*M. Jaramillo, R.L. de Zafra, J. Barrett, L.K. Emmons, P.M. Solomon, and A. Parrish**

State University of New York

Stony Brook, New York 11794, USA

ABSTRACT

Measurements of hydrogen cyanide (HCN) in the springtime antarctic stratosphere were carried out using ground-based mm-wave spectroscopy. We find a volume mixing ratio of 65 $\pm$ 30 pptv at 40 km, and a steeper decrease with altitude than that found at tropical latitudes. This result may be further evidence of substantial vertical subsidence of the antarctic winter stratosphere.

1. OBSERVATIONS AND ANALYSIS

During the National Ozone Expeditions I and II to McMurdo Station, Antarctica (78° S), the Stony Brook team made two measurements of stratospheric HCN, one on the night of September 28, 1986 and one on the same date one year later.

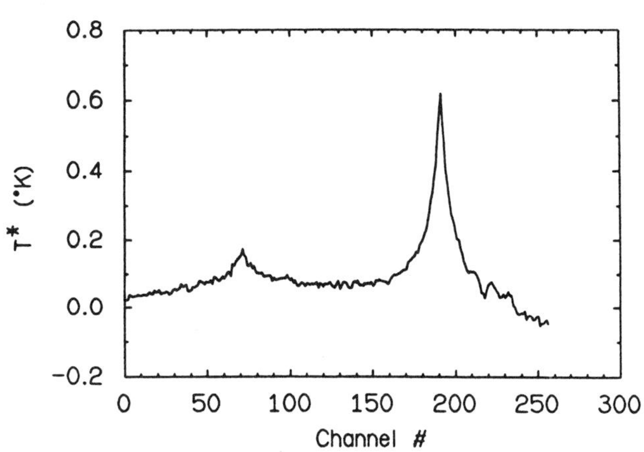

Fig. 1. Averaged spectra from the two antarctic measurements of 1986 and 1987. Vertical scale is relative intensity of emission in units of equivalent Rayleigh-Jeans blackbody temperature (Kelvin). The horizontal scale is frequency, at 1 MHz per channel. The stronger line (right) is the $22_{2,20}$ -> $22_{1,21}$ transition of vibrationally excited ozone at 268.664 GHz; the line towards the left is the HCN signal.

The technique employed, ground-based mm-wave emission spectroscopy, has been described in detail in Parrish et al. (1988a).

The observed HCN spectral line is the J = 3 -> 2 rotational transition at 265.886 GHz. Figure 1 shows the averaged spectra from the two measurements, a total of 13 hours of integration time. The vertical scale in the figure is relative intensity of emission which would be received from the zenith direction, assuming no tropospheric attenuation, and it is marked in units of equivalent Rayleigh-Jeans blackbody temperature (Kelvin). The horizontal scale is frequency and the resolution is 1 MHz per data point.

Two steps were involved in the analysis of the data. The first was to remove a synthesized background due to ozone in order to leave only the net HCN signal (contributions from other molecular species in this region of the spectrum are negligible). The second step is the inversion of the data to obtain the HCN altitude profile information from the line shape, which is determined by pressure broadening from different altitudes. The ozone background subtraction was performed with the aid of data from balloon flights made by the University of Wyoming group (Hofmann et al., 1987 and 1988) to determine line shape contributions from lower altitudes. After the ozone background is removed the spectrum should contain only the HCN signal, but there actually remains a residual baseline curvature of instrumental origin which produces a small asymmetry in the line center. For this reason we have limited our analysis to a bandpass of $\pm$ 70 MHz around the HCN line center. This has the effect of limiting the profile information that we can recover to altitudes above ~ 28 km. (The frequency resolution of the spectrograph [1 MHz] also places an upper limit on the altitude information at about 50 km for the prevailing antarctic pressure profile.)

The retrieval method used to extract the HCN altitude distribution from the line shape (Twomey et al., 1977) requires an initial trial profile from which a synthesized line is generated, compared with the data, and modified accordingly. The process is then repeated iteratively until a given level of fit, dependent on the noise, is reached. The results are not unique, since a multitude of profiles lying within some envelope can give equally good overall fits to the data. The uncertainty inherent to the inversion

*Also at Millitech, Inc., South Deerfield, MA, USA.

ISBN 0-937194-15-8

process, a measure of which is the spread of the different fits, is the main contribution to the experimental error. Calibration of instrument sensitivity and the error in the value of the pressure broadening coefficient contribute an uncertainty of ~ 17% at the 2 sigma (95%) confidence level (Parrish et al., 1988a). The output of the inversion algorithm can be influenced by the initial input profile--this is particularly so at extreme high or low altitudes, where the line shape is relatively insensitive to the mixing ratio.

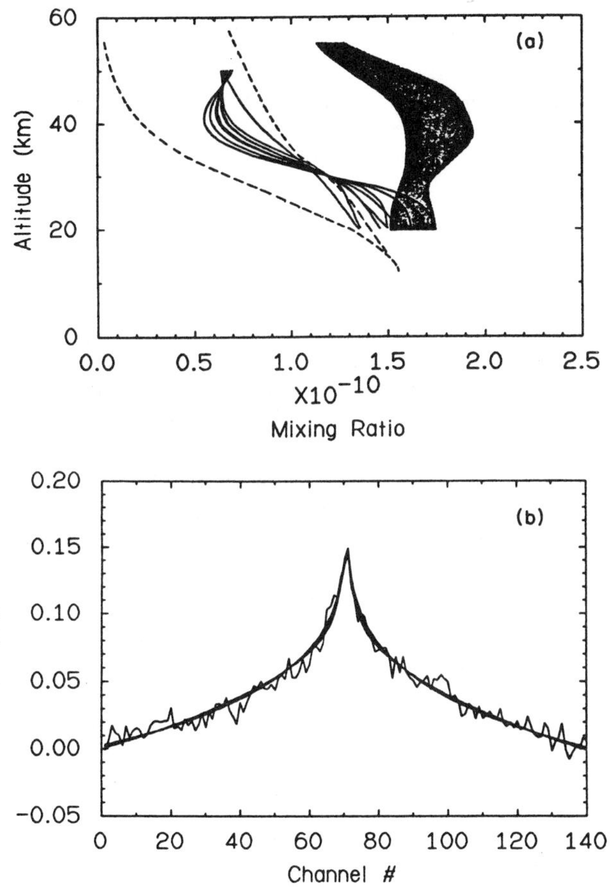

Fig. 2. a: Retrieved HCN profiles, obtained by inversion of the averaged spectra after the removal of the ozone background. Each inversion was made with a different choice of input profile for the deconvolution algorithm. The shaded area represents the most likely profile envelope from our measurements of HCN at 20° N (Jaramillo et al., 1988). The dotted lines correspond to the model of Brasseur et al. (1985) for two cases of HCN photolysis cross section (see text). b: Fits to the spectral HCN data from Antarctica, covering 140 MHz, corresponding to the retrieved profiles of part (a) of the figure.

The solid lines in Fig. 2 (a) are the family of HCN mixing ratio profiles that results from a variety of input distributions, having values at 20 km that correspond roughly to the range of measurements reported in the literature (see "discussion" below). The corresponding fits to the data from the inversions in Fig. 2 (a) are shown in Fig. 2 (b). The two dashed lines in Fig. 2 (a) are the theoretical mid-latitude profiles predicted by Brasseur et al. for the cases of zero photolysis, and HCN photolysis cross section equal to that of HCl. They are shown for reference only, as they were not intended to apply to polar situations (note however that the stratosphere at 79° S is exposed to sunlight for > 17 hrs per day in late September). All but one of the recovered profiles show a pronounced gradient to ~ 40 km, followed by a constant or increasing mixing ratio for h > 40 km, although the latter may be an artifact caused by local noise near the line center.

2. DISCUSSION AND CONCLUSIONS

Atmospheric HCN is believed to be produced at ground level by a variety of natural and anthropogenic sources; it has an average tropospheric mixing ratio of about 160 pptv according to Rinsland et al. (1981) (an earlier measurement by Coffey et al. (1981) gave a column above 12 km corresponding to a mean mixing ratio of 155 ± 20 pptv). Current theory (Cicerone and Zellner, 1983; Brasseur et al., 1985) indicates that the HCN tropospheric mixing ratio should be roughly constant with altitude, with a monotonic decrease in the stratosphere due to reaction with $O(^1D)$, OH, and through photolysis. These facts make HCN a long-lived trace gas, with an atmospheric residence time of a few years. The principal uncertainty claimed in the published models is the lack of information about the photolysis rate.

In previous observations (Jaramillo et al., 1988) made by our group at Mauna Kea, Hawaii (20° N latitude), we found that stratospheric HCN decreases much more slowly with altitude than in the models cited . This discrepancy amounts to 70% at 40 km if zero photolysis is assumed in the model chemistry; if the HCN photolysis rate is taken to be equal to that of HCl, the difference becomes an order of magnitude, rather than 70%. Our Hawaii results are consistent with other recently published measurements, (Abbas et al., 1987; Zander et al., 1988). Hence there is a strong suggestion from the experimental evidence that the rather simple chemistry the models have considered so far may not be complete. This fact complicates the interpretation of the results of the present work.

Our results (Fig. 2 [a]) show that in the antarctic springtime, middle-to-upper stratospheric HCN shows a considerably steeper vertical gradient, exhibiting mixing ratios two to four times lower than those at tropical (Jaramillo et al., 1988) and middle latitudes (Abbas et al., 1987; Zander et al., 1988). At 40 km the value is 65 ± 30 pptv (the estimated uncertainty of ~ 45% includes both calibration and retrieval errors) compared with our corresponding measurement of 190 $\pm$ 40 pptv for Hawaii (shaded area in Fig. 2 [a]). The low stratospheric values measured at McMurdo could be explained at least partly by meridional circulation, with downward transport at high latitudes. At the moment however, a quantitative treatment seems beyond reach, since the special dynamical conditions of the antarctic atmosphere during the winter and spring (namely the polar vortex motion and its consequences) are not adequately incorporated in current chemical/dynamical models. In any case, no photochemical modeling for HCN in the antarctic stratosphere has yet been done, and the lack of success for present 1-D HCN chemical models, when applied to the tropical of mid-latitude conditions they were meant to describe, indicates that some substantial revision must be made before theory can be brought into agreement with experimental data, even in cases where vertical and horizontal transport are thought to be well-modeled.

The present antarctic data leads to a substantially different mixing ratio profile, with a definite and relatively steep negative gradient as a function of altitude, than any of several observational techniques yield for tropical and mid-latitudes (see above). A significant portion this alteration in the HCN profile may be of dynamical origin. In other mm-wave spectroscopic observations at McMurdo Station, we have observed an abnormally low amount of N_2O to be present in the antarctic spring (Parrish et al., 1988b). Since N_2O has an atmospheric lifetime of well over a hundred years, it is an extremely robust tracer of dynamical transport. Its chemistry is similar to that of HCN, being limited to photolysis and to reactions with $O(^1D)$ in the middle and upper stratosphere (Brasseur and Solomon, 1984). The extreme deficit of N_2O registered by mid-September suggests that a prolonged subsidence is taking place over the antarctic continent during the austral winter months (June-August) which would also affect the vertical profile of HCN. Various evidence points to a total downward transport of perhaps as much as 5 km from June to August or early September. It would appear from Fig. 2 (a) that downward transport alone cannot account for the evident difference between tropical and antarctic profiles however, and that further refinement in both the chemistry and dynamics of models will be required to explain the differences.

ACKNOWLEDGEMENTS

This work was funded by the National Science Foundation's Division of Polar Programs, by Grant NAGW-353 from NASA's Upper Atmospheric Research Program, and by Grants FC86-607 and FC87-658 from the Chemical Manufacturer's Association.

REFERENCES

Abbas, M.M., J. Guo, B. Carli, F. Mencaraglia, M. Carlotti and I.G. Nolt, 1987: Stratospheric distribution of HCN from far infrared observations. Geophys. Res. Lett., 14, 531-534.

Brasseur, G., and S. Solomon, 1984: Aeronomy of the Middle Atmosphere. 441 pp., D. Reidel, Dordrecht.

Brasseur, G., R. Zellner, A. De Rudder and E. Arjis, 1985: Is hydrogen cyanide (HCN) a progenitor of acetonitrile (CH_3CN) in the atmosphere? Geophys. Res. Lett., 12, 117-120.

Cicerone, R.J., and R. Zellner, 1983: The atmospheric chemistry of hydrogen cyanide (HCN). J. Geophys. Res., 88, 10,689-10,696.

Coffey, M.T., W.G. Mankin and R.J. Cicerone, 1981: Spectroscopic detection of stratospheric hydrogen cyanide. Science, 214, 333-335.

Hofmann, D.J., J.W. Harder, S.R. Rolf and J.M. Rosen, 1987: Balloon-borne observations of the development and vertical structure of the antarctic ozone hole in 1986. Nature, 326, 59-62.

Hofmann, D.J., J.W. Harder, J.M. Rosen, J.V. Hereford and J.R. Carpenter, 1988: Ozone profile measurements at Mcmurdo Station Antarctica during the spring of 1987. J. Geophys. Res., 93, in press.

Jaramillo, M., R.L. de Zafra, J.W. Barrett, A.Parrish and P.M. Solomon, 1988: MM-wave observations of stratospheric HCN at tropical latitudes. Geophys. Res. Lett., 15, 265-268.

Parrish, A., R.L. de Zafra, P.M. Solomon and J.W. Barrett, 1988a: A ground based technique for millimeter wave spectroscopic observations of stratospheric trace constituents. Radio Science, 23, 106-118.

Parrish, A., R.L. de Zafra, M. Jaramillo, B. Connor, P.M. Solomon and J.W. Barrett, 1988b: Extremely low N_2O concentrations in the springtime stratosphere at McMurdo Station, Antarctica. Nature, 332, 53-55.

Rinsland, C.P., M.A.H. Smith, P.L. Rinsland and G.M. Stokes, 1982: Ground-based infrared spectroscopic measurements of atmospheric hydrogen cyanide. J. Geophys. Res., 87, 11,119-11,125.

Twomey, S., B. Herman and R. Rabinoff, 1977: An extension of the Chahine method of inverting the radiative transfer equation. J. Atm. Sci., 34, 1085-1090.

Zander, R., C.P. Rinsland, C.B. Farmer, J. Namkung, R.H. Norton and J.M. Russell III, 1988: Concentrations of carbonyl sulfide and hydrogen cyanide in the free upper troposphere and lower stratosphere deduced from ATMOS/Spacelab 3 infrared solar occultation spectra. J. Geophys. Res., 93, 1669-1678.

THE CHEMISTRY OF SOME REACTIONS BELIEVED TO BE IMPORTANT IN OZONE DEPLETION OVER ANTARCTICA

Mario J. Molina
Jet Propulsion Laboratory
California Institute of Technology
Pasadena, California 91109, USA

ABSTRACT

The rates and mechanisms of several reactions believed to be important for ozone depletion in the Antarctic stratosphere have been studied in the laboratory, at temperatures around 200 K. Polar stratospheric clouds play a crucial role in this process: our results indicate that they provide an extremely efficient medium to promote the release of photolytically active chlorine from the main inert reservoirs hydrogen chloride and chlorine nitrate. The chemistry and spectroscopy of the dimer of chlorine monoxide has also been studied. This species is most likely involved in an efficient catalytic ozone destruction cycle which does not require atomic oxygen.

1. INTRODUCTION

The abundance of ozone over Antarctica during the spring months has decreased significantly since the late 1970's. Measurements in the laboratory and in the field have suggested recently a chemical origin for this depletion, pointing to the release of man-made chlorofluorocarbons as the primary cause.

The chemical events occur in two steps: in the first one, photolytically active chlorine is liberated from the inert reservoirs HCl (hydrogen chloride) and $ClONO_2$ (chlorine nitrate) through heterogeneous reactions promoted by polar stratospheric clouds (PSC's). In the second step, a catalytic chain reacion involving chlorine free radicals rapidly destroys ozone. Such a cycle is different from the ones operating at mid and equatorial latitudes in that no free oxygen atoms are involved, because of their scarcity over Antarctica. In this paper we exmine those two steps and present some laboratory results related to these chemical mechanisms.

2. HETEROGENEOUS CHEMISTRY

The presence of PSC's is one of the peculiar characteristics of the Antarctic stratosphere. The PSC's consist mainly of water ice or of nitric acid trihydrate. Stratospheric temperatures over the Arctic also reach below -80°C, and hence PSC's are also formed there, although not as frequently as over Antarctica.

2.1 THE HCl + $ClONO_2$ REACTION

Solomon et al. (1986) suggested that the PSC's would promote the reaction between HCl and $ClONO_2$ to produce photolytically active chlorine:

$$HCl + ClONO_2 \rightarrow Cl_2 + HNO_3 \quad (1)$$

However, this suggestion was questioned by others, because it was not considered likely for the two reactants to collide simultaneously with neighboring sites on the surface of a PSC particle. The reaction between $ClONO_2$ and water appeared more likely:

$$H_2O + ClONO_2 \rightarrow HOCl + HNO_3 \quad (2)$$

On the other hand, the reaction with HCl enhances ozone depletion much more efficienlty, given the greater stability, and hence, larger abundance of HCl over $ClONO_2$.

Our laboratory experiments indicated that HCl molecules are efficiently scavenged by ice crystals and that they have a large mobility in these crystals (Molina et al., 1987; Wofsy et al., 1988). Furthermore, our results showed that subsequent collisions of $ClONO_2$ molecules with the HCl containing ice crystals leads with high efficiency to chemical reaction. The product Cl_2 is released immediately to the gas phase, while the other product, HNO_3, remains frozen in the condensed phase. The experimental results of Tolbert et al. (1987, 1988) and of Leu (1988a,b) confirmed our results and showed that other heterogeneous reactions such as the following also occur with high efficiency:

$$HCl + N_2O_5 \rightarrow ClNO_2 + HNO_3 \quad (3)$$

$$H_2O + N_2O_5 \rightarrow 2\ HNO_3 \quad (4)$$

The consequences of these reactions for Antarctic ozone depletion were discussed at length at the Polar Ozone Workshop which took place in May 1988 in Snowmass, Colorado, USA.

2.2 NITRIC ACID TRIHYDRATE

One of the remaining questions has to do with the efficiency of the solid nitric acid trihydrate, $HNO_3 \cdot 3H_2O$, in promoting these reactions. Our original experiments were carried out with regular, hexagonal water-ice crystals (I_h). The structure of the trihydrate crystals is known (Taesler et al., 1975) and is analogous to that of ice I_h from the point of view of being rather open and hence capable of accomodating HCl molecules within the

ISBN 0-937194-15-8

lattice. We have performed some preliminary experiments which indicate that indeed HCl is soluble in nitric acid trihydrate at the several percent level. For atmospheric modeling calculations the most relevant property in this connection should be the HCl vapor pressure over the trihydrate as a function of temperature and of HCl concentration in the solid phase, and we are in the process of measuring these values. The accomodation or sticking coefficients for the collisions of HCl molecules with the trihydrate surface are also of interest, although these coefficients by themselves do not suffice to establish the more relevant vapor pressure values.

2.3 SULFURIC ACID AEROSOLS

It is important to establish whether H_2SO_4 (sulfuric acid) aerosols can promote the type of heterogeneous chemistry which occurs so efficiently with ice in terms of release of active chlorine. We have measured the vapor pressures of HCl dissolved in supercooled, liquid solutions with 40% by weight or more H_2SO_4 in water. For HCl concentrations above about 0.1% by weight the HCl vapor pressures around 200 K are larger than 10^{-3} torr, and hence only trace amounts of HCl are expected to dissolve in the sulfate aerosols prevailing at mid and equatorial latitudes, given that the HCl vapor pressures in the stratosphere are of the order of 10^{-7} torr.

The water in concentrated sulfuric acid solutions is very tightly bound, and is chemically not available. In fact, the hydrolysis of $ClONO_2$ (reaction 2) is an endothermic process with such concentrated solutions. This will affect the competition between HCl and water as reaction partners for $ClONO_2$ (reactions 1 and 2), and it is possible that even trace amounts of HCl will be the more efficient partners.

On the other hand, the sulfuric acid aerosols absorb increasing amounts of water as the temperature approaches the frost point, becoming more dilute and eventually freezing, since supercooling rapidly becomes less probable with dilution. At some point in this process the particles become PSC's, and the question is whether they promote the release of active chlorine even before reaching that point. Carefull laboratory studies will be required to answer this question.

2.4 CONDENSATION OF HF

Measurements carried out over Antarctica during the spring months in 1986 and 1987 indeed show that the abundance of HCl and of HNO_3 in the gas phase is significantly lower than expected in the absence of scavenging by PSC's. In contrast, HF (hydrogen fluoride) appears to be depleted only slightly, or possibly not depleted at all (Toon et al., 1988). The reason is, we believe, that HF has considerably less affinity for ice than HCl or HNO_3, both of which are strong acids. HF is a weak acid, because the H - F bond is very strong and hence the molecule does not dissociate readily. An extrapolation, using the Clausius-Clapeyron equation, of HCl and HF vapor pressures over aqueous solutions to stratospheric temperatures indicates that for similar mixing ratios in the gas phase, the amount of HF in the condensed phase should be significantly smaller than the corresponding amount of HCl; this is a consequence of its smaller enthalpy of solution, which determines the slope of the vapor pressure line in the Clausius-Clapeyron plot.

3. CHEMISTRY OF THE CHLORINE MONOXIDE DIMER

In an earlier publication (Molina and Molina, 1987) we suggested the following cycle as a plausible mechanism for ozone depletion over Antarctica:

$$ClO + ClO \longrightarrow Cl_2O_2$$
$$Cl_2O_2 + h\nu \longrightarrow Cl + ClOO$$
$$ClOO \longrightarrow Cl + O_2$$
$$2\ (Cl + O_3 \longrightarrow ClO + O_2)$$

$$\text{Net:}\quad 2\ O_3 \longrightarrow 3O_2$$

3.1 ULTRAVIOLET AND INFRARED SPECTRA OF THE DIMER

Some evidence for the existence of the ClO dimer came from chemical kinetic studies of the ClO self reaction in which a transient ultraviolet spectrum was observed (e.g., see Hayman et al., 1986). In our earlier work we reported infrared product spectra for systems in which ClO was generated using several different precursors and we concluded that two different isomers of Cl_2O_2 were being formed. We reported the UV spectrum for only one of these "isomers", namely the one formed efficiently with OClO as the excess reactant. Hayman et al. (1986) subsequently obtained a different UV spectrum, with however, a similar absortion tail beyond 300 nm. In more recent work Cox and Hayman (1988) were able to reproduce our UV spectrum, and attributed it to a ClO - OClO adduct, rather than to a ClO dimer.

Recent work in our laboratory indicates that this is indeed the case: the two sets of infrared bands in our earlier studies correspond to the adduct, Cl_2O_3 (bands A, B and E), and to the dimer, Cl_2O_2 (bands C, D and E). The ClO self reaction at room temperature generates OClO, and hence this species was always present in our earlier experiments. In our more recent work, by mixing Cl-atoms with ozone at much lower temperatures (around 200 K) we were able to produce the dimer in the absence of OClO. Only one of the two sets of bands is present under these circunstances (bands C, D and E, at 740, 653 and 562 cm^{-1}), with a corresponding UV spectrum that is in very good agreement with that reported earlier by Hayman et al. (1986) for the dimer, namely with a maximum around 245 nm, as shown in figure 1.

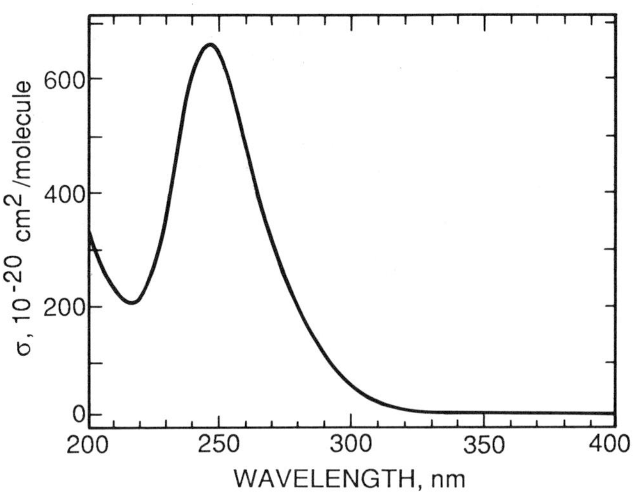

Figure 1. Ultraviolet absorption cross sections of Cl_2O_2 at 200 K.

Additional experiments on the ClO dimer have also been carried out recently by Burkholder et al. (1988) and Friedl and Sander (1988), who report IR and UV spectra, and by DeMore and Roux (1988), who obtained UV spectra. The results of all these experiments are in very good agreement, and there remains little doubt at present about the spectral characterization, and hence, the existence of the dimer.

3.2 STRUCTURE AND PHOTOLYSIS OF THE ClO DIMER

The two most likely structures for the dimer are ClOOCl (the symmetric form) and ClOClO (the asymmetric form). Other dimer structures are also possible, e.g., a central Cl atom bonded to each of the other three. Formation of the asymmetric dimer does not lead to a catalytic cycle with net ozone destruction: if it decomposes to produce OClO, subsequent photolysis of this species yields a free oxygen atom which then regenerates an ozone molecule.

Quantum mechanical calculations strongly suggest that the symmetric form is the most stable one (Toohey et al., 1987; McGrath et al., 1988). Chemical kinetic considerations indicate that this most stable isomer will be formed predominantely, since it will be favored by both enthalpy and entropy factors. In addition, the quantum mechanical calculations predict vibrational frequencies for the symmetric form which match reasonably well the observed dimer IR spectrum.

There is a separate piece of information which points to the symmetric form as the dominant one: the amount of dimer formed, as measured by its UV and IR spectra, depends strongly on the Cl-atom concentration in the system. In fact, it can only be produced in significant quantities in the presence of Cl-atom scavengers, such as O_3 or Cl_2O. This observation can be explained by the following sequence of reactions:

$$Cl + ClOOCl \rightarrow Cl_2 + ClOO$$

$$ClOO \rightarrow Cl + O_2$$

$$\text{Net: } ClOOCl \rightarrow Cl_2 + O_2$$

Thus, a small amount of free Cl atoms catalyzes the destruction of the symmetric dimer. The asymetric form would lead to formation of OClO, which is thermally stable, in contrast to ClOO; no catalytic cycle would be sustained. Furthermore, no OClO formation is observed in the system, around 200 K, even in the absence of Cl-atom scavengers.

Photolysis of the symmetric form may lead, at least in principle, to regeneration of ClO, rather than to the release of free Cl atoms, as needed for the catalytic ozone destruction cycle. However, for this type of species the absorption of UV radiation ussualy promotes an electronic transition involving the non-bonding electrons in the chlorine, and leading to a repulsive upper state which dissociates very rapidly at the Cl atom bond. Experimental studies by Cox and Hayman (1988) indicate that the quantum yield for Cl atom production is unity, as expected. Direct measurements of this quantum yield would be helpful, and we are currently carrying out such measurements.

ACKNOWLEDGEMENT

We thank F.C.-Y. Wang, T.L. Tso and L.T. Molina for their help with these experiments. The research described was performed at the Jet Propulsion Laboratory, California Institute of Technology, under contract with the National Aeronautics and Space Administration.

REFERENCES

Burkholder, J.V., J.J. Orlando, P.D. Hammer, and C.J. Howard, 1988: Measurements of the ClO radical vibrational band intensity and the ClO + ClO reaction product. Paper presented at the Polar Ozone Workshop, Snowmass, Colorado, May 9-13.

Cox, R.A., and G.D. Hayman, 1988: The stability and photochemistry of dimers of the ClO radical and implications for Antarctic ozone depletion. Nature, 322, 796-800.

DeMore, W.B. and E.T. Roux, 1988: Chemistry of the ClO dimer at low temperatures. Paper presented at the Polar Ozone Workshop, Snowmass, Colorado, May 9-13.

Friedl, R.R., and S.P. Sander, 1988: Studies of ClO and BrO reactions important in the polar stratosphere: kinetics and mechanism of the ClO + BrO and ClO + ClO reactions. Paper presented at the Polar Ozone Workshop, Snowmass, Colorado, May 9-13.

Hayman, G.D., J.M. Davies, and R.A. Cox, 1986: Kinetics of the reaction ClO + ClO --> Products and its potential relevance to Antarctic ozone. Geophys. Re. Lett., 13, 1347-1350.

Leu, M.T., 1988a: Laboratory studies of sticking coefficients and heterogeneous reactions important in the Antarctic stratosphere. Geophys. Res. Lett., 15, 17-20.

Leu. M.T., 1988b: Laboratory studies of sticking coefficients and heterogeneous reactions important in the stratosphere. Paper presented at the Polar Ozone Workshop, Snowmass, Colorado, May 9-13.

McGrath, M.P., K.C. Clemitshaw, F.S. Rowland, and W.J. Hehre, 1988: Thermochemical stabilities and vibrational spectra of isomers of the chlorine oxide dimer. Geophys. Res. Lett. 15, 883-886.

Molina, L.T., and M.J. Molina, 1987: Production of Cl_2O_2 by the self reaction of the ClO radical. J. Phys. Chem., 91, 433-436.

Molina, M.J., T.L. Tso, L.T. Molina, and F. C.-Y. Wang, 1987: Antarctic stratospheric chemistry of chlorine nitrate, hydrogen chloride and ice: release of active chlorine. Science, 238, 1253-1257.

Solomon, S., R.R. Garcia, F.S. Rowland, and D.J. Wuebbles, 1986: On the depletion of Antarctic ozone. Nature, 321, 755-758.

Taesler, I., R.G. Delaplane, and I. Olovsson, 1975: Hydrogen bond studies. XCIV. Diaquaoxonium ion in nitric acid tryhydrate. Acta Cryst., B31, 1489-1492.

Tolbert, M.A., M.J. Rossi, R. Malhotra, and D.M. Golden, 1987: Reaction of chlorine nitrate with hydrogen chloride and water at Antarctic stratospheric temperatures. Science, 238, 1258-1260.

Tolbert, M.A., M.J. Rossi, and D.M. Golden, 1988: Heterogeneous chemistry related to Antarctic ozone depletion: Reaction of $ClONO_2$ and N_2O_5 on ice surfaces. Paper presented at the Polar Ozone Workshop, Snowmass, Colorado, May 9-13.

Toohey, D.W., Wm.H. Brune, and J.G. Anderson, 1987: Chemistry of the Antarctic stratosphere: reactions of ClO and BrO. Paper presented at the Eighteenth International Sympoisum on free radicals, Oxford, England, September 10-15.

Toon, G.C., C.B. Farmer, L.L. Lowes, P.W. Schaper, J.-F. Blavier, and R.H. Norton, 1988: Infrared aircraft measurements of stratospheric composition over Antarctica during September 1987. J. Geophys. Res. (submitted).

Wofsy, S.C., M.J. Molina, R.J. Salawitch, L.E. Fox, and M.B. McElroy, 1988: Interactions between HCl, NOx and H_2O ice in the Antarctic stratosphere: implications for ozone. J. Geophys. Res., 93, 2442-2450.

GROUND-BASED MEASUREMENTS OF O_3, NO_2, OClO, AND BrO DURING THE 1987 ANTARCTIC OZONE DEPLETION EVENT

R.W. Sanders, S. Solomon, M.A. Carroll, and A.L. Schmeltekopf
NOAA Aeronomy Laboratory
Boulder, Colorado 80303, USA

1. INTRODUCTION

This paper is a summary of the principal results of ground-based visible and near-uv absorption spectroscopy from McMurdo Station, Antarctica as part of the National Ozone Expeditions (NOZE I and II) in austral spring of 1986 and 1987. More detailed accounts of this work appear in Mount et al. (1987), Solomon et al. (1987b, 1988), Sanders et al. (1987, 1988), and Carroll et al. (1988). The instruments used are 3/8-meter, fixed-grating, crossed Czerny-Turner spectrographs, each with a 1024-element, cooled, diode array detector. The light source is either sunlight scattered from the zenith sky or, at night, direct moonlight. The data is analyzed by least-squares minimization fitting of laboratory spectra collected with the same instruments. [See also Solomon et al. (1987a).]

In 1986 two identical spectrographs were used. Each was capable of operating in either of two spectral bands: 400-450 nm (suitable for measuring NO_2, O_3, and OClO), and 600-680 nm (suitable for NO_3 and H_2O). In 1987, one of the spectrographs was modified to operate only in the spectral band 330-370 nm (suitable for NO_2, O_3, OClO, and BrO).

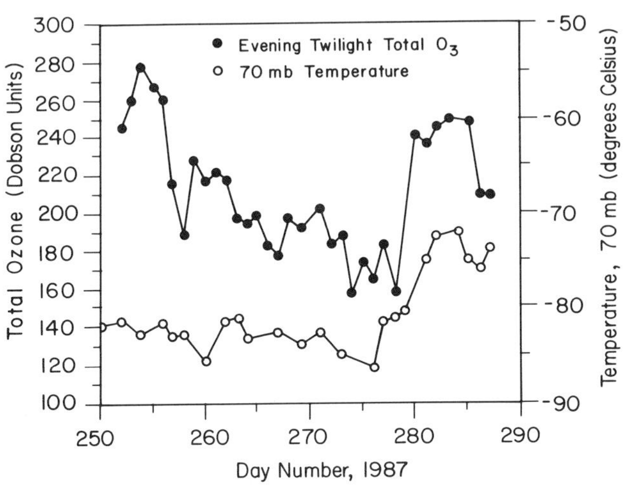

Fig. 1. Results for total column ozone are shown vs. day-of-year from late August through late October, 1987. Temperatures at 70 mb were determined by University of Wyoming balloonsondes.

2. OZONE RESULTS

Figure 1 shows our results for total column ozone from late August through late October, 1987 along with 70 mb temperatures determined by University of Wyoming balloonsondes. The temperature is lower than -80°C throughout most of this period, and ozone shows a steady decline. The sudden temperature increase and corresponding rise in ozone column near day 280 is associated with movement of the polar vortex away from McMurdo and should not be confused with the breakup of the ozone hole, which occurred about eight weeks later. Our measurements of ozone abundance agree very well with ozonesonde measurements by the University of Wyoming and with satellite data from the Total Ozone Mapping Spectrometer (TOMS).

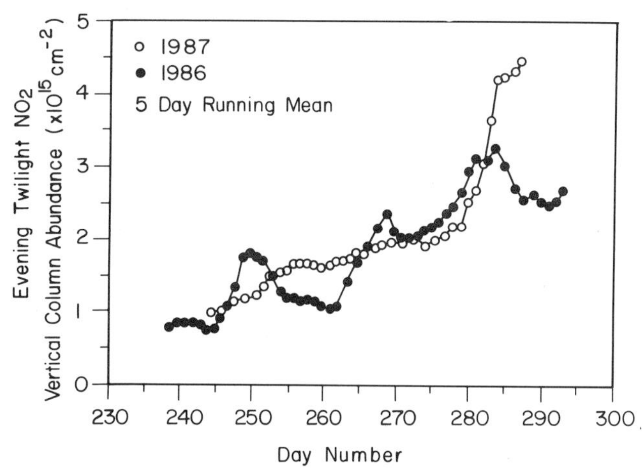

Fig. 2. NO_2 vertical column measured in evening twilight is plotted vs. day number. Open circles are 1987 data, while filled circles represent 1986 data.

3. NO_2 RESULTS

Figure 2 shows the NO_2 vertical column measured in evening twilight. The open circles are 1987 data, while filled circles represent 1986 data. A five-day smooth has been applied to both data sets to make the seasonal trends easier to see. In both years, NO_2 levels are very low in late August and steadily increase with time. As with ozone, the effect of

OZONE IN THE ATMOSPHERE
Rumen D. Bojkov and Peter Fabian (Eds.)

ISBN 0-937194-15-8

displacements of the vortex are readily seen. One large displacement is apparent near day 280 in 1987, while three significant displacements occurred in 1986. In general, the vortex was more stable in '87, and stratospheric temperatures were lower at McMurdo by 5-8 K and less variable.

The overall linear trend in NO_2 is nearly identical in the two years, and probably reflects the seasonal release of NO_2 from its wintertime reservoirs.

We have compared these results to the work of McElroy et al. (1988), in which the seasonal trend in column NO_2 is modelled: (1) for net loss of HNO_3 through heterogeneous reactions on polar stratospheric clouds (PSC's), and (2) for no appreciable loss of HNO_3. Our results lie between the two cases.

Figure 3 compares NO_2 column measurements taken in the morning twilight to those taken in the evening twilight. The evening values are consistently higher than the morning values, and, furthermore, the size of the diurnal variation did not change significantly as total NO_2 increased through the (1987) season. This means that the nighttime reservoir of odd nitrogen is probably a species like HNO_3 , which does not itself vary diurnally, rather than a species like N_2O_5 which quickly photolyzes in daylight.

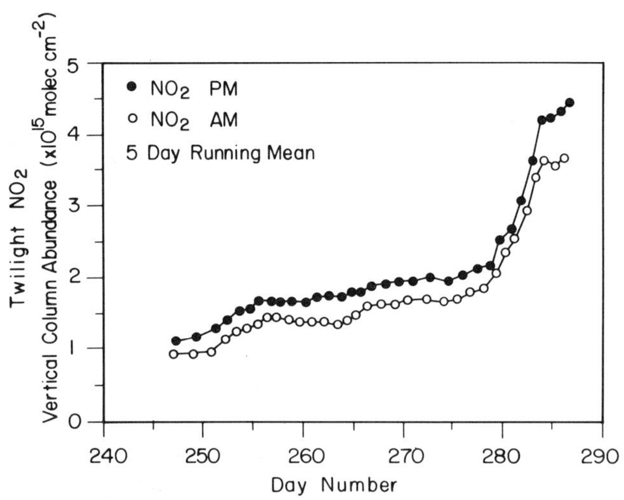

Fig. 3. NO2 column measurements taken in the morning twilight are compared to those taken in the evening twilight.

4. OClO RESULTS

Figure 4 shows the seasonal trend in evening twilight measurements of OClO for the two years. (Again, a 5-day smooth has been applied.) Though both years show a steady seasonal decline, the OClO abundance was significantly higher in '87, probably due to overall colder stratospheric temperatures and more frequent occurrence of PSC's. The rate of decline in the two years was very similar, except that OClO disappeared below our detection limit (indicated by small arrows) later in 1987.

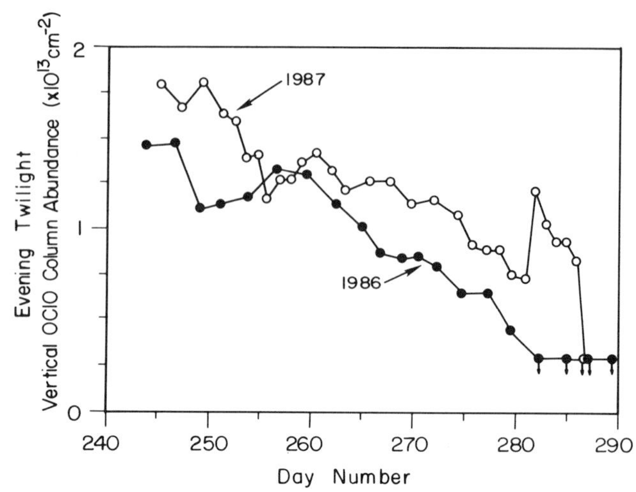

Fig. 4. Evening twilight measurements of OClO are shown for 1987 (open circles) and 1986 (filled circles). A 5-day smooth has been applied. Small arrows indicate values which are upper limits.

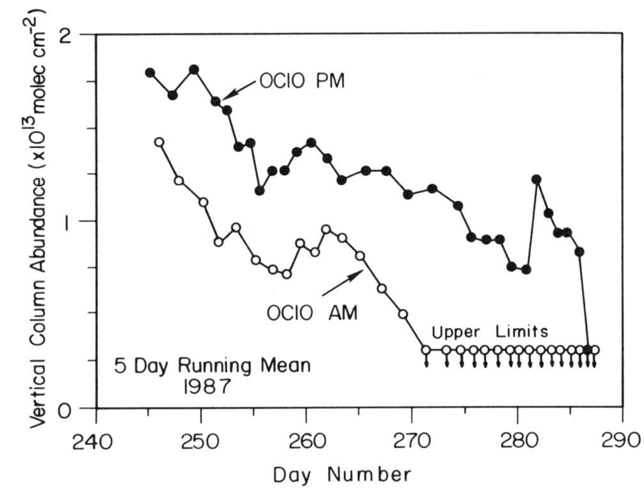

Fig. 5. OClO column measurements taken in the morning twilight are compared to those taken in the evening twilight.

The only established source of atmospheric OClO is the reaction of ClO with BrO. However, it has been suggested that the photolysis of Cl_2O_2 may also produce OClO. If this source of OClO were significant, higher values of OClO would be expected in morning twilight than in the evening. Figure 5 shows the reverse to be true, consistent with OClO formation exclusively by ClO + BrO.

It is further apparent in Fig. 5 that the diurnal variation of OClO is much larger late in the season than in late August. Indeed, by early October, the morning values have dropped below the detection limit. The concurrent, rapid increase in NO_2 abundance suggests that as the available NO_2

increases, active chlorine is removed to form $ClONO_2$. This idea is further supported by Fig. 6, which compares our '86 OClO data with the chlorine nitrate measurements of Farmer et al. (1987). In general, it is seen that the chlorine nitrate abundance increases, while OClO declines.

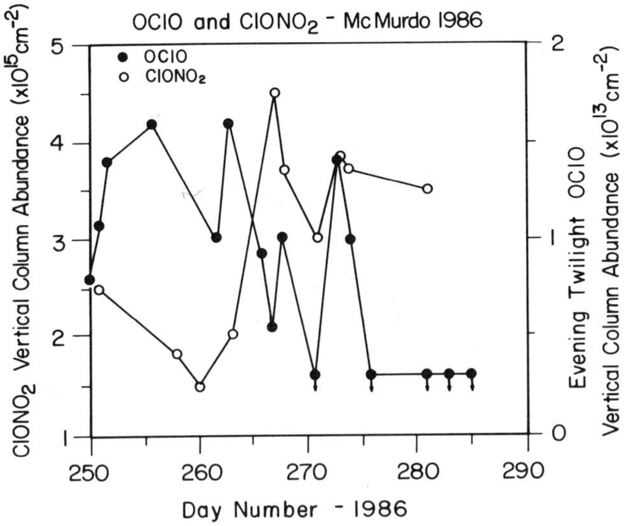

Fig. 6. OClO measurements (filled circles) from 1986 are compared to chlorine nitrate measurements by Farmer et al.

5. BrO RESULTS

Figure 7 shows BrO data from evening twilight on 23 Sep 87, plotted as a function of solar zenith angle (SZA). The slant column measurements are converted to vertical column using airmass factors appropriate to three different layer heights for the BrO. If the abundance of BrO is unchanging-- a good assumption for these SZA's in the evening (but not in the morning!)-- the correct choice of layer height should produce a value for vertical column BrO which does not change with SZA. The data shown indicate a vertical column of about 3 x 10^{13} cm^{-2} centered at an altitude of 15 km. The results for a week of measurements show noticeable day-to-day variation about an average value of approximately 2.5 x 10^{13} cm^{-2}.

TABLE 1. EVENING TWILIGHT BrO MEASUREMENTS

Day of 1987	Slant column at SZA=90° minus slant column at 84° (molec/cm^2 ± 15%)	Vertical column centered at 15 km (molec/cm^2 ± 25%)
262	8.4 x 10^{13}	1.8 x 10^{13}
263	1.3 x 10^{14}	2.8 x 10^{13}
264	1.2 x 10^{14}	2.6 x 10^{13}
265	1.5 x 10^{14}	3.2 x 10^{13}
266	1.4 x 10^{14}	3.0 x 10^{13}
267	1.2 x 10^{14}	2.6 x 10^{13}
268	8.0 x 10^{13}	1.7 x 10^{13}

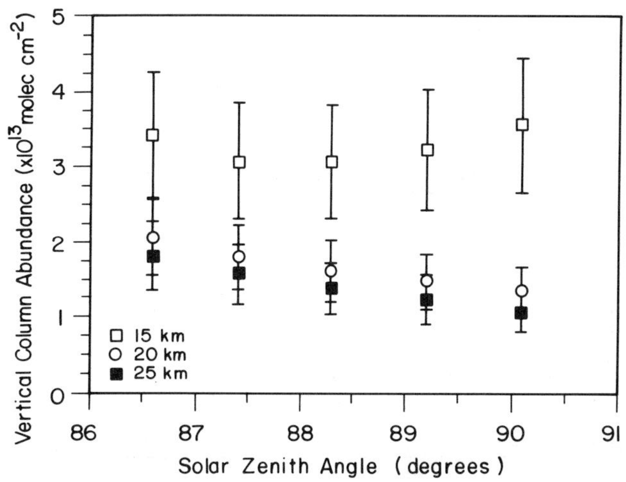

Fig. 7. Plotted vs. solar zenith angle is the evening twilight BrO vertical column inferred from measurements of the slant column, assuming three different values of the layer height.

If ozone loss occurs mainly through the chemistry of the ClO dimer, the rate limiting step is the dimer formation: $ClO + ClO + M \rightarrow Cl_2O_2 + M$. However, McElroy et al. (1986a,b) and Rodriguez et al. (1986) have suggested another mechanism in which the rate limiting step involves coupling between chlorine and bromine chemistry: $ClO + BrO \rightarrow ClO_2 + Br$. The amount of BrO which must be present to make the rate of ozone loss equal for these two mechanisms is dependent on the assumed abundance of ClO; for ClO mixing ratios of 0.6 and 1.0 ppbv, the BrO mixing ratio must be 16 and 24 pptv, respectively.

Figure 8 shows the observed slant column BrO vs. solar zenith angle. The open circles are for day 265, on which the largest BrO abundance was observed; filled circles represent day 264, which is more typical of the week of observations. Also shown is the modeled behavior for BrO mixing ratios of 5 pptv and 10 pptv, assumed to be constant from 150 mb to 20 mb. We conclude that BrO is present at levels between 5 and 15 pptv, which suggests that although loss of ozone due to coupling between bromine and chlorine chemistry is important and cannot be neglected, it does not dominate ozone loss due to chlorine chemistry alone.

BrO also exhibits strong asymmetry between morning and evening values. Figure 9 shows, as a function of zenith angle, the observed change in slant column BrO with respect to a background spectrum taken at SZA = 84 degrees. During the evening (PM) twilight, the BrO abundance is expected to change very little, and the observed increase in slant column simply reflects the increase in airmass factor with increasing zenith angle. The AM values, however, start out low (at SZA = 91°) and go even lower, despite the concurrent decrease in SZA as the sun rises. This implies significant growth of BrO in the morning twilight.

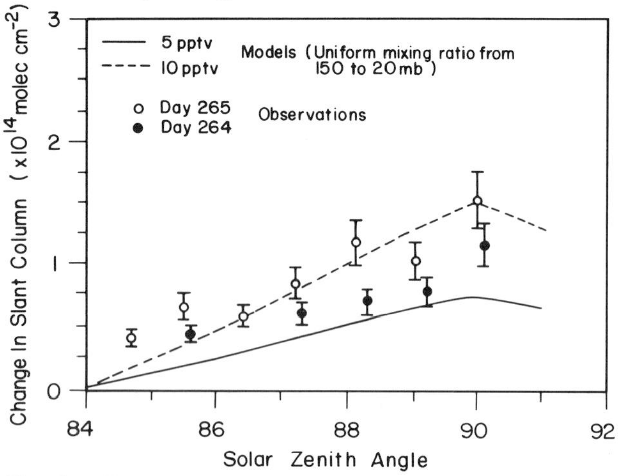

Fig. 8. For two evenings in 1987, the observed BrO slant column is plotted vs. solar zenith angle. Also shown is the modeled behavior for two values of BrO mixing ratio, assumed to be constant from 150 mb to 20 mb.

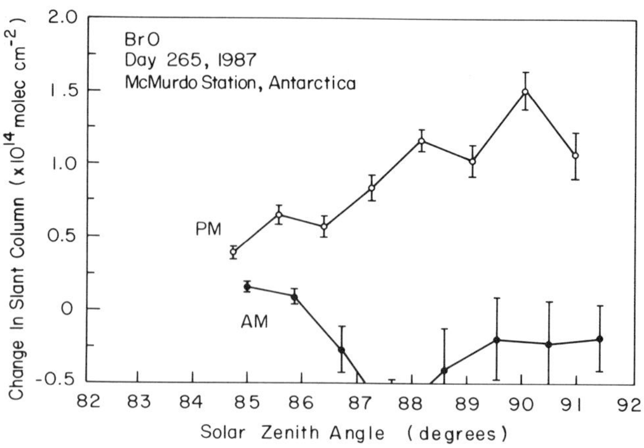

Fig. 9. Plotted vs. solar zenith angle is the observed change in slant column BrO with respect to a background spectrum taken at a solar zenith angle of 84°. Both morning (filled circles) and evening (open circles) twilight data are shown.

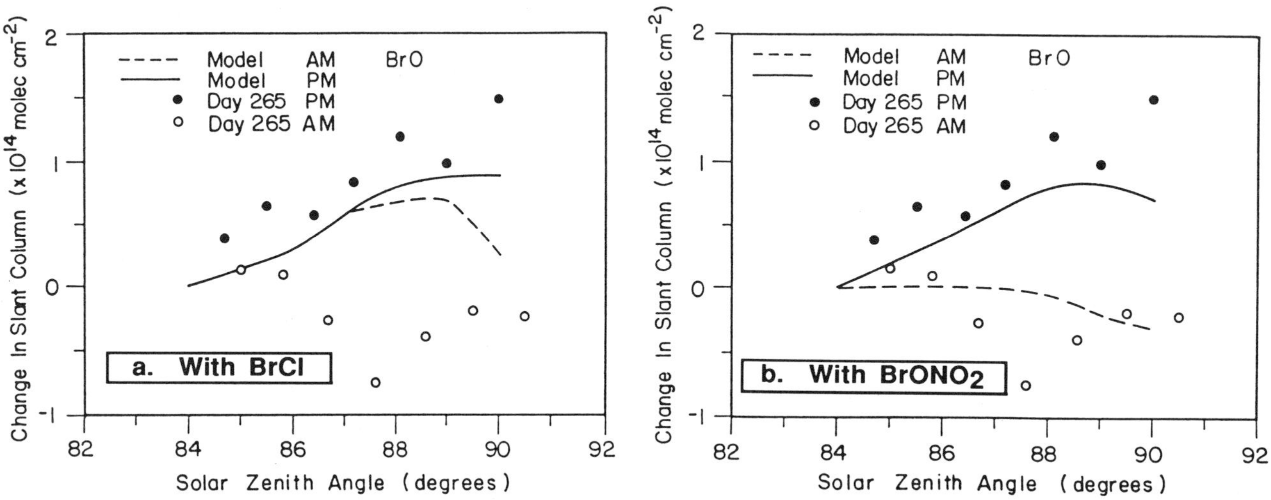

Fig. 10. The diurnal behavior of slant column BrO vs. zenith angle is compared to model predictions in which (a) the nighttime reservoir for BrO is BrCl, and (b) the nighttime reservoir is bromine nitrate.

In Fig. 10 this behavior is compared to the predictions of two models. In one of the models, BrCl is allowed as a nighttime reservoir for BrO, and in the other the reservoir is $BrONO_2$. Both models fit the evening (PM) data reasonably well. However, the morning predictions of the model containing BrCl as the reservoir depart significantly from the observations, because BrCl photolyzes too quickly (on the order of hundreds of seconds). $BrONO_2$ photolyzes on a time scale of thousands of seconds, and the model containing this species agrees better with the observations. From this we conclude that the dominant nighttime reservoir of BrO is not BrCl, but rather bromine nitrate (or some other relatively long-lived species).

REFERENCES

Carroll, M.A., R.W. Sanders, S.Solomon and A.L. Schmeltekopf, 1988: Visible and near-ultraviolet spectro-scopy at McMurdo Station, Antarctica, 5. Observations of BrO. *J. Geophys. Res.*, in press.

Farmer, C.B., G.C. Toon, P.W. Shaper, J.F. Blavier and L.L. Lowes, 1987: Ground-based measurements of the composition of the antarctic atmosphere during the 1986 spring season, I. Stratospheric trace gases. *Nature*, **329**, 126.

McElroy, M.B., R.J. Salawitch, S.C. Wofsy and J.A. Logan, 1986a: Antarctic ozone: reductions due to synergistic interactions of chlorine and bromine. *Nature*, **321**, 759.

McElroy, M.B., R.J. Salawitch and S.C. Wofsy, 1986b: Antarctic O_3: chemical mechanisms for the spring decrease, *Geophys. Res. Lett.*, **13**, 1296.

McElroy, M.B., R.J. Salawitch and S.C. Wofsy, 1988: Chemistry of the antarctic stratosphere. *Planet. Space Sci.*, in press.

Mount, G.H.., R.W. Sanders, A.L. Schmeltekopf and S. Solomon, 1987: Visible spectroscopy at McMurdo Station, Antarctica, 1. Overview and daily variations of NO_2 and O_3 during austral spring, 1986. *J. Geophys. Res.*, **92**, 8320.

Rodriguez, J.M., M.K.W. Ko and M.B. McElroy, 1986: Chemistry of OClO in the antarctic stratosphere: Impact of OClO and Cl_2O_2 and implications for observations. *Geophys. Res. Lett.*, **13**, 1292.

Sanders, R.W., S. Solomon, M.A. Carroll and A.L. Schmeltekopf, 1987: Visible spectroscopy at McMurdo Station, Antarctica, 3. Observations of NO3. *J. Geophys. Res.*, **92**, 8339.

Sanders, R.W., S. Solomon, M.A. Carroll and A.L. Schmeltekopf, 1988: Visible and near-ultraviolet spectro-scopy at McMurdo Station, Antarctica, 4. Overview and daily measurements of NO_2, O_3, and OClO during 1987. *J. Geophys. Res.*, in press.

Solomon, S., A.L. Schmeltekopf and R.W. Sanders, 1987a: On the interpretation of zenith sky absorption measure-ments. *J. Geophys. Res.*, **92**, 8311.

Solomon, S., G.H. Mount, R.W. Sanders and A.L. Schmeltekopf, 1987b: Visible spectroscopy at McMurdo Station, Antarctica, 2. Observations of OClO. *J. Geophys. Res.*, **92**, 8329.

Solomon, S., R.W. Sanders, M.A. Carroll and A.L. Schmeltekopf, 1988: Visible and near-ultraviolet spectro-scopy at McMurdo Station, Antarctica, 6. Observations of the diurnal variations of OClO and BrO. *J. Geophys. Res.*, in press.

ANALYSIS OF OZONE OBSERVATIONS

STATISTICAL ERROR ANALYSIS OF OZONE TRENDS—WINTER DEPLETION IN THE NORTHERN HEMISPHERE

F. Sherwood Rowland and Neil R.P. Harris
Department of Chemistry, University of California
Irvine, California 92717, USA

*Rumen D. Bojkov**
Atmospheric Environment Service, Canada

Peter Bloomfield
Department of Statistics, North Carolina State University
Raleigh, North Carolina, USA

ABSTRACT

Comparison of the average ozone concentrations for the years 1976-1986 versus 1965-1975 shows significant winter-time (DJFM) losses at almost all Dobson stations between 30°N and 64°N latitudes, with smaller ozone changes during the summer (MJJA) months. Multiple regression analyses have been carried out using sets of total ozone data for 1965-1986, provisionally revised to correct for instrument recalibrations at individual stations and then combined to form latitudinal band averages. Separate residual linear trends have been calculated for each calendar month with allowance for the effects of autocorrelation and for the cyclic variations associated with the yearly seasonal cycle, the quasibiennial oscillation (QBO), and the 11-year solar sunspot cycle. The residual linear trends show a measurable decrease in the annual average total column ozone of -1.7% to -3.0% from 1969 to 1986 in three latitude bands from 30°N to 64°N. The decreases are largest during the winter months (-2.3% to -6.2%) of December through March, and contrast with smaller losses (-0.2% to -1.9%) in the summer (May through August). Similar winter-time losses in ozone concentrations are found both in the ODW published data and in the provisionally revised data set, with greater inter-station consistency in the latter.

INTRODUCTION

Continuing concern about possible effects upon stratospheric ozone by chemicals released into the atmosphere requires critical examination of the available time series measurements of total column ozone (NASA/WMO 1988). Ground-based Dobson and filter ozonometer stations provide regular measurements, with many station records covering 20 years or more and therefore useful in determining the nature of any long-term changes in total ozone concetrations. Examination reveals inconsistent features in many of the published ozone data records. In most cases, these can be accounted for by adjusting the data using information about the Dobson instrument recalibrations as recorded and made available by the individual stations.

DATA ANALYSIS

Previous statistical analyses have been chiefly conducted as tests of one-dimensional model calculations of global average ozone concentrations, inherently independent of seasonal or latitudinal variations, and no possibility of seasonal variation in ozone trends was allowed for in these analyses (e.g. Hill et al., 1977; Reinsel, 1981; Reinsel et al., 1981, 1987; St. John et al., 1981; Oehlert, 1986). Examination of the monthly ozone data sets from Arosa and other northern middle latitude Dobson stations indicated substantial differences in the seasonal variation of total ozone changes with much larger losses in the winter months than in the summer (Harris and Rowland, 1987). A statistical approach in which equivalent ozone changes are imposed for all calendar months is inappropriate with a data set which exhibits a pronounced tendency toward seasonal differences in the trends. Appropriate comparisons with the more recent two-dimensional atmospheric model calculations require a statistical approach to the observations which allows seasonal and latitudinal differences to be detected if they exist. Such statistical approach has now been applied.

The second important difference between the previous statistical analyses and those presented here is that the original ozone data as published by the WMO World Ozone Data Center as Ozone Data for the World (ODW) have been thoroughly reexamined for internal consistency.

*Presently with the WMO Secretariat, Geneva

OZONE IN THE ATMOSPHERE
Rumen D. Bojkov and Peter Fabian (Eds.)

ISBN 0-937194-15-8

The statistical techniques have then been applied to two complete sets of data: (1) those published in ODW; and (2) a provisionally revised set of ozone data reevaluated by R.D. Bojkov, used also in the NASA/WMO Ozone Trends Panel Report-1988 and referred to as Bojkov-1987 data set. The methods used to examine the published data for inconsistencies have been described by Bojkov, 1988. They include station calibrations and comparisons with ozone stations from the same region, with the course of the stratospheric temperatures and for last eight years with overpassing TOMS revised data. In the many cases where instrument recalibrations were made at the individual Dobson stations but not reported in ODW or the open literature, corrections to the monthly averages of ozone concentrations have been made using the recalibration information from the stations themselves. These corrections are believed to be approximately correct but are labeled as provisional because more accurate corrections for instrument recalibration can be made on a day-by-day, reading-by-reading basis by the individual stations.

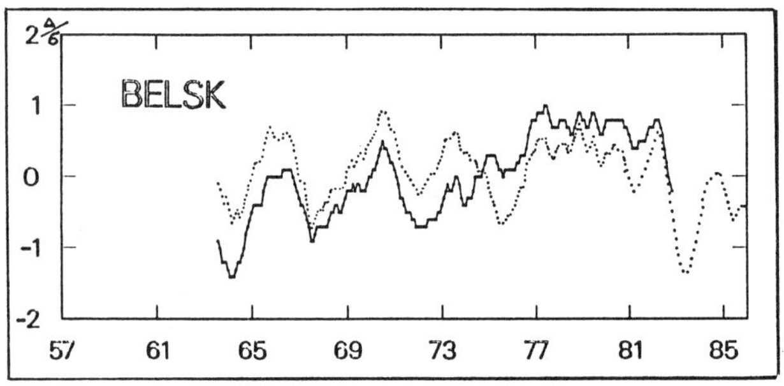

Figure 1. 12-month running means of ozone monthly deviations at Belsk normalized by division with monthly standard deviations. Continuous line old published data. Dotted line re-evaluated data.

A good example of the importance of using corrected total ozone data is given by the record from the station at Belsk, Poland (Figure 1). In the early 1980s the entire ODW ozone record since 1963 was recalculated by the station staff on a reading-by-reading basis using their own recorded instrument calibration history. In an analysis (Reinsel et al., 1981) of the originally published Belsk data a trend coefficient of $+0.83 \pm 0.14\%$, yr^{-1} [one sigma errors] was calculated for the post-1970 part of the period 4/63-12/78. A similar analysis (Reinsel et al., 1987) for the ozone data as recalculated by the station gave a trend of $+0.03 \pm 0.07\%$ yr^{-1} for the period 4/63-12/84, which is in fair agreement with the value of $-0.09 \pm 0.05\%$ yr^{-1} calculated here using the data 1/65 through 12/86. The validity of correcting for recalibration using average monthly corrections to the ODW published data was tested by applying such a correction to the original Belsk data as published prior to the station reanalysis. A very good agreement was found between the two analyses.

WINTER SEASON OZONE LOSSES

All of the above mentioned statistical analyses share the same weakness that the reported trends are based on the use of data "as published" and upon the assumption that equivalent ozone losses must occur in all calendar months. Therefore, here for the first time, a time-series of monthly average ozone concentrations were examined for changes in its seasonal cycle by being split into earlier and later parts which are then compared. The difference between these two long-term means for each calendar month can be easily calculated together with its standard error estimate, and a simple t-test gives some measure of the significance of any calculated changes. Figure 2(a) shows the results of the calculation for ozone data from the Bismarck (U.S.A.) station for the time period 1/63 to 12/86 as published in ODW and divided into two twelve year periods, 1/63-12/74 and 1/75-12/86. The average ozone concentrations are substantially lower for the winter months (DJFM) in the later time period than in the earlier, while the summer (MJJA) ozone values have exhibited smaller changes.

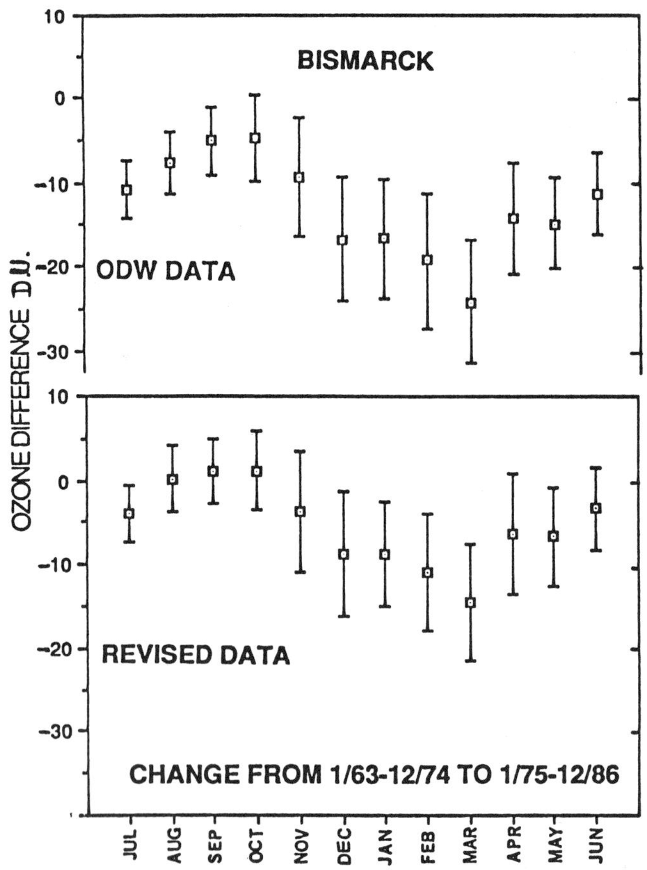

Figure 2. Differences between the monthly means of two time periods (in D.U.) at Bismarck N.D. as discussed in the text.

The results for the same calculation performed on the provisionally revised Bismarck data are shown in Figure 2(b). The different winter/summer response remains, although the magnitudes of the monthly variations are less for the revised data set. In number of stations, the corrections for instrument recalibrations tend to have similar absolute effects on the monthly magnitutudes throughout the year, with little overall influence on inter-monthly comparisons such as the winter/summer differences illustrated in both parts of Figure 2.

Calculations such as those shown for the Bismarck station have been carried out for twenty northern middle latitude Dobson stations, using first the ODW data and then the provisionally revised data sets for the 22-year time period Jan. 1965 to Dec. 1986 divided into earlier and later 11-year periods. For each station, a winter (DJFM) and a summer (MJJA) average ozone change have been calculated for both the ODW and provisionally revised data sets, and the plots are shown on Figure 3 and 4 respectively. Both winter time data evaluations show substantial ozone losses at almost all Dobson stations, while the summer changes tend to be scattered on both the positive and negative sides of zero. No consistent positive or negative change is introduced into the data by the provisional revision of the original ODW data. Instead, the chief effect of the corrections as indicated by comparison of Figures 3 and 4 is that the provisionally revised station data provide a more consistent picture.

STATISTICAL ANALYSIS

The monthly average data series have then been tested statistically. Allowance for the effects of such geophysical phenomena as the quasi-biennial oscillation (QBO) and the solar cycle on total column ozone has been provided using a time series model similar to that of Reinsel et al., (1981; 1987). The general form is:

$$\text{ozone}(i,j) = S(i,j) + X(i,j) + T(i,j) + N(i,j)$$

and in which i and j denote the year and month. S, X, T and N are terms which respectively describe the seasonal cycle, the geophysical phenomena, the possible trend and the autocorrelated noise. The proxy variable used to describe the solar cycle is a 12-month smoothed sunspot series, and that for the QBO is the 50 mbar wind velocity at Singapore. The trend term used in the past has been a "hockey stick" model in which the underlying ozone concentration is assumed to be constant in an early time period after which a year-round linear trend is introduced with its sign and magnitude to be determined from the data. Such a model has been used here, as well as an

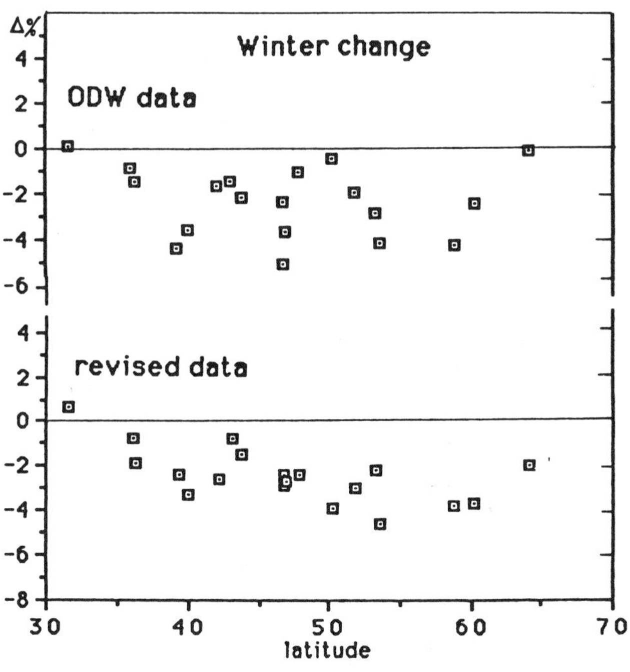

Figure 3. Differences (in %) between the winter season means of two time periods (12/64-3/75 to 12/75-3/86) as discussed in the text.

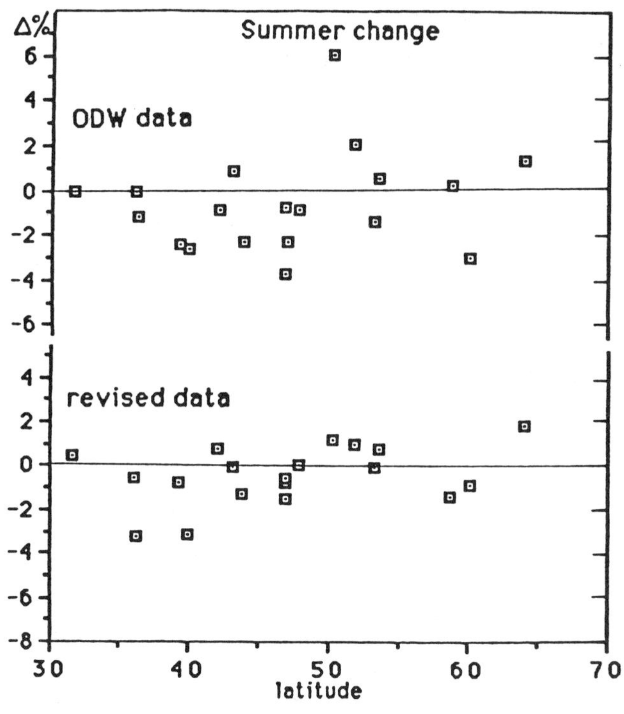

Figure 4. Differences (in %) between the summer season means of two time periods (12/64-3/75 to 12/75-3/86) as discussed in the text.

alternative in which each month is allowed to have a different linear trend ("fan" model). The latter description is more meaningful if seasonally dependent ozone changes are occuring.

Each data set is treated as a whole so that monthly trend coefficients are calculated while including an autocorrelated noise term to allow better estimation of the standard errors. Autocorrelations are generally significant only for one and two month lags at each station, and the independent monthly trends calculated from 12 separate series neglecting autocorrelation are very similar in magnitude to those calculated with inclusion of autocorrelation. To ensure that each month is of equal importance in this least squares fit to the entire data set, the data are weighted by each calendar month's inter-annual standard deviation. (Otherwise, the well-known, much smaller standard deviation for late summer total ozone data at northern middle latitude stations tends to dominate strongly over the winter ozone data with their larger natural scatter). A fuller description of this model appear in NASA/WMO (1988).

The provisionally revised data were examined in two forms: individual stations, and latitudinal band averages. In the statistical analyses presented hereafter the data cover the period 1965 to 1986 and the start of the linear trends has been fixed at 12/1969. Table 1 shows the magnitude of the linear trends for individual months for the period between 1969 and 1986 for three latitudinal band averages: 30-39°N, 40-52°N and 53-64°N. The correlation of ozone change with the QBO is known to be dependent upon latitude and this conclusion is confirmed with the sign change in Table 1. A six month lag has had to be introduced to maintain the QBO-ozone correlation south of about 40°N. The statistically indicated ozone response to the 11-year solar cycle is as large as +2%, increasing from sunspot minimum to maximum. The effects of the QBO on total ozone concentrations sometimes lag behind the equatorial wind direction changes themselves. Tests were made for the phase lag maximizing the apparent QBO/ozone signal, and the trends were evaluated with the lag corresponding to the maximum QBO effect. The calculated ozone decreases are largest at higher latitudes in the winter, e.g. -8.3±2.2% for January at 53-64°N. In the summer, however, the ozone changes are substantially less, with the more significant losses observed at the mid-latitudes. Because of the poor global coverage of Dobson stations south of 30°N latitude,and questionable quality, band averages for the tropics or the southern hemisphere are much less meaningful. Even in the three north temperate latitude bands the longitudinal distribution of Dobson stations is quite uneven.

The equally weighted average of the twelve monthly coefficients indicates a larger ozone loss since 1969 than the calculated year-round coefficient for each of the three bands shown in Table 1. The larger ozone losses occur in the winter months which have the higher inter-annual variation, and a weighted fit to a uniform trend gives more importance to the months with smaller

Table 1

Coefficients of Multiple Regression Statistical Analysis of Re-analyzed Dobson Measurements of Total Ozone Concentrations for the 22 Year Period, 1965-1986. The statistical model assumes a linear change after 1969, with evaluation of a trend coefficient for each month. (Data are expressed in total percent changes for the 17 year period 1969-1986)

Month	Latitude Band		
	53-64°N	40-52°N	30-39°N
January	- 8.3 ± 2.2	- 2.6 ± 2.1	- 2.2 ± 1.5
February	- 6.7 ± 2.8	- 5.0 ± 2.2	- 1.2 ± 1.9
March	- 4.0 ± 1.4	- 5.6 ± 2.3	- 3.5 ± 1.9
April	- 2.0 ± 1.4	- 2.5 ± 1.7	- 1.7 ± 1.3
May	- 2.1 ± 1.2	- 1.3 ± 1.1	- 1.7 ± 0.9
June	+ 1.1 ± 0.9	- 1.8 ± 1.0	- 3.3 ± 1.0
July	+ 0.0 ± 1.1	- 2.2 ± 1.0	- 1.3 ± 1.0
August	+ 0.2 ± 1.2	- 2.4 ± 1.0	- 1.0 ± 1.0
September	+ 0.2 ± 1.1	- 2.9 ± 1.0	- 1.0 ± 0.9
October	- 1.1 ± 1.2	- 1.5 ± 1.5	- 0.9 ± 0.8
November	+ 1.5 ± 1.8	- 2.4 ± 1.3	- 0.1 ± 0.8
December	- 5.8 ± 2.3	- 5.5 ± 1.7	- 2.1 ± 1.1
Average over all months, Percent Change	- 2.3 ± 0.7	- 3.0 ± 0.8	- 1.7 ± 0.7
Winter Average (DJFM),%	- 6.2 ± 1.5	- 4.7 ± 1.5	- 2.3 ± 1.3
Summer Average (MJJA),%	- 0.2 ± 0.8	- 1.9 ± 0.7	- 1.9 ± 0.8
QBO*	- 2.0 ± 0.6	- 1.3 ± 0.6	+ 1.9 ± 0.6
Solar*	+ 1.8 ± 0.6	+ 0.8 ± 0.7	+ 0.1 ± 0.6

* Percent changes per cycle, minimum-to-maximum. All uncertainties are expressed with one sigma statistical significance. The QBO variable is lagged six months in the 30-39°N band.

Calculated ozone change with uniform trend assumed for all months throughout the year in percent change in 17 years.

Percent/17 years	(- 0.7 ±0.6)	(- 2.3 ±0.7)	(- 0.9 ±0.6)

interannual variations which in this case have the smaller losses. This difference between the calculated year-round trend and the averages of the twelve monthly trends is the result of an inappropriate statistical treatment of the physical processes underlying the change rather than an unsatisfactory weighting procedure. A statistical model which posits a constant seasonal cycle and a single, year-round trend cannot adequately describe a situation in which the seasonal changes of the ozone concentrations varies through the years.

The winter-time losses in the two more northerly bands in Table 1 are appreciably larger than can be explained by atmospheric model calculations (NASA/WMO, 1988). A possible reason for this discrepancy is that the winter-time losses are enhanced through heterogeneous chemical processes affecting stratospheric chlorine chemistry, analogous to the processes suspected to cause the Antarctic spring ozone hole. The time and spatial distribution of such effects would be expected to be quite different in the Arctic because of the large differences in meteorology. In contrast to the tight circum-polar vortex in the south which effectively does not permit air exchange over Antarctica all winter long, the Arctic air masses following the Rossbi-wave pattern tend to pass through the polar night and back into sunlight again. Any chemical effects causing ozone depletion in the Arctic winter air would therefore be distributed over a much larger total volume of air than in the constricted Antarctic vortex. Furthermore, enhanced depletion of ozone could be observed throughout the late autumn and winter as air masses emerged into the sunlight, rather than being confined to five-six weeks or so in the late winter/early spring as found in the Antarctic polar vortex (Bojkov, 1986). The much lesser abundance of polar stratospheric clouds in the north, and the shorter exposure time of individual air masses to the polar night would also probably reduce the total amount of ozone depletion induced by such stratospheric chemical reactions.

CONCLUSION

A thorough examination of the available Dobson total column ozone measurements shows that it is necessary to consider the full calibration history of each station instead of using the data "as published" in Ozone Data for the World. Comparison of provisionally revised data with the published ODW data shows that while individual station records might be noticeably affected, on a collective basis there is no large change caused by the revision. A significant winter-time loss is found at latitudes between 30°N and 64°N (between -2.3% and -6.2% over 17 years) together with a smaller summer-time loss (-0.2% to -1.9%). Because the seasonal cycle of ozone concentrations is changing, statistical models based on assumptions of a uniform year-round loss are not suitable.

ACKNOWLEDGEMENT: The authors would like to appreciate Miss Anne Hansson's assistance in the data handling. One of the authors (RDB) further wish to record his thanks to numerous Dobson stations, observers and supervisors through whose thorough record keeping of instruments calibrations, the re-evaluation of the ozone records and the present study became possible.

REFERENCES

Bojkov, R.D., 1988: A critical assessment of published total ozone data and the need for re-evaluation, Proc. Ozone Symposium-1988, Bojkov and Fabian Ed., A. Deepak Publ.

Bojkov, R.D., 1986: Spring-ozone change in Antarctica and the role of the polar vortex, Adv. Space Res., 6/10 pp 89-98.

Harris, N.R.P. and F.S. Rowland, 1987: Trends in total ozone measurements at Arosa, Abstract AGU San Francisco, EOS.

Hill, W.J., P.N. Sheldon, and J.J. Tiede 1977: Analyzing worldwide total ozone for trends, Geophys. Res. Lett., 4, pp 21-24.

NASA/WMO Ozone Trends Panel Report-1988, WMO Ozone Project Report No. 18.

Oehlert, G.W., 1986: Trends in Dobson total ozone-an update through 1983, J. Geophys, Res., 91, pp 2675-2679.

Reinsel, G.C., G.C. Tiao, M.N. Wang, R. Lewis, and D. Nychka, 1981: Statistical analysis of stratospheric ozone data for the detection of trend, Atmos. Environ., 15, pp 1569-1577.

Reinsel, G.C., 1981: Analysis of total ozone data for the detection of recent trends and the effects of nuclear testing during the 1960's, Geophys. Res. Lett., 8, pp 1227-1230.

Reinsel, G.C., G.C. Tiao, A.J. Miller, D.J. D.J.Wuebbles, P.S. Connell, C.L. Mateer and J.J. DeLuisi, 1987: Statistical analysis of total ozone and stratospheric Umkehr data for trends and solar cycle relationship, J. Geophys. Res., 92, pp 2201-2209.

St. John, D.S., S.P. Bailey, W.H. Fellner, J.M. Minor, and R.D. Snee, 1981: Time series search for trend in total ozone measurements, J. Geophys. Res., 86, pp 7299-7311.

HOW MUCH INFORMATION ON ANTHROPOGENIC INFLUENCES ON THE OZONE LAYER CAN BE OBTAINED FROM A SET OF ONE-STATION OBSERVATIONS

Hans U. Dütsch and Johannes Staehelin
Atmospheric Physics ETH
Zürich, Switzerland

ABSTRACT

The Swiss records of total ozone (60 years), of the vertical distribution by the Umkehr method (30 years) and of regular ozone soundings (20 years) are used in an investigation on possible anthropogenic influences on the ozone layer. The correlation between total ozone and temperature on a mountain top station yields some information on the type of ozone variations for the period without soundings. - In the upper stratosphere a significant ozone decrease is found due to the altered photochemical situation. There is considerable but not fully conclusive evidence for changes in the in-situ photochemistry from 24 km into the middle stratosphere, while it is not possible to really distinguish between a change in stratospheric circulation or in the photochemical processes in the tropical source region, as a reason for the strong decrease in the ozone maximum layer; there is, however, an indication for a contribution of the latter. A significant increase of tropospheric ozone is a product of air pollution by man.

1. INTRODUCTION

The 60 year Arosa total ozone record showed between 1970 and 1985 a negative trend of 4.3% after it had before that exhibited large year to year fluctuations around an almost horizontal regression line (+ 0.85% in 45 years). The recent discovery of the Antarctic ozone hole and the suspicion that similar processes could also yield a certain ozone reduction in the Northern hemisphere, let us ask whether at least part of the ozone loss observed at Arosa since 1970 may be of anthropogenic origin.

It is certainly not without danger to base a study on this problem on one-station data only because of the strong influence of the varying large scale weather patterns on total ozone. However, the Arosa record gives the only practically continuous information dating far back into the anthropogenically undisturbed period; only a few other records predate the IGY around which extensive atmospheric nuclear bomb testing, was suspected to have an influence on the ozone layer by NO_x-production. Also because of possible solar cycle effects on total ozone, it is quite desirable to use an observational record longer than 30 years in an investigation on suspected anthropogenic influences.

Additional information can be used in order to reduce the effects of the one-station bias:21 years of regular ozone soundings above Payerne and the 30 year record of measurements of the vertical ozone distribution over Arosa from the indirect Umkehr method. A meteorological parameter (temperature at a mountain top station) was also employed to shed additional light on the processes yielding the observed ozone variations. With this combined material a one-station study with a very long data record should give a useful complement to investigations based on satellite data with world wide coverage over a relatively short time span.

2. MECHANISMS PRODUCING TOTAL OZONE VARIATIONS

One important contribution stems from varying large scale weather patterns which mainly affect the ozone content of the lowest stratosphere - below 16 and to a lesser degree up to 20 km; this influence is of regional character. Variations in the wintertime poleward ozone transport by the stratospheric general circulation have on the other hand a hemispheric effect and are therefore more difficult to separate from changes in photochemical processes especially from those influencing the tropical ozone source region. They affect most strongly the layer of the ozone maximum. If there should be an influence of an Arctic ozone hole type process it would presumably also show up at this level or just below. - Effects of variations in the in-situ photochemistry are expected to be strongest in the upper stratosphere and should also have an influence from the middle stratosphere down to about 25 km.

3. THE AROSA TOTAL OZONE RECORD

Figure 1 shows the annual mean values of total ozone obtained using the single wavelength pair C. The whole record has a sligthly negative trend (2.2 % in 60 years) which is, however, only produced by the accentuated ozone loss since 1970 A much better fit than by a regression line is therefore obtained by a second order parabola which shows a maximum value around 1950. The decrease after 1950 is more pronounced than the increase before that date - giving a certain suggestion that the effect of indicated change in stratosphereric circulation is amplified by alte-

ISBN 0-937194-15-8

red photochemistry, i.e. an anthropogenic influence. - With a few exceptions the QBO does not strongly affect the total ozone record.

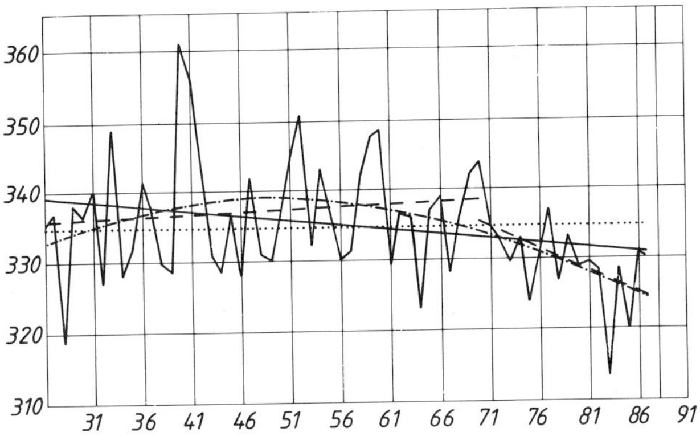

Fig. 1. Arosa total ozone record (annual mean values); regresssion lines for the whole period (full) and from 1926-1970 and 1970-1987(dashed); second order parabola

The main peaks and valleys in the year to year ozone variation in different seasons(fig. 2) do not fully coincide between winter and spring, but they still indicate a strong influence of the wintertime development on spring ozone values; however, the phase shift towards summer and fall becomes considerable and some of the features tend to melt or disappear. The pronounced ozone decrease after 1970 is clearly delayed in summer compared to winter and spring. This indicates that the strong ozone decrease in the 1970ies is due to changes in transport from low latitudes and/or meteorological effects only, while the further fall in the 80ies may have a photochemical component (photochemical in situ processes are strongest at high sun).

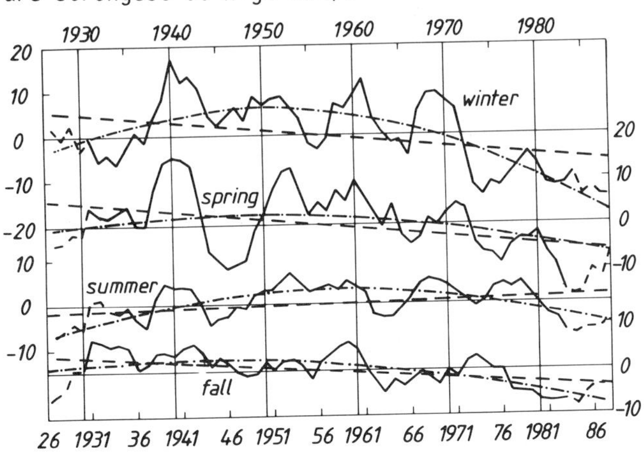

Fig. 2. Arosa total ozone record. Deviation of seasonal values (5-year running means) from the long term average, with regression lines and second order parabola.

4. INFORMATION FROM A METEOROLOGICAL PARAMETER

Cold air advection from higher latitudes under cyclonic conditions is normally connected with high total ozone values as a consequence of large ozone concentrations in the lowest part of the stratospere. This correlation has been used to distinguish between "meteorological" ozone fluctuations tied to large scale weather patterns and those produced by changes in the hemispheric wintertime ozone transport also in the earlier part of the Arosa record for which no concurrent measurements of the vertical distribution are available. The correlation in differenet seasons between mean values of total ozone and of the temperature at the mountain top station Saentis (2500 msl) was empoyed for this purpose.

The regression lines between these two quantities and the corresponding correlation coefficients were computed for each season; further the deviation of the total ozone value from the regression line (Δ y) was determined for each single year. Table 1 shows the slope of the regression lines and the correlation coefficients.

TABLE 1. SLOPE OF REGRESSION LINES (b) AND CORRELATION COEFFICIENTS (r)

	winter	spring	summer	fall
b (D/^{o}C)	- 8.19	- 3.68	- 2.30	- 3.61
r	- 0.704	- 0.327	- 0.282	- 0.530

The slope is steepest and the negative correlation coefficient largest in winter when at least 50% of the year to year ozone variance is explained by meteorological effects; corresponding spring values are already much lower.

In figure 3 the Δ y-line for the 60 year period is compared with that of Δ x (deviation of total ozone from its long term mean). If the whole ozone variation were due to changing large scale weather patterns only, Δ y would be ≛ zero while a coincidence of the 2 curves indicates missing meteorological influence.

The 2 curves differ in winter significantly from each other. The early part of the strong ozone loss after 1970 is at least in its predominant winter contribution a product of meteorological processes, as also shown by the measurements of the vertical distribution, while in the 80ies the decrease becomes more a large scale feature; this developement is in agreement with a possible antropogenic (photochemical) contribution which could not be expected much before 1980

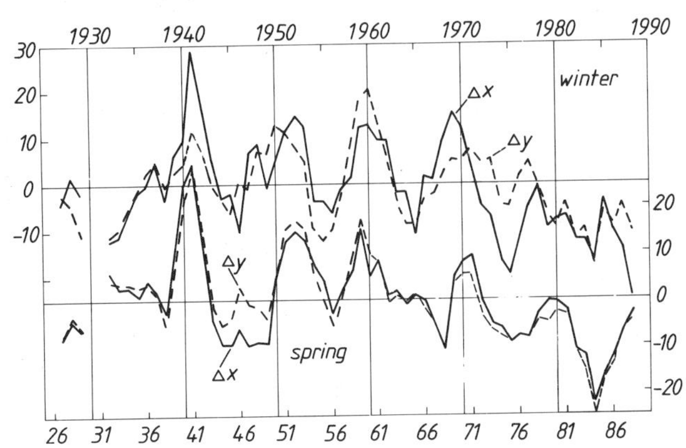

Fig. 3. 3-year running means of (Δ x), full line, and of (Δy), dashed.

In spring there is much less difference between the 2 curves; total ozone is in this season more strongly determined by the ozone accumulation at high latitudes during winter, i.e. by the varying large scale ozone transport, than by weather type influences.

5. VERTICAL DISTRIBUTION

Regular ozone soundings with the Brewer-Mast electrochemical instrument are performed 3 times per week at Payerne (Aerological Station of the Swiss Meteorological Institute) since late 1966. From this material the contribution of different levels to the total ozone loss since 1970 can be determined. For easier discussion the atmosphere up to 10 mb is divided into 13 layers (fig. 4); the mean ozone partial pressure for these layers is used in the following.

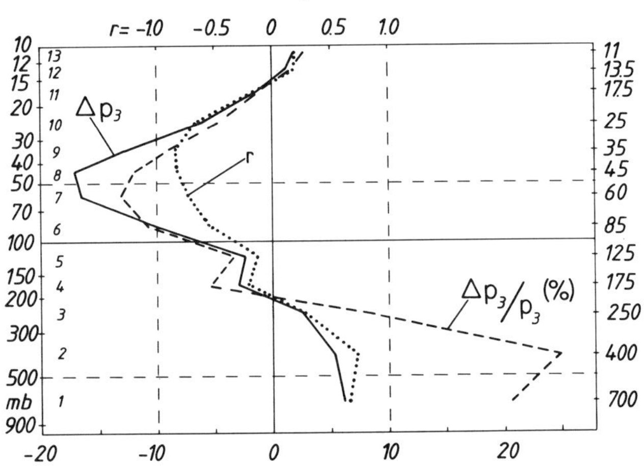

Fig. 4. Changes in mean ozone partial pressure from 1967 to 87 (linear regression) in 13 layers (left side): Absolute in nb and relative in % . Correletion coefficient (r) ozone with time.

Figure 4 shows the absolute and relative trend values and the correlation coefficients of ozone with time. High absolute values of the latter indicate that a regular trend is dominating over random year to year variations.

The most pronounced ozone losses (both absolute and relative) are observed between 100 and 30 mb, with a maximum of 10 - 12 % in a layer where the direct winter time advection from low latitudes is the dominating factor. The mentioned correlation coefficient rises upward from -0.5 to -0.8 showing a rapidly decreasing contribution to the variability from weather type processes. In the layer between 18 and 22 km the ozone loss has been accelerated since the end of the 70ies. This indicates an increasing reduction of the ozone transport from low latitudes but yields no a priori information on a possible contribution of a reduced photochemical production in the tropical source region to this effect.

A significant positive trend is observed in the troposphere (20 - 25 % in 20 years). It can hardly be doubted that this upward trend stems from anthropogenically produced changes in tropospheric photochemistry; considering the concurrent loss in the stratosphere a variation due to an altered flux from this upper level would have the opposite sign.

A correlation coefficient 0.2 absolute shows that in the lowest stratosphere the ozone regime is clearly dominated by random weather type fluctuations and that the negative trend - although 3 - 5 % - is thus rather accidential.

The data interpretation is more difficult and unfortunately not unique in the middle stratosphere, at a level where some information on possible changes of the in-situ photochemistry should be obtainable; this due to considerable observational uncertainties. The sounding quality is reduced a) by the strong influence of baroswitch errors on the ozone value as a function of pressure, and b) by the uncertainty in the at this level rather large pump correction. Thus the slightly positive trend is doubtful especially because the pressure element used in the sounding routine was changed at the end of 1980, followed by a positive excursion of the annual mean from 1980 to 81 of almost 10 % (fig. 5).

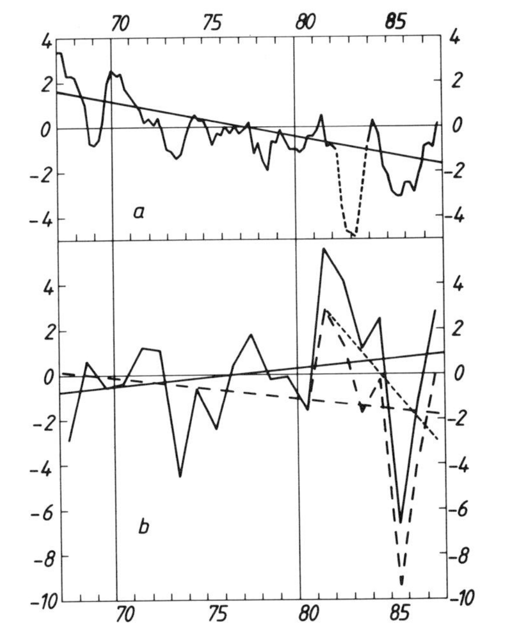

Fig. 5. a: 13 months running means of deviations from the 21 year average of ozone partial pressure in Umkehr layer 6 with regression line (dashed part in 1983 biased by El Chichon aerosol). b: Deviation of annual values from the 21 year mean of ozone partial pressure from soundings in the 10 - 12 mb layer with regression (full lines). Dashed lines: with correction for a possible shift in the record. Short dashed line: regression 1981-87.

The Arosa Umkehr data show in the middle stratosphere (layer 6, 15.6-7.8 mb) from 1980 to 1981 an increase which is only 20 % of that given by the soundings. If this is taken as an indication of a bias in the sounding record beginning in 1981 and thus those data were reduced by only 40% of the 80 to 81 increase the 1967-87 trend becomes negative by 2 %. Between 1981 an 87 there is anyhow a negative trend of 6.5%.

The Umkehr data themselves yield for layer 6 a clearly negative trend of 4.7 % for the 1967-87 period. About 1/4 of this value might be an artefact of the total ozone influence on the concentration in layer 6, introduced by the inversion procedure, leaving a corrected value of -3.5%.

Photochemical processes play a considerable role already in the intermediate 40 to 20 mb layer. Between 40 and 30 mb they contribute significantly to the removal of the surplus ozone accumulated by the wintertime transport and between 30 and 20 mb they even produce a secondary summer maximum after a minimum in May. In the 30 - 40 mb layer the pronounced negative trend in the 80ies has its seasonal maximum between April and June suggesting increased photochemical destruction of the surplus ozone. In the 20 - 30 mb layer the maximum loss occurs already in winter indicating increased photochemical removal already during the accumulation period. Summarizing we may say that the data between 24 km and 33 km suggest a contribution of in-situ photochemical processes to the ozone decrease during the 80ies (fig. 5), however, due to the observational problems in the middle stratosphere they do not supply a definitive proof.

6. NEGATIVE OZONE TREND IN THE UPPER STRATOSPHERE OBTAINED FROM UMKEHR DATA

Trend detection in the upper stratosphere where, according to models, the strongest impact of CFCs is expected, is rendered more difficult by superimposed solar cycle variations and especially by the disturbing effect of strong volcanic eruptions biasing by large particle scattering the Umkehr inversion results for the top layers.

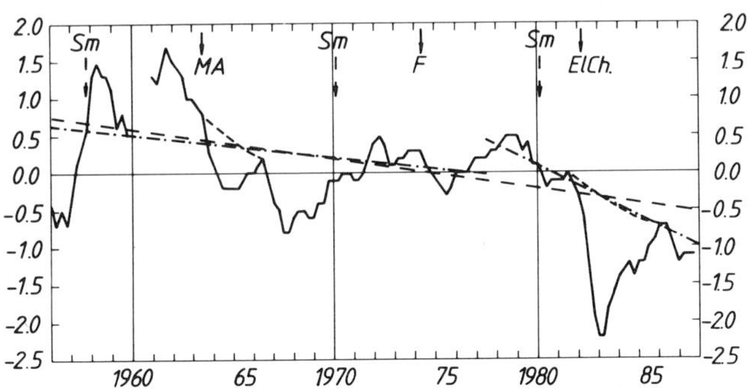

Fig. 6. Deviation of ozone partial pressure from 30 year average in Umkehr layer 9. Regression lines, whole period, full, and 1956-77 and 1977-87, respectively, dashed. Sm = Solar maximum.

Figure 6 shows the development in layer 9 (≈ 45 km) above Arosa (smoothed by 13-months running means). The highest ozone concentrations are connected with the IGY sunspot maximum which was the strongest ever observed. There are 3 major volcanic eruptions disturbing our record (Mt. Agung, 1963, Fuego, 1974 and particullarly El Chichon, 1982). An estimated correction for the volcanic disturbance has been applied (short dashed line in fig. 6) and was used in computing the regression lines.

A negative trend of 20 % in 32 years is obtained which should not be strongly biased by the solar cycle effect (Except may be for the influence of the special IGY maximum) because the data record contains almost exactly 3 cycles. If we divide the record in two pieces containing two and one cycle respectively, we find a much stronger negative trend during the last 10 years (fig. 6), i.e. 21 % and 5 % per decade for the two periods, in accordance with exponentially growing chlorine content in the upper stratosphere.

7. CONCLUSION

Considering all the evidence from total ozone (x) including the ozone-temperature correlation, as well as the development at different levels given by the soundings and the Umkehr measurements, we find that a photchemical (i.e. anthropogenic) contribution to the observed total ozone decrease since 1970 is very probable but can, due to observational problems in the middle stratosphere, not be proved conclusively. Table 2 shows the contribution of changes at different levels and their probable partitioning in photochemical and circulation effects (assuming that photochemical changes in the source region are similar to the alterations in the in-situ photochemistry in mid-latitudes). In the stratosphere alone changes in photochemistry and circulation seem to contribute in equal parts but the former is partly balanced by the increase in the troposphere. We do not try to quantify the uncertainty of the numbers given in table 2.

TABLE 2. ESTIMATED PHOTOCHEMICAL CONTRIBUTION TO OZONE TRENDS

Layer	% of x	trend in % of p_3	trend in % of x	contribution phot.chem.	contribution circ.
troposph.	8	+21	+1.7	+1.7	-
250-100mb	14	-3.0	-0.4	-0.1	-0.3
100-40mb	31	-8.5	-2.6	-0.9	-1.7
40-20mb	24	-5.0	-1.2	-0.7	-0.5
above 20mb	23	-3.5	-0.8	-0.7	-0.1
total	100		-3.3	-0.7	-2.6
in stratosph.	92		-5.0	-2.4	-2.6

POSSIBLE CAUSES OF RECENT TOTAL OZONE ANOMALIES

*Reginald E. Newell and Henry B. Selkirk**
Massachusetts Institute of Technology
Cambridge, Massachusetts 02139, USA

ABSTRACT

Recent total ozone anomalies based on 1963-1979 normals are examined at 32 stations in the surface network. Excess negative anomalies in the 1980's are of two types: those associated with the Antarctic ozone hole and anomalies primarily in middle latitudes. The latter reached a maximum in 1983-85 and are now reduced; the former persist. Changes in atmospheric trace constituents, possibly coupled with atmospheric circulation changes inducing warmer southern oceans in recent years, are thought to govern the Antarctic changes. It is suggested that the middle latitude anomalies are associated with material which ascended into the upper stratosphere and mesosphere following the eruption of El Chichon in 1982; such material would enter the lower stratosphere with a seasonal and latitudinal dependence similar to that of ^{102}Rh, ^{109}Cd and ^{238}Pu from high altitude nuclear debris injections.

1. INTRODUCTION

Coherent patterns of reduced total ozone have been observed since the early 1980's, not only over the Antarctic (Farman *et al.*, 1985), but also over other regions including at least the middle and high latitudes of the Northern Hemisphere (Dütsch, 1985; Angell *et al.*, 1985; Bojkov, 1987; Newell and Selkirk, 1988; Heath, 1988). Two important questions are: Can we understand the reasons for the reductions in total ozone in recent years? Will the reductions continue, strengthen or diminish? In this paper, we examine these questions after updating our previous findings from the surface network.

2. DATA AND RESULTS

Anomalies of total ozone at thirty-two ozone stations in the ground- based network are examined following a station selection procedure used by Newell and Selkirk (1988). Stations are excluded if fewer than 10 years of data are available during the chosen base period 1963-1979 for any month or if one or more of the base period standard deviations (σ) in the summer months June through August exceeds 20 Dobson units (D.U.). The base period was chosen for its statistical stability over a reasonably representative set of stations. In Figure 1 are shown July 1979-December 1987 anomalies relative to the 1963-79 means; the recent trends in the Antarctic and Northern Hemisphere as well as the biennial oscillation show up clearly. [See Newell and Selkirk (*loc. cit.*) for a discussion of problems associated with some of these time series.] The Antarctic decreases commenced in the late 1970's and show few signs of reversing, while the middle latitude anomalies were strongest in the period 1983-86. To assess the likelihood of such anomalies, the number of monthly anomalies exceeding 1.7σ in magnitude were summed in three-month seasonal bins for each year and are expressed in Figure 2 as percentages of the available sample in the northern high latitudes, northern middle latitudes and southern high latitudes. In a normal distribution, the fraction of anomalies with amplitudes greater than 1.7σ over a large sample would tend to 10%. It is clear from Figure 2 that the preponderence of large negative anomalies in the northern middle latitudes as well as in the Antarctic is quite out of the ordinary. In the Antarctic, the negative anomalies have continued to occur, not only in spring, but also at other times of the year, summer in particular. (We note that the winter season numbers are less reliable, as samples are small and based upon fewer observations.) The numbers of large negative anomalies appear to have decreased at the northern middle latitude stations: peak years were 1983 and 1985, while 1987 appears to have fewer than the preceding four years.

3. POSSIBLE EXPLANATIONS

Our analysis suggests that there are two separate phenomena which require explanation: the Antarctic ozone hole and northern hemisphere decreases. In view of the space limitation here we refer to our previous paper (Newell and Selkirk, 1988) for a review of recent ideas on the Antarctic decreases. In that paper we showed (as have others) that Antarctic strato-

*Present affiliation: NASA Ames Research Center
Moffett Field, California

ISBN 0-937194-15-8

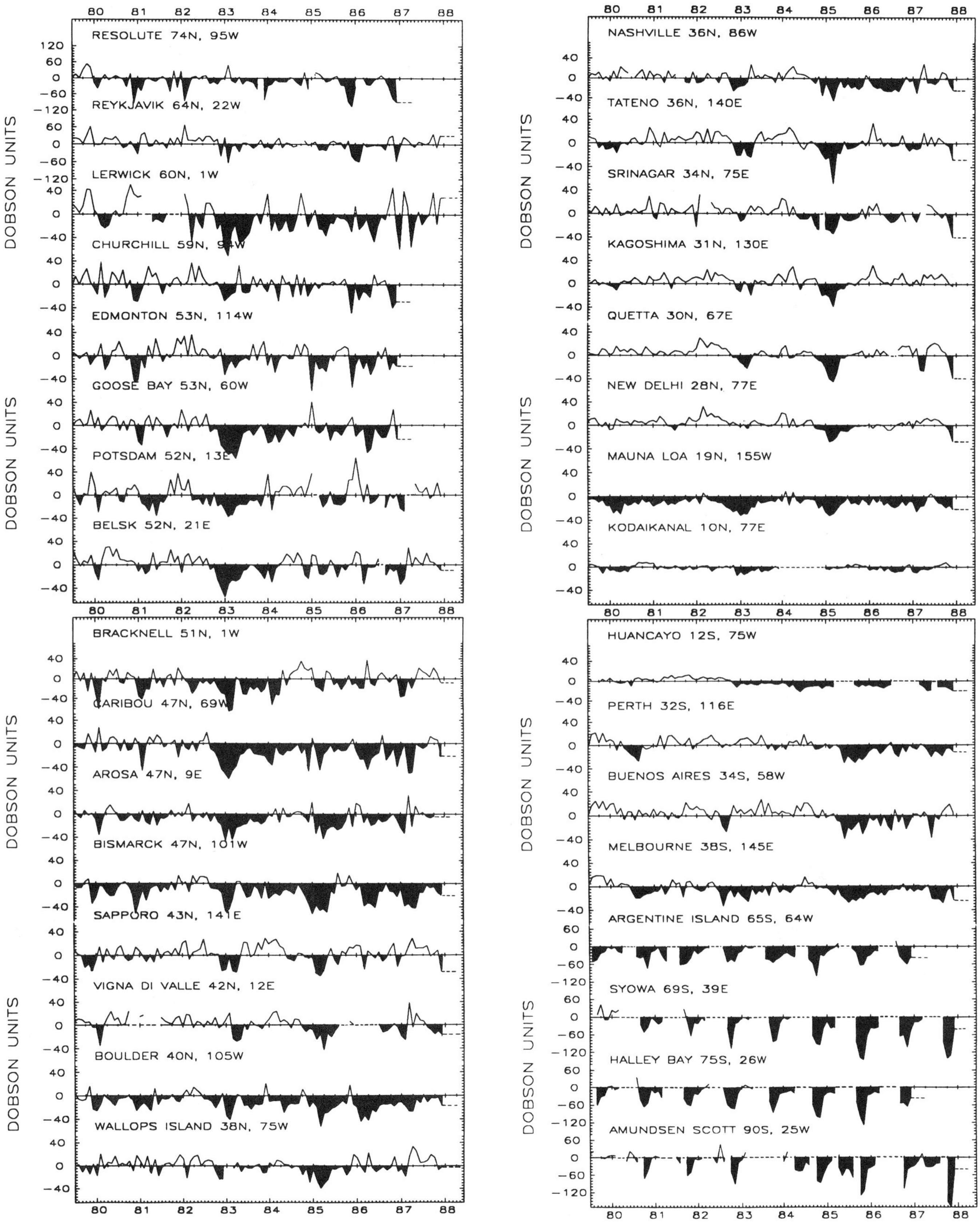

Fig. 1. Total ozone anomalies by station for July 1979 through December 1987 in Dobson units. Negative anomalies shaded for clarity. Missing data indicated by dashed baseline; horizontal dashes at righthand end indicate the final anomaly in each time series.

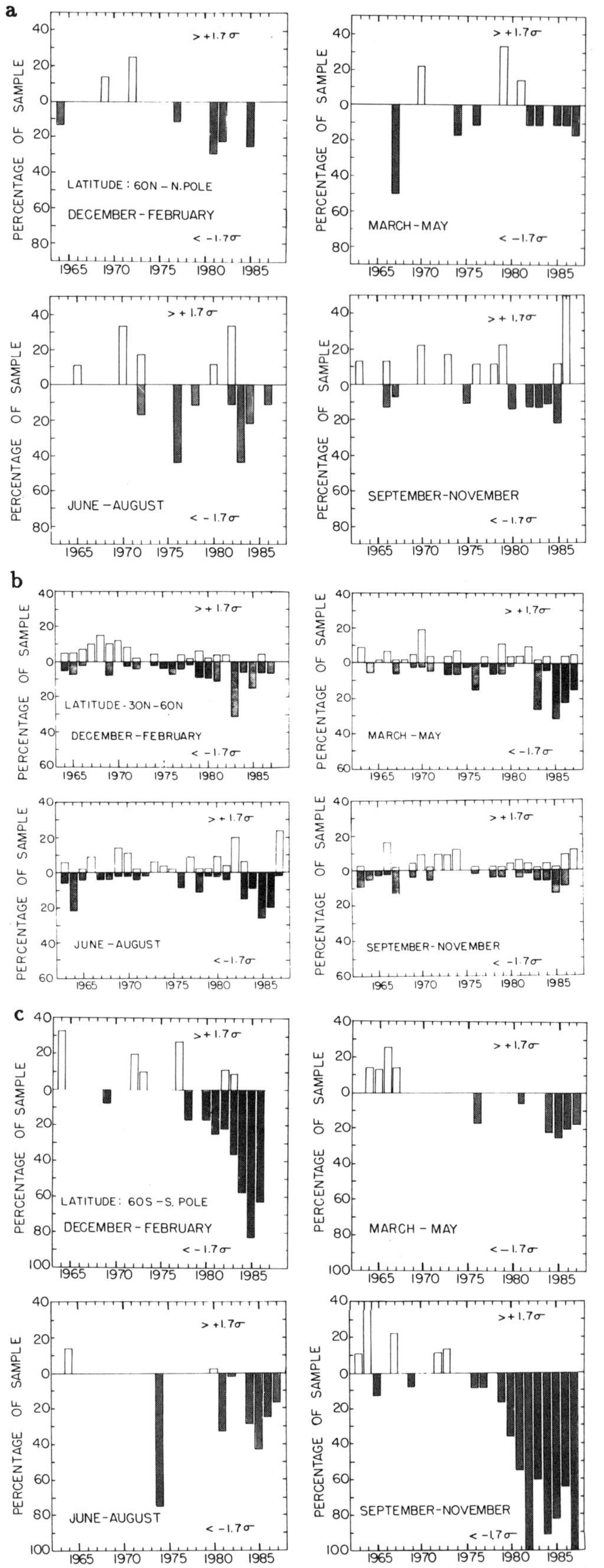

Fig. 2. Percentages by season of total ozone anomalies exceeding $\pm1.7\sigma$: a) 60°-90°N, b) 30°-60°N and c) 60°-90°S.

spheric temperatures have decreased along with ozone. Although lower temperatures may be radiatively caused and thus simply a reflection of chemically-induced reduction of ozone, a simultaneous reduction of both ozone and heat transport into the polar regions is another possible explanation. At the very least a reduction of transport of ozone into polar regions would tend to sustain the effects of chemically-induced reductions which might be occurring there. It has long been argued that both of these transports depend upon the vertical flux of planetary wave energy into the lower stratosphere (Newell, 1963, 1964). This assocation was demonstrated from computations of this flux for the Northern Hemisphere by Newell and Richards (1969) who showed it to be a maximum in the spring at high latitudes. Shiotani and Gille (1987) find the same pattern in the Southern Hemisphere. The annual cycles of total ozone near 60°S and vertical energy flux (calculated using data from Knittel, 1976) show that these quantities vary together as is shown in Figure 3. Thus modulations in the vertical energy flux may have induced recent ozone changes as argued by Mahlman and Fels (1986) and Nagatani and Miller (1987). As we have stresssed elsewhere (Newell and Selkirk, *loc. cit.*), such a change in the atmospheric circulation may have also been responsible for the increase in southern ocean temperatures observed in the same time period (Folland *et al.*,1986). This change, if it exists, has not yet been described in detail, and without further information the termination of this mode clearly cannot be forecast.

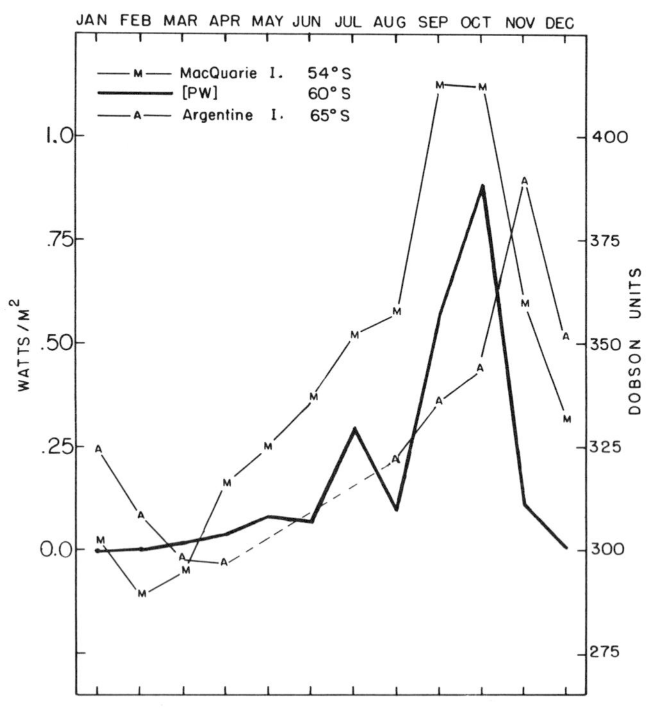

Fig. 3. Annual cycle of planetary wave (Eliassen-Palm) flux at 100 mb, 60°S, compared with total ozone cycle at two southern stations.

We believe that the decreases outside the Antarctic are primarily due to chemical substances injected into the upper troposphere and perhaps lower mesosphere by the eruption of El Chichon in 1982 (Newell and Selkirk, *loc. cit.*). We base this contention on the pattern of the observed anomalies (predominantly in the winter and spring) and on the resemblance of the pattern to that of the concentration in the lower stratosphere of material from very high-altitude nuclear tests and a satellite reactor burn-up (List *et al.*,1966). The El Chichon eruption far exceeded in strength previous volcanic eruptions in the period studied here (for a plot of atmospheric transmission see Newell and Hsiung, 1987). Until several years ago stratospheric trace constituents were monitored regularly by aircraft and balloons (e.g. see Leifer *et al.*, 1981). An illustration of N_2O concentrations appears in Figure 4 for the Northern Hemisphere; N_2O enters the stratosphere through the tropical tropopause and rises into the lower and middle stratosphere. The low values over the pole represent air which has lost N_2O through photodissociation as it subsequently moved polewards and downwards; the difference between the entry values of about 320 ppbv and the high latitude values down to 100 ppbv is found in NO_x compounds which may react with ozone. Anthropogenic chlorine species appear to have a similar distribution (Leifer *et al.*, *loc. cit.*) for the same reasons. Although the substance added by El Chichon is not known (HCl is a possible candidate), it may have acted synergistically with NO_x and chlorofluorocarbon decay products to produce the observed reductions in middle latitudes. If this substance is now passing through the system, it may take 8 years to clear the stratosphere as did ^{238}Pu, although the latter became attached to stratospheric aerosol so that its loss was accelerated. We mention in closing that the effects of this substance may not have been limited to middle latitudes but may also have contributed to changes in the Antarctic after 1982.

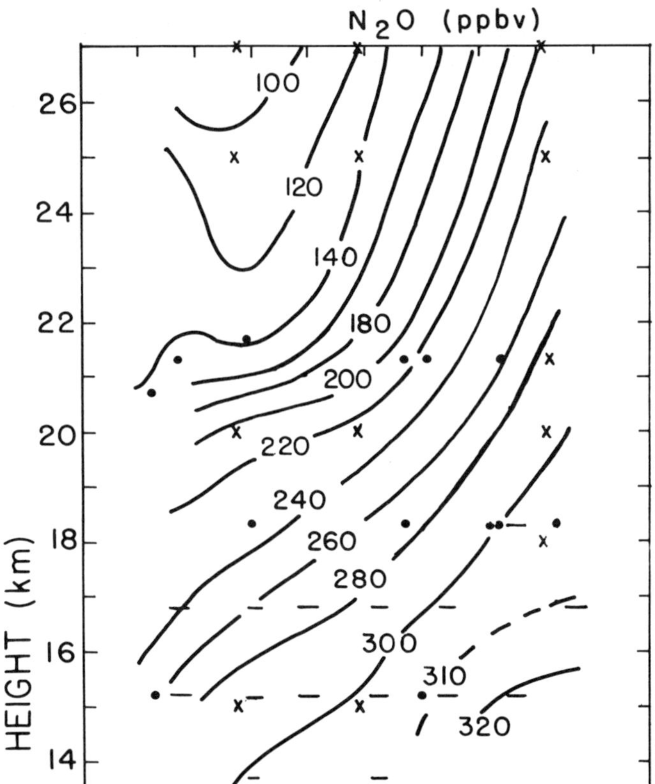

Fig. 4. Meridional cross-section of N_2O mixing ratio (ppbv) drawn from data given by Inn *et al.*, 1977, Krey *et al.*, 1977, Leifer *et al.*, 1981, Schmeltekopf *et al.*, 1977 and Vedder *et al.*, 1978.

4. CONCLUSIONS

Two types of total ozone reductions have occurred in recent years. Those in the Antarctic are thought to be related to changes in trace constituents perhaps accompanied by large-scale atmospheric circulation changes. Elsewhere changes are thought to be due to chemical substances introduced into the stratosphere since the eruption of El Chichon in 1982. The duration of ozone deficits due to circulation changes cannot be predicted, whereas it could be expected that the latter reductions would diminish as the volcanic material clears the stratosphere.

ACKNOWLEDGMENTS

We acknowledge support by the National Science Foundation Climate Dynamics Program under Grant ATM-8517107. The second author is supported at NASA Ames by the National Research Council Associateship Program. Susan Ary organized the data for us and helped compose the figures.

REFERENCES

Angell, J.K., Korshover, J. and Planet, W.G., 1985, Ground-based and satellite evidence for a pronounced ozone minimum in early 1983 and responsible atmospheric layers. *Mon. Wea. Rev.*, **113**, 641-646.

Bojkov, R.D., 1987, The 1983 and 1985 anomalies in ozone distribution in perspective. *Mon. Wea. Rev.*, **115**, 2187-2201.

Dütsch, H.U., 1985, Total ozone trend in the light of ozone soundings, the impact of El Chichon. *Atmospheric Ozone.* Eds. C.S. Zerefos and A. Ghazi. D. Reidel Publishing Co., Dordrecht, 263-268.

Farman, J.C., Gardiner, B.C. and Shanklin, J.D., 1985, Large losses of total ozone in Antarctica reveal seasonal ClO_x/NO_x interaction.*Nature*, **315**, 207-210.

Folland, C.K., Palmer, T.N. and Parker, D.E., 1986, Sahel rainfall and worldwide sea temperatures. *Nature*, **320**, 602-607.

Heath, D.F., 1988, Nonseasonal changes in total column ozone from satellite observations, 1970-1986. *Nature*, **332**, 219-227.

Inn, E.C.Y., Tyson, B.J., and Arvesen, J.C., 1977, Atmospheric halocarbon experiment. *Interhemispheric Survey of Minor Upper Atmospheric Constituents During October-November 1976*, NASA Tech. Memo. X-73630, 79-93.

Knittel, J., 1976, Ein Beitrag zur Klimatologie der Stratosphäre der Südhalbkugel, *Meteorol. Abh. Berlin*, Series A Monographien, Band 2/Heft 1, 112 pp., Maps and Tables.

Krey, P.W., Lagomarsino, R.J., and Schonberg, M., 1977, Correction of: 'N_2O concentrations in the stratosphere in 1975'. *Health and Safety Laboratory, Environmental Quarterly*, U.S. Energy Research and Development Administration, New York, HASL-328, I-15-I-23.

List, R.J., Salter, L.P., and Telegadas, K., 1966, Radioactive debris as a tracer for investigating stratospheric motions. *Tellus*, **18**, 345-354.

Leifer, R., Larsen, R., and Toonkel, L., 1981, Stratospheric distribution, inventories and trends of trace gases in the Northern Hemisphere for 1977-1979. *Environmental Report EML-390*, U.S. Dept. of Energy, New York, I-355-I-406.

Mahlman, J.D., and Fels, S.B., 1986, Antarctic ozone decreases: a dynamical cause? *Geophys. Res. Lett.*, **13**, 1316-1319.

Nagatani, R.M., and Miller, A.J., 1987, The influence of lower stratospheric forcing on the October Antarctic ozone decrease. *Geophys. Res. Lett.*, **14**, 202-205.

Newell, R.E., 1963, Transfer through the tropopause and within the stratosphere. *Q. J. R. Meteorol. Soc.*, **89**, 167-204.

Newell, R.E., 1964, Further ozone transport calculations and the spring maximum in ozone amount. *Pageoph*, **59**, 191-206.

Newell, R.E., and Richards, M.E., 1969, Energy flux and convergence patterns in the lower and middle stratosphere during the IQSY. *Q. J. R. Meteorol. Soc.*, **95**, 310-328.

Newell, R.E., and Hsiung, J., 1987, Factors controlling free air and ocean temperature of the last 30 years and extrapolation to the past. *Abrupt Climatic Change.* Eds. W.H. Berger and L.D. Labeyrie, D. Reidel Publishing Co., Dordrecht, 67-87.

Newell, R.E., and Selkirk, H.B., 1988, Recent large fluctuations in total ozone. *Q. J. R. Meteorol. Soc.*, **114**, 595-617.

Schmeltekopf, A.L., and ten others, 1977, Stratospheric nitric oxide profiles at various latitudes. *J. Atmos. Sci.*, **34**, 729-736.

Shiotani, M., and Gille, J.C., 1987, Dynamical factors affecting ozone mixing ratios in the Antarctic lower stratosphere, *J. Geophys. Res.*, **92**, 9811-9824.

Vedder, J.K., Tyson, B.S., Brewer, R.B., Boitnott, C.A., and Inn, E.C.Y., 1978, Lower stratospheric measurements of variation with latitude of CF_2Cl_2, $CFCl_3$, CCl_4 and N_2O profiles in the Northern Hemisphere. *Geophys. Res. Lett.*, **5**, 33-36.

CALIBRATION OF PRIMARY STANDARD DOBSON SPECTROPHOTOMETER NO. 83 AND TOTAL OZONE TRENDS AT MAUNA LOA OBSERVATORY, HAWAII, AND AMERICAN SAMOA, SOUTH PACIFIC

W.D. Komhyr and R.D. Grass
NOAA/ERL/Air Resources Laboratory
Boulder, Colorado 80303, USA

R.K. Leonard
Cooperative Institute for Research in Environmental Sciences
University of Colorado
Boulder, Colorado 80309, USA

ABSTRACT

The long-term ozone measurement precision of Primary Standard Dobson Spectrophotometer no. 83 has been maintained to within an uncertainity of ±0.5% during 1962-1987. Total ozone data obtained at Mauna Loa Observatory (MLO) since 1963 and Samoa Observatory (SMO) since 1975 have been reprocessed on the scale of Dobson instrument no. 83. On average, total ozone increased at MLO during 1964-1974 at a rate of 0.37 %/yr, but decreased during 1974-1987 at a rate of 0.35 %/yr; at SMO, the ozone decrease rate measured during 1976-1987 was 0.40 %/yr. Unusually low total ozone amounts were observed at SMO at the time in late 1986-early 1987 of the minimum in the tropical Quasibiennial Oscillation in ozone.

1. INTRODUCTION

Dobson spectrophotometer no. 83 was established in 1962 as a standard for total ozone measurements in the United States. In 1980, this instrument was designated by the World Meteorological Organization (WMO) as the Primary Standard Dobson spectrophotometer for the world. Since 1962, 15 domestic and 75 foreign Dobson instruments have been calibrated, many of them several times, either directly or indirectly relative to instrument no. 83.

Absolute calibrations of instrument no. 83 (relative to Vigroux (1953) ozone absorption coefficients for Dobson instrument A and D wavelength pairs) were performed by the Langley slope method (Dobson and Normand, 1962) at Sterling, Virginia, in 1962, and at Mauna Loa, Hawaii, in 1972, 1976, 1978, 1979, 1980, 1981, 1984, 1986, and 1987. Additionally, calibration checks on the instrument have been performed routinely since 1962 with a set of standard lamps. Analysis of results of the two types of calibrations has yielded detailed data on the long-term ozone measurement precision of the instrument (Komhyr et al., 1988). A synopsis of the calibration work is presented here.

Total ozone data obtained with Dobson spectrophotometers at Mauna Loa Observatory (MLO), Hawaii, since 1964, and at American Samoa (SMO), South Pacific, since 1976, have been reprocessed on the calibration scale of Dobson instrument no. 83. Long-term ozone trends at those two sites are presented, based on simple linear regressions of the data.

2. CALIBRATION OF DOBSON INSTRUMENT NO. 83

Calibration of a Dobson spectrophotometer on an absolute scale by the Langley slope method entails obtaining a series of one-half-day total ozone measurements on AD wavelengths of the rising or setting sun, and deducing corrections S_{AD} to the instrument $N_{AD}{}^*$ values, defined below by the fundamental Dobson instrument relation

$$\frac{N_{AD}{}^* - [(\beta-\beta')_A - (\beta-\beta')_D]\, mp/p_o}{\mu} + S_{AD}\left(\frac{1}{\mu}\right) = [(\alpha-\alpha')_A - (\alpha-\alpha')_D]\,x + \frac{[(\delta-\delta')_A - (\delta-\delta')_D]\sec Z}{\mu} \qquad (1)$$

where

$N_A \;=\; log\,\frac{I_\lambda}{I_{\lambda'}} - log\,\frac{I_{o\lambda}}{I_{o\lambda}}\;\overline{\;\;}_{\lambda\lambda' A}$;

$N_D \;=\; log\,\frac{I_\lambda}{I_{\lambda'}} - log\,\frac{I_{o\lambda}}{I_{o\lambda'}}\;\overline{\;\;}_{\lambda\lambda' D}$;

$N_{AD}{}^* \;=\; N_A{}^* - N_D{}^*$;

$N_A{}^* \sim N_A = N_A{}^* + S_A$;

$N_D{}^* \sim N_D = N_D{}^* + S_D$;

$S_{AD} = S_A - S_D$;

I_λ = intensity of the shorter wavelength of the A or D wavelength pair;
$I_{\lambda'}$ = intensity of the longer wavelength of the A or D wavelength pair;
$I_{o\lambda}$ = extraterrestrial value of I_λ;
$I_{o\lambda'}$ = extraterrestrial value of $I_{\lambda'}$;
β,β' = molecular scattering coefficients for the wavelength λ and λ';
m = air mass, the ratio of the actual and vertical path lengths of the solar beam through the atmosphere;
p = mean station pressure;
p_o = sea level pressure;
δ, δ' = particle scattering coefficients for the wavelengths λ and λ';
Z = solar zenith angle;
α,α' = ozone absorption coefficients for the wavelengths λ and λ';
μ = ratio of the actual and vertical path lengths of the solar beam through the ozone layer; and

ISBN 0-937194-15-8

x = total ozone amount in a vertical column of the atmosphere.

For observations made during clear-sky conditions, the last term of (1) becomes negligible, and if the ozone amount is assumed to be constant during the one-half-day's observations, then from (1), a plot of

$$\frac{N_{AD}{}^{*} - [(\beta - \beta')_A - (\beta - \beta)_D]\, mp/p_o}{\mu} \quad vs. \quad \frac{1}{\mu}$$

yields S_{AD}, the correction needed to the provisional $N_{AD}{}^{*}$ values of the Dobson instrument.

The most comprehensive calibration of instrument no. 83 was performed in 1976 at MLO when ozone observations were made during 90 one-half days. The standard error associated with the determination of the mean S_{AD} correction to the instrument $N_{AD}{}^{*}$ values at that time was 0.0006, which corresponds to an uncertainty in the measurement of ozone of 0.07% for $x = 300$ D.U. and $\mu = 2$. This calibration of instrument no. 83 resulted in establishment of an August 26, 1976, Calibration Scale for the intrument that has been used through 1987 in calibrating all domestic and foreign Dobson spectrophotometers.

The August 26, 1976, Calibration Scale was used to establish N values in 1976 for Dobson instrument no. 83 standard lamps 83A, 83B, W, X, Y, Z, 83Q1, and 83Q2, some of which had been run on the instrument since 1962. The lamp data were then used to establish a 1962-1987 Standard Lamp Calibration Scale for instrument no. 83, which, except for the 1976 tie in, is independent of the Langley slope calibrations of the instrument.

Table 1 compares ozone values, x_1, determined with instrument no. 83 during 1962-1987 on the August 26, 1976, Calibration Scale, with similar values, x_2, incorporating Langley slope calibration corrections; with x_3 values measured on the 1976 Standard Lamp Scale; and with x_4 values determined on the 1976 Standard Lamp Scale with Langley slope calibration corrections applied. We draw three conclusions from the results. First, the August 26, 1976, Calibration Scale for Dobson instrument No. 83 has remained unchanged to within an uncertainty of ±0.5% during 1972-1987. Second, because this scale has remained highly stable with time, additional calibration corrections are not needed for domestic and foreign Dobson instruments calibrated since 1976 with Dobson spectrophotometer no. 83. Finally, instrument no. 83 calibration observations made at MLO spanned a solar cycle sunspot maximum and two minima. Because the Langley slope and the standard lamp instrument calibrations gave essentially identical results, we conclude that if the extraterrestrial intensities of the Dobson instrument A and D wavelengths vary during the course of a solar cycle, the variations are not large enough to cause errors in ozone measurements greater than several tenths of a percent.

TABLE 1. STABILITY OF THE AUGUST 26, 1976, CALIBRATION SCALE FOR DOBSON INSTRUMENT NO. 83*

Year	$\frac{100(x_1 - x_2)}{x_2}$ (%)	$\frac{100(x_1 - x_3)}{x_3}$ (%)	$\frac{100(x_1 - x_4)}{x_4}$ (%)
1962	--	-0.7	-1.0
1972	-0.1	-0.3	-0.1
1976	0.1	0.0	0.1
1978	0.0	0.0	0.0
1979	-0.2	-0.1	-0.2
1980	0.2	0.0	-0.4
1981	-0.3	-0.1	-0.7
1984	0.5	-0.1	0.1
1986	0.2	-0.5	-0.5
1987	0.2	-0.2	-0.1

*x_1 ozone values: Based on the August 26, 1976 Calibration Scale and instrument optical wedge calibration of August 26, 1976. x_2 ozone values: As x_1 values, but corrected by the Langley slope calibration method. x_3 ozone values: Based on the Standard Lamp Calibration Scale established in 1976, and updated optical wedge calibrations. x_4 ozone values: As x_3 values, but corrected by the Langley slope calibration method.

3. COMPARISON OF DOBSON INSTRUMENT NO. 83 AND TOMS SATELLITE OZONE DATA AT MLO

The calibration observations made with Dobson instrument No. 83 at MLO on AD wavelengths during 1972-1987 have yielded a unique, highly accurate total ozone data set that is suitable for use in checking on the calibration of satellite total ozone measurement instrumentation. Comparison of the Dobson instrument data with TOMS satellite overpass data at MLO has indicated a nonlinear drift in the calibration of the TOMS satellite ozone sensor of about 3.3% during 1979-1987 (McPeters and Komhyr, 1988).

4. LONG-TERM TOTAL OZONE TRENDS IN HAWAII AND SAMOA

Wavelength settings for Dobson instrument no. 63 used routinely in the past at MLO were determined in late 1963 by the zenith skylight method (Dobson, 1957) rather than with spectral discharge lamps. (The two methods yield short D wavelengths that differ by 0.12 nm). Because, also, a small μ-dependency was inherent in the performance of instrument no. 63, calibration of the instrument in late 1963 by means of standard lamps caused ozone values measured by the instrument to be too high by several percent. Provisional MLO data, routinely archived at the World Ozone Data Centre in Canada, show a step decrease in ozone values in 1976, following recalibration of instrument 83 at that time (Bojkov, 1988). We have reprocessed the MLO total ozone record for 1964-1987 on the August 26, 1976, Calibration Scale of Primary Standard Dobson instrument no. 83. While some shortcuts were taken in reprocessing the data, it is unlikely that more detailed processing will give results over portions of the record that differ on average from current values by more than 1%.

Figure 1 is a plot of ozone anomalies (i.e., deviations of ozone monthly means from monthly normals) at MLO for 1964-1987. Quasibiennial oscillations in ozone are clearly evident in the record. Least-squares linear regression trend lines fitted to the data indicate that ozone at MLO increased on average at a rate of 0.37 ± 0.16%/yr during 1964-1974, but decreased at a rate of 0.35 ± 0.13%/yr during 1974-1987. (Indicated uncertainties are the 95% confidence intervals.) For the record as a whole, the mean ozone trend was -0.07 ± 0.06%/yr.

The SMO Dobson instrument total ozone record was

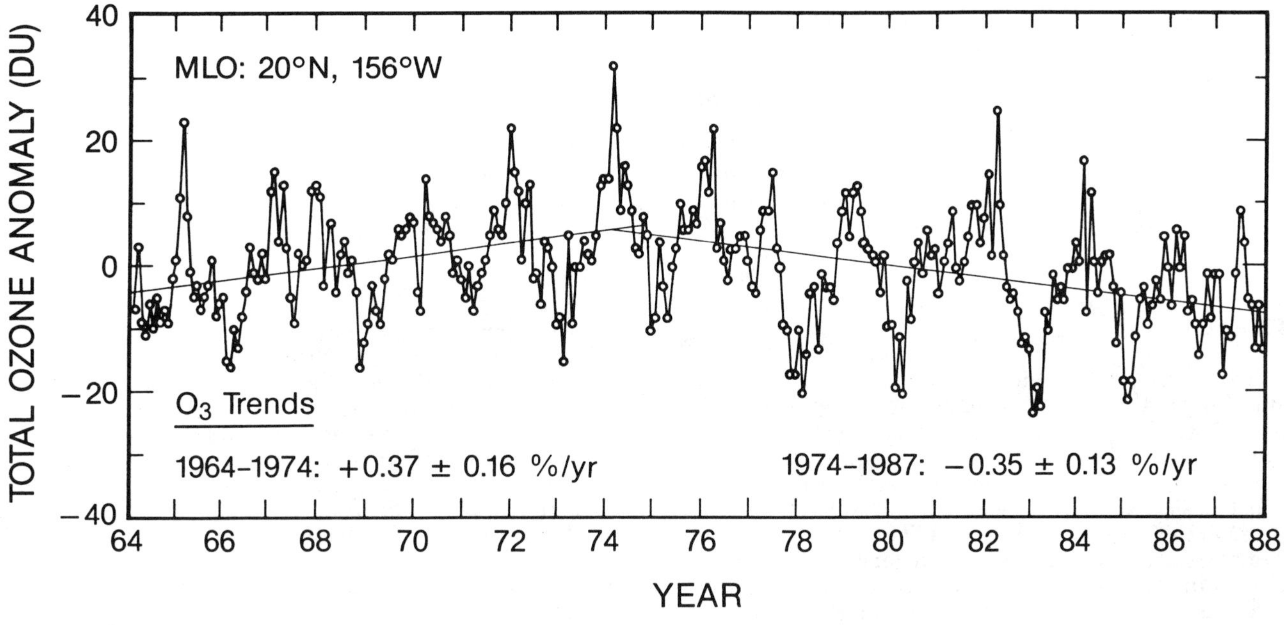

Fig. 1. Deviations of monthly mean total ozone amounts from monthly normals for the period of record at MLO (1964-1987). Least-squares regression trend lines are fitted to the data.

similarly reprocessed. Ozone anomalies for SMO for 1976-1987 are shown in Fig. 2. A least-squares linear trend line fitted to the data shows that, on average, total ozone has been decreasing in Samoa at 0.40 ± 0.11%/yr. (For 1976-1986, the downward trend in ozone was 0.35 ± 0.12%/yr.)

A record low mean monthly total ozone amount of 221 DU was measured at SMO in February 1987, at the time of a deep minimum in the 1986-1987 tropical Quasibiennial Oscillation in ozone. This value was 13% lower than the average total ozone amount for February months of 1976-1987. Although the value is subject to revision, it is unlikely to be too low by more than several percent.

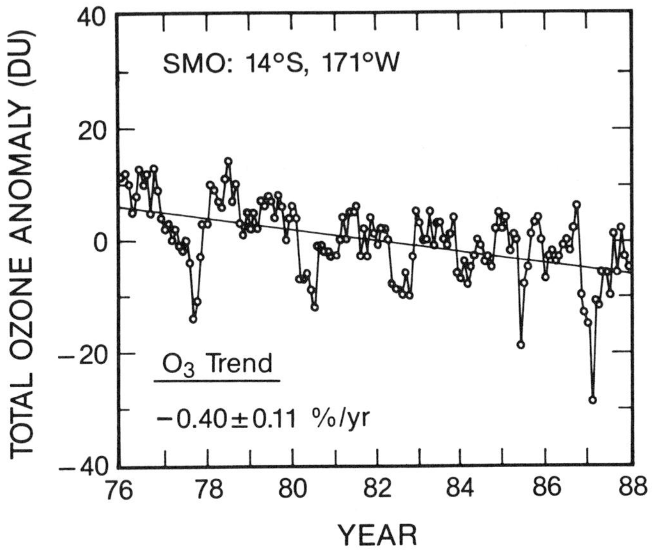

Fig. 2. Deviations of monthly mean total ozone amounts from monthly normals for the period of record at SMO (1976-1987). A least squares regression trend line is fitted to the data.

ACKNOWLEDGMENTS

Appreciation is expressed to the numerous observers who made the Dobson instrument no. 83 calibration observations at MLO, and to the MLO and SMO total ozone observers.

REFERENCES

Bojkov, R.D., 1988: NASA/WMO Ozone trend panel report, *WMO Ozone Project Report No. 18*.

Dobson, G.M.B., 1957: Adjustment and calibration of the ozone spectrophotometer. In *Annals of the International Geophysical Year V, Part I*, Pergamon Press, New York, 90-113.

Dobson, G.M.B., and C.W.B. Normand, 1962: Determination of the constants etc. used in the calculation of the amount of ozone from spectrophotometer measurements and of the accuracy of the results. In *Annals of the International Geophysical Year, XVI, Part II*, Pergamon Press, New York, 161-191.

Komhyr, W.D., R.D. Grass, and R.K. Leonard, 1988: Dobson spectrophotometer no. 83: A standard for total ozone measurements, 1962-1987. Submitted for publication to the *J. Geophys. Res.*

McPeters, R.D., and W.D. Komhyr, 1988: Long-term changes in SBUV/TOMS relative to the world standard Dobson instrument. *Proc. Ozone Symposium-88*, Editors: R.D. Bojkov and P. Fabian, A. Deepak Publishing, Hampton, Virginia.

Vigroux, E., 1953: Contribution a l'étude expérimentale de l'absorption de l'ozone. *Ann. Phys.*, 8, 709-762.

EL NIÑO AND OZONE

Koji Yamazaki and Isamu Yagai
Meteorological Research Institute
1-1 Nagamine Tsukuba Ibaraki 305, Japan

ABSTRACT

The impact of El Niño on total ozone distribution is studied by an atmospheric general circulation model (the MRI·GCM). A simple ozone photochemistry in the stratosphere is included in the model together with an advective process. Results in summer case show the decrease of zonal mean total ozone in the tropics (about 5 D.U.) and the increase in middle latitudes (about 8 D.U.). In summer and winter cases, the significant decrease in total ozone (about 20 D.U.) is simulated over the equatorial central and eastern Pacific where the warm sea surface temperature anomalies lie.

1. INTRODUCTION

Hasebe (1984) studied the interannual variations of total ozone (1970-1977) and found the oscillation with a period of about four years (FYO). This FYO is almost symmetric with respect to the equator showing an out-of-phase relation between the tropics and the extratropics. Its node is located at about 20 °N and 20 °S. The sea surface temperature (SST) in the equatorial eastern Pacific and the FYO of total ozone have a good correlation. During the warm phase of El Niño and just after that, the total ozone in the tropics tends to be low or decreasing and that in the extratropics tends to be high or increasing. Angell (1981) showed that total ozone over North America tends to be a maximum two seasons after the warmest SST in the equatorial eastern Pacific. From these studies, it is suggested that when El Niño takes place, the ozone transport in the stratosphere by the Brewer-Dobson circulation is enhanced and therefore the total ozone increases in the extratropics and decreases in the tropics. In this paper, the impact of El Niño event on total ozone is studied by an atmospheric general circulation model.

2. THE MODEL AND THE EXPERIMENTS

The model used is the Meteorological Research Institute (MRI) · GCM. The model has a horizontal resolution of 5° in longitude and 4° in latitude. Standard physical processes such as solar and terrestrial radiation, penetrative cumulus convection and planetary boundary layer processes are included in the model. A simple ozone photochemical process in the stratosphere is also included in the model together with an advective process (Schlesinger and Mintz, 1979). The photochemical production and destruction of ozone in the stratosphere is comprised of the Chapman reactions and the NO-NO_2 catalytic cycle. The concentration of NO_2 is prescribed to vary only in the vertical direction. The ozone destruction by heterogeneous chemical reactions in the planetary boundary layer is also parameterized in the model. The calculated ozone amount is taken into account for radiation calculation. See Tokioka *et al.* (1984) for details of the model.

In summer case, we used 15 layer version of the model with its top at 10 hPa. Two 3 month integrations are performed from the initial atmospheric condition of 12Z 1 May, 1983. Control run is integrated with the use of climatological sea surface temperatures and the other with the observed sea surface temperatures in 1983. During the period, the 1982/83 El Niño was still active. For the initial condition of ozone, the zonal mean climatological values are used in both runs. In winter case, we used 12 layer version of the model with its top at 1 hPa. Two integrations are performed from the initial atmospheric condition of 12Z 15 December, 1982 through the end of January 1983.

3. SUMMER CASE

Figure 1 shows the July mean and zonal mean total ozone anomaly (El Niño run - Control run). The simulation clearly shows that the total ozone decreases and that in the middle latitude increases. The anomaly is almost symmetric with respect to the equator and its node lies at about 20 °N and 20 °S. The simulated results agree with the observation by Hasebe (1984).

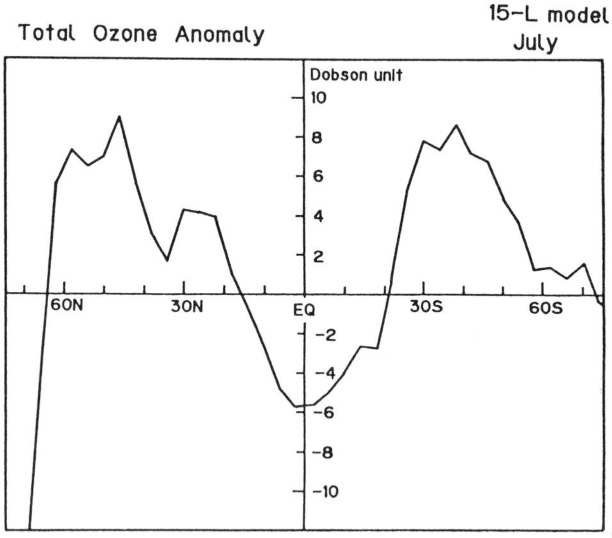

Fig. 1 Simulated July mean and zonal mean total ozone anomaly. Unit is Dobson unit.

The longitudinal distribution of July mean total ozone anomaly is shown in Fig. 2. In the tropics, large east-west variation can be seen. Large negative anomaly with its peak

ISBN 0-937194-15-8

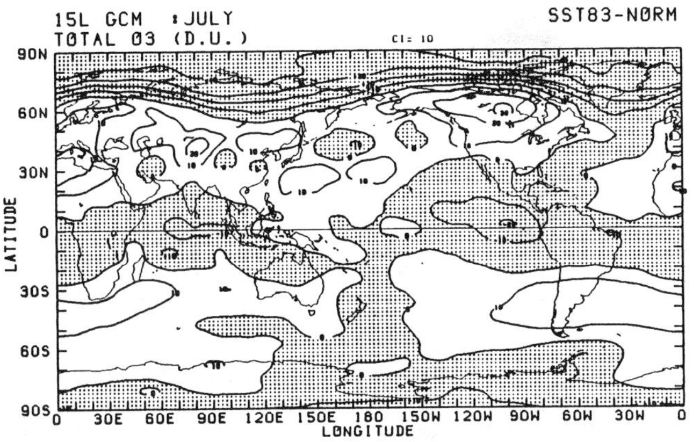

Fig. 2 Simulated July mean total ozone anomaly. Contour interval is 10 D.U.. Negative values are shaded.

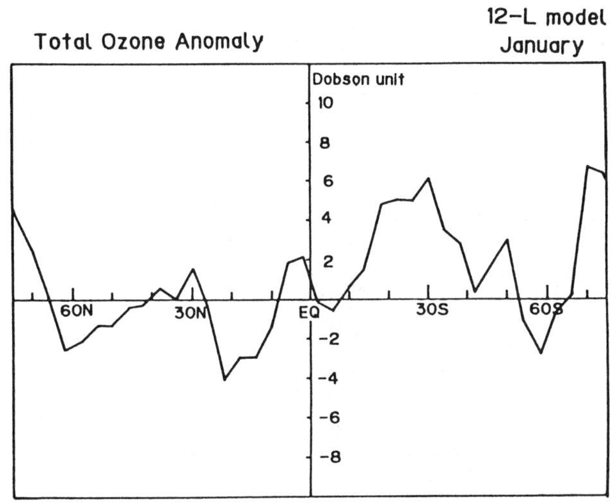

Fig. 3 Simulated January mean and zonal mean total ozone anomaly.

of -20 D.U. is found over the equatorial eastern Pacific where the SST anomaly is large positive and the upward motion anomaly exists. The negative anomaly over there is persistent and is found even in the first month (May). On the other hand, the anomaly over the western Pacific and the maritime continent is very small. Therefore, for detecting the El Niño impact on ozone, the observation over the equatorial eastern Pacific is crucial and the zonal mean values derived from a sparse Dobson network might be misleading.

In the extratropics, anomalies generally increase with time and the pattern also varies with time. In the lower stratosphere, the chemical life-time of ozone is long and the ozone distribution is mainly controlled by dynamical processes. In this El Niño experiment, the main circulation change takes place in the tropics and the effect is expected to propagate gradually poleward. Observations have shown major time-shifting in the appearance of the maximum effect in the extratropical latitudes. Therefore the three month integration might be too short to draw a definite conclusion on the longitudinal anomaly distribution in the extratropics.

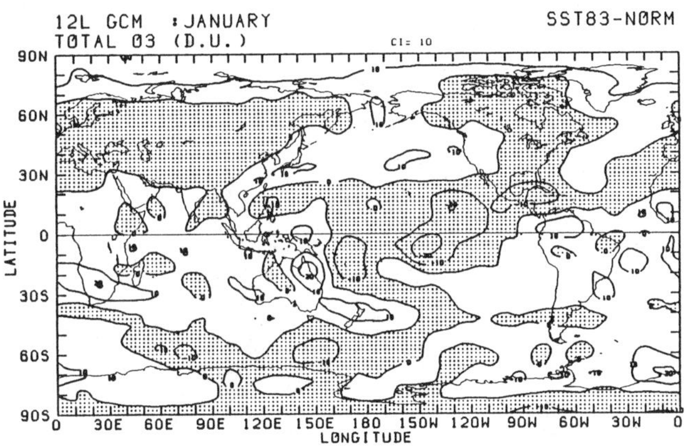

Fig. 4 Simulated January mean total ozone anomaly. Contour interval is 10 D.U.. Negative values are shaded.

4. WINTER CASE

The simulated January mean and zonal mean total ozone anomaly is shown in Fig. 3. The pattern is not so clear as the summer case. While the anomaly is small positive around the equator, it shows moderate negative at 10°N-25°N. At 20°S-40°S, large positive anomaly can be seen. In northern mid-latitudes, the anomaly is small. The El Niño impact does not seem to be so clear in this case.

However, examining the horizontal distribution map (Fig. 4), the El Niño impact is dominant in the tropics. The large negative anomalies (the minimum is less than -20 D.U.) exist over the equatorial central Pacific where the large positive SST anomaly is observed in 1983 winter. Over the maritime continent, moderate positive anomaly appears. These pattern in the tropics is consistent with the change in the Walker circulation associated with El Niño. It is interesting that the anomalies are mostly negative over the Eurasia and North America. In the northern extratropics, the anomaly is generally small probably because the integration is too short for the circulation change to affect the ozone distribution.

The upper stratosphere is included in this case. In the upper stratosphere (above 10 hPa), ozone anomaly is positive in low latitudes and negative in northern middle latitudes. This anomaly is negatively correlated well with the temperature anomaly. The fast photochemical process is accounted for the ozone changes. However, the contribution of the upper stratospheric ozone anomaly to total ozone change is negligible. The change in total ozone is mostly governed by the change in the lower stratosphere.

5. SUMMARY AND DISCUSSION

The change in total ozone due to 1982/83 El Niño was examined by the MRI·GCM for summer and winter cases. Results in summer case showed the decrease of total ozone in the tropics (about -5 D.U.) and the increase in middle latitudes (about 8 D.U.), which was consistent with the intensification of the Hadley cell and the consequent intensification of the Brewer-Dobson circulation. Thus the relationship between El Niño and the total ozone variation observed by Hasebe (1984) is confirmed by the present GCM simulation. In winter case, the integration is continued only one and half month. It is too short to capture the prominent effect in the extratropics because the time-lag effect is important there. Results in winter case is not so clear as far as the extratropics is concerned. The present model is not perfect and has systematic biases such as too strong winter stratospheric jet. When the 12-layer model is integrated long enough (two or three months in winter case), the biases become large. The model must be improved before performing long-term integrations.

In both cases, the significant decrease in total ozone (about -20 D.U.) is simulated over the equatorial central and eastern Pacific where the warm SST anomalies lie. This anomaly is caused by the change in the Walker circulation.

However actually observed total ozone anomaly in 1983 is significantly lower than normal especially in middle latitudes (Bojkov, 1987). Bojkov (1987) pointed out the three candidates for the cause of the total ozone deficiency, i.e., the phase of quasi-biennial oscillation (QBO), the eruption of El Chichón and the El Niño. From the present study, it is suggested that the El Niño is not a main contributor to the deficiency of ozone in 1983. Therefore, the QBO and El Chichón should be considered as the main causes of the observed 1983 ozone deficiency.

REFERENCES

Angell, J.K., 1981: Comparison of variations in atmospheric quantities with sea surface temperature variations in the equatorial eastern Pacific., *Mon. Wea. Rev.*, **109**, 230-243.

Bojkov, R.D., 1987: The 1983 and 1985 anomalies in ozone distribution perspective., *Mon. Wea. Rev.*, **115**, 2187-2201.

Hasebe, F., 1984: The global structure of the total ozone fluctuations observed on the time scales of two to several years. *Dynamics of the middle atmosphere,* Terra Scientific Publishing Company, 445-464.

Schlesinger, M.E. and Y. Mintz, 1979: Numerical simulation of ozone production, transport and distribution with a global atmospheric general circulation model., *J. Atmos. Sci.*, **36**, 1325-1361.

Tokioka, T., K. Yamazaki, I. Yagai and A. Kitoh, 1984: A description of the MRI atmospheric general circulation model (the MRI·GCM-I). *Tech. Rept. of the Meteor. Res. Inst.*, No. 13, 249pp.

CORRELATION BETWEEN TOTAL OZONE AND POTENTIAL VORTICITY

G. Vaughan and D.A. Begum
Physics Department
University College of Wales
Aberystwyth, Dyfed, UK

ABSTRACT

Correlation coefficients have been calculated between total ozone and both tropopause pressure and isentropic potential vorticity as a function of season. The area covered in the study spans the subtropics and extratropics, ranging from 90°E to 160°E and 20°N to 50°N. Potential vorticity was calculated on a number of isentropic surfaces spanning the upper troposphere and lower stratosphere. Gridded assimilation fields from the ECMWF forecast model were used to calculate vorticity and filtered radiosonde profiles used for $d\theta/dp$. A marked seasonal variation is evident in the results. In the subtropics, the correlation is large only around 350K in winter, but in mid-latitudes high correlations follow the level of the tropopause from 300K in February to 370K in summer. Smaller correlations occur at all levels during autumn.

1. INTRODUCTION

The correlation between total ozone and meteorological features has been appreciated since the beginning of research on atmospheric ozone, with sudden increases in total ozone accompanying marked decreases in tropopause height, such as are found during the passage of a cold front or depression. This can be interpreted simply as an increase in the depth of the ozone-rich stratosphere or as a combination of vertical and horizontal advection of ozone (Reed 1950; Schoeberl and Krueger 1983). With the advent of satellite instruments capable of measuring the global distribution of total ozone, the nature of this relationship could be examined more fully. Studies have been reported of the close link between jet streams and gradients in total ozone (Shapiro et al 1982; Reiter and Gao 1982) and of the marked signatures on total ozone maps of features such as tropopause folds and cut-off lows (Uccellini et al 1985; W.M.O. 1985, Chapter 5; Vaughan and Price 1988). However, comparatively little is understood about the extent to which horizontal variability in total ozone simply reflects variability in the height or structure of the tropopause, and the extent to which it reflects variability near the ozone maximum.

One recent study addressing this point was presented by Schubert and Munteanu (1988), in the form of correlation coefficients between TOMS total ozone and gridded tropopause pressures obtained from the NMC global analysis scheme. Many problems were encountered with the model assimilation algorithms, and the correlations reported were restricted to length scales corresponding to wavenumber 6 or greater, i.e. less than 5000 km. Broadly speaking, correlation coefficients >0.6 were found in mid-latitudes of both hemispheres in the summer and equinox seasons, with rather smaller values in winter mid-latitudes and very small values in the tropics and subtropics, where variations in total ozone are themselves small. From this, one might suppose that the influence on total ozone of synoptic-scale variability in the stratosphere proper are most prevalent at mid-latitudes in winter, when planetary wave transport is most active.

In this work, correlations between TOMS total ozone measurements and fields of isentropic potential vorticity (IPV) are examined statistically and compared with similar correlations with tropopause height. In the lower stratosphere and upper troposphere, IPV is strongly correlated with ozone concentration (Danielsen et al 1987; Browell et al 1987), thus a high positive correlation between IPV and total ozone is an indication that variability in total ozone is dominated by the contribution from the relevant isentropic surface. Total ozone should be more sensitive to the IPV distribution than to tropopause height, since the former takes more account of the detailed structure of the atmosphere than the crude definition of the tropopause.

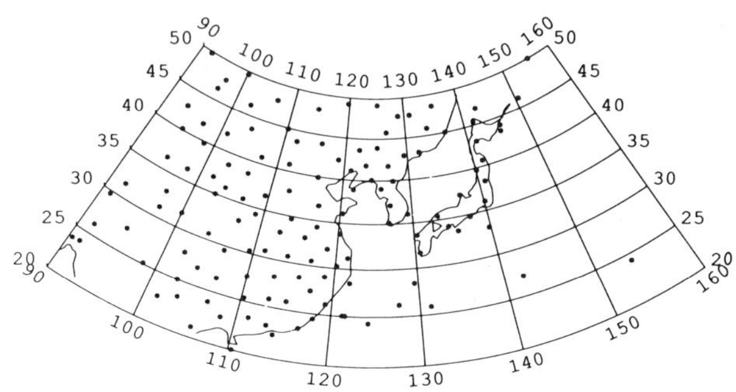

Fig. 1. Area used in this study. Asterisks denote radiosonde stations.

The area chosen for this study spans the entrance region of the East Asian subtropical jet stream, from 20°N to 50°N and from 90°E to 160°E (fig. 1). Choosing such a limited area ensures that the scales of length contributing to variability within it are similar to those considered by Schubert and Munteanu - i.e. predominantly synoptic. Total ozone and meteorological data

OZONE IN THE ATMOSPHERE
Rumen D. Bojkov and Peter Fabian (Eds.)

ISBN 0-937194-15-8

were analysed on a monthly basis from October 1982 to September 1983, to see how the correlation between them varied over the course of one year.

2. METHOD

A straightforward statistical approach was adopted in this study. For each radiosonde station, the tropopause pressure and IPV values on various θ surfaces were obtained for 00 GMT on the first 20 days of each month. For each day, total ozone was interpolated from gridded TOMS data to the station positions. Correlation coefficients were then calculated between O_3 and the other parameters. This pre-supposes a simple linear relationship between the two variables which is likely in practice to be an oversimplification, but which should capture the gross features of any connection. Scrutiny of charts for individual days showing both ozone and tropopause pressure gave the distinct impression that the two tracked well in the synoptic-scale features north of 40°N, but that the abrupt change in tropopause height across the sub-tropical jet-stream was not always matched by a commensurate ozone gradient. The radiosonde stations used in this investigation were therefore divided into 3 groups according to their latitudes - 20-30°N, 30-40°N and 40-50°N - to see whether the correlation between total ozone and tropopause structure was latitude-dependent.

The distribution of radiosonde stations in the chosen area is shown in fig.1; the vast majority are west of 140°E. Since TOMS measures near local noon, the nearest synoptic hour to the ozone measurements in the sector 90-140°E is midnight GMT, with a time displacement of 3-6 hours from the NIMBUS-7 overpass. Meteorological data for 00 GMT were therefore used in this comparison. Synoptic scale meteorological features do not travel very far in this time, so although some error will be introduced by the time displacement, particularly near cold fronts, the statistics should not be markedly degraded.

Daily gridded TOMS data are supplied at intervals of 1° in latitude and 1.25° in longitude. Fleig et al (1986) have reported that recent updates of ozone absorption cross-sections in the Huggins bands mean that the TOMS gridded ozone values supplied for 1982-3 should be increased by about 6%. This correction applies uniformly to the entire dataset used here and therefore should not affect statistical studies. It has not been included in this work.

Values of tropopause pressure were taken from the radiosonde tropopause reports, after quality control to remove spurious values. To calculate IPV, the absolute vorticity on an isentropic surface, ζ , and the stability of the atmosphere, dθ/dp, must be known. The former was calculated from 00 GMT global analysis fields supplied by the European Centre for Medium Range Weather Forecasts (ECMWF). These are provided in the form of wind, height and temperature values on a global 2.5°x2.5° grid on the 100, 200, 300, 500, 700 and 850 mb surfaces. To calculate ζ, northward and eastward wind components were first interpolated linearly with height at each grid-point to the desired isentropic surface, then ζ was determined (as dv/dx - du/dy) at each radiosonde position.

The ECMWF global analysis fields have sufficient horizontal resolution to determine ζ, but lack the vertical resolution for precise determination of dθ/dp. The latter was calculated directly from the radiosonde profiles, after they had been smoothed to remove perturbations on height scales <1km. The smoothing reduces contamination of the stability values by small-scale gravity wave components, although any determination of dθ/dp relying only on single radiosonde profiles cannot guarantee freedom from wave contamination. Derived values of dθ/dp may not therefore be completely representative of an area equivalent to the resolution of the TOMS measurements (about 110 km square). We believe this to be the most important source of error in the IPV calculations, but have no means of estimating its magnitude; it is hoped that smoothing will have removed the worst of the contamination. Correlation coefficients presented here are therefore somewhat underestimated.

3. RESULTS

Correlation coefficients between total ozone and tropopause pressure are presented in fig. 2 and those with IPV in figs. 3-5. The latter are shown as contour plots; individual points were calculated on these plots for the first 20 days of each month at 390, 370, 350, 330, 310 and (northward of 30°N) 300 and 290K. Between 500 and 700 individual points were included in the calculations, but since points adjacent in space and time cannot be considered statistically independent the number of degrees of freedom appropriate to a significance test is considerably smaller than this. With a conservative estimate of 20 for the number of degrees of freedom, correlation coefficients >0.25 should be regarded as significant at the 95% level. In fact, contours are

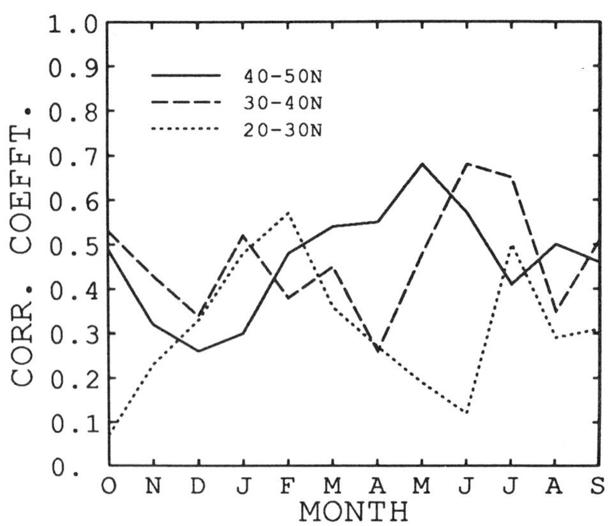

Fig. 2. Correlation coefficients between total ozone and tropopause pressure for each month.

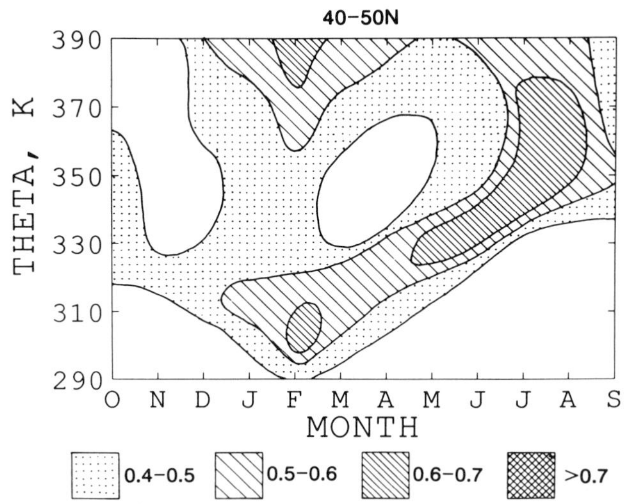

Fig. 3. Contour map of correlation coefficients between total ozone and IPV on different isentropic surfaces for each month, 40-50°N

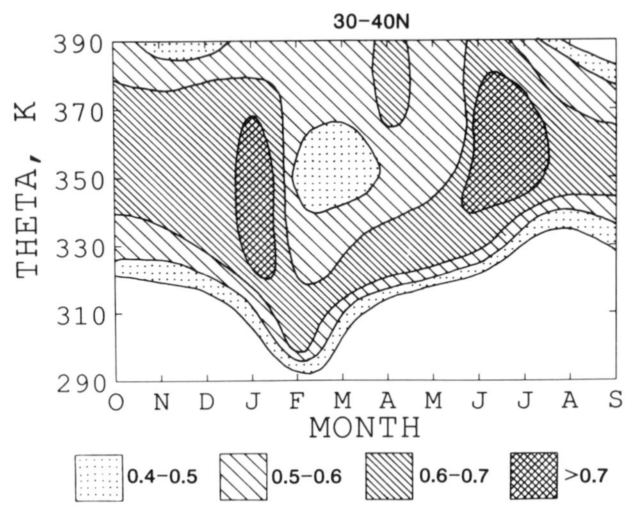

Fig. 4. As Fig. 3, 30-40°N

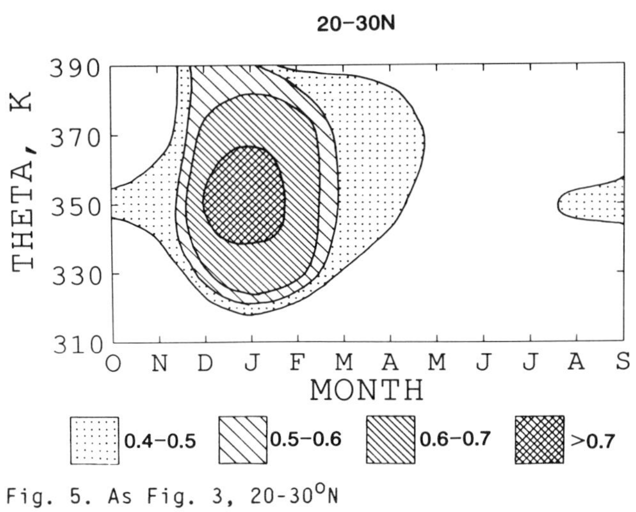

Fig. 5. As Fig. 3, 20-30°N

only shown on figs. 2-5 where the correlation is particularly strong - 0.4 or greater.

4. DISCUSSION

For nearly all months, the maximum value of the coefficient between total ozone and IPV exceeded that with tropopause pressure; the only significant exception is 20-30°N in July. Indeed, in December and January, the maximum coefficient with IPV exceeded that with tropopause pressure by 0.25 or more at all latitudes. This bears out the argument presented in the introduction about the superior diagnostic quality of IPV, but raises an interesting question as to why IPV near the tropopause should correlate so much better with total ozone than tropopause pressure in winter. The supposition that synoptic-scale variability in mid-stratospheric ozone concentrations significantly affects total ozone variability in winter is supported by fig.3, with high correlations at 390K as well as near the tropopause.

The seasonal cycle of the correlation with tropopause pressure in the 40-50°N band is consistent with the findings of Schubert and Munteanu (1988), with broadly similar values. Correlation coefficients in the 20-30° band, however, were found to be much higher than theirs in winter, possibly because of the problems they encountered with the NMC tropopause assimilation in the tropics.

To guide interpretation of figs. 2-5, the range of tropopause pressures observed on each

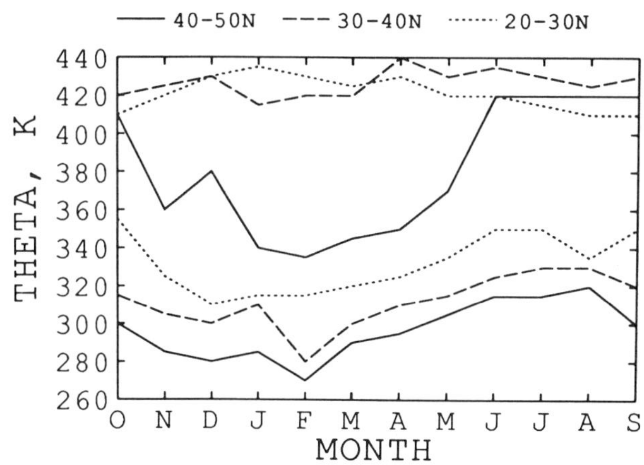

Fig. 6. Range of values reported for tropopause potential temperature for each month.

month in each latitude band is shown in fig.6. This is particularly pertinent to fig. 2, where the isentropic level correlating best with total ozone is seen to lift from 300K in February to 370K in July and August as the range of tropopauses moves to higher potential temperatures. From August to December, the maximum value of the correlation coefficient with IPV is markedly smaller than in the other months. The autumn is a time when latitudinal gradients in total ozone are smallest (Bowman and Krueger 1985) and the lower stratospheric ozone concentration at its minimum (Dutsch 1978) so the correlation calculations would be at their most sensitive to random noise in the stability data. Nevertheless, the marked seasonal asymmetry shown in fig.2 is striking.

An even more striking asymmetry is evident in the sub-tropics (fig. 5). In almost every month, the highest correlations were evident at the 350K level. In winter, the subtropical jet stream is strongest and at its most southerly position, so a strong correlation during these months is not surprising, but the jet stream is almost as strong in March when the correlation of IPV with total ozone drops markedly. During spring, when the subtropical jet stream moves north of the Himalayas, Reiter and Gao (1982) proposed that a considerable amount of tropospheric air is entrained into the lower stratosphere - i.e. the air lies above the stability tropopause but retains tropospheric IPV. Correlation coefficients between total ozone and tropopause pressure fell to very low values during spring - much lower than those with IPV - possibly because of such entrainment.

The pattern in the latitude band 30°-40°N contains the salient features of the two bands either side, but, somewhat surprisingly, correlation coefficients in this band were higher at most levels throughout the year than in the other two bands. Further research is required to understand the reason for this.

5. SUMMARY AND CONCLUSIONS

For the year October 1982- September 1983, correlation coefficients between TOMS total ozone and IPV on isentropic levels were found to be greatest on isentropic levels spanning the tropopause, the only exception being winter mid-latitudes where high coefficients were seen at 390K, well above the tropopause. In almost every case, these correlation coefficients exceeded those calculated with tropopause pressure, establishing the superior diagnostic quality of IPV. In the subtropics, total ozone only correlated well with either parameter in winter, with very low correlation coefficients, particularly with tropopause pressure, in spring and summer. Further study will determine whether this is related to entrainment of tropospheric air on the poleward side of the subtropical jet stream. At all latitudes, correlation coefficients in autumn were smaller than at other times of the year, possibly because of the comparatively low concentration of ozone in the lower stratosphere during that season, causing enhanced susceptibility to random errors in the stability calculations.

We conclude that for most of the year northward of 30°N, and for winter between 20 and 30°N, about half the synoptic-scale variability in total ozone was explained by variations in the flow at tropopause level (taking r^2 as a measure of the fraction of the total variance explained by the correlation). Whether the rest of the variance was caused by random errors in the data, or by variability in the main ozone layer above the levels considered here, could not be resolved in this study.

REFERENCES

Bowman, K. P. and A. J. Krueger, 1985: A Global Climatology of Total Ozone from the Nimbus 7 Total Ozone Mapping Spectrometer. J. Geophys. Res., **90**, 7967-7976.

Browell, E. V., E. F. Danielsen, S. Ismail, G. L. Gregory and S. M. Bleck, 1987: Tropopause fold structure determined from airborne lidar and in situ measurements. J. Geophys. Res., **92**, 2112-2120.

Danielsen, E. F., R. S. Hipskind, S. E. Gaines, G. W. Sachse, G. L. Gre-gory and G. F. Hill, 1987: Three-dimensional analysis of potential vorticity associated with tropopause folds and observed variations of ozone and carbon monoxide. J. Geophys. Res., **92**, 2103-2111.

Dutsch, H. U., 1978: Vertical Ozone distribution on a global scale. Pageoph., **116**, 511-529.

Fleig, A. J., P. K. Bhartia, C. G. Wellemeyer and D. S. Silberstein, 1986: Seven years of total ozone from the TOMS instrument - a report on data quality. Geophys. Res. Lett., **13**, 1355-8.

Reed, R. J., 1950: The role of vertical motions in ozone-weather relationships. J. Meteorol., **7**, 263-7.

Reiter, E. R. and Deng-Yi Gao, 1985: Heating of the Tibet plateau and movements of the South Asian high during Spring. Mon. Weather Rev., **110**, 1694-1711.

Schoeberl, M. R. and A. J. Krueger, 1983: Medium scale disturbances in total ozone during Southern hemisphere summer. Bull. Amer. Met. Soc., **64**, 1358-1365.

Schubert, S. D. and M.-J. Munteanu, 1988: An analysis of tropopause pressure and total ozone correlations. Mon. Weather Rev., **116**, 569-582.

Shapiro, M. A., A. J. Krueger and P. J. Kennedy, 1982: Nowcasting the position and intensity of jet streams using a satellite-borne Total Ozone Mapping Spectrometer, in Nowcasting, ed. K. A. Browning, Academic Press, London.

Uccellini, L. W., D. Keyser, K. F. Brill and C. H. Wash, 1985: The President's day cyclone of 18-19 February 1979: Influence of upstream trough amplification and associated tropopause folding on rapid cyclogenesis. Mon. Weather Rev., **113**, 962-988.

Vaughan, G. and J. D. Price, 1988: Ozone transport into the troposphere in a cut-off low episode. This volume.

WMO, 1985: Atmospheric Ozone 1985. WMO Report no. 16, World Meteorological Organisation, Case Postale No. 5, Geneva.

ACKNOWLEDGEMENTS

We thank Dr. A.J. Fleig, Dr. A.J. Krueger, the TOMS experiment team and the World Data Center-A, Greenbelt, Maryland for supplying us with TOMS data and the staff of the European Centre for Medium-range Weather Forecasts for meteorological data.

ANALYSIS OF THE TEMPORAL VARIANCE OF STRATOSPHERIC OZONE

E.P. Röth and D.H. Ehhalt*
Institut für Atmosphärisch Chemie der
Kernforschungsanlage Jülich GmbH
D-5170 Jülich, FRG

ABSTRACT

The local concentrations of ozone in the stratosphere exhibit a significant temporal variance. Most of that variance is caused by transport, and thus contains information on the transport processes. It is useful to express the variance as 'equivalent displacement height' (EDH). The EDH is defined as the local mean standard deviation divided by the local mean vertical gradient of the ozone mixing ratio. Both, the seasonal and the latitudinal variations of the EDH were determined for ozone and potential temperature - the former from the long-term series of O3 measurements over Mt. Hohenpeißenberg in Bavaria, the latter from the LIMS experiment aboard the satellite NIMBUS 7. The variations of both tracers show similar seasonal and latitudinal trends. They are used to derive stratospheric transport parameters.

INTRODUCTION

The variance of any measured quantity contains information on the cause of the variation. Measurements of stratospheric trace gas concentrations exhibit a substantial temporal variance, which varies with location and season and depends on the species investigated. These variances are by far larger than the analytical precision of the applied method of investigation. This means that there must be other causes and, to elaborate this, Ehhalt and co-workers (1983) introduced a new quantity 'equivalent displacement height' (EDH or). The EDH is defined as the local mean standard variation divided by the local vertical gradient of the mean mixing ratio profile:

$$\Delta = \frac{\sigma}{\partial\mu/\partial z} \qquad (1)$$

The advantage of using the EDH instead of will become apparent, when the variances of different tracers shall be compared. By folding the variance σ by the gradient $\partial\mu/\partial z$, the variances for all species are expressed in the same unit, namely 'height'. Thus, the variances of the different compounds become comparable, and, as has been shown earlier, exhibt similar behavior. It should be emphasized that the expression 'height' does not necessarily mean that the variances are caused by a vertical displacement exclusively. A detailed introduction to the concept of equivalent displacement height is given elsewhere (Ehhalt et al., 1983 and Röth and Ehhalt, 1987).

It could be shown by Ehhalt et al. (1983) that the main cause for the temporal variance of trace gases is a transport effect. Air parcels measured at a certain locality still exhibit the composi-

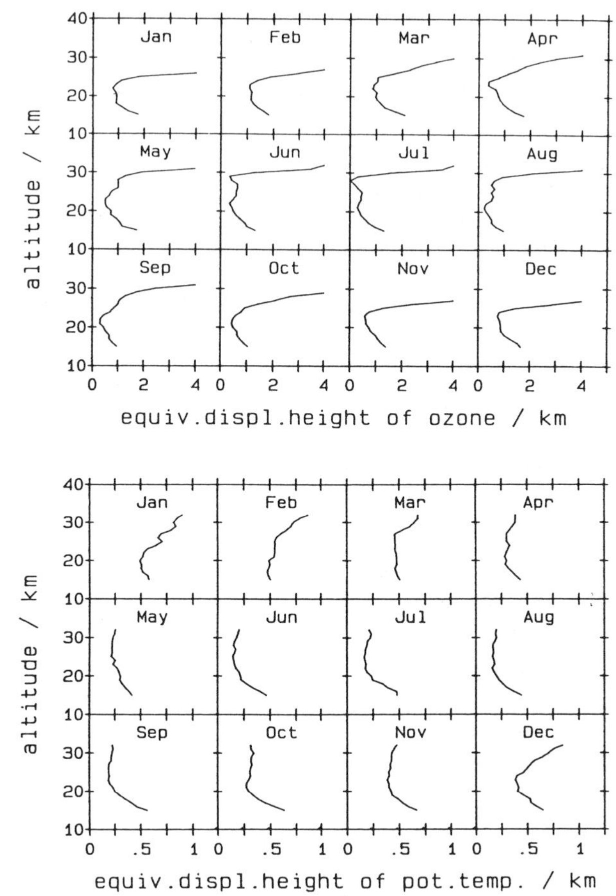

Fig. 1. Monthly mean vertical profiles of the equivalent displacement height of ozone (above) and potential temperature (below), derived from a 18 years' series of measurements over Mt. Hohenpeißenberg.

*) also at Universität Essen Institut für Physikalische Chemie D-4300 Essen 1, FRG

ISBN 0-937194-15-8

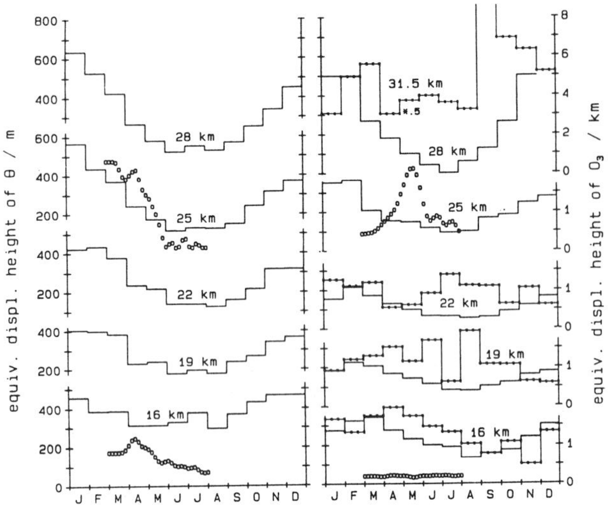

Fig. 2. The seasonal variations of the equivalent displacement height for ozone and potential temperature at five different altitudes. Included are the calculations by Hess and Holton (1985) at 15 and 25 km altitude (o) and the model predicted data of Mahlman et al. (1986) (-x-x-x-).

tion of the point from where they originated. This allows to deduce the vertical (Δz) and horizontal (Δy) displacements of the air parcels from the local temporal variances of long-lived tracer mixing ratios.

As long as chemical reactions don't change the composition of an air parcel, which can be excluded for longlived compounds, and, if longitudinal mixing is neglected, the equivalent displacement height can be expressed by Δy, Δz and the mean meridional gradient β of the surfaces of constant mixing ratio:

$$\Delta \approx \Delta z - \beta \cdot \Delta y \qquad (2)$$

The meridional fields of ozone and potential temperature can be obtained from satellite measurements and, by combining the EDH of both tracers, the vertical and the horizontal displacements can be deduced. Using the mixing length hypothesis of Reed and German (1965) in a meridional plane, the horizontal displacement equals the mixing length and the quotient $\Delta z/\Delta y$ is the mean meridional slope of the transport surface.

In the following, we will present the seasonal variation of the equivalent displacement heights, derived from the long-term series of measurements of ozone and potential temperature over Mt. Hohenpeißenberg in Southern Germany (48° N) (Attmannspacher, 1985) and the latitudinal dependence of the EDH derived from the LIMS experiment aboard the NIMBUS 7 satellite (Gille and Russell, 1984). Together with the equivalent displacement heights, the vertical and horizontal displacements are presented.

2. SEASONAL VARIATION DEDUCED FROM THE HOHENPEISSENBERG DATA

The meteorological observatory of the German Weather Service on Mt. Hohenpeißenberg in Bavaria (48° N latitude) is monitoring stratospheric ozone since November 1966 (Attmannspacher, 1985). The vertical profiles of the O_3 concentration are obtained by balloon-borne electrochemical Brewer/Mast sondes. Temperature and pressure are also measured by balloon-borne instruments. All three quantities are monitored continuously during ascent. From the response time of the O_3 instrument and the ascent rate of approximately 6 m/s, a vertical resolution of the O_3 measurement is estimated to be 200 m. Similar numbers are obtained for pressure and temperature. From temperature and pressure the potential temperature, θ, can be calculated.

Altogether more than 1600 profiles of ozone and potential temperature were evaluated. Over 130 data at each altitude were combined to form the monthly averages and the monthly mean standard deviation. The latter values are plotted as equivalent displacement heights in figure 1a for O_3 and in figure 1b for the potential temperature. The data above 32 km are omitted because most balloons did not reach those altitudes and - as a consequence - the data base becomes sparse. By calculating the EDH of ozone, an experimental error of 6 % is taken into account, i.e. the mean standard deviations are reduced by this amount.

For all seasons, the vertical profiles of Δ_{O3} and Δ_θ show the wellknown C-shape. The values for Δ_θ are lower by a factor of 5 in relation to Δ_{O3} and the upper part of their C-shaped profile is less pronounced.

To examine the seasonal variation of the equivalent displacement heights, the values at 16, 19,

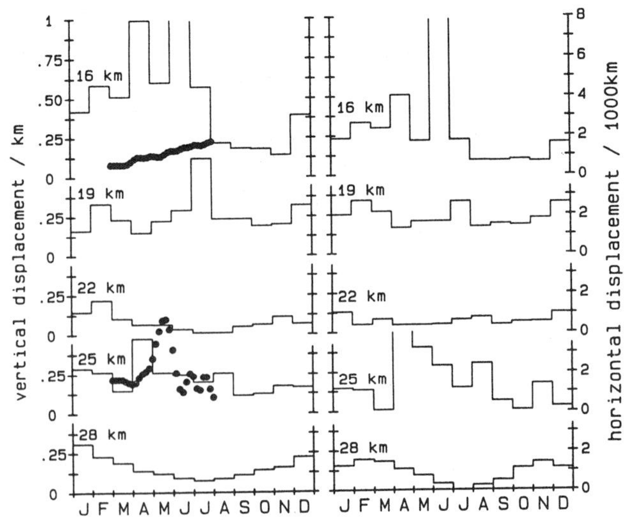

Fig 3. The seasonal variations of the vertical and the horizontal displacements at five different altitudes. The values of Hess and Holton (1985) are indicated by the circles.

22, 25 and 28 km altitude are plotted in figure 2 Both tracers exhibit similar seasonal trends for all altitudes with low Δ-values in summer and high EDH during the winter months. Moreover, the amplitude of the seasonal variation increases with altitude in both cases. There is very little phase shift between the two. It appears, however, that Δ_{O3} peaks in February, whereas Δ_{θ} peaks in January at all altitudes.

These observed seasonal variations can be compared to two model predictions, one by Hess and Holton (1985) for Δ_{θ} and a conservative tracer (N_2O) and the other by Mahlman et al. (1986) for N_2O. Their results are given in figure 2, too.

We first compare our experimental results with the theoretically derived data of Hess and Holton They argued that the remaining turbulence during the 'quiet' summer period are relicts of the final warming. The agreement between prediction and measurement for Δ_{θ} is quite reasonable. The remaining discrepancies may be explained by the

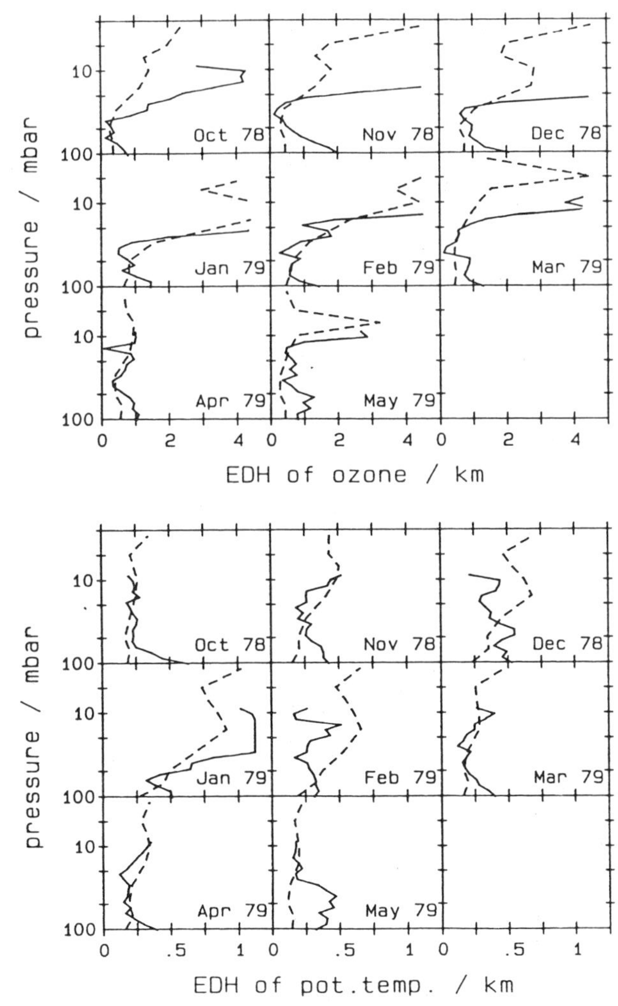

Fig. 4. Monthly averaged equivalent displacement heights of ozone and potential temperature, derived from the Hohenpeißenberg data (——) and from satellite observations (- - -).

fact that the model allows for only one, the final stratospheric warming, whereas the real polar stratosphere experiences several warmings during a winter. And, in addition, in the lower stratosphere wave energy is supplied all the year around from the troposphere, which is not accounted for in the model calculations, but is, of course, present in the observed data.

In contrast to the agreement for the equivalent displacement heights derived from the potential temperature, the modelled EDH for an inert tracer and the observed values for Δ_{O3} differ dramatically, although O_3 is virtually conservative over the observed altitude range. At 16 km the discrepancies may be explained by the above-mentioned limitations in the assumption of Hess and Holton; however, the difference at 25 km is so significant that it is difficult to maintain the assumption that the summer variance is caused by the 'frozen in' final warming. While the model predicts a relatively sharp maximum in Δ_{O3} about 15 days after the zero wind line crossed 25 km altitude, the observed values show a steady decline during May from earlier, higher values. And, as pointed out above, the observed Δ_{O3} and Δ_{θ} show a parallel trend, which is not predicted. From the additional observation that the decay of the higher variance in winter to the lower values in summer proceeds with a time constant of about 80 days, which is much longer than the time constant for the Newtonian cooling of about 20 days, which should dominate the decay of Δ_{θ}, we would argue that the variance is generated throughout the year albeit at lower rates during summer, when tropospheric wave activity is blocked from penetrating into the stratosphere.

Hess and Holton (1985) predicted another parameter, namely the actual vertical displacement Δz, from their model. In the following, we attempt to derive this parameter from the observed EDH using equation (2) given above. We note, however, that there are limitations in using Δ_{θ} because the potential temperature is not a sufficiently inert tracer and, moreover, the difference between the equipotential surface and the surface of transport nearly vanishes, thereby, increasing the relative importance of the unavoidable uncertainties of β_{θ}, the meridional slope of the equipotential surface.

De spite these shortcomings, we plotted the seasonal variations of the vertical and horizontal displacements in figure 3 together with the model predictions of Hess and Holton, to obtain a first guess. Again the same behavior, as for the EDH is observable. The disagreement between the predicted values and the experimental data remains. Due to the large noise in Δy and Δz it is difficult to detect clear seasonal variations in those parameters, except for the highest altitude at 28 km. There Δy and Δz exhibit the same seasonal variation as the equivalent displacement height.

Mahlman et al. (1986) used the Geophysical Fluid Dynamics Laboratory three-dimensional general circulation model to investigate the stratospheric behavior of N_2O. In their figure 9, they gave a one-year model N_2O time series for Laramie,

Wyoming (41° N), which we evaluated to derive the seasonal variation of the EDH in N_2O. These data are also plotted in figure 2.

Considering that the number of points entering each Δ -value is small compared to the number of observational data, the agreement between the model variation and the experimental values 16 km altitude is quite good. The scatter from month to month in the model data is due to the fact that the determination of the local vertical gradient involves large uncertainties since only five points were given for each profile. The observed summer minimum at 19 and 22 km, however, is not predicted by the model at all. Whether this is caused by the limited data base or is an effect of the model itself, remains open until further investigation. Possibly, the model stimulates stratospheric turbulence to too high altitudes.

3. LATITUDINAL VARIATION DEDUCED FROM THE LIMS EXPERIMENT

Aboard the NIMBUS 7 satellite, the LIMS experiment measured, besides other trace constituents, the ozone mixing ratio profile with an altitudinal resolution of 2.5 km and a latitudinal spacing of 4°. The LIMS method is described by Gille and Russell (1984). The experiment operated from October 1978 to May 1979, so that we can compare these data with the Hohenpeißenberg values of the same period. Temperature and pressure were also monitored by the experiment, so that again the potential temperature could be calculated to attain a second tracer for the determination of the vertical and horizontal displacements.

The equivalent displacement heights of the potential temperature, plotted in figure 4, for both data sets agree within their mutual analytical

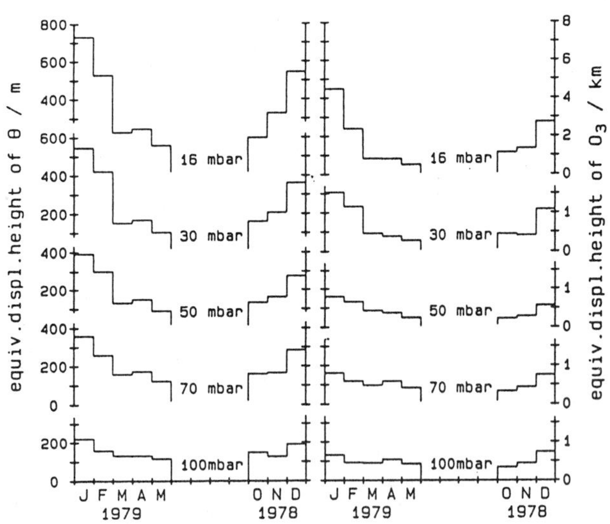

Fig. 5. The seasonal variations of the equivalent displacement height for ozone and potential temperature at 48° N from the LIMS experiment aboard the NIMBUS 7 satellite (Gille and Russell,1984). The observation started in October 78 and ended in May 79

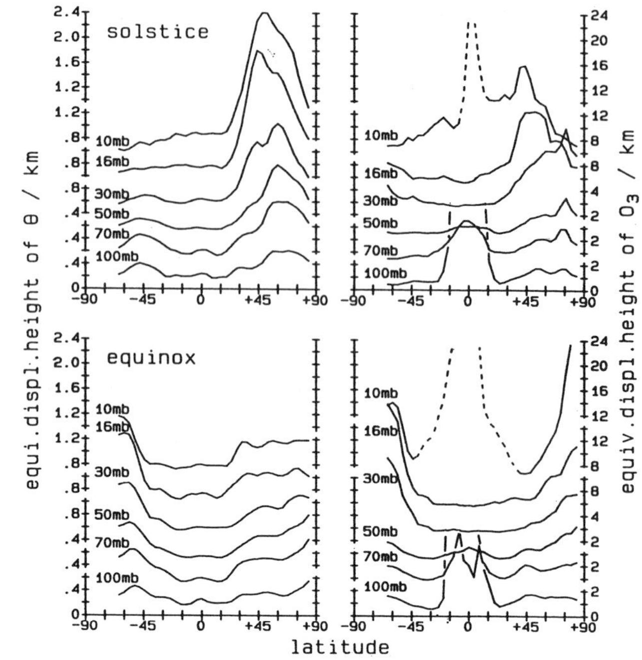

Fig 6. The latitudinal variation of the equivalent displacement height for six different pressure surfaces, derived from potential temperature (left) and ozone-measurements (right) for solstice and equinox conditions.

precision. For the EDH of O_3, the agreement is not satisfactory. The Hohenpeißenberg data are larger and exhibit more scattering than those deduced from the satellite measurements. This can be explained by the fact that the balloon-borne instruments monitor local concentrations, whereas the LIMS experiment integrates over a large observation volume. Nevertheless, the vertical profile of the equivalent displacement heights shows the same pattern in both data sets (fig.5). Thus, we expect that the latitudinal pattern in the EDH is also represented by the satellite data.

The latitudinal variations of the EDH of ozone and potential temperature are given in figure 6 for different pressure surfaces. Again, both tracers show the same latitudinal pattern, but with the amplitude of Δ_θ smaller by a factor of 5. The variance in the tropics is always small and the equivalent displacement height peaks in the polar region. The maximum of Δ_{O_3} at low altitudes in the tropics can be explained by fluctuations in the tropopause height, which can reach above the 100 mbar pressure surface. The tropical maximum at 10 mbar is an artefact caused by the vanishing gradient of the ozone mixing ratio profile at that height. Here, the error bars for the EDH become extremely large.

To demonstrate the seasonal variation of the equivalent displacement height, two weeks of satellite observations were selected to show the latitudinal pattern for solstice (fig.6a) and equinox conditions (fig.6b). During solstice the EDH is largest in the winter hemisphere. The peak of Δ_θ lies at 60° North at 100 mbar and is shifted towards the equator for lower pressures. In the summer hemisphere, a small maximum occurs at 50°S in the lower stratosphere, which disappears above the 16 mbar pressure surface. This pattern holds for both, Δ_θ and Δ_{O3}, but is less apparent for the EDH deduced from ozone measurements.

As expected, the equivalent displacement height is more symmetrical during equinox conditions. For Δ_θ the maxima are less pronounced, whereas for Δ_{O3} the height of the peaks increased above 30 mbar. In the Northern hemisphere, where data were avaible up to 84° North, the maxima seem to occur at the pole. In the Southern hemisphere the maxima are found around 50° South for Δ_θ, while Δ_{O3} is still increasing up to 64° South, the border of the region of observation.

4. CONCLUSION

We have shown the seasonal and latitudinal pattern of equivalent displacement heights deduced from observation. The stratospheric variance of tracer concentrations peaks in winter and is less during the summer months. The latitudinal behahavior exhibits a polar maximum and a quiet tropical zone. This is in accordance with the stratospheric wave activity. But the remaining uncertainties in the EDH and the vertical and horizontal displacement require much more work, in particular the investigation of other conservative tracers would be useful for the interpretation of Δ. The comparison with theoretically predicted EDH shows that there are still discrepancies. A detailed analysis of the temporal trend of the propagation of the EDH from the poles towards the equator may provide further insight in the causes of the variance of stratospheric tracer concentrations.

ACKNOWLEDGMENT

We would like to thank Dr. J. Gille from NCAR for providing us with the LIMS data and for helpful discussions. The programming was carried out by Mrs. H. London, which is gratefully acknowledged. The research project was sponsored by the German Ministry for Science and Technology under contract number 07 045904.

REFERENCES

Attmannspacher, W., 1985: Sonderbeobachtungen des Meteorologischen Observatoriums Hohenpeißenberg. Ergebnisse der aerologischen und bodennahen Ozonmessungen 1966 - 1985.

Ehhalt, D.H., E.P. Röth and U. Schmidt, 1983: On the temporal variance of stratospheric trace gas concentrations, J. Atmos. Chem. 1, 27-51.

Gille, J.C. and J.M. Russell III, 1984: The limb infrared monitor of the stratosphere: experiment description, performance and results, J. Geophys. Res. 89D, 5125-5140.

Hess, P.G. and J.R. Holton, 1985: The origin of temporal variance in long-lived trace constituents in the summer stratosphere, J. Atmos. Sci. 42, 1455-1464.

Mahlman, J.D., H. Levy II and W.J. Maxim, 1986: Three-dimensional simulations of stratospheric N_2O: predictions for other trace constituents. J. Geophys. Res. 91D, 2687-2707.

Röth, E.P. and D.H. Ehhalt, 1987: Seasonal variation in the variance of stratospheric ozone and potential temperature over Hohenpeißenberg, FRG. In: G.Visconti and R.Garcia (eds), Transport processes in the middle atmosphere, p.137-152, by D. Reidel Publ. Comp.

AN OBJECTIVE METHOD OF CORRECTING UMKEHR MEASUREMENTS FOR VOLCANIC AEROSOL ERROR

John J. DeLuisi
NOAA/ERL/ARL/Geophysical Monitoring for Climatic Change
Boulder, Colorado 80303, USA

David U. Longenecker
Cooperative Institute for Research in Environmental Sciences
University of Colorado
Boulder, Colorado 80309, USA

Carlton L. Mateer
Scarborough, Ontario M1W 3T6, Canada

ABSTRACT

An objective method of calculating errors to Umkehr measurements for stratospheric aerosol interference was developed and used on a selected set of Umkehr data. The method involves theoretical calculations that include observations of stratospheric ozone and aerosols. Stratospheric ozone and aerosol profile data used to calculate the errors to the Umkehr measurements are derived from ozonesonde observations and observations provided by five lidar stations in the northern hemisphere middle latitudes. Optical properties of the stratospheric aerosol are deduced from photometric observations and in situ observations of aerosol size distribution. The calculated errors are used to correct Umkehr data for several stations in the Northern middle latitude region. The corrected data display noteworthy variations, one of which is an obvious decrease in ozone concentration in the upper stratosphere during the first half of the 1980 decade. However, the decrease does not appear to be out of the range of variations seen in the long-term data set.

1. INTRODUCTION

It is well known that Umkehr measurements are affected by the presence of aerosols (DeLuisi, 1969, 1979; Dave et al., 1979) that act as additional scatterers and attenuators not included in the present inversion algorithm for obtaining the ozone profiles (Mateer and Dütsch, 1964). Both stratospheric and tropospheric aerosols produce an error to the Umkehr ozone profile that is negative and increases with altitude in the highest levels (30-45 km), and is generally positive in the lowest levels (0-18 km); in between, it can range from slightly negative to zero. The experimentally derived stratospheric aerosol errors from the investigations of DeLuisi (1979), Reinsel et al. (1984), and (see Schnell and Rosson, 1986) are reasonably well matched by the theoretical calculations of Dave et al. (1979) who accounted for multiple scattering and atmospheric sphericity in their work. These investigations have revealed that the tropospheric aerosol error is much less than the stratospheric aerosol error.

The principal aim of the present investigation is the estimation of the stratospheric aerosol error to Umkehr ozone profiles for the initial time period 1979-1986, which includes the stratospheric aerosol enhancement of nearly 2 orders of magnitude caused by the March 1982 eruption of El Chichon. This time period represents one phase of an ongoing effort to investigate the stratospheric aerosol error from the beginning of the Umkehr measurement record.

2. PHYSICAL MECHANISM FOR THE AEROSOL ERROR

At a solar zenith angle near 60° the direct solar rays for the short and long wavelengths penetrate deeply into the atmosphere (below the stratospheric aerosol layer) before they are scattered by the air (and aerosols) vertically downward toward the instrument. Most of this forward scattering occurs in the troposphere. Because the Dobson wavelength pairs are separated by only ~20 nm, the aerosol optical properties are nearly the same, and therefore the optical effect to both intensities is nearly the same as long as the ray paths for both wavelengths are similar. The addition of a stratospheric aerosol layer also slightly enhances the forward-scattering of ultraviolet radiation downward toward the instrument. In this forward-scattering phase, the Umkehr N-value (100 log (I'/I), where the prime is the longer wavelength intensity) ranges from no change to slightly positive depending on the relative shapes of the aerosol and ozone profiles and to some extent the aerosol optical properties.

When the contribution function for the longer wavelength rises over the tropospheric aerosol layer, an aerosol reversal in the N-value will probably occur near a solar zenith angle of 82°. As the solar zenith angle continues to increase further, the rays of the longer wavelength begin to pass nearly tangentially through the stratospheric aerosol

ISBN 0-937194-15-8

layer (rays for the short wavelength are above the layer), thus causing the intensity of the short wavelength to "brighten" relative to the long wavelength; then the N-value further decreases. If the rays of the longer wavelength eventually pass above the stratospheric aerosol layer, another aerosol reversal may occur. As a result of the brightening in the shorter wavelength, the inversion algorithm will place less ozone in the upper layers.

Figure 1 from Dave et al. (1979) illustrates a change to the Umkehr measurement caused by aerosols. The value WN_A is the difference between an Umkehr with aerosol effects and an Umkehr for the same atmospheric conditions, but without aerosol effects. Note the "aerosol Umkehr." In all cases, multiple scattering tends to reduce the amplitude of the primary-scattering error curve.

satellite measurement systems as well as ground-based turbidity measurements.

- Aerosol size distribution from in situ aircraft and dustsonde systems, e.g., Oberbeck et al. (1983).

3.2 THEORETICAL COMPUTATION

The procedure for calculating aerosol errors to the Umkehr profile is as follows:

- Use the Dave et al. (1979) multiple-scattering code with pseudo-spherical atmosphere approximation to calculate an Umkehr profile.

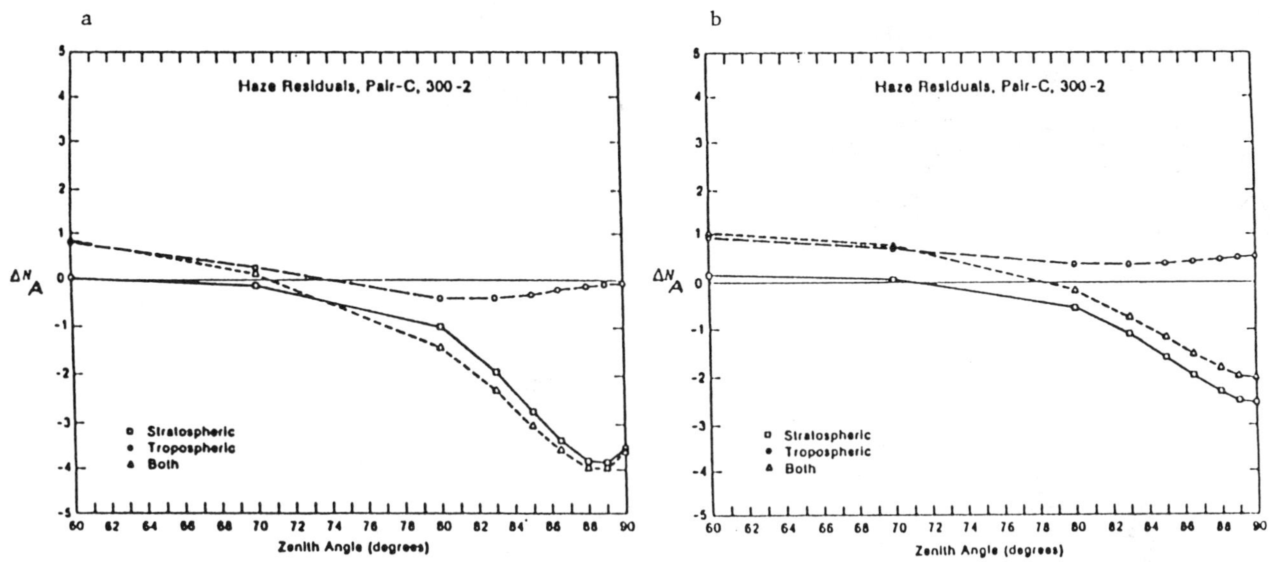

Fig. 1. Typical primary-scattering aerosol errors (a) and (b) multiple-scattering aerosol errors, for tropospheric effects only, stratospheric effects only, and both combined. Stratospheric aerosol optical thickness is 0.0147, tropospheric aerosol optical thickness is 0.188 and aerosol refractive index is 1.5-0.03i. WN_A is the difference between an Umkehr with aerosol effects and one without aerosol effects, for the same atmospheric conditions. Figure from Dave et al. (1979).

An example of theoretically calculated aerosol error to the Umkehr ozone profile is shown in Fig. 2. For ease of comparison, the error is normalized to an optical thickness of 0.01.

3. CALCULATIONS OF THE STRATOSPHERIC AEROSOL ERROR

A schematic diagram that gives the sequence of steps for correcting Umkehr ozone profiles is shown in Fig. 4. The following is a discussion of those steps.

3.1 DATA SOURCES

To calculate the error for a given month, it is necessary to acquire the following data:

- Stratospheric aerosol profile information from lidar sites and from satellites such as in the Stratospheric Aerosol and Gas Experiment (SAGE).
- Optical thickness information from lidar and

- Use monthly average lidar observations of the stratospheric aerosol profile for input to Dave's code (DeLuisi et al., 1988).
- Use monthly average ozonesonde profile data (Tiao et al., 1986) for input to Dave's code.
- Assume surface reflectance = 0 for the present study. The error is small.
- Calculate Umkehr N-values for atmospheres with and without aerosols separately. N_A and N_C stand for N-values with and without clear aerosol effects, respectively. $WN_A = N_A - N_C$.
- Invert the results of calculated Umkehr measurements following the procedure given next.

The monthly average stratospheric optical thickness observed by lidar at five northern middle-latitude sites is shown in Fig. 3 (DeLuisi et al., 1988). Note the abrupt change caused by the eruption of El Chichon. The corresponding vertical profiles to the optical thickness results were used in the theoretical calculations for error.

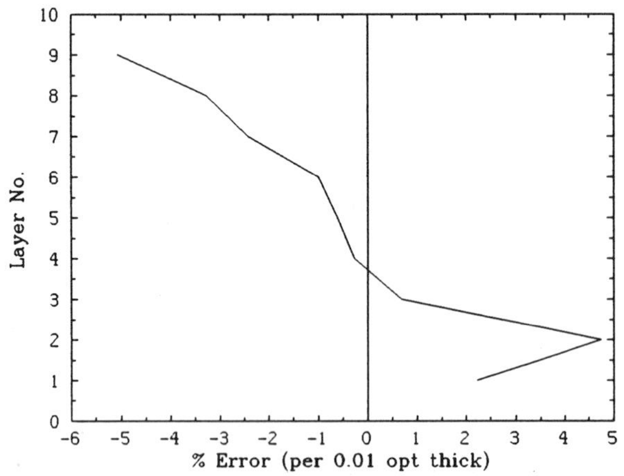

Fig. 2. Umkehr layer number vs. computed error in percent per 0.01 stratospheric aerosol optical thickness for the month of February 1986. The observed stratospheric aerosol optical thickness for that month was 0.005, which results in the actual error being one-half that shown on the plot. Figure from DeLuisi et al. (1988).

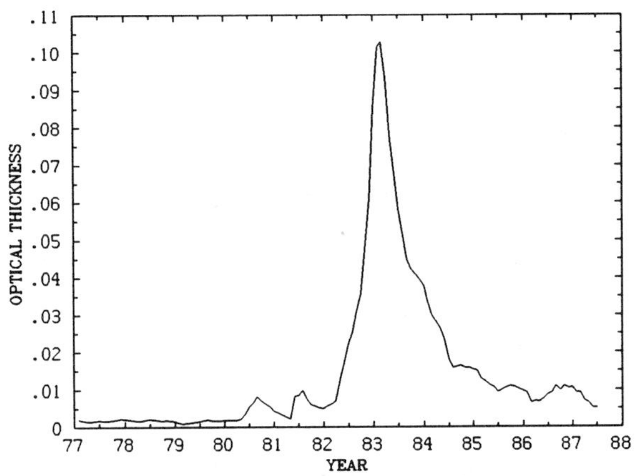

Fig. 3. Monthly averaged (smoothed) stratospheric aerosol optical thickness vs. time observed by lidar at five northern middle latitude sites: Langley, Virginia, U.S.A.; Aberystwyth, Wales; Haute Provence, France; Boulder, Colorado, U.S.A.; and Garmisch-Partenkirchen F.R.G. Figure from DeLuisi et al. (1988).

3.3 INVERSION OF UMKEHR

The procedure for estimating the ozone profile error due to aerosols is as follows:

- Use the WMO Uniform Evaluation System (conventional Umkehr) of Mateer and Dütsch (1964) to invert the Umkehr measurement.
- Use a monthly average actually observed Umkehr profile acquired during conditions of stratospheric quiescence. Invert this Umkehr profile. Then, invert the same Umkehr profile, but with WN_A added to it.
- Calculate the ozone error profile by taking the difference between the two cases (with aerosol and without aerosol) and dividing by the quiescent case.

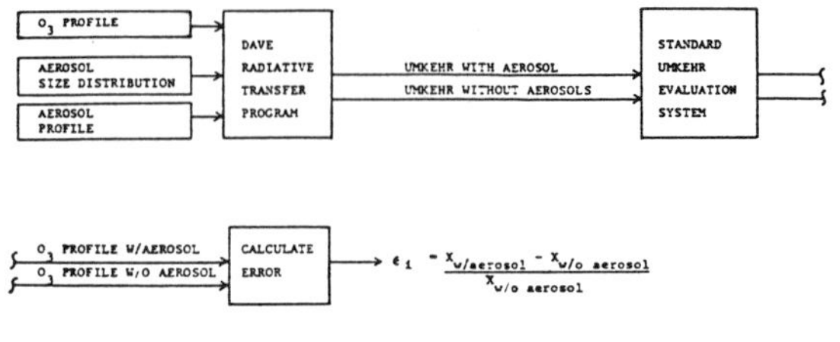

Fig. 4. Scheme of the steps taken to calculate the aerosol error to the Umkehr ozone profile.

4. RESULTS

Figure 5 shows the unweighted average ozone profile, in Umkehr layers, for five Umkehr stations: Belsk, Arosa, Lisbon, Boulder, and Tateno. It is obvious that amounts for layers 5 through 9 were lower than normal during the passage of the major mass of the El Chichon cloud (ref. Fig. 3).

The corrected profiles are shown in Figure 6. We estimate an uncertainty of the calculated correction to be in the range of 10% to 20%. The maximum estimated error was 38% ± 6 % in layer 9 during the winter of 1982-1983. The lower values seen in layer 5 during years 1983 and 1985 and after correction are thought to be due to strong quasibiennial effects (see Bojkov, 1987), and in 1985 possibly some error due to imperfect coincidence of the aerosol and Umkehr data. A 12-month running average was used to smooth the data of Figs. 5 and 6, and a least-squares-linear fit was applied to estimate a linear change, excluding the years 1983 through 1985 and then including these years. The results are shown in Table 1. In the results for corrected Umkehr ozone profiles, there is only a small difference between the use of all data and elimination of the years 1983-1985. The difference is larger for uncorrected profiles. Also, the rate of change for the years 1979-1986 is much larger for uncorrected profiles compared with corrected profiles, due to the error in ozone profiles caused by the El Chichon aerosol. The result for layer 9 must be treated with caution because the inversion algorithm causes layer 9 to be sensitive also to variations in layer 8. Figure 7 shows a plot of ozone concentration in layer 8 vs. time based on a 12-month smoothing of data in Fig. 6 (solid line). Also shown in the diagram is a smoothed plot of the observed solar Mg II line center-to-wing ratio (Heath and Schlesinger, 1986), which is an indicator of the solar ultraviolet emission strength. The two plots seem to correspond quite well and seem to imply a relationship between the two. A continuation of this record through the next solar cycle may reveal a more conclusive relationship.

5. CONCLUSION

We described an objective method for correcting Umkehr observations of ozone profile for the middle latitudes of the northern hemisphere. The method uses stratospheric aerosol profiles observed by lidar, ozonesonde profiles, and rigorous radiative transfer theory similar to that for the Umkehr inversion algorithm. We also used in the calculations stratospheric aerosol optical information obtained directly from observations.

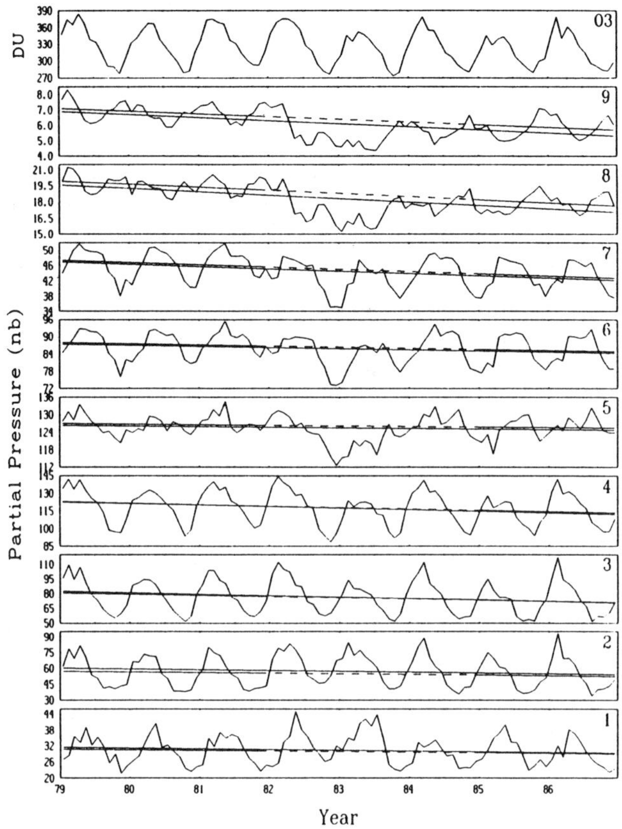

Fig. 5. Monthly average ozone concentration vs. time in Umkehr layers 1-9 for Belsk, Arosa, Lisbon, Boulder, and Tateno. Also shown is a plot of total ozone in Dobson units (DU) (top frame) corresponding to the ozone profile data. Data were supplied courtesy of the WMO World Ozone Data Center at Toronto. These data have not been corrected for stratospheric aerosol error. Note the error effects of El Chichon during the winter of 1982-1983. The lines are the least-squares linear fits for slope. The dashed segments are the fits excluding the years 1982-1984.

After correcting Umkehr data for the years 1979-1986 the apparent error effects of the El Chichon stratospheric aerosol cloud in all layers are no longer obvious.

The corrected data imply a change toward lower ozone concentrations in the upper stratosphere during 1979-1986 (i.e., -1.9% ± 0.26 yr^{-1}, layer 9; -1.1% ± 0.12 yr^{-1}, layer 8; -1.0% ± 0.17 yr^{-1}, layer 7; -0.34% ± 0.12 yr^{-1}, layer 6, and -0.07% ± 0.09 yr^{-1}, layer 5). This change was preceded by an increase toward higher values during the 1970's (see DeLuisi et al., 1988). The lower values do not appear to be extraordinary when comparing them with the Umkehr profile data of the early 1970's. The variations on the order of several years displayed by the data are likely to

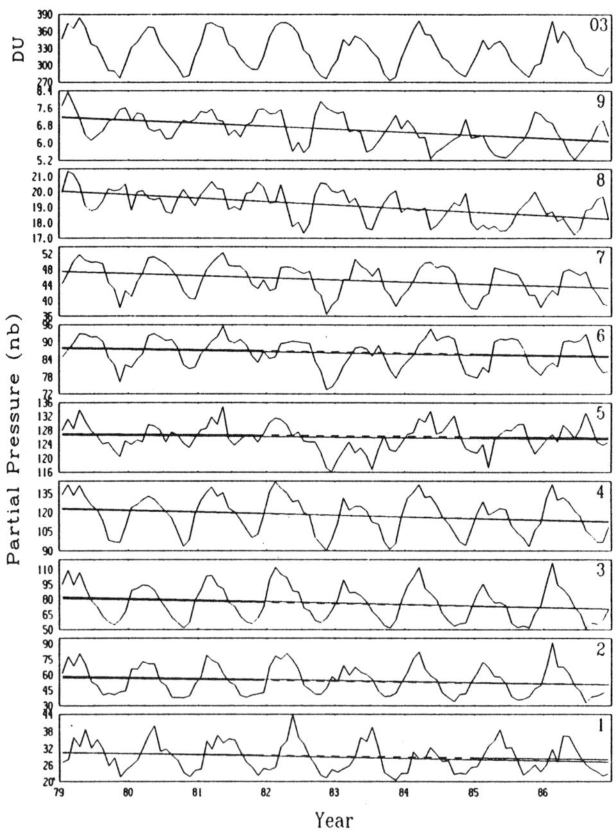

Fig. 6. Monthly average ozone concentation vs. time, as in Figure 5, but corrected for stratospheric aerosol error throughout the entire length of record. Note some scale changes on the ordinate when comparing with Figure 5.

be, at least in part, due to natural forcing from solar variations and perhaps variations in stratospheric circulation, although there is little documentation of such changes in the upper stratosphere because of observational difficulties, especially during the earlier part of the Umkehr record.

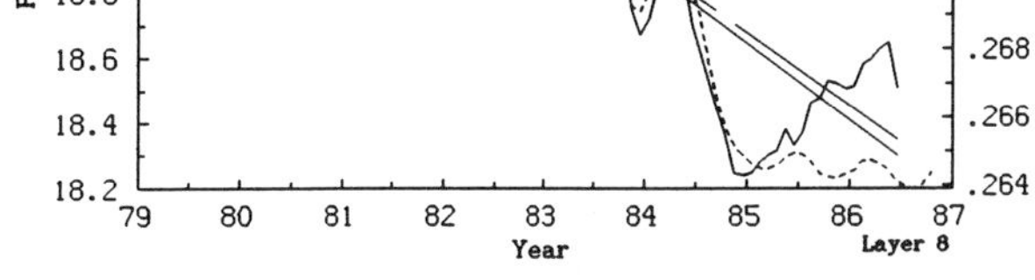

Fig. 7. The 12-month averaged ozone concentration vs. time for layer 8 (solid curve) and an indicator of solar activity given by the core-to-wing (C to W) ratio of the Mg II line from Nimbus-7 satellite data (dashed curve). The straight lines are linear least-square fits for slope. The dashed segments are the fits excluding the years 1982-1984.

TABLE 1. OZONE RATE OF CHANGE FOR 1979-1986 AND 2σ CONFIDENCE FOR FIVE UMKEHR STATIONS.

Layer no.	Altitude (km)	Five-station result* Exclude 83-85 (% yr^{-1})	All (% yr^{-1})
	With aerosol corrections		
9	45.2	-1.9 ± 0.26	-2.1 ± 0.25
8	40.0	-1.1 ± 0.12	-1.2 ± 0.11
7	35.0	-1.0 ± 0.17	-1.0 ± 0.16
6	30.4	-0.34 ± 0.12	-0.34 ± 0.18
5	25.9	-0.07 ± 0.09	-0.08 ± 0.15
4	21.4	-0.61 ± 0.28	-0.63 ± 0.27
3	16.9	-1.0 ± 0.43	-0.99 ± 0.39
2	12.5	-0.77 ± 0.35	-0.77 ± 0.38
1	5.1	-0.70 ± 0.18	-1.1 ± 0.31
	Without aerosol corrections		
9	45.2	-2.51 ± 0.30	-3.11 ± 0.81
8	40.0	-1.45 ± 0.14	-1.72 ± 0.45
7	35.0	-1.11 ± 0.11	-1.16 ± 0.30
6	30.4	-0.37 ± 0.12	-0.38 ± 0.21
5	25.9	-0.11 ± 0.09	-0.14 ± 0.20
4	21.4	-0.65 ± 0.28	-0.69 ± 0.30
3	16.9	-1.01 ± 0.44	-1.02 ± 1.56
2	12.5	-0.42 ± 0.34	-0.26 ± 0.66
1	5.1	-0.33 ± 0.20	-0.41 ± 0.53

*"Excludes 83-85" gives results of analyses that excluded data from the years 1983-1985; "all" gives results of analyses including all years.

ACKNOWLEDGMENTS

A sincere note of thanks is extended to L. Thomas, H. Jaeger, P. McCormick, and G. Megie for generously supplying their lidar data. J. Angell's review and comments are greatly appreciated. D. Donnelly kindly provided the Nimbus-7 Magnesium II line data.

REFERENCES

Bojkov, R. D., 1987: The 1983 and 1985 anomalies in ozone distribution in perspective. *Mon. Wea. Rev.*, 115, 2187-2201.

Dave, J. V., J. J. DeLuisi, and C. L. Mateer, 1979: Results of a comprehensive theoretical examination of the optical effects of aerosols on the Umkehr measurement. *Spec. Environ. Rep. 14*, World Meteorological Organization, Geneva, 15-22.

DeLuisi, J. J., 1969: A study of the effect of haze upon Umkehr measurements. *Q. J. R. Meteor. Soc.*, 95, 181-187.

DeLuisi, J. J., 1979: Umkehr vertical ozone profile errors caused by the presence of stratospheric aerosols. *J. Geophys. Res.*, 84, 1766-1770.

DeLuisi, J. J., D. U. Longenecker, and C. L. Mateer, 1988: An analysis of northern middle latitude Umkehr measurements corrected for stratospheric aerosols for the years 1979-1986, (manuscript to be submitted for review, *J. Geophys. Res.*).

Heath, D. F., and B. M. Schlesinger, 1986: The Mg 280-nm doublet as a monitor of changes in solar ultraviolet irradiance. *J. Geophys. Res.*, 91, 8672-8682.

Mateer, C. L., and H. U. Dütsch, 1964: Uniform evaluation of Umkehr observations from the world ozone network. National Center for Atmospheric Research Report, Boulder, CO, 105 pp.

Oberbeck, V. R., E. F. Danielsen, K. G. Snetsinger, G. V. Ferry, W. Fong, and D. M. Hayes, 1983: Effect of the eruption of El Chichon on stratospheric aerosol size and composition. *Geophys. Res. Lett.*, 10, 1021-1024.

Reinsel, G. C., G. C. Tiao, J. J. DeLuisi, C. L. Mateer, A. J. Miller, and J. E. Fredrick, 1984: An analysis of upper stratospheric Umkehr ozone profile data for trends and the effects of stratospheric aerosols. *J. Geophys. Res.*, 89, 4833-4840.

Schnell, R. C, and Rosson, R. M., 1986: Geophysical Monitoring for Climatic Change, No. 14: Summary Report 1985. NOAA Environmental Research Laboratories, Boulder, CO, pp 29-35.

Tiao, G. C., G. C. Reinsel, J. H. Pedrick, G. M. Allenby, C. L. Mateer, A. J. Miller, and J. J. DeLuisi, 1986: A statistical trend analysis of ozonesonde data. *J. Geophys. Res.*, 91, 13,121-13,136.

LONG-TERM INTERCOMPARISON BETWEEN THE BREWER AND DOBSON SPECTROPHOTOMETERS AT EDMONTON

J.B. Kerr, I.A. Asbridge, and W.F.J. Evans
Atmospheric Environment Service
Downsview, Ontario, M3H 5T4, Canada

ABSTRACT

Results of the intercomparison between direct sun total ozone measurements made simultaneously by Brewer and Dobson spectrophotometers at Edmonton are presented and compared with results of a similar intercomparison at Toronto. After corrections are made to the Dobson data to account for absolute bias due to absorption coefficient uncertainty, SO_2, relative airmass dependence and annual variation, total ozone measurements made by the two Edmonton instruments agree with each other to better than 1%. Also the long term drift between the two instruments was found to be statistically insignificant. An appropriate function is derived to calculate a simulated Dobson ozone value from a Brewer ozone measurement.

1. INTRODUCTION

Automated Brewer spectrophotometers have been installed at seven sites in Canada: Toronto, Edmonton, Churchill, Goose Bay, Resolute, Alert, and Saskatoon. At the five stations excluding Alert and Saskatoon the Brewer instruments are to replace the existing Dobson instruments which have been in operation for more than twenty years. It is essential to carry out the replacement process carefully so that the transition from the Dobson instrument to the Brewer instrument does not cause an erroneous discontinuity in the data record. Prior to the phasing out of the Dobson instrument, both instruments are being operated in parallel at the five sites for a period of at least three years. This is to ensure that any differences between the two instruments are well understood and fully documented and that the ozone record is uninterrupted by and remains continuous during the transition from the Dobson to the Brewer instrument.

Routine Brewer/Dobson intercomparison measurements began at Toronto in February, 1983 and at Edmonton in February, 1984. Results of the intercomparison at Toronto between Dobson instrument # 77 and Brewer instrument # 8 are reported by Kerr et al. (1988). This paper briefly reviews results of the Toronto intercomparison and presents results of the intercomparison at Edmonton between Dobson instrument # 102 and Brewer instrument # 13. Dobson instrument # 102 was last calibrated against the Canadian regional standard in July, 1976 and Brewer instrument # 13 was calibrated against the Brewer standard in October, 1983.

Between February, 1984 and July, 1987, 1460 simultaneous (within one minute) direct sun Brewer and Dobson AD total ozone measurements were made at Edmonton on 350 days. The simultaneous Brewer and Dobson data sets are entirely independent because the calibration of these field Brewer and Dobson instruments are traced to standards which have been absolutely calibrated by independent methods. The two instruments have been measuring the same atmospheric constituent in a field situation for a long time and intercomparison of their results is a valuable aid for evaluating instrument performance.

Results of the Dobson/Brewer intercomparison at Edmonton reported herein follow a similar presentation as results of the intercomparison at Toronto (Kerr et al. ,1988).

2. RESULTS

In order to compare the two sets of data the ratio of total ozone measured by the Dobson instrument to that made simultaneously by the Brewer instrument is considered. The long term mean of this ratio provides a measurement of the systematic difference between the two measurement types. The standard deviation (SD) about the mean is an indication of the random differences which are observed in a field situation. A linear trend with time is fit to the ratio over the intercomparison period in order to determine if there is a statistically significant long term drift between the Brewer and Dobson ozone measurements.

Results of the intercomparison are presented in four stages. The Brewer ozone data are compared with the Dobson data: (i) as measured; (ii) corrected for SO_2; (iii) corrected for SO_2 and an observed airmass dependence; and (iv) corrected for SO_2, airmass dependence, and an observed annual variation.

2.1 REPORTED DATA

All 1460 individual Brewer and Dobson ozone and SO_2 data points are available from the authors on magnetic data medium or in tabular form as functions of date and airmass. These data are also published in the March-April, 1988 issue of Ozone Data for the World. These data are summarized in Fig. 1 where the daily mean ratios of the Dobson to Brewer ozone measurements are plotted. The slope of the best fit line shown in Fig. 1 has a value of -.00026 per year. The mean of all 1460 individual observed ratios is 1.0195 with a SD of .0083. The mean of the 350 daily average values is 1.0188 with a SD of .0076. Table 1 summarizes these statistics.

ISBN 0-937194-15-8

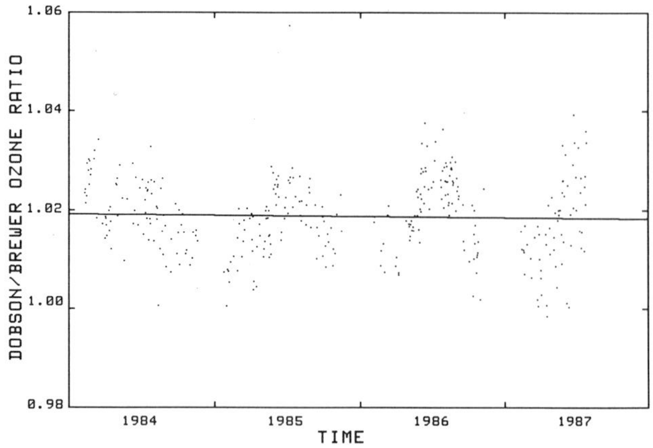

Fig. 1. Plot of the ratio of Dobson to Brewer daily mean total ozone measurements.

TABLE 1. SUMMARY OF STATISTICS FOR THE INTERCOMPARISON DATA AT THE FOUR ANALYSIS STAGES

DOBSON DATA CORRECTIONS	INDIVIDUAL OBS (1460)		350 DAILY AVERAGES		SLOPE (/YEAR)
	MEAN	STD DEV	MEAN	STD DEV	
Uncorrected	1.0195	.0083	1.0188	.0076	-.00026
SO_2	1.0180	.0081	1.0172	.0074	.00050
SO_2,airmass	1.0278	.0083	1.0264	.0078	.00088
SO_2,airmass and annual	1.0255	.0071	1.0254	.0064	.00027

2.2 DATA CORRECTED FOR SO_2

Kerr et al. (1980, 1988) have shown that the presence of SO_2 affects the Dobson AD total ozone measurement. One milli-atmosphere centimeter (matm-cm = 10^{-3} centimeter at STP) of SO_2 increases the Dobson total ozone measurement by 1.06 Dobson Unit (DU). The Dobson data were corrected for SO_2 by subtracting the necessary amount of ozone determined from the value of SO_2 measured by the Brewer instrument. The mean of the 1460 corrected ratios is 1.0180 with a SD of .0081 and the mean of the 350 daily average values is 1.0172 with a SD of .0074. Table 1 summarizes these statistics.

Correcting the Dobson data for SO_2 has reduced the mean value of the daily average Dobson/Brewer ratio from 1.0188 to 1.0172. This reduction of 0.16% is less than the 0.30% reduction observed for the Toronto intercomparison (Kerr et al., 1988). The smaller effect of SO_2 observed at Edmonton is most likely because the rural Edmonton site is cleaner than the suburban Toronto site.

2.3 DATA CORRECTED FOR AIRMASS DEPENDENCE

Previous intercomparisons with Dobson and Brewer instruments have indicated that there can be a relative airmass dependence between the two instruments. Kerr et al. (1988) report a relative airmass dependence between Brewer instrument # 8 and Dobson instrument # 77. Kerr et al. (1985) found that the Australian Dobson instruments # 105 and # 115 had an airmass dependence relative to Brewer instrument # 17 and Köhler (1986) showed that Dobson instrument # 104 at Hohenpeissenberg and Brewer instrument # 17 have a relative airmass dependence.

A relative airmass dependence has also been observed during the Edmonton intercomparison. This is demonstrated in Fig. 2 where the percentage Brewer minus Dobson difference is plotted as a function of airmass using data from only June, July and August. On average the Dobson total ozone is 1.7% larger than that of the Brewer at an airmass value of 1 and at airmass of 3 the Dobson measures 2.8% larger. This airmass dependence is similar to that observed for the Toronto intercomparison where the Dobson is 1.9% larger at airmass 1 and 3.2% larger at airmass 3.

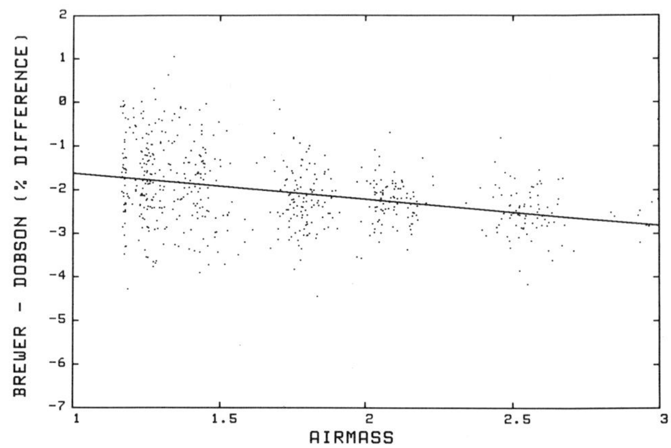

Fig. 2. The percentage Brewer minus Dobson difference as a function of airmass.

The extraterrestrial constant (ETC) for an instrument is used in the calculation of total ozone. In general, an instrument with an erroneous ETC value will exhibit an airmass dependence of total ozone given by:

$$X^* = X + \frac{\varepsilon}{\mu} \tag{1}$$

where X^* is the measured erroneous ozone value, X is the true ozone value, μ is airmass, and ε is a constant proportional to the error in the ETC.

In order to evaluate further the comparison between the two Edmonton instruments, at this stage it is necessary to assume that one instrument is airmass dependent and the other is not. For consistency with the Toronto intercomparison (Kerr et al., 1988) it is assumed that the Brewer instrument is correctly calibrated and the value of ε for the Dobson instrument is determined from the following expression derived from (1) above:

$$X_{Dob} = K{*}X_{Br} + \frac{\varepsilon}{\mu} \tag{2}$$

where X_{Dob} is Dobson total ozone corrected for SO_2 and assumed to be airmass dependent, X_{Br} is Brewer total ozone assumed to be airmass independent, and K is a constant. The constant K is necessary because the Brewer and Dobson absolute ozone values are derived from independent ozone absorption coefficient data.

Equation (2) may be rewritten as follows:

$$\frac{X_{Dob}}{X_{Br}} = K + \frac{\varepsilon}{\mu X_{Br}} \qquad (3)$$

If the intercomparison data are used in a linear regression of X_{Dob}/X_{Br} against $1/\mu X_{Br}$, the intercept is K and the slope is ε. The value for ε was determined using data from June, July and August only so that the range of μ would be similar for all days of the analysis. The resulting value for ε was found to be -5.6 ± 0.6.

The Dobson ozone values corrected for SO_2 were recalculated with an adjustment of $+5.6/\mu$ applied in order to remove the relative airmass variation. This corrected ozone record is plotted in Fig. 3. The slope of the best fit line through the daily mean ratios is .00088 per year. The mean of the Dobson/Brewer ratio for all 1460 data points is 1.0278 with a SD of .0083 and the mean for the 350 daily averages is 1.0264 with a SD of .0078. Table 1 summarizes these statistics.

2.4 DATA CORRECTED FOR ANNUAL VARIATION

The curved line in Fig. 3 is the data smoothed by a triangular filter with a base of 90 days. This smoothed curve has an annual variation with the Dobson/Brewer ratio larger in summer than in winter. Figure 4 is a plot of the daily averages of the corrected Dobson/Brewer ratio as a function of the time of year with the 90 day smoothed curve. There are no data during December and January because the airmass at this time of year is too large. The smooth curve shows the Dobson/Brewer ratio is about 1.2% larger at its maximum in summer than it is at the minimum in early winter with a secondary maximum and minimum in February and March.

Results of this intercomparison have shown that there is relative airmass depencence and annual variation between the Dobson and Brewer ozone measurements. In some situations an erroneous annual variation is observed with an airmass dependent instrument because the airmass in winter is larger than it is in summer. The current intercomparison has avoided a possible interaction between an airmass dependence and an annual dependence because data from the summer

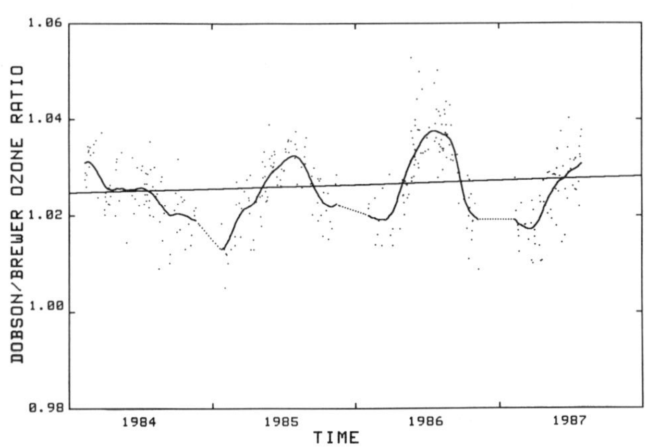

Fig. 3. Same as Fig. 1 except that the Dobson data are corrected for SO_2 and airmass.

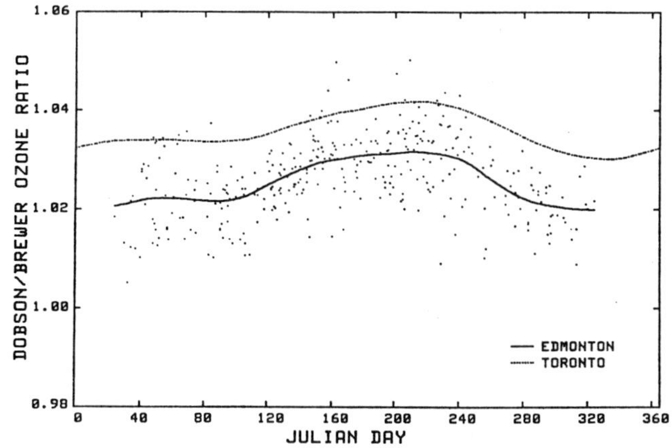

Fig. 4. Daily mean ratios of Dobson/Brewer ozone plotted as a function of time of year with 90 day triangular smoothed curve.

months only were used to determine the airmass dependence. The annual variation shown in Fig. 4 is indeed real and is not an artifact of the correction for airmass.

The annual variation for the Toronto intercomparison (from Kerr et al., 1988) is also shown in Fig. 4. The shape of the two curves is nearly identical including the secondary maximum and minimum. Kerr et al. (1988) suggested that the seasonal variation is most likely due to the annual variation of temperature combined with a different temperature dependence for the ozone absorption coefficients of the Brewer and Dobson instruments. The fact that the Edmonton and Toronto intercomparisons yield the same annual variation confirms that the cause is a real atmospheric phenomenon most likely due to temperature variation.

The data points in Fig. 3 were divided by the normalized annual variation shown in Fig. 4. The resulting data points are plotted in Fig. 5. The linear trend in Fig. 5 is .00027 per year. The mean of all 1460 corrected data point is 1.0255 with a SD of .0071 and the mean of the 350 daily averages is 1.0254 with a SD of .0064. Table 1 summarizes these statistics.

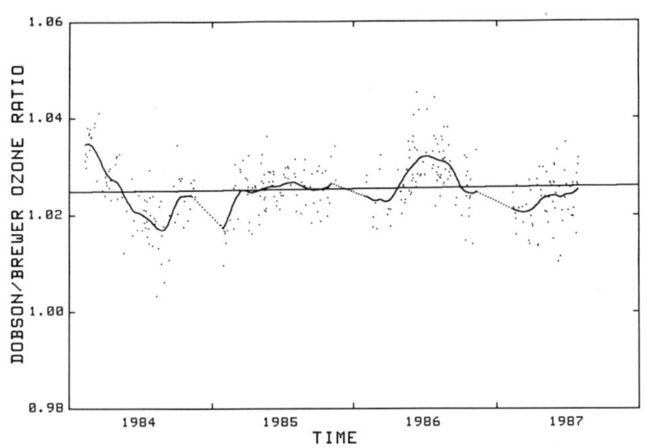

Fig. 5. Same as Fig. 3 except the annual variation is removed by dividing the Dobson data by the normalized curve in Fig. 4.

3. DISCUSSION

The long term average of daily mean total ozone corrected for SO_2 measured by the Edmonton field Dobson instrument is 1.72% larger than that of the Brewer instrument. However, the two instruments have a relative airmass dependence indicating that the ETC for at least one of the instruments is noticeably in error. This error would cause the Dobson instrument to measure 5.6 DU less than the Brewer instrument in the presence of no ozone outside the Earth's atmosphere.

In addition to the ETC calibration difference, the linear response by the two instruments to atmospheric ozone is different. The average of the daily mean Dobson/Brewer ratios corrected for the relative airmass dependence is 1.0264, indicating that the response of the Dobson instrument to ozone is on average 2.64% larger than that of the Brewer instrument. The 2.64% difference is due to the fact that the ozone absorption coefficients are derived from different sources. The Dobson AD coefficient was determined by Vigroux (1967) and the Brewer coefficients are based on those of Bass and Paur (1985) at -45°C. The current intercomparison suggests that the absorption coefficient for Dobson instrument # 102 be increased by 2.64%.

Results of the Toronto intercomparison (Kerr et al., 1988) showed that Dobson instrument # 77 on average measures 2.34% more than Brewer instrument # 8 and, after correcting for airmass, the response of Dobson instrument # 77 is 3.83% larger than that of Brewer instrument # 8. This suggests that the Edmonton Dobson instrument measures 0.6% less than the Toronto Dobson instrument and that the absorption coefficient for Dobson instrument # 77 is 1.19% larger than that for Dobson instrument # 102.

Kerr et al. (1988) showed that the Brewer data could be used to simulate a total ozone measurement made by Dobson instrument # 77. The same procedure may be used to convert a reported Brewer ozone measurement (in micrometers) for Edmonton to a simulated Dobson measurement. This would be done if the past Dobson record is to be linked to the current and future Brewer data record without producing a discontinuity due to instrument changeover. There are basically four adjustments which need to be applied to the Brewer data in order to simulate a Dobson measurement: Dobson/Brewer bias, annual variation, airmass variation, and effect of SO_2. These adjustments are applied using the following transfer function:

$$X_{Sim}=.10265*[1-.005*COS(JD)]*X_{Br}-\frac{5.6}{MU}+.106*SO_2 \quad (4)$$

where X_{Sim} is the simulated Dobson total ozone (in DU), JD is the Julian day, X_{Br} is the Brewer total ozone (in micrometers), MU is the airmass, and SO_2 is the Brewer SO_2 (in micrometers). Here the term of .005*COS(JD) is used to approximate the correction for the annual variation.

Daily mean values of simulated Dobson total ozone were calculated from the Brewer data set by applying Equation (4) above. Figure 6 shows the 90 day smoothed simulated Dobson ozone (from Brewer data) compared with the 90 day smoothed reported Dobson ozone data. The fact that the two curves are nearly equal (within 1.0% at all times) demonstrates that it is possible for a Brewer ozone measurement to simulate a Dobson measurement. The Brewer data in Edmonton can therefore be used with confidence to extend the Dobson data for trend analyses without disturbing the ozone record.

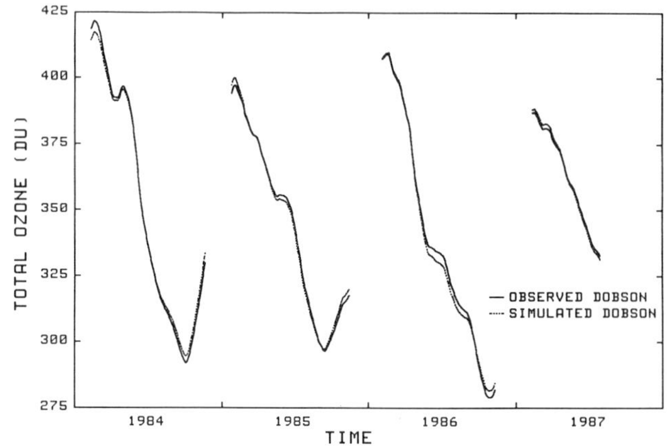

Fig. 6. Measured Dobson ozone data (solid curve) and simulated Dobson data (dotted curve) smoothed by 90 day triangular filter.

REFERENCES

Bass, A.M. and R.J. Paur, 1985: The ultraviolet cross-section of ozone: I. The measurement, Proc. Quadrennial Ozone Symposium, 606-610, Halkidiki, Greece.

Kerr, J.B., I.A. Asbridge, and W.F.J. Evans, 1988: Intercomparison of total ozone measured by the Brewer and Dobson spectrophoto meters at Toronto, accepted by J. Geophys. Res., 93.

Kerr, J.B., W.F.J. Evans, and I.A. Asbridge, 1985: Recalibration of Dobson field spectrophotometers with a travelling Brewer spectrophotometer standard, Proc. Quadrennial Ozone Symposium, 381-386, Halkidiki, Greece.

Kerr, J.B., C.T. McElroy, and R.A. Olafson, 1980: Measurements of ozone with the Brewer ozone spectrophotometer, Proc. Int. Ozone Symposium, 74-79, Boulder, Colorado.

Köhler, U. 1986: , Recalibration of the Hohenpeissenberg Dobson spectrophotometer 104 presuming 'effective absorption coefficients', J. Atmos. Chem., 4, 359-374.

Vigroux, E.,1967: Determination des coefficients moyens d'absorption de l'ozone en vue des observations concernant l'ozone atmospherique a l'aide du spectrometer Dobson, Ann. Phys., 2, 209-215.

SEVERAL RESULTS FROM 20 YEARS OF OZONE OBSERVATIONS AT HOHENPEISSENBERG

Klaus Wege, Hans Claude, and Reinhold Hartmannsgruber
Deutscher Wetterdienst
Meteorologisches Observatorium
Hohenpeissenberg, FRG

ABSTRACT

The correlation between ozone and temperature and furthermore the variability of ozone are discussed on the basis of ozone time series from Hohenpeissenberg. A positive ozone trend is indicated in the troposphere, a negative trend within the ozone maximum.

1. INTRODUCTION

Today there is little doubt that the distribution of atmospheric ozone is affected anthropogenically. Many model calculations predict an ozone reduction in the stratosphere and an increase in the troposphere. The antarctic ozone hole is an indication of the acuteness of this problem. Long-term observations are required to study and register possible changes in ozone concentration. They give information on actual trends and verify model simulations.

To date, 21 years of ozone soundings (1967 - 87) and 17 years of surface ozone observations (1971 - 87) have been recorded at the Meteorological Observatory Hohenpeissenberg.

2. TEMPERATURE - OZONE CORRELATION AND OZONE VARIATIONS

Various studies have been based on the Hohenpeissenberg ozone series (Attmannspacher et al., 1975; 1984a; 1984b; Bojkov, 1983; 1988a; Bojkov et al.; Hartmannsgruber et al., 1985; Köhler et al., 1988; Wege, 1988).

The relationship between ozone concentration and temperature is of special interest. Bojkov (1988b) has investigated correlations between total ozone and various temperatures. The correlation factor between ozone and temperature is plotted as a function of the altitude in Fig. 1. The dashed line shows the correlation of monthly mean values, whereas the solid line represents the correlations of deviations from long-term monthly mean values, i.e. the seasonal variation is eliminated. The dominating effect of seasonal variations is evident in the dashed line. Correlation is very high in the troposphere because the annual variations of ozone and temperature exhibit only a relatively small phase shift. The correlation factor is near 0 below the tropopause, above this minimum a rapid increase to a highly significant positive correlation occurs which is caused by dynamic factors (see below).

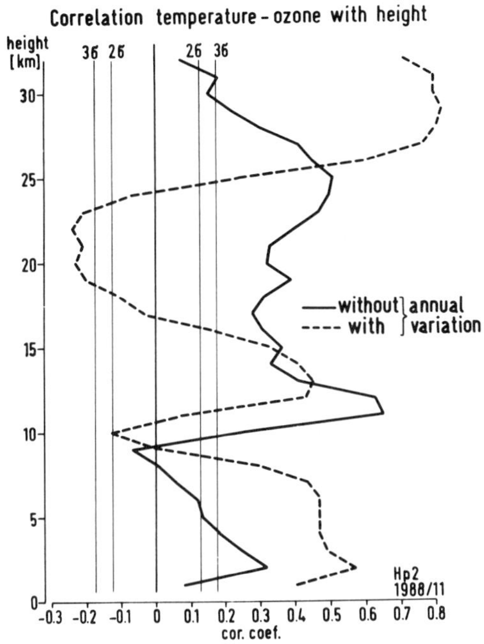

Fig. 1. Correlation temperature - ozone against altitude, Hohenpeissenberg 1967 - 87. Solid line: annual variation eliminated.

The negative correlation in the vicinity of the ozone maximum (Attmannspacher, 1975) is due to the phase difference in the seasonal variation of ozone and temperature. The phase shift is more than 90°, and at 22 km is almost 180°. Correlations become positive again above 25 km, with factors as high as R = 0.82. The solid curve (deseasonalised) is more informative. The positive correlation remains slightly significant in the lower troposphere which is probably due to photochemical processes during increased global radiation. The prominent maximum at 11 km with R = 0.65 is caused by variation in tropopause height, whereby either ozone-rich stratospheric air or ozone-poor tropospheric air is present. Because of the negative correlation between tropopause height and temperature, a positive ozone - temperature correlation is to be expected here (Köhler et al., 1988; Wege, 1988). The correlation decreases at higher levels, but still remains positive and significant due to transport processes. The ozone distribution is strongly coupled with deformations of the vorticity distribution, whereby transport to the pole occurs in narrow tongues which exhibit higher tempera-

OZONE IN THE ATMOSPHERE
Rumen D. Bojkov and Peter Fabian (Eds.)

ISBN 0-937194-15-8

tures (Leovy, 1985). Also Braun (1984) found a positive correlation at these altitudes. In addition, high ozone values are related to the collapse of the polar night low which is accompanied by relatively high temperatures. The correlation factor diminishes substantially above 25 km because transport processes are no longer decisive there. Nevertheless the correlation is still not negative, as would be expected from the photochemistry of the upper stratosphere.

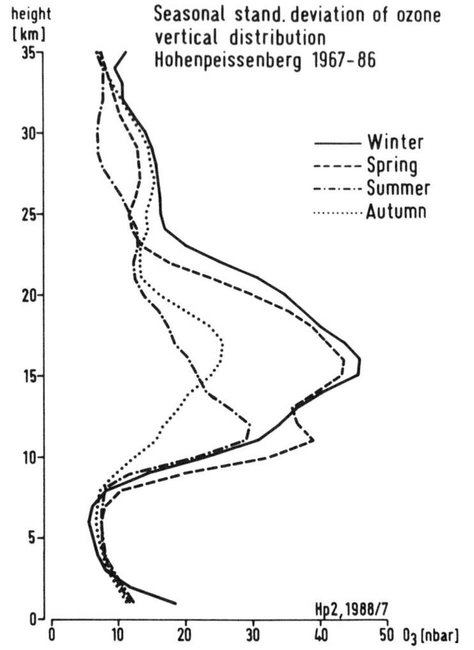

Fig. 2. Standard deviation of ozone against altitude, Hp 1967 - 86, 4 seasons.

The dependency of ozone variability on altitude is interesting in connection with possible trends. The standard deviation is shown in Fig. 2. It is high at the surface, progressively declines towards the ozone minimum below the tropo-pause, and increases thereafter up to 16 km altitude (summer 11 - 12 km). Considerable seasonal differences can be noted. The standard deviation is greatest in spring and winter; the secondary maximum in spring at 12 km is due to frequent occurrences of secondary ozone maxima in this season. The standard deviation decreases substantially further aloft and reveals relatively low values of 10 - 15 nbar above the ozone maximum. If the coefficient of variation is calculated (ratio of ozone standard deviation to mean concentration), another very clear picture results (Fig. 3).

The seasonal dependency is negligible. Large surface values of 60 - 75 % are found, rapidly decreasing with altitude to about 40 % in the middle troposphere and increasing again in the upper troposphere to 75 - 90 % at the tropopause level. This maximum is caused by dynamic factors (see above). A decrease above the tropopause is continuous, a secondary maximum is only present in winter and spring at the level of standard deviation maximum. The coefficient of variation in the ozone maximum and above is very small at 10 - 20 %. The verification of a trend, if present, is difficult at those layers where ozone variability is high, especially near the surface and near the tropopause.

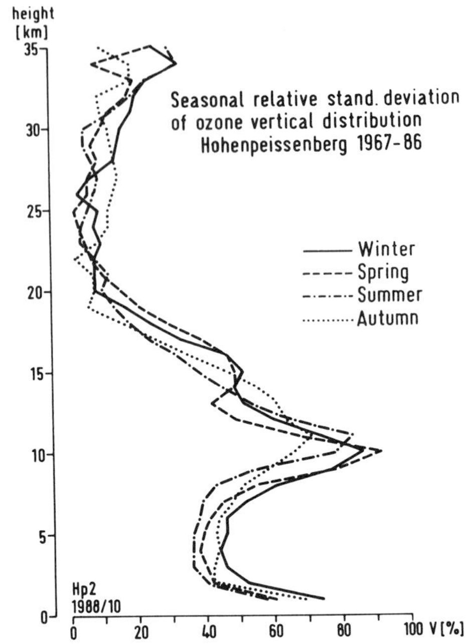

Fig. 3. Coefficient of variation $\sigma/\overline{O}_3$ against altitude. Hp 1967 - 86, 4 seasons.

3. TRENDS

An important question today is whether trends can be deduced from ozone data. The Hohenpeissenberg data have already been examined in this respect. Such investigations have also been performed with other data, especially by Logan (1985), Frederick et al. (1985), Tiao et al. (1986). Here, we have performed the trend test of Cox and Stuart (Schönwiese, 1985) and regression analyses on the Hohenpeissenberg series. The former method utilizes the signs of data changes. Table 1 shows those altitudes where a significant trend is obvious. The trend is positive in the troposphere and negative in the stratosphere around 20 km:

TABLE 1. TREND TEST OF COX AND STUART, 95 % CONFIDENCE LEVEL (TEST VALUE $z \geq 1.96$).

	trend positive	trend negative
year	2 - 9 km	18 - 25 km
spring	2 - 8 km	23 - 24 km
summer	2 - 8 km	--
autumn	2 - 10 km	18 - 23 km
winter	2 - 8 km	22 - 23 km

Figure 4 shows the variation of average annual ozone in the troposphere and the lower stratosphere with the associated regression lines. The slopes of the regression lines were computed in nbar/year by means of regression analyses for altitudes up to 34 km (Fig. 5). The vertical bars indicate the 95 % confidence interval for significant values.

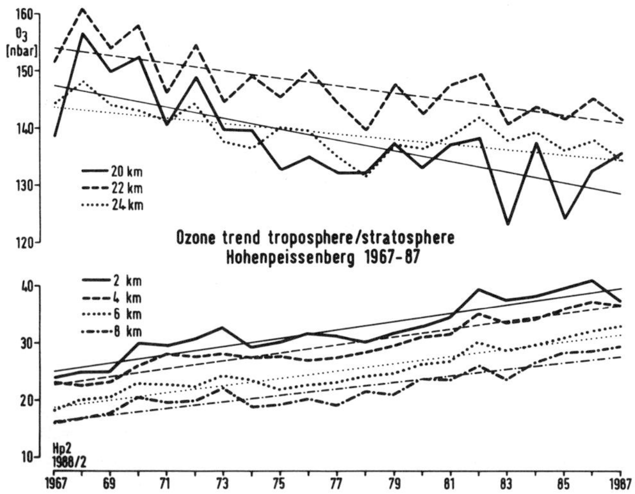

Fig. 4. Ozone trend estimates at different altitudes, Hp 1967 - 87.

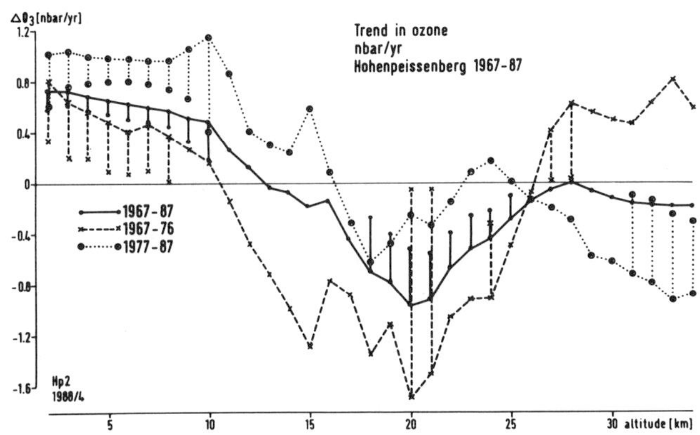

Fig. 5. Ozone trend estimates (nbar/year) against altitude, Hp 1967 - 87. Significant trends have vertical bars showing the 95 % confidence interval (only one half of the bar is plotted).

A highly significant increase is observed in the troposphere up to 10 km with values of 0.6 to 0.7 nbar/year. At the level of the ozone maximum a significant negative trend of up to -1 nbar/year has accured, however the scatter of the data is larger than in the troposphere as can be seen in Fig. 4. Also shown in Fig. 5, the observation series has been divided into 2 periods 1967 - 76 and 1977 - 87. Changes during both periods remain significant in the troposphere, the second period exhibiting the trend more prominently. The negative trend in the lower stratosphere is also present in both periods but is only significant at isolated altitudes. In the interesting region above 30 km, for which model computations predict a negative trend, no decrease is obvious. A weakly significant negative trend is indicated for 1977 - 87, but note that the 1st period shows a (non-significant) rise. A 22-year period of solar activity is possibly indicated here, but this cannot be proven.

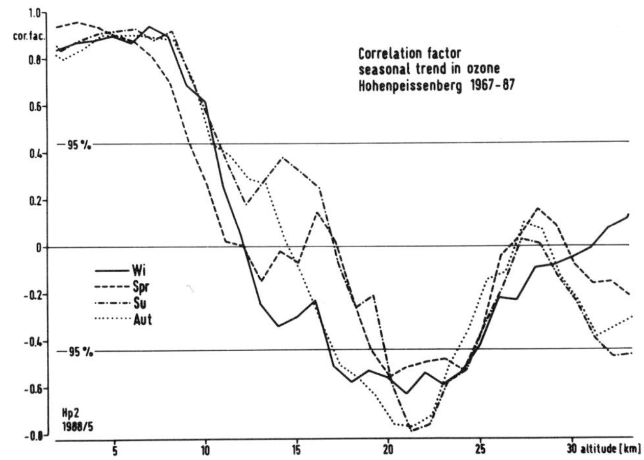

Fig. 6. Correlation between ozone and time against altitude, Hp 1967 - 87, 4 seasons, 95 % level is plotted additionally.

The ozone data were divided into 4 seasons and practically the same ozone - time correlation results (Fig. 6). The significant positive correlation in the troposphere $R \approx 0.90$ is maintained through all seasons, the negative one near 20 km is somewhat less pronounced.

Regression analyses were also performed for individual months. Table 2 gives the number of seasons and months in which the trend is significant for different altitudes. It is noted that the 99 % confidence level is often surpassed.

TABLE 2. NUMBER OF SEASONS AND MONTHS WITH SIGNIFICANT TREND (95 %), + = POSITIVE TREND, - = NEGATIVE TREND.

altitude [km]	surf.	2	4	6	8	10	18	20	22	24
year	+	+	+	+	+	+	-	-	-	-
seasons	+ 2	+ 4	+ 4	+ 4	+ 4	+ 3	- 2	- 4	- 4	- 3
months	+ 3	+12	+12	+12	+12	+ 4	- 4	- 9	- 6	- 4

Tiao et al. (1986) have studied the possible influences of the normalization procedure (Dobson correction) on a trend. The "De Muer" procedure shows no effect, a "constant correction factor" would result in a weakened trend in the troposphere and a negligible change in the stratosphere.

A trend for ozone near the ground was found early in the GDR (Feister and Warmbt, 1987). Attmannspacher et al. (1984), Bojkov (1983, 1988a) and Wege (1988) discussed the Hohenpeissenberg surface ozone data in regard to trends. Annual and winter and spring season trends at Hohenpeissenberg are significant (see Fig. 7). Also trends for the individual months from Jan. to March are significant (see Table 2). The increase in winter is larger than in the other seasons, which could be explained by the arguments of Liu et al. (1987). A seasonal difference in the trend cannot be discerned in the free atmosphere. A partitioning of the surface observation series also shows that a trend is not present in recent years and remains limited to the early part of the series. Figure 7 exhibits the large variations of the yearly and seasonally averaged surface ozone (also compare to Fig. 3). These strong variations and the sink effect at the ground probably diminish the substantial trend that exists in the free atmosphere.

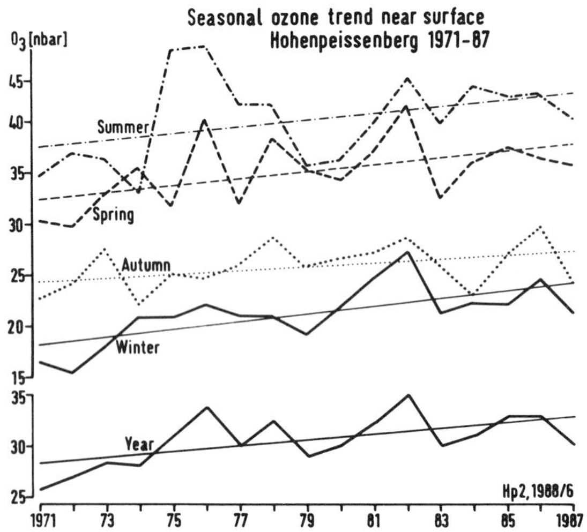

Fig. 7. Trend estimates, surface ozone, seasons and year, Hp 1967 - 87.

4. CONCLUDING REMARKS

An ozone increase in the troposphere at Hohenpeissenberg is apparent even though it does not appear so pronounced near the surface. A significant negative trend near 20 km altitude is also found. Above 30 km the trend is not quite clear; variations in solar activity can play a major role there. All results indicate the importance of long-term observations and the necessity of their continuation.

ACKNOWLEDGMENTS

The authors wish to thank Mr. Weigl for the statistical evaluations, Mr. Steiner for preparing the figures, and Mrs. Kopp and Mrs. Welzmüller for typing the manuscript. The investigation was funded by the Bundesministerium für Forschung und Technologie.

REFERENCES

Attmannspacher, W. and R. Hartmannsgruber, 1975: 6 Jahre (1967 - 72) Ozonsondierungen am Meteorologischen Observatorium Hohenpeißenberg. Berichte des Deutschen Wetterdienstes, 18, 137.

Attmannspacher, W., R. Hartmannsgruber and P. Lang, 1984: Langzeittendenzen des Ozons der Atmosphäre aufgrund der 1967 begonnenen Ozonmeßreihen am Meteorologischen Observatorium Hohenpeißenberg. Meteorologische Rundschau, 37, 193-199.

Attmannspacher, W., R. Hartmannsgruber and P. Lang, 1984: Verbesserung der Grundkenntnisse über die Klimatologie der vertikalen Ozonschicht durch verstärkte Ballonsondierungen. Ges. f. Strahlen- und Umweltforschung, BPT-Bericht, 6/84.

Bojkov, R.D., 1983: Tropospheric ozone, its changes and possible radiative effects. WMO Special Env. Report, No. 16.

Bojkov, R.D., 1988: Ozone changes at the surface and in the free troposphere. Isaksen (ed.), Tropospheric Ozone, 83-96.

Bojkov, R.D., 1988: Ozone variations in the Northern Polar Region. Meteorol. Atm. Phys. 38, 117-130.

Bojkov, R.D. and W. Attmannspacher: Analyses of total and vertical ozone distribution from 20 years observations at Hohenpeissenberg. To be published.

Braun, W., 1984: Nachweis des stratosphärischen Ozontransports in der Winterhemisphäre aufgrund von Beobachtungen und Trajektorienberechnungen. Dissertation ETH, No. 7652.

Feister, U. and W. Warmbt, 1987: Long-term measurements of surface ozone in the German Democratic Republic. J. Atm. Chem., 5, 1-21.

Frederick, J.E. and G.N. Serafino, 1985: The detection of long-term changes in stratospheric ozone: Scientific requirements and current results from satellite-based measurement systems. J. Clim. Appl. Met., 24.

Hartmannsgruber, R., W. Attmannspacher and H. Claude, 1985: Opposite behaviour of the ozone amount in the troposphere and lower stratosphere during the last years. Proc. Quad. Ozone Symp., 770-774.

Köhler, U., K. Wege, R. Hartmannsgruber and H. Claude, 1988: Vergleich und Bewertung von verschiedenen Geräten zur Messung des atmosphärischen Ozons zur Absicherung von Trendaussagen. Gesellschaft für Strahlen- und Umweltforschung, BPT-Bericht, 1/88.

Leovy, C.B., 1985: Transport of ozone in the middle stratosphere: Evidence for planetary wave breaking. J. Atmos. Sci., 42, 3.

Liu, S.C., M. Trainer, F.C. Fensenfeld, D.D. Parrish, E.J. Williams, D.W. Fahey, G. Hubber, and P.C. Murphy, 1987: Ozone production in the rural troposphere and the implications for regional and global ozone distributions. JGR, 92, 4191-4207.

Logan, J.A., 1985: Tropospheric ozone: Seasonal behaviour, trends and anthropogenic influences. JGR, 90 D6, 10463-10482.

Schönwiese, C.D., 1985: Praktische Statistik für Meteorologen und Geowissenschaftler. Available from Gebr. Bornträger Berlin-Stuttgart.

Tiao, G.C., G.C. Reinsel, J.H. Pedrick, G.M. Allenby, C.L. Mateer, A.J. Miller, J.J. De Luisi, 1986: A statistical trend analysis of ozonesonde data. JGR, 91 D12.

Wege, K., 1988: Ozonbeobachtungen am Meteorologischen Observatorium Hohenpeißenberg in der Troposphäre. To be published, Ges. f. Strahlen- und Umweltforschung, BPT-Bericht.

THE INDO-SOVIET COLLABORATIVE EXPERIMENT AT THUMBA TO STUDY DIURNAL VARIATIONS IN THE OZONE VERTICAL DISTRIBUTION OVER THE TROPICS

A. Jayaraman, S. Lal, M. Lal and, B.H. Subbaraya
Physical Research Laboratory
Ahmedabad, India

S.C. Garg, Thomas John, and K.S. Zalpuri
National Physical Laboratory
New Delhi, India

N. Seshadri and C.R. Sreedharan
India Meteorological Department

R. Vijaykumar
Indian Institute of Tropical Meteorology
Pune, India

V.M. Ignatov, G.A. Kokin, S.P. Perov, O.V. Shtrikov, S.V. Tishin, and A.F. Chizhov
Central Aerological Observatory
State Committee for Hydrometeorology and Control of Natural Environment, USSR

1. INTRODUCTION

A rocket programme to study the vertical distribution of ozone at stratospheric and mesospheric altitudes over the tropics was initiated in India in the late seventies. Optical ozonesondes capable of measuring absolute concentrations in the altitude region of 20 km to about 70 km by measuring the attenuation of UV radiation in the 250 nm, 280 nm and 310 nm wavelength bands both during daytime using sun as the source as well as during night time using moon as the source were developed (Subbaraya et al. 1984) and several measurements were made from the Thumba Rocket Launching Station (8.5°N, 76.9°E) using different types of rockets. Even though these measurements were made at different times of the day and in different seasons of the year, it has been possible to construct a mean vertical distribution of ozone for the tropical site, Thumba from the data of nineteen rocket flights (Subbaraya and Jayaraman, 1987). An Indo-Soviet Ozonesonde Intercomparison experiment conducted at Thumba during March 1983 enabled for the first time the collection of a significantly large amount of data on the ozone vertical distribution within a short period of one week. The results are published in Subbaraya et al. (1987). Measurements were made with different types of rocket sensors for intercomparison of sensors and validation of the data and under daytime as well as night time conditions. The different sensors used for daytime measurements could be intercompared and the biases between the different sensors could be estimated. It was found (Acharya et al. 1984) that the Soviet and the PRL optical ozonesondes agree within the stated random errors for these sensors in the altitude region of 25 to 50 km. In this altitude range the absolute values of the difference between the ozone concentrations as measured by the two sensors varried from 0% to 24%. But at altitudes below 25 km there are significant differences from sensor to sensor and from measurement to measurement and no systematic bias could be detected. The observed differences are only partly instrumental. Some of it is believed to be genuine. But it was not possible to separate the two effects from the measurements. At altitudes above 50 km the observed deviations could be partly attributed to the time lapse between the rocket launchings since the photo-chemical time constants at these altitudes is small and also to the somewhat larger measurement errors at these altitudes.

2. DAY TO NIGHT CHANGES IN THE OZONE VERTICAL DISTRIBUTION

Comparison of the daytime measurements with the night time measurements showed a systematic day to night increase in ozone concentration at all altitudes above about 30 km, from all the three sets of rocket measurements, namely 23 March, 28/29 March and 31 March, 1983. The observed changes are significantly above random errors at altitudes above 40 km. Figure 1 illustrates the results. The daytime data is obtained from the mean of several measurements made on that day using different optical ozonesondes. But the night time measurements are single measurements and are made by a different type of

ISBN 0-937194-15-8

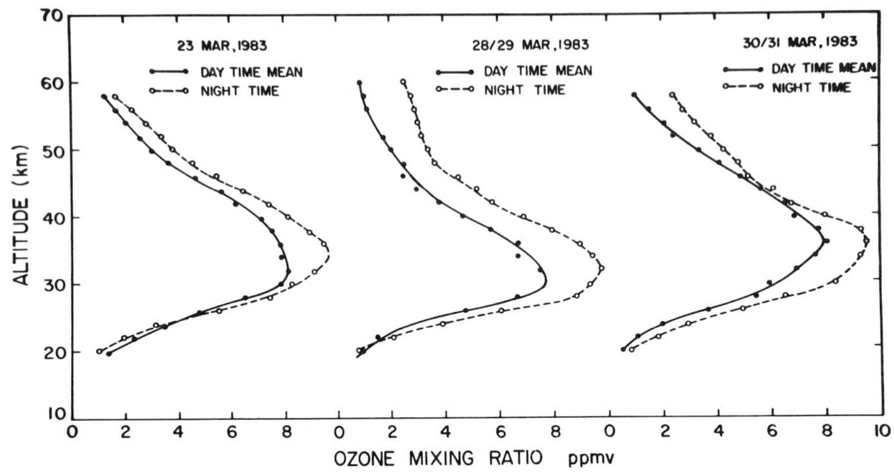

Fig. 1. Day to night changes in ozone concentrations at stratospheric and mesospheric altitudes obtained from the March 1983 Ozonesonde Intercomparison experiment at Thumba.

TABLE 1. NIGHT TIME INCREASE IN OZONE NUMBER DENSITIES AT DIFFERENT ALTITUDES OBTAINED FROM THE MARCH 1983 OZONESONDE INTERCOMPARISON EXPERIMENT AT THUMBA COMPARED WITH EARLIER EXPERIMENTAL RESULTS AND PHOTO-CHEMICAL MODEL CALCULATIONS.

Altitude km.	March 1983 Thumba Expt. Mean of the 3 series.	1980 Day-night series at Thumba	Herman (1979)	Pallister and Tuck (1983)	Vaughan (1984)
58-60	60%	30%	-	-	50%
54-55	54%	15%	24%	-	20%
48-50	42%	((10%	12%	6%	9%
45-46	40%	(	3%	2%	-
40	16%	-	-5%	-4%	-

sensor, the chemiluminescent ozonesonde, one type of sensor whose measurement uncertainties could not be validated by an intercomparison. These results are in qualitative agreement with the results of earlier attempts to study short term variations, such as the day to night, diurnal and solar eclipse induced variations in ozone concentrations in the stratosphere and mesosphere (Herman 1979, Lean 1982, Lal and Subbaraya 1982, Pallister and Tuck 1983, Subbaraya et al. 1984, Vaughan 1984, Herman and McQuillon, 1985). But the observed day to night increases with an average value of 16% at 40 km increasing to a value of 60% in the altitude range of 58 to 60 km are consistently larger than the theoretically predicted values. In fact at 40 km altitude the models predict a slight decrease of ozone from day to night. Table 1 gives a comparison of the March 1983 data with the results of earlier studies. These discrepancies can be attributed to the possible role of eddy diffusion coefficient which can be quite variable. However, since the night time observations were made with the chemiluminescent ozonesonde while the daytime measurements were made with optical ozonesondes and the data from the chemiluminescent instrument needs a calibration, while the optical ozonesondes can in principle yield absolute values of ozone concentrations, and the night time measurements could not be validated by comparison with the data from other sensors, the possibility of a systematic bias between the night time and the daytime sensors can not be ruled out and the results on the day to night increase observed in March 1983 need to be confirmed.

3. THE DECEMBER 1987 OZONE CAMPAIGN AT THUMBA

Short term variations, such as day to night variations are very convenient for verifying the photochemical theories and to study the relative importance of the chemical loss processes and the eddy diffusion coefficient in determining the steady state ozone concentrations. Further, the ozonosphere in the tropics is relatively more stable than in the middle and higher latitudes. Hence a second Indo-Soviet collaborative rocket campaign was undertaken at Thumba during December 1987 with the objectives of

(1) intercomparison of different types of rocket ozone sensors under daytime as well as night time conditions and
(2) to study the diurnal including day to night, variations in ozone concentrations at different altitudes in the stratosphere and mesosphere over the tropics.

Twentytwo Soviet M-100B rockets, nineteen of them equipped with ozonesondes and three equipped with meteorological payloads were launched during the first week of December 1987 in three salvos, a test series on 3 December and the main day-night campaign in two series on 5/6 December and 7/8 December respectively. Twelve of the rocket ozonesondes including five optical sensors for daytime measurements, three chemiluminescent heterophase sensors suitable for night time measurements and four gas-phase chemiluminescent sensors for daytime as well as night time measurements were provided by the Central Aerological Observatory (CAO) of USSR. Seven optical rocket ozonesondes including two daytime and three night time sensors from the Physical Research Laboratory, Ahmedabad and two daytime sensors from National Physical Laboratory, New Delhi were provided by India. The India Meteorological Department conducted six balloon ozonesonde flights from Trivandrum specially for this campaign. The rocket and balloon ozonesonde ascents were supported by ground based observations with the Dobson spectrophotometer, Filter photometer and Surface ozone measurements. Seven out of the twelve Soviet ozonesonde flights and six out of the seven Indian rocket ozonesonde flights were successful and have given good data which is still under analysis. All the six balloon ozonesonde ascents were successful and the data quality is quite good. Due to cloudy weather, Umkehr observations could not be made with the nearby Dobson spectrophotometer at Kodaikanal during this period. When the analysis is completed, nineteen vertical profiles for the ozone vertical distribution over Thumba will be available. These can be used for obtaining a mean ozone vertical distribution over the tropics for the December period and also for study of the diurnal variations. Since the rocket launchings were arranged such that in each group of 2-3 launches the different types of sensors can be intercompared and a statistically significant ozone vertical distribution free from any instrument bias can be obtained, the mean vertical distribution for different parts of the diurnal cycle such as morning, afternoon, evening and night time can be used for a reliable estimation of the diurnal variations of ozone over the tropics. The detailed results of the experiment will be published during 1989.

4. ACKNOWLEDGEMENTS

The organization of the December 1987 Indo-Soviet Collaborative Ozone Campaign at Thumba was possible due to patronage of a number of senior scientists from the two countries. The successful conduct of the experiment was due to the enthusiastic cooperation of the concerned scientific and technical staff at the Thumba Rocket Launching Station, special mention to be made of Messrs A.C. Bahl of RSR and N.P.R. Rao of TERLS, P. Rajaratnam of IMAP-C Office, Bangalore and Mr.K.T. Joseph of the Trivandrum Meteorological Office.

REFERENCES

Subbaraya, B.H. and A. Jayaraman: The vertical distribution of ozone in the equatorial zone. Adv. Space Res. 7, No.9, 119-122, 1987.

Subbaraya, B.H., K.S. Appu, K.P. Chatterjee, A.F. Chizhov, V.D. Grinchinkov, A. Jayaraman, G.A. Kokin, V.A. Kononkov, I.S. Moshnikov, V. Narayanan, S.P. Perov, O.V. Shtrikov, Y.V. Somayajulu, C.R. Sreedharan and K.S. Zalpuri: Scientific results from the Indo-Soviet Ozonesonde Intercomparison experiment at Thumba. Proc. Ind. Academy of Sciences. Earth and Planetary Sciences, 96, 25-40, 1987.

Acharya, Y.B., S.K. Banerjee, A. Jayaraman, B.H. Subbaraya, K.S. Appu, V. Narayanan, K. Chatterjee, C.R. Sreedharan, V.D. Grinchinko, G.A. Kokin, V.A. Kononkov, I.S. Moshnikov, S.P. Perov, A.F. Chizhov, O.V. Shtrikov, K.S. Zalpuri, S.C. Garg, P. Subramaniam, Y.V. Somayajulu, Bh.V. Ramanamurthy, A.M. Selvam, P. Sikka and R. Vijaykumar: The Ozonesonde Intercomparison experiment at Thumba. Advances in Space Research, 4, 59-68, 1984.

Herman, J.R.: Response of the stratospheric constituents to a solar eclipse, sunrise and sunset. Jour. Geophys. Res., 84, 3307-3310, 1979.

Lean, J.L.: Observation of the diurnal variation of atmospheric ozone. Jour. Geophys. Res., 87, 4973-4980, 1982.

Lal, S., B.H. Subbaraya: Solar eclipse induced variations in mesospheric ozone concentrations. Adv. Space Res. 2, No.5, 205-208, 1982.

Pallister, R.C. and A.F. Tuck: Diurnal variation of ozone in the upper stratosphere as a test of photo-chemical theory. Quart. Jour. Royal Met. Soc., 109, 271-284, 1983.

Subbaraya, B.H., Shyam Lal and A. Jayaraman: Rocket measurements of the vertical structure of the ozone field in the tropics. Atmospheric Ozone, Proc. Quadr. Ozone Symposium, Halkidiki, Greece, Sept. 1984. Edited by C.S. Zerefos and A. Ghazi, D. Reidel Pub. Co., 295-299, 1984.

Vaughan, G.: Mesospheric Ozone - Theory and Observations. Quart. Jour. Royal Met. Soc., 110, 239-260, 1984.

Herman, J.R. and McQuillon: Atmospheric chlorine and stratospheric ozone - Nonlinearities and trend detection. Jour.Geophys. Res., 90, 5721-5732, 1985.

PECULIARITIES OF GLOBAL DYNAMICS OF TOTAL OZONE IN ANNUAL COURSE FROM SURFACE MEASUREMENTS

A.N. Grusdev and I.I. Mokhov
Institute of Atmospheric Physics
USSR Academy of Sciences
Moscow, 109017, USSR

ABSTRACT

Amplitude - phase characteristics method is used for an analysis of ground-based measurements of total ozone. Peculiarities of the ozone annual course in different regions of Northern and Southern hemispheres are noted. Additional information on total ozone dynamics is obtained compared with a standard harmonic analysis.

1. INTRODUCTION

There have been many climatological studies of global total ozone (see e.g. Dutsch,1974; London and Oltmans,1978/79; Hilsenrath and Schlesinger,1981; Tolson, 1981; Wilcox et al.,1977; Bowman and Krueger,1985) some of which (Wilcox et al.,1977; Bowman and Krueger,1985) have dealt with a harmonic analysis of an annual course of the global field of total ozone (TO). In the present paper the analysis of an annual course peculiarities of TO is done using the method of the amplitude-phase characteristics (Mokhov,1985; Gruzdev and Mokhov,1988). TO data for the period 1973 - 1985 at 133 Northern (NH) and Southern (SH) Hemisphere stations were obtained from the World Data Center for Ozone, Toronto and from the Main Geophysical Observatory, Leningrad, USSR.

Latitude-longitude peculiarities of TO evolution in annual course were determined: 1) by phase characteristics - by the moments ($t_{\uparrow}$ and $t_{\downarrow}$) of reaching local annual mean values with positive (0-phase, or phase of increase) and negative (π - phase, or phase of decrease) time derivatives and also of achieving local maximum and minimum of TO; 2) by amplitude characteristics - by movement of the boundaries of regions with TO changes by a certain value ΔQ in relation to nonuniform distributions of certain months.

2. PHASE CHARACTERISTICS

The region of essential differences of TO evolution in phases of increase and decrease, of maximum and minimum with estimation of degree of TO annual course asymmetry are noted. So in the equatorial part of the Indian ocean the interval of TO increase $\Delta = t_{max} - t_{min}$ in annual course between minimum and maximum value is close to 8 months. Over India the interval is almost half as long (and consequenty the TO decrease interval is shorter).

Figure 1 presents isochrones (integers correspond to the mid-month) of 0 - phase (a) and π - phase (b). The boundaries of π - phase and TO minimum regimes attainment in annual course over North America are extended on the whole from upper to low latitudes. For the Eurasia the regions with the earliest attainment of the appropriate annual course phases over the North-East Asia and Middle Asia are noted. Peculiarities of these regions are observed also in latitude-longitude evolution of 0 -phase and maximum phase regimes. In the evolution of maximum phase regime the feature of the Europe region is also noted. Over the North America the earlier attainment of maximum phase and 0 -phase is found in the vicinity of the 50s-60s latitudes. Thus 0 -phase and π -phase regimes of TO annual course differ considerably over the North America, although this difference cannot be discovered by a standard harmonic analysis (Wilcox et al., 1977; Bowman and Krueger,1985).

It can also be noted that the attainment of different annual course phases over the Atlantic Ocean is later than over the continents in nontropical NH latitudes. In the Australian SH sector the shift of 0-phase and maximum phase from the region south of Australia to the lower latitudes is noted. The boundaries of -phase and minimum phase regimes are extended from Middle Asia feature in South - East direction to Australia through India. Moreover the feature to the south of India in the Indian Ocean is found. The boundaries of maximum phase and 0 -phase regimes extended from Middle Asia reach 10 S.

ISBN 0-937194-15-8

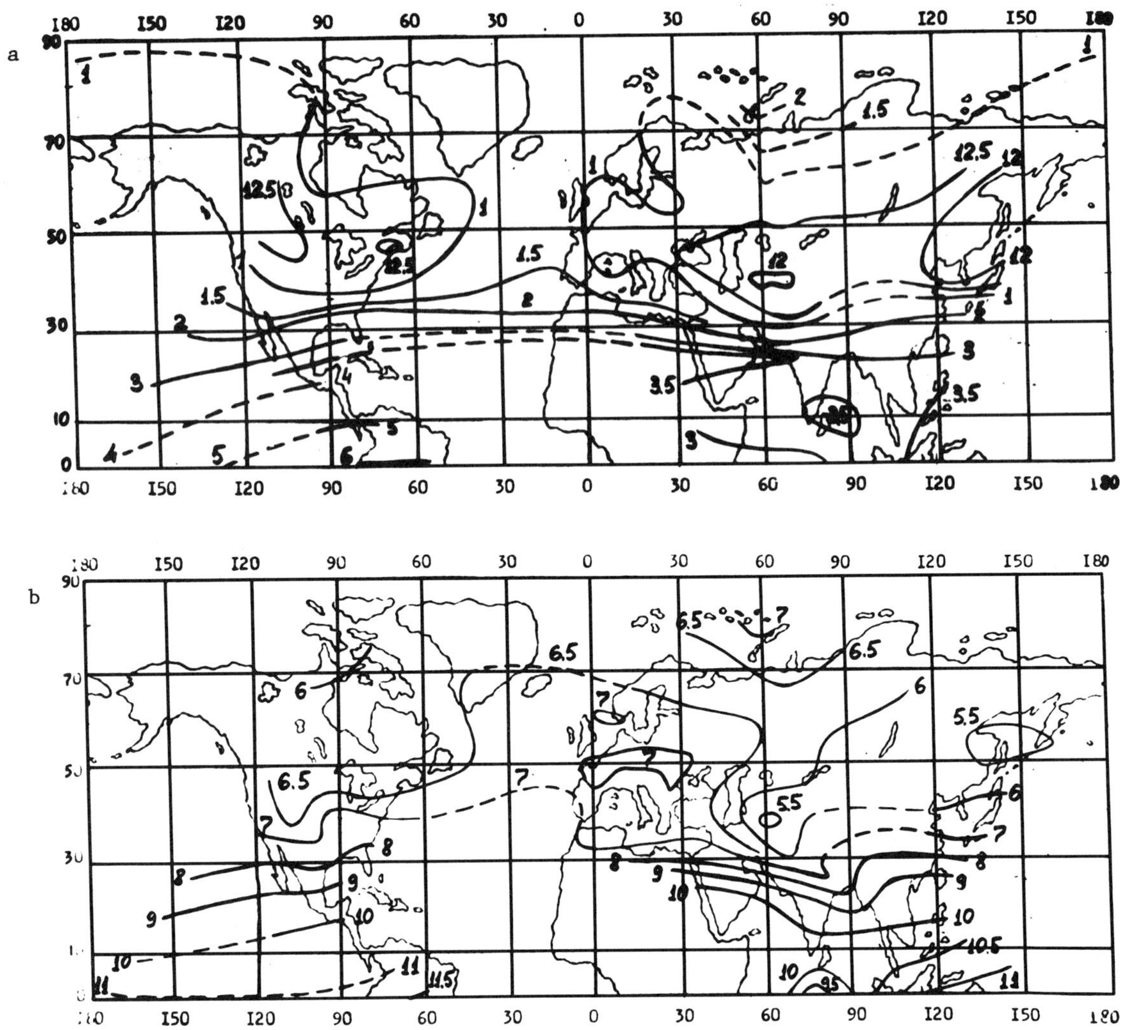

Fig. 1. Isochrones of 0 – phase (a) and π – phase (b) of total ozone annual course in the Northerh Hemisphere (time in months).

3. AMPLITUDE CHARACTERISTICS

Using the amplitude characteristics of TO annual course dynamics the shift of boundaries of regions with TO decrease and increase by ΔQ (20 D.U.and 50 D.U.) against March and September regimes from upper to tropical latitudes is noted (Fig.2). In NH TO inceases to mid – October by ΔQ =20 D.U. against September regime on Fig. 2a over considerable part of the Far East and Okhotsk Sea and to mid-November TO increases over the largest part of Canada. TO increases by ΔQ =50 D.U. in relation to the September regime earlier (in November) over the Okhotsk Sea, to the north of Skandinavia and Kola Peninsular (over the Barens Sea) and by December over Canada. An appropriate analysis of TO annual course evolution against the March regime on Fig.2b has shown that the decrease of TO in addition to spreading from polar latitudes (earlier over the North America) was rather rapid (by ΔQ =20 D.U. by mid-April) over Middle Asia and the Okhotsk Sea. In general, amplitude characteristics demonstrate the features similar to those observed when using phase characteristics (a special TO evolution for regions of Central America and Philippines in NH tropics is noted).

REFERENCES

Bowman, K.P., and A.J. Krueger , 1985: A global climatology of total ozone from the Nimbus 7 total ozone, mapping spectrometer. J. Geophys. Res., 90, 7967-7976.

Dutsch,H.U.,1974: The ozone distribution in the atmosphere. Can. J. Chem., 52, 1491-1504.

Gruzdev, A.N., and I.I.Mokhov, 1988: Diagnostics of stratospheric ozone dynamics in annual course from satellite data. Issledovanie Zemli iz Cosmosa, 2, 3-10.

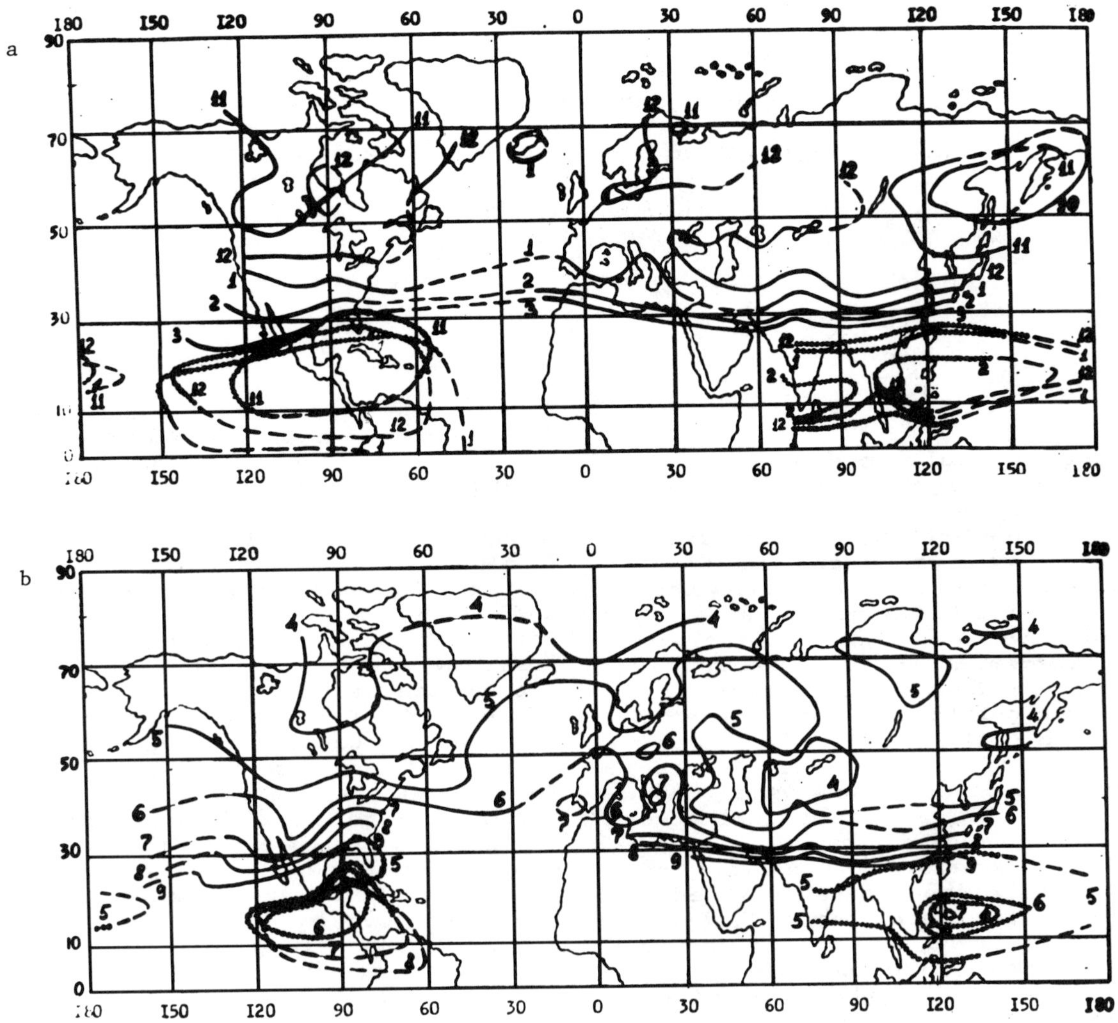

Fig. 2. (a) Isochrone boundaries of regions with the total ozone increase (full curves) and decrease (dotted curves) by 20 D.U. relative to September. Time in months.
(b) The same as in (a) but for March, and full curves denote decrease and dotted curves denote increase.

Hilsenrath,E.,and B.M.Schlesinger, 1981: Total ozone seasonal and inter-annual variations derived from the 7 - year Nimbus 4 BUV data set. J. Geophys. Res., 86, 12086-12096.

London, J., and S.J. Oltmans, 1978/79: The global distribution of longterm total ozone variations during the period 1957-1975.Fure Appl.Geophys., 117,345-354.

Mokhov,I.I.,1985: A method of amplitude-phase characteristics for the analysis of the climate dynamics. Meteorology and Hydrology,5, 80-89.

Tolson, R.H., 1981: Spatial and temporal variations of monthly mean total columnar ozone derived from 7 years of BUV data. J. Geophys. Res., 86, 7312-7330.

Wilcox,R.W.,G.D.Nastrom and A.D.Belmont, 1977: Periodic variations of total ozone and of its vertical distribution. J. Appl. Meteorol., 16, 290-298.

OZONE TRENDS FROM DOBSON STATION AT BELSK, POLAND

Malgorzata Degórska, Janusz Krzyścin, and Bonawentura Rajewska-Wiech
Institute of Geophysics
Polish Academy of Science
Warsaw, Poland

ABSTRACT

The revised total ozone data record and uniformly re-evalueted 25-year Umkehr data set from Belsk, Poland, were analysed for detection of possible trends. An estimate of trend in total ozone for the period 1963-1987 has been obtained as (-0.1 ± 0.4)% per decade. Separate estimates of total ozone trend for winter (December to March) and for summer (June to September) revealed statistically significant negative trends in winter and a positive trend in summer. Statistical significance of the winter trends depends on the analysed data interval, indicating a strong influence of the 1982-1983 ozone deficiency.

Results of statistical analysis of the Umkehr profile ozone data revealed a positive trend in tropospheric layer 1 and a negative trend in stratospheric layer 5 (24-28km). The dependence of ozone content trends in Umkehr layers on the analysed data intervals has been examined. The results indicate that the last decade mainly contributed to the tropospheric ozone increase at Belsk.

1. INTRODUCTION

The recent ozone decrease in Antarctica and reported decrease of total ozone which occured in the Arctic during the northern hemisphere spring season (Bowman,1986) attracted our interest in the long term changes of ozone in the north temperate zone.

The data from long-term operating stations with continuous records are especially important for statistical analyses. The attempt to normalize the Belsk total ozone data record (Dziewulska-Łosiowa et al. , 1983) has resulted in an improved usefulness of these data for trend detection (Dziewulska-Łosiowa, 1983, St. John, 1983, private communication). Therefore, the next step in retrospective improvement of the existing Belsk ozone data was to re-evaluate the Umkehr data set on a uniform basis, i.e., the revised total ozone record for the same period. It was completed in the World Data Centre in Canada due to the kindness of Dr. C. L. Mateer. The re-evaluated Umkehr record from Belsk has been published in the Ozone Data for the World v.28, No. 2 and 3, 1987.

The revised total ozone data record and re-evaluated Umkehr data set from Belsk were analysed for the detection of possible trends in total ozone and in ozone content in different altitude ranges. We have analysed the monthly averages of total ozone and Umkehr data over different periods. According to Reinsel et. al. (1981) concept of the trend estimation, we divide ozone variability into parts to account for ozone trend, periodical pulsations and the noise. Differences between our model and that used by Reinsel et al. (1981) are as follows: long period pulsation are taken into account (2, 4 and 11 years), an annual rate of trend begins since 1963 , an iterative weighted last square procedure is used for calculations (see for the details: Degórska et al. 1988). For to estimate seasonal trend and trend in particular months we exclude the shortest periods (3, 6 and 12 months) in the periodical pulsations term as well as the noise term. In the present study there was no attempt to account for aerosol effect on the Umkehr data, which is especially important for trend estimates in the uppermost Umkehr layers.

This paper presents the results of the statistical analysis, including an investigation of the seasonal variation in the total ozone trend as well as a dependence of trends in total ozone and in its vertical distribution on analysed data intervals. The latter may be connected with increased antropogenic effects on the ozone layer or due to enhanced influence of volcanic aerosol load.

2. TOTAL OZONE VARIABILITY AT BELSK

In Fig. 1. we present deseasonalized total ozone departures from long-term

ISBN 0-937194-15-8

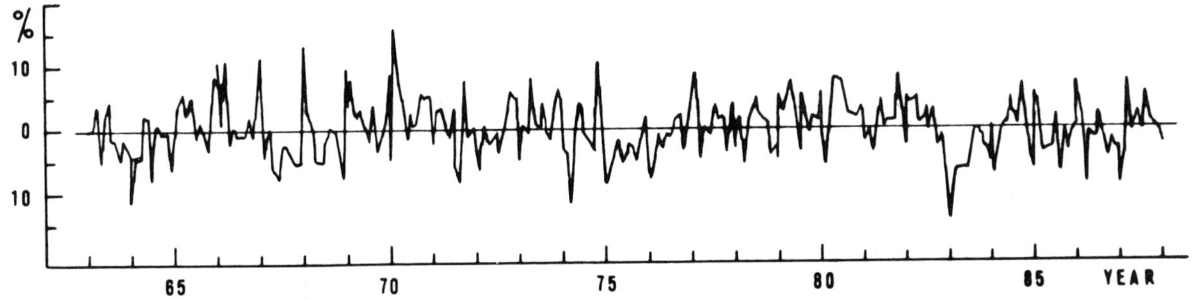

Fig. 1. Total ozone departures from the 1963-1987 averages, at Belsk, Poland

monthly averages in 25-year period at Belsk. A strong ozone decrease observed in the temperate zone in 1982-1983 (Angell et al., 1985, Mantis et al., 1986, Degórska and Rajewska-Więch, 1987, Bojkov, 1987) is clearly demonstrated. Ozone decrease in 1982-1983 was considered to be mainly a conseguence of the quasi-biennial and El Niño/Southern Oscillations circulation effect, and least due to influence of volcanic aerosol load (Bojkov, 1987).

Our estimate of trend in the total ozone over the whole observation period (1963-1987) is (-0.1 ± 0.4)% per decade. However, choosing the shorter trend detection period one may obtain different results. If we perform calculation for the 21 year period 1963-1983, we find statistically significant negative trend in total ozone equal to (-0.8 ± 0.5)% per decade. This is in agreement with the earlier trend estimates (Oehlert, 1986).

We have computed trends in the winter (December-March) and the summer (June-September) total ozone data for two periods, 1963-1987, 1963-1984 (Table 1).

TABLE 1. TOTAL OZONE TREND AND SEASONAL TRENDS (IN % PER DECADE), STANDARD ERRORS ARE SHOWN IN PARENTHESIS

Period	Winter	Summer	Year
1963-87	-1.5(1.0)	1.2(0.6)	-0.1(0.4)
1963-84	-3.3(1.1)	1.6(0.7)	-0.5(0.5)

The most notable features of the table are as follows:

- for the period 1963-1984, statistically significant negative trends in winter and positive in summer are found
- negative trend in winter is less distinct for the period 1963-1987
- positive trend in the summer data is found to be statistically significant at two sigma level.

The effect of ozone deficiency observed at Belsk in 1982-1983 was especially pronounced in the winter trend estimation. We find that negative trend in total ozone in winter is mainly caused by large negative tendency in December (i.e. (-2.4 ± 1.4)% per decade for 1963-1987 and (-4.5 ± 1.7)% per decade for 1963-1984). The positive summer trend is mainly determined by large increasing tendency in July (i.e. (2.1 ± 0.9)% per decade for 1963-1987 and (3.1 ± 1.2)% per decade for 1963-1984).

3. VARIABILITY OF THE OZONE VERTICAL PROFILES

The present study uses 1395 Umkehr profiles of those 1592 performed in the 25-year period (1963-1987). Cases in which the residual exceedes 0.70 are not included in the statistical analysis. The Umkehr data used here are monthly averages of re-evalueted daily recordings. In Figures 2 and 3 we present deseasonalized monthly mean series in Umkehr layer 1 (troposphere) and layers 3-5 (stratosphere). In layer 1 a distinct increase of ozone appears during the last decade. This tropospheric ozone behaviour is in agreement with the previously reported data (see for example Bojkov and Reinsel, 1984) and is confirmed by our trend estimates. Large ozone defficiences are observed at Belsk in layers 3 and 4 in 1982-1983 (Degórska and Rajewska-Więch ,1987) and less pronounced ones in 1985. In layer 5 the negative deviations in 1985 seem to be more significant than those of 1982-1983 (see also Bojkov,1987).

Calculated trends in the Umkehr data are presented in Table 2. The analysis revealed statistically significant positive trend (at the 2σ level) in the troposheric layer 1 and a negative trend in layer 5 (24-28 km). Taking into account the 1σ level of confidence we can see that positive trend has occured in layer 2 and negative one in layer 4. If we estimate trends for the period 1963-1981 (the same as analyzed by Bojkov and Reinsel, 1984) using re-evaluated data, we obtain trend in Umkehr layer 1 equal to (2.5 ± 2.6)% per decade, but for the period 1975-1987 we find trend (18.1 ± 5.4)% per decade. It is clear that

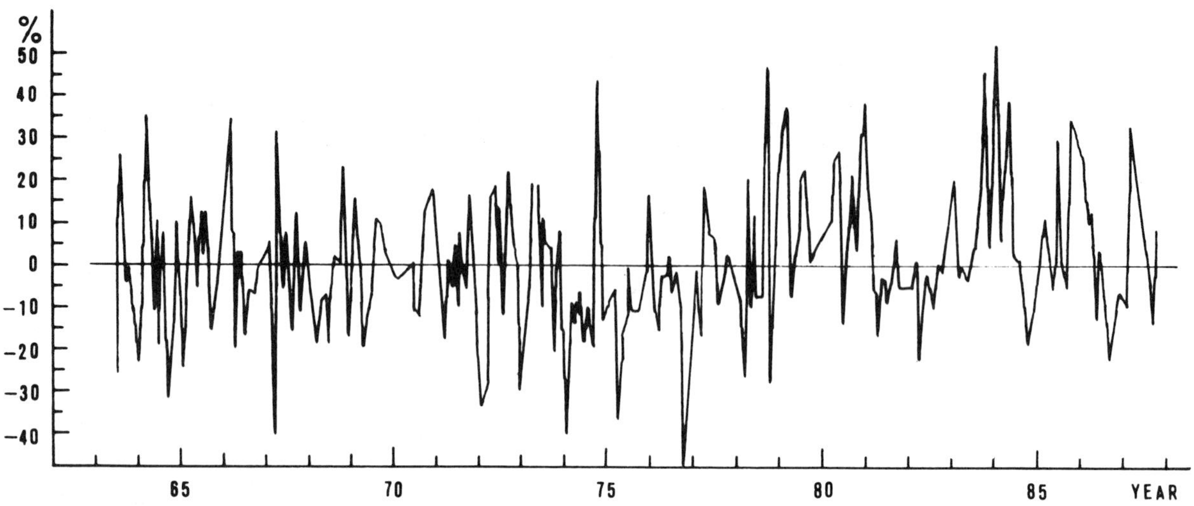

Fig. 2 Time series of monthly mean ozone partial pressure departures from the 1963-1987 monthly averages in Umkehr layer 1; at Belsk, Poland.

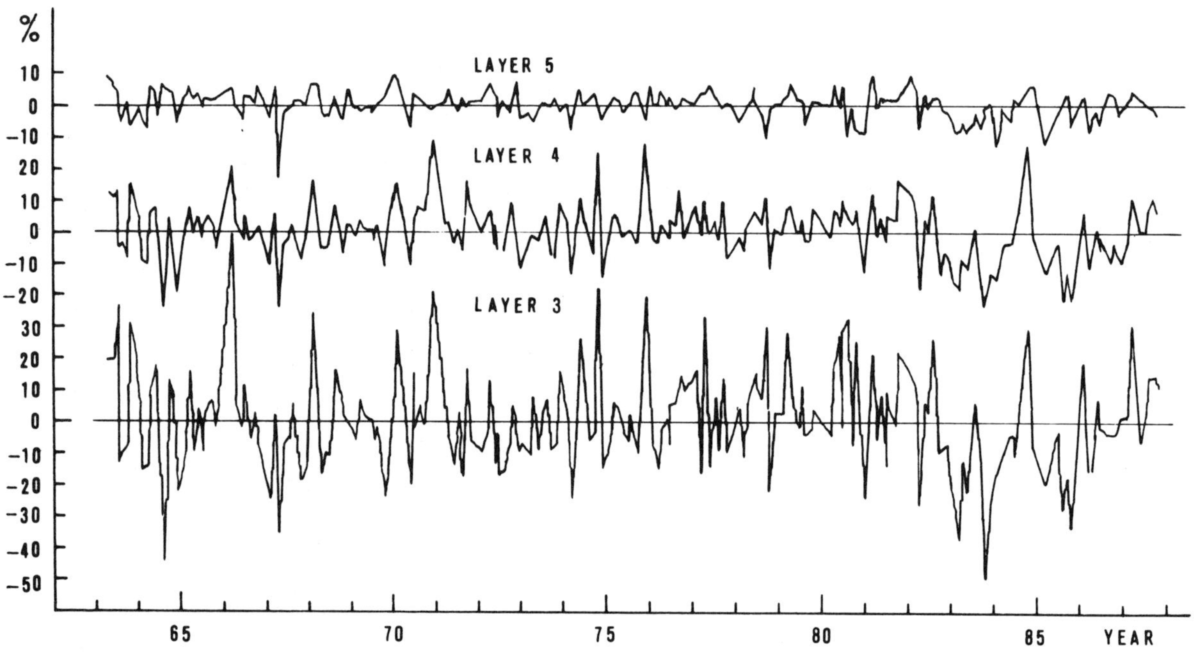

Fig. 3. Same as Fig. 2., but for Umkehr layers 3-5.

last a dozen years has given the main contribution to the tropospheric ozone content increase at Belsk. In layer 5 our results are consistent with previous studies (for example Reinsel et al., 1984) although we used re-evalueted data over the longer period.

We have found no trend in the uppermost Umkehr layers. However, the error of trend estimates in these layers is relatively high. This is mainly due to the observed strong ozone content variability, which is especially pronounced in layer 9 (Degórska et al., 1988). Being aware that the Umkehr data are strongly affected by stratospheric aerosol (in particular in 1982-1984, after the volcanic eruption at El Chichon), we have not performed any detailed analysis of trends in ozone content in the uppermost Umkehr layers.

3. CONCLUSION

An analysis of the 25-year total ozone data set from Belsk has given evidence of seasonal trend, negative for the December to March period and a positive one for the June to September period. It may be of interest to look for mechanisms responsible for the winter ozone defficiency (possibly similar to those acting in Antarctica) and for the summer compensating increase. Short Umkehr data should be more useful for localizing the level at which such seasonal ozone changes appear, as they are more numerous in winter than those of conventional Umkehr and they are less influenced by stratospheric aerosol. Our estimates of the statistically significant positive trend in tropospheric Umkehr layer 1 and a negative one in stratospheric layer 5 (24-28km) seem to be important in connection with their possible climatic impact.

TABLE 2. TREND ESTIMATES (IN PERCENT PER DECADE) IN ALL UMKEHR LAYERS AT BELSK, POLAND, OVER THE PERIOD 1963-1987. ALL UNCERTAINTIES ARE EXPRESSED WITH TWO SIGMA STATISTICAL SIGNIFICANCE

Umkehr Layer								
1	2	3	4	5	6	7	8	9
5.3 ± 3.6	2.9 ± 3.7	-.8 ± 3.2	-1.4 ± 1.6	-1.0 ± 0.8	-.3 ± 0.9	-.7 ± 1.5	-.2 ± 1.6	0.6 ± 2.9

REFERENCES

Angell J.K., J. Korshover, W.G. Planet, 1985: Ground-Based and Satellite evidence for a pronounced total-ozone minimum in early 1983 and responsible atmospheric layers. Mon. Weath. Rev., 113, 4, 641-646.

Bojkov R.D., and G.C. Reinsel, 1984: Trends in tropospheric ozone concentration. Proc. Atmospheric Ozone Symposium 1984., eds. C.S. Zeferos and A. Ghazi, D.Reidel Publ. Co., 775-781.

Bojkov R.D., 1987: The 1983 and 1985 anomalies in ozone distribution in perspective. Mon. Wea. Rev., 115, 10, 2188-2201.

Bowman K. P., 1986: Interannual variability of total ozone during the breakdown of the antartic circumpolar vortex. Geophys. Res. Letters, 13, 12, 1193-96.

Degórska M., and B. Rajewska-Więch, 1987: The ozone minimum observed in the north temperate zone in 1982-1983. Publs. Inst. Geophys. Pol. Acad. Sc., D-28 (211), 93-102.

Degórska M., J. Krzyścin and B. Rajewska-Więch, 1988: Analysis of uniform Umkehr ozone profile data taken at Belsk, Poland, 1963-1987. Publs. Inst. Geophys. Pol. Acad. Sc., D-30(220), 149-162.

Dziewulska-Łosiowa A., 1983: Some remarks on the series of the total ozone measurements. Publs. Inst. Geophys. Pol. Acad. Sc., D-18(169), 101-103.

Dziewulska-Łosiowa A., M. Degórska, and B. Rajewska-Więch, 1983: The normalized total ozone data record, Belsk, 1963-1981. Publs. Inst. Geophys. Pol. Acad. Sc., D-18(169), 23-73.

Mantis H.T., C.S. Zerefos, A. Bais, I. Ziomas, and A. Kelessis, 1986: The Northern Hemisphere ozone minimum in 1982-1983. Arch. Met. Geoph. Biocl., Ser. B 36, 135-145.

Oehlert G.W., 1986: Trends in Dobson total ozone: An update through 1983. J. Geophys. Res., 91, D2, 2675-2679.

Reinsel G., G.C. Tiao, M.N. Wang, R. R. Lewis, and D. Nychka, 1981: Statistical analysis of stratospheric ozone data for the detection of trends. Atmospheric Environment, 15,9, 1569-1577.

Reinsel G., G.C. Tiao, J.J. DeLuisi, C.L. Mateer, A.J. Miller, and J.E. Frederick, 1984: Analysis of upper stratospheric Umkehr ozone profile data for trends and the effects of stratospheric aerosol. J. Geophys. Res., 89, D3, 4833-4840.

TOTAL OZONE AMOUNT IN THE SCANDINAVIAN SECTOR OF THE ARCTIC CHANGES OF OZONE IN CONNECTION WITH THE 30 HPA TEMPERATURE

Soren H.H. Larsen
Department of Physics
University of Oslo
Blindern, 0316 Oslo 3, Norway

ABSTRACT

Measurements of the total amount of ozone during the winter 1987/88 showed large fluctuations. Change in the vertical ozone profile may overestimate the fluctuations when the zenith light observations, with low sun, have to be made. The winter values in Tromsø 1985/86 and 1986/87 are commented. The two winters are very different with ozone regimes wich correspond to significant differences in stratospheric temperatures.

INTRODUCTION

As part of the ground based operational ozone observation network,measurements of the total amount of the atmospheric ozone has been measured in Oslo 60 N, Tromsø 70 N, and Longyearbyen at Spitzbergen, 78 N. Measurements were already made at Tromsø, in 1934 by Dobson and Meetham (1). Bojkov has analysed almost all of these data recently, in a study of the Arctic ozone (2). The result is a manifestation of the large fluctuations wich appear in the ozone data from Tromsø,and it is difficult to establish a contineous, deklining tendency. Bojkov has made an analysis of the data from 12 stations north of the latitude 50 N, and it is interesting to see how the ozone values over Oslo, 60 N,follow the values based on the 12 stations. It seems as if the Oslo measurements are quite representative of the Arctic ozone values. In figure 1, the mean ozone values for March, observed in Tromsø and Oslo are shown in a diagram. In the same diagram the ozone values for October, observed at the Japanese station Syowa, at 69 degrees south, which will be within the polar vortex over Antarctica.

MEASUREMENTS IN TROMSØ THE WINTER 1987/88

As mentioned above,large fluctuations appear in the ozone values from Tromsø, and during the winter season 1987/88 we had some particular events which we will discuss.The ozone amount measured during the winter are given in a diagram shown in figure 2. One has to remember that due to the high latitude, we are dependent on moon measurements and zenith sky measurements. In the diagram one sees the rapid increase in December, from the extraordinary low values measured in November.Both the low values in November, and the large values in December,could partly be affected by the zenith sky method which have been employed to measure total amount of ozone when the sun is under the horizon . If the ozone concentration decreases in middle and upper stratosphere, this will just cause that type of observations , namely: too low ozone value is calculated when deficiency of ozone exists in upper layers, and too high ozone value is calculated if a surplus of ozone at high altitude exsists. The umkehr curves change their shape. This has been calculated by A.Dahlback (3).

TIROS- 10 measurements gave a mean value of 300 DU for November, and this is the same as the moon measurements on December 7. Later in the winter, in the last part of February, low ozone values were observed over Tromsø with the Dobson, but followed by an incrase in the values a few days later.

Our ozone values from Tromsø were sent to R.Neuber. He and his colaborators operated a LIDAR in Kiruna in Sweden at 68 N . He gave us vertical ozone profiles which show interesting changes of ozone concentration in different hight ranges. Their results will be used in our calculations of scattered radiation to see how these change in the vertical ozone profile will alter the ozone amount evaluated from zenith sky observations.

OZONE AND TEMPERATURE

During the winter 1985/86 low total ozone values were observed in February and in March in Tromsø. They are shown in the first diagram in figure 3. The stratospheric temperatures were also low during these months. In this year, 1986, the final stratospheric warming came late .

ISBN 0-937194-15-8

This is indicated in the diagram (fig. 3) with the plotted 30 mb. temperatures. The temperatures are generously provided by Barbara Nujokat (private Communication). The next winter, 1986/87, was different. The final stratospheric warming seems to start in the last part of January. Unfortunately the Dobson instrument was out of order in the most important period, so we have plotted TIROS-9 satellite ozone data from single data point retrieval, carried out at the Tromsø Satellite Station by Jan-Petter Pedersen (4).But the reliability of these TIROS data is limited.
It is seen from the plotted data (second diagram in fig. 3) that the ozone values in February and in March are considerable higher in 1987 than in 1986, in Tromsø . The ozone values for Oslo follow just the same pattern during the initial part of the final stratospheric warming. Such close correspondance between the ozone and stratospheric temperature anomalies over the nothern polar stations are also discussed by Bojkov (1988).

CONCLUSION
Large fluctuation of the ozone values in the Arctic are demonstrated in our measurements. The combinations of ozone and stratospheric temperature measurements, probably indicate transport of ozone rich air when both ozone and temperature increase. The fluctuations may be overestimated if the observations are made on zenith sky. To get more information about the behavior of the ozone, one needs vertical ozone profiles. Next winter it will be carried out ozone sonde ascents in Sodankylæ, 69 N, by Esko Kyrø,and at Spitzbergen Neuber and colaborators will operate the LIDAR in cooperation with our Dobson instrument measurements.

REFERENCES
1) Dobson, G.M.B., A history, App.Opt.,7, 1968.
2) Bojkov,Rumen D.,Ozone variation in the northern polar region, J. Meteorol. Atm. Phys.,Vol.38,117-130,1988.
3) Dahlback,A.,Scattering in a curved,absorbing atmosphe, Cand.real theses, Dept. of Physics, Univ.Oslo,1982.
4) Pedersen,J.-P., Theory, implementation and applikation of a NOAA/TOVS data processing system. Rep. 5003/26, FORUT, P.O. Box 3063 Guleng 9001 Tromsø, 1987.
5) Chubachi,S., Total ozone by lunar observations at Syowa, Antarctica,paper given at Polar Ozone Workshop, Colorado, to be published in JRL,1988.

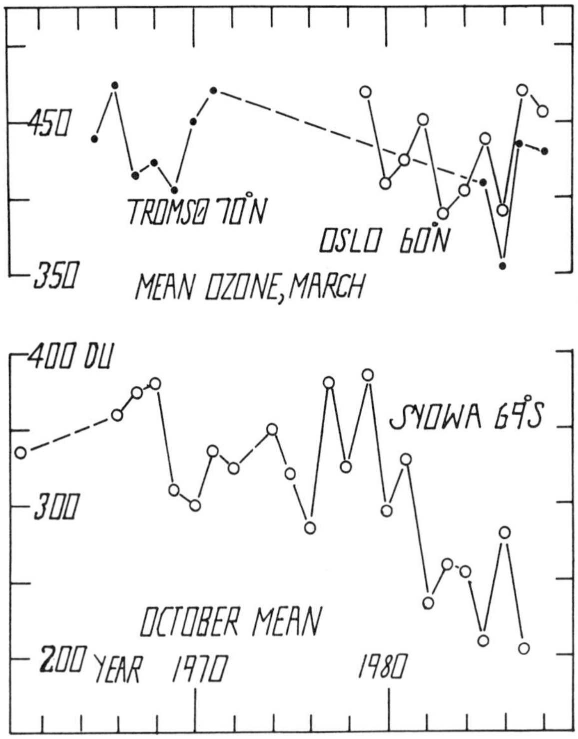

Fig.1: Time series of the mean values of total ozone for March over Tromsø and Oslo,and the mean ozone values for October over Syowa in Antarctica. Ozone amount in DU.

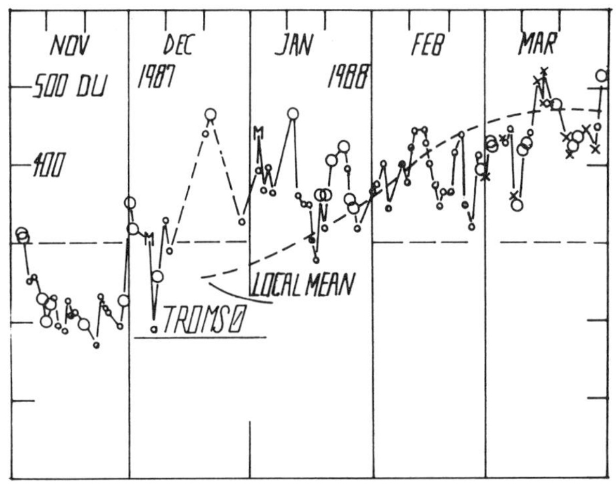

Fig.2: Total amount of ozone observed in Tromsø 70 deg.N, during the winter 1987/88 with the Dobson instrument no 14. X,M: Sun,Moon. o,O: Clear, Cloudy Zenith Sky.

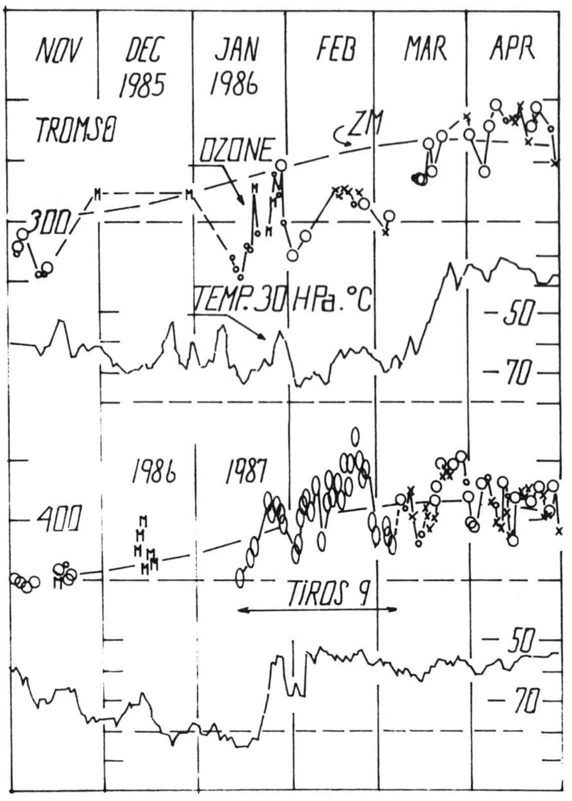

Fig.3: Total amount of ozone observed in Tromsø during the winters 1985/86 and 1986/87. The 30 HPa.temperatures for the same periods are shown in the same diagrams. TIROS measurements have been used since the Dobson instrument was out of order which is indicated. X,M: Sun,Moon. o,O: Clear,Cloudy Zenith Sky. ZM: Zonal Mean for total ozone.

EVALUATION OF VARIOUS FACTORS AFFECTING HOMOGENEITY OF THE TOTAL OZONE TIME-SERIES AT MT. ABU - AHMEDABAD

P.D. Angreji
Physical Research Laboratory and Vedhshala
Ahmedabad, India

ABSTRACT

Mt.Abu/Ahmedabad total Ozone (O_3) data covering 3 decades, 1951-1982, are critically reviewed to delineate various factors causing inhomogenities. Apparent O_3 changes of significant magnitude arising in this time-series due to, for example, revision of O_3 absorption coefficients and "Extra-Terrestrial constants" - N_o, changes in the wavelength settings, variations in opacity of the optical wedges, have been identified and weeded out. The revised, but still preliminary, O_3 data - series shows existence of QBO and larger - period oscillations besides a slow rising from the early 1950's ($\Delta O_3 \approx$ 2% per decade) instead of a three times steeper increase as seen in the earlier published O_3 data.

1. CHRONOLOGY OF THE MT.ABU/AHMEDABAD O_3 TIME-SERIES

Regular daily O_3 measurements on the wavelength pairs CC' (3112/3323 Å and 4536/3323 Å) with Dobson Ozone spectrophotometer No.39 at tropical station Mt.Abu (24°36'N; 72°43'E) in India, began on 25 Oct 1951. This instrument having gelatine-cemented quartz optical-wedges was replaced on 26 May 1957 by instrument No.54 equipped with more stable rhodium coated quartz-wedges cemented by Canadian balsam. It was later transferred twice to Ahmedabad, and was operating alternatively at both places in course of the past four decades (see Table 1). Ahmedabad (23°01'N; 72°39'E) is situated just 176 km south of Mt.Abu, and both these stations are having practically the same climatological characteristics. During the period under review (25 October 1951 - 11 April 1982), the daily O_3 observations and occasional Umkehr observations were taken for a total period of about 21 years from Mt.Abu and 10 years from Ahmedabad. Though the Dobson No.54 was transferred to Ahmedabad in April 1982, it was commissioned from March 1983 only. Except for this major break of about 10 months and another of two months during 1959, the series is almost continuous. Since the distance between these two stations is less than 200 km and are operated by the same agency — Physical Research Laboratory, Ahmedabad — this data - series can be regarded as uninterrupted one.

Chronological and other informations given in Table 1 are self explanatory. The published data upto the end of 1960 are $X_{CC'}$ values (O_3 amount measured with Dobson C and C' pairs and expressed as m atm-cm) and X_{AD} from 1 January 1961 onwards [X_{AD} = O_3 determined from simultaneous measurements on Dobson wavelength pairs A (3055/3254 Å) and D (3176/3398 Å]. These published $X_{CC'}$ and X_{AD} data have been derived, using the following numerical values for effective (decadic) Ozone absorption coefficients (α_λ : units cm^{-1}, base, 10); $(\alpha-\alpha')_C$ = 1.163 due to Ny & Choong (1932) for the data upto 1957, replaced by 0.865 following Vigroux (1953) and recommended for use in the IGY (1957-58) and IGC (1959) (Dobson, 1957). $(\alpha-\alpha')_A - (\alpha-\alpha')_D$ = 1.388 (Dobson, 1957; IOC, 1968) has been used throughout the period 1961-1982.

TABLE 1. MT.ABU/AHMEDABAD O_3 DATA : 1951-1985

Station	Period	Remarks
Mt.Abu	25 Oct.1951- 17 Jan.1960	Dobson Instru.# 54 replaced Instru.# 39 on 26 May 1957.
Ahmedabad	24 Jan.1960- 12 Oct.1969	Reported O_3 values till Dec. 1960 are $X_{CC'}$, and X_{AD} 1961 onwards
Mt.Abu	13 Oct.1969- 11 Apr.1982	Daily O_3 data for 1955-1968 have been published in the Indian J.Met. & Geophy. Vol.6-20.
Ahmedabad	1 Mar.1983 onwards	O_3 and Umkehr data since July 1957 are being published in O_3 Data for the World, Canada

2. VERY LOW O_3 LEVEL OVER INDIAN STATIONS IN 1950'S

A very striking abnormality is noticed upon examinig the published O_3 data for Mt. Abu/Ahmedabad, 1952-1981; it is the very low O_3 level during the 1950's. Same can be seen in case of other Indian Stations as well. Annual means of the published O_3 amounts for Mt.Abu/Ahmedabad have been plotted in Fig.1 alongwith similar O_3 data series for other two Indian stations : New Delhi (28°38'N, 77°13'E; 1953-1980) and Srinagar (34°05'N, 74°50'E; 1957-1960, 1964-1980). In all these time-series of published data $X_{CC'}$ values are shown by crosses (X) and X_{AD} by filled circles (•); and the $X_{CC'}$ values for pre-1957

OZONE IN THE ATMOSPHERE
Rumen D. Bojkov and Peter Fabian (Eds.)

ISBN 0-937194-15-8

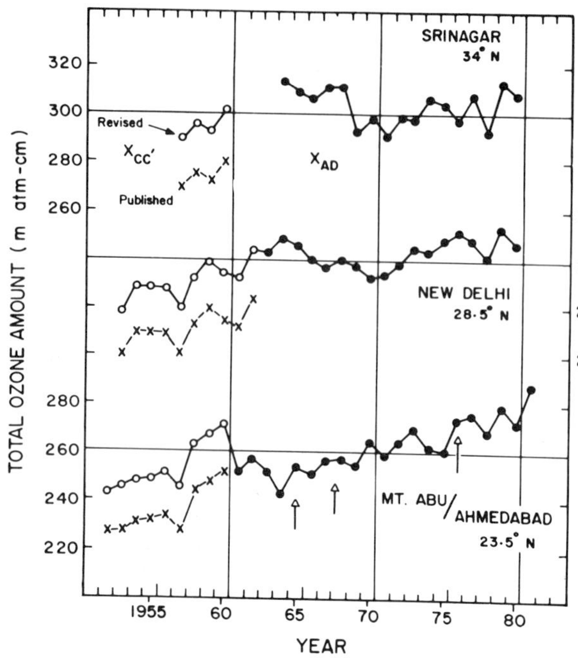

Fig. 1. Annual means of total O_3 amounts observed at the Indian stations during 1952-1981. Revision of the reported $X_{cc'}$ values of 1950's removes sudden jumps in the station O_3 - level with adoption of AD method, at all the stations.

period have been multiplied by 1.33 to take care of the difference in the α_c values of Ny & Choong (1932) and Vigroux (1953) (Dobson, 1957). The station O_3 - levels at all the three stations are conspicuously low during 1950's, increase by about 9-10% in a jump in early 1960's and remain at this higher level during rest of the time. This anomaly stands out very conspicuously when the station O_3 - levels over these Indian stations during 1950's are compared with those during the 1960's and 1970's (Table 2). The same anomaly is found to exist even at Kodaikanal (10° 14'N; 77°28'E)

TABLE 2. O_3 ANOMALY IN THE 1950'S

Station	Period	Mean station O_3-level (m atm-cm)	
Mt.Abu/ Ahmedabad (24°N)	1952-1960	238	
	1961-1981		262
New Delhi (28.5°N)	1953-1962	255	
	1963-1980		280
Srinagar (34°N)	1957-1960	275	
	1964-1980		302
No Obs. during 1961-63			
		$X_{cc'}$	X_{AD}

- another Indian station which was also operating then. Thus this peculiar behaviour, namely extremely low O_3 values seen in 1950's seem to be present throughout India. It may have existed elsewhere also but we don't have such data. If this anomaly of the 1950's is real, it would mean that very low O_3 similar to the values observed in the 1980's were also seen over some stations during the 1950's !

3. O_3 ABSORPTION COEFFICIENTS (α_λ)

It is known that as progressively high precession instruments, more sensitive detectors and larger ranges of low temperature became available for laboratory determinations of the α_λ, previously reported values have been revised from time to time (e.g. Bass and Paur, 1984; Klenk et al., 1984). Since 1957 onwards the "Difference Method"-AD has been adopted at all the O_3 stations (except in India where there was few years' delay) value of the differential O_3 absorption coefficient α_{AD} = 1.388 as determined earlier by Vigroux (1953) and recommended for use during the IGY-IGC period (Dobson, 1957) has remain unaffected at time of issuing a recommendation on the values of α_λ to be adopted for O_3 computation, in 1968 by the International Ozone Commission (IOC) of IAMAP/WMO. The existing X_{AD} time-series hence need no revision. However, opinions favouring an increase in the currently adopted values of α_λ (Komhyr, 1980; Klenk et al; 1984) have been expressed, and this would make result of the ground based O_3 measurements (DOBSON O_3 level) compatible with that of Satellite-borne O_3 measurements (say TOMS); but O_3 revision at all the stations on this count will be taken up in future using, say, Bass and Paur (1984) absorption values. If the repesentative temperature in the stratosphere is taken to be -55°C instead of -44°C (Walshaw et al., 1971) correction to the α_λ values may be necessary on this count also. Based on a new set of α_λ values reported by Vigroux (1967) and results of studies made with Dobson Instrument to determine atmospheric O_3 amount using all the four standard Dobson wavelength pairs A,B,C,D as well as Difference methods such AD,AC,BD,CD etc. (Dobson, 1963; Ramanathan et al., 1965), the IOC recommended use of adjusted α_λ values with effect from January 1968. Partient point to note here is that α_c=0.800 (compatible with α_{AD}=1.388) has been recommended now (IOC, 1968) instead of 0.865, a value which was adopted during 1957-1967. When $X_{cc'}$, reported earlier were increased by 7.5% to accomodate this revision of X_c, the abnormally low O_3 level observed at Indian stations during 1950's is found to vanish (Angreji, 1987).

4. CHANGES IN THE "EXTRA-TERRESTRIAL" CONSTANT : (N_o)

The next question one would like to answer is regarding the O_3 trend in the time-series. There are several possible causes for apparent changes in the station O_3 level which we shall take up one by one. Definition of the so called "Extra-terrestrial" constant $(N_o)_\lambda$ is the ratio of intensity in same wavelength if observed outside the Earth's atmosphere vs. the intensity registered by the instrument. It is determined by plotting the observed intensity ratio $(N)_\lambda$ versus μ (ratio of the observed slant path-length through the ozone layer to the vertical path-length) and then extrapolating it to $\mu = 0$. Hence in ideal situation N_o remains almost constant, the changes in solar spectrum in the observed wavelength band being quite small. But in practice N_o is found to change due to instrumental causes such as change in the spectral response of the detector, the light scattered inside the instrument, etc. Amount and type of the haze, dust and other particulate matter in the lower atmosphere because of differential scattering behaviour, also contribute to changes in $(N_o)_\lambda$ due to non-solar causes. Another very significant contributor to N_o variation is change in opacity gradient of the optical wedges used in the Instrument to equalize intensity of both wavelengths of the pair falling alternately on the detector.

In the case of Mt.Abu-Ahmedabad observations, four significant changes in $(N_o)_\lambda$ have been detected and were duly notified as foot-notes to the published O_3 data in the "Ozone Data for the World" published by WMO's World Ozone Data Center in Canada. Pariticulars regarding the nature and effect of change, the effective date, etc. are given in Table 3. [Times of their occurence have been indicated on Fig.1 by the vertical arrows]. Of the two changes notified in 1965, the second one was re-evaluation. On consulting the station

TABLE 3. CHANGES NOTIFIED IN O_3 DATA FOR THE WORLD

	Change	Effective from	Effect (Apparent increase in O_3-level)
			(m atm-cm)
1.	Revision of N_o values	1 Mar.1965	+ 12
2.	Revision of N_o values	1 Sep.1965	- 5
	(1 + 2 combined)	1 Mar.1965*	+ 7
3.	Revision of N_o & λ -settings	1 Nov.1967	+ 11
4.	Adoption of calibration results	1 Oct.1975**	+ 8

* Revise Effective date as 1 May 1963

** " " " " 9 Oct. 1972

records, it is recommended that the published O_3 be corrected now as follows: Add 7 m atm-cm from 1 May 1963 upto 28 February 1965, and substract 5 m atm-cm from 1 March to 31 August 1965. Similarly the effective date in case of the fourth change may also be revised as 9 October 1972; i.e. 8 m atm-cm be added to the published O_3 data from 9 October 1972 to 18 September 1975.

It may be noted in this connection that the value adopted for L_o (= $\log_{10} I_o/I_o'$ where I_o and I_o' are intensities of the two wavelengths outside the atmosphere) in the routine computations of X_{cc}, was 3.050 during October 1951 to the first half of 1953; 3.070 for the period June 1953 to June 1956; and then 3.080 upto and including February 1957. This progressive change in $(L_o)_c$ is causally related with changing opacity of the optical wedges in use with Dobson No.39.

Besides these N_o variations seem to be due to the instrumental causes, systematic changes in the nature and amount of the haze and dust in the atmosphere seen over the tropics in India are responsible for the seasonal variations in $(N_o)_\lambda$. Variations in case of the wavelength pairs A and B (not shown here) are similar and depict (Fig.2) larger maximum in the winter months and a smaller one in the early summer months. In case of $\lambda\lambda'$ C the amplitudes of these variations are quite small, and only the October minimum is discernable with $\lambda\lambda'$ D. Net effect of these cyclic N_o changes due to the atmospheric conditions, is about 2% and 1% decreases in X_{AD} during December and January respectively.

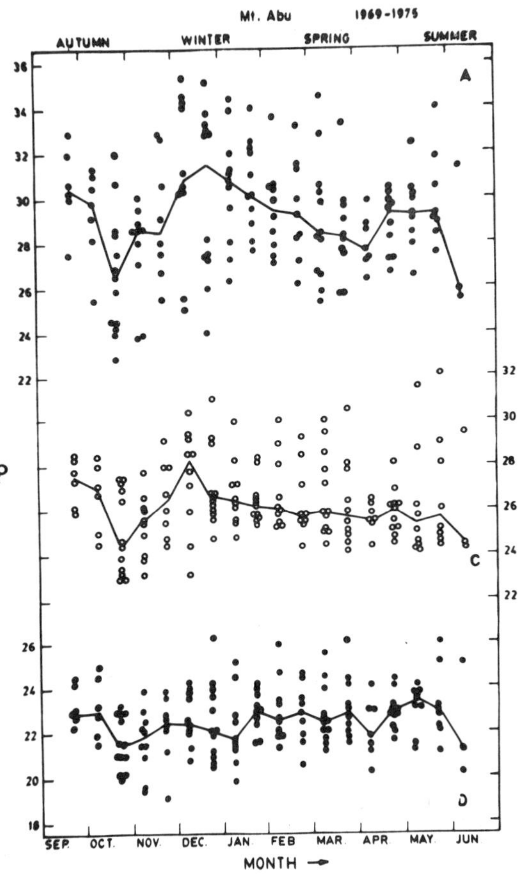

Fig. 2. Seasonal changes in N_o seen at Mt.Abu, 1969-'75 for $\lambda\lambda'$. A, C & D. Each column refers to a fortnight, and median values are shown by continuous curve. Note maxima occuring in the Winter and early Summer months when typical winter time fog and summer haze are seen.

5. DETERIORATION OF THE OPTICAL WEDGES

'Case History' of the optical wedges used in Dobson No.39 and 54 are portrayed in Fig. 3-A and 3-B respectively. The gelatine-divided quartz wedges in Dobson No.39 shows marked changes in the general level as well as presence of localized variations suggestive of phenomenal increase in opacity (seen culminating in 1957 as visible appearance of hair-like growth through the cementing material). Since for the routine observations on wavelength pair C thinner segments of the wedges (R$\leqslant$100°) were required to be used, the situation could be saved; but the C' measurements required for the haze correction and A measurements became totally meaningless during March-May 1957.

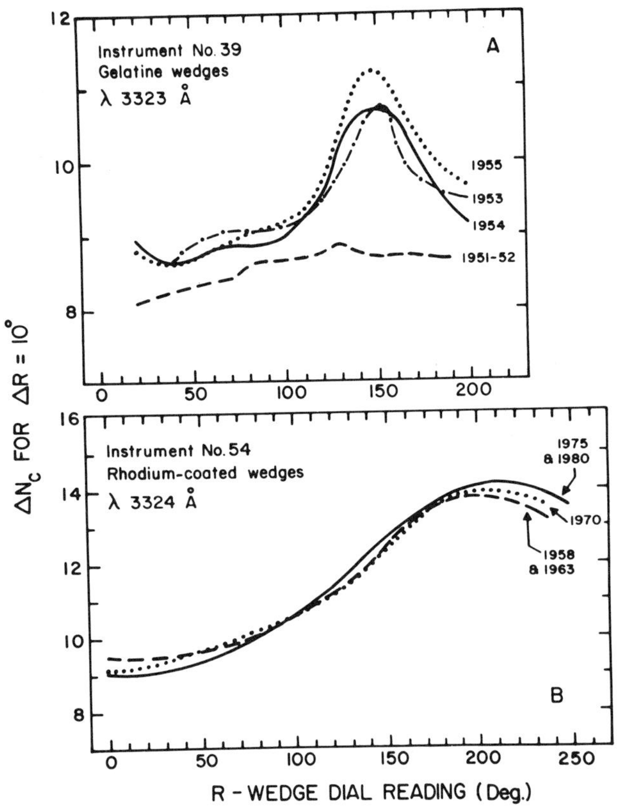

Fig. 3. Changes with time in opacity of the optical wedges of the Dobson Instruments used at Mt.Abu/Ahmedabad. (a) Instrument No.39 (1951-1955); (b) Instrument No.54 (1958-1980).

Another point to be mentioned in the present context is that even in the thinner sections, the change in opacity is of the order of 10-15% between 1951-52 and 1953. This rate of change is smaller than in 1952 but still considerable.

Changes in the rhodium coated optical wedges used in the Dobson No.54, as expected, are much smaller confirming their better quality coating of the Canadian balsam used as cementing agent since 1956 by the manufacturer (Beck Ltd. of England). Still, however, systematic small changes in opposite sense have occured in thinner-end segments (R between 0-100°) and in the thicker-end ones (R between 100-200°) sometime during 1970-75. Considering that the dial ranges used for $\lambda\lambda'$ A and D are R = 100-200° and 40-95° respectively, the resulting errors are additive and do not cancel out. Correction to the published O_3 data on account of this cause is 7 m atm-cm during 1961 to 1968 and 12.5 units during 1969 to September 1975.

6. ERRORS DUE TO THE WAVELENGTH SELECTION & SETTINGS

During the period October 1951 to 1956 the central wavelength of the shorter waveband in case of the C pair was 3112 A. Since January 1957 this nominal wavelength was changed to 3114.5 A. Similarly the value of the shorter wavelength of the A pair which was 3054 A until 1956 end, was changed in January 1957 to 3055 A. Besides these intentional differences in the wavelengths used, labelled as the change due to selection, wavelength settings introducing the wavelength error as much as 2 Å in the C and D pairs and about 1 Å in case of the A pair were found on several occasions. Both types of the wavelengths errors were rectified in this study by using appropriate effective O_3 absorption coefficient obtained using a triangular weighting function (Dobson and Normand, 1962) which resulted in the following corrections:

Period	Correction
1 Jan 1961 - 28 Feb 1963	+ 2 %
22 May 1963 - 31 Oct.1967	3.3
1 Nov 1967 - 31 Oct.1968	1.8
1 Nov 1968 - 22 Oct.1969	+1
23 Oct.1969 - 18 Oct.1975	-1

7. REVISED OZONE TIME-SERIES AT MT.ABU/AHMEDABAD

After all of the above mentioned corrections are applied to the published Ozone data for Mt.Abu/Ahmedabad (1952-1981), the revised time-series plotted in Fig.4 presents a very different picture. Instead of a steep increase ($\Delta O_3 \approx$ 6 percent/decade) seen in the published data what we find now is a slow rise, increase of nearly 2 percent/decade, is in evidence. Now presence of the QBO and the large period variations is discerened very easily. A comparison of the ground based O_3 data with the Satellite-borne O_3 data (TOMS) indicates that the DOBSON O_3 values are 2 to 2.5 percent less than the TOMS values in 1979 and 1980, whereas this difference changes sign, DOBSON O_3 values becoming about 1.3 percent higher than the TOMS values in 1981 (Bojkov, personal communication 1987). Correction due to this discrepancy, however, is not applied in the data shown in Fig.4 and this plot should not be considered as a final but only as a demonstration how important is to introduce all known calibration corrections in the proper places of the ozone record.

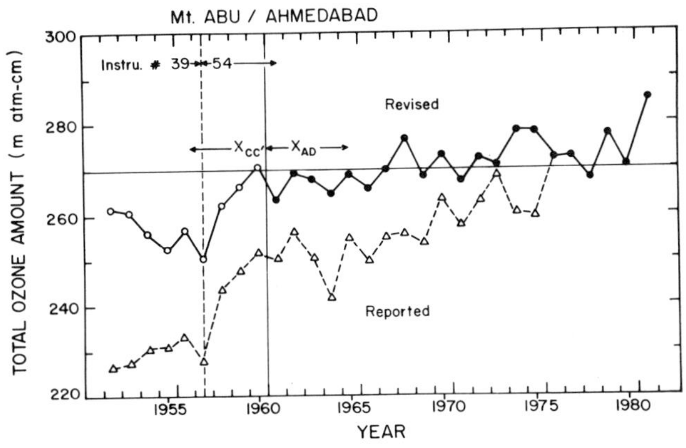

Fig. 4. Revised annual means of the total O_3 amount observed at Mt.Abu/Ahmedabad during 1952-1981. The earlier reported O_3 data upto 1975 have been revised to restore homogenity of this time series.

As a natural consequence of revisions of the total O_3 data series, even the O_3 concentrations at different altitudes will change significantly, because the daily total O_3 value is a parameter involved in evaluation of vertical distribution of O_3 from the umkehr observations.

ACKNOWLEDGEMENTS

It is a pleasant obligation for the author to dedicate this paper to the memory of Prof. K.R.Ramanathan, the initiator and mentor of the ozone measurements in India. The author is much obliged to Dr.Rumen D. Bojkov for inspiring encouragement during the course of this study and for numerous constructive remarks on the earlier manuscript. He is thankful to Prof.B.H.Subbaraya and Dr.D.K. Chakrabarty for their help at various stages of the study. Encouragement and support received from Prof.S.P.Pandya and Prof.R.K. Varma are gratefully acknowledged.

REFERENCES

Angreji, P.D., 1987: Apparent changes in the atmospheric Ozone amounts due to differences in the Ozone absorption coefficients Natl. Space Sci. Symp., Extended Abst., Phys. Res. Lab. Ahmedabad, pp 218-19.

Bass, A.M., and R.J. Paur, 1984: The Ultra-violet cross-sections of Ozone: I. The Measurements, Proc. O_3 Symp., Greece-1984, Eds: Zerefos, C.S., and A.Ghazi, D.Reidel Publ. Co., pp.606-10.

Dobson, G.M.B., 1957: Observer's handbook for the ozone spectrophotometer, Annal. of the IGY, 5, Part III, pp.46-80.

Dobson, G.M.B., 1963: Note on the measurement of ozone in the atmosphere., Quart. J. Roy. Met. Soc., 89, pp.409-11.

Dobson, G.M.B., and C.W.B. Normand, 1962: Determination of constants etc. used in calculation of the amount of ozone from spectrophotometer measurements and an analysis of the accuracy of the results., Annal. of the IGY, 16, Part II, pp.161-91.

I.O.C., IAMAP, 1968: Report of Proc. Publ. IAMAP 14 (IOC, Toronto), p.18.

Klenk, K.F., B. Monosmith, and P.K.Bhartia, 1984: Absorption coefficients of ozone for the backscattered UV Instruments - SBUV, TOMS, and BUV - and for the Dobson instrument., Proc. O_3 Symp. Greece-1984, Eds: Zerefos, C.S., and A.Ghazi, D. Reidel Publ. Co., pp.625-29.

Komhyr, W.D., 1980: Dobson spectrophotometer systematic total ozone measurement error., Geophys. Res. Lett., 7, pp.161-63.

Ny, Tsi-Ze and Shin Piaw Choong., 1932: L' absorption de la lumire par l'ozone entre 3050 et 3400 A (region des bandes de Huggins), Compt Rend Acad. Sci. (Paris), 195, pp.309- .

Ramanathan, K.R., P.D. Angreji., and G.M. Shah, 1965: Ozone absorption coefficients and haze corrections for total ozone measurements with Dobson spectrophotometer, Ind. J. Met. Geophys., 16, pp.671-74.

Vigroux, E., 1953: Contribution a l'etude experimentale de l'absorption de l'ozone. Ann. de Phys, (Paris), 8, p.709- .

Vigroux, E., 1967: Determination des coefficients moyens d'absorption de l'ozone en vue des observations concernat l'ozone atmospherique a l'aide du spectrometere Dobs., Ann. de Phys. (Paris), 2, pp.209-

Walshaw, C.D., G.M.B. Dobson, and B.M.F. McGarry., 1971: Absorption coefficients of ozone for use in the determination of atmospheric ozone, Quart. J.Roy. Met. Soc., 97, p.75-82.

VERTICAL DISTRIBUTION AND VARIATION IN ATMOSPHERIC OZONE OVER LAUDER, NEW ZEALAND

W.A. Matthews
Physics and Engineering Laboratory
Department of Scientific and Industrial Research
Lauder, New Zealand

W. Komhyr, P. Franchois, and S. Oltmans
NOAA/ERL/Air Resources Laboratory
Boulder, Colorado 80303, USA

ABSTRACT

The establishment of the climatology of the vertical ozone distribution over Lauder, New Zealand (45°S) has been attempted by releasing ECC ozone sondes twice weekly through the Southern Hemisphere springs of 1986 and 1987 and at the rate of once per week for the intervening period. Data from this extended series of flights will be presented and discussed.

A feature that is often present in the ozone profiles is a secondary peak just above the tropopause and this is attributed to the influence of the subtropical jet. In the spring of 1987, the ozone concentration appears to be less variable than in the spring of 1986 when on many occasions a "double dip" or "triple peak" phenomenon was observed. Examples of these and other features in the ozone profiles are given.

1. INTRODUCTION

As part of an international effort to study the causes and the ramifications for other latitudes of the "Ozone Hole" phenomenon in Antarctica, an ozone sonde sounding programme was initiated in August 1986 at the Physics and Engineering Laboratory Lauder remote rural site at Lauder, New Zealand, (45°S, 169°E). Small meteorological balloons equipped with an ECC ozone sonde Komhyr (1969) and an hypsometer radiosonde have been launched since that time at a frequency of one sounding per week. An additional sounding per week has been made each austral spring early/summer period.

2. FLIGHT PROCEDURE AND DATA PROCESSING

The ECC ozone sondes were all prepared and preflight calibrated in accordance with the NOAA Technical Memorandum ERL MRL-149 Komhyr (1986). Hydrogen filled 1200 gram rubber balloons were used as the flight vehicles and the average burst height attained was 10 mb. Ozone, air pressure, air temperature and humidity were extracted from the 1680 MHz radiosonde data link at 1 minute intervals from the continuous chart record. A correction to the measured ozone amount was applied to allow for the change in pump efficiency with pressure and also for any change in pump motor speed. The integrated column amount of ozone from each sounding was ratioed with the same day column amount measured by a Dobson spectrophotometer to form a correction factor, and this factor was then applied to all the sonde ozone values. For the 1986 series, the Dobson data was kindly provided by the New Zealand Meteorological Service (S Nichol, private communication) from their Invercargill site, some 170 km SW of Lauder. In January 1987, an automated Dobson was installed at Lauder as one of the seven stations in the "Automated Dobson Network" and since that time the total ozone data has been taken from that instrument.

3. MEASUREMENT UNCERTAINTIES

The ozone measurement accuracy is ±10% in the troposphere and ±5% in the stratosphere up to 10 mb (Hilsenrath et al., 1986). The ozone measurement precision, at the 1-sigma level, is estimated to be ±5% in the troposphere and ±2-3% in the stratosphere up to 10 mb (Komhyr et al., 1985). The agreement between the ozonesonde integrated ozone values and that from the Dobson spectrophotometers was very good with the average correction factor being 1.015 ±0.04 for the 103 flights made in the 1986/87 series. Two of the flights had a correction factor outside the range $0.9 \leq$ corr. factor ≤ 1.25 and these two flights were omitted from the analysis.

ISBN 0-937194-15-8

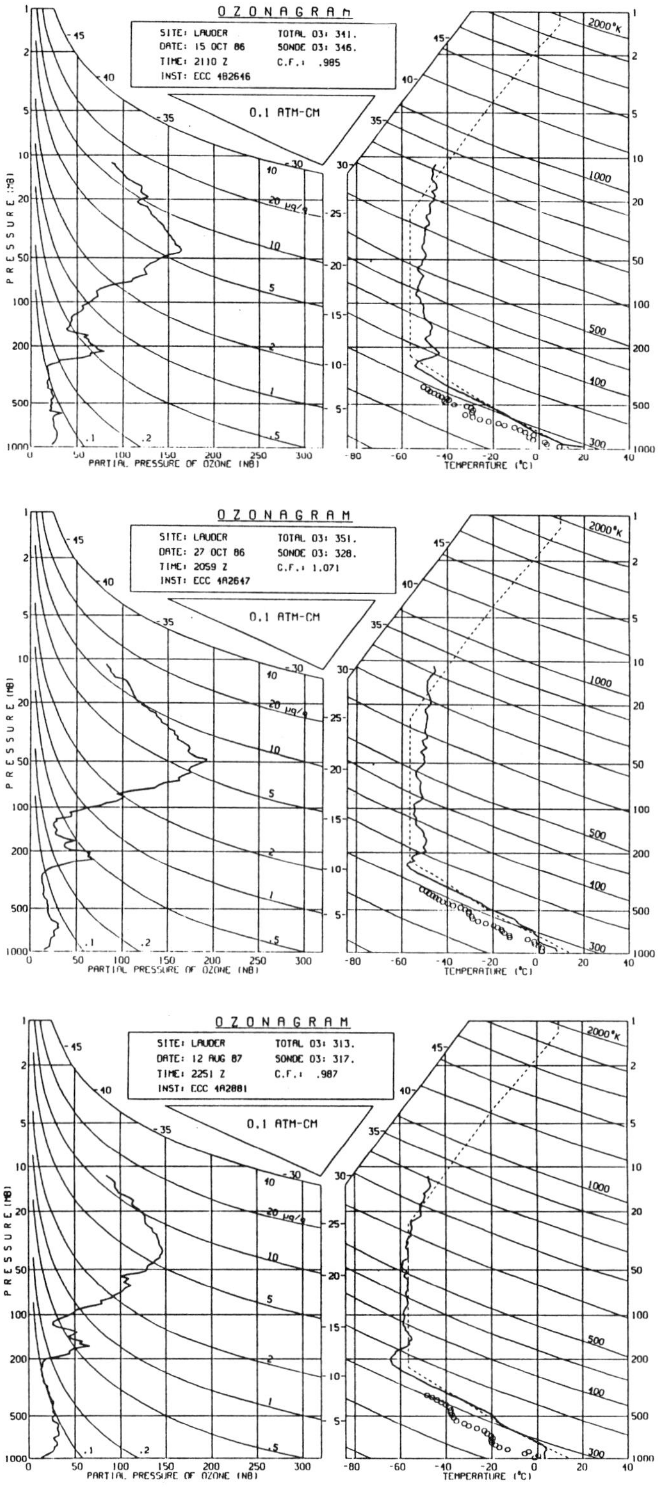

Figure 1. Ozone and temperature profiles slotted as a function of pressure for soundings made over Lauder on October 15, October 27, 1986 and August 12, 1987. The three ozone profiles show the additional peak just above the tropopause.

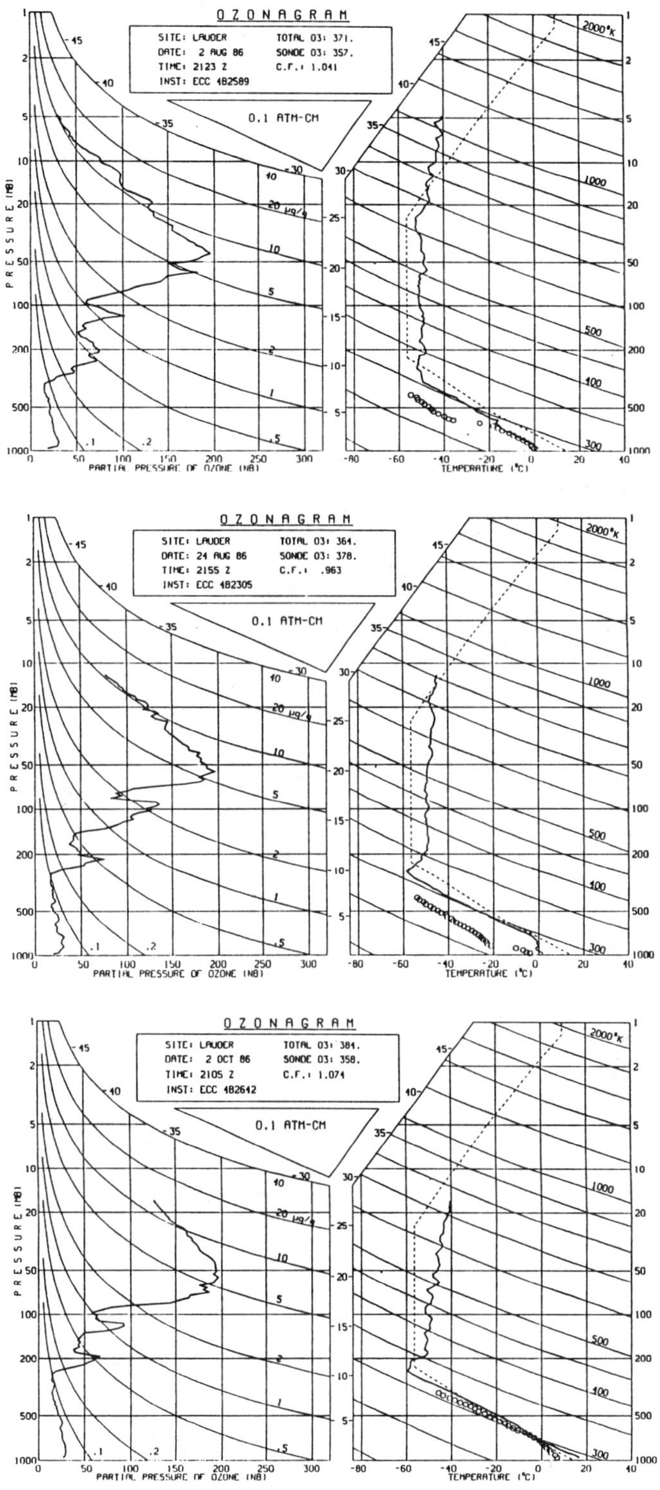

Figure 2. Ozone and temperature profiles for August 2, August 24 and October 2, 1986. These three profiles demonstrate a triple peak" or "double dip" phenomenon.

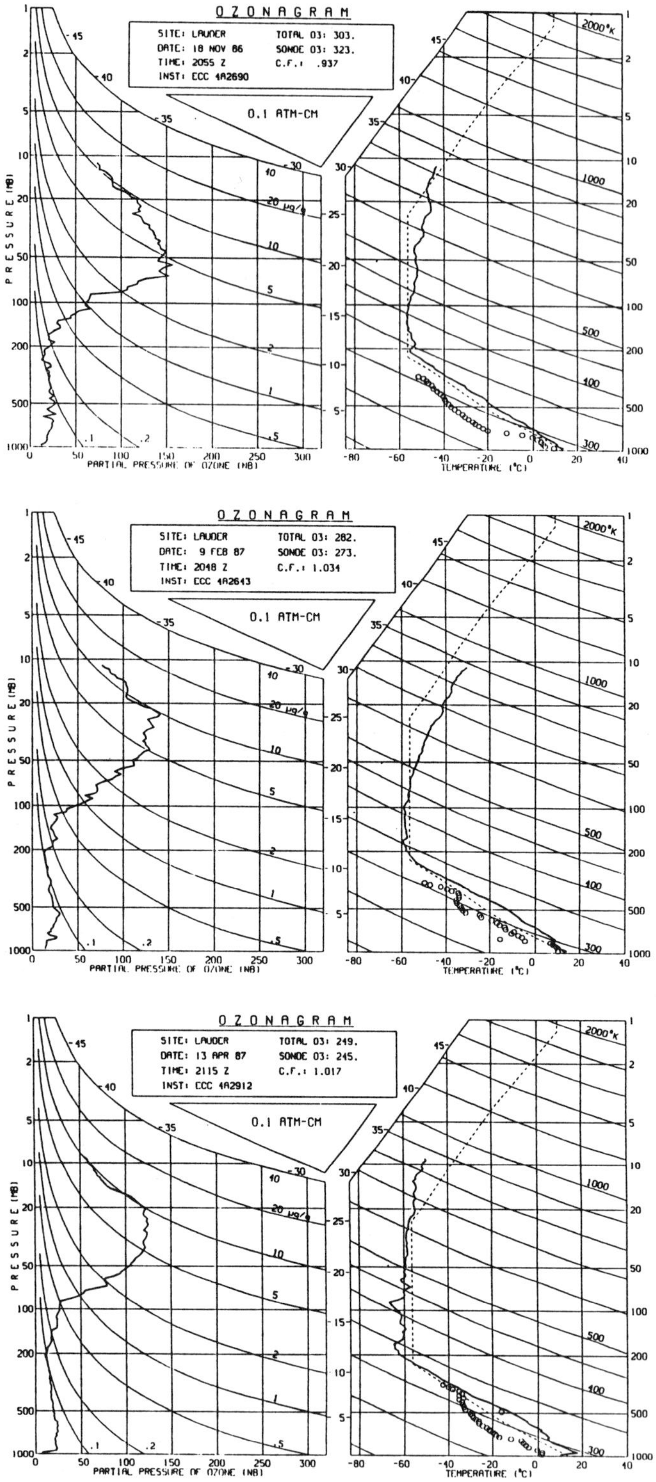

Figure 3. Ozone and temperature profiles for November 18, 1986, February 9 and April 13, 1987. The three profiles in this figure demonstrate the "classical" ozone profile with a single ozone partial pressure maximum in the middle stratosphere.

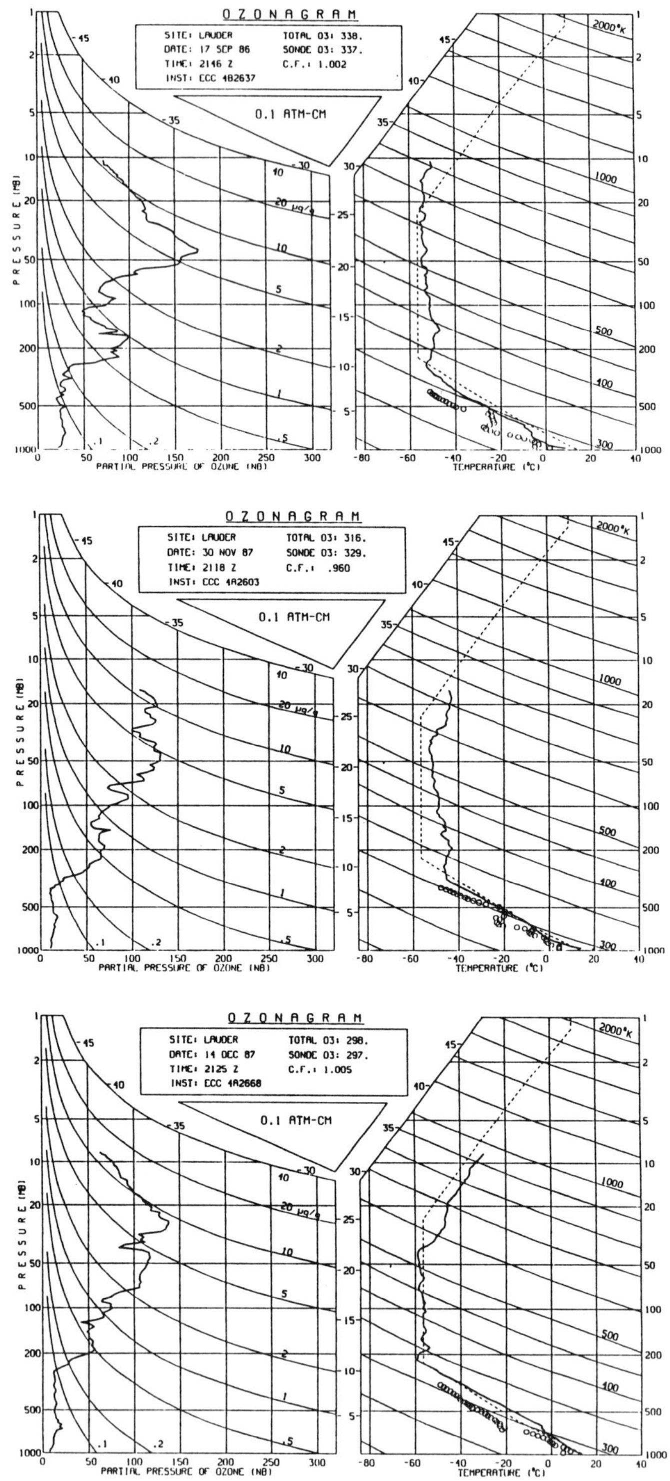

Figure 4. Ozone and temperature profiles for September 17, 1986, November 30, 1987 and December 14, 1987. The ozone profiles in this figure show relatively low ozone partial pressures in regions where normally more ozone is measured.

4. RESULTS

A feature of the ozone profiles over Lauder is the relatively frequent occurrence of a secondary maximum just above the tropopause, examples of which are shown in figure 1 for flights made on October 15, October 27, 1986 and August 12, 1987.

The ambient temperature profile, measured on the same balloon package is given for completeness with each of the ozone profiles presented here. This phenomenon has been reported previously for soundings made at mid-latitudes in the Southern Hemisphere by Hutchings and Farkas (1971) who attributed the cause to mixing associated with the subtropical jet.

On some occasions over Lauder, as is shown in figure 2, for the soundings on August 2, August 24 and October 2, 1986, an additional peak is observed somewhat below the main maximum. However, an equally valid interpretation of this feature of the profile is the interleaving of two regions of relatively ozone depleted air above and below this 100 mb level and therefore interrupting the rather monotonic increase in ozone mixing ratio with height.

Most of the profiles measured in 1987 show the more classical profile shape: a relatively constant mixing ratio in the troposphere up to the tropopause followed by a smooth increase in mixing ratio up to the maximum at approximately 50 mb (21 km) and then a constant mixing ratio above the peak. Examples of such profiles are shown in figure 3 for the ascents made on November 18, 1986, February 9 and April 13, 1987. Indeed, the measurements made in 1987 show in general less variability than those measured in 1986, especially in the spring.

The profile shown in figure 2 for the flight on August 2, 1986 appears to fall in between the "typical" high latitude profile, where the tropopause is around 8 km, and a tropical profile with the tropopause are around 16 km. Figure 4 shows the ozone and temperature profiles retrieved from flights made on September 17, 1986, November 30 and December 14, 1987. These show examples where relatively ozone depleted air is found in the middle stratosphere and poses the question "what transport processes have these slabs of air undergone if this is indeed an effect of dynamics?". This question is currently being addressed.

It is hoped that further measurements will help established the climatology of ozone over this mid-latitude, Southern Hemisphere site.

ACKNOWLEDGEMENTS

The technical support to the balloon sounding program given by Tracy Beck, Des Rowles and Brian McNamara at PEL Lauder is gratefully acknowledged as is the financial support provided for this program by the Chemical Manufactures Association through grants FC86-593 and FC86-609.

REFERENCES

Hilsenrath, E., Attmannspacher, W., Bass, A., Evans, W., Hogemeyer, R., Barnes, R.A., Komhyr, W., Maursberger, K., Mentall, J., Proffitt, M., Robbins, D., Taylor, S., Torres, A., and Weinstock, E., 1986: Results from the balloon ozone intercomparison campaign (BOIC), J. Geophys. Res., **91**, 13137-13152.

Hutchings, J.W. and Farkas, E., 1971: The vertical distribution of atmospheric ozone over Christchurch, New Zealand, Quart. J. of Roy. Met. Soc., **93**, 412.

Komhyr, W.D., 1969: Electrochemical concentration cells for gas analysis, Ann. Geophys., **25**, 203-210.

Komhyr, W.D., Oltmans, S.J., Chopra, A.N., and Franchois, P.R., 1985: Performance characteristics of high-altitude ECC ozonesondes. In Atmospheric Ozone, Proceedings of the Quadrennial Ozone Symposium held in Halkidiki, Greece, 3-7 September 1984. D. Reidel, Dordrecht, Holland, 499-503.

Komhyr, W.D., 1986: Operations handbook - Ozone measurements to 40 km altitude with model 4A electrochemical concentration cel (ECC) ozonesondes (used with 1680 MHz radiosondes). NOAA Technical Memorandum ERL ARL-149, Boulder, Colorado, 49 pp.

VERTICAL DISTRIBUTION OF OZONE OVER HYDERABAD USING A BALLOON-BORNE SUNTRACKING PHOTOMETER

S. Lal, B.H. Subbaraya, Y.B. Acharya, and S.V. Ramani
Physical Research Laboratory
Ahmedabad 380 009, India

R. Borchers and P. Fabian
Max-Planck-Institut für Aeronomie (MPAE)
D–3411 Katlenburg-Lindau, FRG

ABSTRACT

A multi-channel suntracking balloon-borne photometer has been developed and used to measure the vertical distribution of ozone in the 10 to 35 km altitude region. Three high altitude balloon flights have been conducted from Hyderabad (17.5°N, 78.6°E) using this system. Ozone concentrations have been derived from the observed attenuation profiles of solar radiation at one (two) wavelength bands centered at 290 (310) nm. The ozone distribution at Hyderabad peaks at an altitude of 24 km with a value of 4.0 ($\pm$0.5)x10^{-12} mol/cc. This feature is intermediate between the Thumba (8.5°N) profile which peaks at 27 km with a value of 3.46($\pm$0.78)x10^{-12} mol/cc and the mid-latitude profile which peaks at around 21-22 km. The ozone concentration is found to be very variable in the altitude region between 10 to 20 km. Above 25 km, the profiles at Hyderabad merges with the tropical (Thumba) and mid-latitude reference profiles.

1. INTRODUCTION

A balloon-borne multi-channel suntracking photometer has been developed at the Physical Research Laboratory (PRL), Ahmedabad for measuring the attenuation profiles of solar radiation in ultraviolet and visible part of the solar spectrum for measurement of trace gases in the tropical middle atmosphere. The instrument was launched along with a cryogenic air sampler belonging to the Max-Planck-Institute for Aeronomy (MPAE), Lindau from the ISRO/TIFR Balloon Launch Facility at Hyderabad (17.5°N, 78.6°E) on March 26, 1987 to study simultaneously the distribution of trace gases and the attenuation of solar radiation in the tropical middle atmosphere. The experiment was conducted under the ISRO-DFVLR collaborative programme. The vertical distribution of ozone has been derived from the measured attenuation profile of solar UV radiation at 290 nm. This paper presents the results of ozone measurement from this flight along with similar measurements made in 1985.

2. EXPERIMENTAL DETAILS

The multi-channel suntracking photometer operates at six wavelength bands centered at 210, 215, 290, 750, 860 and 950 nm. Each detector consists of a laboratory calibrated interference filter and a suitable photodiode. The detector assembly is mounted on a two axes suntracking platform which tracks the sun in azimuth and elevation and keeps the detectors pointing at the sun within $\pm$1 degree. The amplified signals are digitised and fed to the onboard PCM telemetry. The instrumental details are described elsewhere (Subbaraya et al. 1987).

A 143 300 cubic meter size balloon was launched from Hyderabad on 6 March 1987 carrying the multichannel photometer along with a cryogenic air sampler. The balloon reached the ceiling altitude of 34.5 km. The photometer measured attenuation profiles during ascent as well as during the descent part of the trajectory. The cryogenic air sampler collected 15 air samples during the ascent as well as during the slow apex valve controlled descent of the balloon. The air samples have been analysed for various trace source gases using gas chromatograph techniques at the MPAE. These results are presented separately in this volume.

Ozone concentration values have been estimated for the height region of 10 to 33 km. Since the photometer is continuously tracking the sun within $\pm$1 degree, the accuracy and the height resolution is very good. The derivation of ozone concentration does not need any absolute calibration unlike in the chemiluminescent instrument (eg. Hilsenrath and Kirschner 1980). While the chemical ozonesonde like the ECC have pump efficiency problems (Komhyr et al. 1985), this system does not

ISBN 0–937194–15–8

have any such drawback. The rocket-borne optical system has much lower height resolution as well as it needs aspect corrections which increases the errors (Subbaraya and Lal, 1981). Even though the attenuation profile data is available at lower heights, this is not used below 10 km due to larger Rayleigh and aerosol scattering corrections.

Two additional balloon flights were conducted from Hyderabad using a similar instrument. The details of all the three flights are given in Table 1. The ozone vertical distributions derived from the previous two flights are also presented and compared with the new profile.

Table 1: Details of balloon flights conducted from Hyderabad for the measurement of ozone and related trace constituents.

Balloon flight date	Launch time (IST)	Photometer channels	Channels used for ozone profile derivation
27 March 1985	0608 hrs	210,214,230,290	290 nm 450 and 800 nm
22 Oct. 1985	0611 hrs	214,290,310,450	290 and 310 nm 600 and 800 nm.
26 March 1987	0545 hrs	207,215,291,750	290 nm 860 and 950 nm

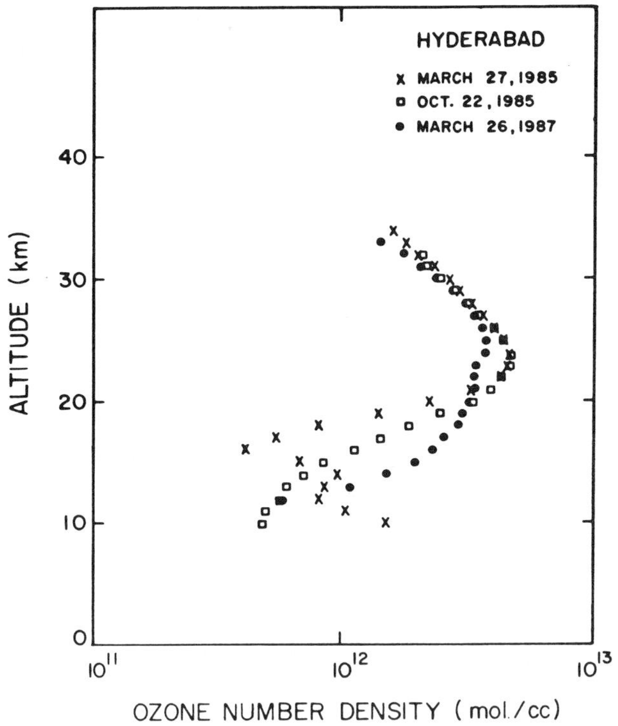

Fig. 1. Vertical distribution of ozone derived from the observed attenuation profile of solar radiation at 290 nm wavelength band. Earlier observations are also shown for comparison.

3. ANALYSIS PROCEDURE

Ozone concentrations are obtained from the measured photometer currents using the Beer-Lamberts' law

$$I_z(\lambda) = I_o(\lambda) exp\{-\tau(\lambda, z)\} \quad (1)$$

and effective values for the ozone absorption $\tau_{eff}(O_3)_z$ and Rayleigh scattering by air molecules $\tau_{eff}(air)_z$ in the pass band of the detector

$$n(O_3)_z = \frac{1}{I_z}\frac{dI}{dz}\frac{1}{F(z,\chi)\tau_{eff}(O_3)_z} - \frac{\tau_{eff}(air)_z}{\tau_{eff}(O_3)_z} n(air)_z \quad (2)$$

Recent measurements of the ozone absorption cross section by Paur and Bass (1985) and the solar flux values of Mount and Rollman (1981) and Nicolet's (1984) empirical formula for Rayleigh scattering are used in the estimation of effective absorption cross sections. The method of anaylses and estimation of errors are given in Subbaraya et al. (1986). Ozone concentrations have been estimated in the altitude region of 10 km to 33 km from the data of three balloon ascents made from Hyderabad in March 85, October 85 and on 26 March 1987. Measurement errors are typically of the order of $\pm$(25-30)% in the 12-15 km altitude region and decrease to a value of $\pm$10% at 20 km altitude. In the 20 km - 30 km altitude region the measurement error are generally less than 5%.

4. RESULTS AND DISCUSSION

Vertical distribution profiles of ozone derived from these measurements are shown in Fig. 1. The March 1985 ozone profiles show a small increase around 14 km. The minimum with a value of 4.0×10^{11} mol/cc appears at 16 km. The ozone concentration increases from 16 km onwards peaking at 23 km with a value of 4.6×10^{12} mol/cc. Above this peak, the ozone concentration decreases steadily reaching a value of 2.7×10^{12} mol/cc at 30 km. The ozone profiles obtained on 22 Oct. 1985 show a different feature in the 10 to 20 km altitude region. The ozone concentration increases from a value of 4.7×10^{11} mol/cc at 10 km to a value of 3.25×10^{12} mol/cc at 20 km. However, above 20 km both these profiles show similar features. The Oct. 1985 ozone profile also peaks at 23 km with the same concentration of 4.7×10^{12} mol/cc. The ozone profile obtained on 26 March 1987 shows larger ozone concentration below 20 km as compared to the two earlier measurements. The ozone concentration on this day increases steadily from a value of 1.05×10^{12} mol/cc at 13 km to a peak value of 3.8×10^{12} at 24 km. Above 26 km, this profile also merges with the other two profiles and shows similar ozone concentrations.

These measurements show large fluctuations in the ozone concentration below 20 km. Since the photochem-

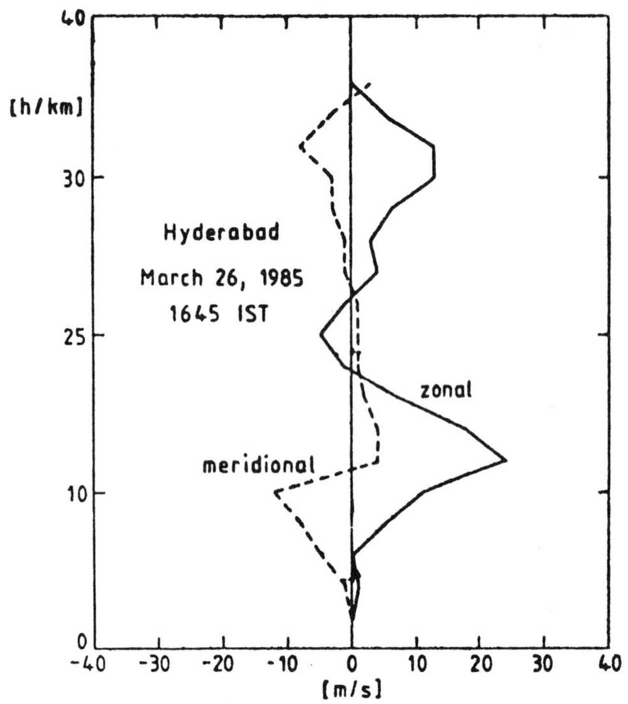

Fig. 2. Wind profile at Hyderabad measured on 26 March 1985.

ical time constants of ozone is much larger than the dynamical time constant, these fluctuations may be caused by dynamical processes. Even though the errors in the estimation of ozone concentration at lower heights are larger, of the order of 20-30% around 14 km, the deviations are much larger and hence they appear to be genuine. Similar large changes from day to day in the ozone concentrations for the altitude region of 10 to 20 km have been reported by Dütsch (1980). The 27 March 1985 ozone profile shows a small peak around 14 km altitude. The meridional wind measured on 26 March 1985 (Fig. 2) shows a change in the direction from southward to northward in the same height region. The observed structure in ozone profile appears to be related to this change in the wind direction. The March 1985 profile shows the lowest values in the 15 to 20 km while the March 1987 profile shows the highest ozone concentrations in the same height region. The wind profiles obtained for these two periods do not show any appreciable change in the zonal component. However, the meridional components differ significantly. The meridional wind measured on 26 March 1985 (assumed representative of 27 March 1985 also) show southward wind peaking around 10 km, but above this height is changes sharply to northward (Fig. 2). In the 15 to 20 km altitude region, the speed is only about 4 m/sec or less. However, the wind profile derived from the 26 March 1987 balloon flight shows a strong northward component with a speed of more than 20 m/sec around 12 km and slowly decreasing above it. Even though the winds on these two periods are different, it is difficult top correlate with changes in ozone concentration. It is needed to have a air-mass trajectory for 2-3 weeks prior to the measurements.

Mean vertical distribution of ozone obtained from the three measurements made from Hyderabad is compared with the reference model ozone profile from Thumba (8.5°N) obtained from the rocket and balloon measurements published by Subbaraya and Jayaraman (1987) and with the mean mid-latitude profile given by Krueger and Minzner (1976) in Fig. 3. The Hyderabad profile deviates from these profiles. The Hyderabad profile shows an ozone peak around 23-24 km with a value of $4.0 \pm 0.5 \times 10^{12}$ mol/cc, while the Thumba profile shows a peak concentration around 26-27 km with a lower value of 3.4×10^{12} mol/cc and the midlatitude profile shows the ozone peak around 21-22 km with a higher value of 4.8×10^{12} mol/cc. Thus the Hyderabad profile shows an intermediate feature as compared to the tropical and mid-latitude region in the 10-30 km altitude region. Above 30 km, there is no appreciable difference between the three profiles.

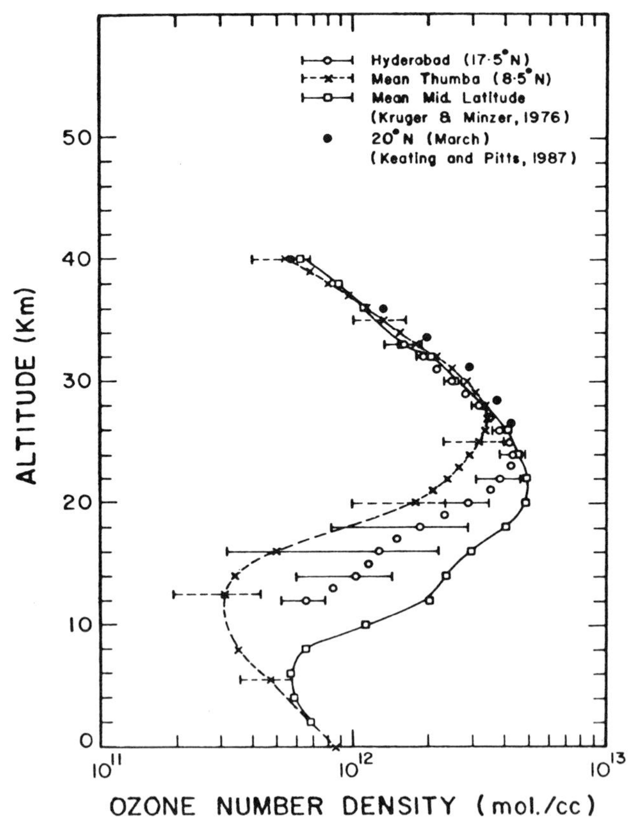

Fig. 3. Mean ozone profile observed at Hyderabad compared with the reference profiles for Thumba (8.5°N) and mid-latitude region. The reference profile for 20°N representative for March month derived from satellite measurements by Keating and Pitts (1987) is also shown.

The tropopause in the Hyderabad region is found to be around 16-17 km with a temperature of -77(±2)° which is not very different from the Thumba value of -80°C around 18 km. Further, the vertical distribution of trace source gases obtained from the analyses of collected air samples from Hyderabad shows tropical features with lower vertical gradient compared with the mid-latitude profiles (Borchers et al., this issue).

The mean ozone concentration over Hyderabad is also compared with the reference model for 20°N proposed by Keating and Pitts (1987) which is based on various satellite measurements. The mixing ratio profile is converted into number density by using the density data published by Sasi and Sen Gupta (1986). These converted values for the month of March are also shown in Fig. 3. The Keating and Pitts model values appear to be slightly higher when compared with the Hyderabad values and also with the Thumba values.

REFERENCES

Dütsch H.U., 1980: Vertical ozone distribution and tropospheric ozone. Proc. NATO Adv. Study Institute on Atmospheric Ozone: Its variation and human influences, Rep. No. FAA-EE-80-20, Ed. A.C. Aikin, p. 7-30

Hilsenrath E. and P.T. Kirschner, 1981: Recent assessment of the performance and accuracy of a chemiluminescent rocket sonde for upper atmospheric ozone measurements. Rev.Sci.Instrum. **51**, 1381-89

Keating G.M. and M.C. Pitts, 1987: Proposed reference models of ozone. Adv. Space Res. **7**, p. 37-48

Komhyr W.D., S.J. Oltmans, A.N. Chopra and P.R. Franchois, 1984: Performance characteristies of high-alttiude ECC ozonesondes. Atmospheric Ozone, Eds C.S. Zerefos and A. Ghazi, Reidel Publishing Co., p. 499-508

Krueger A.J. and R.A. Minzner, 1976: A mid-latitude ozone model for the 1976 U.S. Standard atmosphere. J. Geophys. Res., **81**, 4477-4481

Mount, G.H. and G.J. Rottman, 1981: The solar spectral irradiance 1200-3184 A near solar maximum: 15 July 1980. J. Geophys. Res., **86**, 9193-9198

Nicolet M., 1984: On the molecular scattering in the terrestrial atmosphere: An empirical formula for its calculation in the homosphere. Planet. Space Sci., **32**, 1467-1468

Paur R.J. and A.M. Bass, 1984: The ultraviolet cross-sections of ozone: II. Results and temperature dependence. Atmospheric ozone, Eds. C.S. Zerefos and A. Ghazi, Reidel Publishing Col., 611-616

Sasi M.N. and K. Sen Gupta, 1986: A reference atmosphere for Indian equatorial zone from surface to 80 km. Scientific Rep., No. SPL: SR:006:85, V.S.S.C. Trivadrum, India

Subbaraya B.H. and A. Jayaraman, 1987: The vertical distribution of ozone in the equatorial zone. Adv. Space Res., **7**, 119-122

Subbaraya B.H. and S. Lal, 1981: Rocket measurement of ozone density distsribution in the equatorial strato sphere and mesosphere over Thumba. Proc. Indian Acad. Sci. (Earth Planet. Sci) **90**, 173-187

Subbaraya B.H., A. Jayaraman, Y.B. Acharya, P. Fabian, R. Borchers and S. Lal, 1986: Vertical distribution of ozone over Hyderabad. Indian J. Radio and Space Phys., **15**, 67-70

ANNUAL AND SEMI-ANNUAL OSCILLATION OF THE VERTICAL OZONE DISTRIBUTION OVER CAIRO AND ASWAN, EGYPT

G.K.Y. Hassan
Meteorological Authority of Egypt
Koubry El-Quobba, Cairo, Egypt

Abstract

The monthly averages of the vertical ozone distribution over Cairo (30 05 N), for 1978-1987, and over Aswan (23 58 N) for 1985-1987 are analyzed by harmonic analysis. The variation of the amplitudes and phases of the annual and semi-annual oscillations for each layer are studied and compared to each other.

The results show that, the variation in the vertical ozone distribution over Cairo can be classified "on the basis of the analysis of amplitudes and phases " into three distinguished zones separated by two transition ones.

1. Introduction

This is the first study of the vertical ozone distribution (VOD) over Egypt, using the Umkher observation method (Mateer and Dutsch,1964),at Cairo and Aswan. The computer program of the above method was kindly provided by the Atmosphric Environment Service of Canada. The monthly averages of the ozone mean partial pressures in micro-millibar (umb) of each layer is computed for both Cairo and Aswan (Table 1 and 2 respectively),and the harmonic analysis technique has been applied, The annual oscillations (AO's), and semi-annual oscillations (SAO's),have been studied for each layer of the VOD for both Cairo and Aswan and compared to each other (Tables and Figures are located at end of paper). An analysis for these two oscillations has been also done for Boulder (USA) station for a more comparative study of the general features of the VOD.

2.Data and Analysis

The present study uses 200 and 177 Umkher profiles over Cairo and Aswan respectively. The root mean square (rms) of any profile does not exceeded 0.7 .

The monthly average data of Boulder have been taken from the WMO World Ozone Data publications.

Table 3 shows the average values of the VOD of Cairo,Aswan and Boulder.

Figures 1,2 up to 9 represents the annual cycles of layers 1, 2 up to layer 9 respectively for the three stations.

Figure 10 represents the annual cycle of the total ozone of the corresponding Umkher observations for comparative study.

2.1 Annual oscillations (AO's)

Two maxima in the amplitude of the AO are observed over Cairo, the first one is about 11.5 umb and is observed in layer No. 1 (Fig. 1),with time of maximum occurrence at June, the second one is about 12.6 umb and is observed in layer No. 3 (Fig. 3)with time of maximum occurrence at March.

The maximum AO amplitude over Aswan is observed in layer No. 4 (Fig. 4) and is about 11.7 umb,with phase at May.

The maximum AO amplitude over Boulder is observed in layer No. 3 (Fig. 3),and is about 22.4 umb,with time of maximum occurrence at March. The AO amplitude above layer 4 (Fig. 5) for the three stations is found to be weaker relative to the ozone concentration values.

2.2 Semi-annual oscillations (SAO's)

It is observed that the SAO's amplitudes are much weaker than the AO's amplitudes, their maximum is about 4.7,1.7,and 6.6 umb at Cairo,Aswan,and Boulder respectively. It reaches maximum values at the lower layers and decreases thereafter with height. It is nearly negligible above layer 4. The time of maximum occurrence takes place during Spring.

3. Conclusion

The above results indicate that :

a-The variation in the VOD over Cairo can be classified into three distinguishable zones, the first zone includes layer No. 1, the second zone includes layers 3-4, while the third one includes layers 6-9. Separated by two transitional zones, the first one includes layer No. 2, and the second one includes layer No. 5, the layer of maximum ozone concentration.

b-Layers No. 1 and 6-9 (zone 1 and 3) seem to be affected by the photochemical reactions (according to the vertical distribution of phase of the AO's), since the layers are in phase with the sun cycle.

c-Layer 2 and 3-4 seem to be affected by the circulation in the lower stratosphere.

d-A significant feature of layer No. 5 was observed, out of phase ,at Cairo only which needed more studies for interpretation.

All above findings are quite similar to those by Bojkov (1967,1968a,b), and Maeda (1984).

ISBN 0-937194-15-8

ACKNOWLEDGMENTS

The author wishes to thank Mr.A.M.El-Masry, Chairman of Board of Directors of the Egyptian Meteorological Authority and Mr. M.G. El-Moaty, Director General of Research for their encouragement.

Thank are also due to Dr.Rumen D.Bojkov, Dr.A.M.Ibrahim and Dr.A.A.Hassan for their valuable suggestions,continuous encouraement and advice.

References

Bojkov,Rumen D.,1967:Vertical ozone distribution over South Africa, J. Geophysical Res., 72.

Bojkov,Rumen D.,1968a:Mean pole-to-pole vertical ozone distribution,Proc. of the Seventh Stanstead Seminar, McGill Univ. Publication in Met. No. 90.

Bojkov Rumen D.,1968b: Feature of the vertical ozone distribution,deduced from five years observation at Belsk,Poland.,Acta Geoph.Polonica,Vol.XVI,Nr. 4.

Maeda,K.,1984:Annual and Semiannual oscillation of stratospheric ozone, NASA Publ. Nr.X-961-84-13.

Mateer,C.L. et H.U.Dutsch,1964:Uniform evaluation of Umkher observations from the World Ozone Network,part 1.,Proposed Standard Umkher evaluation Technique,Boulder,National Centre for Atmospheric Research 105 p.

Ozone Data for the World 1986 vol.27. Nr. 6. Published by the Atmospheric Environment Service of Canada

TABLE 1. MONTHLY AVERAGE VERTICAL OZONE DISTRIBUTION OVER CAIRO

MONTH	LAYER OZONE PARTIAL PRESSURE IN MICO MILLIBAR 1	2	3	4	5	6	7	8	9	NO.OF OBS.	TOTAL OZONE
JANUARY	23.5	33.6	52.3	102.6	129.7	87.0	43.1	17.5	5.1	18	286.5
S.D.	7.1	6.6	13.7	14.3	7.5	5.9	4.5	1.7	1.2		22.5
FEBRUARY	25.3	35.8	56.1	107.3	129.6	91.2	48.7	17.6	4.6	20	299.6
S.D.	13.2	8.2	17.9	18.6	8.8	6.5	6.3	2.4	1.6		24.1
MARCH	33.0	38.7	64.3	118.8	129.3	92.8	50.2	18.0	5.3	25	323.2
S.D.	9.5	15.6	16.6	12.6	7.9	6.7	5.1	1.8	1.2		23.5
APRIL	39.4	46.9	69.9	121.7	126.7	93.5	51.3	17.7	5.2	14	338.8
S.D.	11.7	12.6	12.9	8.4	5.7	3.0	2.5	1.0	0.8		16.2
MAY	46.0	37.7	59.3	118.1	129.0	100.5	57.2	17.7	4.5	8	341.7
S.D.	9.5	26.9	13.3	9.5	12.1	8.2	3.9	2.7	2.0		26.4
JUNE	42.6	34.2	49.6	110.3	124.0	94.9	52.3	16.3	3.9	15	316.5
S.D.	6.8	20.9	13.4	4.6	8.1	6.2	5.3	2.9	1.8		13.0
JULY	39.0	26.0	42.0	107.3	126.7	96.2	53.7	18.4	5.1	23	307.3
S.D.	9.4	5.8	8.9	7.5	5.7	3.6	3.0	1.4	1.1		6.7
AUGUST	40.3	29.4	43.8	105.1	123.9	93.1	51.7	17.9	4.9	17	305.9
S.D.	6.7	10.9	10.0	6.5	6.1	6.6	5.3	2.0	1.4		5.8
SEPTMBER	32.3	27.5	37.1	95.5	131.2	94.2	50.5	19.0	5.2	11	292.1
S.D.	7.4	11.1	10.0	11.2	10.5	7.4	3.8	1.5	0.8		7.2
OCTOBER	25.9	29.5	37.1	87.6	129.1	92.1	50.1	20.0	5.6	15	278.8
S.D.	7.0	3.8	10.6	6.2	3.7	3.4	3.5	2.3	1.6		5.5
NOVEMBER	20.0	32.1	51.6	101.6	131.0	88.6	45.5	18.7	5.6	20	285.1
S.D.	7.3	15.9	17.8	9.1	12.0	7.1	3.4	1.9	1.5		17.3
DECEMBER.	19.8	31.0	52.2	102.8	133.7	90.5	46.4	19.0	5.8	14	288.4
S.D.	7.6	10.0	13.5	11.9	8.3	6.1	4.2	1.7	1.1		13.2
AVERAGE OF 200 VALUES	32.5	32.4	50.0	106.4	129.2	93.2	50.0	18.2	5.1	200	305.0
S.D.	8.3	6.1	10.1	9.1	3.0	2.7	3.3	0.8	0.4		18.7

WHEN RMS IS GREATER THAN 0.7 THE VALUE IS NOT COUNTER IN MEAN OR STD DEVNS.

TABLE 2. MONTHLY AVERAGE VERTICAL OZONE DISTRIBUTION OVER ASWAN

MONTH	LAYER OZONE PARTIAL PRESSURE IN MICO MILLIBAR 1	2	3	4	5	6	7	8	9	N OF OBS.	TOTAL OZONE
JANUARY	14.1	19.8	31.7	88.7	130.5	90.2	47.1	18.3	4.7	13	254.0
S.D.	5.7	4.8	8.9	5.0	3.6	4.0	3.0	1.8	1.4		8.0
FEBRUARY	12.5	19.3	38.8	98.4	138.1	94.8	47.8	18.0	4.6	9	267.7
S.D.	3.6	4.5	7.9	7.7	5.8	3.7	2.5	1.5	0.9		16.9
MARCH	14.6	24.5	44.5	98.7	135.9	90.9	46.2	13.5	5.2	11	272.9
S.D.	4.0	5.5	8.2	8.1	5.5	5.7	4.5	1.4	1.2		13.8
APRIL	15.8	24.8	54.6	112.1	139.4	95.0	48.5	17.9	4.8	13	292.2
S.D.	8.0	8.1	15.3	17.5	6.3	3.7	2.9	1.6	0.7		22.4
MAY	15.8	24.3	48.5	106.3	142.3	97.5	49.8	18.1	4.5	4	289.0
S.D.	10.1	2.1	7.5	2.2	3.9	1.3	2.2	0.7	0.5		5.2
JUNE	20.9	25.9	48.0	107.4	137.6	99.0	56.6	18.4	4.4	10	287.7
S.D.	6.6	5.1	9.3	9.9	6.6	2.8	2.2	1.3	0.8		10.3
JULY	22.8	27.1	45.6	103.9	135.5	97.1	52.8	19.0	4.9	13	274.1
S.D.	5.4	3.9	7.8	6.0	4.3	2.4	2.6	1.1	0.8		4.9
AUGUST	23.6	21.0	38.1	97.4	136.9	98.0	51.8	19.9	5.6	13	285.5
S.D.	4.2	3.8	4.1	5.4	2.1	2.3	3.0	1.4	1.0		7.4
SEPTMBER	18.5	19.7	39.1	97.6	137.8	94.9	50.9	20.9	6.4	18	279.2
S.D.	8.4	6.4	9.4	4.2	5.0	5.0	4.5	2.5	1.7		7.1
OCTOBER	16.4	22.4	37.1	91.5	132.2	90.8	48.6	20.0	5.8	21	266.5
S.D.	4.8	3.9	5.9	4.4	3.8	3.1	2.3	0.6	0.5		7.8
NOVEMBER	17.0	21.1	31.5	87.6	129.6	89.2	46.4	19.5	4.9	33	256.1
S.D.	5.2	4.2	8.6	6.4	4.9	4.2	3.5	1.1	0.9		8.4
DECEMBER.	17.8	23.3	33.0	86.7	128.2	88.8	46.1	19.7	5.1	19	257.7
S.D.	2.6	3.4	4.7	5.3	5.3	3.5	3.3	1.7	1.2		7.1
AVERAGE OF 177 VALUES	17.4	22.4	40.7	98.4	135.7	93.9	49.2	18.8	5.1	177	276.0
S.D.	6.3	5.2	10.9	10.8	602	5.0	4.0	1.7	1.2		18.4

WHEN RMS IS GREATER THAN 0.7 THE VALUE IS NOT COUNTER IN MEAN OR STD DEVNS.

TABLE 3. ANNUAL MEAN LAYER VALUES OF OZONE PARTIAL PRESSURE OVER CAIRO, ASWAN AND BOULDER IN U.MB.

LAYER \ STATION		CAIRO	ASWAN	BOULDER
500-250 MB.	MEAN	32.5	17.4	26.8
	S.D.	(8.3)	(3.1)	(3.8)
250-125 MB.	MEAN	32.4	22.4	50.3
	S.D.	(6.1)	(2.5)	(15.7)
125-62.5 MB.	MEAN	50.0	40.7	73.0
	S.D.	(10.1)	(7.6)	(16.5)
62.5-31.2 MB.	MEAN	106.4	98.4	117.4
	S.D.	(9.1)	(8.3)	(12.8)
31.2-15.6 MB.	MEAN	129.2	135.7	129.0
	S.D.	(3.0)	(4.1)	(3.6)
15.6-7.8 MB.	MEAN	93.2	93.9	88.1
	S.D.	(2.7)	(3.3)	(5.5)
7.8-3.9 MB.	MEAN	50.0	49.2	45.6
	S.D.	(3.3)	(2.4)	(4.0)
3.9-1.95 MB.	MEAN	18.2	18.8	18.9
	S.D.	(0.8)	(0.9)	(0.8)
1.95-0.98 MB.	MEAN	5.1	5.1	6.3
	S.D.	(0.4)	(0.6)	(0.6)
TOTAL O3 DU.	MEAN	305.0	276.0	318.0
	S.D.	(18.7)	(15.1)	(27.6)

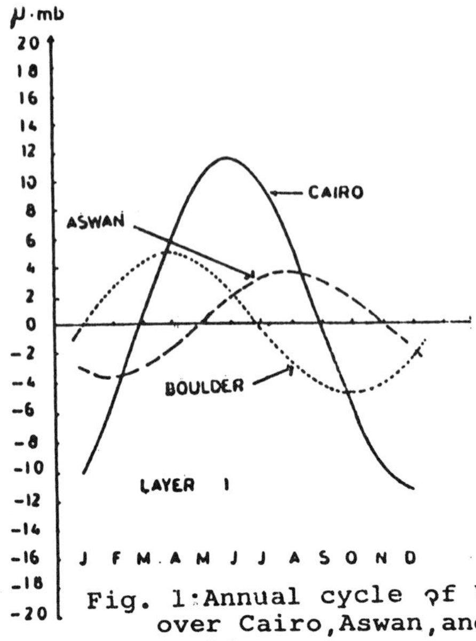

Fig. 1: Annual cycle of VOD over Cairo, Aswan, and Boulder

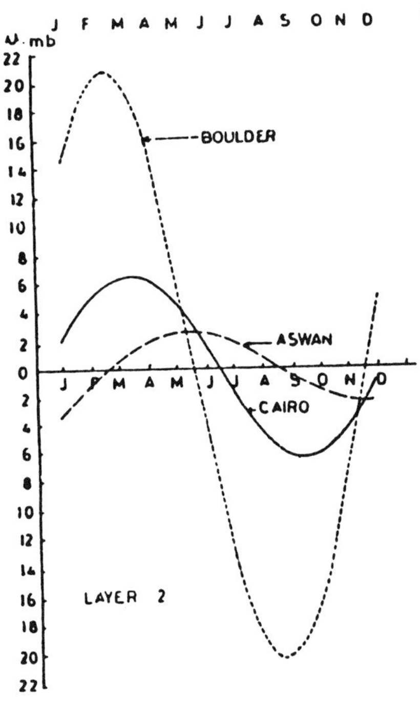

Fig. 2. Same as fig. 1.

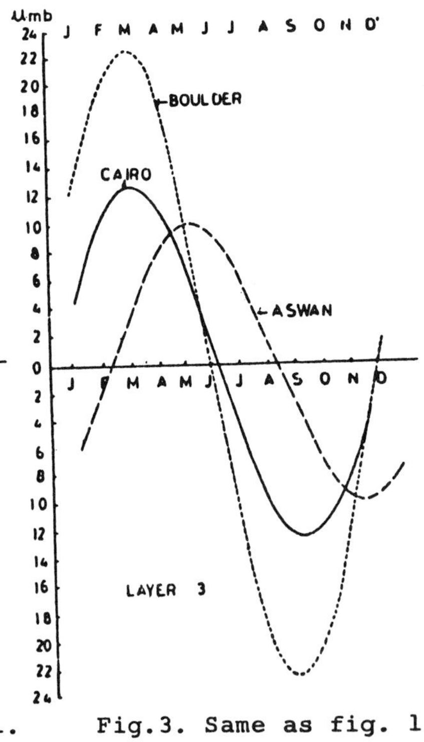

Fig. 3. Same as fig. 1.

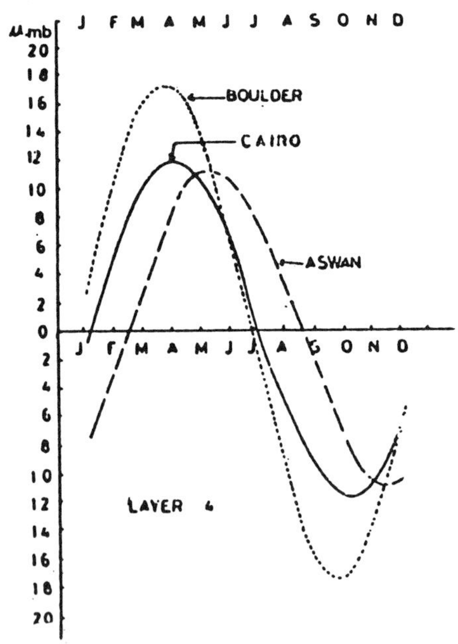

Fig. 4. Same as fig. 1.

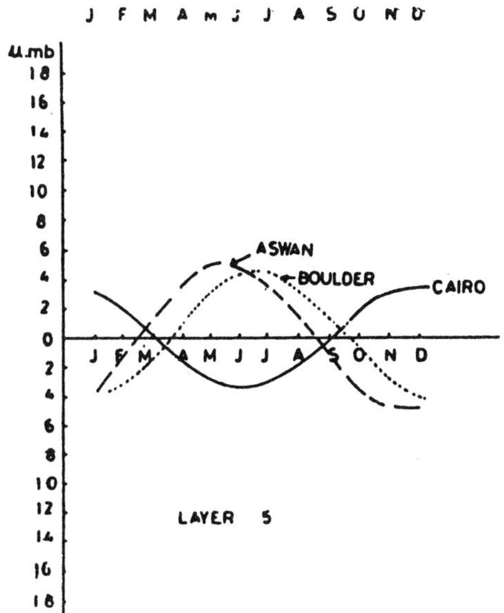

Fig. 5. Same as fig. 1.

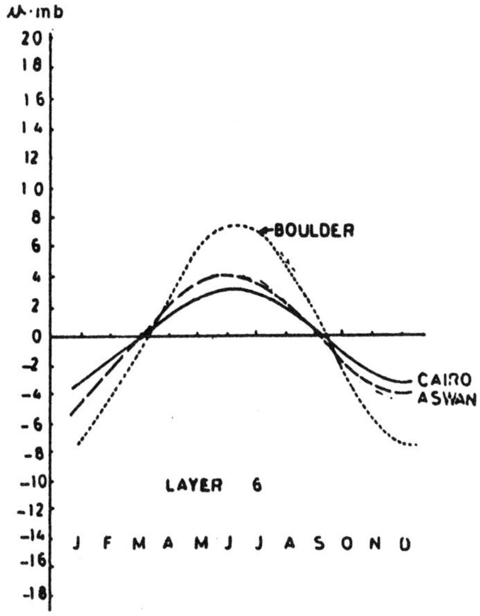

Fig. 6. Same as fig.1.

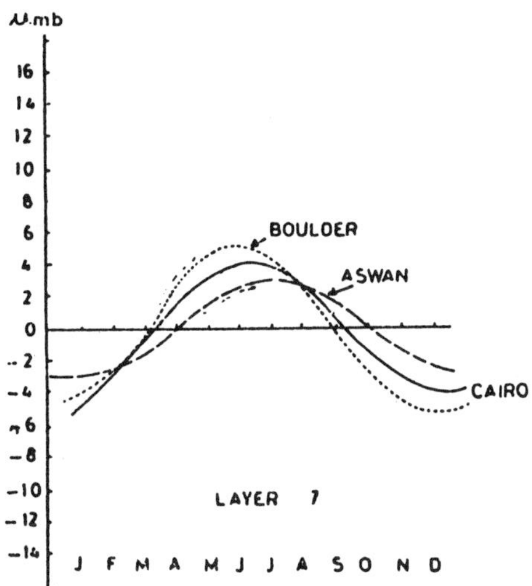

Fig. 7. Same as fig. 1.

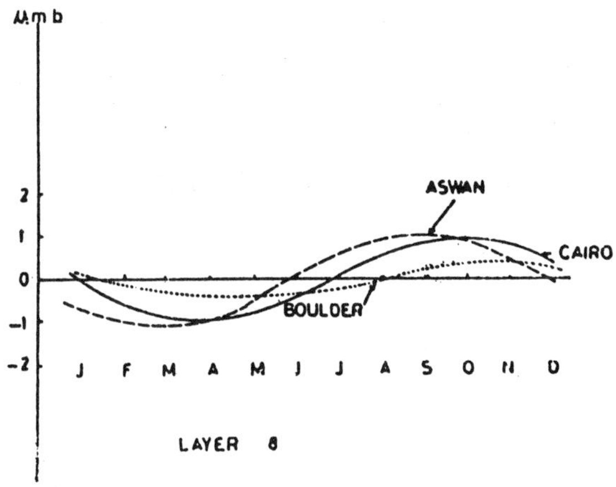

Fig. 8. Same as fig. 1 ,but ordinate is double of 1.

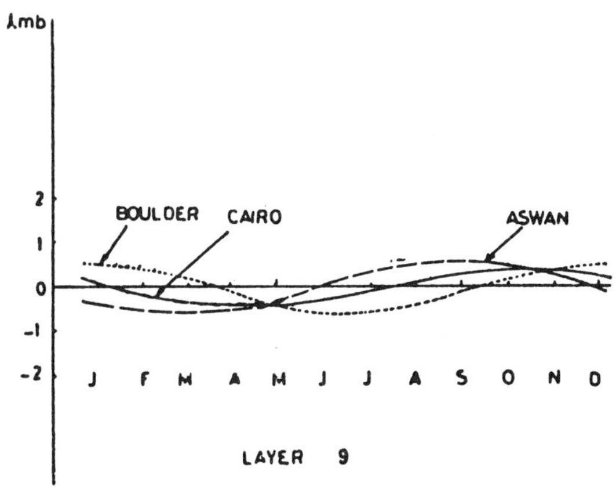

Fig. 9. Same as fig. 8.

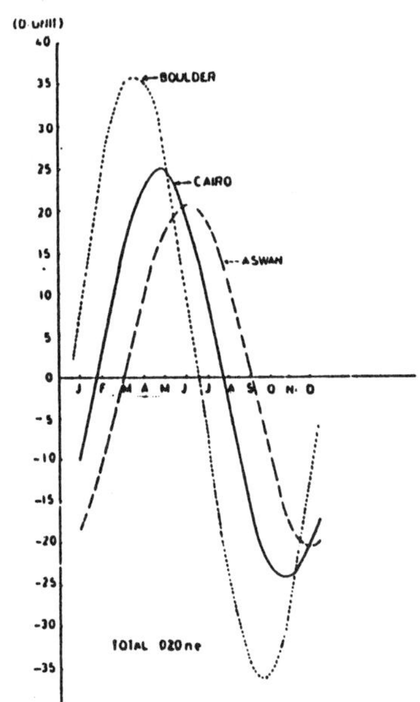

Fig. 10. Annual cycle of total ozone for the three stn;s

TRAVELING STANDARD LAMP CALIBRATION CHECKS OF DOBSON OZONE SPECTROPHOTOMETERS DURING 1985–1987

R.D. Grass and W.D. Komhyr
NOAA/ERL/Air Resources Laboratory
Boulder, Colorado 80303, USA

ABSTRACT

During 1985-1987, calibrated standard lamps were sent from Boulder to check on the calibration level of 95 Dobson instruments in the global total ozone station network. Data received from 81 instruments indicated that 13 were in need of recalibration. Six of the 13 have since been recalibrated by means of direct intercomparison with World Standard Dobson Instrument No. 83.

1. INTRODUCTION

Dobson ozone spectrophotometers serve as a standard for measurements of atmospheric total ozone. World Primary Standard Dobson Spectrophotometer No. 83 is maintained by the U.S. National Oceanic and Atmospheric Administration (NOAA) Geophysical Monitoring for Climatic Change (GMCC) program, in Boulder, Colorado. Since 1962, the long-term ozone measurement precision of instrument No. 83 has been maintained to within an uncertainty of less than ±0.5% (Komhyr et al., this issue). International Dobson instrument comparisons in the past have revealed large calibration errors for many of the instruments (Dziewulska-Losiowa and Walshaw, 1975; Gushchin, 1972); however, the accuracy of the ozone measurements within the global Dobson station network improved markedly following an international Dobson spectrophotometer intercomparison (Komhyr et al., 1981) held in Boulder in 1977. At that time, eight spectrophotometers, designated by the World Meteorological Organization (WMO) as regional secondary standards, were calibrated relative to instrument No. 83 with the intent that they be used to calibrate field instruments within their respective regions. Currently, calibration of virtually all of the instruments in use throughout the world is traceable to either direct or indirect intercalibration with instrument No. 83.

At the 1977 intercomparison, a standard lamp method was devised (Komhyr et al., 1981) for identifying Dobson instruments that have gross absolute calibrations errors. In an ongoing program by NOAA/GMCC to upgrade the calibrations of Dobson instruments throughout the world, seven standard lamp units, each consisting of two calibrated lamps and a stable power supply, were fabricated in 1981 in collaboration with the WMO. The global Dobson instrument network was then divided into seven areas, each containing from 5 to 17 instruments, and a lamp unit was shipped to each area for use in checking the calibrations of the Dobson instruments in each area. Results of the calibration checks were presented at the 1984 Quadrennial Ozone Symposium held in Halkidiki, Greece (Grass and Komhyr, 1985). This paper presents similar results of tests performed with the traveling standard lamps during 1985-1987.

2. DOBSON INSTRUMENT CALIBRATION METHODOLOGY

The standard lamp technique of checking Dobson spectrophometers for gross calibration errors is based on experimental observations (Komhyr et al., 1981) that instruments accurately calibrated on relative and absolute scales yield standard lamp readings

$$N = \log (I/I') + K \tag{1}$$

at the A, B, C, and D Dobson instrument wavelength pairs that are essentially constant provided, also, that the internal light-scattering characteristics of the instruments are similar. (In the above equation, I/I' is the lamp light intensity ratio for a wavelength pair, while K is the associated instrument constant.) Mean absolute calibration error for such instruments, calibrated with standard lamps, is estimated to be ±1.5% at the 95% confidence interval. This error is about twice as large as that which occurs when spectrophotometers are calibrated by direct comparison of instruments. Thus, use of standard lamps for calibrating spectrophotometers on an absolute scale is not generally recommended.

Accurate relative calibration of Dobson spectrophotometers implies accurate optical wedge and wavelength-setting calibrations for the instruments. An instrument with an improperly calibrated optical wedge will, for example, yield wrong N values in most instances for a standard lamp whose N values are accurately known. Because only a limited portion of the optical wedge is used when a standard lamp test is performed, the calibration error determined from correct and observed N-value differences is not, in general, representative of the instrument as a whole. Thus, it is important that spectrophotometer calibration errors determined from tests with standard lamps not be used for correcting ozone data obtained in the past with the improperly calibrated instruments. At most, the lamp tests indicate that an instrument calibration problem exists, and that remedial action is required.

3. DOBSON INSTRUMENT CALIBRATION CHECK RESULTS AND CONCLUSIONS

Standard lamps sent to the Dobson instrument stations were calibrated using World Standard Dobson Spectrophotometer No. 83. Recalibration of the lamps with instrument No. 83 following return of the lamps to Boulder indicated insignificant changes in originally assigned lamp N values. A little more than 2 years were required to circulate the lamp units among all stations. Complete calibration-check data were received from 81 of the 95 instruments. For the 14 remaining instruments, no data were reported for 5, insufficient data were reported for 4, recalibrations were being performed in Boulder for 3, and reconditioning was

ISBN 0–937194–15–8

being performed in Melbourne for 2. Table 1 lists the 14 remaining instruments and the status of these instruments.

TABLE 1. DOBSON INSTRUMENTS FOR WHICH CALIBRATION CHECK DATA ARE NOT AVAILABLE

Station	Inst. no.	Instrument status
Macquarie Is.	6	Isolated station (no data received)
Srinagar	10	Insufficient N-table data
Biscarrosse	11	Unknown (no data received)
Nairobi	18	Unknown (no data received)
Bracknell	41	Recalibrated at Boulder
Cologne	44	No data received
Mt. Abu	54	Insufficient N-table data
Lauder	72	Recalibrated at Boulder
Melbourne	78	Being reconditioned
Natal	93	Recalibrated at Boulder
Casablanca	106	Inoperative since 11/5/85
Melbourne	115	Being reconditioned
Tateno	122	Insufficient N-table data
Sapporo	5702	Insufficient N-table data

Results of the Dobson instrument calibration checks are given in Table 2. Instrument calibration errors shown in the last column of the table are averages derived from

$$\% \text{ Error} = 100\,(\Delta N_{AD})/1.388\mu x \quad (2)$$

for a total ozone amount of 300 m-atm-cm and averaged for equivalent ozone path lengths μ of 1, 2, and 3. The ΔN_{AD} values were computed from the assigned and measured lamp N values.

TABLE 2. RESULTS OF 1985-1987 CALIBRATION CHECKS ON DOBSON SPECTROPHOTOMETERS USING TRAVELING STANDARD LAMPS

Station	Inst no.	Date of calib. check	Inst. calib. error (%)*
Area 1 - North America			
Toronto, Canada	77	850531	-0.29
Resolute, Canada	59	851230	-0.81
Churchill, Canada	60	860212	+1.10
Goose Bay, Canada	62	850713	+0.08
Edmonton, Canda	102	851114	+0.15
Reykjavik, Iceland	50	860610	+0.15
South Pole, Antarctica	82	880202	+0.15
Bismarck, U.S.A.	33	860728	-2.35
Caribou, U.S.A.	86	861006	-0.15
Wallops Is., U.S.A.	38	861010	-0.22
Tallahassee, U.S.A.	58	861231	-0.29
Nashville, U.S.A.	79	870109	+1.17
Poker Flat, U.S.A.	63	870113	-1.10
Mauna Loa, U.S.A.	76	870524	+1.47
American Samoa, South Pacific	42	870715	+1.61
Point Barrow, U.S.A.	91	880129	+0.30
Fresno, U.S.A.	94	870505	+1.18
Boulder, U.S.A.	61	850411	-1.03

TABLE 2. Continued

Station	Inst no.	Date of calib. check	Inst. Calib. error (%)*
Boulder, U.S.A.	65	850411	-1.25
Boulder, U.S.A.	34	880127	+0.22
Boulder, U.S.A.	80	880113	+1.40
Area 2 - South America, Africa			
Mexico City, Mexico	98	850522	+1.32
Huancayo, Peru	87	850401	-0.37
Cachoeira Paulista, Brazil	114	860606	±0.00
Natal, Brazil	93	-	-
Buenos Aires, Argentina	97	851104	-0.52
Buenos Aires, Argentina	99	851106	+0.15
Nairobi, Kenya	18	-	-
Area 3 - Western Europe (1)			
Bracknell, U.K.	41	-	-
Bracknell, U.K.	2	850725	-0.52
Halley Bay, U.K.	123	851228	+3.67
Argentine Is., U.K.	31	860219	+0.37
Seychelles, U.K.	57	850826	-2.50
St. Helena, U.K.	35	850925	-0.59
King Edward Point, U.K.	103	851017	-1.32
Lerwick, U.K.	32	850731	-2.28
Arosa, Switzerland	15	850327	-2.57
Arosa, Switzerland	101	850327	+4.55
Hohenpeissenberg, F.R.G.	104	850422	+2.42
Cologne, F.R.G.	44	-	-
Oslo, Norway	56	850712	-0.66
Tromso, Norway	14	850705	-1.03
Spitzbergen, Norway	8	850703	-0.22
Cambridge, U.K.	73	860909	-2.79
Area 4 - Western Europe (2)			
Aarhus, Denmark	92	850319	+3.60
Uccle, Belgium	40	850405	+1.61
Biscarrosse, France	11	-	-
Haute Provence, France	85	850627	-1.25
Lisbon, Portugal	13	850726	+0.66
El Arenosillo, Spain	120	851008	+0.66
Vigna DiValle, Italy	47	851127	+1.03
Brindisi, Italy	46	851123	+0.88
Sestola, Italy	48	851206	+1.10
Calgiari/Elmas, Italy	113	851116	-0.59
Casablanca, Morocco	106	-	-
Cairo, Egypt	69	860206	-1.25
Cairo, Egypt	96	860211	+1.17
Area 5 - Eastern Europe U.S.S.R.			
Leningrad, U.S.S.R.	108	850402	+0.08
Belsk, Poland	84	850523	+1.24
Hradeč Kralove, Czechoslovakia	74	850626	-4.40
Budapest-Lorini, Hungary	110	850731	-1.54
Bucharest, Romania	121	850905	+2.72
Potsdam, G.D.R.	64	851010	-0.52
Potsdam, G.D.R.	71	851010	+0.66
Area 6 - India			
New Delhi, India	36	850809	+0.73
New Delhi, India	112	850422	-0.52
Srinagar, India	10	-	-
Varanasi, India	55	850822	-1.76
Mt. Abu, India	54	-	-
Poona, India	39	860123	+1.54
Kodaikanal, India	45	850705	+2.42
Quetta, Pakistan	43	860722	+1.40
Quetta, Pakistan	100	860719	+1.47
Bangkok, Thailand	90	861009	-1.17
Singapore	7	861218	+3.06
Manila, Philippines	52	870622	+1.18

TABLE 2. Continued

Station	Inst no.	Date of calib. check	Inst. Calib. error (%)*
Area 7 - Australia, Japan			
Sapporo, Japan	5704	850715	-0.88
Kagoshima, Japan	5705	850618	-0.22
Tateno, Japan	116	850725	-0.34
Naha/Kagamizu, Japan	5706	850629	-0.44
Tateno, Japan	5703	851024	-0.49
Sapporo, Japan	5702	-	-
Tateno, Japan	122	-	-
Xianghe, China	75	870203	+0.66
Kunming, China	3	870616	-0.74
Melbourne, Australia	105	860709	+0.52
Perth, Australia	81	860616	+0.52
Brisbane, Australia	111	860105	+0.22
Hobart, Australia	12	860316	+0.73
Macquarie Is., Australia	6	-	-
Invercargill, New Zealand	17	861017	-1.17
Lauder, New Zealand	72	-	-
Seoul, Korea	124	851130	+1.17
Melbourne, Australia	115	-	-
Melbourne, Australia	78	-	-

*See text for significance of indicated calibration error. Positive error means that instrument yields ozone values that are too large.

Deduced instrument calibration errors of absolute magnitude >2% were taken to signify that an instrument calibration problem exists. According to this criterion, 13 instruments were identified as needing recalibration. Location of the instruments are the following: one in North America; eight in Western Europe; two in Eastern Europe; and two in India, Singapore. Care must be taken not to misinterpret the error data of Table 2 as actual instrument calibration error data, rather than as data indicative of instruments requiring recalibration.

Since the time that the calibration checks were performed on the 13 instruments, 6 have been recalibrated by means of direct intercomparison with Dobson instrument No. 83. The six instruments are Bismarck, No. 33; Arosa, Nos. 15 and 101; Hohenpeissenberg, No. 104; Aarhus, No. 92; and Hradeč Kralove, No. 74.

ACKNOWLEDGMENTS

We thank the numerous Dobson spectrophotometer station personnel who contributed to the Dobson instrument calibration-check data base described in this report.

REFERENCES

Dziewulska-Losiowa, A., and C.D. Walshaw, 1975: The international comparison of ozone spectrophotometers, Belsk, 24 June to 6 July 1974. Publications of the Institute of Geophysics, Polish Academy of Sciences, Vol. 89, 2-59.

Grass, R.D., and W.D. Komhyr, 1985: Traveling standard lamp calibration checks on Dobson ozone spectrophotometers during 1981-83, *Atmospheric Ozone*, Proceedings of the Quadrennial Ozone Symposium held in Halkidiki, Greece, 3-7 September 1984, Reidel, Dordrecht, Holland, 376-380.

Gushchin, G.P., 1972: Comparison of ozonometric instruments. *Actinometry, Atmospheric Optics, Ozonometry* (G.P. Gushchin, editor), Glavnaya Geofizicheskaya Observatoriya Im. A.I. Voeikova, Trudy No. 279, Gidrometeorizdat, Leningrad, 118-126.

Komhyr, W.D., R.D. Grass, and R.K. Leonard, 1981: WMO 1977 international comparison of Dobson ozone spectrophotometers. Proceedings of the Quadrennial International Ozone Symposium, Boulder, Colorado, 4-9 August 1980 (J. London, editor), Vol. I, National Center for Atmospheric Research, Boulder, 25-32.

Komhyr, W.D., R. D. Grass, and R.K. Leonard, this issue: Calibration of Dobson ozone spectrophotometers in the global network during 1962-1987 with World Standard Dobson Instrument 83.

THE LATITUDINAL DISTRIBUTION OF OZONE TO 35 KM ALTITUDE FROM ECC OZONESONDE OBSERVATIONS, 1985-1987

W.D. Komhyr and S.J. Oltmans
NOAA/ERL/Air Resources Laboratory
Boulder, Colorado 80303, USA

P.R. Franchois
Cooperative Institute for Research in Environmental Sciences
University of Colorado
Boulder, Colorado 80309, USA

W.F.J. Evans
Atmospheric Environment Service
Downsview, Ontario, Canada

W.A. Matthews
Physics and Engineering Laboratory, DSIR,
Lauder, New Zealand

ABSTRACT

Electrochemical concentration cell (ECC) ozonesonde observations, made in recent years at nine stations whose locations range from the Arctic to Antarctica, have yielded a self-consistent ozone data base from which the latitudinal ozone distribution to 35 km altitude has been derived. Mean seasonal and annual ozonesonde profiles are presented, together with isopleths of ozone volume mixing ratios to 35 km altitude as a function of latitude. The data should be useful for comparison with model calculations of the global distribution of atmospheric ozone, for serving as a priori statistical information in deriving ozone vertical distributions from satellite and Umkehr observations, and for improving the satellite and Umkehr ozone inversion algorithms.

1. INTRODUCTION

Mean seasonal and annual ozone vertical distribution data are presented for Resolute, Canada, (RES, 74°N, 95°W); Point Barrow, Alaska (BRW, 71°N, 156°W); Edmonton, Canada (EDM, 53°N, 114°W); Boulder, Colorado (BLD, 40°N, 105°W); Hilo, Hawaii (HIL, 19°N, 155°W); American Samoa, South Pacific (SMO, 14°S, 170°W); Lauder, New Zealand (LDR, 45°S, 170°E); Syowa, Antarctica (SYO, 69°S, 39°E); and South Pole, Antarctica (SPO, 90°S, 24.8°W). Most of the observational data used in the analysis were obtained during 1985-1987. Table 1 shows the actual period of record for each station from which the mean data were derived, and the number of soundings made at each station. The data from all stations have been combined to yield isopleths of ozone volume mixing ratios to 35 km altitude as a function of latitude (90°N-90°S). To compare current results with those obtained in the past, see, e.g., Bojkov (1968).

Except at SYO, the ozone observations were made with balloon-borne ECC ozonesondes (Komhyr, 1969). At SYO, carbon-iodine (CI) ozonesondes were used that are predecessors of the ECC sondes and that have performance characteristics similar to those of the ECC instruments.

TABLE 1. PERIODS OF RECORD AND NUMBER OF ECC OZONESONDE SOUNDINGS USED IN DERIVING AVERAGE SEASONAL AND ANNUAL OZONE PROFILES AT NINE STATIONS

		Number of Soundings				
Sta.	Period of Record	DJF	MAM	JJA	SON	ANN
RES	Jan 1982 - Apr 1986	27	33	37	27	124
BRW	Feb 1986 - Apr 1988	16	35	14	17	82
EDM	Jan 1985 - Nov 1987	19	22	21	17	79
BLD	Jan 1985 - Dec 1987	37	43	34	42	156
HIL	Jan 1985 - Dec 1987	33	34	33	35	135
SMO	Apr 1986 - May 1988	16	25	25	19	85
LDR	Aug 1986 - May 1988	24	22	25	39	110
SYO	Jan 1982 - Dec 1986	15	6	4	29	54
SPO	Jan 1986 - Dec 1987	22	19	23	58	122

2. DATA ACCURACY

ECC ozonesondes yield total ozone amounts that agree well with Dobson spectrophotometer ozone values derived from observations on AD wavelengths using ozone absorption coefficients of Vigroux (1953). For 525 soundings made within the NOAA/GMCC network since 1984, over a wide range of operating conditions from the tropics to polar regions, the mean ECC sonde-Dobson instrument normalization factor was 1.009 ± 0.054 (σ).

Relative to the Vigroux ozone absorption coefficient scale, the ozone measurement uncertainty for ECC

ISBN 0-937194-15-8

ozonesondes is estimated to be ±10% in the troposphere, ±5% in the stratosphere to 10 mb, and ±5 to -20% between 10 and 3 mb (Hilsenrath et al., 1986). Although the magnitude of the underestimation of ozone by ECC ozonesondes near 3 mb is subject to question, it is known that the high altitude pumping efficiency characteristics of ECC ozonesondes are variable. Because of possible pumping problems, GMCC ozonesonde profiles are normalized using data to 7.8 mb (~33 km) only. Ozone profile data from non-GMCC stations have also been normalized in this report in a similar manner.

Assuming that ECC-sonde ozone values at 10 mb and above are too low, the following empirical corrections may tentatively be used to correct the 10-3 mb data presented here:

$$y = 1.749 - 0.807\ln x + 0.297(\ln x)^2 - 0.0367(\ln x)^3$$

where y is the correction factor and x is the air pressure in mb at 10 or less mb. Additional research is planned for the near future to verify the adequacy of these corrections.

3. RESULTS AND DISCUSSION

Average December-January-February (DJF), March-April-May (MAM), June-July-August (JJA), September-October-November (SON), and annual (ANN) ozone profiles for the nine stations are plotted in Fig. 1. Mean seasonal and annual ozone partial pressure data, for mandatory pressure levels, are given in Table 2. Three factors affecting the quality of the profiles and data are worth noting: First, because the MAM and JJA average ozone partial pressures for SYO were derived from only six and four ozonesonde soundings, respectively, they are most likely not representative of true conditions at that station during the spring and summer seasons. Second, because of variable balloon-burst altitudes, the seasonal and annual ozone means at the higher altitudes are generally derived from fewer ozone soundings than indicated in Table 1. Finally, SMO data and 1988 data for other stations used herein were not normalized to Dobson spectrophotometer total ozone. Calculated seasonal and annual ozone averages should, however, not be adversely affected since, as indicated earlier,

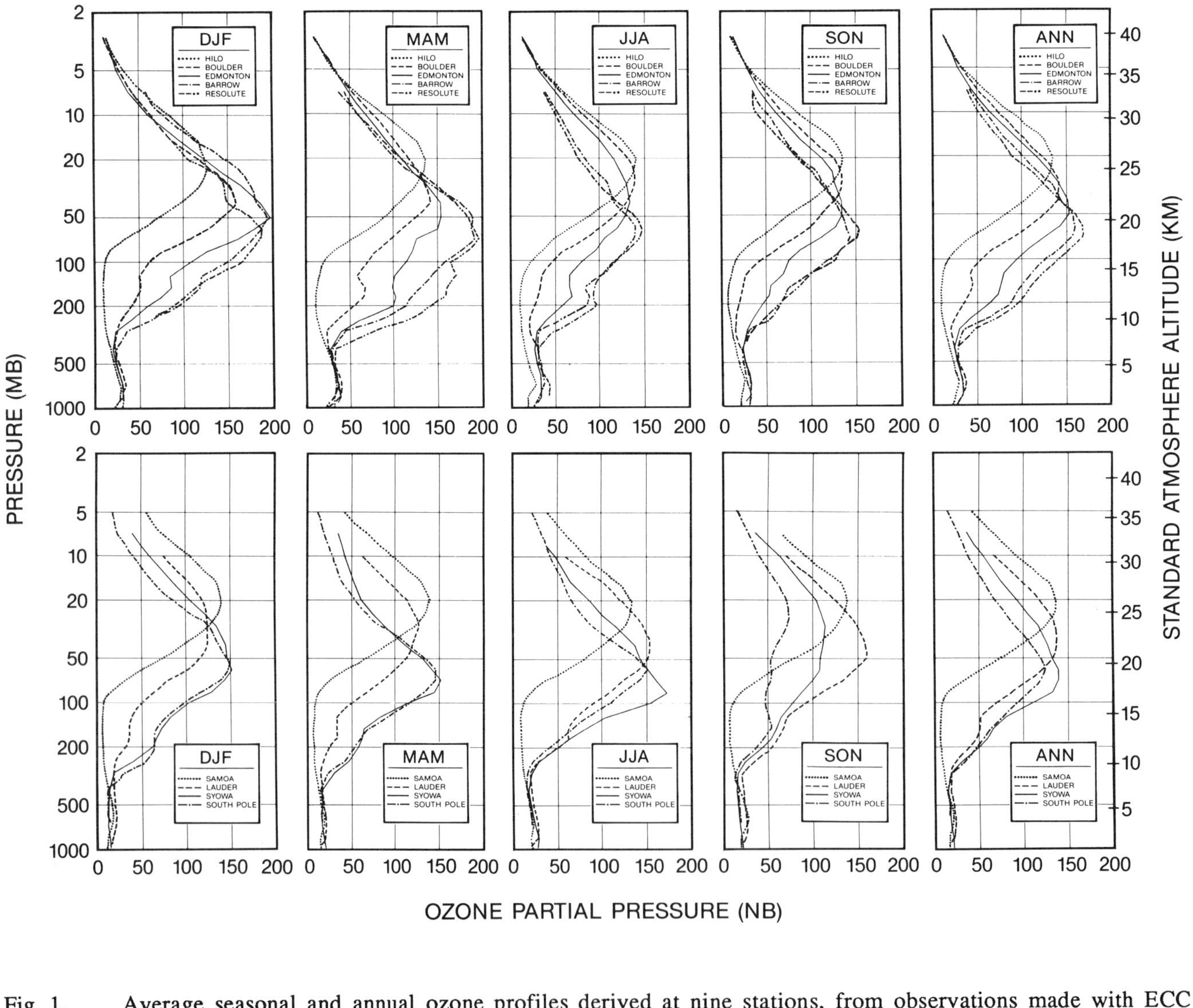

Fig. 1. Average seasonal and annual ozone profiles derived at nine stations, from observations made with ECC ozone-sondes during time periods indicated in Table 1.

TABLE 2. AVERAGE SEASONAL AND ANNUAL OZONE PARTIAL PRESSURED IN NANOBARS FROM ECC OZONESONDE SOUNDINGS AT STATIONS LOCATED BETWEEN 78°N - 90°S LATITUDE*

Pressure (mb)	DJF	MAM	JJA	SON	ANN	DJF	MAM	JJA	SON	ANN	DJF	MAM	JJA	SON	ANN
	Resolute					Point Barrow					Edmonton				
Surface	30.4	21.2	24.8	32.2	27.1	25.4	20.5	22.7	27.7	23.8	24.8	35.5	29.9	26.5	29.2
1000	30.7	22.0	25.9	31.6	27.3	29.2	25.1	24.5	29.9	26.9	-	-	-	-	-
700	30.5	34.0	36.6	30.3	32.8	29.6	36.7	36.8	30.5	33.7	23.3	30.1	29.4	26.4	27.3
500	24.3	31.3	33.0	24.7	28.4	23.3	30.1	33.0	25.5	28.2	22.8	39.6	29.1	28.6	29.9
300	34.4	77.7	46.8	34.1	47.8	37.8	44.8	31.0	30.9	34.8	59.0	99.6	58.0	46.2	64.3
200	87.7	142.4	98.5	65.2	97.4	95.4	116.3	86.4	66.4	88.6	86.3	98.7	66.1	55.2	76.5
150	113.2	160.2	94.4	88.8	112.4	118.6	133.1	84.9	78.4	100.4	108.2	115.3	81.4	75.6	94.8
100	167.3	164.3	116.8	128.6	142.9	149.1	159.4	117.3	120.8	125.7	161.4	126.9	105.8	112.5	127.0
70	186.5	196.3	144.7	152.4	169.1	178.3	192.2	139.6	144.9	156.5	197.8	153.9	128.0	134.8	153.0
50	176.1	183.8	141.0	142.0	160.5	195.1	189.6	131.2	135.3	156.5	164.9	137.5	129.8	126.8	140.2
30	146.3	141.6	98.9	110.7	121.7	175.9	147.9	110.2	114.6	131.4	124.8	109.8	115.4	112.1	115.7
20	106.0	103.6	76.9	83.5	90.6	148.0	107.6	83.2	86.8	98.9	97.1	91.2	100.4	88.3	94.1
15	88.4	82.6	66.2	64.8	74.4	114.5	87.2	70.0	71.1	80.7	60.6	67.2	71.2	62.0	64.3
10	-	57.9	49.8	38.1	49.7	76.8	57.4	53.7	51.3	56.5	38.4	48.2	49.3	42.5	43.8
7	-	43.9	37.2	35.0	39.4	57.7	36.6	37.5	35.8	39.8	25.0	32.7	29.8	29.6	28.6
5	-	-	-	-	-	-	-	-	-	-	20.3	22.9	25.0	22.7	22.5
4	-	-	-	-	-	-	-	-	-	-	13.9	9.0	13.0	13.7	12.5
	Boulder					Hilo					Samoa				
Surface	28.9	39.1	42.5	28.6	34.5	18.0	30.7	20.3	19.4	21.7	11.0	14.5	21.2	18.4	16.5
1000	-	-	-	-	-	22.9	32.1	19.3	20.7	23.5	11.4	15.2	21.2	18.5	16.7
700	32.7	40.8	42.8	32.6	37.3	29.5	37.1	28.5	24.9	29.9	16.7	18.3	22.0	25.9	20.3
500	24.9	30.1	31.1	25.2	27.8	21.8	28.9	20.6	19.8	22.7	15.3	15.6	18.5	22.8	17.9
300	22.9	23.7	20.2	14.6	20.4	13.4	16.7	12.7	11.6	13.5	9.3	8.9	11.7	12.1	10.5
200	41.1	54.3	25.6	18.6	34.9	10.5	11.7	10.1	7.6	9.9	5.9	6.7	8.8	7.4	7.1
150	51.4	67.9	32.7	23.5	44.0	9.4	13.1	12.0	8.8	10.7	5.6	6.3	8.0	6.8	6.7
100	61.1	70.2	46.2	43.7	55.4	11.8	20.8	24.2	15.9	17.9	7.2	8.9	11.0	10.2	9.2
70	106.2	98.2	84.8	84.0	93.7	34.4	55.3	50.1	44.3	45.8	33.5	26.8	36.6	36.8	33.4
50	144.1	128.0	116.0	111.8	125.9	73.3	91.8	88.6	75.4	82.1	72.6	68.0	80.6	77.5	74.5
30	152.0	138.5	138.9	136.6	141.9	120.1	128.4	131.4	127.3	126.7	126.9	125.0	125.2	127.3	126.0
20	116.6	118.2	131.6	125.8	122.9	124.2	137.1	140.9	137.6	134.7	140.3	140.0	133.6	139.3	138.3
15	88.0	101.3	112.7	101.6	100.7	116.5	126.9	125.8	123.4	123.1	134.7	130.7	118.3	131.2	128.4
10	61.6	76.1	80.9	71.3	72.3	83.3	89.9	90.3	86.2	87.4	105.9	96.8	81.5	92.1	94.6
7	42.9	50.7	52.6	48.0	48.5	53.2	54.2	58.3	54.4	55.0	72.3	65.2	57.3	65.3	66.5
5	28.4	31.1	33.0	31.0	30.9	32.3	30.8	34.6	31.3	32.1	55.0	39.3	37.6	-	43.9
4	19.8	20.8	21.8	22.3	21.2	21.3	20.7	24.4	20.7	21.6	-	-	-	-	-
3	10.8	10.2	13.1	14.9	12.1	11.4	10.3	13.0	10.0	11.0	-	-	-	-	-
	Lauder					Syowa					South Pole				
Surface	14.0	15.3	17.8	19.7	16.9	14.3	20.7	26.8	20.9	20.7	14.7	17.1	21.1	18.9	17.8
1000	-	-	-	-	-	11.6	16.5	24.3	18.1	18.3	11.6	14.9	18.1	15.9	15.1
700	20.5	20.1	26.9	26.0	23.3	12.1	17.0	19.5	16.2	17.4	29.7	22.8	15.1	13.4	20.6
500	18.2	17.5	22.1	23.8	20.2	18.2	30.0	20.3	16.3	18.5	64.3	52.9	41.3	37.5	48.9
300	18.0	15.3	18.6	20.2	17.3	62.1	57.3	49.8	43.5	50.5	66.3	67.6	67.3	53.9	64.0
200	33.2	24.3	50.3	55.1	39.7	71.9	63.7	81.0	59.2	66.4	98.6	116.0	111.0	47.1	94.3
150	35.5	33.5	61.6	66.0	49.4	103.1	115.3	153.5	77.3	111.4	140.7	146.9	145.1	53.4	121.7
100	50.7	49.8	91.5	90.0	69.6	143.7	151.7	161.0	99.9	139.4	148.2	133.9	140.6	52.6	118.6
70	81.5	86.3	124.4	135.2	106.9	146.4	131.8	143.0	108.2	132.3	125.0	83.3	89.7	72.4	91.2
50	115.5	114.9	152.4	160.2	134.4	134.3	87.8	116.0	113.9	116.7	84.3	52.5	65.3	71.8	67.2
30	124.0	128.3	146.5	147.6	135.9	104.6	60.6	86.5	104.0	92.3	61.3	37.6	56.3	59.6	53.0
20	117.4	114.4	121.3	126.8	119.4	85.2	52.3	65.2	85.8	75.0	40.5	27.1	41.7	42.5	37.4
15	101.5	92.1	99.9	103.5	98.5	59.0	42.2	47.5	62.9	54.7	22.5	17.5	32.0	28.4	25.7
10	73.9	61.1	55.8	70.3	67.1	40.2	34.8	-	36.0	36.5	17.0	10.5	21.0	16.1	15.4

*Estimated correction factors required to 10, 7, 5, 4, and 3 mb data are 1.017, 1.033, 1.066, 1.103, and 1.172, respectively (see section 2).

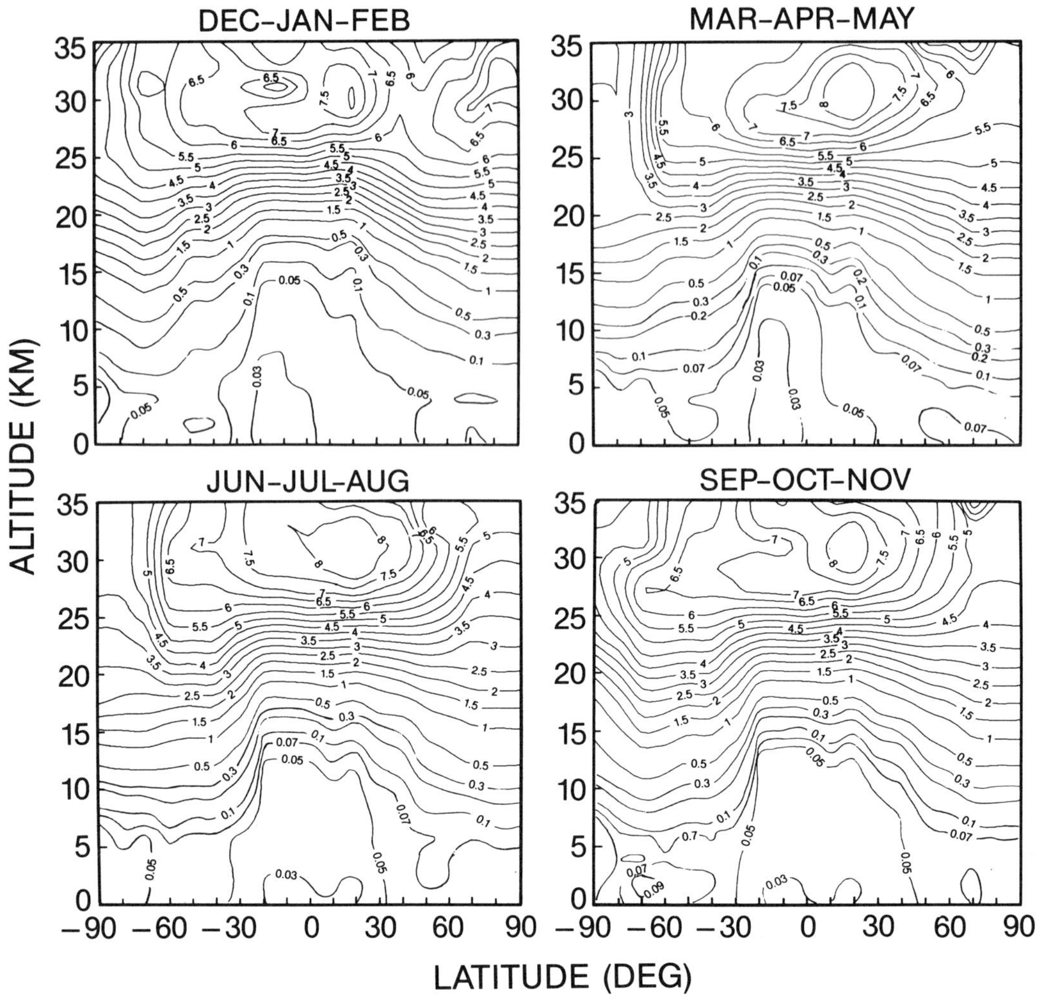

Fig. 2. Average seasonal isopleths of ozone volume mixing ratio to 35 km altitude as a function of latitude, from ECC ozonesonde observations made at nine stations between 74°N and 90°S latitude.

the average normalization factor for the ECC ozonesondes is ~1.0.

Average seasonal isopleths of ozone volume mixing ratio to 35 km altitude, as a function of latitude, are presented in Fig. 2 for the nine stations. SYO ozone data for MAM and JJA were not used in deriving these plots. For 75°-90°N latitude the isopleths were extrapolated, since northernmost ozone data in the Arctic were obtained at RES at latitude 74°N.

ACKNOWLEDGMENTS

Observations at BRW, BLD, HIL, SMO, and SPO were made by personnel of the U.S. National Oceanic and Atmospheric Administration (NOAA), Air Resources Laboratory; at EDM and RES by staff of the Canadian Atmospheric Environment Service (AES); at LDR by members of the New Zealand Physics and Engineering Laboratory, Division of Scientific and Industrial Research; and at SYO by observers of the Japanese Antarctic Expedition, Japan Meteorological Agency (JMA). Funding and logistics support for the various observing programs were provided wholly or in part by NOAA, AES, JMA, CMA (Chemical Manufacturers Association, U.S.), NASA (U.S. National Aeronautics and Space Administration), NESDIS (U.S. National Environmental Satellite, Data, and Information Service), and NSF (U.S. National Science Foundation). Maureen Ballard assisted with data analysis.

REFERENCES

Bojkov, R.D., 1968: Mean pole-to-pole vertical ozone distribution. *Publication in Meteorology No. 90*, Proc. VII Stanstead Seminar, McGill University, 221-240.

Hilsenrath, E., W. Attmannspacher, A. Bass, W. Evans, R. Hagemeyer, R.A. Barnes, W. Komhyr, K. Maursberger, J. Mentall, M. Proffitt, D. Robbins, S. Taylor, A. Torres, and E. Weinstock, 1986: Results from the Balloon Ozone Intercomparison Campaign (BOIC). *J. Geophys. Res., 91(D12)*, 13137-13152.

Komhyr, W.D., 1969: Electrochemical concentration cells for gas analysis. *Annals de Geophysique, 25 (1)*, 203-210.

Vigroux, E., 1953: Contribution a l'étude expérimentale de l'absorption de l'ozone. *Ann. Phys., 8*, 709-762.

UMKEHR OBSERVATIONS AT FIVE AUTOMATED DOBSON SPECTROPHOTOMETER STATIONS IN 1986 AND 1987

R.D. Evans
Cooperative Institute for Research in Environmental Sciences
University of Colorado
Boulder, Colorado 80309, USA

G.L. Koenig and W.D. Komhyr
NOAA/ERL/Air Resources Laboratory
Boulder, Colorado 80303, USA

J. Easson
Bureau of Meteorology
Melbourne, Australia

*P. Marche**
Laboratoire de Physique Moleculaire
University of Reims
Reims, France

ABSTRACT

Umkehr observations have been made with automated Dobson spectrophotometers at Poker Flat, Alaska; l'Observatoire de Haute Provence, France; Boulder, Colorado; Mauna Loa Observatory, Hawaii; and Perth, Australia. The five stations range in latitude from 6°N to 32°. We report on the data obtained only during 1986 and 1987 when adverse effects of El Chichon aerosols on the Umkehr observations were no longer significant. The data set consists of 2199 selected profiles. Averaged seasonal and annual profiles are presented, exhibiting characteristic features related to total ozone amount and station latitude. A brief description of the measurement program, related to data quality, is included.

1. INTRODUCTION

In response to concern about possible detrimental effects on atmospheric ozone from anthropogenic pollutants, a seven-station network of automated Dobson spectrophotometers has been established to provide information on stratospheric ozone. The stations are Poker Flat Research Range, Alaska (POK, 65°N, 148°W); l'Observatoire de Haute Provence, France (OHP, 44°N, 6°E); Boulder, Colorado (BDR, 40°N, 105°W); Mauna Loa Observatory, Hawaii (MLO, 20°N, 156°W); and Perth Airport, Australia (PTH, 32°S, 116°E). Two other stations, Observatorio de Huancayo, Peru (HUA, 12°S, 76°W), and Lauder, New Zealand (LDR, 45°S, 169°E) are operational, but the data from these are not included in this report. Huancayo's instrument was out of operation for much of the time period because of problems with the local electrical power, and Lauder's was not installed until February 1987. The station sites were selected for wide latitudinal coverage, favorable observing conditions, logistical ease, and station proximity to an existing or potential lidar aerosol observing facility. This network was established as a joint EPA/NOAA/WMO project.

The first of these stations (Boulder) was installed in early 1983; the last (Perth) was installed in the summer of 1984. Data for the time period before 1986 have been published by the World Ozone Data Centre in *Ozone Data for the World* (See Section 3 for bibliographic information). Data obtained prior to 1986 are not included in this report because they were adversely influenced by stratospheric aerosols introduced into the atmosphere by the El Chichon volcano eruption in March-April 1982.

The Dobson spectrophotometer (Dobson, 1957) measures the relative intensity of pair wavelengths in the ultraviolet (300-340 nm) spectrum. It is used for observations of both ozone total column content and ozone vertical distribution, the latter by the Umkehr observation technique (Gotz et al., 1934). Details of observational techniques for total ozone column content and Umkehr observations are available in WMO Global Ozone Research and Monitoring Project Report No. 6 (Komhyr, 1980). An automated instrument (Komhyr et al., 1985) was developed because manual Umkehr observations are difficult and tedious to make, and are subject to operator bias.

2. MEASUREMENT PROGRAM

Measurements are made with the Dobson spectrophotometer on zenith skylight while the sun's zenith angle changes between 60° and 90°, in the mornings or afternoons. At appropriate times, the automated system makes the measurements on the Dobson instrument's A (305.5 and 325.4 nm), C (311.45 and 332.4 nm), and D (317.6 and 339.8 nm) wavelength pairs. A plot of these measurements vs. the solar zenith angle has a characteristic reversal (or Umkehr) in its shape that is related to the effective scattering height in the atmosphere of the light being observed at the time of the measurement.

*Present Address: Laboratoire de Vision & Robeotique, University of Orleans, Bourges, France

ISBN 0-937194-15-8

An observer may terminate an Umkehr observation set or select to cancel the next possible Umkehr observation set if observing conditions are poor. Observers also make several daily observations of the ozone total column content, used in the reduction of the Umkehr observations. Calibration checks on the spectrophotometer are made on a bi-weekly basis using spectral and reference lamps. Other calibration checks are done on a longer time scale, both with lamps and by direct comparison with a standard Dobson instrument.

3. DATA PROCESSING

The data presented here are ozone vertical distributions (profiles), retrieved from the observations of the Umkehr effect described above. The number of possible observations at the five stations during 1986 and 1987 was approximately 3410. Many were canceled by an operator because of unfavorable observing conditions. After the data were received in Boulder and examined, more of the observation sets were rejected because of obvious interferences by cloud. After editing the remaining data where necessary, the resultant observation sets were converted to a standard format and sent to the WMO World Ozone Data Centre operated by the Atmospheric Environment Service, Downsview, Canada, for reduction to ozone profiles. There are two methods for reduction of the observations to ozone profiles, the conventional (Mateer and Dutsch, 1964) and short (Mateer and DeLuisi, 1980). Only the profiles retrieved by the conventional method (measurements on C wavelengths) are described in this paper.

An additional criterion was used in selecting the final Umkehr profile data set: only profiles with a root-mean-square difference (the "residual") of less than 0.7 between the observations at the various zenith angles and the observations as reconstructed from the derived profile are included. The number of profiles remaining after editing is 2199.

All data used in this paper have been submitted for publication in *Ozone Data for the World,* available from World Ozone Data Centre,Atmospheric Environment Service, 4905 Dufferin Street,Downsview, Ontario, Canada M3H 5T4. Original data are retained in Boulder, Colorado, USA, at the NOAA/ERL/ARL/GMCC facility.

4. RESULTS AND DISCUSSION

Seasonal and annual ozone partial pressure values in nanobars, for Umkehr layers described in Table 1, are shown in Table 2 for the five automated Dobson stations in 1986 and 1987. Figure 1 displays profiles plotted from the data in Table 1. Annual average data are not given for POK because observations are not possible for the entire year at that latitude.

Year-to-year changes (Fig. 1) in the seasonal and annual average profiles are small. No station shows a year-to-year annual difference in any layer larger than the standard deviations associated with the layer means. This is generally repeated for the year-to-year seasonal differences, except for October-December averages for 1987, an effect of the Quasi-Biennial Oscillation. The profiles of Fig. 1, furthermore, show two generally known features of ozone vertical distribution. The altitude of the ozone maximum decreases as station latitude increases, and the maximum at higher latitude stations is broader than at lower latitude. This broadening is greatest during winter seasons, and is related to total ozone column content amount.

The limitations of the Umkehr technique to accurately depict actual ozone vertical distribution are discussed in detail elsewhere (Bojkov, 1966). These stem primarily from inappropriate first-guess profiles used in the Umkehr data reduction algorithm, as well as from the insensitivity of the Umkehr technique to ozone in layers 1 and 2. With measurements made only during periods of clear skies, consideration must be given to how representative the seasonal and annual averages shown are of the actual average Umkehr profile at the station. At MLO and PTH, the average profiles are clearly representative because of year-round clear skies. At BDR and OHP, the seasonal and annual averages may be biased because profiles from periods of cloudiness do not enter the data set. At POK, the time required to make a complete Umkehr observation can be 8 hours, making it more likely that the true ozone profile will change during the measurement, thereby leading to errors in the deduced profile.

ACKNOWLEDGMENTS

This work was funded by the National Environmental Satellite, Data, and Information Service. The authors thank the observers at the five stations whose work in operation and maintenance of the instruments facilitated collection of the data set.

REFERENCES

Bojkov, R.D., 1966: Differences in the vertical ozone distribution deduced from Umkehr and ozonesonde data at Goose Bay. *J. Appl. Meteor.*, 22, 872-877.

Dobson, G.M.B., 1957:, Observers' handbook for the ozone spectrophotometer. *Annals of the International Geophysical Year,* Vol. 5, Part I, 46-89, Pergamon Press, New York.

Gotz, F.W.P., A.R. Meetham, and G.M.B. Dobson, 1934; *Proc. Roy. Soc.*, London, A, 145, 416.

Komhyr, W.D., 1980: *Operations handbook - Ozone observations with a Dobson spectrophotometer, WMO Global Ozone Research and Monitoring Project Rep. No. 6,* World Meteorological Organization, Geneva, Switzerland.

Komhyr, W.D., R.D. Grass, R.D. Evans, R.K. Leonard, and G.M. Semeniuk, 1985: Umkehr observations with automated Dobson spectrophotometers. *Atmospheric Ozone, Proceedings of the Quadrennial International Ozone Symposium,* 371-375, D. Reidel, Dord recht.

Mateer, C.L., and J.J. DeLuisi, 1980: The estimation of the vertical distribution of ozone by the short Umkehr method. *Proceedings of the Quadrennial International Ozone Symposium, 1,* 64-73, National Center for Atmospheric Research, Boulder, Colo.

Mateer, C.L., and H.U. Dutsch, 1964: Uniform evaluation of Umkehr observations from the World Ozone Network: Part I, Proposed standard evaluation technique. National Center for Atmospheric Research, Boulder, Colo.

TABLE 1. UMKEHR LAYER PRESSURE BOUNDARIES AND LAYER MIDPOINT HEIGHTS

Layer number	Pressure range (mb)	Height (km)
1	1000.0 - 250.0	5.1
2	250.0 - 125.0	12.5
3	125.0 - 62.5	16.9
4	62.5 - 31.2	21.4
5	31.5 - 15.6	25.9
6	15.6 - 7.8	30.4
7	7.8 - 3.9	35.0
8	3.9 - 1.96	40.0
9	1.96 - 0.98	45.2

TABLE 2. SEASONAL AND ANNUAL UMKEHR LAYER AVERAGE OZONE PARTIAL PRESSURES IN NANOBARS AT FIVE UMKEHR OBSERVATORIES DURING 1986 AND 1987

	Layer no.	JFM	AMJ	JAS	OND	Annual
			1986			
POK	9		5.3 ± 0.6	6.1 ± 1.2	6.0	
	8		15.8 ± 0.9	17.1 ± 2.0	15.8	
	7		36.8 ± 1.9	36.1 ± 3.4	31.0	
	6		73.2 ± 2.8	71.9 ± 5.5	65.0	
	5		112.1 ± 4.3	111.4 ± 8.5	100.0	
	4		128.5 ± 6.8	117.0 ± 8.5	115.0	
	3		106.0 ± 12.7	89.4 ± 8.5	91.0	
	2		104.6 ± 17.8	74.3 ± 17.2	89.0	
	1		38.6 ± 8.0	27.3 ± 9.8	30.0	
	No. profiles:		25	16	1	
OHP	9	5.9 ± 0.9	5.4 ± 0.7	5.6 ± 0.7	6.7 ± 1.0	5.9 ± 1.0
	8	17.3 ± 1.4	17.7 ± 1.2	17.7 ± 1.4	18.9 ± 1.9	18.0 ± 1.6
	7	41.8 ± 4.5	49.6 ± 2.4	46.3 ± 3.6	42.2 ± 3.8	45.2 ± 4.7
	6	81.3 ± 6.8	92.7 ± 2.9	86.7 ± 5.1	80.3 ± 3.8	85.5 ± 6.8
	5	120.4 ± 12.6	130.7 ± 4.7	119.7 ± 10.2	119.1 ± 5.2	122.3 ± 9.8
	4	129.5 ± 18.6	134.7 ± 7.2	109.4 ± 15.6	111.5 ± 13.3	119.9 ± 17.9
	3	100.4 ± 22.9	96.5 ± 14.9	65.5 ± 14.2	76.7 ± 15.0	82.6 ± 21.7
	2	93.9 ± 24.3	76.9 ± 22.6	53.0 ± 10.6	58.4 ± 15.8	67.9 ± 23.8
	1	39.6 ± 19.7	33.8 ± 12.4	35.4 ± 18.8	26.6 ± 8.5	33.4 ± 15.9
	No. profiles:	38	50	61	55	204
BDR	9	6.3 ± 0.7	4.9 ± 0.3	5.4 ± 0.5	6.2 ± 0.7	5.8 ± 0.8
	8	17.8 ± 1.0	16.8 ± 0.7	18.1 ± 0.9	18.0 ± 1.3	17.7 ± 1.1
	7	40.6 ± 3.4	47.6 ± 1.5	46.2 ± 2.1	38.2 ± 4.0	42.8 ± 5.0
	6	82.4 ± 3.9	92.2 ± 2.5	92.0 ± 3.2	79.5 ± 5.4	86.0 ± 7.1
	5	127.6 ± 3.9	131.3 ± 4.8	134.3 ± 4.7	125.1 ± 6.0	129.4 ± 6.2
	4	131.8 ± 7.7	126.1 ± 5.8	110.4 ± 8.3	107.8 ± 15.2	117.1 ± 14.5
	3	91.7 ± 10.4	75.1 ± 9.6	58.3 ± 6.2	70.1 ± 17.9	72.3 ± 17.0
	2	56.6 ± 13.1	45.0 ± 13.5	35.4 ± 4.2	47.9 ± 13.1	45.7 ± 13.6
	1	19.6 ± 8.3	24.0 ± 7.9	22.4 ± 6.0	21.0 ± 6.9	21.7 ± 7.3
	No. profiles	34	31	45	51	161
MLO	9	5.4 ± 0.5	5.3 ± 0.4	5.9 ± 0.4	5.5 ± 0.5	5.5 ± 0.5
	8	19.7 ± 0.9	19.3 ± 0.6	20.4 ± 0.6	19.6 ± 0.7	19.7 ± 0.8
	7	53.0 ± 3.0	53.1 ± 1.7	53.0 ± 1.3	51.1 ± 1.7	52.5 ± 2.3
	6	102.7 ± 4.6	104.4 ± 1.8	102.7 ± 1.8	99.2 ± 2.9	102.2 ± 3.7
	5	142.6 ± 6.2	148.0 ± 3.6	146.2 ± 2.1	139.5 ± 4.5	143.7 ± 5.7
	4	92.9 ± 8.4	107.2 ± 5.9	99.7 ± 2.7	91.8 ± 4.6	97.4 ± 8.8
	3	20.7 ± 9.2	40.5 ± 5.9	31.9 ± 5.1	24.6 ± 5.6	28.7 ± 10.4
	2	6.6 ± 5.5	15.9 ± 4.4	9.2 ± 5.2	10.4 ± 4.7	10.4 ± 6.1
	1	17.7 ± 7.8	17.6 ± 6.8	11.0 ± 7.3	16.6 ± 5.8	16.3 ± 7.3
	No. profiles:	110	91	58	97	356
PTH	9	4.8 ± 0.6	6.2 ± 0.5	6.4 ± 0.6	5.0 ± 0.7	5.5 ± 0.9
	8	17.7 ± 0.9	19.0 ± 0.7	18.7 ± 1.0	17.7 ± 1.1	18.2 ± 1.1
	7	45.6 ± 1.5	41.0 ± 2.4	44.1 ± 3.8	49.8 ± 1.7	45.4 ± 4.0
	6	93.8 ± 2.4	84.4 ± 4.1	86.5 ± 4.2	96.0 ± 2.2	90.9 ± 5.8
	5	136.0 ± 3.8	129.0 ± 4.4	128.0 ± 3.8	134.1 ± 5.3	132.5 ± 5.5
	4	91.9 ± 5.3	92.4 ± 7.0	115.0 ± 13.6	113.5 ± 10.4	101.1 ± 13.8
	3	33.0 ± 8.2	45.8 ± 11.0	68.7 ± 14.8	56.2 ± 11.9	47.7 ± 16.3
	2	21.1 ± 4.0	30.7 ± 6.8	41.0 ± 9.0	31.1 ± 6.1	29.1 ± 8.8
	1	18.9 ± 5.0	17.8 ± 4.1	22.2 ± 7.3	24.5 ± 6.6	20.6 ± 6.3
	No. profiles:	75	66	30	69	240

TABLE 2. continued

	Layer no.	JFM	AMJ	JAS	OND	Annual
			1987			
POK	9	8.4 ± 1.0	6.3 ±1.0	5.9 ± 0.7	6.6	
	8	19.7 ± 1.6	17.3 ± 1.3	16.9 ± 1.4	16.4	
	7	36.2 ± 2.2	37.8 ± 3.9	36.3 ± 1.7	30.0	
	6	75.1 ± 3.5	74.6 ± 5.1	70.9 ± 1.6	64.0	
	5	124.4 ± 7.8	115.4 ± 8.2	108.4 ± 3.7	100.0	
	4	148.5 ± 11.5	138.2 ± 12.0	112.5 ± 12.3	115.0	
	3	135.9 ± 16.3	131.2 ± 22.5	87.7 ± 11.0	91.0	
	2	117.2 ± 6.3	129.6 ± 19.5	78.8 ± 15.4	89.0	
	1	38.0 ± 11.9	45.4 ± 15.3	30.1 ± 7.9	33.0	
	No. profiles:	10	33	21	1	
OHP	9	6.1 ± 1.0	5.6 ± 0.6	5.8 ± 0.5	6.5 ± 0.8	5.9 ± 0.7
	8	18.2 ± 1.7	18.2 ± 1.1	18.4 ± 0.9	18.5 ± 1.7	18.3 ± 1.3
	7	45.5 ± 5.2	51.3 ± 2.3	49.5 ± 2.7	39.3 ± 3.5	47.5 ± 5.4
	6	85.6 ± 6.4	93.6 ± 2.9	91.4 ± 3.0	77.2 ± 4.2	88.5 ± 7.2
	5	125.4 ± 7.3	129.7 ± 5.3	125.2 ± 3.7	118.6 ± 6.9	125.4 ± 6.7
	4	131.7 ± 16.6	134.0 ± 8.1	117.1 ± 11.0	108.0 ± 10.9	123.2 ± 15.2
	3	103.8 ± 25.5	99.5 ± 16.1	71.3 ± 11.0	77.4 ± 11.4	86.8 ± 21.2
	2	91.7 ± 31.1	87.7 ± 21.3	53.1 ± 9.8	59.8 ± 15.4	71.7 ± 25.9
	1	38.2 ± 18.9	41.1 ± 10.3	33.4 ± 9.7	25.0 ± 9.9	35.1 ± 13.2
	No. profiles:	28	68	73	40	219
BDR	9	6.4 ± 0.7	5.1 ± 0.6	5.6 ± 0.6	6.2 ± 0.6	5.8 ± 0.8
	8	18.5 ± 1.3	17.1 ± 0.9	18.2 ± 1.1	18.5 ± 1.1	18.1 ± 1.2
	7	42.4 ± 3.7	48.5 ± 1.7	48.2 ± 2.2	40.6 ± 4.6	45.0 ± 4.8
	6	84.6 ± 4.4	93.7 ± 3.0	3.1 ± 2.9	82.9 ± 6.1	88.8 ± 6.6
	5	129.7 ± 6.3	133.2 ± 5.6	131.4 ± 5.0	126.6 ± 6.4	130.1 ± 6.3
	4	123.7 ± 17.6	131.2 ± 8.6	113.1 ± 11.8	103.1 ± 12.7	116.6 ± 16.6
	3	83.8 ± 22.6	80.6 ± 13.1	60.3 ± 11.2	61.8 ± 12.4	69.9 ± 17.9
	2	56.1 ± 20.8	48.8 ± 16.7	38.0 ± 7.9	43.4 ± 8.5	45.4 ± 14.9
	1	21.0 ± 9.9	27.4 ± 9.4	26.4 ± 9.1	23.6 ± 7.6	24.9 ± 9.2
	No. profiles:	38	53	61	62	214
MLO	9	5.3 ± 0.5	5.0 ± 0.3	5.3 ± 0.3	5.3 ±0.4	5.3 ± 0.4
	8	19.5 ± 0.7	18.6 ± 0.5	19.2 ± 0.5	18.7 ± 0.6	19.0 ± 0.7
	7	52.6 ± 1.1	51.9 ± 1.7	51.3 ± 1.3	46.7 ± 1.9	50.4 ± 2.9
	6	100.6 ± 2.5	102.4 ± 2.0	101.7 ± 1.5	92.2 ± 3.7	98.6 ± 5.1
	5	137.9 ± 4.3	144.1 ± 4.9	144.8 ± 2.2	130.2 ± 5.7	138.3 ± 7.5
	4	87.3 ± 5.6	105.4 ± 8.1	99.7 ± 3.0	83.5 ± 6.0	92.3 ± 10.3
	3	17.0 ± 8.3	41.0 ± 9.5	35.1 ± 4.0	22.7 ± 5.8	27.3 ± 11.5
	2	7.9 ± 6.4	19.9 ± 6.0	15.6 ± 4.7	16.9 ± 4.2	14.6 ± 6.9
	1	19.1 ± 7.5	22.6 ± 7.6	16.0 ± 7.0	21.4 ± 6.4	19.6 ± 7.5
	No. profiles:	117	74	101	127	419
PTH	9	4.8 ± 0.5	5.9 ± 0.6	6.4 ± 0.7	5.4 ± 0.7	5.5 ± 0.9
	8	18.0 ± 0.7	18.9 ± 0.7	19.3 ± 1.0	18.3 ± 1.0	18.5 ± 1.0
	7	47.2 ± 1.4	43.7 ± 2.0	46.5 ± 2.6	50.5 ± 1.9	47.4 ± 2.8
	6	95.1 ± 2.5	88.1 ± 3.9	88.5 ± 3.7	95.7 ± 3.1	92.7 ± 4.6
	5	136.2 ± 3.3	131.6 ± 5.1	126.8 ± 4.6	131.7 ± 5.3	132.2 ± 5.7
	4	92.9 ± 6.0	94.1 ± 5.6	108.3 ± 13.3	110.9 ± 13.2	101.2 ± 13.0
	3	34.5± 9.0	46.5 ± 8.0	58.1 ± 13.3	52.2 ± 15.0	46.3 ± 15.2
	2	22.5 ± 5.4	31.8 ± 6.1	36.8 ± 7.1	29.9 ± 5.7	29.1 ± 8.3
	1	19.3 ± 4.0	17.2 ± 4.1	26.2 ± 5.5	29.3 ± 6.4	23.2 ± 6.8
	No. profiles:	105	37	70	67	279

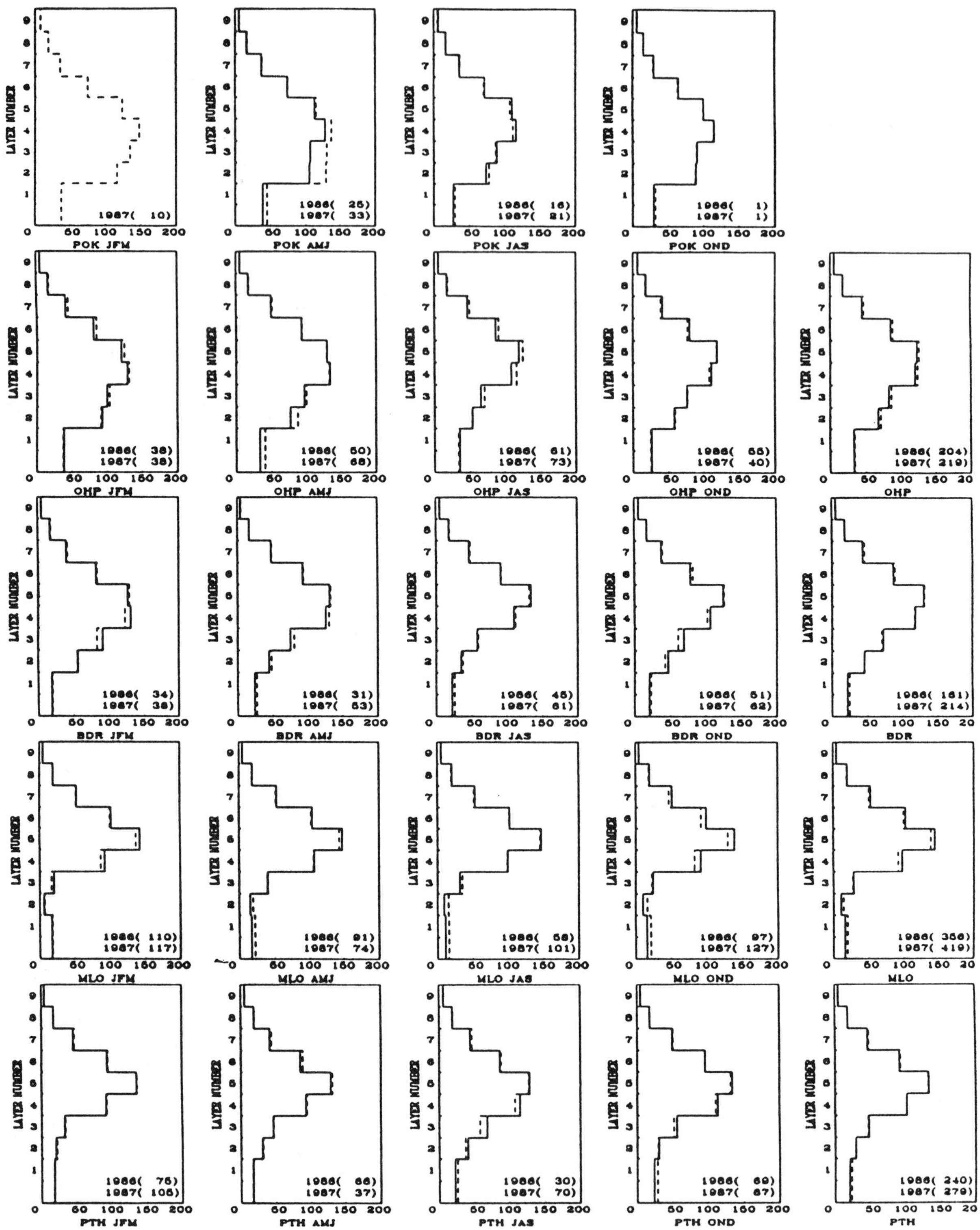

Figure 1. Seasonal and annual Umkehr layer average ozone partial pressures in nanobars at five Umkehr observatories during 1986 and 1987. Solid-line profiles are from 1986; dashed-line profiles are from 1987. The numerals in parentheses after the dates, indicate the number of profiles in the average.

COMPARISON OF ECC OZONESONDE AND UMKEHR PROFILES AT BOULDER AND MAUNA LOA, 1985–1987

P. Franchois
Cooperative Institute for Research in Environmental Sciences
University of Colorado
Boulder, Colorado 80309, USA

G. Koenig
NOAA/ERL/Air Resources Laboratory
Boulder, Colorado 80303, USA

R.D. Evans
Cooperative Institute for Researach in Environmental Sciences
University of Colorado
Boulder, Colorado 80309, USA

W.D. Komhyr
NOAA/ERL/Air Resources Laboratory
Boulder, Colorado 80303, USA

ABSTRACT

Umkehr observations have been made with an automated Dobson spectrophotometer at Boulder, Colorado, since early 1983, and at Mauna Loa, Hawaii, since early 1984. Numerous balloonborne electrochemical concentration cell (ECC-Komhyr) ozonesonde soundings have also been made at Boulder and at Hilo, Hawaii, located near sea level about 60 km from Mauna Loa Observatory. The Umkehr data are compared with quasi-simultaneously obtained ozonesonde data converted to the layer averaged format of the Umkehr profiles. Comparisons are made for seasonally, annually, and long-term averaged data. Results of the comparison are shown to differ according to season and station location but in general indicate agreement to within ±10% between the measurement methods in Umkehr layers 4 to 6.

1. INTRODUCTION

Previous studies (Bojkov,1966; Craig et al., 1967; Dütsch and Ling, 1969; DeLuisi and Mateer, 1971) have compared ozone profile data taken in situ by MAST (Brewer) or chemiluminescent ozonesondes with ozone profile data produced by the indirect Umkehr observation method. A common recommendation of these investigators was that the use of empirical data be increased to improve the first-guess profiles for the Umkehr reduction algorithm. Inaccuracies in early sonde data and low sonde burst heights have limited the usefulness of past Umkehr-sonde comparisons. This present comparative study between standard Umkehr ozone retrievals and profiles from ECC (Komhyr) ozonesondes is concerned mainly with Umkehr layers 3 through 7 (14-40 km), an altitude range for which both Umkehr and ECC (Komhyr) ozonesonde data are available.

ECC (Komhyr) ozonesonde measurement reliability has been enhanced in the last 10 years because of the use of an improved air sampling pump (Komhyr et al., 1985b) and the use of consistent preflight preparation techniques (Komhyr, 1986). Present ozone measurement uncertainty for the ECC (Komhyr) ozonesonde is estimated to be ±10% in the troposphere and ±5% in the stratosphere to 10 mb. Above 10 mb the uncertainty increases, reaching 20% at 3 mb (Komhyr et al., 1987). With the use of 5000 m^3 plastic balloons, burst heights are now typically between 39 and 40 km, through Umkehr layer 7.

Approximately weekly balloon ozone soundings have been made at Hilo, Hawaii (20°N), and Boulder, Colorado (40°N), from January 1985 through the present. Automated Dobson instruments at Mauna Loa and Boulder have operated continuously during that time, making Umkehr measurements when zenith sky conditions were clear. The 3-year data set contains 700 morning and 426 afternoon Umkehr profiles from Mauna Loa and 384 morning and 187 afternoon Umkehr profiles from Boulder. All Umkehr results presented were obtained from the uniform processing (Mateer and Dütsch, 1964) by the WMO-World Ozone Data Centre, Ontario, Canada of observations made on Dobson instrument C wavelengths. Only those Umkehr profiles that had residuals (a measure of data quality) less than 0.7 were used in the comparisons.

In processing the ECC (Komhyr) ozonesonde data, the integrated column ozone amounts were calculated above the burst heights by assuming the mixing ratio to be constant from 7.8 mb to 0.1 mb. The sonde data were then normalized to the Dobson total ozone. The total column ozone for Hilo, Hawaii, was estimated from the Mauna Loa Dobson spectrophotometer total ozone plus sonde ozone data, integrated from Hilo at sea level to the 3.4 km altitude

OZONE IN THE ATMOSPHERE
Rumen D. Bojkov and Peter Fabian (Eds.)

ISBN 0-937194-15-8

data, integrated from Hilo at sea level to the 3.4 km altitude of Mauna Loa Observatory. Only those data from sonde flights with normalization factors of 1.000 ± 0.150 were retained in the comparison group.

For the comparisons, ECC (Komhyr) sonde flight data from December 1984 through December 1987 were matched with the same local day's morning, afternoon, and/or the average morning/afternoon Umkehr observation. Of the 166 total successful sonde flights made at Boulder during the 3 years, 63 were matched with 85 Boulder Umkehr profiles; of the 146 sonde flights made at Hilo, 99 were paired with 145 Mauna Loa Umkehr profiles. Layer comparisons were available for Umkehr layers 1 through 7 (500 to 3.9 mb).

Seasonal and annual layer differences were formed for each of the years and for the 3 years together. The differences are expressed in Tables 1 and 2 and in Figs. 1 and 2 in percentages, calculated as 100 * [(Umkehr - sonde) ÷ Umkehr]. For the percent difference calculations, 95% confidence intervals were computed in order to determine the statistical significance of the Umkehr-sonde differences.

2. RESULTS AND DISCUSSION

2.1 ANNUAL/LONG-TERM DIFFERENCES

Both annual and long-term mean differences in Boulder increase from layer 5 to layer 7: 1-2% in layer 5, 5-10% in layer 6, and 14-18% in layer 7. Likewise, layer 3 has a 0-13% range of annual values with a 6.5% long-term average difference. Conversely, in layer 4 located in the lower stratosphere, ECC (Komhyr) ozonesonde profiles are 10-16% higher annually and over the long-term than the Umkehr method. The net effect on the stratospheric layers in Boulder is that the Umkehr algorithm redistributes an amount of ozone from layer 4 and adds it to layers 3, 5, 6, and 7.

As in Boulder, Umkehr retrievals in Hawaii contain more ozone than ECC (Komhyr) sonde retrievals in layers 5 to 7. Layer 5 has a 3-year 7.6% average difference, about 5% higher than in Boulder, but layer 6 is lower than in Boulder, averaging 4.3%, with annual differences ranging from 2.3-7.1%. Umkehr retrievals in Hawaii are on average 23.5% higher in layer 7, 7.4% higher than in Boulder. If one takes into account that the ECC (Komhyr) sondes may be 5-15% systematically low, the agreement in layer 7 at both stations may be better than 10%.

Unlike the results in Boulder, layer 4 at Mauna Loa agrees well: an annual mean difference of 0-4% and a long-term mean difference of 2.5%. On the other hand, layer 3 is highly variable: annually -6% to 16% from 1985 to 1987, averaging 4.7% over the 3 years. Note that the range of yearly average differences in layer 3 is large, implying that good long-term agreement may not hold for fewer comparisons over a shorter time period.

Annual differences in layer 2 range from 24% to 36% in Boulder and -19 to 31% in Hawaii. But surprisingly, the annual differences for layer 1 in Boulder, 9.6% in 1985, -3.8% in 1986, and 10.6% in 1987, are modest considering that layer 1 is the least well determined by the Umkehr algorithm. At Mauna Loa, except for a -20.2% value in 1985, layer 1 annual differences of -9.0% in 1986 and -2.6% in 1987 are also reasonable. Yet the range of annual differences and the confidence intervals of layers 1 and 2 are the largest of the seven layers compared, challenging the significance of such moderate 5-10% differences in layer 1.

2.2 SEASONAL DIFFERENCES

Mean seasonal differences demonstrate the variability of the Umkehr-sonde comparisons by season. This reflects some of the dependence of the Umkehr retrievals on the limited set of first-guess profiles used in the current standard Umkehr evaluation. In Boulder, more variation from season to season in each layer's Umkehr-sonde difference is found in layers 1 through 4 than in layers 5 through 7, whereas in layers 1 through 4 individual layer differences vary 20-40% over four seasons while layers 5 through 7 vary at most 10%. Spring months (March-May) in Boulder have a range of differences from -6.2% to 11.1% in layers 1 through 6 whereas the autumn months (September-November) have a wider -19.1 to 41.3% range in the same layers.

Layers 4 through 7 in the Mauna Loa comparisons show a 6.3% or less variation in differences through the seasons. Yet the first three layers have a 29.7%, 63.8%, and 26.6% range of differences, respectively, for all seasons. Stratospheric ozone seems to be more consistently measured at both stations than tropospheric ozone, indicated by less seasonal dependence of the Umkehr-sonde comparisons in layers 4 through 7. Differences vary less from season to season for stratospheric ozone (layers 4 to 7) at Mauna Loa than at Boulder. But in the tropospheric layers 1 through 3, Umkehr-sonde differences have less season-to-season variation in Boulder than Mauna Loa possibly because there is more tropospheric ozone over Boulder than over Mauna Loa and in general less percentage uncertainty around the higher mean ozone values.

3. CONCLUSIONS

This study demonstrates that the differences between ozone profiles retrieved by ozonesonde measurements and by Umkehr observations vary according to the season in which the observations are made and the latitude of the measuring site. However, year-to-year variations in seasonal differences between standard Umkehr and ozonesonde measurements are within the boundaries of instrument errors for the most part in layers 4 through 7, confirming that both ECC (Komhyr) ozonesondes and Umkehr observations at Boulder and Mauna Loa/Hilo have yielded consistent data during 1985-1987. Furthermore, it has been demonstrated that the numerous Umkehr profiles obtained at Boulder and Mauna Loa give an adequate climatological vertical ozone distribution in the stratosphere.

Umkehr retrievals could be improved by the compilation of a set of first-guess profiles for each site, based on ozonesonde data. First-guess profiles should be based on total ozone amount, season, and station location. A middle latitude station that has more variability in the total ozone column within a season requires more ozonesonde observations than tropical stations to establish first-guess ozone profiles. The ability of the automated Dobson network to provide numerous Umkehr observations of consistent quality at widely spaced latitudes combined with regular balloon flights at these stations presents a continuing opportunity to relate the two measurement methods.

TABLE 1. ANNUAL, LONG-TERM AND SEASONAL AVERAGES AND PERCENTAGE DIFFERENCES 1985-1987 BY UMKEHR LAYER FOR BOULDER, COLORADO

Layer	Mean Umk.	Std. Dev.	Mean Sonde	Std. Dev.	% Diff.	95% Conf.
1985 Annual (No. U=25 S=20)						
7	42.8	5.1	36.1	2.7	15.7	5.6
6	86.5	8.2	78.7	9.3	9.0	3.0
5	129.0	7.8	126.6	7.7	1.8	2.1
4	113.4	16.0	124.7	12.5	-10.0	5.1
3	69.4	17.9	69.1	23.4	0.4	6.4
2	46.8	16.4	35.5	18.9	24.1	9.8
1	24.6	8.5	22.2	6.3	9.6	12.8
1986 Annual (No. U=27 S=20)						
7	41.0	5.4	33.6	4.8	17.9	6.6
6	83.8	8.0	75.9	11.5	9.5	3.8
5	129.4	6.4	127.6	6.4	1.4	1.8
4	110.7	14.5	128.7	19.4	-16.3	6.5
3	68.2	17.3	59.1	11.6	13.3	7.0
2	41.9	9.4	26.9	14.1	35.6	15.0
1	19.3	6.2	20.0	3.9	-3.8	11.3
1987 Annual (No. U=33 S=23)						
7	44.8	4.7	38.5	4.1	14.2	5.7
6	88.4	6.6	83.6	10.9	5.4	3.0
5	129.7	7.9	128.2	7.0	1.2	2.0
4	114.4	15.5	129.4	14.0	-13.1	6.0
3	68.2	14.8	63.9	16.0	6.2	6.1
2	44.2	12.3	32.6	17.0	26.3	11.1
1	24.5	10.6	21.9	5.6	10.6	12.5
1985 - 1987 Long-term (No. U=85 S=63)						
7	43.0	5.2	36.1	4.4	16.1	3.2
6	86.4	7.7	79.6	10.9	7.9	1.8
5	129.4	7.3	127.5	7.0	1.5	1.1
4	112.9	15.1	127.7	15.4	-13.1	3.2
3	68.5	16.3	64.1	17.8	6.5	3.5
2	44.2	12.8	31.7	16.9	28.3	6.5
1	22.9	9.0	21.4	5.4	6.4	6.8
Dec-Feb Long-term (No. U=18 S=12)						
7	37.3	4.5	31.4	4.0	15.8	6.9
6	77.1	4.1	67.0	8.0	13.0	6.6
5	124.0	4.3	126.7	7.1	-2.1	1.8
4	118.0	11.8	146.6	15.0	-24.2	10.8
3	86.1	10.4	71.5	24.7	16.9	13.0
2	55.1	12.1	38.6	16.0	30.0	15.1
1	19.2	7.3	23.0	6.9	-20.2	20.8
Mar-May Long-term (No. U=14 S=12)						
7	47.6	2.1	38.1	3.7	19.9	10.7
6	92.6	3.3	85.7	2.4	7.5	4.2
5	132.3	4.8	126.6	5.9	4.3	3.1
4	126.0	11.6	133.8	15.3	-6.2	5.7
3	77.4	15.9	72.8	18.6	5.9	7.1
2	50.4	14.0	50.4	15.3	0.1	8.5
1	26.8	12.9	23.8	3.1	11.1	23.0
Jun-Aug Long-term (No. U=22 S=17)						
7	47.3	2.5	39.2	2.7	17.2	6.8
6	93.5	2.8	90.0	4.5	3.7	2.1
5	133.0	7.4	132.0	3.6	0.8	2.2
4	118.5	10.9	120.2	5.7	-1.4	2.9
3	64.1	11.3	58.3	11.9	9.0	4.9
2	37.6	8.8	24.3	13.8	35.2	15.0
1	26.1	8.9	21.9	5.2	16.2	12.3
Sep-Nov Long-term (No. U=29 S=21)						
7	41.4	3.7	35.8	3.5	13.5	5.1
6	84.3	5.9	76.2	8.7	9.6	3.5
5	128.7	8.0	125.3	8.1	2.6	2.1
4	100.2	12.2	119.3	9.9	-19.1	7.1
3	57.0	11.6	60.0	14.4	-5.3	3.8
2	39.3	10.2	23.1	9.9	41.3	14.9
1	21.1	6.7	18.6	4.8	11.8	7.6

TABLE 2. ANNUAL, LONG-TERM AND SEASONAL AVERAGES AND PERCENTAGE DIFFERENCES 1985-1987 BY UMKEHR LAYER FOR MAUNA LOA, HAWAII

Layer	Mean Umk.	Std. Dev.	Mean Sonde	Std. Dev.	% Diff.	95% Conf.
1985 Annual (No. U=38 S=27)						
7	51.9	3.1	40.2	4.0	22.5	6.9
6	101.1	5.4	93.9	7.3	7.1	2.2
5	143.1	9.2	132.2	8.9	7.6	2.3
4	96.3	12.0	92.3	10.4	4.2	2.4
3	27.7	13.3	29.4	8.8	-6.3	8.3
2	8.7	7.3	10.3	7.0	-19.0	17.9
1	12.7	6.3	16.2	6.5	-27.8	16.0
1986 Annual (No. U=51 S=33)						
7	52.5	2.1	39.8	5.7	24.2	6.8
6	102.1	3.8	97.9	6.2	4.1	1.8
5	144.4	6.1	131.6	5.6	8.9	2.4
4	98.0	9.3	93.9	6.7	4.2	2.1
3	29.1	11.3	24.2	8.3	16.8	6.5
2	9.0	5.8	9.7	5.1	-8.6	13.9
1	13.2	5.0	14.4	5.3	-9.0	8.2
1987 Annual (No. U=56 S=39)						
7	50.6	2.7	38.7	5.4	23.5	6.3
6	99.3	5.0	97.0	6.7	2.3	0.8
5	140.1	7.2	131.1	9.3	6.4	2.0
4	93.9	8.9	94.2	11.9	-0.3	1.5
3	29.5	11.0	29.0	10.3	1.6	4.5
2	14.5	7.1	10.0	5.3	31.2	10.9
1	17.0	5.2	17.4	5.2	-2.6	5.9
1985 - 1987 Long-term (No. U=145 S=99)						
7	51.6	2.7	39.5	5.1	23.5	3.7
6	100.8	4.8	96.4	6.8	4.3	0.9
5	142.4	7.6	131.6	8.0	7.6	1.2
4	96.0	10.0	93.6	9.9	2.5	1.1
3	28.9	11.7	27.5	9.5	4.7	3.4
2	11.0	7.2	10.0	5.7	9.4	7.5
1	14.5	5.8	16.1	5.7	-10.7	5.1
Dec-Feb Long-term (No. U=41 S=25)						
7	51.3	2.8	38.6	4.2	24.8	7.9
6	98.5	4.5	93.1	7.4	5.5	2.5
5	136.7	6.8	124.1	6.9	9.3	3.0
4	86.8	7.5	87.3	9.8	-0.6	2.8
3	17.4	8.9	19.6	9.0	-12.8	14.3
2	7.1	5.8	10.1	4.9	-43.2	24.7
1	16.6	4.8	16.5	3.9	0.6	7.7
Mar-May Long-term (No. U=25 S=17)						
7	53.3	1.9	39.1	4.7	26.7	10.4
6	104.1	1.7	100.5	3.9	3.4	1.5
5	146.6	3.6	133.3	7.0	9.1	3.9
4	104.1	10.5	101.3	9.6	2.7	3.5
3	35.8	12.8	30.7	10.9	14.0	8.2
2	13.4	7.6	11.6	7.1	13.6	18.9
1	16.0	5.3	20.7	5.3	-29.1	15.9
Jun-Aug Long-term (No. U=45 S=29)						
7	52.2	1.5	41.5	6.1	20.5	6.5
6	103.0	2.5	98.9	3.6	3.9	1.6
5	147.4	3.3	136.7	4.6	7.3	2.3
4	102.1	3.2	99.4	6.2	2.6	1.8
3	36.2	6.4	34.0	5.4	5.8	3.5
2	12.5	7.1	10.0	5.4	20.1	13.3
1	11.9	5.8	14.7	5.9	-23.9	11.1
Sep-Nov Long-term (No. U=30 S=24)						
7	50.5	2.9	37.9	3.6	24.9	8.8
6	99.5	5.0	95.4	6.8	4.1	2.1
5	141.6	6.8	133.4	5.9	5.7	2.2
4	95.1	5.9	89.6	4.7	5.7	2.3
3	29.1	7.0	26.8	6.7	7.9	8.0
2	11.4	7.2	9.1	5.8	20.6	17.5
1	13.8	5.9	14.1	6.2	-2.3	12.8

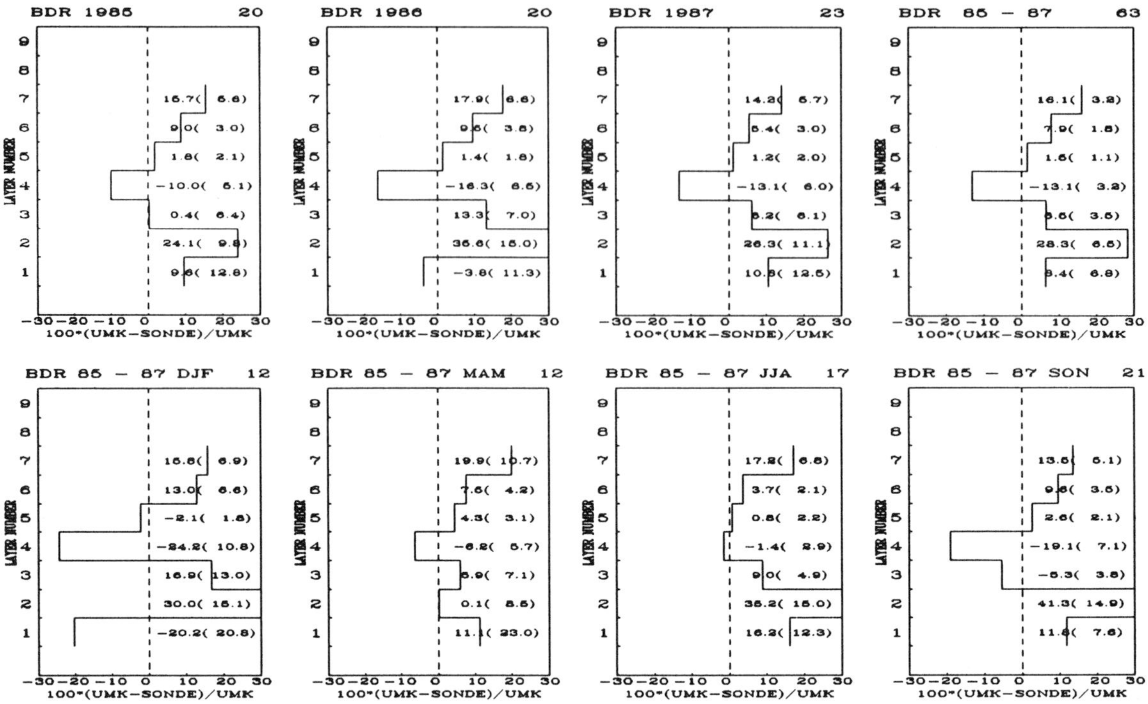

Fig. 1. Annual, long-term, and seasonal percentage differences, 100 * [(Umkehr-sonde) ÷ Umkehr], by Umkehr layer for Boulder, Colorado, 1985 - 1987. Values in parentheses are 95% confidence intervals.

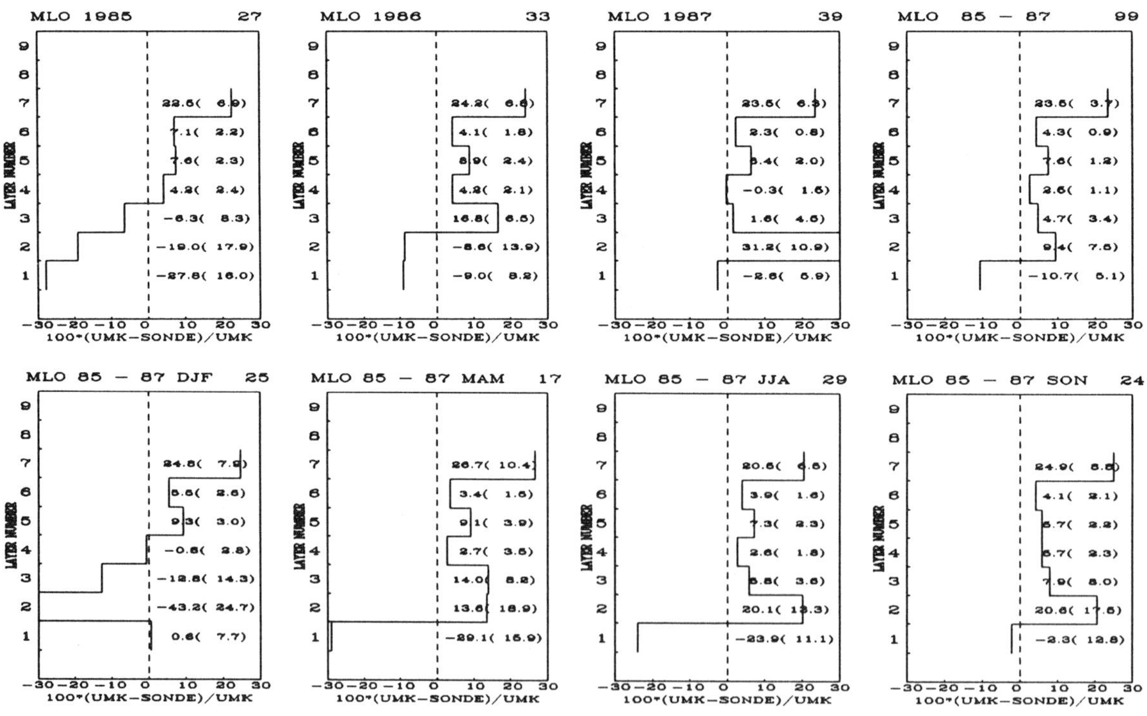

Fig. 2. Annual, long-term, and seasonal percentage differences, 100 * [(Umkehr-sonde) ÷ Umkehr], by Umkehr layer for Mauna Loa/Hilo, Hawaii, 1985 - 1987. Values in parentheses are 95% confidence intervals.

ACKNOWLEDGMENTS

We thank the numerous NOAA/GMCC and CIRES personnel who made the ozonesonde and Umkehr observations at Boulder, Colorado, and at Hilo and Mauna Loa Observatory, Hawaii. Funding for the program was provided by the National Environmental Satellite, Data, and Information Service (NESDIS), of NOAA.

REFERENCES

Bojkov, R, 1966: Differences in the vertical ozone distribution deduced from Umkehr and ozonesonde data at Goose Bay. *J. of Appl. Meteor.,* 5, No. 6, 872-877.

Craig, R.A., J.J. DeLuisi, and I. Stuetzer, 1967: Comparison of chemiluminescent and Umkehr observations of ozone. *J. Geophys. Res.,* 76, No. 6, 1667-1671.

DeLuisi, J.J. and C.L. Mateer, 1971: On the application of the optimum statistical inversion technique to the evaluation of Umkehr observations. *J. of Appl. Meteor.,* 10, No. 2, 328-334.

Dütsch, H.U., and Ch. Ling, 1969: Critical comparison of the determination of vertical ozone distribution by the Umkehr method and by the electrochemical sonde. *Ann. Geophys.,* 25, 211-214.

Komhyr, W.D., 1986: Operations handbook - ozone measurements to 40-km altitude with model 4A electrochemical concentration cell (ECC) ozonesondes (used with 1680-MHz radiosondes). NOAA Technical Memorandum ERL ARL-149, Silver Spring, MD, 49 pp.

Komhyr, W.D., P.R. Franchois, B.C. Halter, and C.C. Wilson, 1987: ECC ozonesonde observations at South Pole, Antarctica, during 1986. NOAA Data Report ERL ARL-11, Silver Springs, MD, 213 pp.

Komhyr, W.D., R.D. Grass, R.D. Evans, R.K. Leonard, and G.M. Semeniuk, 1985a: Umkehr observations with automated Dobson spectrophotometers. *Atmospheric Ozone,* Proceedings of Quadrennial Ozone Symposium, D. Reidel, Dordrecht, 371-375.

Komhyr, W.D., S.J. Oltmans, A.N. Chopra, and P.R. Franchois, 1985b: Performance characteristics of high-altitude ECC ozonesondes. *Atmospheric Ozone,* Proceedings of the Quadrennial Ozone Symposium, D. Reidel, Dordrecht, 499-503.

Mateer, C.L., and H.U. Dütsch, 1964: Uniform evaluation of Umkehr observations from the World Ozone Network: Part I, Proposed standard Umkehr evaluation technique. National Center for Atmospheric Research Report, Boulder, CO, 105 pp.

MEASUREMENTS OF OZONE PROFILES BY A LIDAR METHOD DURING THE ARCTIC WINTER AND SPRING 1988

B.C. Krueger, R. Neuber, and S. El Naggar
Alfred-Wegener-Institut fuer Polar- und Meeresforschung
D-2850 Bremerhaven, FRG

H. Walther
Max-Planck-Institut fuer Quantenoptik
D-8046 Garching, FRG

ABSTRACT

In winter 1987/88 measurements of vertical ozone profiles have been performed in Kiruna, Sweden (67.9 N, 21.0 E) by the differential absorption lidar method. The collected data show rapid changes of the ozone concentration at altitudes in the height range between 16 and 22 km.

1. INTRODUCTION

Investigations of the ozone concentration within the Arctic polar vortex have gained increasing importance since dramatic depletions of stratospheric ozone have been observed over Antarctica during recent years in spring time. Ozone measurements in the Arctic Polar region have been reviewed recently by Bojkov (1988). A first attempt to measure vertical ozone profiles in the Arctic by Lidar was made during the second campaign for Chemistry of Ozone in the Polar Stratosphere (CHEOPS II) from December 20, 1987 to March 18, 1988 at Esrange near Kiruna/Sweden (67.9 N, 21.0 E).

2. INSTRUMENT AND DATA EVALUATION

The applied differential absorption Lidar consists of a light source emitting pulses at two wavelengths, one of which lies in an ozone absorption band. The backscattered radiation is observed by a telescope with a photon-counting detector. A height resolution is achieved by measuring the time between emission and detection of the light pulses. For ozone measurements wavelengths of 308 nm and 353 nm are chosen, since ozone absorbes moderately at 308 nm and is almost transparent for the 353 nm signal.

A XeCl-gas-laser produces light pulses at 308 nm with an energy of 150 mJ/pulse and a repetion rate of 25 Hz. This radiation is partially (20 %) shifted in wavelength to 353 nm by stimulated Raman scattering. This keeps the light collinear and at the same bandwidth (< 15 pm) as the incident light. In the atmosphere the light is scattered mainly by air molecules (Rayleigh scattering) and to a lesser extent by aerosols. Therefore the backscattered intensities decrease logarithmically with height along with the air density.

The backscattered photons are collected by a mirror of 60 cm diameter and focused into a wavelength discriminating detector. The background radiation is reduced by interference filters. The bandwidth of the detector can further be narrowed by up to three pressure tuned Fabry-Perot etalons to less than 20 pm. The combination of a low bandwidth emitter and a low bandwidth receiver allows measurements also during daylight hours.

The registered photons of each pulse are subdivided into 256 height channels. The clock switching the channels is set to a height resolution of 200 m, covering the altitude range from 0 to 51.2 km. Due to the limited dynamic range of the photomultipliers the strong backscattered in-

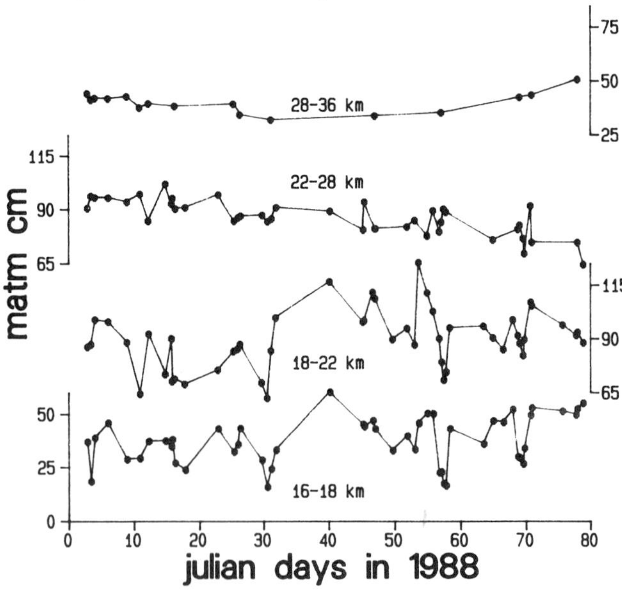

Fig. 1. Ozone columns in different altitudinal ranges calculated from backscatter signals, which have been integrated for up to 16 hours. The statistical errors of these measurements are 3 % in the lowest layer, 1.5 % between 18 km and 22 km, 2 - 5 % between 22 km and 28 km, and 5 - 10 % in the highest layer.

ISBN 0-937194-15-8

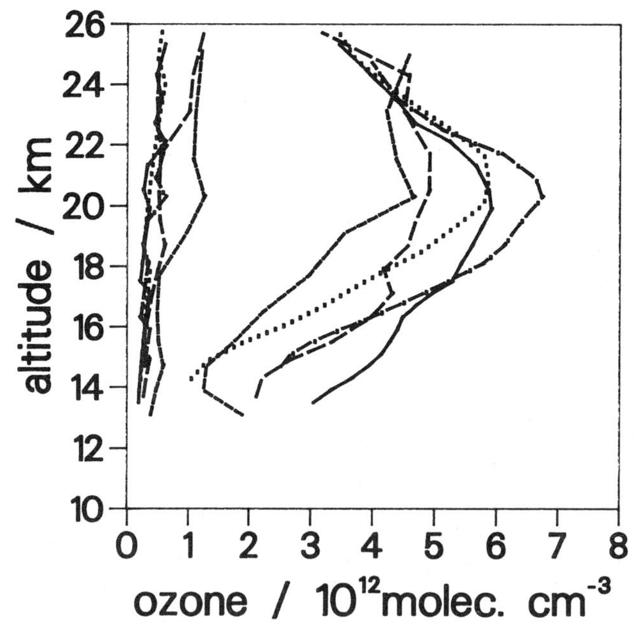

Fig. 2. Temporal development of the ozone concentration from January 25,1988 to January 31, 1988. The curves on the right show the ozone concentration, the curves on the left the calculated statistical error of the ozone concentration for these measurements. All times in UT.
(1)_______: 1/25, 8:23 - 1/25,23:49
(2)_ _ _ _: 1/29,11:06 - 1/29,16:00
(3)-------: 1/30, 9:39 - 1/30,12:50
(4).......: 1/30,22:39 - 1/31, 2:48
(5)_._._._: 1/31, 8:39 - 1/31,19:38

tensities from altitudes below 15 km have to be blocked by a mechanical chopper. The background radiation is determined between two laser pulses and is subtracted from the registered intensities before recording. The full details of the lidar system are described by Steinbrecht et al. (1988).

Due to the ozone absorption the countrates of the 308 nm channel decrease more rapidly than those of the reference channel at 353 nm. Therefore the ozone concentration can be calculated from the difference of the intensity gradients of the two channels [Werner et al.,1983]. The required absorption cross-sections of ozone at the two mentioned wavelengths have been determined by Paur and Bass (1984). The temperature dependence of the cross sections is taken into account by using radiosonde temperatures of a nearby station. The countrates are also corrected for Rayleigh extinction and the dead-time of the photomultiplier tubes. As long as the aerosol concentration above 15 km altitude is small (normal background aerosol), the effect of Mie scattering can be neglected.

The concentration profiles of ozone have been calculated with a variable altitudinal resolution to keep the maximum absolute statistical error of the calculated ozone concentration less than $6 \cdot 10^{11}/cm^3$ at low altitudes and less than $1.2 \cdot 10^{12}/cm^3$ above 25 km. The profiles shown in fig. 2 to 4 present running means over three values. The maximum altitude is limited by scattered sunlight to 27 km at daytime. At night the signals reach up to more than 40 km.

3. RESULTS

Measurements have been performed during all clear sky periods from January to March 1988. Backscatter signals have been collected for 170 hours during January, for 85 hours during February, and for 98 hours during March. The longest continuous measurement lasted for 43 hours while the shortest was less than 1 hour. The sequence of the measurements is displayed by the dots in fig. 1.

We have integrated the raw data to obtain one or two ozone profiles for each day whenever possible. Examples of these profiles are shown in fig. 2 to 4. The temporal developement of the ozone layer during the observing period is shown for four stratospheric sublayers between 16 km and 36 km in fig. 1.

The highest variation in the ozone concentration occurs below 22 km. The uppermost layer shows almost no short term variation, but slowly increasing column densities. Between 22 and 28 km altitude the variation is about 10 % with slowly

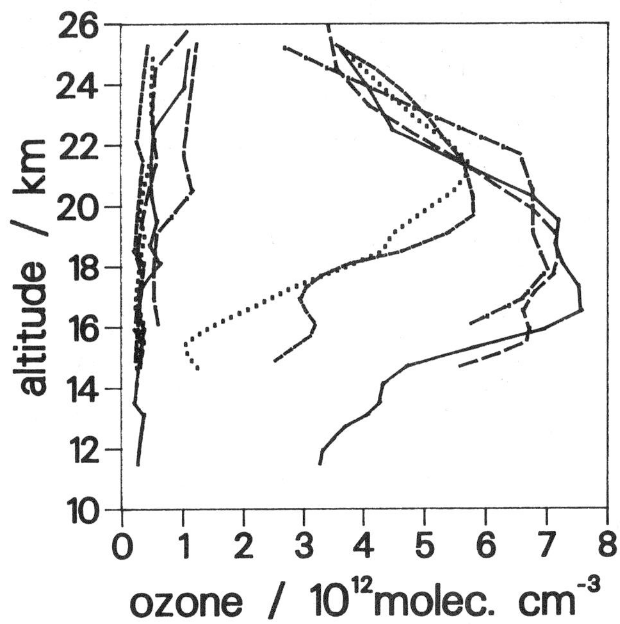

Fig. 3. As Fig. 2 from February 23,1988 (Julian day 54) to February 27,1988
(1)_______: 2/23,14:10 - 2/23,21:49
(2)_ _ _ _: 2/24,13:40 - 2/24,19:21
(3)-------: 2/25,19:47 - 2/26, 2:26
(4).......: 2/26,12:38 - 2/26,18:21
(5)_._._._: 2/27, 6:42 - 2/27, 8:04

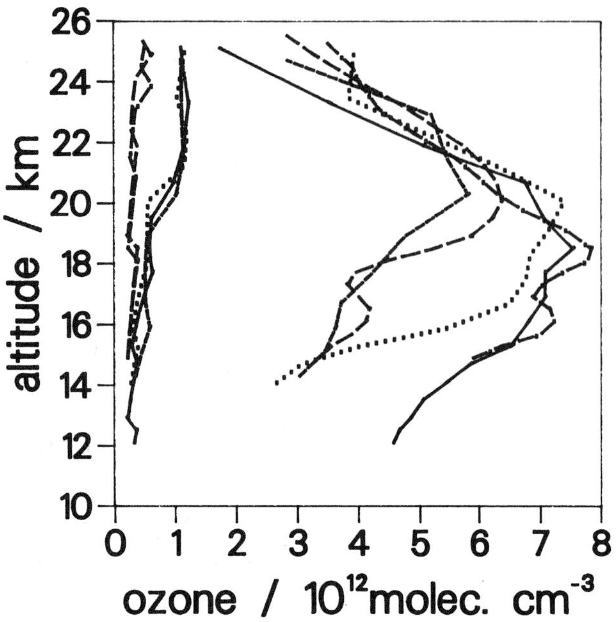

Fig. 4. As Fig. 2 from March 7,1988 (Julian day 67) to March 10,1988.
(1)_______: 3/ 7, 9:11 - 3/ 7,23:49
(2)_ _ _ _: 3/ 8,17:17 - 3/ 9, 0:03
(3)-------: 3/ 9, 6:39 - 3/ 9,13:13
(4).......: 3/10, 8:33 - 3/10,16:28
(5)_._._._: 3/10,16:28 - 3/10,20:45

decreasing values from mid-january to mid-march. On the other hand the column density in the two lower layers varies by more than 50 %. The monthly mean of the ozone column between 16 km and 22 km is increasing from 110 matm cm in January, to 137 matm cm in February, and 140 matm cm in March. Very low total ozone concentrations have been observed on January 17, January 30, February 26, and March 9. These low values result mainly from decreases in the two lower layers. Three of these events are presented in detail in figures 2 to 4.

The daytime measurement from January 30 in fig. 2 (curve 3) shows the lowest ozone concentration of this event. The profiles of the following night (curve 4) and the next day (curve 5) indicate a growing ozone concentration which finally superceeds the values of the first day of this sequence. The 16 to 22 km column density was reduced from 122 matm cm (curve 1) to 80 matm cm (curve 3) and finally reached 133 matm cm.

In the event shown in fig. 3 a fast reduction of the ozone concentration could be observed from the 24th to the 25th of February. A decrease is already obvious from the 23rd to the 24th and again from the 25th to the 26th, however, to a much smaller degree. The 16 to 22 km column densities were 162, 148, 102, and 91 matm cm for curves 1 to 4, respectively. This reduction is followed by a fast increase of the ozone concentration during the night from the 26th to the 27th reaching 150 matm cm in the 16 to 22 km range.

The event at the beginning of March (fig. 4) is marked by a fast decrease from the 7th to the 8th and a slow decrease from the 8th to the 9th. The two profiles from the 10th show a continuous increase of the ozone concentration. The values for the 16 to 22 km column densities for curves 1, 3, and 5 were 152, 112, and 159 matm cm, respectivly.

4. DISCUSSION

The low ozone concentrations observed on February 25 and February 26 coincided with the lowest temperatures measured over northern Europe in the stratosphere this winter. The radiosonde from Esrange on February 26 observed temperatures of -70 C and -82 C at 50 hPa and 15 hPa, respectively. Additionally the wind direction changed from west to north when the ozone concentration decreased and turned back to west when it increased on February 27. The lowest temperatures over the Arctic were located directly north of Kiruna during the period of this event [Naujokat, FU Berlin, priv. comm.]

During the events shown in fig. 2 and 4 the wind direction changed in a similar way, bringing also air from the area with the lowest temperatures towards the measuring site. The temperatures in the stratosphere at these times were about 5 degrees higher than at the end of February. However, the cold air advection is nevertheless an important process also in these cases.

From single station ozone measurements it cannot be conclusively derived whether the observed ozone concentration is modified by transports of different air masses or by local chemical reactions. However, at least the fast increase of ozone in the night from February 26 to 27 is rather likely due to air mass transports as photolytic production of ozone at night is impossible. But it cannot be excluded, that the observed ozone reductions are enhanced by chemical reactions during daytime.

ACKNOWLEDGEMENTS

We are grateful to K.W.Rothe and W.Steinbrecht, University of Munich, who constructed most of the Lidar system and instructed B.C.K. and R.N. how to operate the instrument. Further we appreciate the help of the personel of Esrange/Sweden, who considerably contributed to the success of our work.

REFERENCES

Bojkov,R.D. 1988: Ozone Variations in the Northern Polar Region. Meteorol. Atmos.Phys. 38, 117-130.

Paur,R.J., and A.M.Bass, 1985: The Ultraviolet Cross-Sections of Ozone: II. Results and Temperature Dependence. in: C.S.Zerefos and A.Ghazi (eds.), Atmospheric Ozone, Proc. Quadr. Ozone Symp., Dordrecht, pp. 611-616.

Steinbrecht,W., K.W.Rothe, S. ElNaggar, H. Walther, and E.Augstein, 1988: Measurement of the Stratospheric Ozone Concentration in the Arctic and at the Equator Using a LIDAR System, this symposium.

Werner,J., K.W.Rothe, and H.Walther, ,1983: Monitoring of the Stratospheric Ozone Layer by Laser Radar. Appl.Phys.B 32, 113-118.

LIDAR DAYLIGHT MEASUREMENTS OF STRATOSPHERIC OZONE AT THE NORTH POLE AND THE EQUATOR

*W. Steinbrecht, K.W. Rothe, and H. Walther**
Sektion Physik der Universität München
D-8046 Garching, FRG

S. El Nagger and E. Augstein
Alfred-Wegener-Institut für Polar- und Meeresforschung
D-2850 Bremerhaven, FRG

ABSTRACT

Measurements of atmospheric ozone at altitudes ranging from 5 to 50 km using the differential absorption LIDAR technique (DIAL) are described. For the measurements a XeCl excimer laser with a repetition rate of up to 50 Hz and a wavelength of 308 nm was used. The reference line was produced by stimulated Raman effect in an H_2-cell leading to a wavelength of 353 nm. The time resolved measurement of the backscattered light allows to evaluate the ozone concentration in different altitudes. In the past the ozone LIDAR measurements were only possible at night time. Now a set of specially designed Fabry-Perot filters in the detection system also allows daylight measurements. Thus investigations in polar regions get possible even in the summer periods. First measurements performed in arctic and equatorial regions with the new LIDAR system installed on board of the research vessel 'Polarstern' will be reported.

1. INTRODUCTION

The properties of laser light, such as high pulse energy, low divergence and high spectral purity, allow to perform remote sensing of trace constituents of the atmosphere. A frequently applied technique is the differential absorption LIDAR (DIAL). The feasibility of the DIAL technique for measurements up to distances of several kilometers is already well established (Hinkley (1972)). In the following a LIDAR system suitable for daylight measurements of stratospheric ozone will be described.

2. THE DIAL TECHNIQUE

The differential absorption technique (DIAL) is based on the use of at least two laser wavelengths having different absorption cross-sections for the gas under study: one line is strongly absorbed, the other one is nearly unaffected and serves as a reference. The use of pulsed lasers allows a RADAR-like measurement resulting in a height-resolved information on the backscattering. The comparison of the backscattered intensity at both wavelengths from two adjacent altitudes yields the concentration of the gas under study (e.g. Rothe, Brinkmann, Walther (1974) or Werner, Rothe, Walther (1983)). For this evaluation the absorption cross-sections have to be known. With respect to ozone, high-resolution data of Bass and Paur (1982) are available, which also consider temperature corrections. Especially around 308 nm these data are more precise (2 % or better) than previous measurements (Inn and Tanaka (1953)).

The DIAL technique compares the relative magnitude of backscattered intensities of two adjacent range cells for two wavelengths. It is this relative comparison which excludes any apparative drift during the measurement. Many other techniques for ozone monitoring suffer from instrumental drifts and therefore need corrections.

3. DETAILS OF THE LIDAR SYSTEM

A commercial XeCl excimer laser system (Lambda Physik EMG 150 TMSC) with a pulse energy of 150 mJ, repetition rates of up to 50 Hz, and a linewidth of less than 10 pm serves as the radiation source. The emitted wavelength (308 nm) is located in the wing of a strong ozone absorption band.

Light at a second wavelength (353 nm), used as the reference, is generated by stimulated Raman scattering in a hydrogen gas cell. This Raman scheme proved to be very advantageous since it allows the two wavelengths to be emitted simultaneously into the same field of view; this increases the accuracy of the measurement.

Figure 1 shows the scheme of the LIDAR system. The laser beam is focused into the Raman cell, where the collinear reference beam is generated. The optical receiver consists of a collecting mirror with a diameter of 60 cm and a focal length of 2.4 m. A dichroic filter (D) separates the backscattered intensity at the two wavelengths with minimum losses. Interference filters (IF) and three-stage Fabry-Perot fil-

* also Max-Planck-Institut für Quantenoptik, D-8046 Garching, FRG

ISBN 0-937194-15-8

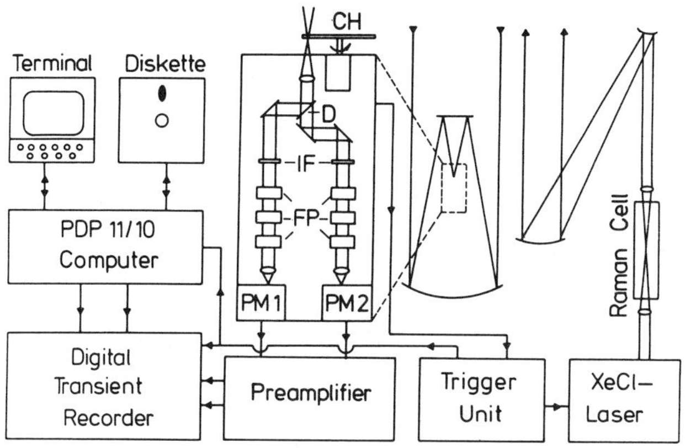

Fig. 1. Scheme of the experimental setup. The transmitter is shown on the right and the receiver on the left part of the figure.

ters (FP) are used to reduce the bandwidth of the receiving system to about 13 pm. The signals are detected by two photomultipliers (PM, EMI 9893) and photoncounters. For the time resolved storage of the photon counts a 512-channel digital transient recorder is used, providing an altitude resolution of 200 m. The signal processing is performed with a PDP 11/10 computer and the data are stored on floppy disks. The complete apparatus was installed in a laboratory container.

Daytime measurements require a reduction of background radiation from the sky. This could be achieved by a spatially and spectrally narrowbanded detection system. In addition a narrowband laser system had to be used. Spatial narrowing was achieved by reducing the field of view of the transmitter and the receiver to about 0.1 and 0.2 mrad respectively. The narrowbanded laser source is an injection locked XeCl laser system, having a bandwidth of less than 10 pm. As described above, the narrow-banded receiving system was equipped with interference filters and three stage Fabry-Perot-interferometers in each detection channel. The details of this system are described by Steinbrecht, Rothe, Walther (1989).

4. DAYTIME LIDAR MEASUREMENTS OF OZONE IN THE ARCTIC

The container with the ozone LIDAR has been installed on board the research vessel and ice-breaker "Polarstern" beginning of May '87. Measurements were performed during the cruise from Bremerhaven to Tromsoe via Spitzbergen (May 14 to July 1, '87). The course of the vessel is shown in Fig. 2. The system was working very reliably, so measurements could be performed whenever there was no overcast. These periods are marked by solid lines in Fig. 2. The main results of these measurements are the following :

In the first part of the cruise the ship was travelling in about one week from 60° N to 80° N. We observed a decreasing ozone concentration at all altitudes with increasing geographic latitude (see Fig. 3 and e.g. Labitzke et al. (1986) for comparison). In the second part of the cruise the ship remained essentially in a region around 75° N. During this period no significant change of the ozone concentration was observed.

The results of Fig. 3 show the advantage of the LIDAR technique providing height resolved data : at 15 km altitude ozone increases from 75° N to 80° N. This increase is probably due to a downward motion of ozone-rich air from higher altitudes, which is caused by the low temperatures in the polar region, marked by the ice-edge around 77° N (see also Plumb et al. (1986)).

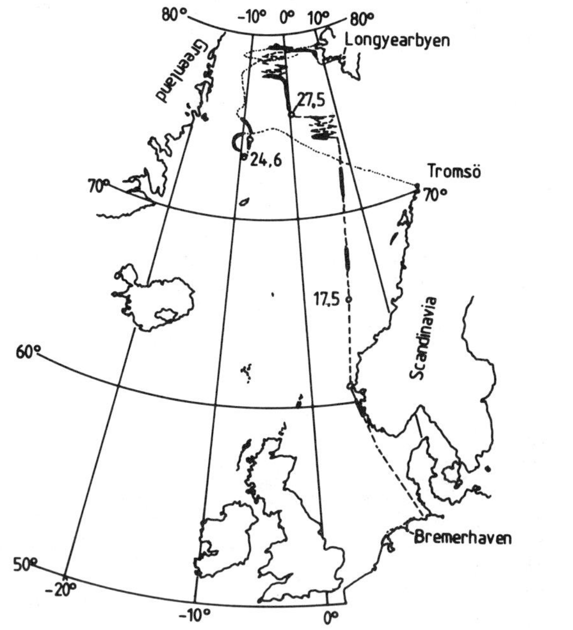

Fig. 2. Course of the ozone LIDAR on board of the ice-breaker and research vessel "Polarstern".

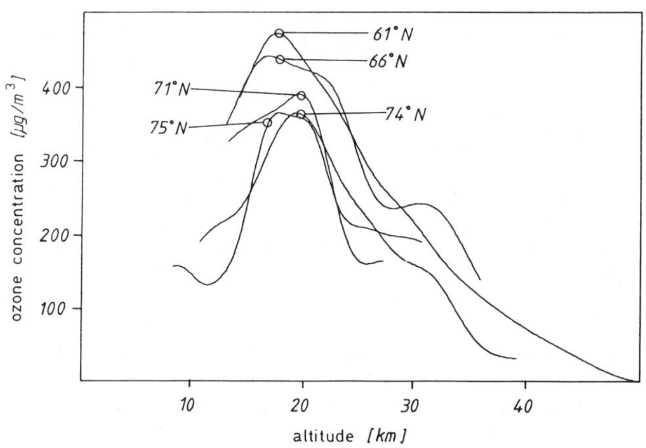

Fig. 3. Ozone concentration as a function of altitude for different geographic latitudes as measured during the polar cruise. The ozone decrease with increasing latitude can clearly be seen.

5. OZONE MEASUREMENTS ON THE "POLARSTERN" CRUISE FROM BREMERHAVEN (WEST-GERMANY) TO RIO GRANDE DO SUL (BRASIL) IN SEPTEMBER/ OCTOBER 1987.

The measurements performed during the trip from Bremerhaven to Rio Grande also show a strong variation of the ozone profiles with geographic latitude. In the equatorial region ozone profiles are shifted towards higher altitudes, ozone peak concentrations are higher and the width of the profiles is smaller than observed in higher latitudes (see Fig. 4). Profiles of around 30° S,

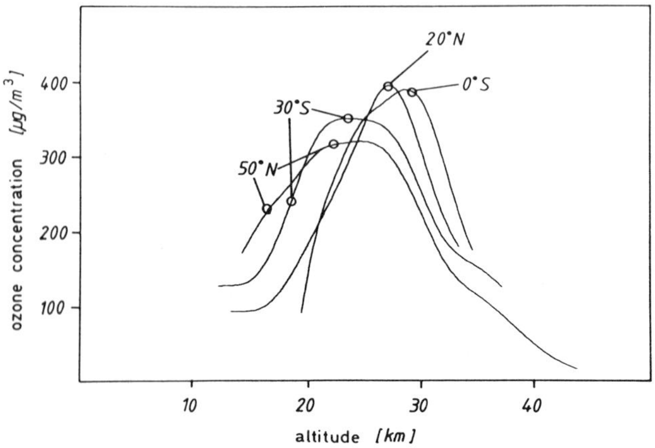

Fig. 4. Ozone profiles measured on the Atlantic cruise. The profiles are averaged over all measurements taken in a 10° intervall around the given latitudes. For the statistical error-bars see Fig. 5.

measured in the end of October, are very similar to profiles measured at 50° N in the end of September (see Fig. 4). These results agree with the general trend following from earlier observations (see e.g. Bowman (1985) or Bhartia et al. (1985)). Towards the equator the ozone content decreases in heights below 23 km whereas it increases in heights above 23 km (see Fig. 5). This behaviour might be caused by two processes : strong solar irradiance in the equatorial regions produces high ozone concentrations in higher altitudes , in addition high temperatures at the earths surface cause an upwards motion of ozone-poor air from the troposphere into the lower stratosphere (see also Plumb et al. (1986)).

Since photochemical reactions and transport processes strongly influence the ozone concentration a diurnal variation of the ozone profile is to be expected. In the mesosphere and upper stratosphere such a variation is well known (e.g. Solomon et al. 1983). To the knowledge of the authors a diurnal variation could not be observed up to now in altitudes around 30 km, as no other technique allows a time-continous, height-resolved observation of the ozone concentration with good precision in this height range.

In order to overcome the problem of natural variations and geographical trends and to minimize the statistical error we separately averaged all daytime and all nighttime measurements taken during the Atlantic cruise. The results are mean

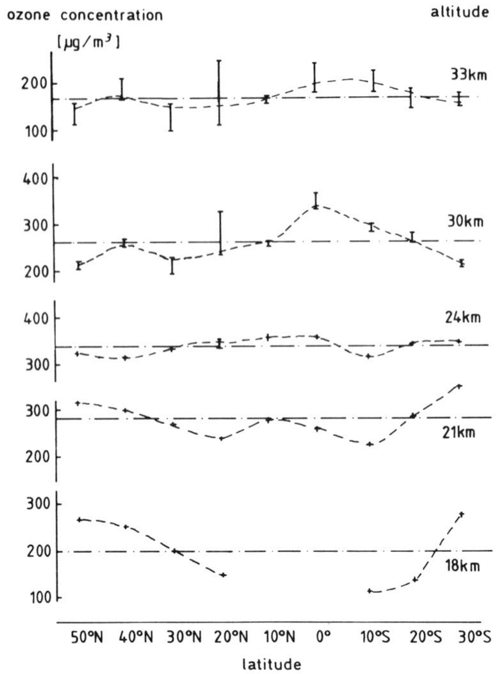

Fig. 5. Variation of the ozone concentration with geographic latitude in different heights measured during the Atlantic cruise.

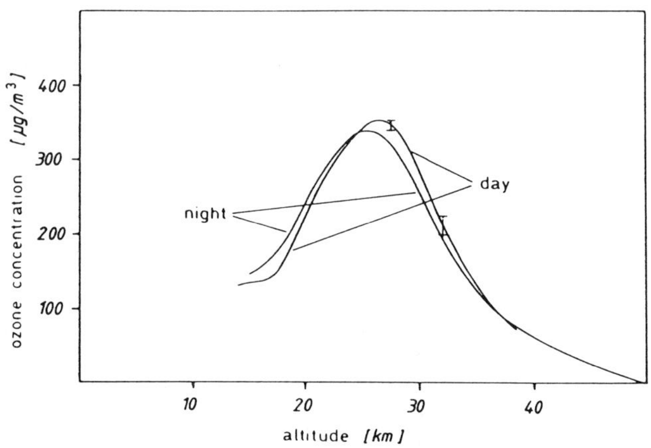

Fig. 6. Mean daytime and mean nighttime profile measured during the Atlantic cruise.

profiles for daytime and nighttime ozone as shown in Fig. 6. Below 25 km the ozone concentration increases during the night and above 25 km it increases during the day, whereby the total ozone between 15 and 38 km remains nearly constant (244 ± 8 DU for daytime and 241 ± 1 DU for nighttime profile). Despite the fact, that the standard deviations of the ozone concentration are very large at these altitudes (about 80 µg/m³ from day to day at 17 km and 40 µg/m³ at 27 km), the observed trend showed up in most of the measurements. The observations thus indicate, that there is a slight diurnal variation in the height range between 15 and 35 km. It might be caused by an upwards motion of the air being warmed up during the day. Such a vertical transport process would leave the total ozone content unchanged.

ACKNOWLEDGEMENT

We especially wish to thank the members of the mechanic and electronic workshops of the Sektion Physik of the University of Munich for their great effort in constructing and setting up the container.

REFERENCES

Hinkley E.D. (Ed.), 1972: Laser Monitoring of the Atmosphere, Springer Verlag, Berlin.

Rothe K.W., U. Brinkman, H. Walther, 1974: Appl. Phys 3, 115 and Appl. Phys 4, 181.

Werner J., K.W. Rothe, H. Walther, 1983: Appl. Phys. B 32, 113.

McPeters R.W., A.M. Bass, 1982: Geoph. Res. Lett. 9, 227.

Inn E.C.Y., Y. Tanaka, 1953: J. Opt. Soc. Am. 43, 870.

Steinbrecht W., K.W. Rothe, H. Walther, 1989: to be published in Appl. Opt.

Bowman K.P., 1985: in Proc. of the Quadrennial Ozone Symposion 1984, Chalkidiki Greece, 363.

Bhartia P.K., D. Silberstein, B. Monosmith, A.J. Fleig, 1985: in Proc of the Quadrennial Ozone Symposion 1984, Chalkidiki Greece, 243.

Labitzke K., J.K. Angell, J.J. Barnett, P. Bowman, M. Corney, G.M. Keating, A.J. Krueger, A.J. Miller, R.M. Nagatani, 1986: in Atmospheric Ozone, WMO-Report No. 16, 1007.

Plumb R.A., D.G. Andrews, M.A. Geller, W.L. Grose A. O'Neill, M. Salby, R.A. Vincent, 1986: in Atmospheric Ozone, WMO-Report No. 16, 311.

Solomon S., D.W. Rusch, R.J. Thomas,R.S. Eckman, 1983: Geophys. Res. Lett 10, 249.

REEVALUATION OF THE AUSTRALIAN TOTAL OZONE DATA RECORD

R.J. Atkinson and J.R. Easson
Australian Bureau of Meteorology
Melbourne, Australia

ABSTRACT

The Australian ozone monitoring network was established during the 1950s. It currently comprises five stations, each equipped with a Dobson ozone spectrophotometer. In recent years the network has been upgraded and the instruments recalibrated by standardised, fully traceable procedures. Some past calibration and maintenance records are being re-examined and the quality of the corresponding total ozone records evaluated. There are discrepancies in these records which would suggest that the published data do not comprise a reliable record for scientific study. Corrected data from Melbourne (1978-1988) and Perth (1980-1987) have been analysed and recent ozone decreases detected. Provisional data from recent months are also presented. The degree of confidence which currently may be attached to recent data from the remaining stations is briefly discussed.

1. INTRODUCTION

The detection of the Antarctic ozone hole, renewed fears of the likelihood of anthropogenic depletion of the global ozone layer and indications from individual records of local decreases in ozone levels have recently sparked a resurgence in attempts at global ozone trend analysis, culminating most recently in that of NASA/WMO (1988). In the course of these recent analyses, criticisms have arisen of the quality of some of the ground-based data sets on which long term trend analysis is possible (e.g. Bojkov et al., 1988), or which might be used to validate remote sensing system data.

The Australian Dobson spectrophotometer total ozone record is one such data set which, extending back to the 1950s, is potentially valuable for trend analysis tasks. We report here on the first results of a reexamination of the calibration and maintenance records of the Australian instruments from the past decade, establishing their traceability of calibration to either the regional or world standard instrument, and suggesting appropriate corrections to be applied to Dobson ozone data currently archived at the World Ozone Data Center (WODC) and published as WMO (1960-1988).

After applying simplified corrections to the archived data, we then perform linear regressions on the total ozone records for Melbourne (1978-1988) and Perth (1980-1987) and compare the resulting trends with those found by NASA/WMO (1988). Finally, having established the credibility of the more recent Australian data, we report on the occurrence of anomalously low total ozone levels over southern Australia during December 1987.

2. AUSTRALIAN NETWORK HISTORY

Routine ozone monitoring, using the Dobson ozone spectrophotometer, commenced in Australia when the Commonwealth Scientific and Industrial Research Organisation (CSIRO) established observations programs at Aspendale (1956), Brisbane (1956) and Macquarie Island (1957). The network was later expanded to include Darwin (1966), Hobart (1967) and Perth (1969), and in support of the Dobson program at Aspendale, weekly balloon-borne ozone sonde flights were commenced there in 1965. The Darwin Dobson program was relocated at Cairns after 1974 before the latter was closed in 1983.

In December 1982 responsibility for the Australian ozone monitoring program was passed to the Bureau of Meteorology. At the same time observations were ceased at Aspendale, the Dobson program being relocated at nearby Melbourne and the sonde program at Laverton, also nearby.

During 1984 Dobson #81 was installed at Perth, as one of several automated instruments in the global network. The instrument is owned by NOAA but maintained by the Bureau of Meteorology, in collaboration with NOAA.

Since 1983 the Bureau of Meteorology has steadily developed the Australian ozone monitoring program. Dedicated staff have been trained in the calibration and maintenance of the Dobson spectrophotometer. Each of the Australian network instruments is now either directly traceable to the world standard #83, (this is the case for #105 and #81) or to the regional standard #105, (for #6, #78 and #111), except for the Hobart instrument, which is indirectly traceable to #105 through standard lamps (#12).

3. INSTRUMENT CALIBRATION HISTORIES

Although the first Australian stations were established during the 1950s, it was not until 1977 that one of the Dobson instruments (#105) was calibrated by side-by-side comparison with a recognised standard (world standard Dobson #83), thus establishing traceability of calibration.

Prior to this the network instruments had been calibrated without reference to a recognised standard, preventing the establishment of their traceability. Also, requisite clear atmospheric conditions were not often experienced during these 'absolute' calibrations which employed the Langley method for extra-terrestrial constant determination, and the perforated gauze method of

ISBN 0-937194-15-8

optical wedge calibration was used, a non-standard technique. Although continued investigation, perhaps necessitating comparison with independent data sets (e.g. stratospheric temperature), might enable reassessment of some of the data from this early period, in this paper further consideration is given only to the period for which traceability of calibration has been established.

From 1978 to 1984 each network instrument continued to be referenced by the Langley method, after having its optical wedge calibrated by the gauze method. During this period intercomparisons with the regional standard were made at the time of referencing, but not used for calibration purposes. Nonetheless, this has enabled the correction of an instrument's data recorded subsequent to its first calibration after 1977.

Since 1984, prior to referencing, most of the Australian instruments have undergone electronic upgrade (as required), optical realignment and wedge calibration by the two lamp method.

No standard lamp corrections have ever been applied to Australian total ozone data prior to their archiving at WODC (apart from those data from Perth since 1984, for which corrections have been routinely applied by NOAA GMCC, the owner of the Perth instrument). This became an established practice because, although devised to keep a check on calibration state, standard lamp tests in the earlier years employed 200 V lamps which were subject to mains variations and their voltage was monitored by poor quality equipment, placing severe doubts on their value for data correction. The practice has been continued since the advent of better quality quartz halogen lamps because lamp test data from Australian stations which have used these standard lamps as reference over a long period of time reveal that although temporal variations at A,C and D wavelengths occur, variations at each wavelength are similar, so that they effectively cancel when total ozone is computed from dual wavelength observations. The only occasions on which significant non-zero standard lamp corrections are warranted for dual wavelength observations is after an instrument has been transferred from its calibration site to an observations site at a significantly different altitude. Where significant non-zero standard lamp corrections have been warranted, the smallness and randomness of the scatter between individual monthly test data indicates that only mean period N_{AD} corrections, rather than monthly corrections, are appropriate.

The instrument calibration histories from each of the Australian stations for the period for which traceability has been established are summarised in Table 1.

As well as the N_{AD} corrections judged appropriate for individual segments of the records from the Australian stations, the right hand side of Column 4 in Table 1 shows approximate mean corrections to be applied to the corresponding total ozone records for those periods and stations. Calculation of precise corrections requires each total ozone observation to be reprocessed, a necessary but time-consuming and as yet incomplete task. Therefore, for the purposes of this paper, we have merely estimated mean period corrections to published data having assumed typical values for ozone optical path length and total ozone involved in the daily calculations.

4. ANALYSIS RESULTS

The revised total ozone data for Melbourne from March 1978 to February 1987, excluding the period from January 1983 to April 1983, have been analysed by least squares linear regression of the deseasonalised monthly means in search of temporal gradients. Similar analysis was performed on the revised data for Perth from January 1980 to October 1987 merged with provisional unpublished data from November and December 1987, for which appropriate corrections have not been applied. The simple analysis performed did not take into account the effects of autocorrelation of the data or natural geophysical variability (e.g. solar cycle, QBO). The resulting temporal gradients of total ozone are -0.59%/year (standard error 0.09%) at Melbourne (latitude 38°S) and -0.76%/year (standard error 0.14%) at Perth (latitude 32°S). These are greater than the TOMS-derived zonal mean trends for the 29S-39S and 39S-53S latitude bands (-2.7% and -4.9% respectively), which were reported by NASA/WMO (1988) for the period from November 1978 to November 1987.

For December 1987, monthly mean total ozone at Melbourne was 288 m.atm.cm., the lowest recorded in the ten year period from 1978 to 1987. Examination of the daily mean total ozone data for the month (see Fig. 1c, solid line) shows that the low monthly mean resulted from a sudden decrease in daily averages which occurred between 11 and 13 December, coinciding with the final breakup of the Antarctic 'ozone hole' (Newman et al, 1988) and which, although displaying subsequent fluctuations, remained low for the remainder of the month. Corroborative evidence which rules out the possibility of instrument malfunction is provided by routine standard lamp test results, which indicate that the instrument calibration was stable at the time. Also, the sudden decrease was reflected in the behaviour of daily 00 UTC 100 hPa temperatures (see Fig. 1c, dashed line), for which the correlation with daily total ozone for the month is 0.77. Having confirmed the credibility of the Melbourne daily total ozone data from the past 10 years, the 1987 data were compared with those from the previous nine years, revealing that the duration of the December 1987 decrease was seasonally anomalous. A comparison of ozone vertical profiles from Mast-Brewer ozone sondes flown at Laverton (17 km SW of Melbourne) a week before and after the decrease, which have been normalised to Melbourne total ozone, reveals that the ozone decrease was confined to the region between 200 and 30 hPa (see Fig. 2). Although the indicated decrease in ozone partial pressures at these levels is comparable to the accuracy of the observational system, support for the truth of the data is indicated by the presence of a temperature decrease with similar vertical structure (see Fig. 2).

Examination of the December 1987 data from the other Australian stations provides an indication of the scale of the event (see Fig.

TABLE 1. DOBSON CALIBRATION HISTORIES AND RECOMMENDED N_{AD} TABLE AND TOTAL OZONE DATA CORRECTIONS

STATION	DATE or PERIOD	INSTR NO.	CORRN N_{AD}	CORRN O_3 #1	COMMENTS
ASPENDALE/ MELBOURNE $\bar{\Omega}$ = 314	08/77	105	+2.3	+9	-Traceability of calibration established for #105 when it was intercompared with #83 to determine back corrections. Not yet known how far back this correction is warranted. #105 thereafter designated as WMO RAV standard.
	08/77-02/78	111	???	??	-Calibration not traceable. Required corrections unknown.
	03/78-12/82, 05/83-11/84	105	-0.5	-2	-Standard lamp correction for period :- mean -0.5 N_{AD} units, std. devn. 0.2 N_{AD} units.
	01/83-04/83	115	???	??	-Calibration not traceable. Required corrections unknown.
	since 12/84	105	0.0	0	-#105/#83 intercomparison 12/84. No additional corrections required. Air wedge fitted to #105, instrument again referenced to #83. Since then standard lamp corrections :- mean 0.0 N_{AD} units, std. devn. 0.2 N_{AD} units.
PERTH $\bar{\Omega}$ = 296	01/80-08/80	12	+1.4	+5	-#12/#105 intercomparison at end of service showed nonlinear differences between +0.5 and +4.0 N_{AD} units. For typical mu/total ozone, average difference was +1.4. Unknown how far back corrn. warranted. Here assumed valid back to 01/80.#2
	09/80-07/84	111	+0.6	+2	-#111/#105 intercomparison 09/80 showed #111 required mean correction of +0.5 N_{AD} units. #111/#105 intercomparison 10/84 showed #111 required mean correction of +0.7 N_{AD} units. No instrument adjustments made during this period. Average period correction assumed as +0.6 N_{AD} units.#3
	since 08/84	81	0.0	0	-#81/#83 intercomparison 05/84. Since 08/84 fortnightly standard lamp corrections routinely applied by NOAA GMCC.
BRISBANE $\bar{\Omega}$ = 285	05/85	6	+4.1	+16	-#6/#105 intercomparison at end of service showed mean corrn. required was +4.1 N_{AD} units. Not yet known how far back this correction is warranted. See also Bojkov et al. (1988).
	since 05/85	111	0.0	0	-#111/#105 intercomparison 04/85. Since then standard lamp corrections :- mean 0.0 N_{AD} units, std. devn. 0.2 N_{AD} units.
HOBART $\bar{\Omega}$ = 315	05/81-07/85	12	0 - +8.0	0 - +31	-#12/#105 intercomparison 04/81 showed large nonlinear corrections required. Detailed investigation incomplete.#4
	since 08/85	12	0.0	0	-Wedge recalibrated by 2 lamp method and referenced by standard lamps 07/85. Confirmation of traceability awaits intercomparison with #105.
MACQUARIE ISLAND#5 $\bar{\Omega}$ = 324	06/80-10/81	115	???	??	-Inappropriate Q-lever adjustments made to instrument. Good data irrecoverable.
	11/81-12/84	78	+1.2	+5	-#78/#105 intercomparison 10/81 used to estimate corrections for entire period.
	01/85-12/86	78	+3.2	+13	-2 lamp wedge calibration and referenced by standard lamps 12/84. #78/#105 intercomparison at end of service used to estimate back corrections for entire period.
	since 12/86	6	+0.2	+1	-#6/#105 intercomparison 11/86. Since then standard lamp corrections :- mean 0.2 N_{AD} units, std. devn. 0.1 N_{AD} units.

#1 Approximate equivalent correction to total ozone data in Dobson Units or m.atm.cm.(see Section 3).

#2 For the period at Perth from 01/80 to 08/80 the precise corrections required to be made to ozone data archived at WODC are not known because the instrument's calibration is not traceable. In order to make available eight completed years of data from Perth for the trend analysis in Section 5 below, we make the assumption here that the calibration of 09/80 may be retrospectively applied to all data from 01/80.

#3 Examination of the 200 V standard lamp test data from the period of #111's service at Perth suggests that a gradual and approximately linear upward trend of +2.7 N_{AD} units occurred between 10/80 and 08/84. Corrections for this drift have not been applied to the data because the more reliable comparisons with #105 carried out immediately before and after the period showed a negligible shift in mean calibration. Additional support for the truth of the intercomparison results is found in data from the two month period at the conclusion of #111's service in Perth. During this period quasi-simultaneous observations were made daily with #111 and #81. Comparison of direct sun total ozone observations made by each instrument indicates that #111 was reading low by approximately 1%. Assuming values for ozone optical path length and total ozone which are representative of typical observing conditions during the period, this equates to a required N_{AD} table correction of +0.6 N_{AD} units, consistent with the two formal intercomparisons.

#4 The large nonlinear corrections that were required to make #12 read as #105 (04/81) were thought to be due to a poorly calibrated optical wedge, so in 1985 the optical wedge was recalibrated by the two lamp method and the instrument referenced by standard lamps. See also Bojkov et al. (1988).

#5 Prior to 1987 total ozone measurements from April to August each year were derived from single C wavelength zenith observations. All winter data require further correction because the zenith correction tables were not individually derived for the instrument, as noted by Komhyr (1980). Since 1987 winter observations have been made on dual CD wavelengths. For the purposes of this paper we confine the analysis to AD total ozone data.

1). A similar sudden decrease is evident in the Perth data, leading in phase the Melbourne event by about two days. Hobart data, although of uncertain quality, as discussed in the preceding section, show signs of the event on 10 December, albeit with a much noisier signal. Macquarie Island data also show a significant decrease on 10 December, but do not exhibit similar marked and persistent features for the remainder of the month. Finally, Brisbane data show no sign of the event.

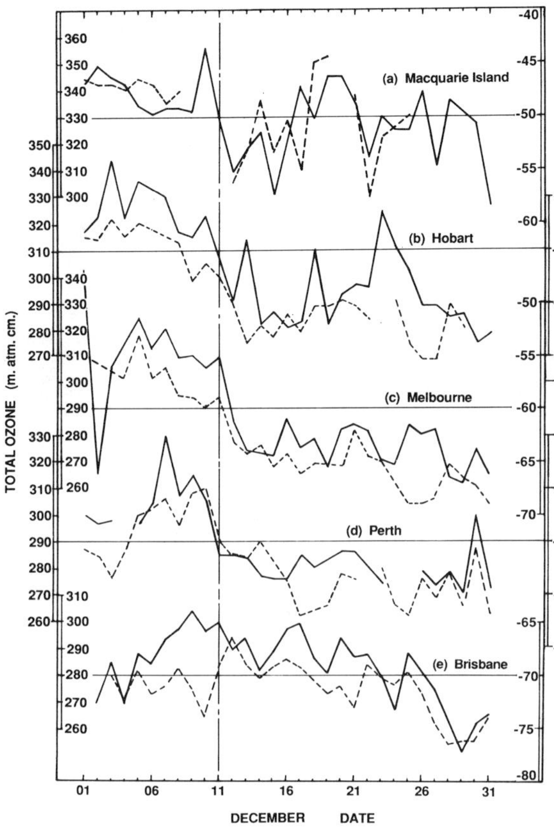

Fig. 1. Daily mean total ozone (m.atm.cm., solid lines, left ordinate) and 00 UTC 100 hPa temperature (^{o}C, dashed lines, right ordinate) vs. December date, 1987.

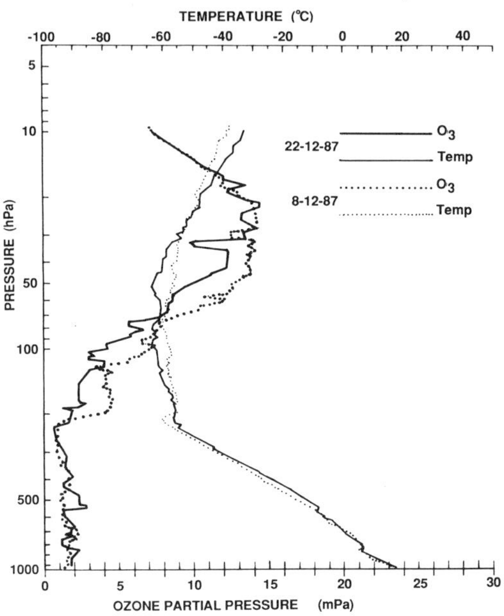

Fig. 2. Temperature (^{o}C) & ozone partial pressure (mPa) vs. pressure (hPa), Laverton.

5. CONCLUDING REMARKS

The reasons for the difference between the 1978-1987 Dobson- and TOMS-derived trends of Section 4 are not clear. It seems unlikely that it is due to instrument error of either system, but this deserves further investigation. More probably it is the result of substantial longitudinal variation in the ozone change of the last decade. Confirmation of this would require a more detailed seasonal analysis of the data, which is beyond the scope of this paper.

The timing of the Melbourne ozone decrease in December 1987 is of particular interest. The data presented here are inadequate to establish the cause of the Melbourne event, and are provided merely to serve as a catalyst for continued investigation. Further clarification of the matter will require more detailed and complex analysis of the meteorological situation by, for example, the use of isentropic back-trajectory or potential vorticity analyses to determine the source of the low-ozone airmass.

We have reassessed and established the credibility of the total ozone records from Melbourne (from 1978) and Perth (from 1980), and have commented explicitly on the quality of data from the remaining stations in the current Australian Dobson network. While further investigation may enable reevaluation of more of the data from these stations, the nature of the calibrations and the documentation of the earlier records are such that this may be impossible without resorting to the application of corrections based on correlations with other geophysical variables.

REFERENCES

Bojkov, R.D., Mateer, C.L., and Hansson, A.L., 1988: Comparison of Ground-Based and Total Ozone Mapping Spectrometer Measurements Used in Assessing the Performance of the Global Ozone Observing System. J. Geophys. Res., 93, 9525- 9533.

Komhyr, W.D., 1980: Operations Handbook - Ozone Observations with a Dobson Spectrophotometer WMO Ozone Proj. Rep. 6, Geneva.

NASA/WMO, (in press) 1988: Report of the NASA/WMO Ozone Trends Panel. WMO Ozone Proj. Rep. 18, Geneva.

Newman, P. A., and Schoeberl, M.R., (in press) 1988: The Breakup of the Southern Hemisphere Spring Polar Ozone and Temperature Minimums from 1979 to 1987. In Proc. Quadr. Ozone Symp. in Gottingen, eds. R.D. Bojkov and P. Fabian, A. Deepak, Hampton, Va.

WMO, 1960 - 1988: Ozone data for the world, Vol.1 No.1 - Vol.28 No.5, Atmospheric Environment Service, Downsview, Ontario, Canada.

SYSTEMATIC LIDAR MEASUREMENTS OF THE STRATOSPHERIC OZONE VERTICAL DISTRIBUTION

S. Godin, S. Mégie, and J. Pelon
Service d'Aéronomie CNRS
91371 Verrières le Buisson, France

ABSTRACT

Lidar measurements of the ozone vertical distribution have been performed on a routine basis at the Observatoire de Haute-Provence starting in October 1986. They provide an unique data base which could be used to derive the seasonal behaviour of ozone in various altitude ranges from 25 to 45 km with a height resolution much higher than previous operational systems. Comparison with Umkehr measurements performed at the same location show a rather good agreement in layers 5 and 6 with unexpected differences appearing in the uppermost layers.

1. INTRODUCTION

Early detection of changes in the stratospheric chemical composition and structure related to human activities is a major issue for the protection of the ozone layer. A recent study (NASA, 1988) shows that satellite observations alone, despite their global coverage of the stratospheric ozone field, are not sufficient to provide an accurate determination of trends mainly because of drifts in instrument calibration. It is thus fully recognized that complementary ground-based measurements are a required component of any global ozone observation system. Such measurements should be conducted routinely from several sites at different latitudes to constitute an effective Network for the Detection of Stratospheric Changes (NDSC-NASA, 1986). As part of the various passive (UV-visible, infrared, microwave spectrometers) and active systems to be implemented, lidars are recognized as being the required instrument to provide the most accurate measurements of the ozone vertical distribution from the ground up to the 45-50 km level.

Since the early development of the ozone lidar at the Observatoire de Haute Provence (OHP - Mégie et al., 1977 ; Pelon et Mégie, 1981) several improvements have been brought to the system which have resulted in succesive increases in the maximum attainable height of the measurements (Pelon and Mégie, 1982 ; Mégie et al., 1985). As a last step, a new excimer laser source has been implemented in 1986, to improve the reliability of the system allowing routine operation starting in October of 1986 (Godin et al., 1986). Such a two years measurements period is obviously too short to provide any indication on potential trends. The data reported here will thus be analysed in terms of seasonal variations and compared to measurements performed at the same location by the Umkehr technique using an automated Dobson spectrophotometer (Marché et al., 1983). Comparison with satellite (SBUV) measurements will be reported in an other paper presented at this Symposium (Godin et al., 1988). A brief description of the system will be given in section 2. Section 3 will emphasize the data processing methods used to retrieve the ozone vertical distribution from the raw lidar signals. Data analysis and results of comparison will be presented in section 3.

2. OPERATIONAL LIDAR SYSTEM

The lidar measurements of the ozone vertical distribution are based on the Differential Absorption Laser (DIAL) technique (Schotland, 1964 ; Mégie and Menzies, 1980) which requires the simultaneous emission of two laser wavelengths. The ozone absorbed wavelength (on line) is provided by a Xenon Chloride (XeCl) excimer laser emitting at 308 nm (Uchino et al., 1978). This laser source (Lambda Physics MG 201) emits an average energy of 0.25 J at a repetition rate of up to 80 Hz. An unstable resonator cavity arrangement is used to reduce the laser divergence down to 0.5 mrad. The spectral emission profile corresponds to two XeCl lines at 307.95 nm and 308.05 nm. It has been carefuly monitored to allow calculation of the effective ozone absorption cross-section which is necessary to determine the ozone number density from the measured local optical thickness (Pelon et al., 1986). The reference wavelength is obtained using the 3rd harmonic (355 nm) of a Nd Yag laser. The output energy is adapted to provide a return signal equivalent to the on-line signal at 40 km altitude. This is done to prevent any bias related to detector dynamics in the whole altitude range.

ISBN 0-937194-15-8

The backscattered signals are collected by a 80 cm diameter telescope and optically separated (308-355 nm) using a dichroïc mirror. Further reduction of the background night sky is obtained using adapted interference filters. The signals are then detected by two photomultipliers tubes and time sampled to retrieve the altitude resolution using simultaneously a 1 MHZ photon counting system (low signal intensity) and a 10 MHz - 10 bits transient waveform recorder (high signal intensity). A LSI 11-23 mini computer is used to control the experiment and for real time data processing and storage.

Details of the experimental procedure have been given in previous publications (Pelon et al., 1986). In today operation, the measurement of the ozone vertical distribution in the 18-45 km range requires an integration time of 2 hours for nightime measurements.

3. DATA PROCESSING

Two types of errors can directly influence the DIAL measurement of the ozone vertical distribution. The statistical error related to the signal to noise ratio defines the limit on the precision of the measurement. Systematic erros due to interferences of molecular and aerosol scattering and extinction, and to temperature dependence of the ozone cross-sections will affect the accuracy of the retrieved ozone profile. Such uncertainties have already been quantified in previous publications (Mégie et al., 1985 ; Pelon et al., 1986) and will not be addressed in details here. In this section however, the emphazis will be set on the possible bias which can be induced by the data processing. Simulations have been performed using a direct-inverse modeling scheme. Given an a priori state of the atmosphere - number density, temperature and ozone profiles - the lidar backscattered signals are calculated using the system parameters as defined in section 1. After the addition of a random noise component adapted to the real return signal characteristics, the inverse algorithms are applied and the retrieved profile is compared to the initial one. The resulting standard deviation and bias related to the data processing are estimated from the average of a large number of profile retrievals in order to obtain statistically significant values. As the ozone vertical distribution is determined from the range derivative of the logarithm of the two return signals ratio, the data processing requires a derivating filter. Besides, a low pass filter with varying cut-off frequency is used to account for the decrease in the signal to noise ratio with altitude. Two combination of filters have finally been selected (Godin, 1987) : a low pass polynomial filter associated with a Taylor derivative filter (combination A) and a derivative polynomial filter (combination B). The results of the simulations are given on figure 1a and 1b. Although the low pass efficiency of the filter A is better, leading to a smaller value of the standard deviation (5 % vs 7 % at 45 km), this filter introduces a larger systematic bias at high altitudes which can reach values up to 4 % at 45 km. Thus if the filter A can be more suitable in terms of measurement precision (short term fluctuations), the study of long term evolution (trends) will certainly require the use of the filter combination B.

4. DATA ANALYSIS

The ozone lidar system at the Observatoire de Haute Provence has been operated on a research mode in 1984-1985 and on an operational mode starting in october 1986. More than sixty ozone profiles have been obtained in the altitude range from 30 to 42 km allowing thus a study of the seasonal variations of the ozone number density at various altitudes. Furthermore, taking advantage of the unique capacity provided by the OHP Geophysical Station to obtain simultaneous measurements of ozone distribution by lidar and Umkehr technique, comparison between these two techniques can be performed on a statistically significant basis. Such comparisons require the conversion of the lidar data (number density vs. height) in units compatible with Umkehr retrieved values (partial pressure in various pressure intervals). Lidar data have first been averaged over the altitude range of the Umkehr layers using pressure-altitude conversion as given by the CIRA 1986 model which includes seasonal and latitudinal variations. The lidar number density ozone values have then been converted to partial pressure using temperature profiles taken again from the CIRA 1986 atmospheric model. Comparison of these temperature data with values obtained by lidar at the OHP, and not necessarily coincident with ozone observations, show that the monthly means do not differ by more than 2 % on the average with a maximum value of 8 % in winter time.

The figures 1a, 1b, 1c represent the seasonal variation of the ozone concentration in the Umkehr layers 5 (23.5 - 25 km), 6(28-33 km) and 7(33-38.5 km) as obtained by lidar measurements over the years 1985-1987. In the layer 5, (fig. 1a), the seasonal variation of ozone is less than 5 % if one excepts the January average. This latter corresponds in fact to only three measurements taken during consecutive days in January 1987 at a time of the year where the planetary wave activity is quite strong and drives the ozone behaviour in this altitude range. They should then not be considered as fully significative. In the above layer 6 (fig. 1b) the amplitude of the seasonal variation is larger (30 %) with a summer maximum in the ozone concentration. The standard deviation of the measurements which can be attributed to a large extent to the natural variability is larger as expected during winter emphasizing again the importance of dynamical processes related to wave propagation. At higher altitudes (layer 7 - fig. 1c), the peak-to-peak variation is again of the order of 35 %, whereas the shape of the variation appears different with a maximum extending through spring and summmer. Here again the day to day variability

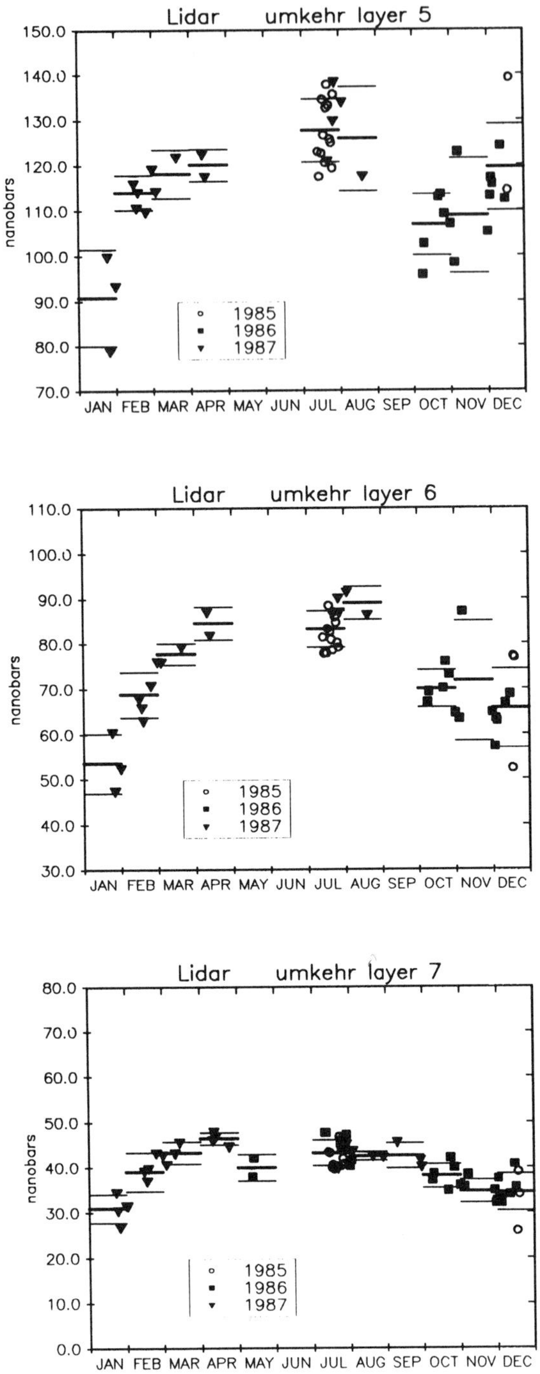

Fig. 1 : Seasonal variation of the ozone content in various Umkehr layers as measured by lidar at the O.H.P..

is somewhat larger during the winter months.

Although the OHP ozone lidar measurements correspond to single point observation and not to zonal means, the seasonal behavior of the ozone concentration as measured reflects the zonally described interaction between dynamical and photochemical processes in an altitude range (25-40 km) where the chemical lifetime of ozone varies by two orders of magnitude. The amplitude of the seasonal variation related to photochemistry alone as calculated by a one dimensional model (Godin, 1987) is less than 2 % below 30 km increasing up to 15 % at 35 km with a maximum in summer.

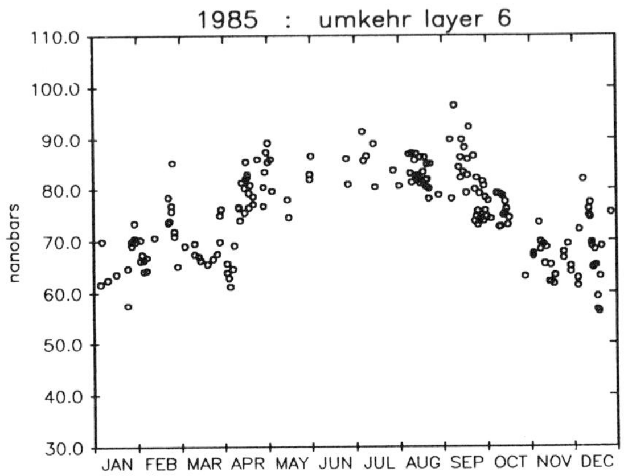

Fig 2 : Seasonal variation of the ozone content in the Umkehr layer 6 as measured by the Dobson spectrophotometer at O.H.P..

The comparison with the Umkehr retrieved profiles show a rather good agreement with the lidar data as far as the general features of the seasonal variation are concerned (fig. 2 for Umkehr layer 6). The major difference lies in the amplitude of the observed variations which are generally 20 to 40 % higher on the lidar data. This effect can most likely be attributed to the non-independance of the various Umkehr layers for Dobson spectrophotometer measurements. Due to the different amplitudes of the ozone variations throughout the altitude range this interdependance results in an averaging effect which reduces the amplitude and modifies the phase of the altitude dependent ozone variations. This can be further assess if one considers the direct lidar measurements at a given altitude (fig. 3 - 35 km $\pm$ 1 km) obtained with a higher vertical resolution which show a larger peak-to-peak amplitude (~ 50 %).

To allow a further comparison of the absolute values of the ozone concentration as retrieved by lidar and by the Umkehr technique at the OHP, all profiles coincident in time have been averaged to provide an accurate estimate of a possible bias. Results are shown on fig. 4 (Umkehr minus lidar relative values) showing a negative bias at lower altitudes (layers 5 and 6) shifting to a positive bias increasing up to

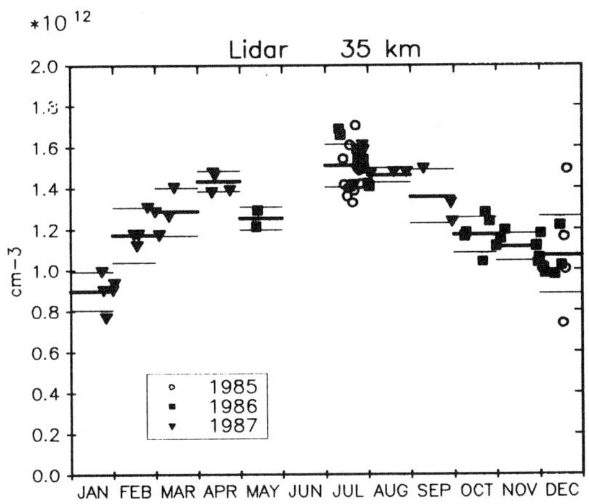

Fig. 3 : Seasonal variation of the ozone number density at 35 ± 1 km as measured by lidar at O.H.P..

10 % in the upper layers 7 and 8. To account for a possible effect of the seasonal variation, comparisons have also been made averaging only summer-time measurements (figure 5) which shows however similar values of the bias. As measurements were taken in the years 1985-1986, one should not expect a large effect of aerosol interferences in the Umkehr measurments as the high stratospheric aerosol load following the El Chichon eruption (Mars-April 1982) had already decreased by one order of magnitude resulting in optical thicknesses of 0.015 as measured by lidar at the same site. This value is compatible with the observed bias in layers 5 and 6 (De Luisi, 1979) whereas it could not explain the positive bias observed in the upper layers and especially the large bias observed in layer 7. Extension of the lidar data base to provide monthly based comparisons will bring further information on a potential seasonal variation in the bias allowing thus to distinguish between the various influencing factors related to data conversion such as temperature, pressure, geopotential height.

5. CONCLUSION

Operational lidar measurements performed at the Observatoire de Haute Provence (1985-1987) confirm the ability of such system to provide a continuous survey of the ozone vertical distribution in the 30-45 km altitude range. Lidar observed seasonal variations of the ozone concentration give evidence for a transition between dynamically controled region at lower altitude and photochemical influence at upper level. If the same features are also observed on Umkehr measurements lidar observations show larger amplitude in the retrieved variations related to the higher vertical resolution of the system. In terms of absolue values, statistically significant comparison show a rather good agreement between lidar and Umkehr retrieved ozone values in layers 5 and 6, although unexpected positive bias corresponding to larger Umkehr values are

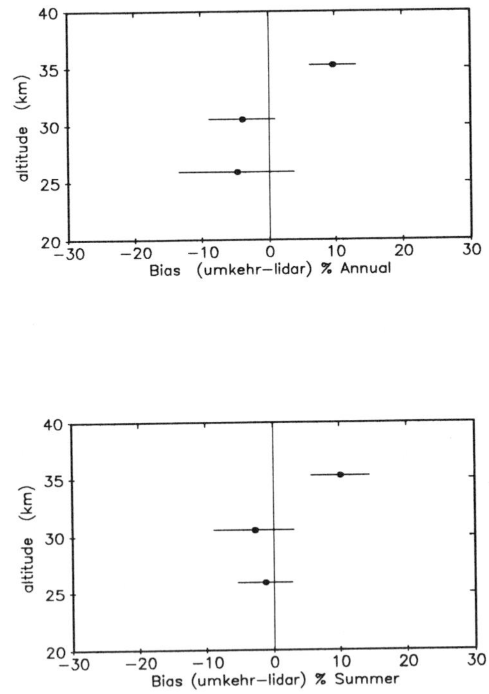

Fig. 4 : Relative bias between Umkehr and lidar measured ozone vertical distribution at the Observatoire de Haute-Provence.

observed in layers 7 and 8.

Continuation of the ozone lidar operation are planned to increase the data base. They are complemented by other lidar measurements of aerosol and temperature, automated Dobson observation of the total ozone content and vertical distribution by the Umkehr technique. Future addition of microwave measurements of chlorine monoxide distribution and UV-visible spectrometer measurements of ozone and nitrogen dioxide is also planned. The global system such implemented at the Observatoire de Haute Provence will thus constitute the basis of an established site of the future Network for Detection of Stratospheric Changes.

6. ACKNOWLEDGMENTS

The Geophysical Station at the Observatoire de Haute Provence is operated with the financial support of the french Centre National de la Recherche Scientifique, Institut National des Sciences de l'Univers, Centre National d'Etudes Spatiales and Ministère de l'Environnement. The operation of the ozone lidar has been supported by a grant from the Chemical Manufacturers Assiociation FPP Pannel. The authors wish to thank particularly Philippe Gildwarg for his contribution to the system operation and data acquisition, and J.P. Schneider and F. Syda for their responsability in the overall maintenance of the station.

7. REFERENCES

De Luisi J.J. (1979) : Umkehr vertical ozone profile errors caused by the presence of stratospheric aerosols. J.Geophys.Res., 84,766.

Godin S., J. Pelon and G. Mégie (1986) :Lidar Stratospheric Measurements at the Observatoire de Haute Provence. 13th International Laser Radar Conference, NASA Conference Publication G2431, 215.

Godin S. (1987) : Etude expérimentale par télédétection laser et modélisation de la distribution verticale d'ozone dans la haute stratosphère. Thèse de Doctorat de l'Université Pierre et Marie Curie, Paris.

Godin S., G. Mégie, J. Pelon, A.J. Miller, R.M. Nagatani and A.J. Fleig (1988) :Comparison of the ozone vertical distribution obtained by lidar and by the SBUV instrument. Quadrennial Ozone Symposium, ed. R. Bojkov and P. Fabian, Götttingen (RFA).

Marché P., Meunier C., Barbe A. and Jouve P. (1983) : Total atmospheric ozone measured by ground-based high resolution infra-red spectra. Comparison with Dobson measurements. Planet. Space Sci., 39, 791.

Mégie G., J.Y. Allain, M.L. Chanin and J.E. Blamont (1977) : Vertical profile of stratospheric ozone (18-28 km) by lidar sounding from the ground. Nature, 270, 329.

Mégie G. and R.T. Menzies (1980) : Complementarity of UV and IR differential absorption lidar for global measurements of atmospheric species. Appl. Opt., 19, 7,1173.

Mégie G., G. Ancellet and J. Pelon (1985) : Lidar measurements of ozone vertical profiles. Appl. Opt., 24, 21,3454-3463.

NASA (1986) :Network for the Detection of Stratospheric Changes. Report of the Workshop, Boulder, Colorado, March 5-7, 1986. Atmospheric Research Program, NASA Headquarters, Washington D.C..

NASA (1988) : Ozone Trends Panel Executive Summary, NASA Headquarters, Washington D.C., March 1988.

Pelon J. and G. Mégie (1981) : Ozone monitoring in the troposphere and lower stratosphere : evaluation and operation of a ground based lidar station. J. Geophys. Res., 87, C7, 4947-4956.

Pelon J. and G. Mégie (1982) :Ozone vertical distribution and total content using a ground based active remote sensing system. Nature, 299, 137-139.

Schotland R.M. (1964) : The determination of the vertical profile of atmospheric gases by means of a ground based optical radar. Proceedings of the 3rd Symposium on Remote Sensing of the Environment, Ann Arbor, U.S.A..

Uchino O., M. Maeda, J. Kohno, T. Shibata, C. Nagasawa and M. Hirono (1978) : Observation of Stratospheric Ozone Layer by XeCl Laser Radar. Appl. Phys. Lett., 33, 807.

THE OBSERVATIONS AND STUDIES OF ATMOSPHERIC OZONE IN BEIJING AND KUNMING, CHINA

Ding-Wen Wei
Institute of Atmospheric Physics, Academia Sinica
Beijing, PRC

ABSTRACT

Two stations for measuring atmospheric ozone have been set up in Beijing and Kunming, China since 1979. A brief review is given on the operations of the stations and corresponding research.

Umkehr method <C> (Wei and Lin, 1964), an inversion technique of Umkehr observations without using a priori climatological data, has been applied in both stations for years. This technique was based on a theoretical work (Wei, 1962), in which, the instability and the non-uniqueness of the solution of the atmospheric remote sensing equation were clear shown. These two fundamental problems in this field, were discovered first from our Umkehr research related to the remote sensing of the ozone vertical profile.

The results given by the Umkehr method <C> show that the secondary maximum concentration existing in ozone vertical profile is a noticable phenomenon especially in spring and summer. The frequency of the secondary maximum has an obvious seasonal variation with a minimum in autumn.

I. OBSERVATIONS AND ANALYSES OF TOTAL OZONE

1.1 OBSERVATIONAL PRECISION

At the end of 1978 and 1979, in Beijing(Shiangher Station, (39° 41'N, 117° 00'E) and Kunming (25° 01'N, 102° 41'E) two atmospheric ozone stations were set up with Dobson instruments No.75 and No.3 respectively. Since then, these two stations have contributed regularly to the WMO global ozone observing system. Before installation, the two instruments had been modernized and intercompared with the regional standard instrument Dobson No.77 in Canada. (Wei and Wen 1980). For ensuring observational quality and developing research, a group comprising members from the Institute of Atmospheric Physics of Academia Sinica and the Department of Geophysics of Yunnan University has been organized

If weather conditions are suitable, five total ozone observations a day are taken using the results of the AD wavelengths with direct sunlight as the formal data. To ensure good data quality, we have not taken cloudy sky observations.

For about nine years, the two instruments have been in quite satisfactory working condition.

In March of 1987, Dobson No.75 was sent to Melbourne, Australia and compared with the regional standard instrument Dobson No.105. Figure 1 shows the results of comparison on April 1, 1987. There were 32 simultaneous observations. The figure shows that the two instruments are in good agreement. The differences for the total ozone are less than 1.0% in general with standard deviation $\sigma = \pm 2.0$ D.U.

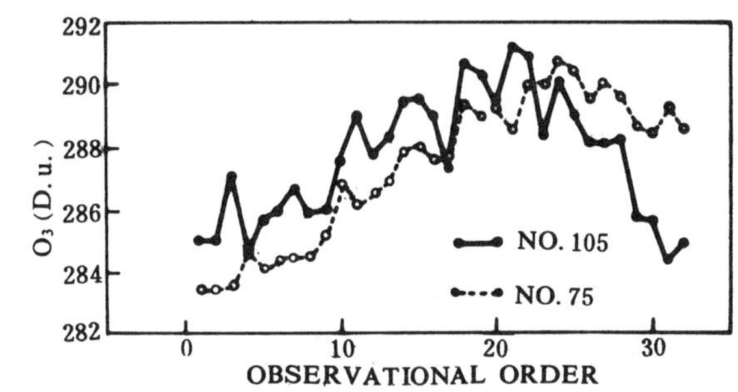

Fig.1. Comparison of Dobson No.75 with Dobson No.105 on April 1, 1987 in Melbourne.

In November of 1987, Dobson instrument No.3 was then shipped to Beijing, and intercompared with Dobson No.75 from Nov. 4 to 7, 1987. In the four clear but breezy days, there were a total of 123 simultaneous observations obtained. The results of Nov.4 are shown in figure 2. The statistical results show that for about 40% of the data, the differences are less than 0.5%; for about 80% of the data, the differences are less than 1.5%; only one observation shows a difference larger than 3%. The standard deviation of the differences was $\sigma = \pm 4.3$ D.U.

Therefore, the results of the intercomparison are basically very satisfactory. Nevertheless, we also noticed that the Dobson No.3 is not as stable as the Dobson No.75, it shows more

ISBN 0-937194-15-8

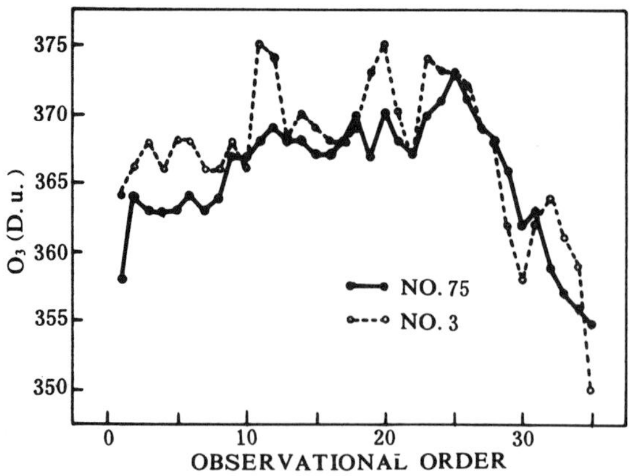

Fig.2. The comparison of Dobson No.3 with Dobson No.75 on Nov. 4, 1987 in Beijing.

small fluctuations.

Just before the beginning of observations at the two stations, No.75 was intercompared with No.77, as mentioned above, and then No.3 was intercompared with No.75. The differences were less than 1.0% and 2.0% respectively(Wei and Wen, 1980). After nine years, we find that the observational precision of the two instruments remains at the same level.

1.2 SOME ANALYSES OF THE TOTAL OZONE

A few of results of analyses of the total ozone observations have been published: the basic states of the atmospheric ozonosphere in both Beijing and Kunming areas; the correlations between the total ozone and atmospheric circulation in Eastern Asia; the analyses and applications of global ozone data, etc. (Wei et al, 1983; Zhang and Guo, 1985; Guo and Wang,1985; Wei, Guo and Zhao,1987). Some interesting phenomena were found. One example is shown as figure 3. It means that in Beijing area, the short period oscillation of total

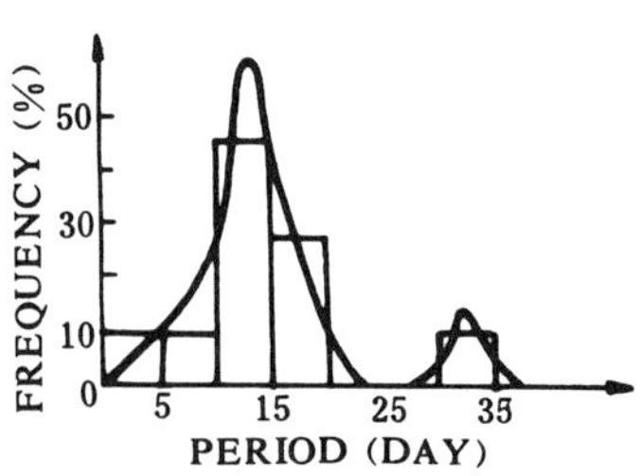

Fig.3. The period of total ozone in Beijing area (from Wei et al, 1983)

ozone, which appears most frequently is the quasi-biweekly oscillation. It was pointed out that this oscillation is associated with the atmospheric circulation patterns in that area.

II. STUDIES ON THE UMKEHR EFFECT AND THE INVERSION TECHNIQUE FOR UMKEHR OBSERVATIONS.

2.1 THEORETICAL STUDIES

Since 1931, when Götz(1931) first observed the umkehr curve, the remote sensing of the vertical distribution of atmospheric ozone by the umkehr method has undergone a substantial development (Götz et al, 1934; Walton, 1957; Ramanathan and dave, 1957; Mateer and Dutsch, 1964; Deluisi and Mateer,1971; Mateer and deluisi,1980). It can be asserted that even today, Umkehr method is still one of the important methods.

Besides, it is interesting that for the modern remote sensing of the atmosphere developed recently, some of the basic problems which they faced are similar to those of Umkehr. This is due to the fact that the basic equations used in the atmospheric remote sensing as well as in the Umkehr method are all Fredholm integral equations of the first kind (Zeng, 1974).

Wei(1962) studied the physical process of the Umkehr effect in great detail. He reached several important conclusions: (1) According to the theory, he demonstrated that the "effective height" of the scattering of $I_{3112Å}$ (C wave lengths) in the Umkehr effect rises with a jump. This means that for some zenith angles, the half width of effective scattering region (weighting function) is very wide and therefore, the measurement of $I_{3112Å}/I_{3223Å}$ is not very sensitive to the vertical distribution of ozone. So, from this principle, he pointed out clearly that the independent information on the vertical distribution of atmospheric ozone, contained in the Umkehr curve, is limited. If the the atmosphere is divided into too many layers, the number of unknowns will be larger than the number of independent equations. Therefore, the solution is non-unique. Later, (Mateer, 1964,1965) further pointed out that the Umkehr curve at most, contains four independent pieces of information about the vertical distribution of atmospheric ozone. (2) He pointed out that small observational errors can produce a very different vertical distribution of atmospheric ozone. (3) As verification, he demonstrated the nonuniqueness of the Umkehr method <B> (Ramanathan and Dave, 1957), an important method at that time for measuring the vertical profile of ozone.

Therefore, according to the similarity mentioned above, two of the most important problems, the instability and nonuniqueness of the solution for the remote sensing equations of the atmosphere in general were discovered first by study of the Umkehr effect for remotely sensing the atmospheric ozone profile (Zeng, 1974).

In the paper of Wei(1962), the normalized curves of spatial and temporal variations of information of $I_{3112\text{Å}}$ were given shown in figure 4 (In which, ΔI_h is the intensity of $I_{3112\text{Å}}$ at height of h, and ΔI_M is the maximum intensity). These kinds of curves are generally called normalized weighting functions or source functions (many atmospheric remote sensing problems involve wavelength instead of zenith angle). It is well known that curves similar to those in figure 4 are fundamental in atmospheric remote sensing problems for discussing the stability and uniqueness of the retrieval, the selection of the channels, as well as the number of channels to should be used, etc. However, figure 4

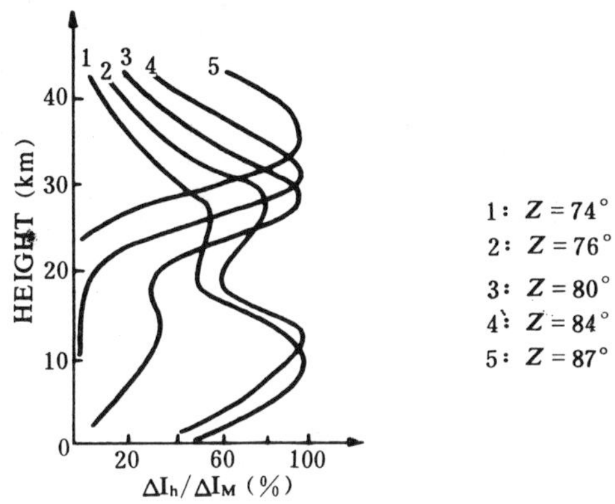

Fig.4. $\Delta I_h/\Delta I_M$ variations with height for different zenith angles (from Wei,1962)

was the first one to appear in the literature for discussing those problems.

2.2 UMKEHR METHOD <C> AND ITS APPLICATIONS

After the work of 1962, we designed umkehr method <C> (Wei and Lin, 1964). In this method, it is coincidental to Mateer's result(1964) that four independent variables, (X_1, X_2, X_1', X_2',) for one Umkehr curve are used to calculate the ozone profile. From the analyses of the distributions and variations of the information of $I_{3112\text{Å}}$, we know that when zenith angle Z>85, the measurement, (I/I') is independent of the vertical distribution of ozone below 28 km. Therefore, we can solve the following two equations numerically first.

$$(I/I')_{88} = f_1\,(X_1, X_2) \qquad (1)$$

$$(I/I')_{90} = f_2\,(X_1, X_2) \qquad (2)$$

After obtained X_1, X_2, the other two equations are then solved numerically for X_1', X_2' below 28 km. Thus the vertical distribution of the atmospheric ozone can be obtained. It can be seen that this inversion technique is different from some other Umkehr methods (Deluisi and Mateer,1971; Mateer and Deluisi, 1980). It does not need the standard profiles of atmospheric ozone below 36 km as a priori information.

TABLE 1. FREQUENCY OF THE SECONDARY MAXIMUM IN DIFFERENT SEASONS.

season	spring	summer	autumn	winter
Number of observations (N)	30	31	52	1
Number of observations with secondary maximum (M)	19	20	19	1
M/N	63%	64%	36%	/

From 1979 to 1986, there were 114 Umkehr observations at Beijing station. The calculated results by method <C> show that the second maximum in ozone vertical distribution in the troposphere is an important phenomenon. The percentages of the second maximum appearing in different seasons are listed in table 1. It can be seen that more than 60% of the observations have secondary maxima in the ozone vertical distribution for both spring and summer. However, the frequency of the secondary maximum appearance decreases very much in the autumn. Therefore, it displayed an obvious seasonal variation. (There are not enough observations in the winter season to be analyzed).

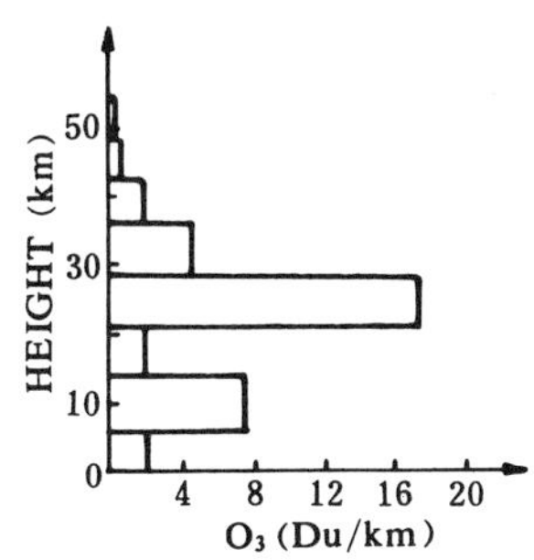

Fig.5 Vertical distribution of ozone in Beijing on March 21,1985.

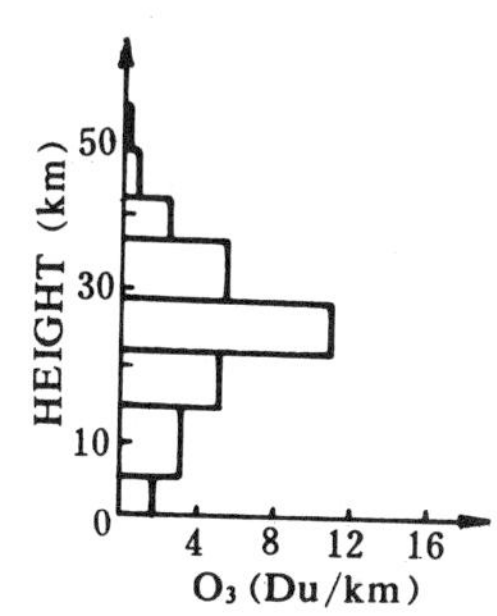

Fig.6 Vertical distribution of ozone in Beijing on Sept. 18, 1986

Figures 5 and 6 show two typical ozone vertical distributions in spring and autumn respectively. In figure 5, the secondary maximum appeared in the region of 5 - 14 km and its value is roughly half of the prime maximum. Figure 6 is a typical and familiar ozone vertical distribution with single maximum located in the region of 21-28 km. In Kunming, the secondary maximum can also appear in spring (Zhang and Guo, 1985)

Overall, the results show that the Umkehr method <C> is a reasonable and useful method. But we also want to point out that this method should be improved because it had been designed in earlier 1960's and the earlier measurements of O_3 absorption coefficient (Ny and Choong, 1933) were used. Especially, it should be compared with the simultaneous observations of balloon sounding in the lower levels as well as the satellite observations in the upper level. Then, probably, we can give a final assessment of this method.

ACKNOWLEDGEMENTS

The author wishes to thank the WMO, Dr. R.D. Bojkov, Prof. Ye Duzheng(T.C. Yeh) and Prof. Zeng Qincun for their support to this work.

REFERENCES

Deluisi, J.J. and C.L. Mateer 1971: On the application of the Optimum statis tical inversion technique to the evaluation of Umkehr observations, J. Appl. Meteor., 2, 328-334.

Götz,F.W.P., 1931: Zum strahlungsklima des spitzbergen sommers, Gerlands Beitrage zur Geophys., 31, 119-154.

Götz,F.W.P., A.R.Meetham and G.M.B.Dobson, 1934: The vertical distribution of ozone in the atmosphere, Proc. Roy. Soc. London, A145,416-446.

Guo SHichang and Wang Weguo. 1985: Some basic charateristics of total ozone variation derived from II stations in Asia, J. of Yunnan University, 4. 498-505.

Mateer,C.L. and J.J.Deluisi, 1980: The estimation of the vertical distribution of ozone by the short Umkehr method, Proc. of Quadrennial Ozone Symposium, Boulder, August 1980, V.1. 64-73.

Mateer,C.L., H,D.Dütsch, 1964: Uniform evaluation of Umkehr observations from the world ozone network, Part 1. Proposed standard Umkehr evaluation technique, Boulder, NCAR, 105 pp.

Mateer,C.L., 1964: A study of the information content of umkehr observations, Doctoral dissertation, University of Michigan, 199 pp.

Mateer,C.L., 1965: On the information content of Umkehr observations, J. Atmos. Sci., 22, 370-381.

Ny Tsi-ze and Choong piaw, 1933: Sur l' absorption ultraviolette de l' ozone, China Journ. Phys., 1, 158.

Ramanathan,K.R. and J.V.Dave, 1957: The calculations of the vertical distribution of ozone by the Gotz Umkehr effect (method B), Annals. of I.G.Y., V.5, 23-45.

Reiter,R. and R.Sladkovic, 1980: on periods of extremely slight and extremely strong fluctuating ozone profiles in the lower stratosphere based on daily ballon radiosonde soundings, Proc. of the Quadrennial ozone symposium, Boulder, August, 1980, 2, 1120-1127.

Walton,G.F., 1957: The Calculation of the vertical distribution of ozone by the Gotz Umkehr-effect (method A), Annals of I.G.Y., V.5, 9-22.

Wei Ding-Wen, 1962: The jump of the effective height of scattering for I in Gotz Umkehr effect and non-uniqueness of the solution of the Umkehr method <B>, Acta Geophysica Sinica, 2, 123-135 (in Chinese); a small part translated into English: Scientia Sinica, 1963, 1065-1070.

Wei Ding-Wen and Lin Chi-Chin, 1964: The Umkehr method <C> for the measurement of vertical distribution of ozone -- An improvement on the umkehr method <A> and <B>, Acta Geophysica Sinica, 4, 260-271; translated into English: Scientia Sinica, 1965, 9, 1339-1351.

Wei Ding-Wen and Wen Chungjia, 1980: Observations and analyses of the total amount of atmospheric ozone in the Beijing region and in the region of Jolmolunma Maitain in Tibet, Proc. of the Quadrennial International ozone symposium, Boulder, August, 1980, V.1, 340-346.

Wei Ding-Wen, Zhang Fangfan, Zhong Xiasping and Guo Shichang, 1983: The basic state of the atmospheric ozonosphere in Beijing area of China and its correlations with other meteorological factors, Scientia Atmospherica Sinica, 4, 375-384.

Wei Ding-Wen, Guo SchiChang and Zhao Yan lian: Atlas of total ozone in Northern Hemisphere: 1963-1985 (to be published), Science press, Beijing, China.

Zeng Qingcun, 1974: The principle of the atmospheric infrared remote sensing, Science Press, Beijing, China.

Zhang Fangfan and Guo Shichang, 1985: On the basic state of the atmospheric ozonosphere in Kunming area, J. of Yunnan University, 1, 125-129.

LONG-TERM VARIATIONS AND RECENT NEGATIVE ANOMALIES OF NEW ZEALAND TOTAL OZONE AND UMKEHR VERTICAL OZONE DISTRIBUTIONS

E. Farkas
New Zealand Meteorological Service
Wellington, New Zealand

ABSTRACT

Total ozone and Umkehr vertical ozone distribution observations made at Invercargill during the 1965-1987 period show unusually persistent negative anomalies since 1985. The long-term total ozone variations have a significant relationship with the quasi-biennial stratospheric wind oscillation (QBWO), in the manner expected from QBWO modulation of the Hadley cell and hence of meridional ozone transport. Lag correlations of the quasi-biennial ozone oscillations (QBOO) at Invercargill and American Samoa with the equatorial QBWO fitted this theoretical expectation. Irregular phase-changes of the QBWO since 1983 were reflected in total ozone variations at both stations during 1983-85.

The possible influence of large Southern Oscillation Index (SOI) anomalies on ozone variations at Invercargill is also considered. Weaker than normal stratospheric westerlies south of Invercargill occurred during midwinter each year since 1984, and on the basis of statistical correlations, these were favourable for lower than normal springtime ozone values.

Umkehr vertical ozone distributions between 1973 and 1987 showed the largest downward linear trends in the highest layers (38-48 km) and in the lower stratosphere below the level of the ozone maximum (10-19 km). The relatively largest layer-content reductions since 1985 occurred during the spring and autumn seasons in the lower stratosphere.

1. INTRODUCTION

Unusually persistent negative total ozone deviations have occurred at Invercargill, New Zealand (46S, 168E) since 1985. The deviations are taken relative to the long-term total ozone series for the period 1965-1987 which has been composited from observations at Wellington (41S, 174E) during 1965-71 and observations at Invercargill from 1971 onwards. The Wellington data were empirically corrected to the latitude of Invercargill by comparisons of average monthly mean values for two non-overlapping six-year periods at the two stations.

The calibration status of the instrument (Dobson No 17) has been regularly maintained, by standard test procedures, and by comparisons with regional standard instruments (Aspendale, 1964, 1978; Toronto, 1981). Two large gaps occurred in the observations, six months in 1974/75, and six months during the second half of 1984 to January 1985. The well-calibrated No 82 Dobson instrument on loan from the Central Ozone Laboratory, Boulder, Colorado, was in operation at Invercargill from February to June 1985, while the New Zealand No. 17 instrument, damaged by floods in January 1985, was undergoing extensive repairs and recalibration. Tests with the Universal Standard Lamps in October 1986 showed that the instrument remained in a steady state of calibration.

Since January 1987 a second Dobson spectrophotometer has been operated in New Zealand, as part of the world automated Dobson network, by the Department of Scientific and Industrial Research at Lauder (45S, 169E), just 180 km from Invercargill. Two series of intercomparisons have shown good correspondence between the Invercargill and Lauder instruments, with consistent differences between their ozone measurements of less than 2%.

2. ANALYSIS OF LONG TERM OSCILLATIONS

2.1 THE RECENT TOTAL OZONE VARIATIONS

Data have been analysed for Invercargill, American Samoa (14S, 170W), and Melbourne (Aspendale) (38S, 145E). The period mean annual variations have been removed from all data, and the residuals have been given in the figures as percentage deviations from these means, or as deviations normalized by their appropriate standard deviations.

The Invercargill monthly mean total ozone values were consistently

ISBN 0-937194-15-8

below normal from February 1985 to September 1987, except during the two slightly above normal months September 1986 and February 1987. The negative departures fluctuated between 2-8% below the mean. Figure 1 shows the 3-month weighted running mean deviations for Invercargill, American Samoa, and

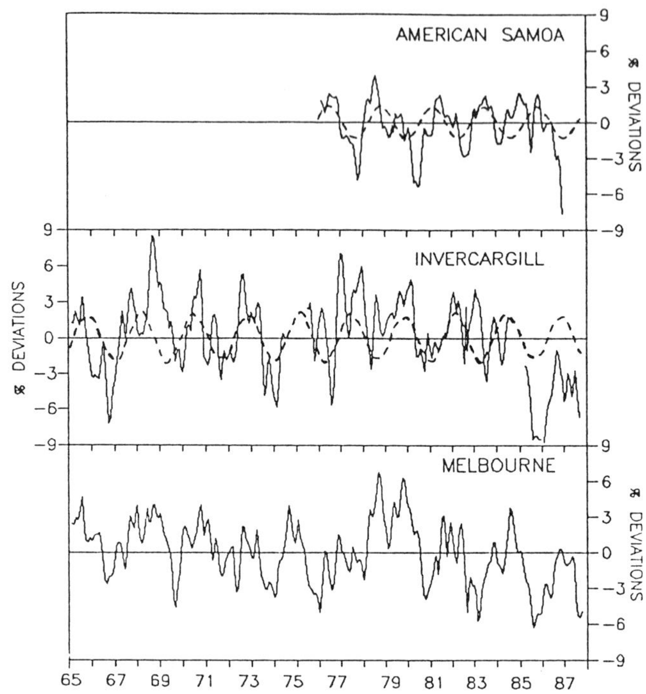

Fig. 1. Monthly mean total ozone deviations for 1965-87. The deviations have been smoothed by weighted 3-term running means. The 28-month QBOO components (dashed) are percentages of the long term monthly means. Ticks on time scale represent January each year.

Melbourne. The largest anomaly, during late 1985, appears at both Invercargill and Melbourne. The decline at Samoa is different, not appearing until late 1986. However, in all three cases, the anomaly coincides with a negative excursion of the quasi-biennial ozone oscillation (QBOO). The mean QBOO is shown in Figure 1 by the dashed line.

2.1 QUASI-BIENNIAL COMPONENTS

Autoregressive moving average analysis of the Invercargill and American Samoa total ozone prior to 1984 have shown quasi-biennial oscillations with periods of 28 months with amplitudes of 5 and 4 matm-cm respectively, accounting for about 12% and 18% of the total variances respectively. These are very similar to the results of Tolson 1981 for the corresponding latitudinal zonal average BUV satellite data.

Quasi-biennial ozone oscillations (QBOO) are thought to be associated with modulations by the quasi-biennial wind oscillations (QBWO) of the Hadley cell circulations, which may affect meridional ozone transport [Newell and Wu, 1978; London and Oltmans, 1978/79]. An analysis of upper air winds in the South-West Pacific made by the author confirms that during the 1965-87 period the amplitude of the QBWO decreases rapidly away from the equatorial latitudes, unlike the QBOO latitudinal variation. Lag correlation analysis between the QBOO and the equatorial QBWO showed phase changes with latitude, with the QBOO at American Samoa positively correlated with a lag of about 6 months, while at Invercargill it was negatively correlated with a lag of about 3 months. Such correlations would be expected to occur in accordance with the Hadley cell modulation theory.

Theory predicts intensification of the Hadley Cell circulations during the easterly phase of the QBWO, with enhanced upward motions at low latitudes and extension of the downward leg of the circulation cell toward middle latitudes, with probably faster poleward and downward ozone transport [Newell and Wu, 1978]. Maximum effects on midlatitude ozone variations would be expected to occur in the years when the easterly phase of the QBWO coincides with the winter phase of the annual variations of the Hadley cell circulation, when transport is poleward in the upper stratosphere [Tolson, 1981].

Time-varying lag correlation analysis of the Invercargill data indeed showed that the largest correlations occurred between values of the QBWO during the Southern Hemisphere midwinter and Invercargill ozone deviations in the following spring (r=-0.6, significant at the 5% level). There was good correlation also between these ozone departures and the quasi-biennial components of the 30 hPa zonal wind departures at Nandi (18S, 178E) (r=-0.67, highly significant.)

The phase of the QBWO was modified during 1983, which was thought to be a consequence of the radiational effects of volcanic dust ejected into the equatorial stratosphere by the eruption of El Chichon in 1982 [Labitzke et al., 1983]. The westerly phase was short during late 1982/early 1983, and the following easterly and westerly phases were longer than normal. Thus conditions would have been favourable, according to the above correlations, for low ozone values to occur in the 1985 spring during the westerly phase of the QBWO.

2.2 CORRELATIONS WITH SOUTHERN OSCILLATION INDEX ANOMALIES

Modulations of the Hadley cell circulations can also be expected to occur in association with variations of the Southern Oscillation Index (SOI) [Kidson, 1975]. According to the theory, during the negative phase of the SOI, modulations of the Hadley cell circulations will be similar in effect to the easterly phase of the QBWO, i.e. a speed-up of the circulations could result, with consequently higher ozone in midlatitudes. The periods and amplitudes of the SOI variations are very irregular, but despite this, our lag correlation analysis gave a high correlation between midwinter values of the SOI and ozone deviations at Invercargill in the following spring, (r=-0.5, significant at 1%).

2.3 HIGH LATITUDE STRATOSPHERIC WIND ANOMALIES

While no detailed studies exist for the Southern Hemisphere of teleconnections between variations of the SOI index and stratospheric winds and temperatures, it can be expected that both the polar night stratospheric jet stream and the subtropical westerlies will be affected to some degree, similarly to the Northern Hemisphere [Van Loon et al., 1982].

For the 1965-86 period, significant positive correlations were found in the present study between the early winter zonal wind deviations at the 30 hPa level over Campbell Island (53S, 179E) and the ozone variations at Invercargill in the following spring (r=0.7, highly significant). During 1983, following the extreme negative SOI index deviations in 1982, the 30 hPa level westerly wind components at Campbell Island were relatively strong, but from 1984 to 1987 they were relatively weak. In general, lower than normal spring-time ozone values at Invercargill would be likely to follow these weaker than normal winter-time stratospheric winds at Campbell Island.

3. LONG-TERM TRENDS IN UMKEHR VERTICAL OZONE DISTRIBUTIONS

Umkehr observations have been made at Invercargill since 1973. Deviations from the long-period seasonal means are shown in Figure 2, for four combined layers, normalized and smoothed by weighted 3-term running means. Seasonal means are used because the frequency of observations is relatively low. The smoothing process reduces the size of the monthly departures by about 15 percent.

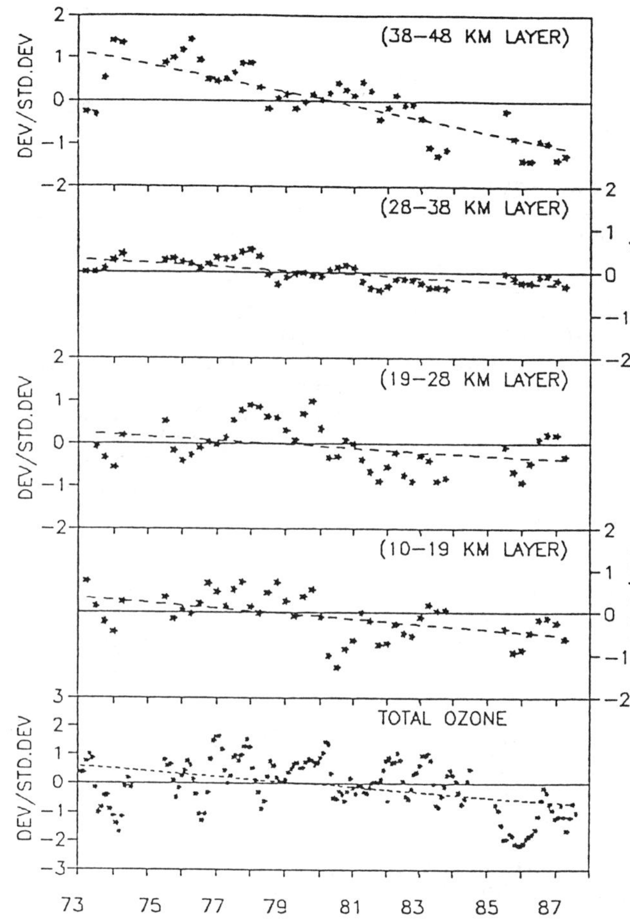

Fig. 2. Normalised monthly mean total ozone and seasonal mean Umkehr layer partial pressure variations during 1973-87 at Invercargill. All series have been smoothed by weighted 3-term running means.

Corrections have been made for the negative errors in the Umkehr evaluations for layers 6 to 9 (28-48 km) due to abnormal statospheric aerosol levels. The percentage reductions relative to changes in stratospheric optical transmission given by Reinsel et al. [1984] were used. Changes in the atmospheric transmission from September 1982 to mid-1983 due to the effects of El Chichon aerosols were estimated on the basis of ground-based optical depth measurements at Invercargill, assuming that the background optical depth of the stratosphere is normally one order of magnitude less than the total depth observed at the ground. The largest corrections amounted to about 50% in October 1982 for layer 9, diminishing rapidly by March 1983 to about 7%. The corrections decrease with decreasing height and were insignificant for layer 6 by the beginning of 1983.

The Umkehr partial pressures appear to drop well below the long-term averages in the second half of 1983 in the top layers (38-48 km) and remain consistently below thereafter. Below-average deviations also persisted in the layers around the level of the ozone maximum (19-28 km) and the layers below (10-19 km), since about 1981. There are suggestions of seasonal patterns in the variations. The relatively largest layer-content reductions occurred in the lower stratosphere centred on layer 3 (15-19 km) in the spring and autumn seasons since 1985. The decreasing linear trends between 1973 and 1987 for the 38-48 km layer, and for the layers in the lower stratosphere, were significant at the 1% level.

Since the total ozone amounts are greatly influenced by the lower stratosphere ozone variations, and these in turn are mostly controlled by atmospheric dynamics at these latitudes, the observations presented here indicate that a large part of the recent negative total ozone departures observed at invercargill is connected with stratospheric circulation anomalies. Non-dynamical causes must be found, however, to explain the significant decreasing trend of ozone found in the highest Umkehr layers.

The ozone observing project at Invercargill was terminated in September 1987. The Invercargill historical series of total ozone and Umkehr vertical ozone distributions can hopefully be extended further by the ozone data from Lauder, after suitable allowances have been made for site diferences.

ACKNOWLEDGEMENTS

Thanks are due to the following N Z Meteorological Service staff: Steve Goulter for helpful discussions on statistical analysis techniques, Sylvia Nichol for supplying the results of intercomparisons between the Invercargill and Lauder instruments, and the technical staff of the observatory at Invercargill who creditably carried out the ozone observations through many years.

REFERENCES

Kidson, J W, 1975: Tropical eigenvictor analysis and the Southern Oscillation, Mon Weath Rev, 103(3), 187-196.

Labitzke, K, B Naujokat, M P McCormick, 1983: Temperature effects on the stratosphere of April 4, 1982 eruption of El Chichon, Mexico, Geophys Res Lett, 10, 24-46.

London, J and S J Oltmans, 1978/79: The global distribution of long-term ozone variations during the period 1957-1975, Pure Appl Geophys, 117, 345-354.

Newell, R E and M F Wu, 1978: A pilot study of concomitant changes in total ozone and atmospheric general circulation, Quart J R Met Soc, 104, 999-1003.

Reinsel, G C, G C Tiao, J J de Luisi, C L Mateer, A J Miller, J E Frederick, 1984: Analysis of upper stratospheric Umkehr ozone profile data for trends and the effect of stratospheric aerosols, J Geophys Res, 89, 4833-4840.

Tolson, R H, 1981: Spatial and temporal variations of monthly mean total columnar ozone derived from 7 years of BUV data, J Geophys Res, 86, 7312-7330.

Van Loon, H, C S Zerefos, C C Repapis, 1982: The Southern Oscillation in the stratosphere, Mon Weath Rev, 110, 225-229.

OZONE OBSERVATIONS FROM SATELLITES

TOTAL OZONE TRENDS FROM TOMS DATA

Richard S. Stolarski, Arlin J. Krueger, and Mark R. Schoeberl
Laboratory for Atmospheres
NASA Goddard Space Flight Center
Greenbelt, Maryland 20771, USA

ABSTRACT

Data from nine and one half years of global total ozone measurements by the TOMS instrument from the Nimbus 7 satellite show a decrease of about 7% between 1979 and 1987 in agreement with previous studies of SBUV total ozone data (Heath, 1988). Because maintenance of precise satellite calibration over long time periods is a significant problem, an alternative version of the TOMS data can be obtained by using the global Dobson network to adjust for any possible small drifts in the calibration. In this paper we deduce how total ozone has changed over the globe from TOMS data which is adjusted in this manner.

1. ADJUSTMENT OF TOMS DATA

Since the Total Ozone Mapping Spectrometer (TOMS) instrument scans across the track of the satellite it is possible to obtain an overpass of each Dobson station on a daily basis where the location separation of the two measurements is less than one degree in longitude and latitude (see Fleig et al., 1988). Data from the overpasses of 41 Dobson stations have been used to obtain an adjustment factor for the global TOMS measurements. Figure 1 shows the difference, TOMS-Dobson divided by Dobson for overpasses averaged over the 41 stations on a daily basis. The data extend from November, 1978 through 1986. The stations used are the those whose record is continuous since 1979 and are listed in Fleig et al. (1988). Also shown on the figure are the differences between TOMS and the world standard Dobson instrument (#83) during the calibration runs of this instrument at Mauna Loa as reported by McPeters and Komhyr (1988). The solid line is an approximation to these data, including an extrapolation through 1988, that has been used to adjust the archival TOMS data set available from the National Space Science Data Center. A study of the D-pair of wavelengths on SBUV reported by Bhartia et al. in this symposium indicates that the drift with respect to Dobson may indeed be in the SBUV and TOMS normalization and lends credence to the use of Dobson as the best overall long-term normalization. In addition, the TOMS instrument began, in 1984, to experience occasional instances of non-synchronization of the chopper wheel which selects the wavelengths (Fleig, et al., 1986). Since that time TOMS has exhibited a slow, latitude dependent drift with respect to SBUV (Wellemeyer, et al., 1988). This drift is near zero at low and mid-latitudes, and negative by nearly 2 percent (total since 1984) close to the poles. To obtain a more accurate, or at least more conservative estimate of total ozone trends from TOMS measurements, this factor has been added to the adjustment of the TOMS observations. As can be seen in the figure 1, the normalization factor changes by nearly 4% from late 1978 to late 1987. A study utilizing 92 stations, including some with incomplete records over this time period, came to similar conclusions concerning the relative drift of TOMS and the ground based network (Bojkov, 1988).

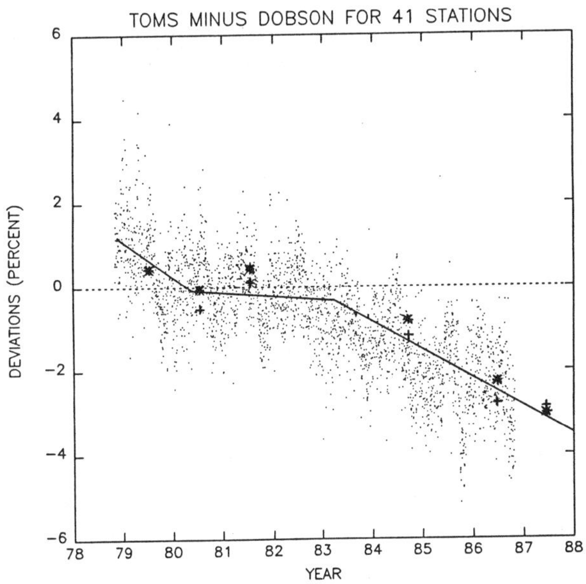

Fig. 1. Daily values of the mean percent deviation of TOMS from 41 Dobson stations shown as TOMS – Dobson divided by Dobson. The asterisks and plusses are the deviations over Mauna Loa during the calibration of the standard instrument number 83 (McPeters and Komhyr, 1988). The solid line is the assumed drift of TOMS with respect to Dobson for the calculations of this paper.

ISBN 0-937194-15-8

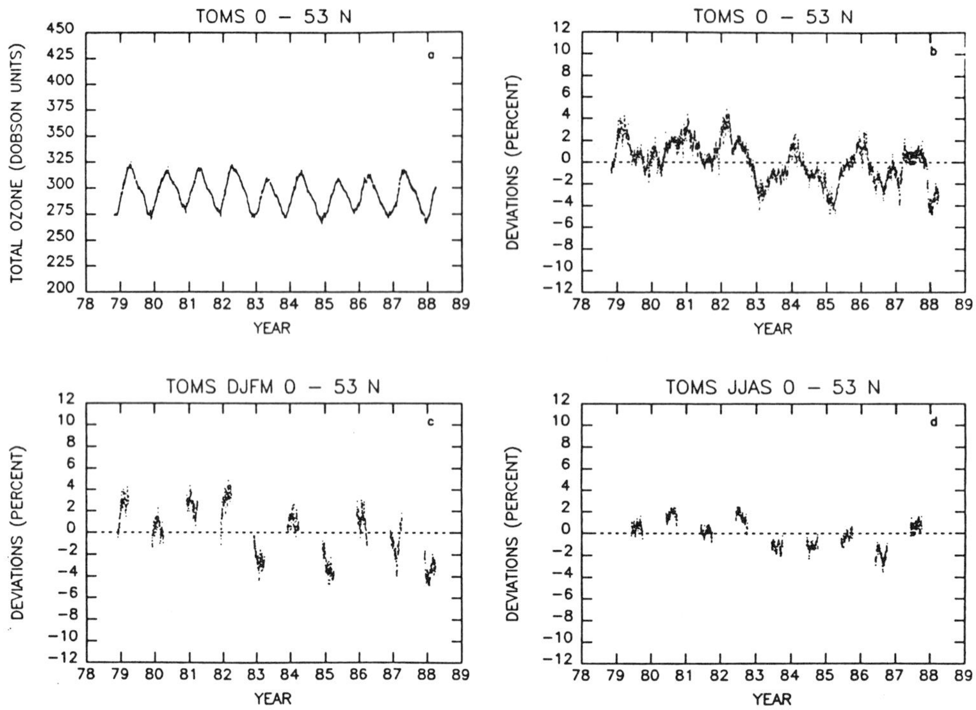

Fig. 2. a) Daily TOMS measurements of total ozone averaged over the latitude band 0 to 53 degrees north. b) Deviations of the daily measurements from the mean taken over 9 plus years for the appropriate week. c) Deviations of the daily measurements from the weekly mean for the winter months only (December, January, February, March). d) Deviations of the daily measurements from the weekly mean for the summer months only (June, July, August, September). All data have been adjusted for the TOMS drift with respect to the Dobson network. Latitude dependence of TOMS drift with respect to SBUV has been used since 1984.

2. TRENDS

Because most of the Dobson stations are located at northern midlatitudes, the adjusted TOMS data at these latitude shows a variation over the last 9 years which looks very much like the analyses of the Dobson network, i.e., dominated by strong negative deviations from climatology in the winters of 82-83 and 84-85. This is shown in figures 2a and 2b which include area weighted data for all northern latitudes from the equator to 53 degrees. There also appears to be a large negative deviation this past winter, i.e. 1987-88. Figure 2, panels c and d show the same deviations broken into two seasons, winter and summer. The winter data exhibit a strong, approximately biennial oscillation. This is despite the fact that the quasi-biennial oscillation in total ozone exhibits a change of phase at about 15 degrees latitude (see e.g. Schoeberl et al., 1988). The summer data show much less variability and are high the first four years, low the next four years, and apparently high again the last year. The summer data is suggestive of a solar cycle signal of a few percent, but this cannot be demonstrated.

Figure 3 shows the results for the Southern Hemisphere. The deviations from seasonal means shown in figure 3b show a strong negative deviation starting in late 1985, continuing into 1986 and another strong negative deviation in 1987. The seasonal breakdown is shown in panels c and d. Winter again shows a strong quasi biennial oscillation while the summer is high at the beginning of the record and lower at the end of the record.

Although the above results show that ozone changes are not simple and monotonic, a global view can be obtained by comparing the early TOMS data with the later data. In figure 4 the difference of the last two complete years of TOMS data (1986-1987) and the first two complete years (1979-1980) is shown as a function of latitude and month. Two years have been used in each case to reduce the variance due to the quasi-biennial oscillation. In the southern hemisphere, strong negative deviations associated with the springtime Antarctic ozone hole are seen. Lesser, but still significant, negative changes are observed throughout the year at southern polar latitudes. In the north polar regions, negative changes are seen in total ozone which are larger than those observed at midlatitudes. They are mostly confined to the early spring when the solar zenith angle is very high. The change in low and mid-latitudes is slightly negative with lesser areas of positive change occurring.

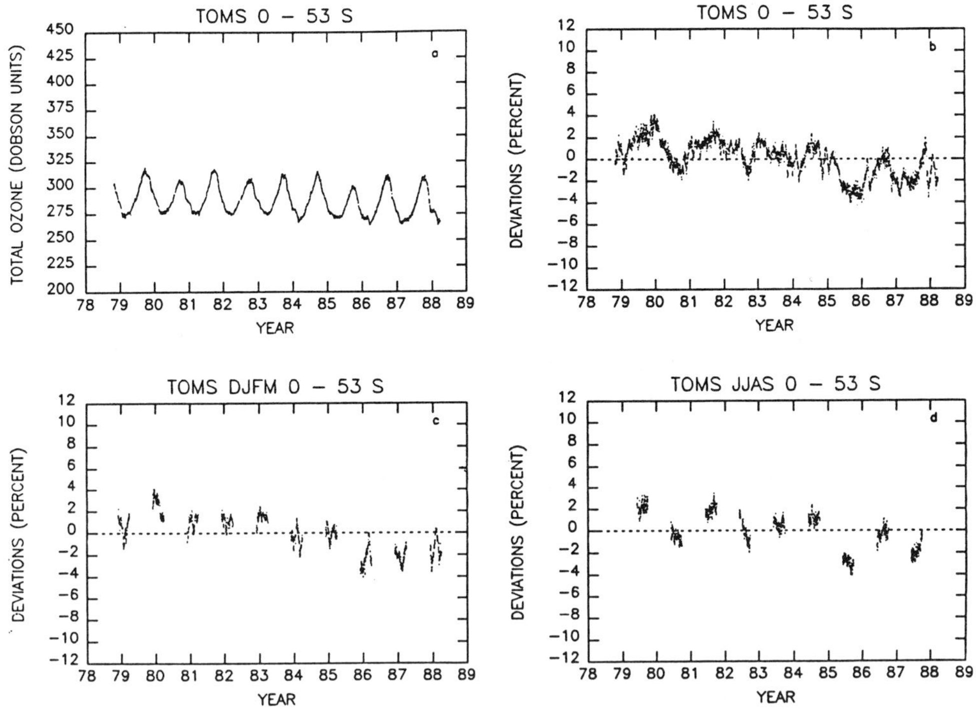

Fig. 3. a) Daily TOMS measurements of total ozone averaged over the latitude band 0 to 53 degrees south. b) Deviations of the daily measurements from the mean taken over 9 plus years for the appropriate week. c) Deviations of the daily measurements from the weekly mean for the summer months only (December, January, February, March). d) Deviations of the daily measurements from the weekly mean for the winter months only (June, July, August, September). All data have been adjusted for the TOMS drift with respect to the Dobson network. Latitude dependence of TOMS drift with respect to SBUV has been used since 1984.

3. CONCLUSIONS

Using the short record of measurements from TOMS, it not possible to separate solar cycle changes from long term trends. There are, however, hints that the summer data show more solar cycle variation than predicted and that the quasi-biennial oscillation dominates in winter.

REFERENCES

Bhartia, P. K., S. Taylor, and A. J. Fleig, Estimation of errors in the long-term change in ozone reported by the SBUV/TOMS instruments-A new way of looking at the old data, Proceedings of the Quadrennial Ozone Symposium, Gottingen, FRG, 1988.

Bojkov, R. D., C. L. Mateer, and A. L. Hansson, Comparison of ground-based and Total Ozone Mapping Spectrometer measurements used in assessing the performance of the Global Ozone Observing System, J. Geophys. Res., 93, 9525-9533, 1988.

Fleig, A. J., P. K. Bhartia, C. G. Wellemeyer, and D. S. Silberstein, Seven years of total ozone from the TOMS instrument-A report on data quality, Geophys. Res. Lett. 13, 1355-1358, 1986.

Fleig, A. J., D. S. Silberstein, C. G. Wellemeyer, R. P. Cebula, and P. K. Bhartia, An assessment of the long-term drift in TOMS total ozone data, based on comparison with the Dobson network, Proceedings of the Quadrennial Ozone Symposium, Gottingen, FRG, 1988.

Heath, D. F., Non-seasonal changes in total column ozone from satellite observations, 1970-86, Nature, 332, 219-227, 1988.

McPeters, R. D. and W. D. Komhyr, Long-term changes in SBUV/TOMS relative to the world primary standard Dobson instrument, Proceedings of the Quadrennial Ozone Symposium, Gottingen, FRG, 1988.

Schoeberl, M. R., L. R. Lait, and P. A. Newman, Quasi-biennial modulation of the Antarctic ozone depletion, submitted to J. Geophys. Res., 1988.

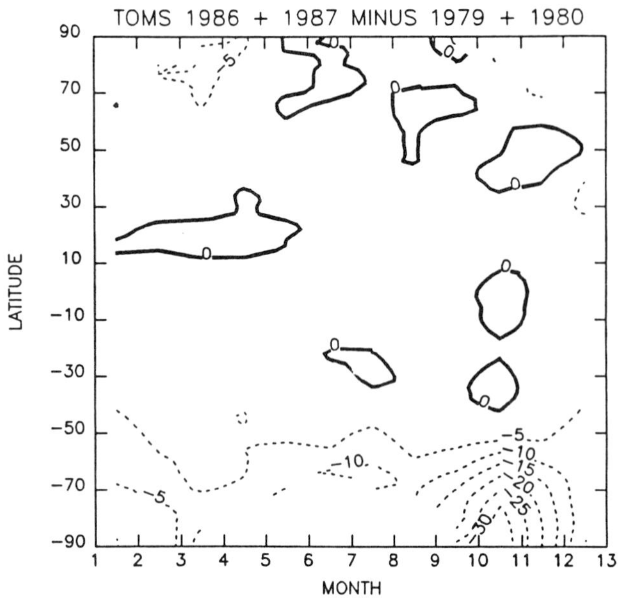

Fig. 4. Difference between last two years of TOMS data and first two years of TOMS data in percent as a function of latitude and month. All data have been adjusted for the TOMS drift with respect to the Dobson network. Latitude dependence of TOMS drift with respect to SBUV has been used since 1984.

Wellemeyer, C. G., A. J. Fleig, and P. K. Bhartia, Internal comparisons of SBUV and TOMS total ozone measurements, Proceedings of the Quadrennial Ozone Symposium, Gottingen, FRG, 1988.

ON THE DETERMINATION OF LONG-TERM TRENDS FROM SBUV OZONE DATA

Robert D. Hudson and Jay R. Herman
NASA Goddard Space Flight Center
Greenbelt, Maryland 20771, USA

George N. Serafino
Applied Research Corporation
Landover, Maryland 20783, USA

1. INTRODUCTION

From October 1976 to the end of 1986, data from the SBUV experiment has shown a global average ozone decrease from 16 to 20% over the pressure altitude range from 4 to 0.4 mb (37.5 to 55 km) (Heath 1988 and Chandra 1988) and an approximately 8% decrease in total ozone (Bowman 1988). These changes are in excess of the reported instrumental errors (Cebula, Park, and Heath, 1988, hereafter referred to as CPH) of approximately 2% over the first six years of the SBUV experiment. This paper argues that the basic model of the SBUV instrument degradation used by CPH is not unique, and can lead to an incorrect estimate of the ozone trend. An alternate analysis of the data is proposed that removes most of the reported ozone decrease and brings the trend into agreement with other data.

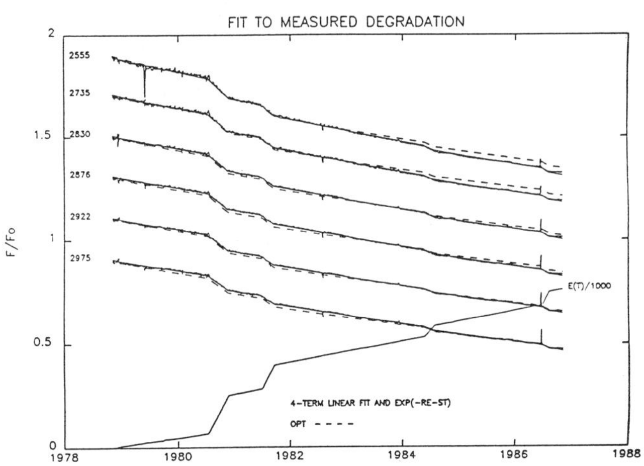

Fig. 1 SBUV instrumental response to the direct solar flux for the first six wavelengths (255.5, 273.5, 283.0, 287.6, 292.2, 297.5 nm). The response at each wavelength has been normalized to 1 at the first data point and displaced by 0.2 units. Also shown is the accumulated exposure time of the diffuser plate in days/1000. The first data point is taken at day number 307 of 1978. The dashed lines represent the fit to the data given by Eq. 6 using the constants obtained by CPH. The smooth solid line is the fit to the data using Eq. 7.

2. DEGRADATION OF THE SBUV INSTRUMENT RESPONSE

SBUV makes observations of the solar radiation backscattered from the earth's atmosphere at 12 fixed wavelengths (255.5 to 339.8 nm) in a nadir view, and a measurement of the solar irradiance for 1 out of 14 orbits per day. The solar irradiance is measured when a ground aluminum diffuser plate is brought out of storage and positioned as the first optical element in front of the depolarizer and spectrometer entrance slits.

Since the spacecraft launch in October 1978, the measured solar irradiance, Fm(t), has been observed to decrease approximately linearly with time at each wavelength (see Fig. 1). These large changes in Fm(t) are not seen in other independent measurements of the solar flux, suggesting that one or more instrumental components are degrading with time. If it is assumed that the degradation of the diffuser plate and the spectrometer are independent, then the measured solar irradiance in terms of the true solar irradiance, Fo(t) is given by

$$Fm(t,E) = Fo(t)D(t,E)S(t,E) \qquad (1)$$

The normalized solar irradiance data shown in Fig. 1 is the result of a daily deployment of the diffuser plate. For brief periods during 1980, 1981, 1984, and 1986 the diffuser plate was exposed once per orbit to the direct solar flux. The periods of increased exposure (about 14:1) appear as abrupt changes in slope in Fm(t)/Fm(0) for each wavelength in exact correlation with the accumulated exposure rate, E(t).

When the diffuser plate is stored, the spectrometer measures the backscattered radiance, Im(t), where

$$Im(t) = Io(t)S(t,E) \qquad (2)$$

Io(t) is the true radiance. From Eqs. 1 and 2 the albedo of the earth's atmosphere is given by:

$$A(t) = \frac{Io(t)}{Fo(t)} = \frac{Im(t)}{Fm(t)} D(t) \qquad (3)$$

ISBN 0-937194-15-8

Thus the albedo, A(t), the quantity that is related to the atmospheric ozone content, can be measured accurately only if the diffuser plate degradation rate, D(t,E), can be accurately specified. However, from Eq. 1, only the product, D(t,E)S(t,E), is actually measured. Since no on-board measurement of D(t,E) was made, it is necessary to construct a model of the degradation rates of the diffuser plate and spectrometer to compare with Fm(t,E(t)).

3. DEGRADATION MODELS

In their paper, CPH assumed that the change in the measured solar irradiance could be modeled for each wavelength by

$$\frac{Fm(t)}{Fm(0)} = P(t)\, e^{-aG(t)}\, e^{-st}\, e^{-rE(t)} \qquad (4)$$

where P(t) is the gain change in the photomultiplier tube. The next term is the long term solar flux variation for each wavelength. In CPH, this was based on a model of Heath and Schlesinger (1986). Next CPH makes the critical assumption that:

$$D(t) = e^{-rE(t)} \qquad (5)$$

and

$$D(t)S(t) = e^{-rE(t)-st} \qquad (6)$$

CPH then determined r and s from the slopes within and just after the 1980 and 1981 periods of enhanced exposure using a least-squares analyses. A comparison of the fit given by Eq. 4 to the entire data record is also shown in Fig. 1.

We suggest an alternate approach to fitting the observed degradation given by:

$$\frac{Fm(t)}{Fm(0)} = A + Bt + CE(t) + DtE(t) \qquad (7)$$

where A, B, C, and D are constants determined for each wavelength by least-squares fitting. An obvious criticism to a fit of this form is that it is wrong for arbitrarily large elapsed time. The implied nearly constant rate of degradation must eventually decrease asymptotically to zero. However, for the 8 years of SBUV operation, this portion of the degradation curve has not been reached and the rate of degradation is approximately linear.

A plot of the fit given by the form in Eq. 7 is also shown in Fig. 1 by the nearly invisible solid line running through the center of the data. The region of worst fit is at the very end of the data just after the last increased deployment rate of the diffuser plate. Except near the end of the data set the error in the overall fit is about 1%.

The terms in the quasi-linear model can be rearranged into a two term product form that can be associated with the separate degradation rates of the diffuser and spectrometer. Of the many product forms possible, two cases are discussed that yield quite different ozone trends.

3.1 CASE 1: MORE OZONE THAN CPH

Starting with Eq. 7, the terms can be rearranged as

$$D(t)S(t) = (1 + kE)\left(1 + \frac{A - 1 + Bt + DEt + hE}{1 + kE}\right) \qquad (8)$$

where $h = C - k$

Let $$k = \frac{C}{1 + f(B/A)t} \qquad (9)$$

then assume the factors can be separated into

$$D(t) = 1 + kE$$

and

$$S(t) = 1 + \frac{A - 1 + Bt + DEt + hE}{1 + kE} \qquad (10)$$

When the scale factor, f, in Eq. 9 is 0.55 the result is about a 15% increase in ozone for the layer between 1 and 2 mb pressure by May 24, 1986 (day 2790) relative to the SBUV-OPT (Solar Backscatter Ultraviolet - Ozone Processing Team) value using the formula of CPH. Increasing this scale factor will increase the amount of ozone relative to the OPT value.

3.2 CASE 2: LESS OZONE THAN CPH

An alternate rearrangement of terms is

$$D(t)S(t) = (A + ht)\left(1 + \frac{CE + DEt + kt}{A + ht}\right) \qquad (11)$$

where $k = B - h$

let $$h = gB \text{ i.e. } k = (1 - g)B \qquad (12)$$

where g is an arbitrary scale factor.

Then assume the factors can be separated into

$$D(t) = 1 + \frac{CE + DEt + kt}{A + ht} \qquad (13)$$

and

$$S(t) = A + ht$$

When g = 1 or larger, case 2 results in an albedo that is larger than predicted by the CPH form, and therefore leads to smaller ozone amounts. Selecting values of g < 1 would lead to albedo values that are smaller than the CPH form, implying more ozone. There is less wavelength dependence for the difference between D(E,t) and exp(-rE) for case 2 than for case 1. This implies that both long and short wavelengths would show a substantial decrease in the calculated albedo, and increases in the ozone amount at all pressure altitudes. A comparison of the resulting amounts of ozone with that measured from SAGE and the Dobson network make this form unlikely for any value of g.

It should be noted that for both case 1 and

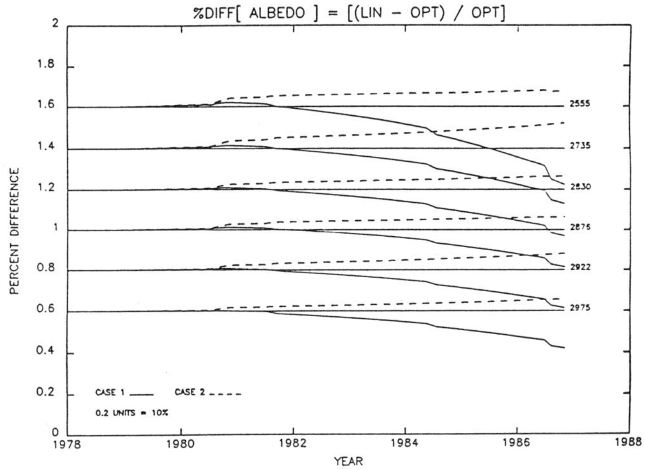

Fig. 2 The percent difference between the calculated albedo using Eq. 5 from CPH and Eq. 10 (Cases 1, solid lines), and Eq. 13 (Case 2, dashed lines). Each curve is displaced 0.2 units. For the wavelengths 273.5, 283.0, and 287.6 the approximation of -1.8% change in ozone for each 1% change in albedo can be used. For longer wavelengths the approximation of constant scale height ratio begins to break down.

case 2 the diffuser and spectrometer terms now depend on both t and E. The case 1 form has properties that may make it useful for SBUV data reduction. When f < 1, the time term in D(E,t) is negligible (less than 1%) until t = 0.01/(fB). After this time, D(E,t) gradually increases relative to exp(-rE). The wavelength dependence of B is such that the difference between D(E,t) and exp(-rE) decreases for increasing wavelength. Using Eq. 3, the different rates of diffuser degradation can be used to calculate the percent change in albedo relative to the CPH formulation, exp(-rE). The

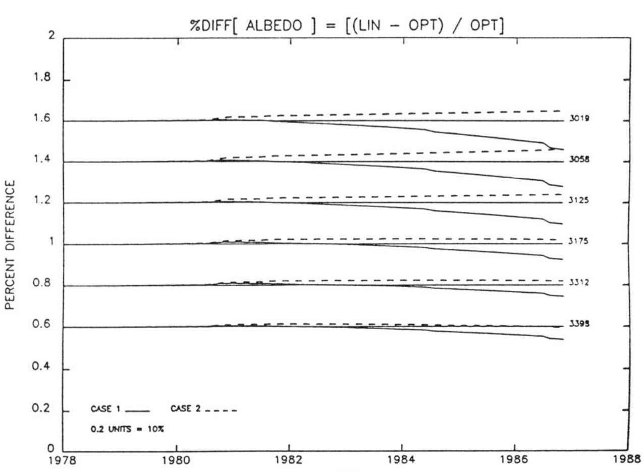

Fig. 3 The same as Fig. 5 except for the last six wavelengths.

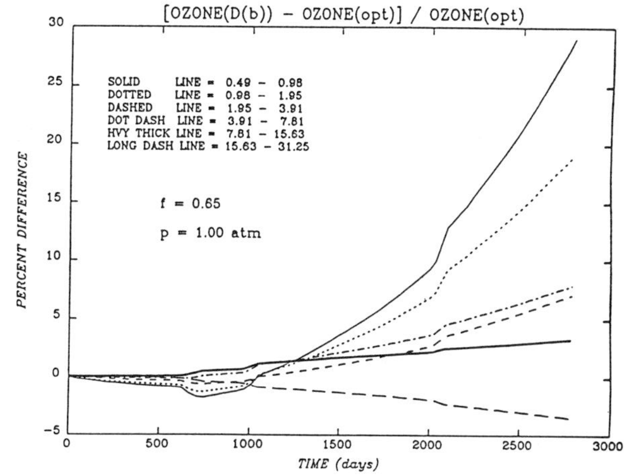

Fig. 4 The time-history of the percent change from the archived SBUV/OPT ozone amounts for the f = 0.65 case 1 correction. The results are shown for six pressure altitudes (0.49-0.98, 0.98-1.95, 1.95-3.91, 3.91-7.81, 7.81-15.63, and 15.63-31.25 mb).

results are shown in Figs. 2 and 3 for all 12 wavelengths used in the ozone retrieval algorithm.

At 273.5 nm the albedo decrease using Eq. 10 relative to Eq. 5 (CPH) is about 8.1% (f=0.55) and 9.7% (f=0.65) (i.e., ozone increases of about 14.6% and 17.5%) by mid 1986 for the region between 1 and 2 mb. Thus, if degradation derived from case 1 of the quasi-linear fit (with f = 0.55 to 0.65) were used instead of that from the CPH formulation, the negative trend of ozone with time in the 1 to 2 mb layer would nearly disappear.

Other than by comparison with other ozone data (Dobson, SAGE, etc.), we have not found a test that might distinguish between the various choices for D(t) and S(t). For example, it might be expected that the inclusion of E(t) in the spectrometer degradation portion of case 1 would lead to structure in the radiance observed at 339.8 nm over the tropical regions of the earth. However, no structure corresponding to the rapid deployment periods is present in any of the three forms of S(t) at 339.8 nm. The magnitudes are also sufficiently close to that from Eq. 5 as to be within the SBUV experimental error.

4. TIME-HISTORY OF THE OZONE TREND CORRECTION

The disagreement between the Umkehr and the satellite results (SAGE and SME) prevents the unequivocal selection of a single value of f with a narrow error range. For purposes of illustration, the ozone trend correction factors for f = 0.65 are shown in Fig. 4 for each of 6 SBUV pressure altitudes (0.49 to 0.98, 0.98 to 1.95, 1.95 to 3.91, 3.91 to 7.81, 7.81 to 15.6, 15.6 to 31.2 mb) and in Fig. 5 for the total ozone.

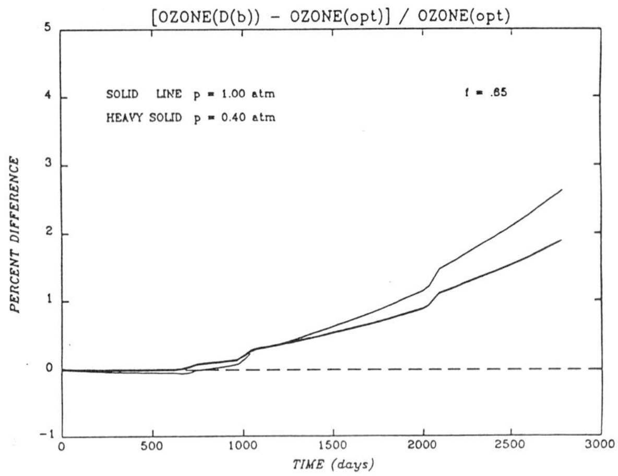

Fig. 5 The time-history of the percent change from the archived SBUV/OPT total ozone amounts for the f = 0.65 case 1 correction. Two lower boundary conditions are used based on a clear atmosphere to the ground at 1000 mb, and a cloud boundary of 400 mb.

The results in the pressure range from 0.49 to 31.2 mb show an almost zero correction factor from launch until 1981. After 1981 the corrected calculation for ozone gives more ozone at all altitudes than indicated by the currently archived SBUV ozone data. The trend and shape of the correction factors correspond to the fact that the CPH diffuser fit (Eqs. 5 and 6) was calculated for the 1980-1981 periods of increased diffuser deployment. After 1981, the decreasing accuracy of the overall fit and the assumption of an exponential form lead to apparent large calculated ozone decreases with time.

5. CONCLUSION

Uncertainty in the model describing the details of the SBUV instrument's degradation between 1978 and 1987 casts doubt on the reported trends in stratospheric ozone. Determining the atmospheric ozone altitude profile is dependent on accurately knowing the change in reflectivity of the SBUV diffuser plate as a function of time. Considered by itself, the SBUV data does not contain sufficient information to construct a unique physical or empirical model that can be used to separate the diffuser plate degradation from that of the spectrometer.

REFERENCES

Bowman, K. P., 1988: Global trends in total ozone, Science, 239, 48-50.

Cebula, R. P., H. Park, and D. F. Heath, 1988: Characterization of the Nimbus-7 SBUV radiometer for the long term monitoring of stratospheric ozone, J. Atmos. Ocean. Tech., 5, 215-227.

Chandra, S., 1988: Recent trends in stratospheric ozone Part I: Implications of El Chichon, solar cycle, and dynamical perturbations, submitted to J. Geophys. Res.

Heath, D. F. and B. M. Schlesinger, 1986: The Mg280 nm doublet as a monitor of changes in solar ultraviolet irradiance, J. Geophys. Res., 91, 8672-8682.

Heath, D. F., 1988: Non-seasonal changes in total column ozone from satellite observations, 1970-1986, Nature, 219-227.

Park, H. and D. F. Heath, 1985: in Atmospheric Ozone Proceedings of the Quadrennial Ozone Symposium held in Halkidiki, Greece 1984, C. S. Zerofos and A. Ghazi, Eds., D Reidel, Dordrecht, 412.

INTERNAL COMPARISONS OF SBUV AND TOMS TOTAL OZONE MEASUREMENTS

Charles G. Wellemeyer
ST Systems Corporation
Lanham, Maryland 20706, USA

Albert J. Fleig
NASA Goddard Space Flight Center
Greenbelt, Maryland 20771, USA

Pawan K. Bhartia
Science Applications Research
Lanham, Maryland 20706, USA

ABSTRACT

The time series of the difference between TOMS and SBUV is examined using the difference of monthly zonal means of total ozone measured by each instrument. The calculation of total ozone from the off-nadir measurements of radiance from TOMS is the major difference in the total ozone algorithms of SBUV and TOMS. When the central samples of the TOMS scan are averaged (nadir-only TOMS) to provide coincidence with the field of view (FOV) of the nadir viewing SBUV, no significant drift (0.0 ± 0.04%/yr) is seen over the majority of the globe. Small differences at high latitudes after 1984 appear to be associated with instrument problems in TOMS reported to commence in April 1984. When the time series of the difference of the nadir and extreme off-nadir total ozone measurements from TOMS is examined, a 2% drift is seen across the eight year data record which is consistent with estimates of the long-term calibration error in TOMS presented elsewhere in these proceedings.

INTRODUCTION

The main goals of this study are to investigate the initial bias and the long-term drift between the total ozone measured by TOMS and SBUV. These parameters have been established through inference from earlier TOMS-Dobson and SBUV-Dobson comparisons [Fleig et al. 1988a], but may be examined in more detail through a direct comparison of the two instruments.

The characterization of the diffuser plate used by SBUV and TOMS in the measurement of solar flux is crucial to the long-term calibration of both measurement systems [Cebula et al., 1988], but the TOMS measures a smaller portion of the diffuser plate than does SBUV. Determination of the long-term drift between the two instruments tests the assumption that a single characterization of the diffuser is appropriate for both instruments.

Instrument problems with TOMS have been noted which have a small relative impact on the TOMS total ozone retrieved from the A-pair (313 nm - 331 nm) and B-pair (318 nm - 331 nm) wavelengths [Fleig et al., 1986a]. Our analysis of solar flux measured by TOMS indicates that the impact of the instrument problems is greatest at 318 nm. The individual results from the A and B-pairs are averaged in the final determination of TOMS and SBUV ozone with a weighting scheme which is solar zenith angle and total ozone dependent such that the B-pair result is given more weight at high latitude [Fleig et al., 1988b]. So, the instrument changes in TOMS may have a latitude dependent impact on the retrieved total ozone.

The differences of monthly averages presented are equivalent to the monthly averages of differences except that the uncertainty in the results is difficult to estimate. The square root of the sum of the squares of the standard errors of

ISBN 0-937194-15-8

the means differenced is nominally 2%. This quantity is helpful in scaling the temporal features in the TOMS-SBUV bias relative to the accuracy with which monthly averages of ozone are determined, but it is an over-estimate of the uncertainty in the determination of the TOMS-SBUV bias.

RESULTS

Shown in Fig. 1 are the results of linear regressions applied to the time series of the difference of the monthly means of the total ozone from nadir-only TOMS and SBUV. The drift is not significantly different from zero over the majority of the globe. This supports the assumption that a single characterization of the solar diffuser common to both instruments is appropriate in spite of the fact that the TOMS views a smaller portion of the diffuser than does SBUV. The negative drifts seen at higher latitudes are apparently due to instrument effects in TOMS unrelated to the diffuser plate. The onset of these changes is relatively sudden so that they are not appropriately characterized, except to first order, by the linear regressions used here.

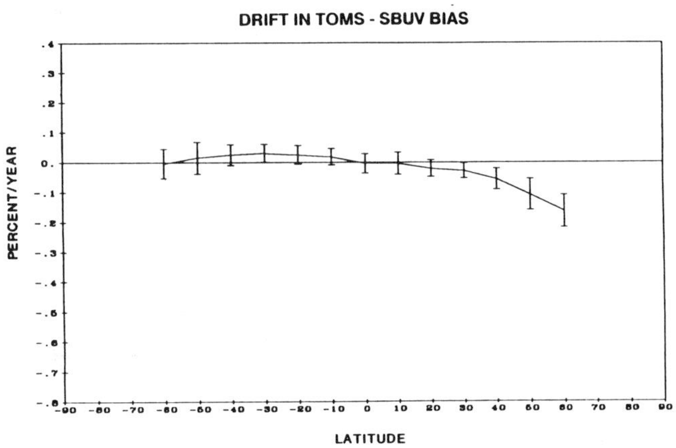

Fig. 1. Slopes in %/year derived by linear regression of percent differences in total ozone (TOMS-SBUV) in 10 degree latitude zones. The central samples from the TOMS scan are averaged to give coincidence with the SBUV FOV. The error bars represent the standard error of the slope.

Shown in Fig. 2 is the nadir-only TOMS-SBUV difference in the monthly means of total ozone in the 10 degree latitude zone from 5 S - 5 N. A nominal 3% bias between TOMS and SBUV is drawn in as a reference. The mean and standard error of the data are 2.5% and 0.04%, respectively. This bias is somewhat smaller than the 2.9% inferred from SBUV-Dobson and TOMS-Dobson comparisons (in which off-nadir data are included) because of a bias between the nadir and off-nadir measurements from TOMS which is discussed further below. Direct TOMS-SBUV comparisons in which the nadir and off-nadir measurements of TOMS are included are consistent with the SBUV-Dobson and TOMS-Dobson results. The initial bias is associated with the uncertainties of the absolute calibration of the TOMS and SBUV instruments [Fleig et al., 1988a].

Seasonal variations of the order of 1% are evident in Fig. 2. The cause of this effect is not known. It may be due to seasonal differences in the reflectivities measured by TOMS at 360 nm and 380 nm and by SBUV at 340 nm.

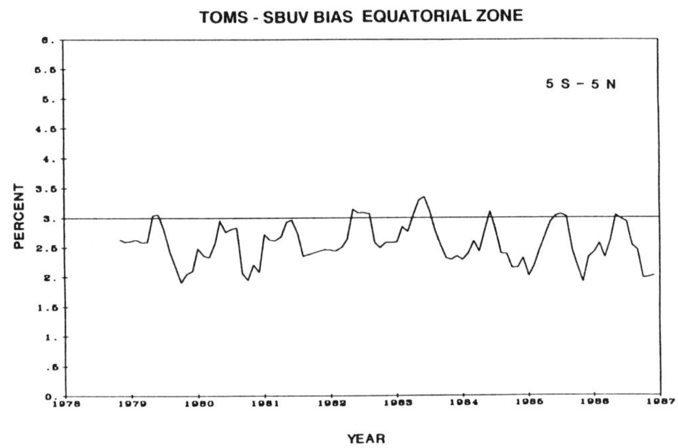

Fig. 2. Percent difference of monthly means of total ozone (nadir only TOMS-SBUV) in the 10 degree latitude zone 5 S - 5 N. The central samples from the TOMS scan are averaged to give coincidence with the SBUV FOV. A nominal 3% bias is drawn for reference.

Shown in Fig. 3 is the nadir-only TOMS-SBUV difference in the monthly means of total ozone in the 10 degree latitude zone from 55 N - 65 N. Evident in these data is a sudden change in the TOMS-SBUV bias around the end of 1984. Table 1 shows the results of modeling the long-term change in the TOMS-SBUV bias as a step discontinuity somewhere in 1984. Significant changes in the TOMS-SBUV total ozone bias are seen south of 60 degrees south and north of 40 degrees north. It is not unreasonable to assume that the step change in the TOMS-SBUV total ozone bias modeled in Table 1 is due to changes in the TOMS total ozone associated with known changes in the TOMS instrument around that time. The latitude dependence implies that the impact of the changes in the TOMS instrument are greatest on the total ozone retrieved from the B-pair or C-pair (331 nm - 340 nm) wavelengths which are given greater weight in the final ozone determination at higher latitudes.

Table 1. Mean of nadir-only TOMS-SBUV percent bias in total ozone for 1 (1/80 - 12/83) and period 2 (1/85 - 12/86). Nominal standard errors mean are 0.1% for period 1 and 0.2% for period 2. Differences represent modeled step change in TOMS total ozone at high latitude.

Latitude	Period 1 (TOMS-SBUV)	Period 2 (TOMS-SBUV)	Difference (Period 2-1)
-70	3.1	1.5	-1.6
-60	2.6	2.5	-0.1
-50	2.6	2.7	0.1
-40	2.6	2.8	0.2
-30	2.3	2.5	0.2
-20	2.2	2.3	0.1
-10	2.4	2.5	0.1
0	2.6	2.5	-0.1
10	2.3	2.2	-0.1
20	1.9	1.7	-0.2
30	2.3	2.1	-0.2
40	2.6	2.2	-0.4
50	2.8	2.0	-0.8
60	3.0	2.0	-1.0
70	2.9	1.1	-1.8

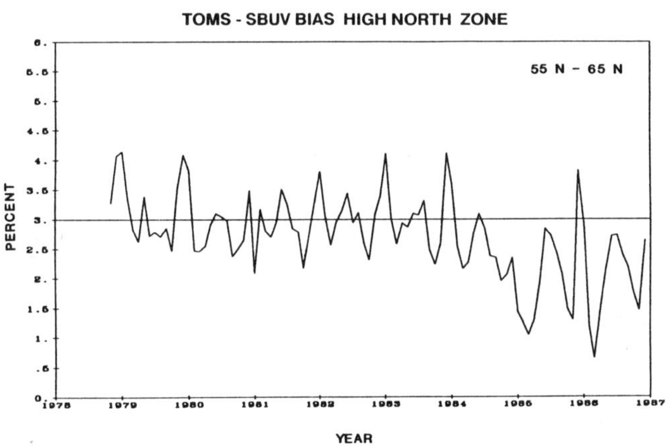

Fig. 3. Percent difference of monthly means of total ozone (nadir only TOMS-SBUV) in the 10 degree latitude zone 55 N - 65 N. Where central samples from TOMS scan are averaged to give coincidence with SBUV swath. Nominal 3% bias is drawn as reference.

Figures 4 and 5 show the nadir-only TOMS-SBUV comparisons in the equatorial zone for the A-pair and B-pair total ozone, respectively. Means (standard error) of the monthly differences plotted indicate that the TOMS A-pair has a 2.8% (0.3%) bias relative to SBUV, and the B-pair has a -0.6% (0.7%) bias.

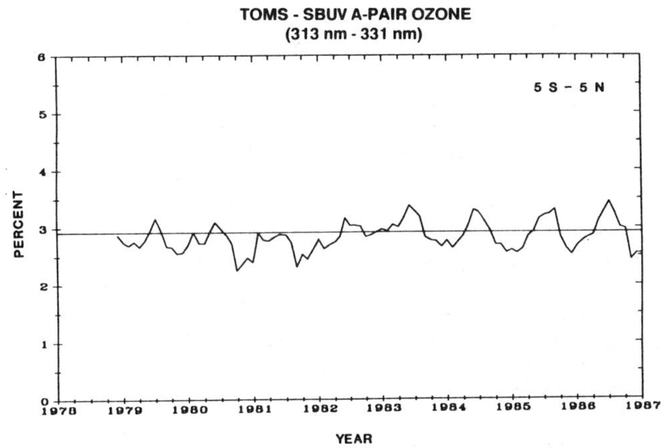

Fig. 4. Percent difference of monthly means of total ozone derived from A-pair wavelengths (TOMS-SBUV) in the 10 degree latitude zone 5 S - 5 N. The central samples from the TOMS scan are averaged to give coincidence with the SBUV FOV. The reference line is the overall mean of the bias.

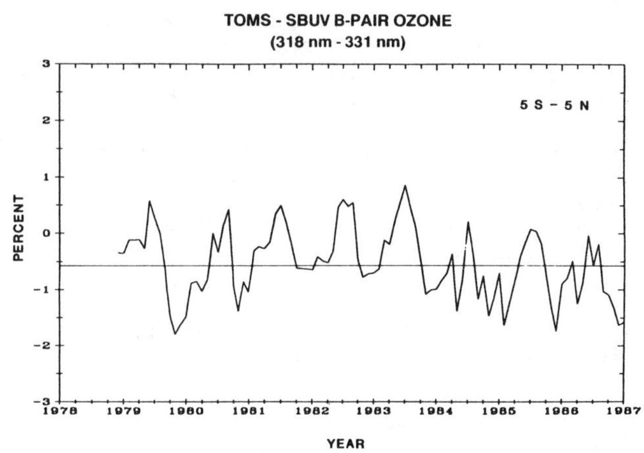

Fig. 5. Percent difference of monthly means of total ozone derived from B-pair wavelengths (TOMS-SBUV) in the 10 degree latitude zone 5 S - 5 N. The central samples from the TOMS scan are averaged to give coincidence with the SBUV FOV. The reference line is the overall mean of the bias.

The C-pair is used only at extremely high solar zenith angles, and the direct TOMS-SBUV C-pair comparison is difficult to perform. The B-pair and C-pair total ozone from SBUV and TOMS are normalized in the mean to the respective A-pair results before the combined total ozone result is computed [Fleig et al., 1988b].

The data in Fig. 2 show a slight downward drift after 1984 (See Fig. 1).

This may be due to a small downward effect on the TOMS B-pair (318 nm - 331 nm) ozone associated with instrument changes in TOMS in early 1984. This small change cannot be resolved in this comparison of total ozone since it is a weighted average of the A-pair and B-pair results, and, in fact, the small seasonal variations persistent in the nadir-only TOMS-SBUV bias tend to obscure these effects even in the individual B-pair comparisons shown in Fig. 4. Some effect is seen, however, where episodic instrument behavior interferes with the seasonal pattern in the nadir-only TOMS-SBUV bias in 1984 and, to a lesser extent, again in 1986. If the step change model considered above is applied to these B-pair data, the mean bias (standard error) is -0.36% (0.05%) for period 1 and -0.83% (0.12%) for period 2.

The calculation of total ozone from the off-nadir measurements of radiance from TOMS is the major difference in the total ozone algorithms of SBUV and TOMS. The comparisons presented above were designed to exclude this difference from consideration. Comparison of the nadir and extreme off-nadir measurements of total ozone from TOMS have also been performed. Figure 6 shows the time series of the difference in the monthly means of total ozone measured by TOMS in the nadir and extreme off-nadir viewing geometries in the 10 degree latitude zone from 5 S - 5 N. Apparent in the plot is an initial bias of approximately -1.5% and a drift of approximately -2% over the eight years plotted. As mentioned above, the initial bias seen is related to the difference between the TOMS-SBUV bias inferred from SBUV-Dobson

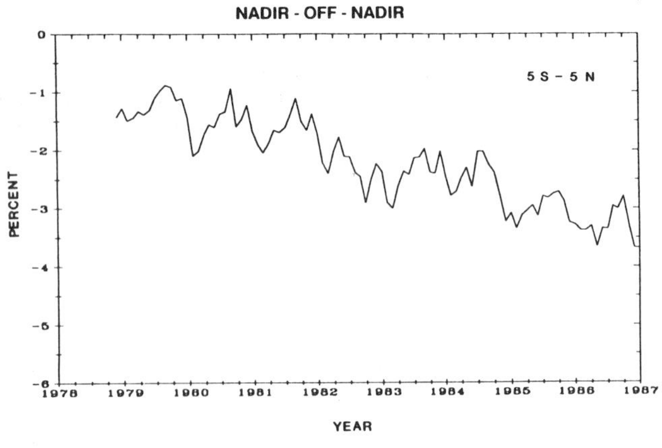

Fig. 6. Percent difference of monthly means of total ozone measured by TOMS in the nadir and the extreme off-nadir in the 10 degree latitude zone 5 S - 5 N. The drift is an indication of long-term calibration error.

and TOMS-Dobson comparisons which include nadir and off-nadir coincidences and the direct comparison of nadir-only TOMS and SBUV. The cause of the nadir - off-nadir bias is not understood. It is probably due to some scan angle dependence in either the instrument calibration or the ozone retrieval algorithm.

The drift seen in the figure is consistent with the drift of the SBUV/TOMS system relative to the Dobson Network first identified through comparisons between SBUV and Dobson [Fleig et al., 1986b]. This is because the impact on retrieved ozone ($\Delta\Omega$) of a given calibration error ($\Delta I/I_{true}$) is path length dependent as illustrated in the equation:

$$\Delta\Omega = -\Delta I/I_{true} \; (1/\alpha(\sec\theta + \sec\theta_o))$$

where α is the ozone absorption cross section, θ is the scan angle and θ_o is the solar zenith angle such that ($\sec\theta + \sec\theta_o$) represents the total path length of the measured radiance. This effect is strongest near the equator where $\sec\theta_o$ is minimum and variations in $\sec\theta$ have maximum impact on the total path length. This effect has been used directly to estimate the long-term calibration error in TOMS [Bhartia et al., 1988].

CONCLUSIONS

The long-term drift between nadir view TOMS total ozone averaged to be coincident with the SBUV swath and SBUV total ozone is 0.0 ± 0.04%/year in the latitude zone from 50 S - 40 N. The absence of a long term drift between the two instruments supports the assumption that a single characterization of the diffuser is appropriate for both SBUV and TOMS. A downward change in the TOMS-SBUV bias is seen at high latitudes, which appears to be the result of a small sudden change in the TOMS total ozone due to unexplained changes in the TOMS instrument which occurred in early 1984. A drift of the extreme off-nadir total ozone measurement relative to the nadir measurement from TOMS, which accumulates to 2% over eight years, appears to be consistent with independent estimates of the long-term calibration error in TOMS [Bhartia et al., 1988].

ACKNOWLEDGMENTS

We wish to thank D.S. Silberstein and R.P. Cebula for their helpful comments, B.A. Raines and J.D. Augustine for their assistance in data analysis, and N.L. Dudley and A.E. Suhari for the final preparation of this manuscript. The SBUV and TOMS data were prepared by the Ozone Processing Team at NASA Goddard Space Flight Center and are archived at the National Space Science Data Center.

REFERENCES

Bhartia, P.K., S. Taylor, A.J. Fleig, 1988: Estimation of Errors in the Long-Term Change in Ozone Reported by the SBUV/TOMS Instruments. A New Way of Looking at the Old Data, these proceedings.

Cebula, R.P., H. Park and D.F. Heath, 1988: Characterization of the Nimbus-7 SBUV Radiometer for the Long-Term Monitoring of Stratospheric Ozone, J. Atm. Oceanic Tech., 5,215.

Fleig, A.J., C.G. Wellemeyer, D.S. Silberstein, R.P. Cebula and P.K. Bhartia, 1988a: An Assessment of the Long-Term Drift in TOMS Total Ozone Data: Based on a Comparison with the Dobson Network, Geophys. Res. Lett., In press.

Fleig, A.J., R.D. McPeters, P.K. Bhartia, Barry M. Schlesinger, Richard P. Cebula, K.F. Klenk, Steven L. Taylor and D.F. Heath, 1988b: Nimbus-7 Solar Backscatter Ultraviolet (SBUV) Ozone Products User's Guide, NASA Reference Publication, National Aeronautics and Space Administration, Washington, D.C.

Fleig, A.J., P.K. Bhartia, C.G. Wellemeyer, and D.S. Silberstein, 1986a: Seven Years of Total Ozone from the TOMS Instrument- A Report on Data Quality, Geophys. Res Lett, 13, 1355.

Fleig, A.J., P.K. Bhartia, and D.S. Silberstein, 1986b: An Assessment of the Long-Term Drift in SBUV Total Ozone Data, Based on Comparison with the Dobson Network, Geophys. Res. Lett., 13, 1359.

COMPARISON OF SBUV/2 AND SBUV OZONE DATA

A.J. Miller and R.M. Nagatani
NOAA/NWS/Climate Analysis Center
Washington, D.C. 20233, USA

W.G. Planet and J.H. Lienesch
NOAA/NESDIS/Office of Research
Washington, D.C. 20233, USA

H.D. Bowman II
NOAA/NESDIS/Office of Satellite Data Processing and Distribution
Washington, D.C. 20233, USA

D.-W. Wei
Academia Sinica, Institute of Atmospheric Physics
Beijing, PRC

Abstract

Data are currently being processed from the SBUV/2 ozone satellite system, the NOAA operational version of the NASA SBUV, beginning in March 1985. Our purpose here is to present comparisons of the SBUV/2 data with that from Dobson, Umkehr and SBUV. The initial focus is on one month, December 1985, and we discuss comparisons of total ozone and the vertical profile results in Umkehr layers 9 through 5. The Umkehr data are adjusted for aerosol effects (DeLuisi, personal communication). The results indicate that SBUV/2 total ozone and Umkehr layer amounts are biased high with respect to SBUV. For total ozone this bias is 3-6% and for Umkehr layer 9 it is 20% decreasing to about 5% in layer 5. In the case of Dobson data, the SBUV/2 data are low by about 1.5%. For Umkehr data the SBUV/2 are high by about 10% in layer 9 decreasing to about 5% in layer 5.

1. Introduction

Along with the significant concern of possible anthropogenic influences on the stratosphere, it is recognized that the anticipated changes in ozone are extremely small. For example, in WMO (1986) it is indicated that the expected decrease of ozone at 45 km is about 3% per decade. Detecting a change of this magnitude places a premium on the radiometric accuracy of the various observing systems, on our understanding of the impact of perturbing influences on the observations such as stratospheric aerosols and on the comparison between data sets from different systems. Concerning the latter, the recently released Executive Summary of the Ozone Trends Panel argues that comparison of ground-based (Dobson and Umkehr) and satellite (Stratospheric Aerosol and Gas Experiment (SAGE)) data does not support previous reports based on Nimbus 7 SBUV (Solar Backscattered Ultraviolet Radiometer) and TOMS (Total Ozone Mapping Spectrometer) data of large global decreases since 1979 (Heath,1988; Bowman,1988).

One other satellite system available during the period of Nimbus 7 observations is the SBUV/2 on-board the NOAA-9 operational meteorological spacecraft. This is a derivative instrument based on the SBUV, and while it has several engineering modifications over the earlier SBUV design, should provide a more direct comparison of ozone determinations from data of the same instrument type. The SBUV/2 was launched in December 1984 and after an initial test and evaluation period, has taken observations continually since March, 1985. As with any new satellite instrument, even one based on proven technology, one must be extremely careful to consider the overall errors and the impacts of the modifications. This is an ongoing collaborative effort between personnel from NOAA, NASA and private industry and we are now in a position to present the first comparisons of data from SBUV. We caution, however, that these data are based on an initial, limited data base and it is quite likely that the results will be refined further as more data become available. For this presentation we focus on one month, December 1985, and discuss comparisons of total ozone and the vertical profile results in Umkehr layers 9 through 5. One final point to be made concerning the Umkehr observations is that they are recognized to be impacted, significantly, by atmospheric aerosols. Of particular concern during the time period of interest are the long-lived effects from the El Chichon eruption in April 1982. These have been examined by John DeLuisi who has kindly provided adjustments to the Umkehr data at 5 stations (Arosa, Belsk, Boulder, Lisbon and Tateno). For December 1985, 3 years after the eruption, the effects are on the order of 5% in layer 9 and reduce to near-zero in layer 5. For each of the stations, a satellite matchup procedure was adopted using a space window of 2 deg latitude by 5 deg longitude and a time window of plus-minus one day.

2. Results

Presented in Figure 1 are the differences of (SBUV/2 minus Dobson) divided by SBUV/2 to give

ISBN 0-937194-15-8

units of percent for the one month as a function of latitude. For these comparisons a 3 deg lat/long space window and a plus-minus 12 hour time window were utilized. One problem immediately apparent is that the inter-Dobson station variability is quite large and tends to make the interpretation of the results very difficult. On average, the SBUV/2 appears lower than the Dobson by about 1.5% which is very similar to the initial SBUV-Dobson comparisons observed after launch (Bhartia et al.,1984). Again, because of the large variability between stations any actual difference with latitude tends to be masked. On this latter point, it is noted that when the two satellite systems are compared, the average difference of the zonal means should agree to about 1% precision (OFCM,1982) because of the large number of observations and any latitudinal variation should be discernible to within this precision. This is shown in Figure 2 which depicts the difference of SBUV/2 minus SBUV (%). The observations (solid line) indicate an overall bias with SBUV/2 values higher than SBUV by 3 to 6% and, in addition, show a large latitudinal variation that is asymmetric about the equator. This, in fact, is anticipated as errors in the calculated albedos from measurements by each instrument are translated through the algorithm to errors in total ozone related to the solar zenith angle, i.e. 1/{1 + SEC(solar zenith angle)}. As the two instruments cross the equator at different local times the solar zenith angles are different and we have utilized the above solar zenith angle relationship in a least-squares regression to investigate the degree to which the latitudinal variation can be accounted for. This result is shown by the model curve (dotted line) and we see that the overall fit is extremely good lending support to the model of the solar zenith angle dependence.

Looking next at the profile information, presented in Figure 3 are the differences, SBUV/2 minus SBUV (%) for Umkehr layers 9 and 7 as a function of latitude and several interesting features are noted. First, the results for layers 9 and 7 appear negatively correlated to each other as a function of latitude. This is a very typical result for a profile algorithm wherein compensation occurs in the vertical to minimize the measurement errors. Second, while there appears to be a slight change with latitude, the overall difference is representative to a few percent. This latter point suggests that we can utilize the overall average values with confidence and compare these results against those obtained from the Umkehr data.

This comparison is presented in Figure 4 which shows the differences of SBUV/2 minus SBUV (solid line) and SBUV/2 minus Umkehr (dotted line) for Umkehr layers 9 through 5. We see that for both, the largest difference is in the top layer with a general decrease to layer 5. Also, the solid curve is to the right of the dotted curve indicating that the bias between SBUV/2 and SBUV is not accounted for by the Umkehr data.

3. Summary

In summary, it has been shown that for the initial comparison period of December 1985, the SBUV/2 total ozone and Umkehr layer amounts (layers 9 - 5) are biased with respect to SBUV with SBUV/2 values higher than those of SBUV. For total ozone this bias is 3-6%. As the SBUV/2 values appear to be lower than the Dobson values by about 1.5%, the net effect suggests that SBUV values are biased low, in general agreement with the results of Bhartia et. al (1984). For the Umkehr layers the SBUV/2 - SBUV bias is about 20% in layer 9 and decreases to about 5% in layer 5. In this case, however, the SBUV/2 - Umkehr bias is of the same sign although generally lower in magnitude indicating that a portion of the bias with SBUV may be due to the SBUV/2. We must caution, however, that these results are for a very limited time period and it is for this reason that we have not focused on the statistical significance of the results. In fact, more recent computations for the months of March and April indicate the difference of SBUV/2 - SBUV in layer 9 is about 10%, suggesting a possible seasonal variation. A critical element of these comparisons will be the consistency with time over the approximate 2 year overlap between SBUV/2 and SBUV. As the SBUV/2 data continue to be processed, these comparisons will be evaluated.

Acknowledgments

The processing of the SBUV/2 data is the result of a team effort of dedicated professionals associated with the NOAA/NESDIS Ozone Science Guidance Committee. We thank, in particular, J. DeLuisi and M. Hill of NOAA, R. Hudson and R. McPeters of NASA and R. Cebula and C. Wellemeyer of STX Corp.

References

Bhartia, P.K., K.F. Klenk, C. K. Wong, D. Gordon and A.J. Fleig, 1984: Intercomparison of the NIMBUS 7 SBUV/TOMS total ozone data sets with Dobson and M83 results. J. Geoph. Res., 89, 5239.

Bowman, K.P., 1988: Global trends in total ozone. Science, 239, 48.

Heath, D.F., 1988: Non-seasonal changes in total column ozone from satellite observations:1970 - 1985. Nature, March 17 issue.

OFCM (Office of Federal Coordinator for Meteorology), 1982: The National Plan for Stratospheric Ozone Monitoring and Early Detection of Change, 1981 - 1986. FCM-P17-1982, 65 pp.

WMO, 1986: Atmospheric Ozone 1985, Assessment of our Understanding of the Process Controlling Its Present Distribution and Change. WMO Global Research and Monitoring Project Report No. 16.

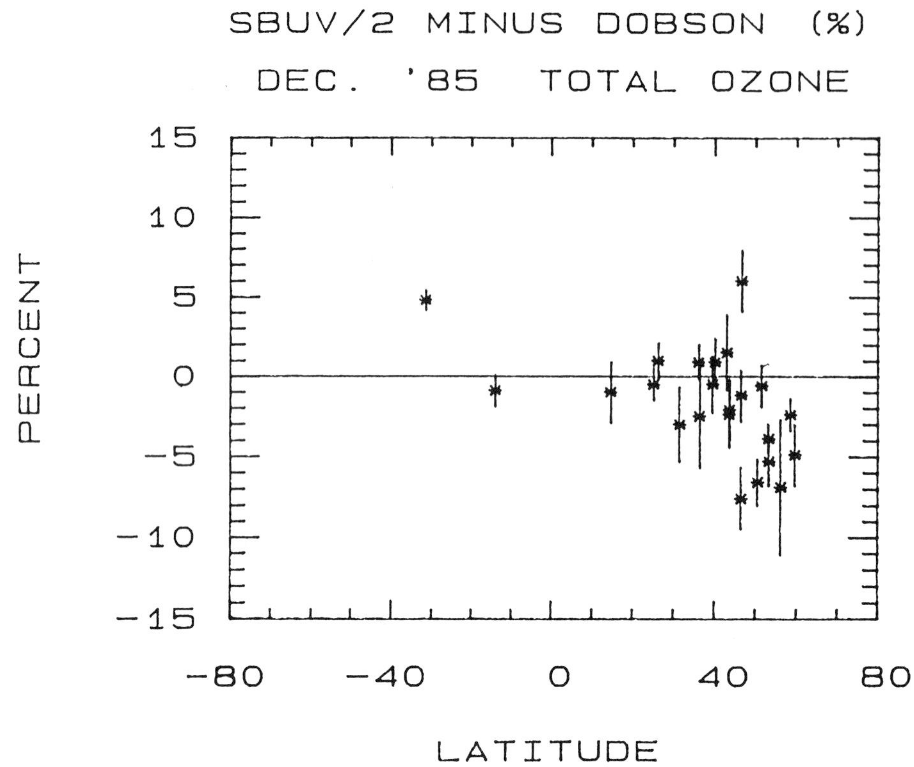

Figure 1. SBUV/2 minus Dobson, % difference, as a function of latitude for December 1985. Vertical lines represent one standard error.

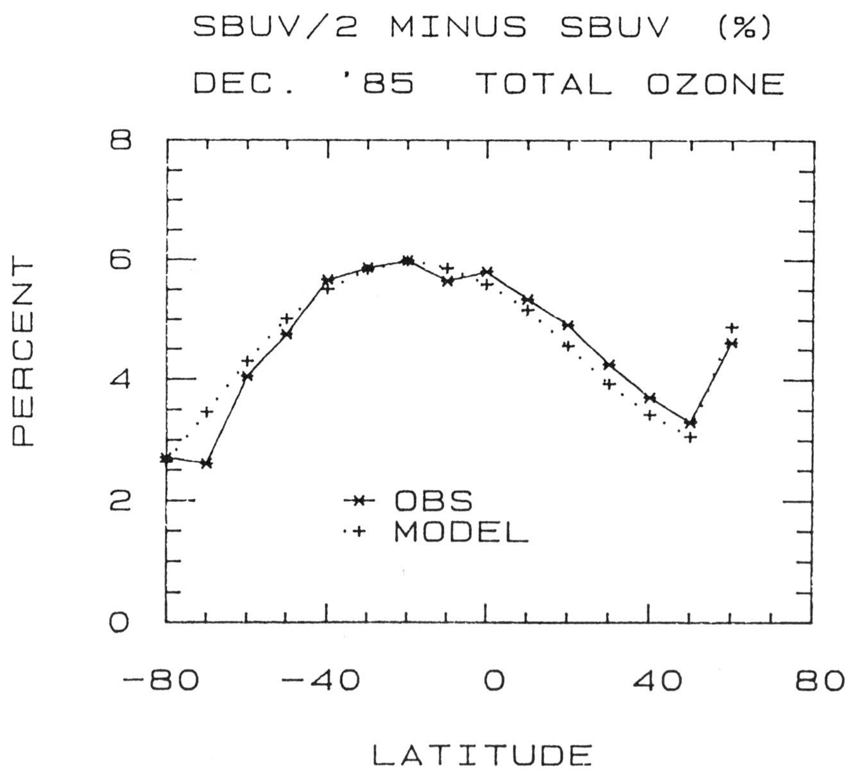

Figure 2. SBUV/2 minus SBUV, total ozone % difference, as a function of latitude for December 1985 (solid line). Dotted line represents model fit to account for solar zenith angle dependence (see text).

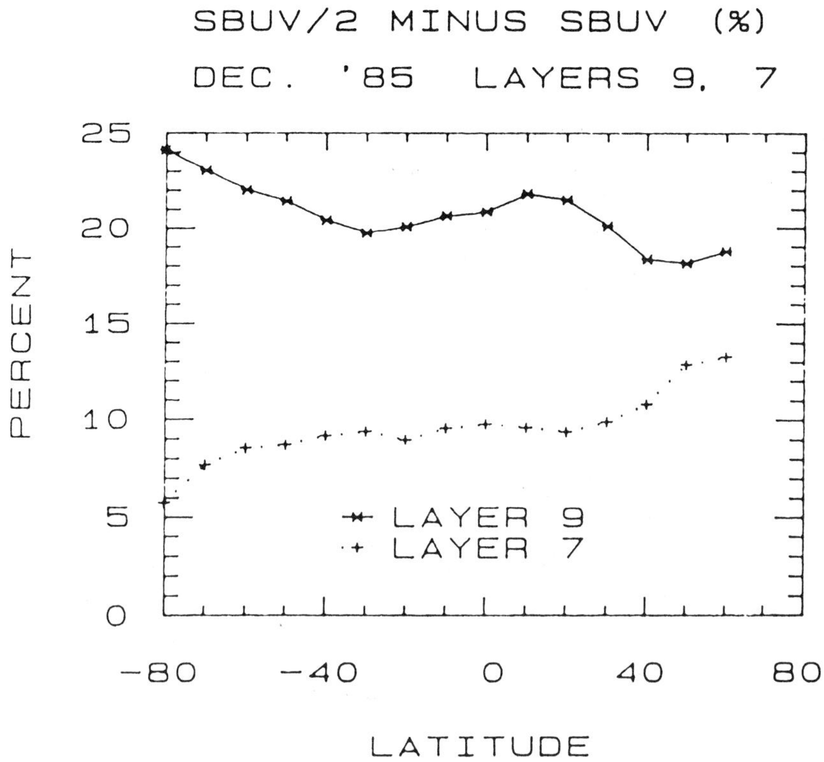

Figure 3. SBUV/2 minus SBUV, % difference, as a function of latitude for December 1985. Solid line is Umkehr layer 9, dotted line is Umkehr layer 7.

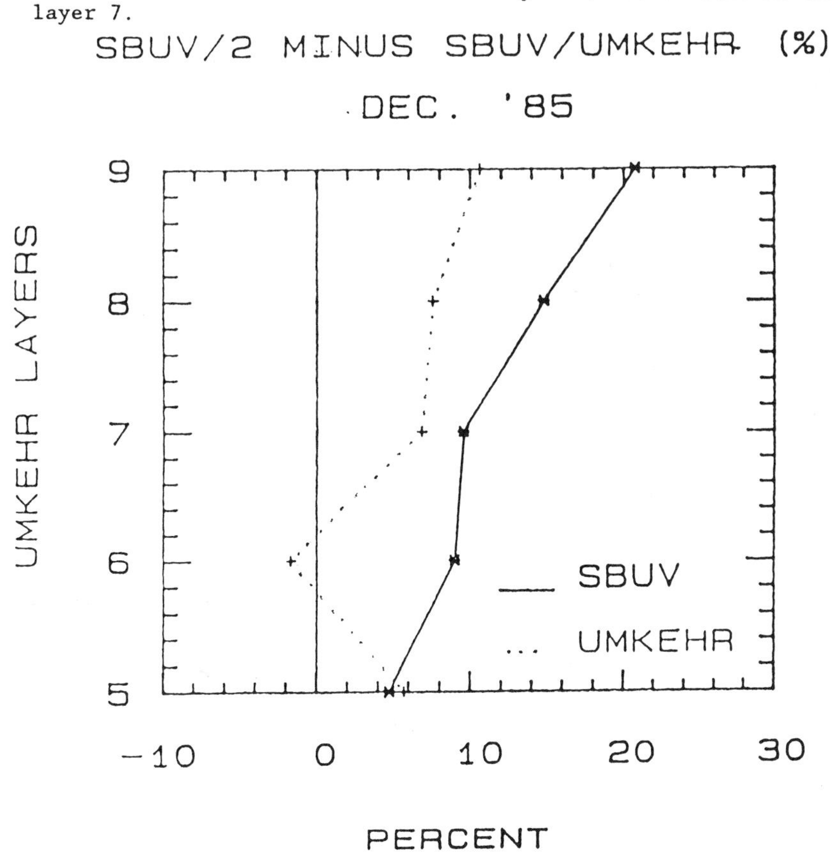

Figure 4. SBUV/2 minus SBUV (solid line) and SBUV/2 minus adjusted Umkehr, % difference, in Umkehr layers 9 through 5 for December 1985.

COMPARISON OF SAGE I AND SAGE II STRATOSPHERIC OZONE MEASUREMENTS

M.P. McCormick, R.E. Veiga[1], *and J.M. Zawodny*
NASA Langley Research Center
Hampton, Virginia 23665, USA

ABSTRACT

Vertical profile measurements of stratospheric ozone concentration measured by the satellite-borne Stratospheric Aerosol and Gas Experiments (SAGE I and SAGE II) are presented and compared for 3-year periods centered in 1980 and 1986, respectively. Time series of the zonal mean ozone show agreement between the two data sets in the temporal and vertical structure of ozone variability. The SAGE I and II profiles were used to estimate changes in ozone over the altitudes from 25 to 50 km. The results show that when averaged over the latitudes 20°N to 50°N and 20°S to 50°S, a maximum decrease in ozone of 3(±4% maximum systematic error between SAGE I and II) at 40 km occurred over the 6 year time period from 1980 to 1986. Essentially no change in ozone occurred at altitudes 28–33 km and 45–48 km for these latitude bands. In addition, for all latitudes from 55°S to 55°N, the analysis shows no change in ozone greater than ±4% at any altitude between 25 and 50 km.

1. INTRODUCTION

SAGE I and II use the solar occultation technique to measure atmospheric aerosol extinction and trace gas concentration [*McCormick et al.*, 1979; *Mauldin et al.*, 1985]. Using this method, the solar spectral measurements attenuated by the Earth's limb during a local satellite sunrise or sunset are normalized to the measured unattenuated solar irradiance. The unattenuated normalizing irradiance measurement is made immediately after the sunrise profile and immediately before the sunset profile. This inherent self-calibration feature and high signal-to-noise ratio, allows accurate long-term changes in atmospheric species to be determined.

Ozone is inferred from these transmission measurements in the 600-nm region at the center of the Chappuis band. The retrieved ozone profiles have a vertical resolution of 1 km, and their range extends from cloud top up to an altitude limited by the signal-to-noise ratio; for SAGE I the upper limit is about 55 km, and for SAGE II it is approximately 65 km. Profiles are smoothed over a 5-km layer at heights above approximately 48 km to increase signal-to-noise at these altitudes.

SAGE I and II instruments utilize essentially the same design with similar optical components. Differences between the two instruments include the number of spectral channels (7 for SAGE II and 4 for SAGE I), the use of narrower spectral bandwidths on SAGE II, and the use of a rectangular field-of-view on SAGE II versus a circular one on SAGE I. These differences do not introduce any significant systematic errors between SAGE I and II ozone measurements as will be discussed below.

The principal sources of error in the measurement of an individual ozone profile are radiometric imprecision, digitizer truncation, and scan mirror pointing errors. These random errors are approximately uncorrelated vertically, and their combined effect on an individual retrieved ozone profile is estimated at each point on the profile from the variance of the measurements from approximately 4-5 scans of the SAGE mirror across the viewing altitude. This uncertainty is provided to the data user in the form of error bars for each ozone profile. Each SAGE profile, however, also possesses an uncertainty in reference altitude of approximately 0.25 km which contributes to uncertainties in profile repeatability [*Cunnold et al.,* 1988].

The principal systematic errors between SAGE I and SAGE II arise from instrument scan-mirror calibration and the uncertainty in our knowledge of the absorption cross-section of ozone at 600 nm convolved with instrument bandpass and solar spectrum. Aerosols can also produce biases in the ozone retrievals, but only at those altitudes where aerosol concentrations are large (i.e., mostly below 25 km altitude). The relative systematic error between the SAGE instruments is estimated to be approximately 4% [*Cunnold et al.,* 1988]. The random error components for both instruments are less than 10% for a vertical resolution of 1 km.

2. ZONAL MEANS

SAGE I and II scan the atmosphere over the Earth's limb at the terminator twice on each of 15 daily orbits. The orbital inclination causes the observation latitude to periodically vary from approximately 75°S to 75°N dependent on season. The latitudes of the SAGE I and SAGE II measurement locations for February 1979 to November 1981 and October 1984 to November 1987, respectively, are displayed for sunset measurements in fig. 1. Due to the SAGE I orbital elements, the coverage does not repeat from one year to the next as it does for the SAGE II measurement location latitudes. Any gaps in the curves of fig. 1 are periods of time during which no satellite occultations occurred. Because of power problems on the spacecraft, only sunset data were taken by SAGE I after July 1979 and additional gaps are observed in the SAGE I curves due to this problem. A complete set of sunrise measurements for SAGE II, however, is available.

For analysis purposes, the SAGE I and II data were spatially grouped into 11 latitude bands. The equatorial latitude band was 15°S to 15°N, and the midlatitude bands were 10° wide centered at ±20°, ±30°, ±40°, ±50°, and ±60°. As the SAGE I or SAGE II observations passed into and subsequently out of one of the latitude bands above, the ozone measurements were averaged to form a zonal mean. In addition, the temporal duration defining a zonal mean was limited to a maximum of 1

[1]This work was done while the co-author held a National Research Council Associateship.

ISBN 0-937194-15-8

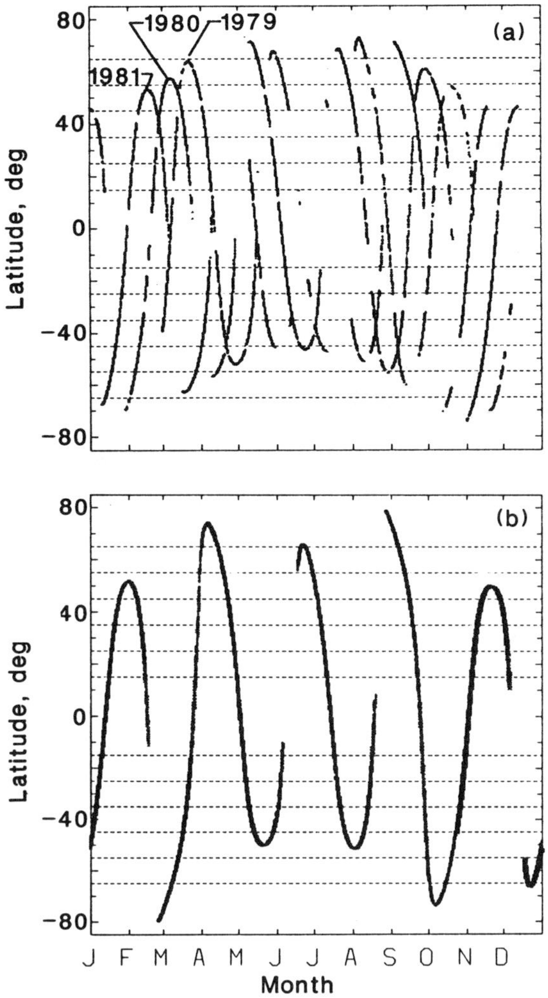

Fig 1. (a) Latitudes of the SAGE I sampling locations for sunset. (b) Latitudes of the SAGE II sampling locations for sunset. Because of the SAGE II orbital characteristics, the latitude coverage appears as a single curve even though 3 years are plotted.

week. Thus, a given latitude band contains more than one zonal mean if SAGE I or SAGE II sampled the latitude band for a continuous period of greater than 1 week duration. These typically occur when the measurement location latitude changes from increasing with time to decreasing with time.

The time series of SAGE I and II ozone zonal means at 30-km altitude are shown in fig. 2a for selected latitude bands. For comparison purposes, they have been time shifted to a common origin. Each time series was Fourier analyzed with a three-component model containing semiannual, annual and 27-month period terms. The model fits to the data, which are intended as a visual aid, are shown as the continuous curves. At 30 km these time series display the expected features of seasonal ozone variability, namely the strong annual variation in the midlatitudes and the semiannual oscillation in the equatorial region. From fig. 2a it is observed in the midlatitudes that both instruments agree in amplitude and phase with no apparent consistent bias. The largest differences occur at the equator and are due to the fact that the long period terms in the model are not in phase between the two data sets.

The vertical distribution of the annual and semiannual component amplitudes between 25 km and 50 km were computed and are shown in fig. 2b. The continuous curves on the plots represent the amplitudes in percentage from the mean. Symbols mark those amplitudes which are statistically significant at the 95% level. A comparison between the SAGE I and SAGE II annual and semiannual component amplitudes shows the similarity in the vertical distribution of amplitude. Both instruments agree exactly on the altitudes at which the midlatitude annual amplitudes effectively vanish, i.e. at 25 km in both hemispheres and at 38 km in the Southern Hemisphere. At 25 km, a phase change occurs between the zonal mean time series below and above this altitude [*Wang et al.,* 1988]. Below 25 km, ozone variability is predominantly controlled by dynamics, while above this altitude, it is primarily driven by photochemistry [*Cunnold et al.,* 1980].

3. CHANGE IN SAGE I AND SAGE II OZONE

The change in stratospheric ozone concentration that occurred between the 33-month lifetime of SAGE I and the first 3 years of SAGE II was estimated in three ways over an altitude range from 25 to 50 km, the region of the atmosphere essentially free of aerosols that could affect SAGE II ozone retrievals.

The first method of estimating the ozone change, and which minimizes the effects of seasonal and spatial sampling biases, is to select sets of profiles where SAGE I and II sampled the same month of the year within the same latitude band. Such a period of time we call an intersection and we define this to be a period of time in the year, of approximate duration 1 month, during which both SAGE I and II sampled the same latitude band. Note that an intersection is independent of the actual year during which measurements were made. The intersection is defined strictly in terms of the month within the year. Within the 11 latitude bands considered, 105 intersections were isolated containing a total of approximately 7300 SAGE I profiles and 19,000 SAGE II profiles. For each intersection, the SAGE I and II profiles were averaged separately.

The second method of estimating the ozone change is to simply compute the zonal mean time series averages and form the percentage difference. This method has the possible disadvantage that seasonal sampling biases may be introduced.

The third method of estimating the ozone change is to use the Fourier coefficients derived from the zonal mean time series and compute the mean value of the series over the time period of the data. This method's disadvantage is the empirical nature of the model.

Using all three methods described above, the percentage difference between the SAGE II mean and the SAGE I mean was computed at each altitude using the SAGE I mean as the reference. The intersection (method 1) percentage differences and their corresponding standard errors are displayed in the upper six panels of fig. 3. For all latitude bands, excluding 60°N and 60°S, methods 2 and 3 gave results that were very similar to those shown in fig. 3 (the limited sampling of SAGE I and II at the extremes of their latitude coverage introduces biases in the results computed by methods 2 and 3). Figure 3 shows that throughout the 11 latitude bands considered, except for 60°S, the ozone changes from SAGE I to SAGE II are on the order of ±4% or less. The larger decreases observed in the Southern Hemisphere near 60°S are heavily driven by the intersection in the austral springtime and are likely due to vortex dynamics.

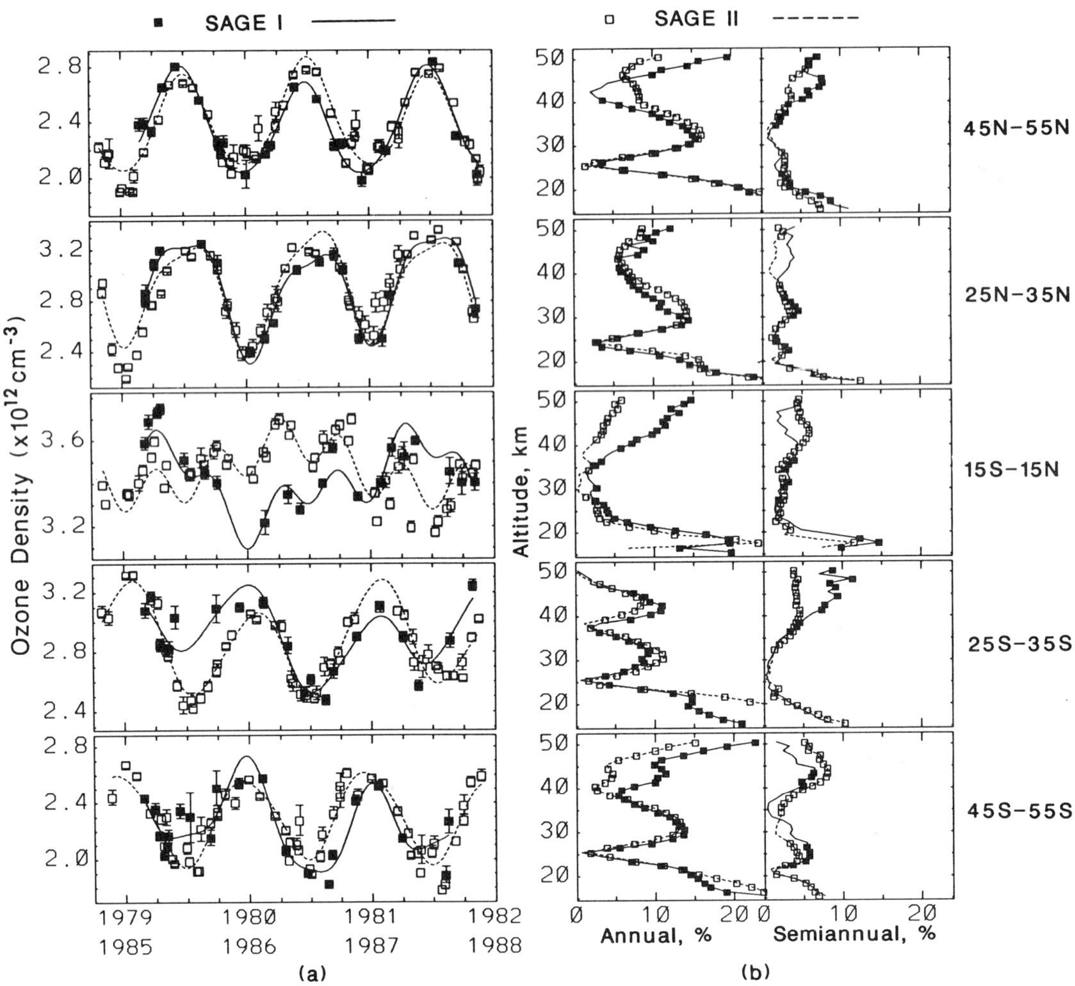

Fig 2. (a) Zonal mean ozone number density at 30 km for SAGE I during the period 2/79–11/81 and SAGE II during the period 10/84–11/87 for the five latitude bands of 45°S–55°S, 25°S–35°S, 15°S–15°N, 25°N–35°N, and 45°N–55°N. The vertical bars are the 95% confidence interval on the zonal mean. Smooth curves are the regression estimates. The data have been time shifted for comparison as indicated on the abscissa. (b) Amplitudes of the seasonal and semiannual components of the SAGE I and SAGE II zonal mean time series as derived from the regression coefficients. Units are in percent relative to the model mean.

The identical analysis (method 1) was performed using two wider latitude bands: 20°N to 50°N and 20°S to 50°S, areas over which there is relatively dense sampling by both instruments. The mean percentage differences appear in the lower panel of fig. 3. The agreement between the two hemispheres is apparent with the magnitude of the ozone change from SAGE I to SAGE II less than ±3% at any altitude. Between 35 and 45 km there is a global decrease in ozone with a maximum ozone reduction of about 3% centered at 40 km. Thus the SAGE I and SAGE II measurements do not support reports that large decreases in global stratospheric ozone have occurred above 25 km [*Heath*, 1987].

4. CONCLUSIONS

SAGE I and SAGE II ozone measurements are shown to display the expected features of seasonal ozone variability at various latitudes. There appears to be no instrumental bias introduced by SAGE I or SAGE II.

On the basis of relative coincidences (method of intersection) and time series means over corresponding 3-year periods centered on 1980 and 1986, comparisons were performed between SAGE I and SAGE II ozone concentration profiles. Over the regions of maximum density of coincidences, 20°N–50°N and 20°S–50°S, these comparisons indicate the following: an ozone decrease between 35 and 44 km with the maximum ozone reduction of 3% occurring at 40 km; a possible ozone decrease of 3% occurring at about 25 km; an essentially zero ozone change at 28-33 km and at 45-48 km. Higher values were found at 60°N and 60°S, however, less sampling occurs at these latitudes. The Southern Hemisphere sampling would weight the decreases to those in the Antarctic springtime (see fig. 1) where peak values of about -10% are observed from about 40 to 45 km.

In view of the estimated relative systematic error of 4% and a 1% standard error of the mean between the SAGE I and SAGE II instruments, the maximum ozone reduction in the midlatitude upper stratosphere indicated by SAGE

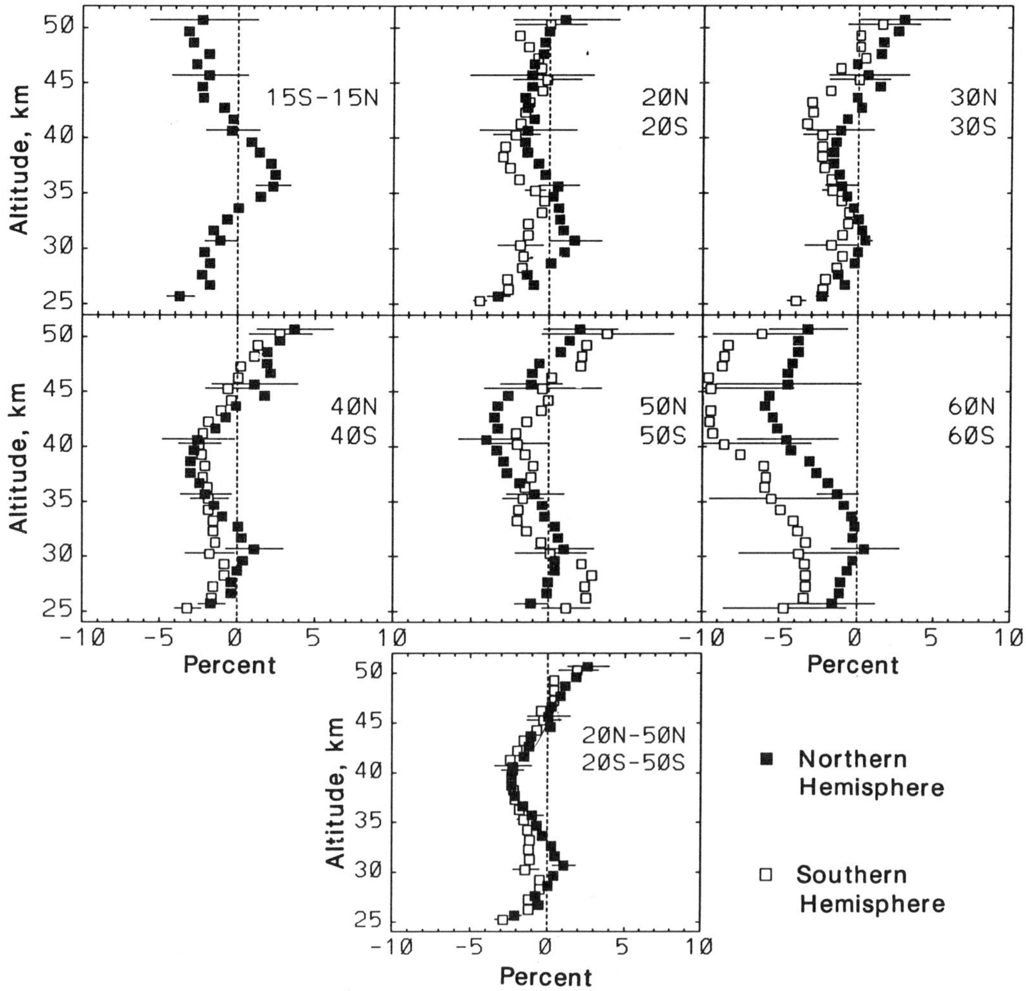

Fig 3. Mean percentage difference between SAGE II and SAGE I versus altitude computed from the intersections of SAGE I with SAGE II. All intersections occurring between 20°N to 50°N or 20°S to 50°S were combined into one sample to produce the results in the lower panel. The horizontal bars, placed at 5 km intervals, represent the standard error of the mean percentage difference. SAGE I is the reference in all percentage calculations.

over the 6-year period is 3 ± 4%. The indicated ozone reduction at 25 km near the peak ozone concentration makes a much larger impact on the total ozone column than the similar percentage effect at 40 km. This apparent decrease in ozone below 25 km is presently being investigated. Among the various techniques of measuring the vertical distribution of ozone, the SAGE I and SAGE II data currently available appear to be the most robust to changes in instrument performance over time [*Watson et al.,* 1988].

REFERENCES

Cunnold, D. M., W. P. Chu, R. A. Barnes, M. P. McCormick, and R. E. Veiga, SAGE II ozone measurements, manuscript accepted by *J. Geophys. Res.*, 1988.

Cunnold, D. M., F. N. Alyea and R. G. Prinn, Preliminary calculations concerning the maintenance of the zonal mean ozone distribution in the northern hemisphere, *Pageoph*, 118, 229–354, 1980.

Heath, D. F., Testimony on Ozone Layer Depletion before the Subcommittee on Health and the Environment, House of Representatives, *Congressional Record*, March 9, 1987.

Mauldin, L. E., N. H. Zaun, M. P. McCormick, J. H. Guy, and W. R. Vaughn, Stratospheric Aerosol and Gas Experiment II instrument: A functional description, *Opt. Eng.*, 24(2), 307–312, 1985.

McCormick, M. P., P. Hamill, T. J. Pepin, W. P. Chu, T. J. Swissler, and L. R. McMaster, Satellite studies of the stratospheric aerosol, *Bull. Am. Met. Soc.*, 60(9), 1038–1046, 1979.

Wang, P-H, M. P. McCormick, L. R. McMaster, S. Schaffner, and G. E. Woodbury, Time-periodic variations in stratospheric ozone from satellite observations, *Proceedings of International Ozone Symposium 1988*, edited by R. D. Bojkov and P. Fabian, published by A. Deepak, Hampton, Virginia.

Watson, R. T. and Ozone Trends Panel, M. J. Prather and Ad Hoc Theory Panel, and M. J. Kurylo and NASA Panel for Data Evaluation, Present State of Knowledge of the Upper Atmosphere 1988: An Assessment Report, NASA Reference Publication 1208, August 1988.

OZONE PROFILES BY UMKEHR, SBUV, AND OZONESONDE: A COMPARISON INCLUDING THE INVERSION ALGORITHMS FOR UMKEHR AND SBUV

J.J. DeLuisi
NOAA/ERL/ARL/Geophysical Monitoring for Climatic Change
Boulder, Colorado 80303, USA

D.U. Longenecker
Cooperative Institute for Research in Environmental Sciences
University of Colorado
Boulder, Colorado 80309, USA

P.K. Bhartia and S. Taylor
System Technology Corporation
Hyattsville, Maryland 20784, USA

C.L. Mateer
Scarborough, Ontario M1W 3T6, Canada

ABSTRACT

The ultraviolet techniques used by the Umkehr and solar backscattered ultraviolet (SBUV) satellite are similar in principle; however, their mathematical constructions differ somewhat. Because of this and because the mathematical radiative transfer calculations in the inversion algorithms for the two techniques may not perfectly simulate true atmospheric conditions, they may not produce identical ozone profiles when simultaneously viewing the same portion of the atmosphere. To determine the Umkehr-SBUV-comparison result that could be "best expected" from an idealized situation, each type of observation was simulated using radiation transfer theory, ozone absorption and Rayleigh scattering coefficients for the Umkehr and SBUV algorithms, and 11 profiles from monthly averaged Stratospheric Aerosol and Gas Experiment (SAGE II) data supplemented by ozonesonde data. Then error noise was added to these simulated measurements and deduced ozone profiles by each technique. The resulting profiles are compared here for bias, mean square deviation (without bias), and correlation. The SAGE II data are included in the comparison. For the real case, observed Umkehr and SBUV profiles, and observed Umkehr and ozonesonde profiles are compared. The final results are interpreted in the context of the "best expected" results, and the combined "effective" atmospheric, instrumental, and algorithm error.

1. INTRODUCTION

Comparisons of ozone profiles derived from the indirect solar backscattered ultraviolet (SBUV) and Umkehr methods reveal differences that are attributable to mathematical algorithm differences, instrument calibration errors, and fundamental limitations of the methods themselves determined by the limited information content of the measurements. The SBUV and Umkehr methods both involve measurements of scattered ultraviolet radiation. Both methods sense ozone profiles reasonably well from layer 4 (20 km) up to layer 8 (40 km). Above layer 8 the SBUV method is still sensitive to ozone, perhaps up to 60 km, but the present Umkehr method begins to suffer a serious loss of sensitivity. Therefore, when ozone profiles from the two methods are compared, for example to validate satellite observations, it is useful to know how well they might agree in the absence of error noise from interfering atmospheric inhomogeneities in time and space, aerosol scattering, instrument calibration errors, and observational inconsistencies in time and space. This paper is concerned with an evaluation of the performance of the Umkehr and SBUV inversion algorithms for a modeled atmosphere, essentially simulating an identical atmosphere for each measurement method. The inversion results for each method are compared with the model ozone profile and with each other. Comparison criteria are bias, standard deviation of the difference, and correlation. To add further insight into the interpretation of this comparison, some results of comparisons between concurrent Umkehr and ozonesonde observations, and concurrent SBUV and Umkehr observations are given and discussed in the context of the algorithm comparison.

2. INTRINSIC PROPERTIES OF THE UMKEHR AND SBUV INVERSION ALGORITHMS

The SBUV and Umkehr methods for estimating the atmospheric ozone profile use measurements of the solar ultraviolet radiation diffusely reflected (SBUV) or diffusely transmitted (Umkehr) by the earth-atmosphere system. A qualitative comparison of what may be achieved by these indirect methods is best accomplished by an examination of

ISBN 0-937194-15-8

their averaging kernels (Rodgers, University of Oxford, UK, manuscript in preparation, 1988). For the current algorithms used with these methods, the averaging kernel may be defined as follows:

$$\ln(\hat{x}/x_o) = \mathbf{A}\ln(x/x_o)$$

where $\hat{x}$ = the retrieved ozone profile vector

x_o = a first-guess ozone profile vector (climatology)

x = the true ozone profile

$\mathbf{A}$ = the averaging kernel matrix.

Clearly, the perfect averaging kernel would be the identity or unit matrix with one on the diagonal and zeroes elsewhere. Examples of the considerably less-than-perfect averaging kernels for the Umkehr are shown in Fig. 1. The imperfections exhibited include kernels with positive response not centered on the correct layer, and negative side-lobes (sometimes quite large) above and/or below the main positive maximum. For the Umkehr ozone retrieval for layer 9, the main response is centered in layer 8, and there are contributions from layers 5, 6, 7, 9 and 10. The SBUV averaging kernels (not shown) for layers 4, 10, 11 and 12 and the Umkehr ones (Fig. 1) for layers 1 and 2 have particularly large negative side-lobes. The largest negative side-lobes may be considerably reduced, and better centering of the main positive response can also be accomplished by changing algorithm parameters, but that discussion is beyond the scope of this paper. Elaboration of other error properties of the present algorithms is given in an about-to-be-published document (International Ozone Trends Panel Report, World Meteorological Oganization, Ozone Research and Monitoring Project, Report No. 18, 1988). When comparing either real or synthetic retrievals, one must keep in mind the basic properties, namely, the averaging kernels for SBUV and Umkehr, and that the first-guess profiles for the two algorithms are different.

3. COMPARISON OF SBUV AND UMKEHR ALGORITHMS

We carried out a synthetic retrieval experiment using the SBUV and Umkehr retrieval algorithms. Monthly average ozone profiles from the second Stratospheric Aerosol and Gas Experiment (SAGE II) for the latitude band 35°-50°N were obtained from a set of over 3000 profiles for 1985-1987. There were no SAGE II profiles for June in this latitude band. Also the profiles extend down only to 12.5 km. Therefore, tropospheric ozone was inserted from the monthly average Boulder ozonesonde data for 1986. Using these SAGE II ozonesonde profiles (as an independent source of profile information), we synthesized SBUV scans and Umkehr observations for the fifteenth of each month for 40°N. These synthesized scans and observations were processed by the respective algorithms. The Umkehr error was ±0.24% and the SBUV error was 0.7%. Results and comparison statistics are given in Table 1.

In discussion of Table 1, we refer to the SAGE II-ozonesonde composite profiles as "Test." The bias, shown as a percent difference of the averages of each technique, is produced by several factors, including the basic modeled physical characteristics of the technique, and the algorithm parameters, including the first-guess profiles. The low value of the Umkehr vs. Test for layers 9 and 8 was an unexpected result. This shows up as a large difference (in percent) in the SBUV vs. Umkehr which is not generally seen in the published comparisons of concurrent observations. Normally, the bias has been only a few percent. We attribute at least part of the cause to the temperature dependence in ozone absorption, which affects the Umkehr observations quite strongly but not the SBUV observations. The column labeled SBUV-Umkehr (temp. effect) accounts for the real temperature bias to the Umkehr and the seasonal variation is also given as the range The ozone absorption coefficients used to provide the temperature effect data were determined by Vigroux (1953). Another possible contributing factor to the bias in layer 9 is the adjustment made to the Umkehr algorithm in 1968 to account for recommended changes to the ozone absorption coefficients. The adjustment assumes the same correction factor to ozone in all layers, and this assumption may not strictly apply to the complex inversion procedure.

The columns under the heading of standard deviation represent the standard deviation of the difference between Umkehr and SBUV retrievals with the average difference (bias) removed. This is a more meaningful measure of the atmospheric and inversion algorithm noise when comparisons are made.

The correlations in Table 1 are artificially high because we are dealing with monthly averages of "measurements" without realistic measurement noise, and most layers exhibit a fairly strong annual cycle. The lowest correlation for Umkehr vs. Test is in layer 5 in which the amplitude of the annual cycle is less than 15%. The relatively high correlation in layer 9 arises because layers 8 and 9 are fairly well correlated in the Test profiles and, as noted earlier, the main contribution to layer 9 retrieval comes from Test ozone changes in layer 8.

For the Umkehr, the largest biases occur in layers 1 and 2, yet these layers also exhibit high correlations with the Test profiles despite the very complex nature of their averaging kernels.

4. UMKEHR VS. OZONESONDE OBSERVATIONS

Table 2 shows the results of a comparison between concurrent Umkehr and ozonesonde observations made at Boulder, Colorado. Certain other papers given at this symposium are also concerned with the same type of comparison, so we have not treated this subject in great detail.

The Umkehr vs. sonde biases (Table 2) agree fairly well with the Umkehr vs. Test biases (Table 1) for layers 1, 2, 4 and 6. The disagreement in layer 7 is very likely due to a deficiency in the sonde pump correction at those pressures, and not to the algorithm. As expected, the correlations in Table 2 are lower than for the synthetic retrieval experiment. As in the experiment, the higher correlations are for layers 2, 3, 6, and 7. The correlation in layer 5 is quite low. The low correlation in layer 1, as opposed to the high correlation in the synthetic experiment, is a manifestation of the fact that layer 1 tends to become a repository for noise and error effects such as stratospheric and tropospheric aerosols, thin clouds, and errors in the observed total ozone that is used as additional information in the inversion algorithm. The high correlations in layers 6 and 7 are quite dramatic and may be partially caused by the small sample size. The vertical pattern of standard deviation follows that of the algorithm tests, but the magnitudes are considerably larger. The layer 7 result suggests noisy ozonesonde data at this level.

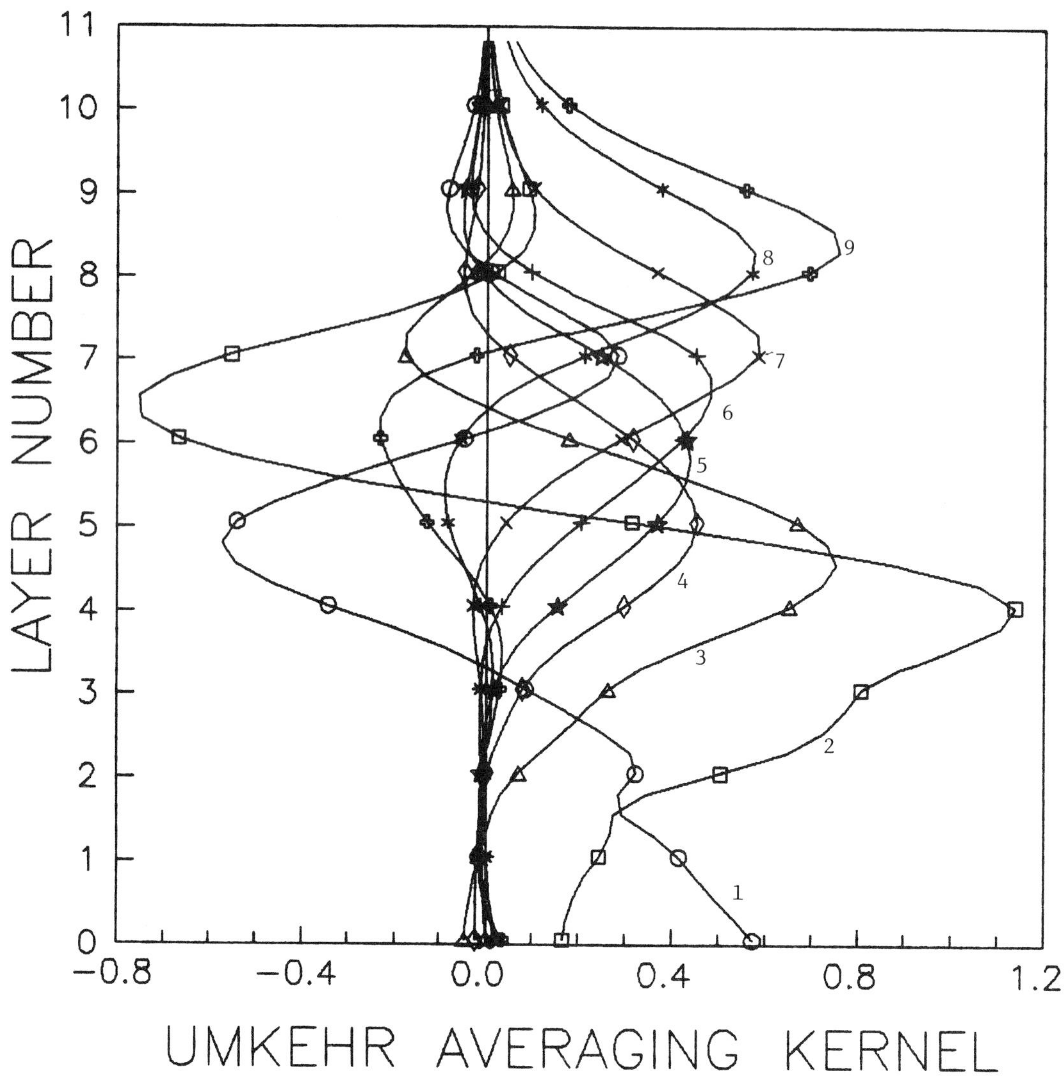

Fig. 1. Umkehr-averaging kernel plots for Umkehr layers 1-9. The numbers on the plots refer to the respective layers.

A review of ozonesonde and Umkehr data comparisons reveals considerable variations among the various investigations (e.g., Bojkov, 1966; Craig et al., 1967; DeLuisi and Mateer, 1971; and others). Factors that contribute to these variations are differences in ozonesonde types, differences in preparation of ozonesondes for flight, differences between the first-guess ozone profiles of the Umkehr inversion algorithm and the actual local ozone profile, differences in the performance and calibration of Dobson instruments at the various sites, and the length of time allowed for a measurement coincidence, i.e., 12 h, 24 h, etc.

5. UMKEHR VS. SBUV OBSERVATIONS

Table 3 shows the results of a comparison between concurrent Umkehr and SBUV observations at Boulder. The comparison is an extended analysis of that done by Bhartia et al. (1984), using the fifth version of the SBUV ozone retrieval algorithm.

The gross features of the Umkehr vs. SBUV biases, seen in Table 3, appear in the algorithm comparison of Table 1. The major exceptions are layers 9 and 3. The unusually low value of layer 9 in Table 3 may be a result of a drift in the satellite calibration and may also be due to error noise from El Chichon, although the Umkehr data were corrected, and data obtained during the major impact of the El Chichon aerosol cloud were not used. In addition, the Dobson Umkehr instrument at Boulder was replaced by an automated version in the winter of 1982-1983, and a slight downward shift in ozone in the highest few layers is seen in the data during a brief overlap period. The rather marked disagreement seen for layer 3 is most likely due to the weak sensitivity of the present Umkehr and SBUV methods to ozone changes in the region below layer 4, as suggested in section 2. The vertical pattern of standard deviation in Table 3 generally follows the pattern of Table 1, except that layer 6 has the minimum in Table 3 and layer 5 has the minimum in Table 1. The highest correlations in Table 3 are for layers 7, 6, 2, and 3. The highest correlations in Table 1 are for layers 9, 8, 7, 6, 4, 3 and 2. Time and spatial coincidence of the

TABLE 1. COMPARISON OF SBUV INVERSION RESULTS WITH UMKEHR INVERSION RESULTS USING SYNTHETIC DATA COMPUTED FROM TEST SAGE II AND OZONESONDE DATA*

a. Bias

Layer No.	Test Profile (annual average) (D.U.)	SBUV-Test (annual average) Difference (%)	Umkehr-Test (annual average) Difference (%)	SBUV-Umkehr (annual average) Difference (%)	SBUV-Umkehr (temp. effect) % ± %
9	2.78	0	-11	12	7.5 ± 2.5
8	8.84	-2	-5	4	4.9 ± 0.9
7	20.4	0	-4	4	3.1 ± 0.8
6	38.7	1	12	-10	0.35 ± 1.1
5	63.0	5	10	-5	-0.75 ± 1.5
4	76.8	8	-8	17	-2.0 ± 1.1
3	57.4	-15	-9	-7	-3.5 ± 1.1
2	25.4	-7	48	-37	-4.0 ± 1.5
1	25.1	9	-27	49	13.6 ± 9.9

b. Standard deviation

Layer no.	SBUV-Test (%)	Umkehr-Test (%)	SBUV-Umkehr (%)
9	2.9	3.6	4.3
8	1.1	3.7	3.9
7	1.0	3.5	3.9
6	1.5	4.1	5.2
5	2.3	3.5	3.3
4	2.9	3.9	5.7
3	7.0	4.7	6.7
2	6.9	9.2	14.4
1	12.8	3.6	11.3

c. Correlation

Layer no.	SBUV-Test	Umkehr-Test	SBUV-Umkehr
9	0.99	0.95	0.98
8	0.99	0.87	0.83
7	0.99	0.90	0.87
6	1.00	0.97	0.96
5	0.80	0.47	0.35
4	0.97	0.92	0.85
3	0.92	0.98	0.92
2	0.99	0.97	0.95
1	0.36	0.97	0.36

*Number of data points is 11.

two types of measurements may be partially responsible for the poorer agreement.

The previous comparisons of Bhartia et al. (1984) and DeLuisi et al. (1985) suggest reasonably good agreement in layer 9 where the Umkehr ozone was 3.8% higher than the SBUV ozone. The correlations were not much different from the present results.

6. DISCUSSION

The algorithm comparison phase of the present study revealed biases in the retrieved ozone profiles when compared with the model ozone profiles. A comparison of the retrieved SBUV and Umkehr profiles revealed a pattern of biases in the profiles that can be attributed to the basic behavior of the two algorithms. An unexpectedly large bias is seen in layer 9, in which the retrieved Umkehr ozone concentration is 11% lower than the modeled profile; the same difference is seen when the Umkehr is compared with the SBUV, which almost exactly reproduces the model ozone concentration in layer 9. In layer 8 the Umkehr ozone is 5% lower than the model ozone. For both methods, the ozone biases in layers 3, 2, and 1 are more serious, but this was expected because of the known low sensitivity of both methods to ozone in these layers.

The algorithm inversion noise, defined as the standard deviation of the difference between the retrieved

TABLE 2. COMPARISON BETWEEN BOULDER OZONESONDE AND BOULDER UMKEHR OZONE PROFILE DATA*

Layer no.	Sonde ozone (D.U.)	No. observations	Umkehr-Sonde Difference %	Standard deviation (%)	Correlation
7	20.0	24	20	10.0	0.82
6	44.1	31	9	7.0	0.93
5	70.6	39	1	6.7	0.24
4	72.0	43	-10	12.3	0.64
3	35.4	43	12	24.2	0.71
2	17.5	43	43	41.1	0.80
1	33.1	43	-21	61.0	0.43

*Number of data points is 43.

and model ozone profile for the SBUV, ranges from about 1 to 3% in layers 4 to 9, inclusive. For the Umkehr, the noise ranges from 3.5 to 4.1% in the same layers. A comparison of the SBUV and Umkehr noise ranges from 3.3 to 5.7% for the same layers.

The algorithm correlations are quite good for both methods, > 0.87 for all layers with the exception of layer 5 (0.80 for the SBUV/Test and 0.47 for the Umkehr/Test), and layer 1 (0.36 for the SBUV/Test). The same pattern of good correlation is seen in the comparison between the SBUV and Umkehr profiles; the exceptions are layer 5 (0.35) and layer 1 (0.36).

A review of the comparisons between Umkehr and ozonesonde observations, and SBUV and Umkehr observations suggests patterns of bias, standard deviations and correlation somewhat similar to the algorithm comparison results. However, as expected, the results do not rank as favorably as the modeled results. The poorer results seen in layers 8 and 9 for the SBUV/Umkehr comparison (correlations 0.58 and 0.66 respectively) may probably be caused by Umkehr noise relating to the relatively greater temperature dependence of the ozone absorption coefficients, than those of the SBUV tropospheric noise from aerosols and thin clouds, and perhaps temporal variations in stratospheric aerosol. Instruments to monitor thin tropospheric clouds and aerosols are now being installed at the seven automated stations. In addition, a world lidar network to monitor stratospheric aerosols has been organized and is beginning to make contributions to the World Ozone Data Center at Toronto for monthly corrections to Umkehr observations.

TABLE 3. COMPARISON BETWEEN OBSERVED SBUV AND BOULDER UMKEHR OZONE PROFILES*

Layer no.	SBUV-Umkehr Difference %	Standard deviation (%)	Correlation
9	-9	14.4	0.66
8	2	7.7	0.58
7	-1	7.0	0.81
6	-11	5.3	0.87
5	-7	6.2	0.39
4	14	6.0	0.66
3	10	14.5	0.87
2	-15	28.1	0.80
1	--	--	--

*Number of data points is 506.

The Umkehr inversion algorithm was developed in 1964 and was changed slightly in 1968 to account for changes in the standard ozone absorption coefficients. An updated retrieval algorithm for producing Umkehr ozone profiles does not yet exist. It will be interesting to see if a new version can improve the correspondence between the SBUV and Umkehr ozone profiles.

ACKNOWLEDGMENTS

We thank Walter Komhyr and his assistants for the Boulder Umkehr and ozonesonde data.

REFERENCES

Bhartia, P. K., K. F. Klenk, A. J. Fleig, C. G. Wellemeyer, and G. Gordon, 1984: Intercomparison of NIMBUS 7 solar backscatter ultraviolet ozone profiles with rocket, balloon, and Umkehr profiles. *J. Geophys. Res.*, 89, 5227-5238.

Bojkov, R. D., 1966: Differences in the vertical ozone distribution deduced from Umkehr and ozonesonde data at Goose Bay. *J. Appl. Meteor.*, 5, 872-877.

Craig, R.A., J. J. DeLuisi, and I. Steutzer, 1967: Comparison of chemiluminescent and Umkehr observations of ozone. *J. Geophys. Res.*, 72, 1667-1671.

DeLuisi, J. J., 1971: Effect of the temperature dependence of ozone absorption on the vertical distribution of ozone deduced from Umkehr observations. *J. Geophys. Res.*, 76, 2131-2137.

DeLuisi, J. J., and C. L. Mateer, 1971: On the application of the optimum statistical inversion technique to the evaluation of Umkehr observations. *J. Appl. Meteor.*, 10, 328-334.

DeLuisi, J. J., C. L. Mateer, and P. K. Bhartia, 1985: On the correspondence between standard Umkehr, short Umkehr, and solar backscattered ultraviolet vertical ozone profiles. *J. Geophys. Res.*, 90, 3845-3849.

Vigrous, E., 1953: Contribution a l´etude expérimentale de l´absorption de l´ozone, *Ann. de Phys.*, 8, 709-762.

HEMISPHERIC DIFFERENCES IN OBSERVED STRATOSPHERIC OZONE

Julius London and Lori M. Perliski
Astrophysical, Planetary and Atmospheric Sciences
University of Colorado
Boulder, Colorado 80309, USA

ABSTRACT

Analyses are presented of the hemispheric differences of time and longitude variations of stratospheric ozone. The present study is based on 8 years (Jan 1979–Dec 1986) of ozone concentration data as derived from the Nimbus 7 SBUV measurements. The results are generally consistent with preliminary reported results based on earlier, limited data sets.

During the winter season, Southern Hemisphere ozone mixing ratio values at 1 mb are considerably higher than those in the Northern Hemisphere. However, during summer, the values are about 15–20 percent lower. The amplitudes of the annual variation in the upper stratosphere are maxima at about 60°N and 50°S but the phases of the maxima are just six months apart (Dec N/Jun S).

At 10 mb the ozone mixing ratio is almost always higher in the Northern Hemisphere, particularly at sub–polar latitudes. In the tropics, however, where the effect of the earth–sun distance is important for ozone photochemistry, the sign of the hemispheric difference alternates from perihelion to aphelion. At lower levels (30–40 mb) the amplitude of the annual harmonic at sub–polar latitudes is slightly higher in the Northern Hemisphere. The phase of maximum at these latitudes in the Southern Hemisphere is retarded about 2 months (compared to the Northern Hemisphere phase), similar to the time variations of total ozone.

Longitude variations at different levels also indicate hemispheric differences. At 1 mb wave #1 is dominant with an amplitude maximum of about 0.5 ppmm during the winter of each hemisphere. These maxima are located at 65°N and 40°S. At lower levels, the maximum amplitude is ~0.8 ppmm at 65°N and ~0.5 ppmm at 50°S. During winter, year–to–year variations of the high latitude amplitude of zonal wave #1 at 1 mb are larger in the Southern Hemisphere but the reverse is found at about 20 mb.

INTRODUCTION

The average time and three-dimensional distribution of stratospheric ozone has been reasonably well documented by ground–based, balloonborne and satellite observations over the past 20 years (see, for instance, WMO, 1985; London, 1985). Ozone mixing ratio is a maximum (~16 ppmm) over the equator at about 7–8 mb. The axis of maximum slopes upward toward the poles, to about 3 mb, in each hemisphere. Above 3 mb the horizontal mixing ratio gradient is from the winter to the summer pole (e.g., Frederick et al., 1983). Average longitude variations are largest at levels below about 10 mb and at most levels at winter polar latitudes.

The ozone concentration in the stratosphere results largely from temperature dependent photochemistry above about 7 mb, interacting photochemical and advective processes in the region 7–25 mb and, dominantly, advection at levels below 25 mb (see, for instance, Miller et al., 1985; Brasseur and Solomon, 1986). To the extent that photochemically relevant trace gases have approximate symmetric hemispheric distributions, the observed asymmetry in the ozone distribution and its time and space variations in the Northern and Southern Hemispheres must result, to first order, from differences in the thermal structure and circulation characteristics of the two hemispheres. These differences, in turn, stem from the basic dissimilarity of their surfaces (e.g., land/water distribution, topography, etc.). Hemispheric differences of the ozone mixing ratio as derived from limited data sets have been discussed by London, 1983; Maeda and Heath, 1983; and others. The following discussion briefly summarizes the observed differences of time and space variations of stratospheric ozone as determined from 8 years of Nimbus–7 SBUV measurements.

DATA SOURCE

Measurements of stratospheric ozone from different satellite platforms have been made almost continuously since Sep 1967 (WMO, 1985). The most extensive set of observations is that made from the Nimbus–7 SBUV satellite (Oct 1978–Apr 1988). Our present analysis makes use of the available SBUV data covering the 8 year period Jan 1979–Dec 1986. A description of the SBUV data base including some error analysis of the derived ozone data is given in Fleig et al., 1984; Bhartia et al., 1984.

The data set is given for 19 pressure levels (0.01–100 mb). The analyses discussed below, however, cover mainly the stratospheric region from 1.0 to 30 mb. We have organized the data set into monthly averages within 5° x 20° lat/long bins. The effect of this time averaging is to damp high frequency time and transient longitude variations. Day–to–day changes of ozone mixing ratio at 30 mb have been discussed, for instance, by Wu et al., 1987. Although the latitude range of the SBUV measurements ±85 deg, the effective limit for annual observations is ±65 deg. The analyzed data were not detrended to account for instrument degradation or possible long term changes in ozone concentration. However, since we are here concerned with hemispheric comparisons, these long term changes should not affect the qualitative results discussed below. Also, because of the relatively short time span of the data set (8 years), hemispheric differences at low latitudes might be affected by such perturbations as El Chichon, ENSO, and the QBO. These influences have also not been filtered out but we

OZONE IN THE ATMOSPHERE
Rumen D. Bojkov and Peter Fabian (Eds.)

ISBN 0–937194–15–8

do not think that their inclusion would alter the main results.

TIME AND LATITUDE VARIATIONS

Hemispheric differences of the ozone mixing ratio are given as a function of latitude and season for 1 mb and 10 mb (Figs. 1a,b). For these diagrams we have taken a 6–month displacement from Northern to Southern hemisphere (i.e., Jan (N) – July (S), etc.) for each latitude. For these two levels the annual phases are approximately 6 months apart but at 30 mb, there is a phase lag of about 2 months at polar latitudes in the Southern Hemisphere. This lag is a result of the delayed breakdown of the circumpolar vortex in the Southern Hemisphere which delays the normal late winter–early spring maximum. Part of the difference shown in Figs. 1a, b is due to latitude variations of ozone maxima or minima, particularly in polar regions. During the winter season, there is about a 15 degree latitude difference at 1 mb between the maximum ozone mixing ratio of each hemisphere (higher latitude in the Northern Hemisphere). An analogous, but smaller, effect occurs at 10 mb. this would tend to slightly exaggerate the absolute values shown in Figs. 1a,b. At 30 mb, the latitude of winter maximum ozone is again displaced about 15–20 degrees equatorward in the Southern Hemisphere.

Average Seasonal Difference of Ozone MR (ppmm) between the Northern and Southern Hemisphere

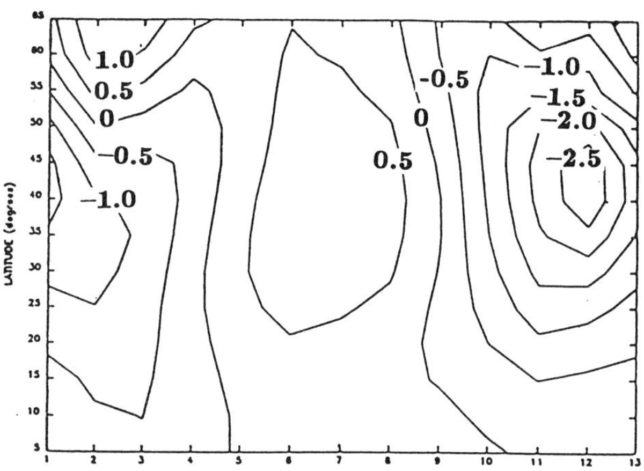

Northern Hemisphere Month (1=Jan(N) -Jul(S), etc.)

Fig. 1a 1 mb

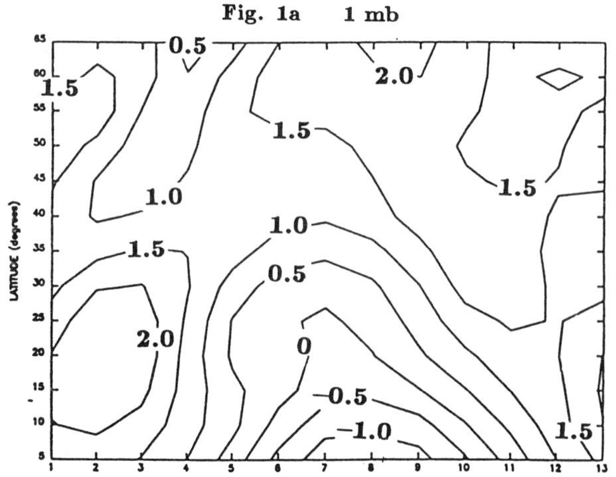

Northern Hemisphere Month (1=Jan(N) -Jul(S), etc.)

Fig. 1b 10 mb

As shown in Fig. 1a, during the winter half–year, the Southern Hemisphere ozone mixing ratio is almost everywhere larger than that in the Northern Hemisphere. The maximum difference, about 3 ppmm occurs between Dec(N) and June(S) at latitude 40–45 degrees. During summer, however, the Northern Hemisphere ozone values are only about 10 percent larger than those found at mid and sub–polar latitudes of the Southern Hemisphere. The large high latitude hemispheric differences at 1 mb during early winter are the combined results of relative latitude displacement of the region of maximum ozone, local temperature dependent photochemical ozone increase associated with colder Southern Hemisphere sub–polar temperature at this level (Fleming et al., 1988), and stronger upward advection along the sloping isentropes.

At levels from 5 to 30 mb, the 6–month differences at latitudes poleward of about 35–40 degrees always show a Northern Hemisphere maximum which varies from about 0.5 to about 2.0 ppmm. At these latitudes there is little seasonal variation in the hemispheric differences at 5 mb. At 10 mb the Northern Hemisphere maximum occurs everywhere except during the summer in the tropics and at 30 mb there is a strong winter maximum, over 2.5 ppmm, in polar regions. This last result stems from the relatively stronger affect of the average eddy poleward ozone transport in the winter Northern Hemisphere (e.g., Wu et al., 1987). In summer, the ozone mixing ratio is, at all levels, higher in the Northern Hemisphere except at equatorial and tropical latitudes above 10 mb.

In the tropics, however, at and above 10 mb, the ozone mixing ratio is under photochemical control and is affected by variations of the earth–sun distance so that there is less ozone during Northern Hemisphere summer (aphelion) than Southern Hemisphere summer (perihelion) (see, for instance, Frederick et al., 1980, 1983). Even so, the so–called winter differences (Feb(N)–Aug(S)) are about twice as large at 10—30 mb as compared with the so–called summer differences.

LONGITUDE VARIATIONS

As mentioned earlier, using monthly averages of the ozone mixing ratio in the present analysis precludes our deriving estimates for high frequency transient longitude variations. Also, year to year rms variations of the phases of waves #1 and #2 are relatively small. Indeed, the variance accounted for by waves #1 and #2 is, in mid and high latitudes of each hemisphere, generally more than 80 percent of the total longitude variance derived from the averaged 8 year data set. At most latitudes wave #1 has the largest amplitude.

The pressure/latitude distribution of the ozone mixing ratio amplitude of longitude wave #1 is shown in Figs. 2a,b for January and July. At 1 mb the winter amplitude maxima are about the same in both hemispheres (~0.5 ppmm) but displaced about 20–25 degrees equatorward in the Southern Hemisphere. In the middle stratosphere at levels 5–10 mb, wave #1 (and 2) are small in both hemispheres. At 30 mb, however, the wave #1 amplitude maximum in winter is 60 percent higher in the Northern Hemisphere. This is consistent with the observed asymmetric winter polar vortex in the Northern Hemisphere stratosphere and the near circumpolar vortex in the Southern Hemisphere. The amplitude maximum for longitude wave #1 is found about 20 degrees farther equatorward in the Southern Hemisphere. Year–to–year rms variations at 30 mb (not shown), in sub–polar regions, are much larger in winter than in summer and, in winter, are larger in the Northern Hemisphere.

Hemispheric differences in the time and longitude variations of stratospheric ozone mixing ratio are evident mostly at the upper (1 mb) and lower (30 mb) stratosphere where temperature dependent photochemistry and quasi–horizontal advection are effective at the

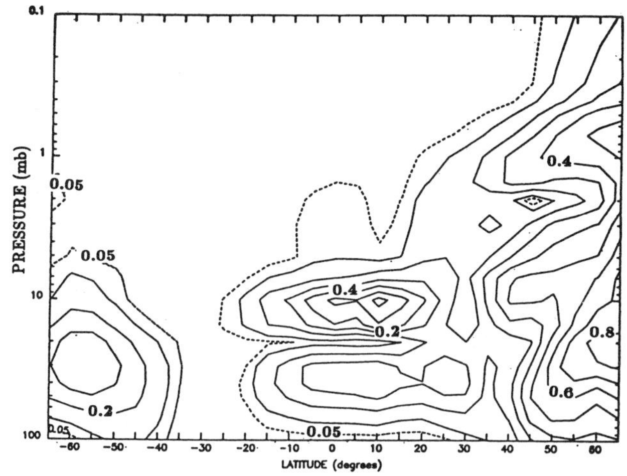

Nimbus–7 SBUV 03 Long Wave #1 AMP MR (ppmm)
Jan 79–86

Fig. 2a

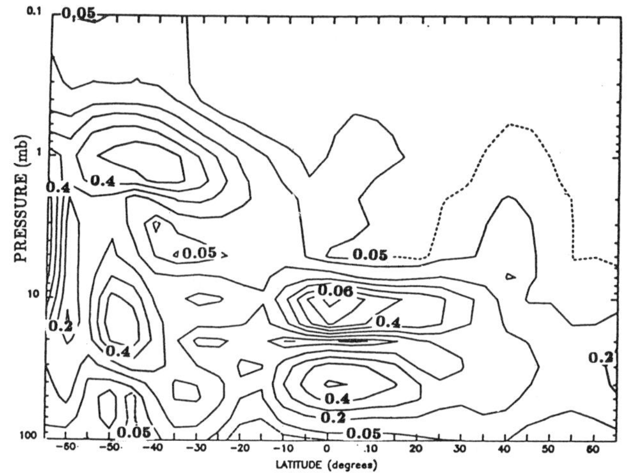

Nimbus–7 SBUV 03 Long Wave #1 AMP MR (ppmm)

Jul 79–86

Fig. 2b

upper and lower levels. These differences reflect, in large part, differences in the nature and form of the underlying boundary surface which, in turn affect hemispheric differences in thermal structure and circulation characteristics of the stratosphere.

ACKNOWLEDGEMENTS

We would like to acknowledge technical assistance provided by Mary S. Sable. This research was supported by grant NSG 5153 from the National Aeronautics and Space Administration. Acknowledgement is made to the National Space Science Data Center for making Nimbus–7 SBUV data available for this study.

REFERENCES

Bhartia, P. K., K. F. Klenk, A. J. Fleig, C. G. Wellemeyer, and D. Gordon, 1984: Intercomparison of Nimbus–7 solar backscattered ultaviolet ozone profiles with rocket, *J. Geophy. Res.*, **89**, 5227–5238.

Brasseur, G. and S. Solomon, 1986: *Aeronomy of the Middle Atmosphere*, second edition, Dordrecht: D. Reidel Publishing Company, 452 pp.

Fleig, A. J., C. Wellemeyer, N. Oslik, K. D. Lee, A. J. Miller and R. Nagatani. 1984: *User's Guide for SBUV/TOMS Ozone Derivative Products, NASA Reference Publication 1116.*

Fleming, E. L., S. Chandra, M. R. Shoeberl, and J. J. Barnett, 1988: Monthly mean global climatology of temperature, wind, geopotential height, and pressure for 0–120 km, *NASA Technical Memorandum 100697.*

Frederick, J. E., R. Dasgupta, and B. Guenther, 1980: An observed annual cycle in tropical upper stratospheric and mesospheric ozone, *Geophys. Res. Lett.*, **7**, 713–716.

Frederick, J. E., F. T. Huang, A. R. Douglass, and C. A. Reber, 1983: The distribution and annual cycle of ozone in the upper stratosphere, *J. Geophys. Res.*, **88**, 3819–3828.

Frederick, J. E., G. N. Serafino, and A. R. Douglass, 1984: An analysis of the annual cycle in upper stratospheric ozone, *J. Geophys. Res.*, **89**, 9547–9555.

London, J., 1983: Periodic and aperiodic ozone variations in the middle and upper stratosphere, *Adv. Space Res.*, **2**, 201–204.

London, J., 1985: The observed distribution of atmospheric zone and its variations, in *Ozone in the Free Atmosphere*, R. C. Whitten and S. S. Prasad, eds., New York: Van Nostrand Reinhold Co., 11–80.

Maeda, K., and D. F. Heath, 1983: Asymmetries of the upper stratospheric ozone distribution between two hemispheres, *J. Atm. Sci.*, **40**, 1353–1359.

Miller, A. J., R. M. Nagatani, and J. E. Frederick, 1985: Ozone–temperature relationships in the stratosphere, in *Atmospheric Ozone*, Hampton: Deepak Publishing Co., 321–324.

World Meteorological Organization, 1985: *Atmospheric Ozone 1985*, Vol. II, Rep. No. 16, 648 pp.

Wu, M.–F., M. A. Geller, J. G. Olson, and E. M. Larson, 1987: A study of global ozone transport and the role of planetary waves using satellite data, *J. Geophys Res.*, **92**, 3081–3097.

OBSERVATION OF VERTICAL OZONE PROFILES WITH THE EXOS-C BACKSCATTERED UV SPECTROMETER

T. Ogawa and M. Koike
Geophysics Research Laboratory
University of Tokyo
Bunkyo-ku, Tokyo 113, Japan

K. Suzuki
Department of Education
Yokohama National University
Hodogaya-ku, Yokohama 240, Japan

ABSTRACT

The EXOS-C/"Ohzora" satellite carried an ultraviolet (UV) spectrometer to measure the Rayleigh scattering of solar UV radiation by the terrestrial atmosphere. The instrumentation and observation aboard the EXOS-C, as well as the data analysis, are described briefly. The intercomparison of the retrieved ozone profiles with those from ground-based and balloon-borne observations shows an agreement within about 10%. We construct the monthly-averaged meridional distribution of ozone mixing ratio. It exhibits a meridional distribution very similar to those from other satellite observations, and agrees with them within $\pm 10 \sim \pm 15\%$ below the height of 1 mb pressure level. The ozone mixing ratio at a fixed pressure level for the period of March 1984 - July 1987 indicates a trend, which is ascribed to the instrumental drift. This implies difficulty in calibrating the spectral sensitivity of the satellite-borne spectrometer against a long-term drift.

1. INTRODUCTION

The solar ultraviolet (UV) radiation backscattered by the terrestrial atmosphere is subjected to absorption by the atmospheric ozone. Since the absorption cross section of ozone varies largely with wavelength in the 260 - 310 nm region, the penetration of solar UV into the atmosphere, and accordingly the height of effective scattering, vary with wavelength. The intensity spectrum of the backscattered UV has information content on the vertical distribution of ozone in the stratosphere and mesosphere. The measurement of the backscattered UV spectrum on board an artificial earth satellite enables us to infer vertical ozone profiles (Singer and Wentworth, 1957; early demonstrations are seen in e.g., Anderson et al., 1969; Iozenas et al., 1969). The BUV experiment aboard the Nimbus-4 satellite successfully obtained the global and systematic data on vertical ozone profiles (Heath et al., 1973; Krueger et al., 1973). This sort of global observation was further developed in late 1970s and 1980s by the Nimbus-7 SBUV, the NOAA-9 SBUV/2 (Heath et al., 1975) and the SME UVS (Rusch et al., 1984).

The EXOS-C satellite (1984-015A) was launched in February 1984 on the orbit of apogee height 850 km, perigee height 350 km, inclination 75° and revolution period 97 min. It carried a UV spectrometer to measure the backscattered UV in the wavelength region 250 - 320 nm. A radiometer was also included to monitor terrestrial albedo at the wavelength 410 nm. This paper will describe the instrumentation and the observation aboard the EXOS-C satellite and will present the results of the data analyses.

2. INSTRUMENTATION AND OBSERVATION

The spectrometer for the backscattered UV made a continuous scan of wavelength in the range 250 - 320 nm with spectral resolution 1 nm. One spectral scan took 0.6 sec and data sampling of 256 points. The field of view of the spectrometer was 10° x 14°. See Suzuki et al.(1985) for instrumental details. The spectrometer was installed in such a way that the center of its field of view was parallel to the main axis of the satellite, which was oriented to the sun within 1° precision. Consequently, the spectrometer always looked at the anti-sun direction. This observation geometry placed a limit on global and extensive observations, but provided us with an advantage in that polarization effect in the spectrometer can be neglected in optical calibration because the Rayleigh scattering does not induce polarization at a scattering angle of 180°.

The observation started in March 1984 and continued until the data recorder went wrong in October 1987. The observations were made intermittently except for interruptions of about 10 days in the middle of February, April, August and October when the spectrometer was unable to look at the atmosphere.

Geographic locations of the observations were sparsely distributed in the latitude range from equator to ~75°N/75°S in summer or to ~50°N/50°S in winter. The horizontal resolution equivalent to the spectrometer's field of view was approximately 200 - 400 km at stratospheric heights for solar zenith angle 60° and for satellite height 600 km.

3. DATA ANALYSIS

In retrieving vertical ozone profiles, which were represented in the coordinates of ozone volume mixing ratio versus atmospheric pressure, we used the spectral intensity data for the wave-

ISBN 0-937194-15-8

length range 260 - 315 nm. We arranged the data by taking the average of 3 spectra and sampling the intensity values at every 1 nm interval. Since the data transmission of one spectrum took 8 sec (32 sec in the case of low telemetry rate), this averaging resulted in degradation of horizontal resolution by ~100 km (~400 km in the low rate case); however, it improved the signal-to-noise ratio of the intensity data.

Unlike the Nimbus-7 SBUV, the EXOS-C spectrometer did not have any diffuser plate to measure solar UV spectrum, accordingly we needed a reference spectrum of solar UV in ozone profile retrieval. We adopted the solar UV spectrum measured by Mentall and Williams (1988) as a reference spectrum. A small correction was made to the absolute sensitivity of the EXOS-C spectrometer in terms of the reference spectrum in such a way that the deviation of the retrieved profile from the observed ones aboard a rocket (above ~35 km; Watanabe and Ogawa, 1987) and balloons (below ~27 km) at Uchinoura, Japan. This resulted in the deviation of the retrieved profile from the rocket

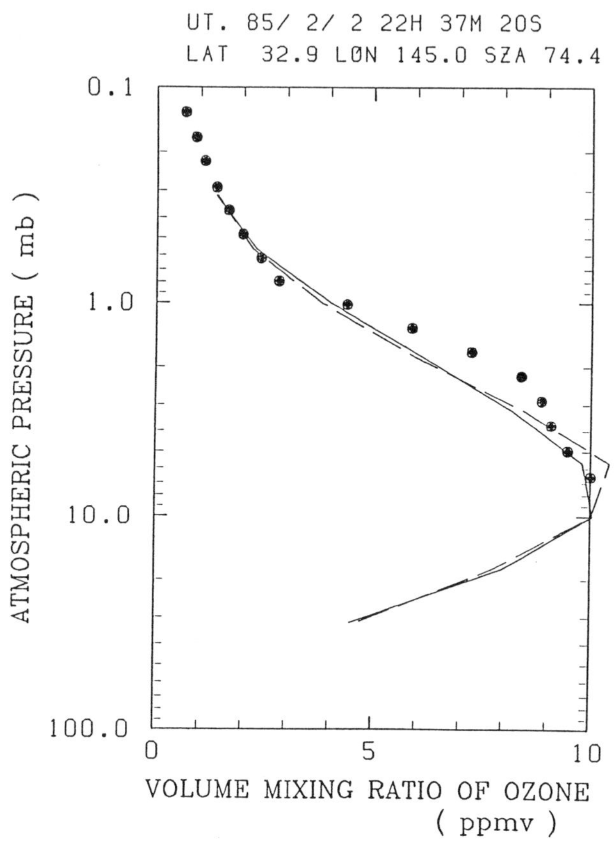

Fig. 1. Calibration of the retrieved ozone profile in terms of the rocket observation (solid circles) at Uchinoura (31°N) on February 7, 1985. Solid and dashed lines represent the retrieved profiles from the satellite overpass observation after correcting spectral sensitivity of the spectrometer: solid line from Twomey's method and dashed line from Chahine's method.

observation profile to be 9% on the average in the height range 5.6 - 0.32 mb (approximately 35 - 57 km; see Fig.1). The average deviation of the retrieved profiles from those obtained with balloon ozonesondes was 16% in the height range 32 - 18 mb (approximately 24 - 27 km). However, because of the uncertainty of efficiency decrease in the ozonesonde pump, the ozonesonde profile gave too small values: 40 - 60% smaller than the retrieved profile at 10 mb level. Accordingly we did not refer to the ozonesonde profiles above the height of ~15 mb.

We tried two different inversion methods in ozone profile retrieval. One is the relaxation method developed by Chahine (1970) and the other is the linear constraint inversion method developed by Twomey (1963,1965). Chahine's method is useful to get a stable solution and to save computer machine time but has a difficulty in evaluating the error of the retrieved profile. Twomey's method applies a linear constraint in solving the matrix equation derived from the integral equation in which the backscattered UV intensity spectrum is represented as a function of ozone mixing ratio with atmospheric pressure as an integration parameter. We need to know the ozone profile to evaluate the matrix elements, and hence, as in the case of Chahine's method, we must set an initial guess to the ozone profile. The matrix elements should be re-evaluated by using the retrieved profile. This procedure was repeated 3 times to get a converged solution.

Figure 1 also shows a comparison of retrieved profiles between the two different inversion methods. The average difference between them is 4% which is about a half of the average difference between the retrieved profile and the ground-based observation profile. In the final version of our retrieval procedure, we employed Twomey's method. The sample data number for one spectrum was 56: sampling with 1 nm interval from 260 nm to 316 nm. The division of the height coordinate was equal in terms of atmospheric pressure: 20 divisions for the height range 1000 - 0.01 mb (approximately 0 - 80 km). We retrieved ozone profiles for the height range 32 - 0.32 mb (approximately 24 - 57 km) to obtain reliable and meaningful values.

4. INTERCOMPARISON

The retrieved ozone profiles should be validated by intercomparisons with those from ground-based and balloon-borne observations. During the MAP GLOBUS Campaign in September - October 1985, we had an opportunity to compare the retrieved profiles with those from balloon-borne UV absorption and chemiluminescence techniques, balloon ozonesondes and ground-based Umkehr and microwave methods. This particular intercomparison implies the average error in the retrieved profiles to be about 10% in the height range 32 - 10 mb (approximately 24 - 31 km; Robbins et al., 1988).

Further intercomparisons of the retrieved profiles were possible during MAP GLOBUS with those from microwave technique at Bordeaux and from Umkehr observations at Haute Provence (de la Noë et al., 1988). We selected several retrieved profiles which were acquired on the same day at the geographic locations within a distance of 1500 km from the ground station. The average deviation was 8% between the retrieved profiles and those measured with microwave technique in the height range 3.2 - 0.32 mb (approximately 39 - 57 km), and 17% between the retrieved profiles and those from Umkehr observations in the height range 32 - 0.32 mb (approximately 24 - 57 km).

5. MERIDIONAL DISTRIBUTION

The retrieved ozone profile is represented in terms of volume mixing ratio at 9 atmospheric pressure levels: 0.316 mb, 0.562 mb, 1 mb, 1.78 mb, 3.16 mb, 5.62 mb, 10 mb, 17.8 mb and 31.6 mb. The data extend for the period from March 1984 to September 1987 except for February and August 1987, but the number of retrieved profiles varies with month, ranging from a few tens to a few thousands. The geographical locations of the observations are distributed over the latitudes between 75°N and 75°S and all over longitudes, but are not distributed uniformly. The distribution density also varies largely from month to month. Despite these insufficient data, we calculated the monthly and zonal average of ozone mixing ratio to construct monthly-averaged meridional distributions.

Figure 2 shows examples of the meridional distribution. They are very similar to those of the proposed CIRA model (Keating and Young, 1985), which represents the average of various satellite observations. The EXOS-C meridional distribution agrees with the proposed CIRA within ± 10 ~ $\pm 15\%$ on the average in the height range 20 mb - 1 mb (approximately 26 - 48 km); however, in the heights above ~0.7 mb (approximately 51 km), the EXOS-C gives smaller ozone mixing ratios by more than 20% than the proposed CIRA. This particular difference may originates from the calibration error in the EXOS-C data.

It is interesting to note in Fig. 2 that ozone mixing ratio seems to have decreased from 1985 to 1986 at 10 - 1 mb (approximately 31 - 48 km). Figure 3 represents time series of ozone mixing ratio at 3 height levels in the latitude zones 20° - 30°N and 20° - 30°S. Seasonal and inter-annual variations are implied, and among others a decrease during 1985 - 1987 is evident at

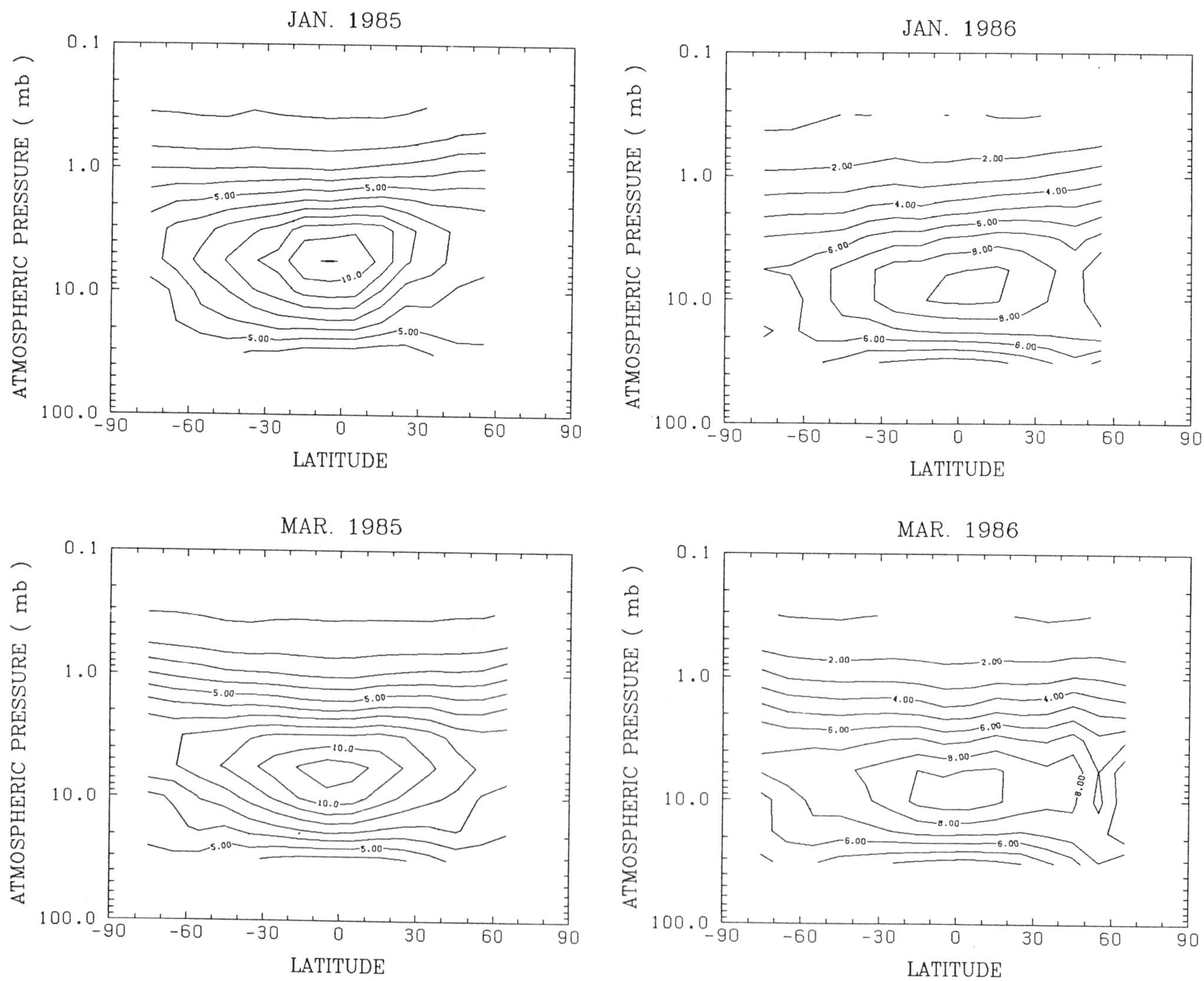

Fig. 2. Monthly-averaged meridional distributions of ozone volume mixing ratio based on the EXOS-C data set. The contours are drawn every 1 ppmv.

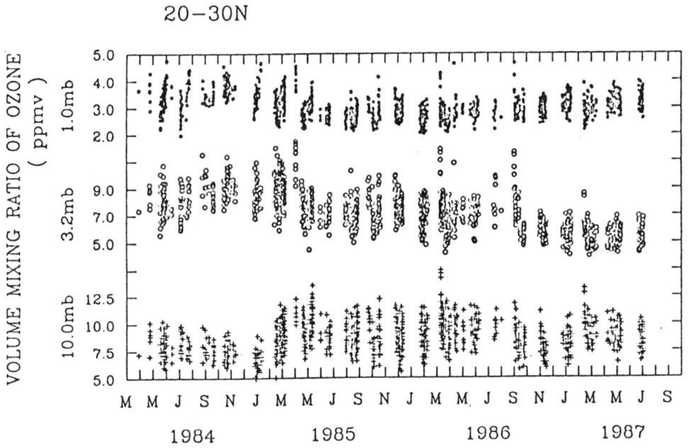

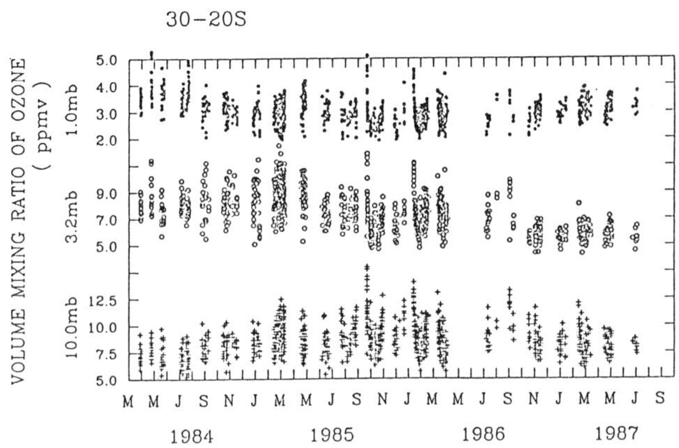

Fig. 3. Variations in the EXOS-C ozone volume mixing ratio at the heights of 10 mb, 3.2 mb and 1 mb pressure levels in latitude zones 20° - 30°N and 20° - 30°S.

3.2 mb level. A preliminary analysis to fit with a linear regression indicates a decrease between 5.6 mb and 1 mb height levels and an increase below 17.8 mb level or above 0.56 mb level. This signature is almost invariant with latitude. The largest decrease rate is -15% per year at ~2 mb level around equator. The increase rate below ~20 mb level is +2 ~ +5% per year, whereas that above 0.56 mb level is more than +20% per year.

A variation of more than 10% per year seems to be statistically significant in our case, and it is too large to be ascribed to the real variation in ozone mixing ratio. We suspect that it is due mainly to the instrumental drift. The most likely source of the drift is a degradation in the spectrometer sensitivity. Since the linear regression shows different signatures at 3 different height regions, the degradation must have been inhomogeneous with wavelength. This implies difficulty in calibrating a satellite-borne UV spectrometer against long-term instrumental drift. We need to establish the calibration system for detecting a probable long-term trend in ozone profiles.

REFERENCES

Anderson, G.P., C.A. Barth, F. Cayla and J. London, 1969: Ann. Geophys. 25, 341-345.

Chahine, T.T., 1970: J. Atm. Sci. 27, 960-967.

de la Noë, J., M. Koike, T.Ogawa and P.Marché, 1988: J. Atm. Chem., submitted.

Heath, D.F., C.L. Mateer and A.J. Krueger, 1973: Pure Appl. Geophys. 106/108, 1238-1253.

Heath, D.F., A.J. Krueger, H.A. Roeder and B.D. Henderson, 1975: Opt. Eng. 14, 323-331.

Iozenas, V.A., V.A. Krasnopolsky, A.P. Kuznetsov and A.I. Lebedinsky, 1969: Izv. Atm. Ocean. Phys. 5, 77-82; 219-223.

Keating, G.M. and D.F. Young, 1985: Handbook for MAP 16, 205-229.

Krueger, A.J., D.F. Heath and C.L. Mateer, 1973: Pure Appl. Geophys. 106/108, 1254-1263.

Mentall, J.E. and D.E. Williams, 1988: J. Geophys. Res. 93, 735-746.

Robbins, D., P. Aimedieu, J. Pélon, J.-P. Pommereau, N. Iwagami, K. Shibasaki, T. Ogawa, M. Koike, P. Marché, J. Naudet, P. Rigaud, D. Huguenin, J. Lenoble and G. Maddrea, 1988: J. Geophys. Res., submitted.

Rusch, D.W., G.H. Mount, C.A. Barth, R.J. Thomas and M.T. Callan, 1984: J. Geophys. Res. 89, 11677-11687.

Singer, S.F. and R.C. Wentworth, 1957: J. Geophys. Res. 62, 299-308.

Suzuki, K., T. Ogawa and S. Kadokura, 1985: J. Geomag. Geoelectr. 37, 225-236.

Twomey, S., 1963: J. Ass. Comp. Mach. 10, 97-101.

Twomey, S., 1965: J. Franklin Inst. 279, 95-109.

Watanabe, T. and T. Ogawa, 1987: Adv. Space Res. 7 (No.9), 123-126.

A STUDY OF THE TIME AND SPATIAL DEPENDENCE OF OZONE NEAR 1.0 MBAR WITH EMPHASIS ON THE SPRINGTIME

D.W. Rusch and R.T. Clancy
Laboratory for Atmospheric and Space Physics
University of Colorado
Boulder, Colorado 80309, USA

ABSTRACT

Models of ozone density near the stratopause are compared with satellite measurements. The models reproduce the observed ozone densities at most times and latitudes except at high latitudes in winter and spring where an additional loss of ozone is required. The transport of aurorally produced *NO* to the stratopause region is postulated as the mechanism for increasing the loss of ozone. Model results including increased high latitude NO_X bring the calculated ozone into agreement with the data. Seasonal and solar cycle variations of ozone near the winter pole that confirm the *NO* transport hypothesis are seen in the satellite data. This measured solar cycle dependence of ozone agrees with previous models in which *NO* transport from the thermosphere to the stratosphere is included. The seasonal and solar cycle signatures evident in the high latitude ozone data are not consistent with ozone changes which may be occurring from CL_X increases.

1. INTRODUCTION

The oxides of nitrogen are known to play an important role in the control of the ozone density in the middle atmosphere of the Earth. The largest source of odd nitrogen in the stratosphere is thought to arise from the oxidation and photolysis of N_2O (Bates and Hays, 1967) which is released at the surface as a result of processes in the soil. An additional source of odd nitrogen, one which has received some serious attention and much speculation in the past, is associated with auroral production of *NO* and its subsequent transport from the thermosphere into the mesosphere and upper stratosphere during the polar night (Garcia et al.,1984; Solomon and Garcia, 1984; Gerard and Deneye, 1984). High densities of odd nitrogen species have been observed in the polar regions during the winter (Russell et al., 1984; Horvath and Frederick, 1985) and the transport time from the thermosphere into the upper stratosphere is of the order of one month. In the sunlit mesosphere *NO* is efficiently destroyed by photolysis and this acts as a barrier to downward *NO* transport.

In this paper, measurements of ozone from the Solar Mesosphere Explorer (SME) (Barth et al., 1983) and the Solar Backscattered Ultraviolet Experiment (SBUV) (Heath et al., 1975) are examined to reveal the possibility of a link between thermospheric production and transport of *NO*, and O_3 near the stratopause. The effect of the aurorally produced odd nitrogen on ozone is expected to be most effective at high latitudes in the springtime. It will be shown that the latitudinal gradient of ozone measured in the winter and early spring can be described by a model which includes larger amounts of NO_X near the poles than at mid and low latitudes.

2. DATA

The data used in this analysis are ozone densities from the Ultraviolet Spectrometer (Rusch et al., 1984) on the Solar Mesosphere Explorer (SME) (Barth et al., 1983) and from the Solar Backscatter Ultraviolet (SBUV) Experiment (Heath et al., 1975; McPeters et al., 1984) on NIMBUS 7. The SBUV data are from the NSSDC and are the result of Level 5 processing. The SME and SBUV data are used in this study to compare to models of ozone in the upper stratosphere and mesosphere, its latitudinal gradients and temporal changes.

3. MODEL DESCRIPTION

The model adopted for this analysis is a slightly modified version of that developed by Eckman (1986) and used by Rusch and Eckman (1986), Clancy et al. (1987), and Rusch and Clancy (1988a,b). The model assumes the background atmosphere of Barnett and Corney (1985), includes measured NMC temperatures (Gelman et al., 1986), and fixed concentration profiles for water vapor, Cl_X, and NO_X. The model can be characterized as a "low odd-hydrogen" model because of the reduced HO_X levels resulting from increased reformation of water by the reaction of $OH + HO_2$. A complete description appears in the aforementioned papers. In this paper we modify the model to include latitudinal gradients of NO_X and compare the results with the data in an attempt to assess the contribution, if any, of the thermosphere to the NO_X density near the stratopause.

4. MODEL/DATA COMPARISONS

In Figures 1a to h, data/model comparisons are presented for monthly averaged ozone mixing ratios at 1.0 mbar for February, March, April, May, August, September, October, and November of 1982. Both the SME and SBUV data are shown. Two model calculations are presented; one for latitudinally invariant *NO*, the other with *NO* changing in a linear fashion from the background density (15 ppbv) at 45 degrees latitude to three times the background amount at 90 degrees.

The agreement between the SME and SBUV data is excellent for the months presented. The two instruments reproduce the magnitude of the ozone mixing ratio and its latitudinal variations extremely well. The agreement is similar for the remaining months of 1982 at this pressure level.

If enhanced *NO* were present in the polar regions during the springtime, we would expect the *NO* invariant model to produce more ozone near the poles than is measured. Indeed we see this to be the case in six of the eight months in the springtime hemisphere at 1.0 mb. The models agree well with the data at other latitudes. Models

ISBN 0-937194-15-8

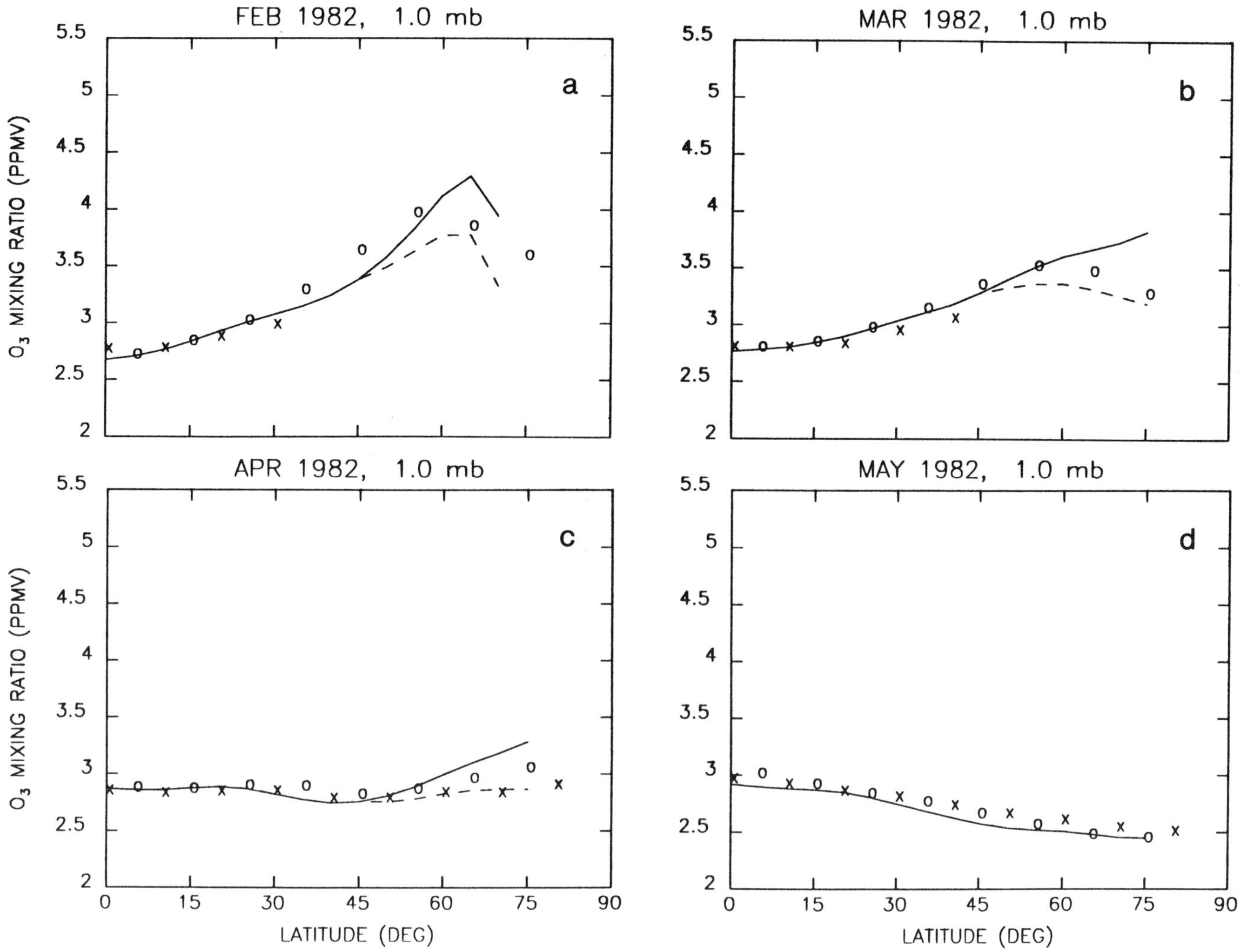

Figure 1. Monthly averaged ozone mixing ratios for six months in 1982, as a function of latitude at 1.0 mbar for SME (x) and SBUV (o) compared with the latitude in varient model (solid line) and model with NOx increasing with latitude (dashed line). See text for a full description. (a) February, (b) March, (c) April, (d) May, (e) August, (f) September, (g) October, (h) November.

containing the latitudinally dependent *NO* results in smaller ozone values and produces ozone gradients similar to the measured ones. It is also interesting to note that the difference between the data and the invariant model decreases with time from February to May and from August to November. The May and November data show excellent agreement with the invariant model indicating that the *NO* has been transported from the polar region and that the flow of *NO* from the thermosphere has been interrupted by the the increasing photolysis rates in the lower thermosphere. These comparisons provide evidence for an increase in ozone destruction at the time of equinox, an increase that can be accounted for by a springtime enhancement in *NO*. This is a similar conclusion to that to that of Solomon and Garcia (1984) and was predicted by Garcia et al. (1984) in their study of the solar cycle variations of minor species. Although the data presented in Figure 1 do not directly reflect on possible solar cycle changes in *NO*, we believe that the comparison of data to model shows that ozone destruction must be increased in the models near the poles in the springtime to agree with the measurements.

Although enhanced *NO* densities do appear to produce the measured effect, other possible scenarios exist which would also increase ozone destruction in the high latitude stratopause. Long-term or high latitude increases in other minor constituents such as chlorine containing compounds could also produce the measured ozone decrease. Temperature, if significantly different from the NMC results used here, would also influence the comparisons. However, springtime increases in *NO* do appear to reproduce the seasonal and latitudinal dependence of the observations and the model/data differences are small at all other times of the year. In particular, there are no significant data/model discrepancies in the summer (see e.g. Rusch and Clancy, 1988b). The relatively short lifetime of *NO* in the upper stratosphere is consistent with the measurements but the expected effect of Cl_X is not as we discuss below.

Solomon et al. (1985) have investigated the latitude dependence of ozone depletion due to increasing Cl_X and have shown that the effects of transport and the distribution of CH_4 play important roles. In particular, near the stratopause, greater losses of ozone occur in the polar regions than at low latitudes and the losses are seasonally dependent as well. Greater depletions appear at times of equinox than solstice. The calculated winter depletion due

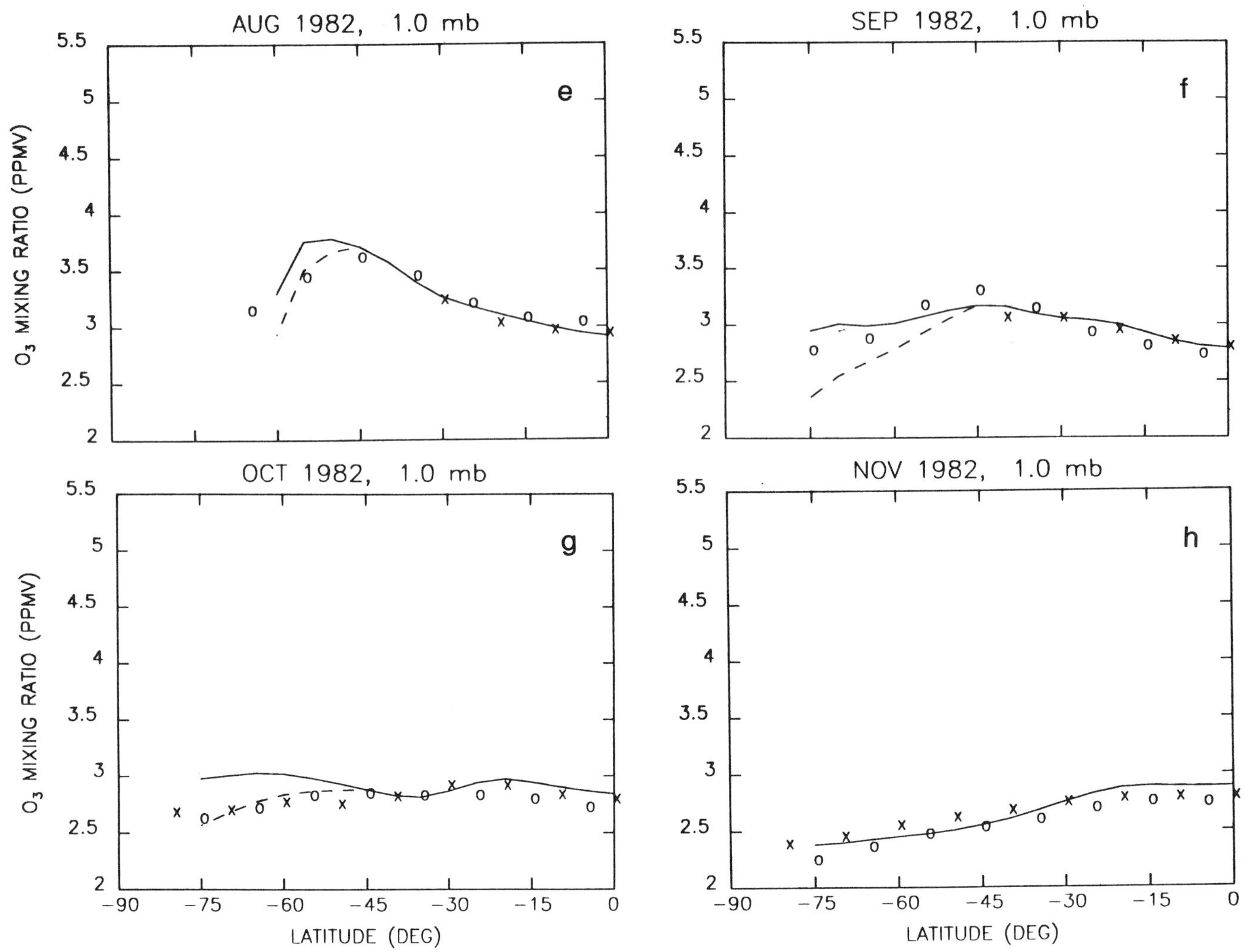

Figure 1. (e-h)

to Cl_X increase is a few percent less than than that in the spring at 70 degrees latitude in the 1.0 mbar region. This effect, although small, would result in an overestimation of the effect of *NO* as Cl_X changes of this nature are not included in our model. The model of Solomon et al. (1985) assumes an increase in Cl_X to 8.0 ppmv. Thus Cl_X dominates the destruction of ozone near 1.0 mbar to a greater degree than in the current atmosphere. However, we do not expect a significant change in the seasonal variation of ozone, in the present day, resulting from Cl_X increases. Rusch and Clancy (1988a) have calculated the change in seasonal variation due to Cl_X increases and have found them to be small compared to that required by the satellite measurements presented here. Thus the effects of Cl_X can be neglected compared to those of *NO* in this study.

We have presented only one scenario for the spatial variation of *NO* in the model as we have no measurements to indicate the true nature of the distribution. Although we could infer this distribution, little would be gained by this exercise and the main conclusion would remain unchanged.

Further indications that the *NO* hypothesis is the correct one stem from analysis of high latitude winter and spring variations in the SBUV data which appear to be anticorrelated with the solar cycle. In Figure 2, we show the normalized, monthly averaged SBUV data for each August from 1979 to 1986 at 2 mbar and 70S. The results for the

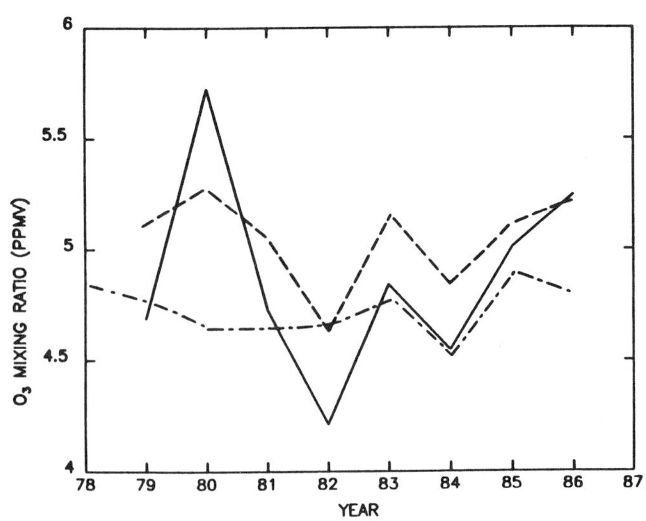

Figure 2. SBUV monthly averaged mixing ratios vs year for August at 70S and 2.0 mbar (solid line), September (dashed line), and October (dot-dashed line). See text for full description.

months of September and October are also shown. The individual data sets are normalized to the equator in 1982 for the month under consideration with the assumption that the solar cycle variation at the equator is small (see Garcia et al., 1984). The equatorial ozone, at the pressures considered here, would respond to solar cycle changes in the solar flux which would be out of phase with the ozone response at high latitude due to *NO* transport from the thermosphere. Therefore, high latitude ozone solar cycle changes will tend to be underestimated. If additional ozone trends exist at the equator at 2 mbar which would cause ozone decreases over time, they will also cause the high latitude trends calculated as described here to be underestimated.

In Figure 2 we see that the minimum ozone mixing ratios for August and September occur in the years 1980 - 1982 which correspond well with the maximum in the solar cycle as determined from F10.7 and sunspot data. The increase in ozone mixing ratio from 1982 to 1986 in Figure 2 is 25% for August, 13% for September, with no significant change for October. The percentage change decreases with altitude becoming 10% at 0.7 mbar for August, and also decreases with decreasing latitude. The increase is 20% at 55S and 15% at 40S. Ozone densities in the northern hemisphere are highly variable due to the high temperature variability of this region. Year to year ozone variability in the north, caused by temperature variations, appears to be as large as any solar cycle changes which may be present in the data.

The ozone solar cycle dependence shown in Figure 2 agrees well with the calculation of Solomon et al. (1984) for the 2 mbar region. Their model includes solar cycle dependent *NO* transport from the thermosphere. The results show an ozone depletion at solar maximum with respect to solar minimum in August at 68S of about 35%, a depletion in September of about 20%, and almost no change in October. The model calculations for July show less depletion than those for August but, at this latitude, July data are unavailable from SBUV. We believe this is a further indication that *NO* is the agent responsible for the ozone decrease in the winter and spring months at solar maximum.

Finally, we show in Figure 3 ozone trends derived from SME data for October at 1.0 mb. The ozone trend is given in %/yr and is calculated for the five years 1982-1986. The solid line is the trend calculated from a model (Rusch and Clancy, 1988a) which includes projected increases in Cl_X, decreases in solar flux (G. Rottman, Private

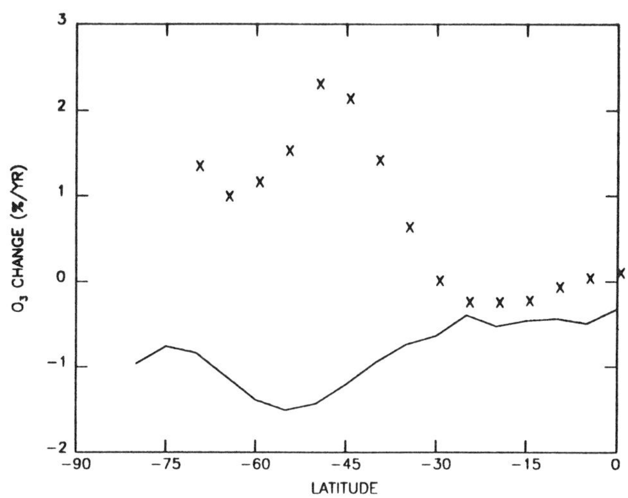

Figure 3. Trends in 1.0 mbar ozone for October as a function of latitude from SME (x's). Model results are also shown.

Communication) as measured by SME, and NMC temperatures. The calculated and measured trends agree very well north of 30S and vary from about 0 to -1%/yr. South of 30S the SME trends rise rapidly to about 2.5%/yr at 50S then decrease but remain larger than the model. The positive trends in ozone at southern latitudes in the spring are consistent with the hypothesis that *NO* is the responsible agent for enhancing ozone destruction in the spring at times of high solar activity with decreasing effect as solar activity decreases.

The ozone trends measured by SME are consistent with Cl_X increases (Rusch and Clancy, 1988b) at most seasons and latitudes. The exceptions occur in the winter and spring near the poles where ozone changes from year to year appear to be controlled by solar cycle effects related to *NO* transport. This effect appears to be significantly larger than long term changes associated with Cl_X increases. Thus, even though models predict that increasing Cl_X will affect high latitude ozone to a greater degree than mid-latitude ozone, the solar cycle and seasonal variations in NO_X density at the stratopause due to thermospheric auroral activity may well preclude any measurement of Cl_X effects in the high latitude stratosphere at least in the winter and spring months. In addition, the solar cycle NO_X effects tend to be sporadic and introduce large geophysical noise into the system from year to year and from cycle to cycle.

5. DISCUSSION

We have found that the transport of *NO* from the thermosphere to the stratopause region is the most likely source for the increase in ozone destruction required to bring the data and model results into agreement in the winter and spring high latitude region. The discovery of a solar cycle effect on ozone in the SBUV and SME data sets is also strong evidence supporting the *NO* hypothesis. The data do not support the hypothesis that a buildup of NO_X is present throughout the solar cycle which would lead to significant enhancements in the global amount of NO_X as suggested by Callis et al. (1986). There is also strong indication from the data that the *NO* is removed from the stratopause region within about one month after the transport of *NO* is photochemically shut down in the lower thermosphere.

6. ACKNOWLEDGMENTS

The authors thank C.A. Barth, A.I. Stewart, and G.E. Thomas for helpful discussions and K. Rosenlof for computing help. The SME mission is sponsored by NASA.

7. REFERENCES

Barth, C.A., D.W. Rusch, R.J. Thomas, G.H. Mount, G.J. Rottman, G.E. Thomas, R.W. Sanders, and G.M. Lawrence, Solar Mesosphere Explorer: Scientific objectives and results, Geophys. Res. Lett., 10, 237-240, 1983.

Barnett, J.J., and M. Corney, Middle atmosphere reference model derived from satellite data, *Middle Atmosphere Program Handbook,* 16, 47-85, July, 1985.

Bates, D.R., and P.B. Hays, Atmospheric nitrous oxide, Planet. Space Sci., 15, 189, 1967.

Callis, L.B., and M. Natarajan, The antarctic ozone minimum:relationship to odd nitrogen, odd chlorine, the final warming, and the 11-year solar cycle, J. Geophys. Res., 91, 10771-10796, 1986.

Clancy, R.T., D.W. Rusch, R.J. Thomas, M. Allen, and R.S. Eckman, Model ozone photochemistry on the basis of Solar Mesosphere Explorer mesospheric observations, J. Geophys. Res., 92, 3067-3080, 1987.

Eckman, R.S., The response of ozone to short-term variations in the solar ultraviolet irradiance. 1. Theoretical model, J. Geophys. Res., 91, 6695-6704, 1986.

Garcia, R.R., S. Solomon, R.G. Roble, and D.W. Rusch, A numerical study of the response of the middle atmosphere to the 11-year solar cycle, Planet. Space Sci., 32, 411-423, 1984.

Gelman, M.E., A.J. Miller, K.W. Johnson, and R.M. Nagatini, Detection of long term trends in global stratospheric temperature from NMC analyses derived from NOAA satellite data, Adv. Sp. Res., COSPAR, Toulouse, France, 1986.

Gerard, J.-Cl., and E.J.F. Deneye, The transport of thermospheric nitric oxide into the mesosphere, Annales Geophysicae, 2, 187-190, 1984.

Heath, D.F., A.J. Krueger, H.A. Roder, and B. Henderson, The solar backscatter ultraviolet and total ozone mapping spectrometer (SBUV/TOMS) for NIMBUS 6, Opt. Eng., 14, 323, 1975.

Horvath, J.J., and J.E. Frederick, In situ measurements of nitric oxide in the high latitude upper stratosphere, Geophys. Res. Lett., 12, 495-498, 1985.

McPeters, R.D., D.F. Heath, and P.K. Bhartia, Average ozone profiles for 1979 from the NIMBUS 7 SBUV instrument, J. Geophys. Res., 89, 5199-5214, 1984.

Rusch, D.W., and R.T. Clancy, Trends in atmospheric ozone: Conflicts between models and SBUV data, J. Geophys. Res., , In Press, 1988a.

Rusch, D.W., and R.T. Clancy, A comparison of ozone trends from SME and SBUV satellite observations and model calculations, Geophys. Res. Lett., In Press, 1988b.

Rusch, D.W., and R.S. Eckman, Implications of the comparison of ozone abundances measured by the Solar Mesosphere Explorer to model calculations, J. Geophys. Res., 90, 12991-12998, 1985.

Rusch, D.W., G.H. Mount, C.A. Barth, R.J. Thomas, and M.T. Callan, Solar Mesosphere Explorer ultraviolet spectrometer measurements of ozone in the 1.0 to 0.1 mbar region, J. Geophys. Res., 89, 11677-11687, 1984.

Russell III, J.M., S. Solomon, L.L. Gordley, E.E. Remsberg, and L.B. Callis, The variability of stratospheric and mesospheric NO_2 in the polar winter night observed by LIMS, J. Geophys. Res., 89, 1984.

Solomon, S., and R.R. Garcia, Transport of thermospheric NO to the upper stratosphere?, Planet. Space Sci., 32, 399-409, 1984.

Solomon, S., R.R. Garcia, and F. Stordal, Transport processes and ozone perturbations, J. Geophys. Res., 90, 12981-12989, 1985.

SATELLITE/ROCKET OZONE COMPARISONS AT NATAL, BRAZIL

Robert A. Barnes
Chemal, Inc.
Wallops Island, Virginia 23337, USA

M. Patrick McCormick
NASA Langley Research Center
Hampton, Virginia 23665, USA

Robert D. Hudson and Chester L. Parsons
NASA Goddard Space Flight Center
Greenbelt, Maryland 20771, USA

ABSTRACT

We present comparisons of satellite, rocket, and balloon ozone profiles near Natal, Brazil (5.9°S, 35.2°W). The low variability of stratospheric ozone at Natal during March and April of 1985 has allowed intercomparisons of reasonably large data sets, rather than a small number of paired satellite/in-situ comparisons. There are sharp differences between the profiles from the SBUV instrument on Nimbus 7 and the in-situ measurements. These results support the conclusions of the NASA Ozone Trends Panel that there is an instrumental cause for the very large changes in upper stratospheric ozone seen by SBUV. Along with other comparisons, these results are being used in a reassessment of the SBUV instrument and its data reduction procedures. The agreement between the ozone profiles from the SAGE II instrument on the ERBS satellite and the rocket values is excellent over the full range of comparisons. Both SAGE II and ROCOZ-A must convert from altitude to pressure for intercomparisons with SME and with SBUV-type instruments. The conversion between pressure and altitude is as important as the ozone measurements, especially in the upper stratosphere where the scale height for ozone is approximately half that for pressure.

1. INTRODUCTION

As part of a support program for NASA and NOAA satellite instruments, measurements with the ROCOZ-A ozonesonde were initiated by NASA in 1983. This program has provided sets of equatorial and northern mid-latitude ozone profiles (Barnes et al., 1987; 1988). In particular, the equatorial measurement series at Natal, Brazil, in March/April 1985 has provided an excellent opportunity for high quality intercomparisons. During the three weeks of the campaign at Natal, stratospheric ozone variability from ROCOZ-A measurements was 2% or less (Barnes et al., 1987). Low variability at a single point for this period implies a comparable uniformity for ozone profiles over a wide area around the measurement site. This condition at Natal relaxes the requirement for exact concurrence between satellite and local ozone measurements, allowing comparisons of larger sets of profiles and improving the precision of the comparisons.

The data set from the in-situ campaign includes profiles of ozone column amount (or overburden), ozone number density, and ozone mixing ratio from the ground to 52 km. The accuracy of ROCOZ-A ozone measurements has been estimated at 5-7% (Barnes et al., 1986). Balloon-based comparisons have shown ROCOZ-A to be 5-10% higher than an in-situ ultraviolet photometer and 4% lower than a mass spectrometer at altitudes near 40 km (Barnes et al., 1987; Watson et al., 1988). This raises concern since both comparison instruments are considered to be absolute standards.

Auxiliary measurements in the rocket series give profiles of atmospheric temperature, pressure, and density from the ground to 52 km. This collection of profiles allows the use of these data as a transfer standard between satellite instruments with different fundamental ozone measurements. The 1985 ozone measurements at Natal were made in conjunction with overpasses from SME, SAGE II, the SBUV on Nimbus 7, and the SBUV/2 on NOAA-9.

2. COMPARISONS WITH SBUV AT NATAL

The SBUV measurements in this analysis come from 32 profiles in 1979 and 38 profiles in 1985. For the first comparison the radiances from the SBUV spectrometer have been converted into ozone column amounts using the algorithm of Thomas and Holland (1977) and measurements from five SBUV wavelengths (Barnes, 1988). There is a distinct skew between SBUV ozone profiles in 1979 and 1985 at pressures from 1 to 5 mbar (fig. 1). There are corresponding changes in the backscattered radiances from the SBUV 'spectrometer for these data sets, with a distinct wavelength dependence for the changes. There is a direct relationship between radiance differences and differences in the resulting ozone column amounts. Using SBUV values alone, it is not possible to separate atmospheric and instrumental changes. ROCOZ-A ozone measurements at Natal in March/April 1985 show reasonable agreement with the 1979 SBUV column amounts there. The SBUV column amounts in 1985 do not agree with the 1985 ROCOZ-A measurements, with ROCOZ-A more than 25% higher at 1.2 mbar (Barnes, 1988).

The primary ozone product from the standard SBUV algorithm is ozone layer amount, which is defined as the ozone column amount within an Umkehr pressure layer (McPeters et al., 1984). The average ozone layer amounts from the 1979 and 1985 SBUV data sets are shown in fig. 2. The differences of the 1985 SBUV profile from 1979 are given in fig. 2b. This figure includes the 95% confidence

ISBN 0-937194-15-8

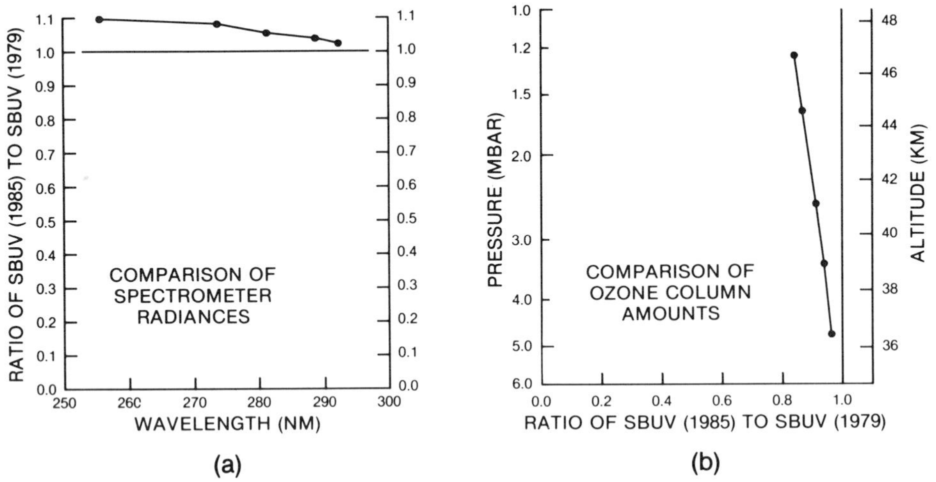

Fig. 1. Relative Changes in SBUV Ozone Measurements at Natal, Brazil, from 1979 to 1985. (a) Changes in the radiances from the SBUV instrument. (b) Changes in ozone column amounts. There is a direct relationship between the wavelength dependent change in the spectrometer values and the pressure dependent change in ozone amounts. After Barnes (1988). Copyright American Geophysical Union.

limits for the differences. At almost all levels the differences between 1979 and 1985 are statistically significant. These changes are representative of the ozone trend reported from SBUV measurements (D.F. Heath, paper presented at the Spring meeting, AGU, Baltimore, MD, 1987).

The profiles from the standard SBUV algorithm can be considered an optimal combination of a first-guess and a measured profile (Cunnold et al., 1984). The measurement region for the SBUV spectrometer is estimated to be from 1 to 16 mbar (McPeters et al., 1986; Barnes, 1988). Above and below these limits the solution profile returns to the first-guess. Care must be taken in the interpretation of the upper stratospheric ozone decrease from the standard algorithm (Umkehr layers 10-12), since much of the reported decrease is outside of the measurement range for the instrument. However, in the pressure range from 1 to 5 mbar, both the Thomas-Holland and the standard SBUV algorithm show the same relative decrease in stratospheric ozone from 1979 to 1985.

Figure 2c shows the comparison of 1985 SBUV ozone layer amounts with concurrent rocket and balloon measurements at Natal. Again, the error bars in figs. 2b through 2d give the 95% confidence limits for the comparisons. In layer 9 there is an 18% difference between ROCOZ-A and SBUV. This is the region where SBUV shows a large decrease in stratospheric ozone from 1979 to 1985. The skew between the 1985 and 1979 SBUV profiles in the upper stratosphere (fig. 2b) is duplicated by the skew between the 1985 SBUV and the 1985 ROCOZ-A results in the same region (fig. 2c).

In Umkehr layers 1 through 4 the SBUV profile shows differences of up to 65% from the rocket and balloon measurements. In this region there is excellent agreement between the ECC ozonesonde measurements and the 6-year climatology at Natal of Logan and Kirchhoff (1986). In these layers the SBUV profile is dominated by the first guess climatology. The agreement of the 1985 SBUV results with the corresponding first guess profiles is shown in fig. 2d. This agreement is better than 5% in layers 1 to 4. Combined with the large differences in fig. 2c this shows the disagreement between the tropical SBUV first-guess and the in-situ measurements in the troposphere and the lower stratosphere.

We have made comparisons between the ECC measurements at Natal and the unpublished climatology of Hilsenrath and Dunn (1979) that is used as a basis for the SBUV first-guess profiles (see Klenk et al., 1983). The low latitude profiles of Hilsenrath and Dunn (1979) cover a latitude range of ±25°, with most of the profiles at higher latitudes. The climatology shows a lower tropopause, which results in considerably more ozone in Umkehr layer 3 than the ECC measurements at Natal (6°S). This can account for most of the 60% difference in layer 3 of fig. 2c.

It should be stressed that the measurement region for the SBUV extends from 1 mbar to about 16 mbar. For pressures outside this range the ozone values from the SBUV archives are combinations of the first-guess climatology and extrapolations of values from inside the measurement region. The SBUV values in layers 1-4 and 10-12 should not be treated as measured ozone amounts. However, because of the coupling of the layers in the SBUV algorithm, an improper first-guess profile can cause problems within the SBUV measurement region (McPeters et al., 1986). The effect of the first-guess on the measurement region of tropical SBUV ozone profiles can be examined by replacing the standard SBUV first-guess in the 1985 data set with the in-situ profiles from Natal.

3. COMPARISONS WITH SAGE II AT NATAL

The SAGE II measurements in this analysis come from 14 profiles in March/April 1985. The intercomparison is made in terms of ozone number density, the primary ozone data product for SAGE II. These measurements also show low variability in stratospheric ozone near Natal. The lower altitude limit for the SAGE II measurements at Natal was created by interference from high tropical cirrus clouds. When these clouds are absent, SAGE II can measure ozone into the troposphere.

The ozone profiles from SAGE II, ROCOZ-A, and ECC are shown in fig. 3a. The differences of SAGE II from ROCOZ-A and ECC are shown in figs. 3b and 3c, respectively. The error bars in figs. 3b and 3c give the 95% confidence limits for the differences. The confidence limits for the SAGE II/ROCOZ-A comparison average 3.5%

For the comparison with ROCOZ-A the relative differences between the satellite and rocket instruments (fig. 3b) average 0.4% (SAGE II lower). The average of the absolute values for the differences between these instruments is 2.4%. For 24 of the 33 comparison altitudes there is no statistical difference between the SAGE II and the ROCOZ-A measurements at the 95% confidence level.

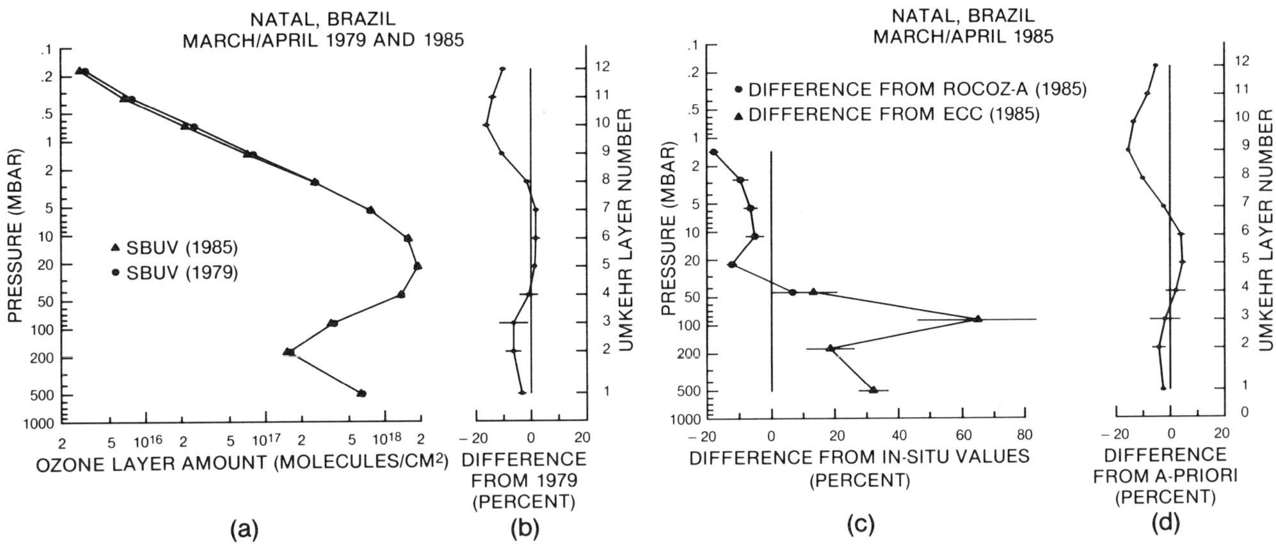

Fig. 2. Comparisons of SBUV Ozone Layer Amounts at Natal, Brazil. (a) SBUV ozone layer amounts in March/April of 1979 and 1985. (b) Percent difference of the 1985 SBUV values from SBUV values in 1979. The error bars give the 95% confidence limits for the comparisons. (c) Percent difference of the 1985 SBUV values from concurrent rocket and balloon measurements. (d) Percent difference of the 1985 SBUV values from their associated first-guess profiles.

Below 20 km there is an increase in the differences of SAGE II from the ECCs, in addition to a dramatic increase in the 95% confidence limits for the differences between the instruments. This reflects increases in the relative standard deviations for the data sets from both the satellite and balloon instruments. Most of this increase is due to increased variability in lower stratospheric ozone, since the variability in the ECC data set (Barnes et al., 1987) is 2 to 5 times greater than the instrument-to-instrument repeatability of the ECCs, themselves (Barnes et al., 1985). The differences between the times and the locations of the satellite and balloon measurements become important, since there is substantial ozone variability in this altitude region.

The temperatures and pressures that are associated with the SAGE II ozone profiles are derived from the NMC gridded analysis. This analysis uses rocket and balloon temperature measurements to provide published corrections for the satellite measurements (Gelman et al., 1986). The equatorial station that provides local measurements for the gridded analysis is Ascension Island.

Temperatures at Natal and Ascension Island are both measured with thermistors, and comparisons will only check the consistency of one measurement type. Since temperatures can be obtained independently from atmospheric density profiles, thermistor-based measurements should also be compared with temperature profiles from rocket-borne falling spheres, from ground-based lidars, and from satellite-based measurements, such as the oxygen sensor proposed for SAGE III. Such comparisons are of prime importance, since temperature is the variable used in the conversion between pressure and altitude.

4. CONCLUDING REMARKS

Both SAGE II and ROCOZ-A measure ozone versus geometric altitude. In terms of the primary data product for SAGE II there is excellent agreement between the two instruments. However, both SAGE II and ROCOZ-A must convert from altitude to pressure for comparisons with SBUV and SME. SBUV ozone profiles are given without information for this conversion (McPeters et al., 1984). For these comparisons the accurate determination of pressure is as important as the ozone measurements, especially in the upper stratosphere where the scale height for ozone is approximately half that for pressure (Barnes, 1988). In this region a 2% discrepancy in pressure creates a 3.5% difference in ozone amount.

For the comparison between SBUV and ROCOZ-A presented above, there is a difference of 18% in Umkehr layer 9. In the upper stratosphere the skew between 1985 SBUV values and 1979 SBUV values duplicates the skew between 1985 SBUV values and 1985 ROCOZ-A ozone amounts. This comparison supports the conclusion of the NASA Ozone Trends Panel that there is an instrumental cause for the very large changes in upper stratospheric ozone seen by SBUV (Watson et al., 1988). The agreement between SAGE II and ROCOZ-A in the upper stratosphere also support this conclusion. The SBUV data set is in the process of reanalysis. Comparisons with other ozone data sets, including the SAGE II and ROCOZ-A sets presented here, will be used as part of an assessment of the rate of long-term changes in the SBUV spectrometer.

REFERENCES

Barnes, R.A., 1988: Changes in SBUV ozone profiles near Natal, Brazil, from 1979 to 1985. J. Geophys. Res., 93, 1704-1717.

Barnes, R.A., A.R. Bandy, and A.L. Torres, 1985: Electrochemical concentration cell ozonesonde accuracy and precision. J. Geophys. Res., 90, 7881-7887.

Barnes, R.A., A.C. Holland, and H.S. Lee, 1986: An improved rocket ozonesonde (ROCOZ-A), 2, Preparation of stratospheric ozone profiles. J. Geophys. Res., 91, 14,521-14,531.

Barnes, R.A., A.C. Holland, and V.W.J.H. Kirchhoff, 1987: Equatorial ozone profiles from the ground to 52 km during the Southern Hemisphere autumn. J. Geophys. Res., 92, 5573-5583.

Barnes, R.A., M.A. Chamberlain, C.L. Parsons, and A.C. Holland, 1988: An improved rocket ozonesonde (ROCOZ-A), 3, Northern mid-latitude ozone

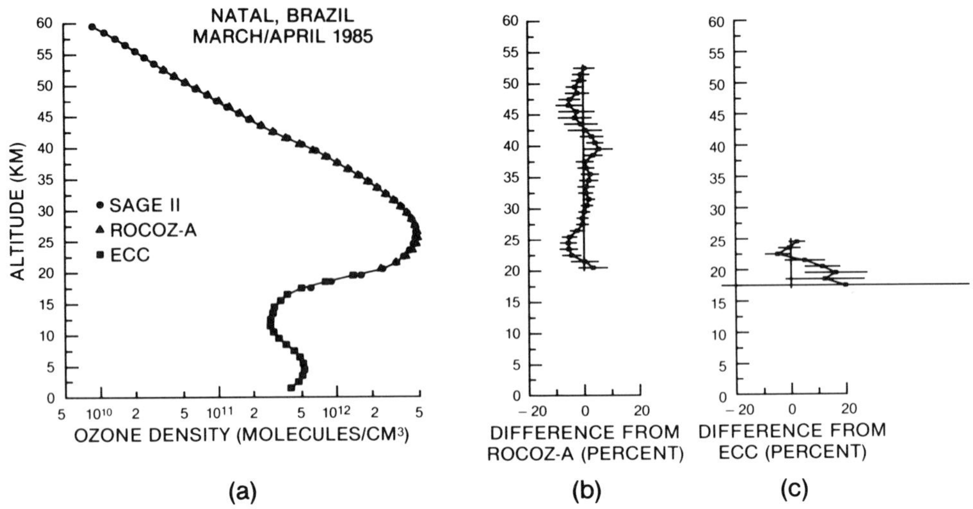

Fig. 3. Comparisons of SAGE II Ozone Number Densities at Natal, Brazil. (a) SAGE II, rocket, and balloon densities in March/April 1985. (b) Percent difference of the SAGE II densities from ROCOZ-A. The error bars give the 95% confidence limits for the comparisons. (c) Percent difference of the SAGE II densities from the ECC's.

measurements from 1983 to 1985. Submitted to J. Geophys. Res.

Cunnold, D.M., M.C. Pitts, and C.R. Trepte, 1984: An intercomparison of SAGE and SBUV observations for March and April 1979. J. Geophys. Res., 89, 5249-5262.

Gelman, M.E., A.J. Miller, K.W. Johnson, and R.M. Nagatani, 1986: Detection of long-term trends in global stratospheric temperature from NMC analyses derived from NOAA satellite data. Adv. Space Res., 6, No. 10, 17-26.

Heath, D. F., 1987: Non-seasonal changes in stratospheric ozone from the Nimbus 7 SBUV instrument, paper presented at the Spring Meeting, AGU, Baltimore, MD.

Hilsenrath, E., and P. J. Dunn, 1979: Standard ozone profiles from balloons and rocket data for satellite and theoretical model input, paper presented at the Spring Meeting, AGU, Washington, DC.

Klenk, K.F., P.K. Bhartia, E. Hilsenrath, and A. J. Fleig, 1983: Standard ozone profiles from balloon and satellite data sets. J. Climate Appl. Meteorol., 22, 2012-2022.

Logan, J.A., and V.W.J.H. Kirchhoff, 1986: Seasonal variations of tropospheric ozone at Natal, Brazil. J. Geophys. Res., 91, 7875-7881.

McPeters, R.D., D.F. Heath, and P.K. Bhartia, 1984: Average ozone profiles from the Nimbus 7 SBUV instrument. J. Geophys. Res., 89, 5199-5214.

McPeters, R.D., R.D. Hudson, P.K. Bhartia, and S.L. Taylor, 1986: The vertical ozone distribution in the Antarctic ozone minimum measured by SBUV. Geophys. Res. Lett., 13, 1213-1216.

Thomas, R.W.L., and A.C. Holland, 1977: Simple relationship between the UV radiation backscattered by the Earth's atmosphere and the vertical ozone profile. Appl. Opt., 16, 2581-2583.

Watson, R.T. and Ozone Trends Panel, M.J. Prather and Ad Hoc Theory Panel, and M.J. Kurylo and NASA Panel for Data Evaluation, 1988: Present state of knowledge of the upper atmosphere 1988: An assessment report. NASA Reference Publication 1208.

SATELLITE TOTAL OZONE CLIMATOLOGY COVERING 16 YEARS

E. Hilsenrath and S. Chandra
NASA Goddard Space Flight Center
Greenbelt, Maryland 20771, USA

ABSTRACT

Satellite total ozone observations have been analyzed from both the BUV on Nimbus-4 and SBUV on Nimbus-7. The former covered the period 1970 to 1977 while the latter covered the period 1979 to 1986. The purpose of this paper is to study the global ozone climatology and seasonal variations. The two data sets show very similar seasonal variations at all latitudes with some exceptions. The total ozone Quasibiennial Oscillation (QBO) is very apparent over the 18 year period with some indication of a 4 year component. The QBO maximum and minimum are best correlated with the West and East components of the QBO wind at 30 mb respectively. The extratropical phase shift in the ozone QBO first seen in the Nimbus-4 data appears similarly in the Nimbus-7 data. Total Ozone eddies and National Meteorological Center temperatures in the lower stratosphere were also analyzed by month and latitude. The ozone eddies correlate well with the temperatures on a month to month basis and over the long term after the seasonal cycles were removed. These long term changes should be nearly independent of instrument drifts and therefore represent a real climatological change.

1. INTRODUCTION

The purpose of this paper is to compare the total ozone climatologies from the Nimbus-4 BUV and Nimbus-7 SBUV data sets. The BUV covers the period 1970 to 1977 where coverage decreased with time, while the SBUV data covered the period 1978 to 1986 and had fairly uniform coverage with time. Both data sets were processed with the same algorithm where the ozone absorbtion cross sections determined by Bass and Pauer (1984) were employed. It is well known that both the BUV and SBUV instruments drifted relative to Dobson observations while in orbit (Flieg et al., 1986; Heath, 1988) and a thorough investigation of the instrument sensitivity changes with time is now under way. Because of these changes, this paper will emphasize total ozone variations that are larger than or independent of instrument problems. Several studies have been performed of total ozone variations employing satellite observations (Hasebe, 1983; Hilsenrath and Schlesinger, 1981; Oltmans and London, 1982; and others). The goal of this paper is to look again at the total ozone climatology using satellite data that covers the period 1970 to 1986. There are no data in 1977.

In addition to the ozone satellite data, we also access the National Meteorological Center (NMC) temperature and wind analyses, (Gelman et al. 1986) that are concurrent with the Nimbus-7 data. The NMC data were then further processed to grids that are coincident with the satellite ozone data. The grid includes monthly mean values in a box size of 10° of latitude and 30° of longitude which makes the NMC data compatible with the ozone data.

Our analysis employs zonal averages and their standard deviation of the 12 values in each latitude zone which we relate to ozone eddies or standing waves. A least square fit to the zonal averages yields annual, semi-annual, and a linear term which includes an ozone trend and instrument degradation. A regression model was derived for each latitude zone which contained the mean value, the amplitude and phase of the periodic components, and the linear term. With this model the data for each zone could be modified to remove the linear trend and/or the seasonal components (de-seasonalized). After the data are deseasonalized, interannual components such as the QBO remained. The data can also be detrended by removing the linear term. If instrument degradation were well characterized and removed, the remainder of the linear term would represent the ozone trend assuming it was also linear.

2. THE ANNUAL CYCLE

That the difference in the evolution of the seasonal variation of ozone in the Northern and Southern Hemispheres is due to circulation differences is well known. Circulation and ozone transport are manifested in the ozone eddies (Wu et al., 1987; Chandra, 1988). In the Northern Hemisphere the eddies are a maximum in winter when ozone begins to build-up. In the Southern Hemisphere the, maximum eddies occur at 70°S and are coincident with the springtime ozone maximum (Chandra, 1988).

An example of the continuity and the difference of total ozone from BUV and SBUV data are shown in Fig. 1. We have selected 50°N, the Equator and 50°S for this example (note scale change for the Equator). As expected the annual wave dominates at higher latitudes and is six months out of phase in the two hemispheres. In the tropics the seasonal cycle appears more complex, because the annual waves in the two hemispheres are superimposed on the QBO. Further discussion on the ozone QBO will follow.

ISBN 0-937194-15-8

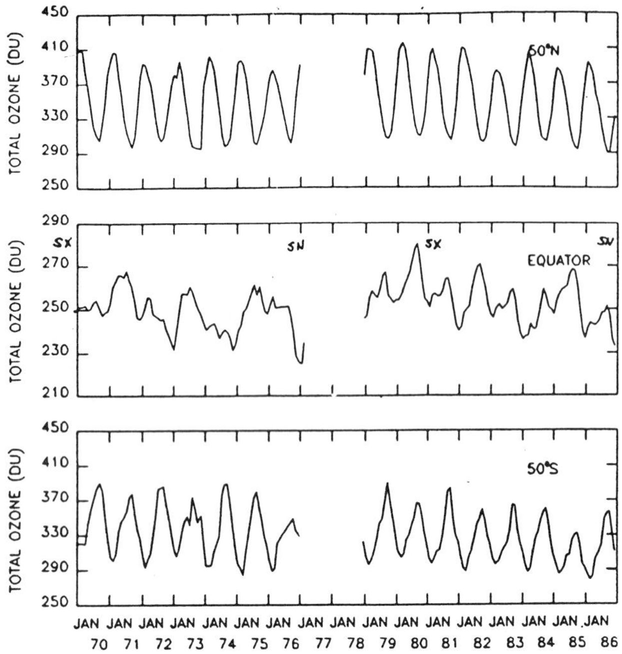

Fig. 1 Zonal mean ozone at three latitudes. BUV and SBUV data are on the left and right sides respectively. SX and SN are times of solar maximum and minimum.

The annual variation in both the BUV and SBUV data sets are very similar when the monthly zonal means are analyzed by a harmonic analysis as employed by Hilsenrath and Schlesinger (1981). This analysis objectively demonstrates that the amplitudes and phases of the annual and semi-annual waves as function of latitude are nearly the same for the two data sets except at high latitudes. In the Southern Hemisphere the amplitude of the annual wave for SBUV decreases with respect to the BUV, where it is 15% lower at 50°S. At higher latitudes, difference in coverage in the two data sets can contribute some uncertainty (5-10%) in the analysis. It is therefore possible that the lower SBUV values in the Southern Hemisphere high latitudes are associated with development of the Antarctic ozone hole which is discussed further below.

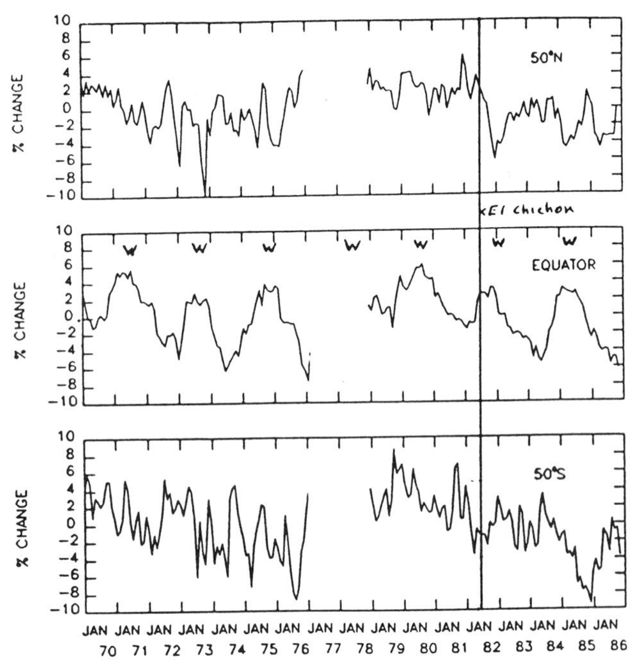

Fig. 2 Zonal mean ozone de-seasonalized. Times of the west wind maximum are shown by W.

3. DE-SEASONALIZED TOTAL OZONE

Figure 2 illustrates total ozone for the same three zones with the annual and semi-annual components removed according to the procedures discussed in the Introduction. Note the scale is percent change from the mean amount. The vertical line through the panels denotes the time for the eruption of El Chichon. At the equator the QBO is clearly seen in both data sets. The period is 28 ±2 months and the best correlation occurs with the tropical west wind maximum at 30 mb which occurs within two months of the ozone maximum. The wind data are taken from Naujokat (1986). There may also be a four year component as suggested by Hasebe (1983) however data from 1977 are missing and El Chichon could be an influence in 1982.

The phase and amplitude of the QBO as a function of latitude has been studied using BUV data as referenced above. Those studies found that both the QBO period and phase had a latitudinal dependence. The QBO observed from BUV and SBUV are illustrated in Figs. 3a and 3b respectively (the contour intervals are 4 Dobson Units). The ozone data employed in these figures were de-seasonalized and detrended leaving only variations larger than one month and less than seven years. In general the two data sets show similar features, namely a phase shift near 20°S and period change at mid and high latitudes. Several attempts have been made to connect the tropical ozone QBO which is clearly associated with the QBO wind with, the ozone QBO at higher latitudes. However the phase shift and the decrease in period has not been satisfactorily explained. Although there are similarities in the BUV and SBUV data, there are striking differences. The most likely cause for these differences is El Chichon and the appearance of the Antarctic ozone hole.

Returning to Fig. 2 we note that at 50°N, El Chichon caused an abrupt and significant decrease in the total ozone zonal average, where the over all decrease is about 6%. Half of this could be instrumental. Additional effects of El Chichon will be discussed later. At 50°S the QBO is apparent but out phase with tropical QBO as discussed earlier. The zonal average decreases about 14% before it seems to have gone up again. Bojkov (1987) has described the 1983 and 1985 events as being triggered by the QBO and

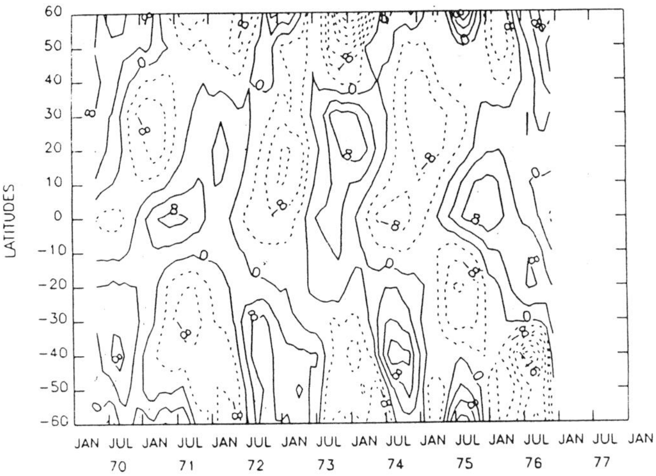

(a)

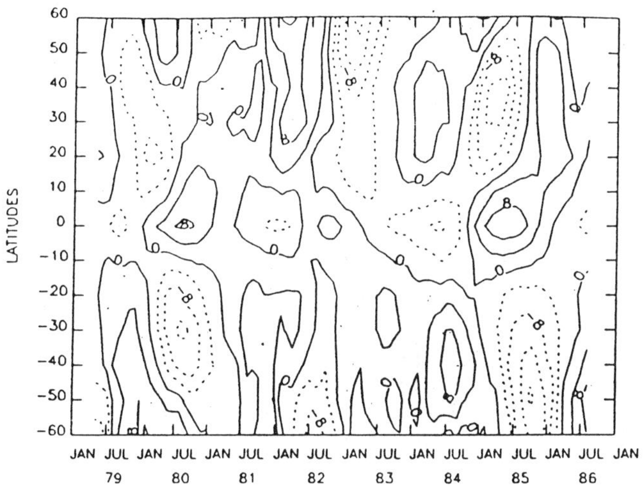

(b)

Fig. 3 Total ozone de-seasonalized, de-trended, and smoothed a) BUV, b)SBUV.

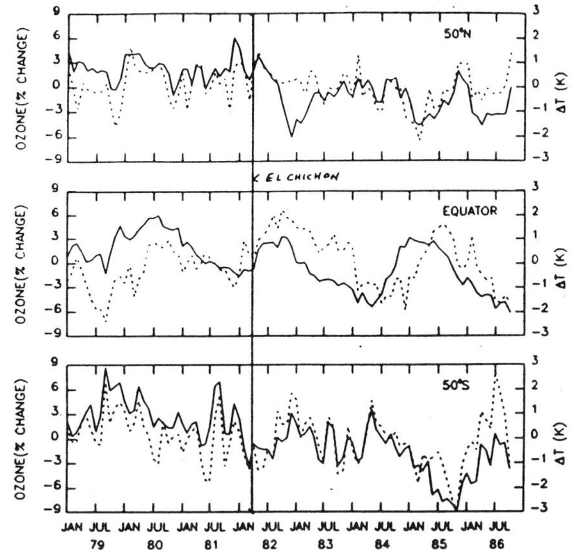

Fig. 4 De-seasonalized total ozone and temperature for the SBUV data. Solid line is ozone and dashed line is temperature.

further augmented by the southern oscillation and El Chichon. It should be noted that almost comparable decreases occurred in the BUV data which are considerably larger than the expected instrument degradation.

4. DE-SEASONALIZED TEMPERATURES

The NMC temperatures were de-seasonalized in the same manner as the total ozone data were in Fig. 2. After de-seasonalizing, the remaining temperature variations represent dynamic activity on monthly and long term time scales. In Fig. 4 temperatures at 70 mb (dashed lines) are shown as departures from the overall mean value and are superimposed on the de-seasonalized ozone data (solid lines) for the SBUV period. At 50°N, El Chichon may have caused some heating, followed by a 3 year decrease before recovering. A QBO signal is clear in the tropics as expected at 70 mb. At 50°S, a small overall decrease in temperature appears between 1978 and early 1984 after which the average drops nearly 4°. There seems to be full recovery in the temperature after early 1986.

When comparing the deseasonlized total ozone and 70 mb temperatures in Fig. 4, we find an in phase correlation, except during the time of El Chichon at 50°N. There is also a concurrent precipitous drop in temperature and ozone at 50°S in early 1984. The in phase relationship depicted here implies that ozone and temperature are following the same dynamical forcing. Nagatani and Miller (1987) also suggested a dynamical origin for simultaneous changes in ozone and temperature at 70°S in the spring months.

5. OZONE EDDIES

Long term changes in ozone eddies may also be indicators of long term climatic changes. To the first order, changes in ozone eddies are independent of instrument sensitivity changes. Fig. 5 illustrates these eddies for the same latitude zones shown earlier. After the first two years of BUV operation coverage decreased to the point where the standard deviations were no longer representative of the standing waves and therefore not shown. For SBUV coverage was continuous. Chandra (1988) demonstrated that the annual cycle of ozone eddies was associated with the annual cycle in ozone. It would not be surprising therefore that there should be corresponding changes in the long term variations of ozone and eddies as well. At the equator these eddies are small as expected. In the Northern hemisphere there is a 5% decrease which may well be real since it is larger than changes that would result from instrument

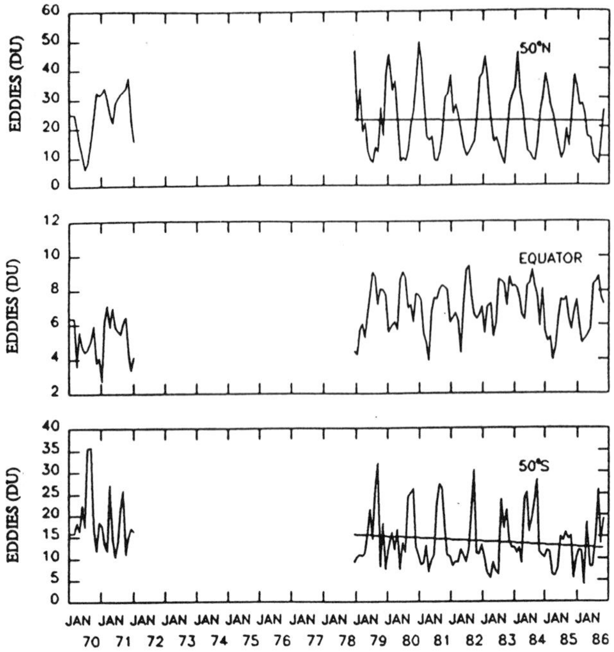

Fig. 5 Ozone eddies for BUV and SBUV data.

degradation which would be no larger than 3%. In the Southern hemisphere the decrease in eddies is nearly 20% with the largest change occurring in early 1985 with an indication of recovery in 1986. The largest decrease appeared in the spring months (October) when ozone eddies are maximum (Chandra, 1988). Wu et al. (1987) found that ozone transport was associated with planetary wave activity derived from geopotential height variability. Therefore we feel a decrease in ozone eddies at 50°S as shown in Fig. 5, most likely represents a weakening of standing waves and transport to higher latitudes which in turn can contribute to the Antarctic ozone hole.

6. OZONE TRENDS

Global and regional changes in ozone are important and formidable scientific and environmental issues. All data must be employed to study this problem with the recognition of the limitations in each measurement. Returning to Fig. 1, we see clearly there is a trend in the data. Using the linear term from the harmonic analysis and after correcting the data sets to Dobson we found that there is 1% downward trend in the BUV data and a 3% downward trend in the SBUV data. Further work is needed to determine the relative biases between the two data sets and the reality of the downward trend in both data sets. This determination is essential since these data cover a major portion of solar cycles 20 and 21. The times for solar maximum and minimum are indicated by the SX and SN respectively in the panel for the Equator in this figure. This comparison implies an ozone minimum and maximum at solar cycle minimum and maximum respectively. This result is consistent with model predictions (Garcia et al., 1981) and observations (Chandra, 1985) who used a more limited data set. More recent studies by the Ozone Trends Panel (1988) indicate that ozone response to solar cycle change may be seasonally and latitude dependent. In the future, we plan to study these effects using this extended data set.

7. SUMMARY

1. The total ozone seasonal and QBO cycles are consistent throughout the 16-year data period. There is some evidence of a four-year component.

2. The total ozone QBO is 28 ±2 months and correlates best with the tropical west wind at 30 mb.

3. After de-seasonalizing lower stratospheric temperatures and total ozone, the monthly variations are in phase except during the time of El Chichon. There is also a significant concurrent drop at 50°S in these to parameters beginning in early 1984 and continuing to 1986 before a recovery began.

4. For the period 1979 to 1986, the ozone standing waves decreased by about 5% at 50°N and nearly 20% at 50°S. These changes can not be attributed to instrument degradation. The decrease at 50°S certainly represents a decrease in transport poleward and likely contributes to the Antarctic ozone hole.

5. There appears to be a solar cycle relationship in the BUV and SBUV data sets, however a more careful study of an instrument induced bias and drift must be performed before a definitive conclusion can be made.

REFERENCES

Bass, A. M. and R. J. Paur, 1984: The ultraviolet cross section of ozone: I. The measurements, in atmospheric ozone, ed. by C. S. Zerefos and A. Ghazi, 606, D. Reidel, Dordrecht.

Bojkov, R. D., 1987: The 1983 and 1985 anomalies in ozone distribution in perspective, Month. Wea. Rev., 115, 2187-2201.

Chandra, S., 1985: Solar induced oscillations in the stratosphere: A myth or reality?, J. Geophys. Res., 90, 2331-2339.

Chandra, S., 1988: Recent trends in stratospheric ozone Part II: Implications of ozone depletion in the Antarctic, submitted to J. Geophys. Res.

Fleig, A. J., P. K. Bartia, D. S. Silberstein, 1986: An assessment of the long term drift in SBUV total ozone data, based on comparison with the Dobson Network, GRL, 13, 1359.

Garcia, R. R., S. Solomon, R. G. Roble, and D. W. Rusch, 1984: A numerical response of the middle atmosphere to the 11 year solar cycle, Planet. Space Sci., 32, 411-423.

Gelman, M. E., A. J. Miller, K. W. Johnson, R. M. Nagatani, 1986: Detection of long term trends in global stratospheric temperatures from NMC analyses derived from NOAA satellite data, Ad. Sp. Res., 6, 17-26.

Hasebe, F., 1983: Interannual variations of global ozone revealed from Nimbus-4 BUV and ground based observations, J. Geophys. Res., 88, 6819-6834.

Heath, D. F., 1988: Non-seasonal changes in total column ozone from satellite observations, 1970-86, Nature, 332, 219.

Hilsenrath, E. and B. M. Schlesinger, 1981: Total ozone seasonal and interannual variations derived from 7 years of Nimbus-4 BUV data, J. Geophys. Res., 86, 12087-12096.

Oltmans, S. J. and J. London, 1982: The quasibiennial oscillation in atmospheric ozone, J. Geophys. Res., 87, 8981-8989.

Report of the Ozone Trends Panel to be published in 1988.

Nagatani, R. M. and A. J. Miller, 1987: The influence of lower stratospheric forcing on the October Antarctic ozone decrease, GRL, 14, 202-205.

Naujokat,. B., 1986: An update of the observed QBO of the stratospheric winds over the tropics, J. Atmos. Sci.

Wu, M.-F., M. A. Geller, J. G. Olson, E. M. Larson, 1987: A study of global ozone transport and the role of planetary waves using satellite data, J. Geophys. Res., 92, 3081-3097.

AN ASSESSMENT OF THE SBUV/TOMS OZONE DATA QUALITY, BASED ON COMPARISON WITH EXTERNAL DATA

Albert J. Fleig
NASA Goddard Space Flight Center
Greenbelt, Maryland 20771, USA

David S. Silberstein, Charles G. Wellemeyer, and Richard P. Cebula
ST Systems Corporation
Lanham, Maryland 20706, USA

Pawan K. Bhartia
Science Applications Research
Lanham, Maryland 20706, USA

John J. DeLuisi
NOAA/ERL/GMCC
Boulder, Colorado 80303, USA

Abstract

The validation effort for the SBUV/TOMS ozone system is summarized. The long term drift of SBUV/TOMS total ozone relative to the Dobson Network over the eight years from launch through October 1986 is presented and discussed. On February 13, 1987 (early in the ninth year of operation), the optical chopper on the SBUV instrument began to fall out of synchronization. As a result, these comparisons comprise the final validation of the in-sync SBUV data set as a whole. The SBUV drift relative to a 41 station Dobson Network is roughly linear with a slope of approximately -0.42 ± 0.09%/yr, but the TOMS drift is better fit by a two-piece linear model with a common intercept around the middle of 1982. The slopes are -0.25 ± 0.17%/yr during the first period and -0.53 ± 0.15%/yr for the second period. Support for the presence of a downward change in the upper level ozone is provided through comparisons of the SBUV profile ozone with data from the Umkehr Network and SAGE II. Correlation analysis of the seasonal behavior of the total ozone with Dobson measurements and profile ozone with Umkehr and balloons is also presented.

TOTAL OZONE COMPARISONS

Additional Dobson total ozone data have been received from the WODC, Toronto, Canada. Dobson comparisons presented earlier for the SBUV [Fleig et al. 1986] and TOMS [Fleig et al., 1988] have been extended to include eight years of data. Methods identical to those used previously gave the results presented in Figure 1 for TOMS and Figure 2 for SBUV. These figures show yearly average percent bias (SBUV or TOMS - Dobson) for the entire Dobson Network (41 selected stations) using two different weighting schemes and monthly Network values computed using a technique called "Locally Weighted Scatter Plot Smoother" (LOWESS) [Cleveland, 1979]. Slight discontinuities apparent in the LOWESS result during the final year of the comparison are due to data dropout from the Dobson Station at Singapore in February 1986 and, to a lesser degree, the Indian Stations at Srinagar, New Delhi, Varanasi and Poona in May 1986. Linear

ISBN 0-937194-15-8

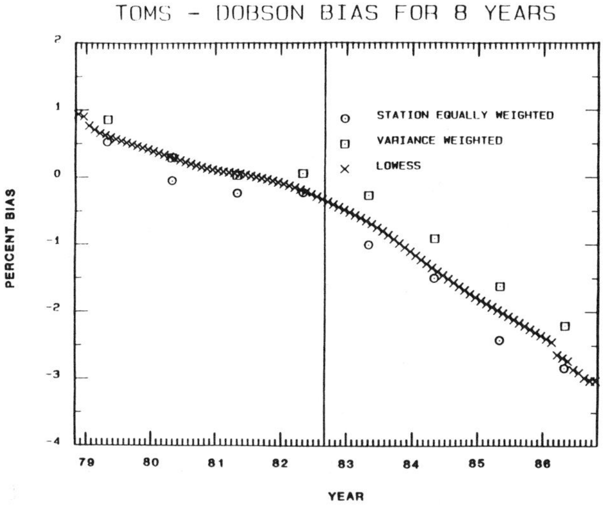

Fig. 1. Yearly and monthly averaged bias between TOMS and the Dobson Network obtained by making different statistical assumptions regarding the quality of the stations. A negative bias indicates the TOMS total ozone values are smaller than those reported by the Dobson stations. Comparisons cover a period of eight years from Nov. 1978 through Oct. 1986.

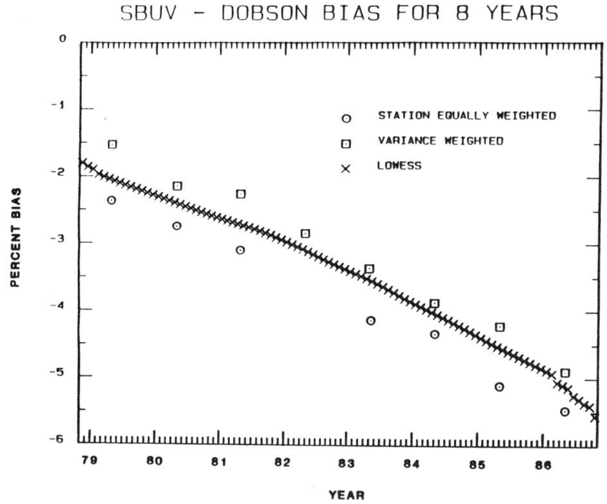

Fig. 2. Yearly and monthly averaged bias between SBUV and the Dobson Network obtained by making different statistical assumptions regarding the quality of the stations. A negative bias indicates the SBUV total ozone values are smaller than those reported by the Dobson stations. Comparisons cover a period of eight years from Nov. 1978 through Oct. 1986.

regression of the SBUV - Dobson bias from the variance weighted comparisons indicates that the SBUV has drifted relative to Dobson with a roughly linear rate of -0.42 ± 0.09% per year. The TOMS - Dobson bias is better fit by a two piece linear model with a change in slope around July 1982. The drift from November 1978 through June 1982 is -0.25 ± 0.17%/yr and from July 1982 through October 1986 it is -0.53 ± 0.15% per year. No significant improvement in fit was achieved in applying the two part model fit to the SBUV - Dobson bias. The results were -0.36 ± 0.19% per year for the first period and -0.46 ± 0.15% per year for the second period. To within the uncertainty of the determination, these drift parameters derived from eight years of data have not changed significantly from those derived earlier from six years of data from SBUV (-0.38 ± 0.13% per year) and seven years of data from TOMS (-0.25 ± 0.17% per year for period one and -0.51 ± 0.21% per year for period two). Further study of the differences in the long term behavior of total ozone derived from SBUV and TOMS is provided through direct comparisons of the two elsewhere in these proceedings [Wellemeyer et al., 1988]. Internal studies of the radiances measured by SBUV and TOMS, also presented in these proceedings [Bhartia and Taylor, 1988], strongly suggest that the drift of SBUV and TOMS relative to the Dobson Network is largely due to a drift in the SBUV/TOMS calibration. We recommend the continued use of drift parameters already published for correction of the long-term total ozone data record from SBUV and TOMS.

Shown in Table 1 are the correlation coefficients of the SBUV and TOMS total ozone measured by each of the 41 stations used in the network comparisons. There is a systematic lowering of the correlations with the equatorial stations probably due to lower amplitude of the seasonal variation in total ozone at low latitudes. Figure 3 shows a sample time series of the total ozone measured by SBUV coincident with the Dobson measurements from the station Arosa which are marked "+". Individual station regression results are presented elsewhere [Fleig et al., 1986a] and [Fleig et al., 1988].

SBUV PROFILE COMPARISONS

Eight years of data from the SBUV were colocated in time and space with all of the Umkehr station and

Table 1. Correlation coefficients of total ozone measured by Dobson Stations with total ozone measured by SBUV and TOMS for the eight year period from 11/78 - 10/86.

STATION NAME	LAT	SBUV-DOBSON # OF COIN	SBUV-DOBSON CORR COEFF	TOMS-DOBSON # OF COIN	TOMS-DOBSON CORR COEFF
Resolute	74.7	254	0.98	1628	0.98
Reykjavik	64.1	270	0.96	1849	0.96
Lerwick	60.1	254	0.95	1869	0.93
Oslo	59.9	219	0.96	1670	0.95
Churchill	58.8	351	0.92	2614	0.94
Aarhus	56.2	295	0.82	2248	0.82
Edmonton	53.6	336	0.97	2746	0.97
Goose	53.3	361	0.95	2772	0.95
Potsdam	52.4	295	0.95	2308	0.95
Belsk	51.8	290	0.97	2330	0.96
Bracknell	51.4	256	0.90	2044	0.91
Uccle	50.8	260	0.97	1937	0.98
Hradec Kralove	50.2	322	0.97	2594	0.96
Hohenpeissenberg	47.8	227	0.97	1732	0.97
Caribou	46.9	287	0.95	2231	0.95
Bismarck	46.8	315	0.96	2420	0.94
Arosa	46.8	273	0.98	2054	0.98
Toronto	43.8	252	0.93	1953	0.93
Sapporo	43.1	327	0.97	2535	0.97
Vigna Di Valle	42.1	282	0.96	2086	0.95
Boulder	40.0	232	0.95	1848	0.97
Shiangher	39.8	255	0.97	1983	0.98
Lisbon	38.8	188	0.95	1504	0.93
Wallops Island	37.9	280	0.94	2094	0.96
Nashville	36.3	266	0.91	2074	0.91
Tateno	36.1	348	0.96	2664	0.96
Srinagar	34.1	237	0.93	1812	0.92
Kagoshima	31.6	295	0.90	2338	0.91
Quetta	30.2	297	0.77	2248	0.80
Cairo	30.1	333	0.94	2551	0.96
New Delhi	28.6	276	0.91	2219	0.94
Naha/Kagamizu	26.2	322	0.93	2513	0.94
Varanasi	25.4	274	0.92	2120	0.92
Mauna Loa Obs.	19.5	231	0.90	1686	0.91
Poona	18.5	291	0.91	2174	0.91
Mahe, Seychelles	4.7	284	0.77	2116	0.73
Singapore	1.3	233	0.85	1767	0.81
Huancayo	-12.1	284	0.55	2206	0.63
Tuituila Island	-14.3	272	0.67	2009	0.64
St. Helena	-15.9	285	0.53	2195	0.53
Perth	-31.9	288	0.96	2183	0.96

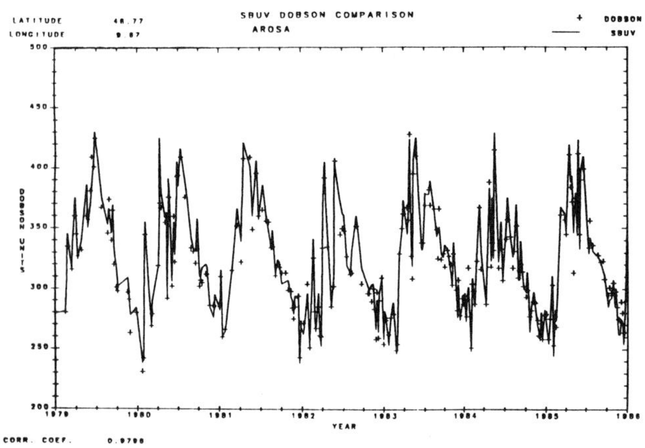

Fig. 3. Sample time series of coincident total ozone measurements from SBUV and the Dobson station at Arosa.

balloonsonde reports available as of July 1987 from the World Ozone Data Center. Criteria for matching were that SBUV and ground station data be from the same day and that the center of the SBUV field of view be within 1° of latitude of the station. Longitudinal separation was varied from 5° for Umkehr layers 1-4 up to a maximum of 10° for layers 5 and above. This variable window, a result of trial and error, represents the best compromise between data volume and comparison noise that results from large longitudinal separation.

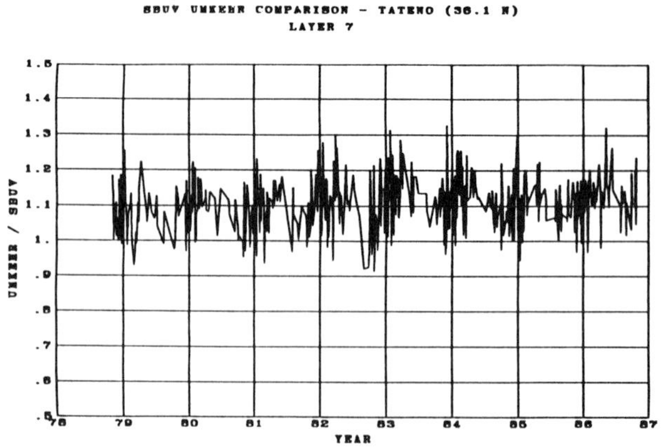

Fig. 4. Sample time series of the ratio of coincident measurements of ozone amount in Umkehr layer 7 (4mb - 8mb) from the ground station at Tateno and SBUV. Slight upward change in this ratio with time is indicative of a downward drift in SBUV ozone amount in layer 7 relative to the Umkehr measurement at Tateno.

The Umkehr data is strongly affected by the presence of aerosols, particularly after the eruption of El Chichon in April 1982. A correction for this effect has recently been developed [DeLuisi et al., 1988] which is in the form of monthly correction factors for stations in the latitude band 30 - 50 degrees north. From this latitude band, a group of four stations have been selected for providing the greatest temporal coverage over the seven-year period. The stations selected were Tateno (36.1 N), Arosa (46.8 N), Boulder (40.0 N) and Lisbon (38.8 N). A sample time series of the corrected Umkehr to SBUV ratio of the ozone amount in Umkehr layer 7 is shown in Figure 4. A seasonal dependence is evident. It is due, mainly, to the temperature dependence in the ozone absorption coefficients which is not taken into account in the Umkehr retrieval [WMO/NASA Ozone Trends Panel Report, 1988]. There are also two transient phenomena in the ratio after the eruption of El Chichon which appear to be associated with the temporal and spatial resolution of the Umkehr correction. Immediately after the eruption the monthly average does not have enough temporal resolution to track the rapid changes in aerosol amounts. Around the end of 1982 there is a slower transient caused by differences in the latitudinal motion of the aerosol cloud over Tateno alone and over the 30 - 50 degree latitude band taken as a whole.

Because of these considerations the first year after the eruption was excluded from the regression analysis performed.

The Balloonsonde Network is comprised of eight stations which were also selected for giving the best temporal coverage over the time period. These stations are Edmonton (53.6 N), Resolute (74.7 N), Goose Bay (53.5 N), Churchill (58.8 N), Hohenpeissenberg (47.8 N), Lindenberg (52.2 N), Legionowo (52.4 N) and Prague (50.2 N). These data are complete through only seven years.

A standard regression model has been used in all of the drift evaluations presented here. In order to remove any residual seasonal variation in the bias and derive the drift between the long term trend of two time series, a linear regression model has been fit to the percent difference SBUV - external data with the addition of annual and semi-annual terms. The overall drift of SBUV relative to a given system is then computed as the mean of the slopes of the individual regressions for each station in the system. A 95% confidence limit is computed based on the standard error of this mean. The results of this method applied to the balloonsonde and Umkehr data are given in Table 2. The Umkehr network used for this comparison had only four stations, and one station had half the number of comparisons as did the other three. A second method of comparison was calculated by finding the regression against all of the Umkehr soundings treated as coming from a single system (Table 2, column 4). This effectively reduced the impact of the Lisbon Umkehr data. The 95% confidence limit in this regression is considerably lower in the upper layers than the one based on the standard error of the network mean.

Table 2. Estimates of long-term drift of layer ozone amounts measured by SBUV relative to other measurement systems expressed in percent per year (95% confidence limits).

Layer # Range mb	SBUV-Balloon	SBUV-Umkehr Standard	SBUV-Umkehr Single Regression	SBUV-SAGE II
9 1-2	-	-0.7 (1.7)	-0.4 (0.4)	-2.5 (1.2)
8 2-4	-	-0.9 (0.5)	-0.8 (0.2)	-1.3 (0.8)
7 4-8	-	-0.5 (0.2)	-0.5 (0.2)	-1.1 (0.7)
6 8-16	-0.7 (1.0)	-0.6 (0.2)	-0.6 (0.1)	-1.6 (0.9)
5 16-32	-0.4 (0.4)	-0.2 (0.3)	-0.3 (0.1)	+2.0 (1.4)
4 32-64	+0.4 (0.7)	0.5 (0.5)	+0.4 (0.2)	+0.5 (1.4)
3 64-125	+0.4 (1.0)	-0.6 (0.6)	-0.6 (0.4)	-
2 125-250	-1.7 (0.9)	-3.9 (1.4)	-3.9 (0.9)	-
1 250-1000	-3.0 (1.9)	-0.8 (3.3)	-0.6 (1.1)	-

Also shown in Table 2 are the results of an SBUV SAGE II comparison. The first two years of SAGE II data (10/24/84 - 11/30/86) were received from the National Space Science Data Center (NSSDC) in Greenbelt, Maryland. These data were analyzed in a fashion analogous to the ground based networks except that instead of individual stations 10 degree latitude zones were considered. The coincidence criteria were the same as above except that the time coincidence was restricted to the nearest half day to avoid incorrect coincidences near the 180th meridian. The same regression model described above was used to derive the drift of SBUV relative to SAGE II in each zone, and the overall drift is the mean across the latitude bands with the 95% confidence limits based on the standard error of the mean.

Figure 5 shows a sample time series of ozone amount in layer 3 measured by SBUV and balloonsonde at Hohenpeissenberg (47.5N). It is evident in this plot that the SBUV system provides ozone information in layer 3. Three stages in the development of that information are represented in the graph. The dashed line is a functional representation of the ozone climatology [McPeters et al. 1984] in layer 3 in the middle latitude region (labelled "climatological profile"). The dotted line is the a priori layer 3 ozone

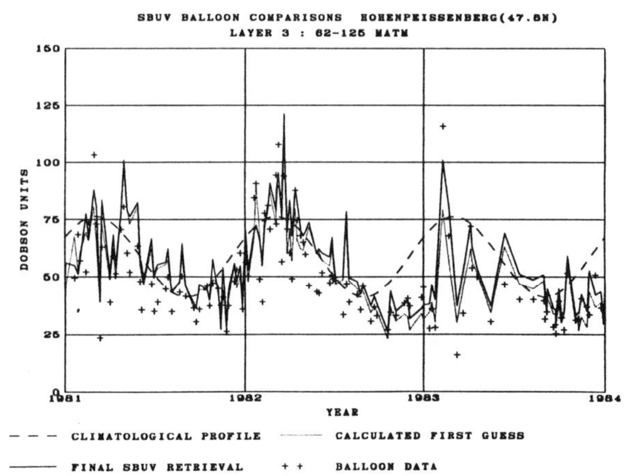

Fig. 5. Sample time series of coincident measurements of ozone amount in Umkehr layer 3 (64 mb - 125 mb) from SBUV and balloonsondes at Hohenpeissenberg. The initial first guess, the final first guess (adjusted by SBUV total ozone measurement) and the final SBUV retrieval illustrate the development of information in the SBUV profile retrieval in the lower layers.

in the sum of layers 2-4 for individual coincidences and monthly averages respectively. The relatively low correlation in layers 5 and 6 is largely due to the low amount of variability in those layers. In fact, the climatology in layer 5 is a straight line so that no correlation is present in the initial first guess. The correlation is low in layer 1 because the SBUV radiances contain little or no information in this layer, and the temporal variations are unpredictable.

Table 4 shows the correlation coefficients of the SBUV layer ozone amounts with the Umkehr measurements from each of the four stations used in the drift determinations. Lisbon is seen to give generally lower correlations than the other three stations. Again, lower correlation coefficients in layer 5 are due to the weaker seasonal signal in that layer. The lower correlations at layers 8 and 9 reflect increased noise in the amount which is an adjustment of the climatology based on the SBUV total ozone measurement (labelled "calculated first guess"). The solid line is the final SBUV retrieval in layer 3, and the "+" mark the balloonsonde measurements. In addition to the albedo measured by SBUV, information about the climatological variability of ozone in layer 3 relative to its neighbors is used to determine the final amount [Klenk et al. 1983]. That information is not necessarily appropriate for the determination of long term trend [WMO/NASA Ozone Trends Panel Report, 1988] since long term changes in ozone may not occur in the same vertical distribution as the shorter term variability represented in the climatology, but it is useful in tracking short term changes in the lower layers. Tables 3a and 3b show the correlation coefficients of the three stages of SBUV information shown in Figure 5 with the Hohenpeissenberg balloonsondes in each of layers 1-6 and

Table 3a. Correlation coefficients of layer ozone amounts measured by balloonsonde at Hohenpeissenberg (47.8 N) with layer ozone amounts from various stages of the SBUV retrieval.

Layer No. Range (mb)	Climatological Profile	Calculated First Guess	Final Retrieval
1 250-1000	0.59	0.18	0.25
2 125-250	0.66	0.82	0.85
3 64-125	0.60	0.83	0.85
4 32-64	0.69	0.77	0.81
Sum of 2-4	0.97	0.99	0.99
5 16-32	-	0.30	0.54
6 8-16	0.29	0.29	0.42

Table 3b. Correlation coefficients of monthly averages of layer ozone amounts measured by balloonsonde at Hohenpeissenberg (47.8 N) with layer ozone amounts from various stages of the SBUV retrieval.

Layer No. Range (mb)	Climatological Profile	Calculated First Guess	Final Retrieval
1 250-1000	0.74	0.19	0.30
2 125-250	0.81	0.88	0.90
3 64-125	0.76	0.88	0.91
4 32-64	0.80	0.85	0.90
Sum of 2-4	0.82	0.93	0.96
5 16-32	-	0.58	0.81
6 8-16	0.73	0.76	0.85

Table 4. Correlation coefficients of Umkehr layer ozone amounts measured by Umkehr and SBUV and comparison population for the eight year period from 11/78 - 10/86.

Umkehr Layer	Tateno Pop.	Tateno Coef.	Arosa Pop.	Arosa Coef.	Boulder Pop.	Boulder Coef.	Lisbon Pop.	Lisbon Coef.
1	287	0.53	346	0.49	257	0.44	140	0.34
2	287	0.79	346	0.83	257	0.80	140	0.73
3	287	0.87	346	0.85	257	0.87	140	0.78
4	287	0.83	346	0.76	257	0.66	140	0.61
5	574	0.44	687	0.48	506	0.39	291	0.16
6	574	0.87	687	0.83	506	0.87	291	0.55
7	574	0.82	687	0.75	506	0.81	291	0.69
8	574	0.57	687	0.59	506	0.58	291	0.25
9	574	0.60	687	0.60	506	0.66	291	0.12

comparison apparent in the uncertainties given in Table 2 and due largely to increased noise in the Umkehr retrieval in the upper layers [Klenk, et al., 1983].

CONCLUSIONS

The long-term drift identified in the SBUV [Fleig et al., 1986] and TOMS [Fleig et al., 1988] total ozone relative to the Dobson Network persists in this extended analysis of eight years of data. Drift parameters cited in the earlier works are still appropriate for adjustment of the long term drift in SBUV and TOMS. They are -0.38 ± 0.13% per year for SBUV and -0.25 ± 0.17% per year for the period 11/78 - 6/82 and -0.51 ± 0.21% per year for the period 7/82 - 10/86 for TOMS. Other work presented in these proceedings [Bhartia et al., 1988] indicates that this drift is largely due to a drift in the long-term calibration of the SBUV/TOMS system. In light of this, it is reasonable to assume that a drift is also present in upper level ozone profile measurement of SBUV. Comparisons with Umkehr and SAGE II presented here indicate that such a drift is present in layer 6 and above, but unexplained differences between

Umkehr and SAGE II make it difficult to quantify this drift.

ACKNOWLEDGMENTS

We wish to thank J.D. Augustine for his help in data analysis, and N.L. Dudley and A.E. Suhari for the final preparation of this manuscript.

REFERENCES

Bhartia, P.K., S.L. Taylor, A.J. Fleig, 1988: Estimation of Errors in the Long-term Change in Ozone Reported by the SBUV/TOMS Instruments - A New Way of Looking at Old Data, these proceedings.

Bhartia, P.K., D. Silberstein, and B. Monosmith, 1984: "Standard profiles of Ozone from Ground to 60 km Obtained by Combining Satellite and Ground Based Measurements" Atmospheric Ozone: Proceedings of the Quadrennial Ozone Symposium held in Halkidiki, Greece, 3-7 September 1984, edited by C.S. Zerefos, and A. Ghazi, pp. 243-247, D. Reidel, Dordrecht Holland.

Cleveland, W.S., 1979: Robust Locally Weighted Regression and Smoothing Scatterplots, J. Amer. Stat. Assoc., 74, 829.

DeLuisi, J.J., D.U. Longenecker, C.L. Mateer, 1988: An Objective Method of Correcting Umkehr Measurements for Volcanic Aerosol Error, these proceedings.

Fleig, A.J., C.G. Wellemeyer, D.S. Silberstein, R.P. Cebula and P.K. Bhartia, 1988: An Assessment of the Long-Term Drift in TOMS Total Ozone Data, Based on Comparison with the Dobson Network, Geophys. Res. Lett., Accepted for publication.

Fleig, A.J., P.K. Bhartia, and D.S. Silberstein, 1986: An Assessment of the Long-Term Drift in SBUV Total Ozone Data, Based on Comparison with the Dobson Network, Geophys. Res. Lett., 13, 1359.

ON THE CONTINUING DISAGREEMENT BETWEEN *IN SITU* AND SATELLITE STRATOSPHERIC OZONE MEASUREMENTS

E.M. Weinstock and J.G. Anderson
Harvard University
Cambridge, Massachusetts 02138, USA

ABSTRACT

The results of the 1984 "BOIC" ozone campaign have clearly shown that properly designed *in situ* ozone photometers do not suffer from "wall loss" and can measure ozone with an accuracy of 5%, even up to 42 km. Accordingly, we utilize a set of *in situ* ozone profiles (mostly represented by recent UV absorption data) to investigate differences with satellite profiles (mostly represented by SBUV data). Results clearly indicate that the satellite data is systematically high above 35 km and the divergence increases with altitude. Using 1-D model calculations and *in situ* data, we provide evidence that modeled ozone at 40 km is about 10 to 15% low, and not 40 to 60% low as has been indicated by other comparisons.

1. INTRODUCTION

Within recent months, based on the most recent conclusions of the 1987 Airborne Antarctic Ozone Expedition and the report of the Ozone Trends Panel, there is a clear and present danger to the ozone layer, with fluorocarbons as the indicted culprit, at least with regard to the ozone loss within the Antarctic polar vortex. The inability of current state of the art photochemical models to completely account for the ozone loss in the stratosphere must again call our attention to the inconsistencies which exist not only between modeled and measured ozone values around 40 km but among the measurements themselves.

The current global data base for stratospheric ozone above 25 km is made up almost entirely of satellite measurements using a solar backscattered ultraviolet instrument (SBUV). While innumerable intercomparisons and validations have been carried out for satellite ozone instruments, systematic disagreements have been found between the *in situ* and SBUV instruments. An early notable example of the divergence of the satellite and *in situ* data occurred on 6 September 1979 on the SABE-2 (Solar Absorption Balloon Experiment) intercomparison in which both the NASA-JSC and NASA Goddard "Dasibi" photometers measured lower ozone than the remote and SBUV instruments for altitudes above 35 km (see Hudson *et al.*, 1981). The shortfall in the photometer data was quickly accounted for by a pressure-dependent wall loss mechanism quantified by Ainsworth *et al.* (1981).

For many years since that time any similar disparity between *in situ* and remote or satellite ozone instruments has been blithely attributed to wall loss. However, the published results of the 1984 balloon ozone intercomparison campaign (BOIC) (Hilsenrath *et al.*, 1986) clearly and irrefutably show that on three independently designed *in situ* photometers there was no evidence of wall loss, even at 42 km. Accordingly, differences between *in situ* and remote measurements can no longer be attributed to wall loss. Additionally, these instruments have demonstrated accuracies of better than 5% up to 42 km.

The second inconsistency concerns the apparent discrepancy at and around 40 km between modeled and measured ozone. This problem is thoroughly analyzed by Froidevaux (1984) and Froidevaux *et al.* (1985) using a 1-D photochemical model with updated chemistry. One of the key conclusions of these publications is that steady state ozone concentrations in the stratosphere are up to 60% smaller than mean observed concentrations. However, the data used for comparison is taken from the IORI 1979 rocket intercomparison and the U.S. Standard Atmosphere. While the apparent agreement between the model and an early NASA JSC Dasibi profile is appropriately dismissed because of the erroneous absorption cell pressure measurement during that flight, the authors neglect to consider the results of more recent *in situ* measurements. They generalize that ozone measurements by *in situ* photometers cannot be trusted above 35 km and therefore are not to be used for meaningful comparisons with model calculations. This approach has a serious impact on their conclusions.

In a recent paper (Weinstock *et al.*, 1986) our 1984 *in situ* ozone data were compared with Harvard 1-D model calculations in order to reinvestigate the 40 km ozone problem. It was concluded that calculated ozone in the 35 to 40 km region is about 10 to 15% low, not 40 to 60%.

Within the same time frame of the publication of the BOIC results and our conclusions using *in situ* data, many authors have continued to ignore the existence of *in situ* profiles. For example, tables of reference ozone profiles for altitudes above 25 km derived solely from satellite data were published (Keating and Young, 1985). Additionally, in the three volume WMO Report No. 16, (1986) there are two separate chapters which address the difference between modeled and measured ozone in the stratosphere. In the chapter on oxygen species, the experimental profiles

OZONE IN THE ATMOSPHERE
Rumen D. Bojkov and Peter Fabian (Eds.)

ISBN 0-937194-15-8

are in fact satellite data composites above 25 km. In the chapter assessing models, the data used are represented as a range of observations taken from a previous publication (see Crutzen and Schmalzil, 1983, and references therein). They include a composite of all profiles from the June 1981 ozone intercomparison in Gap, France, as well as profiles determined from backscattered ultraviolet spectrometers flown on the Ogo and Nimbus-4 satellites. No attempt was made to assess the quality of any of the individual profiles. Data derived from a variety of independent experiments and presented as a mixing ratio range at specific altitudes (as was done in fig. 12-46 of the WMO Report No. 16) can be very misleading even if only one erroneous data set is included. It therefore appears necessary to carefully evaluate the systematic differences between *in situ* and remote ozone measurements in the stratosphere and any conclusions which might be drawn from this study.

In this paper we will investigate the 40 km ozone problem in a systematic way by:

1. Careful evaluation of systematic differences between *in situ* (mostly represented by recent UV absorption photometer data) and remote (mostly represented by SBUV data) ozone profiles.
2. Compilation of recent *in situ* ozone profiles to use as a data base for comparison with published model profiles.
3. Use of previously published results (Weinstock *et al.*, 1986) of the Harvard 1-D photochemical model to vary key chemical species which affect stratospheric ozone, and comparison of the range of these results with *in situ* stratospheric ozone measurements.

2. DATA ANALYSIS

Within the past several years there have been numerous attempts to provide "ground truth" for satellite data. In most, if not all cases, these intercomparisons have verified that the satellite measurements (be it SBUV, LIMS, or SAGE) typically agree to the 10 - 20% level (see Chapter 8, WMO Report No. 16, 1986, and references therein). The limitation of these intercomparisons is the lack of an absolute standard with which to compare. However, following the conclusions of BOIC regarding *in situ* UV absorption photometers, we can take advantage of a series of six balloon-borne, *in situ* measurements which were launched from Palestine, Texas, and either by design or chance have coincided with SBUV overflights. Table 1 summarizes the relevant information for the comparison of the *in situ* and satellite data. Only in the case of the 7 June 1981 flight is the intercomparison questionable because of the one day separating the launch and the overflight. Nevertheless, we include this flight, which shows the smallest difference between the two methods, for completeness.

Figure 1 shows the percent difference between the two methods as a function of altitude, while fig. 2 illustrates the average difference for these six flights. While the spread pictured in fig. 1 is fairly large, it cannot diminish the significance of the systematic trend in the data. Furthermore, this systematic difference between SBUV and *in situ* data strongly suggests the existence of systematic errors in the published ozone profiles often used for model intercomparisons. This, for example, would have a strong effect on the interpretation of the comparison between modeled and measured ozone, as depicted in Chapter 8 of the WMO report (1986) on atmospheric ozone.

While the paper so far has dealt exclusively with SBUV data, it would be worthwhile to investigate any comparable validations of the SAGE ozone data. This is especially important because of the report of the recent intercomparisons of ROCOZ-A and the SAGE instruments (McCormick *et al.*, 1987) which presents data showing excellent agreement from 20 to 53 km. However, ROCOZ-A has measured high when compared to *in situ* instruments both during BOIC (Hilsenrath *et al.*, 1985) and during the 1981 intercomparison in Gap, France (Aimedieu *et al.*, 1983) when an earlier version of the instrument was

TABLE 1. COMPILATION OF *IN SITU* VERSUS SBUV COMPARISON. ALL *IN SITU* DATA TAKEN FROM LAUNCHES FROM PALESTINE, TEXAS, 31.80°N LATITUDE, -95.72° LONGITUDE.

Date	*In Situ* Instrument	Information Relevant to SBUV Data		
		Date	Lat.(°N)	Longitude
7 June 1981	Harvard UV Absorp. Photom.	8 June 1981	32.7	-89.7
22 Sept. 1982	NASA JSC UV Absorp. Photom.	Profile taken from Robbins *et al.* (1988)		
18 July 1983	U. Minn. Mass Spectrometer	Profile taken from Mauersberger *et al.* (1984)		
21 March 1984	Harvard UV Absorp. Photom.	21 March 1984	30.3	-77.6
			32.1	-78.2
			31.5	-104.0
			33.3	-104.6
25 March 1984	NASA JSC UV Absorp. Photom.	Profile taken from Hilsenrath *et al.* (1986)		
14 Sept. 1984	Harvard UV Absorp. Photom.	14 Sept. 1984	31.9	-83.9
			32.2	-109.5

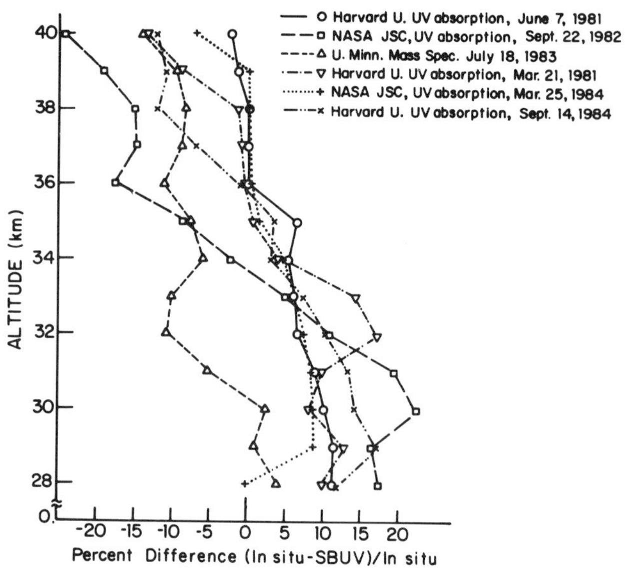

Fig. 1. The percent difference between six *in situ* ozone measurements and SBUV overflight data from 28 to 40 km.

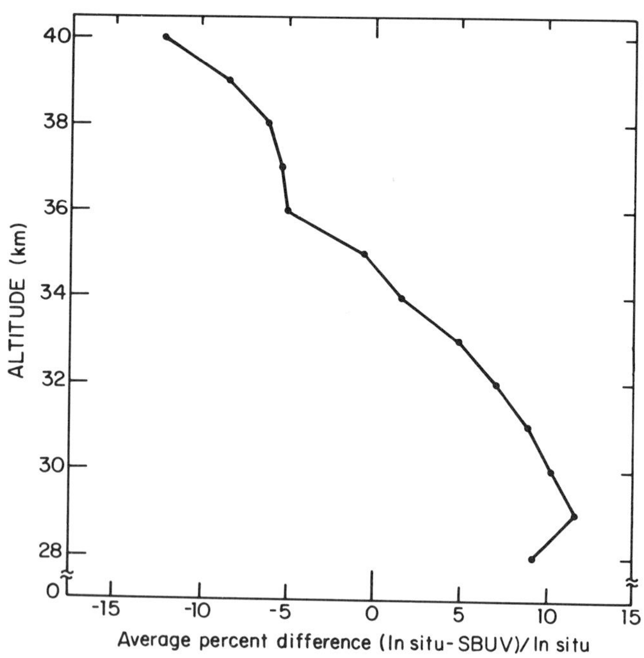

Fig. 2. The average percent difference between six *in situ* ozone measurements and SBUV overflight data from Fig. 1.

used. Also, on 6 April 1979 an *in situ* UV absorption measurement was made on an instrument launched from Palestine, Texas which coincided with a SAGE overflight. The UV absorption data are taken from a LIMS/Dasibi ozone intercomparison (Remsberg *et al.*, 1984) while the SAGE profile is from a SAGE/ECC intercomparison (McCormick *et al.*, 1984). Figure 3 reproduces the SAGE and *in situ* UV profiles and illustrates the significant difference with the two measurements above 28 km. While this is only a single check for the two methods, it is consistent with both the intercomparison of the ROCOZ-A and *in situ* UV instruments and the comparison of SAGE and SBUV data (Cunnold *et al.*, 1984) which, during March and April 1979, shows the zonal mean ozone for 30°N latitude in the 2 to 5 mbar range measured by SAGE to be about 5 to 15% higher than that measured by SBUV. With the use of updated ozone cross sections,the SBUV profiles should be lowered and this difference would be even larger. Additionally, the recent reevaluation of the SAGE data for the Ozone Trends Panel Study might affect these conclusions as well as the SAGE profile in fig. 3, to some extent.

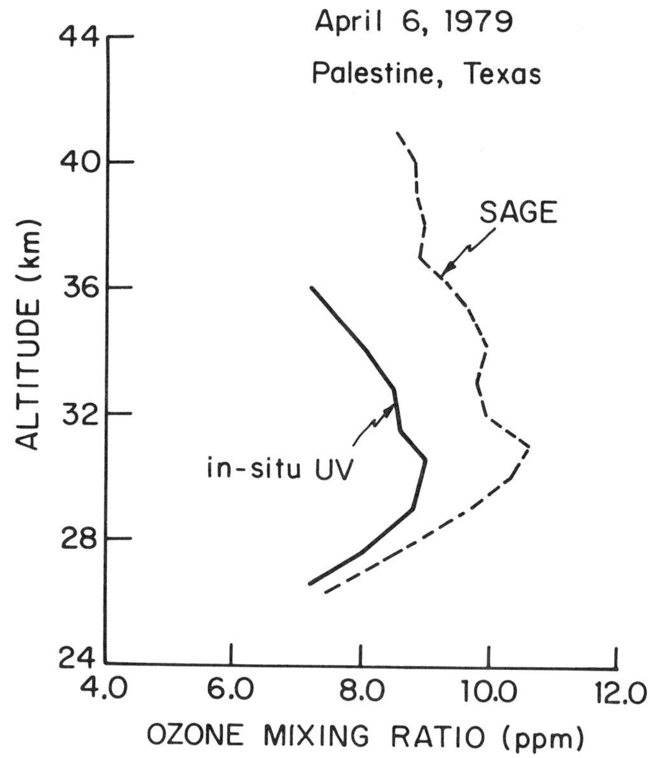

Fig. 3. Comparison of an *in situ* UV absorption profile measured by the NASA JSC "DASIBI" ozone sensor and SAGE overflight data.

Figure 4 shows a comparison between the July 30°N latitude profile used in that figure along with the comparable profile from the Handbook from MAP, Volume 16 (Keating and Young, 1985), and the average of the six *in situ* profiles from Table 1. This figure clearly illustrates the bias in the different measurement techniques.

In any case, we are interested in determining how an averaged *in situ* profile compares with modeled profiles which can provide the flexibility of covering the range of conditions appropriate for the times the measurements were made. We therefore calculate ozone for the extremes of observed stratospheric NO_x and methane mixing ratios with the single constraint of holding the sum of the water

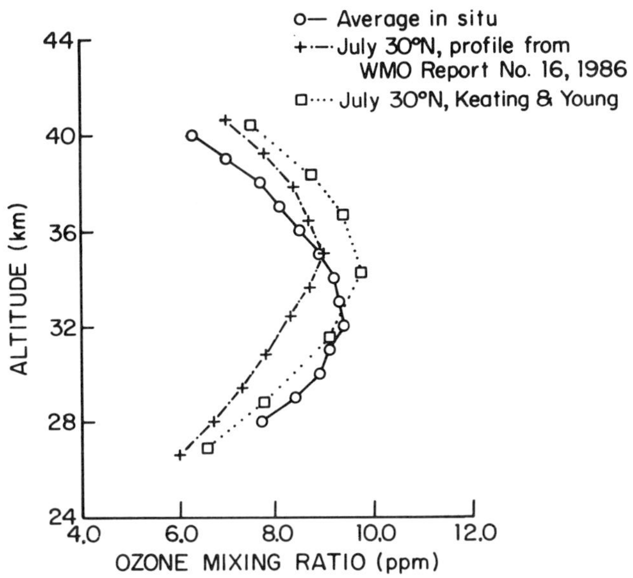

Fig. 4. Comparison of average *in situ* ozone data with July 30°N profiles from the WMO Report No. 16 and the profile published in the MAP Handbook 16 by Keating and Young (1985).

and twice the methane mixing ratios to 6.5 ppm. Figure 5 compares the average *in situ* profile with the same Harvard 1-D model results, as used in our previous paper. This comparison shows again that even with the high NO_x mixing ratios consistent with the LIMS satellite data, the modeled profile is only about 11% low. If, on the other hand, this modeled profile is compared with the average of the six SBUV profiles, it is 21% low. Furthermore, based on the shape of fig. 2, the divergence increases with altitude above 35 km and indicates that the use of satellite based ozone profiles exaggerates the problem in the upper stratosphere.

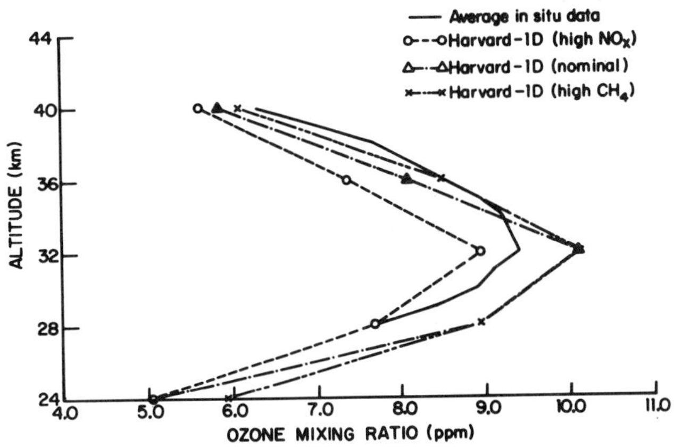

Fig. 5. Comparison of the average *in situ* profile with previously published results of the Harvard 1-D model.

3. CONCLUSIONS

Table 2 summarizes our results by comparing the lowest modeled ozone profile with our average *in situ* profile, the corresponding average SBUV profile, the July 30°N profile from fig. 1, and the comparable Keating and Young (1985) profile. The results clearly indicate that, depending on the experimental data used to evaluate a photochemical model, calculated ozone at 40 km can appear to be 27% low. In fact, if one were to take the percent difference referenced to the modeled ozone mixing ratio, the difference would appear to be 37%. In any case, the data base might be the major factor causing the large discrepancy. In the case of the Keating and Young profile, it is not clear what contributing satellite data set causes the disagreement to be worse than in other cases. A larger

TABLE 2. PERCENT DIFFERENCE BETWEEN MODELED OZONE (HARVARD 1-D HIGH NO_X) AND THREE REFERENCE PROFILES DERIVED FROM EXPERIMENTAL DATA. NUMBERS ARE 100 * (EXP-THEORY)/EXP.

Altitude	*In Situ*	July 30°N WMO Report No. 16	July 30°N Keating and Young	SBUV
40	11.1	23.7	27.1	20.9
36	13.4	16.4	22.8	17.0
32	4.9	-7.7	4.9	2.9
28	0	-23.7	-5.3	-10.6

data base of reliable *in situ* ozone data in the upper stratosphere would be very useful as a means of reinforcing the ideas presented here and clarifying this murky problem.

It is again worth taking note of the history of ozone measurements and the reluctance of the community to accept the accuracy of *in situ* data. The wall loss issue has been resolved (Hilsenrath *et al.*, 1986). For many free radical species, *in situ* data constitutes the data base used for comparisons with models. Many of these species (ClO, OH, and NO, to name a few) are much more difficult than ozone to measure accurately. *In situ* ozone data must be used in the same way.

REFERENCES

Aimedieu, P., A. J. Krueger, D. E. Robbins, and P. C. Simon, 1983: Ozone profile intercomparisons based on simultaneous observations between 20 and 40 km. *Planet. Space Sci.*, 31, 801-807.

Ainsworth, J. E., J. R. Hagemeyer, and E. I. Reed, 1981: Error in Dasibi flight measurements of atmospheric ozone due to wall-loss. *Geophys. Res. Lett.*, 8, 1071-1074.

Cunnold, D. M., M. C. Pitts, and C. R. Trepte, 1984: An intercomparison of SAGE and SBUV ozone observations for March April 1979. *J. Geophys. Res.*, 89, 5249-5262.

Crutzen, P. J., and U. Schmailzl, 1983: Chemical budgets of the stratosphere, *Planet. Space Sci.*, 31, 1009-1032.

Froidevaux, L., 1984: Photochemical modeling of Earth's stratosphere, Ph.D. thesis, Calif. Inst. of Techn., Pasadena, CA.

Froidevaux, L., M. Allen, and Y. L. Yung, 1985: A critical analysis of ClO and O_3 in the midlatitude stratosphere. *J. Geophys. Res.*, 90, 12999-13029.

Hilsenrath, E., J. Ainsworth, W. Attmanspacher, A. Bass, W. Evans, W. Komhyr, K. Mauersberger, A.J. Miller, M. Proffitt, D. Robbins, S. Taylor, and E. Weinstock, 1986: Results from the balloon ozone intercomparison campaign (BOIC). *J. Geophys. Res.*, 91, 13137-13152.

Keating, G. M., and D. F. Young, 1985: Interim reference ozone models for the middle atmosphere, in Handbook for MAP, Vol. 16, edited by K. Labitzke, J. J. Barnett, and B. Edwards, 205-229, SCOSTEP Secretariat, Univ. of Illinois, Urbana.

McCormick, M. P., T. J. Swissler, E. Hilsenrath, A. J. Krueger, and M. T. Osborne, 1984: Satellite and correlative measurements of stratospheric ozone: Comparison of measurements made by SAGE, ECC balloons, chemiluminescent and optical rocket sondes. *J. Geophys. Res.*, 89, 5315-5320.

McCormick, M. P., R. A. Barnes, and M. A. Chamberlain, 1987: A detailed comparison of SAGE II and ROCOZ-A ozone profiles at Natal, Brazil (abstract). *Eos Trans. AGU*, 68, 1232.

Remsberg, E. E., J. E. Russell III, J. C. Gille, L. L. Gordley, P. L. Bailey, W. G. Planet, and J. E. Harries, 1984: The validation of NIMBUS 7 LIMS measurements of ozone, *J. Geophys. Res.*, 89, 5161-5178.

Robbins, D., J. Waters, P. Zimmerman, R. Jarnot, S. Pollitt, W. Traub, K. Chance, N. Louisnard, W. Evans, and J. Kerr, 1988: BIC ozone intercomparison, accepted for publication in *J. Atmos. Chem.*

Weinstock, E. M., C. M. Schiller, and J. G. Anderson, 1986: In situ stratospheric ozone measurements by long path UV absorption: Developments and interpretation. *J. Geophys. Res.*, 91, 5237-5248.

World Meteorological Organization, 1986: Atmospheric Ozone 1985, Global Research & Monitoring Project, Report No. 16, 401-440.

THE RELATIONSHIP BETWEEN 1982–1986 TRENDS IN UPPER STRATOSPHERIC OZONE AND TEMPERATURES

R. Todd Clancy and David W. Rusch
Laboratory for Atmospheric and Space Physics
University of Colorado
Boulder, Colorado

ABSTRACT

Trends for upper stratospheric temperatures and ozone are calculated from the NMC and SBUV data sets, respectively, over the 1982-1986 period. Latitudinal and longitudinal variations in ozone and temperature trends are found to be anticorrelated. The observed temperature-ozone trends relationship appears consistent with ozone sensitivity to temperature dependent kinetic rate coefficients for ozone formation and destruction. The absolute SBUV ozone trends include a constant 4%/yr decrease at 1.0 mbar, most of which may be due to a calibration drift in the SBUV experiment. The absolute NMC temperature trends indicate ~0.5K/yr global decreases in upper stratospheric temperatures over the 1982-1986 period. Such temperature trends would force ~0.5%/yr global increases in upper stratospheric ozone over the same period, based upon the derived relationship between ozone and temperature trend anticorrelations.

1. INTRODUCTION

A number of recent observations of stratospheric ozone trends over the 1979-1986 time period have suggested surprisingly large ozone decreases. In the case of the Antarctic ozone hole, the observed lower stratospheric ozone decreases in austral springtime are far beyond observational uncertainties (e.g., Farman et al., 1985) and appear to indicate heterogeneous photochemical processes unique to the polar winter-spring tropopause region (e.g. Solomon et al., 1986). The observed trend in the extent and depth of the Antarctic ozone minimum is most likely associated with ClOx trends (Solomon et al., 1986) and the effect of stratospheric temperature trends, either through a dynamical relationship with vertical advection (Rosenfield et al., 1988) or through a thermodynamic influence on the formation of polar stratospheric clouds (Sekiguchi, 1986). Possible extensions of winter-spring ozone decreases over the 1979-1986 period have been suggested for the Antarctic upper stratosphere (Aikin and McPeters, 1988), and for the column abundance of ozone in the entire southern hemisphere (Schoerberl, 1986). The inferred amplitudes of these trends are quite small (-1 to -2%/yr) and they remain uncertain. However they significantly exceed predicted ozone trends, and they are not easily explained by the heterogeneous chemistry postulated to explain the Antarctic ozone hole. Observations from the Solar Backscatter Ultraviolet (SBUV) experiment show large negative trends (-3 to -5 %/yr), in global upper stratospheric ozone beginning in 1982, which have been shown to be largely due to instrument degradation [The Ozone Trends Panel Report (OTPR), 1988]. However, it is apparent that significant (1-2 %/yr) latitudinal dependences in the SBUV upper stratospheric ozone trends are related to 1-2K/yr trends in upper stratospheric temperature over the 1982-1986 period (Rusch and Clancy, 1988).

In the following analysis we present comparisons of SBUV 1-5 mbar ozone trends and NMC (National Meteorological Center), (Gelman, et al., 1986) 1-5 mbar temperature trends in order to determine to what degree stratospheric temperature trends over the 1982-1986 period may lead to 2-5% decreases in upper stratospheric ozone over the same period. The inferred temperature-ozone relationship over the relatively long time period of 5 years is compared to the simple photochemical equilibrium relationship between ozone and temperature due to the temperature dependence of kinetic rate coefficients for ozone formation and destruction.

2. OZONE AND TEMPERATURE TRENDS BETWEEN 1982 AND 1986

We have adopted the satellite data bases from SBUV and NMC for analysis of ozone and temperature trends, respectively, over the 1982-1986 period. The 1979-1981 period is excluded from the analysis on the basis of a distinct separation in the behavior of the SBUV calibration between the 1979-1981 and 1982-1986 periods (OTPR). Linear least-squares fits to monthly averages of ozone and temperature between 1982 and 1986 are calculated to determine trends as a function of season (June, September, December, and March), latitude, longitude, and atmospheric pressure (1,2, and 5 mbar). Errors are determined from the formal error analysis of the linear least-squares fitting. The generally large error limits indicate large interannual variations in monthly average temperatures and ozone over the 1982-1986 period.

In figure 1 we present temperature (solid line) and ozone (dotted line) trends for the month of June, at the 1.0 mbar pressure level, at 45 S latitude as a function of longitude. Several aspects of the trends in figure 1 are representative of temperature/ozone trends determined in our analysis. SBUV ozone trends in the 1-5 mbar pressure region exhibit a fairly large constant offset (-3 to -4 %/yr) with significant (1-2 %/yr) longitudinal and latitudinal variations over the 1982-1986 period. The constant offset in the SBUV ozone trends is assumed to be instrumental in nature. Stratospheric temperature trends of -1 to 2 K/yr are generally anticorrelated with the ozone trends in their lon-

ISBN 0-937194-15-8

gitudinal and latitudinal dependences. A correlation coefficient of -.86 is determined for the longitudinal variations of temperature and ozone trends of figure 1. The correlation coefficient (r) between 1.0 mbar temperature and ozone trends in December at 45 N latitude is -.67. Much smaller longitudinal variations of temperature and ozone trends are found for summer months, as shown in the June, 45N latitude results of figure 2. Even here, the very weak longitudinal variations of temperature and ozone trends are anticorrelated (r=-.52).

The latitudinal variations of zonally averaged temperature and ozone trends are also anticorrelated, as shown in figure 3 for the month of June at the 1.0 mbar level (r=-.60). The range of temperature and ozone trends is smaller, as might be expected for a zonal average. In table 1 we summarize the derived correlation coefficients between latitudinal variations of zonally averaged temperature and ozone trends as a function of month, and pressure. Temperature and ozone trends are generally anticorrelated; the maximum anticorrelations are found at the 2.0 mbar level, smaller anticorrelation coefficients are found at the 5.0 mbar level.

3. RELATIONSHIP BETWEEN OZONE AND TEMPERATURE TRENDS

The above anticorrelations of ozone and temperature trends are qualitatively similar to the highly studied anticorrelations of short term (t < month) ozone and temperature variations (e.g., Barnett et al., 1975). Under conditions of photochemical equilibrium (PCE), variations of ozone densities are driven by temperature variations through the temperature sensitivities of kinetic rate coefficients for ozone formation and destruction reactions (most notably, $O + O2 + M \rightarrow O3 + M$ and $O + HO2 \rightarrow OH + O2$). The effects of transport in non-PCE regions, such as winter mid-to-high latitudes, tend to reduce the temperature sensitivity of upper stratospheric ozone (e.g., Rood and Douglass, 1985). The maximum (i.e., PCE) temperature sensitivity of ozone is given by

$$dO_3\,(\%) = -\frac{\Theta}{T^2} \bullet dT\,(K) \bullet 100 \qquad (1)$$

where Θ, the constant of proportionality, is approximately 900-1200 K in the 1-3 mbar region (Froidevaux et al., 1988), and T is the atmospheric temperature in the 1-3 mbar region. A 1K temperature decrease produces approximately a 1.5 % increase in ozone.

We calculate Θ' from the relationship between the zonally averaged SBUV ozone and NMC temperature trends as

$$\Delta O_3\,(\%/yr) = C - \frac{\Theta'}{T^2} \bullet \Delta T\,(K/yr) \bullet 100 \qquad (2)$$

where C is a constant representing degradation of the SBUV instrumental calibration between 1982 and 1986. Tables 2 and 3 present values of C and Θ', respectively, calculated for the months and pressures of table 1. The relative constancy of C versus month (with the exception of December at the 1 and 2 mbar pressure levels) is consistent with the assumption of an instrumental drift in the SBUV calibration. The pressure dependence of C is also consistent with such a calibration drift (Barnes, 1988). However, the values of C must also reflect, to some degree, true global changes in ozone due to solar flux and Clx trends.

Values for Θ' and its one-sigma errors are presented in table 3 for months and pressures at which the

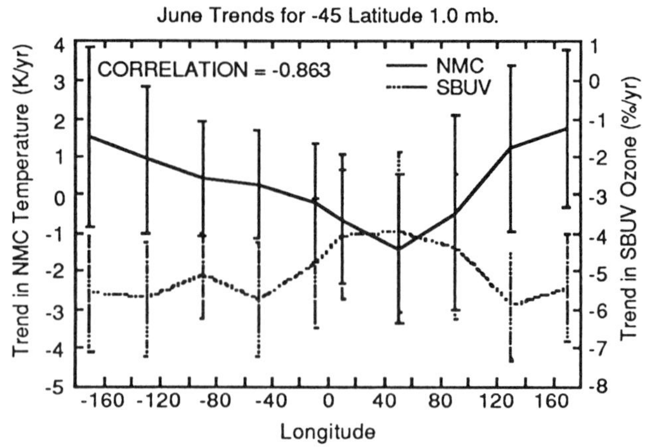

Figure 1. SBUV ozone and NMC temperature trends presented for the 1.0 mbar pressure level over the 1982-1986 period. Trends are calculated versus longitude for the June monthly averages at 45S latitude.

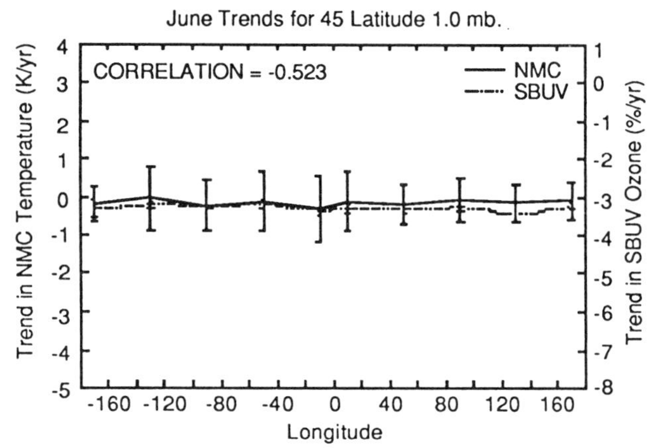

Figure 2. Same as Figure 1 for June trends at 45N latitude.

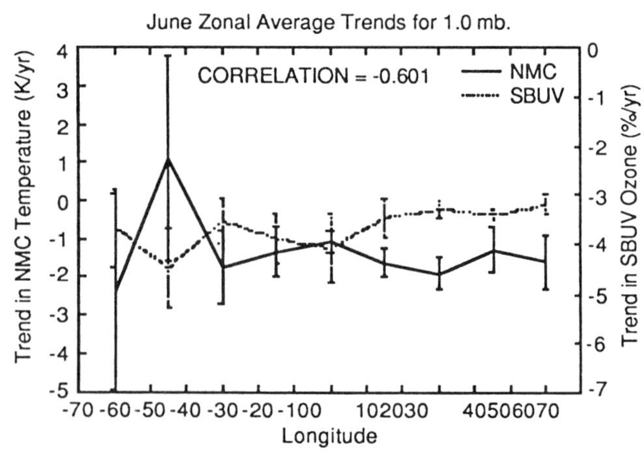

Figure 3. Zonally averaged trends of SBUV ozone and NMC temperatures versus latitude, for June monthly averages between 1982 and 1986 at the 1.0 mbar pressure level.

absolute value of the correlation coefficient between zonally averaged NMC temperatures and SBUV ozone is greater than 0.5 (see table 1). Note that $\Theta' < \Theta_{PCE}$ in all cases, and that Θ'/Θ_{PCE} is generally in the range of 1/2. The relationship between variations in stratospheric temperatures and ozone densities appears to be indicative of the temperature dependence of kinetic rate coefficients, for timescales of temperature and ozone variations as large as 5 years. However, transport related phenomena do appear to dampen the sensitivity of ozone trends to temperature trends. Previous empirical studies of short-term ozone temperature variations

TABLE 1. CORRELATION COEFFICIENTS BETWEEN LATITUDINAL VARIATIONS IN ZONALLY AVERAGED SBUV OZONE AND NMC TEMPERATURE TRENDS BETWEEN 1982 AND 1986.

Pressure (mbar)	December	March	June	September
1.0	.39	-.67	-.60	-.50
2.0	-.79	-.80	-.84	-.66
5.0	-.13	-.71	-.41	-.27

TABLE 2. C-CONSTANT OF SBUV OZONE TRENDS (PERCENT/YEAR).

Pressure (mbar)	December	March	June	September
1.0	-2.5 ± 0.5	-4.0 ± 0.2	-3.9 ± 0.2	-4.0 ± 0.1
2.0	-2.9 ± 0.1	-3.4 ± 0.2	-3.6 ± 0.1	-3.6 ± 0.2
5.0	-1.8 ± 0.2	-2.3 ± 0.2	-2.0 ± 0.2	-1.8 ± 0.1

TABLE 3. Q'-TEMPERATURE SENSITIVITY COEFFICIENT (K) OF OZONE TRENDS BETWEEN 1982 AND 1986.

Pressure (mbar)	December	March	June	September
1.0	—	640 ± 270	600 ± 300	—
2.0	350 ± 100	550 ± 150	820 ± 280	490 ± 200
5.0	—	430 ± 130	—	—

TABLE 4. LATITUDINALLY (60S-60N) AVERAGED NMC TEMPERATURE TRENDS (K/YR) BETWEEN 1982 AND 1986. STATISTICAL ERRORS ARE 0.1K, SYSTEMATIC ERRORS ARE NOT ESTIMATED.

Pressure (mbar)	December	March	June	September
1.0	-.7	-.6	-.3	-.3
2.0	-.3	-.4	-.3	-.4
5.0	-.7	-.7	-.7	-.9

yield values of Θ in the range of 700-1100 K for the 1-3 mbar pressure levels (e.g., Froidevaux et al., 1988). With the exception of the June 2 mbar value, Θ' in Table 3 fall below 800 K, by as much as 4 sigma. Analytical biases due to effects such as the SBUV calibration drift and trends in solar flux and Cl_x (which drive ~1%/yr ozone trends) are minimized by the numerical determination of Θ' from the latitudinal variations of ozone and temperature trends. The latitudinal dependences of ozone trends due to solar cycle and Cl_x trends and instrument biases are second order.

4. GLOBAL UPPER-STRATOSPHERIC TEMPERATURE TRENDS

Due to the presence of a calibration drift in the SBUV ozone data set, no accurate determination of global decreases in upper stratospheric ozone is possible from this study. It is, however, worth noting that the NMC temperature trends imply a general cooling of the upper stratosphere over the 1982-1986 period of 0.5 K/yr. We present NMC temperature trends averaged over the latitude interval 60S-60N versus pressure and month in Table 4.

5. CONCLUSIONS

Examination of SBUV ozone and NMC temperature trends in the upper stratosphere (1-5 mbar) over the 1982-1986 period indicates a strong anticorrelation between the variations of these trends in longitude and latitude. The derived relationship between temperature and ozone trends can be explained by the forcing of ozone trends through the temperature sensitivities of kinetic rate coefficients of reactions for ozone formation and destruction. Ozone trend sensitivities to temperature trends appear to be roughly ≥1/2 of the PCE temperature sensitivity of ozone. Absolute trends in ozone over the 1982-1986 period are not derived from the SBUV data set due to a calibration drift in the SBUV experiment. However, global decreases in upper stratospheric NMC temperatures of 2-3 K over the 1982-1986 period would contribute 2% increases in global upper stratospheric ozone over the same period, given our derived relationship between temperature and ozone trends. The existence of -0.5K/yr trends in upper stratospheric temperatures is not consistent with predicted greenhouse cooling rates of the upper stratosphere (due to long term O_3 decreases and CO_2 increases), which are less than 0.2K/yr. Nor are they consistent with large solar-cycle driven increases (20K) in upper stratospheric temperatures, as reported by Chanin et al. (1987). They are, however, comparable to predicted solar cycle variations of stratospheric temperatures. As the period of this analysis is quite short (5 years), the true timescale of the derived temperature trends is not apparent. In any event, ozone trend analyses must take into account stratospheric temperature trends, which may be quite large over 5-10 year timescales.

6. ACKNOWLEDGMENTS

The authors acknowledge computational support from Nancy Thalman and Karen Rosenlof and assistance in manuscript preparation from Ann Alfaro and Lynn Laubisch. This research was funded under NASA grant NASW-4091.

REFERENCES

Aikin, A.C., and R.D. McPeters, The three-dimensional morphology of the Antarctic ozone minimum, *Geophysical Res. Lett.*, **15**, 413-416, 1988.

Barnett, J.J., J.T. Houghton, and J.A. Pyle, The temperature dependence of the Ozone concentration near the stratopause, *Quart. J. R. Met. Soc.*, **101**, 245-257, 1975.

Barnes, R.A., Changes in SBUV ozone profiles near Natal, Brazil from 1979 to 1985, *J. Geophys. Res.*, **93**, 1704-1717, 1988.

Chanin, M.L., N. Smires, and A. Hauchecorne, Long-term variation of the middle atmosphere at mid-latitude: Dynamical and radiative causes, *J. Geophys. Res.*, **92**, 10, 933-10,941, 1987.

Farman, J.C., B.G. Gardiner, and J.D. Shanklin, Larger losses of total ozone in Antarctica reveal seasonal ClOx/NOx interaction, *Nature*, **315**, 207-210, 1985.

Froidevaux, L., M. Allen, S. Berman, and A. Daughton, The mean ozone profile and its temperature sensitivity in the upper stratosphere and lower mesosphere: An analysis of LIMS observations, *J. Geophys. Res.*, in press, 1988.

Gelman, M.E., A.J. Miller, K.W. Johnson, and R.M. Nagatani, Detection of long-term trends in global stratospheric temperature from MNC analyses derived from NOAA satellite data, *Ad. Sp. Res.*, **6**, 17, 1986.

Ozone Trends Panel Report, Copies of the Executive Summary available from Robert Watson at NASA Headquarters, Washington, DC, 20546.

Rood, R.B., and A.R. Douglass, Interpretation of ozone-temperature correlations, *J. Geophys. Res.*, **90**, 5733-5743, 1985.

Rosenfield, J.E., M.R. Schoerberl, and P.A. Newman, Antarctic-springtime ozone depletion computed from temperature observations, *J. Geophys. Res.*, **93**, 3833-3850, 1988.

Rusch, D.W., and R.T. Clancy, A comparison of ozone trends from SME and SBUV satellites and model calculations, *Geophys. Res. Lett.*, in press, 1988.

Schoerberl, M.R., A.J. Krueger, and P.A. Newman, The morphology of Antarctic total ozone as seen by TOMS, *Geophys. Res. Lett.*, **13**, 1217-1220, 1986.

Sekiguchi, Y., Antarctic ozone change correlated to the stratospheric temperature field, *Geophys. Res. Lett.*, **13**, 1202-1205, 1986.

Solomon, S., R.R. Garcia, F.S. Rowland, and D. Wuebbles, On the depletion of Antarctic ozone, *Nature*, **321**, 755-758, 1986.

TIME-PERIODIC VARIATIONS IN STRATOSPHERIC OZONE FROM SATELLITE OBSERVATIONS

Pi-Huan Wang[1], *M.P. McCormick*[2], *L.R. McMaster*[2], *S. Schaffner*[1], *and G.E. Woodbury*[2]

[1]Science and Technology Corporation, Hampton, Virginia 23666, USA

[2]Atmospheric Science Division, NASA Langley Research Center
Hampton, Virginia 23665, USA

ABSTRACT

The time-periodic variations in stratospheric zonal mean ozone number density are examined with emphasis on annual and semi-annual oscillations. The data base for this analysis is the ozone observation from the SAGE II satellite instrument. A multiple linear regression method is adopted for the analysis. The results show that the amplitudes and phases of the time-periodic ozone variations are functions of altitude and latitude.

1. INTRODUCTION

The time-periodic variations of atmospheric ozone in both total amount and vertical distribution have been studied by many investigators (WMO, 1986). Analyses on their variations have provided useful information toward the understanding of the behavior and controlling mechanism of the atmospheric ozone. In the case of time-periodic variations of the ozone profile, which is directly important to the atmospheric dynamical processes and the Earth's climate, the existing analyses are given in terms of ozone mixing ratio. Studies reveal that variations of total ozone, which is ultimately relevant to the biological structure at the Earth's surface, are highly correlated with ozone variations in the lower and middle stratospheres (i.e., Wang et al., 1983), and that in these regions transport processes play an important role in producing the observed changes in ozone concentration (WMO, 1986). Since the total ozone is related directly to the vertical profile of ozone number density, it is natural to extend the investigation of the ozone variations in terms of ozone number density. The purpose of this study is to examine the time-periodic variations, including annual, semi-annual, and quasi-biennial oscillations, based on the stratospheric ozone number density determined from observations by the satellite instrument of the Stratospheric Aerosol and Gas Experiment (SAGE II).

2. DATA AND APPROACH

The SAGE II satellite instrument is a seven-channel sun radiometer aboard the Earth Radiation Budget Satellite (ERBS) for measuring stratospheric aerosol, ozone, nitrogen dioxide, and water vapor with approximately a 1-km vertical resolution. The vertical distribution of ozone number density is derived from measurements of the attenuated solar intensity profile at a wavelength centered at .60 μm during the spacecraft sunrise and sunset. The latitudinal coverage is from approximately 79°S to 79°N depending on the season. It takes about 1½ months for the SAGE II sunrise/sunset measurements scanning from one latitudinal extreme to the other. The detailed aspects of the SAGE II observation have been described by McMaster (1986) and McCormick (1987). The inversion algorithm of the SAGE II measurements is described in an article by Chu (1986), the ozone data quality by Cunnold et al. (1988).

In this study we have used approximately three years of SAGE II ozone measurements, from October 1984 to November 1987, available to us for the analysis. The ozone data from each of the SAGE II latitudinal scan are first binned into 10° latitudinal bins centered at ±70°, ±60°, ±50°, .., and 0° for the altitude range from 20 to 50 km with 1 km increments. The mean ozone number density is then determined for each bin. Also registered is the corresponding average of the sampling time. These time series means, $n_{O_3}(t)$, obtained from all the SAGE II latitudinal scans are shown in Fig. 1. To study the ozone time-periodic variation a multi-regression curve-fitting approach is applied to the time series data. The ozone variations are estimated by assuming that the time dependence of the mean $n_{O_3}(t)$ can be represented by:

$$n_{O_3}(t) = a_0 + a_1 t + \sum_{i=1}^{3} \Big(b_i \cos(\omega_i t) + c_i \sin(\omega_i t) \Big) + \varepsilon \tag{1}$$

where $\omega_i = \frac{2\pi}{P_i}$, (i = 1,3) represents the annual, semi-annual and biennial components, respectively, P_i is the corresponding period of the oscillation, and ε represents the residual error. With the derived b_i and c_i (i = 1,3), the amplitude and phase of the corresponding harmonics can be easily determined.

3. RESULTS AND DISCUSSIONS

The results of the curve-fitting analysis are displayed in Figs. 1a to 1c. The solid straight lines in Fig. 1 are the results corresponding to the sum of the first two terms,

ISBN 0-937194-15-8

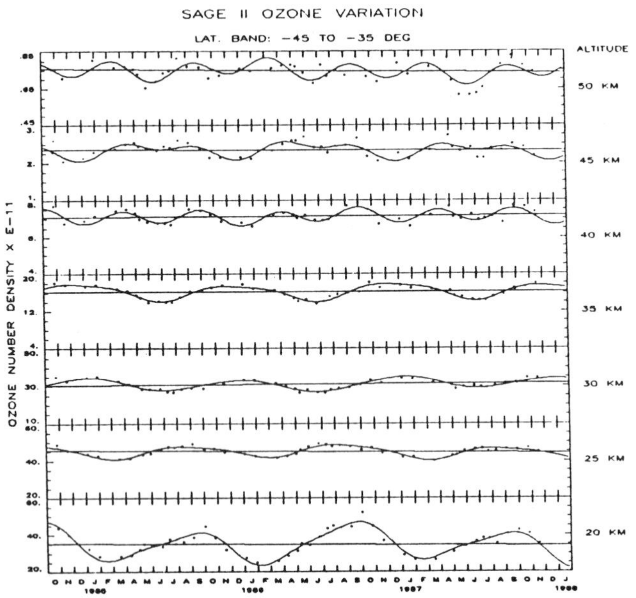

Fig. 1(a). SAGE II time series zonal mean data sets (dots) from October 1984 to November 1987 at 40°S. Solid straight lines represent the linear component of temporal variation. Solid curves are the results of curve-fitting.

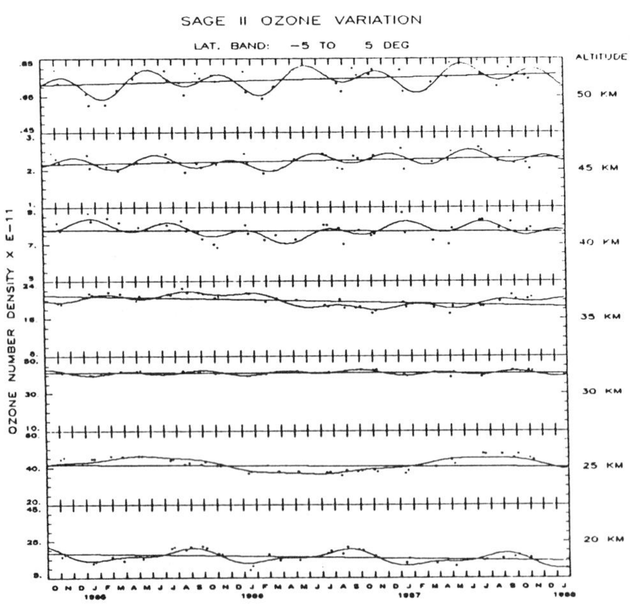

Fig. 1(b). 0°.

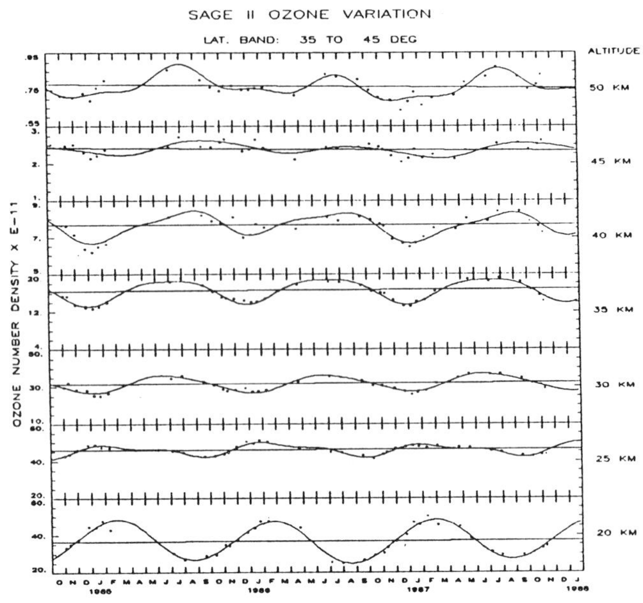

Fig. 1(c). 40°N.

a_0, and a_1t, which appeared in Eq. (1). They represent the linear trend of the ozone for the 3-year period. Since such a limited data set is used in this study, one must interpret these linear trend results with great caution. A different approach to study the long-term trend using both SAGE I and SAGE II ozone data is presented by McCormick et al. (1988). As to the biennial component, there are only 1½ cycles in the data period. As a result, the emphasis of this study is on the annual and semi-annual oscillations. Also shown in Fig. 1 is the sum (solid curves) of all the harmonic terms, the constant and the time linear term. As can be seen, the model results reproduce quite well the mean ozone time series derived from the SAGE II ozone measurements.

a. Annual Oscillation (AO)

The vertical distributions of the amplitude and phase of the ozone AO at latitudes 40°N, 0°, and 40°S are displayed in Fig. 2. The results show a strong AO below altitude ~25 km altitude at 40°N and 40°S. The amplitude of the AO at the 40°N (1.5 x 10^{12} molecule/cm^3) appears to be greater than that at 40°S (.87 x 10^{12} molecule/cm^3). At 40°N this maximum amplitude takes place sometime in early February, and is in early August at 40°S. At 40°N and 40°S, Fig. 2 also indicates a layer of the stratosphere between altitudes 26 and 40 km,

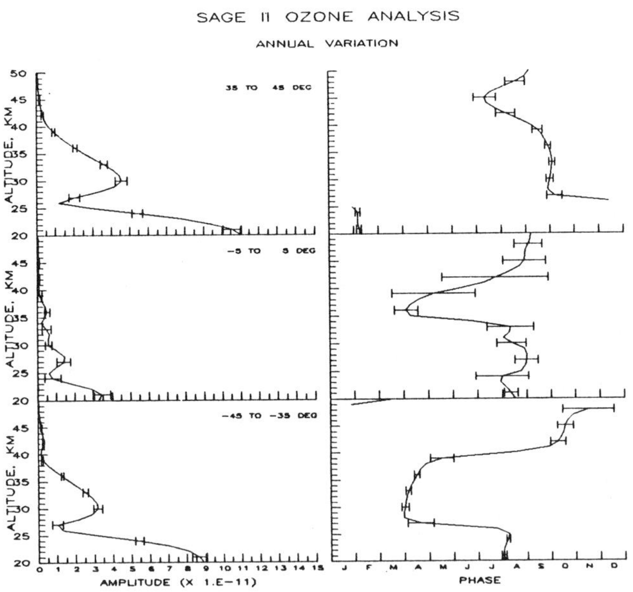

Fig. 2. Amplitude and phase of SAGE II ozone annual oscillation.

which is characterized by a large AO with a maximum of the amplitude located at an altitude about 30 km. The maximum amplitude of this component occurs sometime in early October at 40°N, and is in early April at 40°S. Above altitude 40 km Fig. 2 shows very small AO amplitude at all latitudes. The corresponding AO at the equator is relatively weaker than those at 40°N and 40°S. Figure 2 shows a decrease of the amplitude at the equator with height, from 20 km (the lower boundary of the analysis) to 25 km altitude. Between altitudes 25 and 30 km there is a maximum of the amplitude located at 28 km altitude. As to the phase of the equatorial ozone annual AO, Fig. 2 reveals that it is not as clearly defined as its mid-altitude counterparts. Generally, the maximum of the amplitude occurs during the northern summer months except for the layer between 34 and 40 km. In this layer, the phase results show large variabilities.

The late winter maximum of the midlatitude AO below about 25 km appears to be related to the occurrence of the total ozone build up during the late winter and early spring. The winter planetary wave activities in the lower stratosphere, which transport stratospheric ozone poleward and downward from the source region, are believed to be responsible for this build-up (WMO, 1986). As to the midlatitude AO above 25 km, the plausible mechanism can be the seasonal variation of the middle atmospheric diabatic circulation. Since the midlatitude ozone number density decreases exponentially with height from about 23 km, there would be a positive change in the local ozone concentration at altitudes above ~25 km due to the upward motion associated with the diabatic circulation in the summer months and a negative change due to the downward motion in the winter months. This transport mechanism explains also the appearance of the maximum amplitude of the ozone AO during the late summer months at altitudes above 25 km. It should be mentioned that in the upper stratosphere, ozone, on a seasonal scale, is controlled by photochemical process. Thus, maximum AO amplitude at midlatitude is expected to occur during middle summer. By inspection, one may notice that this situation occurs at altitudes above about 48 km (Fig. 2). However, the amplitude of ozone number density AO is very small in this altitude region.

As to the equatorial ozone AO, Meada (1984) pointed out that there are no obvious mechanisms for the generation of the ozone AO. He attributed this AO feature to: (a) hemispheric asymmetry in the seasonal variation of the planetary wave activities due to the orographic and topographic differences between the Northern and the Southern Hemispheres, and (b) hemispheric asymmetry in the seasonal solar heating in the upper stratosphere due to the earth's orbital eccentricity. Since his analysis was given in terms of ozone-mixing ratio, large AO occurred in the middle and upper stratospheres. In the lower stratosphere the AO was relatively small. But, in terms of ozone number density, only a very small AO component was found in the upper stratosphere (Fig. 2). Therefore, the hemispheric asymmetry in the planetary wave activities is responsible largely for the observed equatorial ozone AO shown in Fig. 2.

b. Semiannual Oscillation (SAO)

The derived amplitude and phase results of the ozone SAO at latitudes 40°N, 0°, and 40°S are displayed in Fig. 3. At the equator, Fig. 3 shows larger equatorial SAO amplitude in two altitude regions. One is located below 25 km with a maximum about 0.13 x 10^{12} molecule/cm^3 at 22 km altitude. The other one is located above 25 km with a peak value of 0.1 x 10^{12} molecule/cm^3 centered around ~30 km . Below about 33 km the maximum amplitude of the equatorial SAO occurs generally in the late equinoxes. Above ~33 km there is a shift of the phase from the late equinox at 33 km to about the late solstice at 45 km. The SAO below ~33 km can be the result of the upward diabatic circulation, which occurs twice a year at the equator during the equinox. This mechanism explains also the occurrence of the maximum amplitude in the late equinoxes. The phase shift in the layer between 33 km and 45 km is indicative of a transition of the controlling mechanism from dynamics to ozone photochemistry associated with the semi-annual modulation of the temperature field. Since the ozone photochemistry is capable of generating a negative correlation between ozone concentration and temperature, it is likely to be responsible for the occurrence of the maximum amplitude during the solstices in the upper stratosphere (above 45 km) at the equator.

At latitudes 40°N and 40°S the ozone SAO is much weaker than that at the equator, except for the altitude regions between 21 and 28 km at

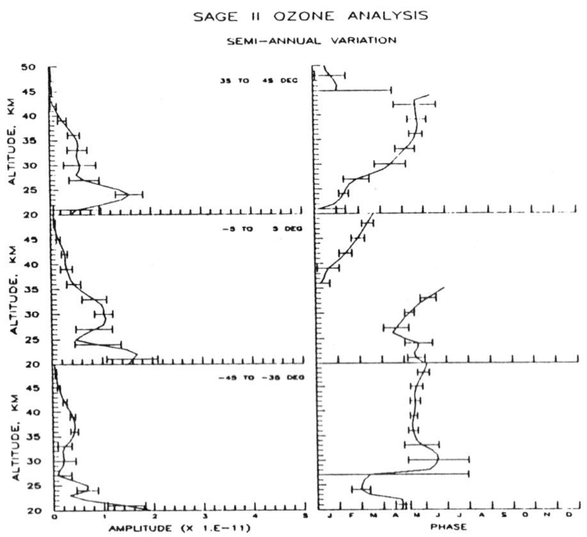

Fig. 3. Amplitude and phase of SAGE II ozone semi-annual oscillation.

40°N, and below altitude 23 km at 40°S. At 40°N the SAO in the layer between altitudes 21 and 18 km is about the same magnitude as that at the equator in the altitude region from 20 to 25 km. This is also the case in the altitude region below altitude 23 km at 40°S. In the middle stratosphere the phase profiles at 40°N and 40°S are generally similar. The maximum amplitude occurs in the late equinoxes. In the lower stratosphere the phase at 40°S shows large vertical variations and uncertainty. At 40°N, on the other hand, there is a phase shift from January at altitude 20 km to about April at 30 km. There is no obvious mechanism for the SAO in the middle and lower stratosphere at the midlatitude. This feature seems to indicate that there is a semi-annual variation in the dynamical processes that transports ozone from the source region to the middle and high latitudes, and that these dynamical processes are characterized by interhemispheric differences.

ACKNOWLEDGMENT

Pi-Huan Wang and S. Schaffner are supported by NASA contract NAS1-18252.

REFERENCES

Chu, W.P., 1986: Inversion of SAGE II Measurements. In extended abstract of the Sixth Conference on Atmospheric Radiation. American Meteo. Soc., Williamsburg, VA, May 13-16, 1986, J49-J51.

Cunnold, D.M., W. P. Chu, R. A. Barnes, M. P. McCormick, R. E. Vaiga, and D. A. Chu, 1988: Error analysis and validation of SAGE II ozone measurements. Proceedings of the Quadrennial Ozone Symposium, R.D. Bojkov and P. Fabian (Eds.), A. Deepak Publishing, Hampton, Virginia.

McCormick, M. P., 1987: SAGE II: An overview. Adv. Space Res., 7, 219-226.

McCormick, M. P., R. C. Viega and J. Zawodny, 1988: A comparison of SAGE I and II stratospheric ozone measurements. Proceedings of the Quadrennial Ozone Symposium, R.D. Bojkov and P. Fabian (Eds.), A. Deepak Publishing, Hampton,

McMaster, L. R., Stratospheric aerosol and gas experiment (SAGE II), 1986: In extended abstracts of the Sixth Conference on Atmospheric Radiation, American Meteo. Soc., Williamsburg, VA, May 13-16, J46-J48, 1986.

Meada, K., 1984: Annual and semiannual oscillations of stratospheric ozone. Atmospheric Ozone, C. S. Zerefos and A. Ghazi (Eds.), D. Reidel Publishing Co., Dordrecht, Halland, 248-252.

Wang, Pi-Huan, M. P. McCormick and W. Chu, 1983: A study on the planetary wave transport of ozone during the late February 1979 stratospheric warming using the SAGE ozone observation and meteorological information. J. Atmos. Sci, 40, 2419-1431.

World Meteorological Organization (WMO), 1986: Atmospheric Ozone 1985: Global Ozone Research and Monitoring Project. Report No. 16.

AN ASSESSMENT OF THE CHOPPER NON-SYNCHRONIZED CONDITION IMPACT ON THE SBUV TOTAL OZONE DETERMINATION

Richard P. Cebula, David S. Silberstein, and Charles G. Wellemeyer
ST Systems Corporation
Lanham, Maryland 20706, USA

Albert J. Fleig
NASA Goddard Space Flight Center
Greenbelt, Maryland 20771, USA

Pawan K. Bhartia
Science Applications Research
Lanham, Maryland 20706, USA

ABSTRACT

Commencing on 13 February 1987 and with increasing frequency thereafter, one of the status monitors aboard the instrument began to indicate that the Nimbus-7 Solar Backscatter Ultraviolet (SBUV) instrument's optical chopper wheel was operating in a non-synchronized (non-sync) condition. By May 1987 essentially 100% of the SBUV data was being flagged for the chopper non-sync condition. The non-sync condition may impact the data stream, increasing the noise in the SBUV total ozone retrieval. More serious, however, would be the introduction of biases into the total ozone determination. Coincident Dobson and simultaneous Total Ozone Mapping Spectrometer (TOMS) total ozone measurements are used to make a preliminary assessment of the quality of the SBUV total ozone measurements in the chopper non-sync environment. At approximately the 2% level there is no evidence of an SBUV chopper non-sync condition induced bias. However, there is an indication that the variance of the SBUV total ozone retrievals increases slightly in the non-sync environment.

1. INTRODUCTION

The SBUV instrument, a subsystem of the Solar Backscatter Ultraviolet and Total Ozone Mapping Spectrometer (SBUV/TOMS) experiment onboard Nimbus-7, has provided continuous global coverage of the terrestrial total ozone amount and stratospheric ozone profile since launch in October 1978. A detailed instrument description may be found in Heath et al. (1975). In brief, SBUV consists of a wavelength scanning tandem monochromator and provides nadir view measurements of the backscattered terrestrial radiance in twelve 1 nm wide wavelength bands spanning the region from 250 to 340 nm. Regular measurements of the incoming solar irradiance are also made. The ratio of the backscattered radiance to the incident solar irradiance is inverted to derive ozone. Only the four longest wavelength bands, 312.6, 317.6, 331.3, and 339.9 nm, are used in the total ozone determination. Light entering SBUV is chopped at 25 Hz by an optical chopper wheel which is driven by a phase-locked loop motor. Synchronously gated up/down counting electronics are used to eliminate the dark current contribution from the output signal; the electronics count up during the chopper open portions of the 1 second integration period, and count down during the chopper closed portions of the integration period. If the optical chopper and counting electronics are not synchronized, then it is conceivable that the instrument output would be biased. The status of the synchronization between the chopper and

ISBN 0-937194-15-8

the counting electronics is continually monitored via the use of a light emitting diode (LED) and photodiode detector which compares its time of illumination by the LED through a machined hole in the chopper wheel to a clock pulse. A latching status test point, telemetered once per scan of data, is used to monitor whether the chopper is synchronized with the counting electronics (in-sync) or not (non-sync) during each scan. The SBUV chopper was never in the non-sync condition during the first 8.25 years of in-orbit operation. However, as of 13 February 1987 and with increasing frequency thereafter, the chopper began to operate in a non-sync condition. Figure 1 shows the percent of the SBUV radiance data flagged for the non-sync condition. By May 1987 essentially 100% of the SBUV data was being flagged.

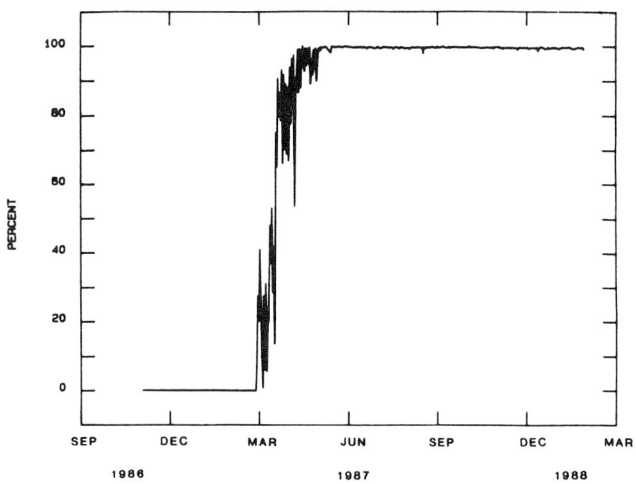

Fig. 1. Percent of the SBUV ozone data flagged for the chopper non-sync condition.

Unless its impact is on the order of tens of percent or more, directly assessing the impact of the non-sync condition on the radiance data is very difficult owing to the substantial geophysical variability of these data. At the SBUV total ozone wavelengths the solar irradiance is stable to within 1% over time scales on the order of days to months (Lean, 1987). Hence, the impact of the non-sync condition is more readily evaluated in these data. Shown in Fig. 2 is the relative SBUV solar irradiance output at 340 nm. No correction for the slow drift in instrument sensitivity has been applied to these data. The most obvious feature of Fig. 2 is that the noise in the 340 nm solar data increased from the pre non-sync level of approximately 0.2% to approximately 2% in the non-sync environment. No clearly discernible impact other than the noise increase is noted. Similar behavior is observed at the other 11 SBUV ozone wavelengths.

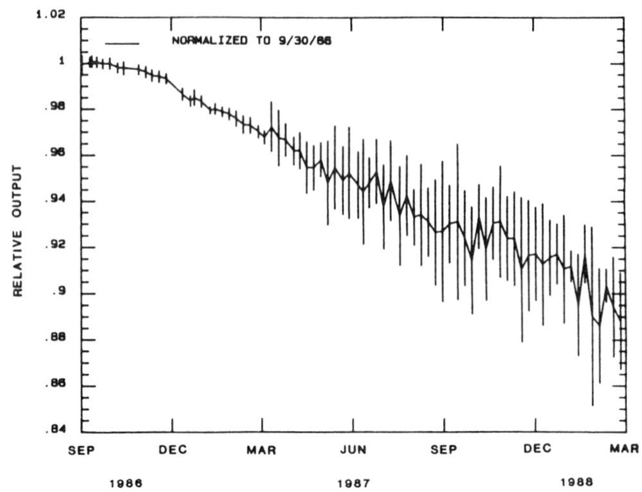

Fig. 2. SBUV 340 nm solar irradiance output, normalized to 30 September 1986. The error bars are ±1 standard deviation.

The impact, if any, of the chopper non-sync condition on the SBUV-derived profile and total ozone measurements is of primary concern. Given the uncertainty in the accuracy of the data in the non-sync environment, the NASA Ozone Processing Team (OPT) believes that SBUV data acquired after 12 February 1987 should be treated as a different data set from the previous data. A decision has recently been made to release the chopper non-sync data to the National Space Sciences Data Center as a continuation of the SBUV data set. Users requesting these data will receive a brief description of the non-sync problem and a summary of potential non-sync impacts with the data. The remainder of this paper presents an initial assessment of the impact of the non-sync condition on the SBUV total ozone measurement; the more complex assessment of the condition's impact on the ozone profile measurement is outside the scope of the present study.

2. COMPARISON WITH SELECTED DOBSON STATIONS

The global Dobson network provides one means of assessing the impact, if any, of the chopper non-sync condition on the SBUV total ozone determination. Previous SBUV-Dobson intercomparisons (Fleig et al., 1986) showed good agreement between the two systems, with the existence of a small bias and a slow long-term drift between SBUV and the Dobson network. The Dobson network consists of more than 100 individual stations. This assessment was limited to the 10 stations that have historically shown the smallest variance with respect to SBUV: Arosa, Cairo, Hradec Kralove, Mahe (Seychelles Is.), Mauna Loa, Naha/Kagamizu, New Delhi, Perth, Poona, and Varanasi. The Dobson data available for this work extends only through September 1987, approximately 5 months after the non-sync condition reached the 100% level. The SBUV and Dobson values were considered coincident if an SBUV measurement was obtained within a 1° latitude by 2° longitude window on the same day as a given Dobson station measurement. Figure 3 presents the comparison of the SBUV and Arosa measurements. No obvious bias or increase in the noise level is observed.

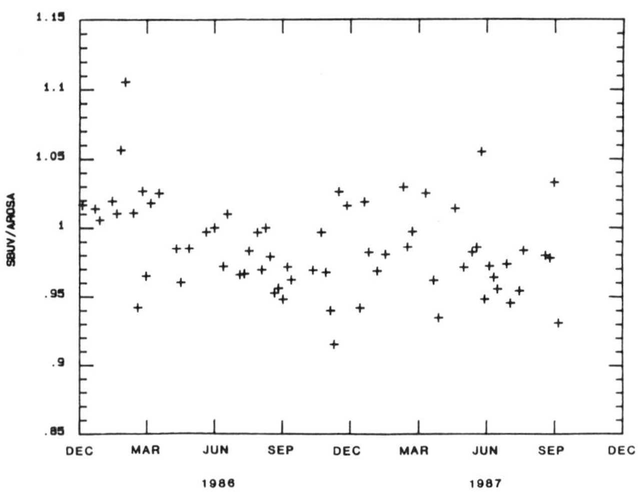

Fig. 3. SBUV/Arosa total ozone comparison.

The Arosa results are typical of all ten comparisons. Historically, intra- and inter-station variabilities have complicated the SBUV-Dobson comparisons. The average ratio of the total ozone from SBUV to the ten selected stations was computed in order to increase the sensitivity of the intercomparison to non-sync effects. However, on average there were coincidences with only 1.9 of the 10 Dobson stations per day, and there were never more than 5 coincidences on any given day. Therefore, the 10 station average, Fig. 4, is nearly as noisy as the individual comparisons. A two week running average was also computed, but again no clear impact of the non-sync condition was observed.

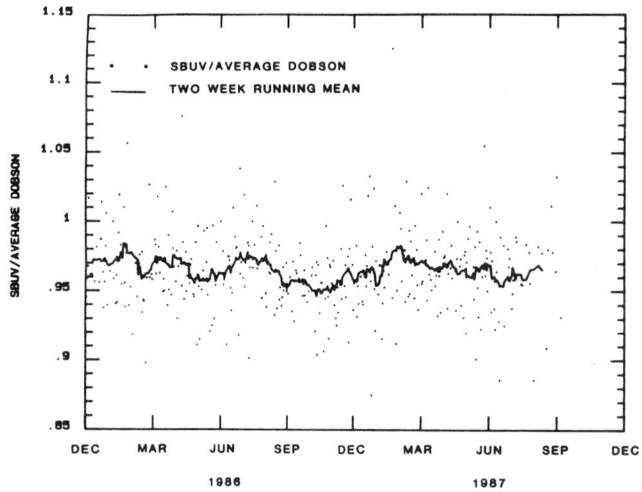

Fig. 4. The average ratio of SBUV to Dobson total ozone for 10 selected stations.

3. COMPARISON WITH TOMS

The TOMS instrument provides simultaneous total ozone measurements which are the best standard by which to assess the impact of the SBUV chopper non-sync condition. While a 2.9% bias, likely reflecting uncertainties in the absolute calibrations of the two instruments, exists between the SBUV and TOMS total ozone values, the two instruments have not drifted substantially relative to one another (Fleig et al., 1986, 1988; Wellemeyer et al., 1988). Similar to SBUV, the TOMS instrument includes an optical chopper wheel to remove the dark current contribution from the output signal (Heath et al., 1975). Commencing in February 1984 the TOMS chopper has exhibited randomly spaced short-term non-sync periods. These periods, during which on the order of 5-10% of the TOMS data are flagged for the chopper non-sync condition, typically last for a few orbits to a few days and are usually separated by several week intervals. The TOMS chopper non-sync data are excluded from the archived data set and were not used in the present SBUV chopper non-sync assessment. While SBUV views only the subsatellite point with a 200 km sq field of view (FOV), TOMS provides daily global coverage via a cross-track scan mechanism. The TOMS FOV at the subsatellite point is 53 km sq. For the present comparison, weekly zonal mean SBUV and nadir TOMS total ozone values, with the TOMS FOV matched to the SBUV FOV, were computed for five ten-degree wide latitude zones centered on -70°, -40°, 0°, $+40^{o}$, and $+70^{o}$.

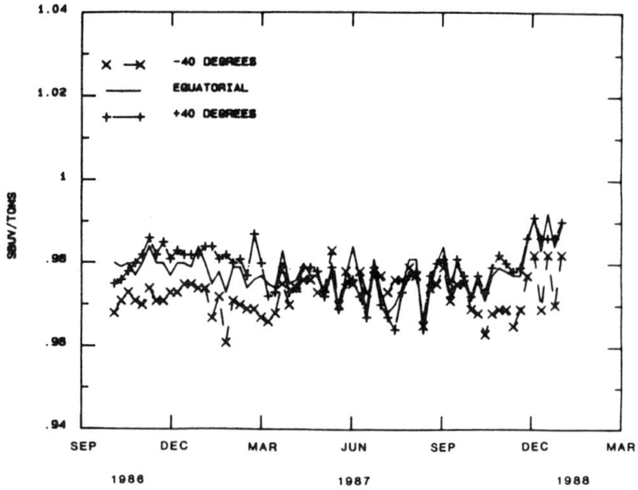

Fig. 5. The ratio of SBUV to TOMS total ozone for three 10° wide latitude zones, centered on -40°, 0°, and +40°.

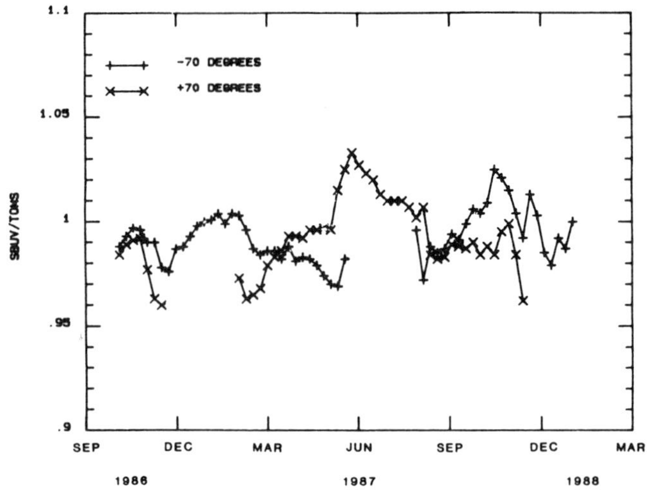

Fig. 6. The ratio of SBUV to TOMS total ozone for two 10° wide latitude zones, centered on -70° and +70°.

Figure 5 presents the ratio of the SBUV to TOMS total ozone values for the three latitude zones centered on -40°, 0°, and +40°. The two high latitude zone comparisons are presented in Fig. 6. At the 1-2% level, there does not appear to be an SBUV non-sync bias introduced into the equatorial or mid-latitude SBUV/TOMS total ozone comparisons. A small (about 1%) upturn in the ratio of SBUV to TOMS total ozone exists in late 1987, however, a longer term comparison is necessary to unambiguously identify if this is a bias or just short-term noise. There is a suggestion that the noise in the comparisons increases slightly coincident with the time that all of the SBUV data began be flagged for the non-sync condition. The high latitude, large solar zenith angle comparisons, Fig. 6, are noisier, however, no features coincident with the SBUV chopper non-sync condition onset are observed at the 3% level. In agreement with the low and mid-latitude comparisons, there appears to be an increase in the high latitude comparisons noise level which is roughly coincident with the onset of the SBUV chopper non-sync condition.

4. SUMMARY

The SBUV optical chopper wheel began to operate in a non-synchronized (non-sync) condition on 13 February 1987. By May 1987 essentially 100% of the SBUV ozone data were being flagged for the chopper non-sync condition. This paper has provided an initial assessment of the impact of the chopper non-sync condition on the SBUV total ozone data. Individual and average comparisons of the SBUV data with ten Dobson stations for the period January 1986 through September 1987 did not reveal any evidence of either a non-sync induced bias or a significant increase in the SBUV total ozone variance. However, variability in the Dobson data could mask such effects if they were smaller than roughly 5 percent. Comparison of SBUV and nadir TOMS measurements were made for five ten-degree latitude zones for the period October 1986 through December 1987. No clear signature of a non-sync induced bias is observed in the SBUV/TOMS comparison at approximately the 2% level. There is, however, a suggestion of a small increase in the variance of the SBUV total ozone retrievals. This is consistent with the approximate 2% increase in the noise in the SBUV irradiance measurements observed coincident with the onset of the chopper non-sync condition. The present work must clearly be considered preliminary. Comparison with a longer term and more complete set of Dobson stations as well as comparison with TOMS 1988 data are planned. Coincident total and profile ozone data from the SBUV/2 instrument on-board the NOAA-9 spacecraft are currently being processed. Comparison of the SBUV and SBUV/2 data for the 3 year overlap is necessary, not only for the assessment of the impact of the SBUV chopper non-sync condition, but also for the evaluation of the long-term calibrations of both instruments.

ACKNOWLEDGMENTS

We wish to thank J.D. Augustine and K. Laamann for their assistance in the data analysis and N.L. Dudley and A.E. Suhari for assistance in the preparation of the manuscript. The SBUV and TOMS data were prepared by the Ozone Processing Team of the NASA Goddard Space Flight Center. This work was supported by the NASA Space and Earth Sciences Directorate.

REFERENCES

Fleig, A.J., P.K. Bhartia, and D.S. Silberstein, 1986: An Assessment of the Long-term Drift in SBUV Total Ozone Data, Based on Comparison with the Dobson Network, Geophys. Res. Lett., 13, 1359-1362.

Fleig, A.J., D.S. Silberstein, C.G. Wellemeyer, R.P. Cebula, and P.K. Bhartia, 1988: An Assessment of the Long-term Drift in TOMS Total Ozone Data, Based on Comparison with the Dobson Network, Geophys. Res. Lett., submitted.

Heath, D.F., A.J. Krueger, H.A. Roeder, and B.D. Henderson, 1975: The Solar Backscatter Ultraviolet and Total Ozone Mapping Spectrometer (SBUV/TOMS) for NIMBUS G, Opt. Eng., 14, 323-331.

Lean, J., 1987: Solar Ultraviolet Irradiance Variations: A Review, J. Geophys. Res., 92, 839-868.

Wellemeyer, C.G., A.J. Fleig, and P.K. Bhartia, 1988: Internal Comparisons of SBUV and TOMS Total Ozone Measurements, these proceedings.

DETERMINATION OF THE OZONE MIXING RATIO PROFILES IN THE ATMOSPHERE FROM TOVS/TIP SATELLITE DATA

L.A. Baranski and K. Rozemski
Institute of Meteorology and Water Management
30–215 Krakow, Poland

ABSTRACT

TOVS/TIP digital data transmitted at VHF-BEACON range from NOAA satellites are receiving and processing at the Institute of Meteorology and Water Management in Krakow (Poland). Receiving station is connected with microcomputer IBM-PC/AT which process TOVS/TIP data for retrieval of vertical profiles of temperature as well as water and ozone mixing ratio in the atmosphere. Preliminary results of spatial distribution of total ozone content over Middle Europe and vertical ozone mixing ratio profile over Krakow are shown. Also results of intercomparison analysis of ozone mixing ratio profiles calculated from TOVS/TIP satellite data vs. ozone sonde measurements are presented.

1. INTRODUCTION

The NOAA satellites are equipped with the TIROS Operational Vertical Sounder (TOVS) that consists of three instruments:

- High Resolution Infrared Radiation Sounder (HIRS),
- Microwave Sounding Unit (MSU),
- Stratospheric Sounding Unit (SSU).

HIRS radiometer is detecting the radiation emitting from Earth's surface and from the atmosphere in one visible and 19 infrared spectral channels. MSU radiometer is detecting the radiation in 4 microwave spectral channels. SSU radiometer is detecting the radiation upwelling from the stratosphere in 3 spectral channels; Schwalb (1978).

The SDRPC is equipped with two independent receiving stations registrated HRPT (High Resolution Picture Transmission) in S-band and TOVS in BEACON-VHF band; Baranski, Struzik (1988).

The data from HIRS radiometer allow the calculation of the vertical distribution of the atmospheric temperature and moisture; Baranski, Rozemski (1988). The vertical profiles of temperature and moisture have been retrieved at the SDRPC since 1985. The intercomparison between several hundreds of the satellite TOVS temperature and moisture retrievals and time-coincident ozone sonde measurements has been performed for seven aerological stations placed in the Middle and South Europe; Baranski, Rozemski (1988). The satellite TOVS data can be used also for retrieval of the ozone mixing ratio vertical profiles as well as total ozone content in the atmosphere.

2. THEORETICAL BACKGROUND

The physical background of the ozone retrieval method based on the phenomenon of the atmospheric emission and absorption of the radiation, is described by the equation of radiative transfer; Smith et al. (1983):

$$I(\nu) = B(T_s,\nu)\,\tau(\nu,p_s) - \int_0^{p_s} B(T(p),\nu)\,\frac{\partial\tau}{\partial p}(\nu,p)\,dp \qquad (1)$$

where:

$I(\nu)$ - the radiance measured in the HIRS channels of central wave number ν,
$B(T,\nu)$ - Planck's function,
$\tau(\nu,p)$ - the atmospheric transmittance function from the pressure level p to the top of atmosphere,
$T(p)$ - the vertical temperature profile,
T_s - Earth's surface temperature,
p_s - the near surface pressure.
$\partial\tau / \partial p$ - named weighting function, which determines the vertical dependance of the sensitivity of HIRS measurements on the atmospheric temperature.

The method for computing the total ozone content in the atmosphere is based on the radiation measurements in 9th channel of the HIRS radiometer (9.7 μm) for which ozone is the main absorbing component. The iterative method for ozone calculation was applied; Ma Xia-Lin et al. (1984). Using the first-order Taylor expansion of the Planck function in terms of temperature and integrating equation (1) by parts we obtain :

$$T_B(\nu) - T_B^{(n)}(\nu) = \int_0^{p_s} \left\{\tau(\nu,p) - \tau^{(n)}(\nu,p)\right\} Q(\nu,p)\,dp \qquad (2)$$

ISBN 0-937194-15-8

where:

$T_B(\nu)$ - is the measured brightness temperature in the 9-th channel of HIRS radiometer,

$T_B^{(n)}(\nu)$ - is the brightness temperature calculated for the n-th estimate of the ozone profile whose corresponding transmittance function is $\tau^{(n)}(\nu,p)$,

$$Q(\nu,p) = \frac{\partial B(T,\nu)/\partial T}{\partial B(T_B,\nu)/\partial T}\,\frac{dT}{dp}$$

Assuming that the true ozone profile is proportional to the n-th estimate and that the absorption coefficient in the 9-th HIRS channel is proportional to the ozone mixing ratio, we obtain the following estimate of transmittance function difference in eq. (2) :

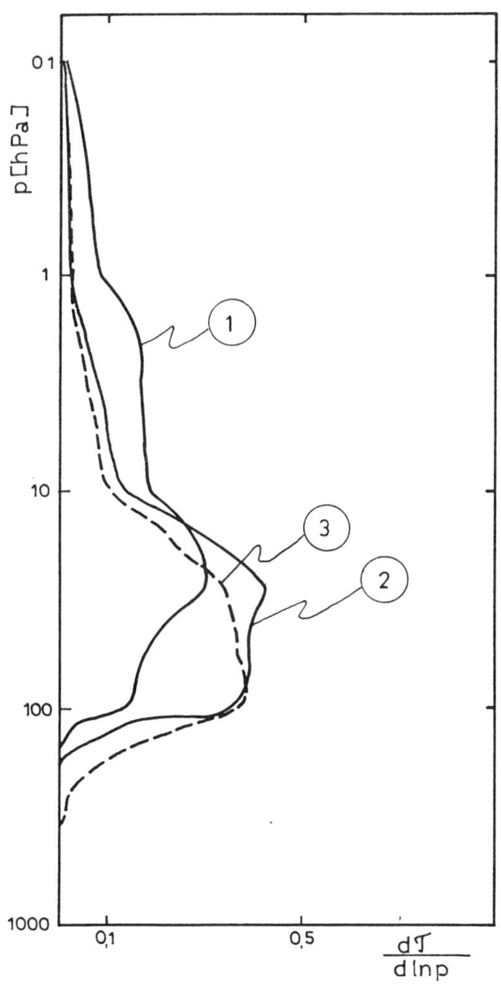

Fig. 1. Weighting functions for channels No 1, 2 and 3 of HIRS radiometer from NOAA-9 satellite.

$$\tau(\nu,p) - \tau^{(n)}(\nu,p) = \tau^{(n)}(\nu,p)\ \ln \tau^{(n)}(\nu,p)\ \ln r \qquad (3)$$

where:

$r = \dfrac{U(p)}{U^{(n)}(p)}$ is the constant scaling factor

$U(p)$, $U^{(n)}(p)$ - are the true and estimated ozone mixing ratio profiles, respectively.

From equations (2) and (3) we get the following iterative formula, which is the basis for the method of ozone mixing ratio retrieval :

$$U^{(n+1)}(p) = U^{(n)}(p)\ \exp\left[\frac{T(\nu) - T_B^{(n)}(\nu)}{\int_0^{p_s} W^{(n)}(\nu,p)\,dp}\right] \qquad (4)$$

where:

$$W^{(n)}(\nu,p) = \frac{\partial B(T,\nu)/\partial T}{\partial B(T_B,\nu)/\partial T}\ \tau^{(n)}(\nu,p)\ \ln \tau^{(n)}(\nu,p)\ \frac{dT}{dp}$$

The iteration is terminated as soon as the difference between the measured brightness temperature and that calculated is less than the noise of measurement level (approximately 0.2K). The total ozone content is obtained by vertical integration of the mixing ratio profile. The first-guess ozone profile is constructed using the brightness temperature measured in the 1,2 and 3 channels of HIRS radiometer. In those three

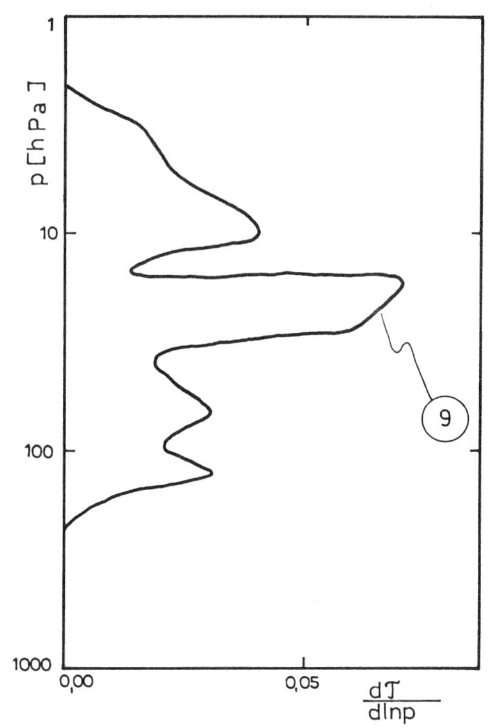

Fig. 2. Weighting function for channel No 9 of HIRS radiometer from NOAA-9 satellite.

channels the radiation from the stratosphere and upper troposphere is measured. Therefore, they characterize the thermal situation in this region, which is strongly correlated with the ozone content. The weighting functions for these channels are presented on Fig.1. The weighting function for 9-th channel if HIRS radiometer is presented in Fig.2. Those graphs were made from the transmittance functions computed for the standard atmosphere using the subroutine TAUHIR of TOVS Export Package; Smith et al. (1983).

3. AN EXAMPLE OF RESULTS

An example of total ozone horizontal distribution obtained from TOVS data received from NOAA-9 satellite over Middle Europe on February 11, 1987 (daily orbit) are illustrated in Fig.3. An example of the ozone mixing ratio vertical distribution over Krakow in Poland (20E, 50N) calculated from TOVS data received from NOAA-9 satellite on February 11, 1987 (daily orbit) are presented in Fig.4.

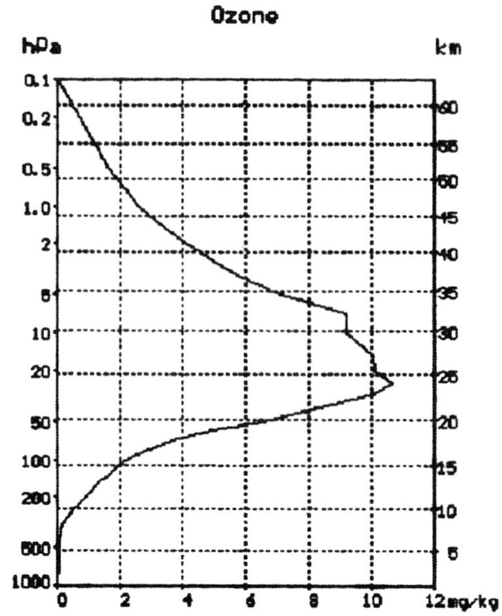

Fig. 4. An example of ozone mixing ratio profile calculated over Krakow (Poland) on February 11, 1987 from TOVS/NOAA-9 satellite data.

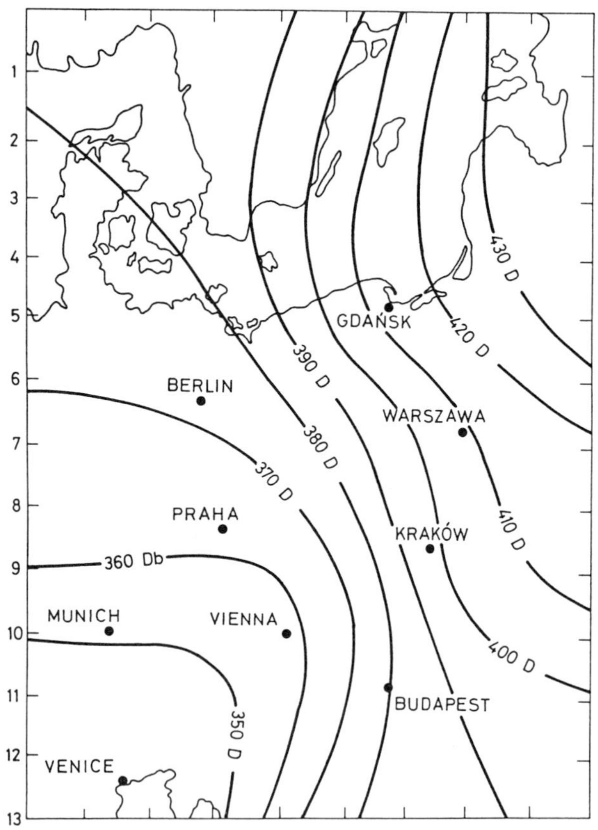

Fig. 3. Results of total ozone content calculations (in Dobson units) from TOVS data obtained from NOAA-9 satellite over Middle Europe on February 11, 1987.

4. INTERCOMPARISON OF SATELLITE AND OZONE SONDE MEASUREMENTS

The intercomparison of 104 ozone retrievals from TOVS data versus ozone sonde measurements from three aerological stations placed in Belsk (Poland), Prague (Czechoslovakia) and Lindenberg (GDR) was performed (February-May, 1987).The ozone sonde data for intercomparison was obtained from World Ozone Data Center in Canada; Atmospheric Environment Service (1987).

Taking into account the strong vertical dependance of ozone mixing ratio, the relative differences rather then absolute one was consider as a measure of discrepancies between TOVS retrieval and ozone sonde measurements. The following statistics were calculated:

- root mean square (RMS) of the relative differences,
- mean relative differences,
- standard relative deviations,
- correlation coefficients.

The statistical material allowed the determination of the vertical dependence of the above mentioned statistics on the range from 500 hPa to 10 hPa. The vertical dependence of the statistics are presented on Fig.5.

On the range 100 hPa through 10 hPa, the relative RMS differences are about 15%. In the region of the highest ozone concentration, the errors of the retrieval method from TOVS data seems to be acceptable. Below about 100 hPa, the differences rapidly grow. This fact can be explained by a poor correlation between HIRS data used here and ozone concentration in this area. The relative mean differences show tendency of TOVS retrievals to overestimate ozone mixing ratio below the level of 35 hPa.

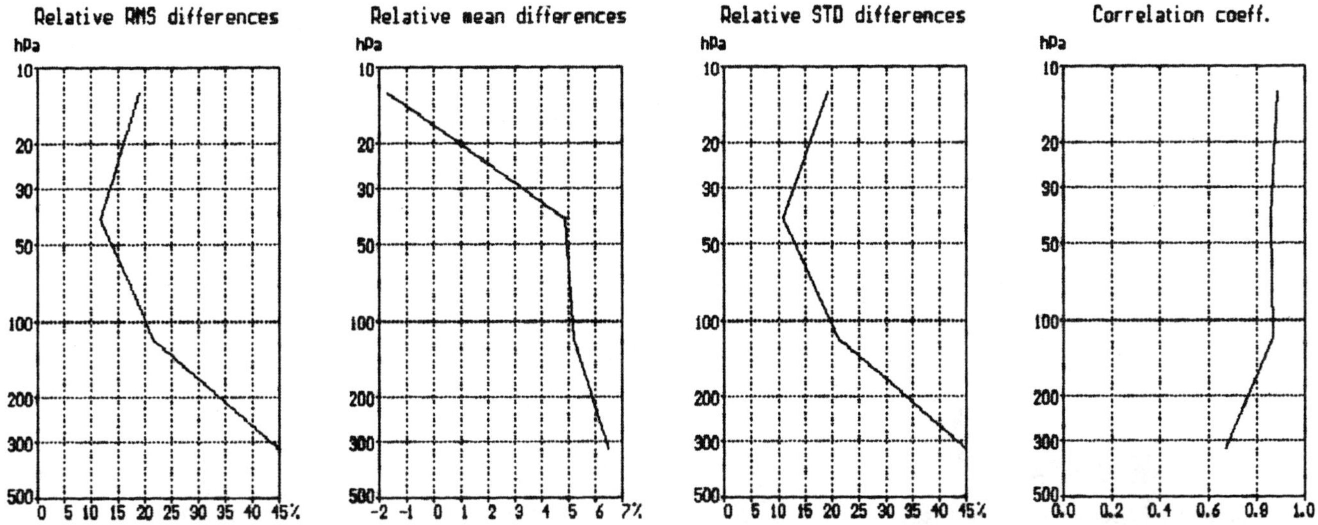

Fig. 5. Comparison statistics for ozone mixing ratio profile.

5. CONCLUSIONS

The satellite TOVS data allow for continous monitoring of three dimensional distribution of ozone content over relatively large areas, within the radio horizon of the receiving antenna. Therefore this kind of ozone measurements is attractive when compared with the ground-based and ozone sonde measurements.

The satellite retrieval method is characterised by an acceptable correspondence to other data in the region of highest ozone concentration. Intercomparison of the results obtained from TOVS data and ground measurements provided from seven European stations for 1987 and 1988 is under preparation. Presented results regarding ozone mixing ratio profiles can be treated as a test of the regression coefficients from TOVS Export Package for calculation of the first guess ozone profile ; Smith W.L. et al, (1983).

The data obtained from Solar Backscatter Ultraviolet (SBUV-2) instrument installed on-board the new series of NOAA satellites will be used in the near future.

REFERENCES

Baranski L.A., Rozemski K., Determination of the temperature and moisture profiles from TOVS satellite transmission, Proc. Symposium on Application of Satellite Measurements in Modelling and Forecasting of the Atmospheric Phenomenons, Prague, Czechoslovakia, in press (April, 1988)

Baranski L.A., Rozemski K., Microcomputer system ATMOSAT for receiving and processing of satellite TOVS/TIP data for vertical sounding of the atmosphere, COSPAR XXVII Planetary Meeting and Associated Activities, Espoo, Finland, in press (18-29 July 1988)

Baranski L.A., Struzik P., Interactive system for registration and processing of digital satellite data using IBM-PC/AT computer, Proc. 16-th International Congress of International Society for Photogrametry and Remote Sensing, Kyoto, Japan, in press (July 1988)

Ma Xia-Lin, Smith W.L., Woolf H.M., Total ozone from NOAA satellites - A physical model for obtaining measurements with high spatial resolution, J. Clim. Appl. Meteor., 29, 9, (1984)

Atmospheric Environment Service, Ozone Data for the World, Ontario, Canada, (1987)

Smith W.L., Woolf H.M., Hayden C.M., Schreiner A.J., Le Marshal I.M., The physical retrieval TOVS Export Package, Techn. Proc. First Intern. TOVS Study Confer., Igls, Austria (1983)

Schwalb A., The TIROS-N/NOAA A-G satellite series NOAA Techn. Memo. NESS 95, Washington D.C., (1978)

QUASI-BIENNIAL AND INTERANNUAL VARIABILITY IN HIGH RESOLUTION TOTAL OZONE DATA (TOMS)

Glen S. Schuster
Department of Meteorology
Florida State University
Tallahassee, Florida 32306, USA

Richard B. Rood and Mark R. Schoeberl
NASA Goddard Space Flight Center
Greenbelt, Maryland 20771, USA

ABSTRACT

The high resolution total ozone data of the Nimbus 7 TOMS has been used to study the equatorial and extratropical QBO from 67°N to 67°S. Emphasis is placed on the Southern Hemisphere. The relationship of the equatorial QBO in zonal winds to the extratropical QBO in high latitude total ozone is investigated. The equatorial QBO periodicity is shown to be statistically significant in the southern hemisphere and implies that there is a relation between equatorial and extraequatorial QBO signals. The relationship becomes weak poleward of 40°S.

1. INTRODUCTION

After removal of the annual cycle significant interannual variability remains in the total ozone record. Near the equator the interannual variability is dominated by the quasi-biennial oscillation (QBO). Outside of the equatorial region significant variance appears in periods of slightly less than two years, but seemingly different from the equatorial QBO. A four year oscillation has also been noted. Interannual variability has been studied by many authors, for instance, Hasebe (1983) Guthrie (1983), Oltmans and London (1982), Hilsenrath and Schlesinger (1981), and Angell and Korshover (1973).

There have been several suggestions that the equatorial QBO has global effects on stratospheric dynamics and hence the global ozone distribution. Most recently Labitzke and van Loon (1988) have suggested a complex relationship between the QBO, the solar cycle, and stratospheric warmings. Garcia and Solomon (1987) and Schoeberl et al. (1988) have investigated the relationship between the QBO and springtime Antarctic ozone trends.

An accurate determination of the interannual variability and its causes is necessary for assessing possible anthropogenic related ozone depletion. This paper investigates the interannual variability of the TOMS Southern Hemisphere total ozone and the possible connection of the equatorial QBO to extratropical signals through the determination of spatial continuity. In particular, the 30 mb zonal wind data at Singapore is used as a definition of the QBO, and the statistical relationship between wind and ozone at various longitude and latitude points is sought. The predominate periodicities are determined as a function of latitude, and the longitudinal coherence of the interannual signal is investigated.

2. DATA AND DATA TREATMENT

Monthly mean TOMS data from the NIMBUS 7 satellite are used. The time series is from November 1978 through April 1987. In order to avoid missing data in the polar night region, only data between 67.5°N and 67.5°S are used. The 50 km resolution data have been placed in 2° latitude by 5° longitude bins to facilitate data handling.

As reported by Bowman (1988; see also Watson and Prather, 1988) and the Ozone Trends Panel (1988) there is a large trend in the TOMS data. Much of this trend is due to instrument degradation. This trend is assumed linear and removed with a least linear squares fit prior to analysis. The seasonal cycle is removed by subtraction of the appropriate time averaged monthly mean from each member of the time series. Three and five point running means have been used to reduce high frequency variability.

3. ZONAL MEAN RESULTS

Figure 1 shows the time series of the zonal mean ozone data as a function of latitude. Figure 1 represents a continuation of the time series of Hasebe (1983) who used ground-based and Nimbus 4 SBUV data. The QBO at equatorial latitudes is visible with a regular period, though with some variability in amplitude. The positive anomaly in 1985 is particularly large. Outside of the equatorial region the variability is less regular and shows a tendency to be out of phase with the equatorial signal, particularly in the Southern Hemisphere.

In both hemispheres the nodal line (where the sign of the anomaly changes) is at approximately 10°. Hasebe (1983) reported the Southern Hemisphere nodal line near 15°. He

ISBN 0-937194-15-8

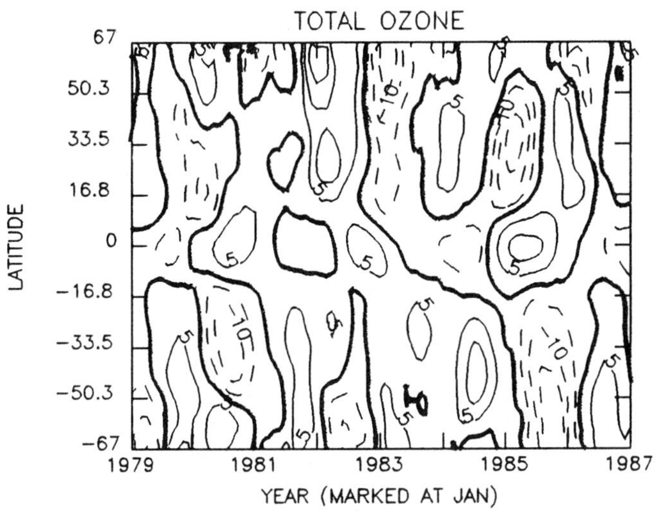

Fig. 1 Deseasonalized, detrended zonal mean total ozone from January 1979 to January 1987. Solid-positive, Dashed-negative, Bold-zero.

cited more variability in Northern Hemisphere nodal line with the location generally more northward at approximately 35°. There are times in both hemispheres where the extraequatorial anomaly is coherent from the nodal line to 67.5°.

Figure 2 shows the zonal mean time series at four selected latitudes. The time series at 1-3°N shows a regular oscillation with the last cycle having somewhat increased amplitude. The time series at 11-13°S is characteristic of the nodal line region and shows little regular variability.

South of the nodal line the signal is generally anticorrelated with the equatorial signal. In particular the positive anomalies at 1-3°N in 1980-81 and 1985 are associated with large negative anomalies at 27-29°S. Otherwise the anticorrelation is not obvious. At 57-59°S the variability is more irregular. The negative anomaly in 1985 is still present, but the negative anomaly in 1980-81 is absent. In its place is a negative anomaly in 1982 that is more or less anticorrelated with a positive equatorial deviation.

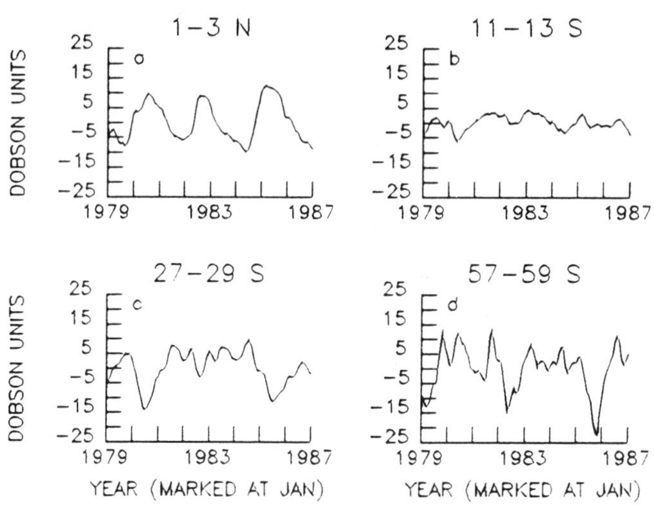

Fig. 2 Deseasonalized, detrended, zonal mean total ozone from January 1979 to January 1987 at selected latitudes.

These time series clearly demonstrate the characteristics of the zonal mean. The variability outside of the tropics is not as regular. While there is a general tendency for the two regions to be anticorrelated, the anticorrelation is not strongly defined. The anticorrelation is frequently determined by coincident specific events rather than a regular signal. These time series do not clearly define a systematic link between the equatorial and polar regions.

Conventional spectral analysis of the zonal mean time series reveal a period at approximately 28 months in the equatorial region (the QBO signal). Just south of the nodal line (11-13°S) the period with the maximum variance shifts from 28 months to a shorter period of 20 to 21 months. This shorter period is statistically significant and generally dominant all the way to 67.5°S. This shift to a shorter period outside of the equatorial region is consistent with previously reported results.

From approximately 20°S to 40°S there is a weaker but significant signal in the 28 month period, the same period as the equatorial QBO. By 50°S this QBO signal has virtually disappeared and the 20-21 month signal is robust.

Hasebe (1983) discussed the four year oscillation (FYO). There is a hint of a FYO at various latitudes, but with such a short time series it is difficult to attach much significance to this finding. The FYO appears primarily as a response to the two times of large depletion seen in late 1980 and 1985-86 in Fig. 1.

4. LONGITUDINAL VARIATIONS

Figure 3 shows the longitude time section of the ozone data at 1-3°N. The QBO shows longitudinal coherence in both amplitude and phase. The large positive deviation noted in the zonal means during 1985 (Fig. 1) occurs at all longitudes.

South of the nodal line in the 17-19°S bin the anticorrelation with 1-3°N between 1979 and 1981 is obvious (Fig. 4). The period of the oscillation at 17-19° appears shorter. After 1981 the oscillation falls apart and the negative cycle that might have been expected in late 1982 is largely missed. The longitudinal coherence also decreases after 1981. A reemergence of both the longitudinal coherence and the anticorrelation occurs in 1985. The behavior in 1982 and 1983 suggests either an El Chichon or El Nino effect. Bojkov (1987) evaluates the relative effects of the QBO, El Nino, and El Chichon on the ground based ozone

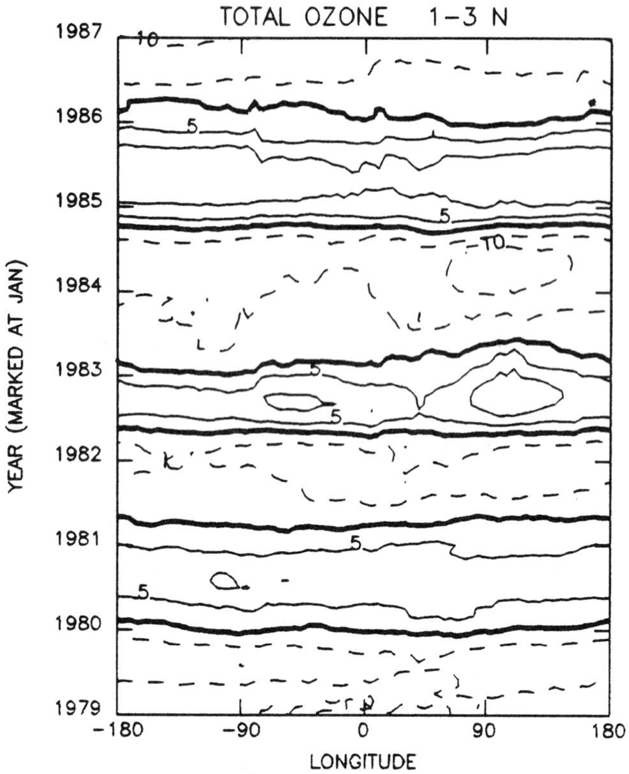

Fig. 3 Longitude versus time, total ozone, 1-3°N.

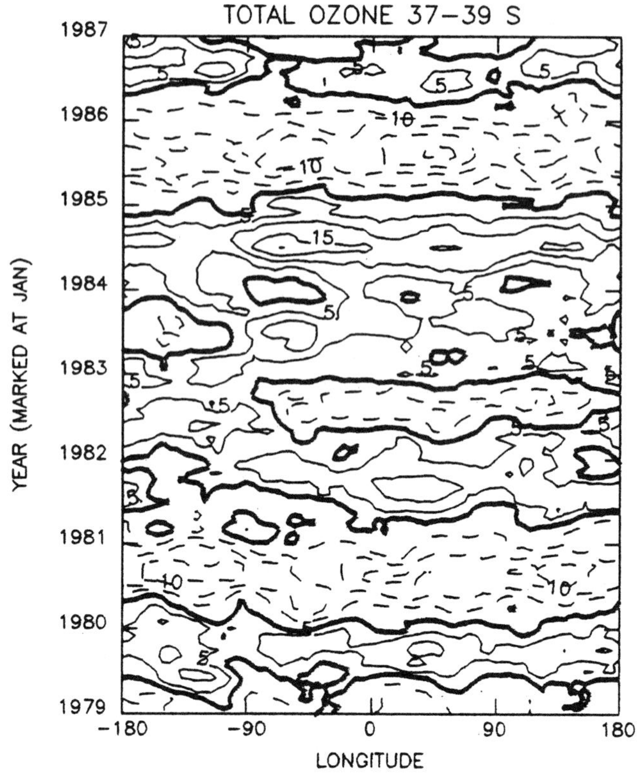

Fig. 5 Longitude versus time, total ozone, 37-39°S.

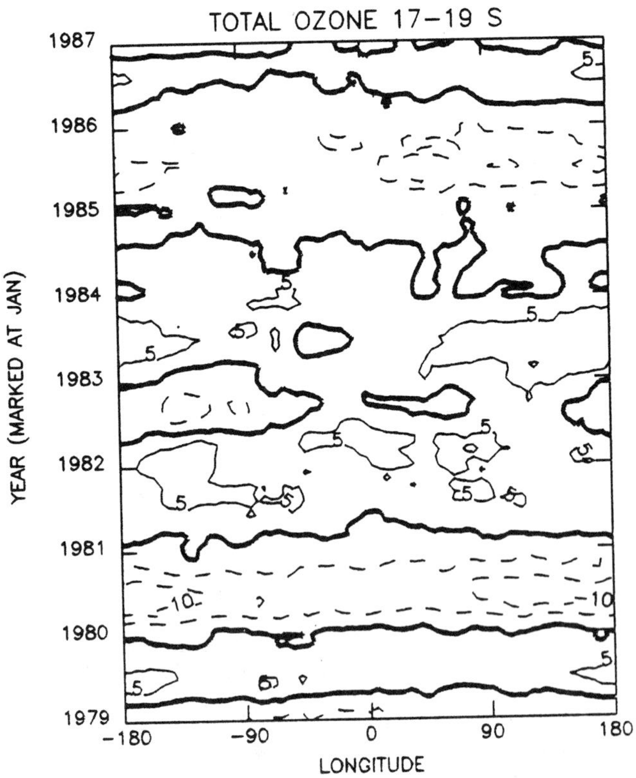

Fig. 4 Longitude versus time, total ozone, 17-19°S.

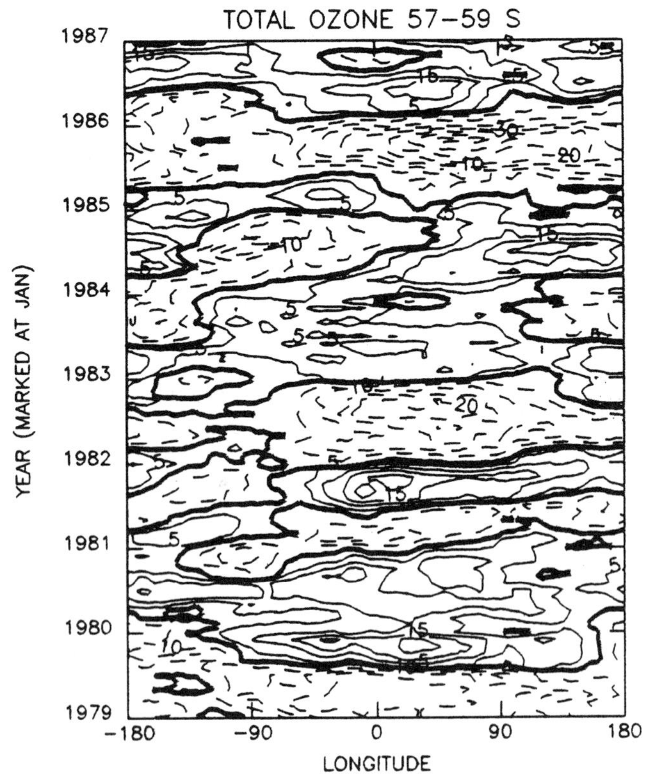

Fig. 6 Longitude versus time, total ozone, 57-59°S.

record.

The data at 37-39°S show an increase in the amplitude of the interannual variability (Fig. 5). There are many features in common with the time series at 17-19°S. A notable difference is that the negative anomaly in late 1982 is somewhat more prominent and confined in longitude between -90° and 180°. Figure 4 shows a small negative anomaly at 17-19°S in 1982 confined to -180° to -90° longitude, out of phase in longitude with 37-39°S data. This might suggest that circulation changes in the Pacific associated with El Nino have significantly altered the ozone distribution.

In Fig. 6 the longitude time series at 57-59°S suggests that the ozone east of -70° to -90° longitude behaves differently from that to the west. This longitude band is just to the west of the Andes. The amplitude of the variability at 0° and just to east is notably larger. This corresponds to the area south of South Africa just north of the Antarctic coast.

To help better determine the relation between ozone at the equator and high latitudes the 30 mb wind field at Singapore has been used to define the QBO (Fig. 7). The 30 mb winds correlate more strongly with ozone than the 50 mb winds. Table 1 summarizes the correlation statistics between the Singapore winds and ozone at particular latitude, longitude bins at zero lag time. The correlation between the Singapore wind and the zonal mean total ozone is also shown as a function of latitude.

The ozone is strongly correlated with the wind at all points in the 1-3°N bin (the latitude of Singapore). The correlation decreases monotonically towards the nodal line, becomes negative, and increases in magnitude. The zonal mean results show the strongest anticorrelation between approximately 20° and 40°S consistent with the spectral analysis discussed above. The most negative correlation

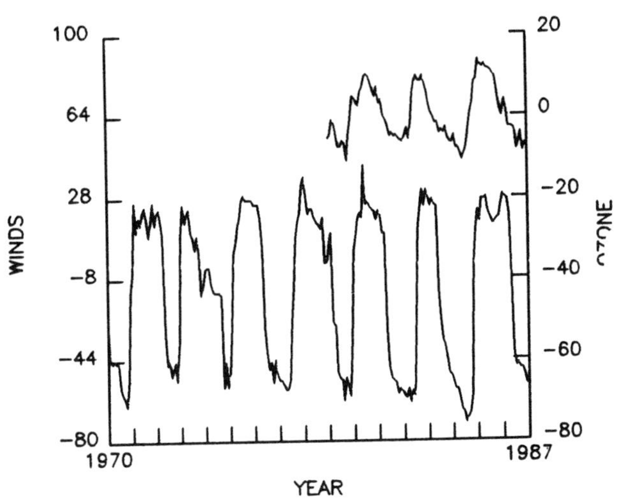

Fig. 7 Zonal wind at Singapore at 30 mb from January 1970 to January 1987, total ozone at Singapore from November 1978 to January 1987.

TABLE 1: Correlation between Singapore 30 mb winds and ozone in specific latitude, longitude bins.

Latitude	Longitude -165°/ -170°	-75°/ -80°	10°/ 15°	100°/ 105°	Zonal Mean
1-3°N	.802	.751	.749	.801	.810
5-7°S	.779	.623	.666	.744	.766
7-9°S	.660	.456	.448	.609	.645
9-11°S	.396	.203	.175	.297	.338
11-13°S	-.048	-.062	-.127	-.045	-.088
17-19°S	-.390	-.487	-.572	-.500	-.606
27-29°S	-.576	-.541	-.518	-.502	-.679
37-39°S	-.470	-.335	-.569	-.522	-.672
47-49°S	-.300	-.101	-.339	-.241	-.404
57-59°S	-.014	-.063	-.154	-.185	-.132
65-67°S	-.164	-.007	-.146	-.124	-.164

is at 34°S. The maximum magnitude of the correlations between Singapore winds and ozone always occurs when lag times are 2 months or less.

South of 50°S the correlations are in general statistically insignificant. They are particularly low in the -75° to -80° longitude bin. More careful analysis reveals that the region just to the west of the Drake Passage, south of South America, is consistently uncorrelated with the tropical winds and with ozone variability in the South Atlantic.

5. DISCUSSION AND SUMMARY

The relationship between equatorial variability and middle and high latitude variability has been investigated using total ozone data from TOMS. The period of the equatorial QBO in both ozone and winds is about 28 months. South of the nodal line most of the variance in the ozone time series appears in the 20-21 month period.

A 28 month oscillation, out of phase with the equatorial signal, is evident from approximately 20-40°S. This signal decreases rapidly poleward of 40°S and the 20-21 month oscillation is dominant. Therefore there is an indication that there is a "tropical" QBO signal in the subtropics and lower middle latitudes, but there is little indication of a direct QBO influence at high latitudes.

An indirect effect based on modulation of planetary wave activity by the phase of the winds at the equator is possible (for instance, Holton and Tan, 1980). This is indeed suggested by Garcia and Solomon (1987) and Schoeberl et al. (1988). A more restrictive analysis based on southern hemisphere winter and/or spring data is necessary to investigate this hypothesis.

An interesting result of this study is the longitudinal variation of the interannual variability at high southern latitudes. For instance the interannual variability is larger and more clearly defined on the Halley Bay side of Antarctica than on the Syowa side. The longitudinal variation suggests possible El Nino effects or an influence by the Andes Mountains.

REFERENCES

Angell, J. K and J. Korshover, 1973: Quasi-Biennial and Long-Term Fluctuations, Mon. Wea. Rev., 101, 426-443.

Bojkov, R. D., 1987: The 1983 and 1985 Anomalies in Ozone Distribution in Perspective, Mon. Wea. Rev., 115, 2187-2201.

Bowman, K. P., 1988: Global Trends in Global Ozone, Science, 239, 48-50.

Garcia, R. R. and S. Solomon, 1987: A Possible Relationship Between Interannual Variability in Antarctic Ozone and the Quasi-Biennial Oscillation, Geophys. Res. Lett., 14, 848-851.

Guthrie, P. D., 1983: Evidence for Quasi-Periodic Components in Dobson Network Total Ozone Records, J. Geophys. Res., 88, 3809-3818.

Hasebe, F. 1983: Interannual Variations of Global Total Ozone Revealed from Nimbus-4 BUV and Ground-Based Observations, J. Geophys. Res., 88, 6819-6834.

Hilsenrath, E., and B. M. Schlesinger, 1981: Total Ozone Seasonal and Interannual Variations Derived from the 7-Year Nimbus-4 BUV Data Set, J. Geophys. Res., 86, 12087-12096.

Holton, J. R., and H. Tan, 1980: The Influence of the Equatorial Quasi- Biennial Oscillation at 50 mb, J. Atmos. Sci., 37, 2200-2208.

Labitzke, K. and H. van Loon, 1988: Associations between the 11-year solar cycle, the QBO, and the atmosphere. Part I: the troposphere and stratosphere in the Northern Hemisphere in winter, J. Atmos. Terr. Phys., 50, 197-206.

Oltmans, S. J., and J. London, 1982: A Model of the Quasi-Biennial Oscillation on a Equatorial Beta-Plane, Quart. J. R. Met. Soc., 108, 335-352.

Ozone Trends Panel, 1988: Ozone Trends Panel Report, NASA, Washington, DC.

Schoeberl, M. R., L. R. Lait, and P. A. Newman, 1988: Quasi-biennial modulation of the Antarctic ozone depletion, submitted to J. Geophys. Res.

Watson, R. T. and M. J. Prather, 1988: Stratospheric Ozone (Letters to Science), Science, 239, 847.

LONG-TERM CHANGES IN SBUV/TOMS RELATIVE TO THE WORLD PRIMARY STANDARD DOBSON INSTRUMENT

Richard D. McPeters
NASA Goddard Space Flight Center
Greenbelt, Maryland 20771, USA

W.D. Komhyr
NOAA/ERL/Air Resources Laboratory
Boulder, Colorado 80303, USA

ABSTRACT

The SBUV and TOMS (solar backscattered ultraviolet and total ozone mapping spectrometer) instruments on Nimbus 7 have measured a decrease in total ozone between 1979 and 1986 of 6%, while the Dobson network reports only a 3% decrease during the same period, bringing the long term calibration of SBUV/TOMS into question. We have examined the time dependence of the calibration by comparing ozone measured by TOMS and SBUV with that measured by Dobson instrument No. 83, which has been maintained since 1962 as a standard for total ozone. Measurements of AD pair ozone made with instrument No. 83 at Mauna Loa observatory in the summers of 1979, 1980, 1981, 1984, 1986, and 1987 were compared with coincident TOMS ozone measurements. The comparison shows that TOMS was stable relative to instrument No. 83 between 1979 and 1982, had decreased by 1% relative to instrument No. 83 by 1984, and had decreased by 3% by 1986 and 1987. A similar time dependence is seen when an ensemble of 41 Dobson stations throughout the world is compared with TOMS over the period 1979-1986. The most likely reason for the relative drift is that the diffuser plate, which is used by both SBUV and TOMS to measure solar flux, has suffered an uncorrected wavelength dependent degradation, with most of the degradation occurring between 1983 and 1986.

1. INTRODUCTION

The basic problem common to all attempts to determine long term trends in ozone is that the instruments themselves are susceptible to drift and it is very difficult to prove that a system has been stable. Only through intercomparison with many independent instruments measuring ozone is it possible to develop confidence that a system such as SBUV/TOMS has accurately measured a trend in ozone. As early as November 1986 Fleig et al. (1986) noted the existence of a long term average drift between the SBUV total ozone and average total ozone for an ensemble of 41 Dobson stations. But individual Dobson stations vary greatly in their comparison with SBUV, so it was possible that any drift could represent a problem with the Dobson measurements.

As an alternative to comparison with a network of stations, comparison with a single Dobson instrument known to have been maintained with an absolutely constant calibration would be sufficient to determine whether SBUV (and TOMS) was stable. Dobson instrument No. 83, which was established in 1962 as the standard spectrophotometer for total ozone measurements in the USA, is arguably the best candidate for such a long term standard instrument. Komhyr et al. (1988) describe the calibration history of this instrument and note that analyses of the calibration data show that the calibration scale for instrument No. 83 has been maintained throughout 1962-1987 to within an uncertainty of ±0.5%. If this instrument has been maintained to this accuracy, it should be possible to clearly identify any instrument drifts in the SBUV/TOMS total ozone.

2. THE OZONE COMPARISON

The comparison is based on a set of observations made at Mauna Loa observatory in the summers of 1979, 1980, 1981, 1984, 1986, and 1987, for the purpose of monitoring the calibration of instrument No. 83. A calibration using the Langley method involves making a number of measurements of ozone on an individual day in order to be able to extrapolate to zero air mass. Thus, multiple Dobson observations are available each day so that self-consistency can be checked and temporal variations detected. Such calibrations with instrument No. 83 were repeated on a number of different days at Mauna Loa over a period of one to two months in order to determine the reproducibility of the calibration.

These observations by instrument No. 83 are compared with TOMS overpass observations. SBUV and TOMS are independent instruments except that they share the diffuser plate that is used for instrument calibration. Total ozone measured by TOMS has been 3.0% greater than that measured by SBUV since launch, possibly because of small wavelength errors in one instrument or the other, but otherwise the two instruments agree extremely well. TOMS data are used here because TOMS, a

ISBN 0-937194-15-8

TABLE 1. DIRECT COMPARISONS OF TOTAL OZONE MEASURED BY INSTRUMENT NO. 83 WITH TOMS OVERPASS TOTAL OZONE (CORRECTED BY 8 DU FOR STATION ALTITUDE)

DATES	DAYS	TOMS AVERAGE	#83 AVERAGE	% DIFF TOMS-#83
June 8 - Aug 14, 1979	34	285.4	(2) 280.0	+1.93% ± .26%
			(3) 280.0	+1.93% ± .26%
June 20 - AUG 1, 1980	24	277.9	(2) 274.0	+1.44% ± .33%
			(3) 275.2	+0.98% ± .33%
June 5 - Aug 12, 1981	48	281.2	(2) 275.9	+1.93% ± .33%
			(3) 276.7	+1.63% ± .33%
Aug 7 - Oct 23, 1984	13	268.3	(2) 266.5	+0.69% ± .29%
			(3) 267.5	+0.32% ± .30%
June 10 - July 11, 1986	11	276.9	(2) 279.1	-0.78% ± .35%
			(3) 280.5	-1.30% ± .37%
May 20 - July 9, 1987	33	287.3	(2) 291.8	-1.53% ± .34%
			(3) 291.3	-1.38% ± .34%

cross-track scanner, observes ozone within a degree of Mauna Loa every day, while SBUV, a nadir viewing instrument, may be on a track that comes no closer than within 13° of Mauna Loa. Ozone correlation begins to decrease with even a few degrees of separation from a Dobson station.

The details of the intercomparison are given in Table 1. In 1979 there were 34 days between June 8 and August 14 on which both instruments measured total ozone. The average ozone measured by TOMS was 293.4 milli-atm-cm (commonly known as the Dobson unit, or DU), while that measured by Dobson No. 83 was 280.0 DU, a difference of 4.8%. Part of this difference is caused by an altitude difference. The TOMS field of view is approximately a 40 km square with average surface altitude near sea level, while the Dobson observation is made at Mauna Loa observatory, altitude 3397 m. Based on ECC sonde observations at Mauna Loa, there are about 8 DU between sea level and station altitude in the summer. Subtracting 8 DU from the TOMS total ozone gives 285.4 DU, which reduces the difference from 4.8% to 1.93%, with a statistical error of 0.26%. Similar comparisons are given for each year. Two different results are given for Dobson each year, labeled (2) and (3), which represent two slightly different calibrations for instrument No. 83. In case 2 the Dobson data are reduced using the 1976 wedge calibration for all years, while for case 3 the data are reduced using updated wedge calibrations. The results for (2) and (3) are shown to give some feel for the level of uncertainty in the instrument No. 83 results.

The percent difference between TOMS and Dobson No. 83 cases 2 and 3 are plotted for each year in Fig. 1. This Figure shows that TOMS was relatively constant with respect to instrument No. 83 in the period 1979-1981, especially if the Dobson results for 1980 are down-weighted because of known problems with the observational technique of the operator that year. But by 1984 the difference was +0.5%, a 1.5% decline relative to the earlier period, and by 1987 the TOMS was 1.4% less than instrument No. 83, a decline of 3.4% from 1979-1981. The overall pattern is that TOMS (and by implication SBUV) was relatively constant relative to instrument No. 83 between 1979 and 1981, declined rapidly between 1981 and 1986, and appears to be declining less rapidly between 1986 and 1987.

This decline relative to Dobson confirms the decline noted earlier relative to many individual

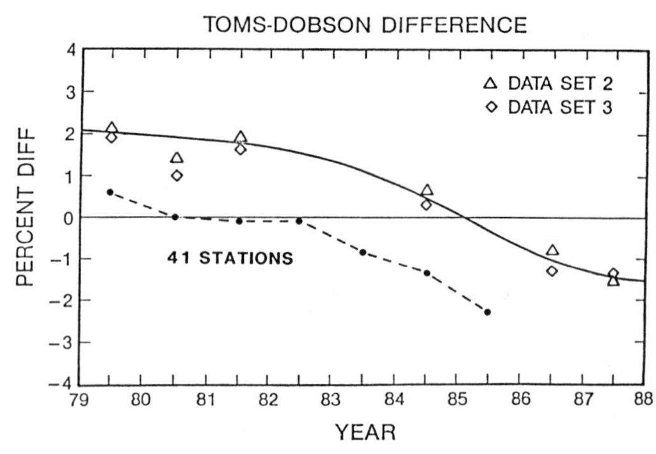

Fig. 1 The average difference each year between TOMS and Dobson instrument No. 83. Difference with an ensemble of 41 Dobson stations also shown.

stations. Also plotted in Fig. 1 (dashed line) is the yearly average TOMS-Dobson trend relative to an ensemble of 41 Dobson stations for the period 1979-1985. These data are the data used by Fleig et al. (1986) to calculate the 0.38% per year trend, but with data for 1985 comparisons added. Statistically the data from the 41 station ensemble could be considered consistent either with a constant linear trend or with the instrument No. 83 result: an initial relatively constant period until 1982 followed by a rapid decline. Very similar results were obtained when Bojkov et al. (1988) compared TOMS with a set of 92 carefully evaluated Dobson and M83 stations.

3. IMPLICATIONS OF A TOMS-DOBSON DRIFT

The conclusion appears to be that there has been about a 3% drift between TOMS and Dobson since 1979 with most of the drift occurring after 1982. The fact that the same pattern, relative stability before 1982 followed by a sharp decline between 1982 and 1986, occurred in the TOMS-Dobson comparisons for so many independent stations was good evidence in itself. But the complete confirmation by comparisons with the standard Dobson instrument No. 83 for both TOMS and SBUV is compelling evidence that this trend relative to Dobson is real.

Accepting the difference as real, could its origin be geophysical and not instrumental? One possibility is that a secular increase in tropospheric ozone has occurred in this time period. Dobson infers ozone by measuring the transmission of sunlight through the atmosphere to the instrument and is equally sensitive to tropospheric and stratospheric ozone. TOMS (SBUV) infers ozone by measuring light that penetrates the atmosphere, Rayleigh scatters in the troposphere or is reflected by clouds or the ground, and is transmitted through the atmosphere again to the satellite. The ozone measurement is accurate because most of the scattering occurs below most of the ozone. But because not all of the light penetrates the lower troposphere, the satellite measurement is only about 50% sensitive to tropospheric ozone; ie., if a 2% change in tropospheric ozone occurs, only about 1% of the change will be detected by TOMS and a TOMS-Dobson difference will result. The observed 3% TOMS-Dobson difference represents 9 DU of ozone, so an 18 DU change in tropospheric ozone between 1981 and 1987 would be required to explain the difference. This large a change has not been observed. A limited set of ECC sonde measurements of the amount of ozone in the lowest 4 km near Mauna Loa averaged 8.3 DU in the summer of 1983 and 7.0 DU in summer 1986, a difference of only 1 DU. Logan (1985) found that in northern Europe tropospheric ozone was increasing between 1979 and 1983, but only by 2% per year at Hohenpeissenberg. Since an increase of more than 100% in tropical tropospheric ozone would be required to explain the observed TOMS-Dobson difference, it seems that this must be discounted as a possible cause of the relative drift.

Since both TOMS and SBUV are changing relative to Dobson, it is most likely that the source of this discrepancy is an error in the characterization of the diffuser plate, the one critical element the two instruments share. This was the conclusion of the Ozone Trends Panel based on all the evidence, including this study and all other comparisons and on analysis that showed that the diffuser calibration characterization that was implemented was non-unique. It should be noted that most of the absolute change in the diffuser plate reflectivity occurred between 1979 and 1982 but was corrected very accurately by the diffuser plate characterization model. Only since 1982 have errors in the characterization become clearly detectable.

ACKNOWLEDGEMENTS

This work was done in support of the Ozone Trends Panel. We thank the members of the Column Ozone Measurements committee headed by F.S. Rowland for helpful discussion and advice.

REFERENCES

Bojkov, R.D., C. Mateer, and A. Hansson, 1988, Comparison of ground-based and TOMS total ozone measurements used in assessing the performance of the global ozone observing system (GO3OS), J. Geophys. Res., accepted in March 1988.

Fleig, A.J., P.K. Bhartia, C.G. Wellemeyer, and D.S. Silberstein, 1986, An assessment of the long-term drift in SBUV total ozone data based on comparison with the Dobson network, Geophys. Res. Lett., 13, 1359-1362.

Komhyr, W.D., R.D. Grass, and R.K. Leonard, 1988, Dobson Spectrophotometer No. 83: a standard for total ozone measurements, 1962-1987, submitted to J. Geophys. Res.

Logan, J., 1985, Tropospheric ozone: seasonal behavior , trends, and anthropogenic influence, J. Geophys. Res., 90, 10463-10482.

OBSERVATIONS OF TRACE CONSTITUENTS RELEVANT TO THE OZONE ISSUE

MAP/GLOBUS NO_x: AN OVERVIEW

J.P. Pommereau
Service d'Aéronomie
91371 Verrières le Buisson, France

ABSTRACT

GLOBUS NO_x is a combined effort of more than 30 international groups to observe nitrogen oxides compounds and the constituents involved in their photochemistry in the stratosphere. Seven multi-instrumented balloons were launched from the Centre National d'Etudes Spatiales balloon base at Aire sur l'Adour in southern France in September 1985. The flights were closely coordinated with four satellite experiments in orbit and with ground-based instruments observations distributed in southern France and Switzerland. The analysis of the data which followed during three years the field campaign itself, has led to a better evaluation of in-situ and remote sensing observations of NO, NO_2, NO_3 and O_3 by various techniques. Several sources of systematic error have been identified and when possible reduced. A consistent data set was issued for comparison with model predictions and previous observations. In the stratosphere, the NO_x budget, the vertical distributions and the diurnal variation of the constituents were found in general agreement with current photochemistry. However, several details differ from what was expected, i e larger ratio between NO and NO_2, and NO_2 reduction during the late afternoon which require further interpretation. In the upper troposphere, the NO_x concentration , larger than predicted by a factor between 3 and 10, indicates an unidentified source.

1. INTRODUCTION

In 1970, Crutzen suggested that man-made NO and NO_2, emitted into the stratosphere by high-altitude aircrafts, would reduce the amount of stratospheric ozone. Since then, many research groups in the world have reported measurements of nitrogen oxides and related species involved in their photochemistry. However, the concentrations measured with different instruments at various locations and seasons have often been very different. The systematic errors of the instruments and natural atmospheric variability were impossible to separate. In order to undestand and perhaps to reduce the existing differences, large multi-instrumented field campaigns have taken place in Europe and the United States in which a wide international community has been involved : the Balloon Intercomparison Campaigns, BIC I and II in 1982 and 1983 (Watson et al. 1988); and MAP/Globus in 1983 (Offermann, 1987).

Taking advantage of the results of the two campaigns, primarily dedicated to instrument evaluations and improvements, a new field experiment for nitrogen oxide studies in the stratosphere was planned : MAP/GLOBUS NO_X. The major objective was to perform simultaneous observations of nitrogen compounds at various local times, in order to provide a combined data set needed for the estimation of total NO_X and the partitioning between the species.

The project was greeted enthusiastically by the international community and more than 30 research groups indicated their intention to participate. Support was given by CNES, NASA and the Fluorocarbon Panel of the Chemical Manufacturers Association (CMA). An operational plan was carefully defined involving US and Japanese satellites, 7 multi-instrumented balloons and 13 experiments on the ground. The campaign took place in September 1985, at the CNES balloon range at Aire sur l'Adour in South Western France. Among the 40 instruments planned, only one total failure occured.

A year later, after data processing, the results of the measurements of each technique were compared and mutually criticized for potential souce of error. Another year of complementary work in the laboratory was needed to reduce the systematic errors which were identified and to produce an almost complete data set of nitrogen compounds, related constituents and parameters involved in the photochemistry. These results are presently under publication and are available for photochemical interpretation. Combined papers on NO_x vertical distributions, diurnal variations and stratospheric or tropospheric budgets are now in preparation.

2. STRUCTURE OF THE CAMPAIGN

Globus NO_X has been centered around seven multi-instrumented balloon gondolas launched in clusters and when possible during or close to a satellite overpass. Measurements from the ground at various altitudes, (Aire sur l'Adour and Bordeaux at sea level, Haute Provence Observatory at 600 m, Pic du Midi at 2800 m and Jungfraujoch Observatory at 3580 m), were performed from late August until early October. Observations from the ground and ozone and temperature soundings were more frequent during the balloon flights.

Table 1 lists the balloon payloads, table 2 the satellite-borne experiments and table 3 the ground-based instruments, together with the measured constituents and the name of the research group or institution operating the experiment.

TABLE 1. BALLOON-BORNE INSTRUMENTS

PAYLOAD/DATE	INSTRUMENT	CONSTITUENTS	INSTITUTION
1-Sept 19, 12 UT	Chemiluminescence	NO,NOx	University of Nagoya PEL,NZ
	Visible Spectrometer	NO2,O3	University of Tokyo
	Visible Spectrometer	NO2	IAS Bruxelles
	Visible Spectrometer	NO2,O3,O4,H2O	Service d'Aeronomie
2-Sept 19, 13:30 UT	IR Interferometer	NO,NO2,HNO3	University of Denver University of Paris VI
	BPMR	NO,NO2,N2O5,HNO3	University of Oxford
3-Sept 20, 01 UT	Chemiluminescence	NO,NOx	MPAE,Lindau
	UV Photometer	O3	NASA,JSFC
	Albedometer/Cameras	Albedo	MPAE,Lindau
4-Sept 23, 22 UT	Visible Spectrometer (nighttime)	NO2,NO3,O3	LPCE,Orleans Geneve Observatory
5-Sept 24, 02 UT	Matrix Isolation/ESR	NO2,HO2	KFA Julich
	Visible Spectrometer	NO2,O3	University of Tokyo
6-Sept 28, 08 UT	UV Vis Radiometers	Direct Solar Flux	WRC Davos
	UV Spectrometers	Direct and Scattered	NASA GSFC
	UV Vis Spectrometer	Direct Solar Flux	IAS Bruxelles
7-Oct 4, 02 UT	Chemiluminescence	O3	Service d'Aeronomie
	Chemiluminescence	NO,NOx	University of Nagoya PEL NZ
	UV Photometer	O3	NASA JSFC

ISBN 0-937194-15-8

TABLE 2. SATELLITE-BORNE INSTRUMENTS

INSTRUMENT	LOCATION	CONSTITUENTS	INSTITUTION
Zeeman Radiometer	Pic du Midi	NO	University of Oxford
Zenith Sky Visible Spectro	ASA/Pic du Midi	NO2	PEL,DSIR,NZ
Visible Photometer	Aire sur l'Adour	O3	University of Graz
Dobson Umkehr	Haute Provence	O3	University of Reims
Microwave	Bordeaux	O3	Bordeaux Observatory
IR Spectrometer	Jungfraujoch	NO,N2O,HCl,HF	University of Liege
TF Spectrometer	Jungfraujoch	NO,NO2,N2O,HCl	University of Liege
Brewer Mast Sondes	Haute Provence	O3	Sevice d'Aeronomie
Lidar	Haute Provence	Temperature	Service d'Aeronomie

TABLE 3. GROUND-BASED INSTRUMENTS

INSTRUMENT	CONSTITUENT	INSTITUTION
SAGE 2	O3,NO2	NASA Langley University of Lille
SME	NO2	LASP Boulder LPCE Orleans
SBUV	NO,O3	NOAA/NASA GSFC
EXOS-C	O3	University of Tokyo

The achieved observations are summarized in table 4. Fig. 1 and 2 give an idea of the altitude and local time distribution of the observations of NO and NO_2. Among the 31 experiments involved in the campaign, only one totally failed, a spectrometer for solar flux measurement, and a balloon orientation system just before sunset, preventing the balloon borne IR interferometer for making NO and NO_2 measurements below the horizon. The balloon flight dedicated for direct and scattered solar flux observations had to be shortened because of the wrong direction of the wind at float. The weather was in general fair throughout the campaign and the ground measurement set is nearly complete. Weather data, radio-soundings, cloud pictures, SSU stratospheric temperature, and pressure charts have been collected, thanks to the weather officer of the balloon range.

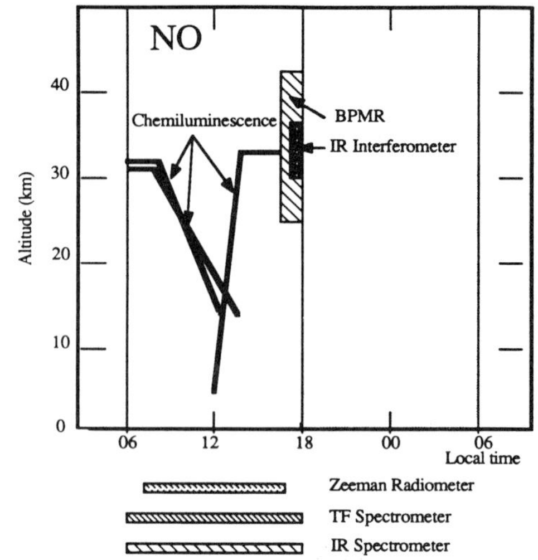

Fig. 1. Vertical distribution of NO observations versus time.

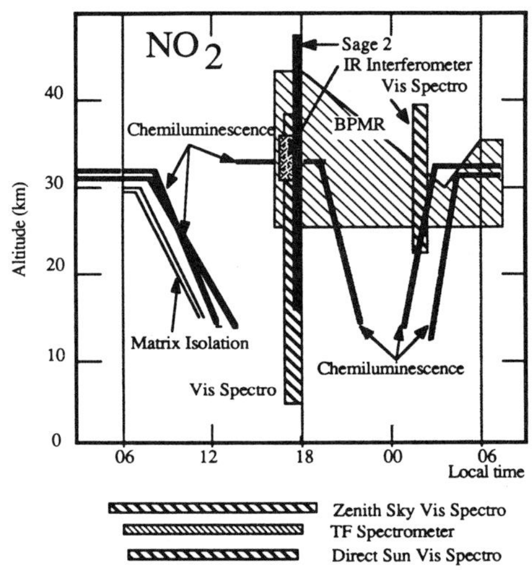

Fig. 2. Same as Fig. 1. for NO_2

TABLE 4. ACHIEVED OBSERVATIONS

		AUGUST						SEPTEMBER																														OCTOBER						
	LOCATION	26	27	28	29	30	31	1	2	3	4	5	6	7	8	9	10	11	12	13	14	15	16	17	18	19	20	21	22	23	24	25	26	27	28	29	30	1	2	3	4	5	6	7
BALLON FLIGHTS	ASA																									1&2&3				4 &5					6						7			
GROUND-BASED																																												
NO2 VIS/Sun TOKYO	ASA																X	X																										
NO2 VIS/Sun CNRS	ASA													X																														
NO2 VIS/Zenith PEL	ASA/PDM			X	X	X	X	X	X	X	X	X	X	X	X	X		X	X	X	X	X	X	X	X	X	X	X	X	X	X	X	X	X	X		X	X	X	X	X			
NO,NO2,N2O IR Specto	Jungfraujoch													X			X	X								X	X			X	X					X								
NO2,N2O,Interf TF	Jungfraujoch													X			X	X	X	X			X		X	X	X			X	X	X	X					X			X			
NO Zeeman Mod Oxford	PDM		X	X	X					X	X	X	X			X	X	X	X	X	X	X	X	X	X	X	X				X				X									
O3 Microwave	Bordeaux															X	X	X	X	X	X	X		X	X	X	X	X	X	X	X	X			X			X						
O3 Umkehr	Haute-Provence			X								X	X		X		X	X	X	X	X	X	X	X				X	X	X	X	X	X	X	X	X	X	X						
O3 Brewer Sondes	Haute-Provence																	X									X							X						X				
Temp Lidar	Haute-Provence								X		X		X			X		X					X	X			X	X		X			X	X		X	X							
SATELLITES																																												
NO2,O3 SAGE 2																							X	X																				
NO2 SME																							Every day within 20° longitude																					
O3 EXOS-C																	X	X				X				X	X		X		X	X	X							X			
NO SBUV																				X																								
O3 SBUV																							Every day within 1000 km																					

3. EXPERIMENTAL RESULTS AND COMPARISONS

Before attempting to use the data for photochemical interpretation, the results were compared, criticized and revised, and their uncertainties assessed. The findings which have been submitted for publication are summarized below.

3.1 NITRIC OXIDE, NO

Balloon-borne, satellite and ground-based remote sensing measurements were found in agreement (McKenzie et al, 1988b). Although the NO column may vary significantly from day to day by up to a factor of 2, these changes were unable to explain a large discrepancy between remote and in-situ (chemiluminescent) measurements, the latter being systematically larger. As a result, the NO chemiluminescence group investigated the problem in the laboratory, where they discovered that an increase of sensitivity with altitude of a mass flow sensor made the NO measurements overestimated at high altitude (Matthews et al., 1988). As a consequence, the NO in-situ data were revised, and the discrepancy with remote-sensing results reduced.

3.2 NITROGEN DIOXIDE, NO_2

NO_2 has been observed by remote sensing instruments from satellites, balloons and the ground and by one in-situ sampler. A comparison between simultaneous observations by three balloon-borne visible spectrometers, at various spectral ranges, has shown that the results are dependent on the spectral range used in the calculations. The use of NO_2 absorption cross-sections measured at low temperature (244K), reconciles all the observations. Spectroscopic measurements in the visible, between 430 and 450 nm without correction for temperature, overestimate the NO_2 concentration in the stratosphere by 15% to 18% (Pommereau et al, 1988). Correcting the data by this factor would have reconciled infra-red and visible occultation measurements during BIC (Roscoe et al, 1988). After revision the balloon data and the satellite SAGE 2 results do not differ by more than 4% in average above 15 km, i.e. well below the error estimations of the two experiments (Lenoble et al., 1988). In the troposphere, interference in the visible spectrum by other absorbing gases has been dicovered, which requires correction before NO_2 can be measured. Interfering constituents have been identified as O_4 (the oxygen dimer) and water-vapor (Pommereau et al, 1988).

Comparison between night-time, morning, evening, remote sensing and in-situ data requires the use of a diurnal model. Naudet et al (1988) have shown that, which such a model, the data set is self-consistent. Ground-based measurements by infra-red, visible direct-sun occultation and visible scattered sunlight at zenith have been compared (Zander et al, 1988). They show a good internal consistency when local contamination in the boundary layer is absent. A more definite assessment between infra-red and visible methods would require simultaneous high altitude observations at the same location.

3.3 OZONE

Numerous observations of ozone have been made during the campaign, coincident in both space and time, by various techniques. At high altitude between 30 km and 55 km, Umkehr, microwave and EXOS-C observations were found to be in better agreement (10% to 20%) than expected from the uncertainties claimed by the authors (de La Noe et al, 1988). In the middle and the lower stratosphere the GLOBUS NO_x results have been discussed together with the findings of previous field campaigns.From the intercomparison, the present status of ozone measurements appears to be (Robbins et al, 1988):

- In-situ UV photometer : 3% total uncertainty ;
- Visible occultation : 5%-10%
- SAGE 2 : 10%
- In-situ gas phase chemiluminescence : 10% ;
- Umkehr : 10% in layers 4 and 5, 15% in layer 6 ;
- SBUV : 10% in average, ozone peak systematically at lower altitude ;
- EXOS-C BUV : 10 - 15% ;
- Brewer sondes : 5 - 10% between 18 and 25 km - 15 - 20% above, occasional failures.

Except for EXOS-C, the assessed uncertainties are worse than those claimed by the authors.

4. PHOTOCHEMICAL RESULTS

The above intercomparisons and evaluation of uncertainties have produced a data set of self-consistent NO_x, related constituents and atmospheric parameters, which can be used for testing photochemical models testing.

Kondo et al., (1988) have shown that the slow NO decrease observed at 32 km at the end of the day after 80° SZA might be interpreted by a progressive reduction of the NO_2 photolysis rate. Naudet et al., (1988) are concluding in the same direction, that sunset conversion of NO into NO_2 begins earlier than predicted by their model. This could be the effect of neglecting multiple scattering in the calculations. In the same paper they show that systematic differences between their model and daytime in-situ observations of NO and NO_2 are apparent. The ratio of NO to NO_2 appears larger than predicted at high altitude (30 km). Matthews et al (1988) analyzing NO vertical profiles have shown that the morning increase of the gas between 26 and 33 km is consistent with the calculated N_2O_5 photolysis.

An analysis of NO_x defined as NO + NO_2 (Naudet et al, 1988) does not show any significant difference from predicted values in the stratosphere. In the upper troposphere, the observed NO_x appears to be larger than those predicted by a factor between 3 and 10. This would imply a source which could be tentatively aircraft emissions or lightning (Pommereau et al, 1988b).

5. CONCLUSION

GLOBUS NO_X has allowed a better evaluation of the uncertainties of measurements of NO_X compounds in the atmosphere. In several cases, it has reduced the systematic errors in these and previous results. When compared to the previous intercomparison campaigns, the quality of the measurements has improved. This is certainly the result of the intercomparisons which forced the workers to better understand their instruments and the interpretation of their data. There are still some remaining inconsistencies and pending questions in the data set, which can be addressed in the near future: is the excess of NO compared to NO_2 significant ? Are all NO measurements self-consistent after correction of the chemiluminescent data ? Can the NO reduction during the afternoon after 80° SZA be explained by the effect of multiple scattering on JNO_2 ? Then, would the predictable NO_2 variation be compatible with the experimental data ? What are the NO_x species measured by chemiluminescence after reduction through a converter ? Are the large NO_x concentrations observed in the upper troposphere compatible with aircraft emissions only or do they need another source? Simple tests in the laboratory or the field would help to improve the results. For example, simultaneous ground-based measurements in the visible and the infra-red at the Jungfraujoch observatory are anticipated for checking the diurnal variation of the NO_2 column. A complete interpretation of GLOBUS NO_x results requires now a modelling effort which has just begun. The data base, including NO_x and related constituents and pertinent atmospheric parameters, is now available for the community.

ACKNOWLEDGMENTS

The campaign was supported by CNES, NASA and the Fluorocarbon Panel of the Chemical Manufacturers Association (Grant FC 85-550) who are gratefully acknowledged, as are the assistance of D. Offermann and R.T. Watson during the program definition phase. The author is indebted to the personnel of the CNES balloon range at Aire sur l'Adour and to his colleagues, scientists and technicians, whose efforts made the campaign successful and wish to express his thanks to H.K.Roscoe for his helpful comments and suggestions.

REFERENCES

de la Noe, J., M. Koike, T. Ogawa and P. Marché, 1988, Comparison of Stratospheric and Mesospheric Ozone Profiles obtained by Ground-based and satellite Observations, Submitted to J. Atm. Chem..

Kondo, Y., W.A. Matthews, P. Aimedieu and D.E. Robbins, 1988, Diurnal Variation of Nitric Oxide at 32 km : Measurements and Interpretation, J. Geophys. Res., 93, 2451.

Lenoble, J., W.P. Chu, D.G. Murcray, O. Lado-Bordowski, C. Camy-Peyret, A. Perrin, J.P. Pommereau, F. Goutail, K. Shibasaki, N. Iwagami and P.C. Simon, 1988, Sage 2 nitrogen dioxide data comparison with Infra-red and visible balloon remote sensing measurements during Globus NO_X, in preparation.

Matthews, W.A., Y. Kondo, P. Fabian and B.C. Krüger, 1988, Nitric Oxide Profiles measured in situ during the Globus 85 Campaign, Submitted to J. Atm. Chem..

McKenzie, R.L., G.D. Peskett and H.K. Roscoe, 1988, Measurements of Stratospheric Columns of NO made with the Zeeman Modulator Radiometer during MAP/GLOBUS 1985, submitted to J. Geophys. Res..

McKenzie, R.L., W.A. Matthews, Y. Kondo, R. Zander, Ph. Demoulin, P. Fabian, D.G. Murcray, F.J. Murcray, O. Lado-Bordowsky, C. Camy-Peyret, H.K. Roscoe, J.A. Pyle and R.D. PcPeters, 1988, Intercomparison of NO Column Measurements during MAP/GLOBUS 1985, Submitted to J. Atm. Chem..

Naudet, J.P., M. Pirre, R. Ramaroson, P. Rigaud, P. Fabian, H. Helten, N. Iwagami, K. Shibasaki, T. Ogawa, Y. Kondo, W.A. Matthews, J.P. Pommereau, F. Goutail, P.C. Simon and W. Peetermans, 1988, Diurnal variation of stratospheric NO and NO_2 from MAP/GLOBUS 1985, Quad. Ozone Symp. , in press.

Offermann, D., 1987, The MAP/GLOBUS Campaign 1983 : Introduction, Planet. Space Sci., 35, 515.

Pommereau, J.P., F. Goutail, N. Iwagami, K. Shibasaki, P.C. Simon, W. Peetermans, J.P. Naudet, P. Rigaud and D. Huguenin, 1988a, Balloon Observations of Nitrogen Dioxide by visible Occultation during Globus NO_X, submitted J. Atm. Chem. .

Pommereau, J.P., F. Goutail, Y. Kondo, W.A. Matthews, and M. Helten, 1988b, An NO_X source in the upper troposphere ?, Quad. Ozone Symp., in press.

Robbins, D.E., P. Aimedieu, J. Pelon, J.P. Pommereau, F. Goutail, N. Iwagami, K. Shibasaki, T. Ogawa, M. Koike, P. Marché, J.P. Naudet, P. Rigaud, D. Huguenin, L. Lenoble, G. Maddrea Jr., 1988, Measurements of Stratospheric Ozone during the 1985 MAP/GLOBUS NO_X campaign, Submitted to J. Geophys. Res..

Roscoe, H.K., B.J. Kerridge, S. Pollitt, N. Louisnard, J.M. Flaud, C. Camy-Peyret, C. Alamichel, J.P. Pommereau, T. Ogawa, N. Iwagami, M.T. Coffey, W. Mankin, W.F. Evans, G.T. McElroy, J.B. Kerr, 1988, Intercomparison of Remote Measurements of Stratospheric NO and NO_2, J. Atm. Chem., In press.

Watson, R.T., H.K. Roscoe, P.T. Woods, 1988, The Balloon Intercomparison Campaigns: An introduction and Overview, J. Atm. Chem. In press.

Zander, R., Ph. Demoulin, G. Roland, W.A. Matthews, P.V. Johnston, J.P. Pommereau, N. Iwagami, K. Shibasaki, 1988, Measurements of Column abundances of Nitrogen Dioxide, NO_2 from the ground during the Globus NO_X Campaign, Quad. Ozone Symp., in press.

NITRIC OZIDE IN THE UPPER STRATOSPHERE

Y. Kondo and A. Iwata
Research Institute of Atmospherics
Nagoya University
Toyokawa, Japan

P. Aimedieu
Service d'Aéronomie, CNRS
Verrieres le Buisson, France

W.A. Matthews
Physics and Engineering Laboratory
Lauder, DSIR, New Zealand

W.R. Sheldon and J.R. Benbrook
Department of Physics
University of Houston
Houston 77004, USA

D.E. Robbins
NASA Johnson Space Center
Houston, Texas 77058, USA

ABSTRACT

A chemiluminescent NO detector was launched, together with in situ ozone instruments, on September 17, 1987 from CNES Balloon Launching Center at Aire sur l'Adour, France (44°N, 0°W). The improvement of the precision of NO measurement in the upper stratosphere involved a precise determination of the sample flow rate at pressures down to 2 mb with a reduced ozone/oxygen flow rate. The balloon reached a float altitude of 40 km (2.9 mb) around 0500 UT, 10 minutes before sunrise. The variation of NO at this altitude was measured for about 5 hours. A very rapid buildup of NO, presumably due to the photodissociation of NO_2, was observed for about 20 minutes. Following this, a much slower increase in NO concentration continued for at least 4 hours. The balloon started a slow descent at 1000 UT. The descent was terminated at an altitude of 24 km around 1300 UT. During the descent, a precise NO profile between 25 and 40 km was obtained. The NO and ozone profiles that were measured simultaneously are compared with previous observations.

INTRODUCTION

In the upper stratosphere, nitric oxide becomes a major species in the oxides of nitrogen family in the daytime. The catalytic destruction of ozone by chlorine species in the upper stratosphere is predicted to depend on the concentration of oxides of nitrogen [e.g., Cicerone et al., 1983; Brasseur and De Rudder, 1987]. The concentration of nitric oxide is therefore an important parameter in predicting any future ozone decrease. Up until now there have been a rather wide spread in the NO data measured in the 30-45 km region [WMO, 1986]. Observations of the diurnal variation of NO in the upper stratosphere are more limited compared to the measurements of NO profiles. Roscoe et al. [1986] observed the diurnal variation of NO between 32 and 43 km with a balloon-borne infrared radiometer. Here we describe a balloon observation to measure the temporal variation of NO from sunrise at 40 km and the NO profile between 40 and 25 km in the late morning.

ISBN 0-937194-15-8

INSTRUMENTATION

A chemiluminescent NO detector described in Kondo et al. [1988a, 1988b] was used to measure the NO concentration. The increase in the sensitivity of the mass flow sensor used to regulate the calibration NO in N_2 gas flow has been taken into account. To increase the precision in the estimate of the sample mass flow rate, the ozone/oxygen flow rate was reduced to about 200 STP cm^3/min. In addition, the sample flow rate at pressures down to 2 mb has been re-calibrated in the laboratory. This new work has made the precision of the NO measurement, even at 40 km, better than 10 %.

For high precision ozone measurements, an in situ UV absorption photometer and a chemiluminescent ozone instrument were flown together [Aimedieu et al., 1987]. Radiometers to measure upward and downward visible radiation were also mounted near the end of the intake tube of the NO detector. A brief description of the radiometer is given in Kondo et al. [1988b].

The NO and ozone instruments are mounted on a gondola as shown in Figure 1. The mixtures of sample air and reactant gases from the chemiluminescent NO and ozone instruments were exhausted through 10 to 15 m long tubes that were directed upward. Inlet tubes for the chemiluminescent NO and ozone instruments are 1.5 m long and are directed horizontally to avoid possible contamination from the gondola.

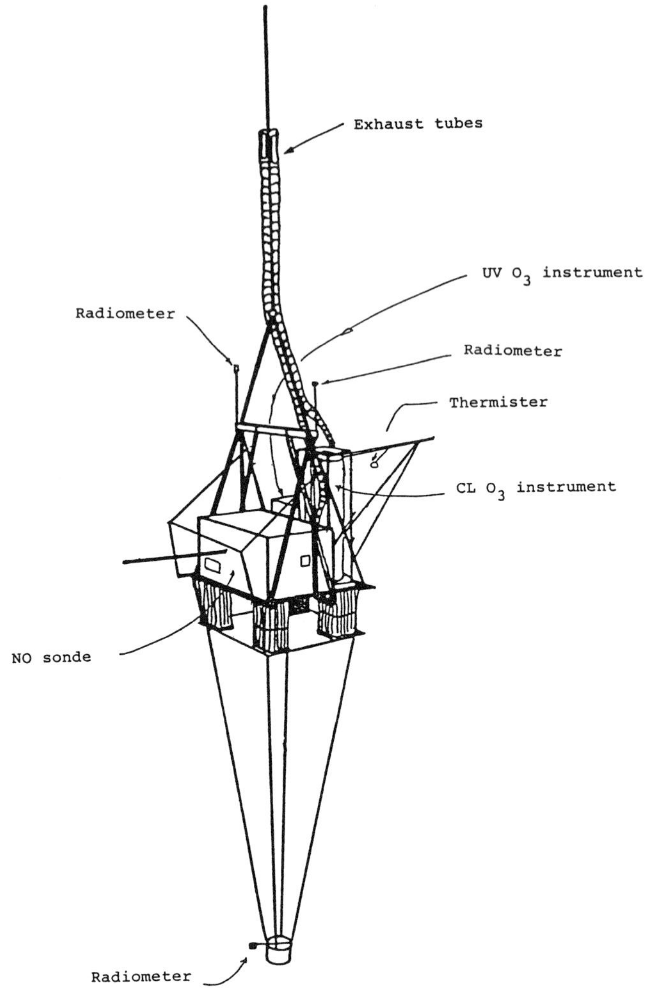

Fig. 1. Configuration of the gondola.

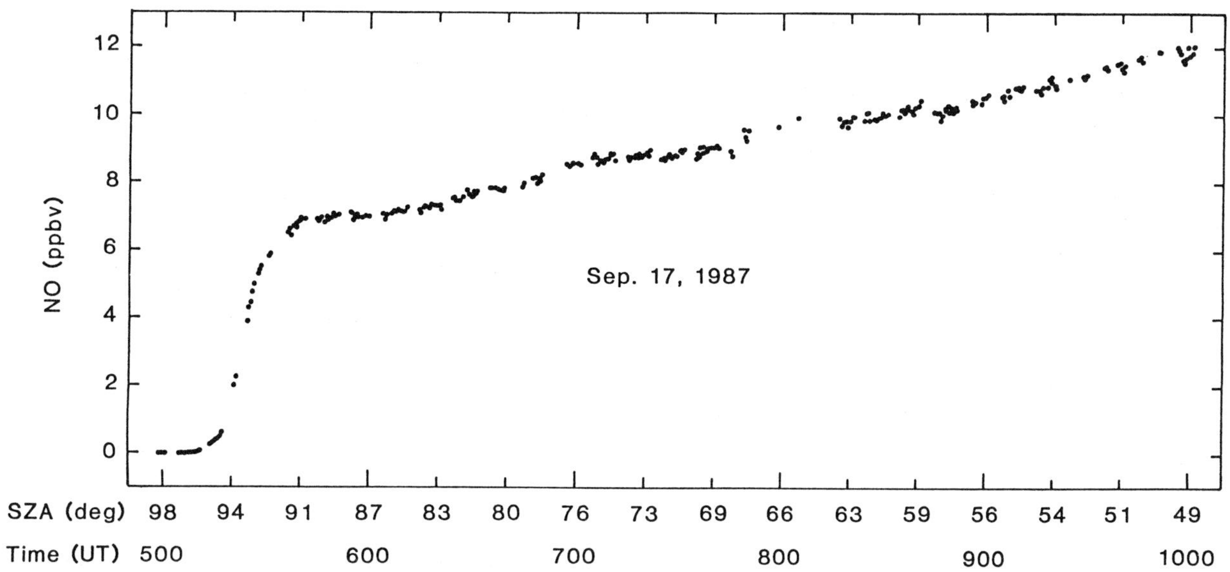

Fig. 2. Temporal variation of NO mixing ratio at 40 km.

BALLOON OBSERVATION

The balloon observation was performed on September 17, 1987 from the Centre National d'Etudes Spatiales (CNES) balloon launching center at Aire sur l'Adour in France (44°N, 0°W). The gondola was suspended 250 m below a 600,000 m^3 balloon. It was launched at 0210 UT and reached a float altitude of 40 km (2.9 mb) around 0500 UT, 10 minutes before sunrise. The photograph taken from Meteosat II at 0930 UT shows that southern France was dotted with little clouds.

The temporal variation of NO measured at the float altitude is shown in Figure 2. Before sunrise, the NO mixing ratio was about 0±0.02 ppbv. It starts to build up rapidly when the solar zenith angle (SZA) reaches 96°. This rapid increase has stopped by the time the SZA decreases to 90°. After 0600 UT, NO continues to increase at a rate of about 1 ppbv/hour for at least a few hours. Kondo et al. [1988a] observed a similar increase in NO at 33 km. The data of Kondo et al. [1988a] has been re-analyzed taking into account the sensitivity change of the mass flow sensor [Matthews et al., 1988]. The rate of NO increase has been reduced to be 0.9 ppbv/hour at 33 km. As the NO/NO_2 ratio will be considerably larger at 40 km than at 33 km, the rate of increase in total NO_x will be smaller at 40 km.

N_2O_5 is believed to be the dominant reservoir of NO_x at night. The decrease in N_2O_5 due to the photodissociation after sunrise can be expressed as $[N_2O_5]/[N_2O_5]_{SR} = \exp(-\int_0 j\,dt)$ where $[N_2O_5]_{SR}$ is the N_2O_5 concentration at sunrise and j is the photodissociation coefficient of N_2O_5 [Kondo et al., 1988a]. The calculated values of $[N_2O_5]/[N_2O_5]_{SR}$ at 40 and 32 km are shown in Figure 3. From this figure, the release of NO_x from N_2O_5 is expected to be almost complete by 1000 UT. The observation at the float altitude was terminated at 1000 UT and therefore it is difficult to be conclusive about this point especially considering that there are some irregular variations seen in the data.

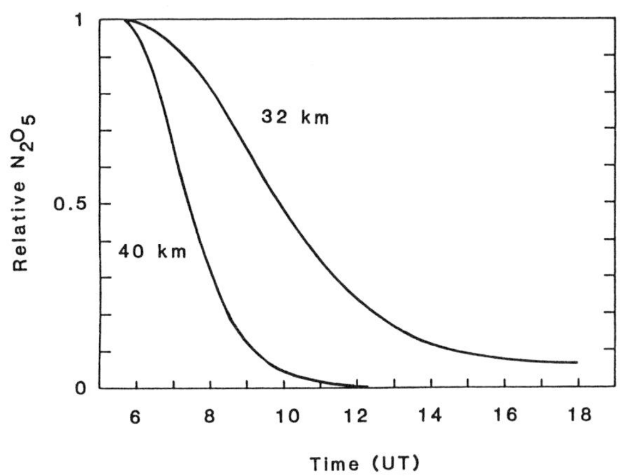

Fig. 3. Calculated decrease of N_2O_5 concentration after sunrise.

At around 1000 UT, the balloon started a slow descent. The NO concentration measured during the descent is shown in Figure 4. It should be noted that the balloon was at 37, 31, and 24 km at 1100, 1200, and 1300 UT respectively. The NO profile obtained here is consistent with our previous measurements in autumn from Aire sur l'Adour [Kondo et al., 1988a; Matthews et al., 1988].

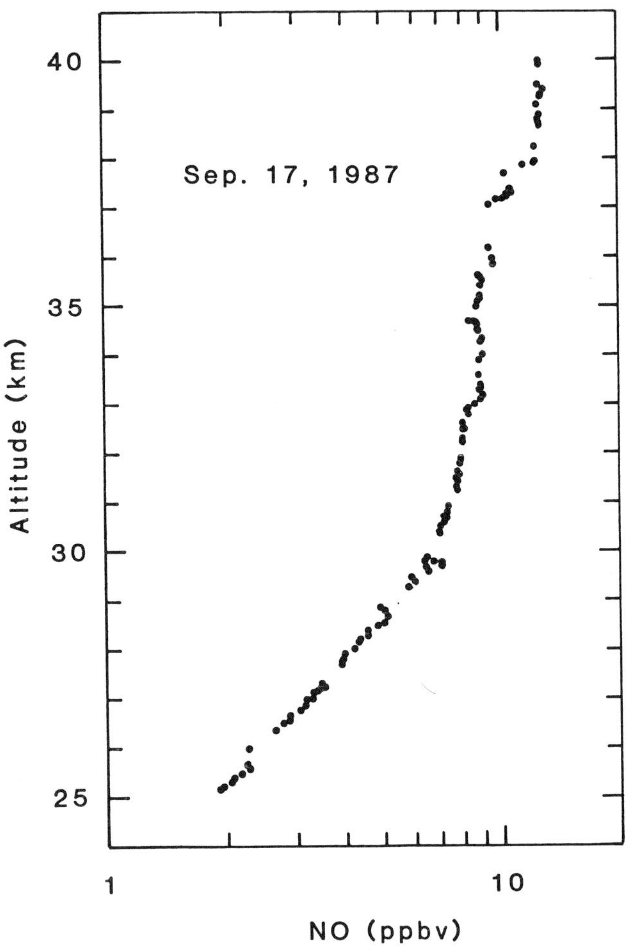

Fig. 4. Profile of NO measured during the slow descent of the balloon.

ACKNOWLEDGEMENTS

We wish to thank Monsieur Faucon, the chief of the Balloon Launching Center of CNES, and his staff at Aire sur l'Adour for the excellent and very professional balloon flight. The balloon flight was funded by CNES through Service d'Aéronomie, CNRS. Technical assistance provided by N. Toriyama, C. Bernard, P. Hulin, and J. Gilbert in preparing the instruments for the balloon experiment is gratefully acknowledged.

REFERENCES

Aimedieu, P., W.A. Matthews, W. Attmannspacher, R. Hartmannsgruber, J. Cisneros, W. Komhyr, and D.E. Robbins, Planet. Space Sci., 35, 563-585, 1987.

Brasseur, G., and A. De Rudder, The potential impact on atmospheric ozone and temperature of increasing trace gas concentrations, J. Geophys. Res., 92, 10903-10920, 1987.

Cicerone, R.J., S. Walters, and S.C. Liu, Nonlinear response of stratospheric ozone column to chlorine injections, J. Geophys. Res., 88, 3647-3661, 1983.

Kondo, Y., W.A. Matthews, P. Aimedieu, and D.E. Robbins, Diurnal variation of nitric oxide at 32 km: Measurements and interpretation, J. Geophys. Res., 93, 2451-2460, 1988a.

Kondo, Y., M. Takagi, A. Iwata, and W.A. Matthews, Diurnal variation of nitric oxide at 26 km, Proc. Int. Ozone Symp., (this issue), 1988b.

Matthews, W.A., Y. Kondo, P. Fabian, and B.C. Krüger, Nitric oxide profiles measured in situ during the GLOBUS 85 campaign, submitted to J. Atmos. Chem., 1988.

Roscoe, H.K., B.J. Kerridge, L.J. Gray, R.J. Wells, and J.A. Pyle, Simultaneous measurements of stratospheric NO and NO_2 and their comparison with model predictions, J. Geophys. Res., 91, 5404-5491, 1986.

VARIABILITY OF STRATOSPHERIC NO_2 PROFILES AT NORTHERN MIDDLE LATITUDES

Kazuo Shibasaki
Kokugakuin University, Shibuya-ku
Tokyo, Japan

Naomoto Iwagami, Kazuyuki Kita, and Toshihiro Ogawa
Geophysics Research Laboratory, University of Tokyo
Tokyo, Japan

ABSTRACT

Five profiles of the stratospheric NO_2 are presented here. All the measurements were made from balloon platforms using long-path visible absorption spectroscopy at sunset. Observation sites were distributed in northern middle latitudes from 32°N to 44°N: Palestine (32°N), Alamogordo (33°N), Sanriku (39°N), Aire sur l'Adour (44°N). A new retrieval technique based on the constrained linear inversion method was applied to our recent data as well as our previous data. Unlike an onion peeling method, this retrieval method allows us to infer the NO_2 distribution above a balloon altitude. A comparison of these five NO_2 profiles reveals a significant variability in lower stratospheric NO_2; however, the vertical column densities are generally consistent with the seasonal behavior observed on the ground. The altitude distribution of the lower stratospheric NO_2 seems to be affected largely by dynamical process as seen in the ozone distribution. We also examined the sensitivity of the temperature dependence of NO_2 absorption cross sections in retrieving the stratospheric NO_2 profile. The inclusion of the temperature effect makes the NO_2 densities reduced by about 20% uniformly with altitude as compared to those deduced by using the cross sections at 298 K.

1. INTRODUCTION

Nitrogen dioxide is one of the key species in stratospheric ozone chemistry. It is, therefore, important to elucidate the variability of the stratospheric NO_2. Several satellites have observed the global atmospheric NO_2 with visible and infrared spectroscopy since late 1970's. Seasonal and latitudinal variations of the stratospheric NO_2 density have been reported by Mount et al.(1983,1984), Russell et al.(1984a,b), Solomon et al.(1984), and Chu and McCormick (1986). However, since the satellite observation method has a limitation in observing the stratospheric NO_2 profiles below ~25 km, these authors discussed the latitudinal variation of the NO_2 profiles only above 25 km by representing zonal mean profiles or total column density.

Dynamical transport plays a dominant role in determining the NO_2 distribution in the lower stratosphere below 25 km where about a half of the total column NO_2 density exists. The balloon-borne measurement is useful to investigate the variability of the lower stratospheric NO_2. In this paper we present five stratospheric NO_2 profiles obtained by balloon-borne measurements with visible absorption spectroscopy in northern middle latitudes, and discuss the variability of the lower stratospheric NO_2 in terms of the dynamical transport effect. We also examine the sensitivity of the retrieved profiles to the temperature effect of NO_2 absorption cross sections adopted.

2. OBSERVATIONS AND DATA ANALYSIS

Our observational technique is based on absorption spectroscopy using the sun as a light source. We selectthe visible wavelength region of about 428 - 452 nm for NO_2 measurment and the ultraviolet wavelength region of about 326 - 339 nm for O_3 measurement. Table 1 gives the locations and the dates of five balloon observations. O_3 measurements were carried out only in the 1985 balloon observations. The instrument used in these observations is a spectrometer combined with a sun tracker, and is slightly modified from observation to observation; however, the same instrument was used in the 1985 observations, and for the observations made from Sankiku, the same instrument was used in 1986 as in 1978.

The slant column densities of NO_2 and O_3 are derived from each whole ratio spectrum using a least squares fitting method (Ogawa et al., 1981). As a reference spectrum, which should be free from NO_2 and O_3 absorptions, we used the solar spectrum obtained at a small solar zenith angle during the same flight. A vertical profile of the stratospheric NO_2 is determined from the variation of slant column density with solar zenith angle using a constrained linear inversion method developed by Twomey (1963). This method has advantage in that errors do not propagate from an upper layer to lower layers unlike the onion peeling method and in that a profile can be

TABLE 1. LOCATIONS AND DATES OF BALLOON OBSERVATIONS

No.	Location		Date
1	Sanriku	39.2°N, 141.8°E	May 29, 1978
5			Oct. 6, 1986
2	Palestine	31.8°N, 95.7°W	June 20, 1983
3	Alamogordo	32.9°N, 106.2°W	Apr. 30, 1985
4	Aire sur l'Adour	43.7°N, 0.3°W	Sep. 19, 1985

ISBN 0-937194-15-8

determined above and below the balloon altitude simultaneously. We adopt the constraint that minimizes the sum of squares of second-order difference plus the sum of squares of error multiplied by unknown constant: a priori assumption that a vertical profile should be smooth.

The precision of deriving slant column density from each ratio spectrum is usually less than $\pm 4\times10^{15}$ cm^{-2}. The precision of the retrieving processes is influenced by many factors such as the uncertainty of balloon position, the signal to noise ratio of the data and the inversion method itself, and it depends on altitude. The uncertainties are less than 20% below the balloon altitude and do not exceed 30% at uppermost regions. The total errors in retrieved vertical profile is estimated to range from 20% to 30%.

Recently, Leroy et al. (1987) reported the temperature dependence of NO_2 absorption cross sections. Their result seems to be important, since the profiles obtained with visible absorption spectroscopy show larger NO_2 densities than those obtained with an infrared technique, as suggested by Ogawa et al.(1981). This difference can probably be attributed to the uncertainty in cross sections adopted for visible absorption spectroscopy. These use Johnston's unpublished data without any temperature dependence (e.g. Harrison, 1979). Later in this paper we dicuss the effect of temperature dependence of absorption cross sections on the retrieved NO_2 profile.

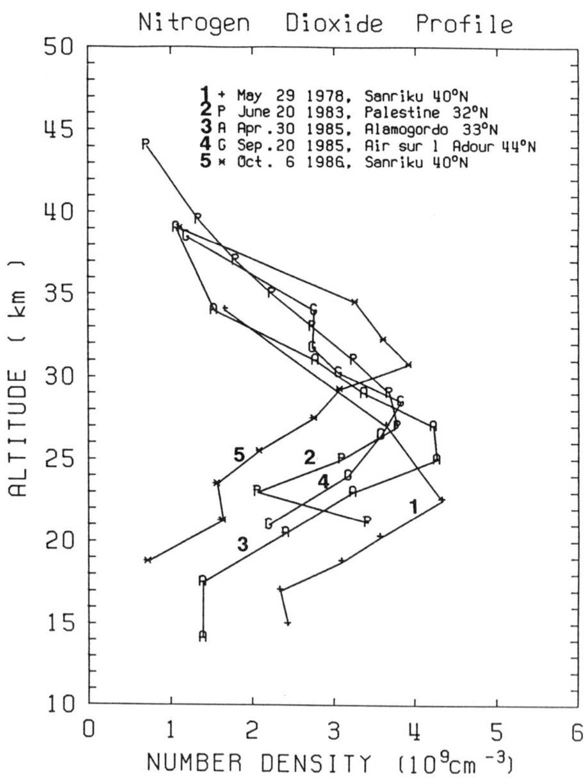

Fig. 1. Derived stratospheric NO_2 profiles in the balloon measurements listed. Each symbol is marked at the center of atmospheric layer. Latitudes denoted in the Figure are those where balloons are located at local sunset.

3. RESULTS AND DISCUSSION

3.1 NO_2 AND O_3 PROFILES

Figure 1 illustrates five NO_2 profiles obtained in the balloon measurements listed in Table 1. The 1983 observation at Palestine was carried out as a part of Balloon Intercomparison Campaign II (BIC II), the 1985 observation at Alamogordo was as a ground truth experiment for SAGE II, and the 1985 observation at Aire sur l'Adour was as a part of MAP/GLOBUS NO_x Campaign. The observation sites and seasons range from 32°N to 44°N and from early summer to mid-autumn, respectively.

Five profiles in Fig. 1 seem to be similar to one another above ~28 km, whereas they are highly variable in the lower stratosphere below ~28 km. The variability is a factor of 4 around 20 km and at most about a factor of 2 above 30km. Two profiles observed at Sanriku in particular show a significant difference in the peak altitude and consequently in the density in the lower stratosphere. Even if excluding these two profiles, the variability of NO_2 is still larger in the lower stratosphere than in the upper stratosphere. Vertical column densities calculated for those five profiles are listed in Table 2. The column densities are in agreement with the column densities at northern middle latitudes observed by Noxon (1979,1980) within the experimental uncertainties.

We concentrate our attention on the two profiles at Sanriku, since they are obtained at the same observation site with the same instrument but in different seasons. Figure 2 illustrates ozone profiles obtained from the corresponding balloon observations. In the 1978 observation, an ECC type ozonesonde on board the balloon gondola measured ozone density during the ascent. In the 1986 observation an ozonesonde was flown from Sanriku Balloon Center at the same time of the balloon sunset observation. Ozonesonde data is also available from the Canadian groups observations during the 1983 BIC II experiment, while for the Alamogordo observation, the simultaneous ozone profile was derived from the solar UV absorption measurement with the same spectrometer that was used in the NO_2 measurement. An ozone profile is available from a Brewer ozonesonde observation at Haute Provence (OHP) on September 20, one day after our balloon experiment for Air sur l'Adour.

Ozone in the lower stratosphere is a good tracer of the air motion because of its long atmospheric lifetime. We can see a significant difference in the two ozone profiles correspond-

TABLE 2. VERTICAL COLUMN DENSITY OF STRATOSPHERIC NO_2

No.*	Vertical column density (10^{15} cm^{-2})
1	7.4 ± 1.2
2	7.0 ± 1.4
3	6.9 ± 1.0
4	6.7 ± 1.2
5	5.5 ± 0.8

* Corresponding to numbers in Table 1.

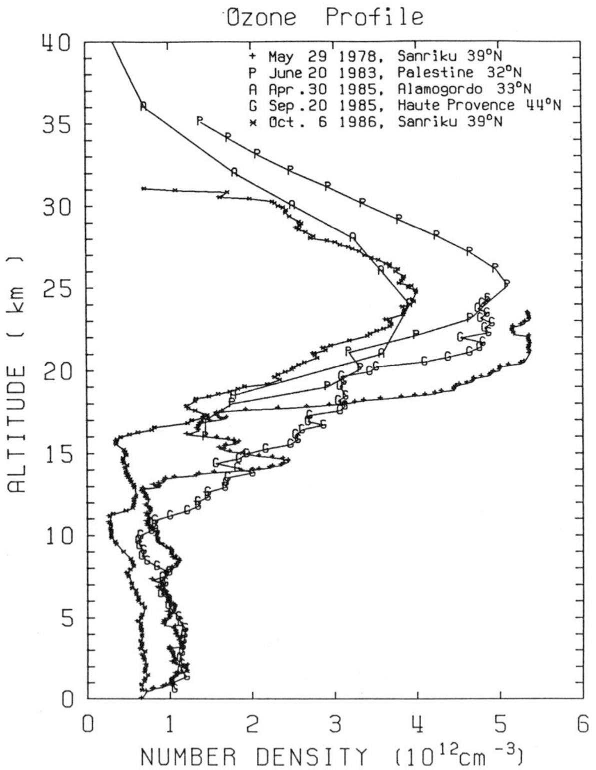

Fig. 2. Ozone profiles obtained at the corresponding balloon observations. Dates and observation sites are given in the Figure. Symbols are the same with corresponding balloon observations.

ing to the two NO_2 observations at Sanriku. A distinct subpeak occurred around 15 km in the 1978 ozone profile, whereas the 1986 profile is fairly smooth. The 1986 profile is representative of those in sub-tropical region. The 1978 profile is representative of those observed during a transient season from winter to summer in middle latitudes, implying that air masses of high latitudinal origin and those from low latitudes are mixed over Sanriku. Consequently a vertical profile of stratospheric NO_2 may be variable according to the origin of air mass.

During the balloon ascent in 1978, the wind direction changed from westerly to easterly/northerly at around 21 km. 30mb and 50mb geopotential height maps (taken from the Berlin Map) and radiosonde data obtained at Akita (39.7°N) of May 28 and 30, 1978 clearly indicate that the origin of air mass is different between 30mb and 50mb levels. The lower stratospheric air mass over Sanriku on the day of observation probably originated from higher latitudes, while the air mass above 30mb originated from similar or lower latitudes. In the 1986 observation the stratospheric wind direction did not change with altitude: the westerlies in the whole stratosphere, as supported by the geopotential height maps of October 6, 1986 (Nikaido, private communication, 1988). The distict difference in NO_2 profiles obtained in different seasons at Sanriku may reflect that difference in atmospheric conditions as above. Our results suggest that the distribution of lower stratospheric NO_2 is strongly controlled by dynamical transport, resulting in complicated and variable structures of stratospheric NO_2.

3.2 EFFECT OF TEMPERATURE DEPENDENCE OF NO_2 ABSORPTION CROSS SECTIONS

Recently Leroy et al.(1987) reported the temperature dependence of NO_2 absorption cross sections in the visible region. This is the region used for the NO_2 measurements discussed here. Their cross-section do not show any significant discrepancy at 298 K from those measured by Johnston and so far used for the NO_2 spectral data analysis. At 235 K, however, absorption decreases from that of 298 K at the band minima and slightly changes at the band maxima.This results in increasing differential absorption between the band maximum and minimum and accordingly in reducing the column density derived from a ratio spectrum.

Four stratospheric NO_2 profiles in Fig. 3 are those retrieved from the slant column density data observed in the 1985 MAP/GLOBUS Campaign using different cross sections. An iterative method is introduced to obtain the final profile (designated as D in Fig. 3). This method is explained in the following.

We define the effective temperature T_e as

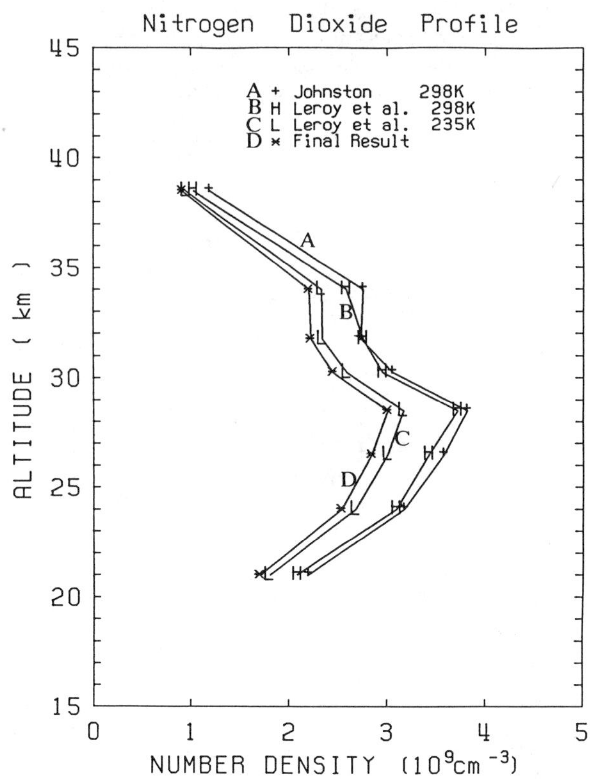

Fig. 3. Stratospheric NO_2 profiles derived using different cross sections. Profiles A,B, C are calculated using cross sections of Johnston at 298 K, of Leroy et al. at 298 K and 235 K, respectively. Profile D is the final one (see text).

$$T_e = \frac{\Sigma n_i T_i l_i}{\Sigma n_i l_i}$$

where n_i is the NO_2 density in the i-th atmospheric layer, l_i is the slant optical pathlength at a given solar zenith angle, and T_i is the atmospheric temperature taken from the U.S.Standard Atmosphere 1966. We assume linear temperature dependence of NO_2 absorption cross sections between of 235 K and 298 K to calculate the cross sections as a function of wavelength and temperature. Adopting the profile A as an initial profile, we calculate the effective temperature and the corresponding cross sections. By using them we derive a slant column density and retrieve a NO_2 profile. The retrieved profile is used to repeat the above procedure until a convergence is reached. The NO_2 densities of the final profile D are reduced by about 20% from those of the initial profile A. The shape of the profile. This fact partly explains the systematic difference in NO_2 profiles between those observed with infrared absorption spectroscopy and those with visible absorption spectroscopy, since room temperature values of the absorption cross sections were used in visible absorption spectroscopy. Note that the latter profiles usually showed larger values.

Five profiles in Fig. 1 are all retrieved using the room temperature values of the NO_2 absorption cross sections measured by Johnston. However inclusion of temperature effect on the cross sections does not affect the shape of the NO_2 profile and accordingly the previous discussion on the variation in the lower stratospheric NO_2.

4. SUMMARY

We present five profiles of the stratospheric NO_2 derived from balloon-borne measurements with visible absorption spectroscopy. Observation sites are located in northern middle latitudes from 32°N to 44°N. The lower stratospheric NO_2 is found to be significantly variable from observation to observation, although the vertical column density agrees with the seasonal behavior observed by Noxon within the experimental uncertainties. Ozone profiles and geopotential height maps during the observa- tion periods suggest that dynamical transport largely affects the distribution of the lower stratospheric NO_2.

Sensitivity is discussed on the retrieved NO_2 profile due to including the temperature dependence of the NO_2 absorption cross sections in visible region. The NO_2 column density is reduced by about 20% due to the inclusion of the temperature effect. As a result, the NO_2 densities so far obtained with visible absorption spectroscopy should be decreased by about 20 % uniformly with altitude, to become closer to those obtained with infrared spectroscopy.

ACKNOWLEDGMENTS

We gratefully acknowledge Prof. J. Nishimura and his colleagues of ISAS for their successful balloon flight operations in 1978 and 1986 at Sanriku. Thanks are also due to Prof. D. G. Murcury of University of Denver for the piggyback balloon flight in Alamogordo, New Mexico. N. Iwagami was financially supported by Nanki-Ikueikai Foundation for the Alamogordo balloon flight. Chemical Manufacturers Association financially supported us to participate in the BIC and MAP/GLOBUS Campaigns (Contract Nos. FC82-391, FC82-431 and FC85-550). The BIC Campaign was a NASA balloon program.

REFERENCES

Chu, W. P. and M. P. McCormick, 1986: SAGE observations of stratospheric nitrogen dioxide, J. Geophys. Res., 91, 5465-5476.

Harrison, A. W., 1979: Midsummer stratospheric NO_2 at latitude 45°S, Can. J. Phys., 57, 1110-1117.

Leroy, B., P. Rigaud and E. Hicks, 1987: Visible absorption cross-sections of NO_2 at 298 K and 235 K, Ann. Geophys., 5A, 247-250.

Mount, G. H. et al., 1983: Measurements of NO_2 in the Earth's stratosphere using a limb scanning visible light spectrometer, Geophys. Res. Lett., 10, 265-268.

Mount, G. H. et al.,1984: Measurements of stratospheric NO_2 from the Solar Mesosphere Explorer Satellite 1. An overview of the results, J. Geophys. Res., 89, 1327-1340.

Noxon, J. F., 1979: Stratospheric NO_2 2. Global behavior, J. Geophys. Res., 84, 5067-5076.

Noxon, J. F., 1980: Correction, J. Geophys. Res., 85, 4560-4561.

Ogawa, T., K. Shibasaki and K. Suzuki, 1981: Balloon observation of the stratospheric NO_2 profile by visible absorption spectroscopy, J. Meteorol. Soc. Japan, 59, 410-416.

Russell, J. M. III et al., 1984 a: Validation of nitrogen dioxide results measured by the Limb Infrared Monitors of the Stratosphere (LIMS) experiment on NIMBUS 7, J. Geophys. Res., 89, 5099-5107.

Russell, J. M. III et al., 1984b: The variability of stratospheric and mesospheric NO_2 in the polar night observed by LIMS, J. Geophys. Res., 89, 7267-7275.

Solomon, S., G. H. Mount and J. M. Zawodny, 1984: Measurements of stratospheric NO_2 from the Solar Mesosphere Explorer Satellite 2. General morphology of observed NO_2 and N_2O_5, J. Geophys. Res., 89, 7317-7321.

Twomey, S., 1963: On the numerical solution of Fredholm integral equations of the first kind by the inversion of the linear system produced by quadrature, J. Assoc. Comp.Mach. 10, 97-101.

NIGHTTIME MEASUREMENTS OF STRATOSPHERIC NO_y FROM BALLOON-BORNE THERMAL EMISSION OBSERVATIONS

Mian M. Abbas
NASA Marshall Space Flight Center
Huntsville, Alabama 35812, USA

Virgil G. Kunde, J.C. Brasunas, W.C. Maguire, and J.R. Herman
NASA Goddard Space Flight Center
Greenbelt, Maryland 20771, USA

Michael J. Glenn
Drexel University
Philadelphia, Pennsylvania 19104, USA

Steven T. Massie
National Center for Atmospheric Research
Boulder, Colorado 80307, USA

William A. Shaffer
Science Systems and Applications, Inc.
Seabrook, Maryland 20706, USA

ABSTRACT

Stratospheric limb thermal emission observations were made on a nighttime balloon flight launched from Palestine, Texas, on September 15–16, 1986, with a cryogenically cooled interferometer spectrometer. Complete limb sequences of the observed spectra obtained in the 600–2000 cm^{-1} region exhibit spectral features of a number of stratospheric constituents controlling the ozone chemistry including important members of the NO_x family. In this paper we present an analysis of the observed spectral data, employing line–by–line radiative transfer calculations, for retrieval of the distributions of NO_2, N_2O_5, HNO_3, and $C\ell ONO_2$. The inferred values of the total odd nitrogen mixing ratios measured directly at nighttime are given for the first time. Comparisons of the total odd nitrogen budget are made with the measurements by the ATMOS experiment and with the predictions of a one–dimensional photochemical model.

1. INTRODUCTION

It is now well recognized that the loss processes in the stratospheric ozone are controlled largely through catalytic cycles involving members of the NO_x, HO_x, and the $C\ell O_x$ families, and to a lesser extent the BrO_x family (WMO, 1985). Of all the chemically active species, the NO_x family plays the dominant role in the catalytic destruction of ozone in the contemporary atmosphere involving an inter–change of the NO and NO_2 free radicals. Active species such as NO and NO_2 are those that directly participate in the catalytic cycles destroying odd oxygen. The active elements of the NO_x family are strongly coupled to the HO_x and $C\ell O_x$ families through an involvement of inactive reservoir species such as HNO_3, HNO_4, N_2O_5, and $C\ell ONO_2$. The efficiency of the O_3 destruction process depends on the lifetimes of the chemically active species compared with those of the inactive species.

For a quantitative knowledge of the processes controlling the ozone chemistry, it is thus important to understand the partitioning between the active and inactive species and the total budget of the nitrogen–containing species. The total odd nitrogen budget of all the nitrogen–containing chemically active species may be defined as $[NO_y] = [NO] + [NO_2] + [NO_3] + [2N_2O_5] + [HNO_3] + [HNO_4] + [C\ell ONO_2]$. A critical test in the understanding and the development of photochemical models thus re–quires simultaneous measurements of these photo–chemically coupled species in the same volume of space and time.

In this paper we present the first simultaneous nighttime measurements of NO_2, N_2O_5, HNO_3, and $C\ell ONO_2$, and the inferred total odd nitrogen mixing ratios. The results are compared with the sunrise measurements by ATMOS (Russell et al., 1988) in the southern hemisphere, and with the predictions of a one–dimensional photochemical model.

2. OBSERVED SPECTRA

The observed spectra analyzed here have been obtained with a liquid nitrogen cooled high resolu–tion stratospheric infrared interferometer spectrometer (SIRIS) capable of making limb thermal emission observations from balloon platforms, and operating in the 600–2000 cm^{-1} region with an unapodized spectral resolution of 0.025 cm^{-1}. The instrument field of view is 0.5°, which gives a 3–km altitude resolution on the limb. The details of the instrument have been given in previous publications (e.g., Brasunas et al., 1986).

ISBN 0–937194–15–8

The stratospheric limb thermal emission spectra analyzed in this paper were obtained on a night–time balloon flight on September 15–16, 1986, launched from Palestine, Texas. After Fourier transformation, the observed spectra were apodized and calibrated to an absolute intensity. An over–view of the observed spectra obtained by SIRIS in a balloon flight in 1984 has been given in detail in a previous publication (Kunde et al., 1987). The stratospheric trace gases identified in the observed spectra to date include: CO_2, O_3, H_2O, CH_4, N_2O, HNO_3, NO_2, NO, N_2O_5, $C\ell ONO_2$, $CFC\ell_3$, $CF_2C\ell_2$, $CHF_2C\ell$, CF_4, and $CC\ell_4$.

A limb sequence of the observed spectra in the 735–740 cm^{-1} region, and 779–781 cm^{-1} region for zenith angles with tangent heights varying from 21 to 40 km are shown in Figs. 1 and 2. The main absorbers in these spectral regions are CO_2, O_3, with the ν_4 band of $C\ell ONO_2$ also occurring at ~ 780.2 cm^{-1}. Figure 3 shows an overview of the observed spectra for three angles with tangent heights from 21 to 28 km indicating the ν_5 and the $2\nu_9$ bands of HNO_3. A portion of the observed spectra from 868 to 872 cm^{-1} employed in the re–trieval of HNO_3 is shown in Fig. 4. An overview of the observed spectra for the 28–km tangent height in the 1575–1625 cm^{-1} channel is shown in Fig. 5 where the main molecular absorber is the ν_3 band of NO_2, along with some strong spectral lines of H_2O (Fig. 6). An analysis of the observed spectra for retrieval of the vertical distribution of stratospheric trace gases is discussed in the follow–ing.

3. ANALYSIS AND RESULTS

A portion of the observed spectra obtained by SIRIS from the balloon flight on September 7–8, 1986, has been analyzed with a focus on calculation of the total nighttime odd nitrogen budget from the simultaneously measured profiles of important mem–bers of the NO_x family. The analysis is based on line–by–line and layer–by–layer calculations and includes curvature and refraction effects. The mo–lecular spectral data employed in the calculations are based on the AFGL compilation HITRAN, 1986 (Rothman et al., 1987). The $C\ell ONO_2$ parameters are taken from the data base of Rinsland et al. (1985) normalized to the laboratory cross sections of Massie et al. (1985). For N_2O_5, room temperature absorption coefficients based on laboratory transmit–tance measurements of Massie et al. were used. The temperature profile used in the analysis was obtained from the soundings made by National Meteorological Center near the balloon track. The inversion technique employed here follows an iterative procedure by successively retrieving mixing ratios from the higher to the lower tangent heights. The relaxation equation used minimizes the weighted RMS difference between observed and the synthetic radiances at selected frequencies around the line centers where the weighting functions are highest. Detailed discussions of the radiative transfer model and the inversion technique employed here have been given in previous publications (Abbas et al., 1985, 1987). The dashed curves in Figs. 1, 2, 4, and 6 represent the spectra calculated with the retrieved gas profiles and indicate the quality of the fit between the observed and the synthetic spectra.

Figure 7 shows the vertical profiles of several members of the NO_x family retrieved from the spectra averaged over several limb sequences ob–tained during the balloon flight from midnight to 5 a.m. on September 15–16, 1986. The pointing angles used in the calculations were retrieved from inversions of a large number of isolated CO_2 lines in the 700–800 cm^{-1} region, with an uncertainty of ~0.03°. The error bars on the mixing ratio profiles indicate the total 1σ uncertainty of the meas–urement representing the quadrature sum of random errors and possible systematic errors. The random errors, given by the standard deviation of the mean mixing ratios determined from different spectral regions or lines, arise from spectral noise, the resid–uals in spectral fitting, and the random component of the pointing errors. The sources of systematic errors are: 5 to 10% errors arising from from a pointing bias of ~2 arc minutes and 10 to 15% errors due to uncertainty in the spectral parameters.

HNO_3: The HNO_3 profile has been retrieved from an inversion of 10 features in the 868–879 cm^{-1} region of the ν_5 band. The HNO_3 features at the

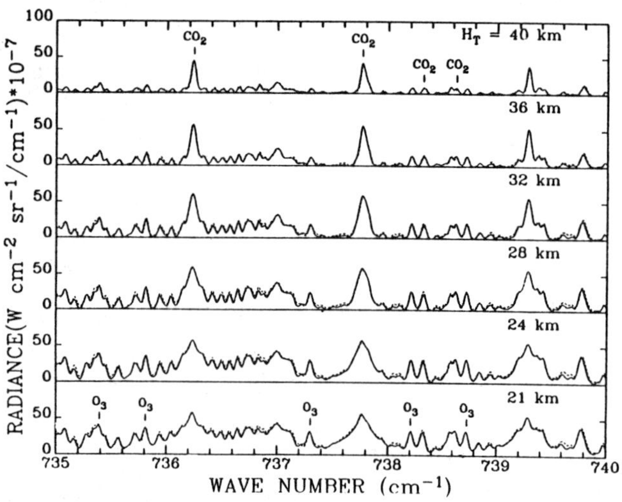

Fig. 1. A complete limb sequence of the observed thermal emission spectra in the 735–740 cm^{-1} region.

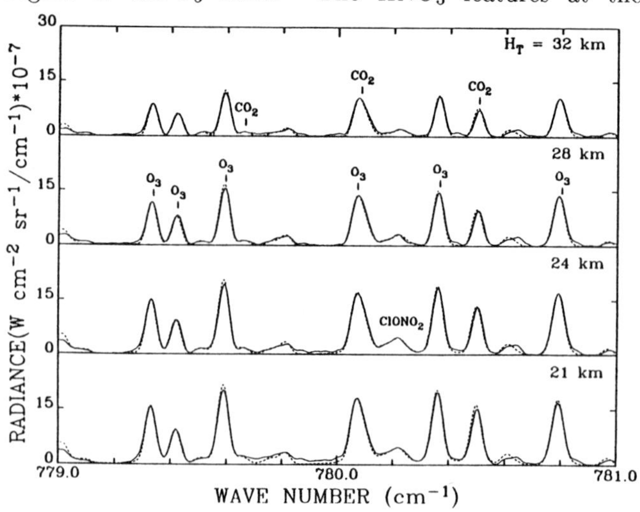

Fig. 2. A limb sequence of the observed thermal emission spectra in the 779–781 cm^{-1} region.

40–km tangent height are too weak to be analyzed, and the mixing ratio at 36 km has been retrieved from the Q branch at 879 cm^{-1} only. The total 1σ uncertainty of the measurements indicated by the error bars is ~16–17% over most of the altitude range.

NO_2: The average profile shown has been retrieved from 67 isolated spectral lines of the ν_3 band in the 1580–1615 cm^{-1} region. The total 1σ uncertainty of the measurements indicated by the error bars varies from 10 to 12% for heights between 37 to 25 km and ~45% at the lowest altitude of 21 km. The large uncertainty at 21 km is due to the small concentrations of NO_2 at lower altitudes.

$C\ell ONO_2$: The profile over the 21–32 km altitude range has been determined from the feature at 780.21 cm^{-1} belonging to the Q branch of ν_4 band. The feature becomes too weak to be analyzed for heights above 32 km. The total 1σ uncertainty

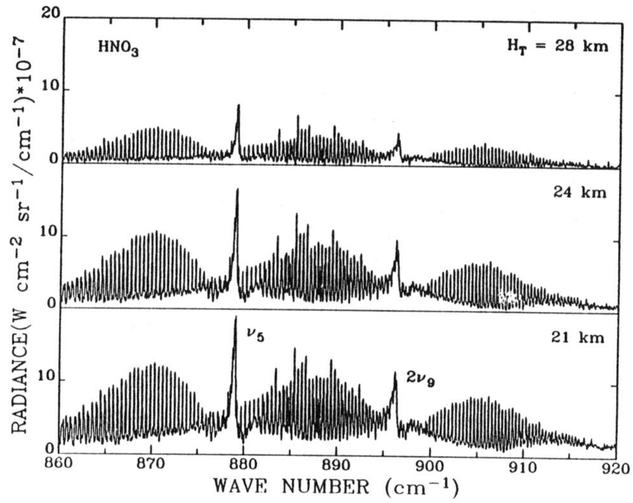

Fig. 3. A limb sequence of the observed spectra in the 860–920 cm^{-1} region.

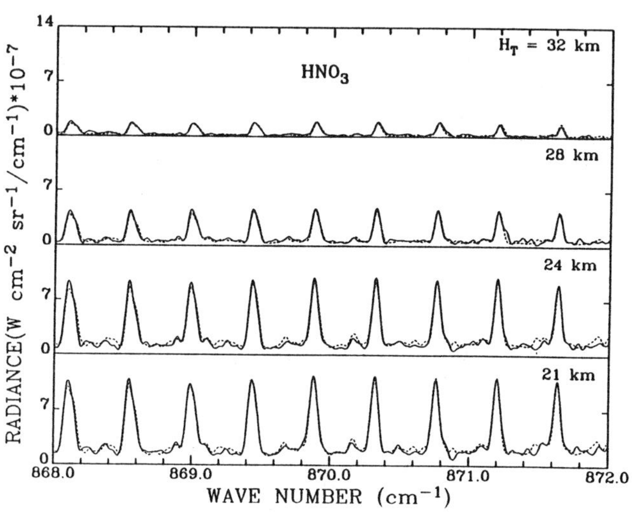

Fig. 4. A limb sequence of the observed spectra in the 868–872 cm^{-1} region.

indicated by the error bars varies from 16 to 19% over the altitude range of the profile and is based on an assumed 15% uncertainty in the line strength.

N_2O_5: The profile shown has been derived from spectral features in the 1250–1260 cm^{-1} region of ν_{12} band with 1σ uncertainty varying from 23% at 22 km to 58% at 38 km. Detailed discussions of the observed N_2O_5 features and the retrieved profile have been given in a separate publication (Kunde et al., 1988).

Total Odd N: The total odd nitrogen concentration $[NO_y]$ is defined here as $[NO_y] = [NO] + [NO_2] + [NO_3] + [2N_2O_5] + [HNO_3] + [HNO_4] + [C\ell ONO_2]$. NO is quickly converted into NO_2 at night and its contribution to the total odd N nighttime budget can be ignored. The distributions of NO_3 and HNO_4 have not been made in the

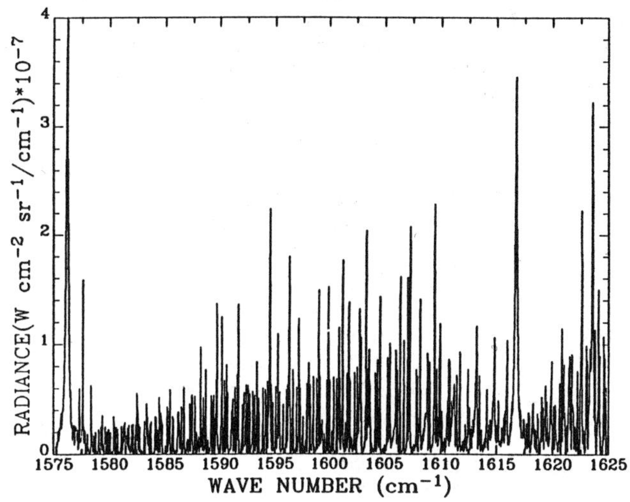

Fig. 5. An overview of the observed spectrum in the 1575–1625 cm^{-1} region.

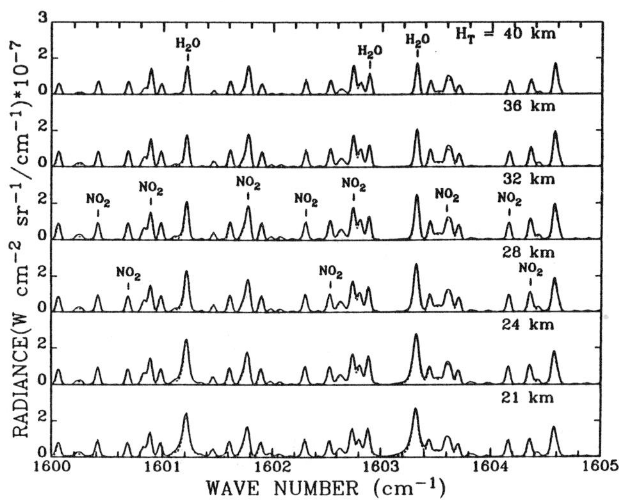

Fig. 6. A limb sequence of a selected portion of the observed spectra in the 1600–1605 cm^{-1} region.

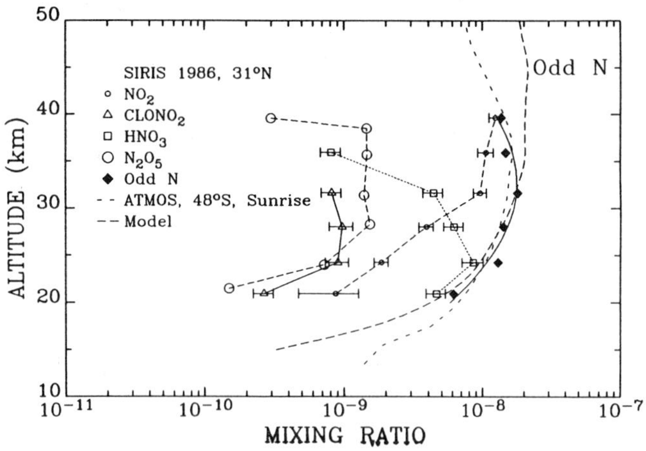

Fig 7. The average volume mixing ratio profiles of HNO_3, NO_2, N_2O_5, and $C\ell ONO_2$ retrieved from the observed nighttime spectra. The inferred total odd nitrogen profile is also shown along with the NO_3 and the HNO_4 profiles taken from a photochemical model. Also shown for comparison are two total odd N profiles obtained by: ATMOS on May 1, 1985, at sunrise at 48° S and a predicted profile obtained from one–dimensional photochemical model calculations for conditions at the time of observations.

present set of measurements by SIRIS. In calculation of the total odd nitrogen budget, predicted profiles based on a one–dimensional photochemical model has been employed for both of these constituents. Although NO_3 has maximum concentrations at night and is important to the production of N_2O_5, it adds less than 0.2% to the total odd N budget. HNO_4 which acts as a reservoir in the NO_x and HO_x chemistry of the lower stratosphere contributes less than 2% to the total budget.

The nighttime total odd N mixing ratios calculated from the observed NO_2, N_2O_5, HNO_3, and $C\ell ONO_2$ profiles, and the model HNO_4 and NO_3 concentrations are shown in Fig. 7. A maximum of $[NO_y]$ of ~18.10 x 10^{-9} is seen to occur at ~30–km altitude. Also shown for comparison is the total odd nitrogen profile obtained by ATMOS (Russell et al., 1988) in the southern hemisphere (48° S) at sunrise. The peak value of the nighttime total odd nitrogen mixing ratio obtained in this work from infrared thermal emission measurements appears to be in fairly good agreement with the ATMOS results and within the uncertainties of the two measurements although the partitioning between the various components of the NO_x family is quite different. In the ATMOS profile for sunrise, the NO contribution to total odd N dominates at altitudes above 30 km, whereas in the nighttime profile presented in this paper NO_2 is the dominant constituent with all NO having been converted into NO_2. Also shown on the Fig. 7 is the predicted odd N profile based on the one dimensional photochemical model of Herman and McQuillan (1985). The model profile shows higher values at upper altitudes and lower values at the lower altitudes with the peak shifted to a higher level of about 45 km. More detailed model calculations are currently being carried out.

ACKNOWLEDGMENT

The work at Drexel University was supported by NASA Grant NAGW–431.

REFERENCES

Abbas, M. M., G. L. Shapiro, B. J. Conrath, V. G. Kunde, and W. C. Maguire, 1985: Thermal emission spectroscopy of the stratosphere from balloon platforms. Advances in Remote Sensing Retrieval Methods, 133–148, A. Deepak Publishing, Hampton, VA.

Abbas, M. M., M. J. Glenn, V. G. Kunde, J. Brasunas, B. J. Conrath, W. C. Maguire, J. Brasunas, and J. R. Herman, 1987: Simultaneous measurement of stratospheric O_3, H_2O, CH_4 and N_2O from limb thermal emissions. J. Geophys. Res., **92**, 8343–8353.

Brasunas, J. C., V. G. Kunde, and L. W. Herath, and H. L. Buijs, 1986: Balloon–borne cryogenic spectrometer for measurement of lower strato–spheric trace constituents. Proc. Soc. Photo–Opt.Instrum. Eng., **619**, 80–88.

Herman, J. R., and C. McQuillan, 1985: Atmospheric chlorine and stratospheric ozone: Nonlinearities and trend detection. J. Geophys. Res., **90**, 5721–5732.

Kunde, V. G., et al., 1987: Infrared spectroscopy of the lower stratosphere with a balloon–borne cryogenic Fourier spectrometer. Appl. Opt., **26**, 545.

Kunde, V. G., et al., 1988: Detection of nighttime stratospheric N_2O_5 in infrared emission spectra. Submitted for publication.

Massie, S. T., et al., 1985: Approximate Absorption crossections of F12, F11, $C\ell ONO_2$, N_2O_5, HNO_3, $CC\ell_4$, CF_4, F21, F113, F114, and HNO_4. Appl. Opt., **24**, 3426–3427.

Rinsland, C. P., et al., 1985: Tentative identification of the 780 cm^{-1} ν_4 band Q branch of chlorine nitrate in high resolution solar absorption spectra of the stratosphere. J. Geophys. Res., **90**, 7931–7943.

Rothman, L. S., et al., 1987: The HITRAN database: 1986 edition. Appl. Opt., **26**, 4058–4097.

Russell, J. M. III, et al., 1988: Measurements of odd nitrogen compounds in the stratosphere by the ATMOS experiment on Spacelab 3. J. Geophys. Res., **93**, 1718–1736.

World Meteorological Organization, 1985: Atmospheric Ozone, Rep. **16**, World Meteorol. Org., Geneva.

HIGH RESOLUTION, BALLOON-BORNE EMISSION SPECTROSCOPY OF TRACE SPECIES IN THE LOWER STRATOSPHERE: N_2O_5, HNO_3

J. Brasunas, J. Herman, V. Kunde, and W. Maguire
NASA Goddard Space Flight Center
Greenbelt, Maryland 20771, USA

L. Herath and W. Shaffer
Science Systems and Applications, Inc.
Seabrook, Maryland 20706, USA

M. Abbas
NASA Marshall Space Flight Center
Huntsville, Alabama 35812, USA

S. Massie
National Center for Atmospheric Research
Boulder, Colorado 80307, USA

ABSTRACT

A liquid-nitrogen cooled Fourier transform spectrometer (SIRIS) measures thermal limb-emission of the stratosphere from a balloon platform at a nominal altitude of 40 km, under night and day conditions, with a 3 km vertical resolution. N_2O_5 and HNO_3 mixing ratios inferred from emission spectra are compared with previous measurements and with the predictions from a one-dimensional photochemical model.

1. INTRODUCTION

The concentration of ozone in the stratosphere is controlled by three families of chemical reactions, nitrogen, chlorine and hydrogen, in addition to the oxygen cycle itself. Each family is composed of active gasses that catalytically destroy ozone, source gasses that break down in the stratosphere to provide the active gasses, and sink gasses (temporary and long-term reservoirs) that remove nitrogen, chlorine and hydrogen compounds from the stratosphere through transport and rain-out. In the present-day stratosphere, the nitrogen-cycle reactions dominate at the altitudes of peak ozone concentration.

The nitrogen family includes the free radicals NO and NO_2, the source gas N_2O, and the reservoir species N_2O_5, HNO_3, $ClONO_2$, and HNO_4. The temporary reservoir N_2O_5 is produced almost entirely at night when the concentration of NO_3 is high. During daytime the N_2O_5 is slowly photolyzed to NO_2. N_2O_5 thus should exhibit a diurnal variation with peak concentrations in the pre-sunrise time period; it is also responsible for the temporary (nighttime) removal of NO and NO_2. HNO_3, a reservoir species without appreciable diurnal variation, is readily incorporated into rain droplets and this leads to removal of ozone-destroying nitrogen species from the stratosphere (Brasseur and Solomon, 1984). Special conditions pertain at high-latitude winter, where models suggest that N_2O_5 is the major component of odd nitrogen, with the conversion of NO_2 to N_2O_5 producing the Noxon cliff (Wuebbles, 1986); and, heterogeneous reactions in the Arctic winter may convert the N_2O_5 to HNO_3.

N_2O_5 was first tentatively identified by a radiometer with a 20 cm^{-1} wide channel operating in limb-emission mode (Roscoe, 1982). This was followed by pre-dawn spectroscopy by Evans (1986) at low resolution using a circular variable filter. The first measurement widely accepted as a definitive spectroscopic identification was the high resolution, absorption-mode ATMOS data (Toon, 1986 and 1987). HNO_3, on the other hand, has been measured by a substantial number of investigators (WMO 1985).

The N_2O_5 and HNO_3 measurements presented here are from a high resolution, balloon-borne, emission-mode Fourier transform spectrometer (SIRIS). This spectrometer has collected data on two balloon flights from Palestine, Texas. The first, on the afternoon of November 6, 1984, produced spectra with tangent heights in the

ISBN 0-937194-15-8

range 14-40 km. The second, during the evening of September 15-16, 1986, produced spectra with tangent heights between 8-40 km. Fifteen molecular species have so far been identified in the data, including members of the nitrogen and chlorine cycles of ozone destruction: HNO_3, NO_2, NO, $ClONO_2$, N_2O_5, CO_2, O_3, H_2O, N_2O, CH_4, CCl_3F, CCl_2F_2, CHF_2Cl, CCl_4 and CF_4.

2. INSTRUMENTATION AND DATA ANALYSIS

The SIRIS spectrometer is cooled to liquid-nitrogen temperature, and the detectors to liquid-helium temperature, to enhance the signal-to-noise ratio (Brasunas et al., 1986, 1988). Five photoconductor detectors cover segments of the spectrum between 700 and 2000 cm^{-1} (5-15 μm) with an unapodized resolution of 0.025 cm^{-1}. SIRIS is a balloon-borne emission-type spectrometer operating in the limb mode; emission measurements allow complete diurnal coverage, and the limb mode enhances the signal of weak features. The vertical resolution is 3 km for altitudes below 40 km.

Figure 1 is an example of a set of limb scans, or limb sequence, from channel 2 for the 1986 flight; the ν_{12} band of N_2O_5 is indicated as a continuum-like feature underlying the strong emission lines of CH_4 and N_2O. The N_2O_5 lines are not resolved at the observed 0.025 cm^{-1} resolution; the continuum-like feature is composed of many unresolved lines. Figure 2 is a

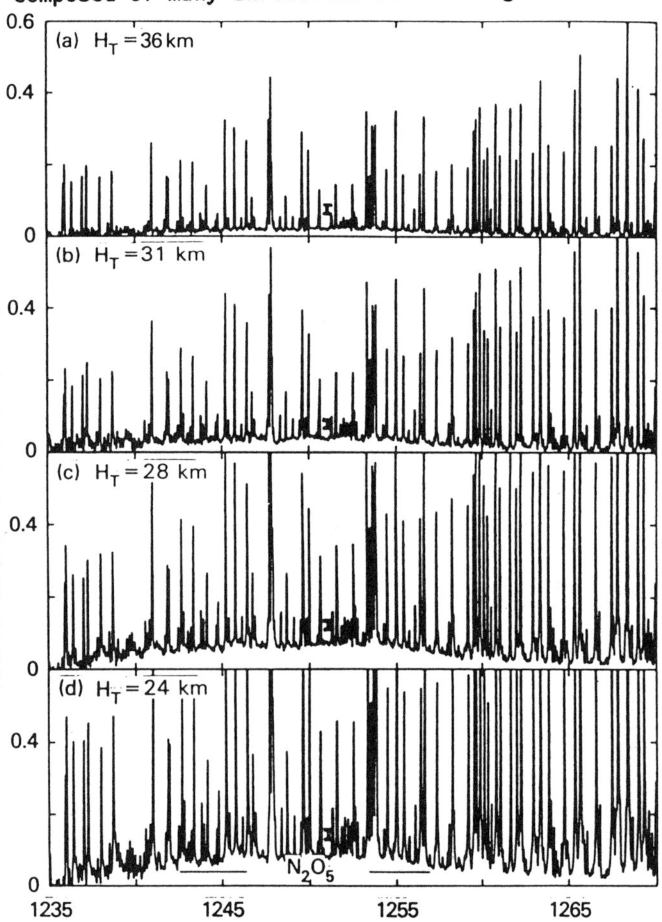

FIG.1 - N_2O_5 LIMB SEQUENCE

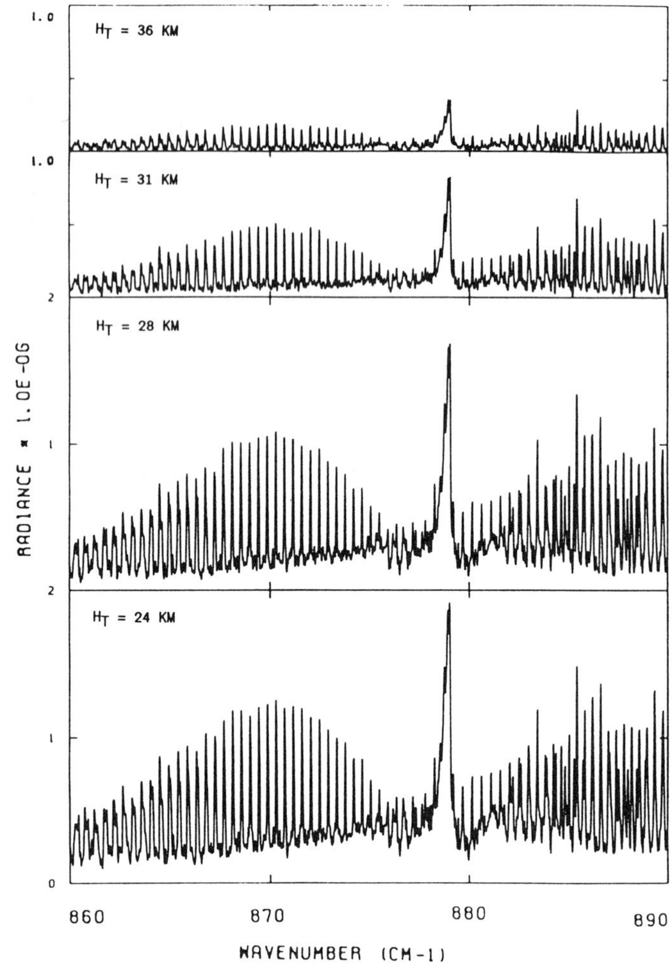

FIG.2 - HNO_3 LIMB SEQUENCE

limb sequence from channel 1 for the 1986 flight, and shows the ν_5 band of HNO_3 at 879 cm^{-1}. Pointing for these data has been verified using CO_2 lines from the 700-810 cm^{-1} region.

Retrieval of constituent mixing ratios from thermal emission measurements requires the inverse solution of the radiative transfer equation. An inversion technique which has been found to be particularly useful for treating the radiative transfer equation is the relaxation method of Chahine (1972). The relaxation method gives unique results for the constituent mixing ratio without any a priori information about the expected solution, and is also able to discriminate between noise and true information without any data smoothing. We have modified the relaxation method to least squares fit several features in the spectrum simultaneously (Shaffer et al., 1988).

If a first guess for the constituent mixing ratio is q_o, the relaxation method determines α such that

$$| I_\nu(\theta_o, Z_o, \alpha q_o) - I_{\nu obs}(\theta_o, Z_o) | < \delta \quad (1)$$

where θ_o is the zenith angle, Z_o the float height of observation, I_ν is the synthetic and $I_{\nu obs}$ the observed spectral intensity, and δ is an arbitrary number small compared to $I_{\nu obs}$. The product αq_o is the "best fit" mixing ratio.

The mixing ratio profile is retrieved from a limb sequence by "onion-peeling". In this method the mixing ratio at each tangent height is retrieved, starting at the highest altitudes and working through successive layers to the lowest altitudes, retaining at all lower altitudes the mixing ratios retrieved at higher altitudes.

Due to the limitations of noise it is desirable to introduce as much information as possible into the retrieval. One way of doing this is to use many spectral lines at each retrieval altitude. If N lines are used then α is found, in the least squares relaxation method by minimizing the residual sum of squares (RSS) given by

$$\mathrm{RSS} = \sqrt{\sum^{N} \left(I_{\nu_j obs}(\theta_o,Z_o) - I_{\nu j}(\theta_o,Z_o,\alpha q_o(Z_o))\right)^2 / N} \qquad (2)$$

where ν_j corresponds to the jth of N line peaks.

For HNO_3, the synthetic radiances are generated by line-by-line techniques using the AFGL line compilation (see Rothman et al., 1987). For N_2O_5, the synthetic radiances are generated from transmission cross-sections, which are derived from room-temperature laboratory measurements at the same spectral resolution (Massie et al., 1985). The peak cross-section is 1.9×10^{-18} cm^2/molecule at 1245 cm^{-1}, which coincidentally is what Davidson et al. (1987) measure at 230K.

Spectral features were chosen which gave optical depths closest to unity in order to optimize the contribution function near the tangent heights of the paths. The contribution functions ($B_\nu(T)W_\nu$, where W_ν is the weighting function in the radiative transfer equation) typical of the lines selected for the retrievals, shown in Fig. 3 for N_2O_5 and Fig. 4 for HNO_3, indicate that the bulk of the path contribution occurs close to the tangent heights. An ideal contribution function would be a delta function.

3. CONCENTRATION RETRIEVALS AND DISCUSSION

Figure 5 presents the SIRIS N_2O_5 retrievals for the midnight and predawn time frames, together with an average over the evening. These data are compared with the output from a one-dimensional photochemical model (Herman, 1979, and Herman and McQuillan, 1985). The error bars represent both random and systematic errors, due to instrument noise, pointing uncertainty, and molecular parameter uncertainty. As discussed in Kunde et al. (1988), there appears to be a trend towards more N_2O_5 later in the night as the models predict, but the error bars rule out a quantitative comparison. The data are generally lower than the model output. Also shown is an upper limit for N_2O_5 at 28 km from the mid-afternoon data of the 1984 flight; the combination of the two flights clearly demonstrates the diurnal variation. Figure 6 compares the SIRIS data with ATMOS results for a sunrise at 47S (Russell et al., 1988); the two data sets appear consistent.

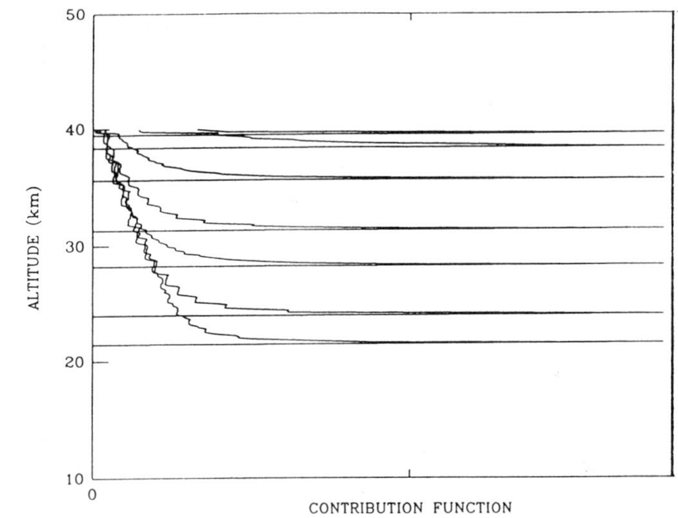

FIG. 3 - N_2O_5

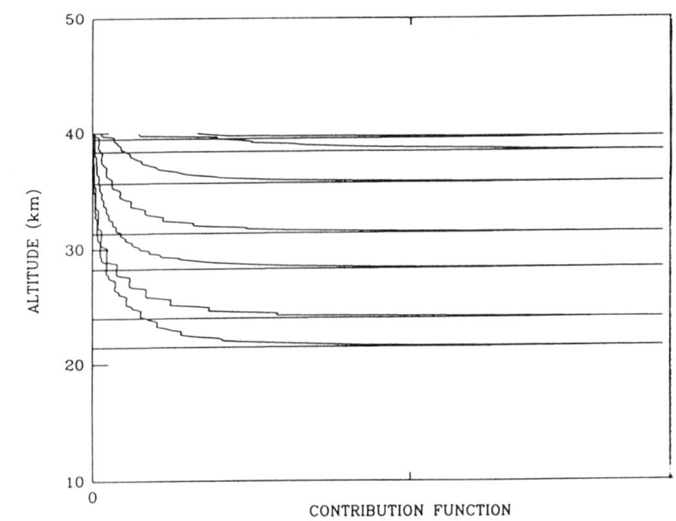

FIG. 4 - HNO_3

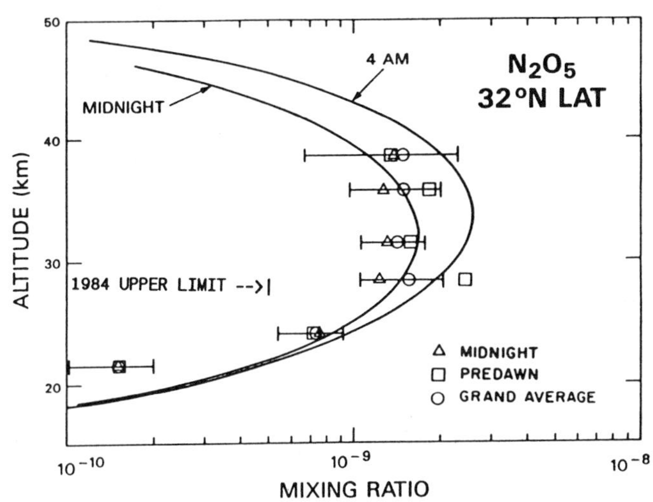

FIG. 5 DIURNAL VARIATION: DATA COMPARED WITH 1D MODEL

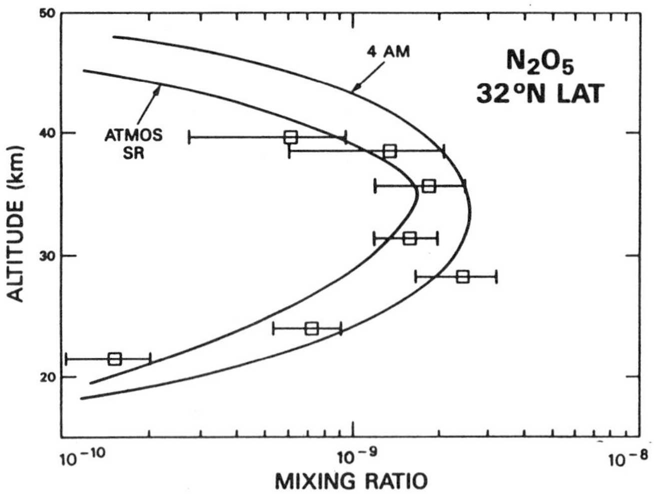

FIG. 6 PREDAWN DATA COMPARED WITH ATMOS AND 1D MODEL

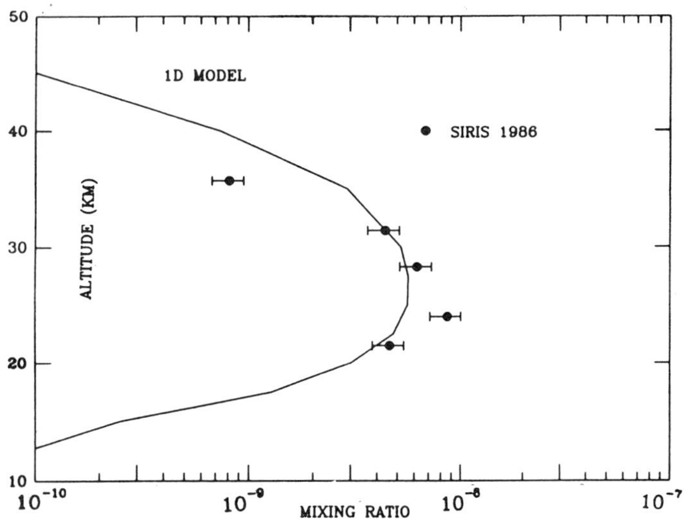

FIG. 7 - MID-LATITUDE HNO_3

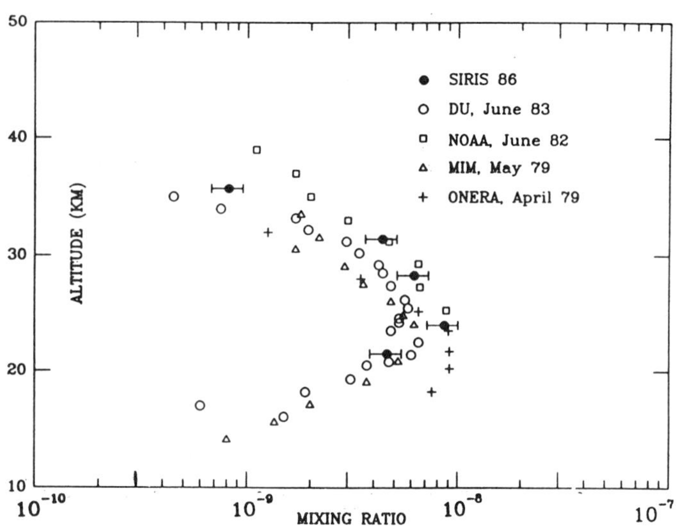

FIG. 8 - MID-LATITUDE HNO_3

Figure 7 compares the SIRIS retrieval of HNO_3 with the one-dimensional photochemical model of Herman (1979). There is general agreement except at higher altitudes and at the peak of the mixing ratio. The tendency of models to overpredict HNO_3 at the higher altitudes is a longstanding problem (WMO, 1985). Figure 8 shows that SIRIS HNO_3 data are within the range of a wide variety of measurements made at mid-latitudes (WMO 1985).

In addition to the direct measurements of N_2O_5 already discussed, the existence of nitrogen reservoir species such as N_2O_5 and $ClONO_2$ can be inferred from the diurnal variations of NO_2 and NO. J.-M. Flaud et al. (1988) have made ground-based observations of the diurnal variation of the NO_2 column density, confirming the existence of a reservoir molecule in apparent agreement with models based on N_2O_5. Kondo et al. (1988) have made in-situ balloon-borne measurements of the diurnal variation of NO at 32 km. From the steady increase of NO after the initial post-sunrise build-up, they infer a sunrise N_2O_5 mixing ratio of 3.9 $\pm$ 1.5 ppbv at 33 km (October, 44N). The SIRIS estimate of 1.2 to 2.2 ppbv is at one-half the amount inferred from the NO decline.

4. CONCLUSION

High-resolution emission spectra of stratospheric species have been recorded from a balloon platform under night and day conditions. The retrieved nighttime N_2O_5 mixing ratios are generally less than a one-dimensional model predicts, however the data qualitatively agree with the model in terms of the midnight to sunrise rate of increase. Comparison of results from night and day flights shows a lower limit of 3 for the N_2O_5 diurnal variation at 28 km. The SIRIS data are in apparent agreement with ATMOS, and both are at one-half the amount inferred from the post-sunrise rate of increase of NO. Retrieved HNO_3 mixing ratios appear consistent with previous mid-latitude measurements, which generally have been less than model predictions above 30 km.

REFERENCES

G. Brasseur and S. Solomon, Aeronomy of the Middle Atmosphere, (D. Reidel, Dordrecht, 1984).

J.C. Brasunas, V.G. Kunde, R.A. Hanel, D. Walser, L.W. Herath, H.L Buijs, J.N. Berube, J. McKinnon, "Balloon-borne cryogenic spectrometer for measurement of lower stratospheric trace constituents," Proc. Soc. Photo-Opt. Instrum. Eng., 619, 80 (1986).

J.C. Brasunas, V.G. Kunde and L.W. Herath, "Cryogenic Fourier spectrometer for measuring trace species in the lower stratosphere," subm. to Appl. Opt. (1988).

M.T. Chahine, "A general relaxation method for inverse solution of the full radiative transfer equation," J. Atmos. Sci., 29, 741 (1972).

J. Davidson, N.C.A.R., Boulder, CO, private comm. (1987).

W.F.J. Evans, "Observation of the 8 μm N_2O_5 thermal emission feature in the stratosphere," Appl. Opt., 25, 1866 (1986).

J.-M. Flaud, C. Camy-Peyret, J.W. Brault, C.P. Rinsland, and D. Cariolle, "Nighttime and daytime variation of atmospheric NO_2 from ground-based infrared measurements," G.R.L., 15, 261 (1988).

J.R. Herman, "The problem of nighttime stratospheric NO_3," J.G.R., 84, 6336 (1979).

J.R. Herman and C. Mcquillan, "Atmospheric chlorine and stratospheric ozone: nonlinearities and trend detection," J.G.R., 90, 5721 (1985).

V.G. Kunde, J.C. Brasunas, W.C. Maguire, J.R. Herman, S.T. Massie, M.M. Abbas, L.W. Herath, and W.A. Shaffer, "Measurement of nighttime stratospheric N_2O_5 from infrared emission spectra," subm. to G.R.L. (1988).

Y. Kondo, W.A. Matthews, D. Aimedieu, D.E. Robbins, "Diurnal variation of nitric oxide at 32 km: measurements and interpretation," J.G.R., 93, 2451 (1988).

S.T. Massie, A. Goldman, D.G. Murcray, J.C. Gille, "Approximate absorption cross sections of F12, F11, $ClONO_2$, N_2O_5, HNO_3, CF_4, F21, F113, and F114," Appl. Opt., 24, 3426 (1985).

H.K. Roscoe, "Tentative observation of stratospheric N_2O_5," G.R.L., 9, 901 (1982).

L.S. Rothman, R.R. Gamache, A. Goldman, L.R. Brown, R.A. Toth, H.M. Pickett, R.L. Poynter, J.-M. Flaud, C. Camy-Peyret, A. Barbe, N. Hudson, C.P. Rinsland and M.A.H. Smith, "The HITRAN database: 1986 edition," Appl. Opt., 26, 4058 (1987).

J.M. Russell III, C.B. Farmer, C.P. Rinsland, R. Zander, L. Friodevaux, G.C. Toon, B. Gau, J. Shaw and M. Gunson, "Measurement of odd nitrogen compounds in the stratosphere by the ATMOS experiment on Spacelab 3," J.G.R., 93, 1718 (1988).

W.A. Shaffer, V.G. Kunde and B.J. Conrath, "Retrieval of constituent mixing ratios from limb thermal emission spectra," accepted by Appl. Opt. (1988).

G.C. Toon, C.B. Farmer and R.A. Norton, "Detection of stratospheric N_2O_5 by infrared remote sounding," Nature, 319, 570 (1986).

G. Toon, "Reply: Detection of stratospheric nitrogen species," Nature, 330, 427 (1987).

WMO, Atmospheric Ozone 1985, Report No. 16 (1985).

D.J. Wuebbles, "Reservoir species discovered in the stratosphere," Nature, 319, 538 (1986).

THE VERTICAL DISTRIBUTION OF SOURCE GASES AT TROPICAL LATITUDES

*R. Borchers, P. Fabian, and O.N. Singh**
Max-Planck-Institut für Aeronomie (MPAE)
D–3411 Katlenburg-Lindau, FRG

S. Lal and B.H. Subbaraya
Physical Research Laboratory (PRL)
Ahmedabad, India

ABSTRACT

The first complete vertical profiles of source gases, in particular halogenated hydrocarbons, up to about 35 km altitude, have been obtained for tropical latitudes. They result from GC- and GC-MS-analyses of whole air samples collected with the balloon-borne cryogenic sampler of MPAE on March 26, 1987, at Hyderabad, India (17.5°N). Except for CH_4, N_2O, CFC-11 and CFC-12, no source gas profiles are as yet documented for the tropical region. We report the first profiles of CFC-10, CFC-13, CFC-14, CFC-22, CFC-113, CFC-114, CFC-115, CFC-116, CH_3Cl and CH_3CCl_3 along with those of the aforementioned species that have been shown before. They all exhibit, as a result of tropical upwelling, volume mixing ratios that decrease less rapidly with height than at midlatitudes.

1. INTRODUCTION

As has been reported earlier, tropical upwelling is the dominating process through which source gases are mixed into the stratosphere [Bischof et al. 1985]. Source gases of anthropogenic origin, in particular halogenated hydrocarbons whose emission is concentrated at northern midlatitudes, thus pass through the tropical troposphere before they enter the stratosphere, carried up by vertical motion. Measurements of source gases at tropical latitudes are thus important as they provide information on the species entering the stratosphere. Since upwelling counteracts the decomposition process in the stratosphere, tropical source gas profiles are different from those measured at midlatitudes in that their mixing ratio falls off less rapidly with height. The comparison of midlatitude profiles of source gases and those measured in the tropical region thus provides a tool for testing two- and three dimensional chemical radiative-dynamical models.

*on leave from the Department of Applied Physics, Institute of Technology, Banares Hindu University, Varanasi, India

2. MEASUREMENTS

Under the auspices of the Indo-German collaboration, a balloon ascent of the MPAE cryogenic whole air sampler was successfully carried out on March 26, 1987, at Hyderabad/India (17.5°N). 14 air samples of about 30 l STP each were collected at different altitudes between 11.4 and 33.5 km. 7 of these were taken during the ascent, at 11.4, 13.9, 15.6, 17.3, 19.4, 21.2, and 23.5 km, while the other 7 were collected during the valve-controlled descent, at 33.5, 31.7, 29.8, 26.6, 25.6, 24.9, and 23.9 km altitude.

Three months later, on June 23, 1987, another balloon flight of the MPAE sampler was carried out at midlatitudes (44°N) and 15 whole air samples were recovered. 5 of these were collected during the ascent, at 10.1, 14.2, 17.2, 19.0 and 21.2 km, while the remaining 10 were sampled during the valve controlled descent at 33.3, 32.1, 30.5, 28.6, 27.1, 25.9, 24.9, 23.8, 22.6, and 21.6 km altitude. The analyses of both sets of balloon samples were carried out together during July of 1987 utilising gas chromatography. Thus we could determine, almost simultaneously and with one common absolute calibration procedure, vertical profiles of source gases at tropical and midlatitudes. Analytical procedures are described in earlier publications, see for instance [Fabian and Gömer, 1984, Fabian 1988]. Table 1 provides information on analytical precision and lowest detection limits. The absolute calibration was achieved, for all constitutents, within less than ± 10%.

3. RESULTS

Vertical profiles of source gases thus obtained for the tropical samples are plotted in figures 1-10 in comparison with midlatitude profiles measured 3 months later. There are two features which all results have in common: 1. As models have predicted, the mixing ratios were found to

ISBN 0–937194–15–8

Table I Analytical details

Substance		Detection	Precision [%]	Lowest detection limit
CH_4		FID	5	10 ppbV
N_2O		ECD	5	2 ppbV
CH_3Cl	(CFC-40)	MS	10	5 pptV
CCl_2F_2	(CFC-12)	ECD	5	0.5 pptV
CCl_3F	(CFC-11)	ECD	5	0.1 pptV
CCl_4	(CFC-10)	ECD	10	0.5 pptV
CH_3-CCl_3	(CFC-140a)	ECD	10	0.5 pptV
$CHClF_2$	(CFC-22)	MS	5	1 pptV
CF_4	(CFC-14)	MS	10	5 pptV
$CClF_2$-CCl_2F	(CFC-113)	ECD	5	0.3 pptV
$CClF_2$-$CClF_2$	(CFC-114)	ECD	5	0.5 pptV
$CClF_2$-CF_3	(CFC-115)	MS	10	0.5 pptV
$CClF_3$	(CFC-13)	MS	10	0.3 pptV
CF_3-CF_3	(CFC-116)	MS	10	0.3 pptV
$CBrClF_2$	(CFC-12B1)	ECD	5	0.03 pptV
$CBrF_3$	(CFC-13B1)	ECD	10	0.1 pptV

decrease with height less rapidly in the tropics than at midlatitudes. 2. All profiles exhibit a pronounced superimposed wave structure. It may be a mere coincidence, but the wave structure of the tropical profiles is very similar to that of the midlatitude profiles except for the heights of the wave minima and maxima, which seem to be higher in the tropics. We do not wish to place any interpretation on this wave structure at this time but rather emphasize that our results show that "real profiles" can differ considerably from "average profiles".

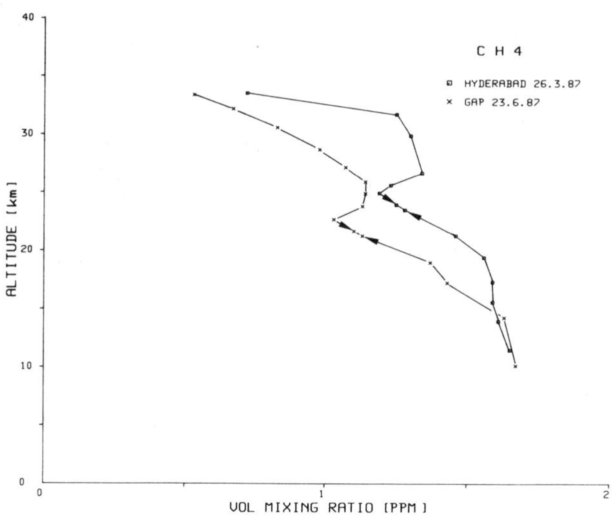

Fig. 1. Vertical profiles of CH_4 measured on March 26, 1987 at Hyderabad/India (17.5°N) and on June 23, 1987 at Gap/France (44°N). Ends of the ascent and the descent sampling phases are marked by arrows.

This wave structure superimposed to the "mean profiles" is best illustrated by CH_4 and N_2O, whose profiles are plotted in the figures 1 and 2 using linear scales. It is still evident in the profiles of CFC-11 and CFC-12 the

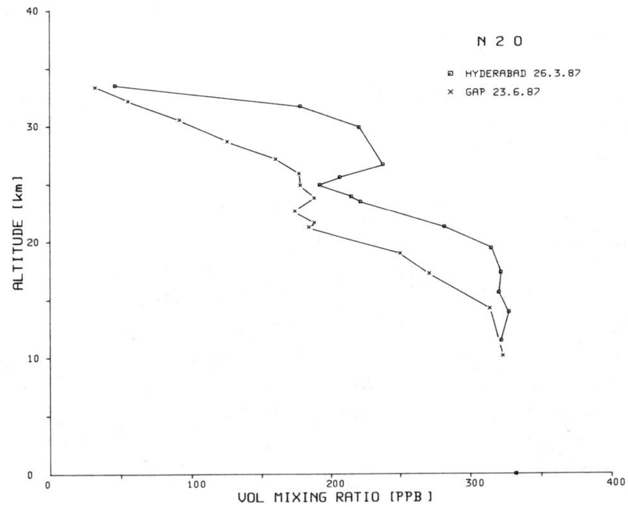

Fig. 2. As fig. 1, but for N_2O

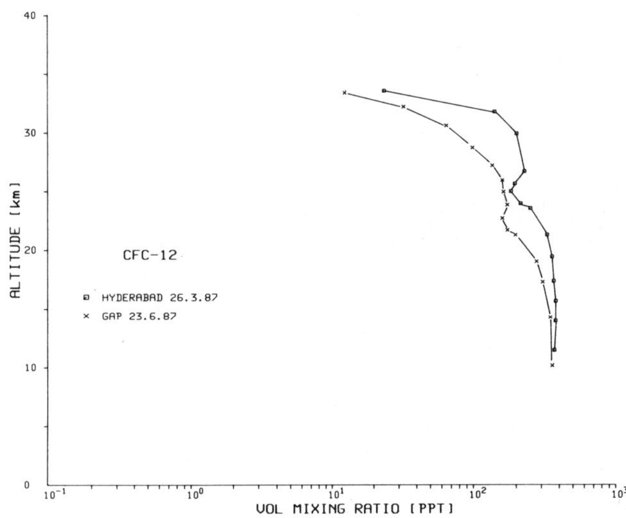

Fig. 3. As fig. 1, but for CFC-12

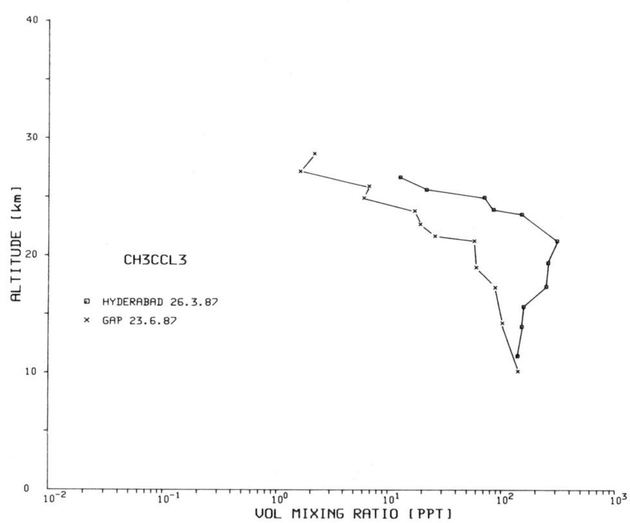

Fig. 4. As fig. 1, but for CH_3-CCl_3

latter is shown in figure 3, although the scale of the abscissa is logarithmic. All these 4 source gases have already been measured, both at midlatitudes and in the tropics, and the latitude dependence shown here is in fairly good agreement with that determined earlier (see for instance Ehhalt et al. [1981]).

All other halocarbons shown in figures 4-10 have, to our knowledge, never been measured in the tropics before. A fairly good data base for northern midlatitudes exists for CH_3Cl, CCl_4, CH_3-CCl_3, $CHClF_2$, $CBrClF_2$, $CClF_2$-CCl_2F, and $CClF_2$-$CClF_2$ (see for instance Fabian [1986], Lal et al. [1985], Fabian et al. [1985], Borchers et al. [1987], Fabian [1988]). Vertical profiles of some of these, determined in 1987 are shown in the figures 4-7. Results for $CHClF_2$ are presented by Fabian et al. [1988]. Reliable midlatitudes data for CF_4 and CF_3-CF_3 have been determined recently [Fabian et a. 1987]. Results of the 1987 cryogenic sampler experiments for these two extremely longlived constituents are consistent with these data. As there is no difference between mid latitudes and the tropics they are not shown here.

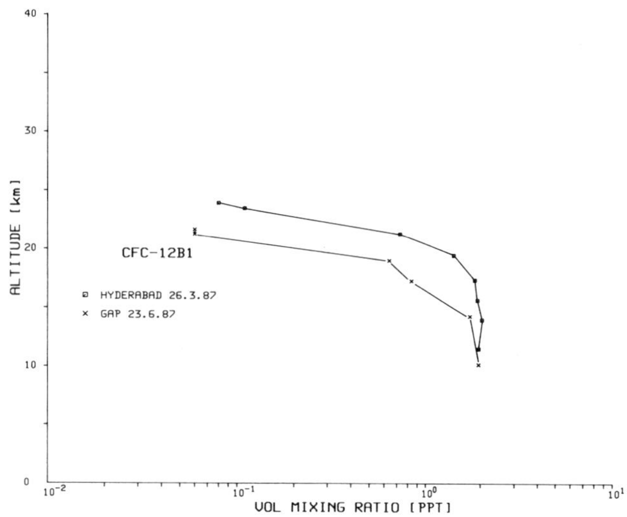

Fig. 5. As fig. 1, but for CFC-12B1

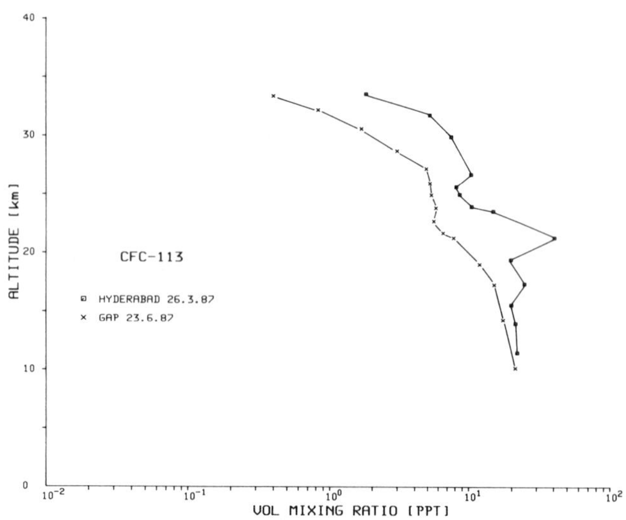

Fig. 6. As fig. 1, but for CFC-113

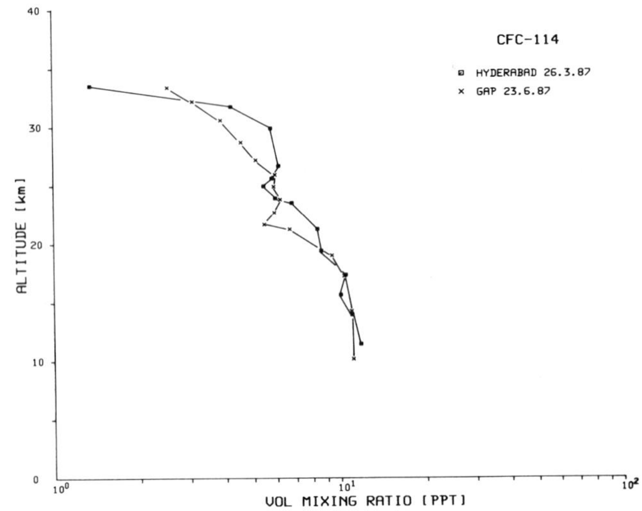

Fig. 7. As fig. 1, but for CFC-114

For $CClF_3$, $CBrF_3$, and $CClF_2$-CF_3, whose profiles are shown in the figures 8-10, only one stratospheric profile had been documented so far [Fabian et al. 1981]. Here, we present new and much more structured midlatitude profiles, together with the first tropical profiles.

It is beyond the scope of this paper to report on 2-D model computations which we have been carried out in collaboration with U. Schmailzl (Mainz) and J.A. Pyle (Cambridge) in order to compute the vertical distribution of all these halocarbons for different latitudes and seasons (a joint report is being prepared at present). Here we wish

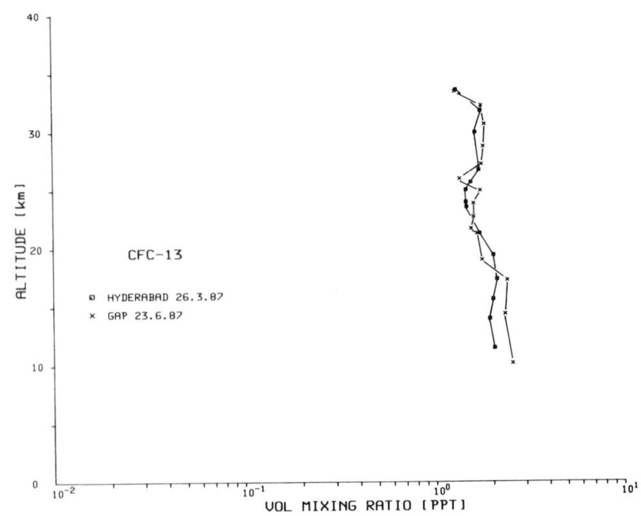

Fig. 8. As fig. 1, but for CFC-13

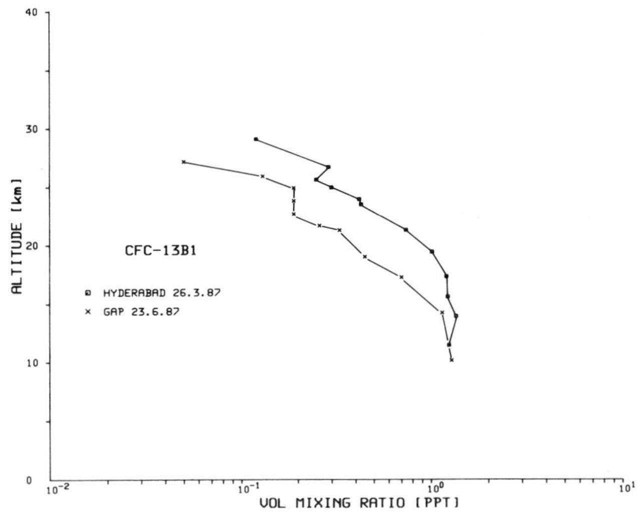

Fig. 9. As fig. 1, but for CFC-13B1

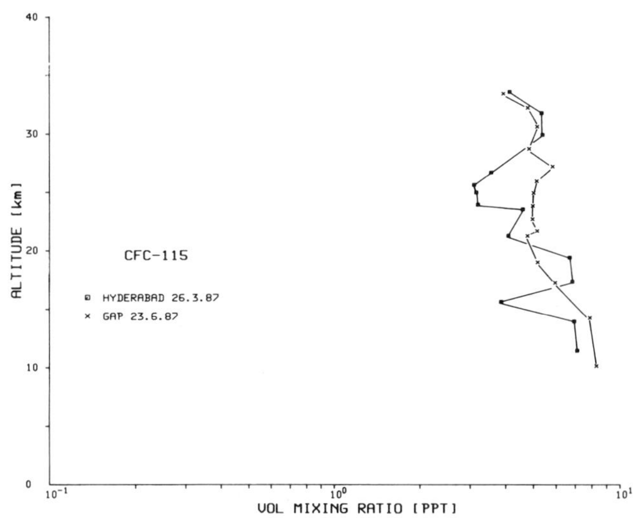

Fig. 10. As fig. 1, but for CFC-115

to communicate the new set of experimental data which may constitute an interesting set of test substances for other models as well.

ACKNOWLEDGEMENTS

Our experimental source gas research program has been sponsored in part by the German Federal Ministry of Research and Technology, Grant No. KBF-MT 04500-070 4563 2. The balloon experiment carried out at Hyderabad was performed unter the auspices of ISRO/DFVLR Indo-German collaboration. We gratefully acknowledge expert support of many Indian scientists, engineers and officials, in particular by R.P. Daniel, S.V. Damle, and R.U. Kundapurkar of the Tata Institute of Fundamental Research, Bombay, R.T. Redkar of the TIFR Balloon Range at Hyderabad, and P. Rajaratnam of the Indian Space Research Organisation (ISRO), Bangalore.

REFERENCES

Bischof, W., Borchers, R., Fabian, P., and Krüger, B.C.: Increased concentration and vertical distribution of carbon dioxide in the stratosphere. Nature **316**, 708, 1985

Borchers, R., Fabian, P., Krüger, B.C., Lal, S., Schmidt, U., Knapska, K., Penkett, S.A.: CFC-113 (CCl_2F-$CClF_2$) in the stratosphere. Planet. Space Sci. **35**, 657, 1987

Ehhalt, D.H.: The observation of long-lived trace gases in the stratosphere. Proc. Int. Ozone Symposium (J. London, Ed.) Boulder, Colo., 728, 1981

Fabian, P., Borchers, R., Penkett, S.A., and Prosser, N.J.: Halocarbons in the stratosphere. Nature **294**, 733, 1981

Fabian, P. and Gömer, D.: The vertical distribution of halocarbons in the stratosphere. Fresenius' Zeitschrift f. Analytische Chemie **319**, 890, 1984

Fabian, P., Borchers, R., Krüger, B.C., and Lal, S.L: The vertical distribution of CFC-114 in the atmosphere. J. Geo- phys. Res. **90**, 13091, 1985

Fabian, P.: Halocarbons in the atmosphere, in the Handbook of Environmental Chemistry (O. Hutzinger, Ed.), Vol. 4A, 24, 1986

Fabian, P., Borchers, R., Krüger, B.C., and Lal, S.: CF_4 and C_2F_6 in the atmosphere. J. Geophys. Res. **92**, 9831, 1987

Fabian, P.: Natürliche und anthropogene Quellgase: Ihre vertikale und globale Verteilung. BPT-Bericht 4/88, GSF München 1988

Fabian, P., Borchers R., Krüger B.C., Singh O., Lal S., Subbaraya B.H.: $CHClF_2$ (CFC-22): Distribution, budget and environmental impact. This Symposium 1988

Lal, S., Borchers, R., and Fabian, P.: Increasing abundance of $CBrClF_2$ in the atmosphere. Nature **316**, 135, 1985

$CHClF_2$ (CFC-22): DISTRIBUTION, BUDGET AND ENVIRONMENTAL IMPACT

*P. Fabian, R. Borchers, H. Duscha, and B.C. Krüger**
Max-Planck-Institut für Aeronomie (MPAE)
D–3411 Katlenburg-Lindau, FRG

S. Lal and B.H. Subbaraya
Physical Research Laboratory (PRL)
Ahmedabad, India

ABSTRACT

Recent measurements of the vertical distribution of CFC-22 carried out at midlatitudes and in the tropics confirm that global abundances of this substance continue to grow by 10%/y. Modelled profiles show lower abundances indicating that published global release rates of CFC-22 have to be scaled by 1.5 in order to fit the observed profiles. Thus present global emission of CFC-22 is about 150 x 10^6 kg/y.

Model computations show that the ODP value of CFC-22 which is 0.05 for the total ozone column may be as large as 0.2 if ozone depletion as a function of height is considered.

1. INTRODUCTION

CFC-22 is exclusively anthropogenic in origin and has primarily been used since the early 1950's, in refrigeration and as a foam blowing agent. Up to 1977, its global release rate into the atmosphere, as assessed by Jesson [1980], has increased by an average of 16%/y. This is reflected in an increase of the CFC-22 abundance of 11.7%/y observed on both hemispheres [Khalil and Rasmussen, 1981]. The main atmospheric sink for the hydrohalocarbon CFC-22 is reaction with OH radicals. This can occur both in the troposphere and the stratosphere leading to an overall atmospheric lifetime of CFC-22 between 12 and 20 years [WMO 1981, Khalil 1981].

$CHClF_2$ was measured in ground-level air for the first time by Rasmussen et al. [1980]. The first stratospheric measurements were made using a U-2 aircraft at altitudes up to about 19 km [Leifer et al. 1981], and first complete stratospheric profiles, up to about 33 km, were obtained by gaschromatographic analyses of air samples collected during 1982/83 by means of balloon borne cryogenic samplers [Fabian et al., 1985]. These show CFC-22 volume mixing ratios decreasing from about 60 pptV at the tropopause to 20 pptV at 33 km (see profile a in fig. 1).

3. MEASUREMENTS

During 1987, more recent data on the distribution of CFC-22 in the atmosphere could be obtained. On June 23, another midlatitude profile was measured at 44°N (profile b in fig. 1) shortly after a first tropical profile had been obtained on March 26, at Hyderabad/India (17.5°N) (profile c in fig. 1, see Borchers et al. [1988]). Moreover, by means of a newly developed rocket borne cryogenic sampler (RASMUS), one midlatitude data point at about 60 km altitude could be measured on May 6, which is also shown in fig. 1 (f, see Duscha et al. [1988]). Experimental details on sampling and analysis are given in the above mentioned publications. The precision of the balloon data is ± 5%, that of the rocket sample is ± 10%. The error of the absolute calibration consistent for the whole data set, is less than ± 10%.

The RASMUS data point at about 60 km is consistent with the midlatitude profile (b) showing clearly that CFC-22, although being partly decomposed in the troposphere, has reached the mesosphere already. The tropical profile (c) falls off less rapidly with height than the midlatitude profile because tropical upwelling counteracts decomposition. Although the tropopause mixing ratio of CFC-22 is about 25% lower at 17.5°N than at 44°N due to the fact that most of this hydrohalocarbon is

*Now at Alfred-Wegener-Institut für Polar- und Meeresforschung, D-2800 Bremerhaven

ISBN 0-937194-15-8

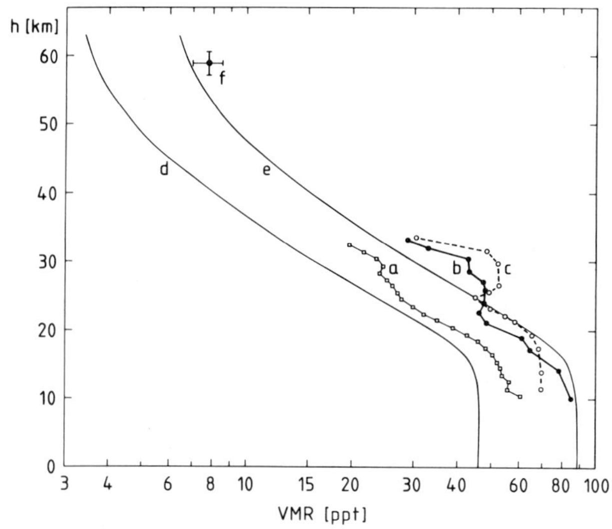

Fig. 1. Vertical distribution of CFC-22
a : Average profile from balloon data for 44° N obtained during 1982/83
b : Profile measured on June 23, 1987 at 44°N
c : Profile measured on March 26, 1987 at 17.5°N
d : Profile computed with a time-dependent 1-D model for end 1984
e : dto for end 1987 but emission scenario scaled by 1.5
f : Rocket sample (RASMUS) of May 6, 1987 at Kiruna/Sweden

emitted at northern midlatitudes, CFC-22 mixing ratios above 25 km are highest at low latitudes. Moreover, both profiles b an c show a huge wave like structure in the profile which is also noticeable in the profiles of other source gases analysed along with CFC-22 [Borchers et al. 1988].

4. BUDGET

The difference of the 1987 profile b as compared to the average 1982/83 midlatitude profile (a) confirms that the abundance of CFC-22 in the atmosphere continues growing. Taking 61 pptV and 85 pptV as the average tropopause levels of CFC-22 for 1982/83 and 1987, respectively, the average growth rate amounts to about 10%/y, slightly less than the 11.7%/y determined earlier [Khalil and Rasmussen, 1981].

We have carried out computations with our time-dependent one dimensional model using global emission rates published by Jesson [1980] for the years up to 1977. We have extrapolated these rates beyond 1977 assuming the growth continued unchanged up to 1980, and emission rates remaining constant thereafter. Thus global emission rates were taken as 77 kt/y, 88 kt/y, and 100 kt/y for the years 1978, 1979, and 1980-1984, respectively. This is a similar emission scenario as that suggested by Wuebbles, MacCracken and Luther [1984], although our emission figures are about 30% higher than theirs. Nevertheless, the profile computed for the end of 1984 (curve d in fig. 1) shows considerably lower values than those measured one to two years earlier (curve c). Thus we have to conclude that our emission scenario based on Jesson's data is considerably underestimated.

Our 1-D model experiments have been accompanied by analogous 2-D model computations, based on the same emission history [B.C. Krüger, P. Fabian, J.A. Pyle, L. Gray, U. Schmailzl, manuscript being prepared]. These show the same discrepancy between modelled and measured profiles. They show good agreement, however for the ratio of midlatitude vs. tropical tropopause mixing ratios of CFC-22, which indicates that the aforementioned discrepancy is not due to unrealistically high OH concentrations but rather underestimated global emission rates.

Thus, as obviously no precise emission figures of CFC-22 are available, we suggest an appropriate scaling to be applied. If we scale the emission history used for our model study by 1.5 we obtain curve e for end 1987 conditions, which is in fairly good agreement with the data measured during 1987. This corresponds to present global emission rates of 150 x 10^6 kg of CFC-22 per year. According to Hoffmann [1988], the global production of CFC-22 was 210 x 10^6 kg in 1985, one third of which was not released but rather used as a starting material for the production of other substances. Thus present release rates into the atmosphere could well amount to about 150 x 10^6 kg/y.

5. ENVIRONMENTAL IMPACT

In an earlier study we assessed the effect of continuously growing emission rates of CFC-22 on the ozone layer. Based on a growth rate of 16%/y, we found for the year 2020 ozone depletions around 40 km altitude due to CFC-22 being comparable to those of CFC-11 and CFC-12 together, for which global emission rates had been kept fixed at the 1976 level [Krüger and Fabian 1986]. It is not unrealistic to assume that CFC-22 emission rates will grow in the future, as this hydrohalocarbon which is not included in the Montreal Protocol, is likely to largely replace the fully halogenated CFC-11 and CFC-12 whose production will be curbed.

It has become common to assess the ozone depletion potential of chemicals with respect to the ozone layer as a whole. Recent calculations of ozone depletion potentials (ODPs) by Hammitt et al. [1987] show that the relative ozone-depletion efficiency of CFC-22 with respect to the total ozone column is 0.05 as compared to CFC-11, i.e. CFC-22 is 20 times less effective than CFC-11. This relatively small ODP of CFC-22 is partly due to the fact that ozone depletion which is largest at about 40 km, is partly compensated by ozone production at about 25 km altitude. This "self-healing" of CFC-22 is, as the paper by Hammitt et al. shows, more effective than that of CFC-11.

In order to assess the ozone depletion potential in more detail, in particular to study the effect of ozone variation as a function of height, we have run the following 1-D model experiments: Anthropogenic halocarbon injection into the atmosphere was reduced to either CFC-11 or CFC-22 emission. A constant global emission rate of 1000 x 10^6 kg/y of either CFC-11 or CFC-22 was taken, which is about the sum of CFC-11, CFC-12 and CFC-22 emission altogether at present. After 150 model years, the model results approached steady state. Relative ozone changes with respect to the initial atmosphere as a function of height thus obtained are shown in figure 2 for CFC-11 (curve a) and CFC-22 (curve b). The largest ozone depletion due to CFC-11 emission of about 50 percent occurs at 40 km, while about 11 percent ozone increase results at 20 km altitude. CFC-22 causes the largest ozone depletion around 40 km, too, although it reaches only 16% the efficiency of CFC-11. The self-healing effect of CFC-22 is different from that of CFC-11 as the maximum ozone increase occurs at higher altitudes. Curve c shows the "relative ozone depletion efficiency" as a function of altitude. It clearly shows values above 5%

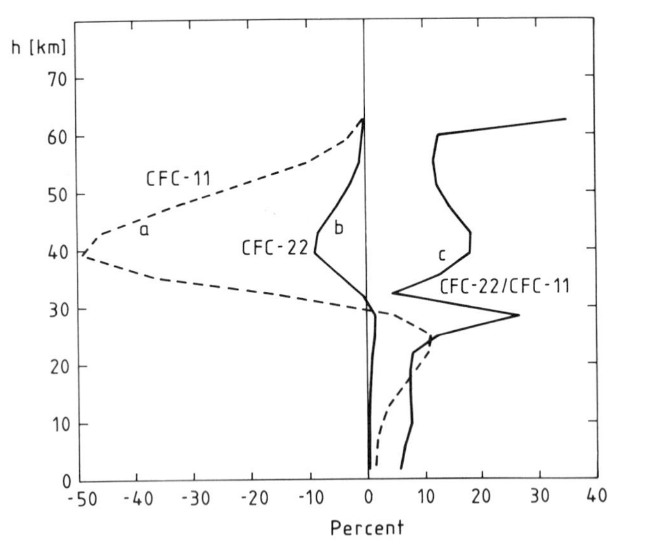

Fig. 2. Steady-state relative ozone change due to CFC-11 (a) and CFC-22 (b) as a function of altitude, based on constant global emission rates of 1000 x 10^6 kg/y.
Curve c: Relative efficiency of CFC-22 with respect to CFC-11

at most altitudes, with highest values even above 20%. Fig. 2 clearly shows that the main impact of CFCs on the ozone layer is not just a change of its overall thickness but, more importantly, a redistribution of the vertical ozone profile, which certainly has implications for global climate.

ACKNOWLEDGEMENTS

This work has been sponsored in part by the German Federal Ministry of Research and Technology through grants KBF-MT 0450-0704 563 2 and 010 E 87016. The balloon flight carried out at Hyderabad was performed under the auspices of the Indo-German collaboration with ISRO-DFVLR sponsorship which is gratefully acknowledged.

REFERENCES

Borchers, R., Fabian, P., Lal, S. and Subbaraya, B.H.: The vertical distribution of source gases at tropical latitudes. Quadrennial Ozone Symposium Göttingen 1988

Fabian, P., Borchers, R., Krüger, B.C., and Lal, S.: The vertical distribution of $CHClF_2$ (CFC-22) in the stratosphere. Geophys. Res. Lett., **12**, 1, 1985

Duscha, H., Borchers, R., Fabian, P.: First results of RASMUS: Source gases in the mesosphere. Quadrennial Ozone Symposium Göttingen 1988

Hammitt, J.K., Camm, F., Connell, P.S., Mooz, W.E., Wolf, K.A., Wuebbles, D.J., and Bamezai, A.: Future emission scenarios for chemicals that may deplete stratospheric ozone. Nature **330**, 711, 1987

Hoffmann, B.: Testimonial during public hearing of the Enquete-Kommission "Vorsorge zum Schutz der Ozonschicht", Berlin 1988

Jesson, J.P.: Release of industrial halocarbons and tropospheric budget, in NATO Advanced Study Institute on Atmospheric Ozone (M. Nicolet and A.C. Aikin, Eds.), Report FAA-EE-90-20, US Dept. of Transportation, Washington, D.C., 373, 1980

Khalil, M.A.K.: Topics in the behaviour of atmospheric trace gases, PhD. dissertation, Oregon Graduate Center, Beaverton, Or. 1079, 1981

Khalil, M.A.K. and Rasmussen, R.A.: Increase of $CHClF_2$ in the Earth's atmosphere. Nature **292**, 823, 1981

Krüger, B.C. and Fabian, P.: Model calculations about the reduction of atmospheric ozone by different halogenated hydrocarbons. Ber. Bunsenges. Phys. Chem., **90**, 1062, 1986

Leifer, R., Sommers, K.C., Guggenheim, S.F.: Atmospheric trace gas measurements with a new clean air sampling system. Geophys. Res. Lett. **8**, 1079, 1981

Rasmussen, R.A., Khalil, M.A.K., Penkett, S.A., and Prosser, N.J.D.: $CHClF_2$ (F-22) in the Earth's atmosphere. Geophys. Res. Lett. **7**, 809, 1980

WMO-Report No 11, The Stratosphere 1981: Theory and Measurements, Geneva 1981

Wuebbles, D.F., Mac Cracken, M.C. and Luther, F.M.: A proposal reference set of scenarios for radiatively active atmospheric constituents. DOE-NBB-0066 TRO 15, Contract No W-7405-Eng-48, US Dept. of Energy, Washington, D.C. 1984

IN SITU OBSERVATIONS OF LONG-LIVED TRACE GASES IN THE ARCTIC STRATOSPHERE DURING WINTER

Ulrich Schmidt, Reimar Bauer, Ahmed Khedim, Erich Klein, Gerhard Kulessa, and Birgit Schubert
Institut für Chemie 3: Atmosphärische Chemie
Kernforschungsanlage Jülich GmbH
D–5170 Jülich, FRG

ABSTRACT

The formation of a cold polar vortex during the winter season is generally associated with large-scale downward air motions over the polar region. Such subsidence should result in substantially increased gradients in the vertical mixing ratio profiles of long-lived trace gases if compared to mid-latitude distributions. Employing a balloon-borne cryogenic whole air sampler, a series of balloon flights were launched over Northern Scandinavia in February 1987 and 1988, respectively. The results obtained for the source gases CH_4, N_2O, $CFCl_3$, CF_2Cl_2, CCl_4, CH_3Cl, CH_3CCL_3, and $C_2F_3Cl_3$ are reported. The present observations confirm the effect of subsidence. Furthermore, the profiles indicate pronounced structures that are probably caused by horizontal transport processes during dynamically active periods in the development of warming events. In the lower arctic stratosphere, the column burden of chlorine bonded in the six halocarbons was found to be lower than at mid-latitudes by about a factor of two.

1. INTRODUCTION

The occurence of large depletions in total ozone in the lower stratosphere over Antarctica during austral spring, the so-called ozone hole, revealed substantial gaps in our knowledge about the physical and chemical processes in the polar stratospheres during winter. When the ozone hole was first described (FARMAN et al, 1985), such an effect had not been predicted by any photochemical model. Since that time various theories were developed in order to explain this regional phenomenon by either chemical or dynamical processes, or by a combination of both (see the review article by SOLOMON, 1988). Interestingly enough, some of the theoretical explanations were purely hypothetical, since no data were available that could have been used to prove or confirm any of the proposed mechanisms. This situation has now changed after several observational campaigns have been performed during the last two years in both the Arctic and the Antarctic.

Observations of long-lived trace gases, especially the source gases, put two important constraints on the theoretical models. First, they provide the actual information about the abundances of these substances, which have not been measured before in the lower stratosphere during winter. Second, they allow a decision as to whether a purely dynamical process, such as the upwelling theory proposed by TUNG et al (1986) could be responsible for the observed depletion in total ozone. The basic argument of this hypothesis suggested that the Antarctic stratosphere was diabatically cooled during the austral winter until radiative equilibrium is established. In spring, this region would be heated again resulting in an upward motion of air that transports tropospheric air with low ozone content into the lower stratosphere.

First observations of a chemically inert, long-lived tracer, N_2O, indicated that upwelling is not a very likely process (PARRISH et al, 1988). They found that the N_2O column was reduced rather than increased. Their ground-based observations have now been confirmed by aircraft-borne in situ observations during the Antarctic Airborne Ozone Expedition in September 1987 (HEIDT et al, 1988). The results indicated that N_2O, as well as other source gases, were strongly depleted at altitudes above about 16 km, especially in the region of the chemically perturbed region.

The winter stratosphere over the Arctic is generally not as cold as over the Antarctic (BARNETT and CORNEY, 1985), because the temperature distribution is affected by large scale dynamic mixing processes. However, the cooling of the polar stratosphere during the absence of sunlight in the polar night also results in large scale subsidence over the Arctic region. Therefore, the stratospheric abundances of long-lived tracers should also be decreased during the Arctic winter. This paper reports on in situ observations of the vertical distribution of variuos source gases in the lower stratosphere, which were performed over Northern Scandinavia during winter.

2. EXPERIMENTAL

We used a balloon-borne cryogenic whole air sampler to collect large air samples at altitudes up to 30 km. So far, three ballon flights have been performed from ESRANGE, near Kiruna in Northern Sweden (67.9°N; 21.1°E) as part of the CHEOPS - Program (CHEOPS = CHEmistry of Ozone in the Polar Stratosphere).

Flight A on 5 February 1987 (CHEOPS I) - Due to a failure of the balloon the ascent was terminated at an altitude of 22.5 km. However, 14 air samples could be collected during parachute descent in the altitude range from 19.3 km to 12.1 km. A complementary set of tropospheric air samples was collected during an aircraft flight.

Flight B on 1 February 1988 (CHEOPS II) - A set of 12 samples was collected during ascent of the balloon in the altitude range from 10.1 km up to 24.2 km. Because of the risk of contamination at higher altitudes no further samples were collected during ascent. Unfortunately the flight had to be terminated after the balloon had reached a maximum altitude at 33.5 km and the uppermost samples had to be collected during the rapid descent on parachute (≈30 m/s). Three further samples could be collected in the altitude range from 33.2 km to 17.9 km, each integrating over height intervals of several kilometers.

Flight C on 10 February 1988 (CHEOPS II) - This flight reached a maximum altitude of 25.8 km. 15 samples were collected during the valve controlled slow descent (2-3 m/s) down to an altitude of 12.0 km.

During Flight A samples had been only collected within time intervals of about 10 seconds in order to assure a height resolution of less than 1 km inspite of the fast descent under parachute. Therefore, the sampled volumes were rather small, and reliable analyses could be only performed for CH_4, N_2O, $CFCl_3$, CF_2Cl_2, and $C_2F_3Cl_3$. The aircraft samples were analyzed for the same set of species. Because of the much larger volumes, all samples collected during Flights B and C could be also analyzed for CCl_4, CH_3CCl_3, and CH_3Cl.

It should be noted that the analyses of these species were calibrated against laboratory standards that can be traced back to those used in our laboratory more than ten years ago (SCHMIDT et al ,1986; VOLZ et al, 1981). This faciliates the comparison with previous observations at mid-latitudes.

ISBN 0–937194–15–8

3. RESULTS

The results obtained for CH_4, N_2O, $CFCl_3$, CF_2Cl_2, $C_2F_3Cl_3$, CCl_4, and CH_3CCl_3 are displayed in the figures 1 through 7 in the form of vertical profiles. Volume mixing ratios are plotted versus pressure altitudes (based on the US-Standard Atmosphere, 1976).

The experimental error for these measurements is generally less than the size of the symbols, unless it is indicated by a horizontal bar. In the case of Flight B the vertical bars indicate the sampling intervals for the three samples that were collected during the rapid descent over altitude intervals of more than 1 km. For all the other samples, the sampling interval is about 500 m or less, i.e. it is comparable to the size of the plotted symbols.

For comparison each of the figures also displays a reference profile. We have chosen to compare the new observations at high latitudes (68°N) against previous observations that were obtained during balloon flights over Southern France (44°N) performed during winter, i.e. in the time period between the fall turn-around and the spring turn-around of the stratospheric circulation. These data were reported by VOLZ et al(1981), FABIAN et al(1981),and SCHMIDT et al(1986).

The reference profiles are average vertical distributions that are calculated as running means over 3-km intervals with steps of 1 km. The horizontal bars represent the 1 σ-variation of all data within the respective 3-km interval. In order to account for the long term trend in the tropospheric mixing ratios of the long-lived trace gases, the individual data were normalized to the date of the 1 January 1987. The trend factors for this correction were taken from WMO (1986).

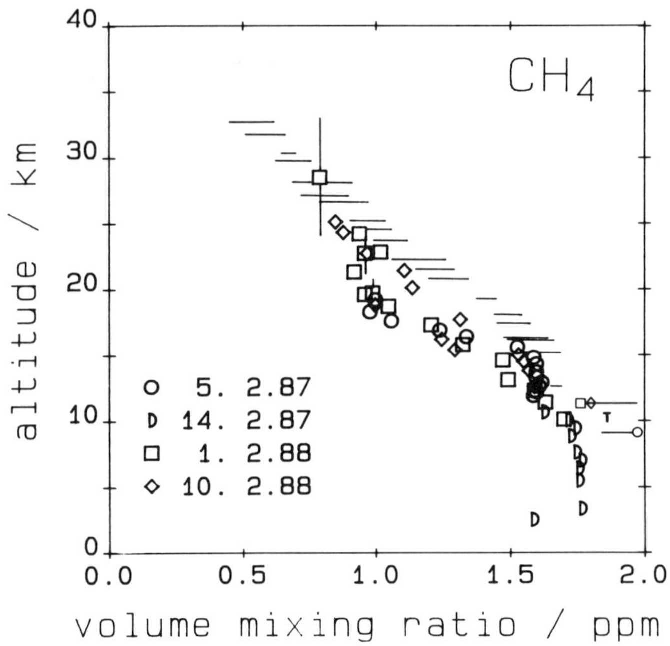

Fig. 1. Vertical distribution of CH_4 at 68°N during winter. The horizontal bars represent the average mid-latitude profile derived from 4 balloon flights during the winter season (VOLZ et al, 1981; SCHMIDT et al, 1986). T = Tropopause

4. DISCUSSION

The most prominent feature in all the plotted profiles is a considerable depletion in the lower stratosphere. The data obtained during Flight A show a strongly enhanced gradient above 16 km altitude and rather constant mixing ratios below that level. A slight difference in the mixing ratios is observed below and just above the tropopause. This feature is most clearly seen for N_2O, CH_4, and CF_2Cl_2, i. e. those species that can be measured with highest precision. The results document, with a high vertical resolution, the very fast decrease of the source gas mixing ratios within a layer about two km thick.

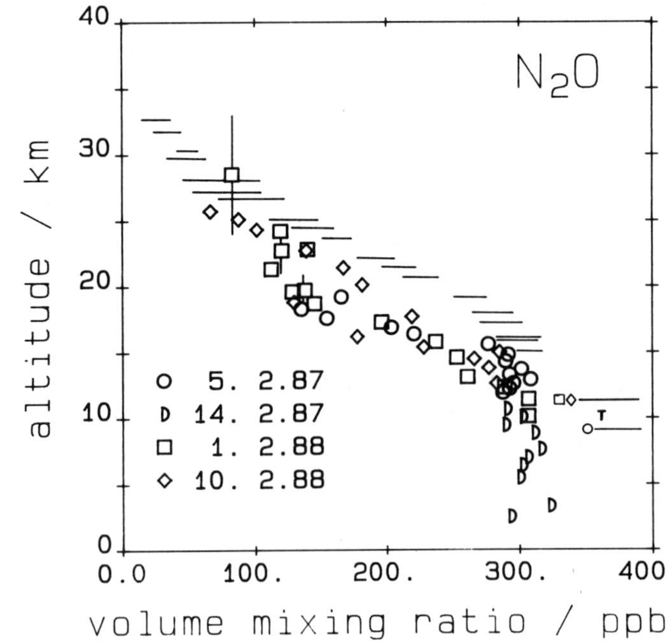

Fig. 2. Same as Figure 1 but for N_2O.

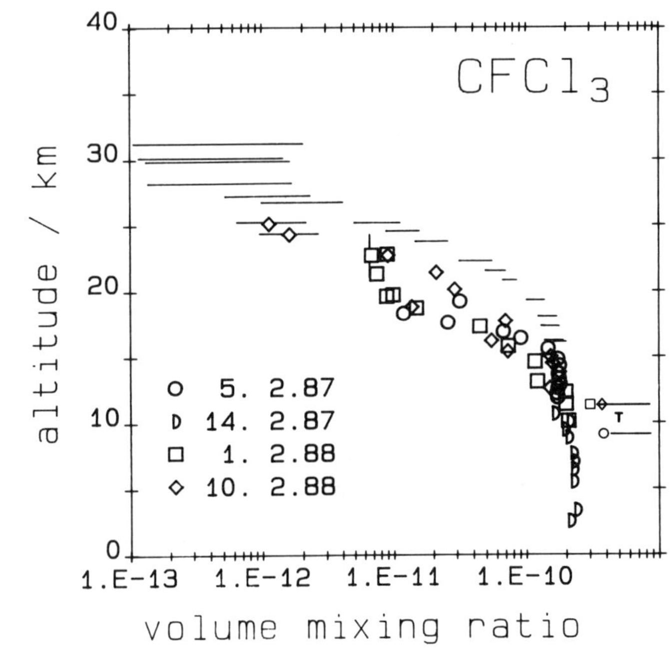

Fig. 3. Same as Figure 1 but for $CFCl_3$.

The results of Flight B confirmed this gradient for the same altitude range. However, above 25 km the measured concentrations are again close to those observed at mid-latitudes. It appears that even the structure around 23 km is quantitatively confirmed by those samples that were collected during parachute descent over much larger altitude intervals.

The Arctic winter of 1987/1988 was unique in the sense that a very strong sudden warming occured rather early, during the first week of December 1987. In fact, it was the strongest mid-winter warming that has ever been recorded since observations of this dynamical phenomenon were begun about 3o years ago. (NAUJOKAT et al, 1988).

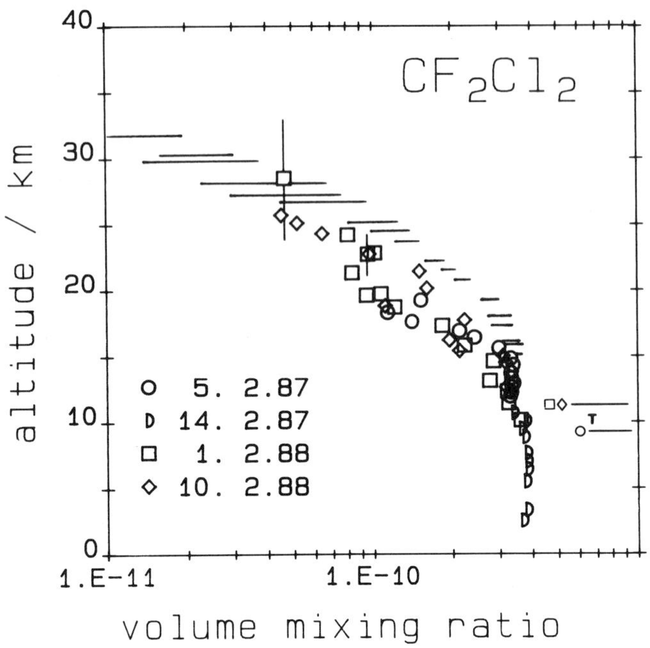

Fig. 4. Same as Figure 1 but for CF_2Cl_2.

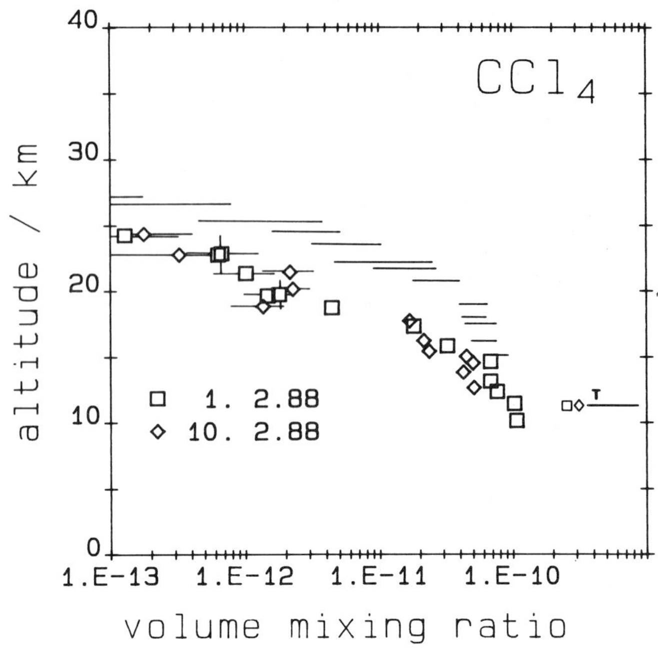

Fig. 6. Same as Figure 5 but for CCl_4.

After this event the temperatures in the lower stratosphere over Northern Scandinavia decreased again only very slowly, although the Arctic vortex reestablished itself and minimum temperatures below -85°C were observed over Greenland at the beginning of February 1988. Some minor warmings developed in the jet region of the elongated polar vortex. At altitudes around 30 km mid-latitude air tended to move towards the pole.

We believe that the reduced mixing ratios in the lower stratosphere reflect the effect of subsidence over the Arctic polar region during winter. The observed deviations from the mid-latitude reference profiles are strongly correlated for all species measured. They are largest around 19.0 km altitude. The mixing ratios observed at this level for the different species correspond to those measured at mid-latitudes during winter at altitudes some 6 km above this level.

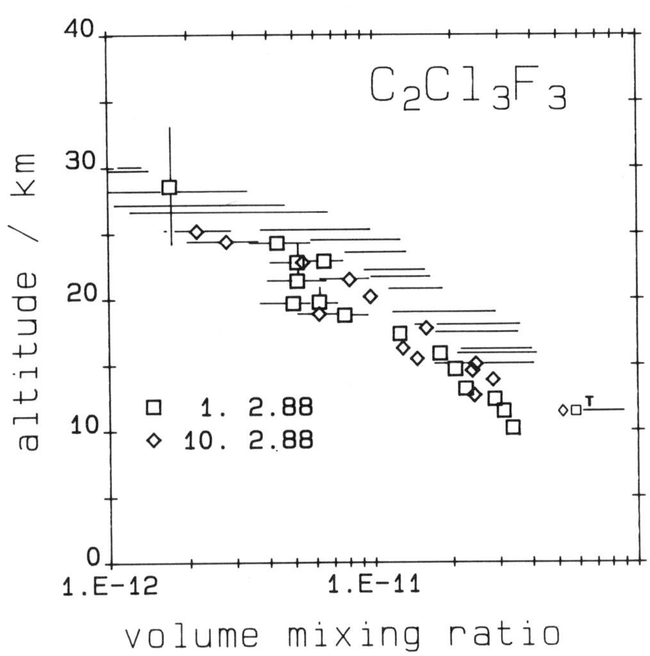

Fig. 5. Vertical distribution of $C_2F_3Cl_3$ at 68°N during winter. The horizontal bars represent the average mid-latitude profile derived from 3 balloon flights during the winter season (SCHMIDT et al, 1986). T = Tropopause.

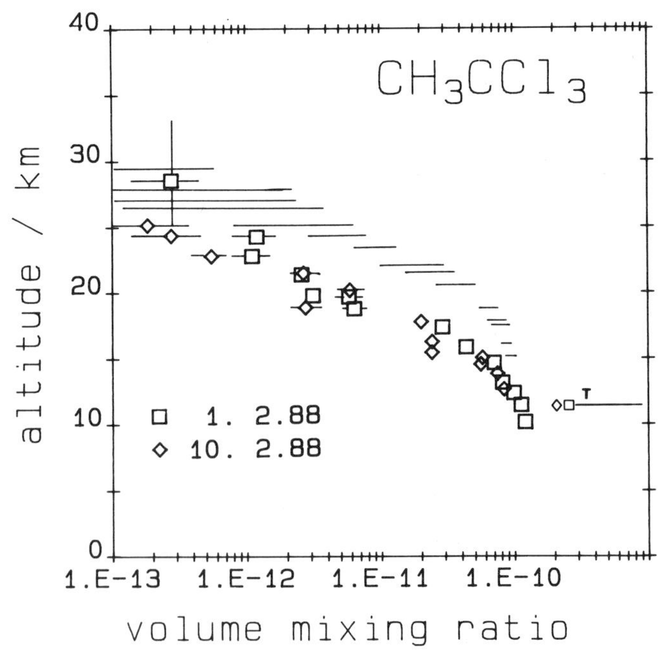

Fig. 7. Same as Figure 5 but for CH_3CCL_3.

The profile observed during Flight C shows much more structure than those measured during the two other flights.Simultaneous measurements of ozone gave variations in the O_3 partial pressure that are highly anticorrelated with the features in the source gas distributions. Furthermore, the temperature of the lower stratosphere was rather variable during the first part of February reflecting the continuing heat flux towards the pole forced by planetary wave activity. The observed structures certainly represent the effect of horizontal mixing as a consequence of the increasing dynamical activity.

Within the altitude range between 15 km and 25 km, the observed distributions over the polar region represent a significant reduction in the column abundance of the various source gases. These observations, therefore, put important constraints on present theoretical models, which assume that the mid-latitude concentration profiles were also representative of the conditions in the polar region. Quantitative information on this column reduction can be estimated from the difference in the average vertical profiles at both latitudes. At 68°N the column burden of CH_4 and N_2O in the layer between 15 and 25 km (120 to 25 hpa) is found to be 19% and 31% lower than at 44°N.

In the case of the chlorine containing source gases such a comparison is more informative when given for the column burden of the total chlorine, available in the six halogenated source gases $CFCl_3$, CF_2Cl_2, CCl_4, CH_3Cl, CH_3CCL_3, and $C_2F_3Cl_3$. Figure 8 shows the vertical distribution of "total chlorine"; it was calculated by adding the mixing ratios of the halogenated hydrocarbons after multiplication by the number of chlorine atoms in the individual compounds. If the local mixing ratio of "total chlorine" is integrated over the altitude interval from 15 km to 25 km we obtain the column burden of chlorine that is available in the six most abundant halocarbons. Based on our measurements the value calculated for northern latitudes is 1.1×10^{15} Cl-atoms/cm^2. This is only about 45% of that derived for mid-latitudes - 2.5×10^{15} Cl-atoms/cm^2.

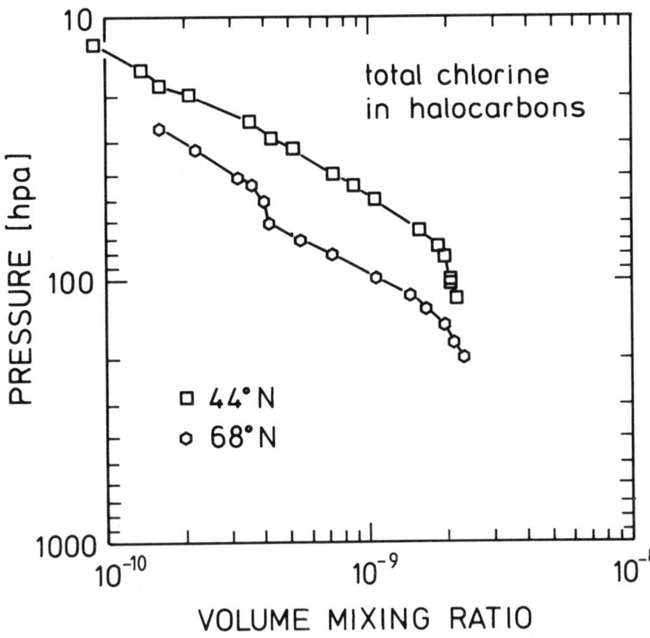

Fig. 8. Vertical distribution of total chlorine bonded in the six most abundant halogenated hydrocarbons (see text).

5. ACKNOWLEDGEMENT

The balloon flights were supported by the German Ministry of Science and Technology (BMFT) through grant No. 07450013.

6. REFERENCES

Barnett, J.J. and M. Corney, 19..: Middle atmosphere reference model derived from satellite data. Handbook for MAP, Volume 16, 47, SCOSTEP-Secretariat, University of Illinois, Urbana, Illinois, USA.

Fabian, P., R. Borchers, G. Flentje, W.A. Matthews, W. Seiler, H. Giehl, K. Bunse, F. Müller, U. Schmidt, A. Volz, A. Khedim, and F.J. Johnen, 1981: The vertical distribution of stable trace gases at mid-latitudes. J. Geophys.Res., 86, 5179.

Farman, J.C., B.G. Gardiner, and J.D. Shanklin, 1985: Large losses of total ozone in Antarctica reveal seasonal ClO_x/NO_x interaction, Nature, 315, 207.

Heidt, L.E., J.F. Vedder, W.H. Pollock, B.E. Henry, and R.A. Lueb, 1988: Trace gas measurements from whole air samples collected over the Antarctic continent, Paper presented at the Polar Ozone Workshop, Snowmass, Colorado, USA, 9-13 May 1988, submitted to J.Geophys.Res.

Naujokat, B., K. Labitzke, R. Lenschow, K. Petzold, R.C. Wohlfahrt, 1988: The stratospheric winter 1978/88: An unusually early major midwinter warming. Beilage z. Berliner Wetterkarte, Nr SO 6/88, Inst.f.Meteorologie, Freie Universität, Berlin, F.R.G.

Parrish, A., R.S. deZafra, M. Jaramillo, B. Connor, P.M. Solomon, and J.W. Barett, 1988: Extremely low N_2O concentrations in the springtime stratosphere at McMurdo Station, Antarctica, Nature, 332, 53.

Schmidt, U., C. Jebsen, F.J. Johnen, A. Khedim, E. Klein, D. Knapska, G. Kulessa, J. Rudolph, G. Schumacher, E. Schunck, 1986: Stratospheric observations of long-lived trace gases at mid-latitudes 1982-1985, Data Report, Jül-Spez.-375, Kernforschungsanlage Jülich GmbH, Jülich, F.R.G.

Solomon, S., 1988: The mystery of the Antarctic ozone "hole", Rev. Geophys., 26, 131.

Tung, K.K., M.K.W. Ko, J.M. Rodriguez, and N.D. Sze, 1986: Are Antarctic ozone variations a manifestation of dynamics or chemistry?, Nature, 322, 811.

Volz, A., U. Schmidt, J. Rudolph, D.H. Ehhalt, F.J. Johnen, and A. Khedim, 1981: Vertical profiles of trace gases at mid-latitudes, Jül-Report 1742, Kernforschungsanlage Jülich GmbH, Jülich, F.R.G.

WMO, Atmospheric Ozone 1985, 1986: Report No 16, World Meteorological Organization, Geneva, Switzerland.

MEASUREMENTS OF COFCl AND $COCl_2$ NEAR THE TROPOPAUSE

Stephen R. Wilson, Gerhard Schuster, and Günter Helas
Max Planck Institute for Chemistry
Mainz, FRG

ABSTRACT

The first measurements of carbonyl fluorochloride, (COFCl) in the atmosphere are reported. COFCl is observed above the tropopause, with a mixing ratio that increases with altitude. In the troposphere the mixing ratio is less than 4 pptv, our lower detection limit. Our measurements extend to approximately 5 km above the tropopause, where a mixing ratio of 18±5 pptv is observed. In the region studied a linear correlation with ozone is observed. This behaviour is consistent with the expected chemistry of COFCl. It is produced primarily from the photo-oxidation of the chlorofluorocarbon F-11 ($CFCl_3$) in the mid-stratosphere. Known loss processes involve hydrolysis (primarily in the troposphere) and photolysis in the middle stratosphere. The comparison with the phosgene mixing ratios determined from the same air samples highlights the importance of tropospheric sources for phosgene in the lower part of the stratosphere.

1. INTRODUCTION

The chlorofluorocarbons have been used for many years in the chemical industry, at least partially because of they are relatively inert. When released into the atmosphere their stability results in very long lifetimes before they are oxidized (As summarized by WMO (1985)). The primary oxidation takes place in the the stratosphere, where they are photo-decomposed, resulting in the release of the halogens.

This photo-oxidation process has been shown to proceed through the halogenated formaldehyde derivative (COXY). Carbonyl fluoride (COF_2) is produced primarily from F-12 (CF_2Cl_2) (Jayanty et al. (1975)) and F-22 (CHF_2Cl). F-11 ($CFCl_3$) produces COFCl (Jayanty et al. (1975)). For phosgene ($COCl_2$) the situation is more complex, with a number of possible tropospheric precursors including the chloroethenes (Singh (1976)). Carbon tetrachloride is believed to be the main source for phosgene in the stratosphere (Crutzen et al. (1978)). The ultimate fate of these compounds is the formation of CO_2 and the respective acid(s) HX.

Of these carbonyl compounds, carbonyl fluoride (COF_2) was the first to be detected in the stratosphere. It has been identified from infrared spectra recorded by the ATMOS experiment (Rinsland et al. (1986), Raper et al. (1987)). Profiles were calculated from the data, which showed a maximum in the mixing ratio of approximately 100 pptv at a height of 32 km. At 17.5 km, the lowest level at which measurements are reported, a mixing ratio of 20 ± 8 pptv has been found.

We have recently identified absorption features in the infrared spectra of air samples due to phosgene (Wilson et al. (1988)). The analysis of these spectra indicates that phosgene has a relatively constant mixing ratio in the upper troposphere of approximately 17 pptv, which then increases in the stratosphere. We report here measurements of the vertical distribution of COFCl in the lower stratosphere, and compare this distribution to that of phosgene.

2. EXPERIMENTAL

The measurements reported here have been made using the Matrix Isolation-Fourier Transform Infrared Spectroscopy technique, which has been described previously (Griffith et al. (1987)). Very briefly, the technique involves the cryogenic trapping of air samples at a temperature of ≤87 K (liquid nitrogen or liquid argon). The material frozen out is primarily CO_2 and water. When appropriately handled, the CO_2 forms a glassy matrix, with the condensable trace gasses embedded in it. The infrared spectrum of this matrix can then be measured; the absorption peaks are characteristic of the specific trace gases and the height or area under each peak gives a quantitative measure of the amount of the trace gas present.

For the measurements described, the samples were collected aboard a Lear Jet (25D) on several flights between Germany and Spitzbergen (50° N - 78° N). The date and approximate tropopause height for each flight is given in table 1. In addition to the measurements of COFCl reported here,

ISBN 0-937194-15-8

the mixing ratio of a number of other compounds were measured, including N_2O, OCS, SO_2, $CFCl_3$, CF_2Cl_2, $CHClF_2$ and CH_3CCl_3. Ozone was measured continuously during the flights with Dasibi ozone monitors. For the ozone measurements the air was sampled from the inlet of the cabin pressurization system. Details of these measurements will be presented elsewhere.

2.1 CALIBRATION

Unfortunately, due to its toxicity COFCl is not commercially available in Europe. We therefore synthesized the small amounts required for calibration using two different techniques (details to be published). First, mixtures of F-21 ($CHFCl_2$), CO_2 and chlorine in synthetic air were irradiated with black lamps. The chlorine atoms so generated extract a hydrogen from the F-21 producing HCl, and the resulting CCl_2F radical then oxidizes to form the required COFCl. The other technique involves calibration mixtures made by irradiating mixtures of either F-11 or F-21 in oxygen with a low pressure mercury lamp. By operating at high concentrations it was found that almost complete conversion could be achieved. These mixtures were then diluted using standard techniques to produce the required mixing ratios.

The calibration of absorption intensity for phosgene has been determined by standard dilution techniques of phosgene in CO_2. It has been observed that the observed absorption intensity is dependent on the amount of water in the matrix. The results discussed here have been corrected for this effect. A dependence on the amount of ozone trapped during sampling has also been observed, and so only results obtained from samples collected with liquid argon have been used.

The spectrum of COFCl in the gas phase has been reasonably well characterized (Nielson et al. (1952), Brown et al., (1987)). For our measurements it was necessary to identify the infrared absorption features of matrix isolated COFCl in CO_2. The spectrum of COFCl in CO_2 (0.5 ppmv = 170 pptv in air) is shown in fig. 1. The strongest absorption feature is the ν_4 absorption at 1088.0 cm^{-1}. The next strongest absorption feature is the ν_2 absorption at 1858.2 cm^{-1}. The frequencies of the absorptions in the CO_2 matrix are summarized in table 2, along with the gas phase frequencies as measured by Nielson et al. (1952).

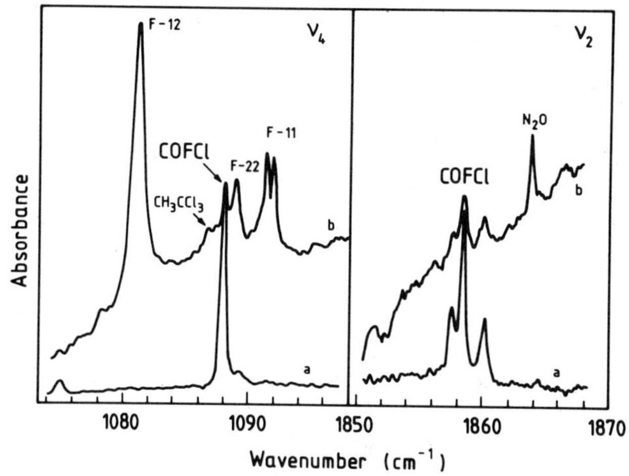

Fig. 1: Spectrum of COFCl in a CO_2 matrix at 77 K, showing both the ν_2 and ν_4 absorption features. a) A spectrum recorded from a mixture of 0.5 ppmv COFCl in CO_2. b) A portion of a spectrum from an air sample collect during the flight LJ8 at 11.8 km. The calculated mixing ratio from this spectrum is 13±3 pptv.

The ν_2 band has a complex structure that cannot be due just to a ^{37}Cl isotope shift. It presumably arises from multiple orientations of COFCl in the CO_2 matrix. Such structure is not visible for the ν_4 band, presumably because the splitting is smaller than the linewidth.

TABLE 1. SUMMARY OF THE DATES AND APPROXIMATE TROPOPAUSE HEIGHTS FOR THE MEASUREMENT FLIGHTS REPORTED.

Flight	Tropopause height (km)	Date
LJ6	12.1	4/12/86
LJ7	9.9	20/1/87
LJ8	7.3	21/2/8
LJ9	9.5	22/4/87
LJ10	9.4	11/6/87

3. RESULTS AND DISCUSSION

The comparison of the reference spectrum with the spectra of flight samples is shown in fig. 1 for both the ν_2 and the ν_4 band. The spectrum used for comparison comes from the flight LJ8, where, due to a very low tropopause, it was possible to collect samples from over 5 km above the tropopause. In both regions it is possible to see very clearly the absorption features due to COFCl. Although the ν_4 band is stronger than ν_2, the

TABLE 2. VIBRATIONAL ABSORPTION FREQUENCIES FOR COFCl. (/cm^{-1})

Gas phase*	CO_2 Matrix	Assignment
776	766.2	ν_1
1095	1088.0	ν_4
1876	1858.2	ν_2
2182	2165.3	$2\nu_4$

* Nielson et al. (1952)

overlap with other absorption features in this region means that it has not been possible, as yet, to use the ν_4 band for quantitative analysis. Therefore, for the results presented here only the absorption due to the ν_2 vibration has been used. It seems unlikely, however, that other (unknown) absorption features could contribute to the absorption assigned to COFCl. Wherever it has been possible to estimate the intensity of the ν_4 band the intensity has been consistent with that of the ν_2 band. Further, the ν_2 absorption feature has a characteristic shape (fig. 1), which can be recognized in all the spectra from air samples except where the absorption feature is weak (<10 pptv).

The errors involved come from two sources. First, the error in the calibration, which includes the error in the quantum yield of production of COFCl for the calibration standards. We estimate that this is of the order of 10%. Secondly, there is the error arising from the sample handling. This can be estimated from the difference in the calculated mixing ratio for air samples collected in parallel. The average difference is 12%, implying an uncertainty of approximately 20% arising from the sampling and analysis.

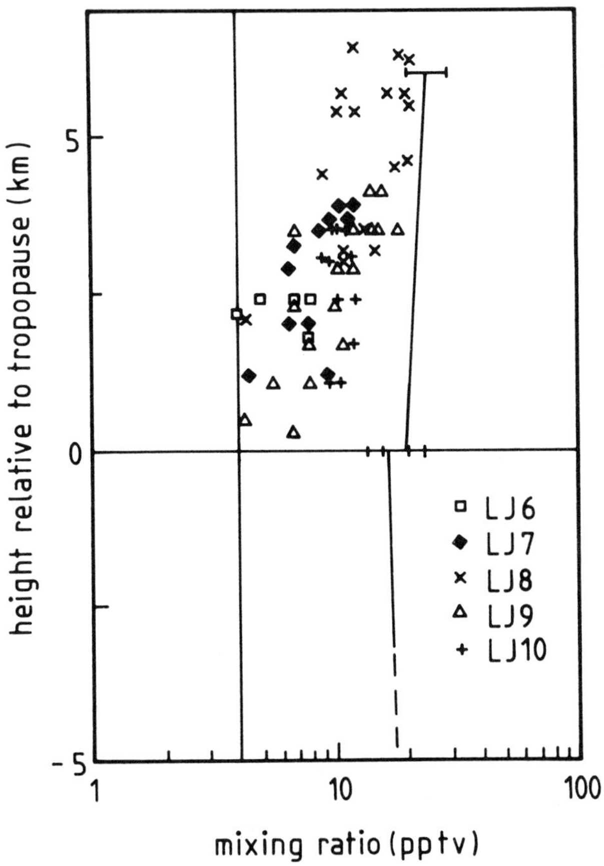

Fig. 2: Vertical profile of the mixing ratio of COFCl versus tropopause height for the flights LJ6-LJ10. The lower detection limit for these measurements is approximately 4 pptv, as indicated by the vertical line. In 34 samples no COFCl could be identified, of which 28 were collected in the troposphere. The average mixing ratio for phosgene is shown as the two thick lines. These lines represent the least squares fit to the measured points, with the error bars representing the estimated uncertainty in the measurements. The lower part of the tropospheric portion is dashed since it is based on relatively few measurements during one flight (LJ6).

The measured COFCl mixing ratios for all flights are shown in fig. 2, plotted against the calculated tropopause level. The tropopause height has been calculated by averaging the radiosonde data. For the samples taken within 1 km of the calculated tropopause the measured ozone mixing ratio has been used to determine whether the sample was taken from below or above the tropopause.

In fig. 2 it can be seen that the mixing ratio of COFCl in the troposphere was always below our lower detection limit of 4 pptv. Once in the stratosphere, the mixing ratio climbs rapidly. This suggests that the primary source of COFCl is photolysis of $CFCl_3$ in the stratosphere and that the troposphere acts as a sink, presumably as a result of hydrolysis.

The scatter in the measured mixing ratio of COFCl comes partially from the detailed structure of the stratosphere at the time of measurement. This can be seen by plotting the COFCl mixing ratio versus that measured for ozone, as shown in fig. 3. In this region of the stratosphere ozone has a very long lifetime, close to the eddy mixing time (Shimazaki (1985)). Therefore the ozone mixing ratio gives a time averaged "height" for the air mass being studied. It can be concluded that in this

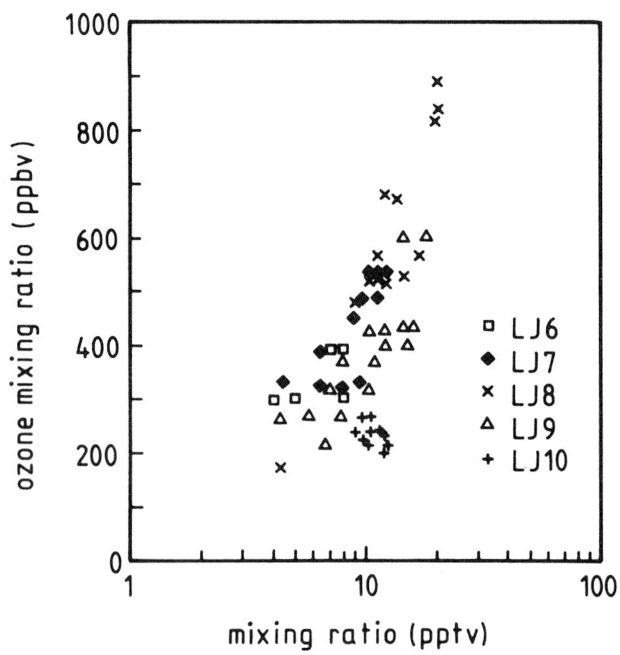

Fig. 3. The mixing ratios of COFCl plotted against ozone mixing ratio (as a measure of "time averaged height", see text).

region of the atmosphere COFCl has a lifetime of some months, and that the simple description of the chemistry given above (photolyisis of F-11 and hydrolysis) is sufficient to describe the behaviour of COFCl in the lower stratosphere.

The mixing ratio profile of phosgene shows distinct differences to that of COFCl. First, there is a significant amount of phosgene in the troposphere, whereas we have not been able to detect COFCl in the same region. Secondly, in the part of the stratosphere that has been investigated the mixing ratio of phosgene is as large or larger than that of COFCl.

The stratospheric behaviour of these two compounds can be discussed in terms of their relative source strengths and decay processes in this region. First, the main known sources for production of $COFCl:COCl_2$ are $CFCl_3:CCl_4$, which have been detected at the tropopause at a ratio of approximately 2:1. The main loss process to be expected is photolysis, which, should be faster for phosgene on the basis of what is known of their respective UV absorption spectra (As summarized in DeMore et al. (1987)). Therefore, if the stratosphere was isolated from the troposphere, the COFCl mixing ratio should be significantly greater than that of phosgene. This has been demonstrated in the only model calculations published on the mixing ratios of these two compounds, which assumed only stratospheric sources (Crutzen et al. (1978)).

The difference between the observations and the expectations presumably arise from the tropospheric behaviour of these two compounds. In particular, it seems likely that the substantial tropospheric source for phosgene is responsible. The resulting increase in tropospheric mixing ratio would cause a net decrease in the diffusion rate from the stratosphere into the troposphere, leading to an increase in the phosgene mixing ratio throughout the height range measured.

The relative hydrolysis rates for these two compounds in the atmosphere are unknown. If the hydrolysis rates are comparable, then the measured tropospheric mixing ratios can be used to set an upper limit on the importance of the stratospheric source for tropospheric phosgene. The data presented here then imply that at most approximately 20% of the tropospheric phosgene is of stratospheric origin.

The halogenated carbonyl compounds are the primary oxidation products of the chlorofluorocarbons, and their chemistry is important in understanding the details of the oxidation of the CFC´s in the atmosphere. However, further theoretical and experimental work is needed, especially with regard to hydrolysis, to fully evaluate the behaviour of these carbonyl compounds in the atmosphere.

ACKNOWLEDGEMENTS

We gratefully acknowledge the helpful advice of Paul Crutzen, and the assistance and expertise of Geert Moortgat and Franz Simon in the study of the oxidation of F-21 and F-22. We wish also to thank Flight Captains Gärtner and Lechner of the FVG for their cooperation during these flights. The flights discussed here were sponsored by the German Federal Ministry of Research and Technology through the grant ATP-88.

REFERENCES

Brown, L.R., C.B. Farmer, C.P. Rinsland and R.A. Toth 1987: Molecular Line Parameters for the Atmospheric Trace Molecule Spectroscopy Experiment. Appl. Optics, **26**, 5154-5182.

Crutzen, P.J., I.S.A. Isaksen and J.R. McAfee, 1978: The Impact of the Chlorocarbon Industry on the Ozone Layer. J. Geophys. Res., **83**, 345-363.

DeMore, W.B., M.J. Molina, S.P. Sander, D.M. Golden R.F. Hampson, M.J. Kurylo, C.J. Howard and A.R. Ravishanka, 1987: "Chemical Kinetics and Photochemical Data for Use in Stratospheric Modelling", Evaluation No 8, JPL publication 87-41, 136.

Griffith, D.W.T. and G. Schuster, 1987: Atmospheric Trace Gas Analysis Using Matrix Isolation-Fourier Transform Infrared Spectroscopy. J. Atmos. Chem., **5**, 59-81.

Jayanty, R.K.M., R. Simonaitis and J. Heicklen, 1975: The Photolysis of Chlorofluoromethanes in the Presence of O_2 or O_3 at 213.9 nm and their Reactions with O (^1D). J. Photochem., **4**, 381-398.

Nielson, A.H., T.G. Burke, P.J.H. Woltz and E.A. Jones 1952: The Infrared and Raman Spectra of F_2CO, FClCO and Cl_2CO. J. Chem. Phys., **20**, 596-604.

Raper, O.F., C.B. Farmer, R. Zander and J.H. Park, 1987: Infrared Spectroscopic Measurements of Halogenated Sink and Reservoir Gases in the Stratosphere with the ATMOS Instrument. J. Geophys. Res., **92**, 9851-9858.

Rinsland, C.P., R. Zander, L.R. Brown, C.B. Farmer, J.H. Park, R.H. Norton, J.M. Russell III and O.F. Raper, 1986: Detection of Carbonyl Fluoride in the Stratosphere. Geophys. Res. Lett., **13**, 769-772.

Shimazaki, T., 1985: "Minor Constituents in the Middle Atmosphere", Vol. 6 in the series "Developments in Earth and Planetary Sciences", D.Reidel Publishing Co., Dordrecht.

Singh, H.B., 1976: Phosgene in Ambient Air. Nature, **264**, 428-429.

Wilson, S.R., P.J. Crutzen, G. Schuster, D.W.T. Griffith and G. Helas Submitted 1988 : Phosgene Measurements in the Upper Troposphere and Lower Stratosphere. Submitted to Nature.

World Meteorological Organization 1985: "Atmospheric Ozone, 1985", report no. 16, Geneva, Switzerland.

HO_2 MEASUREMENTS IN THE LOWER STRATOSPHERE USING MATRIX ISOLATION AND ESR SPECTROSCOPY

M. Helten, H. Fark, H.W. Pätz, and D. Kley
Institut für Chemie 2: Chemie der Belasteten Atmosphäre
Kernforschungsanlage Jülich
D–5170 Jülich, FRG

1. INTRODUCTION

The hydrogen peroxy radical, HO_2, plays an important role in stratospheric photochemistry. As a member of the HO_x radical family it is involved in catalytic reaction chains which destroy ozone (WMO 1985). But only few stratospheric measurements are published (Anderson et al., 1981; Mihelcic et al.,1978; De Zafra et al., 1984; Helten et al., 1984). Pyle et al. (1985, 1988) used the LIMS and SAMS measurements of NO_2 and HNO_3 to calculate HO_2 concentration and its variability together with other radicals.

In this paper we will report on HO_2 measurements from our flights between 1980 and 1987. Also included is the final analysis of results which were previously reported in preliminary form in the WMO Ozone Report 1985 (WMO 1986).

2. SAMPLE COLLECTION AND MEASUREMENT

The experimental technique has been described previously (Helten et al., 1984). A brief desciption will be given here.

During a ballon flight we collect in–situ samples of stratospheric air with a liquid nitrogen cooled cryosampler. Several cold fingers (8, 10 or 14, dependent on the instrument flown) are mounted on a cold plane in a vacuum chamber, which is pumped by a molecular sieve cryopump (Figure 1). Below each cold finger is an orifice, which can be opened and closed by telecommand. The orifice limits the air flow entering the vacuum to a rate below 15 STPl/h for each finger. During the flight the cold fingers are cooled to a temperature in the range between 68 and 73 K. Vapour of D_2O is added to the air stream passing the cold finger during sample collection producing a thin polycristalline ice matrix on the cold finger tip. During one hour, a maximum amount of 15 STPl air is collected for each sample with an ice matrix of 80 μl D_2O. Under these conditions only constituents of stratospheric air that condense at 73 K are trapped in the matrix. Nitrogen, oxygen, and other constituents not condensable at 73 K are pumped into the cryopump. In flight the samples are collected during a valve controlled descent of 1 m/sec.

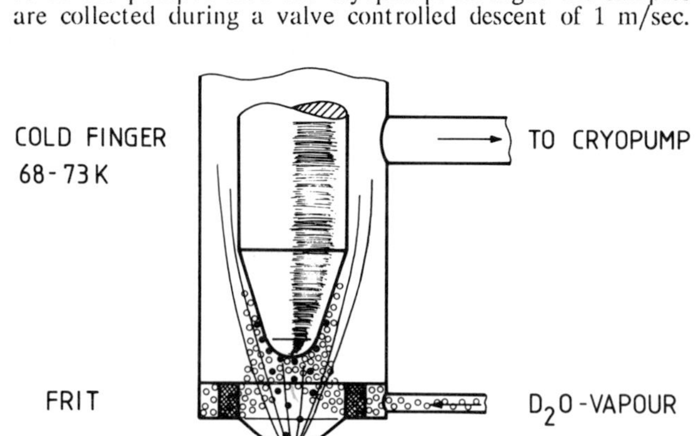

Fig. 1. Principle of sample collection by matrix isolation.

Each sample is collected within one hour, and therefore averages the radical mixing ratios over a 3.6 km height range. Overlapping collection of up to three samples give a better height resolution.

After recovery, the samples are transferred under high vacuum conditions into a liquid nitrogen cooled transport cryostat.

In the laboratory, each sample is transferred into the spectrometer cryostat, which can be cooled with liquid helium to temperatures below 5 K. The Electron Spin Resonance measurements were performed with a Varian ESR–Spectrometer with a 12" magnet, operating in the X–band at 9 GHz.

3. CALIBRATION

For identification of measurable species HO_2, NO_2 and NO_3 the sample spectrum must be compared with standards (Helten et al.,1984). These were prepared using a gas flow and sampling system simulating the balloon borne cryosampler. For the calibration of our previously published HO_2 measurements (Helten et al., 1984), we produced H–atoms by a microwave discharge and mixed them in a flowtube with oxygen for HO_2 production through

$$H + O_2 + M \Rightarrow HO_2 + M .$$

The uncertainty of this method was estimated to be ± 23 %.

We have performed now an independent calibration using a method already employed by Mithelic et al. (1986) for the same radical species. The method used the very stable nitroxide radical ATMPO, (4–Amino–2,2,6,6–tetramethylpiperidine–1–oxyl), to produce calibration samples. ATMPO is commercially available in high purity (97%, Sigma Chemical Comp.).

The double integral of an ESR spectrum is, with a given spectrometer setup, only dependent on the total number of spins in the sample, the absolute temperature, the spin quantum number and Lande's splitting factor of the species. If one has a reliable calibration of a species which is sampled in the same way as the species to be calibrated, one can determine the amount of the species to be calibrated by comparing the double integral of that species with the double integral of a known amout of the other radical with the same spin and Lande–factor. For HO_2, NO_2 and ATMPO the spin is 1/2, and the Lande–factors are identical within a fraction of one percent. The calibration factors should therefore be identical for all three radicals.

From this calibration with ATMPO we could only extract the calibration factor for HO_2. The linearity of the calibration in the concentration range of stratospheric samples, i.e. less than 10^{15} molecules, was not extractable from the measurements. But we believe that the linearity found for the NO_2 measurements in that concentration range (Helten et al., 1984) is valid for the HO_2 measurements as well.

We evaporated ATMPO at temperatures between 50 and 130°C under reduced pressure and mixed it with synthetic air or argon. This was achieved with a small oven in the centre of a small glass vessel, in which a vial (1.5 cm³) with solid ATMPO was inserted. The weight of ATMPO was determined so that the amount of ATMPO evaporated could be calculated from the molecular weight (= 171.26). The

ISBN 0–937194–15–8

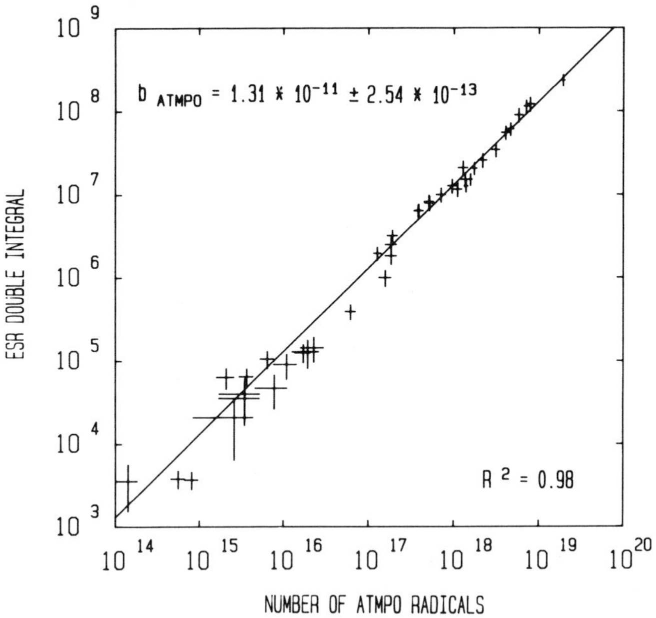

Fig. 2. Double integral of the ATMPO ESR spectrum as function of the ATMPO sampled in the flow system.

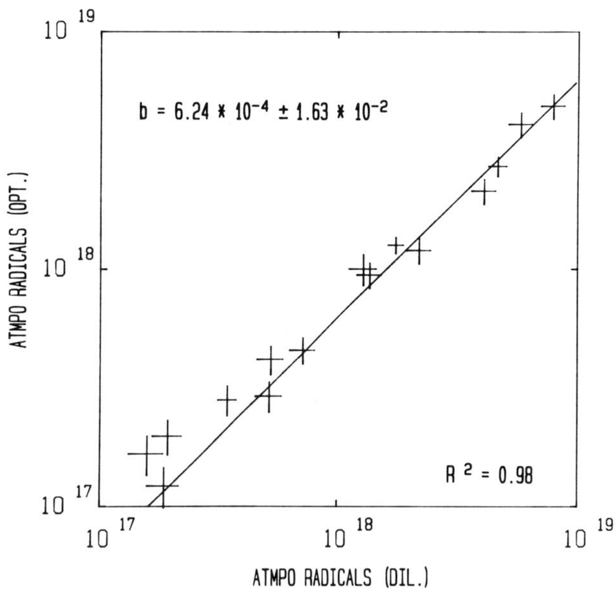

Fig. 3. ATMPO in the sample determined by optical absorption as function of the ATMPO diluted in the system.

dilution gases argon or air were controlled by mass flow controllers. The mixture was then fed into our laboratory sampling system and a matrix sample of the mixture was collected. For very high dilutions a silica capillary was used for a second dilution step prior to sampling.

Each sample was first measured in the ESR spectrometer at 77 K and then warmed up to room temperature. The melt of the matrix was mixed with 1 to 1.5 ml of deionized water. The total amount of ATMPO in this solution was then measured by optical absorption with a spectrophotometer in the 200 to 300 nm wavelength range. The prominent absorption band is centered at 245 nm. The absorption coefficient at 245 nm, determined from a dilution sequence, is 2029±13 [$l \cdot mol^{-1} \cdot cm^{-1}$], in good agreement with the value determined by Mihelcic et al.(1986).

For each sample three measurements of ATMPO could now be compared:
1) ATMPO evaporated and sampled,
2) ATMPO in the matrix measured by ESR and
3) ATMPO in matrix measured by optical absorption.

Figure 2 shows the double integral of the ESR spectrum of each ATMPO sample as function of the ATMPO evaporated and sampled in the flow system. The linear correlation between the data over a concentration range of 5 orders of magnitude is good. The sample with the lowest concentration corresponds to a mixing ratio of 0.5 ppb in the gas mixture. The horizontal error bar gives the estimated accuracy of the diluted amount of ATMPO, the vertical error bar corresponds to the estimated accuracy of the ESR measurements, including the double integration.

Also the optical measurements, shown in figure 3, demonstrate a good linear correlation with the ATMPO evaporated and sampled from the flow system. The detection limit of the optical measurements was above 10^{17} molecules. The slope b of the linear regression indicates that nearly 62%±3% of the sampled ATMPO was collected on the cold finger (Fiqure 3). That means, in each stratospheric sample 62 % of each radical species is collected in the matrix of the cold finger. The remaining 38 % are trapped on cold walls and, finally in the cryopump.

The optical and ESR measurements of the samples, also shows a good linear correlation (Figure 4).

From figure 2 we can derive the new calibration factor $b_{atmpo} = 1.31*10^{-11}$/molecule for our stratospheric HO_2 measurements. This can be compared with the one we used for our old calibration (Helten et al., 1984), $0.91*10^{-11}$ /molecule. We therefore have to correct our published HO_2 data with a factor of 0.69. This correction is within the 37 % range of accuracy given for the old calibration. The accuracy of the new calibration can be derived from the standard error of the calibration factor b_{atmpo} to be ± 2 %. The confidence interval for the 98 % limits is ± 5 %.

The calibration factor of our NO_2 measurements, $b_{no2} = 1.28*10^{-11}$/molecule, determined independently (Helten et al., 1984), is in excellent agreement with $b_{atmpo} = 1.31*10^{-11}$/molecule.

By comparing the slope $2.11*10^{-11}$/molecule of the regression in fiqure 4 with b_{no2} we can independently compute for the joint sampling and detection efficiency of the ESR detection method, and get 61%±12 %. By using the trapping

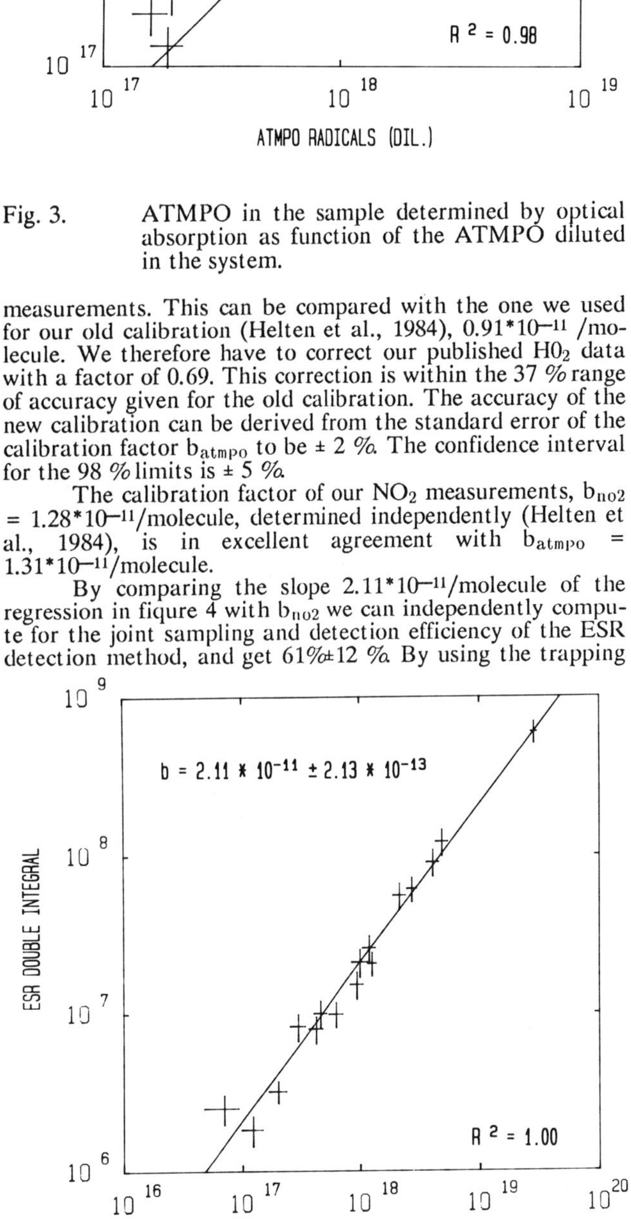

Fig. 4. Double integral of the ATMPO ESR spectrum compared with ATMPO determination of the same sample by optical absorption.

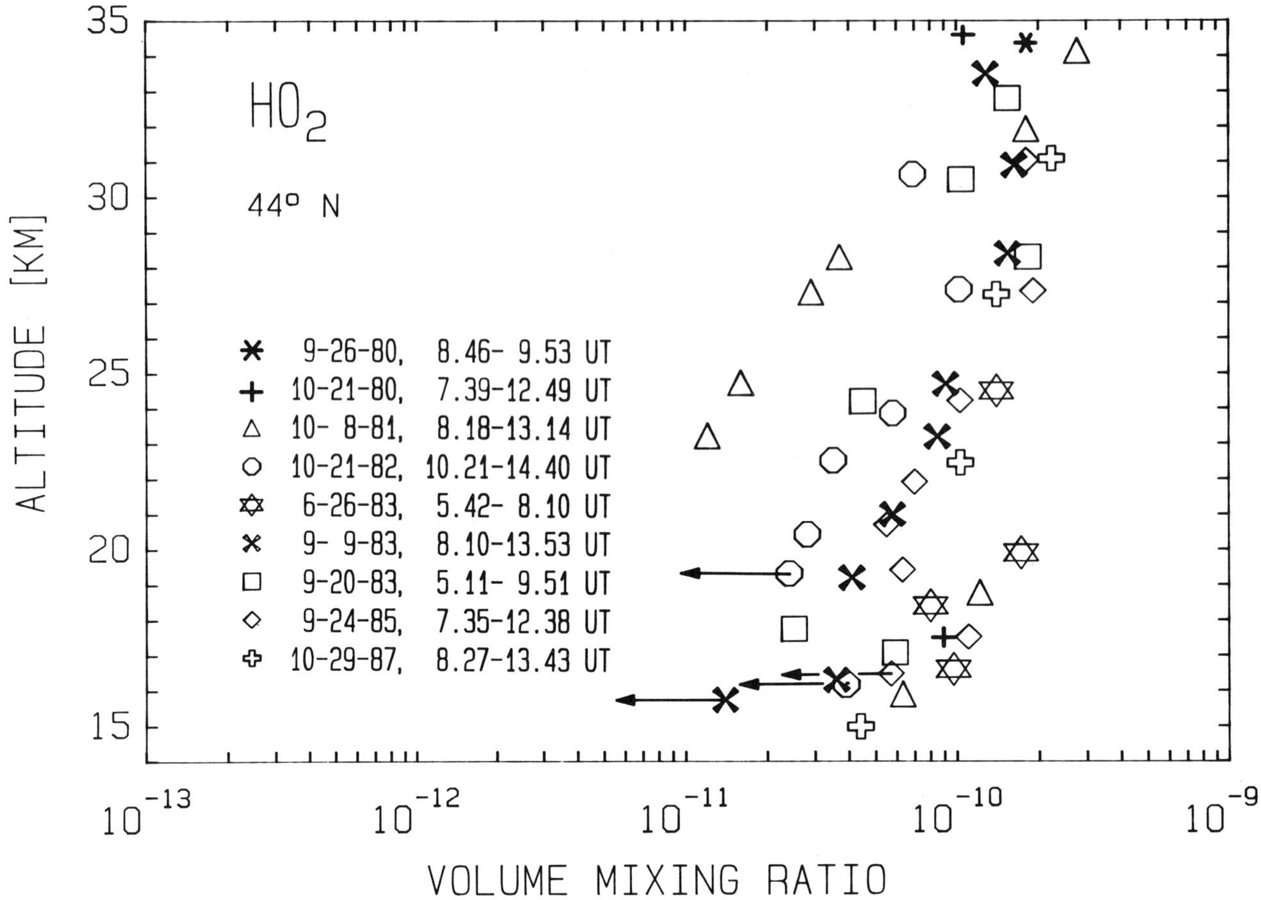

Fig. 5. Altitude profiles of HO_2 measurements from 9 balloon flights over France (44°N).

efficiency of 62 %, as derived from the optical ATMPO measurement, we infer a detection efficiency higher than 90 % for the radicals trapped on the coldfinger and measured by ESR.

With the accuracy of ± 5 % for our new calibration factor, we get a total accuracy of ± 30 % for our HO_2, measurements reported here.

4. HO_2 ALTITUDE PROFILES

Figure 5 presents all HO_2 data we determined from samples of 9 balloon flights over Southern France (44°N) between 1980 and 1987. Included are the three measurements of two flights in 1980, published earlier (Helten et al.,1984), but now with the correction due to the new calibration. NO_2 measurements were executed also on the same samples of all flights. These NO_2 measurements have been published, at least in part, previously (Helten et al., 1984, 1985; Pommereau et al., 1987; Naudet et al., 1987).

The HO_2 mixing ratio profile (figure 5) decreases by a factor 10 between 33 and 15 km. Only the 10–8–1981 profile shows appreciably lower HO_2 mixing ratios than the other profiles in the altitude range between 30 and 20 km. The NO_2 values of that flight (Helten et al., 1984) were also significantly lower in the same height interval than the mean of the other measurements. While there were some technical difficulties (Helten et al., 1984) during the 10–8–1981 flight we have no unequivocal reason to reject the measurements.

HO_2 measurements by other groups using different techniques exist for altitudes only above 28 km (Anderson et al., 1981) or even higher (De Zafra et al.,1984) Anderson's measurements show a different profile slope for each flight. The mean of our HO_2 measurements between 28 and 35 km is somewhat lower than Anderson's data. This could be attributed to the lower latitude and the higher solar zenith angles for Anderson's measurements. The HO_2 profile proposed by De Zafra is significantly lower than our and Anderson's measurements, probably indicating the uncertainty of the profile retrieval in altitudes lower than 35 km due to the noise in the wings of the microwave emission line.

There is a large discrepancy between our and Anderson's measurements, respectively, and different model calculations (WMO 1985). In the 33 km range the model calculations are lower by nearly a factor 2, and at 15 km by more than a factor 10. The mean HO_2 profile calculated from LIMS and SAMS data (Pyle et al., 1986, 1988) is in agreement with the model calculations. The HO_2 variability associated with the satellite data is lower than the variability of our observations, but increases also appreciably with decreasing altitude.

5. DISCUSSION

There remains the question if there could be some contamination of our samples, since this new HO_2 calibration factor is in excellent agreement with that for our NO_2 measurements.

If the cryosampler is collecting air which had contact with the gondola or the instrument itself, this could cause contamination. The inlet valves of the sampler are mounted on the bottom flange, and the sampler itself is hanging on 8 to 12 steel ropes in a gondola of aluminum tubes (diameter = 36 mm), which has a bottom plane with a minimum of four square meters. The orifices of the valves are the lowest points of the instrument. Since samples are collected exclusively in the descent phase, undisturbed air is entering the bottom plane of the gondola, and hits, after a distance of 60 cm, the orifices on the bottom flange of the sampler. This should make contamination impossible.

Another question concerns the possibility of some additional HO_2 produced within the sampler. During two flights we also collected blank samples, i.e. samples for which the orifice was not opened, but which nevertheless collected a D_2O matrix during flight. These samples gave no detectable HO_2 signal. During two flights in June and September 1983 we also collected samples during the night, which allowed for NO_3 and NO_2 measurements. These samples also had no significant HO_2 signal.

The question of long term stability of the samples is addressed by the observation, that samples, sometimes measured up to three months after collection nevertheless fit well within the profile of samples measured several weeks before.

Additionally, we point to the fact that for the HO_2 measurements shown here, corresponding NO_2 measurements exist (Helten et al., 1984, 1985; Pommereau et al., 1987; Naudet et al., 1987), which agree very well with the measurements of other groups and model calculations.

6. ACKNOWLEDGEMENTS

This work was partially funded by the German Ministry for Science and Technology (BMFT) through grants nos. FKW 06 and FKW 25. The 1987 flight and the calibration with ATMPO was also partly funded by the Chemical Manufacturers Association (CMA), Washington, under CMA ref. nos. FC85–571 and FC86–627.

We would also like to thank Prof.D.H. Ehhalt for his continuous support of our work.

We are indebted to the balloon launching team of the French CNES in Aire sur l'Adour for the excellent performance of the balloon flights.

7. REFERENCES

J.G. Anderson, H.J. Shetter, and J.J. Margitan, 1984: HO_2 in the stratosphere: Three in situ observations. Geophys.Res.Lett., 8, 289–292.

R.L. De Zafra, A. Parrish, P.M. Salomon, and J.W. Barrett, 1984: A measurement of stratospheric HO_2 by ground-based millimeter–wave spectroscopy. J.Geophys.Res., 89, 1321–1326.

M. Helten, W. Pätz, M Trainer, H. Fark, E. Klein, and D.H. Ehhalt, 1984: Measurements of Stratospheric HO_2 and NO_2 by matrix isolation and ESR spectroscopy. J.Atm.Chem., 2, 191–202.

M. Helten, W. Pätz, D.H. Ehhalt, and E.P. Roeth, 1985: Measurements of night–time NO_3 and NO_2 in the stratosphere by matrix isolation and ESR spectroscopy.Atmospheric Ozone (Ed. by C.S. Zerefos and A. Ghazi), p.196. D. Reidel, Dordrecht.

D. Mihelcic, D.H. Ehhalt, G. Kulessa, J. Klomfass, M. Trainer, and U. Schmidt, 1978: Measurements of free radicals in the atmosphere by matrix isolation and electron paramagnetic resonance. Pageoph., 116, 530–536.

D. Mihelicic, P. Müsgen, and D.H. Ehhalt, 1985: An improved method of measuring tropospheric NO_2 and RO_2 by matrix isolation and electron spin resonance. J.Atm.Chem., 3, 341–361.

J.P. Naudet, D.W. Rusch, R.J. Thomas, R.T. Clancy, C.A. Barth, J. Wedding, J.M. Zawodny, P. Fabian, and M. Helten, 1987: Stratospheric NO_2 from the Solar Mesosphere Explorer during MAP/GLOBUS 1983. Planet.Space Sci., 35, 631–635.

J.P. Pommerau, P. Fabian, G. Flentje, M. Helten, H.W. Pätz, D.H. Ehhalt, F. Karcher, G. Froment, G. Armand, W.A. Matthews, D. Huguenin, P.C. Simon, W. Peetermans, P. Vandeneede, R. Zander, and G. Roland, 1987: Intercomparison of stratospheric NO_2 and NO_3 measurements during MAP/GLOBUS 1983. Planet.Space Sci., 35, 615–629.

J.A. Pyle, A.M. Zavody, 1985: The derivation of hydrgen–containing radical concentrations from satellite data sets. Quart.J.R.Met.Soc., 111, 993–1012.

J.A. Pyle, A.M. Zavody, 1988: Radical variability determined from LIMS and SAMS satellite data. J.Atm.Chem., 6, 201–220.

WMO, Atmospheric ozone 1985. Global ozone research and monitoring project–report No.16, 1094 pages, Geneva (1986)

NO_2 COLUMN INTERCOMPARISON BETWEEN GROUND-BASED MEASUREMENTS AT LAUDER, NEW ZEALAND (45°S) AND COLOCATED SAGE II SATELLITE MEASUREMENTS

R.L. McKenzie and P.V. Johnston
Physics and Engineering Laboratory, Lauder, New Zealand

J.M. Zawodny and M.P. McCormick
NASA Langley Research Center, Hampton, Virginia 23665, USA

ABSTRACT

NO_2 columns measured at sunset from Lauder, New Zealand (45°S, 169.7°E) are compared with columns derived from colocated profiles measured by the SAGE II satellite instrument. The comparison period covers more than 3 years of data since the launch of SAGE II in October 1984.

The seasonal cycle is well reproduced in the SAGE II data, and there is reasonable agreement between it and the ground based data. Over the period from 1984 to 1988 the SAGE II NO_2 columns show a definite decrease while the Lauder columns remain relatively constant.

1. INTRODUCTION

As part of the SAGE II validation program, NO_2 column measurements made from the ground at Lauder, New Zealand (45°S, 169.7°E) are compared with columns derived from colocated profiles.

NO_2 observations have been made routinely at twilight from Lauder, New Zealand since December 1980. The success rate of observations at sunset is close to 90%. Although the NO_2 columns are dominated by photochemistry, with a large summer maximum, there are day to day variations which can exceed 10% (McKenzie and Johnston, 1982). These short term variations imply that accurately colocated data are desirable in any NO_2 intercomparison studies.

The SAGE II instrument has provided NO_2 profile measurements since Ocotber 1984, and the satellite orbit is such that groups of observations at 45°S occur at regular intervals throughout the year, including periods close to the solstices. This coverage, and the large seasonal signal in NO_2 at 45°S, provide a good opportunity to intercompare the sensitivities of the two techniques.

2. INSTRUMENTATION AND METHODS

The SAGE II instrument and NO_2 retrieval method are described in Mauldin et al. (1985), and Chu (1986). The Lauder instrumentation and retrieval method have been described in McKenzie and Johnston (1982, 1984), and Johnston and McKenzie (1988). Absorption spectroscopy in the spectral range 435-455 nm are used by both, and, since both retrievals use the same NO_2 absorption cross sections (from H. S. Johnston (unpublished)), any spectroscopic errors should be minimized. However, since these cross sections were obtained at room temperature rather than at stratospheric temperatures, differences in the bandpasses sampled may lead to different temperature dependences.

The recommended height limits for SAGE II NO_2 retrievals is 23.5 to 38.5 km. Although the inversion does produce NO_2 profiles down to 15.5 km, the results are of poorer quality. The use of SAGE II data below the recommended 23.5 km is necessary in order to directly compare the two data sets. The ground-based measurements give the column above a height of approximately 15 km (Figure 1), while the SAGE II measurements are number density profiles and must be integrated down to 15 km to provide a comparable quantity. We have investigated the effect of including the poorer data below 25 km by repeating some of the comparisons using a restricted altitude range (h>25 km) for the SAGE II data.

A major difference between the two methods is their viewing geometry. The satellite occultation measurement gives the NO_2 in tangent paths between the satellite and the sun. These are converted to vertical profiles using an onion-peeling technique. The ground-based method uses sunlight scattered from the zenith sky at twilight, and the columns measured at SZA=90 have the altitude weighting function shown in Figure 1. This weighting function was derived from a twilight scattering radiative transfer model developed by J. B. Kerr (see Syed and Harrison, 1980). A similar model has been described by Solomon et al., (1987). The weighting function has a value of 20 in the region 25-30 km where most of the NO_2 is expected, so that typically the slant columns measured from the ground are 20

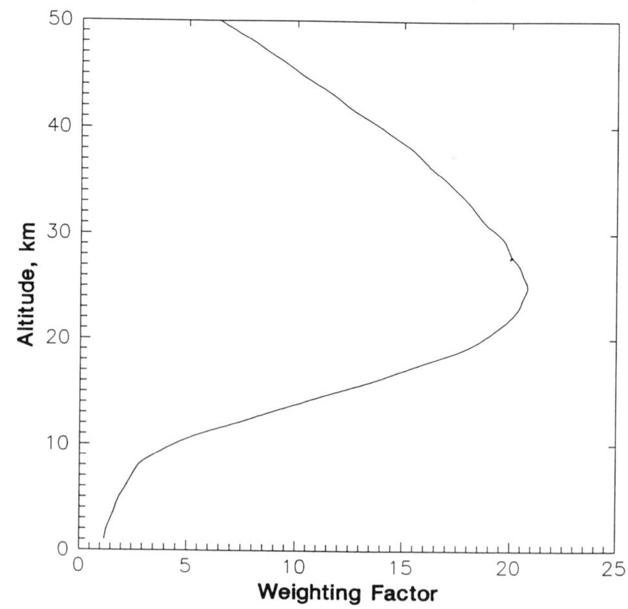

Fig. 1. Effective altitude weighting function for the twilight slant column NO_2 measurements from the ground at Lauder, New Zealand.

ISBN 0-937194-15-8

times the vertical column. These differences in viewing geometry have a further consequence which is potentially important in the measurement of NO_2, which increases markedly after sunset. Whereas SAGE II measures the NO_2 in tangent path that is symmetrical about the terminator (between day and night), the ground-based method gives an effective column displaced to the daylight side by up to 5 degrees (angular displacement varies with height). This difference should lead to systematic underestimations of a few percent in the column measured from the ground compared with that measured by satellite.

3. DATA

For this study a subset of SAGE data was extracted, which included all sunset profiles within 5 degrees of latitude and 10 degrees of longitude from Lauder. This corresponds to a maximum displacement of 950 km, and a maximum time difference of 40 minutes. Using this criterion, 100 colocations with Lauder data were found.

The times that these groupings occur are shown in Figure 2, where individual SAGE colocated columns have been plotted. The two curves show the 1 standard deviation limits in daily variations from monthly means. The measurement period at Lauder started in December 1980, giving a few months of overlap with the earlier SAGE I instrument. A comparison between SAGE I and SAGE II is beyond the scope of this paper.

Sunrise profiles have not been considered in this study, because at the time of writing there were unresolved difficulties with these measurements.

4. COMPARISON OF VERTICAL COLUMNS

For this comparison the Lauder slant columns were divided by 20 to convert to vertical

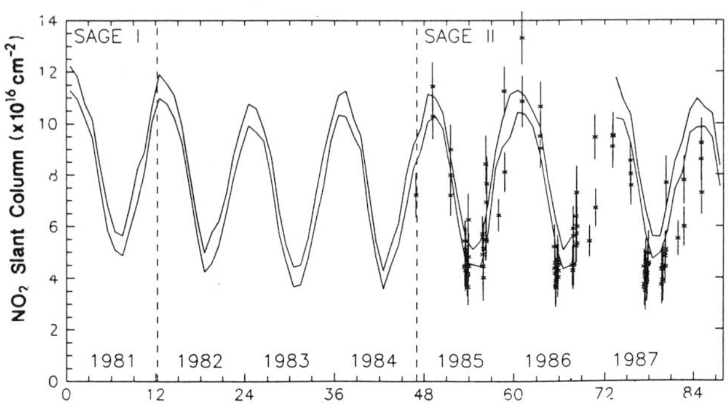

Fig. 2. Monthly mean NO_2 column measurements from Lauder, New Zealand over the period December 1980 to May 1988. The upper curve represents the monthly mean sunset columns plus 1 standard deviation from the day to day variations in that month. The lower curve represents 1 standard deviation below the mean. The individually plotted points show the NO_2 columns with their associated uncertainty, obtained by integrating colocated SAGE II profiles above 15 km.

columns. The SAGE profiles themselves, as well as photochemical considerations (Pyle, unpublished data, 1988) imply that the NO_2 distribution peaks near 25-30 km, so that this enhancement factor of 20 is appropriate (Figure 1).

Figure 3 shows the comparison. Ideally it should be possible to find a regression line to intercept the data points and their associated error bars. If the sensitivities of the two methods are matched the gradient of this line should be unity, and if there are no offsets, the line should pass through the origin. We find that approximately 76% of the points and their associated (1 sigma) errors do lie on the regression line. The gradient of the line indicates that the SAGE II retrievals are 8±2% less than the Lauder. This gradient difference is in the opposite sense from that expected from the geometrical difference discussed above. An enhancement factor of approximately 22 (rather than 20) would be required to bring the two into agreement. A likely cause for the difference is that slightly different height ranges are sampled by the two instruments. The lower boundary in the SAGE II measurements is 15.5 km whereas the Lauder measurements include a small tropospheric contribution (Figure 1). The small y-intercept (-0.09±0.08) shows that any offset differences between the methods are small. Apart from the errors associated with converting Lauder slant columns to vertical columns, the random error in measurement is small compared with SAGE II (typically $\pm 0.1 \times 10^{15}$ molecule cm^{-2}).

However, the SAGE II profiles (as well as unpublished work done at Lauder to measure NO_2 profiles) show that there is a seasonal variation in the shape of the NO_2 profile. In summer the maximum is near 26 km, whereas in winter the peak is near 30 km. A seasonal variation in this

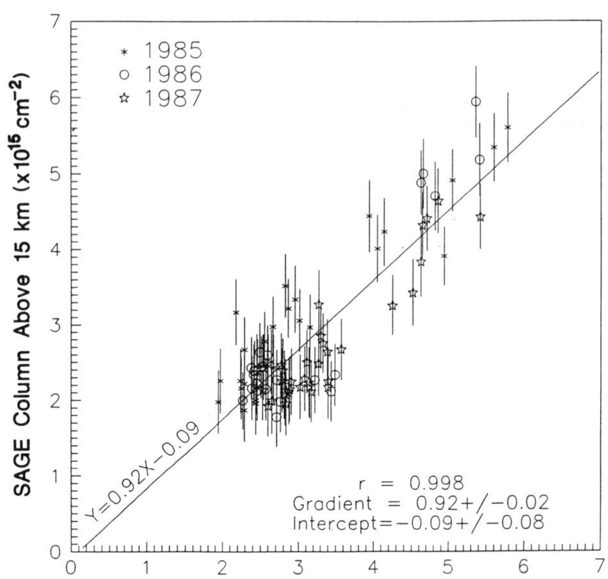

Fig. 3. SAGE II vertical columns above 15 km plotted as a function of the Lauder vertical columns assuming an enhancement factor of 20 (as is appropriate for an NO_2 peak near 28 km (see figure 2).

sense is expected from photochemical considerations. Our fixed enhancement factor will therefore be too large by 5-10% in winter (Figure 1). Thus, the NO_2 vertical columns deduced from the ground measurement may be underestimated by 5-10% compared with the summer columns. Similarly, any day to day differences in the height distribution (which, according to the SAGE II profiles could be as large as ±2 km) lead to errors with this comparison method. These problems are addressed by the approach described in the next section.

5. COMPARISON OF SLANT COLUMNS

The uncertainty in enhancement factors described above can be avoided by comparing the directly measured Lauder slant columns with SAGE II columns derived by scaling the SAGE profiles with the weighting function of Figure 1. Although the resulting columns are less physically meaningful, they should provide a more accurate means of comparison by making use of the SAGE II height information.

Such a comparison was performed and again, the gradient is less than unity, and there is a small negative y-intercept. The correlation coefficient (r) is 0.989 showing that the agreement is similar to that seen in Figure 3 (r=0.998). The lack of improvement may be due to inaccuracies in the SAGE II profiles, and/or the twilight scattering model that was used to generate the weighting function.

6. SENSITIVITY TESTS

No improvement in the scatter in Figure 3 results if the Lauder columns are compared with SAGE II columns integrated above 25 km, a level above which the accuracy is improved. In this comparison the gradient is 0.52, and there is a statistically significant offset due to the differences in the heights sampled by the two methods. Substantial differences in the gradient and intercept of the regression line are expected when the base altitude of the SAGE II columns is changed. As it is increased, less of the total column is included, so smaller gradients are expected. The change in intercept depends upon how the NO_2 varies seasonally above and below the column base altitude.

Surprisingly, in view of the large day to day differences in NO_2, there was little difference in the agreement when the "colocated" data were displaced in time by 1 day. Similarly, there was little improvement if the colocation criteria were tightened to include SAGE data only if they were within 2.5 degrees of latitude and 5 degrees of longitude from Lauder (i.e. <500 km). With these more strict criteria less than 20 colocations were found.

7. LONG-TERM CHANGES

Figure 4a shows a plot of the ratio of the NO_2 columns measured by the two methods. This presentation removes the strong seasonal component of the NO_2 signal, and a difference in trends between the two instruments is immediately apparent, with SAGE II columns decreasing with time. a similar trend is found when the SAGE II data are restricted to altitudes above 25 km (Figure 4b), indicating that the trend is not due to errors in the less precise part of the SAGE II profiles between 15 to 25 km. During 1985 atmospheric aerosol from El Chichon may still be influencing the ground-based measurements, leading to a possible decrease in the measured column, and hence a larger ratio in Figure 4. However, even when the aerosol contamination from El Chichon was at its maximum, the maximum error from this source in the sunset columns measured at Lauder was less than 15% in summer and 20% in winter (Johnston and Mckenzie, 1988). By 1985, the stratospheric aerosol optical depths had already decreased by at least a factor of 2 from this peak (McCormick and Trepte, 1987). Over the 3 years of SAGE II data the aerosol extinction at latitude 45°S decreased by factors of 10, 4 and 2 at 15, 20 and 25 km respectively. The consistency of the ground-based data which includes measurements made near the solar maximum in 1981, and before the atmosphere was contaminated by aerosol from El Chichon (Figure 2) suggests that any long-term changes in NO_2 are small. It would, therefore, be difficult to explain the 20% decrease seen by SAGE II in terms of a solar cycle trend unless its effect has been somehow masked in the Lauder data by changes in aerosol depths.

At present, the reason for the difference in trends is unresolved. Some theories may account for a decreasing NO_2 trend during the decline of the solar cycle (e.g., Callis and Natarajan, 1986) and, therefore, it is important to continue

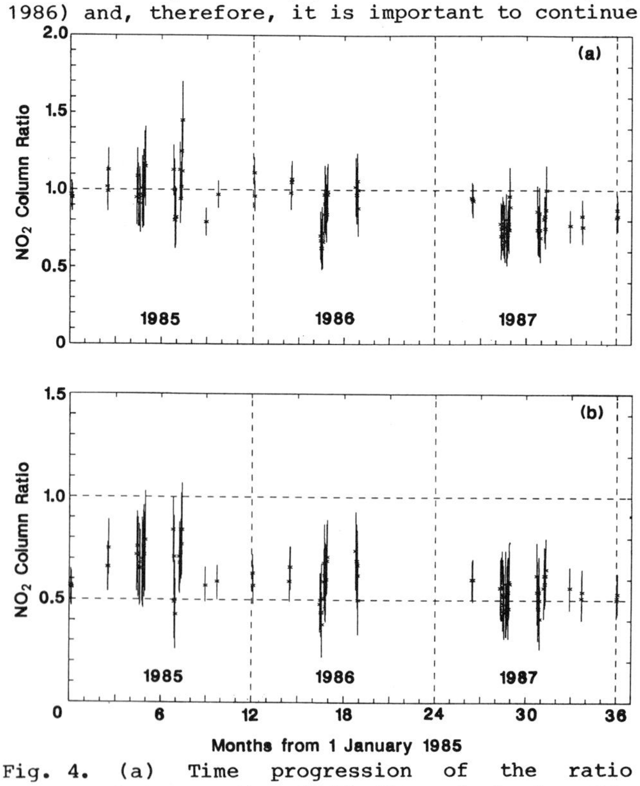

Fig. 4. (a) Time progression of the ratio between the SAGE II and Lauder NO_2 vertical columns above 15 km. The period covered is from the launch of SAGE II to the start of 1988. (b) As for (a), but with altitude range of SAGE II data limited to h>25 km.

to make both sets of measurements over the increasing phase of this solar cycle, to monitor this apparent discrepancy in the long-term trends.

8. CONCLUSIONS

(1.) The seasonal variations are well reproduced in the SAGE II data, though gradients are less than unity. The apparently lower sensitivity in the SAGE II data may however be simply due to differences in the sampling altitudes.

(2.) Averaged over the 3 year period there is reasonable agreement between Lauder data and SAGE II colocated data.

(3.) No improvement in the agreement results if slant columns are compared, or if SAGE II columns are restricted to altitudes above 25 km, or if a more strict colocation criterion is used.

(4.) Measurement errors are such that there is no appreciable difference in agreement even if there is a 1 day error in the colocation time.

(5.) This suggests that a worthwhile comparison would be achieved by comparing monthly zonal means in SAGE II data, with monthly averaged data from Lauder. By these means the statistical advantage will lead to significant reductions in the random measurement error. Furthermore, by averaging over a period as long as a month, many of the day to day variations will be averaged out.

(6.) Over the period 1985 to 1988, the columns measured at Lauder have remained relatively constant, whereas the columns measured by SAGE II have reduced. If approximated by a linear trend, this reduction is 20%, which is larger than the measurement uncertainty.

(7.) Continued measurements are required to resolve the reasons for the differences in long-term trends.

9. REFERENCES

Callis, L. B., and M. Natarajan, 1986: The Antarctic ozone minimum: Relationship to odd nitrogen, odd chlorine, the final warming, and the 11-year solar cycle, J. Geophys. Res., 91, D10, 10771-10796.

Chu, W. P., 1986: Inversion of SAGE II measurements, Proceedings of Sixth Conference on Atmospheric Radiation, Williamsburg, Virginia, May 13-16.

Johnston, P. V., and R. L. McKenzie, 1988: NO_2 observations at 45°S during the decreasing phase of solar cycle 21, J. Geophys. Res., accepted.

Mauldin, L. E., N. H. Zaun, M. P. McCormick, J. H. Guy, and W. R. Vaughn, 1985: Stratospheric aerosol and gas experiment II instrument: a functional description, Optical Engineering, 24 (2), 307-312.

McCormick, M. P., and C. R. Trepte, 1987: Polar stratospheric optical depths observed between 1978 and 1985, J. Geophys. Res., 92, D4, 4297-4306.

McKenzie, R. L., and P. V. Johnston, 1982: Seasonal variations in stratospheric NO_2 at 45°S, Geophys. Res. Lett., 9, 11, 1255-1258.

McKenzie, R. L., and P. V. Johnston, 1984: Stratospheric ozone observations simultaneous with NO_2 at 45°S, Geophys. Res. Lett., 10, 4, 337-340.

Solomon, S., A. L. Schmeltekopf, and R. W. Sanders, 1987: On the interpretation of zenith sky absorption measurements, J. Geophys. Res., 92, D7, 8311-8319.

Syed, M. Q., and A. W. Harrison, 1980: Ground based observations of stratospheric nitrogen dioxide, Can. J. Phys., 58, 788-802.

INTERCOMPARISON ASPECTS OF BALLOON-BORNE MEASUREMENTS OF NO_2

Howard K. Roscoe
Rutherford Appleton Laboratory, Chilton, Didcot
Oxon OX11 0QX, UK

Early measurements of NO_2 showed a wide spread of values, but it was unclear whether this was due to large variability or to differences between instruments (NASA 1979). The Balloon Intercomparison Campaigns (BIC) of 1982 and 1983 were initiated to resolve this question for remote measurements of HCl, HNO_3, NO_2 and NO. Figure 1 shows that values of NO_2 from near-simultaneous measurements by several instruments during BIC differed by a factor 3 below 30km, in excess of known sources of error.

By analysing all balloon-flights prior to 1982, Roscoe (1986) made the case that the variability of NO_2 at mid-latitude and the same season is quite small from day to day or year to year — earlier measurements by Noxon (1979) showing frequent large variability in the NO_2 column at mid-latitudes were rare events probably confined to North America because of frequent excursions of Arctic air (Noxon, private comm., 1984). This is now confirmed by the long series of measurements from New Zealand (McKenzie and Johnston 1982, McKenzie *et al.* 1988). Figure 2 shows the results of the analysis by Roscoe (1986); figure 3 shows the results from BIC2 plotted on the same scale. Note the similarity of the differences in each figure.

Roscoe *et al.* (1988) showed that these differences cannot be explained by radiometric errors in the Balloon-borne Pressure Modulator Radiometer (BPMR), as demonstrated in figure 4. Figure 5 shows that they cannot be explained by errors in the spectroscopic parameters of NO_2 at room temperature. Roscoe and Wells (1988) have now shown that they cannot be explained by uncertainties in the pressure or temperature cycles within the NO_2 pressure modulator (see figure 6). Suspicion must remain that Lorentz-broadened spectroscopic parameters at low temperatures (probably the temperature dependence of half-width) might be the cause of the differences.

However, four recent independent *in-situ* techniques (McFarland *et al.* 1986, Pommereau *et al.* 1987, Webster and May 1987) have measured similarly low NO_2, as shown in figure 7. Furthermore, NO_2 derived from other LIMS and SAMS measurements is a factor two lower than the NO_2 measured by LIMS (Jackman *et al.* 1987, Roscoe 1987). Until the BIC differences are resolved, the absolute concentration of NO_2 in the lower stratosphere must be considered uncertain within at least a factor 2. This makes the process of producing a reference atmosphere for COSPAR (Naudet *et al.* 1986) very difficult.

BIC also highlighted two other discrepencies in NO_2 measurements:

(1) There were differences in results from visible spectrometers which depended on the calibration and analysis procedures used. The differences were greater than 20%, in excess of the error bars. This discrepancy has now been solved as a reult of MAP-GLOBUS/85 and careful work since by the community (Pommereau *et al.* 1988). Figure 8 shows the excellent agreement that can now be obtained between different sensors if the variation in temperature coefficient of NO_2 absorption across the visible band is taken into account. The differences between visible and infra-red occultation sensors in BIC are also accounted for by this temperature coefficient — these differences in figure 1 are within the error bars but were consistent from flight to flight.

(2) There was a systematic difference between measurements of the NO_2 column at 2900cm^{-1}from Kitt Peak and at 1600cm^{-1}from the NCAR airplane. The latter was higher by 20 to 25%, at the limit of the error bars. Coffey *et al.* (1986) have since measured NO_2 from the NCAR airplane at 2900cm^{-1}and 1600cm^{-1}simultaneously and find no difference between results at each wavelength. During MAP-GLOBUS/85 there was a large difference between the NO_2 column measured from the Jungfraujoch at 2950cm^{-1}and that from visible measurements from a balloon and from SAGE II (Zander *et al.* 1988) — similar to the differences during BIC. Simultaneous measurements by infra-red and visible spectrometers at Jungfraujoch are planned in 1989 to help resolve the discrepancy.

ISBN 0-937194-15-8

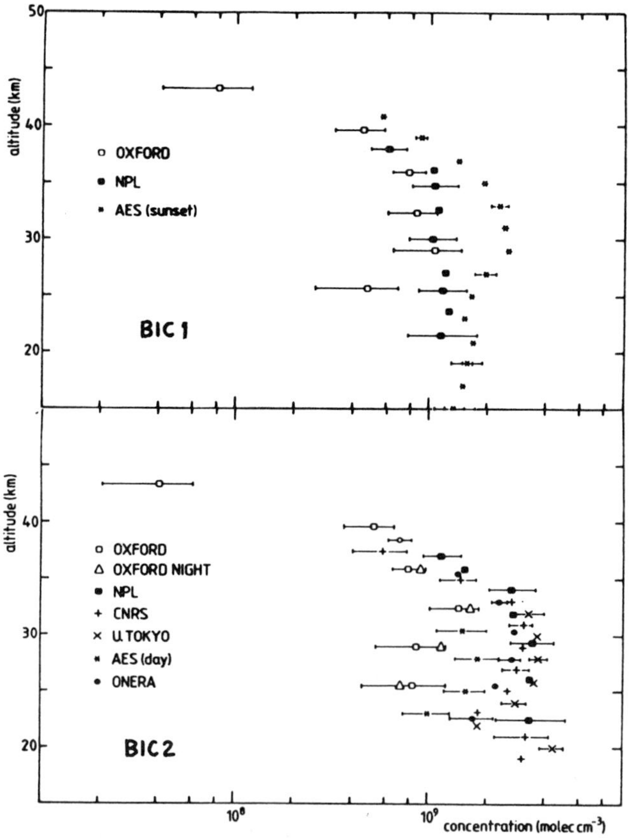

Fig. 1. NO_2 profiles measured during the first and second Balloon Intercomparison Campaigns (BIC).

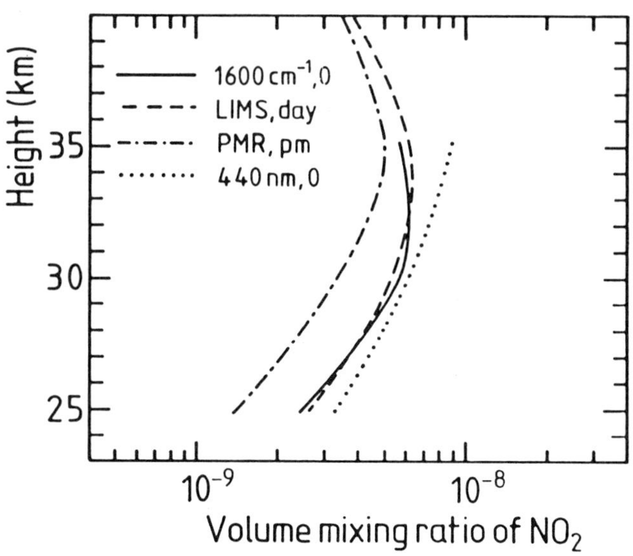

Fig. 2. Mid-latitude profiles of NO_2, close to equinox, measured by various remote sensors prior to BIC. The profiles from balloon-borne sensors are averaged by technique.

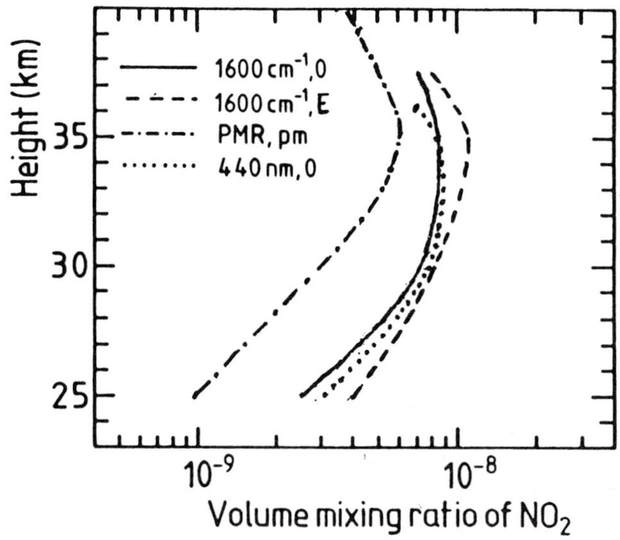

Fig. 3. The results from sensors similar to those of figure 2, during BIC2.

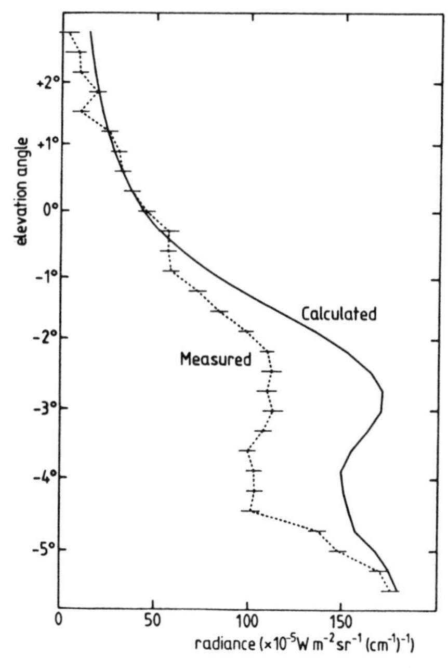

Fig. 4. NO_2 radiances measured by BPMR during BIC2, compared to radiances calculated from the NO_2 profile measured simultaneously by the ONERA spectrometer. The horizontal bars show the random error on the measurements.

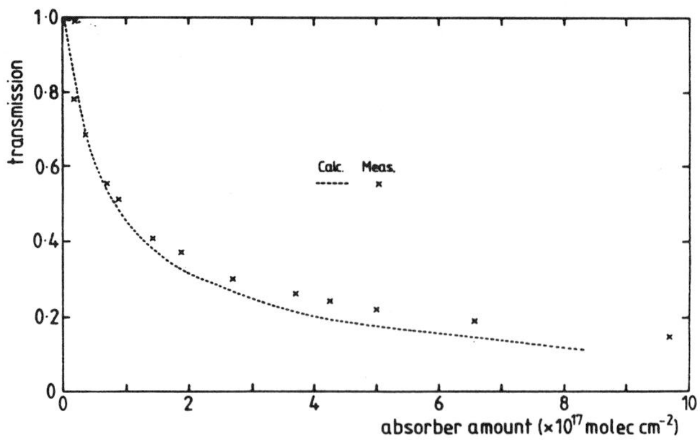

Fig. 5. Laboratory measurements by BPMR of the transmission of a cell containing NO_2 at room temperature, compared to calculated transmissions. Also included are wide-band measurements with simple interference filters. Atmospheric amounts in a limb path are normally less than 10^{17} molec cm^{-2}, where differences between measured and calculated absorptions are less than 5% of the absorption.

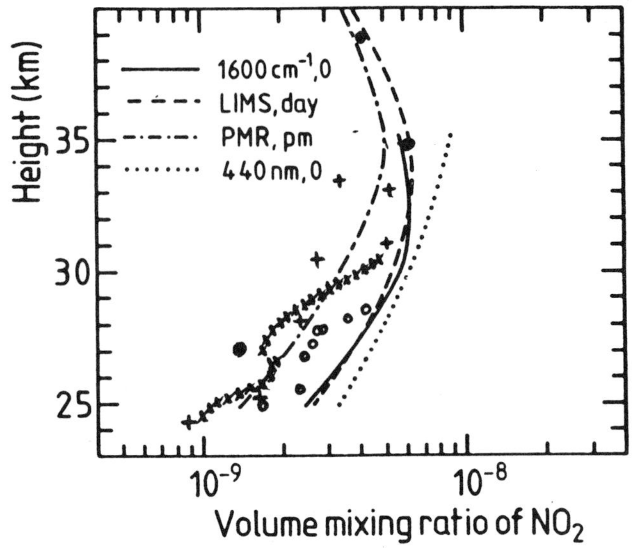

Fig. 7. Volume mixing ratios of NO_2: the historic data at mid-latitudes close to equinox (Roscoe 1986) compared with recent measurements by *in-situ* techniques: (xxx) chemiluminescence, in July 1982 (McFarland *et al.* 1986); (+) chemiluminescence, in September 1983 (Pommereau *et al.* 1987); (o) matrix isolation, in September 1983 (Pommereau *et al.* 1987); (●) BLISS, in November 1985 and October 1986 (Webster and May 1987). The last points have been corrected upward by 10% to take account of new measurements of NO_2 line strengths by BLISS (Webster private comm).

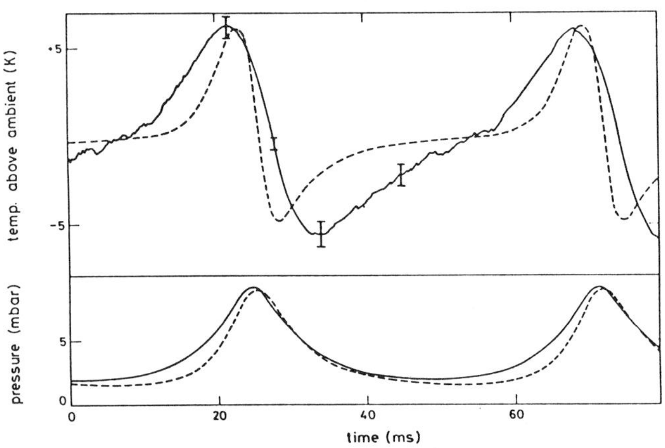

Fig. 6. Calculated temperature and pressure cycles in the balloon-borne modulator at a filling pressure of 4mbar of NO_2 (dashed lines) compared with the measured cycles at the same filling pressure (solid lines). Systematic errors are shown by the vertical bars. Note the excellent agreement between the measured and calculated amplitude and phase of the temperature cycle.

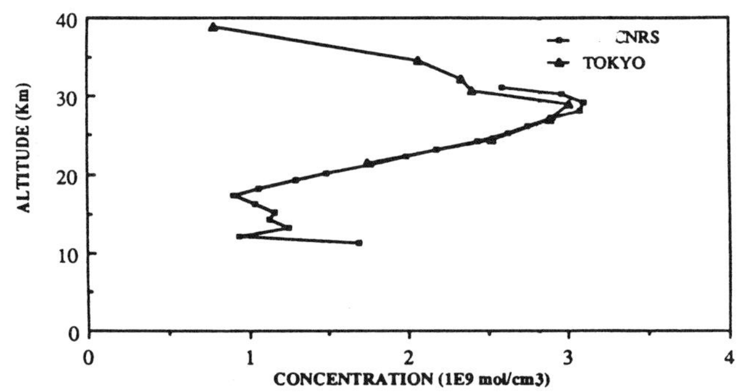

Fig. 8. NO_2 profiles measured by visible spectrometers (430-450nm) during MAP-GLOBUS/85, after correction for the temperature coefficient of the NO_2 spectrum.

REFERENCES

Coffey, M.T., W.G. Mankin, A.Goldman, 1986. "Stratospheric NO_2 retreival from solar absorption spectra in the ν_3 and $\nu_1 + \nu_3$ bands." *Appl. Opt.*, **25**, 2460-2462.

Jackman, C.H., P.D. Guthrie, J.A. Kaye, 1987. "An intercomparison of Nitrogen containing Species in Nimbus 7 LIMS and SAMS data." *J. Geophys. Res.* **92**, 995-1008.

McKenzie, R.L. and P.V. Johnston, 1982. "Seasonal variations in stratospheric NO_2 at 45°S." *Geophys. Res. Lett.*, **9**, 1255-1258.

McKenzie, R.L., P.V. Johnston, M.P. McCormick, 1988. "NO_2 column measurements: comparison between results from 45°S and SAGE II." *Proc. Quad. Ozone Symp., Gottingen 1988* .

McFarland, M., B.A. Ridley, M.H. Proffitt, D.L. Albrittorn, T.L. Thompson, W.J. Harrop, R.H. Winkler, A.L. Schmelterkopf, 1986. "Simultaneous in-situ measurements of nitrogen dioxide, nitric oxide and ozone betwen 20 and 31km." *J. Geophys. Res.* **91**, 5421-5437.

Naudet, J.P., R.J. Thomas, H.K. Roscoe, J.M. Russell III, 1987. "About a possible reference model for stratospheric NO_2", *Adv. Space Res.* **7**, (9)19-(9)23.

Noxon, J.F., E.C. Whipple, Jr., R.S. Hyde, 1979. "Stratospheric NO_2 1. Observational method and behaviour at midlatitude" *J. Geophys. Res.* **84**, 5047-5065.

Pommereau, J-P., P. Fabian, G. Flentje, M. Helten, H.W. Patz, D.H. Ehhalt, F. Karcher, G. Froment, G. Aand, W.A. Matthews, D. Offeann, H. Rippel, P. Rigaud, J.P. Naudet, D. Huguenin, P.C. Simon, W. Peetermans, P. Vandeneede, R. Zander, G. Roland, 1987. "Intercomparison of stratospheric NO_2 and NO_3 measurements during MAP/GLOBUS 1983." *Planet. Space. Sci.* **35**, 615-629.

Pommereau, J-P., F. Goutail, N. Iwagami, K. Shibasaki, P.C. Simon, W. Peetermans, J.P. Naudet, P. Rigaud, D. Huguenin, 1988. "Balloon observations of nitrogen dioxide by visible occultation during GLOBUS NO_x" Submitted to *J. Atmos. Chem.*

Roscoe, H.K., 1986. "Balloon-borne measurements of NO_2: can a comparison of remote techniques be made from the historic data?", *J. Atmos. Chem.* **4**, 479-490.

Roscoe, H.K., 1987. Comment on "An Intercomparison of Nitrogen-Containing Species in Nimbus 7 LIMS and SAMS data", *J. Geophys. Res.* **92**, 14,869-14,870.

Roscoe H.K., B.J. Kerridge, S. Pollitt, M. Bangham, N. Louisnard, C. Alamichel, J-P. Pommereau, T. Ogawa, N. Iwagami, M.T. Coffey, W. Mankin, J-M. Flaud, C. Camy-Peyet, F.J. Murcray, A. Goldman, W.F.J. Evans and C.T. McElroy, 1988. "Intercomparison of remote measurements of stratospheric NO and NO_2." Accepted for publication by *J. Atmos. Chem.*

Webster, C.R. and R.D. May, 1987. "Simultaneous in-situ measurements and diurnal variations of NO, NO2, O3, jNO2, CH4, H2O, and CO2 in the 40-26km region using an open path tunable diode laser spectrometer." *J. Geophys. Res.* **92**, 11,931-11,950.

Zander, R., P. Demoulin, G. Roland, W.A. Matthews, P.V. Johnston, J-P. Pommereau, N. Iwagami, K. Shibasaki, 1988. "Measurements of column abundances of nitrogen dioxide, NO_2, from the ground during the GLOBUS NOX campaign." *Proc. Quad. Ozone Symp., Gottingen, Aug. 1988,*

MEASUREMENTS OF COLUMN ABUNDANCES OF NITROGEN DIOXIDE, NO_2, FROM THE GROUND DURING THE GLOBUS-NO_x CAMPAIGN

R. Zander, Ph. Demoulin, and G. Roland
Institute of Astrophysics, Université of Liège
B-4200, Liège-Ougrée, Belgium

W.A. Matthews and P.V. Johnston
Physics and Engineering Laboratory, DSIR
Lauder, New Zealand

J.P. Pommereau and F. Goutail
Service d'Aéronomie, CNRS
91371 Verrières le Buisson, France

N. Iwagami
Geophysics Research Laboratory, University of Tokyo
Tokyo, Japan

K. Shibasaki
Kokugakuin University
Shibuya-ku, Japan

ABSTRACT

During the Globus NOx campaign of September 1985, the vertical column abundance of nitrogen dioxide, NO_2, has been determined from ground-based observations at three European locations (Jungfraujoch, Switzerland; Aire sur l'Adour and Pic du Midi, France) using different techniques (visible - scanning or fixed diode arrays - spectrometers ; infrared grating and Fourier transform instruments) and methods (direct sun versus twilight). The results derived from these observations are reported here; They show good internal consistency when local NO_2 contamination in the boundary layer is absent. The increase in the column abundance of NO_2 during the day is verified, with a mean daytime value equal to (3.05 ± 0.40) xE15 mol./cm^2, and sunset/sunrise ratios between 1.60 and 1.90. Comparison with other recent ground-based, balloon, and satellite results and with model predictions is included and discussed.

1. DATA BASE AND DATA REDUCTION

1.1 INFRARED DIRECT SUN OBSERVATIONS

Infrared solar observations at the International Scientific Station of the Jungfraujoch (hereafter abreviated by ISSJ) located in the Swiss Alps at 3580 m altitude (latitude : 46.55 deg. N ; longitude : 8.00 deg. E) were made during the September 1985 NOx campaign both with a Fourier transform spectrometer (FTS) and with a double-pass grating spectrometer (DPGS) operated by the Liège group. The FTS instrument was tuned to provide a spectral resolution of 0.005 cm^{-1} unapodized and a signal to noise ratio exceeding 400 over broadband spectral intervals of about 1300 cm^{-1} ; the DPGS achieved 0.02 cm^{-1} resolution and a signal/noise ratio in excess of 150 on tracings averaged from individual fast scans recorded over a few wavenumbers centered on selected target absorption features.

The retrieval of the vertical column abundance of nitrogen dioxide was based on the analysis of absorption features of the v1 + v3 band of NO_2 occuring in the observed 3.4 micrometers region whenever weather conditions permitted. Most of the results presented here have been deduced from the NO_2 absorption located at 2914.65 cm^{-1}. This is the strongest of all NO_2 infrared features observable from the ground, with minimum interferences by other telluric constituents, especially CH_4 in this particular research. Because of its weakness under all observational conditions at ISSJ the weak NO_2 features observed here are not suited for attempting vertical concentration profiles retrievals through root-squares fitting of their observed intensities and envelopes. We have therefore deduced integrated column abundances above ISSJ by measuring the equivalent widths (hereafter abreviated by EQWs) of over 300 pertinent 2914.65 cm^{-1} manifolds and interpolating them among zenith curves of growth computed with a series of typical summer mid-latitude NO_2 concentration profiles scaled relative to each other by a constant throughout the whole altitude span above the ISSJ station. The US. Standard Atmosphere [1966 ; 1976] was used to characterize each 1 km-thick layer in terms of mean pressure, temperature, and density. Linear interpolation between zenith curves of growth was applied, as the observed features verifies linear growth characteristics. Spectroscopic data for both the target and the interfering lines were taken from the AFGL-1986 line parameters compilation [Rothman et al., 1987].

1.2 VISIBLE TWILIGHT OBSERVATIONS

The instrument used by the New Zealand group to make twilight observations of the column abundance of NO_2 and O_3 from Aire sur l'Adour (hereafter abreviated by AsA ; alt.: 50 m : latit.: 44 deg. N; longit.: 0 deg. W) and from Pic du Midi (abreviated hereafter PdM; alt. : 2877 m ; latit. : 42.9 deg.N; longit.: 0.15 deg. E) has been described in detail by McKenzie and Johnston [1982]. It basically consists of an upwards looking 25 cm -f/8 automatic scanning monochromator which analyzes solar radiation scattered from the zenith sky, in the 435 to 450 nm region. The slant column amounts of NO_2 are derived from "ratioed" tracings obtained by dividing twilight spectra with a control spectrum recorded near solar culmination when atmospheric absorption is small. The cross sections filtered to the resolution of the instrument, are least-squares fitted to the Log of the ratio tracings ; two of them account for the observed atmospheric absorbers NO_2 and O_3, and are from H.S. Johnston (they can be found in Syed and Harrison, 1980) ; the third cross section mimics the spectral shape produced by the atmospheric scattering process (i.e., the Ring effect ; see Solomon et al., 1987; Johnston and McKenzie, 1988). After adjustment for the amount of NO_2 in the control spectrum, a NO_2 slant column amount is derived which then allows the vertical column abundance to be calculated using the twilight scattering model

ISBN 0-937194-15-8

discussed by Syed and Harrison [1980]. Further, the twilight results presented here have been corrected for the NO_2 cross section temperature dependence derived recently by Leroy and Rigaud [1987] ; the effect, with respect to earlier visible NO_2 investigations, is to reduce the columns by about 16 to 20% [Pommereau et al., 1988].

1.3 VISIBLE DIRECT SUN OBSERVATIONS

Observations at Aire sur l'Adour were made sporadically during preparation phases of balloon flights by the CNRS and Tokyo groups in the visible region, around 450 nm, where NO_2 exhibits four absorption features. The Tokyo instrument is a 20 cm-focal length concave scanning spectrometer with two exit slits and associated current-mode photomultipliers, producing a spectral purity of 0.15 nm in the 435 to 450 nm spectral interval selected for the present measurements. The CNRS uses a fixed Jobin-Yvon UFS 200-type flat field instrument equipped with a diode array detector assembly, achieving a resolution of 1 nm ; for this investigation it encompasses the 430 to 500 nm region.

The retrievals of slant column abundances of NO_2 are based on least squares fitting the observations over the spectral intervals mentioned above, using the NO_2 absorption cross sections from H.S. Johnston (see Syed and Harrison, 1980). Wavelength calibration is based on positions of solar absorption lines whose own contributions to the absorption are removed by ratioing the ground observations against high altitude high sun spectra ("control" spectra) recorded with the same instruments during balloon flights ; such control spectra were obtained during early afternoon on September 19, 1985, when both instruments were operating at 32 km altitude aboard a balloon payload which was part of the NOx campaign. Instrumentation details, processing methods and errors evaluations for both spectrometers can be found in Ogawa et al. [1981], Roscoe et al. [1988], and Pommereau et al. [1988].

2. RESULTS, DISCUSSIONS AND CONCLUSIONS

The ISSJ results expressed as monthly mean column abundances of NO_2 during September 1985, over discrete zenith angles intervals are reproduced in Fig. 1 by open circles ; the vertical bars correspond to the standard deviation (1-sigma) of the individual column measurements around the mean and represent essentially the variability encountered from day to day during the campaign. A systematic uncertainty of the order of 5-7% remains bound to the line parameters used for treating the absorption of the target molecule NO_2 ; it has not been included in the error bars drawn in Fig 1.

The mean ISSJ results can be summarized as follows :

(i) - the 80 degrees zenith angle sunrise column abundance of NO_2 is equal to (2.25 ± 0.50) x E15 mol./cm^2;

(ii) - the mean culmination value corresponds to (3.0 ± 0.4) x E15 mol./cm^2;

(iii) - the column at sunset (mean zenith angle of 89 deg.) is equal to (4.20 ± 0.7) x E15 mol./cm^2;

(iiii) - the NO_2 column ratio sunset/sunrise is equal to 1.9.

The comparison between the present results and those derived at ISSJ during the MAP-Globus campaign of September 1983 [Pommereau et al., 1987] indicates good agreement, namely mid afternoon columns of (2.75 ± 0.45) x E15 mol./cm^2 in 1983 versus (3.05 ± 0.40) xE15 mol./cm^2 in 1985, and sunset columns of (4.65 ± 0.60) xE15 mol.cm^2 in 1983 versus (4.20 ± 0.70) x E15 mol./cm^2 in 1985 ; the mean 1983 results are noted in Fig. 1 by the dotted area.

The daily NO_2 column abundances derived between August 29 and October 5, 1985 from the twilight measurements at AsA and at PdM are reproduced in Fig. 2. The absence of any appreciable step-like offset when "swithching" sites indicates that the method appears to be insensitive to boundary layer or/and tropospheric biasing ; it provides the column abundance information for the NO_2 burden above about 15 km altitude (Noxon et al., 1979a). Of concern in that respect, is the existence of four measurements corresponding to unusually high column amounts ; whether these are related to stratospheric episodic phenomena or to strong local pollution in the vicinity of the sites of observation, has not been established unambiguously at the present time. However, it is interesting to notice that these maxima correlate satisfactorily with the occurence of thunderstorms activities over northern Spain and the Pyrenees mountains (see the dates marked with full triangles in Fig. 2). The open triangles in Fig. 2 indicate overcast days ; such conditions appear not to influence the results from twilight measurements.

The mean NO_2 column abundances derived from the September 1985 twilight observations are equal to (2.5 ± 0.3) xE15 mol./cm^2 at sunrise, and (4.2 ± 0.3) xE15 mol./cm^2 at sunset, the ratio sunset/sunrise being equal to 1.6. Typical ISSJ columns are indicated in Fig. 2 by horizontal arrows. The twilight results are also reproduced in Fig. 1 by the shaded areas. The morning twilight mean column is definitely larger than the

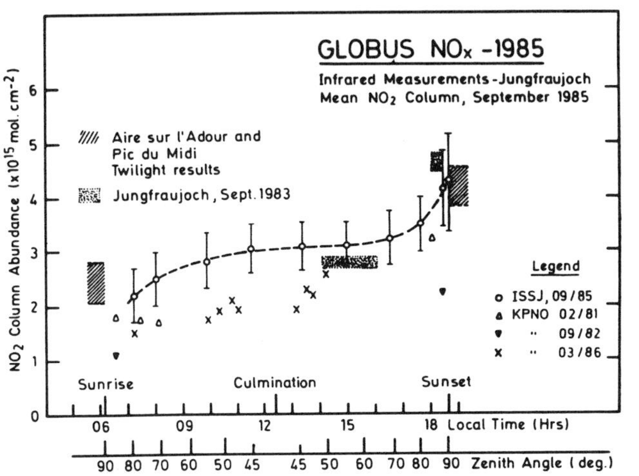

Figure 1 - Mean September 1985 column abundance of NO_2 at discrete zenith angles, derived from the ensemble of the equivalent widths, and showing clearly the column's daytime variation. The error bars correspond to one standard deviation around the mean. Other results are reproduced for comparison purposes.

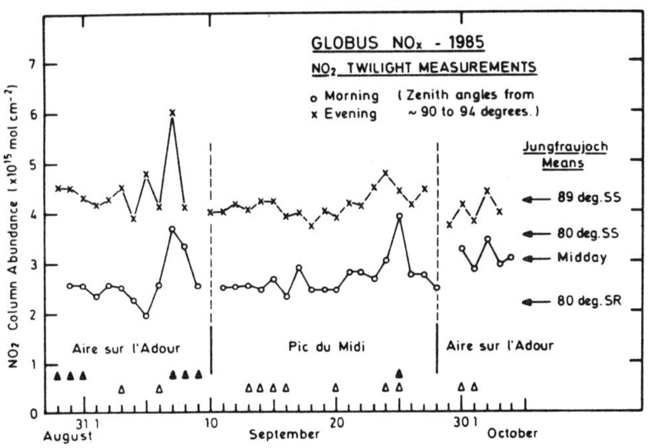

Figure 2 - Evolution of the NO_2 column abundance at Aire sur l'Adour and at Pic du Midi between August 28 and October 5, 1985, obtained by the twilight method at sunrise and at sunset. The full triangles indicate days with thunderstorm activity over northern Spain and the Pyrenees. The open triangles identify days when the sky was overcast.

morning column above ISSJ ; the difference may be the consequence of the twilight measurements being made between about 94 and 90 degrees zenith angle (vs. 80 deg. at ISSJ), when the nighttime column has not yet fully evolved from its late night value to a minimum (factor of about 3 decay) at local sunrise [Brasseur et al., 1987]. During sunset, the conversion of NO into NO_2 is expected to be less abrupt and smoother, explaining the more satisfactory agreement between both results. As the twilight method relates essentially to the amount of NO_2 in the stratosphere, it can be concluded that the tropospheric burden of NO_2 abovre ISSJ is low as compared to the bulk in the stratosphere.

For comparison purposes, series of column abundances of NO_2 above Kitt Peak National Observatory (hereafter KPNO) derived from IR observations made in February 1981 [Camy-Peyret et al, 1983] , September 1982 [Flaud et al., 1983] and March 1986 [Flaud et al., 1988] are given in Figure 1. Lower columns above KPNO are expected in comparison with the Globus-NOx results as a consequence of the latitudinal dependence of the column of NOx, its value increasing with increasing latitude, being also larger in summer than in winter by over a factor 2 at mid-latitudes [WMO Report No. 11, 1982]. Of special interest here is the comparison between the KPNO results of September 20-23, 1982 [Flaud et al., 1983], and the present ones. While the ISSJ and the KPNO sunset/sunrise ratios are equal to ≈ 2, and in good agreement with earlier findings from both series of ground measurements [Noxon et al., 1979b ; Noxon, 1980], from onboard aircraft [Coffey et al., 1981], and by balloons [Evans et al., 1978], the column ratios ISSJ/KPNO for the month of September are also close to 2 ; this is definitely too high when invoking only the latitude effect. Notice that the September 1982 KPNO measurements were part of the BIC1 intercomparison campaign and appear too low when compared to the mean of columns derived above higher altitudes by airplane and balloon observations [Roscoe et al., 1988]. While the various BIC1 measurements occured at similar latitudes (≈ 32 deg.N), some 1000 km in longitude separated the KPNO and the other observations, and it is possible that the airmasses investigated had different NO_2 contents.

Figure 3 reproduces the NO_2 vertical column abundances derived from visible direct sun absorption measurements made sporadically between Sept. 5 - 11 by the CNRS and Tokyo groups at AsA during the Globus-NOx campaign. The scatter among the NO_2 column abundances in Figure 3, as well as their magnitudes in comparison to the twilight results at AsA and the IR results at the ISSJ (shaded areas in Fig. 3), provide clear evidence for the existence of large amounts of NO_2 at ground and/or in the troposphere ; pollution from the Toulouse area is likely to have affected the AsA site, as winds during the campaign were predominantly from the N-E to S-E vector. Notice that the results from the visible observations reported in Fig. 3 have not been corrected in detail for the NO_2 absorption cross-sections temperature dependence suggested by Leroy and Rigaud [1987], because the altitude location of the bulk of the observed column is uncertain ; even with such a correction tentatively applied, the scatter and data amplitudes would not improve the picture significantly. The September 1983 MAP-Globus results obtained with a direct sun-visible instrument operated on the ground at AsA by Matthews gave a mid-day column abundance of NO_2 equal to ≈ 7.5 x E15 mol./cm^2 (see dotted area in Figure 3) ; even with the new proposed 16 to 18% temperature correction applied, that 1983 mid-day column above ASA would still be too high by a factor 2 with respect to most findings regarding mid-day columns around the September months. On the basis of the results from the 1983 and 1985 campaigns, it appears that direct sun visible ground-based measurements of the NO_2 column abundance at AsA are frequently biased by local pollution.

Figure 4 summarizes the Globus-NOx campaign results regarding the column abundance of NO_2 near sunset. The original 1985 results reported in this paper (twilight, direct sun IR, and direct sun visible) are compared (1) to integrated columns of NO_2 derived from weighted averages of data obtained by three visible spectrometers observing simultaneously during sunset in the occultation mode, onboard a balloon platform ; that flight took place on September 19, 1985 at ≈ 32.0 km float altitude over southern France ; (2) to similar results derived from SAGE 2 observations during the sunset occultation of September 17, 1985 which occured about 500 km west of the balloon measurements. Details about these two missions, the data reduction and the evaluation of their uncertainties can be found in Pommereau et al. [1988), for the balloon data base.

While the agreement between the balloon and the satellite results appears excellent, it is equally surprising because the first ones are corrected for the temperature dependence of the cross section while the last ones are not. The mean September 1985 twilight column abundance above ≈ 15 km (square in Fig. 4) is

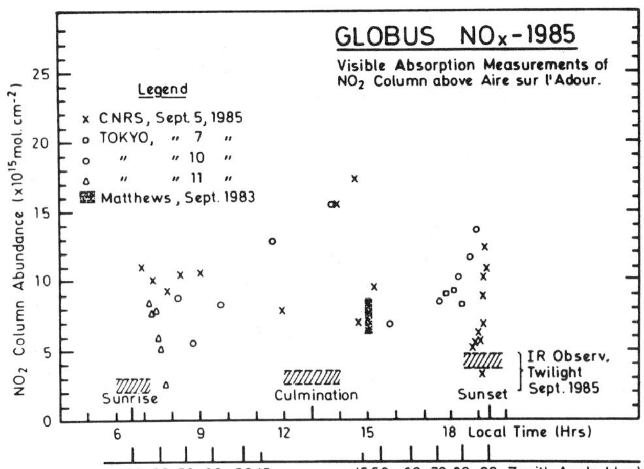

Figure 3 - NO_2 column abundances above AsA derived from direct sun visible observations made between September 5 and 11. The magnitudes of the columns and their variability suggest the occurence of frequent local boundary layer pollution.

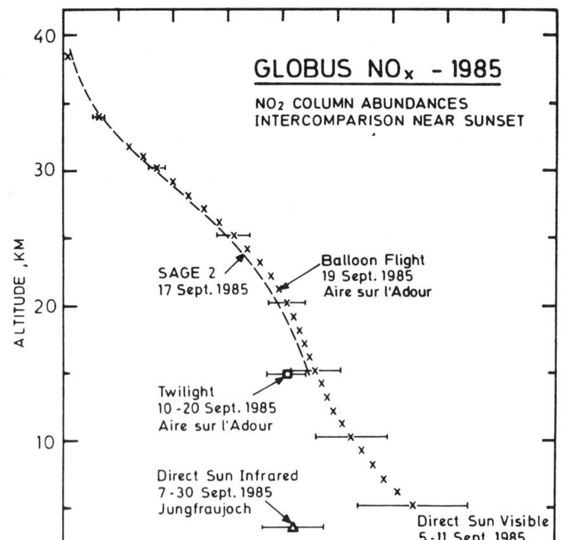

Figure 4 - Comparison of the original near-sunset column abundances presented in this paper with results derived from quasi-simultaneous balloon and satellite observations.

in good agreement with the balloon and the satellite results ; while both high floating balloons and satellite slant paths cross twice the NO_2 rich layers of the atmosphere (double-sided with respect to the tangent height), the columns derived from the twilight observations correspond to less than that double-sided path and consequently weight predominantly the daytime side of the terminator, i.e. always further away in time from the effective shadowing than the other half ; considering the increase of NO_2 at sunset [Brasseur et al., 1987], twilight results are therefore expected to be systematically lower than either satellite or balloon results. Similar arguments invoking observational geometry and partitioning between NO_2 and NO during sunset can also account for the large "apparent" difference between results from ground-to-space or single-sided paths, e.g. the ISSJ sunset observations, and from the double-sided balloon paths ; whether these arguments can explain all or part of the difference between the IR-Jungfraujoch results and the lowermost balloon ones remains to be estiablished quantitatively. Another "a priori" puzzling coincidence among the results is the relative consistency between these low altitude balloon data and the direct sun visible ground-based results at AsA, the magnitude and variability of the last ones implying a significant contribution by local pollution (see earlier discussions) ; however, this does not necessarily imply a strong repercussion of such ground pollution at levels of 5 km and above ; the "coincidence" in this case is misleading, as the balloon half-path on the nighttime side of the terminator crosses layers substantially enriched in NO_2 in comparison to the half-path on the daytime side, i.e. the path along which ground-based observations sense. Only from quantitatively reliable observations covering both and full sunrise and sunset occultations, will it be possible to construct a valid model upon which to satisfactorily quantify differences such as those brought up in this discussion. Meanwhile, however, it is crucial to improve the absolute quality of NO_2 measurements ; owing to the various inconsistencies and unknowns brought up along this paper (and in others, e.g. Roscoe et al., 1988), we believe that clarification of some of these points could and should be addressed through a series of observations with both twilight, infrared and visible direct sun instruments operating simultaneously from a good remote mountain site, i.e. the Jungfraujoch station ; such an easy first step should allow establishment of the relative observational performances of the various types of instruments so heavily involved in today's NO_2 investigations, and help validate retrieval algorithms.

ACKNOWLEDGEMENTS

This research was partly supported by the Fluorocarbon Panel of the Chemical Manufacturers Association (Grant FC 85-550), by the French CNRS and by the Belgian FRFC, which we acknowledge gratefully. We thank C.B. Farmer who allowed us to use the ATMOS computer facility at the Jet Propulsion Laboratory, Pasadena, Calif., for synthetic spectra calculations related to the Jungfraujoch observations, and J. Bosseloirs for the preparation of all figures in this paper.

REFERENCES

Brasseur, G., D. Cariolle, A. De Rudder, L.J. Gray, J.A. Pyle, E.P. Röth, U. Scmaiztl and D.J. Wuebbles, 1987, Odd nitrogen during the MAP/Globus 1983 campaign : theoretical considerations, Planet. Space Sci., 35, 637-645.

Camy-Perret, C., J.M. Flaud, J. Laurent and G.M. Stokes,1983, First measurements of atmospheric NO_2 from the ground, Geophys. Res. Lett., 10 , 35-38.

Coffey, M.T., W.G. Mankin, and A. Goldman, 1981, Simultaneous spectroscopic determination of the latitudinal, seasonal, and diurnal variability of stratospheric N_2O, NO, NO_2 and HNO_3, J. Geophys. Res., 86, 7331-7341.

Evans, W.F.J., H. Fast, J.B. Kerr, C.T. McElroy, R.S. O'Brien, D.I. Wardle,J.C. McConnell and B.A. Ridley,1978, Stratospheric constituents measurements from project Stratophobe, Proc. WMO Symposium on the Geophysical Aspects and Consequences of Changes in the Composition of the Stratosphere, WMO Publ. 511, WMO, Geneva, 55-60.

Flaud, J.M., C. Camy-Peyret, D. Cariolle, J. Laurent and G.M. Stokes, 1983, Daytime variation of atmospheric NO_2 from ground based infrared measurements, Geophys. Res. Lett., 10, 1104-1107.

Flaud, J.M., C. Camy-Peyret, J.W. Brault, C.P. Rinsland, and D. Cariolle, 1988, Nighttime and daytime variation of atmospheric NO_2 from ground-based infrared measurements, Geophys. Res. Lett., 15, 261-264.

Johnston, P.V. and R.L. McKenzie, NO_2 observations at 45 degrees South during the decreasing phase of solar cycle 21, from 1980 to 1987, submitted to J. Geophys. Res., 1988.

Leroy, B.and P. Rigaud, Visible absorption cross-sections of NO_2 at 298 K and 235 K, 1987, Ann. Geophysicae, 5A, 247-250.

McKenzie, R.L. and P.V. Johnston, 1982, Seasonal variations in stratospheric NO_2 at 45 degrees south, Geophys. Res. Lett., 11, 1255-1258.

Noxon, J.F., E.C. Whipple Jr and R.S. Hyde, 1979a, Stratospheric NO_2 - I. Observational method and behvior at mid-latitudes, J. Geophys. Res., 84, 5047-5076.

Noxon, J.F., 1979B, Stratospheric NO_2 behavior, J. Geophys. Res., 84, 5067-5076.

Noxon, J.F., 1980, Correction, J. Geophys. Res., 85, 4560-4561.

Ogawa, T., K. Shibasaki and K. Suzuki ,1981, Balloon observation of stratospheric NO_2 profile by visible absorption spectroscopy, J. Meteorol. Soc. Japan, 59, 410-416.

Pommereau, J.P., P. Fabian, G. Flentje, M. Helten, H.W. Pätz, D.H. Ehhalt, F. Karcher, G. Froment, G. Armand, W.A. Matthews, D. Offerman, H. Rippel, P. Rigaud, J.P. Naudet, D. Huguenin, P.C. Simon, W. Peetermans, P. Vandeneede, R. Zander, and G. Roland,1987, Intercomparison of stratospheric NO_2 and N_3 measurements during MAP/Globus 1983, Planet. Space Sci., 35, 615-629.

Pommereau, J.P., F. Goutail, N. Iwagami, K. Shibasaki, P.C. Simon, W. Petermans, J.P. Naudet, P. Rigaud and D.Huguenin, Balloon observations of nitrogen dioxide by visible occultation during Globus-NOx, submitted to J. Atmos. Chem., 1988.

Roscoe, H.K., B.J. Kerridge, S. Pollitt, N. Louisnard, J.M. Flaud, C. Camy-Peyret, C. Alamichel, J.P. Pommereau, T. Ogawa, N. Iwagami, M.T. Coffey, W. Mankin, W.F.J. Evans, C.T. McElroy, J. Kerr, 1988, Intercomparison of remote measurements of stratospheric NO and NO_2, to appear in J. Atmos. Chem..

Rothman, L.F., R.R. Gamache, A. Goldman, L.R. Brown, R.A. Toth, H.M. Pickett, R.L. Poynter, J.M. Flaud, C. Camy-Peyret, A. Barbe, N. Husson, C.P Rinsland and M.A.H. Smith, The HITRAN data base : 1986 edition, Appl. Opt. 26, 4058-4097, 1987.

Solomon, S., A.L. Schmeltekopf and R.W. Sanders, 1987, On the interpretation of zenith sky absorption measurements, J. Geophys. Res., 92, 8311-8319.

Syed, M.Q., and A.W. Harrison, Ground based observations of stratospheric nitrogen dioxide, Can. J. Phys., 58, 788-802, 1980.

U.S. Standard atmosphere supplements, 1966 and

U.S. Standard atmosphere, 1976, U.S. Printing Office, Washington D.C. 20402, U.S.A.

WMO-Report No. 11, World Meteorological Organization, Geneva 20, Switzerland, 1982.

DIURNAL VARIATION OF STRATOSPHERIC NO AND NO_2 FROM MAP/GLOBUS 1985

J.P. Naudet, M. Pirre, R. Ramaroson, and P. Rigaud
Laboratoire de Physique et Chimie de l'Environnement, CNRS
45071 Orléans, France

P. Fabian
Max-Planck-Institut für Aeronomie, 32411 Lindau, FRG

H. Helten
Institut für Chemie, Jülich Gmbh, 5170 Jülich, FRG

N. Iwagami, K. Shibasaki, and T. Ogawa
University of Tokyo, Geophysical Research Laboratory, Tokyo, Japan

Y. Kondo
Research Institut of Atmospherics, Nagoya University, Toyokawa, Japan

W.A. Matthews
DSIR, PEL Lauder, Central Otago, New Zealand

J.P. Pommereau and F. Goutail
Service d'Aeronomie, CNRS, 91371 Verrières le Buisson, France

P.C. Simon and W. Peetermans
Institut d'Aeronomie Spatiale, Bruxelles, Belgium

1. INTRODUCTION

The sum NO + NO_2 represents the major fraction of odd nitrogen in the mid-latitude stratosphere. The partitioning between NO and NO_2 is mainly determined by the photodissociation of NO_2 at wavelengths shorter than 405 nm:

$$NO_2 + h\nu \rightarrow NO + O \quad (1)$$

and by the catalytic cycle of ozone destruction:

$$NO + O_3 \rightarrow NO_2 + O_2 \quad (2)$$
$$NO_2 + O \rightarrow NO + O_2 \quad (3)$$

Therefore, both species undergo large diurnal variations. At sunset, NO is rapidly converted into NO_2 as solar radiation is fading. During the night, NO_2 is partly converted into a reservoir species believed to be N_2O_5, through the reactions :

$$NO_2 + O_3 \rightarrow NO_3 + O_2 \quad (4)$$
$$NO_3 + NO_2 + M \rightarrow N_2O_5 + M \quad (5)$$

where M represents a third body. At sunrise, NO is rapidly re-formed from the photolysis of NO_2 and then, both NO and NO_2 slowly increases during the day because of the photodissociation of N_2O_5.

This chemical scheme shows that NO, NO_2 and ozone are partners in very powerful temperature dependent photochemical reactions. Therefore, as pointed out by several authors (e.g. Roscoe et al., 1986; Solomon et al., 1986; McFarland et al., 1986; Webster and May, 1987), the measurement of these species through a diurnal cycle is of particular importance in testing photochemical models.

A large set of data including NO, NO_2, ozone and temperature vertical profiles was obtained by various techniques from balloons, from satellites and from ground during the MAP/GLOBUS NO_x campaign of September 1985. The observations were performed at different local times in order to examine the diurnal cycle of NO and NO_2 in the stratosphere.

In this paper, a subset of MAP/GLOBUS NO and NO_2 data taken at pressure levels of 18.1, 10.3 and 7.5 mb from balloons is compared to the predicted diurnal variation. The calculation is made with a time dependent zero dimensional model taking into account the current photochemistry. The model used observed ozone and temperature as fixed parameters.

2. OBSERVATIONS

The experimental NO and NO_2 data used in this work were obtained by different balloon borne instruments. All balloons were launched from the French Centre National d'Etudes Spatiales (CNES) base of Aire-sur-l'Adour (43°42'N, 0°15'W). Table 1 summarizes the characteristics of each flight. Most of the experiments are described in earlier literature. The solar occultation technique at sunset (Pommereau et al., 1987) and the star occultation technique at night (Naudet et al., 1984) were used, in the visible spectral range, to obtain the vertical profile of NO_2. Note that

ISBN 0-937194-15-8

three solar occultation instruments were set-up in the same gondola. Other instruments operated in the in situ mode. A matrix isolation cryosampler (Helten et al., 1984) was used to measure NO_2 during the day, and two chemiluminescent sondes (Fabian et al., 1987; Kondo et al., 1988) measured the mixing ratio of NO and its variation at sunrise and sunset.

Table 1 - Flight characteristics

DATE	TECHNIQUE	EXPERIMENTERS
9/19	NO chemiluminescence	Y. Kondo W.A. Matthews
9/19	NO_2 visible solar occulation	J.P. Pommereau F. Goutail
	NO_2 visible solar occulation	P. Simon W. Peetermans
	NO_2 visible solar occulation	N. Iwagami K. Shibasaki T. Ogawa
9/20	NO chemiluminescence	P. Fabian
9/24	NO_2 visible star occultation	J.P. Naudet P. Rigaud
9/25	NO_2 matrix isolation	H. Helten
10/4	NO chemiluminescence	Y. Kondo W.A. Matthews

Ozone and temperature data are needed for model calculations. The vertical profile of ozone was measured during most of the flights listed in table 1. Additional ozone observations were made by satellite and ground-based experiments. Temperature profiles were derived from meteorological radiosondes, from lidar, from SSU measurements and from measurements made on certain scientific balloon gondolas.

3. MODEL

The calculation presented in this work were made using a time-dependent zero dimensional model (Ramaroson et al., 1987; Pirre et al., 1988) which is well adapted to study the diurnal variation of short lived species such as NO and NO_2. This model neglects dynamics and, therefore, the concentration of the long lived species must be fixed. Also the model includes full spherical geometry, Rayleigh scattering attenuation and mean albedo. The concentration of NO and NO_2 over a 24 hour cycle is calculated at each pressure level and for given ozone and temperature.

The long-lived species are H_2O, CH_4, CO, H_2, and the total amount of nitrogen species NO_y and chlorine species Cl_y. As these were not measured during the MAP/GLOBUS campaign, the mean values available from other observations (WMO, 1985)

Table 2 - Input for model calculations

PRESSURE (mb)	7.5	10.3	18.1
Tmax (K)	238.7	233.4	225.7
Tmin (K)	232.2	226.8	219.0
O_3 max (ppm)	8.6	8.1	6.9
O_3 min (ppm)	7.1	6.5	5.4
NOy (ppb)	21.5	19.0	16.0

were used, except for NO_y that controls NO and NO_2 in the stratosphere. The amount of NO_y, was selected within the range of observations available from WMO (1985) so as to obtain the best agreement between the MAP/GLOBUS observations of NO_2 and the model. However, NO_y may change from day to day because of horizontal transport mechanisms. Therefore, it should be kept in mind that the model represents average conditions of NO_y while the observations may reflect sporadic dynamic effects.

Calculations were made at three pressure levels: 7.5, 10.3 and 18.1 mb that were selected as the best compromise between the number of avaiblable observations and the application range of the model. All available MAP/GLOBUS NO and NO_2 data taken at these levels within an interval of $\pm$ 0.5 mb are used in this work. Also for each level, maximum and minimum values of ozone and temperature as reported by Robbins (private communication, 1988) and Roscoe (private communication, 1988), respectively, were used in the calculations, in order to take into account the temporal variation of ozone and temperature during the MAP/GLOBUS campaign (Table 2).

4. COMPARISON

Figure 1 compares the mixing ratio of NO_2 observed at 7.5, 10.3 and 18.1 mb during MAP/GLOBUS with the corresponding modelled diurnal variation. The model calculations were performed for NO_y and both ozone and temperature values given in table 2, and for an albedo of 0.3. The solar occultation data result from the average of three observations made during the same flight (Table 1). Model and observations agree within error range, apart from the in situ observation made by Helten at 7.5 mb which is about a factor of three lower than the model.

Figure 2 is similar to Figure 1 but for NO mixing ratio. The accuracy of NO measurements is about 15 % (Fabian et al., 1987; Kondo et al., 1988). The values observed by Kondo in the morning at 18.1 and 10.3 mb compare favorably with the model. Other observations indicate that the mixing ratio of NO is larger than that predicted by the model. Also note that the discrepancy, between the model and the observations increases with altitude, and that the increase of NO observed during the day is larger than in the model.

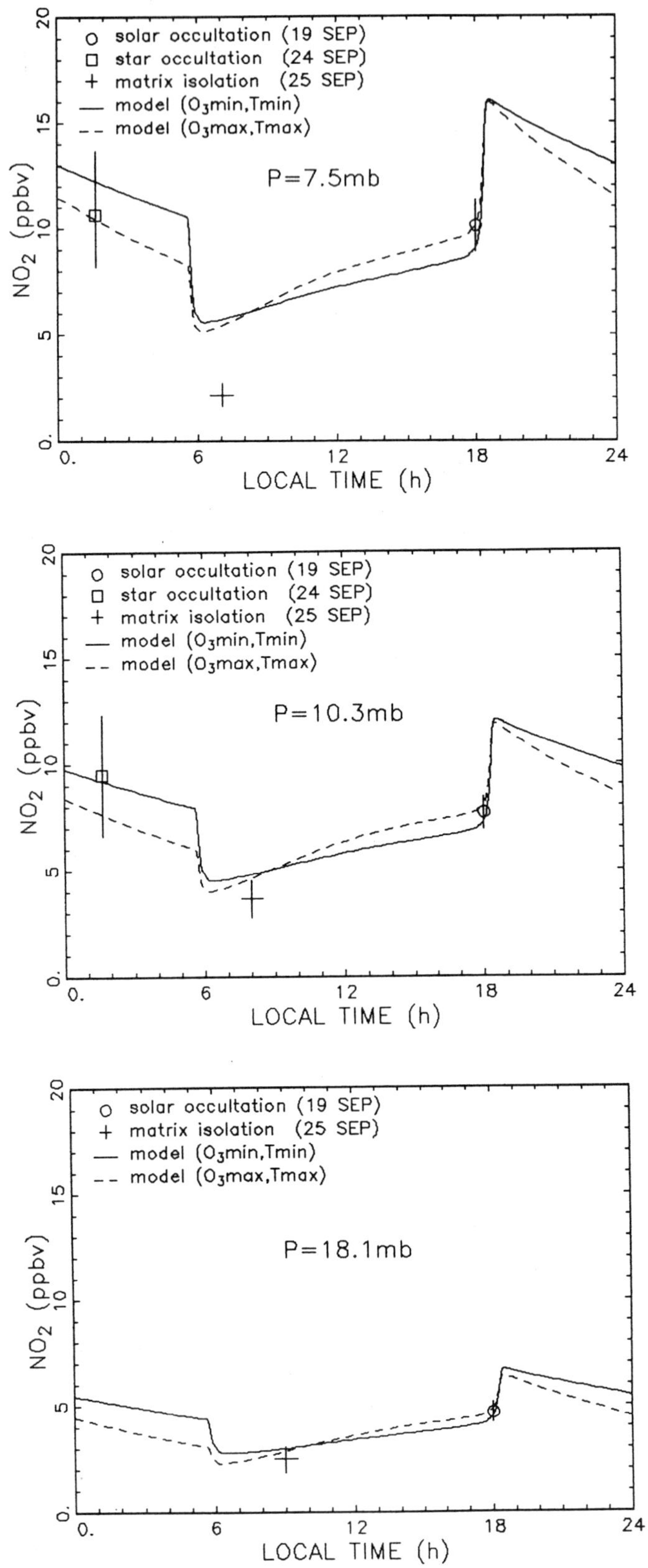

Fig. 1 : MAP/GLOBUS NO_2 mixing ratio compared to the diurnal variation calculated at 7.5, 10.3 and 18.1 mb using NO_y, ozone and temperature from table 2.

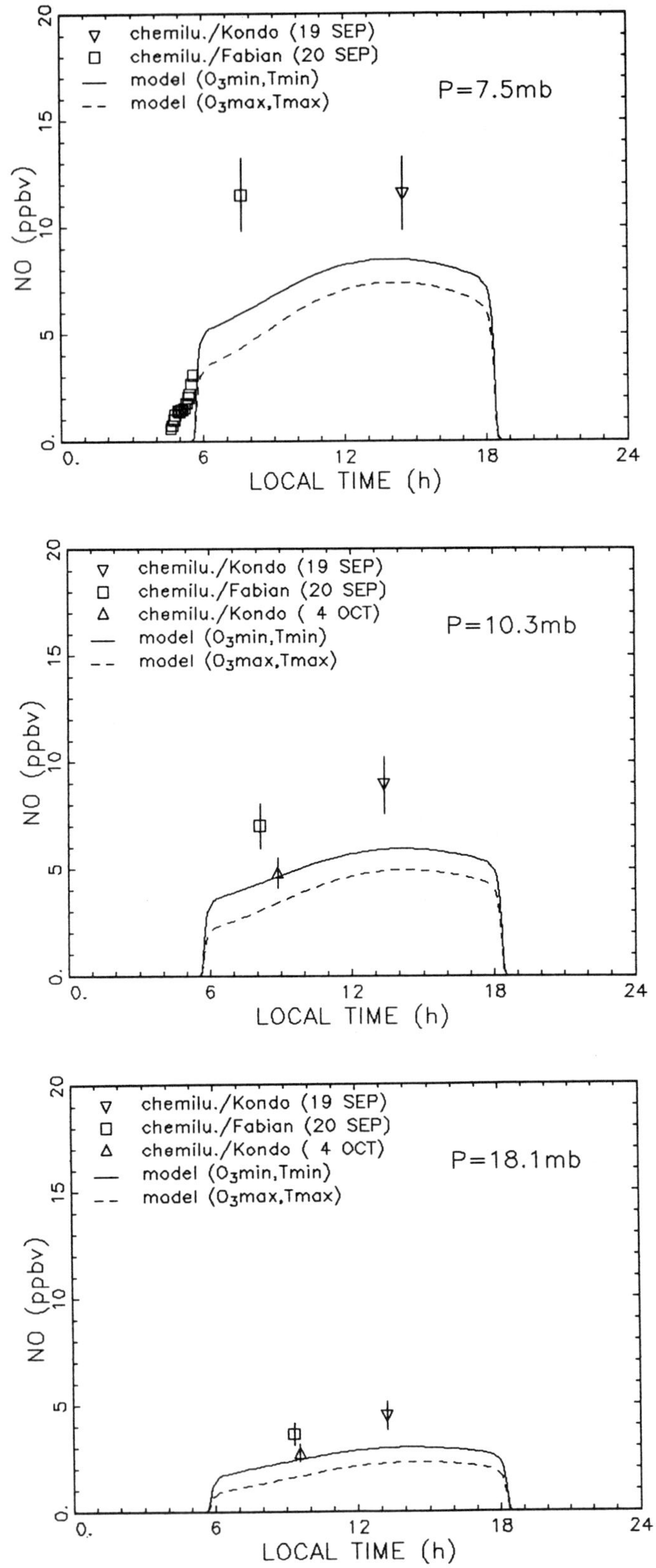

Fig. 2 : Similar to Fig. 1 but for NO.

Model calculations show that, for given NO_y, the changes in NO and NO_2 expected from the temporal variation of ozone and temperature during MAP/GLOBUS 1985 are not large enough to explain the daytime in situ observations. Minimum ozone and temperature lead to maximum NO_2 at night (reaction 4), and maximum NO during the day (reaction 2). Because of the equilibrium between NO and NO_2 (reactions 1 to 3), daytime NO_2 is not very sensitive to the ozone and temperature changes given in table 2.

Since the observations of NO and NO_2 plotted in figures 1 and 2 were not made on the same day, dynamics may be responsible for day to day changes in NO_y leading to changes in NO and NO_2. However, it should be noted that very large changes in NO_y would be necessary to reconcile the model with both NO and NO_2 from MAP/GLOBUS. For example, a value of NO_y as high as 30 ppb would be needed to fit the model to the NO values observed at 7.5 mb in the afternoon. It is also interesting to note that some of the in situ observations of NO and NO_2 can be compared in terms of local time. For example, NO and NO_2 was measured at 7.30 am by Fabian and around 7.00 am by Helten, respectively, for the same pressure level of 7.5 mb. Similar examples can be found at other pressure levels, and the sum of these observations has been plotted on Figure 3 along with the predicted diurnal variation of NO + NO_2. The nighttime value is identical to the NO_2 mixing ratio measured by star occultation (see fig. 1), since NO is entirely converted into NO_2 at sunset. The daytime values derived from the in situ observations of NO and NO_2 compare favorably with the model. Also note that the differences between observations and model seen in Figures 1 and 2 at 7.5 mb tend to cancel each other. Considering that NO + NO_2 directly depends on NO_y and that the model calculations were made using a constant value of NO_y (table 2), the good agreement seen in Figure 3 suggests that no large variation in NO_y occured during the MAP/GLOBUS campaign.

If dynamics cannot explain the discrepancy between in situ observations and model, Figure 1 to 3 indicate that the partitioning between NO and NO_2 differs significantly from that in the model. The balance between NO and NO_2 in the sunlit atmosphere is strongly dependent on the visible radiation field via the photodissociation of NO_2 (equation 1). The photodissociation coefficient J_{NO_2} is a complex function of solar zenith angle and albedo, involving multiple scattering effects (Madronich et al., 1985). The model used in this work takes into account albedo effects but multiple scattering is neglected. Therefore, it is interesting to test the validity of this approximation against direct observations of J_{NO_2}, and also to evaluate the sensitivity of the calculated ratio NO/NO_2 to change in albedo. Figure 4 compares the J_{NO_2} calculated by the model using an albedo of 0.3 with the measurements from Madronich et al. (1985). The good agreement seen on this figure, indicates that the model uses realistic values of J_{NO_2}. An average albedo of 0.3 was selected for the present work, because all balloon flights were made over land during clear sky periods. However the diurnal variation of NO and NO_2 has been re-calculated using a value of 0.6 an overestimate since no clouds were present during the observations. Such an albedo would increase the calculated mixing ratio of NO and decrease that of NO_2 by less than

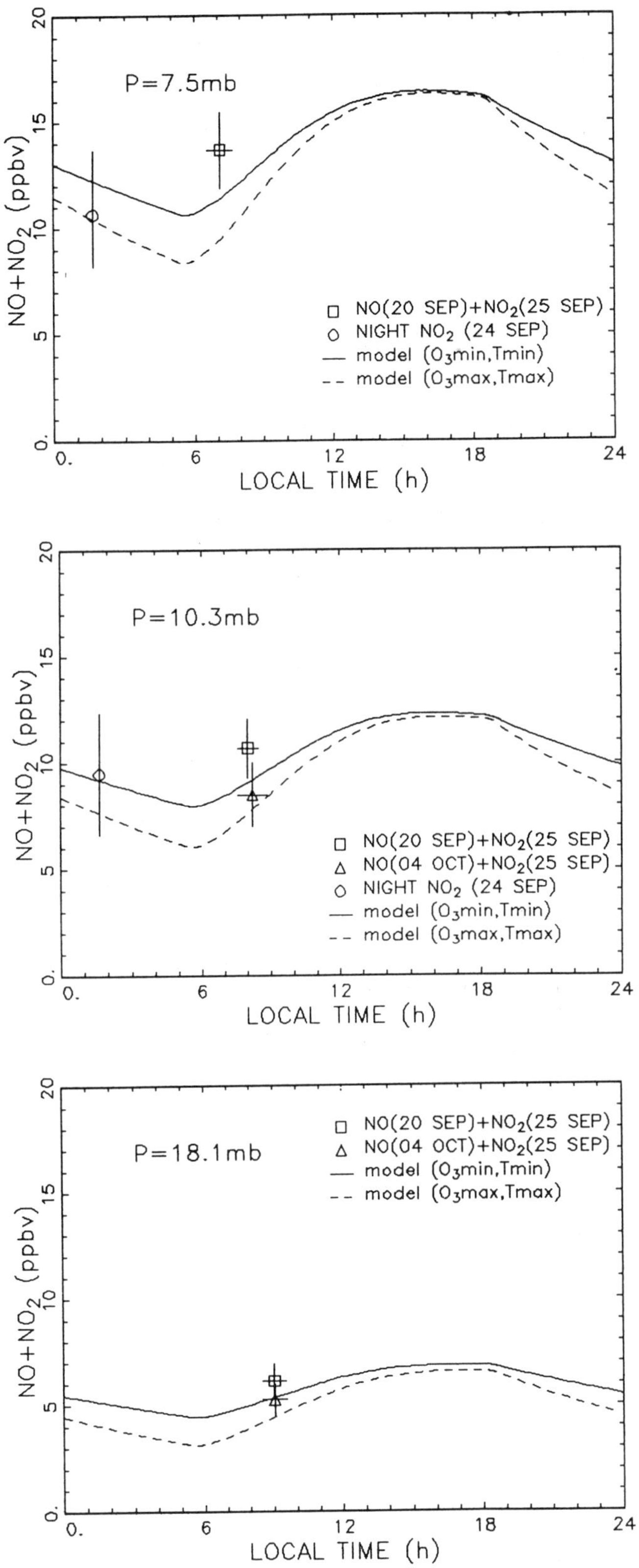

Fig. 3 : Similar to Fig. 1 but for NO + NO_2.

11 %. Therefore changes in albedo cannot explain the difference seen in Figure 1 and 2 between model and in situ observations.

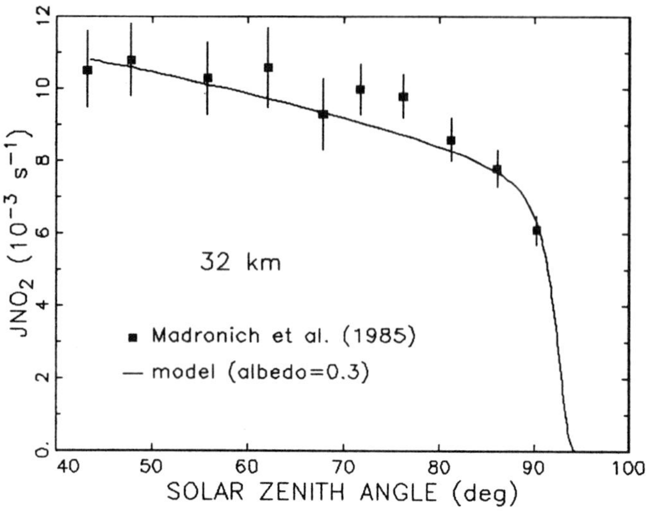

Fig. 4 : Photodissociation coefficient of NO_2 versus solar zenith angle, as observed at 32 km by Madronich et al. (1985) and as calculated by the model using an albedo of 0.3.

Kondo et al. (1988) measured the decay of NO toward sunset during the flight of September 19. They compared the relative variation of NO as a function of solar zenith angle with a simplified calculation based on the photochemical equilibrium between NO and NO_2. Good agreement was found between calculation and observations, but the comparison was limited to solar zenith angles smaller than 90°. Figure 5 extends the comparison toward larger zenith angles using the model of Pirre and Ramaroson. Observations were taken at about 31 km, and the reference value NO day corresponds to that measured at the begining of the afternoon (Kondo et al., 1988). Model calculations were made for 10.3 mb (≃ 31 km) using NO_y, ozone and temperature from table 2. The model generally agrees with the observations. In particular, the model shows the decrease of NO observed during the afternoon. However, a small

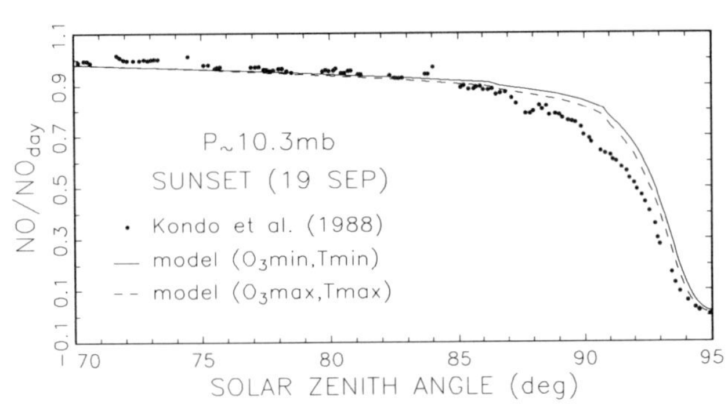

Fig. 5 : NO mixing ratio relative to its daytime value (NO_{day}) versus solar zenith angle, as observed at sunset by Kondo et al. (1988), compared to model calculations using NO_y, ozone and temperature from table 2.

difference between observed and calculated values appears at a zenith angle of 85°, reaching a maximum of 35 % at 91° and then decreasing for larger zenith angles. This suggests that the sunset conversion of NO into NO_2 begins earlier than predicted by the model. Similarly, by using LIMS NO_2 data, Solomon et al. (1986) have shown that, below 5 mb, the buildup of NO_2 at sunset begins earlier than 90°. These authors have concluded that models should take into account multiple scattering processes to fit the observations better. Therefore, as multiple scattering is neglected in the present work, the slight disagreement seen in Figure 5 may partly reflect the solar zenith angle dependence of multiple scattering processes.

5. CONCLUSION

The diurnal variation of NO and NO_2 at 18.1, 10.3 and 7.5 mb have been derived from MAP/GLOBUS data, and compared with a time-dependent zero dimensional model. The model agrees with the observations of NO_2 apart from the in situ observation at 7.5 mb which is lower than the model. The model also agrees with the observations of NO made by Kondo in the morning, but yields lower values than those observed by Fabian in the morning, and by Kondo in the afternoon. These differences between the model and the daytime in situ observations of NO and NO_2 indicates that the calculated partitioning between NO and NO_2 differs from that observed. The examination of the sum NO + NO_2 suggests that no large variation of the total amount of nitrogen species occured at 18.1 mb and above during the MAP/GLOBUS campaign and, therefore, that the comparison presented in this paper is not strongly affected by dynamical processes. It is also shown that the relative variation of NO through sunset favorably compares with the model. Further model calculations should include multiple scattering processes. Also other MAP/GLOBUS observations that are not yet available, such as, for example, the daytime observations of NO_2 by SME would further refine this work.

6. REFERENCES

Fabian, P., G. Flentje and W.A. Matthews, Stratospheric NO profiles measured simultaneously using two chemiluminescent balloon-borne sondes, Planet Space Science, 35, 609, 1987.

Helten, M., W. Patz, D.H. Ehhalt, and E.P. Röth, Measurements of nighttime NO_3 and NO_2 in the stratosphere by matrix isolation and ESR spectroscopy, in Atmospheric Ozone, (Ed. Zerefos C.S. and A. Ghazi), D. Reidel, Dordrech, 196, 1985.

Kondo, Y., W.A. Matthews, P. Aimedieu and D.E. Robbins, Diurnal variation of nitric oxide at 32 km : measurements and interpretation, J.Geophys. Res., 93, 2451, 1988.

Madronich, S., D.R. Hastie, H.I. Schiff and B.A. Ridley, Measurement of the photodissociation coefficient of NO_2 in the atmosphere : II, Stratospheric measurements, J. Atm. Chem., 3, 233, 1985.

McFarland, M., B.A. Ridley, M.H. Proffitt, D.L. Albritton, T.L. Thompson, W.J. Harrop, R.H. Winkler and A.L. Schmeltekopf, Simultaneous in situ measurements of nitrogen dioxide, nitric oxide, and ozone between 20 and 31 km, J. Geophys. Res., 91, 5421, 1986.

Naudet, J.P., P. Rigaud and D. Huguenin, Stratospheric NO_2 at night from balloons, J. Geophys. Res., 89, 2583, 1984.

Pirre, M., R. Ramaroson, J.P. Naudet and P. Rigaud, Altitude distribution of stratospheric NO_3, 2-comparison of observation with model, submitted to J. Geophys. Res., 1988.

Pommereau, J.P. and 19 coauthors, Intercomparison of stratospheric NO_2 measurements during MAP/GLOBUS 1983, Planet Sp. Sc., 35, 615, 1987.

Ramaroson, R., M. Pirre and D. Cariolle, A method adapted to the new generation of supercomputers to model the diurnal variations of the constituents in the middle atmosphere, Terra Cognita, 7, 560, 1987.

Roscoe, H.K., B.J. Kerridge, L.J. Gray, R.J. Wells and J.A. Pyle, Simultaneous measurements of stratospheric NO and NO_2 and their comparison with model predictions, J. Geophys. Res., 91, 5405, 1986.

Solomon, S., J.M. Russell III, and L.L. Gordey, Observations of the diurnal variations of nitrogen dioxide in the stratosphere, J. Geophys. Res., 91, 5455, 1986.

Webster, C.R. and R.D. May, Simultaneous in situ measurements and diurnal variations of NO, NO_2, O_3, jNO_2, CH_4, H_2O and CO_2 in the 40- to 26- km region using an open path tunable diode laser spectrometer, J. Geophys. Res., 92, 11931.

WMO, Atmospheric ozone 1985, report n° 16, world Meteorological Organization, Geneva, Switzerland, 1985.

AN NO_x SOURCE IN THE UPPER TROPOSPHERE?

J.P. Pommereau and F. Goutail
Service d'Aéronomie du CNRS, 91371, Verrières le Buisson, France

Y. Kondo
Research Institute of Atmospherics
Nagoya University
Toyokawa, Japan

W.A. Matthews
Physics and Engineering Laboratory, DSIR
Lauder, New Zealand

M. Helten
Institut für Chemie, Jülich GmbH
5170 Jülich, FRG

ABSTRACT

During the GLOBUS NO_x campaign, NO and NO_2 vertical distributions have been observed from 5 to 31 km from the same balloon, on the same day, by respectively in-situ and remote sensing techniques. The NO_x concentration in the upper troposphere which can be calculated from these measurements, is significantly larger than expected from model predictions. The increase is found to be compatible with an NO_x emission by subsonic aircrafts in a heavy air-traffic area. A contribution from lightning which cannot be estimated, is not excluded.

1. INTRODUCTION

Nitrogen oxides are critical compounds in the troposphere chemistry. Photochemical production of ozone depends on the concentration of NO_x which influence also the concentration of tropospheric OH. Nitric acid formed by reaction of OH with NO_2 is one of the major components of acid precipitation (Logan, 1983). The largest sources are under influence of man principally at ground : combustion of fossil fuels and biomass burning. Natural sources include transport from the stratosphere, microbial activity in soils and lightning in thunderstorms. The importance of this last source strength is difficult to estimate due to the very limited number of observations (Ko et al., 1986). According to current tropospheric chemistry (Logan et al., 1981, Logan, 1983) the lifetime of NO_x is short (less than 1 or 2 days at ground in summer) and therefore natural or man made sources at the surface cannot influence the NO_x budget in the upper troposphere. This budget is under control of transport from the stratosphere, exhaust from aircrafts and ligthning. The few available experimental data have shown that clean tropospheric NO_x mixing ratios vary from 10 pptv in surface air to values of about 200 pptv at the tropopause. Man-made NO_x from ground level sources influences the NO_x profile only in the lower troposphere (Kley et al. 1981).

During the Globus NO_x campaign held in September 1985 in southern France (Pommereau, 1988a), NO and NO_2 have been measured by several independent remote and in-situ methods in the stratosphere. On one occasion, observations of NO in-situ by chemiluminescence and NO_2 by remote visible spectrometry from balloons were obtained down to 5 km on the same day. The analysis of these measurements together with those performed at higher altitude, allows one to calculate the NO_x concentration which will be compared to previously reported data and that resulting from models predictions.

2. EXPERIMENTAL DATA

NO has been measured in-situ from 6 to 30 km around noon during the balloon ascent and NO_2 from 30 to 5 km at sunset by solar occultation in the visible, on the same day. The two instruments were mounted on the same balloon gondola launched on 19 September, 1985 from the Centre National d'Etudes Spatiales (CNES) balloon range at Aire sur l'Adour in south western France.

Nitric oxide was measured by chemiluminescence in-situ. The instrument from the University of Nagoya/PEL Lauder is described in Kondo et al. (1984). The overall accuracy is estimated to be 10% (Kondo et al., 1988). The diurnal variation observed in the stratosphere during this flight at 32 km is discussed in Kondo et al.(1988), and a comparison of vertical profiles obtained during the campaign including a correction for an increase of sensitivity with altitude of the mass flow sensors recently applied, is presented in Matthews et al., (1988). NO_2 was observed by solar occultation in the visible during sunset by the use of a diode array spectrometer. Two other spectrometers, mounted also on the same gondola, have provided stratospheric NO_2 data which compare favorably to the ones of the diode array instrument (Pommereau et al, 1988b). A weighted average profile for the three instruments has been calculated. The total accuracy of these measurements is estimated to be 15% in the stratosphere and 30% below 20 km. The average balloon profile has been found to agree with the NO_2 vertical distribution observed at sunset at nearly the same location by the satellite instrument Sage 2, down to 15 km. NO_2 was also sampled in flight during the morning of 24 September by matrix isolation and analysed later at ground by electron spin resonance (ESR). The instrument is described by Helten et al. (1984). The absolute accuracy of these measurements is estimated to be 25%.

Besides the two NO_x observations which cover the upper troposphere, other data needed for photochemical calculations are available : Ozone was measured in-situ in the stratosphere during the night September 19-20 by the use of an ultra-violet photometer above 15 km with an accuracy of ± 5% (Robbins et al., 1988). In the troposphere, ozone was measured by a Brewer Mast sonde on 20 September at 9 : 25 UT at the Haute Provence Observatory, located 350 km east of Aire sur l'Adour.

Temperatures which will be used in the following discussion were measured with a thermistor mounted on-board the NO/NO_2 gondola. They are found to be in agreement (± 2°C) with the radio-sonde data at Bordeaux at 12 : 00, 100 km north west of the balloon base. Wind speed and direction are derived from tracking the balloon during its ascent early in the afternoon, that is during the NO measurements.

ISBN 0-937194-15-8

3. DATA ANALYSIS

The nitric oxide vertical distribution measured by chemiluminescence during the balloon ascent on 19 September between 12 : 20 and 13 : 25 UT, that is approximately the same solar time, is reported on Fig. 1. Results of NO_2 observations at sunset on the same day by solar occultation in the visible from float level at 31.2 km and of NO_2 in-situ measurements during the slow balloon descent on the morning of 24 September, are shown in Fig. 2.

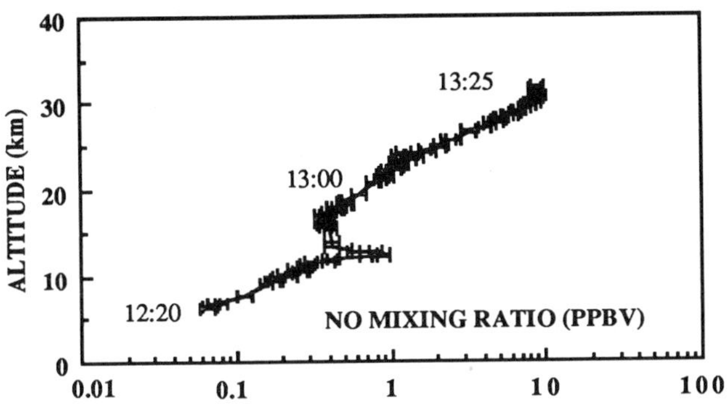

Fig. 1. NO mixing ratio measured by chemiluminescence during the balloon ascent in the early afternoon on 19 September. Time (UT) at which the observations were made are shown on the left of the curve.

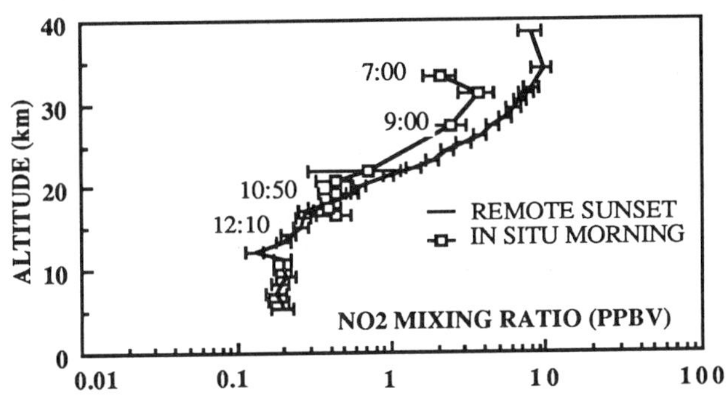

Fig. 2. NO_2 observed by solar occultation at sunset on 10 September and NO_2 sampled in-situ by matrix isolation in the morning of 24 September analysed later at gound by ESR. Time (UT) of the samplings are indicated.

During daytime, and below 25 km where reactions of NO_2 with O and ClO are not effective, the partitioning between NO and NO_2 is controlled by the two reactions:

$$NO_2 + h\nu \rightarrow NO + O \qquad J = 1.2\ 10^{-2} s^{-1} \qquad (1)$$

$$NO + O_3 \rightarrow NO_2 + O_2 \qquad k_2 = 1.8\ 10^{-12} \exp(-1370/T) \qquad (2)$$

At steady state :

$$[NO_2]/[NO] = k_2[O_3]/ J \qquad (3)$$

$$[NO_x] = [NO] + [NO_2] = [NO]\,(1 + k_2\,[O_3]/J) \qquad (4)$$

$$= [NO_2]\,(1 + J/k_2[O_3])$$

At sunset, NO is promptly recombined into NO_2 and therefore at night $[NO_x] = [NO_2]$

The daytime NO_2/NO ratio computed from the actual temperature and ozone concentration is shown in Fig. 3 together with the experimental ratio between NO_2 measured in-situ in the morning of 24 September and NO early in the afternoon on 19 September. Around and below 18 km where the two observations were performed approximately at the same local time, the experimental value agrees with the theoretical one. At higher levels around 30 km, NO_2 was measured early in the morning when N_2O_5 is not completely photolysed. The smaller NO_2/NO ratio observed,compared to the theoritical value at 30 km, is qualitatively explained by the diurnal variation of NO_2. Near 15 km, the experimental ratio appears larger than the theoretical one. JNO2 which was assumed to be constant is in fact reduced by the atmospheric attenuation at low altitudes, and the NO_2/NO ratio is underestimated. Therefore, a calculation of NO_x concentration in the upper troposphere from NO measured early afternoon, through equation (4), should lead to a lower limit of NO_x.

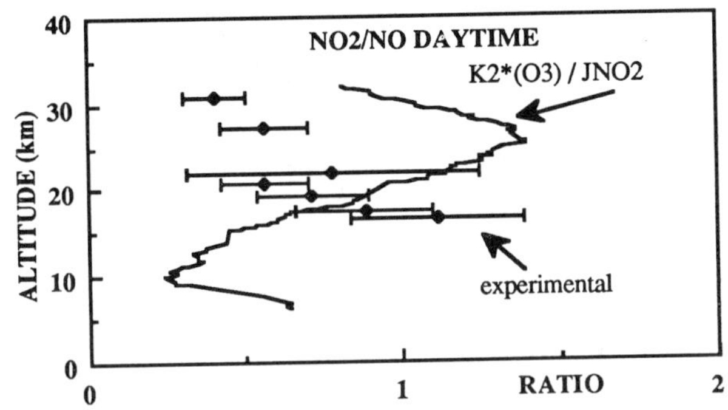

Fig. 3. Daytime NO_2/NO ratio computed from ozone and temperature data assuming a constant NO_2 photolysis rate of $1.2\ 10^{-2} s^{-1}$, compared to the ratio of the two in-situ data.

Fig. 4(A) shows the experimental NO_2 (sunset) ratioed against NO at noon. Fig. 4(B) compares the experimental ratio with a theoritical one calculated for noon. At high altitude, above 25 km, the experimental ratio is larger than expected. There is too much NO compared to NO_2, as already noticed by Naudet et al (1988). Below 25 km, and with the exception of the layer of NO at 12.5 km, the ratio reduces progressively. This may be explained by the reduction of the NO_2 photolysis rate at large solar zenith angles as pointed out by Kley et al (1981). At sunset in the upper troposphere, the NO_2 concentrations measured by solar occultation are close to the night-time values. Since the transmitted solar flux at the balloon level is reduced by more than a factor 1000, NO_2 measured at sunset will represent a lower limit of the NO_x concentration.

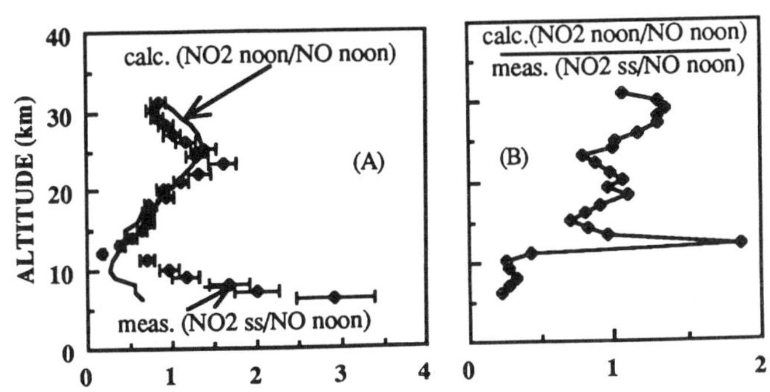

Fig. 4. (A) Observed NO_2 (sunset)/NO (noon) compared to the result of a daytime calculation ; (B) The ratio between the two quantities shows an excess of measured NO compared to NO_2 at altitudes above 25 km. Below, and with the exception of the level of the NO layer at 12.5 km, the ratio decreases.

4. NO_x DISTRIBUTION IN THE UPPER TROPOSPHERE

The two NO_x estimations from NO measured around noon and from NO_2 at sunset between 5 km and 20 km are shown on Fig. 5 where the tropopause level at 13.8 km is also indicated. Also shown are the results of model calculation by Logan et al. (1981) and Kley et al.(1981). Compared to the simulated distributions, the experimental values are larger by a factor of 3 to 10 in the upper troposphere. Moreover, the in-situ data display a sharp NO_x layer centered at 12.5 km, 1 km thick, peaking at more than 1 ppbv. The remote sensing results do not show such a layer but it must be remembered that the solar occultation technique has a low vertical resolution of 2 km and that the region observed is located 400 km away from the balloon itself.

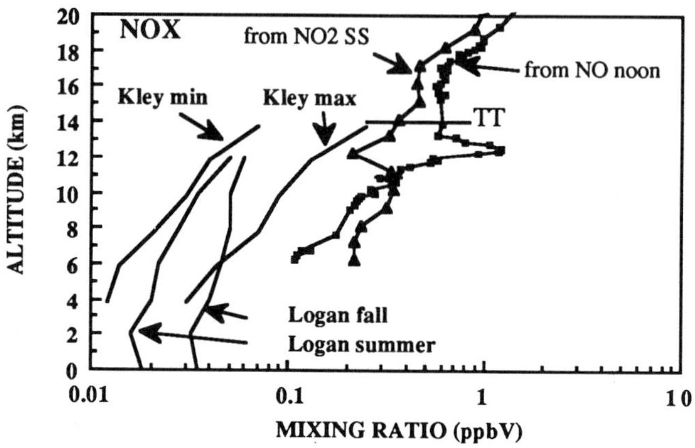

Fig. 5.NO_x estimations from NO measurements at noon (squares) and NO_2 at sunset (triangles) compared to model predictions by Logan et al (1981) and Kley et al (1981)

Fig. 6 shows the observed profiles of temperature, potential temperature and wind speed and direction, measured during the balloon ascent on the same gondola on which NO was measured, and ozone observed by the Brewer-Mast sonde at Haute Provence Observatory early in the morning of the day after. The temperature shows a sharp minimum of - 63.5° at 13.8 km, corresponding to a well defined tropopause, and one can notice on the potential temperature and the ozone profiles the presence of some discontinuity at 11 km. The layer between the two levels corresponds to an horizontal advection from the north isolated between a SE regime in the troposphere and SW in the lower stratosphere. This layer is the one in which the high NO mixing ratios were observed.

5. DISCUSSION

There are few other NO_x experimental data in the upper troposphere for comparison. Kondo et al. (1985) have already observed the NO vertical distribution in the troposphere by in-situ chemiluminescence at Aire sur l'Adour in September 1983. The measurements were made two hours before noon and with the exception of the NO layer, the reported mixing ratios were similar to the ones observed in 1985. Also in summer, Bloxam et al. (1975) have reported large NO_2 mixing-ratios (0.5 ppbv at 10.5 km) at sunset over Canada. Kley et al (1981), have summarized the information available before 1981 which shows the NO_x concentration in winter at mid-latitude or in tropical regions to vary in average from 0.15 ppbv to 0.3 ppbv at 10 km and a positive gradient in the upper troposphere. Ehhalt and Drummond (1987) during a latitudinal survey from an aircraft, have also observed that the daytime NO profiles in the northern hemisphere and the tropics show a positive gradient in the upper troposphere, while the concentrations were lower in the southern hemisphere at the same altitude. In one case, during their southbound flight in June over the Bay of Biscaye, that is close from the site of our

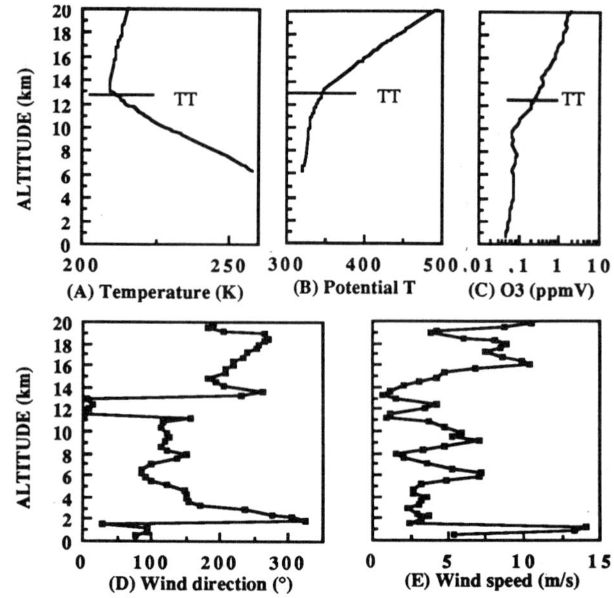

Fig. 6. Temperature (A), potential temperature (B), ozone (C), and wind direction (D) and speed (E) observed during the balloon flight.

observations, they observed NO mixing ratios of up to 3 ppbv around 10 km. Therefore our measurements are consistent with the existing other data at mid-latitudes in the northern hemisphere, at least over continental areas.

Four source processes could contribute to the observed high NOx levels in the upper troposphere: transport from the stratosphere, upwelling from the polluted planetary boundary layer, lightning and exhaust from aircrafts.

The average transport from the stratosphere where NO_x is formed trough N_2O oxidation, is included in the model calculations by Logan et al.(1987) and Kley et al.(1981). Increasing the downward flux at the tropopause level by a factor of between 3 and 10 will not be compatible with the observed ozone distributions. Local transport of ozone cannot be ruled out, but the NO layer observed at 12.5 km does not seem to fit this category. A peak mixing ratio of 1 ppbv would imply a fast downward transport from at least an altitude of 20 km (2 ppmv ozone, 490 K potential temperature) which would have left a signature in the vertical distribution of the two other parameters.

The very short lifetime of NOx compounds at the surface, 2 to 3 days following Kley et al. (1981), would not allow in general the polluted continental PBL to contribute significantly to the upper tropospheric concentration. However, deep convection could locally loft the air fastly from the surface to the upper layers (Chatfield and Crutzen, 1984). The same large convective clouds could also produce NO_x compounds by lightning (see for example Ko et al., 1986).

Ehhalt and Drummond (1987) suggested that their aircraft above the Bay of Biscaye, would have intersected an airmass lofted from the PBL. Simultaneous mesurements of high CO and low O_3 concentrations added to this interpretation.

In our case, the balloon flight was perfomed in anticyclonic conditions and the satellite cloud pictures did not show any cloud above southern France and Spain during the last few days. The only existing features are a perturbation travelling SW-NE over the British Islands and some thunderstorm activity associated with a cold front over southern Mediterannean (Sardigna). The tropospheric winds displayed on Fig.6, were smaller than 5 m/s in average, and therefore these potential sources are located 3 to 4 days upwind. With the exception of the sharp NO layer at 12 km advected from the north, the wind was blowing from the SE in the upper troposphere, that is from the Mediterannean, where the presence of a polluted PBL is unlikely. Lightning remains a potential source but we have no way to evaluate its strength.

The last identified possible NO_x source is the aircraft exhaust.It was proposed at first that the NO layer at 12.5 km could have been the effect of a single aircraft con-trail [Matthews et al. 1988]. Since the separation of the observed half peak values was 1 km, the authors assumed a swath 1 km wide and 1 km deep. A long-haul four engine jet with a fuel flow rate of 5 kg/km, 14% of which will be exhausted as NO_x [Oliver et al, 1977], will then produce an average NO_x mixing ratio, in this swath, of approximately 2.8 ppbv. There are several arguments against this interpretation: (i) The altitude of the layer, 12.5 km, is higher than the cruise altitude of such aircraft; (ii) The diffusion coefficient is not isotropic at a 1 km scale. A layer 1 km deep would be approximately 100 km wide and therefore the NO_x mixing ratio will be diluted by a factor 100; (iii) The NO layer coincides with a wind direction change which indicates that horizontal transport from an NO_x rich region is responsible for the peak. The con-trail hypothesis cannot explain this fact.

If one single aircraft cannot be responsible for the observed NO_x mixing ratio, the thousand jets flying above France each day could perhaps account for it. Hidalgo and Crutzen (1976) have estimated the impact of subsonic aircraft emissions on the upper troposphere NO_x. Their model assumed an average cruise altitude of 10.8 km for the present subsonic type on upper-bound projections to 1990, abitrarily increased by 50 percent to account for short range traffic. The result is an average increase at 10.8 km in the northern hemisphere at 45°N maximum in summer and fall, of the order of 0.12 ppbv. As the air traffic is heavy above western Europe and the light tropospheric winds would have allowed the air to stay for a few days over the continent, it seems reasonable to think that locally, the contribution from aircraft emissions could be multiplied by a factor 2 or 3. The NO layer could be interpreted as resulting from horizontal transport from northern France, one of the heaviest air traffic areas in the world.

6. CONCLUSION

Nitrogen oxide (NO_x) concentrations deduced from in-situ NO measurements using chemiluminescence and from NO_2 observations made by solar occultation in the visible, performed during Globus NO_x campaign, are consistent. In the troposphere, the average NO_x mixing ratio is observed to increase from 0.1 - 0.2 ppbv at 6 km to 0.3 - 0.6 ppbv at the tropopause level. Large structures are present in the vertical profile, in particular, a 1 km deep layer at 12.5 km peaking at 1 ppbv, associated with a well defined wind layer from the north isolated between winds from the SE and the SW directions. The observed mixing ratios in the upper troposphere, consistent with the other data that exist, are larger by a factor between 3 and 10 than than those from model simulations which include downward transport from the stratosphere. A tropospheric source seems to be present. Its required strength appears compatible with aircraft NO_x emissions in a heavy traffic area. The correlation between a high concentration layer and an advection from the north add weight to this explanation. However, a remote contribution from lightning or fast transport from the polluted continental planetary boundary layer cannot be totally excluded.

ACKNOWLEDGMENTS

The authors are indebted to the personnel of the Centre National d'Etudes Spatiales (CNES) balloon launch facilities at Aire sur l'Adour for their support and in particular to the weather officer J. Dedieu for his assistance. This work was supported by the Fluorocarbon Panel of the Chemical Manufacturers Association (Grant FC 85-550), and by CNRS, who are gratefully acknowledged.

REFERENCES

Bloxam, R.M., A.W. Brewer and C.T. Mc Elroy, 1975, NO_2 measurements by absorption spectrophotometer : Observation from the ground and high altitude balloon, Fourth Conference on CIAP, 454, US Department of Transportation, Washington DC.

Chatfield, R.B., and P.J. Crutzen, 1984, Sulfur dioxide in remote oceanic air: cloud transport of reactive precursors, J. Geophys. Res., 89, 7111.

Ehhalt,D.H., and J.W. Drummond, 1987, NOx sources and the tropospheric distribution of NOx during STRATOZ III, Tropospheric Ozone, NATO ASI Series, D.Reidel Pub. Comp.

Helten, M., Patz, W, Trainer, M., Fark, H., Klein E. and Ehhalt D.H. 1984, Measurements of stratospheric HO_2 and NO_2 by Matrix Isolation and ESR Spectroscopy., J. Atm. Chemistry, 2, 191.

Hidalgo H. and P.J. Crutzen, 1976, The tropospheric and stratospheric composition perturbed by NO_X emissions of high altitude aircraft. Proc. of the non-urban tropospheric composition symposium, AGU and AMS, Hollywood Florida.

Kley, D., J.W. Drummond, M. Mc Farland and S.C. Liu, 1981, Tropospheric Profiles of NO_X, J. Geophys. Res., 84, 3153.

Ko, M.K.W., M.B. McElroy, 1986, D.K. Weisenstein and N.D. Sze, Lightning : a possible source of stratospheric odd nitrogen, J. Geophys. Res., 91, 5395.

Kondo Y., A. Iwata, M. Takagi and W.A. Matthews, 1984, Balloon-borne chemiluminescence sonde for the measurement of tropospheric and stratospheric nitric oxide, Rev. Sci. Instrum., 55, 1328.

Kondo,Y., W.A. Matthews, A.Iwata and M.Takagi, 1985, Measurements of Nitric Oxide from 7 to 32 km and its diurnal variation in the stratosphere, J.Geophys.Res., 90,3813.

Kondo Y., W.A. Matthews, P. Aimedieu and D.E. Robbins, 1988, Diurnal variation of nitric oxide at 32 km : measurement and interpretation, J. Geophys. Res., 93, 2451.

Logan, J.A., M.J. Prather, S.C. Wofsy and M.B. Mc Elroy, 1981, Tropospheric chemistry : A global perspective, J. Geophys. Res., 86, 7210.

Logan, J.A., 1983, Nitrogen Oxides in the troposphere : Global and regional budgets, J. Geophys. Res., 88, 785.

Matthews, W.A., Y. Kondo, P. Fabian and B.C. Kruger, 1988, Nitric Oxide profiles measured in situ during the Globus 85 campaign, Submitted to J. Atm. Chemistry.

Naudet, J.P., M.Pirre, R.Ramarson, P.Rigaud, P.Fabian, M.Helten, N.Iwagami, K.Shibasaki, T.Ogawa, Y.Kondo, W.A.Matthews, J.P.Pommereau, F.Goutail, P.C.Simon, and W.Petermans, 1988, Diurnal Variation of Stratospheric NO and NO2 from MAP/Globus 1985, Quad. Ozone Symp., in press.

Oliver R.C., E. Bauer, H. Hidalgo, K.A. Gardner and W. Wasylkiwsky, 1977, Aircraft emissions : potential effect on ozone and climate, US Dept. of Transportation Federal Aviation Agency, Report NO_2 FAA-EQ-77-3

Pommereau, J.P., 1988a MAP/GLOBUS NOX: An Overview, Quad. Ozone Symp.,in press.

Pommereau, J.P., F. Goutail, N. Iwagami, K. Shibasaki, P.C. Simon, W. Peetermans, J.P. Naudet, P. Rigaud and D. Huguenin,1988b, Balloon observations of nitrogen dioxide by visible occultation during Globus NO_X, submitted to J. Atm. Chemistry.

Robbins, D.E., P. Aimedieu, J. Pelon, J.P. Pommereau, F. Goutail, N. Iwagami, K. Shibasaki, T. Ogawa, M. Koike, P. Marché, J.P. Naudet, P. Rigaud, D. Huguenin, J. Lenoble, G. Maddrea Jr, 1988, Measurements of stratospheric ozone during the 1985 Map/Globus NO_X Campaign, Submitted to J. Geophys. Res.

DIURNAL VARIATION OF NITRIC OXIDE AT 26 KM

Y. Kondo, M. Takagi, and A. Iwata
Research Institute of Atmospherics
Nagoya University
Toyokawa, Japan

W.A. Matthews
Physics and Engineering Laboratory, DSIR
Lauder, New Zealand

ABSTRACT

A chemiluminescent NO detector was launched from Uchinoura, Japan (31°N) at 0030 JST on July 29, 1987. The main purpose of the experiment was to measure precisely the diurnal variation of NO in the lower stratosphere using a long distance flight. Upward and downward radiation intensities in the visible region were also monitored on the gondola. The balloon reached a float altitude of 26 km around 0200 JST and stayed at this altitude until 1600 JST when it started a slow descent. During the flight, most of the NO measurement was performed over the East China Sea. At local sunrise, when the solar zenith angle reached 92.7°, the NO concentration started to increase rapidly. This rapid increase in NO continued for about 30 minutes. A slow increase in NO, presumably due to the photodissociation of N_2O_5 continued into the early afternoon.

1. INTRODUCTION

Nitric oxide in the lower stratosphere is predicted to undergo a large diurnal variation by current photochemical models [e.g., Ko and Sze, 1984; Brasseur et al., 1987]. The variation of NO after sunrise is believed to be caused by the photodissociations of NO_2 and the reservoir species such as N_2O_5 formed during the night. The time constant of the photodissociation of NO_2 is of the order of a few minutes, while in the lower stratosphere that of N_2O_5 is of the order of several hours. This is due to the fact that N_2O_5 is photodissociated mainly by solar ultraviolet radiation between 200 and 300 nm, which is strongly absorbed by ozone. In this way, a number of important photochemical processes, such as the production and photodecomposition of reservoir species, are involved in the diurnal variation of NO_x. Consequently the measurements of the diurnal variation of NO should lead to an understanding of these basic photochemical processes. Up until now the temporal variations of NO in the lower stratosphere have been measured by Patel et al. [1974], Burkhardt et al. [1975], and Ridley et al. [1977]. We report a balloon observation of NO made at an altitude of 26 km with a duration of about 12 hours from sunrise on July 29, 1987. The balloon experiment was performed as a part of the Japan-China co-operative balloon program organized by the Institute of Space and Astronautical Science and the Academia Sinica of the Peoples Republic of China. A long distance balloon flight from Japan to China crossing the East China Sea enabled us to observe in detail the temporal variation of NO.

2. INSTRUMENTATION

Detailed descriptions of the chemiluminescent NO detector used for the balloon observation are given in Kondo et al. [1985, 1988]. Purified air was periodically injected upstream of the reaction chamber with mass flow rates twice as large as those of the sample air for 40 s every 10 minutes to measure the zero level of the instrument. In order to improve the precision in the estimate of the sample flow rate, a systematic calibration in the laboatory has been performed. Concentration of the NO in N_2 gas used for the calibration on board the

ISBN 0-937194-15-8

balloon were measured before and after the flight. The uncertainty in the concentration of the calibration NO gas due to a slow decrease in concentration was reduced to within 2 %. Recently we have found that the sensitivity of the mass flow sensor used to control the calibration NO gas increases with the decrease in the ambient pressure [Matthews et al., 1988]. This effect was more important for the mass flow controller used in the present experiment (STEC, SEC-421 MK II) than for that used in our previous experiments (Tylan FC-260). The change in the sensitivity of the mass flow sensor with the ambient pressure was measured in the laboratory and was found to be quite reproducible and is shown in Figure 1. This effect has been taken into account in deriving the NO mixing ratios.

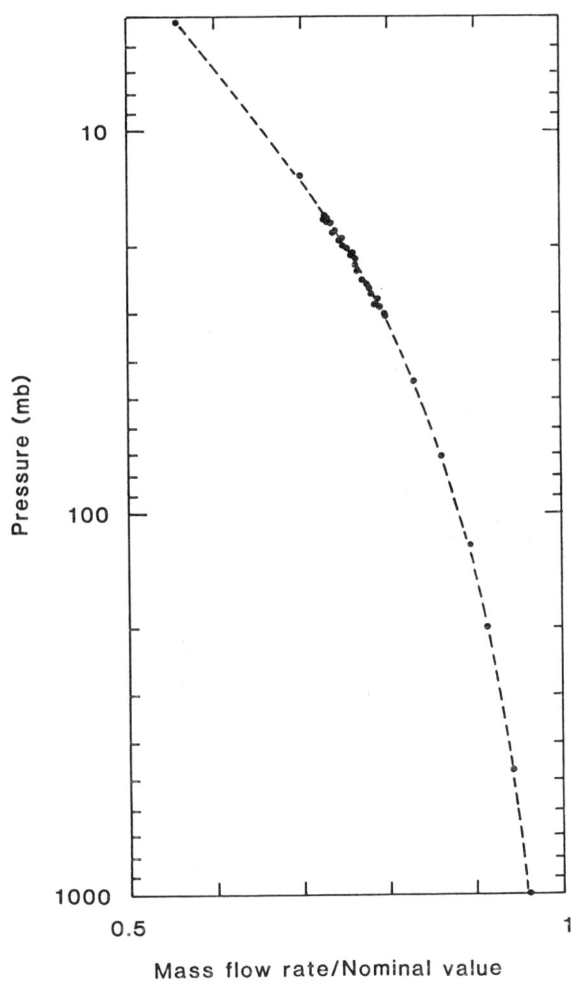

Fig. 1. Change in the relative mass flow rate due to the increase in sensitivity with decreasing ambient pressure.

The top of the radiometer used on the balloon gondola was covered with a plastic hemispherical shell to diffuse incoming radiation. The diffused radiation is passed through an interference filter with a center wavelength of 420 nm and is then detected by a phototube (Hamamatsu R727). The radiometers used to measure upward and downward radiation were essentially identical.

A microcomputer (NEC 98LT) was mounted on the gondola to record all the data during the flight. The data were also transmitted to the ground receiving station by telemetry link.

3. RESULTS

The gondola for NO instrument was suspended about 20 m below a 15 000 m^3 balloon and was launched from Uchinoura, Japan, (31°N, 131°E), at 0030 JST on July 29, 1987. The balloon reached a float altitude of 26 km around 0200 JST and stayed at this altitude until 1600 JST when it started a slow descent. During the flight most of the NO measurement was performed over the East China Sea. Photographs taken from the meteorological satellite show that the balloon flight area was either cloud-free or covered only by a narrow band of clouds.

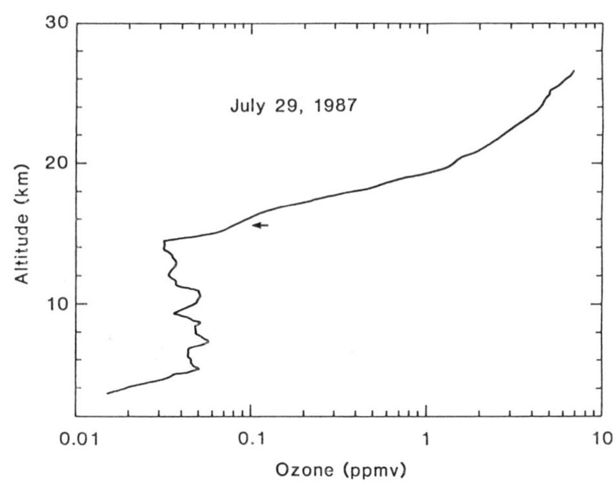

Fig. 2. Ozone profile measured during ascent. The arrow indicates the altitude of the tropopause.

The profile of ozone measured by a KI ozonesonde (Meisei Electric, RSII-KC79) during the ascent is shown in Figure 2. The ozone density between 23.5 and 26.5 km is in good agreement with the ozone profile measured simultaneously by the BUV instrument on board the Japanese satellite EXOS-C (Koike, Ogawa, and Suzuki; private communication). The temperature measured on board the balloon indicated that the altitude and temperature of the tropopause were about 15.5 km and -76°C respectively.

The NO mixing ratio measured at a float altitude is shown in Figure 3. It can be seen that a rapid increase in NO starts at about 05:30 JST when the solar zenith angle reaches 92.7°. This rapid change in NO continues for 20 to 30 minutes and continues thereafter at a much slower rate. These features near sunrise are basically in agreement with those measured previously [Ridley et al., 1977; Kondo et al., 1988]. This increase seems to continue until around 14:00 JST. The NO concentration between 14:00 and 16:30 JST is nearly constant at 2.05 ±0.1 ppbv.

Therefore, NO increases slowly for 7 to 8 hours after sunrise and the daytime NO concentration is 2.5 to 3 times larger than that at 06:00 JST. The slow increase in NO can qualitatively be explained by the photodissociation of N_2O_5. Further quantitative analysis of this effect will be possible by using the ozone profiles that have been measured simultaneously.

The signal of the upward and downward looking radiometer shown in Figure 3 shows that the gondola entered the shadow of the balloon between 11:45 and 14:00 JST. From a calculation taking into account the geometry of the balloon and the gondola, it is expected that the sun as seen from the gondola will be masked by the balloon during this period in accordance with the observation. It seems that some irregular variation in NO is superimposed on the diurnal variation during this period. The reduced visible solar flux might have influenced the NO concentration by shifting the NO_2/NO ratio. A decrease in NO after 16:30 can be considered to be due to a slow descent of the balloon.

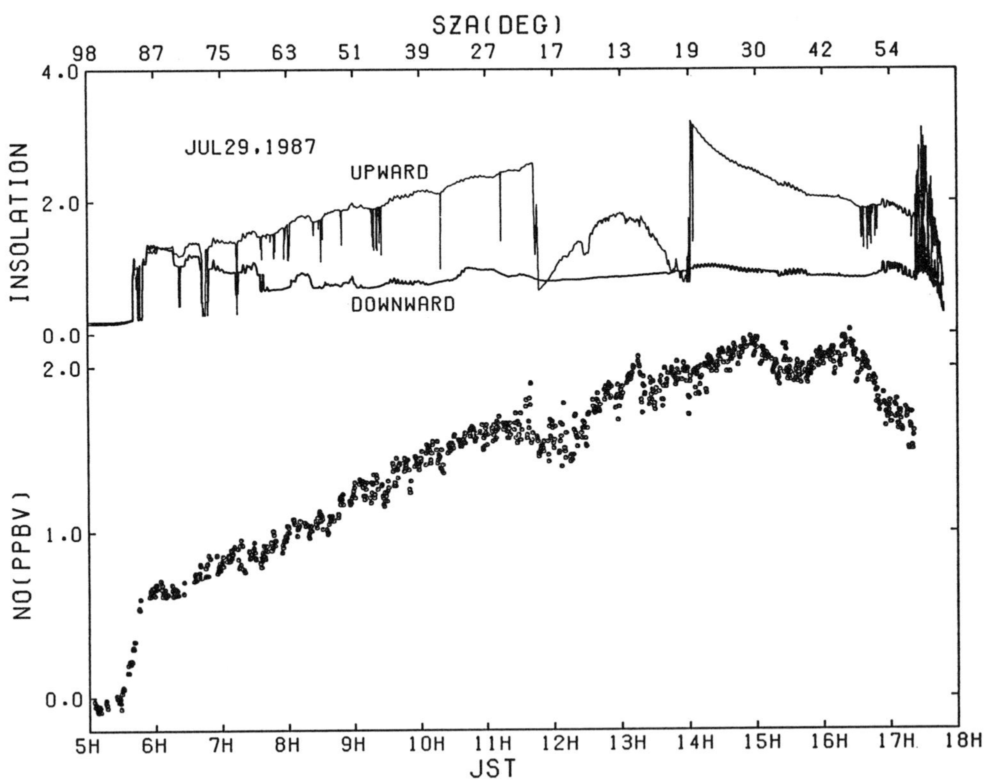

Fig. 3. NO mixing ratios measured at a float altitude. The intensities of radiation measured by upward and downward looking radiometers are also shown. Solar zenith angles (SZA) have been included in this figure for completeness.

ACKNOWLEDGMENTS

We wish to express our thanks to Prof. J. Nishimura and his staff at the Institute of Space and Astronautical Science (ISAS) for the excellent logistical operation of the balloon experiment. The balloon flight was funded by ISAS. The support, including the recovery of the gondola, provided by the Academia Sinica of the Peoples Republic of China is greatly acknowledged. We wish to thank N. Toriyama, H. Jindo, and M. Kanada for their assistance in the preparation of the instruments. This work was supported financially as part of the Middle Atmosphere Program (MAP) by the Ministry of Education, Science, and Culture of Japan.

REFERENCES

Brasseur, G., D. Cariolle, A. De Rudder, L.J. Gray, J.A. Pyle, E.P. Röth, U. Schmailzl, and D.J. Wuebbles, Odd nitrogen during the MAP/GLOBUS 1983 campaign: Theoretical considerations, Planet. Space Sci., 35, 637-645, 1987.

Burkhardt, E.G., C.A. Lambert, and C.K.N. Patel, Stratospheric nitric oxide: Measurements during daytime and sunset, Science, 188, 1111-1113, 1975.

Ko, M.K.W., and N.D. Sze, Diurnal variation of ClO: Implications for the stratospheric chemistries of $ClONO_2$, HOCl, and HCl, J. Geophys. Res., 89, 11619-11632, 1984.

Kondo, Y., A. Iwata, M. Takagi, and W.A. Matthews, Balloon-borne chemiluminescent sonde for the measurement of tropospheric and stratospheric nitric oxide, Rev. Sci. Instrum., 55, 1328-1332, 1984.

Kondo, Y., W.A. Matthews, A. Iwata, and M. Takagi, Measurement of nitric oxide from 7 to 32 km and its diurnal variation in the stratosphere, J. Geophys. Res., 90, 3813-3819, 1985.

Kondo, Y., W.A. Matthews, P. Aimedieu, and D.E. Robbins, Diurnal variation of nitric oxide at 32 km: Measurements and interpretation, J. Geophys. Res., 93, 2451-2460, 1988.

Matthews, W.A., Y. Kondo, P. Fabian, and B.C. Krüger, Nitric oxide profiles measured in situ during the GLOBUS 85 campaign, submitted to J. Atmos. Chem.

Patel, C.K.N., E.G. Burkhardt, and C.A. Lambert, Spectroscopic measurements of stratospheric nitric oxide and water vapor, Science, 184, 1173-1176, 1974.

Ridley, B.A., M. McFarland, J.T. Bruin, H.I. Shiff, and J.C. McConell, Sunrise measurements of stratospheric nitric oxide, Can. J. Phys., 55, 212-221, 1977.

TWO YEARS OF REGULAR STRATOSPHERIC NO_2 MEASUREMENTS AT THESSALONIKI, GREECE USING TWILIGHT PHOTOMETRY

G.T. Amanatidis, C.S. Zerefos, A.F. Bais, I.C. Ziomas, and A.G. Kelessis
Laboratory of Atmospheric Physics (LAP)
University of Thessaloniki
54006 Thessaloniki, Greece

ABSTRACT

A program for ground-based regular measurements of stratospheric NO_2 has been instituted at the Laboratory of Atmospheric Physics of the University of Thessaloniki (40.5°N, 22.9°E) since August 1985 using twilight photometry. The measurements are obtained on clear and unpolluted days with a modified Canterbury filter photometer equipped with narrow band interference filters at wavelengths 437.8 nm, 439.3 nm, 442.0 nm, 444.9 nm and 450.1 nm with half bandwidth of approximately 1 nm.

The optical characteristics of the filters are checked regurarly with two laboratory spectrophotometers and regular standard lamp checks have shown that the photometer had not drifted throughout the two-year period of study. Based on the absorption twilight method, the observational stratospheric NO_2 program includes only unpolluted days with clear skies under favourable meteorological conditions so as to avoid any effects from local pollution sources to the measurements.

The stratospheric NO_2 column over Thessaloniki shows seasonal variation with maximum monthly mean values in summer of about $6x10^{15}$ molec./cm2 and minimum monthly mean values during the winter months of about $2.2x10^{15}$ molec./cm2. A comparison of these measurements of stratospheric NO_2 column at Thessaloniki with earlier measurements shows very good agreement both on the seasonal variation and on the mean values at latitude 40o N. The assymetry of the evening/morning statospheric NO_2 over Thessaloniki was found to be on the average 1.58.

1. Introduction

The importance of NO_2 in the catalytic destruction of ozone in the stratosphere was first pointed by Crutzen (1970). Since then, the seasonal, diurnal and latitudinal behavior of the column abundance of stratospheric NO_2 have been studied using various methods which are summarized in WMO (1985). Except for the satellite observations of NO_2, most of the reported ground-based measurements are sporadic and limited in both time and space coverage. Because of the importance of stratospheric NO_2 for the ozone layer, a program of regular columnar NO_2 observations was instituted at the Laboratory of Atmospheric Physics (LAP) of the University of Thessaloniki (40.5° N) since August 1985 and it continues through the present. The program includes the zenith sky absorption technique (Brewer et al., 1973, Noxon et al., 1979, Syed and Harrison, 1980, Solomon et al., 1987) with a filter photometer, and total column NO_2 measurements with differential absorption photometry in direct sun and focused moon (Amanatidis, 1988).

The observational program is supplemented with total ozone measurements with a Brewer spectrophotometer (Bais et al., 1988) and air pollutants at ground level (Zerefos et al., 1988).

In this paper we present the results of the first two years of stratospheric NO_2 observations (August 1985-August 1987). They include the diurnal and seasonal variation of stratospheric NO_2 column and a comparison with other measurements at similar latitudes.

2. Instrumentation and observations

A modified Canterbury filter photometer (Mathews et al., 1974), initialy developed for total ozone observations, has been used to measure the stratospheric NO_2 column. For this purpose the original ultraviolet interference filters and the attenuator plate were replaced by five new narrow band interference filters with peak wavelengths in the visible region (between 430 nm and 460 nm) ,where the NO_2 absorption spectrum has a significant structure (Jonston, 1974).

The selection of the filters peak wavelengths was based on their correspondance to the NO_2 absorption maxima and minima in this specrtal region. The optical characteristics of the five filters, constructed by Barr Associates Inc. in June 1985, are presented in table 1. Using five wavelengths for the NO_2 observation the effects of other atmospheric constituents, such as ozone absorption and molecular and particulate scattering are minimized as described by Ridley et al. (1984). The interference filters are extremely sensitive to humidity (Furman and Levina, 1971) and temperature variations (Title, 1974) and to the insident radiation (Title, 1974) which shift the peak wavelengths to lower wavelengths and decrese the transmission of the filters. These effects are avoided by placing the filters in a controlled constant temperature housing (about 35 °C) and by placing a cut off filter (transmiting only in the region 420-480 nm) in front of the interference filters. This filter also shields the photomultiplier from the strong visible radiation of the other parts of the spectrum. The dependance of the transmission curve of the filters from the solar zenith angle, known as the Forbes effect, is minimized because of their narrow half-bandwidth (Basher, 1977, Vanier and Wardle, 1969).

The stability of the optical characteristics of the filters is periodically tested with two laboratory spectrophotometers at the University of Thessaloniki and up until now, no significant change is observed. In addition, since October 1986, another stability test is done on a regular basis using a 20 Watt Halogen standard lamp. From the standard lamp test the "standard lamp ratio"

ISBN 0-937194-15-8

TABLE 1. OPTICAL CHARACTERISTICS OF THE NEW INTERFERENCE FILTERS

Wavelength (nm)	Half-bandwidth (nm)	Transmittance (%)
437.8	1.10	36
439.3	0.86	30
442.0	1.00	32
444.9	0.89	30
450.1	0.96	40

,which is the weighted ratio of the five measuring wavelength intensities , is calculated. As it can be seen in figure 1. the "standard lamp ratio" , is very stable through time, with mean value of 0.3482 $\pm$1%.

The method used to calculate the stratospheric NO_2 column was based on taking zenith sky measurements of the scattered solar radiation during sunrise and sunset (zenith angld between 85° and 96°) only on clear skies. At the same time, continuous measurements of ground level NO, NO_2 concentrations using a chemiluminescence analyzer, were taken into account to ensure that the days of the observations were unpolluted, since local lower tropospheric pollution can cause significant errors ($\pm$20%) in the stratospheric NO_2 estimation. Finally, if any distortion of the shape of the calculated absorption curve was observed, these data sets were rejected from our calculations.

3. Results

The mean monthly evening vertical column abundances of NO_2 from August 1985 until August 1987 are shown in figure 2. By applying a harmonic analysis to this time series, it was found that the amplitude of the annual cycle is 1.52x10^{15} molec./cm2. The annual maximum is in July with a value of 6.0x10^{15} molec./cm2, while the annual minimum occures in December with a value of 2.3x10^{15} molec./cm2. The higher summer values of stratospheric NO_2 abundance are probably observed because of the photolysis of the NO_2 species, which occures deeper in the atmosphere due to higher solar elevation. Such reservoir species, converted during winter, are possibly HNO_3, HNO_4, $ClONO_2$ and N_2O_5. Today it is believed that the major winter reservoir species is N_2O_5 (Knight et al., 1982, Solomon and Garcia, 1983) allthough earlier investigators (Noxon et al., 1979, Evans et al., 1982) sugested that NO_2 is depleted during winter due to heterogeneous conversion to HNO_3. It is also believed that HNO_4 and $ClONO_2$ have a negligible participation in wintertime NO_2 conversion because of their low concentrations (Ridley et al., 1984, Solomon and Garcia, 1983).

Having only a short period of observations, long-term trent can not be calculated. However by comparing the three available monthly mean values for August, it appears that there is a small interannual variability of stratospheric NO_2 column.

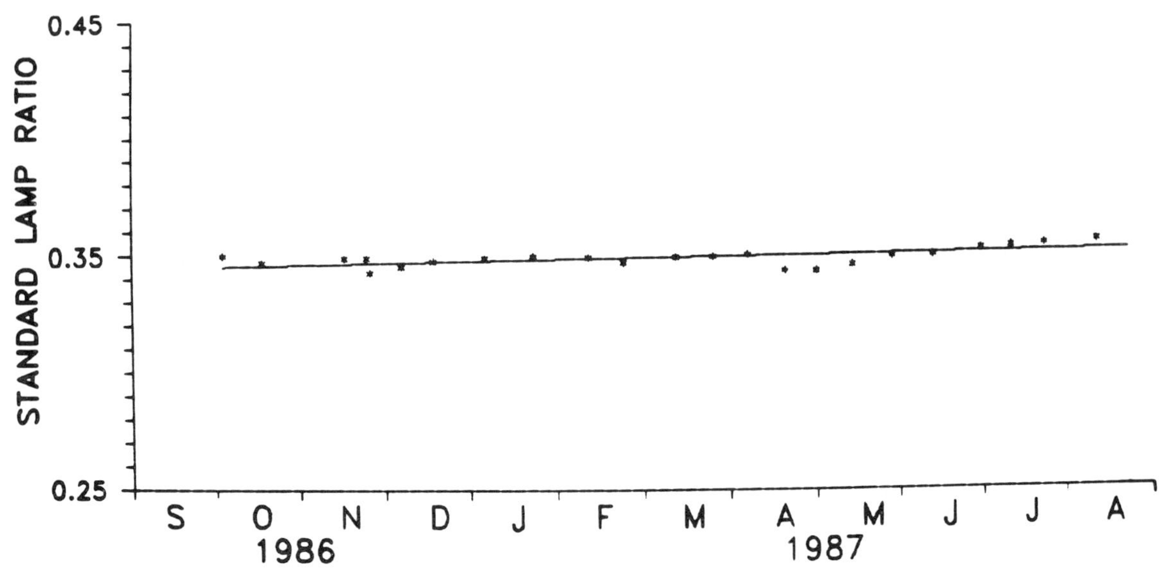

Fig. 1. Standard lamp ratio versus time for the two years of operation.

Our measurements were also compared (fig. 3.) with other made at a similar latitude to this of Thessaloniki, namely at Colorado U.S.A (40o N) during 1974-1977 (Noxon, 1979) and at North Caucasus U.S.S.R (43.7^{o}N) during 1979-1984 (Elansky et al., 1984). The agreement is very good, although there are relatively small diferen-

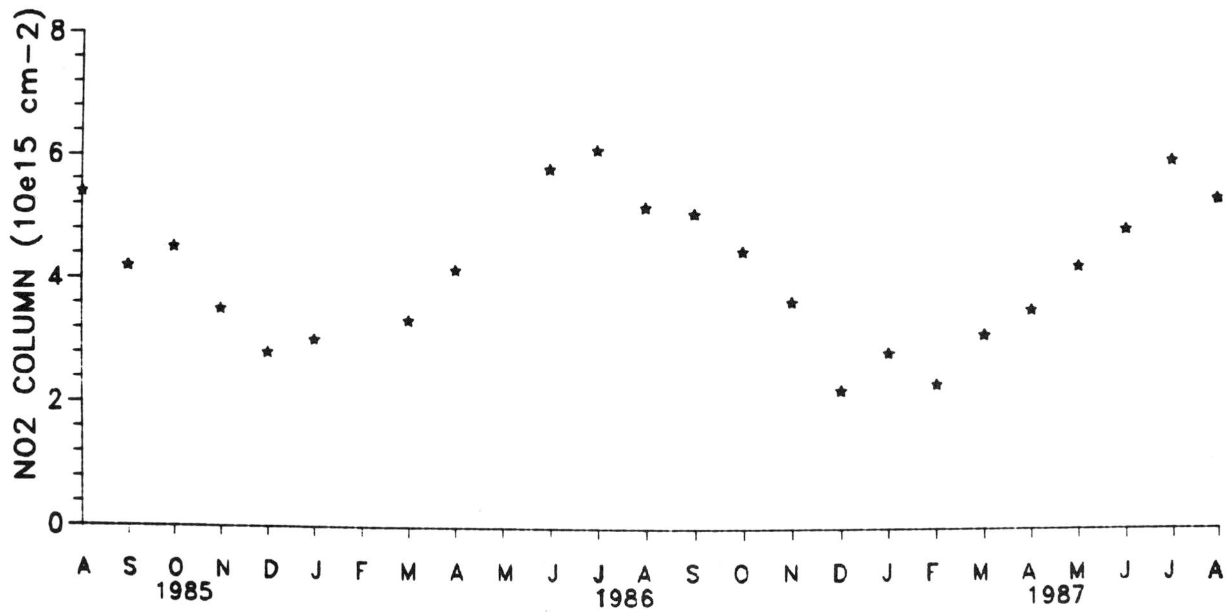

Fig. 2. Monthly means of stratospheric NO2 column at Thessaloniki-Greece.

ces having their origin probably in the zonal assymetry and the interannual variability of NO_2.

The mean of the evening/morning assymetry ratio (i.e. the ratio of the stratospheric NO_2 column measured during sunset to this measured during sunrise) for the examined time period for Thessaloniki is 1.58. This evening/morning assymetry is mainly caused by the photolysis, during daytime, of N_2O_5 and $ClONO_2$ which were formed during the night (W.M.O., 1985). This value of 1.58 when compared with calculated mean evening/morning ratio by other scientists, appears to be in very good agreement with the most of them, as it can be seen in table 2.

4. Conclusions

The analysis of the stratospheric NO_2 column observations made at Thessaloniki-Greece, during the period August 1985 until August 1987, shows a seasonal variation of the stratopspheric NO_2 column with the maximum occuring in July, the minimum in December and amplitude of the annual cycle about 1.52×10^{15} molec./cm2.

By comparing the monthly mean values for August from all available data, it appears that the interannual variability of stratospheric NO_2 column is very small.

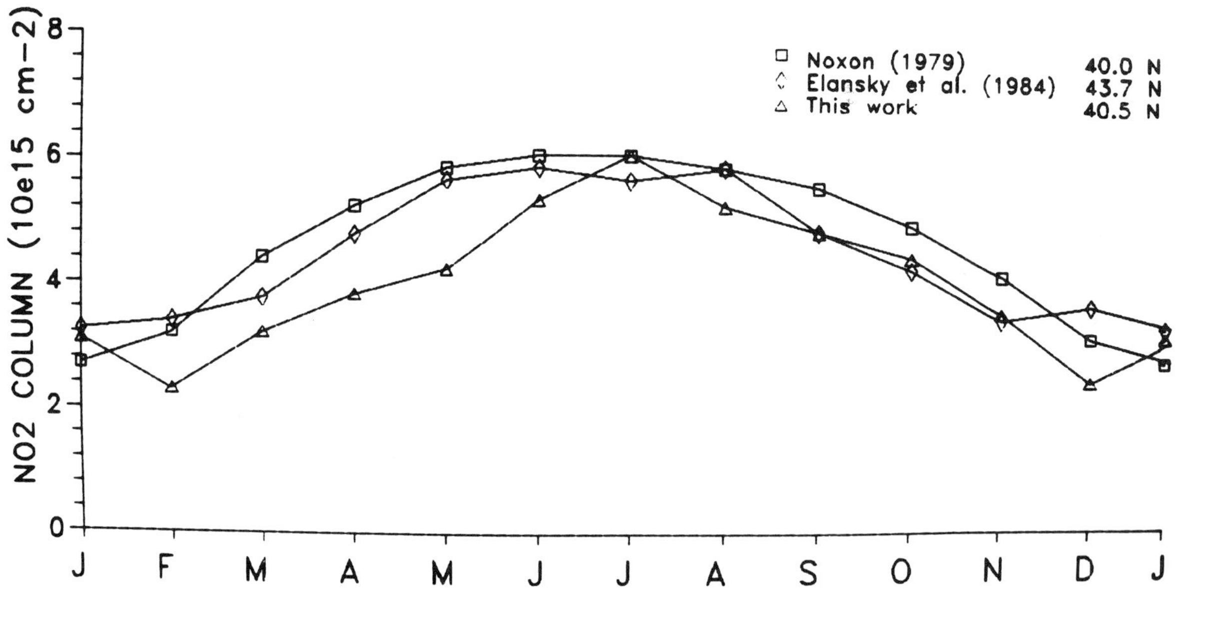

Fig. 3. Comparison of stratospheric NO2 column annual variation at Thessaloniki-Greece with other measurements at similar latitudes.

TABLE 2. EVENING/MORNING ASSYMETRY OF STRATOSPHERIC NO_2 COLUMN

Reference	Season	Latitude	$[NO_2]PM/[NO_2]AM$
Noxon et al. (1979b)	summer	40.0°N	1.70
Ridley et al.(1984)	winter	40.0°N	1.70
Elansky et al.(1985)	all seasons	43.7°N	1.33
McKenzie and Johnston (1982)	all seasons	45.0°S	1.54
Kerr (1988)	all seasons	43.7°N	1.50
This work	all seasons	40.5°N	1.58

There is very good agreement between the annual cycle of stratospheric NO_2 column as observed in Thessaloniki and other observations made at similar latitudes.

The evening/morning assymetry ratio, as calculated from the observations made at Thessaloniki, has a mean value of about 1.58, which is in very good agreement with that calculated by others.

5. References

Amanatidis G.T., 1988: Study of the variations of nitrogen dioxide in the troposphere and stratosphere over Thessaloniki. Ph.D. thesis, University of Thessaloniki.

Bais A.F., A.G. Kelessis, I.C. Ziomas, C.S. Zerefos, G.T. Amanatidis and K.D. Tourpali, 1988: Six years of total ozone observations with a Brewer spectrophotometer at Thessaloniki, Greece. This issue.

Basher R.E., 1977: The effect of badwidth on filter instrument total ozone accuracy. J. Appl. Met., 16, 803-811.

Brewer A.W., J.B. Kerr and C.T.McElroy, 1973: Notrogen dioxide concentrations in the atmosphere.Nature, 246, 129-133.

Elansky N.F., A.Ya. Arabov, A.S. Elskhov and I.A. Senik, 1984: Spatial and temporal variability of the NO_2 total content based on annual observation data.In: Atmospheric Ozone, Quadrenial Ozone Symposium, Halkidiki, Greece, Ed. C.S. Zerefos and A.Ghazi, Reidel, 157-162.

Evans W.F.J., J.B. Kerr, C.T. McElroy, R.S. O'Brien and J.C. McConnel, 1982: Measurements of NO_2 and HNO3 during a stratospheric warmat 54 degrees N in February 1973. Geoph. Res. Lett., 9, 493- 496.

Furman S.A. and M.D. Levina, 1971: Effect of moisture on the optical characteristics of narrow-band interference filters. Opt. Spect., 30,404-408.

Johnston H.S., 1985: Unpublished data, Personal communication.

Kerr J.B., 1988: Ground based measurements of nitrogen dioxide using the Brewer spectrophotometer. This issue. Knight W., D.R. Hastie and B.A. Ridley, 1982: Measurements of nitric oxide during a stratospheric warming. Geoph. Res. Lett., 9, 489-492.

Matthews W.A., R.E. Basher and G.J. Fraser, 1974: Filter ozone spectrophotometer. Pure App. Geoph., 112, 931-938.

McKenzie R.L. and P.V. Johnston, 1982: Seasonal variation in stratospheric NO_2 at 45 degrees S. Geoph. Res. Lett., 9, 1255-1258.

Noxon J.F., E.C. Whipple and R.S. Hyde, 1979: Stratospheric NO_2:1. Observational method and behavior at Mid-latitude. J. Geoph. Res., 84, 5047-5065.

Noxon J.F., 1979: Stratospheric NO_2, 2. Global behavior. J. Geoph. Res., 84, 5067.

Solomon S. and R.R. Garcia, 1983: On the distibution of nitrogen dioxide in the high- latitude stratosphere. J. Geoph. Res., 88, 5229-5289.

Solomon S., A.L. Schmeltekopf and R.W. Sanders, 1987: On the interpretation of zenith sky absorption measurements. J. Geoph. Res., 92, 8311-8319.

Syed M.Q. and A.W. Harrison, 1980: Ground based observations of stratospheric nitrogen dioxide. Canad. J. Phys., 58, 788-802.

Title A.M., 1974: Drift in interference filters, Part II, Radiation effects. Appl. Opt., 13, 2680-2684.

Title A.M., R.P. Pope and J.P. Andelin, 1974: Drift in interference filters, Part I. Appl. Opt., 13, 2675-2679.

Vanier J. and D.I. Wardle, 1969: The effects of spectral resolution on total ozone measurements. Quart. J. Roy. Met. Soc., 95, 395-399.

W.M.O., 1986: Atmospheric Ozone.

Zerefos C.S., I.C. Ziomas, A.F. Bais, G.T. Amanatidis and A.G. Kelessis, 1988: Chemical oxidants in air of Northern Greece in relation to meteorological and solar flux conditions. To be published in Chemosphere.

GROUND-BASED MEASUREMENTS OF NITROGEN DIOXIDE USING THE BREWER SPECTROPHOTOMETER

J.B. Kerr
Atmospheric Environment Service
Downsview, Ontario, Canada, M3H 5T4

ABSTRACT

Atmospheric nitrogen dioxide is measured by a Brewer spectrophotometer which has been modified to measure radiation at visible wavelengths in the NO_2 absorption bands between 430 and 460 nm. Routine ground based NO_2 measurements have been made at Toronto (43°N) since August, 1985 and at Saskatoon (52°N) since January, 1987. The fully automated instrument has been scheduled to obtain zenith sky, direct sun and focused moon measurements. The zenith sky data during sunrise and sunset are used to determine stratospheric NO_2. These measurements show that the column amount of evening stratospheric NO_2 is typically 1.5 greater than the morning values and that the summer values are nearly twice those in the winter. The instrumentation and measurement methods are described. Results of measurements showing diurnal, seasonal and latitudinal variations are presented.

1. INTRODUCTION

Nitrogen dioxide (NO_2) has strong and variable absorption bands at visible wavelengths near 440 nm (Fig. 1). Brewer et al. (1973) demonstrated that these absorption features could be used to measure atmospheric NO_2 by reporting ground based and aircraft based measurements. Ground based measurements of stratospheric NO_2 were made using the zenith sky twilight absorption technique. Since then several other workers have used the ground based visible absorption technique to measure stratospheric NO_2 (Noxon, 1975, 1979; Syed and Harrison, 1980; McKenzie and Johnston, 1982; Solomon et al., 1987). In addition visible absorption spectroscopy has been used to measure the vertical profile of stratospheric NO_2 from balloon platforms (Kerr and McElroy, 1976; Kerr et al., 1982; McElroy, 1988).

The fully automated Brewer ozone spectrophotometer was developed in the early 1980's to measure atmospheric ozone and SO_2 using ultraviolet absorption spectroscopy (Kerr et al., 1983, 1984). In 1985 the instrument was modified to measure atmospheric NO_2 using the automated mode of operation developed for ozone measurements allowing a routine and continuous programmed measurement schedule to be applied to monitoring NO_2.

2. METHOD OF MEASUREMENT

The Brewer ozone spectrophotometer measures ultraviolet light near 310 nm in the Huggins absorption band of ozone with a 1800 line/mm refraction grating used in the second order (Kerr et al., 1984). NO_2 is measured by substituting a 1200 line/mm grating so that second order visible light at about 1.5 times the wavelength (450 nm) falls on the exit slit plane. In addition the broad band nickel sulfate/UG11 order filter was replaced by a broad band blue glass filter to ensure that light from other orders is rejected. Fine tuning of wavelength selection in achieved by rotating the grating with the motor driven micrometer control used for wavelength calibration in normal operation.

Light is measured at five wavelengths simultaneously with a nominal resolution of .85 nm: 431.4 nm, 437.3 nm, 442.8 nm, 448.1 nm and 453.2 nm. The NO_2 absorption coefficients were calculated from data of Johnston (1976). Fig. 1 shows the absorption spectrum for NO_2 at 0.85 nm resolution with the operational wavelengths indicated. The wavelength resolution and NO_2

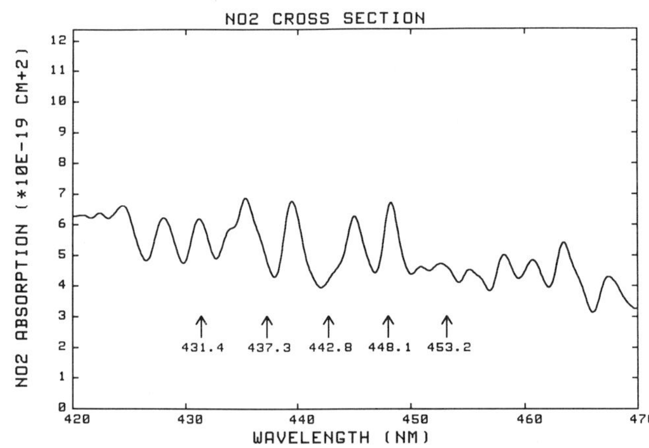

Fig. 1. Absorption spectrum of NO_2 smoothed to .85 nm with the five operational wavelengths indicated.

TABLE 1. NO_2 ABSORPTION COEFFICIENTS AND WAVELENGTH RESOLUTION AT THE FIVE OPERATIONAL WAVELENGTHS OF THE BREWER SPECTROPHOTOMETER

SLIT#	WAVELENGTH (NM)	RESOLUTION (NM)	NO_2 ABSORPTION COEFFICIENT ($*10^{-19}$ cm^2)
1	431.42	.816	6.127
2	437.34	.836	4.749
3	442.82	.868	4.226
4	448.10	.858	6.610
5	453.22	.848	4.612

ISBN 0-937194-15-8

absorption coefficients at the operational wavelengths are summarized in Table 1.

The intensity (I) of solar radiation measured at the ground after passing through the atmosphere and absorbed by NO_2 is given by:

$$I = Io*\exp(-\alpha X \mu - \beta m) \quad (1)$$

where Io is the intensity of light which would be measured outside the Earth's atmosphere, α is the absorption cross section of NO_2 (cm^2/molecule), X is the column amount of NO_2 in the atmosphere (molecule/cm^2), μ (airmass) is the ratio of the slant path through atmospheric NO_2 to the vertical path (approximately the cosecant of the solar zenith angle), β is the Rayleigh scattering coefficient, and m is the ratio of the slant path through the atmosphere to the vertical path.

Light intensities at the five wavelengths of the Brewer NO_2 instrument given by equation (1) above are linearly combined in the following expression:

$$F = Fo - \Delta\alpha X \mu - \Delta\beta m \quad (2)$$

where

F is a linear combination of log(I) terms
$= .1*\log(I_1) - .59*\log(I_2) + .11*\log(I_3) + 1.20*\log(I_4) - .82*\log(I_5)$

Fo is the linear combination of log(Io) terms
$= .1*\log(Io_1) - .59*\log(Io_2) + .11*\log(Io_3) + 1.20*\log(Io_4) - .82*\log(Io_5)$

$\Delta\alpha$ is the linear combination of the NO_2 absorption coefficients
$= .1*\alpha_1 - .59*\alpha_2 + .11*\alpha_3 + 1.20*\alpha_4 - .82*\alpha_5$

$= 2.43*10^{-19}\ cm^2$

$\Delta\beta$ is the linear combination of the Rayleigh scattering coefficients = 0

The weighting coefficient values (whose sum equals zero) are selected so that the linear combination of the Rayleigh scattering coefficients is zero and the linear combination of a function inversely proportional to wavelength is zero. The latter condition reduces effects of particulate scattering. In addition ozone absorbs slightly at these wavelengths and the weighting coefficients given above linearly combined with the ozone absorption coefficients results in zero ozone absorption.

The measured parameter F is linearly proportional to the amount of NO_2 in the optical path and is negligibly affected by the presence of other atmospheric parameters such as clouds, haze and ozone.

The value for Fo is determined by regressing F against the μ value. Equation (2) illustrates that there is a linear relation between F and μ. Fig. 2 is a plot of F versus μ for May 11, 1988. If the amount of NO_2 is constant during the day, the intercept of the regression line in Fig. 2 is Fo and the slope is equal to the amount of NO_2. The value for Fo was determined from regression data on many days when there was little NO_2 present from pollution sources such as May 11, 1988 (Fig. 2). The value for Fo is shown in Fig. 2.

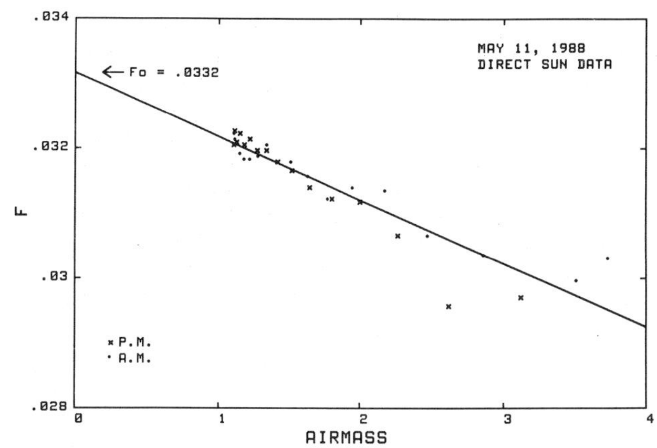

Fig. 2. Example of direct sun measurements made by the Brewer NO_2 spectrophotometer. The value F is linearly proportional to NO_2 in the path along the solar beam.

Equation (2) above may be rewritten in the following form:

$$X = \frac{(F - Fo)}{\Delta\alpha\mu} \quad (3)$$

This equation is used to determine the total column abundance of NO_2 above the station from a direct sun measurement once Fo is known.

Measurements of the value of F for light from the zenith sky during twilight may be used to determine the column abundance of stratospheric NO_2 above a height of 14km. The method by which stratospheric NO_2 is determined by the zenith sky twilight absorption has been extensively discussed by several others (e.g. Noxon, 1979; Syed and Harrison, 1980; Solomon et al., 1987). The increase (decrease) of NO_2 absorption in the zenith sky during sunset (sunrise) is affected primarily by the amount of stratospheric NO_2. Fig. 3 is a plot of Fo-F as a function of solar zenith angle for April 22, 1988 illustrating the enhanced absorption at larger zenith angles near twilight.

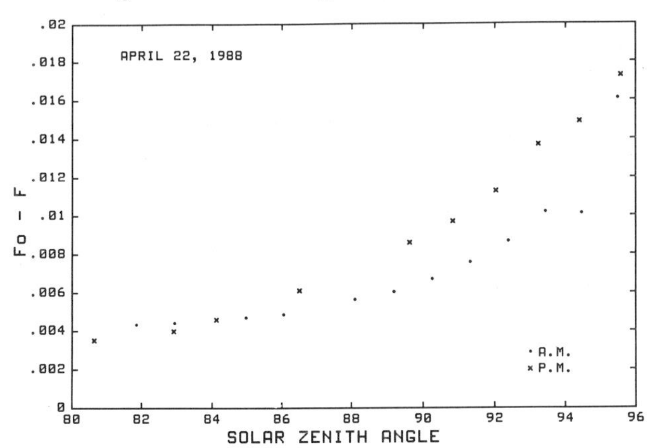

Fig. 3. Example of data showing NO_2 absorption in the zenith sky as a function of solar zenith angle near twilight. The growth of absorption is primarily due to stratospheric NO_2.

Stratospheric NO_2 column abundance is determined by comparing the measured twilight absorption curves with those computed by a multiple scattering model containing NO_2 absorption at different heights in the atmosphere. Fig. 4 shows model results for the "enhancement factor" (the ratio of the effective optical pathlength of light through an absorbing gas in an atmospheric layer to the vertical optical thickness of the gas in the layer) as functions of solar zenith angle and height of the NO_2 layer.

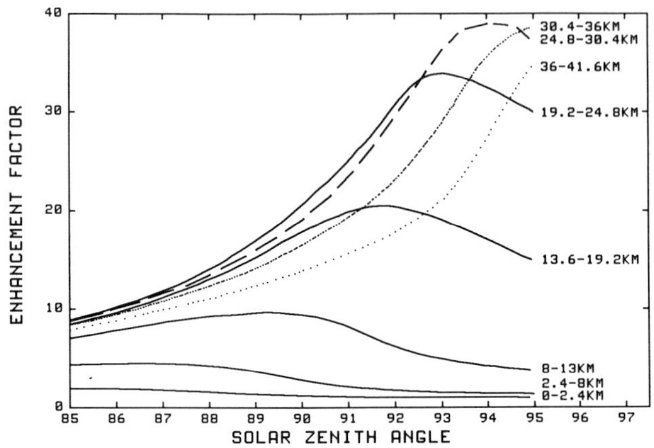

Fig. 4. Model results showing the "enhancement factor" as a function of solar zenith angle and height of the NO_2 layer.

The ground based measurements are affected by NO_2 from pollution sources, particularly at Toronto, and the effects of pollution on the measurements must be considered. In general pollution occurs in the lowest layer in the troposhere. Zenith sky measurements are made vertically upward through this layer and the absorption due to NO_2 is usually constant during the twilight period. The analysis allows for a constant uniform pollution layer so these effects are usually negligible. However, there are rare occasions when the amount of pollution changes during the measurement period and the twilight absorption curve is distorted. When this occurs the fit to the model is rather poor and the measurement is rejected.

In operation twilight absorption measurements are made every sunrise and sunset so morning and evening values for the stratospheric column abundance are obtained each day.

Intercomparison measurements were made with both Brewer NO_2 instruments in Saskatoon in July, 1987. The linear response by the two instruments to atmospheric NO_2 and the morning and evening column abundance values measured by the two instruments agreed to within 3%.

3. RESULTS

Morning and evening stratospheric column abundances are shown in Fig. 5a for Toronto and Fig. 5b for Saskatoon. The lines through the data are the AM and PM values smoothed by a triangular filter with a base of 90 days. For Toronto the scatter of the individual points about the smoothed lines is 20% for AM data and 14% for PM data. For Saskatoon the scatter is 15% for AM and 13% for PM. The larger scatter at Toronto is probably a result of the presence of more pollution at this site than at Saskatoon.

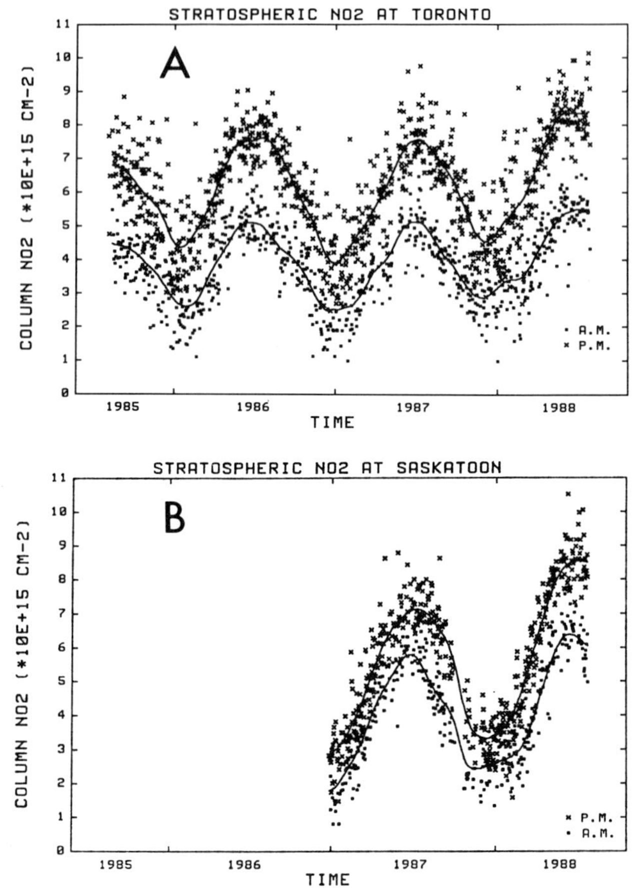

Fig. 5. Morning and evening NO_2 column abundance for Toronto (a) and Saskatoon (b).

The smoothed curves for both sites are shown in Fig. 6. The morning values are larger in the winter and smaller in the summer at Toronto than at Saskatoon. The evening minus morning differences are also shown in Fig. 6. At both sites the maximum difference is in late summer

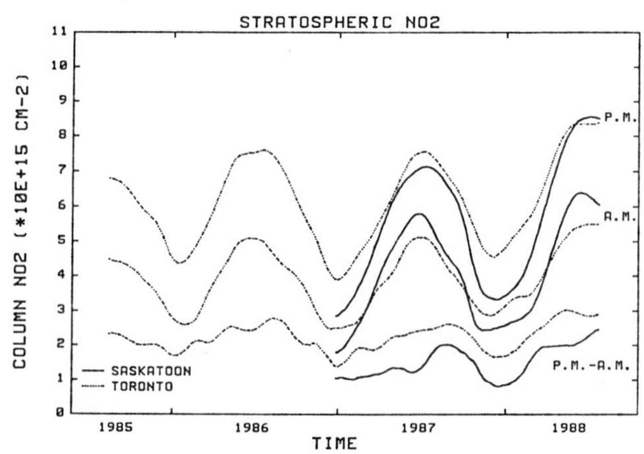

Fig. 6. The smoothed AM and PM data for Toronto and Saskatoon as well as the smoothed PM minus AM differences.

about 8 weeks after the evening and morning maxima in early summer. The evening to morning difference is smaller in Saskatoon than in Toronto for all seasons.

There is a statistically significant increase in stratospheric NO_2 column abundances in 1988 at both Toronto and Saskatoon. The summertime peak for the evening values at Toronto is about 10% larger in 1988 than in 1986 and 1987. At Saskatoon the evening summertime values are about 20% larger in 1988 than in 1987.

4. DISCUSSION

The results presented here are generally 10 to 30 per cent larger than ground based results reported by other workers for comparable situations. For example results from Noxon (1979) with the appropriate correction given by Noxon (1980) show that the evening summer maximum column abundance at 53°N is $6.2*10^{15}$ mol/cm^2, whereas, the summer maximum at Saskatoon (52°N) is about $7.2*10^{15}$ mol/cm^2. Comparison of the Toronto(43°N) data to that reported by McKenzie and Johnston (1982) for 45°S in New Zealand shows a similar difference, although direct comparison between Northern and Southern Hemispherical data may not be conclusive. The evening New Zealand column abundance is $3.2*10^{15}$ mol/cm^2 in winter and $6.4*10^{15}$ mol/cm^2 in summer and the respective values for Toronto are about $4.3*10^{15}$ and $7.5*10^{15}$ mol/cm^2. New Zealand morning winter and summer values are $2.0*10^{15}$ and $4.5*10^{15}$ mol/cm^2 and those for Toronto are $2.5*10^{15}$ and $5.0*10^{15}$ mols/cm^2.

The discrepancies are believed to be a result of differences in the methods by which the absorption data are compared to model results. Direct comparison between a prototype Brewer NO_2 instrument and the spectrometers used by Noxon (1979) at Fritz Peak Observatory in 1979 indicated that the absorption curves measured by the different instruments are consistent to within 5%. One possible cause for the difference is the fact that the results of Noxon (1979) and McKenzie and Johnston (1982) are determined by referencing the twilight absorption spectra to a standard of many noontime spectra averaged over several days. In this work the absorption at twilight is referenced to the zero (extraterrestrial) absorption value. A second possible cause is the difference in assumptions regarding the vertical distribution of NO_2. For this work NO_2 is assumed to be distributed between 14 and 42 km in proportion to a vertical profile based on balloon-based measurements (Kerr et al., 1982). Others apparently assume that the NO_2 is distributed in a thin layer centered at an altitude between 20 and 30 km. Model results shown in Fig. 4 indicate that a thin layer of NO_2 between 20 and 30 km causes more absorption than a more broadly distributed layer between 14 and 42 km.

Comparison of the Toronto and Saskatoon results to those in New Zealand (McKenzie and Johnston, 1982) indicates that there is less variability in the New Zealand data. This is probably because the stratospheric flow is more stable in the Southern Hemisphere than it is in the Northern Hemisphere.

The general seasonal and diurnal variations are consistent with those observed by others. The fact that there appears to be some systematic differences is not too serious for the purposes of long term monitoring at a particular site provided the data aquisition and analysis methods remain unchanged. These long term monitoring records will become more important with time and it is intended to continue the measurements. Questions regarding the cause and importance of the increase of NO_2 in 1988 will be more readily answered with a longer data record. Monitoring at additional sites, particularly in the Arctic, are being planned.

REFERENCES

Brewer, A.W., C.T. McElroy and J.B. Kerr, 1973: Nitrogen dioxide concentrations in the atmosphere, Nature, 246, 129-133, 1973.

Johnston, H.S. and P Graham, 1976: Unpublished absorption coefficients of NO_2 and O_3, Dept. of Chem., University of California, Berkely.

Kerr, J.B. and C.T. McElroy, 1976: Measurements of stratospheric nitrogen dioxide from the AES stratospheric balloon program, Atmosphere, 14, 166-171.

Kerr, J.B., C.T. McElroy and W.F.J. Evans, 1982: Midlatitude summertime measurements of stratospheric NO_2, Can. J. Phys., 60, 196-200.

Kerr, J.B., C.T. McElroy, and W.F.J. Evans, 1983: The automated Brewer spectrophotometer for measurement of SO_2, O_3 and aerosols, Proc. of the WMO/AMS/CMOS Symposium on Meteorological Observations and Instrumentation, 470-472, Toronto.

Kerr, J.B., C.T. McElroy, D.I. Wardle, R.A. Olafson, and W.F.J. Evans, 1984: The automated Brewer spectrophotometer, Proc. Quadrennial Ozone Symposium, pp 396-401, Halkidiki, Greece.

McElroy, C.T., 1988: Stratospheric nitrogen dioxide concentrations as determined from limb brightness measurements made on June 17, 1983, J. Geophys. Res., 93, 7075-7083.

McKenzie, R.L. and P.V. Johnston, 1982: Seasonal variations of stratospheric NO_2 at 45°S, Geophys. Res, Lett., 9, 1255-1258.

Noxon, J.F., 1975: Nitrogen dioxide in the stratosphere and troposphere measured by ground-based absorption spectroscopy, Science, 189, 547.

Noxon, J.F., 1979: Stratospheric NO_2, 2, Global behavior, J. Geophys. Res., 84, 5067-5076.

Noxon, J.F., 1980: Correction, J. Geophys. Res., 85, 4560-4561.

Solomon S., A.L. Schmeltekopf and R.W. Sanders, 1987: On the interpretation of zenith sky absorption measurements, J. Geophys. Res., 92, 8311-8319.

Syed, M.Q. and A.W. Harrison, 1982: Ground based observations of stratospheric nitrogen dioxide, Can. J. Phys., 58, 788-802.

MEASUREMENTS OF COLUMN DENSITIES OF NO_2 AND OTHER SPECIES BY VISIBLE AND UV SPECTROSCOPY AT 68°N

C. Schiller, U. Platt, U. Schmidt, and T. Brauers
Institut für Atmosphärische Chemie
Kernforschungsanlage Jülich
D-5170 Jülich, FRG

ABSTRACT

Measurements of column densities of various stratospheric species were carried out in the Arctic during two periods in winter 1987/88 using direct moon- as well as solar stray light. In the center of the polar vortex NO_2 column densities of less than 6×10^{14} cm^{-2} were observed. Ozone column densities were in the range of 300 to 480 Dobson units. For column densities of OClO and BrO during sunrise and sunset presently only upper limits of 2.5×10^{13} cm^{-2} and 3.5×10^{13} cm^{-2}, respectively, can be given.

1. INTRODUCTION

The springtime depletion of ozone in Antarctica was first reported by Farman et al. (1985). Chemical theories suggest enhanced amounts of halogen species in the stratosphere destroying ozone via catalytic cycles. Possible explanations of the spectacular events in Antarctica center on heterogeneous chemistry, which is believed to play an important role in releasing active halogen species from the reservoirs $ClONO_2$ and HCl (Solomon et al. 1986, McElroy et al. 1986, Crutzen et al. 1986). In particular NO_x should be reduced probably due to chemical transformation into HNO_3 and its incorporation into ice particles. Recent measurements within the Antarctic vortex in fact show very low NO_2 column densities (Mount et al. 1987) together with enhanced halogen amounts (DeZafra et al. 1987, Solomon et al. 1987a, Sanders et al. 1988).

There are good reasons to make measurements also in the northern hemisphere: even if presently larger anomalies in stratospheric chemistry are observed only in Antarctica, there are also indications for smaller deviations in the northern hemisphere (Heath 1988, Rodgers 1988). Moreover, trend measurements will indicate probable developments like those observed in Antarctica in the early stages. Finally the fact that there are currently no dramatic events in the Arctic should provide useful information for the understanding of polar ozone; even different conditions may be a sensitive test of present theories.

For these reasons a campaign exploring the <u>CHE</u>mistry of <u>O</u>zone in the <u>P</u>olar <u>S</u>tratosphere (CHEOPS II) took place in Kiruna/Sweden (68° N, 20° E) during winter 1987/88. In the course of this campaign ground based spectroscopic measurements were carried out to determine column densities of several species relevant to ozone chemistry. Using <u>D</u>ifferential <u>O</u>ptical <u>A</u>bsorption <u>S</u>pectroscopy (DOAS) either direct moon spectra or scattered skylight spectra were analysed for NO_2, O_3, OClO and BrO absorption features. The relevance of OClO as an indicator for ClO and the importance of BrO were pointed out by Rodriguez et al. (1986), McElroy et al. (1986), and Tung et al. (1986).

2. EXPERIMENTAL

The stratospheric vortex of the Arctic is less stable than in Antarctica. This is due to the activity of planetary waves propagating upwards from the troposphere. Meridional transport processes associated with this dynamic activity affect the chemical composition in the stratosphere <u>and</u> the troposphere. Therefore at locations in the vicinity of polluted areas like central Europe column densities derived from ground based observation may be significantly increased by tropospheric trace gas abundances, in particular for NO_2 (Pommereau et al. 1987).

Our experimental setup was designed with regard to these considerations: The instrument was operated in three different modes: First, for direct moon observations a moon tracking system consisting of two mirrors (coelostat) was mounted on the top of the observation building enabling measurements at all moon zenith angles. Second, solar straylight observations were made with the telescope directed to the zenith sky. Third, a xenon arc light source was mounted on a hill at a distance of 4.8 km for measurements of the local tropospheric amount of NO_2 in surface air by horizontal long path absorption spectroscopy (Platt et al. 1979).

For the measurements a spectrograph (Spex 0.5m) with 25 nm scanning range (Platt et al. 1979) was used at various wavelength intervals: NO_2 was analysed from spectra around 440 nm, OClO, BrO and O_3 around 350 nm and O_3 in addition at 320 nm. A relatively low resolution of 1 nm was chosen in order to improve light throughput.

3. DATA EVALUATION

The applied technique has been widely used by other experimentators (Brewer et al. 1973, Noxon 1975, Syed et al. 1980, McKenzie et al. 1982, Solomon et al. 1987). Detailed information about data reduction are given in these papers.

The fundamental equation of DOAS measurements is Lambert-Beer's law. In the form used here it relates the observed slant absorption, expressed by the optical density D, to the vertical column density C_v introducing a geometrical factor L, the 'enhancement factor':

$$D := \log(I/I') = -s * L * C_v \qquad (1)$$

I and I' are the intensities of the measured and the background spectra at a certain wavelength, s denotes the differential absorption cross section and L describes the light path through the atmosphere. For direct moon- or sun light spectra L is simply given by the secant of the corresponding zenith angle. In the case of skylight measurements the calculation of L is more difficult because of the complexity in the geometry. The main features can be described in a simplified model: The light collected on the ground observing exactly the zenith sky at high solar zenith angles has passed along an almost horizontal and long path through high altitude layers; then the light is scattered and passes through the lower air masses vertically with a short absorption path. At high solar zenith angles (85° - 95°) the majority of the received photons are scattered in a relatively thin layer centered at a certain altitude. This altitude depends on the solar zenith angle and the wavelength of the radiation. Therefore this method is most sensitive especially to absorbers located in the altitude layers where scattering is most likely to occur. The enhancement factor L in Eq. (1) now generally becomes a function not only of the zenith angle, but also of the wavelength and the absorber's vertical concentration profile. Therefore some height resolution is possible analysing skylight spectra.

The detection limits and errors of the measurements reported here are determined by the noise in the observed spectra and - in the case of skylight observations - the impossibility of complete removal of Fraunhofer structure due to the 'Ring' effect (Grainger et al. 1962). Of course the method becomes more sensitive with increasing L. For the

ISBN 0-937194-15-8

skylight measurements presented here the enhancement factors are of the order of 20. In that case the main error is determined by the insufficient known enhancement factor and more detailed calculations of L have to be done in the future. Nevertheless the presented data give a good indication for relative changes in the column densities, while the absolute values are less accurate.

4. MEASUREMENTS

Observations were made during two periods: From December 1 to 8, 1987 several series of spectra were obtained using the moon as light source; the observed vertical NO_2 column densities were below the detection limit of 1×10^{15} cm^{-2} until December 6 (fig. 1). During these days the center of the polar vortex was located over northern Scandinavia. A dislocation in the vortex eventually developed the strongest midwinter warming ever observed (Naujokat et al. 1987). With changing air masses around December 7 the NO_2 column density increased up to 4×10^{15} cm^{-2}. During this period the tropospheric measurements did not indicate any significant contribution of NO_2 at lower altitudes (less than 0.5 ppb), and therefore changes in the column densities are solely caused by enhanced stratospheric NO_2.

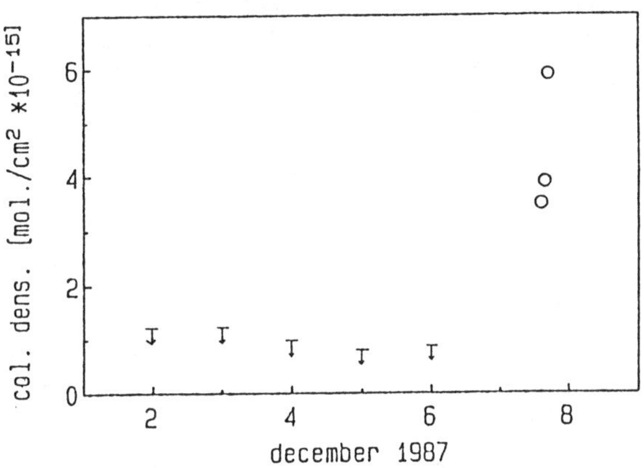

Fig. 1. NO_2 column densities evaluated from moon spectra.

During a second observational period in Kiruna due to cloudy weather only zenith sky measurements were possible. For the same reason tropospheric measurements are not available for every day of that period.

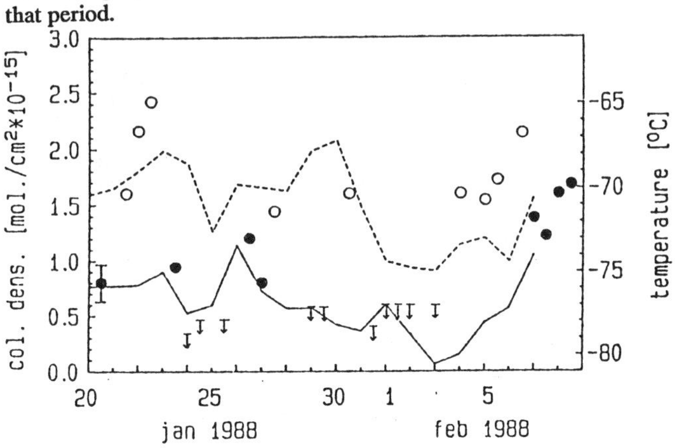

Fig. 2. NO_2 column densities derived from spectra at 440 nm observed close to 90° zenith angle. The filled circles are from measurements without significant tropospheric influence while open circles are values obtained on days with enhanced tropospheric NO_2. The error bar indicates the uncertainty of the absolute values (mostly due to the uncertainty of the enhancement factor), the relative changes of the column densities are considerably less uncertain (see text). Temperatures at 20 hpa (drawn line) and at 50 hpa (dashed line) are included in the figure.

The column densities of NO_2 obtained by zenith sky observation are shown in fig. 2: They are derived from those spectra that are observed close to 90° zenith angle. For a calculation of the absolute column densities L may be assumed to be 19 at these wavelengths for a typical profile of an absorber centered in the middle stratosphere (Syed et al. 1980).

On several days there were winds from southern directions advecting tropospheric air masses from central Europe, and NO_2 mixing ratios up to 3.5 ppb were observed employing the horizontal light path technique. These amounts of NO_2 in the troposphere enhance the optical densities by several 10^{15} cm^{-2}.

On several days no significant NO_2 absorption features in the spectra were observed. Then upper limits for NO_2 column densities of less than 6×10^{14} cm^{-2} were obtained. After the warming in December the polar vortex became stabilized thermally at values of -80° C and less. Due to dynamical impact the center of the vortex was displaced from the pole during the whole period; the temperatures as measured by radiosondes launched from Kiruna for the 20 and 50 hpa layer are also shown in fig. 2: The very low values in NO_2 column densities are well correlated with the lowest temperatures. In the total data set a correlation between temperature as a tracer of the vortex' location and NO_2 in the stratosphere is partly masked by changes in tropospheric NO_2. However, disregarding the values that are obviously influenced by tropospheric pollution, such a correlation becomes more clear.

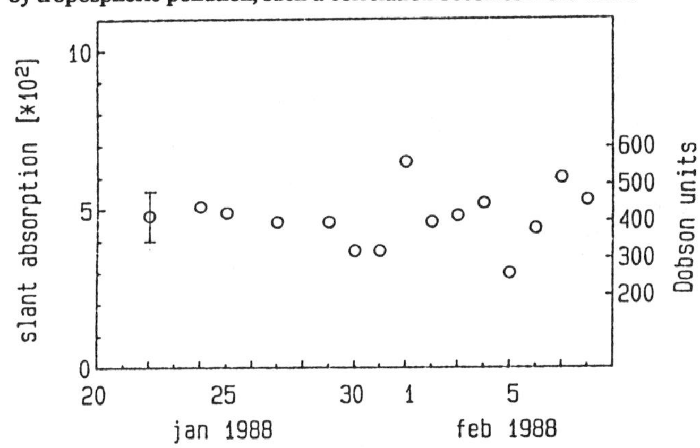

Fig. 3a. O_3 slant absorption derived from spectra around 350 nm obtained near 87° zenith angle and calculated total O_3 column densities.

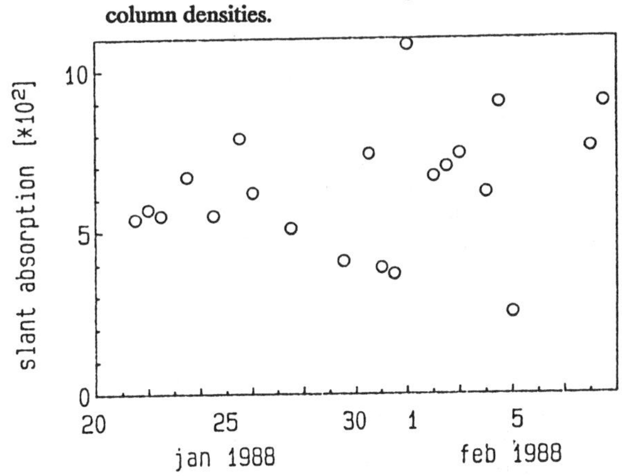

Fig. 3b. O_3 slant absorption derived from spectra around 350 nm obtained near 90° zenith angle. These data are more sensitive to changes at altitudes below the O_3 maximum (see text).

The observed O_3 slant absorption at two different zenith angles observed during this period is given in fig. 3. At a zenith angle of 87° the layer where scattering is most likely to occur (see above) is located below the ozone maximum, so that changes of the total column density are linearely related to the observed slant absorption. The variation of the O_3 columns are in the range from 300 to 480 Dobson units. In contrast to the behaviour at 87° the slant absorption at 90° shows greater amplitudes of variations. A rough estimation shows that most of scattering at this zenith angle occurs around 20 km, the lower part of the ozone layer. As mentioned earlier the observed slant absorption is therefore most sensitive to the O_3 concentration and its changes near this altitude. Comparison of the behaviour at both zenith angles indicates that most of the variation in ozone occurs below the ozone maximum. Clearly more detailed calculations of the absorption path are required since the enhancement factor in Eq. (1) does not adequately resolve the behaviour of the slant absorption and its relation to the vertical column.

Preliminary evaluation of the solar stray light spectra for OClO and BrO shows no significant absorption features. Therefore only upper limits can be given for the column densities: 2.5×10^{13} cm^{-2} for OClO and 3.5×10^{13} cm^{-2} for BrO. Probably these limits may be lowered in a more precise evaluation especially with regard to the treatment of the 'Ring' effect.

5. CONCLUSIONS

Presently only preliminary results can be given. Further evaluation, in particular more detailed calculation of the light path lengths in different atmospheric layers should provide some height resolution and reduction of the uncertainties of the absolute values of the total column densities.

The center of the northern polar vortex is characterized by very low NO_2 columns of less than 6×10^{14} cm^{-2}. These values are comparable to the lowest NO_2 column densities measured in Antarctica (Mount et al. 1987) in spite of slightly higher temperatures in the Arctic stratosphere. Earlier measurements in the northern hemisphere, for instance those made by Noxon (1979) indicate a sharp gradient of NO_2 column densities (the 'Noxon cliff') at high latitude during winter season. During warming events such cliffs were also observed at more southerly locations (McCormick et al. 1979). The measurements reported here demonstrate a similar behaviour: The December column densities are an indication of the cliff moving over northern Scandinavia during the sudden warming, and also the measurements of the second period suggest a NO_2 gradient at the edge of the vortex. The coupling of the NO_2 column densities and the vortex' location demonstrate that one must consider also dynamical processes when interpreting stratospheric NO_2.

The measured total column densities of O_3 vary significantly. The variations are of the same order as those observed in the last years by Dobson spectrometers (Larsen, 1988). The measured changes in ozone column densities may, at least in part, be caused by the dynamics of the vortex. Consequently, for ozone, possible deviations from undisturbed conditions at a smaller scale than observed in Antarctica are difficult to detect by ground based measurements from just a single location. On the other hand an estimation of the altitude where most of ozone variation occures can be given: The measurements at different zenith angles indicate that changes are strongest just below the maximum of O_3 in the stratosphere.

REFERENCES

Brewer, A.W., C.T. McElroy, and J.B. Kerr, 1973: Nitrogen dioxide concentrations in the atmosphere. Nature, 256, 129-133.

Crutzen, P.J., and F. Arnold, 1986: Nitric acid cloud formation in the cold Antarctic stratosphere: a major cause for the springtime 'ozone hole'. Nature, 324, 651-655.

DeZafra, R.L., M. Jaramillo, A. Parrish, P. Solomon, B. Connor, and J. Barret, 1987: High concentrations of chlorine monoxide at low altitudes in the Antarctic spring stratosphere: diurnal variation. Nature, 328, 408-411.

Farman, J.C., and B.G. Gardiner, D. Shanklin, 1985: Large losses of total ozone in Antarctica reveal seasonal ClO_x/NO_x interaction. Nature, 315, 207-210.

Grainger, J.F., and J. Ring, 1962: Anomalous Fraunhofer line profiles. Nature, 193, 762.

Heath, D.F., 1988: Non-seasonal changes in total column ozone from satellite observations, 1970 - 1986. Nature, 332, 219-227.

Larsen, S.H.H., 1988: Ground based ozone measurements at 3 high latitude stations. Paper presented at EGS XIII General Assembly, Bologna, 1988, available from University of Oslo, P.O.Box 1048 Blindern, 0316 Oslo, Norway.

McCormick, M.P., P. Hamill, T.J. Pepin, W.P. Chu, T.J. Swissler, and L.R. McMaster, 1979: Satellite studies of the stratospheric aerosol. Bull. Amer. Meteor. Soc., 60, 1038-1046.

McElroy, M.B., R.J. Salawitch, S.C. Wofsy, and J.A. Logan, 1986: Reductions of Antarctic ozone due to synergistic interactions of chlorine and bromine. Nature, 321, 759-762.

McKenzie, R.L., and P.V. Johnston, 1982: Seasonal variations in stratospheric NO_2 at 45°S. Geophys. Res. Lett., 9, 1255-1258.

Mount, G.H., R.W. Sanders, A.L. Schmeltekopf, and S. Solomon, 1987: Visible spectroscopy at McMurdo station, Antarctica: 1. Overview and daily variation of NO_2 and O_3, austral. spring, 1986. J. Geophys. Res., 92, 8320-8328.

Naujokat, B., R. Lenschow, K. Labitzke, K. Petzold, and R.C. Wohlfart, 1987: First note on the unusually early major stratospheric warming in December 1987. Beil. Berliner Wetterkarte, SO 16, available from Freie Universität Berlin, Dietrich-Schäfer-Weg 6-10, 1000 Berlin, FRG.

Noxon, J.F., 1975: Nitrogen dioxide in the stratosphere and troposphere measured by ground based absorption spectroscopy. Science, 189, 547-549.

Noxon, J.F., 1979: Stratospheric NO_2: 2. Global behaviour. J. Geophys. Res., 84, 5067-5076.

Platt, U., D. Perner, and W. Pätz, 1979: Simultaneous measurements of atmospheric CH_2O, O_3, and NO_2 by differential optical absorption. J. Geophys. Res., 84, 6329-6335.

Pommereau, J.P., P. Fabian, G. Flentje, M. Helten, H.W. Pätz, D.H. Ehhalt, F. Karcher, G. Froment, G. Armand, W.A. Matthews, D. Offermann, H. Rippel, P. Rigaud, J.P. Naudet, D. Huguenin, P.C. Simon, W. Peetermans, P. Vandeneede, R. Zander, and G. Roland, 1987: Intercomparison of stratospheric NO_2 and NO_3 measurements during MAP/Globus 1983. Planet. Space Sci., 35, 615-630.

Rodgers, C., 1988: Global ozone trends reassessed. Nature, 332, 201.

Rodriguez, J.M., M.K.W. Ko, and N.D. Sze, 1986: Chlorine chemistry in the Antarctic stratosphere: Impact of OClO and Cl_2O_2 and implications for observations. Geophys. Res. Lett., 13, 1292-1295.

Sanders, R.W., S. Solomon, M.A. Carroll, and A.L. Schmeltekopf, 1988: Ground based measurements of O_3, NO_2, OClO, and BrO during the 1987 Antarctic ozone depletion event. Paper presented at snowmass meeting, Aspen.

Solomon, S., R.R. Garcia, F.S. Rowland, and D.J. Wuebbles, 1986: On the depletion of Antarctic ozone. Nature, 321, 755-758.

Solomon, S., A.L. Schmeltekopf, and R.W. Sanders, 1987: On the interpretation of zenith sky absorption measurements. J. Geophys. Res., 92, 8311-8319.

Solomon, S., G.H. Mount, R.W. Sanders, and A.L. Schmeltekopf, 1987a: Visible spectroscopy at McMurdo station, Antarctica: 2. Observation of OClO. J. Geophys. Res., 92, 8329-8338.

Syed, M.Q., and A.W. Harrison, 1980: Ground based observations of stratospheric nitrogen dioxide. Can. J. Phys., 58, 788-802.

Tung, K.K., M.K.W. Ko, J.M. Rodriguez, and N.D. Sze, 1986: Are Antarctic ozone variations a manifestation of dynamics or chemistry? Nature, 333, 811-814.

MEASUREMENTS OF OZONE AND OTHER TRACE COMPONENTS NEAR THE TROPOPAUSE OVER NORTHERN EUROPE

G. Helas, G. Schuster, and S.R. Wilson
Max-Planck-Institut für Chemie
Mainz, Germany

ABSTRACT

Grab samples of air have been taken aboard an aircraft flying between Germany and Spitzbergen and have been analyzed by means of a matrix isolation technique developed in our Institute. The sampling height ranged from above cloud level (depending on local conditions between 4 and 7 km) up to 14 km with most of the samples taken above the tropopause. The tropopause height was determined during the flights by measuring the ozone mixing ratio and afterwards by interpolating radiosonde data.

A brief description of sampling and measurement technique is given. Results of the determination of ozone, NO_x, SO_2, and F-11 are discussed.

1. INTRODUCTION

Our understanding of the atmosphere in terms of transport and chemistry is based on both measurements and model calculations. A range of measurement techniques have been developed to determine atmospheric trace components, adapted to the mixing ratios prevailing in the atmosphere. Besides sensitivity, a further criterion for detection methods is the ability to identify and measure a number of trace components simultaneously. The MISST-technique (Matrix-Isolation-Sampling-for-Stratosphere-and-Troposphere) approaches this concept by detecting those compounds which can be frozen out at liquid nitrogen or argon temperatures. In the following we describe results of measurements on samples which have been collected between Germany and northern Scandinavia at heights between 4 and 14 km.

2. EXPERIMENTAL

Ozone was measured photolytically with a Dasibi 1003 or a Dasibi 1008 from air which was taken from the cabin-pressurization system. The mixing ratios thus derived were compared on several occasions with air from the inlet-system used for the grab sampling, which provides outside air directly. Below 10 km height no deviations between the two different sampling ports could be observed. At greater heights the membrane pump of the ozone monitor is, as expected, not able to maintain sufficient gas flow for reliable determination of the ozone mixing ratio at such reduced pressures.

In one flight both ozone instruments were used together for comparison. The inlet pressure of the Dasibi 1003 and cabin temperature were noted separately for later correction of the obtained data, the Dasibi 1008 has its own pressure and temperature transducers and corrects the measured values with a built-in microprocessor. The ozone measurements were conducted during the whole flight.

The air samples were collected through an inlet system of our own design, which passes through a plate bolted on the bottom of the fuselage. Outside air passes through an U shaped tube, with the flow maintained by the ram pressure during flight. The inlet system is kept closed during take-off, and is uncapped only after reaching the cloud-free troposphere.

The air samples were collected cryogenically, so that the atmospheric CO_2 and condensable gases would freeze out. Back in the laboratory the samples were transferred onto a liquid nitrogen cooled mirror, where the sampled CO_2 forms a glassy matrix with the trace components isolated within the solid. The IR spectrum was then measured with a FTIR spectrometer. Usual chromatographic analysis was performed on the IR spectra obtained, where the peak heights and/or areas were measured relative to the amount of CO_2 present. The measurement procedure was calibrated against laboratory standards using the same sample handling as for the flight samples (Griffith and Schuster, 1987).

The aircraft used was a Learjet 25D, with the single modification of a small fuselage-hole for the sampling unit. Normal flight routes were used and weather information provided by the European national Weather services was taken. The sampling flights had a staircase-like flight pattern with ascents after discrete height levels for sampling. Most of the flights were between Tromsö and Spitzbergen. The exceptions were flight LJ6, which was over southern Norway and Denmark, and LJ11, which was over Iceland and the Atlantic ocean due the the general weather situation.

3. RESULTS AND DISCUSSION

3.1 OZONE

Results of the comparison of the two ozone-monitors are shown in Fig. 1a and b. Values were taken at intervals of one minute. The automatic Dasibi 1008 was used during the whole flight, whereas the Dasibi 1003's output was manually recorded, so that some data gaps exist. The

ISBN 0-937194-15-8

difference between both instruments is given by the triangles. It can be seen that differences are within the experimental error of the instruments (±3-5 ppbv), as long as the mixing ratios of ozone vary smoothly. Abrupt changes lead to large deviations. This can be explained by the fact that the microprocessor-controlled Dasibi 1008 shows a moving average making the readout slower than that of the Dasibi 1003. The average difference in the ozone readings in fig. 1a, the flight from Frankfurt to Tromsö on January 19-20, 1987, is -0.9±12.4 ppbv, while in Fig. 1b, the flight from Tromsö via Spitzbergen to Tromsö on

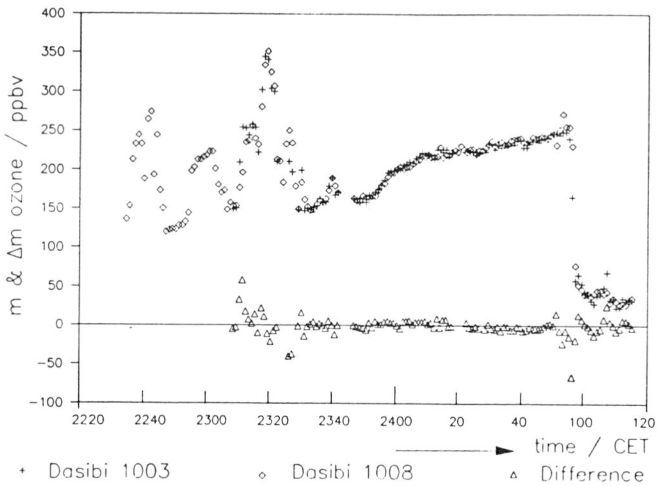

Fig. 1a
Mixing ratios of ozone measured with two different monitors and the differences during flight LJ7a from Frankfurt to Tromsö on Jan. 19-20, 1987.

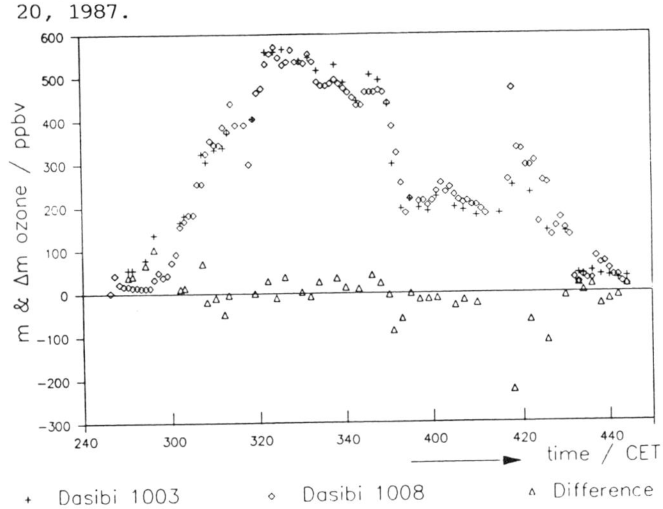

Fig. 1b
Mixing ratios of ozone measured with two different monitors and the differences during flight LJ7b from Tromsö over Spitzbergen and back to Tromsö on Jan. 20, 1987.

January 20, is -4.7±51 ppbv. The large standard deviation derived from the data of fig. 1b stems from the changing of the sampling heights several times with abrupt changes of ozone mixing ratios, while during the flight shown in fig. 1a the flight level was maintained mostly at 12.5 km.

In order to compare the obtained mixing ratios of grab-sampled trace components to ozone concentrations, the ozone mixing ratios have been determined for the individual sampling events. These data are shown in fig. 2 with associated heights relative to the tropopause. Tropopause heights were determined by interpolating radiosonde information. The mixing ratios of ozone exhibit the expected behaviour; they increase above the tropopause. The values obtained, although no direct comparable data exist, fit within the generally accepted range for this region of the atmosphere (WMO-No 16, 1985). The large variation in these data may be understood in terms of the local structure of the atmosphere which is insufficiently described by the tropopause height. Further, it can be assumed that ozone, as a relatively long living species in the lower stratosphere, indicates a time averaged equivalent height. The behaviour of the other trace gases often confirms this assessment.

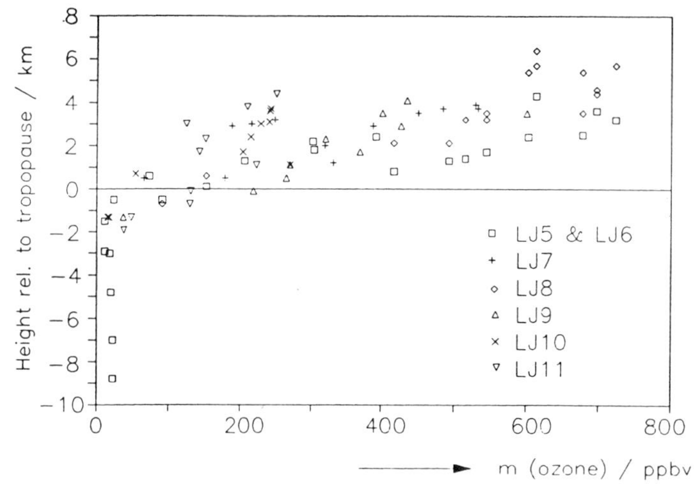

Fig. 2
Mixing ratios of ozone versus height relative to the tropopause. Results shown refer to the individual grab samples taken during the flights.

3.2 NO_x

NO_x is measured as NO_2 with our technique. Under conditions used here NO is rapidly converted to NO_2. Mixing ratios of NO_x found in our samples cover a range of 270 to 4610 pptv as is shown in Fig. 3a. Fig. 3a displays our measurement results versus the height relative to the tropopause, whereas Fig. 3b uses the absolute height in order to compare with published results.The average in our data is 1080 pptv NO_x from a total of 56 samples. There is a slight difference between the mixing ratios found in the troposphere and in the stratosphere with larger values above the tropopause. Comparing these values with results of other measurements show, in Fig. 3b, that our range of data is somewhat larger than is found with other techniques. Chemiluminescent measurement techniques are used by Dickerson (1985), Kley et al. (1981), and Kondo et al. (1987). Only the NO_y data by Dickerson are comparable to ours. The matrix isolation combined with EPR-technique in the troposphere also shows lower mixing ratios (Mihelcic et al., 1985) than we found. Spectroscopic measurements as done by Brewer et al. (data taken from Kley et al., 1981 and refs. therein) and Noxon (1983) show significantly lower mixing ratios for troposphere and stratosphere. Very recent measurements by the TDL method also point to lower values (Harris, private comm.).

Interferences with respect to NO_x in our method are conceivable from those molecules, which produce NO_2 through decomposition during

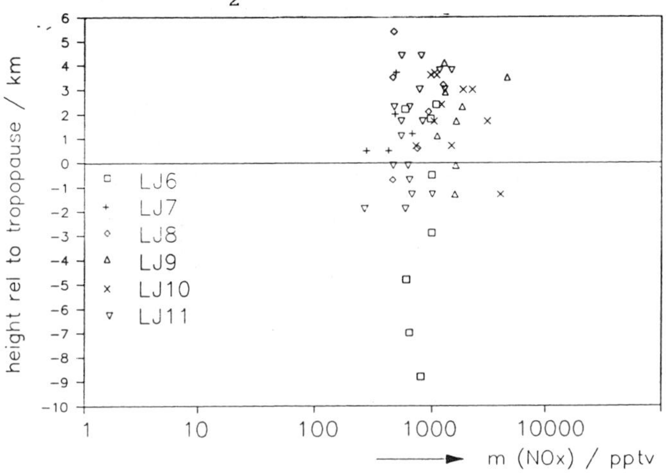

Fig. 3a
Mixing ratios of NO_x versus height relative to the tropopause. The symbols shown indicate different flights.

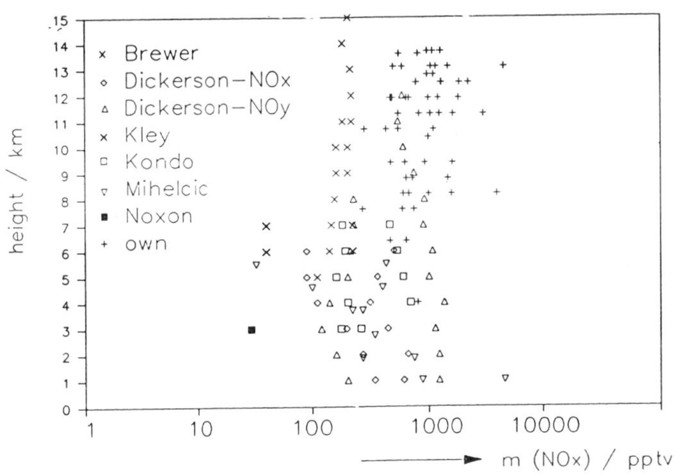

Fig. 3b
Mixing ratios of NO_x versus height. The symbols refering to other groups show minimal and maximal values measured at given heights.

the sample handling. These are predominantly PAN and Pernitric acid. At the moment it is not possible to distinguish such artifacts in our measurements, and therefore our results need to be considered as upper limits.

A slight seasonal behaviour seems to appear in our data with larger variability in mixing ratios during the summer months indicating the presence of higher mixing ratios. This is

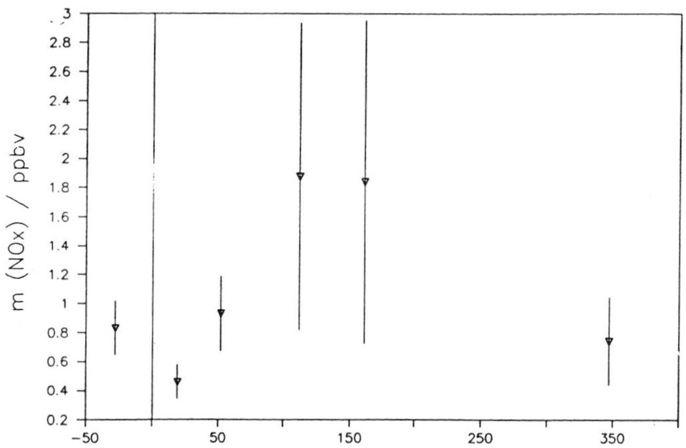

Fig. 4
Mixing ratios of NO_x height averaged for the different flights as a function of the julian date in the course of the year. The vertical lines show the standard deviations of the measurements.

depicted in Fig. 4. A similar tendency is found in the results of Brewer et al.(see Kley et al.,1981). Though the data set is rather scarce for conclusions, this behaviour would be in agreement with model assessments, in which the equilibrium of odd nitrogen compounds is assumed to be shifted to higher nitrogen oxides like N_2O_5 during the extended polar night. Though this compound can be measured with the matrix isolation technique, it may not survive the sample handling procedure.

3.3 SO_2

SO_2 mixing ratios derived by the matrix isolation technique found in the samples range between zero and 24 pptv with three excursions of 64, 177 and 265 pptv, as is shown in Fig. 5a. These three excursions were measured in samples taken in a relatively localized area. These samples were collected at different times during the flight with many measurements in between, ruling out contamination as a cause. This sug-

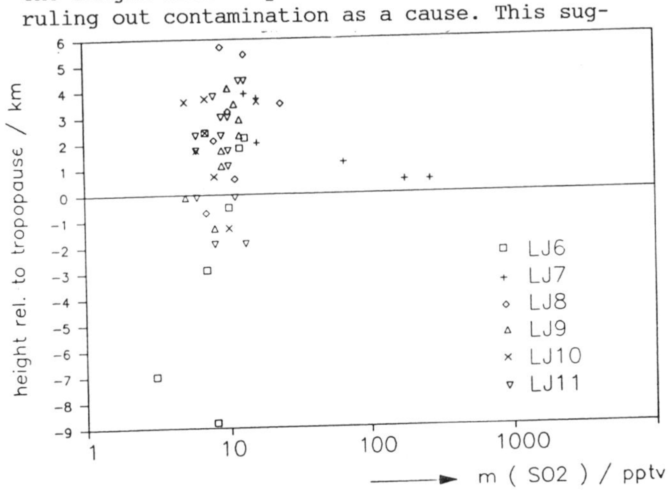

Fig. 5a
Mixing ratios of SO_2 versus height relative to the tropopause. The symbols shown indicate different flights.

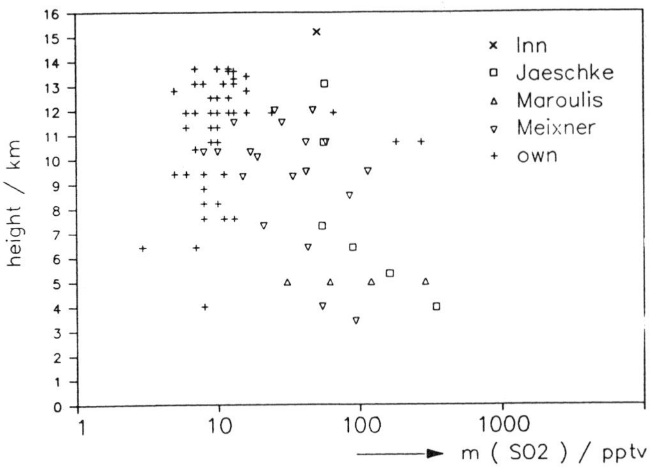

Fig. 5b
Mixing ratios of SO_2 versus height. The values shown of other investigators show individual results, beside those of Maroulis et al, see text.

gests that pollution episodes seem to survive as localized air masses even up to tropopause regions. Excluding this event results average 9 pptv SO_2 in a total of 51 samples. A vertical structure is not observed in our sampling range. Fig. 5b compares our results to previous measurements. The values given by Maroulis et al. are arbitrarily set to a height of 5 km, originally they were designated to the free troposphere. Our results fit to those reported by Meixner (1984). It seems therefore that SO_2 mixing ratios are lower than previously accepted due to measurements of Jaeschke et al. (1976) and Maroulis et al. (1980). The northern high latitudes show for SO_2, in contrast to NO_x, no significant variation during the course of the year.

3.4 F-11 ($CFCl_3$)

As example for the relatively stable halocarbons F-11 is shown in Fig. 6. As expected the mixing ratios remain rather constant in the troposphere and show only a faint decrease with

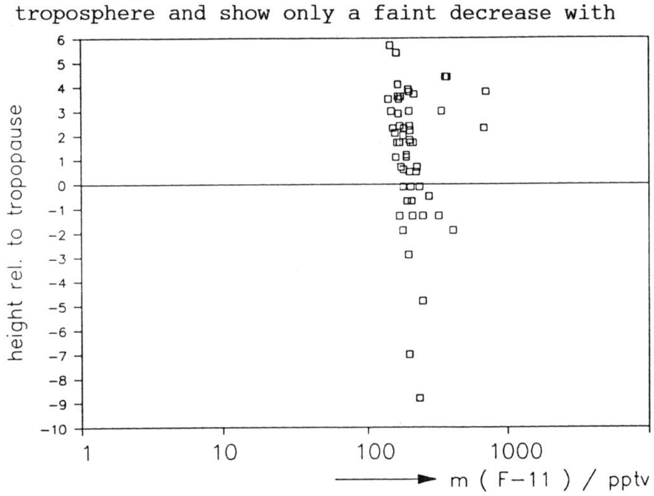

Fig. 6
Mixing ratios of F-11 ($CFCl_3$) versus height relative to the tropopause.

height in the stratosphere up to our maximum sampling height. The tropospheric average is 219 ± 39 pptv,for 16 samples. Here the standard deviation (1 σ) of the measurements is given. This is in reasonable agreement with the reported value of 205 pptv (Rasmussen and Khalil,1983) in 1982 and the observed trend in F-11 (WMO-No 16. 1985). These values may be used to derive an experimental error of 18% of our method. It shows that our method does not achieve the precision of gas-chromatographic measurements, but this technique has the benefit that it can detect a large range of trace components in one sample.

4. ACKNOWLEDGEMENT

This work was in part financially supported by the Bundesministerium für Forschung und Technologie via the grant ATP 0743188, which is gratefully acknowledged. We also thank Flight Cpts. Gärtner and Lechner of the FVG for their help during the flights and Mrs. Paetzhold of the Institut für Meteorolgie der Freien Universität Berlin for meteorological information when preparing for flights.

5. LITERATURE

Dickerson, R. R., J. Geophys. Res. 90, 10739-10743 (1985).

Griffith, D. W. T., and G. Schuster. J. Atmos. Chem. 5, 59-81 (1987).

Inn, E. C. Y., and J. F. Vedder. Geophys. Res. Lett. 8, 5-8 (1981).

Jaeschke, W., R. Schmitt, and H.-W. Georgii. Geophys. Res. Lett. 3, 517-519 (1976).

Kley, D., J. W. Drummond, M. McFarland, and S. C. Liu. J. Geophys. Res. 86, 3153-3161 (1981).

Kondo, Y., W. A. Matthews, A. Iwata, Y. Morita, and M. Takagi. J. Atmos. Chem. 5, 37-58 (1987).

Maroulis, P. J., A. L. Torres, A. B. Goldberg, and A. R. Bandy. J. Geophys. Res. 85, 7345-7349 (1980).

Meixner, F., J. Atmos. Chem. 2, 171-189 (1984).

Mihelcic, D., P. Müsgen, and D. H. Ehhalt. J. Atmos. Chem. 3, 341-361 (1985).

Noxon, J. F. J. Geophys. Res. 88, 11017-11021 (1983).

Rasmussen, R. A., and M. A. K. Khalil. Chemosphere 12, 371-375 (1983).

WMO, Report No. 16, Atmospheric Ozone, 1985.

SOURCE GASES FROM BrO_x IN THE ATMOSPHERE

O.N. Singh, R. Borchers, and P. Fabian*
Max-Planck-Institut für Aeronomie (MPAE)
D-3411 Katlenburg-Lindau, FRG

S. Lal and B.H. Subbaraya
Physical Research Laboratory (PRL)
Ahmedabad, India

ABSTRACT

Almost simultaneously in the first half of 1987, air samples over mid- and tropical latitudes, were collected between 10-35 km altitude by means of balloon borne MPAE neon-cooled cryogenic sampler. After recovery of the payload, by using GC-ECD and GC-MS combinations, various chemical species were analysed. Here, we report the 1987 atmospheric concentrations and profiles of $CBrClF_2$ and $CBrF_3$ over the aforementioned latitudes. From the comparison with the older data we conclude that the atmospheric abundances of $CBrClF_2$ and $CBrF_3$, about 2 pptv and 1.3 pptv at present are growing by $12\%y^{-1}$ and $5\%y^{-1}$, respectively.

1. INTRODUCTION

The possible impact of bromine bearing components on the earth's ozone layer has been causing concern [Watson, 1975; Wofsy et al., 1975; Young et al., 1980; Lal et al., 1985; Singh, 1988]. Catalytic decomposition of ozone by bromine (BrO_x) is more efficient than that by chlorine radicals (ClO_x), and it may even be further enhanced by synergistic effects of BrO_x and ClO_x [McElroy et al., 1985; Pyle and Farman, 1987]. One of the two significant channels of the reaction BrO + ClO leads to a catalytic cycle for ozone destruction which requires neither sunlight nor atomic oxygen, while the other, although leading to a null cycle, is the only gas phase reaction yet demonstrated to produce OClO. Only some 20 ppt by volume of bromine is required to account for the quantities of OClO measured at McMurdo during austral spring [Pyle and Farman, 1987; Tung et al., 1986]. The above mentioned reaction is being advanced as a major contributing factor to the Antarctic springtime ozone column loss of about 50% from 1960-1986 [Pyle and Farman, 1987; Farman et al., 1985; Stolarski et al., 1986; Solomon et al., 1987; Hills et al., 1987]. Only few data are available with regard to the atmospheric concentration of the bromine compounds responsible for the bromine budget in the middle atmosphere, and in several cases they are even considered preliminary [Fabian, 1986; W.M.O., 1985].

The two chief sources of atmospheric bromine are methyl bromide, half of which is manmade and half from marine origins, and halons, chemicals used as nondistructible fire extinguishers, particularly in the electronic industries. The halons $CBrF_3$ (CFC-13B1) and $CBrClF_2$ (CFC-12B1), two of the eight potential ozone depleters (PODs), are supposed to have the maximum ozone depletion efficiency, respectively 11.4 and 2.7 – a few times larger than the relative efficiency of the rest of PODs summed together [Hamitt et al., 1987; Molina et al., 1982]. Like other halons these are of anthropogenic origin and are used as fire retardants. CFC-13B1 has been rapidly adopted as a fire extinguishant in enclosed spaces such as computer rooms, and CFC-12B1 is increasingly used in portable fire extinguishers. These are mostly banked in fire extinguishing systems which have estimated service lifetimes up to 40 years. Some small amounts are emitted annually through leakage, fire and inadvertent discharge, and at disposal. Being fully halogenated hydrocarbons CFC-12B1 and CFC-13B1 have long life times estimated to range between 29-42 y [Lal et al., 1985], and 62-112 y [Fabian, 1986], respectively. Their only sink in the atmosphere is UV photolysis. Since the absorption cross-section of CF_3Br is exceedingly small beyond 290 nm ($\sim 10^{-25}cm^2$) [Molina et al., 1982], it hardly undergoes any photodecomposition in the troposphere and decomposes

*on leave from the Department of Applied Physics, Institute of Technology, Banares Hindu University, Varanasi, India

ISBN 0-937194-15-8

mainly in the stratosphere by UV radiation of 190-285 nm wave-length; $CBrClF_2$ has relatively larger absorption cross-sections beyond 290 nm ($10^{-22}cm^2$) [Molina et al., 1982] and hence its decomposition in the troposphere is not negligible. CFC-12B1 and CFC-13B1, both together contribute 10-30% of the organically bound bromine to the atmosphere [Lal et al., 1985] augmenting the effect that natural bromine has on ozone depletion.

The tropospheric abundance of CFC-13B1 was first measured by Penkett et al. [1981] in 1979 as 0.7 pptv. Its first and, according to our information, the only vertical stratospheric profile, was obtained by Fabian et al. [1981] from air samples collected at 44°N by means of a cryogenic whole air sampler. They found stratospheric $CBrF_3$ mixing ratios decreasing from 1 pptv at 10 km to 0.06 pptv at 25.9 km. First vertical profiles of CFC-12B1 were obtained by Lal et al. [1985] for the same latitude. They found $CBrClF_2$ mixing ratios decreasing from 1.5 pptV measured in 1984 at the tropopause, to about 0.03 pptV at 23 km. They also found, in agreement with surface measurements in the Antarctic [Rasmussen and Khalil, 1984] tropospheric concentrations to increase by 20%/y at that time.

2. EXPERIMENTAL TECHNIQUE

In the first half of 1987, two balloon flights were made – the first over Hyderabad, India (17.5°N), on March 26, the latter on June 23, over Gap, France (44°N). During each of these flights, by means of the MPAE neon cooled cryogenic sampler, [Borchers et al., 1987; Fabian et al., 1979] air samples of 30-50 l volume STP were collected at 15 different altitudes between 10-34 km. Along with various other chemical species of interest the halons $CBrClF_2$ and $CBrF_3$ were analysed after recovery of the payload. $CBrF_3$ was analysed by means of gas chromatography with electron capture detector (ECD) as well as by Dani 6800 GC/Balzers QMG 511 MS combination, whereas for the analysis of $CBrClF_2$ GC-EC was found to be more sensitive. A 3 m x 2 mm id glass column packed with 10% -OV 101 on chromosorb WHP 80-100 mesh was used for separation for EC detection with temperatures programmed between -60°C $\xrightarrow{10^\circ c/min}$ 70° and argon/methane at a flow rate of 30 ml/min being the carrier gas. For GC-MS the separation was achieved by a spherosil XOB 075 column operated at temperatures between -50° and 150° with helium at a flow rate of 30 ml/min as carrier gas. In both cases a cryotrap preconcentrator was used. With this procedure CFC-12B1 and CFC-13B1 could be analysed with a precision of ± 5% and ± 10% or 0.03 pptV and 0.10 pptV, whichever is greater, respectively. The absolute calibration was made by means of a two step static dilution with an accuracy of ±10%. The results are listed in Table 1. The vertical profiles of CFC-12B1 and CFC-13B1 thus obtained are plotted in Fig. 1 and 2. For CFC-13B1 (Fig. 2) we have plotted means of the results obtained by both GC-MS and GC-ECD.

TABLE 1

Volume mixing ratios (VMR) in parts per trillion (10^{-12})

of CFCs-13B1 and -12B1 over 17 and 44°N latitude.

CFC-13B1								CFC-12B1			
Hyderabad (17°N) (26.3.1987)				Gap (44°) (23.6.1987)				Hyderabad (26.3.1987)		Gap (23.6.1987)	
Altitude km	VMR-pptV GC-MS	GC-ECD	mean	Altitude km	VMR-pptV GC-MS	GC-ECD	mean	Altitude km	VMR pptV	Altitude km	VMR pptV
11.44	1.20	1.27	1.24	10.11	1.36	1.20	1.28	11.44	1.96	10.11	1.96
13.91	1.40	1.29	1.35	14.2	1.21	1.06	1.14	13.91	2.04	14.20	1.76
15.57	1.26	1.18	1.22	17.21	0.80	0.59	0.70	15.57	1.93	17.21	0.84
17.29	1.35	1.05	1.20	18.95	0.41	0.48	0.45	17.29	1.87	18.95	0.64
19.37	0.99	1.02	1.01	21.2	0.41	0.25	0.33	19.37	1.44	21.20	0.06
21.22	0.73	0.75	0.74	21.6	0.27	0.25	0.26	21.22	0.73	21.60	0.06
23.42	0.46	0.39	0.43	22.6		0.19	0.19	23.42	0.110		
23.87	0.50	0.33	0.42	23.75	0.20	0.17	0.19	23.87	0.08		
24.88	0.31	0.28	0.30	24.85		0.19	0.19				
25.55	0.27	0.23	0.25	25.85		0.13	0.13				
26.61	0.36	0.22	0.29	27.1		0.05	0.05				
29.0	0.15	0.08	0.12								

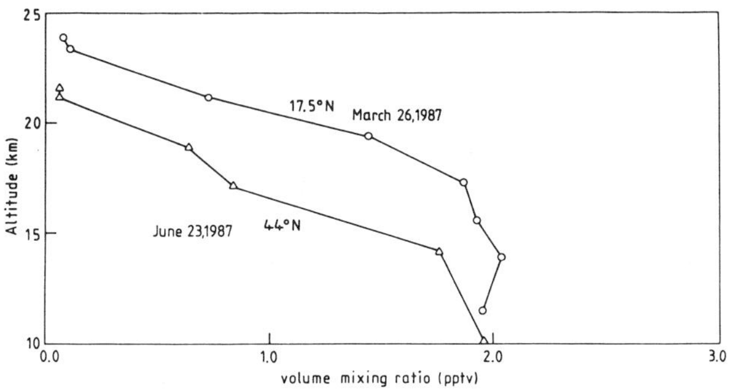

Fig. 1. Vertical profiles of $CBrClF_2$ measured 1987 at tropical and mid-northern latitudes.

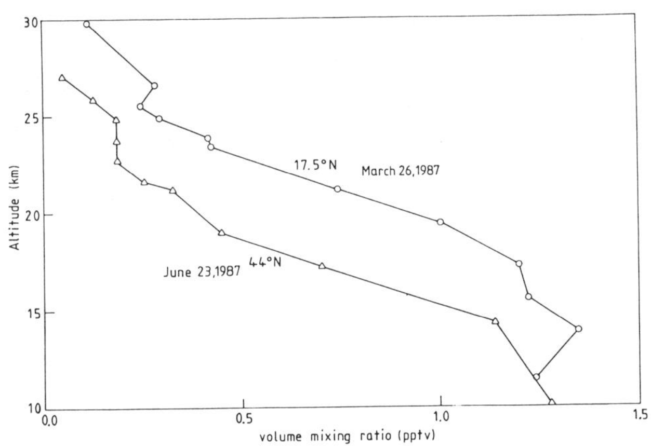

Fig. 2. Vertical profiles of $CBrF_3$ measured 1987 at tropical and mid-northern latitudes.

3. DISCUSSION

By comparing our measurements of the tropospheric concentrations of CFC-12B1 and CFC-13B1 with the values reported earlier, we obtain the annual growth rates of their tropospheric abundances. We are, however, reluctant to claim these to be very accurate, primarily because only a couple of measurements are available in either case and, further, the concentration of these halons is rather small with regard to the measuring capability of the analytical equipments employed. Even a relatively smaller uncertainity in any of the above-mentioned measurements may lead to larger errors in the growth rate estimation. When we compare our tropospheric volume mixing ratios of 1.96 pptv of F-12B1 with the three values of Lal et al. [1985] of Septembers - 1982 & 1983 and October 1984, reported earlier, we get the annual growth rates, respectively to be ~20%, ~13% and ~12%. In our opinion the last value seems to be more realistic. In a similar fashion from the measurements of Penkett et al. [1981] of 1979, Fabian et al. [1981] of 1980 and our present tropospheric concentration of $CBrF_3$ of 1.3 pptv, its annual growth rate is estimated to be little over 4%. This is rather modest compared to the growth rate of $CBrClF_2$. Since $CBrF_3$ has the longest lifetime among all bromine bearing species responsible for the bromine budget of the atmosphere, its gradual build up even at smaller emission rates is more probable and lasting than others. Further, as indicated earlier, CFC-13B1 has the maximum ozone depletion efficiency – over four times larger than even that of F-12B1 which incidently is supposed to have the second largest ozone depletion efficiency among all the CFC's [Hamitt et al., 1987].

No information on worldwide release rates of halons into the atmosphere is available. There even is no information, to our knowledge, on global production figures. If we take 4.5% as the annual increase of CFC-13B1 in the atmosphere, this would correspond to a global release reate of 1.2 kt/y. The global release rate of CFC-12B1 had been determined by Lal et al. [1985], based on 20%/y increase of its abundance in the atmosphere, to amount to 3.6 kt/y. If we accept the more likely growth rate of 12%/y, present global emission rates of CFC-12B1 are likely to range about 2.2 kt/y.

In addition to the mid-latitude vertical profiles of CFC-12B1 and CFC-13B1, we have also measured their vertical distributions over the tropics (17.5°N). These in all probability are the first stratospheric profiles of these two important halons that have been determined at tropical latitudes. As obvious from Fig. 2, for CFC-13B1 the volume mixing ratios at lower altitudes are rather similar over the two aforesaid latitudes, but at higher altitudes its concentration in the tropical air is larger and hence could be measured up to relatively greater heights. In case of F-12B1, although, its tropospheric concentration over the tropics and at mid-latitudes is exactly the same (Fig. 1), its two vertical stratospheric profiles are very different. In tropical air the concentration of $CBrClF_2$ decreases gradually with increasing height, but at midlatitudes there is an abrupt decrease of over 50% at about 15 km altitude (Fig. 1). The above mentioned differences observed between the midlatitude and tropical profiles are mainly due to tropical upwelling. These two halons along with all other species originating from tropospheric sources are carried up to higher altitudes over the tropics through tropical upwelling before they finally decompose and go beyond the detection limit. In most of the profiles we observe a step or wave like structure. This feature is not new and is being observed in the vertical profiles of almost all the trace gases [Fabian et al., 1985].

Although the tropospheric concentration of CFC-13B1, as reported earlier, is considerably less than that of CFC-12B1, $CBrF_3$ could be measured up to higher altitudes both over middle and lower latitudes. This is in confirmity with the absorption cross-sections of these two halons. $CBrClF_2$ has larger absorption cross-section by three orders of magnitude compared to $CBrF_3$, for UV radiation beyond 290 nm wavelength, which is mostly responsible for the photodecomposition of these substances.

ACKNOWLEDGEMENTS

This work was sponsored in part by the German Federal Ministry of Research and Technology, through grant KBF-MT 04500-0704563 2. The balloon experiment carried out at Hyderabad was performed under the auspices of ISRO/DFVLR Indo-German collaboration. We gratefully acknowledge expert suppport of many Indian scientists, engineers and officials, in particular by R.R. Daniel, S.V. Damle, and R.U. Kundapurkar of the Tata Institute of Fundamental Research, Bombay, R.T. Redkar of the TIFR Balloon Range at Hyderabad, and P. Rajaratnam of the Indian Space Research Organisation (ISRO), Bangalore.

REFERENCES

Borchers, R. et al. Planet. Space Sci. **35**, 657-663 (1987).

Fabian, P. et al. J. Geophys. Res. **84**, 3149-3154 (1979).

Fabian, P. et al. Nature **294**, 733-735 (1981).

Fabian, P. et al. Geophys. Res. Lett. **12**, 1-3 (1985).

Fabian, P. Handbook of Environmental Chem. Vol. 4(I), 23-51 edited by O.Hutzinger, Springer-Verlag, NY (1986).

Farman, J.C. et al. Nature **315**, 207-210 (1985).

Hamitt, J.K. et al. Nature **330**, 711-716 (1987).

Hills, A.J. et al. Nature **328**, 405-408 (1987).

Lal, S. et al. Nature **316**, 135-136 (1985).

McElroy, et al. Nature **321**, 759-762 (1985).

Molina, M.J.et al. J. Phys. Chem. **86**, 2672-2676 (1982).

Penkett, S.A. et al. J. Geophys. Res. **86**, 5172-5178 (1981).

Pyle, J.A. and J.C. Farman, Nature **329**, 103-104 (1987).

Rasmussen, R.A. & Khalil, M.A.K.Geophys. Res. Lett. **11**, 433-436 (1984).

Singh, O.N.: Naturwissenschaften 75, 191-193 (1988)

Solomon, S. et al. J. Geophys. Res. **92**, 8329-8338 (1987).

Stolarski, R.S. et al. Nature **322**, 808-811 (1986).

Tung, K.K. et al. Nature **322**, 811-813 (1986).

Watson, R.T. The Natural Stratosphere of 1974 (CIAP Monogr. 1, Sect. 5.7.5, Department of Transportation, Washington DC, 1975).

W.M.O., Atmospheric Ozone - 1985, Global Ozone Research and Monitoring Project, Report No. 16, Vol. II, 636-37 + 643 (1985).

Wofsy, S.C. et al. Geophys Res. Lett. **2**, 215-218 (1975).

Young, Y.L. et al. J. Atmos. Sci. **37**, 339-353 (1980).

TRACE GAS MEASUREMENTS WITH RASMUS IN THE MESOSPHERE

H. Duscha, R. Borchers, and P. Fabian
Max-Planck-Institut für Aeronomie (MPAE)
D–3411 Katlenburg-Lindau, FRG

W. Bischof
University of Stockholm, Sweden

ABSTRACT

RASMUS (Rocket-borne Air Sampler for the Mesosphere and Upper Stratosphere) is a new LHe-cooled cryogenic whole air sampler that has been developed at the MPAE Lindau. It operates at subsonic velocities giving a height resolution of less than 3 km. Here we present the results of our first flight made on May 6th, 1987 from ESRANGE/Sweden. The halocarbons CF_4, $CHClF_2$, $CClF_2$-CF_3, and CF_3-CF_3 have been measured for the first time at a height of about 60 km. Further new mesospheric data of N_2O, CH_4, H_2, CO, and CO_2 are obtained. The results are discussed in comparison with other available experimental data and model predictions.

1. INTRODUCTION

Over the last ten years the mixing ratios of stable trace gases in the atmosphere have been measured at the MPAE Lindau and KFA Jülich from balloon-borne cryogenic whole air samplers obtained from up to 35 km. The mean mixing ratio profiles of a few of these trace gases show only a slight decrease with height in the stratosphere due to their resistance to stratospheric depletion processes. Others are seen to increase in the upper stratosphere due to photochemical sources in the upper mesosphere and lower thermosphere.

To extend the mixing ratio profiles of these interesting substances above 35 km, we have developed a new LHe-cooled cryogenic whole air sampler (RASMUS) at the MPAE Lindau. It operates at subsonic velocities (during a parachute descent) giving a height resolution of less than 3 km. Here we present the results of our first flight from Kiruna/Sweden (68°N) in May 1987.

2. EXPERIMENTAL

The RASMUS-Trajectory (shown in Fig. 1) has an apogee- height of about 65 km. Four samples of about 4 litre STP air can be collected at preprogrammed heights between 65 and 40 km. One can be collected at apogee and three during the descent phase of the flight.

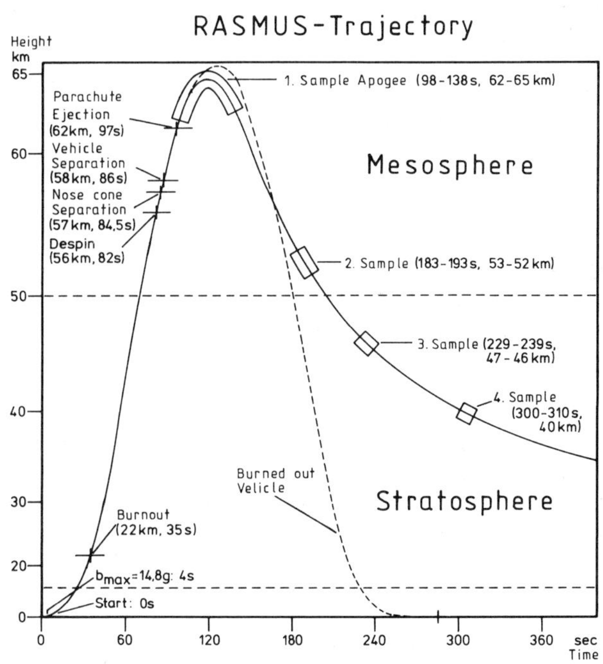

Fig. 1. RASMUS Trajectory: Parachute-breaked payload with sampling intervals 1 to 4 (solid line) and burned-out vehicle (dashed line).

The sampling units consist of gold-plated cold fingers, suspended in electropolished stainless steel containers with liftable caps. They are maintained at the temperature of LHe with a newly-developed umbilical cryocoupling until launch. Their high thermal capacity allows air sampling with temperatures always less than 40 K. After recovery, the trace gas mixing ratios are obtained using the same GC and GC-MS techniques as for the balloon-samples (Fabian et al., 1979; Fabian et al., 1981a). CO_2

ISBN 0-937194-15-8

is analyzed using an IR-Absorption technique (Bischof et al., 1985).

The experiment is not limited to the aforementioned apogee height. We have performed sampling tests on fully exposed cold fingers for the conditions up to 80 km at the MPAE (Duscha, 1988) and plan to extend our sampling height in the future.

3. RESULTS AND DISCUSSION

One sample from the height about 60 km has been successfully analysed. The results are presented in Table 1. Nine substances could be detected, including the halocarbons CF_4 (CFC-14), $CHClF_2$ (CFC-22), $CClF_2$-CF_3 (CFC-115), and CF_3-CF_3 (CFC-116) which have been measured for the first time in the mesosphere. The errors listed in Table 1 include the absolute calibration performed with a static dilution method (Fabian et al., 1981b). For most of the substances the errors are ± 10%. CFC-115 and 116 were close to the detection limit giving a higher error of ± 50%. The accuracy of the CO_2 mixing ratio is better than ± 0.1% (Bischof et al., 1985).

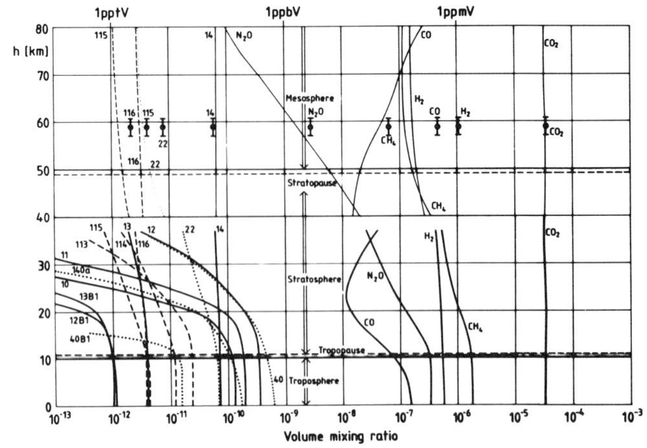

Fig. 2. RASUMUS Results (single dots) compared with measured mean tropospheric and stratospheric trace gas profiles (thick lines) and extension into the mesosphere using 1D-model calculations (thin lines above 40 km height) at northern midlatitudes. Halocarbons are indicated by their trade-name numbers.

TABLE 1

RASMUS Results 1987: Measured trace gas volume mixing ratios at 68°N on May 6th at 5.08 UT for 57.2 to 60.6 km height:

CO_2	342.2 ppm ± 0.1%		
H_2	1046 ppb ± 10%	CF_4	54.6 ppt ± 10%
CO	425 ppb ± 10%	$CHClF_2$	7.74 ppt ± 10%
CH_4	56 ppb ± 10%	$CClF_2$-CF_3	4.13 ppt ± 50%
N_2O	2.8 ppb ± 10%	CF_3-CF_3	2.21 ppt ± 50%

In Fig. 2 the RASMUS results are compared to balloon- measured and modeled trace gas profiles. Below 35 km, these data are mean mixing ratio profiles from data obtained between 1977 and 1984 in northern midlatitude autumn conditions (Fabian, 1987; Fabian et al., 1981a;b). Above 40 km, a few 1D-model results are shown for comparison (Krüger and Fabian, 1986).

We have analyzed the RASMUS sample for those stable halocarbons which either show a slow falloff in the stratosphere such as CFC-14, CFC-22, CFC-13, CFC-15, and CFC-116 or which have significant mixing ratios up to the balloon top-height such as CFC-12 and CFC-40. Four of these could be detected.

The most stable halocarbon CFC-14 shows no significant decrease in the mixing ratio up to 60 km, in agreement with the stratospheric and modelled mesospheric profiles. For the stable CFC-116 we have in the meantime derived new stratospheric profiles (Fabian et al., 1987) with lower mixing ratios as given in Fig. 2 so that the agreement is poorer than suggested here. Our RASMUS result shows that CFC-116 has already reached the mesosphere. The same holds for CFC-115. The high mixing ratio of CFC-22 is important. In conjunction with our upper stratospheric balloon-values and in agreement with the modeled profile, it represents the biggest Cl-source in the upper stratosphere and lower mesosphere. A detailed discussion of CFC-22 is given by Fabian et al. (1988).

The CO_2 mixing ratio of the RASMUS-sample agrees well with the actual ground level mixing ratios and is close to the stratospheric and modeled mesospheric profile. This was expected from the results of the NCAR sampler (Ehhalt et al., 1975) and the University of Pittsburgh sampler (Zipf and Erdman, 1984).

The RASMUS CO mixing ratio agrees well with the increase in the upper stratospheric balloon-profile. This increase due to CH_4 oxidation in the middle atmosphere and CO_2 photolysis in the upper mesosphere and lower thermosphere has also been seen by other investigators such as the ground-based microwave data summarized by Bevilacqua et al. (1985), the spectrometric observations from Spacelab (Lippens et al., 1984), and the zonal mean CO-profile derived from SAMS data (Barnett et al., 1985). Our modeled profile does not contain CO_2 photolysis as a CO-source and therefore is too low.

The RASMUS results for the two long-lived species CH_4 and N_2O confirm the decrease of the stratospheric balloon and modeled mesospheric profiles. This decrease also agrees with the other in-situ rocket results (Ehhalt et al., 1975; Zipf and Erdman, 1984) and the spectrometric observations from onboard Spacelab (Muller et al., 1985). Further, the RASMUS CH_4 mixing ratio is in excellent agreement with the mean reference model derived from SAMS data for May at 70°N (Taylor et al., 1987). However the N_2O mixing ratio measured at 60 km is 65% higher than their model profile for July which predicts the highest values.

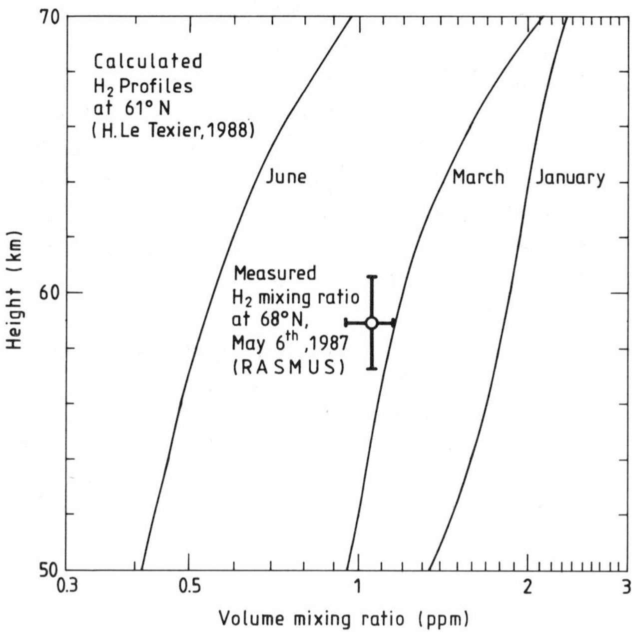

Fig. 3. H_2 volume mixing ratios at high northern latitudes: RASMUS result (dot with height-resolution and error bars) compared with 2D-model calculations by Le Texier et al. (1988) (solid lines).

A surprise has been the high H_2 mixing ratio in the RASMUS sample. From the NCAR sampler results (Ehhalt et al., 1975) and the balloon-profile up to 35 km, we expected a H_2 mixing ratio close to 0.5 ppmv. An explanation for such high H_2 values is given by the 2D-model results of the stratospheric water-vapour budget (Le Texier et al., 1988). In Fig. 3 the RASMUS H_2 mixing ratio at 68°N is compared with calculated H_2 profiles at 61°N deduced from that model (Le Texier, 1988).They propose high H_2 mixing ratios, especially in the high latitude winter mesosphere, due to H_2O photolysis in the upper mesosphere and downward transport by the mean meridional circulation. Between January and June, the mixing ratios decrease due to the slow oxidation of the intruded H_2 and the upward directed mean meridional circulation in summer. The RASMUS May value fits nicely between their March and June profiles.

CONCLUSIONS

The results show that the rocket-borne air sampler RASMUS is suitable for the simultaneous measurement of 9 stable trace gases in the mesosphere. The halocarbons CFC-22, CFC-14, CFC-115, and CFC-116 have been measured for the first time in the mesosphere. CFC-14 and CFC-22 are in good agreement with model profiles and consistent with the balloon data. The new mesospheric data of CH_4, N_2O, CO, and CO_2 are in reasonable agreement with previous measurements and consistent with the balloon data. A high mixing ratio for H_2 in the lower mesosphere has been measured for the first time and this value is in good agreement with a model prediction.

ACKNOWLEDGEMENTS

Project RASMUS has been sponsored in part by the German Federal Ministry of Research and Technology through grant 010E87016. The authors gratefully acknowledge the expert technical support of W. Engelhardt, G. Flentje, H. Hartwig, K.H. Saeger and G. Spitz of MPAE and the DFVLR rocket launching team headed by W. Gabriel and P. Turner.

REFERENCES

Barnett, J.J., Corney, A.K., Murphy R.L., Jones, C.D., Rodgers, R., Taylor W., Williams, E.J. and Vyas, N.M.: Global and seasonal varibability of the temperature and composition of the middle atmosphere. Nature **313**, 439-443, 1985

Bevilacqua, R.M., A.A. Stark, and P.R. Schwartz: The Variability of Carbon Monoxide in the Terrestrial Mesosphere as Determined From Ground-Based Observations of the J = 1 → O Emission Line. J. Geophys. Res. **90**, 5777-5782, 1985

Bischof, W., R. Borchers, P. Fabian, and B. Krüger: Increased concentration and vertical distribution of carbon dioxide in the stratosphere. Nature **316**, 708-710, 1985

Duscha, H.: Quellgasmessungen mit Hilfe der Kryotechnik in der oberen Stratosphäre und unteren Mesosphäre. Thesis Ph.D., Köln, 1988, in press

Ehhalt, D.H., L.E. Heidt, R.H. Lueb and E.A. Martell: Concentrations of CH_4, CO, CO_2, H_2, H_2O and N_2O in the Upper Stratosphere. J. Atmos. Sci. **32**, 163-169, 1975

Fabian P.: Proposed Reference Models for CO_2 and Halogenated Hydrocarbons. Adv. Space Res. **7**, (9)63-(9)72, 1987

Fabian P., R. Borchers, K.H. Weiler, U. Schmidt, A. Volz, d.H. Ehhalt, W. Seiler, and F. Müller: Simultaneously Measured Vertical Profiles of H_2, CH_4, CO, N_2O, $CFCl_3$, and CF_2Cl_2 in the Mid-Latitude Stratosphere and Troposphere. J. Geophys. Res. **84**, 3149-3154, 1979

Fabian P., R. Borchers, S.A. Penkett, and N.J.D. Prosser: Halocarbons in the stratosphere. Nature **294**, 733-735, 1981a

Fabian P., R. Borchers, G. Flentje, and W.A. Matthews: The Vertical Distribution of Stable Trace Gases at Mid-latitude. J. Geophys. Res. **86**, 5179-5184, 1981b

Fabian P., R. Borchers, B. C. Krüger, and S. Lal: CF_4 and C_2F_6 in the Atmosphere. J. Geopys. Res. **92**, 9831-9835, 1987

Fabian P., R. Borchers, H. Duscha, B.C. Krüger, S. Lal, and B.H. Subbaraya: $CHClF_2$ (CFC-22): Distribu tion, budget and environmental impact. Quadrennial Ozone Symposium, Göttingen, 1988

Krüger B.C., and P. Fabian: Model Calculations About the Reduction of Atmospheric Ozone by Different Halogenated Hydrocarbons. Ber. Bunsenges. Phys. Chem. **90**, 1062-1066, 1986

Le Texier H.: Private Communication to P. Fabian, 1988

Le Texier H., S. Solomon, and R.R. Garcia: The role of molecular hydrogen and methane oxidation in the water vapour budget of the stratosphere. Quart. J. Royal Meteorol. Soc. **114**, 281-295, 1988

Lippens C., C. Muller, J. Vercheval, M. Ackerman, J. Laurent, M.P. Lemaitre, J. Besson, and A. Girard, Trace constituents measurements deduced from spectrometric observations on-board spacelab. Adv. Space Res. **4**, (6)75-(6)79, 1984

Muller C., J. Vercheval, M. Ackerman, C. Lippens, J. Laurent, M.P. Lemaitre, J. Besson and A. Girard: Observations of middle atmospheric CH_4 and N_2O vertical distributions by the spacelab 1 grille spectrometer. Geophys. Res. Lett. **12**, 667-670, 1985

Taylor F.W., A. Dudhia, and C.D. Rodgers: Proposed reference models for nitrous oxide and methane in the middle atmosphere. Adv. Space Res. **7**, (9)49-(9)62, 1987

Zipf E.C. and P.W. Erdman, Sounding Rocket Studies of N_2O, NO, and Other Trace Constituents in the Upper Stratosphere and Mesosphere Using Cryogenic Grab Sampling Techniques. NASA Upper Atmosphere Research Program: Research Summaries 1982 - 1983, 1984

COLUMN ABUNDANCE OF HYDROGEN CHOLORIDE ABOVE LAUDER, NEW ZEALAND

W.A. Matthews and P.V. Johnston
Physics and Engineering Laboratory
Department of Scientific and Industrial Research
Lauder, New Zealand

D.G. Murcray, F.J. Murcray, and R.D. Blatherwick
Department of Atmospheric Physics
University of Denver
Denver, Colorado 80210, USA

ABSTRACT

The column abundance of hydrogen chloride over the PEL Lauder site in New Zealand (45°S) has been measured during a series of five campaigns from November 1985 using high resolution Fourier transform absorption spectroscopy. The lack of any diurnal variation or day to day change is demonstrated and the secular increase in HCl over Lauder for the period November 1985 to March 1988 of 10 ±5% per annum is shown. The implication of an annual cycle in the HCl amounts is also discussed.

1. INTRODUCTION

Hydrogen chloride was established as an important trace species in the atmosphere after Molina and Rowland (1974) suggested a catalytic ozone destruction mechanism involving ClO_x. Hydrogen chloride terminates the action of active chlorine in this cycle and its abundance and secular variation provides indirect evidence on this catalytic process provided other sources of hydrogen chloride are small. This is thought to be anthropogenic, through the photolysis of halocarbons (WMO, 1986) and with only minor amounts coming from the reaction of OH with methylchloride released from the oceans and the burning of vegetation. There is also thought to be a small source from the direct injection of HCl into the stratosphere by volcanic eruptions. In contrast with the stratosphere, tropospheric HCl has principally a natural source: the interaction between SO_4 and NO_3 ions with NaCl in ocean spray (WMO, 1986). There is, however, an anthropogenic source from the burning of plastics and certain industrial processes. Consequently, a total column measurement of HCl could be affected by this type of anthropogenic injection. In November 1985, a series of column hydrogen chloride measurements were started at our remote rural site at Lauder, New Zealand (45°S, 169°E). This site is located in the southern centre of the South Island at an altitude of 370 m, and with an annual rainfall of about 350 mm. It is technically classified as the only desert region in New Zealand. The site is 40 km from the nearest "larger" town, Alexandra (pop. 4000) and 150 km from the ocean, and therefore considered a good site to infer stratospheric column abundances of a number of trace gases using high resolution solar absorption spectroscopy. The measurement series has been made during five campaigns using a Fourier transform interferometer. The HCl data obtained in November 1985, June 1986, February and August 1987 and February-March of 1988 form the basis of this paper.

2. INSTRUMENTATION

The instrumentation used to obtain the solar spectra for this study was constructed in the Department of Physics at the University of Denver, for obtaining solar spectra from a balloon platform. As a result, the system is relatively light and robust. The instrumentation consists of a Fourier transform interferometer employing the BOMEM self aligning system which has a nominal non-apodised resolution of 0.01 cm^{-1}, a liquid nitrogen cooled InSb detector, an automatic solar tracking system, and a digital magnetic tape recording system. These digital tapes are read in the field using an IBM/XT computer and this allows the performance of the instrument to be monitoring during a campaign.

3. DATA REDUCTION AND ANALYSIS

The raw data are recorded in the form of interferograms and we have developed software packages that allow this data to be Fourier transformed on an IBM/XT as well

ISBN 0-937194-15-8

as on main frame computers. Under normal operating conditions, an interferogram is recorded every 70 seconds, and so several hundred are generated in a day's operation. Since the processing of the data in the field takes approximately 20 minutes per interferogram, the bulk of the processing is performed, after some preselection and the data has been transferred from magnetic cartridges onto nine-track digital tape, in the NCAR Cray computer in Boulder, Colorado. The analysis of the absorption feature of a particular trace species in the Earth's atmosphere is made easier if there is little or no interference from the absorption of other trace species in the same atmosphere, Zander et al. (1987). For hydrogen chloride observations made from the ground, the most suitable feature is the R1 line of the 1-0 band located at 2925.897 cm^{-1}. This feature is found on the wing of a strong methan absorption feature but sufficiently distant from a water vapour feature at 2925.189 cm^{-1} for water vapour not to effect the analysis – this is particularly important for a relatively low altitude site.

Two sample spectra incorporating the R1 line of HCl, obtained at 0220 UT on June 14, 1986 (winter) and 2017 UT on February 27, 1988 (summer), both with an airmass of approximately 3, are plotted in Figure 1. The vertical column HCl amount has changed from 2.2×10^{15} mol. cm^{-2} on June 14, 1986 to 2.6×10^{15} mol. cm^{-2} on February 27, 1988. It is interesting to note the relatively large change in water vapour content, typical of the summer winter difference, but it is also apparent from this figure that this does not effect the HCl analysis.

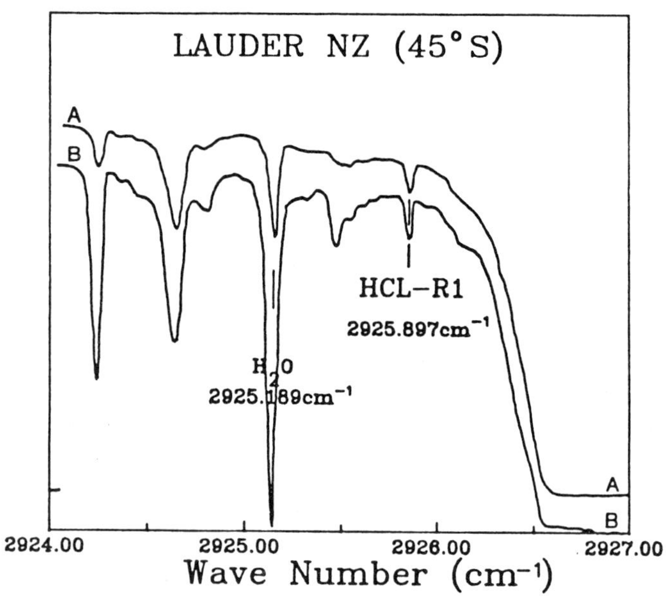

Fig. 1. Two typical sample spectra recorded from the ground at Lauder on June 14, 1986 (A) and February 27, 1988 (B) in the vicinity of the R1 line of the fundamental band of HCl. The strong absorption feature in the right of the figure is due to CH_4 while the 2925.189 cm^{-1} H_2O feature is indicated.

The analysis for this paper was based on equivalent width measurements of the 2925.897 cm^{-1} HCl feature. Spot comparisons of the resultant HCl abundances were made by using the full multicomponent, temperature dependent, non-linear least squares fitting method on the NCAR Cray computer. This method is discussed in Goldman et al (1986). These comparisons showed that the equivalent width analysis agreed with the "full" treatment within the assessed random error of the equivalent width treatment of ±20%. The main sources of these random erors are listed in Table 1.

TABLE 1

Summary of main random error sources in column abundance analysis

Source	% uncertainty
Spectral noise	4
Continuum setting	3
Equivalent width measurements	5
Assumed concentration profile	5
Assumed physical model	4

4. RESULTS

In Figure 2, a sample of the hydrogen chloride abundances measured over Lauder during the period February 19 to March 18, 1988 are plotted as a function of solar zenith angle incorporating both morning and afternoon values. This plot is typical of those obtained for the other four campaigns held at Lauder in as much as the HCl abundance measured in any particular campaign seems to be relatively independent of the time of day. This lack of any diurnal variation indicates that either the HCl abundance does remain constant through the day <u>and</u> the assumed model profile has approximately the correct shape, or within each day, the vertical distribution of HCl changes in such a way so as to compensate for real changes in HCl so the apparent amount measured at the ground as a function of solar zenith angle remains constant. This latter possibility seems somewhat unlikely.

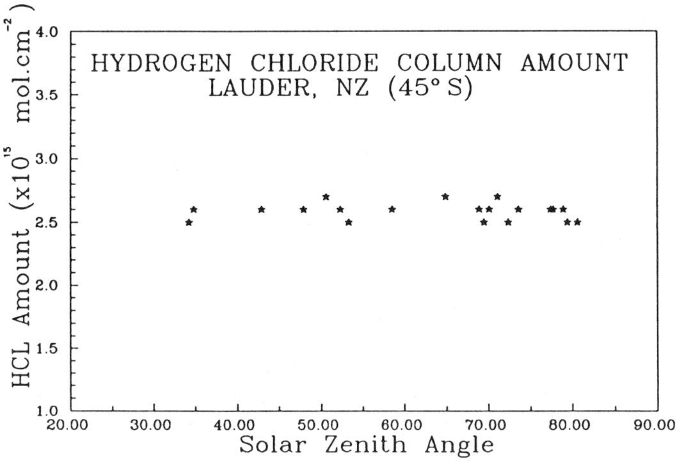

Fig. 2. A sample of the vertical column corrected abundance values of HCl measured at Lauder during the period February 19 to March 18, 1988 shown as a function of solar zenith angle. Both morning and afternoon values are included in this figure.

The daily means of HCl for the February-March 1988 campaign are plotted in Figure 3. This plot shows there is no apparent day to day variation during the campaign and this accords with the findings of Mankin and Coffey (1983) from their aircraft data, but contrasts with the measurements of HCl made by Zander et al (1987) and Marché and Meunier (1983). It is possible that these latter two data sets, both made over the European continent, are being influenced by tropospheric changes.

As has already been stated, the data measured in the four earlier campaigns shows the same solar zenith angle independence, but with the HCl abundance having a different magnitude. Figure 4 displays the secular variation in hydrogen chloride over Lauder, New Zealand during the period November 1985 to March 1988. Each data point displayed represents the mean value for each campaign. The mean values has been formed by taking a sample of the data recorded in each campaign for each day available, and for a range of solar zenith angles. As has been shown in Figures 2 and 3, the lack of a diurnal variation and with no appreciable day to day variability in the Lauder data set enables a mean value to be obtained that is representative of the vertical column abundance for the period. From Figure 4, the secular trend is essentially linear and the best-fit growth rate is 10 ±5% per annum. The lack of seasonal changes in this data set accords with the findings of Mankin and Coffey (1983) but contrasts with the work of Zander et al (1987) who found an annual cycle in HCl over Jungfraujoch with an amplitude of ±10%. An extension to the present data set from Lauder is required to firmly establish the trend that has been observed.

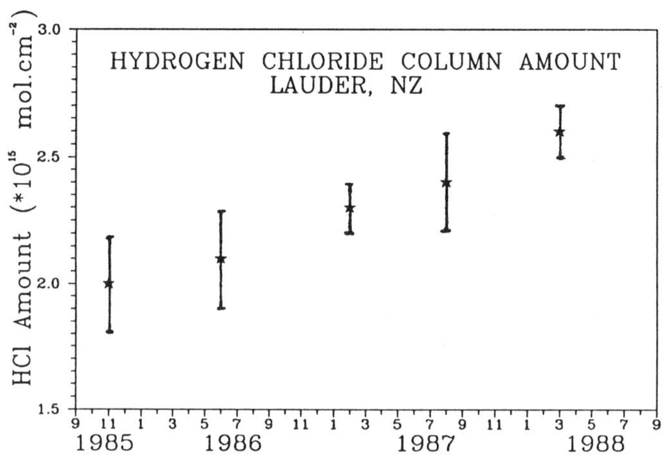

Fig. 4. The secular change of HCl over Lauder New Zealand during the period November 1985 to March 1986. Each data point represents the mean value of sample data set analysed by the equivalent width method for each campaign.

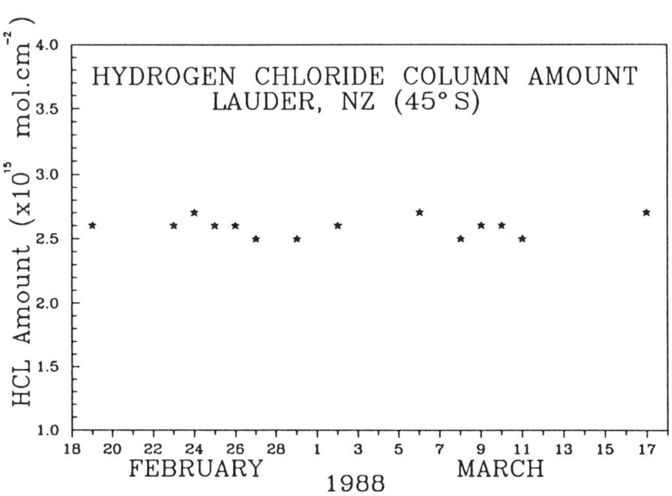

Fig. 3. Daily mean vertical column HCl abundances for the period February 19 to March 18, 1988.

5. CONCLUSION

The hydrogen chloride abundance over Lauder shows essentially no diurnal or day to day variability. The secular trend from the mean values of each of the five campaigns conducted so far shows an annual increase of 10 ±5%.

The data presented here show that a very valuable data set may be obtained by taking high resolution infrared spectra from a relatively low, albeit dry, inland site. The question of the magnitude of the secular change in HCl over this mid-latitude southern hemispheric site is still open until a more continuous data set can be obtained that can be used to access the importance, if any, of a seasonal change.

ACKNOWLEDGEMENTS

We wish to thank Brian McNamara for his dedication in supervising the acquisition of this data, and Gordon Keys and Aaron Goldman for their help with data processing. Financial support was received from the Chemical Manufacturers Association through Grant FC85/575 for which the authors are particularly grateful.

REFERENCES

Goldman, A., Murcray, F.J., Blatherwick, R.D., and Murcray, D.G., 1986: Quantification of HCl from high resolution, ground-based, infrared solar spectra in the 3000 cm^{-1} region, J. Quant. Spectroscl. Radial Transfer, **36**, 375-387.

Mankin, W.G. and Coffey, M.T., 1983: Latitudinal distributions and temporal changes of stratospheric HCl and HF, J. Geophys. Res., **88**, 10776-10784.

Marché, P. and Meunier, C., 1983: Atmospheric trace species measured above Haute-Provence Observatory, Planet. Space. Sci., **31**, 731-733.

Molina, M.J. and Rowland, F.S., 1974: Stratospheric sink for chlorofluoromethanes: chlorine atom catalysed destruction of ozone, Nature, **249**, 810-812.

World Meteorological Organisation, 1986: Atmospheric Ozone 1985: Assessment of our understanding of the processes controlling its present distribution and change, Report 16, World Meteorol. Organ., Global Ozone Res. and Monit. Proj., Geneva.

Zander, R., Roland, G., Delbouille, L., Sauval, A., Farmer, C.B., and Norton, R.H., 1987: Column abundance and the long-term trend of hydrogen chloride (HCl) above the Jungfraujoch Station, J. Atmos. Chem., **5**, 395-404.

MEASUREMENTS OF THE CYCLES OF PRESSURE, TEMPERATURE AND TRANSMISSION IN AN NO_2 PRESSURE MODULATOR

Howard K. Roscoe
Rutherford Appleton Laboratory
Chilton, Didcot, Oxon OX11 0QX, UK

During the Balloon Intercomparison Campaigns (BIC) of 1982 and 1983, differences between simultaneous measurements of NO_2 by the Balloon-borne Pressure Modulator Radiometer (BPMR) and some other remote sensors exceeded a factor three. Roscoe *et al.* (1988) suggested a number of measurements with pressure modulators that might resolve these differences. Foremost amongst these measurements was those of the cycles of pressure, temperature and transmission within the balloon-borne modulator.

Although some measurements of the pressure cycles with low-compression (SAMS) modulators had been made, analysis of data from atmospheric measurements with pressure modulators has relied on computer models of their pressure, temperature and transmission cycles. The possibility had always existed that the temperature amplitude within the high-compression balloon-borne modulator greatly exceeded that predicted by these models, in which case the spectral parameters used in the line-by-line calculation of the transmission functions would be significantly different from those assumed. It is technically difficult to measure the pressure and temperature cycles, but these measurements have now been made with NO_2.

In the pressure modulator, shown in figure 1, a high-speed pressure transducer was fitted to the space above the piston (replacing the plug in figure 1). In order to investigate whether the behaviour of NO_2 differed significantly from that of an ideal gas, pressure cycles were observed with helium and nitrogen, as well as NO_2. Water vapour is known to depart from ideal behaviour, so its pressure cycle was also measured. Figure 2(a) shows that, at ambient temperature, the pressure amplitude for NO_2 is within 93% of that of nitrogen, but the amplitude of H_2O is rather less. This behaviour is characteristic of adsorption and desorption on the surfaces within the modulator. NO_2 and H_2O are known to be strongly adsorbing; figure 2 shows that adsorption and desorption occur within the period of the pressure cycle (about 40ms). Figure 2(b) shows that the effect is greater at lower temperatures, as might be expected. Interestingly, the adsorption and desorption of H_2O is extreme at low temperatures, despite the fact that the scale of saturation temperatures shows that condensation is not occuring.

However, NO_2 does not behave in this extreme way, and at the temperatures of the balloon flights (30°C), computer models of the pressure cycle which ignore adsorption within the cycle will predict pressures accurate within at least 10%.

Figure 3 shows the apparatus for measuring the transmission and temperature cycles. Measurements were made firstly with the modulator off and the vane rotating, to deduce the radiometric gain; then with the vane removed, to deduce the transmission cycle; then with the vane inserted, to deduce the emission and hence temperature cycle; and finally with the vane inserted and nitrogen in the modulator, to measure the signal induced by vibration and electrical interference at the modulator frequency. Because the infrared detector is photoconductive and there is no calibration system between it and the modulator, only the alternating component of the emission signal can be measured. The gas in the cell is at a temperature elevated slightly above ambient (less than 1°C above) because the piston is doing work on the gas; and the value of this elevation affects the shape of the temperature profile deduced from the emission measurements (Roscoe and Wells, 1988). However, the elevation can be deduced by an iterative procedure to find a self-consistent value (Roscoe and Wells, 1988).

Figure 4 compares the measured temperature and pressure cycles at the filling pressure used in the balloon flights to those predicted by the computer model (Kerridge, 1984) used in the data analysis. The figure shows that the measured and calculated amplitudes and phases are in excellent agreement but that some details of the shapes disagree. The amplitude is less than 10K peak-to-peak.

In conclusion, the large differences in NO_2 results during BIC cannot be attributed to uncertainties in the pressure and temperature cycles within the modulator.

ISBN 0-937194-15-8

REFERENCES

Kerridge, B.J., 1984. "Abundances and distribution of Nitrogen Oxides in the atmosphere." *Oxford University D. Phil. thesis,* June 1984.

Roscoe, H.K. and R.J. Wells, 1988. "The variation of pressure, temperature and transmission within a pressure modulator: measurements with a high-compression modulator." Submitted to *J. Quant. Spectrosc. and Radiat. Transfer.*

Roscoe H.K., B.J. Kerridge, S. Pollitt, M. Bangham, N. Louisnard, C. Alamichel, J-P. Pommereau, T. Ogawa, N. Iwagami, M.T. Coffey, W. Mankin, J-M. Flaud, C. Camy-Peyet, F.J. Murcray, A. Goldman, W.F.J. Evans and C.T. McElroy, 1988. "Intercomparison of remote measurements of stratospheric NO and NO_2." Accepted for publication by *J. Atmos. Chem.*

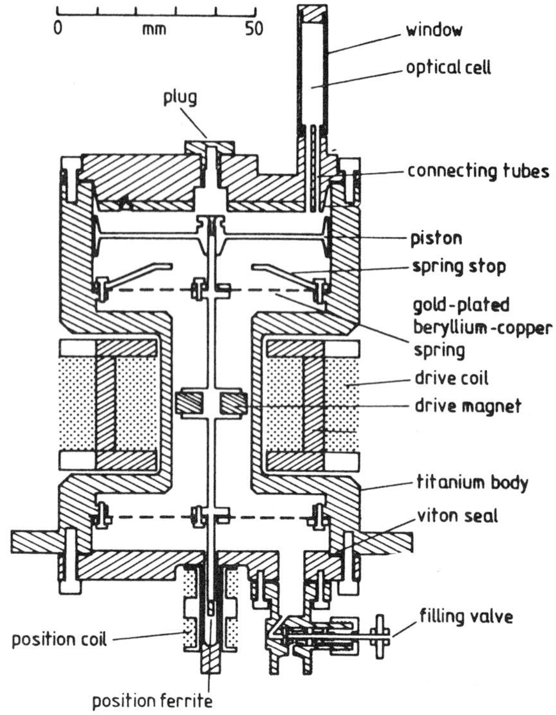

Fig. 1. The balloon-borne modulator used in this study. The piston is driven at its resonant frequency by the moving magnet and external coils, using feedback from the ferrite bead and external transformer, which sense the piston's position. The plug in the cylinder head contained the same volume as the pressure transducer which replaced it when measuring the pressure cycles.

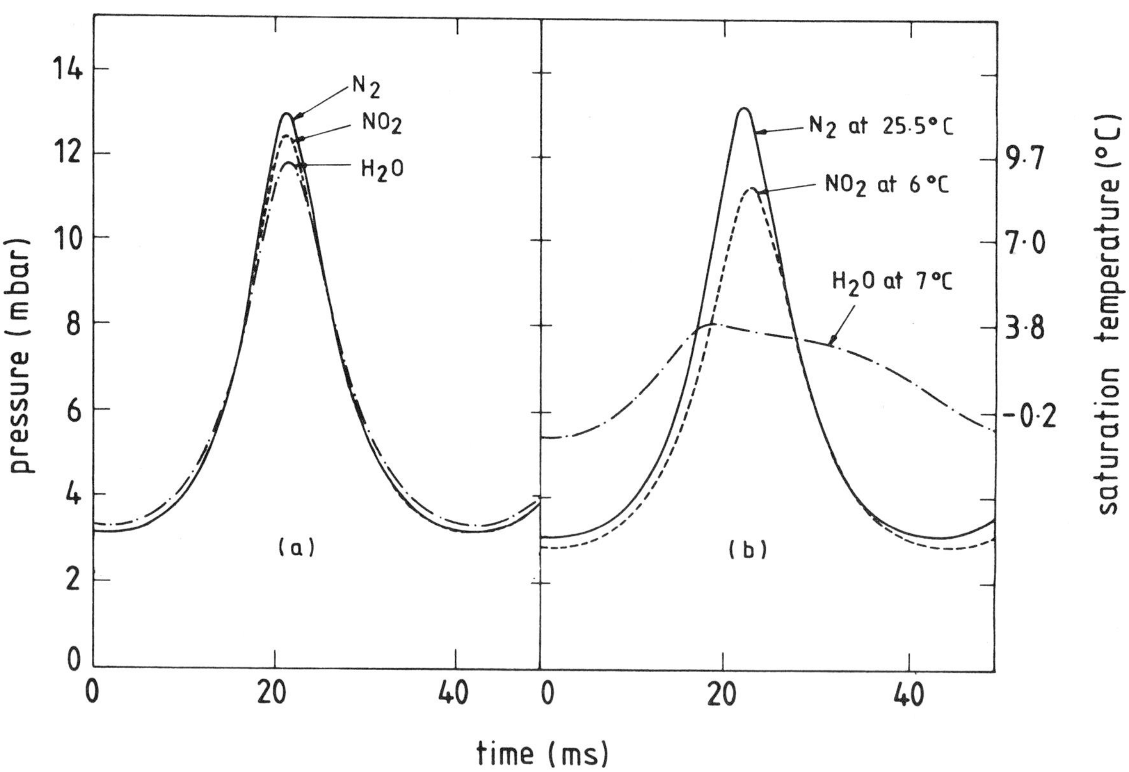

Fig. 2. Pressure cycles at filling pressures of 6mbar of (a) nitrogen, NO_2 and H_2O at 25.5, 24.5 and 23.5°C respectively, and (b) H_2O and NO_2 at 7 and 6°C respectively and nitrogen at 25.5°C. In (a), the compression-ratio of NO_2 and H_2O is less than that of nitrogen, and the lack of compressibility of H_2O is such that the frequency has decreased by 1.6%. The temperatures on the right-hand axis are those at which the saturated vapour pressure of H_2O equals the pressure on the left-hand axis. Note that at no time in curve (b) does the pressure of H_2O exceed the saturation vapour pressure at 3°C.

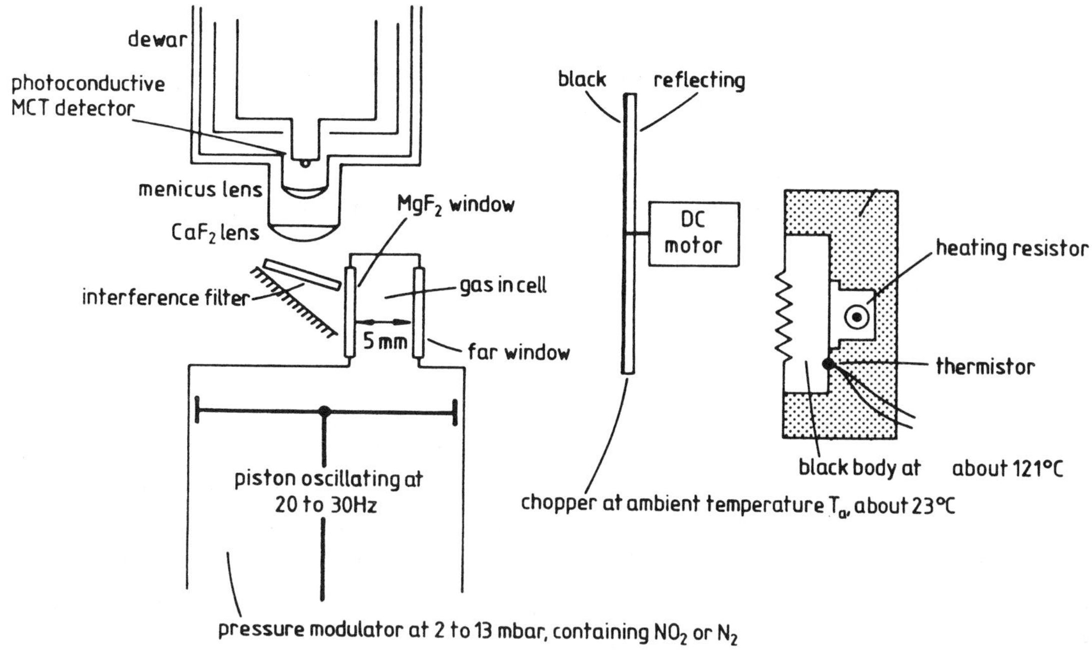

Fig. 3. The experimental arrangement for measuring transmission and emission signals. The emission signal was observed with the ambient vane in place, the transmission signal with the vane removed followed by subtraction of the emission signal. Calibration signals with the vane rotating and the cell empty determined the radiometric gain of the system.

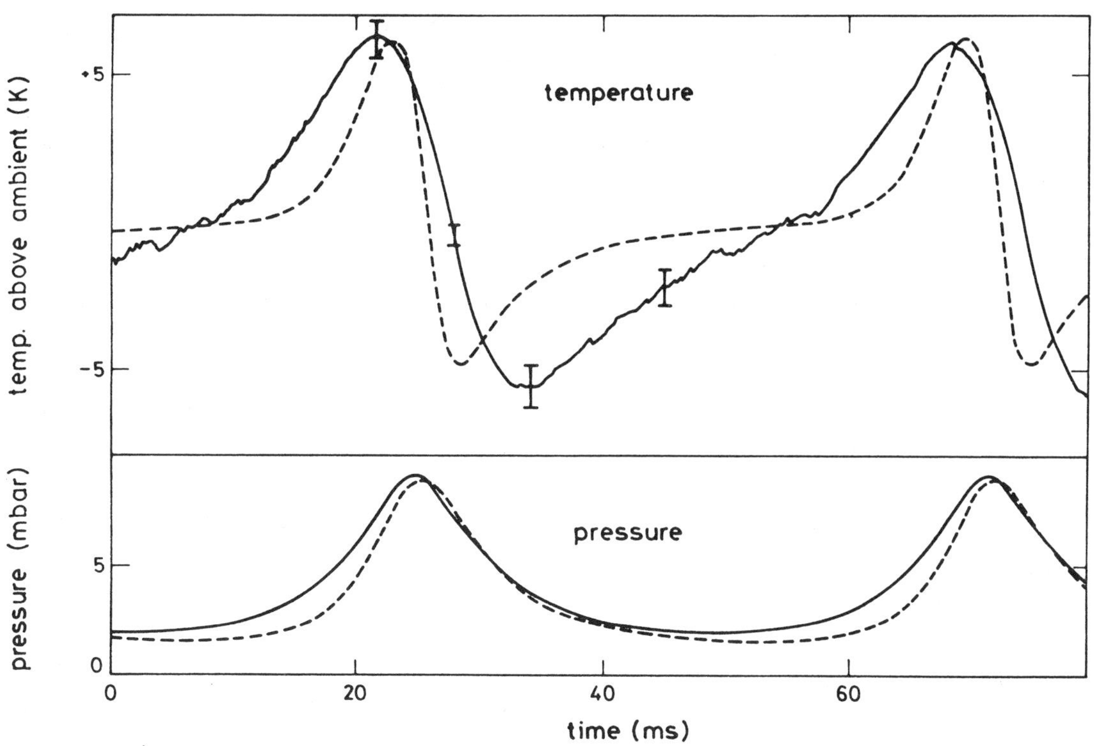

Fig. 4. Calculated temperature and pressure cycles in the balloon-borne modulator at a filling pressure of 4mbar of NO_2 (dashed lines, Kerridge, 1984) compared with the measured cycles at the same filling pressure (solid lines). The linear temperature scale of the calculations is chosen to agree with the slightly non-linear scale of the measurements at $\pm 5K$. Systematic errors are shown by the vertical bars. Note the excellent agreement between the measured and calculated amplitude and phase of the temperature cycle. The agreement in shape is less good.

THE DISTRIBUTION OF SOME HALOGENATED HYDROCARBONS IN THE FREE ATMOSPHERE AND SEA SURFACE LAYER

G.T. Burova, E.A. Chayanova, E.V. Ivanov, and A.G. Katrich
Central Aerological Observatory State Committee for Hydrometeorology
and Control of Natural Environment
Moscow, USSR

ABSTRACT

Presented are the results of measuring five substances $CHCl_3$, CH_3CCl_3, C_2Cl_4, CCl_4, $CFCl_3$ in the atmospheric sea surface layer over the Greenland and Norwegian seas and in the lower troposphere and stratosphere under the continental conditions of East Europe. The measurements were performed using a LHM-80 chromatograph with an electron-capture detector. Air was sampled in the ground layer at 500 m and 800 m levels from a captive balloon, in the lower troposphere at 1,2 and 3 km levels from onboard a helicopter and in the lower stratosphere at 17 to 30 km levels from an automated balloon. The mean concentration values in the stratosphere for $CFCl_3$, CH_3CCl_3, CCl_4 and C_2Cl_4 are 132 ppt, 149 ppt, 127 ppt and 111 ppt respectively. The mean $CHCl_3$ concentration values is 310 ppt. Observations from onboard a research vessel show that in the atmospheric sea surface layer the most inactive halocarbons $CFCl_3$ and CCl_4 are the dispersion around the mean concentration values, which does not exceed 36%, while those for $CHCl_3$, CH_3CCl_3 and C_2Cl_4 are as high as 57, 59 and 86 %.

The atmospheric chemistry of chlorofluorocarbons plays an important role in the assessment of the potential modification by these substances of the atmospheric ozone layer and Earth's climate. The quantitative characteristics of ozone destruction can be obtained by using mathematical models which require the knowledge of the amount of some halogenated minor constituents (CF_2Cl_2, $CFCl_3$, CCl_4, CH_3Cl CH_3CCl_3, $CHCl_3$, C_2Cl_4, etc.) .

To study the vertical distribution of CFC-11 ($CFCl_3$), Chloroform ($CHCl_3$), I,I,I,-Trichloroethane (CH_3CCl_3) tetrachloromethane (CCl_4), and tetrachloroethyene (C_2Cl_4) in the free atmosphere, measurements of these substances were conducted at different altitudes, under the continental conditions of East Europe Rylsk, Kursk Region, 51°N, 34°E in July-August 1986. Air was sampled in the ground layer at 500 m and 800 m levels from a captive balloon , in the lower troposphere at 1,2 and 3 km levels from board a helicopter, and in the lower stratosphere at 17 to 30 km levels from an automated balloon. Air sampling in the troposphere and stratosphere was done by an automated sampling unit developed and constructed at the Central Aerological Observatory for balloon experiments. Air samples with trace halogenated hydrocarbons were analyzed using a chromatograph LHM - 80 with an electron - capture detector; analyzed substances were separated on a column 3 m long, with 10 % squalan on Inerton AW -

ISBN 0-937194-15-8

DMCS, the concentrating of substances performed using hydrophobic sorbent Polysorb - 1.

Mean measurement results for $CFCl_3$, $CHCl_3$, CH_3CCl_3, CCl_4 and C_2CL_4 are given in Table 1. In Fig. 1 summarized are the results of this work in the form of the vertical distributions of halogenated hydrocarbons, which characterize the local properties of the atmosphere during the observation period.

The mean concentration values for CFC-11, tetrachloromethane and I,I,I - trichloroethane in stratospheric air samples, presented here, agree with the results obtained by other authors (Atmospheric Ozone, 1985). The greatest data variability of up to 65 % is for tetrachloroethylene, its amount in 17 to 30 km atmospheric layer being the same as that of CCl_4. Despite C_2Cl_4 emission to the troposphere (660 thousand ton/year) observed in recent years, being considerably greater than that of CCl_4 (50 thousand ton/year), the diffusion to the stratosphere of tetrachloroethylene is less intense. Since C_2Cl_4 is a derivative of a non-limiting compound it is subject to more intense destruction in the troposphere due to its numerous reactions with other minor species contaminating the atmosphere. The amount of chloroform both in troposphere up to 3 km and lower stratosphere at 17 to 24 km, from our data, averages 300 ppt and exceeds the amounts measured by other authors in background atmospheric regions. Mindfull of $CHCl_3$'s possible sink through reactions with tropospheric OH, one should expect a more significant decrease of its concentrations with altitude and lower absolute values in a 17 - 20 km atmospheric layer. However an increase of chloroform production capacities chloroform is used, for example, for the synthesis of CFC-22, a recent

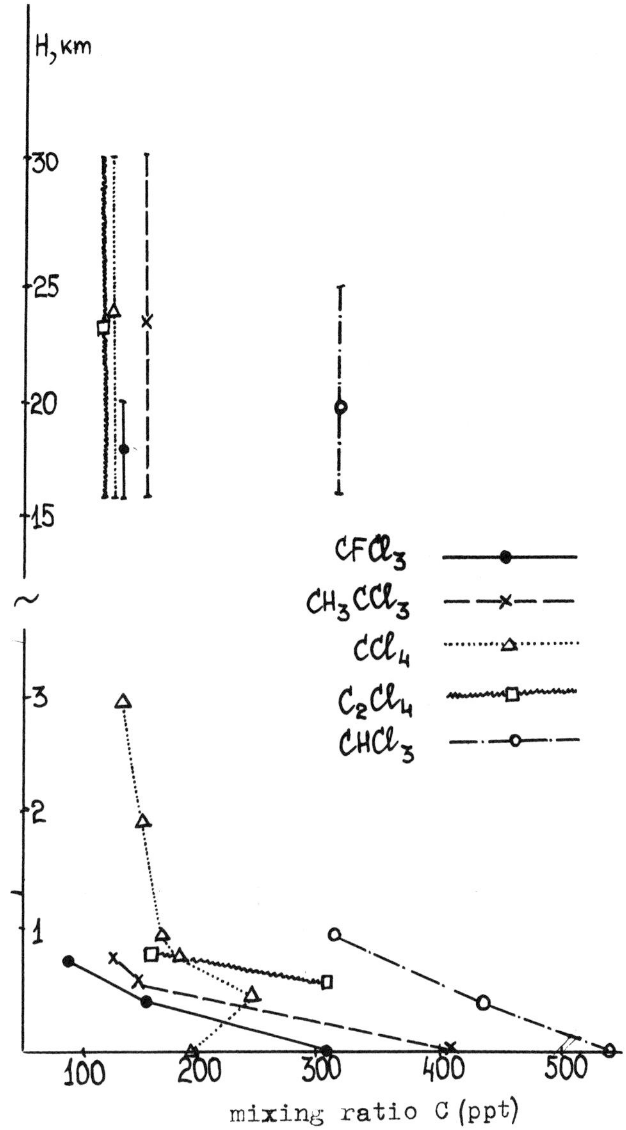

Fig. 1. The vertical distribution of halogenated hydrocarbons in the atmosphere July-August 1986, 51°N, 34°E .

alternative of CFC-12 photochemically active in the stratosphere could cause the accumulation of chloroform in the atmosphere. The scarce data available on $CHCl_3$ content in a background atmosphere refer to sea and seaside regions where a considerable sink of this substance to sea surface is likely. Chloroform solubility in water is better than that of other halogenated hydro-

TABLE 1. MEAN MEASURED CONCENTRATION OF SOME HALOGENATED HYDROCARBONS IN THE FREE ATMOSPHERE

Altitude and type of air sampling	Concentration, ppt				
	$CFCl_3$	$CHCl_3$	CH_3CCl_3	CCl_4	C_2Cl_4
Ground-based H = 0.5 m	297±118	556±171	398±229	187± 81	-
Captive balloon					
H = 500 m	137	432	144	235	303
H = 800 m	82	295	131	167	152
Helicopter					
H = 1 km	-	-	-	166	-
H = 2 km	-	-	-	143	-
H = 3 km	-	-	-	127	-
Autostratostat H = 17-30 km	132±43 H=17-20 km	309±95 H=17-24 km	149±19 H=17-29km	127±46 H=17-30 km	111 ±83 H=17-30 km

carbons analyzed ($CFCl_3$, CCl_4, C_2Cl_4 et al.) .

On the whole, the results obtained on the vertical distributions of halogenated hydrocarbons in the atmosphere confirm the necessity of the regular studies of air composition aimed at the determination of the trends of those substances and prediction of photochemical and climate changes in the atmosphere. Comprehensive monitoring requires observations of halogenated hydrocarbons in different regions, including their latitudional distributions, as well as their altitude distribution up to the upper stratosphere, and such global reservoirs as oceans. The ocean with its large-scale biochemical and chemical processes may play a significant role in the dynamic regulation of the amount of halogen-containing organic substances in the atmosphere of the planet.

The studies of the distributions of CFC-11, chloroform, I,I,I - trichloroethane, tetrachloromethane and tetrachloroethylene in the sea-surface atmospheric layer were conducted by the authors in the regions of the Norwegian and Greenland Seas in August - September 1987, during the 49th cruise of the scientific research vessel "Professor Vieze". These studies were aimed at the acquisition of data on the amount of some halogenated hydrocarbons in the sea surface atmospheric layer in the region of the Arctic Ocean at 61-76°N that has not been studied extensively. The programme aimed at revealing the atmospheric sources and sinks of these substances, and the assessment of their possible sink to the sea surface. $CFCl_3$, $CHCl_3$, CH_3CCl_3, CCl_4, and C_2Cl_4 concentrations were determinated by a method of gas chromatography under the sampling and analysis conditions appropriate for studies of air in the surface layer troposphere and stratosphere. Table 2 presents the distributions of five halogenated hydrocarbons obtained in sea air. These results show that the root-square deviation (RMSD) from the concentrations of analyzed minor constituents is significantly bigger than the relative errors of their gas-chromatographic analysis. Therefore, the variations in the concentrations of halogenated hydrocarbons refer to different air masses and depend on the complex mechanisms of $CFCl_3$, $CHCl_3$, CH_3CCl_3, CCl_4 and C_2Cl_4 transformations in the system "ocean-atmosphere".

Halogenated hydrocarbons CFC-11 and tetrachloromethane, the most inert ones in the atmosphere, are characterized by high stability. The data obtained show the concentration dispersions of 33 and 36 %. The variability of $CHCl_3$, CH_3CCl_3 and C_2Cl_4 concentrations measured by the authors is high, i.e. 57,59 and 86 %. Table 3 presents the data obtained by different authors on $CFCl_3$, $CHCl_3$, CH_3CCl_3, CCl_4 and C_2Cl_4 amounts in the oceanic air of the Northern Hemisphere showing the distribution of these substances in the ocean atmospheric layer. The results obtained by the authors of this paper in the atmospheric studies conducted in an energy-active zone of the ocean (the Norwegian and Greenland Seas) complement the general picture.

TABLE 2. THE DISTRIBUTION OF SOME GALOGENATED HYDROCARBONS IH THE SEA SURFACE LAYER

Substance	The number of measurements	Mean concentrations, ppt	RMSD ppt	RMSD %	Relative error of analysis, %
$CFCl_3$	92	159	± 53	33	16
$CHCl_3$	16	21	± 12	57	15
CH_3CCl_3	55	138	± 81	59	15
CCl_4	37	132	± 48	36	13
C_2Cl_4	55	14	± 12	86	13

TABLE 3. MEAN CONCENTRATIONS OF SOME HALOGENATED HYDROCARBONS IN THE OCEAN ATMOSPHERIC LAYER IN THE NORTHERN HEMISPHERE (PPT)

$CFCl_3$	$CHCl_3$	CH_3CCl_3	CCl_4	C_2Cl_4	Observation region	Authors
159±53	21±12	138±81	132±48	14±12	60° - 76°N	This paper
161	29	117	102	56	35°N	Penkett, 1982 Kirschmer e.al., 1983
206	39	175	151	-	70°N	Rasmussen et.al., 1983
186	21	156	135	29	39°N	Singh et.al., 1983
-	20-50	200±15	140±20	15-30	40° - 60°N	Class et.al., 1986

A comparison of published CH_3CCl_3, $CHCl_3$, and C_2Cl_4 concentrations shows, the discrepancies of 31,85 and 115 %, respectively, which may be accounted for by several facts. Not fully halogenated ($CHCl_3$, CH_2CCl_3) and non-limiting (C_2Cl_4) halogenated hydrocarbons have a significant sink in the atmosphere through reactions which involve the hydroxyl radical OH. Besides, an increased content of $CHCl_3$ and C_2Cl_4 in sea water, 2-10 ng/l and 0.7 ± 0.6 ng/l respectively (Singh et al. 1983, Class et al. 1986) and taking into account their solubility in water and the content in the atmosphere, it provides ground to conclude that sea water samples were supersaturated with chloroform and tetrachloroethylene. This indicates that their most likely sink is the sea surface. Class et.al. 1986 (1986) note a marked effect of $CHCl_3$ washing out by rain (up to 25 ng/l), which is also a sink of this substance from the atmosphere.

Due to local anthropogenic sources, the washing-out effect, natural sources and sinks to the sea surface as well as insufficiently studied biological sources, significant variations of non-stable halogenated hydrocarbons ($CHCl_3$, CH_3CCl_3 C_2Cl_4) are likely in the atmosphere. The variability of the experimental concentrations of some halogenated hydrocarbons, from both our and other authors' data, is determinated by the activity of these substances in the atmosphere. The relative stability of halogenated organic compounds is characterized by their lifetimes in the atmosphere, i.e. 60 (+40/-10) years for $CFCl_3$, 52 years for CCl_4, 9 (+2/-1) years for CH_3CCl_3, 0,6 (+0.2) years for C_2Cl_4 and 0.33 years for $CHCl_3$. The real variability of the amount of these substance in the atmosphere is characterized by experimental data on the amount trace halogenated hydrocarbons in the air in different regions of the globe.

REFERENCES

Atmospheric Ozone, 1985; World Meteorological Organization Global Ozone researth and Monitoring Project Report N 16, v. 1 - 3.

Burova G.T., E.V.Ivanov, E.A.Chayanova, 1988: The vertical distribution of some halogenated hydrocarbons in the atmosphere. Meteorology and hydrology, 5, pp. 112 - 115.

Class Th., K.Ballschmiter, 1986: Chemistry of organic traces in air VI: distribution of chlorinated C_1-C_4 hydrocarbons in air over the northern and southern Atlantic Ocean. Chemosphere, v.15, N 4, pp. 413-427.

Kirschmer P. and K.Ballschmiter, 1983: The Complex Pattern of C_1-C_4 Organohalogens in Continental and Marine Background Air. Int.J.Env.Anal.Chem. v. 14 pp. 275-284.

Penkett S.A., 1982: Non-methane Organics in the Remote troposphere. Atm.Chem. ed. E.D.Goldberg, pp.329-335.

Rasmussen R.A. and M.A.K.Khalil, 1983: natural and antropogenic trace gases in the lower troposphere of the Arctic. Chemosphere, v.12, N3, pp.371-375

Singh H.B., L.J.Salas and R.E.Stiles, 1983: Selected Man-made Halogenated Chemicals in Air and Oceanic Environment. J.Geoph. Research, v.88, N C 6, pp. 3675-3683.

AUTOMATIC SYSTEM FOR MEASUREMENTS OF MINOR SPECIES IN THE ATMOSPHERE

A.S. Elokhov, N.F. Elanśky, O.V. Makarov, and A.S. Smirnov
Institute of Atmospheric Physics
Academy of Sciences of the USSR
Pyzhevsky per. 3, Moscow, USSR

ABSTRACT

A complex of instruments in operation at the research station in Northern Caucasus is now simultaneously producing data on the total content and the vertical distribution of O_3, SO_2, NO_2 and aerosol. The calculations are made in the real-time scale by an iterative method. The operative analysis of the measured parameters and their application for calculation of other values increase the accuracy of measurements and allow us to automatise the functioning of instruments. The recorded variations in ozone and aerosol contents show the orographic influence of air flow over a mountain system on minor species distribution in the troposphere and stratosphere.

1. INTRODUCTION

The determination of minor species content in the atmosphere by means of optical technique is based on a specified optical model. For example, the information about the temperature in the stratosphere, the amount of SO_2 in the air, the spectral optical thickness of aerosol and its vertical profiles is necessary to calculate precise values of the total ozone and its vertical distribution. In their turn, ozone and aerosol data are indispensable for calculation of NO_2, NO_3 contents and evaluation of theire vertical distribution. Theoretically, all these values can be obtained by measuring the spectra of direct and scattered solar radiation with one spectrophotometer under control of microcomputer programming and data processing. The iterative method, applied to produce more accurate values of parameters measured by independent means, may increase the accuracy of determination of minor species contents as compared to the results of traditional methods; this approach may also make measurements less dependent on the changes in the optical properties of the atmosphere during observations. The date can be analysed by the microcomputer processor and the operative regimes of instruments can be automatically selected by programming in accordance with specified problems.

Automatic control of instruments and selection of problems based on the real time-scale processing of incoming data has several advantages, in particular at observation points operating under continuous mountain or polar conditions.

2. MEASUREMENT COMPLEX

An automatic complex of two spectrophotometers and a polarization photometer has been in operation for eighteen mounths at the research station of the Institute of Atmospheric Physics in Northern Caucasus (43.7°N, 42.7°E altitude above sea level 2070 m). The spectrophotometers are recording the direct and scattered solar radiation spectra with differing resolutions in 300-335 and 430-450 nm range. The photometer is measuring the polarization of the solar radiation scattered from zenith at 700 nm wave lenght (Elansky et al., 1987). The microcomputer "Elektronika-60" controls the operation of instruments and data processing.

Figure 1 shows a scheme of iterative process for calculation of the content of gas admixtures and aerosol in tne atmosphere. The order of

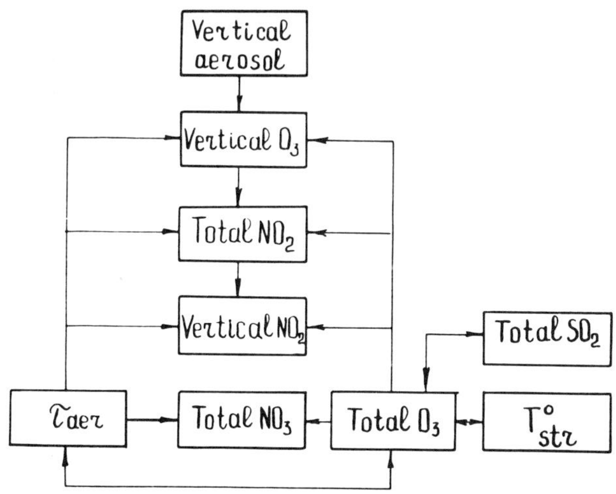

Fig.1. Scheme of iterative process for calculation of minor species content.

ISBN 0-937194-15-8

calculations is as follows: 1) the total amount of NO_2 is determined by the standard method from absorption bands in the UV range of the spectrum (Kerr et al., 1980), and from the same spectrum, recorded with high resolution, the effective temperature of the ozone layer is obtained in the region of its maximum; 2) the coefficient of aerosol scattering and its vertical profile are calculated from the data on spectral measurements in the UV and visible regions and from the data on polarization of scattered radiation (Steinhorst, 1977; Khrgian et al., 1981); 3) the total ozone content is determined by multiwave method (Khrgian et al., 1981); 4) the vertical distribution of ozone is reconstructed from the Umkehr effect by means of statistical regularisation using the autocorrelation ozone matrices (Elansky et al., 1982); 5) the total NO_2 content, its tropospheric part, and the height of maximal NO_2 concentration are estimated from direct and scattered solar radiation in the visible range of the spectrum, with allowance for the data on ozone and aerosol (Syed et al., 1980); 6) the parameters obtained in the day-time are used for the night-time observations of O_3, NO_2, whereas NO_3 content is obtained using the Moon as a light source.

The data on O_3 and NO_3 are applied to reassess the initial parameters of the optical model (SO_2, T_{ef}, aerosol content), and the iteration process continues. Sometimes, when the shortperiod variations of certain values are studied, for example, ozone content, then the iteration cycle can be reduced and for the sake of economy of the time calculations of individual parameters can be simplified.

3. RESULTS

The major purpose of observations at the high-mountain station was to study the possible influence of the orographic lee waves on the content and distribution of admixtures in the disturbed zone. The authors of paper (Elansky et al., 1988) demonstrated that under favourable conditions the lee waves flowing over the Elbrus orogenic system may reach into the stratosphere and influence the ozone content within 1-2% of its total amount. The results of real-time processing of data on short-period ozone variations have confirmed these observations. The content of ozone in the wave zone varies, as a rule, with periods of 23-30, 10-13, 5-7 min. (Fig. 2). Moreover, the long-period oscillations are observed only in the sector of azimuth angles not exceeding ± 15° from

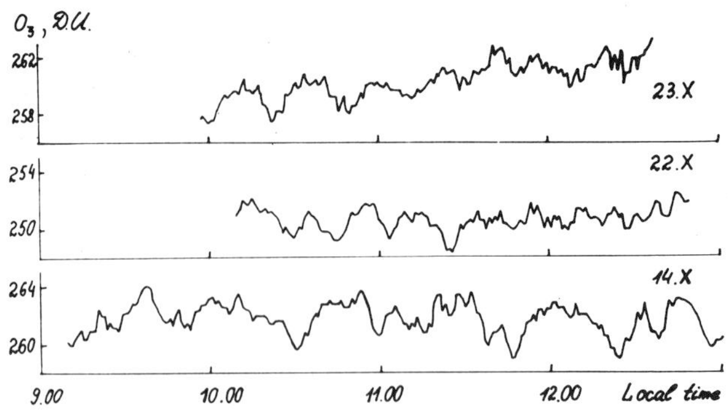

Fig. 2. Variations of the total ozone content in the wave zone of the Caucasus in October 1987 smoothed with 5 min filter.

the normal to the wind direction at the highest point of the range. It is presumed that the long-wave ozone oscillations appear when the Sun scans the stationary wave system in the ozone field which is built up by the stationary lee waves; the 10-13 and 5-7 min oscillations - by nonstationary waves. Short-period ozone oscillations may be connected with short period aerosol variations. Aerosol is able to influence on calculated ozone data if the standard 4 wave method is used. Fig. 3 represents the example of great ozone variations registrated in the morning of 20 April 1988. Ozone measurements were

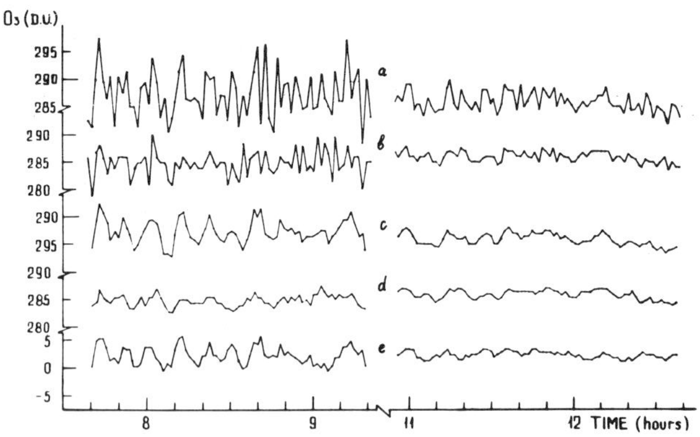

Fig. 3. Temporal variability of total ozone on 20 April 1988 using: a - the standard method of calculations; b - the multiwave method; c and d - smoothed with 5 min filter data; e- difference of c and d data.

carried out in both ways: using standard and multiwave methods. It is shown that the multiwave method which takes into account spectral aerosol attenuation more correctly than standard one do not give 10 min ozone oscillations.

Attempts to find similar variations in NO_2 total content are as yet unsuccessful mostly due to low accuracy of calculations, the total error of NO_2 determination being 15-20% and the mean square deviation 5-8%. Fig. 4 shows the total NO_2 temporal series obtained from direct and scattered radiation.

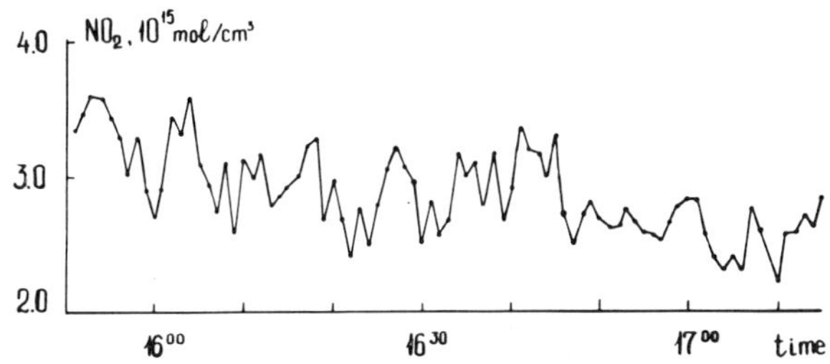

Fig. 4. Variations in NO_2 content from direct solar radiation measurements on 22 March 1988.

In the presence of powerful wave systems in the cloud cover, the polarization of scattered light sometimes also shows periodic variations when the boundary of the Earth's shadow lifts from 0 to 30 km over the observation site. If we reconstruct the vertical aerosol distribution from polarization data by means of method (Steinhorst, 1977), we may then obtain a

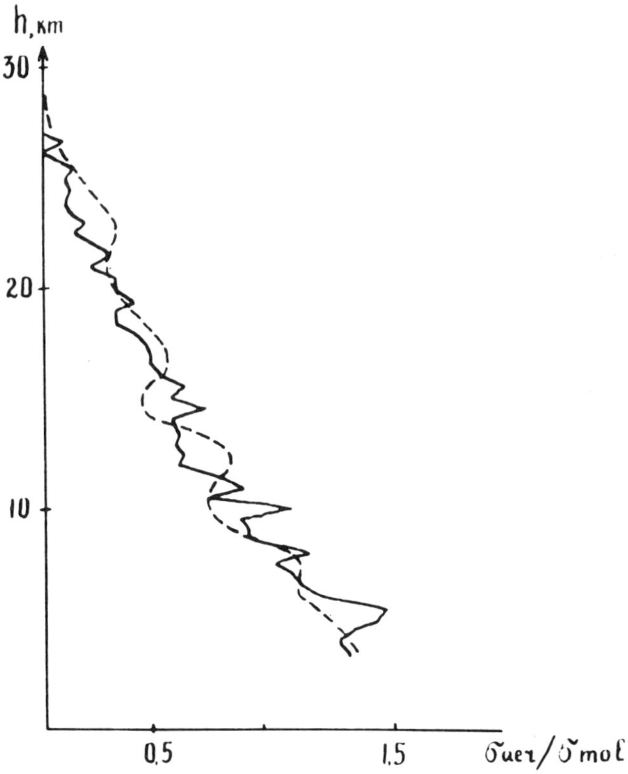

Fig. 5. Ratio of aerosol and molecular scattering coefficient $\sigma_{aer}/\sigma_{mol}$ from polarization measurements (dashed line) and lidar data.

layered structure with the vertical period of several kilometers. Fig. 5 gives a comparison of such reconstructed vertical distribution $\sigma_{aer}/\sigma_{mol}$ with the data of simultaneous lidar observations.

4. CONCLUSION

Simultaneous measurements of a large number of dependent values essentially improve spectrometric data on the contents of admixtures in the atmosphere. These data ensure correct interpretation of the causes of small variations in certain parameters. One of these causes can be the effect of internal gravity waves on the tropospheric and stratospheric composition. Since internal gravity waves are observed not only in the mountains, their influence can be reflected in the ozonometer records on the network of stations. The total ozone fluctuations may be systematic for every station and they should be considered in the determinations of the accuracy of their characteristics.

REFERENCES

Elansky, N.F. and Y.L. Terekhin, 1982: Applications of regularization methods to restoration of vertical ozone profile by observation of the Umkehr effect. Izv. Acad. Sci. USSR. Atm. Oceanic Phys. 18, 944-952.

Elansky, N.F., A.S. Elokhov, E.A. Kadyshevich and S.A. Sitnov , 1987: The influence of volcanic eruptions on polarization of scattered light in the twilight sky. Izv. Acad. Sci. USSR. Atm. Oceanic Phys. 23, 256-263.

Elansky, N.F., I.A. Senik and A.Kh. Khrgian, 1988: The variations of total ozone in the mountain lee waves region. Izv. Acad. Sci. USSR. Atm. Oceanic Phys. 24, in press.

Kerr, J.B., W.F.J. Evans and C.L.Mateer, 1980: Measurements of SO_2 in the Mount St. Helens debris. Proc. Symp. Mt. St. Helens Eruption: Its Atmospheric Effects and Potential Climatic Impact., NASA Conf. Publ. 2240, 219-223.

Khrgian, A.Kh. and G.I.Kouznetzov, 1981: The problem of atmospheric ozone. Moskow, Moskow Univ. Press., 216 p.

Steinhorst, G., 1977: Stratospheric aerosol concentration determined by an iterative method from twilight polarization measurements„ Beitr. zur Physik der Atmosphare, 50, 508-523.

Syed, M.Q. and A.W. Harrison, 1980: Ground-based observations of stratospheric nitrogen dioxide. Can. J. Phys. 58, 788-795.

OZONE AND RADIATION, SOLAR EFFECTS AND MESOSPHERE

RECENT DETECTION OF THE RESPONSE OF THE MIDDLE ATMOSPHERE TO SHORT-TERM SOLAR ULTRAVIOLET VARIABILITY

G.M. Keating
NASA Langley Research Center
Hampton, Virginia 23665, USA

M.C. Pitts
ST Systems Corporation
Hampton, Virginia 23666, USA

G. Brasseur
National Center for Atmospheric Research
Boulder, Colorado 80307, USA

ABSTRACT

The phase and amplitude of the response to short-term solar UV variations has been determined for stratospheric O_3, HNO_3, NO_2, and temperature, and for mesospheric ozone and temperature. The studies were performed using four satellite data sets: Nimbus 7 LIMS, SBUV, and SAMS, and SME 1.27 micron data. Unexpected temperature responses result in unexpected behavior in ozone variations through temperature feedback. Newly detected latitudinal variations in the response of stratospheric ozone and temperature are described.

1. INTRODUCTION

Efforts toward detecting the response of the middle atmosphere to solar activity variations have been pursued for over 50 years (Keating, 1981). Possible 11-year variations in ozone, temperature and other parameters have been difficult to isolate partially because of possible instrument sensitivity drifts and limited instrument lifetimes over such extended periods of time. Also, various climatological variations unrelated to long-term solar variations may be misinterpreted as solar effects. As a result, isolation of 11-year solar cycle variations in the middle atmosphere remains a controversial subject. A new approach has now been undertaken to detect variations related to solar ultraviolet variability. Studies are performed to isolate the response of the atmosphere to short-term solar variations related to the 27-day solar rotation period. Using zonal averages obtained from satellite data, it is possible to obtain the precision necessary to detect the atmospheric response to short-term solar ultraviolet variability. Using this approach, useful statistics on ten 27-day cycles may be obtained with less than one year of satellite data while a similar number of 11-year cycles would require over a century of data. The short-term solar UV variations at 205 nm can exceed five percent, while estimates of the amplitude of 11-year variations at 205 nm are generally less than ten percent. Thus, the amplitudes of short-term and long-term variations seem to be quite comparable, allowing improved estimates of the atmospheric response to the 11-year solar cycle variations. Recent results in this rapidly progressing area have been reviewed (Keating et al., 1987a) and presently the response to short-term variability has been detected in the stratosphere (O_3, HNO_3, NO_2, temperature) and throughout the mesosphere (O_3 and temperature). In addition, Ebel et al. (1986) have found significant spectral coherences between the 10.7-cm solar flux and various harmonic components of planetary waves.

Part of the relevance of these new results is that they provide a method of validation for photochemical models through comparison of observed and predicted amplitudes and phase lags of the response to solar variability. Validation of models and gaining an understanding of the atmospheric response to solar UV variability is crucial for isolation of man-made effects, since the present predicted ozone response to CFC effects and solar UV variations are of the same order.

2. STRATOSPHERIC RESPONSE

The response of stratospheric ozone to short-term solar UV variability can be clearly seen in Figure 1 (Keating et al., 1985). The residuals for solar UV (205-nm solar radiation) and 2-mb O_3 (measured by NIMBUS 7 LIMS and empirically corrected for temperature feedback effects) are five-day running means relative to a 27-day running mean. The detrended residuals are clearly correlated with one another, with ozone responding to each major solar UV change. The 205-nm solar radiation (measured by the NIMBUS 7 SBUV instrument) was chosen as the index of solar UV variability because it is directly involved in the

ISBN 0-937194-15-8

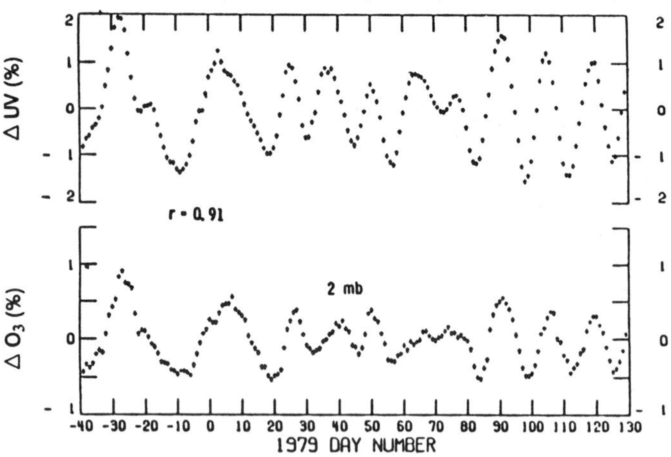

Fig. 1. Response of O_3 (NIMBUS 7 LIMS) to short-term solar UV (205-nm) variability (NIMBUS 7 SBUV). Residuals are relative to a 27-day running mean. Ozone values with temperature effects empirically removed are determined from zonal means between ±40° latitude.

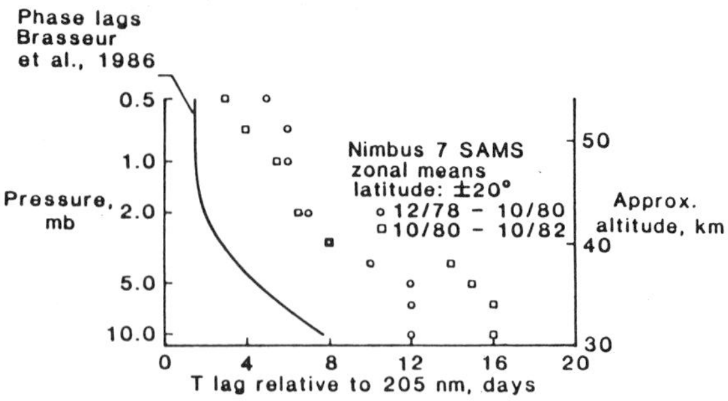

Fig. 2. Observed and predicted temperature response times to short-term solar UV variability. Observations are based on highest quality NIMBUS 7 SAMS data.

ozone production through the photodissociation of O_2, with subsequent production of O_3. We found much higher O_3-solar correlations using 205-nm radiation as opposed to proxy indices such as the 10.7-cm flux (Keating et al., 1985). Note that in this case, the peak to peak ozone variations are less than two percent. In order to obtain the extraordinary precision necessary to provide this clear of a picture of the O_3 response, we averaged zonal means of the LIMS data between ±40° latitude over the entire data set.

To determine the response time, the two time series are simply shifted, relative to one another, to the time giving the highest correlation. For the case in Figure 1, the response time was shorter than one day. Similar studies were performed at different pressure levels to determine the sensitivity (percent change for a one percent increase in 205-nm solar radiation) and response times of various species as a function of pressure level. At levels lower in the atmosphere, the response times were observed to be longer.

Similar studies were performed using four years of the highest quality NIMBUS 7 SAMS temperature data in an effort to detect the response of middle atmosphere temperatures to solar UV variability. Temperatures are affected by a number of short-term dynamical influences independent of solar variations, and thus it is more difficult to isolate the solar signal.

Shown in Figure 2 is the observed response time of temperatures to solar UV variability for two independent sets of NIMBUS 7 SAMS data: 12/78-10/80 and 10/80-10/82. The data were divided into two sets to assure the validity of the results, since the correlations were significant but generally less than 0.4. As may be seen, the independent data sets gave essentially the same response times. Response times near 2 mb (or 43 km) were about seven days. These long response times were consistent with similar studies performed over shorter time intervals using LIMS data (Keating et al., 1987a) and SAMS data (Hood, 1986).

Predicted response times based on diabatic and chemical effects (Brasseur et al., 1987) were much shorter. Perhaps an additional dynamic forcing coherent with the solar variation is required to account for the phase lags of the temperature/UV relation (Hood, 1986; Brasseur et al., 1987).

The unexpectedly long temperature response times have an important effect on the ozone response to solar UV variability through temperature feedback. The effect is shown schematically in Figure 3. Shown on the top of Figure 3 is Case A where we can see the effects of a rapid response of temperature to UV variability. The 27-day variation in solar UV is shown along with the immediate temperature response and the response of ozone before considering the effect of temperature feedback ("No T feedback"). Temperature increases tend to increase the rate of ozone depletion resulting in an inverse relation between temperature variations and ozone perturbations. (See "ΔO_3 from ΔT"). Dynamical effects can also play a role in this inverse relation. The net effect is shown as the curve labeled "T feedback". The effect of temperature feedback for this case is to lower the amplitude of the ozone response to solar UV variability. Shown schematically on the bottom of the Figure (Case B) are the consequences of the observed longer temperature response times. Temperatures are shown lagging the solar UV peak by about seven

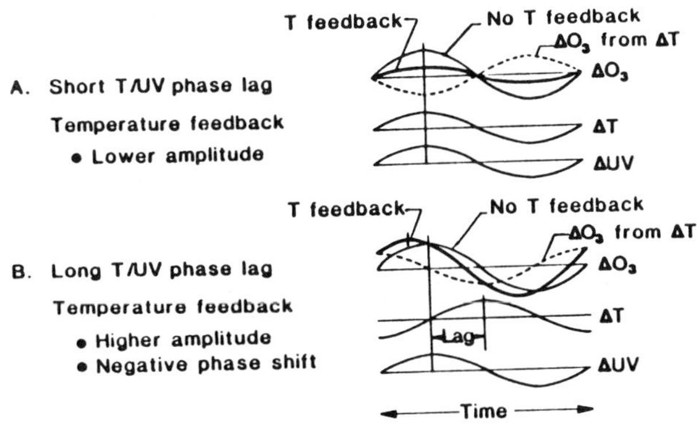

Fig. 3. Schematic of O_3/UV response to different temperature/UV phase lags.

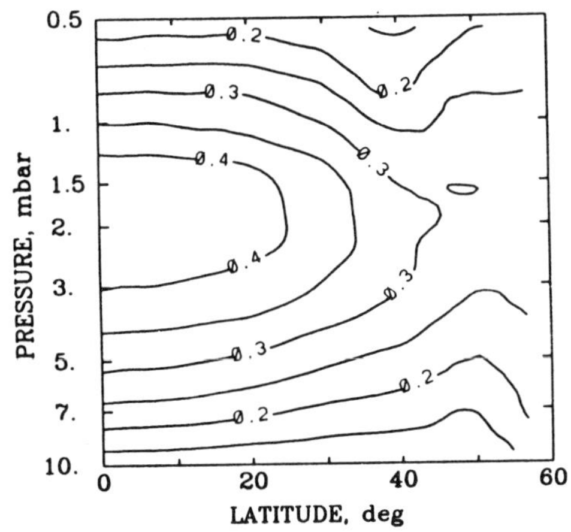

Fig. 4. Mean ozone sensitivity (percent ozone change for a one percent increase in 205 nm radiation) as a function of latitude and pressure, based on first 4 years of detrended NIMBUS 7 SBUV data. Temperature effects are not removed.

days for a 27-day cycle. The effect of temperature feedback on the ozone perturbation is shown again as "ΔO_3 from ΔT". In this case though the net effect (labeled "T feedback") is completely different. As opposed to temperature feedback reducing the amplitude of the O_3/UV effect, the amplitude is increased. Also, the ozone peak occurs before the UV peak. This latter effect occurs essentially because temperature is rising at the time of the UV peak.

Keating et al. [1987a] have demonstrated that these effects are real by comparing cases where temperature effects have been empirically removed with cases where temperature effects are not removed from the ozone data. Brasseur et al., [1987] have demonstrated, using a 1D chemical-radiative time-dependent model, that if the observed temperature response is specified in the model, the resulting ozone response is in accord with Case B. This response pattern can also allow us to understand latitudinal variations in the O_3/UV response.

3. LATITUDINAL VARIATIONS IN RESPONSE

Shown in Figure 4 is the mean sensitivity of ozone in percent for a one percent increase in 205-nm radiation. The sensitivity has been determined as a function of pressure and latitude. The results are based on the first four years of detrended NIMBUS 7 SBUV data. During this period, peak to peak 205-nm variability was at times of the order of five percent, and thus the 0.4 sensitivity contour would correspond in this case to periodic ozone variations of the order of two percent peak to peak. Maximum sensitivities are seen to occur at approximately 2 mb near the equator. The sharp drop in sensitivity with latitude is contrary to model predictions.

Shown in Figure 5 is the observed O_3/UV sensitivity at 2 mb compared to calculations using the Brasseur et al. 2D zonally averaged, time-dependent model with Eulerian residual circulation. The model sensitivities were calculated for a period of one year, and were determined both without temperature feedback and with feedback using specified equatorial temperature/UV sensitivities and long temperature response times. The observed and predicted O_3/UV sensitivities shown in Figure 5 are annual means. The theoretical model with temperature feedback gives higher O_3/UV sensitivities than the model with no temperature feedback, which is consistent with the schematic on long temperature response times (Figure 3, Case B). The sensitivities given by both theoretical model cases show almost no decrease with latitude. This is partially due to latitude independence of the specified temperature sensitivity in the model.

It is hypothesized that the observed sharp drop with latitude of the O_3/UV sensitivity is principally the consequence of a drop in the temperature/UV sensitivity with latitude. Thus there is effectively a transition from sensitivities similar to the upper model curve (temperature feedback) at low latitudes to sensitivities similar to the lower model curve (no temperature feedback) at high latitudes. Of course, there are also differences in the absolute values between predicted and observed sensitivities, but the major discrepancy appears to be the differences in the latitudinal variation.

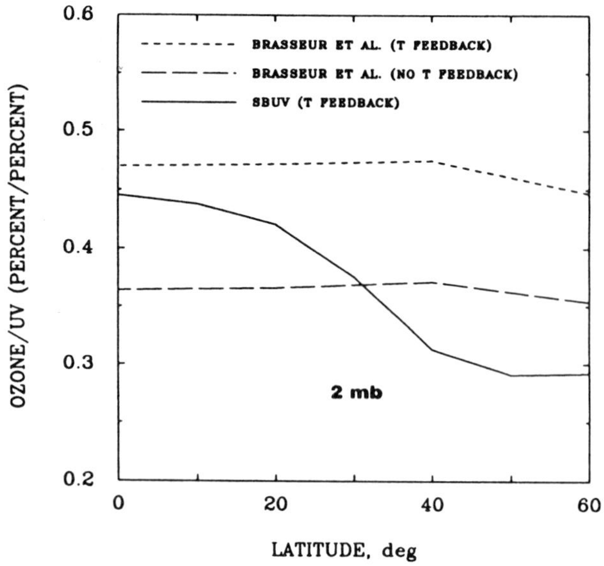

Fig. 5. Observed (SBUV) and predicted ozone/UV sensitivity (percent ozone change for a one percent increase in 205-nm radiation) at 2 mb as a function of latitude.

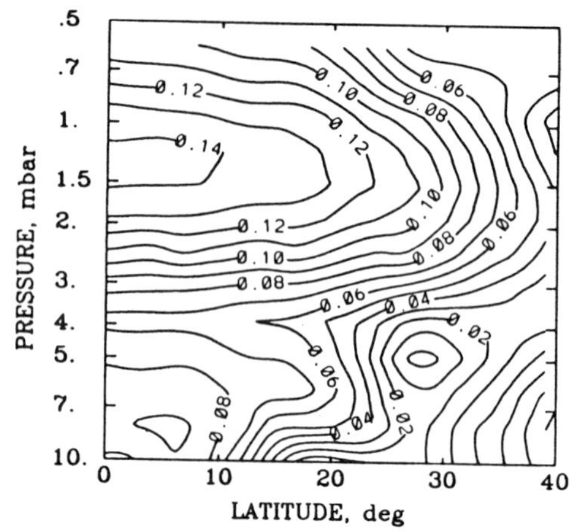

Fig. 6. Observed temperature/UV sensitivity (degrees K change for a one percent increase in 205 nm radiation) as a function of latitude and pressure. Based on all four years of highest quality NIMBUS 7 SAMS data.

To test the above hypothesis, a study was performed in an effort to detect a latitudinal variation in the temperature/UV sensitivity. Temperature data from the entire four-year SAMS data set were analyzed to determine mean temperature/UV sensitivities as a function of latitude. Data were analyzed in $\pm 10^{o}$ latitude bands. The results of the analysis are shown in Figure 6. Sensitivities are given in degrees K for a one percent increase in 205-nm radiation. Maximum temperature/UV sensitivities occur near the equator at approximately 1.5 mb. There is a sharp drop in sensitivity with increasing latitude as was hypothesized. It is clear that the region of highest sensitivity expands in latitude with increasing altitude. At 10 mb, high sensitivities are confined to the equatorial regions. As shown in Figure 2, response times at these altitudes are too long to be explained in terms of conventional theories. This suggests that the high sensitivities at 10 mb may be related to solar modulation of equatorial waves. A study of the temperature/UV response pattern in the Northern and Southern hemispheres (not shown here) shows a striking symmetrical pattern between the two hemispheres at most levels. However, near 1.5 mb, maximum temperature/UV sensitivities occur at about 15°S latitude. A similar study of the O_3/UV response in the two hemispheres also shows the maximum near this same level at approximately the same latitude. Therefore, the asymmetry in the ozone sensitivity pattern may also be explained in terms of temperature feedback.

Thus, the sharp latitudinal variations in O_3/UV sensitivity appear to be largely explained by a sharp latitudinal variation in the temperature/UV sensitivity. Of course, the cause of the temperature/UV sensitivity patterns and response times remains to be fully explained. The nature of the temperature response to 11-year solar variations as a function of the QBO, latitude and season (Labitzke, 1987) is also unexplained by present day theory.

Attention should also be called to the recent detection of the response to short-term solar UV variability of two other stratospheric species which are important to the ozone budget. From a study of LIMS data, HNO_3 at 10 mb is found to be more than twice as sensitive to short-term UV variability as O_3 (Keating et al., 1986). However, HNO_3 decreases with increased solar UV as a result of photodissociation, while O_3 increases. When HNO_3 decreases, NO_2 (a product of HNO_3 photodissociation) is observed to increase (Keating et al., 1986).

4. MESOSPHERIC RESPONSE

Studies have also been performed on the response of mesospheric O_3 and temperature to short-term solar UV variability. Shown in Figure 7 is the scaled response near the equator of the upper stratosphere and the entire mesosphere to a ten percent increase in 205-nm radiation. The ozone response in the mesosphere is based on a study of Solar Mesosphere Explorer (SME) 1.27 micron data. Best correlations were actually obtained using solar Lyman alpha (121.6-nm), also obtained from SME, as the solar index. With short-term increases in solar Lyman alpha (and in 205-nm solar flux), a clear systematic depletion of ozone has been detected peaking near 70 km (Keating et al., 1987a). This depletion is

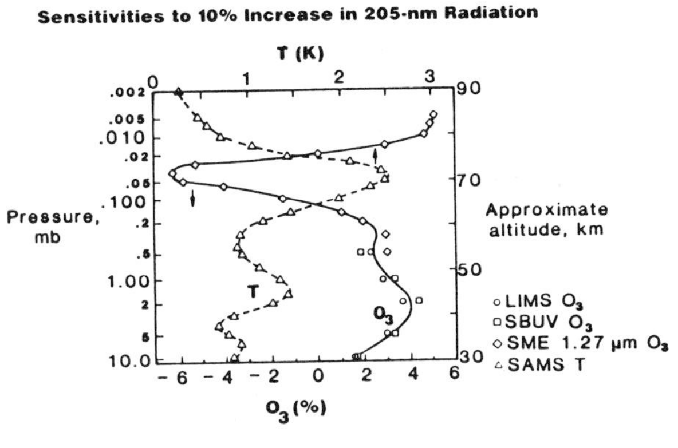

Fig. 7. Scaled response of equatorial stratospheric and mesospheric ozone and temperature to a ten percent increase in 205 nm solar UV radiation based on four satellite data sets.

apparently the result of solar Lyman alpha photodissociating H_2O resulting in the production of HO_x which acts to catalytically destroy O_x and ozone (Frederick, 1977; Garcia et al., 1984; Keating et al., 1987b). More recent studies show an annual mean latitudinal variation in this ozone depletion (Keating et al., 1987b) consistent with the predictions of Garcia et al. [1984]. These studies show promise in checking the detailed validity of present concepts of O_x/HO_x chemistry.

Studies of the temperature/UV response throughout the mesosphere have been performed using the NIMBUS 7 SAMS temperature data. Referring to Figure 7, the peak temperature/UV response also occurs near 70 km. This peak sensitivity, the highest throughout the middle atmosphere, is not predicted with present day theoretical models. The sensitivity peak is strong enough that it also plays a significant role in the ozone depletion near 70 km through temperature feedback. The peak temperature sensitivity near 70 km has also been detected in studies of the temperature response to 11-year solar variations (Mohanakumar, 1985; Chanin et al., 1987). At least part of the peak in temperature response is expected to result from solar Lyman alpha heating (Brasseur et al., 1987).

REFERENCES

Brasseur, G., A. DeRudder, G. M. Keating and M. C. Pitts, 1987: Response of middle atmosphere to short-term solar ultraviolet variations: 2. Theory. J. Geophys. Res., 92, 903.

Chanin, M. L., N. Smires and A. Hauchecorne, 1987: Long-term variations of the temperature of the middle atmosphere at mid-latitude: Dynamical and radiative causes. J. Geophys. Res., 92, 4201.

Ebel, A., M. Dameris, H. Hass, A. H. Manson, C. E. Meek and K. Petzoldt, 1986: Vertical change of the response to solar activity oscillations and periods of 13 and 27 days in the middle atmosphere. Ann. Geophys., 4, 271.

Frederick, J. E., 1977: Chemical response of the middle atmosphere to changes in the ultraviolet solar flux. Planet. Space Sci., 25, 1.

Garcia, R. R., S. Soloman, R. C. Roble and D. W. Rusch, 1984: A numerical response of the middle atmosphere to the 11-year solar cycle. Planet. Space Sci., 32, 411.

Hood, L. L., 1986: Coupled stratospheric ozone and temperature responses to short-term changes in solar ultraviolet flux: An analysis of NIMBUS 7 SBUV and SAMS data. J. Geophys. Res., 91, 5264.

Keating, G. M., 1981: The response of ozone to solar activity variations: A review. Solar Physics, 74, 321.

Keating, G. M., G. P. Brasseur, J. Y. Nicholson III and A. DeRudder, 1985: Detection of the response of ozone in the middle atmosphere to short-term solar ultraviolet variations. Geophys. Res. Lett., 12, 449.

Keating, G. M., J. Nicholson III, G. Brasseur, A. DeRudder, V. Schmailzl and M. Pitts, 1986: Detection of stratospheric HNO_3 and NO_2 response to short-term solar ultraviolet variability. Nature, 332, 43.

Keating, G. M., M. C. Pitts, G. Brasseur and A. DeRudder, 1987a: Response of middle atmosphere to short-term solar ultraviolet variations: 1. Observations. J. Geophys. Res., 92, 889.

Keating, G. M., M. C. Pitts, G. Brasseur, R. Bevilacqua and R. Thomas, 1987b: Detection of middle atmosphere ozone destruction via HO_x modulated by solar Lyman alpha. Proceedings of Sixth AMS Conference on the Dynamics and Chemistry of the Middle Atmosphere, Baltimore, MD.

Labitzke, K., 1987: Sunspots, the QBO, and the stratospheric temperature in the north polar region. Geophys. Res. Lett., 14, 535.

Mohanakumar, K., 1985: An investigation of the influence of solar cycle on mesospheric temperature. Planet. Space Sci., 33, 795.

STRATOSPHERIC RESPONSES TO SOLAR ULTRAVIOLET VARIATIONS: COMPARISONS WITH RADIATIVE-PHOTOCHEMICAL MODELS

L.L. Hood
Lunar and Planetary Laboratory, University of Arizona
Tucson, Arizona 85721, USA

A.R. Douglass
Applied Research Corporation
Landover, Maryland 20783, USA

ABSTRACT

Measurements of 27-day ozone responses to changes in solar ultraviolet flux derived from Nimbus 7 SBUV data are compared to calculations using a parameterized one-dimensional time dependent radiative photochemical model. At levels below about 3 mbar, the model is in agreement with the SBUV ozone response measurements within observational uncertainties. At higher levels, the observed response amplitude is consistently larger than the calculated amplitude. The observed latitude dependence of the ozone response amplitude during the N. H. summer season is also qualitatively consistent with radiative photochemical models only below about 3 mbar. At higher levels, response amplitudes increase with increasing latitude toward the winter hemisphere, suggesting the greater importance of dynamical processes (e.g. a modulation of planetary waves and associated temperature variations) in determining the ozone response. At levels below 3 mbar where approximate agreement between model and observation is obtained, the model is applied to calculate a maximum solar cycle ozone change of approximately 0.39% at 4.5 mbar for every 1% change in the solar flux at 205 nm. Adopting a change in the monthly mean 205 nm flux of ~ 6% between solar maximum in 1980 and solar minimum in 1985 yields a corresponding maximum monthly mean ozone mixing ratio decrease of 2.3% at 4.5 mbar. At other pressure levels, the calculated decreases are smaller, ranging from 2.0% at 3 mbar to 0.8% at 10 mbar. The net decrease during this interval in total ozone at low latitudes resulting directly from these upper stratospheric mixing ratio decreases would be less than 1%.

1. INTRODUCTION

In this paper, we compare short-term ozone responses derived from Nimbus 7 solar backscattered ultraviolet (SBUV) data to calculations using models that consider radiative and photochemical processes but not dynamical processes. At pressure levels and latitudes where model calculations agree with the response measurements, the same model is then applied to estimate long-term (solar cycle) ozone changes that would have been measured with the SBUV instrument during the 1979 to 1985 time period. As motivation for this work, according to the Ozone Trends Panel report, the SBUV and TOMS column ozone data, normalized to ground-based Dobson measurements and averaged between 53° S and 53° N latitude, show a decrease of about 2.5% between late 1979 and late 1985. The report also states that model calculations have predicted a decrease due to solar ultraviolet changes in the same time period (approximately solar maximum to solar minimum) in the range of 0.7 to 2%, depending on the assumed spectral irradiance change and other model uncertainties. The calculated solar-induced ozone changes occur in approximately the same altitude zone in which model calculations predict maximum secular changes due to trace gas release [Herman and McQuillan, 1985]. A more accurate evaluation of solar ultraviolet effects on stratospheric ozone abundance is therefore desirable in order to more reliably constrain possible anthropogenic trends in total ozone at low latitudes using these satellite data sets.

Current model calculations of solar cycle ozone variations are limited in part by uncertainties in measurements of long-term solar spectral irradiance changes [Lean, 1987] but are also limited by theoretical uncertainties and differences in modeling procedures. For example, Brasseur and Simon [1981] have applied a 2-D steady-state radiative photochemical model with parameterized transport to calculate a maximum ozone mixing ratio change of 10% near 1.5 mbar over a solar cycle. They adopted a solar cycle UV flux variation of 22% near 200 nm, decreasing monotonically to zero at 300 nm. On the other hand, Garcia et al. [1984] applied a 2-D model with coupled photochemistry and dynamics to estimate a maximum mixing ratio change of 4.5% near 10 mbar. Their assumed solar flux variation was about 16% near 200 nm, decreasing toward zero at 300 nm. Observational constraints

ISBN 0-937194-15-8

on these long-term model calculations are unfortunately not very restrictive. Although Reinsel et al. [1988] have applied regression methods to estimate long-term total ozone responses to changes in the 10.7 cm flux over the 1978-1985 interval using Nimbus 7 SBUV data, the authors emphasize the difficulty of separating solar UV effects from other natural and anthropogenic trends in this temporally limited data set. Other studies, using ground-based or balloon measurements, have yielded provisional evidence for a nearly in-phase ozone-solar cycle relationship [Angell and Korshover, 1978; Paetzold, 1973], but they extend, at most, over only one or two 11-year cycles, so the statistical significance of the correlations is not large.

To date, responses of stratospheric ozone to changes in solar ultraviolet flux have been clearly detected only on the time scale of the solar rotation period and occur with maximum amplitude near about 40 km altitude at low latitudes [e.g., Hood, 1986; Keating et al., 1987; Chandra, 1986]. The observed short-term ozone responses therefore provide an important test of current model calculations of long-term stratospheric ozone changes in response to the solar activity cycle.

2. SHORT-TERM RESPONSE MEASUREMENTS AND MODEL COMPARISONS

Shown in Figure 1 are 27-day ozone response amplitudes and phase lags derived from two independent 22-month intervals of Nimbus 7 SBUV data [Hood, 1986; Hood and Cantrell, 1988]. These response measurements agree well with an independent analysis by Keating et al. [1987] of the same 4-year time interval. The response amplitude or sensitivity is defined as the mean percent change in SBUV ozone mixing ratio for a 1% change in the solar flux at a reference wavelength of 205 nm. Error bars are standard deviations for values obtained in five separate regression analyses for 10° latitude bands centered on 20° S, 10° S, 0°, 10° N, and 20° N latitude. For general discussions of the short-term ozone and temperature responses in the upper stratosphere, including explanations for the negative ozone lags observed at levels above 3 mbar, see Hood [1986; 1987] and Keating et al. [1987]. Hood and Douglass [1988] have shown that the observed 27-day ozone and coupled temperature responses in the upper stratosphere are consistent with the odd oxygen continuity equation within measurement and model uncertainties.

Brasseur et al. [1987] have previously presented a comparison of 27-day upper stratospheric ozone and temperature responses with calculations using a one-dimensional time dependent radiative photochemical code. In their calculations, diurnal average conditions were assumed for the solar illumination for 30° latitude at equinox. In agreement with earlier analytic calculations by Hood [1986], they concluded that processes in addition to those included in the model (e.g. dynamical processes) must contribute substantially to the temperature response. The solid lines in Figure 1 show the calculated ozone response amplitude and phase lag of Brasseur et al. [1987]. The calculated amplitude is significantly less than the measured amplitude at pressure levels above ~ 7 mbar. However, an important factor in the disagreement may be the use of diurnally averaged photodissociation rates in the model calculations.

In the study of Hood and Douglass [1988], it was found that an important consideration in the photolysis rate calculations is the range of solar zenith angles appropriate for the SBUV ozone observations. At levels below about 10 mbar where the odd oxygen relaxation time becomes comparable to or greater than 1 day, 24-hour diurnal averaging over solar zenith angle is appropriate. However, at higher levels where most response measurements were obtained, the relaxation time is less than 1 day and approaches 1 hour at 0.7 mbar. Thus averaging over a smaller range of angles bracketing the time of observation is necessary. Because the SBUV in-

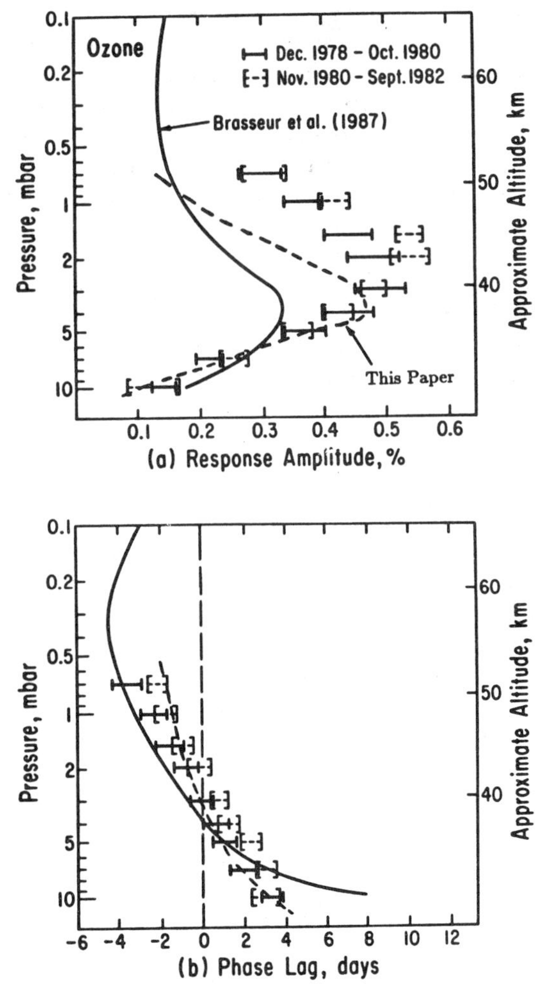

Fig. 1. Comparison between measurements of SBUV ozone responses to short-term solar ultraviolet variations and radiative photochemical model calculations. The measurements are taken from Hood and Cantrell [1988].

strument obtained measurements at low latitudes near local noon, the photodissociation rate changes at these levels are significantly larger (by a factor of nearly 2) than would have been calculated using diurnal averaging.

We have carried out calculations using a parameterized perturbation-order radiative photochemical model but using a variable solar zenith angle range in the photodissociation calculations that is more appropriate for comparison with the SBUV measurements. Photochemical parameters are taken from the model of Stolarski and Douglass [1985] for the equator at equinox. The resulting ozone response amplitude and phase lag are shown in Figure 1 as dashed lines. The increased photodissociation rate changes result in a higher amplitude for the ozone response that is more nearly in agreement with the observed amplitude at levels below about 3 mbar.

To provide an alternate approach toward evaluating the adequacy of a radiative photochemical model formulation for simulating the observed 27-day ozone responses to solar ultraviolet variability, it is of interest to consider the latitude and seasonal dependences of the ozone responses. These dependences are diagnostic of the importance of dynamical processes in contributing to the measured responses. During the past year, we have begun a program to calculate these dependences for both ozone and temperature responses using the nearly 4 years of simultaneous Nimbus 7 SBUV and stratospheric and mesospheric sounder (SAMS) data sets. Preliminary ozone response results for the northern hemisphere summer season are shown in Figure 2 at latitudes from 40° S to 60° N (a range determined by availability of the SAMS data). During this season, correlation coefficients between the solar 205 nm flux and SBUV ozone mixing ratio (detrended to remove periods greater than about 35 days) are consistently positive ($\leq$ 0.75) allowing the determination of response amplitudes and phase lags with reasonable confidence. Phase lags (in days) are indicated in parentheses beside each response amplitude value.

At the highest altitudes (0.7 and 1.5 mbar), the response amplitude of Figure 2 increases with increasing latitude toward the *winter* hemisphere. This is opposite to expectations based on solar zenith angle considerations and indicates that purely radiative and photochemical processes are not responsible for this latitude dependence. However, such a latitude dependence (with increasing amplitude toward the winter hemisphere) could be explained by a phase modulation of planetary waves with periods comparable to the solar rotation period. At lower levels (especially between 3 and 5 mbar), the response amplitude maximizes in a broad band centered approximately on the equator and is therefore more consistent with purely photochemical and radiative processes. The reduced importance of dynamical effects at these levels is explained in part by the lower temperature response amplitudes compared to the larger ozone response amplitudes. This results in a smaller percentage temperature feedback contribution to the ozone response. Since the ozone response without temperature feedback is dominantly photochemical, dynamical effects are minimized. We conclude that the observed summer season latitude dependence is support-

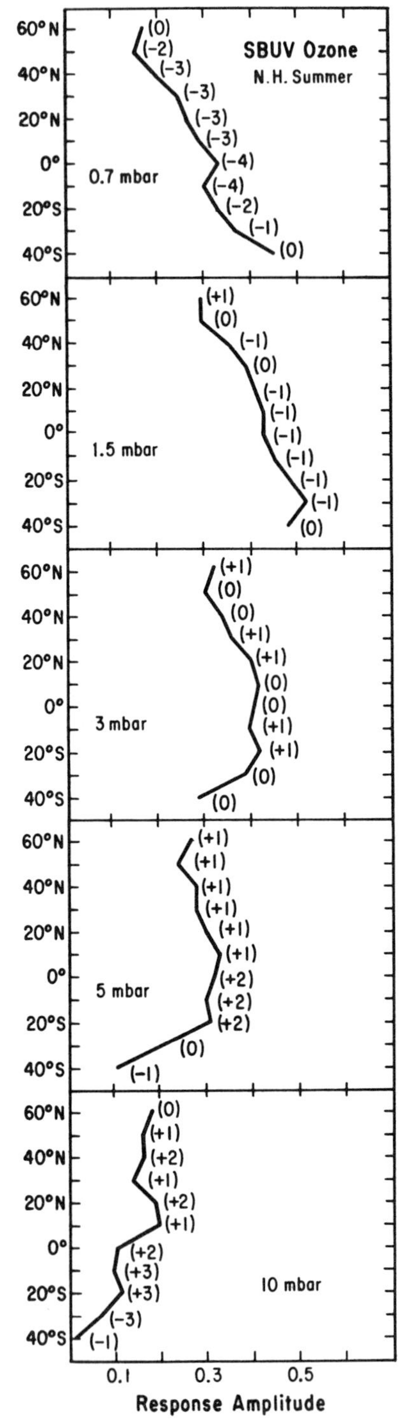

Fig. 2. Dependence on latitude of the mean 27-day ozone response amplitude and phase lag during the N. H. summer season (defined as May to September). Phase lags in days are indicated in parentheses for each 10° latitude band.

ive of the ability of radiative photochemical model calculations to simulate the observed mean 27-day ozone response at pressure levels below about 3 mbar.

3. EXTENSION TO THE SOLAR CYCLE TIME SCALE

The same radiative photochemical code used to produce the dashed line model of Figure 1 for the 27-day time scale was applied in order to calculate the expected long-term variation of ozone mixing ratio at levels between 3 and 10 mbar and at equatorial latitudes during the period of operation of the SBUV instrument. Results show a maximum amplitude of 0.39% at 4.5 mbar ($\sim$ 37 km altitude) for every 1% change in the solar 205 nm flux. At other levels, the calculated change is smaller, ranging from 0.34% at 3 mbar to 0.14% at 10 mbar.

Although the actual solar spectral irradiance change between 1979 and 1985 remains controversial, some convergence of estimates has occurred in recent years. Using the Mg II core index of Heath and Schlesinger [1986], a change in the monthly mean value of the SBUV 205 nm flux between solar maximum in 1980 and solar minimum in 1985 of $\sim$ 6% may be estimated. (If peak-to-peak monthly variations are also considered, the maximum change during the same time period is $\sim$ 9%.) Using the SME 200-205 nm flux, a monthly mean change between early 1982 and early 1985 of $\sim$ 5% may be estimated although instrument error sources are still being evaluated [G. Rottman, private communication, 1988]. Since the Mg II core index shows little or no monthly mean change between 1980 and 1982, the SME and Mg II index estimates of the 205 nm flux change between solar maximum and minimum may be consistent within instrumental uncertainties.

Adopting a change in the monthly mean solar 205 nm flux of $\sim$ 6% between solar maximum in 1980 and solar minimum in 1985 yields corresponding monthly mean ozone mixing ratio decreases of 2.0% at 3 mbar, 2.3% at 4.5 mbar, and 0.8% at 10 mbar. Such decreases are small in comparison to seasonal variations and would be difficult to detect relative to other natural sources of variability such as the El Chichon volcanic eruption. The net change in the ozone column at low latitudes resulting directly from these upper stratospheric mixing ratio decreases would be less than 1%. It is therefore concluded that the total ozone trends of $\sim$ 2.5% reported by the Ozone Trends Panel for the same time interval based on SBUV and TOMS data can not be explained as due to changes in solar spectral irradiance alone. The remaining $\sim$ 1.5% may represent, in part, an anthropogenic trend due to release of chlorofluoromethanes. However, additional natural effects such as those due to El Chichon aerosols must be quantitatively evaluated before an accurate low-latitude anthropogenic trend can be determined from existing satellite data sets.

REFERENCES

Angell, J.K., and J. Korshover, Global ozone variations: An update into 1976, *Mon. Weather Rev., 106*, 725-737, 1978.

Brasseur, G., A. De Rudder, G. M. Keating, and M. C. Pitts, Response of middle atmosphere to short-term solar ultraviolet variations: 2. Theory, *J. Geophys. Res., 92*, 903-914, 1987.

Brasseur, G. and P. C. Simon, Stratospheric chemical and thermal response to long-term variability in solar UV irradiance, *J. Geophys. Res., 86*, 7343-7362, 1981.

Chandra, S., The solar and dynamically induced oscillations in the stratosphere, *J. Geophys. Res., 91*, 2719-2734, 1986.

Garcia, R., S. Solomon, R.G. Roble, and D.W. Rusch, A numerical response of the middle atmosphere to the 11-year solar cycle, *Planet. Space Sci., 32*, 411-421, 1984.

Heath, D. F. and B. M. Schlesinger, The Mg 280-nm doublet as a monitor of changes in solar ultraviolet irradiance, *J. Geophys. Res., 91*, 8672-8682, 1986.

Herman, J.R. and C.J. McQuillan, Atmospheric chlorine and stratospheric ozone nonlinearities and trend detection, *J. Geophys. Res., 90*, 5721–5732, 1985.

Hood, L. L., Coupled stratospheric ozone and temperature responses to short-term changes in solar ultraviolet flux: An analysis of Nimbus 7 SBUV and SAMS data, *J. Geophys. Res., 91*, 5264-5276, 1986.

Hood, L. L. and S. Cantrell, Stratospheric responses to solar ultraviolet variations: Reproducibility of low-latitude response measurements, *Annales Geophysicae*, in press, 1988.

Hood, L. L. and A. R. Douglass, Stratospheric responses to solar ultraviolet variations: Comparisons with photochemical models, *J. Geophys. Res., 93*, 3905-3911, 1988.

Keating G. M., J. Y. Nicholson III, D. F. Young, G. Brasseur, and A. De Rudder, Response of middle atmosphere to short-term solar ultraviolet variations: 1. Observations, *J. Geophys. Res., 92*, 889-902, 1987.

Lean, J., Solar ultraviolet irradiance variations: A review, *J. Geophys. Res., 92*, 839-868, 1987.

Paetzold, H.K., The influence of solar activity on the stratospheric ozone layer, *Pure Appl. Geophys., 106-108*, 1308-1312, 1973.

Reinsel, G. C., G. C. Tiao, S. K. Ahn, M. Pugh, S. Basu, J. DeLuisi, C. L. Mateer, A. J. Miller, P. S. Connell, and D. J. Wuebbles, An analysis of the 7-year record of SBUV satellite ozone data: Global profile features and trends in total ozone, *J. Geophys. Res., 93*, 1689-1703, 1988.

Stolarski, R. S., and A. R. Douglass, Parameterization of the photochemistry of stratospheric ozone including catalytic loss processes, *J. Geophys. Res., 90*, 10,709-10,718, 1985.

ON THE INFLUENCE OF INTERNAL GRAVITY WAVES ON RADIATION FIELD AND PHOTODISSOCIATION RATES OF GASES OF THE ATMOSPHERE

M.Yu. Danilin and G.I. Kouznetsov
Department of Atmospheric Physics
Moscow State University
Moscow, USSR

Internal gravity waves (IGW) modulate air density and temperature in the region of their propogation. This fact has been established both experimentally and by theoretical calculations(Chanin et al,1981,1984; Philbrick,1981;Chunchuzov,1978). Change of air density due to IGW leads to the corresponding changes of the radiation field and trace gas concentrations.

The main aim of this study is to calculate changes of the radiation field and photodissociation rates due to this mechanism. We used the following data for modelling: solar flux at the top of the atmosphere(Mount et al,1985),cross section and quantum yield of molecules (Baulch et al,1984;Nicolet et al,1986), Rayleigh scattering(Nicolet et al,1986), ozone vertical distribution(Krueger et al,1976). Light transfer equation was solved in height region 0-88km in spectral region L_α -750nm. It was assumed that solar radiation was reduced by ozone(O_3) and oxygen(O_2) absorption and Rayleigh scattering. Spectral step $\Delta\lambda$ was irregular,i.e. $\Delta\lambda$=10nm(λ>320nm) and $\Delta\lambda$=500cm^{-1} in Schumann-Runge bands of O_2.

The solution of light transfer is the following:

$$I(\lambda,z,\chi)=I_0(\lambda)\exp\left(-\sum_{O_2,O_3,RS}\sigma_i(\lambda)\int_z^\infty n_i(h)\,dh/\cos\chi\right) \quad (1)$$

where $I(\lambda,z,\chi)$-solar radiation at height z at wavelenth λ,

$I_0(\lambda)$-the same at the top of the atmosphere,

χ-solar zenith angle.

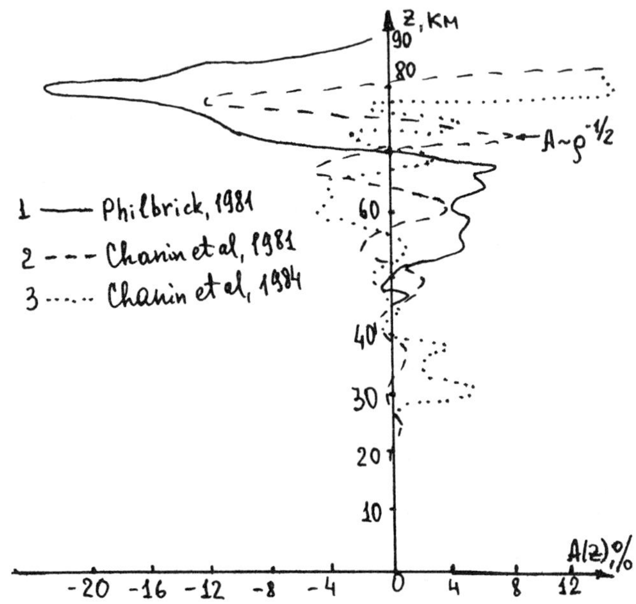

Fig.1. Modulation of air density by IGW.

Experimental data characterized the air density modulation by IGW are still rare. Results of (Chanin et al,1981,1984) and (Philbrick,1981) are shown in Fig.1. Dashed line corresponds to 2% air density deviation at 50km due to IGW with λ_z=12km (this value is typical for atmosphere).

Both stationary and travelling IGW have been detected in the atmosphere (Vincent,1984). Let us consider,first, stationary plane IGW(SPIGW). In the general case(Fig.2) wave azimuth angle α, angle between nadir line and wave plane equals β. It may be shown that SPIGW modificates formula (1) to (2):

ISBN 0-937194-15-8

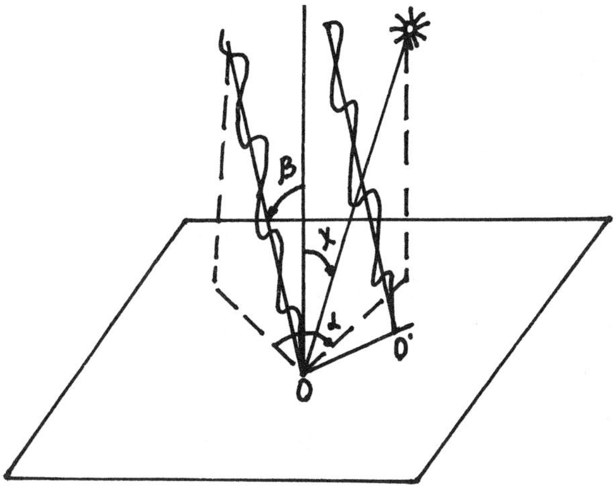

Fig.2. Possible mutual Sun and wave localisation.

$$I^W(\lambda,z,\chi)=I_0(\lambda)\exp\Big(-\sum_{O_2,O_3,RS}\sigma_i(\lambda)\int_z^\infty n_i(h)\cdot\Big(1+A(h)\frac{\cos 2\pi\left(\frac{t_0}{T}-\frac{h\cos\beta}{\lambda^*}\right)}{\cos(\gamma-\beta)}\Big)\,dh/\cos\chi\Big) \qquad (2)$$

where T-wave period, $A(h)=\Delta n(h)/\bar{n}(h)$,

$\beta=\tan^{-1}(\lambda_z/\lambda_x)$, $\lambda^{*2}=(\lambda_x^{-2}+\lambda_z^{-2})^{-1}$,

$\gamma=\sin^{-1}(\sin\chi\cos\alpha)$.

Radiation field change by IGW was denoted in terms of $\xi(\lambda,z,\chi)$:

$$\xi(\lambda,z,\chi)=(I^W(\lambda,z,\chi)-I(\lambda,z,\chi))/I(\lambda,z,\chi)$$

Consider the following situation: $t_0=0$, $\alpha=\chi=0^\circ$, $\lambda_z=12$km, $\lambda_x=108$km for SPIGW with air density modulation for case 2. Radiation field changes were shown for this scenario in Fig.3. It is important that $\xi(\lambda,z,\chi)$ has wave-like structure with phase lag 90° with respect to SPIGW phase. Maximum values of radiation field modulation for chosen wavelength are the following: $\xi(L_\alpha,52\text{km})=16.5\%$, $\xi(180,46\text{km})=-11.5\%$, $\xi(205,16\text{km})=0.39\%$, $\xi(250,28\text{km})=3.7\%$, $\xi(320,22\text{km})=-0.08\%$. It is important to stress the dependence of these results on t_0, i.e. for $t_0=T$ the sign of $\xi(\lambda,z,\chi)$ changes, but the absolute value remains approximately the same.

The effect of solar radiation redistribution (ESRR) due to IGW is the important result of our work. ESRR' origin was demonstrated in Fig.4.

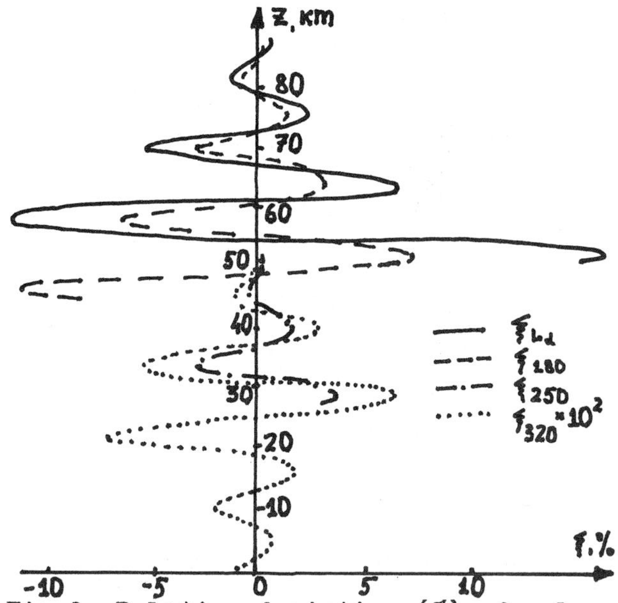

Fig.3. Relative deviations(%) of solar radiation at L_α, 180nm, 250nm and 320nm due to SPIGW for case 2.

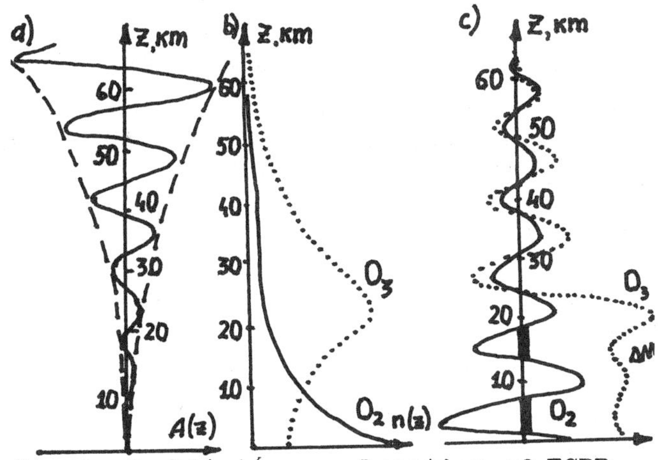

Fig.4. Qualitative explanation of ESRR origin: a) modulation of air density by IGW ($\lambda_z=12$km); b) vertical ozone and oxygen distributions; c) vertical distributions of $\Delta N_3(z)$ and $\Delta N_2(z)$. Solid broad lines at vertical axes show height region of ESRR, i.e. $\Delta N_3(z)$ and $\Delta N_2(z)$ have opposite signs here.

Deviatins of column ozone $\Delta N_3(z)=\int_z^\infty n_3(h)A(h)\cos 2\pi h/\lambda_z\,dh$ and oxygen $\Delta N_2(z)=\int_z^\infty n_2(h)A(h)\cos 2\pi h/\lambda_z\,dh$ were shown in Fig.4c. $\Delta N_2(z)$ and $\Delta N_3(z)$ have opposite signs below the maximum in O_3

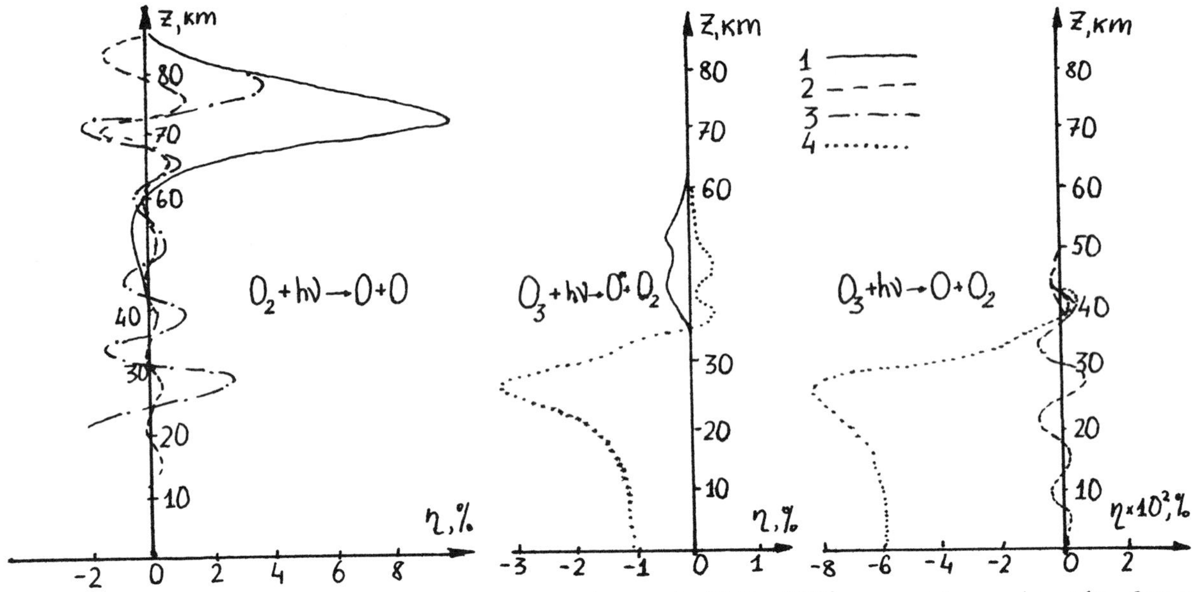

Fig.5. Responses of photodissociation rates of (R1)-(R3) reactions to air density modulation by SPIGW ($\alpha=0°$): 1- $\chi=0°$,Philbrick; 2- $\chi=0°$,Chanin et al,1981; 3- $\chi=75°$,Chanin et al,1981; 4- $\chi=0°$,Chanin et al,1984.

concentration and in the mesosphere,where scale heights of the atmosphere and ozonosphere differ considerably. It means that $\xi(\lambda,z,\chi)$ may have opposite signs at the same height in such wave condition,that implies solar energy transport within the spectrum in the atmosphere. ESRR becomes more obvious with growth of solar zenith angle. For example,for air deviation profile (Chanin et al,$t_0=0,\alpha=0°$) at z=18km $\xi(290nm,0°)=-0.10\%$, $\xi(335nm,0°)=0.02\%$; but when $\chi=75°$: $\xi(290nm,75°)=-0.51\%$, $\xi(335nm,75°)=0.05\%$.

Situation for travelling IGW(TIGW) is more complicated,because is not a conservative trace gas in this case due to its small photochemical relaxation time $\tau(h,\chi)$ in the upper stratosphere. For TIGW ozone oscillation are less by factor $\exp(-T/\tau(h,\chi))$ and ozone phase lag $t_3=T\cdot\tan(T/\tau(h,\chi))$ (Gruzdev et al,1983). So the solution of light transfer equation for TIGW takes the following form:

$$I^W(\lambda,z,\chi)=I_0(\lambda)\exp\Big(-\sum_{O_2,RS}\sigma_i(\lambda)\int_z^\infty n_i(h)\Big(1+A(h)\frac{\cos 2\pi\left(\frac{t_0}{T}-\frac{h\cos\beta}{\lambda^*}\right)}{\cos(\gamma-\beta)}\Big)\frac{dh}{\cos\chi}-\sigma_3(\lambda)\int_z^\infty n_3(h)\Big(1+A(h)\exp\Big(-\frac{T}{\tau(h,\chi)}\Big)\cdot\frac{\cos 2\pi\left((t_0-t_3)/T-z\cos\beta/\lambda^*\right)}{\cos(\gamma-\beta)}\Big)dh/\cos\chi\Big)$$

Let us choose TIGW parameters so that TIGW produces the same air density modulation as SPIGW does. Then we compare radiation field modulation by SPIGW and by TIGW,we may conclude that oxygen-active spectral region(i.e.$\xi(L_\alpha,z)$ or $\xi(180nm,z)$) are the same at all heights,but $\xi(205nm,z)$,$\xi(250nm,z)$ and $\xi(320nm,z)$ decrease in the stratosphere by factors 1.1.1.6,2.1 respectively. ESRR is not stationary and exists in wider height intervals for TIGW.

Parameter $\eta_i(z,\chi)$ is defined in the following form:

$\eta_i(z,\chi)=(J_i^W(z,\chi)-J_i(z,\chi))/J_i(z,\chi)$, where $J_i^W(z,\chi)=\int_0^\infty I^W(\lambda,z,\chi)\sigma_i(\lambda)\varphi_i(\lambda)\,d\lambda$ and $J_i(z,\chi)=\int_z^\infty I(\lambda,z,\chi)\sigma_i(\lambda)\varphi_i(\lambda)\,d\lambda$ are the disturbed and undisturbed photodissociation rates respectively, $\sigma_i(\lambda)$ is molecular cross-section.

This parameter was used as response of i-gas photodissociation rate to IGW action.

Changes of photodissociation rates : $O_2+h\nu=O+O$ (R1)

$$O_3 + h\nu = O^* + O_2 \quad (R2)$$
$$O_3 + h\nu = O + O_2 \quad (R3)$$

(i.e.reactions of 1-st,2-d and 3-d type (Luther and Gelinas,1976)) are presented in Fig.5 for SPIGW. When SPIGW and TIGW modulate air density equally,(R2)-(R3) responses decrease by factor 1.5 and 2.0. (R1) response do not change. Photodissociation reaction responses increase with the growth of such TIGW characteristics as T, λ_z,vertical energy flux F_z.

It is obvious that photodissociation rate responses have the wave-like structure. Lines 1 and 2 in Fig.5 show large change (R1) and (R2): η_{O_2}(72,0)=9.7% η_{O_3}(26km,0°)=2.9%. This "resonance" takes place at such a height due to the fact that height of maximum of the air density deviation and that of the absorption of the spectral region,which makes the greatest contribution to the photolys rate of this gas,coincide. Responses of photodissociation rates decrease by a factor of a 1000 when the number of spectral types of the reaction grows from 1 to 3. So both considerable differences in the photolys rate responses and ESRR can lead to local perturbation (or thin structure) of vertical distribution of gases. Further work should consider radiation field modulation by spaced-limited wave sources and changing the wave phase along line OO'(Fig.2).

R E F E R E N C E S

Baulch D.L..R.A.Cox,R.F.Hampson Jr.,J.A. Kerr,J.Troe,R.T.Watson.,1984: Evaluated kinetic and photochemical data for atmospheric chemistry.Supp.2. J.Phys.Chem.Ref.Data,13,1259-1378.

Chanin M.-L.,A.Hauchecorne,1981: Lidar observation of gravity and tidal wave in the stratosphere and mesosphere. J.Geophys.Res.,86,9715-9721.

Chanin M.-L.,A.Hauchecorne,1984; Lidar studies of temperature and density using Rayleigh scattering.-Handbook for MAP,13,87-98.

Chunchuzov E.P.,1978: On energetical characteristics of internal gravity waves registrated on hydroxil emission near mesopause.-Izv.AN USSR, Phys.Atmosph.Ocean,14,1094-1097.

Gruzdev A.N.,N.F.Elansky,Yu.L.Trutse, 1983: IGW influense on ozone and other trace gases in the atmosphere. In:Atmosph.Ozone(ed.N.F.Elansky & A.Kh.Khrian),Moscow,Nauka,18-27.

Krueger A.J.,R.A.Minzner,1976:A midlatitude ozone model for the 1976 US standard atmosphere. J.Geophys. Res.,81,4477-4481.

Luther F.M.,R.J.Gelinas,1976: Effect of molecular multiple scattering and surface albedo on atmospheric photodissociation rates. J.Geophys. Res. ,81,1125-1132.

Mount G.H.,G.J.Rottman,1985: Solar absolute spectral irradiance 118-300 nm:July 25,1983. J.Geophys.Res.,90, 13031-13036.

Nicolet M.,R.Kennes,1986: Aeronomic problem of the molecular oxygen photodissociation.1.The O_2 Herzberg continuum. Planet.Space Sci.,34,1043-1059.

Philbrick C.R.,1981: Measurements of structural features in profiles of mesospheric density. Handbook for Map, 2,333-340.

Vincent R.A.,1984: Gravity-wave motions in the mesosphere,J.Atmos.Terr.Phys., 46,119-131.

MESOSPHERIC OZONE BEHAVIOR DETERMINED FROM SATELLITE DATA

Arthur C. Aikin
NASA Goddard Space Flight Center
Greenbelt, Maryland 20771, USA

ABSTRACT

Sunset equatorial ozone profile data obtained by observing ultraviolet solar occultations with the UVSP instrument on the SMM spacecraft are reported for the 1985-1988 period. Between 20°N and 20°S an annual as well as a semiannual oscillation is evident in the lower mesospheric ozone data, with greatly reduced amplitude at 20°S. These UVSP data are compared with SBUV measurements of ozone mixing ratios at 0.4 mb, and both the SBUV and UVSP data sets agree in the latitude distribution of the equatorial oscillations. The SBUV ozone concentrations are as much as 16% less than those observed by UVSP/SMM. The SBUV data exhibit an downward trend of approximately 2% per year since 1980 with the largest drop occuring since 1984. Comparison of individual sunset SAGE II profiles with UVSP/SMM shows good agreement to 60 km. Above 55 km UVSP shows that sunrise ozone is greater than sunset with profiles diverging as altitude increases.

1. INTRODUCTION

The Solar Maximum Mission, SMM, spacecraft carries a high resolution ultraviolet spectrometer polarimeter, UVSP, which has been employed since 1980 to measure ozone by the technique of solar occultation. Since 1985 all observations have been carried out at 2788 Å, a wavelength for which unit optical depth at twilight is 60 km. Useable ozone profiles are obtained between 50 and 70 km with the greatest accuracy centered at 60 km. Employing a 1 x 180 arc second^2 slit to observe the sun yields a height resolution of 11 meters when the long axis of the slit is parallel to the limb of the earth and 2.1 km for a 90° rotation of the slit. In the present analysis there was no sorting according to slit orientation. Due to the 28.5° inclination of the spacecraft orbit occultation data are only available between ±40°. The results presented here are based on more than 15000 profiles obtained between August 1985 and May 1988.

During occultation the intensity of radiation observed by the instrument is given by

$$I_\lambda(h_T) = I_\lambda(h_o)\tau_\lambda(h_T) \qquad (1)$$

where $I_\lambda(h_T)$ and $I_\lambda(h_o)$ are observed and unattenuated intensities respectively at the tangent height h_T and the height h_o where there is no attenuation. The attenuation factor τ_λ is defined as

$$\tau_\lambda(h_T) = \exp\left(-2\sigma_\lambda \int_{h_T}^{h_o}[O_3]dl\right) \qquad (2)$$

where σ_λ is the ozone absorption cross section at 2788 Å and dl is the path element along the tangent height line between the instrument and the sun. The total amount of ozone between the sun and the satellite is equivalent to twice the integrated amount of ozone between the tangent height h_T and the sun. By taking the intensity ratio $I_\lambda(h_T)/I_\lambda(h_o) = \Phi(h_T)$ an integral equation can be written for the ozone concentration

$$\Phi(h_T) = (1/2\sigma_\lambda)\ln[1/\tau_\lambda(h_T)] = \int_{h_T}^{h_o}[O_3]dl \qquad (3)$$

The solution of this equation for the ozone concentration as a function of altitude is given by Aikin et al. (1982).

Examples of the profiles are shown in Fig. 1, where a sunset and a sunrise ozone concentration versus altitude profile are plotted. These profiles were both obtained on 18 December 1985 at 2° latitude. Two prominent differences between the profiles are the magnitude of the concentrations and the scale height of the distribution. In the present case the sunrise and sunset profiles agree at altitudes below 55 km and diverge at higher altitudes. At 65 km the sunrise ozone

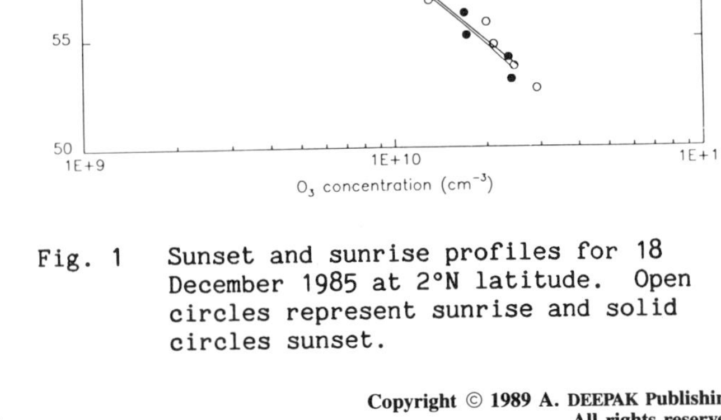

Fig. 1 Sunset and sunrise profiles for 18 December 1985 at 2°N latitude. Open circles represent sunrise and solid circles sunset.

ISBN 0-937194-15-8

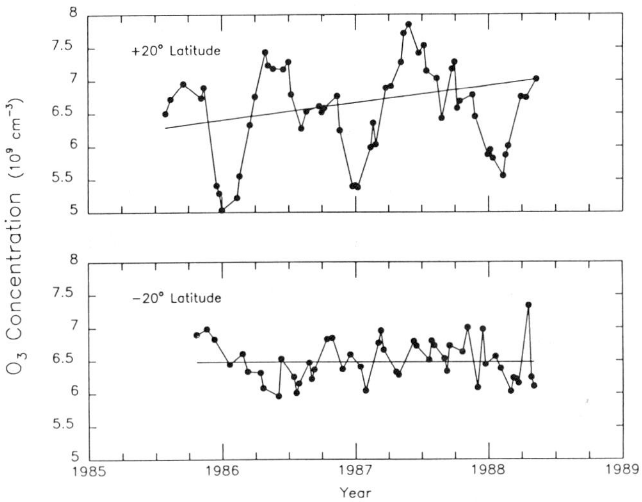

Fig. 2 The long term behavior of ozone concentration at 60 km for 20°N, top panel, and 20°S, bottom panel. Solid line represents the trend line.

concentration is 55% greater than at sunset. Between 60 and 65 km scale heights are 4 km for sunset profiles and 5 km for sunrise profiles.

Only sunset profiles are included in the analysis of long term behavior. Individual UVSP profiles have been compared with SAGE II and found to agree to within 10% for altitudes below 60 km.

2. ANNUAL VARIATIONS

In order to examine the data for long-term behavior, profiles have been binned into weekly means. Because of the occultation nature of the experiment there is no uniform distribution of data at a given latitude. Nevertheless, all data in a latitude zone of ±5° are used to obtain a zonal mean centered at latitudes 0, ±10°, and ±20° for the purpose of the present study. Figure 2 displays weekly averaged ozone concentrations for an altitude of 60 km. Mean ozone is displayed for 20°N and 20°S.

The ozone at 20°N exhibits a strong semiannual and annual oscillation. At 20°S the same oscillation occurs but with greatly reduced amplitude. Equatorial annual and semiannual waves have also been observed in temperature and wind data and are present in the SBUV ozone data at lower altitudes. The waves exhibit a seasonal asymmetry whose exact origin is unclear. These waves are probably of the type assigned to the stratosphere as opposed to the waves which originate near the mesospause. Such waves should be present in upper mesospheric ozone data. Because of the severe damping of the waves at 20°S, it is not possible to derive the phase. However, at 10°S the ozone concentrations are in phase with 20°N. The largest wave amplitude occurs at the beginning of the year in the northern hemisphere.

3. SECULAR CHANGES

Trend analysis of the weekly averaged data was carried out be fitting the data with a function containing an oscillatory as well as a linear term. The slope of the linear term is a measure of the trend in the data. A trend line is drawn through the 60 km data in each case of Fig. 2. For 20°N the trend is 4.2% in 2.75 years or an increase of 1.5% per year between 1985 and 1988. In contrast, at 20°S there is almost no change. Analysis of the ozone data at all latitudes indicates that there is no downward trend. An upward trend is determined for northern latitudes particularly 20° and greater. Interpretation is complicated by the short span of data and the presence of the quasibiennial oscillation. Sunspot cycle 22 started in September 1986. If the observed northern hemisphere ozone increase is associated with this increase in solar activity, then it appears that wave activity masks solar activity detection at 20°S. Data over a longer time span is required to clarify this point.

Trends observed by the UVSP are different from the Nimbus 7 SBUV data at 0.4 mb. This is illustrated in Fig. 3. The SBUV ozone mixing ratio data at 0.4 mb have been converted into concentration by use of NMC temperature data. The resulting zonally averaged concentrations at 20°N and 20°S are shown for the period 1980 to 1987. A downward trend of 2% per year is deduced from this SBUV data. Also displayed is the UVSP data at 0.4 mb. In 1987 there is a difference of 16% between UVSP and SBUV.

4. AVAILABILITY OF THE DATA SET

The data set consists of more than 15000 ozone profiles between ±40°. It is planned to

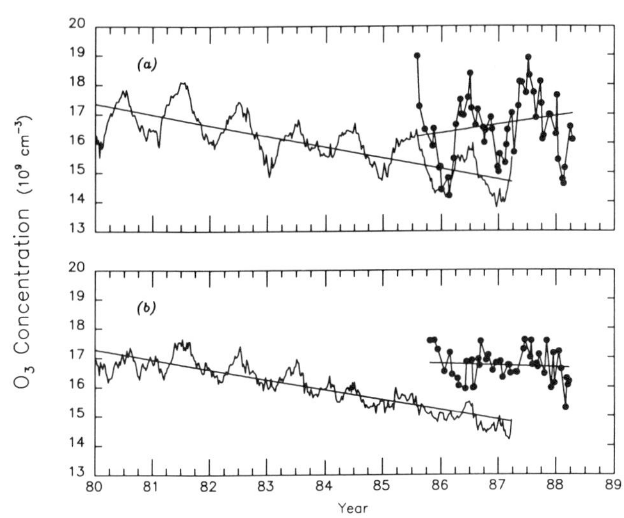

Fig. 3 The long term behavior of UVSP ozone and SBUV ozone concentration at the altitude equivalent to 0.4 mb. Solid lines represent the trend. The upper panel represents 20°N and the bottom panel 20°S.

continue observations during the next solar maximum between 1988 and 1991. Data will be archived with other solar data from the SMM spacecraft and made available to the scientific community.

5. CONCLUSIONS

Twilight ozone profiles exhibit as expected an increased amount of ozone at sunrise as compared to sunset above 55 km. Hemispherical differences at low latitudes are one of the most surprising aspects of the UVSP ozone data. There is a great reduction in the oscillations at 20°S compared to 20°N at both 55 and 60 km. Reduction of the annual and semiannual variation at 20°S is observed at 0.4 mb by both the UVSP and the SBUV instruments. Since 1985, mesopheric ozone concentrations, as measured by UVSP, have either remained nearly constant or increased slightly. Wave activity masks the degree to which solar control manifests itself in the ozone data at 20°S.

ACKNOWLEDGEMENTS

The author would like to thank Dr. William Henze for his contributions to wavelength determination and development of the necessary corrections for limb darkening. Thanks are also due to Dr. H. J. P. Smith and D. Kendig for development of the algorithm and data reduction. Finally, thanks are due to the entire UVSP team particularly Dr. J. Gurman for making this experiment a success.

REFERENCE

Aikin, A. C., B. E. Woodgate, and H. J. P. Smith, 1982: Atmospheric ozone determined by solar occultation using the UV spectrometer on the solar maximum mission, Appl. Opt., 21, 2421.

ATMOSPHERIC OZONE AND SOLAR ULTRAVIOLET RADIATION: THE TROPICAL SCENE

Mohammad Ilyas
School of Physics
University of Science of Malaysia
Penang, Malaysia

ABSTRACT

Solar ultraviolet radiation measurements made at the equatorial location of Penang (5.2°N, 100°E) were used to establish the cloud cover effect on the incoming solar ultraviolet radiation. This together with long term global cloud cover data is used to evaluate the annual ultraviolet-B radiation received at the surface under unperturbed (present) ozone level conditions at the tropical latitudes as well as the relative latitudinal increase in UV-B dosage due to a depletion in global ozone content. The annual cloud cover and incoming radiation combine to produce a nearly constant increase in the UV-B dosage for any depletion in the ozone layer. Also, in absolute terms, at the tropical latitudes this increase alone will be more than the total radiation input at the high latitudes. Therefore, the protection of ozone layer is very important from tropical perspectives. The paper discusses these aspects together with specific health problems enchanced by the presence of high UV radiation levels.

1. INTRODUCTION

For quite a while, we had taken the atmospheric ozone as a natural protective shield against the damaging solar ultraviolet radiation. In more recent years, we recognized the danger which some of human activities may pose to the ozone layer and intensive efforts are now being made to evaluate the degree of danger to the ozone layer and its adverse climatic, biologic and environmental effects to the humans, living organisms, plants and aquatic systems.

The best understood and most direct effect of any ozone-layer depletion is the damage to human skin due to an increase in the erythemal solar UV-B dosage. Under the prevailing natural equilibrium condition, there are only small dosages of UV radiation reaching the surface to which populations, living organisms, plants and aquatic systems have generally become adapted. The harmful effects are thus considered to be generally very small except perhaps in the not so well studied geographical regions of the tropics and the equatorial belt.

The surface dosage of natural UV increases substantially from high and mid latitudes to the equator. Generally people living in the high dosage tropical belt have a darker skin tan which provides greater protection to the skin against ultraviolet radiation penetration. As a result, this high dosage situation has not been given much serious attention in the context of adverse health and environmental effects that might occur in these regions. An equally important aspect which needs our consideration is the effect of any ozone layer depletion on the surface-level UV flux at the tropical latitudes and how this increase may enhance the radiation related adverse effects.

So far, very little experimental work has been done in measuring the base levels of atmospheric ozone and ultraviolet radiation in the tropics. However, a detailed study involving measurements of surface level UV radiation and other related parameters including surface ozone, vertical ozone and other meteorological parameters over the past 10 years has been in progress at the equatorial station of Penang (5.2°N, 100°E), Malaysia. The long-term UV data together with local and global cloud cover have enabled us to examine the UV-radiation climatology and its modification due to any ozone depletion. These results are discussed in the context of the tropical region.

2. DATA ANALYSIS

The latitudinal and seasonal distribution of incoming solar UV flux at the surface can be easily calculated for clear (cloudless) weather conditions (Johnson et al., 1976; Mattingly, 1976). For the present study, the work of Mattingly (1976) provides a good summary of yearly-erythemal (UV-B) dosage for clear weather (Fig. 1). The increase in total UV-B dosage due to a ten-percent ozone depletion (using a simple (average) basis that for each one-percent ozone decrease, there will be a two-percent UV-B increase) is also shown in Fig. (1).

Warren et al. (1986) have presented a detailed study of global cloud cover on the basis of world-wide data (1971-81). As can be expected, the cloud cover and cloud types vary with season and latitude. The data on land zonal average as a function of latitude for four different seasons provides a good basis to evaluate the average surface UV dosage. Using this cloud cover data, an average annual cloud cover curve was constructed. This is shown in Fig. (2). The bars on the curve reflect the extent of seasonal variability of the original data.

ISBN 0-937194-15-8

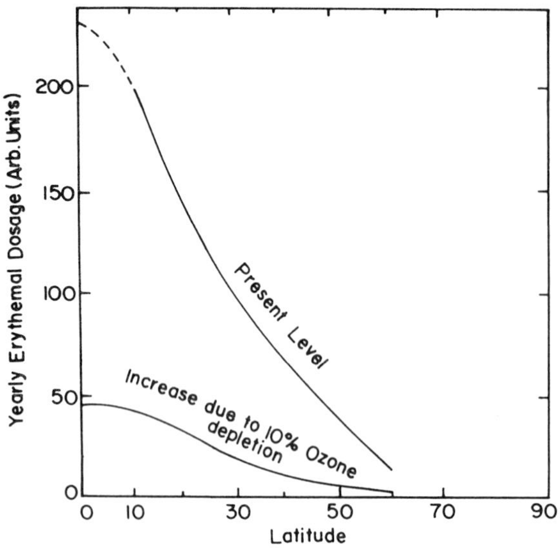

Fig. 1. Latitudinal variation of yearly erythemal solar ultraviolet dosage (upper curve) and the average increase in this flux due to a 10% ozone depletion (lower curve).

Using 10-year long cloud cover data and 4-year long measured UV data, both at Penang, Ilyas (1987) showed the effectiveness of an empirical relationship to estimate the average cloud cover effect on the UV radiation transmission to surface:

$$F_s/F_o = (1 - 0.056C) \quad (1)$$

where F_o and F_s are radiation flux reaching the surface for clear weather and cloud cover C respectively (C is the number of tenths of the sky covered by the clouds).

This relationship (eqn. 1) enables us to evaluate the radiation modulation factor as a result of average global cloud cover shown in Fig. (2). The resultant modulation effect (Fig. 2) allows us to evaluate the average surface dosage of UV flux using the available clear weather data (Fig. 1).

3. RESULTS

The average yearly surface dosage of erythemal UV-B radiation after modulation (reduction) due to average cloud cover is shown in Fig. (3) at different latitudes. We also show the average increase in this flux as a result of 10% reduction in ozone using the simple 2:1 magnification scheme.

The average yearly flux (Fig. 3) shows a slight hemispherical asymmetry as a result of the cloud cover effect being slightly different in the two hemispheres (Fig. 2). The results in Fig. (3) indicate a sharp increase in the surface level radiation flux towards the equator. However, due to the cloud cover effect, the overall magnitude of the difference between equator and mid latitudes (say 50°) has been somewhat reduced compared to the radiation flux under clear weather condition (Fig. 1).

The relative increase in the UV-B flux due to a 10% ozone decrease remains practically same over the equatorial/tropical region. In absolute terms, this resultant increase alone would be more than the present level of radiation flux at mid latitudes (50°).

4. DISCUSSION

The high ultraviolet flux condition prevails at the lower latitudes (due to low ozone content and high incoming UV flux) in comparison to the low UV flux at mid and high latitudes. Besides, the high (air) temperatures and humidities in the tropics occur throughout the year and may couple together with the high UV radiation in producing more severe biological and physiological effects including tropical diseases and medical problems. Lack of education, understanding and proper awareness of UV-induced damages, skin's false sense of protection, increased outdoor activities including young school children being exposed to relatively large amounts of ultraviolet radiation (under hot and humid conditions) due to inadequate clothing and/or outdoor work (like Physical Education lessons and games) at wrong times of the day may all add up to the seriousness of the situation.

Although skin cancer and other related problems, hitherto the prime concern in Western populations, may not be of sufficient importance in the tropics, more serious effects on humans such as viral infections (herpes, hepatitis), eye damage (cataracts), damage to the immune system and life expectancy are important avenues for future work. Also, what kind of adverse effects the very high dosages may have on tropical plants and aquatic organisms and the exposed materials would also need to be studied. Obviously, the effect of any increases in the UV dosages as a result of ozone-layer modification on these tropical problems needs serious consideration. Would these further increases in UV dosage affect the threshold limit - to which the tropical system (may?) have become adapted?

5. CONCLUSIONS

It is clear that the total surface dosages of erythemally effective UV radiation are very high at the tropical latitudes even after the dampening effect of relatively high cloud cover in this region. The average surface dosages presented in this study represent a kind of lower limit. This is because at many tropical locations, actual sky conditions may be found to be clearer. Any increases in UV flux due to possible ozone depletion would have maximum effect at the tropical latitudes. Unfortunately, there is a serious deficiency in the area of basic data acquisition on atmospheric ozone and ultraviolet radiation pertaining to the tropics. There are a few locations where limited studies on this matter have been made. Our own work during past 10

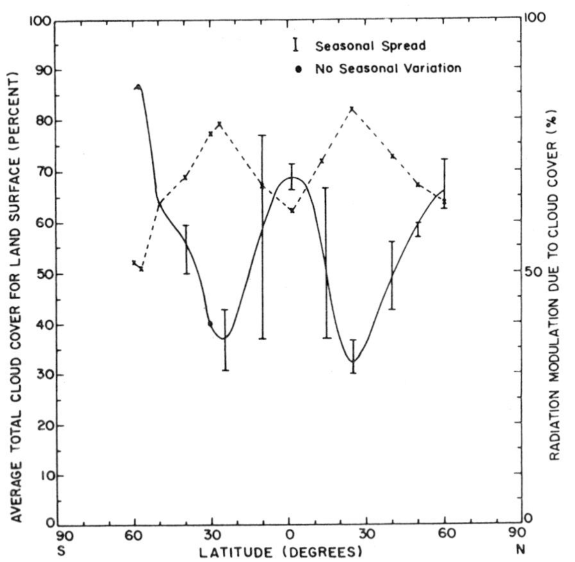

Fig. 2. Yearly average total cloud cover for land surface (left scale) and its effect on the incoming radiation; represented as percent of incoming radiation with no cloud cover (dotted curve).

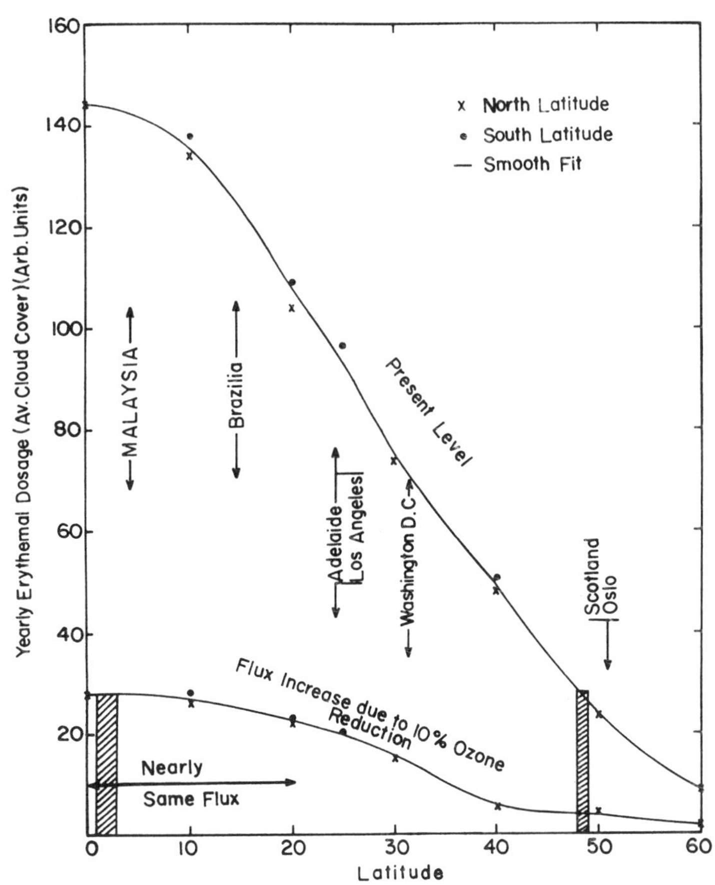

Fig. 3. The surface level erythemal ultraviolet radiation dosage after reduction due to average cloud cover (top curve) and the increase in this flux due to a 10% ozone reduction (lower curve).

years, although covering several inter-related aspects - vertical ozone (Ilyas, 1979; Ilyas et al., 1983; Ilyas, 1984a), surface ozone (Ilyas, 1984b), UV flux (Ilyas and Barton, 1983; and Ilyas et al., 1988) and other parameters - needs further enhancement especially on the total and vertical ozone study.

Recently we have initiated the establishment of a new astro-geophysical observatory which is expected to be functional by mid 1989. We plan to establish it as a S.E. Asian Ozone station and initiate a series of soundings for the measurements of atmospheric ozone and other important minor constituents. Hopefully, this place may also be used as an equatorial station by others to make similar measurements so that, in due course a comprehensive understanding on tropical ozone and related aspects may be achieved. It will be ideal to intensify this effort through similar work at other tropical locations. Undoubtedly, there is a serious need to give the tropical region a special consideration with respect to ozone layer - ultraviolet study.

ACKNOWLEDGEMENTS

A partial grant from the Organizing Committee for Quad. Int'l Ozone Symposium enabling my participation in the meeting is gratefully acknowledged.

REFERENCES

Ilyas, M., 1979: Sains Malaysiana, 8, 13.

Ilyas, M., 1984a: In: Atmosph. Ozone (D. Reidel: Dordrecht), 247.

Ilyas, M., 1984b: In: Atmosph. Ozone (D. Reidel: Dordrecht), 791.

Ilyas, M., 1987: Atmosph. Env., 21, 1483.

Ilyas, M. and I.J. Barton, 1983: Atmosph. Env., 17, 2069.

Ilyas, M., K.K. Chow and T.Y. Ho, 1983: J. Phys. Malaysia, 4, 71.

Ilyas, M., D.A. Aziz and M.R. Tajuddin, 1988: Int'l J. Derm., 21.

Johnson, F.S., T. Mo and A.E.S. Green, 1976: Photochem. Photobiol., 23, 179.

Mattingly, S.R., 1976: Atmosph. Env., 10, 935.

Warren, S.G., C.J. Hahn, J. London, R.M. Chervin and R.L. Jenne, 1986: Global Distribution of Total Cloud Cover and Cloud Type Amounts Over Land, NCAR Technical Note 273, National Center for Atmospheric Research: Boulder.

SHORT-TERM SOLAR RADIATION VARIATIONS AND ATMOSPHERIC OZONE

M.Yu. Danilin, T.V. Matveeva, and G.I. Kouznetsov
Department of Atmospheric Physics
Moscow State Unversity
Moscow, USSR

The 27- and 13-day solar ultraviolet irradiance variation (SUVIV) influence on trace gases and the thermal regime of the middle atmosphere was established experimentally (Ebel et al, 1986). This problem attracted scientists' attention (Keating et al,1985,1986,1987; Hood,1986; Chandra,1986; Kouznetsov, 1987),but its solution has not been completed as yet. So the main purpose of our research is to develop an experimental and theoretical approach to study the response of the ozonosphere to SUVIV periodicities.

The mechanism for this response is the following: SUVIV modulate photodissociation rates of atmospheric gases, hence they change trace gas contents and the temperature of the atmosphere.

We used total ozone (TO) observations by 5 tropical stations: Manila,Singapore,Kodaikanal,Mahe and Huancayo (The Ozone Data for the World, 1978-1982). Since the 205nm radiation is responsible for ozone production by O_2 photodissociation,the Sun-ozone relation was studied using I_{205} as the solar index (Keating et al,1985; Hood,1986). The relation between TO and I_{205} was analysed for the period 1978 to 1982.It was established that TO has a good 27-day periodicity when such periodicity was present in I_{205} (for example, October 1979-January 1980). Comparison between detrended 205nm solar flux(Hood, 1986) and 7-day running average and linear interpolated TO data for given stations was shown in Fig.1a. Correlation coefficients between I_{205} and TO equal 0.51,0.60,0.66 and 0.71 for 0-,1-,2- and 3-day lag respectively. Similar results were obtained in "MONEX-79"(Danilin and Kouznetsov,1987).

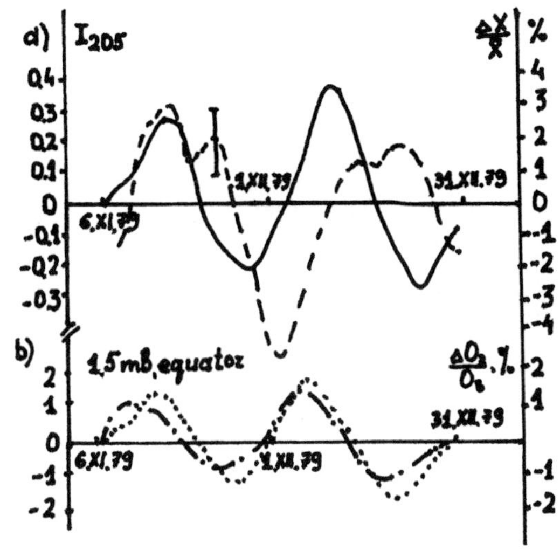

Fig.1. A comparison of detrended 205nm solar flux data I_{205} and TO response during November-December 1979(a);ozone response accordingly (Hood,1986) and our model results at 1.5mb (equator) during the same period (b).

The one-dimentional photochemical model was developed to explain the ozone (O_3) behaviour. Our model covers the height region 10-60km with 2 km height step and includes 68 reactions of O_x-, HO_x-,NO_x- and ClO_x-cycles and 17

ISBN 0-937194-15-8

photodissociation reactions. **We used** the following data for modelling(Baulch et al,1984;Brasseur and Solomon,1984;Mount and Rottman,1985;Nicolet and Kennes, 1986). Concentrations of $O,O^*,O_3,H,OH,HO_2,H_2O_2,N,NO,NO_2,NO_3,N_2O_5,HNO_3$ were calculated using "family" techniques; concentrations of $O_2,N_2,H_2,H_2O,CH_4,HCl,Cl,CO,ClO,CHO,CH_2O,CH_3O_2$ were fixed (Danilin and Kouznetsov,1987).

The 27-day spectral variability of solar irradiances adopted in the model was the same as that(Brasseur et al, 1987). Solar irradiance variability for 13-day cycles is reduced at all spectrum by factor 0.9, i.e. $\Delta I_{205}/\bar{I}_{205}=2.30\%$.

Disturbed photodissociation rate at the height z $J'(z,\chi,t)$ was calculated by:

$$J'(z,\chi,t)=\int_0^\infty I_o(\lambda)\left(1+\frac{\Delta I(\lambda)}{\bar{I}(\lambda)}\sin\frac{2\pi t}{27}\right)\sigma_i(\lambda)\varphi_i(\lambda)\tau(\lambda,z,\chi)d\lambda \quad (1)$$

where $I_o(\lambda)$- the solar flux at the top of the atmosphere,

$\sigma_i(\lambda)$-cross-section of i-molecula,

$\varphi_i(\lambda)$-quantum yield,

$\tau(\lambda,z,\chi)$-optical depth of the atmosphere.

Temperature disturbation during 27-day SUVIV was taken according to (Hood,1986). Ozone response was described in terms of sensitivity",i.e.:

$$S(z)=\frac{\Delta O_3(z)/\bar{O}_3(z)}{\Delta I_{205}/I_{205}(z)}$$

We discovered that duirnal ozone variations considerably masked ozone response. For example,O_3 response at midnight decreases by factor 1.8 in the mesosphere in comparison with that at midday. So we considered the scenario when the maximum of 27(13)-day sinusoid coincided with model midday. In this case ozone response is a maximum. Comparison of experimental (Hood,1986) and our theoretical ozone responses at 1.5mb on the equator was shown in Fig.1b. Their good agreement demonstrates the validity of the description of the mechanism,that governed ozone behaviour in this phenomenon.

Lines 1 and 2 in Fig.2 show ozone responses to 27-day sinusoidal SUVIV with and without temperature feedback at equator in June. Temperature feedback reduced O_3 response in the mesosphere and increased in the stratosphere, reaching maximum value S(z)=0.63(i.e. 1.6% ozone concentration) at heights z=44÷48km. Line 3 in Fig.2 shows ozone response to 13-day SUVIV. Assumption concerning the absence of temperature disturbances at night did not change ozone response at midday.

It is necessary to stress that the advance of ozone response with reference to I_{205} at height z>40km was due to the amplitudinal and phase thermal response characteristics. Below 40km ozonosphere disturbances decreased and delayed deep into atmosphere monotonously. Such ozone behaviour could be explained by decrease SUVIV,increasing ozone concentration and increased O_3 photochemical relaxation time deep into the atmosphere. Possible discrepancy between our results and data obtained by (Brasseur et al,1987) should be explained by the fact that our results correspond those at midday(i.e.maximum) response,but Brasseur' results are daily averaged.

We analysed the ozone response dependence on spectral variability of the solar irradiance. Being increased by factor 1.56 (i.e. $\Delta I_{205}/ I_{205}=4\%$) proportionally at all wavelength SUVIV led to the same values of ozone sensitivity S(z) at all heights and TO response increased from 0.6% to 0.94%.

This results showed linearity of the ozone response to proportional change of the spectrum of SUVIV.

When we changed variability only in wavelength region (L_α -200nm) proportionally by factor 4 (Lean,1987) ozonosphere response increased by factor 1.1 at the height 60 km and remained the same below 44 km. This fact stress that I_{205} was chosen as solar index for the solar-ozone relations correctly.

When we studied seasonal and

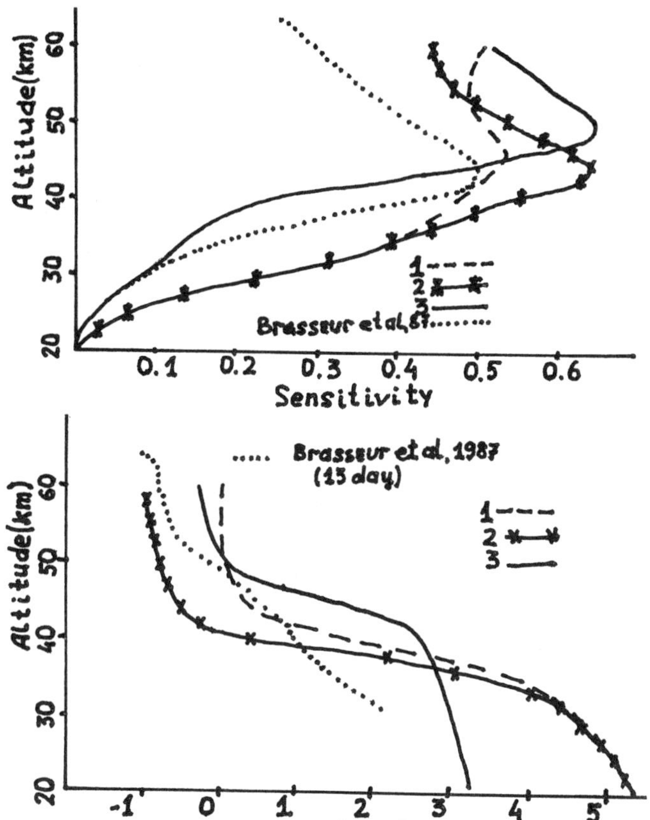

Fig.2. Ozone sensitivity (percent/percent) S(z) (a) and phase lag (days); (b) for following scenario 1 and 2- 27-day SUVIV without and with temperature feedback; 3- 13-day SUVIV with temperature feedback.

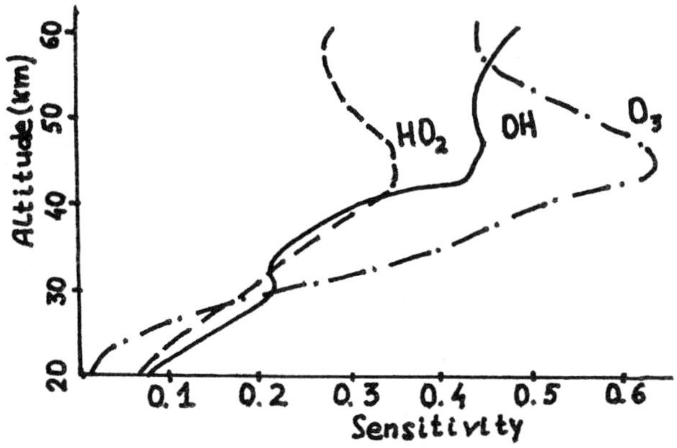

Fig.3. Sensitivities of OH,HO_2 and O_3 to 27-day SUVIV with temperature feedback.

latitudinal dependence of the ozone response we used the same temperature dependence taken by Chandra (1986). It was concluded that ozone response became maximum at equator in summer. Here the solar zenith angle was minimum, so disturbances of the solar radiation penetrated deeper into atmosphere. For example, summer ozone response at 60°N was twice as great as the winter response.

Fig.3 illustrates OH,HO_2 and O_3 sensitivity for 27-day SUVIV. The growth of OH and HO_2 contents was due to increased photodissociation rates of H_2O_2,H_2O and HNO_3. OH and HO_2 responses were experimentally undetectable due to small values of their sensitivities.

"MONEX-79" experimental data (Kouznetsov,1987) and our theoretical results of the TO responses to a real profile of SUVIV are shown in Fig.4 for April-June 1979. The 2-3 day delay TO responses with references to I_{205} are obvious.

Quantitive discrepancy between experimental and theoretical TO responses could be explained since we have

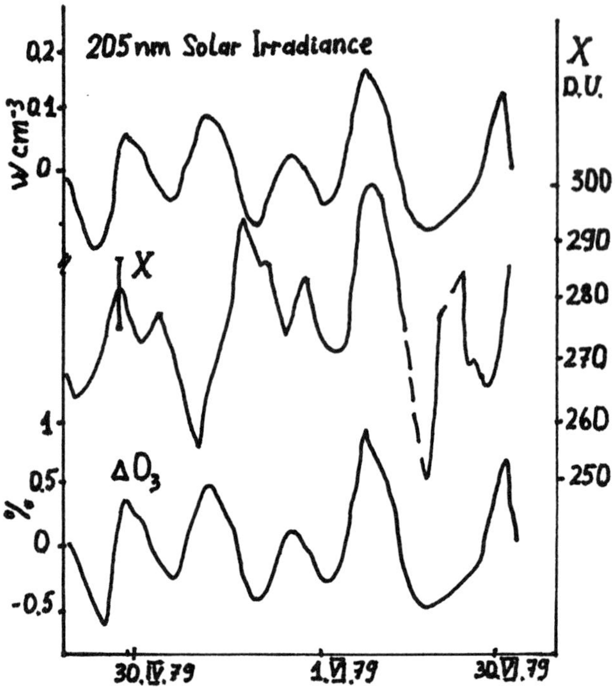

Fig.4. A comparison of detrended I_{205} data, TO amount (in Dobson Units) X, accordingly "MONEX-79" and ozone response O_3(%) for April-June 1979.

neglected dynamic effects (Chandra, 1986). In our future research we shall take into account the above effect.

REFERENCES

Baulch D.L. et al, 1984: Evaluated kinetic and photochemical data for atmospheric chemistry. Suppl. II. J. Phys. Chem. Ref. Data, 13, 1259-1378.

Brasseur G., S. Solomon, 1984: Aeronomy of the middle atmosphere.-D. Reidel Pub. C.

Brasseur G., A. DeRudder, G.M. Keating, M.C. Pitts, 1987: Response of middle atmosphere to short-term solar UV variations: 2. Theory. J. Geophys. Res., 92, 903-914.

Chandra S., 1986: The solar and dynamically induced oscillations in the stratosphere.-J. Geophys. Res., 91, 2719.

Danilin M.Yu., G.I. Kouznetsov, 1987: On the influence of short-term solar radiation variations on atmospheric ozone. Izv. AN USSR. Phys. Atm. Oc., 23, 830-838.

Ebel A. et al, 1986: Vertical change of the response of solar activity oscillations with period around 13 and 27 days in the middle atmosphere. Ann. Geophys., 4A, 271-280.

Hood L. Coupled stratosperic ozone and temperature responses to short-term changes in solar ultraviolet flux: An analysis of Nimbus-7 SBUV and SAMS data -J. Geophys. Res., 91, 5264-5276, 1986.

Keating G.M. et al, 1985: Detection of the response in the middle atmosphere to short-term solar ultraviolet variations. Geophys. Res. Lett. , 12, 449.

Keating G.M. et al., 1986: Detection of stratospheric HNO_3 and NO_3 response to short-term solar uv variability. Nature , 322, 43-46.

Keating G.M. et al., 1987: Response of middle atmosphere to short-term solar ultraviolet variations: 1. Observations. J. Geophys. Res., 92, 889-902.

Kouznetsov G.I., 1987: Research of the connection between ozone and some solar and geophysical factors. Atmos. Ozone. Proc. 6th All-Union Ozone Symp., 209-217, Leningrad, Gidrometeoizdat.

Mount G.H., G.J. Rottman, 1985: Solar absolute spectral irradiance 118-300nm: July 25, 1983. J. Geophys. Res., 90, 13031.

Nicolet M., R. Kennes, 1986: Aeronomic problems of the molecular oxygen photodissociation. 1. The O_2 Herzberg continuum. Planet. Space Sci. , 34, 1043-1059.

DIURNAL VARIATIONS OF OZONE IN THE MESOSPHERE

S.P. Perov and S.V. Tishin
USSR State Committee for Hydrometeorology
Central Aerological Observatory
Dolgoprudnij Moscow Region, USSR

ABSTRACT

The results of several tens flights of meteorological rockets with optical and chemiluminescent ozonesondes, carried out in different regions and different seasons, in the day time and at night have been analyzed. Diurnal effect comes in to being at the 50 km altitude and at the 60 km is 1,8 to 3 depending on various data groups. In the tropical region diurnal effect comes into being at 40 km. According to the data obtained with gas-phase ozonesondes the greatest variability of the diurnal effect is observed in the range of 74...82 km. The results obtained are compared with the photochemical theory conclusion from various models.

The investigation of atmospheric ozone behavior during day and night in the region of ozonosphere photochemical equilibrium (30...80 km) must help the better understanding of photochemical processes in this region and their relations with the dynamics of atmosphere. Considerable diurnal variation in ozone is noticed in many atmospheric layers: the most noticeable amplitude of ozone oscillations near by Earth's surface, in the region of 6 .... 11 km (evidently, episodic), in the upper stratosphere (40 ... 50 km), in mesosphere, where the amplitude of oscillations reaches its maximum value near by mesopause (80 ...85 km) and in the lower thermosphere. As a rule the night values of ozone concentration exceed the day ones with the exception of the altitude region between 40 and 50 km where the reverse situation can be observed.

The results of preliminary analysis of data, based mainly on meteorological rockets M-100B launches with the ozonometric apparatus of three types: heterophase chemiluminescent (Perov and Hrgian, 1980),(Kononkov, Kon'kov and Perov, 1981), gasphase chemiluminescent (Perov and Tishin, 1986), and optical (Brezgin et al., 1984), are presented in this report. During analysis the data of ozone concentration measurements by optical rocket ozonometers of India (Subbaraya, 1987) and the USA have also been used.

The method of comparison of night and day ozone concentration values in stratosphere and mesosphere based on the groups of one or two measurements technique from our point of view presents the interesting data in spite of the difference in time of separate launches because: firstly, day/night variations reach a factor of IO at mesospheric levels according to theoretical evaluations and secondly, there is very little data on this subject in the literature. Consequently it is interesting to compare the different data sets even if the systematic errors differ. According to some evaluations the errors are essentially less than the variations in ozone content in the mesosphere, especially at levels below 55 to 60 km. In some cases in using data obtained with the gas-phase chemiluminescent technique the normalization of the output vertical distribution to the value of signal at definite altitude (in the region of 40....50 km), where the diurnal variations of ozone are not large, was performed. The gas-phase chemiluminescent technique (Perov and Tishin, 1986) in spite of its high sensitivity and little hysteresis has systematic errors that have not been extensively analysed.

The results of ozone concentration ratios at night time and day time, obtained for two latitude zones: midlatitude (30°-50°) and equatorial zone (0°.. ..8°) are summarised in Table I.

ISBN 0-937194-15-8

Table 1. The results of comparison of daytime and nighttime ozone concentrations in the stratosphere and mesosphere according to different measurements data and models (ratios n_{night}/n_{day}).

No	Authors, sources	Technique	Conditions of experim., number of launches	Latitude, time	40 km	45 km	50 km	55 km	60 km	65 km	70 km
			a) mid latitudes								
1	Kononkov Kon'kov Perov, 1981 Perov, Hrgian, 1980	Ch getero-phase	night 7 launches	48°N 1975-81	0,8	0,9	1,1	1,4	1,8		
	Krueger Minzner, 1976	optical model	day 17 launches	45°,15° 1976							
2	Kononkov Kon'kov Perov, 1981	Ch getero-phase	night 7 launches	48°N 1975-81		0,9	1,0	1,7	1,4		
	Chizhov Brezgin et al., 1984	optical	day 4 launches	48°N 1975-78							
3	Perov Tishin present paper	Ch. gas-phase	night 2 launches	28°N 25.02.85 36°S 21.06.85			1,0	1,0	1,9	3,0	
	Perov Tishin present work	Ch. gas-phase	day 2 launches	48°N 28.04.82 02.08.83							
			b) equatorial zone								
4	Chizhov Kononkov Perov	Ch getero-phase	night 3 launches	8°N 03.1983	1,0	1,3	1,4	1,7	2,3		
	Subbaraya et al., 1987	optical	day 3 launches	8°N 03.1983							
5	Subbaraya, 1987	optical, solar eclipse	day 3 launches	8°N 16.02.80	1,1	1,3	1,4	1,8	2,5	3,4	
6	Chizhov Kononkov Perov	Ch getero-phase	night 3 launches	8°N 03.1983	1,2	1,4	1,3	2,0	3,3		
	Subbaraya et al., 1987	optical	day 8 launches	0°N 05.1979							
7	Perov Tishin present paper	Ch gas-phase	night 1 launch	0° 02.07.85			1,0	1,7	3,0		
	Brezgin Chizhov et al., 1988	optical	day 1 launch	0° 03.07.85							

Table 1.

No	Authors, sourses	Technique	Conditions of experim., number of launches	Latitude, time	40 km	45 km	50 km	55 km	60 km	65 km	70 km
8	Perov Tishin present paper	Ch gas-phase	night 2 launches	8°N 06.12.87, 08.12.87			1,1	2,3	3,3	4,1	
	Perov Tishin present paper	Ch gas-phase	day 2 launches	8°N 03.,05. 12.87							

CONCLUSIONS:

1. Diurnal variation in the ozone concentration becomes noticeable at around 50 to 55 km, reaching the values of 2 ... 3 in the 60 to 65 km in midlatitude region.
2. Diurnal variation in ozone concentration in the equatorial region becomes noticeable at around 45 to 50 km, reaching the values of 2.5 ... 3.4 in the 60 to 65 km region.
3. Such a difference in ozone concentration behaviour at different latitudes is caused, evidently, by following reasons:
   a) there is more powerful influence of sun radiation on ozonosphere in tropics in comparison with the midlatitude zone;
   b) there is less influence of ozone-active components of atmosphere on ozone concentration in tropics in comparison with the midlatitude zone;
   c) there is different influence of the dynamical factors in two latitude zones.
4. Modern photochemical theories (Vaughan G., 1984) on ozone diurnal variations can not explain, for the present at least, some important features of ozone behavior shown in these experimental data.
5. Dynamic factors (planetary and other waves) have influence on the character of the diurnal ozone variations and they have to be taken into consideration when creating photochemical models of ozonosphere.

Note: Ch - chemiluminescent, in a) launches were carried out in s. Volgograd (48°N), also in the Pacific Ocean (28°N) and in the Indian Ocean (36°S), in b) at Thumba launch station (8° N) and in the Indian Ocean (0°... 6°N).

REFERENCES

Acharya Y.B., Chizhov A.F., Kononkov V. Perov S.P., et al. Indo-USSR Ozone - sonde Intercomparison Experiment at Thumba - part II. - SR ISRO-IMAP-SR--24-85, ISRO, Bangalore, 1985, pp. 1-50.

Brezgin N.I. et al. Racketnye izmereniya ozona i opticheskih harakteristik aerozolya s pomoschyu photometrov USSR i GDR v srednih shirotah i tropikah. Issledovaniya verhney atmosphery Zemly. Trudy III Mezdunarodnogo Simposiuma po Kosmitcheskoy Meteorologii.- - L.: Hidrometeoizdat, 1984, pp.73-80.

Brezgin N.I. et al.- Meteorologiya and hydrologiya, 1988, No 5. Issledovanie verticalnogo raspredeleniya ozona v Juznom polusharii opticheskimi raketnimy ozonometrami.

Chizhov A.F., Kononkov V.A., Perov S.P. et al. Results from the Indo-USSR ozonosonde intercomparison experiment. - Proc. Indian Acad. Sci.(Earth Planet. Sci.), 1987, v.96, N.1, pp.25 -40.

Kononkov V.A., Kon'kov V.I., Perov S.P. Experiment employing chemiluminescent method for measurement of atmospheric ozone on board M-100B Meteorological rockets. - Symp. on Investigations of Atmospheric Ozone, TBILISI, 23-27 Nov. 1981, Material from reports published in METSNIEREBA, TBILISI, 1982, pp.78-82.

Krueger A.T. and Minzner R.A. A midlatitude ozone model for 1976 U.S. standard atmosphere. - J. Geoph. Res., v.81, No 24, 1976, pp. 4477-4481.

Perov S.P. and Hrgian A.H., Sovremennie problemy atmosphernogo ozona. L., Hydrometeoizdat, 1980.-287p.

Perov S.P. and Tishin S.V., A rocket gas-gas chemiluminescent technique for atomic oxygen and ozone concentration in the 15-95 km region. - Atmospheric ozone. Proc. Quadr. Ozone Symp., Halkidiki, Greece, 3-7 Sept., 1984, O.Reidel Publ. Comp. 1986, pp. 527-531.

Subbaraya B.H. Vertical distribution of minor constituents in the tropical middle atmosphere. - Indian Journ. of Radio and Space Physics, 1987, pp.25-35.

Vaughan G. Mesospheric ozone-theiry and observation. - Quart.J.R. Met. Soc., 1984. 110, pp. 239-260.

CALCULATIONS OF WATER VAPOR IN THE MESOSPHERE FROM THE OZONE CONCENTRATION

D.V. Bromberg and S.P. Perov
USSR State Committee for Hydrometeorology
Central Aerological Observatory
Dolgoprudnij Moscow Region, USSR

ABSTRACT

The role of chemical reactions with odd oxygen and odd hydrogen was estimated for photochemical equilibrium of odd oxygen. Water vapor mixture ratios were determinated in the mesosphere in heights interval of 55-70 km, the τ_{H_2O} distributions were functions of τ_{O_3}, photolytic rate constants of O_2, O_3, H_2O and temperature factors. Seasonal and latitudinal variations of τ_{H_2O} were developed using ozone data from the ozone reference model (Keating et al.,1985) and P, T, ρ, data from the reference atmosphere model

Lower mesosphere at equator and middle latitudes is relatively dry ($\sim$ 5 ppmv), in the northern hemisphere τ_{H_2O} gradient for 0-40° is close to zero, gradient for 40°-70°N is small $\sim$ 1%/1°N. The model results were compared with the experiment results for variations and height distributions of τ_{H_2O} in the mesosphere.

Water vapor is one of the most importent minor constituents of the atmosphere. Photodissociation of H_2O or its reactions with atoms produce odd hydrogen HO_x (OH + HO_2 + H) and the HO_x catalytic cycle controls the ozone destruction at heights above 55 km. Mesospheric H_2O behaviour is important for the problem of noctilucent clouds and other aeronomic problems. There are some experimental data for concentrations of mesospheric H_2O obtained by different techniques (Bevilacqua, 1983; Yushkov, 1983), but it is single measurements at local latitudes and the obtained H_2O concentrations differ to the factor 10 or more. The non-satellite data are unable to obtain global distribution of H_2O for different latitude and season. Satellite data are obtained only at the stratospheric altitudes lower than 45km. There are many one- and two-dimensional model results for the vertical distribution of H_2O (Solomon, 1984; Stolarski, 1985), but there are many input parameters in this models, and these values must be confirmed by experimental data. The central problem in models is the treatment of eddy terms, there are various approximations of these terms and these introduce large uncertanties in model results.

Another method of calculations of minor constituents (Perov, 1980;Callis, 1985) is based on calculations using photochemical region of atmosphere the constituents of which influence the ozone balance and the experimental data for ozone and the thermodynamical parameters of the atmosphere. In this paper latitudinal-seasonal distributions of day-time water vapor concentrations are calculated at heights between 55-70 km from the values of ozone concentrations and thermodynamical parameters (T, P, ρ). Lifetime for transport for all constituents, which influence on O_3 balance at heights between 55 and 70 km is more than 5 days. Photochemical lifetimes (τ) are different for different constituents typically are $\lessapprox$ 1 hour. The behaviour of these constituents is not dependent on dynamical conditions, the concentrations are steady-state values and are not time dependent if the input parameters (photodissociation factor, reaction rates, neutral constituents concentrations) are constant for time intervals equal to the photochemical lifetimes. The estimates of the role of various reactions allow to extract key reactions and constituents (the reactions whose role is less than 10% of the key reactions are not take into account).

Key reactions for O_3 steady state

$$O_3 + h\nu \xrightarrow{J_{O_3}} O_2 + O$$
$$O + O_2 + O_2 \xrightarrow{K_{O-O_2-O_2}} O_3 + O_2 \qquad \tau < 1 \text{ min}$$
$$O + O_2 + N_2 \xrightarrow{K_{O-O_2-N_2}} O_3 + N_2$$

ISBN 0-937194-15-8

Key reactions for O steady state

$$O_2 + h\nu \xrightarrow{J_{O_2}} O + O \qquad \tau < 1 \text{ hour}$$
$$O + OH \xrightarrow{K_{OH-O}} H + O_2$$

Key reactions for OH steady state

$$HO_2 + O \xrightarrow{K_{O-HO_2}} OH + O_2 \qquad \tau < 10 \text{ c}$$
$$OH + O \xrightarrow{K_{O-OH}} H + O_2$$

Key reactions for HO_x steady state

$$O(^1D) + H_2O \xrightarrow{K_{O(^1D)-H_2O}} 2\, OH$$
$$H_2O + h\nu \xrightarrow{J_{H_2O}} OH + H \qquad \tau < 1 \text{ hour}$$
$$OH + HO_2 \xrightarrow{K_{OH-HO_2}} H_2O + O_2$$
$$H + HO_2 \xrightarrow{K_{H-HO_2}} H_2 + O_2$$

Key reactions for O (^{1}D) steady state

$$O_3 + h\nu \xrightarrow{J_{O_3}} O(^1D) + O_2$$
$$O(^1D) + O_2 \xrightarrow{K_{O(^1D)-O_2}} O + O_2 \qquad \tau < 1 \text{ min}$$
$$O(^1D) + N_2 \xrightarrow{K_{O(^1D)-N_2}} O + N_2$$

Taking into consideration all these reactions we can derive the analytical expression for water vapor mixing ratios

$$R_{H_2O} = \frac{J^2_{O_2}}{J^2_{O_3}} \frac{N(O,2)^4}{R^3_{O_3}} \frac{A^2 K_{OH+HO_2}}{K_{O+OH} K_{O+HO_2}\{B+J_{H_2O}\}} + \frac{J^2_{O_2} [O_2]^2 A K_{H+HO_2}}{J_{O_3} K_{H+O_2+M} N^2[O_3] K_{O+HO_2} (B+J_{H_2O})} \quad (1)$$

$$A = K_{O+O_2+O_2} [O_2] + K_{O+O_2+N_2} [N_2]$$

$$B = \frac{K_{O(^1D)+H_2O} J_{O_3} [O_3]}{(K_{O(^1D)+O_2} [O_2] + K_{O(^1D)+N_2} [N_2])}$$

J_A - photodissociation rate of A-molecules,
K_{A+B+C} - kinetic rate constant A+B+C —— products,
[A] - concentration of A-molecules (cm^{-3}),
R_A - molecule mixture ratio (PPMV),
N - air density (cm-3).

Kinetic rate constants come from ozone 85 (WMO, 1985). Photodissociation factor for O_2 is sum of photodissociation factor in the Schumann-Runge bands (calculated using the Allen and Frederick approximation (Allen,1982)) the photodissociation factor in the Herzberg continuum and photodissociation factor in L_α (calculated with Nicole approximation (Nicolet, 1984). Solar flux at the top of the atmosphere and O_2,O_3 cross sections come from (WMO, 1985). Thermodynamical zonal mean parameters come from (Ivanovski, 1984) zonal mean mixture ratio of O_3 come from (WMO, 1985). Figures 1,2 show the results of calculations.

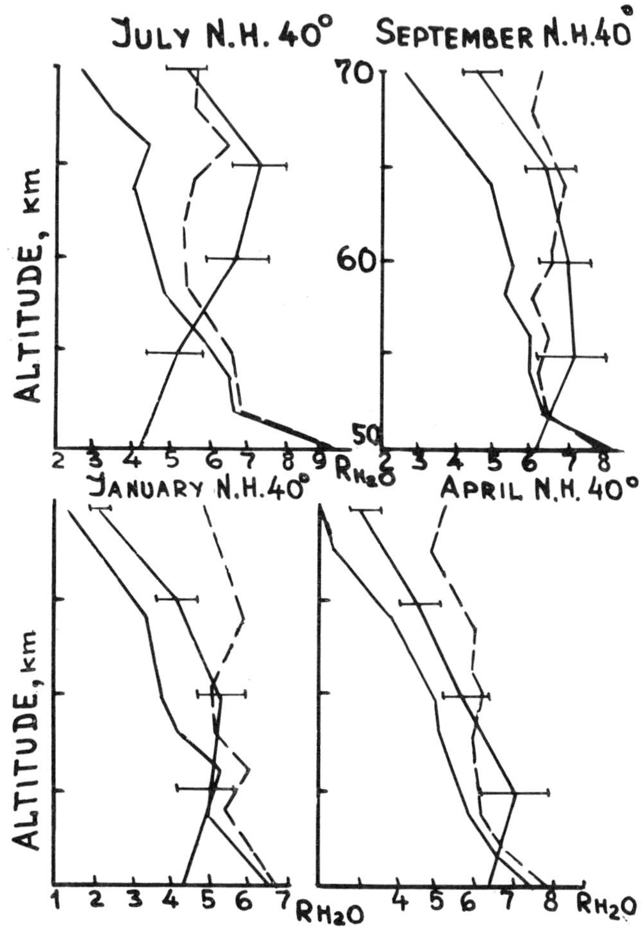

Fig. 1 Altitude distribution of H_2O mixture ratios on 4 days is four different seasons: winter - 15.01, spring - 15.04, summer - 15.07, fall - 15.09.
------ K_{H+HO_2} come from JPL-84 /6.0 · 10^{-12}/
- - - K_{H+HO_2} come from Ozone-85 /7.4 · 10^{-11}/

Fig. 1 shows the altitude distributions of H_2O mixture ratio on 4 different days representing the four seasons: winter - 15.01, spring - 15.04, summer - 15.07, autumn - 15.09. There are also experimental data in this fig. 1 that come from microwave spectroscopic technique (Bebilacqua, 1983).

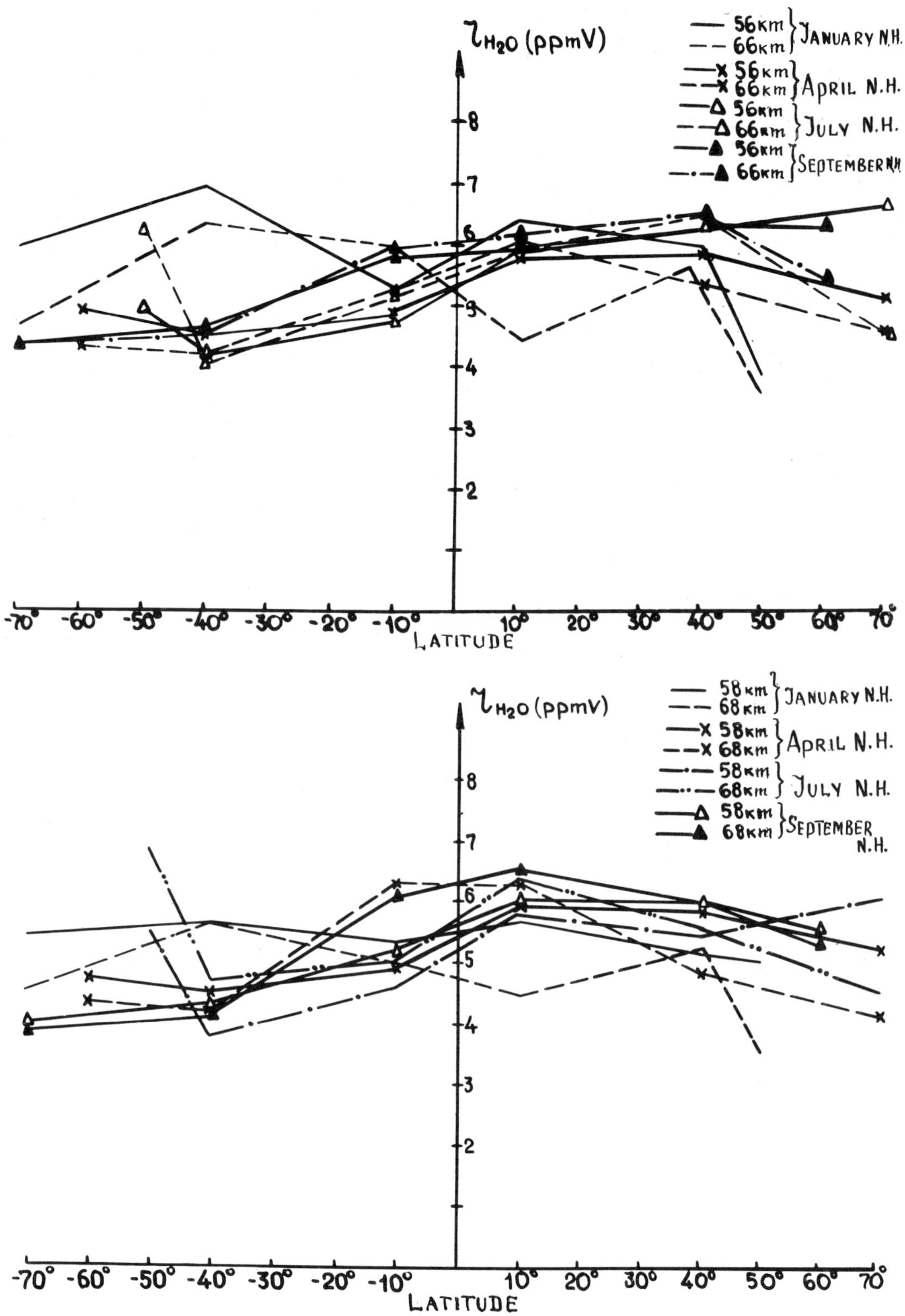

Fig.2 Latitudinal distribution of H_2O mixture ratios in 4 different seasons at altitudes 56 and 66 km, 58 and 68 km.

There are other experimental data (Yushkov, 1983; Martinkevich, 1969) that give concentrations as large as ~ 100 ppmv at 55-70 km heights, but in recent comparisons of 10 different techniques of H_2O measurements (WMO, 1985), the H_2O mixing ratio of 2-8 ppbV were obtained, so the mixture ratio ~ 100ppmV may be incorrect. Microwave spectroscopic techniques are the only techniques that have a sufficient data base of H_2O measurements at the altitudes > 55-70 km and there are experimental data for all modelled seasons, so it is appropriate to compare this data with results of calculations.

There is good agreement between modelled and measured H_2O at fig. 1, the agreement would be even better if the kinetic rate constant (K_{H+HO_2}) would be the average of the value in JPL-84 and Ozone-85. Fig.2 show the latitudinal distribution of r_{H_2O} in 4 seasons at altitudes 56 and 66 km, 58 and 68 km. All mixing ratio at all latitudes and seasons are in the range from 4.5 to 6.5 ppmV. The variations are small < 40%. This conclusion agrees with microwave data and LIMS satellite data for the lower altitudes (< 45 km).

REFERENCES

Allen M. and Frederick J.E., Effective photodissociation cross sections for molecular oxygen and nitric oxide in the Schuman-Runge bands. J. Atm. Sci., v.39, No 9, 2066-2074, 1982.

Bevilacqua R.M., Olivero J.J., et al. An observational study of water vapour in the mid-latitude mesosphere using ground-based microwave techniques, J.G.R., 88, 8523-8534, 1983.

Callis L.B. and Natarajan M., Stratospheric photochemical studies using Nimbus 7 data 2. Development of infer-red trace species distributions, J.G.R. D91, 1167-1197, 1985.

Ivanovsky A.I., Fedorov V.V., Glazkov V.N., An analytic approach to modelling the structure and circulation of the middle atmosphere. Preprint of paper presented at XXI COSPAR, Graz, Austria, 1984.

Martinkevich G.M., Meteorologiya i Gidrologia, No 12, 3-8, 1969.

Nicolet M. Photodissociation of molecular oxygen in the terrestrial atmosphere. Simplified numerical relations for the spectral range of Schuman-Runge bands. J.G.R., D89, 2573-2582, 1984.

Perov S.P. and Hrgian A.N., Sovremennie problemi atmosphernogo ozona. L., Gidrometizdat, 1980, 288 p.

Solomon S., Garsia R.R., On the distribution of long-lived trasers and chlorine species in the middle atmosphere, J.G.R., 89, 11633-11644, 1984.

Stolarski R.R. and Douglas A.R., Parametrisation of the photochemystry of stratospheric ozone including catalytic loss processes, J.G.R., 90, 10709-10818, 1985.

WMO, Global ozone research and monitoring project - report No. 16, Atmospheric ozone 1985, v. 1-3, 1985.

Yushkov V.A. Proceedings of the Indo-Soviet symposium on space research held during 21-26 Feb. 1983/ISRO, Bangalore/, 1983, 7.04-1-7. -7.

SOLAR IRRADIANCE AND OZONE COLUMN DENSITY IN POLAR REGIONS DERIVED WITH AN UV SENSITIVE DOUBLE MONOCHROMATOR

K. Henriksen and K. Stamnes
The Auroral Observatory, University of Tromsø
N-9001 Tromsø, Norway

ABSTRACT

Spectral measurements of solar radiation in the Arctic during the summer of 1987 were carried out with a double monochromator calibrated in absolute units. Under clear sky conditions the measured spectral irradiances are closely reproduced by radiative transfer calculations. The column density of atmospheric ozone is derived by fitting calculated UV irradiances to the measurements using a least squares procedure. The derived ozone values are in basic agreement with the ozone amount inferred by a nearby Dobson spectrometer and thus demonstrate the feasibility of determining simultaneously atmospheric ozone abundances and spectral ultraviolet irradiances with an UV-sensitive spectrometer.

1. INTRODUCTION

The solar radiation powers all life and activity on earth, and the solar constant and its variability have been determined with relatively high accuracy. However, the spectral solar irradiance reaching the surface of the earth under different atmospheric conditions is insufficiently investigated. At present with increasing pollution of the atmosphere it is more important than ever to perform such ground-based spectral measurements. Only then will it be feasible to quantify the effects of natural and anthropogenic "pollution" on the UV and solar radiation received by the biosphere, and provide data for a variety of studies in biology, dermatology, climatology, and geophysics.

The spectral measurements and the derivation of ozone column densities from the data are described in some detail elsewhere (Stamnes et al., 1988; Henriksen et al., 1988a; Henriksen et al., 1988b). Here we provide a summary of the salient features of these studies.

2. INSTRUMENTATION AND OPERATION

The instrumentation consisted of an UV-sensitive, double monochromator, H10D, produced by Jobin-Yvon, Fig. 1. A resolution of 20 Å was used, and the monochromator was scanning in the first order from 2900 to 8000 Å. No precaution was taken for second order contributions at wavelengths longer than 6000 Å, since the sensitivity of the instrument in second order was reduced by two orders of magnitude.

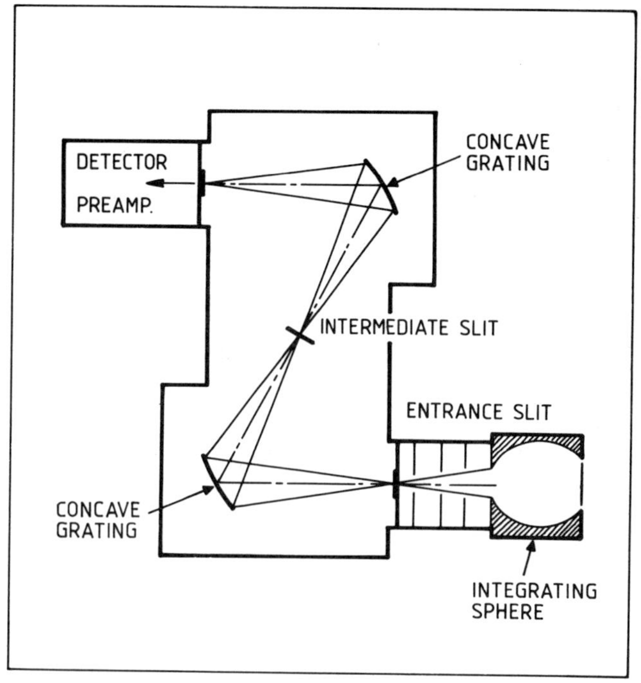

Fig. 1. The double monochromator, H10D.

A UV-sensitive photomultiplier was used and the pulses counted. The operation of the monochromator and the recording of the data were performed with a microcomputer.

The monochromator was placed inside a light-tightened, thermo-regulated box. When measuring the global radiation, an integrating sphere was placed in front of the entrance slit, and the diffuse radiation was obtained by screening off the direct solar radiation. Direct solar measurements were carried out without the integrating sphere, by pointing the spectrometer towards the sun. The field of view was $2^{\circ} \times 2^{\circ}$.

3. SPECTRAL MEASUREMENTS

A sample of spectral measurements are given in Fig. 2, and diurnal UVB global irradiances from a two-week observation period in Longyearbyen, Svalbard are shown in Fig. 3. The period extended from July 23 to August 5, 1987 with the sun above the horizon all the time. The solar elevation varied from 5° to 30°, and the

ISBN 0-937194-15-8

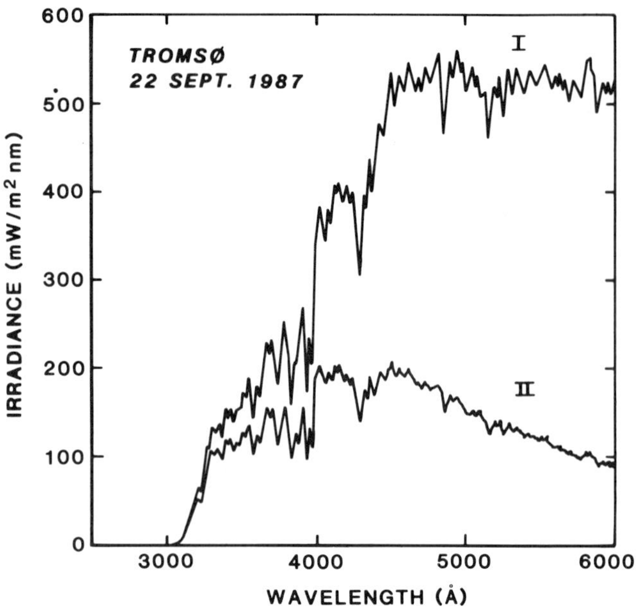

Fig. 2. Spectra of solar irradiance at 20° elevation.
I Global horizontal irradiance.

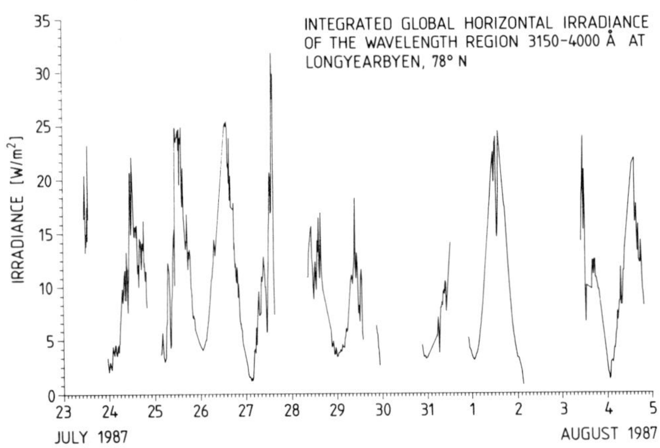

Fig. 3. UV-B from July 23rd to August 4th, 1987. Measurements were stopped during rain and data gaps occur.

measurements were carried out continuously both under clear and cloudy sky conditions. Tabulated and analog representations of the data were presented by Henriksen et al. (1988a).

The averaged integrated UVB, UVA and full-range irradiances are shown in Fig. 4, and we see that the maximum irradiances are about 0.15, 15, and 250 W/m^2, respectively.

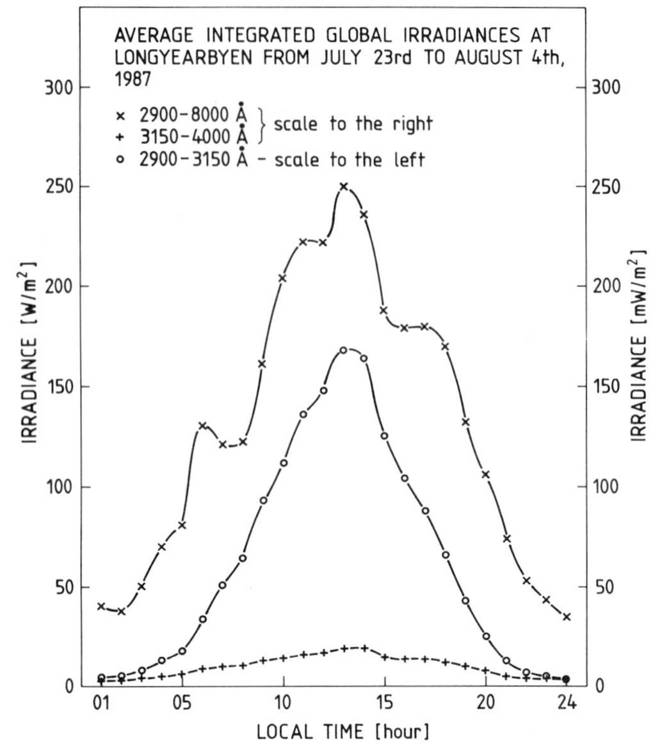

Fig. 4. Averaged diurnal variations.

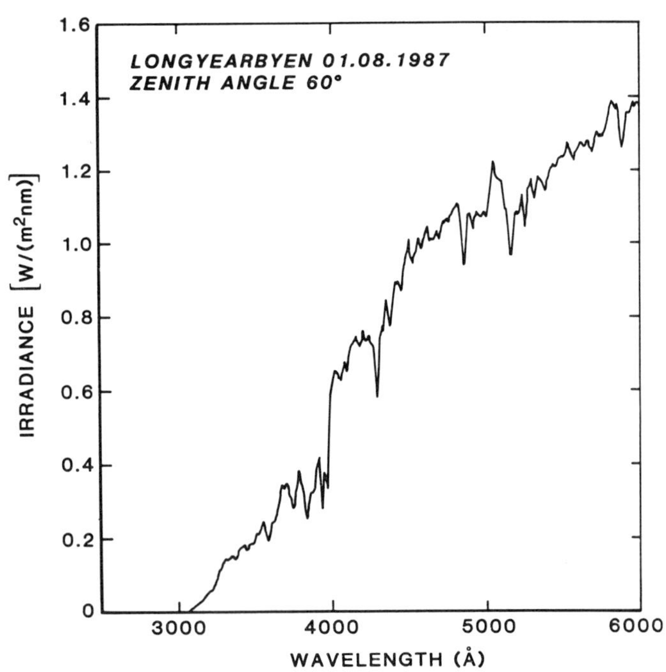

Fig. 5. Spectrum of direct solar irradiance.

4. OZONE COLUMN DENSITY

The ozone content is derived from direct solar measurement and a sample of a direct solar spectrum is given in Fig. 5. Using such data, the ozone density can be derived by applying Beer's law. Six data points can be defined in the region from 3030 to 3300 Å, and used to provide spectral wavelength pairs in a manner similar to the Dobson technique. We have used the extraterrestrial spectrum as a reference instead of extrapolating the measurements to zero airmass. For each spectral scan, the ozone density can be derived by a least squares fit, using the six data points, and results from a limited amount of data are given in Table 1.

A more detailed account of this procedure is given by Stamnes et al. (1988).

TABLE 1. COMPARISON OF DOBSON AND SPECTROMETER OZONE MEASUREMENTS

Observational Site Day Month Year	UT	Zenith Angle (°)	Amount ± s.d. (D.U.)
Tromsø 69.7°N, 18.9°E			
16 06 1987	12:32	48.9	367±25
	12:44	49.6	359±22
	12:56	50.3	357±19
	15:08	51.0	350±19
	15:20	51.1	352±21
	15:32	52.3	318±21
Longyearbyen 78.2°N, 15.6°E			
26 07 1987	10:00	60.3	291±16
	10:12	60.0	307±19
	10:24	59.7	317±30
	12:31	59.5	308±18
	12:39	59.4	331±21
	12:47	59.2	318±21
Longyearbyen			
01 08 1987	12:10	60.0	334±16
	12:22	60.0	319±16
	12:37	60.1	345±19
	12:45	60.2	338±19
	12:52	60.3	348±20
	13:00	60.4	345±20
	13:08	60.5	357±19
Longyearbyen			
01 08 1987	16:13	66.7	365±17
	16:25	67.3	375±17
	16:37	67.9	356±20
	15:49	68.5	380±18
	17:01	69.1	372±15

	Dobson measurements (D.U.)	Averaged spectrometer measurements (D.U.)
16.06	390	355
26.07		312
27.07	357	
01.08		340
03.08	355	377

5. UV RADIATION AND HEALTH RISKS

From the spectral measurements the relation between UVB and ozone column density can be derived, and such measurements should be carried out for varying atmospheric conditions in order to quantify the effects of clouds and anthropogenic "pollution" (leading e.g. to ozone depletions) on the transmitted UV and visible radiation. Ozone depletion should, if everything else affecting UV radiation in the atmosphere is unchanged, result in an increase in UVB irradiance. In a recent study by Scotto et al. (1988) show that UVB irradiances have decreased inside USA during the time period that there has been a decrease in atmospheric ozone above Canada (NASA, 1988). This indicates that other environmental parameters such as atmospheric turbidity and cloud amount (which are at present poorly documented) may play an important role in UV attenuation.

Our limited data give numbers for polar regions. The UVB irradiance is about an order of magnitude lower than at the equator, and for humans the risk of getting skin cancer is not overwhelming. Small changes in UVB might have much higher impact on vegetation and wild life.

The extraterrestrial spectrum together with the calculated spectral irradiance at the equator is plotted in Fig. 6. Below 3300 Å the attenuation effect of ozone dominates and shields the biosphere for most of the dangerous UVB and all the UVC radiation. In Table 2 measured and calculated UVB and UVA irradiances are placed together.

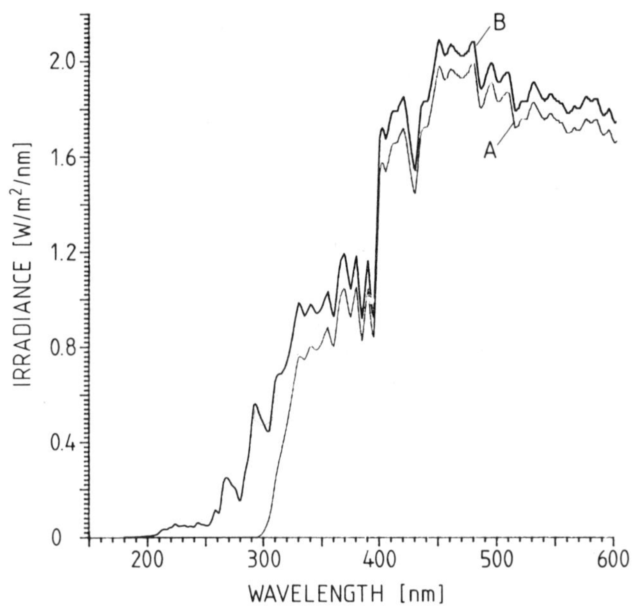

Fig. 6. Comparison of
A: Irradiance calculated at ground with 0° zenith angle, 0.3 ground albedo, and 230 D.U. of ozone.
B: Measured extraterrestrial spectrum WMO (1988).

In addition to ozone, other important modifiers are solar elevation, ground albedo and turbidity. The numbers in Table 2 indicate that the maximum UVB irradiance in Tromsø is an order of magnitude lower than at Equator. Dermatologists believe that the reason for increased frequency of skin cancer in Scandinavia is mostly due to increased vacation travels to tropical and subtropical areas and use of solar studios, rather than decreasing atmospheric ozone. It can be added that also UVA may have a significant impact on the frequency of skin cancer. UVA is little affected by ozone, while clouds and aerosols have a large effect.

TABLE 2. MEASURED AND CALCULATED UVB AND UVA IRRADIANCES

Place	Mode	SEA (deg)	Weather	UVB (mW/m^2)	UVA (W/m^2)	Albedo	Ozone (D.U.)
TR	M	20	Clear	120	14	~0.3	~280
	C	20	"	110	16	0.3	280
LYR	M	30	Clear	240	30		~350
	C	30	"	230/190	30/27	0.8/0.3	350
	C	30	"	110/450	27/28	0.3	450/200
LYR	M	6	Clear	2	3	~0.3	~350
LYR	M	30	Cloudy	100	10	0.3	~350
LYR	M	6	Cloudy	4	5	~0.3	~350
LYR	C	35	Clear	280/360	33/44	~0.3/0.8	~350
Equator	C	90	Clear	2100/2650	68/83	0.3/0.8	230
Extra-terrestrial	M			17000	86		

Mode: C - Calculated
M - Measured

An albedo value of 0.3 indicates bare ground, whereas 0.8 pertains to a snow-covered ground.

6. CONCLUDING REMARKS

The most disturbing aspect of a stratospheric ozone depletion is the resulting increase in UVB irradiance at the ground. Measurements are urgently needed to document the relationship between UV radiation received by the biosphere and atmospheric ozone content, mainly because the effects of atmospheric turbidity (including clouds) are at present largely unknown. In addition to decreasing plant production an increase in UV radiation can generate skin cancer. Our study is comparable to the work of Ilyas (1988) and show that there is a large latitudinal variation in the UVB irradiance, which is enhanced by the lowered ozone content towards the Equator.

Dütsch and Oeschger (1988) argue that the increase in UVB irradiance at ground caused by decreased atmospheric ozone content, may enhance the greenhouse effect. However, increased anthropogenic pollution of the troposphere may decrease UVB. Since increased UVB irradiance can be a key factor of an increasing greenhouse effect, a global monitoring system for UVB should be established.

REFERENCES

Dütsch, H.U., and H. Oeschger, 1988: Another possible link between the ozone layers and the greenhouse effect. Presented at the Quadrennial Ozone Symposium.

Henriksen, K., K. Stamnes and P. Østensen, 1988a: Solar radiation at 78°N. Submitted to Atmospheric Environment.

Henriksen, K., K. Stamnes, G. Volden and E. Falk, 1988b: UV irradiances at high latitudes and risks of skin diseases. Submitted to Photodermatology.

Ilyas, M., 1988: Atmospheric ozone and solar Ultraviolet radiation: The Tropical Scene. Quadr. Ozone Symposium - 1988, Göttingen, Bojkov and Fabian, Eds., A. Deepak Publishing, USA.

NASA Reference Publication 1208, August 1988. Present state knowledge of the upper atmosphere 1988: An assessment report.

Scotto, J., G. Cotton, F. Urback, D. Berger, and T. Fears, 1988: Biologically effective ultraviolet radiation: Surface measurements in the United States, 1974 to 1985. Science 239, 762.

Stamnes, K., K. Henriksen and P. Østensen, 1988: Simultaneous measurements of UV radiation and total ozone amount. Geophys. Res. Lett. 15, 784.

World Meteorological Organization, 1985: Global ozone research and monitoring project. Rep. No. 16, Atmospheric ozone 1985, Vol. 1.

THE DISTRIBUTION OF TOTAL OZONE, STRATOSPHERIC OZONE, AND TROPOSPHERIC OZONE AT LOW LATITUDES DEDUCED FROM SATELLITE DATA SETS

Jack Fishman
Atmospheric Sciences Division
NASA Langley Research Center
Hampton, Virginia 23665-5225, USA

Catherine E. Watson
Planning Research Corporation, Hampton, Virginia 23666, USA

Jack C. Larson
STX Corporation, Hampton, Virginia 23666, USA

ABSTRACT

A preliminary analysis of more than 6000 ozone profiles from SAGE I (1979-1981) and SAGE II (1984-1986) at low latitudes (30° N to 30° S) shows that the distribution of the integrated amount of ozone above 100 mb is considerably different from the distribution of total ozone obtained using concurrent TOMS measurements. From these data, we are able to calculate a "tropospheric residual" which represents the integrated amount of ozone between the surface and 100 mb. The tropospheric residual is highest over the eastern tropical Atlantic Ocean where integrated amounts of tropospheric ozone are more than twice as high over this region than at the same latitudes in the western tropical Pacific Ocean. A distinct seasonality is also observed for the tropospheric residual with the highest values being found during the dry season over western Africa and eastern South America (July through October), supportive of the premise that this ozone is likely the result of widespread biomass burning. Our analyses agree well with the limited amount of in situ tropospheric ozone data that are available in the Tropics.

1. INTRODUCTION

Building upon methodologies described in Fishman and Larsen (1987) and Fishman (1988), we present a preliminary analysis of 6704 ozone profiles obtained from SAGE (Stratospheric Aerosol and Gas Experiment) I and II in the Tropics to show that the variability of the total ozone field (i.e., the integrated amount of ozone in the stratosphere and the troposphere) is primarily a result of the variability observed in the troposphere, even though the tropospheric component accounts for only 5-15% of the total signal. Whereas the data discussed in these earlier two studies covered only one year of concurrent SAGE I profiles and total ozone measurements from TOMS (Total Ozone Mapping Spectrometer) between 15° N and 15° S, this study uses all the concurrent measurements from the reprocessed SAGE I (February 1979-November 1981) data base and the first 26 months (November 1984-December 1986) of the SAGE II data base between 30° N and 30° S. Because of this expanded data base, we are able to derive a 2-dimensional (latitude and longitude) depiction of the total ozone and integrated stratospheric ozone (above 100 mb) fields. Although we have examined many of the potential errors that may be associated with the use of satellite data for studies of tropospheric ozone (e.g., see Fishman et al., 1986; 1987b; Fishman and Larsen, 1987), we consider the analyses presented herein to be preliminary because we have not taken into account the additional consideration of the varying height of the tropopause, which may occasionally fall below 100 mb at the higher latitudes within this expanded analysis domain.

2. TOTAL OZONE, STRATOSPHERIC OZONE, AND TROPOSPHERIC OZONE DISTRIBUTIONS

In Fig. 1, we present the distributions of integrated ozone above 100 mb, total ozone, and the calculated tropospheric residual between 30° N and 30° S. The units on the contours in these plots are in Dobson Units (DU, where 1 DU=2.69 x 10^{16} mol. O_3 cm^{-2}). In general, both the total ozone and the stratospheric ozone have their lowest values near the equator and gradually increase poleward. The primary difference between these two depictions is the pronounced longitudinal gradient seen in the total ozone field close to the equator (especially between 15° N to 20° S), whereas the integrated stratospheric ozone between these two latitudes exhibits virtually no longitudinal gradients (Fishman and Larsen, 1987). As a result of these differences, the derived tropospheric residual distribution is similar to the one for the total ozone field (with a constant value subtracted) derived from the TOMS data.

The tropospheric residuals are highest along the northern border of this depiction (30° N). Without a knowledge of the heights of the tropopause for each individual SAGE profile, it is not possible to determine how much this analysis was biased by the fact that we have arbitrarily set the top of the troposphere at 100 mb. If, during some of the times of the SAGE observations, the tropopause was located significantly below 100 mb, then this analysis would erroneously include contributions from the lower stratosphere, which

ISBN 0-937194-15-8

would show up in this analysis as higher tropospheric residuals. On the other hand, the presence of relatively higher tropospheric ozone concentrations in the northern hemisphere would be consistent with the observed distribution of tropospheric ozone (e.g., Seiler and Fishman, 1981) and our understanding of the tropospheric ozone budget (e.g., see Fishman et al., 1979; Logan et al., 1981).

Closer to the equator, the tropospheric residual values are greatest off the southern west coast of Africa with annually averaged values of 31 DU. The lowest tropospheric residual values are found over New Guinea with annually averaged values of 13 DU. This is the same location where the lowest annually averaged TOMS total ozone values (255 DU) are found. Interestingly, we note that the integrated ozone above 100 mb is 244 DU at these two tropical locations where the tropospheric residuals are a maximum and a minimum, respectively.

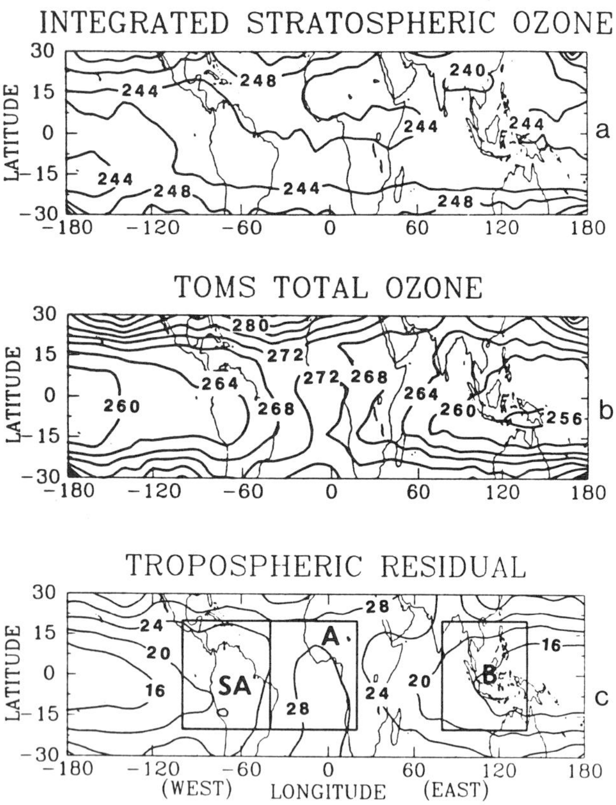

Fig. 1. The distribution of integrated ozone above 100 mb (a), total ozone (b), and the resultant troposperic residual between the surface and 100 mb (c) are depicted from an analysis of 6704 SAGE profiles. The units on the contours are Dobson Units (DU), with a contour interval of 4 DU. The boxes in Fig. 1c define the regimes over which the seasonal cycles shown in Fig. 3 were calculated.

3. ANNUAL CYCLE OF THE TROPOSPHERIC RESIDUAL

In Fig. 2, the distribution of the tropospheric residual shown in the bottom part of Fig. 1 is divided into two different time periods. The top figure depicts the distribution for the four months between July and October. During this time of the year, prevailing monsoonal circulations bring about the "dry" season in both central Brazil in South America and in southern and central Africa. Because of the relative dryness, biomass burning is more prevalent over these regions during this time of the year than at other times of the year (Batchelder, 1967). As a result, photochemically generated ozone becomes much more abundant (e.g., see Browell et al., 1988; Cros et al., 1988) over South America and southern Africa. This seasonal difference is clearly seen in Fig. 2a, where a pronounced maximum of nearly 40 DU is found during the dry season just downwind (west) of the Congo coast. The integrated tropospheric ozone amount is nearly 15 DU higher in this region during the "burning" (dry) season than the amount during the remainder of the year (Fig. 2b). For the most part, the tropospheric residuals are higher everywhere in these depictions in the dry season, except over a relatively small region in the tropical western Pacific Ocean, where residual amounts are about 4 DU lower in July-October than during the remanider of the year. It is noteworthy that Malingreau et al. (1985) describe widespread burning taking place over Borneo in the February-May time frame.

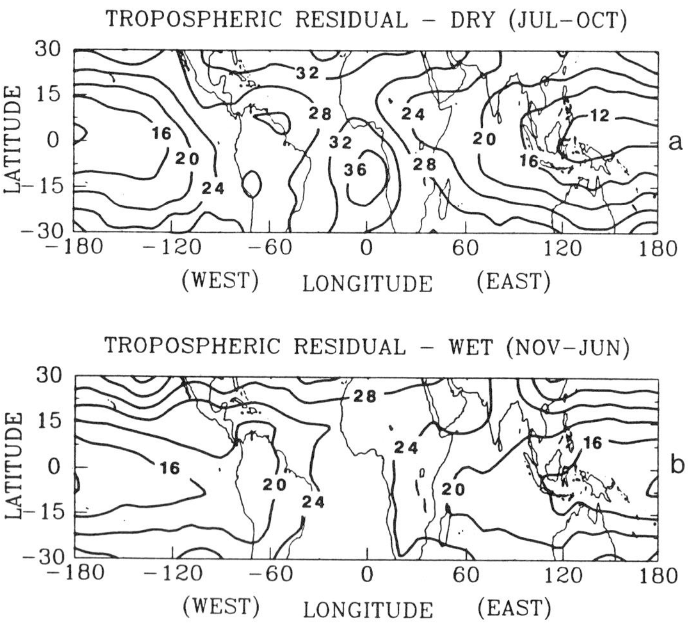

Fig. 2. The distribution of the tropospheric residual from Fig. 1c is shown for two different seasons of the year: the months of July through October (a), and the remainder of the year (b). The units on the contours are Dobson Units (DU), with a contour interval of 4 DU.

The seasonal cycles of the tropospheric residual over three tropical regions between 20° N and 20° S are examined in detail in Fig. 3. The solid thick (top) curve depicts the seasonal cycle over the region bounded by 20° N, 20° S, 40° W, and 20° E (box A, in Fig. 1c), which incorporates western Africa and the eastern tropical Atlantic Ocean. This curve peaks in September at 34 DU and minimizes in April with a value 22 DU. The box adjacent to it (labeled SA) incorporating most of tropical South America exhibits a very similar seasonal cycle , but shifted downward by ~5 DU. In contrast, the dashed curve, representing the area labeled "B" in Fig. 1c shows a pronounced maximum in late May, consistent with the discussion in Malingreau et al. (1985) describing maximum burning in Borneo between February and May. It should be noted that the particular case studies descibed by Malingreau et al. occurred primarily in 1983, in between the SAGE I and SAGE II data. Nonetheless, this period of greater biomass burning seems to be reflected by the average seasonal of the tropospheric residual. The secondary maximum seen in the dashed curve in late November may be the result of the climatological east-to-west transport in the tropics associated with the earlier maxima seen in September for box "A" and in October for Box "SA". A more rigorous analysis of the climatological circulation patterns in the tropics is necessary to verify such a hypothesis.

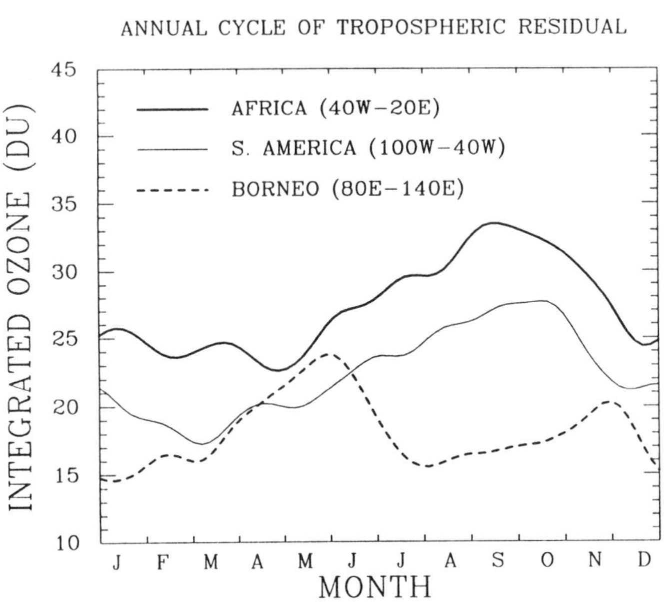

Fig. 3. The annual cycles of tropospheric ozone for three geographic areas between 20° S and 20° N are shown. The thick solid line is the averaged tropospheric residual (TOMS minus SAGE) from 40° W to 20° E (Africa), the thin solid line from 100° W to 40° W (S. America), and the dashed line from 80° E to 140° E (Borneo). The boxes representing these areas can be seen in Fig. 1c.

4. COMPARISON WITH OTHER MEASUREMENTS

In Fishman (1988), it was shown that the integrated amount of tropospheric ozone obtained from these measurements is consistent with the amount of ozone measured in situ in the tropical troposphere using the profiles reported by Seiler and Fishman (1981), Delany et al. (1985), Fishman et al. (1987a), and others. In a recent study comparing the amount of ozone over the Amazon Basin during the dry season and wet season, Kirchhoff (unpublished data, presented at the Spring 1988 AGU Meeting) reported integrated ozone amounts obtained from special ozonesonde launches of 29 and 24 DU between the surface and 200 mb for the dry (1985) and wet (1987) seasons, respectively. Assuming a representative ozone distribution between 200 mb and 100 mb, the integrated amount of ozone between the surface and 100 mb for the dry and wet seasons would be 32 and 26 DU, respectively. By comparison, the depictions in Figs. 2 and 3 show integrated tropospheric residual values of 27 and 21 DU, for the corresponding two seasons at the same location as Kirchhoff's ozonesonde profiles. Thus, the tropospheric residual values derived from the satellite data show approximately the same difference between the two seasons, but are lower in both seasons by ~20%. This finding is consistent with the direct comparison presented in Fishman and Browell (1988), which showed that the tropospheric residual values obtained on the same day as two special ozonesonde profiles in July 1985 were also ~20% lower than the in situ measurements. Thus, it appears that the use of satellite-derived tropospheric residual data accurately depicts spatial and temporal variations of the tropospheric ozone distribution, but tends to underestimate the actual amount of ozone present in the troposphere. The magnitude of this underestimate (15-20%) would be consistent with the study of Klenk et al. (1982), which indicated that the TOMS instrument can underestimate the amount of ozone in the planetary boundary layer and that the amount of this error was dependent on the surface albedo (see Fishman et al., 1987b).

TABLE 1. CORRELATION COEFFICIENTS

Latitude Band	Data Points	TOMS/SAGE	TOMS/TROP
5 S - 5 N	1067	-.06	.94
10 S - 10 N	2057	.07	.97
15 S - 15 N	3073	.03	.98
20 S - 20 N	4176	.23	.97
30 S - 30 N	6704	.47	.90

Table 1. Calculated correlation coefficients for the TOMS and SAGE data and the TOMS and Tropospheric residual, for five latitude bands. The number of data points used in each calculation is also given.

5. STATISTICAL ANALYSIS OF TOTAL OZONE, STRATOSPHERIC OZONE, AND TROPOSPHERIC OZONE

In Table 1, we summarize the correlation coefficients between the amounts of total ozone (from TOMS), integrated stratospheric ozone above 100 mb (from SAGE), and the derived tropospheric residual (denoted as TROP in the Table). To compute these correlation coefficients, the data are put into 5°-longitude bins (see Fishman and Larsen, 1987). Particularly noteworthy is the fact that there is a very strong positive correlation between the calculated troposheric residual and the total ozone data, especially in the latitude bands closest to the equator. Between 15° N and 15° S, on the other hand, there is absolutely no correlation between the TOMS and SAGE data. As the region of interest is expanded from the equator, we note that a positive correlation between the TOMS and SAGE data increases to significant levels, although the calculated correlation coefficient is still considerably smaller than the one for TOMS and the tropospheric residual. When we compare the data for only the latitude band between 20° N and 30° N (not shown in Table 1), we note that the correlation coefficient between the TOMS total ozone and the integrated stratospheric ozone from SAGE is 0.91, whereas the correlation coefficient between the TOMS total ozone and the derived tropospheric residual is 0.82. This high correlation between TOMS and SAGE indicates that the variations in ozone column at these higher latitudes are strongly influenced by the amount of ozone in the stratosphere. A similar finding is also observed between 20° S and 30° S.

6. SUMMARY AND CONCLUSIONS

For the first time, a 2-dimensional depiction of the distribution of tropospheric ozone derived from satellite information has been presented. The depictions shown in Figs. 1 and 2, together with the correlations summarized in Table 1 are extremely enlightening because they suggest that TOMS data, which have provided daily maps of total ozone since 1978, can be used with an extremely high degree of reliabilty to investigate the distribution of tropospheric ozone in regions where in situ measurements are severely lacking. Our ongoing research efforts will concentrate on these relationships to provide more definitive quantitative information on the potential use and shortcomings of satellite data for tropospheric studies. An examination of the annual cycles of the tropospheric residuals over Africa, South America, and Borneo shows that the peak values found over each of these tropical regions correlates strongly with the respective times at which widespread biomass burning is most pronounced at each locale.

ACKNOWLEDGEMENTS

This study has benefited from discussions with W.P. Chu, M.P. McCormick and R.K. Seals, Jr., of NASA LaRC and was funded in part through NASA's Tropospheric Chemistry Program.

7. REFERENCES

Batchelder, R.B., 1967: Spatial and temporal patterns of fire in the tropical world, Proc. 6th Annual Tall Timber Fires Ecology Conf., 171-191.

Browell, E.V., G.L. Gregory, R.C. Harriss, and V.W.J.H. Kirchhoff, 1988: Tropospheric ozone and aerosol distributions across the Amazon Basin, J. Geophys. Res., 93, 1431-1451.

Cros, B., R. Delmas, D. Nganga, B. Clairac, and J. Fontan, 1988: Seasonal trends in equatorial Africa: Experimental evidence of photochemical formation, J. Geophys. Res., 93, 8355-8366.

Delany, A.C., P.J. Crutzen, P. Haagenson, S. Walters, and A.F. Wartburg, 1985: Photochemically produced ozone in the emission from large-scale tropical vegetation fires, J. Geophys. Res., 90, 2425-2429.

Fishman, J., 1988: Tropospheric ozone from satellite total ozone measurements, Tropospheric Ozone, D. Reidel Publishing Co., 111-123.

Fishman, J. and E.V. Browell, 1988: Comparison of satellite total ozone measurements with the distribution of tropospheric ozone obtained by an airborne UV-DIAL system over the Amazon Basin, Tellus (in press).

Fishman, J., and J.C. Larsen, 1987: Distribution of total ozone and stratospheric ozone in the tropics: Implications for the distribution of tropospheric ozone, J. Geophys. Res., 92, 6627-6634.

Fishman, J., S. Solomon, and P.J. Crutzen, 1979: Observational and theoretical evidence in support of a significant in situ photochemical source of tropospheric ozone, Tellus, 31, 432-446.

Fishman, J., P. Minnis, and H.G. Reichle, Jr., 1986: The use of satellite data to study tropospheric ozone in the tropics, J. Geophys. Res., 91, 14451-14465.

Fishman, J., G.L. Gregory, G.W. Sachse, S.M. Beck, and G.F. Hill, 1987a: Vertical profiles of ozone, carbon monoxide, and dew-point temperature obtained during GTE/CITE I, October-November 1983, J. Geophys. Res., 92, 2083-2094.

Fishman, J., F.M. Vukovich, D.R. Cahoon and M.C. Shipham, 1987b: The characterization of an air pollution episode using satellite total ozone measurements, J. Clim. Appl. Meteor., 26, 1638-1654.

Klenk, K.F., P.K. Bhartia, A.J. Fleig, V.G. Kaveeshvar, R.D. McPeters, and P.M. Smith, 1982: Total ozone determination from the Backscattered Ultraviolet (BUV) experiment, J. Appl. Meteor., 21, 1672-1684.

Logan J.A., M.J. Prather, S.C. Wofsy, and M.B. McElroy, 1981: Tropospheric chemistry: A global perspective, J. Geophys. Res., 86, 7210-7254.

Malingreau, J.P., G. Stevens, and L. Fellows, 1985: Remote sensing of forest fires: Kalimantan and North Borneo in 1982-83, Ambio, 14, 314-320.

Seiler, W., and J. Fishman, 1981: The distribution of carbon monoxide and ozone in the free troposphere, J. Geophys. Res., 86, 7255-7266.

OZONE TRANSPORT INTO THE TROPOSPHERE IN A CUT-OFF LOW EVENT

G. Vaughan and J.D. Price
Physics Department, U.C.W.
Aberystwyth, UK

ABSTRACT

TOMS ozone maps and meteorological data have been studied for a cut-off vortex in the upper tropospheric flow which developed over the eastern United States in April 1983. It is shown that the propagation of jet streaks around such a vortex can a) induce tropopause folding on its western flank and b) replenish the vortex with air high in potential vorticity and ozone. During a period when jet streaks were not active, total ozone inside the vortex area decreased as convection eroded the low tropopause, mixing tropospheric air into the lower stratosphere. It is estimated that during this period the rate of transfer of ozone from stratosphere to troposphere amounted to about 16% of the average rate for the whole of the Northern hemisphere. This, combined with the induction of folds, means that cut-off lows are potent regions of stratosphere-troposphere exchange.

1. INTRODUCTION

In recent years concern about increases in tropospheric ozone has been growing because of ecological damage caused by pollution events (Logan 1985) and also because of possible climatic effects (Ramanathan et al 1985). To asses reliably the anthropogenic component of free tropospheric ozone, the nature of stratosphere-troposphere exchange must be well understood. Until recently it was believed that the main transfer of lower stratospheric ozone occurs in tropopause folding events associated with rapid cyclogenesis (Danielsen 1968; Reiter 1975; Keyser and Shapiro 1986; Browell et al 1987). However, observations near a cut off low in the upper troposphere (Bamber et al 1984) suggest that such features may also contribute to the removal of ozone from the stratosphere. This is noted as a major uncertainty in the WMO report on atmospheric ozone (1985).

In this paper, a case study of a cut off low in the upper troposphere is presented, with a view to assessing the potential for cross-tropopause transport of ozone and estimating any amounts transferred to the troposphere during the lifetime of the event. The vortex occurred over the north-eastern United States and eastern Canada at the same time as the cut-off low investigated by Bamber et al (late April 1983). Use is made in this study of gridded TOMS total ozone maps, global assimilation fields from the ECMWF operational forecast model and radiosonde observations. The dense network of radiosonde stations over North America enables a more detailed case study than would be possible for a cut-off low occurring over the ocean, and was the reason for choosing the present example.

2. DEVELOPMENT OF CUT-OFF LOW SYSTEM

The cut off system ensued from a large southward extension of the polar jet over the eastern United States during the 16th and 17th April 1983. The trough deepened further and cut off on the 18th, with the beginnings of a tropopause fold apparent on TOMS total ozone maps as a hook shape in the total ozone contours. The system was then 1900km across and was at its most southerly extension.

By the 19th the fold had intensified and appeared as a marked feature on the western side of the trough (fig.1). At the same time a large area of the vortex had broken away and was moving northwards, leaving the main low much reduced in size. Visible in the low on the ozone map for the 19th is a large crescent shaped area of very high ozone (up to 480 DU) and low tropopause starting in the south east and curving through north east to the north west of the low. Upwind adjacent to the folding there was very high wind shear with winds of $40ms^{-1}$ only 300-400 km from the centre of the low. The absolute vorticity in this shear zone was at least twice the planetary value, so in order to conserve its potential vorticity (PV) a parcel of air reaching the jet exit had either to turn cyclonically, diverge horizontally or both. The crescent of high ozone at the base of the low strongly suggests that air at the jet exit did turn cyclonically, taking high PV into the cut-off vortex. This new air originated at high latitudes, where the planetary vorticity was large and the tropopause low, and it is this low tropopause which caused an enhancement of the total ozone. This event may also have been responsible for displacing air originally in the vortex, causing it to move poleward and re-join the main zonal flow.

A very similar case to the present example is shown in WMO (1985, p.205) for the cut-off low studied by Bamber et al (1984). Once again, a jet streak propa-

ISBN 0-937194-15-8

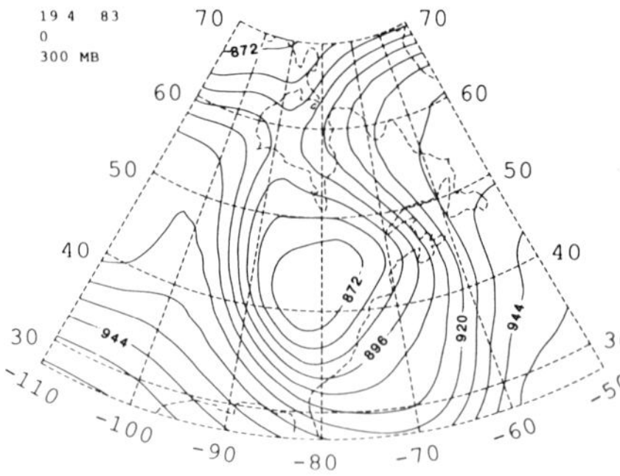

FIG 1a

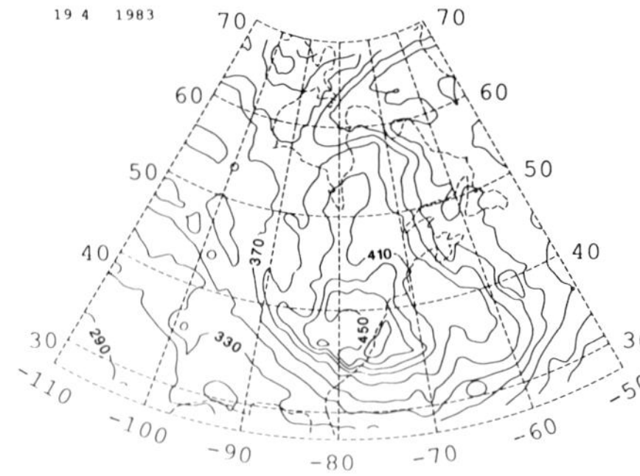

FIG 1b

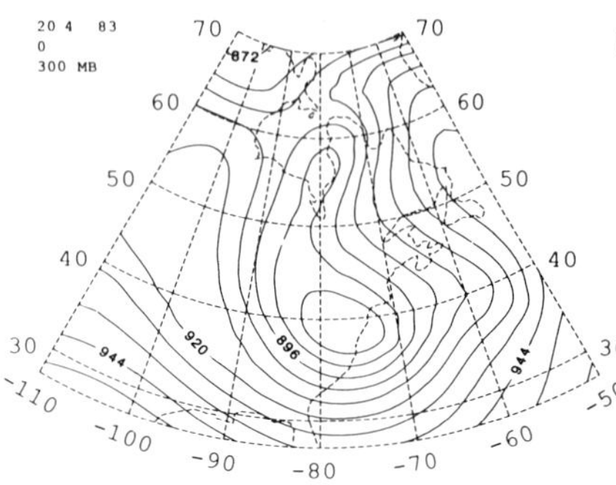

FIG 2a

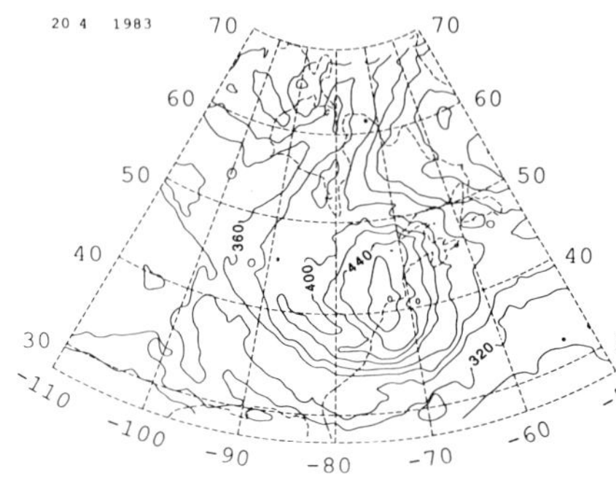

FIG 2b

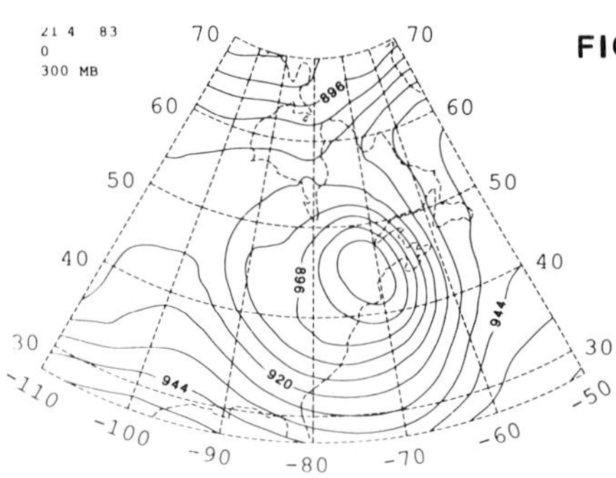

FIG 3a

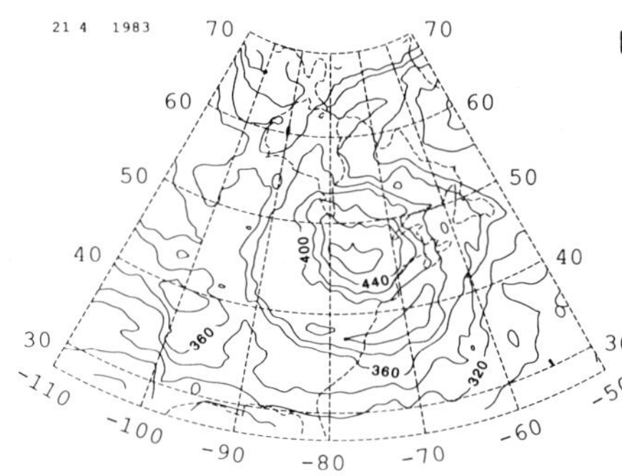

FIG 3b

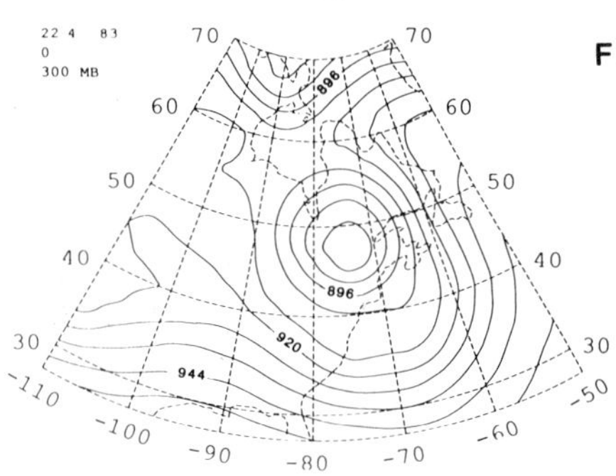

FIG 4a

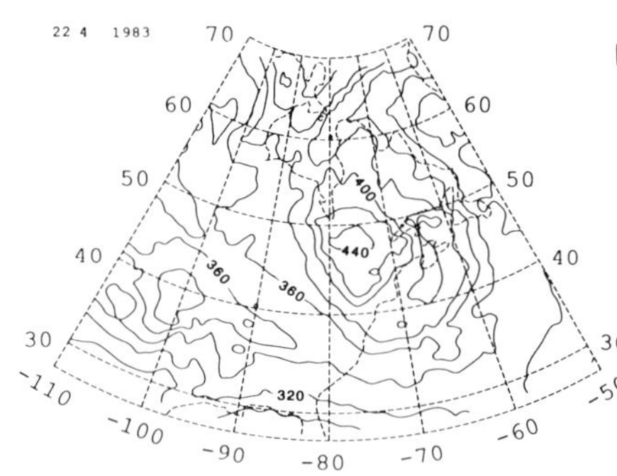

FIG 4b

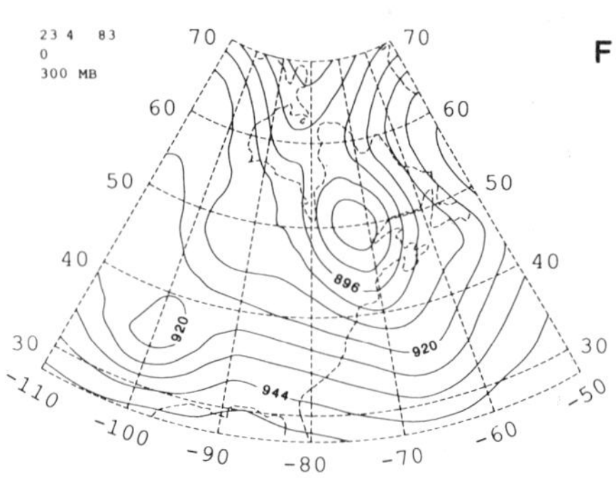

FIG 5a

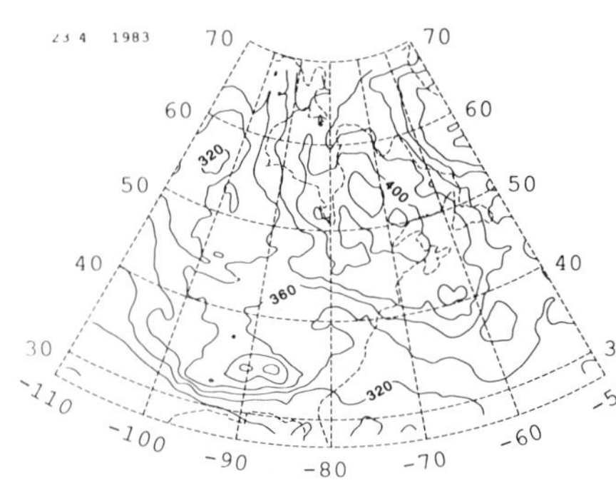

FIG 5b

FIGURES 1–5. CHARTS FOR THE PERIOD 19–23 APRIL 1983.
a) 00 GMT, 300mb GEOPOTENTIAL HEIGHTS, Dm b) TOMS TOTAL OZONE, D.U., c. 1700 GMT

gated around the low, a tropopause fold developed on its western flank and an area of enhanced total ozone moved from the base of the trough (jet exit) northwestwards into the vortex.

These observations are important as they strongly suggest that jet streaks propagating around a cut-off low are capable of injecting high PV into the vortex and prolonging its life. We can also note that a weak jet streak following a stronger one that has penetrated the low cannot displace it as its reduced shear vorticity cannot generate the radius of curvature necessary to enter the low. We might then postulate that only air of similar or higher PV than that already in the vortex can penetrate it.

3. FURTHER DEVELOPMENT

By the 20th (fig. 2) the system was more settled with the cut off region more circular and the highest ozone in the centre of the low, which had drifted northeastwards along the eastern coast of America. From the 20th - 23rd the low drifted slowly northwards (figs. 2-5) and reduced in size while remnants of the fold tracked around it. At the same time total ozone values in the vortex decreased. This quiescent period represents the time when the vortex could be regarded as a distinct mass of air, for which the ozone transfer could be estimated.

On the 24th total ozone in the vortex had increased again, with ECMWF geopotential maps showing that the cut off region had expanded as a new jet streak propagated around it. This again illustrates exchange of air between the low and its surroundings. The system then lay further north over snow fields with little potential for any large scale convection, and by the 26th at 1200 hours it had re-joined the main zonal flow.

Included in the total ozone data is an implicit addition to compensate for the tropospheric component below cloud (Fleig et al 1982). This includes Nimbus 7 THIR data to determine reflectivity and cloud top height. The magnitude of this correction is about 5DU, and since the mean deviation in the TOMS reflectivity over the period considered was less than 4%, it may be expected to remain nearly constant during this period. It is therefore not considered here.

4. EFFECT OF CONVECTION

Deep convection reaching the tropopause may be expected to warm the upper troposphere and, by overshooting the tropopause, cool the lower stratosphere and lift the tropopause. Radiosonde ascents in the centre of the vortex were examined from the 20th-22nd to see if this had occurred. The majority of tropopauses on the 22nd were warmer than on the 20th, with a slightly more stable upper troposphere (>500mb). The average tropopause rose by 20mb. Figure 6 shows the average

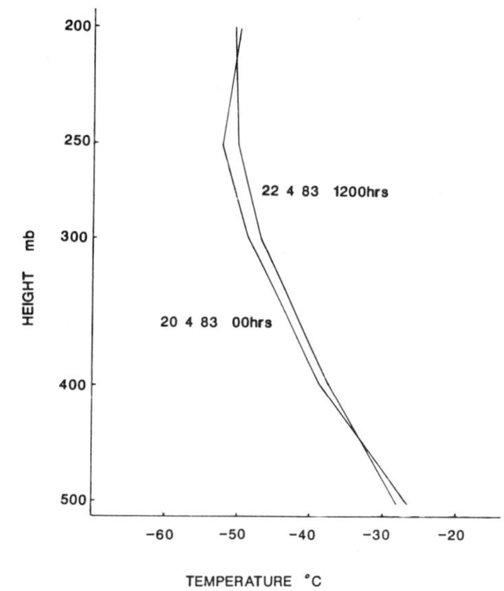

Fig. 6. Average temperature profiles in the centre of the low calculated from standard level radiosonde data. (Note that the tropopause is not shown in this figure)

lapse rates for the two days and illustrates the tropospheric warming as well as showing evidence for stratospheric cooling above 225mb. It must be emphasised that the range of values encountered in temperature and tropopause height on the two days were larger than the differences between the averages, so that the differences may not be significant. The range of values encountered, however, was noticably smaller on the 22nd suggesting more thorough mixing by that time.

5. ESTIMATE OF CROSS TROPOPAUSE FLUX OF OZONE

Total ozone maps (figs.2-5) show ozone values increasing within the vortex from the 16th to the 19th April as the system developed, then decreasing from the 20th to the 23rd, followed by a small increase on the 24th before decaying on the 25th and 26th. At the same time total ozone in the surrounding area changed little, demonstrating that the observed changes occurred in the vortex itself, at the base of the stratosphere, rather than higher up in the main ozone layer. By determining the size of the cut off region from ECMWF charts, the total amount of ozone in a vertical column above the vortex was calculated from the TOMS maps. Changes in this total were then taken to relate to changes within the vortex; in particular the decrease from the 20th - 23rd was interpreted as transfer of air from stratosphere to troposphere.

It is presumed in this analysis that the decrease in TOMS total ozone within the vortex is caused by clouds penetrating the stratosphere, reflecting sunlight back to the satellite at a higher level. Thus the 'missing' ozone is transferred below cloud top, out of sight of the instrument, and into the troposphere. The transfer therefore represents the tropopause being re-established above ozone-rich air originally in the stratosphere, not direct transfer of ozone into the lower troposphere, such as occurs in tropopause folds (although some ozone may be so transferred in convective downdraughts). It is clear from the TOMS maps that a cut-off low can induce folding events along its flank (fig.1) thus further contributing to stratosphere-troposphere exchange, but this effect is not included here.

Since it is difficult to determine precisely from the charts which air parcels were trapped in the vortex and which were part of the flow around its flank, upper and lower limits for the transfer of ozone were estimated by considering maximum and minimum probable cut off areas. The amount of ozone transferred as a consequence of raising the tropopause within the low from the 20th-23rd was estimated to lie between 7.4×10^{33} and 1.1×10^{34} molecules. An error for the lower value was estimated by displacing the cut off regions slightly over the ozone maps and noting changes in the total. This was found to be typically 5%. The increased uncertainty in the larger estimate will make its error larger than this.

A further point can be made by considering the amount that the tropopause will rise in order to transfer 7.4×10^{33} molecules. If we assume a simple cylindrical model for the vortex and a partial pressure for ozone of 75 nb in the lower stratosphere, we can estimate the rise assuming that the resulting tropospheric air will be an equal mixture from the two regions. This gives an average rise of the tropopause of about 0.6km. That observed over the period 20th-23rd April was 0.45km. Considering the limitations of the model and the uncertainties in the measured tropopause height, these two results appear to be consistent.

6. SUMMARY AND CONCLUSION

From the case given here, the potential for cross-tropopause flow in cut off cyclones is demonstrated. Estimates of the total cross-tropopause flux have been made by Danielsen (1968) and Danielsen and Mohnen (1977) using radioactivity, ozone and potential vorticity. The latter paper gives a cross-tropopause flux of 7.8×10^{10} ozone molecules $cm^{-2}s^{-1}$ for the northern hemisphere. Later, general circulation model estimates (Mahlman et al 1980; Gidel and Shapiro 1979; Levy et al 1985) suggest a smaller flux, between 4 and 7×10^{10} $cm^{-2}s^{-1}$ in the northen hemisphere (and a factor of 2 smaller in the southern hemisphere). If the cross-tropopause flux is 7×10^{10} $cm^{-2}s^{-1}$, the transfer of ozone during the present episode of 2.5×10^{33} molecules per day amounts to some 16% of the average daily transfer in the northern hemisphere. This is a significant proportion of the total.

ACKNOWLEDGEMENTS

We thank Dr. A.J. Fleig, Dr. A.J. Krueger, the TOMS experiment team and the World Data Center-A, Greenbelt, Maryland for supplying us with TOMS data and the staff of the European Centre for Medium-range Weather Forecasts for meteorological data.

REFERENCES

Bamber, D. J., P. G. Healey, B. M. Jones, S. A. Penkett, A. F. Tuck and G. Vaughan, 1984: Vertical profiles of tropospheric gases - chemical consequences of stratospheric intrusions. Atmos. Environ., **18**, 1759-1766.

Browell, E. V., E. F. Danielsen, S. Ismail, G. L. Gregory and S. M. Bleck, 1987: Tropopause fold structure determined from airborne lidar and in situ measurements. J. Geophys. Res., **92**, 2112-2120.

Danielsen, E. F., 1968: Stratospheric-tropospheric exchange based on radioactivity, ozone and potential vorticity. J. Atmos. Sci., **25**, 502-518.

Danielsen, E. F. and V. A. Mohnen, 1977: Project Dustorm report: Ozone transport, in situ measurements and meteorological analyses of tropopause folding. J. Geophys. Res., **82**, 5867-5877.

Fleig, A. J., K. F. Klenk, P. K. Bhartia and D. Gordon, 1982: Users guide for the TOMS instrument first year ozone-T data set. Goddard Space Flight Center, Green Belt, Maryland.

Gidel, L. T. and M. A. Shapiro, 1980: General circulation model estimates of the net vertical flux of ozone in the lower stratosphere and the implications for the tropospheric ozone budget. J. Geophys. Res., **85**, 4049-4058.

Keyser, D. and M. A. Shapiro, 1986: A review of the structure and dynamics of upper level frontal zones. Mon. Weather Rev., **114**, 452-499.

Levy, H. B., J. D. Mahlman, W. J. Moxim and S. Liu, 1985: Tropospheric ozone: the role of transport. J. Geophys. Res., **90**, 3753-3771.

Logan, J. A., 1985: Tropospheric ozone: seasonal behaviour, trends and anthropogenic influence. J. Geophys. Res., **90**, 10463-10482.

Mahlman, J. D., H. B. Levy and W. J. Moxim, 1980: Three-dimensional tracer structure and behaviour as simulated in two ozone precursor experiments. J. Atmos. Sci., **37**, 655-685.

Ramanathan, R. J., R. J. Cicerone, H. B. Singh and J. T. Kiehl, 1985: Trace gas trends and the potential role in climate change. J. Geophys. Res., **90**, 5547-5566.

Reiter, E.R., 1975: Stratospheric-tropospheric exchange processes. Rev. Geophys. Space Phys., **13**, 459-474.

THE LATITUDINAL AND VERTICAL DISTRIBUTION OF TROPOSPHERIC OZONE OVER THE ATLANTIC OCEAN IN THE SOUTHERN AND NORTHERN HEMISPHERES

Herman G.J. Smit, Dieter Kley, Stuart McKeen, Andreas Volz, and Stefan Gilge
Institut für Chemie 2: Chemie der Belasteten Atmosphäre
Kernforschungsanlange Jülich, D–5170 Jülich, FRG

ABSTRACT

Vertical ozone soundings were made during a cruise of the research vessel "Polarstern" in March/April 1987 on the Atlantic between Puerto Madryn (42°S) and Bremerhaven (54°N). Electrochemical O_3 sondes (ECC) were used in combination with radio sondes and a novel telemetry that allowed the simultaneous measurement of O_3, P, T, and RH. In addition, the O_3 concentration was continuously monitored aboard the ship with an UV–photometer.

Tropospheric O_3 in the southern hemisphere was characterized by low values near the surface and a large vertical gradient. The lowest ozone mixing ratios were 8 ppbv, measured at approximately 10°S. Photolysis of ozone in the lower troposphere, followed by reaction of $O(^1D)$ with water vapor, provides the major sink for ozone in the tropics. In the northern hemisphere, large mixing ratios up to 55 ppb were found near the surface. The vertical gradient was very weak. Strong influences by stratospheric/tropospheric exchange processes were observed around 30°N.

1. INTRODUCTION

Although ozone is certainly the most important trace gas in the troposphere, there is a dearth of information on its distribution with respect to altitude/latitude. Ozone is most frequently measured at or very close to ground. In fact, there are hundreds of ground-based measuring sites in the northern hemisphere (NH). Tropospheric profiles of ozone have mostly been obtained at stations that use electrochemical ozone sondes. Traditionally, their principal purpose is the collection of stratospheric ozone data. In Central Europe the most significant station that has reported tropospheric ozone profiles is Hohenpeißenberg, operated by the Deutscher Wetterdienst [Attmannspacher et al., 1984].

Tropospheric cross sections of ozone over the Central Atlantic have not been reported up to the present. This fact is significant since maritime ozone profiles are needed for several reasons. Profiles obtained over the North Central Atlantic would give information on maritime tropospheric background levels. By comparison with continental ones over the American and the European continent, respectively, they could be used to assess questions in regard to transport and photochemical production. Finally, cross sections over the Atlantic are important to establish upwind boundary conditions for model studies of O_3 over Europe. In this article we report a cross section of ozone between 40°S and 45°N, obtained mostly along the 30°W meridian.

2. EXPERIMENTAL DETAILS

The vertical soundings were performed aboard the German research vessel "Polarstern", operated by the Alfred Wegener Institut, Bremerhaven, during cruise ANT/V5 from Puerto Madryn (42°S) to Bremerhaven (54°N) between March 19 and April 18, 1987. Figure 1 gives the daily position of the "Polarstern" with the vertical soundings marked by circles. Double circles denote simultaneous launches of two sondes. In addition, the O_3 concentration was continuously measured aboard the ship about 20 m above sea level by means of a UV–photometer (Dasibi).

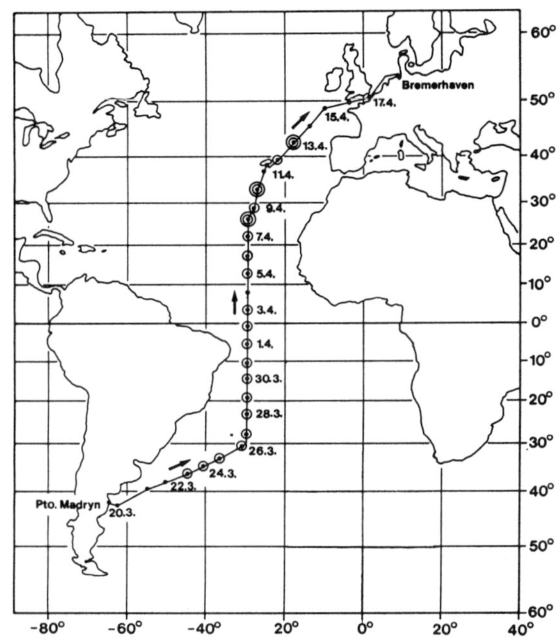

Fig. 1. Cruise track of the Polarstern. The dots give the daily position of the ship at noon. Circles denote vertical soundings, double circles the launch of twin flights.

For the vertical O_3 soundings, electrochemical concentration cells (ECC, Scientific Pump) after Komhyr (1969) were used. They were deployed together with RS–80 radiosondes (Vaisala, Helsinki) on 600 g Totex balloons. Normally, the signal from the ECC is fed into the radiosonde and the transmitted O_3 data replace those on relative humidity. We used a microcomputer controlled interface board (TMAX, Boulder) that allows the simultaneous transmission of ECC current, temperature of the air sampling pump, as well as the data from the radio sonde, e.g. pressure, temperature, and relative humidity. All the parameters are digitized in-flight by the interface. The data stream consisting of ASCII hexadecimals is then modulated as a two–tone signal on the 403 MHz carrier frequency of the RS–80 sonde and transmitted to the ship. It is demodulated by a conventional FM–receiver and fed into a personal computer via a modem. Every 5 s, a full data frame is transmitted.

Data were taken during the ascent of the balloon. Under the experimental conditions (1050 g payload, 1100–1300 g free lift), the ascent velocity was around 5 m/s and the burst altitude of the balloon near 26 km. The 5 s sampling intervals thus rendered a height resolution of 25 m. As the inherent time constant of the ECC is about 20 s, the effective height resolution is only about 100 m. The meteorological data obtained by the radiosonde, namely temperature, pressure and relative humidity were recorded with the same time/height-resolution as O_3. Altogether, 24 flights were made during the cruise of which 3 were twin flights.

ISBN 0–937194–15–8

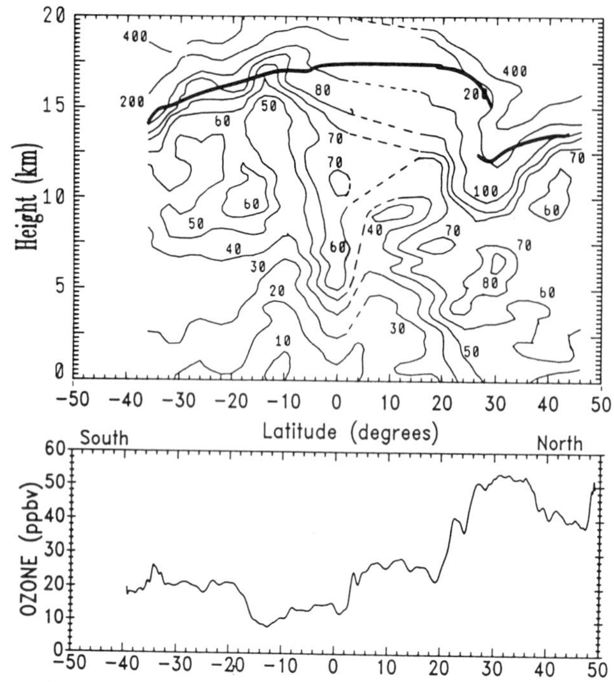

Fig. 4. a) Tropospheric meridional cross section of ozone. Regions of missing observations are interpolated by dashed lines. Tropopause heigts are indicated by the thick line.
b) Ozone levels near the surface (20 m above sea level), measured by UV–photometry.

Vertical O_3 profiles over the Central Atlantic have not been reported so far. There are, however, data from the STRATOZ III mission (Marenco, 1988) that went along the eastern coastline of the South American continent to Natal, crossed the Atlantic between 5°S and 15°N, and finally went from Dakar over the Canary Islands to Portugal and France. Although the two data sets were collected three years apart as well as during slightly different times of the year (STRATOZ III took place in June 1984) they seem to be worth comparing with regard to general differences. Unfortunately, the STRATOZ experiment was restricted to altitudes below 12 km and thus did not sample the lower stratosphere, except at high latitudes. In addition, as the aircraft usually landed at the larger cities, local photochemical influences could have been present in the ABL. The latter is most evident at mid latitudes in the SH, where the STRATOZ data show high O_3 values around 60 ppb in the ABL. They are most likely due to photochemical production in the large metropolitan plumes of Buenos Aires and Rio de Janeiro. In this context it must be noted that the STRATOZ data do not represent a meridional cross section in the sense of representing typical conditions of either maritime or continental air masses. There are, however, two large scale features seen in both data sets, namely a strong horizontal gradient between 20–30°N and almost no vertical gradient north of 30°N.

The most prominent large scale features seen in Figure 4 are the low O_3 concentrations around 10°S extending from the bottom almost throughout the whole troposphere and the high values at 30–40°N penetrating from the lower stratosphere into the upper and mid troposphere. The low concentrations at 10°S suggest the existence of a sink in that region (Liu et al., 1983). The accompanying high concentrations of water vapor measured by the radiosondes provide a photochemical explanation: Destruction of ozone is provided by photolysis, leading to $O(^1D)$, followed by reaction of $O(^1D)$ with water vapor.

O_3	$+ h\nu$	$\longrightarrow$	$O(^1D)$	$+ O_2$	$J(O^1D)$
$O(^1D)$	$+ M$	$\longrightarrow$	$O(^3P)$	$+ M$	k_1
$O(^1D)$	$+ H_2O$	$\longrightarrow$	$2\,OH$		k_2

The relative O_3 loss rate is given by (1).

$$-\frac{1}{[O_3]}\cdot\frac{d[O_3]}{dt} = J(O^1D)\cdot\frac{[H_2O]\cdot k_2}{[H_2O]k_2+[M]k_1} \quad (1)$$

It increases nearly proportionally with increasing water mixing ratio. We will, at a later time, be in a position to provide a complete cross section of the water mixing ratio. At present, we still need to evaluate this type of sensor for its accuracy in the upper troposphere. However, there is no question concerning the sensor's performance in the lower troposphere. From these data, we have calculated isolines of relative O_3 loss rates (normalized to $J(O^1D)$,) for the lower troposphere in the region 30°S to 30°N, as shown in Figure 5. According to equation 1, they are almost proportional to the isolines of the water vapor mixing ratio. By comparing Figures 4 and 5 it is seen that the loss rate is largest where the lowest O_3 concentrations are observed. In first approximation, the relative loss rates can be converted to absolute 24 h averages of the photochemical O_3 destruction by multiplying the isolines in Figure 5 with the respective average of $J(O^1D)$. With a value of $7\cdot10^{-6}\ s^{-1}$ a life time of O_3 around 10^6 s against photolysis is obtained in surface air at 10°S. In comparison, O_3 loss by dry deposition to the ocean surface yields a lifetime of $3\cdot10^6$ s, as obtained with a deposition velocity of $0.05\ cm\cdot s^{-1}$ (Lenschow et al., 1982; Galbally and Roy, 1980) and an ABL height of 1.3 km. The latter was derived from the potential temperature. Bearing in mind that the photochemical loss was estimated conservatively, that is without the additional loss due to the reaction of O_3 with HO_2 radicals, it follows that photochemistry provides the major sink for O_3 in the ABL at 10°S.

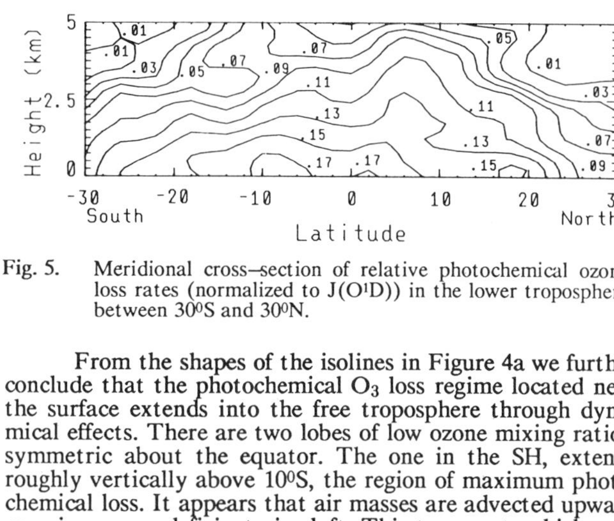

Fig. 5. Meridional cross–section of relative photochemical ozone loss rates (normalized to $J(O^1D)$) in the lower troposphere between 30°S and 30°N.

From the shapes of the isolines in Figure 4a we further conclude that the photochemical O_3 loss regime located near the surface extends into the free troposphere through dynamical effects. There are two lobes of low ozone mixing ratios, symmetric about the equator. The one in the SH, extends roughly vertically above 10°S, the region of maximum photochemical loss. It appears that air masses are advected upward carrying ozone deficient air aloft. This transport, which may be identified with the rising branch of the tropospheric Hadley cell, extends into the upper troposphere, perhaps even into the stratosphere. The fact that ozone mixing ratios increase from 10 ppb at ground to values around 50 or 60 ppb near tropopause indicates horizontal confluence in addition to vertical advection. Its companion branch in the NH does not seem to extend above 10 km of altitude. It is tempting to explain this compression as being caused by ozone–rich air that was brought down from higher altitudes. While in the SH ozone mixing ratios between 70 and 80 ppb prevail at tropopause, the values are substantially larger in the NH; the 70/80 ppb isolines lie several km below tropopause height. At 28°N we observe the well known tropopause break, a phenomenon associated with the subtropical jet stream. The shape

The accuracy of the ECC sondes was checked aboard the ship by comparison with the UV–photometer prior to launch. Only sondes that agreed within ± 3 ppb were used. In addition, the precision of the sondes was checked several times during the cruise by performing twin flights. For this purpose two ECC's were prepared and launched simultaneously. Each ECC had its own radiosonde. The 403 MHz–carrier frequency of one sonde was slightly detuned with respect to the other and receiving was achieved with our own and the shipboard FM receiver, respectively.

The results of a twin flight on 10th of April at 33.5° N, 27.2° W are shown in Figure 2. Some small scale structure of the ozone mixing ratio was observed. Both instruments tracked these structures well, which proves that they are atmospheric features. Above 5 km the two sondes gave slightly different values which we treat as a measure of in–flight precision. Defined in this way the precision for a tropospheric sounding was around ± 5%. Similar results were obtained from the two other twin flights.

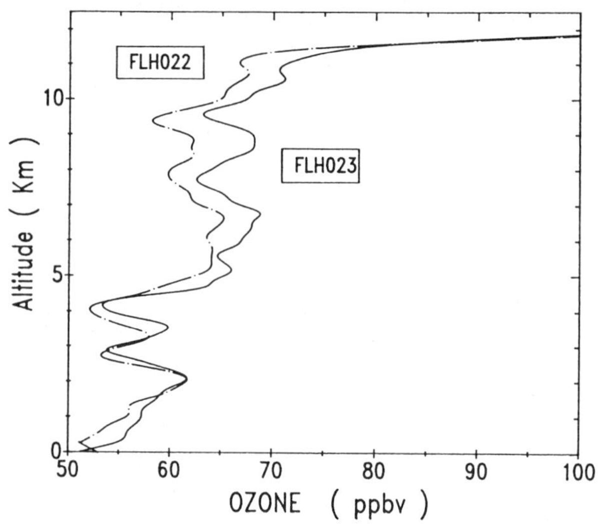

Fig. 2. Twin–ozone sounding at 33.5°N, 27.2°W (10–4–1988). Note the suppression of the x–axis.

Although the performance of the ECC sondes was found to be good, more rigorous checks will be made in the future in an environmental chamber that allows the simulation of ozone, pressure and temperature changes as they occur during balloon flights.

On a few occasions, the balloons drifted through the ship's engine exhaust plume after launch. This caused transient erratic readings that were discarded. The ozone sensor recovered, however, within 1–2 minutes after passage through the exhaust fumes. Finally, a Saharian dust storm was encountered in the upper troposphere between 3°N and 15°N causing erratic response of the ECC's. During this time O_3 data above 10 km were not obtained.

3. RESULTS

Presentation of all the individual vertical O_3 profiles including temperature and humidity would be beyond the scope of this article. They will be published as a technical report. Figure 3 shows two individual profiles from 9.5°S and 41°N, respectively, as typical examples. In the southern hemisphere, very low O_3 concentrations were observed close to the surface and a relatively steep vertical gradient throughout the troposphere. At 9.5°S, the O_3 concentration was 10 ppb at shipboard increasing to 80 ppb near the tropopause. In the northern hemisphere, on the other hand, the O_3 distribution was characterized by fairly high concentrations and weak vertical gradients. The example given in Figure 3 shows almost constant mixing ratios around 60 ppb.

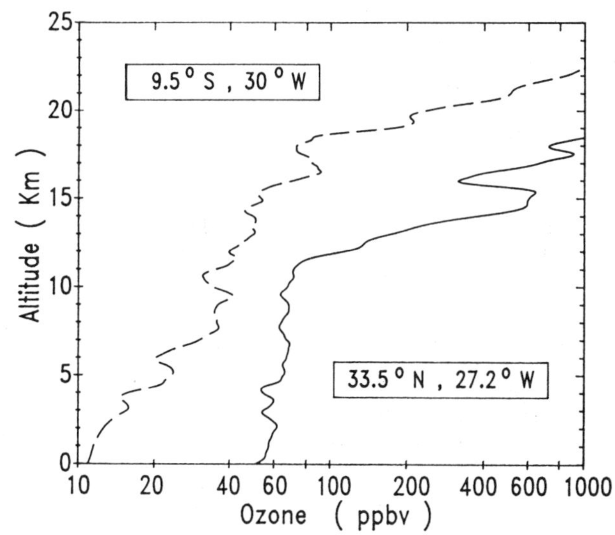

Fig. 3. Ozone soundings at 9.5°S, 30°W (31–3–1987) and at 33.5°N, 27.2°W (10–4–1987).

From the individual profiles the tropospheric meridional cross section of ozone shown in Figure 4a was derived by using the following procedure:
First, the height resolution of the individual profiles was reduced to obtain equidistant data with a vertical spacing of 250 m by linear interpolation. Secondly, equidistant profiles were generated at a 2 degree spacing by linear interpolation in the horizontal direction. Thirdly, a low-pass filter was used to smooth structures with length scales smaller than 500 m vertical and 4 degrees horizontal.

The meridional cross section is based on 4000 independent measurements and is fairly uniform both horizontally and vertically. In the horizontal, we have a resolution of about 3–4° of latitude. In the vertical, as said before, the resolution is, in principal, about 100 m which is much higher than the width of the observed features. The distribution plotted in Figure 4a is, therefore, well sampled and represents the large scale tropospheric ozone distribution over the Central Atlantic ocean at the time of March/April 1987. Tropopause heights determined from potential temperature are also given in Figure 4a. Values of the ozone mixing ratio by the UV–photometer at shipboard height are given in Figure 4b.

The general features of the meridional cross section are as follows:
Atmospheric boundary layer (ABL): Ozone at and close to the surface showed extremely low mixing ratios in the SH. The lowest values of ~ 8 ppb were measured between 10°S and 15°S. In the NH, north of 20°, we observed much larger mixing ratios. There was a sharp horizontal gradient between 20° and 30°N with an increase of ozone to values around 55 ppb at 30° to 40°, followed by slightly lower values north of 40°.

Free Troposphere: Ozone aloft had a strong vertical gradient in the SH, with a 5 fold increase between surface and upper troposphere. Conversely, north of 20° in the NH, there was hardly any vertical gradient.

4. DISCUSSION

Measurements of O_3 in the ABL over the Atlantic were reported by Winkler (1988) from a series of cruises during the last decade. The general features of the latitudinal variation given by Winkler are in good agreement with that shown in Figure 4b for spring 1987. It has the same low values between 10 and 20 ppb in the SH and a strong gradient between the equator and 20°N. The O_3 concentrations found by us around 30°N are higher than the climatological mean given by Winkler.

of the ozone isolines implies transport of stratospheric ozone into the troposphere near this latitude. The actual transfer is thought to occur during tropopause folding events [Danielsen, 1968], and is usually connected with large scale wave amplification and the formation of rapid cyclogenesis. Our measurements, extending over a 6 day time period between 20^0N and 40^0N, do not show evidence for a recent folding event, since the isolines of potential temperature followed closely an undisturbed pattern characteristic of the normal zonal mean distribution [Danielsen, 1983]. Nevertheless, the ozone–rich air extending deep into the troposphere at 30^0N shows a clear signature of recent stratospheric origin. This is inferred by our simultaneous measurements of relative humidity. Although qualitative at this time, it shows extremely dry air associated with the largest ozone mixing ratios. Ozone and water vapor are not only negatively correlated, they are inversely proportional to each other. The island of maximum ozone mixing ratio ($\simeq$ 90 ppb at 7 km, 30^0N) corresponds to the lowest water mixing ratio that was measured in the troposphere during this cruise.

In the ongoing debate on the origin of tropospheric ozone a major question is related to the amount of ozone produced by photochemistry from natural and anthropogenic precursors on the one side and by stratospheric intrusion on the other.

The large ozone mixing ratios between 40 and 55 ppb we observed north of 30^0 could be indicative of photochemical production. However, as seen from the tropospheric cross section in Figure 4a, the situation requires a detailed discussion of the relative role of transport versus photochemical production. There is no doubt that the island of ozone with mixing ratios $\geq$ 70 ppb sitting at an altitude of 7 km near 30^0N was of stratospheric origin. Since the measurements of water vapor showed very dry air even at altitudes as low as 3.5 km, there is strong evidence that the large values of ozone underneath this island may have been advected from higher up as well.

5. CONCLUSIONS

The ozone sonde and surface measurements taken aboard the 1987 Polarstern cruise provide an excellent view of latitudinal ozone variations over the central Atlantic Ocean. Southern hemispheric ozone in the middle and lower troposphere has a pronounced vertical gradient, with surface concentrations much lower than corresponding latitudes in the north. Much of the latitudinal and vertical ozone variation within the troposphere can be explained in terms of transport from regions of ozone sources or sinks. A distinct ozone minimum (< 10 ppbv) was found at $\sim 10^0$S, coinciding with a region of high water vapor concentrations and accompanying high photochemical loss. The effects of this loss region markedly influence the ozone distribution up to tropopause heights through upward transport within an 8^0 latitude band.

At 30^0N, an air mass characterized by an O_3 maxium with values up to 90 ppb was observed at an altitude of 7 km. Large values of surface ozone around 55 ppbv were found directly below this. Based on simultaneously measured water vapor profiles, the maximum at 7 km was of recent stratospheric origin and extended down to 3.5 km. This suggests the influence of the stratospheric O_3 source in the lower troposphere. However, the possibility of anthropogenic influences cannot be discounted in light of hydrocarbon measurements also obtained during the cruise (Rudolph, priv. comm.).

In order to examine the seasonal differences in the latitude–height ozone cross sections and the relative influence of tropospheric source and loss regions, we plan to repeat the ozone sonde program on the FS–"Polarstern" in September/October of 1988.

6. ACKNOWLEDGEMENT

The authors wish to thank the Alfred Wegener Institut, Bremerhaven, FRG, for the opportunity to participate on the ANT/V5 expedition. Also, we wish to thank the crew of the research vessel FS–"Polarstern" for their expert assistance.

7. REFERENCES

Attmannspacher, W., R. Hartmannsgruber and P. Lang, 1984: Long period tentencies of atmospheric ozone based on ozone measurements started in 1967 at the Hohenpeissenberg Meteorological Observatory. Meteorol. Rdsch., 37, 193–199.

Danielsen, E.F., 1968: Stratosphere–troposhere exchange based on radioactivity, ozone and potential vorticity. J. Atmos. Sci., 25, 502–518.

Danielsen, E.F., 1983: An Atlas of objectivity analyzed atmospheric cross–sections 1973–1980, NASA report, prepared for Ames Research Center under cooperative agreement NCC 2–73 (Space Sciences Division, NASA Ames Research Center, Moffet Field, CA 94035, USA).

Galbally, I.E. and C.R. Roy, 1980: Destruction of ozone at the earth's surface. Quart. J. Roy. Met. Soc., 106, 599–620.

Komhyr, W.D., 1969: Electrochemical concentration cells for Gas analysis. Ann. Geoph., 25, 203–210.

Lenschow, D.H., R. Pearson, Jr. and B.B. Stankov, 1982: Measurements of Ozone Vertical Flux to Ocean and Forest. J. Geophys. Res., 87, 8833–8837.

Liu, S.C., M. McFarland, D. Kley, O. Zafiriou and B. Huebert, 1983: Tropospheric NO_x and O_3 Budgets in the Equatorial Pacific. J. Geophys. Res., 88, 1360–1368.

Marenco, A., 1988: Large scale distributions of O_3, CO and CH_4 in the troposphere from scientific aircraft measurements (STRATOZ III) In: Tropospheric Ozone – Regional and Global Scale Interactions, edited by I.S.A. Isaksen, NATO ASI Series C Vol. 227, 73–81, D. Reidel Publ. Comp.

Winkler, P., (1988): Surface Ozone over the Atlantic Ocean. J. Atmos. Chem., in press

MERIDIONAL DISTRIBUTION OF SURFACE OZONE OVER THE ATLANTIC (83°N–76°S)

Peter Winkler
German Weather Service
Meteorological Observatory
Frahmredder 95, D–2000 Hamburg 65

SUMMARY

Surface ozone has been measured on board various research vessels since 1977. Available data from the Atlantic ocean between 83°N and 76°S show the ozone concentration in the NH to be about twice as high as in the SH. Secondary maxima indicate latitudes at which stratospheric-tropospheric exchange preferentially occurs. There is evidence that in the SH around 60°S this exchange is much smaller than in the respective northern latitude. In the ITCZ pronounced maxima have been observed in about half of all trans-sections which are attributed to stratospheric-tropospheric exchange due to deep convection. - Seasonal variations as function of latitude also exhibit pronounced differences between the hemispheres. In the NH a maximum isfound in May which becomes sharper and higher when going northward. North of 60°N this maximum decreases in height and shifts to June. In the SH, the spring maximum is present only north of 30°S.

Southwards of 30°S a broad minimum appears during the summer month. - Comparisons with concurrent Kr85 measurements by other groups allows for the conclusion that in some cases photochemically produced ozone from polluted air masses is transported over the North Atlantic.

1. INTRODUCTION

Tropospheric ozone has experienced increasing interest recently, because trends have been documented (e.g. Attmannspacher et al. 1984; Feister and Warmbt 1985; Logan 1986). Ozone as a greenhouse gas is influencing the radiation budget and as consequence the climate. Presently there is much discussion on the relative contributions of the stratospheric, the tropospheric and man made source to the tropospheric ozone concentration (e.g. Levi et al. 1985). In all these debates the necessity to a better observational data base is stressed. In this paper which is discussed in detail in Winkler (1988) ozone measurements from ship cruises are reported. Because the surface destruction over the ocean is roughly one tenth of that over land, surface measurements over the ocean are more representative for the lower troposphere than the observations from land.

2. LATITUDINAL DISTRIBUTION

Meaurements are made with the KI method. Times where the method was disturbed by pollution from the ship itself were excluded from evaluation. The program was started in 1977 and is still continuing. Up till now the results of 32 cruises have been evaluated covering the latitudes between 83°N and 76°S. The measurements from all cruises have been averaged for each 5° latitude belt. The latitudinal distribution with the standard deviation for each 5° section is depicted in fig. 1. Because the cruises do not cover all seasons with the same frequency the distribution in fig. 1 slightly underestimated the yearly average in the southern hemisphere by 3-5 ppb, while in the northern hemisphere the curve is fairly representative. The standard deviation shows maxima at 3 latitudes (60-65°, 20-25°N, and 30-35°S). These latitudes correspond to the average latitudinal positions of the subtropical and polar jet streams where stratospheric-tropospheric exchange frequently occurs. In a latitude where stratospheric influx occurs the ozone concentration can expected to be quite variable resulting in a high standard deviation.

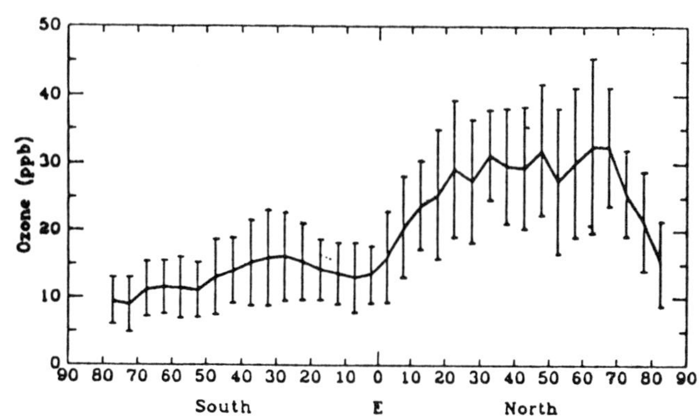

Fig. 1: Latitudinal distribution of ozone with average scattering bars. Due to incomplete data-coverage of all seasons the concentrations southwards of 40°S should be increased by 3-5 ppb to obtain yearly averages.

ISBN 0–937194–15–8

The most pronounced feature of the latitudinal distribution is the concentration difference between the hemispheres. The question whether the higher concentration in the NH is due to a higher exchange rate between troposphere and stratosphere or due to photochemical production in consequence of a higher pollution level cannot be answered satisfactorily. Gidel and Shapiro (1980) and Mahlmann et al. (1980) derived a hemispheric difference in the exchange of the order of a factor of 2 which is in agreement with our measurements. Arpe et al. (1980) derived from a long series of atmospheric data, produced by the ECMWF model, the turbulent kinetic energy (KE) distribution with respect to high, latitude, and season. They could show that KE hat two maxima in the NH-summer but only one maximum during austral summer at 30°S. This is also in agreement with our observation indicating that stratospheric-tropospheric exchange in of minor importance at the SH polar front. The main reason for this difference can perhaps be found in the distribution of land masses or high mountainous regions which stimulate stratospheric-tropospheric exchange processes.

A special observation was made in the ITCZ which usually forms over the Atlantic between 3°N in winter and 8°N in summer. In the ITCZ a pronounced maximum was observed in about half of all transsections. In a quarter of the transsections minima were observed while in another quarter a simple increase from low SH values to high concentrations was found. Stallard et al. (1975) attributed the maximum to deep convection which penetrates the tropical tropopause and which causes exchange from the stratosphere to the troposphere by the sinking motion compensating the mass balance. In all cases where we observed maxima or minima, cumulonimbus clouds were reported, however, since no information on the height of cloudtops is available, we cannot offer a different interpretation. Nevertheless, maxima in the ITCZ seem to occur regularly. It may be of interest to mention that in cases where ITCZ-maxima of ozone were observed, the SH ozone concentration was lower by 2-5 ppb (at least in the region ITCZ - 20°S) as compared to the cases with ITCZ minima.

3. SEASONAL VARIATIONS

Since the cruises to the Arctic take place preferentially during summer and to the Antarctic during austral summer our material is incomplete. We have calculated seasonal variations as far as possible for latitudinal sections each 10 degrees wide. In the figure the NH-seasons are shifted by 6 months to the SH-seasons so that the yearly cycles can be compared directly. Thick full lines represent the NH-, thick dashed lines the SH-concentrations. Thin continuous and thin dashed lines correspond to landbased measurements in the NH and SH, respectively. (Nat = Natal, Brasil; NDJ = Ndjamena, WHK = Windhoek, HUS = Hermanus (all in Africa); BRW = Barrow; FPK = Fritzpeak; MLO = Mouna Loa; HP = Hohenpeissenberg; Qui = Quickborn near Hamburg; GVN = Georg-von-Neumayer-Station, Antarctica).

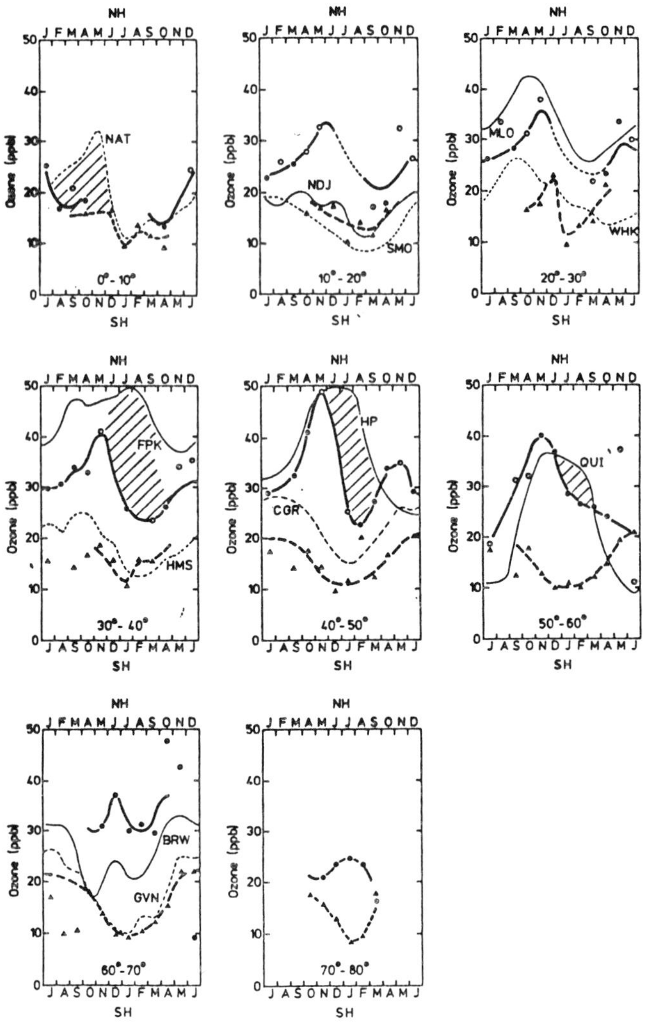

Fig. 2: Yearly cycles of ozone concentration, averaged over 10° latitude. Open and full symbols indicate that less resp. more than 200 values were available. Thick full lines represent the NH, thick dashed lines the SH. Note that NH and SH seasons are shifted by 6 month. Thin lines indicate land based measurements (NAT = Natal, NDJ = Ndjamena, WHK = Windhoek, HMS = Hermanus, BRW = Barrow, FPK = Fritzpeak. MLO = Mouna Loa. SMO = Samoa, CGR = Cape Grim, HP = Hohenpeißenberg, Qui = Quickborn near Hamburg, GVN = Georg-von-Neumayer-Station. Hatched areas visualize periods with photochemical production over continents.

From fig. 2 we can derive the following general behaviour of ozone over the Atlantic:

(1) In the NH a spring maximum is observed in all latitudes between 0° and 80°N. For the latitudes between 0° and 60°N the peak appears during May while north of 60° the peak shifts to June and even to July (70°N). The minimum concentration occurs during fall or winter in

most latitudes. North of the polar front (60^oN) there is an indication for higher winter concentrations.
The spring peak increases in height from the ITCZ to 40°-50^oN and reduces when going northward.

(2) In the SH a spring maximum is indicated only in the latitudinal range between 0° and 30^oS. South of 40^oS a broad summer maximum appears from November through March. Although the oceanic data base is incomplete the seasonal variation is similar to the GVN-Station (Antarctica) where a continuous record of the ozone concentration is available since 1982. This congruent seasonal variation indicates that the troposphere seems to be well mixed between 40°-70^oS.

(3) Comparing oceanic with continental measurements we find in the NH the spring peak extending to summer for sites in industrialized regions. As indicated by hatching this extending maxima are most probably due to photochemical production.
At Natal in Brazil a pronounced maximum is observed which according to Logan and Kirchhoff (1986) is attributed to burning activities.

The marked difference in seasonal behaviour between NH and SH confirms the conclusion, mentioned earlier, that in the SH the stratospheric-tropospheric exchange is most probably less effective than in the NH.

4. LONG RANGE TRANSPORT

During five cruises Kr85 has been measured along with ozone (Weiss, 1982; Weiss et al., 1983). Kr85 originates from nuclear reprocessing plants which are all situated in the NH. Because Kr85 has no sink except radioactive decay, high relative concentrations indicate transport from industrialized areas. During some of the cruises a positive correlation between ozone and Kr85 was found in the NH, during other cruises the correlation was always negative. Because in the stratosphere the Kr85 level is low, negative correlation means that ozone originated from the stratosphere in this case. In the case of positive correlation we have evidence for long range transport of photochemically produced ozone from industrialized regions. In one case the weather pattern clearly indicated that european sources were responsible for high ozone levels over the Atlantic, in the other case a more detailed investigation would be necessary to find the sources.

5. CONCLUSIONS

Concentrations of ozone in the NH are nearly twice as high as in the SH. The higher concentrations in the NH may partly be caused by photochemical anthropogenic production as is demonstrated by occasionally positive correlation between ozone and Kr85.

A pronounced spring maximum in the NH has no respective counterpart in the SH. It is supposed that a reduced stratospheric-tropospheric exchange in the SH is responsible for this difference.

Another explanation would require an increased destruction of ozone in the SH during austral summer probably due to biogenic emissions, however, not observational evidence is presently available for such a conclusion.

In the region of the ITCZ ozone exhibits frequently pronounced maxima but also minima have been observed. The maxima are most probably caused by compensating downdrafts being connected with deep convection.

LITERATURE

Arpe, K., C. Brancovic, E. Oriol, P. Speth: Beitr. Phys. Atm. 59 (1986) 321-355.
Attmannspacher, W., R. Hartmannsgruber, P. Lang: Meteorol. Rdsch. 37 (1984) 193-199.
Feister, U., W. Warmbt: Proc. Quadr. Ozone, Symp. Halkidiki, Reidel, Dordrecht (1985) 782-787.
Gidel, L.T., M.A. Shapiro: J. Geophys. Res. 85 (1980) 4049-4058.
Levi II, H., J.D. Mahlmann. W.J. Moxim, S.C. Liu: J. Geophys. Res. 90 (1985) 3753-3772.
Logan, J.A.: J. Geophys. Res. 90 (1985) 10.463-10.482.
Logan, J.A., W.V.J.H. Kirchhoff: J. Geophys. Res. 91 (1986) 7875-7881.
Mahlmann, J.D., H. Levi II, W.J. Moxim: J. Atm. Sci. 37 (1980) 655-685.
Stallard, R.F., J.M. Edmond. R.E. Nevell: Geophys. Res. Letters 2 (1986) 289-292.
Weiss, W., H. Stockburger, H. Sartorius, K.O. Münnich: Ber. Polarforschung 6 (1982) 72-73.
Weiss, W., A. Sittkus, H. Stockburger, H. Sartorius, K.O. Münnich: J. Geophys. Res. 88 (1983) 8574-8577.
Winkler, P.: J. Atm. Chem. (1988) in press.

A COMPARATIVE STUDY OF TROPOSPHERIC OZONE IN THE AMAZONIAN RAINFOREST, THE CERRADO, AND A COASTAL SITE

Volker W.J.H. Kirchhoff
Instituto de Pesquisas Espaciais (INPE)
São José dos Campos, São Paulo, Brazil

ABSTRACT

Ozone measurements have been obtained at three sites representative of very different remote ecosystems in Brazil: the Amazonian rainforest (site 1, 3°S, 60°W), the savannah-like cerrado (site 2, 16°S, 58°W), and a site at the Atlantic coast (site 3, 6°S, 35°W). The results show significant differences, each site with its own characteristics. Ozone concentrations obtained at about 1.5 m above the surface vary from a few ppbv to about 40 ppbv at these sites. No diurnal variation is seen at site 3, while the continental sites show strong cycles. At site 1, the nocturnal concentrations are very low, often zero, and daytime maxima are not larger than 15 ppbv. The largest values are seen at site 2, with a consistent diurnal variation. The low concentrations in the forest seem to be related to an abundant foliage, which increases the effective surface area, plus a strong convective activity characteristic of the equatorial region. Site 2 is strongly influenced by seasonal biomass burning activities, when occasionally ozone highs of up to 180 ppbv have been observed. The seasonal variation is strong. For sites 2 and 3 this is true in the whole troposphere, concentrations being twice as large in October as compared to April. Ozonesonde profiles show that less O_3 is present at site 1, compared to site 3, in the whole troposphere, which is probably the result of strong mixing at site 1.

1. INTRODUCTION

Brazil is the largest country in South America extending east to west over almost 4000 km, from 35°W to 74°W, and north to south over an almost identical length from 5°N to 34°S. The Brazilian territory is covered by very distinct and sui generis vegetation types: the equatorial rainforest in the northwest; the savannah (campo cerrado) of central Brazil; the semi-arid (caatinga) of the northeast; the marshlands of the Pantanal; and the long shoreline at the Atlantic coast. These different Brazilian environments are shown in Fig. 1.

In November 1978, in collaboration with NASA, a long-term program was started to obtain equatorial ozone profiles using balloon sondes. The launch site chosen was Natal (6°S, 35°W) beacause of the availability of the nearby Brazilian rocket range Barreira do Inferno. The ozone data obtained has shown a strong seasonal variation of tropospheric ozone (Kirchhoff et al., 1981, 1983; Logan and Kirchhoff, 1986). On the basis of previous work this result was rather unexpected and stimulated measurement programs and field expeditions to other sites. These revealed a strong influence of the local surface environment on the ozone concentration observed near (1.5 m above) the surface. It was decided, therefore, to pursue observations at sites with different characteristics (Fig. 1 and Table 1).

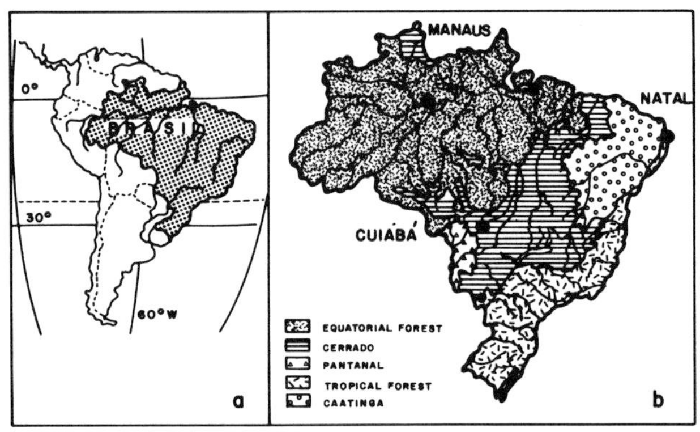

Fig. 1. a) The position of Brazil in South America; b) The geographic map of Brazil, showing its different natural surface environments, and the location of three sampling sites: Manaus (3°S, 60°W), Cuiabá (16°S, 58°W), and Natal 6°S, 35°W).

TABLE 1. SITE DESCRIPTION

Nº	Name	Lat.	Long.	Location
1	Manaus	3°S	60°W	Rainforest
2	Cuiabá	16°S	58°W	Savannah
3	Natal	6°S	35°W	Coast

2. SITE DESCRIPTION

Fig. 1 shows the location of the three sampling sites. The site near Manaus (3°S, 60°W) is characteristic of the equatorial rainforest; sampling was performed 30 km north of Manaus, in a protected environment, the Ducke forest preserve owned by INPA, the Institue for Amazonian Studies.

ISBN 0-937194-15-8

The site in the Brazilian savannah is Cuiabá (16°S, 58°W), where sampling is performed 7 km northeast of the downtown area. The coastal site Natal (6°S, 35°W) is located at the northeast extreme of the Brazilian Atlantic coast. The prevailing winds are easterlies and, therefore, this is a marine station that samples Atlantic air masses.

3. OBJECTIVE OF PAPER

The objective of this paper is to describe and compare the O_3 results that were obtained at the sampling sites, shown in Fig. 1, covering at least one seasonal cycle. We concentrate on diurnal and seasonal variations. The excellent correlation between O_3 and CO seasonal variations is used to argue that phytomass burning in the northern part of the state of Mato Grosso, is very likely responsible for the large concentrations of O_3 and CO observed in September.

4. MEASUREMENT TECHNIQUE

Ozone sensors that use the ultraviolet absorption photometric technique have been used at the three sites. The instruments have been inter-calibrated at regular time intervals. Although these instruments renew their readings at intervals of about 20 seconds, we have not analysed short term variations. For the surface ozone records we have concentrated on hourly averages for the study of diurnal variations and these are then compared in different seasonal periods.

For the ozone concentration variations in the upper troposphere ECC ozonesondes launched on balloons were used for Natal and Manaus. No ECC sondes have been launched so far at Cuiabá.

In order to investigate possible photochemical effects triggered by phytomass burnings, it was decided to measure CO as well. The basic CO measurement technique used is chromatographic separation followed by mercury vapor detection, which has excellent resolution, precision, and linearity. At around 100 ppbv, the typical precision is 0.2%.

5. RESULTS

1- Amazonia. Ozone measurements near the surface were obtained during field expeditions in 1985, in July and early August, during the local dry season, and in 1987, during the month of April and May, representative of the local wet season. The difference in observed concentrations is quite remarkable. Very low concentrations were measured in both seasonal periods, but in the wet season, concentrations were about half the values observed in the dry season (Kirchhoff, 1988).

The dry-wet seasonal periods of the Amazonian region in terms of O_3 concentrations is shown in Fig. 2. The diurnal averages are shown, as well as the maxima and minima found during the measurement periods. Larger O_3 concentrations are found during the daytime. Concentrations at night are very low, often zero. The top panel in Fig. 2 shows the dry season result; the lower panel shows the wet season concentrations, about one half of the dry season values.

The whole troposphere in Amazonia undergoes a strong variation between wet and dry seasonal periods as shown by ozonesonde measurements made in the same periods previously mentioned. The average O_3 profile for the dry season of 1985, was obtained from 14 ECC sonde launches, and the 1987 wet season average profile, was obtained from 20 ECC launches. Both profiles are shown in Fig. 3. Evidently, the whole troposphere is depleted of O_3 in the wet season.

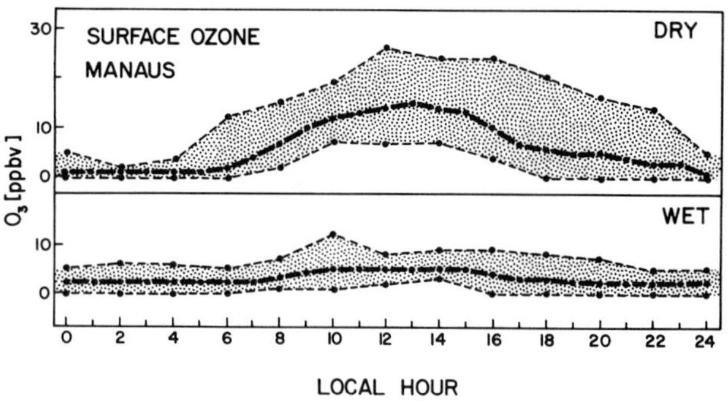

Fig. 2. Surface ozone diurnal variation in the Amazonian rainforest. Top panel shows diurnal averages during the dry season; lower panel shows the wet season average.

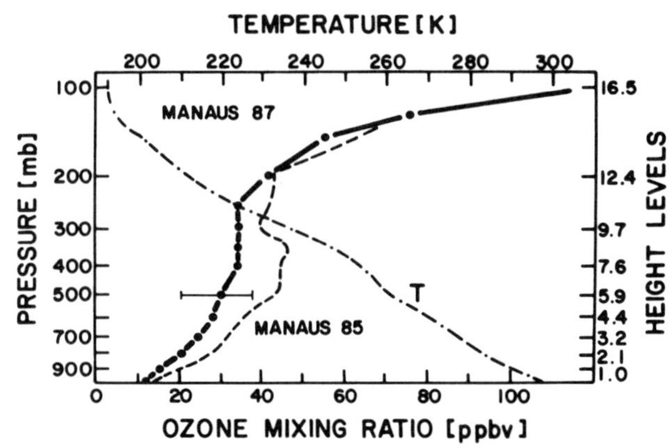

Fig. 3. Average ozone profiles in the troposphere of the Brazilian equatorial rainforest. The O_3 concentrations are larger in the dry season in the whole troposphere.

2- The Savannah. The O_3 concentrations at this site are the largest seen at a remote site in Brazil. The daytime maxima are of the same magnitude as values observed at western sites of the USA. A large diurnal and seasonal variation can be seen at this site, as shown in Fig. 4. The upper panel shows the diurnal variation during the dry seasonal period (August-September-October), as well as maxima and minima, whereas the lower panel shows the wet season (January-February-March) average. Daytime maxima near 50 ppbv are common in the dry season, but only 15 ppbv can be seen in the wet season.

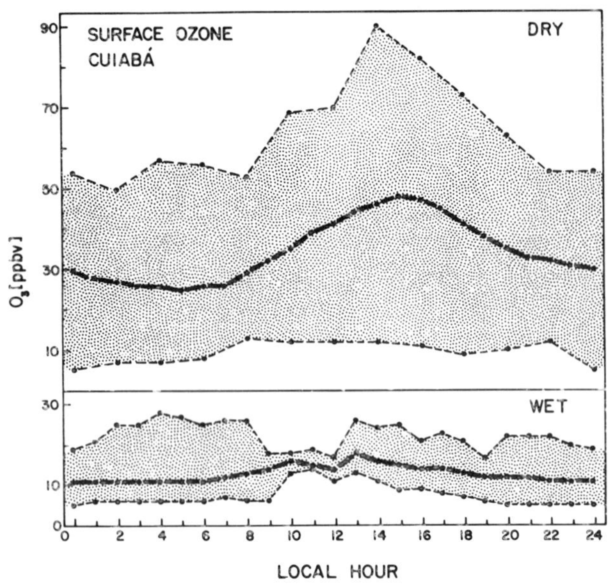

Fig. 4. Surface ozone diurnal variation in the Brazilian savannah. Top panel shows the dry season average and lower panel shows the wet season average.

3- The marine site. The ozone concentrations observed at the Atlantic coast, near Natal (6°S, 35°W) do not show a clear diurnal variation, in contrast to all other sites tested in Brazil. There is, however, a strong seasonal variation for the O_3 concentrations at Natal. The diurnal averages are shown in Fig. 5, for the dry (September-October) and wet (February-March) periods. There is no indication of a diurnal cycle present in the higher concentration record of the September-October data, shown in the upper panel of Fig. 5. In the lower panel, there might be a small diurnal oscillation present, but not a strong signal as was the case for the two continental sites, Manaus and Cuiabá.

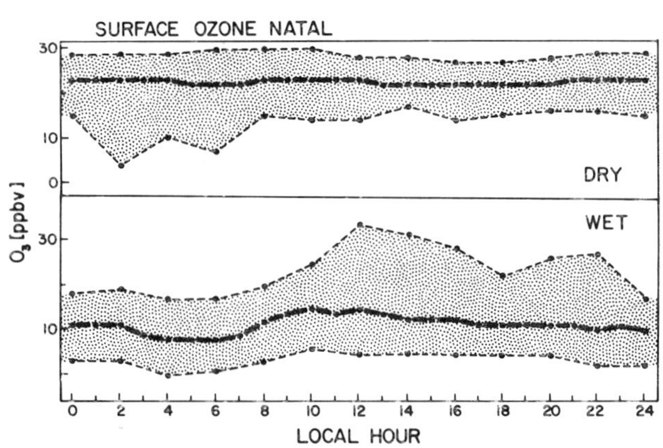

Fig. 5. Surface ozone diurnal variation measured at the Atlantic coast. Top panel shows the September-October period, and lower panel shows the February-March period.

The seasonal variation of the O_3 concentrations is present at all levels of the marine troposphere, a result obtained from ozonesoundings at Natal (Logan and Kirchhoff, 1986). The largest oscillation is seen near the surface, with an amplitude of about a factor of 2, which decreases towards higher altitudes. Other aspects of the Natal ECC data have been discussed elsewhere (Kirchhoff, 1981, 1983; Logan and Kirchhoff, 1986).

6. DISCUSSION

It appears that our results for Manaus and Cuiabá are both strongly influenced by local and remote phytomass burning activities in the dry season period. The Cuiabá site is close to the burning activity and often the smoke in the air is so intense that the Cuiabá airport must be closed for aircraft operation. This is also true for other airports of the region. In order to monitor this activity a long term program has been started to obtain systematic samples for the measurement of ozone and carbon monoxide. Fig. 6 shows the sequence of a complete seasonal cycle between the 1987 dry season maximum and the 1988 wet season minima in the monthly mean concentrations of O_3 and CO, for the Cuiabá site. Evidently, there is a high degree of correlation between the CO and O_3 variations. Note that the factor between minima and maxima is about 4.

Previous work on O_3 and CO analyses in Brazil by Crutzen and co-workers (Crutzen et al., 1985, Delany et al., 1985) used an aircraft platform that sampled over the rainforest and the savannah, in the dry season, showing the effects of phytomass burnings. Ozone concentration bulges were seen in the mixing layer in association to local biomass burnings. Further associations between biomass burnings and associated O_3 and CO correlations have been reported by Browell et al. (1988) and Fishman et al. (1986).

The evidence in favor of the strong influence of burnings on the CO and O_3 concentrations in Cuiabá becomes even stronger if one compares our values to observations at marine remote sites, for example, in Australia where only 44 ppbv of CO is reported by Fraser et al. (1986). Our own results for Natal, using the same absolute CO calibration scale, give about 100 ppbv. For a site in South Africa, Seiler et al. (1984) report between about 50 and 80 ppbv of CO, with a seasonal variation less than a factor of 2. It may be concluded that the burning activity in central Brazil more than doubles the amplitude of the natural seasonal variation of atmospheric CO. The ozone concentration variation seems to follow this behavior very closely, as shown in Fig. 6.

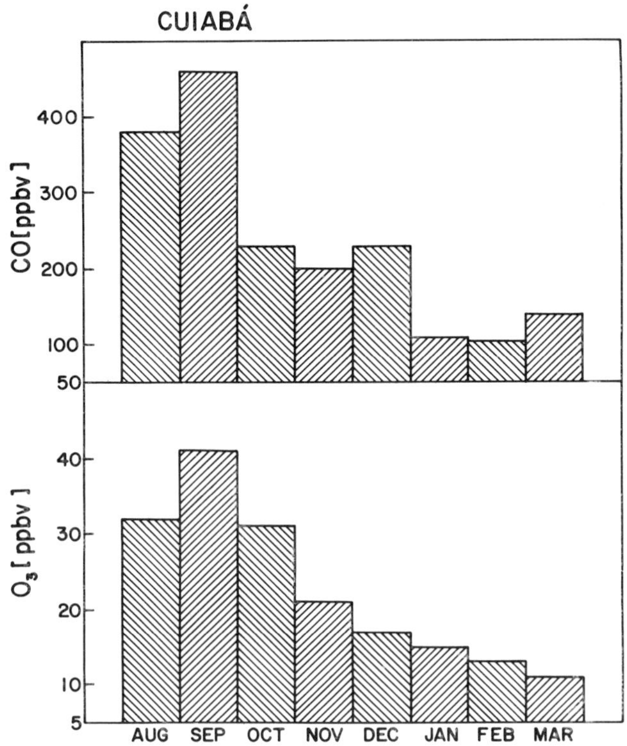

Fig. 6. Seasonal variation of surface O_3 and CO at Cuiabá (16°S, 58°W), showing monthly means.

7. CONCLUSIONS

Ozone concentrations in the troposphere were measured at three ecologically distinct sampling sites in Brazil: the equatorial rainforest, the savannah, and the Atlantic coast. The highlights of this comparison are:

1- The continental sites Manaus and Cuiabá show strong diurnal cycles in the surface ozone concentrations. In contrast, the marine site shows no evident diurnal variation.

2- The seasonal variation is strong at all sites. It is of the order of a factor of 2 for Natal and Manaus. At Cuiabá the concentration maximum during the dry season period is about four times as large as the concentration during the wet season period.

3- The strong seasonal variation of O_3 at Cuiabá is also seen in the CO concentration measurements made at this site, which is thought to be the result of intense phytomass burning activities in central Brazil during the dry season, when the CO concentrations are about 4 times larger.

4- The strong seasonal variation is present at all levels of the troposphere, but decreases slightly upwards, as shown by ozonesoundings made at Natal and Manaus.

ACKNOWLEDGEMENTS

I thank my colleagues at the remote sites for their effort in obtaining high quality data. I also thank NASA for suport dur.ng the ABLE-2B expedition to Amazonia, and Arnold Torres, NASA-Wallops Island, for his longstanding support to the Natal ECC program. Steven Wofsy has been of help many times by lending equipment and providing replacement parts, and Rei Rasmussen has provided CO calibration gas.

REFERENCES

Browell, E.V., G.L. Gregory, R.C. Harriss, and V.W.J.H. Kirchhoff, 1988: Tropospheric ozone and aerosol distribution across the Amazon basin. J. Geophys. Res., 93, 1431-1451.

Crutzen, P.J., A.C. Delany, J. Greenberg, P. Haagenson, L. Heidt, R. Lueb, W. Pollock, W. Seiler, A. Wartburg, and P. Zimmermann. 1985: Tropospheric chemical composition measurements in Brazil during the dry season. J. Atmos. Chem., 2, 233-256.

Delany, A.C., P. Haagenson, S. Walters, A.F. Wartburg, and P.J. Crutzen. 1985: Photochemically produced ozone in the emission from large-scale tropical vegetation fires. J. Geophys. Res., 90, 2425-2429.

Fishman, J., P. Minnis, and H.G. Heichle, Jr. 1986: Use of satellite data to study tropospheric ozone in the tropics. J. Geophys. Res., 91, 14451-14465.

Fraser, P.J., P. Hyson, R.A. Rasmussen, A.J. Crawford, and M.A.K. Khalil. 1986: Methane, carbon monoxide, and methylchloroform in the Southern Hemisphere. J. Atmos. Chem., 4, 3-42.

Kirchhoff, V.W.J.H., Y. Sahai, and A.C. Mota. 1981: First ozone profiles measured with ECC sondes at Natal (5.9°S, 35.2°W). Geophys. Res. Lett., 8, 1171-1172.

Kirchhoff, V.W.J.H., E. Hilsenrath, A.G. Mota, Y. Sahai, and R.A. Medrano-B. 1983: Equatorial ozone characteristics as measured at Natal (5.9°S, 35.2°W). J. Geophys. Res., 88, 6812-6818.

Kirchhoff, V.W.J.H. 1988: Surface ozone measurements in Amazonia. J. Geophys. Res., 93, 1469-1476.

Logan, J.A., and V.W.J.H. Kirchhoff. 1986: Seasonal variations of tropospheric ozone at Natal, Brazil. J. Geophys. Res., 91, 7875-7881.

Seiler, W., H. Giehl, E.C. Brünke, and E. Halliday. 1984: The seasonality of CO abundance in the Southern Hemisphere. Tellus, 36, 219-231.

OZONE MEASUREMENTS IN EQUATORIAL AFRICA DURING DECAFE 88

G. Helas and M.O. Andreae
Max-Planck-Institut für Chemie
Mainz, Germany

J. Fontan
Laboratoire d'Aérologie, Université Paul Sabatier
Toulouse, France

B. Cros and R. Delmas
Laboratoire de Physique de l'Atmosphère, Université Marien Ngouabi
Brazzaville, People's Republic of the Congo

ABSTRACT

Measurements of ozone and some other trace components were carried out in the tropical forest region of the Congo near Impfondo (2°N, 18°E) in the dry season (February) within the framework of the DECAFE-experiment. Observations were made both on the ground and aboard an aircraft. Mixing ratios of ozone are compared to recent measurements in the same region during the rainy season (Cros et al., 1987) and to other reported ozone measurements in tropical areas.

1. INTRODUCTION

The presence of ozone in the troposphere is still under discussion in terms of local production from precursor oxidants and of transport from aloft (Logan, 1985). The tropical regions near the ITCZ ozone mixing ratios are of particular interest because there is both local production of ozone by precursors of anthropogenic or biogenic origin, and effective atmospheric transport due to strong vertical exchange events.

The experiment DECAFE (Dynamique et Chimie de l'Atmosphère en Forêt Equatoriale) has been designed to study the behaviour of natural trace components emitted by vegetation and soil. Included were also meteorological measurements in order to define mass fluxes between biosphere and atmosphere.

2. EXPERIMENTAL

The ozone measurements were made with a ThermoElectron 49 and a Dasibi 1008, and an Environnement S. A. instrument, which has the same basic design as the Dasibi 1003. All the instruments use UV absorption. In contrast to the Dasibi types, the ThermoElectron is a dual-beam instrument. The TE data are used whenever possible for the data reported here since this instrument has the fastest response, but all three instruments were in good agreement. The instruments performed well during most of the flight conditions encountered. Comparison of data obtained with and without the pressure and temperature correction circuit showed that the correction worked satisfactorily.

Measurements were carried out in a tropical forest region near Impfondo (2°N, 18°E) in the northern part of the People's Republic of the Congo during February, which is the "local" dry season. For the experiments described here measurement sites were located within the forest, one on a ca. 100 x 250 m clearing approximately 25 km from Impfondo, and the other on a large clearing approximately 2 km in diameter about 5 km from Impfondo. The latter clearing is polluted, for this area is inhabited and is connected to the outskirts of the village nearby. This latter clearing was the base for our operations. The airborne measurements have been performed on a Britten-Norman Islander aircraft, which operated within a range of 100 km around Impfondo and was used up to 4 km height. For logistical reason no flights were possible during darkness.

3. RESULTS

Measurements within the canopy layer of the forest area showed that ozone was essentially absent. Cross calibrations of the instruments used repeatedly supported these findings. Also, in the small clearing, the ozone mixing ratios were zero during the night. During the day, however, the ozone mixing ratios started to increase after sunrise and then varied between 10 and 20 ppbv.

The ozone mixing ratios measured in the large clearing were also about 20 ppbv during the day time, as can be seen in Fig. 1, which shows the situation of Feb. 19, 1988. During the night the values sometimes decreased to the lower detection limit of the instrument.

ISBN 0-937194-15-8

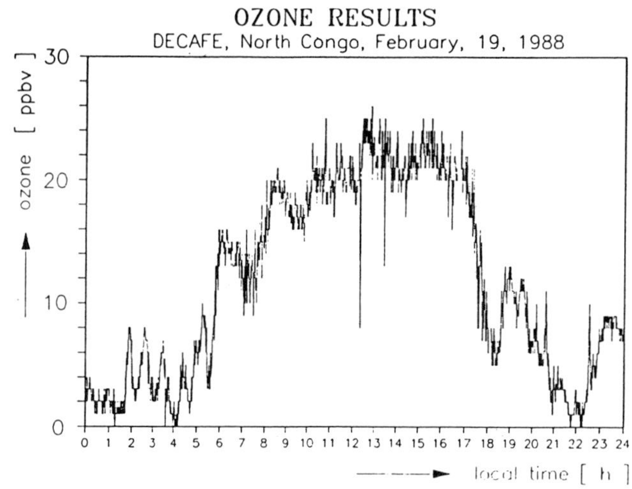

Fig. 1: Ozone mixing ratios as function of time on Feb. 19, 1988. Measurement site: large clearing, sampling height: 2m.

This behaviour was also found during the course of the day in the aircraft measurements at levels from the ground up to approximately 0.3 km. Above this height, up to approximately 3 km, mixing ratios ranged between 20 and 70 ppbv with highest amounts around 2 km. Above 3 km ozone concentrations decreased again to values typically found in the mid-troposphere. This is depicted in Fig. 2, which shows the range of mixing ratios found during the flights. A large persistent maximum centered near 2 km height was found. This maximum coincides with a CO maximum (J. Rudolph, unpublished results) and also with

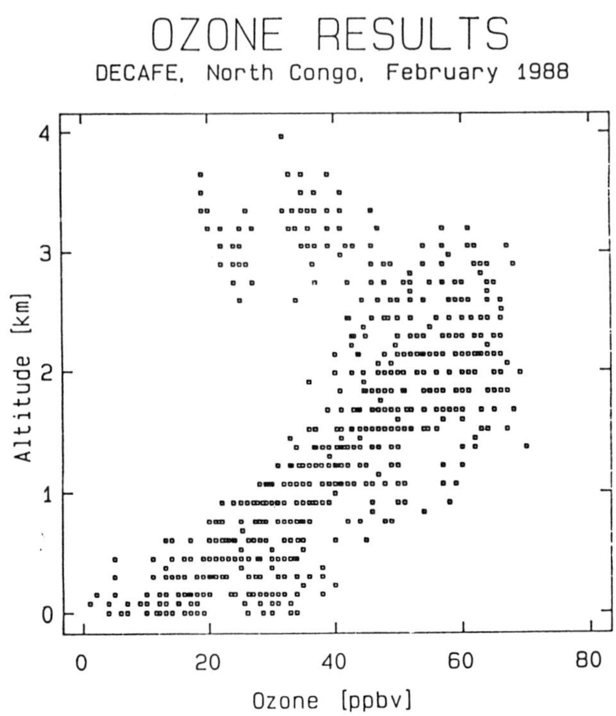

Fig. 2: Ozone mixing ratios as a function of altitude above the tropical forest area around Impfondo.

high values of CO_2 (J. Rudolph) and organic acids.

4. DISCUSSION

The variation of the ozone mixing ratios near the ground may be discussed in terms of small-scale transport and chemistry. Except for local pollution episodes, which usually occur on a relatively short time scale, the amount of ozone present obviously depended on the stratification of the air. This could be derived from CO_2 measurements, which were been conducted in parallel (unpublished results). During the night, when the wind speed remained low, there was a distinct build-up of CO_2 concentrations. This coincided with low values of ozone. Consequently a sink process, either physical adsorption or chemical reaction, exists near the ground. The measurements carried out in the small clearing in the forest suggest chemical reactions may be important. Here, mixing ratios of NO showed the opposite behaviour to those of ozone, being lower during day time (unpublished results). Therefore the NO, which is emitted from the tropical environment, is probably keeping the ozone concentrations low through chemical interaction, as long as there is no effective mixing within the boundary layer. Measurements in the Amazonia region indicate a similar behaviour (Kaplan et al., 1988; Kirchhoff, 1988). Kaplan et al. (1988) also indicate that surface deposition and reaction with NO are sinks for ozone in tropical forests.

The large persistent maximum of ozone in the mixing layer is not yet fully understood. It appears to be of photochemical origin. The source of the precursor compounds, however, is not clear. The slope of the CO/CO_2 regression is about 0.004. This is at least a factor of ten smaller than expected for biomass burning (Andreae et al, 1988, and references therein). Consistent with the low CO/CO_2 ratio, the O_3/CO slope of 0.15 is much higher in these data compared to those obtained from the biomass burning plumes studied over the Amazon (Andreae et al., 1988). This suggests that the observed ozone is mostly the product of the photooxidation of non-methane hydrocarbons. It implies, that the ozone mixing ratios found should not be restricted to the dry season, as would be the case if biomass burning were the main source.

The range and variation of the ozone mixing ratios prevailing in Impfondo do not significantly differ from those observed in Brazzaville or near Pointe-Noir (Cros et al., 1987). Therefore in these sites the presence of ozone predominantly depends on the biologically dictated environment. Excursions of high ozone episodes, as described by Cros et al. (1987), may be explained, in the terms of the results presented here, as downdrafts within the mixing layer and not necessarily from well above it.

The diurnal variation in tropical regions near the surface on the continents usually is similar to that found here (Crutzen et al., 1985; Sanhueza et al., 1985). Kirchhoff (1988) shows in detail a similar pattern for a forest clearing of the Amazonia basin. Only for Samoa the opposite behaviour has also been described (Oltmans, 1981). Marine air masses from this region of the Atlantic do not show a diurnal pattern (unpublished results), indicating the absence of local sink processes.

High ozone mixing ratios in the upper mixing layer do not always seem to be present. The data of Crutzen et al. (1985) and Delany et al. (1985) show no ozone maximum for the humid forest area of Brazil, but indicate one over savanna regions during the dry season. Gregory et al. (1986) have no evidence for enhanced ozone values above the Guyana forest during the wet season. Equally, during the dry season there seems to be no ozone maximum in the boundary layer over Amazonia (Gregory et al., 1988). However, Kirchhoff et al. (1983) deduce a minor maximum from ozone soundings approximately 3 km above Natal, Brazil. So, comparing the tropical forest areas of Central Africa and Amazonia, differences in source qualities for the precursors leading to the ozone maximum within the mixing layer must exist.

5. LITERATURE

Andreae, M. O., E. V. Browell, M. Garstang, G. L. Gregory, R. C. Harriss, G. F. Hill, D. J. Jacob, M. C. Pereira, G. W. Sachse, A. W. Setzer, P. L. Silva Dias, R. W. Talbot, A. L. Torres, and S. C. Wofsy. J. Geophys. Res. 93, 1509-1527, 1988.

Cros, B., R. Delmas, B. Clairac, and J. Loemba-Ndembi. J. Geophys. Res. 92, 9772-9778, 1987.

Crutzen, P. J., A. C. Delany, J. Greenberg, P. Haagenson, L. Heidt, R. Lueb, W. Pollock, W. Seiler, A. Wartburg, and P. Zimmerman. J. Atmos. Chem. 2, 233-256, 1985.

Delany, A. C., P. Haagensen, S. Walters, A. F. Wartburg, and P. J. Crutzen. J. Geophys. Res. 90, 2425-2429, 1985.

Gregory, G. L., R. C. Harriss, R. W. Talbot, R. A. Rasmussen, M. Garstang, M. O. Andreae, R. R. Hinton, E. V. Browell, S. M. Beck, D. I. Sebacher, M. A. Khalil, R. J. Ferek, and S. V. Harriss. J. Geophys. Res. 91, 8603-8612, 1986.

Gregory, G. L., E. V. Browell, and L. S. Warren. J. Geophys. Res. 93, 1452-1468, 1988.

Kaplan, W. A., S. C. Wofsy, M. Keller, and J. M. DaCosta. J. Geophys. Res. 93, 1389-1395, 1988.

Kirchhoff, V. W. J. H.. J. Geophys. Res. 93, 1469-1476, 1988.

Kirchhoff, V. W. J. H., E. Hilsenrath, A. G. Motta, Y. Sahai, and R. A. Medrano-B.. J. Geophys. Res. 88, 6812-6818, 1983.

Logan, J. A.. J. Geophys. Res. 90, 10463-10482, 1985.

Oltmans, S. J.. J. Geophys. Res. 86, 1174-1180, 1981.

Sanhueza, E., K. H. Octavio, and A. Arrocha. J. Atmos. Chem. 2, 377-385, 1985)

SEASONAL BEHAVIOR OF THE TROPOSPHERE OZONE IN RURAL JAPAN

Haruo Tsuruta
Yokohama Research Institute for Environmental Science
Yokohama 235, Japan

Kozo Shinya
Japan Weather Association
Tokyo 102, Japan

Tsuguo Mizoguchi
National Institute for Environmental Studies
Ibaraki 305, Japan

Toshihiro Ogawa
Geophysics Research Laboratory
University of Tokyo
Tokyo 113, Japan

ABSTRACT

Surface ozone has been measured at three rural sites in Japan with Dasibi ultraviolet photometers from June 1985 - December 1987. Monthly mean ozone concentration clearly shows the seasonal variation with a summer minimum and a spring maximum. The summer minimum is a feature different from those observed in Europe and North America where an autumn minimum occurs, whereas the spring maximum is a common feature at middle latitudes in the northern hemisphere. Ozone concentration is less than 20 ppbv when oceanic air masses prevail over Japan Islands in summer, but it is higher than 30 ppbv when continental air masses coming from the northwest prevail in winter. Due to these climatic characteristics, the seasonal behavior of tropospheric ozone in Japan is quite different from those in mid-latitude Europe and North America.

1. INTRODUCTION

Interest in the tropospheric ozone has recently arisen from a possible climatic impact due to the increase of tropospheric ozone concentration, which may be a result of the long-range transport of anthropogenic precursor gases for ozone.

Surface ozone concentrations at rural sites have been measured in different latitudes (e.g., Oltmans, 1981), representing a seasonal cycle with a spring maximum and a fall minimum at middle latitudes (e.g., Chatfield and Harrison, 1977; Fabian and Pruchniewicz, 1977; Singh et al., 1978). This type of seasonal variation is similar to those in the lower stratospheric ozone at middle latitudes and indicates a close coupling between the lower stratosphere and the troposphere.

In East Asia, according to the analysis of the oxidant data acquired from Japanese air pollution monitoring station network (Ogawa and Miyata, 1985), the behavior of tropospheric ozone seems to exhibit a summer minimum in Japan which is located at the edge of the Asian continent and the Pacific Ocean. However, the data base used in their analysis may have been erroneous because of a degradation of the instrument sensitivity during summer. It had been already demonstrated in a laboratory experiment that the sensitivity of the ozone instruments based on colorimetry or coulometry of KI solution decreases at higher temperature, but they were unable to make a correction because data on the instrumental temperature were not available.

In June 1985 we began to measure surface ozone using Dasibi ultraviolet photometers at three stations whose locations are more remote in Japan than the monitoring stations mentioned before. In this paper we describe the seasonal and latitudinal variations of the tropospheric ozone observed in June 1985 - December 1987, and discuss the temporal variations with regard to meteorological conditions.

2. OBSERVATION SITES

Three observation sites are Naze (28°23' N, 129°33'E; 298 m a.s.l.), Odaigahara (34°14'N, 136°11'E; 1006 m a.s.l.), and Hachikoda (40°42'N, 140°52'E; 1324 m a.s.l.).

The site of Naze is located in the northern part of Amamioshima Is. The island is of an area of 700 km^2, surrounded by the Pacific Ocean, and lies 300 km south of Kyushu Is. and 800 km east of the Chinese coast. On the shore 6 km southwest from the observation site there is a small city of Naze whose population is 50,000,

ISBN 0-937194-15-8

but the city is not windward of the site under normal local wind conditions.

The site of Odaigahara is located at the west plateau of the Odaigahara mountain region, whose highest peak is 1695 m a.s.l. It is surrounded by the forest covered with snow in winter. The urban and industrial area which may affect the atmospheric environment of this observation site is Osaka, which has a population of about 3,000,000, and is 80 km away to the northwest.

The site of Hachikoda is located in the middle of the northwestern slope of Mt. Hachikoda whose peak is 1585 m a.s.l. The surroundings are the shrub forest covered with snow in winter and spring. There are two local cities, Aomori (population of 300,000) 20 km to the north and Hirosaki (population of 180,000) 40 km to the west. The observation site may be downwind of these cities in winter, but the site is free from pollutions because of the strong winter westerlies.

3. DIURNAL VARIATION

The amplitude of diurnal variation in surface ozone concentration is relatively small at these stations unlike that at the air pollution monitoring stations in Japan. The diurnal average, however, shows a significant seasonal change as shown in Fig. 1. We will discuss this point later.

Ozone concentration at Naze is constant at ~10 ppbv all through the day in summer when the southerlies are dominant, whereas it shows a slight but broad minimum in the daytime of spring and autumn. The daytime minimum seems to be due to sea breezes which carry ozone-poor air in the planetary boundary layer over the sea. Further evidence for a ozone concentration in marine air masses was obtained with a Dasibi photometer on board a research vessel in Sagami Bay 60 km south of Tokyo during summer. Ozone concentration was ~15 ppbv and did not show any diurnal variation in that experiment (Tsuruta, 1983). These results indicate that the diurnal variation of surface ozone does not occur over the tropical Pacific Ocean.

Any diurnal variation is not seen at Odaigahara in winter when the strong westerlies prevail. In other seasons, however, we see the diurnal variation with a maximum in the early evening; the maximum is most pronounced in May through September. According to the data from the air pollution monitoring station network, the ozone photochemically produced in the urban and industrial area of Osaka was probably transported by the extended sea breezes (the northwesterlies) to Odaigahara in the afternoon. Consequently, the pollution may be partly responsible for the fact that the summertime ozone concentration is higher at Odaigahara than at Naze.

A diurnal variation is also observed at Hachikoda, exhibiting a maximum at night and a broad minimum around noon. This diurnal pattern seems to be closely associated with the mountain and valley winds as seen in the case of Mauna Loa (Oltmans, 1981). Since ozone in the air carried by the valley winds is subjected to the destruction due to plants on the mountain slope, ozone concentration at the top of the mountain shows a minimum during the day. However, at Hachikoda, the day-to-night difference of ozone concentration is most distinct in winter when the mountain is covered with snow, on which ozone destruction is believed to be small. This result is quite different from those measured at other mountain top stations such as Mt. Norikura in the mountainous region of central Japan; ozone does not exhibit any diurnal variation in winter. At this time we are unable to find a reasonable explanation for the daytime minimum in winter at Hachikoda.

4. SEASONAL VARIATION

The surface ozone concentration clearly shows a seasonal variation with a spring maximum and a summer minimum as seen in Fig. 2, being in agreement with the previous analysis (Ogawa and Miyata, 1985). A similar seasonal cycle is also found at low latitude sites under similar climatic condition (Logan, 1985). The spring maximum is a common feature at middle latitudes

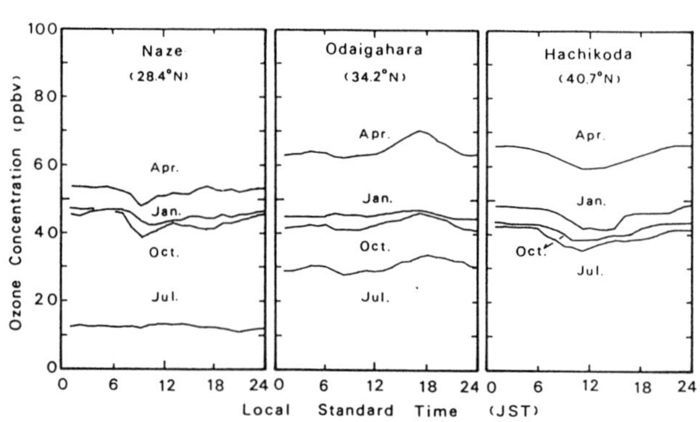

Fig. 1. Diurnal variations of surface ozone concentration at three stations in rural Japan. Representative monthly average values for January, April, July and October in 1985 - 1987.

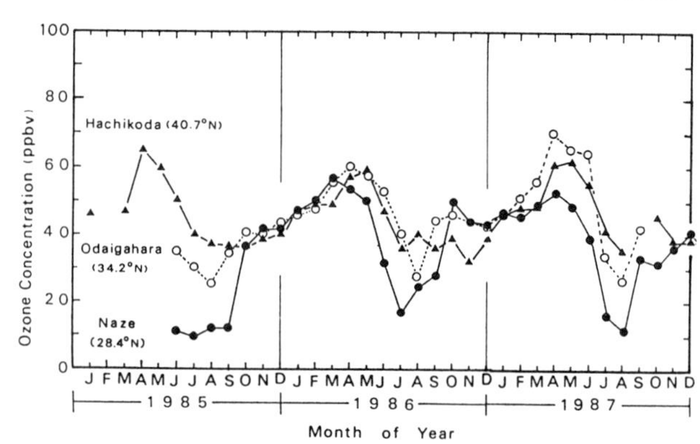

Fig. 2. Monthly average concentration (based on 24 hours hourly values) of surface ozone at three stations in Japan.

in the northern hemisphere, and can be interpreted in terms of ozone transport from the lower stratosphere where ozone concentration attains its maximum in spring.

The summer minimum is smaller at lower latitudes, and the duration of the summer decrease is longer at lower latitudes. The summer decrease extends over four months from June to September at 28.4°N latitude (Naze), whereas it is only for two months from July to August at 34.2°N latitude (Odaigahara). The data from Hachikoda (40.7°N) represent a decrease in summer and autumn, but the summer concentration is much higher than that at the other two sites. Accordingly, the seasonal behavior at Hachikoda is very similar to that observed at middle latitudes in Europe and North America.

Ozone concentrations in other seasons seem to be homogeneous with latitude over Japan. Actually, latitudinal difference in monthly mean ozone concentrations is less than 10 ppbv during the period of October through March.

The above seasonal characteristics are closely related to the climatic cycle in Japan. In summer, tropical oceanic air masses prevail over Japan Is. and the dominant winds are the southerlies. In winter, polar continental air masses are brought by the strong westerlies or northwesterlies, which sometimes carry ozone-rich air from the lower stratosphere. Consequently, the origin of air mass over Japan is quite different in summer to that of winter. We will discuss this point in detail in the following section.

5. OZONE CONCENTRATIONS IN CONTINENTAL AND OCEANIC AIR MASSES

The seasonal change in atmospheric condition over Japan is explained in terms of the configuration of continental and oceanic air masses. The continental air mass originates from the lower stratosphere and/or the upper troposphere in high latitudes, and is relatively dry and cool. The oceanic air mass originates from the lower troposphere in the tropics and is warm and humid accordingly. In the boundary of these two air masses a front is formed. This front is fairly stable at the beginning and the end of summer, extending from the Chinese coast to the east of Japan Is. roughly along a latitude slightly south of Japan Is. This particular front is called the "Baiu" frontal zone (BFZ) and the "Akisame" frontal zone (AFZ) for the periods of mid-June through mid-July and of September, respectively. The frontal zone moves toward the north, as the tropical oceanic air mass grows in summer. It is located at a latitude of ~30°N in June, and at ~40°N in late July - August; or it mostly disappears. In September as the continental air mass grows, the frontal zone is again located at ~30°N, south of Japan Is.

Fig. 3 shows that the exchange of air masses occured at Naze in mid-June and mid-September. In the period between mid-June and mid-September or in summer, Naze is affected by tropical oceanic air masses as characterized by high temperature and high humidity. The diurnal mean ozone concentration during this period is ~10 ppbv, which is significantly lower than those before and after this period. This level of ozone concentration is almost the same as at Samoa in January - April (Oltmans, 1981), and may be representative of those in tropical oceanic air masses. Tropical oceanic air should be poor in ozone, because it is independent of the air which comes from the lower stratosphere or the upper troposphere, and because the sources of precursor gases for photochemical ozone production such as NO_x, CO, methane and non-methane hydrocarbons are inert over the ocean.

It is interesting to note that ozone concentration sometimes exceeds 20 ppbv in summer at Naze. These events are associated with the arrival of continental air as indicated by the trajectory analysis.

Hachikoda is located north of the Baiu frontal zone. As shown in Fig. 4, it is affected

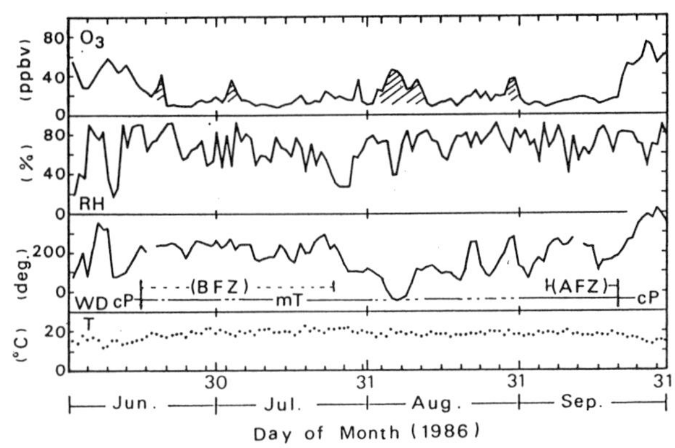

Fig. 3. Day-to-day variations of surface ozone concentration (O_3) and meteorological parameters such as relative humidity (RH), wind direction (WD) and temperature (T) at the 850 mb level. All the data are obtained at Naze. cP, mT, BFZ and AFZ denote continental air mass, tropical oceanic air mass, Baiu frontal zone and Akisame frontal zone, respectively.

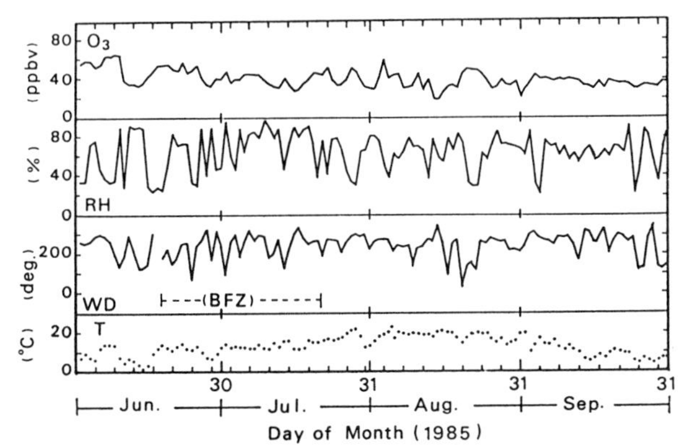

Fig. 4. Similar to Fig. 3 except for the ozone concentration at Hachikoda and meteorological parameters at Misawa located 40 km away to the east.

by continental air brought by the westerlies even after the disappearance of the front. At this site and probably in northern Japan, diurnal mean ozone concentration does not decrease below ~30 ppbv in summer. However, when a typhoon moved north and carried with it a tropical air, the ozone concentration decreased to as low as ~20 ppbv at Hachikoda.

Summertime day-to-day variations of surface ozone are more complicated at Odaigahara than at other sites. The Baiu frontal zone stays in close proximity, with its position relative to Odaigahara moving from day to day. If the front is located south of Odaigahara, then the atmospheric condition over Odaigahara is affected by continental air which is cool and dry, and if the front is located north of Odaigahara, then it is affected by oceanic air which is warm and humid. A negative correlation is found between ozone concentration and relative humidity at the 850 mb level over Odaigahara.

6. YEAR-TO-YEAR VARIATION

Surface ozone measurement began in April 1983 at Hachikoda, and the data are available since then. The tropospheric ozone concentration at mid-latitudes depends on the magnitude of the downward ozone flux from the stratosphere, and accordingly any long-term trend in tropospheric ozone may be associated with the variation in air exchange strength between the stratosphere and the troposphere. Of interest also is the impact of growing anthropogenic sources on the tropospheric ozone trend.

Monthly average concentration based on 24 hours hourly values, as shown in Fig. 5, does not show any significant trend during these four years. The variation seems to be a year-to-year fluctuation.

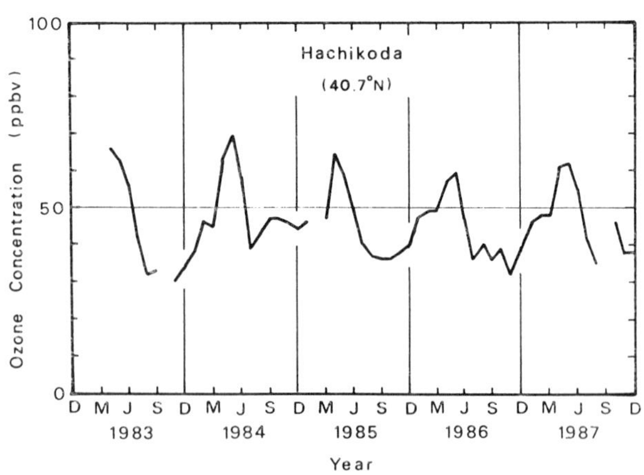

Fig. 5. Monthly average concentration (based on 24 hours hourly values) of surface ozone at Hachikoda for 1983-1987. M, J, S and D represent March, June, September and December, respectively.

7. SUMMARY

Tropospheric ozone has been measured at three rural sites in Japan, and it shows a seasonal behavior different from that seen in Europe and North America. The spring maximum observed at all three sites is a common feature at mid-latitudes, but the summer minimum appearing more conspicuously at lower latitudes in Japan is different from the autumn minimum in other sites at mid-latitudes. The summer minimum can be found at the site where tropical air mass prevails in summer. Our present result indicates that the ozone concentration in tropical oceanic air mass is ~10 ppbv without any distinct diurnal variation.

At higher latitudes in Japan, the seasonal ozone behavior exhibits a minimum from late summer to fall, and it becomes closer to those in mid-latitude Europe and North America. Since continental air carried by the westerlies prevails over northern Japan even in summer, the ozone concentration of 30-40 ppbv is more typical of that found at mid-latitudes in summer.

In fall and winter, the atmospheric condition is affected by polar continental air mass covering Japan Is. Monthly mean ozone concentration therefore exhibits almost the same values at different latitudes in Japan. This feature is very similar to those of middle latitudes in other continents.

The exchange of air masses between continental and oceanic ones is clearly seen at Naze (28.4°N) in accordance with the occurrence of the Baiu frontal zone and the disappearance of the Akisame frontal zone. At these events, surface ozone concentration is found to manifest dramatic changes from the state of continental air mass to that of oceanic air mass, and the reverse.

ACKNOWLEDGEMENT

The authors thank personnel of Aomori Prefectural Center for Environmental Protection for providing the ozone data at Hachikoda.

REFERENCES

Chatfield, R. and H. Harrison, 1977: J. Geophys. Res., 82, 5969-5976.

Fabian, P. and P.G. Pruchniewicz, 1977: J. Geophys. Res., 82, 2063-2073.

Logan, J.A., 1985: J. Geophys. Res., 90, 10463-10482.

Ogawa, T. and A. Miyata, 1985: J. Meteorol. Soc. Japan, 63, 937-946.

Oltmans, S.J., 1981: J. Geophys. Res., 86, 1174-1180.

Singh, H.B., F.L. Ludwig and W.B. Johnson, 1978: Atm. Env., 12, 2185-2196.

Tsuruta, H., 1983: Spec. Rep. of Yokohama Research Institute for Environmental Science No. 47 (in Japanese).

DIURNAL AND SEASONAL VARIATIONS OF THE TROPOSPHERIC OZONE IN TROPICAL ASIA

T. Ogawa and Ninong-Komala
Geophysics Research Laboratory, University of Tokyo
Tokyo 113, Japan

ABSTRACT

Tropospheric ozone near the ground surface has been observed at Watukosek (7°35'S, 112°40'E), a village in East Java since December 1986. We analyzed the ozone data for December 1986 - March 1988 along with meteorological data. Ozone concentration exhibits a significant diurnal variation: maximum of ~40 ppbv before noon, minimum of ~10 ppbv at night and diurnal average of ~20 ppbv. A preliminary measurement indicates NO_2 concentration to be 4 - 6 ppbv. These results demonstrate that the ozone concentration is controlled basically through photochemical production and removal. However, the diurnal maximum occurs at ~10 LT in dry months and at 11 - 12 LT in other months. This early diurnal maximum should be interpreted in terms of transport due to sea breezes.

1. INTRODUCTION

The global distribution of tropospheric ozone is one of the important issues in atmospheric science. It is related with global air quality and photochemistry and with impact on atmospheric thermal radiation field. Tropospheric ozone concentration in the tropics is believed to be low as compared with those in middle latitudes, because of inactive transport from the stratosphere to the troposphere (Fabian and Pruchniewicz, 1977; Routhier et al. 1980). However, tropospheric ozone is supplied by photochemical reactions as well as transport from the stratosphere. We expect high photochemical activity in the tropics where intense solar insolation is available. Photochemical productivity of ozone also depends on the concentrations of NOx, CO, methane and non-methane hydrocarbons (NMHC). The land area provides potential sources of these species in the tropics (Crutzen et al.,1985), but the ocean does not seem to be the sources (McFarland et al., 1979; Helas and Warneck,1981). Tropospheric ozone data implies that the marine air is lean in ozone (Stallard et al., 1975; Routhier et al., 1980; Oltmans, 1981; Piotrowicz et al., 1986).

Tropical Asia, in particular Indonesia, is a region of islands and seas. Its surrounding nature may be oceanic on the average, but the tropical land is favorable for vegetation and agricultural activity, providing the sources of NOx, CO, methane and NMHC. Consequently, it is interesting to investigate the tropospheric ozone behavior in tropical Asia (Ilyas, 1987) comparing with those in tropical America and Africa (Sanhueza et al., 1985; Logan and Kirchhoff, 1986; Cros et al.,1987; Kirchhoff, 1988). The University of Tokyo has started a program to measure atmospheric ozone in Indonesia in collaboration with the Aerospace Research Center of Indonesian LAPAN. This paper presents an initial result from the surface ozone measurements in East Java.

2. INSTRUMENTATION AND OBSERVATION SITE

The ozone instrument used in this study is a Dylec model 1006 AHJ, which is based on ultraviolet absorption technique and produced under a license of Dasibi Inc. The cycle time of measurement is about 12 sec. The measured concentration of ozone is represented in units of ppbv with the resolution 1 ppbv and recorded on a strip chart. In December 1986, we set up this instrument at Watukosek (7°35'S,112°40'E) in Java Is., Indonesia, and we have been continuing the ozone observation since then. We have also been acquiring meteorological data such as wind speed and direction as well as solar insolation.

The observation site is located in the agricultural village area of East Java. It is 50 m high above sea level and on the eastside slope of a low hill cultivated for a cashew plantation. The air intake is 4 m high above the ground, and 1 - 2 m above the canopy of cashew trees. The surroundings are agricultural areas mainly consisting of rice paddies. Two roads of considerable car traffic run. One is on the northwest foot of the hill and the other is about 10 km to the east. Small biomass burnings are seen here and there in the paddy fields and in the villages. They probably provide the sources of NOx, CO and hydrocarbons (HC). The seashore is about 20 km away to the east.

Climate in Java Is. is of tropical nature characterized by dry and wet seasons. In the dry season which prevails in June - September, it is affected by Australian continental air mass, whereas in the wet season which prevails in December - March, it is affected by Asian continental and Pacific oceanic air masses. Consequently, the monsoon wind is easterly in the dry season and westerly in the wet season. However, in East Java, rainfall is not so significant as in a typical tropical rain zone like West Java. The transition of seasons is sometimes vague in Watukosek and its timing also fluctuates from year to year.

The local wind system at Watukosek is characterized by a diurnal pattern of land and sea breezes. The land breeze comes from the west and

ISBN 0-937194-15-8

the sea breeze comes from the east. The dominant wind direction seems to be determined as a result of a combination of the local wind system and the large-scale monsoon wind. Westerlies prevail all through the day of the wet season and easterlies prevail in the daytime of the dry season. There is no urban area toward these winds.

3. RESULTS AND DISCUSSION

The chart record of ozone concentration exhibits fluctuations of time scales of 10 sec - 10 min probably due to dynamical and chemical small-scale inhomogeneity in the planetary boundary-layer atmosphere. In particular, concentration fluctuations in the order of 1 min are significant in the morning and in the early afternoon. In this study we will concentrate on the hourly values of ozone concentration calculated in terms of one hour average.

3.1 DIURNAL VARIATION

Figure 1 shows diurnal variations of the ozone concentration observed at Watukosek in typical months of wet and dry seasons. The average feature of the diurnal variation is described as follows: Ozone concentration begins to increase at sunrise, attains its maximum before noon, subsequently decreases gradually until sunset or several hours after sunset, and remains almost constant after midnight. This average feature is regularly seen day after day, noting that the standard deviations are small ($1\sigma < 10$ ppbv on the average).

Larger ozone concentration in the daytime than at night is explained in terms of photochemical process and/or vertical mixing process. Ozone is produced through NOx-CO-HC reactions in the daytime and removed through NO reactions at night. On the other hand, the planetary boundary layer is stabler in vertical mixing at night than in the daytime. Inactive vertical mixing at night, combined with chemical removal, would result in low ozone concentration. Vertical mixing normally becomes active just after sunrise, and accordingly an increase in ozone concentration occurs within a few hours after sunrise. If the NOx concentration is low enough (less than 1 ppbv), as seen in the rural region in middle latitudes, the daytime ozone concentration is almost constant, representing of the tropospheric concentration (Singh et al.1978; Fehsenfeld et al.,1983; Ogawa and Miyata, 1985).

The ozone increase in the morning at Watukosek seems to be too slow to be ascribed to the vertical mixing. The daytime ozone variation at Watukosek is very similar to those in mid-latitude urban areas, and accordingly suggests active photochemical ozone production. However, the diurnal maximum occurs earlier at Watukosek than in the urban atmosphere, and the nighttime concentration is larger at Watukosek (around 10

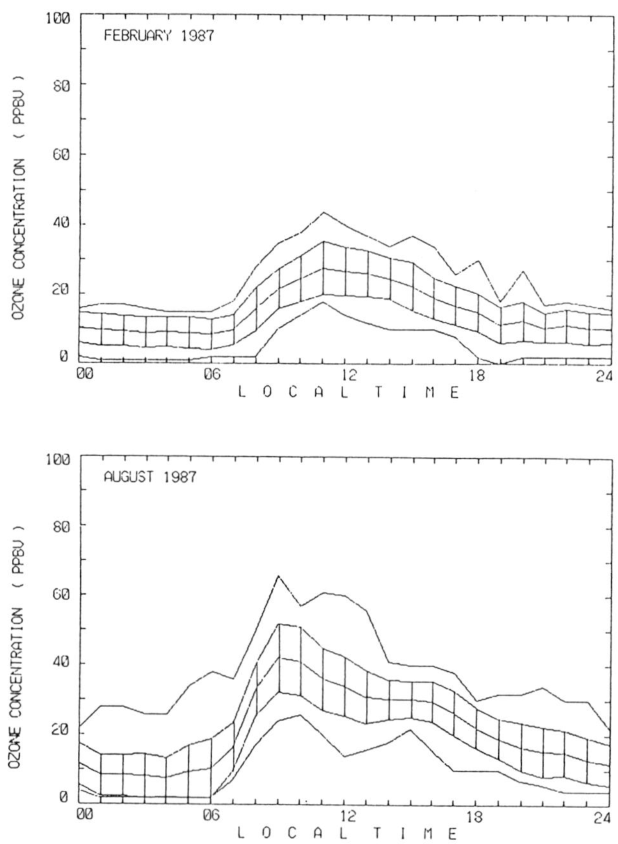

Fig. 1. Diurnal variations of surface ozone concentration at Watukosek in wet (February) and dry (August) months. The panel represents monthly maximum, minimum and average with a vertical bar for the standard deviation at each local time.

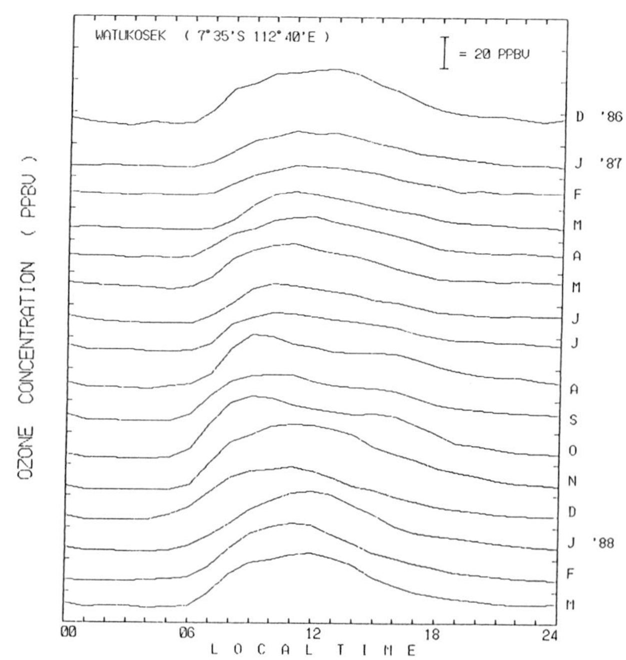

Fig. 2. Monthly-averaged diurnal variations at Watukosek for the period December 1986 - March 1988.

ppbv) than in the urban atmosphere where ozone concentration is less than ~1 ppbv at night.

We carried small NOx sampling capsules, which contained sampling filters, to Watukosek and made samplings on February 22 - 24, 1988. The chemical analysis of sample filters exhibits the average NO_2 concentration to be 3.5 ppbv in the daytime and 5.7 ppbv at night. These values are smaller than those measured in the urban atmosphere but seem to be large enough to yield ozone through photochemical reactions. For further study on photochemical activity at Watukosek, we are planning systematic measurements of NOx, CO, methane and NMHC along with ozone measurement.

3.2 SEASONAL BEHAVIOR

Figure 2 exhibits monthly-averaged diurnal variations for the period of December 1986 - March 1988. We can see month-to-month variations in some characteristics of diurnal variation such as diurnal maximum, the local time at which diurnal maximum occurs, diurnal minimum (or nighttime level) and the amplitude of diurnal variation.

The diurnal maximum of ozone concentration may depend on solar insolation because of photochemical ozone production. However, according to the global solar radiation data observed at Watukosek, the daily sum of insolation does not very significantly with month or between dry and wet seasons. In fact, any systematic seasonal trend of diurnal ozone maximum is not found in the data for December 1986 - March 1988. The diurnal amplitude (difference between maximum and minimum) varied in the same way as the diurnal maximum.

Diurnal ozone minimum or the nearly constant level at night is thought to be affected by the chemical removal and vertical mixing of ozone, accordingly NO concentration and atmospheric stability. The ozone data implies that diurnal minimum is smaller in wet months than in dry months. Since the timing of turnaround between dry and wet seasons is likely to fluctuate from year to year at East Java, further data acquisition is necessary to demonstrate that implication clearly.

A seasonal variation is clearly seen in Fig. 3 exhibiting the local time when diurnal ozone maximum occurs. The diurnal maximum occurs at ~10 LT in May - October, whereas it occurs at 11 - 12 LT in other months. Moreover, it is significant in Fig. 3 that monthly standard deviation is smaller in May - October than in other months. Figure 4 shows that the ratio of ozone concentrations at 12 LT and 09 LT is lower in May - October than in other months. Since solar insolation does not very significantly with day and between dry and wet months in East Java which is located outside the monsoon rain zone, those seasonal variations shown in Figs. 3 and 4 should not be ascribed to solar insolation. It is highly probable that those seasonal variations are related with the development of sea breezes in the dry season.

The typical wind direction in the daytime of the dry season at Watukosek is easterly: the monsoon winds and sea breezes, which carry marine air. Since ozone concentration is low in the tropical marine air (e.g. Stallard et al., 1975; Ogawa and Miyata,1985; Logan, 1985), the easterlies are able to suppress ozone increase in the daytime. Consequently, the diurnal ozone maximum at ~10 LT and the subsequent gradual decrease can be interpreted in terms of the development of sea breezes.

Figure 5 demonstrates the differences in diurnal variations between the days of well-developed easterlies and of weak winds. The post-sunrise increase begins earlier on the days of the well-developed easterlies than on those of weak winds. This is probably due to active vertical mixing on the days of the easterlies. The daytime concentration is roughly 10 ppbv smaller on the days of the well-developed easterlies than on those of weak winds. This difference represents the effect of the sea breezes which are lean in ozone.

In the wet season, the monsoon westerlies prevail and the typical wind direction is westerly all through the day, although sometimes even in the wet season the easterlies of sea breezes are evident around noon and in the afternoon. The westerlies coming to Watukosek may carry land air

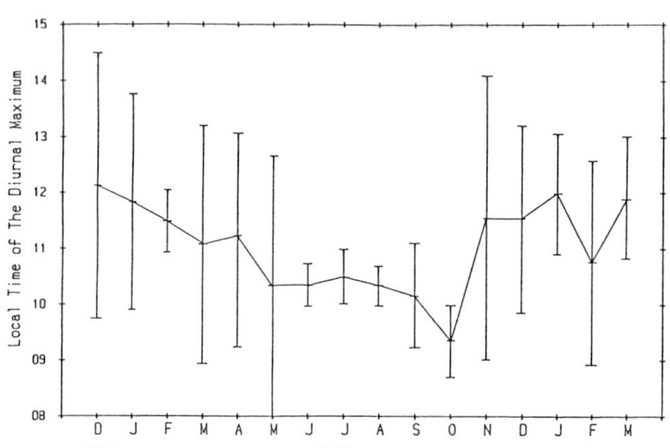

Fig. 3. Month-to-month variation in the local time when ozone concentration attains its diurnal maximum. The panel exhibits monthly averages and standard deviations.

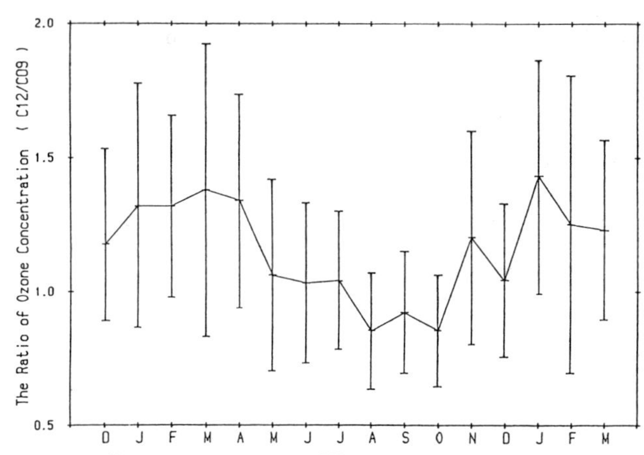

Fig. 4. Month-to-month variation in the ratio of ozone concentrations at 12 LT and 09 LT. The panel exhibits monthly averages and standard deviations.

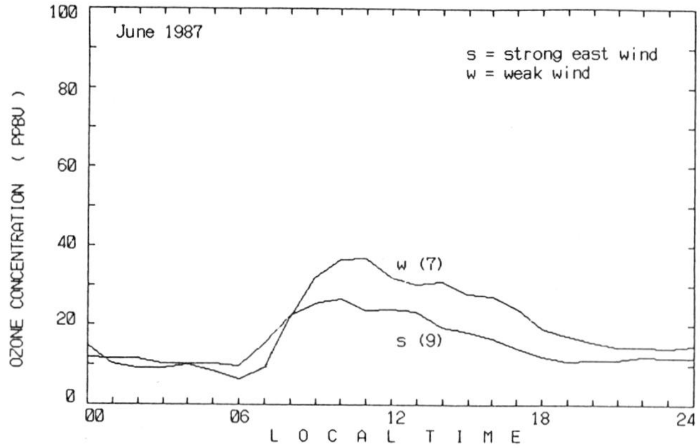

Fig. 5. Difference in diurnal variations due to wind speed. The "strong east wind" day means that the easterlies persist throughout the daytime and the maximum of hourly values is not smaller than 4 m s^{-1}. The "weak wind" day means that the hourly values of wind speed do not exceed 2.5 m s^{-1} irrespectively of wind direction. Numerals in the parentheses are for the number of days.

mass of significant NOx, CO and HC concentrations, providing the potential of ozone production. Consequently, the diurnal variation and the wind effect of ozone concentration are complicated in the wet season.

4. CONCLUDING REMARKS

The data on the surface ozone concentration observed at Watukosek exhibit a significant diurnal variation. Diurnal maximum, average and minimum are ~40 ppbv, ~20 ppbv and ~10 ppbv on the average, respectively. The concentration varies from day to day and from month to month, but does not show any regular seasonal behavior. The diurnal variation itself is a consequence of photochemical ozone production, because NOx concentration is moderately high according to our preliminary observations (NO_2 concentration: 4 - 6 ppbv). However, diurnal maximum occurs at ~10 LT in the dry season and 11 - 12 LT in the wet and transition seasons. The early diurnal maximum is thought to be closely related with the development of sea breezes which carry ozone-lean air. For a detailed analysis on dynamical and chemical effects, we need a simultaneous data set for NOx, CO, methane and NMHC together with ozone.

REFERENCES

Cros, B., R. Delmas, B. Clairac and J. Loemba-Ndembi, 1987: J. Geophys. Res. 92, 9772-9778.

Crutzen, P.J., A.C. Delany, J. Greenberg, P. Haagenson, L. Heidt, R. Lueb, W. Pollack, W. Seiler, A. Wartburg and P. Zimmerman, 1985: J. Atm. Chem. 2, 233-256.

Fabian, P. and P.G. Pruchniewicz, 1977: J. Geophys. Res. 82, 2063-2073.

Fehsenfeld, F.C., M.J. Bollinger, S.C. Liu, D.D. Parrish, M. McFarland, M. Trainer, D. Kley, P.C. Murphy, D.L. Albritton and D.H. Lenschow, 1983: J. Atm. Chem. 1, 87-105.

Helas, G. and P. Warneck, 1981: J. Geophys. Res. 86, 7283-7290.

Ilyas, M., 1987: Atm. Env. 21, 1799-1803.

Kirchhoff, V.W.J.H., 1988: J. Geophys. Res. 93, 1469-1476.

Logan, J.A., 1985: J. Geophys. Res. 90, 10463-10482.

Logan, J.A. and V.W.J.H. Kirchhoff, 1986: J. Geophys. Res. 91, 7875-7881.

McFarland, M., D. Kley, J.W. Drummond, A.L. Schmeltekopf and R.H. Winkler, 1979: Geophys. Res. Lett. 6, 605-608.

Ogawa, T. and A. Miyata, 1985: J. Meteorol. Soc. Japan 63, 937-946.

Oltmans, S.J. , 1981: J. Geophys. Res. 86, 1174-1180.

Piotrowicz, S.R., D.A. Boran and C.J. Fischer, 1986: J. Geophys. Res. 91, 13113-13119.

Routhier, F., R. Dennett, D.D. Davis, A. Wartburg, P. Haagenson and A.C. Delany, 1980: J. Geophys. Res. 85, 7307-7321.

Sanhueza, E., K.H. Octavio and A. Arrocha, 1985: J. Atm. Chem. 2, 377-385.

Singh, H.B., F.L. Ludwig and W.B. Johnson, 1978: Atm. Env. 12, 2185-2196.

Stallard, R.F., J.M. Edmond and R.E. Newell, 1975: Geophys. Res. Lett. 2, 289-292.

SURFACE OZONE AND SOLAR RADIATION

U. Feister and K.-H. Grasnick
Meteorological Service of the GDR
Meteorological Main Observatory
Telegrafenberg, Potsdam, GDR–1561

G. Jakobi
Meteorolgical Service of the GDR
Meteorological Observatory
Dresden-Radebeul, GDR–8122

ABSTRACT

A substantial portion of tropospheric ozone is produced by photochemical reactions within the troposphere in the presence of solar ultraviolet radiation. Measurements of surface ozone and solar UV radiation, which is mainly determined by the amount of clouds and sun angle, have been made at Potsdam. Similar variations of both parameters with maximum values in summer have been observed. Due to other processes such as vertical ozone transport from the stratosphere and horizontal ozone transport, the varying amount of constituents destroying ozone, ozone destruction at the surface in dependence of soil type as well as the variation in the concentration of ozone precursors (NO_x, hydrocarbons), the relationship between ozone concentration and solar radiation is not clearly discernible at times. Trend analyses of total global radiation and surface ozone have shown similar, though sometimes non-significant increases or decreases of both parameters at the stations in the period 1972 to 1987. Thus it seems to be important not to neglect variations of solar radiation due to varying cloudiness in model calculations that simulate the long-term changes of photooxidants in the lower atmosphere.

1. INTRODUCTION

Solar UV radiation does not only affect the biosphere and damages plastics and paints, but also plays an important role in the photochemistry of the lower atmosphere. According to estimates made by Bojkov (1983) $1.85 \cdot 10^{12}$ kg per year ozone is produced by photochemical processes in the troposphere. This amount, which is about three times the estimated amount injected from the stratosphere, nearly corresponds to more than 50% of the global ozone mass or about 16 times the amount of tropospheric ozone. As proposed by Frenkiel (1959), the main source reaction for photochemically produced ozone is photodissociation of NO_2 by solar UV radiation with wavelengths $\lambda < 400$ nm

$$NO_2 + h\nu \longrightarrow NO + O(^3P), \quad \lambda < 400 \text{ nm}$$

which provides the needed O atoms for ozone production. Recognizing the significant importance of solar UV radiation, WMO (1977) recommended monitoring of UV radiation at stations where atmospheric ozone is measured. The paper will discuss first results of UV measurements and surface ozone measurements that have been taken at the Potsdam Observatory since 1984/85.

2. INSTRUMENTS USED FOR MEASUREMENTS OF SURFACE OZONE AND UV RADIATION

Photocell instruments using glass or metal interference filters to select spectral regions of the ultraviolet spectrum are used to monitor UV radiation A detailed description of the instruments and their calibration will be given by Feister and Grasnick (1989). Three spectral intervals with halfwidths from 311-320 nm, 315-391 nm and 296-371 nm are measured at Potsdam. Surface ozone is measured by the electrochemical potassium iodide method using instruments of the type OZONOGRAPH (Warmbt 1983). A description of station network and results of long-term observations of surface ozone was given by Feister and Warmbt (1987). The wet chemical method has to be applied with great care to provide reliable results. Fig. 1 shows some surface ozone measurements carried out at different stations in the Antarctic during the fifties and sixties. Our wet chemical method was applied at the station Mirny (Kolbig and Warmbt 1978),

ISBN 0-937194-15-8

while the other stations used Regener's chemical ozone recorder (Little America, Hallett station, Amundsen Scott) and Brewer Mast type recorders (Hallett station, Base King Baudoin).

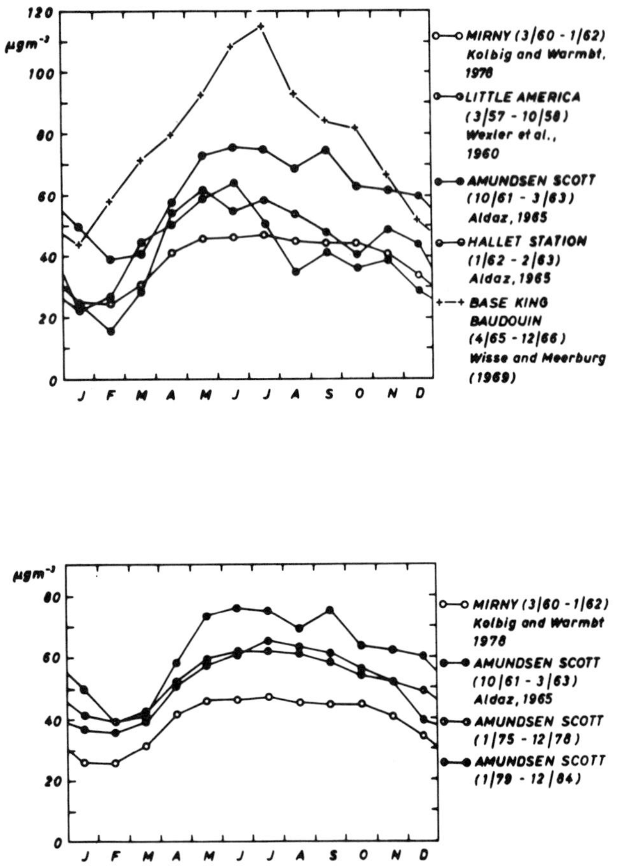

Fig. 1. Seasonal variation of surface ozone measured at the Antarctic stations
Mirny (66°33' S, 93°01' E, 30 m)
Little America (78°11' S, 162°10' W, 44 m)
Hallett Station (72°18' S, 170°18' E, 15 m)
Amundsen Scott (89°59' S, 24°48' W, 2835 m)
Base King Baudoin (7o°26' S, 24°19' E, 38 m)

Though the general seasonal ozone variation is similar at all stations with a minimum in January/February and a broad winter maximum, there are considerable differences between the ozone records that can be ascribed partly to differences in space and time of the measured data and to a greater extent to measurement errors. A comparison of the old measurements made at Mirny and at Amundsen Scott with later measurements at Amundsen Scott (Fig. 1b), which were made by the ECC method (1/75 - 12/78) and by Dasibi UV photometric ozone recorders (1/79 - 12/84) (Oltmans and Komhyr 1976) and which were determined from data published by the AES Toronto (Ozone data for the World), shows that there is closest correspondence in the seasonal variation between the Amundsen Scott ozone data (1975 - 1984) and the old Mirny data. For the conversion of units, climatological data of pressure and temperature were used (Orvig 1970). The bias between these curves may be caused by the more southward location (2600 km distance) and the higher elevation of Amundsen Scott (2835 m) as compared to Mirny (30 m). This result may serve to illustrate that also wet chemical measurements can be considered reliable provided that they are well taken.

3. RELATIONSHIP BETWEEN SURFACE OZONE AND SOLAR UV RADIATION

Tichy (1939), who made concurrent measurements of both surface ozone and UV radiation on cloudless days in spring 1937 and 1938 at Schreiberhau (700 m asl) noticed a similar daily variation of both parameters (Fig. 2).

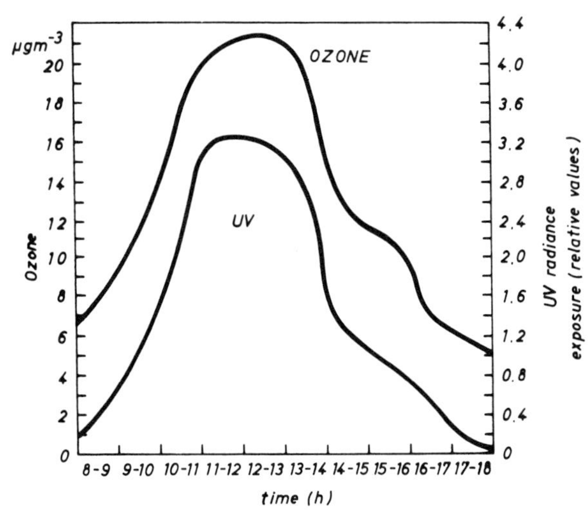

Fig. 2. Average daily variation of surface ozone (Cauer method) and solar UV radiation (UV dosimeter of IG Farben) as determined from measurements at Schreiberhau (700 m asl) on 13 cloudless days in spring 1937 (Tichy 1939)

However, he ascribed the ozone variations to vertical transport, which generally intensifies from morning to afternoon. Similar diurnal variations have been observed at Potsdam on cloudless days. Fig. 3 shows the daily radiance exposure of global UV radiation (296 - 371 nm) and global total radiation (400 - 2500 nm) measured at Potsdam in June 1986.

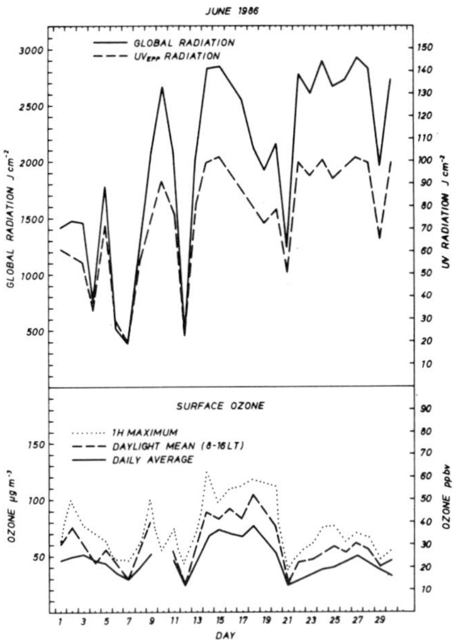

Fig. 3. Daily radiation exposure of global UV radiation (296 - 371 nm) and global total radiation (400 - 2500 nm) as measured at Potsdam in June 1986

The UV portion is about 3-4% of total global radiation. Strong variations of UV radiation from day to day by about 20 J cm^{-2} to more than 140 J cm^{-2} can be seen, which are mainly due to changing cloudiness. A similar pattern occurs in surface ozone, which is shown in the lower part of Fig. 3. The highest amplitude can be observed in the 1 h maximum ozone data with values ranging from 40 µg m^{-3} to 125 µg m^{-3}. A somewhat lower amplitude between 25 and 105 µg m^{-3} occurs in the daylight mean values (8 - 16 local time) of surface ozone. Due to the dependence of the surface ozone concentration on further parameters such as ozone precursors and other constituents or vertical and horizontal ozone transport, the variation of solar radiation from day to day is not in each case reflected by a similar ozone variation.

Daily totals of UV radiation and 1 h maximum values of surface ozone for the periods April 1985 to July 1986 and April to September (summer) 1985 show correlation coefficients of 0.64 and 0.48, respectively, which are well above the 1% significance level. Similarly, Pearson's correlation ratios (Taubenheim 1969) which can generally be applied, if the data are not necessarily normally distributed, also show a significant relationship between daily UV radiance exposure and daily maximum surface ozone. To eliminate the possible autocorrelation of data, only ozone and UV radiation values of every third day were used in the calculation of correlation coefficients. This reduced data set did also show a significant correlation coefficient of 0.64.

4. LONG-TERM CHANGES OF SURFACE OZONE AND SOLAR RADIATION

A significant long-term increase of surface ozone by 1 - 3% per year from the mid-fifties to about 1980 was observed in the ozone records of the Meteorological stations in the GDR (Feister and Warmbt 1987). Since about 1980 the ozone values seem to decrease at some stations. Table 1 shows trends of surface ozone at some stations in the 16 years' period from 1972 to 1987. While up to the beginning of the eighties most stations showed a significant ozone increase, the addition of recent years' periods provides no significant trends any more at most stations. Possible causes for that effect as discussed by Feister and Warmbt (1987) are a changed ratio of nitrogen oxides to volatile organic compounds or the change of meteorological conditions such as ozone transport and varying cloudiness. In addition, Table 1 and Figure 4 show seasonal averages of total global radiation at the three stations Arkona, Dresden-Radebeul and Fichtelberg. In spite of the high interannual variability, which shows no direct relationship between annual totals of solar radiation and surface ozone, trends of both parameters are either positive (Dresden, Fichtelberg) or negative (Arkona). Although most of these trends are non-significant, they could be an indication that changing cloudiness was indeed one influencing factor contributing to the long-term change in surface ozone.

TABLE 1. SURFACE OZONE (95 PERCENTILES OF HOURLY AVERAGES) AND GLOBAL SOLAR RADIATION TRENDS IN THE PERIOD 1972 TO 1987 (SUMMER: APRIL - SEPTEMBER, WINTER: JANUARY - MARCH, OCTOBER - DECEMBER)

STATION	PARAMETER	AVERAGE (µg m^{-3} and J cm^{-2})	TREND	SIG. LEVEL
ARKONA (42 m)	ozone summer	87.2	-0.58±1.05	
	solar radiation[1)] summer	1620	-0.72±0.35	5%
	ozone winter	66.1	-0.39±0.99	
	solar radiation[1)] winter	344	-0.49±0.43	

STATION	PARAMETER	AVERAGE ($\mu g\ m^{-3}$ and $J\ cm^{-2}$)	TREND	SIG. LEVEL
DRESDEN-RADEBEUL (246 m)	ozone summer,	90.6	0.67±0.55	
	solar radiation, summer	1450	0.07±0.45	
	ozone winter,	51.4	1.43±0.53	1%
	solar radiation, winter	443	0.98±0.45	5%
FICHTEL-BERG (1213 m)	ozone summer[2],	101.0	3.44±1.05	1%
	solar radiation, summer	1370	0.38±0.53	
	ozone winter,	86.2	0.30±1.36	
	solar radiation, winter	452	1.23±0.74	

1) Global radiation at Arkona was determined from measured sunshine duration (see Feister and Warmbt, 1985)

2) Period 1972 - 1986

An increase in surface ozone on days with high relative sunshine duration (> 80%) was found for the summers 1979 - 1984 in the ozone record of Garmisch-Partenkirchen (740 m) by Reiter et al. 1987.

5. SURFACE OZONE AND WIND

An analysis of the surface ozone measurements and wind data taken at the station Schmücke (937 m) in Thuringia for the years 1985 - 1986 has shown that higher ozone values occur with definite prevailing wind directions. The prevailing wind direction for each day is defined as the centre of a 90° sector with at least 75% of the individual 1 h wind direction values. For 2/3 of the days between April and September 1985 and 1986, a prevailing wind direction was determined. As an example, the average daily 1 h ozone maximum values for different wind directions during the summer months (April to September) are drawn in Fig. 5. The dashed curves designate directions with less than 5% of all cases. While the lengths of the lines give the average 1 h ozone maximum for the wind direction, the numbers at the end of the lines give the absolute ozone maximum ($\mu g\ m^{-3}$) with that wind direction. It is obvious that the highest ozone maxima occur when the wind comes from southwest, and lowest values occur with northerly winds. Ozone values are highest for a wind velocity of 3 ms^{-1}, but there is a broader secondary maximum with wind speeds between 2 and 6 ms^{-1}. The highest average ozone maximum of 132 $\mu g\ m^{-3}$ occurs with winds from the south. However, this wind direction is

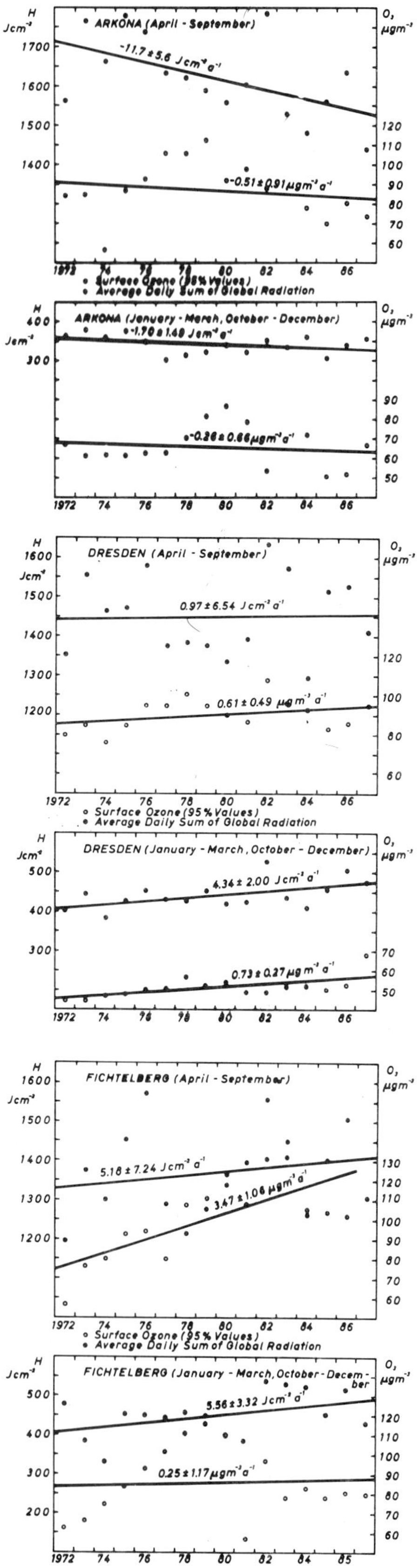

Fig. 4. Annual and seasonal averages of surface ozone (95 percentiles and global total radiation)

observed in only 1.7% of all cases.

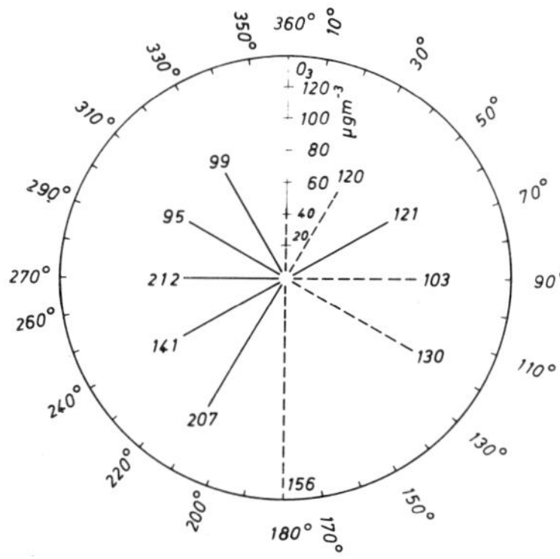

Fig. 5. Daily averages of 1 h maximum values of surface ozone with different prevailing wind directions at the station Schmücke (937 m) in the period April - September 1985 and 1986

Much more frequently, in 60% of all cases, the wind comes from southwest and west at the station. We are aware that surface wind is only a rough approximation of horizontal transport of ozone and precursors. The situation becomes even more complicated, because solar radiation should play an important role in the photochemical ozone production. Fig. 6 shows the daily average sunshine duration classified for wind directions at the station Schmücke in the summer months of 1985 and 1986.

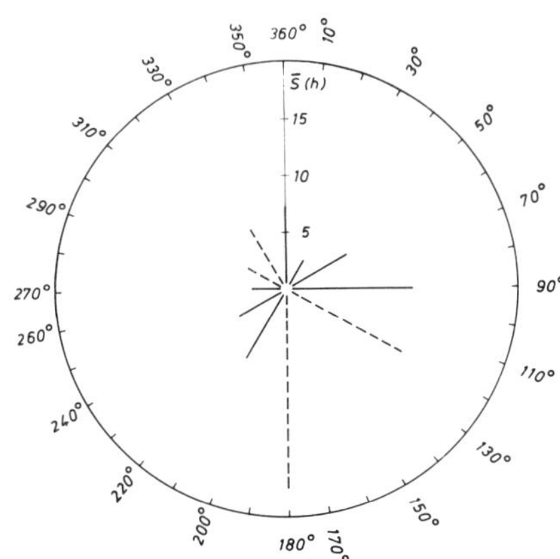

Fig. 6. Daily averages of sunshine duration with different prevailing wind directions at the station Schmücke in the periods April - September 1985/ 1986

A distinct maximum of sunshine duration of 17.5 h occurs, if the wind comes from the south. However, again this direction represents only 1% of all cases and characterizes some episodes with high ozone values. Secondary maxima of sunshine duration can be recognized with winds from E, N and SSW. Figures 5 and 6 should illustrate that both transport of ozone and/or precursors and solar radiation are important parameters that determine the ozone concentration near the ground.

6. CONCLUSION

The ozone concentration near the ground is to a great extent determined by the variability of solar UV radiation. As a rough approximation, radiation with wavelengths $\lambda < 400$ nm contributes about 5-7% to the global solar radiation reaching the lower atmosphere. On the other hand, on days with cloud cover the daily total of radiance exposure is reduced to about 20% - in extreme cases down to 5% - of the radiance received on days with clear sky. This reduction in daily radiance exposure by clouds and aerosols nearly corresponds to the average seasonal variation of radiance exposure at mid-latitudes, i.e. the daily radiance exposure in December is about 10% of that in summer. In addition to the variability of transport, both the short- and long-term variability of cloud cover and/or - if relevant - atmospheric aerosol-, or in other words radiation reaching the lower atmosphere, should be taken into account, when models are applied to simulate photochemical processes in the boundary layer. Solar radiation and surface ozone measurements taken at stations of the Meteorological Service of the GDR seem to indicate that varying amounts of cloudiness may have contributed to the longer-term variation of surface ozone.

REFERENCES

Aldaz, L., 1965: Atmospheric ozone in Antarctica. J. Geophys. Res., 70, 1767-1773.

Bojkov, R.D., 1983: Tropospheric ozone, its changes and possible radiative effects. WMO Special Environmental Report No. 16. WMO Technical Conference on Observation and Measurement of Atmospheric Contaminants (TECOMAC), Vienna, 17-21 October 1983.

Feister, U., Grasnick, K.-H., 1989: Solar UV radiation measurement at Potsdam (52^o22'N, 13^o5'E), (in preparation).

Feister, U., Warmbt, W., 1985: Long-term surface ozone increase at Arkona (54.68°N, 13.43°E). Atmospheric Ozone, Reidel Publishing Company. 1985, 782-787.

Feister, U. Warmbt, W., 1987: Long-term measurements of surface ozone in the German Democratic Republic. J. Atmosph. Chemistry, 5, 1-21.

Frenkiel, F.N., 1959: Tropospheric ozone. IUGG Symposium on Atmospheric Ozone, Oxford, 20-25 July 1959. IUGG Monograph No. 3, 33-36.

Kolbig, J., Warmbt, W., 1978: Messungen des bodennahen Ozone in Mirny - Antarktika. Zeitschr. Meteorol., 28, 274-277.

Oltmans, S.J., Komhyr, W.D., 1976: Surface ozone in Antarctica. J. Geophys. Res. 81, 5359-5364.

Orvig, S., 1970 Ed.: World Survey of Climatology, vol. 14. Climate of the Polar Regions. Elsevier Publishing Company, Amsterdam - London - New York.

Reiter, R., Sladkovic, R., Kanter, H.-J., 1987: Concentration of trace gases in the lower troposphere, simultaneously recorded at neighbouring mountain stations: Part II: Ozone. Meteorol. Atmosph. Phys., 37, 27-47.

Taubenheim, J., 1969: Statistische Auswertungen geophysikalischer und meteorologischer Daten. Akadem. Verlagsges. Leipzig 1969.

Tichy, H., 1939: Gleichzeitige Messungen von Ultraviolett und bodennahem Ozon. Der Balneologe. 6. Jg., 125-130.

Wexler, H., Moreland, W.B., Weyant, W.S., 1960: A preliminary report on ozone observations at Little America, Antarctica. Monthly Weather Rev., 88, 43-54.

Wisse, J.A., Meerburg, A.J., 1969: Ozone observations at Base King Baudoin in 1965 and 1966. Arch. Met. Geophys. Biokl., Ser. A, 18, 41-54.

WMO 1977: Global Ozone Research and Monitoring Project.
Report of the meeting of experts on UV-B monitoring and research. Geneva, 16-20 May 1977, Report No. 3.

CORRELATION OF OZONE AND SOLAR RADIATION AT MONTSOURIS AND HOHENPEISSENBERG: INDICATIONS FOR PHOTOCHEMICAL INFLUENCE

Andreas Volz, Heiner Geiss, Stuart McKeen, and Dieter Kley
Institut für Chemie 2: Chemie der Belasteten Atmosphäre
Kernforschungsanlage Jülich
D–5170 Jülich, FRG

ABSTRACT

At the end of last century, ozone concentrations at Montsouris, Paris, showed no variation with time of year, except for a slight maximum in spring possibly due to enhanced downward flux from the stratosphere. Today, a strong seasonal component with a summer maximum is seen in most data sets from continental stations in the northern hemisphere. An analysis of the measurements made since 1971 at Hohenpeißenberg yields a clear correlation between O_3 and sunshine on time scales below, at, and above one year.

A more detailed analysis of the typical meteorological patterns observed at Hohenpeißenberg demonstrates the local nature of this phenomenon: The sunny periods are usually associated with easterly winds advecting polluted air from Munich. Photochemical production of O_3 is observed during these periods. They are followed by the passage of a front, associated with both cloudiness and much lower O_3 concentrations, but also with a change in wind direction to NW. The extremely low values of $\lesssim$ 10 ppb observed during these events suggest chemical destruction of O_3 (and NO_x) in the air travelling behind the cold front, possibly due to NO_3 and N_2O_5 formation at night and heterogeneous removal.

1. INTRODUCTION

There has been an extensive discussion about the sources of O_3 in the troposphere, particularly with regard to the relative contributions of flux from the stratosphere and photochemical in–situ production (Fishman and Crutzen, 1978; Liu et al., 1980; Logan, 1985, Levy et al., 1985). While it is quite clear, that the budget of O_3 in polluted air masses is dominated by photochemistry, only little is known about the O_3 budget in the free troposphere and in rural areas (Vukovich et al., 1985; Parrish et al., 1986; Liu et al., 1987). It has been argued that O_3 levels over the continents of the northern hemisphere have more than doubled since the beginning of industrialization (Volz and Kley, 1988; Crutzen, 1988) and increased during the past decades at a rate of approximately 0.3 ppb/yr (Attmannspacher et al., 1984; Feister and Warmbt, 1987; Bojkov, 1988; Kley et al., 1988).

The marked change in the seasonal cycle of O_3 as suggested from the modern and the few historic records also points towards photochemical production being the major source of tropospheric O_3 today (Attmannspacher et al., 1984; Logan, 1985; Volz and Kley, 1988), however, it is not clear wether the observations reflect changes in the whole northern hemisphere or are representative of areas in the vicinity of pollution sources only.

In this paper, we compare the data gathered last century at the Observatoire de Montsouris (in the following labeled **M**) with the series of ground–based measurements made since 1971 at the Hohenpeißenberg Observatorium (labeled **HPB**) with regard to solar radiation correlations. We also show evidence for O_3 at **HPB** being influenced by the Greater Munich area.

2. THE DATA

The O_3 measurements made at **M** have been analyzed for interferences by SO_2 from the nearby city of Paris (Volz and Kley, 1988). It was concluded that only those measurements made during meteorological situations with south westerly winds are free of such interferences. We have used here the daily O_3 concentrations as selected by Volz and Kley together with the concurrent daily averages of the irradiance measured at **M** (Bull. Hebd. Stat. Municip., 1876–1886).

For **HPB**, we used the hourly mean O_3 concentrations (Sonderbeobachtungen des Observatoriums Hohenpeißenberg, 1971–1984; Attmannspacher et al.,1984). Since irradiance measurements at **HPB** were not available to us, we used the cumulative hours of sunshine per month at that site (European Meteorological Bulletin). In general, these two different quantities exhibit a similar seasonal behaviour, with the irradiance having a somewhat larger amplitude due to the dependence on solar zenith angle.

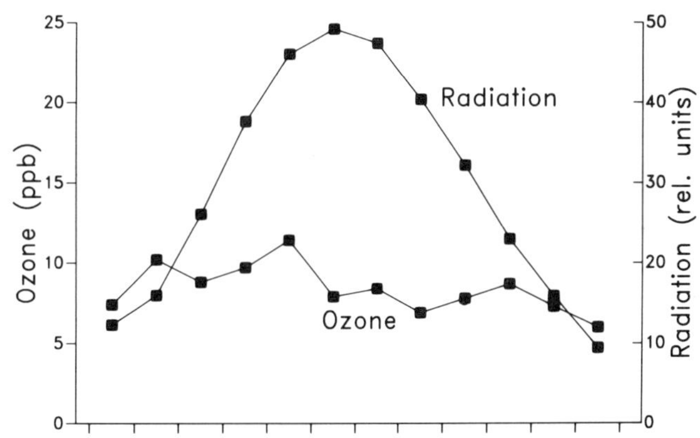

Fig. 1. Average seasonal variation of O_3 and solar radiation at Montsouris, 1876–1886. Data from Bull. Hebd. Stat. Munic. Paris, SW-sector only (c.f. Volz and Kley, 1988).

The average seasonal cycle of solar radiation and O_3 for the periods 1876-1886 and 1971–1984, respectively, is reproduced in figs. 1 and 2. While the irradiance at **M** and the cumulative hours of sunshine (CHS) at **HPB** are quite similar, the seasonal cycle of O_3 is quite different. At **M**, the concentrations are rather constant around 8 ppb throughout the year. At **HPB**, the O_3 concentrations are, on the average, a factor of 3–4 greater than

OZONE IN THE ATMOSPHERE
Rumen D. Bojkov and Peter Fabian (Eds.)

ISBN 0–937194–15–8

they were at **M** and show a pronounced summer maximum. The amplitude in the annual cycle of O_3 at **HPB** is about 20 ppb.

A statistical analysis performed with the daily O_3 concentrations and the corresponding irradiance at **M** gave a small and statistically nonsignificant correlation coefficient of 0.3. The same result was obtained for the correlation of monthly averages, which is in line with the absence of a summer maximum at **M**.

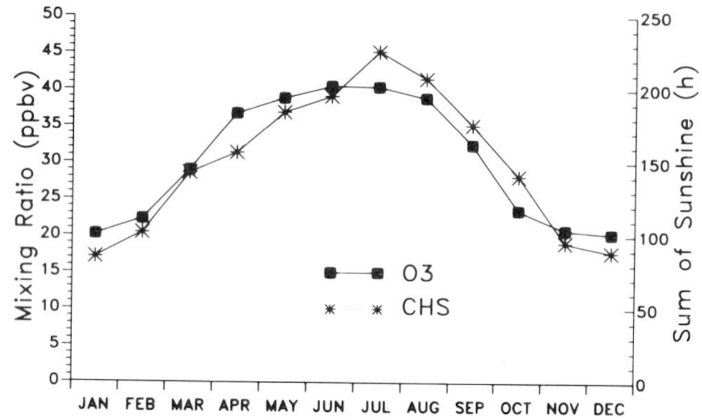

Fig. 2. Average seasonal variation of O_3 and average monthly sum of sunshine at Hohenpeißenberg, 1971–1984. Data from German Wether Service.

The analysis was then performed with the monthly mean O_3 concentrations and CHS at **HPB**. Figure 3 gives the correlation diagram for the individual monthly averages. The correlation coefficients for the individual years were large, between 0.67 and 0.96, and the slope of the linear regression ranged from 0.1 to 0.2 ppb/CHS. The solid line is a linear regression to the entire 14 year period.

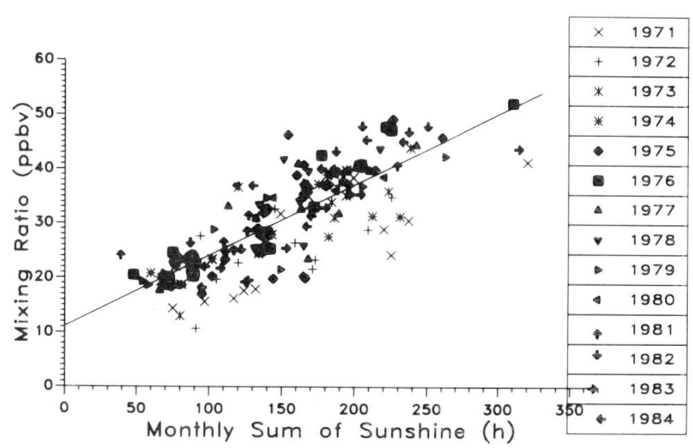

Fig. 3. Correlation between monthly averages of O_3 and monthly sum of sunshine at HPB.

Figure 4 gives the time series of monthly O_3 and sunshine for **HPB**. Again, the seasonal cycle is the most prominent feature. At a closer look, correlations on time scales below and above one year are evident. In particular, the month to month variation seen in the sunshine hours is also reflected in the O_3 concentrations. This is emphazised in fig. 5 for the year 1979. The two sunshine maxima in May and September are clearly reflected in the O_3 concentrations.

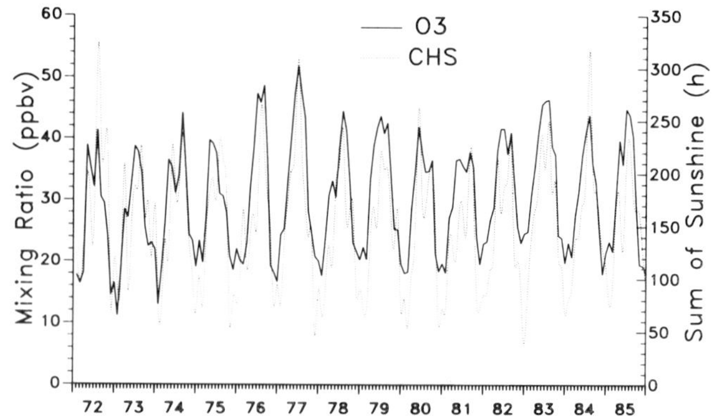

Fig. 4. Monthly O_3 concentrations and monthly sum of sunshine at HPB for the years 1971–1984.

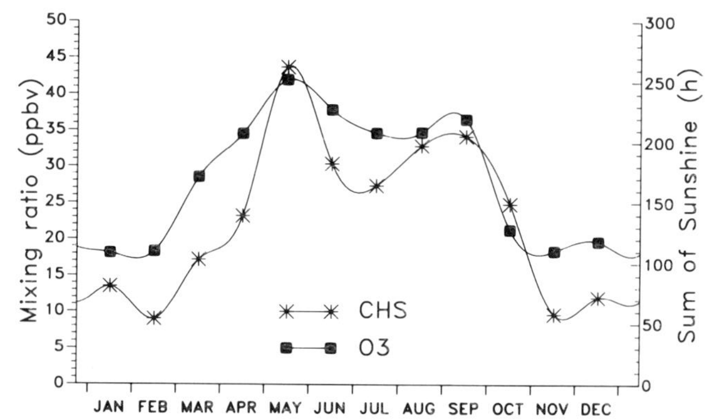

Fig. 5. O_3 and sunshine at HPB during 1979.

3. DISCUSSION

The relation between O_3 and NO_x has been subject of a number of in–situ observations and model calculations (Kelly et al., 1984; Parrish et al., 1986; Volz et al., 1988; Liu et al., 1987). In the absence of NO_x in the troposphere, photochemistry serves to destroy O_3 via reactions (R1–R4). At large NO_x concentrations, O_3 is formed via reactions (R5–R6). The efficiency of the production depends on the concentration of natural and anthropogenic hydrocarbons (Liu et al., 1987; McKeen et al., this volume).

O_3	$+ h\nu$	$\longrightarrow$	$O(^1D)$	$+ O_2$	R1
$O(^1D)$	$+ H_2O$	$\longrightarrow$	2 HO		R2
HO + CO + O_2		$\longrightarrow$	HO_2	$+ CO_2$	R3a
HO + RH + O_2		$\longrightarrow$	RO_2	$+ H_2O$	R3a
HO_2	$+ O_3$	$\longrightarrow$	HO	$+ O_2$	R4a
RO_2	$+ O_3$	$\longrightarrow$	HO	$+ O_2$	R4b
HO_2	+ NO	$\longrightarrow$	HO	$+ NO_2$	R5a
RO_2	+ NO	$\longrightarrow$	RO	$+ NO_2$	R5b
$NO_2 + h\nu + O_2$		$\longrightarrow$	NO	$+ O_3$	R6

As a consequence, enhanced radiation causes lower O_3 concentrations in the absence of NO_x, while it produces higher O_3 concentrations in the presence of NO_x. As was derived by means of a simple photochemical model described later in the text, the effect is mainly due to the increased NO/NO_2 ratio at higher radiation intensities (R6). At NO_x mixing ratios around 0.1 ppb, O_3 concentrations are fairly insensitive to changes in the radiation.

The absence of a seasonal cycle in the O_3 data and the absence of a correlation with solar radiation at **M** suggest that NO_x concentrations in continental air during that period were around 0.1 ppb . The strong correlation between sunshine and O_3 found at **HPB**, on the other hand, is consistent with much higher concentrations of NO_x around 1 ppb, as they are indeed observed in rural areas of Germany today (Monatsberichte des Umweltbundesamtes).

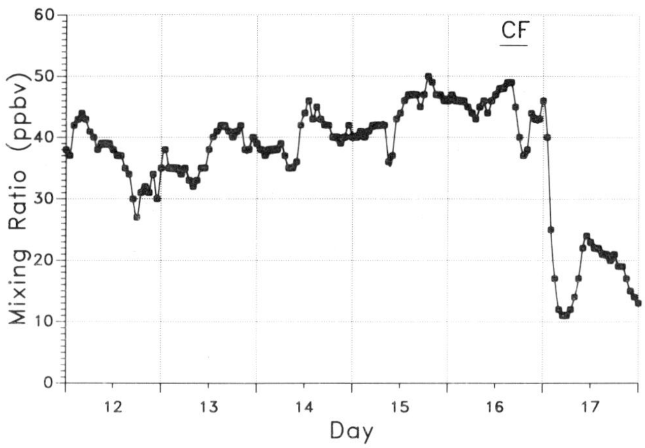

Fig. 6. Hourly O_3 mixing ratios at HPB during a typical episode in 1979. High pressure persisted until April,16. CF denotes passage of the cold front at HPB.

In order to investigate the spatial extent of the photochemical O_3 production implied by the **HPB** data, we have investigated the day to day variations of O_3 at **HPB** for accompanying changes in the meteorological situation. Guided by fourier analysis, a fairly regular pattern with time scales on the order of 5–10 days was detected in both O_3 and sunshine. Sunny periods with higher O_3 levels were frequently followed by a sudden drop in O_3 associated with cloudiness. A typical example is shown in fig. 6. Between April 12 and 16, 1979, ozone increased slowly for 4 days and then suddenly dropped to 10 ppb within a few hours. Similar patterns were observed to occur during all years, mainly in spring, summer and fall. High and low O_3 concentrations were usually correlated, respectively, with sunshine and cloudiness.

The synoptic situation for the example in fig. 6 was as follows: a high pressure system resided over Europe during April 12–16. At **HPB**, winds were from the east at 1–4 m/s, advecting polluted air from the Greater Munich area to **HPB**. Photochemical O_3 production is indicated by both the general increase as well as the diurnal cyle during that period. The sudden drop on April 16 is accompanied by the passage of a cold front associated with clouds and, most important, a change in wind direction to NW with velocities around 4–5 m/s. As shown in fig. 7, this cold front had originated around Scotland on April 15 and had moved over England, The Netherlands and West Germany to **HPB** within less than 2 days.

We have used a box model (height: 1 km, chemical scheme from Liu et al., 1987) to simulate the gross features seen in the O_3 concentrations at **HPB** on occasions as is shown in fig. 6. Three different fluxes of NO_x

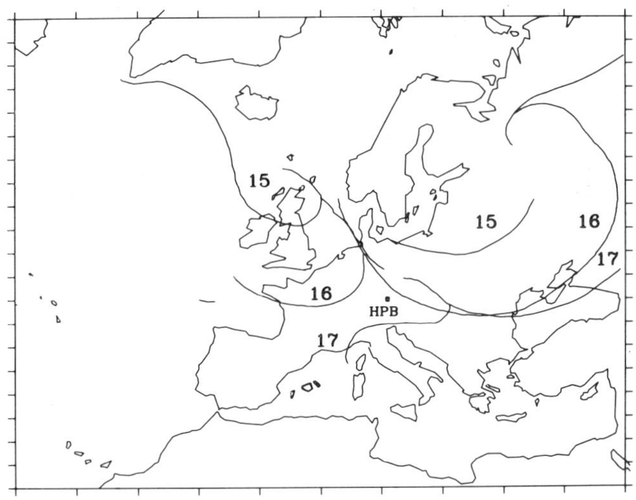

Fig. 7. Map of Europe with the positions of the cold front on April 15, 16, and 17 at noon.

in the range $3–30 \cdot 10^{10}$ molec./cm^{2}s were used. The results are given in fig. 8. For sunny conditions, e.g. the situation during April 12–16, the model simulates the increase in O_3 and the diurnal cycle quite well. The best agreement was obtained with NO_x fluxes around $1 \cdot 10^{11}$ molec./cm^{2}s.

The simulations were repeated for cloudy conditions, as they are representative for the air mass travelling with the cold front on its way from Scotland to **HPB**. Two different cases were investigated: a slow and a fast heterogeneous removal rate for water soluble species, like H_2O_2, N_2O_5, NO_3, and HNO_3. The slow removal rate (5 day lifetime) is typical for average conditions used in chemical models (Logan et al., 1981), while the faster

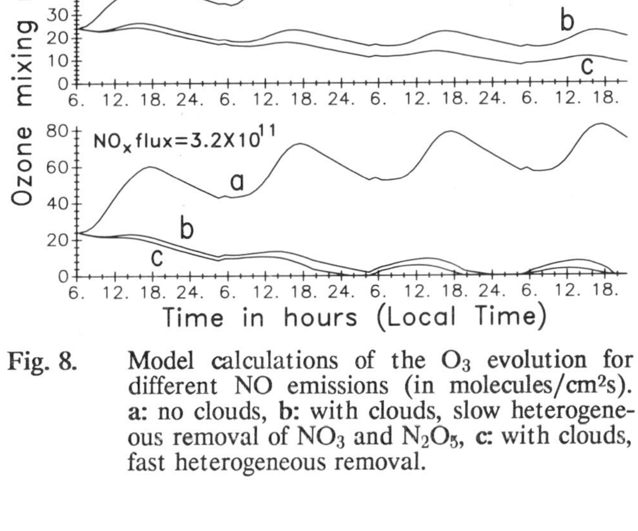

Fig. 8. Model calculations of the O_3 evolution for different NO emissions (in molecules/cm^{2}s). **a:** no clouds, **b:** with clouds, slow heterogeneous removal of NO_3 and N_2O_5, **c:** with clouds, fast heterogeneous removal.

rate (1 hr lifetime) applies to conditions during a moderate shower (Levine and Schwartz, 1982). The optical depth of the cloud was adjusted so that the daily averaged photolysis rate $J(O^1D)$ was reduced by 80% (Chang et al., 1987). The model simulates the following picture: As the air mass travels across Europe, it is exposed to emissions of NO from anthropogenic sources. O_3 is removed by dry deposition. In addition, reactions R7–R9 followed by heterogeneous removal of the water soluble species NO_3 and N_2O_5 serve to remove 1.5–2 O_3 molecules per NO molecule oxidized.

$$O_3 + NO \longrightarrow NO_2 + O_2 \qquad R7$$
$$O_3 + NO_2 \longrightarrow NO_3 + O_2 \qquad R8$$
$$NO_2 + NO_3 \longleftrightarrow N_2O_5 \qquad R9$$
$$NO_3 \text{ (het.)} \longrightarrow NO_3^- \qquad R10$$
$$N_2O_5 \text{ (het.)} \longrightarrow 2\,NO_3^- \qquad R10$$

As is seen in fig. 8, O_3 decreases under cloudy conditions. Only little is reformed during daytime because of the cloud cover. Including the fast heterogeneous removal further reduces O_3. The nonlinear behaviour of the chemical system with respect to NO_x can also be inferred from fig. 8. At the lower NO flux, the system is NO_x limited and the inclusion of a fast heterogeneous removal has only little effect. At the medium NO flux, the heterogeneous removal is most efficient, while the system starts to be O_3 limited at the highest NO flux due to titration (R7). With fluxes around $1–2\cdot10^{11}$ molec./cm^{2}s, as they are representative for Europe (McKeen et al., *this volume*), the O_3 concentration drops from initially 25 ppb to 12 ppb or less during the 2 days before it reaches **HPB**.

It can be concluded that both the high and the low concentrations of O_3 observed at **HPB** can be explained with similar NO_x emissions if the changes in the meteorological situation are taken into account. Photochemical production is effective during the persistence of the high pressure system. The low concentrations suggest chemical destruction of O_3 by NO_x to occur in the air mass passing **HPB** with the cold front. It is also clear that the high concentrations at **HPB** cannot be considered representative of Europe, as they are usually associated with easterly winds and thus may only reflect the situation in the Greater Munich area.

4. CONCLUSIONS

The absence of a summer maximum in the O_3 concentrations at Montsouris 100 years ago and the absence of a correlation with solar radiation can be explained by much lower NO_x emissions at that time. The summer maximum observed at Hohenpeißenberg today as well as the good correlation with sunshine, on the other hand, are in line with much higher NO_x emissions, most likely because of anthropogenic activities.

The good correlation with sunshine is, at least partially, due to concurrent changes in the meteorological situation. The results indicate that the high O_3 concentrations found at **HPB** are strongly influenced by the vicinity of the Greater Munich area and thus do not necessarily reflect a change of O_3 in the whole northern hemisphere. A network of stations distributed over Europe in combination with vertical soundings must be used in conjunction with mesoscale models in order to better understand the formation, destruction and transport of O_3 on a European scale.

As important are the extremely low concentrations observed at **HPB** upon passage of a front. Night time chemistry and heterogeneous processes can explain the low O_3 levels in this air mass after being exposed to NO emissions for 2 days under cloudy conditions. The situation of a moving front crossing Europe provides an excellent opportunity for aircraft experiments designed to studying the balance of O_3 in polluted air masses and its relation to NO_x.

ACKNOWLEDGMENTS

We thank K. Wege and R. Hartmannsgruber of the Hohenpeißenberg Observatorium, who made the data tapes with the hourly O_3 concentrations available to us.

REFERENCES

Attmannspacher, W., 1971–1984: *Sonderbeobachtungen des Meteorologischen Observatoriums Hohenpeißenberg,* (Deutscher Wetterdienst, Hohenpeißenberg, FRG)

Attmannspacher, W., Hartmannsgruber, R. & Lang, P., 1984: *Meteorol. Rdsch.* 37, 193–199.

Bojkov, R.D., 1986: *J. Clim. Appl. Met.* **25**, 343–352.

Bojkov, R.D., 1988, in: Isaksen, I.S.A. (ed.) *Tropospheric Ozone, Regional and Global Scale Interactions, NATO ASI Series* **C 227**, 83–96 (Reidel, Dordrecht, NL)

Bull. Hebdomadaire Statistique Municicipal, 1876–1886 (Paris, France)

Europ. Meteorol. Bull., 1971–1984

Crutzen, P.J., 1988 in: Isaksen, I.S.A. (ed.) *Tropospheric Ozone, Regional and Global Scale Interactions, NATO ASI Series* **C 227**, 3–32 (Reidel, Dordrecht, NL)

Chang, J.S., et al., 1987: *J. Geophys. Res.* **92**, 14,681–14,700.

Feister, U. & Warmbt, W.J., 1987: *J. Atmos. Chem.* **5**, 1–21.

Fishman, J. & Crutzen, P., 1978: *Nature* **274**, 855–858.

Kley, D., et al., 1988 in: Isaksen, I.S.A. (ed.) *Tropospheric Ozone, Regional and Global Scale Interactions, NATO ASI Series* **C 227**, 67–72 (Reidel, Dordrecht, NL)

Levine, S.Z. & Schwartz, S.E., 1982: *Atmos. Env.* **16**, 1725–1734

Liu, S.C., et al., 1987: *J.Geophys. Res.* **92**, 4191–4207.

Liu, S.C., et al., 1983: *J.Geophys. Res.* **88**, 1360–1368.

Liu, S.C., 1988 in: Isaksen, I.S.A. (ed.) *Tropospheric Ozone, Regional and Global Scale Interactions, NATO ASI Series* **C 227**, 303–318 (Reidel, Dordrecht, NL)

Logan, J.A., et al., 1981: *J.Geophys. Res.* **86**, 7210–7254.

Logan, J.A., 1985: *J.Geophys. Res.* **90**, 10463–10482.

Monatsberichte aus dem Messnetz, 1984–1987 (Umweltbundesamt, Berlin, FRG)

Parrish, D.D. et al., 1986: *J. Atmos. Chem.* **4**, 63–80.

Volz, A. & Kley, D., 1988: *Nature* **332**, 240–242.

Volz, A. et al. 1988 in: Isaksen, I.S.A. (ed.) *Tropospheric Ozone, Regional and Global Scale Interactions, NATO ASI Series* **C 227**, 293–302 (Reidel, Dordrecht, NL)

Vukovich, F.M., Fishman, J. & Browell, E.V., 1985: *J.Geophys. Res.* **90**, 5687–5698.

RELATIONSHIPS BETWEEN TROPOSPHERIC OZONE AND GLOBAL SOLAR RADIATION IN GERMANY

Manfred Schmidt
Max-Planck-Institut für Aeronomie (MPAE)
D–3411 Katlenburg-Lindau, FRG

ABSTRACT

Theories on the origin of tropospheric ozone, published in recent years, show that this ozone mainly is created by UV radiation < 310 nm. These UV wavelengths (UV-B) are a component of the global solar radiation, especially in summer and in higher altitudes (mountains). The annual variation of the total UV-B runs parallel to that of the global radiation. In the last ~ 15 years, ozone stations in Germany recorded increased concentrations of ozone in summers with increased solar radiation, i.e. in dry and sunny summers. The paper shows the average relationships between ozone mixing ratios and the global radiation in summer for several stations situated in different altitudes between the coast of the North Sea and the peaks of the Alps, including rural and industrial areas. Particularly the following details are discussed: time series of both parameters, altitude and North-South-dependences in Germany, and correlation coefficients between ozone and radiation. From the ozone soundings of the Observatory Hohenpeißenberg, we have calculated mean ozone profiles up to the lower stratosphere for days with high and low global radiation in the summer months. The results yield prominent and significant differences, which reflect the mean altitude region and the intensity of ozone production, depending on the altitude, during times of strong radiation. The correlation between the ozone concentration and global solar radiation is high on the summit of Hohenpeißenberg (988 m a.s.l.), at similar mountain stations and at places with extensive industry (Mannheim), low in Hamburg and other places in the North German lowlands. There, it seems, other processes (transports, destruction of ozone etc.) are more significant. Since ozone is an important pollutant with respect to the alarming forest decline in Europe, especially for forests in higher altitudes, i.e. hills and mountains, the results are important for these problems too.

1. INTRODUCTION

In the recent years we have made some investigations about the possible influences of dry, sunny and warm summers on the one hand, and overcasted and wet summers on the other hand on the forest disease in Central Europe. The summers 1976, 1982 and 1983 in Germany have been some of the dryest at all and -vice versa- the summers 1979 to 1981 were summers with overcasted sky during long times, these of 1980 and 1981 have been very wet. Ozone, likely produced in large quantities by radiation in sunny summers, is one of the pollutants, suspected as a poison-gas with respect to the forest decline. For that reason we have extended our investigations to the ozone content of the air too.

2. RESULTS

Comparisons of the ozone concentration and the global solar radiation at some stations in Germany in the summers of the recent 10 to 16 years show the interesting result, that there are localities with obviously high correlations between the ozone content and the global solar radiation, and other stations with low ones.

Three examples are shown in fig. 1. On the top of Hohenpeißenberg, an isolated mountain in 990 m a.s.l. and close to the Alps, both curves run very good parallel. The same is true at the recording stations of Mannheim, ~ 100 m a.s.l. in the Rhine area of South Germany, a large city of ~ 500 000 inhabitants (with the nearby city of Ludwigshafen) and with extensive industry, mainly chemical

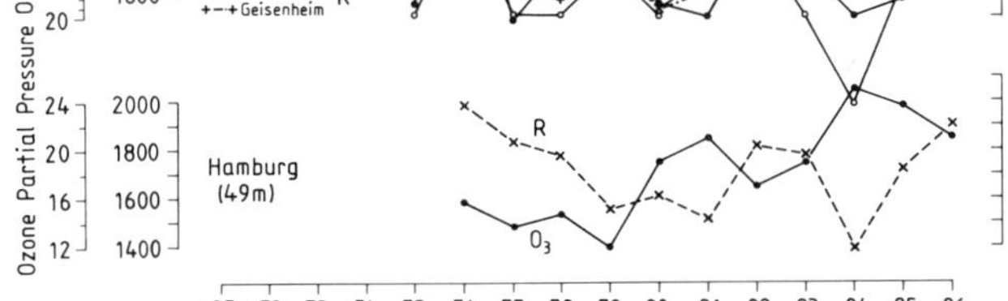

Fig. 1. Ozone concentration O_3 and global solar radiation R (average May-July) at three places in Germany

ISBN 0-937194-15-8

factories. On the other hand, the correlation is low at the Meteorological Observatory of Hamburg in the outskirts of the large city of Hamburg, ~ 50 m a.s.l. and ~ 90 km distant from the coast of the North Sea. The two curves of Hamburg in fig. 1 run nearly antiparallel. At a first sight according to fig. 1, there seem to exist dependences of both parameters on the height above sea level, South to North in Germany, and wheather the locality is in a rural or industrial area.

In a next step we have calculated correlation coefficients r between the ozone content O_3 and the global radiation R, considering the particular monthly means. The results of two summer periods are plotted in fig. 2 together with mean values of the ozone and radiation. In this figure, seven stations with 10 years of available records and one station with 8 years have been considered. Fig. 3 shows the same, but for the summer period April to September only and for an extended number of 16 stations with 6 or 7 years of records. All stations are arranged according to decreasing height above sea level, and roughly in the direction South to North in Germany. The name of a nearby station with global solar radiation records, used for the correlations, is printed in brackets. In a general view, both parameters O_3 and R decrease with decreasing height until ~ 100 m a.s.l., as it can be expected. There exists one station with unusual high O_3 (Waldhof), the reason is still unknown. The similar high O_3 mean value of Westerland (island of Sylt) is certainly due to the position of the station at the coast of the North Sea with relatively pure air, and to the higher solar radiation at the Isle of Sylt. The lowest concentrations of O_3 have been found in Mannheim, Hamburg, Quickborn and Neuglobsow, the first and second one being large cities with extended industry, the third place is in a distance of merely 20 km from Hamburg.

In both figures 2 and 3 the correlation between O_3 and R is highest at the Observatory Hohenpeißenberg. In fig. 2 with the stations of ~ 10 years records, r is similar high at the South German stations Zugspitze and Mannheim, and low at the North German lowland stations, especially at those nearby or immediately at the sea coast. But these results must be modificated because of fig. 3, with a double number of stations considered, but with shorter duration of the records. Here, r is highest at Hohenpeißenberg (r=0.65), an isolated mountain near the Alps (see also fig. 1). But the r values of the three Mannheim stations in an industrial and large city are also high with values between 0.47 and 0.57, they are similar high as on the top of Zugspitze (2962 m, r=0.52) and higher than at 3 further stations in the German high-

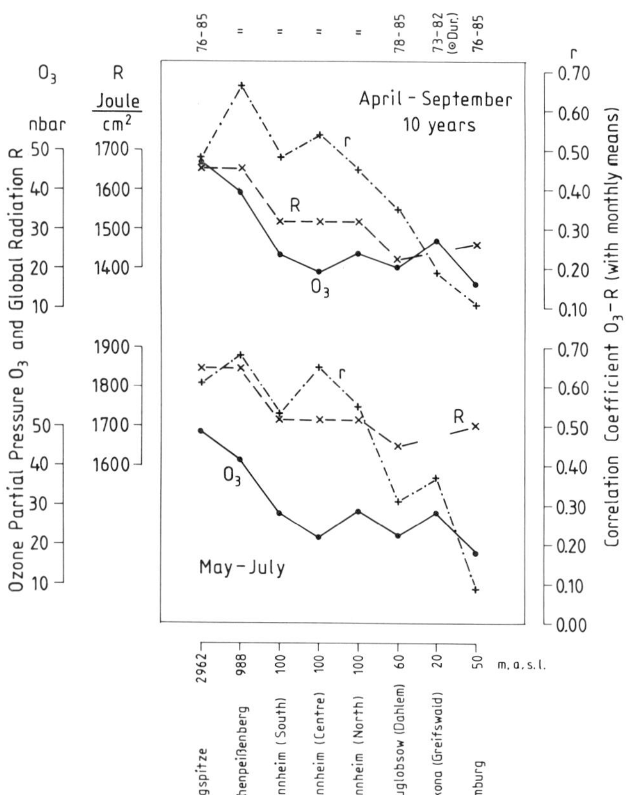

Fig. 2. Ozone concentration O_3, global solar radiation R and correlation coefficient r of both at eight stations in Germany. Two periods in the summers of ten years considered.

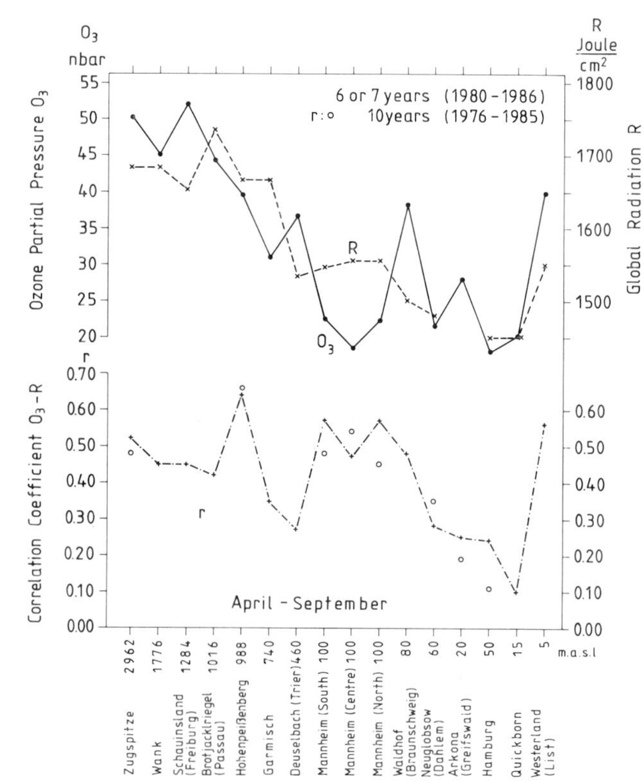

Fig. 3. Ozone concentration O_3, global solar radiation R, and correlation coefficient r of both at 16 stations in Germany. April to September of six or seven years considered, names in brackets = nearby stations with radiation measurements.

lands in $\sim$ 1000 to 1800 m a.s.l. (r $\sim$ 0.44). The very high value at Westerland is probably due to the increased values of O_3 and R. The lowest r values ≤ 0.25 have been found again at the three North German stations near the coast, with exception of Westerland.

The relatively high correlation coefficients between ozone and global radiation of r $\sim$ 0.35 to 0.65 at a number of these stations (South German stations situated higher than 700 m a.s.l., Mannheim, Westerland/Sylt) point to the fact, that ozone will be produced in considerable amounts by the shortwave part of solar radiation < 310 nm, especially during the summer months. Before we will discuss the circumstances and the different efficiency of this ozone production, we will show some more facts.

Fig. 4 presents results of evaluations of ozonesonde measurements above Hohenpeißenberg/Bavaria with respect to high and low global solar radiation. We have considered all ascents during the summer season April to August, if the global solar radiation on the day before and the day of the ascent had exceeded values of $R \geq 2000$ Joule·cm^{-2} (fine days) on the one hand, and $R \leq 1000$ Joule·cm^{-2} (overcasted days) on the other hand. For all days of these conditions we have calculated mean values of the ozone concentrations at standard pressure levels, and the results are plotted in diagrams like fig. 4. We have calculated the mean values for the single summers of all years between 1976 and 1985, and the 10-year averages for all single months April to August. Fig. 4 shows some examples, 1976 and 1983 have been sunny and dry summers, 1979 and 1980 overcasted and wet summers. Generally, the diagrams show a similar picture. The ozone

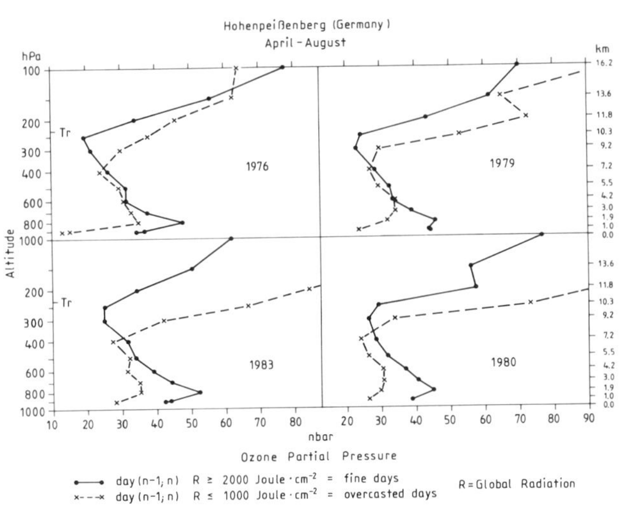

Fig. 4. Average vertical distributions of ozone concentration in the period April-August calculated from the soundings of Hohenpeißenberg, examples of four years. Selections: ——— fine days only, the day before and the day of the ascent with $R \geq 2000$ Joule·cm^{-2}, - - - overcasted days, both days with $R \leq 1000$ Joule·cm^{-2}.

concentrations in the surface layer up to $\sim$ 2 km height at the sunny days are by a factor of 1.5 to 2 higher than at the overcasted days. This maximum difference in $\sim$ 2 km height decreases with altitude. In most of these cases this difference obtains 0 at $\sim$ 400 hPa or $\sim$ 7 km height. Above of this altitude the ozone partial pressures at the fine days are lower than these of the overcasted days. If the assumption of an ozone production by the UV part of the global radiation is correct, it can be concluded from this and similar figures, that the O_3 production area would extend in the mean from the surface up to $\sim$ 7 km height in the troposphere, with a maximum in $\sim$ 2 km, and decreasing efficiency between 2 and 7 km.

3. DISCUSSIONS

In all diagrams of these Hohenpeißenberg evaluations, the maximum of ozone at fine days is in the level of 800 hPa $\hat{=}$ 1.9 km, the concentrations increases from the surface in 988 m a.s.l. to this altitude. The same is valid for the overcasted days, in these cases the much lower maxima vary between 1.9 and 4 km in most of the cases. This could be explained with the average height of the mixing layer. According to Schneider-Carius [1953], the average top heights of the different types of mixing layers are found between 1.1 and 1.4 km above the surface, with temperature inversions, characterizing the top of these layers. This is in agreement with maps of the average mixing heights in the morning and afternoon in the USA, published by G. Holzworth [1972]. The mixing layer normally is enriched with haze, aerosols and other tracers, able to destruct ozone. The height of the ozone maximum in fig. 4 is in the average well and immediately above the top of the mixing layer and this is probably the reason for the maximum.

These facts are certainly one of the reasons for the decreasing average ozone content with decreasing altitude down to $\sim$ 100 m a.s.l. in figs. 2 and 3. But we have to consider the radiation and the concentrations of precursors too. These are air pollutants: nitrogen oxides, hydrocarbons and carbon monoxide, produced by human activity. According to figs. 2 and 3 the global radiation increases from North to South in Germany, but this could be in a certain respect due to the election of the stations. If we inspect maps of the average daily sunshine duration in summer in Germany, measured at much more stations than the radiation, we find an increase of this duration from lowest values in Middle and Northwest Germany in both directions, to the North and to the South. The highest values are at the coast of the North Sea (Isle of Sylt) and the Baltic Sea (Isle of Fehmarn) in the North direction, and in Bavaria and Württemberg in the South direction.

According to fig. 3, the global radiation is highest in the summer months in altitudes higher than $\sim$ 700

m. The UV-B radiation, important for the formation of ozone, is a small part of the global radiation, and increases considerably with height. Measurements of the UV radiation, made by R. Reiter et al. [1972] during the years 1964 to 1971 demonstrate this increase clearly. The UV intensity in the summer seasons May - July and April - September has been in the average at Garmisch (740 m a.s.l.) 210 (190) units, at the peak of Wank (1776 m) 325 (290) units and on the top of Zugspitze (2962 m) 615 (530) units. On the other hand, the concentrations of the above mentioned gaseous precursors decrease with altitude. Fig. 5 shows an example, with average values of measured NO_2 concentrations in summer 1985, together with the simultaneously measured global radiation R, and with the UV-B values of Garmisch and Wank, extrapolated to the further altitudes. The NO_2 values of the two large cities Hamburg (33 $\mu g \cdot m^{-3}$) and Mannheim are very high. Considering the increase of the UV radiation and the decrease of the precursors (e.g. NO_2) with height, there must be an optimum of ozone formation in rural areas, like in South Germany far of large cities and industrial areas. This presumed optimum probably is the reason for the high correlation coefficient between O_3 and R at the top of Hohenpeißenberg (figs. 2 and 3) and at stations in similar altitudes. The relatively high ozone concentration and correlation coefficient at the peak of Zugspitze (figs. 2 and 3) could be due to the very high UV radiation intensity there, nevertheless the concentrations of the precursors are already low. The reasons for the high correlation coefficients of the Mannheim stations probably are the high concentrations of gaseous precursors in that area with extensive chemical industry, in connection with

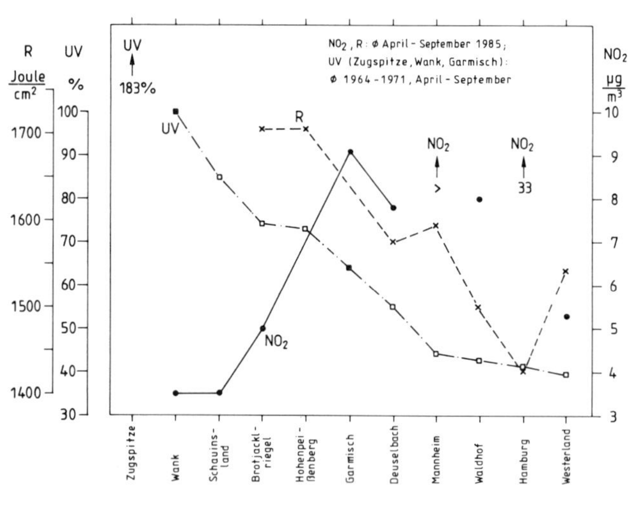

Fig. 5. UV-B values (average April-September, Wank = 100%), global radiation R and NO_2 concentration (averages April-September 1985) at several German stations. UV-B: ■ = measured, □ = extrapolated values. R partly from a nearby station.

still sufficient high intensity of radiation, fig. 3. In the large city of Hamburg with likewise high concentrations of precursors, caused by extended industry and traffic, the correlation coefficients between ozone and global radiation are low. There both parameters, the ozone concentration and the global radiation, are low in the average. This could be a reason for the low correlation coefficients. On the other hand it should be considered, that the average wind velocity increases from South to North in Central Europe. The wind velocity is lowest at surface stations in South Germany and highest at the coast areas of the North Sea and Baltic Sea. That means, the ventilation there very often removes and dilutes the produced gaseous precursors. These two factors, the low intensity of global and (within it) UV-radiation, and the much higher ventilation, compared with the conditions in South Germany, could be the reason of the low ozone concentrations and the low correlation coefficients between O_3 and R at the stations in North Germany considered. Single exceptions (e.g. Waldhof) are perhaps due to local conditions and have to be investigated separately. The same holds for the station Westerland, situated immediately at the seashore, but there the global radiation is much higher than in most of the other North German lowlands.

REFERENCES

Bayerisches Landesamt für Umweltschutz: Lufthygienischer Monatsbericht. Several annual reports

Deutscher Wetterdienst, Meteorologisches Observatorium Hamburg: Ergebnisse von Strahlungsmessungen in der Bundesrepublik Deutschland sowie von speziellen Meßreihen am Meteorologischen Observatorium Hamburg. Several annual reports.

Environment Canada, Atmospheric Environment Service: Ozone Data for the World. Several Volumes

Feister, U. and W. Warmbt: Long-term surface ozone increase at Arkona. In: Atmospheric Ozone, Reidel Dordrecht, 782-787, 1985

Holzworth, G.: Mixing heights, wind speeds, and pollution throughout the continguous United States. Environmental Protection Agency, Research Triangle Park, North Carolina, Jan. 1972

Landesanstalt für Umweltschutz Baden-Württemberg: Ozondaten der Mannheimer Stationen 1975-1986

Munzert, K., R. Reiter, H.-J. Kanter and K. Pötzl: Effect of stratospheric intrusions on the tropospheric ozone. In: Atmospheric Ozone, Reidel Dordrecht, 735-739, 1985

Reiter, R., W. Carnuth und R. Sladkovic: Ultraviolettstrahlung in alpinen Höhenlagen, Wetter und Leben **24**, 231-247, 1972

Schneider-Carius, K.: Die Grundschicht der Troposphäre. Akad. Verlags-Ges. Geest & Portig KG, Leipzig 1953

Umweltbundesamt Berlin: Monatsberichte aus dem Meßnetz. Several annual reports

ON THE EXTENT OF PHOTOCHEMICAL OZONE PRODUCTION OVER THE UK

K.J. Weston and P.J.A. Kay
Department of Meteorology
Edinburgh University, EH9 3JZ, UK

D. Fowler
Institute of Terrestrial Ecology
Bush Estate, Midlothian EH26 OQB, UK

A. Martin
Scientific Services Department, CEGB
Nottingham, NG11, OEE, UK

J.S. Bower
Warren Spring Laboratory, Stevenage
Hertfordshire, SG1 2BX, UK

ABSTRACT

During late April 1987 an ozone episode occurred during which mid-afternoon hourly ozone concentrations reached between 80 and 100 ppbV at many sites in the UK during a 5-day period. Such an occurrence is exceptional for April. Data from the recently-established UK network of ozone-monitoring stations is used to estimate ozone concentrations in the mixed layer of the lower troposphere as air moves across the UK. Detailed analyses of individual episode days has enabled estimates to be made of the contribution of 'local' photochemical ozone production to the observed concentrations at various UK sites.

For the period 24-30 April 1987 air trajectories in the lower troposphere were calculated from objectively-analysed, 6-hourly wind observations made at routine upper-air stations in Britain, Ireland, N. France, Belgium and the Netherlands. Ozone concentrations measured near the surface are representative of the deep daytime boundary layer during periods of active convection, usually late morning and early afternoon. Ozone budgets were derived by following air columns between such well-mixed phases on consecutive days along trajectories, taking into account the variation of trajectories with height. Surface deposition was estimated using a diurnally-varying deposition velocity and observed surface ozone concentrations along the trajectory. Net photochemical production was deduced as a residual, knowing ozone concentrations at the end points of the trajectory.

The results from 10 budget calculations show a clear relationship between the apparent generation of ozone in the mixed layer and the sunshine experienced along the respective trajectory. A net ozone generation of about 4 ppbV per hour of bright sunshine is indicated, implying maximum daily generations of about 45 ppbV in April and about 60 ppbV in mid-summer.

1. INTRODUCTION

The availability of data from the new UK ozone monitoring network, and from several other sites where routine ozone measurements are made, presents new opportunities for studies of ozone episodes over the UK. In late April 1987, an episode occurred during which mid afternoon measurements of hourly average ozone concentrations ranged from 80 to 100ppbV at many UK sites. The data allow concentrations of ozone in the mixing layer to be observed as the air moved between the East Midlands of England and Central Scotland with winds generally from the south-east under the influence of an anticyclone over Eastern Europe. This flow of warm, dry continental air gave remarkably hazy conditions over most of the UK, particularly during the period 25-28 April, when day-time visibilities between 2 and 5 km were widespread. Long-term April maximum temperature records were exceeded in several places.

In this contribution a detailed analysis of individua episode days during 24-30 April 1987, considering the trajectory of the air and depletion at the ground by dry deposition, has enabled estimates to be made of the contribution of 'local' photochemical ozone production to the observed concentrations at various UK sites.

2. ANALYSIS

2.1 GENERAL APPROACH

In this study photochemial ozone production is estimated from a knowledge of the ozone mass in an air column on successive afternoons and from an estimate of surface deposition during the period using observed ozone concentrations along the trajectory and an assumed value for deposition velocity. Fig 1 shows the basis adopted for the budget calculations. The air column shown at site C represents the mixed layer at the time of maximum observed ozone concentration (mid afternoon) at the site. Air forming this column will have followed a range of paths, but in the budget calculation these are represented by two trajectories. Line AC is a low-level trajectory, the 300m level is used, chosen to represent air trajectories for the lower portion of the column, and line BC represents air trajectories for the upper portion of the column, for which a level of 900m is used. A and B are points on the two trajectories at the time of maximum (hence well-mixed) ozone concentration measured on the previous afternoon. At the end and start times of these trajectories the ozone concentration measured at the surface is considered to

ISBN 0-937194-15-8

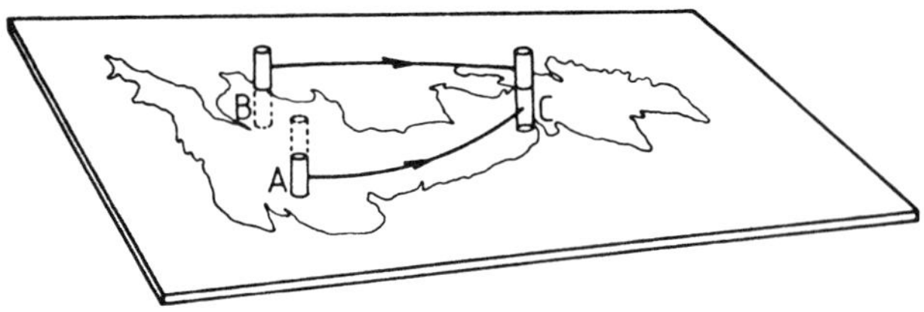

Fig. 1. Schematic view of the basis of the budget calculation. Air forming the lower part of the mixed layer at location C originates from location A, while the upper part of the mixed layer originates from B.

represent the value throughout the boundary layer since it is well-coupled by convective mixing. Thus initial and final ozone concentrations for the budget calculation can be inferred from values measured at monitoring sites.

During the episode studied here, unusually high turbidity of the lower atmosphere and relatively low sun angle (April) are likely to ensure a relatively short period of deep convection (typically 1200-1600 GMT). Thus soon after the start point of these trajectories, and until a few hours before the end point, the upper portion of the column (following track BC) is decoupled from the surface by low-level stability, so that this air does not suffer depletion due to surface deposition. Air following track AC, however, will experience surface deposition from its lower levels. In the budget calculation it is assumed that losses due to dry deposition occur only from the lower portion of the column and at a rate which can be estimated from a knowledge of the ozone concentration along the trajectory and a realistic deposition velocity.

From these components of the budget, the net ozone generation by chemical and photochemical processes during the budget period is given by: Concentration change due to net generation = final concentration − initial concentration + decrease in concentration due to surface deposition

This budget equation is expressed, for convenience, in terms of concentration (ppbV), but the budget calculation is, of course, in terms of ozone mass.

2.2 TRAJECTORIES

Wind data were used from routine upper-air stations in Britain, Ireland, N. France, Belgium and the Netherlands for each of the days during the episode at all observation times (00, 06, 12 and 18 GMT) to determine trajetories. These data were supplemented with data over a wider area of Europe for each 00 GMT data time. In analysing a field of wind observations for use in medium-scale transport studies, it is desirable to introduce some smoothing to reduce local influences, while retaining the larger-scale flow pattern. The analysis technique used here is to fit polynomial functions in two dimensions to the fields of west-east and south-north wind components by using orthogonal polynomials as base functions (Dixon et al., 1972).

In general, the larger the area to be analysed, the higher the order of the polynomials required to adequately represent the synoptic-scale flow pattern. The area considered has charateristic dimensions of 1000-1500km, for which polynomials of second or third order are sufficient to give good representations. Back trajectories were determined from the objectively-analysed wind fields by integrating backwards in time and using winds linearly interpolated between adjacent times of analysis, six hours apart

2.3 CALCULATION OF SURFACE DEPOSITION

Rates of turbulent deposition of ozone onto vegetation and soil have been reported by Galbally (1971) and by Weseley et al (1982). These measurements show that vegetation and soil are the major sink for ozone in the boundary layer except in urban areas with large nitric oxide concentration. With typical rates of deposition expressed as a deposition velocity (where deposition velocity (v_g) at height z is equal to the vertical flux divided by the concentration of ozone at height z) of about 10 mm s^{-1}, the half-life of ozone molecules in a well-mixed 1 km deep boundary layer is approximately 1 day.

For the photochemical ozone episode in April 1987 the trajectories described traverse mainly lowland, agricultural districts of Britain with some areas of moorland at higher elevations. The agricultural areas during late April comprise mainly grassland and cereal crops growing in districts which at this stage in the season have not developed water shortage and with the warm bright weather was growing rapidly. In these conditions the canopy conductance for water vapour and for ozone exchange show large diurnal cycles with a daytime maximum and a nocturnal minimum.

To parameterize the dry deposition of ozone for this study it is necessary to generalize over distance scales of 500 km and over irregular terrain. In the absence of detailed information on the surfaces along the trajectory it is only possible to justify day and night values (0600-2000 hrs, 2000-0600 hrs respectively). The values chosen to represent day and nighttime deposition velocities over land of 10 mm s^{-1} and 5 mm s^{-1} respectively are consistent with appropriate experimental data. These values are larger than some published values, Galbally (1971) and Garland and Derwent (1979) but the smaller values in the literature are associated with either slow growing, water stressed or senescent vegetation or bare soil. Other published values for rapidly growing, unstressed vegetation show day maximum deposition velocities $\geq$10 mm s^{-1}.

Over water the deposition velocity is likely to be small (Galbally and Roy, 1980), and since only some of the trajectories cross the sea, and then for relatively short periods, the value used is not critical. A value of zero is assumed.

2.4 MIXED LAYER DEPTH

Radiosonde ascents made at mid-day at the U.K. network of upper-air stations (8 stations) were used to estimate the depth of the well-mixed boundary layer. Shallow convection imposes characteristic lapse rates in the boundary layer: immediately above the surface is a shallow superadiabatic layer in which potential temperature falls with height by about 1 or 2 deg. C between screen level (1.2m) and the top of the layer; above this is an adiabatic layer in which the potential temperature is almost constant; and this is normally capped by a strongly-stable layer, separating the well-mixed layers below from the subsiding atmosphere above (Ludlam, 1966). The depths of mid-afternoon convection at the beginning and end points of trajectories were estimated using the afternoon maximum temperatures observed at Meteorological Office Stations in the areas and assuming a decline of 2 deg C between screen level and the top of the superadiabatic layer and with a dry adiabatic lapse rate above. The intersection of this profile with the environment curve provides a reliable estimate of the top of the mixed layer under anticyclonic conditions.

3. RESULTS

Ten trajectories were found to be suitable for budget calculations. The monitoring site at Bush, near Edinburgh in Eastern Scotland, was particularly well-positioned, with flow generally from the south-east and with a daily travel distance of about 400km, the Bush station lay downwind of monitoring sites in central England on four of the days. In all, ten trajectories were found to be suitable for budget calculations.

Fig. 2(a) shows trajectories arriving at Bush at 1600 GMT, 24 April 1987, at both 300m and 900m. Surface ozone concentrations, derived from monitoring sites along the air trajectory at 300m, are shown in Fig. 2(b) as a continuous line. Using these, surface deposition rates from the moving column were calculated, and these are shown as a histogram. On the 23 April maximum ozone concentrations in the vicinity of the trajectories at 300 and 900m were generally observed at 16 GMT. Ozone concentrations for the lower and upper parts of the boundary layer column (equivalent to positions A and B in Fig. 1) were estimated from the pattern of maximum ozone concentrations, giving values of 78 and 72 ppbV for the lower and upper columns reflectively; these are entered in Fig 2(b). The integrated

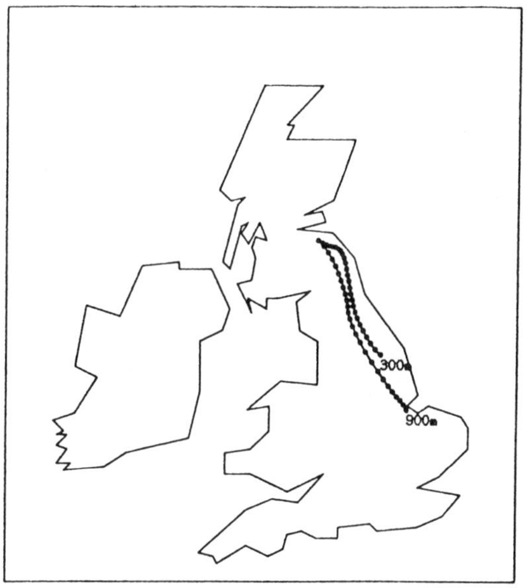

Fig. 2a. Trajectories for air arriving at the Bush monitoring site at 16 GMT 24 April 1987 at 300m and 900m. The start time for both trajectories is 15 GMT 23 April, the time of maximum ozone concentration at the locations concerned. Circles indicate positions at successive hours.

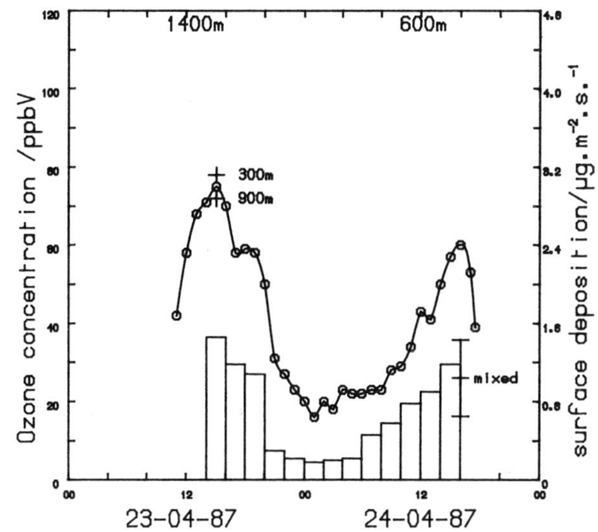

Fig. 2b. Ozone concentrations and surface deposition along the 300m trajectory shown in Fig. 2a. The continuous line with open circles shows ozone concentrations at the surface for air moving along the 300m trajectory, derived from observations at monitoring sites lying along the trajectory. Assuming a deposition velocity of $1cms^{-1}$ between 06 and 20 GMT and $0.5cms^{-1}$ between 20 and 06 GMT, these concentrations imply surface deposition values shown by the histogram. Mixed layer depths are shown at midday at the top of the diagram. Crosses indicate initial ozone concentations for the lower and upper parts of the budgeted column (labelled 300m and 900m) and the average of these, depleted by surface deposition, is labelled 'mixed' at the end of the budget period.

surface deposition during the 24 hour period would have depleted the average ozone concentration from its initial value of 75 ppbV by 49ppbV to 26 ppbV. On the following day the depth of the mixed layer at the Bush site was relatively small, inferred to be only 600m. This means that surface deposition effects a considerable decrease in ozone concentration in the boundary layer, since it has not extended far into the layer of ozone-rich air of the previous day's boundary layer above. If deposition velocities are increased or decreased by 20%, the depleted value changes by 10ppbV. The value of 26±10ppbV is plotted in Fig. 2(b) and compares with the measured ozone value for the column of 60ppbV. These results imply that the net increase in ozone concentration of the column due to chemical and photochemical means was 34±10ppbV.

Reliable estimates of sunshine hours experienced by budgeted air columns can be inferred from reports of sunshine hours from Meteorological Office stations, AVHRR satellite cloud images and synoptic weather charts. For air arriving at the Bush station on 24 April, this value was 9 hours.

The results of this budget calculation, together with those for nine more trajectories, are shown in Table 1.

4. DISCUSSION

The principal objective of the analysis presented here was to obtain estimates of the extent of 'pollutant' ozone production over the UK during an episode and to show the influence of meteorological variables on the amount produced.

The estimates of production of ozone (Table 1) show variability with observed insolation along the trajectories, the extent of production increasing with sunshine hours (Fig. 3). The slope of the line of best fit to these data implies an ozone production of about 4 ppbV per hour of sunshine. The regression line is not constrained to pass through zero, since the net generation includes chemical, as well as photochemical processes. These changes may be non-zero even in the absence of sunshine. Thus the maximum daily ozone production would be limited to about 45 ppbV in April and about 60 ppbV in midsummer, consistent with UK observations to date. The average length of trajectories in the boundary layer for a 24-hour period was about 500 km and this would be typical of anticyclonic situations during ozone episodes. Thus a value of 45-60 ppbV represents a reasonable value for ozone generation over the UK under cloudless conditions in the summer half of the year and with adequate precursor gas concentrations.

TABLE 1. BUDGET VALUES.

Date	End Point	Mixed Layer Depth of Column(m)		Average ozone concentration /ppbV					Sunshine hours
		at 12 Z	at 12 Z prev day	Initial	Surface depletion	Depleted value	Observed value	Apparent Generation	
24.04.87	Bush	600	1400	75	49	26	60	34 ± 10	9
25.04.87	Lough Navar	1100	1650	74	27	47	78	31 ± 6	7
26.04.87	Bush	1325	1150	82	24	58	99	41 ± 5	12.5
27.04.87	Bush 12GMT	1150	1200	85	13	72	83	11 ± 3	4.5
	16GMT			85	24	61	84	23 ± 5	8
27.04.87	Aston Hill	1300	1150	56	20	36	71	35 ± 4	6.5
28.04.87	Lough Navar	1500	1500	76	19	57	79	22 ± 4	8
28.04.87	Strath Vaich	1350	1750	68	12	56	63	7 ± 2	3
28.04.87	Bush	1825	1450	77	20	57	98	41 ± 4	9.5
29.04.87	Bush	800	1400	77	22	55	66	11 ± 4	3
30.04.87	Bush	1050	900	44	11	33	43	10 ± 2	1.5

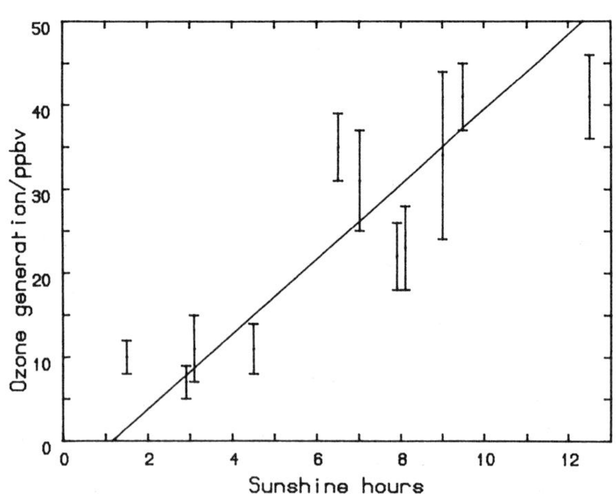

Fig.3. Ozone generation, deduced from budget calculations, versus sunshine hours along the air trajectory. The least squares regression line is indicated.

ACKNOWLEDGEMENT

This work was supported by the UK Department of the Environment.

REFERENCES

Dixon, R., Spackman, E.A., James, I. and Francis, A. (1972). The global analysis of meteorological data using orthogonal polynomial base functions. J. Atmos. Sci., 29, 609-622.

Galbally, I. (1971). Ozone profiles and ozone fluxes in the atmospheric surface layer. Quart. J.R. Met. Soc., 97, 18-29.

Galbally, I.E. and Roy, C.R. (1980). Destruction of ozone at the earth's surface. Quart. J Royal Met. Soc., 106, 599-620.

Garland, J.A. and Derwent, R.G. (1979). Destruction at the ground and the diurnal cycle of concentration of ozone and other gases. Quart. J. Royal Met. Soc.,105, 169-183.

Ludlam, F.H. (1966). Cumulus and cumulonimbus convection. Tellus, 18, 687-698.

Weseley, M.L., Eastman, J.A. Stedman, D.H. and Yawac, E.D. (1982). An eddy-correlation measurement of NO_2 flux to vegetation and comparison to O_3 flux. Atmospheric Environment 16, 815-820.

MEASUREMENTS OF NO AND NO_y OBTAINED DURING TROPOZ I

F. Rohrer and D.H. Ehhalt
Institut für Atmosphärische Chemie
Kernforschungsanlage Jülich GmbH
D-5170 Jülich, FRG

ABSTRACT

NO and NO_y profiles were measured during the TROPOZ I mission from December 11. to December 22., 1987. The flight track from Paris - Lisbon - Las Palmas -Dakar - Abidjan - Douala - Pointe Noir - Brazzaville covered latitudes between 50°N to 5°S. Vertical profiles were obtained between 0.3 and 10 km altitude. The flights included extensive level flights at 300 meters above forest, savanna and ocean. We will present here the data obtained in the tropical region, which are strongly affected by brush fires and the meteorological situation.

1. INTRODUCTION

Nitrogen oxide plays a controlling role in the chain of catalytic reactions which produce tropospheric ozone. This role is most easily illustrated by the oxidation of carbon monoxide. In the presence of sufficient NO it takes the path,

$$CO + OH \rightarrow H + CO_2 \quad R1$$
$$H + O_2 + M \rightarrow HO_2 \quad R2$$
$$HO_2 + NO \rightarrow OH + NO_2 \quad R3$$
$$NO_2 + h\nu \rightarrow NO + O \quad R4$$
$$O + O_2 + M \rightarrow O_3 + M \quad R5$$
$$\text{-------------------------}$$
$$CO + 2O_2 \rightarrow CO_2 + O_3$$

which results in a net production of O_3. In the absence of NO, the oxidation of CO takes a different path:

$$CO + OH \rightarrow H + CO_2 \quad R1$$
$$H + O_2 + M \rightarrow HO_2 \quad R2$$
$$HO_2 + O_3 \rightarrow OH + 2\ O_2 \quad R6$$
$$\text{-------------------------}$$
$$CO + O_3 \rightarrow CO_2 + O_2$$

In this case, ozone is destroyed. At the "critical" volume mixing ratio of NO of 10 pptv, production (by R1 to R5) and destruction (by R1, R2, R6) are equal. Above this value, the oxidation of CO leads to a photochemical production of ozone.

To estimate the global tropospheric production of O_3, the global background concentration of NO is required. We have therefore recently started to measure the NO volume mixing ratio on a global scale (Drummond et al., 1988).

Here, we report vertical profiles of the NO and NO_y volume mixing ratios made aboard an aircraft (Caravelle 116) over the equatorial region of West Africa. To investigate the role of brush fires in the production of nitric oxide, we have chosen two different areas: the area around Brazzaville (Congo) with tropical forest, and the area near Abidjan (Ivory Coast) with savanna.

A number of other components (CO, CO_2, CH_4, O_3, non methane hydrocarbons) were measured by other groups during the flights.

2. EXPERIMENTAL

NO and NO_y were measured by a chemiluminescence detector described earlier (Drummond et al., 1985). (We define here NO_y as the sum of NO, NO_2, NO_3, 2 x N_2O_5, HNO_2, HNO_3, HNO_4, PAN and nitrate-aerosol (Fahey et al., 1985)).

The instrument flown is shown schematically in Fig. 1. It consisted of two detectors operated in parallel. NO was detected by the chemiluminescence reaction between NO in the sample air and ozone added from an ozone generator.

$$NO + O_3 \rightarrow NO_2^* + O_2 \quad R7$$
$$NO_2^* \rightarrow NO_2 + h\nu \quad R8$$

NO_y was measured after conversion to NO in a gold converter (Fahey et al., 1985). The converter consisted of a heated (400°C) gold tube (l = 0.3 m, d = 6 mm). A low amount of CO (0.1 %) was added to the sample air for a complete reduction of the various NO_y-components.

The fluorescence was detected by a photomultiplier tube (EMI 9658 R, cooled with dry ice) through a red filter (RG 610). To increase the sensitivity, the reaction chamber consisted of gold-plated aluminium.

ISBN 0-937194-15-8

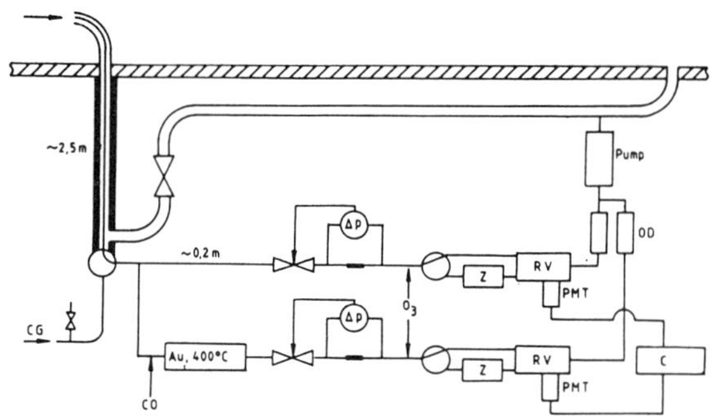

Fig. 1. Schematic drawing of the chemiluminescence detector system
RV reaction vessel
Z zeroing volume
D ozone destroyer
C computer
CG calibration gas

The dark current of the PMT and the interfering chemiluminescence reactions were measured in a zero-mode by passing the sample air-ozone mixture through a teflon-coated relaxation volume. The NO-signal was calculated from the difference between the measure-mode (without passage through the relaxation volume) and the average of the two neighboring zero modes. The signal from the PMT's was fed to photon counters, which were controlled by a computer. The counters were read out each 10 seconds with a mode switching time of 80 seconds.

The sensitivity of the two detectors was determined prior to each flight with a calibrated NO/N_2-mixture (AIRCO) and synthetic air. The calibration mixture was prepared by using two thermostatted mass flow controlers (Advanced Semiconductor Materials), which had been calibrated by soap bubble flow meters. The NO-detector showed a sensitivity of 4 counts s^{-1} ppt^{-1} (NO), the NO_y-detector 0.5 counts s^{-1} $pptv^{-1}$ (NO_y). The calibrations showed day to day variations of less then 5 %.

The detection limit was calculated from the frequency distribution of the differences of the measured and calculated zero mode signals. These distributions were gaussian and showed standard deviations of 8 ppt NO and 40 ppt NO_y (for a 10 second integration time).

The reaction mixture was pumped through the reaction vessel at 10 mbar pressure and a volume flow rate of 100 l min^{-1}. The flow was controlled by measuring the pressure drop across a capillary in the inlet. That signal which was obtained with the help of differential pressure gauges, was used to drive a Teflon needle valve which regulated the flow to a constant rate. The pressure drop was calibrated against the flow controlers of the gas calibration system prior to each flight. With this equipment, the flow rate could be held constant to better than 2 % even in a fast descent of the aircraft.

The inlet system consisted of a 2.5 m long Teflon tube with an inner diameter of 4 mm, which was cooled by flowing outside air over its outer surface. The air intake was located 0.20 m above the hull of the aircraft to prevent contamination.

Auxiliary meteorological information namely altitude, wind direction, wind speed, aircraft position, outside temperature and pressure was supplied by the aircraft navigation system. The temperature and pressure was used to calculate the potential temperature.

3. RESULTS

The flight tracks are shown in Figure 2. The flight data are summarized in Figure 3, which gives the vertical profiles of NO, NO_y, potential temperature , wind direction and of the calculated ratio NO_y/NO. For the local flights all data falling into an altitude band of 330 m were averaged. For the transfer flight, the data for the ascent and descent will be shown seperately.

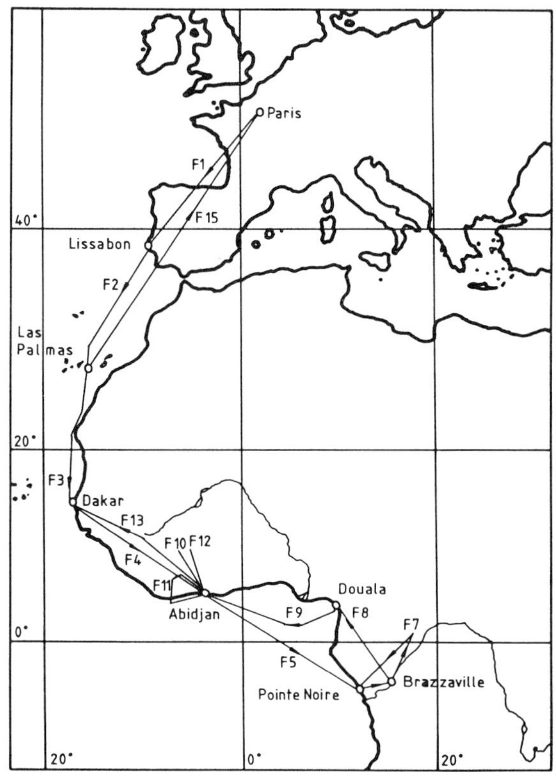

Fig. 2. Flight tracks during TROPOZ I.

The local flight at Brazzaville, F7, was carried out on December 15, 1987 between 13:30 UT and 17:00 UT. The flight track was triangular, starting at Brazzaville, going in the NNE-direction to the equator, from there in SW-direction to the coast at Point Noir, and then back to Brazzaville. The flight covered altitudes between 0.3 km and 6.1 km.

The transfer flight from Douala to Abidjan took place on December 16, 1987. The aircraft was climbing from 0.3 km to 9.5 km between 14:00 UT and 14:45 UT at 2°S, 6°E over the ocean, 300 km apart from the coast (F9 ascent). The aircraft descended at Abidjan between 15:40 UT and 16:15 UT (F9 descent).

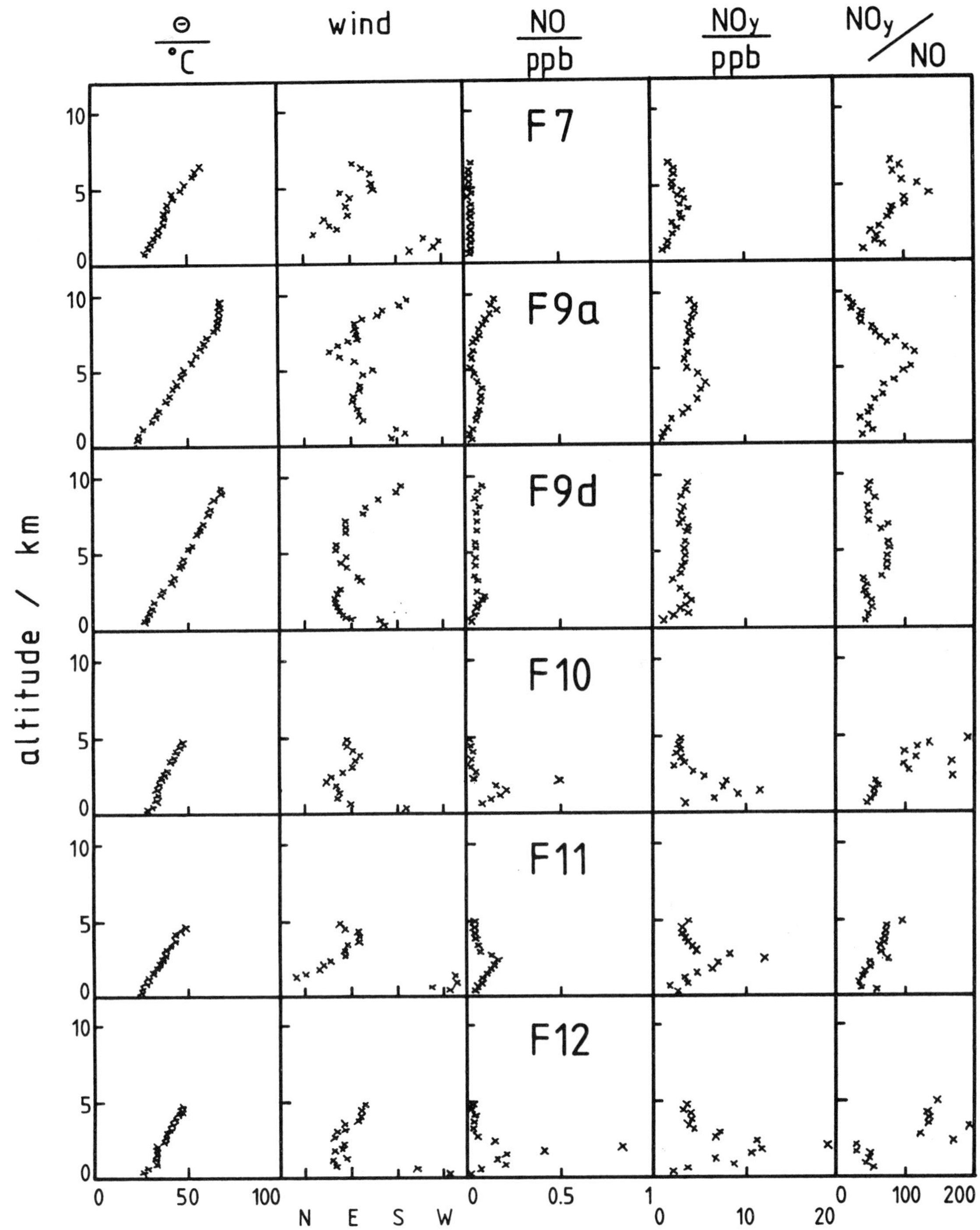

Fig. 3. Vertical profiles of the NO and NOy mixing ratios, potential temperature, wind direction and the ratio NOy/NO.
F7 : local flight at Brazzaville, 15.12.1987,13:30 - 17:00 UT
F9a: ascent at 2°S, 6°E, 16.12.1987,14:00 - 14:45 UT
F9d: descent near Abidjan, 16.12.1987,15:40 - 16:15 UT
F10: local flight at Abidjan, 17.12.1987,15:00 - 18:00 UT
F11: local flight at Abidjan (coast) 18.12.1987,09:30 - 12:30 UT
F12: local flight at Abidjan, 19.12.1987,15:00 - 18:00 UT

Two of the three local flights at Abidjan (F10 and F12) had a flight track going from Abidjan 600 km in the NNW-direction over the savanna and then back to Abidjan. During these flights, a pattern of black circles could be seen in the savanna, originating from bush fires. This pattern was the only hint to bush fires; no burning fire could be detected. These flights took place on December 17 and 19, 1987 between 15:00 UT and 18:00 UT.

The third flight, F11, on December 18, 1987, 9:30 UT - 12:30 UT followed a different flight track: 350 km along the coast in WSW-direction, then 300 km NNE and back to Abidjan in the SE-direction. The flight track, F11, barely touched the border of the savanna.

All the flights shown here had a similar profile of wind direction. At low altitudes up to 1 km, the wind direction was S to SW (sea breeze). At altitudes above 1 km, the wind direction changed sharply to NE (continental breeze). For altitudes over 7.5 km, the wind changed back to S.

The NO and NO_y profiles can be divided into three groups. The flights F7 at Brazzaville and F9 ascent near Douala show profiles of low mixing ratios of NO (20-100 ppt) and NO_y (1-5 ppb) with a shallow maximum at 3-4 km altitude. The profile of the ratio NO_y/NO shows a V-shape with values between 20 (at 1 km and 9 km altitude) and 130 (at 5 km altitude).

The profiles of flight F9 descent and F11 at the coast near Abidjan give a different picture. Here, the NO and NO_y mixing ratios peak at 2 km altitude. The ratio of the mixing ratios NO_y/NO is constant for altitudes lower than 3 km with a value of 50 and is than sharply changing to 80. At higher altitudes over 6.5 km, this ratio is sharply changing back again to 50.

The third group of flights consists of flights F10 and F12 over the savanna. Here, the NO mixing ratio varies between 20 ppt and 800 ppt showing a sharp maximum at 2 km altitude. The NO_y-mixing ratio has a similar behavior. It changes between 2 ppb and 20 ppb, again with a sharp peak at 2 km altitude. The ratio NO_y/NO is constant for altitudes lower than 2.5 km with a value of 50. At 2.5 km altitude, somewhat higher than the NO and NO_y peaks, the ratio is changing sharply to 150.

4. SUMMARY AND DISCUSSION

The NO and NO_y distribution obtained during TROPOZ I at the west coast of Africa in the tropics exhibits marked regional differences in the altitude profiles. The profiles seem to be strongly affected by the wind field. In the lower and upper mid-troposphere, the air masses influenced by the "sea breeze" during daytime show relatively low mixing ratios of NO and especially of NO_y. In the lower mid-troposphere, there is a layer of polluted air caused by NO_y-sources at the ground, which is not affected by the "sea breeze". This layer is quite sharply defined in the area near Abidjan, probably because of nearby NO_y-sources (brush fires). In the region near the equator, and more removed from NO_y-sources, this layer shows a less elevated NO concentration transiting smoothly into the "clean" air mass condition above and below.

ACKNOWLEDGEMENTS

The authors wish to thank the staff of the Centre d'Essais en Vol (CEV), Bretigny for the opportunity to participate in the TROPOZ I mission. The research was sponsored by the German Ministry for Research and Technology under grant 0744102, project "Einfluß anthropogener Kohlenwasserstoffe auf die Chemie der Troposphäre in außerstädtischen Gebieten".

REFERENCES

Drummond,J.W., Ehhalt,D.H., and Volz,A. Measurements of nitric oxide between 0 - 12 km altitude and 67°N - 60°S latitude obtained during STRATOZ III. Submitted to J. Geophys. Res. 1988.

Drummond,J.W., Ehhalt,D.H., and Volz,A. An optimized chemiluminescence detector for tropospheric NO measurements. J. Atmos. Chem. 2, 287-306, 1985.

Fahey,D.W., Eubank,C.S., Hübler,G., and Fehsenfeld,F.C. Evaluation of a catalytic reduction technique for the measurement of total reactive odd-nitrogen NO_y in the atmosphere. J. Atmos. Chem. 3, 435-468, 1985.

PROFILES OF PHOTOCHEMICALLY ACTIVE TRACE GASES IN THE TROPOSPHERE

R.R. Dickerson, B.S. Gockel, W.T. Luke, D.P. McNamara, L.J. Nunnermacker, and K.E. Pickering
Departments of Meteorology and Chemistry, University of Maryland
College Park, Maryland 20742, USA

J.P. Greenberg and P.R. Zimmerman
National Center for Atmospheric Research
Boulder, Colorado 80307, USA

ABSTRACT

The relative strength of the two main sources of O_3 in the troposphere - transport from the stratosphere and local photochemistry - stands as one of the major uncertainties in atmospheric chemistry. This paper describes results from a series of research flights over the central United States during which O_3, CO, NO, NOy, nonmethane hydrocarbons (NMHC), and meteorological parameters were measured. In summer, convective processes and planetary boundary layer (pbl) dynamics rapidly mix O_3 precursors from the surface layer into the free troposphere. In spite of intense convective mixing, O_3 was often highly stratified all the way from the surface to the tropopause. To deduce the origins of O_3 in these strata, we examined chemical and meteorological variables to calculate correlation coefficients. Positive correlations between O_3 and CO or H_2O indicate local photochemical O_3 production. Correlations between NOy and O_3 were positive for profiles showing O_3 of either stratospheric or tropospheric origin. These analyses and results from a simplified photochemical model indicate that although regions of pristine air are discernable, photochemistry tends to produce O_3 throughout much of the troposphere over the central U.S. Calculations of the rate of summer column O_3 production suggest that deep mixing can amplify the effects of photochemistry and significantly increase tropospheric O_3.

1. INTRODUCTION

Ozone acts not just as a local pollutant, but also as an integral part of the Earth's climate system. An understanding of global tropospheric O_3 is essential to understanding the chemical and radiative properties of the atmosphere. In what was probably the first attempt to calculate a global budget of tropospheric O_3, Junge (1962) suggested that O_3 is transported from the stratosphere into the troposphere where it is essentially inert until it is destroyed by contact with the Earth's surface. This paradigm can explain O_3 distributions under many circumstances, but as more observations became available, Crutzen (1973) and Chameides and Walker (1973) realized that additional production and destruction terms must be included in the budget. The relative roles of transport and photochemistry in the tropospheric O_3 budget remain as one of the central questions in atmospheric science.

If all pollutants were uniformly mixed into the troposphere, O_3 would soon increase on a global scale. Several important precursors of O_3 (NO, NO_2, and some hydrocarbons), however, are short-lived, and usually remain in the pbl where most emission occurs. In modeling studies, Chatfield and Crutzen (1984), among others, pointed ou that convection is a noteworthy exception to the general case of slow vertical mixing. Experiments have since demonstrated high concentrations of O_3 precursors in the outflow regions of convective clouds (e.g. Dickerson et al., 1987). In contrast, Pickering et al. (1988) demonstrated that some deep convective clouds do not transport pbl air to the upper troposphere. This paper will examine results from several flights over the central U.S. to try to begin to assess general trends.

Several investigators, including Fishman et al. (1980) have used the correlative nature of profiles of trace gases in the troposphere to determine the origin of O_3. Here we compare correlations among profiles of O_3, CO, NO, H_2O, and NOy (total reactive nitrogen) to calculations of the rate of O_3 production or destruction using observed concentrations and a simplified photochemical model.

2. EXPERIMENTAL TECHNIQUES

The analytical techniques and research aircraft, the NCAR Sabreliner, were described by Dickerson et al. (1987) and Pickering et al. (1988).

NO was measured by chemiluminescence (Dickerson et al. 1984) with a detection limit of about 10 ppt (parts per 10^{-12} by volume) for a 20 s integration time. Total reactive nitrogen, NOy, is detected by reduction to NO and is defined as the sum of NO, NO_2, NO_3, $2xN_2O_5$, HONO, HNO_3, HO_2NO_2, PAN (peroxy acetyl nitrate), RONOx (organo nitrates and nitrites), and small nitrate particles. For earlier work, a ferrous sulfate converter was used to measure NOx, which is usually defined as the sum of NO and NO_2. This converter, however, suffers an interference from PAN, and for this paper "NOx" will be defined as the sum of NO, NO_2, and PAN (Fehsenfeld et al., 1987).

Carbon monoxide was measured by IR absorption (Dickerson and Delany, 1988). Carbon monoxide, along with CH_4 and NMHC, was also measured in grab samples taken during the flights. These samples were later analyzed by gas chromatography with a flame ionization detector (Greenberg and Zimmerman, 1984).

The NO_2 and O_3 photolysis rate coefficients were derived from Eppley UV radiometer signals and overhead O_3 abundances (Dickerson et al., 1982).

ISBN 0-937194-15-8

3. RESULTS

Three flights (Figure 1) are investigated in detail. The first occurred on June 13, 1985, in cloud-free air; the second shortly after noon on June 22, 1985 under partly sunny skies behind a squall line, and the third also on June 22, but in cloud-free air well after the rains stopped. Both soundings (altitude profiles) and constant-altitude legs were flown.

A complete listing of the data would be too lengthy for this paper, and the reader is referred to Pickering (1987), Pickering et al. (1988) and McNamara (1988). An example of NO, CO, and O_3 mixing ratios measured at constant altitude on flight 1 appears on Figure 2. The general trend of high concentrations in the lower troposphere, relatively clean air in the mid troposphere and high concentrations again in the upper troposphere was frequently encountered under clear-sky conditions.

To facilitate a quantitative measure of correlation, best fit regression lines were drawn through the profiles so that the soundings could be "detrended." The correlative nature of the fine structure of these profiles is

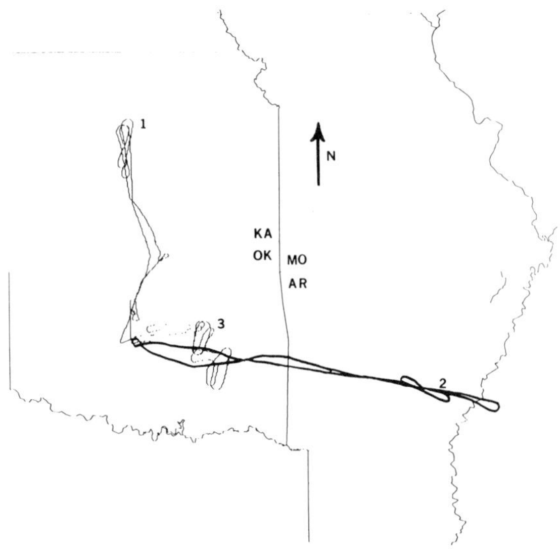

Fig. 1. Flight tracks: North-south track is flight 1, thick line is flight 2 and thin east-west track is flight 3.

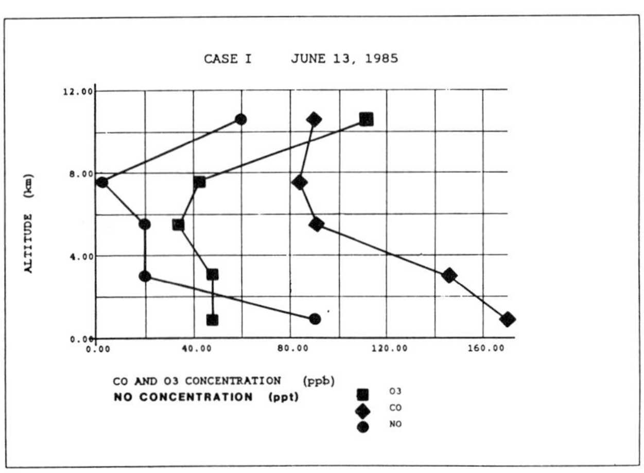

Fig. 2. Trace gas concentrations from constant altitude legs of fight 1.

then calculated in a manner similar to that of Fishman and Seiler (1983). A normalized correlation coefficient of > 0.5 is represented by "+" and called positive, < -0.5 is represented by a "-" and called negative, with all other coefficients represented by a "0". Table 1 shows the correlations of the trace gases measured on all three flights. In an independent test, simple correlations coefficients showed that correlations called positive or negative were generally statistically significant at the 90% level even after accounting for autocorrelation (Quenouille, 1952). On Table 1 "NOy" is written when NO, NOx, or NOy was measured.

TABLE 1. TRACE GAS PROFILE CORRELATIONS

FLIGHT	ALTITUDE (km)	SPECIES	CORRELATION*
1	0.9 to 4.2	H2O/O3	+
1	0.9 to 4.2	CO/O3	+
1	0.9 to 4.2	NOy/O3	+
1	0.9 to 4.2	CO/NOy	+
1	5.5 to 7.6	H2O/O3	+
1	5.5 to 7.6	CO/O3	+
1	5.5 to 7.6	NOy/O3	+
1	5.5 to 7.6	CO/NOy	0
1	7.6 to 10.6	H2O/O3	0
1	7.6 to 10.0	CO/O3	-
1	10.0 to 10.6	CO/O3	0
1	7.6 to 10.6	NOy/O3	0**
1	7.6 to 10.6	CO/NOy	+
2	0.9 to 10.6	H2O/O3	0
2	0.9 to 10.6	CO/O3	0
2	0.9 to 10.6	NOy/O3	0
2	0.9 to 10.6	CO/H2O	0
3	0.9 to 10.6	H2O/O3	0
3	0.9 to 10.6	CO/O3	+
3	0.9 to 10.6	NOy/O3	+
3	0.9 to 10.6	CO/H2O	+

* See text. ** Positive before detrending.

Twelve grab samples for hydrocarbons analysis were collected on this project. Meaningful correlations with other trace gases would require even more data, but these samples do provide information on the potential for photochemical O_3 formation. To summarize this large data set, we have calculated the "effective methane" concentration that would produce approximately the same effect on HO_x chemistry. Effective methane, $[CH_4]_{eff}$, is defined as follows:

$$[CH_4]_{eff} = [CH_4] + \sum_{n=2}^{10} n^*[C_nH](k_{CnH}/k_{CH4})$$

where k represents the rate constant for reaction with OH, square brackets represent concentration and n represents the number of carbon atoms. These values (Table 2) were used in a simplified photochemical model described below.

TABLE 2. EFFECTIVE METHANE CONCENTRATIONS FROM GRAB SAMPLES

SAMPLE	ALTITUDE (km)	TOTAL NMHC (ppbC)	$[CH_4]_{eff}/[CH_4]$	[CO] (ppb)	$[CH_4]$ (ppm)
1011	10.7	7.2	1.8	85	1.69
1022	5.5	9.8	4.8	103	1.68
1043	.9	18.9	12	122	1.71
1013	5.8	18.9	19	134	1.71
1049	2.4	15.7	21	94	1.65
1012	10.7	29.6	25	123	1.73
201	.9	20.6	27	87	1.68
1006	.9	26.7	57	202	1.96
200	10.1	5.8	64	108	1.66
1000	7.6	53.0	150	103	1.72
202	10.7	11.5	167	97	1.68
1042	6.7	80.3	549	118	1.68

4. DISCUSSION

With the goal of identifying the sources of O_3 layers in the troposphere, we consider the correlations of trace gases. Water and CO both have primary sources at the Earth's surface, while reactive nitrogen compounds can originate from combustion and biological activities at the surface, lightning throughout the troposphere, or downward transport from the stratosphere. If a particular stratum in the atmosphere has high O_3 accompanied by high CO and H_2O, an origin near the surface is inferred, and photochemistry is implicated. If photochemistry is the source of this high O_3 then there *must* also be (or have been) high NO or NOx in this stratum, because without NOx photochemical reactions destroy O_3. If, on the other hand, a layer of relatively high O_3 is accompanied by low CO and H_2O, a stratospheric source for this O_3 is inferred. Local photochemistry may still have produced the O_3 in such a layer if the stratosphere supplied sufficient NOx (Liu et al., 1980; Kley et al., 1981).

For flight 1, in the lowest 4.2 km of the troposphere O_3 is positively correlated with H_2O, CO, and NOy (Table 1) - clear evidence of photochemical O_3 production in these layers. The layers can be identified as the pbl or an old pbl, but the air was not highly polluted. NOx levels ranged from 0.1 to 1.0 ppb.

In the mid troposphere correlations were again generally positive but not as strong as in the lower troposphere, and the correlation between CO and NOy is lost. Between 7.6 and 10.6 km the correlations are generally not significant, suggesting a combination of O_3 sources. CO and O_3 are negatively correlated between 7.6 and 10.0 km, but the correlation becomes positive (although not significant) above 10 km. This implies that the nature of the air suddenly changes at 10.0 km, and in fact an isentropic back-trajectory traces the air to a deep convective cell near Denver on the previous morning (Pickering, 1987).

For flight 2 none of the correlations is significant. This is at first surprising in view of the plentiful supply of O_3 precursors; [NO] in the free troposphere ranged from 40 to 900 ppt. The vertical structure of the atmosphere on this occasion was highly disturbed by the recent passage of a squall line. When mixing is so thorough, there is insufficient time for photochemical reactions to produce O_3 in distinct layers. In addition, during precipitation, H_2O can be lost from an air parcel without loss of O_3 or CO. The lack of correlation among the trace gases is more a demonstration of the limitations of such analysis techniques than an indication of the origins of the O_3, for as the next section shows, this air has a potential to produce copious quantities of O_3 *in situ*.

Flight 3 occurred later in the day over generally the same area, but in air that did not experience frontal passage. Positive correlations are restored among trace gases from the surface to near the tropopause showing that local photochemistry dominates O_3 production in this air mass. The correlation between O_3 and H_2O was positive but < 0.5.

In order to calculate the rate of O_3 production a box model with 18 species participating in 26 reactions (McNamara, 1988) was solved with the Gear method. Species not measured directly were assumed to be in photochemical steady state. The net O_3 production, $P(O_3)$, was calculated from the following equation:

$$P(O_3) = k_1 [NO][HO_2] - k_2 [O_3][HO_2] - k_3 [O_3][OH] + k_4 [NO][CH_3O_2] - k_5 [O(^1D)][H_2O]$$

The model was first run for flight 1 for the five altitudes of level flight. To determine the effect of NMHC, we ran the model with $[CH_4]_{eff}$ of 1.7, 3.0 and 110 ppm to represent CH_4 only, the minimum observed NMHC and the approximate mean NMHC. Figure 3 shows the rapid production of O_3 calculated for all altitudes except between 6.5 and 10 km where the very low NO concentration (Figure 2) allowed net O_3 destruction. The influence of NMHC can also be seen, especially in the pbl where the inclusion of NMHC increases $P(O_3)$ by about a factor of ten.

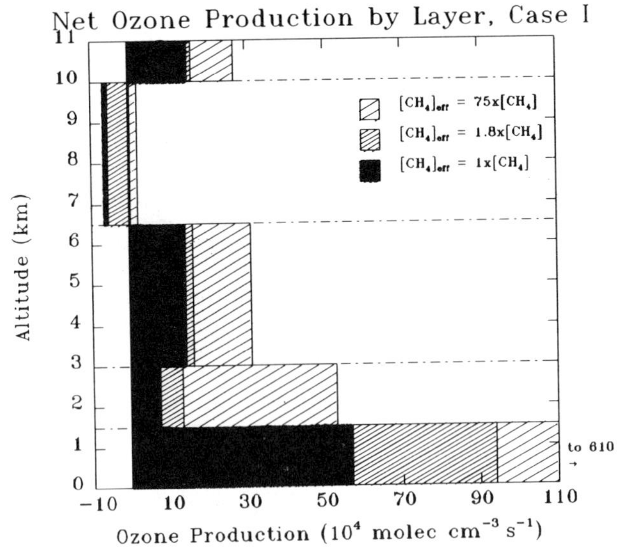

Fig. 3. Calculated ozone production rates for five layers in the troposphere.

Column O_3 production rates for all three flights are compared (Figure 4) to calculations by Chameides et al. (1987) for the remote Pacific where NMHC were assumed unimportant. The higher CO and NOx levels found over the continental U.S. cause O_3 to be produced at about 100 times the rate in clean marine air. Liu et al. (1987) calculated O_3 production for average summer conditions over the continental U.S.; they assumed that NMHC were limited to the pbl. Flight 1 shows results similar to those of Liu et al., but calculations for flights 2 and 3 with representative NMHC concentrations predict about four times as much O_3 production.

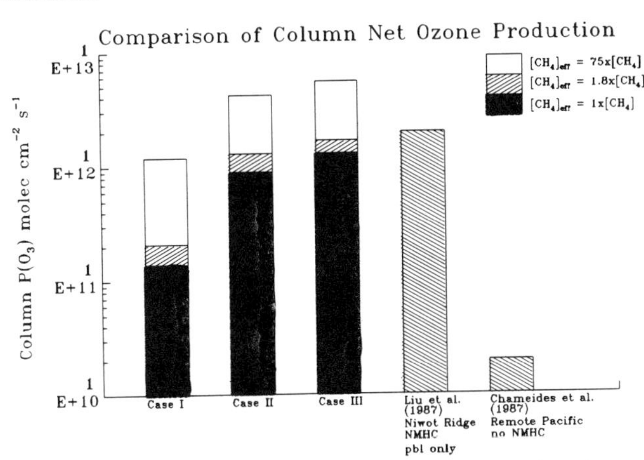

Fig. 4. Calculated column ozone production for three flights compared to previous estimates.

The rate of column O_3 production is higher for the well mixed troposphere described here even though the pbl NOx concentration are lower than the U.S. mean calculated by Liu et al. The number of O_3 molecules produced per NOx molecule is much greater at higher altitude and lower NOx concentrations (e.g. Liu et al. 1987). Thus in the long run, the increase in column O_3 production would be even greater.

5. CONCLUSIONS

Correlations of O_3, CO, NO, H_2O, and NOy profiles for all cases tested indicate that in the lower troposphere the dominant source of O_3 is local photochemistry. In the mid and upper troposphere both surface and stratospheric sources are suggested, but the correlations cannot distinguish between direct injection of stratospheric O_3 and *in situ* formation from NOx of stratospheric origin.

A photochemical model showed the dominance of *in situ* formation of O_3 in the troposphere with the maximum rate of production in the pbl and the minimum in the mid troposphere where the NO level can be low enough for net O_3 destruction. Ozone production in the upper troposphere is often amplified by pollution from deep convective cells hundreds of km upwind.

Particularily rapid O_3 production was calculated for air behind a squall line where the troposphere was well mixed from the surface to 10.6 km over a horizontal distance of at least 700 km. Correlations studies were of little use in this case due to the loss of H_2O to precipitation and the lack of time to produce O_3.

In the summer over the Great Plains of the U.S., the deep pbl and frequent thunderstorms cause efficient vertical mixing of NOx and NMHC and increases the rate of photochemical O_3 production in that area to such an extent that these processes should be considered in regional or hemispheric budgets of tropospheric O_3.

ACKNOWLEDGMENTS

This work was supported by NSF Grant ATM-86-19491, and the DOE PRECP program and NOAA/NSF PRESTORM programs. NCAR's RAF provided aircraft support. NCAR is sponsored by NSF. The U. MD Office of Graduate Studies and Research supported travel.

REFERENCES

Chameides, W., and J.C.G. Walker, 1973: A photochemical theory of tropospheric ozone, J. Geophys. Res., 78, 8751-8760.

Chameides, W.L., D.D. Davis, M.O. Rodgers, J. Bradshaw, S. Sandholm, G. Sachse, G. Hill, G. Gregory, and R. Rasmussen, 1987: Net ozone photochemical production over the eastern and central North Pacific as inferred from GTE/CITE 1 observations during fall 1983, J. Geophys. Res., 92(2), 2131-2152.

Chatfield, R.B., and P.J. Crutzen, 1984: Sulfur dioxide in remote oceanic air: Cloud transport of reactive precursors, J. Geophys. Res., 89(5), 7111-7132.

Crutzen, P.J., 1973: A discussion of the chemistry of some minor consitituents in the stratosphere and troposphere, Pure Appl. Geophys., 106, 1385-1399.

Dickerson, R.R., D.H. Stedman, and A.C. Delany, 1982: Direct measurements of ozone and nitrogen dioxide photolysis rates in the troposphere, J. Geophys. Res., 87(7), 4933-4946.

Dickerson, R.R., 1984: Measurements of reactive nitrogen compounds in the free troposphere, Atmos. Environ., 18(12), 2585-2593.

Dickerson, R.R., A.C. Delany, and A.F. Wartburg, 1984: Further modification of a commercial NO_x detector for high sensitivity, Rev. Sci. Instr., 55(12), 1995-1998.

Dickerson, R.R., G.J. Huffman, W.T. Luke, L.J. Nunnermacker, K.E. Pickering, A.C.D. Leslie, C.G. Lindsey, W.G.N. Slinn, T.J. Kelly, A.C. Delany, J.P. Greenberg, P.R. Zimmerman, J.F. Boatman, J.D. Ray and D.H. Stedman, 1987: Thunderstorms: An important mechanism in the transport of air pollutants. Science, 235, 460-465.

Dickerson, R.R., and A.C. Delany, 1988: Modification of a commercial gas filter correlation CO detector for enhanced sensitivity, J. Atmos. Oceanic Tech., 53, 424-431.

Fehsenfeld, F.C., R.R. Dickerson, G. Huebler, W.T. Luke, L. Nunnermacker, E.J. Williams, J. Roberts, J.G. Calvert, C. Curran, A.C. Delany, C.S. Eubank, D.W. Fahey, A. Fried, B. Gandrud, A. Langford, P. Murphy, R.B. Norton, K. Pickering, and B. Ridley, 1987: A ground-based intercomparison of NO, NO_x, NO_y measurement techniques, J. Geophys. Res., 92(12), 14710-14722.

Fishman, J., W. Seiler, and P. Haagenson, 1980: Simultaneous presence of O_3 and CO bands in the troposphere, Tellus, 32(5), 456-463.

Fishman J., and W. Seiler, 1983: Correlative nature of O_3 and CO in the troposphere: Implications for the tropospheric O_3 budget, J. Geophys. Res., 88, 3662-3670.

Greenberg, J.P., and P.R. Zimmerman, 1984: Non-methane hydrocarbons in the remote tropical, continental and marine atmospheres, J. Geophys. Res., 89(3), 4767-4778.

Junge, C.E., 1962: Global ozone budget and exchange between stratosphere and troposphere, Tellus, 14, 363-377.

Kley, D., J.W. Drummond, M. McFarland, and S.C. Liu, 1981: Tropospheric profiles of NO_x, J. Geophys. Res., 86, 3153-3161.

Liu, S.C., D. Kley, M. McFarland, J.D. Mahlman, and H. Levy II, 1980: On the origin of tropospheric O_3, J. Geophys. Res., 85, 7546-7552.

Liu, S.C., M. Trainer, F.C. Fehsenfeld, D.D. Parrish, E.J. Williams, D.W. Fahey, G. Huebler, and P.C. Murphy, 1987: Ozone production in the rural troposphere and the implications for regional and global O_3 distributions, J. Geophys. Res., 92(4), 4191-4207.

McNamara, D.P., 1988: Origin of Tropospheric Ozone over Central U.S., M.S. thesis, Dep. of Meteo. Univ. of Maryland, College Park, MD.

Pickering, K.E., 1987: Observations of Tropospheric Trace Gases and Techniques for Assessing Regional Transport of Air Pollutants, Ph.D. dissertation, Dep. of Meteo. Univ. of Maryland, College Park, MD.

Pickering, K.E., R.R. Dickerson, G.J. Huffman, J.F. Boatman, and A. Schanot, 1988: Trace gas transport in the vicinity of frontal convective clouds, J. Geophys. Res., 93(1), 759-773.

Quenouille, M.H., 1952: Associated Measurements, Butterworth Scientific Publications, London.

OZONE, CO AND NO_x DISTRIBUTION IN THE TROPOSPHERE DURING STRATOZ III

Ph. Gerhardt and D. Poppe
Institut für Atmosphärische Chemie
Kernforschungsanlage Jülich
D-5170 Jülich, FRG

A. Marenco
Laboratoire d'Aérologie
(CNRS-LA 354) Université Paul Sabatier
F-31062 Toulouse Cedex, France

ABSTRACT

The distributions of O_3, CO, and NO, measured between 0 and 12 km altitude and 67°N-60°S latitude during the STRATOZ III campaign in June 1984, are analysed. Altogether 42 profiles were taken. The concentrations exhibit a considerable variability with altitude and latitude. Correlations between the O_3 and the CO distributions are discussed in connection with a chemical in situ production of tropospheric ozone. The results show that ozone distributions are considerably influenced by dynamic processes.

1. INTRODUCTION

Little is known about the detailed global distribution of trace gases with lifetimes between a day and several months. This is also true for those trace gases which are of particular importance for atmospheric chemistry such as CO (atmospheric lifetime $\tau \approx 2$ months [VOLZ et al. 1981]), O_3 ($\tau \approx 3$-4 months [GALBALLY et al. 1980, SEILER et al. 1981]), and NO_y (the sum of odd nitrogen, $\tau \approx 1$ day [EHHALT 1987, EHHALT and DRUMMOND 1982]). All of these gases affect the tropospheric OH concentration and thus the lifetimes of other reactive trace gases [e.g. LOGAN et al. 1981].

The first experimental surveys in the troposphere with considerable latitudinal and altitudinal resolution were made between 1964 and 1967 over North America. The ozone measurements of these campaigns were analysed by CHATFIELD and HARRISON [1977]. ROUTHIER et al. [1980] and SCHIFF et al. [1979] described measurements of ozone and tropospheric aerosols (H_2O) and NO, obtained during flights over the central and eastern Pacific Ocean (20°S - 37°N) and the North American continent (40°N - 60°N) during flights in 1977 and 1978.

In 1974 the distributions of H_2, CO_2, CO, and O_3 were measured during the ascents and descents of 39 consecutive aircraft flights over the North Atlantic and the American continent between 57°S and 67°N. The data have been discussed by U. SCHMIDT [1978], R. SCHMITT [1976], and SEILER et al.[1981]. As a result of these studies FISHMAN et al. [1983] derived interrelations between CO and O_3 in the vertical profiles, which they discuss in the light of an in situ chemical production of O_3 from CO under stable atmospheric conditions, when transport including that from the stratosphere [FIEDLER 1982, SHAPIRO 1980] could be neglected. The chemical production of O_3 is promoted by the reaction sequence

$$RO_2 + NO \longrightarrow RO + NO_2$$
$$NO_2 + h\nu \longrightarrow NO + O$$
$$O + O_2 \longrightarrow O_3$$

(which is part of the net cycle $CO + 2\,O_2 \longrightarrow CO_2 + O_3$) and should be taken into account, whenever the NO_x concentration is high enough. In time resolved experiments at Pic du Midi between 1981 and 1984 there was clearly a correlation between CO and O_3, indicating a maximum O_3 production in the summer afternoon, when maximum CO concentrations and photolysis rates of NO_2 are reached [MARENCO 1986].

In this paper we present global two-dimensional distributions of O_3, CO, and NO_x, which are derived from simultaneous measurements during the STRATOZ III flight in June 1984.

2. DESCRIPTION OF MEASUREMENTS

The STRATOZ III mission was carried out jointly by the Etablissement d'Etudes et de Recherche Meteorologiques (EERM), Toulouse, and the Kernforschungsanlage Jülich, using the Caravelle 116 aircraft of the Centre d'Essais en Vol, Bretigny, France. The flight track (see Figure 1 in the paper of POPPE et al., this symposium) was similar to the flight described by SEILER et al. [1981], with the exception of the return (northbound) flight, which led over the eastern coast of South America and the western coast of North Africa. In 21 ascents and descents a number of vertical trace gas profiles were measured in addition to O_3, CO, and NO, some reported elsewhere [DRUMMOND et al. 1988, RUDOLPH et al. 1987,1988, KARCHER et al. 1988]. Also, meteorological data (wind speed and direction, temperature, pressure) and photon fluxes for NO_2 photolysis were continuously recorded during the flight. Results for O_3 and CO are reported by MARENCO [this symposium], the NO measurements were given by DRUMMOND et al. [1988], their interpretation by EHHALT and DRUMMOND [1988].

3. DATA ANALYSIS

3.1 NO_x CONCENTRATIONS

During STRATOZ III only NO was measured reliably. To derive the NO_x concentration, which is a

ISBN 0-937194-15-8

Table 1: Main reaction paths for NO_2 evaluation
Rate constants are in units cm^3/s (second order) and s^{-1} (first order); z is the solar zenith angle; j_2 is derived from a photon flux model and checked against the measured upwelling UV flux.

Number	Reaction	Rate coefficient
1	$O_3 + NO \longrightarrow NO_2 + O_2$	$k_1 = 1.8E{-}12 \exp(-1370\ K / T)$
2	$NO_2 + h\nu \longrightarrow NO + O_3$	j_2
3	$O_3 + h\nu \longrightarrow O(^1D) \xrightarrow{H_2O} 2\ OH$	$k_3 j_3 = 1/(1 + 0.13\ [M]/[H_2O]) * 0.0116 \exp(-1.58/\cos z) * j_2$
4	$OH + CO \longrightarrow HO_2 + CO_2$	$k_4 = 1.52E{-}13$
5	$OH + NO_2 \xrightarrow{M} HNO_3$	$k_5 = 1.13E{-}11 * p/bar$
6	$HO_2 + O_3 \longrightarrow OH + 2\ O_2$	$k_6 = 1.42E{-}14 \exp(-580\ K / T)$
7	$HO_2 + NO \longrightarrow OH + NO_2$	$k_7 = 3.7E{-}12 \exp(240\ K / T)$
8	$HO_2 + HO_2 \longrightarrow H_2O_2 + O_2$	$k_8 = 4.2E{-}12 + 3.1E{-}34 \exp(2980\ K / T) * [H_2O]/cm^{-3}$
9	$HO_2 + OH \longrightarrow H_2O + O_2$	$k_9 = 4.8E{-}11 \exp(250\ K / T)$
10	$OH + CH_4 \longrightarrow CH_3O_2 + H_2O$	$k_{10} = 3.1E{-}18 * (T/K)^{2.1} \exp(-1240\ K / T)$
11	$CH_3O_2 + NO \longrightarrow HCHO + HO_2 + NO_2$	$k_{11} = 1.4E{-}11 \exp(-180\ K / T)$
12	$HCHO + OH \longrightarrow HO_2 + CO + H_2O$	$k_{12} = 1.015E{-}11$
13	$ACO_3 + NO \longrightarrow CH_3O_2 + NO_2 + CO_2$	$k_{13} = 0.974E{-}11$
14	$ALD + OH \longrightarrow ACO_3 + H_2O$	$k_{14} = 3.87E{-}11 \exp(-260\ K / T)$
15	$PAN \xrightarrow{M} ACO_3 + NO_2$	$k_{15} = 6.3E15 \exp(-13160\ K / T)$
-15	$ACO_3 + NO_2 \xrightarrow{M} PAN$	$k_{-15} = 6.08E{-}12$

more conservative quantity, the NO_2 concentration has to be estimated. This was done using a simple chemical reaction scheme taking the main reaction paths for NO_2 into account. The reactions are summarized in Table 1. Steady state is assumed for the radicals ACO_3 (Acetylperoxyde), CH_3O_2 (Methylperoxyde), HO_2, and OH. Mean PAN (Peroxyacetyl nitrate) concentrations are taken from the simultaneous measurements during STRATOZ III [RUDOLPH et al. 1987, EHHALT and RUDOLPH 1987]. Constant HCHO and ALD (Aldehydes other than HCHO) concentrations are assumed. With steady state for NO the reaction scheme leads to an improved photostationarity for $[NO]/[NO_2]$, since the influence of the peroxy radicals is included.

3.2 GLOBAL TWO-DIMENSIONAL DISTRIBUTIONS OF O_3, CO, AND NO_x

In Fig. 1 the meridional cross-sections of the mixing ratios of CO and NO_x averaged over the north- and southbound paths of STRATOZ III are presented. The data for continental and marine air do not differ significantly within the available spatial resolution. But one has to keep in mind, that NO_x could have much different concentrations in marine air far from the continental sources, because of its short lifetime. CO concentrations at northern mid latitudes are more than a factor of 2 higher than in the southern hemisphere. NO_x shows larger variations and is also a factor of 2 higher in the northern hemisphere. Exceptionally high concentrations are found in the planetary boundary layer of polluted areas: Rio de Janeiro (23°S), Las Palmas (28°N), and western Europe (38°N - 50°N). Over northern Spain (43°N, 5°W) very high NO mixing ratios of up to 3 ppb were recorded between 9.6 and 10.6 km altitude when the flight passed over an area of cumulus clouds. In this area relatively low ozone concentrations (40 - 50 ppb) were found, only half of the concentrations found in the adjacent regions. Highest O_3 concentrations (up to 180 ppb) were found at the transition zone, where NO declined sharply to normal values.

In Fig. 2 the mixing ratios of O_3 are shown in a two-dimensional plot, divided for continental and marine air. The differences between the two plots are not significant, indicating that high O_3 concentrations can be maintained in marine air. On the other hand, the concentration profiles show a clear distinction between polluted areas (northern mid latitudes) with values mostly in the range between 40 and 60 ppb and small gradients in the free troposphere, and clean areas (southern hemisphere) with mean mixing ratios of about 25 - 35 ppb. In addition, positive vertical gradients in the equatorial belt (tropical rain forests in South America) are clearly visible, resulting most likely from the high deposition rates in this area.

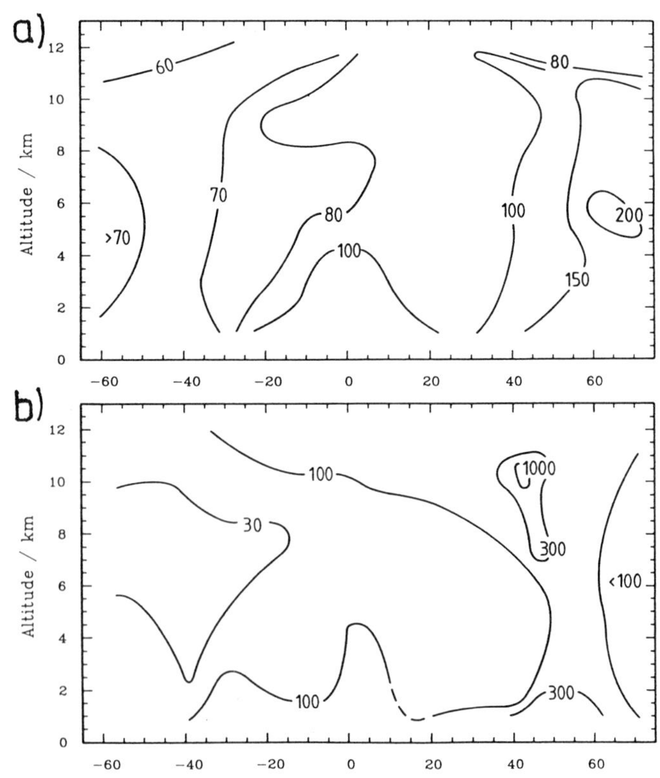

Fig. 1. a) Global CO distribution from STRATOZ III. Isolines indicate mixing ratios in ppb
b) Global NO_x distribution (ppt units)

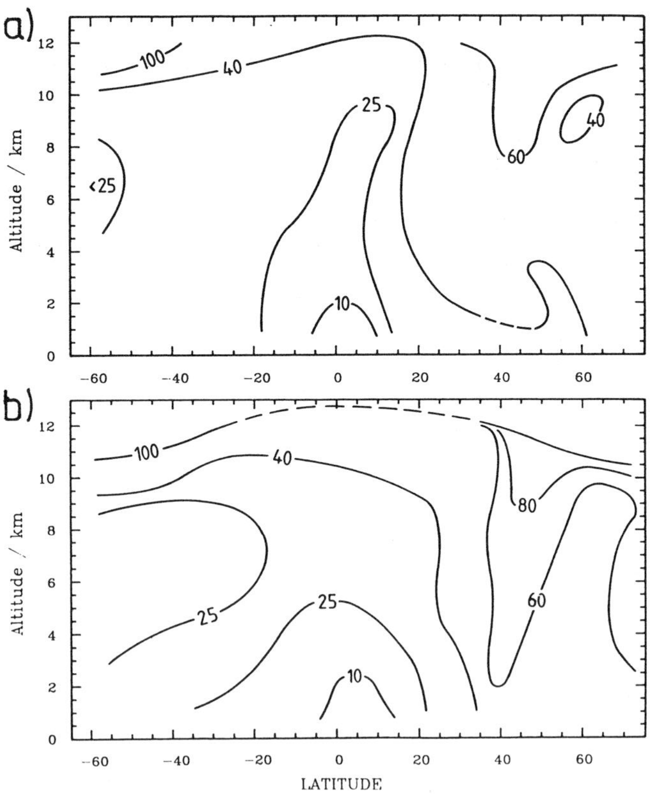

Fig. 2. Global ozone distribution (ppb units) a) marine air, b) continental air

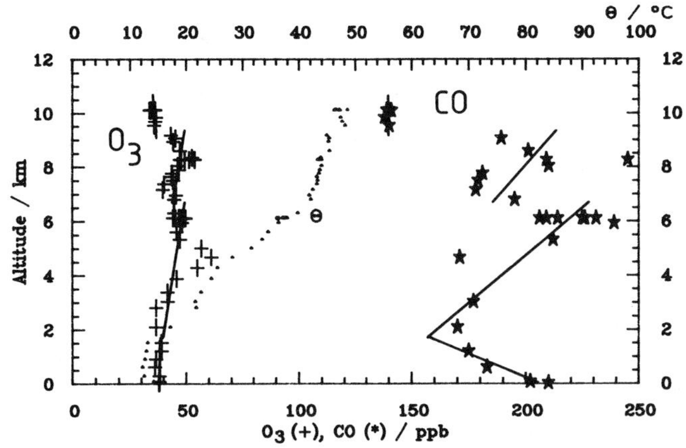

Fig. 3. Profiles of O_3 and CO mixing ratios for the ascent from Prestwick. Different air masses are observed between 0-2 km, 2-6.7 km, 6.7-9.3 km, and 9.3-10.3 km, indicated by the straight lines, representing the local gradients (not used in this evaluation)

3.3 CORRELATIONS BETWEEN O_3, CO, AND NO_x

To get more insight into the ozone distributions, the concentration profiles of all the ascents and descents were analysed in a way which is similar to the procedure described by FISHMAN et al. [1983]. The relative deviations from linear profiles of the mixing ratios of O_3 on the one hand and of CO or NO_x on the other hand were computed and correlated. In an attempt to achieve a better separation of transport and chemistry we modified Fishman's treatment by dividing each profile into subprofiles belonging to different air masses. The latter are identified using the wind field and the vertical mixing rates which are related to the potential temperatures, Θ. The divisions according to different air masses has the important advantage that all data points within the air mass have similar origins and the same chemical history. This makes a statistical analysis much more meaningful and eventually allows for a separation of transport and in-situ chemical processes. With Fishman et al. we assume that there is an overall linear dependence of the whole mean profile with altitude, given by the gradient g. Transport for an air mass effects therefore only its mean mixing ratio while the gradient is still characterized by g.

Accordingly, correlation coefficients, C, are calculated for the expressions $x_i(h_i)$ - [$\langle x \rangle$ + $g \cdot (h_i - \langle h \rangle)$], x_i being the mixing ratios measured at height h_i, and $\langle\ \rangle$ denoting the average within each different air mass.

The procedure has been applied to all 42 profiles. As an illustration (Fig. 3) we show the profile for the ascent from Prestwick (55°N - 60°N). Four different air masses can be distinguished. The straight lines indicate linear approximations to the profiles of the different air masses. Correlation coefficients of -0.1, 0.2, 0.9, and -0.5 were calculated, whereas the mean correlation coefficient without a subdivision is 0.70. This shows that some of the apparently significant correlation between O_3 and CO is lost, if a subdivision is made. The uppermost part of the profile obviously would lead to a larger C in Fishman's treatment, but is due to transport of less polluted air.

To discuss a possible in situ chemical production of ozone from CO in the troposphere we have constructed a two-dimensional plot of the mean correlation coefficients (Fig.4). One expects a strong positive correlation in a region where the NO_x and the OH concentrations are high (northern mid latitudes). There are indeed areas in which O_3 and CO seem to be strongly correlated. However, these places are not identical with those expected (cf. Fig. 1). In particular, areas with the highest NO_x concentrations are anticorrelative in CO and O_3. That anticorrelation arises from transport of polluted air. Over the clouded area in northern Spain cited

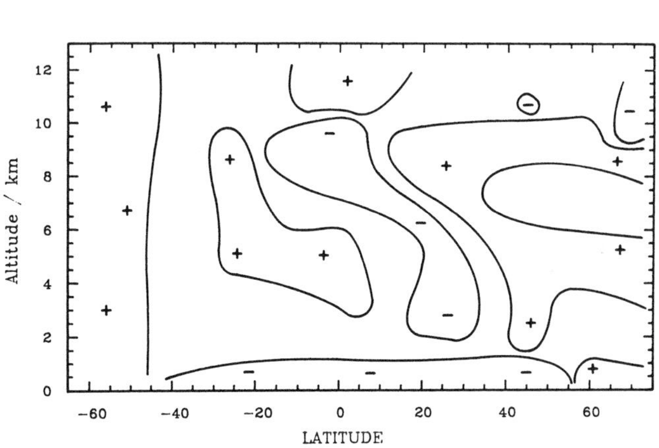

Fig. 4. Distribution of O_3 and CO correlations. Positive signs indicate regions with $C > 0.5$, negative signs those with $C < -0.5$.

above, the CO mixing ratio shows a similar enhancement as NO_x from 70 ppb in the boundary region to about 120 ppb in the center, whereas O_3 is decreasing.

In addition we have tried to make a plot of the correlation of O_3 and NO_x. Again, no structure was found for the global distribution of those correlation coefficients.

4. DISCUSSION

The data presented above confirm the findings of FISHMAN et al. [1983] in the sense that there are regions of strong correlations of O_3 and CO concentrations, but their locations are quite different. In our data out of 82 subdivisions of the 42 profiles, free from obvious transport effects, 20 are positively ($C > 0.5$) and 14 are negatively correlated ($C < -0.5$). However, the explanation is probably not a simple in situ chemical production of O_3 from CO (or other hydrocarbons which have similar distributions), as perceived by those authors.

In the case of positive C, the correlation can arise from:
- similar sources, e.g. formation in polluted air,
- transport of polluted air with positive C into higher altitudes,
- in situ chemical production of O_3 from CO, if $[NO_x]$ and [OH] are high.

Anticorrelations can arise from:
- different sources and sinks, e.g. dominating O_3 deposition,
- transport of air, e.g. from the stratosphere with high O_3 but low CO concentrations.

The absence of extensive areas with strong positive correlations of O_3 and CO concentrations does not necessarily contradict a significant contribution of ozone by local production in the troposphere. It merely underlines the fact that the local variations of O_3 and CO concentrations are dominated by transport processes and the large scale gradients of these gases. Any local production of O_3 is therefore masked and hard to detect in such a low resolution approach.

REFERENCES

CHATFIELD, R.B., and H. HARRISON, 1977: Tropospheric ozone, variations along a meridional band. *J. Geophys. Res.*, **82**, 5969-5976

DRUMMOND, J.W., D.H. EHHALT and A. VOLZ, 1988: Measurement of Nitric Oxide between 0 and 12 km altitude and 67°N - 60°S latitude obtained during STRATOZ III. *J. Geophys. Res. (submitted)*

EHHALT, D.H., 1987: Free radicals in the atmosphere. *Free Rad. Res. Comms.*, **3**, 153-164

EHHALT, D.H., and J.W. DRUMMOND, 1982: The tropospheric cycle of NO_x. In: *'Chemistry of the unpolluted and polluted troposphere', H.W. Georgii, W. Jaeschke (Eds.), D. Reidel Publ. Corp.*, p. 219-251

EHHALT, D.H., and J.W. DRUMMOND, 1988: NO_x sources and the tropospheric distribution of NO_x during STRATOZ III. In: *'Tropospheric ozone, regional and global scale interactions', NATO ASI Series C, vol. 227, I.S.A. Isaksen (Ed.)*, p. 217-237

EHHALT, D.H., and J. RUDOLPH, 1987: Die globale Verteilung von Stickoxid und Kohlenwasserstoffen in der Troposphäre. *Rep. Kernforschungsanlage Jülich*

FIEDLER, F., 1982: Atmospheric circulation. In: *'Chemistry of the unpolluted and polluted troposphere', H.W. Georgii, W. Jaeschke (Eds.), D. Reidel Publ. Corp.*, p. 119-154

FISHMAN, J., and W. SEILER 1983: Correlative nature of ozone and carbon monoxide in the troposphere: implications for the tropospheric ozone budget. *J. Geophys. Res.*, **88**, 3663-3670

GALBALLY, I.E., and C.R. ROY, 1980: Destruction of ozone at the earth's surface. *Q.J.R. Meteorol. Soc.*, **106**, 599-620

KARCHER, F., M. AMODEI, G. ARMAND, C. BESSON, B. DUFOUR, G. FROMENT and J.P. MEYER, 1988: Simultaneous measurements of stratospheric HNO_3, NO_2, HCl, O_3, N_2O, CH_4, H_2O and CO from the STRATOZ III flights. *This symposium*

LOGAN, J.A., M.J. PRATHER, S.C. WOFSY and M.B. McELROY, 1981: Tropospheric chemistry: a global perspective. *J. Geophys. Res.*, **86**, 7210-54

MARENCO, A., 1986: Variations of CO and O_3 in the troposphere: evidence of O_3 photochemistry. *Atmospheric Environment*, **20**, 911-918

POPPE, D., J. DRUMMOND and D.H. EHHALT, 1988: Tropospheric ozone budget during STRATOZ III, a model study. *This symposium*

ROUTHIER, F., R. DENNETT, D.D. DAVIS, E. DANIELSEN, A. WARTBURG, P. HAAGENSON and A.C. DELANY, 1980: Free tropospheric and boundary layer airborne measurements of ozone over the latitude range of 59°S and 70°N. *J. Geophys. Res.*, **85**, 7307-7321

RUDOLPH, J., B. VIERKORN-RUDOLPH and F.X. MEIXNER: Large-scale distribution of peroxyacetyl nitrate results from the STRATOZ III flights. *J. Geophys. Res.*, **92**, 6653-6661

SCHIFF, H.I., D. PEPPER and B.A. RIDLEY, 1979: Tropospheric NO measurements up to 7 km. *J. Geophys. Res.*, **84**, 7895-7897

SCHMIDT, U., 1978: The latitudinal and vertical distribution of molecular hydrogen in the troposphere. *J. Geophys. Res.*, **83**, 941-946

SCHMITT, R., 1976: Globale Untersuchungen des Kohlendioxid-Konzentrationsfeldes mittels eines neuentwickelten Messgerätes. Dissertation Univ. Frankfurt

SEILER, W. and J. FISHMAN, 1981: The distribution of carbon monoxide and ozone in the free troposphere. *J. Geophys. Res.*, **86**, 7255-7265

SHAPIRO, M.A., 1980: Turbulent mixing within tropopause folds as a mechanism for the exchange of chemical constituents between the stratosphere and troposphere. *J. Atmos. Sci.*, **37**, 994-1004

VOLZ, A., D.H. EHHALT and R.G. DERWENT, 1981: Seasonal and latidudinal variation of ^{14}CO and the tropospheric concentration of OH radicals. *J. Geophys. Res.*, **86**, 5163-5171

THE RELATIONSHIP BETWEEN OZONE AND HYDROGEN PEROXIDE: FIELD OBSERVATION AND MODEL EVALUATION

Menachem Luria
CIRES, University of Colorado, Boulder, Colorado 80309, USA
The Hebrew University, Jerusalem 91904, Israel

Charles C. Van Valin and Joe F. Boatman
GMCC, Air Resources Laboratory, NOAA
Boulder, Colorado 80303, USA

John D. Ray
CIRES, University of Colorado
Boulder, Colorado 80309, USA

ABSTRACT

Continuous measurements of ozone and hydrogen peroxide were made using the NOAA King Air research aircraft. The atmosphere was sampled over the United States through the latitude range between 29° and 41°N, during seasonal research periods. The concentration of O_3 showed relatively small seasonal variation; flight averages ranged between 49 and 70 ppb. The concentration of H_2O_2 showed strong seasonal and intermittent latitudinal effects. Flight averages of H_2O_2 ranged from 0.2 ppb during winter to >6 ppb during the summer. Although both H_2O_2 and O_3 are produced photochemically, no clear interdependence between their concentrations was found. The flight data further indicated that increased H_2O_2 levels are heavily dependent upon humidity. A computer model for the photochemical reaction system was used to simulate the production of H_2O_2 and O_3 under conditions similar to the experimental observations. The model results were in close agreement with the averages and the extremes of the seasonal values of the field data.

1. INTRODUCTION

Ozone and hydrogen peroxide are present in the atmosphere as a result of photochemical activity. The production of both substances requires the presence of free radicals; however, the production of these substances proceeds via two different branches of the free radical-oxidation mechanism. Tropospheric O_3 formation follows the photodissociation of nitrogen dioxide:

$$NO_2 + h\nu \rightarrow NO + O(^3P) \quad (1)$$

$$O(^3P) + O_2 \rightarrow O_3. \quad (2)$$

The conversion of NO to NO_2 proceeds with the aid of hydrocarbons (RH) and the hydroxyl radical (HO):

$$HO + RH \rightarrow H_2O + R \quad (3)$$

$$R + O_2 \rightarrow RO_2 \quad (4)$$

$$RO_2 + NO \rightarrow NO_2 + RO \quad (5)$$

$$RO + O_2 \rightarrow R'CHO + HO_2 \ (R{=}R'CH_2) \quad (6)$$

$$HO_2 + NO \rightarrow HO + NO_2. \quad (7)$$

This mechanism indicates that maximum O_3 formation will occur at an optimal RH to NO_x ratio (Dunker, 1984). When this ratio is not maintained, for example when NO levels are low, an increase in the concentration of the hydroperoxy radical (HO_2) is predicted. This radical cannot promote ozone production directly; however its recombination leads to the production of H_2O_2:

$$HO_2 + HO_2 + M \rightarrow H_2O_2 + O_2 + M. \quad (8)$$

According to the scheme of Atkinson and Lloyd (1984), the rate of this reaction is much higher when $M{=}H_2O$ than if M represent any other constituent.

The Central U.S. RADM Test and Assessment INtensives (CURTAIN) experiment (in which RADM is the Regional Acid Deposition Model) provided the opportunity to evaluate the concentrations of O_3 and H_2O_2 during the four seasons and over the latitude range of 29° to 41°N. In this report we discuss the relationship between these two quantities and the influence of other environmental quantities on their production. The analysis was assisted by the atmospheric photochemical model of Luria and Meagher (1988), previously used for the analysis of data from the 1985 Western Atlantic Ocean Experiment (Luria et al. 1987; Van Valin and Luria 1988).

2. EXPERIMENTAL DESCRIPTION

Air was sampled during flights from Little Rock, AR (35°N, 92°W), along a north-south track near 91.5°W between 29° and 41°N. The aircraft used was the National Oceanic and Atmospheric Administration (NOAA) instrumented King Air C-90. Measurements were carried out during seasonal research periods: February 5-10, April

ISBN 0-937194-15-8

22-26, July 22-26, and October 22-28, 1987. The general operation plan for each period included two daytime northward flights, two daytime southward flights and two nighttime flights within 250 km north and south of Little Rock. Most flights were made at two altitudes of about 1600 and 2600 m above sea level. Further details about the experiment were given by Boatman et al. (1988a).

The instruments carried by the King Air were described by Wellman et al. (1988) and by Boatman et al. (1988b). During the flights, flask samples were collected for CO and C_1-C_5 hydrocarbon analysis and filter samples were collected for anion analysis. Water vapor density and solar radiation were monitored. Continuous measurements included temperature, wind speed, wind direction, altitude, heading and position, and concentrations of SO_2, NO_x (summer only), O_3, and H_2O_2.

In this report we discuss some factors affecting the concentrations of O_3 and H_2O_2. Ozone was measured with a photometric monitor (Thermo Instruments Inc., Model 49) and the H_2O_2 with a continuous analyzer employing the enzyme catalyzed dimerzation of p-hydroxyphenylacetic acid (Lazrus et al., 1985).

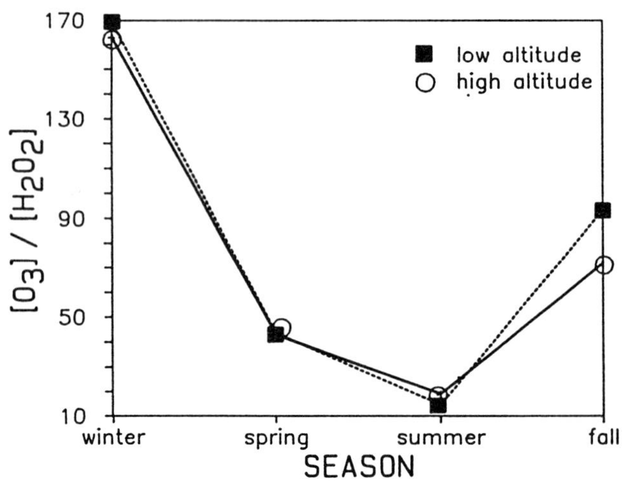

Fig. 1. The ratio between the concentrations of O_3 and H_2O_2 as a seasonal function. The closed squares represent the average of all lower altitude measurements; the open circles represent the average of all higher altitude measurements.

3. EXPERIMENTAL RESULTS

Average values for the main quantities of interest, i.e., O_3, H_2O_2, H_2O, temperature, and solar radiation, were calculated for each seasonal study period at both the lower and higher altitudes. The seasonal means and standard deviations are presented in Table 1. The average concentrations of both O_3 and H_2O_2 were usually slightly higher at the lower sampling altitude, especially during the summer, but the change with altitude was within the standard deviation. All of the lower altitude measurements and many of the higher altitude measurements were performed within the planetary boundary layer.

As expected, maximum concentrations of O_3 and H_2O_2 were observed during the summer; however, whereas the O_3 maximum was 40% greater than the minimum, the extreme concentrations of H_2O_2 differed by a factor of more than 10. The ratio between O_3 and H_2O_2 concentrations at both altitudes is presented in Fig. 1. The quantities that are expected to influence the production rate of the peroxides (O_3 and H_2O_2) are solar radiation, hydrocarbons, NO_x and water vapor (Table 1). Of these quantities, the solar radiation increases by 50-90% from the winter minimum to the summer maximum, non-methane hydrocarbons decrease by a factor of more than 2, and water vapor content increases by a factor of nearly 5.

TABLE 1. SUMMARY OF AVERAGE METEOROLOGICAL DATA AND CONCENTRATIONS OF QUANTITIES OBSERVED DURING THE FOUR SEGMENTS OF THE CURTAIN EXPERIMENT

	Lower Altitude				Higher Altitude			
Quant.	Win.	Spr.	Sum.	Fall	Win.	Spr.	Sum.	Fall
O_3, ppbv	51 (3)	70 (3)	63 (10)	51 (3)	49 (3)	69 (6)	61 (7)	55 (5)
H_2O_2, ppbv	0.3 (0.1)	1.6 (1.2)	4.1 (1.2)	0.54 (0.4)	0.3 (0.1)	1.6 (1.3)	3.3 (1.2)	0.76 (0.5)
H_2O, %	0.34 (0.20)	0.78 (0.15)	1.60 (0.75)	0.47 (0.36)	0.24 (0.14)	0.54 (0.10)	0.87 (0.40)	0.28 (0.3)
NMHC, ppbvC	19.0 (15.0)		9.0 (5.0)		11.5 (11.0)		5.3 (3.9)	
NO_x, ppbv			1.0 (1.0)					
SRD, W m^{-2}	434 (131)	492 (59)	821 (195)	327 (72)	660 (56)	527 (73)	991 (150)	496 (102)
Temp., °C	5.6 (2.2)	11.2 (3.2)	18.0 (2.2)	7.2 (1.1)	1.9 (2.0)	7.1 (1.7)	11.6 (0.8)	3.4 (1.7)

Note: Arithmetic means are presented for each quantity and standard deviations are given below the means, in parentheses; NMHC--nonmethane hydrocarbons (C_2-C_5); SRD--solar radiation.

4. MODEL DESCRIPTION

The atmospheric photochemical model employed in this study was previously used to simulate the sulfur chemistry in the remote marine atmosphere (Luria and Meagher, 1986), the transformation of anthropogenic pollutants during transport over the Atlantic Ocean (Luria and Meagher, 1988) and the relationship between dimethyl sulfide and particulate sulfate over the mid-Atlantic Ocean (Luria et al., 1988). The numerical model is a two-level Lagrangian scheme. One level (the mixed layer, ML) is for detailed chemical simulation; the other level (the free troposphere, FT) serves as a reservoir that is available for exchange. The model assumes that the ML is homogeneous in the vertical and horizontal dimensions. The model simulates the diurnal solar cycle and the diurnal change of the ML, using the parameterization of Fishman and Carney (1984).

The chemical mechanism employed is based on the scheme of Atkinson and Lloyd (1984), with modification to include ethane and methane chemistry. Photolysis rates were calculated for the geographical central point (Little Rock, AR) for each of the four sampling seasons. Deposition velocities were taken from Chatfield and Crutzen (1984) and supplemented by those used by Fishman and Carney (1984). Model calculations were carried out for a period of 5 days to achieve a complete equilibrium among the various species. Integration of the chemical equations was carried out on a Digital Equipment Corp. VAX/780 computer system, using the Gear (1971) algorithm. Computation time for simulating 5 days was about 15 min.

5. MODEL RESULT

Initial conditions for the model runs for summer and winter were based on the chemical and meteorological observations made at Little Rock during the summer and winter research periods; these are referred to as the standard summer and winter runs. A summary of the initial concentrations for both seasons is given in Table 2, and initial meteorological quantities in Table 3. In the calculations the model assumed that the concentrations of hydrocarbons and water vapor were maintained constant at the average observed values for the duration of the calculation. The model assumed a constant source of NO that was sufficient to provide a daily average NO_x ($NO + NO_2$) concentration of approximately 1 ppbv. The introduction of a constant source for NO is based on the fact that NO chemistry is relatively rapid and it is converted to NO_2 and/or other NO_x species at a rate dictated by the time of day and the concentration of the other reactive species. For the standard conditions, and during both seasons, the model assumed that 80% of the maximum potential solar radiation is available for photochemistry. The 20% reduction was to account for clouds above flight altitude and light scattering by aerosols; all flights were conducted in clear air. In the present study the influence of four major quantities on O_3 and H_2O_2 concentrations were examined. Simulations were performed by maintaining all but one quantity constant and varying that quantity by a factor of 3-4 greater or less than the averaged observed value. A summary of the model runs, showing the input and output values, is presented in Table 4. The model predicted a near-sinusoidal diurnal cycle for both O_3 and H_2O_2, having a maximum level at sunset (between 1800 and 2000 h) and a minimum before sunrise (between 0500 and 0700 h). The O_3 and H_2O_2 concentrations listed in Table 4 represent the daily extreme predicted levels.

TABLE 2. A SUMMARY OF THE INITIAL CONCENTRATIONS USED FOR THE STANDARD MODEL SIMULATIONS

Compound	Average concentration (ppbv) winter	summer
H_2O_2	0.1	3.0
O_3	50	60
H_2O	3.4E+06	1.7E+07
C_2H_4	0.35	0.15
C_2H_6	3.2	1.4
C_3H_6	0.4	0.16
C_3H_8	2.0	1.0
C_4H_{10}	1.2	0.6
NO(ER)	0.25*	0.25*

* No emission rate; ppt min^{-1}

TABLE 3. METEOROLOGICAL QUANTITIES USED AS MODEL INPUT

Quantity	Winter	Summer
Cloud coverage, %	20%	20%
Solar radiation expressed as NO_2 photolysis rate-- $J(NO_2)_{max}$	0.44	0.54
Mixed layer height		
Lower	1000	1000
Higher	2500	2500

TABLE 4. SUMMARY OF COMPUTER MODEL RUNS

	Model input				Model output			
Run ID	H_2O ppmv	HC ppbvC	SRD (%)	NO(ER) ppt/min	$[H_2O_2]$, ppbv Min.	Max.	$[O_3]$, ppbv Min.	Max.
Runs with increased water vapor concentration--summer								
S22	4000	9	80	0.25	0.8	1.6	74	84
S9	8000	9	80	0.25	1.6	3.0	73	82
S6*	16000	9	80	0.25	2.9	5.3	69	77
S7	24000	9	80	0.25	4.0	7.3	66	74
S8	32000	9	80	0.25	4.9	8.9	63	70
Runs with increased water vapor concentration--winter								
W3	1400	19	80	0.25	0.09	0.11	47	55
W2#	3400	19	80	0.25	0.12	0.25	49	60
W1	4700	19	80	0.25	0.14	0.35	49	61
W4	6000	19	80	0.25	0.17	0.45	50	63
W5	9000	19	80	0.25	0.21	0.66	50	63
Runs with increased NO emission rate--summer								
S10	16000	9	80	0.1	2.7	4.8	64	66
S11	16000	9	80	0.2	2.9	5.3	68	75
S6*	16000	9	80	0.25	2.9	5.3	69	77
S12	16000	9	80	0.5	2.7	5.4	74	88
S13	16000	9	80	1.0	1.7	4.4	72	98
Runs with increased NO emission rate--winter								
W10	3400	19	80	0.1	0.16	0.43	50	57
W11	3400	19	80	0.2	0.15	0.32	49	60
W2#	3400	19	80	0.25	0.12	0.25	49	60
W12	3400	19	80	0.5	0.07	0.09	42	48
W13	3400	19	80	1.0	0.06	0.08	25	39
Runs with increased HC concentration--summer								
S14	16000	2.5	80	0.25	2.6	4.7	64	71
S15	16000	4.8	80	0.25	2.7	4.9	66	74
S6*	16000	9	80	0.25	2.9	5.3	69	77
S16	16000	18	80	0.25	3.3	5.9	74	84
S17	16000	27	80	0.25	3.6	6.4	77	88
Runs with increased HC concentration--winter								
W14	3400	4.9	80	0.25	0.08	0.10	46	51
W15	3400	9.8	80	0.25	0.09	0.13	47	53
W2#	3400	19	80	0.25	0.12	0.25	49	60
W16	3400	38	80	0.25	0.18	0.51	51	66
W17	3400	54	80	0.25	0.26	0.77	52	71
Runs with decreased solar radiation--summer								
S18	16000	9	100	0.25	3.7	6.9	71	80
S6*	16000	9	80	0.25	2.9	5.3	69	77
S19	16000	9	60	0.25	2.1	3.8	66	74
S20	16000	9	40	0.25	1.3	2.3	62	69
S21	16000	9	20	0.25	0.6	1.0	55	61
Runs with decreased solar radiation--winter								
W6	3400	19	100	0.25	0.15	0.40	51	65
W2#	3400	19	80	0.25	0.12	0.25	49	60
W7	3400	19	60	0.25	0.10	0.15	47	54
W8	3400	19	40	0.25	0.09	0.11	45	50
W9	3400	19	20	0.25	0.08	0.09	43	47

NOTES: The symbols * and # represent the standard model conditions for summer and winter respectively; SRD--solar radiation; NO(ER) NO--emission rate and HC--hydrocarbons.

6. DISCUSSION

The annual cycle and absolute concentrations of O_3 observed in this study are consistent with those reported for rural locations at similar geographical latitude, such as in Missouri (Logan, 1985) and Israel (Lifschitz et al., 1988). The levels are also close to those observed in our early study off the coast of Virginia and near Bermuda (Van Valin and Luria, 1988). Hydrogen peroxide concentrations were usually in the range of concentrations from other measurements performed at similar rural locations (Lazrus et al., 1986; Olszyna et al., 1988). The peak summer concentration is, however, the highest yet reported.

The computer model was successful in predicting the concentrations of H_2O_2 and O_3 for both the winter and summer seasons. A comparison between the observations and the model is shown in Fig. 2. This figure illustrates the agreement between model predictions for various water vapor concentrations and the lower altitude experiment average at the observed humidity. As shown in this figure, the concentration of H_2O_2 is strongly dependent on the absolute humidity. Even when all other parameters are maintained constant the model predicts that the concentration of H_2O_2 will increase by a factor of >3 when the relative humidity is increased from 20 to 70% (from 4000 to 16000 ppm). Likewise, the winter H_2O_2 is predicted to change from a daily maximum value of 0.11 to 0.66 ppbv when the relative humidity changes from 0.2 to 1% at the corresponding winter temperature (from 1400 to 9000 ppm). The strong effect of the humidity on H_2O_2 production is the result of reaction (8). Although increased H_2O increases free-radical formation by means of:

$$H_2O + O(^1D) \rightarrow 2\ HO, \tag{9}$$

the model indicates that the enhancement is insufficient to make up for the loss of free radicals by reaction (8), thus resulting in a slight decrease of O_3 production (see summer O_3 plot in Fig. 2).

The differences between the summer andwinter humidities do not account for the entire difference in the production of oxidant. The model indicates that for the atmospheric composition observed during the CURTAIN experiment, oxidant production increases with increasing solar radiation. As shown in Table 4, most summer and winter simulations were carried out with 80% of the radiation available for each season. However, the total daily solar radiation dosage for the summer is 83% greater than that for winter, on the basis of differences in the NO_2 photolysis rate ($J(NO_2)$) and the number of daylight hours. Thus, the change in solar radiation between the seasons provides a significant supplementary avenue for the change observed in the oxidant level.

For the conditions of the present study, an increase in the hydrocarbon concentration also enhances oxidant formation. However, here the hydrocarbons (HC) served to diminish the

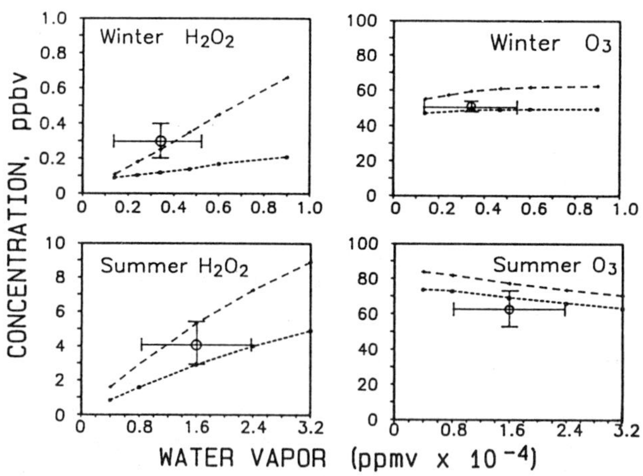

Fig. 2. Experimental observation and model simulation of the mixing ratios of H_2O_2 and O_3 for the winter and summer sampling seasons as a function of the water vapor mixing ratio. The circles represent the observed averages for the season at the lower altitude. The lines connect the model values predicted for various humidities; the attached error bars represet the standard deviations. The lower line represents the minimum value observed before sunrise and the upper line represents the maximum value observed at sunset.

summer-winter gap because winter HC levels were more than double the summer levels. In fact, when the observed winter HC concentrations were used in the model with summer conditions for the other quantities, the concentrations of both O_3 and H_2O_2 increased by 7-14% (Table 4).

As mentioned earlier, NO_x data are available only for the summer season; the standard model conditions assume that similar levels were present other seasons. For summer, the model predicts that a fourfold increase in NO would increase O_3 maximum levels by almost 30%. However, a further increase would result in O_3 depletion. For H_2O_2 the optimal HC/NO is different; the model predicts maximum concentration near the experimentally observed HC to NO ratio. The winter simulation showed a different picture. In this case the effect of increased NO from the levels assumed for the standard conditions resulted in decreased oxidant levels.

7. CONCLUSION

The small seasonal variations of O_3 (maximum to minimum ratio of 1.4) and large seasonal variations of H_2O_2 (maximum to minimum ratio of 14) observed over the rural south-central United States were duplicated using a homogeneous photochemical model. For the conditions of the experimental observations the model indicated that the increase in solar radiation from winter to summer contributed to the enhancement of the oxidant level; a similar enhancement is derived from increased absolute humidity. The model also predicted that maximum production of H_2O_2 occurs at a higher HC to NO_x value than the one required for maximum production of O_3.

REFERENCES

Atkinson, R., and A. Lloyd, 1984: Evaluation of kinetic and mechanism data for modeling of photochemical smog. J. Phys. Chem. Ref. Data, **13**, 315-444.

Boatman, J.F., D.L. Wellman, M. Luria, C.C. Van Valin and R.C. Schnell, 1988: In-flight intercomparison of some aircraft meteorological and chemical measurement techniques. Global Biogeochem. Cycles, **2**, 1-11.

Boatman, J.F., D.L. Wellman, C.C. Van Valin, R.L. Gunter, J.D. Ray, Y. Kim, S.W. Wilkison and M. Luria, 1988b: Airborne sampling for validation of the Regional Acid Deposition Model. J. Geophys. Res. (submitted).

Chatfield, R.B., and P. Crutzen, 1984: Sulfur dioxide in remote oceanic air: Cloud transport of reactive precursors. J. Geophys Res., **89**, 7111-7132.

Dunker, A.M., S. Kumar and P.H. Berzines, 1984: A comparison of chemical mechanisms used in atmospheric models. Atmos. Environ., **18**, 311-326.

Fishman, J., and T.A. Carney, 1984: A one dimensional photochemical model of the troposphere with planetary boundary layer parameterization. J. Atmos. Chem., **1**, 351-376.

Gear, C.W., 1971: The automatic integration of ordinary differential equations. Commun. Assoc. Comput. Mach., **14**, 1179.

Lazrus, A.L., G.L. Kok, S.N. Gitlin, J.A. Lind and S.E. McLaren, 1985: Automated fluorimetric method for hydrogen peroxide in atmospheric precipitation. Anal. Chem., **57**, 917-922.

Lazrus, A.L., G.L. Kok, J.A. Lind, S.N. Gitlin, B.G. Heikes and R.E. Shetter, 1986: Automatic fluorimetric method for hydrogen peroxide in air. Anal. Chem., **58**, 594-597.

Lifschitz, B., M. Peleg and M. Luria, 1988: The influence of long range transport on O_3 levels at a rural site in Israel. J.

Atmos. Chem. (in press).

Logan, J., 1985: Tropospheric ozone: Seasonal behavior, trends, and anthropogenic influence. J. Geophyss. Res., **90**, 10463-10482.

Luria, M., and J. F. Meagher, 1986: Computer simulations of the oxidation and removal of natural sulfur compounds in the marine atmosphere. Proceedings of the 7th World Clean Air Congress, Sydney, Australia, **Vol I**, 295-302.

Luria, M., C.C. Van Valin, J.F. Boatman and R.F. Pueschel, 1987: Sulfur dioxide flux measurements over the western Atlanlic Ocean. Atmos. Environ. 21, 1631-1636.

Luria, M., and J.F. Meagher, 1988: Computer simulation of the boundary layer oxidation and removal of atmospheric pollutants over the western Atlantic Ocean. Atmos. Environ., 22, 307-316.

Luria, M., C.C. Van Valin, W.C. Keene, D.L. Wellman and J.F. Boatman, 1988: The relationship between dimethyl-sulfide and particulate sulfate in the mid-Atlantic Ocean atmosphere. Atmos. Environ. (in press).

Olszyna, K.J., J.F. Meagher and E.M. Bailey, 1988: Gas phase cloud and rain water measurements of hydrogen peroxide at a high elevation site. Atmos. Environ. (in press).

Van Valin, C.C., and M. Luria, 1988: O_3, CO, hydrocarbons and DMS over the western Atlantic Ocean. Atmos. Environ. (in press).

Wellman, D.L., M. Luria, C.C. Van Valin and J.F. Boatman, 1988: The use of an airborne air sampling platform for regional air quality studies. National Oceanic and Atmospheric Administration, Technical Report (in press).

OBSERVATIONS OF OZONE PROFILES IN THE DEVELOPING CONVECTIVE BOUNDARY LAYER

*Carmen J. Nappo and Jerold A. Herwehe**
Air Resources Laboratory, NOAA
Oak Ridge, Tennessee 37831, USA

Anne M. Thompson
NASA Goddard Space Flight Center
Greenbelt, Maryland 20771, USA

ABSTRACT

Simultaneous profiles of meteorological variables and ozone concentrations were obtained using a tethered-balloon sounding system. Soundings were made approximately once each hour, from before sunrise until about mid-day. Presented here are some sample observations and qualitative interpretations of the temporal variation of ozone profiles in developing convective planetary boundary layers.

1. INTRODUCTION

This note reports on some initial results of a long-term program of observations of ozone in the planetary boundary layer (PBL). Here, we present some sample observations and qualitative interpretations of the temporal variation of ozone profiles in developing convective boundary layers.

2. INSTRUMENTATION AND FIELD SITES

Simultaneous profiles of temperature, wind speed, wind direction, water vapor, and ozone concentration are made with a tethered-balloon sounding system. Ozone concentrations are measured using a lightweight, wet-chemistry sensor. The sensor is calibrated before each experimental period. The balloon ascent rate is kept between 0.25 and 0.5 m/s, and observations are taken only during the ascent.

The observations were made at two sites near Oak Ridge, Tennessee, which lies near the western side of the Tennessee River Valley. This is a region of complex terrain consisting mostly of parallel ridges and valleys running from southwest to northeast. The ridges are typically 30 to 75 m high and 1000 m wide, and the valleys between the ridges are typically 1-2 km wide. The area surface cover is mostly deciduous forest. The winds are generally from the SW, up-valley, during the day, and from the NE, down-valley, during the night. A 950 Mw coal-fired power plant with a 244 m exhaust stack lies NE of the observation sites.

The observations were made on 29 July 1987 and 19-20 November 1987. The July observations were made at a site about 7.2 km SW of the power plant and in the same valley containing the plant. The ground elevation of this site is the same as that of the power plant. The November observations were made at a site about 13.6 km SW of the plant, but in a valley adjacent to the July site and with a ground elevation about 40 m higher. During the November observations, several controlled burns of forest material were made about 3 km NE of the observation site.

3. RESULTS

29 July, 1987: This day began with a light fog which dissipated by 08:00 EDT; however, a heavy dew remained on the surface vegetation until about 10:00 EDT. Clear skies prevailed throughout the morning. Profiles of wind speed, wind direction, potential temperature, and ozone concentration during the morning hours are presented in Fig. 1. A strong, ground-based inversion extended upward to about 400 m AGL during the first sounding, Fig. 1(c); however, by 10:00 EDT this inversion had been destroyed by convection, resulting in a near-neutral thermal structure. Winds were from the northeast throughout the morning, Fig. 1(b). A jet located at about 300 m AGL is seen in the 08:37 EDT sounding, Fig. 1(a); however, it is gone by 10:00 EDT, and for the remainder of the day wind speeds were variable at about 3 m/s throughout the PBL. From these profiles, we can conclude that convection and entrainment are the primary processes affecting the meteorology in the PBL.

Profiles of ozone concentration throughout the morning are shown in Fig. 1(d). An interesting feature in this figure is the

*Oak Ridge Associated Universities, Oak Ridge TN. On permanent assignment to ATDD.

ISBN 0-937194-15-8

concentration minimum between 280 and 470 m in the 08:37 EDT profile. We attribute this "hole" in the ozone profile to the effects of emissions from a power plant. The strong buoyancy of this plume can produce a plume rise of a few hundred meters within a few kilometers of the plant. As shown in Fig. 1(b), winds were from the northeast up to at least 600 m, and so the plume was most likely over our site throughout the morning hours. Van Valin and Ganor (1987) report on the effects of a power plant plume on PBL ozone concentrations measured near Boulder, Colorado. They show that during the night, ozone concentrations within the power plant plume are low due to scavenging of ozone by NO.

By 10:07 EDT, ozone concentration is essentially vertically uniform below 480 m. The 11:39 EDT profile shows no trace of this ozone depletion. The heavy broken-line segments in Fig. 1(d) represent conjectures as to what the profiles might look like in the absence of the power plant plume. These profiles, show large increases in PBL ozone throughout the morning. The average ozone flux between the ground surface and 350 m is estimated from these profiles to be about -0.89 ppbv m/s, which is much greater than the -0.19 ppbv m/s measured by Lenschow et al. (1981) over eastern Colorado. Using a mass balance, they concluded that the major source of ozone in the PBL is photochemical production. In the present case, vertical transport of ozone into the PBL by entrainment from above is important.

19 November, 1987: Profiles of wind speed, wind direction, potential temperature, and ozone concentration are shown in Fig. 2. Until about 10:00 EST, winds were steady and from the northeast; after this time, winds became increasingly variable and light, especially below 300 m. The growth of the convective PBL is shown in the evolution of the potential temperature profiles in Fig. 2(c).

The wind speed, wind direction, and potential temperature profiles shown in Fig. 2 are typical of those found in a developing

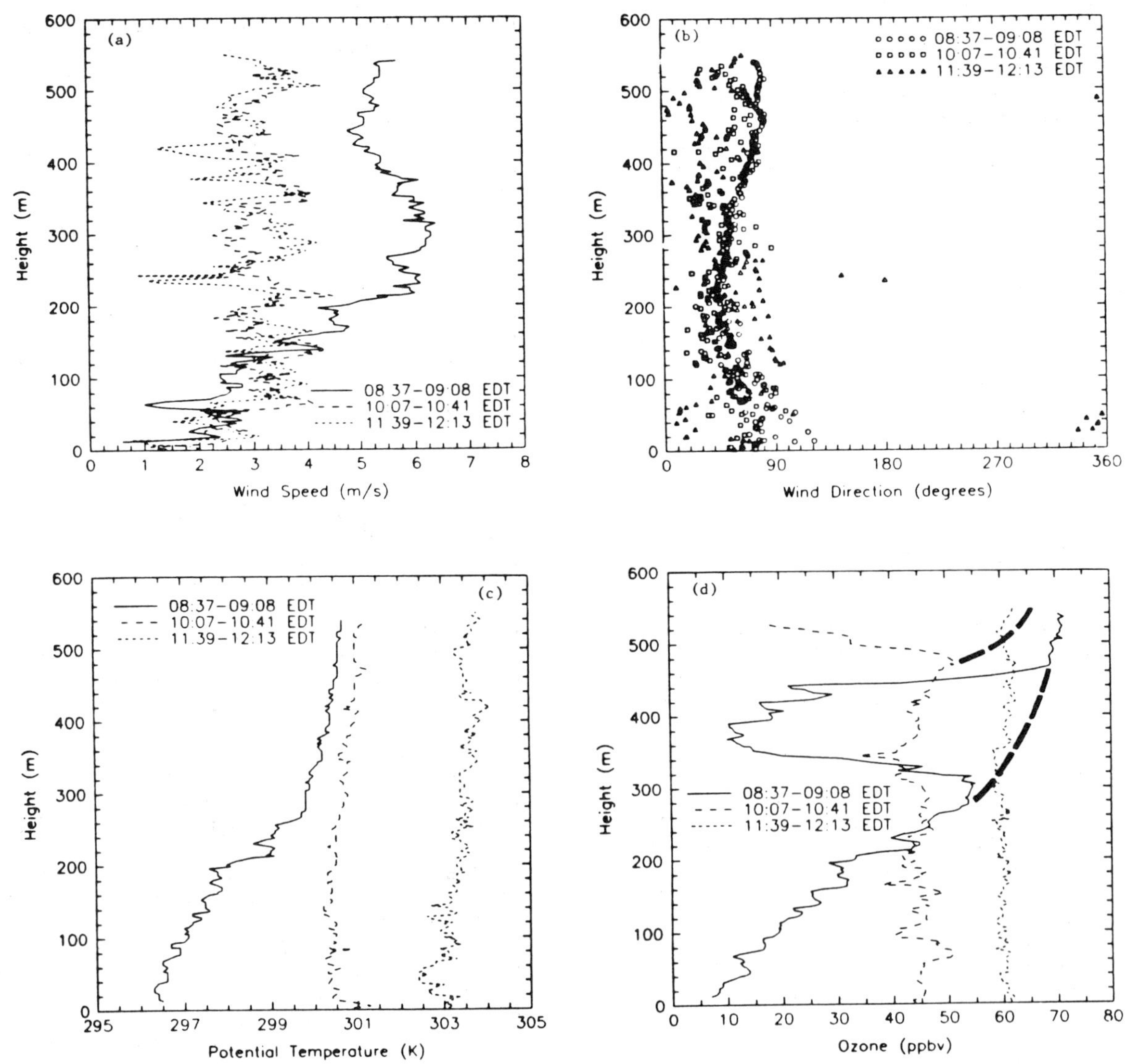

Fig. 1. Profiles from 29 July, 1987: (a) wind speed, (b) wind direction, (c) potential temperature, and (d) ozone concentration.

convective PBL. However, the ozone profiles presented in Fig. 2(d) indicate a behavior which is quite different from that shown by a passive constituent of the PBL. The scavenging of ozone by the power plant plume is again seen in the 09:21 EST and 10:05 EST soundings. The 10:53 EST sounding does not show a zone of diminished ozone; however, such a zone centered at about 500 m is seen in the 12:06 EST profile. The temporal changes of ozone concentration at all heights appear almost random. In the lower PBL, say below 200 m, the fluctuations may be due to the brush fires only a few kilometers upwind of the site.

20 November, 1987: On this day, scattered clouds and high winds associated with an approaching front retarded the development of a convective PBL. Instead, a near-neutral boundary layer extended from the ground surface upward to about 350 m, and this persisted up to at least 11:00 EST when soundings had to be terminated due to extreme winds and turbulence. The meteorological profiles are presented in Fig. 3(a,b,c). The winds were from the west and relatively steady above about 300 m.

The ozone concentration profiles shown in Fig. 3(d) contrast strongly with those observed on the previous day, Fig. 2(d). Notably absent are the elevated zones of diminished ozone concentration. This we attribute to the absence overhead of the power plant plume. The plume on this day would have been to the east, and hence directed about 135° away from the observation site. Also, the plumes from the controlled fires, which may have affected ozone concentrations near the ground surface on the previous day, were directed toward the northeast on this day. This in part can explain the vertical uniformity of the ozone profiles in Fig. 3(d). Ozone concentration increases throughout the PBL in a generally uniform way, and this increase we attribute primarily to photochemical production.

4. CONCLUSION

We have presented some initial results of

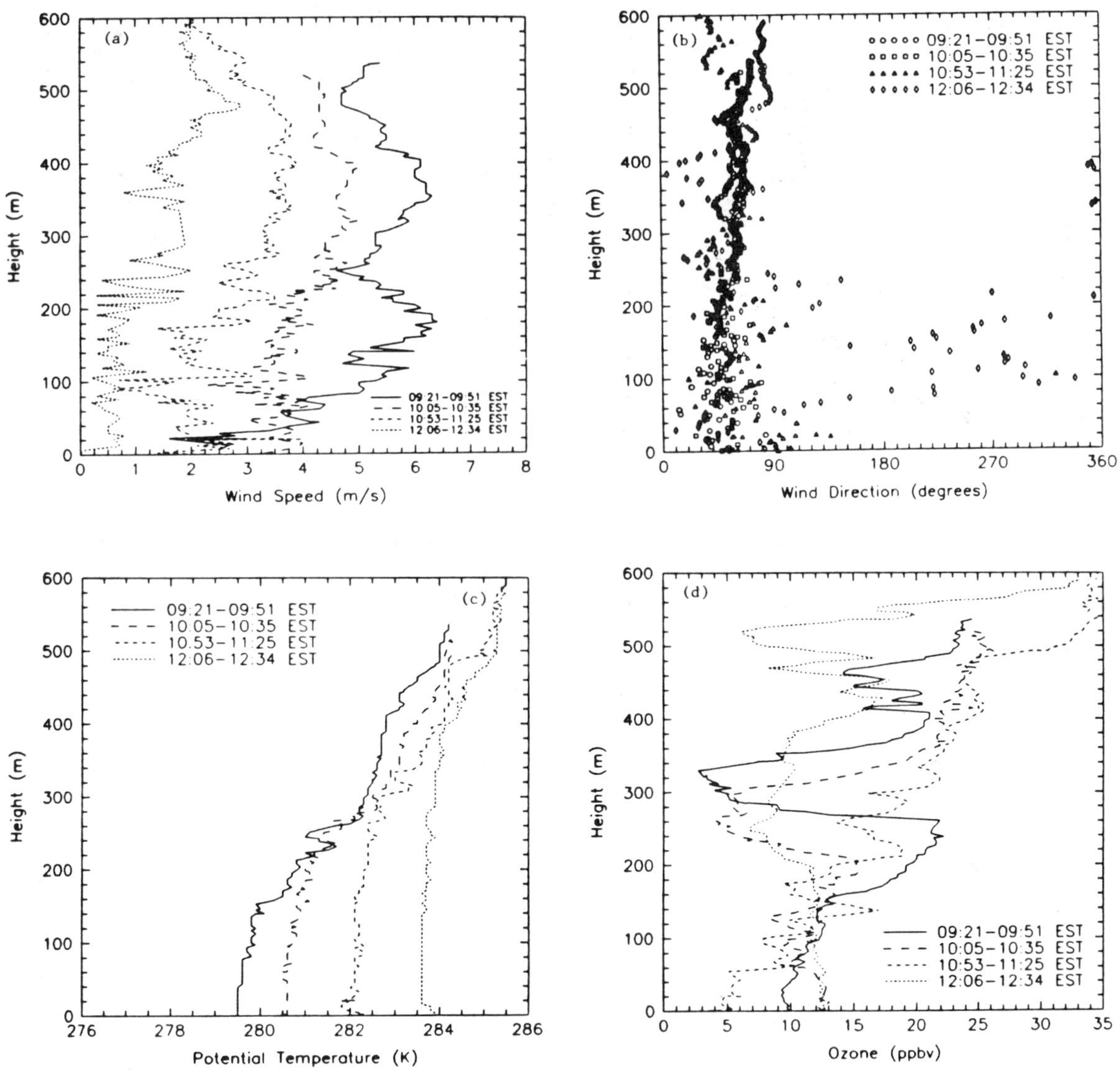

Fig. 2. Same as Fig. 1, but from 19 November, 1987.

a long term program of ozone measurements through the PBL. The results should be representative of ozone in a mid-latitude forested continental region of moderately complex topography subject to both natural and anthropogenic sources of NOx, hydrocarbons, and CO. Nearby sources of pollution can significantly affect the vertical distribution of ozone. Although the destruction of ozone by these pollutants is expected to occur primarily at nighttime, our measurements show that the effects of a power plant plume continue well into the morning hours.

ACKNOWLEDGEMENTS

This work was conducted under an agreement between the U.S. Department of Energy and the National Atmospheric and Oceanic Administration, and with the support of the NASA Tropospheric Chemistry program.

REFERENCES

Lenshow, D. H., R. Pearson, Jr., and B. B. Stankov, 1981: Estimating the ozone budget in the boundary layer by use of aircraft measurements of ozone eddy flux and mean concentration. J.G.R., 86, 7291-7297.

Van Valin, C. C., and E. Ganor, 1987: Air pollution measurements at the Boulder Atmospheric Observatory. Water, Air and Soil Pollution, 35, 357-372.

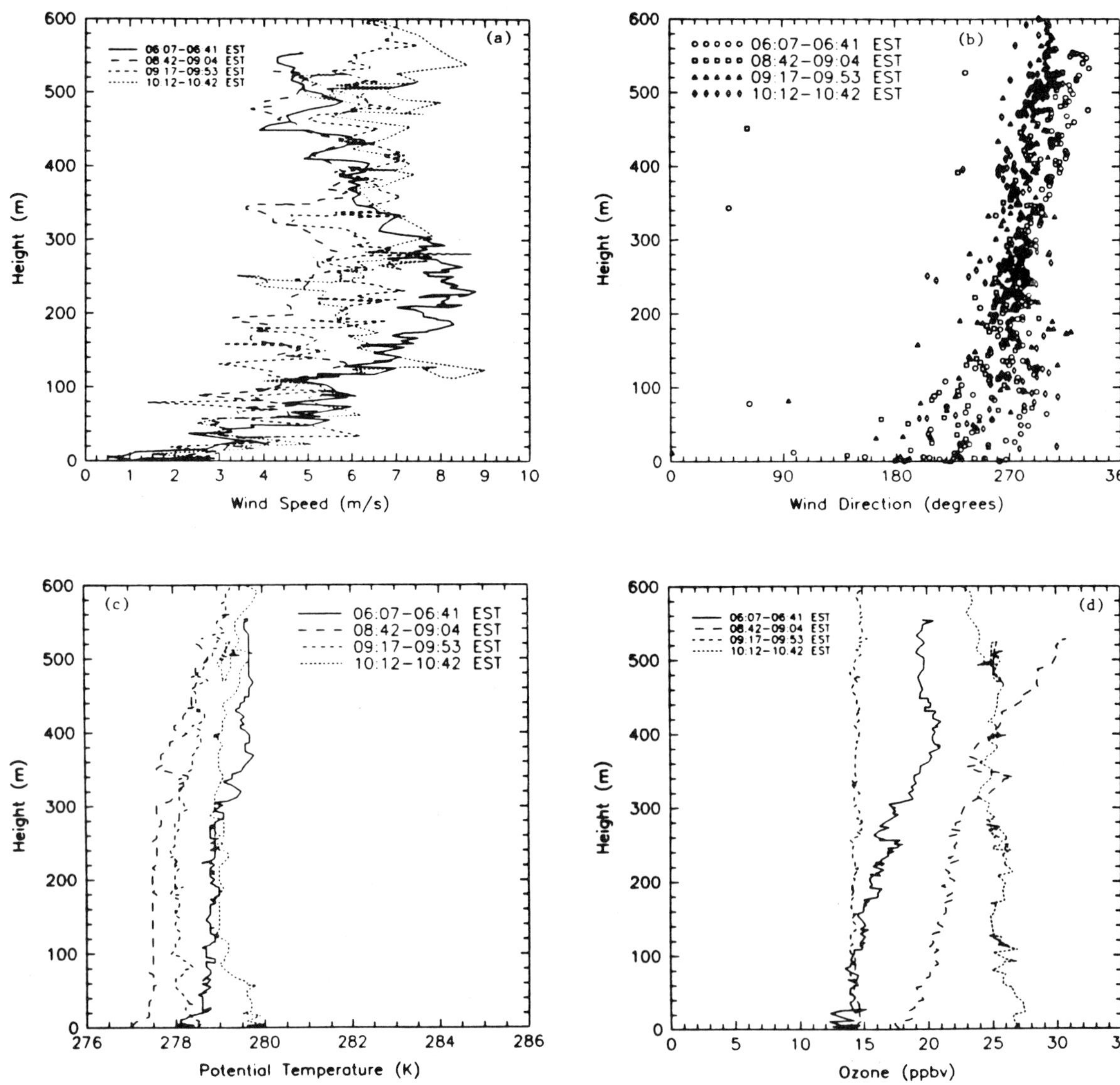

Fig. 3. Same as Fig. 1, but from 20 November, 1987.

OZONE CONCENTRATION BEHAVIOUR AT MARINE AND FORESTED SITES IN FINLAND

H. Hakola, P. Taalas, S. Joffre, and H. Lättilä
Finnish Meteorological Institute
00810 Helsinki, Finland

We present results of two years of ozone measurements, carried out in Utö, a small island in the Baltic Sea, and one and a half years in Ähtäri, a forested site in Central Finland. The annual variation shows two peaks in Utö, one in March and the other in July. In Ähtäri there is only one peak, in spring (fig. 1).

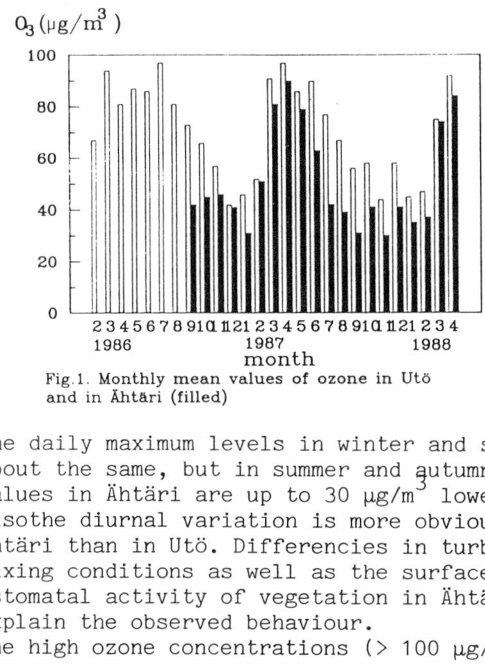

Fig.1. Monthly mean values of ozone in Utö and in Ähtäri (filled)

The daily maximum levels in winter and spring are about the same, but in summer and autumn the values in Ähtäri are up to 30 µg/m^3 lower (fig.2) Alsothe diurnal variation is more obvious in Ähtäri than in Utö. Differencies in turbulent mixing conditions as well as the surface sinks (stomatal activity of vegetation in Ähtäri) explain the observed behaviour.
The high ozone concentrations (> 100 µg/m^3) were found to occur when the wind direction is 90°-279° in Ähtäri, and 90°-240° in Utö, which shows that long-range transport is usually the cause for enhanced ozone concentrations. The smallest diurnal amplitudes occur with wind directions corresponding to high ozone concentrations. This is observed in spring and in summertime (fig.3).

The yearly variation of ozone concentration departures at Utö in connection with meteorological/chemical factors has been studied by a multi-correlation method. The monthly ozone concentrations were corrected for the annual trend. The effectiveness of various ozone sinks/sources seems to have an observable yearly variation, while the effectiveness of surface sink (sea/ice) remains approximately constant. We note that in spring ozone correlates positively with Be-7.

The positive correlation with the global radiation is only observed in summer. In winter ozone correlates negatively with relative humidity and SO_2. This would indicate that photochemical production and stratospheric intrusions influence ozone concentration behaviour in spring and summer while chemical sinks seem to be important in winter and autumn.

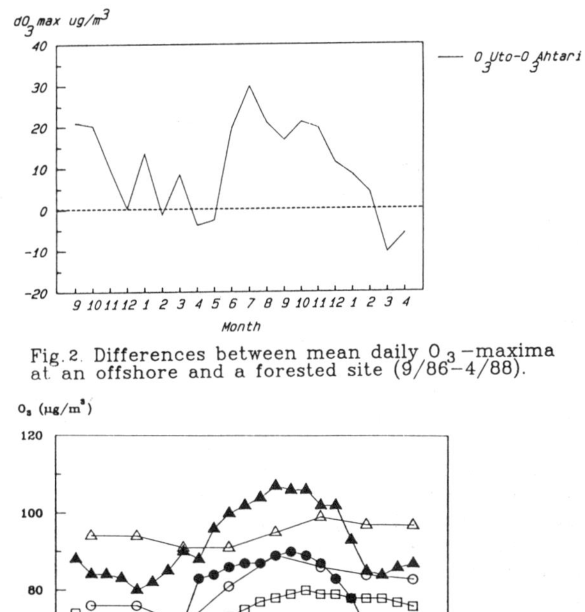

Fig.2. Differences between mean daily O_3-maxima at an offshore and a forested site (9/86-4/88).

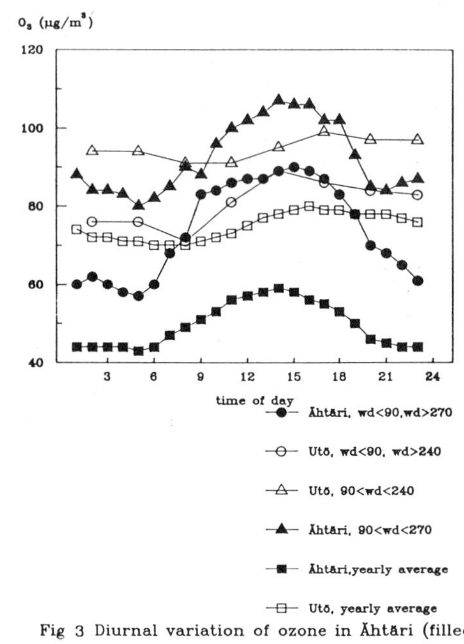

Fig 3 Diurnal variation of ozone in Ähtäri (filled) and in Utö in different wind direction sectors in spring and the yearly averages.

ISBN 0-937194-15-8

NOCTURNAL OZONE CONCENTRATIONS AT A RURAL SITE IN NORTHWEST ENGLAND

I. Colbeck
Institute of Aerosol Science, University of Essex
Colchester CO4 3SQ, UK

ABSTRACT

Occasions when daily ozone maxima occurred at night time have been investigated. They may be classified into two groups; one due to the breakdown of the nocturnal inversion, by an increase in turbulence; and the other the result of high windspeeds and approximately constant ozone levels through the day. Vertical profiles of ozone and temperature have been used to infer deposition velocities. These show that dry deposition is adequate to account for the observed diurnal variation of ozone.

1. INTRODUCTION

The diurnal variation of ozone in rural areas has been well described during the last fifty years (Auer, 1939; Garland and Derwent, 1979; Colbeck and Harrison, 1985a; Meagher *et al.*, 1987). In the summer months, a maximum in the afternoon and a minimum in the early morning generally occurs. The warm, sunny anticyclonic conditions which bring high daytime ozone levels are also conducive to the formation of nighttime inversions, due to radiative cooling at the surface. Above this inversion there are only limited chemical sinks for ozone and since vertical mixing through the stable layer is inhibited a reservoir of ozone is formed. Below the inversion, there are also limited chemical sinks, but deposition will be possible as a slow mass transfer takes place towards the ground. Concentrations at ground level remain low until the following morning as daybreak causes the breakup of the nocturnal inversion allowing downward mixing of ozone-rich air from aloft. Photochemical production produces an ozone maxima in the early to late afternoon.

A contrasting cycle of ozone is often observed at mountain tops (Stasiuk and Coffey, 1974; Wolff *et al.*, 1987) with a nocturnal maxima. However, at some mountain sites, with good exposure, little diurnal variation has been observed (Galbally et al., 1986).

2. NOCTURNAL MAXIMA

Ozone has been measured at rural sites near Lancaster. The data analysed in this study was collected between 1981 and 1984. Details of the monitoring sites, equipment, and a summary of the data are given elsewhere (Colbeck and Harrison, 1985b). Figure 1 shows histograms for the time of occurrence of the daily maximum and minimum concentration for the summer months (May to September). A similar diagram for the winter months (November to March) is shown in Figure 2. It is clear that while the summer months exhibit a large diurnal variation, the winter months show little variation. In the winter, periods of strong winds or overcast weather occur and the formation of the nocturnal stable layer is inhibited. Alternatively, in clear winter weather, the stable layer may persist throughout the day and night and concentrations may remain low for days on end.

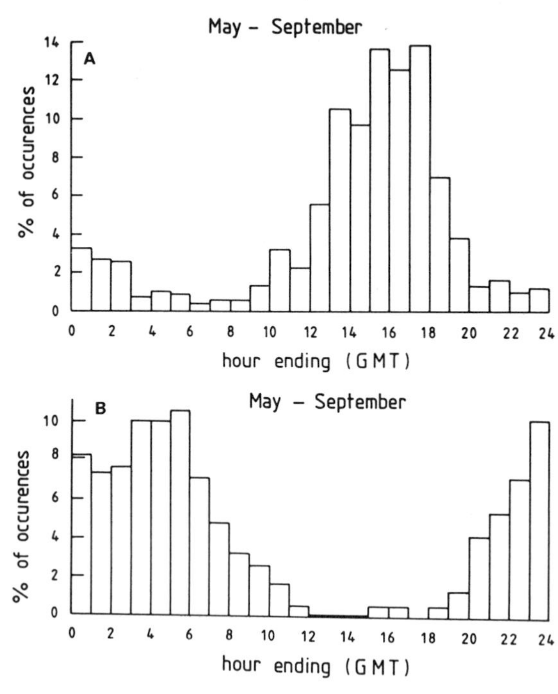

Fig. 1 Percentage of occurrences of the daily (A) maximum and (B) minimum hourly ozone concentrations for each hour of the day: May-September.

ISBN 0-937194-15-8

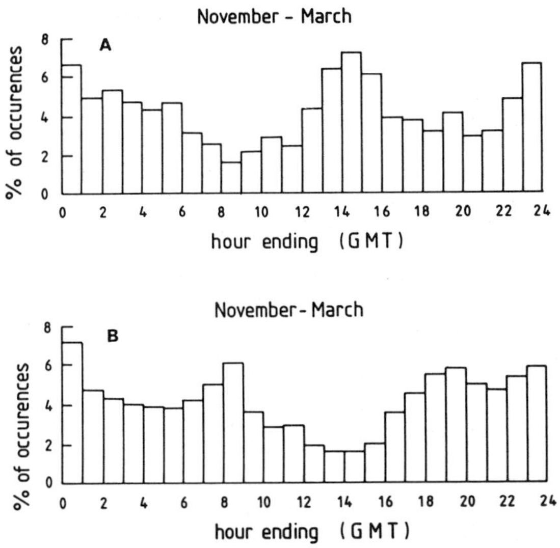

Fig. 2 Percentage of occurrences of the daily (A) maximum and (B) minimum hourly ozone concentrations for each hour of the day: November-March

It is evident from Figure 1 that minimum ozone levels rarely occur during the afternoon in the summer, as expected. However, the daily maximum concentration occurred during the night (hour ending 2300-0040 GMT) approximately 11% of the time. Each occurrence of nighttime maxima was then examined to investigate the processes involved in nighttime ozone maxima.

Average ozone concentrations between 1300-1800 GMT are normally much greater than any nocturnal hourly average concentration. The nocturnal maxima were compared with the previous afternoon's average and approximately 50% of the maxima exceed the afternoon average by 8 ppb or less (Colbeck, 1988). On all these occasions, ozone levels were approximately constant throughout the day, (Figure 3) windspeeds were high and the meteorological conditions were generally not conducive to ozone formation. In this case nocturnal maxima are purely random and not due to any systematic phenomena. Of the remaining occasions, those with concentration differences in excess of 8 ppb, a change in the stability of the boundary layer occurred. This change gives rise to two different nocturnal profiles: one where the windspeed falls, (Figure 4) hence an increase in atmosphere stability and a gradual decline in ozone; and one (Figure 5) where the windspeed increases resulting in enhanced vertical mixing. The former type of nocturnal maxima accounted for 15% of the occurrences, while just over 30% of the maxima could be correlated with an increase in windspeed. There were also several occasions when rainfall was recorded simultaneously with increased windspeed and ozone.

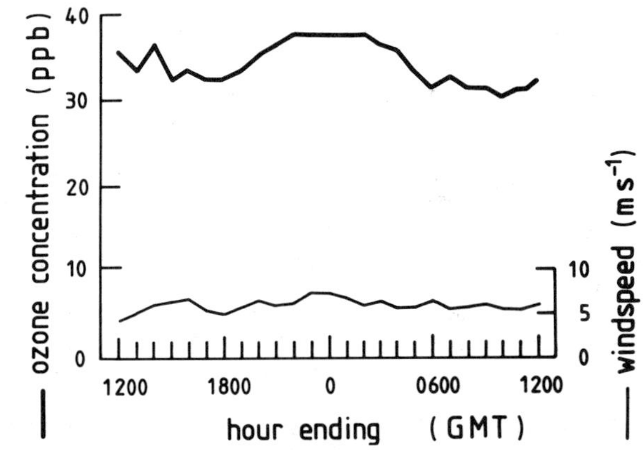

Fig. 3 Ozone concentration and windspeed measured on 27-28 May 1983.

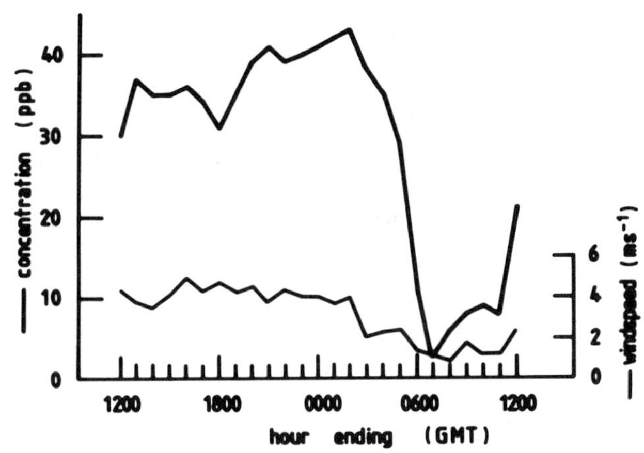

Fig. 4 Ozone concentration of windspeed measured on 28-29 September 1981 (Colbeck, 1988).

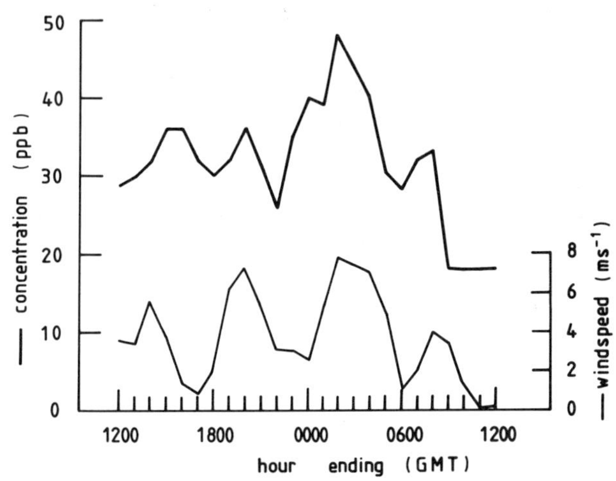

Fig. 5 Ozone concentration and windspeed measured on 24-25 September 1983 (Colbeck, 1988).

3. VERTICAL PROFILES AND DRY DEPOSITION

Vertical profiles of ozone, during the summer, have shown that the nocturnal ozone depletion is limited to the lowest tens of metres of the atmosphere (Colbeck and Harrison, 1985a; Varey *et al.*, 1988). Concentrations show little vertical variation from about midday through to the late afternoon, consistent with a well mixed atmosphere. During the night the ozone concentration is only depleted in the lowest layer of the atmosphere, the height of which corresponds to a surface-based radiation inversion.

Examination of the ozone records during photochemical episodes showed that after 1700-1800 GMT the concentration generally declines in a roughly exponential fashion, (Figure 6). The concentration, c, after a time, t, may be expressed as:

$$c(t) = c(0)\exp(-\lambda t) \quad (1)$$

where λ is the decay constant.

Garland and Derwent (1979) studied this decay in some detail and gave an expression relating the time constant to the deposition velocity, V_g, referenced to 10 m above ground level and the height of the boundary layer, h:

$$\lambda = V_g/h. \quad (2)$$

From a knowledge of windspeed and cooling rate they predicted the height of the mixed layer. Colbeck and Harrison (1985a) calculated the nocturnal mixing height, according to this method, for various photochemical episodes.

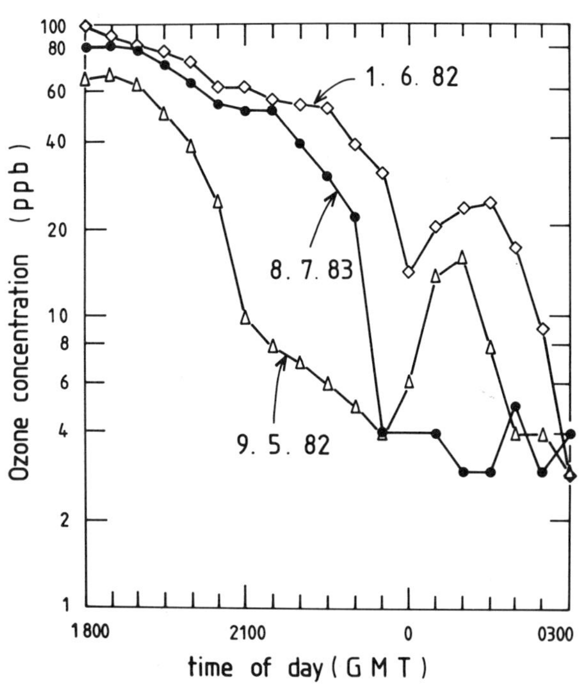

Fig. 6 Decrease in ozone concentration with time of day

Estimated mixing height ranged from 22m to 326m. From this and the observed decay constants it is possible to calculate the deposition velocity. The results are summarized in Table 1. The deposition velocities ranges from 0.06 to 1.18 cms^{-1} with a mean value of 0.48 cms^{-1}.

On three nights Colbeck and Harrison (1985a) also measured vertical profiles of ozone and temperature. On the night of 8-9th July 1983 the observed decay constant was 0.61 hr^{-1}, the height of the inversion obtained from the vertical profile, was approximately 30m. Hence, from the above expression, the deposition velocity is 0.51 cms^{-1}. For the night of 11-12th July 1983 a deposition velocity of 0.2 cms^{-1} may be obtained. On night of 10-11th July 1983 the measured mixing height was 24m, compared with the calculated height of 46m. Using the measured mixing height gives a deposition velocity of 0.16 cms^{-1}, compared with 0.24 cms^{-1} assuming the height calculated after Garland and Derwent. This discrepancy appears to result from the model over-estimating the boundary layer depth.

TABLE 1 Estimated deposition velocities at 10m

Date	Observed decay constant (h^{-1})	Estimated mixing height (m)	Deposition velocity (cms^{-1})
29.8.81	0.21	33.7	0.20
4.9.81	0.06	33.3	0.06
7.9.81	0.20	96.8	0.54
10.5.82	0.81	22.6	0.51
11.5.82	0.05	172.9	0.24
12.5.82	0.13	326.1	1.18
30.5.82	0.11	144.0	0.44
1.6.82	0.33	50.1	0.46
3.6.82	0.32	43.3	0.38
4.6.82	0.40	31.3	0.35
30.7.82	0.19	53.1	0.28
5.8.82	0.69	18.4	0.35
17.9.82	0.21	53.2	0.31
31.5.83	0.37	54.4	0.56
19.6.83	0.12	98.7	0.33
9.7.83	0.27	78.0	0.59
10.7.83	0.19	45.9	0.24
16.7.83	0.35	85.9	0.84
18.8.83	0.54	42.6	0.64
19.8.83	0.22	175.0	1.07

All the deposition velocities calculated in this paper have been referenced to 10m above the ground. The ground cover at the sites was grass. Previous workers have measured nighttime deposition velocities over grass and obtained values of 0.29 cms^{-1} at 1m (2) and 0.24 cms^{-1} at 2m (Galbally and Roy, 1980). Since the concentrations decreases towards the ground, deposition velocities at 10m are expected to be lower than those at 1 or 2m. Deposition velocities calculated from two of the vertical profiles agree with this. However, deposition velocities calculated by equation (2) give relatively high values. This may be the result of

(1) the mixing height being over-estimated, as on 10-11th July; or
(2) that some of the assumptions in the calculation, such as horizontal homogeneity, were not fulfilled. The largest values occur at high wind speeds suggesting that the model of Garland and Derwent is inappropriate in such cases. It is doubtful whether inversion height and mixing conditions, as assumed in their model, remain constant.

4. CONCLUSION

Occasions when the daily ozone maxima occurs between the hours of 2200 and 0400 GMT, in the summer months, can be classified into two groups: one when ozone levels are approximately 'constant' throughout the day, average windspeeds are above 5 ms^{-1} and atmospheric conditions are not conducive to photochemical ozone formation; and another due to a change in the atmospheric stability. Hence ground level ozone concentration is a good indicator of the stability of the surface layers of the atmosphere during the night.

Deposition velocities calculated according to the method of Garland and Derwent (1979) are higher than those measured by other workers over similar ground cover. This may well be the result of the assumptions in the model were not fulfilled in the atmosphere. While equation (2) is a good approximation in an idealised inversion model, it is probably a crude approximation in most real circumstances. Deposition velocities calculated from measurements of the decay constant and mixing height are in much better agreement.

REFERENCES

Auer, R. 1939: Über der täglichen Gang des Ozonehaltes der bodennahen Luft. Beiträge zur Geophysik, 54, 187-145.

Colbeck, I. 1988: The occurrence of nocturnal ozone maxima at a rural site in NW England. Environ. Tech. Lett., 9, 75-80.

Colbeck, I. and R.M. Harrison 1985a: Dry deposition of ozone - some measurements of deposition velocity and of vertical profiles to 100 metres. Atmos. Environ., 19, 1807-1818.

Colbeck, I. and R.M. Harrison 1985b: The frequency and causes of elevated concentrations of ozone at ground-level at rural sites in NW England. Atmos. Environ., 19, 1577-1587.

Galbally, I.E. and C.R. Roy, 1980: Ozone and nitrogen oxides in the southern hemisphere. Proc. International Ozone Symposium, Boulder, Co.

Galbally, I.E., A.J. Miller, R.D. Hoy, S. Ahmet, R.C. Joynt and D. Attwood 1986: Surface ozone at rural sites in the Latrobe Valley and Cape Grim, Australia. Atmos. Environ., 20, 2403-2422.

Garland, J.A. and R.G. Derwent 1979: Destruction at the ground and the diurnal cycle of concentration of ozone and other gases. Quart. J. R. Met. Soc., 105, 169-183.

Meagher, J.F., N.T. Lee, R.J. Valente and W.J.Parkhurst 1987: Rural ozone in the southeastern United States. Atmos. Environ., 21, 605-615.

Stasiuk, W.N. and P.E. Coffey 1974: Rural and urban ozone relationships in New York State. JACPA, 24, 564-568.

Varey, R.H., D.J. Ball, A.J. Crane, D.P.H. Laxen and F.J. Sandalls 1988: Ozone formation in the London plume. Atmos. Environ., in press.

Wolff, G.T., P.J. Lioy and R.S. Taylor 1987: The diurnal variations of ozone at different altitudes on a rural mountain in the eastern United States. JAPCA, 37, 45-48.

COMPARISON OF METHODS TO MEASURE OZONE DRY DEPOSITION

I. Colbeck and A. Simmons
Institute of Aerosol Science
University of Essex
Colchester CO4 3SQ, UK

ABSTRACT

Deposition velocities for ozone to grassland have been measured using the profile method and the box method. From a preliminary analysis of the results it appears that the two techniques give consistent results. The variation of deposition velocity with friction velocity is also investigated.

1. INTRODUCTION

The destruction of ozone at the earth's surface plays an important role in the budget of tropospheric ozone. Dry deposition is the direct absorption/destruction of a gas at a surface. This distinguishes it from removal by precipitation, although water surfaces as well as vegetation and soil may participate. In studies of dry deposition the rate of transfer, or deposition velocity, v_g of a substance from the atmosphere to a surface may be expressed by

$$v_g = F/c(z)$$

where F is the downward flux of depositing species and c its concentration at height z, close to the surface. The deposition velocity varies over different types of terrain and vegetation, and, because of its dependence on concentration, its value will depend on the height at which the concentration is measured.

Dry deposition velocities have been measured by several techniques. The main methods are: (1) The box method, (Aldaz, 1969; Regener and Aldaz, 1969; Galbally and Roy, 1980) which consists of an open-bottomed box covered inside with an inert material. Ozone is injected into the box and its rate of decay is measured. The rate of decrease of ozone within the system is equal to its destruction rate at the surface being studied; (2) The profile method, (Garland and Derwent, 1979; Delany and Davies, 1983; Colbeck and Harrison, 1985) equates the ozone flux with the product of ozone gradient and the associated eddy diffusivity. It uses the hypothesis that, under appropriate conditions, the transfer processes of ozone and momentum are identical, so that eddy diffusivity theory is valid; and (3) Eddy correlation methods, (Welsey et al., 1978 & 1982; Droppo, 1985) which involve measuring concentration variations, c', over a period of 1 second or less. The covariance between c' and fluctuations in the vertical windspeed w', yield the vertical flux i.e.

$$F = -\overline{c'w'}.$$

In this paper, we describe the advantages and disadvantages of the box and profile methods for determining ozone dry deposition velocities. Some preliminary results, obtained by both methods, are reported. Finally, using data from Colbeck and Harrison (1985) the variation of deposition velocity with friction velocity is examined.

2. THE BOX METHOD

As stated earlier, this consists of a box lined with a material inert to ozone, with an open base which is placed upon the test surface. A small amount of ozone is injected into the box, and rate of decrease of ozone within the system is monitored. If the concentration is uniform throughout the volume V of the box and if A is the exposed area at the base of the box then the rate of change of ozone is given by

$$\frac{d\,[O_3]}{dt} = -\frac{K'A\,[O_3]}{V}$$

where K' is a constant. Galbally and Roy have shown K' is equivalent to the deposition velocity. The above equation implies that the ozone in the box decays exponentially with the time constant V/K'A. Since some ozone is destroyed on the walls of the box some allowance must be made for it. This may be determined by placing the box on the inert liner and measuring the decay. Figure 1 shows first order plots for ozone destruction over grass and FEP Teflon. From these the half life ($t_{½}$) for ozone decay may be determined. With adjustment for wall losses it is found (3)

$$(t_{½}\text{ surface})^{-1} = (t_{½}\text{ measured})^{-1} - \frac{5}{6}(t_{½}\text{ inert base})^{-1}$$

This value of $t_{½}$ surface may then be used to calculate v_g. Deposition velocities obtained by this method are 0.6 cm/s over grass, 0.04 cm/s over water and 0.02 cm/s over snow (Aldaz, 1969). Alternatively, others (Galbally and Roy, 1980) have obtained 0.63 cm/s over grass during the day, 0.25 cm/s over grass during the night, 0.1 cm/s over snow and 0.1 cm/s over water.

ISBN 0-937194-15-8

The main advantage of this method is that it enables the surface resistance to ozone uptake to be measured. Other techniques calculate this value by difference between measured and calculated parameters. Other advantages are that it can be used over various surfaces without the 'fetch' limitations of profile methods, a detailed knowledge of the atmospheric boundary layer is not required, apart from the gas analyser the remainder of the equipment is relatively cheap and additionally the measurements may be made relatively quickly.

Measurements made by this method are useful, but do not represent conditions that actually occur in nature. It has been pointed out that artifical stirring within the box creates a loss rate that is higher than occurs naturally (Fabian and Junge, 1970). Stirring ensures a uniform ozone density within the volume of the box whereas in the atmosphere, a sharp vertical ozone gradient exists in the lowest metre.

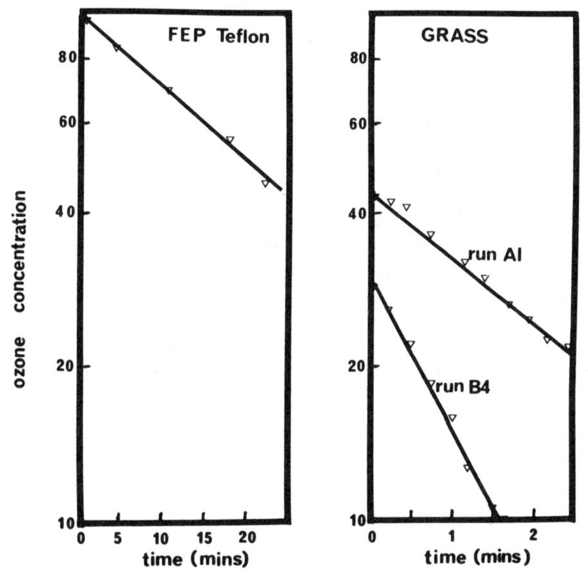

Fig. 1. Decay in ozone concentration over FEP Teflon and grass.

3. THE PROFILE METHOD

This method is based on the calculation of the ozone flux from the vertical gradients of concentration, windspeed and temperature. The techniques for calculating the flux and deposition velocity are described in detail elsewhere (Garland, 1977; Nicholson, 1988). Normally, the vertical flux, F is related to the concentration gradient dc/dz and eddy diffusivity K by

$$F = K\, dc/dz$$

Assuming a log linear wind profile and correcting for the effects of convection or damping the flux is given by

$$F = k^2 \frac{\dfrac{du}{d\ln(z-d)} \; \dfrac{dc}{d\ln(z-d)}}{\Phi_M \, \Phi_H}$$

where k is von Karman's constant, d is zero plane displacement, u is wind velocity and Φ_M and Φ_H are the dimensionless shears of momentum and heat respectively. Typically a reference height of 1 m is taken. This ensures that the concentration gradient is measured in the constant flux part of the turbulent boundary layer.

The advantage of this method over eddy correlation is that fast response analysers are not required. The windspeed and temperature gradients used to determine the eddy diffusivity may be used to infer atmospheric resistances which describe the processes which control deposition.

This method requires an extensive, uniform area to ensure a constant flux layer. Therefore, this area should extend upwind for a fetch of at least a hundred times the maximum observation height. Local sources or the advection of a different air mass during sampling may give rise to spurious vertical fluxes. Concentration differences are normally less than 10% (Garland, 1977) and should be made at two or more heights with a resolution of 1% or less. By allowing measurements at several heights by the same detector systematic errors are eliminated. Ideally, the meteorological conditions should not change significantly during the period of measurement.

4. COMPARISON OF METHODS

Ozone deposition velocities to grassland were measured using the box method and the profile method. The box consisted of a wooden frame (0.5 m per side) internally lined with FEP Teflon film. The air inside the box was stirred by a Teflon fin powered by an external electric motor. Air was sampled from the box at a rate of 0.8 l/min and analysed by a Monitor Labs 8810 photometric analyser. The equipment used in the profile studies is similar to that described by Colbeck and Harrison (1985) with the chemiluminescent analyser being replaced by a photometric analyser. A schematic diagram of the profile apparatus is shown in Figure 2.

The results are summarised in Table 1. On run A the profile method gave a deposition velocity of 0.4cm s^{-1} and the box method gave 0.22cm s^{-1}, whereas run B resulted in similar values but those for each method have interchanged. On this day, the first three box runs gave similar results ($v_g \approx 0.4$cm s^{-1}), while the final box run showed an increase in v_g to 0.63cm s^{-1}. It was noted that during these measurements, it became sunny and the temperature increased. The first two profile measurements for run C show low values of r_s, and the resistance is dominated by r_a. At the end of the second profile run the box method showed that v_g had fallen to 0.18cm s^{-1}. The following profile run gave a similar result. In

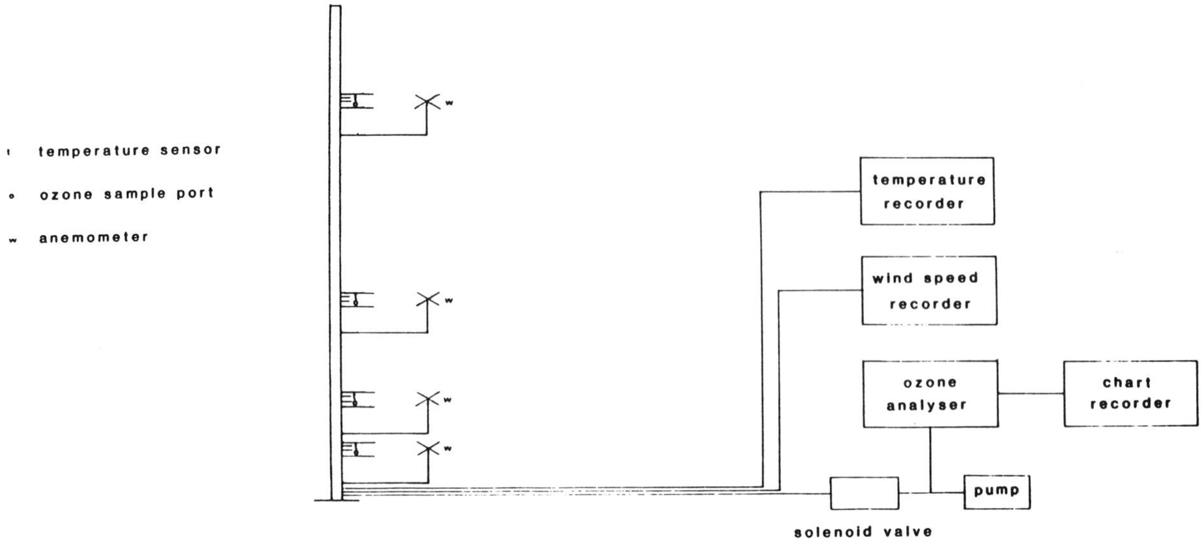

Fig. 2. Schematic diagram of experimental setup for profile measurements

TABLE 1 Comparison of Box and Profile Measurements

	Method	Local time	u_* (cm s^{-1})	v_g (cm s^{-1})	r_a (s cm^{-1})	r_b (s cm^{-1})	r_s (s cm^{-1})
A	Profile	1145-1250	26.3	0.40	0.52	0.26	1.74
	Box	1225		0.22			
		1504		0.14			
B	Profile	1151-1309	27.8	0.22	0.52	0.23	3.87
	Box	1315		0.41			
		1330		0.33			
		1344		0.36			
		1355		0.63			
C	Profile	1029-1147	19.0	0.81	0.74	0.36	0.16
		1148-1314	23.7	0.73	0.59	0.27	0.51
		1342-1454	23.5	0.22	0.59	0.27	3.69
	Box	1324		0.18			
		1504		0.52			
		1519		0.44			
		1530		0.57			
		1540*		0.24			

*air inside box not stirred

this case r_a and r_b were the same as profile 2 but r_s had increased by a factor of 7. It is possible that towards the end of profile 2 the conditions controlling deposition were changing, but by sampling over 90 minutes these changes were not sufficient to alter significantly the final result. Later box runs found v_g to be approximately 0.5cm s^{-1}.

From a preliminary analysis of the results it appears that the box method gives consistent results. Before definite conclusions can be drawn between the profile and box methods the duration of the profile method needs to be reduced so that changes in ground and atmospheric conditions are not averaged out. Work is now in hand to enable r_s to be measured by the box method. This will then enable direct comparison of r_s obtained by both methods.

5. CALCULATIONS OF v_g FROM FRICTION VELOCITY

To relate the deposition velocity to transport processes in the surface boundary layer, the concentration at the effective surface must be known. To study transfer mechanisms close to the ground the reciprocal of deposition velocity, the resistance, may be divided into various components. The total resistance r to transport from some height to the surface can be written as the sum of the resistances from each of the processes involved:

$$r(z) = r_a(z) + r_b + r_s$$

where r_a is the aerodynamic resistance, r_b relates to transport to the surface through air close to surfaces where microphysical or molecular processes are dominant, and r_s is the resistance controlled by the surface itself.

It has been shown that (Galbally and Roy, 1980)

$$r_a = u(z)/u_*^2$$

$$r_b = 1/Bu_*$$

where u_* is the friction velocity and B the dimensionless Stanton number. Only rather scanty evidence is available of the values of B for transport of a vapour to vegetation in the field. There are various formulations of B that can be used, but in this study

$$B^{-1} = 10.2\, u_*^{1/3}$$

is used where u_* is in units of m s^{-1}. For a surface roughness in the range 1 to 2 cm values of $u(z)/u_*$ at 1 m, for various sites, have an average value of 12 (Galbally, 1970). This is very close to the mean value of $u(5)/u_*$ found by Garland and Derwent (1979). Colbeck and Harrison (1985) showed an inverse relationship between r_s and u_*. Earlier (Garland and Derwent, 1979) it had been suggested that such a variation could be explained by diffusion through stagnant air among the grass. Previous work where r_s and u_* have been measured has been reviewed and we find that r_s depends on the inverse variation of u_*^{δ}, where δ is constant, $0<\delta<1$. Therefore we find

$$v_g = \left| \frac{12}{u_*} + \frac{10.2}{u_*^{2/3}} + \frac{r_s'}{u_*^{\delta}} \right|^{-1}$$

Figure 3 shows a plot of u_* against v_g using the data from Colbeck and Harrison (1985). Also plotted are curves generated by the above equation for various values of r_s' and δ. Galbally and Roy (1980) found that the best fit to their data, for soil and grassland during the daytime, occurred when r_s' = 100 s m^{-1} and δ = 0. None of the curves in Figure 3 gives a good fit to the data as a whole. When grouped according to surface condition it appears that $r_s' = 50/u_*^{½}$ is the appropriate value, while for wet conditions $r_s' = 200/u_*^{½}$ gives a good fit.

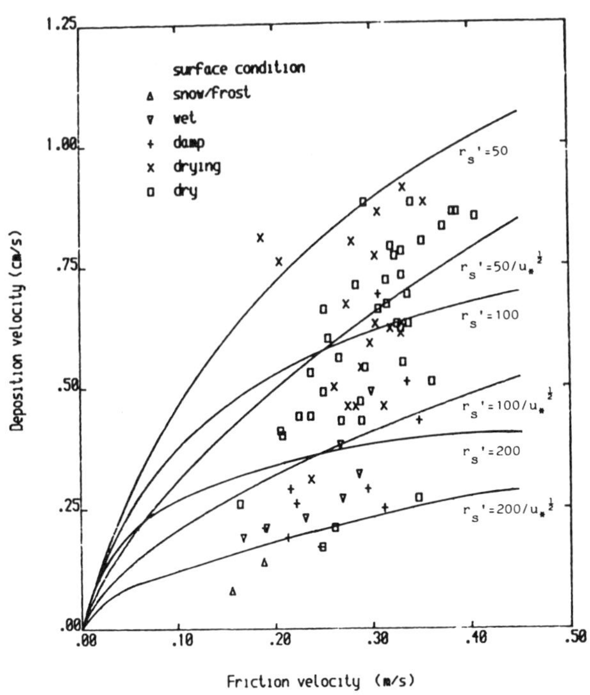

Fig. 3. Variation of deposition velocity with friction velocity

Therefore, we may conclude that the above equation is a good approximation to use to determine daytime ozone deposition when only the friction velocity and surface conditions are known. The equation gave rise to some success for some measurements and some large discrepancies for others.

REFERENCES

Aldaz, L. 1969: Flux measurements of atmospheric ozone over land and water. J. Geophys. Res., 74,6493-6946.

Colbeck I. and R.M. Harrison 1985: Dry deposition of ozone - some measurements of deposition velocity and of vertical profiles to 100 metres. Atmos. Environ., 19, 1807-1818.

Delany, A.C. and T.D. Davies 1985: Dry deposition of NO_x to grass in rural East Anglia. Atmos. Environ., 17, 1391-1394.

Droppo, J.G. 1985: Concurrent measurements of ozone dry deposition using eddy correlation and profile flux methods. J. Geophys. Res., 90, 2111-2118.

Fabian, P. and C.E. Junge 1970: Global rate of ozone destruction at the earth's surface. Arch. Met. Geoph. Biokl., 19, 161-172.

Galbally I.E.: 1971: Ozone profiles and ozone fluxes in the atmospheric surface layer. Quart. J. Roy. Met. Soc., 97, 18-29.

Galbally I.E. and C.R. Roy 1980: Destruction of ozone at the earth's surface. Quart. J. Roy. Met. Soc., 106, 599-620.

Garland J.A. 1977: The dry deposition of sulphur dioxide to land and water surfaces. Proc. Roy. Soc., 354, 245-268.

Garland J.A. and R.G. Derwent 1979: Destruction at the ground and the diurnal cycle of concentration of ozone and other gases. Quart. J. Roy. Met. Soc., 105, 168-183.

Nicholson K.W. 1988: The dry deposition of small particles - a review of epxerimental measurements. Atmos. Environ., in press.

Regener V.H. and L. Aldaz 1969: Turbulent transport near the ground as determined from measurements of the ozone flux and the ozone gradient. J. Geophys. Res., 74, 6935-6942.

Wesely M.L., J.A. Estman, D.R. Cook and B.B. Hicks 1978: Daytime variations of ozone eddy fluxes to maize. Boundary-Layer Met., 15, 361-373.

Wesely M.L., J.A. Eastman, D.H. Stedman and E.D. Yalvac 1982: An eddy-correlation measurement of NO_2 flux to vegetation and comparison to O_3 flux. Atmos. Environ. 16, 815-820.

ATMOSPHERE-SURFACE EXCHANGE OF OZONE TO A SPRUCE CONIFER FOREST: MEASURED AND INFERRED

D.R. Matt and J.D. Womack
Atmospheric Turbulence and Diffusion Division
Oak Ridge, Tennessee 37830, USA

ABSTRACT

Ozone fluxes to a spruce fir forest were measured using eddy correlation technique. The measured concentrations, fluxes and deposition velocities exhibited diurnal cycles as expected. The deposition velocities ranged from $v_d < 0.1$ cm/s during nighttime to daytime values of approximately 0.5 cm/s. Measured results are compared to inferred fluxes and deposition velocities derived from the Dry Deposition Inferrential Model. An analysis of measured and inferred deposition velocities showed that v_d was overestimated because the inferred cuticular resistance is too small for nighttime-dry canopy conditions. During the daytime, the canopy resistance is underestimated in the afternoon when the trees are expected to show water stress.

1. INTRODUCTION

The transfer of ozone to common surfaces in the Eastern Half of the United States has been summarized in a paper by Wesely (1983) in which he discusses measurements made to a variety of surfaces, both biologically active and dormant. In this communication we present the results of measuring the flux of ozone to a spruce-fir forest in Howland, Maine from September 24-29, 1987. Measured deposition rates are compared to the inferred fluxes estimated by the Dry Deposition Inferential Model (DDIM) developed by the Air Resources Laboratory Atmospheric Turbulence and Diffusion Division (ATDD) of National Oceanic and Atmospheric Administration (NOAA). The comparison of the measured to inferred results is important because a version of the DDIM (or very similar model) will be used by the government and private organizations in estimating dry deposition of pollutants in the National Acidic Precipitation Assessment Program (NAPAP) and elsewhere (Hale et al. 1987 and Baer and Nester, 1987).

The DDIM is a resistance analog model discussed in detail by Hicks et al. (1987). In the model the transfer of material (ozone) is represented by

$$F = V_d * [C] \tag{1}$$

where F is the flux of material to the surface, V_d is the deposition velocity, and C is the concentration. The deposition velocity is the inverse of the total resistance to transfer, R, which may be separated into three components in series:

$$1 / V_d = R = R_a + R_b + R_c \tag{2}$$

where R_a is the aerodynamic resistance, R_b is the quasi-laminar boundary layer resistance, and R_c is the surface resistance. R_a characterizes the ability of atmospheric turbulence to transport material to the surface and thus is dependent on atmospheric stability. R_b characterizes the transfer of material through the quasi-laminar boundary layer and is related to the molecular diffusivity of the material and shape of surface. For ozone over biologically active surfaces, the controlling resistance is R_c. Although usually referred to as a canopy resistance, it is the equivalent of parallel resistances to stomata, cuticle, soil, water, etc. A detailed discussion of DDIM's technique for determining R_c including references and sensitivity analysis is discussed by Baldocchi et al. (1987).

2. EXPERIMENTAL MEASUREMENTS

The experimental site is a mixed species coniferous forest with less than 10% composed of deciduous species. Spruce, white pine, and hemlock are the dominant conifer species. In this experiment, anemometry, hygrometry, thermometry, and chemical sensors were all mounted at 24.5 m on a walkup tower extending through the forest canopy. Mean tree height was 20 m thus the instrumentation is mounted at approximately 10 m above the zero plane displacement height (estimated as 0.6 - 0.7 mean tree height).

The fluxes of heat, momentum, sulfur dioxide and ozone were measured by eddy correlation technique following the procedures discussed by Matt et al. (1987). The experimental setup is shown in Figure 1. Ozone and sulfur dioxide were sampled through a draft tube extending to within 25 cm of the sonic anemometer. Measured sulfur dioxide concentrations were below 2 ppb for all but 6

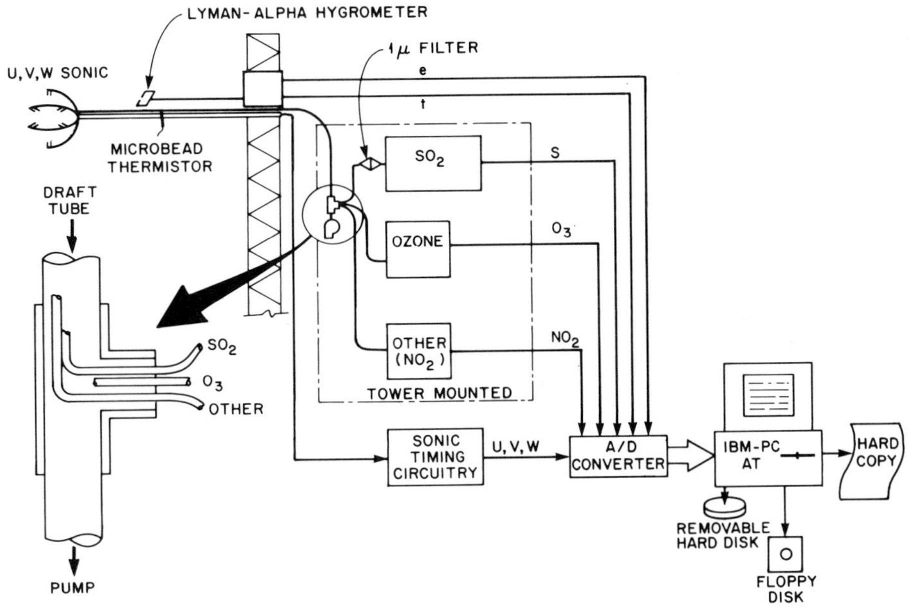

Fig. 1. Schematic representation of the experimental setup. Chemical sensors, anemometry, hygrometry, and thermometry are all mounted about ten meters above zero plane displacement height of the forest.

hours on the last day of sampling thus only the results from the ozone measurements will be discussed here. Data acquisition and processing was done via a portable personal computer using software discussed by McMillen (1986, 1988). This software obtains deviations from the mean using a 200 s digital filter, forms all covariances, and rotates the coordinate system at the end of the averaging period (usually 30 minutes). The vertical exchange rates are interpreted as turbulent fluxes perpendicular to the mean streamline flow. The data are recorded on magnetic media and hard copy. Additionally, this software permits scaling, offsets, and switch selectable continuous data retention for post processing.

3. RESULTS AND DISCUSSION

Time series of some measured micrometeorological variables are shown in Fig. 2. The mean wind, u, was generally above 2 m/s, well above the criteria for application of the eddy correlation technique. The friction velocity, u_*, had a diurnal cycle, and increased during the day to around 20% of u. The energy fluxes are shown in Fig. 2(b). Both sensible heat, H, and latent heat, LE, were measured during the daytime, while only sensible heat was measured at night. Net radiation almost always exceeded the sum of sensible and latent heat suggesting that the forest volume is capable of storing a significant amount of the available energy (between 15-30%). The forest biomass (undried) was determined to be 50 kg/m^2 by plotting available energy versus change in air temperature.

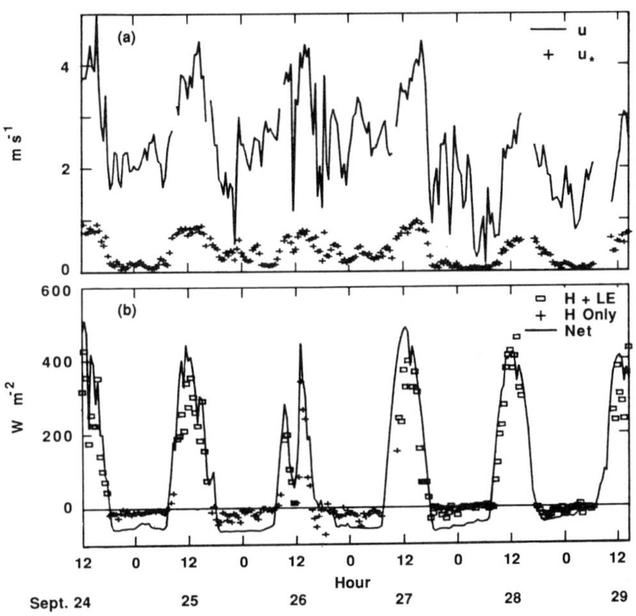

Fig. 2. Time series of (a) winds and (b) energy fluxes for Sept. 24-29, 1987. The forest site is situated on micrometeorologically simple terrain providing excellent fetch in all directions.

The time series of ozone and sulfur dioxide concentrations are shown in Fig. 3(a). The mean concentrations of both sulfur dioxide and ozone are measured by separate instruments than those used for flux measurements, since the flux instruments tend to drift as the temperature proceeds through the diurnal cycle. The ozone concentrations varied from 5 ppb in the early morning to 50 ppb during midday, exhibiting a strong diurnal photochemical cycle. The measured and inferred ozone fluxes are shown in Fig. 3(b). Positive fluxes indicate deposition to the forest surface. Significant deviation between measured and inferred fluxes was observed on Sept. 26 and Sept. 27. On Sept. 26 skies were partly cloudy with some brief scattered showers around noon and again in the evening. The cause of the difference between the measured and inferred fluxes during the afternoon is not known although some of the variation is due to stomatal component of the surface resistance which is controlled by visible light intensity. Overall, the inferred fluxes approximated the measured fluxes reasonably well for all days except Sept. 26 and Sept. 27.

Combining the concentrations with the measured fluxes (Equation 1) produces the measured deposition velocity, Fig. 3(c). Also shown is the inferred V_d derived from DDIM. In general, the inferred V_d again closely approximated the measured V_d. Differences between measured and inferred V_d were the largest during the nighttime periods.

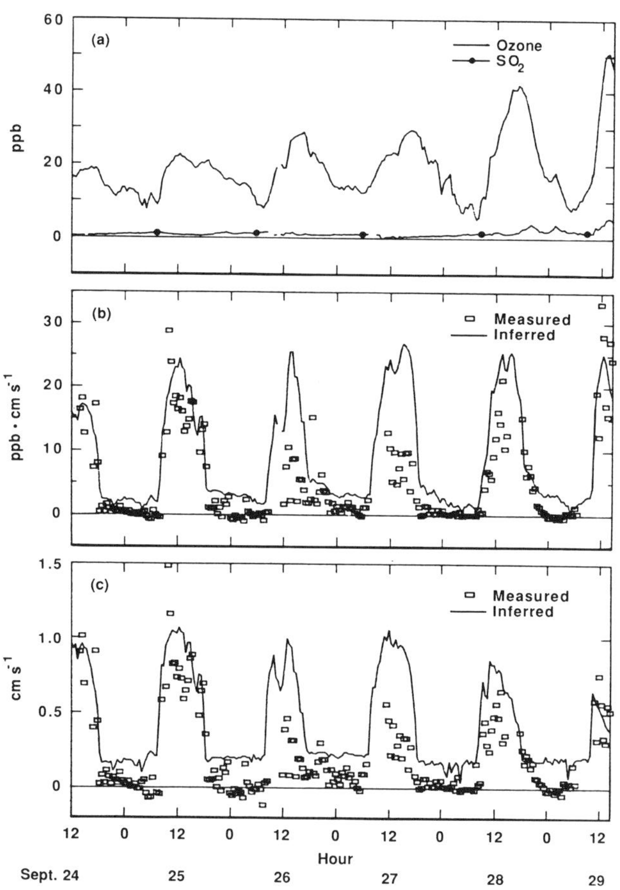

Fig. 3. Time series of (a) ozone and sulfur dioxide concentration, (b) measured and inferred ozone fluxes and (c) measured and inferred deposition velocities.

Again the Sept. 26-27 period showed the largest deviations. Although the ozone concentration and the ozone flux each exhibit a diurnal cycle, they are not proportional by a constant value. That is, $\overline{V_d}$ also exhibits a diurnal cycle.

Figure 4 shows the averaged diurnal cycle of measured and inferred deposition velocity for the total time period for which data were taken. Results show that the measured V_d were smaller then the inferred. Although the nighttime differences are very large, when expressed in percent of measured, the effect on calculated fluxes is not significant since the concentrations during those times is low.

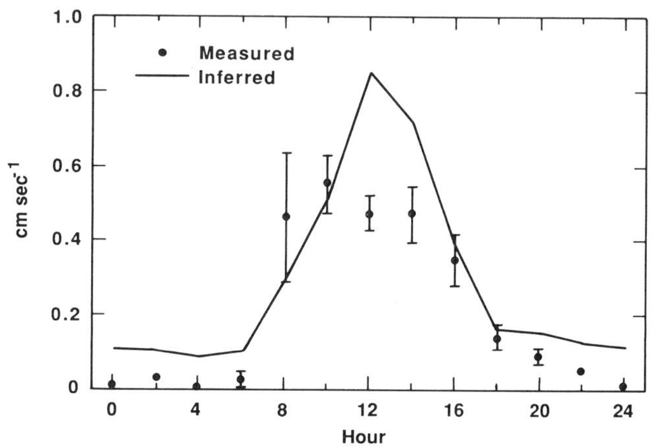

Fig. 4. Average diurnal cycle for measured and inferred deposition velocity for ozone.

In order to determine the cause of the differences between the DDIM and the measured resistances, it is informative to look at the individual resistance components. Fig. 5 shows the relative magnitudes of the three measured resistance components. As can be seen, the surface resistance dominates throughout the diurnal cycle even though R_a and R_b increase during the night. The inferred values for R_a and R_b are not included on the plot but have same magnitudes as the measured values and would fall within the estimated error for both. The point labeled $R_{stomata}$ is the leaf area index (LAI = 5) scaled value for the stomatal resistance to water vapor transport measured on Sept. 25 using a diffusion porometer. These measurements were made on white pine and hemlock on both the forest floor and at the top of the canopy. The R_{cut} shown is the residual of the resistance for the same time period and is a measure of the resistance to ozone uptake by the cuticle and all other surfaces (e.g. soil). Thus it does not include those controlled by the stomata. This value of R_c has the same magnitude as that obtained for nighttime average resistance, confirming our results. Wesely (1983) reports 12 s/cm for nighttime R_c to loblolly pine measured in North Carolina.

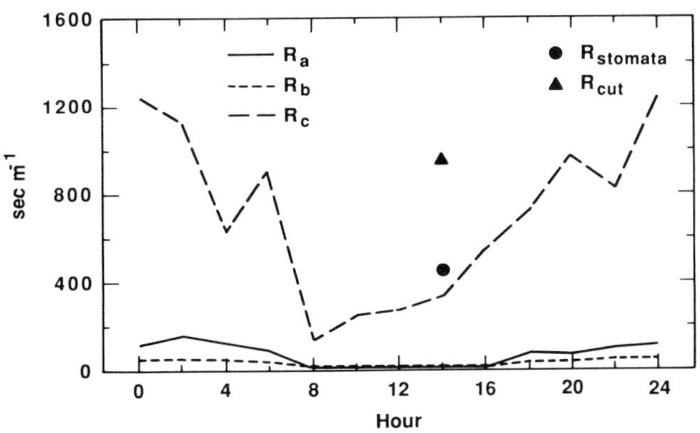

Fig. 5. Diurnal cycle of the measured resistance components.

The diurnal cycle for measured and inferred surface resistance is shown in Fig. 6, with the standard deviations of each for the total time period when valid flux data were collected. The DDIM underestimated R_c. During the daytime, measured R_c was 300 s/cm while inferred R_c was approximately 100 s/cm. At nighttime when the surface is wet with dew (the 4 am and 6 am data) the DDIM estimate of R_c more closely approximates measured R_c then it does when the surface is dry (8 pm till 2 am for this experiment). Thus the current version of DDIM has too small parameters for the cuticular component of R_c, especially for dry surfaces.

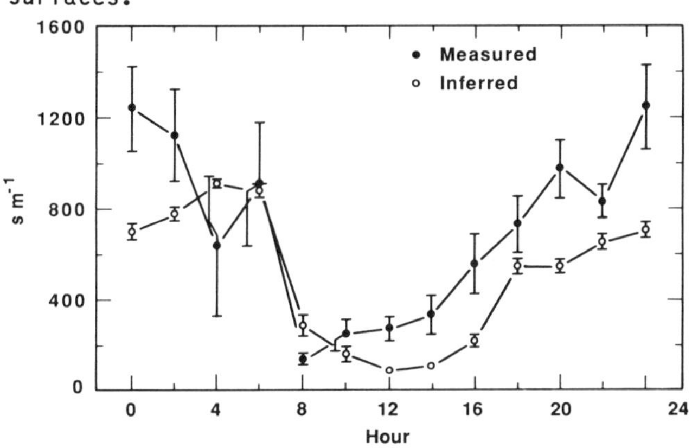

Fig. 6. Comparison of diurnal cycle of measured and inferred surface resistance component.

4. CONCLUSIONS

The main purpose of this experimental program is to measure fluxes of pollutants to various types of surfaces and then to compare these measurements with estimated fluxes from the Dry Deposition Inferential Model. The DDIM is currently being used by NOAA to estimate dry deposition at 13 sites throughout the United States, predominantly in the north eastern quadrant. The measured fluxes over various surfaces are used as benchmark tests for the DDIM. In this experiment, the fluxes measured to the conifer forest showed that current parameterization of the "cuticular + soil" resistance is too small. Thus the modeled fluxes are in variance with measured fluxes, causing an overestimate of total dry deposition flux by generally 50% for this experiment. Parameterizations of the DDIM are reviewed on an annual basis and improvements are made to the model, reflecting the current state of science.

5. ACKNOWLEDGMENTS

This work was conducted with the support of the U. S. Department of Energy and the National Oceanic and Atmospheric Administration. This work is a contribution to the National Acid Precipitation Assessment Program, under the auspices of Task Group 2: Atmospheric Chemistry. The authors wish to thank Dr. S. M. Goltz, Forest Scott, and Evan Hillman, University of Maine, Orono for hosting this experiment and providing excellent facilities and assistance during its duration. Also, the authors wish to thank R. T. McMillen for giving up some of his vacation time to present this paper at the Tropospheric Ozone Workshop.

6. REFERENCES

Baldocchi, D. D., B. B. Hicks, and P. Camara, 1987: A canopy stomatal resistance model for gaseous deposition to vegetated surfaces. Atmos. Environ., 21, 91-101.

Baer, M. and K. Nester, 1987: Simulation of concentration, dry deposition and fluxes of SO_2 during TULLA. Ann. Geophysicae, 5B, 497-502.

Hale, J. M., B. B. Hicks, and J. M. Miller, 1987: The role of research measurement-networks as contributors of federal assessments of acid deposition. Bull. Amer. Meteor. Soc., 68, 216.

Hicks, B. B., D. D. Baldocchi, T. P. Meyers, R. P. Hosker, Jr. and D. R. Matt, 1987: A preliminary multiple resistance routine for deriving dry deposition velocities from measured quantities. Water, Air, Soil Poll., 36, 311-330.

Matt, D. R., R. T. McMillen, J. D. Womack, and B. B. Hicks, 1987: A comparison of estimated and measured SO_2 deposition velocities. Water Air, Soil Poll., 36, 331-347.

McMillen, R. T., 1986: A basic program for eddy correlation in non-simple terrain. NOAA Tech. Memo., ERL ARL-147.

McMillen, R. T., 1988: An eddy correlation technique with extended applicability to non-simple terrain. Bound.-Layer Meteor., 43, 231-245.

Wesely, M. L., 1983: Turbulent transport of ozone to surfaces common in the eastern half of the United States, in Trace Atmospheric Constituents: Properties, Transformations, and Fates, edited by S. E. Schwartz, John Wiley & Sons, Inc., New York, 345-370.

INVESTIGATIONS OF MESOSCALE DYNAMICS PROCESSES OVER MOUNTAIN REGION BY DEFORMATION OF OZONE FIELD

N.F. Elanśky
Institute of Atmospheric Physics
Academy of Sciences of the USSR
Pyzhevsky per. 3, Moscow, USSR

ABSTRACT

In July-August 1985, an experiment was carried out with the purpose of studying the peculiarities in the air flows streaming over the Ukrainian Carpathians. The observational results of the distribution of ozone concentration within the troposphere were used to determine the parameters of lee-waves of different scale. The observations were conducted on board of two aircraft. The results indicate that the lee-waves originate over the system of mountain ranges of the Ukrainian Carpathians at practically any direction of the overflowing current, provided there are favourable conditions for vertical wind and temperature distribution. The most intervise wave system is formed by the south-western current flowing normally to the ridges. The system involves the entire tropospheric layer. The typical lengths of waves of these overflows are 7-9 and 20-30 km. The vertical air movement amplitudes reach several hundreds of meters.

1. INTRODUCTION

The orographic disturbances of air currents, when flowing over such large mountain massifs as the Carpathians, notably affect the circulation regime and climate of the adjoining regions. For examle, the moisture regime in the Carpathians depends on wave activity. In 1985, in the Ukrainian Carpathians, a complex experiment was carried out to study the orographic disturbances in the air current overflowing the mountain massif. The method based on observations of the ozone field deformations in the disturbed zone was used in the experiment for the determination of lee-wave parameters (Gruzdev, A.N., and N.F. Elansky, 1984 a,b). Under the influence of the waves, at every fixed altitude, the air particles are substituted by the particles arriving from above or from below. During this process, the periodic changes of ozone concentration may occur in the horisontal direction along the current overflowing the mountain range; these changes are connected with the vertical gradient of concentration in the undisturbed oncoming flow.

Ozone in the troposphere is adequately conservative as regards processes lasting several tens of minutes. Its horisontal distribution in the undisturbed area is, as rule, almost homogeneous, whereas the vertical distribution often has large gradients. Owing to these characteristics, the ozone can be used as a convenient tracer, in particular, when temperature measurements produce little information because of dense cloud cover of almost adabatic temperature stratification.

2. RESULTS

Ozone concentrations were measured in the period from 19 July to 7 August 1985 on board of two aircraft Yak-40 specially equipped for meteorological observations. The flights started from Lvov and covered the Ukrainian part of the Carpathians Besides ozone concentration, which was measured with electrochemical cells, the observations included the UV radiation flows, total O_3 and NO_2 contents, temperature pulsations, and wind velocity component. The variations of UV radiation were also recorded from helicopters, their altitudes ranging from the Earth's surface up to 3 km. These measurements and the wind regime control were also conducted from ground-based stations.

The aircraft flights were synchronous and followed two schemes: 1) at different altitudes on one course at right angles to the range, Bobrka - Sredney - Bobrka; 2) at two altitudes over a large area of the Carpathians along the ranges and across the ranges in different places (Fig. 1).

The structure of ozone distribution in the area of the flights was complicated and corresponded to the complicated system of mountain ranges of the Ukrainian Carpathians. In this area, the lee-waves appear at practically any direction of the air flow, provided the vertical wind and temperature profiles are favourable. In many cases the wave pattern can be interpreted as superposition of waves from small local sources and of waves

ISBN 0-937194-15-8

Fig. 1. The topography of the Ukrainian Carpathians and two schemes (1 and 2) of the flught tracks

associated with the overflow of ranges.

Figure 2 shows examples of profiles of ozone concentration when the south-western air current flowed at right angles to the major ranges.

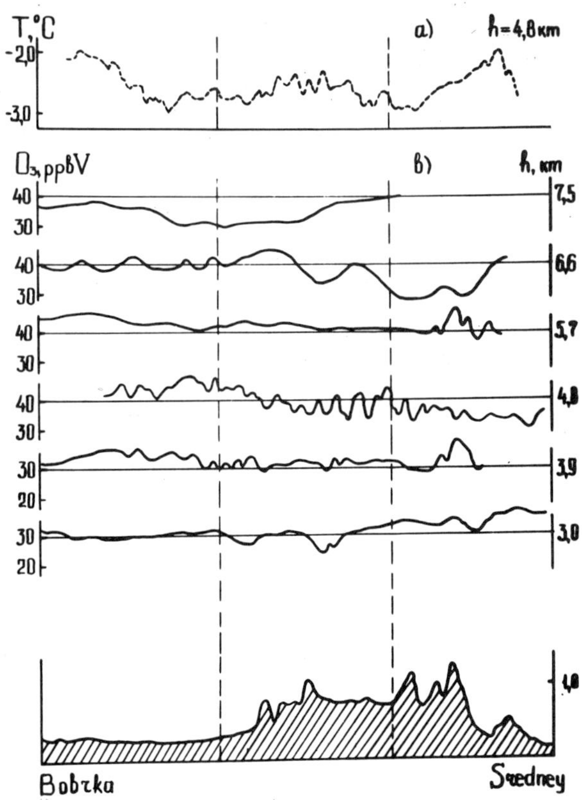

Fig. 2. Aireraft measurements of ozone and temperature variations for horizontal flight Bobrka-Srcdney on 7 August 1985

Periodic oscillations with 8 km wave length are recorded in the relatively thin layer at h = 4.8 km. These oscillations are not observed above and below this layer, though the ozone gradients were rather high there. The vertical air displacements reached $\pm$ 90 m in the isolated layer with favourable conditions for the appearance of waves. The vertical gradient of temperature in the layer 4,5-5,0 km is minimal and equals 4,2 km^{-1}. The conditions of existence trapped waves are faourable in this area. These waves apear also in variations of T that were registrated during the flight Bobrka - Sredney. The amplitudes of vertical air displacements from temperature data are about 100m and approximately equal to quantities obtained from ozone data.

The similar ozone and T variations are observed on 2.08.On this day trapped waves with wavelengths 7,5-8,0 km and amplitudes 80-90 m were found in the layer 4,5-6,0 km (Fig. 3).

During the observation period, the most typical were the waves 7-9 km and 15-20 km long. The most frequent waves 7-9 km long occurred in the layer whose height is no more than 2,0-2,5 km. The position of layers with periodic ozone oscillations were compared with the height interval where favourable

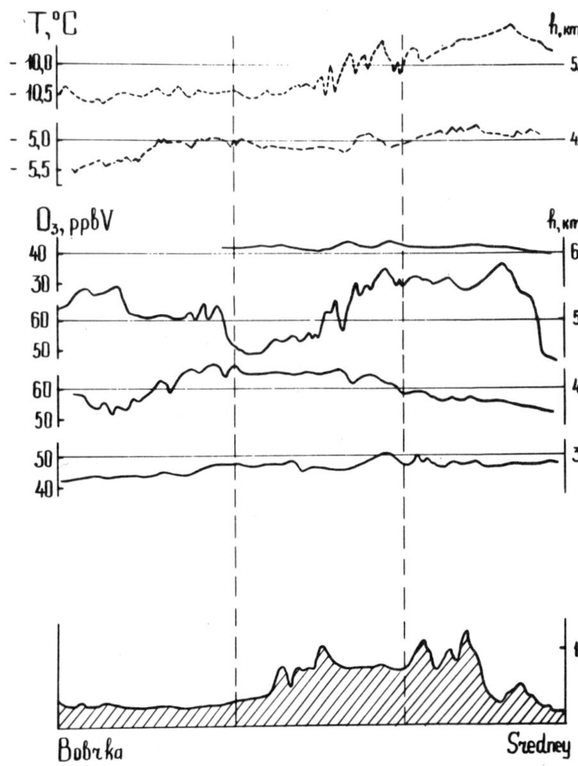

Fig. 3. Aireraft measurements of ozone and temperature variations for horizontal flight Bobrka-Sredney on 2 August 1985

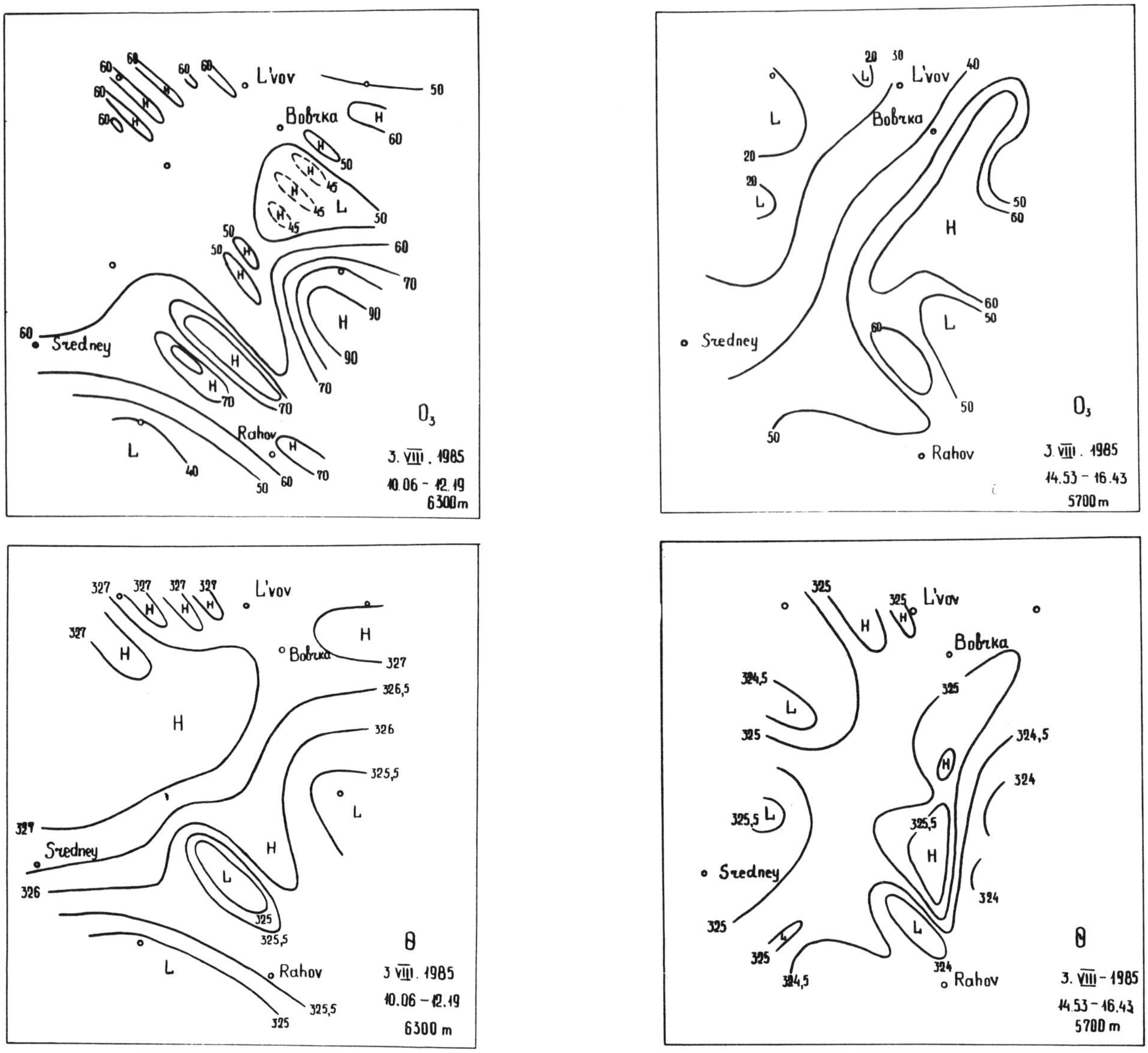

Fig. 4. The ozone (a,b) and potential (e,d) temperature analysis for two fliqht levels 5,7 and 6,3 km

conditions for lee-waves formation are indicated by computed values of the Scorer parameter. On the whole, both kinds of data are in agreement, but the numerical analysis of meteorological information, as a rule, shows larger by height areas of the possible appearance of waves than the areas observed from aircraft. This result can be apparently attributed to insufficient space-time resolution of the wind and temperature field from balloon sounding data which does not allow us to reveal the fine vertical structure of air flows over the mountain region.

Over the highest part of topography, the horisontal fields of ozone concentration have areas of higher or lower concentration values. The changes of concentration are recorded on the boundary between the mountain and the plain, thus implying vertical air movements over the mountains during quasistatic overflow. Individual peculiarities in the distribution, however, cannot be otherwise interpreted except as evidence of manifestation of lee-waves of different scale, from 4-5 to 30-40 km (Fig. 4).

These waves are stable for several hours. They are displaied not only in ozone distribution but in potential temperature distribution as well. For the vertical ozone and temperature profiles, the evaluations of vertical displacement produced values in the order of several hundreds of meters. Maximal displacements of 500-600 m were recorded over the highest part of the Ukrainian Carpathians. In the area of

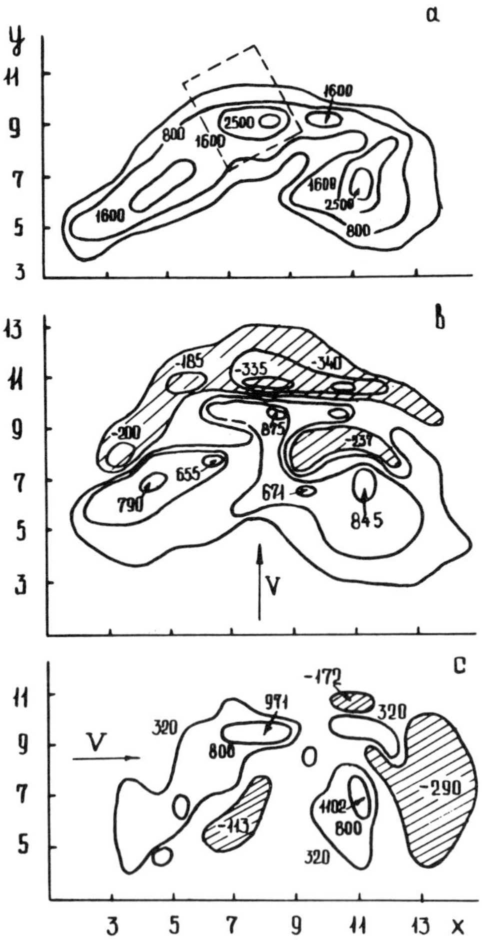

Fig. 5. The topography of the Carpathians used in the numerical simulations (a) and the isolines of vertical displacement (in meters) of the isopycnic surface 2 km for the south-west air flow (b) and for the north-east air flow (c)
$V=V_o(1+kh), V_o=10m\ c^{-1}, K=10^{-3}c^{-1}$

most frequently observed waves 7-10 km long, the vertical displacements varied from several tens of meters to 150 m. In the interval of longer waves (15-30 km) these displacements increase to 200-300 m.

Large-scale inhomogeneities of the ozone field, recorded from aircraft, were compared with the results of mathematical modelling of the air streaming over the Carpathians (Granberg, I.G., 1987). A solution of the stationary problem with real topography and real values of vertical displacement of the oncoming air flow velocity was undertaken. The comparison of overflow patterns for identical meteorological conditions reveals that mathematical models can identify the major large-scale peculiarities of these patterns (Fig. 5). Computer interpretation of processes in the order of several kilometers is apparently as yet an unattainable problem for such a complicated system as the Carpathians.

The observation of ozone concentration at the Earth's surface and of its total content in the vertical column showed a number of impressive examples of interval waves propagation towards the Earth and into the stratosphere. Distinct temporal ozone oscillations with periods of 10 and 30-40 min were recorded at the surface, and periodic variations of the total content were measured from aircraft. The amplitude of periodic variations reached 1-2%, which corresponds approximately to the values recorded in the Caucasus (Elansky, N.F. et al., in press.).

3. CONCLUSION

The application of ozone concentration data determination of parameters of lee-waves proved to be effective. The obtained distributions of vertical air displacement in the wave zone almost invariably coincided with analogous distributions compiled from the potential temperature field data. To conclude, the ozone remote sensing measurements by lidars or by satellite optical techniques can be successfully used for the study of meso-scale dynamic processes.

Acknowledgments the author extends sincere thanks to R.V.Girgzhdene and A.Yu.Girgzhdis for their assistance in the organisation of ozone concentration measurements from aircraft.

4. REFERENCES

Granberg, I.G. 1987:Numerical simulation of the South-West air floms over the Carpathians. Izv. Acad. Sci. USSR Atm. Oceanic Phys.,23,235-241

Gruzdev, A.N., and N.F. Elansky, 1984: Estimate of lee-wave influence on the distribution c o n t r n t of qaseous tropospheric species. Izv. Acad. Sci. USSR Atm. Oceanic Phys.

Gruzdev, A.N. and N.F. Elansky, 1984: Ozone observations in mountain lee-waves. Izv. Acad. Sci. USSR Atm. Oceanic Phys., 20, 705-714

Elansky, N.F. I.A. S e n i k and A.Kh. Khrgian: The total ozone variations in the mountain lee-waves region. Izv. Acad. Sci. USSR Atm. Oceanic Phys., 24, in press.

SPRINGTIME SURFACE OZONE FLUCTUATIONS AT HIGH ARCTIC LATITUDES AND THEIR POSSIBLE RELATIONSHIP TO ATMOSPHERIC BROMINE

S.J. Oltmans[1], *P.J. Sheridan*[2], *R.C. Schnell*[3], *R.E. Peterson*[4], *J.W. Winchester*[5], *L.A. Barrie*[6], *J.D. Kahl*[2], *and W.D. Komhyr*[1]

[1]NOAA/Geophysical Monitoring for Climatic Change
Boulder, Colorado 80303, USA

[2]NOAA/GMCC, Boulder, Colorado 80303, USA

[3]CIRES, University of Colorado
Boulder, Colorado 80309, USA

[4]Chemistry Department, University of Washington
Seattle, Washington 98195, USA

[5]Department of Oceanography, Florida State University
Tallahassee, Florida 32306, USA

[6]Atmospheric Environment Service
Downsview, Ontario M3H 5T4 Canada

ABSTRACT

Ozone and filterable bromine measurements in the high Arctic during the spring return of solar radiation suggest a rapid concurrent destruction of ozone and the conversion of gaseous to particulate bromine. Multi-year observations show that this pattern is an annual feature of ozone measured near the surface at Barrow, Alaska and other Arctic locations. Aircraft measurements show low ozone amounts and high filterable bromine concentrations beneath the surface temperature inversion over ice throughout the Arctic. A wintertime buildup of the gaseous organic compound bromoform and its rapid depletion in the spring may be a link between the episodic ozone depletion events and the concomitant rise in filterable bromine.

At high Arctic stations such as Barrow, Alaska, springtime near-surface ozone amounts fluctuate between the highest and lowest values seen during the course of the year (Oltmans, 1981). Episodes when the surface ozone concentration is essentially zero last up to several days during this time of year (Fig. 1). In the Arctic Gas and Aerosol Sampling Program (AGASP-I and AGASP-II) in 1983 and 1986, it was found that ozone concentrations often showed a very steep gradient in altitude with very low values near the surface (Herbert et al., 1988; Bridgman et al., 1988). The cold temperatures, and snow-covered ground make it unlikely that the surface itself would rapidly destroy significant amounts of ozone (Galbally and Roy, 1980).

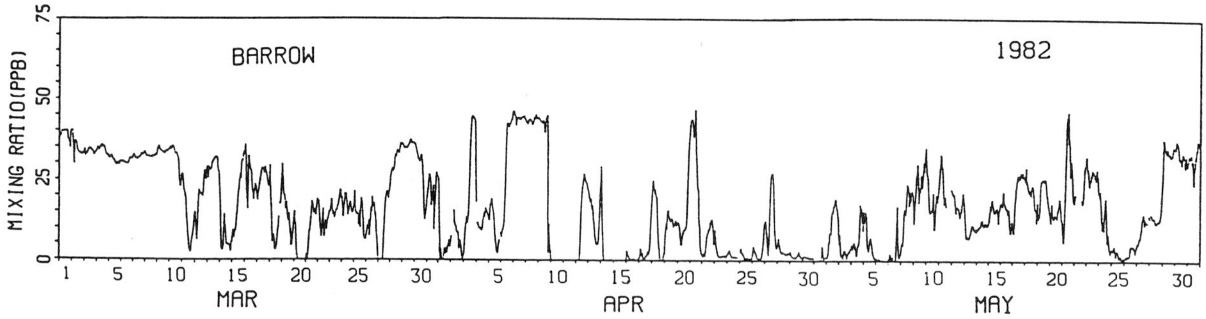

Fig. 1. Hourly surface ozone concentrations at Barrow for Mar-May 1982 showing large fluctuations and periods of near-zero values.

ISBN 0-937194-15-8

Fifteen years of ozone observations near the surface at Barrow, Alaska (71°N, 157°W) show (Fig. 2a) a late autumn and winter maximum, a sharp springtime drop; a modest recovery to a secondary maximum in June; a small dip during the summer; followed by rising concentrations during autumn. Although ozone concentrations in Fig. 2a show a great deal of variability which ranges in scale from days to years, the pattern described above is in evidence each year. A very similar pattern is seen in the ozonesonde record at Resolute, NWT (75°N) for observations near the surface. Such a seasonal pattern is unique to the high Arctic and is not seen at other relatively high latitude stations (e.g. Churchill at 59°N) outside the Arctic Basin. Ozonesonde observations at both Barrow and Resolute show that this behavior is confined to the boundary layer.

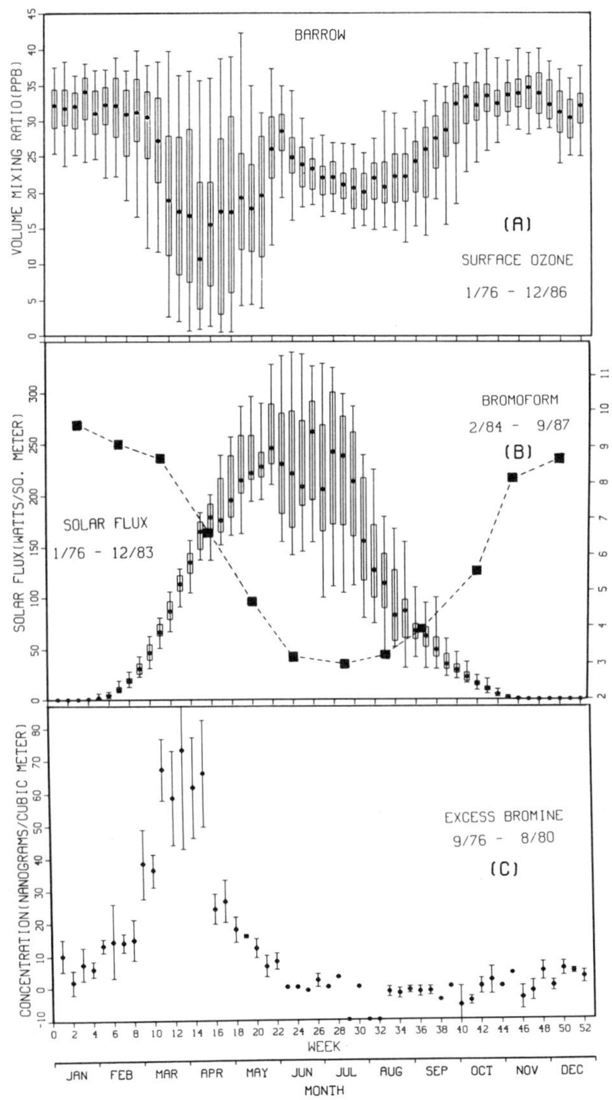

Fig. 2. (a) The seasonal variation of ozone at Barrow, AK for 1973-1987. Box and whisker plots at weekly intervals are from daily means. The median (dot), the upper and lower quartiles (ends of the box) and upper and lower 5th percentiles (ends of lines) are shown; (b) Total global flux of solar radiation at Barrow for 1976-1983 and mean monthly bromoform mole fraction (Cicerone et al., 1988); and (c) weekly mean and standard deviations of the excess filterable bromine at Barrow, AK (Berg et al., 1983).

The AGASP aircraft measurements that found low ozone concentrations in the lowest layers of the troposphere also found that filterable excess bromine (the amount of bromine in excess of the sea salt component) in samples collected wholly or partially beneath the temperature inversion had higher bromine concentrations than other tropospheric samples. Figure 3 shows the excess filterable bromine in the aerosol above that of seawater for all non-stratospheric filter samples collected during AGASP-I and AGASP-II covering the Artic from Alaska to Greenland to Norway. Samples were classified as containing significant marine aerosol contributions (M), when collected at low altitude over open ocean or as having been collected over the pack ice in tropospheric air (T) at least 200 km from open ocean. Samples were plotted according to the weighted mean ambient pressure during sample collection. Pressure ranges are also shown for the T samples.

M samples with a lower percentage of time spent in the marine boundary layer (represented with a lower mean ambient pressure) show slightly higher excess bromine concentrations probably because of the sampling of mid-tropospheric air. T samples collected wholly in the free troposphere show intermediate excess bromine concentrations (~0.5-2. ng/m^3). The remaining T samples in figure 3 had significant aerosol contributions from below the temperature inversion layer. Most had sampling time near the surface at pressures greater than 1000 mb. These samples contained elevated excess bromine concentrations. Three of the four largest T sample concentrations were collected over pack ice north of Alert, N.W.T.

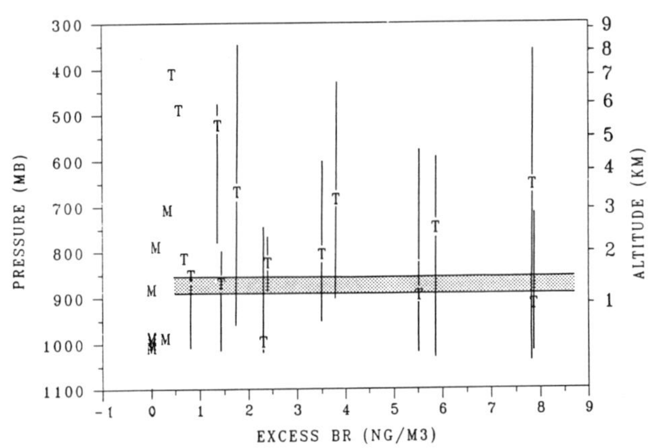

Fig.3 Excess filterable bromine in the Arctic from AGASP aircraft samples. M = over ocean, T = over ice, tropospheric sample. the M and T symbols are plotted at the weighted mean ambient pressure of the sample with the vertical 'ars indicating the sampling altitude range. The shaded section indicates the mean position of the temperature inversion over ice observed in the Arctic springs 1983 and 1986.

Surface measurements of excess filterable bromine (Fig. 2c) at Barrow (Berg et al., 1983) as well as at three other stations (Fig. 4) in the high Arctic (Sturges and Barrie, 1988) show a strong seasonal dependence. Concentrations rise dramatically in March, remain high through April, then decline in May and remain low during the remainder of the year. There is a definite tendency for the large concentrations of filterable bromine to appear earlier at more southerly locations. This tendency is most apparent when looking at the most southerly site (Igloolik - 69°N) and most northerly location (Alert - 82°N). The approximate 4-week lag in the appearance of the peak bromide concentrations at Alert suggests that the appearance of sunlight is an important factor in the bromide formation.

The abrupt appearance of high filterable bromine concentrations corresponds to the dramatic reductions in ozone seen at Barrow (Figs. 1 & 2). Recently, Barrie et al. (1988) found that at Alert, NWT ozone mixing ratios during April 1986 were anti-correlated at a level of better than 0.95 with filterable bromine concentrations.

A potential source for the link between the episodes of low surface ozone concentration and high filterable bromine amounts is the presence of organic bromine gases in the Arctic (Cicerone et al., 1988). Bromoform, one of these gases shows a dramatic seasonal dependence (Fig. 2b). The levels rise sharply beginning in the autumn, reach their peak during the winter, fall rapidly in the spring, and reach the lowest values in the summer. Winter amounts are at least three times greater than in the summer. Figure 2b also shows the seasonal march in the solar energy received at the surface at Barrow. After week 11 when the total daily flux of solar radiation reaches approximately 50-100 watts/m^2, the bromoform concentration drops dramatically. This is the beginning of the period of large ozone and bromide fluctuations. The excess filterable bromine data from Barrow (Fig. 2c) are plotted with the ozone, bromoform, and solar radiation information emphasizing the timing of the spring events.

The importance of solar radiation in the springtime ozone loss and high bromide concentrations is also suggested by the autumn events which do not duplicate those seen in the spring. In the autumn, bromoform concentrations do not begin to rise significantly until solar radiation falls to about 50 watts/m^2 at around week 37. This is six months after bromoform concentrations began to fall. Thus there is very little bromoform available during the period of declining solar energy. This also suggests that the bromoform is easily photolyzed in the presence of sunlight.

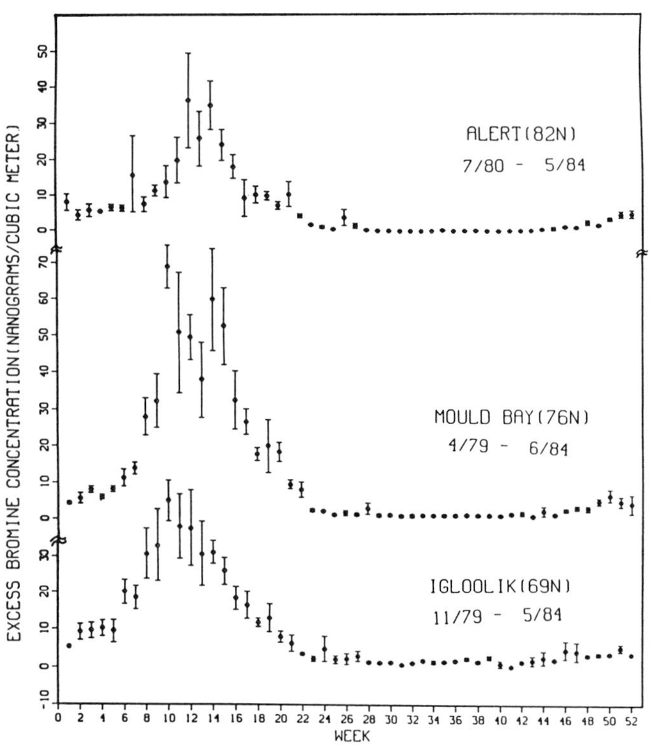

Fig. 4. The weekly mean and standard deviations of the excess filterable bromine at Igloolik, Mould Bay, and Alert, NWT (Sturges and Barrie, 1988).

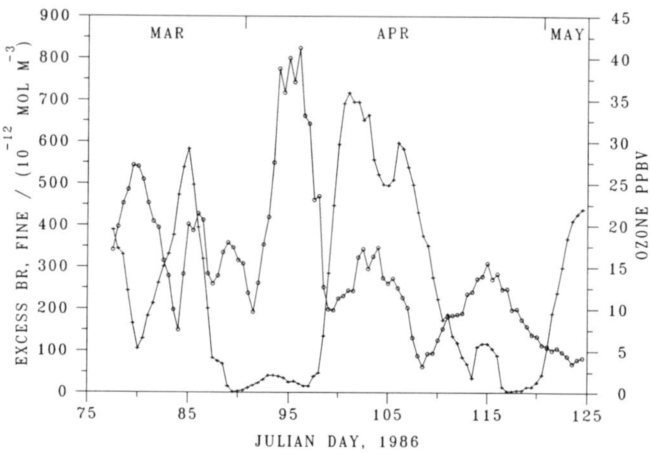

Fig. 5. Surface ozone mixing ratio (+) and excess filterable bromine (o) during spring 1986 at Barrow for the fine particle (<1µm) fraction. Points represent 60-hour running means of hourly ozone and 12-hourly bromine measurements.

The springtime minimum in the seasonal cycle of surface ozone concentrations at high Arctic latitudes appears to be linked to the destruction of ozone in the boundary layer by reactions with bromine compounds (Wofsy et al., 1975; Barrie et al., 1988). Organic bromine compounds emitted from the predominatly ice-covered ocean appear to react with ozone in the presence of adequate sunlight to destroy ozone and produce filterable bromine. On the AGASP flights, three of the four lowest minimum ozone concentrations were associated with the highest bromine enrichments. Simultaneous measurements of filterable bromine and ozone at the surface were made at Barrow during the spring of 1986. The broad features of the relationship between excess filterable bromine concentrations and ozone mixing ratio are shown in Fig. 5. A 60-hour running mean has been applied to these data to smooth out small scale fluctuations. Although the significance of the correlation is difficult to assess because of the smoothing, it appears that during periods of high filterable bromine amounts, the ozone concentration is low and vice versa.

There are some differences between the ozone/bromide relationship as seen at Barrow and Alert. At Alert, the high anti-correlation between ozone and filterable bromine results from the alternating presence of air from off the frozen ocean below the inversion and air from above the inversion mixed to the surface by winds blowing over the rougher and higher terrain of Ellesmere Island (Barrie et al., 1988). At Barrow, on the other hand, the regimes under which above-inversion and below-inversion air is sampled seem to be less easily defined. The temperature inversion at low levels is present most of the time at Barrow. During the period in 1986 when ozone and bromide measurements were made, there did not appear to be a significant relationship between the strength of the inversion and high or low ozone or bromide amounts. The meteorological conditions under which ozone is depleted are not easily segregated from those showing concentrations more typical of the free troposphere above the boundary layer. Work is currently underway to see if different air flow regimes reaching Barrow are associated with depleted ozone values. If, as expected, the ocean is the main source of the bromoform, then air that has traveled over the ocean is likely to have lower ozone concentrations and higher excess filterable bromine than air coming from continental regions. Inspection of surface winds at Barrow during spring 1986 does show that southerly winds are not usually associated with lower ozone amounts. Winds with a significant southerly component are not prevalent during this time of year, however. It seems likely that at Barrow existing measurements will not produce the same detailed relationship between ozone and bromide seen at Alert.

The multi-year measurements of surface ozone, solar radiation, organic bromine gases and filterable bromine compounds at Barrow, Alaska suggest that springtime ozone is depleted in the Arctic boundary layer as a result of photochemical processes brought on by the appearance of sunlight in the presence of bromine. Detailed measurements of the quantities discussed here as well as measurements from the ocean, and of other relevant atmospheric gases such as nitrogen oxides will be required to properly model the observed ozone depletion. Attention may also have to be paid to possible heteorogenous reactions on ice crystals as has been suggested as a cause of the dramatic springtime ozone depletion in the stratosphere in Antarctica (Solomon et al., 1986; McElroy et al., 1986).

REFERENCES

Barrie, L. A., J. W. Bottenheim, R. C. Schnell, P. J. Crutzen, and R. A. Rasmussen, 1988: Ozone destruction and photochemical reactions at polar sunrise in the lower arctic atmosphere, *Nature,* in press.

Berg, W. W., P. O. Sperry, K. A. Rahn, and E. S. Gladney, 1983: Atmospheric bromine in the Arctic, *J. Geophys. Res.,* 88, 6719-6736.

Bridgman, H. A., R. C. Schnell, G. A. Herbert, B. A. Bodhaine, and S. J. Oltmans, 1988: Meteorology and haze structure during AGASP-II, part 2: Canadian Arctic flights, 13-16 April 1986, *J. Atmos. Chem.*, in press.

Cicerone, R. J., L. E. Heidt, and W. H. Pollock, 1988: Measurements of atmospheric methyl bromide (CH_3Br) and bromoform ($CHBr_3$). *J. Geophys. Res.*, 93, 3745-3749.

Galbally, I.E., and C.R. Roy, 1980: Destruction of ozone at the earth's surface. *Q. J. R. Meteorol. Soc.*, 106, 599-620.

Herbert, G. A., R. C. Schnell, H. A. Bridgman, B. A. Bodhaine, S. J. Oltmans, and G. E. Shaw, 1988: Meteorology and haze structure during AGASP-II, part I: Alaskan Arctic flights, 2-10 April, 1986. *J. Atmos. Chem.*

McElroy, M.B., R.J. Salawitch, S.C. Wofsy, and J.A. Logan, 1986: Reductions of Antarctic ozone due to synergistic interactions of clorine and bromine. *Nature*, 321, 759-762.

Oltmans, S. J., 1981: Surface ozone measurements in clean air. *J. Geophys. Res.*, 86, 1174-1180.

Solomon, S., R.R. Garcia, F.S. Rowland, and D.J. Wuebbles, 1986: On the depletion of Antarctic ozone. *Nature*, 321, 755-758.

Sturges, W. T., and L. A. Barrie, 1988: Halogens in Arctic aerosols. *Atmos. Env.*

Wofsy, S. C., M. B. McElroy, and Y. L. Yung, 1975: The chemistry of atmospheric bromine. *Geophys. Res. Lett.*, 2, 215-218.

O_3 - AND NO_x - MEASUREMENTS IN FOREST AND URBAN AREAS

G. Baumbach, K. Baumann, and F. Dröscher
Institut für Verfahrenstechnik und Dampfkesselwesen
Universität Stuttgart
D-7000 Stuttgart 80, FRG

ABSTRACT

Horizontal and vertical profiles of ozone (O_3) as well as diurnal variations of the ozone concentration were investigated at different locations in Baden-Württemberg based on continuous measurements at or near ground level. The influence of nitrogen oxides and light intensity on the ozone concentration is discussed in this paper.

Ozone concentrations are depressed in the vicinity of NO_x emission sources like urban areas or heavily used motorways. These interrelationships are studied from measurements carried out at the city of Freudenstadt/Black Forest and a forest area four kilometres away.

On days with high solar radiation, the average afternoon ozone levels are equal in town and forest. At night, when no photochemical ozone production takes place, NO sources in the town cause an immediate ozone reduction, while ozone levels in the forest decrease only slightly.

With a special measuring technique, vertical concentration gradients are determined in forest stands. At ground level inside the forest stand, the degradation of ozone is observed during night time which may be caused by NO exhalation from the soil. During sunny day hours, especially in summer, ozone is produced above the tree canopy.

Special investigations with NBR elastomers point to a positive correlation between the deterioration of test samples exposed at different locations and the measured ozone concentrations at those same sites.

1. INTRODUCTION

To further the understanding of the causes of forest damage, measurements of air pollution are carried out by our institute in two forest areas in Baden-Württemberg. One measuring station - in the following context named "Betzenberg"- is situated on a hill in the forest area Schönbuch 25 kilometres south of downtown Stuttgart at an elevation of 500 metres above sea level (asl). The other one-named "Schöllkopf" - lies on a hill near Freudenstadt in the northern Black Forest at an elevation of 840 m asl. At both stations, the gaseous pollutants SO_2, NO, NO_2 and O_3, pariculate and wet deposition, as well as various meteorological parameters, are continuously measured. The database is represented by half-hour mean values.

The specially developped measuring technique allows gas concentration measurements above the canopy and inside the forest stand in a quasi-continuous manner. This technique is described in detail elsewhere (Baumbach and Käß, 1985). Not only the concentration of gaseous pollutants in the forests is registered, but also conclusions about the absorption by the trees of such pollutants can be drawn. Figure 1 illustrates the arrangement of a forest measuring station. The Betzenberg station has been operating since December 1983, the Schöllkopf station since June 1984. Some of the results concerning the ozone measurements are presented here.

ISBN 0-937194-15-8

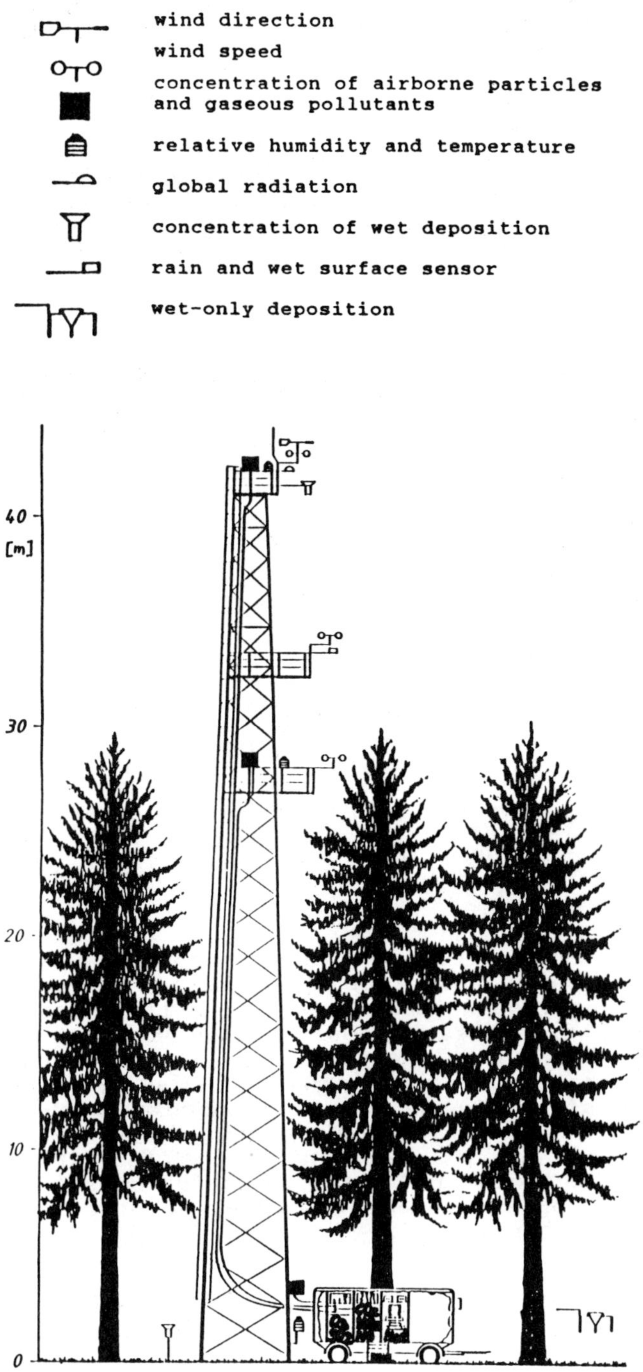

Fig. 1. Arrangement of the forest measuring station operated by the IVD

2. COMPARISON OF THE POLLUTION LOADS MEASURED AT DIFFERENT FOREST SITES

Under the influence of low pressure systems with westerly winds, only very low concentrations of gaseous pollutants are measured at both locations. Under stable weather conditions, however, especially during good weather periods with northerly winds, increased NO_x concentrations are often detected at the Betzenberg station. This relation of NO_x concentrations to northerly wind directions points to the influence of the metropolitan area of Stuttgart with its centre 25 km north (Baumbach, 1986a). During such periods of clear sky and sunshine Stuttgart itself is often covered by a brown veil of haze which indicates the accumulation of pollutants.

In late autumn and winter time the forests of Baden-Württemberg are beset by SO_2 on a large scale when north-easterly winds prevail (Baumbach et al., 1988). A long distance transport of SO_2 takes place only when temperature inversions built up at particular altitudes (800-1500 m asl) predominate. Then maximum ground level SO_2 concentrations are observed both at the Betzenberg and the Schöllkopf stations, although the difference in altitude is 340 m and the distance between the stations is about 40 km. The high SO_2 levels observed at the northeast station Betzenberg decrease as the "SO_2 cloud" moves further to the southwest station Schöllkopf due to dispersion and deposition.

Figure 2 (top) shows average diurnal variations of ozone measured at three differently located forest sites during July and August 1987. The "Hornisgrinde" is the highest mountain of the northern Black Forest area and lies approximately 10 km north of the Schöllkopf hill. For this comparative analysis, the behaviour of ozone must be especially examined:

During stable weather periods ground inversions build up as the ground surface cools down due to the radiation energy loss at clear nights. Right after sunrise, the ground warms up and the inversion layer may become elevated and disappears by noon. At higher altitudes, an inversion layer exists which acts as a main barrier moving between 1500 and 2500 m asl during summer (Baumbach, 1986b). During late afternoon and night, primary air pollutants remain in the ground inversion layer.

Hence, above this layer, in ranges between 800 and 1600 m, a certain "ozone reservoir" (Paffrath, 1988) can be built up, since anthropogenic NO_x cannot reach this reservoir. The ozone profile measured at the Hornisgrinde (1140 m asl) shows clearly that the top of this mountain is in the ozone reservoir all the time. The Schöllkopf station (860 m asl) may be in this ozone reservoir sometimes. This can be assumed to be due to the fact that the measured ozone levels at both stations do not differ much as is shown in Figure 2.

In the case of increased solar irradiation during summer months, the photochemical conversion of the nitrogen oxides and hydrocarbons emitted in the metropolitan area of Stuttgart 10 km away (border line) may also contribute to the increased ozone concentrations in the forest at the Betzenberg. Ozone is produced on the way from Stuttgart to the forest site (Baumbach et al., 1987a).

Although it is difficult to define individual concentration limits above which certain air pollutants may cause phytotoxical effects, a presentation of the statistical variation of air pollutant concentrations can be helpful in describing the general pollution situation of the investigated area. In Figure 3, the relative excessive frequencies of NO and O_3 are plotted versus the respective concentrations measured at the forest sites Betzenberg and Schöllkopf. It is clearly visible that, during a measuring period of three years, higher ozone concentrations are detected more frequently at the higher situated station Schöllkopf (860 m asl) than at the Betzenberg(520 m asl).

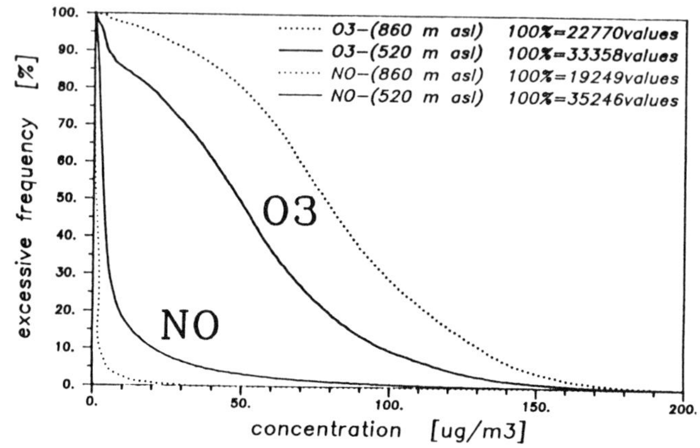

Fig. 3. Excessive frequencies of NO and O_3 as percentage of their total of measured half-hour mean values for given concentrations (1985-1987)

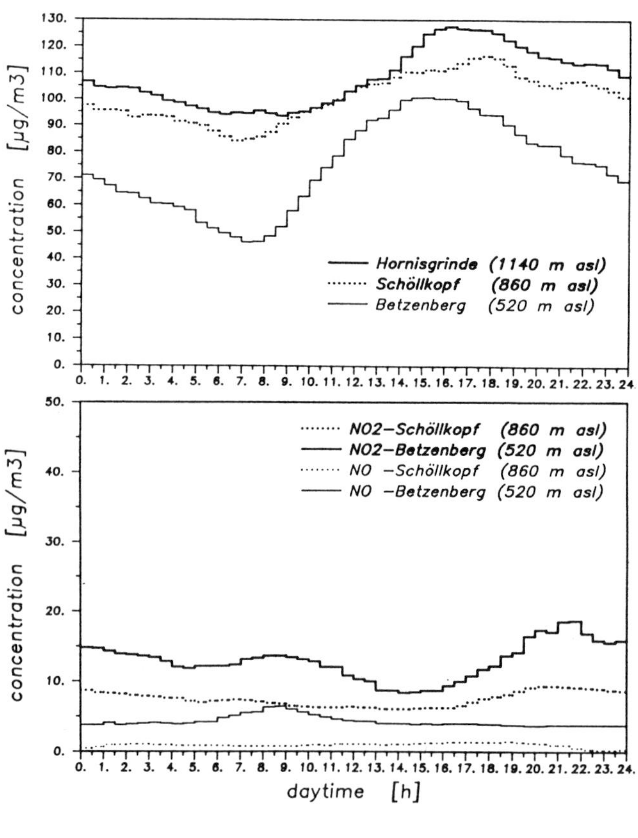

Fig. 2. Comparison of average diurnal variations of NO, NO_2, and O_3 (top) concentrations measured at different altitudes in July and August 1987

This kind of graphic representation of the statistical distribution permits comparison with any preventive limits, but the awareness of the related time scale controls the interpretation. It is important to note that data reduction by means of statistic methods and calculations is also accompanied with a loss in information content. Interpretation of overall reflections like these is stated correctly only by knowledge of short-time processes and observations.

3. CONCENTRATION DIFFERENCES BETWEEN THE AIR LAYERS ABOVE AND BELOW THE CANOPY

The specially developped measuring technique -as is illustrated in Figure 1- allows quasi-continuous measurement of air pollutants above and below the canopy, two metres above the ground. The detected concentration differences between these two horizontal air layers depend on different factors which cause specific effects on each individual air pollutant.

Inside the forest stand, the NO_x concentrations temporarily exceeded the values measured above the canopy at the same time. Evidently, small NO_x concentrations were emitted by the forest soil.

To further investigate these initial observations, average diurnal variations of the nitrogen oxides and ozone were calculated for the measuring site Betzenberg, and an example is plotted in Figure 4 for June 1986. Nighttime ground inversions prevent primary air pollutants to be transported from the adjacent valleys to the Betzenberg hill. For the same reason, the NO concentrations exhaled by the forest soil are not diluted by fresh air, and therefore become detectable. Only with an elevated ground inversion after sunrise, the primary air pollutants rise up from the valley floor and couse high NO concentrations above the canopy during the late morning hours.

With increasing turbulant air exchange, which causes the mixing with the air below the canopy, the NO ground level concentrations increase also, but do not reach the values from above so that there exists an NO concentration gradient until sunset. It is obvious that the NO soil exhalation is not detectable during daylight hours.

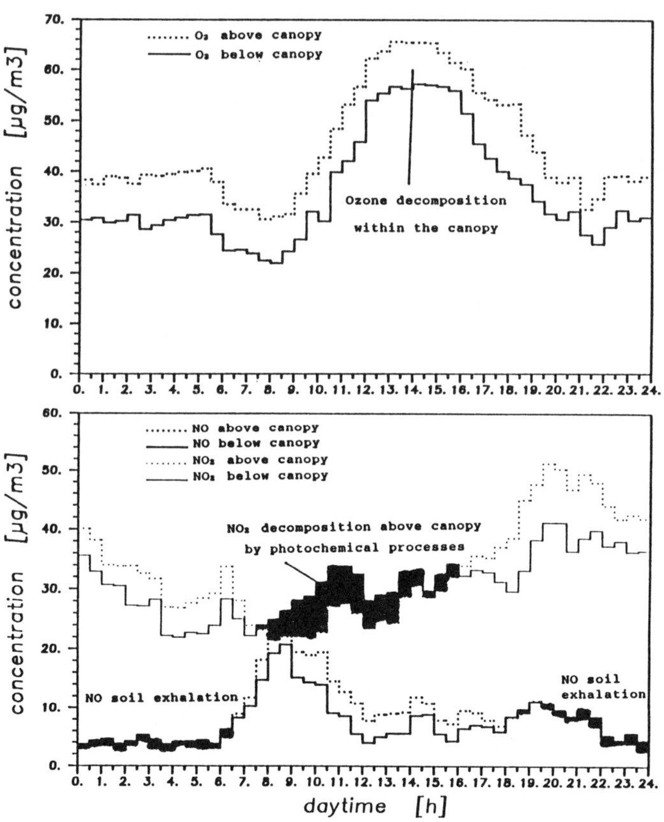

Fig. 4. Ozone and NO_x concentrations above the canopy and inside the forest stand (2 m above the ground) as average daily cycles in June 1986

For NO_2, on the other hand, a distinct concentration gradient towards the ground level can be seen during nicht time between dusk and dawn. However, around midday intense solar irradiation causes the photolysis of NO_2 above the canopy corresponding to the photostationary equilibrium: The increasing ozone concentration during this time of day is evident, see Figure 4 (top). Due to less intensive solar radiation in the shaded area below the canopy, less NO_2 can be decomposed there. The relatively high amount of turbulent air exchange and maximal photochemical interactions from above may be the reason for the parallel course of the ozone ground level concentration during the early afternoon.

At dusk, more NO_2 is produced above the canopy than below due to more intensive mixing with fresh ozonic air. At night, when the ozone concentrations are low, no vertical air exchange takes place due to the ground temperature inversion, so that the consumed ozone at ground level cannot be renewed: The O_3 concentration differences between the ambient air above the canopy and the air below the canopy inside the forest stand, are large.

4. COMPARISON OF MEASURED OZONE CONCENTRATIONS BETWEEN URBAN AND FOREST AREAS

From November 1985 until October 1986,comparative measurements were taken in the Black Forest town of Freudenstadt at an elevation of 810 m asl on top of the town hall tower and at the Schöllkopf hill station 3.6 km away, that has been mentioned already (Baumbach et al., 1987b). Some of the results are summarised here.

For the month of July 1986, a comparison of average daily cycles of NO, NO_2 and O_3 concentrations between the investigated urban and forest areas is presented in Figure 5. It is evident that the NO and NO_2 levels measured in the town exceed those measured in the forest due to the vicinity of the traffic emissions. The early morning rush-hour peak may be seen in the respective NO and NO_2 levels. A further increase of these air pollutants during the day is prevented by an increasing turbulent air exchange due to stronger winds. The traffic-dependent course of NO and NO_2 is not noticeable at the forest site.

Due to the above-mentioned overlapping processes of turbulent air mixing and photochemical reactions, an intensive ozone production takes place in the urban area where the ozone values reach those measured in the forest area, see Figure 5 (top). At night, when the photochemical influence is diminished, the ozone concentration decreases due to the close vicinity of NO sources.

Some conclusions can be drawn from comparative measurements taken at a distance of 60 m from a heavily used motorway and those from the Betzenberg hill, 15 km away, during a period of 10 days in July 1985 (Baumann, 1987). This is shown in Figure 6. A comparison of the average diurnal variations of NO_2 and O_3 during this period are shown. Since the traffic density on the investigated three-lane motorway, and therefore the NO emission rate, is extremely high, the secondary decomposition which initializes the ozone production at a distance of 60 m results in very high ozone levels, exceeding even those detected at the forest site on the Betzenberg hill.

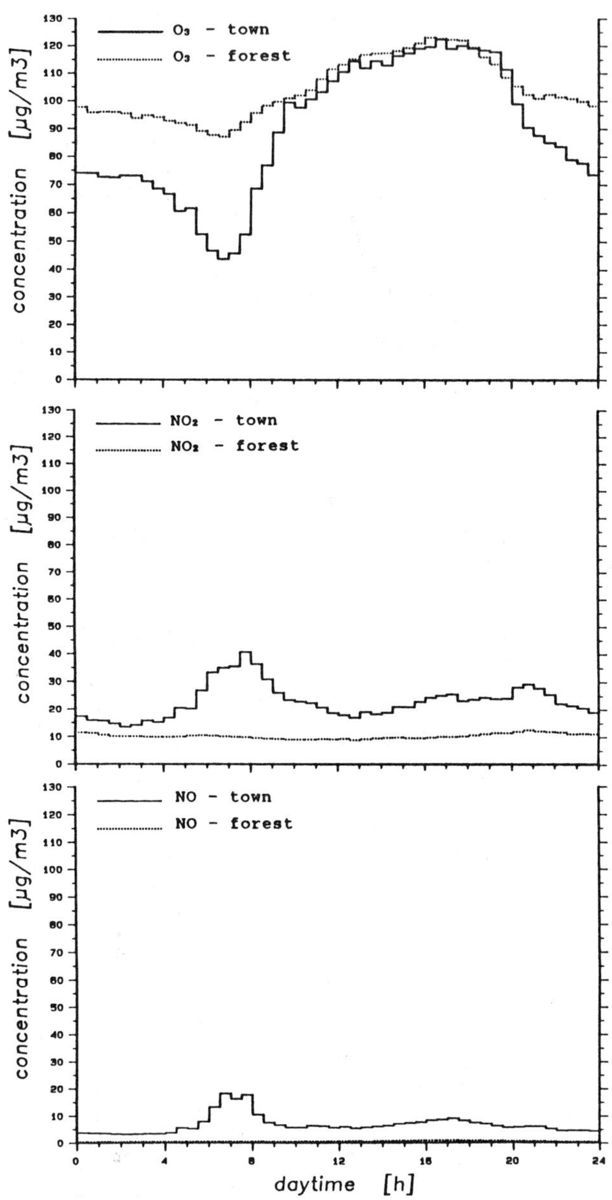

Fig. 5. Comparison of average diurnal variations of NO, NO_2, and O_3 concentrations between the town of Freudenstadt (810 m asl) and the forest area around Schöllkopf (860 m asl) 3.6 km away in July 1986

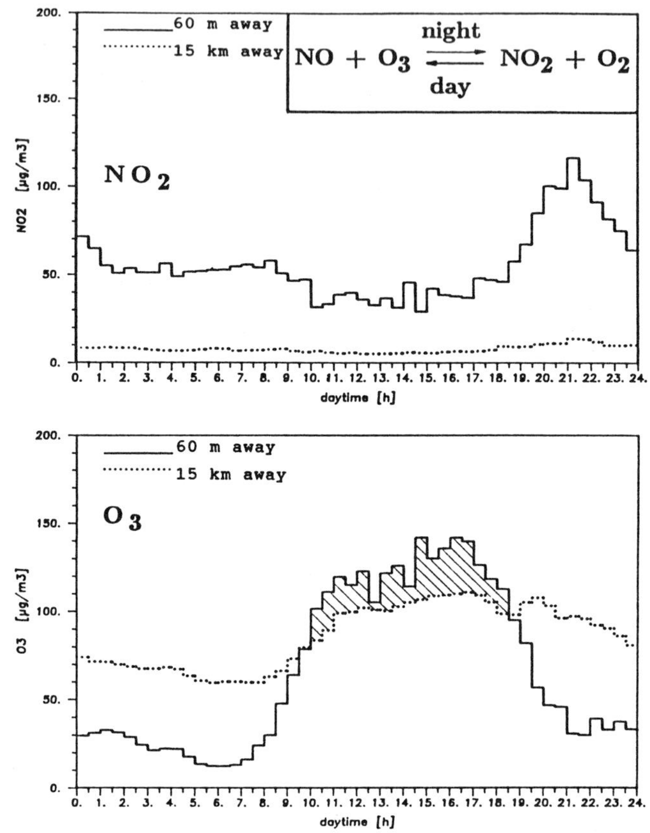

Fig. 6. Average daily cycles of NO_2 and O_3 concentrations at a distance of 60 m from a heavily used three-lane motorway compared to the conditions at the forest area at the Betzenberg hill from July 22 to July 31, 1985

In exposition experiments of plastic indicators at different locations a strong correlation of ozone concentration levels and the deterioration of the rubber test samples was established (Schliep, 1986; Rudolph, 1986). Thus elastomer rubber can serve as useful ozone indicator. Figure 7 shows standardized ultimate tensile strength (uts) of "Acrylnitril-Butadien-Rubber" after 70 days of (shaded) exposure time. The uts variation of four respectively exposed test samples is depicted at the top of each light bar.

Furthermore, the overall means of ozone concentrations measured during the respective exposure time at each location are depicted as broken bars. It is clearly visible that the deterioration increases with the ozone concentration. Locations in the vicinity of a heavily used motorway are compared with urban areas and remote forest sites so that an equivalent correlation may be seen between the measured ozone concentrations and the deterioration of forests. However, more detailed investigations are necessary in the future to verify this correlation.

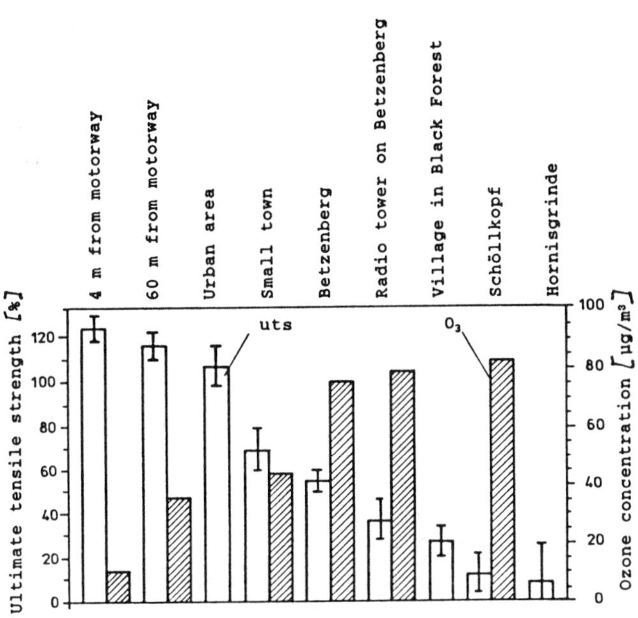

Fig. 7. Ultimate tensile strength of NBR-elastomere after an exposure time of 70 days with respective O_3 mean concentrations

5. ACKNOWLEDGEMENTS

This report is generally based on a research program sponsored by the German government through the Bundesminister für Forschung und Technologie (BMFT), under contract no. 0339112B. Special investigations on the despersion of air pollutants in the vicinity of a motorway presented here, were sponsored by the Autobahnamt Stuttgart (federal motorway department).

6. REFERENCES

Baumann, K., 1987: Schadstoffausbreitung im Nahbereich einer Autobahn, Institut für Verfahrenstechnik und Dampfkesselwesen, University of Stuttgart, Dept. Reinhaltung der Luft, Report No. 8

Baumbach, G., M. Käß, 1985: Ermittlung von vertikalen Schadstoffkonzentrations-Profilen in Waldbeständen in Baden-Württemberg - Meßtechnik und erste Ergebnisse, Staub-RdL 45 No. 6, 274-278

Baumbach, G., 1986a: Occurance of gaseous pollutants in forest stands, in: Georgii, H.W. (ed.): Atmospheric pollutants in forest areas, Reidel Publ., Dordrecht, 177-187

Baumbach, G., 1986b: Luftverunreinigungen in Mittelgebirgen durch Anreicherungen unter atmosphärischen Sperrschichten, in: Atmosphärische Prozesse, Beiträge zum IMA-Querschnittsseminar, Berlin, January 22-24, Texte des Umweltbundesamtes Berlin

Baumbach, G., K. Baumann, F. Dröscher, 1987a: Luftverunreinigungen in Wäldern, Institut für Verfahrenstechnik und Dampfkesselwesen, University of Stuttgart, Dept. Reinhaltung der Luft, Report No. 5

Baumbach, G., K. Baumann, F. Dröscher, 1987b: Luftqualität in Freudenstadt, Institut für Verfahrenstechnik und Dampfkesselwesen, University of Stuttgart, Dept. Reinhaltung der Luft, Report No. 6

Baumbach, G., K. Baumann, F. Dröscher, 1988: Behaviour of air pollutants under inversion weather conditions, in: K. Grefen and J. Löbel (eds.): Environmental Meteorology, Kluver Academic Publ., Dordrecht, 115-124

Paffrath, P., 1988: Untersuchung über die Verteilung und Bildung von Ozon im Bereich der Alpen aus Flugzeugmessungen, presentation at the International Symposium on Distribution and Effects of Photooxidants in the Alpine Environment, held at Garmisch-Partenkirchen, April 11-15

Rudolph, E., 1986: Verwendung eines Vergleichs-Elastomers als Wirkungsobjekt für Ozon, Staub-RdL 46 No.9, 393-395

Schliep, M., 1986: Kunststoffe als Indikatoren für Luftverunreinigungen, Institut für Verfahrenstechnik und Dampfkesselwesen, University of Stuttgart, Dept. Reinhaltung der Luft, Report No. 4

IN SITU AND REMOTE MEASUREMENTS OF NO_2 IN THE TROPOSPHERE AT LAUDER, NEW ZEALAND

W.A. Matthews and N.B. Jones
PEL Lauder, DSIR
Lauder, New Zealand

and

Y. Kondo, A. Iwata, and M. Takagi
Research Institute of Atmospherics
University of Nagoya
Toyokawa, Japan

ABSTRACT

An in situ measurement programme of nitric oxide, nitrogen dioxide, ozone and aerosols in a number of size ranges commenced at the PEL Lauder site (45°S) in August 1986. NO_2 data from the in situ chemiluminescent instrument are presented and comparisons drawn with remote measurements of NO_2 made at the same site using a scanning monochromator and a distant light source. Under conditions when the wind strength exceeds 15 knots for periods greater than 1 hour, the lower troposphere is shown to be sufficiently well mixed that not only do the NO_2 values obtained from the remote instrument remain relatively constant and low in magnitude, but the chemiluminescent in situ data tracks this finding in magnitude and phase. Examples of this intercomparison are given from this the first phase of an oxides of nitrogen budget analysis experiment.

1. INTRODUCTION

The Global Tropospheric Panel of the US National Research Council stated in their 1984 publication that night-time budget experiments of the oxides of nitrogen and humidity need to be performed to provide a more complete understanding of the many facets of nitrogen oxide chemistry. These experiments they state should be performed after adequate instrument intercomparisons and cross calibrations have been performed.

Two techniques have been employed to measure the concentration of nitrogen dioxide in the free troposphere as a first stage in such a budget analysis experiment at the Physics and Engineering Laboratory Lauder's relatively dry and inland site at Lauder, New Zealand, (45°S, 169°E). The purpose was to cross-check the experimental precision of each technique and determine the meteorological conditions that need to prevail to ensure that the measured data could be taken as representing the concentration in the free troposphere. The purpose of this paper is to present this first step in the experiment.

The next step will be to use the NO and NO_2 data obtained from the chemiluminescent detector in conjunction with simultaneous NO_3 and humidity measurements from the remote absorption technique, made under the "right" climatic conditions and perform a budget analysis.

2. INSTRUMENTATION

2.1 CHEMILUMINESCENT DETECTOR

The chemiluminescent detector used in this experiment was purpose built by the Research Institute of Atmospherics of the University of Nagoya for deployment at Lauder, New Zealand. It incorporates essentially the design of the instruments developed for our balloon experiments, described in Kondo et al (1984), and that used for aircraft measurements, Kondo et al, (1987). Ambient air is pulled through a teflon tube, the orifice of which is 7 m above ground level, into a reaction chamber where it is mixed with ozone generated by silent discharge and the resulting chemiluminescent signal is measured with a photomultiplier. Under the condition of excess ozone, this signal is proportional to the amount of NO in the atmosphere sampled. By monitoring a variety of housekeeping parameters such as the pressure and temperature of the ambient air in the reaction chamber, the mass flow in the system and by calibrating the efficiency of the reaction system using a series of standard mixtures, the NO mixing ratio can be obtained.

The ground based equipment at Lauder, has in addition, a 500 W mercury lamp converter system that is included in series upstream of the reaction chamber. The ambient air is then irradiated with intense ultraviolet before entering the reaction chamber. This irradiation process occurs at preset intervals and for the irradiation band

OZONE IN THE ATMOSPHERE
Rumen D. Bojkov and Peter Fabian (Eds.)

ISBN 0-937194-15-8

selected, nitrogen dioxide is photolysed to nitric oxide. The resulting increase in the chemiluminescent signal is taken as a measure of the amount of NO_2 in the air sample. NO_2 calibrations are performed by introducing known amounts of NO_2 from a commercial elution source into the ambient air and observing the change in chemiluminescent signal. The precision of the NO_2 measurement is 15 pptv.

2.2 LONG PATH ABSORPTION

The long path absorption method currently used at Lauder to measure the tropospheric concentration of NO_2 is similar to that already reported by McKenzie and Johnston (1984). White light from four 55 W Hella Rally 2000 quartz halogen lamps is collected by an F4 320 mm paraboloid. This is fed via a lens at approximately F15 onto the slit of a monochromator.

The path between the lamps and the mirror is 9.2 km over pasture land and the mean height of the path is < 30 m. The four lamps are spaced 0.5 m apart vertically with the bottom lamp of the chain 4 m above the ground. The monochromator, a McKee-Peterson 1018B, is optimally fed at F8 and incorporates a 1180 g/mm ruled grating blazed at 500 nm. A beam splitter just before the exit slit feeds 4% of the light onto the broad exit slit (15 mm width). At the conventional exit slit (0.5 mm width) an EMPI 9558B photomultiplier tube responds to both the NO_2 spectral signature and atmopsheric turbulence, termed here as flicker. However, behind the broad slit an identical photomultiplier tube responds to the atmospheric flicker only.

A ratio is formed from these two analog signals after simultaneously sampling at 1 kHz for a period of 1 second at each wavelength of the scan. Diffusers are used in front of both photomultipliers to minimise photocathode inhomogeneities. Any residual errors are further reduced by coadding successive scans and the total integration time used is 30 minutes.

A 'control' spectrum is obtained by using the same system but with the atmospheric path reduced to just a few metres. This spectrum is used to remove the instrumental function giving a residual noise level of less than 0.05%.

A detection threshould of approximately 25 pptv is obtained by using the least squares reduction technique on a synthetic spectrum as employed by McKenzie and Johnston (1984) in their earlier work.

3. RESULTS

It has been found that the NO_2 values measured by both techniques are particularly sensitive to wind strength, but relatively insensitive to wind direction. This is shown in Figure 1 where, with the onset of a NW wind at 1800

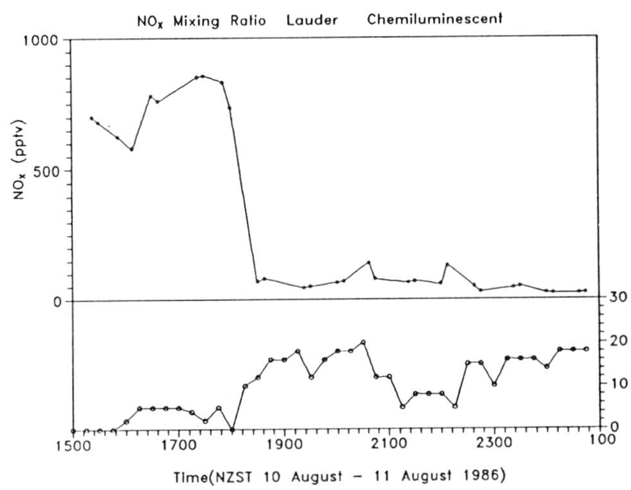

Fig. 1. In situ concentration of NO_x at Lauder as a function of New Zealand Standard Time for the night of August 10/August 11, 1986. The wind speed measured nearby is plotted in the same figure. At 1800 NZST the wind strength increases and the NO_x concentrations fall to consistently low values indicating the free troposphere is being sampled.

on 12 August 1986, almost immediately the NO_x values measured with the chemiluminescent detector dropped from 750 pptv to 60 pptv. We have numerous examples where under light wind conditions, the NO_2 values are highly variable, but this variability virtually disappears when the wind strength remains over 15 knots for an hour or two. The NO_2 values under these conditions are seen to drop to consistently low values with very little scatter.

At Lauder, the condition of winds consistenlty over 15 knots for periods exceeding 1 hour seems to be a sufficient criterion for selecting data representative of the free troposphere. The wind direction seems to have little effect on the data. This finding is perhaps not surprising because of Lauder's relative isolation. The data is however biased towards winds from the north-west since the climatology at Lauder is such that the wind strength exceeds 15 knots for more than 1 hour the wind tends to be blowing on most occasions from the NW sector.

We have therefore used this 15 knot minimum wind strength for periods of 1 hour as our data selection filter and for the intercomparison of the two measurement systems used data that satisfies this criterion.

In figure 2 the volume mixing ratios of NO_2 measured in situ (open triangles) are plotted as a function of New Zealand standard time. In the same figure the 2 sigma bounds of the 30 minute integrated data from the remote technique are also shown. Finally, the wind

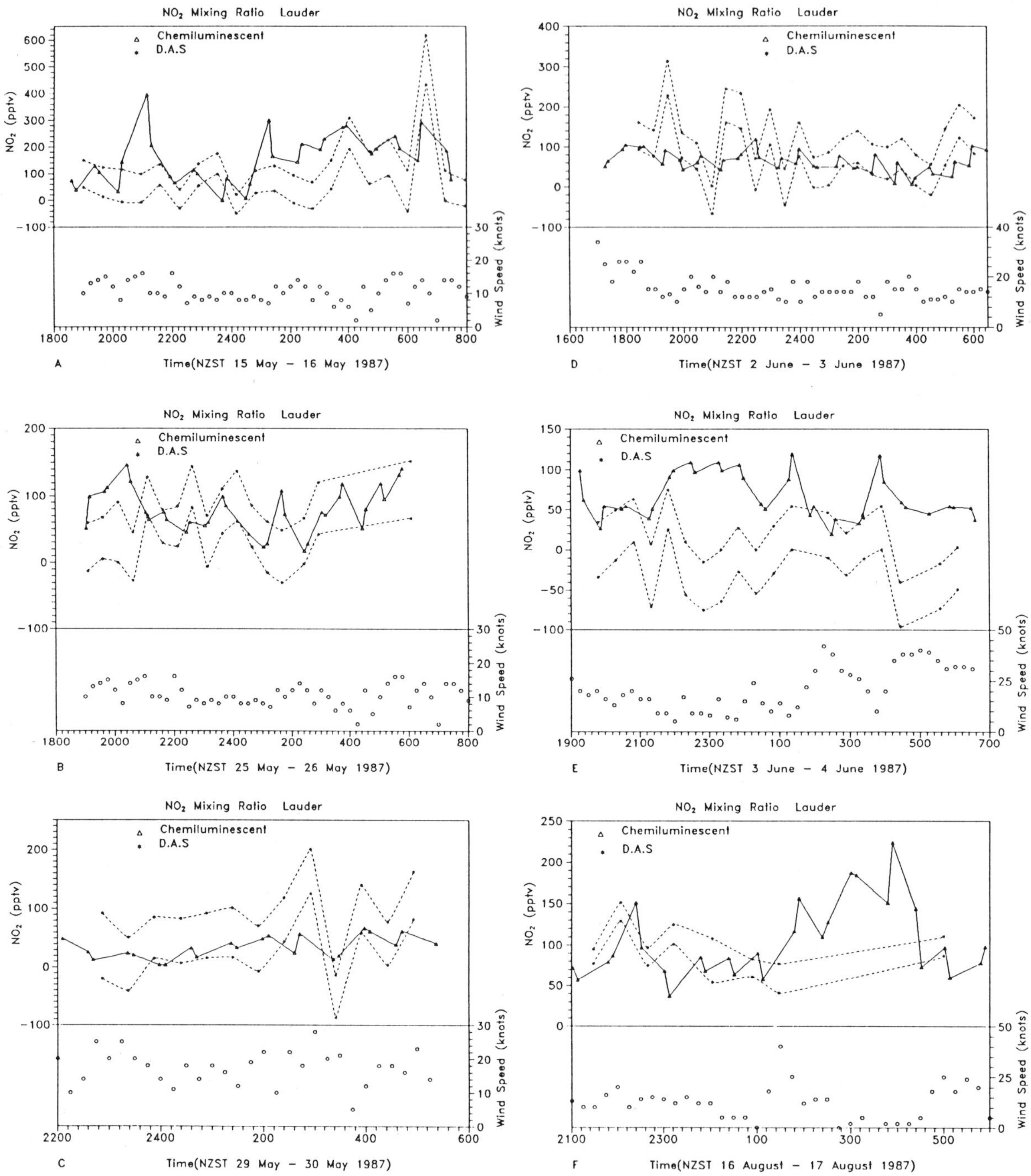

Fig. 2. Examples of volume mixing ratios of NO_2 measured in situ (open triangles) are plotted as a function of New Zealand Standard Time. The 1 sigma bounds of the 30 minute integrated data from the remote technique are also shown as is the wind strength. The agreement between the two NO_2 data sets is excellent provided the wind strength exceeds approximately 15 knots for periods larger than 1 hour.

strength has been added to each section of the figure for completeness. The anemometer was located near the chemiluminescent system and at one end of the long path for the remote system, the monochromator and chemiluminsecent system being housed in the same laboratory.

In general, the in situ measurements track inside the 1 sigma bounds of the remote measurements provided the wind strength, as measured at one end of the remote path exceeds about 15 knots. On the night of 15/16 May 1987, the wind started to decrease around midnight and the discrepancy between the in situ sample and the 9.2 km path average grows. The wind strength increased around 5am and the NO_2 values from both systems dropped to similar levels.

From the data given in figure 2, the mean NO_2 mixing ratio in the night, under conditions of a well mixed lower troposphere appears to be approximately 40-60 pptv in the winter-early spring of 1987.

4. CONCLUSION

The remote visible absorption technique is species specific for atmospheric NO_2 since the retrieval process identifies the effects of any interfering species. There is excellent agreement both in phase and absolute magnitude between the two NO_2 data sets provided data is compared when the wind strength exceeds 15 knots. This implies that under these conditions, the lower troposphere is sufficiently mixed over the 9 km path that the in situ system was also observing well mixed free tropospheric air.

The agreement in absolute amount implies that not only does no NO_2 like species other than NO_2 itself contribute appreciably to the NO_2 values measured with the chemiluminescent system, but also the conversion efficiency of the converter in the field maintains its laboratory calibration. This is an important finding as the validity of this technique for measuring oxides of nitrogen in the atmosphere has been the subject of some discussion in the past.

With the successful completion of this first stage of the experiment, the budget analysis phase has commenced with simultaneous NO_3 and humidity measurements also being made and this data will be discussed in a subsequent publication.

ACKNOWLEDGMENTS

The diligent technical support given to this programme by Brian McNamara and Alan Thomas at PEL Lauder is greatefully acknowledged as is the support of the Japan Society for the Promotion of Science.

REFERENCES

Global Tropospheric Chemistry Panel of National Research Council of United States of America, 1984: 'Global tropospheric chemistry: a plan for action', National Academy Press, Washington DC.

Johnston, P.V. and R.L. McKenzie, 1984: Long-path absorption measurements of tropospheric NO_2 in rural New Zealand, Geophys. Res. Lett., 11, 69-72.

Kondo, Y., A. Iwata, M. Takagi and W.A. Matthews, 1984: Balloon-borne chemiluminescent sonde for the measurement of tropospheric and stratospheric nitric oxide, Rev. Sci. Instr., 55, 1328-1332.

Kondo, Y., W.A. Matthews, A. Iwata, Y. Morita and M. Takagi, 1987: Aircraft Measurements of Oxides of Nitrogen along the Eastern Rim of the Asian Continent: Winter Observations, J. Atmos. Chem., 5, 37-58.

OBSERVATION OF NITRATE RADICAL CONCENTRATIONS AND LIFETIMES IN TROPOSPHERIC AIR

Ulrich Platt[1], *Dieter Perner*[2], *and Sabine Semke*[1]
[1]Institut für Chemie 3: Atmosphärische Chemie
Kernforschungsanlage Jülich GmbH
D-5170 Jülich, FRG
[2]Max-Planck-Institute für Chemie, Abt. Luftchemie
D-6500 Mainz, FRG

ABSTRACT

A total of 25 simultaneous night time series of NO_2, NO_3, and O_3 were observed between February 1983 and May 1985, about half of which were recorded during strong fall- or wintertime inversions isolating the observed air masses from the ground. NO_3-lifetimes derived from those data range from a few minutes to several hours. While usually the NO_3 appears to be limited by loss of N_2O_5, some of the nocturnal profiles of NO_3 and NO_2 are only consistent with direct loss of NO_3, possibly due to the reaction of NO_3 with biogenic hydrocarbons.

1. INTRODUCTION

Following the experimental detection of nitrate radicals (NO_3) in the troposphere by Platt et al. [1980] and Noxon et al. [1980] it was suggested by several authors [Ehhalt and Drummond, 1982; Richards, 1983; Heikes and Thompson, 1983, 1984; Calvert and Stockwell, 1983] that the NO_3 radical might play an important role in the non-photochemical conversion of NO_x to HNO_3. In addition, NO_3 is a strong oxidizing agent and initiates the nighttime removal of atmospheric trace species such as olefines, aromatic hydrocarbons, or organic sulfur compounds [Winter et al., 1984; Atkinson et al., 1984a,b,c, 1985; Mac Leod et al., 1986].

The atmospheric lifetime of NO_3 during sunlight hours is known to be of the order of seconds due to efficient photolysis by visible light as well as the reaction with NO. At night, however, its lifetime can be orders of magnitude longer. The lifetime can be derived by measuring the nocturnal variation in the NO_3 concentration simultaneously with its precursers O_3 and NO_2 and then solving the simple system of differential equations based on to the following reactions:

$$NO_2 + O_3 \rightarrow NO_3 + O_2 \qquad (1)$$

$$(-2) \qquad NO_2 + NO_3 \leftrightarrow N_2O_5 \qquad (2)$$

$$NO_3 \rightarrow loss \qquad (3)$$

$$N_2O_5 \rightarrow loss \qquad (4)$$

In an uniform environment, i.e. if there is no change in air masses during the time of observation the NO_3-lifetime simply becomes [Platt et al., 1984]:

$$\tau_{NO3} = 1/k_3 = [NO_3] / ([NO_2]*[O_3]*k_1) \qquad (5)$$

It should be noted that the NO_3-lifetime given by eq. (5) can also be limited by an irreversible loss of N_2O_5 ($k_4 > 0$) because both species are in a rapidly established thermodynamic equilibrium (reactions 2, -2). However, the shape of the NO_3 profiles during the course of the night is likely to be quite different: An NO_3-loss caused by removal of NO_3 itself will (under steady conditions) result in a constant lifetime of NO_3, while on the other hand, if the NO_3-loss is caused by removal of N_2O_5, it will be proportional to the N_2O_5/NO_3 ratio, which in turn is a function of the NO_2 concentration and temperature.

The exact nature of the NO_3 sink mechanism has remained unclear to date. In cases where high humidity prevails during the night, heterogeneous loss of N_2O_5 (to form HNO_3) should be efficient in removing NO_3. Reactions of NO_3 with hydrocarbons can certainly play a role. For measurements near the ground dry deposition and reaction of NO_3 with soil emitted NO can also be ruled out.

Despite the theoretical interest in nighttime NO_x chemistry few experimental investigations of tropospheric NO_3 and its lifetime have been reported to date. All those measurements of NO_3 were performed by long path absorption spectroscopy using the molecule's strong visible absorption lines. While part of the investigations used the sun as a light source to obtain total column densities being dominated by stratospheric NO_3 [Noxon et al., 1978], horizontal light paths between an artificial light source and the spectrometer have been used to determine the NO_3 concentration typically very close to the ground. Moreover, only few different types of air masses were investigated: Heavily polluted air [Platt et al., 1980, 1981; Pitts et al., 1984], moderately polluted air in Germany [Platt et al., 1981], background to slightly polluted air [Noxon, 1981; Noxon et al., 1980, 1983; Platt et al., 1984].

The NO_3-lifetimes observed in these studies range from a fraction of a minute in polluted areas over a few minutes in background- or slightly polluted air masses up to, but never exceeding one hour.

ISBN 0-937194-15-8

2. EXPERIMENTAL AND DATA REDUCTION

NO_3, NO_2, and O_3 were measured by the long-path differential light absorption technique (DOAS) presented before [Platt et al., 1979; Platt and Perner, 1983] using a high-pressure Xe-arc lamp as light source. The light beam was operated between two mountain tops in the German Black Forest (Feldberg and Schauinsland, 1220 to 1480 m above see level), its average distance to the ground being 500 m (Fig. 1). It had a path length of 8.6 km. The light was received by a Newton type telescope and was focussed on the entrance slit of a spectrograph (SPEX 1870, f = 0.5 m, 600 grooves/mm grating blazed at 400 nm and operated in first order). A cutoff filter in front of the spectrograph blocked the light below 600 nm from interfering via higher orders. The resulting spectral resolution was 0.3 nm. Nighttime spectra at 662 nm were ratioed against daytime spectra for the determination of NO_3. Spectra recorded in the 321 - 349 nm spectral interval were evaluated for NO_2 and O_3 by deconvolution as described by Platt and Perner [1983]. Minimum optical densities of $(1\text{-}2)*10^{-4}$ (base e) were obtained within integration times of 15 minutes for NO_3 of 30 minutes for NO_2 and O_3. The differential absorption cross section for the 662 nm band of NO_3 was taken at $1.77*10^{-17}$ cm^2, as given by Graham and Johnston [1978]. This number is not far from recent determinations, at least for the temperatures that prevailed during our measurements (-8° to +16°C; see for instance Sander [1986], Cantrell et al. [1987]). The use of this value allows direct comparison of the present data with data already published. Detection limits were $\approx 10^7$ cm^{-3} for NO_3, $\approx 10^9$ cm^{-3} for NO_2, and $\approx 3*10^{10}$ cm^{-3} for O_3. The sunlight intensity was monitored by a pyranometer (Kipp and Zonen CM6), temperature and relative humidity (R.H.) were recorded by standard meteorological instrumentation. The majority of the observations were made during periods of strong inversion, thus probing free tropospheric air.

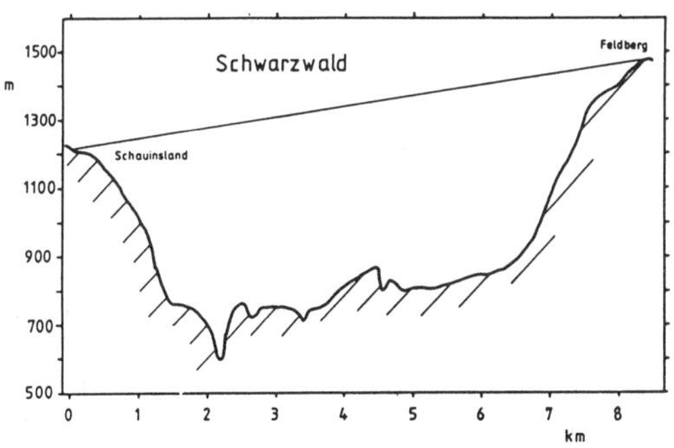

Fig. 1. Light-path and terrain profile in the Black Forest. The DOAS light beam is emitted at the Feldberg (1480 m above sea level) and received at the Schauinsland (1220 m).

3. MODEL CALCULATIONS

A simple box model was used to gain some understanding of the night time series of NO_2 and NO_3. The employed reactions are those given by Graham and Johnston [1978], except for the rate of thermal decay of N_2O_5 (k_{-2}), which was adopted from Malko and Troe [1982]. In a series of model runs either k_3 (describing irreversible NO_3-loss) or alternatively k_4 (describing irreversible N_2O_5-loss) were varied to obtain an optimum fit between measured and calculated night time series of NO_3 and NO_2.

4. RESULTS

A total of 25 simultaneous night time series of NO_2, NO_3, and O_3 were observed between February 1983 and May 1985, twelve of which were made in fall or winter (Oct., Dez., Jan.), when strong inversions isolated the observed air masses from influence from the ground.

Table 1 gives an overview of the observed night time series of NO_3, NO_2, and O_3. In contrast to earlier measurements in Germany [Platt et al., 1981] NO_3 levels always exceeded the detection limit with peak concentrations ranging from $\approx 8*10^7$ to $1.2*10^9$ cm^{-3}. Nitrogen dioxide concentrations, ranged from below detection limit ($\approx 10^9$ cm^{-3}) up to $8*10^{10}$ cm^{-3}. Ozone levels did not deviate much from the average value of $1.25*10^{12}$, except for a few extreme values ranging from 0.5 to $1.8*10^{12}$ cm^{-3}. In most cases (15 out of 25) a broad maximum in the NO_3 concentration is reached, typically during the second half of the night, which permits the approximation of the NO_3-lifetime by eq. (5). In Table 1 those data are marked with 'S'. In case when conditions differed due to change of air masses in the course of the night, maximum NO_3 and corresponding NO_2 concentrations are given. Also given are average temperature and R.H. values together with the calculated values for τ_{NO3}.

The last column in Table 1 gives the ratios of $(NO_2)_1/(NO_2)_0$, where the indices 0, 1 denote the conditions before the NO_3 concentration reaches it's plateau level and after sunrise, respectively.

TABLE 1. Concentrations of NO_3, NO_2, and O_3 measured in the Black Forest.

Date		NO_3 $10^8 cm^{-3}$		NO_2 $10^{10} cm^{-3}$	O_3 $10^{12} cm^{-3}$	Temp. °C	R.H. %	τNO_3/min	τN_2O_5	$\frac{(NO_2)_1}{(NO_2)_0}$
Feb. 83,	18/19	1.08	S	≤0.11	1.47	- 2	52	74	77	-
	19/20	3.8	S	≤0.11	1.35	- 3	72	293	355	-
	22/23	2.0	S	≤0.14	1.07	- 6	23	170	415	-
Feb. 84,	14/15	1.7		2.00	1.20	- 8	86	-	-	-
June 84,	15/16	1.7		3.40	0.95	7	70	-	-	-
	18/19	12.3	S	3.30	1.50	13	63	17	61	0.46
	19/20	8.6	S	1.43	1.8	15	55	22	26	0.32
	20/21	3.2	(S)	1.10	1.3	16	56	14	11	1.0
	26/27	1.9	S	1.00	1.4	9	75	11	20	0.26
Oct. 84,	19/20	4.0		0.25	1.2	4	67	-	-	-
	20/21	3.8	S	1.00	0.9	5	80	37	125	≈0.5
Dec. 84,	06/07	2.3	S	0.50	1.0	3	76	43	97	1.0
	07/08	7.1		1.10	1.2	5	51	-	-	-
	10/11	1.2		0.07	1.2	4	54	-	-	-
Feb. 85,	20/21	6.6		0.86	1.5	- 7	29	-	-	-
	21/22	7.0		3.00	1.1	- 7	80	-	-	-
	22/23	1.1		2.10	1.5	- 7	80	-	-	-
May 85,	03/04	0.83	S	2.70	1.3	0	94	2.5	46	0.5
	04/05	1.9		4.10	1.9	1	98	-	-	-
	05/06	4.6	S	2.00	1.2	5	79	16.8	113	0.39
	06/07	7.5	S	6.00	0.8	(5)		14	277	0.65
	12/13	5.8	S	1.10	1.3	(5)		36	132	0.33
	13/14	10.0		8.00	0.5	(5)		-	-	-
	14/15	3.2	S	0.90	1.2	(5)		21.6	79	0.46
	15/16	3.4		0.60	1.5	(5)		33.3	67	0.13

Three representative examples of nighttime NO_3 behaviour are presented in Figs. 2 to 4 and will be discussed in somewhat more detail here:

1. During Feb. 19 to 23, 1983 existed a strong inversion, so that pollution from below could not rise to the level of the light path. The ground at the measuring site was snow covered. The visibility was about 250 km. In the night of Feb. 22/23 the NO_2 concentrations were below the detection limit between 23:00 and sunrise, while the NO_3 concentration gradually rose to $\approx 2*10^6$ cm^{-3} (Fig. 2). The best fit between observed and calculated NO_3 is reached assuming a NO_3-lifetime of 3.5 hours and an initial NO_2 concentration of $1.2*10^9$, just below the detection limit for NO_2 for those particular measurements.

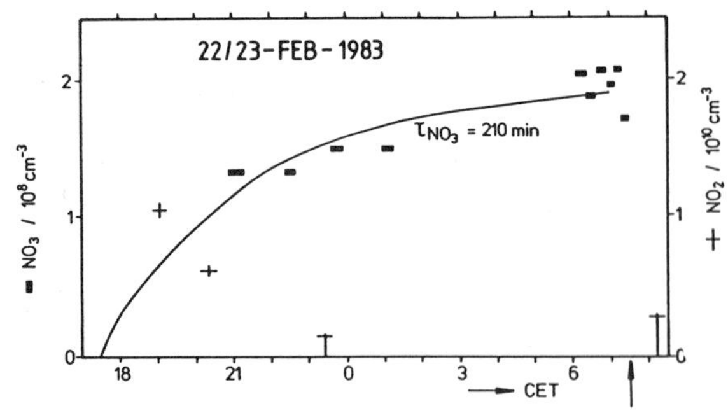

Fig. 2. Measured night time series of NO_2 and NO_3 concentrations Feb. 22/23, 1983, -6°C, 22 % R.H., $[O_3] = 1.07 \cdot 10^{12}$ cm^{-3}. Drawn line: calculated NO_3 assuming τ_{NO3} = 210 min.

2. Considerably higher NO_2 levels were found during the night of June 18/19, 1984 (see Fig. 3), as to be expected from the summertime meteorological situation, when inversion heights during daytime exceeding 2000 m above sea level allowed NO_x transport to those altitudes [Platt et al., 1988]. The NO_3 concentration reached a plateau level of about $1.5 \cdot 10^9$ cm^{-3} after midnight. Beginning at 2:00 in the early morning a slight decrease was observed, which turned into an abrupt decrease at the photolysis during dawn. The best match between model and observation is obtained assuming a NO_3-lifetime of 17.5 min, lifetimes of 16 or 19 min already yield considerably worse fits. The most interesting feature, however, is the observation that the assumption of a N_2O_5-loss leads to a much poorer fit of the calculated NO_3 profiles. Moreover, a N_2O_5-loss would lead to a loss of NO_2 twice as large as that actually observed.

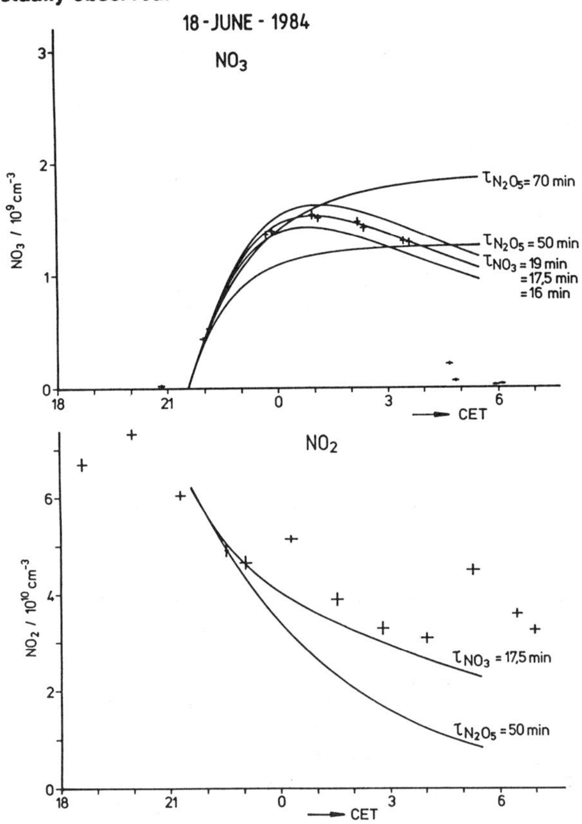

Fig. 3. Measured night time series of NO_3 (upper panel) and NO_2 (lower panel) concentrations June 18/19, 1984, +13°C, 63 % R.H., $[O_3] = 1.5 \cdot 10^{12}$ cm^{-3}. Drawn lines: calculated NO_3 and NO_2 profiles assuming τ_{NO3} = 17.5 min.

3. During the night of October 20/21, 1984 (see Fig. 4) measured NO_3 concentrations agree well with a calculated curve assuming a NO_3-lifetime of 27 min until midnight. Later, air masses apparently containing more NO_2 and NO_3 are advected. Here the observed decrease of NO_2 is somewhat steeper than expected from NO_3 loss alone, thus N_2O_5 loss may have occured as well.

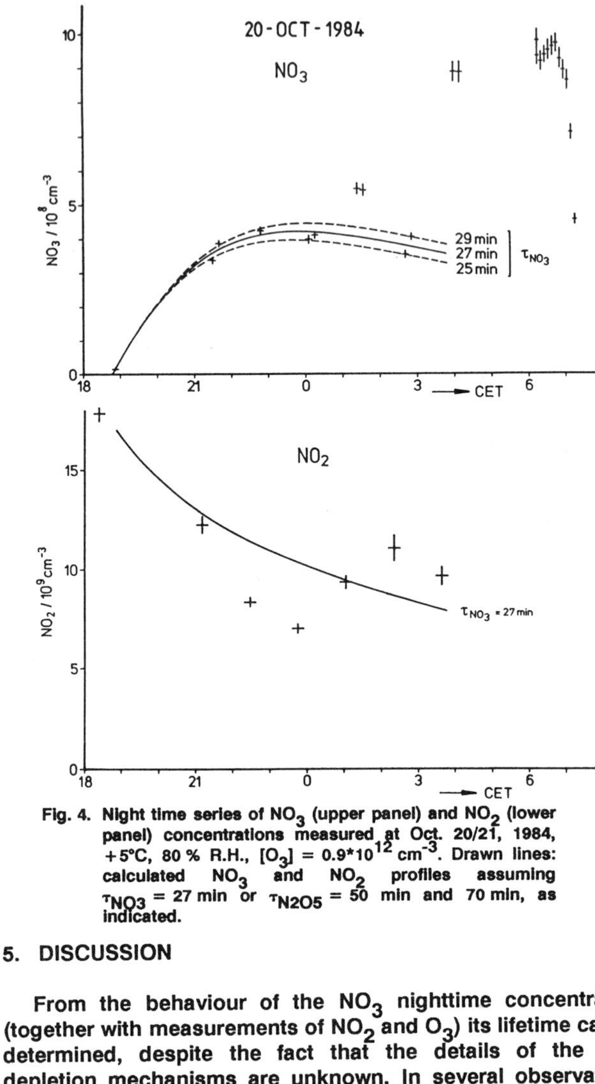

Fig. 4. Night time series of NO_3 (upper panel) and NO_2 (lower panel) concentrations measured at Oct. 20/21, 1984, +5°C, 80 % R.H., $[O_3] = 0.9 \cdot 10^{12}$ cm^{-3}. Drawn lines: calculated NO_3 and NO_2 profiles assuming τ_{NO3} = 27 min or τ_{N2O5} = 50 min and 70 min, as indicated.

5. DISCUSSION

From the behaviour of the NO_3 nighttime concentration (together with measurements of NO_2 and O_3) its lifetime can be determined, despite the fact that the details of the NO_3 depletion mechanisms are unknown. In several observations above strong wintertime inversions NO_3-lifetimes exceeding one hour are infered.

Note, that in the three night time series presented above, the NO_3-lifetimes derived from fitting the observed NO_3 profiles to the model calculation agree quite well with the lifetimes derived from simple stationary state considerations (e.g. eq. (5)) as given in Table 1.

While in the majority of the cases reported in Table 1 N_2O_5-loss limits the lifetime of NO_3, in some cases, especially during summertime, the data can only be explained by a direct loss of NO_3. Likely mechanisms removing NO_3 are reactions with organic compounds like olefines, terpenes, or aromatic species. For instance, of the order of $3 \cdot 10^7$ terpene molecules per cm^3 (roughly 1 ppt) are required to limit the NO_3-lifetime to 20 min [Winer et al., 1984]. While such levels of terpene concentrations can easily by reachen, [Roberts et al., 1985] the fact that the lifetime of NO_3 seems to vary little throughout the night would require a steady addition of terpenes into the observed air mass. Therefore, a reaction of NO_3 with less reactive and more abundant species appears to be more likely.

6. ACKNOWLEDGEMENTS

The authors gratefully acknowledge the help of W. Weiss, H. Stockburger, and H. Sartorius from the Institute of Atmospheric Radioactivity, Freiburg in siting the DOAS spectrometer at the Schauinsland. The crew of the Feldberg weather station was very helpful in allowing us to set up our Xe-arc light source. We wish to thank the members of the Umweltbundesamt station Schauinsland for providing meteorological data.

7. REFERENCES

Atkinson, R., C.N. Plum, W.P.L. Carter, A.M. Winer, and J.N. Pitts, 1984a: Rate constants for the gas-phase reactions of NO_3 radicals with a series of organics in air at 298±1 K. J. Phys. Chem., 88, 1210-1215.

Atkinson, R., C.N. Plum, W.P.L. Carter, A.M. Winer, and J.M. Pitts, 1984b: Rate constants for the gas-phase reactions of nitrate radicals with a series of organics in air at 298±1 K. J. Phys. Chem., 88, 4446.

Atkinson, R., C.N. Plum, W.P.L. Carter, A.M. Winer, and J.M. Pitts, 1984c: Kinetics of the gas-phase reactions of NO_3 radicals with a series of alkanes at 296±2 K. J. Phys. Chem., 88, 2361-2364.

Atkinson, R., S.M. Aschmann, A.M. Winer, and J.N. Pitts, 1985: Kinetics and atmospheric implications of the gas-phase reactions of NO_3 radicals with a series of monoterpenes and related organics at 294±2 K. Environ. Sci. Technol, 19, 159-163.

Calvert, J.G., and W.R. Stockwell, 1983: Acid generation in the troposphere by gas-phase chemistry. Environ. Sci. Technol., 17, 428A-443A.

Cantrell, C.A., J.A. Davidson, R.E. Shetter, B.A. Anderson, and J.G. Calvert, 1987; The temperature invariance of the NO_3 absorption cross section in the 662 nm region. J. Phys. Chem., 91, 5858-5863.

Ehhalt, D.H., and J.W. Drummond, 1982: The tropospheric cycle of NO_x: Chem. Unpoll. Poll. Troposphere, Proc. NATO Adv. Study Inst., Corfu, Greece, 28. Sept. - 10. Oct., 1981, H.W. Georgii, W. Jaeschke (Eds.), Reidel, Dordrecht, Boston, London, 219-251.

Graham, R.A., and H.S. Johnston, 1978: The photochemistry of NO_3 and the kinetics of the N_2O_5-O_3 system. J. Phys. Chem., 82, 254-268.

Heikes, B.G., and A.M. Thompson, 1983: Effects of heterogeneous processes on NO_3, HONO, and HNO_3 chemistry in the troposphere. J. Geophys. Res., 88, 10883-10895.

Heikes, B.G., and A.M. Thompson, 1984: Correction to "Effects of heterogeneous processes on NO_3, HONO, and HNO_3 chemistry in the troposphere". J. Geophys. Res., 89, 11829.

Japar, S.M., and H. Niki, 1975: Gas-phase reactions of the nitrate radical with olefins. J. Phys. Chem., 79, 1629-1632.

Johnston, H.S., C.A. Cantrell, and J.G. Calvert, 1986: Unimolecular decomposition of NO_3 to form NO and O_2 and a review of N_2O_5/NO_3 kinetics. J. Geophys. Res., 91, 5129-5172.

Levy, H., J.D. Mahlman, and W.J. Moxim, 1980: A stratospheric source of reactive nitrogen in the unpolluted troposphere. Geophys. Res. Lett., 7, 441-444.

Mac Leod, H., S.M. Aschmann, R. Atkinson, E.C. Tuazon, J.A. Sweetman, A.M. Winer, J.N. Pitts, 1986: Kinetics and mechanism of the gas-phase reactions of the NO_3 radical with a series of reduced sulfur compounds. J. Geophys. Res., 91, 5338-5346.

Malko, M.W., and J. Troe, 1982: Analysis of the unimolecular reaction $N_2O_5 + M \rightarrow NO_2 + NO_3 + M$. Int. J. Chem. Kin., 14, 399-416.

Noxon, J.F., R.B. Norton, and W.R. Henderson, 1978: Observation of atmospheric NO_3. Geophys. Res. Lett., 5, 675-678.

Noxon, J.F., 1981: NO_x in the mid-Pacific troposphere. Geophys. Res. Lett., 8, 1223-1226.

Pitts, J.N., Jr., H.W. Biermann, R. Atkinson, and A.M. Winer, 1984: Atmospheric implications of simultaneous nighttime measurements of NO_3 radicals and HONO. Geophys. Res. Lett., 11, 557-560.

Platt, U., D. Perner, and H. Paetz, 1979: Simultaneous measurement of atmospheric CH_2O, O_3, and NO_2 by differential optical absorption. J. Geophys. Res., 84, 6329-6335.

Platt, U., D. Perner, A.M. Winer, G.W. Harris, and J.N. Pitts, 1980: Detection of NO_3 in the polluted troposphere by differential optical absorption. Geophys. Res. Lett., 7, 89-92.

Platt, U., D. Perner, J. Schroeder, C. Kessler, and A. Toenissen, 1981: The diurnal variation of NO_3. J. Geophys. Res., 86, 11965-11970.

Platt, U., and D. Perner, 1983: Measurements of atmospheric trace gases by long path differential UV/visible absorption spectroscopy: Optical and Laser Remote Sensing, D.K. Killinger and A. Mooradian (Eds.), Springer Ser. Optical Sci., 39, 95-105.

Platt, U.F., A.M. Winer, H.W. Biermann, R. Atkinson, and J.N. Pitts, 1984: Measurement of nitrate radical concentrations in continental air. Environ. Sci. Techn., 18, 365-369.

Platt, u., M. Rateike, W. Junkermann, J. Rudolph, and D.H. Ehhalt, 1988: New tropospheric OH measurements. J. Geophys. Res., 93, 5159-5166.

Richards, L.W., 1983: Comments on the oxidation of NO_2 to nitrate - day and night. Atmosph. Environ., 17, 397-402.

Roberts, J.M., C.J. Hahn, F.C. Fehsenfeld, J.M. Warnock, D.L. Albritton, and R.E. Sievers, 1985: Monoterpene hydrocarbons in the nighttime troposphere. Environ. Sci. Technol., 19, 364-369.

Sander, S.P., 1986: Temperature dependence of the NO_3 absorption spectrum. J. Phys. Chem., 90, 4135-4142.

Winer, A.M., R. Atkinson, and J.N. Pitts, 1984: Gaseous nitrate radical: possible nighttime atmospheric sink for biogenic organic compounds. Science, 224, 156-159.

THE OZONE FORMING POTENTIAL OF THE BIOGENIC α-PINENE IN COMPARISON WITH ANTHROPOGENIC HYDROCARBONS

F. Nolting and C. Zetzsch
Fraunhofer-Institut für Toxikologie und Aerosolforschung
D-3000 Hannover 61, FRG

ABSTRACT

A new method of evaluation of smog chamber kinetic data is presented to describe the degradation of olefines by OH and ozone simultaneously. The scheme has been validated with several olefines and five monoterpenes. In the dark reaction of ozone with a surplus of olefines, the formation of OH radicals can be detected. In experiments under irradiation, the formation of ozone in the presence of α-pinene, cis-butene-(2) and propene is compared at different hydrocarbon/NO_x ratios. The lowest ozone forming potential is found for the monoterpene. Propene yields twice as much, whereas cisbutene-(2) with a factor of 1.5 is in between.

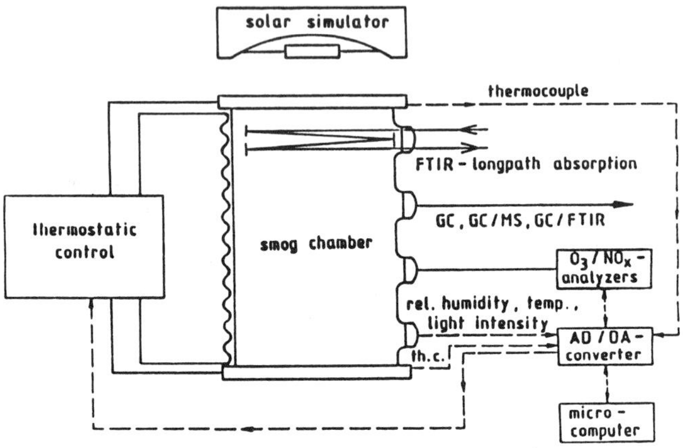

Fig. 1. Schematic diagram of the smog chamber with peripheral instrumentation

1. INTRODUCTION

There has been a long and continuous debate about an apparent discrepancy between the high emission rates of the biogenic hydrocarbons (especially the terpenes) and the rather low concentrations observed (Went, 1960, and Kohlmaier et al., 1984). Many contributions deal with the photochemical activity of these compounds in view of their importance in the discussion of forest decline, where phytotoxic transients might be formed in the direct vicinity of the plants at high concentrations.

On the other hand, only little information is available about the portion of the biogenic terpenes (and their primary degradation products) in the photochemical reactions taking place together with the anthropogenic airborne pollutants that are found in forest air. e.g. NO_x and hydrocarbons (e.g. Lloyd et al., 1983).

Using a smog chamber, experiments have been performed to study the photochemical degradation of the terpenes themselves and the degradation of their transient products. In order to examine the ozone forming potential of α-pinene, comparative measurements have also been undertaken with propene and cis-butene-(2).

2. EXPERIMENTAL

Figure 1 shows the smog chamber, a cylinder made of borosilicate glass (Duran, Schott), with a volume of 700 l. Irradiation is provided with a solar simulator at about 50% of the intensity of the sun at noon of a summer day in middle europe. The lid and the walls of the chamber are thermostated at room temperature, the botttom is kept at a slightly higher temperature (by 3°C) to improve mixing. The chamber can be heated up to 60°C to study the photochemical degradation of compounds with low volatility (e.g. n-hexadecane). The regulation of the temperature, the control of the analyzers for NO_x, O_3, and SO_2, and the recording of these data and of the humidity and light intensity are performed by means of a microcomputer.

The hydrocarbons are measured automatically every 35 minutes by cryofocussing 20 ml air samples with capillary gas chromatography (Nolting et al., 1988), and selective and simultaneous information on the actual concentrations is obtained down to the lower ppbv level. By employing two columns with different polarity, the degradation of the primary unpolar compounds and the formation of polar transients can be monitored simultaneously with high reproducibility.

3. RESULTS

3.1. DEGRADATION BY OZONE AND DETECTION OF OH FORMATION IN THE DARK

The precise and simultaneous monitoring of several hydrocarbons enables us to apply a new method for the determination of the rate constants for the reaction of hydrocarbons with ozone, originally developed by Behnke et al. (1987), where Cl atoms have been monitored in a smog chamber in the presence of OH. The decay of a

OZONE IN THE ATMOSPHERE
Rumen D. Bojkov and Peter Fabian (Eds.)

ISBN 0-937194-15-8

compound, S, is described by the reaction with OH radicals and with ozone and by the dilution of the chamber contents (from the gas consumption of the analyzers). The dilution can be eliminated by monitoring the decay of inert compounds (e.g. hexafluorobenzene). In integrated form the following expression is obtained:

$$-\ln([S]_t/[S]_o) = k_{OH}\int[OH]dt + k_{O_3}\int[O_3]dt$$

The integrals $\int[O_3]dt$ and $\int[OH]dt$ can be determined from the decay pattern of 4 - 6 reference compounds with known rate constants using the following relations:

$$-\frac{\ln([S]_t/[S]_o)}{k_{OH}} = \int[OH]dt + \frac{k_{O_3}}{k_{OH}}\int[O_3]dt, \text{ or}$$

$$-\frac{\ln([S]_t/[S]_o)}{k_{O_3}} = \int[O_3]dt + \frac{k_{OH}}{k_{O_3}}\int[OH]dt$$

From plots of $\ln([S]_t/[S]_o) + k_{OH}\int[OH]dt$ vs. $\int[O_3]dt$ the rate constant of the reaction with O_3 can be calculated as slope. The internal consistency of the evaluation is checked in a final balance of the overall decay. The rate constants obtained in this way are summarized in table 1 in comparison with literature values.

Table 1. RATE CONSTANTS ($k_{O_3} \pm 2\sigma$)/10^{-17} cm^3s^{-1} FOR THE REACTIONS OF OLEFINES AND TERPENES WITH OZONE (NOLTING ET AL., 1988, REFERENCED TO CIS-BUTENE-(2) AS INTERNAL STANDARD AT 297 K), NUMBER OF DETERMINATIONS, N, AND COMPARISON WITH LITERATURE VALUES OF ATKINSON AND CARTER (1984)

Compound	this work	(N)	lit.
cis-3-Methylpentene-(2)	47.2 ± 1.7	(10)	45.6
2,3-Dimethylbutene-(2)	125.0 ± 6.0	(6)	116.0
cis-Butene-(2)	13.0		
trans-Butene-(2)	21.1 ± 1.0	(21)	20.0
i-Butene	1.42 ± 0.1	(33)	1.21
Propene	1.42 ± 0.3	(14)	1.13
Hexene-(1)	1.16 ± 0.4	(5)	1.17
Heptene-(1)	0.98 ± 0.1	(7)	1.15
Octene-(1)	0.91 ± 0.2	(8)	0.80
Nonene-(1)	1.5 ± 0.7	(5)	1.45
Cyclopentene	62.4 ± 3.5	(19)	96.9
Cyclohexene	7.8 ± 0.5	(21)	10.4
Cycloheptene	28.3 ± 1.5	(20)	31.9
Isoprene	1.17 ± 0.1	(43)	1.43
α-Pinene	8.6 ± 1.3	(6)	8.4
β-Pinene	1.4 ± 0.2	(6)	2.1
Limonene	15.3 ± 3	(7)	64
d3-Carene	5.6 ± 9	(4)	12
Tricyclene	<0.1	(6)	-

In spite of complete darkness during these experiments, OH radicals are observed as a product of the reaction between ozone and olefines with considerable contribution to the degradation of the compounds.

3.2 THE OZONE FORMING POTENTIAL OF THREE OLEFINES IN COMPARISON

Three olefines have been chosen that are observed in forest air: α-pinene as a representative of biogenic hydrocarbons (a frequently observed terpene), cis-butene-(2) as anthropogenic olefine (with a reactivity similar to that of α-pinene), and propene a predominantly anthropogenic olefine (a standard compound of many studies of atmospheric photochemistry).

Under irradiation with simulated sunlight in mixtures with NO_x, these hydrocarbons lead to the formation of ozone. The ozone forming potential can be defined as the increase of ozone with different ratios of hydrocarbons/NO_x compared to experiments in the absence of hydrocarbons.

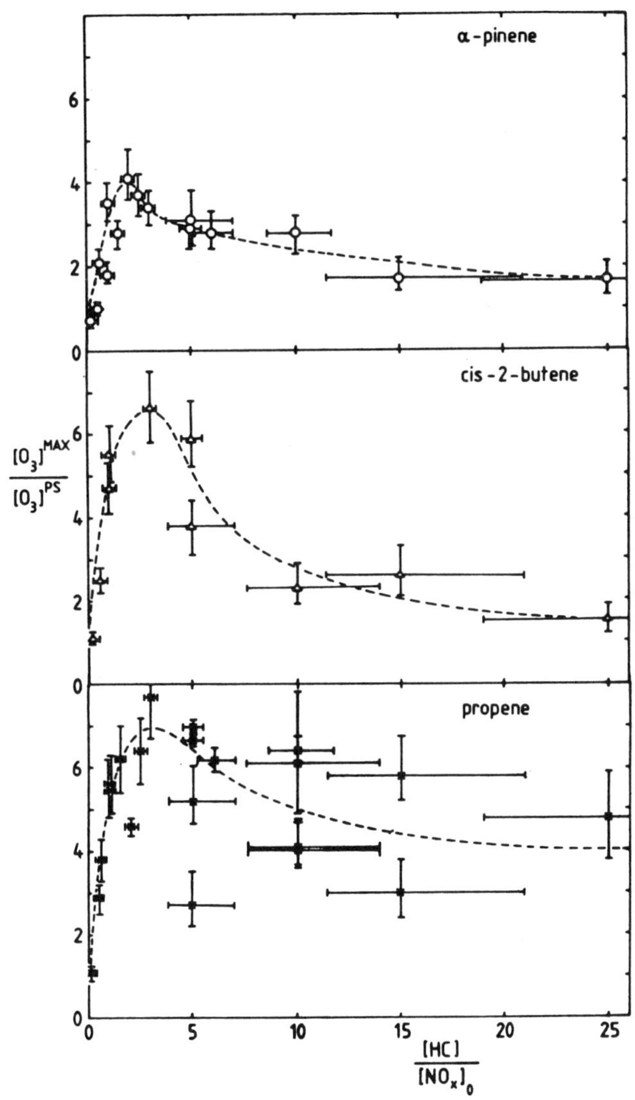

Fig. 2. Relative ozone maxima of three olefines at different hydrocarbon/NO_x ratios. The biogenic α-pinene yields only half of the increase of ozone observed for propene.

The observed concentrations of ozone are running through a maximum, because the hydrocarbons are largely degraded in the course of the irradiation and - especially in the case of terpenes - their transient products are forming aerosols. The particles are deposited together with nitrogen containing compounds formed from NO_x. The results of these experiments are plotted in fig. 2 in dependence of the hydrocarbon/NO_x ratio.

4. CONCLUSIONS

The present relative method for the determination of reactivities against ozone allows us to use several internal standards simultaneously and to evaluate rate constants with consideration of possible side reactions selectively and precisely.

Even during nighttime the degradation by ozone leads to the formation of OH radicals with considerable contribution to the transformation of hydrocarbons.

In comparative experiments with the biogenic α-pinene and the predominatly anthropogenic olefines cis-butene-(2) and propene the lowest ozone forming potential is observed for the monoterpene, whereas propene yields twice as much ozone. Cis-butene-(2) occupies a medium position with a factor of 1.5.

With respect to forest decline (as far as the results obtained with α-pinene are transferable to all terpenes) one might expect a rather shielding influence on forests. For a more detailed description, the aspect of time has to be considered for the rate of formation of ozone with special impact on the actual reactions (and concentrations) in the direct vicinity of the plants.

ACKNOWLEDGEMENT

This work was supported by the Projekt Europäisches Forschungszentrum für Maßnahmen zur Luftreinhaltung (PEF), Land Baden-Württemberg, in Karlsruhe.

REFERENCES

Atkinson, R., and W.P.L. Carter, 1984: Kinetics and mechanisms of the gas-phase reactions of ozone with organic compounds under atmospheric conditions. Chem. Rev., 84, 437-470.

Behnke, W., and C. Zetzsch, 1987: The simultaneous determination of two active species in smog chambers: evaluation of rate constants for Cl atoms in the presence of OH, 3rd French - German Workshop: Tropospheric Chemistry by Laboratory Studies, 6. - 8. October, Leinsweiler, Report No. 15, Fachbereich 9, Physikalische Chemie, (eds. K.H. Becker and P. Carlier), pp. 59-60 (Available from K.H. Becker, Universität Wuppertal).

Kohlmaier, G.H., H. Bröhl and E.O. Siré, 1984: Über die mögliche lokale Wechselwirkung anthropogener Schadstoffe mit den Terpen-Emissionen von Waldökosystemen, Allg. Forst- u. Jagdzeitung, 155, 197-199.

Lloyd, A.C., R. Atkinson, F.W. Lurmann and B. Nitta, 1983: Modelling potential ozone impacts from natural hydrocarbons-I. Development and testing of a chemical mechanism for the NO_x-air photooxidations of isoprene and α-pinene under ambient conditions, Atmospheric Environment, 17, 1931-1950.

Nolting, F., W. Behnke and C. Zetzsch, 1988: A smog chamber for studies of the reactions of terpenes and alkanes with ozone and OH, J. Atmos. Chem., 6, 47-59.

Went, F.W., 1960: Organic matter in the atmosphere and its possible relation to petroleum formation, Proc. Acad. Sci., 46, 212-221.

SMOG CHAMBER INVESTIGATIONS OF THE INFLUENCE OF NaCl AEROSOL ON THE CONCENTRATION OF O_3 IN A PHOTOSMOG SYSTEM

W. Behnke and C. Zetzsch
Fraunhofer-Institut für Toxikologie und Aerosolforschung
D-3000 Hannover 61, FRG

ABSTRACT

A strong enhancement of ozone production is observed in a simulated photochemical smog in the presence of NaCl aerosol. The increase in ozone is accompanied by a formation of atomic Cl. The production of Cl atoms requires humidity and probably not NO_x (NO_x was less than 1 ppb), and it is catalyzed by the glass chamber walls (borosilicate glass, Duran 50, Schott) or in a teflon bag chamber by the presence of quartz aerosol (AEROSIL 200, DEGUSSA). Cl_2 is identified to be the main photochemical precursor of Cl; the formation of Cl_2 is presently not understood from known reaction mechanisms. Heterogeneous reactions of O_3 and OH radicals are postulated to explain this effect, that might have atmospheric relevance.

1. INTRODUCTION

NaCl is considered to be the most important aerosol world-wide in the form of sea spray with a source strength of the order of 10^9 t/a. A certain portion of this mass is transported across the continents, where an intrusion of the acids H_2SO_4 and HNO_3 occurs in the presence of SO_2 (Meinert and Winchester, 1977, Hitchcock et al., 1980, Clegg and Brimblecombe, 1985, and NO_x (Cadle and Robbins, 1960, Martens et al., 1973). The reaction of these acids with aerosols releases the highly volatile HCl. In Western Europe the direct emission of HCl from the stack gas of coal fired power plants contributes largely to this source (Lightowlers and Cape, 1988), and the atmospheric photooxidation of chlorinated hydrocarbons like tri- and perchloroethene may contribute on a global scale (because of the rather low reactivity of these compounds against OH). Depending on humidity, number density and composition of the aerosol, a portion of the atmospheric HCl will be adsorbed on the surface of the aersosol or will remain in the gas phase (Cocks and McElroy, 1984). It has been postulated that atmospheric emissions of chlorine (in the form of Cl_2 or HCl) may enhance the photochemical smog by photochemical oxidation cycles leading to the formation of (the extremely reactive) atomic Cl (Hov et al., 1985). A preliminary study in our laboratory showed that the presence of NaCl aerosol leads to the formation of Cl atoms, which can contribute considerably to the consumption of alkanes (Behnke et al., 1984) and may explain discrepancies in field determinations of tropospheric OH from the rate of disappearance of hydrocarbons (Zetzsch, 1987).

In the present study, photochemical oxidation cycles of Cl^- (in the form of HCl in the gas phase or bound in aerosols) are investigated under simulated tropospheric conditions in the presence of NaCl aerosol, NO_x, ozone and hydrocarbons. The presence of atomic Cl might add to the tropospheric photochemical production of ozone, induced by the attack of OH or Cl to hydrocarbons.

2. EXPERIMENTAL

The simulation experiments are performed either in a 2.3 m^3 glass chamber (made of borosilicate glass, Duran 50, Schott) or in teflon bags (PTFE, Dupont) with 0.6 m^3 volume. The glass chamber is optimized for long residence times of aerosols (Holländer et al., 1984). The spectrum of the solar simulator is adjusted to the spectrum of the sun at ground level (Behnke et al., 1987), its intensity corresponds to the twice of an European summer day at noon. The spectrum of the solar simulator can be limited to wavelengths larger than 360 nm by using a cut-off filter (Plexiglass), the analytics of the trace gases O_3, NO_x (by gas analyzers) and the hydrocarbons (by capillary gas chromatography with a preconcentrator) is described elsewhere (Behnke et al. 1988, Nolting et al., 1988). The aerosol is produced by atomizing aqueous solutions of NaCl; initial aerosol mass densities of 3 to 8 mg m^{-3} are achieved at mean particle diameters of 0.5µm. The aerosol is analyzed quantitatively from filter samples of at least 100 µg by ion chromatography and microgravimetry. The hydrocarbons are injected as a mixture of 6 compounds into the smog chamber at a total mixing ratio of 160 ppb; the components are selected to cover a wide range of relative reactivities against Cl and OH from toluene (k_{Cl}/k_{OH} = 10) to 2,2,3,3-tetramethylbutane (k_{Cl}/k_{OH} = 140). Pentafluorobenzene (or in some experiments hexaflourobenzene) serves for an almost inert dilution standard. The concentration time profiles of the hydrocarbons are utilized to compute the profiles of Cl and OH simultaneously and independently according to a procedure described by Behnke and Zetzsch (1987).

ISBN 0-937194-15-8

3. RESULTS AND DISCUSSION

Figure 1 shows the disappearance of the various components of this mixture versus the irradiation time. The decrease of pentafluorobenzene is almost exclusively caused by the dilution of the chamber contents by the gas consumption of the gas analyzers, and the influence of aerosol sampling (taking a larger portion of gas) can be seen as sudden steps. Besides this (in constant volume chambers inherent) dilution, the other compounds can be seen to react markedly with more than one active species (presumably OH and Cl).

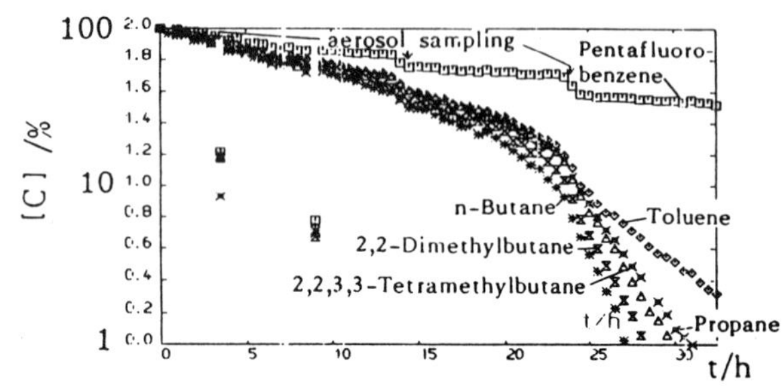

Fig. 1. Gas chromatographic monitoring of the indicator compounds used for the simultaneous detection of OH and Cl during a smog chamber run in the presence of NaCl aerosol

The initial decrease of toluene in figure 1 is similar to the decrease of 2,2,3,3-tetramethylbutane, but after 23 h the decay of this alkane accelerates much faster than the decay rate of toluene. The concentrations of the radicals OH and Cl can be evaluated quantitatively from this behaviour (Behnke and Zetzsch, 1987). For a discussion of the source strengths it may be more useful to consider the rates of consumption of these radicals. In the initial phase of the irradiation, these radicals will react mostly with educts, and the rate of consumption of Cl can be determined from known rate constants and concentrations, yielding the expression for Cl:

$$d[Cl]/dt = - \sum_i [HC]_i \; k_{Cl_i} [Cl](t) + \text{Cl-prdn.} \quad (I)$$

In the steady state approximation ($d[Cl]/dt = 0$) the Cl consumption will be equal to the Cl production.

Figures 2a,b show a comparison of the concentration profiles of NO, NO_2, O_3 and OH in the absence and presence of NaCl aerosol. At the beginning of each experiment the hydrocarbons and 300 ppb NO_2 were injected and the irradiation by the solar simulator was started (full solar spectrum). In a few seconds the photostationary state is established, yielding 30ppb O_3 and a concentration of OH of 2×10^6 cm^{-3} in both experiments. The initial contribution of Cl atoms can be neglected in the presence of NaCl aerosol. After 4h NO is added continously at 1ppb/min in both experiments to replenish the NO_x loss to the chamber walls, and the slow decrease of O_3 (due to the consumption of the hydrocarbons) is not influenced in the absence of aerosol.

In the presence of NaCl aerosol, O_3 shows a slow increase from the beginning and a marked increase after 5h, yielding a maximum much larger than 1.5 ppm, outside the range of the O_3 analyzer. The increase of the radicals OH and Cl can be detected even earlier, after 3h, and it can be monitored up to 9h, where the hydrocarbons are consumed to a level below 1ppb (the detection limit of our gas chromatography). Then the level of Cl has reached 15% of the level of OH, and reactions of Cl will dominate in the consumption of hydrocarbons because of the higher rate constants of Cl. After 10h, NO_x is decreasing in the presence of NaCl aerosol in spite of the continuous addition of NO.

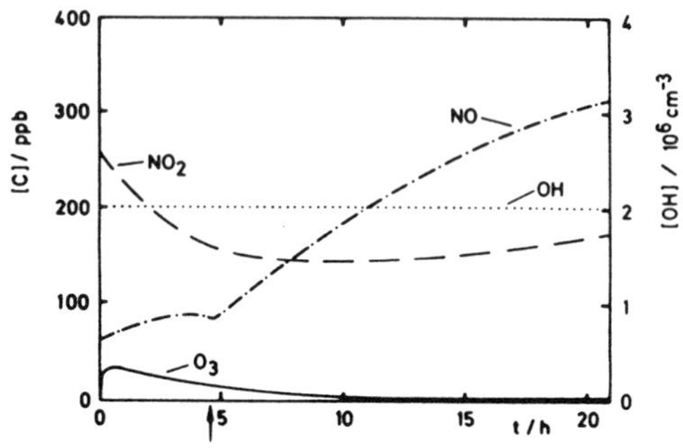

Fig. 2a. Behaviour of NO, NO_2 and O_3 during a smog chamber run in the absence of aerosol. The arrow marks the beginning of a continuous addition of NO at 1 ppb min^{-1}. The concentration of OH radicals is constant (to within ± 20%) at the indicated level of 2×10^6 cm^{-3}.

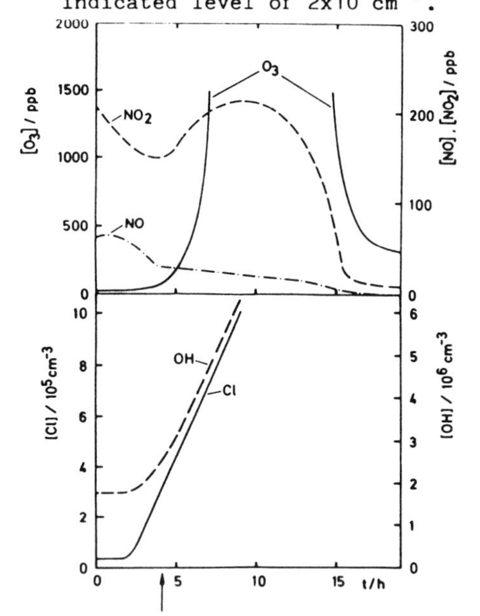

Fig. 2b. Behaviour of NO, NO_2 and O_3 during a smog chamber run in the presence of NaCl aerosol at 7 mg m^{-3}. Although NO is added at the same rate as in fig. 2a, O_3 increases rapidly until the NO_x is mostly consumed (top). The concentration profiles of Cl and OH are also shown (bottom).

Further experiments were performed to understand if the O_3 production is only enhanced by the presence of Cl (initiating O_3 formation by attack to the hydrocarbons, similar to the attack of OH) or if the production of atomic Cl is caused by the presence of O_3 or of OH. Ozone was injected into the chamber at the ppm level in the presence of NaCl aerosol, and in the absence of NO_x (at levels below 1 ppb) no formation of Cl was observed under irradiation. But in the presence of NO_x, Cl is observed at a level of $6x10^5 cm^{-3}$ very soon and even increasing linearly, as shown in figure 3 (top). After 4h, the hydrocarbons are consumed to a level below 1ppb, and Cl levels increase to above $2x10^6 cm^{-3}$. Under these extreme conditions the OH attack to the hyrocarbons can even be neglected, compared to the attack of Cl. On the other hand the Cl-consumption, shown in figure 3 (bottom) demonstrates that the increased concentration of Cl is not caused by an increasing source strength but by a decreasing sink (the hydrocarbons, the main sink, are mostly consumed).

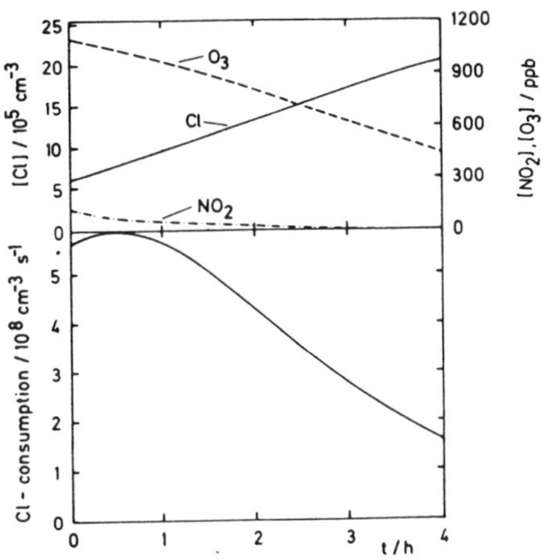

Fig. 3. Concentration profiles of O_3, NO_2 and Cl atoms (top) and consumption rate of Cl according to eq. I (bottom) observed in the photochemical smog in the presence of NaCl aerosol.

In corresponding dark experiments, molecular Cl_2 was identified to be formed from NaCl aerosol as well, by applying a chemical method (the bleaching of methyl orange solutions in impingers, a technique capable of detecting Cl_2 at levels above 1ppb according to Kettner and Forwerg, 1969).

Analysis of the aerosol by ion chromatography revealed that a portion of the Cl^--contents of the aerosol has been converted to NO_3^-. This intrusion of NO_3^- is expected to lead to a corresponding release of HCl.

The loss of 5-10% of the total Cl^-- contents from 7mg m^{-3} aerosol corresponds to a release of 150-300ppb HCl. This HCl release corresponds roughly to the additional loss of NO_x observed in the presence of aerosol. On the other hand, only a small portion (less than 10ppb) could be detected in the gas phase (by using a denuder technique and ion chromatography). Hence, it is assumed that HCl is adsorbed immediately in the (inherent) water layer on the glass walls of the smog chamber. Oxidation reactions of Cl^- by O_3 in the water layer (Hoigné et al., 1985) might be a reason for the formation of atomic Cl via Cl_2 and subsequent photolysis. According to a reaction mechanism proposed by Yeatts and Taube (1949), Cl_2 may be formed in the presence of O_3 in aqueous solutions of HCl, according to the rate law

$$d[Cl_2]/dt = k_a[O_3][Cl^-] + k_b[O_3][Cl^-][H^+]$$

with the rate constants:

$k_a = 2x10^{-3}M^{-1}s^{-1}$ and $k_b = 2x10^{-2}M^{-2}s^{-1}$ at room temperature.

Further experiments were performed to check this mechanism by replacing the NaCl aerosol with the molar equivalent of HCl gas (300ppb) directly. The solar simulator was modified by a UV cut-off filter at 360nm (in order to lower the direct photolysis of Cl_2), and O_3 was injected for the first 20 min. Figure 4 shows observed profiles of O_3 and Cl_2 and of the consumption rate of Cl atoms according eq. I. Although Cl_2 is still photolyzed under these conditions at a rate of $k_{phot} = (3 \pm 0.5)x10^{-4} s^{-1}$, Cl_2 increases rapidly to 17ppb with a maximum at the same time with O_3. Later on both components show a slower, exponential decrease.

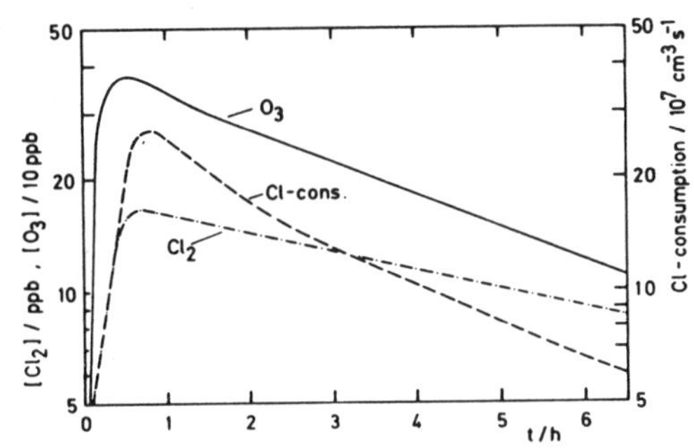

Fig. 4. Time profiles of O_3 and Cl_2, observed in a smog chamber run in the presence of HCl, using a cut-off filter at 360 nm. Also shown is the consumption rate of atomic Cl according to eq.I

A crucial question for the quantitative determination of Cl_2 by the method of Kettner and Forwerg (1969) is the possibility of interferences (the interference by O_3 was corrected for, it limits the detection of Cl_2 in the presence of a large surplus of O_3, but other unknown components like HOCl or its precursor, Cl_2O, might bleach methyl orange as well). Hence, the consumption rate according to eq. I was compared to the photolytic production rate of Cl from the Cl_2 levels observed in the experiment, according to the simple equation

$$d[Cl]/dt = 2 k_{phot} [Cl_2] \quad (II)$$

The results of this comparison are listed in table 1. The agreement of the values in the initial phase is striking; later on the estimated consumption becomes slower than the calculated production by up to 50%. It should be kept in mind that the reactions of the products with Cl have been neglected in this simple estimate and that these species and their reactivities are mostly unknown. Hence, the agreement must be considered satisfactory.

TABLE 1. COMPARISON OF THE CONSUMPTION OF Cl, ESTIMATED FROM THE CONSUMPTION OF THE HYDROCARBONS BY Cl USING KNOWN RATE CONSTANTS (EQ. I) AND THE PRODUCTION OF Cl ATOMS FROM THE PHOTOLYSIS OF Cl_2 (EQ. II), IN UNITS OF $10^8 cm^{-3}s^{-1}$

Irradiation time (min)	Consumption of Cl	Production of Cl
40	2.5	2.55 ± 0.43
70	2.4	2.03 ± 0.34
120	1.6	1.95 ± 0.33
180	1.25	1.88 ± 0.32
240	1.00	1.80 ± 0.30
360	0.65	1.35 ± 0.23

The role of the glass walls was further investigated by some preliminary experiments in teflon bags. Briefly, these experiments revealed that the recovery of HCl is much higher in teflon bags (about 50% at the 100 ppb level). No additional source of atomic Cl was detected, besides the simple, well-known gas phase reaction of HCl with OH (DeMore et al., 1985), and no correlation of the rate of consumption of Cl with the concentration of OH was observed in teflon bags. If quartz aerosol (AEROSIL 200, DEGUSSA) is introduced into the teflon bags at 50% relative humidity, the additional (obviously heterogeneous) production of Cl is observed, it is negligibly small at low relative humidities below 5%.

4. POSSIBLE SOURCES OF Cl ATOMS

Let us first consider the mechanism of Yeatts and Taube (1949) for an explanation of the formation of Cl_2. At a mixing ratio of O_3 in the gas phase of 400ppb, the equilibrium concentration of O_3 in the water layer on the smog chamber walls can be estimated from Henry's law to be $5x10^{-9}$ M (adopting a Henry constant of $1.3x10^{-2}$ M atm^{-1} at 298 K, according to Briner and Perrottet (1939). The concentration of HCl in the water layer on the walls of the chamber can be estimated from experiments, where the amount of water necessary to achieve 50% relative humidity was determined. A surplus of 10ml was needed over the amount calculated from the volume of the chamber. If 300 ppb HCl are absorbed completely from the gas phase in this amount of water, the Cl^-- concentration will be $3x10^{-3}$ M, yielding a Cl_2- production according to the mechanism of Yeatts and Taube (and in agreement with the rate constant for the elementary reaction of O_3 with Cl^-, determined by Hoigné et al., the acid catalyzed term in the expression of Yeatts and Taube contributes less than 10%) of $3x10^{-14}$ M s^{-1} from the 10 ml of solution. If this Cl_2 evaporates from 10 ml of solution into the gas phase, the corresponding source strength of Cl_2 molecules in the 2.3 m^3 smog chamber will be as low as 90 $cm^{-3}s^{-1}$. The experimentally observed mixing ratio of 17ppb Cl_2 after 40 min can be converted to a source strength of $1.8x10^8$ $cm^{-3}s^{-1}$, more than 6 orders of magnitude higher than estimated from the mechanism under consideration.

Another possible mechanism is the oxidation of Cl^- by OH in the liquid phase (Jayson et al., 1973) according to the reactions:

$$OH + Cl^- \rightarrow ClOH^- \qquad k = 4.3x10^9 M^{-1}s^{-1}$$
$$ClOH^- + H^+ \rightarrow Cl + H_2O \qquad k = 2.1x10^{10} M^{-1}s^{-1}$$
$$Cl + Cl^- \rightarrow Cl_2^- \qquad k = 2.2x10^{10} M^{-1}s^{-1},$$

and according to Wagner et al., 1986, Cl_2^- disproportionates rather rapidly:

$$Cl_2^- + Cl_2^- \rightarrow Cl_2 + 2\,Cl^- \qquad k = 7x10^7 \; M^{-1}s^{-1}.$$

At the low pH of our experiments Cl_2 (which is in equilibrium with HOCl) would volatilize. On the other hand, the present experiments give no support to the hypothesis that OH radicals from the gas phase are a precursor of Cl, since the Cl production was not observed to correlate with the OH concentration.

A further, heterogeneous reaction mechanism may be considered, similar to the H_2O_2 production from O_3 in impingers, observed by Heikes (1984). Transition metal cations in the glass surface are postulated form various active species: O^-, O^{2-}, O_2^-, O_2^{2-}, O_3^- and others. A formation of OH might then proceed in the liquid phase according to the reactions:

$$O_3 + O_2^- \rightarrow O_3^- + O_2$$
$$O_3^- \rightleftharpoons O^- + O_2$$
$$(k_f = 3.3x10^3 s^{-1}, \qquad k_r = (2.5\text{-}4)x10^9 M^{-1}s^{-1})$$
$$OH + OH^- \rightleftharpoons O^- + H_2O$$
$$(k_f = (3\text{-}120)x10^8 M^{-1}s^{-1}, \qquad k_r = (1.5\text{-}2)x10^6 s^{-1})$$

5. CONCLUSIONS

Cl atoms are formed at high production rate in the presence of Cl^- (in the form of NaCl aerosol or HCl), O_3 and simulated sunlight. Only a small portion of this production rate of atomic Cl can be explained from the gas-phase reaction of OH radicals with HCl. It can be shown that a heterogeneous reaction is responsible for the major portion of the production of atomic Cl, where Cl_2 is the main precursor. Although in the stratosphere, ozone is consumed by Cl, in the troposphere Cl produces ozone by enhancing the photosmog.

ACKNOWLEDGEMENTS

This work was supported by the Bundesminister für Forschung und Technologie, helpful discussions with Prof. W. Stöber and the aid of G. Pfahler with the ion chromatography are gratefully acknowledged.

REFERENCES

Behnke, W., F. Nolting and C. Zetzsch, 1984: Rate constants of chlorine atoms with hydrocarbons determined by an aerosol smog chamber technique. Abstr. 16th Informal Conference on Photochemistry, Harvard.

Behnke, W., and C. Zetzsch, 1987: The simultaneous determination of two active species in smog chambers: Evaluation of rate constants for Cl atoms in the presence of OH, Proceedings of the 3rd French-German Workshop on Tropospheric Chemistry by Laboratory Studies, 6. -8. October, Leinsweiler Hof, Report No 15 of Fachbereich 8, Physikalische Chemie, pp. 59-60, Universität Gesamthochschule, Wuppertal.

Behnke, W., F. Nolting and C. Zetzsch, 1987: A smog chamber study on the impact of aserosols on the photodegradation of chemicals in the troposphere. J. Aerosol Sci., 18, 65-71.

Behnke, W., W. Holländer, W. Koch, F. Nolting and C. Zetzsch, 1988: A smog chamber for studies of the photochemical degradation of chemicals in the presence of aerosols. Atmos. Environ., 22 (in press).

Briner, E., and E. Perrottet, 1939: Détermination des solubilités de l'ozone dans l'eau et dans une solution aqueuse de chlorure de sodium: calcul des solubilités de l'ozone atmosphérique dans les eaux. Helv. Chim. Acta, 22, 397-405.

Cadle, R.D., and R.C. Robbins, 1960: III. Physical and chemical properties: Kinetics of atmospheric reactions involving aerosols. Discuss. Faraday Soc., 30, 155-161.

Clegg, S.L., and P. Brimlecombe, 1985: Potential degassing of hydrogen chloride from acidified sodium chloride droplets. Atmos. Environment, 19, 465-470.

Cocks, A.T., and W.J. McElroy, 1984: The absorption of hydrogen chloride by aqueous aerosols. Atmos. Environ., 18, 1471-1483.

DeMore, W.B., J.J. Margitan, M.J. Molina, R.T. Watson, D.M. Golden, R.F. Hampson, M.J. Kurylo, C.J. Howard and A.R. Ravishankara, 1985: Chemical kinetics and photochemical data for use in stratospheric modeling. Evaluation Number 7, JPL Publication 85-37, Jet Propulsion Laboratory, Pasadena, CA.

Heikes, B.G., 1984: Aqueous H_2O_2 production from O_3 in glass impingers. Atmos. Environ., 18, 1433-1445.

Hitchcock, D.R., L.L. Spiller and W.E. Wilson, 1980: Sulfuric acid aerosol and HCl release in coastal atmospheres: Evidence of rapid formation of sulfuric acid particulates. Atmos. Environ., 14, 165-182.

Hoigné, J., H. Bader, W.R. Haag and J. Staehelin, 1985: Rate constants of reactions of ozone with organic and inorganic compounds in water-III. Water Res., 19, 983-1004.

Holländer, W., W. Behnke, W. Koch, G. Pohlmann, 1984: Design and performance of an aerosol reactor for photochemical studies. Proceedings of the 3rd European Symposium on Physico-Chemical Behaviour of Atmospheric Pollutants (G. Versino and B. Angeletti eds.), pp. 309-311, Reidel, Dordrecht.

Hov, Ø., 1985: The effect of chlorine on the formation of photochemical oxidants in southern Telemark, Norway. Atmos. Environ., 19, 471-485.

Jayson, C.G., B.J. Parsons and A.J. Swallow, 1973: Some simple, highly reactive chloride derivates in aqueous solutions. J. Chem. Soc. Faraday Trans. I, 69, 1597-1607.

Kettner, K., and W. Forwerg, 1969: Bestimmung geringer Chlorkonzentrationen nach der Methylorange-Methode. Atmos. Environ., 3, 215-220.

Lightowlers, P.J., and J.N. Cape, 1988: Sources and fate of atmospheric HCl in the U.K. and Western Europe. Atmos. Environ., 22, 7-15.

Martens, C.S., J.J. Wesolowski, R.C. Harris and R. Kaifer, 1973: Chlorine loss from Puerto Rican and San Francisco bay area marine aerosol. J. Geophys. Res., 78, 8788-8792.

Meinert, D.L., and J.W. Winchester, 1977: Chemical relationship in the north atlantic marine aerosol. J. Geophys. Res., 82, 1778-1782.

Nolting, F., W. Behnke and C. Zetzsch, 1988: A smog chamber for studies of the reactions of terpenes and alkanes with ozone and OH. J. Atmos. Chem., 6, 47-59.

Wagner, I., J. Karthäuser and H. Strehlow, 1986: On the decay of the dichloride anion, Cl_2^-, in aqueous solution. Ber. Bunsenges. Phys. Chem., 90, 861-867.

Yeatts, R.B., and H. Taube, 1949: The kinetics of the reactions of ozone and chloride ion in aqueous solution. J. Am. Chem. Soc., 71, 4100-4105.

Zetzsch, C., 1987: Simulation of atmospheric photochemistry in the presence of solid airborne aerosols. Formation, Distribution and Chemical Transformation of Air Pollutants (ed. R. Zellner), Dechema monograph 104, 187-212.

MODEL CALCULATIONS OF THE TROPOSPHERIC OH MIXING RATIO AND COMPARISON WITH FIELD DATA

Dirk Poppe, Ulrich Platt, Hans-Peter Dorn, and Dieter H. Ehhalt
Institut für Atmosphärische Chemie
Kernforschungsanlage Jülich
D-5170 Jülich, FRG

ABSTRACT

Tropospheric OH concentrations were calculated for chemically well characterized airmasses. The required data were taken from a series of campaigns performed in W. Germany from 1980 to 1987 where photolysis rates J_{O3}, J_{NO2} and trace gas concentrations (NO_x, O_3, H_2O, CO, CH_4, NMHC, SO_2, CH_2O) were obtained. This approach allows the use of a zero dimensional (only time dependent) model modified to incorporate a variable height of the planetary boundary layer. The chemical mechanism from RADM (regional acid deposition model) is used. The influence of several longer lived precursors on the OH concentration is investigated. It is also demonstrated that the diurnal variation of the mixed layer is of large influence on many trace gas concentrations. The model results are compared with the actual OH concentrations measured by long path absorption spectroscopy.

1. INTRODUCTION

The OH radical plays a very important role in the chemistry of the troposphere. For many atmospheric trace gases of natural or anthropogenic origin the radical initiates chemical reaction chains, which oxidize the species and finally lead to their elimination from the atmosphere. The chemical lifetime of OH and the related peroxiradicals are short (< 1 min) so that the concentrations are approximately determined by the steady state values which depend on the mixing ratios of longer lived chemical precursors.

Here we report on model calculations which investigate

(a) the influence of several atmospheric constituents on the OH concentrations,
(b) the role of peroxiacetyl nitrate (PAN) as a temporary chemical source or sink for radicals,
(c) the impact of the diurnal variation of the depth of the planetary boundary layer.

A zerodimensional (time dependent) model is used which incorporates source and sink terms modified to include the variation of the height of the mixed layer. The chemical reaction mechanism includes only gas phase reactions and is based on the chemical mechanism of RADM as published by Stockwell (1987).

The calculations are compared to experimental OH data which were obtained in field studies in West Germany together with the simultaneous measurements of the photolysis rates J_{O3}, J_{NO2}, and trace gas concentrations (NO_x, O_3, H_2O, CO, CH_4, NMHC, SO_2, and CH_2O).

2. RESULTS

(1) In Fig. 1 a schematic overview of the chemical mechanism for the formation and destruction of the OH radical and related species is shown. The OH radical is primarily formed in the reaction of H_2O with $O(^1D)$ which is produced in the photolysis of O_3. The chemical dynamics is largely governed by a very efficient recycling of radicals between OH, HO_2, MO_2 (methyl peroxy radical), ACO_3 (acetyl peroxy radical), and HCP (other organic peroxy radicals). Destruction of radicals is provided by the reactions

$$OH + HO_2 \rightarrow H_2O + O_2$$

$$HO_2 + HO_2 \rightarrow H_2O_2 + O_2$$

$$OH + NO_2 \xrightarrow{M} HNO_3$$

the latter being the dominant loss term in case of high NO_2 (≥ 0.2 ppb) concentrations.

(2) Various production and destruction processes that determine the steady state concentrations of the radicals. The resulting dependence of the OH abundance is shown in Fig. 2 for several trace gases.

(3) The chemical life time of PAN is much larger (≈ 30 min at T = 298 K) than the life time of the radicals and depends strongly on the temperature and the NO_x concentration. Therefore the PAN reactions

$$PAN \rightarrow ACO_3 + NO_2$$

$$ACO_3 + NO_2 \xrightarrow{M} PAN$$

provide temporary sources or sinks for radicals depending on whether the momentary PAN concentration is above or below its steady state limit (Fig. 3). Nitrous acid is coupled in a similar way to the radical chemistry by the reactions

OZONE IN THE ATMOSPHERE
Rumen D. Bojkov and Peter Fabian (Eds.)

ISBN 0-937194-15-8

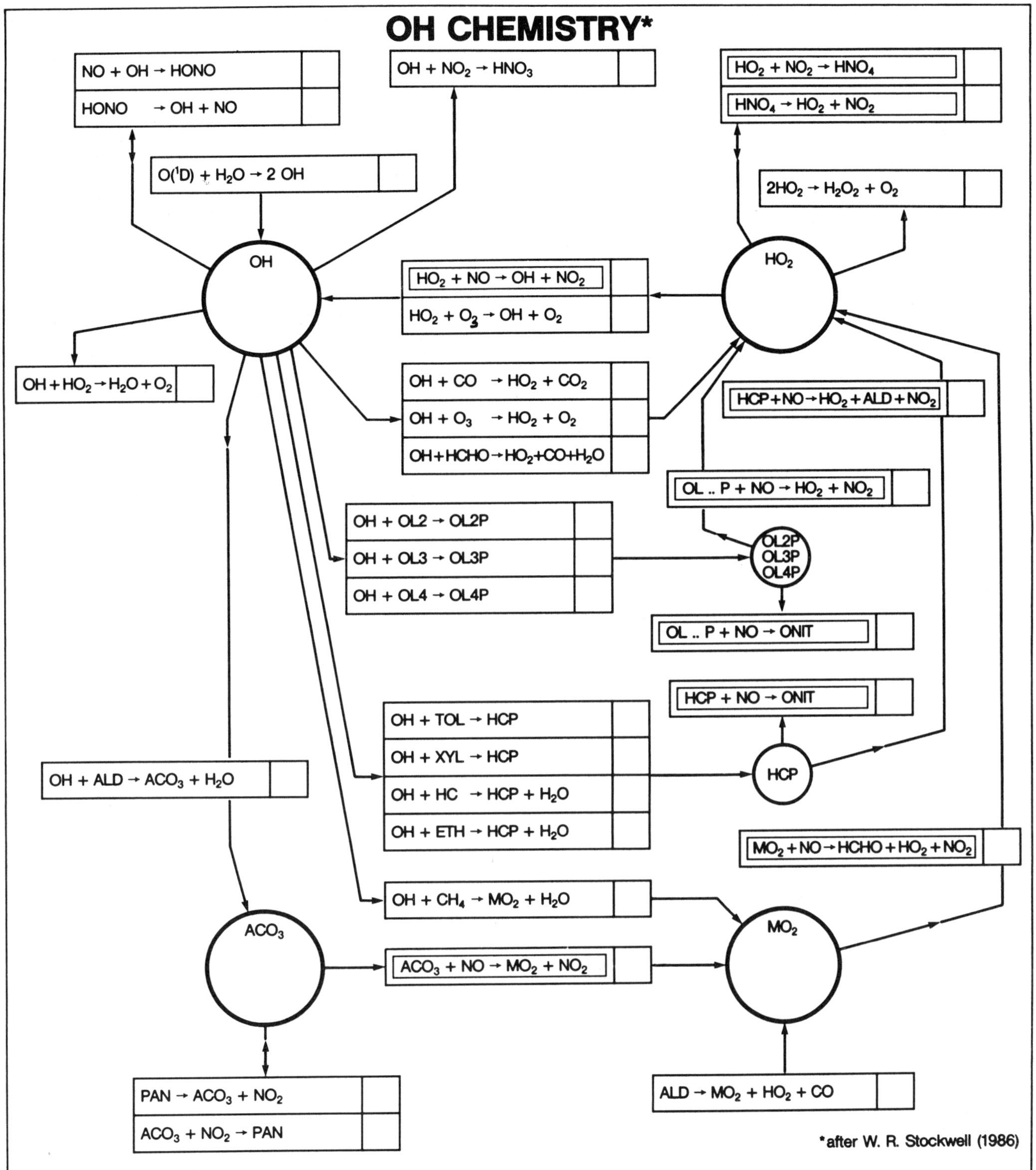

Methane	CH_4	Slow aromatics	TOL	HO-ethylene peroxy radical	OL2P
Ethane	ETH	Fast aromatics	XYL	HO-propylene peroxy radical	OL3P
Other alkanes	HC	Formaldehyde	HCHO	HO-higher olefin peroxy radical	OL4P
Ethylene	OL2	Other aldehydes	ALD	Acetyl peroxy radical	ACO3
Propylene	OL3	Methyl peroxy radical	MO2	Organic nitrates	ONIT
Other olefins	OL4	Organic peroxy radical	HCP	Peroxyacetyl nitrates	PAN

Fig. 1. Chemistry of the OH radical after Stockwell (1986).

$$HONO \rightarrow OH + NO$$

$$OH + NO \xrightarrow{M} HONO$$

However, due to the short life time of HONO both processes are balanced within minutes so that HONO has no significant influence on OH, as long as heterogeneous production of HONO is excluded.

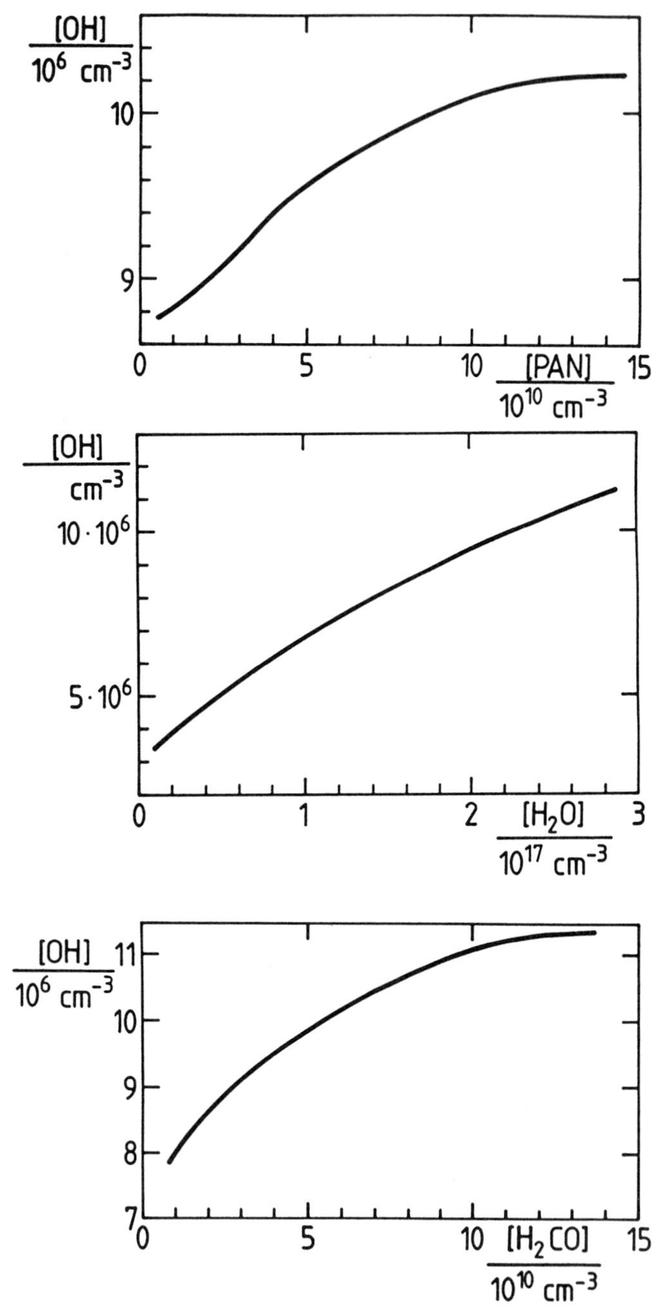

Fig. 2. Theoretical results for the OH abundance as function of several trace gases. T = 283 K, P = 850 mb, $[O_3]$ = 49, $[NO_2]$ = 1, $[CH_4]$ = 1820, [ETH] = 2.05, [OL2] = 0.76, [OL3] = 0.2, [OL4] = 0.04, [TOL] = 0.28, [XYL] = 0.23 (in ppb).

(4) The increase of the height of the mixed layer during the day constitutes a virtual sink for many of the trace gases due to dilution. For comparison with the field data taken in Jülich (FRG) at July 14, 1987 a height increase from 100 m in the morning to 1100 m in the afternoon was assumed in the model in agreement with the meteorological observations. Fig. 4a demonstrates that particularly the diurnal variation of the precursor gases NO_2 and O_3 is influenced. Comparison with field measurements (Fig. 4b) for both species shows qualitative agreement. Equally important for the OH radical is the effect of the dilution on the temporary radical reservoirs, PAN, HCHO, H_2O_2. Since not all of these have been measured in the earlier experiments, the precise prediction of these concentrations by the model is important for the correct interpretation of the measured OH concentration. This result is reflected in the comparison of measured and calculated OH concentrations. While the asymmetry of the diurnal cycle with respect to local noon is well reproduced, there are still differences in the concentrations pointing to unsufficiently accurate OH precursor abundances.

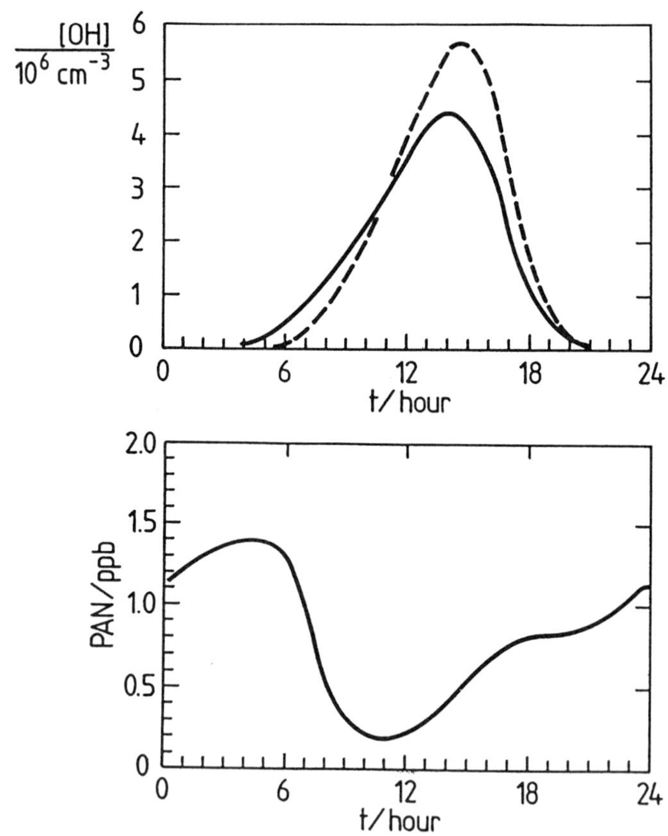

Fig. 3. Diurnal variation of OH and PAN including PAN reactions (solid line) and without PAN reactions (dashed line).

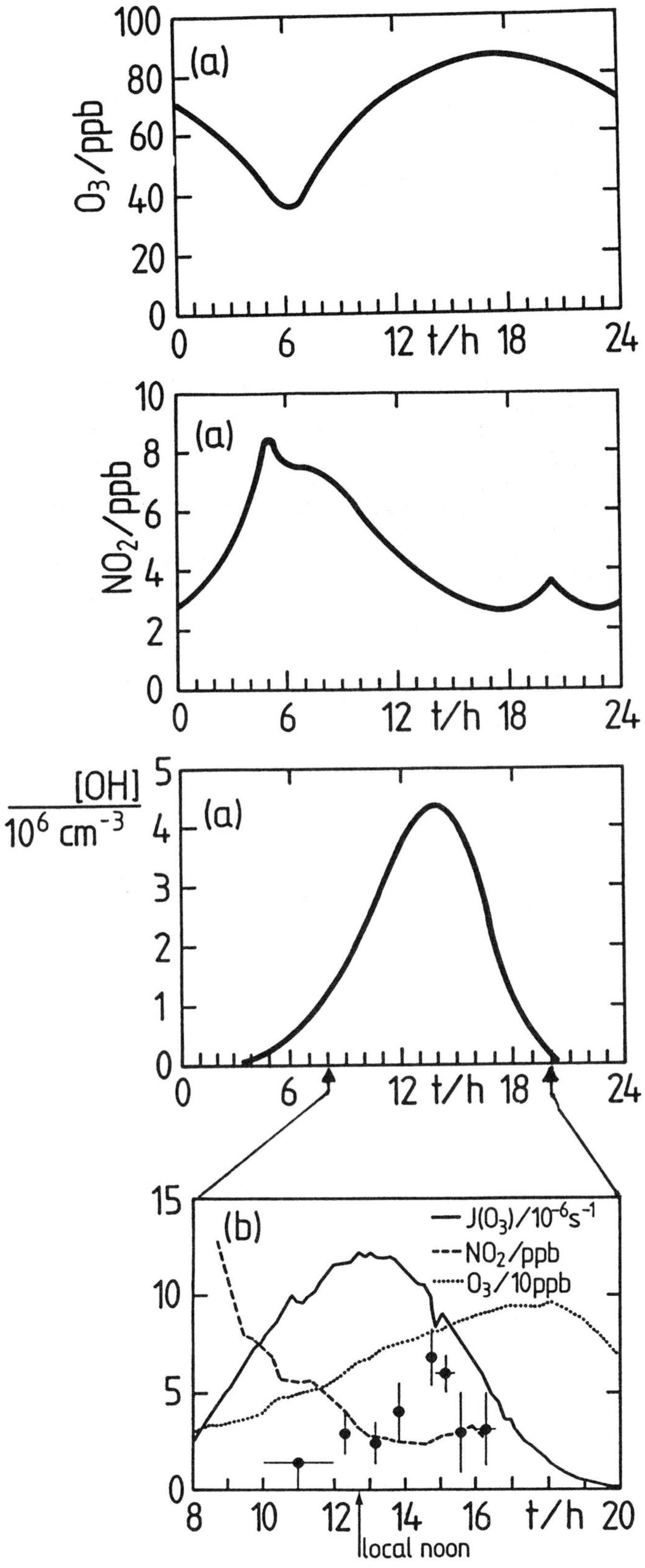

Fig. 4: Diurnal variation of NO_2, O_3 and OH. The latter are given by the • in unit of 10^6 cm^{-3}.
(a) calculations with an increase of the height of the mixed layer from 100 m in the morning to 1100 m in the late afternoon.
(b) experimental values by H.-P. Dorn et al. (1988) taken in Jülich at July 14, 1987.

REFERENCES

Callies, J., H.-P. Dorn, U. Platt, and D.H. Ehhalt, 1988: Tropospheric OH concentration measurements by laser long-path absorption spectroscopy. Proceedings of the International Ozone Symposium, Göttingen, 1988 (this volume).

Dorn, H.-P., J. Callies, U. Platt, and D.H. Ehhalt, 1988: Measurement of tropospheric OH concentrations by laser long-path absorption spectroscopy. Accepted for publication in Tellus.

Stockwell, W.R., 1986: A homogeneous gas phase mechanism for use in a regional and deposition model. Atmos. Envir. 20, 1615.

MEASUREMENTS OF OZONE CONCENTRATION IN SELECTED POINTS OF KATOWICE PROVINCE

Weronika Markiewicz and Apoloniusz Sztyler
Institute of Environmental Protection
Katowice, Poland

ABSTRACT

The paper presents results of 30 min. ozone concentration measurements in near-ground layer of the air. Measurements were carried out with chemiluminescence ozone analyzers /model 8002 Bendix/in five towns of the Katowice Province. Simultaneously with ozone concentration comparison between average ozone values and meteorological factors will be presented. Also will be presented the dependency between standardized daily ozone concentrations, in ground layer of the air, and the total amount of atmospheric ozone.

MEASUREMENTS OF OZONE CONCENTRATION IN SELECTED POINTS OF KATOWICE PROVINCE

The Katowice Province is a highly industrialized and urbanized region. It is observed that the natural environmental equillibrium is disturbed. In the view of the number and variety of existing emission sources also the quantity of air pollutants is very big. Furthermore the pollutants emitted into the atmosphere undergo many physical and chemical conversions. The paper presents ozone concentrations measured in selected points of the Katowice Province. The measurements were carried out with the aid of ozone analyzer produced by Bendix model 8002. The analyzer makes use of fotometric chemiluminescence detection principle caused by non-flame reaction of gas phase ethylene and ozone. Air sample through teflon filters and control valve gets ethylene. Ozone in air sample reacts with ethylene. As a result of non-flame reaction photons are produced, which in photo-duplicator converted into electricity are reinforced. The amount of photons produced and voltage of photo-duplicator are directly proportional to the ozone amount in the air sample. Reported 30-minute ozone concentrations were calculated to obtain average ozone concentrations at the given point. The measurements were carried out in 4 towns located in the south-east part of the Katowice Province and Katowice itself located in centre of the Province. Markiewicz W. et. al., 1981). The measurements were carried out starting from April 1980 till July 1981 so they covered both the summer time and the heating period. The influence of industrial plants burning considerable amounts of coal was taken into the consideration while selecting the measurement points. For this area coal combustion is the largest pollution source. 30-minute average ozone concentrations, refering to individual measurement point, were compared with the Polish standard value amounting $0.07 mg/m^3$. The analysis of measurement data in comparison with the standard value allowed to draw following conclusions:

- measurement point in Jaworzno - 30-minute ozone concentrations were between $5.8 - 78.0 ug/m^3$ in the heating period and $4.0 - 216.0 ug/m^3$ non-heating period /15 April - 15 October. During the whole measurement period 37.3 % of measured average 30 minutes exceeded the standard value for $0.07 mg/m^3$, out of which 16.9 % exceeded the standard level twice and more. In the heating period only 7.1 % of measured average 30-minute ozone concentrations slightly exceeded the level of $0.07 mg/m^3$. In non-heating period 52.3 % of measured average 30-minute ozone concentrations exceeded the standard level. 23.7 % exceeded the standard twice and more.
- measurement point in Myślachowice In the heating period the concentrations were between $16.0 - 62.0 ug/m^3$ and in non-heating period between $28.0 - 128.0 ug/m^3$. During the whole measurement period 37.8 % of measured 30-minute ozone concentrations exceeded the standard, including which exceeded the Polish standard twice or more. In the heating period the standard was not exceeded. In non-heating period 63.6 % of measured average 30-minute ozone concentrations were above the standard including 4.5 % exceeded the standard twice and more.
- measurement point in Luszowice - In the heating period 30-minute ozone

OZONE IN THE ATMOSPHERE
Rumen D. Bojkov and Peter Fabian (Eds.)

ISBN 0-937194-15-8

concentrations were recorded between 6.4. - 40.0 ug/m^3, in non heating period between 40.0 - 160 ug/m^3. During the whole measurement period 25 % measurements exceeded the standard value. In the heating period the concentrations did not exceed the standard value but in non-heating period about 33 % of measurements were higher than 0.07 ug/m^3.

- measurement point in Libiąż - 30-minute concentrations were in the ranges from 7.6 - to 88.0 ug/m^3 in the heating period and 70-200.0 ug/m^3 in non heating period 44.6 % of measurements exceeded the standard value and 16.4 % exceeded the standard value twice and more. In the heating period 18.2 % of measurements were above the value 0.07 mg/m^3, in non-heating period already 60.6 %. 27.3 % exceeded the standard twice and more.
- measurement point in Żurada - 30-minute ozone concentrations were in the ranges from 30.0 - to 300 ug/m^3 in the heating period and 50.0 - 192.0 in non heating period 72.1 % of the measurements were above the standard value and 27.9 exceeded the standard twice and more. In non-heating period 84.6 % of measurements were above the value 0.07 mg/m^3. 31 % of the results exceeded the standard twice and more. Remaining 30-minute results were on the level below 0.07 mg/m^3.
- measurement point in Katowice - In non-heating period 30-minute ozone concentrations were in the ranges 5.6 - 164 ug/m^3 and in the heating period from 4.0 - 36 ug/m^3. 22.1 % of the measurements exceeded the standard value 25 % exceeded the standard twice and more. In the heating period the excess of standard was not observed. In summer time 39.1 % of results exceeded the value 0.07 mg/m^3. 4.3 % were higher twice or more.

From the results in measurements of ozone concentrations one could draw a conclusion that definetely higher ozone concentrations in the air in the Katowice Province are in non-heating period/April - October/. During this period frequent excesses of standard value of permissible concentrations were recorded which for 30-minute measurements were 0.07 mg/m^3. The main objective of the paper was to provide some information on ozone concentration levels in the Katowice Province area.

Information on concentration levels were simulataneously compared with meteorological conditions existing in the measurement area /based on data gathered by the Meteorological Institute/.
This comparison /except Katowice city/ is presented graphically in Figures 1-4. Comparison of average ozone concentrations with meteorological parameters /wind direction, overcast, visibility/ existing on a given day of measurement concentration performance it becomes evident that ozone concentration level is influenced by overcast and wind direction. Generally, higher ozone concentration levels were observed at a small overcast and west winds. The results of ozone concentrations measurements helped to determine the correlation between ozone concentration in air-ground layer with total amount of atmospheric ozone. Earlier, the presented data on ozone concentrations levels in individual measurement points pointed at considerable local differentiation of ozone concentrations at ground-air layer. Determination of ground layer ozone concentration /C_3/ from the total amount of atmospheric ozone /Ω/ in the reference area, required independency from the described differentiation. One should point, however, that data on total amount of atmospheric ozone was obtained from Geophysical Observatory of the Polish Academy of Sciences in Belsk. Publ. Inst. Geoph. /1981/ /1982/. Furthermore, independency from local ozone concentration differentiation in air-ground layer was achieved, relating concentrations from ozone measurements to average values /$\bar{C}_3$/ at given measurement point. This was the way to get "standardized" daily values of ozone concentrations as a relation $C_3/\bar{C}_3$. The dependency between standardized daily ozone concentrations in ground layer of the air /$C_3/\bar{C}_3$/ and the total amount of atmospheric /Ω /$\bar{\Omega}$/ in the south east part of the Katowice Province presents Fig.5.

The magnitude of disperison of points from simple regression line /Fig. 5/ can be interpreted as a result of:

a/ local conditions of ozone generation or ozone transport in air ground layer
b/ intensity of vertical air exchange in the atmosphere

Extreme dispersion magnitude from average conditions determined above represent values marked in the spaces I and II in Fig. 5. One can note that concentration values in space I /deviation in minus/ might be caused by blocking effect of strong higher inversions in the troposphere, making difficult inflow of big ozone portions into lower air layers. It could also be the result of a very quick increase of ozone in the higher atmosphere which beginning fall on the period of ground observations.
After eliminating in calculation data covering space I and II in Fig. 5 correlation coefficient between magnitude $C_3/\bar{C}_3$ and Ω /$\bar{\Omega}$ grew from 0.31 to 0.64. It was assumed that ozone concentration dependence in air-ground layer from total

amount of atmospheric ozone is strong when data dispersion value from simple regression line /Fig. 5/ does not exceed the value of one standard deviation. According to that, however, quite arbitary criterion significant correlation of ozone concentration /C_3/ from its total amount in the atmosphere /Ω/ took place in about 60 % of cases. Presented data in Table 1 is the best confirmation. In spite of this-area differentiation of ozone concentrations in air-ground layer is /as it was pointed earlier/ Quite big. It says of significant impact of different local processes on ozone concentrations in the Katowice Province.
Neglecting the quantitative aspect of ozone concentrations - comparison between measurement points in Katowice /the lowest concentrations/ and Żurada /the highest concentrations/ in the view of relatively small number of measurements and the qualitative picture of area decay concentrations seems to be not incidental. According to actinometric measurements carried out in Katowice. Sztyler A./1988/ and Żurada. Sztyler A./1981/ reduction of direct utraviolet radiation /λ = 0 - 385 nm/, as result of scattering and absorption by aerosols, amounts in one year relatively 75 % and 50 % /measurements performed with the use of Eppley's pyrheliometer/. These values should be understood on one hand as a general comparison of pollution levels in both places - on the other hand these data say of considerable difference of ultraviolet radiation intensity active in the processes of ozone generation. Perhaps it can partly explain some difference of ozone concentrations in Katowice and Żurada. Search for other cases of ozone concentration differences in the area under study requires further measurements and observations.

CONCLUSIONS

- Carried out measurements pointed at area differentation of ozone concentration reflecting an important role of factors and local processes in formation of ozone concentration levels.
- In the area under study standard of ozone concentrations are exceeded particulary in summer time /non heating period/ and during strong impact of sun energy /low overcast/.
- There are reasons to say that in highly polluted regions ozone concentrations can reach lower values.
- In the area under study there is a quite clear correlation of ozone concentrations in the ground layer of the air from total amount of atmospheric ozone.

REFERENCES

Markiewicz W., and Przybyla E., 1981: Measurements of ozone concentration in the ground layer part of the atmosphere.
Report of WHO Programme POL/RCE-003, Katowice,

Publ.Inst.Geophys. Pol.Acad. Sc., 1981, D-13/149/, PWN, Warszawa

Publ. Inst. Geophys.Pol. Acad. Sc., 1982, D-15/157, PWN, Warszawa.

Sztyler A., 1981: Changes of direct solar radiation intensity in spectral bands important for biological and energetical processes in urban conditions;
Report of WHO Programme POL/RCE-003, Katowice

Sztyler A., 1988: Influence of air pollution on solar radiation in choosen spectral wavelength;
Report Inst. of Enwin. Protec. Katowice

TABLE 1. DATA CONCERNING AVERAGE TOTAL AMOUNT OF ATMOSPHERIC OZONE /Ω/ AND AVERAGE OZONE CONCENTRATIONS IN NEAR GROUND LEVEL OF THE AIR /C_3/

Place	Average ozone concentrations /C_3/ /ugm^{-3}/	Average total amount of atmospheric ozone /Ω/
Jaworzno	84.3	362.1
Libiąż	81.7	338.6
Katowice	52.6	357.3
Myślachowice	64.8	354.2
Żurada	115.4	365.3

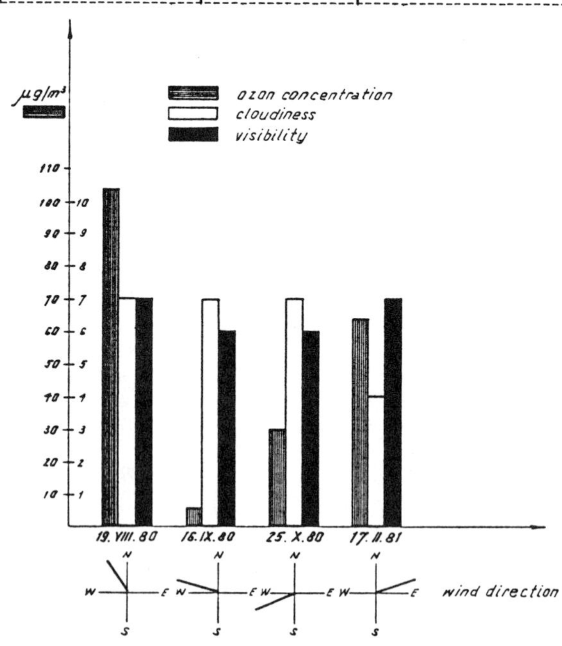

Fig. 1. Comparison between average ozone concentration and meteorological factors - Jaworzno Station.

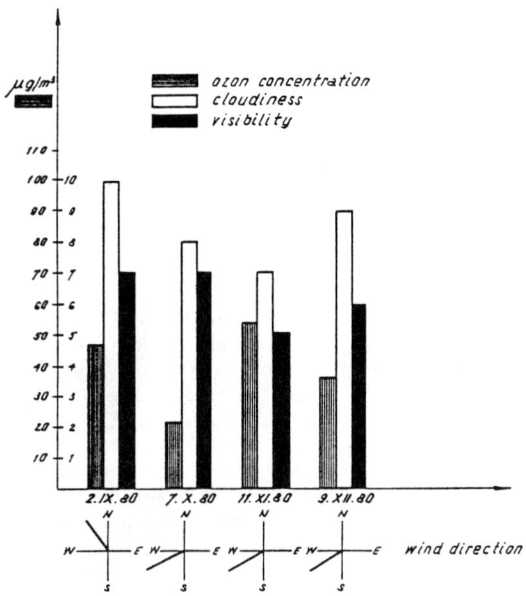

Fig 2. Comparison between average ozone concentration and meteorological factors - Myślachowice Station.

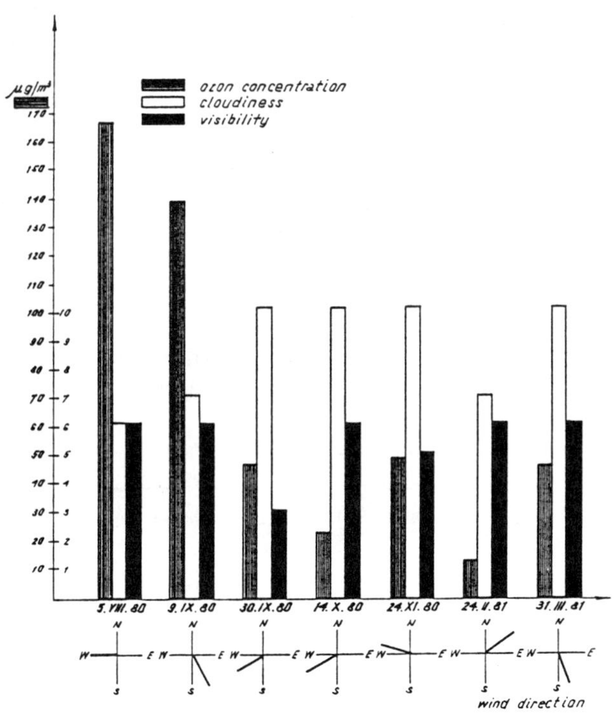

Fig. 3. Comparison between average ozone concentration and meteorological factors - Libiąż Station.

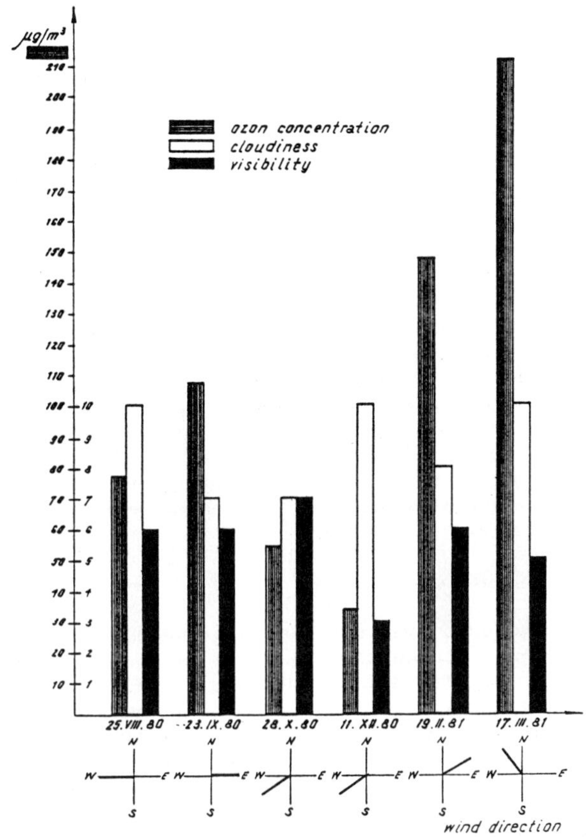

Fig. 4. Comparison between average ozone concentration and meteorological factors - Żurada Station.

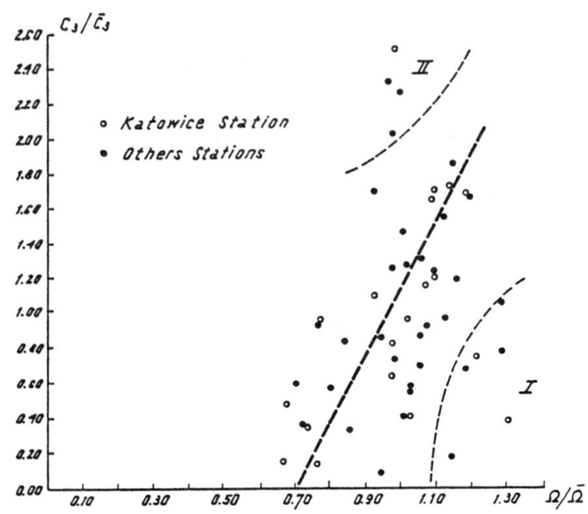

Fig. 5. The dependency between standardized daily ozone concentration in ground layer of the air $/C_3/\bar{C}_3/$ and the total amount of atmospheric ozone / /.

OZONE VARIABILITY AT A SEMIRURAL SITE IN FRANCE

M. Rahbar
Laboratoire de Physique et Chimie de L'Environnement, CNRS
45071 Orleans Cedex 2, France

ABSTRACT

The ground-level concentration of ozone has been monitored at Orléans (France) since 1985 using an electro-chemical sensor.

The characteristics of the instrument will be described. Also simultaneous measurements were made with a DASIBI allowing a direct comparison between both instruments. Results from both instruments are in excellent agreement. The main features observed during the three-year period of measurements are described.

1. INTRODUCTION

Two principal sources of tropospheric ozone have been proposed; the first one is natural ozone transported from the stratosphere and the other is photochemical production during transport of pollutants on regional scales or generated from local anthropogenic. The contribution of the pollutants would be dependent on the meteorological conditions (Junge, 1962; Vassy, 1965; Chameides and Walker, 1973; Fishman, 1984).

For improving understanding of the coupling between both sources, surface ozone concentration (mixing ratio) have been continuously measured between 1985 and 1988 near Orléans (France) using a modified Brewer bubbler ozone sensor (Brewer and Milford, 1960; Rahbar, 1986). The site of observation is semi rural, (47°50N, 1°56E), located at 10 and 120 km south of Orleans and Paris, respectively. Therefore this site may be under the influence of the atmospheric pollution.

2. DESCRIPTION OF SURFACE OZONE SENSOR

Figure 1 shows the main parts of surface ozone sensor. The measuring cell (bubbler) consists of a tube (B) with two electrodes into a buffered potassium iodide solution. The cathode is a platinum gauze (about 10 sq cm in area) and the anode a piece of silver wire. The bubbler is kept inside a reservoir (R) with a communicating 2 mm diameter hole (h) at the bottom of the bubbler.

Bubbler and reservoir are made of pyrex glass having a diameter of about 2 and 11 cm, respectively. Air from the atmosphere is sucked into the bubbler through a short tube of teflon. The output from the bubbler is connected to a recorder. A volume of 150 cm^3 of air is bubbled through the cell in one minute. A 10 mm diameter hole in the wall of bubbler prevents the solution from rising in case of high air flow rate. The sensor is kept inside a weatherhood made of wood which is coverd by an aluminium sheath to protect the sensor from light, heat and electrical perturbations. The reservoir solution keeps nearly constant the concentration of the solution and its level in the bubbler. The volume of solution in the bubbler is small enough to obtain a response of 30 s.

Simultaneous measurements were made with a DASIBI allowing a direct comparison between both instruments. The results from both instruments are in exellent agreement.

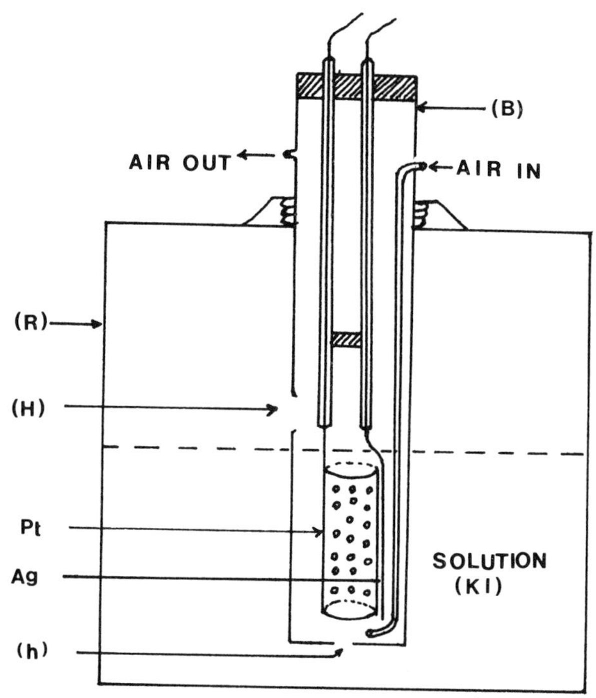

Fig. 1 Main parts of surface ozone sensor.

3. MEASUREMENTS AND RESULTS

Figure 2 illustrates the diurnal variation of ozone concentrations for 10 consecutive days in June 1986.

The first continuous measurements of surface ozone concentration at Orléans indicate large concentration near the ground.

The surface ozone concentration shows a pronounced diurnal variation with a minimum early

ISBN 0-937194-15-8

in the morning and a maximum in the afternoon. The concentration usually varies between 0-30 ppb in winter and 20-50 in summer. Sometimes, a secondary maximum of ozone occurs during the night in winter. In summer, the ozone concentration increases at sunrise and decreases before sunset.

Figure 3 shows the diurnal variation of ozone concentration and temperature during June 1986. Error bars represents the standard deviation from the monthly mean.

The variation of ozone concentration follows the diurnal variation of surface temperature.

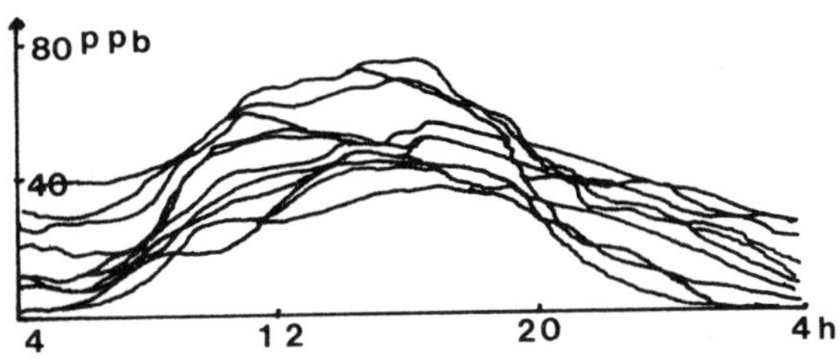

Fig. 2 Diurnal variation of ozone mixing ratio for 10 consecutive days.

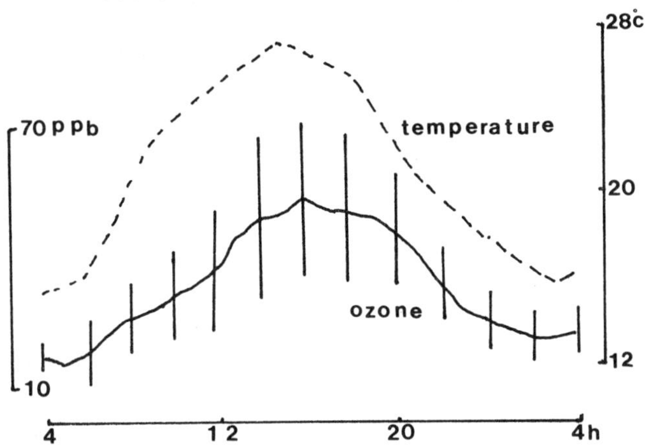

Fig. 3 Diurnal variation of ozone mixing ratio and temperature during June 1986.

Figure 4 presents the maximum values of ozone and temperature observed for each day in May and June 1988.

The concentration of ozone is highly variable from day to day. These variations are usally associated with changes of daily maximum temperature.

Usually northerly winds bring air pollutants from urban and the industrial areas of Orleans and Paris. The very high concentration of ozone (100 ppb) occurred in periods of high temperature, northerly flows and high insolation. Also some of the largest concentrations of ozone were observed during periods of a high pressure system . On the other hand, low pressure and temperature are always associated with low ozone.

Figure 5 shows ozone concentration as a function of month. The concentration is minimum in winter and reaches a maximum during late spring and early summer.

Some transient effects like showers and insolation variation can modify the normal diurnal character of the surface ozone concentration.

For example, local rainshowers lead to a fast decrease of ozone concentration. Sometimes, ozone remains at low level for few hours after the showers end (Figure 6). This does not support earlier studies (Dobson et al, 1956; Vassy, 1969) that indicated that thunderstorms could lead to high surface O_3 concentration.

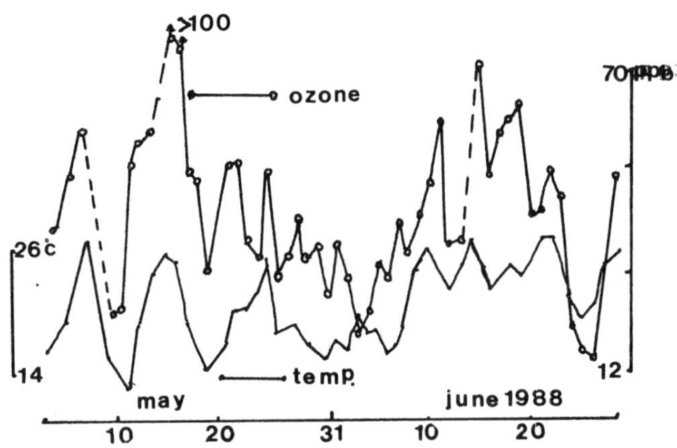

Fig. 4 Maximum values of ozone mixing ratio and temperature observed for each day in May and June 1988.

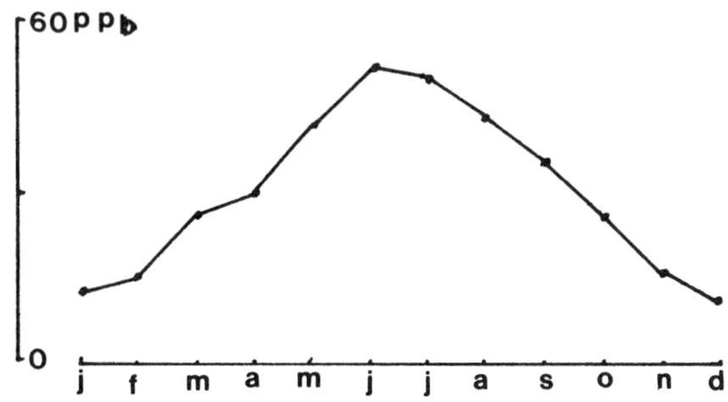

Fig. 5 Monthly mean variation of ozone mixing ratio.

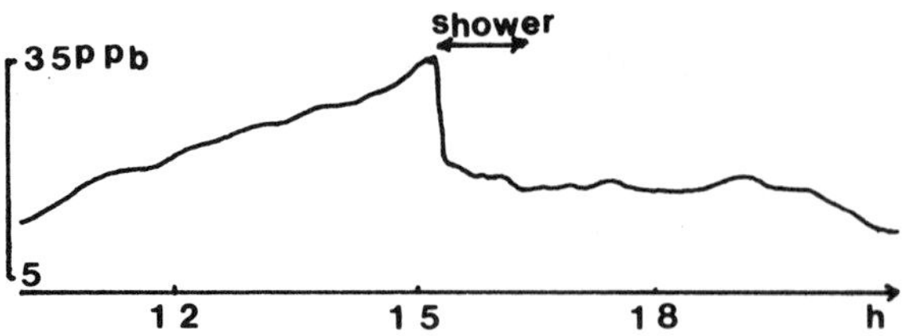

Fig. 6 Fast decrease of ozone mixing ratio caused by rainshower.

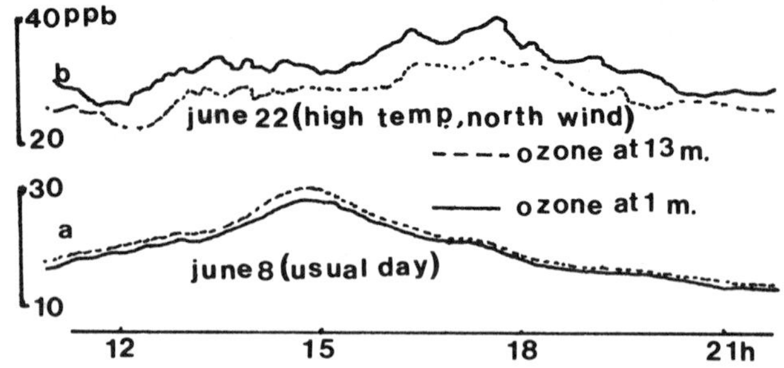

Fig. 8 Ozone mixing ratio observed at 1 and 13 m above the ground during usual days and during northerly wind and high temperature period.

In many cases, the sporadic changes in ozone concentration were also associated with high changes in sunshine. Figure 7 presents variations of ozone concentration and insolation observed on May 26, 1988.

In order to study the variation of surface ozone concentration with height in the boundary layer, two sensors were installed at 1 and 13 m, above the ground.

Usually, there are weak positive gradients of ozone. But on a few occasions, during northerly wind and high temperature, the higher ozonometre shows lower ozone values (Figure 8).

Figure 8a presents the ozone concentration as usally observed by both sensors, Figure 8b gives an example of ozone concentration observed for northerly wind and for temperature higher than usual. The large values and the rapid variations of ozone concentration seen in figure 8b are likely related to pollutants from urban sites.

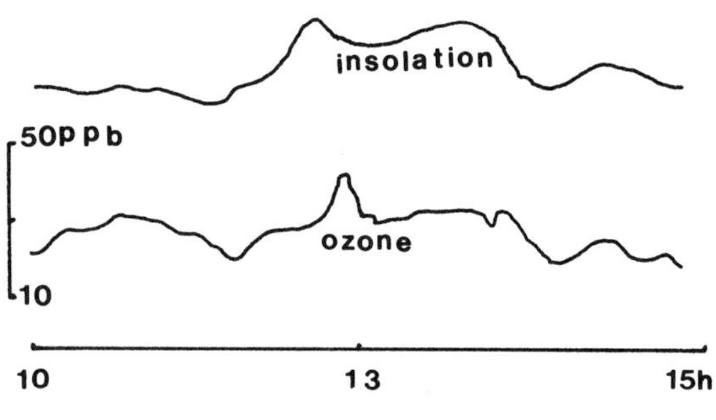

Fig. 7 Variations of ozone mixing ratio and isolation observed on May 26, 1988.

4. CONCLUSION

1. A net diurnal variation of ozone with a minimum early in the morning and a maximum in the afternoon.
2. Seasonal variation of ozone with a minimum in winter and a maximum in early summer.
3. Fast decrease of ozone concentration during showers.
4. Sporadic changes in ozone related to changes in insolation.
5. Peak of ozone when northerly wind brings pollution from Paris and Orléans.

5. REFERENCES

Brewer A.W., and J.R. Milford, 1960: The oxford kew ozone sonde, Proc. Roy. Sco., A 256.

Chameides W.L., and J.C.C. Walker, 1973: A photochemical Theory of tropospheric ozone, J.G.R. 78, 8751-8760.

Dobson G.M.B., et al, 1956: Meteorology of the lower stratosphere, Proc. Roy. Soc., A 185, 144-175.

Fishman J., 1984: O_3 in the free atmosphere. Ed. Whitten and Prasad, Van Norstrand/Rheinhold, NYC.

Junge C.E., 1962: Global ozone budget and exchange between stratosphere and ionosphere. Tellus 14, 363-377.

Rahbar M., 1986: Mesure de l'ozone au niveau du sol par un appareil simple, CNRS Orléans, DT/1034/LPCE.

Vassy A., 1965: Advances in Geophysics, Vol. 11, 115-173.

Vassy A., 1969: Production d'ozone atmosphérique par les décharges silencieuses, Ann. de Geophys. 25, 153-155.

TRENDS IN TROPOSPHERIC OZONE AND OZONE TRANSPORT FROM THE STRATOSPHERE

H. Muramatsu
Disaster Prevention Research Institute
Kyoto University
Kyoto 611, Japan

ABSTRACT

Ozone partial pressure in the upper troposphere (500-250 mb) obtained from the umkehr observations over Japan shows different long term trends at Sapporo(43° N,141° E),Tateno(36° N,140° E) and Kagoshima(31° N,130° E) for the period 1960-1985. At Sapporo and Kagoshima ozone trend (μ mb/year) is positive ,while it is negative at Tateno in winter.The trend of ozone in the upper troposphere has the correlation with the trend of ozone transport from the lower stratosphere over the individual station.

Upward trend at Sapporo in the upper troposphere is closely correlated with the upward trends in both frequency of the intrusion events and the ozone concentration in the lower stratosphere. At Tateno downward trend in ozone is due to the decrease in the intrusion events.At Kagoshima the ozone trend in the troposphere is scarcely related to the intrusion events.

1. INTRODUCTION

Angell and Korshover(1983)have shown that in the tropospheric 2-8km layer of north temperate latitudes the ozonesonde data suggest a 12% increase in ozone between 1970 and 1981,but little ozone change in 8-16km layer. Bojkov and Reinsel (1984) have shown that the weighted average trend estimated for umkehr layer-1 is 0.65± 0.69 % per year during 1970 and 1981. The increase in the free troposphere(850-400mb) detected in the ozone sonde data over the northern hemisphere is 1.12± 0.54% per year during the past 15 years. Logan (1985) has shown good evidence for an increase in ozone in the middle troposphere over Europe during past 15 years and weaker evidence for a similar increase over North America and Japan.

The importance of photochemical ozone production in the troposphere has been recognized. Liu et al.(1980) report 7-15 % ozone increase should have occurred in the northern hemisphere middle and upper troposphere due to subsonic aircraft during the period from 1968 to 1978 and this is not inconsistent with the ozonesonde data which show an increase of about 8%. Crutzen and Gidel (1983) have shown by calculation that only the simultaneous inclusion of NO_x lightning and ground level pollution sources seems to give reasonable agreement between the calculated and observed tropospheric concentrations of O_3 and HNO_3. Isaksen and Hov (1987) showed in their 2-D model that if the emissions of CO, nitrogen oxides and nonmethane hydrocarbons were increased by 3% per year and methane by 0.5% per year,the concentration of ozone increased by about 1% per year over a 30-year period.

Observed trends in tropospheric ozone are not always consistent with the suggested trends by photochemistry in the troposphere. There are no apparent long-term trends in surface ozone at the 4 GMCC stations(Oltmans,1981).The increase of ozone in the 2-8 km in south temperate latitudes (Aspendale) is similar in magnitude to that in north temperate latitudes.According to most theories this increase should be considerably greater in the northern hemisphere where industry and commerce are concentrated (Angell and Korshover ,1983).Payerne(Switzerland) has exhibited little change in tropospheric ozone through 1980 even though nearby Hohenpeissenberg showed increase (Angell and Korshover,1983).There is the evidence for an increase in tropospheric ozone and it is consistent with theory in general, but because of some contradictions stated above firm conclusions regarding trends and their causes are not elucidated.

2. METHOD FOR ESTIMATION OF OZONE TRANSPORT

The intrusion of the stratospheric air and ozone into the troposphere around the polar jet stream is one of the important processes in the stratospheric-tropospheric exchanges (Fabian and Pruchniewicz,1977; Danielsen,1985; Shapiro,1980; Muramatsu et al.,1984a).

We have made several observations from an aircraft and estimated the ozone transported by such events(Muramatsu et al.,1984b). We induced from observational study a model to estimate the long-term variation of ozone transport (Fig.1). When the tropopause (solid curve) descends below 350mb (mean altitude of the polar jet stream) in the cyclonic side of the jet stream,stratospheric ozone is transported through the frontal zone(within the broken curves) into the troposphere. Downward flux of ozone ,$M(O_3)$,over any place with area dS is estimated from

$$M(O_3) = \overline{[O_3]}\overline{V_z}P_r dS \quad \cdots\cdots\cdots\cdots\cdots (1)$$

where $\overline{[O_3]}$ and $\overline{V_z}$ are monthly or seasonal mean values of ozone density and downward vertical velocity,respectively,and $P_r dS$ is the probability of occurrence of the intrusion event over any place. P_r can be obtained from Aerological Data of Japan. Viezee et al.(1983) estimated the down-

ISBN 0-937194-15-8

ward flux of ozone intrusion events by similar method using horizontal area of intrusion instead of PrdS. Our method is applicable to the statistical estimation of the trend.

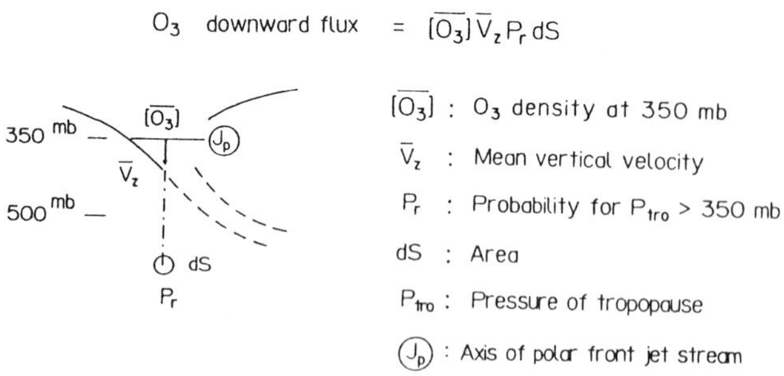

Fig.1 Transport model to estimate the downward flux of ozone.

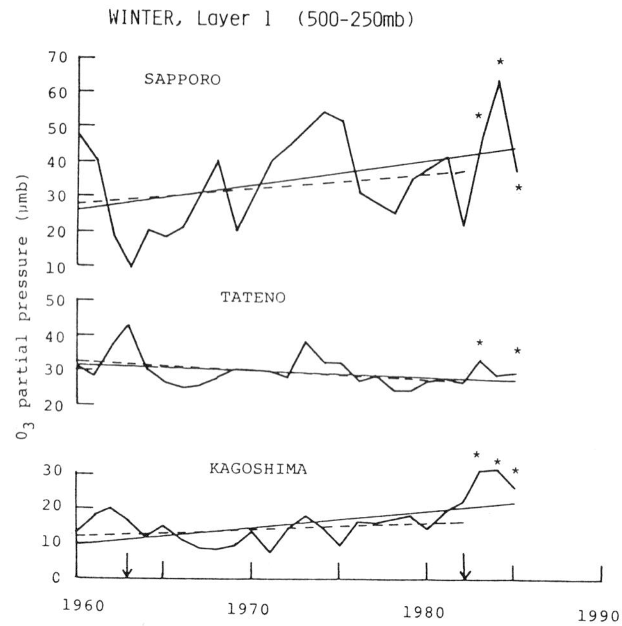

Fig.2 Year-to-year variation of ozone partial pressure in layer-1 in winter.Vertical arrows show the volcanic eruptions of Agung(1963) and El Chichon(1982).

3. TRENDS IN OZONE IN THE UPPER TROPOSPHERE

Year-to-year variations of ozone partial pressure of umkehr layer-1 in winter(Dec., Jan., Feb.) are shown in Fig.2 to see the trends in ozone in the upper troposphere. It should be recognized that at Kagoshima layer-1 is totally in the troposphere, but at Sapporo and Tateno uppermost part of layer-1 is in the stratosphere.

Changes in ozone for 1960-1985 as estimated from linear regression(solid lines in Fig.2) are 0.74, -0.15 and 0.50 μ mb per year at Sapporo, Tateno and Kagoshima, respectively.The confidence coefficients 1-α (α is significance level) are 0.94, 0.78 and 0.99, respectively. So it is said that tropospheric ozone has been increasing at Sapporo and Kagoshima while decreasing at Tateno. It is known that stratospheric aerosols have an effect on the umkehr ozone profile in upper layers and indirect effect in lower layers (DeLuisi, 1979). As the eruption of El Chichon in 1982 may have produced anomalous ozone content in layer-1, changes in ozone are estimated from linear regression (broken lines in Fig.2) excluding the data after 1983 (shown by stars).In this case changes in ozone are 0.45μ mb per year(0.67), -0.24 μ mb per year(0.90) and 0.21μ mb per year(0.89) at Sapporo,Tateno and Kagoshima, respectively.(Figures in parentheses are confidence coefficients.) Upward trends at Sapporo and Kagoshima decreased while greater downward trend is given at Tateno.

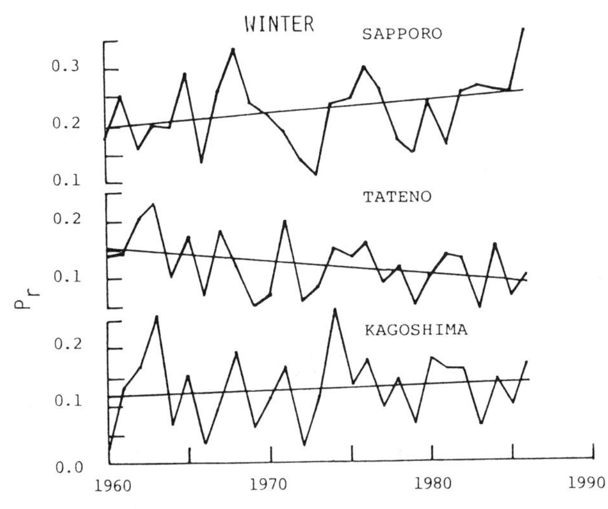

Fig.3 Year-to-year variation of the probability in ozone intrusion events.

4. TREND IN OZONE INTRUSION

Pr is evaluated making use of Aerological Data from the frequency that the altitude of the tropopause lowers below 350 mb. Fig.3 shows the changes in the probability Pr of ozone transport events. Trends in Pr for 1960 -1986 as estimated from linear regression (shown by solid lines) are 1.0, -2.0 and 0.56 % per year at Sapporo, Tateno and Kagoshima, respectively. Corresponding confidence coefficients are 0.85,0.94 and 0.34,respectively. At each station the trend in Pr coincides with that in ozone in layer-1.Trends in ozone and intrusion are positive at Sapporo and Kagoshima, while they are negative at Tateno. The trend in intrusion at Kagoshima is not significant statistically.

5. TREND IN OZONE IN THE LOWER STRATOSPHERE

Changes in ozone in umkehr layer-2 (250-125 mb) are shown in Fig.4. At Sapporo and Tateno they show the trends in ozone in the lower stratosphere,while at Kagoshima layer-2 is in the upper troposphere.

Trends in ozone for 1960-1985 estimated from linear regression(solid lines in Fig.4) are 1.12, 0.055 and 0.041 μ mb per year with corresponding confidence coefficients of 0.86, 0.30 and 0.28 at Sapporo,Tateno and Kagoshima,respectively. Trends in ozone for the period 1960-1982 (data affected by the eruption of El Chichon are excluded) are 0.833, -0.150 and -0.135μ mb per year with confidence coefficients of 0.69, 0.67 and 0.62 at Sapporo,Tateno and Kagoshima,respectively.

From the facts that the confidence coefficients are low and that the sign of trends in ozone changes at Tateno and Kagoshima when the data affected by the eruption of El Chichon are excluded, it is concluded that trends in ozone in layer-2 are not significant at Tateno and Kagoshima. On the other hand the upward trend at Sapporo is significant though the confidence coefficient is lowered by excluding the data affected by El Chichon.

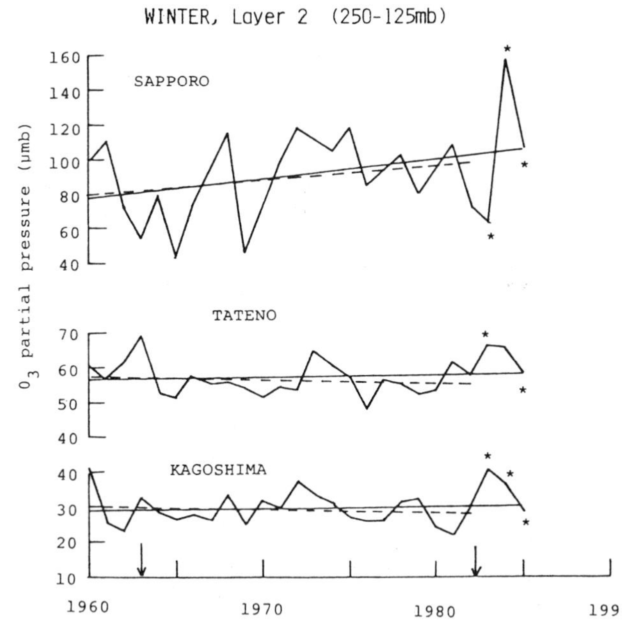

Fig.4 Same as Fig.2,but for ozone in layer-2.

TABLE 1. TRENDS (% PER YEAR) OF OZONE AND INTRUSION FREQUENCY IN WINTER WITH 95 % CONFIDENCE LIMITS

Station	Period	Trend(%/year)	Period*	Trend(%/year)
Ozone Layer 1 (500 -250 mb)				
Sapporo	1960-1985	2.1± 2.2	1960-1982	1.4 ± 3.0
Tateno	1960-1985	-0.51± 0.78	1960-1982	-0.81 ± 0.98
Kagoshima	1960-1985	3.1± 1.7	1960-1982	1.5 ± 1.8
Ozone Layer 2 (250 -125 mb)				
Sapporo	1960-1985	1.2 ± 1.7	1960-1982	0.94 ± 1.98
Tateno	1960-1985	0.096± 0.49	1960-1982	-0.27 ± 0.57
Kagoshima	1960-1985	0.14 ± 0.97	1960-1982	-0.48 ± 1.1
Intrusion frequency				
Sapporo	1960-1986	1.0 ± 1.4		
Tateno	1960-1986	-2.0 ± 2.1		
Kagoshima	1960-1986	0.56 ± 2.5		

* :Data affected by the stratospheric aerosols from the eruption of El Chichon are omitted.

6. DISCUSSION

To estimate downward flux of ozone from the stratosphere to the troposphere from eq.(1), we must evaluate $[O_3]$, $\overline{V_z}$, and P_r. P_r is obtained as shown in section 4. Ozone density in the lower stratosphere is used for $[O_3]$. As for $\overline{V_z}$ we suppose as suggested by Pruchniewicz(1973) $\overline{V_z} \propto (1-cV_h)$,where c is constant and V_h is the mean horizontal wind velocity at 350 mb.

Changes in horizontal wind velocity at 350mb in winter for 1960-1986 are estimated. Trends based on linear regression are -0.20± 0.73,0.02± 0.41 and 0.02± 0.37 %per year with 95% confidence limits at Sapporo, Tateno and Kagoshima, respectively. So we suppose for the first approximation that $\overline{V_z}$ has been unchanged for 1960-1986. Then the trend in downward flux of ozone is determined from $[O_3]\cdot P_r$. Trends in ozone in layer-1,layer-2 and trends in ozone intrusion events are shown in Table 1 by % per year with 95% confidence limits.

As for the relation between the trends in downward flux of ozone and the ozone in the upper troposphere the following results can be deduced from the preceding description.

(1)At Sapporo the ozone density in the upper troposphere has upward trend over last 25 years. The trends in both ozone intrusion events and the ozone density in the lower stratosphere are also positive. So, the increase in ozone in the upper troposphere is intimately connected with the increase in the ozone intrusion ,but being unable to determine of which factor is more closely related, ozone intrusion event or ozone density in the lower stratosphere.

(2) At Tateno the ozone density in the upper troposphere has the downward trend.The ozone density in the lower stratosphere has insignificant trend, while the frequency of the ozone intrusion from the stratosphere has the significant downward trend. So the downward trend in ozone in the troposphere is induced by the decrease in the frequency of ozone intrusion from the stratosphere.

(3)At Kagoshima ozone in the troposphere(500-250 mb) shows the upward trend. The frequency in the ozone intrusion has the upward trend which is insignificant statistically. The change of the ozone density that is transported to the tropospheric layer of 500-250 mb is not recognized if it is from the layer-2 above Kagoshima (in the troposphere) or Tateno(in the stratosphere). So, the positive trend in ozone in the tropospheric layer of 500-250 mb has not obvious relation with the ozone intrusion events.

The spatial distribution of the trend (% per year) in probability P_r of the ozone intrusion events in winter over Japan is shown in Fig.5. It is seen that in the northern and southern part of Japan trend in P_r is positive, while it is negative in the central part. Sapporo (SAP) and Kagoshima(KAG) are in the positive region, while Tateno(TAT) is in the center of the negative region.In spring the region of positive trend in P_r occupies the northern half of Japan and that of negative trend occupies the southern half.Changes in ozone in the troposphere agree with the trends in P_r at Sapporo and Tateno.

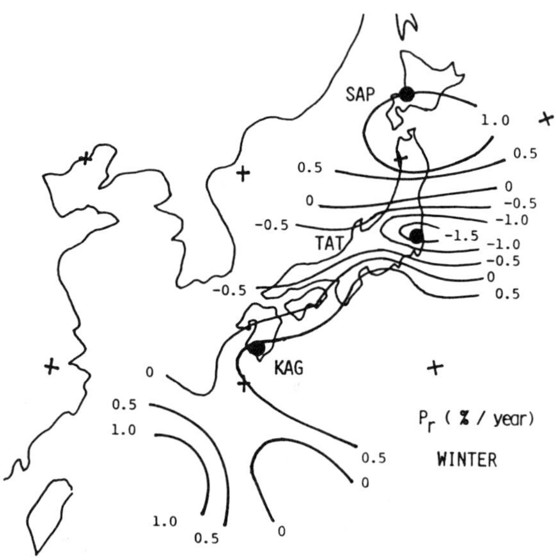

Fig.5 Spatial distribution of the trend in P_r (% per year) in winter.

REFERENCES

Angell,J.K. and J.Korshover,1983: Global variation in total ozone and layer-mean ozone: An update through 1981.J.Climate and Appl.Meteor. 22,1611-1627.

Bojkov,R.D.and G.C.Reinsel,1984: Trends in tropospheric ozone concentration. Pro.Ozone Symp. Halkidiki,Reidel,pp.775-781.

Crutzen,P.J.and L.T.Gidel,1983: A two-dimensional photochemical model of the atmosphere 2 : The tropospheric budgets of the anthropogenic chlorocarbons,CO, CH_4, CH_3Cl and the effect of various NOx sources on tropospheric ozone.J. Geophys.Res.,88,6641-6661.

Danielsen,E.F.,1985: Ozone transport in "Ozone in the Free Atmosphere"R.C.Whitten and S.S.Prasad eds.,Van Nostrand Reinhold Co.,N.Y.,pp.123-159.

DeLuisi,J.J.,1979: Umkehr vertical ozone profile errors caused by the presence of stratospheric aerosols. J.Geophys.Res.,84,1766-1770.

Fabian,P. and P.G.Pruchniewicz,1977: Meridional distribution of ozone in the troposphere and its seasonal variations.J.Geophys.Res.,82, 2063-2073.

Isaksen,I.S.A. and O.Hov,1987: Calculation of trends in the tropospheric concentration of O_3, OH,CO,CH_4 and NO_x.Tellus,39B,271-285.

Liu,S.C., D.Kley, M.McFarland, J.D.Mahlman and H. Levy II,1980: On the origin of tropospheric ozone. J.Geophys.Res.,85,7546-7552.

Logan,J.A.,1985: Tropospheric ozone: Seasonal behavior,trends and anthropogenic influence.J. Geophys.Res.,90,10463-10482.

Muramatsu,H.,T.Sasaki,M.Hirota and Y.Makino,1984a: An aircraft observation of an intrusion process of stratospheric ozone into the troposphere. Papers in Meteor.and Geophys.,35,1-10.

Muramatsu,H.,Y.Makino,M.Hirota and T.Sasaki,1984b: Intrusion process of stratospheric ozone into the troposphere over Japan. Inter.MAP Symp, Kyoto 1984.

Oltmans,S.J.,1981: Surface ozone measurements in clear air.J.Geophys.Res.,86,1174-1180.

Pruchniewicz,P.G.,1973: The average tropospheric ozone content and its variation with season and latitude as a result of the global ozone circulation. Pure Appl.Geophys.,106-108, 1058-1073.

Shapiro,M.A.,1980: Turbulent mixing within tropopause folds as a mechanism for the exchange of chemical constituents between the stratosphere and troposphere.J.Atmos.Sci.,37,994-1004.

Viezee,W.,W.B.Johnson and H.B.Singh,1983: Stratospheric ozone in the lower troposphere-II. Assessment of downward flux and ground-level impact.Atmos.Env.,17,1979-1993.

TROPOSPHERIC OZONE: VARIATIONS FROM SURFACE AND ECC OZONESONDE OBSERVATIONS

S.J. Oltmans and W.D. Komhyr
Air Resources Laboratory, NOAA/ERL
Boulder, Colorado 80303, USA

P.R. Franchois
CIRES, University of Colorado
Boulder, Colorado 80309, USA

W.A. Matthews
Physics and Engineering Laboratory, DSIR
Lauder, New Zealand

ABSTRACT

A self-consistent set of ozonesonde soundings from an eight station network stretching from 75°N to 90°S confirms that there is more ozone in the NH troposphere than in the SH. There is a very strong enhancement at mid-tropospheric levels in ozone at Samoa (14°S) during the austral spring. This is also the period of enhanced tropospheric ozone over tropical South America and Africa.

Fifteen years of surface measurements at the four Geophysical Monitoring for Climatic Change observatories show increasing ozone concentrations at the two NH sites (Barrow and Mauna Loa). At the two SH locations (Samoa and South Pole), ozone amounts have decreased dramatically (~2.0%yr^{-1}) during the austral summer (DJF) which is also the time of the seasonal minimum.

1. INTRODUCTION

The study of tropospheric ozone, whose role has often taken a back seat to stratospheric ozone in atmospheric chemistry studies, has emerged as an important constituent in our understanding of tropospheric chemistry (NAS, 1984), atmospheric dynamics (Junge, 1962), and climatic change (Ramanathan et al., 1985). Trends in tropospheric ozone have been inferred from existing observations (Angell and Korshover, 1983; Logan, 1985; Bojkov, 1985). Evidence appears to be convincing that ozone in the troposphere has been increasing over continental regions of the Northern Hemisphere (Logan, 1985) probably as a result of increasing anthropogenic emissions into the atmosphere (Logan, 1985, Liu et al., 1987). In the more remote troposphere, long-term changes have also been detected with less conclusive evidence as to possible causes (Oltmans and Komhyr, 1986, Bojkov, 1985 , Logan, 1985).

A self-consistent set of ozonesonde data are used to determine the seasonal cycle and latitudinal variation throughout the troposphere in a band stretching from the high Arctic to the South Pole. These include observations never previously reported on from Hawaii, Samoa, and New Zealand.

In addition, long-term (up to 15 years) surface-based measurements from the NOAA Geophysical Monitoring for Climatic Change (GMCC) locations are used to search for trends in tropospheric ozone behavior.

2. SEASONAL AND LATITUDINAL VARIATIONS FROM OZONESONDES

The basis for the seasonal and latitudinal patterns derived here is a network of eight stations listed in Table 1. In addition, data from Syowa on the coast of Antarctica are used. The Carbon-Iodine ozonesonde used by the Japanese is an early version of the ECC ozonesonde with similar characteristics, although it is not known whether the procedures used are the same as at the other stations. An earlier intercomparison campaign (Attmanspacher and Dütsch, 1978) found that a basic characteristic that is consistent among instruments of this type is the correction factor of close to one when comparing with Dobson column amounts. Results described here are consistent with the conclusion that the Syowa data are compatible with the other eight sites.

The ECC sonde is capable of making ozone measurements on an absolute scale. Comparisons with the total column amount give correction factors to the Dobson very close to unity. For over 525 soundings made within the NOAA/GMCC ozonesonde network since 1984, the correction factor is 1.009 with a standard deviation of 0.054. This represents the behavior over a wide range of operating conditions from the Antarctic to the tropics. Comparisons with a number of ECC sondes flown together along with more sophisticated UV photometers confirm that the sondes are capable of reasonable precision (standard error of 7%) and accuracy (5%) in the troposphere (Komhyr et al., 1984). Because of the good absolute accuracy of the ECC sonde, high quality ozone soundings can be obtained when an independent measurement of total column ozone is not available. This is particularly important at high latitudes where a Dobson or other column measurement is often not possible.

ISBN 0-937194-15-8

Table 1. ECC Ozonesonde Network

Station (Abbreviation)	Period of Observation	Latitude	Longitude	Elevation (meters)	Correction Factor (standard deviation)	Number of Soundings	Agency
Resolute, NWT (RES)	DEC79-APR86	74°43'N	94°59'W	64	1.022 ± 0.074	202	AES/Canada
Barrow, AK (BAR)	FEB86-APR88	71°19'N	156°36'W	11	1.006 ± 0.057	80	NOAA-GMCC/USA
Edmonton, ALB (EDM)	AUG79-SEP86	53°33'N	114°06'W	766	1.022 ± 0.066	245	AES/Canada
Boulder, CO (BLD)	MAR79-DEC87	40° 00'N	105°15'W	1743	1.007 ± 0.056	222	NOAA-GMCC/USA
Hilo, HI (HIL)	SEP82-DEC87	19°43'N	155°04'W	11	1.001 ± 0.052	170	NOAA-GMCC/USA
Pago Pago, AS (SAM)	APR86-APR88	14°15'S	170°34'W	15	-	85	NOAA-GMCC/USA
Lauder, NZ (LAU)	AUG86-APR88	45°00'S	170°00'W	400	1.022 ± 0.054	110	DSIR/New Zealand
Syowa, ANT (SYW)	MAR66-DEC84	69°00'S	39°35'E	21	1.013 ± 0.088	187	NIPR/Japan
South Pole, ANT (SPO)	JAN86-DEC87	90°00'S	-	2835	1.022 ± 0.053	124	NOAA-GMCC/USA

The seasonal cycle is a prominent feature of the tropospheric ozone distribution (Oltmans, 1981; Levy et al., 1985; Logan, 1985). Both the phasing and variability in this cycle can be useful in explaining its origin at a given location. Figure 1 a,b show the average vertical ozone profile during two seasons of the year at four locations in the Northern Hemisphere (NH). The two seasons Mar.-Apr.-May (MAM) and Sep.-Oct.-Nov. (SON) represent the seasons of maximum and minimum ozone at most locations and altitudes in the NH troposphere. This is not uniformly the case and the deviation from this pattern will be discussed later. Despite the wide variation in latitude (20°N-75°N), it appears that at middle tropospheric levels, ozone amounts during the MAM season are close to the same. This encompasses levels from above the boundary layer up to levels which are not encompassed by the stratosphere during a portion of the year. This is roughly the 700-400 mb layer (only to 500 mb at high Arctic locations with lower springtime tropopauses). The major exception to this is Boulder where there appears to be strong enhancement from the surface to 700 mb during this season. During the SON season, there is a strong gradient between Resolute and Hilo.

In the Southern Hemisphere (SH) the vertical profiles shown for each season at four stations (figure 2 a-d) present a very different picture than in the NH. During all seasons at Lauder, there is more ozone than at higher and lower latitudes. In fact, even during the seasonal minimum at Lauder (MAM) mixing ratios equal or exceed the amount (averaged over a season) for any season at all other latitudes in the SH from about 700 to 400 mb Although longitudinal gradients appear to be quite weak in model studies (Levy et al., 1985) we do know that strong local sources in the tropics of South America (Crutzen et al., 1985; Logan and Kirchhoff, 1986; Cros et al., 1987) at low levels in the troposphere may alter this picture. At Lauder through much of the year, there is also a very strong ozone gradient between the upper and lower troposphere. This is weaker at high latitudes and even reversed during parts of the year at Samoa. These data are consistent with the idea that ozone injected into the troposphere from the stratosphere at middle latitudes of the SH and transported equatorward and poleward plays a major role in determining the seasonal behavior. Although latitudinal gradients are weaker in the SH, the clear picture is of ozone decreasing away from mid-latitudes. A clear exception to this is during the month of October where there is substantially more ozone at Samoa than Lauder at mid-tropospheric levels (500 mb). More will be said about this later.

This set of ozone soundings make the difference in the amount of ozone in the NH and SH very clear (Fishman et al., 1979). On an annual basis (figure 3), all locations in the NH have more ozone than any location in the SH. Thus, Hilo which has the lowest annual average ozone amount in the NH has about 15% more ozone between 700 and 500 mb than does Lauder, the site with the most tropospheric ozone in the SH. Edmonton has 30% more ozone than Lauder in the region between 850 and 400 mb. Below 500 mb Boulder has more ozone than any other location. There is ample evidence that ozone is produced in the troposphere in the Boulder region during summer months (Oltmans, 1981; Parrish et al., 1986). At Samoa, the layer between 700 and 500 mb has more ozone than at Antarctic latitudes.

Both Hilo and Samoa show mid-tropospheric secondary maxima. At Hilo this occurs most prominently during the MAM seasonal maximum, while at Samoa, this is most marked during SON. High latitudes of the NH have roughly twice as much ozone as similar latitudes in the SH.

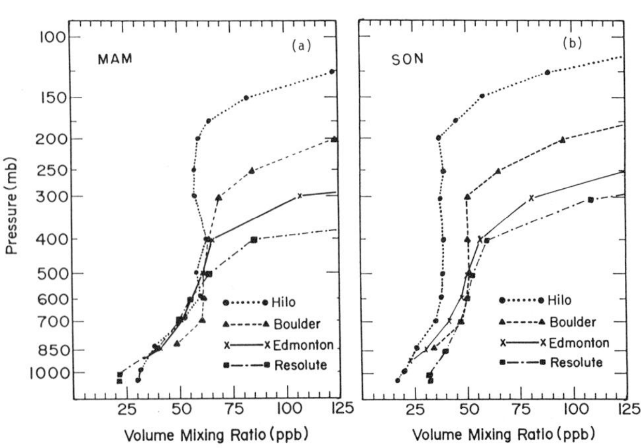

Fig.1. Ozone mixing ratio profiles from 4 stations in the NH for (a) Mar.-Apr.-May (MAM) and (b) Sep.-Oct.-Nov. (SON).

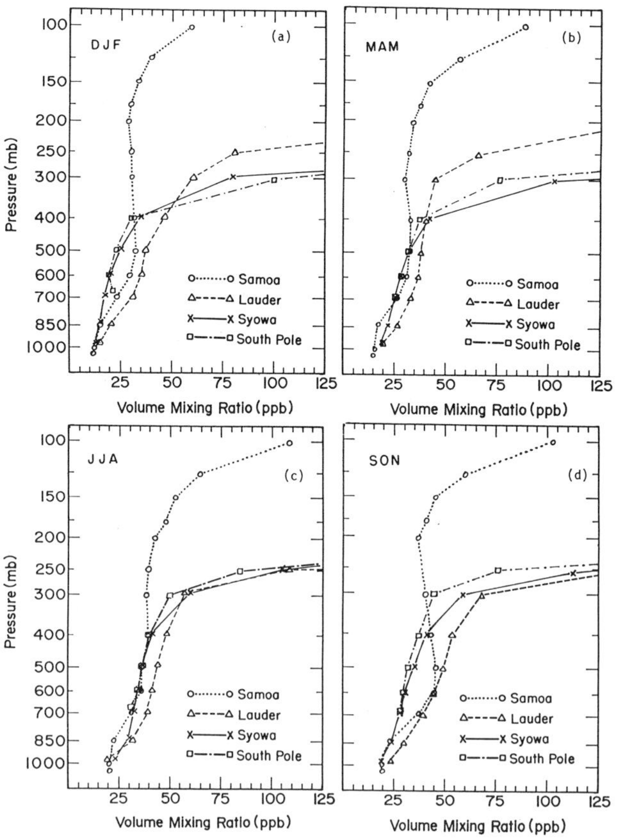

Fig. 2. Ozone mixing ratio profiles from 4 stations in the SH for (a) Dec.-Jan.-Feb. (DJF), (b) Mar.-Apr.-May (MAM), (c) Jun.-Jul.-Aug. (JJA) and (d) Sep.-Oct.-Nov. (SON).

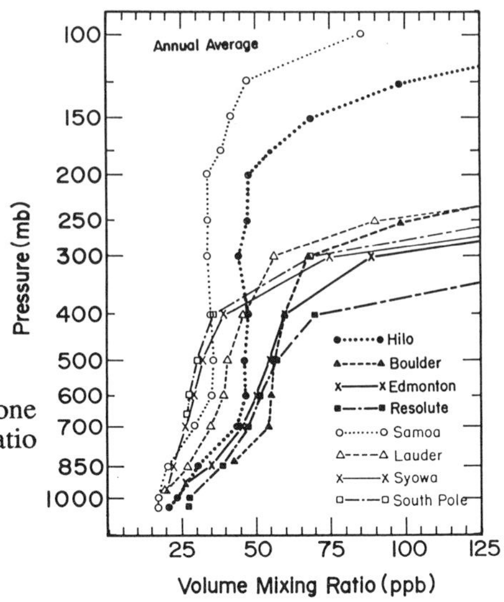

Fig. 3. Average annual ozone mixing ratio profiles.

While it is clear that these data support the idea of more ozone in the NH than the SH, it is less clear what the source of this difference is. Photochemical production of ozone through oxidation of nonmethane hydrocarbons in the presence of nitrogen oxides (Crutzen, 1988) is not inconsistent with the profile at Boulder where strong mid-tropospheric enhancements of ozone are seen. This is in contrast to Edmonton where ozone increases with increasing altitude in all seasons. Injection of ozone into the mid-latitudes of the SH appears to be an important source of tropospheric ozone in that hemisphere, but the high values at Samoa in October in the mid-troposphere raise questions as to other possible sources. In the NH, injection from the stratosphere and horizontal advection appear to be important, but there is clear indication of important regional contributions from tropospheric production as seen at Boulder. The contribution to the hemispheric difference through the oxidation of methane to carbon monoxide which should lead to ozone production in the nitric oxide rich environment of the NH and ozone loss in the low nitric oxide SH (Crutzen, 1988) is not easily detected in examining the character of the profiles alone.

The times of the seasonal maximum and minimum seem to be well-preserved with altitude through the troposphere at a given location, at least if one considers seasonal groupings of three months duration. The major exception to this is at the surface in the high Arctic where the seasonal maximum occurs in SON and the minimum in MAM. A potential mechanism, that has recently been suggested, is ozone destroying reactions with organic bromine gases at the time of the return of sunlight in the spring (Barrie et al., 1988; Oltmans et al., 1988). The preservation of the phasing of the seasonal cycle with altitude (within a couple of months) suggests that the forcing mechanism is the same with altitude but again does not necessarily distinquish between tropospheric production and transport from the stratosphere.

In Antarctica the gradient above the surface is very weak indicating very weak loss at the surface in this region. There are stronger gradients between the surface and the lower troposphere in Hawaii than in Samoa. This is especially marked during the seasonal maximum (MAM) when at Hawaii the levels between 700 and 200 mb have very large ozone amounts relative to other times of the year. In general, the gradient between the surface and 700 mb is stronger at the NH locations than those in the SH. This might be excepted in light of the fact that the SH sites are either away from major continental areas (Samoa and Lauder) or in Antarctica where the surface is often snow-covered (Galbally and Roy, 1980). Since this contrast between the NH and SH is in general the case, it is likely that ozone destruction at the surface is substantially larger in the NH as evidenced by these data.

3. SURFACE OZONE OBSERVATIONS

Surface observations can be made nearly continuously and at relatively low cost, but give data only at the lowest level of the troposphere which is complicated by boundary layer effects. At some higher altitude locations, it appears that data representative of the lower free troposphere can be obtained (Oltmans and Komhyr, 1986). It is also important to understand tropospheric ozone behavior near the surface. Surface data reported on here fall in both categories: data strongly influenced by boundary layer effects (Barrow, AK), and a station clearly representative of air above the boundary layer (Mauna Loa, HI).

Instrument operation and calibration methods for the surface ozone data discussed here are described in Oltmans (1981) and Oltmans and Komhyr (1986). Basic features of the ozone distribution such as diurnal, day-to-day and seasonal variability were also discussed in these references.

Trend information was also reported based on approximately 10 years of observations. At that time significant positive trends were reported in the NH while no significant changes over this time period were detected in the SH.

The basis of this analysis is the record of monthly mean surface ozone amounts at each of the 4 GMCC observatories (figure 4 a-b). The Mauna Loa Observatory is located at the 3400-meter level. Ozone concentrations are strongly influenced by the diurnal wind flow regime (Oltmans, 1981). Ozone measured during the nighttime downslope flow appears to be representative of air in the free troposphere at the level of the observatory (Oltmans and Komhyr, 1986) rather than surface boundary layer air. As in the past, the data from the downslope regime (00-08 LST) are used for this analysis. At the other locations, data from all hours not contaminated by nearby local sources are used. As mentioned earlier, surface data at Barrow is decoupled from the behavior at higher tropospheric levels; thus, is not representative of the layers above the surface. In the seasonal pattern, surface data at the South Pole are very similar to what is seen at other tropospheric levels. The conclusion drawn earlier that the processes responsible for producing the annual cycle are operative through broad altitude regions of the troposphere, lends weight to the idea that long-term changes in the lower troposphere can be monitored at the surface at South Pole.

At Barrow a linear trend of nearly 0.8%/year is found for the departures from the monthly normals (figure 4 a). This trend is more than twice as large (1.9%/year) during the period from June-September. Ozone is also increasing significantly at the time of the seasonal maximum (November). Decreases are seen only during February, March and April. The addition of 3 years of observations to the record at Mauna Loa has seen the magnitude of the overall trend drop markedly from 1.37%/yr through 1984 to .78%/yr at the end of 1987. The record through 1984 was strongly influenced by the very large amounts recorded in 1983. Recent years have seen near normal amounts.

Although no significant overall trend at Samoa is apparent, a significant decrease has taken place during the seasonal minimum (Dec.-Mar.) of 1.85%/yr. This is similar to what was found during the same time of the year at the South Pole (Komhyr et al., 1988). Earlier it was suggested the ozone at high and low latitudes of the SH reflects the transport from middle latitudes. Komhyr et al. (1988) gave evidence for a summertime weakening of transport into Antarctica producing the lower seasonal minimum at that location. This is the first indication of a possible related weakening in equatorward transport or of weakened transport from the stratosphere source during these months. During much of this period, springtime Antarctic lower stratospheric ozone amounts have also shown a dramatic decline (Komhyr et al., 1988). Increasing methane amounts may also lead to tropospheric ozone decreases when nitric oxide amounts are very low (Crutzen, 1988).

4. DISCUSSION AND CONCLUSIONS

Several new features of the tropospheric ozone distribution have appeared from a set of ozonesonde soundings, many of which were made where such measurements have not previously been made. The mid-latitude troposphere of the SH contains more ozone than does the tropics or high latitudes. The tropics of both hemispheres show marked relative ozone maxima during the

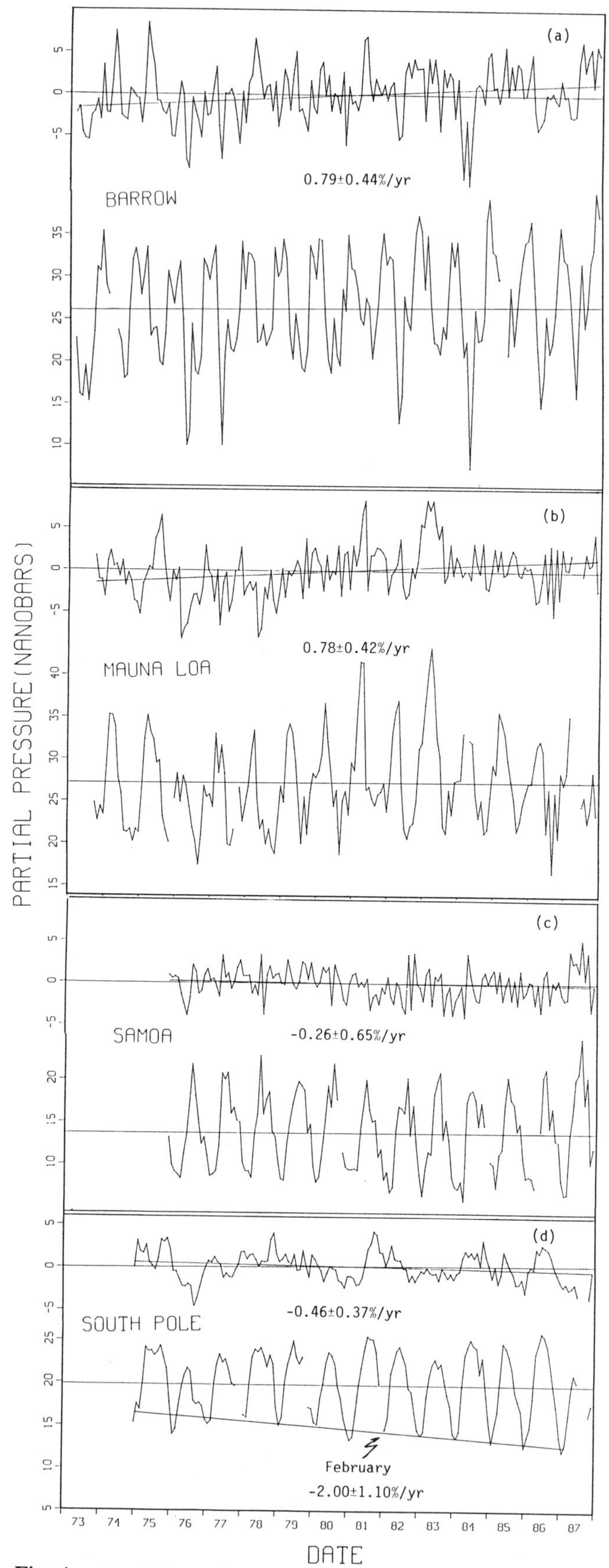

Fig. 4. Monthly surface ozone partial pressures and anomalies for (a) Barrow, AK, (b) Mauna Loa, HI, (c) Samoa and (d) South Pole.

spring in the region between 700-300 mb. Because this data set can be considered to represent broad climatological regions, they lend strong support to the presence of more ozone in the NH than in the SH. The soundings from Lauder, Samoa, and Hilo give a much better determination of the latitudinal gradient than was previously possible.

The larger ozone amounts in the NH raise the question as to whether this condition is primarily a result of relatively recent increases due to human activity or are a result of differing fluxes from the stratosphere in the two hemispheres. The model of Levy et al. (1985) in which there is no tropospheric ozone production gives greater amounts of ozone in the NH than in the SH. The hemispheric patterns and seasonal cycles presented here seem to replicate many of the model features lending some credence to a hemispheric difference in ozone based on differences in the flux of ozone from the stratosphere into the troposphere.

The ozonesonde data at Samoa invite comparison with the other important set of SH tropical ozone profiles at Natal, Brazil (Logan and Kirchhoff, 1986). Because of differences in the concentration of the sensing reagent, the Natal values are reported on a scale approximately 4% higher than for the Samoa results. The differences discussed here far exceed this difference. The two sets of data are also for different periods of time and include about 90 profiles at each site. At the surface during the austral autumn minimum, both locations have comparable concentrations while the spring maximum which occurs a month or two later at Natal is about 25% higher there. At 700 mb this difference is near 50% while at 500 mb and 300 mb, it is 30%.

An important question that arises from this comparison is whether the very large amounts of ozone found over Brazil are a source of ozone at far distant Samoa. We have already noted that the mid-tropospheric concentration during October at Samoa exceeds that at mid-latitudes of the SH. Of course, higher tropospheric altitudes (300 mb) at mid-latitudes do have such concentrations but such a source seems unlikely given the profiles shown in figure 2d. The very low NO_x values expected in the western Pacific (Logan, 1983) suggest that this region should be a net sink rather than source for tropospheric ozone. At this point, the biomass burning on the South American and African continents appears to be the only identified source of what may be an overall enhancement of SH tropical tropospheric ozone.

REFERENCES

Angell, J., and J. Korshover, 1983: Global variations in total ozone and layer mean ozone: An update through 1981, *J. Climate Appl. Met.*, 22, 1611-1626.

Attmanspacher, W., and Dütsch, H.U., 1981: 2nd international ozonesonde intercomparison at the Observatory Höhenpeissenberg 5-20 April 1978, *Berichte des Deutschen Wetterdienstes NR.* 157, Offenbach am Main, FRG, 65 pp.

Barrie, L.A., J.W. Bottenheim, R.C. Schnell, P.J. Crutzen, and R.A. Rasmussen, 1988: Ozone destruction and photochemical reactions at polar sunrise in the lower Arctic atmosphere, *Nature*, 334, 138-141.

Bojkov, R.D., and G.C. Reinsel, 1985: Trends in tropospheric ozone concentration, in *Atmospheric Ozone, Proceedings of the Quadrennial International Ozone Symposium*, pp 775-781, D. Reidel, Dordrecht.

Cros, B., R. Delmas, B. Clairae, J. Loemba-Ndembi, and J. Fontan, 1987: Survey of ozone concentrations in an equatorial region during the rainy season, *J. Geophys. Res.*, 92, 9772-9778.

Crutzen, P.J., A.C. Delany, J. Greenberg, P. Haagenson, L. Heidt, R. Lueb, W. Pollack, W. Seiler, and P. Zimmerman, 1985: Tropospheric chemical composition measurements in Brazil during the dry season, *J. Atmos. Chem.*, 2, 233-256.

Crutzen, P.J., 1988: Tropospheric ozone: An overview in *Tropospheric Ozone*, pp 3-32, D. Reidel, Dordrecht.

Fishman, J., S. Solomon, and P.J. Crutzen, 1979: Observational and theoretical evidence in support of a significant in-situ photochemical source of tropospheric ozone, *Tellus*, 31, 432-446.

Galbally, I.E., and C.R. Roy, 1980: Destruction of ozone at the earth's surface, *Q. J. R. Meteorol. Soc.*, 106, 599-620.

Junge, C.E., 1962: Global ozone budget and exchange between stratosphere and troposphere, *Tellus*, 14, 363-377.

Komhyr, W.D., S.J. Oltmans, A.N. Chopra, and P.R. Franchois, 1985: Performance characteristics of high-altitude ECC ozonesondes, in *Atmospheric Ozone, Proceedings of the Quadrennial International Ozone Symposium,* 796-802, D. Reidel, Dordrecht.

Komhyr, W.D., S.J. Oltmans, and R.D. Grass, 1988: Atmospheric ozone at South Pole, Antarctica, in 1986, *J. Geophys. Res.*, 93, 5167-5184.

Levy, H., II, J.D. Mahlman, W.J. Moxim, and S.C. Liu, 1985: Tropospheric ozone: The role of transport, *J. Geophys. Res.*, 90, 3753-3772.

Liu, S., M. Trainer, F. Fehsenfeld, D. Parrish, E. Williams, D. Fahey, G. Hubler, and P. Murphy, 1987: Ozone production in the rural troposphere and implications for regional and global ozone distributions, *J. Geophys. Res.*, 92, 4191-4207.

Logan, J.A., 1983: Nitrogen oxides in the troposphere: Global and regional budgets, *J. Geophys. Res.*, 88, 10785-10807.

Logan, J.A., 1985: Tropospheric ozone: Seasonal behavior, trends and anthropogenic influence, *J. Geophys. Res.*, 90, 10463-10482.

Logan, J.A., and V.W.J.H. Kirchhoff, 1986: Seasonal variations of tropospheric ozone at Natal, Brazil, *J. Geophys. Res.*, 91, 7875-7882.

NAS, 1984: Global tropospheric chemistry: A plan for action, National Academy Press, Washington, DC, 194 pp.

Oltmans, S.J., 1981: Surface ozone measurements in clean air, *J. Geophys. Res.*, 86, 1174-1180.

Oltmans, S.J., and W.D. Komhyr, 1986: Surface ozone distributions and variations from 1973-1984 measurements at the NOAA Geophysical Monitoring for Climatic Change baseline observatories, *J. Geophys. Res.*, 91, 5229-5236.

Oltmans, S.J., R.C. Schnell, P.J. Sheridan, J.W. Winchester, R.E. Peterson, J.D. Kahl, and L.A. Barrie, 1988: Seasonal surface ozone in the high Arctic, submitted to *Atmos. Environ.*

Parrish, D.D., D.W. Fahey, E.J. Williams, S.C. Liu, M. Trainer, P.C. Murphy, D.L. Albritton, and F.C. Fehsenfeld, 1986: Background ozone and anthropogenic ozone enhancement at Niwot Ridge, Colorado, *J. Atmos. Chem.*, 4, 63-80.

Ramanathan, V., R. Cicerone, H. Singh, and J. Kiehl, Trace gas trends and their potential role in climate, 1985: *J. Geophys. Res.*, 90, 5547-5566.

MODEL STUDIES OF FREE TROPOSPHERIC OZONE FORMATION FROM POLLUTION SOURCES

Ivar S.A. Isaksen[1], *Jan E. Jonson*[1], *Claire E. Reeves*[1,2],
Sverre Solberg[1], *and Robert B. Chatfield*[3]
[1]Institute for Geophysics
University of Oslo, Blindern
0315 Oslo 3, Norway
[2]On leave from School of Environmental Science
University of East Anglia
Norwich, NR4 7TJ, UK
[3]National Center for Atmospheric Research
Boulder, Colorado 80307, USA

ABSTRACT

A "channel model" has been constructed to study the development of chemically active trace species over continental and oceanic regions at mid latitudes (30N-60N). Particular emphasis is given to the distribution and changes of ozone (O_3), hydrogen peroxide (H_2O_2) and hydroxyl (OH). A situation representing spring (April) is chosen. Ozone is found to be photochemically formed when air masses are transported from industrialized continents to oceans where the loss process is slow. Maximum ozone values of almost 60 ppbv are obtained in surface air over the mid Atlantic. OH and H_2O_2 are found to be anti-correlated: OH has its maximum over industrialised areas where NO_x levels are high, while H_2O_2 are highest over remote unpolluted areas where its concentration exceeds 1 ppbv. Ozone is found to have increased substantially since the turn of the century as a result of man-made releases of pollutants (NO_x, HC, CO).

1. INTRODUCTION

It is becoming evident that human activities are affecting the ozone distribution, not only on a local but also on a regional scale throughout large parts of the northern hemisphere. Measurements over Europe and the U.S. in the last two to three decades indicate that there has been a continuous increase in regional ozone levels during this time period, (Logan, 1985). Recent analysis of ozone data as far back as the end of the last century indicate that the northern hemispheric troposphere has experienced a substantial increase in ozone levels due to increased human activity. Bojkov (1986), and Kley et. al. (1988), have published observations from the U.S. and Europe during the last decades of the nineteenth century, and found that surface ozone levels are now twice as high.

2. MODEL DESCRIPTION

Tropospheric ozone for April at northern mid latitudes is studied applying the zonal channel model described by Solberg et. al. (1988). The objective is to simulate ozone generation during its transport and that of its precursors from polluted continental regions (e.g. the U.S. and central Europe) to less polluted oceanic and continental- background regions. Of particular interest is to study the zonal gradients for ozone and other secondary species, like hydroxyl (OH) and hydrogen peroxide (H_2O_2). Both species are important to the tropospheric chemistry.

Some modifications have been made to the model. Diffusion across latitudinal boundaries is omitted, as well as K_{xz} and K_{zx}. The convective cloud transport is distributionally connected to the lightning activity in addition to the monthly mean precipitation. In this way more weight is given to transport in highly convective areas as opposed to areas of frontal precipitation.

The chemical species included in the model (47 altogether) are the same as used in previous studies (Isaksen and Hov, 1987). In addition to photochemical processes, heterogeneous removal through wet and dry deposition, and emissions of ozone precursors from natural and anthropogenic sources are included. Emissions are shown in figure 1. Also downward fluxes of O_3, NO and HNO_3 are assumed into the top layer.

The following deposition velocities for surface removal over land are adopted (cms^{-1}): HNO_3 - 0.7, PAN - 0.2, O_3 and NO_2 - 0.5. HNO_3 has the same deposition velocity over ocean as over land, while for O_3 and NO_2 the deposition velocities are reduced by a factor of 10 over the ocean, and for PAN by a factor of 4. Reversible and irreversible washout are included as a function of height, and of precipitation (Deutsche Wetterdienst, 1976), and cloud cover (Gordon et. al., 1984), respectively. HNO_3 is assumed to be washed out reversibly. Several species are washed out irreversibly, among them H_2O_2 and CH_2O. An additional sink, in the lowest 10 km, is loss to aerosols. This removal rate reduces exponentially with height and is a factor of 10 lower over sea due to the lower particle concentration.

The dissociation rates (Isaksen et. al., 1977) are calculated with an average cloud cover of 50%, extending between a

ISBN 0-937194-15-8

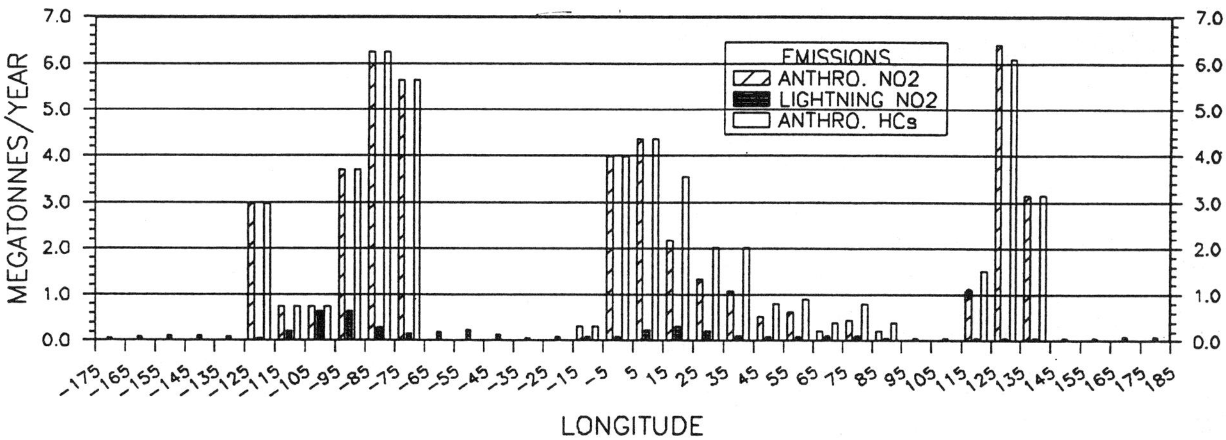

Fig. 1. *Anthropogenic emissions of NO_x as (NO_2) and hydrocarbons (HC). NO_x produced from lightning is also shown, with zonal distribution given by Turman and Edgar (1982). It is assumed that each flash produces 10^{26} molecules of NO_2 and that this source is distributed vertically up to 9 km. Natural sources for HC and NO_2 are assumed to be $3.3 * 10^{-8}$ $gm^{-2}s^{-1}$ and 10^{-9} $gNm^{-2}s^{-1}$ over land. HC emissions are distributed internally as 20% C_2H_4, 20% C_2H_6, 10% C_3H_6, 20% C_4H_{10}, 10% C_6H_{14} and 10% m-xylene. The remaining 10%, assumed to be methane, is not included, as methane is kept constant at 1.6 ppmv. Emissions of CO are assumed as 9 times that of the HC emissions.*

base at 1 km to a top at 4.5 km, assuming an albedo of 0.5.

The distribution of NO_x in the troposphere is important for the ozone formation (and loss). Together with CO and hydrocarbons, which are the sources of the free hydrogen peroxy radical, it provides the source for ozone through reactions such as:

$$OH + CO \rightarrow CO_2 + H \quad (1)$$

$$H + O_2 + M \rightarrow HO_2 + M \quad (2)$$

and:

$$NO + HO_2 \rightarrow NO_2 + OH \quad (3)$$

$$NO_2 + h\nu \rightarrow O + NO \quad (4)$$

$$O + O_2 \rightarrow M + O_3 \quad (5)$$

A major loss term for NO_x is a reaction with OH producing HNO_3. The latter component is lost mainly through rainout and ground deposition. However, particularly in winter at high latitudes, additional loss can occur efficiently through the night time chemistry of NO_3 and N_2O_5.

$$NO_2 + O_3 \rightarrow NO_3 + O_2 \quad (6)$$

$$NO_3 + NO_2 + M \rightleftharpoons M + N_2O_5 \quad (7)$$

Observations (Platt 1986) have shown that, under favourable conditions for removal, the lifetime of NO_3 - N_2O_5 could be of the order of a few minutes only. This is attributed to reactions, probably on the surface of deliquesent aerosols, producing particulate nitrate. The limiting process is probably reaction 6.

Several 30 day model simulations are performed:

- Standard run simulating the present spring-time (April) atmosphere.
- Anthropogenic emissions reduced by a factor of 10, simulating conditions at the turn of the century.
- Anthropogenic emissions reduced by 50% over the North American continent only.
- Without the night-time chemistry of NO_3-N_2O_5, producing particulate nitrogen.

3. RESULTS

In the present spring-time lower troposphere ozone is built up over the oceans, with a peak surface concentration of almost 60 ppbv in the mid-Atlantic (figure 2a). This demonstrates the chemical development in air masses, which have been exposed to pollution releases over continents. During transport over sea surfaces, where deposition is low, ozone builds up gradually providing sufficient NO_x is present, and not oxidized to HNO_3. A net chemical destruction only takes place over the most remote oceanic areas near the surface. (In this case, the eastern parts of the oceans, since the prevailing zonal winds are westerly.) Over the eastern U.S. and western Europe surface concentrations of around 50 ppbv are found. These values, although appearing to be high, are still within the upper limits of observed average concentrations (Becker et. al. 1985, Logan 1985). As our lowest level values are average concentrations up to 500m, they are expected to be higher than the observed surface concentrations.

The distribution of OH and H_2O_2 are depicted in figure 2b and 2c respectively. It is clearly seen that they are anti-correlated in the lower troposphere. OH has its maximum values over industrialised continents, where NO_x levels are high as a result of reaction 3. H_2O_2 has its maximum values in remote regions. This is explained by its dependence on the HO_2 concentration, as production occurs via the reaction:

$$HO_2 + HO_2 \rightarrow H_2O_2 + O_2 \quad (8)$$

Hence, when NO is low, HO_2 loss through reaction 3 is slow, favouring high HO_2 and H_2O_2 levels. OH concentrations, over the source areas, exceed 10^6 molecules cm^{-3}. There is also a distinct maximum in the upper troposphere over the east-

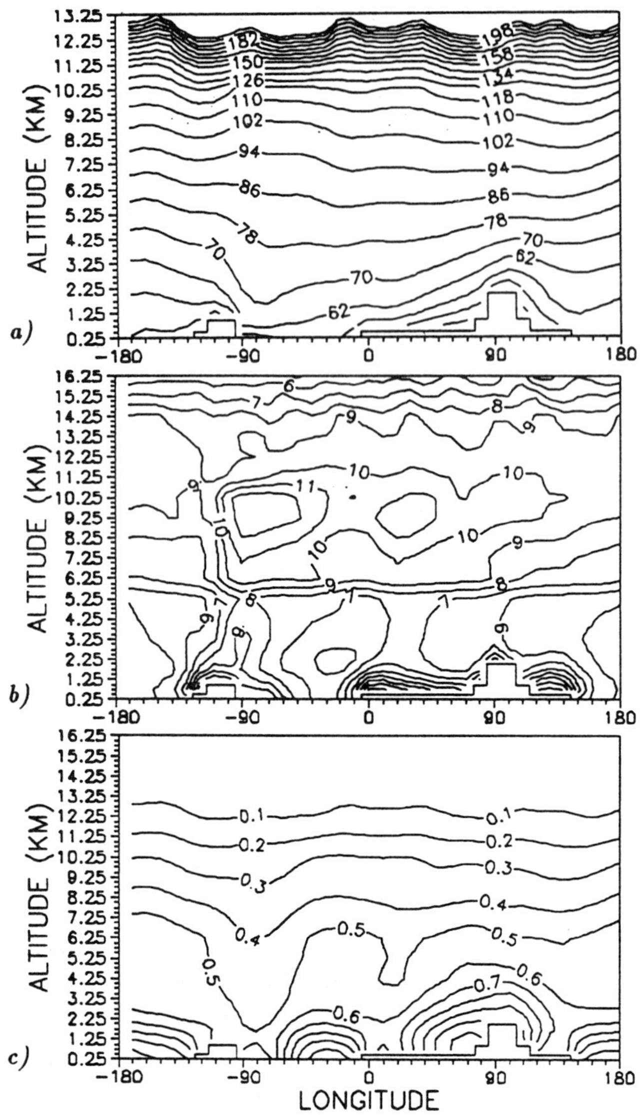

Fig. 2. *Daily average concentrations of a) ozone in ppbv, b) OH in molecules cm*$^{-3} * 10^5$ *and c)* H_2O_2 *in ppbv. The stepped outlines represent elevated ground level, but low level areas may extend further.*

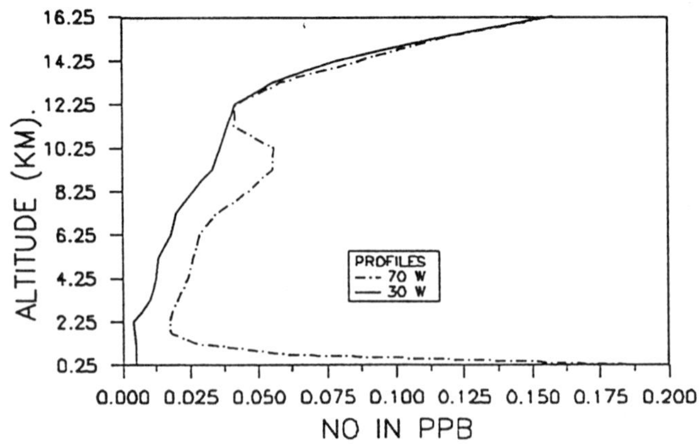

Fig. 3. *Vertical profiles of daily average NO concentrations for two sites: a continental source region at 70W, and a remote oceanic area at 30W.*

ern North American continent, reflecting enhanced NO levels. The marked increase with height around 6 km. is a result of enhanced photodissociation over clouds. Maximum H_2O_2 concentrations, which are found in the lower troposphere over the remote areas, exceed 1 ppbv.

Short-lived components, such as non-methane hydrocarbons (NMHC), PAN, NO_x, and HNO_3, show distinct maxima near the sources. In figure 3, a vertical profile for NO is shown for two longitudes. 70W represents a continental source region, whereas 30W is a remote oceanic site. The vertical profile for the former area decreases sharply from surface values of approximately 200 pptv to less than 25 pptv in the lower part of the free troposphere. Surface values are dramatically reduced over the remote oceanic regions, where the concentrations are only a few pptv. The free tropospheric NO values are also enhanced over the continental source region compared to the remote area. This is partly due to the high lightning activity, and partly due to increased vertical transport by convective cloud activity.

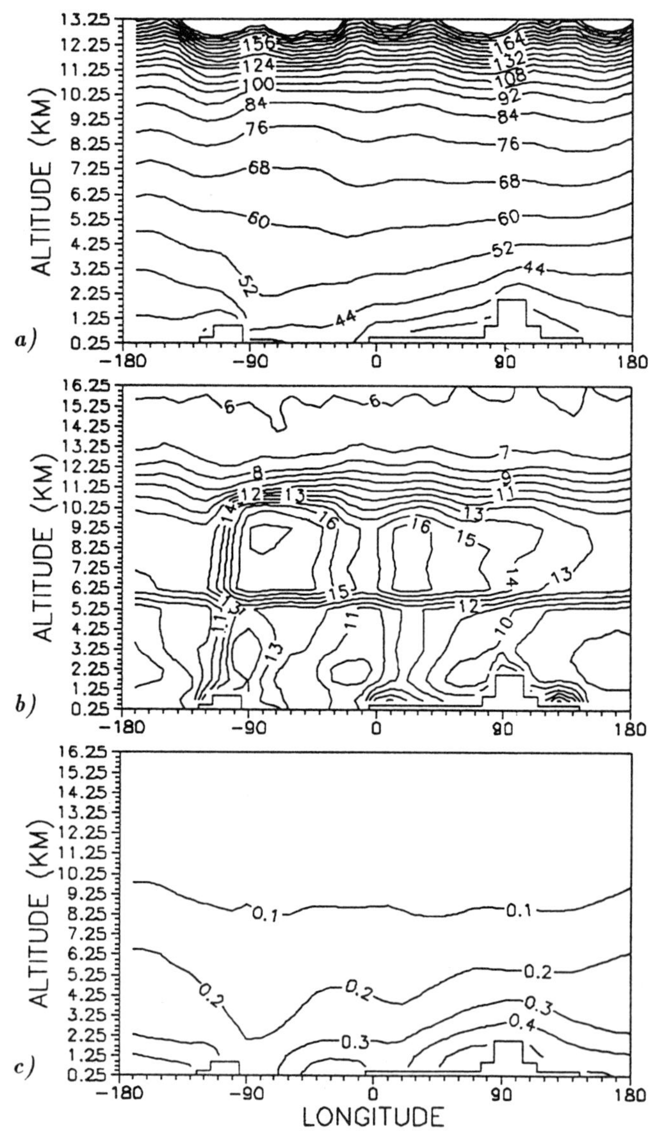

Fig. 4. *As figure 2, but with anthropogenic emissions reduced to 10%.*

Running the model with reduced emissions indicates that there has been an increase of tropospheric ozone of more than 50% in the northern mid latitudes since the turn of the century. We calculate previous surface ozone concentrations to have been approximately 30 - 35 ppbv in Western Europe and the Eastern U.S. (figure 4a). Peak values in the mid Atlantic are slightly above 40 ppbv. NO_x concentrations over polluted areas appear to have increased by approximately a factor of 7, somewhat less than the increase in emissions. However, natural emissions are assumed unchanged. NMHC and PAN show similar changes. OH concentrations (figure 4b) appear to have decreased since the turn of the century, except for polluted areas, where concentrations are unchanged or slightly higher. The decrease in OH concentrations are probably due to an increase in methane and CO, which break down OH. H_2O_2 concentrations (figure 4c) were about a factor of 2 lower, with peak values over the central Asian continent, instead of the mid Atlantic.

Reducing the anthropogenic emissions over the North American continent by 50% will result in a substantial decrease in pollutants over the region (figures 5 and 6). The reduction downwind depends on the lifetime of the species in question. For a relatively shortlived component like NO_2, a reduction of about 5% may be found over the eastern part of the Atlantic. For HNO_3 a reduction can be traced even further, extending into the European continent. The ozone reduction is 8% regionally and small over Europe, 2-3% (figure 6).

A major uncertainty in model studies is the night-time chemistry of NO_3-N_2O_5, which could be a major sink of nitrogen oxides (NO_x) through particulate nitrate formation. The assumption that no nitrate formation occurs leads to an increase in NO_x levels of more than 100% over remote areas, but far less near the source areas. NO_x is broken down less efficiently, and thus transported further. An increase in NO_x levels also leads to an increase in ozone, which goes up by 10-20%. The major source of HNO_3 in this case is the reaction of NO_2 with OH, with very little particulate nitrate being produced from NO_x. The consequence is that gaseous HNO_3 levels are more than 150% higher in remote areas, and more than 50% higher over polluted continents, than in the case where night-time NO_3-N_2O_5 is converted to particulate nitrate.

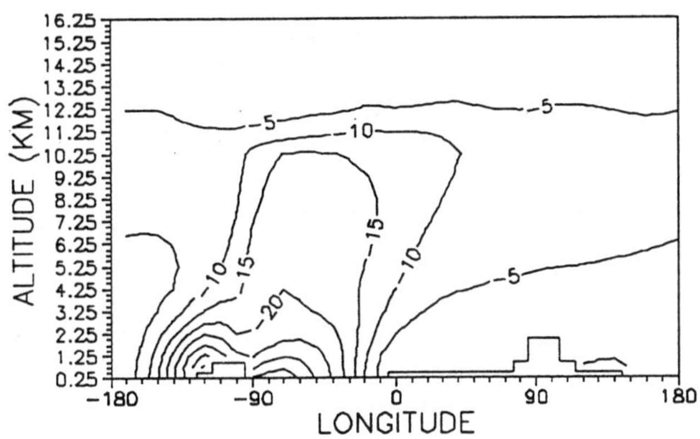

Fig. 5. *Percentage change in HNO_3 concentration due to a decrease in anthropogenic emissions of 50% over the U.S.*

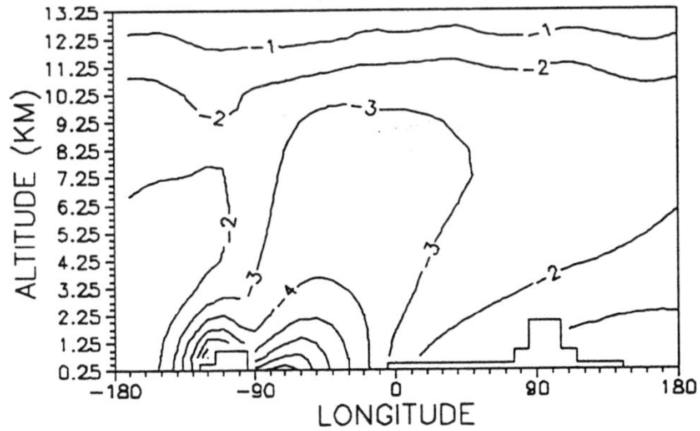

Fig. 6. *Percentage change in ozone concentration due to a decrease in anthropogenic emissions of 50% over the U.S.*

REFERENCES

Becker K.H., Fricke W., Lobel J., Schurath U. 1985, Formation, transport, and control of photochemical oxidants. Air pollution by photochemical oxidants. (ed. Guderian R.), Springer, Berlin, pp. 11-67.

Bojkov R.D. 1986, Surface ozone during the second half of the nineteenth century. J. Clim. Appl. Meteor., Vol. 25, pp. 343-352.

Deutcher Wetterdienstes monatskarten des niederslags für die ganze erde. 1976, Berichte nr. 139, band 18, Offenbach.

Gordon C.T., Hovanec R.D., Stern W.F. 1984, Analyses of monthly mean cloudiness and their influence upon model-diagnosed radiative fluxes. J. Geophys. Res., vol. 89, no. D3, pp. 4713-4738.

Isaksen I.S.A., Crutzen P.J., Midtbø K.H., Sunde J. 1977, A simplified method to include molecular scattering and reflection in calculations of photon fluxes and photodissosiation rates. Geophysica Norwegica, Vol. 31, no. 5.

Isaksen I.S.A., Hov Ø. 1987, Calculations of trends in the tropospheric concentrations of O_3, OH, CO, CH_4 and NO_x. Tellus, Vol 39B, pp. 271-283.

Kley D., Volz A.,Mülheims F. 1988, Ozone measurements in historic perspective. Tropospheric ozone, regional and global scale interactions. (ed. Isaksen I.S.A.) NATO ASI Series C, vol. 227, pp. 63-72, D. Reidel Publishing Company, Dordrecht.

Logan J.A. 1985, Tropospheric ozone: seasonal behavior, trends, and anthropogenic influence. J. Geophys. Res. vol. 90, no. D6, pp. 10463-10482.

Platt U. 1986, The origin of nitrous and nitric acid in the atmosphere. Chemistry of multiphase atmospheric systems. (ed. W. Jaeschke), NATO ASI Series G, vol.6, pp. 299-319, Springer-Verlag, New York.

Solberg S., Chatfield R.B., Isaksen I.S.A. 1988, Design of a channel model to assess mid-latitude pollution effects including increase of global tropospheric ozone. This Workshop.

Turman B.N., Edgar B.C. 1982, Global lightning distributions at dawn and dusk. J. Geophy. Res., Vol. 87, pp. 1191-1206.

DESIGN OF A CHANNEL MODEL TO ASSESS MID-LATITUDE POLLUTION EFFECTS

Sverre Solberg and Ivar S.A. Isaksen
Institute of Geophysics
University of Oslo
Blindern, 0315 Oslo 3, Norway

Robert B. Chatfield
National Center for Atmospheric Research
Boulder, Colorado 80307, USA

ABSTRACT

In this paper we have described the construction of a meridionally averaged 2D model for mid latitudes for the study of reactive species in the troposphere. The eddy coefficients were calculated from isentropic trajectories, which gave Kzz-values of $5 \cdot 10^4$ cm^2/s in the troposphere. The fact that in most 2D models the Kzz-values are larger, supports the idea that a significant fraction of the vertical motion is driven by release of latent heat which is characteristic of rapid convection as parameterized in this work. Model calculations for radon show marked zonal variations up to 4 km, above 8 km these are smoothed out. Even with our highly simplified parameterization of cloud transport, we find a better agreement with observations than with only eddy diffusion.

1. INTRODUCTION

Based on the success with zonally averaged 2D-models for the stratosphere, similar tropospheric models have been developed, and have shown to represent well the global distribution of long lived trace gases. Tropospheric ozone, however, is known to be controlled in a nonlinear way by reactive species like NOx and HO_2 with lifetimes much less than the characteristic time for zonal mixing. This makes the basic assumptions necesarry for these models more weak and uncertain when applied to the study of ozone and short lived components.

In order to examine the deviations from zonal symmetry of these gases, we have constructed a two dimensional model in altitude and longitude, that is a meridionally averaged model, describing the industrial belt from 30° to 60°N, and extending around the globe starting at 180°W. This paper describes the formulation of the model with advective and diffusive transport. Of major importance to the transport is the model description of vertical exchange, and the flux across the northern and southern boundaries. For the vertical transport we have looked at the effects when a simple cloud parameterization is included. Finally the transport is tested with radon as a tracer. A more detailed description of the model is given by Solberg (1988).

2. THE CONTINUITY EQUATION

To express the average mixing ratio between 30° and 60°N, we define a quantity, $\hat{\chi}$, averaged meridionally over n discrete datapoints as :

$$\hat{\chi} = \frac{\sum_{i=1}^{n} \chi_i \rho_i \cos\phi_i}{\sum_{i=1}^{n} \rho_i \cos\phi_i}$$

where ρ is the air density. When this averaging procedure is applied to the whole gas continuity equation, we get in geographical coordinates:

$$\begin{aligned}\frac{\partial \hat{r}}{\partial t} = & -\frac{1}{\sum(\rho_i \cos\phi_i)} \frac{\partial}{a\partial\lambda}[\hat{r} \sum (\rho_i u_i)] - \frac{1}{a\triangle\phi \sum(\rho_i \cos\phi_i)}[\rho r v \cos\phi]_{\phi_1}^{\phi^n} \\ & -\frac{1}{\sum(\rho_i \cos\phi_i)} \frac{\partial}{\partial z}[\hat{r} \sum (\rho_i w_i \cos\phi_i)] \\ & + \frac{1}{\tilde{\rho} a \cos\phi^*} \frac{\partial}{\partial\lambda}[\tilde{\rho} Kxx \frac{1}{a\cos\phi^*} \frac{\partial \hat{r}}{\partial\lambda} + \tilde{\rho} Kxz \frac{\partial \hat{r}}{\partial z}] \qquad (1) \\ & + \frac{1}{a\triangle\phi \sum(\rho_i \cos\phi_i)}[\rho Kyy \cos\phi \frac{\partial r}{a\partial\phi}]_{\phi_1}^{\phi^n} \\ & + \frac{1}{\tilde{\rho}} \frac{\partial}{\partial z}[\tilde{\rho} Kzx \frac{1}{a\cos\phi^*} \frac{\partial \hat{r}}{\partial\lambda} + \tilde{\rho} Kzz \frac{\partial \hat{r}}{\partial z}] \\ & + \frac{1}{\tilde{\rho}} P - \frac{1}{\tilde{\rho}} L\tilde{c} + \left(\frac{\partial \hat{r}}{\partial t}\right)_{cloud}\end{aligned}$$

where r = mixing ratio, c = concentration, P = chemical production, L = chemical loss, , ϕ^* = an angle between 30° and 60° (chosen to 45°), $\triangle\phi$ = distance between datapoints, and :

$$\tilde{\rho} = \frac{\sum_{i=1}^{n} \rho_i \cos\phi_i}{\sum_{i=1}^{n} \cos\phi_i}$$

$$\tilde{c} = \frac{\sum_{i=1}^{n} c_i \cos\phi_i}{\sum_{i=1}^{n} \cos\phi_i}$$

To arrive at this equation we had to assume that the components are reasonably well mixed from 30° to 60°N.

3. SOLUTION OF THE ADVECTIVE EQUATION

Numerically the model area was divided into 37*20 gridboxes, 37 gridsteps of 10° longitude, and 20 vertical levels as specified by Isaksen and Rodhe (1978). At the lateral boundaries we calculated zonal averaged concentrations in the latitude belts

ISBN 0-937194-15-8

from 0° to 30°N and 60° to 90°N at each vertical level from the mean and turbulent fluxes across 30° and 60°N , giving 2*20 extra gridpoints, assuming that zonal variations are smoothed out as the components are transported out of the gridbelt.

For the advective parts of the transport we used seasonally averaged u-, v-, and w-fields (Oort, 1983). In this work we use data for the spring season. To get massconservation in each gridbox, we redefined new meridional velocities satisfying the mass balance equation. This was done by multiplying the v-velocities at 30° and 60° with the pair of factors closest to 1, which gave a mass conserving, three dimensional velocity field. At most gridpoints the new v-velocities where quite close to the original ones.

The advection equation was solved numerically by an up-wind scheme with correction for numerical dispersion, given by Smolarkiewicz (1983). Based on numerical tests we found that due to the smaller number of grid points and the larger gradients expected over one grid distance compared to that described by Smolarkiewicz, the numerical solution could be markedly improved by repeating the correction, or by increasing the scaling factor Sc. For stability and computer cost reasons we used Sc = 1.5 and one correction.

4. SOLUTION OF THE EDDY TERMS

There are few references to zonal eddy diffusivities, Kxx and Kxz. For the vertical eddy diffusion, Kzz, there are a variety of references, but also a large variation in the values, ranging from $4 \cdot 10^4$ cm^2/s (Machta, 1974) to $7 \cdot 10^5$ cm^2/s (Liu et al., 1984). One possible explanation for this uncertainty is that traditionally all kinds of vertical movement (except large scale advection) have been parameterized with K-theory, mixing together very different transport phenomens, like rapid convection in clouds and the much slower diffusive process that is due to statistical variations in winds. The eddy diffusion applies most properly when the modeled transporting process is composed of many events which move material a small distance compared to the depth of the troposphere (Mactha, 1974).

Another way of parameterizing eddy transport is by using an integral transport technique (Chatfield and Crutzen, 1984; Fiedler and Moeng, 1985). Through this description the modeled quantity is redistributed throughout the entire tropospheric column at one location over a time step.

In this paper we propose that modeling large scale, quasi-isentropic motions with diffusion and small scale, rapid latent-heat driven motions with an integral transport technique, will capture basic averaged transport behaviour of the atmosphere.

To do so we attempted to estimate the eddy coefficients for the isentropic movement only, based on isentropic trajectories. The trajectories were precalculated, and are partially described by Chatfield and Haagenson (1987). For each of 6 stations over the North Atlantic there where 24 trajectories for each of 3 height levels, with arrival points in approximately 1000, 850, and 500 mb. The (x,y,p)-positions were given at 1, 36, 72, and 84 hours before arrival.

We calculated the eddy coefficients from the expression:

$$\aleph = \begin{pmatrix} Kxx & Kxz \\ Kzx & Kzz \end{pmatrix} = \frac{1}{2}\frac{\partial}{\partial t}\begin{pmatrix} \overline{\xi^2} & \overline{\xi\zeta} \\ \overline{\xi\zeta} & \overline{\zeta^2} \end{pmatrix} \tag{2}$$

where $\overline{\xi^2}$, $\overline{\zeta^2}$ and $\overline{\xi\zeta}$ are respectively the variance of the x- and z-trajectorie postions, and the covariance terms.

Equation 2) is obtained from theory of the stratosphere (Plumb and Mahlman 1987). We have neglected terms which we expect to bee small in the troposphere.

Using Eulerian time derivative, we get K-values averaged for all 6 stations for the time periods 1-36h (A), 36-72h (B), and 72-84h (C). For the actual values used, we averaged A- and B-values (fig. 1), interpolated up to low values in the stratosphere (Luther, 1974), and assumed constant values with longitude.

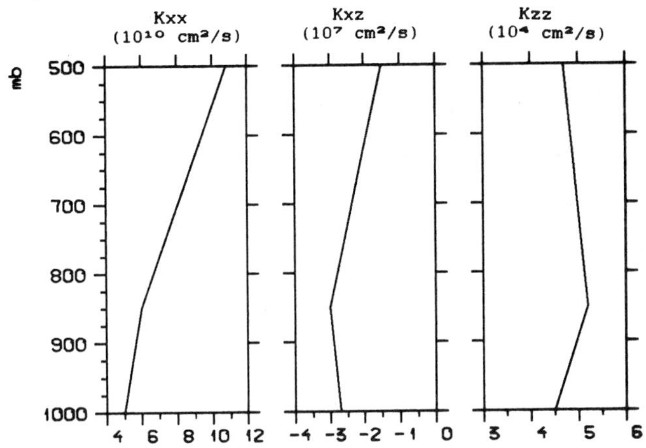

Fig. 1. Calculated eddy coefficients.

Fig. 1) shows that Kzz $\sim 5 \cdot 10^4 cm^2/s$, which is similar to the lowest ones referenced (Machta, 1974). The Kxx-values are increasing with height in agreement with general ideas that Kxx is proportional to the variance in zonal wind $\overline{u'u'}$. Due to the correlation between eastward and sinking motion, Kxz is negative (Chatfield and Haagenson, 1987).

To account for the difference in diffusion over land and ocean, and the fact that most of the trajectories were over the North Atlantic, we added a time varying boundary layer diffusion over land, adopted from Hov (1983), giving maximum Kzz-values of 10^6 cm^2/s in the lowest 1 km from noon to 6 pm.

5. CLOUD TRANSPORT

As already pointed out, we wanted to separate the eddy terms into a slow, isentropic motion described by diffusion, and a rapid, non-isentropic motion, described by the integral technique. From Chatfield and Crutzen (1984) we get :

$$\left(\frac{\partial c_k}{\partial t}\right)_{cloud} = \sum_{l=1}^{top}[P(z_k|z_l)\frac{\rho_k}{\rho_l}c_l] - \sum_{l=1}^{top}[P(z_l|z_k)c_k] \tag{3}$$

where

c_k = concentration in height level k

$P(z_k|z_l)$ = a statistical function describing the probability of transport from z_l to z_k.

To obtain this expression we assumed that the cloud transport is strictly vertical, and that we can separate the chemical terms. We stress that the last assumption is valid only for components with lifetimes longer than the characteristic time of the cloud transport.

Eq. 3) is a very general expression, allowing exchange between all vertical levels at a specified longitude. The physical characteristics of convection leads us to the further assumption that this process consists mainly of two mechanisms, a direct transport from the cloud base to any layer up to the cloud top, and an accompanying slower subsidence to the nearest layer below (Chatfield and Crutzen, 1984; Fiedler and Moeng, 1985). In z-coordinates we have from Chatfield and Crutzen:

$$\rho_l w_{k,l} = P(z_k|z_l)\rho_k \triangle z_k$$

where $\rho_l w_{k,l}$ = mass transport from layer l to layer k (in absolute values).
We then can write eq. 3) :

$$\left(\frac{\partial c_k}{\partial t}\right)_{cloud} = \frac{1}{\triangle z_k}[\rho_b w_{k,b} r_b + \rho_{k+1} w_{k,k+1} r_{k+1} - \rho_k w_{k-1,k} r_k]$$

Here b denotes the cloudbase.

The values $w_{k,b}$ can be expressed as a fraction of the total mass flux from the cloudbase :

$$\rho_b w_{k,b} = \epsilon(k|b)\,\rho_b W_b$$

$$W_b = \sum_{k=b+1}^{top} w_{k,b}$$

$$\sum_{k=b+1}^{top} \epsilon(k|b) = 1$$

Then the subsidence is given by mass conservation when W_b and ϵ are specified.
For tropical areas there have been extensive work to estimate these cloud parameter values, for midlatitude the corresponding references are sparse. A detailed description of W_b and the distribution function ϵ varying with longitude, would require a study beyond the scope of this work. Instead we suggest some hypothesis about ϵ and W_b, giving preliminary, first guess values.

For the mass flux $\rho_b W_b$ we assume :

$$\rho_b W_b = \frac{Prec}{f \cdot r_{sat,b}}$$

where
Prec = precipitation per time and area,
f = rain efficiency,
$r_{sat,b}$ = saturation mixing ratio of water at cloud base.

The function $\epsilon(k|b)$ was solved by assuming it is dependent only of land or ocean. Over land we assumed equal flux input:

$$\epsilon(k|b) = \frac{\rho_k \triangle z_k}{\sum_{l=b+1}^{top}(\rho_l \triangle z_l)}$$

Over oceans we adopted values of "region 3" from Chatfield and Crutzen (1984), adjusting them to our height levels. This gave more mass transport to the lowest kilometers than above. This is a crude representation of the deeper and stronger convection usually found over the continents, and the higher frequency of shallow, low clouds over oceans (fig. 3).

Values for the precipitation were read from the Deutsche Wetterdienstses (1976) for April, and the rain efficiency, f, was set equal to 0.65 based on comparisons between observed and calculated precipitation. These model calculations were similar to that described by Isaksen and Rodhe (1978), except that in addition to taking out as precipitation the fraction of the water vapor exceeding a critical value, we assumed a constant conversion rate from vapor to precipitation in each grid point based on the reversible scavenging parameters. Fig. 2 and 3 show the resulting values of W_b and ϵ.

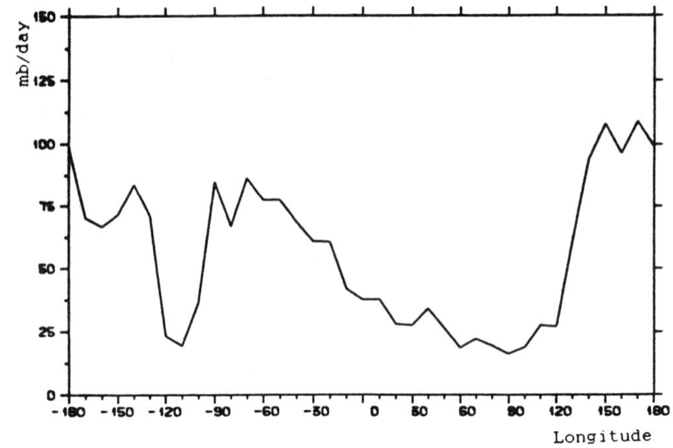

Fig. 2. The mass flux from cloudbase (mb/day).

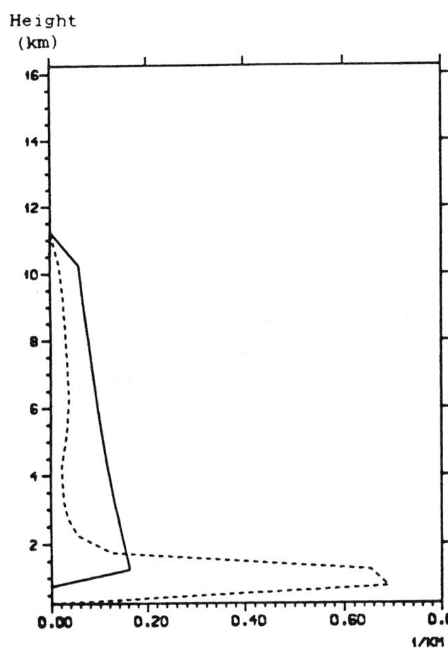

Fig.3.

Vertical distribution of the mass flux from cloud base over land and ocean. Solid line = land, dashed = ocean.

The area inside each curve is equal to 1, so that the fraction of the base mass flux transported to a layer between z_1 and z_2 is equal to the area inside the curves and z_1 and z_2.

We assumed a cloud base at 500 m over oceans, 1 km over land, and a cloud top height at 10 km. In addition the mass flux from cloud base was given a daily variation over the continents, following a sine function with a maximum at 4 pm. and an amplitude of 50 % of the average value. This is according to the observed behaviour of Cb-clouds at northern mid latitudes (NCAR, 1986).

6. ESTIMATES OF RADON DISTRIBUTION

To give an idea of how the model works, we chose radon as a tracer, with emission 1.2 atom/cm^{2}s over land, and a chemical lifetime 5.5 days (Turekian et al., 1977; Liu et al., 1984). Fig. 4 shows the calculated values at steady state. Fig. 5 shows observed values (Liu et al., 1984) together with average calculations over land, with and without cloud transport. From fig. 4 we see that there are marked zonal variations up to 4 km, showing the separation between continental, radon rich, and marin, radon poor air. Above 8 km these differences are smoothed out. We found that the cloud transport was critical for maintenance of the zonal variations in the lowest 4-5 km. Over source regions the effect of the cloud transport is most pronounced in the upper troposphere (fig. 5), where the radon concentration increases by one order of magnitude due to this transport compared to the eddy transport alone.

The observed profiles with the characteristic smaller lapse rate between 3 and 8 km, seem to indicate that cloud transport is an important mechanism for transport of pollutants

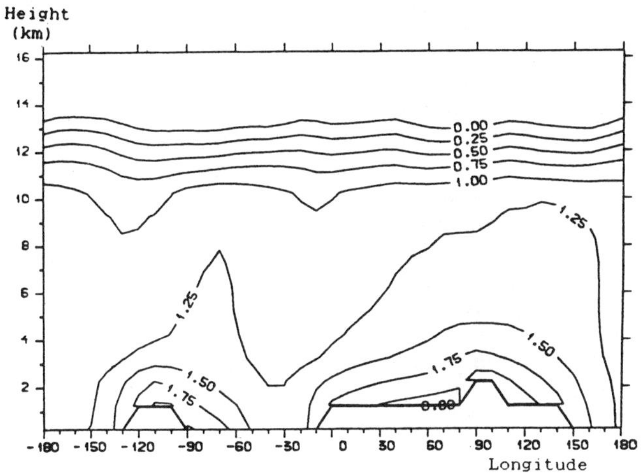

Fig. 4. Calculated mixing ratio of radon. Units : log(pCu/m^3).

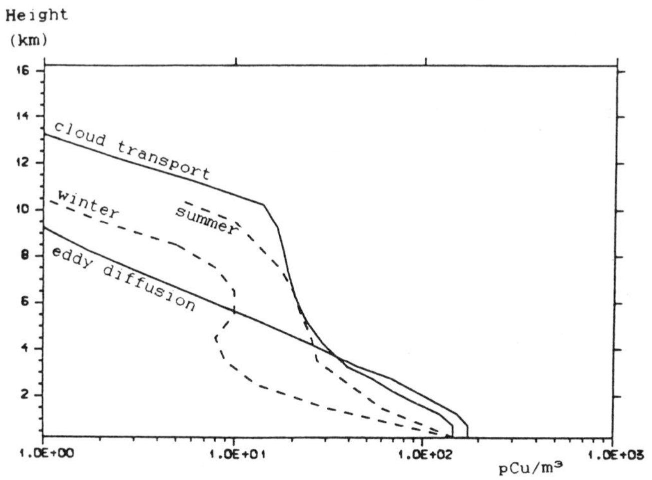

Fig. 5. Average of calculated vertical distribution of radon over land, with and without convection (solid lines).
Observations for summer and winter are dashed. Unit : pCu/m^3.

out of the boundary layer. However, there are indications that our adopted cloud top height at 10 km probably is too high. Compared to the ordinary eddy diffusion model, however, we believe that the cloud model is an improvement, especially for spring and summer conditions.

REFERENCES

Chatfield,R.B., and Crutzen,P. , Sulfur dioxide in remote oceanic air: Cloud transport of reactive precursors, J. Geophys. Res., 89, 7111-7132 (1984).

Chatfield,R.B., and Haagenson,P.L. , Sampling mid-tropospheric air at aeroce surface sites. NATO Workshop on tropospheric ozone: Regional and global ozone and its enviromental consequenses, D. Reidel (in press, 1987).

Deutsche Wetterdienstes , Monatskarten des Niederschlags fur die ganze Erde. Berichte nr. 139, band 18, Offenbach (1976).

Fiedler,B.H., and Moeng,C.H. , A practical integral closure model for mean vertical transport of a scalar in a convective boundary layer, J. of Atm. Sci., 42, 359-363 (1985).

Hov,Ø. , A one-dimensional vertical model for ozone and other gases in the atmospheric boundary layer, Atm. Envir. vol 17, 535-562 (1983).

Isaksen,I.S.A., and Rodhe,H. , A twodimensional model for the global distribution of gases and aerosol particles in the troposphere. Report AC-47. Dep. of Meteorology, University of Stockholm (1978).

Liu,S.C., McAfee,J.R., and Cicerone,R.J. , Radon 222 and tropospheric vertical transport, J. of Geophys. Res. 89, 7291-7297 (1984).

Luther,F.M. , The natural stratosphere of 1974, CIAP Monograph 1, 6.31 -6.39 (1975).

Machta,L. , Global scale atmospheric mixing, Adv. Geophys., 18B (1974).

NCAR Technical notes , Global distribution of total cloud cover and cloud type amounts over land, (1986).

Oort,A.H. , Global atmospheric circulation statistics, 1958-1973, NOAA professional paper 14 (1983).

Plumb,R.A., and Mahlman,J.D. , The zonally averaged transport characteristics of the GFDL general circulation transport model, J.of Atm. Sci., 44, 298-326 (1987).

Smolarkiewicz,P.K. , A simple positive definite advection scheme with small implicit diffusion, Am. Met. Soc. 111, 479-487 (1983).

Solberg,S. , Hovedfagsoppgave i meteorologi, institutt for geofysikk, Universitetet i Oslo (1988).

Turekian,K.K., Nozaki,Y., and Benninger,L.K. , Geochemistry of atmospheric radon and radon products, Annu. Rev. Earth Planet Sci., 5, 227 (1977).

THE HISTORICAL TREND OF TROPOSPHERIC OZONE OVER WESTERN EUROPE: A MODEL PERSPECTIVE

Stuart McKeen, Dieter Kley, and Andreas Volz
Institut für Chemie 2: Chemie der Belasteten Atmosphäre
Kernforschungsanlage Jülich
D–5170 Jülich, FRG

Abstract

Annual mean ozone mixing ratios in the late nineteenth and early twentieth centuries measured near Paris, France were between 5 and 15 ppbv (Volz and Kley, 1988), while present day average mixing ratios are typically greater than 25 ppbv for remote locations in western Europe. The increase of NO_x and non-methane hydrocarbon (NMHC) emissions from anthropogenic activity over this time period is known to be a determining factor in regard to this increase.

In this study, a one-dimensional model of the troposphere is used to construct a time history of surface ozone for conditions applicable to summertime in western Europe. The model incorporates a high resolution, diurnally varying planetary boundary layer, stratospheric sources of ozone and NO_y, and surface emissions from natural and anthropogenic sources of NO_x and NMHC. Results of the model are reasonably consistent with recently observed increases in surface ozone at two European stations. The low concentrations observed in the beginning of this century are reproduced accurately only if biogenic emissions of NMHC are included, and if currently recommended mechanisms of the NMHC oxidation process are invoked. Surface fluxes and concentrations of NO_x that, if exceeded, lead to net tropospheric ozone production are also determined, and are also found to be sensitive to natural NMHC emissions and assumed oxidation pathways. Formation of ozone and other photochemical oxidants from estimates of present day NO_x and NMHC emissions are also discussed.

Introduction

A reexamination of ozone determined by the arsenometric method used at Montsouris France from 1876 until 1910 (Volz and Kley, 1988) shows a summer average ozone mixing ratio of about 10 ppbv, which is nearly a factor of 3 below present day background values in Europe. As pointed out by Crutzen (1988), such persisently low average O_3 values are currently observed over the equatorial oceans and at maritime locations of the southern hemisphere mid-latitudes. Smit et al. (this volume) give a photochemical explanation for the anticorrelation between low O_3 ($\sim$10 ppbv) and water vapor measurements over the equatorial Atlantic. Although an extensive number of Schönbein measurements were made between 1855 and 1930 (Bojkov, 1986), there are quantitative uncertainties associated with the technique (Kley et al., 1988). The most reliable O_3 data set prior to 1910 is undoubtedly the Montsouris series. Assuming the pre-1900 air chemistry at Montsouris was not unique for the surrounding region, an explanation for such low summertime O_3 mixing ratios is required before the time history of the European O_3 increase can be addressed.

From continuous observations of tropospheric ozone over the last 18 to 30 years, an ozone increase of roughly 1% per year has been deduced for western Europe (Logan, 1985; Tiao et al., 1986), with the largest increases in O_3 values occuring during the summer months. Several studies (Crutzen and Gidel, 1983; Dignon and Hameed, 1985; Isaksen and Hov, 1987) have illustrated the importance of anthropogenic NO_x compared to other perturbations in terms of tropospheric O_3 increases over a global scale. The importance of NO_x to regional O_3 concentrations is both theoretically expected, and quantitatively observed (Liu et al., 1987, and references therein).

Two versions of a one-dimensional planetary boundary layer (PBL) model described by Trainer et al. (1987a) are used to investigate the O_3 dependence on NO_x and NMHC. The first version is used to assess the influence of natural NMHC emissions on ozone concentrations before the widespread introduction of anthropogenic NO_x and NMHC emissions. The second version considers a particular 1000 × 1000 km^2 area of western Europe with significantly higher NO_x fluxes.

Formulation of the Model

The model described by Trainer et al. (1987a) has been extended to an assumed tropopause height of 12.5 km. The model uses an unequally spaced vertical grid with 2 meter resolution at the surface, increasing to about 800 meters at the top of the domain. Below 6 km, the temperature and humidity are taken from the original model. Above 6 km, U.S. Standard Atmosphere (1976) values of temperature and water vapor are used. Model calculations are carried out until near steady-state conditions, which typically requires 20 days of simulation. The reaction scheme used is extended to include the heterogeneous losses of the soluble species HNO_3, H_2O_2, CH_3OOH, and CH_2O according to Logan et al. (1981). All reactions, photolysis cross-sections and quantum yields have been updated to JPL (1985) values.

The reaction set incorporates the CH_4-CO-HO_x-NO_x-O_x chemistry described by Liu et al. (1980). The chemistry of C_2 to C_4 alkanes and the alkenes C_2H_4 and C_3H_6 are modeled according to Atkinson et al. (1982). Natural NMHC emissions in western Europe are expected to be dominated by monoterpenes

ISBN 0-937194-15-8

and isoprene (Jüttner, 1988; Hov et al., 1983). However, the chemistry of monoterpene oxidation is not understood nearly as well as that of isoprene (Lloyd et al., 1983). For this reason, isoprene is used as a surrogate for natural NMHC, with mechanisms and rates according to Lloyd et al. (1983). HO_2 reactions with RO_2 are also considered, with an RO_2 independent rate constant of 3×10^{-12} cm^3-s^{-1} according to Atkinson and Lloyd (1984).

Vertical transport within the model is parameterized in terms of a vertical diffusion coefficient (K_z). Within the PBL, the adopted K_z parameterization is that of the original model. Above the PBL, the K_z determined from summertime Radon 222 observations (Liu et al., 1984a) is used. In both model versions, the stratospheric flux of O_3 is taken to be the globally averaged value of 5×10^{10} molec.-cm^{-2}-s^{-1} (Mahlman et al.,1980). A downward NO_y flux of 3×10^8 (Kley et al., 1981) is also assumed, and split evenly between NO_x and HNO_3. In all simulations, the initial distribution of O_3 is taken to be 3 ppbv in the lowest 3 km, linearly increasing to 50 ppbv at 10 km.

Pre-Anthropogenic Ozone

The photochemistry of the CH_4-NO_x-HO_x-O_x system has been extensively studied (e.g. Fishman et al., 1979; Logan et al., 1981; Crutzen and Gidel, 1983). In summary, at low NO_x mixing ratios (less than 12 to 35 pptv), photochemical ozone production by the conversion of NO to NO_2 via reactions with peroxy and methylperoxy radicals (folowed by NO_2 photolysis) is slower than the ozone loss by the reactions of O_3 with HO_2 and OH and the reaction of O^1D with water. For NO_x concentrations higher than about 35 pptv, ozone production (and concentrations) increase linearly with NO_x, until NO_x reaches about .5 ppbv, while the loss rate for odd oxygen remains nearly constant.

The influence on O_3 by naturally emitted NMHC for conditions of low NO_x have been mentioned in previous studies (Lurmann et al., 1983; Trainer et al. 1987b; Jacobs and Wofsy, 1988). For isoprene, the Lloyd et al. (1983) oxidation mechanism yields a direct O_x loss through reaction with O_3:

$$C_5H_8 + O_3 \rightarrow .05HO_2 + .2CO + CARB + CRIE \quad [1]$$

where CARB denotes carbonyls and CRIE denotes Criegee biradicals. On the other hand, O_x (as NO_2) is produced during the oxidation of isoprene by OH via peroxy radical formation:

$$C_5H_8 + OH \xrightarrow{O_2} RO_2 \quad [2]$$

and the subsequent NO reaction:

$$RO_2 + NO \xrightarrow{O_2} (1-\alpha)[NO_2 + HO_2 + \mathrm{CARB}] + \alpha RONO_2 \quad [3]$$

where $\alpha RONO_2$ represents the fraction of organic nitrate formation. The fractional yield α is usually assumed to be .1 (Lurmann et al., 1986). However, the kinetics of organic nitrate formation, especially for ambient background NO_x concentrations, is still uncertain. Also, organic nitrates can be expected to undergo photolysis and reactions with OH (Liu et al., 1984b). The fraction of NO_x returned would then reduce the net NO_x loss. For the purpose of this discussion, we therefore treat α as a free parameter.

To investigate pre-industrial O_3 we have incorporated the following scenario. CO and CH_4 are fixed at 100 and 1000 ppbv respectively (Rasmussen and Khalil, 1984). Stratospheric fluxes of O_3 and NO_y are prescribed as stated above. The height profiles for all species are calculated for surface NO_x emissions ranging from 1×10^8 to 3.2×10^{10} molec.-cm^{-2}-s^{-1}. These emission rates span the NO_x emissions for uncultivated grasslands according to Williams et al. (1987) (1×10^9 and 1×10^{10} at about 10°C and 20°C, respectively). Isoprene emission rates are according to Lamb et al. (1985) for a deciduous forest, with emissions occuring from 8:30 am until 5:30 pm local time, yielding a diurnally averaged emission rate of 1.2×10^{11} molec.-cm^{-2}-s^{-1}. For an NO_x flux of 1×10^{10} in the model, this isoprene flux yields a maximum isoprene mixing ratio of 2 ppbv at 5 m above ground, consistent with summertime measurements in the U.S. and Japan (Lamb et al., 1985; Yokouchi and Ambe, 1988).

The diurnally averaged O_3 and NO_x height profiles for an NO_x flux of 1×10^{10} are shown in figure 1, along with the profile of an inert tracer having the same stratospheric flux and deposition velocity as ozone. The addition of NMHC causes a reduction of NO_x and O_3 in the upper troposphere due to the formation of PAN (and PAN homologues of methacrolein and methyl vinyl ketone) in the isoprene oxidation mechanism. As the conversion of NO_x to organic nitrate is allowed to increase, significant changes in the vertical profiles of O_3 and NO_x are observed. The depletion of NO_x occurs mainly between 1 and 6 km as a consequence of the $RONO_2$ formation channel. It affects the entire column of O_3 due to the coupling of vertical transport and chemistry. For this particular NO_x flux, the photochemical O_3 production and loss integrated over the column are nearly equal for the case where $\alpha=.1$.

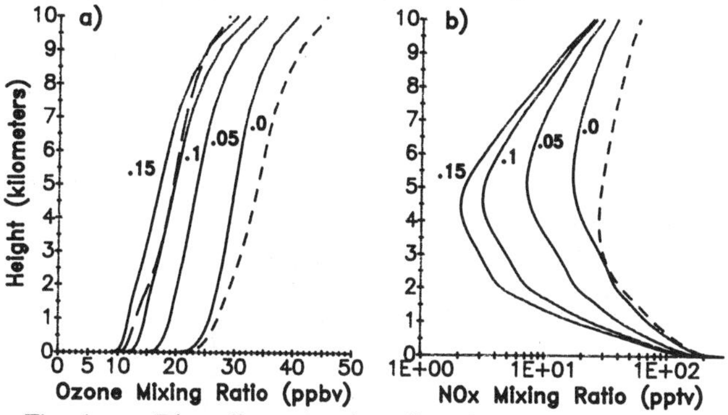

Fig. 1. : Diurnally averaged profiles of O_3 (a) and NO_x (b) for different values of α and NO_x flux = 1×10^{10}. The short dashed line is the case with no natural NMHC emissions. The long dashed line in figure 1a is for an inert tracer with the same stratospheric flux and deposition velocity as O_3

Figure 2 shows the O_3 and NO_x mixing ratios at 5 m as a function of the NO_x flux. The horizontal dashed line is the division between net O_3 production or loss at 5 m. With no NMHC emissions, O_3 is lower than the inert species only for very small NO_x fluxes, with a corresponding NO_x mixing ratio of 19 pptv. This critical value dividing O_3 loss and production is in agreement

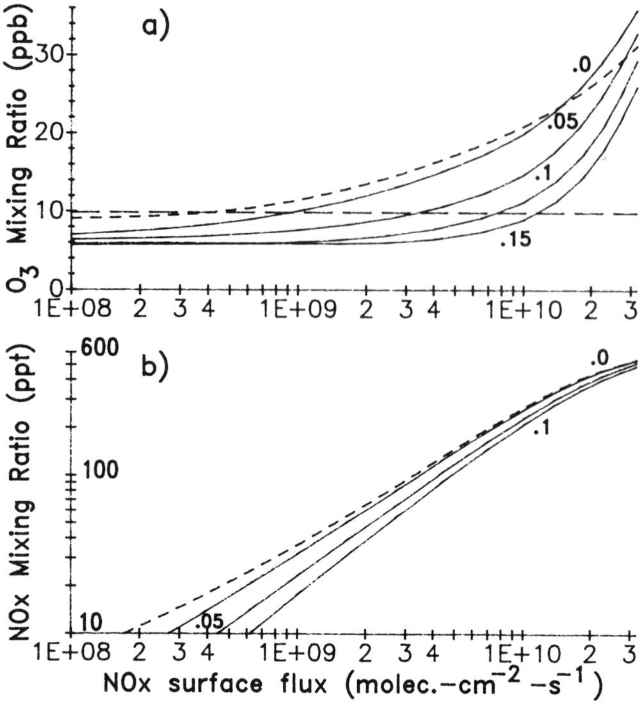

Fig. 2. : Diurnally averaged O_3 (a) and NO_x (b) at 5m above ground as a function of the NO_x surface emission and for different values of α. The dashed lines are the same as in figure 1.

with the 24 pptv reported by Fishman et al. (1979). However, this is certainly fortuitous since their calculations were for the global troposphere and used fixed O_3 mixing ratios. The inclusion of NMHC has a twofold effect on the predicted O_3. Less NO_x is available for a given NO_x flux due to reaction 3 and formation of PAN. From figure 1b, it is obvious that this NO_x loss is not so apparent at the surface, but that it significantly affects NO_x (and hence the O_3 formation) above 500 m. At low NO_x concentrations, the RO_2 created from reaction 2 will not efficiently regenerate HO_x from reaction 3, but will form organic peroxides from reaction with HO_2, providing an additional HO_x sink. When HO_x is reduced, isoprene increases, and also the direct O_3 loss from reaction 1. The decreased O_3 production in the middle troposphere and the increase in O_3 loss below 2 km combine to increase the surface NO_x concentration for which net O_3 production begins to occur (~150 pptv for $\alpha = .1$).

Although several assumptions and approximations are used in this study, our results give a plausible explanation for the low summertime O_3 concentrations ($\leq$10 ppbv) observed at Montsouris 80 to 110 years ago. Using isoprene as a surrogate for NMHC emissions, figures 1 and 2 imply that natural NMHC have a significant effect on regional photochemistry under conditions that prevailed prior to widespread anthropogenic intervention. The 8 to 10 ppbv reported as the average summertime O_3 level at Montsouris (Volz and Kley, 1988) was actually derived using data when the wind was blowing from the southwest sector. Air masses from this sector would have to travel about 300 km across the continent before reaching the site. It is therefore expected that NMHC emissions from natural vegetation had sufficient time to influence these air masses. Low maritime O_3 concentrations could also have contributed to the low O_3 values observed at Montsouris, but estimates of pre-industrial maritime O_3 levels are not available.

One assumption in our study is the assignment of isoprene as the only biogenic NMHC. Coniferous forests typically emit less isoprene than is assumed here, but measurements of monoterpenes in European forests (Hov et al., 1983) are up to a factor of 10 higher (by carbon number) than those predicted by the model, depending on forest type (Jüttner, 1988). The oxidation pathway for the dominant monoterpene, α-pinene, is not known (Lloyd et al., 1983), and until such information is available (in particular the organic nitrate formation channel), it is impossible to assess the importance of α-pinene to O_3 loss under conditions of low NO_x. However, a measured shortfall between total NO_y and the sum of nitrogen containing species in a coniferous U.S. forest has lead Fahey et al. (1987) to conclude organic nitrates could be a significant fraction of NO_y.

Several assumptions within the reaction scheme have been discussed by Trainer et al. (1987a). These include the treatment of organic peroxides and uncertainties with the photolysis rates of carbonyls formed by reactions 1 and 3. We have also neglected the effects of cloudiness, which is known to affect isoprene emissions, but could also affect the below cloud photochemical balance in favor of reduced O_3 (Volz et al., this volume). We also examined the influence of the adopted diffusion coefficients by varying the coefficients in the PBL and above the PBL separately by a factor of 2. A 50% decrease in the above PBL coefficients yields the greatest influence. In this case, the O_3 and inert tracer curves of figure 2a are reduced by 2.1 ppbv, while the NO_x curves of figure 2b are increased by ~13%. Thus, our conclusions are not significantly affected by a factor of 2 shift in the diffusion coefficient profile.

Anthropogenic Ozone

Current annually averaged NO_x emission rates for western Europe average about 2×10^{11} molec-cm^{-2}-s^{-1} (OECD, 1987a). Such large fluxes yield unrealistically high NO_x and O_3 concentrations in the previously described one-dimensional model due to the assumption of horizontal homogeniety. In order to apply the 1-D to a specific area with higher emission rates than the global average, advection of background air into the column, and of polluted air out of the column, must be considered. We assume that regionally averaged conditions are only influenced by vertical transport processes and horizontal advection.

The influence of horizontal advection on the concentration of a particular species (q) is parameterized by:

$$\left(\frac{\partial q}{\partial t}\right)_{advection} = \frac{-\,u(z)(q(z) - q_o(z))}{1 \times 10^6 meter},$$

where z is height, u(z) is the 8 year climatic mean summertime zonal wind speeds for 50°N of Danielsen

(1983) and $q_o(z)$ is the upwind vertical profile of the species considered. The zonal winds vary from 15 to 2.5 $m\text{-}s^{-1}$ at 11 and 1.5 km respectively, and are fixed at 2.5 $m\text{-}s^{-1}$ below 1.5 km. The upwind concentrations are obtained separately with the 1-D model adjusted for oceanic conditions. The diffusion coefficients are also adjusted so the mid-tropospheric O_3 profile matches that found by Smit et al. (this volume) in the southern hemisphere at mid-latitudes. The O_3 and NO_x profiles so derived are meant to represent an extremely clean air mass upwind of the perturbed region. Anthropogenic emissions of the C_2 - C_4 hydrocarbons within the polluted column are scaled to the NO_x emissions according to Hov et al. (1983). CO and CH_4 are fixed at 150 ppbv and 1.6 ppmv, respectively.

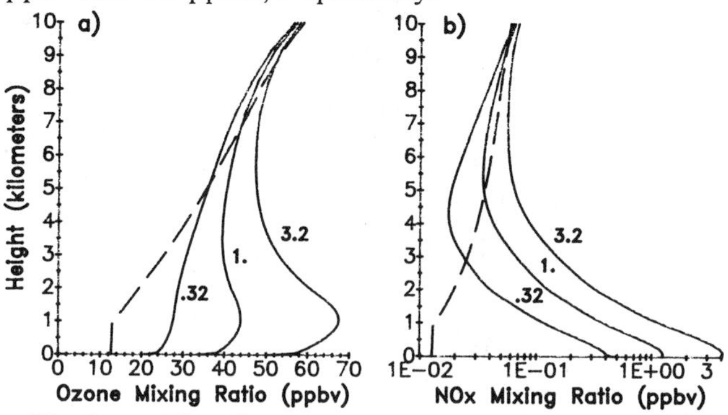

Fig. 3. : Diurnally averaged O_3 (a) and NO_x (b) for different values of the NO_x surface flux ($\times 10^{11}$ molec.-cm^{-2}-s^{-1}). The dashed line is the upwind concentrations used in the horizontal advection parameterization.

Figure 3 shows the O_3 and NO_x predicted for three NO_x emission cases as well as the upwind values used in the advection parameterization. Above 5 km, the respective concentrations are largely controlled by horizontal advection. The O_3 vertical gradient below 2 km is a consequence of diurnal averaging and the effects of surface deposition occuring throughout the night. Mid-day O_3 exhibits a constant vertical gradient from ground to 2 km. Essentially no diurnal variation is observed above 2 km.

To relate the calculated O_3 mixing ratios to NO_x emissions for western Europe, we choose a 1000×1000 km^2 region with corner points at Birmingham England, Bordeaux France, the northwestern tip of the Adriatic Sea, and the southern most tip of the Baltic Sea. This region approximately encompasses a fractional area of the following countries: Great Britian(.3), France(.6), Germany(1.), Austria(.25), Belgium(1.), Holland(1.), Luxembourg(1.), and Switzerland(1.). Using the OECD (1987a) NO_x emissions for each of these countries weighted by the fractional area, the temporal variation of NO_x from 1970 to 1985 can be derived as shown in figure 4. Values for years prior to 1970 are scaled to the U.S. NO_x emission survey of Gschwandtner et al. (1986). These values surely become more uncertain as one goes back in time, and are only meant to serve as a reference for discussion. Emission estimates after 1985 are from projected energy demands (OECD, 1987b) with the 1985 NO_x emission to energy consumption ratio used as a scaling factor for each country.

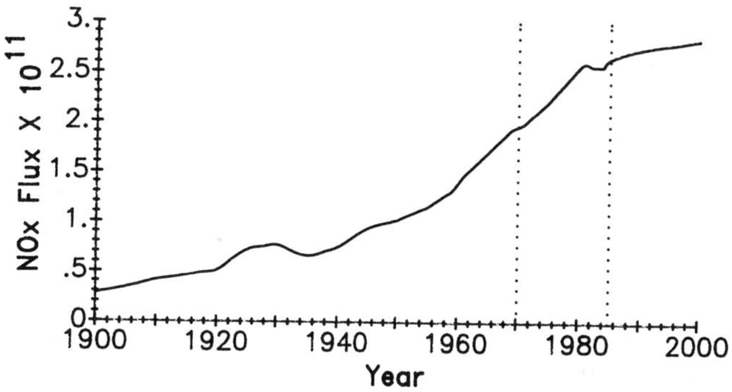

Fig. 4. : Estimated temporal variation of the NO_x flux (in molec.-cm^{-2}-s^{-1}) for the 1000×1000 km^2 region. See text for details.

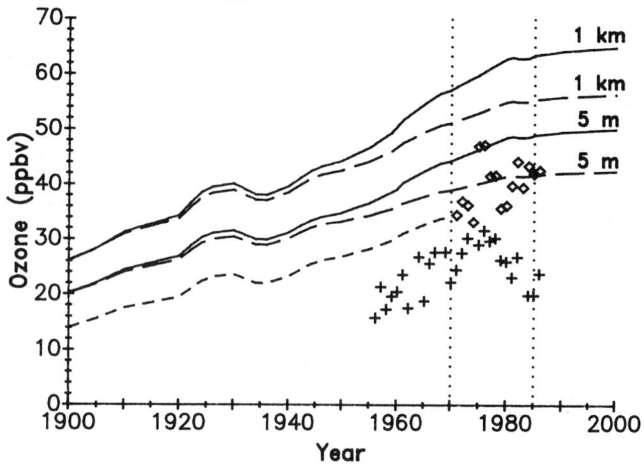

Fig. 5. : Model estimates and observations of surface O_3 during summertime. Diamonds-Hohenpeißenberg, plus signs-Arkona, solid lines- anthropogenic plus natural NMHC emissions included, long dashed lines-natural NMHC emissions only, short dashed line- the 5 m case with anthropogenic NMHC included but NO_x flux reduced by 50%.

Figure 5 shows the results of combining the temporal NO_x emission increases of figure 4 with the model calculations depicted in figure 3. Also shown are summer-time average O_3 values for Hohenpeißenberg (e.g. Attmannspacher et al., 1984), and Arkona (Feister and Warmbt, 1987). A comparison between our model results and observations at a particular location is not very meaningful. However, assuming the temporal increase at each station is due to a regional influence, a comparison between absolute O_3 increases (in units of ppbv per year) provides a check on the consistency of the model and input data with the observations.

When linear regressions are performed on the Hohenpeißenberg (975m ASL) and Arkona (42m ASL) data, summertime O_3 increases of .40 and .22 ppbv/year are derived, respectively. The same regression for the model results from 1956 to 1986 yields .43 and .34 ppbv/year at the 1 km and 5 m levels, respectively. The agreement is certainly fortuitous, but illustrates that the NO_x emission increase in combination with the NMHC scaling of Hov et al. (1983) is consistent with O_3 increases at Hohenpeißenberg. The decrease in summertime O_3 at Arkona between 1978 and 1986 suggests that O_3 at this station is not entirely controlled by precursor emissions over a regional scale. Strong correlations between hourly sunshine amount and O_3 suggest other influences related to synoptic weather patterns, such as cloudiness and humidity, that affect O_3 at a particular location (see e.g. Volz et al., this volume).

Figure 5 also illustrates the effect of anthropogenic NMHC upon the predicted O_3 increase. The rate of increase drops by 35% when only natural NMHC emissions are included in the model. This is a consequence of the non-linear behavior of O_3 formation under conditions of HO_x and RO_2 limited oxidation as discussed by Trainer et al. (1987a). Also from figure 5, even reducing the anthropogenic NO_x flux by 50% does not reduce the predicted O_3 for 1900 below 14 ppbv. The model results in figure 5 are also larger than the reported observations, which undoubtedly results from the simplifying assumption of a uniform anthropogenic NO_x flux over the entire 1000×1000 km^2 area.

Summary

Estimates of natural NO_x and NMHC emissions and currently recommended mechanisms for isoprene oxidation suggest that natural NMHC emissions could have played an important part in regulating summertime O_3 before anthropogenically derived NO_x began to influence the regional chemistry. Provided the organic nitrate formation channel during NMHC oxidation is significant, and organic nitrates do not rapidly reconvert to NO_x, biogenic NMHC can influence the entire tropospheric O_3 column over time scales of a few weeks. For NO_x fluxes exceeding that of expected natural emissions, the presence of biogenic NMHC strongly affects the O_3 dependence on the NO_x emissions.

Estimates of regional O_3 levels in Europe prior to 1970 are uncertain due to a lack of knowledge concerning NO_x emissions. However, using NO_x emission estimates from 1970 to 1985, along with currently accepted NMHC scaling, good agreement is found with the long term increase of O_3 observed during summer months at Hohenpeißenberg.

Acknowledgments

The authors wish to thank Michael Trainer and Shaw Liu of the NOAA Aeronomy Lab, Boulder, Colorado for useful discussions and ideas that contributed to this paper.

References

Atkinson, R., et al. 1982: *Atmos. Environ.*,16,1341-1355.
Atkinson, R., and A. C, Lloyd, 1984: *J. Phys. Chem. Ref. Data*,13,315-444.
Attmannspacher, W., et. al., 1984: *Meteorol. Rdsch.*, 37,193-199.
Bojkov, R. D., 1986: *J. Clim. Appl. Met.*, 25, 333-352.
Crutzen, P. J., and L. T. Gidel, 1983: *J. Geophys. Res.*, 88, 6641-6661.
Crutzen, P. J., 1988: in: Isaksen, I.S.A. (ed.) *Tropospheric Ozone: Regional and Global Scale Interactions, NATO ASI Series* C 227, 3-32, (Reidel, Dordrecht).
Danielsen, E. F., 1983: *An Atlas of Objectively Analyzed Atmospheric Cross Sections: 1973-1980*, NASA/ARC Report, Contract No. NCC2-73.
Dignon, J., and S. Hammed, 1985: *J. Atmos. Chem.*, 3,491-506.
Fahey, D. W., et al., 1986: *J. Geophys. Res.*, 91,9781-9793.
Feister, U., and W. Warmbt, 1987: *J. Atmos. Chem.*, 5,1-20.
Fishman, J., et al., 1979: *Tellus*,31,432-446.
Gschwandtner, G., et al., 1986: *J. Air. Pollut. Control Assoc.*,36,139-149.
Hov, Ø., et al., 1983: *J. Geophys. Res.*,88,10679-10688.
Jacobs, D. J., and S. C. Wofsy, 1988: *J. Geophys. Res.*, 93, 1477-1486.
JPL, 1985: *Chemical Kinetics and Photochemical Data for Use in Stratospheric Modeling*, Eval. No. 7, JPL Publ. 85-37, JPL/NASA, CIT, Pasadena, CA.
Jüttner, F., 1988: *Chemosphere*,17,309-317.
Isaksen, I. S. A., and Ø. Hov, 1987: *Tellus*,39B,271-285.
Kley, D., et al., 1981: *J. Geophys. Res.*,86,3153-3161.
Kley, D., et al., 1988: in: Isaksen, I.S.A. (ed.) *Tropospheric Ozone: Regional and Global Scale Interactions, NATO ASI Series* C 227, 63-72, (Reidel, Dordrecht).
Lamb, B., et al. 1985: *J. Geophys. Res.*,90,2380-2390.
Liu, S. C., et al., 1980: *J. Geophys. Res.*,85,7546-7552.
Liu, S. C., et al., 1984a: *J. Geophys. Res.*,89,7291-7297.
Liu, S. C., et al., 1984b: *Eos Trans. AGU*, 65, 833.
Liu, S. C., et al., 1987: *J. Geophys. Res.*, 92, 4191-4207.
Lloyd, A. C., et al., 1983: *Atmos. Environ.*, 17, 1931-1950.
Logan, J. A., et al., 1981: *J. Geophys. Res.*, 85, 7546-7552.
Logan, J. A., 1985: Tropospheric ozone: *J. Geophys. Res.*, 90, 10463-10482.
Lurmann, F. W., et al., 1983: , *Atmos. Environ.*, 17, 1951-1963.
Lurmann, F. W., et al., 1986: *J. Geophys. Res.*, 91, 10905-10936.
Mahlman, J. D., et al., 1980: *J. Atmos. Sci.*, 37, 655-685.
OECD, 1987a: *OECD Environmental Data Compendium, 1987*, OECD, Paris, France.
OECD, 1987b: *Coal Information 1987*, IEA-OECD publications, Paris Cedex 16, France.
Rasmussen, R. A. and M. A. K. Khalil, 1984: *J. Geophys. Res.*, 89, 11599-11605.
Tiao, G. C., et al., 1986: *J. Geophys. Res.*, 91, 13121-13136.
Trainer, M., et al., 1987a: *J. Geophys. Res.*, 92, 11879-11893.
Trainer, M., et al., 1987b: *Nature*,329, 705-707.
U.S. Standard Atmosphere, 1976: U.S. Government Printing Office, Washington D. C..
Volz, A. and D. Kley, 1988: *Nature*, 332, 240-242.
Williams, E. J., et al., 1987: *J. Geophys. Res.*, 92, 2173-2179.
Yokouchi, Y., and Y. Ambe, 1988: *J. Geophys. Res.*, 93, 3751-3759.

CHANGES IN TROPOSPHERIC OZONE: A SIMPLE MODEL EXPERIMENT

Øystein Hov
Norwegian Institute for Air Research
N–2001 Lillestrøm, Norway

ABSTRACT

The number of ozone molecules generated per NOx molecule emitted is calculated using a box model of the troposphere and integrating to steady state. From about 1 to more than 50 O_3 molecules are formed per NOx molecule with urban emissions and global average emissions of NOx and volatile organic compounds (VOC), respectively. The effect of changing the VOC emission is very slight at low NOx emissions.

1. INTRODUCTION

The episodic increase in the concentration of ozone in the atmospheric boundary layer in the summer time over Europe, North America and other industrialized parts of the world, is well documented (e.g. Grennfelt et al., 1987). Convincing evidence of a growing concentration of ozone in the remote northern hemisphere troposphere has also been published, based in particular on European measurements (Angell and Korshover, 1983; Attmannspacher et al., 1984; Feister and Warmbt, 1987; Volz and Kley, 1988). The free tropospheric concentration in the northern hemisphere may have doubled in 100 years, and there has been a 1-3%/a increase in ozone over the last 30 years with a particularly steep rise during the summer months.

Anthropogenic emissions of nitrogen oxides (NOx) and volatile organic compounds (VOC) are important both for the episodic occurrence of elevated ozone concentrations during the summer in the atmospheric boundary layer over industrialized continental areas, and for the long term increase in tropospheric ozone in the northern hemisphere (Hov et al., 1978; Isaksen and Hov, 1987; Dignon and Hameed, 1985).

Modelling studies of the episodic occurrence of elevated ozone concentrations in the atmospheric boundary layer in general show that VOC emission reduction is more efficient in the abatement of ozone than NOx emission reduction. For European conditions it has been shown that in some situations the emissions of NOx are so high that they may suppress the ozone formation (Eliassen et al., 1982; Hov, 1988; Builtjes and Stern, 1988).

In the remote troposphere where NOx is very low, model calculations and the simultaneous measurement of ozone and precursors at remote locations, show that the production of ozone is limited by NOx (Liu et al., 1987; Parrish et al., 1986). Dignon and Hameed (1985) found using a one-dimensional latitudinal model of the troposphere that a significant portion of the northern hemisphere ozone increase can be attributed to the increase in global anthropogenic NOx emissions during the last decades, while the contribution of methane (CH_4) to the increase is quite small. Isaksen and Hov (1987), applying a global meridional 2-d model of the troposphere, arrived at similar conclusions.

In this paper the relationship between emissions of VOC and NOx, and the efficiency of formation of tropospheric ozone will be investigated in a simple model calculation where atmospheric transport processes are neglected. A numerical integration of the photochemical formation and destruction of tropospheric ozone is carried out until steady state is reached where the net photochemical formation of ozone due to emissions of NOx and VOC, is balanced by dry removal at the surface.

2. MODEL DESCRIPTION

The troposphere was approximated by a single box of height 10 km and extending infinitely in the horizontal. At the surface there were constant emissions of VOC and NOx, and dry removal of ozone, NO_2, HNO_3 and PAN (0.6, 0.1, 1.0 and 0.2 cm/s, respectively. The value 0.1 cm/s for NO_2 is smaller than the majority of measured values which are closer to 0.5 cm/s, but this has no consequences for the results reported here). The emissions of VOC were approximated by a mixture of ethene, propene, ethane, n-butane, m-xylene, formaldehyde and carbon monoxide, while the concentration of methane was fixed (1.6 ppm). The composition of the VOC emissions and the formulation of the photochemistry of ozone formation were similar to what is used in the extended EMEP model with boundary layer chemistry (Eliassen et al., 1982; Hov, 1988).

The continuity equation for each of about

TABLE 1. EMISSIONS OF NOx.

	kt N/a	Molecules/ cm^2 x s
United Kingdom (1979)	500	$2.8x10^{11}$ (1.00)
United Kingdom (1985)	550	$3.1x10^{11}$ (1.10)
Europe (EMEP) (1985)	6 340	$5.5x10^{10}$ (0.20)
Global average (1980)	30 000	$8x10^{9}$ (0.03)
Northern hemisphere average (1980)	25 000	$2.3x10^{10}$ (0.08)
Ruhr (1985)	200*	$1.2x10^{12}$ (4.32)

* Maximum emission averaged over an EMEP grid square (150 x 150 km^2).

ISBN 0–937194–15–8

40 components in the model was integrated for 80 d leaving sufficient time for ozone to reach steady state. The emissions of NOx and VOC were varied over a wide range, from less than global average fluxes to fluxes which may be found over heavily industrialized areas in Europe, see Table 1.

The emissions of VOC were set equal to the emission of NOx measured as NO_2 on a compound basis. This is a reasonable approximation for the EMEP region where it is estimated that in 1985, the emissions of NOx were about 21 Mt (NO_2) and VOC 17 Mt.

3. RESULTS

The number of ozone molecules generated per NOx molecule emitted (O_3/NOx) was calculated from the expression

$$(*) \quad O_3/NOx = \left[O_{3,t=80d} - O_{3,t=0} + \int_0^{80d} v_g\, O_3 dt\right] \Big/ \int_0^{80d} \frac{E_{NOx}}{H}\, dt$$

In Figure 1 is shown the results for NOx emissions ranging from 0.01 to 10 times the UK average for mid summer conditions. Also shown are the results when the VOC/NOx mass ratio in the emissions is changed from 1 to 2 or 0.5 (vertical bars). The number of O_3 molecules generated per NOx molecule emitted ranges from 1.2 to more than 50, and the effect of changing the VOC/NOx ratio is very slight for NOx emissions comparable to or smaller than the UK average.

At 10% NOx emission of the UK average, increasing VOC/NOx by a factor of 4 only increases the ratio (*) by 3.4%, while an increase of NOx emissions by a factor of 4 changes the ratio (*) by a factor of 2.22 (122%). At the UK average NOx emission, increasing VOC/NOx by a factor of 4 increases the ratio (*) by 18%, while for 10 times UK NOx emissions the corresponding increase is 67%. This shows that for low NOx emissions there is very strong NOx control of the formation of ozone, while for NOx emissions comparable to or higher than the UK average, VOC is also important. It should be remembered that the numbers shown in Figure 1 are steady state numbers, which do not necessarily reflect the behaviour of the chemical system in a real situation. In particular in the atmospheric boundary layer the air masses are exposed to emissions which may vary significantly in time and space implying that steady state conditions are rarely met close to the ground.

From Figure 1 it can be seen when the ozone concentration may drop in an air mass as a result of NOx and VOC emissions and when it may increase, given steady state conditions. If for instance there is 40 ppb of ozone in the air and UK average emissions of NOx, the ground removal flux with v_g = 0.6 cm/s is $6x10^{11}$ molecules/(cm^2xs) and the net production of O_3 is UK average NOx flux ($2.8x10^{11}$ molecules/cm^2xs) times 4.5 (number of O_3 molecules produced per NOx molecule emitted), i.e. the flux of O_3 produced is $12.6x10^{11}$ molecules/cm^2xs implying that O_3 is increasing.

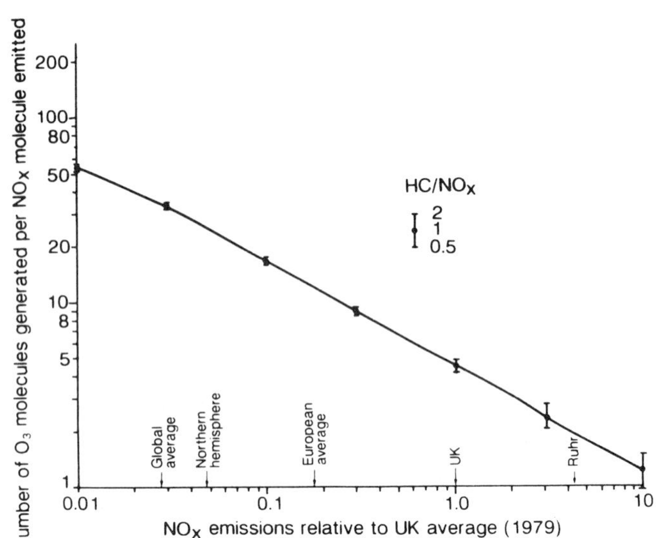

Fig. 1. Number of O_3 molecules generated per NOx molecule emitted as a function of the emission of NOx. Results indicated for VOC/NOx ratios in the emissions equal to 1, 2 and 0.5, for 50^0N latitude and mid summer.

With 90 ppb ozone in the air, the flux to the ground would be $13.5x10^{11}$ molecules/cm^2xs, implying a decrease in the O_3 concentration.

The relationship between changes in emissions of VOC and NOx, and the production of O_3 demonstrated in Figure 1, is in a good qualitative agreement with the findings in more elaborate episodic, continental and long term, global ozone calculations (Hov, 1988; Isaksen and Hov, 1987).

From Figure 1 some estimates of tropospheric ozone lifetime can be made: As a global, annual average there is approximately $1x10^{18}$ molecules/cm^2 of O_3 in the troposphere. The global NOx emissions of 30 Tg N/a give rise to approximately 30 O_3 molecules per NOx molecule, or a flux of $2.5x10^{11}$ molecules O_3/cm^2xs which corresponds to a tropospheric lifetime for ozone of 50 d. For the northern hemisphere the lifetime is calculated to be 29 d, for European emissions 19 d (mid summer, taking 50^0N latitude as representative).

On an annual basis approximately 50% of the NOx emissions in Europe (including USSR west of Ural) are deposited within Europe through dry or wet removal processes (EMEP results to be published). The rest of the emissions are transported out of Europe to the free troposphere or in the atmospheric boundary layer, mainly as NO_2 and to a lesser extent as sum nitrate or PAN. If it is assumed that one half of what leaves Europe, is deposited outside of Europe, and the rest is exchanged into the free troposphere, 1/4 of European NOx emissions enter the free troposphere in some form. If it is assumed that 1/2 of this quarter is found as NOx (NO_2), the flux of NOx to the free troposphere over Europe is 1/8 of the surface emissions, giving rise to an O_3 flux of $1.1x10^{11}$ molecules/cm^2xs (mid summer).

In Figure 2 are shown the same results as in Figure 1, but as a function of the steady state concentrations of NO_2 and NO_y (daily mean), also for mid summer. In Table 2 is shown the steady state budget for compounds derived from emission of NOx for the case with UK average NOx emission, VOC/NOx = 1 and mid summer (v_{d,NO_2} = 0.5 cm/s in this calculation). The budgets are quite similar also for other emissions.

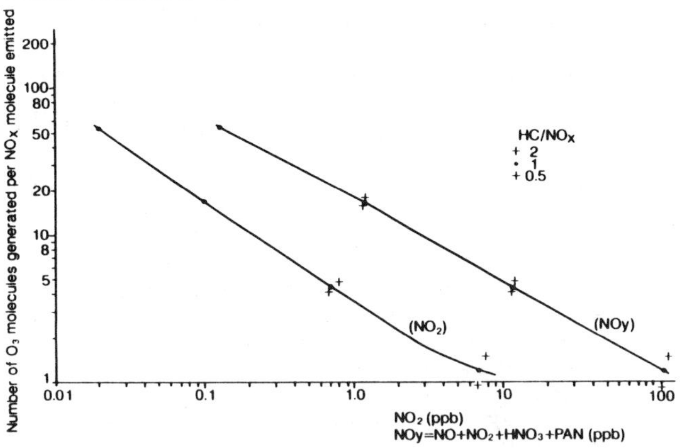

Fig. 2. Number of O_3 molecules generated per emitted NOx molecule as a function of the steady state concentration of NO_y (= NO + NO_2 + HNO_3 + PAN) and NO_2, for VOC/NOx ratios in the emissions equal to 1, 2 and 0.5, for 50^0N latitude and mid summer.

In Figure 3 is shown the number of O_3 molecules produced per NOx molecule emitted as a function of season at 50^0N. It is interesting to note that for a solar declination less than about -17.5^0 (before 1 February and after 15 November) there is very little ozone formed, while for solar declinations larger than -10^0 (after 20 February and before 20 October) the ozone formation is fairly independent of the solar declination with a slight maximum around equinox (21 March). The concentration of ozone

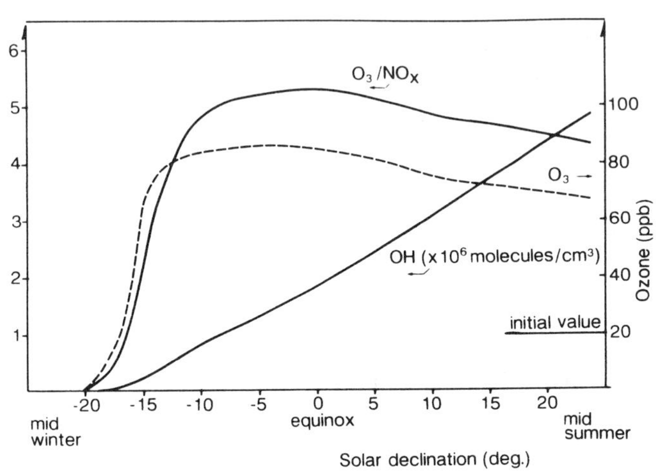

Fig. 3. Number of O_3 molecules generated per emitted NOx molecule at 50^0N latitude, as a function of time of the year (expressed by the solar declination). Also shown is the diurnal average OH concentration as a function of time of the year.

TABLE 2. STEADY STATE NOx BUDGET; NOx EMISSIONS EQUAL TO UK AVERAGE, MID SUMMER, 50^0N. TOTAL NOx EMISSION AFTER 80 D IS 75 PPB (AT GROUND AIR DENSITY).

Species	Airborne						Dry deposition		
	NO	NO_2	Sum nitrate	PAN	NO_3	N_2O_5	NO_2	Sum nitrate	PAN
Fraction of NOx emissions (%)	0.08	0.94	14.0	0.21	0.00	0.00	1.74	82.7	0.34

(in ppb of surface air density) is seen to have a spring maximum as well.

The results in a qualitative sense reproduce the observed spring maximum in tropospheric ozone. The box model gives a similar maximum at the fall equinox which is not in line with observations, indicating that transport processes, stratospheric exchange etc. of course are important for the tropospheric ozone distribution.

In Figure 3 is also shown that the diurnal average steady state OH concentration is steadily increasing towards mid summer, in response to the increase in the $O(^1D)$ formation due to O_3 photolysis. The mid summer daily average value of 4.8 x 10^6 molecules/cm^3 corresponds to a maximum hourly value of about 9 x 10^6 molecules/cm^3. Up to 8.7 x 10^6 molecules/cm^3 has been measured in Schauinsland in the Black Forest in June 1984 (Platt et al., 1988).

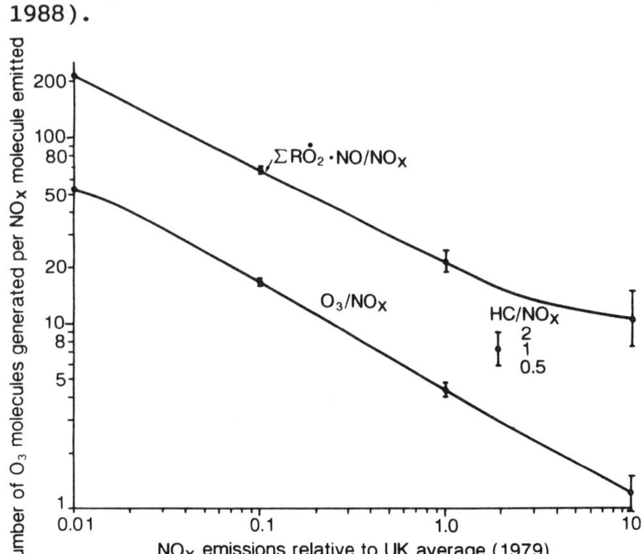

Fig. 4. Number of O_3 molecules produced per emitted NOx molecule as a function of the emission of NOx for VOC/NOx = 1, 2 and 0.5 (upper curve). Also shown is the same curve as in Figure 1 (50^0N latitude, mid summer).

In Figure 4 is shown the net production of O_3 per NOx molecule emitted as a function of the NOx emissions. It is calculated as the time integrated product of the sum of peroxy radicals and NO divided by the time integrated emissions of NOx. It can be seen that the distance from the O_3/NOx curve (cpr. Figure 1) increases with increasing NOx emissions, indicating that the net ozone loss reactions (mainly $O_3 + HO_2 \rightarrow 2O_2 + OH$, $O_3 + OH \rightarrow HO_2 + O_2$ and $O(^1D) + H_2O \rightarrow 2OH$) gain importance with higher NOx emissions, leaving less net O_3 (and consequently a lower O_3/NOx ratio).

In Figure 5 is shown the steady state sum of peroxy radicals (HO_2 + the organic peroxy radicals), the concentration of airborne plus deposited O_3 and the concentration of O_3 as a function of the emission of NOx. The sum of peroxy radicals is remarkably constant with increasing NOx emission, reflecting that both the production (OH + VOC-reactions) and loss ($R\dot{O}_2 + NO \rightarrow R\dot{O} + NO_2$) increase with increasing NOx emission. The sum is independent of a change in VOC/NOx at low emissions, but very different at high emissions, pointing to the main reason for the change in the O_3/NOx ratio with the VOC/NOx ratio in the emissions as the emissions increase.

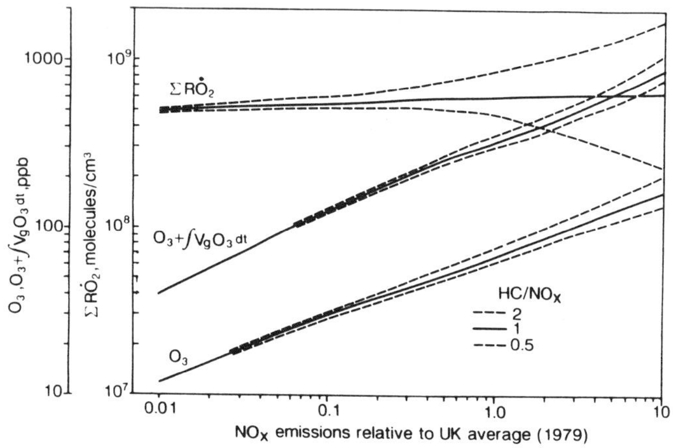

Fig. 5. The sum of HO_2 and organic peroxy radicals (denoted $\Sigma R\dot{O}_2$), total ozone generated (sum of airborne steady state concentration minus initial value plus deposited amount), and the airborne O_3 steady state concentration as a function of NOx emission, for 50°N latitude mid summer. Results are shown for VOC/NOx in the emissions equal to 1 by volume (full lines), 2 (dashed, upper curves) and 0.5 (dashed, lower curves). (Mixing ratios refer to surface air density.)

4. ACKNOWLEDGEMENT

The work reported here is supported by a research grant from the Royal Norwegian Research Council for Science and Technology.

5. REFERENCES

Angell, J.K. and Korshover, J. (1983) Global variation in total ozone and layer-mean ozone: An update through 1981. J. Clim. Appl. Meteor., 22, 1611-1627.

Attmannspacher, W., Hartmannsgruber, R. and Lang, P. (1984) Langzeittendenzen des Ozons der Atmosphäre aufgrund der 1967 begonnenen Ozonmessreihen am Meteorologischen Observatorium Hohenpeissenberg. Meteorol. Rdsch., 37, 193-199.

Builtjes, P. and Stern, R.M. (1988) Photochemical processes: Emissions and atmospheric transport. Paper presented at the European Convention on Combustion, pollution, reduction; new techniques in Europe. Hamburg 16-17 May.

Dignon, J. and Hameed, S. (1985) A model investigation of the impact of increases in anthropogenic NOx emissions between 1967 and 1980 on tropospheric ozone. J. Atmos. Chem., 4, 491-506.

Eliassen, A., Hov, Ø., Isaksen, I.S.A., Saltbones, J. and Stordal, F. (1982) A lagrangian long-range transport model with atmospheric boundary layer chemistry. J. Appl. Meteor., 21, 1645-1661.

Feister, U. and Warmbt, W. (1987) Long-term measurements of surface ozone in the German Democratic Republic. J. Atmos. Chem., 5, 1-21.

Grennfelt, P., Saltbones, J. and Schjoldager, J. (1987) Oxidant data collection in OECD-Europe 1985-87 (OXIDATE). Lillestrøm (NILU OR 22/87).

Hov, Ø., Hesstvedt, E. and Isaksen, I.S.A. (1978) Long range transport of tropospheric ozone. Nature, 273, 341-344.

Hov, Ø. (1988) Photochemical oxidant episodes, acid deposition and global atmospheric change. The relationship with emission changes of nitrogen oxides and volatile organic compounds. Lillestrøm (NILU OR 12/88).

Isaksen, I.S.A. and Hov, Ø. (1987) Calculation of trends in the tropospheric concentration of O_3, OH, CO, CH_4 and NOx. Tellus, 39B, 271-285.

Liu, S.C., Trainer, M., Fehsenfeld, F.C., Parrish, D.D., Williams, E.J., Fahey, D.W., Hübler, G. and Murphy, P.C. (1987) Ozone production in the rural troposphere and the implications for regional and global ozone distributions. J. Geophys. Res., 92, 4191-4207.

Parrish, D.D., Fahey, D.W., Williams, E.J., Liu, S.C., Trainer, M., Murphy, P.C., Albritton, D.L. and Fehsenfeld, F.C. (1986) Background ozone and anthropogenic ozone enhancement at Niwot Ridge, Colorado. J. Atmos. Chem., 4, 63-80.

Platt, U., Rateike, M., Junkermann, W., Rudolph, J. and Ehhalt, D.H. (1988) New tropospheric OH measurements. J. Geophys. Res., 93, 5159-5166.

Volz, A. and Kley, D. (1988) Evaluation of the Montsouris series of ozone measurements made in the nineteenth century. Nature, 332, 240-242.

THE DISTRIBUTION OF LIGHT NONMETHANE HYDROCARBONS IN THE TROPOSPHERE AND THEIR POTENTIAL IMPACT ON PHOTOCHEMICAL OZONE FORMATION

Jochen Rudolph, Ralf Koppmann, Franz-Josef Johnen, and Ahmed Khedim
Institut für Chemie 3: Atmosphärische Chemie
Kernforschungsanlage Jülich GmbH
D–5170 Jülich, FRG

ABSTRACT

Several measurement series during the past years have shown that the mixing ratios of light nonmethane hydrocarbons in the remote troposphere are very low, in general only a fraction of a ppb, never exceeding a few ppb. Consequently only those NMHC which react quickly with OH radicals or ozone can significantly contribute to the chemistry of the atmosphere. Naturally these very reactive NMHC, such as alkenes, have only rather short atmospheric lifetimes, several hours to a few days. It is, therefore, not surprising that these species can only be observed at more than marginal levels in the vicinity of their sources. However, during a flight mission (STRATOZ III) in June 1984 we observed surprisingly high ethene and propene mixing ratios (exceeding 1 ppb) over the west coast of Africa between 15°N and 35°N at altitudes up to 12 km. Nimbus 7 satellite pictures from the "Coastal Zone Colour Scanner" showed that during the time of the measurements the phytoplankton concentration in the coastal regions at the African coast were very high. Later measurements at a different season (March) with lower plankton concentrations in this region of the Atlantic showed significantly lower ethene and propene mixing ratios in the troposphere above. The ethene and propene mixing ratios seldom exceeded 1 ppb and were generally only a few tenths of a ppb. These results indicate that, depending on the seasonal variability of the sources in remote areas, the impact of the light unsaturated hydrocarbons may vary significantly. Since the emission of ethene and propene during periods of high phytoplankton concentrations in ocean water seems to be substantial, the possible impact of oceanic emissions of light alkenes may contribute significantly to seasonal variations in the atmospheric chemical reaction cycles such as ozone production or destruction.

1. INTRODUCTION

The photochemical oxidation of carbonmonoxide, methane and nonmethane hydrocarbons in the atmosphere leads to the production of ozone, provided sufficient NO is present in the atmosphere. The photochemical ozone formation is not only important for urban or industrialized areas but also for the remote troposphere (cf. Crutzen and Gidel, 1983; Crutzen, 1974). Methane and carbonmonoxide have atmospheric lifetimes of several years and a few month resp. and their large scale tropospheric distributions are reasonably well known (cf. Schmidt et al., 1982). However, the tropospheric abundance of most of the NMHCs is not well known as yet and most of the NMHCs show a considerable variability in their tropospheric mixing ratios (cf. Rudolph, 1988). The NMHC mixing ratios in the remote troposphere are in general less than one ppb, practically never exceeding a few ppb. Compared to methane and carbonmonoxide the NMHC levels are roughly a factor of 1000 or 100 resp. lower. However several of the NMHCs, especially the alkenes, are by several orders of magnitude more reactive with the OH radical than carbonmonoxide or methane (cf. Rudolph and Khedim, 1985). This can compensate for the much lower mixing ratios of the NMHC compared to methane and carbonmonoxide. Even in the remote troposphere, ethene and propene levels have been observed which indicate that the atmospheric reactions of these species can significantly contribute to the photochemical reaction cycles in the atmosphere (cf. Bonsang and Lambert, 1985; Bonsang et al., 1988; Rudolph and Ehhalt, 1981). Rudolph (1988) presented measurements of the latitude/altitude distribution of several light NMHCs, including ethene and propene. The ethene and propene mixing ratios varied considerably, ranging from some 10 ppt up to a few ppb. Surprisingly enough, alkene levels exceeding 1 ppb were not only observed over regions which are subject to substantial man made hydrocarbon emissions, but also during the flights along the west coast of Africa alkene mixing ratios exceeding 1 ppb were observed. These alkene concentrations are high enough to be of considerable importance for the atmospheric photochemistry in these regions.

In order to get a better understanding of the possible impact of the light alkenes on the chemistry of the remote troposphere we estimate their potential for ozone production from their latitude-altitude distributions measured during the STRATOZ III flights. We will also present the results of ethene and propene measurements made during two further flight missions over the west coast of Africa, the region where the highest alkene levels were observed during the STRATOZ III flights in June 1984.

2. THE FORMATION OF OZONE FROM THE PHOTOCHEMICAL OXIDATION OF ALKENES IN THE ATMOSPHERE

In order to estimate the potential ozone formation from alkenes in the atmosphere we use a simplified reaction scheme, only considering the main reaction pathways. As an example, the oxidation of ethene is shown here.
The first step is the reaction of ethene with an OH-radical,

$$C_2H_4 + \bullet OH \rightarrow \bullet CH_2CH_2OH \qquad (1)$$

followed by the rapid addition of molecular oxygen.

$$\bullet CH_2CH_2OH + O_2 \rightarrow \bullet O_2CH_2CH_2OH \qquad (2)$$

In the presence of sufficient NO concentrations the peroxyradical reacts with NO;

$$\bullet O_2CH_2CH_2OH + \bullet NO \rightarrow \bullet OCH_2CH_2OH + \bullet NO_2 \qquad (3)$$

again followed by rapid reaction with oxygen:

$$\bullet OCH_2CH_2OH + O_2 \rightarrow 2\,CH_2O + \bullet HO_2 \qquad (4)$$

Formaldehyde (CH_2O) is destroyed by reaction with OH-radicals:

$$CH_2O + \bullet OH \rightarrow \bullet CHO + H_2O \qquad (5)$$

or photolysis:

$$CH_2O + h\nu \begin{cases} \rightarrow \bullet CHO + \bullet H \\ \rightarrow H_2 + CO \end{cases} \qquad (6)$$

OZONE IN THE ATMOSPHERE
Rumen D. Bojkov and Peter Fabian (Eds.)

ISBN 0–937194–15–8

Both the hydrogen atom (•H) and the formylradical (•CHO) react with oxygen and form peroxyradicals:

$$H\bullet + O_2 \rightarrow HO_2\bullet \qquad (7)$$

$$\bullet CHO + O_2 \rightarrow HO_2\bullet + CO \qquad (8)$$

In the presence of NO the HO_2-radical is recycled to OH via:

$$HO_2\bullet + \bullet NO \rightarrow \bullet OH + \bullet NO_2 \qquad (9)$$

if the NO/O_3 ratio is lower than $2\cdot 10^{-4}$ (Crutzen and Gidel, 1983) the reaction

$$HO_2\bullet + O_3 \rightarrow \bullet OH + 2\,O_2 \qquad (10)$$

is becoming more important than reaction (9), and the alkene oxidation would result in ozone destruction instead of production.
NO is photolytically recycled from NO_2:

$$NO_2 + h\nu \rightarrow NO + O \qquad (11)$$

followed by the formation of ozone:

$$O + O_2 + M \rightarrow O_3 + M \qquad (12)$$

Based on this simplified reaction scheme we can estimate that for each reacting ethene molecule on the average 3.8 ozone molecules are produced, provided sufficient NO is present.

The atmospheric oxidation of propene proceeds via very similar reactions. Due to the presence of 3 carbon atoms in the molecule, the number of reaction steps involved in the degradation of propene is larger and more ozone molecules can be formed (on the average 6.8 instead of 3.8). During the photooxidation of propene, acetaldehyde is formed as an intermediate. This allows additional photochemical reactions (e.g. the formation of peroxyacetylnitrate) which are not accesible from the oxidation of methane.

During the oxidation of both ethene and propene, two molecules of carbonmonoxide are formed for each oxidized alkene molecule. The photo-oxidation of each CO molecule yields another molecule of ozone, however, the atmospheric lifetime of CO is in the order of month and we therefore did not include this step in our estimates of the potential in situ ozone production from alkenes.

3. THE POTENTIAL OZONE FORMATION FROM ALKENES ESTIMATED FROM THE STRATOZ III MEASUREMENTS

In order to determine the production rate of ozone we assume that the NO concentrations are large enough to make reactions (3) and (9) the dominant reaction pathways for the peroxyradicals. Under these conditions we can as a first approximation calculate the potential ozone formation rate simply by multiplying the rate of the reaction of the individual alkenes with OH radicals by the number of ozone molecules produced for each oxidized alkene molecule (i.e. 3.8 for ethene and 6.8 for propene).

$$P_{O3} = k_{OH}\,[OH][Alkene] \cdot N_{O3}$$

P_{O3}: potential O_3 production from oxidation of the alkene
k_{OH}: rate constant for the reaction of the alkene with OH radicals
[OH]: concentration of OH radicals
[Alkene]: concentration of the individual alkene (i.e. ethene or propene)
N_{O3}: number of ozone molecules produced for each alkene molecule oxidized

For our estimates we use the measured atmospheric ethene and propene concentrations from STRATOZ III (Rudolph, 1988), the two dimensional OH distribution calculated by Volz et al. (1981) and the reaction rate constants recommended by Atkinson (1984).

The potential ozone production was calculated for each box of a two dimensional grid with a resolution of 10° latitude and 1 km altitude. The calculated ozone production from ethene and propene oxidation is shown in Fig. 1 in the form of isolines.

The ozone production varies between a tenth of a ppb per day and several ten ppb per day. It should be noted that the ozone formation from propene oxidation is by about a factor of 2-3 higher than from ethene oxidation.

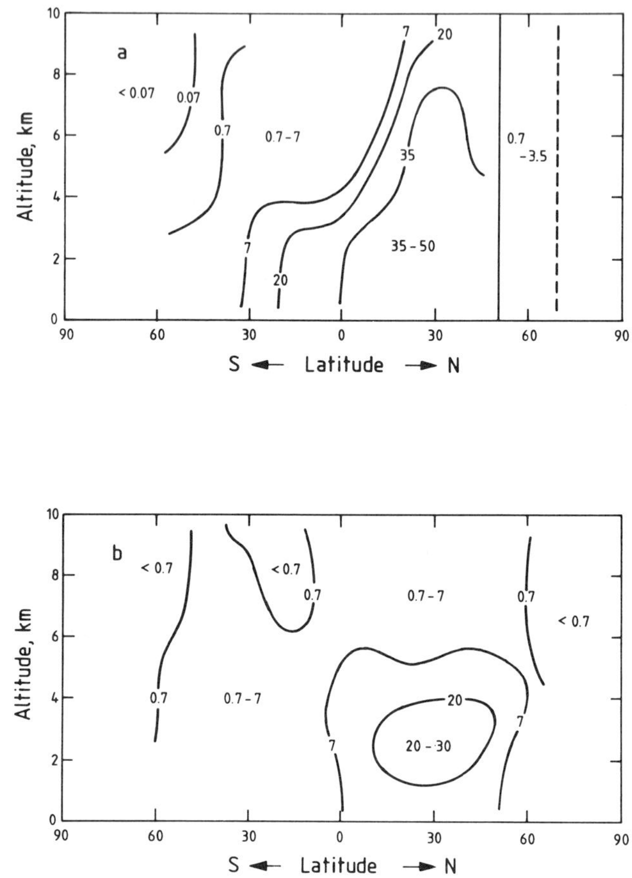

Fig. 1. Potential ozone formation by photooxidation of ethene and propene in the troposphere, calculated from the alkene measurements made during the STRATOZ III experiment.
a: northbound flights
b: southbound flights

The potential ozone production has a maximum at 30° - 40°N during the southbound flights and approximately between 40°N and the equator for the northbound flights.

The southbound flights at 30 to 40°N were made in the vicinity of the North American coast and thus the high calculated ozone production rates probably reflect the impact of anthropogenic hydrocarbon sources. Indeed, the observed trace gas pattern suggests a considerable level of man made hydrocarbons in this region (Rudolph, 1988). However, the ozone production rates of up to 50 ppb over the African westcoast are very surprising. We can compare the ozone production rates from alkenes with the potential ozone formation from methane and carbonmonoxide. For the region above the African westcoat we measured methane mixing ratios of 1.65 - 1.7 ppb and CO levels in the range of 100 - 150 ppb. If we calculate the ozone production from CO and methane oxidation in a similar way as for the alkenes, we find that for the region above the African westcoast the ozone formation potential of ethene and propene exceeds that of methane plus carbonmonoxide by more than a factor of two. These findings are only valid for a very restricted time period and the question is whether and to what extend they can be generalized.

4. ALKENE MEASUREMENTS OVER THE AFRICAN WEST-COAST IN MARCH 85 AND DECEMBER 87

During two further large scale aircraft experiments in March 1985 (STRATOZ IIIS) and December 1987 (TROPOZ I) we had the opportunity to measure the mixing ratios of light hydrocarbons, including ethene and propene in the troposphere above the African west-coast between the equator and about 30°N. The sampling and measurement procedure was the same as used during the STRATOZ III experiment (Rudolph, 1988). We can compare the results from these flights with the findings from STRATOZ III in order to estimate the potential impact of alkenes on the photochemistry of this region on a better basis.

The results of the ethene and propene measurements made in March 85 and December 87 are shown in Figure 2a - d. The vertical profiles show the results from several airplane flights made within 5 days (STRATOZ IIIS) and 10 days (TROPOZ I) respectively. The average atmospheric lifetime of ethene and propene is estimated to be less than one day at low latitudes (Rudolph and Ehhalt, 1981). Consequently we can expect a large variability, which at least to a large extent should show no apparent systematic trends. Indeed, the vertical profiles of the ethene and propene mixing ratios (Fig. 2) cover a range of more than an order of magnitude. However, these profiles give no indication of the fast decrease in mixing ratios with increasing altitude which is predicted by one dimensional model calculations for short-lived hydrocarbons (Chameides and Cicerone, 1978; Brewer et al., 1983; Fishman and Carney, 1984; Kasting and Singh, 1986). The vertical profiles of both ethene and propene show no systematic altitudinal gradients, but more or less a layered structure, sometimes with maximum mixing ratios in the middle or upper troposphere. Examples are the increased mixing ratios at about 10 km altitude observed during the flights between Las Palmas and Dakkar in March 85 (Fig. 2a) and the flights Las Palmas - Dakkar and Dakkar - Abidjan in December 87 (Fig. 2c). These elevated ethene and propene mixing ratios are about 1 - 2 ppb and 0.5 - 1 ppb respectively. Compared to the results of the STRATOZ III experiment of more than 2 ppb of ethene and more than 1 ppb of propene at comparable latitudes over Africa these values are lower by roughly a factor of 2. Another layer of elevated ethene and propene mixing ratios was found between 6 and 7 km altitude over equatorial Africa in December 87 (Fig. 2d). With up to roughly 5 ppb, the ethene and propene mixing ratios reached surprisingly high values. These observations of pronounced layers with enhanced trace gas mixing ratios in the free troposphere supports the assumption that the vertical profiles of these shortlived trace gases are determined by fast exchange processes between the boundary layer and the free troposphere such as in convective cloud systems (cf. Gidel, 1983; Chatfield and Crutzen, 1984; Ehhalt et al., 1985).

For both experiments (STRATOZ IIIS and TROPOZ I), the ethene and propene mixing ratios in the middle and lower troposphere are higher over continental equatorial regions (Fig. 2b and d) than at slightly more northern latitudes (Fig. 2a and c). This indicates that substantial ethene and propene emissions are occuring in equatorial Africa, probably from biomass burning and from vegetation. The ethene mixing ratios over equatorial Africa (Fig. 2b and d) average around 0.3 - 0.5 ppb with the exception of the layer of increased concentrations at 6 - 7 km altitude (Fig. 2d) and several of the measurements made below 0.5 km. The corresponding propene values are on the average a factor of 1.5 - 2 lower. The ethene mixing ratios measured at slightly higher latitudes (10° - 30°N) are significantly lower, 0.1 - 0.3 ppb. Again the propene data are about a factor of 1.5 - 2 lower than the ethene values. These values measured in December 87 and March 85 are a factor of 3 - 10 lower than the ethene and propene mixing ratios measured during the STRATOZ III flight in June 84 over the same area (Rudolph, 1988). This agrees with the hypothesis that the high ethene and propene values found in June 1984 over the African westcoast were caused by high oceanic alkene emissions associated with the occurence of very high phytoplankton concentrations in the coastal areas of Westafrica during this period, Esaias et al. (1986).

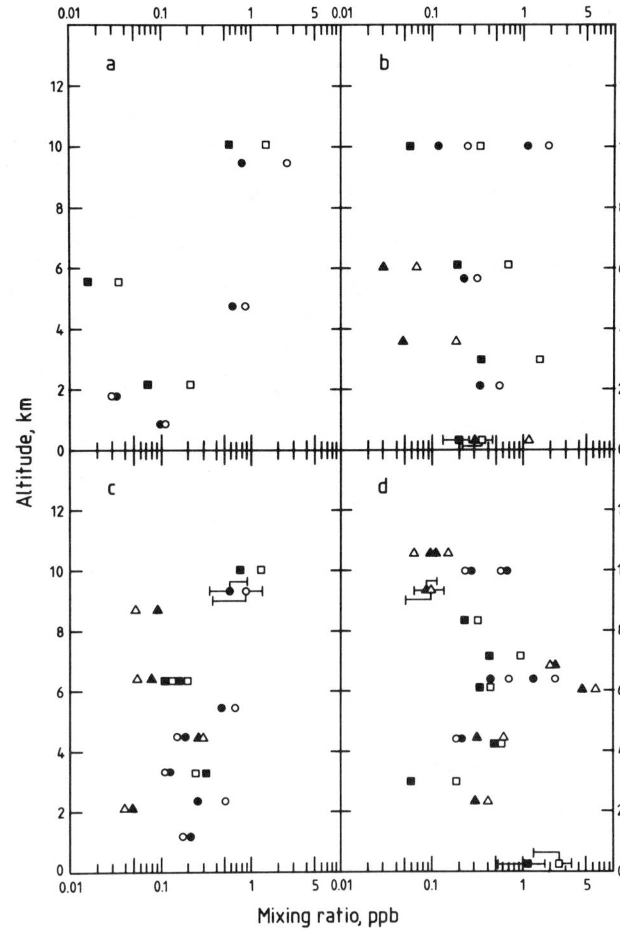

Fig. 2. Vertical profiles of ethen (open symbold) and propen (full symbols) measured in March 1985 (STRATOZ IIIS) and December 1987 (TROPOZ I) over the coast of Westafrica.

a:	Las Palmas - Dakkar	(03/18/85),	circles
	Dakkar - Las Palmas	(03/22/85),	squares
b:	Dakkar - Abidjan	(03/19/85),	circles
	Abidjan - Abidjan	(03/21/85),	squares
	Abidjan - Dakkar	(03/22/85),	triangles
c:	Las Palmas - Dakkar	(12/11/87),	circles
	Dakkar - Las Palmas	(12/21/87),	triangles
	Dakkar - Abidjan	(12/12/87),	squares
d:	Abidjan - Point Noir	(12/12/87),	circles
	Brazzaville - Douala	(12/16/87),	squares
	Abidjan - Dakkar (via Bamako)	(12/21/87),	triangles

The data points with error bars represent the average and its statistical error of three or more measurements made at the same altitude during one flight.

5. CONCLUSIONS

The ethene and propene mixing ratios measured over the west coast of Africa during three flight missions in June 84 (STRATOZ III), March 85 (STRATOZ IIIS) and December 87 (TROPOZ I) cover a range of more than an order of magnitude. The highest values are in the range of a few ppb, lowest values are several tenth's of a ppt. In general the propene mixing ratios are by a factor of 1.5 - 2 lower than the ethene values. In contrast with published one dimensional model calculations, the data show no or only a very weak decrease with increasing altitude, but often a layered structure with increased mixing ratios at higher altitudes is observed. This supports the assumption that the vertical profiles of the alkenes are determined by fast upward transport mechanisms such as convective cloud systems.

The alkene mixing ratios measured in June 84 are considerably higher than in March 85 and December 87. This agrees with the hypothesis that these high alkene concentrations were caused by oceanic emissions during a phytoplankton bloom off the coast of Westafrica. However, biomass burning and emissions from vegetation in tropical regions are probably important sources for atmospheric alkenes as indicated by the substantial ethene and propene mixing ratios over continental equatorial Africa found in December 87. The ethene and propene levels are mostly high enough to allow a considerable production of tropospheric ozone from their photochemical oxidation, provided sufficient NO is present. However the high variability of the light alkenes both with time and space makes it very difficult to give a representative picture of their large scale distribution. The data base that is currently available is still limited. Nevertheless, the presented data strongly suggest that ethene and propene are important components in the chemistry of the troposphere, even over areas with little industrial or urban air pollution and outside the planetary boundary layer.

6. ACKNOWLEDGEMENTS

The authors thank F. Karcher from EERM (Toulouse) and A. Marenco from the University of Toulouse who organized the STRATOZ and TROPOZ flights. We also thank the airplane's crew from the Centre d'Essais en Vol at Bretigny sur l'Orge. This work was supported financially by the Bundesminister für Forschung und Technologie of the Federal Republic of Germany.

7. REFERENCES

Atkinson, R., and A.C. Lloyd, 1984: Evaluation of kinetic and mechanistic data for modeling of photochemical smog. J. Phys. Chem. Ref. Data, 13, 315-444.

Bonsang, B., and G. Lambert, 1985: Nonmethane hydrocarbons in an oceanic atmosphere. J. Atmos. Chem., 2, 257-271.

Bonsang, B., M. Kanakidou, G. Lambert, and P. Monfray, 1988: The marine source of C_2 - C_6 aliphatic hydrocarbons. J. Atmos. Chem., 6, 3-20.

Brewer, D., A. Augustson, and J.S. Levine, 1983: The photochemistry of anthropogenic nonmethane hydrocarbons in the troposphere. J. Geophys. Res., 88, 6683-6695.

Chameides, W.L., and R.J. Cicerone, 1978: Effects of nonmethane hydrocarbons in the atmosphere. J. Geophys. Res., 83, 947-952.

Chatfield, R.B., and P.J. Crutzen, 1984: Sulfur dioxide in remote oceanic air: Cloud transport of reactive precursors. J. Geophys. Res., 83, 947-952.

Crutzen, P.J., 1974: Photochemical reaction initiated by and influencing ozone in unpolluted tropospheric air. Tellus, 26, 47-57.

Crutzen, P.M., and L.T. Gidel, 1983: A two-dimensional photochemical model of the atmosphere 2: The tropospheric budgets of the anthropogenic chlorocarbons, CO, CH_4, CH_3Cl and the effect of various NO_x sources on tropospheric ozone, J. Geophys. Res. 88, 6641-6661.

Ehhalt, D.H., J. Rudolph, F. Meixner, and U. Schmidt, 1985: Measurements of selected C_2 - C_5 hydrocarbons in the background troposphere: Vertical and latitudinal variations. J. Atm. Chem., 3, 29-52.

Esaias, W., G. Feldman, C.R. Milain and J.A. Elrod, 1986: Monthly satellite-derived phytoplankton pigment distribution for the north Atlantic Ocean basin, EOS Transactions 67, 835-837.

Fishman, J., and T.A. Carney, 1984: A one-dimensional photochemical model of the troposphere with planetary boundary layer parametrization. J. Atm. Chem., 1, 351-376.

Gidel, L.T., 1983: Cumulus cloud transport of transient tracers. J. Geophys. Res., 88, 6587-6599.

Kasting, J.F., and H.B. Singh, 1986: Nonmethane hydrocarbons in the troposhpere: Impact on the odd hydrogen and odd nitrogen chemistry. J. Geophys. Res., 91, 13.239-13.256.

Rudolph, J., and D.H. Ehhalt, 1981: Measurements of C_2 - C_5 hydrocarbons over the North Atlantic. J. Geophys. Res., 86, 11.959-11.964.

Rudolph, J., and A. Khedim, 1985: Hydrocarbons in the non-urban atmosphere: Analysis, ambient concentrations and impact on the chemistry of the atmosphere. Intern. J. Environ. Anal. Chem., 20, 265-282.

Rudolph, J., 1988: The two dimensional distribution of light hydrocarbons, results from the STRATOZ III experiment. J. Geophys. Res., in press.

Schmidt, U., A. Khedim, F.J. Johnen, J. Rudolph, and D.H. Ehhalt, 1982: Two dimensional meridional distributions of CO, CH_4, N_2O, $CFCl_3$, and CF_2Cl_2 in the remote troposphere over the Atlantic ocean, Preprint Volume: Second Symposium on the Composition of the Nonurban Troposphere, May 25-28, 1982, Williamsburg, Va., 52-55, American Meteorological Society, Boston, Mass.

Volz, A., D.H. Ehhalt, and R.G. Derwent, 1981: Seasonal and latitudinal variation of ^{14}CO and the tropospheric concentration of OH-radicals. J. Geophys. Res., 86, 5163-5171.

TROPOSPHERIC OZONE BUDGET DURING THE STRATOZ III: A MODEL STUDY

Dirk Poppe, John Drummond, and Dieter H. Ehhalt*
Institut für Atmosphärische Chemie
Kernforschungsanlage Jülich
D-5170 Jülich, FRG

Alain Marenco
Laboratoire d'Aerologie
Université Paul Sabatier
F-31062 Toulouse Cedex, France

ABSTRACT

Vertical profiles of various trace gases including NO, CO, CH_4, O_3 and several organic compounds were determined between 0 - 12 km altitude and 67°N - 60°S latitude during the STRATOZ III campaign in June 1984. A zero dimensional model with a complex chemical reaction mechanism is used to investigate the ozone chemistry. Using the measured trace gas concentrations the local chemical production (or loss) of ozone is calculated. The results show a considerable variability of the O_3 gain (or loss) depending on the NO mixing ratios and the hydrocarbon concentrations. Particularly the light alkenes, ethene and propene are demonstrated to be of significant influence.

1. INTRODUCTION

There are two main sources of O_3 in the troposphere, namely a flux from the lower stratosphere and a chemical production within the troposphere. Intense experimental and theoretical work over the last years have provided evidence that the net chemical formation of ozone is influenced by human activities, mainly by the enhanced emissions of hydrocarbons and NO. The rate of chemical O_3 formation and its dependence on chemical precursors, geographical and meteorological parameters is still a matter of research. Despite many efforts there are still large uncertainties in our knowledge about the global distribution of the ozone mainly because of the lack of experimental data.

It is the purpose of this paper to estimate the net chemical O_3 formation in the troposphere based on experimental data which were obtained during the STRATOZ III flight campaign (Drummond et al., 1988; Ehhalt et al., 1987; Rudolph et al., 1987; Marenco, 1986; Marenco, 1988; Rudolph, 1988). The flights took place in June 1984 and covered latitudes between 70°N and 60°S and altitudes between 0 and 13 km (see Fig. 1).

Fig. 1. Flight track of the STRATOZ III mission in June 1984.

Vertical profiles as well as latitudinal distributions for NO, NO_2, O_3, CO, CH_4 and the non methane hydrocarbons, propene, and ethene were determined. The photolysis rate for NO_2 was also measured. These experimental data are used as input for a model to find the ozone production. Since we restrict ourselves to the chemical formation and take concentrations of the ozon precursors from measured data fields a zero dimensional calculation suffices to model the ozone production and destruction.

The paper is organized as follows: In section 2 the main features of the gas phase ozone chemistry are briefly summarized. The results for

*Unisearch Associates Inc., 222 Scidercroft Rd., Concord, Ontario 14K 1D5, Canada

ISBN 0-937194-15-8

the ozone generation as function of latitude and altitude for three different scenarios are presented and discussed in section 3.

2. OZONE CHEMISTRY

Since the tropospheric chemistry of ozone has been discussed in the literature (see for example Logan et al., 1981 and Logan, 1985) it is sufficient to briefly recall the main features:

(1) Ozone molecules are produced via the production of RO_2 radicals and subsequent transfer of an O-atom to NO to form NO_2. The photodissociation of NO_2 leads to generation of O_3.

(2) The production of RO_2 depends largely on the presence of OH radicals. These stem mainly from the photodissociation of O_3 to give $O(^1D)$ which reacts with water molecules to give OH radicals. This path constitutes the main sources for OH radicals during the day.

(3) O_3 attacks the alkenes producing formaldehyde and higher aldehydes. These processes tend to reduce the O_3 concentration and form the dominant O_3 loss during the night.

(4) The O_3-production is further increased by photolysis of the aldehydes leading to production of HO_2 which recycles to OH.

All these reaction paths are coupled in a complex manner. They are part of the gase phase chemistry of the troposphere. For our numerical calculations we employed the gas phase chemistry mechanism of the Regional Acid Deposition Model (RADM) which we used as published by Stockwell (1986).

3. RESULTS

The STRATOZ III data were averaged over longitude and divided into boxes of 10° in latitude and 1 km in altitude. Photolysis rates are scaled to the experimental values for J_{NO2}. In order to simulate an average over the daylight part of the diurnal cycle the zenith angles are fixed to their values at 10^h local time. A zero-dimensional model is applied to each box in order to calculate the net ozone production $d[O_3]/dt$.

Results were obtained for three different scenarios A, B, and C, respectively (see Tab. 1 and Fig. 2). Scenario A neglects the influence of all hydrocarbons. The net chemical O_3 production varied between -0.3 ppb/h and +2.9 ppb/h apart from the very high value of 23 ppb/h at 0° - 5° latitude and 2 - 3 km altitude. It is caused by the polluted air of Bogota and will not be concedered in the following. Large positive values are concentrated at high altitudes in high northern latitude where also the NO concentrations are found to be high (Ehhalt et al., 1987). There are also areas mainly on the southern hemisphere where there is a small net ozone gain or even a loss. In scenario B the influence of ethene and

TABLE 1.

Species			
NO, CH_4, CO, O_3	STRATOZ III		
NO_2	photostationarity		
PAN OH, RO_2 and other shortlived species	stationarity		
	Scenario		
	A	B	C
HCHO	0	0	0.3 ppb
other aldehydes	0	0	"
higher alkenes	0	0	"
ethane	0	0	"
higher alkanes	0	0	"
slow aromates	0	0	"
fast aromates	0	0	"
ethene	0	STRATOZ III	STRATOZ III
propene	0	"	"

propene is included. Compared to A there is in general a considerable increase in O_3 gain with the highest value of 3.0 ppb/h. The production on the southern hemisphere is still small for two reasons. Firstly the photolytic activities are small due to large zenith angles. Secondly the alkene concentrations are found to be rather low. The influence of an additional background of hydrocarbons is illustrated in scenario C. Here we find in general further increase of the ozone production rate because of enhanced photolysis of the aldehydes and the oxidizing of the alkanes.
We used the data speculate about the chemical ozone formation. Taking an average over altitude and latitude we find:

Scenario	ozone production ($cm^{-2}s^{-1}$)
A	6.2×10^{11}
B	9.5×10^{11}
C	9.3×10^{11}

These values are rather large compared to the hemispherically averaged data compiled in by Marenco (1986). This is so for basically two reason. Firstly, the flight tracks sample mainly coastal lines leading to hydrocarbon mixing above the global average. Secondly, the NO_x concentrations along the flight rates are in general larger then on a global average.

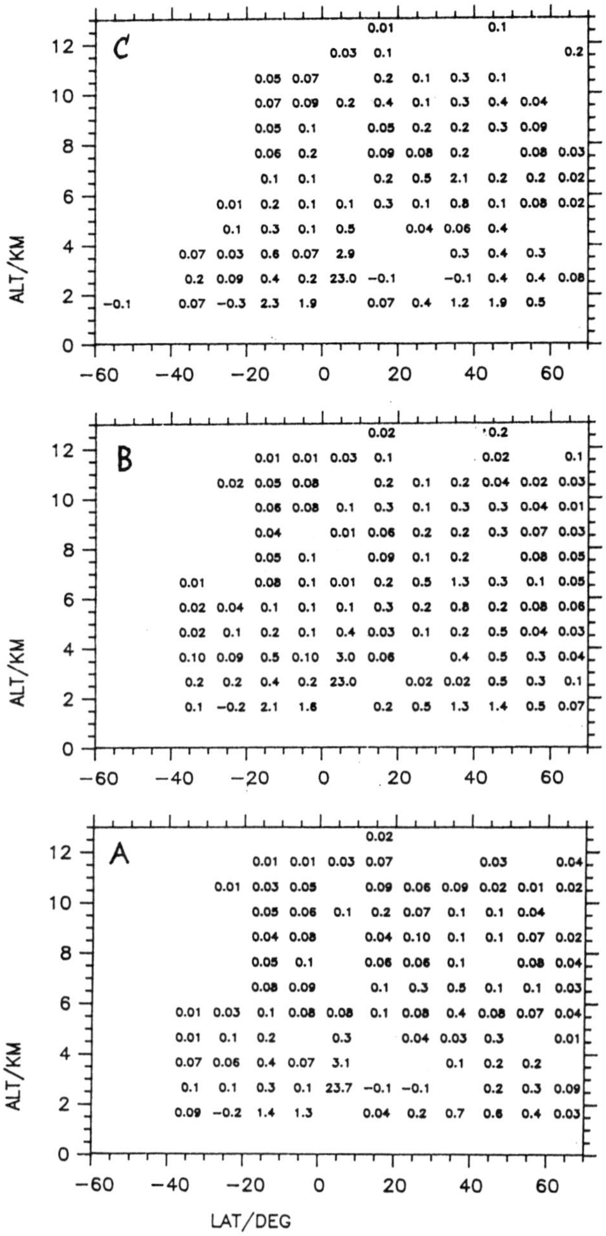

Fig. 2. Latitudinal and altitudinal distribution of the chemical ozone production (in ppb/h) as calculated by a zerodimensional model. The type of scenario is indicated. No entry has been made for boxes which have $|d[O_3]/dt|<0.01$ ppb/h or which are not covered by the flight route.

REFERENCES

Drummond, J.W., D.H. Ehhalt and A. Volz, 1988: Measurements of nitric oxide between 0 - 12 km altitude and 67°N - 60°S latitude obtained during STRATOZ III. Submitted to J. Geophys. Res.

Ehhalt, D.H., and J.W. Drummond, 1987: NO_x sources and tropospheric distribution of NO_x during STRATOZ III, in: Ivar Isaksen (ed.), Tropospheric ozone, regional and global scale interactions, Reidel Publ. Comp., Dordrecht/Holland.

Logan, J.A., M.J. Prather, St.C. Wofsy, and M.B. McElroy, 1981: Tropospheric chemistry: A global perspective, J. Geophys. Res. 86 (no. C8), 7210-7254.

Logan, J.A., 1985: Tropospheric ozone: Seasonal behaviour, trends, and anthropogenic influence, J. Geophys. Res. 90, (No. D6), 10.463-10.482.

Marenco, A., 1986: Variations of CO and O_3 in the troposphere: evidence of O_3 chemistry, Atmos. Environ. 20, 911-918.

Marenco, A., 1988: Study of tropospheric ozone in the tropical belt from STRATOZ campaigns. Proceedings of the International Ozone Symposium, Göttingen, 1988 (this volume).

Rudolph, J., B. Vierkorn-Rudolph, and F.X. Meixner, 1987: Large-scale distribution of peroxyacetylnitrate results from the STRATOZ III flights. J. Geophys. Res. 92, 6653-6661.

Rudolph, J., 1988: The two dimensional distribution of light hydrocarbons, results from the STRATOZ III experiment, J. Geophys. Res., in press.

Stockwell, W.R., 1986: A homogeneous gas phase mechanism for use in a regional acid deposition model, Atmos. Envir. 20, 1615-1632.

THREE-DIMENSIONAL MODELING OF THE OZONE BUDGET OVER THE NORTH ATLANTIC OCEAN

R.A. Brost, R.B. Chatfield, J.P. Greenberg, P.L. Haagenson, B.G. Heikes, S. Madronich, B.A. Ridley, and P.R. Zimmerman
National Center for Atmospheric Research
Boulder, Colorado 80307, USA

ABSTRACT

We used a synoptic scale meteorological model and a separate chemical model to simulate the chemical evolution of about 33% of the northern hemisphere troposphere for one month. A mesoscale meteorological model simulated April 1982; this month of meteorology in turn drove a chemical model, which calculated advection by horizontal and vertical winds, subgrid scale transfer by turbulence and by clouds, surface emission and deposition, and gas phase and aqueous phase chemistry.

In the budget of ozone, the temporal change in the total mass of ozone in the domain equaled the sum of the gains minus the sum of the losses. For the entire month and this large domain, the budget reduced to a net horizontal influx being dry deposited at the surface. Chemical reactions produced large production and destruction rates, but the net chemical change was much smaller and was negative. Generally, the net chemical effect was production in source regions, and destruction in regions without sources. For smaller regions and shorter periods, the net chemical production or destruction and the temporal change could become important. The transit time across the domain (1-4 weeks) limited the chemical time scale that could be simulated. Because the chemical time scale of ozone destruction could be one month or longer, only a global or perhaps a hemispheric model could calculate reliable tropospheric budgets of O_3.

1. INTRODUCTION

Pollutant emissions of nitrogen oxides and hydrocarbons fuel photochemical reactions producing ozone (O_3), which then blends into the hemispheric background of ozone from the stratosphere (Crutzen, 1988; Liu et al., 1987; Logan, 1985). To study this we used an Eulerian modeling system that included the mesoscale meteorological model of Anthes and Warner (1978) and the regional chemical model of Chang et al. (1987; henceforth C87). The domain ran from the Pacific Ocean westward to Eastern Europe and from near the North Pole to about the Equator (Fig. 1). For a smaller domain, Brost (1988) found that simulated ozone was sensitive to lateral boundary conditions.

For the same meteorology, domain, and modeling system, Brost and Chatfield (1988; hereafter BC88) tested cloud parameterizations and advection schemes by simulating the evolution of a decaying tracer, radon 222. Likewise, for the same meteorology, domain, and modeling system, Brost et al. (1988c; hereafter B88c) calculated the anthropogenic fluxes of sulfur and nitrogen to the Atlantic Ocean and compared with observations and other model studies. The current study is an improvement over B88c in that we use more realistic initial and boundary conditions, and analyze the results for the entire month rather than just the last two weeks. In addition, we calculate budgets that were not done before. We have simulated the month without cloud transport, and using two models of cloud transport. Based on the the results of BC88, however, here we normally use a cloud parameterization (labeled as pc in BC88) with less vertical transport than that used in B88c. Many details applicable to this study are given in B88c and BC88.

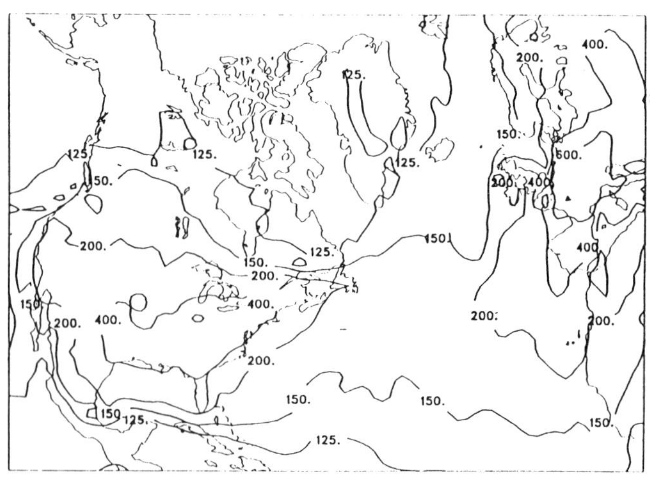

Fig. 1. The 15-d average CO (ppb) for z=160-290 m, in the absence of clouds.

ISBN 0-937194-15-8

2. MODELING SYSTEM

We simulated the meteorology of April 1982 using 10 consecutive runs of 3-day duration with the mesoscale model. We used simulated rather than observed meteorology because the temporal resolution was higher and the spatial coverage of precipitation data was more complete. Each model simulation departed from observations, so we returned to the observed state of the atmosphere every third day. The vertical had 15 levels from the surface to 100 mb. The horizontal resolution was about 190 km, with 46 by 61 grid cells. The vertical resolution was variable and terrain following. We saved three-dimensional fields of winds, temperature, and water vapor mixing ratio, and also saved two-dimensional fields of surface pressure, total rain, and ground temperature. The chemical model used these meteorological fields to diagnose precipitating and nonprecipitating cloudiness, vertical velocity, and vertical eddy diffusivity. B88c and BC88 discuss the meteorology in more detail.

In the chemical transport model, advection was handled by the iterated upstream finite difference scheme of Smolarkiewicz (1983), which was tested by Brost et al. (1988a,b). The vertical turbulence parameterization in Brost et al. (1988a,b) was used. The surface fluxes of heat and momentum were calculated following Businger et al. (1971), and dry deposition of chemical species was estimated using a resistance formulation (Walcek et al., 1986).

The model had 24 transported and 11 non-transported species (Stockwell, 1986; C87). The photolysis rate constants varied with latitude, height, hour angle, and cloudiness (C87). Walcek and Taylor (1986) parameterized the aqueous chemistry and the vertical redistribution of chemicals in precipitating clouds. We used only anthropogenic emissions of sulfur, nitrogen oxides, reactive hydrocarbons, and ammonia from the United States, Canada, and Europe (B88c).

Table 1 shows the initial and boundary conditions used here. We integrated the chemical model for two months. The first month spun up the chemical model; we only analyzed results from the second month.

TABLE 1. INITIAL AND BOUNDARY CONDITIONS

Species that are constant.
Sulfur dioxide=10 ppt, acetaldehyde=1 ppt, formaldehyde=0.2 ppb, butane and higher alkanes=0.5 ppb, ethane=1 ppb, CO=120 ppb, NO_3=N_2O_5=sulfate=organic acid=0.

Species that decrease linearly with latitude.
H_2O_2=2-0 ppb, organic peroxide=1-0 ppb, peroxyacetic acid=0.5-0 ppb.

Exponential decrease, $C_0 \exp\{-z/Z_0\}$, with height z in km.
[Species, C_0 (ppb), Z_0 (km)] = [ammonia, 0.5, 5], [ethylene, 0.2, 3], [propylene, 0.1, 3], [butene, 0.01, 3], [toluene, 0.03, 3], [xylene, 0.03, 3]

Linear increase with height.
$O_3 = (20 + 50\, z / H)$ ppb, where z=0-13.9 km and H=13.9 km.

Linear interpolation between concentrations at 1-km intervals for z=0, 1, 2, ..., 14 km.

NO_2 = 6, 9, 15, 18, 21, 27, 33, 45, 48, 51, 57, 72, 75, 75, 75 ppt.
NO = NO_2 / 3.
HNO_3= 30, 35, 40, 60, 100, rest are 120 ppt.
PAN = 4, 6, 10, 20, 80, rest are 120 ppt.

3. BUDGETS

Table 2 give the terms of the ozone budget in parts per billion by volume per day (ppb/d). Budgets were calculated for each three-day period for 30 days. We also give the standard deviation of the three-day periods about the 30-day averages. Sometimes we only show the budgets for the last three-day period.

TABLE 2. OZONE BUDGET (PPB/D)

	$\partial O_3/\partial t$	Net Chemistry	Div. of Mean Flux	Div. of Turb. Flux	Cloud Effects
Domain 0-15 km 30-d average	-0.001±0.20	-0.074±0.07	0.29±0.13	-0.22±0.01	0.0
Domain 0-15 km last 3 days	-0.35	-0.18	-0.02	-0.24	0.0
Domain 0-15 km last 3 days no clouds	-0.34	-0.15	-0.04	-0.23	0.0
Remote areas 0-15 km last 3 days	-0.41	-0.55	0.24	-0.15	0.0
North Amer. 0-15 km last 3 days	-0.29	1.5	-0.96	-0.80	0.0
North Amer. 0-1 km last 3 days	-0.90	10.2	-5.53	-7.01	1.4

These budgets result from post-processing of meteorological and chemical data stored at 6-h intervals. It would have been prohibitively expensive to record each budget term at each location during the original integration of the chemical model. In addition, post-processing is more flexible because we can later decide to calculate new budgets or do the calculations for new areas. The penalty, however, is that the calculation procedure introduces truncation error in some budget terms, which are suggested later. In general, unless a term is known precisely, the accuracy with which the term is known depends on its size, i.e., the larger the term, the more reliable.

There are five rates of interest. The temporal change of the total ozone is always accurately known, and should equal the sum of all the other terms. For a three-day period, the time change can be large, but it is often small averaged over 30 days. Our post-processing gives the least reliable results for the net effect of chemical production minus chemical loss. For species with fast chemistry or large diurnal changes, this term was calculated by difference to make the budget balance, and hence might accumulate the truncation errors from the other terms. The horizontal and vertical divergences of the fluxes by resolved (mean) winds are approximately known. If chemical concentration varied linearly in time between 6-h save files, the post-calculation of the term would be exact. If the volume extends from the surface to 15 km, then there is no net vertical flux into or out of the volume, and the term is the net horizontal divergence. If the budget volume includes the full vertical domain (0-15 km), the divergence of the vertical turbulent flux is exactly known and equal to the dry deposition. Otherwise, the term is only approximately known. Cloud effects include aqueous chemistry, wet deposition, and vertical transport, and are not well resolved. Aqueous chemistry and wet deposition are unimportant for ozone, but vertical redistribution can be important when the domain does not extend from 0-15 km. Of all important processes, the dry deposition of ozone changes the least between three-day periods.

The balance in the total domain budget for the month is the net import of ozone from the lateral boundary equals the large loss from dry deposition and the small loss from chemistry. This supports the basic picture of stratospheric ozone injected into the troposphere, transported, and then dry deposited on continental surfaces (Levy et al., 1985). The other ozone budgets we calculated are for the final three-day period. For the total domain, the individual budget terms for the final three-day period may depart considerably from the month-long budget.' In the final three days the individual budget terms hardly differ with and without cloud transport and aqueous chemistry. About 80% of the domain does not have anthropogenic emission from the underlying surface. This "remote" region is largely the Atlantic Ocean (Fig. 1). Compared to the entire domain, the remote region has a larger net chemical destruction and a greater net import of ozone by the mean winds. Clearly, results in remote regions are sensitive to boundary conditions because of the critical role of nitrogen oxides in the photochemical production or destruction of ozone (Liu et al., 1987; Logan, 1985; Fishman et al., 1979). The troposphere above the source regions in North America has a net chemical production of ozone and a large export of ozone. The lowest 1 km above sources in North American has a net chemical production of 10.2 ppb/d, an export of -5.53 ppb/d, and a loss to dry deposition of -7.01 ppb/d.

The average net chemical production rate is often small because it changes sign in the vertical and horizontal. Horizontally averaging the net chemical production rate across the domain we typically find chemical production for $z < 0.5$ km and $z > 8$ km, and destruction for $z=0.5$-8 km. Likewise, there is production in and near source regions and destruction in remote regions.

The simulated absolute ozone concentration and the ozone budget of this confined source region are probably insensitive to initial and boundary conditions. As the continental plume moves out over the Atlantic Ocean, the absolute concentrations become less reliable. The mean concentrations are least reliably simulated in middle and upper troposphere regions that are strongly affected by boundary conditions, and weakly affected by continental sources. The results suggest that, aside from isolated planetary boundary layers in source regions, ozone budgets may not be useful on less than a hemispheric or global scale.

For comparison, Table 3 shows budgets of other species. Here we use parts per trillion by volume per day (ppt/d). The chemical lifetime of CO is long compared to transport time across the domain, so that CO is sensitive to lateral boundary conditions. To first order the budget of CO is horizontal advection of CO off the lateral walls balances the net chemical loss (by reaction with OH) in the domain. It seems implausible that the large emissions of CO from North America and Europe are not comparable to or larger than the net mean transport into the domain. Probably the boundary condition of CO was too large or the average hydroxyl was too large. It is also possible that this simulated budget is correct because large sources of CO outside this domain, say in the Amazon and Asia, would maintain the the large CO on the lateral boundaries and in the domain.

The chemical time scale of hydrogen peroxide is sufficiently faster than that of O_3 or CO that the lateral inflow is less important directly. Here the net chemical production in the domain balances losses from the net mean advection from the domain, from cloud effects, and from dry deposition. The budget of nitric acid is similar to that of H_2O_2, but dry deposition is more important and export less important for nitric acid. PAN has a simple budget in which chemically produced PAN is advected from the domain. In the budget of SO_2, anthropogenic emission balances losses from all other processes, and every process is important.

TABLE 3. DOMAIN TOTAL, 0-15 KM, 30-D AVERAGE BUDGETS FOR VARIOUS SPECIES

$\partial CO/\partial t$ (ppt/d)	Net CO Chemistry	Div. of CO Mean Flux	Div. of CO Turb. Flux	Cloud CO Effects	Emissions of CO
-120±130	-1590±60	1160±100	0.0	0.0	310
$\partial H_2O_2/\partial t$ (ppt/d)	Net H_2O_2 Chemistry	Div. of H_2O_2 Mean Flux	Div. of H_2O_2 Turb. Flux	Cloud H_2O_2 Effects	Emissions of H_2O_2
16±12	200±15	-87±5	-15±2	-80±12	0.0
$\partial HNO_3/\partial t$ (ppt/d)	Net HNO_3 Chemistry	Div. of HNO_3 Mean Flux	Div. of HNO_3 Turb. Flux	Cloud HNO_3 Effects	Emissions of HNO_3
1±10	69±5	-16±4	-24±5	-28±2	0.0
$\partial PAN/\partial t$ (ppt/d)	Net PAN Chemistry	Div. of PAN Mean Flux	Div. of PAN Turb. Flux	Cloud PAN Effects	Emissions of PAN
0.8±1.6	4.6±1.3	-4.0±0.7	0.0	0.0	0.0
$\partial SO_2/\partial t$ (ppt/d)	Net SO_2 Chemistry	Div. of SO_2 Mean Flux	Div. of SO_2 Turb. Flux	Cloud SO_2 Effects	Emissions of SO_2
-3±12	-46±4	-29±7	-44±3	-19±6	135

4. CONCLUSIONS

It is probably more difficult to calculate the budget than the concentrations, because, for example, the net inflow or outflow can depend on small differences in concentrations at inflow and outflow grid cells.

In calculating budgets for a limited domain model, ideally we want sources and sinks in the domain to be about equal; net lateral flux to or from the domain should be unimportant. (A net export is not a problem unless the exported mass returns to affect our inflow boundary conditions. Net import, however, means that our results might be sensitive to assumptions about the lateral boundary conditions.) In addition, we should do simulations with different lateral boundary conditions to verify that our conclusions are not dependent on boundary conditions.

CO and O_3 are sufficiently long-lived chemically that the domain total budgets of both species are sensitive to poorly known lateral boundary conditions. Aside from the confined PBL in source regions, budgets of these quantities should be done on hemispheric or global scales. The budget of anthropogenic SO_2 (and sulfur in general) is probably the best simulated because three budget terms are more important than the net export of SO_2. Besides, it is probable that anthropogenic SO_2 should be exported from this domain, and the exported mass could be depleted in the other 2/3 of the northern hemisphere before reentering this domain. For similar reasons, HNO_3 is almost as well simulated as SO_2. It is unclear whether it is reasonable that the largest loss term of H_2O_2 should be lateral export, as found here. Our domain is so large, yet not hemispheric, that we wonder whether the large exported mass of PAN can circle the world and reenter our domain.

ACKNOWLEDGMENTS

The National Center for Atmospheric Research is sponsored by the National Science Foundation.

REFERENCES

Anthes, R.A. and Warner, T.T., 1978: Development of hydrostatic models suitable for air pollution and other mesometeorological studies. Mon. Wea. Rev., **106**, 1045-1078.

Brost, R.A., 1988: The sensitivity to input parameters of atmospheric concentrations simulated by a regional scale chemical model. J. Geophys. Res., **93**, 2371-2387.

Brost, R.A., Haagenson, P.L. and Kuo, Y.-H., 1988a: The effect of diffusion on tracer puffs simulated by a regional scale Eulerian model. J. Geophys. Res., **93**, 2389-2404.

Brost, R.A., Haagenson, P.L. and Kuo, Y.-H., 1988b: Eulerian simulation of tracer distribution during CAPTEX. J. Appl. Meteor., **27**, 579-593.

Brost, R.A., Chatfield, R.B., Greenberg, J.P., Haagenson, P.L., Heikes, B.G., Madronich, S., and Ridley, B.A., 1988c: Three-dimensional modeling of transport of chemical species from continents to the Atlantic Ocean. Tellus B, in press.

Brost, R.A., and Chatfield, R.B., 1988: Transport of radon in a three-dimensional synoptic scale model. J. Geophys. Res., submitted.

Businger, J.A., Wyngaard, J.C., Izumi, Y. and Bradley, E.F., 1971: Flux profile relationships in the atmospheric surface layer. J. Atmos. Sci., **28**, 181-189.

Chang, J.S., Brost, R.A., Isaksen, I.S.A., Madronich, S., Middleton, P., Stockwell, W.R. and Walcek, C.J., 1987: A three-dimensional Eulerian acid deposition model: Physical concepts and model formulation. J. Geophys. Res., **92**, 14,681-14,700.

Crutzen, P.J., 1988: Tropospheric ozone: An overview. In **Tropospheric Ozone**, ed. by I. S. A. Isaksen, D. Reidel Publishing Company, 3-32.

Fishman, J., Solomon, S., and Crutzen, P.J., 1979: Observational and theoretical evidence in support of a significant in-situ photochemical source of tropospheric ozone. Tellus, **31**, 432-446.

Levy II, H., Mahlman, J.D., and Moxim, W.J., 1985: Tropospheric ozone: The role of transport. J. Geophys. Res., **90**, 3753-3772.

Liu, S.C., Trainer, M., Fehsenfeld, F.C., Parrish, D.D., Williams, E.J., Fahey, D.W., Hubler, G., and Murphy, P.C., 1987: Ozone production in the rural troposphere and its implications for regional and global ozone distributions. J. Geophys. Res., **92**, 4191-4207.

Logan, J.A., 1985: Tropospheric ozone: Seasonal behavior, trends, and anthropogenic influence. J. Geophys. Res., **90**, 10,463-10,482.

Smolarkiewicz, P.K., 1983: A positive definite advection scheme with small implicit diffusion. Mon. Wea. Rev., **111**, 479-486.

Stockwell, W.R., 1986: A homogeneous gas phase mechanism for use in a regional acid deposition model. Atmos. Environ., **20**, 1615-1632.

Walcek, C.J. and Taylor, G.R., 1986: A theoretical method for computing vertical distributions of acidity and sulfate production within growing cumulus clouds. J. Atmos. Sci., **43**, 339-355.

Walcek, C.J., Brost, R.A., Chang, J.S. and Wesely, M.L., 1986: SO_2, sulfate and HNO_3 deposition velocities computed using regional landuse and meteorological data. Atmos. Environ., **20**, 949-964.

HOW LIGHTNING AND HUMAN ACTIVITY DETERMINE TROPICAL OZONE LEVELS

Robert B. Chatfield
National Center for Atmospheric Research
Boulder, Colorado 80307, USA

ABSTRACT

A first analysis of a coarse three dimensional photochemical-transport model suggests that ozone variation in the global tropical troposphere is only partially explained by simulated natural processes. The model is compared to satellite data which suggest that ozone appears to vary in both time and space in a complex manner, but also with clear large-scale features. It shows a distinct maximum spatially centered near Africa, the "continental tropics", and a minimum west of the Date Line, the "oceanic tropics". The ozone maxima depend on season. While various data indicates anthropogenic pollution causing local ozone increases, there are powerful natural processes helping determine ozone in the tropical belt. The model was used first to measure the magnitude of thes conjectured effects on the very largest scales in the tropics. This paper investigates (1) the role of meteorological transport from the mid-latitudes and also (2) the effect of natural NOx emissions in producing these ozone variations. In the modeled troposphere, the effects of transport and lightning-enhanced production during the summer seasons are counterbalanced by the large effect of greater middle tropospheric water vapor in destroying ozone. An effect of lightning can be discerned in one season, but appears to have spatial effects, horizontal and vertical, that do not produce the regions of highest ozone. The global spread of tropical pollution or other sources on NOx appear reasonable, but much more data analysis and modeling are required to demonstate this.

1. INTRODUCTION: VARYING LEVELS OF TROPICAL O_3

1.1 CHANGING VIEWS OF TROPICAL OZONE

The phenomena of ozone in the tropical troposphere were for years considered very passive responses to some very strong inputs occuring in midlatitudes. When some of the pioneering, and still very interesting, measurements of tropical ozone were made [Fabian, 1973], they were interpreted in terms of mid-latitude stratospheric injection, transport to the tropics, "deposition" to the earth's various surfaces. When Levy [1972] showed how the reaction $O(^1D) + H_2O \longrightarrow 2\,OH$ produced ample tropospheric OH, Crutzen [1972] stressed its role as a sink for tropospheric ozone.The introduction of this sink globally required an equally strong source, which was identified as the smog-like chemical production of tropospheric ozone [Crutzen, 1972, Chameides and Walker 1973, Fishman *et al.*, 1979] *i.e.*, the photochemical oxidation of CO and methane, *in the presence of sufficient quantities of NO*. Currently relatively recent analyses of the global ozone budget [Crutzen and Gidel 1983, Logan *et al.*, 1981 Chameides *et.al.*, 1987a] portray the tropics as largely an area of ozone destruction, due to low NO levels in the lower troposphere. Several studies of the tropics have recently suggested rather intense ozone production over the Brazil [Delany *et al.*, 1985, Logan and Kirchoff, 1986] and presumably in similar, more densely inhabited, areas. The precursor nitrogen oxides and hydrocarbons originate from tropical agricultural burning. Sources and sinks of ozone very nearly counterbalance in the remote atmosphere [Chameiedes *et al.* , 1987a]. In describing the whole tropics, there is still a view that meteorology and transport from mid-latitudes are dominant in determining tropical O_3 [Levy, 1988].

A Puzzling Variation: Recently, satellite measurements of the total column-integrated density of atmospheric ozone have been interpretated to highlight rather localized "hot-spots" of intense ozone production over the inhabited regions of South America and Africa. Consequently, Fishman and Larsen [1987] undertook the more delicate process of interpreting total ozone patterns throughout the tropics as *primarily* reflecting variations in tropospheric ozone. They argued that while the stratospheric contribution to the ozone column is often nearly ten times large than the tropospheric effect, the stratospheric variation is rather small, and may be removed by sufficient (month-long) temporal averaging of the data. Their paper is recommended as a preparation for this one. Fishman and Larsen suggested very large-scale variation of tropospheric ozone in the tropics (see their Figure 1), with a maximum in all seasons over the half of the tropics centered over Africa, and a minimum centered west of the international date line. Furthermore, the amplitude of the variation depended on season, with small peak-trough amplitudes in February, approximately 10 Dobson units, and larger in September, approximately 25 Dobson units. The zonal mean values varied also, being 30 Dobson units higher in September. (10 Dobson units of ozone spread evenly through the troposphere is approximately 14 ppb of ozone. When "average tropospheric" ozone mixing ratios are mentioned below, they are defined as the ratio of the O_3 column integrals of ozone and total air.) There was also frequently noticeable a general variation in ozone between the northern and southern tropics, as described in their Figure 4. Jack Fishman has kindly supplied maps of the TOMS averages for various months, and there is frequently a clear, small (just over 10 Dobson units) variation in many months, with the division drawn near the local ITCZ over the oceans.

It is tempting to relate these large-scale variations in ozone to the intense but very local production portrayed by Fishman *et al.* [1986], but that relation involves a large extrapolation, and also implies a very large effect of humans on natural ozone levels, one beyond that which has ever been attributed to such a significant trace gas. Anthropogenic ozone increases could potentially or actually affect the natural environment. However, it is wise to avoid the tempting fallacy that measurements made in the most remote areas of the world represent the state of nature before human influence, when weather, vegetation, natural emissions, and other factors are very strongly correlated with the areas where people live.

1.2 TWO NON-POLLUTION HYPOTHESES

Accordingly, this paper investigates two alternative hypotheses about tropical ozone which are more conservative and describe natural processes. One is that the ozone maxima seen in the tropics is due to transport from the midlatitudes,

ISBN 0-937194-15-8

which have a summer maxima in tropospheric ozone whose nature has occasioned some debate [Fishman, *et al.*, 1979, Chatfield and Harrison, 1978]. Preferential ozone transfer to the tropics might occur during winter and summer monsoons, but especially in the latter, since these move ozone southward at high altitudes, where ozone mixing ratios are typically elevated. This is an idea which traces to the stimulating early work of Junge and associates [Junge, 1962, Fabian 1973, Pruchniewicz, 1973]. Furthermore, the longitudinal variation might well be explained by the following argument. The oceanic tropics frequently divide into two segments at the ITCZ, with Hadley cell circulation on each side. Over the continental tropics, however, the ITCZ travels widely following the sun creating a single powerful monsoonal cell [Oort, 1983]. A glance ahead to Figure 2 indicates a very reasonable hypothesis about the import of ozone. It may be imported from the northern mid-latitudes, where stratospheric injection, aided by uncertain amounts of natural and pollutant production, make for high ozone levels.

Lightning NOx hypothesis: Another hypothesis is that lightning provides an effective source of NOx to catalyze ozone production [Liu *et al.*, 1983]. Lightning activity is strongly concentrated over the continents, and also maximizes in the summer hemisphere of the tropics. (See Turman and Edgar [1982].) Both of these hypotheses suggest reasonable correlations with longitude, season, and hemisphere, and both act on a very large scale. Perhaps these natural differences influencing ozone are indeed correlated with what makes regions habitable. These hypotheses clearly require somewhat detailed calculation of both transport and chemical effects, especially since ozone chemistry is self-modifying ("non-linear").

2. MODEL DESCRIPTION:

This search for wavenumber-one variations in ozone over seasonal timescales makes reasonable the use of a very coarse three dimensional model. It is called a 2.5-d model, since it resembles 2-d models in approach more than 3-d, as it uses climatically averaged winds and transport parameters and relatively few grid points. It is in spherical geometry with pressure coordinates in the vertical. It describes 18 grid areas, 9 describing the tropics from 0 to 30 N, 9 for the tropics from 0 to 30 S. Boundary conditions are described for important and long-lived species. Ten tropospheric levels were described in the vertical. The model integrates the integro-differential equations of air composition. Horizontal transport has the standard advective and diffusive terms. Mean motions are taken from the monthly summaries of Oort [1983]. Mean motions dominate circulations over most of the tropics [e.g., Riehl, 1979], but a weak diffusion, proportional to the wind speed, was allowed, with a magnitude $K_x \sim 10^6$ m s^{-1} u, $K_y \sim 10^6$ m s^{-1} v. This diffusion had little effect. Boundary conditions for ozone and carbon monoxide were taken from Chatfield and Harrison [1977] and Logan *et al.*, [1979], those for the NOx species were taken from Fehsenfeld *et al.*, [1988].

Vertical motion was described by transport kernels, as described in Chatfield and Crutzen [1984]. The kernels were estimated by a new technique. Perhaps the best-known database that is related to convective vertical motion in the tropics is rainfall. The oceanic analyses of Jaeger [1976] was used to estimate the cloud base mass flux of air into updrafts by a rainfall efficiency argument. Rainfall "storm" efficiencies are frequently quoted in the cloud physics literature to describe the ratio of those water molecules that fall to the earth as precipitation to the number that are moving upward through cloud base. (Note that the rain water may originate from water molecules entrained above cloud base by this definition.) The transfers of other species may be ratioed to this (as described in Chatfield and Crutzen, [1984]). Efficiencies of 50% were used in this work [Riehl, 1979]. The advantages of this parameterization are its simplicity, the availability of input data, and the fact that the location of convection is directly tied to observation, rather than to a model that may or may not correctly locate convection and give an appropriate intensity. The material rising through cloud base was allowed to detrain in a simple manner, *i.e.*, twice as much at the top of the troposphere as at cloud base, with an interpolation between linear in altitude. This profile allows for more lower tropospheric detrainment, and may be more consistent [Grell, 1988] with life-cycle descriptions of cloud transport than those that motivated the profiles of Chatfield and Crutzen [1984].

Lightning Data were digitized from the maps presented by Turman and Edgar [1982]. These data define a horizontal pattern for dawn and dusk for the months selected, Februrary and August.

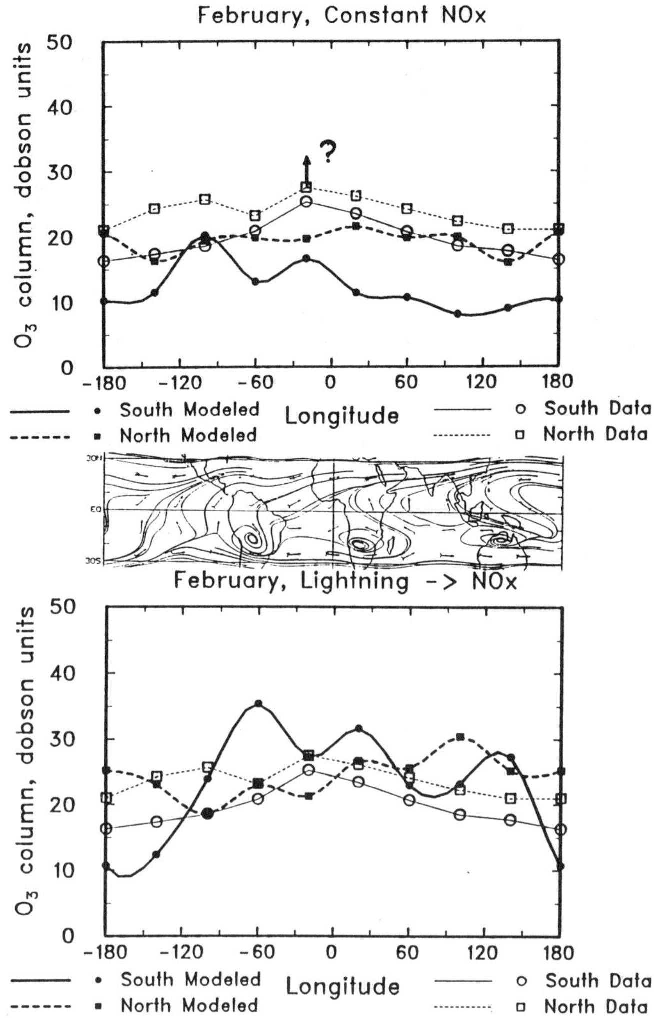

Fig. 1. February simulations of tropospheric ozone. (a) Horizontally constant source of NOx: this simulation highlights the effects of the wind systems alone in transporting ozone. The satellite data shown is highly averaged in time, November–June The arrow indicates a very approximate uncertainty in its interpretation. In some cases it appears to be ~20% low. (b) Lightning-input simulations, with a nitrogen source rate of 1.2 Tg(N) yr^{-1}. More ozone is produced over the continental tropics, and agreement with the averaged observations might be considered "fair".

A harmonic interpolation was made for all other hours. Chameides *et al.*, [1987b] indicate a relatively low efficiency for production of NO from lightning, and I cut down the global figure to obtain a tropical input rate, using the figures of Turman and Edgar to estimate how much was in mid-latitudes. Finally, there is needed a vertical profile descibing the "virtual source" of NO after the molecule leaves the immediate environment of its thunderstorm. This is very poorly understood. For the studies reported, I took a source profile parabolic in altitude, $S = -6\zeta^2 + 6\zeta$ where ζ is the altitude normalized by the tropopause height, 16 km. Results not reported here indicate that this source profile produced NOx profiles in differing regions of the model that were in reasonably good accord with observations [Ridley *et al.*, 1987, Feshsenfeld *et al.*, 1988, Ehhalt and Drummond, 1988].

Chemical Reaction Rates were taken from the review of Baulch *et al.*, [1982], with updates using the review of DeMore *et al.*, [1988] (which focuses on rates of stratospheric interest) for reactions such as HOO + HOO (+ H_2O). A very complete set of relevant inorganic species and methane-derived species was integrated. The lack of PAN and associated organic precursors is likely the greatest limitation to this study, but since NO and NO_2 are simulated at values near those observed, the effect of this omission upon O_3 is likely not very great. Temperatures and water vapor concentrations were taken from the longitudinally and latitudinally varying data of Oort [1983]. The ozone deposition velocity, a bulk value for the boundary layer, was taken as 0.1 cm s^{-1} everywhere; if higher values were simulated in proportion to land, and lower values in proportion to ocean, our conclusions would be stronger. The model does not attempt the complexities of boundary layer chemistry either; that is left as a next step. Since it omits ozone reaction with rain-forest and other terpenes, it likely over-represents tropospheric ozone in the lower troposphere over the continents [Browell *et al*, 1988]. This destruction should more than compensate for various natural surface sources of NOx that lead to ozone production. If so, even stronger conclusions than those given below are again in order.

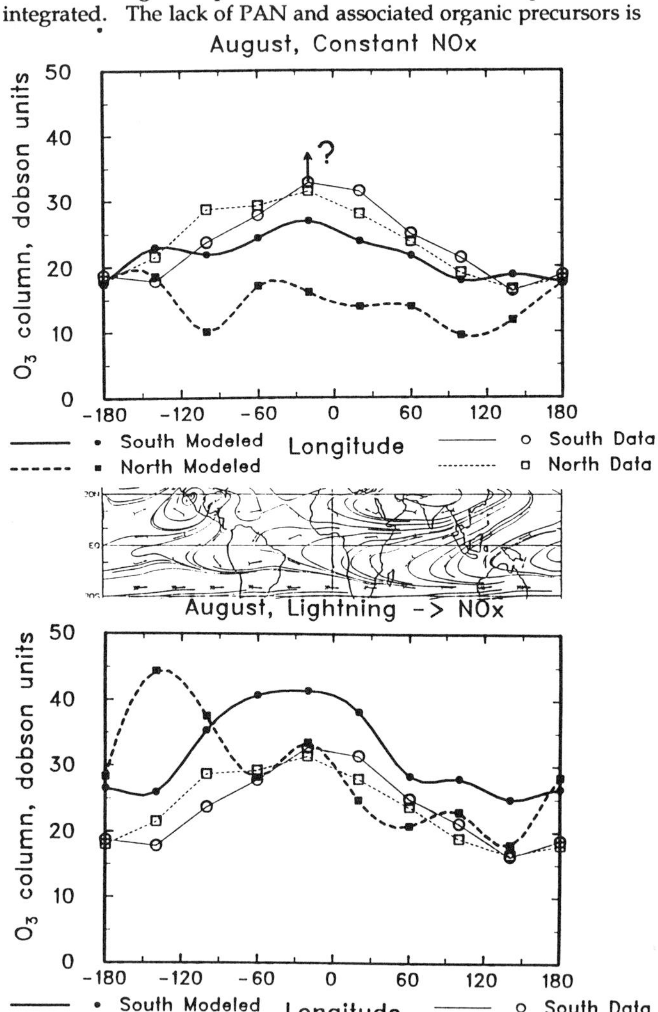

Fig. 2. August simulations of tropospheric ozone. (b) Horizontally constant source of NOx: this simulation highlights transport by the summer monsoon of ozone to the continental tropics. Effects are small.. The arrow again indicates ~20% uncertainty in the July–October data. (b) Lightning-input simulations with a rate of 1.5 Tg(N) yr ^{1}. Comparison to the observed ozone data here is good for the Southern Hemisphere (possibly from the co-incidence of land, lightning, and burning), poor for the Northern.

3. RESULTS AND DISCUSSION

Simulations Highlighting Meteorological Transport: Figures 1a and 2a show results of simulations which maintain a constant amount of NOx in all grid boxes at a given level. The vertical distribution of NOx was very close to that reviewed in Fehsenfeld *et al.*, [1988]. A suggested distribution of at least a large part of the column of ozone for the months Novemer–June is shown in Figure 1a to compare with the February simulation, and distribution for July–October is shown to compare with August in Figure 2a. These are abstracted from the data presented by Fishman, Watson, and Larsen in this symposium volume. Data presented for the months of February and September in Fishman and Larsen [1987] and also maps of ozone for individual months of 1985 (Fishman, personal communication) seem to indicate more spatial variability than the data shown, which is reasonable given the long averaging period. Large-scale East–West and North–South differences appear more defined in these data, but the quantification of tropospheric residual amount is considerably more difficult. Additionally, as Fishman, Watson, and Larsen indicate, there appears a difference between their troposopheric residual and data measured *in-situ*. The satellite residual appeared to be 20% low, but the degree of shortfall is not certain. An arrow in the diagram indicates this uncertainty. For their part, the simulations show point-to-point variability not necessarily considered within the capability of the model to simulated. The model was intended to capture larger-scale wavenumber 1 and 2 longitudinal variations and interhemispheric differences.

The agreement of the constant–NOx data with observations is not good. The simulations are both rather constant with longitude, while the data suggests a definite "bump" in the continental tropics.. The data suggest that there is a tendency for higher ozone in the summer hemisphere, while both simulations, Figures 1a and 2a suggest higher ozone is simulated in the winter hemisphere. An examination of the model runs indicated that production was O_3 similar in both hemispheres, and that transport added as expected, but that O_3 loss was always higher in the summer hemisphere, due to the much larger amount of water vapor there [Oort, 1983]. Ozone was lost to produce OH, and under these low-NOx conditions, little ozone was returned from the relatively increased OH radicals.

Effects of Lightning Activity: Since lightning should also maximize with convective activity in the summer hemisphere, it would seem that these could produce the summer ozone maximum. Figures 1b and 2b show the simulations based on lightning ozone input. The February simulation is almost acceptable, showing comparable concentrations in both hemispheres and a definite ozone "bump" in the continental tropics. The August simulation is very poor for the northern hemisphere, showing a bump in the Northern East Pacific, not over Africa. Conceivably, this is due to weather variations causing anomalies in lightning or winds in the year sampled by Turman and Edgar [1982], in the winds (a the year sampled by Turman and Edgar [1982], in the winds (a climatology was used), or in ozone (see above). Simulations of other months are planned. The southern hemisphere ozone shows a bump in the right area, and given the uncertainties in the tropospheric residual ozone and the simulations, is remarkably close. Individual monthly maps indicate remarkably distinct features of high total ozone observed in the African region, a behavior very different from other areas and other seasons.

Nitrogen Photochemistry and Ozone: The model as constructed so far likely overestimates ozone for two reasons. The lack of detail for deposition and near-surface chemistry has been mentioned. Chatfield and Crutzen [1983], indicate that the source region for lightning NOx, cumulonimbus clouds, is also simulated to be a region of high OH. The coarse resolution of the model does not resolve the NO_2 destruction by OH, and so NOx spreads too far into the troposphere. Likely, then, the simulations err by giving O_3 too high, the satellite-deduced data may be too low, and there is considerably room for anthropogenic pollution process to add to tropical ozone.

4. CONCLUSIONS:

Tropical tropospheric ozone shows large-scale variation with season and location. A coarse model of the global tropical circulation and chemistry can produce some of the very large, wavenumber-one variations that are deduced from in-situ and TOMS satellite observations. Transport of ozone from mid-latitudes does not seem to have a major effect in producing the observed tropical ozone distribution. Water vapor in the atmosphere has a larger control on the ozone destruction rate.

Lightning-produced NOx appears to have a significant role in producing the observed ozone distribution, as Liu *et al.*, 1983] suggested, but its effects appear too weak, and may be incorrectly located. Simulations not reported here suggest that if the lightning NOx input goes more to the lower middle troposphere, somewhat more ozone is created, while the NOx levels go significantly above observations. Several conclusions need to be tested with a finer mesh model including organic nitrate chemistry. It is a reasonable hypothesis that human-produced pollution, namely large-scale biomass burning, contributes importantly to ozone in the continental half of the tropics. Simulations of these effects are important to substantiate this conclusion, and are left as important, future work.

Acknowledgments: I am grateful to Jack Fishman for discussions and for providing data, and to Brian Ridley for reading an early draft. NCAR is funded by the U. S. National Science Foundation. This is publication number ACD-1905-88-5 of the NCAR Atmospheric Chemistry Division.

REFERENCES

Browell, E.V. G.L. Gregory, R.C. Harriss, and V.W.J.J Kirchoff, Tropospheric ozone and aerosol distibutions across the Amazon Basin, *J. Geophys. Res., 93* , 1431–1451, 1988.

Baulch, D.L, R.A. Cox, P.J. Crutzen, R.F. Hampson, Jr., J.A. Kerr, J. Troe, R.T. Watson, Evaluated kinetic and photochemical data for atmospheric chemistry, *J.Phys. Chem. Ref. Dat., 11,* 327-496, 1982.

Chameides, W., and J.C.G. Walker, A photochemical theory of tropospheric ozone, *J. Geophys. Res., 7,* 8751– 8760, 1973.

Chameides, W.L., D.D. Davis, M.O. Rodgers, J. Bradshaw, and S. Sandholm, Net ozone photochemical production over the eastern and central north Pacific as inferred from GTE/CITE 1 observations during fall 1983, *J. Geophys. Res., 92,* 2131–2152, 1987a.

Chameides, W.L., D.D. Davis, J. Bradshaw, M. Rodgers, S. Sandholm, and D.B. Bai, An estimate of the NOx production rate in electrified clouds based on NO observations from the GTE/CITE 1 fall 1983 field operation, *J. Geophys. Res., 92,* 2153–2156, 1987b.

Chatfield, R.B. and P. J. Crutzen, Sulfur dioxide in remote ocean air: Cloud transport of reactive precursors, *J. Geophys. Res., 89,* 7111–7132, 1984.

Chatfield, R.B., and H. Harrison, Tropospheric ozone, 2, Variations along a meridional band, *J. Geophys. Res., 82,* 5969–5976. 1977.

Crutzen, P.J., and L.T. Gidel, A Two-dimensional photochemical model of the atmosphere. 2: The tropspheric budgets of the antropgenic chlorcarbons, CO, CH4, CH3Cl, and the effects of various NOx sourcs on tropospheric ozone, *J. Geophys. Res., 88,* 6641-6661, 1983.

Delany, A.C, P.J. Crutzen, P. Haagenson, S.Walters, and A.F. Wartburg, photochemically porduced ozone in the emission from large-scale tropical vegetation fires, *J. Geophys. Res., 90,* 2425–2429, 1985.

DeMore, W.B., M.J. Molina, S.P. Sander, D.M. Golden, R.F. Hampson, M.J. Kurylo, C.J. Howard, A.R. Ravishankara, *Chemical Kinetics and Photochemical Data for Use in Stratospheric Modeling,* JPL Publication 87-41, 1987.

Fabian, P., A theoretical investigation of tropospheric ozone and stratospheric–tropospheric exchange processes, *Pure Appl. Geophys. 106–108,* 1044–1056, 1973.

Fehsenfeld, F.C., D.D. Parrish, and D.W. Fahey, The measurement of NOx in the non–urban troposphere, I.S.A. Isaksen (ed.), *Tropospheric Ozone,* 185–215, D. Reidel Publishing Company, 1988.

Fishman, J., S. Solomon, and P.J. Crutzen, Observational and theoretical evidence in support of a significant in–situ photochemical source of tropospheric ozone, *Tellus, 31,* 432–446, 1979.

Fishman, J., and J.C. Larsen, Distribution of total ozone and stratospheric ozone in the tropics: implications for the distribution of tropospheric ozone, *J. Geophys. Res., 92,* 6627–6634, 1987.

Jaeger , L., Monadskarten des Niederschlags für die ganze Erde, *Berichte des Deutschen Wetterdienstes, 139,* 1996.

Junge, C.E., Global ozone budget and exchange between stratosphere and troposphere, *Tellus, 14,* 363–377, 1962.

Levy, H., Normal atmosphere: Large radical and formaldehyde concentrations predicted, *Science, 173,* 141–143, 1971.

Levy, H., II, Global transport of ozone, I.S.A. Isaksen (ed.), *Tropospheric Ozone,* 319–325, D. Reidel Publishing Company, 1988.

Logan, J.A., M.J. Prather, S.C. Wofsy, and M.B. McElroy, Tropospheric chemistry: A global perspective,, *J. Geophys. res., 86,* 7210–7254,1981.

Logan, J.A., and V.W.J.H. Kirchoff, Seasonal variations of tropospheric ozone at Natal, Brazil, *J. Geophys. Res., 91,* 7875–7882, 1986.

Liu, S.C., M. McFarland, D. Kley, O. Zafiriou, and B. Huebert, Tropospheric NOx and O_3 budgets in the equatorial Pacific, *J. Geophys. Res., 88,* 1360–1368, 1983.

Liu, S.C., Model studies of background ozone formation, I.S.A. Isaksen (ed.), *Tropospheric Ozone,* 303–318, D. Reidel Publishing Company, 1988.

Oort, A. H., Global *Atmospherhic Circulation Statistics, 1958–1973,* NOAA Professional Paper 14, Rockville, MD, 1983.

Pruchniewicz, P.G., The average tropospheric ozone content and its variation with season and latitude as a result of the global ozone circulation, *Pure Appl. Geophys, 106–108,* 1973.

Riehl, H., *Climate and Weather in the Tropics,* Academic Press, London, 1979.

Turman, B.N., and B.C. Edgar, Global lightning distributions at dawn and dusk, *J. G eophys. Res., 87,* 1191–1206, 1982.

ESTIMATES OF PAST AND FUTURE TROPOSPHERIC OZONE CHANGES FROM CHANGES IN HUMAN RELEASED SOURCE GASES

Ivar S.A. Isaksen, Terje Berntsen, and Sverre Solberg
Institute of Geophysics, University of Oslo
Blindern, 0315 Oslo 3, Norway

ABSTRACT

A global 2-D model is used to study the distribution of chemical species in the troposphere and their changes with time. The studies are focused on O_3 and OH, the two major compounds in the gas phase oxidation process in the troposphere, over the last as well as future chemical development.

The calculations indicate that a substantial increase in global tropospheric ozone (≈25%) has occurred since the turn of the century. Increases are most pronounced at northern mid-latitude surface levels where ozone is estimated to have doubled. OH concentrations show a small decrease in the same time period. Nearly all future scenarios which were adopted gave smaller ozone changes. From 1985 to 2100 the global ozone change is in the range -5% to +25 %. Global average OH changes are always less than 10%.

1.INTRODUCTION

Observed changes in the large scale distribution of some of the main ozone precursors as well as in ozone itself, indicate that human activities over the last decades are changing the chemical composition and thereby probably also the oxidizing capacity of the global troposphere (Isaksen, 1988). The Northern Hemispheric tropospheric ozone may have increased by as much as 1% pr. year during the last two decades. The complicated interaction between a large number of gaseous species of both natural and anthropogenic origin which determines the formation of ozone and other secondary gaseous products in the troposphere, makes it difficult to evaluate the significance of the release of pollutants based only on observed changes. Thus model simulations of future gaseous changes are a helpful tool in evaluating the atmospheric significance of pollution releases. In this study we use the global model previously described (Isaksen and Hov, 1987) to perform model runs with different assumptions for anthropogenic releases of nitrogen oxides (NO_x), carbon monoxide (CO), non-methane hydrocarbons (NMHC) and methane (CH_4). We will also simulate the global change in ozone and other secondary gaseous compounds due to changes in anthropogenic emissions over the last 100 years.

2.EMISSIONS

Emission-data of NO_x, CO and CH_4 for the future scenarios were provided by EPA (Environmental Protection Agency). The emission was divided into 16 geographical regions. In our model the emissions must be given at each 10° of latitude. To accomplish this the contribution from each region to the different gridpoints at the ground was estimated. Emissions of the NMHCs were held constant at an estimated 1985-level, except for one model run where they were increased simultanously with the NO_x emissions. NO_x-emissions from aircraft were also provided by EPA. Total emissions and a height profile were given, while the meridional distribution was the same as the anthropogenic distribution given by Isaksen and Rodhe (1978).The total emissions of NO_x from aircraft were held constant at 0.35 TgN/year throughout the calculations.

The calculations cover the time period 1985-2100. Except for methane where emission rates were changed every year, changes were made every 10th year from 1985 to 2025, and every 25th year from 2025 to 2100. The model was run for one year with complete chemistry after emission rates were changed. After one year all components except methane were assumed to be close to equilibrium. During the intermediate years the model was run with only methane calculations, using a linear change in the methane emissions. Every time the calculations of a new scenario were started, new 1985 values were calculated.

Scen.nr	1	2	3	4	5	6	7	8
CH_4	L	L	L	L	H	H	H	H
CO	L	L	H	H	L	L	H	H
NO_x	L	H	L	H	L	H	L	H

Table 1: Combinations of high (H) and low (L) emissions for CO ,CH_4 and NO_x giving the 8 scenarios.

Time development of the emissions are given in figures 1 and 2 for the high and low cases. The scenarios are combinations of these curves.

3.ADJUSTMENT OF METHANE EMISSIONS

Recent measurements indicate that the present CH_4-concentration is about 1.7 ppmv in the troposphere (WMO,1985). To obtain this as an initial condition we adjusted the 1985 CH_4 emission every time a new run was started. Because of the feedback from OH, this was solved by an iterative method, where a correction factor for the CH_4-emission for each scenario was calculated. Table 2 shows the correction factors with which the given CH_4 emission rates were reduced.

ISBN 0-937194-15-8

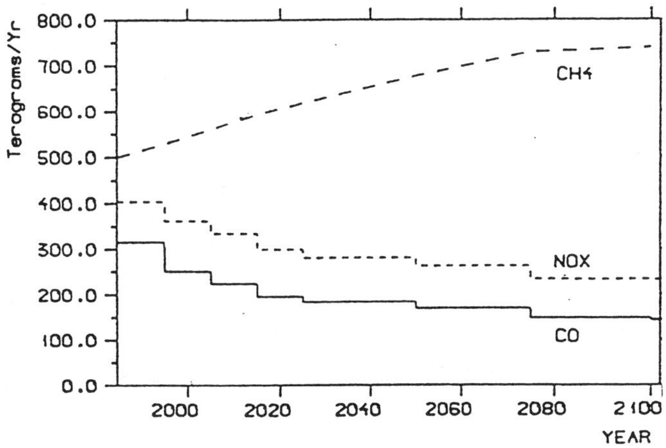

Figure 1. Time development of low emission rates for CO (Tg C/year), CH_4, (Tg CH_4/year) and NO_x (0.1 TgN/ year).

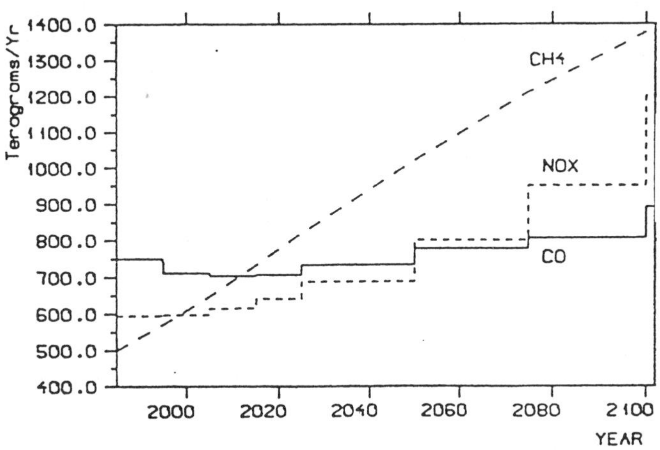

Figure 2. Same as figure 1 for high emission rates.

Scenario nr.	1,5	2,6	3,7	4,8
Correction factors for CH_4 emissions	1.44	1.13	1.65	1.28

Table 2: Correction factors for the given CH_4 fluxes

4.CHANGES IN THE PAST

Changes over the last 100 years were studied with two model runs ending up with 1985-emissions as given by EPA scenarios 1 and 4 respectively. Natural emissions were held constant while the changes in the anthropogenic emissions (except for methane), were estimated to follow the world's energy consumption (Gates, 1985), which increased with 1.8% pr. year until 1950, and with 4.5% pr. year afterwards. Methane concentrations were not calculated, but were given fixed values which increased by 0.34% pr. year before 1950 and 1.1% afterwards (NASA, 1988).For the other species new emission rates were adopted each 25th year until 1950, and each 10th year from 1950 to 1980. One year with complete chemistry was calculated each time the emission rates were changed.

Figure 3 shows global mean concentrations of O_3 and OH from 1900 to 1980. The increase in the calculated global mean O_3- concentration is approximately 25% in both cases. The increase is largest between 50° and 60° N at ground level, where O_3-concentrations rise by almost 100% in both simulations. Global OH concentrations are calculated to have decreased over the last 100 years. The decrease is approximately the same in both hemispheres with as much as 25% between 4 and 8 km height at low latitudes. Global average change is more moderate, of the order of 5-10%.

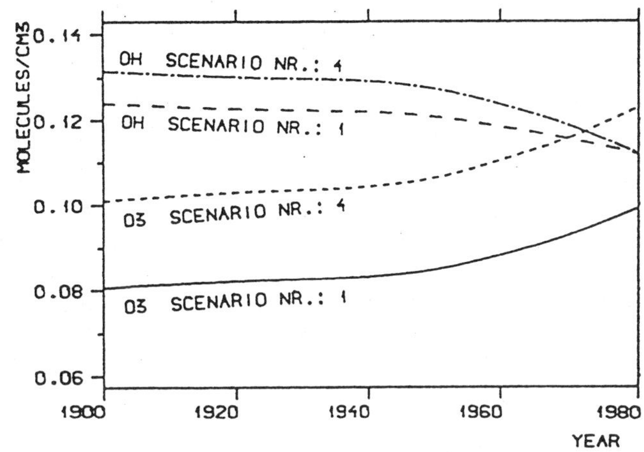

Figure 3. Time development of O_3 and OH from 1900 to 1980. Concentrations are multiplied by $10^{-13}(O_3)$ and 10^{-7} (OH).

The calculations indicate that there have been changes in the photochemical activity in the troposphere, particularly in the distribution of ozone, and to a lesser degree in OH. This can be attributed to the increased anthropogenic activities over the last decades. The changes are, however, highly dependent on height and latitude. Ozone increases are confirmed by measurements (Bojkov, 1986; Kley et al, 1988).

5.SCENARIOS 1985-2100 - RESULTS AND DISCUSSIONS

O_3 and OH are key species in the tropospheric oxidation. We will therefore concentrate our discussion on their calculated distribution and future changes. We want to look at how the different anthropogenic emissions affect both the global mean values and the distribution of these key components in the troposphere.

5.1 Ozone changes

Compared to the changes in emissions, global mean O_3 changes are relatively small for most scenarios. All changes in 2100 are within -5% to +25% of their 1985-values. Note, however, that there are large differences in the 1985- values, (and also 2100 values).

Global average O_3 concentrations vary between 1.1 and 1.8 x $10^{12} molecules/cm^3$ in the Northern Hemisphere (N.H).

	Scenario nr.							
	1	2	3	4	5	6	7	8
O_3 ^{x}	1.13	1.52	1.24	1.66	1.20	1.71	1.29	1.81
OH^{xx}	1.43	1.87	1.02	1.39	1.35	1.80	1.00	1.39

$^x 10^{12} molec./cm^3$, $^{xx} 10^6 molec./cm^3$

Table 3: Calculated average tropospheric O_3 and OH concentrations for the year 2100 in the Northern Hemisphere.

Figure 4 shows the calculated time development of global average O_3-concentrations in the troposphere for 4 scenarios.

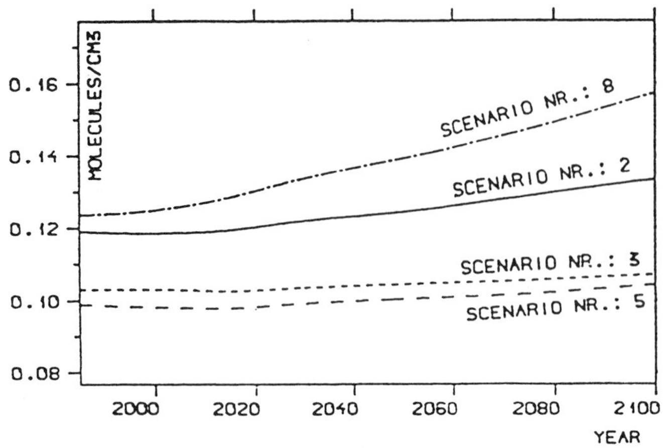

Figure 4. Time development of the global mean concentration of O_3 for 4 different scenarios. Concentrations are multiplied by 10^{-13}.

The highest increase in ozone with time (approximately 25%), is obtained in scenario 8 where all source gases have high release rates, the average increase in tropospheric concentration of NO_x, CO and CH_4 are 40%, 50% and 155% respectively.

Figures 5 and 6 show a latitudinal-height cross section of the O_3 changes between 1985 and 2100 in percent for scenario 2 and 8. High emissions of CO, NO_x and CH_4 give maximum ozone increase of 50-60% in the latitudinal belt 10° -30° N in the lowest level of the troposphere (fig.6). The emissions of NO_x which increase by a factor 2.5-3 in this area are largely responsible for this. With a decrease in the CO emissions and a slow increase in the methane emissions (fig.5) we get the same pattern, but the maximum increase in ozone is now reduced to 20-30%. With low CO and CH_4 and high NO_x emissions we see that even though there is a general increase in ozone by 10-20% in the whole troposphere , we get a small decrease in areas with the largest NO_x emissions. This leads us to conclude that NO_x is a dominant compound in the ozone chemistry. Increases in NO_x will generally lead to ozone increases on a large scale. In more NO_x rich environments, increases in NO_x will lead to reduced ozone formation and reduce ozone levels. One reason for this is that the loss reaction for NO_x:

$$NO_2 + OH \rightarrow HNO_3$$

removes NO_2 which otherwise through photodissociation would lead to ozone formation. Changes in CO and CH_4 emissions will strongly modify the ozone trends. Increased emissions enhance the ozone formation.

One model run equal to scenario 2, except that we increased the NMHC emissions simultaneously with the increases in NO_x emissions, showed that the ozone development is affected. While the difference in the ozone change was small in most of the troposphere, it was pronounced at latitudes with the largest emissions. This indicate that the ozone levels are sensitive to NMHC releases in polluted regions, while they are less sensitive in background areas of the troposphere.

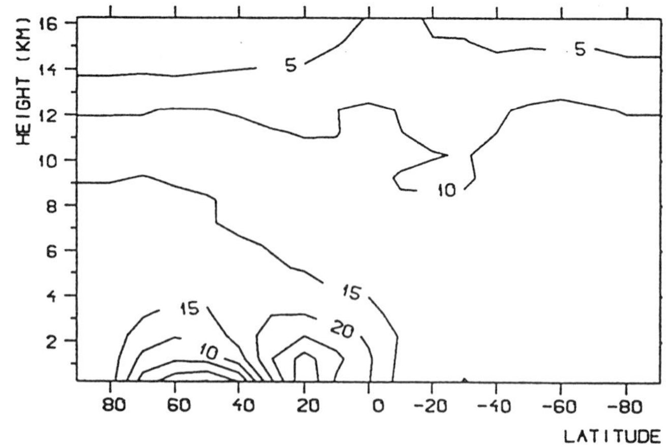

Figure 5. Local change of O_3 concentrations from 1985-2100 for scenario 2 in percent.

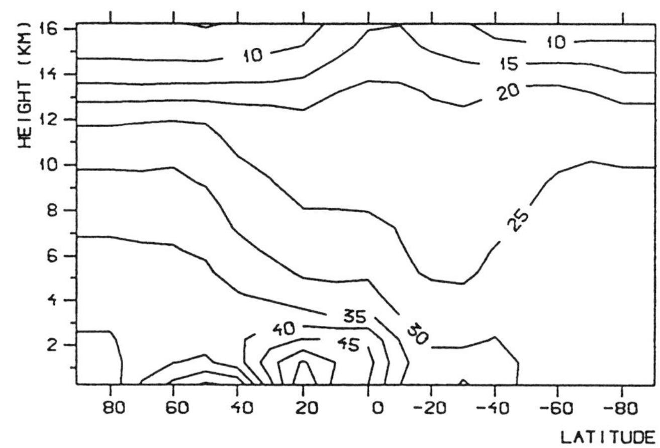

Figure 6. Same as figure 5 but for scenario 8.

In order to look at how CO, CH_4 and NO_x simultaneously affect O_3, we computed differences in average tropospheric O_3 concentrations produced by these three components separately and in combination. We find that both in the combined $CO + NO_x$ (with low methane) and the $CH_4 + NO_x$ (with low CO) scenarios, the amount of O_3 produced was larger than the sum of the separate effects. Combination of CO and CH_4 gave less O_3 than the sum of each, especially in the case with high NO_x.

5.2 OH changes

The OH radical plays a key role in this picture. It is increased by high NO_x while it is reduced by oxidation with CO and CH_4. Its behaviour will therefore be quite different from the O_3 development as NO_x, CO and CH_4 changes.

Figure 7 shows the calculated time development of average tropospheric OH concentrations for some scenarios. The change with time for each scenario is surprisingly small, and mostly within +/-10%. The fact that the differences between the individual scenarios are quite larger than the changes with time within each scenario (table 3), suggest that the average OH concentrations are determined by emissions which are not varying with time, but have different values in the scenarios. The only natural source which is changed from one scenario to the other is the lightning production of NO_x.We have adopted two values;1 TgN/year in cases with low NO_x scenarios and 10 TgN/year in the high NO_x scenario cases. Although lightning contributes only 16-17% to the total NO_x emissions at most, and only 10% at the most by the year 2100, it emits NO_x directly into the free troposphere. Because of the short lifetime of NO_x, we find that the lightning produced NO_x is highly significant for the free tropospheric NO_x budget, and thereby for OH which is produced by the reaction:

$$NO + HO_2 \rightarrow NO_2 + OH$$

From the discussion it seems clear that the tropospheric NO_x through its effect on OH has a substantial effect on the oxidation of species like CO and CH_4.

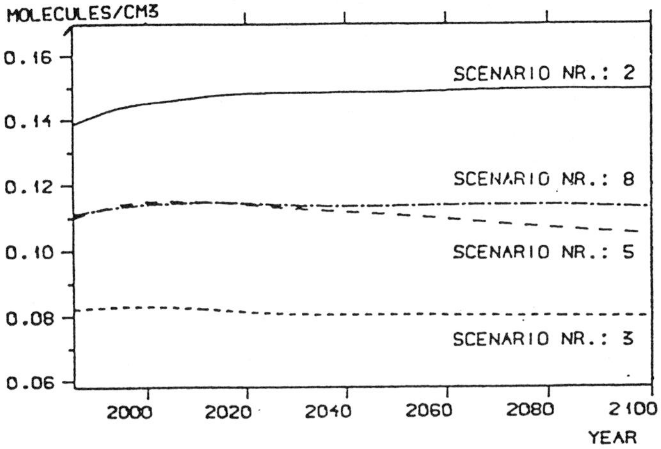

Figure 7. Time development of mean global OH-concentrations for 4 different scenarios. Concentrations are multiplied by 10^{-7}

What is said above does not hold for the whole troposphere, but first of all in all areas where the ratio of NO_x are large compared to hydrocarbons. Fig.8 shows the distribution of change in the calculated OH concentration in percent from 1985 to 2100 for scenario 6. OH increases up to 100% north of 30° N. This is due to the increase in NO_x concentrations of about the same order. South of 30° N is OH generally decreased by ≈ 20%. Here are NO_x levels much smaller and the decrease in OH is a result of the reaction with CO and CH_4. A striking feature shown in figure 8 is the large percentage buildup of OH above 12 km, specially towards the poles. The concentrations are, however, low, and it has little significance for the chemistry on a global scale.

The distribution of OH and its development with time when the emissions of its precursors, NO_x, CH_4 and CO are changed, show in many respect a behavior which is similar to ozone. It is highly dependent on the combination of precursor emissions, and on the background levels, particularly on NO_x. Its variation differs from that of ozone in most regions of the atmosphere. NO_x changes have an opposite effect on OH than CH_4 and CO have. For this reason most scenarios tend to give relatively small OH changes over the time period 1985 to 2100. In the background atmosphere the effect of NMHC was small, NO_x was highly important for the ozone production, both directly and by the enforcement of O_3 production from CO and CH_4.

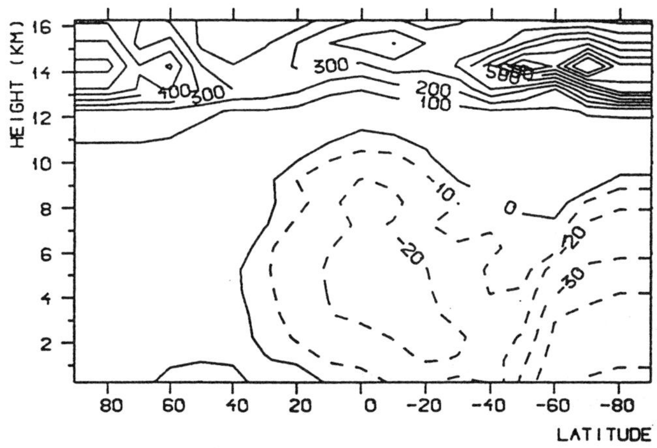

Figure 8. Same as figure 5, but for the OH concentration of scenario 6. Note the different contour interval for negative and positive values.Dashed line indicates a decrease.

ACKNOWLEDGEMENTS

This work has been sponsered by the US Environmental Protection Agency (EPA) and the Norwegian Research Council for Sciences and the Humanities (NAVF).

REFERENCES:

Bojkov R.D. Surface ozone during the second half of the nineteenth century. J.Clim.Appl.Meteor., Vol.25,pp. 343-352, 1986.

Gates, D.M. Energy and ecology, Sinnaner Assoc., Inc.Sunderland Massachusetts, 1985.

Isaksen,I.S.A. Is the oxidizing capacity of the atmosphere changing? The Changing Atmosphere pp.141-157, ed.F.S. Rowland and I.S.A.Isaksen, John Wiley & Sons. Ltd., 1988.

Rodhe,H and I.S.A.Isaksen , A two-dimensional model for the global distribution of gases and aerosol particles in the troposphere, Report AC-47. Dept.of Meteorology, Univ. of Stockholm 1978.

Isaksen, I.S.A. and Hov,Ø. Calculations of trends in the tropospheric concentrations of O_3, OH,CO, CH_4 and NO2x. Tellus, Vol 39B,pp.271-283,1987.

Kley,D., Volz A.,Mülheims F. , Ozone measurements in historic perspective. Tropospheric ozone, regional and global scale interactions. (ed.Isaksen I.S.A.). NATO ASI Series C, vol.227,pp.63-72, D.Reidel Publishing Company, Dordrecht, 1988.

Ozone Trend Panel , Chapter on trends in source gases, NASA Report (1988, in preparation). WMO-Report No.16, Atmospheric Ozone (1985).

ATMOSPHERIC CHEMICAL AND CLIMATE CHANGE: POSSIBLE EFFECTS ON TROPOSHERIC OZONE

Anne M. Thompson
NASA Goddard Space Flight Center
Greenbelt, Maryland 20771, USA

ABSTRACT

We have evaluated the sensitivity of tropospheric ozone to increases in CH_4, CO, and NO_x and predicted changes in stratospheric O_3, temperature and moisture with a one-dimensional photochemical-transport model. Sensitivities in tropospheric ozone and OH are different in urban and nonurban areas, in marine and continental regions and in the boundary layer and free troposphere.

The model is used with projections of CO, NO_x, and CH_4 to predict tropospheric ozone and OH changes over the next fifty years. Generally, CH_4 and CO increase O_3 and suppress OH. Similar calculations assuming stratospheric ozone depletion and climate warming show moderation of OH loss and tropospheric ozone enhancement.

1. INTRODUCTION

Changes in atmospheric levels of certain pollutants (CO, NO_x, and nonmethane hydrocarbons [NMHC]) have the potential to affect the rate of tropospheric O_3 formation and the abundance of the free radical oxidant OH on a global scale. Perturbations to stratospheric O_3 and climate (temperature and moisture) can also influence the rates of key photochemical processes affecting tropospheric O_3 and OH. Since OH and O_3 are the chief tropospheric oxidants, these changes may alter the oxidizing capacity of the atmosphere.

Modeling studies have demonstrated that increases in global CH_4 and CO are consistent with a loss in global OH and an increase in tropospheric ozone [Levine et al, 1985; Thompson and Cicerone, 1986]. Ozone increases over the past century have been reported for a few Northern Hemisphere locations [Bojkov, 1986]. Besides CH_4 and CO, NMHC and NO_x emissions affect oxidant levels. The lifetimes of these compounds are much shorter than those of CH_4 and CO (minutes to hours vs. months to years for CH_4) which means their direct effects on O_3 and OH are more localized. This suggests that geographical areas that are distinct in general chemical characteristics will respond differently to changes in CH_4/CO/NO/NMHC emissions over time.

These geographical distinctions motivate the present study of the effects of trace gas and climate change on tropospheric O_3 and OH. Calculations have been made with a one-dimensional photochemical model [Thompson and Cicerone, 1986] to quantify O_3 and OH dependence on changing NO, CH_4, and CO emissions and stratospheric ozone and climate change for a given set of conditions, each one characteristic of a different "chemically coherent region." Sensitivities of different regions can be compared when they are quantified in terms of coefficients of the type $\partial \ln[y]/\partial \ln[x]$. Specific scenarios of expected CH_4, CO, NO, and climate changes have been included in model runs to estimate how OH and O_3 will change in these areas.

2. EVALUATION OF SENSITIVITY COEFFICIENTS

Evaluation of sensitivity coefficients and the assignment of "chemically coherent regions" are described in detail in Thompson et al [1988]. Calculations are performed by running the model to represent perturbed conditions in these areas, i.e. vary the influx of CO, NO, or CH_4 at the lower boundary, reduce the stratospheric O_3 column in calculating photolysis rates, fix water vapor at a higher amount (20% more water vapor at the ground if the temperature increases 3K; Ramanathan et al, 1985). Results are shown here for conditions

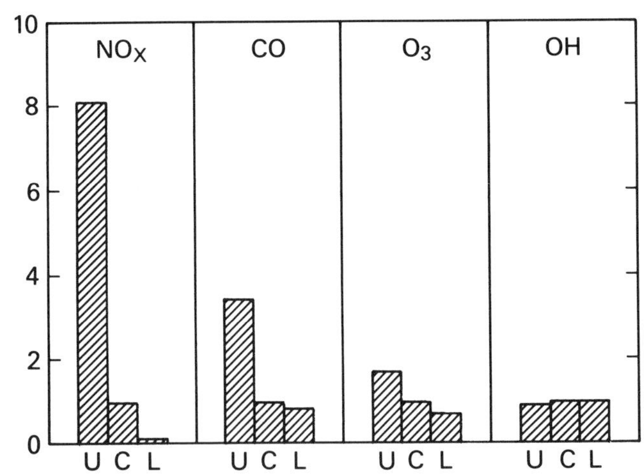

Fig. 1 Surface mixing ratios of NO_x, CO, O_3, and tropospheric OH (column-integrated, 0-15 km) for chemically coherent regions. U = urban mid-latitude; C = clean continental mid-latitude; L = low latitude. Values are normalized to continental: NO_x = 0.20 ppbv; CO = 130 ppbv; O_3 = 39 ppbv; OH = 6.1 (11) cm-2.

ISBN 0-937194-15-8

corresponding to: (1) low latitudes (15N); and (2) two "prototype" regions characteristic of the mid-latitude Northern Hemisphere, "urban" and "clean continental." Surface mixing ratios of NO, CO, O_3 and tropospheric OH (integrated from 0-15 km) for the three regions are shown in Fig. 1.

2.1 OH SENSITIVITY TO CHEMICAL AND CLIMATE CHANGE

Sensitivity coefficients for OH variation as a function of changing CO, NO, CH_4 emissions are given in Table 1. The coefficients can be understood by considering a simple form of the photochemical equilibrium equation governing OH:

$$[OH] = \frac{2k\,[O(^1D)][H_2O]}{k[CO] + k[CH_4] + k[NMHC]}$$

where the k's signify reaction rate coefficients and brackets [] denote equilibrium concentrations of CO, CH_4, etc. The equilibrium level of OH is determined by two opposing tendencies. Increases in CH_4, CO (or NMHC) suppress OH by increasing the denominator relative to the numerator. However, adding these gases to the atmosphere initiates photochemical production of O_3 (see below) which adds to the numerator and gives back OH via:

$$O_3 + h\nu \rightarrow O(^1D) + O_2$$
$$O(^1D) + H_2O \rightarrow 2OH$$

Ozone formation requires substantial levels of NO. The level of NO_x essentially determines whether OH increases or decreases in a given region as CO or CH_4 is added. In Table 1 note the differences among the three chemically coherent regions illustrated. The sensitivity functions (from which the coefficient is derived) are nonlinear [Thompson et al, 1988].

Table 1: OH Sensitivity Coefficients for Unperturbed ("Present Day") Conditions

$(\partial \ln[OH]/\partial \ln x)$ at $S_{NO}(1.0)$, $S_{CO}(1.0)$, $S_{CH_4}(1.0)$

x=	S_{NO}	S_{CO}	S_{CH_4}	Strat.* O_3	H_2O
Urban Mid-Lat. (45N, UML)	+0.05	-0.48	-0.04	+1.0	+0.88
Clean Continental Mid-lat. (CCML)	+0.60	-0.41	-0.40	+0.68	+0.56
Low Latitude (15N, LL)	+0.27	-0.17	-0.43	+0.62	+0.47

* Positive coefficient signifies increase in OH when stratospheric O_3 decreased.

Table 1 also shows values of coefficients for OH change when the stratospheric O_3 column is reduced. The value is positive for O_3 depletion because more uv penetrates to the troposphere and $O(^1D)$ and OH are produced. The coefficient is uniform over a range of stratospheric O_3 loss from 0 to 15%, which is what stratospheric models predict as likely to occur over the next 50-100 years. Similarly, if a climate warming is assumed to proceed with no change in relative humidity, water vapor will increase about 10-20% at the ground in a 50-100 year time frame. The model calculates an OH response similar to that for perturbed stratospheric O_3. A summary of how OH will change with perturbed chemical inputs can be described: generally CO and CH_4 emissions depress tropospheric OH, NO adds to it up to a point; global warming or stratospheric ozone depletion will tend to increase OH.

Table 2: O_3 Sensitivity Coefficients for Unperturbed ("Present Day") Conditions

$(\partial \ln[O_3]/\partial \ln x)$ at $S_{NO}(1.0)$, $S_{CO}(1.0)$, $S_{CH_4}(1.0)$

x=	S_{NO}	S_{CO}	S_{CH_4}	Strat.* O_3	H_2O
Urban Mid-Lat. (45N, UML)	+0.11	+0.37	+0.14	+0.14	-0.18
Clean Continental Mid-lat. (CCML)	+0.10	+0.19	+0.24	-0.08	-0.21
Low Latitude (15N, LL)	+0.07	+0.07	+0.28	-0.34	-0.40

* Positive coefficient signifies increase in tropospheric O_3 when stratospheric O_3 decreased.

2.2 O_3 SENSITIVITY TO CHEMICAL AND CLIMATE CHANGE

The model calculations described above were analyzed to derive sensitivity coefficients for tropospheric O_3 (Table 2). As with OH there are significant differences among O_3 sensitivities for different chemically coherent regions. A difference between OH and O_3 responses, however, is that increased emissions of CH_4 and CO in all three regions increase O_3, giving positive coefficients. The reason is that CO, CH_4 and NO contribute to photochemical ozone formation as the reactions show:

$$OH + CO\ (+ O_2) \rightarrow CO_2 + HO_2$$
$$HO_2 + NO \rightarrow NO_2 + OH$$
or
$$OH + CH_4\ (+ O_2) \rightarrow CH_3O_2 + HO_2$$
$$CH_3O_2 + NO \rightarrow NO_2 + CH_3O$$

$$NO_2 + h\nu \rightarrow NO + O$$
$$O + O_2 + M \rightarrow O_3 + M$$

Stratospheric O_3 loss (and increased uv flux to the troposphere) mean photochemical destruction of tropospheric O_3, except that at high NO_x (urban) levels more OH (and higher NO_2/NO ratio) cause photochemical O_3 formation: positive coefficient from stratospheric O_3 loss. More H_2O vapor is associated with an ozone loss, through conversion of $O(^1D)$ to OH. Then OH, at low NO_x levels as in a marine, low-latitude environment, actually destroys O_3 through a catalytic cycle with HO_2.

3. OH AND OZONE CHANGES FROM PERTURBED CH_4/CO EMISSIONS

Model calculations have been used to estimate O_3 and OH changes resulting from

Table 3: Scenarios for 50-Year Runs (Starting year is 1980)

No.	CH_4	CO	Strat. O_3	H_2O
1	+ 1%/yr mix. rat.	+ 0.5%/yr flux	Constant	Constant
2	+ 1%/yr mix. rat.	+ 0.5%/yr flux	- 10% at 50 yr	Constant
3	+ 1%/yr mix. rat.	+ 0.5%/yr flux	- 10% at 50 yrs	+ 12% at 50 yrs

several alternative scenarios based on extrapolations of recent CO and CH_4 trends. The scenarios are summarized in Table 3. Calculations are made with a steady-state version of the model that is run with perturbed conditions prescribed at 5-year intervals from 1980 to 2030. The unperturbed conditions shown in Fig. 1 are assumed to represent 1980. Methane levels are prescribed, corresponding to a 1%/year increase in mixing ratio, taken to be 1.58 ppmv in 1980. CO influx increases 0.5%/year. The specification of 0.5%/yr reflects CH_4-CO-OH feedback (less than 1%/year increase in source will produce the 1%/yr increase that is observed) and also the fact that only about half the global CO source is anthropogenic.

As the sensitivity analysis suggests (positive coefficients in Table 2), ozone increases in all three regions at constant NO flux (Fig. 2a). The increase is greatest for the urban case, where NO_x values are highest and photochemical formation of O_3 is most efficient. Also shown in Fig. 2a is a curve corresponding to a 0.3% /yr increase in tropospheric O_3, that is a rough mean for the mid-latitude increase in tropospheric O_3 that has been deduced by the 1988 NASA Ozone Trends Study Panel. None of the calculated increases in O_3 is as great but the urban one is not very different. Surface mixing ratio increases in O_3 are higher than the total tropospheric changes:

Urban: Total: + 0.25%/year 0 km: + 0.35%/yr
Cont. ML: Total: + 0.21%/year 0 km: + 0.26%/yr
Low-lat.: Total: + 0.15%/year 0 km: + 0.19%/yr

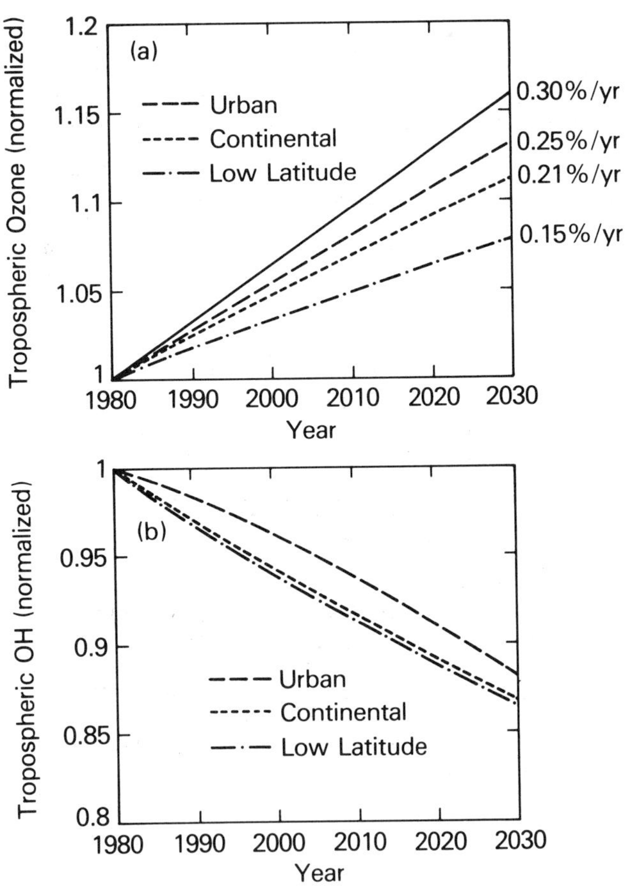

Fig. 2 Calculated changes (relative to 1980) for (a) tropospheric ozone; (b) tropospheric OH, CH_4, and CO changes follow Scenario 1 (Table 3).

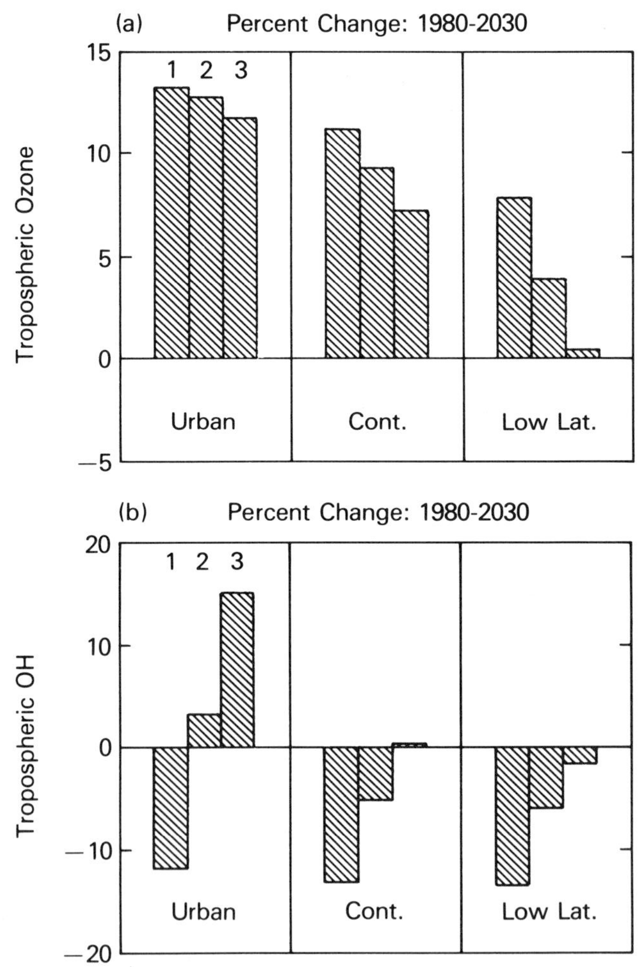

Fig. 3 Calculated changes (relative to 1980) for (a) tropospheric ozone; (b) tropospheric OH. Scenarios 1-3 are followed. (Table 3)

This means that one cause of large-scale tropospheric O_3 increases may be increasing atmospheric CH_4 (which has doubled since about 1800) and possibly CO. Although it has not been included in these scenarios, increasing NO_x may also contribute to increasing tropospheric O_3 [Dignon and Hameed, 1985]. Although the NO_x increases remain localized, O_3 from urban areas can live long enough to be transported far from source regions [Liu et al, 1987; Fishman, 1987] and its consequences may therefore appear in the recent trends. Figure 2b shows the losses in OH that accompany the O_3 increases in Fig. 2a because CO and CH_4 suppress OH levels. The relative magnitudes of OH loss correlate with the trends of the sensitivity coefficients (Table 1).

Figure 3 shows fractional O_3 increases and OH decreases calculated with all three scenarios at the 50-year point. In Scenario 2 the possibility was added of a 15% depletion in stratospheric O_3 [WMO, 1986] taking place at the end of 50 years. Ozone would be photolyzed more rapidly with more uv penetrating from the upper atmosphere to the troposphere and the $O(^1D)$ formed by photolysis reacts with H_2O to give more OH. In nonurban areas, this leads to some destruction of ozone, but superimposed on the specified CO and CH_4 increases, there is still a net increase in O_3 but not as great as before. Looking at the effect of these changes on OH (Fig. 3b) it is seen that in the urban case, the uv effect actually cancels out the loss of OH that CH_4 and CO increases bring about. In the nonurban situations they counteract the OH loss substantially (cut it in about half) but do not completely cancel it.

The effects of Scenario 3 on O_3 and OH after 50 years are also given in Fig. 3. In addition to the CH_4 and CO increases of Scenarios 1 and 2, and the stratospheric O_3 loss in Scenario 2, it is assumed in this calculation that there will be about 12% more H_2O vapor in the troposphere. This means more OH and less O_3 due to $O(^1D) + H_2O \rightarrow 2OH$. The effect is almost no net OH change in the nonpolluted atmosphere. Accompanying increases in O_3 are smaller than with the chemical perturbations alone.

4. SUMMARY AND CONCLUSIONS

A photochemical model has been used to calculate sensitivity of the tropospheric oxidants OH and O_3 to $CO/CH_4/NO_x$ emissions and to possible changes in stratospheric O_3 and tropospheric H_2O. Sensitivity coefficients provide a quantitative comparison of O_3 and OH changes in different regions.

Specific scenarios of $CO/NO/CH_4$ in a photochemical model simulating perturbed conditions of different regions show that CO and CH_4 increases suppress OH and enhance O_3 in nonpolluted areas. In urban regimes increases in CO and CH_4 add to OH. Stratospheric O_3 loss increases uv flux and tropospheric OH and decreases O_3 except in polluted regions. Increasing water vapor tends to increase OH and decrease O_3.

Actual changes in global tropospheric O_3 and OH will be a composite of regional $CO/NO/CH_4$ changes. Inasmuch as these calculations have not been done with a coupled chemistry-global transport model, they are of limited value in estimating global oxidant change. However, they are useful in pointing out regions of greatest sensitivity and uncertainty in 2- and 3-D models that are being developed for studying chemical change [Isaksen and Hov, 1987; Douglass et al, 1988]. They can also be used to test model parameterizations of chemistry based on simplified chemical schemes.

REFERENCES

Bojkov R. D., 1986: Surface ozone during the second half of the nineteenth century. J. Clim. Appl. Meteor., 25, 343-352.

Dignon J. and S. Hameed, 1985: A model investigation of the impact of increases in anthropogenic NO_x emissions between 1967 and 1980 on tropospheric ozone, J. Atmos. Chem., 3, 491-506.

Douglass, A. R., C. H. Jackman, R. S. Stolarski, 1988: Comparison of model results transporting the odd nitrogen family with results transporting separate odd nitrogen species, submitted to J. Geophys. Res.

Fishman J., 1987: Tropospheric ozone measurements from satellite total ozone measurements, Tropospheric Ozone, ed. by I. S. A. Isaksen, pp 111-123, D. Reidel, Dordrecht.

Isaksen I. S. A. and O. Hov, 1987: Calculation of trends in the tropospheric concentration of O_3, OH, CO, CH_4, and NO, Tellus, 39B, 271-285.

Levine J. S., C. P. Rinsland, and G. M. Tennille, 1985: The photochemistry of methane and carbon monoxide in the troposphere in 1950 and 1985, Nature, 318, 254-257.

Liu S., M. Trainer, F. C. Fehsenfeld, D. D. Parrish, E. J. Williams, D. W. Fahey, G. Hubler, and P. C. Murphy, 1987: Ozone production in the rural troposphere and the implications for regional and global ozone distributions, J. Geophys. Res., 92, 4194-4207.

Ramanathan V., R. J. Cicerone, H. Singh, and J. T. Kiehl, 1985: Trace gas trends and their potential role in climate change, J. Geophys. Res., 90, 5547-5566.

Thompson A. M. and R. J. Cicerone, 1986: Possible perturbations to atmospheric CO, CH_4 and OH, J. Geophys. Res, 91, 10853-10864.

Thompson A. M., R. W. Stewart, M. A. Owens, and J. A. Herwehe, 1988: Sensitivity of tropospheric oxidants to global chemical and climate change, submitted to Atmos. Environ.

CHEMICAL-RADIATIVE-DYNAMICAL MODEL STUDIES

ON THE TWO-DIMENSIONAL MODELLING OF THE STRATOSPHERE

Martyn P. Chipperfield
Department of Physical Chemistry
Cambridge, CB2 1EP, UK

Richard S. Eckman
NASA Langely Research Center
Hampton, Virginia 23665, USA

Lesley J. Gray
Rutherford Appleton Laboratory, Chilton, Didcot
Oxon, OX11 OQX, UK

Joanna D. Haigh
Centre for Remote Sensing, Imperial College
London, SW7 2BZ, UK

John A. Pyle
Department of Physical Chemistry
Cambridge, CB2 1EP, UK

ABSTRACT

The role of 2-D models is discussed. We consider the formulation ('classical Eulerian' versus 'diabatic') and argue that the most important consideration is that the physics and chemistry are included consistently. We show results from a classical Eulerian model which agree well with observations. One area of poor agreement is the polar lower stratosphere. The implications for studies of the Antarctic ozone hole and the future perturbed stratosphere are discussed. Finally, some calculations relevant to Antarctic ozone depletion are presented.

1. INTRODUCTION

In the past few years two-dimensional models have been developed by a number of groups so that they have now become the main tool for stratospheric assessment studies. Investigations into the treatment of eddy processes and the formulation of the mean circulation (see NASA/WMO, Chapter 12 for detailed discussion) have advanced the theoretical basis of these models. Two generic forms of model can now be identified: the classical Eulerian models which use the traditional Eulerian circulation with, for example, an indirect cell in the high latitude winter, and models which employ some kind of residual or diabatic circulation, after the theoretical work of Andrews and McIntyre (1976) and Dunkerton (1978). In the former, the diffusion treatment follows the work of Reed and German (1965) with higher diffusion coefficients than used in the residual or diabatic models. This has led to a misconception that diffusion is the most important factor in determining trace gas distributions in the classical Eulerian models and that the constituent fields calculated in these models will tend to be flatter. While these observations may be true for certain specific models, there is no reason for this to be generally true. Note, for example, the sensitivity of results in the model of Ko et al (1985), a diabatic model, to horizontal diffusion (see NASA/WMO figure 12.15).

The residual models arise from a reformulation of the dynamical equations. This reformulation has some advantages but contains no new physics. As long as the physics and chemistry are formulated correctly, either type of model should produce useful results. Rather than discussing the relative merits of different types of 2-D models, perhaps we should be concentrating on a question raised by Houghton (1978) in his Presidential Address to the Royal Meteorological Society, namely, why do 2-D models, despite their inherent limitations, achieve a successful balance between advection and diffusion, thus producing reasonable simulation of observed constituent fields?

In the next section, the important details of our 2-D model are briefly presented. Results from the model are then compared with observations. The agreement in general is good. An important area of discrepancy is the Antarctic lower stratosphere. We discuss the implications of this. Finally, some calculations specifically relevant to high latitude ozone depletion are shown.

2 THE MODEL

The model used in this study is the 2-D model developed by Harwood and Pyle (1975). The model extends from pole to pole and from the ground to ~90km with a resolution of ~9.5° in the horizontal and ~3.5km in the vertical. The model calculates not only the mixing ratios of a variety of constituents, but also the zonal mean winds and temperature.

The chemical scheme is included up to 60km and includes what are currently thought to be the most important O_X, NO_X, HO_X, ClO_X, BrO_X and CHO_X reactions using current photochemical data (NASA/JPL 1987). The levels of ClO_X and BrO_X are calculated using realistic surface emission rates for their source gases. Heterogeneous processes, when included, are treated as equivalent gas phase reactions, as discussed in more detail later.

The model includes a detailed scheme for radiative transfer throughout the stratosphere and mesosphere (Haigh 1984). For example, both solar u.v. heating and radiative transfer in the 9.6μm ozone band depend on the modelled ozone fields, allowing an important feedback between chemistry, radiation and dynamics. This radiative scheme is a particularly important advance in the lower stratosphere over some previous schemes used in the model.

ISBN 0-937194-15-8

3. SOME RESULTS FOR THE CONTEMPORARY ATMOSPHERE

In this short paper it is not possible to discuss the model performance in detail. Instead some relevant results are presented with some indication as to reasons for the model performance. A more detailed analysis will be given in a paper in preparation.

Figure 1 shows the latitude versus time section of the ozone column calculated for the mid-1980's. Apart from the failure to model the ozone hole (see later) the distribution agrees well with observations.

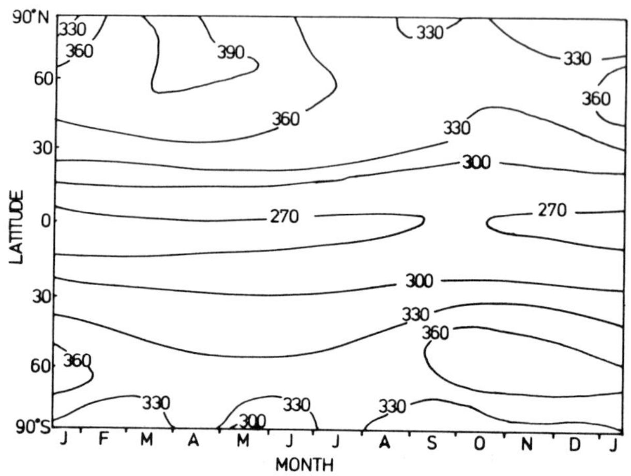

Fig. 1. Latitude versus time section of column ozone from the 2-D model calculated for the mid 1980's.

Figure 2 shows the ozone field in March compared with LIMS observations (Russell et al, 1986). The agreement is very good in terms of both magnitude and morphology. The model underestimates the observations by only a small amount in the upper stratosphere. We believe that the excellent temperature structure (fig. 3) in the model is the reason for such a small discrepancy.

Figure 3 shows modelled and observed temperatures. The agreement is very good. Compared with some earlier model calculations the stratopause temperature is lower and more realistic, as is the stratospheric lapse rate. These improvements result from the improved radiative schemes, as shown first by Haigh (1984).

The modelled field of total odd nitrogen agrees well with the field derived from LIMS data (see figure 10.68 NASA/WMO). In NASA/WMO, comparisons between different 2-D models and the NO_y fields were made. No model produced good agreement at all heights and latitudes. In particular, in our previous calculations

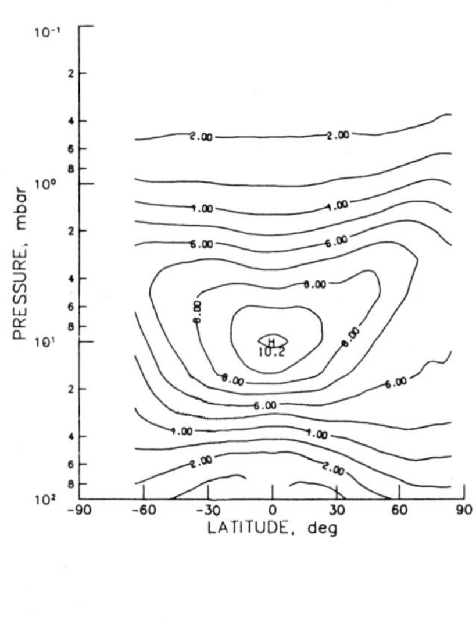

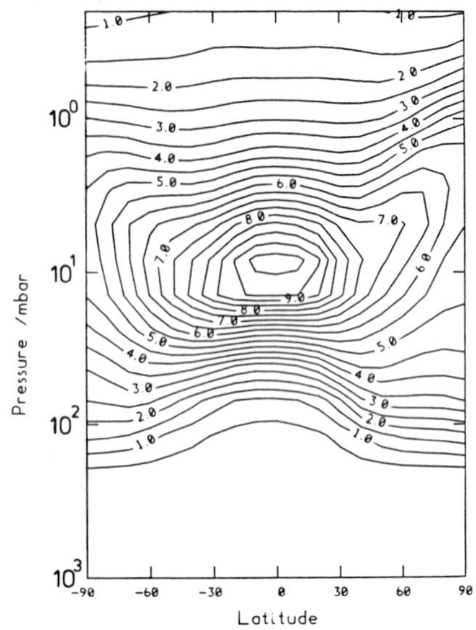

Fig. 2. a) LIMS monthly zonal mean ozone cross section (ppmv) for March 1979 (Russell et al, 1986) compared with b) 2-D model calculation of latitude versus height section of ozone for March.

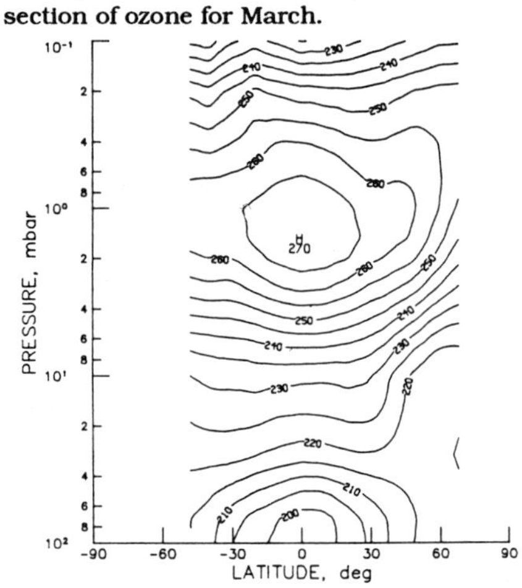

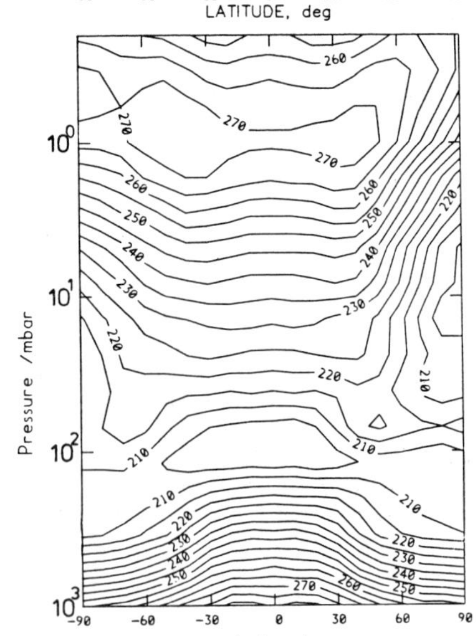

Fig. 3. a) SAMS monthly zonal mean temperature cross section in K) for March 1979 (Russell et al, 1986) compared with b) 2-D model calculation of latitude versus height section of temperature for March.

(see fig. 4 'Pyle') the fields were too flat, attributed by some to the fact that classical models are too diffusive. With the new radiative scheme and updated chemistry, fig. 4 shows excellent agreement for both magnitude and horizontal and vertical gradients. The effect of the radiative scheme is to strengthen the stratospheric extension of the equatorial Hadley cell. This consequently sharpens the horizontal gradients as well as advecting more N_2O into the middle stratosphere, thus producing higher NO_y mixing ratios. This strengthened circulation also produces the deepened equatorial minimum in total ozone found in these runs.

We conclude that the model is performing well in a number of respects and achieves a satisfactory balance between mean and eddy transports.

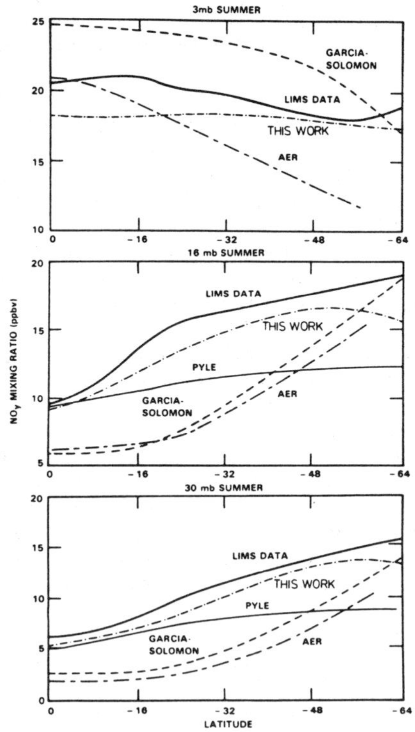

Fig. 4. Latitudinal gradients of NO_y observed by LIMS, and calculated in various 2-D models, at the 3mb, 16mb and 30mb levels. Figure adapted from NASA/WMO (1985) figure 10-69.

4 THE HIGH LATITUDE LOWER STRATOSPHERE

Figure 5 shows the modelled HCl cross section for September. HCl is a reasonable proxy for total active chlorine in both the upper stratosphere and the high latitude stratosphere below about 20km, in the absence of heterogeneous chemical processes. Notice the low mixing ratios of HCl in the polar lower stratosphere, a feature also shown in some 'diabatic' models (see figure 12.39 from NASA/WMO). Results from the recent Airborne Antarctic Ozone Expedition (Brune et al, 1988) and the NOZE Expedition (de Zafra et al, 1987) indicate that the levels of active chlorine in this region are very high and that the air is much depleted in the chlorine source gases. This can only be achieved by a downward transport from an area of strong photolysis in the upper stratosphere. 2-D models of whatever formulation do not appear to reproduce this transport. This is one reason why figure 1 shows no sign of an ozone hole. It casts doubt on the results of 2-D models for assessment studies. Furthermore, if the models underestimate active chlorine levels in the Antarctic lower stratosphere, it can

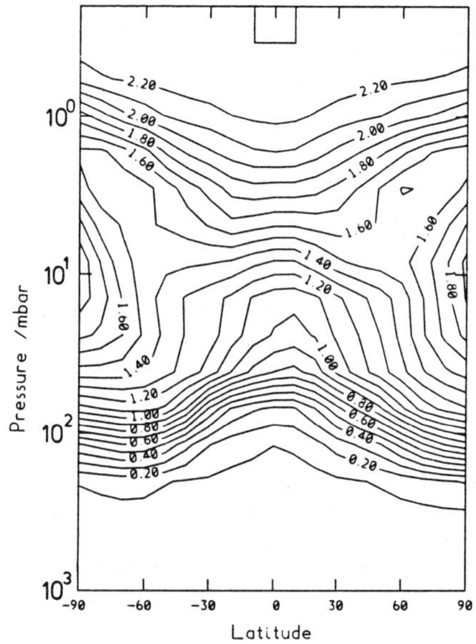

Fig. 5. Modelled latitude - height cross section of HCl (in ppbv) for September.

be anticipated that they are also deficient in chlorine in the middle latitude lower stratosphere. Since the lower stratosphere is the key to determining the ozone column change for simultaneous perturbation by CFCs, CO_2, CH_4, N_2O etc., this failure casts yet more uncertainty on 2-D calculations of ozone change into the next century.

To illustrate the necessity of bringing down high mixing ratios of active chlorine, a photochemical box model was used to test the sensitivity of the modelled ozone depletion to ClO_x levels. The model was run at 76°S and 17km, for 100 days from 19th July. The model has a sophisticated photolysis routine with a full diurnal cycle and spherical geometry. The following heterogeneous reactions were included (Solomon et al 1986)

$$N_2O_5 + H_2O \rightarrow 2HNO_3 \quad (1)$$
$$ClONO_2 + H_2O \rightarrow HOCl + HNO_3 \quad (2)$$
$$ClONO_2 + HCl \rightarrow Cl_2 + HNO_3 \quad (3)$$

These reactions were included in the model using equivalent first order rates of $3.23 \times 10^{-6} s^{-1}$, $6.45 \times 10^{-6} s^{-1}$ respectively for reaction 1 and 2 and a second order rate of 2.0 x

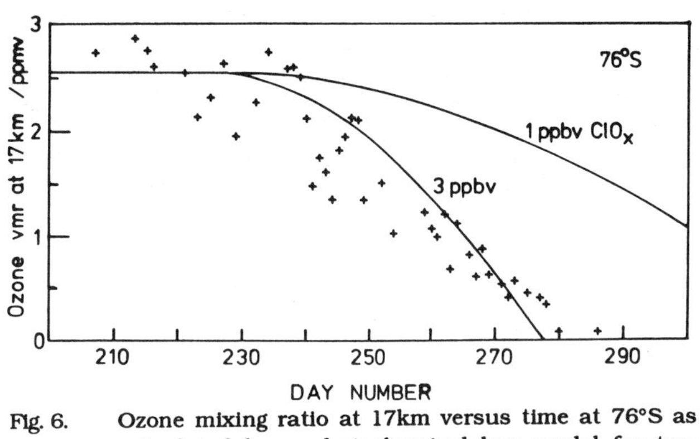

Fig. 6. Ozone mixing ratio at 17km versus time at 76°S as calculated by a photochemical box model for two levels of active chlorine. The + marks are measurements from Halley Bay in 1987. Day 244 is September 1st.

10^{-14} molecules^{-1} cm^{3}s^{-1} for reaction 3. These reactions serve to keep ClO levels high and allow for ozone destruction via the following cycle (Molina and Molina, 1986)

$$ClO + ClO + M \rightarrow Cl_2O_2 + M$$
$$Cl_2O_2 + h\upsilon \rightarrow ClO_2 + Cl$$
$$ClO_2 + M \rightarrow Cl + O_2 + M$$
$$\underline{2(Cl + O_3 \rightarrow ClO + O_2)}$$
$$\text{net } 2O_3 \rightarrow 3O_2$$

The ozone mixing ratio at 17km versus time is shown in fig. 6 for this model along with measurements from Halley Bay in 1987 (Gardiner and Farman, 1988). The difference in modelled ozone depletion between levels of ClO_x of 1ppbv and 3ppbv is striking.

Hence in order to produce an Antarctic ozone hole in a 2-D model not only must the relevant physico-chemical processes, both heterogeneous and homogeneous, be included but at some time of year downwelling must occur over Antarctica to produce high levels of ClO_x in the lower stratosphere. It appears necessary that this downwelling should occur before the onset of polar night while there is still the opportunity of photochemical destruction of the high ozone mixing ratios which will also be brought down. Descent during the winter would lead to large O_3 columns at the end of polar night. Anderson et al (1988) saw no evidence for higher O_3 columns at the start of the spring. The descent will also produce low levels of some gases such as CH_4 in the vortex. As noted by Salawitch et al (1988), this will help to maintain high ClO levels due to inhibition of the reaction:

$$Cl + CH_4 \rightarrow HCl + CH_3$$

Finally, we note that even when the relevant physical and chemical processes outlined above are included in our 2-D model, along with high ClO_x mixing ratios in the lower Antarctic stratosphere, a realistic ozone hole can only be produced when an isolated vortex is imposed on the model. This can be achieved by reducing drastically the eddy diffusion coefficients at ~70°S (Chipperfield and Pyle, 1988).

5. CONCLUSIONS

Results from a classical Eulerian 2-D model have been presented, showing excellent agreement with observations of O_3, temperature and satellite derived NO_y. It is concluded that this good agreement, better than in our previous calculations, arises from the use of a newer radiative scheme and updated chemistry, and that the model achieves a satisfactory balance between mean and eddy transports. This balance supports our conclusion that either classical Eulerian or diabatic models may be used to study stratospheric processes and that their successes will depend on the inclusion of the correct physics and chemistry within the individual models.

This 2-D model (and others) does not reproduce the high levels of active chlorine found in the lower polar stratosphere . This has serious implications for studies of both polar ozone depletion and future perturbations of the ozone layer.

ACKNOWLEDGEMENTS

The work at the University of Cambridge was supported by the U.K. Department of the Environment. MPC thanks NERC for a studentship.

REFERENCES

Anderson, J.G., W.H. Brune and M.J. Proffit, Ozone Destruction by Chlorine Radicals within the Antarctic Vortex: The Spatial and Temporal Evolution of ClO-O_3 Anticorrelation Based on In Situ ER-2 Data, J. Geophys. Res., (submitted), 1988.

Andrews, D.G., and M.E. McIntyre, Planetary waves in horizontal and vertical shear: The generalized Eliassen-Palm relation and the mean zonal acceleration, J. Atmos. Sci., **33**, 2031-2048, 1976.

Brune, W.H., J.G. Anderson and K.R. Chan, In Situ observations of ClO in the Antarctic: ER-2 Aircraft Results from 54°s to 72°s Latitude, J. Geophys. Res., (submitted), 1988.

Chipperfield, M.P. and J.A. Pyle, Two-dimensional modelling of the Antarctic lower stratosphere, Geophys. Res. Lett., (submitted), 1988.

Dunkerton, T., On the meridional mass motions of the stratosphere and mesosphere, J. Atmos., Sci., **35**, 2325-2333, 1978.

Gardiner, B .G. and J.C. Farman, Results of the 1987 Ozonesonde Programme at Halley Bay, Antarctica, British Antarctic Survey Report, 1988.

Haigh, J.D., Radiative heating in the lower stratosphere and the distribution of ozone in a two-dimensional model, Quart. J.R. Meteorol. Soc., **110**, 167-185,. 1984.

Harwood, R.S., and J.A. Pyle, A two-dimensional mean circulation model for the atmosphere below 80km, Quart. J.R. Meteorol. Soc., **101**, 723-748, 1975.

Houghton, J.T., The stratosphere and mesosphere, Quart. J.R. Meteorol. Soc., **104**, 1-29, 1978.

Ko, M.K.W., K.K. Tung, D.K. Weinstein and N.D. Sze, A zonal-mean model of stratospheric tracer transport in isentropic coordinates: Numerical simulations for nitrous oxide and nitric acid, J. Geophys. Res., **90**, 2313-2329, 1985.

Molina, L.T., and M.J. Molina, Production of Cl_2O_2 from the self-reaction of the ClO radical, J. Phys. Chem., **91**, 433-436, 1986

NASA/JPL Chemical kinetics and photochemical data for use in stratospheric modelling, Evolution No. 8, JPL Publication 87-41, 1987.

NASA/WMO, Atmospheric Ozone, Report No. 16, 1985.

Reed, R.J. and K.E. German, A contribution to the problem of stratospheric diffusion by large-scale mixing, Mon. Wea. Rev., **93**, 313-321, 1965.

Russell, J.M., S. Solomon, M.P. McCormick, A.J. Miller, J.J. Barnett, R.L. Jones and D.W. Rusch, Middle Atmosphere Composition Revealed by Satellite Observations, Handbook for MAP, volume 22, 1986.

Salawitch, R.J., S.C. Wofsy and M.B. McElroy, Influence of Polar Stratospheric Clouds on the Depletion of Ozone over Antarctica, Geophys. Res. Lett., (submitted), 1988.

Solomon, S., R.R. Garcia, F.S. Rowland and D. Wuebbles, On the depletion of Antarctic ozone, Nature, **321**, 755-758, 1986.

de Zafra, R.L., M. Jaramillo, A. Parrish, P. Solomon, B. Connor and J. Barrett, High concentrations of chlorine monoxide at low altitudes in the Antarctic spring stratosphere: diurnal variation, Nature, **328**, 408-411, 1987.

THE ROLE OF PARAMETERIZED EDDY MIXING IN MODELING THE STRATOSPHERIC CIRCULATION AND TRACER DISTRIBUTIONS

Arthur Y. Hou, Hans-Rainer Schneider, and Malcolm K.W. Ko
Atmospheric and Environmental Research, Inc.
Cambridge, Massachusetts 02139, USA

ABSTRACT

The effects of parameterized quasigeostrophic eddy mixing on the zonally averaged circulation and tracer slopes are examined in the context of idealized and realistic models of the stratosphere.

As a result of nonlinear advection of relative angular momentum, the effect of reduced eddy mixing in the stratosphere is to contract the meridional width of the residual circulation (thereby limiting its poleward transport) and to intensify the circulation locally, resulting in a steeper meridional tracer slope relative to the isentrope. This is characteristically different from the behavior of a geostrophic circulation, which is constrained to vanish as the eddy mixing tends to zero.

Model simulations with a realistic solar heating cycle suggest that meridional contraction of the residual circulation resulting from weaker wave activities in the Southern Hemisphere can provide a simple explanation for the asymmetry in the observed column ozone maxima between the Northern and Southern springs.

1. INTRODUCTION

The purpose of this work is to investigate the response of the zonal-mean transport circulation to a parameterized mixing by quasigeostrophic waves in the stratosphere, with a special emphasis on the meridional extent of the circulation and meridional tracer slopes.

Recently, Hou and Ko (1988) showed that as the planetary wave forcing weakens, the residual circulation in a nonlinear, ageostrophic channel model does not vanish, so that the solution does not approach the theoretical 2-D radiative equilibrium state, unlike the linear, geostrophic solution of Holton (1986). Using an idealized, steady-state model of the stratosphere, they showed that the effect of reduced eddy mixing of potential vorticity is to contract the width of the residual circulation and to steepen the meridional slope of a long-lived tracer relative to the isentrope. A brief summary of these results is given in section 2.

In section 3, we verify these findings in a realistic simulation of the column ozone distribution using a high-resolution version of the interactive radiative-chemical-dynamic model of Schneider et al. (1988). Specifically, we seek to determine whether the asymmetry in the observed ozone maxima between the Northern and Southern springs can be explained in terms of a meridional contraction of the residual circulation due to low wave activities in the Southern Hemisphere. We summarize our conclusions in section 4.

2. COMPARISON OF GEOSTROPHIC AND AGEOSTROPHIC CIRCULATIONS

Some of the key issues concerning the effect of eddy mixing on the mean transport circulation are: Does the mean circulation vanish in the absence of eddy mixing? Does the stratosphere approach radiative equilibrium as the eddy mixing decreases? What happens to tracer slopes? Results from geostrophic model studies (e.g. Holton, 1986) suggest that as the eddy mixing vanishes, the mean circulation vanishes, as do tracer slopes relative to the isentropes. However, from the work of Held and Hou (1980) on nonlinear Hadley cells, it would appear that the answer depends on whether the nonlinear advection of relative angular momentum becomes important compared with the Coriolis torque as the eddy mixing decreases.

To investigate the characteristic response of linear, geostrophic and nonlinear, ageostrophic circulations to varied amounts of mixing of potential vorticity by quasigeostrophic waves, Hou and Ko (1988) adopted a system of equations for an idealized channel flow identical to that of Holton (1986), except for adding the nonlinear ageostrophic terms (in brackets) in the zonal momentum equation; i.e.,

$$[\overline{V u_y} + \overline{W u_z}] - f\bar{V} = \nabla\cdot(\mathbf{V}\bar{M}) = \sigma_o\overline{v'q'} = -\sigma_o K\bar{q}_y, \qquad (1)$$

where the overbar denotes the zonal average; $\bar{u}$ is zonal velocity, $\mathbf{V} = (\bar{V},\bar{W}) = \sigma_o(\bar{v},\bar{w})$ is the meridional mass flux vector in isentropic coordinates ($\bar{v}$ and $\bar{w}$ are the horizontal and "vertical" velocities, respectively), $\bar{q}_y = \bar{u}_{yy} - (F/N)^2(\sigma_o\bar{u}_z)_z/\sigma_o$ is the quasigeostrophic mean potential vorticity gradient on a constant f-plane (neglecting the variation of f with latitude does not alter the basic physics), and K is a constant eddy mixing coefficient; $\sigma_o(z) = \sigma_s\exp(-z/H)$ is the mean density, σ_s is a reference density, $z = Hc_pR\ \ln\theta$ is the vertical coordinate, $H = RT_*/g$ is a mean scale height, R is the gas constant, T_* is a mean stratospheric

ISBN 0-937194-15-8

temperature, g is the gravitational acceleration, c_p is the heat capacity of air at constant pressure, and θ is the potential temperature. $\bar{M} = \bar{u} - fy$ is the absolute angular momentum per unit mass divided by the earth's radius, y is latitude, and f is the Coriolis parameter.

In addition to (1), the model consists of the thermal wind equation, the thermodynamic equation and the mass continuity equation (see Holton, 1986). The behavior of this system may be understood by examining the limiting cases of strong and weak eddy mixings. In the case of strong eddy mixing (which acts to destroy the gradients of zonal velocity, as will be shown), the ageostrophic advection is small compared with the Coriolis term (i.e. $\bar{u}_y \ll f$) and (1) reduces to the geostrophic approximation,

$$f\bar{V} = \sigma_o K\bar{q}_y. \qquad (2)$$

As $K \to 0$, the meridional velocity is directly proportional to K (since $\bar{u}$ is bounded). The result that the departure from radiative equilibrium being proportional to the mixing of potential vorticity by planetary waves is thus a necessary consequence of the geostrophic approximation.

In the limit of small eddy mixing, (1) becomes,

$$\nabla\cdot(\mathbf{V}\bar{M}) \simeq 0 \qquad (3)$$

which states that the absolute angular momentum is effectively conserved. As $K \to 0$, (3) may be satisfied in either of two ways: The first is that $\mathbf{V} \to 0$ (implying that the net diabatic heating tends to zero and the system approaches the radiative equilibrium solution). The second is that $\bar{M} \to$ constant, while the meridional circulation and the net diabatic heating remain finite; an example of this is the nonlinear Hadley cells obtained by Held and Hou (1980) for the troposphere.

To determine which limit is appropriate for the circulation in the stratosphere, we solved both the geostrophic (2) and ageostrophic (1) versions of the model for the following boundary conditons: $\bar{w}(y,D) = \partial\bar{u}/\partial z(y,D) = 0$ (where D is model height), and $\bar{u}(y,0)$ = zonal wind at the tropopause, in conjunction with an open meridional mass flux at z = 0.

Figure 1 shows the calculated streamfunctions. As the value of K decreases, the ageostrophic circulation undergoes a pronounced meridional contraction, while the geostrophic circulation weakens proportionally with K without significant structural changes. The inviscid limit to which the ageostrophic circulation contracts may be determined from the analytic model of Held and Hou (1980). For the model parameters used here, the limiting width of the inviscid meridional cell is y/L = 0.45. As the circulation contracts, the maximum sinking motion (not shown) is displaced significantly equatorward: it is located at y/L = 1 for K = 10^5 m^2/s and y/L = 0.4 for K = 10^4 m^2/s.

Figure 2 compares the results for the meridional slope at the mid-channel for a long-lived tracer with a photochemical loss rate

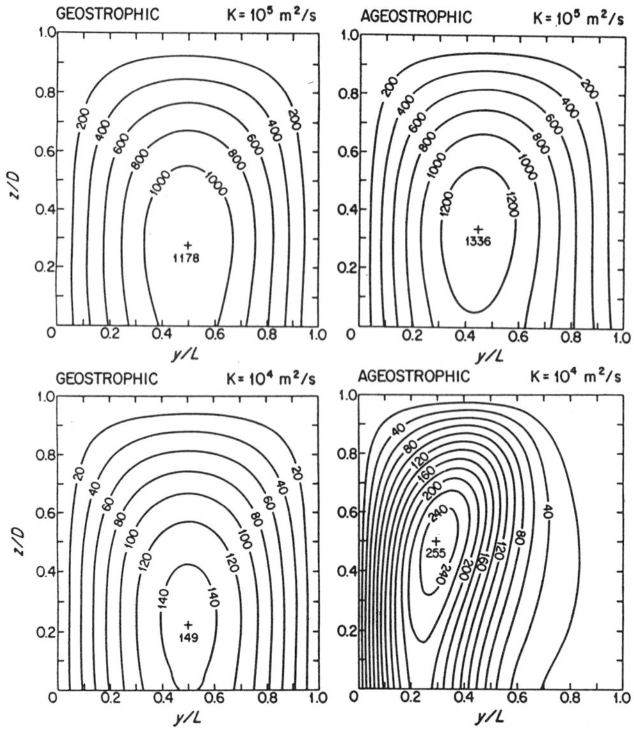

Fig. 1. Calculated meridional streamfunctions in units of m^2/s for the channel model. As K decreases from 10^5 to 10^4 m^2/s, the residual circulation of the geostrophic model decreases roughly linearly with K, without significant strctural changes, while that of the ageostrophic model undergoes a pronounced meridional contraction, with the downward branch moved equatorward.

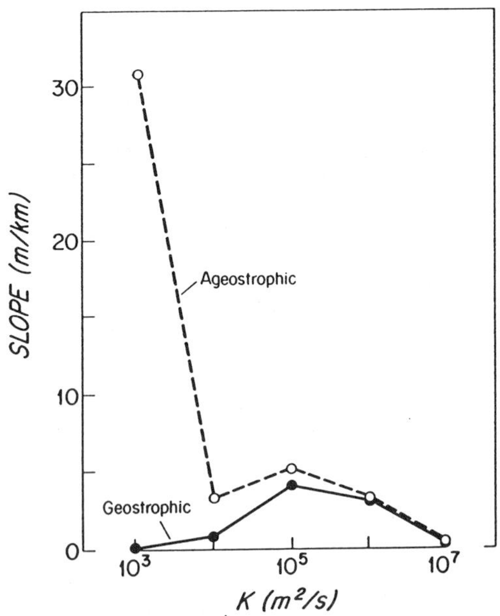

Fig. 2. Meridional tracer slopes between the maximum rising and sinking motions for a long-lived tracer (see text for details).

comparable to that of N_2O at 35 km. The response of tracer slopes to reduced eddy mixing is qualtitatively different for the two models.

3. REALISTIC SIMULATIONS WITH AN ANNUAL CYCLE

To test whether the simple model results of section 2 remain valid for a realistic 2-D model with an annual solar heating cycle, we performed a 2-year simulation using a high-resolution (3° latitude) version of the interactive radiative-chemical-dynamical model of Schneider et al. (1988). The model consists of a primitive-equation zonally-averaged dynamics module, a full radiative transfer scheme, and a simplified HO_x ozone model with adjusted loss terms. The model can be run with very little diffusion, making it possible to study the model behavior in the inviscid limit.

We performed a simulation with the eddy diffusion coefficient set to 3×10^5 m^2/s in the Northern Hemisphere and 10^4 m^2/s in the Southern Hemisphere. The same coefficients were used for the potential vorticity flux in the zonal momentum equation and the eddy mixing in the ozone transport equation. These different eddy mixing values for the Northern and Southern Hemispheres are the only imposed hemispheric asymmetry in the model.

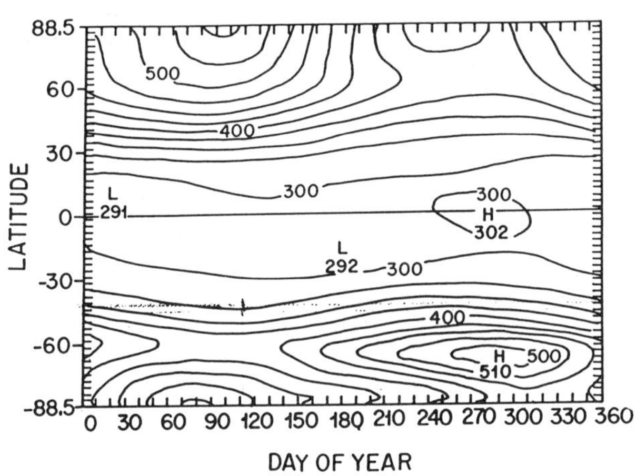

Fig. 3. The column ozone map for model year 2. The values are in Dobsons, and the time in Julian days.

Figure 3 shows the calculated ozone column abundance for model year 2. The equatorial ozone amounts are slightly higher than observed, but the asymmetry between the amounts at high latitudes in the two hemispheres bear a direct resemblance to observations (Bowman and Krueger, 1985).

Note also in Fig. 3 that the meridional slope of the column ozone at the high latitudes (45°-60°) is steeper for the Southern spring (Day 270) than for the Northern spring (Day 90); this steepening in the presence of small eddy mixing was also anticipated by the simple model (cf. Fig. 2).

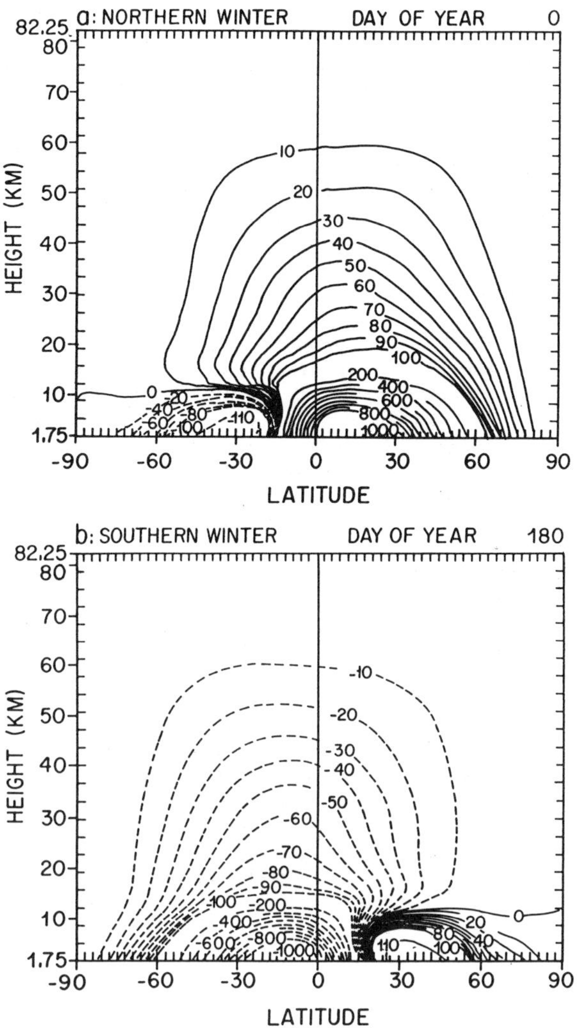

Fig. 4. Meridional mass streamfunctions for (a) Northern winter (Day 0), (b) Southern winter (Day 180). The values are scaled by the surface density and given in units of m^2/s.

Figure 4 shows the calculated mass streamfunctions for two selected days. During the Northern winter (Day 0, Fig. 4a) the sinking branch of the residual circulation extends further into the pole region, as compared with that for the Southern winter (Day 180, Fig. 4b), as is evident from the equatorward displacement of the streamlines. In the 15-20 km region, the 50 m^2/s contour for the sinking branch is at 45°S in Fig. 4b, compared with 60°N in Fig. 4a. These calculations show that, in the presence of the smaller eddy mixing in the Southern Hemisphere, the downward branch of the residual circulation is displaced away from the Southern winter pole and ozone accumulates most effectively around 60°S until the onset of spring. By contrast, during the Northern winter the maximum subsidence occurs over the pole to build a column ozone maximum at the pole.

We have also performed a calculation with an eddy mixing coefficient which is 10^4 m^2/s everywhere except for a 40° latitudianal band centered around 50°, where it increases to 5 x 10^5 m^2/s, more in keeping with the structure suggested by data (Newman et al., 1986). The Southern spring total ozone maximum still occurs away at subpolar latitutes, suggesting that ageostrophic effects can be important in controlling the global structure of the residual circulation if the areas of large eddy mixing are localized.

4. CONCLUSIONS

The main conclusions of this study are:

As eddy mixing decreases, the zonally averaged stratospheric circulation approaches an effectively inviscid, thermally controlled, nonlinear state, instead of the theoretical, 2-D radiative equilibrium solution.

The effect of reduced eddy mixing is to contract the meridional width of the residual circulation (thereby limiting its poleward transport) and to confine the circulation to the maximum heating region. The circulation can intensify locally, resulting in steeper tracer slopes relative to the isentropes, within the domain of the circulation.

Numerical simulations with an annual solar cycle show that the displacement of the maximum sinking motion away from the pole as a result of reduced wave mixing can produce a subpolar maximum in the column ozone. Given that the planetary wave forcing is generally weaker in the Southern Hemisphere, this could account for the asymmetry in the observed column ozone maxima between the Northern and Southern springs.

ACKNOWLEDGMENTS

This work was supported by NASA Contract NASW-4080 and by the Fluorocarbon Program Panel of the Chemical Manufacturers Association.

REFERENCES

Bowman and Krueger, 1985: A global climatology of total ozone from the Nimbus 7 total ozone mapping spectrometer. J. Geophys. Res., **90**, 7967-7976.

Held, I. M., and A. Y. Hou, 1980: Nonlinear axially symmetric circulations in a nearly inviscid atmosphere. J. Atmos. Sci., **37**, 515-533.

Holton, J. R., 1986: Meridional distribution of stratospheric trace constituents. J. Atmos. Sci., **43**, 1238-1242.

Hou, A. Y., and M. K. W. Ko, 1988: The influence of nonlinear advection of zonal momentum on the residual circulation and tracer distribution in the stratosphere. Submitted to J. Atmos. Sci.

Newman, P. A., M. R. Schoeberl, and R. A. Plumb, 1986: Horizontal mixing coefficients for two-dimensional chemical models calculated from National Meteorological Center data. J. Geophys. Res., **91**, 7919-7924.

Schneider, H.-R., M. K. W. Ko, N. D. Sze, G.-Y. Shi, and W.-C. Wang, 1988: The effect of eddy diffusion on the meridional circulation and ozone distribution in an interactive 2-D model. Submitted to J. Atmos. Sci.

THE EFFECT OF PARAMETERIZED FRICTION AND EDDY DIFFUSION ON OZONE PERTURBATION CALCULATIONS IN A 2-D MODEL

Hans R. Schneider and Malcolm K.W. Ko
Atmospheric and Environmental Research, Inc.
Cambridge, Massachusetts 02139, USA

ABSTRACT

A variety of ozone perturbation calculations have been performed with a 2-D model of dynamics, radiation and chemistry. Interactive calculations with various values of friction and eddy diffusion are compared with each other and with offline calculations where a predetermined streamfunction was used for the transport. A simplified HOx ozone model is used and increased chemical ozone loss terms are simulated by increasing the water vapor content in the stratosphere. It is shown that the ratio of the perturbed to the unperturbed ozone amounts is remarkably insensitive to the parameters used. The radiative dynamical feedback is shown to have a relatively small effect, amounting to about one third compensation in local mixing ratio differences in the upper stratosphere and one tenth in the ozone column amounts. The results are interpreted with the aid of stationary linear models for the diabatic circulation.

1. INTRODUCTION

Predictions of 2-D stratospheric models for a chemically perturbed environment are difficult to interpret because most models do not include all the feedback mechanisms between the radiatively active trace gases and the transport circulation. Many assessment calculations have been made with models that use a predetermined temperature and circulation without any feedback (e.g. WMO, 1985).

In recent years, efforts have been made to include more and more of the feedback between chemistry and dynamics into 2-D models. It is now state of the art to use coupled models of zonal mean dynamics, chemistry and radiative transfer. These interactive models are quite expensive to run and form a complex system in themselves. The different parts of the model often interact with each other in non-trivial ways. Furthermore, the results of these models are very sensitive to the parameterizations of gravity wave drag and the diffusion induced by large scale Rossby waves (Schneider et al.,1988, hereafter referred to as SKSSW).

If a change in the ozone distribution occurs, the solar heating in the stratosphere changes and, as a result, an adjustment in temperature, meridional and zonal winds takes place. The purpose of this paper is to provide an estimate of the importance of the radiative dynamic feedback on ozone perturbation calculations and to investigate the sensitivity of assessment calculations on the choice of eddy diffusion coefficients in an interactive 2-D model.

Results of interactive and offline calculations are compared with each other. It is shown that, for the cases studied, the feedback effect is comparatively small and it is unlikely that it will increase substantially by allowing for a change in the eddy diffusion coefficients as a function of the zonal mean state. With the adopted formulation used in this 2-D model, although the absolute amounts of ozone calculatd in the model depend on the parameters used, the percent changes of ozone in the perturbation calculations are quite insensitive to the parameters.

The perturbation experiments with different parameters and model configurations are described in Section 2. In section 3, the model results are interpreted with the aid of linear stationary models of the 2-D circulation. Conclusions are summarized in Section 4.

2. OZONE PERTURBATION CALCULATIONS

The 2-D model used for this study has been described in detail in SKSSW. It consists of a primitive equation zonal mean dynamics module, a full radiative transfer scheme and a simplified HO_x ozone model with adjusted loss terms. The flux of eddy (quasigeostrophic) potential vorticity is assumed to be proportional to the mean gradient of the potential vorticity with the same coefficient that is used for the eddy diffusion in the tracer transport equation. It was shown in SKSSW that ozone distributions generated with the model are very sensitive to the amount of mechanical damping and the exact value of the eddy diffusion coefficients used in the stratosphere.

In order to simulate increased ozone losses with the simplified chemical model used here, the (fixed) water vapor content in the stratosphere was increased from 6ppm for the baselines to 12 ppm for the perturbation runs.

Fig. 1 shows the calculated ozone column amounts for the baseline and the percentage difference due to increased ozone loss terms obtained with an interactive calculation using a frictional timescale of 30 days and an eddy diffusion coefficient of $4x10^5$ m^2/sec in the

ISBN 0-937194-15-8

region between 10 and 40 km. The eddy diffusion coefficients, K_{yy}, are reduced in the tropics and vary with the time of the year as desribed in SKSSW. The global ozone reduction amounts to about 11%.

Fig. 2 represents results of an off-line calculation where streamfunction and temperatures were specified independent of the calculated O_3 distributions.

Several other cases with different combinations of friction and K_{yy} have been studied. The base-line ozone concentrations for these cases are significantly different from each other (see also SKSSW). Nevertheless, the percentage differences in ozone amounts between a perturbed and an unperturbed atmosphere seems to be relatively stable within three percentage points for a large range of parameters.

Fig. (3a) shows the January ozone mixing ratios obtained in the interactive run with normal water vapor content and the percentage difference for the high water vapor content. The ozone reduction at high altitudes is about 15%. In the winter polar regions of the stratosphere mixing ratios drop by about 25%. A self healing effect of about 6% due to increased UV-penetration occurs in the ozone source region above the tropical tropopause. Temperature reductions (not shown) in the upper stratosphere are about 4 deg., in the middle and lower stratosphere they are of the order of 1 deg. The differences in the zonal mean wind between the two runs with regular and increased ozone loss terms is very small, about 2 m/sec in the polar night jets and less than .5 m/sec everywhere else. Therefore, feedback on K_{yy} is not expected to be of great significance.

In order to estimate the radiative-dynamical feedback, the increased water vapor run was repeated in the off-line mode. The meridional streamfunction and temperatures of the baseline were used to transport ozone and calculate the chemical reaction rates for the increased water vapor conditions. The ozone difference to the interactive perturbation calculation is shown in Fig. 4. The radiative feedback amounts to an increase of about 5% in ozone in the upper stratosphere and about 2% or less in the lower layers of the atmosphere. The high altitude gain is due to the reduced temperature in the interactive run, the differences in the lower layers are about equally divided between the temperature and the streamfunction change. In the total column, the feedback amounts to 2 to 3% at high latitudes.

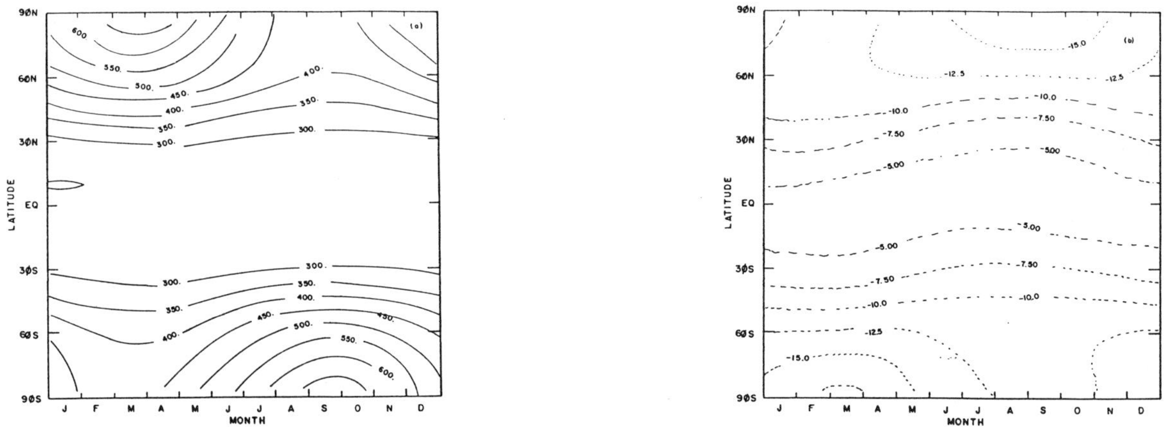

Fig. 1. (a) Ozone column amounts (Dobson units), obtained in an interactive run.
(b) the percentage differences in the column amounts due to increased chemical loss terms (Run 1b) (see text).

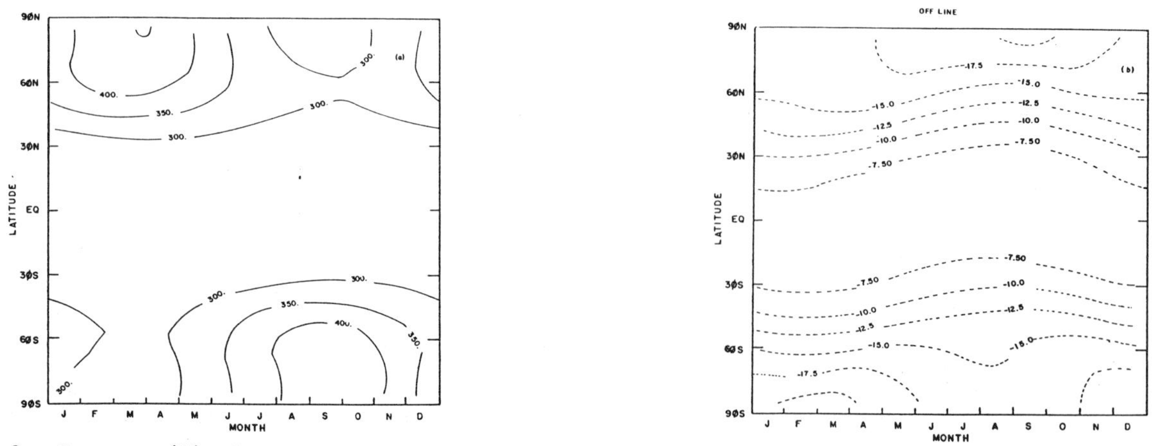

Fig. 2. Same as (1), but off-line calculations with a predetermined streamfunction for both baseline and perturbation.

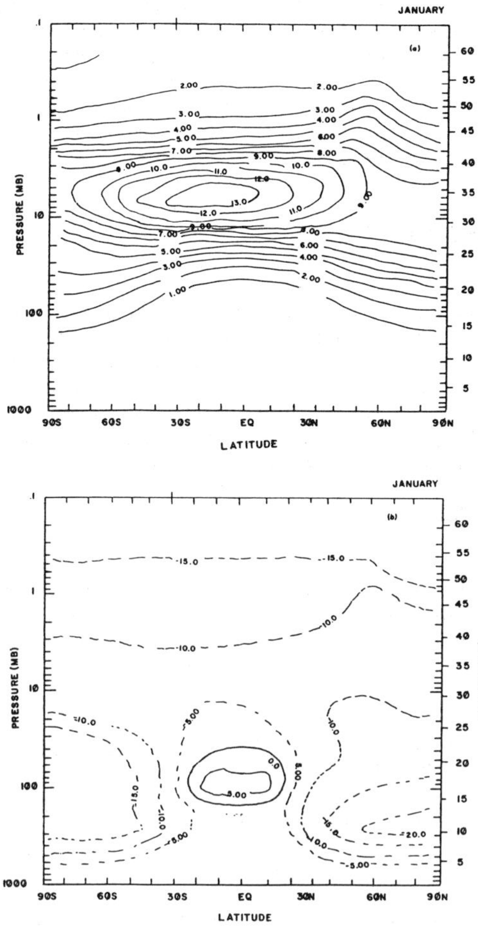

Fig. 3. (a) January ozone mixing ratios (ppm) for the interactive baseline calculation.
(b) Percentage differences in mixing ratio for the interactive perturbation.

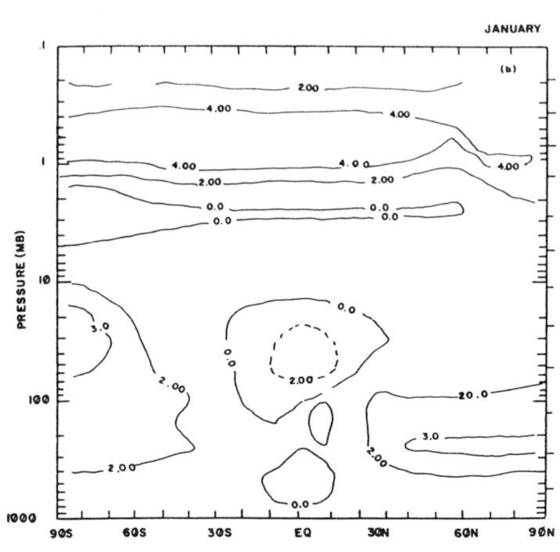

Fig. 4. Percentage difference in January mixing ratios between the interactive perturbation run and an off-line perturbation run.

3. LINEAR MODELS OF OZONE PERTURBATION

Assuming Rayleigh friction and Newtonian cooling, the zonal mean fields are to 0th. order governed by (e. g. Fels et al., 1980; Garcia, 1987)

$$f\bar{v} = \beta_R \bar{u} \quad (1)$$

$$\bar{w}S = \bar{J} - \alpha\bar{T} \quad (2)$$

together with the continuity equation and the thermal wind equation which are also linear.

Here, $\bar{u}$, $\bar{v}$ and $\bar{w}$ are the zonal, north-south, and vertical wind components, f is the Coriolis parameter, β_R the Rayleigh friction coefficient, $\bar{J}$ the input heating, S the static stability, α the Newtonian cooling coefficient. We allow α and β_R to be functions of z, the static stability is assumed constant.

Many authors have commented on parameter ranges where the system is radiatively or dynamically dominated depending on the ratio of α and β_R. It is worth noting, however, that as far as perturbations are concerned the distinction between radiative or dynamic response of the above system dissappears. If we have a solution to a given forcing $\bar{J}$, and if $\bar{J}$ is replaced by $(1+\delta)$ $\bar{J}$, the new solution is obtained by multiplying all fields by $(1+\delta)$. This is true for all values of $\alpha(z)$ and $\beta_R(z)$ as long as δ is a constant with respect to z. For arbitrary functions $\delta(z)$ it can be shown that the response to a solar heating perturbations is limited to a height range of less than two scale heights.

Numerical solutions for altitude dependent α and β_R show that the changes in all variables remain nearly proportional to each other for $\bar{J}$ - perturbations of sufficiently large vertical scale. As long as the atmosphere is in linear balance, its response to solar heating perturbations is neither radiative nor dynamical. All fields respond proportionally to the forcing perturbation. These results also hold if the Rayleigh friction in (1) is replaced by a quasigeostropic potential vorticity flux as in Holton (1986). Details of the response of linear systems will be presented elsewhere.

To understand the response of the full model we have looked at the differences between the solar UV heating for the perturbed and the unperturbed ozone for the interactive runs. In the lower stratosphere, they amount to 5 to 8% in the poleward half of the winter hemisphere, i.e. in the downward branch of the Brewer Dobson circulation which determines the column abundances. The full model responds with a 3 to 4% change in the net heating and, therefore, the transport circulation. This is the same magnitude as the difference in the column density changes observed between offline and interactive calculations.

4. CONCLUSIONS

Ozone perturbation calculations have been performed in both offline and fully interactive model setups. The values of the mechanical damping time and the eddy diffusion have been varied across a range of numbers consistent with previous modeling experience and observations.

Despite the sensitivity of the baseline runs to the precise value of the dissipation/diffusion parameters, the perturbation calculations exhibit roughly the same percentage change in the ozone colums for a variety of very different baseline distributions consistent with the response expected from a more or less linear system.

In a linear system, dynamical quantities respond to an ozone change with the same factor by which the solar input heating changes as result of the chemically induced change in ozone. It seems that this factor is an upper bound for the radiative feedback on the ozone distributions in an interactive 2-D model. In most calculations presented in this paper, the actual temperature and circulation change was about one half the ratio of the solar input heating differences divided by the unperturbed solar heating.

Although a simplified chemistry was used, the results presented here are still relevant. The ozone reductions obtained in our experiments were of the order 20 to 25% in the upper stratosphere and the polar regions of the lower stratosphere which is the range of changes expected in many assessment scenarios. Column ozone decresed from about 5% in the tropics to about 17% at high latitudes. The radiative dynamical feedback amounted to about 5% in local mixing ratios and 3% in ozone column densities at high latitudes. The temperature effect was dominant in the upper stratosphere, circulation and temperature changes were about equally important in the lower stratosphere and are reflected in the column amounts.

In the present study, eddy diffusion coefficients were specified externally and not modified as a function of the mean flow changes. The zonal mean flow changes observed in the model are very small and, at present, no parameterized description of the eddy mixing is good enough to be triggered by such small changes.

The perturbation calculations presented here illustrate the behavior of 2-D models of the stratosphere, not necessarily the behavior of the atmosphere. The validity of the results must await comparison of the simulated ozone trend using a model with full chemistry with the observed ozone trend.

ACKNOWLEDGEMENTS

This work was sponsored by the National Aeronautics and Space Administration (NASA-4080) and by the Fluorocarbon Panel Program of the Chemical Manufacturers Association. Thanks are due to to Arthur Y. Hou for many helpful discussions during the course of this work and to Debra K. Weisenstein for her help with the ozone model and the graphics software.

REFERENCES

Fels, S. B., J. D. Mahlman, M. D. Schwarzkopf and R. W. Sinclair (1980): Stratospheric sensitivity to perturbations in ozone and carbon dioxide: radiative and dynamical response. J. Atmos. Sci., 37, 2265-2297.

Garcia, R. R. (1987): On the mean circulation of the middle atmosphere. J. Atmos. Sci., 44, 3599-3605.

Holton, J. R. (1986): Meridional distribution of stratospheric trace constituents. J. Atmos. Sci., 43, 1238-1242.

Schneider, H. R., M. K. W. Ko, G.-Y. Shi, N. D. Sze and W.-C. Wang (1988): The effects of eddy diffusion, drag and nonlinear terms on the meridional circulation and ozone distributions in an interactive 2-D model. Submitted to J. Atmos. Sci..

WMO 1985: Atmospheric ozone. World Meteorological Organization, Global Ozone Research and Monitoring Project - Report No. 16.

SOLAR RADIATIVE HEATING DUE TO ABSORPTION BY OZONE

K.P. Shine and J.A. Rickaby
Department of Atmospheric, Oceanic and Planetary Physics
University of Oxford
Oxford OX1 3PU, UK

ABSTRACT

The radiative heating due to the absorption of solar radiation by ozone is calculated using recent measurements of the absorption cross-section of ozone and the extraterrestrial solar irradiance. It is shown that the difference between two sets of absorption cross-sections leads to only very minor changes in the heating rate; the effect of the temperature dependence of the cross-section is also shown to be small. The heating rates are compared with those calculated using two parameterisations. Differences in diurnally-averaged heating rate can exceed 1 K/day in the upper stratosphere. A new parameterisation is derived which fits the more detailed calculations to within 0.1 K/day in diurnally-averaged heating rate over a wide range of conditions. This parameterisation is shown to give a globally-averaged net heating rate in the upper stratosphere considerably closer to zero than earlier calculations.

1. INTRODUCTION

A number of computationally fast schemes exist for the calculation of the heating rate due to the absorption of solar radiation by ozone (e.g. Lacis and Hansen 1974; Strobel 1978). These are widely used in numerical models of the middle atmosphere. In recent years a number of high resolution, measurements of the absorption cross-section of ozone (including its temperature dependence) have been performed (e.g. Bass and Paur 1985; Molina and Molina 1986) and improved measurements of the extraterrestrial solar irradiance have become available (e.g. WMO 1986). This paper compares radiative heating rates calculated using recent absorption cross-sections and irradiances and assesses the importance of the temperature dependence. These heating rates, calculated with a high spectral resolution, are then compared with earlier parameterisations. The differences are sufficiently large to necessitate the derivation of a new parameterisation.

2. CALCULATIONS USING RECENT MEASUREMENTS OF OZONE ABSORPTION CROSS-SECTIONS

WMO (1986) presented a reference spectrum of extraterrestrial solar irradiance, together with absorption cross-sections for diatomic oxygen and ozone for wavelengths between 175.439 nm and 852.5 nm. An important part of the reference compilation in the Hartley and Huggins bands was derived from recent measurements by Bass and Paur (1985) and included a temperature dependence.

Molina and Molina (1986) also presented measurements of the absorption cross-section of ozone in the Hartley and Huggins bands for temperatures in the range 225-298 K.

Radiative heating rates were calculated using the WMO(1986) reference solar irradiances and two sets of absorption cross-sections: the first was as given by WMO (1986); the second set replaced the WMO values by the Molina and Molina measurements for wavelengths up to 350 nm. The calculations were performed for wavelengths longer than 206.186nm. The overlap between the Herzberg continuum due to O_2 and the ozone bands between 206.186 and 246.918 nm is included.

The radiative heating calculation is straightforward (see e.g. Strobel 1978) given the extraterrestrial irradiance, the absorption cross-sections and the vertical profile of absorber. The temperature dependence in the cross-sections was incorporated for the local cross-sections by using local temperature, and for column integrated absorber amounts by calculating a temperature weighted by absorber amount. Ozone amounts for these calculations were derived from the Keating *et al.* (1987) climatology, and a three-year average (1979-1981) of reprocessed SBUV data. Temperatures were derived from Barnett and Corney (1985). Diurnal averages were calculated using a three-point Gaussian integration over half a day. Heating due to reflection from the troposphere is ignored for simplicity until Section 6.

Figure 1 shows the heating rates at 87.5°S for mid-January using the WMO cross-sections incorporating temperature dependence, together with the heating rate differences for two other cases. The first neglects the temperature dependence and the second uses the Molina and Molina cross-sections at wavelengths where they are available. The Molina and Molina cross-sections generate smaller heating rates in the mesosphere which reach 0.12 K/day at 87.5°S, or about 1% of the heating rate. Such agreement is encouraging. Given a radiative relaxation time of about 5 days at these levels, such a heating rate difference would generate changes in the radiative equilibrium temperature differences of no more than 0.6 K. The neglect of the temperature dependence of the cross-sections is of the same order. Assuming a temperature of 250 K at all heights and latitudes, errors of up to 0.15 K/day at 87.5°S are generated in the upper stratosphere.

For the purposes of radiative heating calculations in the middle atmosphere, the above results indicate that the differences amongst compilations of cross-sections and the effects of incorporating temperature-dependent cross-sections are relatively small; compared to the uncertainties in the calculation of other components of the diabatic heating, such differences are negligible.

3. COMPARISON WITH PREVIOUS PARAMETERISATIONS

Two parameterisations of the absorptance of ozone are those of Strobel (1978) and Lacis and Hansen (1974).

ISBN 0-937194-15-8

Strobel's parameterisation is based on the ideas of Lindzen and Will (1974) whereby the ultra-violet and visible spectral regions are split into a number of broad bands; the absorption cross-section in these bands is represented either by a constant value or by some simple function of wavelength. An important update to Strobel's coefficients is given by Apruzese *et al.* (1982).

Lacis and Hansen (1974) derived analytical fits to the ozone absorptance in two intervals. This scheme is computationally very fast, but lacks some of the flexibility of Strobel's scheme.

Figure 2 shows a comparison of the heating rates obtained using the high-resolution calculations (including temperature dependence) and these two parameterisations. The difference between the high-resolution calculations and the updated Strobel scheme is in excess of 1 K/day between about 35 and 40 km at 87.5°S; this amounts to more than 10% of the total solar heating due to ozone. The bulk of the error in the mid-stratosphere is in the Huggins band. However, changes in both the solar irradiances and absorption cross-sections means that differences in other spectral regions are present. The Chappuis band was found to underestimate the heating by about 8%; the Hartley band overestimates by about 1.4% at the stratopause; in Strobel's Herzberg band, the heating rates are an overestimate by about 10% in the upper mesosphere and an underestimate of about 6% at the stratopause. Rosenfield *et al.* (1987) have noted similar differences.

It must be emphasized that Strobel's original coefficients give a serious discrepancy in heating rate; at 87.5°S the heating rate is overestimated by 2.9 K/day at the stratopause, or more than 20% of the total heating.

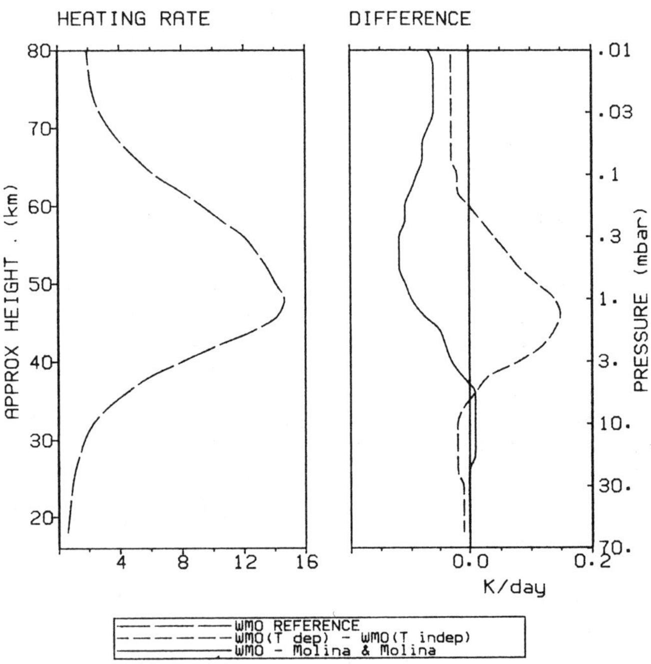

Fig. 1. The first frame shows the solar radiative heating (K/day) calculated at 87.5°S in mid-January using the WMO(1986) reference ozone cross-sections and extraterrestrial irradiance. The second frame shows the change in heating rate when the temperature dependence of the cross-sections is ignored, and when the WMO cross-sections are replaced by the measurements of Molina and Molina (1986).

The Lacis and Hansen ozone absorptance shows better agreement than the updated Strobel scheme in the upper stratosphere, but in the lower mesosphere differences reach 0.8 K/day.

4. A NEW PARAMETERISATION

The errors in the parameterisation schemes shown in Fig.2 are sufficiently large that the derivation of new coefficients was considered to be desirable. Strobel's method was followed. The same spectral regions were used although the form of some of the fits differs slightly. For example, in the Huggins band, Strobel uses an exponential variation of cross-section with wavelength to derive a fit. Using the new cross-sections, such a method was found to generate unacceptably large errors.

In the following description, the heating rate at some level z is given by Q, $[O_2]$ and $[O_3]$ are the number concentrations at z (in cm^{-3}) of diatomic oxygen and ozone respectively, and N_2 and N_3 are the corresponding total column densities (in cm^{-2}) integrated along the path of the direct solar beam to level z.

To obtain the heating rate in K/sec, the values of Q below must be multiplied by $10^2/(\rho c_p)$ where ρ is the air density and c_p is the specific heat of air at constant pressure; 10^2 converts the flux divergence from cm^{-1} to m^{-1}. Note, however, that the determination of number density from an absorbing constituent's mixing ratio requires multiplication by ρ. Hence it is unnecessary to calculate the air density if it is omitted from the calculation of both the number density and the factor converting to K/sec.

The names used for the bands follow those used by Strobel (1978). The errors in the fits refer to a comparison with the WMO(1986) coefficients for a typical ozone profiles and a solar zenith angle of 30°.

Herzberg Band, (206.186 - 243.902 nm, WMO (1986) bands 18-32): In this spectral region there is significant overlap in the absorption by ozone and diatomic oxygen. The new fit is

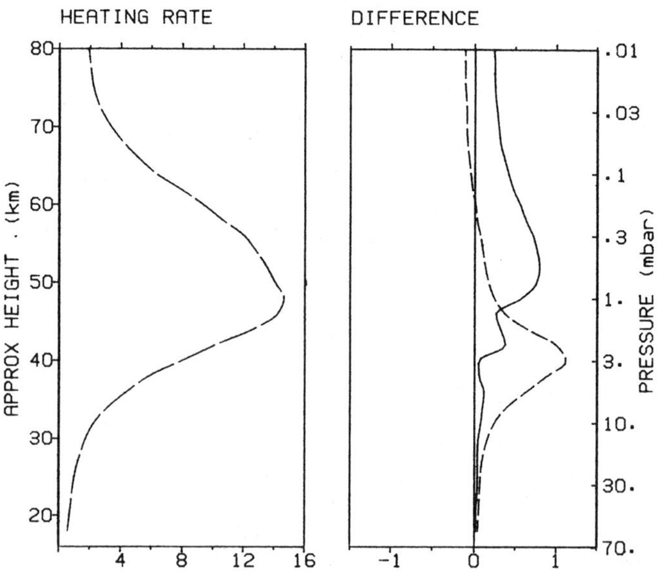

Fig. 2. As Fig. 1, but now the differences are between the WMO(1986) reference compilation and two computationally fast parameterisations.

$$Q_{Hz} = F_1(\sigma(O_2)_1[O_2] + \sigma(O_3)_1[O_3])$$
$$\times exp(-(\sigma(O_2)_1 N_2 + \sigma(O_3)_1 N_3))$$
$$+F_2(\sigma(O_2)_2[O_2] + \sigma(O_3)_2[O_3])$$
$$\times exp(-(\sigma(O_2)_2 N_2 + \sigma(O_3)_2 N_3)),$$

where $\sigma(O_2)_1 = 5.50 \times 10^{-24} cm^2/molecule$,
$\sigma(O_3)_1 = 6.22 \times 10^{-18} cm^2/molecule$,
$\sigma(O_2)_2 = 1.34 \times 10^{-26} cm^2/molecule$,
$\sigma(O_3)_2 = 1.66 \times 10^{-18} cm^2/molecule$,
$F_1 = 0.913\ W/m^2$
and $F_2 = 0.631\ W/m^2$.

At levels of any appreciable heating, the error in the heating rate is less than 2%; throughout the mesosphere the error is less than 1%.

Hartley Band, (243.902 - 277.778 nm, WMO (1986) bands 33-42): The fit in this region is

$$Q_{Ha} = [O_3]\sum_{i=1}^{2} \sigma_i F_i exp(-\sigma_i N_3),$$

where $\sigma_1 = 9.65 \times 10^{-18} cm^2/molecule$,
$\sigma_2 = 4.79 \times 10^{-18} cm^2/molecule$,
$F_1 = 3.68\ W/m^2$
and $F_2 = 1.0\ W/m^2$.

At the stratopause and above, the error is less than 0.2%; the error increases to about 1% in the mid-stratosphere, but by these heights this band is of little importance.

Huggins Band, (277.778 - 362.5 nm, WMO (1986) bands 43-60): The fit in this region is

$$Q_{Hu} = [O_3]\sum_{i=1}^{3} \sigma_i F_i exp(-\sigma_i N_3),$$

where $\sigma_1 = 8.94 \times 10^{-19} cm^2/molecule$,
$\sigma_2 = 1.09 \times 10^{-19} cm^2/molecule$,
$\sigma_3 = 3.07 \times 10^{-18} cm^2/molecule$,
$F_1 = 6.11\ W/m^2$,
$F_2 = 13.25\ W/m^2$
and $F_3 = 2.28\ W/m^2$.

At levels of maximum heating (the mid-stratosphere) errors are less than 1%. The error at the mesopause increases to 6 to 10%. This is slightly less accurate than Strobel's scheme, with respect to the WMO reference coefficients. However, the error with respect to the Molina and Molina coefficients is less, being about 3 to 5%. In any case, the contribution of this band to the total heating at these levels is relatively small. Errors also increase into the lower stratosphere, where the heating rates are again small.

Chappuis Band, (407.5 - 852.5 nm, WMO (1986) bands 70-158): The fit here is

$$Q_{Ch} = [O_3]\sigma F exp(-\sigma N_3),$$

where $\sigma = 3.16 \times 10^{-21} cm^2/molecule$
and $F = 364.9\ W/m^2$.

The error is less than 0.01% at all levels.

5. LATITUDE-HEIGHT CROSS-SECTIONS OF HEATING RATE

Figure 3 shows a latitude-height cross-section of heating rate using the high resolution calculations for January along with the difference between this heating rate and that calculated using the new parameterisations. The percentage error is generally less than 2% everywhere; the absolute error never exceeds 0.1 K/day. At latitudes close to the edge of the polar night, where the Sun stays low in the sky for much of the day, errors can reach 10% in the lower stratosphere; the absolute error in heating rate is, however, no more than 0.025 K/day.

6. IMPACT OF USING THE NEW SCHEME

There are a number of simple ways of assessing the impact of the new parameterisation on the representation of the atmosphere. These include the effect on radiative-equilibrium temperatures (e.g. Shine 1987) and the effect on dynamical quantities diagnosed using a radiation scheme and the observed distributions of temperature and ozone (e.g. Shine 1988).

The calculations in this section used Strobel's parameterisation for the heating due to the Schumann-Runge bands and continuum of O_2, and include the heating due to reflection from the troposphere. The thermal infrared heating was calculated using the scheme described by Shine (1987).

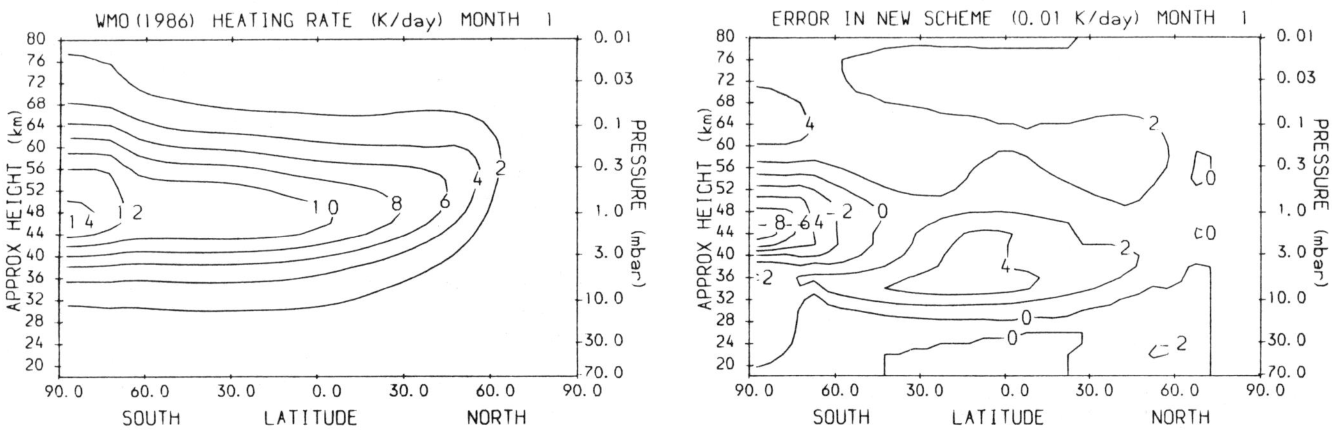

Fig. 3. Latitude-height cross-sections of the solar radiative heating (K/day) for January using the WMO(1986) reference compilation and the difference between this heating rate and that calculated using the new parameterisation (in 0.01 K/day).

The radiative equilibrium temperature was calculated for mid-January at 87.5°S using the new fits, along with the differences between this calculation and one performed using the updated Strobel fits. As would be anticipated from Fig. 2, the peak changes occur between 35 and 40 km; the new scheme leads to a warming by 6 K at 87.5°S, and all underlying stratospheric levels are warmed by at least 1 K.

A number of recent studies (e.g. Rosenfield *et al.* 1987; Shine 1988) have used radiative transfer schemes, together with observed distributions of temperature and ozone to derive the residual or diabatic circulation for a variety of purposes. One problem that pervades such studies is that in principle the radiation schemes, used in conjunction with the atmospheric data, should yield globally-averaged net heating (on a pressure surface) close to zero. As is discussed by Shine (1988) none of the recent works achieve such a balance, and a correction must be made to the heating rates; the need for this correction limits the confidence that can be placed in diagnosed quantities. The lack of balance is probably due to a combination of errors in the radiative calculations and the atmospheric data.

Figure 4 shows the global average thermal infrared cooling for January along with the net heating (solar heating minus thermal infrared cooling) using the new scheme and the updated Strobel scheme. In the mesosphere there is little difference between the two schemes, but in the upper stratosphere the new scheme reduces the departure from global balance by up to 0.5 K/day, or by one-third. At around 35 km at low latitudes, the old scheme gave a net cooling (not shown), but the new fits now give a region of net heating. Hence the new parameterisation has led to a significant amelioration of a previous problem.

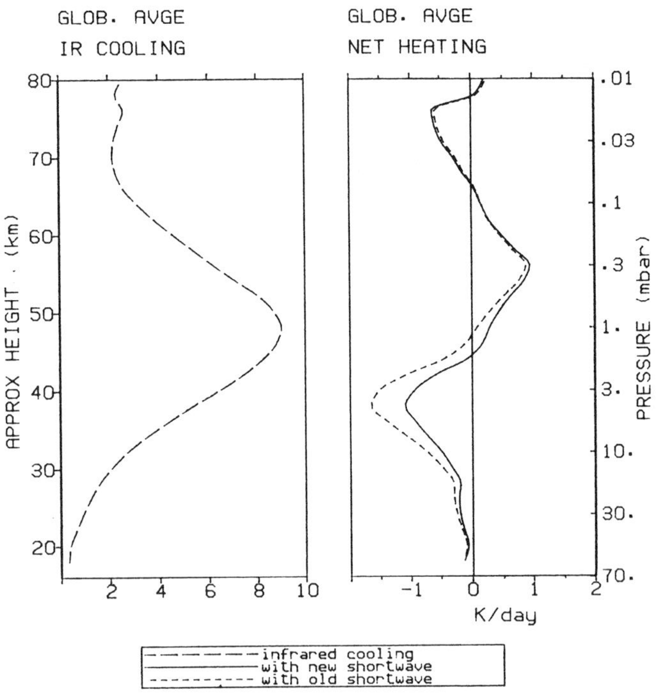

Fig. 4. The first frame shows the globally-averaged thermal infrared cooling rate (K/day) for January; the second frame shows the net heating (solar heating minus infrared cooling) using the new parameterisation and the Strobel/Apruzese *et al.* ('old') scheme.

7. CONCLUSIONS

Differences in recent compilations of absorption cross-sections for ozone in the visible and ultra-violet, and the temperature dependences of those absorption coefficients, are shown to have a relatively small effect on calculations of the radiative heating in the middle atmosphere.

However, substantial differences exist between the heating rate calculated using these recent compilations and earlier parameterisations of the ozone absorption. A new, computationally fast, parameterisation was derived and this yields errors of less than 0.1 K/day for a wide variety of conditions.

The new scheme substantially improves previous calculations of the globally-averaged heating rate in the upper stratosphere.

ACKNOWLEDGEMENTS

This work was supported by the UK Natural Environment Research Council.

REFERENCES

Apruzese, J.P., Schoeberl, M.R. and Strobel, D.F., 1982: Parameterization of IR cooling in a middle atmosphere dynamics model. 1. Effects on the zonally averaged circulation. *J. Geophys. Res.*,**87**, 8951-8966.

Barnett, J.J. and Corney, M., 1985: 'Middle atmosphere reference model derived from satellite data'.Pp 47-85 in Handbook for MAP, **16:***Atmospheric structure and its variations in the region 20-120 km: Draft of a new reference middle atmosphere.* Edited by K.Labitzke, J.J.Barnett and B.Edwards.

Bass, A.M. and Paur,R.J., 1984: The ultraviolet cross-sections of ozone: I The measurements. Pp 606-610 in *Atmospheric Ozone.* Edited by C.S.Zerefos and A.Ghazi. D.Riedel.

Keating, G.M., Young, D.F. and Pitts, M.C., 1987: Ozone reference models for CIRA.*Adv. Space Res.*,**7**, 105-115.

Lacis, A.A. and Hansen,J.E., 1974: A parameterization for the absorption of solar radiation in the Earth's atmosphere.*J. Atmos. Sci.*,**31**, 118-133.

Lindzen, R.S. and Will, D.I., 1973: An analytic formula for heating due to ozone absorption.*J. Atmos. Sci.*,**30**, 513-515.

Molina, L.T. and Molina,M.J., 1986: Absolute absorption cross sections of ozone in the 185 to 350-nm wavelength range.*J. Geophys. Res.*,**91**, 13501-14508.

Rosenfield, J.E., Schoeberl, M.R. and Geller, M.A., 1987: A computation of the stratospheric diabatic circulation using an accurate radiative transfer model. *J. Atmos. Sci.*,**44**, 859-876.

Shine, K.P., 1987: The middle atmosphere in the absence of dynamical heat fluxes.*Quart. J. Royal Meteorol. Soc.*, **113**, 603-633.

Shine, K.P., 1988: Sources and sinks of zonal momentum in the middle atmosphere diagnosed using the diabatic circulation. *Quart. J. Royal Meteorol. Soc.* (to appear).

Strobel, D.F., 1978: Parameterization of the atmospheric heating rate from 15-120 km due to O_2 and O_3 absorption of solar radiation.*J. Geophys. Res.*,**83**, 6225-6230.

WMO, 1986: Atmospheric ozone 1985.*WMO Global Ozone Research and Monitoring Project No. 16*, WMO, Geneva.

THE EFFECTS OF STRATOSPHERIC CHANGES DUE TO CO_2 DOUBLING ON THE CLIMATE OF THE TROPOSPHERE

W.-C. Wang, G. Molnar, H.R. Schneider, and M.K.W. Ko
Atmospheric and Environmental Research, Inc.
Cambridge, Massachusetts 02139, USA

ABSTRACT

Increases in atmospheric CO_2 have a strong effect on the earth's radiative energy balance. The net effect of a CO_2 increase is a warming of the troposphere-surface system and a cooling of the stratosphere. Associated with the stratospheric cooling are increases in O_3. Here we present a study on the effects of stratospheric changes due to CO_2 increases on the tropospheric climate. The study is based on a 2-D seasonal climate model which includes stratospheric levels. Dynamical heat fluxes and O_3 for the stratospheric levels are taken from runs of an interactive stratospheric 2-D model. The results suggest that the combined effects of changes in dynamical heat fluxes and O_3 in the stratosphere decrease the tropospheric warming and increase the stratospheric cooling resulting directly from CO_2-induced radiative perturbation.

1. INTRODUCTION

Model calculations indicate that increases of atmospheric CO_2 lead to substantial decreases of temperatures in the mid to upper stratosphere through increased thermal radiative cooling. The decreased temperatures will cause O_3 to increase in that region through the temperature dependence of the reaction rates. The increased UV absorption associated with O_3 increase can not fully compensate the thermal radiative cooling effects. There is also a change in the meridional overturning in the stratosphere. The circulation feedback is, however, small compared to the effect of the temperature reduction. It is interesting to note in this context that the stratospheric cooling due to CO_2 increase may partially compensate O_3 losses expected from simultaneous increases in CFC emissions (Eckman et al., 1987).

Increases of CO_2 can directly warm the troposphere-surface system through the perturbation to its thermal radiative energy balance. Our previous 2-D model studies of the effects of increasing CO_2 on tropospheric climate assumed radiative equilibrium with no change of O_3 in the stratosphere, i.e., the solar radiative heating balances the thermal radiative cooling (Wang and Molnar, 1985; Wang et al., 1988). In so doing, the effects of dynamical heat fluxes and their changes as well as O_3 changes in the stratosphere were ignored. Since the perturbation to the thermal radiative energy balance depends not only on the CO_2 amount but also on the stratospheric temperature, the CO_2-induced indirect effects of changes in O_3 and dynamical heat fluxes discussed above may further modify the direct warming effect.

Here, we present a study of the effects of stratospheric changes due to a CO_2 doubling on the tropospheric climate. The study is based on a 2-D seasonal tropospheric climate model with stratospheric levels. The changes in stratospheric temperature, O_3 and dynamical heat fluxes were calculated with a 2-D interactive dynamical-chemical model of the stratosphere. The same radiative transfer scheme is used in both models. The effect of the stratosphere on tropospheric climate is studied by comparing CO_2-induced temperature changes calculated with and without taking into account the changes in the stratospheric dynamical heat fluxes and O_3 associated with a doubling of CO_2 from current level of 345 ppmv (1 x CO_2).

2. MODELS

2.1 TROPOSPHERIC CLIMATE MODEL

The 2-D seasonal climate model is an extension of the annual model (Wang et al., 1984; Wang and Molnar 1985) and includes three latitudinal zones representing the tropical latitudes (30°S-30°N), southern hemispheric extratropical latitudes (30°S-90°S), the southern zone and northern hemispheric extratropical latitudes (30°N-90°N), the northern zone. Because of the large difference in thermal inertia between land and oceans, which is important in simulating the seasonal cycle, both land and ocean sectors are included within each latitude zone. The seasonal model computes the seasonal and vertical distribution of temperature from the heat balance for the atmosphere, and for the subsurface (ocean/land) layer.

Rather than consider the moisture budget explicitly we simply prescribe the climatological relative humidity. Ice and snow albedo feedbacks are parameterized for both hemispheres. Empirically derived horizontal (meridional and zonal) heat flux parameterizations are used while the vertical dynamical heat flux is parameterized through dynamical adjustment processes using both the moist adiabatic lapse rate and the critical lapse rate

ISBN 0-937194-15-8

for baroclinic adjustment (Wang et al., 1984). In order to simulate the current climate, we need to include the seasonal variation of ocean heat transport and storage, which for present climate is derived from the model surface energy balance with (fixed) observed seasonal ocean surface temperature. During climate change experiments, the sum of the ocean heat transport and storage is assumed to remain the same as that of the present climate, while the ocean surface temperature is calculated. This approach has been used in general circulation models (see Chapter 4 of DOE, 1985).

For northern and tropical zones, the model calculated seasonal variation of the surface temperature and lapse rates as well as the radiation budget at the top of the atmosphere are in good agreement with observations. However the model calculates a much warmer southern land surface temperature.

2.2 STRATOSPHERIC MODEL

The 2-D stratospheric model consists of three modules for dynamics, radiative transfer, and O_3 chemistry, respectively. The vertical level spacing corresponds to about 3.5 km with the lowest level centered at about 700 mb and the top at approximately 80 km height. Nineteen grid points are used in the horizontal direction from pole to pole. The chemical module used is a simple HO_x model. As the O_3 removal reactions involving nitrogen, chlorine and bromine species are not included, calculated O_3 mixing ratios are too high. In order to avoid excessive heating rates in the upper stratosphere and to bring O_3 closer to observed levels, the calculated loss frequency is scaled up by a factor of 1.5. This simple chemistry model was chosen because O_3 becomes annually periodic after about five years of integration. Models with chlorine and nitrogen cycles must include long lived source gases such as N_2O and CFCs and would require much longer integration times to achieve periodicity. For the CO_2 doubling experiment, tropospheric temperature changes were neglected. In so doing, the effect of tropospheric warming on O_3 changes at lower stratospheric levels is not considered. The objective of this study is to find out to what extent the stratospheric changes affect the tropospheric climate. Ultimately we plan to make the two models interactive.

The model is able to simulate the seasonal cycle of winds, temperatures and O_3 reasonably well compared with observations. Also, the effects of the CO_2-doubling compare well with other model calculations, using a more complete chemistry (e.g. Haigh and Pyle, 1982). As far as the simulation of dynamical quantities is concerned, the temperatures in the polar region of the lower and middle stratosphere are about 10°C cooler than observations, the upper tropical stratosphere is about 10°C warmer. This latter problem is directly related to the simplified chemistry which leads to too large O_3 amounts at high altitudes. Details of the model can be found in Schneider et al. (1988).

3. RESULTS

3.1 EFFECT OF STRATOSPHERIC INTERACTIONS ON TROPOSPHERIC TEMPERATURES

As noted earlier, the tropospheric climate model has coarser latitudinal resolution. Therefore the stratospheric model calculated dynamical heat fluxes Q_D and O_3 were averaged over the three zones.

Q_D, defined as the dynamical heat flux divergence, was used in the stratosphere of the tropospheric climate model as an additional heating to the solar radiative heating Q_{SR} and thermal radiative cooling Q_{IR}. In the present study, we run the experiments corresponding to perpetual January condition. Therefore, in the stratosphere, the temperatures are calculated based on $Q_{SR} + Q_{IR} = 0$ for radiative equilibrium (RE) and $Q_{SR} + Q_{IR} + Q_D = 0$ for radiative dynamical equilibrium (RDE). Figure 1 shows the tropospheric climate model calculated January temperature distributions for 1 x CO_2 with and without Q_D. Note here that we used in both calculations the O_3 calculated from the stratospheric model. Note also that the temperatures with Q_D are nearly identical to those calculated from the stratosphere model.

For the northern zone, the stratosphere without Q_D is colder and nearly isothermal between 15 and 40 km. On the other hand, for the tropical zone, the Q_D provides a warming to the mid and upper stratosphere but a cooling to the lower stratosphere. Similar feature but with smaller magnitude is calculated for the southern zone. The effects of Q_D on the surface temperature suggest that the northern zone is slightly warmer while the tropical and southern zones are slightly colder, reflecting the difference in the lower stratospheric temperatures between the cases with and without Q_D (see Fig. 1).

3.2 EFFECT OF STRATOSPHERIC INTERACTIONS ON WARMING OF TROPOSPHERE CAUSED BY CO_2 DOUBLING

We performed two types of experiments to illustrate the importance of including an interactive stratosphere in studying the cooling of the stratosphere and warming of the troposphere caused by a doubling of CO_2. First, we use the 2-D tropospheric climate model with a radiative equilibrium stratosphere to calculate the temperature changes. Secondly, the stratospheric dynamical heating and the O_3 changes associated with the increases of CO_2 calculated from the 2-D stratospheric model are included in the stratosphere of the 2-D tropospheric climate model in CO_2 doubling experiments.

Figure 2 shows the stratospheric model calculated changes of O_3 vertical distribution resulting from a doubling of CO_2. All three zones show an increase of O_3 except for the tropical zone between 12 and 25 km, where the "self-healing" effect is evident. For the three zones, O_3 increase peaks around 40-50 km with a magnitude of 18-22 %.

Given the changes in O_3 (shown in Fig. 2)

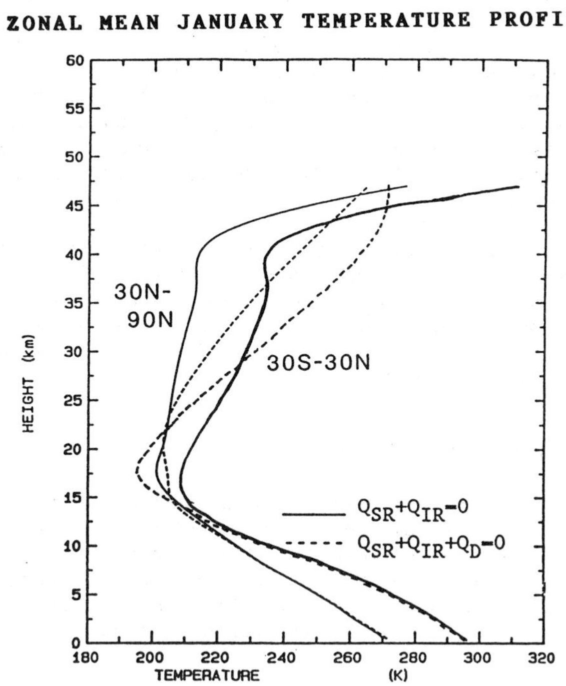

Fig. 1 Two-dimensional climate model calculated radiative equilibrium and radiative dynamical equilibrium January temperature distributions for 30°N-90°N and 30°S-30°N. Q_{SR}, Q_{IR} and Q_D are solar radiative, thermal radiative and dynamical heatings, respectively. Note that here Q_D are calculated from stratospheric model corresponding to $1xCO_2$.

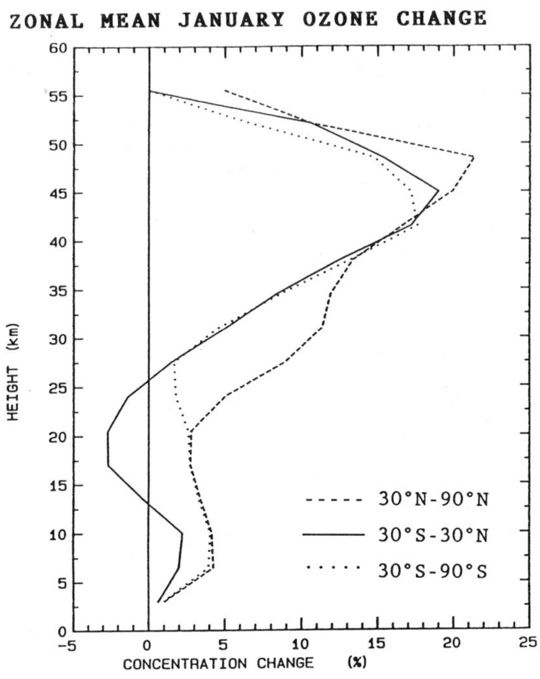

Fig. 2. Two-dimensional stratospheric model calculated changes in January O_3 distributions due to CO_2 doubling.

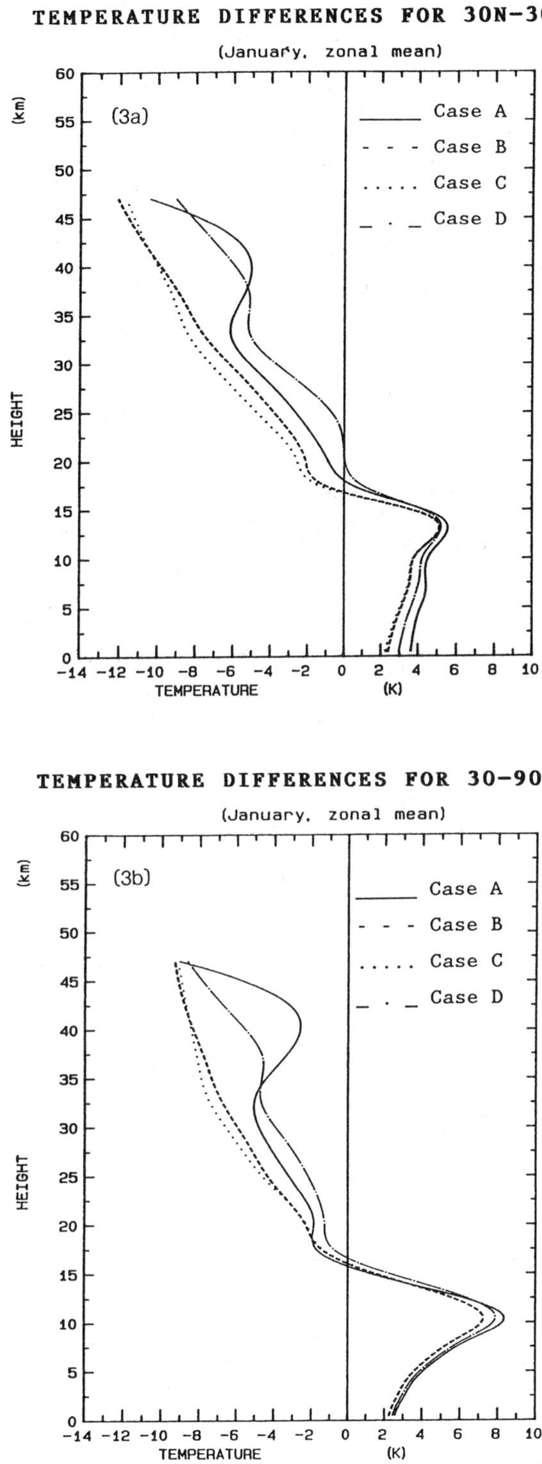

Fig. 3. Two-dimensional climate model calculated changes in January temperature distribution caused by CO_2 doubling for (a) 30°N-90°N and (b) 30°S-30°N. Four cases are shown: (A) $Q_{SR}+Q_{IR}=0$ without O_3 changes, (B) $Q_{SR}+Q_{IR}+Q_D=0$ with O_3 change and Q_D correspond to $2xCO_2$, (C) $Q_{SR}+Q_{IR}+Q_D=0$ without O_3 change and Q_D correspond to $2xCO_2$ and (D) $Q_{SR}+Q_{IR}+Q_D=0$ with O_3 change and Q_D correspond to $1xCO_2$.

and the Q_D calculated from the stratospheric model when CO_2 concentration is doubled, we used the tropospheric climate model to compare the difference in the temperature distributions between the cases with and without changes in Q_D and O_3. Note that the model calculated Q_D are different between 1 x CO_2 and 2 x CO_2 cases.

Figure 3 shows the changes in RE and RDE temperature distributions caused by a doubling of CO_2 while the surface temperature changes are summarized in Table 1. In addition, the individual effect of change in Q_D and O_3 is also investigated. Note that, for Case A, the smaller cooling around 40 km region for the three zones when the Q_D are not considered reflects the temperature feature of the model calculated 1 x CO_2 temperature distribution shown in Fig. 1.

In general, the inclusion of the effect of the stratospheric interactions is to cause more cooling in the stratosphere and less warming of the troposphere and surface when CO_2 is doubled (Case A versus Case B). In the stratosphere, between 18 and 42 km, O_3 increases tend to provide a warming effect (Case B versus Case C) while the change of Q_D caused by CO_2 increase is to cool the temperature (Case B versus Case D). In the troposphere, the effect of stratospheric interactions calculated to be much larger for the northern zone. For example, the changes in the surface temperature shown in Table 1 is 2.81 °C and 2.08 °C for RE and RDE, respectively, while the corresponding changes in southern zone is 2.36 °C and 2.09 °C. The calculated January globally averaged surface temperature is decreased from 2.5 °C for RE to 2.1 °C for RDE. Note here that for RE condition

TABLE 1. MODEL CALCULATED ZONAL MEAN JANUARY SURFACE WARMING (°C) CAUSED BY CO_2 DOUBLING.

	30°S-90°S	30°S-30°N	30°N-90°N	Globe
A. Q_D = 0 without O_3 change				
	2.36	2.49	2.81	2.54
B. Q_D with O_3 change				
	2.09	2.17	2.08	2.13
C. (B-A)				
	-0.27	-0.32	-0.73	-0.41

the contrast in the warming between the southern and northern zones is caused mainly by the warming during wintertime in the northern land which is equally important as oceans in determining the zonal mean warming. During wintertime over land, the vertical sensible and latent heat fluxes are very small and thus the surface temperature is influenced primarily by radiative energy balance. Therefore, when CO_2 is increased, the vertical sensible and latent heat fluxes, which usually increase during surface warming and thus provide a negative feedback to surface warming, is not in effect.

It is interesting to find that larger warming at northern zone calculated for RE is no longer present in the RDE case. This is mainly caused by a larger RDE calculated stratospheric cooling caused by CO_2 doubling (see Fig. 3), which will result in a smaller CO_2-induced increase of downward thermal radiation from stratosphere to troposphere-surface and thus less surface warming. However, this feature needs to be further investigated using the stratospheric model with full chemistry and the tropospheric model with higher latitudinal resolution.

In summary, the interactions in the stratosphere involving the changes in the dynamical heat fluxes and O_3 caused by CO_2 increase can influence both the warming in the troposphere and cooling in the stratosphere. The effect is particular large in northern high latitudes during wintertime.

ACKNOWLEDGEMENT

This research was supported by the Climate Dynamics Program of the National Science Foundation (ATM-8805908), the Carbon Dioxide Research Division of the Department of Energy (DE-FG02-86ER60485) and the Fluorocarbon Program Panel of the Chemical Manufacturers Association.

REFERENCE

DOE, 1985: Projecting the climatic effects of increasing carbon dioxide. MacCracken, M. C. and Luther, F. M. (eds.), Report ER0239, U.S. Department of Energy, Washington, DC.

Eckman, R.S., J.D. Haigh, and J.A. Pyle, 1987: An important uncertainty in coupled chlorine-carbon dioxide studies of atmospheric ozone modification. Nature, **329**, 616-619.

Schneider, H.R., M.K.W. Ko, N.D. Sze, G.Y. Shi, and W.-C. Wang, 1988: The effects of eddy diffusion, drag, and nonlinear terms on the meridional circulation and ozone distributions in an interactive 2-D model. Submitted to J. Atmos. Sci.

Haigh, J.D. and J.A. Pyle, 1982: Ozone perturbation experiments in a two-dimensional model. Quart. J. Roy. Met. Soc., **108**, 551-574.

Wang, W.-C, and G. Molnar, 1985: A model study of the greenhouse effects due to increasing atmospheric CH_4, N_2O, CF_2Cl_2 and $CFCl_3$ J. Geophys. Res., **90**, 12971-12980.

Wang, W.-C., G. Molnar, M.K.W. Ko, S. Goldenberg, and N.D. Sze, 1988: Atmospheric trace gases and global climate: A two-dimensional seasonal model study. Submitted to Tellus.

Wang, W.-C., G. Molnar, T.P. Mitchell, and P.H. Stone, 1984: Effects of dynamical heat fluxes on model climate sensitivity. Geophys. Res., **89**, 4699-4711.

A TWO-DIMENSIONAL MODEL STUDY OF PAST TRENDS IN GLOBAL OZONE

Donald J. Wuebbles and Douglas E. Kinnison
Lawrence Livermore National Laboratory
University of California
Livermore, California 94550, USA

ABSTRACT

Emissions and atmospheric concentrations of several trace gases important to atmospheric chemistry are known to have increased substantially over recent decades. Solar flux variations and the atmospheric nuclear test series are also likely to have affected stratospheric ozone. In this study, the LLNL two-dimensional chemical-radiative-transport model of the troposphere and stratosphere has been applied to an analysis of the effects that these natural and anthropogenic influences may have had on global ozone concentrations over the last three decades. In general, model determined species distributions and the derived ozone trends agree well with published analyses of land-based and satellite-based observations. Also, the total ozone and ozone distribution trends derived from CFC and other trace gas effects have a different response with latitiude than the derived trends from solar flux variations, thus providing a "signature" for anthropogenic effects on ozone.

1. INTRODUCTION

At the Quadrennial Ozone Symposium in 1984, we presented a theoretical analysis, using the LLNL one-dimensional model, on the past trends in ozone and temperature expected from emissions of chlorofluorocarbons and other trace gases (Wuebbles, 1984). In this paper, we examine these trends using the LLNL two-dimensional model, which provides much more information on latitudinal and seasonal effects. Emissions and atmospheric concentrations of several trace gases important to global atmospheric chemistry, including CO_2, CH_4, N_2O and several chlorocarbons, are known to have increased substantially over recent decades. In addition, variations in ultraviolet radiation during the 11-year solar cycle are likely to have influenced upper stratospheric photochemistry. We also consider the effect of nitrogen oxides (NO_x) produced from the atmospheric nuclear test series of the late 1950s and early 1960s.

The LLNL two-dimensional chemical-radiative-transport model of the troposphere and stratosphere has been applied to analyzing the effects that these natural and human-related influences may have had on global ozone concentrations over the last three decades. Similar calculations were included in a recent international study of global trends (Watson, 1988); however, results presented here are based on an updated version of the model.

2. THE LLNL TWO-DIMENSIONAL MODEL

The LLNL zonally-averaged two-dimensional chemical-radiative-transport model currently determines the atmospheric distributions of 31 chemically active atmospheric trace constituents in the troposphere and stratosphere. The model domain extends from pole to pole, and from the ground to 0.56 mb (approximately 0 to 54 km). Approximately 95 chemical and photochemical reactions are included in the model. Reaction rates, solar flux data, absorption cross-sections, and quantum yields are based on the latest NASA panel recommendations (Demore et al., 1987, except for NO_2 + O which is based on prior recommendations). Photodissociation rates, including the effects of multiple scattering, are computed as a function of time at each zone, with optical depths consistent with calculated species distributions.

The diabatic circulation for the ambient atmosphere is determined using net heating rates calculated in an internally consistent way with the derived species distributions. The technique for deriving the diabatic circulation is similar to that used by Solomon et al. (1986): the vertical velocity is determined from the zonally averaged residual Eulerian thermodynamic equation, while the horizontal velocity is determined using the equation for mass continuity. The net heating rates are determined using accurate solar and infrared radiative models. The solar model includes absorption and scattering effects of O_3, O_2, and NO_2 at ultraviolet and visible wavelenghts, and for H_2O, CO_2, and O_2 in the near infrared. The longwave emission and absorption by O_3, CO_2, and H_2O are included in the infrared submodel.

Temperatures for the atmosphere vary continuously over the annual cycle based on the reference data set of Barnett and Corney (1985). The derived diabatic circulation depends strongly on the temperature distribution; by using observed temperatures for the ambient atmosphere, an accurate representation of the diabatic circulation can be derived. The model determined net radiative heating rates, and resulting diabatic circulation, compare well with those derived from LIMS data (Kiehl and Solomon, 1986; Rosenfield et al., 1987; Solomon et al., 1986).

For all of the model calculations presented here, the diabatic circulation is assumed to be unchanged from that derived for a reference atmosphere corresponding to the mid-1980s. At the conference, we plan to discuss internally consistent perturbations to stratospheric temperatures; unfortunately results from this version of the model were not available by the time we had to write this paper.

3. HISTORICAL EMISSIONS AND INFLUENCES

Modeling past trends require estimates of the historical changes in trace gas emission and concentrations, variations in solar ultraviolet radiation, and NO_x produced from nuclear tests. Possible effects from other influences such as the El Chichon eruption, the Antarctic ozone hole, or possible changes in climate or atmospheric dynamics, were not included. As much as possible, available measurements and emissions evaluations were used in the development of the historical scenarios.

The historical emissions for the chlorocarbons $CFCl_3$, CF_2Cl_2, CCl_4 and CH_3CCl_3 were based on the expressions

A. DEEPAK Publishing
ISBN 0-937194-15-8

developed by Wuebbles et al. (1984) from available Chemical Manufacturing Association data and other databases. The time history of CH_4 concentrations were based on measurements made since 1978 indicating a 1% per year increase in methane concentrations (Blake and Rowland, 1988; WMO, 1985) and data for earlier periods (e.g., spectral data and ice core data). Historical changes in N_2O surface mixing ratios were based on the data and analysis of Weiss (1981). The expression for increase in CO_2 concentrations developed by Wuebbles et al. (1984) was used; this was based on observations at Mauna Loa since 1958 and assumes a pre-industrial concentration of 270 ppmv.

Many uncertainties remain regarding the variations in ultraviolet radiation during the 11-year solar cycle. The variations assumed here are derived from the analyses of Heath and Schlesinger (1984, 1986). Their analyses, based on SBUV data, assume that the spectrum of solar rotation induced variations are preserved over the solar cycle and that the irradiance variations are linearly related to the ratio of the core to wing emissions of the MgII line. Their analyses produced variations in the ultraviolet flux from solar minimum to solar maximum of 9% at 205 nm, 4% at 250 nm, and 1% at 270 nm. These variations are much smaller than used in prior published modeling studies. For the model studies presented here, the solar cycle flux changes are assumed to have a sinusoidal variation with time. Magnitudes of respective solar cycles are based on F10.7 solar radio flux data (Heath and Schlesinger, 1984), with the magnitude of solar cycle 21 (1974–1985) taken to be 1.0 (i.e., same solar flux variations from solar min to solar max as given above).

Past modeling studies (e.g., Chang et al., 1979; Wuebbles, 1983) have indicated that nitrogen oxides produced from atmospheric nuclear tests could have had significant effects on stratospheric ozone during the early 1960s. In the calculations presented here, all reported tests are included, with the timing and yield of the various test devices based on Bauer (1979). The altitude for the bottom and top of the stabilized cloud are taken from the empirically-based analyses of Peterson (1970). Likewise, the distribution of NO_x formed within the cloud is based on Peterson (1970). There still remains uncertainty regarding the amount of NO produced per megaton of explosive energy; we assumed 0.67×10^{32} NO molecules/Mt based on the discussion in Chang et al. (1979).

4. RESULTS AND DISCUSSION

Figure 1 shows the model-determined annual and globally averaged changes in the total ozone for the period 1950–1986 relative to the total ozone amount calculated for 1960. Shown are three cases corresponding to the effects of trace gas emissions only, trace gases plus solar cycle, and finally, with nuclear test effects also included. These calculations indicate that trace gas emissions could have decreased total ozone by about one percent since 1960 (note that effects due to the Antarctica ozone hole, which are not included here, would tend to cause larger decreases). Prior calculations with the LLNL 1-D model suggest that the effects of temperature feedback, including stratospheric cooling due to the past growth in the CO_2 concentration, would tend to reduce the magnitude of the trend over this period. Inclusion of solar flux variations result in significant year-to-year changes in the total ozone amount. Although dependent on the particular cycle, the model estimates changes in total ozone of about 1.8% from solar minimum to solar maximum. These calculated global variations are consistent with available statistical analyses attempting to determine the solar cycle induced changes in total ozone from Dobson and satellite data (Reinsel et al., 1987, 1988; Oehlert, 1986). Likewise, the global trend in total ozone determined from trace gases is also consistent with published analyses of Dobson data for the 1970s and 1980s. (Angell and Korshover, 1983; Reinsel et al., 1987).

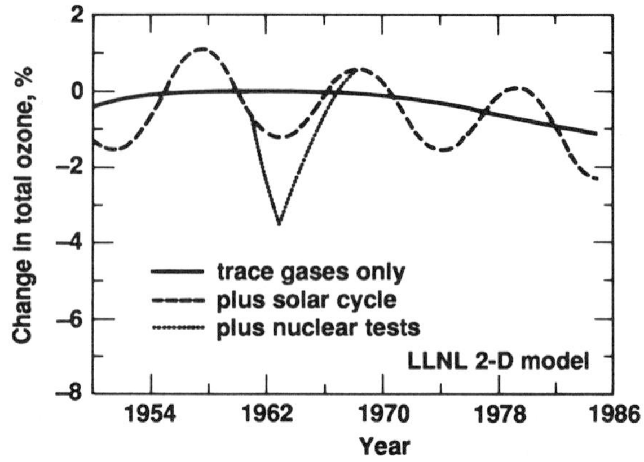

Fig. 1. Calculated global- and annual-averaged changes in total ozone from 1950 to 1986 relative to total ozone for 1960 for three cases: (1) trace gases only; (2) trace gases plus solar cycle variations; and (3) with atmospheric nuclear test effects also included.

Figure 2 shows the model calculated distribution of total ozone for the year 1985. A comparison of this diagram with Dobson and satellite data indicates good general agreement. The model simulates the latitudinal contrast well, with maxima generally occurring approximately during early spring at high latitudes. The estimated high latitude minima are also in good agreement with observations. However, there is too much ozone calculated for the tropics.

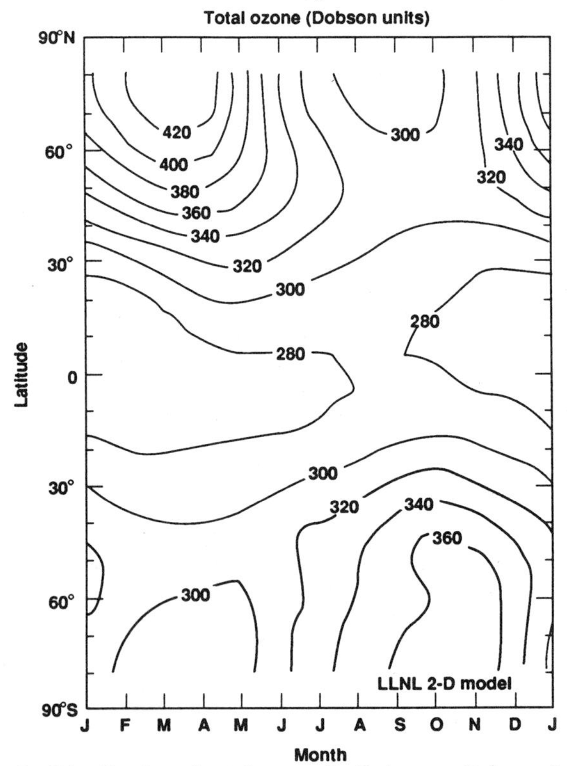

Fig. 2. Distribution of total ozone (in Dobson units) as a function of latitude and month calculated for model year 1985.

In agreement with past 1-D model studies, a significant decrease in total ozone (see Fig. 1) was calculated due to the atmospheric nuclear tests in the early 1960s. The maximum effect of the nuclear tests on globally and annual averaged ozone occurred in 1963, with a maximum change of −2.3%. This is consistent with available analyses of Dobson data for this period (Reinsel, 1981; Chang et al., 1979). A solar minimum also occurs during this same time period. The recovery of ozone concentrations from the nuclear tests occurred over the next five years. Including the effects of solar cycle variations, the calculated total ozone increase from 1963 to 1970 is about 3.8%.

Figure 3 shows the model calculated changes in total ozone due to the nuclear tests for model year 1962 relative to 1960. Large ozone decreases are calculated in late 1962 at high northern latitudes, with a maximum decrease of 12.5% occurring poleward of 70°N in November 1962. This is consistent and showed a high correlation with a recent analysis of data from Northern Hemisphere Dobson stations (R. Bojkov, private communication, 1988). The largest changes occur at this time because of a significant number of high latitude (~75°N) tests by the USSR (eleven tests >5 Mt) and several tests at lower latitudes (2–17°N) by the US (two tests >5 Mt).

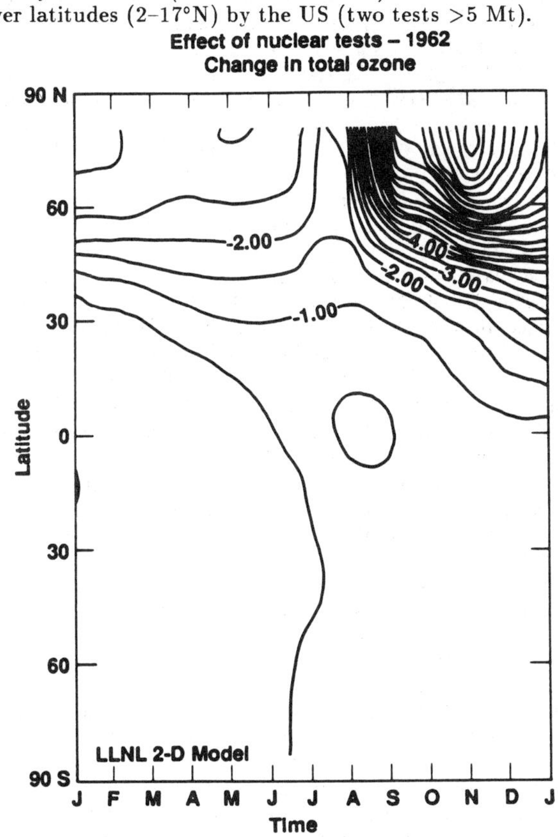

Fig. 3. Changes in total ozone calculated due to the effects of atmospheric nuclear tests for model year 1962.

Atmospheric nuclear testing was banned after 1962, although, as indicated earlier, the maximum global effect on total ozone occurred in 1963. Throughout the year, ozone at high northern latitudes continues to recover from the initial odd-nitrogen loading. Odd-nitrogen is also being transported to lower latitude and into the Southern Hemisphere, resulting in a 2% decrease in S.H. total ozone by late 1963. Later years continue to show gradual recovery of ozone concentrations at all latitudes.

The period from 1979 to 1986 is of particular interest because of the availability of satellite data. The trace gas emissions scenario gives a 0.6% decrease in globally-averaged total ozone for this time period. A much larger decrease in total ozone, −2.6%, is determined when solar cycle effects are included, reflecting the transition from solar maximum to solar minimum conditions. Analyses of SBUV data (Bowman, 1988; Reinsel et al., 1988) for the 1979–1986 period give much larger trends, but these analyses do not adequately account for apparent instrumental degradation over the period. The Reinsel et al. (1988) analysis does indicate a drift of data from SBUV relative to the Dobson data; when this drift is accounted for they determine a change of −0.35 ± 0.28% per year or −2.5% over this time period. This is in excellent agreement with the model determined trend, implying that most of the ozone decrease observed over this time period is related to solar cycle effects. In turn, an increase in total ozone from solar cycle effects would be expected over the next five years, until solar maximum is reached in 1990.

Figure 4 shows the change in total ozone and in local ozone (for January) calculated over the 1979–1986 period for the trace gas only scenario. A similar set of diagrams is shown in Fig. 5 for the trace gas plus solar cycle case. In each case, the largest changes in total ozone are found at the high latitudes in early spring, and the smallest changes occur in the tropics. However, these latitudinal and seasonal distributions of the changes are related primarily to the variations in trace gas concentrations. Solar cycle variations alone induce very little gradient with season or latitude. This implies that global measurements of total ozone showing such gradients are indicative of effects expected to be induced by from CFCs and other trace gases, and perhaps providing a "signature" for trace gas influence on ozone.

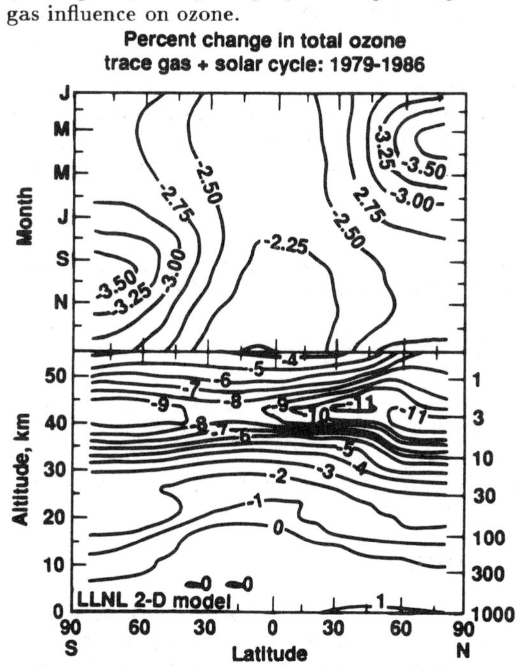

Fig. 4. Changes in total ozone (top diagram) from 1979 to 1986 for trace gases only, and changes in local ozone vs. latitude and altitude (bottom diagram) in January for trace gas only scenario.

Similarly, the largest percentage changes in ozone as a function of latitude and altitude occur in the upper stratosphere, near 3 mb, and show a gradient with the largest changes at higher latitudes and minimum changes in the tropics. While the maximum solar effect also occurs near 3 mb, the calculated gradient with latitude is small. Therefore, measured latitudinal gradients in ozone change near 3 mb would be primarily indicative of CFC and other trace gas effects.

These calculated changes in stratospheric ozone also compare extremely well with available Umkehr data at mid-northern latitudes for layer five through eight (J. DeLuisi, private communication, 1988). There are indications, however, that the model may underestimate decreases in ozone in the lower

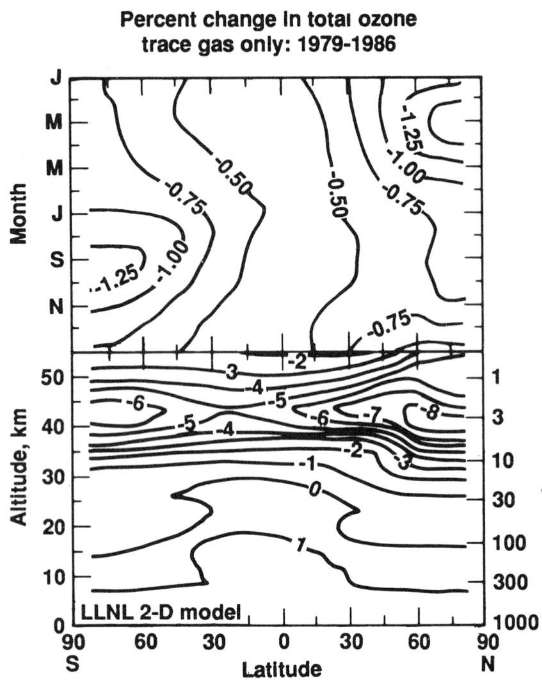

Fig. 5. Changes in total ozone (top diagram) from 1979 to 1986 for trace gases only, and changes in local ozone vs. latitude and altitude (bottom diagram) in January for trace gas only scenario for trace gases plus solar cycle scenario.

stratosphere as implied by measuremeasurements near 20 km from both Umkehr and ozonesonde data.

ACKNOWLEDGMENTS

This work was performed by the Lawrence Livermore National Laboratory under the auspices of the U.S. Department of Energy under contract W-7405-ENG-48 and was supported in part by the Department of Energy's Carbon Dioxide Research Division.

REFERENCES

Angell, J.K., and J. Korshover, 1983: Global variation in total ozone and layer-mean. Ozone: an update through 1981. *J. Climate and Appl. Meteor. 22,* 1611-1627.

Barnett, J.J., and M. Corney, 1985: Middle atmosphere reference model derived from satellite data. Middle Atmosphere Program Handbook, Vol. 16.

Bauer, E., 1979: A catalog of perturbing influences on stratospheric ozone, 1955–1975. *J. Geophys. Res., 84,* 6929–6940.

Blake, D., and F.S. Rowland, 1988: Continuing worldwide increase in tropospheric methane, 1978 to 1987. *Science 239,* 1129–1131.

Bowman, K.P., 1988: Global trends in total ozone. *Science 239,* 48–50.

Chang, J.S., W.H. Duewer, and D.J. Wuebbles, 1979: The atmospheric nuclear tests of the 1950s and 1960s: a possible test of ozone depletion theories. *J. Geophys. Res., 94,* 1755–1765.

Demore, W.B., J.J. Margitan, M.J. Kurylo, C.J. Howard, and A.R. Ravishankara, 1987: Chemical kinetics and photochemical data for use in stratospheric modeling. Jet Propulsion Lab., Pasadena, CA: JPL Publication 87-41.

Heath, D.F., and B.M. Schlesinger, 1986: The Mg 280-nm doublet as a monitor of changes in solar ultraviolet irradiance. *J. Geophys. Res., 91,* 8672–8628.

Heath, D.F., and B.M. Schlesinger, 1984: Temporal variability of uv solar spectral irradiance from 160–400 nm over periods of the evolution and rotation of active regions from maximum to minimum phases of the sunspot cycle. In *IRS '84: Current Problems in Atmospheric Radiation,* edited by G. Fiocco, A. Deepak Publishing, Hampton, VA.

Kiehl, J.T., and S. Solomon, 1986: On the radiative balance of the stratosphere. *J. Atmos. Sci., 43,* 1525–1534.

Oehlert, G.W., 1986: Trends in Dobson total ozone: an update through 1983. *J. Geophys. Res., 91,* 2675–2679.

Peterson, K.R., 1970: An empirical model for estimating world-wide deposition from atmospheric nuclear detonations. *Health Physics, 18,* 357–378.

Reinsel, G.C., 1981: Analysis of total ozone data for the detection of recent trends and the effects of nuclear testing during the 1960s. *Geophys. Res. Lett., 8,* 1227–1230.

Reinsel, G.C., G.C. Tiao, A.J. Miller, D.J. Wuebbles, P.S. Connell, C.L. Mateer, and J.J. De Luisi, 1987: Statistical analysis of total ozone and stratospheric Umkehr data for trends and solar cycle relationship. *J. Geophys. Res., 92,* 2201–2209.

Reinsel, G.C., G.C. Tiao, S.K. Ahn, M. Pugh, S. Basu, J.J. DeLuisi, C.L. Mateer, A.J. Miller, P.S. Connell, and D.J. Wuebbles, 1988 An analysis of the 7-Year record of sbuv satellite ozone data: global profile features and trends in total ozone. *J. Geophys. Res., 93,* 1689–1703.

Rinsland, C.P., J.S. Levine, and T. Miles, 1985: Concentration of methane in the troposphere deduced from 1951 solar spectrum. *Nature, 318,* 245–249.

Rosenfield, J.E., M.R. Schoeberl, and M.A. Geller, 1987: A computation of the stratospheric diabatic circulation using an accurate radiative transfer model. *J. Atmos. Sci., 44,* 859–876.

Solomon, S., J.T. Kiehl, R.R. Garcia, and W. Grose, 1986: Tracer transport by the diabatic circulation deduced from satellite observations. *J. Atmos. Sci., 43,* 1603–1617.

Watson, R., 1988: Executive summary of the ozone trends panel. NASA.

Weiss, R.W., 1981: The temporal and spatial distribution of tropospheric nitrous oxide. *J. Geophys. Res., 86,* 7185–7195.

World Meteorological Organization, 1985: *Atmospheric ozone 1985: assessment of our understanding of the processes controlling its present distribution and change.* Global Ozone Research and Monitoring Project-Research No. 16.

Wuebbles, D.J. 1983: A theoretical analysis of the past variations in global atmospheric composition and temperature structure. Ph.D. Thesis, University of California.

Wuebbles, D.J., M.C. MacCracken and F.M. Luther, 1984: A proposed reference set of scenarios for radiatively active atmospheric constituents. U.S. Department of Energy, Carbon Dioxide Research Division, Technical Report 015.

Wuebbles, D.J., 1984: Trends in ozone and temperature structure: comparison of theory and measurements. In *Atmospheric Ozone,* edited by C.S. Zerefos and A. Ghazi, D. Reidel Publishing, Dordrecht, Holland.

MODEL STUDIES OF STRATOSPHERIC OZONE DEPLETION

Björg Rognerud, Ivar S.A. Isaksen, and Frode Stordal
Institute for Geophysics
University of Oslo
Blindern, 0315 Oslo 3, Norway

ABSTRACT

Future effects, on ozone depletion for different control strategies for fully halogenated CFCs and halons are studied in a 2-D model. Five scenarios which cover the range from no control to 90% reduction in future CFC are used. Of particular interest is the scenario based on the Montreal agreement for future CFC and halon reductions. The calculations are time dependent and are extended to 2035.

The results indicate that the "Montreal" scenario could lead to substantial reductions in total column ozone in the next century. A shift to $F-22$ on the other hand could lead to a substantial reduction in the depletion as $F-22$ has a much smaller ozone depletion potential (ODP) than the fully halogenated CFCs. Bromine compounds released from man made halons in the stratosphere could contribute noticeably to future ozone depletion (20% of total depletion). This depletion occurs predominantly in the lower stratosphere at high latitudes as a result of reactions between bromine and chlorine species. As there are large uncertainties connected to the chlorine/bromine chemistry in these regions, there is a possibility that the ozone depletion could be larger than calculated.

1. INTRODUCTION

Model studies of future ozone changes due to the release of man made source gases (e.g. CFCs and Halons) are strongly dependent on the scenarios used for releases. As a result of the signing of the Montreal protocol limiting future releases of fully halogenated chlorocarbons ($CFC-11$, $CFC-12$ and $CFC-113$) and halons we expect future releases to be substantially less than what could be expected if no control measures are taken. In addition there are concerns that the control of release imposed by the Montreal agreement is not sufficient for limiting odd chlorine (Cl_x) in the stratosphere to give a sufficient protection to the ozone layer. There are suggestions to reduce the release of fully halogenated chlorocarbons with up to 85-90%. Another point which has been raised is the possibility of replacing fully halogenated CFCs with less stable, and thereby less efficient species with regard to ozone depletion (e.g. $CFC-22$). These different aspects of future ozone depletion will be studied using our 2-D model which has been used previously to similar ozone perturbation studies (Stordal et al.1985; Isaksen and Stordal 1986; Stordal and Isaksen 1987). The model has also been used in a recent intercomparition study (NASA in preparation) of ozone variation over the last 30 years. We have also included bromine chemistry in the model to study the impact of future bromine increases on the ozone depletion in the stratosphere. Although the contribution from man made bromine compounds (e.g. $H1211$ and $H1301$) to the odd bromine level (Br_x) in the stratosphere is small in the present atmosphere, the increase of the man made bromine compounds is rapid, and could significantly increase bromine levels in the future.

2. OZONE DEPLETION POTENTIAL (ODP) FOR THE SOURCE GASES

The impact certain chlorine or bromine containing species has on stratospheric ozone is critically dependant on how efficient the compound is transported into the stratosphere and where Cl_x and Br_x is released. This will to a large extent depend on the atmospheric lifetime of the source gases. Table 1 shows the atmospheric lifetime and ODP of man made chlorine and bromine source gases of importance for stratospheric ozone.

The ODPs are obtained through steady state calculations with chlorine and bromine source gas releases giving Cl_x and Br_x levels of 8.6 ppbv and 15 pptv respectively in the stratosphere. ODP from Halons are also calculated with Cl_x levels of 4.5 ppbv. The ODP is obtained by a small change in the release of one of the source gases at a time and running the model until a new steady state is obtained. Among the chlorine compounds CH_3CCl_3 and $F-22$ have substantially shorter atmospheric lifetimes than the other chlorine gases as they are oxidized in the troposphere by the hydroxyl radical. This leads to a substantially smaller ODP than for most other chlorine species (fully halogenated CFCs). The extremely stable chlorine species (e.g. $CFC-115$) are only partially broken down in the stratosphere, but rather transported to higher altitudes where the chlorine

ISBN 0-937194-15-8

Comp.	Lifetime	ODP	ODP
CH_3CCL_3	6.2	0.13	
F22	17.0	0.038	
F11	57.7	1.0	
F12	96.4	0.86	
CCl_4	50.8	1.19	
F113	103.0	0.93	
F114	247.0	0.82	
F115	548.0	0.12	
H1211	17.7	(1) 3.94	(2) 3.02
H1301	72.1	(1) 9.82	(2) 7.78
H2402	23.3	(1) 6.55	(2) 4.95

(1) Total chlorine in the stratosphere: 8.6 ppbv
(2) Total chlorine in the stratosphere: 4.5 ppbv

Table 1: Ozone depletion potential for the different chlorine species. The comparison is made for the same mass released of each compound with the efficiency of CFC-11 set each to 1.

effect on ozone is small. ODPs are therefor dependant on how efficient gases are oxidized in the troposphere, and where in the stratosphere the main oxidation occurs. For this reason ODP for $CFC-115$ is substantially less than for $CFC-11$ and $CFC-12$. $CFC-11$ and $CFC-12$ have lifetimes which are sufficiently long for the species to be transported efficiently to the stratosphere, but not stable enough to survive the shortwave radiation penetrating down to the middle and lower stratosphere.

The ODP of bromine species is not as uniquely defined as for chlorine species, but depends markedly on the chlorine level adopted in this study. This is because a large fraction of the ozone depletion occur via the reaction

$$BrO + ClO \rightarrow Br + ClOO \qquad (1)$$

in the lower stratosphere. The estimates show that when chlorine levels are increased by approximately 90% (stratospheric chlorine levels increasing from 4.5 to 8.6) ODP for the bromine compounds increases by approximately 30%. The ODP for bromine source gases are all larger than the ODP for chlorine compounds with $H1301$ being the most efficient bromine compound. Its ODP is almost 10 times as large as that of the most efficient chlorine compound ($CFC-11$) for a stratospheric Cl_x level of 8.6 ppbv. It should be mentioned that the ODP for chlorine species to a certain extent depends on the Br_x level in the stratosphere due to the interaction of BrO and ClO through the reaction above.

3. SCENARIOS FOR FUTURE OZONE CHANGES

The 2-D model is used to study the future development in stratospheric and tropospheric ozone as the release of trace gases changes. In addition to changes in CFCs and the Halons changes in the release rates of CH_3CCl_3 (MCF), CCl_4, N_2O and CH_4 were included. N_2O and CH_4 are assumed to increase at their present rates (0.20% and 0.017ppmv pr. year) in all five scenarios which were adopted.

a) A no control scenario.
The assumed release rates are given in Table 2. The growth rate is specific for each compound up to the year 2000, and is 2.5% pr. year for all compounds after year 2000.

b) A "Montreal" protocol scenario with restrictions in use of CFC and halons.
The protocol case simulates emissions based on the requirements of the recently signed Montreal Protocol on Substances that Deplete the Ozone Layer. CFCs -11, -12, and -113 undergo a phased reduction to 50 percent of their 1986 use levels (in the protocol also $CFC-114$ and -115, however, these species are not included in the present calculations), and Halons 1211 and 1301 are frozen at 1986 levels. The various exceptions to the requirements (e.g., the 10-year lag for developing countries) were also reflected. It was assumed that there would be 100 percent compliance in the U.S., about 94 percent compliance in other developed nations, and 65 percent compliance among developing nations. Because less than 100 percent global compliance is assumed, total weighted CFC emissions do not fall to 50 percent of the 1986 values (Table 3).

c) The "Montreal" release without increase in bromine species.

d) A scenario where the release of fully halogenated CFCs are immediately replaced by the same amount of $F-22$. Release rates are thereafter held constant.

Year	F11	F12	F22	F113	CCl_4	MCF	H1211	H1301
1985	278.3	363.7	73.8	150.4	87.3	813.0	1.4	2.0
1990	383.4	451.0	119.3	184.9	110.8	689.2	3.2	3.4
1995	458.5	544.7	156.6	224.6	127.4	788.5	6.0	4.3
2000	544.9	668.4	192.8	272.8	146.6	902.1	9.6	5.1
2005	638.8	791.0	227.2	308.7	165.9	1020.6	14.5	6.1
2010	724.9	899.2	258.0	349.3	187.7	1154.8	19.2	7.3
2015	820.9	1019.9	292.4	395.2	212.3	1306.5	23.7	8.5
2020	929.2	1154.7	330.9	447.1	240.2	1478.2	28.7	9.8
2025	1051.3	1306.5	374.4	505.8	271.8	1672.4	33.3	11.3
2030	1189.5	1478.2	423.6	572.3	307.5	1892.2	38.6	13.0
2035	1345.8	1672.4	479.3	647.5	348.0	2140.8	44.6	14.9

TABLE 2: Scenario a (No control,EPA)

Year	F11	F12	F22	F113	CCl_4	MCF	H1211	H1301
1985	278.3	363.7	73.8	150.4	87.3	813.0	1.4	2.0
1990	366.8	437.9	119.3	154.1	104.2	689.2	3.2	3.4
1995	363.2	438.2	156.6	129.5	93.0	788.5	5.6	3.9
2000	315.4	415.5	192.8	99.4	72.8	902.1	8.5	4.4
2005	318.6	421.2	227.2	92.6	70.2	1020.7	12.3	4.9
2010	295.1	401.6	258.0	89.4	65.7	1154.8	15.5	5.5
2015	279.7	391.7	292.4	91.1	66.8	1306.5	16.9	6.0
2020	258.6	386.2	330.9	93.0	67.8	1478.2	18.6	6.1
2025	260.6	354.6	374.4	94.6	69.0	1672.5	19.1	6.3
2030	263.4	319.7	423.6	97.0	70.2	1892.2	19.7	6.3
2035	269.3	324.1	479.3	99.1	71.4	2140.9	20.0	6.3

TABLE 3: Scenario b ("Montreal",EPA)

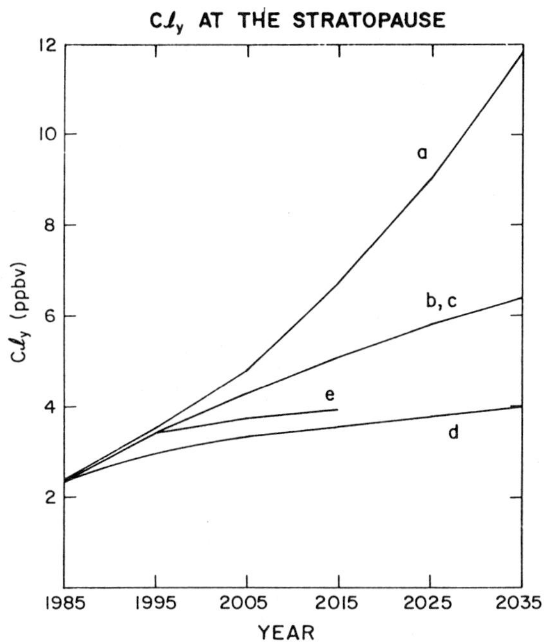

Fig. 1: Time development between 1985 and 2035 of total chlorine in the upper stratosphere in the five model cases. Substitution of all fully halogenated CFCs with F-22 lead to lowest future chlorine levels in the stratosphere.

e) A scenario where the release of fully halogenated CFCs are reduced to 50% in 1991 and to 10% in 1995.

Except for case e) where the time dependent calculations are run to 2015, all cases are run to the year 2035.

Depletion of ozone depends on the amount of Cl_x which is added to the stratosphere in the future. Fig. 1 shows the global average upper stratospheric future chlorine levels. In all cases, even those where we assume a drastic reduction in future CFC releases, Cl_x is estimated to increase due to the long lifetime of the CFCs.

In the case of the "Montreal" scenario the chlorine content in the stratosphere is expected to increase to more than 6 ppbv in the next 50 years. It is seen that a change to $F-22$, due to its low ODP, would have a substantial impact on the future build up of Cl_x. The corresponding changes in the global average total ozone is shown in Fig. 2. The "Montreal" scenario, although limiting chlorine build up as compared to a no control scenario, will have a substantial effect on the future ozone levels. Average global ozone depletion will exceed 3% in the next 50 years.

The bromine chemistry is calculated to have a noticeable effect on future ozone levels. We estimate the depletion by bromine to be approximately 20% of the total depletion. We see that a switch to $F - 22$ from fully halogenated chloroflourocarbons will lead to a pronounced reduction in the ozone depletion in the future.

Local ozone depletion in the upper stratosphere,

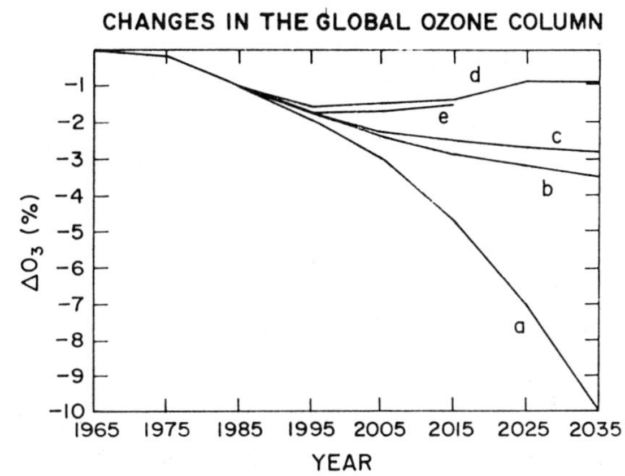

Fig. 2: Time development of global average ozone depletion from 1965 until 2035 for the five selected scenarios.

particularly at high latitudes are pronounced in the "Montreal" case. Asmuch as 50% reduction in the upper stratospheric ozone is calculated to develop over the next 50 years (Fig. 3a). This will have a marked effect on the total column ozone depletion which is in the range 7-10% at high northern latitudes during the winter months (Fig. 3b).

The effect of bromine chemistry (assuming no change in chlorine) is shown in Fig. 4 a) and b). The ozone depletion from bromine reactions (Fig. 4a) is particular significant in the lower high latitude stratosphere as a result of the reaction where both chlorine (ClO) and bromine (BrO) are involved. This is in contrast to the chlorine chemistry where the main loss occurs in the middle and upper stratosphere. It is also seen that the self healing effect at low latitude, in the lower stratosphere is negligible.

Changes in total column ozone from bromine reactions (Fig. 4b) show ozone depletions which increase markedly towards high latitudes. However, in contrast to the depletion due to chlorine, ozone depletion is large throughout the year. This is again a result of the ozone depletion occuring at high latitudes and at lower altitudes. The significance of the ozone depletion by bromine/chlorine reactions is that it occurs at high latitudes in the region of maximum ozone values where there still are large uncertainties (e.g. heterogeneous chemistry).

ACKNOWLEDGEMENTS

This research was supported by the Norwegian Council for Science and the Humanities (NAVF) and the Norwegian Department of Environment (MD).

REFERENCES

Isaksen, I.S.A. and Stordal, F., 1986: Ozone pertur-

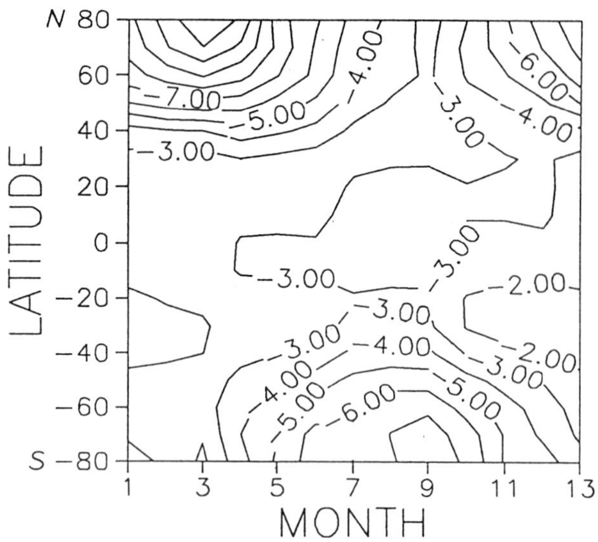

Fig. 3a: Column ozone depletions between 1965 and 2035 for scenario b. The depletions are given as a function of season and latitude.

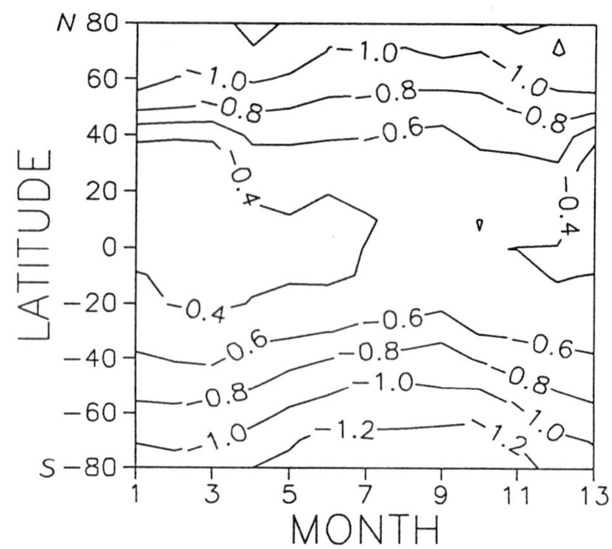

Fig. 4a: Column ozone depletion between 1965-2035 due to releases of bromine in the "Montreal" scenario (scenario c vs scenario b). The depletions are given as a function of season and latitude.

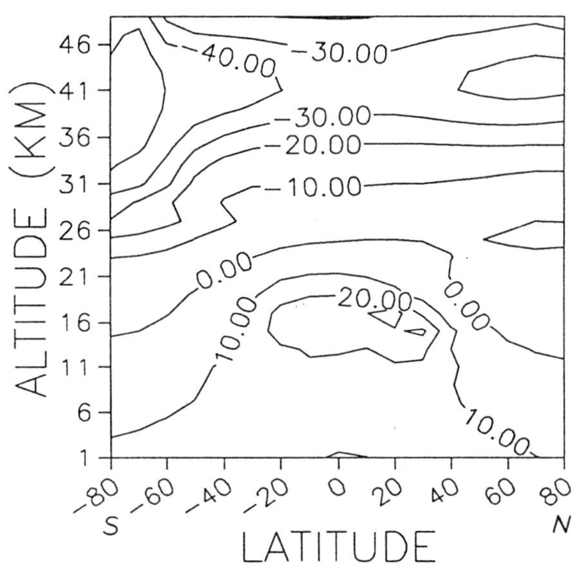

Fig. 3b: Ozone changes between 1965 and 2035 as a function of height and latitude for scenario b. The Figure represent summer months in the northern hemisphere.

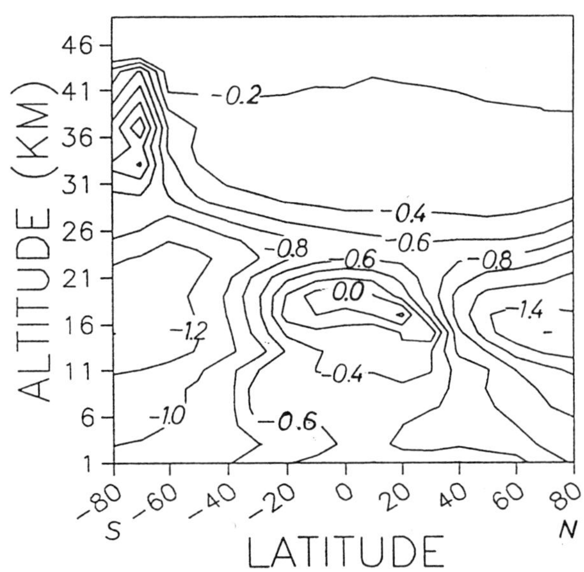

Fig. 4b: Ozone depletion between 1965 and 2035 as a function of height and latitude due to releases of bromine in the "Montreal" scenario (scenario c vs scenario b). The Figure represent the summer months in the northern hemisphere.

bations by enhanced levels of $CFCs$,N_2O and CH_4: A two-dimensional diabatic circulation study including uncertainty estimates. J. Geophys. Res., 91, 5249-5263.

Stordal, F., Isaksen, I.S.A. and Horntveth, K., 1985: A diabatic circulation two-dimensional model with photochemistry: Simulations of ozone and ground-released tracers, J. Geophys.Res., 90, 5757-5776.

Stordal, F. and Isaksen, I.S.A., 1987: Ozone perturbations due to increases in N_2O, CH_4 and chlorocarbons: two-dimensional time-dependent calculations, Tellus, 39B, 333-353.

WMO, Atmospheric ozone, 1985 : Assessment of our understanding of the processes controlling its present distribution and change, World Meteorological Organization, WMO Rep.No.16 ,1986.

TWO-D MODEL SIMULATION OF OZONE CLIMATOLOGY AND YEAR-TO-YEAR VARIATIONS

Hu Yang, Ka-Kit Tung, and Eduardo P. Olaguer
Clarkson University,
Potsdam, New York 13676, USA

ABSTRACT

Using a coupled 2-D model where all transport parameters are determined interactively from radiative transfer calculations using the observed NMC temperature input for the individual years in the period 1979–1986, we simulated the seasonal and year-to-year changes in the column ozone. Comparing the result with the zonally averaged data from TOMS, we find that the earlier climatology based on the first four years of TOMS data is well simulated, including the timing, magnitude and location of the spring maxima, and also the weak ozone hole that appeared. The year-to-year changes in column ozone at high latitudes follow the Quasi-Biennial Oscillation (QBO), as pointed out by several authors. This feature is also reproduced by the model.

1. INTRODUCTION

It has been well known since 1930s (Dobson, 1930) that the observed latitudinal and seasonal distributions of ozone abundance are almost opposite to that predicted by photochemical theory of ozone. "This gave rise to what has been known as the *ozone problem*, which is still not entirely resolved today and which represents a problem of considerable scientific interest" (Hunt, 1969). Considerable progress has been made towards the resolution of the ozone problem during the last two decades including a number of three-dimensional model studies (e.g. Hunt and Manabe, 1968; Hunt, 1969; London and Park, 1973; Newson, 1974; Cunnold *et al.*, 1975; Schlesinger and Mintz, 1979; Mahlman *et al.*, 1980). With the maturing of the 2-D models (see WMO, 1986), longer-term runs with coupled radiation, dynamics and chemistry are becoming feasible.

Today, we are interested not only in explaining the *sign* of the latitudinal and seasonal gradients, but also in quantitatively accounting for the drastic differences between the climatological ozone distribution between the Northern and Southern Hemispheres and in accounting for their year-to-year variations. The present work presents our first attempt in using a coupled third generation model with radiative feedbacks incorporated to study the ozone problem.

2. MODEL DESCRIPTION

Our 2-D model with coupled dynamics, radiation and chemistry has been documented in Yang (1988). The dynamical equations can also be found in Tung and Yang (1988b). The difference between the earlier version of the model (Tung and Yang, 1988b) and the present version lies mainly in the "comprehensive" radiative transfer code used in the present version and the additional modes of radiative-dynamical coupling that are now incorporated.

3. OZONE CLIMATOLOGY

The climatology of zonal mean column ozone, based on Nimbus 7 satellite data for a period of four years (from October, 1978 to September, 1982), can be found in Bowman and Krueger (1985). Model simulated ozone climatology is plotted in Fig. 1, which is an average of ozone values for model years from 1979 to 1982 when we have temperature data.

Important features present in both observed and simulated ozone climatology include the following: (1) the persistent minimum in the tropics, where the photochemical ozone production is most efficient with plenty of sunshine; (2) the spring maximum and fall minimum in high latitudes in NH; (3) the spring maximum in *mid*-latitudes in SH; (4) the springtime minimum in southern high latitudes (or the Antarctic ozone "hole"); (5) And the "filling up" of the ozone hole in November. We note that these features are robust, and can be found in observational data sets made in the early 1970s (see Hilsenrath *et al.*, 1979; Hilsenrath and Schlesinger, 1981; Prabhakara *et al.*, 1976). In particular, we suspect that the springtime high latitude minimum is a feature that persists every year, although the level in earlier years is higher (above or about 300 Dobson units). If this is true even for the pre-CFC era, then the *formation* of the ozone minimum is probably

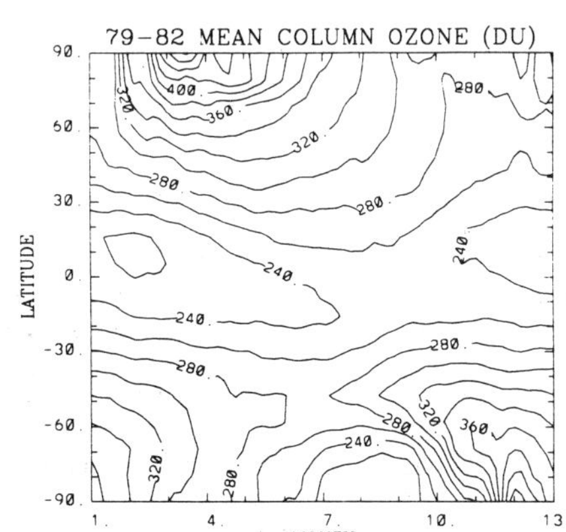

Fig. 1. Calculated time-latitude cross-section of the zonal mean total ozone (DU) averaged over model years 1979—1982.

ISBN 0-937194-15-8

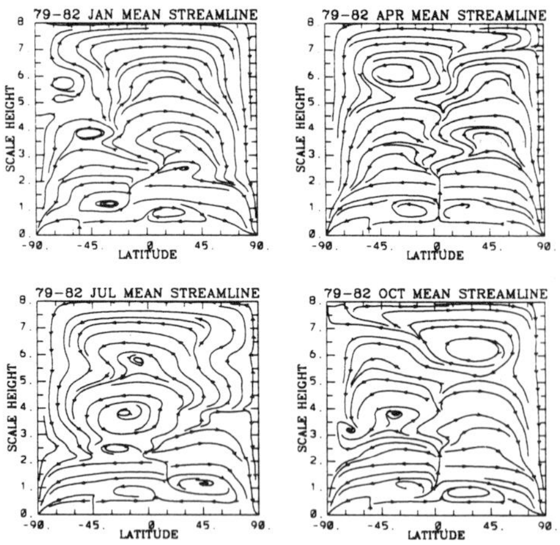

Fig. 2. Monthly mean streamlines of diabatic circulation for January, April, July and September averaged over model years 1979-1982.

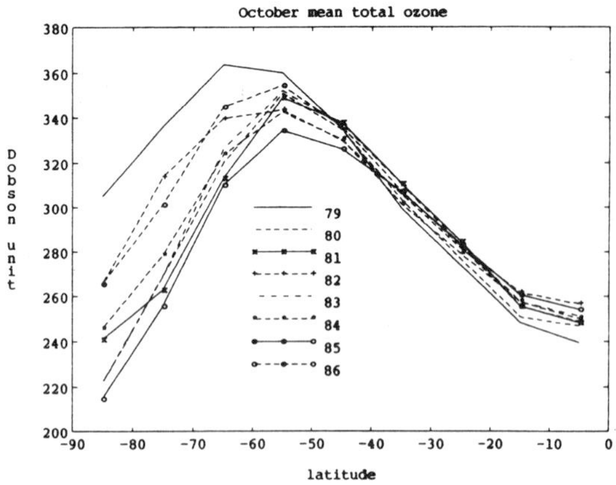

Fig. 3. Calculated monthly mean column ozone amount as a function of southern latitude for October of 1979—1986.

attributable to dynamics, although the *intensification* of the ozone hole in recent years may have an important chemically-induced component. Major departures of the simulated ozone climatology from observations include: (a) an underprediction of the column ozone amount during the southern polar night; and (b) values near the south pole after the final warming are too high and last too long.

The climatological column ozone distribution can be understood by examining the diabatic circulation calculated using our model. In Fig. 2, we plot the streamlines of diabatic circulation for January, April, July and October averaged for model years from 1979 to 1982. The upward motion near the tropics throughout the year constantly lifts ozone-poor air from the troposphere to replace ozone-rich air in the lower stratosphere, where ozone is produced by photochemistry, creating a minimum in the equatorial region. When the air rises, it cools through adiabatic expansion, resulting in the very cold equatorial tropopause, a region with abundant sunshine. This cold region traps most of the moisture reaching it and prevents it from entering the stratosphere, which explains the dryness of the stratosphere (WMO, 1986). In fact, the same kind of circulation, now known as the Brewer-Dobson circulation, has been inferred before from observed tracer distributions including water-vapor and ozone (Brewer, 1949; Dobson, 1956).

Higher values of ozone column amount are generally found in high latitudes due to poleward transport. During the northern winter when the photochemical time scale is very long in the polar night region, the ozone transported from lower latitudes accumulates in the polar region, leading to a maximum in spring. However, the situation during the southern polar night is quite different. The strong polar vortex prevents the transported ozone from reaching the polar region. The transport largely stops at the edge of the polar vortex so that a maximum is reached in mid-latitudes in the southern spring, while low values of column ozone amount are maintained during the polar night in the polar region. After spring in both hemispheres, the seasonal reversal in the upper circulation takes place and the lower circulation cell weakens, which together with enhanced photochemical destruction leads to the ozone column amount decreasing and later reaching a minimum in the fall. In our simulated ozone distribution, the fall minimum is not present in SH due to the lack of the increase phase in late fall and winter, which also leads to underprediction of ozone column amount in August and September.

In SH, the strong polar vortex also prevents effective heat transport into the polar night region, so that this region is very close to radiative equilibrium, especially in late winter. After the sun returns, increased heating due to solar absorption by ozone (and/or PSC which we have not included in the model) leads in our model to a net positive heating so that a reverse circulation develops in the lower stratosphere. This reverse circulation, as first suggested by Tung *et al.* (1986), will reduce the column ozone amount by replacing relatively ozone-rich air with ozone-poor air from below, just as in the tropics, thus helping to "deepen" the Antarctic ozone hole. After the final warming, the hole is "filled in" by ozone transported into the polar region when the polar vortex breaks down.

4. YEAR-TO-YEAR VARIATION

It is observed that the level of ozone in the Antarctic region has decreased dramatically in recent years. The observed zonal mean, monthly mean column ozone amount as a function of latitude for October of 1979 to 1985 is reported by Schoeberl *et al.* (1986). The October mean column ozone amount near the south pole in 1985 is more than 40% lower than that in 1979. The 1986 level is a little higher, but the 1987 level is more than 50% lower than that of 1979 (Schoeberl, personal communication).

In Fig. 3, the calculated column ozone amount is plotted as a function of latitude for October of 1979 to 1986. The simulated October mean column ozone amount compares favorably with the observations. In high latitudes, the column amount is well simulated by the model for the early four years, while for the later four years, the calculated values are higher than the observations. At 85S, the simulated ozone decline from 1979 to 1985 by the model is about 90 DU, 72% of the observed decrease, which is about 125 DU. This result should be compared with the 100 DU reported by Tung and Yang (1988b) using the Newtonian cooling approximation without the feedback between ozone and the radiative heating rates. At 65S, the model produced about a 50 DU decline from 1979 to 1985, same as in Tung and Yang (1988b), while the observed change in the maximum region is about 80 DU.

The non-monotonic nature of ozone year-to-year variation is evident from Fig. 4. Note that the up-and-down pattern of year-to-year column ozone variation produced by the model matches that of observations almost perfectly. This is because

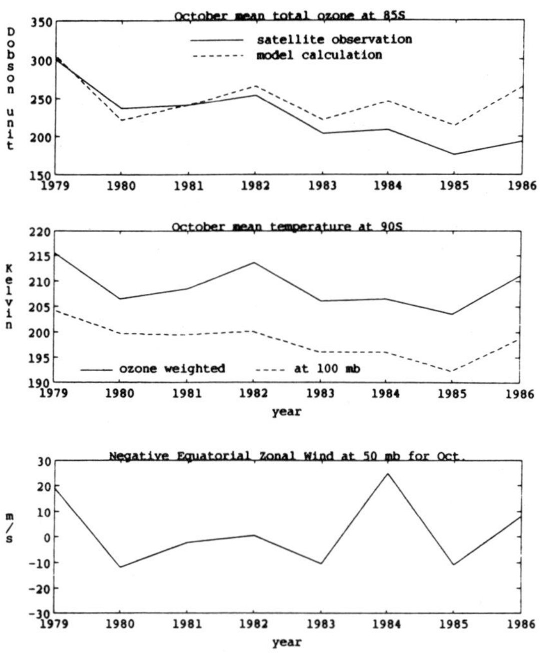

Fig. 4. Correlation between column ozone amount, temperature and equatorial zonal wind for October of 1979—1986. Top: the observed and calculated monthly mean ozone at 85S; middle: the ozone partial pressure weighted and 100 mb monthly mean temperatures at 90S; bottom: equatorial zonal wind (data taken from Garcia and Solomon, 1987).

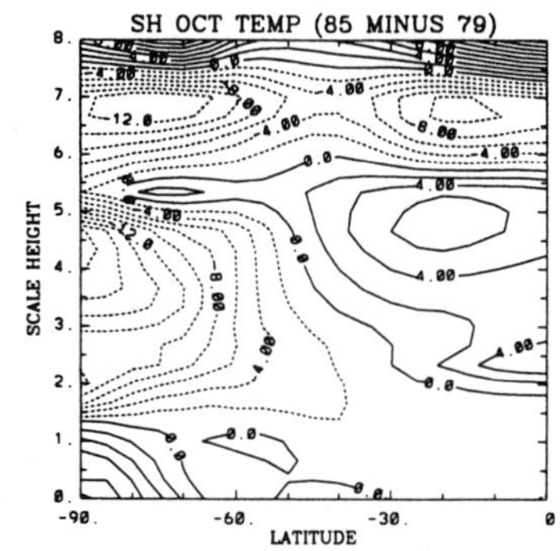

Fig. 5. Difference in Southern Hemisphere October mean temperatures (Kelvin) between 1985 and 1979.

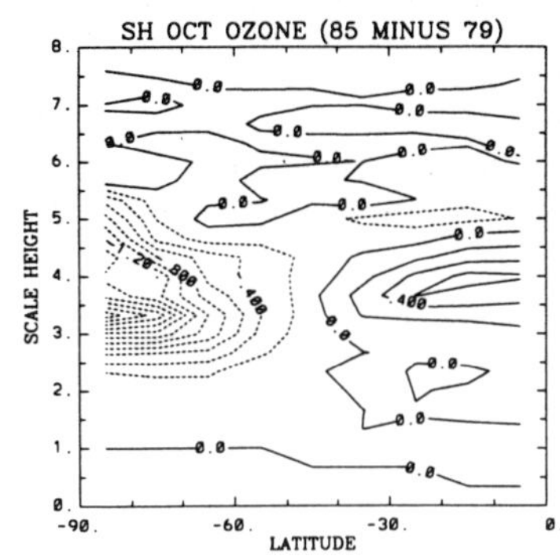

Fig. 6. The difference of the simulated October mean ozone mixing ratio (ppmv) between 1985 and 1979.

both the model simulated ozone and the ozone in the real atmosphere are correlated with the observed temperature. In the middle panel of Fig. 4, the ozone partial pressure weighted and 100 mb temperatures are plotted for October. One can see that the temperature variations in October match those of ozone changes.

It has been pointed out by many authors (Sekiguchi, 1986; Newman and Schoeberl, 1986; Chubachi, 1986; Angell, 1986; Mahlman and Fels, 1986) that the total ozone column amount is well correlated with the temperature of the lower stratosphere, with low column ozone amount correlated with low temperature and high column ozone amount correlated with high temperature. The mechanism of this correlation was explored by Tung and Yang (1988a). Cooling of the lower stratosphere induces a transport taking ozone away from the cooled region and it was shown analytically that a 1% temperature change weighted by ozone partial pressure produces a 7% change in column ozone amount. One can see that 1985 is colder than 1979 at the south pole by 12°K when temperature is weighted by ozone partial pressure. For a 12°K (about 5.5%) temperature change from 1979 to 1985, the simple analytic solution appears to give an estimate of the right magnitude. If a second-order correction is added to the first-order solution, the ozone decrease should be 36% or 108 DU for a 5.5% temperature change from 1979 to 1985. This should be compared to the simulated 90 DU decrease, and the observed 125 DU decrease.

The correlation of column ozone amount and temperature does not tell which one is the cause. Temperature change due to dynamical variation can influence the ozone transport, while chemically induced ozone change would also affect the temperature due to altered ozone absorption. The present work seems to suggest that an October mean temperature change of the order of 12°K can induce as much as 72% of the observed ozone depletion. However the work of Kiehl *et al.* (1988) suggests that the radiative effect of ozone reduction on temperature can be a cooling of the October mean temperature of about 3°K (see Tung and Yang, 1988b). Even if one assumes that *all* of the ozone-induced temperature change is due to chemistry and subtracts it from the total temperature change, the temperature change left (presumably due to dynamics) can still account for about 54% of the observed ozone depletion, according to the present work. However, there are still large uncertainties in the model calculations (e.g., associated with radiative transfer codes).

As pointed out by Garcia and Solomon (1987), the phase of QBO may have important effects on the observed ozone and temperature variations. One can see from Fig. 4 that warm (cold) years and high (low) ozone are correlated with the easterly (westerly) phase of the QBO. If the link between QBO and temperature/ozone variation is established, then the temperature and ozone changes in October are likely to be strongly influenced by dynamical causes, since QBO is a tropical phenomenon.

In Fig. 5, the difference of October mean temperature between 1985 and 1979 is plotted. In Fig. 6, we plot the difference of October mean ozone concentration between 1985 and 1979. Comparing Fig. 6 with Fig. 5, one sees the striking *spatial* correlation between the temperature and ozone, except, of course, in the upper stratosphere where ozone is controlled by chemistry, which is fixed in our model for every year. The decrease in high latitudes is accompanied by an *increase* in low latitudes. The spatial correlation between the *total* ozone pattern and the distribution of temperature in the lower stratosphere has been pointed out by many authors before (Stolarski *et al.*, 1986; Newman and Schoeberl, 1986; Sekiguchi, 1986; Chubachi, 1986).

One of the important features of the Antarctic ozone hole

is that the seasonal depletion occurs mainly in a layer from about 12 to 20 km (Chubachi, 1985; Hofmann *et al.*, 1987; Gernandt, 1987; Komhyr *et al.*, 1988). Examining Fig. 2 indicates that a reverse diabatic circulation (Tung *et al.*, 1986) exists in the same region where the ozone amount is observed to decrease. However, the reverse circulation based on the present study is quite weak, which suggests that in recent years (after 1983), the accelerated deepening of the Antarctic ozone hole cannot be explained by dynamics alone. When accelerated chemical depletion rates are added into our model to bring the model Antarctic values to the observed ones in 1986, we find that the heating rates are further reduced to close to radiative equilibrium values even in spring in the presence of increasing solar radiation (but not enough ozone to absorb it). The absence of upwelling in 1986 in our model and in the observed data (Hofmann *et al.*, 1987) does not imply that the dynamical mechanism is incorrect. One can argue that if the vertical velocity is close to zero for the 1986 condition, then in an earlier year when the ozone amount in the lower stratosphere is higher, and/or the temperature is lower, the vertical velocity may well be positive and play an important role in the ozone seasonal decline in those early years.

ACKNOWLEDGMENT

The research was supported by NASA under grant # NAGW-910.

REFERENCES

Angell, J. K., 1986: The close relation between Antarctic total-ozone depletion and cooling of the Antarctic low stratosphere. *Geophys. Res. Lett.*, **13**, 1240-1243.

Bowman, K. P., and A. J. Krueger, 1985: A global climatology of total ozone from the Nimbus 7 total ozone mapping spectrometer. *J. Geophys. Res.*, **90**, 7967-7976.

Brewer, A. W., 1949: Evidence for a world circulation provided by the measurement of helium and water vapour distribution in the atmosphere. *Quart. J. Roy. Meteorol. Soc.*, **75**, 351-363.

Chubachi, S., 1985: A special ozone observation at Syowa station, Antarctica from February 1982 to January 1983, in *Proceedings of the Quadrenniel Ozone Symposium, Halkidiki, Greece, 1984*, ed. by C. S. Zerefos and A. Ghazi, Reidel Pub. Co.

——, 1986: On the cooling of stratospheric temperature at Syowa, Antarctic. *Geophys. Res. Lett.*, **13**, 1221-1223.

Cunnold, D., F. N. Alyea, N. Phillips, and R. G. Prinn, 1975: A three-dimensional dynamical chemical model of atmosphere ozone. *J. Atmos. Sci.*, **32**, 170-194.

Dobson, G. M. B., 1930: Observation of the amount of ozone in the Earth's atmosphere and its relation to other geophysical conditions. *Proc. R. Soc. London, Ser. A*, *129*, 411.

——, 1956: Origin and distribution of polyatomic molecules in the atmosphere. *Proc. Roy. Soc. London*, **A236**, 187-193.

Garcia, R. R., and S. Solomon, 1987: A possible relationship between interannual variability in Antarctic ozone and the quasi-biennial oscillation. *Geophys. Res. Lett.*, **14**, 848-851.

Gernandt, H., 1987: The vertical ozone distribution above the GDR research base, Antarctica in 1985. *Geophys. Res. Lett.*, **14**, 84-86.

Hilsenrath, E., D. F. Heath, and B. M. Schlesinger, 1979: Seasonal and interannual variations in total ozone revealed by the Nimbus 4 backscattered ultraviolet experiment. *J. Geophys. Res.*, **84**, 6969-6979.

——, and B. M. Schlesinger, 1981: Total ozone seasonal and interannual variations derived from the 7 year Nimbus 4 data. *J. Geophys. Res.*, **86**, 12087-12096.

Hofmann, D. J., J. W. Harder, S. R. Rolf, and J. M. Rosen, 1987: Balloon-borne observations of the development and vertical structure of the Antarctic ozone hole in 1986. *Nature*, **326**, 59-62.

Hunt, B. G., 1969: Experiments with a stratospheric general circulation model, III: Large-scale diffusion of ozone including photochemistry. *Mon. Weather Rev.*, **97**, 287-306, 1969.

——, and S. Manabe, 1968: Experiments with a stratospheric general circulation model, II: Large-scale diffusion of tracers in the stratosphere. *Mon. Weather Rev.*, **96**, 503-539.

Kiehl, J. T., B. A. Boville, and B. P. Briegleb, 1988: Response of a general circulation model to a prescribed Antarctic ozone hole. *Nature*, **332**, 501-504.

Komhyr, W. D., S. J. Oltmans, and R. D. Grass, 1988: Atmospheric ozone at South Pole, Antarctica, in 1986. *J. Geophys. Res.*, **93**, 5167-5184.

London, J., and J. Park, 1973: Application of general circulation models to the study of stratospheric ozone. *Pure Appl. Geophys.*, **106-108**, 1611-1617.

Mahlman, J. D., and S. B. Fels, 1986: Antarctic ozone decreases: A dynamical cause? *Geophys. Res. Lett.*, **13**, 1316-1319.

——, H. B. Leovy II, and W. J. Moxim, 1980: Three-dimensional tracer structure and behavior as simulated in two ozone precursor experiments. *J. Atmos. Sci.*, **37**, 655-685.

Newman, P. A., and M. R. Schoeberl, 1986: October Antarctic temperature and total ozone trends from 1979-1985. *Geophys. Res. Lett.*, **13**, 1206-1209.

Newson, R. L., 1974: An experiment with a tropospheric and stratospheric three-dimensional general circulation model, in *Proceedings Third Conference on the Climatic Impact Assessment Program, DOT-TSC-OST-74-15*, edited by A. J. Broderick and T. M. Hard, pp. 461-474, Dept. of Transp., Washington DC.

Prabhakara, C., E. B. Rodgers, B. J. Conrath, R. A. Hanel, and V. G. Kunde, 1976: The NIMBUS 4 infrared spectroscopy experiment, 3: observations of the lower stratospheric thermal structure and total ozone. *J. Geophys. Res.*, **81**, 6391-6399.

Schlesinger, M. E., and Y. Mintz, 1979: Numerical simulation of ozone production, transport and distribution with a global general circulation model. *J. Atmos. Sci.*, **36**, 1325-1361.

Schoeberl, M. R., A. J. Krueger, and P. A. Newman, 1986: The morphology of Antarctic total ozone as seen by TOMS, *Geophys. Res. Lett.*, *13*, 1217–1220.

Sekiguchi, Y., 1986: Antarctic ozone change correlated to the stratospheric temperature field. *Geophys. Res. Lett.*, **13**, 1202-1205.

Stolarski, R. S., A. J. Krueger, M. R. Schoeberl, R. D. McPeters, P. A. Newman, and J. C. Alpert, 1986: Nimbus 7 SBUV/TOMS measurements of the springtime Antarctic ozone decreases. *Nature*, **322**, 308-311.

Tung, K. K., M. K. W. Ko, J. M. Rodriguez, and N. D. Sze, 1986: Large Antarctic ozone variation: a manifestation of dynamics or chemistry? *Nature*, **322**, 811-814.

——, and H. Yang, 1988a: Dynamical variability of column ozone. *J. Geophys. Res.*, in press.

——, and ——, 1988b: Dynamical component of seasonal and year-to-year changes in Antarctic and global ozone. *J. Geophys. Res.*, in press.

WMO, 1986: *Atmospheric Ozone 1985: Assessment of our understanding of processes controlling its present distribution and change.* World Meteorological Organization.

Yang, H., 1988: *Numerical Simulation of Global and Antarctic Ozone Using a 2-D Model in Isentropic Coordinates.* Ph. D. thesis, MIT.

SIMULATION OF THE NORTHERN HEMISPHERE OZONE DISTRIBUTION WITH A GENERAL CIRCULATION MODEL

D. Cariolle, M. Déqué, and P. Simon
Direction de la Météorologie
Centre National de Recherches Météorologiques
31057 Toulouse, France

ABSTRACT

A spectral General Circulation Model is used to study the dynamics and the ozone distribution in the Northern hemisphere middle stratosphere. Results from two successive winter simulations are discussed.

Each winter shows the development of minor warmings which, during the second winter, are followed by a major warming in late February. Maps of potential vorticity and ozone on isentropic surfaces at different stage of the warming illustrate the important role played by the irreversible mixing due to planetary wave-breaking, and the wave damping by radiative processes and by momentum deposition due to the breaking of orographic gravity-waves. These results are compared to similar diagnostics obtained from satellite observations.

1. MODEL DESCRIPTION.

The model is derived from the French operational weather forecasting hemispheric model "Emeraude" (Coiffier et al,1987).The global version used in this study has 30 vertical levels extending from the ground to the mesosophere. A T21 resolution is used corresponding to the explicit horizontal resolution of the dynamical structures up to wavenumber 21 in the zonal and meridional directions. Nonlinear terms, including physical and chemical parameterizations are computed on the "Gaussian" grid with 32 latitude circles and 64 equidistant points on each latitude. A complete physical package is used to parameterize the main physical processes: radiative transfer, rainfall and latent heat release, convection. In addition a parameterization has been developed for the momentum deposition due to wavebreaking of sub-grid scale orographic gravity waves (Cariolle and Rochas,1987).

The ozone mixing ratio is treated as a fully interactive variable, advected according to the model wind fields. Its concentration is also used to compute the ultraviolet and infrared heating rates. The ozone photochemical production and loss rates are parametrized following the linear scheme developed by Cariolle and Déqué (1986).

2. MODEL SIMULATION AND QUASI-STATIONARY PLANETARY WAVE ACTIVITY

The model has been integrated for two years starting from the situation on the 17 December of 1978. Thus two successive winter simulations can be compared, giving an initial impression of the model interannual variations. The geopotential heights fields at 10 mb, averaged over periods of 15 days, are presented in fig. 1 for both winters in January and February.

Both late Januaries are characterized by the development of minor warmings with the buildup of a small geopotential maximum at high latitudes near the Aleutian Islands. The polar vortex is slightly displaced towards Europe with an elongation over the American continent. This picture is typical of a wavenumber one disturbance superposed to the mean zonal flow. As a result of a more pronounced tropospheric Pacific ridge, the Aleutian anticyclone is more intense at 10 mb during the second winter.

During the second halves of February, the two winter simulations exhibit very different situations. An elongated vortex develops during the first year with the build up of two anticyclonic votices, one over the pacific and the second one over Europe. At midlatitudes the wavenumber two pattern dominates whereas at high latitudes the flow is still very zonal despite that the vortex center is slightly deplaced towards Europe.

By contrast the second winter is much more perturbed with the development of a major warming. The vortex is much smaller in size and displaced over Scandinavia. A large tongue of air extracted from the

ISBN 0-937194-15-8

vortex and rotating clockwise around the Aleutian anticyclone is well developed. A second anticyclonic center is growing over the Himalayas. The synoptic picture is very close to what has been observed in the atmosphere during late January 1987. This rather perturbed situation cannot be simply described as a superposition of waves on the mean flow.

A 10 mb 16-30 JANUARY YEAR 1

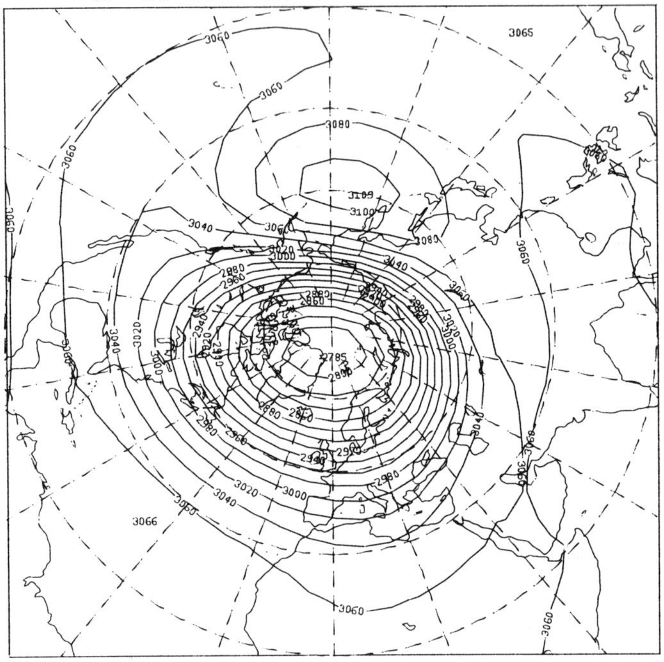

B 10 mb 16-29 FEBRUARY YEAR 1

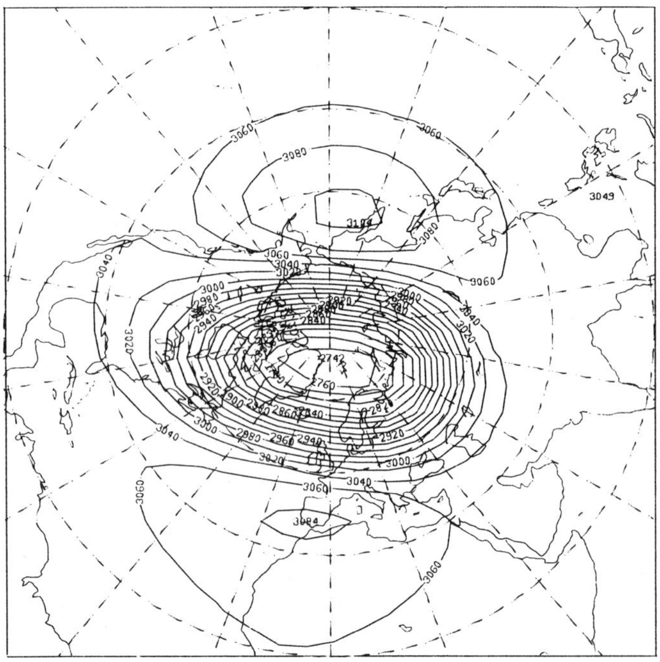

C 10 mb 15-29 FEBRUARY YEAR 2

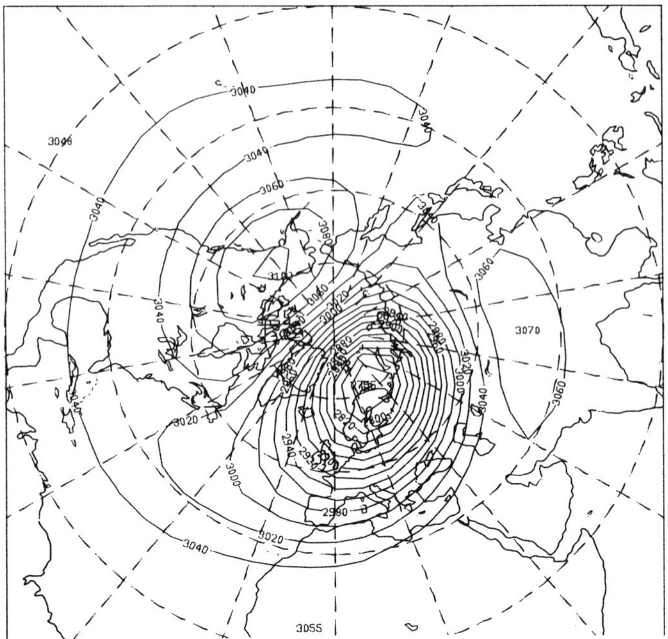

Fig. 1. Geopotential height fields at 10 mb averaged over periods of two weeks:a) 16-30 January of year 1 ,b) 16-29 February of year 1, and c) 15-30 February of year 2.

3. THE DISTRIBUTION OF OZONE AND POTENTIAL VORTICITY

Ertel's Potential Vorticity (EPV) is a quasi-conservative quantity in the stratosphere. Its distribution and variations with time give very good information on the transport of air masses. In particular, during sudden warmings, it is possible to distinguish the polar air from the air that originated at lower latitudes since the potential vorticity has much higher values inside the vortex. Palmer and McIntyre (1983) have shown that, based on EPV maps calculated from SSU data, the vortex is continually eroded during winter due to the presence of large amplitude planetary waves. Since our model reproduces also episodes with large amplitude planetary waves, it is important to see if the concept of "wave breaking" described by Palmer and McIntyre applies to the model simulation.

The equation for the time evolution of EPV (Haynes and McIntyre,1987; equ.4.3) may be written as:

$$d(\rho EPV)/dt = -\, div\{\, \rho\, EPV\, \vec{U} - \dot{\theta}\, (Rot\, \vec{U} + 2\vec{\Omega}) - \vec{F} \wedge \vec{\nabla\theta}\, \} \qquad (1)$$

where $\dot{\theta}$ represents the diabatic heating rate and $\vec{F}$ the frictional forces, mostly dissipation and drag due to gravity wave breaking in the model simulation.

Figure 2a,b shows the maps of the 10 mb geopotential heights and the EPV on the 850 K iso-θ surface on the February 16, during the warming of the second year. The vortex is clearly defined by the large gradients surrounding the region with high EPV content. The air with relatively high EPV content is extracted from the main vortex and is advected towards lower latitudes around the high geopotential center. During the

GEOPOTENTIAL 10 hPa FEBRUARY 16

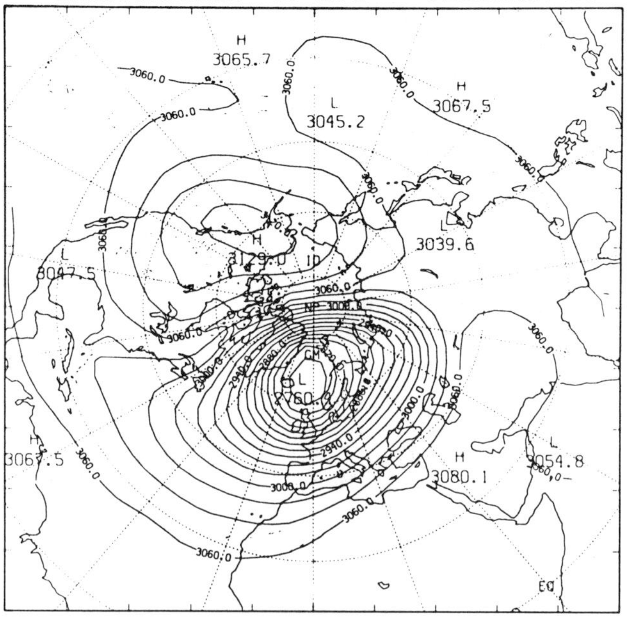

POTENTIAL VORTICITY 850K FEBRUARY 16

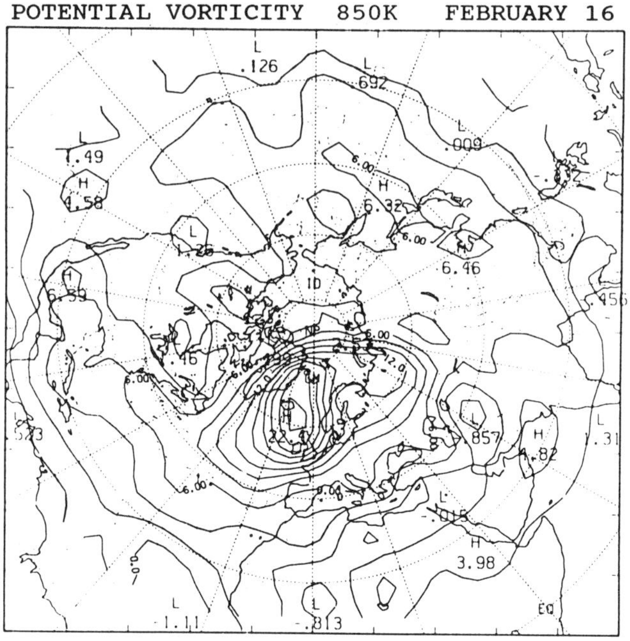

Fig. 2. Geopotential height field at 10 mb (a), EPV (b) and ozone mixing ratio (c) on the 850 K isentropic surface for the 16 of February of year 2 during the major warming.

OZONE (ppmv) 850 K FEBRUARY 16

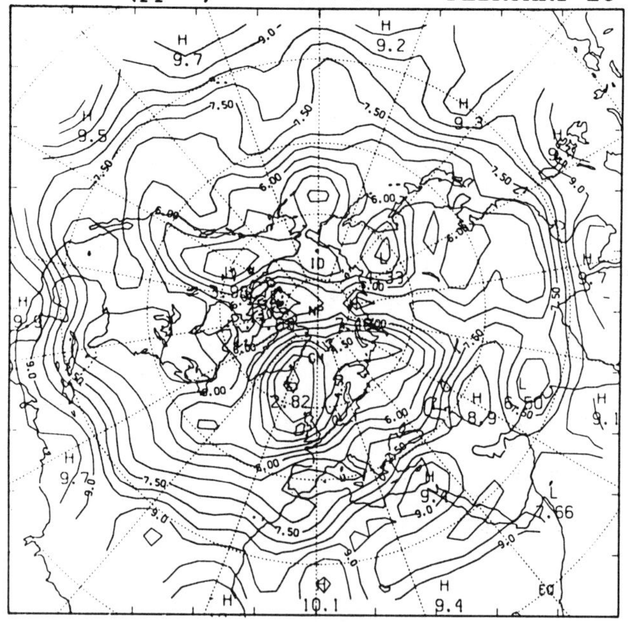

mature stage of the warming, air from the vortex is encountered at midlatitudes over the Pacific and Asia. This region, as well as the one located inside the anticyclonic area, is dominated by smaller scale structures in the EPV maps and was named the "surf zone" by McIntyre and Palmer. Also noticeable is the tongue of low latitude air with low EPV content extending from North Africa, Central Europe to the north pole.

There is a generally good anti-correlation between EPV and the ozone mixing ratio (fig. 2c). In particular low ozone contents are found inside the main vortex, whereas the tongue of low latitude air has a high ozone mixing ratio. This means that in those regions EPV and ozone are relatively well conserved and act as pure tracers.

By contrast, in the region of the "surf zone", EPV and ozone are progressively decorrelated. In particular, low ozone mixing ratios are found inside the anticyclone without being associated with high EPV. This particular feature is consistent with the one observed during the January 1979 major warming. Leovy et al.(1985) have presented for January 22, an ozone map using LIMS data similar to our fig. 2c , with a distinct ozone maximum in the anticyclonic region; and McIntyre et al. (1983) have shown an EPV map for January 23 without any particular structure in the same region. Thus the model calculation and observations are completely consistent, showing very good anticorrelation between ozone and EPV in the vortex and in the air rapidly advected around the vortex, and a progressive decorrelation in the "surf zone".

This result provides a good illustration of the important damping that occurs in the "surf zone". Since the ozone photochemical sources and sinks are mainly latitudinally dependant and are in any case very weak at high latitudes in January and February, the origin of the decorrelation between EPV and ozone must arises from EPV production and destruction. From equation (1) it is clear that the main physical processes responsible for EPV production and destruction are the radiative damping and the momentum deposition by gravity wave breaking. Both processes are important in the model simulation. The radiative cooling of the low latitude air crossing over the dark pole can reach several degrees per day and the deceleration of the polar vortex due to gravity wave breaking sometimes exceeds several tens of meters per day during the warming.

4. CONCLUDING REMARKS

The winter simulations of our model bear many similarities with the atmosphere. In particular, the succession of several minor warmings before the development of a major warming is very consistent with observations. As a result of different tropospheric large scale wave amplitudes from one year to an other,the model also shows a significant year to year variability in planetary wave activity.

The ozone and EPV behaviours on isentropic surfaces illustrate the different characteristics of planetary wave breaking as has already been discussed by McIntyre and Palmer. In our simulation, during the major warming of the second year, the regions of the main vortex and the surf zone are clearly identified. The latter being characterized by an important decorrelation between ozone and EPV. This result is consistent with satellite data and clearly shows the importance of the planetary wave damping by radiative processes and by momentum deposition due to gravity wave breaking.

Although our results agree with those already obtained with a model with lower horizontal resolution (Cariolle and Rochas,1987), it is important to evaluate the influence of the dissipation due to horizontal truncation on the model simulation. Mahlman and Umscheid (1987) have found a dramatic improvement of their grid point model results when the resolution was increased from 5°x5° to 1°x1°. We will soon repeat the present calculations with T42 and T79 resolutions to evalute the sensitivity of our spectral model to horizontal resolution.

The exact relationship between tropospheric and stratospheric planetary-scale disturbances still need to be clarified. It is hoped that further diagnostics from model simulations will help to solve those remaining issues.

5. REFERENCES

Cariolle,D. and M. Déqué, 1986: Southern Hemisphere medium scale waves and total ozone disturbances in a spectral general circulation model. J.Geophys.Res.,91,825-84.

Cariolle,D. and G. Rochas, 1987: L'influence des ondes de gravité d'origine orographique sur le developpement des réchauffements soudains stratosphériques.C.R.Acad.Sci.,305, 189-192.

Coiffier,J. Y. Ernie, J.F. Geleyn, J. Clochard, J. Hoffman and F. Dupont,1987: The operational hemispheric Model at the French Meteorological Service. J.Meteor.Soc.Japan,337-345.

Haynes,P.H. and M.E. McIntyre,1987: On the evolution of isentropic distribution of potential vorticity in the presence of diabatic heating and frictional or other forces. J.Atmos.Sci.,828-841.

Leovy,C.B.,C.-R. Sun, M.H. Hitchman, E.E. Remsberg, J.M. Russel, L.L. Gordley,J.C. Gille, L.V. Lyjak,1985: Transport of ozone in the middle stratosphere: evidence for planetary wave breaking.J.Atmos.Sci.,42,230-244.

McIntyre, M.E. and T.N. Palmer,1983: Breaking planetary waves in the stratosphere, Nature,305,593-600.

Mahlman, J.D. and L.J. Umscheid,1987: Comprehensive modeling of the middle atmosphere: the influence of horizontal resolution. Transport Processes in the Middle Atmosphere,G. Visconti and R.Garcia (Eds.),251-266.

THE BEHAVIOR OF OZONE AND OTHER TRACE GASES IN THE STRATOSPHERE DURING DYNAMICALLY PERTURBED SITUATIONS: A THREE-DIMENSIONAL MODEL

Guy Brasseur[1], *Klaus Rose*[2], *and Charles M. Smythe*[1]
[1]National Center for Atmospheric Research
Boulder, Colorado 80307, USA*
[2]Free University of Berlin
Berlin, FRG

ABSTRACT

A mechanistic three-dimensional dynamical model with chemistry is used to study the behavior of chemically active tracers in the stratosphere during periods disturbed by planetary-scale waves (winter). Reactions affecting oxygen, nitrogen, hydrogen and chlorine compounds are explicitly included in the model. The circulation is driven by a forcing specified at the lower boundary of the model. The relative roles of chemical and dynamical processes are described and the largely irreversible character of meridional transport during a stratospheric warming is shown.

1. INTRODUCTION

During winter, when the stratosphere becomes significantly disturbed by the propagation and the dissipation of large-scale planetary waves, meridional transport of trace species becomes important and greatly affects the distribution of these species. In the present paper we intend to study the behavior of ozone and other chemically active species in the stratosphere during a simulated warming event. The model which is used in this study is a mechanistic three-dimensional model (Rose, 1983) to which a chemical scheme has been added. The model, which extends from 10 to 80 km altitude, is forced at its lower boundary by a wave like perturbation that propagates in the stratosphere and mesosphere and affects the dynamics and consequently the distribution of tracers.

2. BRIEF DESCRIPTION OF THE MODEL

The model is based on the primitive equations which are expressed in spherical coordinates and solved on a three-dimensional hemispheric grid with a resolution of 22.5 degrees in longitude, 5 degrees in latitude and about 3 km in altitude. Diabatic processes are expressed by a Newtonian cooling term such that, in the absence of forced disturbances, the model relaxes towards a climatological rather than the radiative equilibrium temperature. The (radiative) dissipation time constant is of the order of 23 days at 15 km, 13 days at 30 km, 5 at 50 km, and 4.5 days at 70 km.

The model is forced at its lower boundary by a perturbation in the geopotential (maximum amplitude of 300 gpm and 100 gpm for the first and second harmonic respectively). The phase is zero degree for both waves.

The three-dimensional distribution of the trace species is obtained by solving continuity equations in which the transport is expressed as a function of the winds derived from the dynamical scheme, and the source/sink terms explicitly include the most important chemical and photochemical reactions occuring in the middle atmosphere. Eight constituents or families ($O_x=O+O_3$; $NO_x=N+NO+NO_2+NO_3+HO_2NO_2+ClONO_2$; HNO_3; N_2O_5; $Cl_x=Cl+ClO+HOCl+ClONO_2$; HCl; H_2O; CH_4) are transported in the model. The partition between family members is performed assuming photochemical equilibrium conditions between these species. It should be noted that since the chemical coupling of HNO_3 and N_2O_5 with the other odd nitrogen compounds is weak, these species are treated independently from the NO_x family. HCl is also isolated from the remaining odd chlorine compounds. The chemical rate constants adopted in this study are taken from the NASA/JPL compilation (De More et al., 1985).

The prognostic equations are solved by the leapfrog scheme (see Rose, 1983) with a timestep of 450 seconds (7.5 min). Linear stability is assured by filtering out the high-frequency waves near the pole using a Fourier filtering method. In order to avoid non-linear instability, a 8th order filter is applied to all prognostic variables in the meridional as well as in the zonal direction.

3. MODEL RESULTS AND DISCUSSION

The model simulation is started with zonally symmetric fields of all dependent variables. The distributions of the trace gases are constructed from both

* The National Center for Atmospheric Research is funded by the National Science Foundation.

ISBN 0-937194-15-8

observations and results of a two-dimensional model (Brasseur et al., 1988). As the wave applied at the lower boundary of the model propagates, the dynamical fields in the stratosphere and mesosphere become disturbed and meridional transport of heat, momentum and chemical constituents takes place. In this discussion, we will restrict our analysis mainly to the situation of the middle stratosphere on the 11.6 mb level (about 31 km). As shown in Figure 1, the polar vortex is displaced from the pole so that the circulation may transport air parcels (e.g., with high ozone concentration) effectively from the subtropical zone to the polar region. The formation of an anticyclone at the edge of the vortex indicates the beginning of a sudden stratospheric warming during the model run. The amplitude of the geopotential wave at 60 degrees is of the order of 625 gpm. The air in the displaced vortex remains cold but, in the vicinity of the high pressure cell, it is adiabatically heated and, on day 15 for example it is about 40 K warmer than inside the vortex.

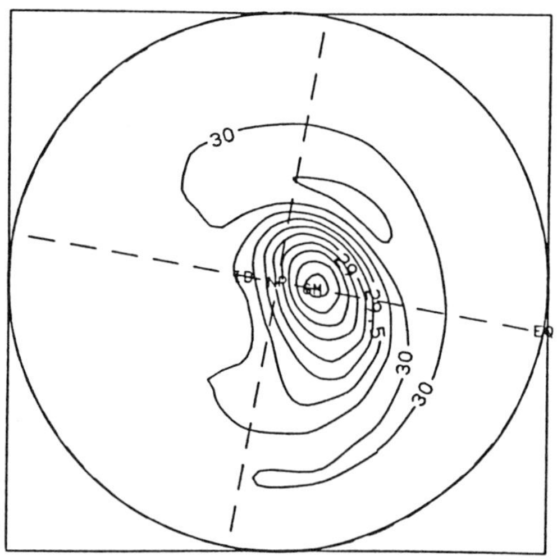

Fig. 1. Distribution of the geopotential height (in km) on the 11.6 mb pressure surface on day 15 of the model integration.

Figures 2a-d show a sequence of ozone distributions on the 11.6 mb surface during the development of the planetary-scale disturbance. The air with low ozone concentration, initially located over the pole, is displaced together with the vortex which thus behaves as a material entity. At the same time a tongue of rich-ozone air is produced at the edge of the displaced vortex and indicates that the meridional transport of tracers in winter does not occur on straight South-North trajectories (as described by two-dimensional models) but results essentially from advection around the polar vortex. The irreversibility of the transport depends on the strength of dissipation, transience and non-linearity as well as on the existence of chemical eddies. The highly non-linear process of wave breaking leads to irreversible mixing of chemical constituents that have meridional gradients on isentropic surfaces (Andrews et al., 1987). The model shows that, as time evolves, smaller structures become apparent in the ozone distribution, pointing out the transport to be a non-linear process leading to an amplification of higher wave numbers. Un-

a

b

c

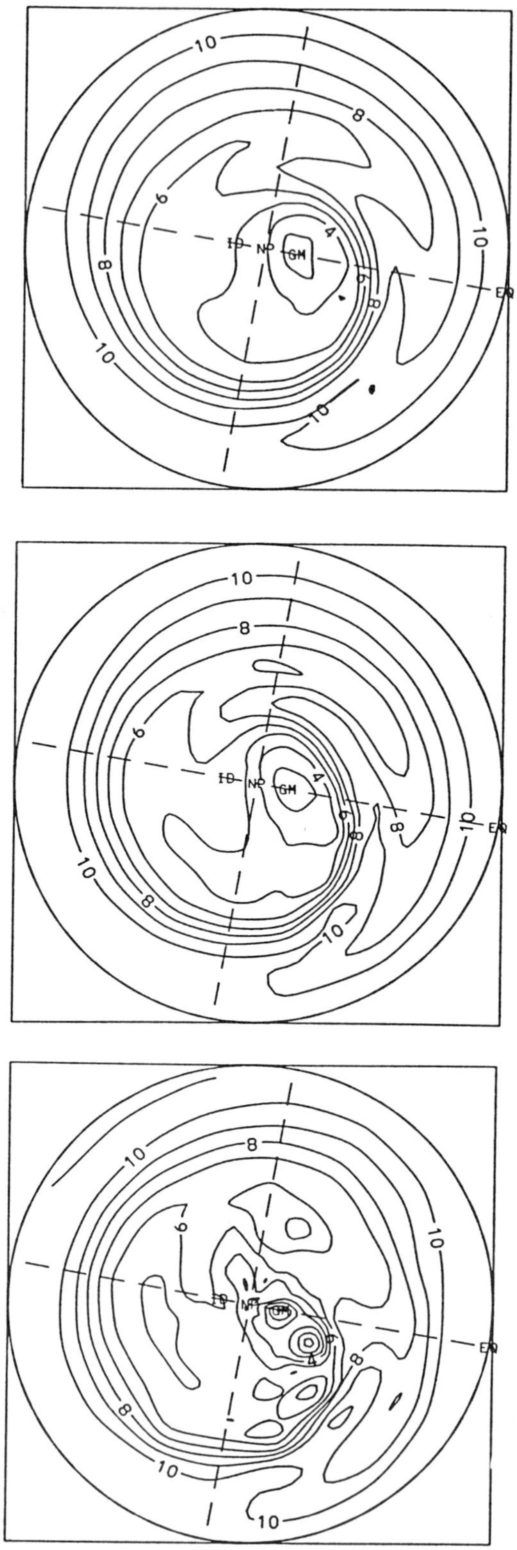

Fig. 2.a-c Ozone distributions (ppmv) on the 11.6 mb level on days (a) 12, (b) 15 and (c) 18 of the model integration.

d

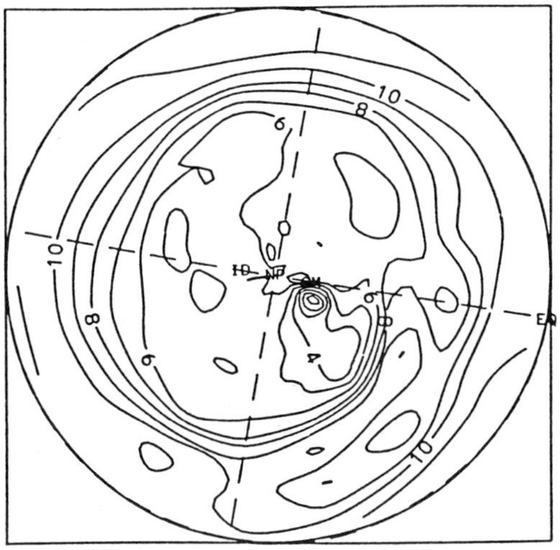

Fig. 2.d Same as Fig. 2.a-c but for day 23.

fortunately, the resolution of the model is too coarse to study these processes in detail. But there is indication that a large area with weak concentration gradients is formed at high latitude outside the region of the polar vortex, consistently with the idea of the "surf zone" developed by McIntyre and Palmer (1984). Furthermore the model shows a substantial erosion of the vortex during the warming event, which produces efficient mixing near the vortex.

The distribution of odd nitrogen species is more difficult to analyze since it is determined by equally important chemical and dynamical processes. Figure 3a shows the distribution of nitric acid on the 11.6 mb level on day 15. The morphology of the concentration of this constituent, which behaves as a quasi-conservative tracer, is similar to that of ozone except that the high mixing ratio values are in this case located inside the vortex and the low values are in the tropics. The distribution of N_2O_5 on the 11.6 mb level (Fig. 3b) appears to be influenced more substantially by chemistry. Despite the displacement of air parcels together with the vortex, high N_2O_5 mixing ratios remain present over the pole as a result of continuous conversion of NO_2 to N_2O_5 in the region of the polar night. The spatial variation of the N_2O_5 density inside the winter polar cap provides information on the length of time during which air parcels have been in the dark. The diurnal variation of the nitrogen species is clearly visible on the maps. Nitric oxide (Fig. 3c) is only present during daytime and is rapidly converted into nitrogen dioxide (Fig. 3d) immediately after sunset. As seen from Fig. 3d, the NO_2 mixing ratio is a factor of 2 larger during nighttime (about 10 ppbv at mid-latitude) than during daytime (about 5 ppbv). The relatively slower conversion between NO_2 and N_2O_5 is also noticeable. At a given latitude outside the polar night cap, the concentration of N_2O_5 increases as a function of local time in the nighttime region and decreases in the daytime region. The opposite is found for $NO + NO_2$. An examination of the distribution of NO_2 and NO_2 shows that transport affects these species particularly at high latitudes. A tongue of air with high NO_x density is present at the Southern edge of the displaced vortex and strong latitudinal gradients in the mixing ratio are observed (a factor 2 variation over 10 degrees latitude) in the vicinity of the tongue. Such rapid variations in the density of nitrogen dioxide have been observed and are referred as the "Noxon cliff" (Noxon, 1979). The concentration of nitrogen oxides is thus expected to be highly variable versus time and space as a result of simultaneous effects of photochemistry and dynamics. Interpretation of observations of these species is therefore difficult and requires an inspection of the dynamical fields as well as knowledge of the solar elevation during the time of the measurement.

The model simulation suggests that the distribution of other long-lived trace gases is also significantly perturbed during warming events. Figure 4 shows, as an example, the calculated distribution of methane on the 11.6 mb pressure level on day 15. A mixing ratio of less than 0.6 ppmv is found inside the displaced vortex and a tongue of air with concentration values larger than 0.9 ppmv originates from lower latitudes. There are few global observations of source gases such as methane in the stratosphere but the data from the SAMS instrument on Nimbus 7 (Jones, 1984) shows clear evidence of material deformation during planetary wave breaking.

4. CONCLUSIONS

The distribution of minor constituents in the stratosphere is substantially affected by the propagation and breaking of planetary waves. The transport of these species is therefore three-dimensional in nature, especially in winter. In the region near 10 mb, source-gases, ozone, nitrogen and chlorine compounds are affected by both chemistry and dynamics. Their behavior should preferably be studied using three-dimensional transport/chemistry models.

Acknowledgment

The authors thank Professor K. Labitzke and Dr. J. C. Gille for helpful discussions. This work was supported in part by NASA under grant W162215. G. Brasseur was supported by the Belgian National Fund for Scientific Research and is on leave from the Belgian Institute for Space Aeronomy, Brussels, Belgium.

References

Andrews, D. G., J. R. Holton, and C. B. Leovy, 1987: Middle atmosphere dynamics. Academic Press, Orlando, Florida, 489 pp.

Brasseur, G., M. Hitchman, M. Dymek, M. Pirre, and E. Falise, 1988: An interactive chemical-dynamical radiative two-dimensional model of the middle atmosphere. In preparation.

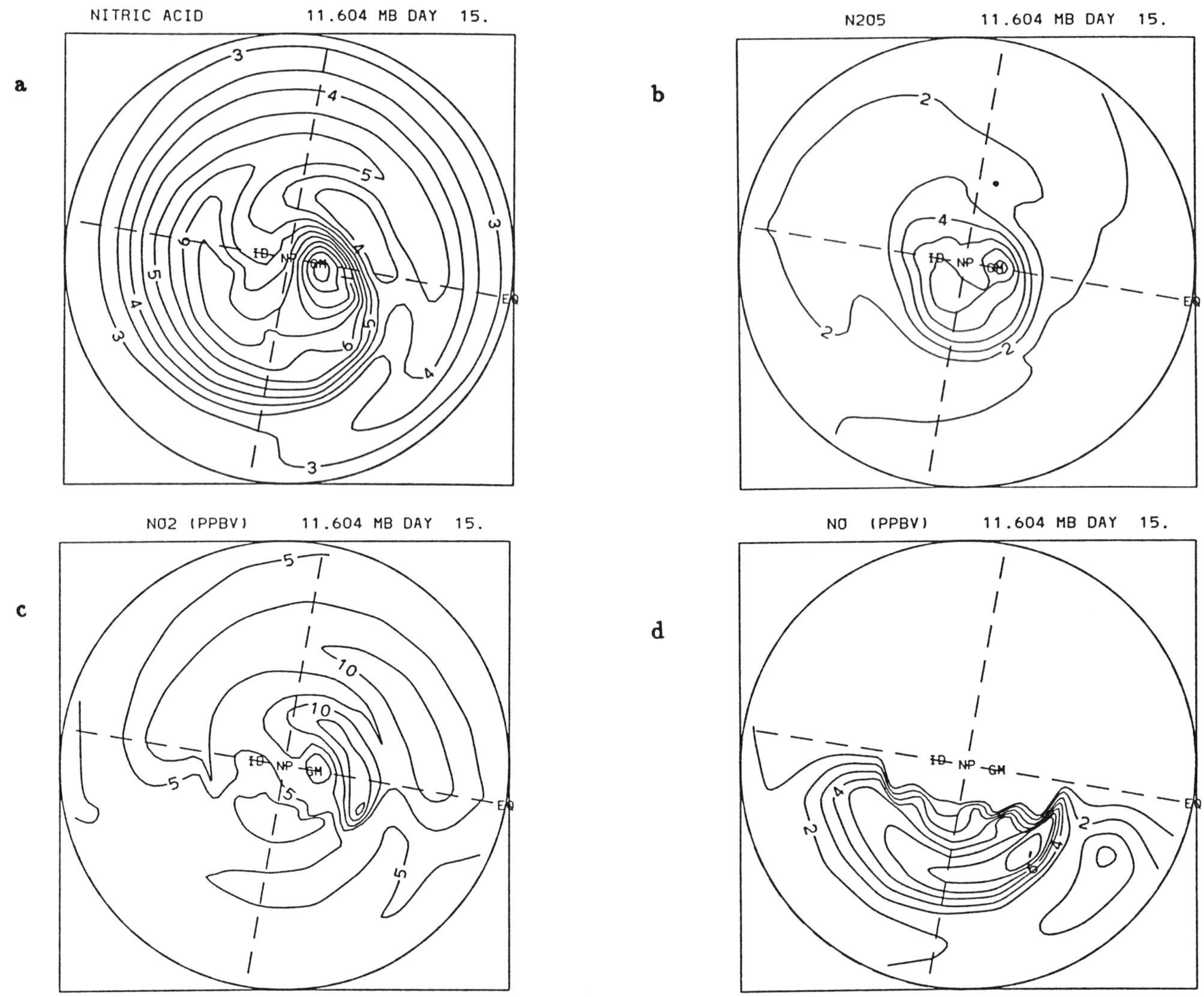

Fig. 3. Distribution on the 11.6 mb pressure surface of odd nitrogen species (mixing ratio expressed in ppbv): (a) HNO_3; (b) N_2O_5; (c) NO_2; (d) NO.

DeMore, W.B. *et al.*, 1985: Chemical kinetics and photochemical data for use in stratospheric modeling. NASA, JPL publication 85-37.

Jones, R.L., 1984: Satellite measurements of atmospheric composition: Three years' observations of CH_4 and N_20. *Adv. Space Res.*, **4**(4), 121-130.

McIntyre, M.E., and T.N. Palmer, 1984: The "surf zone" in the stratosphere. *J. Atmos. Terr. Phys.*, **46**, 825-849.

Noxon, J. F., 1979: Stratospheric NO_2, 2. Global behavior. *J. Geophys. Res.*, **84**, 5067-5076.

Rose, K., 1983: On the influence of nonlinear wave-wave interaction in a 3-D primitive equation model for sudden stratospheric warmings. *Beitr. Phys. Atmosph.*, **56**, 14-31.

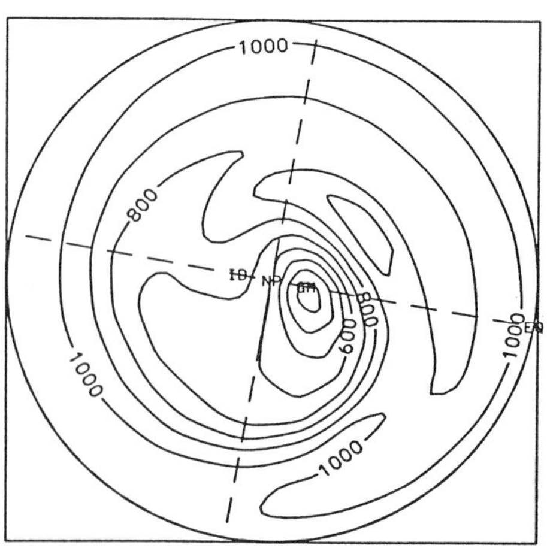

Fig. 4. Distribution of the methane mixing ratio (ppbv) on the 11.6 mb pressure surface on day 15.

THE USE OF WINDS AND TEMPERATURES FROM A STRATOSPHERIC ASSIMILATION MODEL IN THREE-DIMENSIONAL CONSTITUENT TRANSPORT STUDIES

Richard B. Rood and Jack A. Kaye
NASA Goddard Space Flight Center
Greenbelt, Maryland 20771, USA

Dale J. Allen
Applied Research Corporation
Landover, Maryland 20783, USA

Wayman E. Baker
National Meteorological Center
Washington, DC, USA

David J. Lamich
Centel Federal Services Corporation
Rockville, Maryland 20850, USA

ABSTRACT

An assimilation model that incorporates stratospheric rawinsonde data, rocketsonde profiles, and satellite temperature soundings into a general circulation model at 6 hour time intervals has been developed at NASA/GSFC. The model extends from the ground to 0.4 mb. The assimilation model generates wind fields that are balanced with the mass field through the model dynamics and physics. Wind fields obtained from an assimilation model are the best estimates of stratospheric winds that can be obtained in the absence of direct wind measurements.

Winds and temperatures from our assimilation model have been used in our 3-dimensional transport and chemistry model. Comparisons with results from stratospheric forecast models and data derived geostrophic winds, demonstrate significant improvement in attempts to simulate the time evolution of LIMS constituent fields (O_3, HNO_3, NO_2, H_2O). However, notable discrepancies remain between the observations and the simulations. In particular, the meridional transport in the planetary wave systems is overestimated.

1. INTRODUCTION

Calculations of budgets with satellite constituent data and geostrophic winds during wintertime reveal serious imbalances. Most notably the observed changes in the fields are much smaller and spatially smoother than those predicted by the calculation of the advection terms (see Rood et al. 1988). The reasons for these discrepancies in budget calculations are not explicitly clear. Possible explanations include errors in either the constituent or wind data, deficiency of the geostrophic approximation, or simply that the satellite data do not contain sufficient information to calculate budgets on a daily basis.

This paper reports on a continuing parallel modeling and data analysis effort to investigate the information content of satellite data. In particular, the use of dynamical data acquired from a data assimilation procedure to study constituent transport is discussed. In the next section the data assimilation is described and compared to the National Meteorological Center (NMC) analysis. This section is followed by a brief discussion of the constituent transport model. Then experiments using winds and temperatures from the assimilation to calculate stratospheric constituent fields are described. Finally the results and their relevance to analysis of stratospheric constituent observations are discussed.

2. DATA ASSIMILATION

Data assimilation is a form of analysis that has been used extensively in numerical weather prediction. First a short term prediction (6 hrs) is performed with a general circulation model. The model output is then passed to the analysis procedure that combines the model output with data from rawinsondes, rocketsondes, and satellites (see Baker et al., 1987). This produces an analysis that is constrained by data where data are available, and where data are not available the model provides estimates of the meteorological fields. This analysis is then used as the initial condition for the next short term forecast.

ISBN 0-937194-15-8

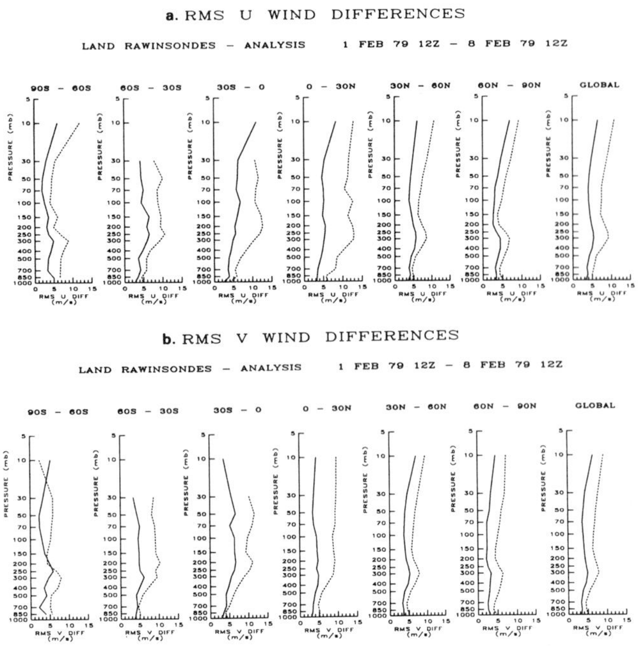

Fig. 1 RMS differences between rawinsonde winds and NMC geostrophic winds (dashed) and data assimilation winds (solid), a) zonal b) meridional.

The assimilation procedure produces a wind, temperature, and height analysis that is constrained by observations. The impact of the winds analysis is particularly great in the stratosphere because there are few direct wind observations. Winds are traditionally estimated using geostrophic balance in the stratosphere.

Elson (1986), Boville (1987), and others have shown that the geostrophic approximation is not very good during winter. The deficiencies are particularly large at the pole. Of course, the geostrophic approximation breaks done in the tropics, and wind estimates in the tropics are particularly suspect. The data assimilation technique conceptually replaces the geostrophic balance by a balance defined by the primitive equations of the model.

Winds from the assimilation, therefore, are in proper dynamical balance globally. Furthermore, since the divergent component of the horizontal wind field has been estimated, the vertical velocity field can be derived. The wind fields from the assimilation are potentially useful in studies of constituent budgets.

The increased accuracy over geostrophic winds of the data assimilation winds is shown in Fig. 1. Both the geostrophic wind from the NMC analysis, and the winds from the data assimilation are differenced from rawinsonde observations. The RMS differences of the assimilation product are smaller than the geostrophic values. Figure 2 shows the isotachs and horizontal wind vectors on the 10 mb surface for the data assimilation and the NMC analysis. Regions of strong winds are shaded, with dark shading for winds greater than 100 m/sec and light shading for winds between 40 and 100 m/sec.

Significant differences are seen with the NMC analysis showing more area with both scales of shading. The magnitude of the cross polar

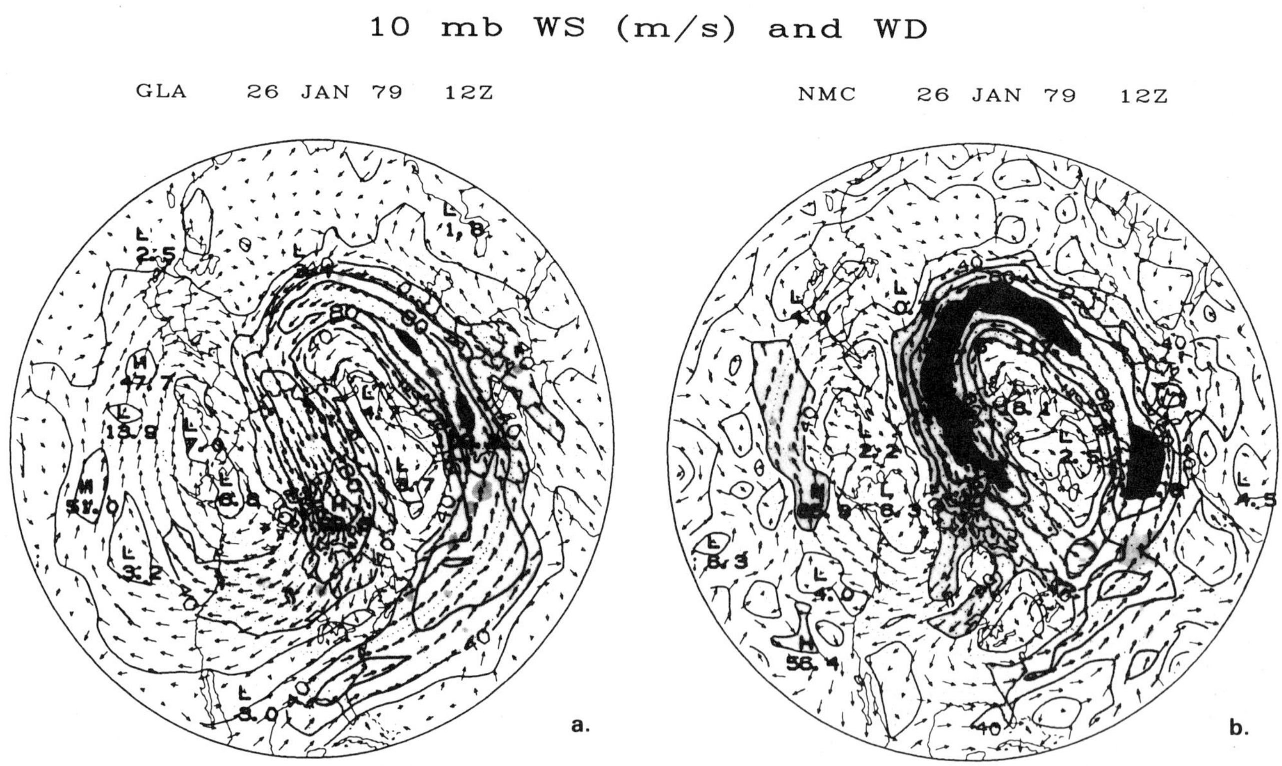

Fig. 2 10 mb isotachs and wind vectors for 26 January 1979. Dark shading indicates speeds in excess of 100 m/sec. Light shading indicates speed between 40 m/sec and 100 m/sec.

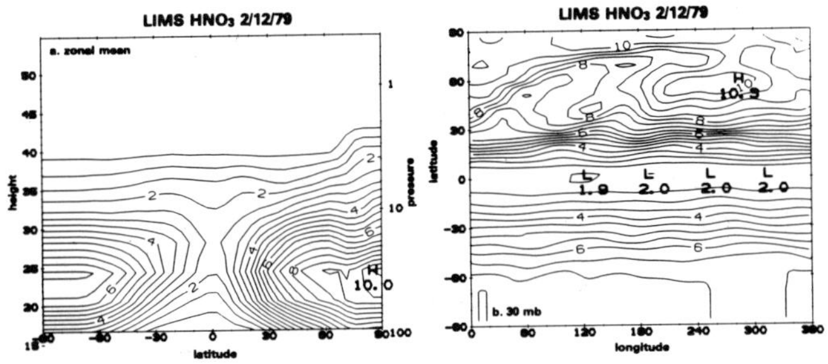

Fig. 3 LIMS data on 2/12/79, the initial condition, a) Zonal mean, b) 30 mb. (ppbv)

flow is significantly reduced in the assimilation. More important than the changes in wind speed are changes in the circulation pattern. The assimilation shows a well defined jet which would transport parcels from the region over the eastern United States, around the cyclone that is displaced to the European side of the pole, across the pole, and finally around the Aleutian anticyclone. The inflow region over the eastern United States and the outflow region over Canada are clearly separated in the assimilation. This separation is not as clear in the geostrophic analysis from the NMC data. In the tropics and subtropics very profound circulation differences are indicated.

Since there is not a known wind field for validation, it is not definitive which of these analyses is better. It is obvious that the transport characteristics of the two fields will be different. The calculation of constituent transport with the assimilated winds and the subsequent comparison to observations offers an independent check of the quality of the analysis.

3. CONSTITUENT TRANSPORT MODEL

The wind and temperature data from the assimilation are interfaced with our offline transport and chemistry model. The results presented here use a simple nitric acid chemistry based on Limb Infrared Monitor of the Stratosphere (LIMS, Gille et al., 1984) observables. No attempt has been made to model the missing polar night source of nitric acid (for instance, Jackman et al., 1987). This chemistry is the same as Rood et al. (1987), and because of its simplicity many experiments can be run. The transport is the same as Rood et al. (1988).

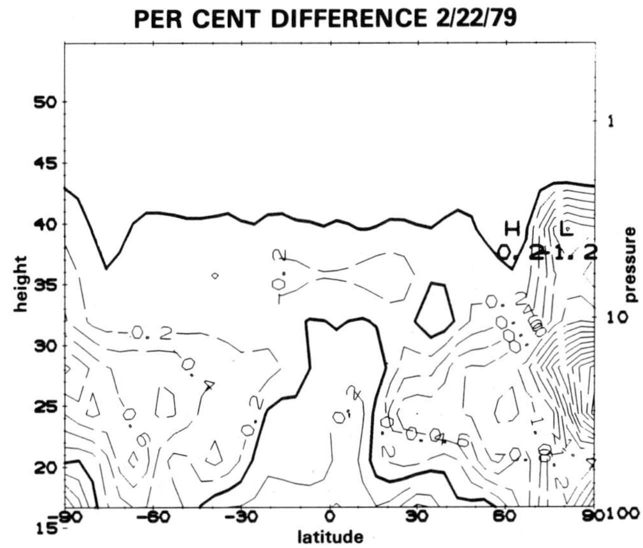

Fig. 5 Zonal mean HNO_3 difference between computation and LIMS data. (ppbv)

The assimilated winds have also been interfaced with our more complete chemistry package (Kaye and Rood, 1988; Kaye et al., this issue). The expense of this model prevents extensive experimentation, but when transport is provided by winds from the assimilation, the most realistic representation possible of three dimensional dynamics and chemistry is obtained. Some results from this model are described in Kaye et al. in this volume.

4. RESULTS

Figure 3 shows the initial condition for an experiment that uses with LIMS nitric acid fields and winds from the assimilation on February 12, 1979. The experiment is run for 15 days and includes a wave 2 major warming. In the zonal mean field note that the contours below the nitric acid maximum poleward of 70° north are relatively flat.

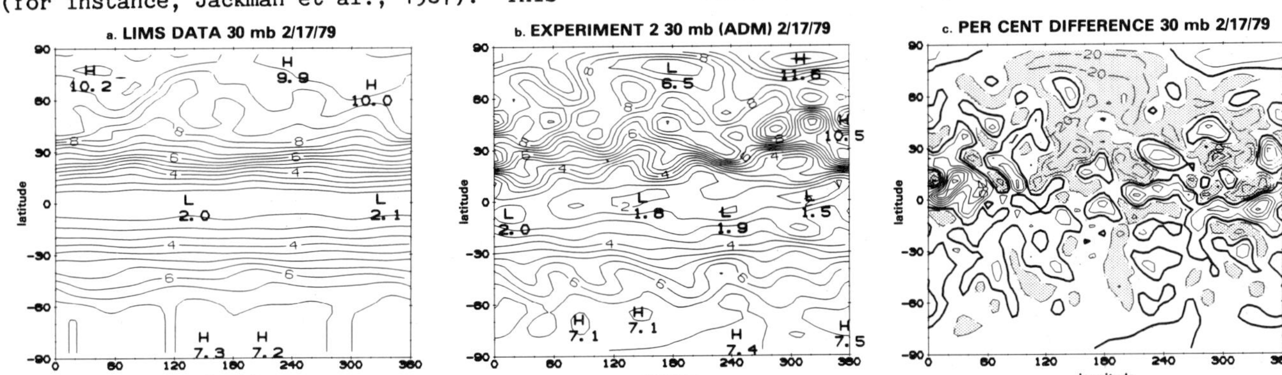

Fig. 4 30 mb charts on Feb. 17, 1979, a) LIMS data, b) Computations, c) per cent difference between parts a and b, shading < -10%, dashed positive. (ppbv)

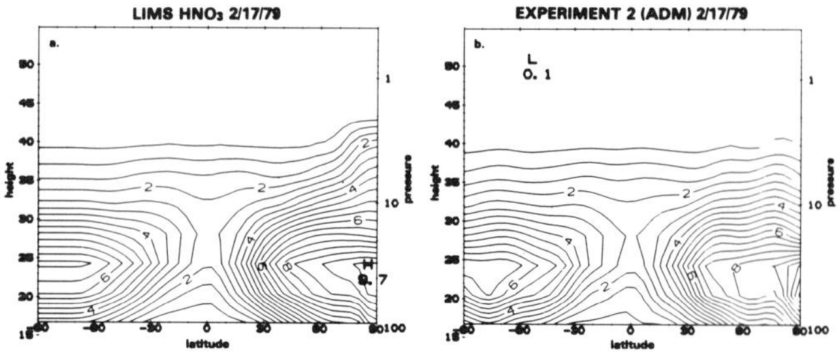

Fig. 6 Zonal mean HNO_3 on 2/17/79, a) LIMS data, b) Computation. Note the kink at 75°N between 16 and 20 km. (ppbv)

Figure 4 shows the computed, the observed, and the per cent difference field on February 17. The computed field contains much more structure than the observations. This is due in part to the higher spatial resolution of the model and in part to apparent excessive horizontal transport in the calculation. Possible explanations of the excessive meridional transport include errors in the wind field, neglect of important subscale processes, and incompatibility of the initial dynamical and nitric acid fields.

The difference field in Fig. 4 shows a general deficiency of nitric acid in extratropical northern latitudes. Part of this deficiency may be due to the transport of too much low nitric acid air from low to high latitudes. Part of the deficiency may be due to the 'missing' nitric acid source that has been postulated in the winter hemisphere. Figure 5 shows the difference of the computed and observed zonal mean fields on February 22 after 10 days of integration. The northern hemisphere shows a generalized deficit that maximizes in the polar night. It can be concluded that the chemistry in the model cannot maintain the observed nitric acid distribution in the presence of realistic wintertime planetary wave dynamics.

Significant problems have been noted in the quantitative simulation of nitric acid concentrations. Comparisons of the time behavior of the computed and observed concentrations are remarkably consistent. Figure 6 shows the zonal mean fields on February 17. Poleward of 70° below the nitric acid peak a sharp kink is observed in the LIMS data. The computed field also has this structure. This kink was not present in the initial condition (Fig. 3). The transport and chemistry model forms and maintains this kink on the same time scales that it is observed to occur.

Above the nitric acid peak at Arctic latitudes the instantaneous plot in Fig. 6 does not show such good agreement. The time series presented in Fig. 7 indicate, however, that the computed and observed variance are in general agreement. There is an initial bias generated at the start of the integration. Neglecting the first 2 days, the correlations between the remaining time series is 0.89.

The wind fields from the assimilation appear to accurately represent the variance of the nitric acid field after the initial

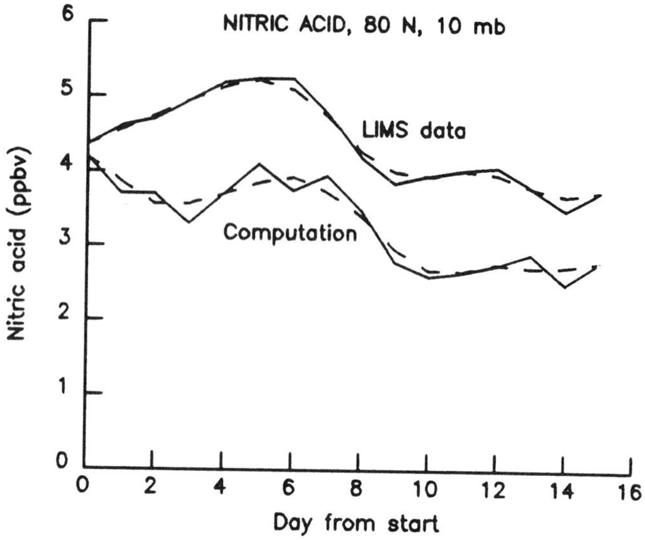

Fig. 7 Time series of HNO_3 at 80°N, 10 mb (32 km) from computation and LIMS data. The correlation between the model and the data (excluding the first three points) is 0.89 for the raw data and 0.95 for three point smoothed data. (ppbv)

transience. As discussed in more detail in Rood et al. (1988) these wind fields are a great improvement over geostrophic winds derived from the operational NMC analysis. The quality of the simulation at high latitudes suggest that this technique provides a mechanism for accurately assessing constituent variability in polar regions.

5. SUMMARY AND CONCLUSIONS

Transport calculations using winds from our assimilation model suggest that these winds are superior to geostrophic estimate. The variance of the constituent fields is accurately represented by the model. Because of this success the model is currently being used to assess the impact of localized heteorogeneous processes on global ozone fields.

REFERENCES

Baker, W. E., S. C. Bloom, J. S. Wollon, M. S. Nestler, E. Brin, T. W. Schlatter, and G. W. Brantstator, 1987: Experiments with a three-dimensional statistical objective analysis scheme using FGGE data, Mon. Wea. Rev., 115, 272-296.

Boville, B. A., 1987: The validity of the geostrophic approximation in the winter stratosphere and troposphere, J. Atmos. Sci., 44, 443-457.

Elson, L. S., 1986: Ageostrophic motion in the stratosphere from satellite observations, J. Atmos. Sci., 43, 409-418.

Gille, J. C., J. M. Russell III, P. L. Bailey, E. E. Remsberg, L. L. Gordley, W. F. J. Evans, H. Fischer, B. W. Gandrud, A. Girard, J. E. Harries, and S. A. Beck, 1984: Accuracy and precision of the nitric acid concentrations determined by the Limb Infrared Monitor of the Stratosphere experiment on Nimbus 7, J. Geophys. Res., 89, 5179-5190.

Jackman, C. H., P. D. Guthrie, and J. A. Kaye, 1987: An intercomparison of nitrogen-containing species in Nimbus 7 LIMS and SAMS data, J. Geophys. Res., 92, 995-1008.

Kaye, J. A., and R. B. Rood, 1988: Chemistry and transport in a three dimensional stratospheric model: Chlorine species during a simulated stratospheric warming, to appear J. Geophys. Res.

Kaye, J. A, R. B. Rood, and D. J. Allen, 1988: Variability of chlorine containing species as revealed by three-dimensional stratospheric transport and chemistry models, this volume.

Rood, R. B., J. A. Kaye, J. E. Nielsen, M. R. Schoeberl, and M. A. Geller, 1987: Nitric acid forecast experiments, Physica Scripta, 36, 337-354.

Rood, R. B., D. J. Allen, W. E. Baker, D. J. Lamich, and J. A. Kaye, 1988: The use of assimilated data in constituent transport calculations, to appear, J. Atmos. Sci.

A THREE-DIMENSIONAL MODEL SIMULATION OF OZONE IN THE STRATOSPHERE

R.E. Turner, W.L. Grose, W.T. Blackshear, and R.S. Eckman
Atmospheric Sciences Division
NASA Langley Research Center
Hampton, Virginia 23665, USA

ABSTRACT

A three-dimensional atmospheric circulation/transport model has been developed to study dynamics and transport processes and to simulate the distribution of ozone and other chemically active trace constituents. The model extends from the surface to 60 km and incorporates a comprehensive treatment of chemistry using the "family" technique. Some results from a simulation of the seasonally varying evolution of ozone and other constituents are discussed. Emphasis will be placed on the ozone and odd-nitrogen species, including comparisons with observations and other models. The coupling between photochemical and dynamical processes for a disturbed period during Northern Hemisphere winter is illustrated using diagnostic calculations on isentropic surfaces.

1. INTRODUCTION

Understanding the complex coupling between photochemistry and the transport of ozone has been the focus of a considerable amount of research in recent years. The current concern over Antarctic ozone depletion (Solomon, 1988) emphasizes the current requirement for understanding the interplay between radiative, chemical, and dynamical processes in determining the morphology of atmospheric ozone.

Improved observational evidence in recent years has made it evident that the stratospheric circulation often departs significantly from zonal symmetry. Particularly during winter, air parcels can undergo substantial meridional and vertical excursions in a matter of days (Leovy et al., 1985). Such evidence supports the rationale for three-dimensional modeling efforts (WMO, 1985). However, it is currently not feasible to attempt such studies using a comprehensive, fully interactive formulation of radiation, chemistry, and dynamics. This restriction is a joint consequence of deficiencies in understanding both dynamical and chemical processes and the prohibitive computational burden that such a formulation would entail. Such restrictions have motivated the more simplified approach described herein.

2. DESCRIPTION OF THE MODEL

A global, primitive equation, general circulation model (GCM) (Blackshear et al., 1987 and Grose et al., 1987) was used for the studies described herein. The model extends from the surface to approximately 60 km, including the effects of forcing by orography and land-ocean thermal contrast. Time integration is accomplished with a semi-implicit technique using a 30-minute time step. Transport model simulations are performed with an "off-line" technique (Mahlman and Moxim, 1978) in which wind and temperature fields are generated separately by the GCM and then used as input to a set of mass continuity equations. A scale-selective, biharmonic diffusion term is included to parameterize horizontal, subgridscale processes. Evaluation of the net chemical source (sink) term required during the integration requires consideration of a large number of species and chemical reactions (39 and 115, respectively, in this version of the model).

The complexity of the chemistry makes explicit transport of each constituent an unwieldy and expensive option. Also, the extremely fast chemistry characteristic of some species would dictate integration time steps so small as to be impractical for conducting long-time simulations. To circumvent these difficulties, the "family" approach has been adopted in which related species which undergo relatively fast chemical transformations are grouped into a family with a characteristic chemical lifetime much greater than for the component members--a concept frequently used in one- and two- dimensional models (Solomon and Garcia, 1983). In the present version of the model, mixing ratios of $O_x[O_3 + O(^3P) + O(^1D)]$, $NO_y[NO + NO_2 + NO_3 + N + HNO_2 + HNO_4]$, $Cl_x[Cl + ClO + ClONO_2 + ClO_2 + HOCl + HCl]$, HNO_3, and N_2O_5 are determined by explicit integration of their respective mass continuity equations. Partitioning of the families into individual species is accomplished through various equilibrium relationships. Photolysis rates are evaluated by integration with respect to wavelength of the products of photodissociation cross section, quantum efficiency, and photon flux. Certain chemical species (e.g. N_2O, CH_4, and fluorocarbons) exhibit extremely long

ISBN 0-937194-15-8

lifetimes in the atmosphere. These species are specified as functions of latitude, longitude and season (as appropriate).

3. RESULTS AND DISCUSSION

The transport model simulation discussed in this section is a portion of an annual cycle integration. Zonally symmetric initial conditions for the chemical constituents were inferred either from December monthly mean data of the Limb Infrared Monitor of the Stratosphere (LIMS) experiment (Gille and Russell, 1984) or taken from two-dimensional model simulations (Garcia and Solomon, private communication). Wind and temperature fields used in the transport model simulation were obtained from a multi-year integration of the circulation model.

Subsequent discussion will focus on a highly disturbed period with strong transients during Northern Hemisphere winter. Results from a single day (February 5) have been selected for the purpose of this discussion. The dynamical evolution which characterized this period is discussed in some detail in Blackshear et al. (1987). Some idea of the strongly disturbed and rapidly changing synoptic situation can be gained from the Northern Hemisphere ozone mixing ratio distribution (10-mb level) shown in Fig. 1. Both the qualitative features and the mixing ratio levels are in good agreement with LIMS observations for disturbed wintertime conditions (Leovy et al., 1985). For time scales on the order of a week, (at this pressure level in high latitudes) ozone is approximately conserved and sufficiently correlated with the dynamical tracer, Ertel's potential vorticity, to reveal considerable insight into the ongoing transport processes. The accentuated 6 ppmv contour illustrates the region of low ozone concentration within the polar vortex which is being drawn out and "wrapped" about the Aleutian anticyclone. The polar vortex is highly elongated from the Pole towards 90 degrees west with the axis of the vortex displaced from the Pole. The elongated vortex is flanked on the left by the Aleutian anticyclone and on the right by a secondary Atlantic anticyclone (see Blackshear et al., 1987). The darkened arrows indicate large material tongues that are alternately drawn in between the main vortex and the flanking anticyclones, first, directly across the pole and secondly, from the Tropics near 90 degrees west into the high latitudes, consistent with observations (Leovy et al., 1985 and Clough et al., 1985). Note the magnitude of this phenomenom--with the elongated region of low ozone extending from Northern Siberia over the polar region into the middle of the United States.

In this highly disturbed atmospheric state, ozone with high mixing ratios (≈10 ppmv) is transported from low to high latitudes in a few days, while ozone with low mixing ratios is moved equatorward. This cycle will be repeated throughout the winter with maximum ozone accumulating in the high latitudes, reaching peak values in spring (Grose et al., 1987). Similar behavior is apparent in the satellite observations during winter/spring 1979 (Leovy et al., 1985).

A comparison of simulated and observed vertical profiles of hydroxyl (OH) at 30°N is shown in Fig. 2 (redrawn from WMO, 1986, Fig. 12-30, page 693). The LaRC model results are daytime averages. The solid lines bound a range of 2-D model results. The 3-D model values lie near the lower range of the 2-D models, but are well within the range of measurements.

A similar comparison of simulated and observed vertical profiles of chlorine nitrate ($C\ell ONO_2$) at 30°N is shown in Fig. 3 (redrawm from WMO, 1986, Fig. 12-43, page 706). Also shown in the figure are results from the Atmospheric Trace Molecular Spectroscopy (ATMOS) experiment (Zander et al., 1986). The peak mixing ratio values occur near the 10- to 20-mb level and decrease sharply into the troposphere and into the upper stratosphere. The LaRC model profile is in good agreement with the NOCAR model results. The ATMOS values are somewhat higher, but uncertainties of 54 percent are quoted for these observations (Zander et al., 1986).

The simulated distribution of nitrogen pentoxide (N_2O_5) at the 10-mb level shown in Fig. 4 is strongly influenced by the diurnal photochemical cycle in the lower latitudes. The local mixing ratio maximum of 2.5 ppbv at about 40 degrees north and 40 degrees west occurs near local sunrise. Proceeding eastward along a constant latitude circle, photolysis acts to destroy N_2O_5 until local sunset, which occurs near 140 degrees east. Continuing eastward, N_2O_5 is being reformed in the nighttime. In the polar latitudes, the situation is complicated by transport into and out of the polar night. As an air parcel is drawn in between the main vortex and the Aleutian anticyclone approaching the polar night from 90 degrees east,it has a low N_2O_5 mixing ratio (presumably because of daytime photolysis). Parcels approaching the Pole, moving into polar night and circulating about the Aleutian anticyclone, experience increased production of N_2O_5 during darkness (Solomon and Garcia, 1983). Those parcels moving along the flank of the main vortex away from the Pole, recirculating about the polar vortex out of polar night, experience destruction of N_2O_5 by photolysis. The large N_2O_5 maximum (4.25 ppbv) shown in Fig. 4, extending from 90 degrees east to about 270 degrees west in the

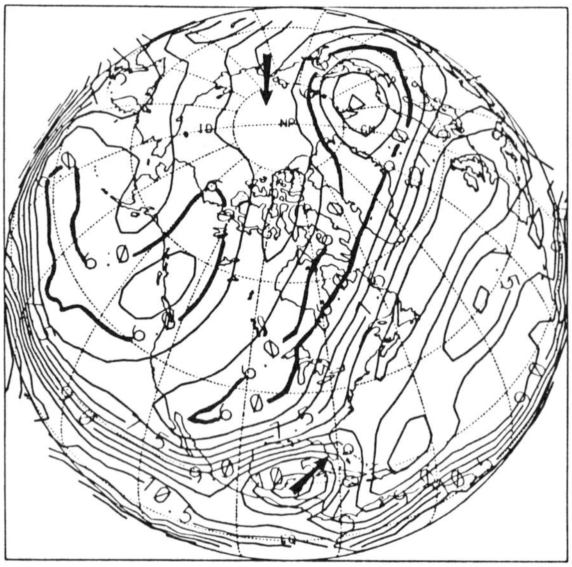

Fig. 1 Simulated O_3 mixing ratio, 10 mb, Feb. 5, Northern Hemisphere perspective.

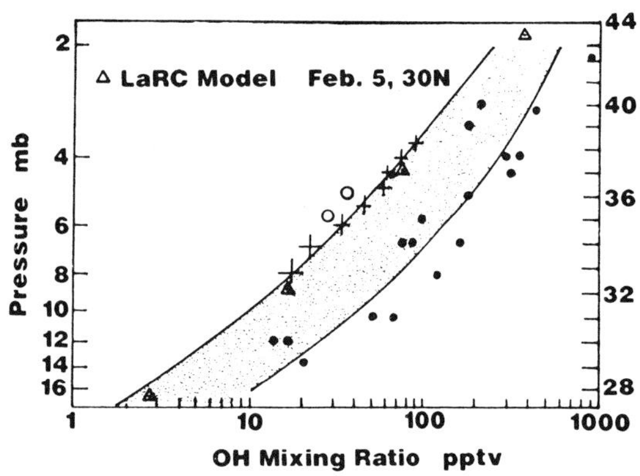

Fig. 2 Vertical profiles of OH mixing ratio.

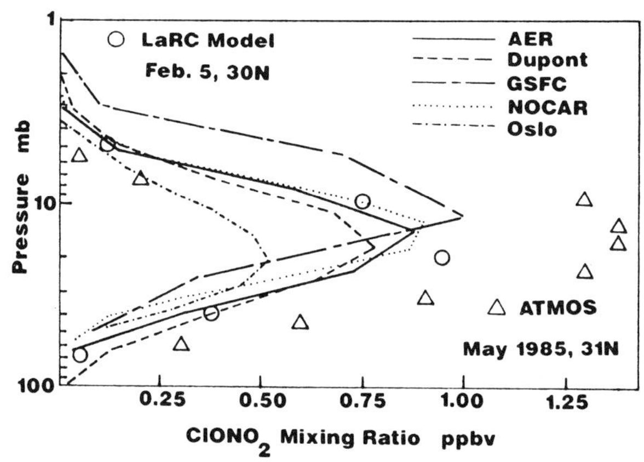

Fig. 3 Vertical profiles of $ClONO_2$ mixing ratio.

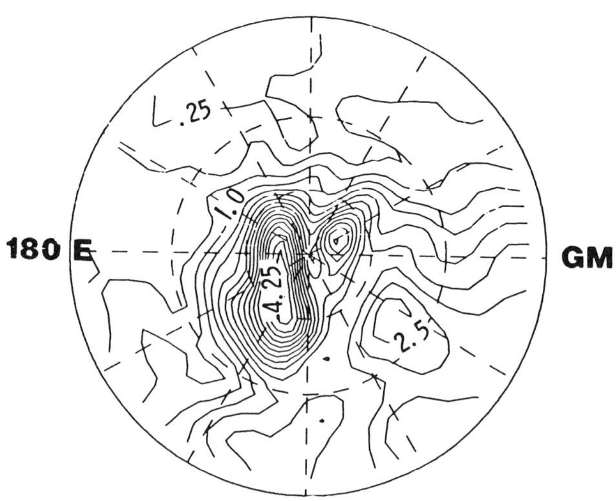

Fig. 4 Simulated N_2O_5 mixing ratio, 10 mb, Feb. 5, Northern Hemisphere.

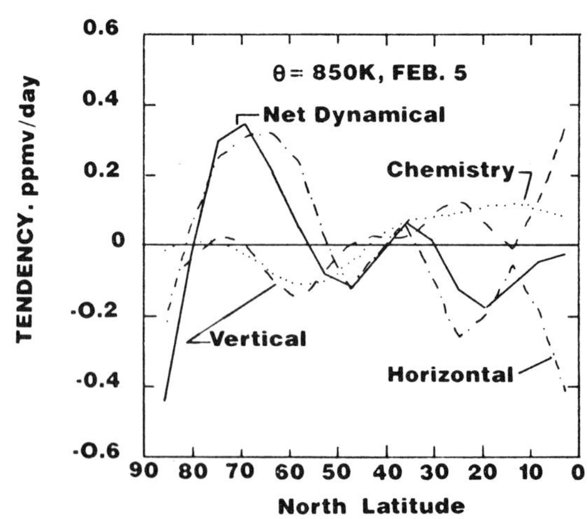

Fig. 5 Zonal mean O_3 tendency, 850 K, Feb. 5, Northern Hemisphere.

polar region is consistent with this complex process. A secondary polar maximum near the Greenwich meridian is influenced by the same process. The extent to which photochemical processes may influence a constituent field can be seen by comparing the N_2O_5 distribution in Fig. 4 with the ozone distribution shown in Fig. 1. High latitude ozone (which is approximately conserved) tends to align itself along the parcel trajectories, while N_2O_5 varies because of chemical transformations along similar parcel trajectories.

To illustrate the interplay between the various processes which control the ozone distribution, the components which contribute to the zonal mean ozone tendency on an isentropic surface (850°K, nominally about 10 mb) are shown in Fig. 5. The net dynamical tendency (principally the horizontal component) acts to increase ozone in the high latitudes from 50°N to about 80°N, whereas chemistry is relatively weak, acting in opposition. In the lower latitudes the net dynamical transport generally acts to remove ozone from the Tropics towards higher latitudes along approximately constant potential temperature surfaces (over timescales of about a week or less) by the complex, irreversible process termed wave-breaking (Mcintyre and Palmer, 1983 and McIntyre and Palmer, 1984). Three-dimensional diagnosis of the ozone tendency (not shown here) indicates that the high latitude increases are dominated by quasi-horizontal (on isentropic surfaces) transport which is largest in the regions depicted by arrows in Fig. 1. The process is strongly non-linear in these regions when the wave-breaking phenomenom is occurring (Grose et al., 1987). The importance of three-dimensional diagnosis is strikingly evident during disturbed periods and provides fuller insight than displayed in the zonal mean context.

4. CONCLUDING REMARKS

Some representative results from a three-dimensional atmospheric simulation model have been presented. The results are generally encouraging and compare favorably with observations. The results also suggest that the off-line transport approach is a useful and reasonable alternative to a fully coupled three-dimensional model. However, the reader should note that model/ data comparisons are difficult, at best, because of interannual variability and the scarcity of simultaneous, global observations of key constituents. Much further research is obviously required.

5. REFERENCES

Blackshear, W. T., W. L. Grose, and R. E. Turner, 1987: Simulated sudden stratospheric warming; Synoptic evolution. Quart. J. Roy. Meteor. Soc., 113, 815-846.

Brasseur, G., and S. Solomon, 1984: Aeronomy of the Middle Atmosphere. D. Reidel and Co., Dordrecht, Holland.

Clough, S. A., N. S. Grahame, and A. O'Neill, 1985: Potential vorticity in the stratosphere derived using data from satellites. Quart. J. Roy. Met. Soc., 111, 335-358.

Gille, J. C., and J. M. Russell III, 1984: The limb infrared monitor of the stratosphere experiment description, performance, and results. J. Geophys. Res., 89, 5125-5140.

Grose, W. L., J. E. Nealy, R. E. Turner, and W. T. Blackshear, 1987: Modeling the transport of chemically active constituents in the stratosphere. Transport Processes in the Middle Atmosphere, Visconti and R. Garcia (Eds.), D. Reidel Publishing Co., 229-250.

Leovy, C. B., C.-R. Sun, M. H. Hitchman, E. E. Remsberg, J. M. Russell III, L. L. Gordley, J. C. Gille, and L. V. Lyjak, 1985: Transport of ozone in the middle stratosphere: Evidence for planetary wave breaking. J. Atmos. Sci., 42, 230-244.

Mahlman, J. D., and W. J. Moxim, 1978: Tracer simulation using a global general circulation model: Results from a midlatitude instantaneous source experiment. J. Atmos. Sci., 35, 1340-1374.

McIntyre, M. E., and T. N. Palmer, 1983: Breaking planetary waves in the stratosphere. Nature, 305, 593-600.

McIntyre, M. E. and T. N. Palmer, 1984: The "surf zone" in the stratosphere. J. Atmos. Terr. Phys., 46, 825-850.

Solomon, S. and R. Garcia, 1983: On the distribution of nitrogen dioxide in the high-latitude stratosphere. J. Geophys. Res., 88, 5229-5239.

Solomon, S., 1988: The mystery of the Antarctic ozone "hole." Rev. of Geophys., 26, 131-148.

World Meteorological Organization, 1986: Atmospheric Ozone 1985: Global Ozone Research and Monitoring Project. Report No. 16.

Zander, R., C. P. Rinsland, C. B. Farmer, L. R. Brown, and R. H. Norton, 1986: Observations of several chlorine nitrate ($C\ell ONO_2$) bands in stratospheric infrared spectra. Geophys. Res. Lett., 13, 757-760.

MODELING

SENSITIVITY STUDY OF GLOBAL OZONE TO NO_x EMISSIONS FROM AIRCRAFT

Douglas Kinnison and Harold Johnston
Lawrence Berkeley Laboratory
University of California
Berkeley, California 94720, USA

Donald J. Wuebbles
Lawrence Livermore National Laboratory
University of California
Livermore, California 94550, USA

ABSTRACT

There recently has been renewed interest in the development of faster and more efficient aircraft for intercontinental passenger flights (Ott, 1988). Such aircraft would probably spend a large fraction of their flight time in the stratosphere, perhaps as high as 35 km. As a natural progression from studies that were done in the early 1970s (Johnston, 1971; CIAP, 1975; WMO, 1985), this study reinvestigates the sensitivity of stratospheric ozone to NO_x emissions based on current understanding of atmospheric chemical and physical processes.

The LLNL one-dimensional and new two-dimensional chemical-radiative-transport models of the troposphere and stratosphere are used in this investigation. The two-dimensional model provides latitudinal, altitudinal, and seasonal resolution; and the computationally faster one-dimensional model is used for sensitivity studies. Because of uncertainties in possible future emissions, it is necessary to examine the model sensitivity to a wide range in magnitude, altitude, and latitude of assumed NO_x emissions. As an initial study, the models are used in simulation of a typical scenario from the CIAP era, and results are compared with those of a published two-dimensional model, which includes some three-dimensional features (Alyea et al., 1975; Cunnold et al., 1977). This investigation lays the ground work for future studies, including different background amounts of ClX and for other aircraft emission scenarios.

1. ATMOSPHERIC MODELS

1.1 DESCRIPTION OF LLNL TWO-DIMENSIONAL MODEL

The LLNL zonally-averaged two-dimensional chemical-radiative-transport model currently determines the atmospheric distributions of 31 chemically active atmospheric trace constituents in the troposphere and stratosphere. The model domain extends from pole to pole, and from the ground to 0.56 mb (approximately 0 to 54 km). The sine of latitude is used as the horizontal coordinate with intervals of about 10°. The vertical coordinate corresponds to the natural logarithm of pressure ($z^* = H_0 \ln(p/p_0)$, where H_0 is the assumed scale height of 7.2 km, and p_0 is the surface pressure, 10^{13} mb). The vertical resolution is $\ln(p/p_0) = 0.417$ or about 3 km. Ninety-five chemical and photochemical reactions are included in the model. Reaction rates, solar flux data, absorption cross-sections, and quantum yields are based on the latest NASA panel recommendations (DeMore et al., 1987), except that the previous value is used for the rate of the reaction, $O + NO_2 = O_2 + NO$. Photodissociation rates, including the effects of multiple scattering, are computed as a function of time at each zone, with optical depths consistent with calculated species distributions.

The diabatic circulation for the ambient atmosphere is determined using net heating rates calculated in an internally consistent way with the derived species distributions. The technique for deriving the diabatic circulation is similar to that used by Solomon et al. (1986): the vertical velocity is determined from the zonally averaged residual Eulerian thermodynamic equation, while the horizontal velocity is determined using the equation for mass continuity. The net heating rates are determined using accurate solar and infrared radiative models.

Temperatures for the ambient atmosphere vary continuously, over the annual cycle, based on the reference model of Barnett and Corney (1985). The derived diabatic circulation depends strongly on the temperature distribution; by using observed temperatures for the ambient atmosphere, a more accurate representation of the diabatic circulation can be derived.

For the perturbed atmosphere, a perturbation form of the thermodynamic equation is solved for the changes in stratospheric temperatures resulting from changes in the distribution of ozone and other radiatively active constituents. Using this approach, the diabatic circulation is assumed to be unchanged in the perturbed atmosphere from that calculated for the ambient.

Turbulent eddy transport is parameterized through diffusion coefficients K_{yy} and K_{zz}. In the current version of the model, a value of K_{yy} of 2×10^9 cm^2 s^{-1} is assumed at all stratospheric altitudes and latitudes; and values of 1×10^{11} are assumed in the troposphere, with a transition region at the tropopause. Values of K_{zz} are 1×10^3 cm^2 s^{-1} in the lower stratosphere, increasing slowly with altitude based on gravity wave modeling studies.

The continuity equation for each individual species is solved using a variable time-step, variable order, implicit technique for solving stiff numerical systems with strict error control. Advection terms are treated using the two-dimensional

A. DEEPAK Publishing
ISBN 0-937194-15-8

transport algorithm of Smolarkiewicz (1984). The diurnal-averaged concentrations for each species at each zone are calculated at each time step. Diurnal calculations are used to derive time-varying factors for each chemical and photochemical reaction included in the diurnal-averaged version of the model.

1.2 ONE-DIMENSIONAL MODEL

The LLNL one-dimensional eddy-diffusion model of the troposphere and stratosphere was also used in this study. The details of this model are discussed in Luther et al. (1979) and Wuebbles et al. (1983), with photochemical rate coefficients from DeMore et al. (1985, 1987). Briefly the 1-D model contains 37 species. There are 119 chemical and 46 photochemical reactions included in the model. The 1-D model atmosphere is divided up into 44 vertical layers which extend from the surface up to 56 km. Transport of trace species is controlled by an empirical eddy diffusion function (K_z) which was derived to match observed distributions of long-lived atmospheric tracers (N_2O, CH_4, CFC-11, CFC-12, and CH_3Cl).

The boundary values of the reference atmosphere were selected to reproduce the composition of 1988. The mixing ratios and altitudes of various families of species are: 3.1 ppbv of ClY at 55 km, 13 pptv of BrY at 55 km, and 19.4 ppbv of NOY at 37.5 km.

2. RESULTS

Hemispherical and global average ozone changes from 2-D model calculations are given in Table 1.A; and results of one-dimensional model calculations at similar altitudes are given in Table 1.B. The standard injection is taken to be 1.8 million metric tons (10^9 kg) of NO_x (as NO_2) per year (CIAP, 1975; Alyea et al., 1975; Cunnold et al., 1977). For this standard NO_2 injection, spread uniformly at 19.5 ± 1.5 km and between 37 and 49 degrees north, the calculated percentage changes of the ozone vertical column are presented as a function of latitude and season in the top panel of Fig. 1, and the local ozone percentage changes are given as a function of latitude and altitude in the lower panel of Fig. 1. For the same NO_2 injection at 31.5 ± 1.5 km, the column and local percentage changes of ozone are shown in Fig. 2.

With the less expensive 1-D model, a wide range of calculations was made, and these results are summarized as a contour plot of percentage ozone-column change as a function on injection rate and altitude of injection, in Fig. 3. In the 1-D model the injection rate of 1500 molecules $cm^{-3}s^{-1}$ over a one kilometer vertical band corresponds to a global injection of 1.8 MT yr^{-1} of NO_2, so that Fig. 3 spans injection rates from zero to twice the standard rate. In terms of aircraft, these altitudes roughly correspond to current commercial aircraft (12 km), to the French-British Concorde (17 km), to the 1971 American-planned supersonic transport (20 km), and to possible future hypersonic aircraft (30 to 34 km).

3. DISCUSSION OF RESULTS

For the standard NO_x injection at low altitude (19.5 ± 1.5 km) in the 2-D model, the maximum ozone column change of -9% occurs in the polar region of the Northern Hemisphere; the maximum local concentration change is -16% near the injection corridor (Fig. 1); the global average ozone change is -3.3%; and the ratio of total ozone change in the Northern to Southern Hemisphere is 3.1 (Table 1). For injections at high altitude (31.5 ± 1.5 km), the 2-D model gives a maximum ozone column change of -32% in the winter polar regions; the maximum local ozone change is -55% (Fig. 2); the global average ozone change is -8.0%; and the ratio between Northern and Southern Hemispheres for ozone change is 6.9 (Table 1). In comparison, the older three/two-dimensional calculations for a 20 km injection altitude gave a global average ozone change of -12% and a hemispherical ratio of 2/1 (Cunnold et al., 1977). There have been many changes in photochemical parameters since 1977, and these comparisons are primarily of historical interest.

The changes in ozone vertical column calculated in the 1-D model are compared with 2-D results for total global ozone (Table 1). For the standard injection centered near 20 km, the calculated global-average ozone change is -3.3% for the 2-D model and -8.7% for the 1-D model, using the same photochemistry. For the same injection near 30 km, the ozone changes are -8.0% for the 2-D model and -12.5% for the 1-D model. For an NO_2 injection of 1.8 MT yr^{-1} at 34.5 ± 1.5 km, the 2-D global-average ozone change is -7.2% when the injection is made between 37 and 49 degrees north,

PERCENT CHANGE IN OZONE INJECTION AT 19.5 KM

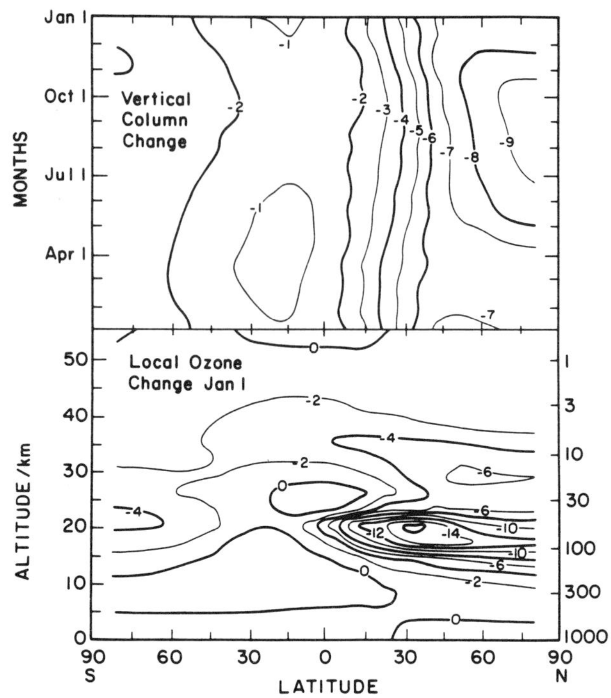

Fig. 1. Percent change in ozone for injection of 1.8 MT yr^{-1} of NO_2, where the latitude of injection is between 37 and 49°N and the altitude of injection is between 18 and 21 km. The upper panel shows the change of the ozone vertical column as a function of latitude and month of the year. The lower panel shows the change of local ozone (as of January 1) as a function of latitude and altitude. Photochemical coefficients are from DeMore et al. (1987), except for use of older value (DeMore et al., 1985) for reaction O + NO_2.

TABLE 1. CALCULATED CHANGES IN GLOBAL AND HEMISPHERICAL OZONE VERTICAL COLUMNS AS A RESULT OF NO_x INJECTIONS BY STRATOSPHERIC AIRCRAFT.

A. LLNL two-dimensional model. The boundary values include 300 ppbv of nitrous oxide and enough chlorofluorocarbons to give 2.8 ppbv of stratospheric chlorine at 50 km. JPL 1987,##.

NO_x Injection as NO_2			% Ozone Change			t(NOY)*
km	Lat. N	MT/yr	Global	N.H.	S.H.	yr
16.5	37–49	1.8	+0.01	−0.02	+0.04	0.3
19.5	37–49	1.8	−3.3	−5.0	−1.6	1.2
31.5	37–49	1.8	−8.0	−13.8	−2.2	1.8
34.5	37–49	1.8	−7.2	−12.4	−1.8	1.8
34.5	37–49	1.8	−7.7**	−13.1	−2.3	
34.5	N.H.	1.8	−7.1	−11.3	−2.9	1.8
34.5	Global	1.8	−8.0	−7.5	−9.6	1.9
34.5	37–49	0.9	−4.0	−6.9	−1.0	
34.5	37–49	0.45	−2.1	−3.6	−0.5	

B. One-dimensional model results. 3.1 ppbv ClX.**

		JPL (DeMore et al.)			t(NOY)*
km	MT yr^{-1}	1985†	1987†	1987	yr
12	1.8		+0.2		
17	1.8	−2.9	−2.9	−3.5	1.5
20	1.8	−8.6	−8.7	−10.0	2.7
30	1.8	−12.1	−12.5	−13.4	3.9
34	1.8	−10.1	−10.5	−11.4	3.6

* Stratospheric residence time of excess NOY calculated as ratio of steady-state excess inventory of NOY over injection rate of NO_x.

** Temperature feedback is not included in this case.

† Old, slow rate for reaction O + NO_2.

−7.1% when spread uniformly over the Northern Hemisphere, and −8.0% when spread uniformly over the globe; the corresponding ozone change calculated by the 1-D model is −10.5%. It is interesting that the 2-D model gives approximately the same global-average ozone change regardless of the latitudinal spread of the injection band for the high altitude injection. When the injection is all in the Northern Hemisphere, the average ozone reduction in the Northern Hemisphere is four to seven times that in the Southern Hemisphere.

For the standard NO_x injection at 34.5 km, one calculation was made omitting temperature feedback, and the global average ozone change was −7.2% with temperature feedback and −7.7% without temperature feedback, a small difference. For NO_x injections of 0.45, 0.9, and 1.8 MT yr^{-1}, the global-average ozone changes are −2.1%, −4.0%, and −7.2%, respectively, which is a slightly slower change than linear.

The 1-D model gives larger ozone reductions than the 2-D model: by a factor of 1.5 when injections occur at 30 km and by a factor of 2.6 for injections at 20 km. Reasons for these differences are being investigated, and it appears to be related to differences in transport rates in the two models.

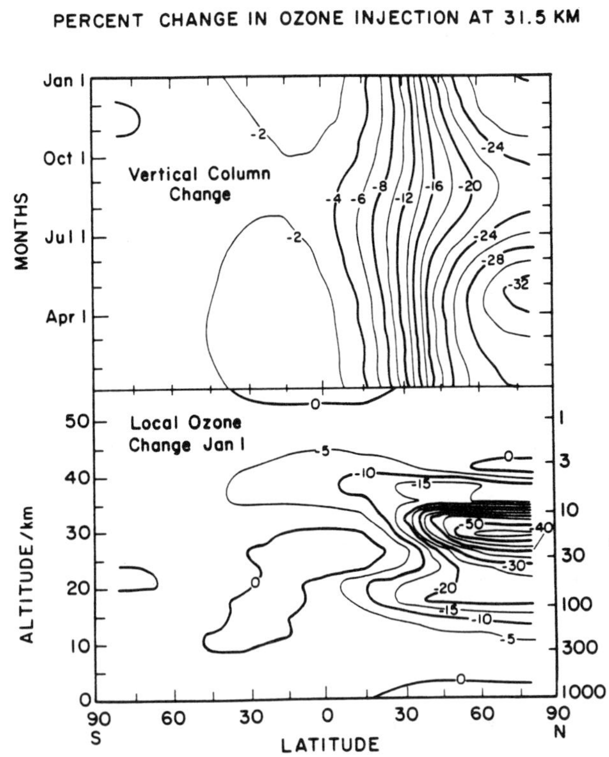

Fig. 2. Same as Fig. 1, except the injection altitude is between 30 and 33 km, nominally 31.5 km.

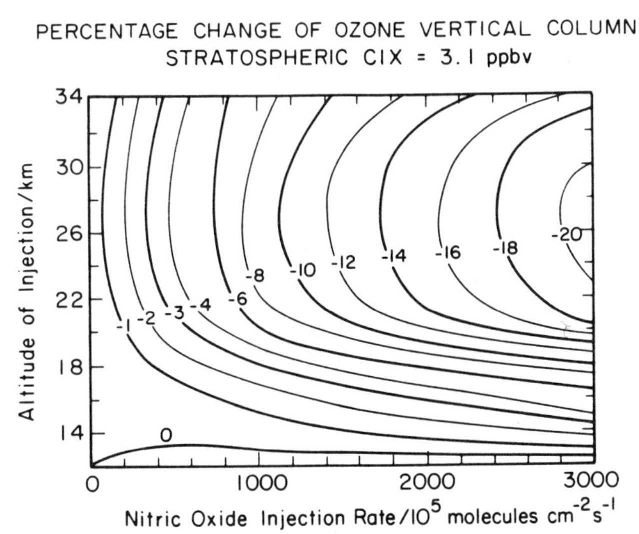

Fig. 3. Percentage change of the ozone vertical column as calculated by the one-dimensional model as a function of altitude of NO_x injection and magnitude of NO_x injection in units of molecules cm^{-3} s^{-1} over a one kilometer band centered at the stated altitude. The stratospheric ClX is assumed to be 3.1 ppbv, which is approximately the 1988 value. An injection of 1500 in these units corresponds to a global injection of 1.8×10^{12} g yr^{-1} (as NO_2). Photochemical coefficients are from DeMore et al. (1985).

The lifetime of excess NO_x injected into the stratosphere was calculated from the relation

$$\text{lifetime} = t(\text{NOY}) = \frac{\text{global excess NOY at steady state}}{\text{global NO}_x\text{ injection rate}}$$

for both 1-D and 2-D models, and these values are included as the last column in Table 1. For NO_x injection at 30 to 35 km, the 1-D lifetime of excess NOY is twice that for the 2-D model; at 17 km injection altitude the 1-D lifetime is almost five times as long as that for the 2-D model. As can be seen from Figs. 1 and 2, the 2-D model sweeps NOY out of the Northern Hemisphere faster than it transports NOY to the Southern Hemisphere. Studies are underway testing these models against atmospheric carbon-14 data to see which model gives the more nearly correct stratospheric lifetime for an inert tracer. The calculated ozone reduction is slightly greater (10 to 20%) when we use the new (DeMore et al., 1987) temperature dependent value for the rate coefficients for the reaction $O + NO_2$, relative to that for the constant value, 9.3×10^{-12} cm^3 s^{-1}, Table 1-B.

The ozone-column changes calculated with the 1-D model for altitudes between 12 and 34 km and for injection rates between 0 and 3000 molecules $cm^{-3}s^{-1}$ over a 1 km band (0 to 3.6 MT yr^{-1} of NO_2) are shown as a contour plot in Fig. 3. These calculations were made with the 1985 photochemical coefficients (DeMore et al. 1985) before the 1987 values were published, and stratospheric ClX is taken to be 3.1 ppbv. According to Fig. 3, NO_x injected below 13 km in any amount increases the ozone column. Above 13 km, injected NO_x decreases the ozone column, with magnitude of the decrease increasing with rate of injection and with altitude. Above 30 km, the reduction of the ozone column decreases with altitude for a given injection rate; this feature is also given by the 2-D model. Further investigation of this problem is in progress, including study of the effects of large changes in ClX.

ACKNOWLEDGMENTS

The work at the University of California and the Lawrence Berkeley Laboratory was supported by the Director, Office of Energy Research, Office of Basic Energy Sciences, Chemical Sciences Division of the U.S. Department of Energy under Contract No. DE-AC03-76SF00098. Work at Lawrence Livermore National Laboratory was performed under the auspices of the U.S. Department of Energy under Contract W-7405-ENG-48 and was supported in part by NASA's Upper Atmospheric Research Program.

REFERENCES

Alyea, F.N., D.M. Cunnold, and R.G. Prinn, 1975: Stratospheric ozone destruction by aircraft-induced nitrogen oxides. *Science, 188,* 117–121

Barnett, J.J., and M. Corney, 1984: A middle atmosphere temperature reference model from satellite measurements. *Adv. Space Res., 5,* 125–134.

CIAP, 1975: Monograph 2, Propulsion effluent in the stratosphere; Monograph 3, The stratosphere perturbed by propulsion effluent. U.S. Department of Transportation, Washington, D.C.

Cunnold, D.M., F.N. Alyea, N.A. Phillips, and R.G. Prinn, 1977: Relative effects on atmospheric ozone of latitude and altitude of supersonic flight. *AIAA Journal, 15,* 337–345.

Demore, W.B., J.J. Margitan, M.J. Molina, R.T. Watson, D.M. Golden, R.F. Hampson, M.J. Kurylo, C.J. Howard, and A.R. Ravishankara, 1987: Chemical kinetics and photochemical data for use in stratospheric modeling. Jet Propulsion Lab., Pasadena, CA: JPL Publication 85-37 and 87-41.

Johnston, H.S., 1971: Reduction of stratospheric ozone by nitrogen oxide catalysts from supersonic transport exhaust. *Science, 173,* 517–522.

Luther, F.M., J.S. Chang, W.H. Duewer, J.E. Penner, R.L. Tarp, and D.J. Wuebbles, 1979: Potential environmental effects of aircraft emissions. Lawrence Livermore National Laboratory Report UCRL-52861.

Ott, J., 1988: High-speed transport study focuses on lower Mach range. *Aviation Week & Space Tech., 60–61,* Feb. 29.

WMO (World Meteorological Organization), 1985: Atmospheric ozone: Assessment of our understanding of the processes controlling its present distribution and change. Global ozone research and monitoring project. Report No. 16.

Smolarkiewicz, P.K., 1984: A fully multidimensional positive definitive advection transport algorithm with small implicit diffusion. *J. Comp. Phys., 54,* 325–362.

Solomon, S., J.T. Kiehl, R.R. Garcia, and W. Grose, 1986: Tracer transport by the diabatic circulation deduced from satellite observations. *J. Atmos. Sci., 43,* 1604–1617.

Wuebbles, D.J., 1983: A theoretical analysis of the past variations in global atmospheric composition and temperature. Lawrence Livermore National Laboratory Report UCRL-53423.

A TWO-DIMENSIONAL MODELING STUDY OF THE SENSITIVITY OF OZONE TO RADIATIVE FLUX UNCERTAINTIES

Keith E. Grant and Donald J. Wuebbles
Lawrence Livermore National Laboratory
University of California
Livermore, California 94550, USA

ABSTRACT

Radiative processes strongly effect equilibrium trace gas concentrations both directly, through photolysis reactions, and indirectly through temperature and transport processes. We have used the LLNL 2-D chemical-radiative-transport model to investigate the net sensitivity of equilibrium ozone concentrations to several changes in radiative forcing. Doubling CO_2 from 300 ppmv to 600 ppmv resulted in a temperature decrease of 5 K to 8 K in the middle stratosphere along with an 8% to 16% increase in ozone in the same region. Replacing our usual shortwave scattering algorithms with a simplified Rayleigh algorithm led to a 1% to 2% increase in ozone in the lower stratosphere. Finally, modifying our normal CO_2 cooling rates by corrections derived from line-by-line calculations resulted in several regions of heating and cooling. We observed temperature changes on the order of 1 K to 1.5 K with corresponding changes of 0.5% to 1.5% in O_3. Our results for doubled CO_2 compare favorably with those by other authors. Results for our two perturbation scenarios stress the need for accurately modeling radiative processes while confirming the general validity of current models.

1. INTRODUCTION

The time and space dependent evolvement of trace gas concentrations within a 2-D chemistry-transport model is strongly affected by the effects atmospheric radiative processes have on temperatures, transport, and photolysis reactions. Changes to shortwave and infrared fluxes perturb model chemistry and transport calculations via a number of interconnecting mechanisms. Shortwave flux densities control photolysis rates, directly affecting trace gas concentrations. Changes to either short wave or infrared fluxes perturb net heating rates, changing the diabatic circulation and temperature profiles calculated by the model. Transport changes directly change the latitude-altitude distribution of space and time varying trace gas concentrations, in turn feeding back upon radiative absorption. Temperature changes perturb chemical reactions rates, also modifying trace gas concentrations and feeding back upon radiative fluxes.

We have investigated the net sensitivity of equilibrium ozone concentrations, estimated using the LLNL two-dimen sional chemical-radiative-transport model, for several changes in radiative forcing. The radiative scenarios used were chosen to be representative of physically motivated modeling concerns from our continuing model development, rather than sensitivity studies based on statistical treatment of individual parameter uncertainties.

2. DESCRIPTION OF LLNL TWO-DIMENSIONAL MODEL

The LLNL zonally-averaged two-dimensional chemical-radiative-transport model currently determines the atmospheric distributions of 31 chemically active atmospheric trace constituents in the troposphere and stratosphere. The model domain extends from pole to pole, and from the ground to 0.56 mb (approximately 0 to 54 km). The sine of latitude is used as the horizontal coordinate with uneven increments corresponding to about 10 in latitude. The vertical coordinate corresponds to the natural logarithm of pressure with resolution of approximately 3 km.

Approximately 95 chemical and photochemical reactions are included in the model. Reaction rates, solar flux data, absorption cross-sections, and quantum yields are based on the latest NASA panel recommendations of Demore et al. (1987), with the exception of the rate for NO_2 + O, which is based on prior recommendations. Photodissociation rates, including the effects of multiple scattering, are computed as a function of time at each zone, with optical depths consistent with calculated species distributions.

The diabatic circulation for the ambient atmosphere is determined using net heating rates calculated in an internally consistent way with the derived species distributions. The technique for deriving the diabatic circulation is similar to that used by Solomon et al. (1986): the vertical velocity is determined from the zonally averaged residual Eulerian thermodynamic equation, while the horizontal velocity is determined using the equation for mass continuity.

The net heating rates are determined using accurate solar and infrared radiative models. The solar model includes absorption and scattering effects for O_3, O_2, and NO_2 at ultraviolet and visible wavelengths, and for H_2O, CO_2, and O_2 in the near infrared. The solar model for visible and UV wavelengths uses the Sagan and Pollack (1967) two-stream model to calculate reflection and transmission operators for scattering of diffuse incident radiation by a single layer. Scattering from the solar beam is calculated for each layer using the delta-Eddington technique (Joseph et al., 1976). Merging of individual layers, including multiple-scattering, is accomplished via a flux formulation of the adding technique (Harshvardhan et al., 1987). The longwave emission and absorption by O_3, CO_2, and H_2O are included in the infrared submodel.

Temperatures for the ambient atmosphere vary continuously, over the annual cycle, based on the reference model of Barnett and Corney (1985). The derived diabatic circulation depends strongly on the temperature distribution; by using observed temperatures for the ambient atmosphere, a more accurate representation of the diabatic circulation can be derived. The model determined net radiative heating rates, and resulting diabatic circulation, compare well with those derived

A. DEEPAK Publishing
ISBN 0-937194-15-8

from LIMS data (Kiehl and Solomon, 1986; Rosenfield et al., 1987; Solomon et al., 1986).

For the perturbed atmosphere, a perturbation form of the thermodynamic equation is solved for the changes in stratospheric temperatures resulting from changes in the distributions of ozone and other radiatively active constituents. Using this approach, the diabatic circulation is assumed to be unchanged in the perturbed atmosphere from that calculated for the ambient.

Turbulent eddy transport is parameterized through diffusion coefficients K_{yy} and K_{zz}. In the current version of the model, a value of K_{yy} of 2×10^9 cm^2 s^{-1} is assumed at all stratospheric altitudes and latitudes, and values of 1×10^{11} in the troposphere, with a transition region at the tropopause. Values of K_{zz} are 1×10^3 cm^2 s^{-1} in the lower stratosphere, increasing slowly with altitude based on gravity wave modeling studies.

The continuity equation for each individual species is solved using a variable time-step, variable order, implicit technique for solving stiff numerical systems with strict error control. Advection terms are treated accurately using the two-dimensional transport algorithm of Smolarkiewicz (1984). The diurnal-averaged concentrations for each species at each zone are calculated at each time step. Accurate diurnal calculations are used to derive time-varying factors for each chemical and photochemical reaction included in the diurnal-averaged version of the model.

3. RADIATIVE PERTURBATION SCENARIOS

As a baseline for comparison, we have used a year's seasonal cycle for a chemically ambient atmosphere. A CO_2 mixing ratio of 300 ppmv was chosen for this baseline. Although this is lower than the ambient CO_2 mixing ratios in current general use, it was compatible with available data for line-by-line CO_2 cooling rates (Fels and Schwarzkopf, private communication, 1987). These cooling rates were used in deriving one of our perturbation scenarios. For both the ambient and radiatively perturbed runs described below, the 2-D model was run for sufficient model years that the ozone concentrations had reached steady-state, apart from a seasonal cycle. All scenarios were run for clear-sky conditions, with seasonally varying, latitude dependent ground albedos.

As a first radiative perturbation scenario, the LLNL model was run to radiative-transport equilibrium for doubled CO_2 (600 ppmv), while allowing only stratospheric temperatures to respond to the radiative forcing. A second doubled CO_2 scenario was then run in which full chemical-radiative-transport equilibrium was allowed to occur. Since doubling CO_2 is a relatively standard radiative perturbation, these scenarios allowed comparison between the response of the LLNL model and the responses of other 2-D models.

A third scenario was created by modifying the present shortwave multiple-scattering submodel to simulate the simpler Rayleigh multiple-scattering formulation of Luther (1980). Luther's model was, until recently, used in LLNL models to calculate photolysis rates. Similar models are being used in other 2-D models. In current LLNL models, we have switched to the more rigorous scattering model, described above, which is able to treat clouds and aerosols as well as being more accurate at larger zenith angles. For this scenario, the radiation scattered out of a layer was approximated as the total single-scattering extinction minus the corresponding gaseous absorption. Diffuse radiation was assumed to have an effective zenith angle of approximately 53°, and scattered radiation was partitioned equally in the forwards and backwards directions. This scenario was run without temperature feedback, so the changes in equilibrium ozone were forced entirely by changes in photolysis rates.

For the final scenario in this study, we added corrections to the heating rates calculated by the longwave submodel internal to the LLNL 2-D model. These corrections were based on line-by-line calculations, which are expected to give a more accurate representation of cooling rates than those obtainable from any band model. An earlier version of the longwave submodel used by the LLNL 2-D model was described in Harshvardhan et al. (1987). The current version has improved treatments of absorption by CO_2 and doppler broadening of H_2O in the upper stratosphere (Harshvardhan, private communication, 1987). Using this improved longwave submodel we calculated CO_2 cooling rates for four vertical profiles with corresponding line-by-line calculations of infrared cooling rates for 300 ppmv CO_2. The two sets of cooling rate profiles were differenced to obtain cooling rate corrections for the McClatchey et al. (1972) midlatitude summer, midlatitude winter, tropical and subarctic winter profiles. After averaging the temperature and cooling-rate-correction profiles to the LLNL 2-D model vertical grid, singular value decomposition was used with these data to develop a linear model that related vertical profiles of cooling rate corrections to model generated temperature profiles. Cooling rate corrections generated by this linear model were then added to the cooling rates calculated by the longwave submodel in the LLNL 2-D model during model execution. This scenario was forced by temperature feedback to chemical-radiative-transport equilibrium.

4. RESULTS

Figure 1 shows the total column ozone as a function of month and latitude as calculated by the LLNL 2-D model for an ambient atmosphere with 300 ppmv CO_2. This distribution generally compares well with observations, although it shows too much total ozone in the tropics. The ozone distributions for all the other scenarios discussed, will be relative to the ozone for this atmosphere.

In Figure 2, we show the change in temperature for July, resulting from doubling CO_2 to 600 ppmv and allowing only the stratospheric temperatures to come into radiative-transport equilibrium (i.e. no chemistry feedback). No changes were made to tropospheric or surface temperatures— the temperature changes shown are forced only by changes in longwave cooling. The observed cooling increases with altitude with little latitudinal variation in the lower and middle stratosphere. In the upper stratosphere, there is slightly stronger (2 K) cooling at the winter pole and weaker cooling in the tropics. These results are in general agreement with those of Fels et al. (1980).

In Figure 3, we show (for July) the temperature change and percent ozone change, resulting from doubling CO_2 with full chemistry feedback. Allowing for somewhat different lower boundary conditions and a change in CO_2 of 300 ppmv versus 350 ppmv, these results compare closely with those of Brasseur and Hitchman (1988). Our temperature changes for doubled CO_2 with chemistry feedback also compare closely with those reported by Haigh (1984). Relative to Figure 2, Figure 3a shows less cooling (3 K) in the upper stratosphere, except slightly greater cooling (1 K) at the winter pole. These changes are explained by increased infrared cooling due to increased ozone, overshadowed by increased shortwave heating except at the winter pole. The changes in ozone in Figure 3 are, in general anticorrelated with the temperature changes. In the lower stratosphere the ozone increase shows a much greater latitudinal gradient, being smallest in the tropics and largest at the summer pole. The ozone increase peaks around 45 km at 18 to 22 percent.

Figure 4 shows the percent change in ozone for July caused by changing the 2-D model scattering algorithm to the simple Rayleigh treatment described above. In summary, an increase

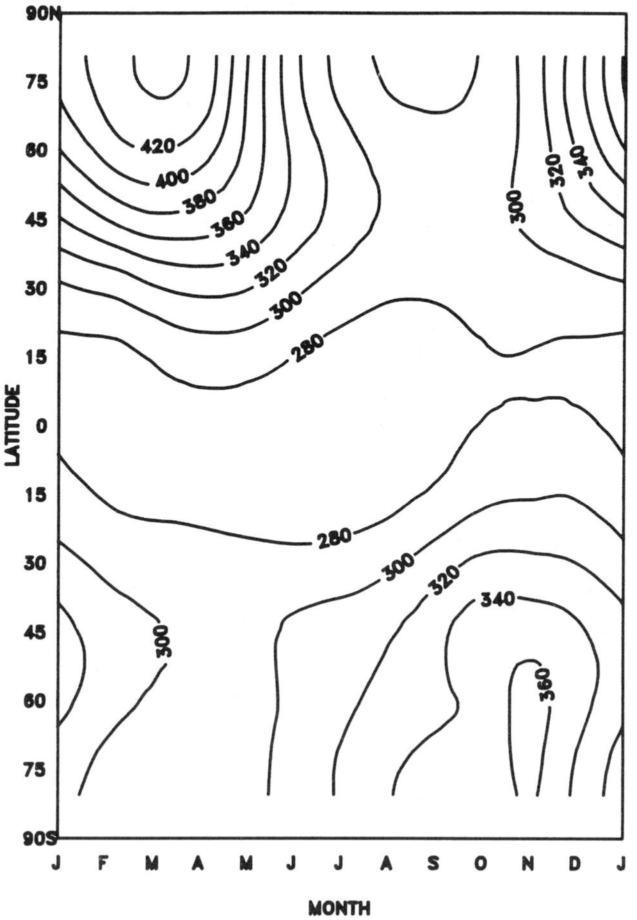

Fig. 1. Total ozone in Dobson units for an ambient atmosphere with 300 ppmv CO_2.

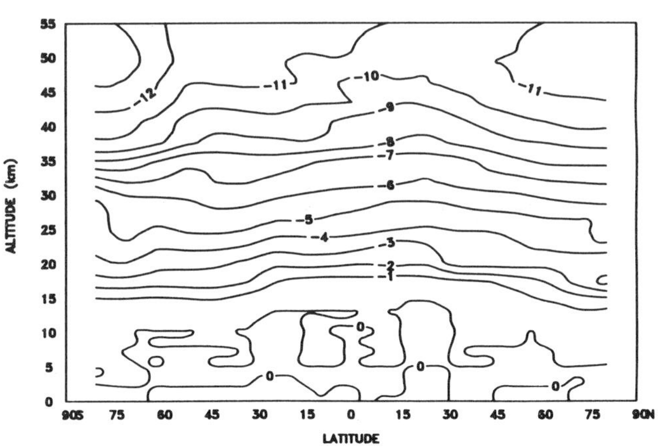

Fig. 2. Temperature change for July in response to doubling CO_2 without chemistry feedback.

in ozone is observed, peaking in the lower stratosphere at 1.5 to 2.5 percent and becoming largest in the summer hemisphere for small solar zenith angles. This increase is attributable to a increase in the photolysis of O_2. This case displays the importance of scattering algorithm choice, even under the simplest clear-sky conditions. It differs from the other scenarios presented here, in that it is driven solely by change in photolysis rates without accompanying temperature feedback.

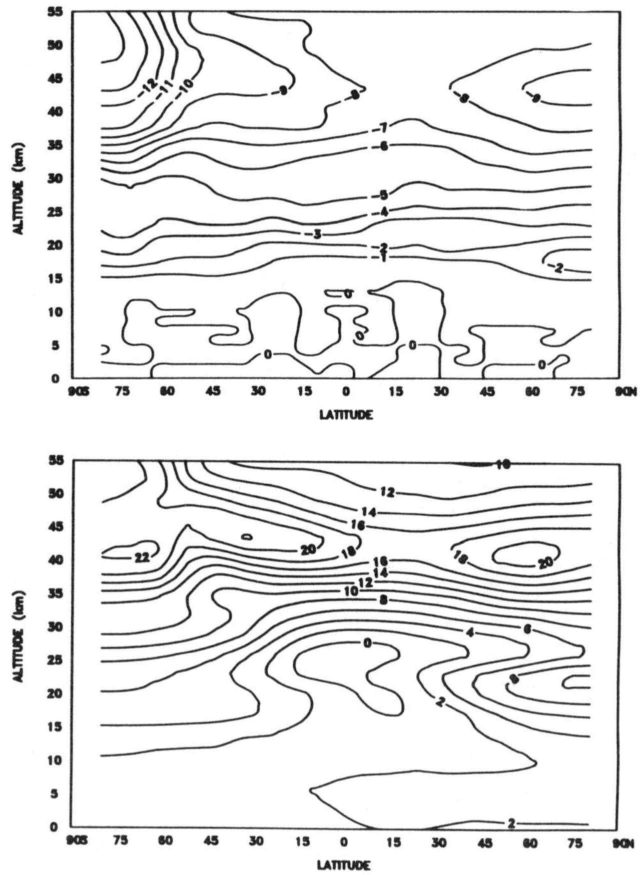

Fig. 3. (a) Temperature change for July in response to doubling CO_2 with chemistry feedback (b) Ozone percent change for July in response to doubling CO_2 with chemistry feedback.

radiation package. As described above, these correction terms were generated as a function of temperature profile from a linear model based on 300 ppmv CO_2 line-by-line calculations. The general features created by this correction include: slight warming (1 K) just above the polar tropopause, a region of slight cooling (1.0 to 1.5 K) between 20 km to 35 km, warming (0.5 K) between 35 km and 45 km, and cooling (1 to 3 K) above 45 km. A region of stronger cooling (greater than 3 K) occurs above 45 km at the winter pole. The ozone response in Figure 5 is anticorrelated with the temperature change, and of the order of one percent for each degree.

5. CONCLUSIONS

The studies described here should not be construed as being quantitatively predictive of errors which would be obtained in estimating ozone changes for chemical perturbation scenarios. For instance, the radiative perturbations of scenarios three (Figure 4) and four (Figure 5) would be expected to have a smaller effect on relative ozone than on absolute ozone concentrations. Nor are these studies in any sense comprehensive. Rather, they form part of our ongoing process of studying and evaluating the sensitivity of the LLNL 2-D model to changes and unavoidable tradeoffs in its radiative parameterizations (e.g. the most accurate and physically correct radiation models are far too slow for inclusion in chemical-radiative-transport models). However, The good agreement of our results for doubled CO_2 with those for other 2-D models suggests that the magnitude and form of the effects we saw for small radiative changes is generally valid. The response of

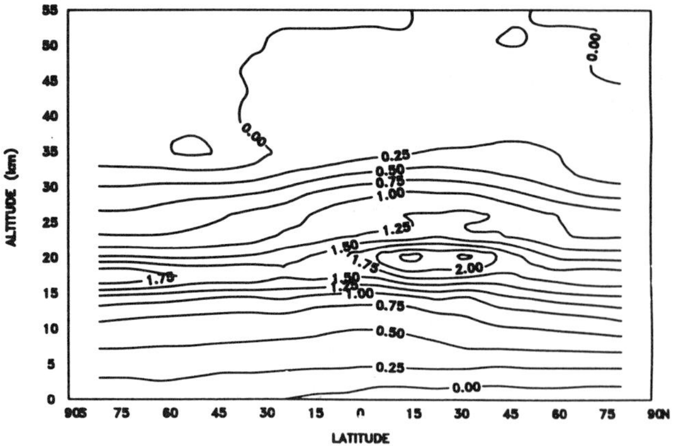

Fig. 4. Percent ozone change for July in response to replacing the shortwave scattering treatment with a simplified Rayleigh scattering algorithm.

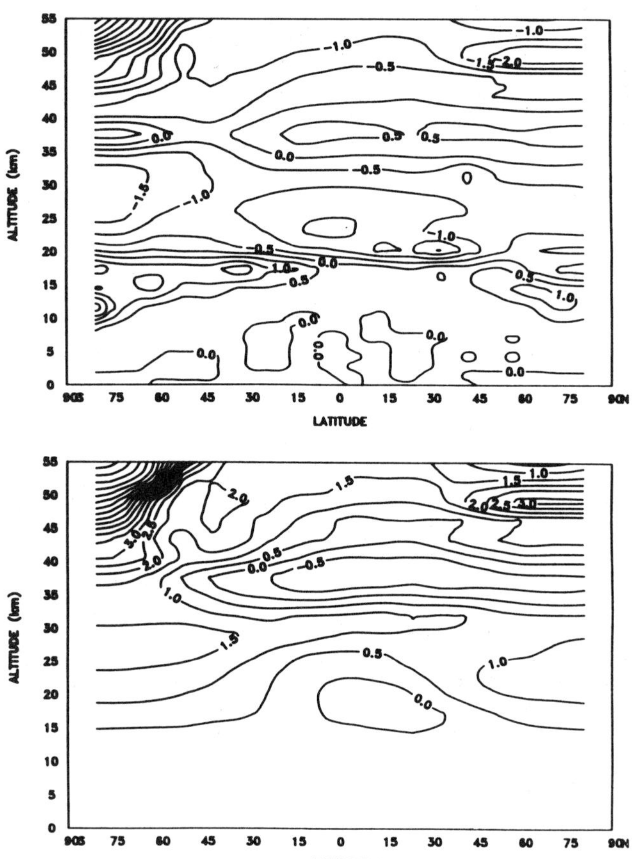

Fig. 5. (a) Temperature change for July in response to adding corrections to the normal CO_2 longwave cooling rates. These corrections were derived using line-by-line calculations for 300 ppmv CO_2. (b) Ozone percent change for July in response to adding above corrections.

ozone and temperature changes to the small radiative perturbations of scenarios three and four, relative to the responses for doubled CO_2 with chemical feedback in scenario two, provide an incomplete yet valid benchmark of the magnitude of error in ozone concentration predictions attributable to current radiative transfer approximations. Our results stress the importance of accurately modeling radiative processes when accurate distributions of trace gas concentrations are required. They also indicate, however, that model predictions for total ozone are not unduly susceptible to the reasonable approximations required for computationally efficient radiation models.

ACKNOWLEDGMENTS

The authors wish to express their thanks to Harshvardhan for providing radiation models and advise that have contributed greatly to the development of the radiation submodels used in the LLNL 2-D model, to Steven Fels and Dan Schwarzkopf for their tape of line-by-line calculations, and to Ray Tarp and Mary Ann Mansigh for their extensive and ongoing programming assistance.

This work was performed under the auspices of the U.S. Department of Energy by the Lawrence Livermore National Laboratory under Contract W-7405-Eng-48.

REFERENCES

Barnett, J.J., and M. Corney, 1985: Middle atmosphere reference model derived from satellite data. *Middle Atmosphere Program Handbook, vol. 16.*

Brasseur, G., and M.H. Hitchman, 1988: Stratospheric response to trace gas perturbations: changes in ozone and temperature distributions. *Science, 240,* 634-636.

Demore, W.B., J.J. Margitan, M.J. Kurylo, C.J. Howard, and A.R. Ravishankara, 1987: Chemical kinetics and photochemical data for use in stratospheric modeling. Jet Propulsion Lab., Pasadena, CA, JPL Publication 87-41.

Fels, S.B., J.D. Mahlman, M.D. Schwarzkopf, and R.W. Sinclair, 1980: Stratospheric sensitivity to perturbations in ozone and carbon dioxide: radiative and dynamical response. *J. Atmos. Sci., 17 (10),* 2265-2297.

Haigh, J.D., 1984: Radiative heating in the lower stratosphere and the distribution of ozone in a two-dimensional model. *Quart. J. R. Met. Soc., 110,* 167–185,

Harshvardhan, R. Davies, D.A. Randall, and T.G. Corsetti, 1987: A fast radiation parameterization for atmospheric circulation models. *J. Geophys. Res., 92 (D1),* 1009-1016.

Joseph, J.H., W.J. Wiscombe, and J.A. Weinman, 1976: The delta-eddington approximation for radiative flux transfer. *J. Atmos. Sci., 33,* 2452-2459.

Kiehl, J.T., and S. Solomon, 1986: On the radiative balance of the stratosphere. *J. Atmos. Sci., 43,* 1525-1534.

Luther, F.M., D.J. Wuebbles, W.H. Duewer, and J.S. Chang, Effect of multiple scattering on species concentrations and model sensitivity. *J. Geophys. Res. 83 (C7),* 3563-3570.

Luther, F.M., 1980: *Annual Rept. of Lawrence Livermore National Laboratory to the FAA on the High Altitude Pollution Program - 1980, Sec. 2.6: Accuracy of the LLNL Multiple Scattering Algorithm,* UCRL-50042-80, Lawrence Livermore National Laboratory, 94 pp.

McClatchey, R.A., R.W. Fenn, J.E.A. Selby, F.E. Volz, and J.S. Gering, 1972: *Optical Properties of the Atmosphere,* 3rd ed., AFCRL-72-0497, Air Force Cambridge Research Laboratories, Mass, NTIS AD-753 075.

Rosenfield, J.E., M.R. Schoeberl, and M.A. Geller, 1987: A computation of the stratospheric diabatic circulation using an accurate radiative transfer model. *J. Atmos. Sci., 44,* 859-876.

Sagan, C., and J.B. Pollack, 1967: Anisotropic nonconservative scattering and the clouds of Venus. *J. Geophys. Res., 72,* 469-477.

Smolarkiewicz, P., 1984: A fully multidimensional positive definite advection transport algorithm with wmall implicit diffusion. *J. Comput. Phys., 54,* 325-362.

Solomon, S., J.T. Kiehl, R.R. Garcia, and W. Grose, 1986: Trends transport by the diabatic circulation deduced from satellite observations. *J. Atmos., Sci., 43,* 1603-1617.

ASPECTS OF CFC RELATIVE OZONE DESTRUCTION EFFICIENCES DETERMINED IN THE LLNL 2-D MODEL

P. Connell, K. Grant, and D. Wuebbles
Lawrence Livermore National Laboratory
University of California
Livermore, California 94550, USA

ABSTRACT

We have investigated the efficiency of surface emission of several individual chlorofluorocarbons (CFCs) and hydrogen-containing chlorofluorocarbons (HCFCs) for reducing the calculated steady state total ozone column in both one- and two-dimensional models, relative to the effects of CFC-11 ($CFCl_3$) emssion. We find that CFC and HCFC relative efficiencies can vary over two orders of magnitude, mostly as a result of atmospheric lifetimes. For CFCs and HCFCs with long stratospheric lifetimes the calculated efficiency is also reduced by the effects of lower and mid stratospheric photochemistry. Ozone in this region changes according to the net effect of changing rates of oxygen photolysis, direct chlorine-catalyzed ozone loss and interference by active chlorine species in the nitrogen oxide-catalyzed ozone loss process. In these cases, the relative efficiencies show a marked latitudinal dependence, being larger at high latitudes and smaller at lower latitudes. Additionally in these cases, the one-dimensional model result appears to overestimate the globally and annually averaged result that is obtained from the two-dimensional model. These considerations should be taken into account when quantities similar to the relative efficiency defined here are applied to CFC production and emission decisions.

1. INTRODUCTION

Chlorofluorocarbons (CFCs) as a class of compounds exhibit a variety of physical properties that make them useful in many technological applications. CFCs that are emitted into the atmosphere are also implicated by photochemical theory, as well as atmospheric measurements in polar regions, to reduce the average ozone abundance in the stratosphere. Although the direct interaction with ozone photochemistry occurs only after the CFCs are destroyed by photolysis or reaction with OH or $O(^1D)$, releasing the photochemically active chlorine atom, differences in CFC ultraviolet (uv) spectral properties and reactivity can strongly influence the extent to which a given emission rate of any particular CFC will affect stratospheric ozone. This is the basis for investigating substitution of currently produced CFCs with other CFCs or hydrogen-containing compounds (HCFCs) with appropriate properties. An understanding of these differences and their relationship to stratospheric photochemistry may also be useful in cost-benefit analysis and production decisions for individual CFCs and their applications.

"Relative efficiencies" for ozone depletion by a number of CFCs based on 1-D model results have been reported by Wuebbles (1981). Hammitt et al. (1987) combined estimates of "relative depletion efficiency" for six CFCs with their economic analysis of CFC production and emission trends in calculating the associated probability distribution of change in the global average ozone abundance.

Several recent studies have reported the results of model calculations of the effects of stratospheric chlorine introduced by CFCs on ozone distributions in two-dimensional models (for example, Brasseur and Hitchman, 1988). A consistent result in these studies is that larger depletion is calculated at higher latitudes in both hemispheres and in the respective winter seasons. We report here preliminary work, using the LLNL 2-D model, on the extent to which these latitudinal and seasonal effects are controlled by the assumed individual CFC source compound abundances, with the important proviso that the heterogeneous processes currently thought to be of great importance in the Antarctic ozone hole phenomenon are not treated in the model. We also briefly discuss some aspects of the concept of ozone depletion efficiency from the standpoint of one-dimensional calculations of global-annual average ozone abundance.

2. DESCRIPTION OF LLNL 2-D MODEL

The LLNL zonally-averaged two-dimensional chemical-radiative-transport model determines the distributions of 31 atmospheric trace constituents in the troposphere and stratosphere, including CCl_4, $CFCl_3$ (CFC-11), CF_2Cl_2 (CFC-12) and CH_3CCl_3 (methyl chloroform). The model domain extends from pole to pole, and from the ground to 0.56 mb (approximately 0 to 54 km). The horizontal resolution is about 10° in latitude. The vertical resolution is approximately 3 km. Reaction rate constants, solar flux data, absorption cross-sections, and quantum yields are based on the latest NASA panel recommendations (DeMore et al., 1987), except for the reaction of NO_2 with O, which was left unchanged from the earlier recommendation (DeMore et al., 1985). Accurate diurnal calculations are used to derive time-varying factors for each chemical and photochemical reaction included in the diurnal-averaged version of the model.

The diabatic circulation for the ambient atmosphere is determined using net heating rates calculated in an internally consistent way with the derived species distributions. Temperatures for the ambient atmosphere vary continuously over the annual cycle, based on the

A. DEEPAK Publishing
ISBN 0-937194-15-8

reference model of Barnett and Corney (1985). Turbulent eddy transport is parameterized through diffusion coefficients K_{yy} and K_{zz}. A value of K_{yy} of 2×10^9 cm^2 s^{-1} is assumed at all stratospheric altitudes and latitudes, with a value of 1×10^{11} in the troposphere. Values of K_{zz} are 1×10^3 cm^2 s^{-1} in the lower stratosphere, increasing slowly with altitude.

The effects of changes in the abundances of individual CFCs on ozone are calculated with reference to a converged steady state mid-1980s model. In the perturbed atmosphere, the diabatic circulation is assumed to be unchanged from the reference state. In the results presented here the temperature field is also held fixed.

The 1-dimensional model used has been described previously (Hammitt et al., 1987 and references therein). The results quoted here are again for a fixed temperature profile. Global and annual average results presented here from the 1-D model are for the most part based on the 1985 NASA Data Panel recommendations (DeMore et al., 1985). Rate constants for HCFC reactions with OH are taken from DeMore et al. (1987).

For the 1-D calculations of the effects of individual CFCs on ozone, the surface CFC emission rate was specified to produce a total column ozone change of about 1%, and the model was integrated time-dependently until a steady state was approached (typically three or four atmospheric lifetimes of the relevant CFC). The reference atmosphere was taken to be free of CFCs and CH_3CCl_3.

3. THE CONCEPT OF RELATIVE EFFICIENCY

The "relative efficiency" for ozone depletion on a global basis, as defined in this study, is the ratio of CFC-11 mass emission rate to the mass emission rate of the other CFC, such that at steady state in each case the global ozone burden is reduced by 1% relative to the reference atmosphere. The choice of a 1% depletion in the global ozone burden is made to avoid any nonlinear effects that might occur if Cl photochemistry were to become the dominant ozone loss process throughout the stratosphere. We also consider, for the 2-D model, seasonal and latitudinal dependences of "relative efficiency", defined as the ratio of the local seasonal and latitudinal change in total column ozone for a CFC perturbation relative to the change calculated for CFC-11, at steady state. Many other definitions of "relative efficiency" are possible, i.e. the ratio of ozone perturbation given a constant fixed mass emission rate, or the relative calculated ozone depletion on a molar rather than a mass basis of CFC emission. Relative efficiencies may also be functions of the magnitude of the calculated ozone change.

The relative efficiency concept we will discuss is based on steady state perturbation calculations. For CFCs such as CH_3CCl_3 with lifetimes shorter than the lifetime of the reference compound, CFC-11, will appear initially, in a time-dependent integration, to have larger relative efficiencies, while the reverse is true of long-lived species such as CFC-115. Relative efficiencies defined at steady state are thus not complete indicators of the near term effects of individual CFC emission rates on ozone.

4. RESULTS AND DISCUSSION

Relative efficiencies for several CFCs and HCFCs, derived from 1-D model results, are given in Table 1. Most of the spread in efficiencies can be attributed to differences in atmospheric lifetime, number of Cl atoms and molecular weight. Figure 1 shows the ozone depletion efficiencies relative to CFC-11 plotted against the product of lifetime and number of Cl atoms divided by molecular weight, again relative to CFC-11. Were all sources of stratospheric Cl equivalent in the photochemical sense once they reach the stratosphere, the CFCs would all lie along the line with unit slope. In fact, several lie below the line, that is, their relative efficiencies are lower than expected, while none lie above it.

This behavior of the 1-D model results follows for three major reasons. First, several of the HCFCs have a sufficiently short lifetime, by virtue of reaction with OH, that they are not well-mixed in the troposphere at steady state in the 1-D model. The HCFC mole fraction in air entering the stratosphere then falls below that predicted on the basis of emission rate and total atmospheric lifetime alone, causing the stratospheric Cl and ozone perturbations to be smaller than predicted. Second, mono- and some bi-chlorinated CFCs and HCFCs tend to have higher energy $n \rightarrow \sigma^*$ transitions than poly-chlorinated CFCs, corresponding to smaller uv cross sections around 200 nm and longer stratospheric lifetimes with respect to photodissociation. At steady state in the region of the stratosphere where Cl photochemistry is important for determining ozone abundance, these CFCs are incompletely dissociated, again with the effect of decreasing the Cl and ozone perturbations below the simple predictions based on total atmospheric lifetime. This effect is somewhat independent of total atmospheric lifetime.

TABLE 1. 1-D RELATIVE OZONE DEPLETION EFFICIENCIES AT STEADY STATE

NAME	FORMULA	LIFETIME (years)	#Cl/MW CFC-11	EFF. CFC-11
	CCl_4	67.	1.19	1.10
CFC-11	$CFCl_3$	75.	1.00	1.00
CFC-12	CF_2Cl_2	148.	0.76	1.03
CFC-21	$CHFCl_2$	2.5	0.89	0.020
CFC-22	CHF_2Cl	20.	0.53	0.062
CFC-31	CH_2FCl	1.8	0.67	0.011
CFC-113	$CFCl_2CF_2Cl$	89.	0.74	0.80
CFC-114	CF_2ClCF_2Cl	185.	0.54	0.75
CFC-115	CF_3CF_2Cl	380.	0.30	0.42
CFC-123	CF_3CHCl_2	2.4	0.60	0.015
CFC-124	CF_3CHFCl	7.6	0.34	0.018
CFC-132b	CH_2ClCF_2Cl	5.7	0.68	0.048
CFC-133a	CF_3CH_2Cl	25.	0.39	0.084
	CH_3CCl_3	7.7	1.03	0.11
CFC-142b	CH_3CF_2Cl	27.	0.46	0.076

Finally, differences among CFCs in absorption cross sections and reactivity with OH and $O(^1D)$ produce differences in the profile of the stratospheric chlorine perturbation and, subsequently, varying effects on ozone-related photochemistry. In the region of the ozone maximum around 25 km, the dominant ozone loss process depends on NO and NO_2 as catalysts. Increased ClO levels have the effect of decreasing ozone loss catalyzed by NO_2 by reacting with NO_2 to form $ClONO_2$. Another effect of the vertical profile of ozone loss is increased penetration of uv in the 200-240 nm region following overhead column ozone decrease. An upper stratospheric ozone decrease is accompanied by greater photolysis of oxygen in the lower stratosphere, providing a negative feedback for the calculated total column change.

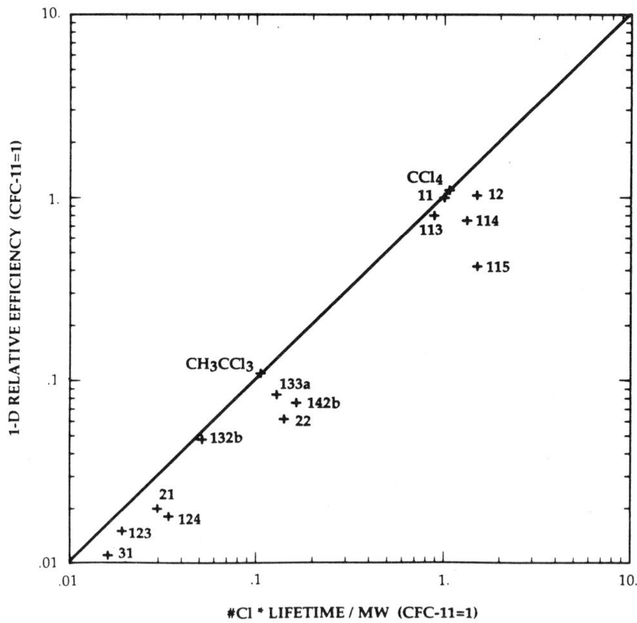

Fig. 1. Calculated 1-D efficiencies for ozone depletion relative to CFC-11 plotted against the product of number of Cl atoms and lifetime, divided by mol. wt., again relative to CFC-11.

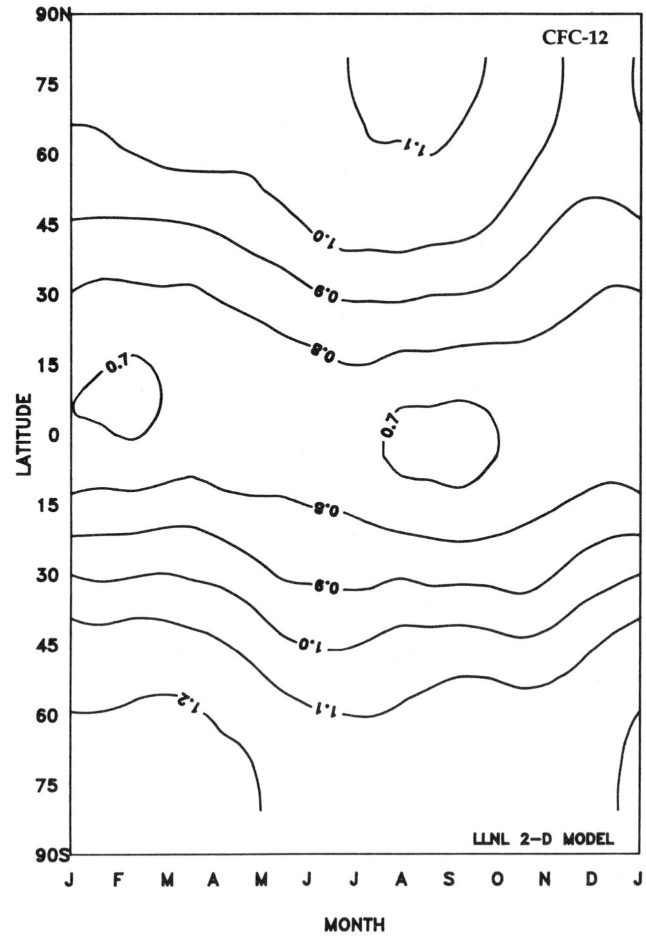

Fig. 2. Ozone depletion efficiency for CFC-12 relative to CFC-11.

4.1 TWO-DIMENSIONAL MODEL RESULTS

The limitations and difficulties of 1-dimensional model representations of the troposphere and stratosphere are well known. Two considerations arise in applying a 2-dimensional model to the calculation of CFC relative efficiencies. First, does the 1-D model adequately predict CFC atmospheric lifetimes and globally and annually averaged CFC profiles and photochemistry? And second, what are the seasonal and latitudinal dependences of relative efficiencies and are there inherently multi-dimensional aspects to the CFC/ozone photochemistry which can not be represented in a 1-D model?

The larger losses for incremental CFC-11 emission calculated at higher latitudes and in the respective hemispheric spring seasons are consistent with other 2-D studies of CFC perturbations (Brasseur and Hitchman, 1988). The calculated CFC-11 lifetime of 77 years is essentially equal to the 1-D result. The CFC-11 emission rate required to produce the globally and annually averaged 1% ozone decrease is about 8% smaller than the corresponding 1-D value, probably associated with several minor differences in rate constant recommendations between the 1985 and 1987 NASA Data Panel documents (DeMore et al., 1985 and 1987). Lifetimes calculated for CFC-12 and CH_3CCl_3 from the 2-D model were 150 years, again essentially equal to the 1-D result, and 5.6 years, respectively.

The seasonal and latitudinal dependence of the effective ozone column depletion for CH_3CCl_3 relative to CFC-11 is small. The global average relative efficiency of 0.085 compares well with the 1-D value of 0.11, taking into account the shorter lifetime for CH_3CCl_3 in the 2-D model. There is a small decrease with increasing latitude in both hemispheres. This is probably the result of relatively faster loss of CH_3CCl_3 during the equator to pole diabatic transport producing a relative depletion of active chlorine in the polar mid stratosphere.

The result for CFC-12 relative to CFC-11 is quite different (Fig. 2). The globally averaged efficiency of 0.92 is about 10% smaller than the corresponding 1-D result, although the lifetimes are essentially equal. A strong latitudinal gradient is also calculated, resulting from the superposition of several effects. CFC-12 differs from CFC-11 and CH_3CCl_3 in that it is not effectively completely dissociated at steady state in the lower stratosphere in the region of the maximum ozone abundance. The abundance of active Cl species in the lower stratosphere relative to the upper (around 40 km) is thus lower for a CFC-12 perturbation than for CFC-11 or CH_3CCl_3.

At lower latitudes, introduction of Cl causes a decrease in ozone in the upper stratosphere and an increase in the lower (Figs. 3a and b). The net change in ozone-controlling reaction rates around 28 km at lower latitudes is largely the sum of ozone-increasing changes in oxygen photolysis (increase) and the reaction of NO_2 with O atoms (decrease), and an increase in the rate of ClO + O, with the effect of reducing ozone (Figs. 4a and b). For CFC-11 and CH_3CCl_3 the behavior is very similar; for CFC-12, lower Cl abundances in the lower stratosphere reduce the importance of changes in the rates of NO_2+O and ClO+O, allowing the increase in oxygen photolysis to dominate and result in greater negative feedback to the overall column ozone change. This regime persists into the midlatitudes. At higher latitudes, though, transport from lower latitudes begins to dominate the lower stratosphere and reduced penetration of solar uv slows lower stratospheric photochemistry. The higher Cl abundance in the upper

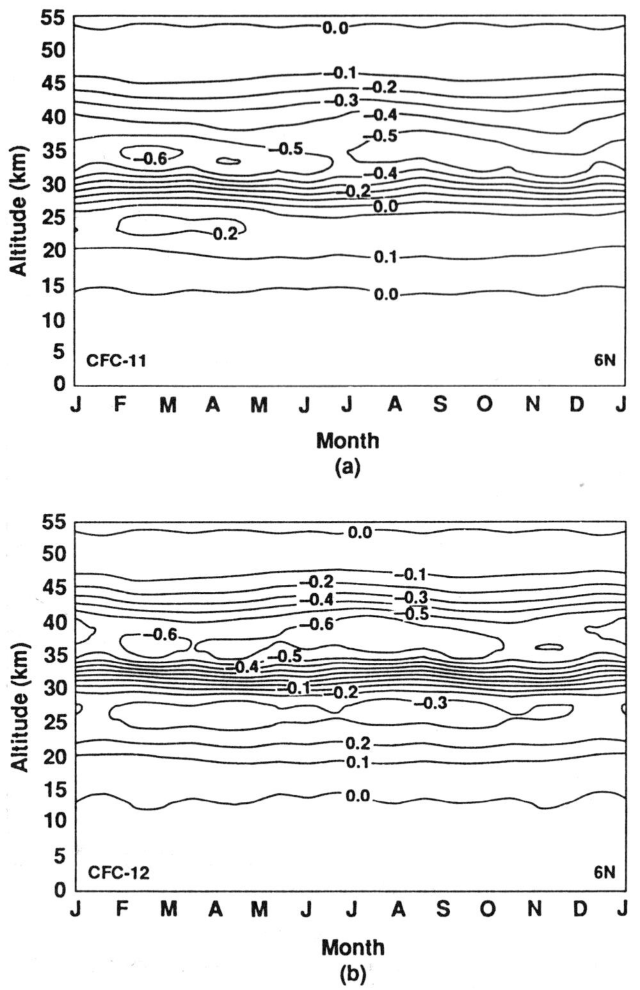

Fig. 3. a) Absolute local ozone changes for CFC-11 at 6°N in units of 10^{11} mol cm-3; b) Ozone changes for CFC-12.

stratosphere for the CFC-12 perturbation results in larger column ozone depletion relative to CFC-11.

5. SUMMARY

These preliminary 2-dimensional calculations show that 1-D model results for global average CFC relative efficiencies for column ozone depletion are representative, at least for CFCs with lifetimes of several years. However, differences in the profiles of the abundances of active Cl species, resulting mostly from differences in CFC uv absorption coefficients, can introduce substantial latitudinal dependence into the efficiency relative to CFC-11 for some CFCs and HCFCs. The results of 2-D model calculations of the effect of assumed future CFC abundances will then depend on the prescribed mix of individual CFC abundances as well as on the total stratospheric Cl perturbation produced. Proper theoretic treatment of the HCFCs with the shortest lifetimes will also depend on the model's treatment of OH in the troposphere.

ACKNOWLEDGEMENTS

This work was performed under the auspices of the U.S. Department of Energy by the Lawrence Livermore National Laboratory under contract No. W-7405-Eng-48.

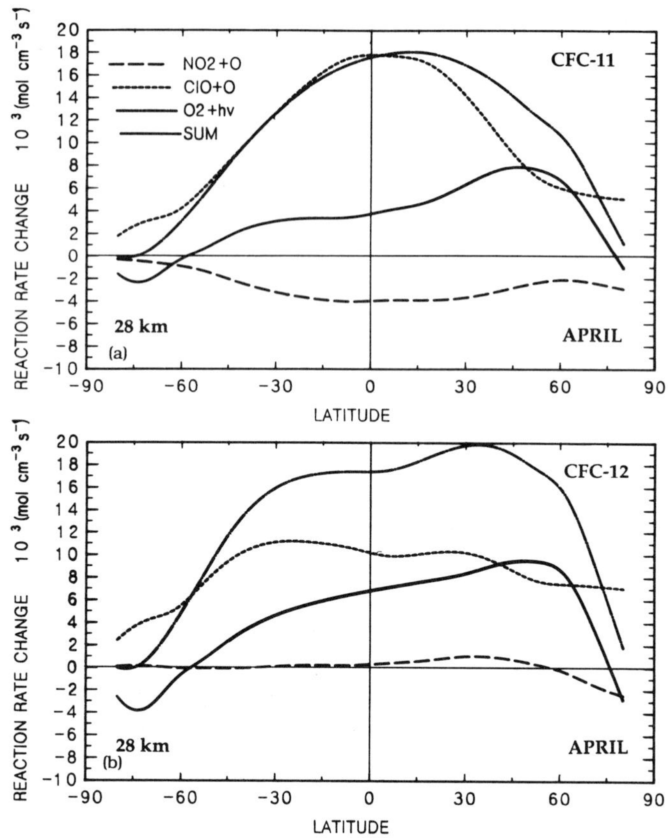

Fig. 4. a) Changes in ozone-controlling rate processes for CFC-11 at 28.5 km in April; b) Changes for CFC-12.

REFERENCES

Barnett, J.J., and M. Corney, 1985: Middle Atmosphere Reference Model Derived from Satellite Data, Middle Atmosphere Program Handbook, Vol. 16.

Brasseur, G., and M.H. Hitchman, 1988: Stratospheric Response to Trace Gas Perturbations: Changes in Ozone and Temperature Distributions. Science, **240**, 634-636.

Demore, W.B., J.J. Margitan, M.J. Molina, R.T. Watson, D.M. Golden, R.F. Hampson, M.J. Kurylo, C.J. Howard, and A.R. Ravishankara, 1985: "Chemical Kinetics and Photochemical Data for Use in Stratospheric Modeling", Jet Propulsion Lab., Pasadena, CA, JPL Publication 85-37.

Demore, W.B., J.J. Margitan, M.J. Kurylo, C.J. Howard, and A.R. Ravishankara, 1987: "Chemical Kinetics and Photochemical Data for Use in Stratospheric Modeling", Jet Propulsion Lab., Pasadena, CA, JPL Publication 87-41.

Hammitt, J.K., F. Camm, P.S. Connell, W.E. Mooz, K.A. Wolf, D.J. Wuebbles and A. Bamezai, 1987: Future emission scenarios for chemicals that may deplete stratospheric ozone. Nature, **330**, 711-716.

Wuebbles, D.J., 1981: The relative efficiency of a number of halocarbons for destroying stratospheric ozone. pp. 11, LLNL Report UCID-18924.

EFFECT OF COMPUTED HORIZONTAL DIFFUSION COEFFICIENTS ON TWO-DIMENSIONAL OZONE ASSESSMENT STUDIES

Charles H. Jackman[1], *Anne R. Douglass*[2], *Paul A. Newman*[2], *and Paul D. Guthrie*[1]
[1]NASA Goddard Space Flight Center
Greenbelt, Maryland 20771, USA
[2]Applied Research Corporation
Landover, Maryland 20783, USA

ABSTRACT

The sensitivity of the distributions of several source and photochemically active gases, including ozone, to changes in the horizontal diffusion coefficients (K_{yy} and K_{yz}) used in our two-dimensional photochemical model has been explored. We also investigated the effects of horizontal diffusion on ozone perturbations caused by increases in atmospheric chlorofluorocarbons, nitrous oxide, and methane.

The use of these computed K_{yy} and K_{yz} values in the model compared to a base case using minimal K_{yy} and K_{yz} values showed several differences. Total ozone was increased at the polar latitudes and decreased at mid-latitudes with input of these K_{yy} and K_{yz} values; maximum changes were apparent in the high latitudes because of large variations in the values of K_{yy} in the lower stratosphere. Most other species including Cl_x, odd nitrogen, CH_4, N_2O, and the chlorofluorocarbons also showed an enhanced transport to higher latitudes, especially in the winter because of large K_{yy} values in the middle to upper stratosphere at higher latitudes.

An ozone perturbation scenario was investigated in which Cl_x was increased from 2.5 ppbv to 8.2 ppbv, N_2O was increased by 20%, and CH_4 was doubled. The total amount of ozone depletion was about the same in the minimal horizontal diffusion model experiment compared to the computed horizontal diffusion model experiment, however, the latitude of maximum total ozone depletion was changed somewhat, especially in the middle to high latitudes.

1. INTRODUCTION

The representation of the atmosphere in a zonally averaged sense has gone through many changes in the past several years. Early two-dimensional (2D) models were represented in an Eulerian framework with zonal mean winds and large eddy diffusion representing the dynamics (Pyle, 1980; Whitten et al., 1981; Vupputuri, 1978/79). In recent years, however, most 2D models have been formulated with a residual mean or diabatic circulation (see, e.g., Miller et al., 1981; Garcia and Solomon, 1983; Guthrie et al., 1984; Stordal et al., 1985). It was thought (Tung, 1982; Kida, 1983) that low horizontal eddy diffusion (~2 X 10^9 cm^2 S^{-1}) was appropriate for the residual mean or diabatic circulation 2D models, however, recent work using NMC data (Newman et al. 1986, 1988) has shown that horizontal diffusion has large temporal and spatial dependence and that values larger than 4 X 10^{10} cm^2 S^{-1} are possible in the upper stratosphere at high latitudes in the winter.

We investigated the influence of these temporally and spatially dependent K_{yy} and K_{yz} values in our 2D model on the N_2O distribution (Jackman et al., 1988). We found that the model predictions for N_2O were in better agreement with SAMS data when these larger K's were used.

Two-dimensional models have been used in recent years for predicting changes that could be caused by Cl_x increases due to anthropogenic changes in source gases (WMO, 1986). The prediction of ozone changes is important to the understanding of the impact of anthropogenic trace gases on the environment, and since the diffusion changes were shown to cause a significant change in N_2O (Jackman et al., 1988), we performed model computations in this study to derive the impact of diffusion changes on ozone depletion predictions.

2. MODEL DESCRIPTION

We use the 2D model described in Douglass et al. (1988) which has a horizontal extent from -85°S to +85°N and a vertical extent from the ground to about 60 km in log pressure coordinates. The horizontal grid is spaced every 10° and the vertical grid spacing is approximately every 2 km. We transport 13 species or families and compute 34 species including the members of the O_x, NO_x, HO_x, and Cl_x families as well as all the methane-related hydrocarbons. The computation includes over 100 photochemical reactions with a one day model time-step.

3. MODEL COMPUTATIONS

We performed two separate steady-state model prediction computations (model experiments) in which Cl_x was increased from 2.5 to 8.2 ppbv, N_2O

OZONE IN THE ATMOSPHERE
Rumen D. Bojkov and Peter Fabian (Eds.)

ISBN 0-937194-15-8

was increased by 20%, and CH_4 was doubled. The Cl_x changes were all reflected in the ground boundary conditions for the Cl_x source gases. The ground boundary condition of CH_3Cl was fixed at 700 pptv, CCl_4 was fixed at 100 pptv, CH_3CCl_3 was fixed at 100 pptv, $CFCl_3$ was increased from 170 pptv to 800 pptv, and CF_2Cl_2 was increased from 285 pptv to 2200 pptv.

The advection field was computed using the residual mean approximation. The heating rates above 100 mbar were taken from Rosenfield et al. (1987) and the heating rates below 100 mbar were taken from Dopplick (1974, 1979).

For the first model experiment we use low K's everywhere in the stratosphere. The K_{yy} values are set to 2×10^9 cm^2 S^{-1} everywhere above the tropopause, being increased gradually in the troposphere to reach 1×10^{10} cm^2 S^{-1} near the ground. For this model experiment we set K_{yz} = 0.0 everywhere. The K_{zz} values are fixed in both model experiments to be 2×10^3 cm^2 S^{-1} everywhere above the tropopause, being increased gradually in the troposphere to reach 1×10^4 cm^2 S^{-1} near the ground.

We compare all our model results to SBUV data. The SBUV instrument measures total ozone, as does TOMS, but also provides profile information. We were given the SBUV data by Richard D. McPeters (NASA/GSFC, private communication, 1988). These data are presented in Fig. 1A and are a three year average (1979-81) of the SBUV data. The SBUV data used in this study differs slightly from that given in McPeters et al. (1984) because the algorithm used to derive the ozone from radiances was corrected using recently measured ozone cross sections (Bass et al., 1985; Paur et al., 1985). The yearly global average amount of total ozone from the SBUV data was computed to be 298 D.U.

The first model experiment base run (simulated year 1980) gives a total ozone model distribution which is represented in Fig. 1B. The yearly integrated amount of total ozone (334 D.U.) is about 12% higher than that observed, however, the qualitative distribution is similar. The northern hemisphere high latitudes in both the data and model show the large winter buildup in total ozone. Even though the 440 D.U. contour is evident in both Figs. 1A and 1B, indications are that the model total ozone is too low in the northern hemisphere high latitudes late winter and early spring time period.

The total ozone phase is correct in the tropical latitudes although the quantitative amounts of ozone are incorrectly predicted by the model. The ozone amount follows the solar zenith angle: smaller amounts of ozone are predicted by the model during mid-winter and larger amounts of ozone are predicted by the model during mid-summer. The southern hemisphere off-pole maximum is predicted by the model, however, the model forecasts the onset of the maximum at an earlier time period than is evidenced in the data.

The prediction for ozone change due to anthropogenic increases of trace gases is given in Fig. 2A. The minimal ozone change is indicated near the Equator with larger changes at the higher latitudes. The peak ozone change in both hemispheres is computed to occur in the late

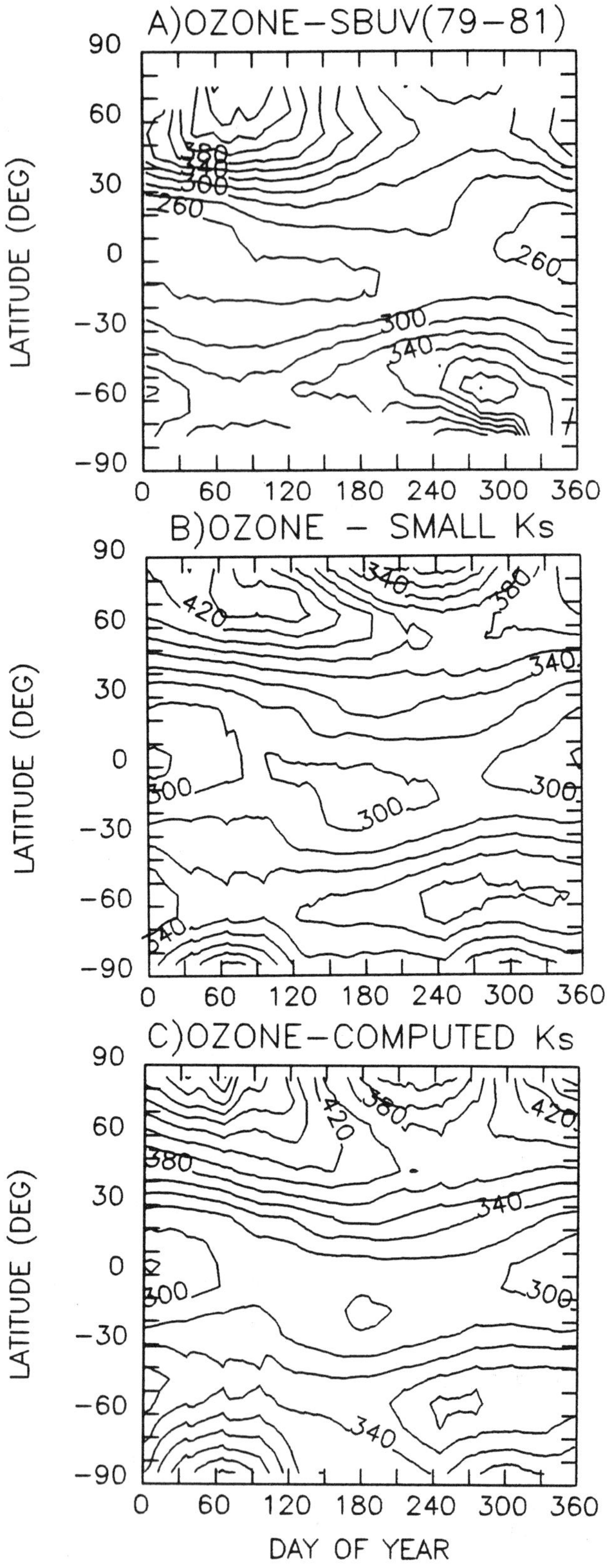

Fig. 1. Total yearly ozone distribution in Dobson units from A) SBUV data (1979-81); B) first model experiment - low horizontal diffusion; and C) second model experiment - computed horizontal diffusion.

spring, early summer time period.

Our second model experiment uses the computed K_{yy} and K_{yz} values from Newman et al. (1988) [see Jackman et al. 1988 for method of inclusion into the 2D model]. The total ozone from the base run of this experiment is indicated in Fig. 1C. More ozone is transported to the poles leading to better qualitative agreement with SBUV data. The global yearly average ozone amount is even larger (343 D.U.) in this model experiment than in the first one. More ozone is transported to higher latitudes via the transport induced by the larger K's to regions of the atmosphere in which the photochemical lifetime of ozone is longer. The major difference, other than the problem with the magnitude of the total ozone predicted, is that the model predicts the ozone maximum amounts at high latitudes for time-periods that are earlier than indicated in the data.

The prediction for the anthropogenically-produced ozone change for this model experiment is given in Fig. 2B. The ozone change in the low latitude region is very similar to the first experiment, however, the peak ozone change in both hemispheres has been moved to the early spring time period. Thus horizontal eddy diffusion changes can cause shifts of a couple of months in the maximum ozone depletion predicted.

It should be noted that these predictions for anthropogenically-produced ozone change do not take into account the possible feedbacks on dynamical parameters that can be caused by changes in the ozone distribution. Some of these feedback effects on perturbations studies have been discussed by Eckman et al. (1987) and Guthrie et al. (1988).

4. CONCLUSIONS

The yearly average global ozone change predicted was approximately the same in both the first and second experiments: -3.2%. The latitude dependence of the depletion was different, however, between the two model experiments especially at northern hemisphere high latitudes (compare Figs. 2A and 2B).

Since the larger K's (second experiment) are more consistent with the dynamical formulation (Newman et al. 1986, 1988) and give a more reasonable ozone distribution when compared to SBUV data, we expect the predictions for the ozone changes from the second model experiment to be more realistic. Even though certain species such as N_2O (see Jackman et al. 1988), CH_4, and other source gases can be changed dramatically with different K's, total ozone and, especially, perturbation studies only show minimal differences.

ACKNOWLEDGEMENTS

We thank Richard D. McPeters (NASA/GSFC) and Joan Stokes (Science Applications Research) who kindly provided updated SBUV data.

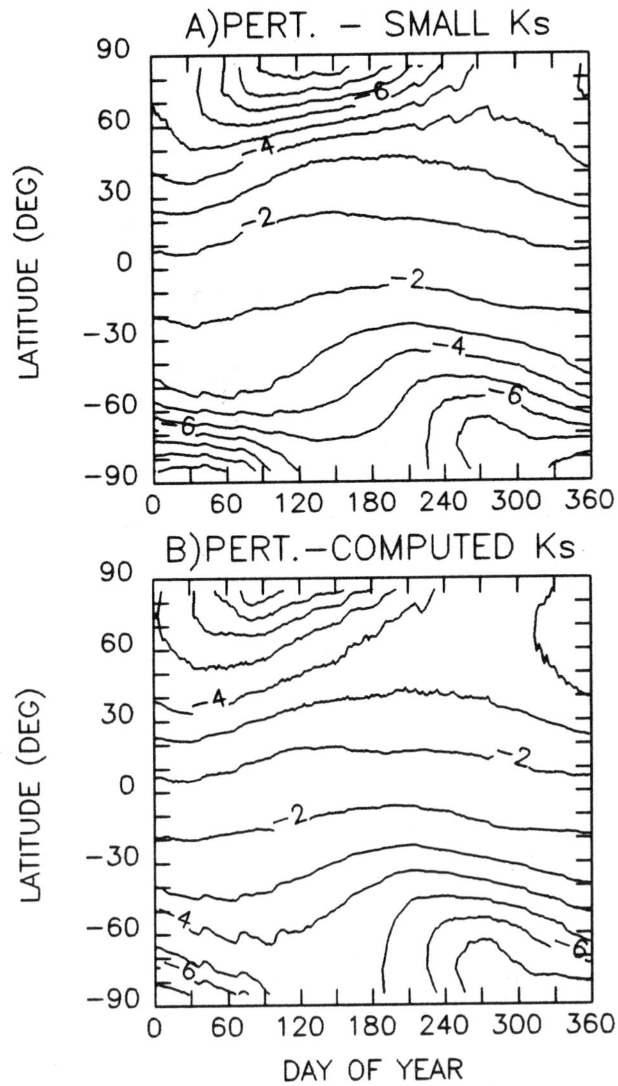

Fig. 2. Total percentage ozone change predicted from a perturbation study in which Clx increases from 2.5 to 8.2 ppbv, N_2O increases by 20%, and CH_4 doubles for A) first model experiment - low horizontal diffusion and B) second model experiment - computed horizontal diffusion.

REFERENCES

Bass, A. M. and R. J. Paur, 1984: The ultraviolet cross-sections of ozone: I. The measurements, Atmospheric Ozone, 606-610, Proceedings of the Quadrennial Ozone Symposium held in Halkidiki, Greece, Sep. 3-7.

Dopplick, T. G., 1974: The heat budget, in The General Circulation of the Tropical Atmosphere and Interactions With Extratropical Latitudes, vol. 2, chap. 7, edited by R. E. Newell, J. W. Kidson, D. G. Vincent, and C. J. Boer, The MIT Press, Cambridge, Mass.

Dopplick, T. G., 1979: Radiative heating of the global atmosphere: Corrigendum, J. Atmos. Sci., 36, 1812-1817.

Douglass, A. R., C. H. Jackman, and R. S. Stolarski, 1988: Two-dimensional model comparisons of odd nitrogen family chemistry with separate calculations for the odd nitrogen species, submitted to J. Geophys. Res..

Eckman, R. S., J. D. Haigh, and J. A. Pyle, 1987: An important uncertainty in coupled chlorine-carbon dioxide studies of atmospheric ozone modifications, Nature, 329, 616-619.

Garcia, R. R., and S. Solomon, 1983: A numerical model of the zonally averaged dynamical and chemical structure of the middle atmosphere, J. Geophys. Res., 88, 1379-1400.

Guthrie, P. D., C. H. Jackman, J. R. Herman, and C. J. McQuillan, 1984: A diabatic circulation experiment in a two-dimensional photochemical model, J. Geophys. Res., 89, 9589-9602.

Guthrie, P. D., T. L. Kucsera, and C. H. Jackman, 1988: On the latitude dependence of ozone depletion predictions, Proceedings of the Quadrennial Ozone Symposium held in Gottingen, West Germany, Aug. 8-13.

Jackman, C. H., P. A. Newman, P. D. Guthrie, and M. R. Schoeberl, 1988: Effect of computed horizontal diffusion coefficients on two-dimensional N_2O model distributions, J. Geophys. Res., 93, 5213-5219.

Kida, H., 1983: General circulation of air parcels and transport in the stratosphere and troposphere derived from CCM, 1, Mean mass flow in the lower stratosphere, J. Meteorol. Soc. Jpn., 61, 171-186.

McPeters, R. D., D. F. Heath, and P. K. Bhartia, 1984: Average ozone profiles for 1979 from the NIMBUS 7 SBUV instrument, J. Geophys. Res., 89, 5199-5214.

Miller, C., D. L. Filken, A. J. Owens, J. M. Steed, and J. P. Jesson, 1981: A two-dimensional model stratospheric chemistry and transport, J. Geophys. Res., 86, 12039-12065.

Newman, P. A., M. R. Schoeberl, and R. A. Plumb, 1986: Horizontal mixing coefficients for two-dimensional chemical models calculated from National meteorological Center data, J. Geophys. Res., 91, 7919-7924.

Newman, P. A., M. R. Schoeberl, R. A. Plumb, and J. E. Rosenfield, 1988: Mixing rates calculated from potential vorticity, J. Geophys. Res., 93, 5221-5240.

Paur, R. J. and A. M. Bass, 1984: The ultraviolet cross-sections of ozone: II. Results and temperature dependence, Atmospheric Ozone, 611-616, Proceedings of the Quadrennial Ozone Symposium held in Halkidiki, Greece, Sep. 3-7.

Pyle, J. A., 1980: A calculation of the possible depletion of ozone by chloroflurorcarbons using a two-dimensional model Pure Appl. Geophys., 118, 355-377.

Rosenfield, J. E., M. R. Schoeberl, and M. A. Geller, 1987: A computation of the stratospheric diabatic residual circulation using an accurate radiative transfer model, J. Atmos. Sci., 44, 859-876.

Stordal, F., I. S. A. Isaksen, and K. Horntveth, 1985: A diabatic circulation two-dimensional model with photochemistry: Simulations of ozone and ground released tracers, J. Geophys. Res., 90, 5757-5776.

Tung, K. K., 1982: On the two-dimensional transport of stratospheric trace gases in isentropic coordinates, J. Atmos. Sci., 39, 2330-2355.

Vupputuri, R. K. R., 1978/79: The structure of the natural stratosphere and the impact of chlorofluoromethanes on the ozone layer investigated in a 2-D time dependent model, Pure Appl. Geophys., 117, 448-485.

Whitten, R. C., W. J. Borucki, H. T. Woodward, L. A. Capone, C. A. Riegel, R. P. Turner, I. G. Poppoff, and K. Santhanam, 1981: Implications of smaller concentrations of stratospheric OH: A two-dimensional model study of ozone perturbations, Atmos. Environ., 15, 1583-1589.

World Meteorological Organization, 1986: Atmospheric Ozone 1985, Rep. No. 16, Global Ozone Research and Monitoring Project, Geneva.

TWO-DIMENSIONAL MODEL COMPARISONS OF ODD NITROGEN FAMILY CHEMISTRY WITH SEPARATE CALCULATIONS FOR THE ODD NITROGEN SPECIES

Anne R. Douglass
Applied Research Corporation
Landover, Maryland 20783, USA

Charles H. Jackman and Richard S. Stolarski
NASA Goddard Space Flight Center
Greenbelt, Maryland 20771, USA

ABSTRACT

Two dimensional models may be used to calculate possible effects of changes in trace gas concentrations. Although these models are fast compared to a three dimensional model with full chemistry, and it is possible to calculate the effects of long term changes in trace gases (e.g., chloro-fluorocarbons, methane, nitrous oxide), many two dimensional models with full chemistry are too slow to examine a large number of scenarios or sensitivities to the dynamical assumptions which are required in their development. Simplifying assumptions may be made in order to develop a faster model. Here we consider family chemistry approximations in which certain species are grouped together and transported as one species. These assumptions are tested by comparing results for the family transport model (FTM) in which odd N = N + NO + NO_2 + NO_3 + 2 N_2O_5 + HNO_3 + HO_2NO_2 + $ClONO_2$ with results for the separate transport model (STM) in which N_2O_5, $ClONO_2$, HNO_3 and HO_2NO_2 are transported separately from NO_x = N + NO + NO_2 + NO_3. Two cases are considered: 1) a current atmosphere annual cycle; and 2) a typical scenario for increases in fluorocarbons, methane and nitrous oxide. Although there are differences in odd nitrogen species partitioning, especially at high latitudes, the calculated O_3 distributions are nearly identical. For the perturbation scenario, although the column ozone change is 0.5% less for STM than FTM, the major features of the distribution of ozone losses are reproduced.

1. INTRODUCTION

It has long been recognized that photochemical equilibrium relationships control the concentrations of some of the trace species that are important in stratospheric photochemistry (e.g., Nicolet, 1965; Crutzen, 1971). In a model, a family of related species is transported and acted on by the net photochemical source, and the member species are calculated from equilibrium relationships. Practical gains of this approach include a reduction in the number of continuity equations to be solved, an increased photochemical time scale, and elimination of a diurnal cycle for the transported family. Because of the increased photochemical time scale, the diffusion term which represents the effects of non-zonal processes for conservative tracers is a better description of family transport than of individual member transport (Pyle and Rogers, 1980).

Although these approximations are commonly used, the photochemical lifetime in the lower stratosphere during high latitude winter becomes long for species such as HNO_3, and the family approximation fails. The total family concentration may be calculated correctly, but the partitioning among the species may be in error. We have examined the family chemistry approximation for odd nitrogen using a fast two-dimensional residual circulation stratospheric model. Calculations of the ambient atmosphere and of the atmosphere resulting from a standard perturbation scenario are compared for the family transport model (FTM) in which Odd N = N + NO +NO_2 + NO_3 + HNO_3 + $ClONO_2$ + $2N_2O_5$ + HO_2NO_2 and a separate transport model (STM) in which HNO_3, N_2O_5, $ClONO_2$ and HO_2NO_2 are transported separately from NO_x = N+NO+NO_2+NO_3.

2. MODEL DESCRIPTION

A brief description of the model used in this study is found in Jackman et al., 1988; details are given in Douglass et al., 1988. The aspects of the model which are important to this study are given here.

The central calculation of this and other models is solution of the longitudinally averaged species continuity equation

$$\frac{\partial \chi}{\partial t} = - \vec{v} \cdot \vec{\nabla} \chi + P - L\chi - D(\chi) \qquad (1)$$

Here χ represents a species concentration, $\vec{v}$ is the wind field, and the first term on the right is advection by the circulation. The photochemical source term is approximated by a constant term P and a loss term that is proportional to the species concentration $L\chi$; these are dependent on reaction rates, photolysis rates and other species concentrations. The diffusion term D represents small scale mixing and also includes the effects of planetary waves and gravity waves.

Changes in species concentration for one time step are assumed small, and therefore the photochemical terms in (1) are evaluated using the previous time step's species concentrations.

ISBN 0-937194-15-8

Results using this assumption compared well with results of a model in which species solutions are obtained simultaneously (Guthrie et al., 1984), and model computer requirements decreased substantially. Implementation of process splitting, i.e., assuming that the advective forcing in (1) can be separated from photochemical forcing (McRae et al., 1983), also increased the speed. The continuity equation (1) is written in difference form

$$\frac{\chi^t - \chi^o}{\Delta t} = \frac{\chi' - \chi^o}{\Delta t} + (P - L\chi^t) - D(\chi') \qquad (2)$$

where χ^t is the result, χ^o the initial condition, and χ' is the advected field. The advected constituent field is calculated using an accurate, efficient, non-dispersive transport scheme (Prather, 1986). The final constituent values are calculated by solving (2) for χ^t

$$\chi^t = \frac{(\chi' + (P - D(\chi'))\Delta t)}{(1 + L\,\Delta t)} \qquad (3)$$

3. FAMILY APPROXIMATIONS

There are three chemical families which are transported in the basic model; these are $O_x = O + O_3 + O(^1D)$, odd $N = N + NO + NO_2 + NO_3 + 2\,N_2O_5 + HNO_3 + HO_2NO_2 + ClONO_2$ and $Cl_x = Cl + ClO + HCl + HOCl + ClONO_2$. The general strategy for using family chemistry in this model is outlined using the odd N family as an example. The net photochemical production for odd N is reaction of N_2O with $O(^1D)$. The only stratospheric loss, reaction N + NO, is written as a function of odd N using the ratios N/NO, NO/NO_2 and NO_2/odd N:

$$\text{Loss} = k_{N,NO}\frac{[N]}{[NO]}\left[\frac{[NO]}{[NO_2]}\right]^2\left[\frac{[NO_2]}{[\text{Odd N}]}\right]^2[\text{Odd N}]^2 \qquad (4)$$

The ratios N/NO and NO/NO_2 are formulated simply using the production and loss reactions, e.g.,

$$\frac{[N]}{[NO]} = \frac{J_{NO}}{k_{N,NO}[NO] + k_{N,O_2}[O_2]} \qquad (5)$$

The expression for NO_2/odd N is derived from the definition of odd N using ratios of each species to NO_2. The ratios N_2O_5/NO_2 and $ClONO_2/NO_2$ account for their diurnal cycles. After calculating the updated value of total odd N from (3), the individual species concentrations are calculated using these ratios.

4. DESCRIPTION OF EXPERIMENT

In the separate transport model (STM), the reactions by which HNO_3, HO_2NO_2, N_2O_5 and $ClONO_2$ are produced and destroyed are loss and production for NO_x. The transport equation is modified to account for the diurnal effects.

Total odd N should be the same for both calculations if the family approximations are valid, and the net production and loss of odd N must be unchanged. Five separate continuity equations must be written for the four transported species and NO_x. Because the transport and diffusion operators are linear, the sums of the transported terms and the diffusion terms from STM should equal the transport and diffusion terms for total odd N in the family transport model (FTM).

Equality of net production and loss requires cancellation among terms, illustrated by considering the production and loss of NO_x from HNO_3. The production term $P(NO_x)_{HNO_3} = \{J_{HNO_3} + k_{OH,HNO_3}[OH]^o\}[HNO_3]^o$ must balance the loss of HNO_3, $\{J_{HNO_3} + k_{OH,HNO_3}[OH]^o\}[HNO_3]^t$. Here J and k, the photodissociation and reaction rates, are subscripted by the reacting species. Cancellation of these terms requires $[HNO_3]^t \approx [HNO_3]^o$, i.e., that the day to day changes in the species concentrations are small. This is important because this assumption is central to our use of the previous day's values to calculate the photochemical terms throughout the model.

5. RESULTS

5.1 AMBIENT ATMOSPHERE

The FTM and STM were run for 20 years to obtain a repeating annual cycle. At equinox, the two models produce similar results. HNO_3 calculated from FTM compares well with the HNO_3 from STM, as indicated by the ratio HNO_3/NO_2 (Figure 1). The only differences are seen at high latitude, and in the troposphere. Total odd nitrogen values are found to compare well at all latitudes and seasons, suggesting the validity of the assumption of slow day to day changes in species concentrations for FTM.

In Figure 2, the difference in HNO_3 calculated by the two formulations is given at 85° N as a function of time. The winter differences, which become significant about 20 days past fall equinox and persist until about 10 days past spring equinox, are largely attributed to failure of the photochemical equilibrium approximation. Between 100 and 10 mb, STM HNO_3 peak values are

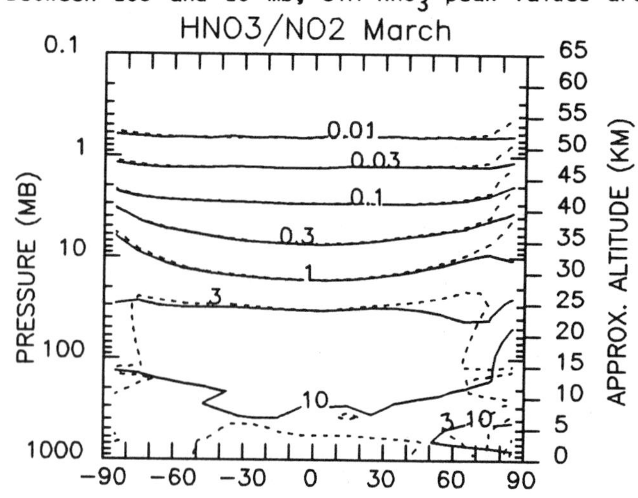

Fig. 1. Equinox ratio HNO_3/NO_2 from STM (solid) and FTM dashed)

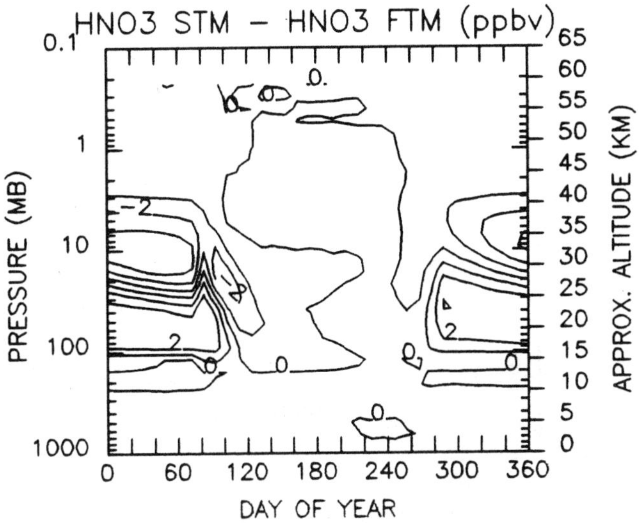

Fig. 2. Difference in HNO_3 at 85° N

transported to lower altitudes during winter. The low FTM HNO_3 values reflect the decrease in NO_2 as solar insolation decreases and the photochemical equilibrium assumption. Neglecting transport, HNO_3 is given by

$$[HNO_3]^t = \frac{P}{L}(1 - e^{-L\Delta t}) + [HNO_3]^o e^{-L\Delta t} \quad (6)$$

where $P = k_{OH,NO_2,M}[OH][NO_2][M]$ and $L = J_{HNO_3} + k_{OH,HNO_3}[OH]$. FTM assumes that HNO_3 is given by its photochemical equilibrium limit $[HNO_3]^t = P/L$ although for long lifetime ($e^{-L\Delta t} \sim 1$), $[HNO_3]^t$ is more closely approximated by its initial value $[HNO_3]^o$. These very low values of HNO_3 in FTM are accompanied by large wintertime values of N_2O_5. Agreement between FTM and STM could be improved by accounting for relaxation of species such as HNO_3 to their photochemical equilibrium values.

Total $Cl_x = Cl + ClO + ClONO_2 + HCl + HOCl$ is nearly identical in both models. In STM, as for HNO_3, $ClONO_2$ is transported to the lower stratosphere during the winter (Figure 3). Unlike HNO_3, $ClONO_2$ differences persist at spring equinox (Figure 4). STM maintains higher values of $ClONO_2$ (and lower values of ClO) in the lower stratosphere through much of the year.

The calculated ozone distributions are similar throughout the year. (Figures 5 and 6). The differences in NO_2 and ClO are confined to the high latitide lower stratosphere where photochemical forcing of ozone is minimal.

5.2 PERTURBED ATMOSPHERE

The percentage changes in local ozone for a perturbation in which Cl_x increases from 2.5 to 8.2 ppbv, N_2O increases by 20% and CH_4 is doubled are given at each altitude and latitude for March and June in Figures 7and 8. Above 25 km, FTM and STM produce the same ozone responses. There are larger differences in the lower stratosphere, which are observed because more odd chlorine is tied up in $ClONO_2$ in the separate transport case (Figures 3 and 4), and there is a smaller concentration of ClO in the lower stratosphere. The difference in ozone response is least in the winter hemisphere, as seen by comparing the southern and northern high latitudes in Figure 8. Because STM has smaller values of ClO than FTM, the global ozone change is less (2.7% compared to 3.2%). However, the latitude, altitude and seasonal behavior of the ozone response is similar in both FTM and STM. This pattern of ozone response, with the largest ozone depletions at high latitudes, is also produced by other models, and has been previously been discussed (WMO, 1986).

6. CONCLUSIONS

High latitude lower stratosphere differences in odd nitrogen and odd chlorine species partitioning resulting from family approximations have little impact on the calculated ozone or on the response of ozone to a standard perturbation of chlorine, nitrous oxide, and methane.

Agreement between the total odd nitrogen distributions for FTM and STM indicates no

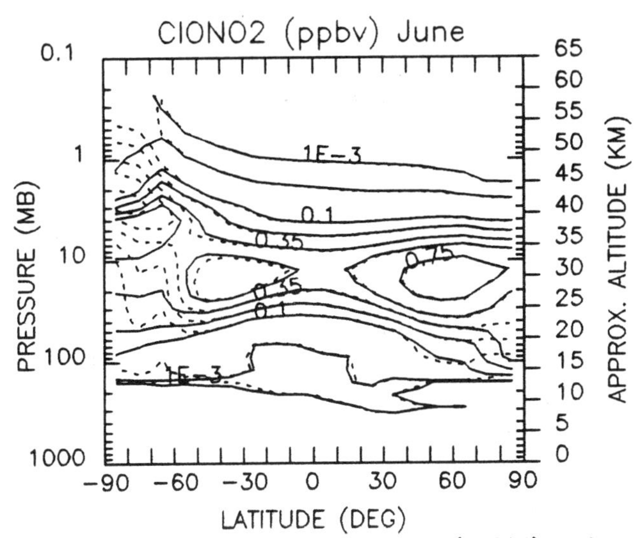

Fig. 3. Solstice $ClONO_2$ from STM (solid) and FTM (dashed)

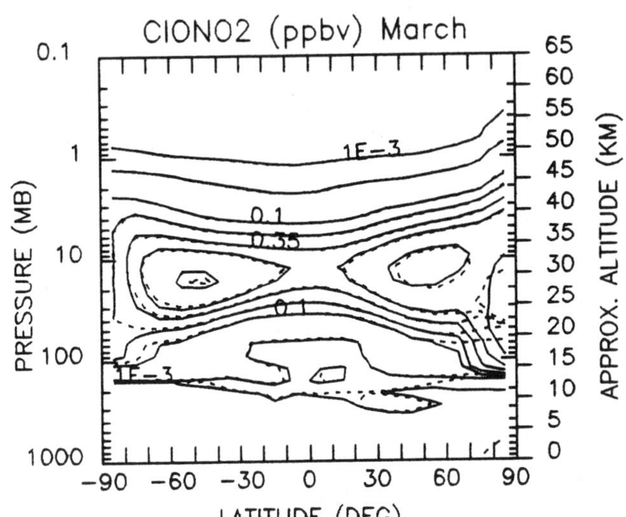

Fig. 4. Equinox $ClONO_2$ from STM (solid) and FTM (dashed)

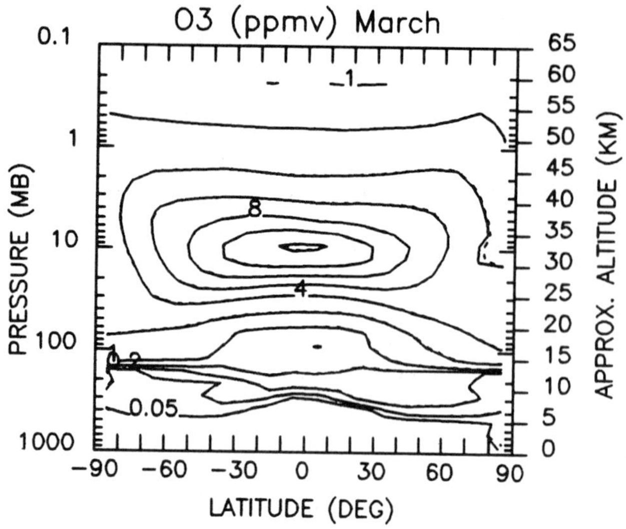

Fig. 5. Equinox O_3 from STM (solid) and FTM (dashed).

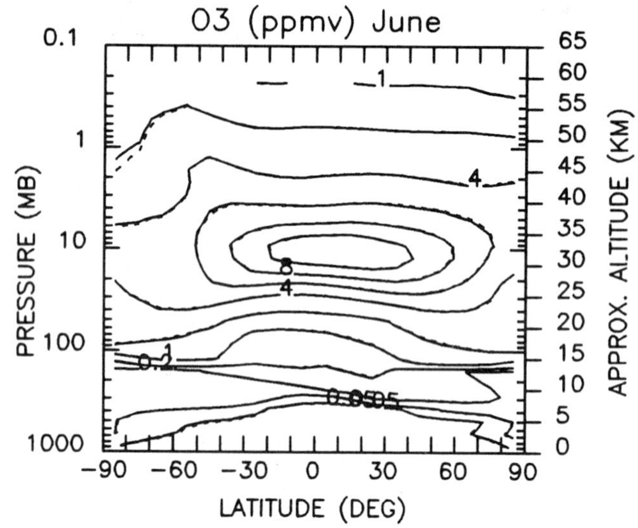

Fig. 6. Solstice O_3 from STM (solid) and FTM (dashed)

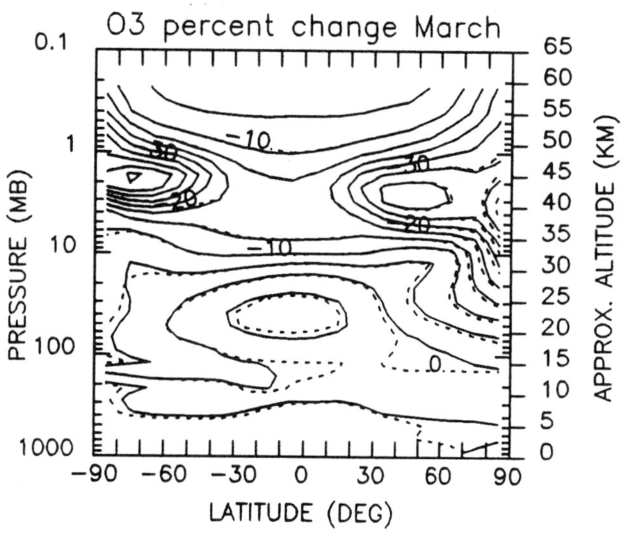

Fig. 7. Equinox percent change in O_3 due to perturbation STM (solid) FTM (dashed)

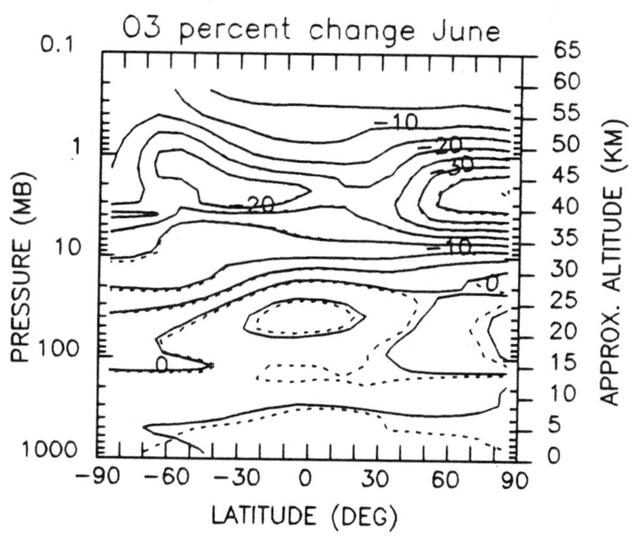

Fig. 8. Solstice percent change in O_3 due to perturbation STM (solid) FTM (dashed)

differences in the net production and loss of odd nitrogen and supports the model assumption of FTM that photochemical forcing terms can be evaluated using previous timestep species concentrations.

REFERENCES

Crutzen, P. J., 1971,: Ozone production rates in the oxygen-hydrogen-nitrogen oxide atmosphere, J. Geophys. Res., 76, 7311-7327.

Douglass, A. R., C. H. Jackman and R. S. Stolarski, 1988: Comparison of model results transporting the odd nitrogen family with results transporting separate odd nitrogen species, submitted to J. Geophys. Res..

Guthrie, P. D., C. H. Jackman, J. R. Herman, and C. J. McQuillan, 1984: A diabatic circulation experiment in a two-dimensional photochemical model, J. Geophys. Res., 89, 9589-9602.

Jackman, C. H., A. R. Douglass, P. A. Newman and P. D. Guthrie, 1988: Effect of computed horizontal diffusion coefficients on two-dimensional ozone assessment studies, this volume, Proceedings of International Ozone Symposium.

McRae, G. J., W. R. Goodin, and J. H. Seinfeld, 1982: Numerical solution of the atmospheric diffusion equation for chemically reacting flows, J. Comput. Phys., 45, 1-42.

Nicolet, M., 1965: Nitrogen oxides in the chemosphere, J. Geophys. Res., 70, 679-689.

Prather, M. J., 1986: Numerical advection by conservation of second-order moments, J. Geophys. Res., 91, 6671-6681.

World Meteorological Organization, 1986: Atmospheric Ozone 1985, Rep. No. 16, Global Ozone Research and Monitoring Project, Geneva.

ON THE LATITUDE DEPENDENCE OF OZONE DEPLETION PREDICTIONS

P.D. Guthrie and C.H. Jackman
NASA Goddard Space Flight Center
Greenbelt, Maryland 20771, USA

T.L. Kucsera
Applied Research Corporation
Landover, Maryland 20783, USA

ABSTRACT

Assessments of the effects of chemical perturbations on the ozone column have now been carried out using a number of 2-D models. Until recently most such studies showed a characteristic variation of fractional depletion with latitude, having minimum depletion in the tropics with greater depletion at middle and high latitudes. Two studies have shown a "reversed" latitude gradient (maximum depletion in the tropics) when temperature and circulation feedbacks were included in the models, and a combined perturbation (increased chlorine and doubled CO_2) was imposed. We report here a series of studies with a model with coupled chemistry, temperature, and circulation which constrains the lower stratospheric circulation less than previous models. These studies indicate that models may produce a reversed gradient in ozone depletion for both a chlorine-only perturbation and the combined perturbation.

1. INTRODUCTION

Until recently there has been a broad consensus among modelers as to the latitude dependence of ozone depletion in two-dimensional model simulations of chlorine perturbations of the stratosphere. The basic picture showed minimal percentage depletion of the ozone column in the tropics, increasing poleward. This picture has been accepted at the policy-making level (EPA, 1987). While the magnitude of the model-predicted depletion for today's atmosphere has been called into question (1988 Ozone Trends Panel Report) the sense of the latitude gradient predicted by the models has appeared to be correct. This is likely to be fortuitous, however, since the models do not include the heterogeneous processes thought to lead to the Antarctic and possible Arctic ozone "hole" phenomena, which dominate the observed column ozone changes at high latitude.

The models used for the assessment studies cited above used specified winds to transport constituents in the model atmosphere. Because the winds and temperature distributions in the stratosphere are largely determined by radiative heating due to ozone, changes in the ozone distribution may in principal lead to significant changes in temperature and transport, which may in turn further modify the ozone distribution. A new generation of models is being developed which attempt to include these transport and temperature feedback processes in ozone perturbation assessments. Preliminary results published by Eckman et. al. (1987) and Brasseur and Hitchman (1988) indicated that the latitude gradient of predicted depletion could in some cases be reversed when these feedback processes were included.

In this paper we report the results of perturbation studies in a model with coupled chemistry, radiation and transport. The model is conceptually similar to that of Brasseur and Hitchman, but with many differences in detail, particularly in the determination of the transport properties of the lower stratosphere. We find that, when feedbacks are included for both the base and perturbed cases, the model predicts maximum percentage depletion of column ozone in the tropics ("reversed" latitude gradient) even for a chlorine-only perturbation, whereas without the feedbacks (using specified transport) the predicted depletion agrees with the consensus latitude gradient. The latitude gradient appears to be quite sensitive to model treatment of the lower stratosphere.

2. MODEL DESCRIPTION

A detailed description of the model and its performance in simulating the present atmosphere appear in Guthrie et al. (1988b). Briefly, the model domain is from -85° to +85° in latitude with 10° latitude resolution and from the ground to ~60 km in height using log pressure coordinates. The model computes the radiative heating and cooling everywhere using the radiative transfer code described by Rosenfield et al., (1987). An additional "latent heating" component is specified below a latitudinally varying tropopause height, and dominates the tropospheric transport, along with a large specified tropospheric eddy diffusion coefficient. Temperature is computed prognostically using the residual mean formulation for the thermodynamic equation for potential temperature. Momentum forcing is obtained through a closure approximation based

ISBN 0-937194-15-8

on a Taylor expansion of the time derivative of the zonal momentum equation and the zonal acceleration obtained by differencing the zonal winds obtained for the new and previous temperatures distributions. The eddy diffusion coefficients are obtained from the momentum forcing by treating quasigeostrophic potential vorticity as a conserved tracer (see, e.g. Newman et al. 1986). The net heating and momentum forcing are used with the residual mean stream function equation to determine new vertical and meridional winds, which in turn are used to update the transport terms in the constituent continuity equations. These continuity equations are solved as in Guthrie et al. (1984) without family groupings. The temperature, transport, and ozone distributions are thus consistent with one another above the tropopause, while the temperature and heating rates are constrained below the tropopause.

3. THE MODEL EXPERIMENTS

The experiments reported here consist of a base case, representing the atmospheric trace gas content around 1980; a chlorine only perturbation, representing the approximate chlorine level for steady state with CFC emissions at 1980 fluxes; and a combination of the chlorine perturbation and doubled CO_2. All were carried out with the same chemical kinetic dataset (JPL 1983). While this data is not the most current, by using it for all scenarios we avoid confusion from changing chemistry and can compare our results directly to previous work. The primary aim of this paper is to examine model behavior when feedbacks are included. None of the experiments presented here should be interpreted as predictions of actual atmospheric ozone distributions.

The perturbations were carried out by initializing with the base case, introducing the perturbation, and running the model through several calendar years until it converged to a repeating annual cycle in temperature, ozone and trace gas distributions. Typically convergence takes about 3-4 years of model time. Convergence is much faster in the model with feedback than in the same photochemical model without feedback (specified temperature and transport). In the latter, convergence of trace gas distributions after a change in transport data typically requires 7-10 years of model time.

The percentage change in the ozone mixing ratio for the chlorine-only perturbation is shown in Fig. 1 as a function of height and latitude for July. As in uncoupled model calculations, the largest percentage depletion occurs at about 3 mbar and at high latitude. The percentage depletion here is greater than in most uncoupled models, largely because of a decrease in methane in the upper stratosphere. This decrease is due to increased hydroxyl (OH) and decreased upward transport in the middle and lower stratosphere.

Unlike in most uncoupled model results, the ozone mixing ratio increases in two regions at

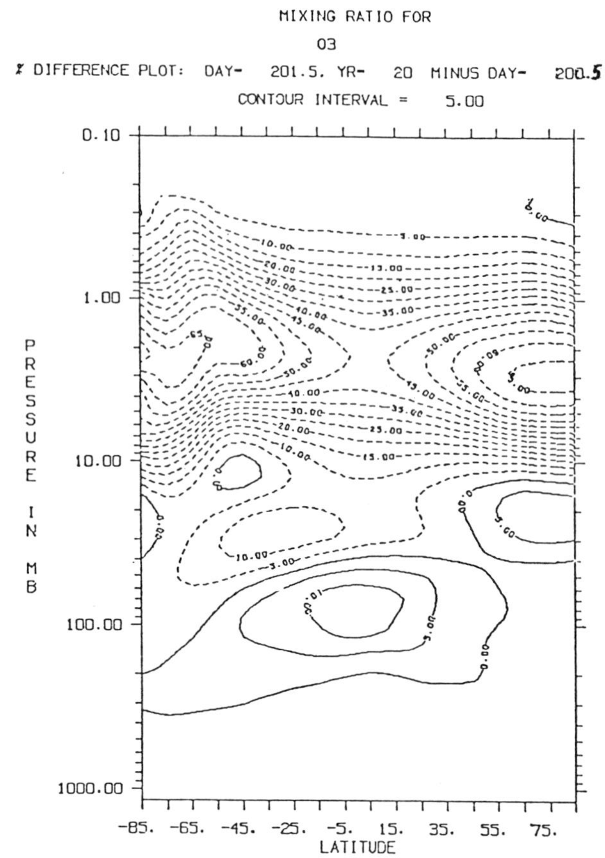

Fig. 1 Percentage change in ozone mixing ratio for Cl_x-only perturbation (July).

about 20 mbar at high latitude, with a greater increase in the summer hemisphere than in the winter. This effect appears to be due to a combination of slightly increased poleward transport of ozone from the tropics and a reduction in the catalytic efficiencies of the HO_x and NO_x cycles due to interference with the chlorine cycle.

More importantly for the distribution of column ozone with latitude, the region of "self-healing" in the tropics occurs at lower altitude than in uncoupled models. This is due to the change in the ratio of O_2 to O_3 photolysis stemming from increased depletion of ozone in the upper stratosphere in the coupled model, and to changes in the temperature distribution affecting the ozone formation rate. Even a slight downward shift of the self-healing region can have a substantial effect on the ozone column, since it represents greater destruction near the heart of the ozone layer.

The region of enhanced ozone in the lower stratosphere extends all the way to the winter pole between 200 and 300 mbar in the coupled model. This represents the effect of transport by a circulation which is more poleward than downward in this region. Because such a large fraction of the high latitude ozone column resides in this region of the lower stratosphere, the overall effect is a local minimum in the fractional column depletion at

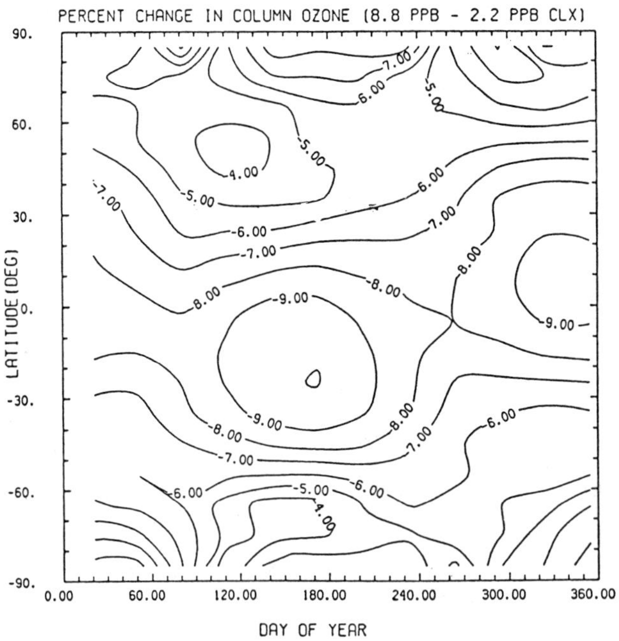

Fig. 2 Percentage change in ozone column for Cl_x-only perturbation.

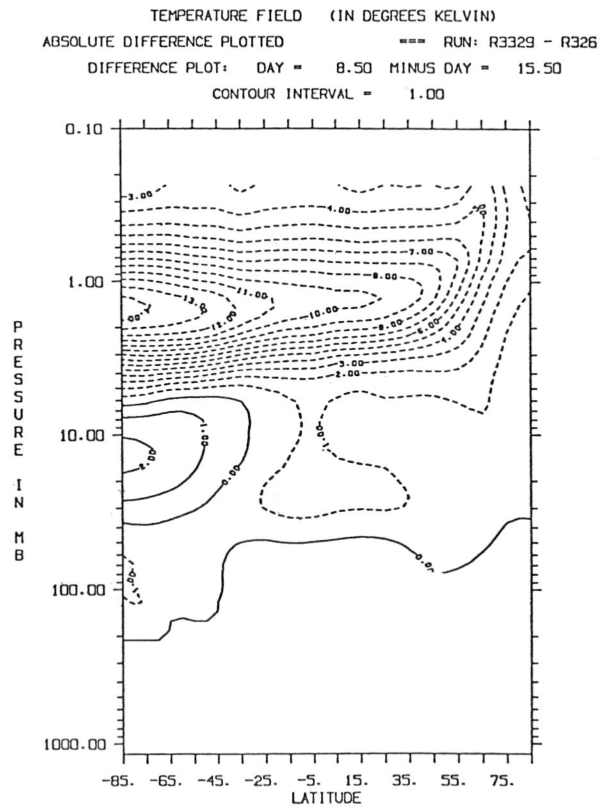

Fig. 3 Change in temperature distribution for Cl_x-only perturbation (Jan.).

high latitude in winter.

The percentage change in the ozone column as a function of latitude and season is shown in Fig. 2 for the chlorine only perturbation. One sees the distinct reversed latitude gradient, with maximum fractional depletion in the tropics, a relatively constant smaller depletion in midlatitudes, and a strong seasonal variation at high latitudes with maximum depletion there occurring in late spring. There is a substantial difference between the base case transport obtained in the coupled model and the transport obtained for observed ozone and temperatures (used in most model studies with specified transport). This has two effects on comparison of this study with others. First, since the overall circulation is much weaker in the coupled model, the region of photochemical dominance extends further down in the model stratosphere, and the strength of the intense downward transport at high latitudes is reduced as compared to results from previous studies. This is probably a major reason for the lessened high latitude column depletion noted here. A second effect is that, because the base case circulation is weak, a small change in temperature or net heating may constitute a larger fractional change than would be the case using a circulation such as that obtained by Rosenfield et al (1987) based on NMC temperatures. The coupled model net ozone transport is thus likely to be more sensitive to changes in ozone distribution than might be expected based on uncoupled model experiments. As noted by Jackman et al. (1988) and explored further in Guthrie et al. (1988a, b) the tracer distributions obtained in an uncoupled 2-D photochemical model using the Rosenfield et al. (1987) circulation are not satisfactory. In particular, the ozone distribution produced in the photochemical model is very different from that used to generate the circulation.

The pattern of column ozone depletion obtained here for a Cl_x-only perturbation is reminiscent of that obtained by Brasseur and Hitchman (1988) for combined Cl_x and CO_2 perturbations. It should be recalled that the two perturbations are very different. Cl_x affects ozone directly through its own catalytic cycle, and through interference with the HO_x and NO_x cycles, whereas CO_2 affects ozone only indirectly through changes in temperature distribution which affect the catalytic efficiency of all cycles and may affect the global circulation.

The temperature perturbations produced in our Cl_x-only and combined Cl_x and CO_2 perturbations are shown in Figs. 3 and 4, respectively, for January. The tropospheric changes in Fig. 4 are imposed as a schematic representation of the CO_2-induced tropospheric warming. As in other models, increased CO_2 produces stratospheric cooling, but in this model the zero net change contour occurs at higher altitude than in the results of Fels et al. (1980), for example. This is due in part to small differences in the radiation codes used, but also to interactive changes in ozone and temperature.

The percentage depletion in column ozone

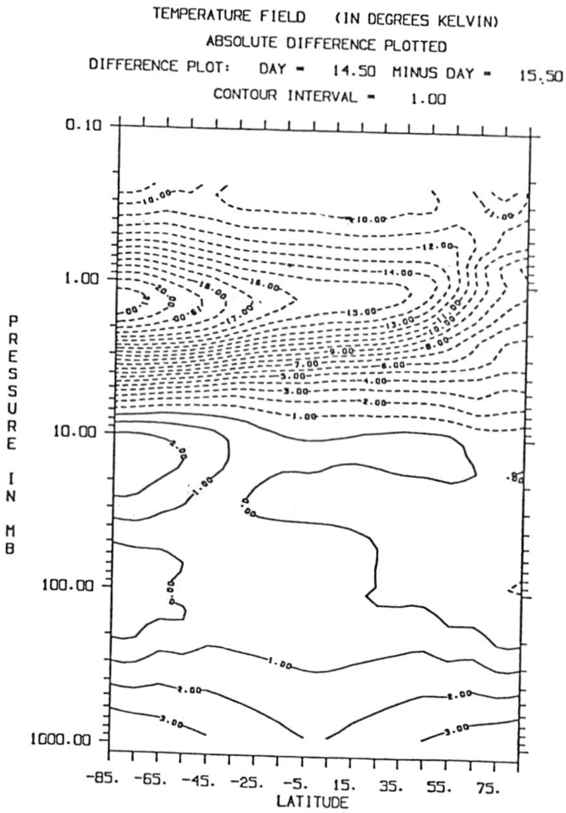

Fig. 4 Change in temperature distribution for combined Cl_x and CO_2 perturbations (Jan.).

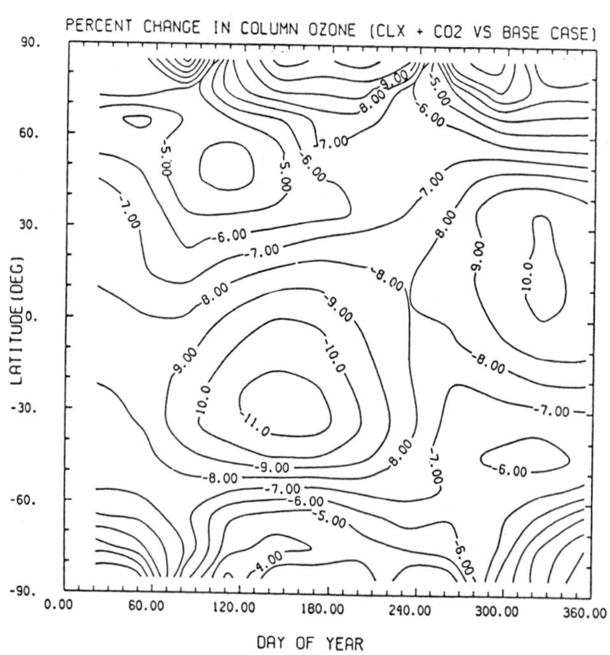

Fig. 5 Percentage change in ozone column for combined Cl_x and CO_2 perturbations.

with respect to the base case for the combined perturbations is shown in Fig. 5. The reversed gradient is retained, with the depletion increasing roughly uniformly by approximately 1% of the base case. The amplitude of the high latitude seasonal cycle increases.

4. DISCUSSION

The difference between our results and others lies mainly in the distribution of ozone in the lower stratosphere. This is the region of greatest sensitivity of the ozone column to differences in transport, and the region in which the net radiative heating which drives the transport is most difficult to calculate due to the near cancellation of large heating and cooling terms. The similarity of our combined perturbation results to those of Brasseur and Hitchman suggests that when feedbacks are included in the perturbation run, the temperature and circulation patterns are forced toward a new equilibrium resulting in the reversed gradient of ozone depletion. The difference between this model and others for the Cl_x-only perturbation may derive from differing treatments of net heating and transport in the lower stratosphere. Eckman et al. (1987) made the assumption of radiative equilibrium in this region. Brasseur and Hitchman (1988) employed a dynamical boundary condition on the stream function at 15 km altitude. Both of these approaches would tend to reduce the sensitivity of the model lower stratospheric transport to changes in the ozone distribution. If this interpretation is correct it would imply that coupled models with weak base case circulations will predict a reversed gradient for ozone depletion due to temperature and circulation changes for some combination of Cl_x and CO_2 perturbations, depending on the formulation of the model transport in the lower stratosphere. In this case, model predictions of the latitude gradient must be regarded as highly uncertain.

REFERENCES

Brasseur, G. and M. H. Hitchman, 1988: Stratospheric response to trace gas perturbations: Changes in ozone and temperature distributions, Science, 240, 634.

Eckman, R. S., J. D. Haigh, and J. A. Pyle, 1987: An important uncertainty in coupled chlorine-carbon dioxide studies of atmospheric ozone modification, Nature, 329, 616.

EPA, 1987: Assessing the risks of trace gases that can modify the stratosphere, U.S. E.P.A. 400/1-87/001.

Fels, S. B., J. D. Mahlman, M. D. Schwarzkopf, and R. W. Sinclair, 1980: Stratospheric sensitivity to perturbations in ozone and carbon dioxide: Radiative and dynamical response, J. Atmos. Sci., 37, 2265.

Guthrie, P. D., C. H. Jackman, T. L. Kucsera, and J. E. Rosenfield, 1988a: On the sensitivity of residual circulation models to temperature data, submitted to J. Geophys. Res.

Guthrie, P. D., T. L. Kucsera, C. H. Jackman, and J. E. Rosenfield, 1988b: A prognostic 2-D model with coupled chemistry, radiation and transport 1: The base case, submitted to J. Geophys. Res.

Guthrie, P. D., C. H. Jackman, J. R. Herman, and C. J. McQuillan, 1984: A diabatic circulation experiment in a two-dimensional photochemical model, J. Geophys. Res., 89, 9589.

Jackman, C. H., P. A. Newman, P. D. Guthrie, and M. R. Schoeberl, 1988: Effect of computed horizontal diffusion coefficients on two-dimensional N_2O model distributions, J. Geophys. Res., 93, 5213.

JPL, 1983: Chemical kinetics and photochemical data for use in stratospheric modeling, Evaluation no. 6, Jet Propulsion Laboratory, 83-62.

Newman, P. A., M. R. Schoeberl, and R. A. Plumb, 1986: Horizontal mixing coefficients for two-dimensional chemical models calculated from National Meteorological Center Data, J. Geophys. Res., 91, 7919.

Rosenfield, J. E., M. R. Schoeberl, and M. A. Geller, 1987: A computation of the stratospheric diabatic residual circulation using an accurate radiative transfer model, J. Atmos. Sci., 44, 859.

MODEL STUDY OF OZONE PHOTOCHEMISTRY USING DATA FROM ATMOS AND SAGE-II EXPERIMENTS

Murali Natarajan
ST Systems Corporation
Hampton, Virginia 23666, USA

Linwood B. Callis
Atmospheric Sciences Division
NASA Langley Research Center
Hampton, Virginia 23665, USA

ABSTRACT

The ATMOS experiment aboard Spacelab - 3 measured vertical profiles of various odd nitrogen and odd chlorine species in the stratosphere at 30N and 48S latitudes during April-May, 1985. For the same time period, O_3 and NO_2 distributions have been measured by the SAGE II experiment. These data sets have been used in the present work to study the consistency of the photochemistry of O_3 in the stratosphere. It is found that current recommendations of the kinetic rate data, along with the constraints provided by the observations, result in good agreement between theory and observations. At 30N latitude the calculated O_3 mixing ratio in the stratosphere is within 12% of the measured value. We discuss in this paper the changes in the model input parameters and their influences on the model results.

1. INTRODUCTION

The balance between the photochemical production and destruction of stratospheric O_3 has been the subject of research for many years. With the advent of satellite based experiments, global scale measurements of various trace species have become available. Quite predictably, many of the theoretical model studies have made use of the large data sets, such as that from the Limb Infrared Monitor of the Stratosphere (LIMS) experiment. [Natarajan et al., (1986), Callis et al., (1986), Jackman et al., (1986), Schmailzl and Crutzen (1984)]. A major conclusion from all the studies which used the Nimbus 7 data set has been that the theoretical models incorporating the contemporary reaction rate data underestimate the upper stratospheric O_3 by nearly 30%. The discrepancy between theory and observation has been all the more puzzling because these models, in general, have constrained the levels of important trace species by making use of the available data. Most of the attempts to explain these differences have focussed on possible uncertainties in key reaction rate constants, and errors in the observations. While modifications to certain rate constants could bring the model results closer to the data, these changes were not necessarily within the quoted limits of uncertainties in the rate data.

The ATMOS and SAGE II experiments have provided yet another opportunity to study the question of the O_3 photochemistry in a self consistent manner. The ATMOS experiment, launched on Spacelab 3, obtained vertical profiles of a number of trace gases important to the odd nitrogen, odd hydrogen, and odd chlorine families at 30N and 48S latitudes during April - May, 1985. (Farmer et al., (1987), Raper et al., (1987), Russell et al., (1988)). The solar occultation measurements included simultaneous observations of species such as NO, NO_2, HNO_3, $ClONO_2$, HNO_4, N_2O_5, HCl, and O_3. Measurements of O_3 and NO_2 for nearly the same latitude region and time period were made by the SAGE II experiment, which was launched in October, 1984. Together they form a comprehensive set of measurements of stratospheric trace species. The purpose of the present study is to take a further look at the O_3 photochemistry using this recent data set.

2. DESCRIPTION OF DATA AND MODEL

The approach used for this work follows closely that of our earlier study with the LIMS data (Natarajan et al., 1986). The model considers the following trace species: $O(^3P)$, O_3, NO_2, NO, HNO_3, OH, H_2O_2, NO_3, N_2O_5, HCl, Cl, ClO, $ClONO_2$, HOCl, HNO_4, HO_2, CH_2O, and CH_3O_2. H_2O data from LIMS and CH_4 from SAMS experiments are used in the present study. Even though the ATMOS measurements include profiles of H_2O and CH_4, these results have not been released yet. The NMC temperature profile, which forms a part of the SAGE II data set, is used to obtain the pressure and total number density at the various altitudes. Since ATMOS data consist of all the major

ISBN 0-937194-15-8

odd nitrogen species and also the upper stratospheric HCl, the levels of total odd nitrogen and odd chlorine are known quantities. The photochemical equilibrium relations used in the computations of unmeasured trace species are similar to those given in Callis et al., (1986). The solar occultation experiment gives data corresponding to sunset or sunrise conditions. However, since the total family concentrations do not have any diurnal variation, one can still estimate the noon time concentrations of the short-lived trace species using equilibrium relations. The chemical reaction scheme does not include any source or sink for either odd nitrogen or odd chlorine; therefore, the constraints imposed by the observed data can be retained and proper partitioning of the species family concentration at different times of the day can be obtained by the use of a diurnal model. The altitude range considered is from 34 to 52 km. Solar UV flux measured by the Solar Mesosphere Explorer (SME) experiment, O_3 from SAGE II, and temperature from NMC data are used in the computation of the photodissociation rate constants. We have adopted the kinetic rate data recommended in the recent JPL publication (JPL ,1987). As in the past, advances in laboratory experiments have led to new recommendations for many reactions. The modification that is most critical to the ozone photochemistry in the stratosphere appears to be the change in the rate constant for the reaction of OH with HO_2. This reaction represents the major sink for odd hydrogen throughout the upper stratosphere. The current recommendation gives nearly a 30% increase in this rate constant at 250 K.

3. RESULTS AND DISCUSSION

In this section we compare the model results with observed data. Mixing ratios and other derived quantities corresponding to the appropriate time of measurement have been obtained from the diurnal model. ATMOS measurements were made at sunset conditions in the northern hemisphere (30N), and sunrise conditions in the southern hemisphere (48S). The latitudinal and temporal coverage of the SAGE II experiment were limited. SAGE II made sunset measurements around 30N latitude on April 27, 1985, and sunrise measurements around 48S latitude on April 30, 1985. We have not had access to the ATMOS O_3 data and this has necessitated the use of SAGE II data for this study.

Figure 1 shows the comparison of the O_3 mixing ratio as a function of altitude for 30N. Though SAGE II O_3 values have been used to initialize the model runs, this comparison is meaningful because O_3 is treated as a variable in the diurnal model. The reaction scheme includes all the major reactions involved in the production and destruction of odd oxygen in the stratosphere. Since the chemical lifetime of odd oxygen in

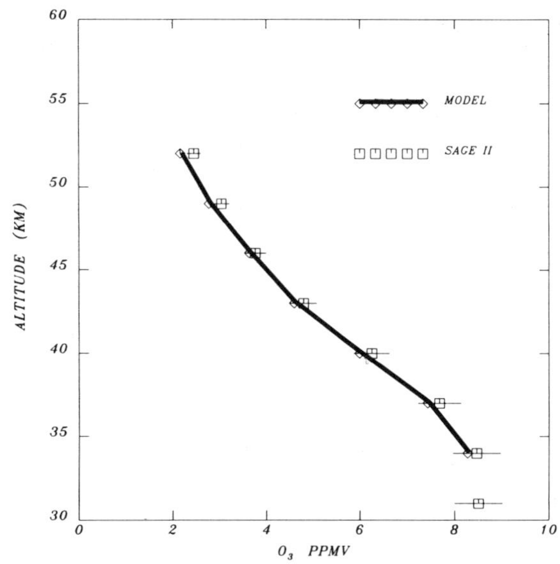

Fig. 1. Calculated and observed ozone mixing ratio during sunset at 30N latitude.

the upper stratosphere is less than a day above 40 km at 30N in May, the O_3 level could change from the initial data and reach a steady state depending on the production and loss rates over the course of five diurnal cycles. The agreement between the calculated and observed O_3 is within 12%. This result is quite significant because most of the studies based on the Nimbus 7 data set had indicated a discrepancy of more than 30% in O_3, with the models giving lower O_3 than the observations. A combination of different factors could explain this change in the model results. Chief among

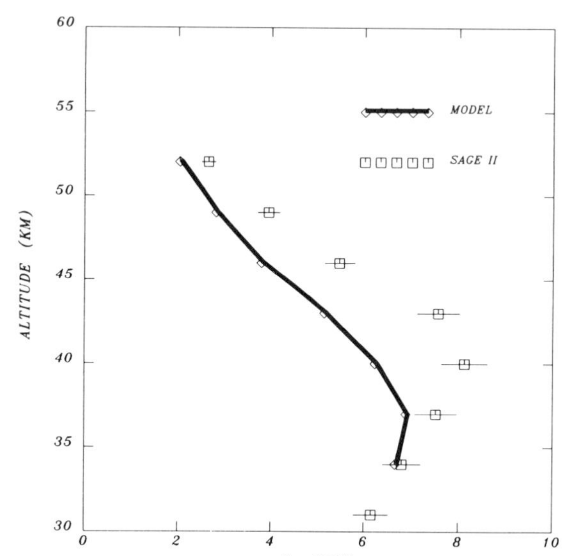

Fig. 2. Calculated and observed ozone mixing ratio during sunrise at 48S latitude.

them is the revised rate constant for the reaction of OH with HO_2, which results in a reduction in the HO_x level. This factor alone could account for about 5-10% increase in O_3 at the upper altitudes. Another parameter that affects the model results is the level of odd chlorine used in the calculations. The asymptotic level of HCl measured by ATMOS at 30N is 2.6 ppbv, and this is a reliable value for the total upper atmospheric burden of chlorine in 1985 (Raper et al. 1987). The total odd chlorine in most of the earlier model studies ranged from 2.5 to 3.0 ppbv [Natarajan et al., (1986), Jackman et al., (1986)]. Considering the ATMOS value of HCl as the reference and assuming the average rate of increase in atmospheric odd chlorine to be 4% per year, one would obtain for the 1979 time period, when LIMS measurements were made, a level of odd chlorine of about 2.0 ppbv. Though there are no global scale data on upper stratospheric odd chlorine for 1979, it appears that the chlorine estimates used in the earlier model studies were too high. This leads to an overestimation of the loss of odd oxygen due to chlorine reactions. Finally, it should be noted that the O_3 concentrations around 40 km measured by SAGE II in April 1985 at 30N are lower than those measured in 1979 for the same location and season by LIMS and SAGE.

Figure 2 shows the comparison of the O_3 mixing ratio for sunrise conditions at 48S latitude. It is clear that the agreement between the calculated and measured O_3 is not as good as in the northern hemisphere. Dynamical effects play a significant role in determining the O_3 level in the winter stratosphere. The scope of our present study does not permit any consideration

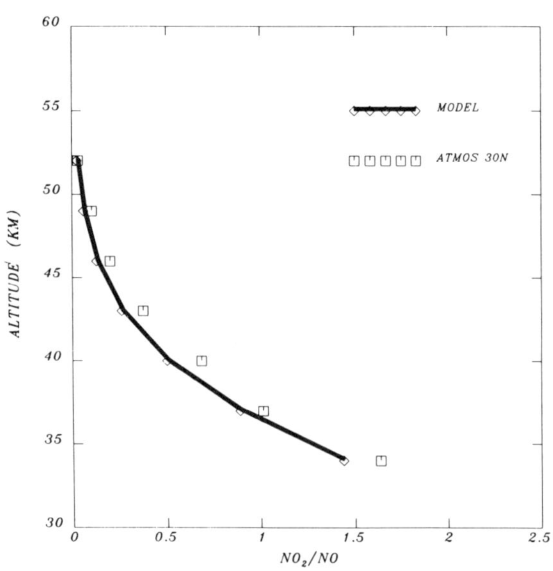

Fig. 3. Calculated and observed ratio of NO_2 to NO during sunset at 30N latitude.

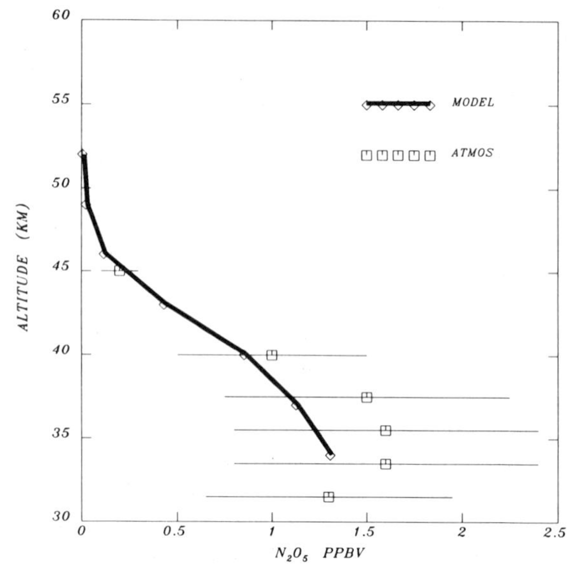

Fig. 4. A comparison of the calculated profile of N_2O_5 with the observation at 48S latitude during sunrise.

of transport. The chemical lifetime of odd oxygen below 40 km is greater than 3 days at 48S latitude in May. Therefore a purely photochemical model cannot be used to compute the O_3 mixing ratio at these altitudes.

ATMOS data set also presents an ideal opportunity to check the photochemical partitioning of the odd nitrogen species. Figure 3 shows a comparison of the ratio of NO_2 to NO at 30N latitude. The agreement shown between the data and the calculated value for sunset conditions reflects the consistency of the description of the odd nitrogen photochemistry in the model. One of the most significant results provided by the ATMOS experiment is the measurement of N_2O_5. A signal corresponding to N_2O_5 was present in the sunrise spectra at 48S latitude, but not in the sunset spectra at 30N latitude (Toon et al., 1986). A comparison of the calculated and observed N_2O_5 for sunrise conditions at 48S is shown in figure 4. The conversion of NO_2 into N_2O_5 during nightime seems to be well simulated by the model. Figure 5 shows the calculated and observed vertical distribution of HCl at 30N latitude. The total odd chlorine in the model is based on the ATMOS measurement of HCl near the stratospause. The conversion of this odd chlorine into other reactive forms such as ClO, $ClONO_2$, and HOCl is responsible for the decrease in HCl at lower altitudes.

The fractional contribution of various catalytic cycles to the destruction of odd oxygen in the stratosphere is illustrated by figure 6. The HO_x reactions are dominant in controlling the ozone level near 50 km, so the calculated O_3 is sensitive to the H_2O used in the model. We have used the LIMS data for H_2O in this

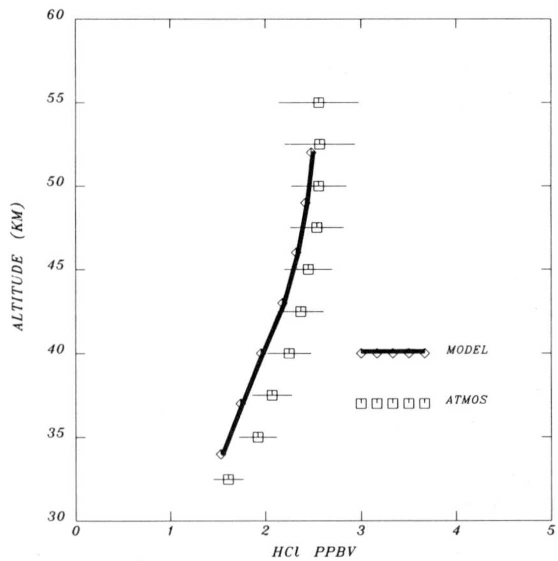

Fig. 5. Vertical profile for HCl. Calculated vs observed at 30N latitude.

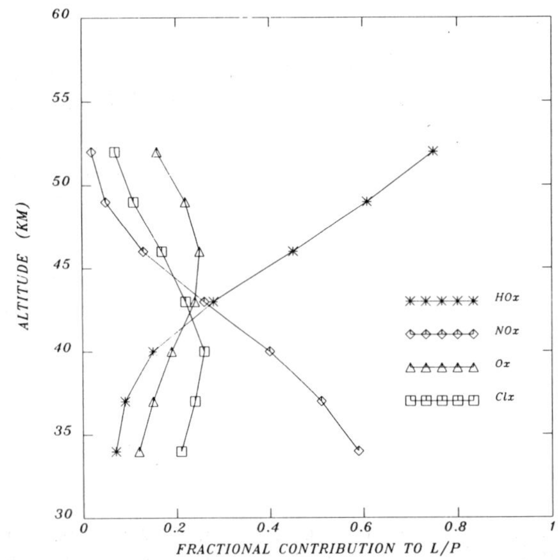

Fig. 6. Fractional contribution of the various catalytic cycles to the destruction of odd oxygen (24-hour averaged destruction from the model results, normalized to the odd oxygen production) as a function of altitude.

work, and it would be of interest to study whether ATMOS H_2O showed any appreciable difference. Around 43 km, all the four cycles are equally important in the destruction of odd oxygen. At lower altitudes the NO_x cycle becomes dominant. Since the total odd nitrogen and odd chlorine have been constrained in this study by the observations, the agreement in midlatitude summer hemisphere O_3 between theory and data seems to validate the photochemical description that is currently in use.

REFERENCES

Callis, L. B., M. Natarajan, R. E. Boughner, J. M. Russell III, and J. D. Lambeth, 1986: Stratospheric photochemical studies using Nimbus 7 data. 2. Development of inferred trace specie distributions. *J. Geophys. Res.*, 91, 1167–1197.

Farmer, C. B., O. F. Raper, and F. G. O'Callaghan, 1987: Final report on the first flight of the ATMOS instrument during the Spacelab 3 mission, April 29 through May 6, 1985. *JPL Publication 87-32.*

Jackman, C. H., R. S. Stolarski, and J. A. Kaye, 1986: Two- dimensional monthly average ozone balance from limb infrared monitor of the stratosphere and stratospheric and mesospheric sounder data. *J. Geophys. Res.*, 91, 1103–1116.

Jet Propulsion Laboratory, 1987: Chemical kinetics and photochemical data for use in stratospheric modeling, Evaluation number 8. *JPL Publication 87-41.*

Natarajan, M., L. B. Callis, R. E. Boughner, J. M. Russell III, and J. D. Lambeth, 1986: Stratospheric photochemical studies using Nimbus 7 data. 1. Ozone photochemistry. *J. Geophys. Res.*, 91, 1153-1166.

Raper, O. F., C. B. Farmer, R. Zander, and J. H. Park, 1987: Infrared spectroscopic measurements of halogenated sink and reservoir gases in the stratosphere with the ATMOS instrument. *J. Geophys. Res.*, 92, 9851–9858.

Russell, J. M., III, C. B. Farmer, C. P. Rinsland, R. Zander, L. Froidevaux, G. C. Toon, B. Gao, J. Shaw, and M. Gunson, 1988: Measurements of odd nitrogen compunds in the stratosphere by the ATMOS experiment on Spacelab 3. *J. Geophys. Res.*, 93, 1718–1736.

Schmailzl, U. and P. J. Crutzen, 1984: Budgets of stratospheric trace gases from 2-D model calculations and satellite observations. *Atmospheric Ozone*, Proceedings of the Quadrennial Ozone Symposium.

Toon, G. C., C. B. Farmer, and R. H. Norton, 1986: Detection of stratospheric N_2O_5 by infrared remote sounding. *Nature*, 319, 570–571.

RADIATIVE HEATING RATES AND THE DISTRIBUTION OF TEMPERATURE AND OZONE IN A TWO-DIMENSIONAL MODEL OF THE STRATOSPHERE

J.D. Haigh
Blackett Laboratory, Imperial College
London SW7 2BZ, UK

J.A. Pyle
Department of Physical Chemistry
Cambridge, CB2 1EP, UK

ABSTRACT

An assessment is made of some of the approximations made in the calculation of thermal infrared radiative cooling rates in the stratosphere. Large differences in cooling rates arise and the effect of errors of such magnitude on temperature and ozone fields is studied using a two-dimensional model with full representation of photochemical and radiative processes. Serious uncertainties in modelled ozone and temperature are found for errors of order $1Kd^{-1}$ in cooling rate but uncertainties of order 1K in temperature and 2% in ozone concentration for errors of order $0.1Kd^{-1}$. An experiment in which the model dynamics are inhibited from responding to changes in temperature confirms the necessity for interactive transport-chemistry-radiation in ozone perturbation calculations.

1. INTRODUCTION

An estimate of net diabatic heating rate is essential for any numerical model of the middle atmosphere. The gases providing contributions to this heating and the spectral properties of these gases are, for the most part, well known and, should sufficient computing resources be available, accurate estimation of the heating rates using line-by-line calculations would be possible in the model. In practice such computing power is not freely available and parameterisations of the gaseous absorptions need to be made. Errors in these approximations will directly affect the model's temperature structure and, in the case of models including photochemical calculations, uncertainties in temperature will be reflected in reaction rates and ultimately in the fields of chemical constituents. Ozone in particular is sensitive to temperature in the upper stratosphere. Modifications to the temperature structure will also affect the wind fields and thus tracer transport.

The work described in this paper represents an attempt to estimate the accuracy to which net diabatic heating rates must be calculated to give an acceptable uncertainty in the modelled fields of temperature and ozone where the definition of "acceptable" will depend on the context of the calculations.

The discussion of heating rate accuracy in the next section concentrates on thermal infrared parameterisations as, in models including photochemistry, the solar ultraviolet/visible component may be accurately estimated as a by-product of photodissociation rate calculations.

2. APPROXIMATIONS IN COOLING RATE CALCULATIONS

2.1 EXCLUSION OF VARIOUS GASES OR ABSORPTION BANDS

Figure 1 shows the contributions of the most significant gases/absorption bands to the infrared cooling rate in the stratosphere near the summer pole. The contributions due to the CO_2 15μm band, the O_3 9.6μm band and the H_2O bands were calculated as described by Haigh (1984). The CH_4 and N_2O estimates used the same framework for flux divergence calculation and the approximate transmissions of Green *et al* (1964) and the O_3 14μm used a wide band model with parameters estimated from the AFCRL line data compilation.

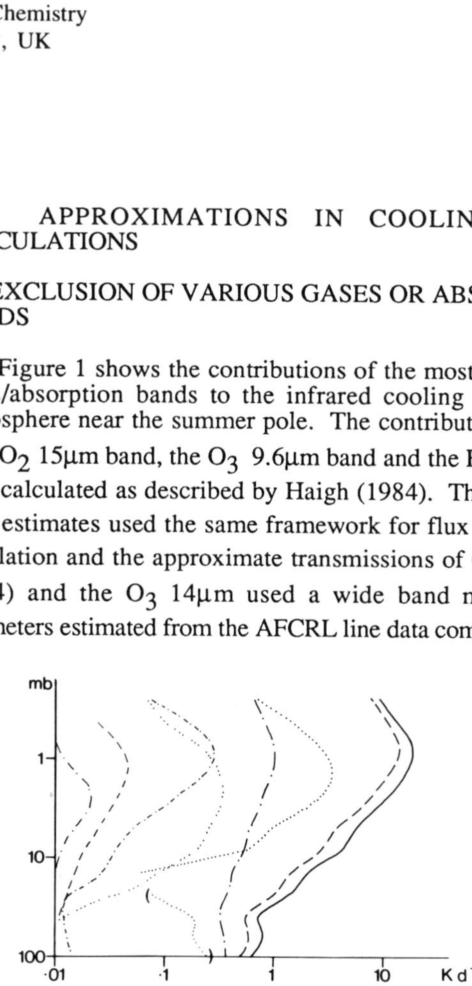

Fig. 1. Contributions to infrared cooling rate (Kd^{-1}) as a function of height due to various gases/spectral bands. (Note abcissa has logarithmic scale).

	gas	spectral region (cm^{-1})
– – – –	CO_2	550 - 750
—·—·—·	H_2O	520 - 820 overlap
··· ··· ···	O_3	500 - 840 overlap
...........*	O_3	960 - 1080
—·—·—	H_2O	0 - 550
-·-·-·-·-	H_2O	1080 - 2200
- - - - - -	CH_4 & N_2O	1200 - 1400 overlap

The designation "overlap" indicates the cooling provided by that gas/band in addition to that of the major contributor in that spectral region. *(Section in parentheses indicates warming rather than cooling).

ISBN 0-937194-15-8

The CO_2 15μm band provides approximately 70% of the total throughout this region and the O_3 9.6μm band up to 20% in the upper stratosphere. These bands are usually included in stratospheric models. The H_2O rotation bands, sometimes neglected, provide about $1Kd^{-1}$ in the upper stratosphere but a more significant $0.8Kd^{-1}$, being around half the total, in the lower stratosphere. It is interesting to note that the O_3 14μm band contribution, invariably omitted because it is only a few tenths of $1Kd^{-1}$, is responsible for about 5% of the cooling near 10mb. Other H_2O bands, CH_4 and N_2O and other gases not shown here, including CFC11 and CFC12, provide very minor contributions.

2.2 BAND MODELS

A detailed discussion of the approximations made in various band models is not appropriate here. Other authors have conducted investigations into various aspects of these approximations; in particular Kiehl and Ramanathan (1983) and Crisp *et al* (1986) have provided analyses of several aspects of the parameterisations used for the CO_2 15μm band.

The stages at which approximations are made will be briefly summarised so that their effects on cooling rates may be demonstrated:

2.2.1 Lineshapes

In the transition between between the pressure-broadening in the lower stratosphere and Doppler-broadening in the upper stratosphere the lineshape may be described by the Voigt function. An analytical expression for the integrated Voigt line is not possible so approximate expressions are needed (see e.g. Fels 1979).

2.2.2 Line overlap

In the upper stratosphere the ratio of line-spacing to line-width is , in most cases, sufficiently large that the lines may be considered independent. Lower down the degree of overlap depends on position and spacing but needs to be represented in any approximation.

2.2.3 Band width

Two different approaches have been adopted: use of several narrow spectral bands and use of one wide band. The assumptions that can be made about line distribution within a band depends on its width.

2.2.4 Temperature dependence of band strengths

Transmittances calculated for one temperature profile will not be exact for any other. Fig.2 shows CO_2 15μm cooling rates calculated using various of these assumptions. The solid line indicates a calculation using a Goody random band model in thirty bands, each of width $10cm^{-1}$, between 520 and $820cm^{-1}$. Doppler broadening is included using an expression for equivalent width as a function of absorber amount which is a polynomial curve fitted to accurate numerical integrations. Transition between pressure and Doppler broadening is accomplished using a technique similar to that derived by Rodgers and Williams (1974) for individual spectral lines. Temperature dependence of bandstrengths is incorporated using the method of Rodgers and Walshaw (1966). The short-dashed line represnts a calculation identical but lacking the temperature dependence. The dotted line represents the temperature-dependent calculation using the Fels (1979) approximation for Voigt lineshape. The method represented by the long-dashed line used the Ramanathan (1976) wideband parameterisation. For the dash-dotted line spectral lines were

Fig.3(a) shows the cooling rates calculated for the H_2O rotation bands using a Goody random model in one band (dashed line) and fifty-five $10cm^{-1}$ bands (solid line). Fig.3(b) indicates the difference in the O_3 9.6μm band between using the Ramanathan (1976) and Rodgers (1968) methods.

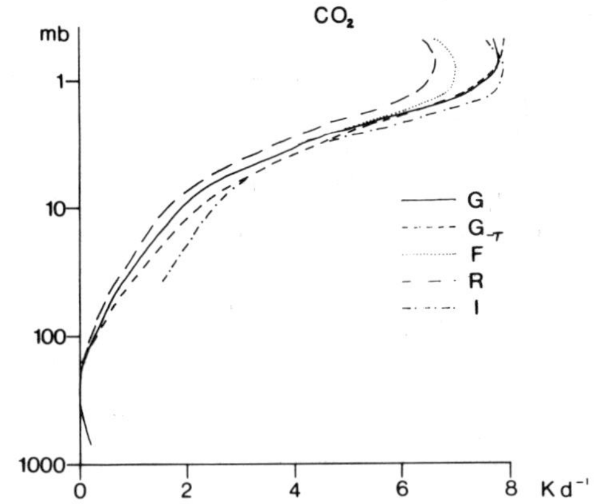

Fig. 2. Cooling rate (Kd^{-1}) as a function of height in the CO_2 15μm band for standard atmosphere using different band models. See text for details.

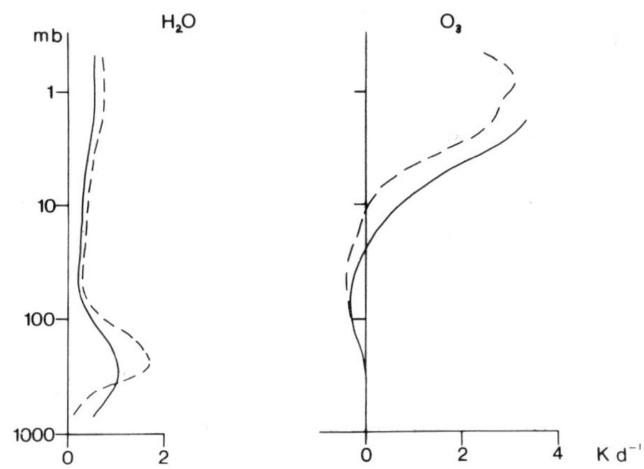

Fig. 3(a). Cooling rate (Kd^{-1}) as a function of height due to H_2O rotation bands (0-$550cm^{-1}$) calculated using a Goody random model in 55 bands (_______) and in one band (_ _ _ _ _).

(b). Cooling rate (Kd^{-1}) as a function of height due to O_3 9.6μm band using band model of Rodgers (1968) (________) and Ramanathan (1976) (_ _ _ _ _).

Any of these techniques might legitimately have been used in a numerial climate model and the extent of the variation in cooling rates is thus used as a guide for the experiments described in the next section.

3. RESPONSE OF OZONE AND TEMPERATURE TO HEATING RATE APPROXIMATIONS

The accurate calculation of net diabatic heating rates is of fundamental importance for a realistic simulation of atmospheric temperature structure. Changes in temperature will affect the dynamical structure and thus the advection of tracers by the wind and will also directly affect the chemical structure through temperature-dependent reaction rates. The correlation between O_3 and temperature in the stratosphere has been the subject of many studies both from observational (e.g. Keating *et al* 1985) and theoretical (e.g. Haigh and Pyle 1982) viewpoints. In the upper stratosphere, where photochemical time constants are short, there is a negative correlation between O_3 and temperature. In the lower stratosphere O_3 behaves effectively as an inert tracer and is positively correlated with temperature due to the simultaneous transfer of heat and O_3 from low to high latitudes.

In order to assess the sensitivity of model fields to changes in heating rates a two-dimensional model with fully interactive dynamics, radiation and detailed photochemical scheme has been used (see Chipperfield *et al* 1988). Two experiments have been performed in which the radiation schemes have been amended and the resultant temperature and O_3 fields compared with a control run.

In Run A the net heating rate was increased by $1Kd^{-1}$ everywhere. This gross change was suggested by the order of the approximations shown in Figs. 1-3 and, although it offers a pessimistic estimate of available accuracy in the lower stratosphere, probably represents a conservative view of potential disagreement near the stratopause. The experiment was performed in two modes: A1 in which the dynamical fields were allowed to respond freely to the changing temperatures and A2 in which the wind structure was fixed at its control run values. Run A2 is thus using an approximation very similar to that described as Fixed Dynamical Heating by Fels *et al* (1980); the difference lies in the advection terms for in this work changing temperature gradients might affect their contribution. For a uniform change in heating rate this difference will actually be very small.

In Run B the CO_2 15μm band cooling rate was amended in line with the errors in the Ramanathan (1976) parameterisation as suggested by Crisp *et al* (1986, see their Fig.17). The error varies in sign and increases in magnitude above the stratopause (see Table 1).

TABLE 1. Percentage error in the cooling-to-space part of the cooling rate incurred by use of Ramanathan (1976) parameterisation (after Crisp *et al* 1986) and typical errors in a winter mid-latitude cooling rate profile associated with these. (Minus sign indicates that cooling rate is underestimated i.e. net heating will be too high).

Approx pressure (mb)	Percent error	Typical error (Kd^{-1})
0.261	-35	-2.45
0.431	-27	-2.08
0.710	-18	-1.40
1.17	-8	-0.58
1.93	-2	-0.11
3.18	4	0.16
5.25	5	0.14
8.65	2	0.04
14.3	-3	-0.05
23.5	-7	-0.08
38.8	-8	-0.06
63.9	-7	-0.03
105.	3	0.01

4. RESULTS OF RUN A

The difference in temperature between RunA1 and the control run are shown in Fig.4. Large increases are indicated throughout the region with over 30K in the equatorial lower stratosphere decreasing to around 5K at the summer polar stratopause. As might be expected with a uniform $1Kd^{-1}$ increase this structure strongly resembles that of the Newtonian damping time (see e.g. Kiehl and Solomon 1986). The percentage difference in O_3 concentration is shown in Fig.5.

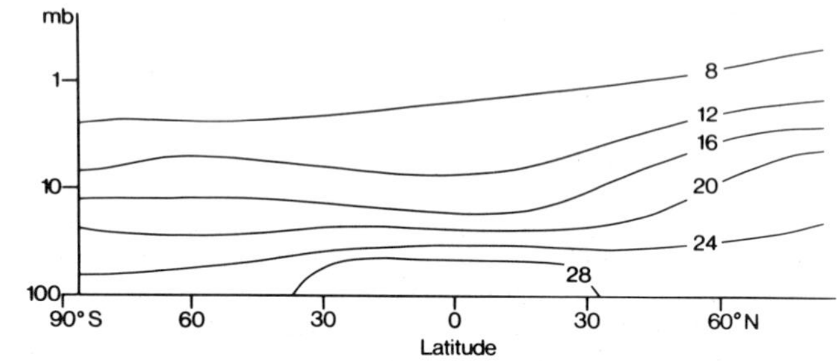

Fig. 4. Difference in temperature (K) as a function of latitude and height at the December solstice between Run A1 (see text for details) and the control run.

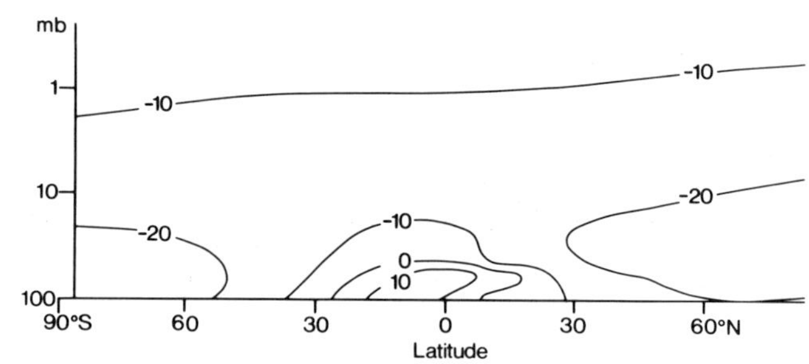

Fig. 5. Percentage increase in O_3 (other details as Fig.4)

Decreases of between about 6 and 20% in the upper stratosphere reflect the increase of temperature here while in the lower stratosphere there are larger decreases in high latitudes with a region of increased concentration near the equator. This corresponds not only to a 'self-healing' effect of ultraviolet radiation in low latitudes but also to a weakened mean circulation and thus a reduced transportation of O_3 out of this area. The global decrease of O_3 in Run A1 with respect to the control was 14.5%.

It is instructive to compare these results with those of Run A2 in which the dynamical fields remain unaffected by the changing temperature structure. The temperature field in Run A2 is very similar (within 2K throughout most of the stratosphere) to that of A1 discussed above confirming that the Fixed Dynamical Heating approximation is a good one for studying the effect of radiative perturbations on temperature. The ozone field, however, is quite different.

The difference in ozone concentration between Run A2 and the control run is shown in Fig.6. In the upper stratosphere the values are very similar to those in Fig.5 but in the lower stratosphere, and thus in the column and global totals, the values are small reflecting the fixed transport in A2. This has important implications for those types of two-dimensional model in which the mean meridional circulation is pre-calculated particularly for calculations in which some form of perturbed situation is studied. The inability of thes model dynamics to respond to changes in the temperature fields may seriously limit the validity of their results.

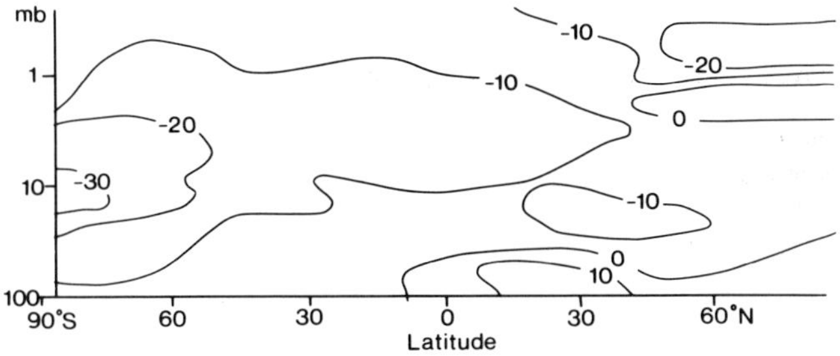

Fig.6. Percentage increase in Run A2 relative to control run.

5. RESULTS OF RUN B

The difference in temperature between Run B and the control run is shown in Fig.7. Corresponding to the larger heating rate changes at higher altitudes there are decreases of 6-8K near the stratopause but differences of less than 1K throughout most of the stratosphere with small regions of warming near 5mb corresponding to the structure of the heating rate changes (see Table 1). The difference in O_3 concentration between Run B and the control run are shown in Fig.8. Except near the stratopause, where the temperature decreases have resulted in O_3 increases of around 10%, values are small. The change in temperature structure has slightly strengthened the meridional circulation with a resultant increase in transport of O_3 from low to high latitudes but changes in the lower stratosphere are of order 1% only. The resultant increase in global total O_3 is 0.7%.

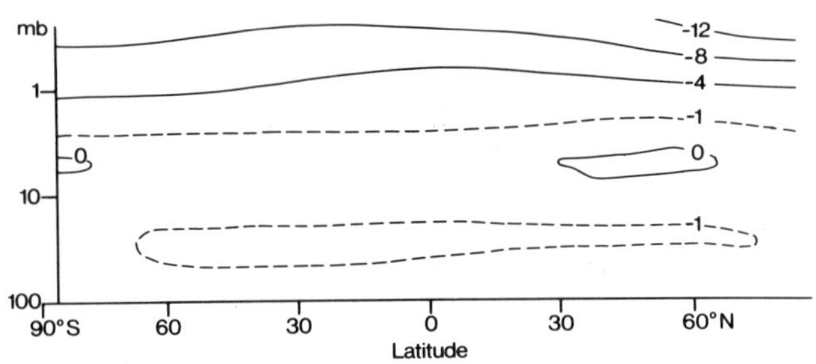

Fig. 7. Difference in temperature between Run B and control run.

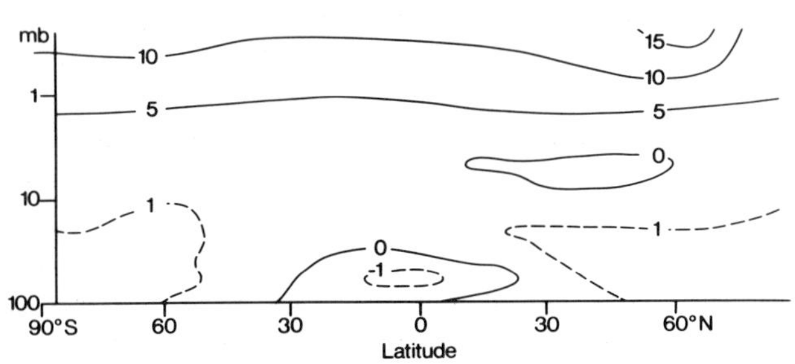

Fig. 8. Percentage increase in O_3 in Run B relative to control run.

6. CONCLUSIONS

Between 15 and 40km errors of less than 1K in temperature and 2% in ozone concentration are incurred as a result of using the Ramanathan (1976) wide-band parameterisation for absorption in the CO_2 15μm band. Above 40km larger errors are likely but the accurate incorporation of Doppler line-broadening effects, such as included in recent versions of the 2-D model used in this work, will considerably ameliorate these. Systematic errors of more than of the order of $0.1Kd^{-1}$ in heating rates will lead to serious errors in both temperature and ozone fields.

Chemical or radiative perturbations to the stratosphere will result in changes in net heating rate and the results of predictive models in which the dynamics are prohibited from adjusting to these changes must be assessed with caution.

ACKNOWLEDGEMENTS

The assistance of Dr.R.Brugge with the computing and of Mrs.E.Modin with the diagrams is gratefully acknowledged.

REFERENCES

Crisp D., Fels S.B. and Schwarzkopf M.D., 1986: Approximate methods for finding CO_2 15μm band transmission in planetary atmospheres. *J. Geophys. Res.* ,**91**, 11851-11866.

Chipperfield M.P., Eckman R.S., Haigh J.D., Gray L.J. and Pyle J.A.,1988: On the two-dimensional modelling of the stratosphere. *This volume.*

Fels S.B., 1979: Simple strategies for inclusion of Voigt effects in infrared cooling rate calculations. *Appl. Opt.*,**18**, 2634-2637.

Fels S.B., Mahlman J.D., Schwarzkopf M.D. and Sinclair R.W. , 1980: Stratospheric sensitivity to perturbations in ozone and carbon dioxide: radiative and dynamical response. *J. Atmos. Sci.,* **37**, 2265-2297.

Green A.E.S., Lindenmeyer C.S. and Griggs M., 1964: Molecular absorption in planetary atmospheres. *J. Geophys. Res.*, **69**, 493-504.

Haigh J.D., 1984 : Radiative heating in the lower stratosphere and the distribution of ozone in a two-dimensional model. *Quart. J. Roy. Meteorol. Soc.,* **110**, 167-185.

Haigh J.D. and Pyle J.A., 1982: Ozone perturbation experiments in a two-dimensional circulation model. *Quart. J. Roy. Meteorol. Soc.,* **108**, 551-574.

Keating G.M., Brasseur G.P., Nicholson J.Y.III and DeRudder A., 1985: Detection of the response of ozone in the middle atmosphere to short-term solar ultraviolet variations. *Geophys. Res. Lett.* ,**12**, 449-452.

Kiehl J.T. and Ramanathan V., 1983: CO_2 radiative parameterization used in climate models: comparison with narrow band models and with laboratory data. *J. Geophys. Res.* , **88**, 5191-5202.

Kiehl J.T. and Solomon S., 1986: On the radiative balance of the stratosphere. *J. Atmos. Sci.* ,**43**, 1525-1534.

Ramanathan V., 1976: Radiative transfer within the earth's troposphere and stratosphere : A simplified radiative-convective model. *J. Atmos. Sci.* ,**33**, 1330-1346.

Rodgers C.D., 1968: Some extensions and applications of the new random model for molecular band transmissions. *Quart. J. Roy. Meteorol. Soc.* ,**94**, 99-102.

Rodgers C.D. and Walshaw C.D., 1966: The computation of infrared cooling rates in planetary atmospheres. *Quart. J. Roy. Meteorol. Soc.*, **92**, 43-54.

Rodgers C.D. and Williams A.P., 1974 : Integrated absorption of a spectral line with the Voigt profile. *J. Quant. Spectrosc. Radiat. Trans.*, **29**, 279-281.

GLOBAL BUDGETS OF ODD NITROGEN AND CHLORINE SPECIES

Frode Stordal
Institute of Geophysics
University of Oslo
Blindern, 0315 Oslo 3, Norway

ABSTRACT

The global budgets of odd nitrogen and chlorine have been investigated by using a truncated 2-D model of the troposphere and the stratosphere. The total amount of inorganic chlorine, Cl_y, in the stratosphere - produced from $CFCl_3$ and CF_2Cl_2 in this model - has a mixing ratio in the upper stratosphere which equals the mixing ratio of the chlorine contained in the source gases in the troposphere. The Cl_y content of the stratosphere therefore depends on parameters that determine the turnover of the source gases in the atmosphere. It is shown to be more sensitive to changes in stratospheric transport than to variations in the photochemical loss rates of the source gases. The budget of odd nitrogen, NO_y, in the stratosphere is more complex since only a fraction of the source gas N_2O ends up as NO_y in the stratosphere. Besides, there are other sources, such as lightning in the troposphere, as well as a sink in the upper stratosphere. The sensitivity of stratospheric NO_y to variations in the strength of the various sources and sinks is studied. Variations in the stratospheric tracer transport parameters show that not only transport of NO_y, but also transport of N_2O, plays an important role for the stratospheric NO_y budget.

1. INTRODUCTION

Odd nitrogen and chlorine are known to influence the distribution as well as the trend of stratospheric ozone. Various measurements of odd nitrogen species show a considerable scatter, and there are still quantitative uncertainties in our understanding of the turnover of source gases for stratospheric chlorine (WMO, 1986). Various models also predict significantly different distributions of odd nitrogen and chlorine species in the stratosphere. More accurate mesurements as well as careful theoretical budget studies for e.g. these components are therefore needed.

In this work the sensitivity of the turnover and distribution of odd nitrogen (NO_y) and chlorine (Cl_y) to the variation in various physical and chemical model parameters have been studied. The purpose of these sensitivity tests are threefold. The tests are useful firstly when model results are compared with observations, and secondly in model intercomparisons. Thirdly they can be used to evaluate changes in e.g. odd nitrogen and chlorine in perturbed atmospheres, which have undergone physical or chemical changes.

2. MODEL EXPERIMENTS

The present work is based on experiments performed with the highly truncated 2-D model of Holton (1986a), in the way it was implemented by Stordal and Garcia (1987). This model, which is often denoted as a 1 1/2-D model, is inexpensive compared to a full 2-D model, and allows a large number of sensitivity tests to be performed without being computer costly.

The governing equation for the mixing ratio of a trace species in this model is

$$\frac{\partial X_0}{\partial t} - \frac{1}{\rho_0}\frac{\partial}{\partial z}\left(\rho_0 K \frac{\partial X_0}{\partial z}\right) = -\frac{1}{\rho_0}\frac{\partial}{\partial z}(\rho_0 \gamma) + S_0 \qquad (1)$$

where

$$K = W_2^2 \langle P_2^2 \rangle \frac{1}{\left(\tau_d^{-1} + l_0 + \epsilon l_2\right)} \qquad (2)$$

Here $\langle\rangle$ denote global integration at a certain altitude level, $P_2 = \frac{1}{2}(3\mu^2 - 1)$ is the second-degree Legendre polynomial, μ is the sine of the latitude, X_0, S_0 and l_0 are the global mean at a certain altitude of the mixing ratio, the net chemical source term and chemical loss rate respectively, $\rho_0(z) = \rho_{00} * \; exp(-z/H)$, ρ_{00} is the density at a reference level, z is the log-pressure altitude, H is a standard atmospheric scale height, W_2 is the projection of the vertical residual meridional velocity onto the P_2 mode, $\tau_d = a^2/D$ is the dynamic timescale for meridional diffusion, D is the diffusion coefficient used to parameterize the horizontal flux divergence of transient, dissipative eddies and a is the radius of the earth.

Essentially the model takes the form of a one-dimensional model, with individual vertical diffusion coefficients K for each component, depending on its photochemical loss rate. Equation (1) includes an advection term, which appears after taking latitudinal variations in the chemical

ISBN 0-937194-15-8

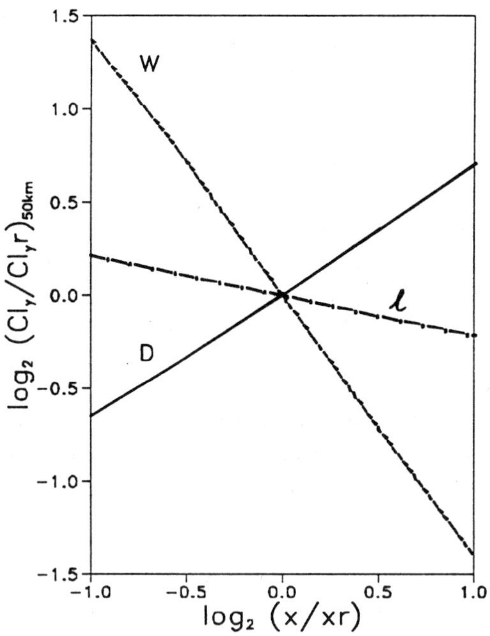

Fig. 1: Changes in Cl_y at 50 km as a function of changes in W_2, D and the loss rate (l) for the source gases. x/x_r is the ratio of the values of these parameters to their values in the reference case.

terms into account. The advective flux γ is defined and explained in Stordal and Garcia (1987). The model can also solve for the projection onto P_2 of the tracer mixing ratio, allowing latitudinal variations to be computed. An important constraint of the model is that, due to the high truncation excluding asymmetric components, it cannot describe any interhemispheric differences, and is therefore applicable only to yearly averaged conditions.

In the present experiments the model has been run in the base case adopting the same parameter values as in Stordal and Garcia (1987). They demonstated that the response of this model to e.g. variations in vertical residual meridional winds and horizontal diffusion coefficients is similar to that of a full two-dimensional model. Sensitivity experiments have been performed where the dynamical parameters, W_2 and D, as well as chemical source and sink terms have been multiplied by 2^q, where q=-1.0, -0.5, 0.0, 0.5, and 1.0.

Photochemical calculations have not been performed in this study. As in Stordal and Garcia (1987) the photochemical terms are taken from experiments with the 2-D model of Stordal et al. (1985).

3. THE CHLORINE BUDGET

The chlorine species described in the following model experiments are the source gases $CFCl_3$ and CF_2Cl_2 as well as Cl_y, which is the total amount of chlorine produced from these source gases. Figure 1 shows that Cl_y at 50 km, a quantity which is of large importance for CFC induced ozone reductions, is very sensitive to changes in the transport parameters W_2 and D, whereas the changes in the chemical loss rates of the CFCs are of less importance. The results in Figure 1 are obtained when a constant flux has been used as a lower boundary condition for the CFCs. As seen from Table 1, variations in Cl_y in the upper stratosphere is a mirror of the variations of the tropospheric mixing ratios of the CFCs. This is because the two CFCs chosen here are reduced to extremely low mixing ratios in the upper stratosphere, and are therefore totally decomposed to Cl_y within the stratosphere. The behaviour of Cl_y in the upper stratosphere can therefore be understood by analyzing the variations in the source gases.

Component	W_2	D	l
$CFCl_3$, surface	2.69^{-1}	1.64	1.15^{-1}
CF_2Cl_2, surface	2.54^{-1}	1.62	1.16^{-1}
Cl_y, 50km global average	2.64^{-1}	1.63	1.16^{-1}
Cl_y, 25km equator	5.15^{-1}	2.06	1.01

Table 1: Changes in the mixing ratios of Cl_y and the source gases when W_2 ,D and l (eq.1 and 2) are increased by a factor 2. The experiments reported are performed with variations in W_2 ,D and l individually. The changes in the mixing ratios of the chlorine species are given as the ratios X/Xr, where X and Xr denote the mixing ratios in the perturbed and the reference case respectively.

Globally averaged the amount of source gases is governed by the equation

$$\{\rho\mu l\}\tau = \{\rho\mu\} \quad (3)$$

where {} denote global volume integretion, and τ is the residence time of the source gas (see e.g. Stordal et al., 1985). An increase in W_2 (D) results in faster (slower) transport of the source gases from the troposphere to the stratosphere where they are decomposed. The residence time is thereby decreased (increased). Since the experiments are performed with constant flux input for the CFCs, the integral on the left hand side is kept constant (it equals the the surface flux), the mixing ratio of the CFC will change in the troposphere (where the bulk of the CFCs reside) to compensate for the change in the residence time. The effect of changing W_2 is larger than in the case of D, as would be expected due to the quadratic dependence in K from W_2 (see equation 2).

The content of CFCs in the troposphere, and therefore also Cl_y in the upper stratosphere is less sensitive to variations in the photochemical loss rate of the CFCs (see Figure 1). The residence time of a CFC does not respond linearly to a change in the loss rate. This can also bee seen from the fact that the dissociation rate of $CFCl_3$ is 6-7 times higher in the whole stratosphere than the one of CF_2Cl_2, while the ratio between their residence times is less than 2. Because the vertical gradient of the loss rate l is very steep, an increase in l by a factor 2 only implies that the decomposition of the CFCs takes place at a slightly less altitude. The residence time is therefore only weakly decreased, leading to a slight decrease in the upper stratospheric Cl_y.

The Cl_y resulting from releases of $CFCl_3$ and CF_2Cl_2 will exhibit constant mixing ratios with altitude in the upper stratosphere, above the level where the source gases vanish. In the middle and lower stratosphere, the Cl_y mixing ratios will decrease towards lower altitudes. These altitude gradients will also give rise to latitudinal gradients, through the effect of the meridional circulation (see e.g. Holton,1986b). The changes in Cl_y at these altitudes, which are of major importance for the perturbations in the total ozone column, are therefore expected to be influenced by other factors than the tropospheric mixing ratios of the source gases. The Cl_y mixing ratio at the 25 km altitude at the equator is chosen to demonstrate this effect. As seen from Table 1, Cl_y at 25 km at equator is more sensitive to variations in W_2 and D than Cl_y in the upper stratosphere. This is the case since an increase in W_2 (D), in addition to lowering (increasing) the suface mixing ratio of the CFCs, will increase (decrease) the equatorial mixing ratios of the CFCs in the lower stratosphere, leaving less chlorine available for Cl_y in this region.

4. ODD NITROGEN

The nitrogen studies reported here involve N_2O and the total amount of odd nitrogen, NO_y. The photochemical source of NO_y (from the reaction $N_2O+O(^1D)$) is allowed to vary as N_2O varies. The amount of $O(^1D)$ is however kept constant in the calculations. Another source for NO_y is lightning in the troposphere. In the present calculations 4MT of NO_y per year has been used as a source strength. NO_y has been released between 4 and 15 km uniformly at all latitudes (for a discussion of the strength and distribution of the source of NO_y from lightning, see Ko et al., 1986). NO_y is removed by wet deposition in the troposphere as well as by photodissociation of NO in the upper stratosphere.

Sensitivity studies have been performed by varying all the above mentioned sources and sinks of NO_y, as well as the two dynamical parameters, in the same manner as described in the previous section for chlorine. Table 2 summarizes the sensitivity of NO_y at two altitudes (20 and 36 km, global avarages) to the various parameters. The numbers give the ratio of the relative change in NO_y to the relative change in each parameter, a number denoted here as the sensitivity factor s, i.e.

$$s = \frac{\delta NO_y}{NO_y} / \frac{\delta p}{p} \qquad (4)$$

where p is the value of one of the parameters. The sensitivity factors given in Table 2 are based on experiments where each parameter has been changed individually. In each parameter study the model has been run with p given the standard value as well as a factor $2^{0.5}$ higher. In case of a linear relationship between NO_y and p, the sensitivity factor will be 1 (-1 in case of an inverse relationship), while a quadradic relationship would yield the factor 2.

For the chemical terms, the results reflect the influence of each process at the different altitudes. Near the altitude of maximum production of NO_y from N_2O, at 36 km, the relation between NO_y and the production rate is nearly linear, since the sensitivity factor is 0.92. At 20 km the factor is reduced to 0.77. Even at 20 km NO_y is more sensitive to changes in the production from N_2O (s=0.77) than the production from lightning (s=0.13).

The two loss processes, dissociation of NO and rainout, will influence the stratospheric NO_y in different ways. The former process will have the main influence in the upper stratosphere (s is reduced in magnitude from -0.27 to -0.16 from 36 to 20 km). The latter process will have the largest influence in the lower stratosphere, but the sensitivity factor is only -0.06 at 20 km. This only implies that the transport of NO_y from the stratosphere to the troposphere is the limiting process. Once NO_y is transported into the troposphere, the deposition is, within the limits of the present variations, fast enough to remove NO_y.

Changes in the dynamical parameters W_2 and D will influence the stratospheric NO_y through impact on the transport of NO_y as well as the transport of N_2O. The direct transport effect on NO_y is studied in experiments where N_2O has been kept unaffected by variations in W_2 and D. A higher W_2 (D) will lead to faster (slower) transport of NO_y from the source region (the middle stratosphere) to the sink regions (the upper stratosphere and the troposphere), leading to decrease (increase) in the NO_y mixing ratios in the entire stratosphere. As in the chlorine case, the impact of changes in W_2 is stronger because K depends quadratically on W_2 and only linearly on D.

When also transport effects on N_2O is taken into account, this picture is modified. Firstly we will consider a case where transport changes will be allowed to modify the distribution of N_2O except at the ground (fixed mixing ratio boundary condition for N_2O). An increase in W_2 (D) will then increase (decrease) the transport of N_2O from the troposphere to the stratosphere, giving higher (lower) concentrations of N_2O in the stratosphere. This will enhance (lower) the production of NO_y. This effect will therefore counteract the direct transport effect on NO_y. At the 36 km level it dominates the direct effect, while at 20km the total effect is still a weak decrease (increase) in NO_y due to increased W_2 (D).

Secondly, we allow N_2O to change also at the surface (fixed flux condition). As in the chlorine case, an increase in W_2 (D) will then shorten (lengthen) the residence time of N_2O, yielding lower mixing ratios in the troposphere as well as in the stratosphere compared to the case with a fixed surface mixing ratio condition. Like the direct transport effect on NO_y, this will decrease (increase) the

Parameter	s_{20km}	s_{36km}
Rate of production from N_2O	0.77	0.92
Rate of production from lightning	0.13	0.14
Rate of loss by dissociation of NO	-0.16	-0.27
Rate of loss by tropospheric deposition	-0.06	-0.02
W_2	[a]-0.83	[a]-0.81
W_2	[b]-0.27	[b]0.08
W_2	[c]-0.87	[c]-0.76
D	[a]0.58	[a]0.46
D	[b]0.15	[b]-0.14
D	[c]0.64	[c]0.38

[a] N_2O constant mixing ratios in the entire domain
[b] N_2O constant mixing ratios at the lower boundary
[c] N_2O constant flux at the lower boundary

Table 2: Sensitivity factors s (equation 4), describing the sensitivity of NO_y mixing ratios at 20 and 36 km (globally averaged) to various parameters.

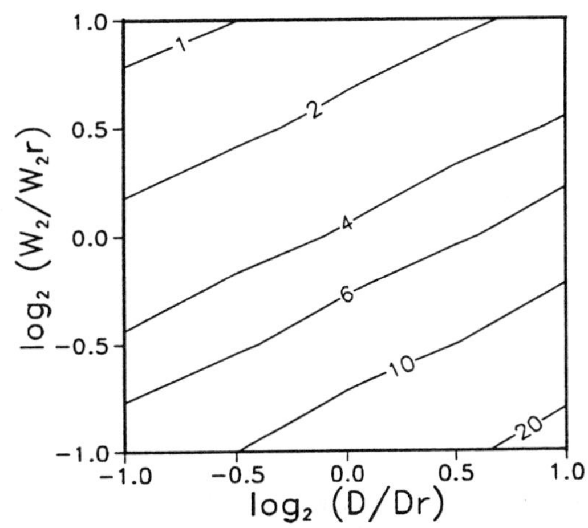

Fig. 2: Residence time for NO_y at 20 km (years) in experiments with different values for W_2 an D. W_2r and Dr denote the values of the parameters in the reference case

content of NO_y as W_2 (D) increases. In fact, the impact on NO_y from variations in N_2O now nearly vanishes. The total impact on NO_y is nearly the same as in the case when N_2O was kept constant.

In the experiments where transport parameters have been varied, also the decay time for NO_y has been studied. Figure 2 gives the decay time at 20 km altitude (for globally averaged NO_y) as a function af changes in W_2 and D. In these experiments the model has been run for 10 years without production terms for NO_y. The decay time, defined as the time needed to reduce the NO_y concentrations to 1/e of the initial values, have been estimated based on the changes during these 10 years. The standard set of transport values give a decay time of about 4 years at 20 km. An increase in W_2 will give a shorter decay time, since the reservoir of stratospheric nitrogen is more efficiently transported to the loss regions. Increased horizontal mixing (D) will on the other hand lengthen the response time. In this case NO_y transported downward at high latitudes will mix equatorwards more efficiently. NO_y will then enter low latitudes where it is transported the long way through the Brewer-Dobson cell, thereby lowering the efficiency of the decay of NO_y in the troposphere. Figure 2 shows that, due to uncertainties in our quantitative description of the transport in the stratosphere, considerable uncertainty exists for a resulting decay time for NO_y in the lower stratosphere. For example, as a response to variations in an upper stratospheric source one could expect a variation in lower stratospheric NO_y which is delayed by a time which could be from only about one year up to the length of one solar cycle.

5. ACKNOWLEDGEMENTS

This research was supported by the Norwegian Council for Science and the Humanities (NAVF) and the Norwegian Department of Environment (MD).

6. REFERENCES

Holton, J.R., A dynamically based transport parameterization for one-dimensional photochemical models of the stratosphere, *J.Geophys.Res.*, 91, 2681-2686, 1986a.

Holton, J.R., Meridional distribution of stratospheric trace gas constituents, *J.Atmos.Sci.*, 43, 1238-1242, 1986b.

Ko, M.K.W., McElroy, M.B., Weisenstein, D.K. and Sze, N.D., Lightning : A possible source of stratospheric odd nitrogen, *J.Geophys.Res.*, 91, 5395-5404, 1986.

Stordal, F. and Garcia, R.R., Sensitivity studies and a simple ozone perturbation experiment with a truncated two- dimensional model of the stratosphere, *J.Geophys. Res.*, 92, 11909-11918, 1987.

Stordal, F., Isaksen, I.S.A. and Horntveth, K., A diabatic circulation two-dimensional model with photochemistry: Simulations of ozone and ground-released tracers, *J.Geophys.Res.*, 90, 5757-5776, 1985.

WMO, Atmospheric ozone 1985 : Assessment of our understanding of the processes controlling its present distribution and change, World Meteorological Organization, *WMORep.No.*16, 1986.

SOME CALCULATIONS OF OZONE CONTENT IN THE STRATOSPHERE

Geng-Xiong Chen and Kang-Kun Tschu
Institute of Geophysics
Academia Sinica
Beijing, PRC

ABSTRACT

This paper consists of the following 3 parts: (1) Atmospheric photochemistry chain relating to stratospheric ozone, (2) Turbulent diffusion equation and one dimensional theoretical models of stratospheric minor constituents, (3) Effects on production and depletion of stratospheric ozone, e.g. due to atmospheric nuclear explosion. Photochemical reactions of the relevant constituents in the oxygen-hydrogen-nitrogen atmosphere are listed and computational results are depicted in figures and briefly discussed.

1. STRATOSPHERIC CHEMISTRY OF OZONE

The most important trace constituent of the stratosphere is ozone. Although present only in a few ppm, ozone, nevertheless, is responsible for shielding the earth from ultraviolet (UV) radiation that is harmful to life. The amount of ozone in the stratosphere is maintained as the result of a dynamic balance between formation and destruction processes. According to the original Chapman (1930) scheme the formation and destruction of stratospheric O_3 is as follows
(a) Chapman or "O_2 only" scheme

$$\begin{aligned} &O_2 + h\nu \ (UV) \rightarrow O + O \\ &O + O_2 + M \rightarrow O_3 + M \\ &O_3 + h\nu \ (UV \text{ or Visible}) \rightarrow O_2 + O \\ &O_3 + O \rightarrow O_2 + O_2 \end{aligned} \qquad (1)$$

The equations above account for the production of all the atmospheric O_3, but for the removal of only about 20% of the O_3. About 10% of the destruction is caused by catalytic cycles involving hydrogen-containing species: free hydrogen atoms (H), hydroxyl (OH), and hydroperoxyl (HO_2), which can achieve the same effect as reaction (1) without being themselves removed. For example, we have
(b) HO_x cycle (M. Nicolet, 1975)

$$\begin{aligned} &O + HO_2 \rightarrow OH + O_2 \\ &\underline{OH + O_3 \rightarrow HO_2 + O_2} \\ \text{Net: } &O + O_3 \rightarrow O_2 + O_2 \end{aligned} \qquad (2)$$

These hydrogen-containing species are produced by the reaction of naturally occurring water vapor and methane with excited oxygen atoms, $O(^1D)$, from O_3 photolysis

$$\begin{aligned} &O(^1D) + H_2O \rightarrow 2OH \\ &O(^1D) + CH_4 \rightarrow CH_3 + OH \end{aligned}$$

Similarly, we may have the following (c), (d), (e), and (f) cycles, among which the process involving NO and NO_2 provides the most important destruction process (nearly 70%) for ozone.
(c) NO_x cycle 1 (P.J. Crutzen, 1971)

$$\begin{aligned} &O + NO_2 \rightarrow NO + O_2 \\ &\underline{NO + O_3 \rightarrow NO_2 + O_2} \\ \text{Net: } &O + O_3 \rightarrow O_2 + O_2 \end{aligned} \qquad (3)$$

(d) NO_x cycle 2

$$\begin{aligned} &NO + O_3 \rightarrow NO_2 + O_2 \\ &NO_2 + O_3 \rightarrow NO_3 + O_2 \\ &\underline{NO_3 + h\nu \ (\text{Visibl}) \rightarrow NO + O_2} \\ \text{Net: } &2O_3 + h\nu \rightarrow 3O_2 \end{aligned} \qquad (4)$$

(e) Cl_x cycle (H.S. Johnston, 1975)

$$\begin{aligned} &O + Cl\,O \rightarrow Cl + O_2 \\ &\underline{Cl + O_3 \rightarrow ClO + O_2} \\ \text{Net: } &O + O_3 \rightarrow O_2 + O_2 \end{aligned} \qquad (5)$$

(f) Bromine gas (R.P.Turco, 1985)

$$\begin{aligned} &2(Br + O_3 \rightarrow Bro + O_2) \\ &\underline{Bro + Bro \rightarrow Br + Br + O_2} \\ \text{Net: } &2O_3 \rightarrow 3O_2 \,. \end{aligned} \qquad (6)$$

2. ONE DIMENSIONAL THEORETICAL MODELS OF STATOSPHERIC MINOR CONSTITUTES

2.1 RELEVANT FORMULAS

In this paper, relevant data such as chemical reactions, thermal constants and photodissociation rates are taken from WMO report (1982), H.S. Johnston et. al. (1978) and P.J. Crutzen (1971). Let [A] be the concentration of minor constituent A, it can be derived to the following 3D general equation

$$\frac{d[A]}{dt} = \frac{\partial}{\partial x}\left(K_x [M] \frac{\partial \mu_A}{\partial x}\right) + \frac{\partial}{\partial y}\left(K_y [M] \frac{\partial \mu_A}{\partial y}\right) + \frac{\partial}{\partial z}\left(K_z [M] \frac{\partial \mu_A}{\partial z}\right) + P(A,\vec{x},t) - L(A,\vec{x},t) \qquad (7)$$

where $P(A,\vec{x},t)$ and $L(A,\vec{x},t)$ are production and loss velocities of minor constituent A, $K=K_x\vec{i}+K_y\vec{j}+K_z\vec{k}$ is the coefficient of turbulent transfer, $\mu_A=[A]/[M]$ is the mixing ratio of A, relative to the density of the unit volume of air. Here we consider the one-dimensional case only, that is, assuming the minor constituent to be horizontally uniform and neglecting the

ISBN 0-937194-15-8

vertical transport term due to the small vertical velocity. Equation (7) can be written as

$$\frac{\partial [A]}{\partial t} = \langle \mu_A \rangle + P(A,z,t) - L(A,z,t) \quad (8)$$

where $\langle \mu_A \rangle = \frac{\partial}{\partial z}\left(K_z [M] \frac{\partial \mu_A}{\partial z}\right)$.

Furthemore, the daily mean values are used for all the quantities, we have

$$\langle \mu_A \rangle + P(A,z) - L(A,z) = 0 \quad (9)$$

We now define two time constants. One is the time constant for vertical turbulent diffusion T_{ED}

$$T_{ED} = H^2/K_z \quad (10)$$

where H is the scale height of the atmosphere. In the stratosphere, K_z is of the order 10^3--10^5 $cm.s^{-1}$, and $H \approx 8km$, therefore T_{ED} is of the order of 10^6--10^8 s. The other time constant is the one for photochemical reaction $\tau(A,z)$

$$\tau(A,z) = [A]/L(A,z) \quad (11)$$

When $T_{ED} \gg \tau$, the controlling equation will be

$$P(A,z) - L(A,z) = 0 \quad (12)$$

When $T_{ED} \ll \tau$, turbulent diffusion should be included, i.e. the controlling equation (8) above must be used.

2.2 VERTICAL DISTRIBUTION BETWEEN 20KM AND 50KM

In the stratosphere among trace constituents $O(^1D)$. O. NO. NO_2. NO_3. H. OH. HO_2 H_2O_2. HNO_3. N_2O the time constant of the former 9 elements is far smaller than the respective T_{ED}, whereas for the last two the reverse is the case, i.e. $T_{ED} \ll \tau$. Using profile data from U.S. Standard Atmosphere (1976) and Hunten's K_z value, we have made numerical calculations for different constituents. All results are depicted as shown in Fig. 1-5.

Fig.1 shows the vertical distribution of O, $O(^1D)$, and O_3. In view of the prodominance of O_3 in the stratosphere it is quite resonable to use $O_x = O(^1D) + O + O_3$ approximating the O_3 content. Thus we have $d[O_x]/dt = d[O_3]/dt + d[O]/dt + d[O(^1D)]/dt \approx d[O_3]/dt$; in essence, $d[O(^1D)]/dt \approx 0$ and $d[O]/dt \approx 0$.

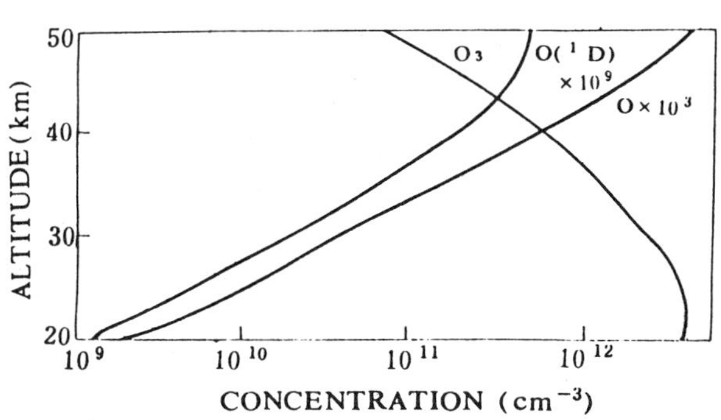

Fig.1. Vertical distribution of O, $O(^1D)$, O_3.

The vertical distributions of N_2O and NO_x are shown in Fig.2, where $NO_x = (N + NO + NO_2 + NO_3 + HNO_3)$, so grouped for computational convenience and the turbulent transfer term is also included. The major natural source of NO_x in the stratosphere is provided by the oxidation of nitrous oxide (N_2O), which is produced by bacteria in soil and water. We note that $O(^1D) + N_2O \rightarrow 2NO$, so $P(NO_x) = 2k_{40}[O(^1D)][N_2O]$. Fig.2 shows that the NO_x-curve is increasing with height, whereas the N_2O curve is decreasing rapidly with height.

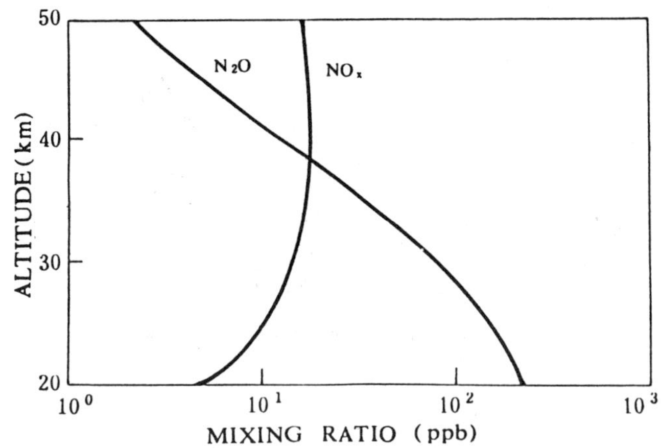

Fig.2. Vertical distribution of N_2O and NO_x.

Fig.3 and Fig.4 illustrate respectively the vertical distributions of nitrogen-containing compounds and hydrogen- containing compounds. At lower altitudes of the stratosphere concentrations of NO, NO_2, and HNO_3 are of the same order, but at higher altitudes the prodominance of NO is clearly shown in Fig.3. As regards hydrogen-containing compounds Fig.4 reveals the interesting fact that below 40km, H_2O_2 is the most abundant species in this family while OH is the least. But toward 40km and higher altitudes, the reverse is the case, i.e. the abundance of OH and the rapid decay of H_2O_2.

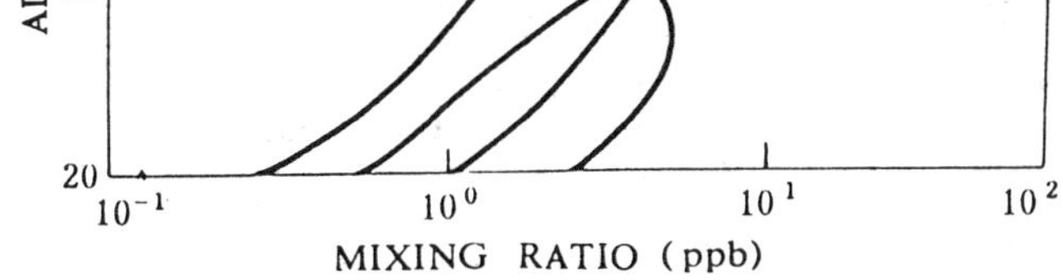

Fig.3. Vertical distribution of nitrogen-containing compounds.

Fig.5 shows photochemical reaction time constants as a function of height for different species, calculated according to the formula (11). It can be seen that lifetimes of NO_x, N_2O, HNO_3 below 30km are of same order or greater than the time constant for vertical turbulent diffusivity T_{ED} (eq.10), thus turbulent diffusion term should be included in computation (eq.9). Also, N_2O due to its longer life may be taken as a major source species together with N_2 and O_2 in the stratosphere.

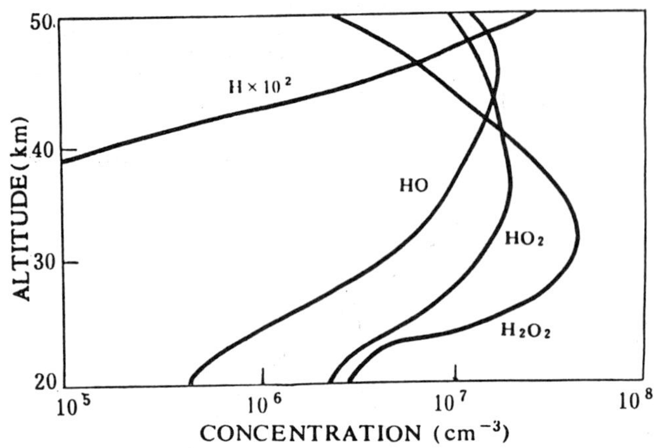

Fig.4. Vertical distribution of hydrogen-containing compounds.

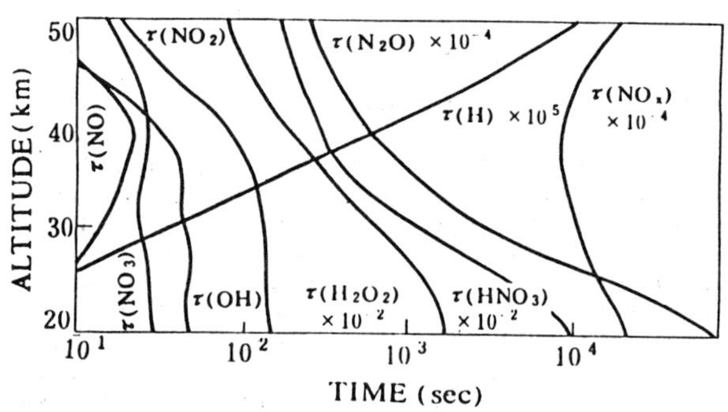

Fig.5. Photochemical reaction time constants as a function of height for different species.

3. EFFECTS ON PRODUCTION AND DEPLETION OF STRATOSPHERIC OZONE

R.P. Turco (1985) in a detailed review has summarized stratospheric ozone perturbations due to natural and man-made agents. Natural physical events affecting ozone include solar variations (11-year cycle and daily variation), solar proton events, aurora, meteor,supernova and volcanoes, among which volcanic injections of H_2O and HCl can reduce ozone chemically and large injections of sulfur gases (SO_2 and H_2S) can suppress the HO_x cycle. For anthropogenic perturbations of stratospheric O_3 it may be divided into 2 subgroups: ozone depletion and ozone enhancement. In the case of ozone depletion the perturbations agents are SST-NO_x, subsonic aircraft, CFMS (F-11, and F-12), CH_xCl_y, N_2O doubling and nuclear explosions. The ozone enhancement may arise due to CO_2 doubling and CH_4 doubling of the atmosphere.

H.S. Johnston et.al. (1978) demostrated a general differential equation for ozone may be transferred to an equivalent equation that dirctly identifies the catalytic cycles and sequences. This method of treating stratospheric photochemistry properly partitions ozone destruction rates among O_x, NO_x and HO_x families of reactions. Fig.6 is one example of our calculated results, showing the reaction rates in the formation and distribution processes of ozone as a function of height. Here, we have

$$S(O_3) = P(O_3) - \{L_O(O_3) + L_H(O_3) + L_N(O_3)\} \quad (14)$$

where

$$P(O_3) = 2J_3[O_2] \quad (14a)$$

$$L_O(O_3) = 2k_2[O][O_3] + 2k_{10}[O(^1D)][O_3] \quad (14b)$$

$$L_H(O_3) = 2k_{12}[O][HO_2] + 2k_{14}[O_3][H] + k_{18}[O_3][HO_2] + 2k_{19}[OH][HO_2] + 2k_{29}[HO_2][HO_2] \quad (14c)$$

$$L_N(O_3) = 2k_{32}[O][NO_2] + 2k_{53}[O][NO] + 2k_{39}[O(^1D)][N_2O] + 2k_{40}[O(^1D)][N_2O] + 2J_{50}[NO_3] \quad (14d)$$

This figure shows the production of ozone and the loss term respectivity due to O_x, HO_x, and NO_x species. Easily seen that below 40km the nitrogen-containing species is most important for destruction of ozone, nearly above 40% of all. Between 40-45km heights all three loss terms are of the same order. Above 45km the loss term due to NO_x becomes rapidly decreasing, being 11% of all, whereas the loss term due to HO_x becomes more important.

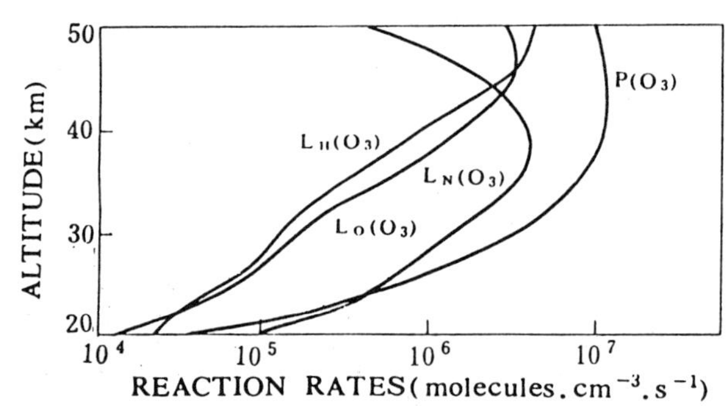

Fig.6. Reaction rates in the format destruction processes of ozone function of height.

In the following part we will make a simple approach to explain the depletion of ozone when the nitric oxide is injected into the atmosphere. Suppose the concentration of ambient ozone is in chemical equilibrium. From equation (14d) we know that the first term in the right side is the main term, being above 90% in the reactions. Neglecting all small terms due to $L_O(O_3)$, $L_H(O_3)$, $L_N(O_3)$ and keeping the first term of (14d) only. We may arrive getting the variations of ozone due to the injection of nitric oxide (symbolized by super-script M, while N indicating nature-existed) as follows

$$d[O_3]^M/dt = -2k_{32}[O]^N[NO_2]^M = -2k_{32}[O]^N\left(\frac{k_{42}[O_3]^N[NO]^M}{J_{48} + k_{32}[O]^N} + \frac{k_{42}(NO]^N + [NO]^M)}{J_{48} + k_{32}[O]^N}[O_3]^M\right) \quad (15)$$

The initial condition is given by

$[O_3]^M=0$ when $t=0$.

Thus the solution of (15) will be

$$\frac{[O_3]^M}{[O_3]^N} = \frac{[NO]^M}{[NO]^M + [NO]^N} \times \left\{\exp\left(-\frac{2k_{32}k_{42}[O]^N([NO]^M + [NO]^N)}{J_{48}+k_{32}[O]^N}t\right)-1\right\} \quad (16)$$

It can be seen that so long as nitric oxide is injected by human activity, the stratospheric ozone will be decreased. Furthermore, as $t\rightarrow\infty$, the decrease of ozone will be

$$[O_3]^M = \frac{[O_3]^N[NO]^M}{[NO]^N + [NO]^M} \quad (17)$$

as depicted in Fig.7. Thus, the more nitric oxide is injected, the more ozone is depleted. When the amount of nitric oxide injections is 99 times higher than the background, only 0.01% of the original amount will remain in the upper atmosphere.

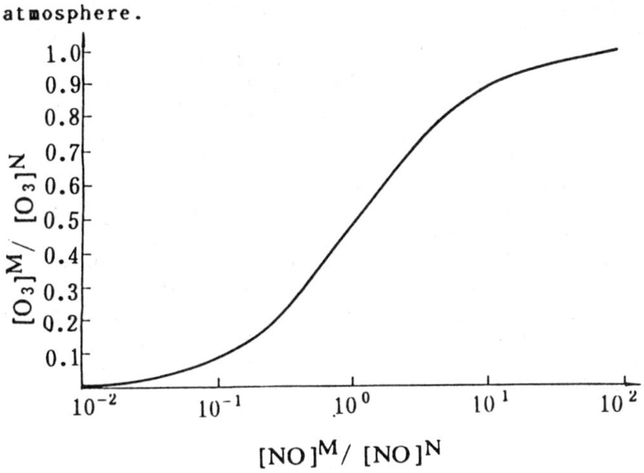

Fig.7. Decrease of $[O_3]^M/[O_3]^N$ when $t\rightarrow\infty$ as a function of $[NO]^M/[NO]^N$.

As well known to scientific community the nitric oxide can be produced by atmospheric nuclear explosions. Here take one example, suppose that 10MT explosion will produce so much nitric oxide molecules by the production rate of 10^{32} molecules/MT, and the distribution will be homogeneous in a cone of radius 2km between 20 to 30km. In the equation (16) let $[O_3]^M=0.99[O_3]^N$, so we solve for time t as

$$t = \frac{J_{48} + k_{32}[O]^N}{2k_{32}k_{42}[O]^N([NO]^M + [NO]^N)} \times \mathrm{Ln}\left\{1 - 0.99\frac{[NO]^M + [NO]^N}{[NO]^M}\right\} \quad (18)$$

Fig.8 is the calculation result according to this formula. It can be seen that the total ozone may be depleted in the fireball, before the injected nitric oxide due to atmospheric nuclear explosion is transported and diffused.

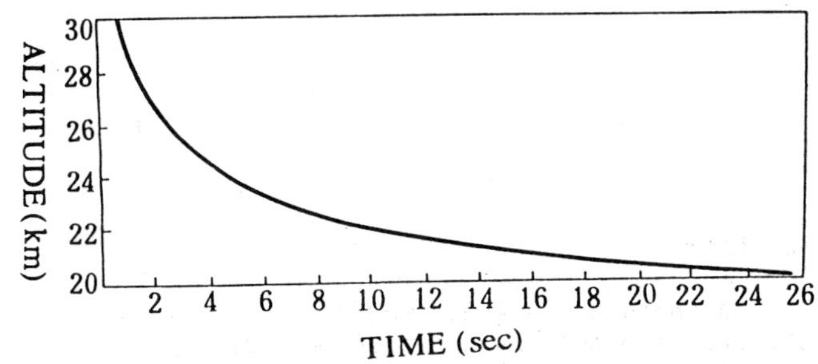

Fig.8. The time needed for $[O_3]^M=0.99[O_3]^N$ as a function of altitude.

REFERENCES

Chapman, S., 1930: A theory of upper atmospheric ozone. Mem. Roy. Meteorol. Soc., 3, 103.

Crutzen, P.J., 1971: Ozone Production rates in an oxygen-hydrogen-nitrogen oxide atmosphere. J. Geophy. Res., 76, 7311-7327.

Crutzen, P.J., and U. Schmailzl, 1983: Chemical budgets of the stratosphere. Planet. Space Sci., 31, 1009-1032.

Johnston, H.S., 1975: Global ozone balance in natural stratosphere. Rev. Geophys. Space Sci. 13, 637-649.

Johnston, H.S., 1976: Use of excess carbon 14 data to calibrate models of stratospheric ozone depletion by supersonic transports. J. Geophys. Res. 81, 368-380.

Johnston, H.S. and J. Podolske, 1978: Interpretations of stratospheric photochemistry. Rev. Geophys. Space Phy., 16, 491-519.

Nicolet, M., 1975: Stratospheric ozone: an introduction to its study. Rev. Geophys. Space Sci., 13, 593-636.

Turco, R.P., in Chapter 5 of the book: Ozone in the Free Atmosphere. ed. by R.C. Whitten and S.S. Prasad, (1985).

U.S. Standard Atmosphere 1976: Pull. NOAA-S/T-1562, U.S. Government Print Office, Washington. D.C..

WMO Report 1981: The stratosphere 1981 theory and measurements. report NO.11.

VARIABILITY OF CHLORINE CONTAINING SPECIES AS REVEALED BY THREE-DIMENSIONAL STRATOSPHERIC TRANSPORT AND CHEMISTRY MODELS

Jack A. Kaye and Richard B. Rood
NASA Goddard Space Flight Center
Greenbelt, Maryland 20771, USA

Dale J. Allen
Applied Research Corporation
Landover, Maryland 20783, USA

ABSTRACT

A three-dimensional model of the combined effects of chemistry and transport in the Earth's stratosphere is described together with its application to the period of the February 1979 major stratospheric warming. The model includes 12 transported constituents and/or families, including those containing odd oxygen, nitrogen, hydrogen, and chlorine, and is designed to simulate the time dependence of diurnally varying constituents and to correctly represent the behavior of constituents near the terminator. Results presented include the zonal mean concentrations of chlorine-containing species for conditions corresponding to local noon and local midnight, and latitude-longitude distributions of the chlorine-containing constituents at 10 mb. The way in which the non-zonal dynamics are reflected in the distributions of reservoir constituents is discussed.

1. INTRODUCTION

The role of reservoir species in stratospheric photochemistry is well-established. These species, whose photochemical lifetime is of the same magnitude as the time scales for transport, play a crucial role in tying up those elements (H, N, Cl, Br) which in other chemical forms are capable of catalytically destroying stratospheric ozone. They therefore serve to reduce the efficiency of these trace elements in ozone destruction.

Only recently has substantial information about the full three-dimensional distribution of these reservoir species become available, however. Data from the Limb Infrared Monitor of the Stratosphere (LIMS) instrument on the Nimbus 7 satellite provided three-dimensional distributions of both nitric acid (HNO_3) and nitrogen dioxide (NO_2) over most of the stratosphere for a seven month period from October 1978 to May 1979 Gille and Russell, 1984). HNO_3 is long-lived in much of the stratosphere and thus behaves essentially as a "tracer" of stratospheric motions. In daytime NO_2 is in photochemical equilibrium with NO and thus can be considered as a photochemically "active" species; at night NO_2 is a true reservoir species, being converted only slowly to N_2O_5.

There are no such detailed data available about the distribution of other reservoir species, including those containing chlorine ($ClONO_2$, HCl, and HOCl, primarily), as well as several others (N_2O_5, HNO_4, H_2O_2). Indeed, with the exception of data from the Atmospheric Trace Molecule Spectroscopy (ATMOS) experiment, which flew on the space shuttle as part of Spacelab 3 (Raper et al., 1987) there are no space-borne data on any of the chlorine-containing constituents. Until such data on these become available as a result of several instruments to be flown on the Upper Atmosphere Research Satellite (UARS) in 1991, models must be used to derive information about the three-dimensional distribution of these constituents. This paper includes the results of calculations we have performed to simulate the concentrations of stratospheric reservoir species.

2. THE MODEL

The "off-line" model used here includes both the effects of dynamics and chemistry on stratospheric composition. The winds used for the model are those obtained from the stratospheric data assimilation effort at the Goddard Space Flight Center (Baker et al., 1987). Its usefulness in producing wind fields suitable for stratospheric transport calculations has been described by Rood et al. (1988). Model extent, resolution, and transport are as described by Rood et al. (1988). The chemistry portion of the model, which includes 12 transported species (constituents and/or familes) represents an enhancement over that described by Kaye and Rood (1988), in which only 8 species were included ($O_x = O_3 + O + O(^1D)$, $NO_z = NO + NO_2 + NO_3$, HNO_3, N_2O_5, $Cl_z = Cl + ClO$, $ClONO_2$, HCl, HOCl). The new transported constituents in the present calculations are H_2O_2, HNO_4, H_2O, and CH_4 (the latter two were held fixed in the earlier calculations).

The current calculations are the first in which the "full chemistry" package described by Kaye and Rood (1988) is used together with the winds from the data assimilation project. They

ISBN 0-937194-15-8

should present a much more accurate picture of dynamically-induced variations in constituent concentrations than our former work (Rood et al., 1987; Kaye and Rood, 1988) due to the improved wind fields from the assimilation. We consider here the same time period simulated in our former work, beginning on 2/17/79, just prior to the onset of the major stratospheric warming of that winter.

3. RESULTS AND DISCUSSION

Zonal mean fields of the chlorine-containing constituents after six days of model simulation, corresponding to 2/23/79, are shown in Fig. 1 for both local noon and local midnight conditions. These were obtained by averaging the constituent mixing ratios at the model longitude closest to local noon and local midnight, respectively, every hour between 5.5 and 6.5 days of model integration. By retaining data in this way, we obtain fields which are similar to those which would be obtained from a polar-orbiting satellite in its ascending and descending nodes, respectively.

The diurnal variation of these constituents and its dependence on latitude and altitude are clear from the panels of this figure. Cl_z, which equals ClO at night and is almost entirely ClO during daytime because of the small concentration of Cl, is much larger during the day than at night at all but the highest altitudes, where the nighttime conversion of ClO to $ClONO_2$ is unimportant. $ClONO_2$, on the other hand, is more abundant at night than in the day, although the difference is very small at low altitudes where $ClONO_2$ photolysis is slow. HCl has essentially no diurnal variation, as befits a moelcule with a long lifetime, and HOCl has a small and altitude-dependent diurnal variation. In general, the mixing ratio of HOCl is smaller during the day in the lower and middle stratosphere, and has little diurnal variation above there. Above 50 km, nighttime HOCl tends to be slightly larger than daytime. All of these results are consistent with those seen in one-dimensional model calculations (Ko and Sze, 1984).

A synoptic plot of the latitude-longitude distribution of $ClONO_2$ (Fourier decomposed and reexpanded to include only six zonal wave numbers) at the model level closest to 10 mb after six days of model integration is shown in Fig. 2 (left panel). Both dynamical and photochemical influences on the distribution can be seen. Photochemical influences are clearest in the tropics and at mid-latitudes in the southern hemisphere. At the model time for which this plot was calculated, the sun was over the 0°E longitude line, which corresponds to local noon, so low $ClONO_2$ values are expected there. The highest values are seen just east of the 180°E longitude line, corresponding to local midnight. At mid-latitudes in the southern hemisphere, the largest $ClONO_2$ amounts are seen between approximately 120°E and 210°E, reflecting the dominant diurnal photochemical variation. In the northern hemisphere, the influence of dynamics is much stronger. There is a strong wave in the $ClONO_2$ field, and low $ClONO_2$ air can be seen reaching high northern latitudes at approximately 150°E longitude, while high $ClONO_2$ values are found at approximately 210°E. Another area of high $ClONO_2$ is seen just west of 0°E, but it does not extend nearly as far southward as the one near 150°E.

The dynamical and photochemical influences on the $ClONO_2$ distribution may be more clearly seen by examination of the 10 mb maps obtained for local noon and local midnight conditions,

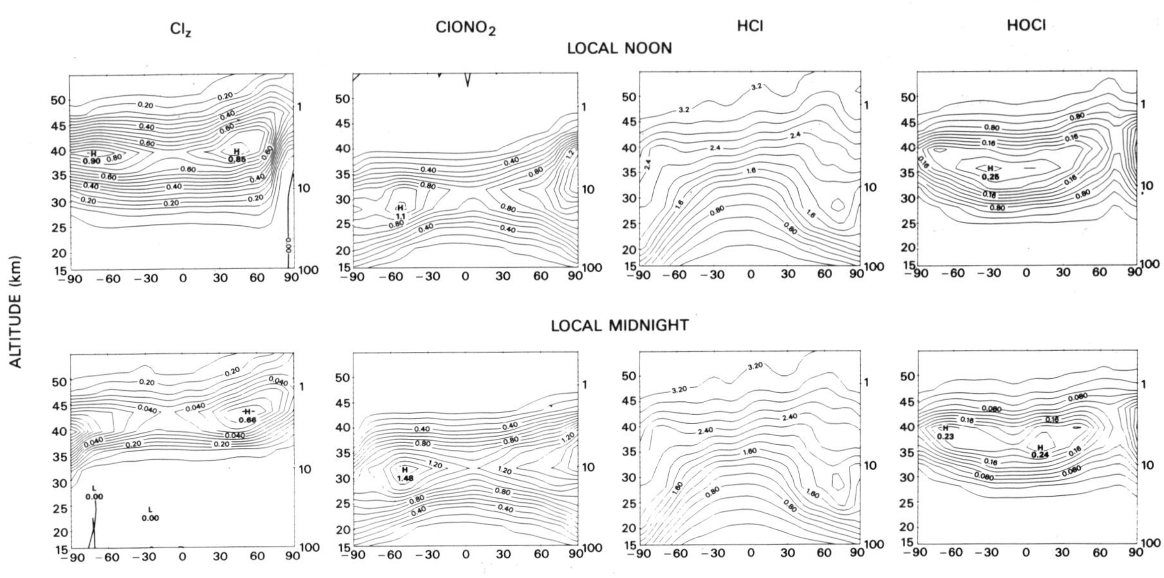

Fig. 1 Plot showing zonally averaged mixing ratios of chlorine-containing constituents between 5.5 and 6.5 days of model integration for local times corresponding to noon (top) and midnight (bottom). From left to right, the constituents are Cl_z, $ClONO_2$, HCl, and HOCl.

also shown in Fig. 2. As pointed out above, these figures use data taken between 5.5 and 6.5 days of the model simulation, and some dynamically-induced variation between the two figures is expected. The larger nighttime values of $ClONO_2$ in the tropics and southern hemisphere are clear. There is also far less zonal variation in the tropics in the noon $ClONO_2$ concentrations than in the midnight values.

At high northern latitudes, both the dynamical influences and the effects of considering chemistry and transport simultaneously become evident. For example, in the single-time figure (left panel of Fig. 2), the 1.2 ppbv contour extends nearly to the equator at 230°E and then goes northward and eastward, reaching 60°N at 300°E. Much of the large southward extent of the high $ClONO_2$ air near 230°E is dynamical in origin. This may be seen more clearly from the local noon figure (center panel), in which the 1.2 ppbv contour extends as far south as 30°N and slopes north and west near about 210°E. The presence of high $ClONO_2$ air in the region near 60°N and 270°E is, however, due more to photochemical effects associated with the large nighttime concentrations of $ClONO_2$ (see right panel of Fig. 2) than it is to dynamical effects. As a result, the strong meridional gradients in $ClONO_2$ seen in the left panel of Fig. 2 near 300°E are mostly photochemical in origin, while those near 180°E are dynamical.

Latitude-longitude distributions of the other chlorine-containing constituents at local noon are shown in Fig. 3. Values of $Cl_z \approx ClO$ are small at high northern latitudes, where the limited sunlight makes $ClONO_2$ photolysis unlikely. There is some evidence for non-zonal structure in the ClO distribution, especially between 120° and 200°E, but it is not nearly as prominent as that in $ClONO_2$. The species with the largest extent of non-zonal structure is HCl, for which tongues of low HCl air extend nearly to the pole, while high (mixing ratio > 2 ppbv) HCl air is seen as low as approximately 45°N in latitude. The zonal variability can be substantial; near 70°N the HCl values at this pressure level range from below 1.5 ppbv to above 2.5 ppbv.

For HOCl, there is little zonal variability at high northern latitudes; what variation there is seems to be anti-correlated with that of HCl, although the correlation is by no means perfect. The largest zonal variation in the local noon field, at least in an absolute sense, is seen at low latitudes in the southern hemisphere, where values from between 0.19 and 0.24 ppbv are seen.

4. CONCLUSIONS

The results of this model give an indication as to the magnitude of the zonal variation in the chlorine-containing constituents in the stratosphere, as well as of their diurnal variation, especially at high northern latitudes. Dynamical influences are seen both on the concentration of the relatively long-lived HCl and the shorter lived $ClONO_2$, although the influences are much stronger for HCl.

The large zonal variation in the HCl field at 10 mb suggests that there might be important zonal (and thus temporal) variability in the HCl column, for which there are extensive ground-based (Zander et al., 1987) and aircraft-based (Mankin and Coffey, 1983) measurements. In particular, it has been shown (Marche and Meunier, 1983) that there can be large changes in the HCl column over a period of several days. A model such as this, which consistently includes the effects of chemistry and dynamics, can play an important role in predicting and interpreting such variations.

The model also accurately simulates the evolution of the three-dimensional distributions of other long-lived and reservoir species; this will be shown in detail elsewhere. By consideration of the results of additional runs of this model, we should be able to build up useful information about the three-dimensional nature, temporal variability, and dynamical

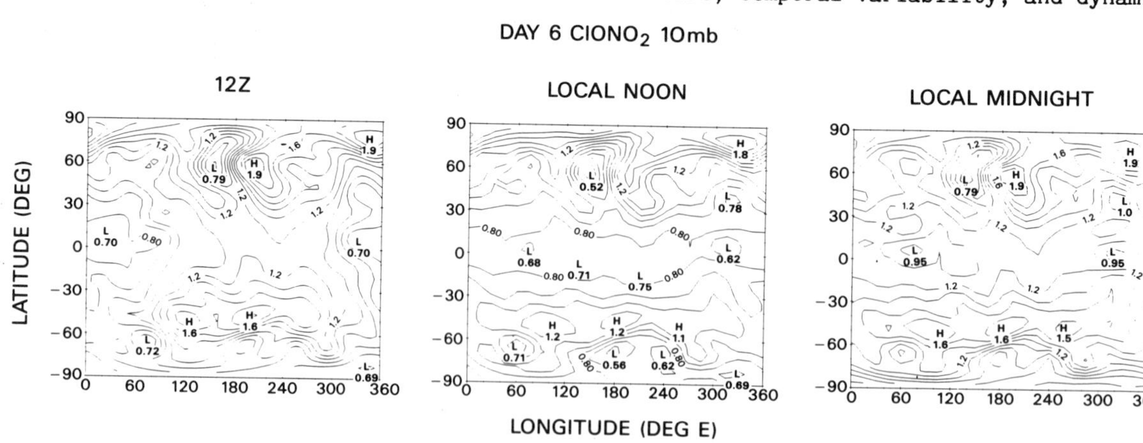

Fig. 2 Plots showing latitude-longitude distribution of $ClONO_2$ field at model level at 10.4 mb. Calculated fields have been Fourier decomposed and reexpanded to include only up to wave number six prior to plotting. a: (Left) Field after 6.0 days of model integration. b: (Middle) Local noon field calculated between 5.5 and 6.5 days of model integration. c: (Right) Local midnight field calculated between 5.5 and 6.5 days of model integration.

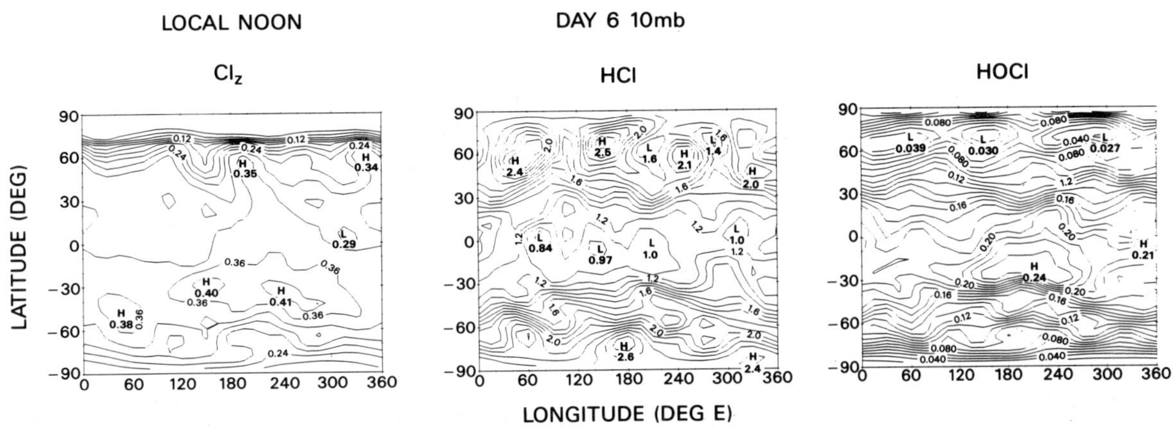

Fig. 3 Latitude-longitude distributions at 10.4 mb model level calculated between 5.5 and 6.5 days of model integration for Cl_z (left), HCl (middle), and HOCl (right). Calculated fields have been Fourier decomposed and reexpanded to include only up to wave number six prior to plotting.

influences on the distribution of these compounds. Such information should prove useful in attempting to interpret data on these constituents to be obtained by UARS, especially in the understanding of the limitations of asynoptic measurements for reservoir species (Salby, 1987).

ACKNOWLEDGMENTS

We thank Anne Douglass, Marvin Geller, Charles Jackman, and Richard Stolarski for helpful discussions and comments on this manuscript. We also thank Pamela G. Redmond for her assistance with the preparation of this manuscript. This is contribution number 48 of the Stratospheric General Circulation with Chemistry Modeling Project (SGCCM) at NASA/GSFC.

REFERENCES

Baker, W. E., S. C. Bloom, J. S. Woolen, M. S. Nestler, E. Brin, T. W. Schlatter, and G. W. Brantstator, 1987: Experiments with a three-dimensional objective analysis scheme using FGGE data, Mon. Wea. Rev., 115, 272-296.

Gille, J. C. and J. M. Russell III, 1984: The limb infrared monitor of the stratosphere: Experiment description, performance, and results, J. Geophys. Res., 89, 5125-5140.

Kaye, J. A. and R. B. Rood, 1988: Chemistry and transport in a three dimensional stratospheric model: Chlorine species during a simulated stratospheric warming, submitted to J. Geophys. Res.

Ko, M. K. W. and N. D. Sze, 1984: Diurnal variation of ClO: Implications for the stratospheric chemistries of $ClONO_2$, HOCl, and HCl, J. Geophys. Res., 89, 11,619-11,632.

Mankin, W. G. and M. T. Coffey, 1983: Latitudinal distributions and temporal changes of stratospheric HCl and HF, J. Geophys. Res., 88, 10,776-10,784.

Marche, P. and C. Meunier, 1983: Atmospheric trace species measured above Haute-Provence observatory, Planet. Space Sci., 31, 731-733.

Raper, O. F., C. B. Farmer, R. Zander, and J. H. Park, 1987: Infrared spectroscopic measurements of halogenated sink and reservoir gases in the stratosphere with the ATMOS instrument, J. Geophys. Res., 92, 9851-9858.

Rood, R. B., J. A. Kaye, J. E. Nielsen, M. R. Schoeberl, and M. A. Geller, 1987: Nitric acid forecast experiments, Physica Scripta, 36, 337-354.

Rood, R. B., D. J. Allen, W. Baker, D. Lamich, and J. A. Kaye, 1988: The use of assimilated stratospheric data in constituent transport calculations, submitted to J. Atmos. Sci.

Salby, M., 1987: Irregular and diurnal variability in asynoptic measurements of stratospheric trace species, J. Geophys. Res., 92, 14,781-14,805.

Zander, R., G. Roland, L. Delbouille, A. Sauval, C. B. Farmer, and R. H. Norton, 1987: Column abundance and long-term trend of hydrogen chloride (HCl) above the Jungfraujoch station, J. Atmos. Chem., 5, 395-404.

MODEL ASSESSMENT OF OZONE AND OZONE ACTIVE ATMOSPHERIC GAS CONTENT CHANGES DUE TO ANTHROPOGENIC INFLUENCE AND TO SOLAR ACTIVITY VARIATIONS

I.L. Karol, A.A. Kiselev, and E.V. Rosanov
Voeikov Main Geophysical Observatory
Leningrad, 1940I8, USSR

ABSTRACT

Changes of the gas composition and temperature regime of the atmosphere due to atmospheric pollution and to the UV intensity variations in the 11 year solar activity cycle are estimated using the one-dimensional radiative photochemical model and the actual data for the 21th solar cycle. Contribution of both these factors in ozone concentration changes are quantitatively comparable, having the opposite sign in 1975-1980 and the same sign in 1980-1986 with maximal ozone concentration decrease at about 40 km height level.

1. INTRODUCTION

The modeling and analysis of mostly satellite based observations have been widely used to study the influence of spectral intensity variations in UV radiation due to solar activity cycles on the photochemical transformations & gas composition in different atmospheric layers (Ramanathan, 1982; Keating et al., 1987). In comparing simulated ozone and other radiatively active trace gas (RATG) content variations and their data during the eleven year solar activity cycle, the effect of other (mostly anthropogenic) sources of RATG variations is generally neglected. Among these sources, the observed increase of concentrations of methane CH_4, of carbon oxide CO, of nitrous N_2O and nitric oxide NO, NO_2 and especially of halocarbons in the global lower atmosphere may be considered.

The paper deals with model studies of the influence of measured anthropogenic release of above gases into the atmosphere in 1976-1986 on the RATG and mostly on the ozone atmospheric content during the 21th cycle of solar activity. The possible variations of these effects in the first phase of 22nd cycle (1986-1991) are also considered.

2. MODEL INITIAL PARAMETERS

Special studies of the UV variations in the solar activity cycle using the two-dimensional photochemical model (PCM) with seasonally varying parameters reveal that the latitudinal and seasonal variations of these effects on the ozone and other RATG content are not significant. In view of that and due to necessity of model calculations projected for several decades, the one-dimensional RPM with the annually averaged parameters from (Karol et al., 1986) has been used.

Atmospheric gas composition and temperature variation of atmospheric layer up to 50 km level in the temperate latitudes of the Northern hemisphere are calculated for several numerical experiments:

1) For continuously varying solar UV radiation intensity I(λ) in 1975-1986 period, excluding the anthropogenic releases - the solar dynamical case.

2) For stationary states with the constant maximal (minimal) I(λ) values in the 21 solar cycle, excluding again the anthropogenic releases-the solar static case.

3) For the increased anthropogenic release in 1975-1986 and constant I(λ), excluding the I(λ) variation in the 11-year solar activity cycle.

4) For real simultaneous I(λ) variations and anthropogenic releases in the 1975-1991 period, using the extrapolation of I(λ) from the 21 cycle to the 1986-1991 period and of release rate trend observed in 1980-1985.

The choice of I(λ) variation in 21 cycle for extrapolation is caused by its better coverage by satellite observations. Our estimates of 22 cycle effects will possibly present the upper boundary of expected values due to probable lower intensity of 1991 solar maximum compared to 1980 solar maximum. The I(λ) values for λ > 180nm and their variations during the 21 solar cycle have been taken from (Lean, 1987) and assumed to be proportional to the solar radiofluxes at the 10.7 cm wavelength according to data presented in (Lean , 1987).

The anthropogenic release estimates of RATG in 1970-1986 have been based on concentration measurements in background conditions in the surface air of the temperate latitudes of the Northern

OZONE IN THE ATMOSPHERE
Rumen D. Bojkov and Peter Fabian (Eds.)

ISBN 0-937194-15-8

hemisphere: for the CO_2, N_2O, F-11, 12- the Geophysical Monitoring for Climatic Change (GMCC) data (GMCC,1985-1987); for CO- its measurements in Moscow environment (Dvoryashina et al.,1986). The ground level releases of NO and NO2 assigned in RPM, according to (Hameed, Dignon, 1988) have been increasing with 4% annual rate before 1974 and with 1.5% rate after this year. In 3 and 4 numerical experiments the 1966 has been chosen as the initial year with the zero F-11,12 concentrations in the surface air, but these of CH_3CL and CCl_4 have been taken as equal to 0.7 ppbv and to 0.1 ppbv and remained constant during the period of interest. F-11,12 surface air concentrations in the temperate latitudes of the Northern hemisphere have been extrapolated into the 1966-1974 period from the measured values of 1975, proportionally to estimates of F-11,12 production in the US in the previous year. These and similar CH_3CCl_3 concentration data have been taken from (Prinn et al. 1983).

3. MODEL RESULTS

The period of 21th solar activity cycle is divided into two sub-periods: 1975-1980 and 1980-1986 with growing and decreasing UV flux intensity I(λ). The model changes of RATG concentrations and of air temperature T are examined in these subperiods and in 1986-1991 separately.

Fig.1 presents the model vertical profiles of ozone (O_3) concentration deviations for the above subperiods. The I (λ) increase with time in 1975-1980 and in 1986-1991 leads to (O_3) growth with maximums in the upper-stratosphere (up to 5.6% in the "static" and up to 2.7% in the "dynamic" experiments). The rise of the anthropogenic releases leads to ozone decrease at levels higher than 30 km, up to 3% in 1975-1980 and up to 5% in 1986-1991 at 40 km level. In these subperiods the ozone concentration changes in the upper stratosphere due to anthropogenic pollution and due to 11-year solar cycle effects have the same order of magnitude, but the opposite signs. The tropospheric ozone increase by 3-4% in each subperiod is due to anthropogenic pollution. In 1980-1986, when UV intensity was decreasing, model (O_3) at the 40 km level diminished by 4.5% for "dynamical" experiment 1 and by 8% for "static" experiment 2 due to both forcings.

Fig.2 shows that the gradual increase of anthropogenic NO_x release leads to a significant growth of tropospheric concentrations of (NO_x) and (O_3). As (HNO_3) concentration prevails over the concentrations of other components of NO_x in the troposphere, the profiles of (NO_x) and (HNO_3) deviations are coincident there. Over the 30 km level the significant (up to 6-8%) (HNO_3) decrease during the I (λ) increase in 1975-1980 is due to a rise of HNO_3 photodissociation rate, which is the principal photochemical HNO_3 sink in this layer. The anthropogenic content of HNO_3 is increasing in the same way as NO and NO_2 and higher than 40 km level amounts to 1% per year.

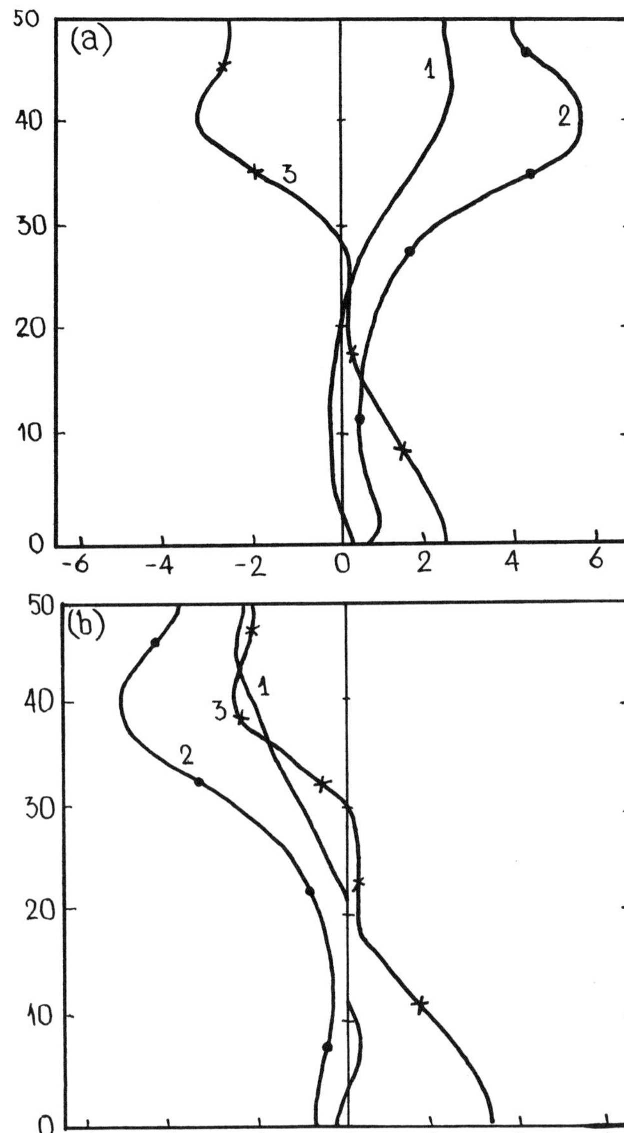

Fig.1. Vertical profiles of ozone concentration deviations for subperiods: 1975-1980 (a); 1980 - 1985 (b) due to solar activity in "static"-1 and "dynamic"- 2 experiments and due to anthropogenic pollution-3.

In 1975-80 and in 1986-1991 subperiods the UV intensity increase has dest -

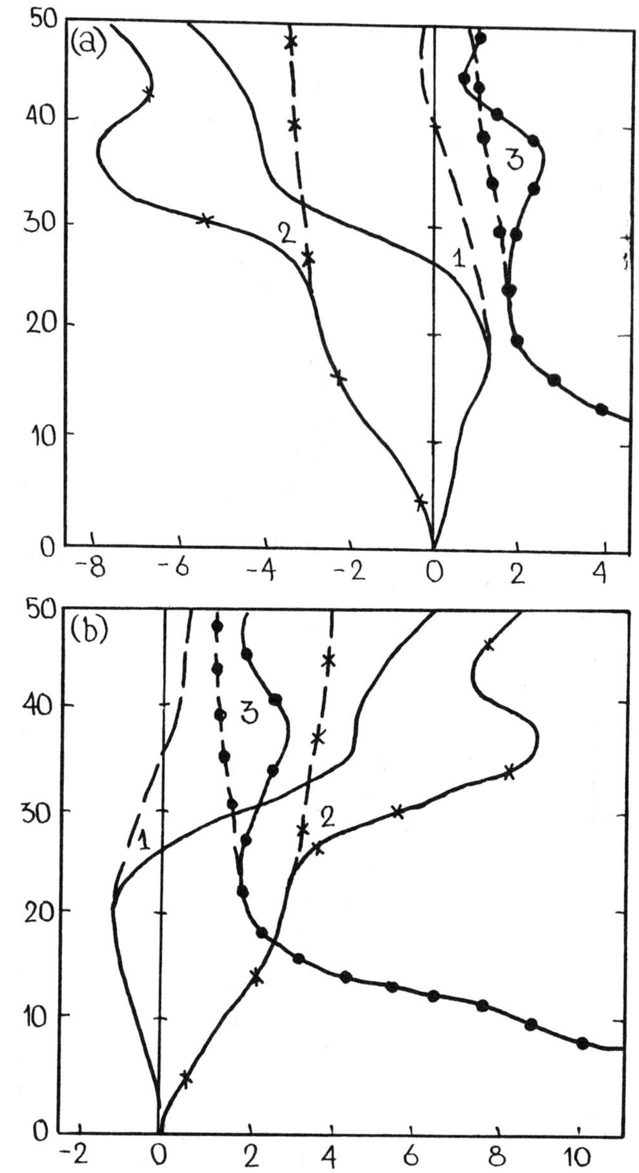

'Fig.2. Vertical profiles of concentration deviations of HNO_3 (solid line) and NO_x (dashed line)for subperiods:1975-1980 (a); 1980-1985 (b) due to solar activity in "dynamic-1 and "static"-2 experiments and due to anthropogenic pollution-3.

royed most intensively in the upper stratosphere, the gases having only ground level sources and photochemical sinks determined by their photodissociation (N_2O,halcarbons); e.g.over 45km in experiment 2 (CF_2Cl_2) have diminished by 20%. But the anthropogenic increase of (CF_2Cl_2) during 1975-1980, with account of its accumulation in the atmosphere, amounts to 30% almost homogeneously throughout the layer. Therefore, the UV acceleration of CF_2Cl_2 (and of other halocarbons) destruction is almost entirely exceeded by their anthropogenic input. In 1980-1986 period the anthropogenic growth of (CF_2Cl_2) by 27-30% in the entire 0-50 km atmospheric layer have been supplemented by its 10% rise at 36 km level and its 28% rise at 48 km level due to UV decrease in the solar cycle.

Though stratospheric NOx change in all periods is insignificant (excepting the "static" experiment 2), the (HNO_3) variations in the middle and upper stratosphere due to solar activity amounts to 4-6% for 5-6 years in the "dynamical" regime and exceeds noticeably the anthropogenic (HNO_3) increase during the same period. Therefore, the (HNO_3) monitoring may be recommended for the studies of solar cycle effects on the gaseous atmospheric composition.

The above changes in the atmospheric gas composition have comparatively weak influence on its thermal regime.In the 40-50km atmospheric layer of maximum temperature change ΔT in each of subperiods the anthropogenic cooling does not exceed 0.3-0.6 K, but T rises approximately by the same amount due to UV intensity growth in 1975-1980 and 1986-1991. Such small change of radiatively induced T and close RCM and PCM(photochemical model) $\Delta(O_3)$ values (maximum $\Delta(O_3)$ difference in RCM and PCM near the stratopause is less 1.0-1.5%) allows to use the PCM without the air temperature-gas composition feedback for the study of forcing effects in the current period.

4. DISCUSSION OF RESULTS

The above numerical experiments reveal that the global variations of stratospheric gas composition and temperature affected by the 21th solar activity cycle have the same magnitude as their changes induced by anthropogenic releases during the same period. The effects of these small disturbances are additive. The growth of ozone and other RATG content in the troposphere and lower stratosphere is almost entirely caused by the anthropogenic pollution.
The (O_3) growth in the middle and upper stratosphere due to I (λ) increase in 1975-1980 is compensated by its anthropogenic decrease (see Fig.1). Both the solar and anthropogenic forcings have reduced (O_3) at 40 km level to about 2% each during 1981-1986.

Nearly the same was the change not only of ozone concentration in the 30-40 km layer, but also the total ozone according to surface and satellite data in that period (Deluisi et al.,1987). The appreciable drop of these values began in 1980 before the famous El-Chichon volcano eruption in Mexico, occurred in April 1982 at 17°N. According to several studies, the eruption products caused a sharp reduction in total ozone and its content in the tropical lower stratosphere in the middle of 1982 and at other latitudes from the end of 1982 and in the beginning of 1983. However, these products have not affected the upper stratospheric ozone, and its content decreased by 3-4% during 1980-1985 (Bojkov, 1987). This agrees well with the above model estimates.

TABLE. ANNUALLY AVERAGED TOTAL OZONE CHANGES IN THE SOLAR CYCLE PHASES AND ANTHROPOGENICALLY CAUSED

Period	Due to ΔI (λ) in experiments		Anthropogenic		Total	Observed *
	Statical	Dynamical	PCM	RPM		
1	2	3	4	5	6	7
1975-1980	1.61	0.63	2.03	1.85	2.48	3.1
1980-1986	-1.58	-0.56	1.14	0.78	0.22	-6.2
1986-1991	1.6	0.6	0.19	-0.11	0.5	-

*According to (De Luisi et al.1987) for the temperate zone of the Northern Hemisphere

The table compares the modeled and observed changes of annually averaged total ozone X in the periods under consideration. After their good agreement in 1975-1980, a strong drop of X in the 1980-1985 exceeds considerably the model estimates. Besides the El-Chichon eruption effect, the significant contribution to this decrease has been produced by the sharp drop in X in the early 1985, caused by the stratospheric circulation changes (Bojkov, 1987) not accounted for in the models used.

5. CONCLUSION

With condition of keeping the current rate of global anthropogenic RATG releases into the atmosphere, the effects of UV variations in the 22th solar cycle in 1986-1997 on ozone content in the upper stratosphere will be overshaded by the anthropogenic forcing, particularly of halocarbon releases. Model estimates for 1986-1991 period, based on the extrapolation of ΔI(λ) from the 21th cycle and of anthropogenic release trends in the 1980-s, show, that total ozone X will remain almost unchanged and at the 40 km level the relative change in ozone will not exceed 0.2%, (Fig.1).

Therefore ozone, as an indicator of UV change effects on the stratospheric gas composition and photochemistry, may be replaced by the reliably measured HNO_3, as this has been proposed by Keating et al.(1987). For the detection of solar activity variation effects, measurements of gases in the lower mesosphere up to 70 km level may be recommended together with model studies of atmospheric chemistry of anthropogenic RATG in this layer.

REFERENCES

Boikov, R., 1987: The 1983 and 1985 anomalies in ozone distribution in perspective. Month. Weath. Rev., 115, 2187-2201

Deluisi, J.G., C. Tiao, G.C. Reinsel, C.L. Mateer, P.S. Connell, D.T.Wuebbles, 1987: Statistical analysis of total ozone and stratosphere Umkehr data for trend and solar cycle relationship. J.Geoph. Res., 92, 2201-2209.

Dvoryashina, E.V., V.I.Dianov-Klokov et al., 1986: The results of spectrometric measurements of methane content in the atmosphere of Northern Hemisphere. Izv. AN SSSR. Atm. and Ocean. Phys., 22, 87-89 (in Russian).

Geophys. monitoring for climatic change. 1986: Summ. Rep., 1985, №14; 1987: Summ.Rep. 1986, №15, Boulder CO., 146 p.; 155 p.

Hameed S., J.Dignon. 1988: Changes in geographical distribution of global emissions of NOx and SOx from fossil fuel combustion between 1966 and 1980. Atm.Envir., 22, 441-449.

Karol I.L., (ed.), Babanova et al. 1986: Radiative-photochemical models of the atmosphere. (In Russian), Leningrad, Gidrometeoizdat, 192 p.

Keating G.M., Brasseur G. et al. 1987: Response of middle atmosphere to short-term solar ultraviolet variations. 1. Observations. 2. Theory. J. Geophys. Res., 92, 889-902; 903-914.

Lean J. 1987: Solar ultraviolet irradiance variations: a review. J. Geoph. Res., 92, 839-868.

Prinn, et al. 1983.

Ramanathan. 1982.

REACTION KINETICS AND LABORATORY STUDIES

UV ABSORPTION CROSS SECTIONS OF SULFUR HEXAFLUORIDE AND ACETONITRILE

Cornelius Zetzsch
Fraunhofer-Institut für Toxikologie und Aerosolforschung
D-3000 Hannover 61, FRG
and
Physikalische Chemie I, Ruhr-Universität Bochum

ABSTRACT

UV absorption cross sections are determined at room temperature for SF_6 in the gas phase in the wavelength range from 115 to 170 nm and for CH_3CN from 155 to 185 nm. For an extension of the wavelength range, the cross sections of CH_3CN are determined in the liquid phase from 190 to 240 nm for comparison. The cross sections are lower than previous literature data and are discussed with respect to the fate of these molecules.

1. INTRODUCTION

The fate of long lived trace gases released from the surface of the earth may influence the chemistry of the stratospheric ozone layer if the gases are consumed by photolysis or reactions at these altitudes. A long atmospheric residence time of a trace gas leads necessarily to its accumulation. This may have an effect on climate by the greenhouse effect, especially if the infrared absorptions occur in the atmospheric window. Hence, an investigation of the chemical and photolytical destruction rates is desired for at least all anthropogenic compounds.

CH_3CN, besides applications as a polar and inert solvent, is an unwanted side product in the synthesis of acrylonitrile (Hagedorn and Gelbke, 1979). It is also emitted from natural fires as well (Becker and Ionescu, 1982). It has been detected in the stratosphere (Arnold et al., 1978, Arijs et al., 1980, Böhringer and Arnold, 1981, and Brasseur et al., 1983) at levels around 3 ppt at 25 km, decreasing to less than 0.3 ppt at 45 km altitude. It has been observed at ground level by Becker and Ionescu (1982) at 2-7 ppb in the troposphere but only at 56 ppt by Snider and Dawson (1984). The suggestion that HCN might be a precursor of CH_3CN (Murad et al., 1984) has been rejected (Cicerone and Zellner, 1983, and Brasseur et al., 1985) as a result of model calculations.

SF_6 is probably an anthropogenic compound. It is currently used as an insulating gas in industrial high voltage applications and as an inert tracer for atmospheric field experiments in science. One might suggest it as a substitute for fluorochlorocarbons in several applications, e.g. closed cell foams, since this molecule contains no Cl atoms, which might (after photolytic release) consume stratospheric ozone. It is used as a rather UV transparent, inert and efficient third body in kinetic experiments, and quantitative data on its absorption cross sections are required to estimate potential interference of photolytically formed F atoms in VUV photolysis experiments. On the other hand, the F atoms have a high bond energy, and SF_6 has accordingly been observed to be distributed rather uniformly in the troposphere; average mixing ratios of 0.31 ppt have been observed in the northern hemisphere and of 0.27 ppt in the southern hemisphere (Singh et al., 1979). In the stratosphere, mixing ratios of 0.4 ppt have been observed by Leifer et al. (1981). Such low gradients are an indication for a long atmospheric residence time of the order of 100 years.

Data on UV absorption cross sections of acetonitrile in the wavelength region relevant for stratospheric chemistry are still sparse, and only the study of Suto and Lee (1985) appears to cover the desired wavelength range. Similarly, the weak UV absorption of SF_6 in the stratospheric range has been studied only once, Bastien et al. (1985).

2. EXPERIMENTAL

The VUV cross sections in the gas phase are determined by using an inexpensive single beam apparatus, consisting of a 200 W deuterium lamp (D200F-SO, Heraeus-Hanau, equipped with a MgF_2 window), a 0.2 m Seya-Namioka-type monochromator with folding mirrors (Jobin Yvon H 20 UV-V, equipped with a holographic grating with 1200 grooves/mm and 100 nm blaze, operating at a slit width of 0.05 mm), an absorption cell, made of borosilicate glass (either 10 cm path length with MgF_2 windows, Harshaw, or 1 m path length with quartz windows, Suprasil, Heraeus) and a solar blind photomultiplier (EMI Gencom G 26 H 315) with a CsTe photocathode. The components of the apparatus are connected by vacuum-tight flanges. The monochromator is made leak-tight simply by adhesive tape, and all accessible parts of the light path of the apparatus are flushed continuously by N_2 gas evaporating from a liquid N_2 reservoir. A schematic of the optical path is shown in fig. 1.

ISBN 0-937194-15-8

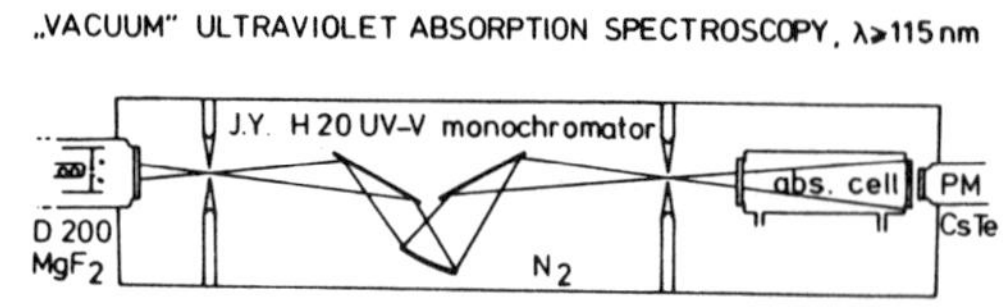

Fig. 1. Schematic of the inexpensive, N_2-flushed single beam arrangement for the determination of vacuum-UV absorption cross sections.

Spectra of the D_2 lamp, observed with MgF_2 and SiO_2 windows after flushing the apparatus with dry N_2 overnight, and of the stray light portion, observed at total absorption by CCl_3F in the 1 m cell, are shown in fig. 2.

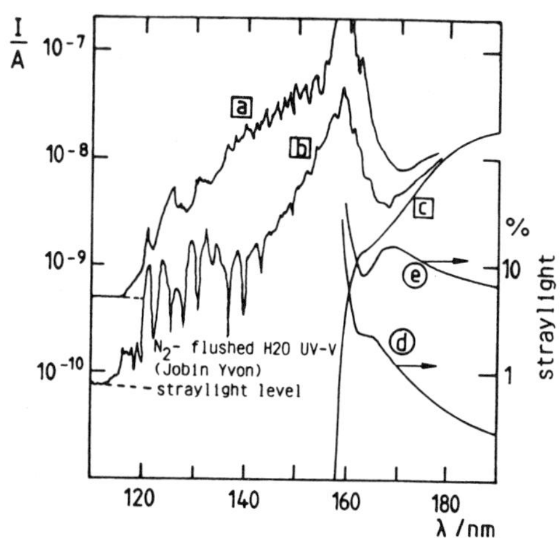

Fig. 2. Spectra of the D_2 lamp with MgF_2 windows taken with an evacuated Seya-Namioka instrument (Scharping et al., 1985, 1987) (a), and with the N_2 flushed H20 UV-V- monochromator and with MgF_2 windows on the 10 cm cell (b), or with SiO_2 windows on the 1 m cell (c). The portions of stray light background at total absorption by CCl_3F, observed in the 1 m cell with the H 20 UV-V-monochromator (d) and with a 0.5m Czerny-Turner instrument (e, Hubrich, 1979) are shown in the right hand scale for comparison.

An evacuable vuv monochromator became available later in our laboratory (Scharping and Zetzsch, 1985, and Scharping et al., 1987), and a spectrum obtained with evacuated light path is shown for comparison. The spectra of liquid CH_3CN are determined in a 1 cm quartz cell (Suprasil, Heraeus) by using a double beam spectrophotometer (Hitachi) and an empty reference cell.

Since the SF_6 sample (Linde, 99.8 %) contained a considerable impurity of O_2 (500 ppm), it was passed through a purification cartridge (Oxisorb, Messer Griesheim). Especially for SF_6 it was essential to flush the 1 m absorption cell very carefully with SF_6 gas at 1 atm for extended periods (several hours) to remove traces of other strongly absorbing molecules adsorbed on the walls, e.g. H_2O. Three different samples of CH_3CN were used, per analysis (Merck), UVASOL (Merck) and a special sample of high purity, prepared by J. Heinze at the University of Freiburg for electrochemical purposes. The CH_3CN samples have to be stored in the dark and air excluded by using inert gas (e.g. N_2) in order to preserve their UV transmitting properties. The samples are transferred into an evacuated glass bulb, where they were thoroughly degassed and their vapour was subsequently used. CCl_3F (Hoechst, 99.99%) was used without further purification. The gases are introduced into the evacuated absorption cell by stepwise release through a needle valve and are evacuated stepwise again to check the reproducibility of intensities and possible influences of leaks and adsorption/ desorption hysteresis, similar to previous work of Hubrich et al. (1977) on chlorofluorocarbons. The total pressures are determined by capacitance manometers (MKS Baratron). The intensities are recorded by an electrometer amplifier (Keithley) and a strip chart recorder.

3. RESULTS

Absorption cross sections, σ, are determined from the slopes of semilogarithmic plots of the background-corrected intensities, I, on the molecular number densities, N, at fixed absorption path length, l. The Beer Lambert law, $\ln (I/I_o) = - \sigma N l$, was obeyed throughout by the data. Figure 3 shows the observed absorption cross sections in a semilogarithmic diagram in comparison with other data from the literature.

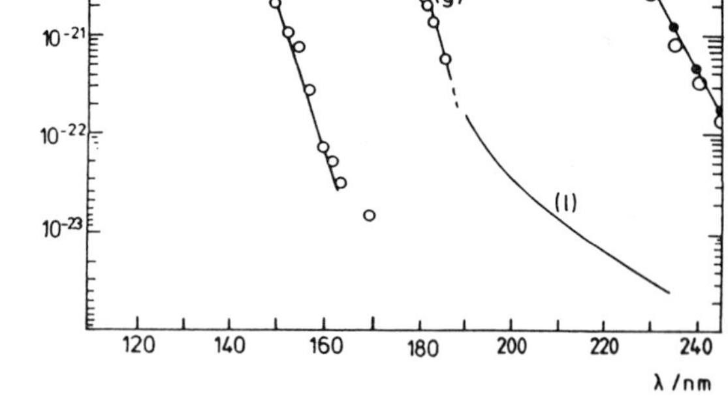

Fig. 3. Absorption cross sections of SF_6, CH_3CN and CCl_3F, obtained in the present study, and comparison with literature data.

4. DISCUSSION

4.1 PROPERTIES OF THE EQUIPMENT

Although the recent, previous studies on SF_6 and CH_3CN used synchrotron radiation as a light source, the D_2 lamp spectra of fig. 2 show clearly that enough intensity is available from this classical light source, even with the inexpensive, N_2 flushed arrangement, where no interfering absorptions of O_2 or H_2O can be detected. In addition, the stray light properties of the monochromator have been checked and compared with those of a 0.5m Czerny-Turner monochromator, equipped with a conventionally ruled grating (Minuteman 305M, grating with 1200 l/mm, 150nm blaze, see Hubrich, 1979). It can be seen in fig. 2 that the instrument with the holographic grating yields superior straylight properties, increasing to an advantage of more than an order of magnitude with increasing wavelength.

4.2 COMPARISON OF ABSORPTION CROSS SECTIONS

The absorption cross sections observed for CCl_3F are in excellent agreement with the data of Hubrich et al. and with the recommendations of the NASA-evaluation by DeMore et al. (1985). The absorption cross sections of SF_6 of the present study are lower than the previous determination by Bastien et al., 1985. The ratio to the present values increases from a factor of 1.5 at 125 nm to almost 100 at 160 nm. It should be noted that the SF_6 of the present study was purified by an Oxisorb cartridge, that is specified to remove O_2 and H_2 down to less than 0.1 ppm. The true absorption cross sections above 160 nm may still be lower since they may contain a portion of Rayleigh scatter and should hence be considered as upper limits.

The absorption cross sections of CH_3CN determined in the present study are again lower than the previous data of Suto and Lee (1985), which are higher by a factor of 1.8 at 160 nm up to a factor of 6 at 182 nm. Since the decrease of the absorption cross sections of the present study is steeper, a simple wavelength shift by approximately 4 nm is not an acceptable explanation for this discrepancy. In addition, the absorption cross sections of the liquid phase agree (in accordance with the manufacturer's specifications of the UV transmitting properties of UVASOL, Merck) favorably with those of the gas phase, extending the accessible range of absorption cross sections by almost 4 orders of magnitude below the lower limit attained by Suto and Lee (1985).

The purity of the compounds was observed to be extremely important in the present study, as can be expected from the large range of cross sections of more than 5 orders of magnitude covered by the measurements in the gas phase. This can demonstrated from the measurements of liquid phase absorption spectra of different CH_3CN samples in fig. 4.

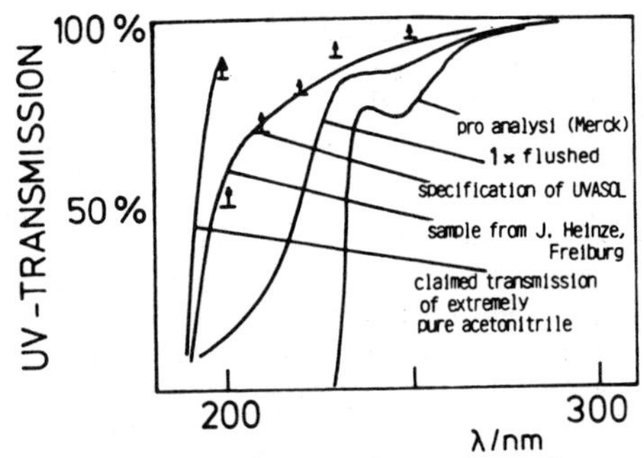

Fig. 4. UV transmission of various samples of liquid CH_3CN for 1 cm absorption path.

The spectrum of a p.a. sample from the stock shows an absorption band at 245 nm and a cutoff at 230 nm. After flushing the cell once with a freshly purified sample from the University of Freiburg, the cutoff lowers to less than 200 nm; the best spectrum is obtained after flushing more than 10 times, and this spectrum meets the specifications of Merck company for the UVASOL quality approximately. On the other hand, information about an attainable transmission of more than 90% at 190 nm can be found in the literature about UV transparent solvents. Hence, the absorption cross sections of CH_3CN obtained in the present study should again be considered as upper limits above 200 nm.

4.3 ATMOSPHERIC FATE OF SF_6 and CH_3CN

The absorption of SF_6 in the stratospheric region is extremely weak, and photolysis and probably chemical reactions in the troposphere and stratosphere can be neglected. Hence, SF_6 is expected to be similarly inert as was observed for CF_4 and C_2F_6 by Fabian et al. (1987). Since SF_6 has not been produced for very long it might again be possible to determine the anthropogenic portion from the height profile in the upper stratosphere. Although acetonitrile attains comparable absorption cross sections at 30 nm longer wavelengths, its absorption is still weaker than that of CO_2, where the increase through human activities has been deduced from height profiles (Bischof et al., 1987). Both compounds will be photolyzed slowly by the extraterrestrial radiation and especially SF_6 is expected to escape to a major extent into space.

Although acetonitrile contains weak C-H-bonds (with hydrogen atoms which might be attacked by radicals like OH, O or Cl) and a triple bond (that might be attacked by ozone) its reactivity against these radicals and molecules is unexpectedly low. For the reaction with OH radicals at room temperature values between 1.9 and $5x10^{-14}$ cm^3s^{-1} have been determined (Zetzsch, 1981-1983, Fritz et al., 1982, Kurylo and Knable, 1984, Poulet et al., 1984, and Harris et al. 1981), and a slight pressure dependence of the reaction (indicating again an addition mechanism) has been detected below 30 mbar by Zetzsch (1983) in contrast to the results of Poulet et al. (1984). For the reaction with Cl atoms, a rate constant of of $2\text{-}9x10^{-15}$ cm^3s^{-1} (Kurylo and Knable, 1984, and Poulet et al., 1984) and for O atoms a rate constant of $2.3x10^{-16}$ cm^3s^{-1} has been observed by Bonanno et al., 1977, and addition has again been claimed to be an important reaction path with both atoms. Adopting the average global tropospheric concentrations reported by Graedel (1978), [OH]= $4.1\text{x}10^{5}$ cm^{-3}, [Cl] = $2\text{x}10^{4}$ cm^{-3} and [O]= $2.5\text{x}10^{4}$ cm^{-3}, the atmospheric residence time of acetonitrile will be limited to 4, 800, and 5500 years by each of the active species, respectively (at room temperature, neglecting the activation energies and using the lowest rate constants). Although the upper limit of $1.5\text{x}10^{-19}$ $cm^{3}s^{-1}$ for the reactivity with ozone, stated by Harris et al. (1981), would yield a lifetime of the order of 2.5 months, it is anticipated that the true reactivity of acetonitrile with ozone is much lower than the stated upper limit. Hence, all these processes together are too slow to explain the gradient in the mixing ratios between the stratosphere and ground level, that have been explained by Brasseur et al. (1983) and Hamm et al. (1984) as a contribution of rainout and uptake of this water soluble compound by the ocean with a resulting tropospheric residence time (including the reaction with OH) of less than 2 years.

The CH_3CN absorption is weaker than $CHClF_2$ (F22) above 155 nm, and above 175 nm weaker than $CClF_3$ (F13). Hence its photolytic lifetime is expected to exceed 100 years, and correspondingly low gradients are expected in the upper atmosphere. Since the reaction with OH is the main sink in the stratosphere, profiles of OH may be deduced from profiles of CH_3CN and the known rate constant. For further sinks one might speculate about the reactions with $O(^1D)$ and about a possible influence of adsorption on the photolysis process, as suggested by Krüger et al. (1987) for the the fluorocarbons.

For SF_6, no chemical sinks are known. Since it absorbs less than CO_2 (similar to CF_4 and C_2F_6, see Sandorfy, 1976), the atmospheric lifetime is expected to exceed 10000 years. In the upper atmosphere, the profiles of both gases may still reflect the history of their anthropogenic production.

ACKNOWLEDGEMENTS

The experimental part of this study work was supported by the Umweltbundesamt and by the Fonds der Chemischen Industrie and it was performed at the University of Bochum, where the author thanks H. Richtering and F. Stuhl for the loan of equipment and for support. The author is indebted to M. Ramdohr (deceased), U. Lutze, R. Vogelsang, B. Weß and N. Wingender for experimental help. A highly purified sample of acetonitrile was delivered from the University of Freiburg by J. Heinze.

REFERENCES

Arnold, F., H. Böhringer, and C. Henschen, 1978: Composition measurements of atmospheric positive ions. Geophys. Res. Lett. 5, 653-656.

Arijs, E., D. Nevejans and J. Ingels, 1980: Unambiguous mass determination of major stratospheric positive ions. Nature, 288, 681-686.

Bastien, F., P.A. Chatterton, E. Marode, and J.L. Moruzzi, 1985: Photoabsorption measurements in SF_6, SF_6 mixtures and some fluorocarbon gases. J. Phys. B: Appl. Phys., 18, 1327-1337.

Becker, K.H., and A. Ionescu, 1982: Acetonitrile in the lower atmosphere. Geophys. Res. Lett., 9, 1349-1351.

Bischof, W., B. Borchers, P. Fabian and B.C. Krüger, 1985: Increased concentration and vertical distribution of carbon dioxide in the stratosphere. Nature, 316, 708-710.

Böhringer, H., and F. Arnold, 1981: Acetonitrile in the stratosphere- implications from laboratory studies. Nature, 290, 321-322.

Bonanno, R.J., R.B. Timmons, L.J. Stief and R.B. Klemm, 1977: The kinetics and mechanisms of the reactions of O (^3P) atoms with CH_3CN and CF_3CN. J. Chem. Phys., 66, 92-98.

Brasseur, G., E. Arijs, A. De Rudder, D. Nevejans and J. Ingels, 1983: Acetonitrile in the atmosphere. Geophys. Res. Lett., 10, 725-728.

Brasseur, G., R. Zellner, A. De Rudder and E. Arijs, 1985: Is hydrogen cyanide a progenitor of acetonitrile in the atmosphere? Geophys. Res. Lett., 12, 117-120.

Cicerone, R.J., and R. Zellner, 1983: The atmospheric chemistry of hydrogen cyanide. J. Geophys. Res., 88, 10689-10696.

DeMore, W.B., J.J. Margitan, M.J. Molina, R.T. Watson, D.M. Golden, R.F. Hampson, M.J. Kurylo, C.J. Howard and A.R. Ravishankara, 1985: Chemical kinetics and photochemical data for use in stratospheric modeling. Evaluation Number 7,. JPL Publication 85-37, Jet Propulsion Laboratory, Pasadena, CA.

Fabian, P., B. Borchers, B.C. Krüger and S. Lal, 1987: CF_4 and C_2F_6 in the atmosphere. J. Geophys. Res., 92, 9831-9835

Fritz, B., K. Lorenz, W. Steinert and R. Zellner, 1982: Laboratory kinetic investigations of tropospheric oxidation of selected industrial emissions. Proc. 2nd Europ. Sympos. on the Physico-Chemical Behaviour of atmospheric Pollutants, Varese, (eds. B. Versino and G. Angeletti), pp. 192-202, Reidel, Boston.

Graedel, T.E., 1978: Chemical compounds in the atmosphere. Academic Press, New York.

Hagedorn, F., and H.P. Gelbke, 1979: Nitrile, aliphatische. Ullmanns Encyklopädie der technischen Chemie, 17, pp. 323-332, Verlag Chemie, Weinheim

Hamm, S., J. Hahn, G. Helas and P. Warneck, 1984: Acetonitrile in the troposphere. Geophys. Res. Lett., 11, 1207-1210.

Harris, G.W., T.E. Kleindienst, J.N. Pitts, Jr., 1981: Rate constants for the reactions of OH radicals with CH_3CN, C_2H_5CN and CH_2=CHCN in the temperature range 298-424K. Chem. Phys. Lett., 80, 479-483.

Hubrich, C., 1979: UV-Absorptionsspektren von Chlor-Fluor-Derivaten des Methans und Ethans sowie von Distickstoffmonoxid im Wellenlängenbereich von 160 bis 375 nm bei den Temperaturen 298K und 208K und ihre spektroskopische Interpretation. Dissertation, p. 17, Universität Bochum.

Hubrich, C., C. Zetzsch and F. Stuhl, 1977: Absorptionsspektren von halogenierten Methanen im Bereich von 275 bis 160 nm bei Temperaturen von 298 und 208 K. Ber. Bunsenges. Phys. Chem., 81, 437-442.

Krüger, B.C., G.Q. Wang and P. Fabian, 1987: The antarctic ozone depletion caused by heterogeneous photolysis of halogenated hydrocarbons. Geophys. Res. Lett., 14, 523-526.

Kurylo, M.J. and G.L. Knable, 1984: A kinetics investigation of the gas phase reaction of $Cl(^2P)$ and OH $(X^2\Pi)$ with CH_3CN: Atmospheric significance and evidence for decreased reactivity between strong electrophiles. J. Phys. Chem., 88, 3305-3308.

Leifer, R., K. Sommers and S.F. Guggenheim, 1981: Atmospheric trace gas measurements with a new air sampling system. Geophys. Res. Lett., 8, 1079-1081.

Murad, E., W. Swider, R.A. Moss and S. Toby, 1984: Stratospheric sources of CH_3CN and CH_3OH. Geophys. Res. Lett., 11, 147-150.

Poulet, G., G. Laverdet, J.L. Jourdain and G. Le Bras, 1984: Kinetic study of the reactions of CH_3CN with Cl and OH radicals. J. Phys. Chem., 88, 6259-6262.

Sandorfy, C., 1976: UV absorption of fluorocarbons. Atmos. Environment, 10, 343-346.

Scharping, H., and C. Zetzsch, 1985: Substituent effects in the vuv spectra of monochlorobenzene, and o-, m-, and p-dichlorobenzene. J. Mol. Spectrosc., 112, 8-17.

Scharping, H., C. Zetzsch and H.A. Dessouki, 1987: The vuv spectra of the trichlorobenzenes and the higher chlorinated benzenes in the gas phase and in n-hexane solution. J. Mol. Spectrosc., 123, 382-391.

Singh, H.B., L.J. Salas, H. Shigeishi and E. Scibner, 1979: Atmospheric halocarbons, hydrocarbons, and sulfur hexafluoride: Global distributions, sources and sinks. Science, 203, 899-903.

Snider, J.R., and G.A. Dawson, 1984: Surface acetonitrile near Tucson, Arizona. Geophys. Res. Lett., 11, 241-242.

Suto, M., and L.C. Lee, 1985: Photoabsorption cross section of CH_3CN: Photodissociation rates by solar flux and interstellar radiation, J. Geophys. Res., 90, D7, 13037-13040.

Zetzsch, C., 1981: Rate constants for the reactions of OH with nitriles: acetonitrile, benzonitrile, and acrylonitrile. Abstr. NATO Advanced Study Institute on Chemistry of the Unpolluted and Polluted Troposphere, Corfu, Greece (ed. W. Jaeschke), Frankfurt.

Zetzsch, C., 1981: Abbauverhalten von Acetonitril aus seiner Reaktivität gegen OH-Radikale. Abstr. 9. Bunsenkolloquium der Bunsen-Gesellschaft, Frankfurt.

Zetzsch, C., 1983: Bestimmung absoluter OH-Geschwindigkeitskonstanten mit gepulster Photolyse und Resonanzfluoreszenz. Abstr. 16. Bunsenkolloquium der Bunsen-Gesellschaft, Frankfurt.

OXYGEN PHOTOLYSIS AT 214 nm AND 25°C

A. Horowitz, G. von Helden, W. Schneider, F.G. Simon, P.J. Crutzen, and G.K. Moortgat
Max-Planck-Institut für Chemie
D-6500, Mainz, FRG

ABSTRACT

Ozone formation for the 214 nm photolysis of oxygen at pressures ranging from 380 to 1300 Torr was investigated. The rates of ozone formation increase with the square of oxygen pressure mirroring the pressure dependence of the absorption cross sections of oxygen. This observation demonstrates the importance of the collision induced changes in oxygen absorbance and points to a need to examine the role of oxygen dimer $(O_2)_2$ in reactions initiated by its photodissociation. Because of the high dilution of ozone its formation was found to be linear with time even though, on the average, its rate of photolysis exceeded the rate of its generation from oxygen. The quantum yield of ozone formation was 1.86 ± 0.17 independent of O_2 pressure. From this result the quantum yield of the primary process: $O_2 + h\nu$ (214 nm) -> $2O(^3P)$ is estimated as being equal to 0.93 ± 0.08(2σ). Based on the observed linear buildup of ozone and on model calculations, an upper limit of about 0.03 is estimated for O_3 photolysis via a channel which leads to $O(^3P)$ and an excited O_2 species that can regenerate ozone in its reaction with oxygen.

1. INTRODUCTION

The importance of ozone in the chemistry of air pollution, its vital role in attenuating solar u.v. radiation as well as fundamental interest in its photochemistry lead to extensive studies which have been recently reviewed (Wayne 1987). In particular the understanding of the processes that govern the photochemical behaviour of ozone and oxygen, is of major importance for the precise assessment of the chemistry of the atmospheric ozone layer.

In view of the imbalance observed in the stratospheric ozone budget it is rather unfortunate that ozone photochemistry below 248 nm has not been as extensively studied as at the longer wavelengths. Current status of our knowledge does not seem to rule out the possibility that, with the increase in the available energy, ozone photodissociation might occur along additional pathways. Conceivably, reactions of oxygen, with species formed via these additional photodissociation channels, could result in an enhanced regeneration of ozone and thus could constitiute an additional source of "odd" oxygen in the stratosphere (Saxon and Slanger 1986, Crutzen 1986). This attractive possibility has not been investigated sofar.

Finally, it is worth noting that, in all atmospheric model calculations related to ozone formation and destruction, the quantum yield of O_2 photodissociation in the Herzberg continuum is assumed to be unity. This is a very sound postulate which does not appear however to be based on a direct determination.

The present study of oxygen photolysis at 214 nm addresses itself to some of the problems related to the ozone balance in the atmosphere. In particular we were interested in the determination of the quantum yields of ozone formation under conditions where its secondary reactions can be minimized.

2. EXPERIMENTAL

The photolysis of oxygen was carried out in a quartz cross shaped flushable cell that has been previously used in O_3 photolysis (Moortgat and Warneck 1975). The photolytic light source was a custom made 1 kW Sb lamp (Hereaus) from which the wavelength band centered near 214 nm was isolated. Ozone concentration was determined from the absorbance of 254 nm light generated by a low pressure Hg lamp. Decomposition of ClNO, monitored at 254 nm was used for actinometry assuming that the quantum yield of its photodecomposition is 2.

3. RESULTS

The formation of ozone was determined at several O_2 presures ranging from 380 to 1300 Torr and as a function of incident light intensity. At the small conversions at which ozone formation

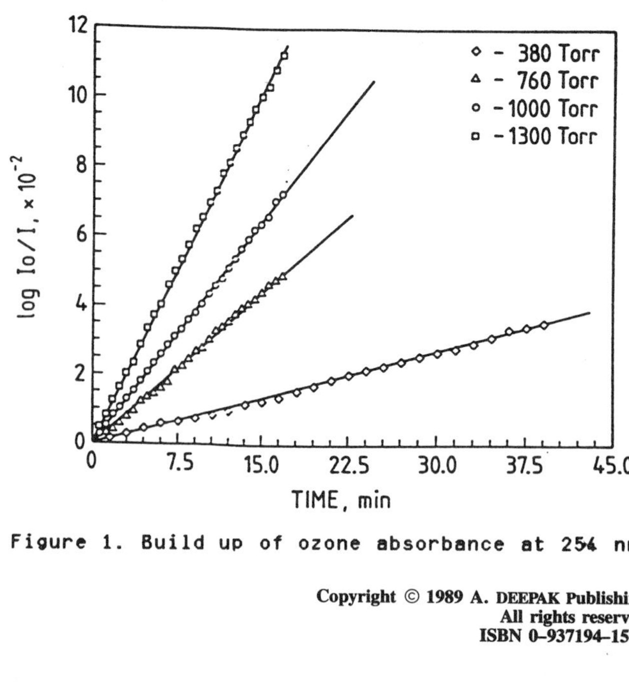

Figure 1. Build up of ozone absorbance at 254 nm.

ISBN 0-937194-15-8

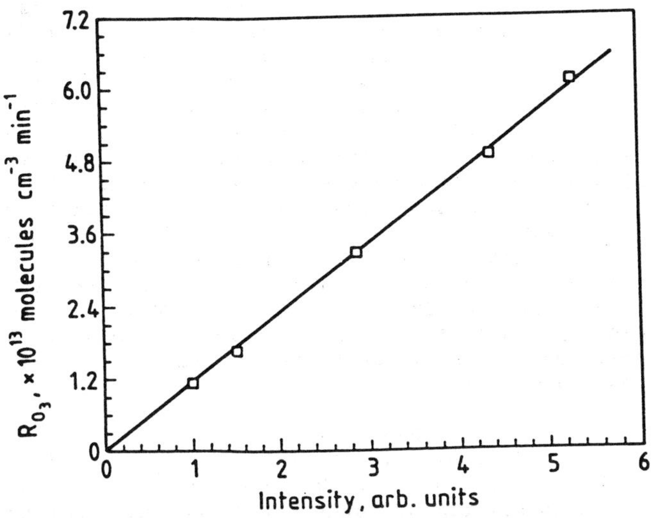

Figure 2. Variation of the rate of ozone formation with incident light intensity. O_2 = 1300 Torr.

was studied, no departure from linear variation of its concentration with time was observed, see Fig. 1, although on the average O_3 absorbed considerably more light than oxygen. Even in the presence of about 2 ppm of water no departure from this linear behaviour was observed. As expected, rates of ozone formation, R_{O_3}, were found to be proportional to the light intensity as shown in Fig. 2. Rather unexpectedly though, an increase of R_{O_3} with the square of O_2 was observed as shown in Fig. 3.

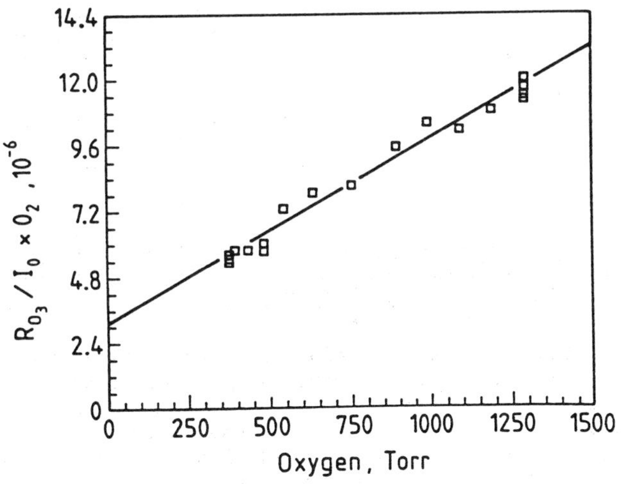

Figure 3. Plot of R_{O_3}/P_{O_2} vs. P_{O_2}.

4. DISCUSSION.

4.1 THE ROLE OF OXYGEN DIMER

In principle, the observed increase of the rates of ozone formation with the square of oxygen pressure, can be attributed either to some unknown photochemical and photophysical processes that result in an increase of Φ_{O_3} or to non-linear effect in the absorption of O_2 in the Herzberg continuum. Our choice of the latter alternative is based on the fact that an increase of O_2 absorption cross section with increasing oxygen pressure indeed has been observed and is well documented (Cheung et al. 1986, Johnston et al. 1984). This assumption is further supported by the observation that linear least squares fit of the data plotted in Fig. 3 gives $R_{O_3}/P_{O_2}I_0 = (3.09 \pm 0.40)x10^{-6} + (6.60 \pm 0.43)x10^{-9}P_{O_2}$. The slope to intercept ratio obtained from this relation of $(2.14 \pm 0.31)x10^{-3}$ $Torr^{-1}$ compares very well with the slope/intercept ratio derived from the pressure dependence of the absorption cross section of O_2 , σ_{O_2} (Cheung et al. 1986). These authors used the following expression for the pressure dependence

$$\sigma_{O_2}(\lambda) = \sigma_o(\lambda) + \alpha(\lambda)P_{O_2} \qquad (1)$$

where P_{O_2} is in Torr and $\sigma_o(\lambda)$ is the pressure independent absorption cross section. They found, at 296 - 300°K, that $\alpha(\lambda)/\sigma_o(\lambda)$ = (2.05 ± 0.17) $x10^{-3}$, $(2.00 \pm 0.18)x10^{-3}$ and $(2.13 \pm 0.24)x10^{-3}$ $Torr^{-1}$ at 210, 215 and 220 nm, respectively. The agreement with the results of Johnston et al. (1984) is also quite good. The slope/intercept ratios, in this case are: $(1.74 \pm 0.17)x10^{-3}$, $(2.34 \pm 0.23)x10^{-3}$ and $(1.80 \pm 0.21)x10^{-3}$ $Torr^{-1}$ for the same wavelengths as before.

The pressure dependence of oxygen absorption cross sections in the Herzberg continuum is usually ascribed to the absorbance of the weekly bound dimer, $(O_2)_2$ or O_4, which is in an equilibrium with oxygen, reaction 2.

$$2O_2 \;\longleftrightarrow\; (O_2)_2 \qquad (2)$$

Accordingly, the pressure dependent term in exp.1, can be considered as being equal to $K_2.\sigma_{(O_2)_2}.P_{O_2}$. Using the detailed spectral data (Cheung et al. 1986) and integrating it over the spectrum of the photolysing light it can be shown that 45 percent of the light absorbed at O_2 pressure of 380 Torr is due to $(O_2)_2$ absorption. Furthermore, for most of the pressures employed, the dimer absorbes over half of the absorbed light and this fraction reaches 75 percent at 1300 Torr. Hence, before any conclusion can be drawn from the present results, the possible consequencies of this observation, which, to a large extent, depend on the photochemical behaviour of the O_2 dimer and the nature of its bonding, have to be examined.

Absorption of oxygen in the near u.v., the visible and the infrared is known for more than a century. It appears that, in as much as the nature of the bonding in $(O_2)_2$ is concerned, the most informative work is the study in which the infrared absorption of gaseous O_2 from 1480 to 1800cm^{-1} and between 77 to 300°K was examined (Long and Ewing 1971). These authors observed in the O_2 absorption spectrum a structureless band down to 90°K where a doublet appears. Absorption in both regions increased with the square of O_2 concentration and hence they rationalized their findings in terms of two distinct phenomena: collision induced absorption resultant in structureless band and formation at low temperatures of an $(O_2)_2$ bound by van der Waals interactions. For the latter species they estimated, from the temperature dependence of the doublet absorption, an activation energy of formation equal to 530 cal mol^{-1}. The corresponding value for the temperature dependence of the collision induced spectrum was found to be lower by more than an order of magnitude. Johnston et al.(1984) also arrived at a similar conclusion. They determined O_2 cross sections in the Herzberg continuum from 206 to 327°K. and derived the temperature dependence of K_2 from the macroscopic virial coefficient B, given by exp. 3 and the excluded volume b in the microscopic equation of state, exp. 4

$$PV/nRT = 1 + nB/V + \cdots \quad (3)$$

$$(V-b) = (n_2 + n_4)RT \quad (4)$$

where n_2 and n_4 are the molar concentrations of oxygen and its dimer, respecively. The values of K_2 thus obtained increased by a factor of about two from 327 to 206°K, in marked contrast to the experimental findings that, within the experimental error, the term reflecting $(O_2)_2$ absorbance did not show any detectable variation with temperature. Based on this observation and the parallelism between the O_2 and $(O_2)_2$ spectra these authors concluded that the pressure dependence of O_2 spectrum is caused by a perturbation of the wave function of an oxygen molecule by the force field of nearby O_2 rather than by stable dimer formation.

Our results appear to be a first demonstration of the correspondence between the pressure dependence of O_2 absorption in the Herzberg continuum and its photochemistry. Nevertheless, in view of the rather convincing evidence, we tend to accept the opinion that $(O_2)_2$ and O_4 reflect a convenient notation more than an existence of a real species that could differ from molecular oxygen in its primary photodissociation processes.

4.2 MECHANISM AND QUANTUM YIELDS

Based on the conclusion reached in the previous section and the observation that in the 170 - 237 nm region the quantum yield of the energetically possible dissociation channel leading to $O(^1S)$ does not exceed 0.001 (Lee et al. 1980), photodissociation of O_2 under our experimental conditions appears to be fully accounted for by reaction 5

$$O_2 + h\nu \longrightarrow 2\,O(^3P) \quad (5)$$

However, even in the present system in which the average O_2/O_3 concentration ratio was as high as $(2.35 \pm 0.5)x10^4$, photolysis of ozone could not be completely suppressed. It can be seen from the results presented in Table 1, that, at the end of all runs, the amount of light absorbed by ozone considerabely exceeded the amount absorbed by oxygen. The value of $I_{abs}(O_3)/I_{abs}(O_2)$ at the end of the runs was close to 3 at highest pressure and about 6 at the lower one. Accordingly the following reactions have to be also considered

$$O_3 + h\nu \longrightarrow O(^1D) + O_2(^1\Delta g) \quad (6a)$$

$$\longrightarrow O(^3P) + O_2(^3\Sigma_g^-) \quad (6b)$$

Subsequenly, the initially formed species can react either with ozone or with oxygen according to the following reaction scheme:

$$O(^3P) + O_2 \longrightarrow O_3 + O_2 \quad (7)$$

$$O(^3P) + O_3 \longrightarrow O_2 + O_2 \quad (8)$$

$$O(^1D) + O_2 \longrightarrow O(^3P) + O_2(^1\Sigma_g^+) \quad (9)$$

$$\longrightarrow O(^3P) + O_2 \quad (10)$$

$$O(^1D) + O_3 \longrightarrow O_2 + O_2 \quad (11)$$

$$\longrightarrow O_2 + 2O(^3P) \quad (12)$$

$$O_2(^1\Delta_g) + O_2 \longrightarrow O_2 + O_2 \quad (13)$$

$$+ O_3 \longrightarrow O_2 + O_2 + O(^3P) \quad (14)$$

$$O_2(^1\Sigma_g^+) + O_2 \longrightarrow O_2 + O_2 \quad (15)$$

$$+ O_3 \longrightarrow O(^3P) + O_2 + O_2 \quad (16)$$

In terms of this reaction scheme it is quite clear that decomposition of ozone in its mixtures with oxygen will take place only if the amount of oxygen is not sufficient to suppress completely the reactions of $O(^1D)$, $O(^3P)$, $O_2(^1\Delta_g)$ and $O_2(^1\Sigma_g^+)$ with ozone. The situation becomes more complex however when traces of water are present in the system. In this case, generation of the hydroxyl radical via reaction 17, can lead to a chain decomposition of ozone propagated mainly by reactions 18 and 19

$$O(^1D) + H_2O \longrightarrow OH + OH \quad (17)$$

$$OH + O_3 \longrightarrow HO_2 + O_2 \quad (18)$$

$$HO_2 + O_3 \longrightarrow OH + O_2 + O_2 \quad (19)$$

Decomposition of ozone should be indicated by a decrease in its rate of formation which in turn has to be reflected in the quantum yields, Φ_{O_3}. From our results we obtain $\Phi_{O_3} = 1.86 \pm 0.17$ (2σ) with no indication of pressure dependence, which would be expected to exist if the reaction of $O(^1D)$ with water or ozone would have taken place. Also, in line with this argument, presence of small amounts of water did not affect the O_3 quantum yield and within the experimental error, no departure from linearity of ozone growth with time was detected. These observations seem to provide a strong support for the assumption that, under the experimental conditions in this work, $O(^1D)$ atoms react only with oxygen. In this case, it can be inferred fom our results that the primary quantum yield of oxygen photodissociation is close to unity.

TABLE 1 Model Computed Values of Ozone Formation in the 214 nm Photolysis of Oxygen[a].

O_2, x10[19] [b]	H_2O, x10[13]	$\Phi(A^3\Sigma_u^+)$	$O_{3,1m}$, x10[13] [c]	$O_{3,f}$, x10[13] [d]
1.23	0	0	0.928	3.617(3.617)
	3.7	0	0.928	3.598(3.671)
	3.7	0.03	0.934	4.048(3.642)
	3.7	0.1	0.937	5.420((3.654)
2.46	0	0	2.677	4.498(4.498)
	7.4	0	2.677	4.488(4.498)
	7.4	0.03	2.687	4.720(4.514)
	7.4	0.1	2.709	5.325(4.551)
4.21	0	0	6.583	11.06(11.06)
	12.6	0	6.583	11.06(11.06)
	12.6	0.03	6.606	11.59(11.10)
	12.6	0.1	6.662	13.08(11.19)

[a] All concentations in units of molecules cm^{-3}, [b] The corresponding pressures in Torr are: 380, 760 and 1300. [c] ozone concentration after 60 s.[d] Ozone concentations after $2.34x10^3$, $1.08x10^3$ and $1.08x10^3$ s at O_2 pressures of 380 ,760 and 1300 Torr, respectively; in parenthesis ozone concentrations at the same times extrapolated linearily from the values at 60 s in previous column.

As a final test of the accuracy and meaning of the quantum yields determined in this work model calculations were carried out. In these calculations all possible reactions, in the dry as well as the water containing system were included in order to examine any possible cause for non linearity of R_{O_3} and reduction of Φ_{O_3}.

Potential formation of ozone in reactions 20 and 21 involving O_2 in the $A^3\Sigma_u^+$ state was also tested by the model calculations. We assumed conditions that would maximize ozone formation from this source, namely formation of $O_2(A^3\Sigma_u^+)$ at the expense of $O(^1D)$ and neglecting the possibility that reaction 22 can also take place. In fact, only its overall rate constant was determined (Kenner and Ogryzlo 1980).

$$O_3 + h\nu \longrightarrow O_2(A^3\Sigma_u^+) + O(^3P) \quad (20)$$
$$O_2(A^3\Sigma_u^+) + O_2 \longrightarrow O_3 + O(^3P) \quad (21)$$
$$\longrightarrow O_2 + O_2 \quad (22)$$

Turning now to the results of the model calculations summarized in Table I, it should be emphasized that the simulation includes an assumption that the photodissociation of oxygen proceeds with a quantum yield of unity. Consequently, linearity with time of the rate of ozone formation, should be considered as a confirmation of the procedures used in the experimental determination of ϕ_{-O_2}. It should also be mentioned that, as an additional precaution in the linearity test, the water concentration was set at 3 ppm, the upper limit specified by the manufacturer. This value is by an order of magnitude higher than the measured water content. It can be seen that, even under these rather extreme conditions, the maximum expected deviation from linearity, at 380 Torr, is less than one percent. We conclude thus that ϕ_{-O_2} is indeed very close to unity as indicated by the experimentally determined value of 0.93 ± 0.08, the error limits of which reflect the scatter of the results but do not take into account systematic errors. Inaccuracies in the absorption spectra of O_2 and O_3 appear to be the main causes of such errors.

The model calculations seem also to exclude the possibility that, under our experimental conditions, $O_2(A^3\Sigma_u^+)$ has a major role in the mechanism of ozone formation. Since within the experimental time scale the presence of water does not reduce the rate of ozone formation, the ozone generating reactions of molecular oxygen in this state should result in an upward curvature in the time dependence of ozone build up. This assumption is indeed verified by the model calculations. However, even if as a measure of safety we assume that, failure to detect 10 percent deviation from linearity was possible, the model calculations show that the quantum yield of the primary photodissociation step in which $O_2(A^3\Sigma_u^+)$ is formed cannot exceed 0.03.

4.3. ATMOSPHERIC CONSEQUENCES

Since it is generally assumed that photodissociation of O_2 in the Herzberg continuum proceeds with a quantum yield of unity, the present results do no seem to have any direct bearing upon the known chemistry of ozone formation. Indirectly however, we were able to show that, at least at 214 nm, $O_2(A^3\Sigma_u^+)$ cannot be an important source of ozone in the atmosphere. This conclusion applies to any other states of O_2 formed in the photolysis of ozone at 214 nm. The suggestion that the A and A' states, potentially formed upon ozone photolysis below 230 nm, could generate significant amounts of ozone in their reactions with ground state oxygen (Saxon and Slanger 1986, Crutzen 1986), thus does not seem to be borne out by the experimental findings at high O_2 pressures. Conceivably, such reactions could become important at lower pressures and shorter wavelength. This possibility is presently investigated by us.

The clear demonstration that $O_2 - O_2$ interactions are reflected in the photodissociation of oxygen is quite important in as far as extrapolation of laboratory results to the real atmosphere is concerned. For instance, a rather trivial error can occur when pressure effects on various reactions initiated by O_2 photodissociation are investigated and the experimental results are interpreted assuming a constant absorption cross section. Isotope enrichement studies represent another case where missinterpretation of the results is possible in particular if the O_2 dimer exhibits isotope selectivity in its photodissociation reactions.

ACKNOWLEDGEMENT

We thank T. G. Slanger (SRI) for helpful suggestions. A. H. gratefully acknowledges the support under the MPI Visiting Scientist Fellowship scheme.

REFERENCES

Cheung, A. S. C., K. Y. Yoshino, W.H. Parkinson, S.L.Guberman and D. E. Freeman, 1986: Absorption cross section measurement of oxygen in the wavelength region 195-241 nm of the Herzberg continuum. *Planet. Space Sci.*, 34, 1007-1021.

Crutzen P. J., 1986 in " Proceedings of the CMA Sponsored Meeting on Stratospheric Chemistry." Boulder 1986, CMA - FPP - 85 - 559, 157.

Johnston, H. S., M. Paige and F. Yao, 1984: Oxygen absorption cross sections in the Herzberg continuum and between 206 and 327 K *J.Geophys. Res.*, 89, 11661 - 11665.

Kenner, R. D. and E. A. Ogryzlo 1980: Deactivation of $O_2(A^3\Sigma_u^+)$ by O_2, O and Ar. *Int. J. Chem. Kinet.*,12, 501-508.

Lee, L. C., G. Black, R. L.Sharpless and T. G. Slanger 1980: $O(^1S)$ yield from O_3 photodissociation at 1700-3100A. *J. Chem. Phys.*, 73, 256-258.

Long, C. A. and G. E. Ewing, 1971: The infrared spectrum of bound state oxygen dimers in the gas phase. *Chem. Phys. Lett.*, 9, 225-229.

Moortgat, G. K. and P. Z. Warneck, 1975: Relative $O(^1D)$ quantum yields in the near uv photolysis of ozone at 298 K *Naturforsch.* 30a, 835-843.

Saxon, R. P.and T. G. Slanger, 1986: Molecular oxygen absorption continua at 195-300 nm and O_2 radiative lifetimes. *J. Geophys. Res.*, 91, 9877-9879.

Wayne, R. P., 1987: The photochemistry of ozone *Atmospheric Environment*, 21, 1683-1694.

INFRARED MATRIX ISOLATION STUDIES OF PHOTOCHEMICAL PROCESSES OF ATMOSPHERIC INTEREST WITHIN WATER, SO_2 AND OZONE AGGREGATES

A. Schriver, L. Schriver, C. Barreau, D. Carrère, and J.P. Perchard
Université Pierre et Marie Curie
75252 Paris Cedex 05, France

K. Jaeger and O. Schrems
Alfred Wegener Institute for Polar and Marine Research, Columbusstrasse
D-2850 Bremerhaven, FRG

ABSTRACT

Matrix isolation coupled with IR spectroscopy has been applied in laboratory studies in order to simulate the formation of sulfuric acid in the troposphere by photooxidation of sulfur oxides. In the first step, 1:1 (H_2O-O_3) and 1:2 (O_3-H_2O-O_3) complexes have been identified for the system water/ozone in solid argon and UV-photolysis of these samples generated hydrogen peroxide. In the SO_2/O_3 system irradiation led to the formation of SO_3 and SO_4. The photoreaction products in the system $SO_2/H_2O/O_3$ were H_2O_2, SO_3 and SO_4.

1. INTRODUCTION

Among the most significant atmospheric pollutants involved in acid rain formation is sulfur dioxide which mainly enters the atmosphere from anthropogenic sources like fossil fuel burning and the smelting of metals. The primary pollutant SO_2 may be transformed by a manifold of reactions giving sulfuric acid as the eventual product. Pathways for the oxidation of SO_2 in the atmosphere are complex and have not yet been fully understood down to the smallest detail. A number of mechanisms have been proposed to explain the oxidation of atmospheric sulfur dioxide to sulfuric acid and sulfates. These may be classified into two types of processes:
(a) homogeneous gas-phase oxidation
(b) heterogeneous oxidation in fog and cloud droplets or aerosol particles.

The mechanisms of the reactions under conditions excisting in clouds are poorly understood. The SO_2 oxidation rate is obviously much enhanced by the presence of oxidants or oxidant precursors carried along with polluted air (Warneck 1988). In the presence of clouds and fogs, SO_2 may dissolve to some extent in the water droplets, where it forms HSO_3^- and SO_3^{2-}. These ions are subject to oxidation not only by molecular oxygen, but also by the highly reactive trace constituents ozone and hydrogen peroxide that are the products of secondary atmospheric photochemical reactions. The recognition that ozone is an important oxidant in clouds is due to Penkett (1972) and has been supported by further work (Penkett et al. 1979, Maahs 1983a, 1983b, Nahir et al. 1987). They showed in laboratory experiments that ozone rather rapidly oxidized sulfur dioxide absorbed in artificial fog droplets. It was demonstrated that the SO_2 oxidation rate was proportional to the mixing ratio of O_3 in the air (Maahs 1983).

The relative importance of the various reaction pathways in the SO_2 oxidation depends on factors such as relative humidity, aerosol loading and intensity of solar radiation. As Sodeau and Lee (1980) pointed out, the existence of molecular donor-acceptor complexes or molecular clusters may also be important in the photo-oxidation of SO_2. Studies of complexes or clusters are most relevant to conditions in which gas-phase reactions dominate acid production and gas-to-particle conversion proceeds through nucleation (Keese and Castleman Jr. 1987). Sulfur dioxide is thought to be a major source of sulfate cloud condensation nuclei, which constitute a significant control on climate via their strong influence on cloud droplet size spectra (Chatfield and Crutzen 1984).

Concerning the area of mechanisms, it seems well established that oxidation by OH· radicals is the dominant gas-phase reaction in the conversion of SO_2 to H_2SO_4. The HSO_3· radical, a key intermediate in the oxidation of SO_2 has been identified and characterized by matrix-isolation infrared spectroscopy (Nagase et al. 1988). It has only been recently, that the detection of the gaseous HSO_3· radical has been reported (Egsgaard et al. 1988). In a gas-phase and matrix-isolation study Bhatia and Hall (1987) found that the reaction SO_2 + OH yields the $HSO_3 \cdot H_2O$ complex as the final product in the matrix.

ISBN 0-937194-15-8

It is well known that ozone dissolved in water is a potential source of hydroxyl radicals.

$$O_3 \xrightarrow{h\nu} O_2 + O(^1D); \quad \lambda<320 \text{ nm} \qquad (1)$$

$$O(^1D) + H_2O \longrightarrow H_2O_2 \qquad (2)$$

The hydrogen peroxide serves both as an oxidant as a source of solution hydroxyl radicals when it is photolyzed:

$$H_2O_2 \longrightarrow HO\cdot + HO\cdot; \quad \lambda<380 \text{ nm} \qquad (3)$$

In the first part of a study devoted to photochemical processes of atmospheric relevance involving H_2O, SO_2 and O_3 we have been led to examine in detail the infrared spectra of $(H_2O)_m(SO_2)_n$ complexes trapped in solid argon and and nitrogen (Schriver et al. 1988). The present work gives a survey of bi- and tri-molecular photoreactions involving H_2O/O_3, SO_2/O_3, $SO_2/H_2O/O_3$ and SO_3/H_2O trapped in solid argon. The aim of such studies is the identification of the successive reaction steps in the generation of atmospheric H_2SO_4.

2. EXPERIMENTAL

The gas mixtures were prepared by standard manometric techniques and sprayed through either a single jet or dual jet system onto a CsI window kept below 20 K in a liquid helium cryostat (TBT). Ozone was prepared by tesla coil discharge of oxygen at low pressures in Pyrex finger partially immersed in liquid nitrogen. Water was doubly destilled and outgassed through several freeze and thaw cycles. SO_2 (Matheson) was dried over P_2O_5 and destilled at low temperature. The photolysis experiments were performed with unfiltered and filtered light of a 100 W medium pressure mercury arc. The spectra were recorded on a Perkin-Elmer Model 580 IR spectrometer. The resolution varied according to the spectral region within 1.8 cm^{-1} at 3700 cm^{-1} and 0.9 cm^{-1} in the range 1600 - 1000 cm^{-1}.

3. RESULTS AND DISCUSSION

3.1 THE H_2O/O_3 SYSTEM

3.1.1 Identification of complexes

The spectral analysis of water/ozone doped argon matrices leads to the identification of two complexes with different stoichiometries. In both cases vibrational perturbations are weak but significant enough to elucidate some structural properties.

The 1:1 complex: At low concentrations of H_2O and O_3 ($H_2O/O_3/Ar$ = 1/5/1000) in the host matrix new features which are not present in the spectra of the parent meolecules are identified in the stretching and bending regions of water. The corresponding frequencies are summarized in table 1 for three isotopic species of water.

TABLE 1. FREQUENCIES (cm^{-1}) OF THE H_2O ABSORPTIONS IN THE H_2O-O_3 COMPLEX TRAPPED IN SOLID ARGON AT 10 K.

	H_2O		D_2O		HOD	
	obs	calc	obs	calc	obs	calc
ν_1	3632.4	3630.1	2653.1	2654.2	2697.0	2700.0
ν_2	1592.5	1592.3	1176.5	1174.9	1399	1400.8
ν_3	3726.8	3726.2	2765.6	2766.0	3691	3691.3

Despite of the weakness of the frequency shifts with respect to the monomeric H_2O species, isotopic decoupling observed for HOD shows that the two OH oscillators are not equivalent. Force field calculations corrected from anharmonicity proves this non-identity. A possible structure of the complex is shown below:

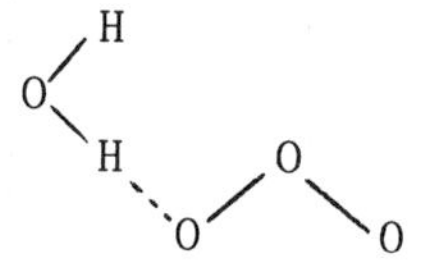

1:1 complex

The 1:2 complex: In the presence of a large excess of ozone with respect to water, new bands are observed in the water spectral regions (table 2). The most significant result is, as in the case of the 1:1 complex, the asymmmetrical decoupling in the HDO molecule, leading to the conclusion that the two OH oscillators are not equivalent. A possible structure is shown below:

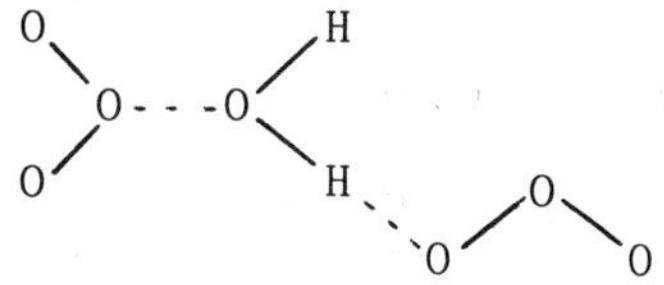

1:2 complex

TABLE 2. EXPERIMENTAL FREQUENCIES (cm^{-1}) OF THE H_2O ABSORPTIONS IN THE H_2O-(O_3) COMPLEX TRAPPED IN SOLID ARGON AT 10 K.

	H_2O	D_2O	HDO
ν_3	3706.9	2740.4	3671
ν_1	3603	2635.4	2681
ν_2	1599	1180.1	1403

3.1.2 Irradiation Experiment

The photolysis of the H_2O/O_3 doped argon matrices were performed either with unfiltered or filtered UV light (medium pressure mercury arc) in various time intervals. Irradiation at $280 < \lambda < 360$ nm led to the formation of H_2O_2 with simultaneous disappearance of the 1:1 complex. Figure 1 shows the depletion of the absorption bands of H_2O and O_3 involved in the H_2O-O_3 complex and the increase of the ν_6 band assigned to the reaction product H_2O_2.

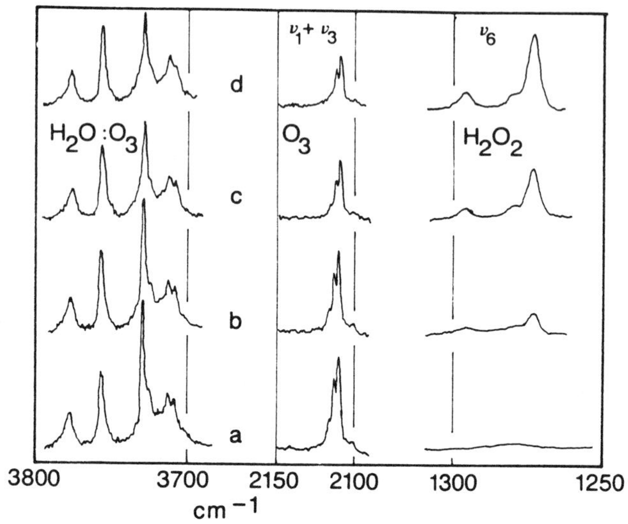

Fig. 1 . Evolution of the spectrum of a $H_2O/O_3/Ar$ = 1/6/600 matrix with the time of irradiation (band pass filter: $280 < \lambda < 360$ nm). Irradiation time: a = 0, b = 30, c = 40, d = 150 min.

Accordingly we suggest the following mechanism for the formation of hydrogen peroxide:

$$H_2O{\cdot}O_3 \xrightarrow{h\nu} H_2O{\cdot}O(^1D) + O_2 \longrightarrow H_2O_2 \qquad (4)$$

The first step is the photodissociation of O_3 producing oxygen in the ^1D state. The formation of the $H_2O{\cdot}O(^1D)$ complex can be considered as transition state. Finally the oxygen atom reacts with the H_2O molecule of the original complex. No diffusion of O within the matrix is required for the reaction to proceed, which makes this process very efficient. The diffusion of O would imply $O(^1D) \rightarrow O(^3P)$ conversion due to the short life time of $O(^1D)$.

3.2 THE SO_2/O_3 SYSTEM

Mixtures of SO_2/Ar and O_3/Ar have been codeposited in order to form SO_2/O_3 complexes within the matrix. However, the IR spectra obtained from matrices prepared in this way do not definitely lead to the observation of absorption bands assignable to SO_2/O_3 complexes, probably because of the weakness of the frequency shifts with respect to monomers upon complexation.

Irradiation of the samples with visible and/or UV light generates SO_3 characterized by by one strong absorption band at about 1390 cm^{-1}, With proceeding time much weaker lines appear at 1442 and 1272 cm^{-1} which are assigned to the SO_4 species (figure 2).

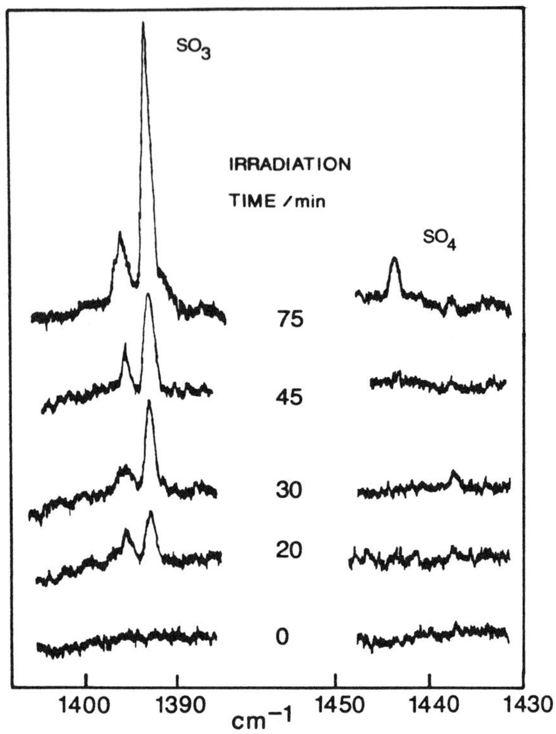

Fig. 2. Evolution of the spectrum of a $SO_2/O_3/Ar$ = 1/5/500 matrix with the time of irradiation (band pass filter: $296 < \lambda < 360$ nm).

3.3. THE $H_2O/SO_2/O_3$ SYSTEM

The experimental findings for the H_2O/O_3 and the SO_2/O_3 systems suggest that for the ternary system $H_2O/SO_2/O_3$ conversion of SO_2 to H_2SO_4 might take place either via the reactions described in equation 5 or equation 6 and 7:

$$SO_2 \xrightarrow{h\nu,\ O_3} SO_3 \xrightarrow{H_2O} H_2SO_4 \qquad (5)$$

$$H_2O \xrightarrow{h\nu,\ O_3} H_2O_2 \qquad (6)$$

$$SO_2 + H_2O_2 \longrightarrow H_2SO_4 \qquad (7)$$

The photoreaction products which we have observed in triply doped ($H_2O/SO_2/O_3$) matrices are the same as those obtained in the $H_2O/O_3/Ar$ and $SO_2/O_3/Ar$ experiments, namely H_2O_2 and SO_3 + SO_4. In fact, under the experimental conditions used no trace of H_2SO_4 was detected.

3.4. THE SO_3/H_2O SYSTEM

The results of the photolysis of the $H_2O/SO_2/O_3$ system prompted us to reexamine the SO_3/H_2O system (Bondybey et al. 1985) in either Ar or N_2 matrix.

H_2O/SO_3 interactions have also been studied by Tso and Lee (1984) who observed the 1:1 complex of SO_3 and H_2O in the photooxidation of H_2S in in solid O_2. The interaction of SO_3 with H_2O and the corresponding potential surface are of considerable interest. Hofmann-Sievert and Castleman, Jr. (1984) report a rapid isomerization of the adduct $SO_3 \cdot H_2O$ to H_2SO_4 in the gas phase (barrier of isomerization is a maximum of 13 kcal/mol).

The results of our experiments show, that the vibrational perturbations of SO_3 by H_2O in the 1:1 complex are scarcely observable and noticeably matrix dependent. This is different to what has been observed for SO_3 complexes with stronger oxygen containing bases (Sass and Ault, 1986). For example, in Ar matrix we found that the ν_2 mode of SO_3 is red shifted (blue shift for stronger oxygen bases) and the ν_3 mode is very weakly blue shifted (red shift for stronger bases). Thus, these observations suggest the formation of a weak hydrogen bond OH···O, rather than a charge transfer complex between the oxygen atom of H_2O and the electron acceptor sulfur atom of SO_3 as in the case of stronger oxygen bases.

In the experiments which we have performed so far we have not yet detected sulfuric acid as result of the expected isomerization of the $SO_3 \cdot H_2O$ complex to H_2SO_4.

ACKNOWLEDGEMENT

K. J. thanks the Gottlieb-Daimler and Karl-Benz Stiftung for fellowship support during a stay at UPMC Paris.

REFERENCES

Bhatia, S.C., and J.H. Hall,Jr., 1987: Spectroscopic identification of products formed in the gas-phase reaction of OH with atmospheric sulfur compounds, in: The chemistry of acid rain, R.W. Johnson and G.E. Gordon (eds) American Chemical Society, Washington D.C.

Bondybey, V.E. and J.H. English, 1985: Infrared spectra of SO_3 polymers and complexes in rare gas matrices. J. Mol. Spectr. 109, 221-228.

Chatfield, R.B., and P.J. Crutzen, 1984: Sulfur dioxide in remote oceanic air: cloud transport of reactive precursors. J. Geophys. Res. 89, 7111-7132.

Egsgaard, H., L. Carlsen, H. Florencio, Th. Drewello and H. Schwarz, 1988: Experimental evidence for the gaseous HSO_3 radical.The key intermediate in the oxidation of SO_2 in the atmosphere. Chem. Phys. Lett., 148 537-540.

Hofmann-Sievert, R., and A.W. Castleman,Jr., 1984: Reaction of SO_3 with water clusters and the formation of H_2SO_4. J. Phys. Chem., 88, 3329.

Keese, R.G., and A.W. Castleman,Jr., 1987: Acid clusters, in: The chemistry of acid rain, R.W. Johnsoon, and G.E. Gordon (eds.), American Chemical Society, Washington D.C.

Maahs, H.G., 1983 a: Measurements of the oxidation rate of sulfur (IV) by ozone in aqueous solution and their relevance to SO_2 conversion in non-urban tropospheric clouds. Atmos. Environ., 17, 341.

Maahs, H.G., 1983 b: Kinetics and mechanism of the oxidation of S (IV) by ozone in aqueous solution with particular reference to SO_2 conversion in non-urban tropospheric clouds. J. Geophys. Res., 88, 10721-10732.

Nagase, S., S. Hashimoto and H. Akimoto, 1988: $HOSO_2$ and $HOSO_4$ radicals studied by ab initio calculation and matrix isolation technique. J. Phys. Chem., 92, 641-644.

Nahir, T.M., and G.A. Dawson, 1987: Oxidation of sulfur dioxide by ozone in highly dispersed water droplets. J. Atmos. Chem., 5, 373-383.

Penkett, S.A., 1972: Oxidation of SO_2 and other atmospheric gases by ozone in aqueous solution. Nature, 240, 105-106.

Penkett, S.A., B.M.R. Jones, K.A. Brice and A.E.J. Eggleton, 1979: The importance of atmospheric ozone and hydrogen peroxide in oxidising sulfur dioxide in cloud and rainwater. Atmos. Environ., 13, 323-337.

Sass, C.S., and B.S. Ault, 1986: Matrix isolation infrared spectra of the 1:1 molecular complexes of sulfur trioxide with selected oxygen-containing bases. J. Phys. Chem., 90, 4533.

Schriver, A., L. Schriver and J.P. Perchard, 1988: Infrared matrix isolation studies of complexes between water and sulfur dioxide: Identification and structure of the 1:1, 1:2 and 2:1 species. J. Mol. Struct., 127 125-142.

Sodeau, J.R., and E.K.C. Lee, 1980: Photooxidation of sulfur dioxide in low-temperature matrices J. Phys. Chem., 84, 3358-3362.

Tso, T.-L., and E.K.C. Lee, 1984: Formation of sulfuric acid and sulfuric trioxide/water complex from photooxidation of hydrogen sulfide in solid O_2 at 15 K. J. Phys. Chem., 88, 2776-2781.

Warneck, P., 1988: Chemistry of the Natural Atmosphere., Academic Press, Inc., San Diego.

OZONOLYSIS OF ALKENES UNDER ATMOSPHERIC CONDITIONS

O. Horie and G.K. Moortgat
Max-Planck-Institut für Chemie
D-6500 Mainz, FRG

ABSTRACT

Ozonolysis of C_2H_4, C_3H_6, and *trans*-2-C_4H_8 was carried out under atmospheric conditions in a continuously-stirred tank-reactor coupled with molecular-beam sampling and matrix-isolation FTIR spectroscopy, and the effect of the addition of SO_2, HCHO and CH_3CHO was examined. A new intermediate, hydroxyethyl formate $CH_3CH(OH)$-O-CHO, was formed as a major reaction product in the C_3H_6 and *trans*-2-C_4H_8 ozonolysis. This compound completely disappeared with the addition of SO_2. In both systems, CH_3CHO, HCHO, CO_2, CO, CH_3OH and CH_4 were major products, and CH_2CO, HCOOH and CH_3COOH were minor products.

1. INTRODUCTION

Accurate kinetic and mechanistic data on the ozonolysis of alkenes are essential in the modelling of the tropospheric atmosphere. However, except for C_2H_4, no detailed quantitative analyses of main reaction products under atmospheric conditions have been made (Atkinson and Lloyd 1984). This is largely due to the complexities of the elementary reactions involved, in particular of the formation and subsequent reactions of intermediate products such as hydroxymethyl formate (HMF) formed in the O_3/C_2H_4 system (Su et al. 1980, Kan et al. 1981, Niki et al. 1981).

The purpose of this work was to carry out the ozonolysis of C_2H_4, C_3H_6 and *trans*-2-C_4H_8 under atmospheric conditions, and to analyze the main reaction products in the presence and absence of added SO_2, HCHO, and CH_3CHO. A continuously-stirred tank-reactor was constructed for this purpose, which enabled us to analyze reaction intermediates as well as stable prodcuts via molecular-beam sampling and matrix-isolation IR spectroscopy.

2. EXPERIMENTAL

A continuously-stirred tank-reactor (CSTR) was used in this study. The apparatus together with experimental procedures will be described in detail separately (Horie and Moortgat 1988). The reactor was made of a 2 *l* Pyrex flask with 2.14 *l* reaction volume equipped with a Teflon stirrer, two opposingly placed quartz windows, inlet ports for reactants and the carrier gas, and a pumping and sampling port. The reaction mixtures were continuously withdrawn through a molecular-beam sampling unit and deposited on a cold finger maintained at 7±1 K. The matrix-isolated IR spectra were recorded with a FTIR spectrometer (Bomem DA03.01).

Ozone was prepared in an electric discharge of pure O_2 (Linde 99.999%). The ozone concentration was determined by the absorption at 254 nm upstream of the reactor. C_2H_4 (Messer-Griesheim, 99.95%), C_3H_6 (Matheson, CP), and *trans*-2-C_4H_8 (Messer-Griesheim, 99%) were metered through a needle valve and their concentrations were determined by a pressure-drop method. SO_2 (4.8% in synthetic air), CH_3CHO , and HCHO which was generated by pyrolyzing trioxane were diluted in Ar. The mixtures of Ar and O_2 were used as the carrier gas in which O_2 was maintained at 20±1% during a run. Actual sampling was performed by opening a shutter in front of the cryostat chamber for 5 to 60 min.

The range of the experimental conditions were as follows: $T/K = 294–301$, $p/\text{Torr} = 745–768$, $t/s = 39–60$, $[O_3]_0/10^{15}\text{cm}^{-3} = 6–11$, $[C_2H_4]_0/10^{15}\text{cm}^{-3} = 2–10$, $[C_3H_6]_0/10^{15}\text{cm}^{-3} = 7$, $[trans\text{-}2\text{-}C_4H_8]_0/10^{15}\text{cm}^{-3} = 2–18$, $[HCHO]_0/10^{15}\text{cm}^{-3} = 3–16$, $[CH_3CHO]_0/10^{15}\text{cm}^{-3} = 16–70$, and $[SO_2]_0/10^{15}\text{cm}^{-3} = 2–11$.

3. RESULTS AND DISCUSSION

3.1 C_2H_4

Results are summarized in Fig. 1 and in Table 1. In the absence of the added compounds, HCHO, CO, CO_2 and H_2O (though not quantitatively analyzed) were the major stable products and HCOOH was a minor product. Hydroxymethyl formate $CH_2(OH)$-O-CHO, abbreviated to HMF hereafter, was formed as a dominant intermediate product. This compound was first observed by Su et al.(1980) in a static ozonolysis using long-path FTIR spectroscopy and further studied by Niki et al.(1981) in a similar experiment. The initial reactions in the O_3/C_2H_4 system produce, via the primary ozonide, HCHO and excited Criegee radicals CH_2OO^*, which either decompose unimolecularly to CO and CO_2 among other products or yield the collisionally stabilized Criegee radicals CH_2OO

ISBN 0-937194-15-8

(Atkinson and Lloyd 1984). The stabilized CH_2OO radicals then react with HCHO to form HMF:

$$CH_2OO + HCHO \rightarrow CH_2(OH)\text{-}O\text{-}CHO \quad [1]$$

Most of the present results are consistent with the results of the static experiments (Su et al. 1980, Kan et al. 1981, Niki et al. 1981). However, formic anhydride $(HCO)_2O$ was observed as a trace component here. This is explained by much shorter reaction time (60 s) used in this study as compared to longer reaction times (up to 60 min) in the static experiments, where HMF was observed to decay to $(HCO)_2O$.

The results of the effect of SO_2 are shown in Table 1. The complete disappearance of HMF and a large increase in [HCOOH] are consistent with previous observations (Su et al. 1980, Hatakeyama et al. 1986). However, an increase in [HCHO] was not observed, in contrast to about 30% increase found by Hatakeyama et al.(1986). With about 100-fold larger reactant concentrations used in this study, SO_2 may react with HMF as well as with CH_2OO. In fact, in the matrix-isolation FTIR spectra, several small absorption peaks in the wavelength range between 1000 and 1200 cm^{-1} were observed, which is the characteristic absorption range of the S=O stretching vibration of sulphoxides, sulphites or sulphones (Socrates 1980). Also, an ion chromatographic analysis of the aqueous solution obtained by rinsing the reactor with distilled water after the SO_2 -experiment indicated the presence of at least three ionic species besides H_2SO_4. They might be due to the reaction products between HMF and SO_2.

One interesting finding with the added SO_2 is the very efficient HMF-removal capability of the aerosol-coated reactor wall which had been exposed to the $O_3/C_2H_4/SO_2$ reaction mixture (Table 1, last column). This effect persisted even after an overnight pumping of the reactor. As indicated above, this aerosol is not simply sulfuric acid but contains complex reaction products including those between SO_2 and HMF. Further studies on these phenomena are underway.

TABLE 1. THE EFFECT OF SO_2 ON THE O_3/C_2H_4 SYSTEM

run	I	II	III[b]
$[O_3]_0^{a)}$	9.38	9.27	9.31
$[C_2H_4]_0^{a)}$	8.99	9.29	8.91
$[SO_2]_0^{a)}$	0	10.7	0
O_3 conv/%	38.8	47.4	37.2
C_2H_4 conv/%	29.5	34.9	30.0
product yield relative to the converted C_2H_4/%			
HCHO	59.9	66.3	61.8
CO	24.6	18.5	24.4
CO_2	16.9	13.8	20.0
HCOOH	3.8	9.2	13.5
HMF[c]	64.5	0	0

T/K = 295–299, p/Torr = 752–758, t/s = 60. a) The concentrations are in $10^{15}cm^{-3}$ units. b) Run II preceded run III. c) The absolute concentration of HMF was estimated on the basis of a 100% carbon balance.

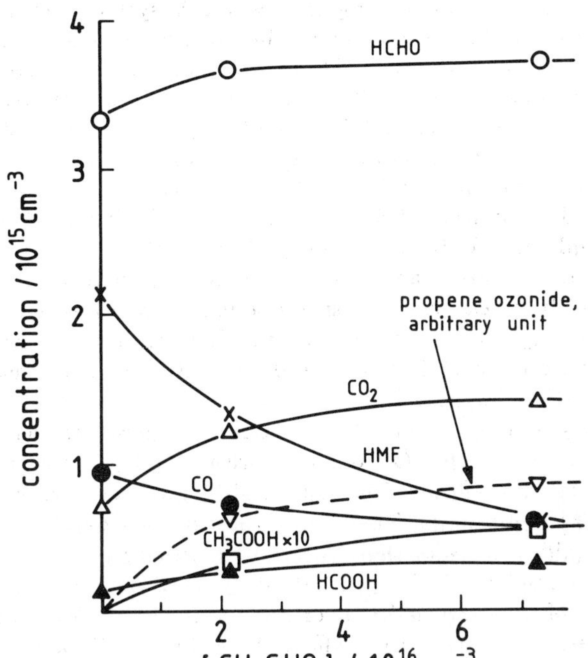

Fig. 1. The effect of the addition of CH_3CHO on the O_3/C_2H_4 system. T/K=298, p/Torr=755, t/s=55, $[O_3]_0/10^{15}cm^{-3}$=10.0, $[C_2H_4]_0/10^{15}cm^{-3}$=10.0.

The effect of the added CH_3CHO is noted by the decrease in [HMF] and the appearances of propene ozonide (POZ) and of CH_3COOH as shown in Fig. 1. Although the formation of propene ozonide in the $O_3/C_2H_4/CH_3CHO$ system was previously observed (Su et al.1980, Hatakeyama et al.1986), the decrease in [HMF] and the formation of CH_3COOH had not been observed before. The present results can be explained by the competitive reaction of HCHO and CH_3CHO toward CH_2OO radicals to form the propene ozonide, which decomposes either to CH_3CHO and HCOOH or to HCHO and CH_3COOH:

$$CH_3CHO + CH_2OO \rightleftharpoons POZ \rightarrow CH_3CHO + HCOOH \quad [2]$$
$$\rightarrow HCHO + CH_3COOH \quad [3]$$

3.2 *trans*-2-C_4H_8

Main results are summarized in Table 2. In the absence of the added compounds, CH_3CHO, HCHO, CO, CO_2, CH_3OH, CH_4, and H_2O (not quantitatively determined) were formed as the major stable products. Ketene and HCOOH were minor products, and CH_3COOH and propene ozonide were detected as trace components. The initial reactions produce CH_3CHO and excited Criegee radicals CH_3CHOO^*. The unimolecular decomposition of CH_3CHOO^* leads to CH_4+CO_2, $CH_3+CO+OH$, CH_3+CO_2+H, or CH_3O+HCO (Atkinson and Lloyd 1984). Further oxidations of CH_3 and CH_3O radicals produce HCHO.

The most interesting finding in this system was the formation of a new reaction product which was tentatively described as hydroxyethyl formate $CH_3CH(OH)$-O-CHO, abbreviated to HEF. The main absorption lines of HEF in cm^{-1} are: 850 and 854 (doublet), 931, 1116, 1160, 1193, 1132 and 1336 (doublet), and 1745. Three sets of experiments were performed to characterize this product.

The addition of SO_2 to this system eliminated HEF completely. At the same time, CH_3COOH and HCOOH increased more than 10 and 2 times, respectively. The total disappearance of this compound in the presence of SO_2 suggests that this is a product between the primary initial products CH_3CHOO and CH_3CHO, as in the case of HMF in the C_2H_4 system.

The addition of HCHO should therefore decrease [HMF] since HCHO competes with CH_3CHO toward CH_3CHOO radicals. The added HCHO however increased [HEF] significantly. The increase in propene ozonide was insignificant. Note that [HCOOH] increased more than 10 fold which can be attributerd to the secondary reactions of HO_2 and HCHO (Horie and Moortgat 1988). The added CH_3CHO did not affect the yield of HEF as seen in Table 2. These results suggests that this compound is a product between the stabilized CH_3CHOO radicals and HCHO having a structure similar to that of HMF:

$$CH_3CHOO + HCHO \rightarrow CH_3CH(OH)\text{-O-}CHO \quad [4]$$

Although unexpected, all the experimental evidence, including that on the O_3/C_3H_6 system (see below) indicates that HCHO and not CH_3CHO is the reactant in the intermediate formation.

There are several unidentified small absorption peaks in the matrix isolation FTIR spectra, but the total concentration of these products was estimated to amount to less than 5% of the total prodcut yield. The absolute concentration of HEF listed in Table 2 was estimated by assuming a 100% carbon balance in the O_3/*trans*-2-C_4H_8 runs in the absence of the added compounds.

In the matrix isolation FTIR spectra of the reaction mixtures with the added SO_2 , much less and much smaller new peaks were observable than in the C_2H_4 case. One reason may be that reaction products between HEF and SO_2 were rapidly converted to aerosols and little fraction remained in the gas phase after 40 s reaction times. It should be also noted that the increases in [CH_3CHO] and [HCHO] in the presence of SO_2 were not found, contrary to the earlier observations for higher alkenes by Cox and Penkett (1972). The same argument used in the C_2H_4 system should apply here; namely due to higher concentrations of SO_2 and the reactants, SO_2 reacted with HEF as well as CH_3CHOO radicals.

The present results give the first quantitative analysis of the majority of reaction products including the intermediate HEF in the ozonolysis of 2-butenes carried out under atmospheric conditions. The only partially quantitative analysis of main reaction products is due to Niki et al.(1977). In the presence of 10 ppm HCHO, the initial reaction mixture consisting of 10 ppm *cis*-2-C_4H_8 and 5 ppm O_3 gave, at 50% conversion of C_4H_8, about 13% CH_4, 5% HCOOH, 8% CH_3OH, and 37% CO_2 relative to the reacted C_4H_8. The low pressure study of Atkinson et al. (1973) using photo-ionization mass spectroscopy listed HCHO, CH_3CHO and CH_2CO as major, and $(CHO)_2$, HCOOH, CH_3OH and H_2O_2 as minor products. CO and CO_2 were not detectable by the method used. The low pressure study of O_3/*cis* and *trans*-2-C_4H_8 systems by Martinez et al. (1981) did not give quantitative results. Kühne et al. (1980) observed in a study of the O_3/*cis*-2-C_4H_8 system with the reactant concentrations of about $1\times10^{18}cm^{-3}$ (about 100 times larger concentrations than used in this study) in their product spectra butene ozonide and formic acetic anhydride. Neither of them were detected in this study.

TABLE 2. THE OZONOLYSIS OF *trans*-2-C_4H_8

expt.	SO_2 effect		HCHO effect		CH_3CHO effect	
$[O_3]_0^{a)}$	8.41	8.32	9.18	9.20	8.56	8.57
$[C_4H_8]_0^{a)}$	6.62	6.58	6.93	6.76	6.79	6.65
$[added]^{a)}$	0	7.18	0	5	0	16
O_3 conv/%	96.3	96.5	97.0	96.4	97.0	97.9
C_4H_8 conv/%	90.6	85.6	85.6	85.2	83.8	81.4
product yield relative to the converted C_4H_8/%						
CH_3CHO	75.8	80.8	91.1	99.4	96.7	$394^{b)}$
CO_2	33.9	35.8	38.3	39.0	40.3	52.3
CO	20.2	19.5	21.9	32.0	23.0	22.0
HCHO	10.7	7.5	14.1	$109^{b)}$	14.7	20.9
CH_3OH	9.4	5.5	10.2	13.4	11.8	16.0
CH_4	10.0	12.1	11.5	13.1	12.0	12.0
CH_2CO	1.6	0	1.7	1.7	1.8	2.0
CH_3COOH	0.4	4.5	0.4	0.8	0.4	1.4
HCOOH	2.5	5.6	2.4	25.8	2.5	4.1
HEF	59.3	0	59.2	102.0	63.0	59.6

T/K = 295–299, p/Torr = 745–765, t/s = 39. a) The concentrations are in 10^{15} cm^{-3} units. b) the total concentration (= added + formed).

None of the previous studies seem to have indicated the formation of the intermediate HEF. Also, previous studies in which SO_2 was added to the reaction system seem to have payed little attention to organic reaction products, in particular to organic acids (Cox and Penkett 1972, Hatakeyama et al. 1984). The increase in [HCOOH] and [CH_3COOH] were first observed in this study. The mechanism for the formation and increase in the concentration may be closely connected to the formation of the intermediate and its disappearance in the presence of SO_2. Further work is planned.

3.3 C_3H_6

Experiments with C_3H_6 were carried out briefly to see if both of the intermediatees HMF and HEF were formed in this system. The results are summarized in Table 3. Both HMF and HEF were formed as the major products comparable to HCHO and CH_3CHO. The concentrations of HMF and HEF were estimated from the calibration determined in the O_3/C_2H_4 and O_3/*trans*-2-

TABLE 3. THE EFFECT OF SO_2 ON THE O_3/C_3H_6 SYSTEM

run	I	II	III	IV[b)]
$[O_3]_0^{a)}$	8.82	8.98	9.02	9.17
$[C_3H_6]_0^{a)}$	7.13	7.19	7.10	7.07
$[SO_2]_0^{a)}$	0	2.24	7.18	0
O_3 conv/%	71	76	74	74
C_3H_6 conv/%	53	52	53	53
product yield relative to the converted C_3H_6/%				
CH_3CHO	43.3	40.8	44.8	41.9
HCHO	57.5	47.3	45.7	44.5
CO_2	32.0	29.7	32.9	33.1
CH_3OH	4.7	2.7	0	3.7
CH_4	9.6	10.5	11.0	11.1
CH_2CO	1.5	0.1	0	0.1
CH_3COOH	0.2	0.8	1.7	0.6
HCOOH	6.2	7.4	14.0	8.5
HEF	46.0	0	0	0
HMF	43.0	0	0	0

T/K = 295–299, p/Torr = 752–758, t/s = 39. CO was not included due to a high CO impurity level in C_3H_6. However, in run I, [CO] produced was about $0.8\times10^{15}cm^{-3}$ or 21% relative to the converted C_3H_6. a) The concentrations are in $10^{15}cm^{-3}$ units. b) Run III preceded run IV.

C_4H_8 system, respectively. The carbon balance in the O_3/C_3H_6 system based on these estimates was satisfied within 30%, which indicates that the formation of both HMF and HEF is consistent with the following initial reaction yielding both CH_3CHOO and CH_2OO radicals:

$$O_3 + C_3H_6 \rightarrow \text{primary ozonide} \quad [5]$$
$$\text{primary ozonide} \rightarrow CH_2OO^* + CH_3CHO \quad [6]$$
$$\rightarrow CH_3CHOO^* + HCHO \quad [7]$$

A fraction of these excited Criegee radicals will give stabilized Criegee radicals (Atkinson and Lloyd 1984).

It is interesting to note that only the formates HMF and HEF were found, although CH_3CHO could form the analogous products, hydroxymethyl acetate and hydroxyethyl acetate, with the stabilized Criegee radicals. It is to be noted in this connection that in the $O_3/C_2H_4/CH_3CHO$ system, no product between CH_3CHO and CH_2OO radicals except propene ozonide was formed, despite a large decrease in [HMF] (Fig.1). Only HCHO may be able to form products with Criegee radicals CH_2OO or CH_3CHOO. In other words, for HCHO the HMF and HEF require lower potential barriers than to form corresponding secondary ozonides. The fact that the addition of HCHO to the O_3/*trans*-2-C_4H_8 system had little effect on the propene ozonide yield while addition of CH_3CHO to the O_3/C_2H_4 system produced a large amount of the same ozonide may be consistent with the above presumption.

The addition of SO_2 reveals a further interesting aspect of the reaction. The formation of HMF and HEF were already completely suppressed at lower [SO_2], while [HCOOH] did not increase. The effectiveness of the reactor wall which had been exposed to the $O_3/C_3H_6/SO_2$ system in eliminating HMF and HEF is also apparent from Table 3.

ACKNOWLEDGEMENT

We would like to thank U. Deister, Max-Planck-Institut für Chemie, for the ion-chromatographic analysis.

REFERENCES

Atkinson, R., B. J. Finlayson and J. N. Pitts, Jr., 1973: Photoionization mass spectrometer studies of gas phase ozone–olefin reactions. *J. Am. Chem. Soc.*, **95**, 7592–7599.

Atkinson, R. and A. Lloyd, 1984: Evaluation of kinetic and mechanistic data for modeling of photochemical smog. *J. Phys. Chem. Ref. Data*, **13**, 315–444.

Cox, R. A. and S. A. Penkett, 1972: Aerosol formation from sulphur dioxide in the presence of ozone and olefinic hydrocarbons. *J. Chem. Soc., Faraday Trans.* 1, **68**, 1735–1753.

Fu Su, J. G. Calvert and J. H. Shaw, 1980: A FTIR spectroscopic study of the ozone–ethene reaction mechanism in O_2-rich mixtures. *J. Phys. Chem.*, **84**, 239–246.

Hatakeyama, S., H. Kobayashi and H. Akimoto 1984: Gas-phase oxidation of SO_2 in the ozone–olefin reactions. *J. Phys. Chem.*, **88**, 4736–4739.

Hatakeyama, S., H. Kobayashi, Zi-Yu Lin, H. Takagi, and H. Akimoto, 1986: Mechanism for the reaction of CH_2OO with SO_2 . *J. Phys. Chem.*, **90**, 4131–4135.

Horie, O. and G. K. Moortgat, 1988: The formation of intermediate products in the gas-phase ozonolysis of alkenes, to be submitted.

Kan, C. S., Fu Su, J. G. Calvert and J. H. Shaw, 1981: Mechanism of the ozone–ethene reaction in dilute N_2/O_2 mixtures near 1-atm pressure. *J. Phys. Chem.*, **85**, 2359–2363.

Kühne, H., M. Forster, J. Hulliger, H. Ruprecht, A. Bauder, and H.-H. Günthard, 1980: Linear reactor infrared matrix and microwave spectroscopoy of the *cis*-2-butene gas-phase ozonolysis. *Helv. Chem. Acta*, **63**, 1971–1999.

Martinez, R. I., J. T. Herron and R. E. Huie, 1981: The mechanism of ozone–alkene reactions in the gas phase. A mass spectroscopic study of the reactions of eight linear and branched-chain alkenes. *J. Am. Chem. Soc.*, **103**, 3807–3820.

Niki, H., P. D. Maker, C. M. Savage and L. P. Breitenbach, 1977: Fourier transform IR spectroscopic observation of propylene ozonide in the gas phase reaction of ozone–*cis*-2-butene–formaldehyde. *Chem. Phys. Lett.*, **46**, 327–330.

Niki, H., P. D. Maker, C. M. Savage and L. P. Breitenbach, 1981: A FT IR study of a transitory product in the gas-phase ozone-ethylene reaction. *J. Phys. Chem.*, **85**, 1024–1027.

Socrates, G., 1980: *Infrared characteristic group frequencies*, John Wiley and Sons.

MEASUREMENTS OF OZONE ABSORPTION CROSS SECTIONS IN THE UV AT WAVELENGTHS USED AS REFERENCE IN OZONE DIAL SYSTEMS

Marco Cacciani, Alcide di Sarra, Giorgio Fiocco, and Antonella Amoruso
Department of Physics
University "La Sapienza"
00185, Rome, Italy

ABSTRACT

Absolute measurements of the ozone absorption coefficient in the Huggins bands at different temperatures have been carried out.

Ozone is produced with an electrical discharge and stored cryogenically; differential absorption measurements are subsequently obtained in a slowly evolving mixture of ozone and molecular oxygen. The radiation from a Xe lamp is focused on the slit of a Jobin-Yvon U 1000 double monochromator. The monochromatic beam, split into two channels for differential absorption measurements, is collected on a photomultiplier for photon-counting detection. The total pressure is controlled by means of a MKS Baratron pressure transducer, and the temperature by Pt 100 resistance detectors.

High resolution (to 0.012 nm) measurements cover a spectral range (339-360 nm) where the ozone absorption shows a strong dependance on temperature. Comparisons with existing data sets are possible between 339 and 345 nm, where our results agree within experimental errors with Brion, Daumont and Malicet (1984). Outside this spectral region our measurements appear to be the only recent systematic results available.

Results at temperatures between 220 and 293 K will be reported; they are particularly interesting in view of the utilization of this spectral region as a low absorption reference for the observation of atmospheric ozone profiles by active probing techniques (DIAL). In this respect, three wavelengths around 351 nm, 353 nm and 355 nm can be obtained as fundamental of the Xe F excimer laser, by H_2 Raman shifting of a XeCl excimer laser output and as third harmonic of the Nd YAG laser fundamental.

1. EXPERIMENTAL SET UP

In fig. 1 a description of the instrumentation is schematically shown. The radiation source is a Xenon lamp, housed in a lantern filled with molecular nitrogen and cooled by water circulation. The radiation is focused by quartz lenses on the entrance slit of a Jobin Yvon U 1000 double monochromator, with 1 m focal length.

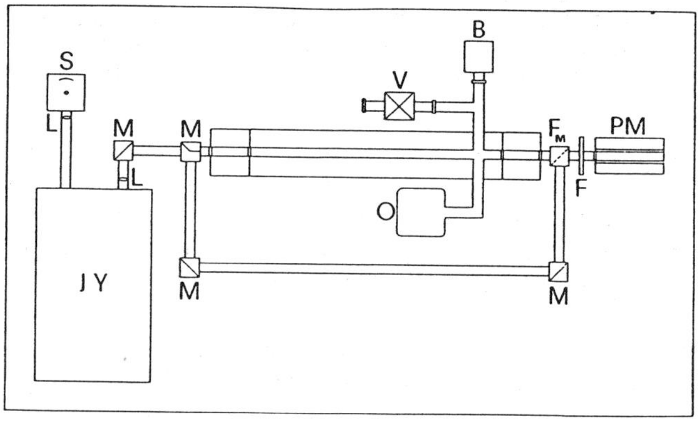

Fig. 1. The optical system: S, source of radiation, xenon lamp; L, lens; M, mirror; JY, Jobin-Yvon monochromator; O, ozonizer; B, MKS Baratron pressure gauge; V, valve, connection to the vacuum system; F_m, flipping mirror; F, bandpass filter; PM, photomultiplier. The upper chamber, with the cooling system, is the absorption cell, the lower is the reference.

The main characteristics of the monochromator are the low stray light level and the high resolution, usually kept at 0.012 nm. The light output from the monochromator is collimated by a lens and is split into two channels (absorption and reference chamber respectively). The optical path in the gas sample is about 1 m. Situated at the extremities of the absorption chamber are two cells that can be emptied to avoid condensation of atmospheric water vapour on the windows during measurements at low temperatures.

The beams pass through the two chambers, reach a flipping mirror that alternately selects the light emerging from either cells and sends it to the

ISBN 0-937194-15-8

detector. The detector consists of a photomultiplier followed by a photon counting system. The temperature is regulated by a cooling system (ethyl alcohol flows in an enclosure surrounding the absorption chamber) and is controlled by four resistance detectors which are placed around the chamber. The total pressure in the chamber is measured by a MKS Baratron gauge, whose output signal, with the temperature signal, is digitized and is acquired by a microprocessor during the measurement process. The process runs almost automatically, controlled by the microprocessor.

The absorption coefficient determination is based upon the application of the Beer Lambert's law:

$$k = P\,T_{O3}\,d\,/\,L\,T\,P_{O3};\quad d = \mathrm{Log}_{10}(I_0/I)$$

Independent determination of partial ozone pressure (P_{O3}), temperature (T_{O3}), optical path (L) and optical density (d) are needed to obtain an absolute value of the absorption coefficient. P and T are natural pressure and temperature, I_o and I light intensities in the reference and in the absorption channel respectively.

Much of the difficulty in the measurement consists in the ability to estimate the instantaneous value of the ozone partial pressure.

Ozone is produced by a crown discharge cooled by liquid nitrogen, from UPP molecular oxygen, purified by passing through a liquid nitrogen cold trap. The produced ozone is pure and is stored cryogenically.

The pure ozone in the chamber progressively undergoes a slow temperature-dependent transformation back to molecular oxygen. The ozone decay is principally due to a first order heterogeneous reaction at the walls, being the thermal conversion much slower than the heterogeneous reaction.

In fig. 2 we have a graph of the total pressure of the gas sample inside the chamber as a function of time. After the initial period, during which the liquid ozone evaporates and the evolving gas mixture reaches the thermal equilibrium, the only transformation is the O_3 to O_2 conversion, following the law

$$P_t = P_i\ (\ 1 - 1/3\ e^{-t/\mu})$$

The behaviour of P_t with time, according to this expression, is shown by the solid line: the crosses are the experimental total pressure values. P_i is the oxygen pressure converted to ozone, P_t is the total pressure at time t. μ is about 10^5 -10^6 s, depending on gas temperature and pressure. The ozone partial pressure is determined by measuring P_t and computing the P_i asymptotic value of the oxygen pressure, by means of a fit of the total pressure in function of time. The ozone partial pressure is

$$P_{O3} = 2\ (P_i - P_t)$$

The absorption chamber is constructed of stainless steel and glass; Viton o-rings and fluorinated stopcock grease are used to minimize chemical reaction at the walls. The main characteristics of the instrumentation are:

- the ability to produce relatively large quantities of pure liquid ozone (up to 800 torr of oxygen converted to ozone);
- the optical system, with the monochromator placed between the radiation source and the absorption chamber (the gas sample is hit by a monochromatic beam), to minimize photolysis of ozone;
- the use of material inert to ozone, to minimize heterogeneous chemical reactions.

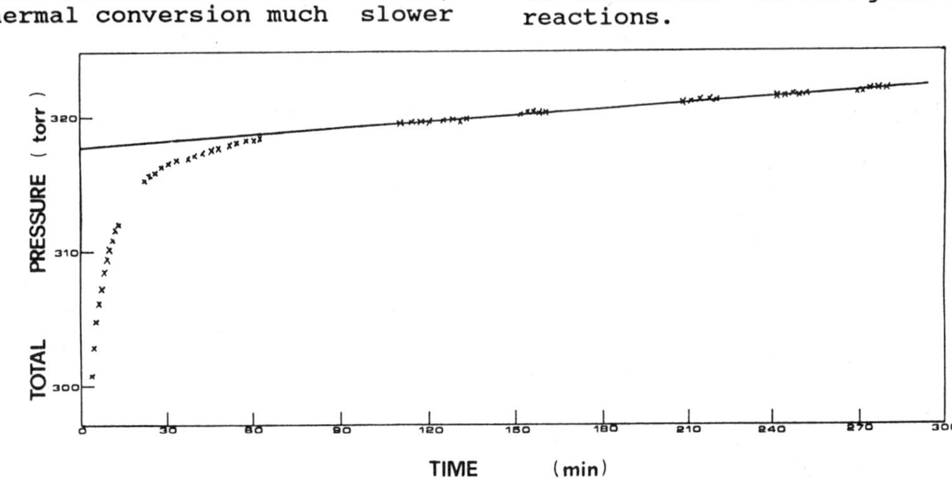

Fig. 2. Behaviour of the total pressure vs time at 293 K.

2. RESULTS

The Huggins bands, that begin at 300 nm and are situated at the longer wavelength end of the Hartley bands, are characterized by a strong wavelength dependance coupled with a strong temperature dependance. The cross section has a generally diminishing trend with wavelength, and discrete spectral features appear. The bands rise from the superposition of two systems:

1) Intense bands (cold bands) with weak temperature sensitivity (in fig. 3 they correspond to the arrows). These are due to transitions from a vibrational ground state to an upper electronic state in which three vibrational frequencies can be excited.

2) Strongly temperature sensitive bands (hot bands), attributable to transitions from excited vibrational levels in the ground state.

These spectral features are currently used for differential absorption measurements of ozone atmospheric profiles, for example by the Dobson instruments, by many satellite instruments (TOMS, BUV, SBUV) and by Lidar DIAL systems.

DIAL systems for ozone in the UV utilize:

Werner et al., 1983
Absorption channel 308 nm
Reference channel 353 nm or 338 nm

Pelon et al., 1986
Absorption channel 308 nm
Reference channel 355 nm

Shibata et al., 1987
Absorption channel 313 nm
Reference channel 360 nm

Sugimoto et al., 1988
Absorption channel 308 nm
Reference channel 351 nm.

These reference wavelengths are obtainable as follows:

351.1 nm - Xe F excimer laser output
353.2 nm - First Stokes line of a Xe Cl laser output (308 nm)
354.7 nm - Third harmonic of a Nd:YAG laser output (1064 nm)
360 nm - Third Stokes line of H_2 Raman shifting of a Kr F laser output (248 nm)

In fig. 3 we report the results in the 339-355 nm range. These curves (curve (a) refers to 293 K, curve (b) refers to 220 K) are obtained each by averaging at least four independent cross-section determinations, obtained at steps of 0.05 nm. Curve (c) is the ratio the absorption cross section at 220 and 293 K. This curve shows how the "cold spectrum" bands are less temperature sensitive than the "hot spectrum" bands.

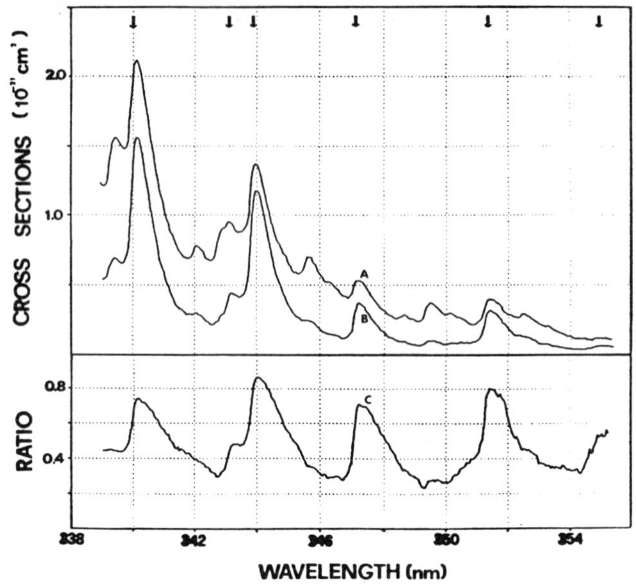

Fig. 3. Upper figure: absorption cross sections of ozone in the region 390-355 nm. Curve a, T=293 K; curve b, T=220 K. The heads of the "temperature insensitive" bands are indicated by arrows. Curve c: ratio of the absolute cross section at 220 K to the cross section at 293 K.

These results are in good agreement with the measurements of Brion, Daumont and Malicet (1983) and those of Bass and Paur (1984), at least in the 339-345 nm wavelength region, which has been investigated by these authors.

Most of the wavelengths used as reference in DIAL systems are in the proximity of the 350 nm band, that is a "cold spectrum" band, with a relatively weak temperature dependance.

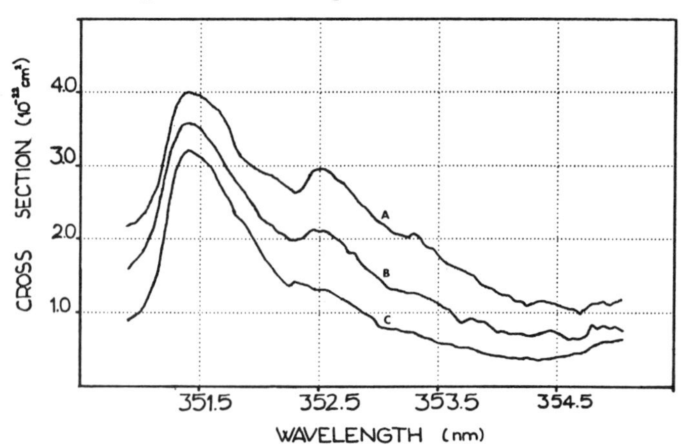

Fig. 4. Expanded portion of the spectrum 350-355 nm. a) T=293 K; b) T=273 K; c) T= 220 K.

We have studied this region more in detail, and we have performed measurements also at 273 K. The results are reported in fig. 4. At each wavelength the T dependence can be fitted by a quadratic law

$$\sigma = C_o + C_1 T + C_2 T^2$$

In table 1 the coefficients C_o, C_1 and C_2 are reported, for 351.1 nm, 352.3 nm and 354.7 nm. In table 2 the absorption cross sections at 293 K are reported; these values can be compared with the cross section at 308 nm (Bass et al., 1984), wavelength currently used as absorption channel in DIAL systems.

TABLE 1

Wavelength	C_0	C_1	C_2
351.1 nm	7.56	-6.47×10^{-2}	1.61×10^{-4}
353.2 nm	21.39	-0.178	3 . 8 1 5
354.7 nm	12.21	-9.97×10^{-2}	2.104×10^{-4}

Cross sections are in units of 10^{-22} cm^2

In the current analyses of DIAL data the absorption cross section of the reference channel is usually considered negligible. This assumption does not involve a considerable error in the ozone measurement, but future corrections may be useful where high accuracy is desirable.

TABLE 2
T = 293 K

Wavelength (nm)	Cross section (cm^2)
308	1.2×10^{-19}
351.1	2.27×10^{-22}
353.2	2.05×10^{-22}
354.7	1.05×10^{-22}

3. CONCLUSIONS

We have carried out absolute ozone absorption cross sections measurement in a spectral region that has been poorly investigated in the past. The results are in good agreement with recently published work.

The results are particularly interesting for eventual corrections in the determination of ozone profiles by active probing technique in the near UV.

REFERENCES

Bass, A.M., and R.J. Paur, The ultraviolet cross sections of ozone, in Atmospheric Ozone, Zerefos and Ghazi eds., 1984.

Brion, J., D. Daumont, and J. Malicet, New measurements of the absololute absorption cross sections of ozone at 294 and 223 K in the 310-350 nm spectral range, J. Phys. Lettres, 45, 57-60, 1984.

Pelon, J., S. Godin, and G. Megie, Upper stratospheric (30-50 km) lidar observations of the ozone vertical distribution, J. Geophys. Res., 91, 8667-8671, 1986.

Shibata, T., T. Fukuda, T. Narikiyo, and M. Maeda, Evaluation of the solar blind effect in ultraviolet ozone lidar with Raman lasers, Appl. Opt., 26, 2604-2608, 1987.

Sugimoto, N., Y. Sasano, S. Hayashida-Amano, H. Shimizu, N. Takeuchi and H. Akimoto, Multi wavelength ozone lidar for stratospheric and tropospheric measurements, XIV International Laser Radar Conference, S. Candido, Italy, 1988.

Uchino, O., M. Maeda, J. Kohno, T. Shibata, C. Nagasawa, and M. Hirono, Observation of stratospheric ozone layer by a Xe Cl laser radar, Appl. Phys. Lett., 33, 807-809, 1978.

Werner, J., K.W. Rothe, and H. Walther, Monitoring of the stratospheric ozone layer by laser radar, Appl. Phys., B 32, 113-118, 1983.

TEMPERATURE DEPENDENCE OF ULTRAVIOLET ABSORPTION CROSS SECTIONS OF BROMINATED METHANES AND ETHANES

D. Gillotay, P.C. Simon, and L. Dierickx
Institut d'Aéronomie Spatiale de Belgique
B-1180 Bruxelles, Belgium

ABSTRACT

Absorption cross-sections of five brominated methanes (CH_3Br, CH_2Br_2, CF_3Br, CF_2Br_2 and CF_2BrCl), and of two brominated ethanes (CF_2Br-CF_2Br and CF_3-$CHBrCl$) have been measured between 172 and 300 nm with a classical single beam system, for temperature ranging from 210 to 295 K. These data are compared with previous determinations made at room temperature (Robbins, 1976 and Molina et al., 1982).

Numerical values of absorptions cross-sections are given and the temperature effects on the photodissociation coefficients of these molecules discussed.

1. INTRODUCTION

Recent studies (Prather et al., 1984; Rodriguez et al., 1986; Yung et al., 1980) show that brominated compounds can play a significantly role in the ozone catalytic destruction in the stratosphere. RBr species concentrations have been measured in the atmosphere by several authors (Fabian et al., 1981; Lal et al., 1985; Cicerone et al., 1988; Berg et al., 1984; Penkett et al., 1985; Class et al., 1986) confirming their possible impact on the ozone layer.

The reliability of models is strongly dependent on the photodissociation pattern adopted for these bromocarbons. Until now, one set of measurements of absorption cross-sections is available at room temperature for most of the bromocarbons (Molina et al., 1982), a second set of data being available for CH_3Br (Robbins, 1976). The purpose of this paper is to report a new investigation of ultraviolet absorption cross-sections of five brominated methanes (CH_3Br, CH_2Br_2, CF_3Br, CF_2Br_2 and CF_2BrCl), and of two brominated ethanes (CF_2Br-CF_2Br and CF_3-$CHBrCl$) measured between 172 and 300 nm, for temperatures between 295 and 210 K. Temperature dependence of absorption cross-section is clearly demonstrated. Photodissociation coefficients are calculated and their temperature dependence is discussed.

2. EXPERIMENTAL

The absorption measurements has been performed by means of a single beam experimental device previously discribed in Gillotay and Simon, (1988), including a Deuterium or a Tungsten filament light source, a 1 m McPherson 225 monochromator, a 2 m thermostatic absorption cell and a EMR type 542 P-09-18 solar blind photomultiplier. The temperature is regulated down to 210 K by circulation of cooled methylcyclohexane through double jacket around the absorption cell and is determined with a precision better than 1% at 210 K. The pressures ranging from 2×10^{-3} to 1000 torr are measured with three capacitance manometers MKS Baratron with a precision better than 0.1%.

Determination of absorption cross-sections is made after at least fifteen sequential recordings of the incident and absorbed fluxes measured for identical temperature conditions, using the Beer-Lambert's law.

3. RESULTS

Numerical values of absorption cross-sections for selected wavelengths between 168 and 304 nm (2 nm intervals) are given in Tables 2-8. The absorption spectra are illustrated in Figures 1 for the bromo-methanes.

Measurements have been performed at working pressures given in Table 1. In all cases, Beer-Lambert's law was verified for absorptions ranging from 10 to 85%. In such conditions, and according to the error budget recently published (Simon et al., 1988), the absorption cross- sections given in Tables 2-8 are determined with an accuracy of $\pm$ 2% at room temperature and of $\pm$ 3 to $\pm$ 4% at the lowest temperature.

Brominated compounds display a continuous absorption in the 172-300 nm range, with absorption cross-sections ranging from 10^{-17} to 10^{-25} cm^2 $molecule^{-1}$, with a maximum around 200 nm for the monobrominated carbons and around 225 nm for the dibrominated ones. The results are compared with the previous available values (Molina et al., 1982; Robbins, 1976). Disagreements up to 20% are observed between our set of measurements and the values proposed by Molina et al., (1982), the latter being lower than ours mainly in the region of high absorption.

Absorption cross-section values change with temperature by a factor which depends on both the wavelength and the chemical composition of the compound. For each wavelength, an exponential dependence of the absorption cross-section versus temperature is clearly shown, with a decrease of absorption cross-sections in the region of low absorptions (up to 80% at 310 nm and 210 K) and an increase (up to 20% at 210 K) near the maximum of absorption.

ISBN 0-937194-15-8

TABLE 1. : PRESSURE RANGES FOR ABSORPTION CROSS-SECTIONS MEASUREMENTS.

Compound \ Temp. (K)	PRESSURE (torr) 295 K	255 K	210 K
CF_3Br (*)	$722 - 6x10^{-1}$	$597 - 6x10^{-1}$	$384.6 - 2x10^{-2}$
CF_2Br_2 (*)	$234 - 3x10^{-2}$	$67 - 1x10^{-1}$	$3.4 - 2x10^{-2}$
CF_2BrCl (*)	$453 - 1x10^{-1}$	$80 - 8x10^{-2}$	$10.3 - 9x10^{-2}$
CH_3Br (**)	$22 - 4x10^{-2}$	$19 - 7x10^{-2}$	$9.6 - 9x10^{-2}$
CH_2Br_2	$36 - 4x10^{-2}$	$1 - 5x10^{-2}$	
$C_2F_4Br_4$	$80 - 2x10^{-2}$	$67 - 5x10^{-2}$	$11.2 - 2x10^{-2}$
C_2F_3HBrCl	$147 - 4x10^{-2}$	$20 - 4x10^{-2}$	$9x10^{-1}-4x10^{-2}$

(*) : Gillotay and Simon, 1988a.
(**) : Gillotay and Simon, 1988b.

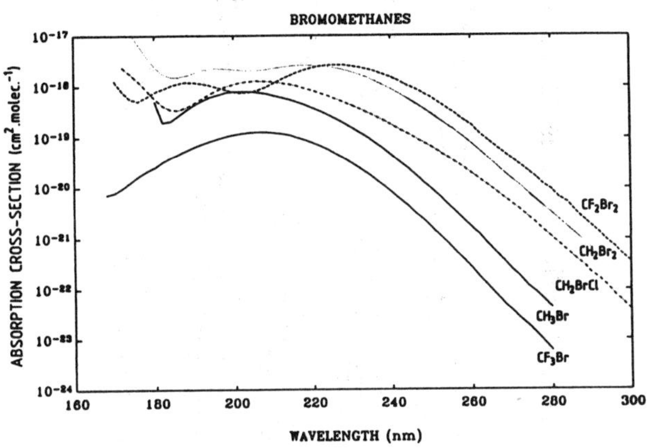

Fig. 1. Ultraviolet absorption cross-sections of bromomethanes with respect to wavelength at 295 K.

TABLE 2. ABSORPTION CROSS-SECTIONS [$\sigma(\lambda) \times 10^{21}$ (cm^2 $molec.^{-1}$)] OF CH_3Br (Gillotay and Simon, 1988b)

λ(nm)	295K	270K	250K	230K	210K
180	500	500	500	500	500
182	198	198	198	198	198
184	208	208	208	208	208
186	271	271	271	271	271
188	346	346	346	346	346
190	439	439	439	439	439
192	529	529	529	529	529
194	620	620	620	620	620
196	691	691	691	691	691
198	760	760	760	760	760
200	791	791	791	791	791
202	797	797	797	797	797
204	793	793	793	793	793
206	767	767	767	767	767
208	727	727	727	727	727
210	666	666	666	666	666
212	614	613	613	613	613
214	557	557	556	556	556
216	493	493	492	492	492
218	440	438	436	434	432
220	378	374	372	369	367
222	325	321	318	316	313
224	278	273	268	264	260
226	231	226	222	218	214
228	190	187	184	181	179
230	155	152	149	147	144
232	125	122	120	117	115
234	99.2	96.1	93.7	91.3	89.0
236	77.3	74.2	71.7	69.3	67.1
238	59.7	56.6	54.3	52.0	49.9
240	44.8	42.7	41.0	39.4	37.9
242	33.3	31.6	30.2	28.9	27.7
244	25.1	23.3	22.0	20.8	19.6
246	18.2	16.9	15.9	15.0	14.1
248	13.0	12.0	11.2	10.5	9.81
250	9.57	8.65	7.98	7.37	6.80
252	6.89	6.09	5.52	5.01	4.54
254	4.83	4.26	3.85	3.48	3.14
256	3.39	2.93	2.61	2.32	2.07
258	2.40	2.02	1.75	1.52	1.32
260	1.70	1.43	1.24	1.07	0.932
262	1.23	0.946	0.800	0.678	0.573
264	0.810	0.638	0.529	0.439	0.364
266	0.557	0.429	0.348	0.282	0.229
268	0.384	0.287	0.227	0.180	0.142
270	0.255	0.191	0.147	0.114	0.0875
272	0.183	0.127	0.0953	0.0713	0.0533
274	0.118	0.0847	0.0614	0.0444	0.0322
276	0.0880	0.0567	0.0394	0.0275	0.0193
278	0.0616	0.0375	0.0253	0.0170	0.0114
280	0.0434	0.0251	0.0162	0.0104	0.00672

TABLE 3. ABSORPTION CROSS-SECTIONS [$\sigma(\lambda) \times 10^{21}$ (cm^2 $molec.^{-1}$)] OF CH_2Br_2

λ(nm)	295K	270K	250K	230K	210K
174	11709				
176	6624	6352	6143	5940	5744
178	3772	3594	3458	3326	3200
180	2410	2331	2270	2211	2153
182	1784	1714	1660	1608	1557
184	1544	1469	1412	1357	1303
186	1535	1487	1450	1414	1379
188	1661	1646	1634	1623	1611
190	1870	1879	1887	1894	1902
192	2093	2115	2133	2151	2170
194	2225	2270	2306	2343	2381
196	2283	2322	2354	2387	2419
198	2260	2291	2317	2342	2368
200	2201	2209	2216	2223	2230
202	2143	2133	2126	2119	2111
204	2101	2087	2075	2064	2052
206	2112	2101	2092	2084	2075
208	2175	2186	2195	2205	2214
210	2269	2307	2339	2370	2402
212	2428	2530	2614	2702	2792
214	2548	2706	2839	2979	3126
216	2622	2825	2998	3182	3377
218	2647	2879	3080	3294	3524
220	2621	2868	3082	3311	3558
222	2547	2792	3005	3234	3481
224	2430	2659	2859	3073	3304
226	2276	2479	2656	2844	3046
228	2094	2264	2411	2566	2732
230	1893	2026	2140	2261	2388
232	1682	1779	1861	1946	2035
234	1470	1533	1585	1639	1695
236	1264	1297	1324	1352	1380
238	1070	1079	1086	1094	1101
240	892	883	876	868	861
242	733	711	694	678	662
244	594	565	542	521	500
246	475	442	417	394	372
248	375	342	317	294	273
250	293	261	238	217	197
252	226	197	176	158	141
254	172	147	129	114	100
256	130	109	93.9	81.2	70.1
258	97.8	79.7	67.7	57.5	48.8
260	72.7	58.0	48.5	40.5	33.8
262	53.7	42.0	34.5	28.3	23.2
264	39.4	30.2	24.4	19.7	15.9
266	28.8	21.6	17.2	13.7	10.9
268	20.9	15.4	12.1	9.47	7.42
270	15.2	11.0	8.48	6.55	5.06
272	11.0	7.81	5.95	4.53	3.45
274	7.93	5.55	4.17	3.14	2.36
276	5.73	3.95	2.93	2.18	1.62
278	4.15	2.82	2.07	1.52	1.12
280	3.01	2.02	1.47	1.07	0.773
282	2.19	1.45	1.04	0.750	0.539
284	1.60	1.05	0.747	0.532	0.379
286	1.18	0.764	0.540	0.381	0.269
288	0.877	0.562	0.393	0.275	0.193
290	0.656	0.416	0.290	0.201	0.140

TABLE 4. ABSORPTION CROSS-SECTIONS [$\sigma(\lambda) \times 10^{21}$ (cm^2 $molec.^{-1}$)] OF CF_3Br (Gillotay et al., 1988a)

λ(nm)	295 K	270K	250K	230K	210K
168	5.17	5.09	5.04	4.98	4.93
170	6.96	6.89	6.84	6.79	6.73
172	9.28	9.24	9.21	9.17	9.14
174	12.2	12.3	12.3	12.3	12.3
176	16.0	16.1	16.2	16.3	16.4
178	20.5	20.8	21.1	21.3	21.6
180	26.1	26.6	27.1	27.5	28.0
182	32.6	33.5	34.3	35.0	35.8
184	40.2	41.6	42.7	43.9	45.1
186	48.8	50.8	52.4	54.1	55.9
188	58.2	60.9	63.2	65.5	67.9
190	68.4	71.9	74.8	77.8	81.0
192	78.8	83.2	86.9	90.7	94.7
194	89.3	94.6	99.0	104	108
196	99.4	105	110	116	121
198	109	115	121	127	133
200	116	123	129	136	142
202	122	130	136	142	149
204	126	133	139	146	152
206	128	134	140	146	152
208	127	132	137	142	148
210	123	128	132	136	140
212	117	121	124	127	130
214	110	112	114	116	118
216	101	102	102	103	104
218	90.6	90.3	90.1	89.9	89.7
220	79.9	78.6	77.6	76.6	75.6
222	69.2	67.1	65.4	63.8	62.3
224	58.8	56.1	54.1	52.1	50.1
226	49.1	46.1	43.8	41.6	39.5
228	40.3	37.1	34.7	32.5	30.5
230	32.4	29.3	27.0	25.0	23.0
232	25.7	22.8	20.7	18.8	17.1
234	20.0	17.4	15.6	13.9	12.4
236	15.3	13.1	11.5	10.1	8.88
238	11.6	9.66	8.35	7.22	6.24
240	8.62	7.04	5.98	5.09	4.33
242	6.33	5.06	4.23	3.54	2.96
244	4.59	3.59	2.95	2.43	2.00
246	3.28	2.52	2.04	1.65	1.33
248	2.32	1.75	1.39	1.11	0.881
250	1.63	1.20	0.941	0.738	0.579
252	1.13	0.818	0.632	0.489	0.378
254	0.777	0.554	0.422	0.322	0.246
256	0.530	0.373	0.281	0.212	0.160
258	0.360	0.250	0.186	0.139	0.104
260	0.243	0.167	0.123	0.0911	0.0674
262	0.164	0.111	0.0817	0.0600	0.0440
264	0.110	0.0742	0.0542	0.0396	0.0290
266	0.0736	0.0496	0.0362	0.0264	0.0192
268	0.0494	0.0333	0.0243	0.0177	0.0129
270	0.0331	0.0225	0.0164	0.0120	0.00882
272	0.0223	0.0153	0.0112	0.00829	0.00611
274	0.0151	0.0105	0.00779	0.00580	0.00432
276	0.0103	0.00725	0.00547	0.00414	0.00313
278	0.00706	0.00509	0.00392	0.00301	0.00232
280	0.00489	0.00363	0.00286	0.00225	0.00177

TABLE 5. ABSORPTION CROSS-SECTIONS [$\sigma(\lambda) \times 10^{21}$ (cm^2 $molec.^{-1}$)] OF CF_2Br_2 (Gillotay et al., 1988a)

λ(nm)	295 K	270 K	250 K	230 K	210 K
170	1245	1108	1000	891	782
172	781	696	629	560	493
174	553	551	550	549	548
176	495	523	546	568	590
178	603	641	671	700	730
180	750	791	824	857	890
182	866	939	997	1056	1114
184	1009	1086	1148	1210	1272
186	1118	1204	1272	1341	1409
188	1180	1262	1327	1392	1458
190	1168	1252	1319	1387	1454
192	1109	1185	1245	1306	1366
194	1022	1085	1135	1185	1235
196	920	973	1016	1058	1100
198	825	869	905	941	976
200	748	792	827	862	898
202	716	758	791	825	858
204	735	778	812	846	880
206	810	853	887	921	955
208	936	989	1033	1076	1119
210	1110	1173	1223	1273	1323
212	1431	1408	1466	1525	1583
214	1586	1670	1738	1806	1873
216	1841	1935	2011	2086	2162
218	2081	2186	2270	2353	2437
220	2281	2394	2482	2577	2676
222	2477	2585	2675	2767	2863
224	2540	2643	2729	2817	2909
226	2545	2640	2718	2800	2883
228	2493	2577	2646	2717	2790
230	2390	2461	2519	2578	2639
232	2246	2302	2347	2394	2442
234	2068	2109	2143	2177	2212
236	1869	1897	1918	1941	1964
238	1659	1674	1686	1698	1710
240	1448	1452	1455	1459	1463
242	1243	1238	1235	1232	1229
244	1050	1039	1031	1023	1015
246	875	860	848	836	825
248	718	701	687	674	661
250	582	564	549	535	521
252	466	447	433	419	406
254	369	351	337	325	312
256	289	272	260	248	237
258	224	209	198	189	178
260	172	159	150	141	132
262	130	119	112	104	97.3
264	98.7	89.6	82.9	76.7	71.0
266	74.0	66.5	61.0	56.0	51.3
268	55.1	48.9	44.5	40.5	36.8
270	40.8	35.9	32.3	29.1	26.3
272	30.1	26.1	23.3	20.8	18.6
274	22.1	18.9	16.7	14.8	13.1
276	16.1	13.7	12.0	10.5	9.17
278	11.7	9.81	8.51	7.38	6.39
280	8.52	7.04	6.04	5.18	4.44
282	6.17	5.03	4.27	3.62	3.08
284	4.47	3.59	3.02	2.53	2.13
286	3.23	2.56	2.13	1.77	1.47
288	2.34	1.83	1.50	1.23	1.01
290	1.69	1.30	1.06	0.856	0.695
292	1.22	0.928	0.744	0.596	0.478
294	0.876	0.663	0.525	0.415	0.329
296	0.645	0.474	0.370	0.290	0.226
298	0.470	0.340	0.262	0.202	0.156
300	0.343	0.244	0.186	0.142	0.108
302	0.252	0.176	0.132	0.0993	0.0746
304	0.185	0.1274	0.0944	0.0699	0.0518

TABLE 6. ABSORPTION CROSS-SECTIONS [$\sigma(\lambda) \times 10^{21}$ (cm^2 $molec.^{-1}$)] OF CF_2BrCl (Gillotay et al. 1988a)

λ(nm)	295 K	270 K	250 K	230 K	210 K
170	3230	3200	3180	3160	3150
172	2342	2300	2285	2250	2227
174	1760	1720	1680	1660	1660
176	1209	1180	1175	1170	1160
178	847	840	834	830	825
180	581	580	579	579	578
182	419	419	418	418	418
184	350	353	356	359	362
186	341	347	353	359	366
188	389	405	420	437	456
190	474	500	524	548	573
192	584	626	666	707	748
194	722	768	816	866	922
196	845	919	974	1020	1078
198	990	1041	1090	1139	1190
200	1197	1244	1284	1324	1366
202	1230	1283	1328	1374	1422
204	1244	1302	1350	1400	1452
206	1239	1299	1350	1402	1457
208	1216	1277	1328	1381	1435
210	1177	1237	1286	1337	1391
212	1124	1180	1227	1276	1326
214	1060	1111	1154	1198	1245
216	986	1032	1070	1110	1151
218	907	947	980	1014	1049
220	824	858	885	914	943
222	741	768	790	813	836
224	659	680	697	714	732
226	580	595	608	620	633
228	505	516	524	533	541
230	436	442	447	452	457
232	373	376	378	380	382
234	316	316	316	316	316
236	266	264	262	260	259
238	222	218	215	212	210
240	183	179	175	172	168
242	150	145	141	138	134
244	123	117	113	109	106
246	99.2	94.0	90.0	86.1	82.5
248	79.7	74.7	70.9	67.3	63.9
250	63.7	59.0	55.5	52.2	49.1
252	50.5	46.2	43.1	40.2	37.5
254	39.8	36.0	33.3	30.7	28.3
256	31.2	27.9	25.5	23.3	21.3
258	24.3	21.4	19.4	17.5	15.9
260	18.8	16.4	14.7	13.1	11.7
262	14.5	12.4	11.0	9.73	8.61
264	11.1	9.39	8.21	7.18	6.28
266	8.46	7.05	6.09	5.26	4.54
268	6.42	5.26	4.49	3.83	3.26
270	4.84	3.90	3.29	2.77	2.33
272	3.63	2.88	2.39	1.99	1.65
274	2.71	2.11	1.73	1.42	1.16
276	2.01	1.54	1.24	1.00	0.810
278	1.48	1.11	0.887	0.706	0.562
280	1.09	0.803	0.629	0.493	0.386
282	0.796	0.576	0.444	0.342	0.264
284	0.579	0.410	0.311	0.236	0.179
286	0.419	0.290	0.216	0.161	0.120
288	0.301	0.204	0.149	0.109	0.0802
290	0.215	0.143	0.103	0.0737	0.0530
292	0.153	0.0990	0.0698	0.4927	0.0348
294	0.108	0.0683	0.0472	0.0327	0.0226
296	0.0761	0.0468	0.0317	0.0215	0.0146
298	0.0532	0.0318	0.0211	0.0140	0.00929
300	0.0369	0.0215	0.0139	0.00905	0.00587
302	0.0255	0.0144	0.00914	0.00579	0.00367

TABLE 7. ABSORPTION CROSS-SECTIONS [$\sigma(\lambda) \times 10^{21}$ (cm^2 $molec.^{-1}$)] OF $C_2F_4Br_2$

λ(nm)	295K	270K	250K	230K	210K
170	509	539	564	590	618
172	564	594	620	646	674
174	623	653	678	705	732
176	685	715	740	765	791
178	751	779	802	826	851
180	818	844	866	888	911
182	886	909	929	949	969
184	953	973	990	1007	1024
186	1018	1035	1048	1062	1076
188	1080	1092	1102	1112	1122
190	1135	1143	1149	1155	1161
192	1184	1186	1188	1190	1192
194	1223	1221	1219	1217	1214
196	1252	1245	1238	1233	1227
198	1269	1257	1247	1237	1228
200	1273	1257	1244	1231	1218
202	1264	1244	1228	1212	1197
204	1241	1218	1200	1182	1165
206	1205	1180	1160	1141	1122
208	1157	1131	1110	1090	1070
210	1099	1072	1051	1030	1010
212	1031	1004	983	963	943
214	957	931	910	890	871
216	877	852	833	814	795
218	795	772	753	735	718
220	712	690	673	657	641
222	630	611	595	580	566
224	551	534	520	506	493
226	477	461	449	437	425
228	408	394	383	372	362
230	345	332	323	314	305
232	288	278	269	261	253
234	239	229	222	215	208
236	195	187	181	175	169
238	158	151	146	140	135
240	127	121	116	111	107
242	101	95.5	91.3	87.4	83.6
244	79.5	74.8	71.2	67.8	64.5
246	62.1	58.0	54.9	52.0	49.2
248	48.1	44.5	41.9	39.4	37.1
250	37.0	33.9	31.6	29.5	27.6
252	28.2	25.6	23.7	21.9	20.3
254	21.4	19.2	17.6	16.1	14.7
256	16.1	14.2	12.9	11.7	10.6
258	12.1	10.5	9.38	8.38	7.49
260	9.04	7.70	6.77	5.96	5.24
262	6.72	5.61	4.85	4.20	3.63
264	4.99	4.06	3.45	2.93	2.49
266	3.69	2.93	2.43	2.02	1.68
268	2.73	2.10	1.71	1.39	1.13
270	2.01	1.50	1.19	0.942	0.745
272	1.49	1.07	0.825	0.634	0.488
274	1.10	0.763	0.569	0.424	0.316
276	0.816	0.542	0.390	0.281	0.203
278	0.607	0.385	0.267	0.185	0.129
280	0.454	0.273	0.182	0.121	0.0809

TABLE 8. ABSORPTION CROSS-SECTIONS [$\sigma(\lambda) \times 10^{21}$ (cm^2 $molec.^{-1}$)] OF CF_3-CHBrCl

λ(nm)	295K	270K	250K	230K	210K
170	7026	7014	7005	6996	6987
172	6146	6058	5988	5919	5851
174	4968	4849	4756	4665	4575
176	3798	3682	3592	3504	3418
178	2811	2715	2640	2567	2496
180	2061	1988	1932	1877	1823
182	1533	1481	1441	1401	1363
184	1184	1148	1120	1093	1067
186	971	948	929	912	894
188	866	853	842	832	822
190	906	905	903	902	901
192	985	988	991	994	997
194	1052	1061	1069	1076	1084
196	1105	1120	1132	1145	1157
198	1143	1163	1180	1197	1214
200	1164	1190	1210	1232	1253
202	1169	1199	1223	1247	1272
204	1158	1190	1217	1244	1272
206	1132	1166	1194	1223	1253
208	1092	1127	1156	1186	1216
210	1041	1076	1104	1134	1164
212	981	1014	1042	1070	1099
214	913	945	971	998	1025
216	842	871	894	919	944
218	768	794	815	837	859
220	694	716	734	753	773
222	621	640	655	671	687
224	551	566	579	592	605
226	484	496	506	516	527
228	422	431	439	447	454
230	365	372	377	383	388
232	313	318	321	325	329
234	267	269	272	274	276
236	226	227	228	229	230
238	190	190	190	190	190
240	159	158	157	156	155
242	132	130	129	128	126
244	109	107	105	103	102
246	89.2	86.9	85.1	83.4	81.7
248	72.7	70.4	68.5	66.8	65.0
250	58.9	56.6	54.8	53.1	51.4
252	47.5	45.3	43.6	42.0	40.4
254	38.1	36.1	34.5	33.0	31.6
256	30.4	28.6	27.2	25.8	24.5
258	24.2	22.5	21.2	20.1	18.9
260	19.1	17.6	16.5	15.5	14.6
262	15.0	13.7	12.8	11.9	11.1
264	11.7	10.6	9.85	9.12	8.45
266	9.11	8.21	7.55	6.94	6.38
268	7.05	6.30	5.75	5.26	4.80
270	5.42	4.81	4.36	3.96	3.59
272	4.16	3.65	3.29	2.97	2.68
274	3.17	2.76	2.47	2.21	1.98
276	2.40	2.07	1.85	1.64	1.46
278	1.80	1.55	1.37	1.21	1.07
280	1.35	1.15	1.01	0.892	0.785
282	1.01	0.852	0.746	0.652	0.571
284	0.745	0.627	0.545	0.475	0.413
286	0.548	0.458	0.397	0.344	0.298
288	0.401	0.333	0.287	0.247	0.213
290	0.291	0.240	0.206	0.177	0.152

4. DISCUSSION

Photodissociation coefficients of these molecules have been calculated, neglecting effects of multiple scattering, for given altitude (z), zenith angle (χ) and wavelength interval according to the relations :

$$J(z) = \sigma_\lambda \, q_\lambda(z) \; ; \; q_\lambda(z) = q_\lambda(\infty) \, e^{-\tau_\lambda(z)}$$

$$\tau_\lambda(z) = \int_z^\infty [n(O_2)\, \sigma_\lambda(O_2) + n(O_3)\, \sigma_\lambda(O_3) + n(air)\sigma_{scatt}] \sec\chi \, dz \qquad (1)$$

where

z is the altitude,
σ_λ are the absorption cross-sections,
$q_\lambda(z)$ and $q_\lambda(\infty)$ are the solar irradiance at altitude z or extraterrestrial ($z = \infty$),
n is the number of particles per volume unit,
for solar zenith angles of 0° and 60° (sec χ = 1 and 2),
taking into account the values of $\sigma_\lambda(O_2)$ and $\sigma_\lambda(O_3)$ from WMO and Kockarts (1976), σ_{scatt} from Nicolet (1984) and the values of $q_\lambda(\infty)$ from WMO (1985) and taking into account the actual values of cross-sections corresponding to the temperatures conditions at each altitude. Stratospheric photodissociation coefficients (for altitude ranging from 15 to 50 km) calculated with the temperature dependent absorption cross-sections, are about 15% greater than those calculated with the room temperature values in the 20-35 km region, due to the increase in the absorption cross-sections in the 200 nm region. The photodissociation coefficients of CH_3Br and $C_2F_4Br_2$ seem to be temperature independent. Tropospheric photodissociation coefficients for all these molecules are very low (between 10^{-8} and 10^{-11} sec^{-1}) and are reduced down to 20% of their room temperature values, using the temperature dependent cross- sections values. Therefore, the theoretical tropospheric residence times are increased, and the production of Br radicals in the troposphere by photodissociation of brominated species may be considered as negligible.

In conclusion, this work presents a new set of experimental data on the absorption cross-sections of brommocarbons in atmospheric temperature conditions and also highlights the non negligible temperature dependence for stratospheric studies.

4.1. REFERENCES

BERG, W.W., L.E. HEIDT, W. POLLOCK, P.D. SPERRY, R.J. CICERONE, and E.S. GLADNEY, 1984, Brominated organic species in the Arctic atmosphere, Geophys. Res. Lett., **11**, 429-432.

CLASS, Th., R. KOHNLE, K., BALLSCHMITER, 1986, Chemistry of organic races in air VII : Bromo- and Bromochloromethanes in air over Atlantic Ocean, Chemosphere, **15**. 429-436.

CICERONE, R.J., L.E. HEIDT and W.H. POLLOCK, 1988, Measurements of atmospheric Methyl bromide (CH3Br) and Bromoform (CHBr3), J. Geophys. Res., **93**, 3745-3749.

FABIAN, P., R. BORCHERS, S.A. PENKETT and N.J.D. PROSSER, 1981, Halocarbons in the stratosphere, Nature, **294**, 733-735.

GILLOTAY, D. and P.C. SIMON, 1988a, Ultraviolet absorption spectrum of trifluro-bromo-methane, difluoro-dibromomethane and difluoro- bromo-chloromethane in the vapor phase, submitted for publication to J. Atmos. Chem.

GILLOTAY, D. and P.C. SIMON, 1988b, Ultraviolet absorption cross-sections of methyl bromide at stratospheric temperature, Ann. Geophysicae, **6**, 211-215.

KOCKARTS, G., 1976, Absorption and photodissociation in the Schumann- Runge bands of molecular oxygen in terrestrial atmosphere, Planet. Space Sc., **24**, 589-604.

LAL, S., R. BORCHERS, P. FABIAN and B.C. KRUGER, 1985, Increasing abundance of $CBrClF_2$ in the atmosphere, Nature, **316**, 135-136.

MOLINA, L.T., M.J. MOLINA and F.S. ROWLAND, 1982, Ultraviolet absorption cross sections of several brominated methanes and ethanes of atmospheric interest, J. Phys. Chem., **86**, 5172-5178.

NICOLET, M., 1984, On the molecular scattering in the terrestrial atmosphere : an empirical formula for calculation in the homosphere, Planet. Space Sc., **32**, 1467-1468.

PENKETT, S.A., B.M.R. JONES, M.J. RYCROFT and D.A. SIMMONS, 1985, An interhemispheric comparison of the concentrations of bromide compounds in the atmosphere, Nature, **318**, 550-553.

PRATHER, M.J., M.B. McELROY and S.C. WOFSY, 1984, Reductions in ozone at high concentrations of stratospheric halogens, Nature, **312**, 227-231.

ROBBINS, D.E., 1976, Photodissociation of methyl chloride and methyl bromide in atmosphere, Geophys. Res. Lett., **3**, 213-216.
Erratum in Geophys. Res. Lett., **3**, 757, 1976.

RODRIGUEZ, J.M., M.K.M. KO and N.D. SZE, 1986, Chlorine chemistry in the Antarctic stratosphere : Impact of OClO and Cl_2O_2 and implications for observations, Geophys. Res. Lett., **13**, 1292-1295.

SIMON, P.C., D. GILLOTAY, N. VANLAETHEM-MEUREE and J. WISEMBERG, 1988, Ultraviolet absorption cross-sections of chloro and chloro- fluoro-methanes at stratospheric temperatures, J. Atmos. Chem., **6**, in press.

WMO, Atmospheric ozone 1985, WMO global ozone research and monitoring project, Report 16, 1985.

YUNG, Y.L., J.P. PINTO, R.T. WATSON, and S.P. SANDER, 1980, Atmospheric bromine and ozone perturbations in the lower stratosphere, J. Atmos. Sci., **37**, 339-353.

NEW ABSOLUTE ABSORPTION CROSS-SECTION OF O_3 IN THE 195–350 nm SPECTRAL RANGE

Dominique Daumont, Alain Barbe, Jean Brion, and Jacques Malicet
Spectrométrie Moléculaire et Atmosphérique
U.F.R. Sciences
Reims, France

ABSTRACT

New absolute cross-section values of ozone have been determined at ambient temperature (295 K) in the 195-300 nm spectral range. A comparison with the more recent data have been done, with a particular attention to the problems of wavelength calibration and normalisation of the relative data. In addition to continuous measurements, the temperature effect at the reference mercury line (254 nm) have been studied. A difference in the cross-section values measured at 295 and 228 K, of - 1% is observed and is in good agreement with the other available published values.

1. INTRODUCTION

On account of its importance in atmosphere, numerous works have been carried out to determine the cross-section values for O_3. Earlier works are those of INN and TANAKA (1953), VIGROUX (1953-1967), HEARN (1961), GRIGGS (1968)... However, discrepancies are observed between these different data. The main reason is connected to the difficulty of measuring the ozone concentration, due to the spontaneous ozone decomposition.

Recently, two new sets of data have been published, giving more accurate values : the absolute values determined by MOLINA (1986) and the relative values of BASS and PAUR (1984), calibrated to the Hearn's value ($\sigma = 114.7 \times 10^{-19}$ cm^2) at the mercury line at 254 nm.

The comparison, between these two sets shows a relatively good agreement. Nevertheless, recents mass spectrometry measurements of MAUERSBERGER (1986) have shown, that the reference value taken by BASS and PAUR is 1% too large. Furthermore a small shift in the wavelength scale is shown in the vibrational structure of the Huggins bands between the two sets.

So, we have performed new measurements in this region, in order to study these two problems.

2. CONTINUOUS MEASUREMENTS AT ROOM TEMPERATURE-EXPERIMENT AND RESULTS

The measurements were done using a high-resolution monochromator of 1.5 m of focal length. The wavelength scale was calibrated by emission lines of mercury, cadmium and iron. The accuracy is about 0.01 nm.

Ozone was prepared by a method described by GRIGGS (1968) and the gas mixture (O_2 + O_3 containing about 90% of O_3) filled the quartz absorption cell. For the determination of the absolute total ozone concentration in the cell, we have developped a method using pressure measurements. At constant volume and temperature, the partial pressure of O_3 is given by a simple relation

$$P_{O_3} = 2\,[(P_{O_2})_i - P_T]$$

between the initial pressure of oxygen $(P_{O_2})_i$ before ozonisation and the total pressure of the gas mixture (O_2 + O_3) which was recorded simultaneously with the absorption spectrum.

The present absolute cross-sections are determined at room temperature in the spectral region 195-350 nm and represented on the Fig.1. The comparison has been made with the two more recent sets of measurements.

For the cross-section values, we have a relatively good agreement. With Molina's values, in the whole region above 235 nm, our measurements are lower from 1 to 3%. Below this wavelength (235 nm) our values are exactly the same (Fig.2). This agreement shows in particular, that the previous values of INN and TANAKA (1953), still used in this region in the modelling calculations are wrong. This is important because in this region corresponding to the atmospheric window at 210 nm lying between the ozone absorption (Hartley band) and the oxygen absorption (Schumann-Runge Bands and Herzberg continuum) few measurements are available. Furthermore there are in this region some discrepancies between the calculated and observed solar flux.

Compared to the BASS's values measured at the maxima and minima, our results are lower. The difference between the two sets is very small within 0% and - 2.0%. This difference is similar to the difference (- 1%) observed by MAUERSBERGER at the reference value at 254 nm. So, if we normalize the Bass's data to this new reference, the difference between the two sets of values still remains below 1%, in the limits of the experimental errors.

The comparison has been made also in the wavelength scale. We may observe a small shift of about 0.04 nm between our values and those of BASS. This shift is not constant and is maximum at 320 nm. It corresponds to the shift observed between BASS and MOLINA's values (Fig.3). So, we may think that this difference is due to the BASS wavelength calibration.

3. TEMPERATURE EFFECT AT THE REFERENCE MERCURY WAVELENGTH (254 nm)

In addition to these continuous scanning measurements we have studied the temperature effect at 254 nm. Up to now, the cross-section value, at this wavelength corresponding to the maximum of the absorption curve, was considered as independent

OZONE IN THE ATMOSPHERE
Rumen D. Bojkov and Peter Fabian (Eds.)

ISBN 0-937194-15-8

with the temperature. Our measurements have shown that the variation is really small. But we found that the cross-section increase significantly of about 1% ± 0.5% when the temperature decrease from room temperature (25°C, σ = 112.6 x 10^{-19} cm^2)*, to atmospheric temperature (- 45°C, σ = 113.9x$10^{-19}$$cm^2$)

To measure this small variation, we have developed a new technique of measurement using the recent determination by MAUERSBERGER (1985) of the liquid ozone vapor pressure at the normal boiling point of argon (- 185.9°C). During the experiment, liquid ozone is kept at the liquid argon temperatures. So the ozone pressure (about 4.05x10^{-2} Torr) remains constant independently of the temperature of the cell where is made the absorption measurements. The experimental error (1.5%) is mostly due to the uncertainty on the determination of the ozone vapor pressure. However this error does not affect the determination of the variation $\Delta\sigma$ = (- 1.3 ± 0.6) x 10^{-19} cm^2. This value is in agreement with that of MOLINA (- 0.9 x 10^{-19} cm^2) and with theoretical calculation made by ATABEK (1986) ($\Delta\sigma$ = - 1.1 x 10^{19} cm^2).

CONCLUSION

We think that it is now possible to normalize the relative Bass data to the new reference determined by MAUERSBERGER at 254 nm and take account of the variation of this reference with the temperature. With an appropriate wavelength calibration for Bass data, we may obtain three coherent sets of data in the acceptable limits of the experimental errors.

* On the continuous measurements σ = 113.0 $x^{-19}$$cm^2$.

REFERENCES

ATABEK, O., M.T. BOURGEOIS and M. JACON, 1986 : Three dimensional analytical model for the photodissociation of symmetric triatomics. Absorption and fluorescence spectra of ozone, J. Chem. Phys. 84, 6699-6711.

BASS, A.M., and R.J. PAUR, 1984 : The Ultraviolet cross-section of ozone : the measurements, in Atmospheric Ozone, Proceedings of the Quadriennal Ozone Symposium in Halkidiki, Greece, 3-7 Sept. 1984, 606-610D, Reidel, Hingham, Mass and private communication.

GRIGGS, M., 1968 : Absorption coefficients of ozone in the ultraviolet and visible region, J. Chem. Phys. , 49, 857-859.

HEARN, A.G., 1961 : The absorption of ozone in the ultraviolet and visible region of the spectrum, Proc. Phys. Soc., 78, 932-940.

INN, E.C.Y., and Y. TANAKA, 1953 : Absorption coefficients of ozone in the ultraviolet and visible regions, J. Opt. Soc. Amer., 43, 870-873.

MOLINA, L.T., and M.J. MOLINA, 1986 : Absolute absorption cross-sections of ozone in the 185-350 nm wavelength range, J. Geophys. Res., 91, 14501.

MAUERSBERGER, K., D. MANSON and J. MORTON, 1985 : A new ozone standard : the vapor pressure of ozone at liquid argon temperatures, Geophys. Res. Lett., 12, 89-92.

MAUERSBERGER, K., J. BARNES, D. HABSON and J. MORTON, 1986 : Measurements of the ozone absorption cross-section at the 253.7 nm mercury line, Geophys. Res. Lett., 13, 671.

VIGROUX, X., 1953 : Ann. Phys., 8, 709 ; 1967 : Ann. Phys., 2, 209.

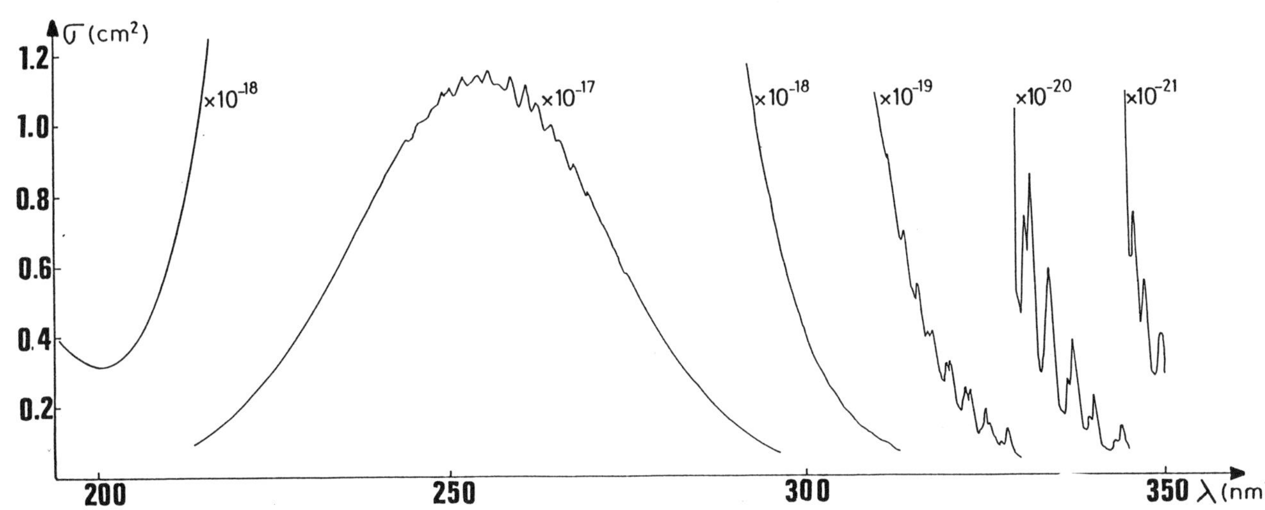

Fig.1. UV absorption cross-sections of O_3 : Hartley and Huggins bands.

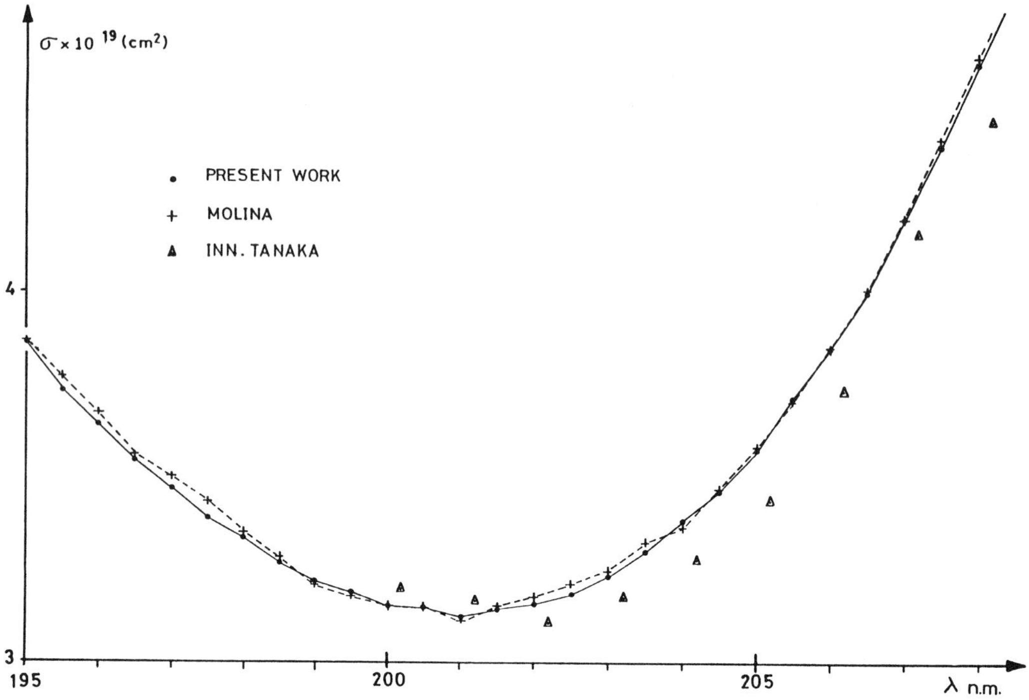

Fig.2. Comparison of the O_3 cross-section values in the 200 nm atmospheric window.

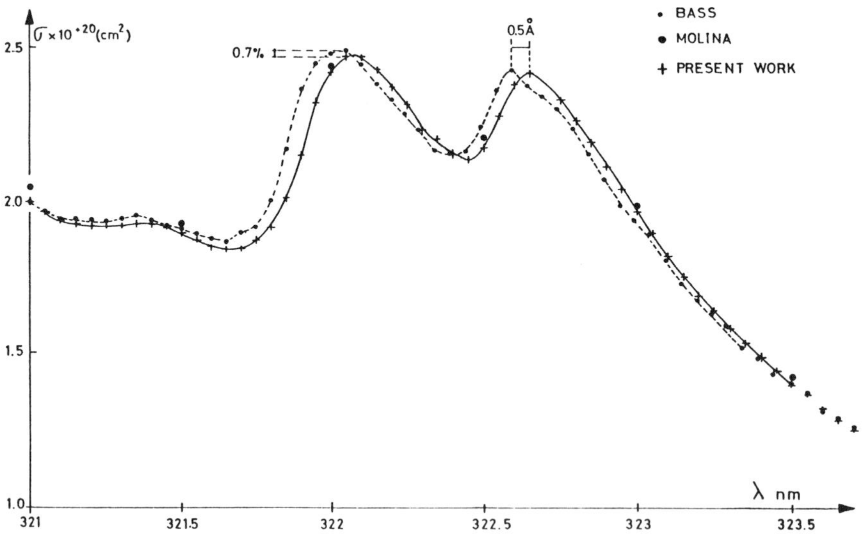

Fig.3. Huggins bands : observed λ shift.

DEVELOPMENTS IN OBSERVATIONAL TECHNIQUES

SIX YEARS OF TOTAL OZONE OBSERVATIONS WITH A BREWER SPECTROPHOTOMETER AT THESSALONIKI, GREECE

A.F. Bais, A.G. Kelessis, I.C. Ziomas, C.S. Zerefos, G.T. Amanatidis, and K.D. Tourpali
Laboratory of Atmospheric Physics (LAP)
University of Thessaloniki
54006, Thessaloniki, Greece

ABSTRACT

Total ozone measurements with the Brewer spectrophotometer #005 have been instituted at Thessaloniki (40.5°N, 22.9°E) since March 1982 and they continue to present. Prior to delivery the Brewer #005 had been calibrated with Brewer #001 and Dobson #077 spectrophotometers at AES in Canada. Since then, Brewer #005 was repeatedly intercompared against Brewers #008 and #017 (secondary standard), as well as against Dobsons #076 and #083 (standard) in Toronto, Thessaloniki and Arosa. The standard deviation of the differences between the Brewer #005 and the above mentioned instruments ranges from about ±0.6 matm-cm to ±3.2 matm-cm of total ozone. In addition to these intercomparisons on single days, the long-term stability of the Brewer #005 is tested by performing daily standard lamp and mercury calibration tests throughout the period of record. Several wavelength scans with Hg lamp showed no shift in the measuring wavelengths of Brewer #005 throughout the period 1982 to present. Also the uncertainty of the instrument in total ozone measurements was found to be between ±0.5% and ±1.5%, based on the daily standard lamp tests, which have shown no significant drift until present. On the other hand from Arosa intercomparison it follows that no μ-dependance can presently be established for Brewer #005. Having at hand one of the longest available continuous series with Brewer total ozone data, some results are olso reported concerning two periods with extreme total ozone values at Thessaloniki. They include the period of widespread ozone minima in 1983 and 1985, as well as March 1987 with extreme total ozone maxima over Europe.

1. INTRODUCTION

Concern for detection of the ozone trends calls attention to high precision measurements of total ozone. Since the Brewer total ozone network is more or less rapidly expanding and supplements the existing network of total ozone observing stations, the accuracy and general performance of the Brewer instruments is important for the ozone issue.

The first commercially available Brewer instrument (#005) was installed at the laboratory of Atmospheric Physics (LAP) of the University of Thessaloniki (40.5°N, 22.9°E, near sea level) in March 1982 and a regular total ozone observing program was instituted since then. The ozone data used in this work are only from direct sun observations. The performance of Brewer #005 during its first six years of continuous operation is discussed together with important features of the ozone variability during the period of study.

2. THE OPERATIONAL PROGRAM

All measurements with the Brewer #005 are taken preferably near the local apparent noon and

TABLE 1. INTERCOMPARISON SUMMARY OR BREWER #005 VS OTHER INSTRUMENTS

DATE	LOCATION	BREWERS			DOBSONS		
		#001	#008	#017	#076	#077	#083
FEB 1983	TORONTO	0.5 ±0.6				-1.0 ±1.3	
AUG 1983	TORONTO		0.3 ±1.4		-0.9 ±3.2		
SEP 1984	GREECE			0.1 ±1.2			
AUG 1986	AROSA			-3.0 ±1.3			1.4 ±1.6

ISBN 0-937194-15-8

when possible throughout daytime. Several tests are daily performed on the instrument, the most important of which are the standard lamp test and the mercury calibration test.The mercury test is performed several times during the day to ensure the proper measuring wavelength settings of the spectrophotometer, while the standard lamp test provides information on the stability of the instrument. Other tests namely the Deadtime, High voltage, Run/stop, Slitmask timing and Scan on direct sun tests are also made frequently following the directions of the manufacturer (Kerr et al. 1984, SCI-TEC 1985).

The results from the standard lamp tests can be seen in figure 1., which shows the ozone ratios, i.e. the weighted ratio of the light intensity logarithms at the measuring wavelengths, plotted versus time. As it appears from this figure the Brewer #005 was stable throughout the six years period with an error in total ozone ranging from ±0.5 to ±1.5% depending on the solar zenith angle. The solid line in this figure is a linear best fit which shows that the overall drift of the instrument is no more than 0.03% per year. This drift is negative, small and acceptable. It is thus necessary in the operational mode of the Brewer spectrophotometers to apply regularly thorough service of the instrument to assure its high performance.

In addition to the calibration and testing program discussed above, the Brewer #005 was intercalibrated with several other spectrophotometers (Brewers and Dobsons). Prior to delivery it was calibrated against the Brewer #001 and the Dobson #076 at AES and since then three more intercomparisons were made. The results of these intercomparisons are shown in Table 1. The data shown are daily means calculated from many simultaneous measurements of the Brewer #005 and the corresponding other spectrophotometers. Thus the individual observations during the intercomparison campains cover a large range of solar zenith angles with the μ ranging from 1.1 to 3.0.

From the intercomparisons shown here, it appears that the Brewer #005 compares well with other instruments, especially with the Brewer #017 which is considered as the secondary traveling standard instrument. Moreover, during the Arosa intercomparison campain,in which the largest data volume was collected, it appeared from comparisson of Brewer #005 with Brewer #017 that #005 is not μ-dependent. Brewer #017 is considered to be well calibrated and it has been used for calibration of the Hohenpeissenberg's Brewer #010 (Kohler 1986).

3. THE TOTAL OZONE RECORD AT THESSALONIKI

Figure 2. shows the time series of monthly mean total ozone at LAP (40.5°N, 22.9°E). The linear best fit which was calculated for the period March 1982 to March 1988 shows an insignificant change of total ozone, of about +0.02% per year. However it shows the important features of short-term periods with total ozone minima that characterized ozone in the northern hemisphere; the sharpest low ozone period which lasted from November 1982 to March 1983, the second lower ozone period from March through May

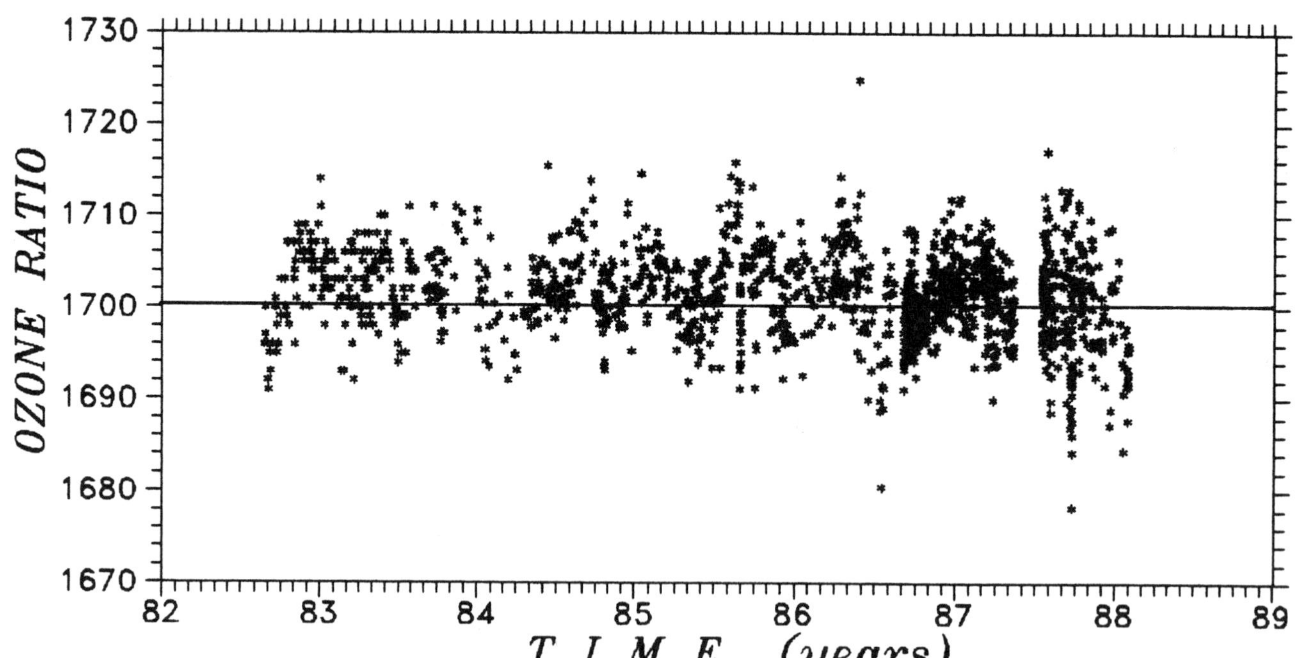

Fig. 1. The "Ozone Ratio" of the Brewer #005 standard lamp tests versus time, for the six years of operation.

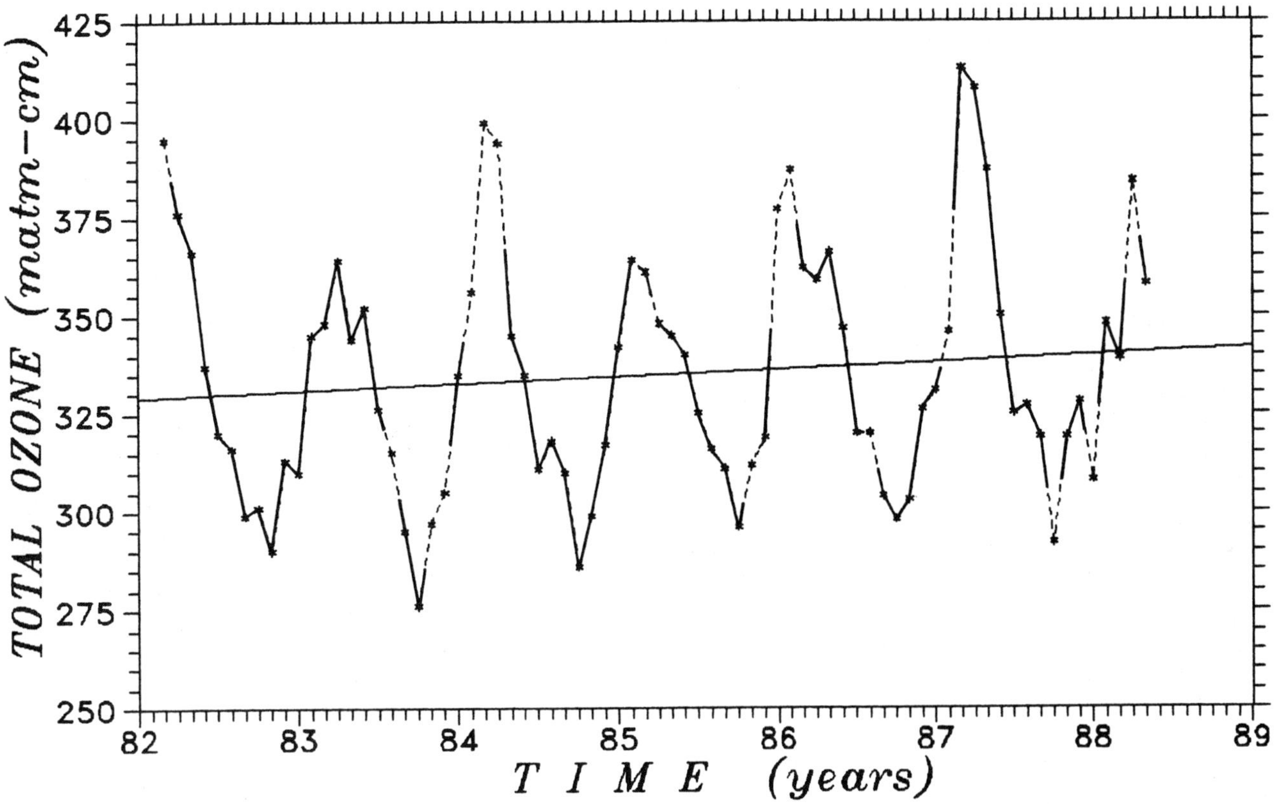

Fig. 2. Monthly means of total ozone at Thessaloniki-Greece. Dashed lines denote months with less than 12 days of observations.

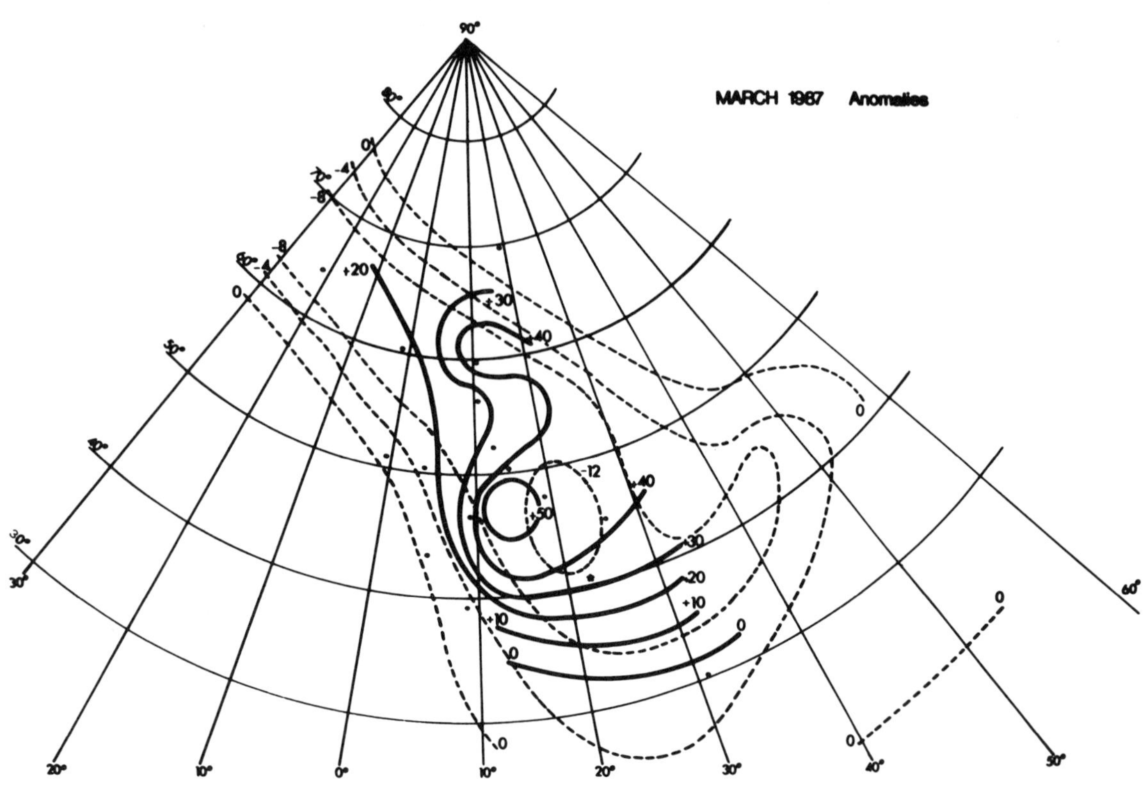

Fig. 3. March 1987 total ozone anomalies (solid line) and 300 mbar isobaric surface anomalies (dashed line) over Europe.

1985 and the highest record value of March 1987. The periods with low ozone have been extensively discused in the literature (Bais et al. 1985, Mantis et al. 1986, Bojkov 1987). As to their cause they have probably been caused by a combination of the QBO, ENSO and disturbed following the El Chichon volcanic eruption (Bojkov 1987).

In March 1987 record high values of total ozone exceeding 500 D.U. have been observed at LAP (40.5°N). This month was exceptionaly cold in the Balkans with persistent upper air troughs over central Europe and the peninsula, throughout the first half of the month. During the first week of March total ozone climbed at Thessaloniki by more than 120 D.U., peaking between the 6th and the 8th of March with more than 500 D.U. of total ozone. Total ozone was transported in the lower stratosphere by the mean flow in central Europe, as can be seen from the ozonsonde data at Hohenpeissenberg (47.5°N) and Lindenberg (52.1°N) as well as by comparing total ozone anomalies with the anomalies of the 300 mbar isobaric surface (Meteorologishe Abhandlugen 1987) shown in figure 3. The coinsidence of the negative height anomalies with these of increasing total ozone is as expected. The synoptic conditions were such, that March of 1987 was positively abnormal in total ozone over most of the european stations. Seeking for long-term trends by season in total ozone especially during the cold part of the year, should take into consideration extreme months that can be simply prodused by climatological fluctuations of persisting synoptic upper air patterns.

REFERENCES

Bais A.F., C.S. Zerefos, I.C. Ziomas, N. Zoumakis, H.T. Mantis, D.J. Hofmann and G. Fiocco. Decreases in the Ozone and the SO_2 Columns Following the Appearence of the El Chichon Aerosol Cloud at Midlatitude. In: Atmospheric Ozone, Proc. Quadrennial Ozone Symposium, Halkidiki, Greece, Sep 1984, Reidel, 353-356.

Bojkov R.D. (1987). The 1983 and 1985 Anomalies in Ozone Distribution in Perspective. Monthly Weather Review, 115,10,2187-2201.

Kerr J.B., C.T. McElroy, D.I. Wardle, R.A. Olafson and W.F.J Evans (1984). The Automated Brewer Spectrophotometer. In: Atmospheric Ozone, Proc. Quadrennial Ozone Symposium, Halkidiki, Greece, Sep 1984, Reidel, 239-401.

Kohler U. (1986). Recalibration of the Hohenpeissenberg Dobson Spectrophotometer 104 Presuming "Effective Absorption Coefficients. Journal of Atmospheric Chemistry, 4, 359-374.

Mantis H.T., C.S. Zerefos, A.F. Bais, I.C. Ziomas and A.G. Kelessis (1986). The Northern Hemisphere Ozone Minimum in 1982-1983. Arch. Met. Geoph. Biocl. Ser. B 36, 135-145.

Meteorologishe Abhandlugen (1987). Frei Universitat Berlin, Institut fur Meteorologie, Ser.B, Band 55, Heft 3.

SCI-TEC Instruments Inc. (1985). Brewer Ozone Spectrophotometer (Operator's manual) OM-BA-C05. Available from SCI-TEC Instruments Inc. Saskatoon, Saskatchewan, Canada.

ISOTOPIC ABUNDANCES OF STRATOSPHERIC OZONE FROM BALLOON-BORNE OBSERVATIONS OF HIGH RESOLUTION INFRARED SOLAR SPECTRA

A. Goldman, F.J. Murcray, D.G. Murcray, and J.J. Kosters
University of Denver, Denver, Colorado 80210, USA
C.P. Rinsland
NASA Langley Research Center, Hampton, Virginia 23665, USA
C. Camy-Peyret and J.-M. Flaud
Pierre and Marie Curie University, 75252 Paris Cedex 05 France
A. Barbe
Reims University, Reims France

Abstract. Infrared solar absorption spectra at 0.002 cm^{-1} resolution spectra obtained in the 10 μm region during a recent balloon flight have been analyzed for isotopic ozone quantification in the stratosphere. The analysis has utilized recent updates of O_3 line parameters for both the normal and isotopic ozone species. Preliminary results show isotopic enhancements of 1.20 ± 0.14 and 1.41 ± 0.18 for $^{16}O^{18}O^{16}O/^{16}O^{16}O^{16}O$ and $^{16}O^{16}O^{18}O/^{16}O^{16}O^{16}O$ respectively above 37 km altitude.

Introduction

Since the initial study of Cicerone and McCrumb [1980], much interest has developed in chemistry and dynamics of the heavy ozone enhancement in the stratosphere, [cf Bates, 1988]. The contribution of the isotopic and "hot" bands O_3 lines to the atmospheric infrared spectrum was first recognized from 0.5 cm^{-1} resolution long path balloon-borne spectra [Goldman et al, 1970]. However, the low resolution and the limited spectroscopic information available for the O_3 lines at the time did not permit detailed spectroscopic analysis of the isotopic lines in the atmospheric data. Balloon-borne spectra obtained later at 0.02 cm^{-1} resolution allowed spectroscopic update of several of the important atmospheric O_3 bands, but the spectroscopy of the isotopic lines remained crude [Rothman et al, 1987]. Subsequent high resolution laboratory spectra of isotopic enhanced O_3 allowed spectroscopic update of many atmospheric isotopic O_3 transitions in the far infrared, millimeter wave region and middle infrared and the identification and quantification of several isotopic lines in the atmospheric spectrum [Rinsland et al, 1985; Abbas et al, 1987].

During a balloon flight performed on November 18, 1987 from Ft. Sumner, New Mexico, infrared solar spectra at 0.002 cm^{-1} resolution were recorded with the University of Denver balloon-borne Fourier transform spectrometer. Measurements were obtained during part of the ascent and from float altitude of 37 km, and cover the 750-1350 cm^{-1} region with medium optical path through the ozone layer. The spectra are rich in new detailed information on stratospheric ozone lines in the ν_3, ν_1 region, with numerous well isolated isotopic ^{18}O lines and several "hot" bands lines of $^{16}O_3$. The recent updates of the line parameters allow assignments and detailed analysis of stratospheric lines of $^{16}O_3$, $^{16}O^{18}O^{16}O$ and $^{16}O^{16}O^{18}O$. In the present paper, the equivalent width method for single, isolated lines and the spectral least-squares fitting method for narrow intervals are used for the quantification of the observed isotopic enrichment, at high altitudes, of $^{16}O^{18}O^{16}O$ and $^{16}O^{16}O^{18}O$ relative to $^{16}O_3$.

Data Analysis

Figures 1-2, plotted on the same wavenumber scale, show samples of balloon flight spectra obtained at 0.02 (apodized) resolution with an older interferometer system [Goldman et al, 1987] and at 0.002 cm^{-1} (unapodized) resolution with the new interferometer system. The spectral details gained by the 0.002 cm^{-1} resolution are quite apparent. Line-by-line calculations with the HITRAN database [Rothman et al, 1987] over the 970-1200 cm^{-1} region showed that even though only medium optical path was obtained during this flight, <u>all</u> of the O_3 bands in this region require updating for proper spectral simulations of the new balloon flight data. The new line positions and intensities of ν_3, ν_1, $\nu_3 + \nu_2 - \nu_2$, $\nu_1 + \nu_2 - \nu_2$, $2\nu_3 - \nu_3$, $\nu_3 + \nu_1 - \nu_1$ of $^{16}O_3$ and ν_3, ν_1 of $^{16}O^{18}O^{16}O$ and $^{16}O^{16}O^{18}O$ are based on the recent work of Flaud et al [1986, 1987], Camy-Peyret et al [1986], Malathy Devi et al [1987] and Rinsland et al [1988]. New ozone air-broadened halfwidths have been derived from the experimental data of Smith et al [1988], which agree very well with recent halfwidth calculations [Hartman et al, 1988]. The new line parameters provide much improved and more consistent agreement with the balloon data.

Figure 3 shows sample spectra and analysis of the CO_2 R(14) 10.4 μm line near 971.930 cm^{-1} The thermal core and the pressure broadened wings are quite distinct in these spectra. Spectral least-squares fitting is used to derive the atmospheric temperature profile.

Figure 4 shows a spectral least-squares analysis of the 993 cm^{-1} interval. From such analysis and from equivalent width analysis of weak lines, O_3 isotopic enhancements have been determined.

In the weak line approximation, the equivalent width W is proportional to the gas amount U, by $W(cm^{-1}) = S(cm^{-1}/molec\ cm^{-2})U(molec/cm^2)$, where S is the intensity of the spectral line.

ISBN 0-937194-15-8

Considering first the ratios of the equivalent widths of a selected set of, say, $^{16}O^{18}O^{16}O$ lines to a selected set of $^{16}O^{16}O^{16}O$ lines in the ground-based spectrum, and then the corresponding ratios in the float-altitude spectrum, and then averaging the ratios of the above ratios, we obtain an isotopic abundance (relative to $^{16}O^{16}O^{16}O$) ratio from the two spectra. Since about 90% of the ozone amount is below the float altitude, an estimate for the upper altitudes isotopic enhancement is obtained.

Similar to Rinsland [1985], we have selected ten $^{16}O^{18}O^{16}O$ lines in the 980-994 cm^{-1} interval, seven $^{16}O^{16}O^{18}O$ lines in the 1042-1045 cm^{-1} interval, and nine $^{16}O^{16}O^{16}O$ lines in the 1116-1145 cm^{-1} interval, each with medium ground state energy to minimize the temperature dependence. Spectral least-squares analysis was performed for several intervals, to verify the validity of the weak line approximation and to include more O_3 lines in each interval. From the equivalent width double ratios described above, values of 1.20 ± 0.14 and 1.41 ± 0.18 were obtained for $^{16}O^{18}O^{16}O/^{16}O^{16}O^{16}O$ and $^{16}O^{18}O^{16}O/^{16}O^{16}O^{16}$ respectively.

The present results indicate smaller enhancement of heavy ozone in the stratosphere compared to previous results of sampling and spectroscopic measurements [Mauersberger, 1987; Abbas et al, 1987]. While most theoretical calculations [Kaye, 1986] predict negligibly small isotopic enhancement, it appears that the processes involved are not fully understood [Bates, 1988]. The method reported in this paper has the high resolution advantage for isolating isotopic O_3 lines, and the quantification is insensitive to the line intensities. Solar spectra over longer optical path will produce more information on stratospheric ozone and its isotopes.

We have also used tunable diode laser heterodyne spectrometer to generate extremely high resolution (0.0005 cm^{-1}) ground-based solar spectra in narrow intervals of the 9.6 micron ozone region [McElroy et al, 1988]. It is apparent that despite the pressure broadening of the lower stratosphere O_3 lines, resolution better than 0.002 cm^{-1} is desired for more complete spectral characterization.

Acknowledgments. This work was supported in part by the National Aeronautics and Space Administration, and in part by the National Science Foundation. Acknowledgment is made to the National Center for Atmospheric Research, which is supported by the National Science Foundation, for computer time used in this research.

References

Abbas, M.M., J. Guo, B. Carli, F. Mencaraglia, M. Carlotti, and I.G. Nolt, Heavy ozone distribution in the stratosphere from far-infrared observations, J. Geophys. Res., 92, D11, 13231-13239, 1987.

Bates, D.R., Suggested explanation of heavy ozone, Geophys. Res. Lett., 15, 13-16, 1988.

Camy-Peyret, C. J.-M. Flaud, A. Perrin, V. Malathy Devi, C.P. Rinsland, and M.A.H. Smith, The hybrid-type bands, ν_1 and ν_3 of $^{16}O^{16}O^{18}O$: line positions and intensities, J. Molec. Spectrosc., 118, 345-354, 1986.

Cicerone R.J., and J.L. McCrumb, Dissociation of isotopically heavy O_2 as a source of atmospheric O_3, Geophys. Res. Lett., 7, 251-254, 1980.

Flaud, J.-M., C. Camy-Peyret, V. Malathy Devi, C.P. Rinsland and M.A.H. Smith, The ν_1 and ν_3 bands of $^{16}O^{18}O^{16}O$: line positions and intensities, J. Molec. Spectrosc., 118, 334-344, 1986.

Flaud, J.-M., C. Camy-Peyret, V. Malathy Devi, C.P. Rinsland and M.A.H. Smith, The ν_1 and ν_3 bands of $^{16}O_3$: line positions and intensities, J. Molec. Spectrosc., 124, 209-217, 1987.

Goldman, A., R.D. Blatherwick, F.J. Murcray, J.W. Van Allen, F.H. Murcray, and D.G. Murcray, New atlas of stratospheric IR absorption spectra, September 1987 Edition, Vol. I. Line positions and identifications; Vol.II. The spectra, Department of Physics, University of Denver, September 1987.

Goldman, A., T.G. Kyle, D.G. Murcray, F.H. Murcray, and W.J. Williams, Long path atmospheric ozone absorption in the 9-10-μ region observed from a balloon-borne spectrometer, Appl. Opt., 9, 565-580, 1970.

Hartman, J.M., C. Camy-Peyret, J.-M. Flaud, J. Bonamy, and D. Robert, New accurate calculations of ozone line-broadening by O_2 and N_2 submitted to J. Quant. Spectros. Radiat. Transfer, 1988.

Kaye, J.A., Theoretical analysis of isotope effects on ozone formation in oxygen photochemistry, J. Geophys. Res., 91, 7865-7874, 1986.

Malathy Devi, V., J.-M. Flaud and C. Camy-Peyret, C.P. Rinsland and M.A.H. Smith, Line positions and intensities for the $\nu_1 + \nu_2$ and $\nu_2 + \nu_3$ bands of $^{16}O_3$, J. Molec. Spectrosc., 125, 174-183, 1987.

Mauersberger, K., Ozone isotope measurements in the stratosphere, Geophys. Res. Lett., 14, 80-83, 1987.

McElroy, C.T., A. Goldman, and D.G. Murcray, Heterodyne spectrophotometry of ozone in the 9.6 micron band using a tunable diode laser, Appl. Opt., in press, 1988.

Rinsland, C.P., V. Malathy Devi, J.-M. Flaud, C. Camy-Peyret, M.A.H. Smith, and G.M. Stokes, Identifications of ^{18}O-isotopic lines of ozone in infrared ground-based solar absorption spectra, J. Geophys. Res., 90, D6, 10719-10725, 1985.

Rinsland, C.P., M.A.H. Smith, J.-M. Flaud, C. Camy-Peyret, and V. Malathy Devi, Line positions and intensities of the $2\nu_3$, $\nu_1 + \nu_3$, and $2\nu_1$ bands of $^{16}O_3$, J. Mol. Spectrosc., in press, 1988.

Rothman, L.S., R.R. Gamache, A. Goldman, L.R. Brown, R. A. Toth, H. Pickett, R. Poynter, J.-M. Flaud, C. Camy-Peyret, A. Barbe, N. Husson, C.P. Rinsland, and M.A.H. Smith, The HITRAN database: 1986 edition, Appl. Opt., 26, 4058-4096, 1987.

Smith, M.A.H., C.P. Rinsland, V. Malathy Devi, D.C. Benner, and K.B. Thakur, Measurements of air-broadened and nitrogen-broadened halfwidths and shifts of ozone lines near 9 μm, J. Opt. Soc. Am., B, 5, 585-592, 1988.

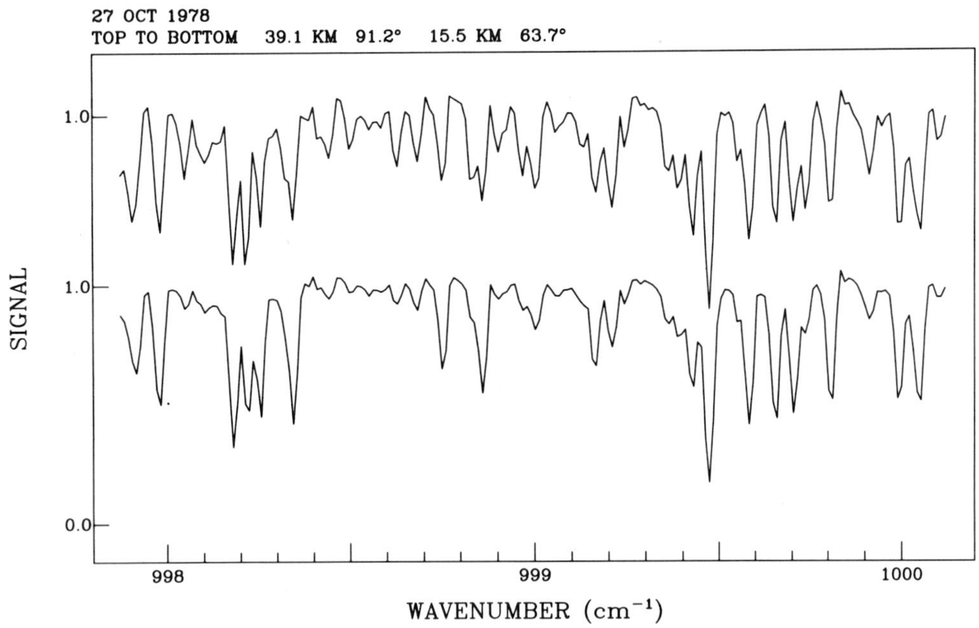

Figure 1. Balloon-borne spectra in the 998-1000 cm^{-1} region obtained on October 27, 1978, from Alamogordo, New Mexico, with a 0.02 cm^{-1} resolution interferometer.

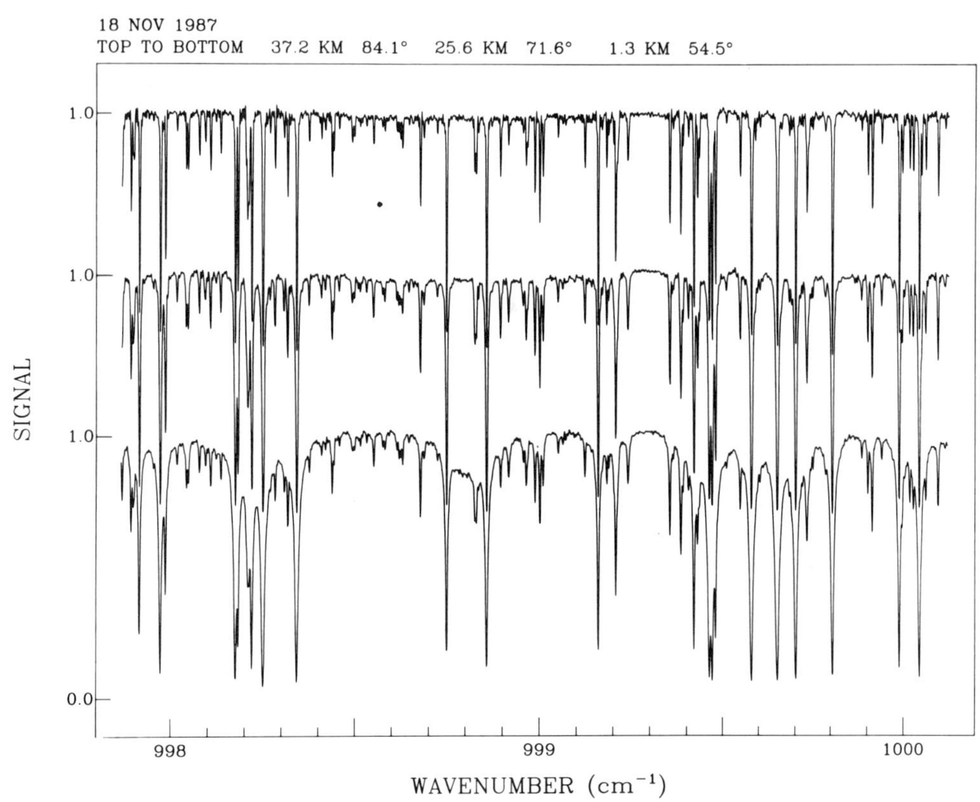

Figure 2. Ground-based and balloon-borne spectra in the 998-1000 cm^{-1} region obtained on November 18, 1987, from Ft. Sumner, New Mexico, with a 0.002 cm^{-1} resolution interferometer.

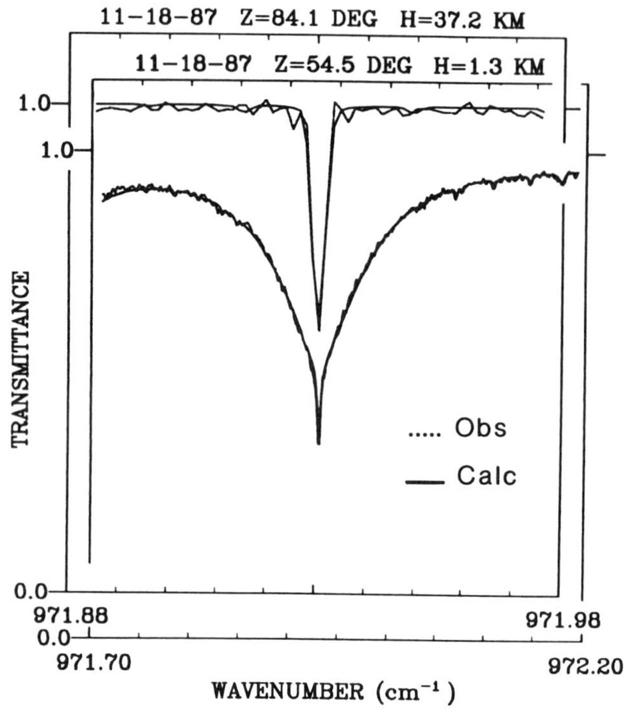

Figure 3. The CO_2 R(14) 10.4 μm line at 971.9303 cm^{-1} (E"=1470.1192 cm^{-1}) as observed during the November 18, 1987 balloon flight from Ft. Sumner, New Mexico. Spectral least-squares is used to derive the atmospheric temperature profile.

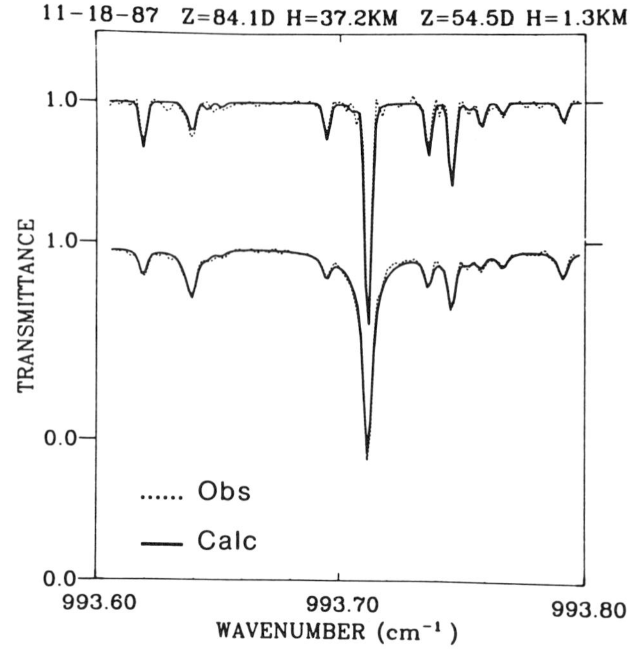

Figure 4. Spectral least-squares fitting of ground-based and balloon-borne solar spectra in the 993.60-993.80 cm^{-1} region. The spectra were obtained during a balloon flight on November 18, 1987, from Ft. Sumner, New Mexico, with the University of Denver 0.002 cm^{-1} resolution interferometer.

SUGGESTION ON THE POSSIBILITY TO SOUND ATMOSPHERIC OZONE WITH A RADIOSONDE EQUIPPED WITH TWO TEMPERATURE SENSORS, SENSITIVE AND NON-SENSITIVE TO THE LONG-WAVE RADIATION

T. Kitaoka
158 4–25–10, Tamagawa, Setagaya-ku
Tokyo, Japan

ABSTRACT

The 5-year mean report of upper air soundings by radiosonde in Japan for the period 1981 to 1985 was published by the Japan Meteorological Agency in 1987. In analyzing data in this report, a conspicuous singular increase in the temperature differences between day and night, $\Delta T_1 = T_9 - T_{21}$, was found at the 40-50 mb level for the period from February to May.

This increase was observed over the Northern area of Japan at about 44°N, while a small decrease was observed for the same period over the Southern area at about 25°N.

Over the N-area, the temperature difference ΔT_1 reaches a maximum of about 1.2°C in March and gradually decreases from spring to autumn and is nearly zero in autumn. On the contrary, it remains at almost zero throughout the year over the S-area, except for a small decrease of about 0.6°C in spring. The temperature sensor used in these radiosonde soundings was a naked thermistor.

On the other hand, this singular annual trend of temperature differences was not observed in any record of the 5-year means of 1966-70, 1971-75 and 1976-80, in which the temperature sensor was bimetal. It showed only a monotonic trend of annual variation, affected by the insolation put in the ventilation duct of the radiosondes.

Results of this analysis was supposed to provide supporting evidence that a naked thermistor is sensitive to the long-wave radiation, and that the singular annual trend in temperature differences might be caused by a change in the long-wave radiation emitted by atmospheric ozone based on some kind of behavior in emission processes.

In the following description, results of the above analysis are described in detail, and further evidence, which shows the effect of long-wave radiation on the temperature measured by thermistor sensor, is shown. As a result, the possibility of sounding the atmospheric ozone by using radiosondes equipped with two sensors is suggested.

1. INTRODUCTION

By analyzing data of four 5-year mean reports on upper air soundings in Japan which were published by Japan Meteorological Agency every 5 years, it was made clear that the thermistor sensor of radiosonde, which is covered by white paint and exposed to the air flow with naked form, is sensitive for the long-wave radiation emitted not only from the earth surface, cloud and water vapor but also from the atmospheric ozone by some measurable degrees. The temperature sensor of the Japanese radiosonde has been changed from bimetal to thermistor in March 1981, so that the last 5-year mean report is on the monthly mean results of soundings by thermistor sensor, the other three ones being those taken by bimetal. The investigation was first made on the differences of elements measured by day (0900) and by night (2100) and then made on the results of nighttime soundings when no consideration is necessary for the insolation error. The following phenomena which would have some relation with the atmospheric ozone were found.

1. A conspicuous increase of temperature difference, $\Delta T_1 = T_9 - T_{21}$, recognized in the record of 5-year means from 1981 to 85 of temperature measurements by radiosonde, which were equipped with thermistor sensor in the period from February to May, giving a maximum of about 1.2°C in March at $40 \sim 50$ mb level.
   This increase is observed over the northern area of Japan at about 44°N, while a small decrease of about 0.6°C was observed for the same period over the southern area at about 25°N.
   On the other hand, those singular annual trend of temperature difference is not observed in any record of 5 year mean of 1966-70, 71-75 and 76-80, measured by bimetal sensor, non sensitive for the long wave radiation .
2. The seasonal change of the turning levels from negative values in the lower layer to positive ones in the upper layer are observed in the vertical changes of temperature difference ΔT_1 by thermistor.
   The level is lowest in spring and highest in autumn over both areas.

The temperature differences ΔT_1, and $\Delta T_2 = T_{Th}$ - T_{Bi} depend not only on the amount of ozone existing in the relevant layer, but also on the temperature difference between the thermistor and the ozone layer. The temperature difference ΔT_2 would appear to be due to the difference in the lag coefficients between two sensors, but it was made clear by further detailed analysis and discussions that it can not be explained only by the lag difference. As a result, it is concluded that the temperature measured by thermistor at the upper layers produced an effect on atmospheric ozone of a few measurable degrees. It was suggested that some very useful additional informations on atmospheric ozone could be obtained by developing adequate sounding techniques; hardware and software and by applying them to the normal system of upper air soundings by radiosondes, without high additional costs.

2. DATA USED FOR THE INVESTIGATION AND INFORMATION RELATING TO MEASURING ACCURACY

In Japan, before 1981, radiosondes equipped with bimetal sensors were used for upper air soundings. In March 1981, however, the temperature sensor was changed from bimetal to thermistor. The statistical results of soundings are published in 5-year mean reports by Japan Meteorological Agency. The accuracy of temperature soundings has become stable and relatively good since about 1965. For that reason, three reports of 5-year means, 1966-70, 1971-75 and 1976-80, in which the monthly means of sounding results by bimetal are included, and one report of one 5-year mean for 1981-85, by thermistor, were used for this investigation.

Bimetal and thermistor sensors used for the soundings are installed in the radiosonde body as shown in Fig. 1: the former protected in the ventilation duct, the latter exposed to the air flow outside the body of the radiosonde itself.

ISBN 0–937194–15–8

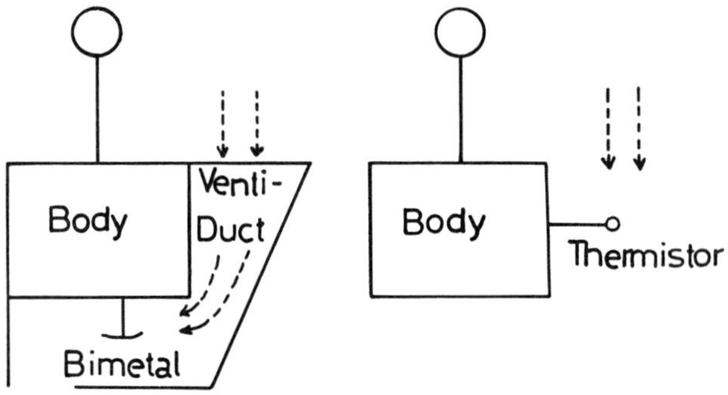

Fig. 1 General view of radiosonde temperature sensor and body

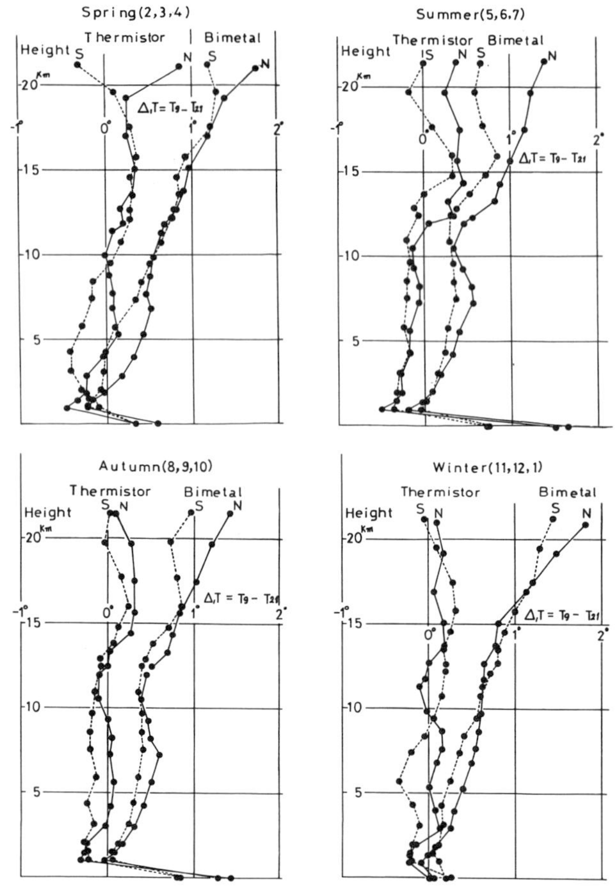

Fig. 2 Results on the difference between day and night, $\Delta T_1 = T_9 - T_{21}$

The difference of lag coefficient of the two sensors was found by another investigation based on the results of 41 comparative soundings between the two kinds of radiosonde, made by hanging two radiosondes in the same train under one balloon. This difference is shown in Table 1.

TABLE 1

Press. Range (mb)	Lag Diff. (Sec)
1024-788	1.08
838-613	0.88
709-462	0.96
585-339	0.54
472-240	0.85
397-165	1.04
333-117	1.31

However, the lag difference at upper levels above about 200 mb is not yet clear.

3. RESULTS OF INVESTIGATION

In order to avoid any confusion resulting from the difference in the lag coefficient and of other characteristic performances between the two sensors, the investigation was first made on the differences of elements measured by day (0900) and by night (2100). It was then made on the results of soundings at night when no consideration was necessary for the insolation error.

Systematic differences were found in temperatures by the two sensors. Fig. 2 shows the vertical change of temperature between day and night, $\Delta T_1 = T_9 - T_{21}$, for four seasons and over the two areas. Values over the Northern and Southern areas were the means of three stations in each area, whose mean latitudes were about 44° and 25°N, respectively. They were deduced from the 5-year means of temperatures (or thicknesses at upper levels) by thermistor in the period 1981-85, and from the 15 year means by bimetal sensor in the period, 1966-80.

The temperature differences between day and night run almost parallel with height over both areas for values by thermistor and bimetal, the values of thermistor being shifted negatively by about 0.5°C, as shown in Fig. 2, except for somewhat of an increase in the values from the bimetal sensor at upper levels due to its insolation error increasing with height above about 300 mb level (about 10 km) as described below.

However, the temperature differences of thermistor at lower levels show larger negative values (warmer by night) in higher humid seasons and region, compared with those in lower humid seasons and region.

In summer, the humid season, the temperature differences are negative up to the level of about 12 km over both areas. Over the S-area which is more humid, the differences are negative at lower levels up to about 9 km for all seasons.

These features could be explained by the differences of net flux of long-wave radiation from thermistor sensor by day and by night, which were calculated mainly with the vertical profile of relative humidity and temperature with some assumption on the relation between cloud and relative humidity.

At upper levels above 9-10 km, the temperature differences of thermistor also become positive. The turning level from negative to positive is lower in spring and winter when ozone is more abundant and higher in autumn and summer when there is less ozone. The positive value at upper levels is largest in summer, smallest in winter and almost the same in spring and autumn. The value of temperature differences also depends on the vertical profile of temperature at upper levels, giving larger positive values in the season of larger increases in temperature with height, and smaller positive values in the season and region of smaller increases in temperature or almost isothermal with height.

This shows that the value of the temperature differences depends not only on the amount of atmospheric ozone, but also on the temperature changes with height. This results from the fact that the amount of radiation emitted by ozone varies with its amount and temperature and that the net flux of radiation from the thermistor sensor is equal to the difference between the black body radiation emitted by the thermistor sensor itself and the radiation coming from the above and below.

In Fig. 3, the annual variation of temperature measured by the two sensors and those of the temperature differences between day and night, ΔT_1, at upper level at 50-40 mb, are shown for both thermistor

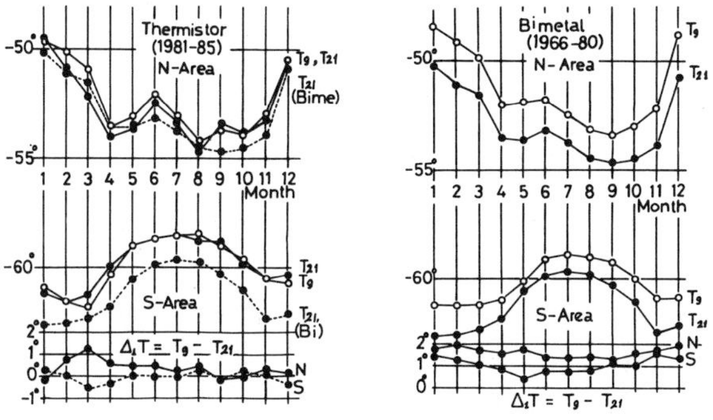

Fig. 3 The annual variation of temperature measured by the thermistor and bimetal sensor, $\Delta T_2 = T_{Th} - T_{Bi}$.

and bimetal sensor over the same areas as above.

In the case of bimetal sensors, the effect of insolation energy coming into the ventilation duct was not perfectly corrected at upper levels, above 300 mb, giving a maximum difference of day values from night ones at the Winter solstice, the time of greatest distance between the earth and the sun and a minimum at the Summer solstice, the time of the shortest distance.

The values of ΔT_1 by bimetal sensor over both areas, which are given at the right side of Fig. 3, run almost parallel throughout the year. However they show somewhat irregular changes, caused by a lack of uniformity of the number of observations in each month, because of the balloons bursting or other causes. Such accidents happened more frequently at the beginning of the sounding program.

The larger value of the insolation effect over the N-area compared to the S-area cannot be real, but is probably due to the correction technique for the insolation error.

In the case of the thermistor sensors, the effect of insolation is almost eliminated by the white paint covering on the sensor. However, the paint works like a black body for long-wave radiation and the characteristic feature of the annual trend of ΔT_1, as shown at the left side of Fig. 3, is supposed to be caused by this effect.

The increase and decrease of the temperature difference between day and night over the N- and S-areas in the period from February to April are supported mainly by the increase and decrease in thermistor temperatures by day as shown in Fig. 3. Because the temperatures measured by the bimetal sensor and by the thermistor at night run almost parallel through the year the temperatures from the thermistor during the day show such a singular change of ΔT_1, as shown in the left lower side of Fig. 3.

The cause of these singular changes of thermistor temperatures may be explained by the change in the net flux of long-wave radiation emitted from the thermistor by day, based on some behavior of ozone in the period of abundant ozone. However, this problem should be solved in future by more information related to this problem.

The vertical changes in temperature differences between the thermistor and the bimetal sensor, $\Delta T_2 = T_{Th}$ - T_{Bi}, are given in Fig. 4, for the values by night only, when no consideration is necessary for the insolation error, over the N- and S-areas, respectively.

From Fig. 4-1, in which the values of yearly means are shown, we can see that differences are negative at lower levels and positive at upper levels over both areas. However, the negative values are larger and the positive values are smaller over the N-area than over the S-area and the turning level from negtive to positive is lower over the N-area than over the S-area. This shows the same sense as above.

From Fig. 4-2, in which the differences of temperatures measured by thermistor from those by bimetallic sensor at upper levels above 200 mb are shown for separate seasons, the same feature can be clearly seen.

4. DISCUSSION AND CONCLUSION

The difference between temperatures measured by thermistor and bimetal sensor by night would appear to be due to the difference in the lag coefficients between the two sensors. But, if we assume that the temperature difference between the thermistor and bimetal sensor comes from the difference in lag coefficient, the lag difference must be almost 50-60 seconds at upper levels. Such a large difference in the lag coefficient could not be expected even at the upper levels,

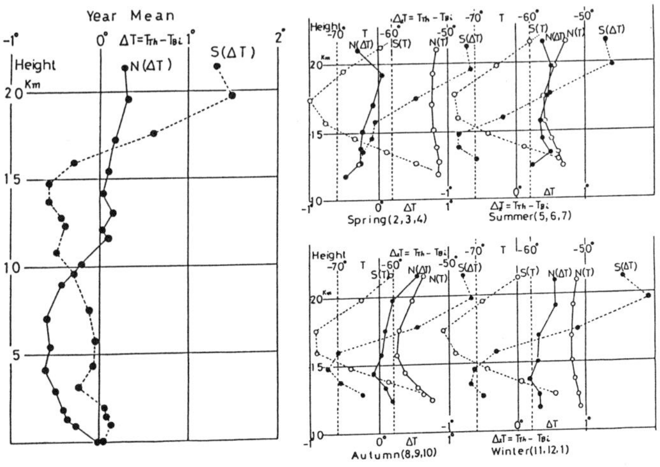

Fig. 4 The vertical changes in temperature differences between the thermistor and the bimetal sensor $\Delta T_2 = T_{Th} - T_{Bi}$ for the values by night; 4-1. values for yearly means, 4-2. values for the seasons

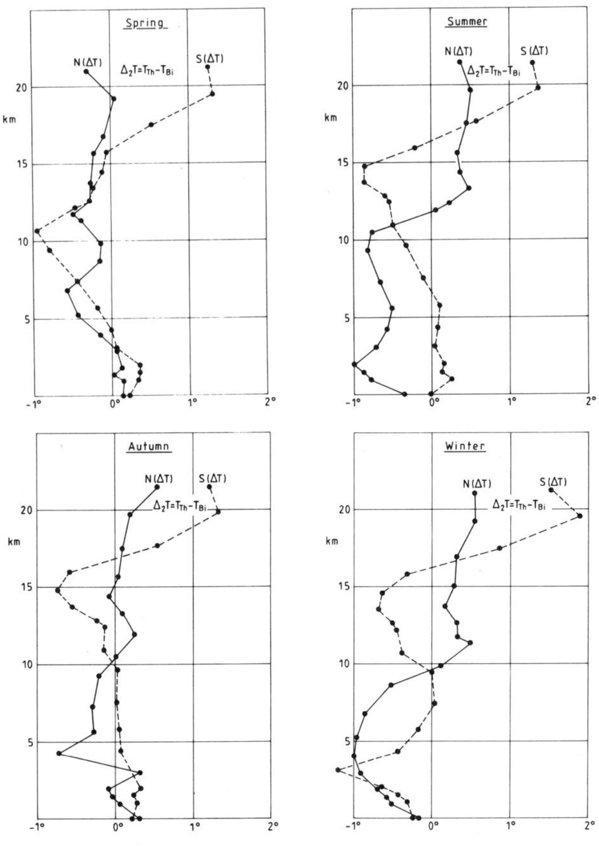

Fig. 5 The temperature difference between temperatures measured by thermistor and bimetal in the whole layers down to the ground for four seasons.

from such a small difference of less than one second at lower levels, as found by the other investigation and shown in Table 1. Moreover, the following analysis also shows that the temperature difference ΔT_2 can not be explained only by the lag differences.

In Fig. 5, the difference between the temperatures measured by thermistor and by bimetal in the whole layer are shown for four seasons separately.

In Fig. 6, the annual changes of surface temperature which is the most important element for warming the thermistor sensor from below are shown. In this figure, effective surface temperatures are also entered. On the right side of this figure, the seasonal means by the effective surface temperatures and the differences of ground average temperatures in two periods, 1981-85, and 1966-1980, which are the periods sounding by thermistor and by bimetal respectively, are also given by numerical formula.

Here, the temperature of effective surface has the following meaning: The radiosonde balloon is flown by the westerlies, therefore the balloon moves to the East during the sounding by radiosonde. The Japan Island is surrounded by the Ocean, so the most of the balloons come over the Ocean soon after their ascents. The temperature of sea surface is different from the one of the ground surface, especially during the cold season, being pretty warmer in general.

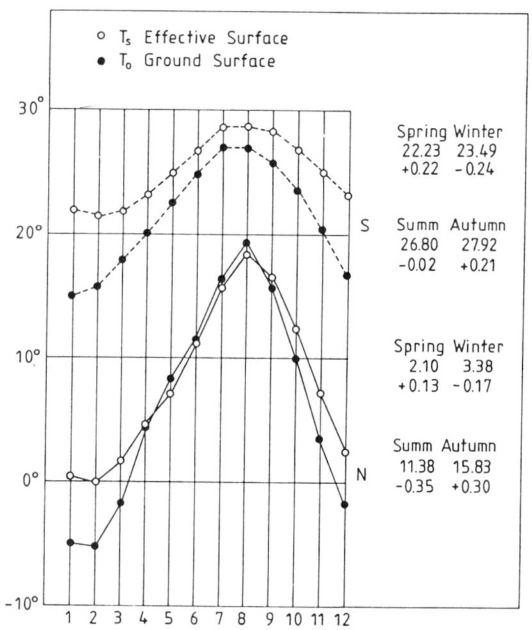

Fig. 6 The annual changes of surface temperature.

Considering these circumstances, some adequate averages of temperatures of ground and of sea surfaces should be taken as the effective surface temperature effective for warming the thermistor sensor in upper levels, instead of the temperatures of the ground themselves. The differences in average ground temperatures in the two periods are considered as some indications of the climatic differences. The higher the effective surface temperature, the warmer the thermistor will get. When the difference of average ground temperature is negative, that is, the average temperature in the period 1981-85, is lower than the one in the period 1966-80, the climate in the first period is considered clearer than that in the second period. Because the average ground temperature by night decreases at clearer nights more than at clouded ones due to radiation cooling.

The temperature differences, ΔT_2 are larger over the S-Area in spite of the smaller amount of ozone. The warming effect of ozone on the thermistor sensor changes not only with the ozone existing in the relevant layer but also with the temperature difference between the layer and the thermistor. The temperature increases conspicuously with height over S-Area, whereas it increases by smaller degree or nearly isothermal with height over N-area. The radiation exchange between ozone and thermistor is, of course, zero in the special case of isothermal layer, even at very heigh ozone concentrations. This is one reason for the fact mentioned above.

Comparing the temperature differences ΔT_2 over N-Area in summer and in autumn, we can observe that the difference in autumn is smaller than in summer, in spite of having almost the same vertical profile of temperature. The similar but contrary features can be observed in comparison between spring and winter, which have also the same vertical profile of temperature. In the case of the first comparison, the smaller temperature difference in autumn seem to be explained by smaller ozone concentrations, but in the case of the second comparison, the temperature difference ΔT_2 is smaller in spring with more abundant ozone than in winter. This seems to be in the contrary direction. However, this contradiction can be solved by taking into account the differences of effective surface temperatures between two seasons and the differences in climate difference of two periods between two seasons. It is as follows. The effective surface temperature in winter over N-area is higher (about 1-3°C) than in spring. If the effective temperature in spring was the same as in winter, the temperature difference, ΔT_2, in spring would become larger. On the other hand, the difference of average ground temperature of two periods is negative in winter and positive in spring. If the difference in spring was so negative as it is in Winter, the temperature difference ΔT_2 in spring would become larger too. These two factors act in the same sense and magnify the temperature difference in spring, or make it smaller in winter. How much degree should be corrected by this factors is not yet estimated, but the large difference of ΔT_2 in lower levels below 9 km of two seasons suggests us that the temperature difference ΔT_2 in the upper level in winter should be more reduced and on the contrary, ΔT_2 in spring should be increased by smaller degree. As a result, ΔT_2 in spring must become larger than in winter, however it is only a qualitative estimation. Comparing summer and autumn, the effective surface temperature over N-Area in autumn is also warmer about 45°C than in summer. But the difference of average ground temperature in two periods is positive in autumn and negative in summer and in the contrary sense against the above case. So, in this case, the effective surface temperature makes the tempeatures difference ΔT_2 larger in summer and the temperature difference of average ground temperature makes ΔT_2 smaller in summer. So these two factors act in the opposite direction and neutralized each other. Therefore, the relative feature of the values of temperature difference ΔT_2 in the upper level would not change even through the two correction factors. Similar discussions as above could be made on the temperature differences ΔT_2 in four seasons over S-Area and come to the same results.

The results of analysis in the above mentioned comparisons between two sensors show us that the temperature difference ΔT_2 in the upper level is determined not only by the vertical profile of temperature and cannot be explained only by the difference of lag coefficients of the two sensors. It show us that the magnitude of ΔT_2 changes mainly with ozone existing in the upper layer above the thermistor sensor, when the vertical profile of temperature is the same. It is concluded that the temperature measured by thermistor is affected by atmospheric ozone of a few measurable degrees. On the other hand, the temperature increase of the thermistor sensor from the temperature measured by bimetal sensor by night at upper levels shown in Fig. 4-1 and the singular differences between the northern and southern areas can be explained by the effect of long-wave radiation emitted from atmospheric ozone, if we take 2.5 x 10^8 x σ as the energy of net flux for 1°C increase or decrease in thermistor temperature. By developing adequate of both hardware and software sounding techniques and by applying them to the normal system of upper air soundings by radiosonde, we could get very useful additional information on atmospheric ozone without high additional costs. In developing software, consideration must be given to the effect of long-wave radiation by water vapor and clouds too. In developing hardware, it is most desirable to use two sensors having almost the same lag coefficients but quite different characteristic performances for long-wave radiation.

UMKEHR OBSERVATIONS MADE WITH THE BREWER OZONE SPECTROPHOTOMETER

C.T. McElroy, C.L. Mateer, J.B. Kerr, and D.I. Wardle
Atmospheric Environment Service
Downsview, Ontario, M3H 5T4, Canada

ABSTRACT

The Brewer ozone spectrophotometer has now been in service at the Edmonton station of the Canadian ozone network for four years. During the years 1985 and 1986 the instrument was programmed to make automatic Umkehr observations. The 28 profiles from these years have been analyzed to evaluate the performance of the Brewer instrument in making these observations and to examine the accuracy of the inversion program which was developed for their analysis. It was found that the Umkehr observations taken with the Brewer instrument produced vertical profiles of ozone which were in good agreement with those taken using the Dobson short Umkehr technique and with the ozonesonde data collected at Edmonton during the period of the observations. At levels of interest Brewer Umkehr measurements produce results that are in agreement with the sonde data to better than 10%, a performance which is essentially identical to that of the short Umkehr technique using the Dobson instrument.

1. INTRODUCTION

It is of critical importance that the amount of ozone in the upper layers of the stratosphere be measured reliably in order to monitor the future rate of depletion of the ozone layer, and to improve man's understanding of the chemical and dynamic processes which control the ozone layer. In the past, monitoring of the upper levels of the ozone layer has been done predominantly through the analysis (inversion) of Dobson ozone spectrophotometer Umkehr data (Mateer and Dütsch, 1964).

Dobson data are analyzed principally by two methods. One, which has been in use longest, is called the standard Umkehr analysis method (Mateer and Dütsch, 1964). In this technique, measurements made on the Dobson "C" wavelength pair (at angles between 60° and 90°) are inverted to produce a vertical profile of ozone. A more recently developed method is also in use. This is called the "short Umkehr technique" (Mateer and Deluisi, 1980) because it utilizes measurements from a smaller range of zenith angles (80° to 89°) than the standard analysis method, and therefore requires less total observation time. The short Umkehr method uses observations on three wavelength pairs (the Dobson A, C, and D wavelengths) instead of the usual one (C pair). The extra information thereby obtained permits the smaller range of observing angles to be used.

The Brewer Umkehr observations (Mateer et al., 1984) comprise intensity measurements in eight wavebands centred at 306.3, 310.1, 313.5, 316.8,320.0, 323.2, 326.4, and 329.5 nm. The instrument is equipped with a polarizer. For Umkehr observations the polarization is set so that the component of the zenith radiation with the electric vector in the plane containing the sun and the local vertical is rejected.

2. DATA AND ANALYSIS

Throughout the years 1985 and 1986, a number of automatic Umkehr observations using the Brewer instrument were obtained at the Stony Plain upper air station near Edmonton, Alberta. During 1985 the Brewer was operated in a semi-automatic mode and a total of 30 Umkehr measurements were made. Of these, 23 included observations over a sufficient range of zenith angles to be analyzed and had good ozone total column measurements associated with them. In 1986 the collection of Umkehr data was fully automatic. A total of 59 Umkehrs were collected and 46 of these were processed.

In processing the data for analysis, the raw spectral intensity data are first fitted as functions of the zenith angle using cubic splines, and the scatter of the data about the best fit spline curve is used as a measure of the quality of the observations. The results are plotted and the plots used to accept or reject data for specific days. In order to process Umkehr data it is essential to have a good estimate of the total ozone amount above the station, and so, only days for which good ozone measurements were available near 70° solar zenith angle are accepted for analysis.

After the Umkehrs were analyzed it was found that 22 of the 1985 observations were well matched by the inversion procedure, but only 6 taken during 1986 were acceptable. To be an acceptable fit, the modelled observations must have an RMS difference from the observed data of less than 0.7 N units. This is the same fitting criterion routinely used with both the Dobson standard and short Umkehr methods. The cause of the large number of rejections in 1986 is a malfunction in the instrument control software peculiar to the automated Umkehr mode. It resulted in the long wavelength set of observations being made occasionally at an incorrect wavelength setting. It has since been corrected and in future a much larger proportion of observed Umkehrs should be acceptable for analysis.

The preprocessed data are analyzed by an inversion algorithm to generate vertical distributions of ozone. The inversion algorithm is a linear inversion method which has been

ISBN 0-937194-15-8

previously described by Mateer and Deluisi (1971). The inversion process is considered in detail in the review paper of Rodgers (1976), which particularly addresses the proper inclusion of a priori information. The Umkehr inversion method makes use of a climatological mean distribution appropriate to the total column amount of ozone measured and to the location where the observations were made. The solution sought is one which balances off the contribution of the uncertainty of the observations against the degree to which the first guess profile is believed to be a good representation of the profile measured on any particular day. The covariance matrix which expresses the uncertainty of the first guess profile, and the first guess profile itself are input data for the inversion algorithm. These data are largely based on the analysis of actual ozone profiles collected over a long period of time by ozonesondes.

In addition to the Brewer data, both ozonesonde profiles and Dobson Umkehr observations were regularly collected. A total of 22 useful comparison observations were made in 1985 and six in 1986 using the Brewer instrument. Umkehr data were collected only in the morning during 1985. Both morning and evening Umkehrs were obtained during 1986.

TABLE 1: COMPARISON OF THE TWO-YEAR MEAN PARTIAL PRESSURE PROFILES OF OZONE AS DETERMINED BY OZONESONDES, THE BREWER OZONE SPECTROPHOTOMETER AND THE DOBSON INSTRUMENT AT THE EDMONTON UPPER AIR STATION DURING 1985 AND 1986.

Layer Number	Brewer (28)	Dobson Short (52)	Dobson Std. (38)	Sonde (71)	B-S %	B-D %	B-DS %
0	25.2	25.1	—	33.3	-24	—	0
1	25.3	27.0	32.0	28.5	-11	-21	-6
2	67.0	73.5	64.6	66.1	1	4	-8
3	109.7	115.5	83.4	103.2	6	32	-5
4	142.5	138.8	119.8	146.7	-3	19	3
5	114.2	108.3	121.9	120.1	-5	-6	5
6	71.4	69.0	81.6	73.2	-2	-13	3
7	42.0	41.1	41.6	32.6	29	1	2
8	18.7	18.1	17.3	—	—	8	3
9	6.26	6.00	5.69	—	—	18	4
10	1.83	1.73	—	—	—	—	6
11	0.51	0.48	—	—	—	—	6
12	0.14	0.13	—	—	—	—	8

1)B-S (Brewer-Sonde)/Sonde %.
2)B-D (Brewer-Dobson Standard)/(Dobson Standard) %.
3)B-DS (Brewer-Dobson Short)/(Dobson Short) %.
4)Numbers in parentheses are the number of samples averaged for each column of the table.

3. RESULTS

Since only a small number of vertical profiles from the different methods were actually coincident in time (one day had all three types of observations and a further three had Brewer observations and one other type), a two-year mean of profiles has been chosen to make a comparison between the methods. Table 1 shows the two-year mean values of the Brewer data, both types of Dobson data and the results of ozone sonde flights from the Stony Plain upper air station. As can be seen from the table, the Brewer data agree extremely well with the data from the Dobson short Umkehr analysis. This is not surprising since the Brewer and Dobson measurements span the same wavelength region in the near UV. The Brewer Umkehr and short Dobson data are better estimates of the ozone profile than the standard Dobson data as indicated by the comparison with the mean sonde data. The superiority of the short over the standard Umkehr has also been demonstrated by Mateer and Deluisi (1985).

Table 2 presents data for April 3, 1985, the only day when profiles by all three methods were available for comparison. Here again, the ozone sonde profile is better estimated by the Brewer or Dobson short Umkehr methods than by the standard Umkehr analysis. There are over and under estimates of the ozone amount at various levels by the standard Umkehr method which are consistent with the mean data in Table 1.

TABLE 2: RESULTS OF A SONDE FLIGHT AND BREWER AND DOBSON PROFILES FOR APRIL 3, 1985 (units nb).

Layer Number	Brewer	Dobson Short	Dobson Standard	Sonde
0	31.2	30	—	38.4
1	37.8	39	58	35.9
2	121.8	128	117	128.3
3	159.4	165	115	128.7
4	152.6	147	134	144.1
5	108.5	103	119	115.4
6	64.9	65	79	—
7	42.4	43	42	—
8	20.5	22	18.0	—
9	6.90	7.1	6.4	—
10	1.95	2.0	—	—
11	0.52	0.51	—	—
12	0.14	0.13	—	—

1)Only one day's data were available with all types of observations.

4. CONCLUSIONS

The Brewer Umkehr evaluation algorithm is already providing data whose quality is essentially the same as the Dobson short Umkehr method. The Brewer instrument is more thoroughly automated than any other accurate ozone spectrophotometer. Consequently, Umkehr observations at the Brewer stations in the World Ozone Network could easily be instigated and would provide improved spatial and temporal coverage for global monitoring of the ozone profile. Satellite measurement intercomparison also would benefit from the higher density of the ground-based data.

A routine evaluation method for Brewer Umkehr evaluation is now being finalized to be used for the operational processing of Brewer data at the World Meteorological Organization World Ozone Data Centre in Toronto.

5. ACKNOWLEDGMENT

The authors would like to express their appreciation for the great skill with which Mr. I.A. Asbridge has operated the Canadian ozone network for many years.

REFERENCES

Mateer, C.L., and H.U. Dütsch, 1964: Uniform evaluation of Umkehr observations from the World Umkehr Observation Network, Part 1, National Center for Atmospheric Research, Boulder, Colorado, 105pp.

Mateer, C.L., and J.J Deluisi, 1980: The estimation of the vertical profile of ozone by the short Umkehr method, Proc. Quad. Ozone Symp., Int'l Ozone Comm., 64–73.

Mateer, C.L., and J.J Deluisi, 1984: A comparison of ozone profiles derived from standard Umkehr and short Umkehr measurements from fifteen stations, Proc. Quad. Ozone Symp., Reidel, 290–294.

Mateer, C.L., J.B. Kerr and W.F.J. Evans, 1984: Ozone profiles derived from Umkehr observations obtained with the Brewer ozone spectrophotometer, Proc. Quad. Ozone Symp., Reidel, 407–411.

Rodgers, C.D., 1976: Retrieval of atmospheric temperature and composition from remote measurements of thermal radiation, Rev. Geophys. and Space Phys., 14, 609–624.

OBSERVATIONS OF ARCTIC TOTAL OZONE WITH THE BREWER SPECTROPHOTOMETER DURING THE POLAR WINTER USING THE MOON AS A LIGHT SOURCE

J.B. Kerr
Atmospheric Environment Service
Downsview, Ontario, M3H 5T4, Canada

ABSTRACT

Fully automated Brewer ozone spectrophotometers were installed for routine operation in the Canadian ozone network at Resolute and Alert. A major objective of these installations is to obtain winter measurements of total ozone during the Arctic night using the moon as a light source. An algorithm which calculates the moon's position in the sky has been applied to the automatic operation of the Brewer instrument. The Brewer pointing system is capable of positioning an image of the moon on the entrance slit with an accuracy of ±0.25 lunar diameters. Measurements are made automatically without the need of an operator. Good quality ozone measurements have been made using this technique at both Resolute and Alert during the winter of 1987-1988. The technique used to measure ozone with the moon is described and the measurement method is evaluated. Results of the Arctic winter measurements are presented.

1. INTRODUCTION

Ozone has been observed to be decreasing over the Antarctic during spring and the available data suggests that a similar but smaller decrease in ozone is occurring over the Arctic during late winter. These concerns have prompted an increased effort to monitor total ozone over the Canadian Arctic, particularly during winter. Fully automated Brewer ozone spectrophotometers were installed at Resolute (74°N, 95°W) in August, 1987 and Alert (82°N, 62°W) in November, 1987. Total ozone direct moon measurements were made during the polar winter of 1987-1988 at the two sites.

The fully automated Brewer ozone spectrophotometer and the method by which it measures total ozone using the sun as a light source have been described by Kerr et al. (1983, 1984). The instrument operates from a predetermined schedule which is programmed by the operator. Part of the observation schedule includes measurement of moonlight at times when the sun is sufficiently below the horizon so that there is no scattered solar radiation. The instrument automatically points at the moon and measurements are made without requiring the presence of an observer.

The method used to measure ozone using moonlight is described and evaluated. Results of the measurements from Resolute and Alert are presented.

2. METHOD OF MEASUREMENT

The theory and data analysis used to measure total ozone with the moon as a light source is similar to those used for the direct sun measurement. Intensity measurements of ultraviolet light at wavelengths in the Huggins absorption bands of ozone are measured after it has passed through and been absorbed by ozone in the atmosphere. The direct sun total ozone measurement method applied to the Brewer spectrophotometer is described by Kerr et al. (1980).

In operation it is necessary to direct moonlight into the instrument to measure its intensity. This is done by the same pointing optics used to measure sunlight. The celestial position of the moon is calculated with a correction for parallax applied and the azimuth and zenith pointing controls for the Brewer instrument are adjusted to the appropriate positions. Pointing requirements are more crucial for direct moon measurements than for direct sun measurements. This is because the low light intensity of the moon requires the moon's image to be focussed on the entrance slit, whereas, the direct sun uses a ground quartz diffuser to reduce light intensity.

A measurement of total ozone using the direct moon method is the average of five individual observations each lasting about two minutes. The Brewer instrument points slightly in advance of the moon's position at the beginning of the observation. During the two minute observation period the image of the moon traverses across the entrance slit. This ensures that light from all parts of the moon's surface are used for all observations. The mean of the five observations is used as the total ozone measurement and the standard deviation of the individual observations about the mean is used to indicate the quality of the measurement. Should the standard deviation be too large, the measurement is assumed to be of poor quality (usually because of clouds) and is not used to determine the daily average value.

3. EVALUATION OF DIRECT MOON MEASUREMENTS

Routine automatic direct moon measurements of total ozone are made at Toronto with Brewer instrument # 15 and these measurements are used to evaluate the quality of the moon data.

ISBN 0-937194-15-8

Between January, 1987 and May, 1988, 706 acceptable quality direct moon measurements were made on 91 nights. The total number of measurements made over this period as a function of the day relative to the full moon is shown in Figure 1. Measurements were made up to ± 7 days of the full moon.

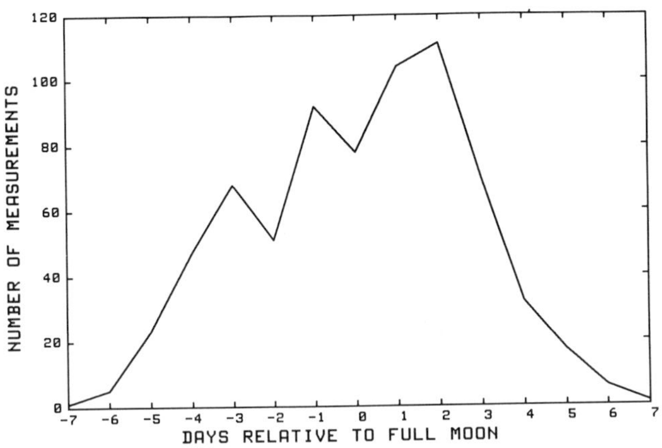

Fig. 1. Plot of the number of direct moon observations as a function of time relative to the full moon.

The nightly direct moon total ozone measurements are compared to direct sun total ozone measurements made on the days before and after. There are 71 instances when there are direct sun measurements on both the day preceeding and following a night with direct moon data. The average of the daily mean direct sun ozone values for the two adjacent days is compared with the nightly mean direct moon value in Figure 2 where the nighttime data is plotted against the daytime data. The slope of this line is 1.03 and the intercept is -12 DU with respective 95% confidence estimates of ± .11 and ± 43 DU. The standard deviation of data poir... about the best fit line is ± 20 DU. A slope of 1.0 for this comparison would indicate that the ozone absorption coefficient used for moon measurements is the same as that used for sun measurements. An intercept of 0 DU would indicate that the extraterrestrial

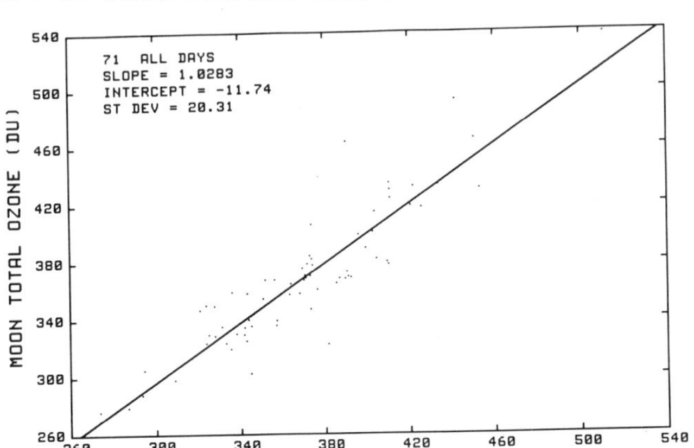

Fig. 2. Plot of nightly mean direct moon total ozone values as a function of the average of the direct sun daily mean values for the to adjacent days.

value used for the moon is the same as that for the sun. The comparison between sun and moon data suggests that both the extraterrestrial constant and ozone absorption coefficient are the same for the sun and moon to within the observed uncertainty.

The observed variability of ± 20 DU (5.5%) between nightly direct moon measurements and daily direct sun measurements could result from two possible causes. There could be a real variation of atmospheric total ozone during the twelve hours difference between night and day or there may be a random error for the direct moon measurements. Random measurement error for the daily mean direct sun measurement is less than ± 0.5% (Kerr et al., 1988). Results of the moon measurements in Toronto indicate that the precision of an individual observation is ± 9 DU (2.4%). The variability of a nightly mean direct moon value is observed to be ± 5.5 DU (1.5%). These values suggest that only a small part of the 5.5% daytime/nighttime variability is due to random measurement error for the direct moon measurements.

The cause of the daytime to nighttime variability was evaluated further by considering direct moon data and direct sun data on days 36 hours after the night of the moon measurements in addition to daytime data on adjacent days. There are 49 instances between January, 1987 and May, 1988 when direct sun measurements were made on 0.5 days before (D1), 0.5 days after (D2) and 1.5 days after (D3) a night (N1) during which direct moon measurements were made. The mean differences and standard deviations of measurements made 0.5 days, 1 day and 2 days apart are determined and summarized in Table 1. The mean of the 98 daytime minus nighttime differences separated by 12 hours (D1-N1 and D2-N1) is 0.5 DU with a standard deviation of 25.1 DU. The mean of the 98 differences between consecutive days (D1-D2 and D2-D3) is 2.5 DU with a standard deviation of 35.5 DU. The mean of the 49 differences of ozone on days separated by two days (D1-D3) is 4.9 DU with a standard deviation

TABLE 1. MEANS AND STANDARD DEVIATIONS FOR DIFFERENCES BETWEEN SUN AND MOON DATA SEPARATED BY .5, 1.0 AND 2.0 DAYS

DATA	DELTA TIME (DAYS)	NUMBER OF OBS	MEAN (DU)	STANDARD DEVIATION (DU)
D1 - N1 and D2 - N1	0.5	98	0.5	25.1
D1 - D2 and D2 - D3	1.0	98	2.5	35.5
D1 - D3	2.0	49	4.9	50.4
$\frac{D1+D2}{2}$ - N1	0.5	49	0.5	17.5
$\frac{D1+D3}{2}$ - D2	1.0	49	-3.5	25.1

of 50.4 DU. The mean of the 49 differences between the average of D1 and D2 minus N1 is 0.5 DU with a standard deviation of 17.5 DU. The mean of the 49 differences between the average of D1 and D3 minus D2 is -3.5 DU with a standard deviation of 25.1 DU. These statistics indicate that the variability of total ozone measured on adjacent days is on average ± 35 DU (10%) and that this variability increases for data separated by two days and decreases for data separated by 0.5 days. It is therefore concluded that the observed daytime/nightime variability is caused mostly by real atmospheric changes in total ozone and that the nightly mean direct moon total ozone values are accurate to within ± 1.5%.

The effects of the lunar phase on the direct moon measurements were investigated. The data were divided into two categories depending on whether the measurements were made on days before (-ve days) or after (+ve days) the full moon and days on the full moon were not included. Figure 3a show the plot of the 35 pre full moon observations as a function of the average direct sun total ozone values from the days before and after and Fig. 3b is the plot for the 31 post full moon values. Results of these regressions are similar to those using all the data (to within 4%) and indicate that to within the observed uncertainty the absorption coefficients and extraterrestrial constants are the same before and after the full moon.

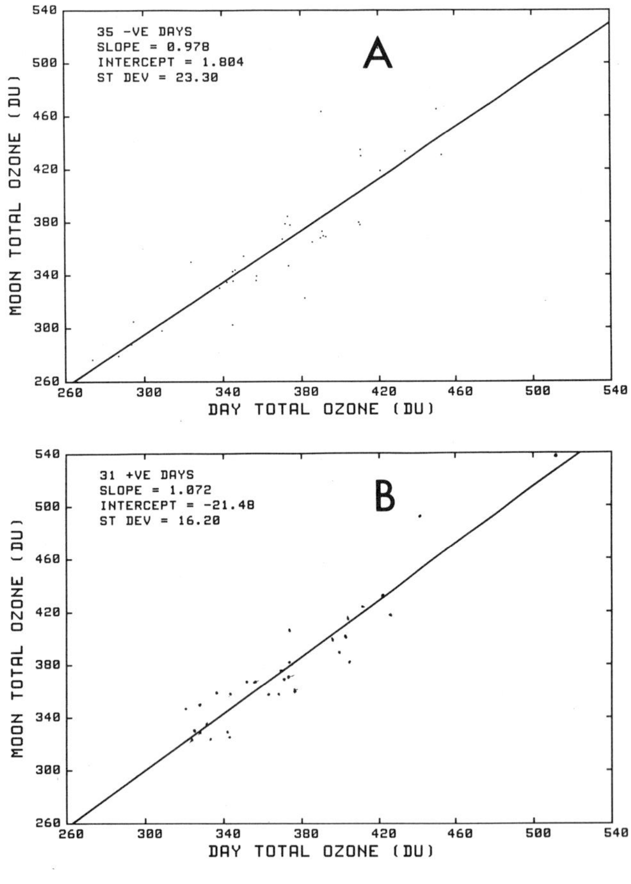

Fig. 3. Same as in Fig. 2 except using data only from days before the full moon (a) and after the full moon (b).

Results of the differences between data separated by 0.5 days, 1 day, and 2 days are summarized in Table 2 for the pre full moon data and in Table 3 for the post moon data. These results are again similar to those using all the data. It is therefore concluded that a direct moon measurement does not have a statistically significant systematic dependence on the phase of the moon.

TABLE 2. SAME AS TABLE 1 EXCEPT USING DATA FROM DAYS BEFORE THE FULL MOON

DATA	DELTA TIME (DAYS)	NUMBER OF OBS	MEAN (DU)	STANDARD DEVIATION (DU)
D1 - N1 and D2 - N1	0.5	38	4.9	23.4
D1 - D2 and D2 - D3	1.0	38	0.4	32.5
D1 - D3	2.0	19	0.9	46.7
$\frac{D1+D2}{2}$ - N1	0.5	19	4.9	17.5
$\frac{D1+D3}{2}$ - D2	1.0	19	-5.0	22.6

TABLE 3. SAME AS TABLE 1 EXCEPT USING DATA FROM DAYS AFTER THE FULL MOON

DATA	DELTA TIME (DAYS)	NUMBER OF OBS	MEAN (DU)	STANDARD DEVIATION (DU)
D1 - N1 and D2 - N1	0.5	50	-4.3	26.8
D1 - D2 and D2 - D3	1.0	50	3.0	40.0
D1 - D3	2.0	25	5.9	57.9
$\frac{D1+D2}{2}$ - N1	0.5	25	-4.3	17.9
$\frac{D1+D3}{2}$ - D2	1.0	25	-2.8	28.1

Evaluation of direct moon measurements made at Toronto between January, 1987 and May, 1988 have indicated that the accuracy of a nightly mean direct moon measurement using the Brewer spectrophotometer is about ± 1.5% and that there is no significant dependence on lunar phase.

4. RESULTS OF ARCTIC MEASUREMENTS

The results of direct moon total ozone measurements during the polar night in the Arctic are summarized in Fig. 4 for Alert and Fig. 5 for Resolute. Nightly mean values of total ozone using the direct moon method are shown.

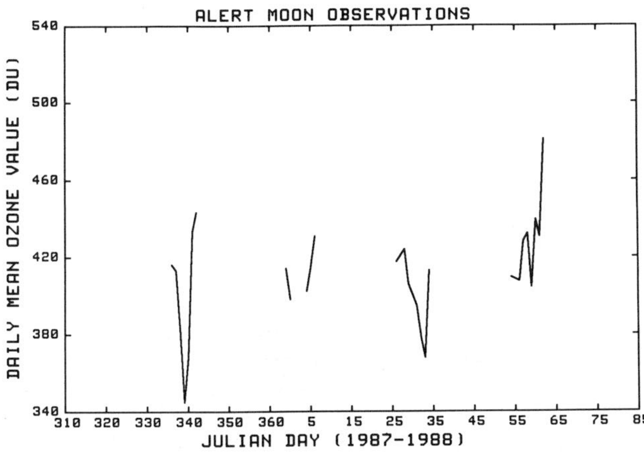

Fig. 4. Nightly mean direct moon total ozone at Alert. There are 279 measurements made on 30 nights.

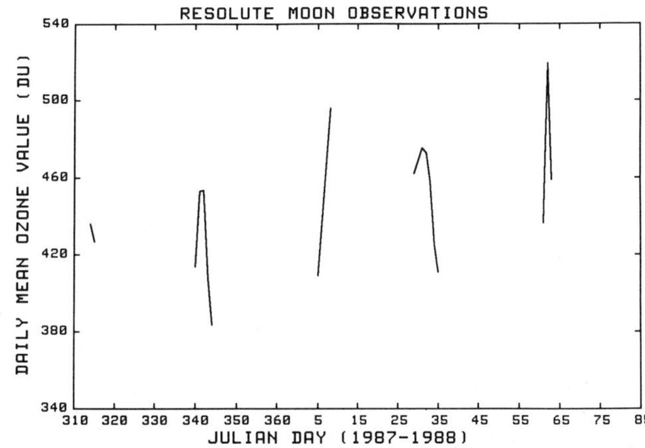

Fig. 5. Nightly mean direct moon total ozone at Resolute. There are 158 measurements made on 19 nights.

REFERENCES

Kerr, J.B., I.A. Asbridge, and W.F.J. Evans, 1988: Intercomparison of total ozone measured by the Brewer and Dobson spectrophotometers at Toronto, accepted by J. Geophys. Res., 93.

Kerr, J.B., C.T. McElroy, and W.F.J. Evans, 1983: The automated Brewer spectrophotometer for measurement of SO_2, O_3 and aerosols, Proc. of the WMO/AMS/CMOS Symposium on Meteorological Observations and Instrumentation, 470-472, Toronto.

Kerr, J.B., C.T. McElroy, and R.A. Olafson, 1980: Measurements of ozone with the Brewer ozone spectrophotometer, Proc. Int. Ozone Symposium, 74-79, Boulder, Colorado.

Kerr, J.B., C.T. McElroy, D.I. Wardle, R.A. Olafson, and W.F.J. Evans, 1984: The automated Brewer spectrophotometer, Proc. Quadrennial Ozone Symposium, 196-401, Halkidiki, Greece.

TUNABLE DIODE LASER HETERODYNE SPECTROPHOTOMETRY OF OZONE

P.F. Fogal[1], C.T. McElroy[2], A. Goldman[1], and D.G. Murcray[1]
[1]University of Denver
Denver, Colorado 80208, USA
[2]Atmospheric Environment Service
Downsview, Ontario, M3H 5T4 Canada

ABSTRACT

Tunable diode laser heterodyne spectrophotometry (TDLHS) has been used to make extremely high resolution (< 0.0005 cm^{-1}) solar spectra in the 9.6 micron ozone band. Observations have shown that a signal-to-noise ratio of 95:1 (35% of theoretical) for an integration time of 1/8 second can be achieved at a resolution of 0.0005 wavenumbers. The spectral data have been inverted to yield a total column amount of ozone, in good agreement with that measured at the nearby National Oceanographic and Atmospheric Administration (NOAA) ozone monitoring facility in Boulder, Colorado.

1. INTRODUCTION

Tunable Diode Laser Heterodyne Spectroscopy (TDLHS) has the ability to produce very high resolution atmospheric spectra. The concept of heterodyning and its applications to instruments for atmospheric measurement have been discussed previously by a number of authors, including Fujii et al. [1978], Menzies et al. [1981], Glenar et al. [1983], and Kostiuk and Mumma [1983]. In particular, Frerking and Meuhlner [1977] have applied TDLHS to the study of atmospheric ozone, with a resolution of 0.0067 cm^{-1} and a signal-to-noise ratio of 40. At the University of Denver, the instrument described by Allario et al., [1979, 1980] has recently produced spectra having a resolution of 0.0005 cm^{-1} or better, at wavelengths in the region of the 9.6 μm ozone band. Some of these spectra have been inverted to provide column ozone amounts which were found to be in good agreement with values obtained at the National Oceanographic and Atmospheric Adiministration (NOAA). A number of the line positions have been compared to the Hitran database [Rothman et al., 1987], with updated ozone lines [Goldman and Murcray, 1988] and in the majority of the cases the frequencies are found to match to better than 0.001 cm^{-1}.

2. INSTRUMENTATION

The theory behind heterodyne detection is well known, and its application to the TDLHS has been discussed by McElroy et al. [1988]. The voltage signal-to-noise ratio S_V [Blaney, 1975], can be arrived at by considering our photoconductor to be a square-law detector and if we consider the noise bandwidth of the output filter to be on the order of $1/\tau$, where τ is the post-detection integration time constant of the system. Therefore,

$$S_V = \frac{\eta p_s}{2h\nu B_{IF}(\quad 1 + \eta\delta_B\beta \quad)} \sqrt{2B_{IF}\tau}$$

where η is the quantum efficiency of the photoconductor, ν is the photon frequency, B_{IF} is the bandwidth of the intermediate frequency, $\delta = [\ \exp(h\nu/kT_B) - 1\ ]^{-1}$, where T_B is the black-body temperature of the source, and $\beta = 1$ for a thermal source that completely fills the field-of-view. If values typical to this instrument and atmospheric spectroscopy are used, that is, $\eta = 0.38$, $T_B = 5500$ K, $B_{IF} = 3.1$ MHz, $\tau = 0.125$ s, $\nu = 3 \cdot 10^{13}$ Hz, $p_s = 2h\nu\delta\beta B_{IF}$, and we apply the quantum limit $h\nu >> kT_B$, then S_V is approximately 1200:1. Other losses such as chopper efficiency (50%) and losses due to the optics (45%) result in a final figure of approximately 260:1.

The University of Denver TDLHS was constructed by the NASA research group in Langley, Virginia. The instrument concept was described by Allario et al. [1979, 1980]. During the past year of its life at the University of Denver, the instrument has been considerably modified resulting in a reduction in temperature tuning errors and a consequent improvement in wavelength stability. The optical arrangement is more or less unchanged, except for the addition of

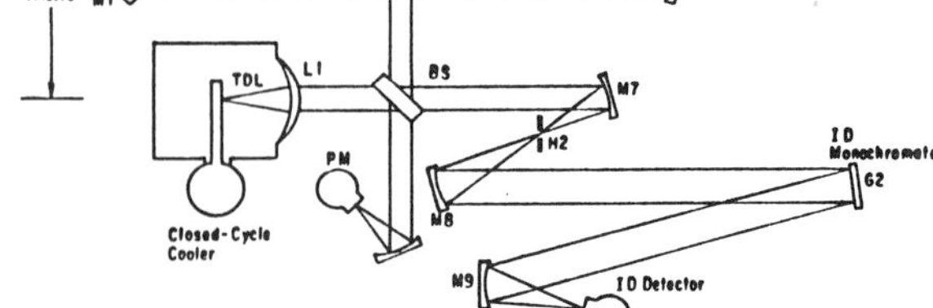

Fig. 1. The optical schematic diagram of the instrument. Mirrors are indicated by the letter M, diffraction gratings by G, the beamsplitter by BS and the lens by L1.

ISBN 0-937194-15-8

focussing capability at the detector, and is illustrated in Figure 1. Solar radiation is directed into the instrument by means of a clock-driven mirror. The solar tracker directs the light to mirror M1, which reflects it to an off-axis paraboloid M6, which produces an image of the sun at the pin-hole located immediately before the chopper. The light exiting the pinhole is collimated by mirror M5, dispersed by grating G1, and focussed onto the exit pinhole H1 by mirror M4. The light exiting H1 is collimated by mirror M3 and directed towards the beamsplitter BS by mirror M2. Light from the TDL local oscillator is collected by lens L1 and it then falls on the beamsplitter where it combines with the solar beam with phase matching. After the beamsplitter, the composite beam is focussed upon a high-speed mercury-cadmium-tellurium (HgCdTe) photomixer biased as a photoconductor. The heterodyne signal, which contains the spectral information, is then separated from the DC component. The amplified RF signal is then bandpass filtered to set the desired instrumental resolution. Currently, there are 7 different bandpass options available. The highest resolution is derived from a channel having a bandpass that ranges from 1 to 4.1 MHz. After the bandpass filtering, the signal is square-law detected by a broadband RF detector (0.1 to 1 GHz) and subsequently phase-sensitive detected using a dual slope integrator. The integrator is ramped up during the open cycle of the chopper, and then ramped down again during the closed cycle. The phase-sensitive detection process provides a large rejection ratio for sources of noise such as detector thermal noise and other background sources which are unmodulated.

It should be noted, that the signal from the detector RF amplifier is proportional to the square root of the solar intensity, or equivalently, to the solar field strength. After the signal is square-law detected, a signal which is proportional to the solar intensity as well as the local oscillator intensity results. This is the signal which is phase-sensitive detected and then recorded as spectral data.

3. SPECTRAL ANALYSIS

The spectra are viewed and manipulated by means of software written for an IBM PC/XT. All of the spectra presented here have had a 3 point (FWHM) triangular filter applied to them. A wavelength calibration is generated through the application of calculated dispersion coefficients. The dispersion coefficients are generally calculated by taking the current scan values for selected lines and then performing a linear or quadratic regression against line positions as determined for the Bomem DA3 interferometer [Murcray et al., 1988], which is currently demonstrating an unapodized resolution of 0.002 cm^{-1} . For a linear regression, this method generally provides agreement between the wavelength calibration of the spectra of the order of 10^{-4} cm^{-1}. In all cases, only fairly narrow, well-defined lines were used in these regressions, one of which is required for each current scan. It is necessary to provide a wavelength scale in this manner, because a sufficiently accurate independent method has not

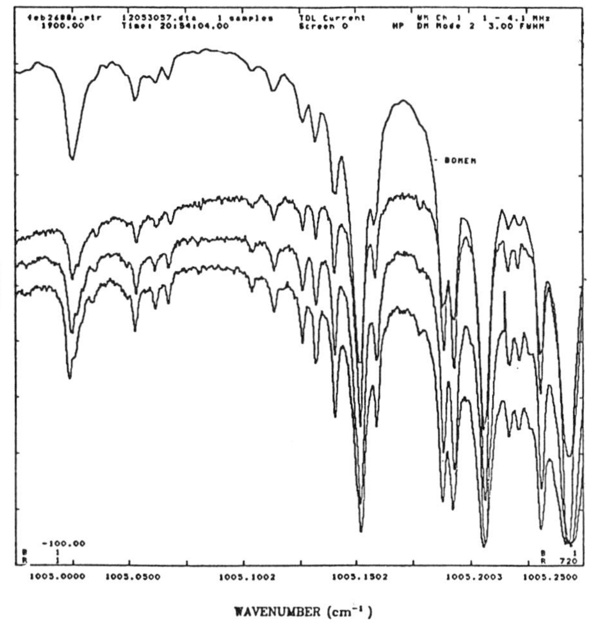

Fig. 2. Comparison of a Bomem solar spectrum to 3 TDLHS solar spectra between 1005.0 and 1005.25 cm^{-1} . The Bomem spectrum was taken on Oct. 21, 1987 at a zenith angle of 66.5°. The TDLHS spectra were taken on Feb. 26, 1987 at zenith angles of 54°, 70°, and 77° respectively.

been incorporated into the TDLHS as yet.

The signal-to-noise ratios quoted here were calculated by determining the standard deviation of a number of points in a region where the spectral intensity appeared to be relatively constant. This method could well yield a noise level greater than the actual one, but it is unlikely to yield one that is less. The theoritical signal-to-noise ratio for the 4.1 MHz channel is 260:1, while the observed signal-to-noise ratio calculated by this method is 95:1, or 35% of theoretical.

TDLHS spectra have been compiled in 4 regions. They are as follows: 996 - 997 cm^{-1}, 1004.9 - 1005.3 cm^{-1}, 1009.6 - 1010.2 cm^{-1}, 1012.0 - 1012.5 cm^{-1}. The observed TDLHS spectra have been compared to the Hitran database, and to Bomem DA3 spectra taken during October and November of 1987. In each case the higher resolution and signal-to-noise of the TDLHS can be seen. Figure 2 shows three scans taken on February 26, 1988 from 1005.0 to 1005.25 cm^{-1}. Each scan was calibrated independently as described earlier. The higher resolution can be seen in that the TDLHS can better resolve a number of lines that the Bomem interferometer spectra show as slope changes in the sides of lines. Figure 3 shows a comparison between Bomem and TDLHS spectra in the region 1009.6 cm^{-1} to 1009.8 cm^{-1}. The TDLHS resolves a line at 1009.61880 cm^{-1}, while the Bomem spectrum indicates that there may be a feature at that frequency. A similar phenomenon is seen with the line at 1009.70334 cm^{-1}. All the interferometer data were collected by scans of 4 minute duration.

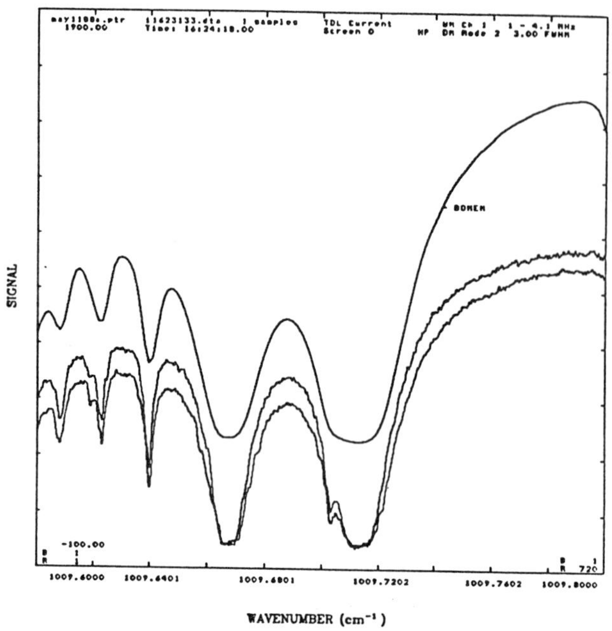

Fig. 3. Comparison of TDLHS solar spectra to a Bomem solar spectrum between 1009.6 and 1009.8 cm^{-1}. The Bomem is as described in Figure 3, while the TDLHS spectra were taken on May 11, 1988 at zenith angles of 56.5° and 63.5° respectively

Figure 4 illustrates a spectrum taken January 21, 1987 at the University of Denver (altitude 1.6 km, 39.5°N 105°W). This spectrum was taken while the instrument was in its original configuration, and it has a signal-to-noise ratio of 45:1, or 17% of theoretical. The curves in Figure 5 compare an expanded portion of the observed spectrum taken at 0.0016 cm^{-1} resolution, to a simulation using line parameters which are the same as those on the Hitran compilation, with the exception of the line positions, which were adjusted to agree with the observed data. The climatological vertical distribution of ozone at mid-latitude was taken as a starting point for a non-linear spectral least squares fitting procedure which scaled the whole profile until the best possible agreement between the two curves was obtained [Goldman et al., 1983]. It can be seen that the three lines shown are not all fit with the same precision. This difference is due to the combined effects of inaccuracies in the compiled line intensities and halfwidths, and the fact that the observed lines result from states of rather different, relatively high, energies and are therefore affected differently by changes in the local temperature.

In order to investigate the possibility that the observed and simulated spectra might be brought into better agreement if the temperature profile of the model atmosphere were changed, the fitting program was run in such a way that the temperature profile could be scaled to achieve the best agreement between the data and the model results. When this was done a slightly different total column amount of ozone was obtained, with no apparent improvement of the fit, with the difference between the two amounts being about

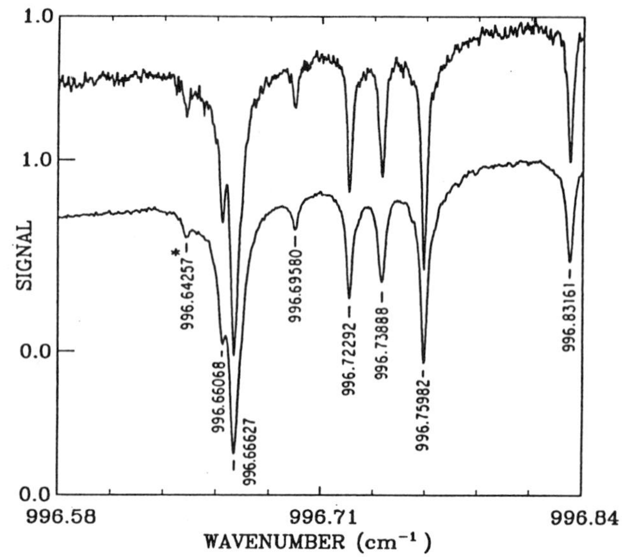

Fig. 4. A solar spectrum taken by the TDLHS on January 21, 1987, at a zenith angle of 60.1°. The upper curve shows data from the 1 to 4.1 MHz channel, while the lower one plots the signal from the 10 to 20 MHz bandpass channel.

8%. The column amounts obtained in these calculations are 344 and 317 m-atm-cm for the fixed and adjusted temperature profiles, respectively. The ozone total column amount determined by the Dobson ozone spectrophotometer at NOAA at Boulder [private communication from W. Komhr and R. Evans, 1987], is 355 m-atm-cm.

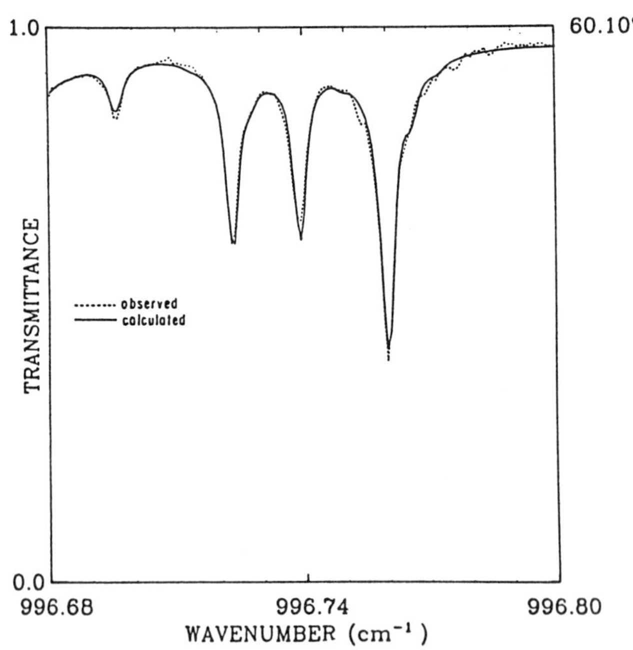

Fig. 5. This figure shows a comparison of the observed data and the results of a spectral simulation over a narrow region containing several ozone lines. The ozone amount was adjusted to give the best fit, and the spectrum shown was calculated with 344 m-atm-cm of ozone in the model atmosphere.

The Dobson total ozone amount was adjusted by 3.9% before being compared to the infrared results, with the reported 369 Dobson unit Boulder measurement corrected to 355 m-atm-cm. This adjustment is needed because the Dobson values are now known to read high compared to ultraviolet and chemical measurements of ozone [private communication from J.B. Kerr, Atmospheric Environment Service, 1987].

4. CONCLUSIONS

The TDLHS is just now beginning to realize its potential. It is hoped that a better selection of TDL's will become available in the not to distant future, thereby increasing the number of spectral regions which can be investigated. The higher resolution of the TDLHS provides the potential to better resolve the spectral lines of absorbers whose lines overlap those of other atmospheric constituents. Higher spectral resolution will provide better resolution of constituent vertical profiles which would benefit line shape inversion studies.

ACKNOWLEDGEMENTS

Acknowledgement is made to the National Center for Atmospheric Research, which is sponsored by the National Science Foundation, for computer time used in this research. Funding to support this work was provided by NASA/Langley.

REFERENCES

Allario, F., J. M. Hoell, S. J. Katzberg, J. C. Larsen, 1979: An Experiment Concept to Measure Stratospheric Trace Constituents by Laser Heterodyne Spectroscopy, Proc. NASA Conf. on High Resolution Infrared Spectroscopy Techniques for Atmospheric Measurements, Publ. No. 2134, 57-79.

Allario, F., J.M. Hoell, and S.J. Katzberg, 1980: An experiment concept to measure stratospheric trace constituents by laser heterodyne spectroscopy, Appl. Phys., 23, 47-56.

Blaney, T. G., Signal-to-Noise Ratio and Other Characteristics of Heterodyne Radiation Receivers, 1975: Space Science Reviews, 17, 691-702.

Frerking, M. A. and D. J. Muehlner, 1977: Infrared Heterodyne Spectroscopy of Atmospheric Ozone, Appl. Opt., 16, 526-528.

Fujii, Y., J. Yamashita, S. Shikata, S. Saito, 1978: Incoherent Optical Heterodyne Detection and its Application to Air Pollution Detection, Appl. Opt., 17, 3444-3449.

Glenar, D., T. Kostiuk, D. E. Jennings, D. Buhl, and M. J. Mumma, 1981: Tunable Diode-Laser Heterodyne Spectrometer for Remote Observations near 8μm, Appl. Opt., 21, 253-259.

Goldman, A., F. G. Fernauld, F. J. Murcray, F. H. Murcray, and D. G. Murcray, 1983: Spectral Least Squares Quantification of Several Atmospheric Gases From High Resolution Infrared Solar Spectra Obtained at the South Pole, J. Quant. Spectrosc. Radiat. Transfer, 29, 189-204.

Goldman, A. and D. G. Murcray, 1988: High Resolution Studies of Atmospheric Infrared Emisssion and Absorption Spectra, Proceedings of the Spie Conference.

Kostiuk, T., and M. J. Mumma, 1983: Remote Sensing by IR Heterodyne Spectroscopy, Appl. Opt., 22, 2644-2654.

McElroy, C. T., A. Goldman, and D. G. Murcray, 1988: Heterodyne Spectrophotometry of Ozone in the 9.6 Micron Band Using a Tunable Diode Laser, submitted to Appl. Opt..

Menzies, R. T., C. W. Rutledge, R. A. Zanteson, and D. L. Spears, 1981: Balloon-Borne Laser Heterodyne Radiometer for Measurements of Stratospheric Trace Species, Appl. Opt., 20, 536-544.

Murcray, F. J., J. J. Kosters, and R. D. Blatherwick, 1988: A Very High Resolution Fourier Transform Spectrometer for Balloon-Borne Applications, Submitted to Appl. Opt.

Rothman, L. S., R. R. Gamache, A. Goldman, L. R. Brown, R. A. Toth, H. M. Pickett, R. L. Poynter, J. -M. Flaud, C. Camy-Peyret, A. Barbe, N. Husson, C. P. Rinsland, and M. A. H. Smith, 1987: The Hitran Database: 1986 edition, Appl. Opt., 26, 4058-4097.

ZENITH-SKY CLOUD DETECTION FOR AUTOMATED DOBSON UMKEHR MEASUREMENTS

J.J. DeLuisi, E.G. Dutton, and G.L. Koenig
NOAA/ERL/ARL/Geophysical Monitor for Climatic Change
Boulder, Colorado 80303, USA

D.U. Longenecker and P.J. Reddy
Cooperative Institute for Research in the Environmental Sciences
University of Colorado
Boulder, Colorado 80309, USA

ABSTRACT

Seven Dobson instruments have been automated by the Geophysical Monitoring for Climatic Change program (GMCC) for unattended operation to obtain short and conventional Umkehr measurements. The large-particle scattering condition of the zenith sky, primarily clouds, is manifested in the Umkehr measurement, which is accepted or rejected on the basis of the smoothness of the Umkehr curve. A sensitive photometer (zenith sky cloud detector, ZSCD) was designed and developed by GMCC to observe approximately the same portion of the zenith sky as the Dobson instrument does during operation, but at a different wavelength (862nm). The ZSCD information is used as a means to assess the quality of each Umkehr observation in addition to aiding the diagnosis of a malfunctioning instrument.

1. INTRODUCTION

The seven automated Dobson instruments of the Geophysical Monitoring for Climatic Change (GMCC) program operate in the Umkehr mode without the presence of an operator (Komhyr et al., 1984). In the absence of an on-site observer it is necessary to have an independent measure of zenith-sky conditions because the large-particle scattering condition (primarily clouds) of the zenith sky is manifested in the Umkehr measurement, which is accepted or rejected on the basis of the smoothness of the Umkehr curve. Erroneous Umkehr data could be due either to such cloud interference or to instrumentation malfunction. Large-particle scatterers such as cirrus clouds present the most serious problem because of their optical thickness and variability over short periods of time. Atmospheric aerosols in the troposphere are less serious because they are more homogeneously distributed in the lower atmosphere where the Umkehr sensitivity is low and their optical thickness is considerably less than that of cirrus clouds.

To monitor the zenith-sky brightness as an indicator of cloud presence during an Umkehr measurement, GMCC constructed a special photometer--a zenith sky cloud detector (ZSCD)-- to view essentially the same sky as the as the Dobson instrument does. This paper briefly describes the ZSCD and the results of measurements made at two Dobson sites: Boulder, Colorado, and Mauna Loa, Hawaii.

2. INSTRUMENTATION

Figure 1 is a drawing of the design of the ZSCD. The lens and aperture fix the field of view at 0.060 steradians. The filter bandpass is 10 nm centered at 862 nm. The columnator tube is protected with a dust- and rain-resistant glass cover, and the shade tube prevents the sun's direct rays from reaching the glass cover. Water drain holes are located in the block holding the shade tube. A stable EG and G silicon photodiode is used with an amplifier that is adjusted to give an output of 0 to 10 V. The system dark current is obtained each time the hatch cover of the Dobson housing dome is automatically closed between observation times. The output is adjusted to give a voltage reading of a few volts for a cloud-free zenith sky. All field instruments were compared simultaneously and amplifiers were adjusted to give the same readings.

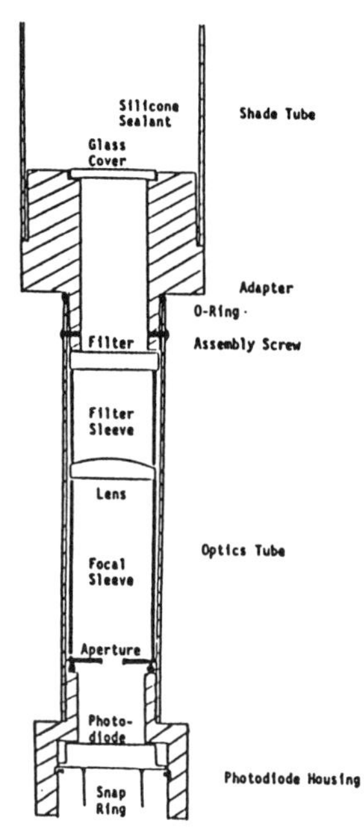

Fig. 1. A schematic of the zenith-sky cloud detector.

ISBN 0-937194-15-8

At the automated Dobson field sites the output of the ZSCD was fed to the data acquisition system. During the data processing procedure which was carried out at Boulder, the digitized cloud detector signal and a plot of the signal strength versus time were automatically recorded.

3. ZENITH-SKY CLOUD INFORMATION

Figure 2 shows an example of zenith-sky signals received under conditions completely free of clouds on the morning of 21 March 1988, during the time that the Umkehr observation (solid line) was made. Note the low value of the ZSCD signal, and the very smooth continuity (dashed line) in the fitted curve. Figure 3 shows an example of very slight cloud interference at a solar zenith angle near 65° on 19 March 1988. On this same day, during the afternoon Umkehr measurement, clouds interfered severely. The interference effects are shown in Fig. 4.

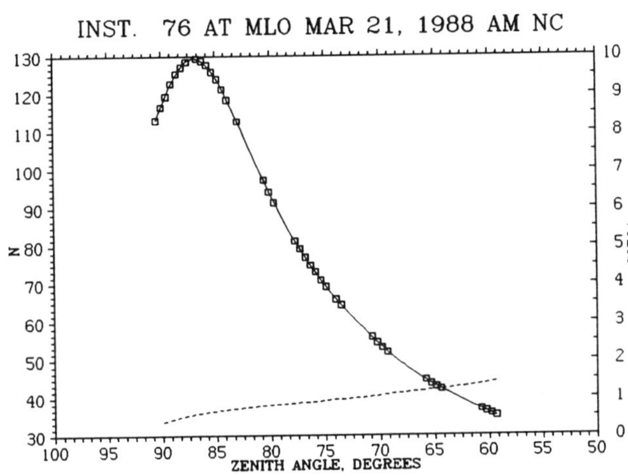

Fig. 2. Umkehr observation N-values versus solar zenith angle (solid line), taken at Mauna Loa, Hawaii GMCC observatory. The dashed line represents the zenith-sky signal observation from the ZSCD versus zenith angle at the same time as the Umkehr observation. These results obtained during the morning of 21 March 1988, are typical of completely cloud-free conditions.

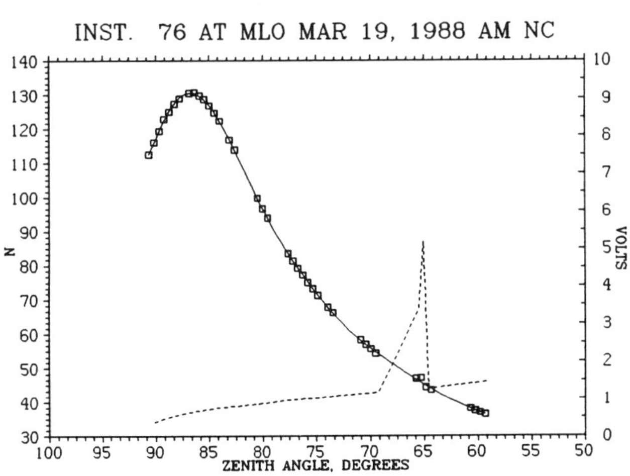

Fig. 3. Same as Fig. 2 except for the morning of 19 March 1988 and under conditions of slight cloud interference near 65° solar zenith angle.

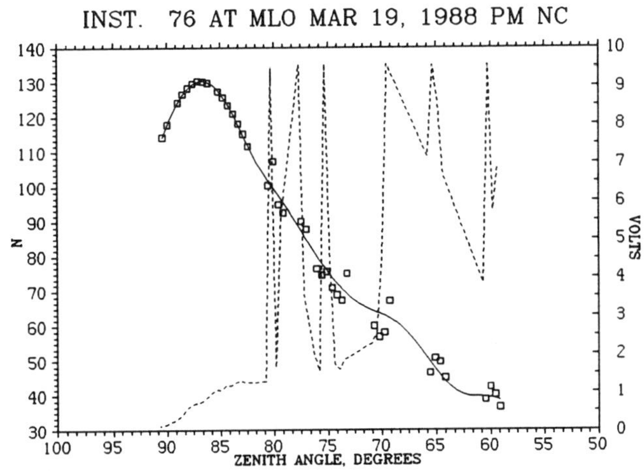

Fig. 4. Same as Fig. 2 except for the afternoon of 19 March 1988 and under conditions of severe cloud interference for part of the observation.

These examples serve to illustrate the sensitivity of the zenith-sky cloud observations and the ability to quantify the quality of the zenith-sky conditions during an Umkehr measurement. The Umkehr, being a ratio measurement, does not clearly display direct relationships between cloud presence and error, other than to produce a noisy observation. However, in general, thin clouds will cause the Umkehr observation N-values to become erroneously high at the smaller zenith angles. At the larger zenith angles, interference effects become less than at the smaller zenith angles.

4. ZENITH-SKY AEROSOL EFFECTS

Aerosols increase the brightness of the zenith sky by increasing the downward scattering of solar radiation. At 862 nm the zenith sky is relatively dark from Rayleigh scattering because the optical depth is only 0.016. The aerosol optical depth is generally greater than that for Rayleigh and therefore the signal can be easily detected. Aerosol optical depth measurements are routinely made using sunphotometers at Boulder, where one of the automated Dobson instruments is in operation. To show a relationship, we plotted the zenith-sky brightness from the ZSCD versus aerosol optical depth from the sunphotometers for a solar zenith angle of 70° and for cloud-free conditions only. The result is shown in Fig. 5. The scatter is due to noise in the observations because of noncoincident observation times and because of variations in the aerosol forward-scattering phase function, which is closely associated with variations in aerosol size distribution. Note in the figure the nearly linear relationship; this is expected because aerosol scattering is directly proportional to the aerosol optical mass. The y-intercept from a least-squares fit is approximately 2.2, which is the signal for Rayleigh scattering only.

5. DISCUSSION

The zenith-sky cloud detector described in this paper represents an independent and highly sensitive observation of

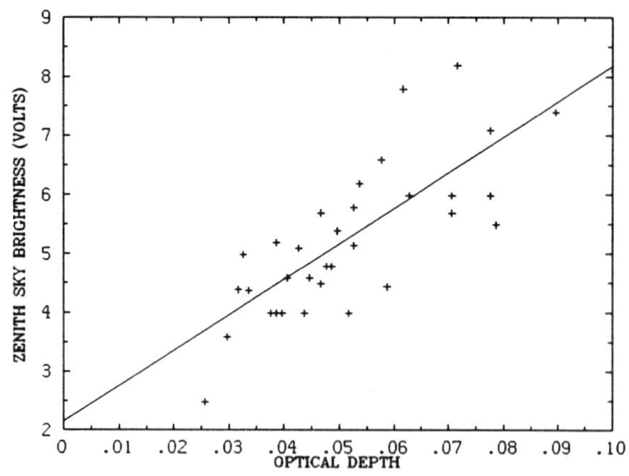

Fig. 5. Plot of cloud-free zenith-angle signal from the ZSCD versus aerosol optical thickness from the sunphotometers, for a sample of 33 Umkehr observations.

the zenith-sky conditions during an automated Dobson Umkehr observation. In the absence of an on-site observer where automated Dobson instruments operate, the ZSCD provides valuable ancillary information for quality control. Seven automated Dobson instruments are now being operated by GMCC. The information from the ZSCD can be used to assess the quality of observing conditions at a site, to diagnose possible instrument problems, and to rank a long-term set of Umkehr observations at a particular site according to the quality of zenith-sky conditions.

An additional potential application is to provide cloud information that can be applied to the validation of certain satellite measurements of ozone profiles, such as those derived from the Solar Backscatter Ultraviolet (SBUV) satellite, and from the Stratospheric Aerosol and Gas Experiment (SAGE) satellite. Both of these satellite systems sense clouds and use this information in the interpretation of their measurements. A situation of particular importance occurs when very thin clouds are present. Threshold conditions may exist wherein the signal-to-noise capability of the satellite cannot clearly discern cloud interference. The ZSCD can provide such valuable additional information.

6. REFERENCE

Komhyr, W. D., R. D. Grass, R. D. Evans, R. K. Leonard, and G. M. Semeniuk, 1984: Umkehr observations with automated Dobson spectrophotometers. *Atmospheric Ozone, Proc. Quadrennial International Ozone Symp.*, 3-7 September 1984, Halkidiki, Greece, C. S. Zerefos and A. Ghazi, Eds., Reidel, Dordrecht, 371-375.

THE HOHENPEISSENBERG LIDAR: FIRST RESULTS AND INTERCOMPARISON WITH BREWER/MAST SOUNDINGS

Hans Claude and Klaus Wege
Deutscher Wetterdienst
Meteorologisches Observatorium
Hohenpeissenberg, FRG

ABSTRACT

A lidar system operating on Hohenpeissenberg since autumn 1987 measures ozone profiles up to 50 km. A short description of this system is given including an error consideration. The ozone profiles obtained with the lidar are compared with Brewer/Mast soundings and total ozone measurements with Dobson/Brewer spectrophotometers. The results of these intercomparisons indicate a good agreement among the ozone observations. Short-term measurements in a close sequence show unexpected intensive and rapid variations in the ozone vertical distribution.

1. INTRODUCTION

Ozone measurements have been regularly performed on Hohenpeissenberg for over 20 years without interruption. In use are Brewer/Mast sondes to measure the vertical distribution, different conventional ozone analyzers for ozone measurements at surface level and a Dobson and a Brewer spectrophotometer for total ozone measurements. The long-term series of measurements at Hohenpeissenberg are repeatedly used for trend reports (Logan, 1985; Bojkov, 1987). Due to technical reasons, the sonde operates only up to 32 - 35 km altitude. No profile data is available for higher levels. However, just because ozone decrease is expected to be stronger at altitudes around 40 km (Brasseur and De Rudder, 1987), exact measurements are urgently required also for these levels. Ozone profile measurements up to 50 km altitude can be carried out from ground level with lidar equipment (Pelon et al., 1986; Werner et al., 1983).

Following the successful application of an ozone lidar on the Zugspitze (2965 m) (Werner, 1984), a second improved unit was set up on the Hohenpeissenberg in cooperation with the Max-Planck-Institut f. Quantenoptik, Garching and the Sektion Physik of the Universität München, and was put into operation last autumn.

2. EXPERIMENTAL SET UP

The Hohenpeissenberg lidar (Fig. 1) is basically the same as the earlier ozone lidar on the Zugspitze, with a more powerful laser and an improved optical design (Geh, 1987). In spite of the lower location (980 m to 2965 m) and the associated weakening of the signal due to aerosols and higher air density, a factor 2 gain in the backscattered signal could be attained.

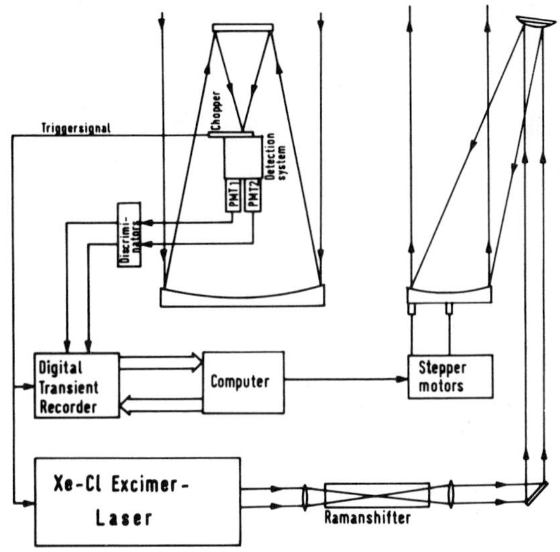

Fig. 1. Schematic diagram of the ozone lidar on the Hohenpeissenberg.

A Xe-Cl excimer laser delivers pulses with 300 mJ energy and a length of 30 ns at a repetition rate of 20 Hz. The emitted light with 308 nm wavelength (absorption line) is focussed in a hydrogen-filled Raman cell where the collinear reference wavelength beam of 353 nm (1st stokes line) is simultaneously generated. Upon leaving the Raman cell, both beams are reflected, expanded and vertically emitted into the atmosphere.

The backscattered light is sent over a concave mirror (60 cm Ø) and a folding mirror to the visual field iris of the detection system. The detection system itself is housed in a solid aluminium detector block, which is located in the optical axis between the receiving mirror and the folding mirror. Here the two wavelengths are separated with a dichroic mirror and sent through a narrow band interference filter to two low noise photomultipliers. The backscatter count rates of both wavelengths are first buffered in high speed digital transient recorders. The gate time is 2 µs which corresponds to a height resolution of 300 m. The computer reads the buffer every 32 laser pulses and stores the data on hard disk.

ISBN 0-937194-15-8

3. ERROR CONSIDERATION

The main source of error in lidar measurements would be the statistical error of the backscattered photons (Pelon and Mégie, 1982). However, because a relatively high statistical accuracy can be attained with the described apparatus, other error sources must also be considered, e.g. a not exactly known ozone absorption cross section, possible interferences and the effects of aerosols. The statistical error predominates in higher altitudes firstly because of the exponential drop off of the backscattered signal (Fig. 2).

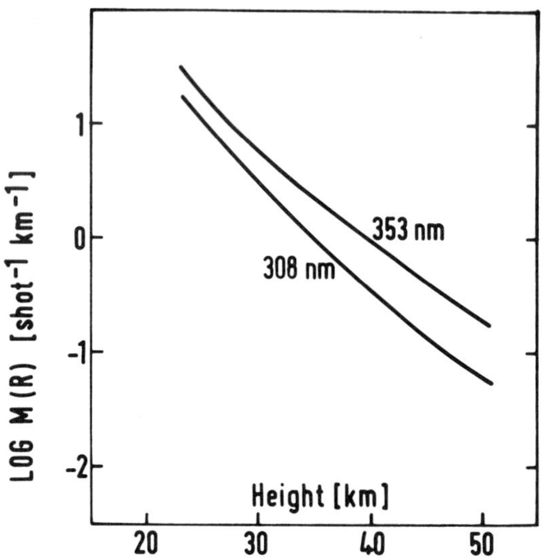

Fig. 2. Logarithmic graph of backscattered count rate versus altitude.

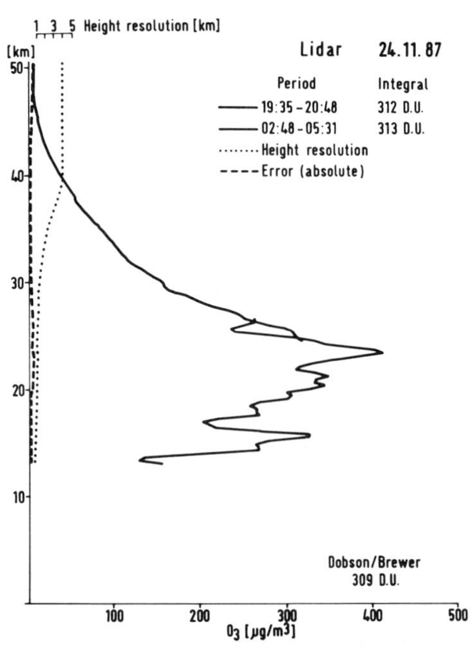

Fig. 3. Vertical ozone distribution with error curve and height resolution. The discrepancy in the overlap range is due to the time lag of the measurements.

Although several 10^5 laser shots are emitted in a typical night measuring period, only a few 10^3 counts per layer are retrieved for evaluation from a height of 50 km. On the other hand, there the ozone content drops to about 1/100 of its maximum value, which also places extreme demands on the detection limit of the system. The statistical error can be reduced by combining several layers, which decreases the effective height resolution. Figure 3 illustrates this relationship on the basis of the measurement from Nov. 24, 1987. The evaluation program determines the height resolution depending on the number of counts per layer. The effective height resolution (...) thus alters with increasing altitude from ± 1 to ± 4 km. It is nevertheless unavoidable that the absolute statistical error (---) corresponds to the strongly decreasing ozone profile above 45 km, i.e. the error is just as large as the measured value itself.

TABLE 1. INTERDEPENDENCE OF HEIGHT RESOLUTION ΔR, INTEGRATION TIME t AND ACCURACY OF THE LIDAR MEASUREMENT IN RELATION TO ALTITUDE.

H (km)	ΔR (km)	t (δN/N = 2 %)	δN/N (t = 12 h)
15	± 0.4	7.1 min	
20	± 0.6	2.6 min	
25	± 1.0	3.2 min	
30	± 1.4	16 min	
35	± 1.9	1.2 h	< 1 %
40	± 2.5	8.7 h	1.7 %
45	± 3.1	89 h	5.7 %
50	± 3.9	955 h	19 %

Table 1 illustrates the close connection between altitude, height resolution and integration time on the one hand and accuracy on the other. It shows that measurements up to 30 km altitude can be made in a very short period (< 60 min) with high accuracy. In altitudes above 40 km however, either the integration time or the height resolution or the error increases to an unacceptably high value. As will be later shown, the temporal and local changes in the ozone content at this range are by far not so extreme as below the main maximum. For this reason, the height resolution can be reduced in favor of the measuring accuracy, or several night measuring periods can be combined.

4. COMMENTS ON THE COMPARISON BETWEEN BREWER/MAST SONDE, LIDAR AND SPECTROPHOTOMETERS

Theoretical error estimates are always necessary when introducing a new operational measurement system. It is just as important to perform direct comparisons with conventional systems in field use, whereby the specific equipment characteristics should not be left out of consideration. The Brewer/Mast sonde displays some typical features of all wet chemical sondes. These have effects on the measured ozone profile. The sonde drifts with the air flow; the drift sometimes reaches several 100 km. Compared with the lidar method, the ozone measurement seldom conforms to the ozone distrubution over the station.

The balloon ascent lasts approx. 90 minutes. During this period "point measurements" are performed along the ascent route which do not fully reproduce extremes in the vertical distribution and lead to a smoothing of the profile (Claude et al., 1987). In addition, the sonde exhibits several systematic errors which are well known from international sonde comparisons (Attmannspacher and Dütsch, 1981; Hilsenrath et al., 1986).

The measured ozone profile is improved to a certain degree with the so-called Dobson correction, which is applied by all operational ozone measurement stations (WMO, 1980). With this method, the measured raw profile is compared to a nearly simultaneous total ozone measurement by spectrophotometer (Dobson/Brewer) and corrected accordingly. The relative error of the spectrophotometer measurement lies typically between 1 - 2 %. The ten year mean of the correction factor for the ozone soundings on Hohenpeissenberg is 1.097 ± 0.090, i.e. the ozone raw data delivered by the Brewer/Mast sonde are ≈ 10 % too low in long term mean. Errors depending on altitude and pressure cannot be corrected with this procedure. The decreasing pump efficiency at altitudes above 28 - 30 km belongs to this error category, and leads to a reduction in the measured values in the upper portion of the ozone profile.

A comparison between lidar measurement and total ozone measurement with a spectrophotometer is also quite informative. Because the correction factors are commonly used as a quality standard for sounding profiles, an analogous procedure could well be used for the lidar measurements. Before the total ozone value of the lidar measurement and the corresponding pseudocorrection factor can be determined by integrating the entire profile, the missing portion (~15% of total ozone) must be completed. Although only seasonal mean profiles from Hohenpeissenberg sounding data are used for this purpose, the calculated pseudocorrection factors are very good:

All lidar measurements from September 1987 - May 1988 (33 cases), correlated with the spectrophotometer measurement closest in time to it, produce a pseudocorrection factor of 0.997 ± 0.07 with a correlation coefficient of 0.82.

5. RESULTS

The profiles in Fig. 4 represent two lidar profiles that were consecutively measured on October 2, 1987, the lower part (16 - 32 km) with a reducer, the upper part (20 - 48 km) without. This partitioning is always necessary because the very high backscattered count rate from below 20 - 23 km would overload the photomultipliers without reducers. In spite of the time shift, the profiles correspond well in the overlap range, the deviation amounts to max. 10 %. The ozone distribution measured by the Brewer/Mast sonde on the following morning is also in excellent agreement with the lidar profiles. The mean day value with the Dobson is 306 DU. The total ozone value for the lower lidar measurement is 307 DU (+ 0.3 %), and 311 DU for the upper measurement (+ 1.6 %).

In contrast to soundings, profile measurements with a lidar system can be made in short

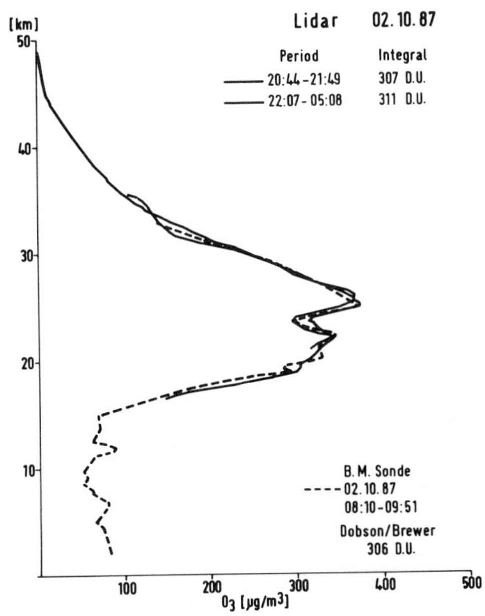

Fig. 4. Comparison of ozone profiles measured with lidar, Brewer/Mast sonde and total ozone values.

sequence. Such measurements show that especially in late winter/spring rapid ozone change can occur in the part of the ozone profile which is dominated by dynamic processes. It is also typical for this season that strong secondary maximums appear below the main maximum.

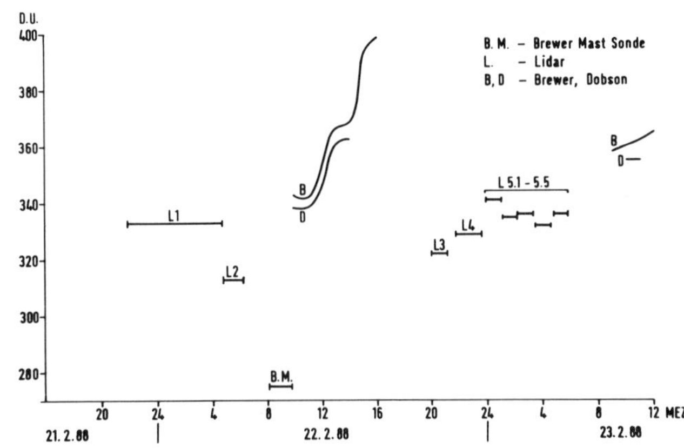

Fig. 5. Temporal representation of a series of ozone measurements with lidar, Brewer/ Mast sonde and Dobson/Brewer spectrophotometers including corresponding total ozone values.

The measurements from Feb. 22/23, 1988 demonstrate this case. The sampling periods are represented in Fig. 5. The total ozone values resulting from the profile measurements, or the total ozone measurements themselves can be read in Dobson Units.

The initial situation (Fig. 6) shows a distinguished secondary ozone maximum below the main maximum between 12 and 15 km altitude. The sounding and lidar correspond quite well. The first

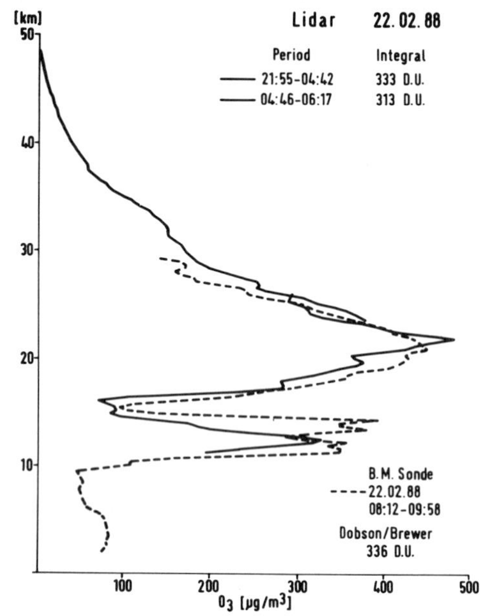

Fig. 6. Ozone vertical distribution on Feb. 22, 1988 at the beginning of the series (Lidar and Brewer/Mast sonde).

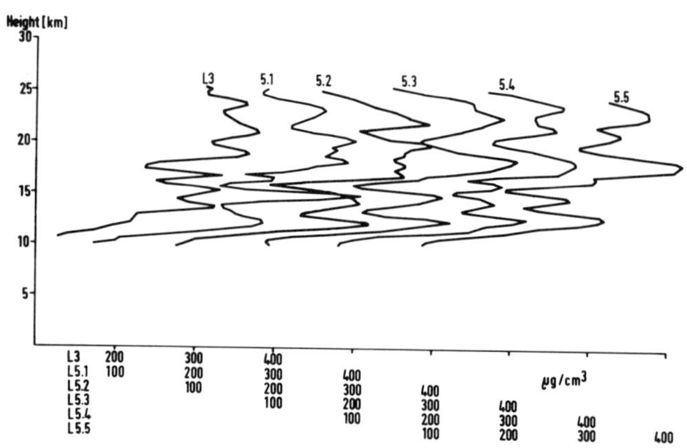

Fig. 7. Rapid ozone variations in the lower stratosphere observed with 1 hour lidar measurements.

lidar measurement (Fig. 7) on the following evening (L3) indicates that the growing indistinctness of the secondary maximum measured before noon has progressed, and also the main maximum no longer forms a peak. The significant minimum at 15 - 16 km has dissolved completely. Five further 1 hour lidar measurements were performed between about midnight and 6 a.m. (L5.1 - L5.5). They indicate how the ozone distribution in the lower stratosphere continues to change in very short periods, whereby the concentration or dissipation of individual maximums can be observed.

6. CONCLUSION

The previous exerience gained with the lidar system shows the outstanding capability of this modern system to measure ozone profiles up to altitudes above 40 km. Measurements in short periods and in close sequence can be made in the lower stratosphere, which indicate unexpected intensive and rapid changes in the ozone distribution. In spite of these rapid changes and basic differences between lidar and balloon soundings, the measured profiles mostly exhibit a high correspondence. A remarkable agreement is also yielded in comparison with the total ozone measurement using the spectrophotometers (Dobson/Brewer), although these measurements were never performed simultaneously. The major restriction on the lidar measurement is its dependence on clear nights, which are rare in middle Europe. They amount to 15 % on Hohenpeissenberg.

REFERENCES

Attmannspacher, W. and H.U. Dütsch, 1981: 2nd international ozone sonde intercomparison at the Observator. Hohenpeissenberg, April 1978. Ber. d. Deutschen Wetterdienstes, 157, 1.

Bojkov, R.D., 1987: The 1983 and 1985 anomalies in ozone distribution in perspective. Mon. Wea. Rev., 115, 2187-2201.

Brasseur, G. and A. De Rudder, 1987: The potential impact on atmospheric ozone and temperature of increasing trace gas concentrations. JRG, 92 D9, 10903-10920.

Claude, H., R. Hartmannsgruber, U. Köhler, 1987: Measurement of atmospheric ozone profiles using the Brewer/Mast sonde. WMO Global Research and Monitoring Project, Rep. No. 17.

Geh, B., 1987: Aufbau einer automatischen Lidar-Apparatur zur Messung der Stratosphärischen Ozonkonzentration Diplomarbeit, Sektion Physik Univ. München.

Hilsenrath, E., W. Attmannspacher, A. Bass, W. Evans, R. Hagemayer, R.A. Barnes, W. Komhyr, K. Mauersberger, J. Mentall, M. Proffitt, D. Robbins, S. Taylor, A. Torres and E. Weinstock, 1986: Results from the balloon ozone intercomparison campaign (BOIC). JGR, 91 D12, 13137-13152.

Logan, J.A., 1985: Tropospheric ozone: seasonal behaviour, trends, and anthropogenic influence. JGR, 90 D6, 10463-10482.

Pelon, J., Mégie, G., 1982: Ozone monitoring in the troposphere and lower stratosphere. JGR 87 C7, 4947-4955.

Pelon, J., S. Godin and G. Mégie, 1986: Upper stratospheric (30 - 50 km) lidar observations of the ozone vertical distribution. JGR, 91 D8, 8667-8671.

Werner, J., K.W. Rothe und H. Walther, 1983: Monitoring of the stratospheric ozone layer by laser radar. Appl. Phys., B32, 113-118.

Werner, J., 1984: Messung der stratosphärischen Ozonkonzentration mit Hilfe eines Laser-Radar-Systems. Dissertation, Fakultät f. Physik, Universität München.

WMO, 1980: Report of the meeting of experts on assessment of performance characteristics of various ozone observing systems. WMO Global Research and Monitoring Project, Rep. No. 9.

MULTIPLE-WAVELENGTH DIAL AND A NEW ANALYSIS TECHNIQUE TO DEDUCE THE OZONE PROFILE WITHOUT SYSTEMATIC ERRORS DUE TO AEROSOL EFFECTS

Yasuhiro Sasano, Hideaki Nakane, Sachiko Hayashida-Amano, Nobuo Sugimoto, and Ichiro Matsui
The National Institute for Environmental Studies
Tsukuba, Ibaraki 305, Japan

ABSTRACT

The present paper describes a new analysis technique for a multiple-wavelength differential absorption lidar (DIAL). Numerical integration is applied to the simultaneous differential equations derived from the basic lidar equation for each wavelength. The present method can simultaneously determine aerosol and gas distribution profiles as well as wavelength dependence of aerosol extinction and backscatter coefficients. Results of numerical simulation are presented for stratospheric ozone measurements with an excimer-laser-based DIAL system.

1. INTRODUCTION

Differential absorption lidar (DIAL) has been widely used for measuring atmospheric gaseous constituents. Recent studies concerning stratospheric ozone depletion due to anthropogenic gases such as chlorofluorocarbons have suggested more important roles for the DIAL technique. Considerable effort has been made to establish an operational lidar network for ozone monitoring and early detection of stratospheric change.

The DIAL technique has great potential for measuring gas concentration profiles with high spatial and temporal resolution as well as with high accuracy. However, it must be recognized that, under certain conditions, the aerosol distribution produces systematic errors in the derived gas concentration profiles. These errors are most notable when the difference in laser wavelengths (on-line and off-line) is so large that the effects of wavelength dependence in aerosol optical properties can not be neglected for heavy aerosol loadings (Schotland, 1974).

Browell et al. (1985) proposed a correction method for the systematic errors due to aerosols by solving the off-line lidar signal for the aerosol distribution, and substituting it in the DIAL equation to estimate the aerosol backscatter and extinction terms. They also investigated sensitivity of their analysis to aerosol optical properties. Other investigators (Uchino et al., 1983; Pelon et al., 1986) have suggested, but have not yet been able to implement, the use of a third wavelength laser source for obtaining information on aerosol optical properties.

This paper proposes a new analysis technique and describes simulation results for solving multiple-wavelength DIAL data. The National Institute for Environmental Studies has recently introduced a multiple-wavelength DIAL system for ozone measurements in the troposphere and the stratosphere. The system has three lasers (KrF, XeCl and XeF) and three Raman shifters using H_2 and D_2 to generate multiple wavelength laser pulses. Among them, the 277, 292 and 313 nm are used for lower altitude (1.5 - 15 km) measurements and the 308, 339 and 351nm are for higher altitude (15 - 50 km) measurements. Details of the system will be described elsewhere.

2. ANALYSIS TECHNIQUE

The present method (Sasano, 1988) employs a numerical integration technique to solve the simultaneous differential equations which are derived from the lidar equations for two wavelengths. The lidar equation for a wavelength is written as,

$$X(R)=P(R)R^2 = K\{\beta_1(R)+\beta_2(R)\} \exp\left[-2\int_0^R \{\alpha_1(r)+\alpha_2(r)+\sigma N(r)\}\,dr\right] \quad (1)$$

where P(R) is the received signal at the range R, K the system constant, β and α the volume backscatter and extinction coefficients, σ the absorption cross section for the gas, and N the gas molecule concentration. The subscripts 1 and 2 denote the aerosol and molecular quantities, respectively.

The differential equations to be solved are derived from eq.(1) and are written for the wavelengths λ and λ^* as

$$\frac{d\ln y(R)}{dR} - 2S_1 y(R) - 2\sigma N(R) = \frac{d\ln X(R)}{dR} - 2(S_1-S_2)\beta_2(R) \quad (2)$$

$$\frac{d\ln y^*(R)}{dR} - 2S_1^* y^*(R) - 2\sigma^* N(R) = \frac{d\ln X^*(R)}{dR} - 2(S_1^*-S_2)\beta_2^*(R) \quad (3)$$

ISBN 0-937194-15-8

where y(R) is the total backscatter coefficient defined by $y(R)= \beta_1(R) + \beta_2(R)$. Given constant extinction-to-backscatter ratios S_1 (aerosol) and S_2 (air molecules), we assumed the following relationship between extinction and backscatter,

$$\alpha_1 = S_1 \beta_1 \quad \text{and} \quad \alpha_2 = S_2 \beta_2$$

The quantities without (with) an asterisk represent those for wavelength λ (λ^*).

Assuming the widely used power-law relationship between extinction and backscatter coefficients and the wavelength, we obtain

$$\alpha_1^* = \alpha_1(\lambda^*/\lambda)^{-\gamma}, \qquad \beta_1^* = \beta_1(\lambda^*/\lambda)^{-\delta},$$

$$S_1^* = S_1(\lambda^*/\lambda)^{-(\gamma-\delta)},$$

$$\beta_2^* = \beta_2(\lambda^*/\lambda)^{-4}$$

When we assign appropriate values for γ and δ, we can rewrite the above expression as $\beta_1^*=A\beta_1$, $\beta_2^*=B\beta_2$ and $S_1^*=CS_1$ with constants A, B and C, and we then can obtain the following expression:

$$y^*(R)=Ay(R)+(B-A)\beta_2(R) \qquad (4)$$

With eqs.(2), (3) and (4), we now have the system of differential equations for the unknowns y(R), $y^*(R)$ and N(R). Solving the equations is straightforward by a numerical integration technique such as the Runge-Kutta scheme with an appropriate boundary condition $y(R_0)$ at $R=R_0$.

When the absorption term $\sigma^* N$ is negligible for the wavelength λ^*, eq.(3) is reduced to

$$\frac{d\ln y^*(R)}{dR} - 2S_1^* y^*(R) = \frac{d\ln X^*(R)}{dR} - 2(S_1^* - S_2)\beta_2^*(R) \qquad (5)$$

which is the so-called Bernoulli equation (Fernald, 1984; Browell et al., 1985). In this case, the present method becomes identical to that by Browell et al. (1985), although they use the analytical expression for the solution of eq.(5).

Note that the parameters S_1, γ and δ must be assigned for the integration. In addition we need $\beta_2(R)$, σ and σ^*. The absorption cross-sections can be determined from the literature. X(R) and $X^*(R)$ are given from the measurements. The advantage of solving these simultaneous differential equations is that the aerosol distribution can be directly obtained even if the absorption term is not negligible in the off-line lidar signal.

If we have measurements with a third wavelength λ^{**}, it is possible to infer the values of S_1, γ and δ by minimizing D_1 and/or D_N defined by

$$D_1 = \int \{ \beta_1(r) - \beta_1'(r) \}^2 dr$$

$$D_N = \int \{ N(r) - N'(r) \}^2 dr$$

when the upper and lower limits for the integration are appropriately chosen. Here β_1 and N are the solutions obtained from the combination of the measurements with λ^* and λ; β_1' and N' are from the combination of λ^{**} and λ. In principle, β_1 and β_1' (and N and N') should coincide with each other under the assumption that the power-law relationship with the same values of γ and δ holds between the three wavelengths. A numerical method for searching for the best estimates of S_1, γ and δ is employed.

The conventional DIAL equation (Schotland, 1974) can be derived from eqs.(2) and (3) as follows:

$$N(R) = -\frac{1}{2(\sigma - \sigma^*)}\left[\frac{d}{dR}\ln\frac{X(R)}{X^*(R)} - \frac{d}{dR}\ln\frac{y(R)}{y^*(R)} + 2(\alpha_1 - \alpha_1^*) + 2(\alpha_2 - \alpha_2^*)\right]$$

The second and third terms in the right hand side are due to the wavelength-dependence of aerosol backscatter and extinction coefficients, and are neglected in conventional DIAL analysis. The last term can be estimated from appropriate meteorological measurements or atmospheric models and included in the present analysis.

3. NUMERICAL SIMULATION

Numerical simulation was carried out to show the effectiveness and limitations of the present method. Fig.1 shows the model profiles for backscatter coefficients for air molecules and the aerosol and for the ozone number density. The number densities of air molecules and ozone are from the US Standard atmosphere. The profiles of aerosol backscatter coefficient were generated by modifying Elterman's model (Elterman, 1968). The aerosol peak around the altitude of 20 km is enhanced to represent the condition of high (a) and moderate (b) aerosol loadings such as after major volcanic eruptions. Parameters for the aerosol optical properties S_1, γ and δ are assumed as 50.0, 1.0 and 1.0, respectively. The lidar signals $\underline{X}(R)$, $\underline{X}^*(R)$ and $\underline{X}^{**}(R)$ were generated as

$$\underline{X}(R_i) = \frac{1}{\Delta R}\int_{R_i - \frac{\Delta R}{2}}^{R_i + \frac{\Delta R}{2}} X(r)dr$$

where ΔR is the range resolution and $R_i = i\,\Delta R$. The integration was replaced by a summation of X(R) at ten equally-divided discrete points in the range ΔR.

The ozone concentration was estimated with the conventional DIAL equation and the errors of the derived ozone concentration relative to the modelled one at each altitude is shown in Fig.2 for the wavelength pair of 308 nm and 339 nm. The lidar signals were produced with a range resolution of 50 m and values of N(R) were calculated at every 50 m in altitude. Distinct distortion was found in the derived profiles, especially in case (a).

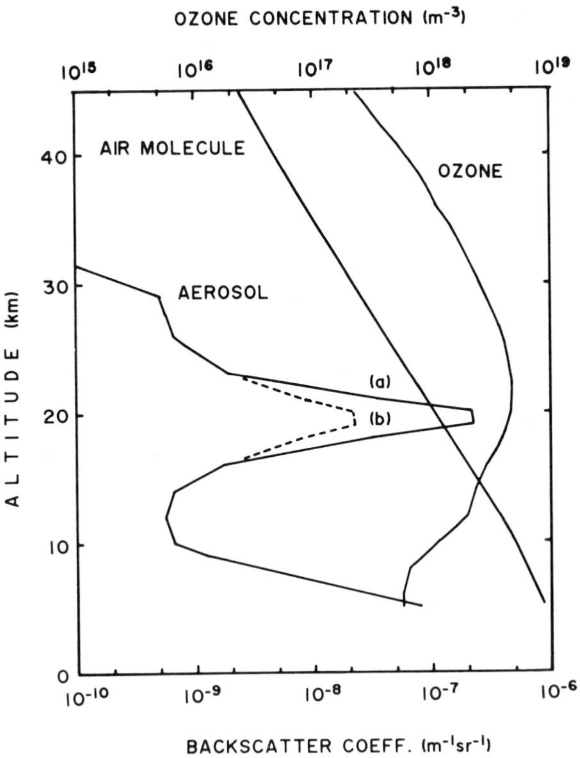

Fig.1 Backscatter coefficients for 308nm wavelength for the aerosol and air molecules, and the ozone number density used for the simulation.

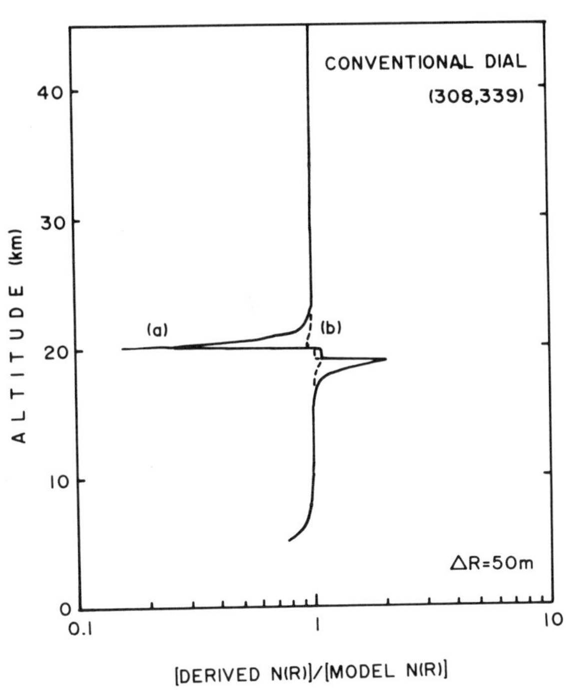

Fig.2 Relative error in the ozone concentration profile obtained by conventional DIAL analysis (solid curve for case (a) and dashed curve for case (b)). Range resolution is 50m.

The results of analysis by the multiple-wavelength DIAL are shown in Fig.3 in the same way as in Fig.2. As a boundary condition, $\beta_1 = 0.0$ was assigned at the altitude of 45km. The derived ozone profile for case (b) almost coincides with the model profile. In case (a), two small spikes appear at altitudes of 19 km and 20 km where the aerosol profile in the model has sharp inflections.

The optical parameters of aerosol S_1, γ and δ were estimated by minimizing D_1 in the present analysis and are summarized in Table 1. Much better estimates were made for case (a) than case (b).

Fig.4 and 5 show the results of conventional DIAL analysis and of multiple wavelength DIAL analysis, respectively, with range resolution of 500m. Due to the smoothing effect in the calculation of the received signal, the magnitude of distortion in the derived profiles were made smaller than the estimation with the range resolution of 50m. The multiple-wavelength DIAL analysis, however, does not give significantly better solutions. The relative error is less than 10 %.

Even when the correct values (50.0, 1.0 and 1.0) were given to the parameters S_1, γ and δ ,

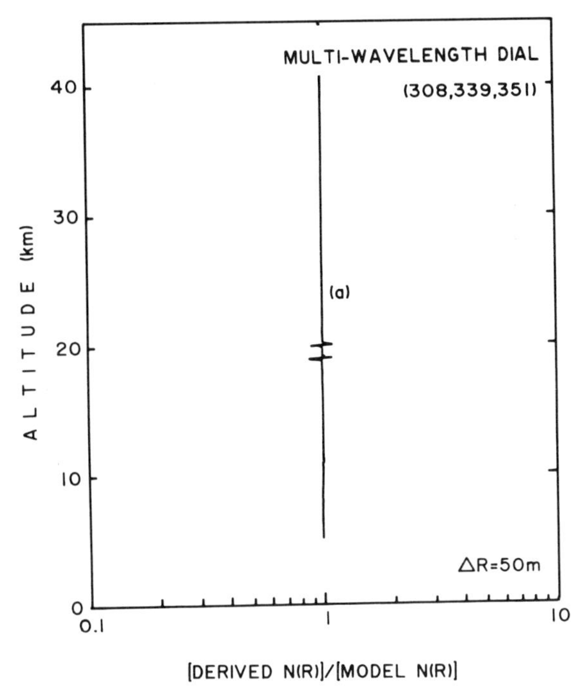

Fig.3 Relative error in the ozone concentration profile obtained by the multiple-wavelength DIAL analysis. Range resolution is 50m.

Table 1 OPTICAL PARAMETERS S_1, γ AND δ ESTIMATED BY THE MULTIPLE WAVELENGTH DIAL ANALYSIS

	Model	Case(a)		Case(b)	
	R=	50m	500m	50m	500m
S_1	50.0	50.5	48.2	41.2	39.1
γ	1.0	1.05	1.15	1.16	1.68
δ	1.0	1.01	1.02	1.10	1.16

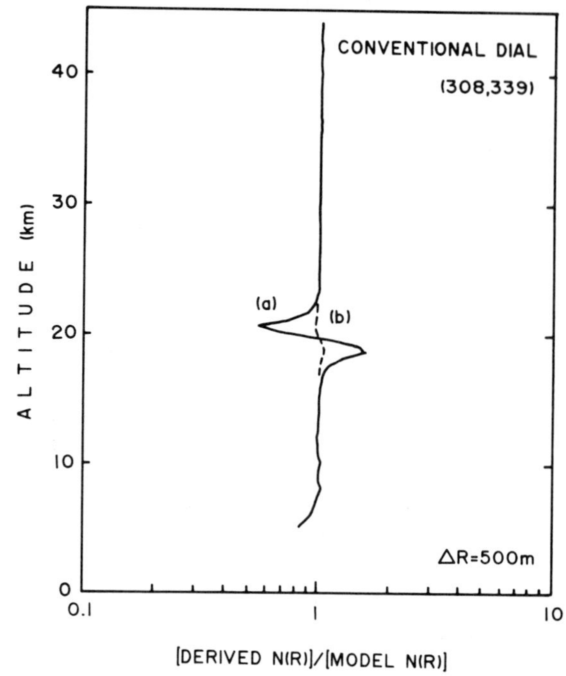

Fig.4 The same as Fig.2 except that the range resolution is 500m.

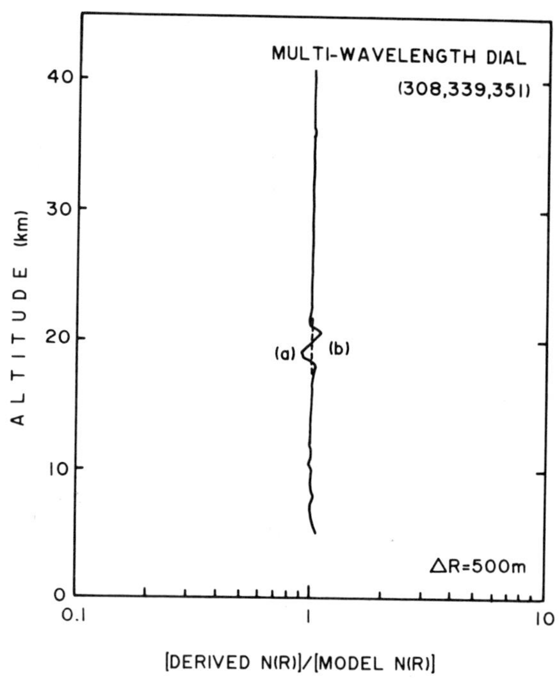

Fig.5 The same as Fig.3 except that the range resolution is 500m.

the same degree of distortion were found in the ozone profile in the simulation with range resolution of 500m. This means the distortion in these cases is not due to the inaccurate estimation of the optical parameters but due to the smoothing effect.

4. CONCLUDING REMARKS

Even with the condition of heavy and quite localized aerosol loadings, the present analysis method produces correct ozone profiles when the range resolution is sufficiently high. Case (a) employed in the simulation corresponds to the condition after a major volcanic eruption such as El Chichon.

It should be noted that, due to the smoothing effect, artificial structures appear in the derived profiles with low range resolution data when the aerosol exists in a narrow layer, even when the multiple-wavelength analysis is applied.

More detailed analysis for realistic situations with measurement errors is needed, which is, however, out of the scope of the present paper.

References

Browell, E. V., S. Ismail and S. T. Shipley, 1985: Ultraviolet DIAL measurements of O_3 profiles in regions of spatially inhomogeneous aerosols. Appl. Opt., **24**, 2828-2836.

Elterman, L., 1968: UV, visible and ir attenuation for altitudes to 50 km. Air Force Cambridge Research Laboratories, AFCRL-68-0153, Environmental Research Paper, No.285.

Fernald, F. G., 1984: Analysis of atmospheric lidar observations: some comments. Appl. Opt., **23**, 652-653.

Pelon, J., S. Godin and G. Megie, 1986: Upper stratospheric (30-50 km) lidar observations of the ozone vertical distribution. J. Geophys. Res., **91**, D8, 8667-8671.

Sasano, Y., 1988: Simultaneous determination of aerosol and gas distribution by DIAL measurements. Appl. Opt., **27**, 2640-2641.

Schotland, R. M., 1974: Error in the lidar measurement of atmospheric gases by differential absorption. J. Appl. Meteorol., **13**, 71-77.

Uchino, O., M. Tokunaga, M. Maeda and Y. Miyazoe, 1983: Differential-absorption-lidar measurement of tropospheric ozone with excimer-Raman hybrid laser. Optics Letters, **8**, 347-349.

A NOVEL FAST RESPONSE CHEMILUMINESCENCE SONDE FOR ROUTINE SOUNDINGS OF STRATOSPHERIC OZONE UP TO 1.5 mb

W. Speuser, S. Sahand, and U. Schurath
Institut für Physikalische Chemie der Universität
D–5300 Bonn 1, FRG

ABSTRACT

A novel balloon-borne surface-chemiluminescence sonde is described which can measure ozone in the stratosphere up to 1.5 mb (ca. 45 km). In the development phase 10 sondes were launched from Hohenpeissenberg Meteorological Observatory. The ozone profiles obtained with the new sonde are compared with profiles of BREWER-MAST sondes which were launched in tandem or in succession. A preliminary comparison with a LIDAR profile is also presented.

1. INTRODUCTION

With the advent of inexpensive light-weight electrochemical ozonesondes, balloon soundings have become feasible on a routine basis. Recent model calculations predict that anthropogenic perturbations of the stratosphere should cause the largest relative change in ozone mixing ratio at an altitude of about 40 km. Wuebbels et al. (1983) calculate that, up to the year 2000, northern hemispheric ozone will decrease by 12 % at this altitude. For the period 1979-1985 SAGE I and II data suggest a maximum ozone reduction of 3 % at 40 km, while the reduction could be as high as 7 % according to SBUV satellite data (Rodgers 1988). Unfortunately this critical altitude range is inaccessible to routine sounding with wet chemical ozone sondes which freeze at the triple point (6.1 mb for pure water, ≙ ca. 35 km altitude). Ozone detection by surface chemiluminescence with dyes like Rhodamine B, which was first introduced by Regener (1964), is not subject to this altitude limitation, and has in fact been utilized for rocket soundings (Hilsenrath and Kirschner 1980; Krueger 1986). These sondes are expensive and complicated and not suited for balloon soundings.

2. DEVELOPMENT OF THE SONDE

2.1 TESTING OF DYES AND DYE SUBSTRATES

In a preliminary laboratory study the most suitable dye for ozone detection was selected. Thin layer chromatography (TLC) sheets were treated with solutions of 13 laser dyes, and dried in a vacuum oven at 60 ^{o}C. Circular discs were cut and tested after 3 h exposure to about 100 ppb ozone in a small wind tunnel ("preozonization") by measuring the chemiluminescence intensity as a function of ozone mixing ratio in ambient air at fixed flow rates (usually 60 l/h). The chemiluminescence was measured with a cheap side-window photomultiplier (Hamamatsu 931B). Its high stability bialkali photocathode is most sensitive in the 400-500 nm range, showing little overlap with the red fluorescence of Rhodamin B. Under these conditions pale yellow Coumarin 47 was found to be the most efficient and stable dye. Its fluorescence spectrum coincides with the quantum yield function of the photomultiplier. In contrast to most other dyes, addition of gallic acid had little effect on the efficiency and stability of Coumarin 47. However, addition of a commercially available hydrophobing emulsion was found to enhance the sensitivity. The stability of the dye could be further enhanced at the expense of sensitivity (which was much higher than required anyway) by rinsing the dyed sheets with a solution of silicon oil in CCl_4 and drying again. Neither this nor other coatings (e.g. treatment with hexamethyldisilazane) had a significant effect on the water vapour dependence of the sensitivity, in contrast to the findings of other authors (Hodgeson et al. 1970). The humidity effect was evaluated by measuring the sensitivity for ozone in dry synthetic air and in laboratory air (ca. 30 % r.h. at 25 ^{o}C). The sensitivity was roughly a factor 2 higher in laboratory air, independent of the dye studied. Humidity effects are irrelevant for stratospheric applications.

We have also tested Coumarin 47 on different substrates. TLC sheets coated with polyamide 11, silica gel 60, cellulose, aluminium oxide 60, and Kieselguhr were investigated. Although the chemiluminescence was stronger on polyamide, silica gel was preferred because of its weaker temperature dependence.

In the upper stratosphere interference by oxygen atoms must be considered, because above 40 km the O : O_3 ratio may be as high as 1 : 1000 during the day. We have compared the chemiluminescence yields in a low pressure flow system. Oxygen atoms were produced by titration of N atoms (from a microwave discharge in nitrogen) with NO. The surface chemiluminescence yield of oxygen atoms with Coumarin 47 was found to be a factor 25 less than that of ozone. Oxygen atom interference can thus be entirely neglegted.

Further details of our laboratory work have been published elsewhere (Sahand et al. 1987). The results reported in this paper pertain to discs of 25 mm diameter. They were cut from TLC sheets (0.2 mm silica gel 60 on aluminium), dyed with a solution of 5 g Coumarin 47 and 16 g gallic acid in one liter methanol mixed with the same volume of a commercial hydrophobing emulsion. The vacuum dried

ISBN 0–937194–15–8

sheets were coated with silicon oil as described, and stored in a desiccator.

2.2 DESCRIPTION OF THE SONDE

Surface chemiluminescence exhibits a flow rate dependence, Fig. 1, because the reaction rate at low flow is limited by turbulent transport of ozone to the surface. At very high flows chemiluminescence is independent of the flow, and molecular diffusion of ozone into the porous dye support becomes reaction rate limiting. This gives rise to a pressure dependence in the flow rate independent regime (cf. Fig. 3 above 20 mbar).

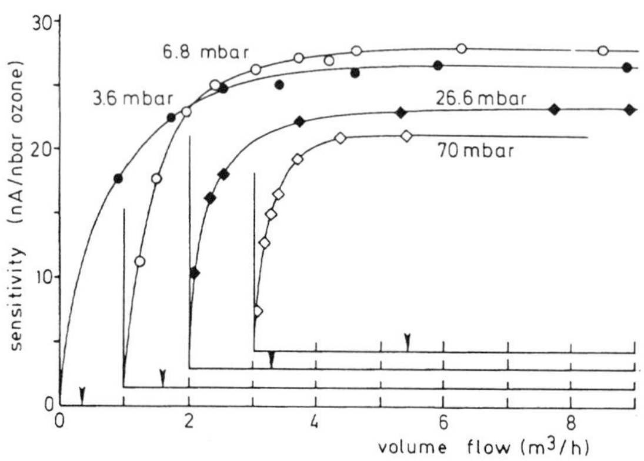

Fig. 1. Flow rate dependence of the signal at different total pressures.

We have constructed a sonde which as far as possible works in the flow independent regime: air is forced past a chemiluminescent surface by means of a high performance miniature fan (Micronel V361 L, 12 V). The power consumption decreases from 1 W at ambient pressure to 0.15 W at 30 mbar. The flow rate through the sonde, 18,000 liter/h at ambient pressure, decreases to 240 liter/h at 3 mbar.

The sonde body, which houses the photomultiplier and the chemiluminescent disc, was turned from a block of black PVC. The photomultiplier base, high voltage power supply (RCA, PF1042), and amplifier are encapsuled in a gas-tight box. All other electronics and the lithium batteries (15 à 1200 mAh, 4 à 750 mAh) are contained in a styrofoam box. The batteries ensure at least 5 h operation under stratospheric conditions. Air is sucked into the sonde through corrugated teflon tubing (20.5 mm i. d., 550 mm long) wound to form a light trap. The electric fan is attached downstream of the photomultiplier. The exhaust is a spiralled corrugated aluminium tube of 40 mm i.d. and 600 mm length. The weight of the sonde (ca. 1.3 kg including batteries) can be significantly reduced by miniaturization.

The time resolution of the sonde was measured by switching an ozone producing uv lamp on and off directly in front of the inlet tube. The signal was measured with a fast electrometer (Keithley 610C). Response was practically instantaneous, yielding a time constant of 0.1 s or less. This enables the sonde to measure ozone on descent also, if required. A time constant of 1 s was imposed under flight conditions. At ambient temperature and pressure a detection limit of 10 ppt was readily achieved.

The intensity I of the chemiluminescence is a function of the surface temperature. The dependence in the range 250 - 295 K is approximately Arrhenius-like,

$$I = A \exp(-T^*/T),$$

with $T^* = 1540$ K. Therefore the aluminium base of the chemiluminescent disc had to be electronically thermostated at + 5 $^{\circ}$C. A test flight was arranged to measure air temperature outside and inside the sonde and the surface temperature of the chemiluminescent disc in the stratosphere. The results are plotted in Fig. 2, showing that the surface temperature remained constant. The temperature in the styrofoam box never dropped below 10 $^{\circ}$C.

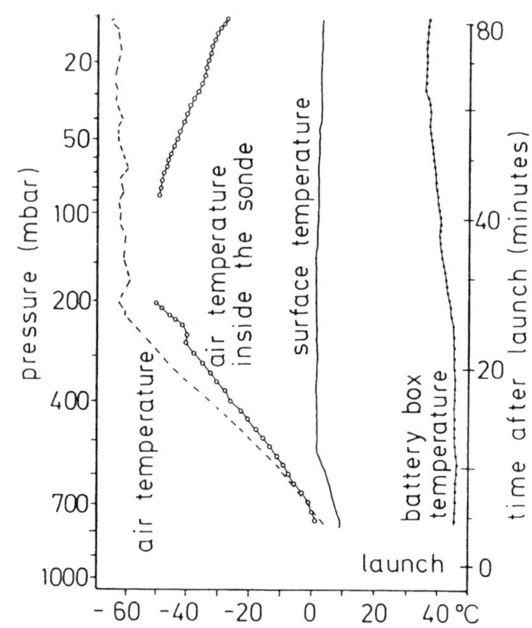

Fig. 2. Temperatures inside and outside the sonde during a stratospheric flight.

We have measured the sensitivity of the sonde in ambient air over periods of several days, in parallel with a commercial ozone analyzer. E.g., at 30 $^{\circ}$C surface temperature the sensitivity of a "preozonized" disc exposed to 50 ppb ozone remains constant for 30 h. Then the sensitivity starts to drop linearly, reaching half its plateau value after another 55 h. This is typical of a zero order reaction. The time interval of constant sensitivity was in fact halved when the ozone mixing ratio was doubled. At lower surface temperatures the stability is enhanced.

3. RESULTS

3.1 PROPERTIES AND CALIBRATION OF THE SONDE

The ozonesonde was developed and tested in cooperation with the Meteorological Observatory at Hohenpeissenberg, where 10 sondes were launched up to date. Before launching a sonde the sensitivity had to be determined in the laboratory. Calibration in the pressure range 3 - 1000 mbar was at first accomplished by mounting the entire sonde in an evacuable smog chamber. The chamber was filled with various pressures of synthetic air, ozone was added, and the concentration measured by uv absorption (3 m path length) at 254 nm. This procedure, apart from being extremely time consuming, risks to exhaust the disc by exposing it to very high ozone concentrations.

We have recently constructed a more practical calibration system operating in the 1.5 - 200 mb range. A cylindrical glass chamber of 40 liter volume is pumped by a powerful Roots blower and rotary pump. Ozone in synthetic air is metered into the chamber by electronic flow controllers. The ozone mixing ratio (0 - 10 ppm) is measured upstream at ambient pressure. Most of the air flowing through the chamber bypasses a dummy sonde through a glass tube. When the electric fan of the sonde is activated, a small fraction of air is aspired upstream and exhausted downstream of the tube. Since the pressure drop along the tube (i.d. 80 mm) is negligible, the flow rate through the sonde is entirely controlled by the electric fan, simulating flight conditions. The chemiluminescent disc, which is cooled to 5 ^{o}C, can be interchanged from outside without opening the chamber. Fig. 3 shows a calibration curve, which overlaps with a high pressure calibration in the smog chamber. The sensitivity increase between 1000 and 20 mb is due to enhanced molecular diffusion, as noted above. Below 20 mb the sensitivity decreases because the flow rate drops below the plateau region. Vertical arrows in Fig. 1 indicate approximate flow rates under flight conditions.

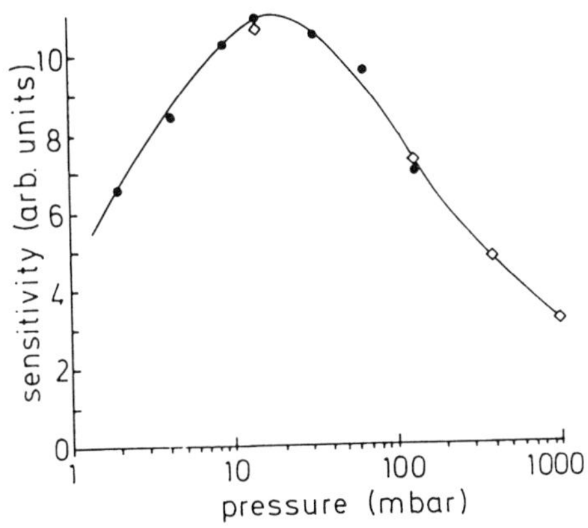

Fig. 3. Calibration curve of the ozonesonde. Squares refer to smog chamber calibration.

The sensitivities of "preozonized" discs cut from the same TLC sheet differ by ± 5 %. We are confident that this scatter can be reduced by improved preparation techniques.

The photomultipliers in the sondes differ in quantum yield and amplification. Therefore the photomultipliers are now calibrated in situ by replacing the chemiluminescent disc of the sonde with a heavily attenuated LED. The intensity of the LED is exactly reproducible by comparison with a standard photomultiplier unit operated with a precision power supply and electrometer.

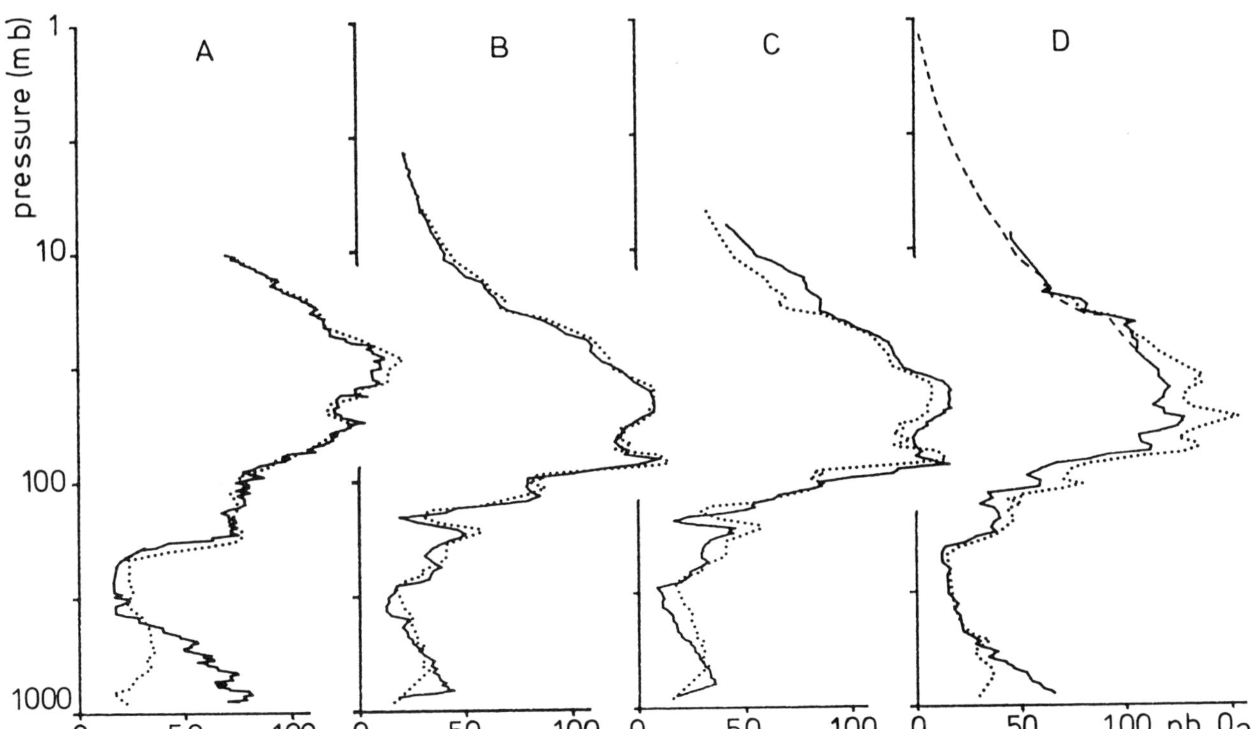

Fig. 4. Flight of chemiluminecence sonde (solid line) and BREWER-MAST sonde (dotted line). A, tandem flight of 4 August 1987; B, tandem flight of 2 December 1987; C, chemiluminescence sonde profile, 5 h after 4B (dotted BREWER-MAST profile of 4B shown for comparison); D, BREWER-MAST profile of 4 December 1987 (launch at 06:53 UTC), and profile of chemiluminescence sonde launched 3 h later. The dashed line represents a LIDAR profile from the previous night.

3.2 INTERCOMPARISON FLIGHTS

The results of four more recent intercomparison flights of our sonde (solid lines) with BREWER-MAST sondes (dotted lines) are shown in Fig. 4. Figs. 4A and 4B represent tandem flights. Both show good agreement above the tropopause. Note the resolution of fine structure in the solid profile. The solid profile in Fig. 4C was measured 5 h after the flight shown in Fig. 4B, which had been launched at 8:01 UTC. The BREWER-MAST profile from the previous flight, dotteld line in Fig. 4C, is shown for comparison. Fig. 4D represents the last flight. The BREWER-MAST sonde was launched at 6:53 UTC, followed by the chemiluminescence sonde about 3 h later. The difference between the two profiles seems somewhat too large for a 3 h interval, and a calibration error cannot be excluded. The dashed line at high altitude is an ozone profile measured by the Hohenpeissenberg LIDAR during the previous night (H. Claude, private communication). It is in reasonable agreement with both profiles in the overlap region, considering the averaging effect and limited altitude resolution of the LIDAR technique.

ACKNOWLEDGEMENTS

The authors gratefully acknowledge the continuing help and encouragement by the DWD team of Hohenpeissenberg, and the expertise and efficiency of our electronics workshop. The project is supported by the Bundesminister für Forschung und Technologie.

REFERENCES

Hilsenrath, E., and P.T. Kirschner, 1980: Recent assessment of the performance and accuracy of a chemiluminescent rocket sonde for upper atmospheric ozone measurements. Rev. Sci. Instrum., 51, 1381-1389

Hodgeson, J.A., Krost, K.J., O'Keefe, A.E. and R.K. Stevens, 1970: Chemiluminescent Measurement of Atmospheric Ozone, Response Characteristics and Operating Variables. Anal. Chem., 42, 1795-1802

Krueger, A.J., 1986: 2. Rocket Techniques for Measurement of Ozone and Related Neutral Minor Constituents. In Handbook for MAP (R.A. Goldberg, Ed.), 19, 29-55

Regener, V.H., 1967: Measurement of atmospheric ozone with the chemiluminescent method. J. Geophys. Res., 69, 3795-3800

Rodgers, C., 1988: Global ozone trends reassessed. Nature, 332, 201

Sahand, S., Speuser, W. and U. Schurath, 1987: A Battery-Powered Light-Weight Ozone Analyzer for Use in the Troposphere and Stratosphere. In Physico-Chemical Behaviour of Atmospheric Pollutants, G. Angeletti and G. Restelli (Eds.), D. Reidel, Dordrecht, 33-44

Wuebbels, D.J., Luther, F.M. and J.E. Penner, 1983: Effect of coupled anthropogenic perturbations on stratospheric ozone. J. Geophys. Res., 88, 1444-1456

GROUND-BASED MICROWAVE SENSOR TO RETRIEVE STRATOSPHERIC OZONE PROFILES

G. Rubin and K.F. Künzi
Institute of Applied Physics
University of Bern, Switzerland

ABSTRACT

A microwave radiometer for measuring ozone profiles in the stratosphere is described with emphasis on the techniques used to extract information in the lower stratosphere. The ozone profile is determined by observing a pressure broadened ozone absorption line at 142 GHz and inverting the radiative transfer equation. Until now this method was limited to the altitude range from 25 km to 75 km. By extending the bandwidth of the receiver from 200 MHz to 1200 MHz it was possible to include the lower stratosphere (15 km to 25 km) in the observed region. Measurements made with the broadband radiometer show a good correlation between the microwave measurements and balloon sonde data from the nearby meteorological station. An agreement to 0.5 ppmV was obtained for simultaneous microwave and balloon observations at altitudes between 15 km and 30 km. Comparison with Umkehr data also shows good agreement.

1. INTRODUCTION

Since 1981 microwave radiometry has proved to be a reliable method of monitoring ozone in the upper stratosphere and mesosphere. Comparison with other methods show good agreement between the microwave and Umkehr, Lobsiger et al. (1984) and SBUV data, Zommerfelds et al. (1988). A drawback of this method was the limited altitude range from 25 km to 75 km which allows only for 40% of the total column ozone to be measured. Because the ozone concentration peaks at 22 km, about 60% of the total atmospheric ozone was not accounted for by conventional microwave remote sensing. There was no possibility to make estimates of the total ozone column based on microwave observations. By extending the receiver bandwidth it became possible to include the lower stratosphere (15 km to 25 km) in the observed region thus increasing the amount of ozone covered by microwave measurements to approximately 85%. By combining balloon sonde and microwave data it is possible to estimate total atmospheric ozone.

2. PRINCIPLE OF OPERATION

A complete description of the experimental set-up operated at the University of Bern, and the retrieval method used is given by Lobsiger (1987). Lobsiger describes the radiometer used for ozone measurements at altitudes ranging from 25 km to 75 km, and in Lobsiger et al. (1986) he deals with observations of diurnal variations of mesospheric ozone. This work will focus on the ozone observations from 15 km to 25 km.

The emission of the thermally excited 142.175 GHz rotation line of ozone is detected with a heterodyne microwave radiometer (Fig. 1). Singlesideband system noise temperature is 700 K.

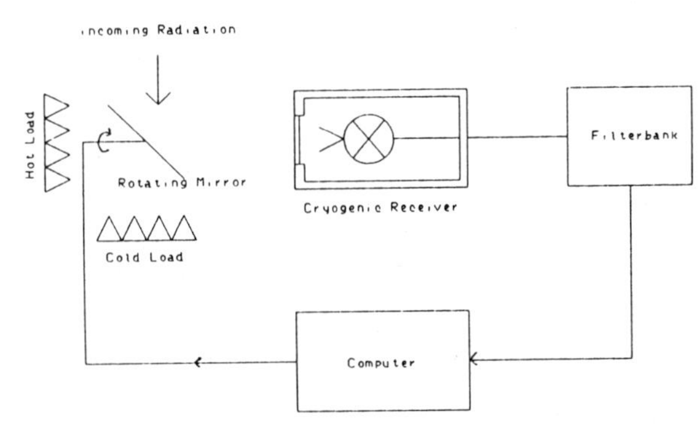

Fig. 1 Block diagram of the radiometer. The microwave radiation enters the receiver via the rotating mirror situated between the hot and cold calibration loads.

After several downconversion steps a realtime spectrum analysis is performed using conventional filterbanks having a resolution from 100 MHz at the linewing to 0.2 MHz at the linecenter. Calibration is performed by looking periodically at cold (80 K) and hot (310 K) absorbers, which act as blackbodies of known

ISBN 0-937194-15-8

temperature. Correction for tropospheric effects is performed assuming a isothermal troposphere as described in by Lobsiger et al. (1984).

The dominant broadening mechanism in the altitude range from 15 km to 75 km is pressure broadening. This allows a retrieval of the ozone profile from the measured lineshape by inverting the radiative transfer equation. The inversion algorithm used is the Chahine relaxation method adapted by Randegger (1980). For the inversion the brightness temperature differences of neighbouring channels are used. The altitudes corresponding to channel pairs are obtained by calculating a simulated line with an ozone profile that is varied stepwise over the entire altitude range. The altitude where a change in the ozone profile has the largest impact on the temperature difference for a given channel pair becomes the inversion altitude associated with this channel pair. Thus the sensitivity functions of the radiometer channels are obtained.

The lower altitude limit is determined by the total system bandwidth. The latter increases very rapidly with decreasing altitude. It is not feasible to use an arbitrary large bandwidth because the signal to noise ratio at the linewings becomes too small for a meaningful inversion. Model calculations with a simulated line have shown that a bandwidth of 1100 MHz seems to be an upper limit, giving a brightness temperature difference of a few tenths of degree Kelvin for a channel pair at the wings. This results in an altitude of 13 km corresponding to the outermost channel pair.

Table 1 shows the broadband channels that have been added to the radiometer described by Lobsiger (1987) and the corresponding maxima of the sensitivity functions. The halfwidth of the functions at half power is approximately 4 km and neighbouring functions overlap so that the information of neighbouring channel pairs is not independent and the best altitude resolution to be expected is about 4 km. The outermost channel is used for reference purposes and is not considered for the inversion, so the lowest altitude is 15 km.

TABLE 1 Frequencies of channels used for retrieval of profile in the lower stratosphere. The bandwidth of the channels is 100 MHz. The altitude of the maximum sensitivity of channel pairs is shown on the right.

Channel pair number	Frequency offset from linecenter [MHz]	Maximum sensitivity [km]
	± 550	
1		13
	± 450	
2		15
	± 350	
3		17
	± 250	
4		19
	± 150	
5		22
	± 50	

3. MEASUREMENTS

From May 26th, 1987 to June 9th, 1987 a series of observations were made. For comparison balloon sonde data from the nearby Payerne meteorological station (distance 40 km) and from the Arosa Umkehr station (distance 170 km) were obtained. Due to the close location of the observation sites no significant deviations of the ozone profiles are expected. Although there exist microwave data for the entire period, only those days for which simultaneous balloon observation were available will be considered for the comparison. In Table 2 averaged data for seven simultaneous measurements are shown. Since the Umkehr method depends on a clear sky, there is just one day (May 26th) for which observations with all three techniques exist.

The balloon sondes are released at 12.00 GMT and they take about three to four hours to reach maximum altitude. Therefore the microwave data taken from 12.00 to 18.00 hours are averaged for the comparison. Due to the lack of Umkehr data taken simultaneously with microwave and balloon measurements the Umkehr data from June 6th and June 11th were averaged and compared with the microwave and balloon data from June 9th.

Fig. 2 and 3 show measured brightness temperature differences and the corresponding profiles from May 26th, 1987 and June 9th, 1987 together with Umkehr and balloon sonde data. Both the Umkehr and the microwave method have an altitude resolution which is far too coarse to resolve the fine details observed with the balloon sonde. Agreement between the microwave data and Umkehr and balloon observations is quite good above 20 km. The database is too small, however, to conduct a significant statistical analysis.

The total ozone column can be estimated by combining microwave data from 15 km to 75 km (approximately 85% of total ozone) with balloon sonde data from ground to 15 km. This results in 366 DU for the observation period in May and June which does agree well with the Arosa value of 359 DU for the same period.

TABLE 2 Comparison of microwave and balloon sonde data for period from May 26th to June 9th, 1987. N_m and N_b denote number of microwave and balloon measurements respectively.
The difference is (Conc.micr-Conc.ball)*100/Conc.micr

Altitude [km]	N_m	Conc.micr. $[10^{-18}\ m^{-3}]$	N_b	Conc.ball. $[10^{-18}\ m^{-3}]$	Difference %
15	7	3.57	7	3.03	+15.1
17	7	2.98	7	3.28	-10.1
19	7	3.00	6	2.89	+3.6
23	7	4.04	5	4.06	-0.5
26	7	3.78	4	3.57	+5.6

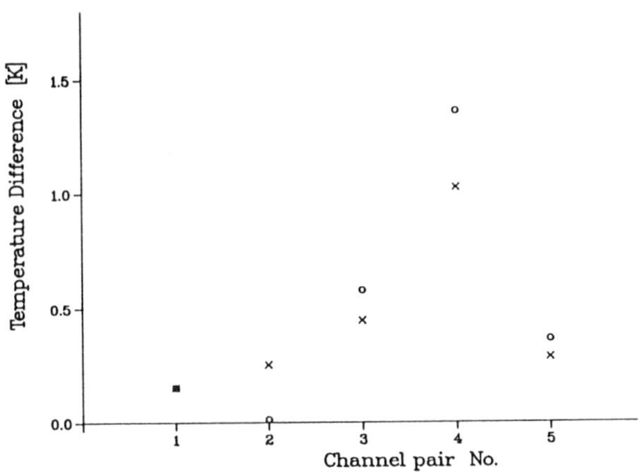

Fig. 2 Measured brightness temperature differences for May 26th, 1987 (x) and June 9th, 1987 (o). Channel pair numbers as in table 1.

4. ERROR ANALYSIS

The ozone profile error can be divided into a statistical part and a systematic part. The statistical part is a measure for the repeatability of the measurement. Its main contribution is the system noise and fluctuations in the calibration load temperature. The error resulting from the system noise is $\Delta T = T_{sys}/\sqrt{Bt}$ with B=100 MHz and t=6 hours giving 0.005 K. The contribution from changing hot and cold load temperatures is 0.1 K. The statistical error of the retrieved profile is 15%.

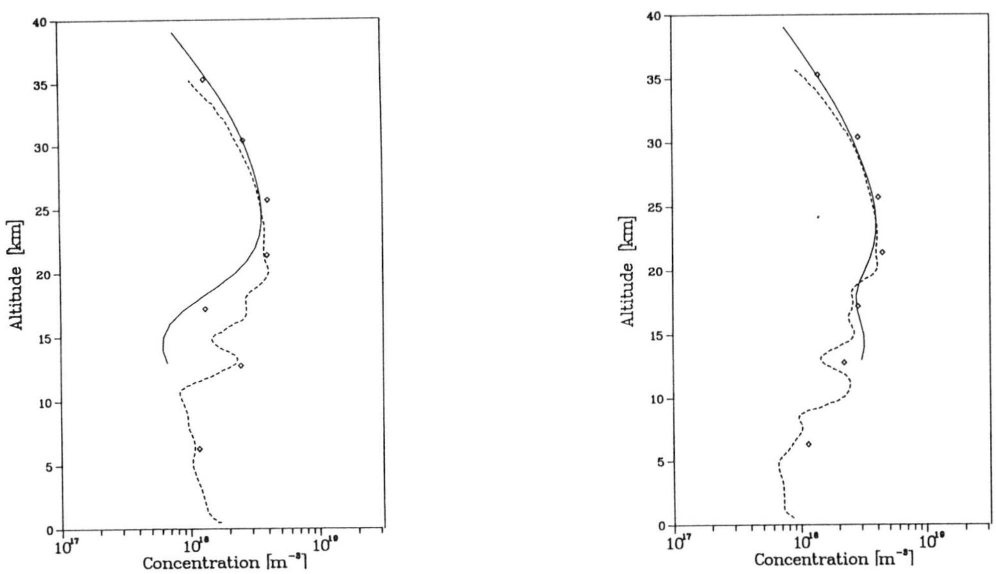

Fig. 3 Ozone profile for May 26th, 1987 (left) and June 9th, 1987 (right, see text details of averaging the data). The solid line is the microwave profile, dashed is the balloon sonde profile and diamonds denote Umkehr layers 1 to 7.

The systematic error stems mainly from baseline problems. They are the result of reflections between the receiver and the calibration loads which produce standing waves. To reduce this effect we use a reciprocating mirror as proposed by Gustincic (1977). The remaining baseline was determined by looking at a blackbody at 235 K and taking the deviation form this temperature at each channel as the baseline which was subtracted from the measured ozone spectrum. A further reduction of the baseline effect was achieved by correcting for the slight slope (0.4 K/GHz) in the tropospheric absorption. A symmetric lineshape was then assumed and the line folded at the centre before further processing. Errors affecting the absolute value of the brightness temperature, such as imageband response of the mixer or assumed temperatures for tropospheric correction, are negligible, since only temperature differences are used for the inversion of the profile.

The total error in measuring the brightness temperature is 0.15 K, which is nearly 40% of the temperature difference at the linewing. Performing the inversion with input data varying 40% it is seen that the resulting profiles have a variation of approximately ±1 ppmV. This corresponds to an error in the concentration from $\pm 10^{17}\ m^{-3}$ at 35 km to $\pm 3 \cdot 10^{18}\ m^{-3}$ at 15 km. Assuming the balloon data in Fig. 3 to be correct we can estimate the uncertainty to be approximately 10% at 35 km increasing below 20 km to close to 100% at 15 km altitude.

Agreement is better at higher altitudes with improving accuracy of the microwave method due to better signal-to-noise-ratio.

5. CONCLUSIONS

It has been shown that microwave radiometry can determine ozone profiles down to 15 km. Although longtime observations do not yet exist, the first measurements indicate a good correlation between microwave radiometry and established techniques such as Umkehr and balloon sonde observations, in particular for the altitude range > 20 km. Thus microwave radiometry seems suited to complement existing ozone observation station on an operational basis since the microwave sensor is a reliable automated instrument not requiring a skilled operator.

REFERENCES

Gustincic J.J., 1977: A_quasi-otical receiver design, Proc. IEEE Conf. on MTT, San Diego, 99-100.

Lobsiger E., Künzi K.F., Dütsch H.U., 1984 : Comparison of stratospheric ozone profiles retrieved from microwave radiometer and Dobson-spectrometer data, J. atmos. terr. Phys. **46**, 799-806, 1984

Lobsiger E., Künzi K.F., 1986: Night-time increase of mesospheric ozone measured with a ground-based microwave radiometer, J. atmos. terr. Phys. **48**, 1153-1158.

Lobsiger E., 1987: Ground-based radiometry to determine startospheric and mesospheric ozone profiles, J. atmos. terr. Phys. **49**, 493-501.

Randegger A.K., 1980: On the Determination of Atmospheric Ozone Profile for Ground Based Microwave Measurements, Pure appl. Geophys. **118**, 1052-1065.

Zommerfelds W., Künzi K.F., Gille J.: Comparison of ozone profiles obtained by ground-based microwave radiometry, Umkehr and SBUV, to be published

A NEW MULTIPASS OZONOMETER

G. Moreau, C. Robert, and P. Rigaud
Laboratoire de Physique et Chimie de l'Environnement/CNRS
45071 Orleans Cedex 2, France

ABSTRACT

A balloon-borne in situ ozonometer was built, which is based on the absorption of a laser beam by ozone in the Hartley band. The first flight took place on 4 May 1988 at Aire-sur-l'Adour (France), to obtain an ozone profile during ascent and checking the accuracy of the instrument with the small ozone variation already observed after sunrise, during a two-hour float. Though laser and electronics were correctly functioning, mechanical failures during ascent prevented ozone measurements with the accuracy expected. However, it was possible to derive an ozone profile, up to 36 km, from data obtained before the failure, data obtained above this level being not reliable. For long-path absorption, a multipass cavity was used composed of two concave mirrors; this type of instrumentation had never before been used in balloon experiments.

1. INTRODUCTION

It is necessary to dispose of accurate instruments for in situ ozone concentration measurements at altitudes of 40 km and higher. Two main reasons justify this need :. 1) to study ozone variations in connection with other parameters ; 2) to calibrate satellite measurements Such instruments must have absolute accuracies of 1-2 %, which can be obtained when two conditions are fulfilled. The first is to avoid contamination in the instrument by gases exuding from the gondola walls, which cause discrepancies in usual instruments that are difficult to estimate. The second concerns measurements without ozone, which must be made easily, to provide the zero offset.

In recent years, UV absorption, using Dasibi instruments or derived instruments (1), has given the most reliable values. This is why UV absorption instruments are used as references for other instruments, in intercomparison campaign (2). We, too, have chosen UV absorption measurements, but with a laser beam traveling in a long multipass-cell composed of two identical concave mirrors. This cell geometry, in conjonction with the laser source, differs fundamentally from the well-known White cell used with normal lamp which has been extensively used in the past for the detection of ozone, and other absorbing gases, in balloon-borne instrument (3).

To comply with the first condition given above (avoiding contamination), that is rarely met, widely opened shutters allow natural air circulation in the absorption zone, which is far away from the walls. When these shutters close the absorption zone, neutral gas is introduced, expelling the confined air, permitting zero measurement and providing absolute optical transparency of the instrument. This second condition is fulfilled several times during a flight.

The laser beam is absorbed over a long optical path, which can be varied from 200 to 400 m, depending only upon geometrical parameters that are preset for the flight. A reference beam (before entering the absorption zone), and an analysis beam (after travelling the multipass cavity), are detected by two fast vacuum diodes enabling a high laser rate (several hundreds of shots per second). For the first flight a low was selected frequency value of 0.25 Hz was selected.

2. DESCRIPTION OF INSTRUMENT

The instrument is shown in figure 1. It consists of a bench made of light materials on which are fixed the different components. They consist of, the laser, the two mirrors of the multipass cavity at each end of the bench extremity, the detector assembly and the electronics. An excimer laser emits at 248.3 nm wavelength a low divergence ($5\ 10^{-4}$ radian), short pulsed (30 ns) beam. Two small mirrors direct the laser beam coming to the multipass cavity, exactly to a 1 cm diameter mirror (mirror "a" on figure 1), situated near the cell mirror M1. After this, the beam travels inside the cavity, following principles first given by Herriott (4), and exits by the same mirror "a", the two cavity mirrors being identical concave mirrors of focal length f. In the cavity the beam is reflected a certain number of times N, N/2 on each mirror. Reflection impacts are on circles, with the impact number 0 and the impact number N confused on mirror "a". In this particular configuration, the figures not being ellipses as would be in the general configuration, the impacts are equally spaced. Impacts are on a circle of 14 cm diameter (20 cm is the diameter of each mirror M1 and M2), that does not authorize a large injection mirror "a".

ISBN 0-937194-15-8

The number of reflections N is simply drawn from the two relations

$$N=2\ k\pi/\vartheta \qquad (1)$$

$$\vartheta=\mathrm{Arc}\cos\,(1-d/2f) \qquad (2)$$

being a geometrical characteristic of the cavity, with the restriction that N/2 and k, an integer must be prime numbers. An optical path no shorter than 200 m was imposed, with a distance between cavity mirrors d about 4 m. In all, 50 passes or reflections are present, and for a given exact focal length for each mirror f=7.342 m, the d value is exactly 3.980 m. We deduce a total optical path of f=194,04 m. There are other d values for which the two relations (1) and (2) are still obeyed, even close to the chosen d value and giving different total path lengths. However, there is no doubt about the number of reflections because it is easy to discriminate between different geometrical configurations of the cavity.

Distance d is not changed during the flight, but it is possible to translate mirror M1, so as to get other d values and, consequently, N and L values. The cavity can be pre-set with L up to about 400 m; in this limit configuration we have exactly d=4.136 m, N=98 reflections and L=405.3 m.

Before the beam enters in the multipass cavity it goes through a beam splitter (a flat silica window at 45°) which deflects the reference beam. After travelling in the cavity, the whole beam reaches the analysis detector D2, thanks to an an optical system conjugating it with the mirror "a". There, in the two fast vaccum photodiodes, used as detectors, resulting photo electron are collected in 50 pF condensators polarised by 10 volts maximum. Since we know optical losses and quantum yield of photodiodes, only $2\ 10^{11}$ photons are needed in the laser beam (or 170 nJ). For this reason the beam should be attenuated just after the laser, which provides 500 J energy. A variable optical attenuator made of four thick flat silica windows, combined with a spatial filter give the beam the right energy. We measure the condensator polarisations V1 and V2 respectively for D1 and D2 detectors about 100 S after a laser pulse, a lapse of time long enough after electrical perturbations, caused by the laser, thus ensuring a very low noise.

The mechanical stability of the bench is very important. It is made of honeycomb carbon-fiber plates and carbon fiber tubes. The carbon fiber material was selected for its low density, good rigidity, and low temperature heat expansion factor. Therefore, distance d does not vary more than 2 mm within the 100 degrees of temperature variations. We calculated the shift of reflection impacts inside the cavity in relation with mirror mounts misalignment. A 2 cm edge was allowed for around the reflection impact circle, to permit up to 6 minutes in angular variations of the M1 and M2 mirrors. Under this conditions, it was concluded that stability of the cavity has to stay within easily attainable limits (5).

3. UNCERTAINTIES AND ACCURACY

The Beer law applies to the absorption of a short pulsed laser beam by ozone, providing two non linearities are absent. The first is related to the depletion of ozone molecules along the beam. The low beam energy renders this phenomenon negligible. The medium is then optically thin. The second concerns detectors : we selected vacuum photodiodes to measure laser fluxes, so as to minimize electronic space charge effects.

Several uncertainties affect the absolute accuracy of ozone densities derived from the Beer law :

$$V1/V0 = k_o \exp\,(-\ \sigma N L) \qquad (3)$$

1. The absorption cross section of ozone at 248,3 nm, whose value is extracted from (5) $\sigma=1{,}08510^{-17}\ cm^2$, with an uncertainty generally considered to be less than 1 %.

2. The optical path length L is know with the very good accuracy of 0.1 %.

3. The optical transparency k_o of multipass cavity is subject to variations, difficult to estimate, but one can assume it to be constant for at least a short time before and after its determination. The first cause of variation is in the reflectivity of mirror coatings which may evolve with time, though this variation is certainly small, since the outer coating is in silica, that normally is unreative to ozone or other reactants. The second cause is linked to the shifting of reflection impacts with mirror misalignment. Large surface coatings techniques do not allow sufficient homogeneity of the coatings and therefore of the reflectivity. It results that any angular mirror change in shifting impacts, will produce a variation of total reflectivity and thus of k_o. When ordering the mirrors, we specified that their maximum reflectivity be on the impact circles. Indeed, calculations show that, with the coating maximum reflectivity on the impact circle, there will be a negligible influence on k_o. A k_o constant was measured before launch. Its value close, to 1.20, maximizes accuracy on the V2/V1 ratio.

4. The effects of gondola degassing are eliminated. Thanks to the widely open shutters, and because the gondola velocity is generally higher than 1 m/s relative to air, there is an efficient washing out of the absorption zone.

5. The temperature, as a consequence, is assumed to be the same, outside and inside with open shutters.

So from the different incertainties contributing to the accuracy of N the ozone molecule density, we deduce that a value of 2 % or less can be achieved at the altitude of 40 km. The ratio V2/V1 is measured within $\langle$ 0,1 %; without taking into account accuracy on , and, with a total absorption length of the order of 400 m, 1 % accuracy can be achieved.

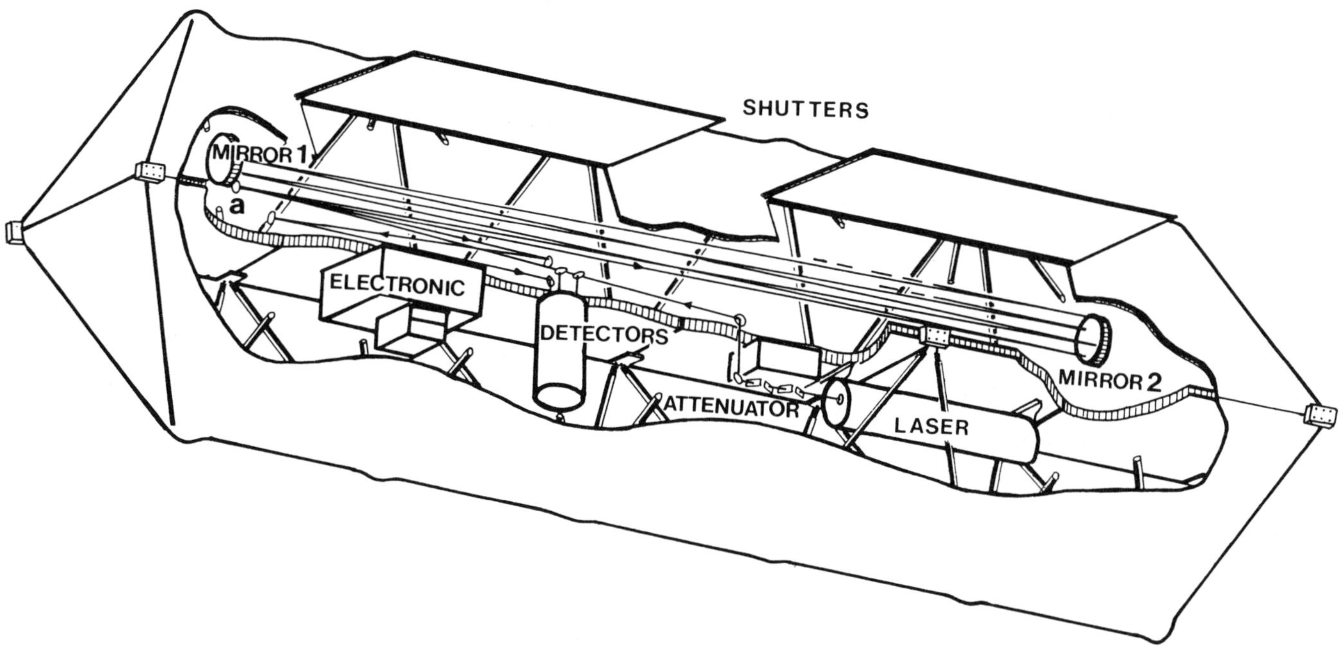

Fig. 1. An exploding view of the multipass ozonometer inside the triangular structure. The different components are attached to a bench, made of carbon fiber honeycomb boards. The laser beam is attenuated, and deflected by two mirrors, to the small mirror "a" by which the beam enters and quits the cavity. Two large identical concave mirrors of 20 cm diameter, in a stable laser configuration, form the multipass cavity. The adjustment of distance d gives automatically the numbers of reflections and the optical path length. Because of the losses in the optics, and the different sensitivities of the detectors, the reference beam (9 % of the incident beam) and the analysis beam have nearly equal energies. The instrument sits in a triangular metallic structure (not shown), designed to absorb shock at landing.

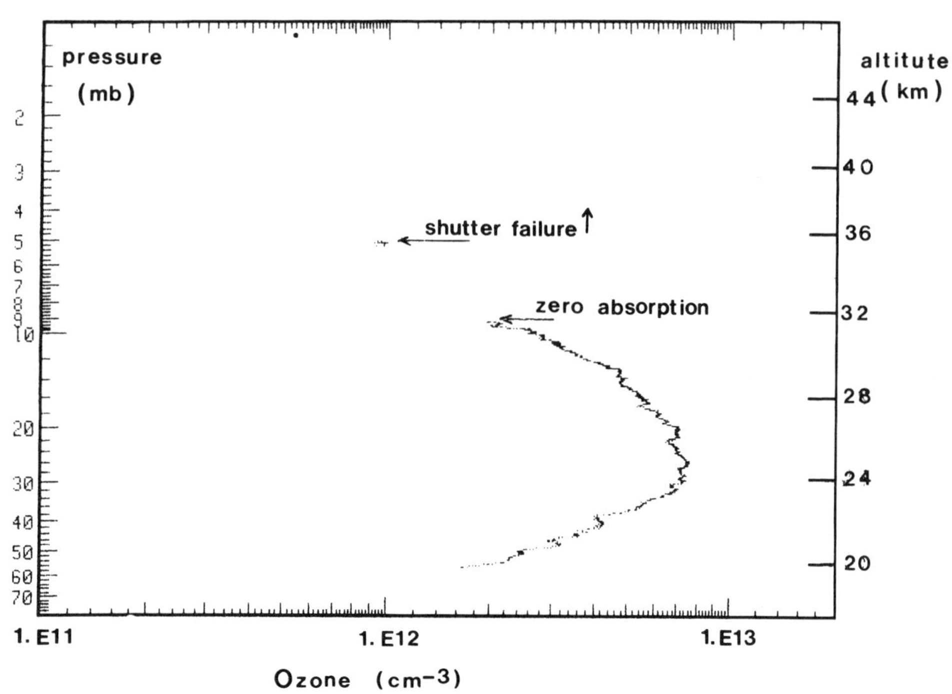

Fig.2. Smoothed ozone profile obtained during ascent at AIRE SUR L'ADOUR (FRANCE) on 4 May 1988. Noticeable events are indicated. The high value at the maximum agrees with earlier, experimental observations made in spring.

4. RESULTS

The first flight occurred on 4 May 1988. Ascent was planned for the end of the night with floating commencing half an hour before sunrise. Total float time had to be at least 2 hours at 3.2 mb pressure (or 39 km altitude). It was planned to determine an ozone profile during ascent and during floating to test the instrument with a small ozone variation of less than 1 % after sunrise predicted by models and already observed.

Measurements began at 18 km during ascent, after shutters opened. Unfortunately, it became immediately apparent, that the laser impact on the small mirror "a" had shifted a few millimeters, resulting in a V2 signal lower than expected. This was caused by the two mirrors, just after attenuator, being unstable with cold (-60°C at the tropopause). It was possible to save measurements because the increase in instability was relatively constant. It was then decided to measure k_o by closing the cavity and introducing pure nitrogen inside, even though this phase has been intended to start during float, just before sunrise. Another failure occurred, which at first, partially damaged shutter opening system; above 36 km the shutter opening system had become completely unserviceable. Thanks to zero absorption performed at about 34 km altitude, it was possible to derive the smoothed profile of ozone molecule density versus altitude (figure 2). This profile is only indicative with an estimation of accuracy not less than 10 % but certainely no more than 20 %, because of large fluctuations of beam reflections energies on the incriminated mirror "a". The obtained profile is in fairly good agreement with standard ozone profiles, but some remarks must be added. At low attitude it underestimates ozone, the maximum concentration is certainly too high with $7.3\ 10^{12}$ molecules/cm^3 at 25 km. However, this observation agrees with the fact that ozone concentration is in spring.

The flight ended with a non-functioning instrument though laser and electronics were still operational. In particular it was impossible to test its performance and accuracy at float. Surprisingly, when temperature increased at the end of the flight, after sunrise, the laser-pulse impacts on mirror "a" reverted to their initial position; but irreversible shutter failure prevented reliable absorption measurements.

The gondola landed with no domage at all, which showed that choices on the triangular structure, the table and the materials used for them, had been correct.

5. CONCLUSION

This is the first time that the Herriott multipass cavity array, composed of two identical mirrors was carried in a **balloon** gondola. The array behaved satisfactorily over a large range of temperature, even with a very long distance of 4 m between the mirrors. It is obvious that this array possesses good applicability on long-distance absorption, up to 1 km, thanks to its easy mounting and alignment even though it requires a laser source.

The low temperatures encountered during ascent were responsible of the two failures, on deflection mirrors and on shutters. Almost certainly failures should not have happened with a flight programmed during day time.

An ozone profile was derived from 20 km to 36 km, but it is only indicative, because of mechanical failures. Above 36 km, where the instrument had to provide an ozone concentration accuracy of less than 2 %, the measurements were not successful. However, theses flaws are identified, they will be corrected before the next flight planned for September 1988.

6. REFERENCES

Proffitt, M.H., and R.J. McLaughlin, 1983: Rev. Sci. Instrum., 54, 1719-1728.

Robbins, D., et al.: Measurements of stratospheric Ozone during the 1985 MAP/GLOBUS NO_x Campaign.

Weinstock, E.M., C.M. Schiller, and J.G. Anderson, 1986: Journal of Geophys. Res., 91(D4), 5237-5248.

Herriott, D., H. Wogelnik, and R. Kompener, 1984: Applied Optics, 3(4), 523-526.

Moreau G., C. Robert, 1985: J. of Optics, 16(4), 177-183.

Molina, L.T., M.J. Molina, 1986: Journal of Geophys. Res., 91(D3), 14501-14508.

THE EFFECT OF STRAY LIGHT ON TOTAL OZONE MEASUREMENTS AT BELSK, POLAND

M. Degórska and B. Rajewska-Wiech
Institute of Geophysics
Polish Academy of Science
Warsaw, Poland

ABSTRACT

Series of total ozone measurements using the Dobson spectrophotometer No. 84 were taken out to high airmass, and the results were compared with the two parameter stray light model presented by Basher (1982). These data show that the value of $\mathcal{R}o$, the model parameter which is dependent on the quality of the instrument, is close to that suggested to be typical for most Dobson instruments. The values of the other parameter, a, which is the atmosphere's relative attenuation coefficient of the stray light band to the desired band, were found to vary over Belsk from one day to another, and to have only a weak dependence on ozone amount. Preliminary examination of atmospheric transparency data indicates a relationship of the parameters with aerosol scattering.

1. INTRODUCTION

A number of factors influence the Dobson spectrophotometer's · ozone accuracy. One of the major sources of measurement error is the stray light contribution to, principally, the shorter wavelength band of a Dobson bandpair. As the value of airmass μ becomes greater, this error becomes progressively greater, which significantly limits the airmass range of the Dobson instrument (e.g. see Fig. 1.).

The "stray light", as a general term, describes the stray radiation scattered by the atmosphere in the instrument's field of view as well as the unwanted radiation scattered within the instrument. G.M.B. Dobson (in Walshaw, 1975) examined both sources of stray light, describing a detailed investigation of errors in ozone measurements made with a low sun. Basher (1982) made an useful distinction between homochromatic and heterochromatic stray light. The stray light effect varies noticeably among instruments (Olafson and Asbridge, 1981).

One of the methods of testing for stray light is taking measurements when the value of μ is more than about 3.0 and comparing the results with a stray light model. The measurements of Dobson No. 84 have been compared with the heterochromatic stray light model presented by Basher (1982). The two parameters of this model are: $\mathcal{R}o$, which is reported to range from 10^{-5} to 10^{-3}, depending (inversely) on the quality of the instrument's stray light rejection, and a , the atmosphere's relative attenuation coefficient of the stray light band, which ranges from about 0.7 to 1.2, depending on ozone amount.

The Belsk instrument No. 84 has been believed to have a stray light problem. At the very beginning of ozone measurements at Belsk, the usable range of airmasses for A, D, C bandpairs were established (Dziewulska-Łosiowa, 1965). The change in the extraterrestrial constant values and the extended range of airmasses resulting from the use of $NiSO_4$ filter were determined for the CD band combination (Łękawska-Degórska, 1972). After the International Comparison of Ozone Spectrophotometers at Belsk in 1974, corrections arising from a dependence on μ were deduced from the intercomparison with the Canadian instrument No. 77. These corrections were applied to the total ozone observation at Belsk in the years 1975-1981. However, in the republished 19-year revised total ozone data (Dziewulska-Łosiowa et al., 1983), the corrections have not been applied.

2. COMPARISON OF THE EXPERIMENTAL DATA WITH BASHER'S STRAY LIGHT MODEL

For many years the airmass dependence of Belsk instrument has been observed to vary noticeably even from one day to another. In some cases the effect appears in the AD wavelength pairs as early as μ = 2.2, while in other cases the onset may not occur until approximately μ = 3.0. Figures 1 and 2 can illustrate this variation in the airmass dependences. Thus, it is of

ISBN 0-937194-15-8

interest to find what factor it is that is the cause of the dependence and its variation.

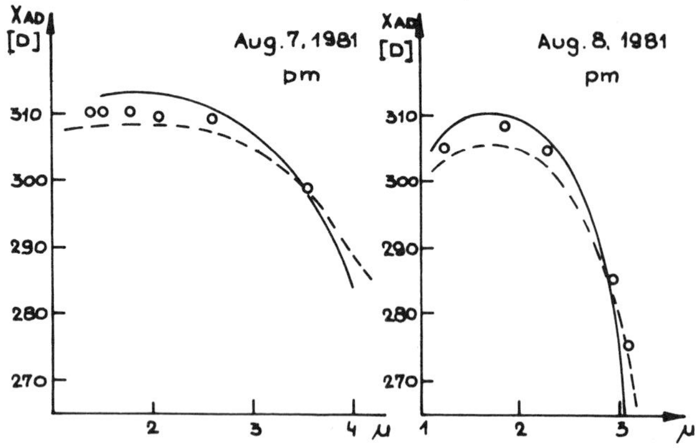

Fig. 1. Comparison of experimental data with stray light models: — $\mathcal{R}o = 10^{-3.8}$, -- $\mathcal{R}o = 10^{-4.0}$, a = 0.9 (left), a = 1.2 (right).

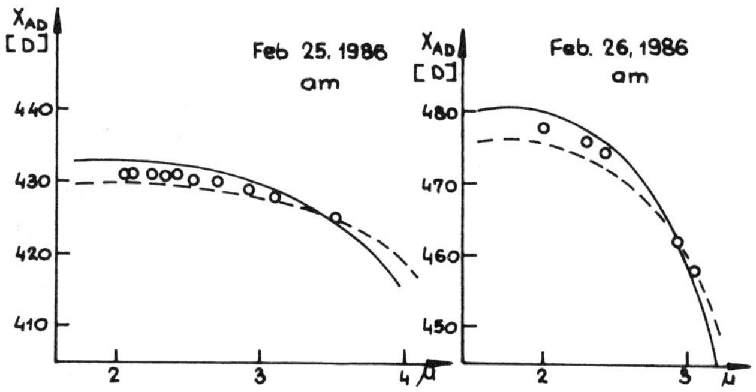

Fig. 2. Same as Fig. 1., but for: a = 0.8 (left) and a = 1.1 (right).

During recent years, a substantial number of observational data have been accumulated at Belsk with some days extending out to μ = 4.0. These data were partly published in the issues of the Publications of the Institute of Geophysics Polish Academy of Sciences and are available at the Belsk Observatory on request. Several series of the AD measurements have been selected to be compared to the values calculated by Basher's stray light model.

Our computations suggest that the Belsk instrument's $\mathcal{R}o$ value is about $10^{-4.0}$ to $10^{-3.8}$ for all the analysed measurement series, which is similar to the value reported by Basher (1982) as typical of most Dobson instruments. The parameter a ranged from 0.8 to 1.2 which is also reasonably consistent with the a values given by Basher. However, the values of parameter a show only a weak relationship with the total ozone (Table 1., Figs. 1 and 2), which is not as expected from the stray light model. A comparison of 850 AD measurements taken at the airmass of about 2.0 with those made beyond the airmass 2.5 also does not show the expected dependence of the differences on the total ozone. This may indicate the presence of interfering absorbers other than ozone, the concentrations of which appear to be highly variable. Komhyr and Evans (1980) showed that the most important of these are SO_2 and NO_2.

TABLE 1. VALUES OF MODEL PARAMETER a, TOTAL OZONE AND ATMOSPHERIC TRANSPARENCY DATA: T - LINKE'S TURBIDITY FACTOR, T_λ - SPECTRAL TURBIDITY FACTOR (392 nm EFFECTIVE WAVELENGTH), p - TRANSPARENCY, p_λ - SPECTRAL TRANSPARENCY

Date	a	X [D]	T	p [%]	T_λ	p_λ [%]
1981.02.20	1.0	402	-	-	-	-
1981.07.08	1.1	350	5.3	63	1.5	50
1981.08.07	0.9	310	3.3	75	0.6	74
1981.08.08	1.2	306	4.9	65	1.1	62
1981.09.07	1.0	310	5.0	69	1.4	53
1981.10.09	1.0	305	-	-	-	-
1981.10.29	1.0	320	4.0	72	1.0	65
1984.07.11	1.1	325	4.5	68	1.4	56
1985.06.04	1.1	361	3.2	76	1.1	69
1985.06.05	1.1	365	2.9	78	1.0	71
1986.02.25	0.8	430	3.1	77	1.2	60
1986.02.26	1.1	475	3.5	74	1.3	58
1986.03.17	0.8	308	-	-	-	-
1987.02.26	1.0	411	2.4	82	0.9	69
1987.02.27	1.0	398	2.3	83	0.9	68
1987.03.12	1.0	412	3.1	77	1.1	63
1987.06.18	1.2	403	6.6	54	2.4	35
1987.07.07	1.0	339	3.3	77	1.1	63
1987.07.08	1.1	353	4.0	70	1.3	57
1987.07.15	0.9	353	2.6	80	0.8	71

Total NO_2 amount measurements were made using the Dobson spectrophotometer at Belsk in 1977-1979 (Dziewulska-Łosiowa and Rajewska-Więch, 1980). The obtained values ranged from 5 to 18 m atm-cm. Such total NO_2 amounts have been estimated to cause the Dobson spectrophotometer ozone measurement error of only about 1% (Komhyr and Evans, 1980). Special attention should be paid to SO_2, the other atmospheric gas absorber that can affect ozone measurements, and which is the principal precursor of H_2SO_4 aerosols.

Furthermore, the scattering by aerosols, especially under very hazy conditions, may also contribute to the discrepancy in the model.

The atmospheric turbidity data presented in Table 1 suggest a relationship between the aerosol scattering and the values of the attenuation parameter a analysed in terms of the Basher approach.

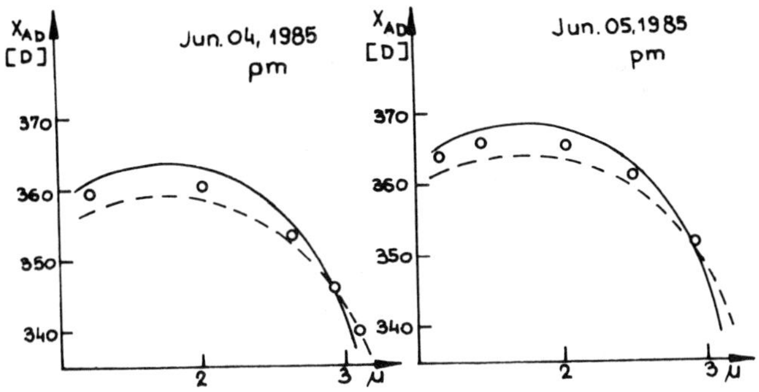

Fig. 3. Same as Fig. 1., but for: a = 1.1

Figure 3 presents the airmass dependences of total ozone AD measurements taken on two consecutive days. Values of model parameters are the same for both data series. Table 1 shows that the atmospheric transparency values are nearly the same on each of the days. However, it may be seen in Figs. 1 and 2 as well as in Table 1 that when the atmospheric transparency values decrease, the a parameter values increase.

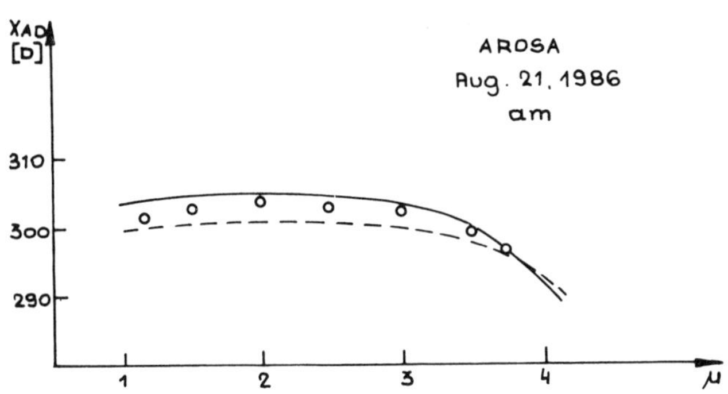

Fig. 4. Same as Fig. 1., but for : a =0.7

Figure 4 shows one of the series of the total ozone measurements made with the Dobson instrument No. 84 during the International Comparison of Dobson Spectrophotometers at Arosa, Switzerland, in 1986. Values of the atmospheric parameter a range from 0.7 to 0.8 for all the series of ozone measurements performed at Arosa. Correspondingly, the instrument's operating range of airmass extended beyond the airmass 3.0. These few series of measurements taken at clear air in high mountains seem to support the results obtained from the ozone measurements performed at Belsk.

Concluding, it may be suggested that, because of stray light and high aerosol amounts, in polluted regions the Dobson ozone values obtained beyond the airmass 2.2 should be treated with caution. These may lead to errors in ozone records which could cause fictitious negative trends.

ACKNOWLEDGEMENTS.

Our sincerest gratitude is due to Dr. Reid E. Basher for his valuable comments, to Anna Michałowska-Smak for providing us the atmospheric transparency data and to Dr. Regina Hryniewicz for helpful discussions.

REFERENCES

Basher R.E., 1982: Review of the Dobson spectrophotometer and its accuracy. WMO Global Ozone Research and Monitoring Project. Rep. No. 13, World Meteorological Organization, pp. 90.

Dziewulska-Łosiowa A., 1965: Pomiary całkowitej zawartości ozonu atmosferycznego w Obserwatorium Geofizycznym w Belsku i ich analiza. Publs. Inst. Geophys. Pol. Acad. Sc.(Materialy i Prace), 5, 65 -74.

Dziewulska-Łosiowa A., Degórska M., Rajewska-Więch B., 1983: The normalized total ozone data record - Belsk, 1963-1981. Publs. Inst. Geophys. Pol. Acad. Sc., D-18 (169), 23-73.

Dziewulska-Łosiowa A., Rajewska-Więch B., 1980: The results of the three-years series of the measurements of total NO_2 amount at Belsk. Publs. Inst. Geophys. Pol. Acad. Sc., D-11 (141), 105-110.

Komhyr W.D., Evans R.D., 1980: Dobson spectrophotometer total ozone measurement errors caused by interfering absorbing species such as SO_2, NO_2, and photochemically produced O_3 in polluted air. Geophys Res. Lett., 7, 2, 157-160.

Łękawska-Degórska M., 1972: An analysis of the accuracy of atmospheric ozone observations made at Belsk Observatory, 1971. Publs. Inst. Geophys. Pol. Acad. Sc. (Materialy i Prace), 56, 55-58.

Olafson R.A., Asbridge I.A., 1981: Stray light in the Dobson spectrophotometer and its effect on ozone measurements. Proc. Quad. Int. Ozone Symp., J. London (ed.), Boulder, Colorado, USA, vol. 1, 46 - 47.

Walshaw C.D. (ed.), 1975: Papers of Professor G.M.B. Dobson F.R.S. Publs. Inst. Geophys. Pol. Acad. Sc., 89, 61-115.

BACKSCATTERED UV OZONE AND AEROSOL MEASUREMENTS FROM THE GEOSTATIONARY ORBIT: FEASIBILITY AND INTERPRETATION

C. Muller and P.C. Simon
Institut d'Aéronomie Spatiale de Belgique
B-1180 Bruxelles, Belgium

ABSTRACT

The implications of the geostationary geometry on the feasibility of a backscattered U.V. monitoring instrument are studied at wavelengths ranging from 250 nm to the blue region of the spectrum. Simulations of the data show a coverage from the equator for 80 degrees of latitude, the technique is shown to be very effective for the monitoring of upper stratospheric and total ozone, however total aerosols, NO_2 and tropospheric ozone will be difficult to survey due to the strong effects of clouds and varying ground albedo. Specific channels must however be included for the study of these phenomena so that the obtained data can be fed back in order to refine the ozone inversion process.

INTRODUCTION

Up to now, remote sensing of stratospheric ozone has been made from low Earth orbit (altitude lower than 1200 km) which limits both the geographical area that could be observed at any time (200 km for SBUV) and the time scale during which one scene can be continuously observed. The use of the geostationary orbit will allows to obtain synoptic views of both total ozone contents and its vertical distribution above 35 km over one half of the Earth surface on time scales of a few hours. This type of data is crucial to any studies of ozone distributions related to atmospheric dynamics. The high negative correlation that exists between total ozone content and tropopause heights can be used to study short-term variations in the dynamics of the upper troposphere. Synoptic observations of the distribution of total ozone content will permit to identify regions of large gradient which are related to the location of the polar jet stream. Therefore ozone observations from a geostationary orbit will complement the meteorological data obtained from the same platforms.

The observation of the back scattered solar ultraviolet radiation is well adapted for monitoring ozone levels from a geostationary orbit; it is a well established remote sensing technique, based on physical processes, which have been extensively studied. In addition, with the experience of previous back scattered radiation instruments used in space to drawn in, a very good assessment of the technique is possible producing a clear identification of instrumental problems.

ASSESSMENT OF THE IMPACT OF THE GEOSTATIONARY ORBIT ON THE BACK SCATTERED ULTRAVIOLET OZONE OBSERVATIONS

The geostationary orbit is very different in respect of backscattered light observation from the nadir looking geometries which have been the rule in previous ozone observation instruments. Figure 1 gives the different angular parameters which are important for such an observation, in the case of 12h U.T., the satellite being positionned on the Greenwich meridian, the angle of incidence of solar radiation on an observed ground location is then given by its latitude and the solar declination of the day. For example, at the summer solstice, the solar elevation is equal to the sum of 23° and the difference between 90° and the latitude. An other important point is the latitude where the observation is no longer possible. It is expressed by :

$$\phi_{max} = \text{arcos } (R/z + R)$$

where R is the Earth radius and z the satellite altitude which in the case of the geostationary orbit is 81°. It means that the Antarctic ozone "hole" which has been discovered at Halley Bay (76°S) and Syowa (69°S) can be followed from the geostationary orbit as well as the high ozone belts which surround the Antarctic vortex in the Spring and Summer.

A simulation program has been endeavored in order to determine the radiation levels expected from a BUV instrument in the meridian case given in figure 1. The atmosphere is divided in 1 km layers from the ground to the altitude of 100 km and the radiation is computed in the direction of the satellite taking into account, as a source, the Rayleigh scattering of UV solar radiation and, as sinks, the losses by ozone absorption, Rayleigh scattering and losses by aerosol extinction and other absorbing gases. This permits to compare the intensities in the different wavelengths with extraterrestrial solar energy and gives the necessary input to instrument designers, the maxima of the Rayleigh phase functions being in forward and backscattering, this meridian geometry gives the maximum of radiation possible. The results are indicated on figure 2 for the wavelengths of 250 nm (corresponding to the ozone maximum) and 280 nm. The ozone values have been varied above the altitude of 40 km to cover the ranges from 100% of current value down to 20%. It is seen that ozone variations are well observed from 0°

ISBN 0-937194-15-8

to 80° latitude using this technique, however the sensitivity decreases at high latitudes and, as expected, latitudes higher than 67° cannot be observed at the Winter solstice.

The following step was to introduce in the computation the effects of albedo, NO_2 and aerosols at these two wavelengths (250-280) and at 300 nm, 350 nm, 435 nm and 440 nm. The wavelength of 300 nm corresponds to a range typically used for total ozone observations while 350 nm is at the extreme limit of the ozone spectrum measured in the laboratory. The two last wavelengths are a couple which can be used for NO_2 determinations. The albedo is represented here by a reflecting layer of 0.4 efficiency, corresponding to a value for a snow cover or complete cloud cover. At 250 nm, no effects are obtained while at 280 nm, this reflecting layer has to be elevated up to 20 km to produce a significant difference in received signal. At 300 nm, the results for a 10 km and a ground layer are shown in figures 3 and 4. One important result is that ozone variations above 40 km still have an influence on the results of this channel. In the 350 nm case, ozone absorption does not affect the results anymore and this wavelength can already be used as albedo monitor, the results being evidently as good for this purpose at wavelengths as 360 and 400 nm which are known to be completely free of ozone absorption.

The effects of NO_2 have been considered at 435 and 440 nm, the computation performed at 440 nm used two NO_2 vertical distributions, one was as determined by the Spacelab 1 grille spectrometer and the other one was a factor of ten higher. The results are compared with the NO_2 = 0 value and unfortunately, the observation seems to require a much higher sensitivity than the one necessitated for ozone. This can only be achieved through correlation between a NO_2 minimum and maximum, absolutely requiring two channels (figure 5).

Aerosols were introduced in the form of a unimodal distribution of particles of a scattering efficiency of 2 (corresponding grosso-modo to 0.12 μm radius stratospheric aerosols at 440 nm). Three distributions were introduced : first a uniform distribution of 2.5 particles/cc, second, 10 particles/cc and finally 10/cc plus a stratospheric layer of 100 particles/cc between 17 and 23 km. The results on figure 6 show an effect at 440 nm which is much more significant than the one of NO_2, fortunately the aerosol extinction and scattering wavelength dependence is weak compared to the NO_2 and even ozone spectral variations. At high latitudes, especially, in the Summer solstice case, the aerosol effects tend to cancel itself, as both a source and an absorber of light. Their importance makes them a critical factor of all inversion algorithms, as their phase function is critically dependant on their size distribution, this one will have to be determined first and this

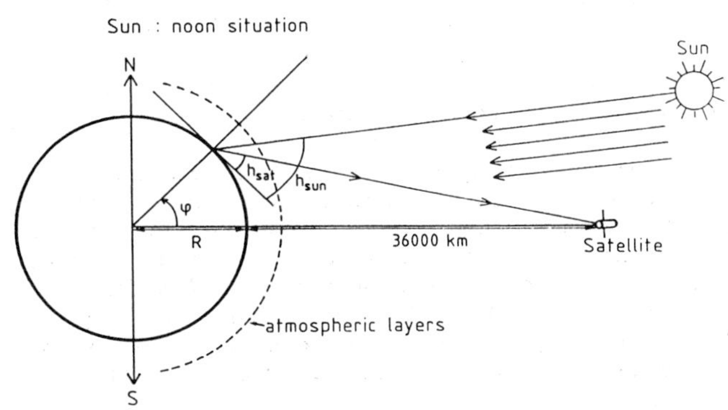

Fig. 1. Geometrical GBUV situation, the satellite at geostationary altitude, sees at noon the backscattered solar radiation over an entire hemisphere.

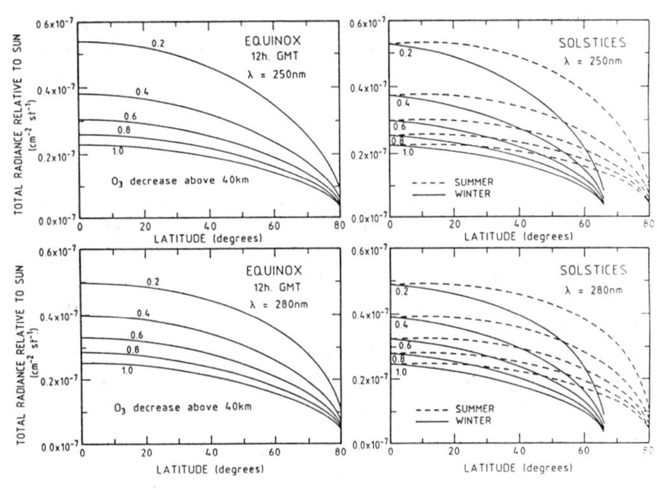

Fig. 2. Computation of the radiation received at the geostationary altitude in terms of solar radiance at 250 and 280 nm, the different curves correspond to an 03 variation above the 40 km altitude, the lower curve corresponding to present and the upper curve corresponding to an eighty percent ozone decrease. The indicated latitudes show that the Antarctic ozone hole can be monitored using this technique.

requires several wavelengths, preferably as far apart as possible and decoupled from ozone absorptions. Unfortunately, due to the importance of the albedo parameter, nadir sounding will always yield coarser results on aerosol distributions than limb sounding as exemplified by the SAGE satellite and aerosol models deduced from simultaneous specialised observations will always have to be tested against the GBUV aerosol data in order to check its accuracy. However, GBUV will provide unique data on long term trends and exceptional events as volcanic explosions.

A further computation must include multiple scattering effects, realistic albedo phase functions and aerosols distributions and should also include variations of the hour angle. However the present computation, if insufficient to provide a data base for the testing of inversion algorithms, provides the range of possible values for light received at the satellite.

This analysis shows that a range from 1% of direct solar radiation down to 10^{-8} has to be measured with accuracies of better than 1%. The albedo of the ground and cloud cover are found to be a major component of the signal above 280 nm as well as aerosols, which, as was already well known from previous visible NO_2 observations, act both as a source of scattered light and absorber. Ozone, especially above 40 km, can be efficiently monitored by the technique using simple algorithms where the layers below 20 km are crudely modelled while monitoring of ozone below the maximum and minor constituents (NO_2) will demand an extensive study of aerosols and albedo data from the observation itself before a final version.

CONCLUSIONS

As several new geostationary meteorological satellite programs are now in the course of evaluation within the large space agencies, attention should be given to the use of these platforms for monitoring of long term climatic parameters and especially the ozone. The back-scattered U.V. technique permits a large number of possible designs for spectral imaging radiometers using the new developments of detector arrays. One of these instruments used from the geostationary point would generate maps of total ozone and vertical distributions on an entire hemisphere, ideally completing the lower-altitude present sounders which are limited by their ground-track coverage.

Aerosol and other gases measurements are more difficult with this instrument than by using a limb sounder due to the strong albedo signal present outside the ozone absorption, however they can and must be achieved using the same instrument in order to refine the ozone inversion algorithms, by including all possible perturbations.

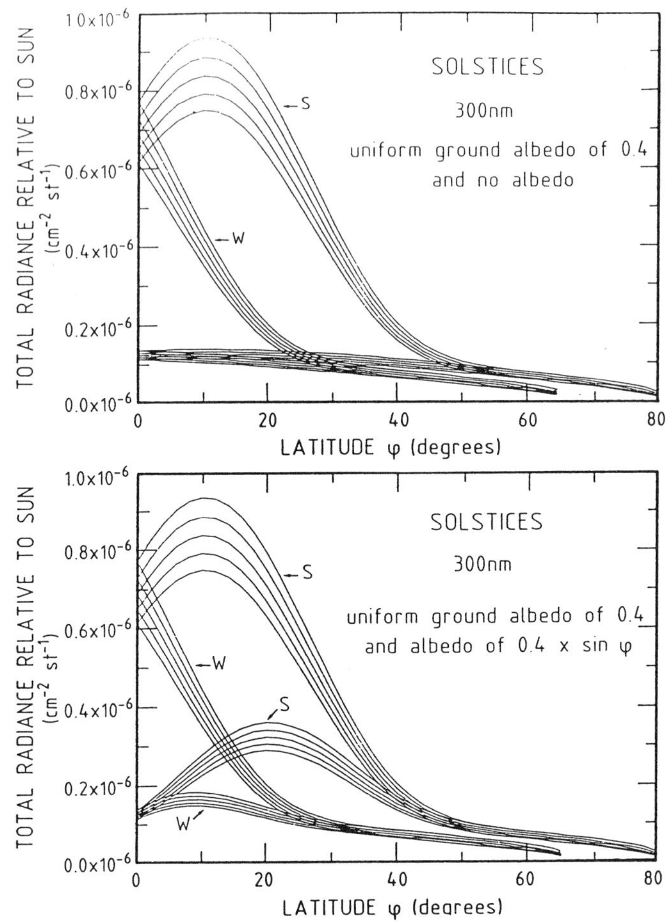

Fig. 3. Effect of a 10 km altitude reflecting layer of albedoes of 0.4 and 0.1 on the signal level at 300 nm, the different curves correspond to an ozone variation over 40 km and to the Summer and Winter equinoxes as in figure 2.

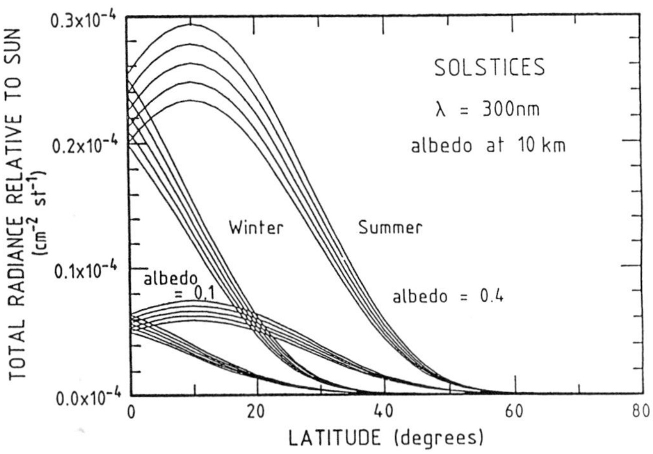

Fig. 4. Effect of a ground reflecting layer on the signal level at 300 nm, two computations are shown in order to determine the incidence of an albedo latitude variation. The other indications correspond to Fig. 2.

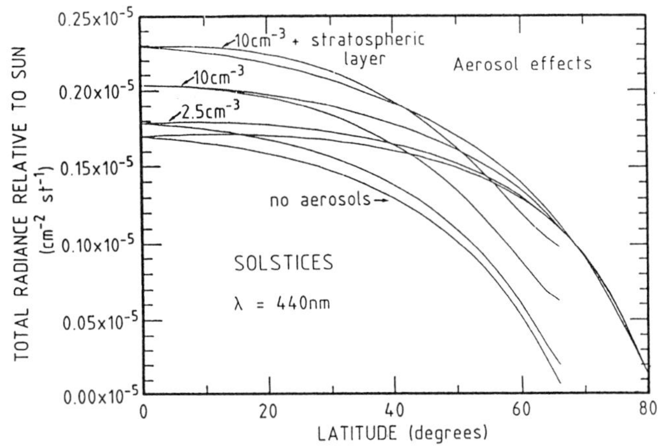

Fig. 6. Aerosol effects for three different aerosol models, unfortunately, the effects tend to cancel themselves at high latitudes, while at low latitudes outside the ozone spectral region, the aerosol variations, if optically thin are difficult to distinguish from the ground albedo effects.

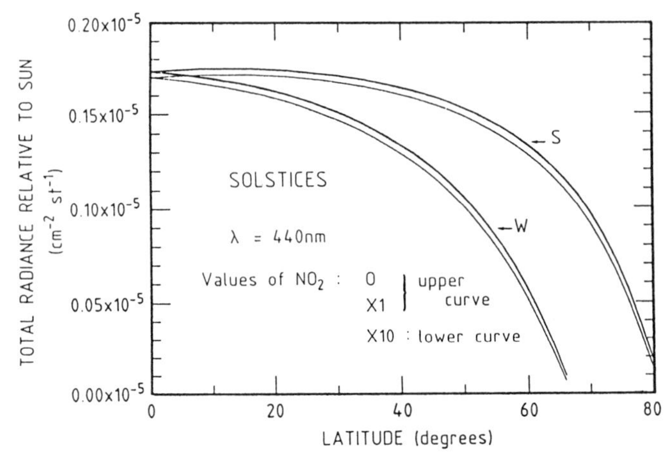

Fig. 5. Effect of a NO_2 variation of a factor of 10 on the received signal, this effect, in absolute, is negligeable compared to albedo variations and differential measurements are necessary in order to observe NO_2.

AN AUTOMATED SPECTRAL COMPLEX FOR MEASUREMENTS OF ATMOSPHERIC OZONE AND UV RADIATION

V.M. Dorochov, V.U. Khattatov, T.N. Klimova, A.V. Pankratov, and V.A. Torgovichev
Central Aerological Observatory
The State Committee of the USSR for Hydrometeorology
Moscow, USSR

ABSTRACT

Presented are the characteristics of newly developed automated spectral complex for atmospheric ozone and UV-radiation monitoring. The results of measuring ozone with this instrument are intercompared with simultaneous Dobson spectrophotometer No 107 measurements. The theoretical and experimental studies conducted show that the new instrument and multiwave technique make it possible to determine total ozone with accuracy not worse than that of Dobson spectrophotometers. Presently, these new instruments have became commercially available.

1. INTRODUCTION

Currently, the ground-based ozonometric network is equipped with Dobson spectrophotometers and filter ozonometers M-124. The Dobson instruments having stopped to be produced, the available ones aging, and there being the need to extend the geography of ozonometric observations. In Canada there have been developed and made commercially available automated Brewer spectrophotometers (Kerr et. al., 1984). Now there are 45 Brewer instruments operating in 12 countries. To determine total ozone from spectrometry observation data obtained by network ozonometers, there are used the calculation algorythms based on the difference technique using one or a combination of several wavelength pairs in 305-340 nm waveband. One of the basic disadvantage of this technique is the inaccuracy of taking into account of the selective properties of radiation attenuation by atmospheric aerozols. A number of recent paper (Saunders et.al., 1984, Dorokhov et.al., 1985) have shown that this drawback could be almost completely removed and total ozone observation accuracy increased through measurements of direct sun radiation spectrum in a shorter wavelength range and using multi-wave spectrogram processing algorythms.

In this report are presented the characteristics and results of testing a prototype automated spectral complex (Gikkel et al., 1987) for determination of the following optical atmospheric parameters: total amount and vertical distribution of ozone; levels, spectral distribution and biologically effective values of natural UV-radiation; total amount of some trace gases with specific absorption spectrum in the range of 300 - 600 nm (SO_2, NO_2 etc.); spectral atmosphere transparence and characteristics of aerosol turbidity. Simultaneous determination of these pa-

ISBN 0-937194-15-8

rameters is necessery for an early detection of changes in the ozone layer caused by natural and anthropogenic factors.

2. AUTOMATED STATIONARY UV-SPECTROPHOTOMETER SUVS

SUVS is constructed out of three main parts an optico-mechanical block (OMB), a control panel, and a microcomputer. SUVS is also equipped with a set of alignment and control and testing instruments.

Fig. 1. The general view of the automated UV-spectrophotometer SUVS

OMB (see Figure 1) is based on a double scanning monochromator (f/6, 1200 lines/mm) in a thermostabilized casing. A block of standard slits provides the discrimination of a spectral intervals of 0.35 to 1.7 nm in the range of 250 .. 600 nm with a discrete scanning step of 0.5 nm. OMB has three switched optical inputs. Direct sun radiation is received by a telescope (ø 100 mm)with a automated two-mirror tracker (accuracy 0,5). To observe the spectral fluxes of scattered solar radiation there is a second optical channel with a cosine integrating head (ø 200 mm) at the input, having a flap shading screen. Also important in the OMB is a built-in reference light source, a block of light filters, a modulation chopper with as a photomultiplier block with a feedback system. A/D control panel scheme provides the pre-setting and visual control of the OMB operation regime. The control panel also performs preliminary signal processing and presentation in the form fit for input to the microcomputer "Electronica-60".

To conduct observations OMB is placed on an open site and controlled by an operator from a laboratory room, using a two-way communication line.

Preliminary processing a photomultiplier signal consists of synchronous detection with accumulation and spectrum intercomparision of measured solar radiation with the radiation of a stable reference source at each step of monochromator scanning. This makes it possible to achieve high sensitivity in the detection of week light sources characteristic of the short-wave range of solar radiation spectrum and facilitates the setting of the instrument spectral sensitivity stabilization. The principle of spectrum intercomparison facilitates the task of determing the absolute values of the energy illumination spectral density of UV-radiation. With such a measurement method used basic metrological spectrophotometer characteristics are depended only on the reproducibility of the spectrum form and radiation intensity of the built in reference source.

3. THE TECHNIQUE OF DETERMING TOTAL OZONE

To calculate total ozone from SUVS measurement data on direct sunlight, an improved multiwave technique(Dorokhov VM.

et al. 1985) is now used by the authors:

$$X = \frac{\sum_{1}^{N}(f_i - f_p)(\alpha_p - \alpha_i)}{\mu \sum_{1}^{N} (\alpha_p - \alpha_i)^2} \qquad (1)$$

$$f_i = \ln E_i - \ln E_i^{o} - \beta_i m$$

The index "i" or "p" indicates current wavelength (λ_i) or reference wavelength λ_p .

X - total ozone in the atmosphere,

E, E^{o} - the intensity of measured direct sun radiation and its value beyond the atmosphere.

α - ozone absorption coefficient

β - optical density of Rayleigh atmosphere

μ, m - slant optical ozone and Rayleigh masses.

The parameters E_i^{o} and α_i are calculated from published data (Bass et al., 1984) through their reduction by the instrument spectral transmission function.

The main difference of our technique of total ozone determination from earlier ones (Saunders et al., 1984; Kuznetsov, 1975) consists in the application of a least square method only for the minimization of the root-mean-square deviation of one, the most informative parameter (X) in solving a system of linear logorithmic equations. This improves significantly the resistance of the equation system solution system solution to the measurement track noise.

The minimization of an aerosol effect on the results of total ozone determination is attained by using the spectrum range 295-315 nm where ozone absorption of solar radiation considerably exceeds aerosol attenuation. Figure 2 shows the results of calculating the aerosol correction, ΔX_a to X depending on the selection of an operation wavelength range position.

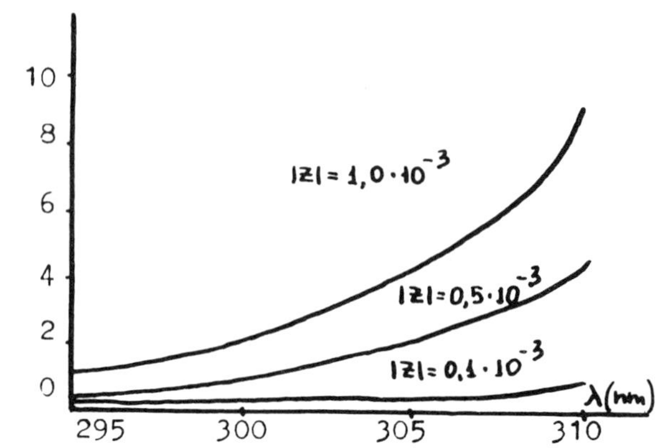

Fig. 2. The dependence of the total ozone aerosol error on the position of the operation wavelength range the width of the range of 15 nm, $\lambda = \lambda_p + 15$ nm

Here the same width of this range is supposed 15 nm, while the wavelength correspond to the begining of scanning the reference wavelength being at the long-wave end of the range $\lambda_p = \lambda + 15$nm. It can be seen that the error of total ozone determination using (1) does not exceed 0.5% for the scanning spectrum range shorter than 315 nm. To lessen this error caused by SO_2 and NO_2 selective absorption in the atmosphere, calculations were made to select an optimum set of the wavelengths of this range. At a certain choise of λ_i , the errors caused by SO_2 can be reduced to 1.5 % even under heavy atmospheric pollution and made, on the average, by an order of magnitude less than similar errors for Dobson spectrophotometers (Basher, 1982). Total ozone errors values measured by SUVS are erroneous due to NO_2 two times less than those measured by Dobson instruments. Besides, the SUVS errors are of different sign with SO_2 errors, which contributes to their partial compensation. The intercomparison of calculated errors for SUVS and Dobson instruments in the determination of other parameters affecting the accuracy of determining total ozone is presented in Table 1.

TABLE 1. INTERCOMPARISON OF THE CONSTITUENTS OF RANDOM ERRORS IN TOTAL OZONE DETERMINATION BY DOBSON AND SUVS INSTRUMENTS

	Dobson	SUVS 88
1. The error of measuring radiation intensity	0,5%	1%
2. The error of extraterrestrial constants determination	1%	2%
3. The error of air and ozone masses determination	0,5%	0,5%
4. The ozone absorption coefficient uncertainty	0,5%	0,5%
5. The reproducibility of setting the wavelength scale	0,05nm	0,8nm
$\Delta X/X$	1,6%	0,8%

The estimates of the components of these errors for Dobson spectrophotometers are borrowed from Basher (1982) while for SUVS presented are the experimental data on the metrological characteristics of the prototype instrument.

4. THE RESULTS OF TESTING SUVS

The technical characteristics of SUVS instruments in the recording of spectra in the shortest wavelength part of the UV - B range and the accuracy of total ozone determination were obtained in experimental studies at the ozonometric station "Moscow" in the spring and summer of 1987.

During these experiments there were performed simultaneous measurements of total ozone in the atmosphere, using Dobson spectrophotometer No 107, network filter ozonometer M-124 and a SUVS prototype instrument.

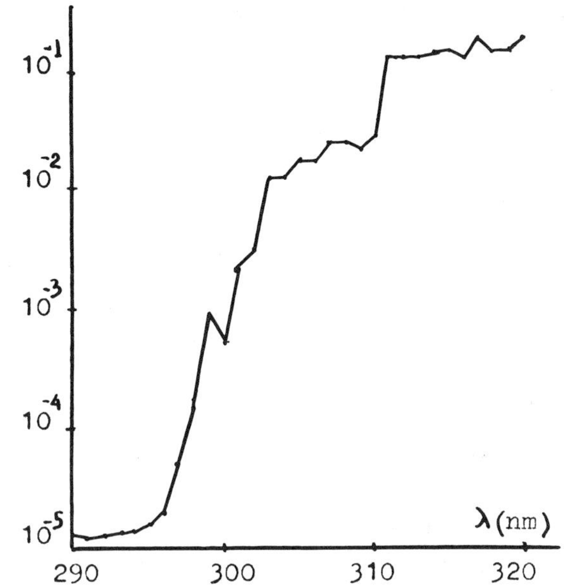

Fig. 3. The SUVS spectrogram of direct sun radiation $X = 360$ DU, $m = 1,8$

For example, Figure 3 shows sun radiation spectrogram obtained by SUVS. Figure 4 gives the results of total ozone determination using formula (1), depending on the shift of the initial wavelength (λ) of the scanning range $\Delta\lambda$ = 15 nm for which total ozone was estimated. Here also are given the similar results calculated by using G.I.Kuznetsov's formula (Kuznetsov, 1975).
The vertical segments in these graphs indicate the magnitude of $\overline{X}$ values dispersion from the results of intensified observation series (10 spectrograms). From the analysis of these measurements it might be concluded, which agrees with the relevant model calculation, that when the scanning range shifts to a short-wave spectrum range the accuracy and reproducibility of total ozone data improves considerably, i.e. three-

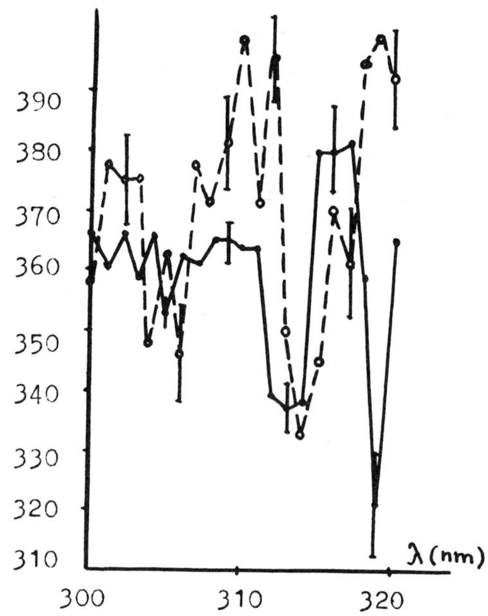

Fig. 4. Intercomparison of total ozone data for different spectral ranges calculated by formula (1) and G.I.Kuznetsov's (1975) formula. 1 - X values by Kuznetsov's formula, 2 - X values by formula (1) of this paper.

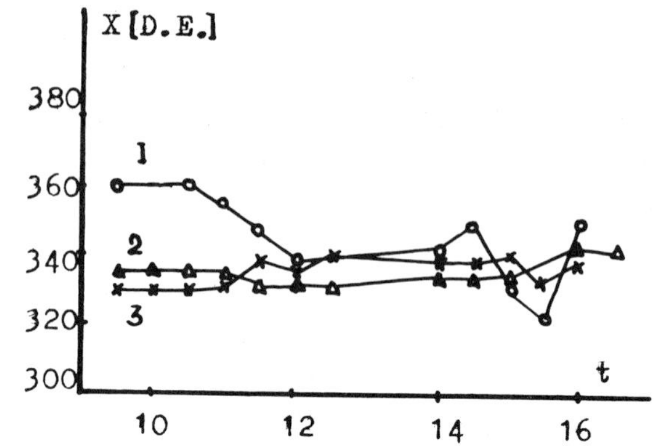

Fig. 5. Total ozone obtained by Dobson spectrophotometer (1), SUVS (2), and filter ozonometer M-124 (3) throughout the day.

fold with the shift from a 310...325 nm spectrum range to a 300...325 nm. Also observed is a trend of total ozone undertimation through calculations by formula (1) using long-wave spectrum range data. This seems to be connected with the effect of an aerosol correction and agrees well with the theoretical data(Figure 2). Total ozone values calculated by the formula from Kuznetsov (1975) are almost independent to the effect of an aerosol error, the latter being automatically taken into account. However, these data have a considerable spread depending on the selection of a wavelength range. This is due to the instability of this algorithm of a linear equation system solution.

Figure 5 illustrates the compatibility of total ozone data obtained by Dobson instrument No 107 using an AD pair, M-124 (an integral method), and SUVS by formula (1) for λ =306...315nm and N = 10. It can be seen that the random deviations of single total ozone observations from the mean total ozone value, obtained from Dobson spectrophotometer and SUVS data, do not exceed 1 % with confidence probability of 0.95. M-124 mean diurnal totao ozone values are close to those of the other two instruments, but the single observations conducted throughout the day are made by a strong variation, particularly for large air masses. It will also be noted that the SUVS total ozone data in Figure 5 were obtained after the correction of its extraterrestrial constants calculated from the results of its intercomparison with Dobson spectrophotometer No 107 at the same station. This correction was necessitated by the earlier 3 % difference between readings of the two instruments. However, it is not at all imposible that this systematic constituent of the error is connected with the difference between the effective coefficients of ozone absorption,assumed for the two instruments. Note also that the extraterrestrial constants of the Dobson spectrophotometer used were earlier verified by the authors by plotting Bouguer straight lines for stable optico-atmospheric conditions of a Pamir high mountain region (H = 3,2 km). The other metrological parameters of Dobson spectrophotometer No 107 were verified by the technique adopted by the world Ozonometric Network. As a result of these studies Dobson spectrophotometer No 107 was metrologically certified by the USSR State Standards Committee as a reference instrument for total ozone measurements with an accuracy of not worse that ±11DU.

5. CONCLUSION

The conducted theoretical and experimental studies show that newly developed automated spectral complex and multiwave technique make it possible to determine total ozone in the atmosphere with an accuracy not worse than that of Dobson spectrophotometers. The application of SUVS instruments at ozonometric stations of an observatory type is promising for the improvement of the ground-based observation network and control of the ozonosphere state. Presently, these instruments have become commercially available.

REFRENCES

Basher, R.E., 1982: Review of the Dobson spectrophotometer total ozone measurement accuracy. Report No 13,WMO.

Bass, A.M., Paur R.J., 1984: The ultraviolet cross-section of ozone Proc. Quadr.Ozone Symp., Greece, 1984.

Dorokhov, V.M., Klimova T.N., V.A.Torgovichev, Khattatov V.U, 1985: On a possibility to improve the accuracy and informativity of ozonometric observations using UV-spectrophotometers. Abstracts, IIIrd All-Union Conference on Aerology, Dolgoprudny, 102-103.

Gikkel, A.A., V.A.Maizlik, G.P.Maximov, L.N.Mishin, G.S.Revsin, O.N.Semchenko, V.M.Dorokhov, T.N.Klimova, V.A.Torgovichev, V.U.Khattatov, 1985: A Ground-based spectral Complex for the Ozonosphere State Observations. In: "Atmospheric Ozone" Proceedings. VIth All-Union symposium. Leningrad, Gidrometeoizdat, 1987, 36-43.

Kerr, J.B., C.T.McElroy, D.J.Wardle, R.A.Olafson, W.F.J.Evans, 1984: The automated Brewer spectrophotometer. Proc.Quadr.Ozone Symp., Greece, 396-401.

Kuznetsov, G.I., 1975: A multiwave technique and apparatus to study atmosphere ozone and aerosol. Izv. USSR Acad. of Sci. "Physics of Atmosphere and Ocean, 11, 6, 647-651.

Saunders, R.D., H.J.Kostkowski, A.E.S. Green, J.F.Ward and C.H.Popenoe, 1984: High-precision atmospheric ozone measurements using wavelength between 290 and 315 nm, J.Geophys. Res., 89, D4, 5215-5226.

TROPOSPHERIC OH CONCENTRATION MEASUREMENTS BY LASER LONG-PATH ABSORPTION SPECTROSCOPY

J. Callies, H.-P. Dorn, U. Platt, and D.H. Ehhalt
Kernforschungsanlage Jülich GmbH
Institut für Atmosphärische Chemie
D-5170 Jülich, FRG

ABSTRACT

Tropospheric hydroxyl (OH) radical concentrations were measured in Jülich during July 1987 by broadband laser long-path absorption (laser-LPA) spectroscopy observing the $Q_1(2)$, $Q_{21}(2)$, $Q_1(3)$, $Q_{21}(3)$, and $P_1(1)$ rotational absorption lines around 308 nm.
The OH absorption spectra were recorded using a high resolution multichannel spectrometric detector.
Simultaneous to the OH measurements the concentrations of NO_2, HCHO, and SO_2 were monitored by a second LPA spectrometer. O_3, aerosol concentrations, meteorological parameters, and absolutely calibrated photolysis frequencies of O_3 and NO_2 were also recorded. Carbon monoxide (CO), methane (CH_4), and light hydrocarbon concentrations were determined by gas chromatographic analysis.
Three measured OH diurnal profiles reached peak values of $(5 - 8) \cdot 10^6$ OH/cm^3. Two of the profiles are asymmetric with respect to the course of the O_3 photolysis frequency, which is a measure for the primary production of tropospheric OH radicals. This asymmetry can be qualitatively explained by a strong decrease in the NO_2 concentration during the morning hours.

1. INTRODUCTION

The reaction of the OH radical with atmospheric trace gases is one of the most important chemical processes in the troposphere. For many species this reaction initiates a chain of oxidation reactions, of which it forms the rate limiting step. Therefore, the OH concentration determines the atmospheric life time and thus the concentration of a large variety of trace gases. The high reactivity of tropospheric OH has two consequences: First, its concentration is very low with noontime peak values rarely exceeding 10^7 molecules/cm^3. Therefore, the measurement of OH radicals in the ambient atmosphere makes high demands on the detection methods. Second, the steady state OH concentration is established within minutes (Perner et al., 1987). Thus, OH chemistry takes place on a *local* scale. Its transport can be neglected. Due to the sensitivity of the OH concentration to changing atmospheric parameters (for example: trace gas concentrations, photolysis frequencies) OH measurements are ideally suited for testing and improving photochemical reaction models (*box models*). Meaningful comparison of calculated and measured hydroxyl radical concentration is, however, only possible, if the parameters that determine the actual OH concentration in ambient air are measured simultaneously to OH.

2. THE OH-SPECTROMETER

The tropospheric OH radical concentration was measured by laser long-path absorption spectroscopy observing several of the well resolved strong rotational lines ($Q_1(2)$, $Q_{21}(2)$, $Q_1(3)$, $Q_{21}(3)$, and $P_1(1)$ of the $A^2\Sigma^+(v'=0)$ - $X^2\Pi(v''=0)$) transition in the vicinity of 308.0 nm. The method and the laser system have been described in detail by Hübler et al., 1984, thus only the recent improvements, which concerned mainly the receiving system, will be described in more detail. The light source consists of a frequency doubled picosecond dye laser, synchronously pumped by a modelocked argon ion laser (fig. 1a). The separated ultraviolet light is expanded by a telescope to reduce its divergence and to keep the level of self-generated OH radicals well below the detection limit. The beam is reflected back to the laboratory by a plane reflector 2.9 km away (total light-path length 5.8 km). The received laser light is focused onto the entrance slit of a high resolution double spectrometer (fig. 1b).
The spectrum is recorded by a linear silicon photodiode array (Reticon RL 1024 S) and transferred to a microcomputer for signal averaging and further data evaluation. In order to adapt the dispersion of the spectrometer to the physical dimensions of the detector it is necessary to magnify the spectrum about ten-fold (fig. 1b).
The *detection limit* of the instrument is determined by the noise in the spectrum. The fundamental limit is set by the photoelectron noise (shot noise) of the detector. Using a specially developped (Callies, 1988; Dorn et al., 1988) low noise electronics for control and read-out of the detector information, the observed noise level is reduced to only 20% above the shot noise limit in laboratory experiments. Under atmospheric conditions the noise is

ISBN 0-937194-15-8

about 3 times higher. Part of that excess noise appears to be due to air turbulences, which cause a varying illumination of the diode array (optical noise). This noise determines the lower detection limit (LDL) for OH radicals. With the present light path length of 5.8 km a LDL of $1.5 \cdot 10^6$ OH/cm^3 is reached at 60 min integration time observing the $Q_1(3)$ line and $1.3 \cdot 10^6$ OH/cm^3 observing the $Q_1(2)$ line.
The multichannel detector provides a scanning range of 0.3 nm, which is sufficient to cover all of the OH rotational lines mentioned above. Because of the limited emission-bandwidth of the laser system (0.15 nm), we could either observe the $Q_1(2)$, $Q_{21}(2)$ or $Q_1(3)$, $Q_{21}(3)$, $P_1(1)$ line groups.
The increased spectral range allows a more reliable identification of interfering gases like SO_2. This is important, since in industrialized regions SO_2 is roughly 10^5 times more abundant (typ. $\simeq$ 1 - 50 ppbv) than OH during daytime. Although its absorption coefficient is about 4 orders of magnitude lower, in polluted air the atmospheric absorption spectra can be dominated by SO_2. CS_2 and HCHO which also have absorption lines in this wavelength region are of minor influence because of their lower absorption coefficients and concentrations in ambient air. These well identified interferences can be

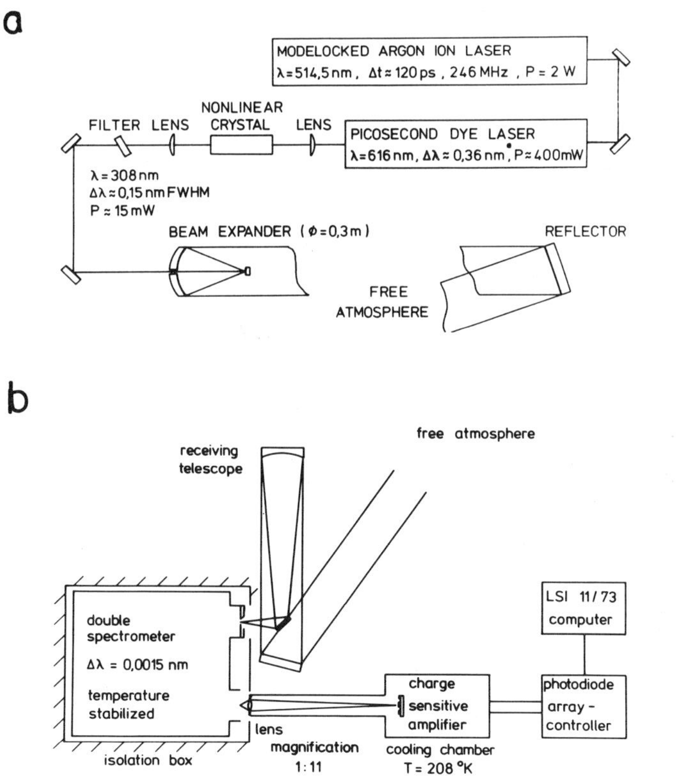

Fig. 1 a,b: Experimental setup for the OH radical detection by laser long-path absorption spectroscopy. The light path length in Jülich was 5.8 km.

removed from the spectra mathematically (see Dorn et al., 1988).
However, in several cases additional absorption features were recorded in the spectra. Some of these appeared at the same wavelength positions, but at varying strengths, other lines varied in both, strength and position. The latter are likely artefacts of the spectrometer system, the former are probably due to absorption by - yet unknown - atmospheric species. Current work at our laboratory has shown that, for example, naphthalene has some strong narrow-band absorption lines around 308 nm.

3. SUPPORTING MEASUREMENTS

Simultaneously to OH we measured a set of parameters (see table 1) that determine the actual OH concentration in ambient air and therefore are essential as input parameters for model calculations. The trace gases NO_2, SO_2, and HCHO were measured by their characteristic *vibrational* band structures, again using differential optical absorption spectroscopy (DOAS). This second spectrometer operates with a Xe-arc as a light source. Both LPA-instruments sampled the same air volume, and both used the same reflecting mirror. (For further details of the DOAS instrument see Platt et al., 1979; Platt and Perner 1983).

Table. 1: Most important parameters necessary for a characterization of the air mass and corresponding measurement technique

species	measurement technique
SO_2	DOAS[a]
HCHO	DOAS
NO_2	DOAS, Chemiluminescence
NO	Chemiluminescence
O_3	UV-Photometer (DASIBI)
H_2O	Psychrometer
Aerosol	Nephelometer
CO CH_4 Hydrocarbons	Gas Chromatography
$J(O_3)$ $J(NO_2)$	Absolute calibrated photoelectric sensors
Temperature Air pressure Windspeed Winddirection	Standard meteorologic instrumentation using a computerized data recording system

[a] DOAS = **D**ifferential **O**ptical **A**bsorption **S**pectroscopy

4. RESULTS

In July 1987 measurements were made in Jülich (51°N), a relatively polluted site owing to its proximity to the industrial regions of Belgium and the Rhein-Ruhr valley. The diurnal profiles of the OH concentration recorded on July 5 (observing the $Q_1(2)$ line group), July 11, and July 14 (observing the $Q_1(3)$ line group) are plotted in figs. 2a,3a,4a together with the profiles of NO_2, O_3, and ozone photolysis frequency, $J(O_3)$. Diurnal profiles of HCHO, SO_2, $J(NO_2)$, and the aerosol light scattering signal, β, (corresponding to the effective aerosol surface per unit volume) are plotted in figures 2b,3b, and 4b. On July 5 and July 14 a comparison of the diurnal profile of OH with that of $J(O_3)$ - which is a measure for the primary production of tropospheric OH radicals - shows, that the increase of the OH concentration is delayed with respect to the rise of $J(O_3)$. The peak value in OH (7/5/87: $4.7 \cdot 10^6$ OH/cm^3; 7/14/87: $6.8 \cdot 10^6$ OH/cm^3) appears roughly two hours after the maximum $J(O_3)$ was reached (figs. 2a, 4a). Between 10:00 am and 11:15 am on July 5 (and 12:00 pm on July 14), no absorption bands could be determined in the spectra, so only upper limits (sum of signal and σ error) are given.

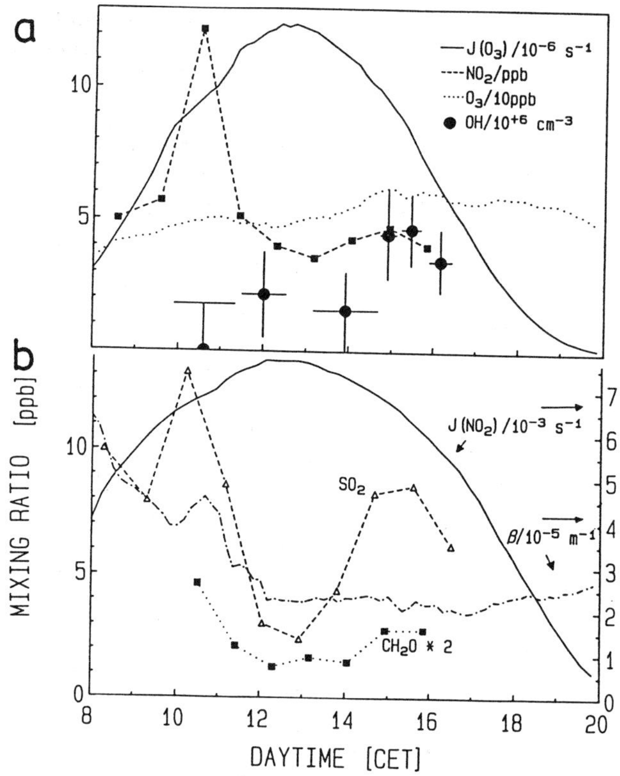

Fig. 2 a: Diurnal profile of the OH concentration (filled circles), O_3 photolysis frequency, $J(O_3)$, (drawn line), NO_2 mixing ratio (dashed line), and O_3 mixing ratio (dotted line) in Jülich on July 5, 1987. Horizontal bars indicate integration times, vertical bars the σ errors of the OH concentration. The data point in the morning corresponds to the upper limit of the OH concentration. b: Diurnal profiles of SO_2 (triangles), NO_2 photolysis frequency, $J(NO_2)$(drawn line, right scale), HCHO(squares), and aerosol light scattering signal, β (dashed line, right scale) on July 5, 1987.

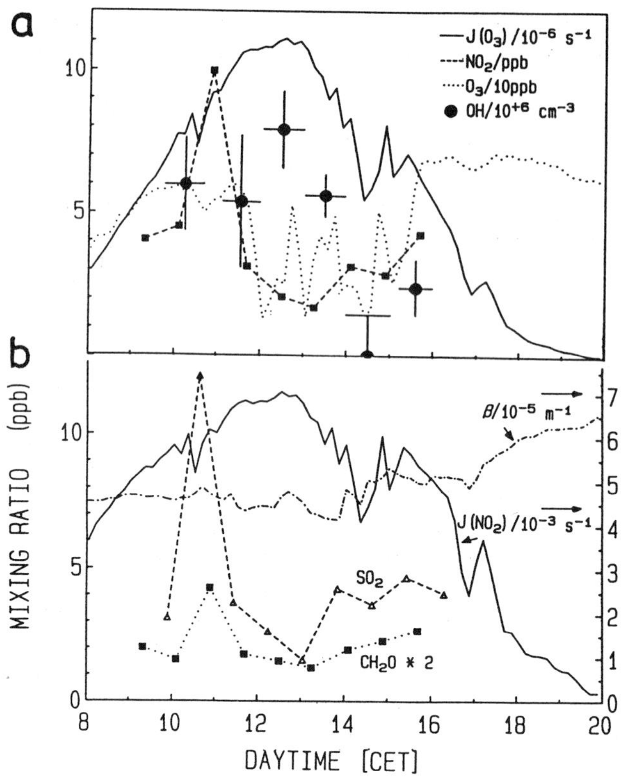

Fig. 3 a,b: Diurnal profiles in Jülich on July 11, 1987 (same symbols as in figs. 2a,b)

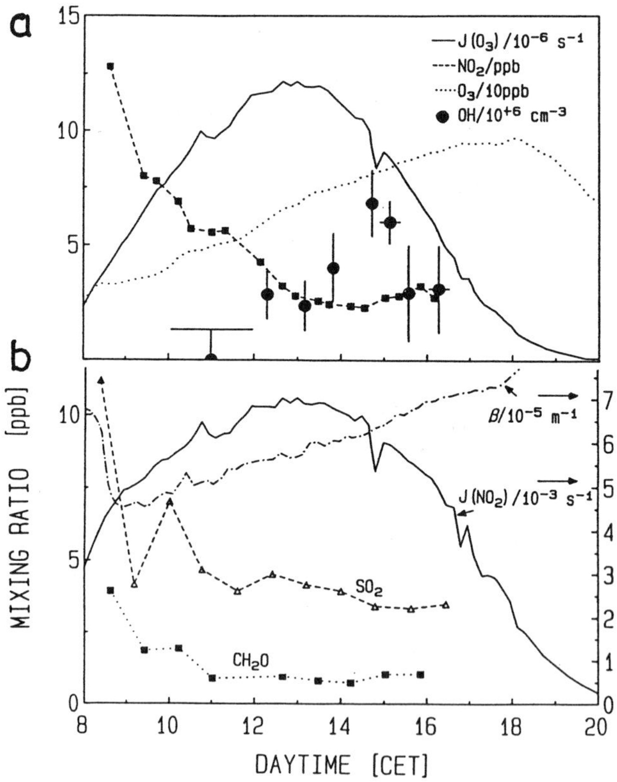

Fig. 4 a,b: Diurnal profiles in Jülich on July 14, 1987 (same symbols as in figs. 2a,b)

A look at the profile of the NO_2 concentration (figs. 2a, 4a) shows that during the morning hours NO_2 concentrations higher than 10 ppb were present. At NO_2 concentrations of a few ppb the reaction of OH with NO_2 forming nitric acid (HNO_3) is the major and net sink for OH radicals. Therefore, the expected OH concentration remained below our detection limit as long as the NO_2 concentration was above $\simeq$5 ppb. Indeed we observed no OH absorption at those conditions.
With the decrease in the NO_2 concentration at noon to values below 4 ppb the destruction reaction slowed down sufficiently to allow the OH concentration to increase above our detection limit. The primary production of OH at that time also increased due to the increasing ozone concentration at nearly constant photolysis frequency (on July 14). In addition, on July 5, the formaldehyde concentration showed an about two-fold increase in the afternoon. The photolysis of HCHO leads to an increase of HO_2 radicals, which are rapidly converted to OH by the reaction with NO and thus increases the OH source. In the later afternoon - at low NO_2 - the decay of the OH concentration is governed by the decrease of the primary source process.
On July 11 the peak concentration in OH coincided with the noontime maximum in $J(O_3)$. However, the observed air mass was strongly influenced by the emission of a nearby coalburning power-plant, as indicated by substantial variations in the ozone concentration (fig. 3a). Therefore definite conclusions are difficult to draw in this case.

5. CONCLUSION

We demonstrated the usefulness of a diode array detector for the measurement of tropospheric hydroxyl radical concentrations in polluted air. The broadband excitation and detection technique allows reliable identification of interferences.
We have shown the importance of supporting measurements performed during the time of OH measurements. Only a sufficiently complete chemical characterization of the air mass can provide the basis for meaningful model calculations to interprete the measured OH concentration profiles.
Ongoing research to improve the OH measurements concentrates in three areas: changes in the laser system in order to broaden the UV bandwidth, changes in the optics of the receiving system to reduce optical noise. Furthermore we are trying to identify other atmospheric species that have absorption lines near 308 nm and might appear in sufficient concentration in ambient air to interfere with OH measurements.

6. REFERENCES

Callies, J., 1988: Absorptionsspektroskopischer Nachweis von Hydroxyl–Radikalen in der Troposphäre. PhD thesis, University of Cologne, FRG.

Dorn, H.-P., J. Callies, U. Platt and D.H. Ehhalt, 1988: Measurement of tropospheric OH concentrations by laser long-path absorption spectroscopy. Tellus B in press

Hübler, G., D. Perner, U. Platt, A. Tönnissen, and D.H. Ehhalt, 1984: Groundlevel OH Radical Concentration: New Measurements by Optical Absorption. J. Geophys. Res. **89**, 1309-1319

Perner, D., U. Platt, M. Trainer, G. Hübler, J.W. Drummond, W. Junkermann, J. Rudolph, B. Schubert, A. Volz, D.H. Ehhalt, K.J. Rumpel, and G. Helas. 1987: Tropospheric OH concentrations: A comparison of field data with model predictions. J. Atmosph. Chem. **5**, 185-216

Platt, U., D. Perner, and H. Pätz, 1979: Simultaneous measurement of atmospheric HCHO, O_3, and NO_2 by differential optical absorption. J. Geophys. Res. **84**, 6329-6335

Platt, U., and D. Perner, 1983: Measurements of atmospheric trace gases by long path differential UV/visible absorption spectroscopy. In Springer Ser. Opt. Sci. 39, Optical and Laser Remote Sensing (ed. D.K. Killinger and A. Mooradian). Heidelberg: Springer, 95-105

RESULTS OF INTERNATIONAL DOBSON SPECTROPHOTOMETER AND BREWER SPECTROMETER CALIBRATIONS, AROSA, SWITZERLAND, 1986

W.D. Komhyr and R.D. Grass
NOAA Air Resources Laboratory
Boulder, Colorado 80303, USA

R.D. Evans and R.K. Leonard
Cooperative Institute for Research in Environmental Services
University of Colorado
Boulder, Colorado 80309, USA

ABSTRACT

An international comparison of Dobson spectrophotometers and Brewer spectrometers was conducted in Arosa, Switzerland, in August 1986. Countries participating with Dobson instruments were Belgium, Czechoslavakia, Denmark, Egypt, the Federal Republic of Germany, France, the German Democratic Republic, Norway, Poland, Portugal, Spain, Switzerland, and the United States. The reference standard instrument for the comparisons was World Primary Standard Dobson spectrophotometer no. 83 maintained by the United States. The mean difference in ozone measured by the 15 participating Dobson instruments relative to Dobson instrument no. 83, for observations in the air mass range 1.15-3.2, was -0.5 ± 1.5(1σ) %. Countries participating with Brewer spectrometers were Canada, Greece, and the Federal Republic of Germany. The Brewer instruments yielded ozone values lower on average than Dobson instrument no. 83 values by 0.3 ± 0.6(1σ) %.

1. INTRODUCTION

In an ongoing program by the Dobson Spectrophotometer Central Laboratory, NOAA Air Resources Laboratory, Boulder, Colorado, to upgrade the quality of total ozone observations made throughout the world, an international comparison of Dobson spectrophotometers and Brewer spectrometers was held in Arosa, Switzerland, August 3-25, 1986, under auspices of the WMO Global Ozone Research and Monitoring Project. A total of 16 Dobson instruments from 13 countries participated (Table 1). Participating Brewer spectrometers were no. 017 from Canada, no. 010 from the Federal Republic of Germany, and no. 005 from Greece. In addition to the instrument calibrations at Arosa, two Dobson spectrophotometers (nos. 71 and 104) were modernized electronically and fitted with new optical wedges. Following optical realignment, optical wedge and wavelength-setting calibrations were performed on the instruments. Optical wedge and wavelength-setting calibrations were performed, also, on Dobson instruments nos. 13, 96 and 101.

2. DOBSON SPECTROPHOTOMETER CALIBRATIONS

The reference spectrophotometer used at the Arosa Dobson and Brewer instrument intercomparisons was World Primary Standard Dobson spectrophotometer no. 83 of the Dobson Spectrophotometer Central Laboratory in Boulder. The long-term (1962-1987) ozone measurement precision of instrument no. 83 has been maintained to within an uncertainty of ±0.5% (Komhyr et al., 1988). Total ozone data obtained at Arosa with Dobson instrument no. 83 were initially processed on the August 26, 1976, Calibration Scale of the instrument. Final results, presented herein, have been processed on the July 10, 1987, Calibration Scale for instrument no. 83. This scale was established at Mauna Loa

TABLE 1. INSTRUMENTS AND PARTICIPANTS OF THE 1986 AROSA DOBSON AND BREWER INSTRUMENT CALIBRATIONS

Instrument No.	Country	Participants
Dobson Instruments		
13	Portugal	None
15	Switzerland	H. Dütsch
40	Belgium	D. DeMuer
51	Switzerland	K. Aeschbacker
56	Norway	S. Larsen
64	German Democratic Republic	U. Feister
71	German Democratic Republic	P. Plessing
74	Czechoslavakia	V. Vanicek
83	United States	W. Komhyr, R. Grass, R. Evans, K. Leonard
84	Poland	M. Degórska, B. Rojewska-Wiech
85	France/United States	A. Bekkadour
92	Denmark	P. Christensen, J. Bjerglind
96	Egypt	G.K.Y. Hassan
101	Switzerland	A. Röch
104	Fed. Republic of Germany	R. Hartmannsgruber, H. Claude
120	Spain	J. Cacho
Brewer Instruments		
05	Greece	I. Ziomas, A. Bais
10	Fed. Republic of Germany	U. Köhler
17	Canada	J. Kerr, K. Lamb

ISBN 0-937194-15-8

TABLE 2. OZONE VALUES IN DOBSON UNITS FOR VARIOUS AIR MASSES μ AT AROSA ON AUGUST 15, 1986

Inst. No. 83 Calib. Scale	Air Mass (μ) 1.15-1.5	1.5-2.0	2.0-2.5	2.5-3.2
August 26, 1976	301	302	303	304
July 10, 1987	302	303	304	304

Observatory in 1987 by the Langley plot method (Dobson and Normand, 1962), following calibration of the optical wedge of instrument no. 83 in 1987. As shown in Table 2, the two calibration scales yield ozone data that, on average over the μ range 1.15 to 3.2, agree to within 0.3% (μ is a quantity closely related to air mass).

Procedures followed in intercalibrating the instruments were virtually identical to ones used at the 1977 International Comparison of Dobson Ozone Spectrophotometers held in Boulder, Colorado, July 25-August 20, 1977 (Komhyr et al., 1980): Several instruments including no. 83 were used to make simultaneous observations on direct sun and A, C, and D wavelength pairs during mornings as μ varied between 1.15 and 3.2 or more. At high μ, with the sun rising rapidly in the sky, observations were made on each wavelength pairs at 4-minute intervals. As the rate of solar ascent decreased, observations were made less frequently.

Data were processed by computer to yield A, C, and D wavelength-pair N values as a function of μ for the various instruments. ($N = \log I_o/I_o' - \log I/I'$ where I_o and I_o' are Dobson instrument solar radiation wavelength pair intensities outside the earth's atmosphere, and I and I′ are corresponding intensities at the instrument location.) Spectrophotometer calibration corrections, namely, values of ΔN_A, ΔN_C, and ΔN_D, were determined from comparisons of instrument no. 83 N values and N values of the instruments being calibrated. ΔN values were determined for the μ ranges 1.15-1.5, 1.5-2.0, 2.0-2.5, 2.5-3.2, 3.2-4.0, and the overall standard observing range of $1.15 \le \mu \le 3.2$. Included in the output data were comparison total ozone amounts x for the pair instruments. Values of x_A, x_C, and x_D were calculated using standard equations (Dobson, 1957) for processing Dobson spectrophotometer data; x_{AD} values were, however, derived from

$$x_{AD} = \frac{N_A/\mu_A - N_D/\mu_D}{(\alpha-\alpha')_A - (\alpha-\alpha')_D} - \frac{[(\beta-\beta')_A - (\beta-\beta')_D](m_A + m_D)p}{[(\alpha-a')_A - (\alpha-\alpha')_D](\mu_A + \mu_D)p_o},$$

which is a modified form of the standard equation for deducing ozone from Dobson instrument observations on pair wavelengths. In this equation α, α′ and β, β′ are the ozone absorption and air molecule scattering coefficients, respectively; p is the mean station pressure; and p_o is sea-level pressure. This equation is applicable when observations of N_A and N_D are made close together but not coincidentally in time. A similar equation was used in computing values of x_{CD}. Also included in the computer output were various types of wavelength-dependent N, ΔN, and x vs. μ plots. These were useful in identifying observer and optical wedge calibration errors, and in providing information on various instrument performance characteristics such as sensitivity, stability, light scattering properties, etc.

Results of preliminary calibrations of the various Dobson spectrophotometers are shown in Table 3. Preliminary calibrations assess the "as is" status of the instruments, yielding information that may be used in correcting total ozone data obtained in the past. The last column in Table 3 shows mean percent differences in ozone measured by the various instruments within the air mass range 1.15-3.2, compared with ozone values measured by Dobson instrument no. 83. Columns 4-8 of Table 3 present

TABLE 3. RESULTS OF PRELIMINARY CALIBRATIONS AT AROSA OF VARIOUS DOBSON SPECTROPHOTOMETERS RELATIVE TO WORLD PRIMARY STANDARD DOBSON INSTRUMENT NO. 83

Dobson Inst. Number	Country	Calibration Date	Total Ozone Difference in Percent* μ= 1.15-1.5	μ= 1.5-2.0	μ= 2.0-2.5	μ= 2.5-3.2	μ= 3.2-4.0	μ= 1.15-3.2
13	Portugal	25-08-86	1.01	-0.33	-2.61	-5.82	-17.90	-1.94
15	Switzerland	15-08-86	-2.17	-2.55	-2.05	-3.04	-7.20	-2.45
40	Belgium	15-08-86	1.93	0.97	0.68	0.94	0.43	1.13
51	Switzerland	15-08-86	-1.28	-2.12	-1.83	-1.86	-1.81	-1.77
56	Norway	15-08-86	-1.51	-1.44	-0.83	-0.84	-1.23	-1.15
64	G.D.R.	21-08-86	-0.50	0.10	0.07	-0.19	-0.32	-0.13
71	G.D.R.	15-08-86	1.19	1.25	2.49	2.30	0.74	1.81
74	Czechoslovakia	15-08-86	-2.20	-1.77	-1.38	-1.37	-0.84	-1.68
83	U.S.A.		Reference standard instrument					
84	Poland	21-08-86	2.05	1.50	1.24	0.74	-0.12	1.38
85	France/U.S.A.	21-08-86	-0.67	-0.73	-0.63	-0.44	-0.03	-0.62
92	Denmarck	15-08-86	1.61	0.55	0.35	-0.23	-1.46	0.57
96	Egypt	15-08-86	0.85	0.08	0.06	-1.14	-2.95	-0.04
101	Switzerland	15-08-86	-3.07	-3.09	-2.98	-3.54	-4.95	-3.17
104	F.R.G.	06-08-86	3.06	0.82	-0.34	0.00	-0.68	0.89
120	Spain	25-08-86	0.67	-0.92	-1.62	-1.67	-1.43	-0.89

*Negative percent differences denote low ozone values compared with Dobson instrument no. 83 data.

TABLE 4. RESULTS OF FINAL CALIBRATIONS AT AROSA OF VARIOUS DOBSON SPECTROPHOTOMETERS RELATIVE TO WORLD PRIMARY STANDARD DOBSON INSTRUMENT NO. 83

Dobson Inst. Number	Country	Calibration Date	Total Ozone Difference in Percent*					
			μ= 1.15-1.5	μ= 1.5-2.0	μ= 2.0-2.5	μ= 2.5-3.2	μ= 3.2-4.0	μ= 1.15-3.2
13	Portugal	20-10-87	-	1.03	-0.10	-0.51	-0.61	0.14
15	Switzerland	15-08-86	2.03	0.63	0.24	-1.07	-5.65	0.45
40	Belgium	15-08-86	0.28	-0.28	-0.23	0.18	-0.18	-0.01
51	Switzerland	15-08-86	1.65	0.10	-0.24	-0.52	-0.74	0.24
56	Norway	15-08-86	0.14	-0.19	0.07	-0.08	-0.62	-0.01
64	G.D.R.	21-08-86	-0.32	0.24	0.17	-0.11	-0.25	0.00
71	G.D.R.	25-08-86	0.28	0.10	-0.07	-0.21	-0.82	0.02
74	Czechoslovakia	15-08-86	0.34	0.15	0.01	-0.20	0.09	0.07
83	U.S.A.		Reference Standard Instrument					
84	Poland	21-08-86	0.13	0.03	0.22	-0.15	-0.79	0.06
85	France/U.S.A.	25-08-86	-0.54	-0.34	-0.11	0.43	0.42	-0.14
92	Denmarck	15-08-86	1.08	0.15	0.06	-0.47	-1.67	0.20
96	Egypt	21-08-86	0.27	0.13	0.05	-0.35	-2.85	0.03
101	Switzerland	21-08-86	-0.70	-0.09	0.48	0.08	-0.95	-0.05
104	F.R.G.	21-08-86	0.20	0.16	-0.26	-0.12	0.22	-0.01
120	Spain	25-08-86	2.32	0.36	-0.67	-0.90	-0.81	0.28

*Negative percent differences denote low ozone values compared with Dobson instrument no. 83 data.

similar data for narrower air mass observing ranges.

Final calibration data, expressed as percent differences in ozone measured by the various instruments compared with Dobson instrument no. 83 values, are given in Table 4. These results were obtained after calibrations of the different instruments relative to instrument no. 83 were completed. As in Table 3, performance of the instruments in Table 4 is characterized as a function of air mass.

3. BREWER SPECTROMETER CALIBRATIONS

Results of preliminary calibrations of the three participating Brewer spectrometers with Dobson instrument no. 83 are shown in Table 5. The spectrometers of the Federal Republic of Germany (no. 010) and Greece (no. 005) were calibrated in the past by Canadian Brewer spectrometer no. 017. Calibration of Brewer instrument no. 017 is traceable to Canadian Dobson spectrophotometer no. 77, which was calibrated relative to Dobson instrument no. 83 in Boulder, in 1977. A calibration check on Dobson instrument no. 77, using the Langley plot calibration method, was performed by the Canadian Atmospheric Environment Service at Mauna Loa Observatory in 1980 (A. Asbridge, private communication, 1980).

4. CONCLUSIONS AND DISCUSSION

Preliminary intercomparison data (Table 3) show that four instruments measured ozone values that were lower in the mean than Dobson instrument no. 83 values by about 2% or more. Portugal instrument no. 13 ozone values exhibited an extreme μ dependency, but with values on average in the μ range 1.15-3.2 being only 1.94% lower than instrument no. 83 values. The μ dependency was largely eliminated (Table 4) after electronic modernization, optical alignment, and recalibration of the instrument in Boulder in 1987. Switzerland instruments nos. 15 and 51 measured ozone values on average too low relative to instrument no. 83 by

TABLE 5. RESULTS OF PRELIMINARY CALIBRATIONS AT AROSA OF SEVERAL BREWER SPECTROPHOTOMETERS RELATIVE TO WORLD PRIMARY STANDARD DOBSON INSTRUMENT NO. 83

Brewer Inst. Number	Country	Calibration Date	Total Ozone Difference in Percent*					
			μ= 1.15-1.5	μ= 1.5-2.0	μ= 2.0-2.5	μ= 2.5-3.2	μ= 3.2-4.0	μ= 1.15-3.2
005	Greece	21-08-86	-0.4	-0.5	-0.6	-1.1	-0.3	-0.7
010	F.R.G.	15-08-86	-0.4	-1.1	-1.3	-1.1	-0.7	-1.0
010	F.R.G.	21-08-86	-0.1	-0.2	-0.4	-0.7	0.2	-0.4
017	Canada	21-08-86	0.9	0.4	0.3	0.1	0.8	0.4

*Brewer minus Dobson instrument ozone differences are shown. The Dobson instrument values are derived from observations on AD wavelengths.

2.45% and 1.77%, respectively. Both instruments exhibited a μ dependency in the observations that may be due to light scattering, or perhaps some optical alignment problem. Switzerland instrument no. 101 ozone values were also lower on average than corresponding instrument no. 83 values, by 3.17%. The μ dependency exhibited by instrument no. 101 in the calibration data of August 15, 1986, with calibration corrections applied (not shown) was, however, significantly reduced following recalibration of the optical wedge of the instrument and final calibration (Table 4). Ozone values on average 1.68% too low, indicated by Czechoslovakian instrument no. 74, appear to have resulted from instrument calibration drift; final intercomparison data in Table 4 show no significant μ dependency.

Instrument no. 71 of the German Democratic Republic initially measured ozone values too high on average by 1.81%, while Federal Republic of Germany instrument no. 104 ozone values were too high by 3.06% in the μ range 1.15-1.5 (Table 3). After "air-space" optical wedges were installed into the two instruments, and the instruments were optically aligned and recalibrated, both measured ozone values that closely approximated those of Dobson instrument no. 83 (Table 4).

Dobson instrument intercomparison data obtained following final instrument calibrations (Table 4) show that in the mean, for observations made in the standard μ observing range of 1.15-3.2, all instruments yielded ozone amounts that agreed with those of instrument no. 83 to within several tenths of a percent. At high sun elevations ($1.15 \leq \mu \leq 1.5$), however, several instruments (nos. 15, 51, 92, and 120) measured ozone values too high by up to about 2%. Such ozone measurement errors should be taken into account when processing data from routine ozone observations.

The Brewer spectrometers measured total ozone amounts at all μ intervals that agreed to within about ±1% of corresponding Dobson instrument no. 83 values. On average, the Brewer instrument ozone values were lower by $0.3 \pm 0.64(1\sigma)$ %. Considerably less μ dependency in the ozone data was exhibited by the small sample of Brewer instruments than that shown by the Dobson spectrophotometers.

ACKNOWLEDGMENTS

The 1986 Arosa International Calibration of Dobson Spectrophotometers and Brewer Spectrometers was conducted under auspices of the WMO Global Ozone Research and Monitoring Project. Appreciation is expressed to H. U. Dütsch of the Laboratory for Atmospheric Physics ETH, Zürich, Switzerland, for providing the excellent Arosa Observatory facility in which the instrument comparisons were conducted.

REFERENCES

Dobson, G.M.B., 1957: Observers' handbook for the ozone spectrophotometer. *Annals of the International Geophysical Year*, V, Part I, Pergamon Press, New York, 46-89.

Dobson, G.M.B., and C.W.B. Normand, 1962: Determination of constants etc. used in the calculation of the amount of ozone from spectrophotometer measurements and the accuracy of the results. *Annals of the International Geophysical Year*, XVI, Part II, Pergamon Press, New York, 161-191.

Komhyr, W.D., R.D. Grass, and R.K. Leonard, 1980: WMO 1977 international comparison of Dobson ozone spectrophotometers. *Proc. of the Quadrennial International Ozone Symposium*, Boulder, Colorado, August 4-9, National Center for Atmospheric Research, Boulder, 25-32.

Komhyr, W.D., R.D. Grass, and R.K. Leonard, 1988: Calibration of primary standard Dobson spectrophotometer no. 83 and total ozone trends at Mauna Loa Observatory, Hawaii, and American Samoa, South Pacific. *Proc. International Ozone Symposium-88*, R.D. Bojkov and P. Fabian, Eds., A. Deepak Publishing, Hampton, Virginia.

MODERNIZATION OF THE CANADIAN OZONE MONITORING NETWORK

W.F.J. Evans, J.B. Kerr, D.I. Wardle, and C.T. McElroy
Atmospheric Environment Service
Downsview, Ontario, M3H5T4 Canada

ABSTRACT

The Canadian ozone network has been improved over the last 10 years so that it is equal to the task of monitoring the small trends predicted by ozone depletion scenario models. The Dobson instrument has been replaced by the Brewer instrument in the Canadian ozone monitoring network, the ECC ozonesonde has been substituted for the Mast sonde, a new chemical luminox ozonesonde for nitrogen dioxide is under development and an ozone DIAL system for ozone profiling is planned for Toronto.

1. INTRODUCTION

In the 1970s, it was evident from scenario model predictions that, in the 1980s, one would have to be able to detect ozone trends in total ozone of the order of the order of 0.1% per year or absolute changes of 0.5% over 5 years. It was considered unlikely that the Dobson network could detect such small changes, particularly after large discrepancies between instruments were revealed at the Belsk Dobson intercomparison in 1974 (Dziewulska-Losiowa and Walshaw, 1975). To meet the challenge of detecting small trends, a new instrument called the Brewer spectrophotometer was developed with the aim of improved long-term stability and operational effectiveness. The Brewer instrument has now been deployed in the Canadian network and a calibration system has been set up.

2. DEVELOPMENT OF THE BREWER SPECTROPHOTOMETER

The Brewer spectrophotometer was designed and built at the AES laboratories in Toronto in the early 1980s and the technology was transferred to SCITEC Instruments Ltd. of Saskatoon for manufacturing. More than 15 Brewers are available in 14 countries in addition to the 10 Brewers operating in Canada. Brewer serial number 48 is now under construction at SCITEC. The concept of the Brewer ozone calibration system is shown in Fig. 1 (Evans et al., 1987).

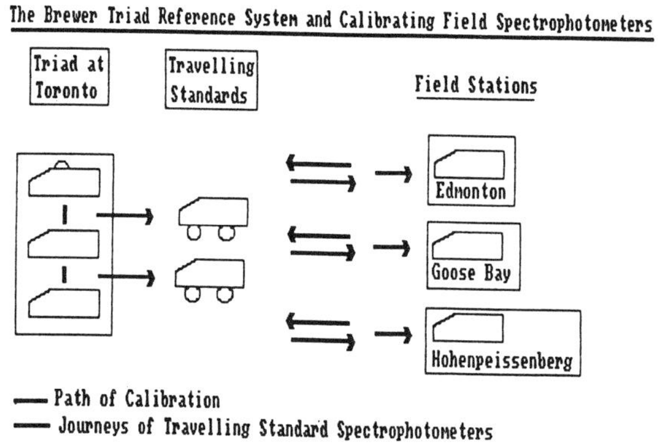

Fig. 1. The concept of the Brewer ozone monitoring system.

A set of three reference instruments forms a triad standard while two travelling standard instruments are used to transfer the calibration to the network instruments in the field. The calibration of the travelling standard is checked against the triad before and after the field calibration trip. The travelling standard means that the station instrument does not have to be moved for calibration with attendent risk of calibration shift.

3. DEPLOYMENT OF THE BREWER

The map of Canada in Fig. 2 shows the stations in the Canadian ozone monitoring network; the old network has been expanded to include new stations at Saskatoon and Alert. A new station to extend Canadian longitudinal coverage will be put into operation at Victoria in early 1989.

ISBN 0-937194-15-8

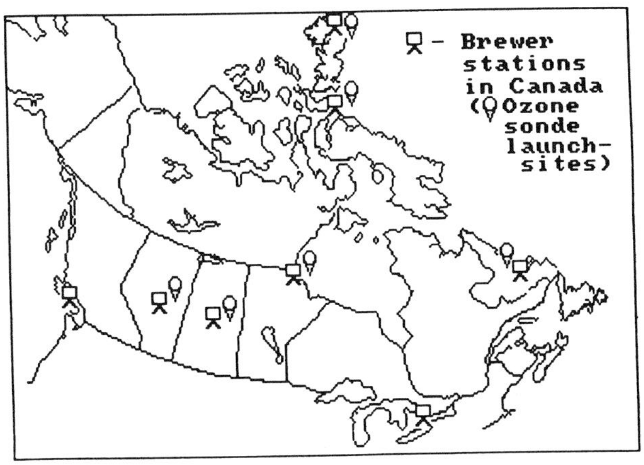

Fig. 2. The location of Canadian ozone stations.

Ozone profiles are measured with ozonesondes at Alert, Resolute, Goose Bay, Churchill and Edmonton on a routine weekly basis. At each station, Brewers are being run in parallel with Dobson instruments for a period of 3 years before the Dobsons are retired. The Edmonton Dobson was retired in August, 1988 and the Goose Bay Dobson is scheduled for retirement in October, 1988. The Dobson instruments which have been taken out of service will be carefully preserved so that further tests or comparisons can be made if the need should arise in the future. The Edmonton Dobson is now on display in the AES museum of meteorological instruments at AES in Toronto.

In the replacement process, the Dobson instruments are characterised by a transfer function and then synthetic Dobson data are calculated from the Brewer data so that there will be no discontinuity in the ozone data record (Kerr et al., 1988a, 1988b). Only direct sun Dobson data is used. The success of this process is demonstrated in Fig. 3, which shows the trend data from Toronto using synthetic data in the upper panel and the actual Dobson data in the lower panel over the last five years. The comparison shows no apparent difference over the last 5 years -- there would have been a larger difference if we had placed the Edmonton Dobson in operation at Toronto, because there is typically a difference of greater than 1% between Canadian network Dobson instruments.

A problem in the use of Dobson data for the determination of ozone trends has been found which results from the inclusion of sky measurements in a trend analysis. Figure 4 shows a comparison of direct sun measurements with zenith sky measurements at Toronto.

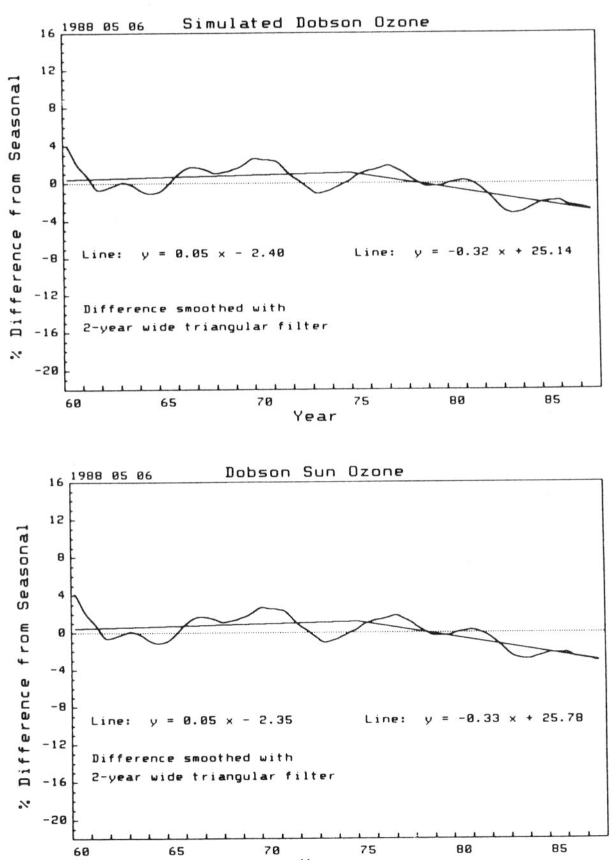

Fig. 3. A comparison of ozone trends at Toronto determined from synthetic Dobson data and from original Dobson data.

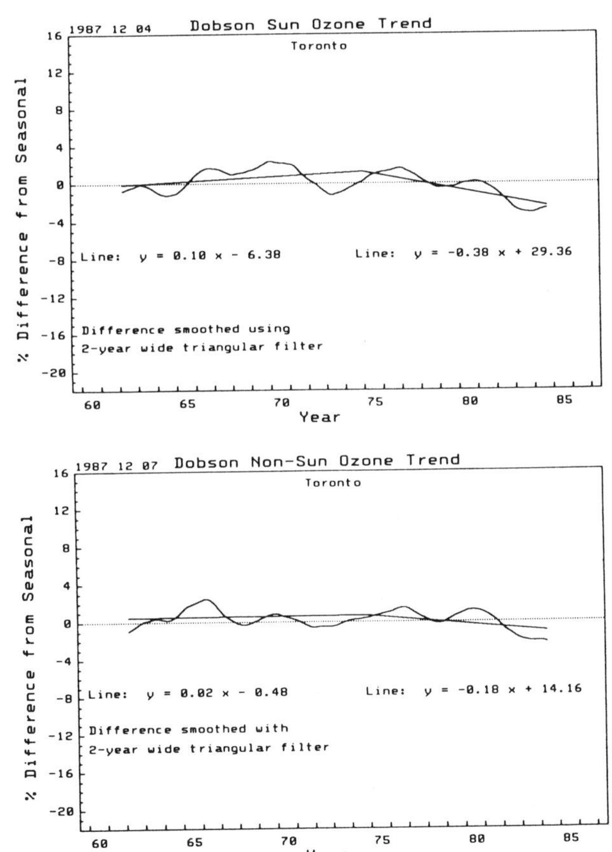

Fig. 4. A comparison of ozone trends determined from direct sun and sky ozone measurements at Toronto from 1960 to 1987.

The direct sun measurements have been used for the upper panel, while the non-sun or sky measurements (zenith blue and zenith cloud) have been used for the lower panel. The direct sun measurements indicate a downward trend since 1975 of 0.38% per year. However, the sky measurements indicate a much smaller trend of 0.18% per year. A similar problem was found in the sky data from other Canadian stations. This demonstrates that sky measurements from the Dobson data in the Canadian network may not be reliable for trend studies. Preliminary results indicate that the Brewer sky data may not suffer from this problem which seems to be associated with the Dobson wavelength bandpass function. In any case, the automatic operation of a Brewer instrument provides more days (about 20 per month in the Canadian network) of observations which include at least one good direct sun determination using the direct sun method (Evans et al., 1987).

4. DATA MANAGEMENT FOR THE BREWER SPECTROPHOTOMETER

Management of ozone data for the Canadian ground-based ozone network has been modernised by the development of the Brewer data base management system (BDMS). The aim is to ensure that calibration and operational characteristics of all field instruments in Canada and elsewhere are recorded and documented in a standard format in a centralised computer data base. A particular field instrument is operated by one of several models of Commodore or IBM PC compatible microcomputers which are programmed to schedule and execute the observations as well as to analyse, record and print out the data. About one week of data are stored on either 3.5 or 5.25 inch floppy diskettes which are duplicated with one copy retained at the field station while the other is sent to the AES in Toronto where the BDMS is located. The BDMS has been developed to process all Brewer data by a common analysis process at a central location and to present the data in a format suitable for publication in a tabular or graphical format. Ozone measurements are accessed by instrument number and a date range. Statistical analyses are carried out on the data and calibration corrections are applied if necessary. It is expected that a summary of the data will ultimately be available to Brewer users via telephone lines.

5. REPLACEMENT OF MAST OZONESONDES

In the 1970s we noticed a severe problem with the Mast ozonesondes as flown in the Canadian network; this problem is demonstrated in Fig. 5.

At Goose Bay, the scale correction factors of integrated ozone against ground based total ozone were biased at a mean of 1.29 with an rms deviation of 22%; the actual data points are demonstrated in Fig. 5. The Kohmyr ECC sonde was known to have a much smaller correction

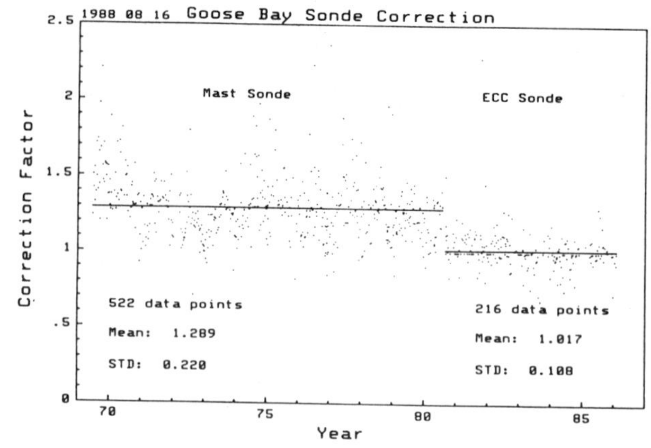

Fig. 5. Correction factors for ozonesondes at Goose Bay in the Canadian network.

factor and scatter, and based on field tests carried out by AES at Palestine, Texas, the Mast sonde was replaced by the ECC sonde in the Canadian network. After 1980, the correction factor dropped to a mean of 1.017 and the standard deviation dropped to 10% at Goose Bay as demonstrated in Fig. 5. A similar improvement was found at the other Canadian stations. At Edmonton, the mean dropped from 1.26 with a deviation of 31% to 1.02 with a deviation of 13% after the changeover in 1980. Although this substitution of sondes may have had an effect on trends from the ozonesonde profiles for some investigations, it was obviously an great improvement, because the Mast ozone concentration profiles were also skewed with height due to this large correction factor.

Since it would desirable to have a more precise ozonesonde, particularly at levels above 30 km, a new chemical ozonesonde is currently under development by UNISEARCH, a Toronto firm in cooperation with AES. This is based on the use of the new LUMINOX (liquid luminol) technology and potentially could provide a further improvement in performance of ozonesondes. With a simple change in dye, the same sonde will also measure nitrogen dioxide which will enhance reseach programs on tropospheric and stratospheric chemistry in the world. The monitoring of nitrogen dioxide profiles with the new sonde will complement the ground based Brewer measurements at Toronto and Saskatoon.

6. OZONE DIAL LIDAR

In cooperation with the new Institute for Space and Terrestrial Science at York University, a Canadian company named OPTECH is developing an ozone lidar for operation at Toronto.

The functional block diagram is shown in Fig. 6. The system employs a xenon chloride ultraviolet laser together with a methane Raman shifting cell to generate beams at 308 nm and 338 nm. It will use a 100 cm telescope for collection of the return beam. This system is

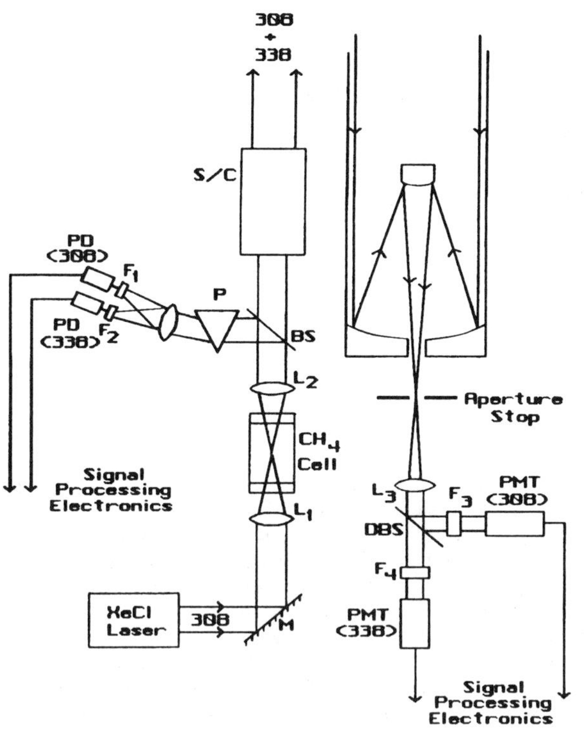

Fig. 6. Block diagram of the planned ozone DIAL LIDAR for ozone profiling at Toronto.

intended to measure ozone profiles with a precision of better than 5% up to 30 km on an hourly basis. Operating the system at Toronto would be particularly advantageous because of local air traffic density which prevents the flying of ozonesondes; this will add profile measurements on a regular basis to supplement the total ozone measurements.

7. CONCLUSIONS

The Canadian ozone monitoring network has been upgraded in order to be equal to the task of measuring the small trends predicted by models. The purpose of this report is to encourage other countries to upgrade their networks using the same technologies, particularly the Brewer spectrophotometer for total ozone measurement. The Canadian network will be further upgraded in the future with the DIAL ozone LIDAR and the new nitrogen dioxide sonde. At this time, when ozone depletion looks ominous, the current world ground based network may not be accurate enough to properly monitor the predicted rate of ozone depletion in order to test scenario models. The scientific community has been forced to rely on the present ground network to recalibrate the satellite instruments due to instrumental drift problems in SBUV and TOMS. What is needed is a total system for monitoring ozone; individual instruments are not adequate whether they are on the ground or in space. Ozone can only be properly monitored with a well-designed system which includes a sophisticated calibration methodology and automatic instrument performance testing.

ACKNOWLEDGEMENTS

W.A. Clark, I.A. Asbridge, J.J. Bellefleur and L. Morrison have contributed for many years to the excellent operation of the Canadian network.

REFERENCES

Dziewulska-Losiowa, A. and C.D. Walshaw, 1975: The International Comparison of Ozone Spectrophotometers at Belsk, 24 June - 6 July 1974. Publs. Inst. Geoph. Pol. Ac. Sci., Vol. 89, pp. 3-123. (Available from: Instytut Geofizyki Polskiej Akademii Nauk ul. Pasteura 3, 02-093 Warszawa, Poland)

Evans, W.F.J., H. Fast, A.F. Forester, G.S. Henderson, J.B. Kerr, K. Vupputuri and D.I. Wardle, 1987: Stratospheric Ozone Science in Canada, An Agenda for Research and Monitoring, Environment Canada Internal Report ARD-87-3. (Available from: AES, 4905 Dufferin Street, Downsview, ONTARIO, M3H5T4)

Kerr, J.B., I.A. Asbridge and W.F.J. Evans, 1988a: Long Term Intercomparison of Total Ozone Measured by the Brewer and Dobson Spectrophotometers at Toronto, 1988: J. Geophys. Res., 93, in press.

Kerr, J.B., I.A. Asbridge and W.F.J. Evans, 1988b: Long Term Intercomparison Between the Brewer and Dobson Spectrophotometers at Edmonton, Proc. Quad. Ozone Symp., this volume.

Kerr, J.B., D.I. Wardle, C.T. McElroy and W.F.J. Evans, The Automated Brewer Spectrophotometer, 1984: Proc. Quad. Ozone Symp., Haldiki, Greece, pp 396-401, (1984).

Komhyr, W.D., R.D. Grass, R.D. Evans and R.K. Leonard, 1988: Results of International Dobson Spectrophotometer and Brewer Intercomparisons, Arosa, Switzerland, 1986, Proc. Quad. Ozone Symp., this volume (1988).

OPERATIONAL GROUND-BASED OZONE SENSOR (OGOS)

W. Zommerfelds and K.F. Künzi
Institute of Applied Physics
University of Bern, Switzerland

ABSTRACT

Successful ground based microwave measurements of stratospheric constituents have been made by various research groups. Based on this experience an automatic, operational ground based ozone sensor (OGOS) was developed. Ozone profiles from 15 to 75 km are retrieved from measurements of the thermal emission of a rotational O3 transitions at 142 GHz typically once every hour.

1. INTRODUCTION

The discussion on drift and accuracy of satellite instruments showed the importance of ground based ozone sensors as a reference and calibration network for satellite observations.

A microwave radiometer is well suited to measure ozone profiles in the entire middle atmosphere (stratosphere and mesosphere) [Lobsiger and Künzi 1986, Lobsiger 1987, Rubin and Künzi 1988]. Microwave transmission in the atmosphere is not affected by stratospheric aerosols (e.g. from volcanic eruptions), and only little attenuated by clouds. Only precipitation or very high tropospheric water vapour concentrations limit the measuring capability of microwave sensors. From data published by Clark et al. [1984] we conclude, that the tropospheric total precipitable water should not exceed ~2.7 cm, this value corresponds to a tropospheric transmittance of ~50%, which still allows an ozone profile retrieval. From the data collected in Bern (47°N, 7°E, 500 m Alt.) we conclude that more than 80% of observing time in winter and more than 30% in summer is available. By going to a higher and therefore dryer observing site, observing time will rapidly increase.

Since microwave radiometers measure the thermal emission from rotational transitions of atmospheric constituents, such sensors can operate equally well during night- and day time. Microwave spectroscopic parameters for the O_3 molecule are sufficiently well known, to retrieve the altitude Ozone profile from high spectral resolution measurements of the shape of pressure broadened emission lines. For the selected 142 GHz line this allows to obtain O_3 information up to ~75 km altitude.

Ground based microwave observations of stratospheric ozone have been made in Bern since 1981. Annual and diurnal variations of the ozone profile have been observed successfully and the results show [Lobsiger et al. 1984, Lobsiger and Künzi 1986, Lobsiger 1987], that microwave radiometry is a very useful and reliable way to monitor stratospheric ozone. Based on this experience, an automatic "Operational Ground-based Ozone Sensor" (OGOS) was designed, built and tested successfully in Bern.

2. INSTRUMENT AND OBSERVATION METHOD

Fig. 1 shows a block diagram of OGOS. The atmospheric microwave radiation at 142.175 GHz is received by the receiver front end, where it is converted down to 3.7 GHz IF and amplified. This particular line was chosen, because it is a fairly strong line, located in a "window region" with little interferences from lines of other atmospheric constituents. In three steps, depending on the frequency resolution required, the signal is amplified further and converted to lower intermediate frequencies. With a total of 75 filters we obtain the power spectrum of the line within a total bandwidth of 1200 MHz and a resolution of 100 MHz at the line wings increasing to

OZONE IN THE ATMOSPHERE
Rumen D. Bojkov and Peter Fabian (Eds.)

ISBN 0-937194-15-8

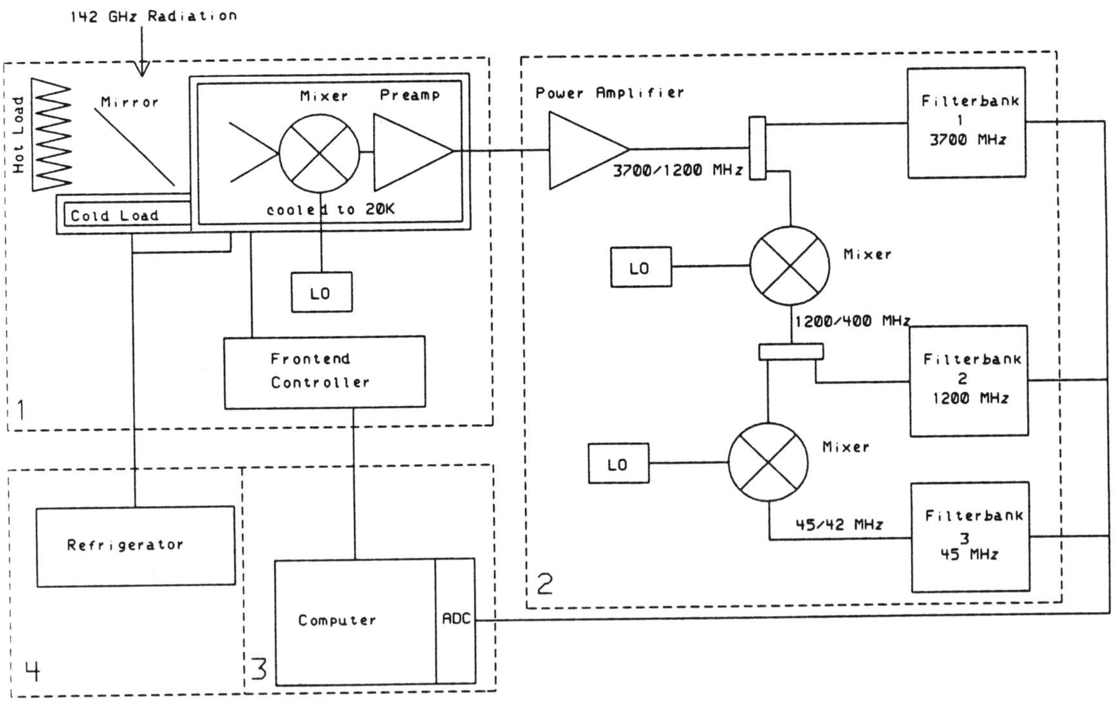

Fig.1: OGOS block diagram; (1) Receiver front end, (2) Filterbanks, (3) Computer, (4) cooling system.

0.05 MHz at the line centre. Detectors convert the filter power output into a voltage that is read by an ADC into the computer.

In the microwave frequency range the Raleigh-Jeans approximation is valid and the brightness temperature of a black body is equivalent to its physical temperature. By rotating the mirror, the black body radiation of a hot (300 K) and a cold (150 K) reference absorber and the atmospheric emission is measured in a 10 second cycle. By comparing the measured sky emission with the radiation of two references, the sky brightness temperature is determined. With this calibration the effect of gain fluctuations of the radiometer very much reduced. A two step refrigerator cools down the cold reference absorber, and also the front end mixer and the first FET preamplifier to ~20 K to reduce the receiver noise. An IBM-AT compatible computer controls the radiometer, positions the rotating mirror, reads the filtered spectra, calibrates and integrates the measurements and keeps track of housekeeping data. A complete algorithm, described in the following section, to retrieve stratospheric ozone profiles from the measured spectra, is also installed on the computer.

3. SOFTWARE AND RETRIEVAL ALGORITHM

The OGOS software package includes a complete set of programs to operate the radiometer, to measure ozone spectra and to calculate the corresponding stratospheric ozone profiles.

The radiometer program controls the whole radiometer operation such as positioning of the rotating mirror, scan of ADC to read the filter output, calibration and integration of measured spectra. The measured spectra are saved on hard disk. A set of preprocessing routines prepares the data for the inversion. The ground based measured sky brightness temperature T_{BG} consists of the stratospheric signal T_{Bstr} but also from contributions of the troposphere:

$$T_{BG} = T_{Bstr} \cdot A + T_{tr} \cdot (1-A) \qquad (1)$$

We approximate the troposphere by a single layer with a temperature T_{tr} taken from the CIRA model atmosphere at 1000 m altitude. Using standard profiles we calculate the expected stratospheric brightness temperature of the ozone line wings (±550 MHz off centre), where the ozone contribution is only very small), compare it with the observed temperature T_{BG} and get the tropospheric transmittance A.

The rotation transition is symmetric around the line centre and baseline fluctuations are reduced by folding the line around the centre. Finally with a spline fit the line is smoothed and we get the brightness temperatures at the desired frequencies for the inversion.

By inverting the atmospheric radiative transfer integral with an iterative algorithm [Randegger, 1980] we obtain the O_3 volume mixing ratio (VMR) profile from the measured line shape. We start from an initial guess for the ozone profile of $q_1(z)=1$ ppm VMR and calculate the expected brightness temperatures T_{BC}. The differences $\Delta T_{BC,i,n}$ of the temperatures at neighbouring filter channels (labelled by i) are compared with the measured brightness temperature differences $\Delta T_{B,i}$. The profile is modified in n iterations at the iteration altitudes z_i, using the relaxation formula

$$q_{n+1}(z_i) = q_n(z_i) \cdot \frac{\Delta T_{B,i}}{\Delta T_{BC,i,n}} \qquad (2)$$

as derived by Chahine [1972]. After usually 7 iterations the differences between calculated and observed ΔT_B become small compared to the data error.

Figure 2 shows a typical example of a measured emission line, and Figure 3 the corresponding retrieved ozone profile. The data in Fig. 2 was obtained with an earlier version of OGOS.

4. SYSTEM SPECIFICATIONS AND ERROR CONSIDERATIONS

The altitude resolution was determined by calculating the sensitivity of the brightness temperature differences to an increase of 1 ppm VMR in a 1 km layer. At higher altitudes the sensitivity functions become very wide as Doppler dominates over pressure broadening and the signal rapidly decreases.

With sufficient measurement time errors introduced by receiver noise can be neglected. The reference absorbers are not perfect black bodies and we estimate a systematic error in our calibration of ~5%. In agreement with Lobsiger et al. [1984], we estimate a statistical error introduced by tropospheric effects of ~5%. The major source of errors are baseline ripples on the measured line shape. Some are caused by standing waves in the radiometer front-end. This effect is reduced by reciprocating the mirror, as proposed by Gustincic [1977]. Fluctuations smaller than 0.15 K, also caused by nonlinearities of the system, however remain on the measured spectra. From simulated inversions we estimate a systematic error of 1 ppm VMR, due to a systematic baseline ripple and calibration errors, and a statistical error of ~15% caused by tropospheric variations, baseline fluctuations and receiver noise.

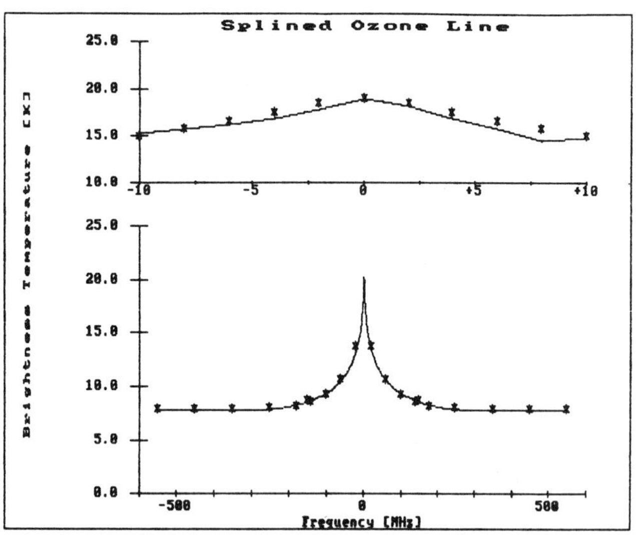

Fig.2: Measured ozone line.

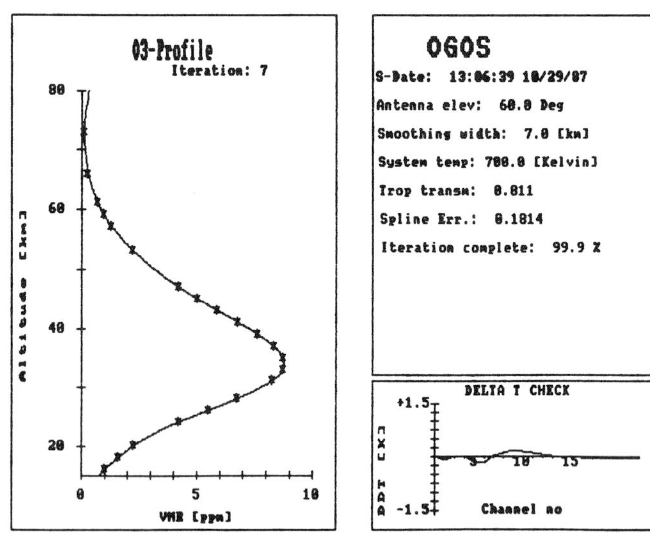

Fig.3: Corresponding retrieved ozone profile.

5. SUMMARY OF THE OGOS CHARACTERISTICS

PHYSICAL:	
Power consumption .... :	2.5 kW @ 220 V
Operating temperature .... :	15 - 25°C
Dimensions	
Main unit .... :	55 cm x 75 cm x 180 cm
Compressor .... :	60 cm x 70 cm x 50 cm
Computer .... :	100 cm x 50 cm x 50 cm
Weights	
Rack .... :	200 kg
Compressor .... :	50 kg
Computer .... :	20 kg
Service interval .... :	6 months (Compressor)
OPERATIONAL:	
System noise temperature .... :	<1000 K
Total bandwidth .... :	1200 MHz
Frequency resolution .... :	0.05 MHz at line centre 100. MHz at line wing
Measurement time .... :	0.5 - 3 h
OZONE PROFILE:	
Altitude range .... :	15 - 75 km
Altitude resolution .... :	~8 km (below 50 km alt.) 10 - 18 km (50-80 km alt.)
OZONE VMR ERROR:	
Absolute (systematic) .... :	± 1 ppmV
Statistical .... :	± 15%

REFERENCES

M.T.Chahine, A general relaxation method for inverse solution of the full radiative transfer equation, J.Atmos.Sci, 29, 741-747, 1972.

W.Clark, J.E.Miller, P.H.Richardson, Sky brightness temperature measurements at 135 GHz and 215 GHz, IEEE TAP, AP-32, 9, 928-932, 1984.

J.Gustincic, A quasi-optical receiver design, Proc. IEEE conf. on MTT, San Diego, 99-100, June 1977.

E.Lobsiger, K.F.Künzi, H.U.Dütsch, Comparison of stratospheric ozone profiles retrieved from microwave radiometer and Dobson-spectrometer data, J. of Atm. and Terr. Phys., 46,9,799-806, 1984.

E.Lobsiger, K.F.Künzi, Night-time increase of mesospheric ozone measured with a ground-based microwave radiometer, J. of Atm. and Terr. Phys, 48, 11-12, 1153-1158, 1986.

E.Lobsiger, Ground based radiometry to determine stratospheric and mesospheric ozone profiles, J. of atm. and terr. Phys, 49, 5, 493-501, 1987.

A.Randegger, On the Determination of Atmospheric Ozone Profile for Ground Based Microwave Measurements, Pure appl. Geophys., 118, 1052-1065, 1980.

G.Rubin, K.F.Künzi, Ground-Based Microwave Sensor to retrieve stratospheric Ozone Profiles, Proc. of the 1988 Ozone Symposium, Göttingen, FRG.

APPENDIX 1

WORKSHOP ON TROPOSPHERIC OZONE August 4-6, 1988

QUADRENNIAL OZONE SYMPOSIUM August 8-13, 1988

List of Participants

1.	**Ackerman, Marcel**	Inst. d'Aeronomie Spatiale, Circulaire 3, B-1180, Brussels, Belgium
2.	**Adriani, Alberto**	FA–CNR, Frascati, Italy
3.	**Aikin, Arthur**	NASA/GSFC, Code 616, Greenbelt, MA 20771, USA
4.	**Aimedieu, Patrick**	CNRS/SA, BP 3, 91371 Verrieres-le-Buisson, France
5.	**Albritton, Daniel L.**	NOAA/ERL, R/E/AL, 325 Broadway, Boulder, CO 80303-3328, USA
6.	**Amanatidis, George T.**	University of Thessaloniki, BP 149, 54006 Thessaloniki, Greece
7.	**Amoruso, Antonella**	Universite di Roma "La Sapienza", Roma, Italy
8.	**Ancellet, Gerard**	CNRS, Service d'Aeronomie, B.P. 3, 91371 Verrieres le Buisson, France
9.	**Ansmann, Albert**	MPI f. Meteorologie, Bundesstr. 55, 2000 Hamburg, F.R. Germany
10.	**Arijs, Etienne**	Belgian Institute for Space Aeronomy, Ringlaan 3, Brussels, Belgium
11.	**Atkinson, Roger**	Bureau of Meteorology, Post Box 1289K, 3001 Melbourne, Victoria, Australia
12.	**Bais, Alkiviadis**	University of Thessaloniki, Lab. of Atmospheric Physics, P.B. 149, 54006 Thessaloniki, Greece
13.	**Barański, Leslaw**	Institute of Meteorology and Water Management, Piotra Borowego str. 14, 30-215 Krakow, Poland
14.	**Barbe, Alain**	Universite de Reims, Reims, France
15.	**Barnes, Robert**	Chemal. Inc., P.O.Box 44, Wallops Island, VA 23337, USA
16.	**Basher, Reid**	New Zealand Meteorological Service, P.O.Box 722, Wellington, New Zealand
17.	**Bauer, Helmut**	Gesellschaft für Strahlen- und Umweltforschung, Ingolstädter Landstr. 1, 8042 Neuherberg, F.R. Germany
18.	**Beagley, Stephen**	Cambridge University, Department of Physical Chemistry, Lensfield Road, Cambridge CB2 1EP, Great Britain
19.	**Beck, Jeannette**	Institute of Environmental Health and Protection, P.O.Box 1, 3720 Bilthoven, Netherlands
20.	**Beckers, Werner**	Westfälische Wilhelms Universität, Dodostr. 21, 4400 Münster, F.R. Germany
21.	**Behnke, Wolfgang**	Fraunhofer Institut, Nikolai-Fuchs-Str. 1, 3000 Hannover 61, F.R. Germany
22.	**Berntsen, Terse**	Inst. of Geophysics, University of Oslo, P.B. 1022, 0315 Blindern Oslo 3, Norway
23.	**Bhartia, Pawan K.**	Science Applications Research, 4400 Forbes Blvd., Lanham MD 20706, USA
24.	**Biggs, Peter**	Physical Chemical Laboratory, Oxford University, South Parks Road, Oxford OX1 3Q2, Great Britain
25.	**Bisling, Peter**	GKSS-Forschungszentrum, Postfach 1160, 2054 Geesthacht, F.R. Germany
26.	**Bodenmann, Patrick**	Institute for Applied Physic, University Berne, Sidlerstr. 5, 3000 Berne, Switzerland
27.	**Bojkov, Rumen**	WMO, Case Postale No. 1, 1211 Genf, Switzerland
28.	**Bösenberg, Jens**	MPI f. Meteorologie, Bundesstr. 55, 2000 Hamburg 13, F.R. Germany
29.	**Borchers, Reinhard**	Max-Planck-Institut f. Aeronomie, Postfach 20, 3411 Katlenburg-Lindau, F.R. Germany
30.	**Bräutigam, Hans**	Kali-Chemie AG, Postfach 220, 3000 Hannover 1, F.R. Germany
31.	**Brasseur, Guy**	Inst. d'Aeronomie Spatiale, 3, Avenue Circulaire, 1180 Brussels, Belgium
32.	**Brasunas, John**	NASA/GSFC, Code 693.2, Greenbelt MD 20771, USA

33.	**Brost, Richard**	NCAR, P.O.Box 3000, Boulder, CO 80303-3000, USA
34.	**Browell, Edward**	NASA/Langley Research Center, Hampton, VA 23665-5225, USA
35.	**Brueckner, Guenter**	Naval Research Laboratory, Code 4160, Washington, D.C. 20375, USA
36.	**Brühl, Christoph**	MPI f. Chemie, Postfach 3060, 6500 Mainz, F.R. Germany
37.	**Brüning, Dirk**	ICH3, KFA Jülich, Postfach 1913, 5170 Jülich, F.R. Germany
38.	**Cacciani, Marco**	Universite di Rome "La Sapienza", Rue Aldo Moro 2, 00185 Roma, Italy
39.	**Cacho, Javier**	INTA, Carretora de Gialuir, 28890 Torrejon, Madrid, Spain
40.	**Calander, Karin**	Swedisch Environm. Res. Inst., Box 470 86, 402 58 Göteborg, Sweden
41.	**Callis, Linwod B.**	NASA Langley, Hampton, VA, USA
42.	**Camy-Peyret, Claude**	CNRS, Tour 13, 4 Place Jussieu, 75255 Paris Cedex 05, France
43.	**Cariolle, Daniel**	EERM/Centre National de Recherches Met., 42 Av. Cariolis, Toulouse 31957, France
44.	**Carlotti, Massimo**	Universita di Bologna, Dipartimento di Chimica Fisica e Inorganica, 40136 Bologna, Italy
45.	**Carnuth, Walter**	Fraunhofer Institut, Kreuzeckbahnstr. 19, 8100 Garmisch-Partenkirchen, F.R. Germany
46.	**Carroll, Mary Anne**	NOAA/Aeronomy Laboratory, R/E/Al6, 325 Broadway, Boulder, CO 80303, USA
47.	**Cebula, Richard**	STSystems Corporation, 4400 Forbes Blvd., Lanham MD 20706, USA
48.	**Chakrabarty, D. K.**	Physical Research Laboratory, Navrangpura Ahmedabad, 380 009, India
49.	**Chandra, Sushil**	NASA/Goddard Space Flight Center, Code 616, Greenbelt, MD 20771, USA
50.	**Chang, Julius**	Atmospheric Sciences Center, State University of New York at Albany, 100 Fuller Road, Albany, NY 12205, USA
51.	**Chatfield, R.B.**	National Center for Atmospheric Research, P.O.Box 3000, Boulder, CO 80307, USA
52.	**Chipperfield, Martyn**	Cambridge University, Dept. of Physical Chemistry, Lensfield Road, Cambridge CB2 1EP, United Kingdom
53.	**Chubachi, Shigeru**	Meteorological Research Institute, 1-1 Nagamine, Tsukuba, Ibaraki 305, Japan
54.	**Cieslik, Stanislaus**	Joint Reserach Center, Eurotom, Ispra, Italy
55.	**Cisneros, Juan M.**	Instituto Noc.de Met., Ceudad Universitaria, 28071 Madrid, Spain
56.	**Clancy, R. Todd**	LASP, University of Colorado, Box 392, Boulder, CO 80309, USA
57.	**Clarkson, Thomas**	New Zealand Met. Service, P.O.Box 722, Wellington, New Zealand
58.	**Claude, Hans**	Deutscher Wetterdienst, Met. Obs. Hohenpeißenberg, Albin-Schwaiger-Weg 10, 8126 Hohenpeißenberg, F.R. Germany
59.	**Clough, Toni**	AER, Inc., Cambridge, MA 02139, USA
60.	**Coffey, Michael T.**	National Center for Atmospheric Research, Boulder, CO 80307, USA
61.	**Colbeck, Ian**	University of Essex, Dept. of Chemistry, Wivenhoe Park, Colchester CO4 3SQ, Great Britain
62.	**Colombo, Tiziano**	Italian Met. Service, Via Spatiale Ovest No. 5, 41029 Sestola, Italy
63.	**Connell, Peter**	Lawrence Livermore National Laboratory, P.O.Box 808 L-262, Livermore, CA 94550, USA
64.	**Cox, Richard A.**	Harwell Laboratory, Engineering Sciences Division, B 551, Harwell Laboratory, Didcot, Oxfordshire OX11 ORA, United Kingdom
65.	**Cunnold, Derek**	Georgia Tech. School of Geophysical Sciences, Atlanta, GA 30332, USA
66.	**Dameris, Martin**	Universität zu Köln, Institut f. Geophysik und Meteorologie, 5000 Köln 41, F.R. Germany

67.	Davis, G.R.	Mullard Space Science Lab., Holmbury Street Mary, Dorking, Surrey, RH5 6NT, Great Britain
68.	de Bakker, Carl-Peter	Universität Wuppertal, Gaußstr. 20, 5600 Wuppertal, F.R. Germany
69.	Degórska, Malgorzata	Polish Academy of Science, Inst.of Geophysics, Pasteura 3, P.O.Box 155, 00-973 Warsaw, Poland
70.	de la Noë, Jérôme	Observatoire de Bordeaux, BP 21, 33270 Floirac, France
71.	Deluisi, John	NOAA/ERL/ARL/GMCC, 325 Braodway, Boulder, CO 80303, USA
72.	de Muer, Dirk	Koninklijk Met. Institut, Ringlaan 3, 1180 Brussels, Belgium
73.	de Zafra, Robert	State Univ. of New York, Inst.for Atmospheric Sciences, Physics Bld., Stony Brook, NY 11794, USA
74.	Diament, P.	Meteorological Office, London Road, Bracknell, Berkshire RG12 2SZ, Great Britain
75.	Dickerson, Russell	University of Maryland, Department of Meteorology, College Park, MD 20742, USA
76.	di Girolamo, Paolo	Dipartimento di Fisica, Universita "La Sapienza", Via della Porretta No. 5, Roma, Italy
77.	Dinelli, Bianca Maria	Inst. di Spettroscopia Moleculare, Via de Castagnoli 1, 40126 Bologna, Italy
78.	Ding-Wen Wei	Inst. of Atmospheric Physics, Acadmia Sinica, Beijing, China
79.	Dirmhirn, Inge	Institut f. Meteorologie, Universität f. Bodenkultur, Gregor-Mendel-Str. 33, 1180 Wien, Austria
80.	di Sarra, Alcide	Dipt. di Fisica, Universite di Roma "La Sapienza", Via A. Moro 5, 00185 Roma, Italy
81.	Dlugi, Ralph	Universität München, Meteorologisches Institut, Theresienstr. 37, 8000 München, F.R. Germany
82.	do Nascimento, M.F.	CNRS, Laboratoire d'Aerologie, 118, Route de Narbonne, 31062 Toulouse Cedex, France
83.	Dorn, Hans-Peter	KFA Jülich, ICH-3, Postfach 1913, 5170 Jülich, F.R. Germany
84.	Douglass, Anne	Applied Research Corp., Atmospheric Chemistry and Dynamics Branch, NASA/GSFC, Code 616, Greenbelt, MD 20771, USA
85.	Duscha, Harald	Max-Planck-Institut für Aeronomie, Postfach 20, 3411 Katlenburg-Lindau, F.R. Germany
86.	Dütsch, Hans Ulrich	Kantstr. 11, CH-8044 Zürich, Switzerland
87.	Ebel, Adolf	University of Cologne, Institute for Geophysics and Meteorology, 5000 Cologne 41, F.R. Germany
88.	Ehhalt, Dieter H.	Institut für Chemie der KFA Jülich, Postfach 1913, 5170 Jülich, F.R. Germany
89.	Elansky, Nikolay F.	Institute of Atmospheric Physics, Academy of Sciences USSR, Pyzhevsky 3, 109027 Moscow, USSR
90.	El Naggar, Saad	Alfred-Wegener-Institut, Columbusstr., 2850 Bremerhaven, F.R. Germany
91.	Enders, Gerhard	Lehrstuhl f. Bioklimatologie u. Angewandte Meteorologie, Universität München, Amalienstr. 52, 8000 München 40, F.R. Germany
92.	Erisman, Jan-Willem	Institute of Public Health and Environmental Protection (RIVM), P.O.Box 1, 3720 BA, Bilthoven, The Netherlands
93.	Evans, Robert	University of Colorado, NOAA/ERL/ARL/GMCC, Mail Code R/E/AR4, 325 Boradway, Boulder, Colorado 80303, USA
94.	Evans, Wayne F.J.	Atmospheric Environment Service, 4905 Dufferin Street, Ontario M3H 5T4, Canada
95.	Fabian, Peter	Max-Planck-Institut für Aeronomie, Postfach 20, 3411 Katlenburg-Lindau, F.R. Germany
96.	Fark, Horst	KFA Jülich, Postfach 1913, 5170 Jülich, F.R. Germany
97.	Farkas, Edith	N.Z. Meteorological Service, 80 Ranui Cres, Wellington 4, New Zealand
98.	Feichter, Johann	Max-Planck-Institut für Chemie, Postfach 3060, 6500 Mainz, F.R. Germany

99.	**Feister, Uwe**	Main Meteorological Observatory, Telegrafenberg, 1561 Potsdam, G.D.R.
100.	**Fiocco, Giorgio**	Department of Physics, University "La Sapienza", P.le A. Moro 5, 00185 Rome, Italy
101.	**Fischer, Herbert**	Institut für Meteorologie und Klimaforschung, KFZ Karlsruhe, Postfach 3640, 7500 Karlsruhe 1, F.R. Germany
102.	**Fishman, Jack**	NASA Langley Research Center, Mail Stop 401A, Hampton, VA 23665-5225, USA
103.	**Fleig, Al**	NASA/GSFC, Code 600, Greenbelt, MD 20771, USA
104.	**Flentje, Günther**	Max-Planck-Institut für Aeronomie, Postfach 20, 3411 Katlenburg-Lindau, F.R. Germany
105.	**Fogal, Pierre**	University of Denver, Dept. of Physics, University Park, Denver, CO 80208-020, USA
106.	**Franchois, Paul**	CIRES/University of Colorado NOAA/GMCC, R/E/AR4, 325 Boradway, Boulder, CO 80303, USA
107.	**Fries, Jochen**	DFVLR, Institut für Optoelektronik, 8031 Wessling, F.R. Germany
108.	**Froidevaux, Lucien**	JET Propulsion Laboratory, 4800 Oak Grove Drive, Pasadena, CA 91109, USA
109.	**Gandrud, Bruce W.**	Atmospheric Chemistry Division, NCAR, Boulder, CO 80307, USA
110.	**Gansert, Dirk**	Systemat.-Geobot. Institut Göttingen, Untere Karspüle 2, 3400 Göttingen, F.R. Germany
111.	**Garber, Wolf**	Umweltbundesamt, Bismarckplatz 1, 1000 Berlin 33, F.R. Germany
112.	**Garcia, Maria**	Columbia University, GISS, 2880 Broadway, New York, NY 10025, USA
113.	**Gardiner, Brian G.**	British Antarctic Survey, Madingley Road, Cambridge CB3 OET, Great Britain
114.	**Gerhardt, Philipp**	KFA Jülich, ICH-3, Postfach 1913, 5170 Jülich, F.R. Germany
115.	**Ghazi, Anver**	Commission of the European Communities, Rue de la Loi 100, 1049 Bruxelles, Belgium
116.	**Gil, Manuel**	INTA, Torrejon de Ardoz 28850, Spain
117.	**Gillotay, Didier**	Institute d'Aeronomie Spatiale de Belgique, Avenue Circulaire, 3, B-1080 Brussels, Belgium
118.	**Godin, Sophie**	JPL/NASA, P.O.Box 367, Wrightwood, CA 92397, USA
119.	**Gömer, Dagmar**	Umweltbehörde des Senats Hamburg, Neuer Kemp 23-25, 2000 Hamburg, F.R. Germany
120.	**Goldman, Aaron**	University of Denver, Dept. of Physics, Denver, Colorado 80208, USA
121.	**Goutail, Florence**	Service d'Aeronomie du CNRS, BP 3, 91371 Verrieres le Buisson, France
122.	**Granier, Claire**	Service d'Aeronomie du CNRS, BP 3, 91371 Verrieres le Buisson, France
123.	**Grant, Keith**	Lawrence Livermore National Laboratory, P.O.Box 808, L-262, Livermore, CA 94550, USA
124.	**Grasnik, Karl-Heinz**	Meteorologisches Observatorium, Telegrafenberg, Potsdam, GDR
125.	**Grass, Robert D.**	NOAA/ERL/ARL/GMCC, 325 Broadway, Boulder, CO 80303, USA
126.	**Guicherit, Robert**	TNO Division of Technology for Society, P.O.Box 217, 2600 AE Delft, The Netherlands
127.	**Gupta, Surya**	Physical Research Laboratory, Navarangpura, 380 009 Ahmedabad, India
128.	**Guthrie, Paul**	NASA/GSFC, Code 616, Greenbelt, MD 20771, USA
129.	**Haigh, Joanna**	Centre for Remote Sensing, Imperial College, London SW7 2BZ, Great Britain
130.	**Hakola, Hannele**	Finnish Meteorological Institute, Sahhajankatu 22 E, 00810 Helsinki, Finland
131.	**Haleem, Noor M.**	Geophysical Centre, Pakistan Met. Deptartment, P.Box 2, Quetta, Pakistan
132.	**Hanson, David**	University of Minnesota, 116 Church St. SE, Physics Building, Minnesota, Minneapolis 55455, USA

133.	**Hansson, Anne L.**	Atmosph. Environ. Service ARPP, 4905 Dufferin Street, Downsview, Ontario M3H ST4, Canada
134.	**Harris, Neil**	University of California, Department of Chemistry, Irvine, CA 92717, USA
135.	**Hartmann, Gerd**	Max-Planck-Institut für Aeronomie, Postfach 20, 3411 Katlenburg-Lindau, F.R. Germany
136.	**Hartogh, Paul**	Max-Planck-Institut f. Aeronomie, Postfach 20, 3411 Katlenburg-Lindau, F.R. Germany
137.	**Harvey, George**	NOAA, 4301 Rickenbacker Causeway, Miami, Florida 33149, USA
138.	**Hassan, Galal K.Y.**	Egyption Met. Authority, P.O.Box 11784, Cairo, Egypt
139.	**Hausamann, Dieter**	DFVLR, Institut für Optoelektronik, 8031 Oberpfaffenhofen, F.R. Germany
140.	**Hauswald, Horst**	Bund für Umwelt und Naturschutz, Erbprinzenstr. 18, 7753 Allensbach 1, F.R. Germany
141.	**Hayashida-Amano, S.**	National Institute f. Environ. Studies, Onogawa, Tsukuba, Ibaraki 305, Japan
142.	**Hayman, Garry David**	Harwell Laboratory, Environmental and Medical Science Division, Didcot, Oxon OX11 ORA, Great Britain
143.	**Heaps, William**	NASA/Goddard Space Flight Center, Code 615, Greenbelt, MD 20771, USA
144.	**Heath, Donald**	NOAA/Environmental Research Laboratory, 325 Broadway, Boulder, CO 80303, USA
145.	**Helas, Günter**	Max-Planck-Institut f. Chemie, Dept. Biogeochemie, Postfach 3060, 6500 Mainz, F.R. Germany
146.	**Helten, Manfred**	KFA Jülich, Postfach 1913, 5170 Jülich, F.R. Germany
147.	**Henderson, Grant**	Atmosph. Environ. Service, 4905 Dufferin St. York, Ontario M3H 5T4, Canada
148.	**Henriksen, Kjell**	The Auroral Observatory, University of Tromso, 9000 Tromso, Norway
149.	**Heydtmann, Horst**	Universität Frankfurt, Inst. f. Physik und Theoretische Chemie, Postfach 11 19 32, 6000 Frankfurt, F.R. Germany
150.	**Hilsenrath, Ernest**	NASA/Goddard Space Flight Center, Code 616, Greenbelt MD 20771, USA
151.	**Hingane, Laxman**	Indian Institute of Tropical Meteorology, Shivajinagar, 411 005 Pune, India
152.	**Hirschberg, Michaela**	Max-Planck-Institut für Aeronomie, Postfach 20, 3411 Katlenburg-Lindau, F.R. Germany
153.	**Hoefken, Klaus D.**	GSF München, Projektträger, Ingolstädter Landstr. 1, 8042 Neuherberg, F.R. Germany
154.	**Hoffmann, José A.J.**	Consejo Nacional de Invest. Cientificas y Tecnicas, 25 de Moyo 658, 1002 Buenos Aires, Argentina
155.	**Hofzumahaus, A.**	KFA Jülich, Postfach 1913, 5170 Jülich, F.R. Germany
156.	**Hood, Lon L.**	Lunar and Planetary Lab., University of Arizona, Tucson, Arizona 85721, USA
157.	**Horie, Osamu**	Max-Planck-Institut f. Chemie, Postfach 3060, 6500 Mainz, F.R. Germany
158.	**Horowitz, Abraham**	Deptartment of Rad. Chemistry, Yavne 70600, Israel
159.	**Hou, Arthur**	Atmospheric and Environmental Research, 840 Memorial Drive, Cambridge, MA 02139, USA
160.	**Hough, A. Michael**	Harwell Laboratory, Environmental and Medical Sciences, Didcot, Oxfordshire, Great Britain
161.	**Hov, Øystein**	Norwegian Institute for Air Research, P.O.Box 64, 2001 Lillestram, Norway
162.	**Hudson, Robert**	NASA/Goddard Space Flight Center, Code 616, Greenbelt, MD 20771, USA
163.	**Hübler, Gerhard**	NOAA - Aeronomy Laboratory, R-E-AL7, 325 Boradway, Boulder, CO 80303, USA
164.	**Hutchinson, John**	BASF AG, HUM/EB, Bau F500, 6700 Ludwigshafen, F.R. Germany
165.	**Ilyas, Mohammad**	School of Physics, University of Science of Malaysia, 118000 USM Penang, Malaysia
166.	**Isaksen, Ivar**	University of Oslo, Institute of Geophysics, P.O.Box 1022, Blindern 0315 Oslo, Norway

167.	**Jackman, Charles**	NASA/Goddard Space Flight Center, Code 616, Greenbelt, MD 20771, USA
168.	**Jaramillo, Mauricio**	S.U.N.Y. at Stony Brook, Dept. of Physics, Stony Brook, NZ 11794, USA
169.	**Joffre, Sylvain**	Finnish Met. Institute, Sahaajankatu 22E, SF-00810 Helsinki, Finland
170.	**Jones, Roderic Lewis**	Meteorological Office, London Road, Bracknell, Berkshire RG12 2SZ, Great Britain
171.	**Jonson, Jan Eiof**	Institute of Geophysics, University of Oslo, P.O.Box 1022, Blindern 0315 Oslo 3, Norway
172.	**Jørgensen, Torben**	Danish Met. Institute, Division of Geophysics, Lyngbyvej 100, 2100 Copenhagen, Denmark
173.	**Josefsson, Weine**	SMHI, Swedish Met. and Hydrol. Institute, S-601 76 Norrköping, Sweden
174.	**Kämpfer, Niklaus**	Institute of Applied Physics, Sidlerstr. 5, CH-3012 Berne, Switzerland
175.	**Karcher, Fernand**	Centre National de Recherches Met., 42m, Avenue Coriolis, 31057 Toulouse-Cedex, France
176.	**Karbach, Hans-Josef**	MPI f. Chemie, Abtl. Luftchemie, Postfach 3060, 6500 Mainz, F.R. Germany
177.	**Kawahira, Kohji**	Toyama National College of Technology, 13 Hongo-cho, Toyama 939, Japan
178.	**Keating, Gerald M.**	NASA Langley Research Center, MS 401 B, Hampton, VA 23665, USA
179.	**Kelder, Hennie**	Royal Netherl. Met. Inst., P.O.Box 201, 3730 AE de Bilt, Netherlands
180.	**Kelessis, Apostolos**	University of Thessaloniki, Lab. of Atmospheric Physics, P.O.Box 149, Thessaloniki 54006, Greece
181.	**Keppler, Erhard**	Max-Planck-Institut für Aeronomie, Postfach 20, 3411 Katlenburg-Lindau, F.R. Germany
182.	**Kerr, James B.**	Atmosph. Environ. Service, 4905 Dufferin Street, Downsview, Ontario M3H 5T4, Canada
183.	**Khattatov, Y.**	Central Aerological Observatory, State Committe of Hydrometeorology, Moscow I23376, USSR
184.	**Kinnison, Douglass**	Lawrence Livermore Laboratory, P.O.Box 808, L-262, Livermore, CA 94550, USA
185.	**Kirchhoff, Volker**	INPE-CP 515, 12201 S. Jose Dos Campos, Sao Paulo, Brazil
186.	**Kitaoka, Tatsumi**	FHL, 158-4-25-10 Tamagawa, Setagaya-ku, Tokyo, Japan
187.	**Klenk, Kenneth**	SAR/National Space Science Data Center, 4400 Forbes Blvd., Lanham, MD, USA
188.	**Kley, Dieter**	KFA Jülich, Postfach 1913, 5170 Jülich, F.R. Germany
189.	**Koenig, Gloria**	NOAA/ERL/ARL/GMCC, Mail Code R/E/AR4, 325 Broadway, Boulder, CO 80303, USA
190.	**Köhler, Axel**	Institut der Deutschen Wirtschaft, Postfach 510669, 5000 Köln 51, F.R. Germany
191.	**Kondo, Yutaka**	Nagoya University, Research Institute of Atmospherics, 3013 Honohara, Tokyokawa, Aichi, Japan
192.	**Komhyr, Walter D.**	NOAA/R/E/AR4, 325 Broadway, Boulder, Colorado 80303, USA
193.	**Koppmann, Ralf**	KFA Jülich, Postfach 1913, 5170 Jülich, F.R. Germany
194.	**Kosmus, Walter**	Inst. f. Analytical Chemistry, Universitätsplatz 1, A-8010 Graz, Austria
195.	**Kourtidis, Kostas**	Max-Planck-Institut für Aeronomie, Postfach 20, 3411 Katlenburg-Lindau, F.R. Germany
196.	**Krueger, Arlin**	NASA/Goddard Space Flight Center, Greenbelt, MD 20771, USA
197.	**Krüger, Bernd C.**	Alfred-Wegener-Institut, Columbusstr., 2850 Bremerhaven, F.R. Germany
198.	**Künzi, Klaus**	University of Berne, Sidlerstr. 5, 3012 Berne, Switzerland
199.	**Kundu, Namita**	National Physical Lab., Radio Science Division, New Delhi 110 012, India
200.	**Kyrölä, Erkki**	Finnish Met. Inst., P.O.Box 503, SF-00101 Helsinki, Finland
201.	**Lahoz, William**	Edinburth University, Department of Meteorology, Kings Building, Edinburgh EH9 3J2, Great Britain

202.	Lal, Shyam	Physical Research Laboratory, Navrangpura 380 009, Ahmedabad, India
203.	Lamb, Ken	SCI-TEC Instruments Inc., 1526 Fletcher Road, Saskatoon, S1M5MI, Canada
204.	Langematz, Ulrike	Met.Institut der FU Berlin, Dietrich-Schäfer-Weg 6-10, 1000 Berlin 41, F.R. Germany
205.	Larsen, Sören H.H.	University of Oslo, Dept. of Physics, Box 1048, Blindern 0316 Oslo 3, Norway
206.	Lary, David John	Dept. of Physical Chemistry, Cambridge University, Lensfield Road, Cambridge, Great Britain
207.	Law, Katharine	Dept. of Phys. Chemistry, Cambridge University, Lensfield Road, Cambridge, CB7 2EP, Great Britain
208.	Levy II, Hiram	GFDL/NOAA, P.O.Box 303, Princeton NJ 08540, USA
209.	Legrand, Michel	Lab. de Glaciologie et Geophys. de l'Environ., BP 96, 38402 Saint Martin d'Heres Cedex, France
210.	Lloyd, Steven	Harvard University, Atmospheric Research Project, 40 Oxford St., Cambridge, Mass. 02138, USA
211.	Loidl, Alfred	Max-Planck-Institut für Aeronomie, Postfach 20, 3411 Katlenburg-Lindau, F.R. Germany
212.	London, Julius	University of Colorado, Boulder, CO 80303, USA
213.	Longenecker, David	NOAA/CIRES, R/E/AR4, 325 Broadway, Boulder, CO 80303, USA
214.	Low, P.S.	University of East Anglia, Climatic Reserach Unit, Norwich NR4 7TJ, United Kingdom
215.	Lu, Da-ren	Acadmia Sinica, Inst. of Atmospheric Physics, Beijing 100 011, China
216.	Luria, Menachem	The Hebrew University, Dept. of Environmental Sciences, Jerusalem, Israel
217.	Makino, Yukio	Meteorological Research Institute, 1-1 Nagamine, Tsukuba, Ibaraki-ken 305, Japan
218.	Mani, Anna	Raman Research Institute, Bangalore 560 080, India
219.	Mankin, William G.	NCAR, P.O.Box 3000, Boulder, CO 80307, USA
220.	Manning, Amanda J.L.	University of Oxford, Clarendon Laboratory, Parks Road, Oxford OX1 3PU, Great Britain
221.	Markiewicz, Weronika	Environmental Pollution Abatement Centre, ul. Kossutha 6, 40-832 Katowice, Poland
222.	Mateer, Carl	1210-255 Bamburgh CIR, Scarborough, Ont MIW 3T6, Canada
223.	Matthews, W. Andrew	PEL Lauder, DSIR, Lauder, Central Ontago, New Zealand
224.	Mauersberger, Konrad	University of Minnesota, 148 Physics Bldg, 116 Church Str. SF, Minneapolis, MN 55455, USA
225.	McConnell, John C.	Observatoire de Besancon, L'Avenue de L'Observatoire, 25044 Besancon Cedex, France
226.	McElroy, C. Thomas	Atmosph. Environ. Service, 4905 Dufferin Street, Downsview, Ontario, Canada
227.	McMillen, Bob	NOAA/ATDD, P.O.Box 2456, Oak Ridge, TN 37831, USA
228.	**McKeen, Stuart**	**NOAA RE/AL4, 325 Broadway, Boulder, Colorado 80303, USA**
229.	McKenzie, Richard L.	PEL Lauder, PB Omakau, Central Otago, New Zealand
230.	McPeters, Richard D.	Goddard Space Flight Center, Code 616, Greenbelt, MD 20771, USA
231.	Megie, Gérard	Service d'Aeronomie-CNRS, BP 3, 91371 Verrieres-le-Buisson, France
232.	Miller, Alvin	National Weather Service, NOAA, 5200 Auth Road, Room 805, Washington, D.C., USA
233.	Molina, Mario J.	Jet Propulsion Laboratory, M/S 183-301, 4800 Oak Grove Drive, Pasadena, CA 91109, USA
234.	Moortgat, Geert	MPI f. Chemie, Abtl. Luftchemie, Postfach 3060, 6500 Mainz, F.R. Germany
235.	Moreau, Guy	LPCE/CNRS, 34 Avenue de la Recherche Scientifique, 45071 Orleans Cedex 2, France
236.	Muller, Christian	Belgian Institute for Space Aeronomy, Avenue Circulaire 3, B-1180 Brussels, Belgium
237.	Muramatsu, Hisafumi	Kyoto University, Disaster Prevention Research Institute, Gokasho, Uji, Kyoto 611, Japan

238.	**Murphy, Daniel**	Aeronomy Lab/NOAA/ERL, 325 Broadway, Boulder, CO 80303-3328, USA
239.	**Nagatani, Ronald**	NOAA/Climate Analysis Center, W/NMC53, 5200 Auth Road, Washington, DC 20233, USA
240.	**Nastrom, Gregory**	Department of Earth Sciences, St. Cloud State University, St. Cloud, MN 56301, USA
241.	**Natarajan, Murali**	ST Systems Corporation, 28, Reserach Drive, Hampton, Virginia 23666, USA
242.	**Naudet, Jean-Pierre**	LPCE/CNRS, 34 Avenue de la Recherche Scientifique, 45071 Orleans, France
243.	**Neuber, Roland**	Alfred-Wegener-Institut, Columbusstr., 2850 Bremerhaven, F.R. Germany
244.	**Newchurch, Michael**	Teledyne Brown Engineering, Cummings Research Park, Huntsville, Alabama 35807, USA
245.	**Newell, Reginald E.**	M.I.T., 54-1522, Dept. of Earth, Atmospheric and Planetary Sciences, Cambridge, Mass. 02139, USA
246.	**Newman, Paul A.**	Applied Research Corporation, NASA/GSFC, Code 616, Greenbelt, MD 20771, USA
247.	**Nicolet, Marcel**	Inst. d'Aeronomie Spatiale, Avenue Circulaire 3, 1180 Brussels, Belgium
248.	**Njau, Leonhard**	Kenya Meteorological Department, Dagoretti Corner, Ngong Road, P.O.Box 30259, Nairobi, Kenya
249.	**Nolting, Friedrich**	Fraunhofer Institut, Nikolai-Fuchs-Str. 1, 3000 Hannover 61, F.R. Germany
250.	**Nouchi, Isamu**	National Institute of Agro-Environ. Sciences, Kannondai 3-1-1, Tsukuba, Ibaraki 305, Japan
251.	**Oelhaf, Hermann**	Institut f. Meteorologie und Klimaforschung, KFZ Karlsruhe GmbH, Postfach 3640, 7500 Karlsruhe 1, F.R. Germany
252.	**Offermann, Dirk**	Bergische Universität Wuppertal, Gaußstr. 20, 5600 Wuppertal, F.R. Germany
253.	**Ogawa, Toshiro**	University of Tokyo, Geophysics Research Laboratory, Hongo, Bunkyo-ku, Tokyp 113, Japan
254.	**Oltmans, Samuel**	NOAA/GMCC, R/E/AR4, 325 Broadway, Boulder, CO, USA
255.	**Paetzold, Hans-Karl**	Institute of Geophysics and Meteorology, Volksgartenstr. 10, 5000 Cologne, F.R. Germany
256.	**Papayannis, A.**	Service d'Aeronomie, CNRS, B.P. No.3, 91371 Verrieres le Buisson, France
257.	**Parr, Douglas**	Oxford University, Physical Chemistry Laboratory, South Parks Road, Oxford, Great Britain
258.	**Pelon, Jacques**	CNRS - Service d'Aeronomie, Universite Paris 6, TIS-ES, 4 Place Jussien, 75030 Paris Cedex 05, France
259.	**Perner, Dieter**	Max-Planck-Institut f. Chemie, Saarstr. 23, 6500 Mainz, F.R. Germany
260.	**Peter, Reto**	Institute for Applied Physic, Sidlerstr. 5, CH-3000 Berne, Switzerland
261.	**Perov S.P.**	Central Aerological Observatory, State Committee of Hydrometeorology and Control of Natural Environment, I23376 Moscow, USSR
262.	**Pfeilsticker, Klaus**	Alfred-Wegener-Institut, Columbustr., 2850 Bremerhaven, F.R. Germany
263.	**Pickett, Herbert**	JPL, Calif. Inst. of Technology, 4800 Oak Grove Drive, Mailstop 183-601, Pasadena, CA 91109, USA
264.	**Piotrowicz, Stephen**	NOAA/ AOML, 4301 Rickenbacker Causeway, Miami, Florida 33149, USA
265.	**Pirre, Michel**	LPCE/CNRS, 34 Avenue de la Recherche Scientifique, 45071 Orleans Cedex, France
266.	**Platt, Ulrich**	KFA Jülich, Postfach 1913, 5170 Jülich, F.R. Germany
267.	**Podolske, James**	NASA Ames Research Center, MS 245-5, Space Science Division, Moffett Field, CA 94703, USA
268.	**Pommereau, J.-P.**	CNRS - Service d'Aeronomie, 91371 Verrieres-le-Buisson, France
269.	**Poppe, D.**	KFA Jülich, Postfach 1913, 5170 Jülich, F.R. Germany
270.	**Prather, Michael**	NASA/GISS, 2880 Broadway, New York, NY 10025, USA

271. **Price, Jeremy D.** Physics Dept., University of Wales, Penglais Campus, Aberystwyth, Dyfed, Wales, SY23 3BZ, G.B. (please send 2 proceedings)

272. **Proffitt, Michael** NOAA Aeronomy Lab. and CIRES, R/E/AL6, 325 Broadway, Boulder, CO 80303, USA

273. **Pueschel, R.F.** NASA Ames, Mail Stop 245-5, Moffett Field, CA 94035, USA

274. **Puliafito, Enrique** Max-Planck-Institu für Aeronomie, Postfach 20, 3411 Katlenburg-Lindau, F.R. Germany

275. **Puliafito, Carlos** Max-Planck-Institut für Aeronomie, Postfach 20, 3411 Katlenburg-Lindau, F.R. Germany

276. **Putz, Erich** Inst. f. Met. und Geophysik, Universität Graz, Halbärthgasse 1, A-8010 Graz, Austria

277. **Rajewska-Wiech, B.** Polish Academy of Sciences, Institute of Geophysics, P.O.Box 155, Pasteura 3, 00-973 Warsaw, Poland

278. **Ramanamurty, Y.V.** National Physical Laboratory, Dr. K.S. Krishnan Road, New Delhi 110 012, India

279. **Ramaroson, R.A.** LPCE/CNRS, 3A, Avenue de la Recherche Scientifique, 45071 Orleans Cedex 2, France

280. **Rasmussen, Jean** Greenpeace, Thomas Lanbs Gade 11-13, 2100 Kobenhavn, Denmark

281. **Reeves, Claire** Universitet Oslo, Institut for Gedophysic, P.O.Box 1022, Blindern, Oslo 3, Norway

282. **Regener, Victor H.** VHR Systems, 7200 Jefferson NE, Albuquerque, NM 87109, USA

283. **Remedios, John** Clarendon Laboratory, Department of Atmospheric, Oceanic and Planetary Physics, Oxford University, Parks Road, Oxford OX1 3PU, Great Britain

284. **Rigaud, Pierre** LPCE/CNRS, 3A, Avenue de la Recherche Scientifique, 45071 Orleans Cedex 2, France

285. **Robert, Claude** LPCE/CNRS, 3A, Avenue de la Recherche Scientifique, 45071 Orleans Cedex 2, France

286. **Rodriguez, Jose M.** Atmospheric and Environmental Research, Inc., 840 Memorial Drive, Cambridge, MA 02139, USA

287. **Roeder, Herbert A.** Perkin-Elmer ASD, 2771 N. Garey Ave., Pomona, CA 91761, USA

288. **Röth, Ernst-Peter** Universität Essen, FB8-Physikalische Chemie, Universitätsstr. 5, 4300 Essen 1, F.R. Germany

289. **Rognerud, Bjørg** University of Oslo, Institute for Geophysics, PB 1022, Blindern 0315 Oslo 3, Norway

290. **Rohrer, Franz** KFA Jülich, ICH-3, Postfach 1913, 5170 Jülich, F.R. Germany

291. **Rood, Richard** NASA/GSFC, Code 616, Greenbelt, MD 20771, USA

292. **Roscoe, Howard K.** Heriot Watt University, Rutherford Appleton Laboratory, Chilton, Didcot, Oxon OX1 OQR, Great Britain

293. **Rosenbauer, Helmut** Max-Planck-Institut für Aeronomie, Postfach 20, 3411 Katlenburg-Lindau, F.R. Germany

294. **Rossi, Michael** Department of Chemical Kinetics, Chemical Physics Laboratory, SRI International, Menlo Park, CA 94025, USA

295. **Rowland, F. S.** University of California, Department of Chemistry, Irvine, CA 92717, USA

296. **Rubin, Georg** University of Berne, Institute of Applied Physics, Sidlerstr. 5, CH-3012 Berne, Switzerland

297. **Rudolph, Jochen** KFA Jülich, ICH-3, Postfach 1913, 5170 Jülich, F.R. Germany

298. **Rusch, David** Laboratory for Atmospheric and Space Physics, University of Colorado, Campus Box 392, Boulder, CO 80309, USA

299. **Saiz de Aja, Maria J.** INTA, Dept. Programy Espaciales, Torrejon de Ardoz, 28850 Madrid, Spain

300. **Sanders, Ryan** NOAA Aeronomy Laboratory, R/E/AL8, 325 Broadway, Boulder, CO 80303, USA

301. **Seckmeyer, Gunther** Gesellschaft f. Strahlen- und Umweltforschung mbH, Ingolstädter Landstr. 1, 8042 Neuherberg, F.R. Germany

302. **Seiler, Wolfgang** Fraunhofer Institut f. Atmosphärische Umweltforschung, Kreuzeckbahnstr. 19, 8100 Garmisch-Partenkirchen, F.R. Germany

303. **Sekiguchi, Yoshiro** Japan Weather Association, Kaiji-Center Bldg., 4-5 Kojimachi, Chiyoda-ku, Tokyo 102, Japan
304. **Shibasaki, Kazuo** Kokugakuin University, 10-28 Higashi 4-chome, Shibuya-ku, Tokyo 150, Japan
305. **Shine, Keith P.** Dept. of Atmospheric Physics, University of Oxford, Parks Road, Oxford OX1 3PU, Great Britain
306. **Simidchiev, Dimitri** Institute of Hydrology & Meteorology, Chervena Rosa Str. Bl.36 A, Apt. 4, 1404 Sofia, Bulgaria
307. **Singh, Onkar N.** Benares Hindu University, Department of Applied Physics, Institute of Technology, Varanasi 221 005 , India
308. **Simon, Paul C.** Inst. d'Aeronomie Spatiale, 3 Av. Circulaire, B-1180 Bruxelles, Belgium
309. **Smit, Herman G.J.** KFA Jülich, ICH-2, Postfach 1913, 5170 Jülich, F.R. Germany
310. **Smith, Stuart** P.C.L. Oxford University, South Parks Road, Oxford, Oxon Great Britain
311. **Solberg, Sverre** Institute of Geophysics, University of Oslo, PB 1022, Blindern 0315, Oslo 3, Norway
312. **Subbaraya, B.H.** Physical Research Laboratory, Navrangpura, 380 009 Ahmedabad, India
313. **Summers, Michael E.** Naval Research Laboratory, Mail Code 4140, Washington, D.C. 20375, USA
314. **Schmidt, Dieter** Am Klingenberg 14, 8729 Sulzdorf, F.R. Germany
315. **Schmidt, Manfred** Max-Planck-Institut für Aeronomie, Postfach 20, 3411 Katlenburg-Lindau, F.R. Germany
316. **Schmidt, Ulrich** KFA Jülich, ICH-3, Postfach 1913, 5170 Jülich, F.R. Germany
317. **Schneider, Hans R.** Atmospheric & Environmental Research, Inc., 840 Memorial Drive, Cambridge, MA 02139, USA
318. **Schrems, Otto** Alfred-Wegener Institut, Postfach 120161, 2850 Bremerhaven, F.R. Germany
319. **Schurath, Ulrich** University of Bonn, Institute Physical Chemistry, Wegelerstr, 12, 5300 Bonn, F.R. Germany
320. **Speuser, Winfried** Universität Bonn, Institut f. Physikalische Chemie, Wegelerstr. 12, 5300 Bonn 1, F.R. Germany
321. **Staehlin, Johannes** Laboratorium f. Atmosphärenphysik, ETH-Hoenggerberg, 8093 Zürich, Switzerland
322. **Stefanutti, Leopoldo** I.R.O.E of C.N.R., Via Panciatichi 64, Florenz, Italy
323. **Steinberger, Eva H.** The Hebrew University, Dept. of Atmospheric Sciences, 91904 Jerusalem, Israel
324. **Steinbrecht, Wolfgang** Universität München, Sektion Physik, Am Coulombwall 1, 8046 Garching, F.R. Germany
325. **Stephens, Sherry** University of Colorado, NCAR, Campus Box 449, Boulder, CO 80309, USA
326. **Stief, Gisela** Strada di Presciano, 6, 53010 Presciano-Siena, Italy
327. **Stolarsky, Richard** NASA/Goddard Space Flight Center, Code 616, Greentbelt, MD 20771, USA
328. **Stordal, Frode** University of Oslo, Institute of Geophysics, P.B. 1022, Blindern 0315, Oslo 3, Norway
329. **Strahan, Susan** NASA/Ames Research Center, SSG:245-5, Moffett Field, CA 94035, USA
330. **Stranz, Dietrich** Am Hainbuchenfeld 10, 2121 Mechtersen, F.R. Germany
331. **Stuart, Derek** National Physical Laboratory, Queens Road, Teddington, Middx TW11 OLW, Great Britain
332. **Taalas, Petteri** Finnish Meteorological Institute, Sahaajankatu 22 E, 00810 Helsinki, Finland
333. **Thomas, Ronald** University of Colorado, LASP, Box 392, Boulder, CO 80309-0392, USA
334. **Thompson, Anne** NASA/GSFC, Code 616, Greenbelt, MD 20771, USA
335. **Thro, Pierre** MPI f. Meteolologie, Bundesstr. 55, 2000 Hamburg 13, F.R. Germany
336. **Kang-Kun Tschu** Institute of Geophysics, Academia Sinica, P.O.Box 928, Beijing, China
337. **Toumi, Ralf** Cambridge University, St. Catharine's College, Lensfield Road, Cambridge CB2 1EP, Great Britain

338.	**Turner, Richard**	NASA Langley Res. Center, Mail Stop 401B, Hampton, VA 23665-5225, USA
339.	**Varotsos, Costas**	University of Athens, Laboratory of Meteorology, Knossou 36, Ano Glyfada 16561, Athens, Greece
340.	**Vassy, Arlette**	110 Quai Louis Bleriot, 75781 Paris Cedex 16, France
341.	**Vaughan, Geraint**	Physics Department, University of Wales, Aberystwyth, Great Britain
342.	**Visconti, Guido**	Dipartimento di Fisica, Universita degli Studi - L'Aquila, 67100 L'Aquila, Italy
343.	**Vogel, Heike**	Kernforschungszentrum Karlsruhe, Postfach 3640, 7500 Karlsruhe 1, F.R. Germany
344.	**Vupputuri, Rama**	Canadian Climate Center, 4905 Dufferin Street, Downsview, Ontario M3H 5T4, Canada
345.	**Wang, Gengchen**	Acadmia Sinica, Institute of Atmospheric Physics, Beijing, China
346.	**Wang, GuiQin**	Peking University, Department of Geophysics, Beijing, China
347.	**Wang, Pi-Huan**	Science & Technology Corporation, 101 Research Drive, Hampton, VA 23666, USA
348.	**Wang, Wei-Chyung**	Atmospheric & Environment Research, Inc., 840 Memorial Drive, Cambridge, MA 02139, USA
349.	**Warmbt, Wolfgang**	An der Wilhlemshoehe 9, 8122 Radebeul, G.D.R.
350.	**Wayne, Richard P.**	University of Oxford, Physical Chemistry Laboratory, South Parks Road, Oxford OX1 3QZ, Great Britain
351.	**Wege, Klaus**	Deutscher Wetterdienst, Met. Observatorium, Albin-Schwaiger-Weg 10, 8126 Hohenpeissenberg, F.R. Germany
352.	**Weigl, Elmar**	Deutscher Wetterdienst, Met. Observatorium, Albin-Schwaiger-Weg 10, 8126 Hohenpeißenberg, F.R. Germany
353.	**Weinstock, Elliot**	Harvard University, Atmospheric Research Project, 40 Oxford St. Cambridge, MA 02138, USA
354.	**Weitkamp, Claus**	Institut für Physik, GKSS Forschungszentrum, Postfach 1160, 2054 Geesthacht, F.R. Germany
355.	**Wellemeyer, Charlie**	ST Systems Corporation, 4400 Forbes Blvd., Lanham, MD 20706, USA
356.	**Wendt, Jürgen**	Max-Planck-Institut für Aeronomie, Postfach 20, 3411 Katlenburg-Lindau, F.R. Germany
357.	**Wilson, Stephen R.**	Max-Planck-Institut für Chemie, Postfach 3060, 6500 Mainz, F.R. Germany
358.	**Wölcken, G. C.**	Universidad del Salvador, Arenales 2518, RA-1602 Florida Buenos Aires, Argentina
359.	**Wuebbles, Donald J.**	Lawrence Livermore National Laboratory, P.O.Box 808, L-262, Livermore, CA 94550, USA
360.	**Zalpuri, Kunwar S.**	National Physical Laboratory, Radio Science Division, New Delhi 110 012, India
361.	**Zander, Radolphe**	Inst. of Astrophysics, University of Liege, 5, Avenue de Cointe, 4200 Liege-Ougree, Belgium
362.	**Zawodny, Joseph**	NASA Langley, Mail Stop 475, Hampton, VA 23665, USA
363.	**Zellner, Reinhard**	Universität Göttingen, Institut für Physikalische Chemie, Tammannstr. 6, 3400 Göttingen, F.R. Germany
364.	**Zerefos, Christos**	University of Thessaloniki, Laboratory of Atmospheric Physics, 540 06 Thessaloniki, Greece
365.	**Zetzsch, Cornelius**	Fraunhofer Institut, Nikolai-Fuchs-Str. 1, 3000 Hannover 61, F.R. Germany
366.	**Zimmermann, P. H.**	MPI f. Chemie, Abtl. Luftchemie, Postfach 3060, 6500 Mainz, F.R. Germany
367.	**Ziomas, Ioannis**	University of Thessaloniki, Laboratory of Atmospheric Physics, Campus Box 149, 540 06 Thessaloniki, Greece
368.	**Zommerfelds, W.C.**	University of Berne, Institute of Applied Physics, Sidlerstr. 5, CH-3012 Berne, Switzerland

APPENDIX 2

International Association for Meteorology and Atmospheric Physics • International Ozone Commission

WORKSHOP ON TROPOSPHERIC OZONE August 4-6, 1988

QUADRENNIAL OZONE SYMPOSIUM August 8-13, 1988

University of Göttingen, Fed. Rep. of Germany

Scientific Program

August 1988

WORKSHOP	Thursday, 4 Aug.	
	09:30 - 13:00	Tropospheric Ozone observations - global and regional aspects
	14:00 - 17:30	Solar radiation and ozone
	Friday, 5. Aug.	
	09:00 - 13:00	Tropospheric Ozone and Precursors
	14:00 - 16:00	Tropospheric Ozone and Precursors
	16:00 - 18:00	Happy Hour Poster Session (Tropospheric Ozone)
	Saturday, 6 Aug.	
	09:00 - 10:30	Tropospheric Ozone Trends
	11:00 - 13:00	Tropospheric Modelling
SYMPOSIUM	Monday, 8 Aug.	
	09:00 - 11:00	Formal Opening
	11:30 - 12:30	Reviews on Ozone Trends and Polar Ozone
	13:30 - 16:30	Polar Ozone
	16:30 - 18:00	Happy Hour Poster Session (Polar Ozone)
	20:00 - 22:00	IOC first (private) meeting
	Tuesday, 9 Aug.	
	09:00 - 10:20	Polar Ozone
	10:50 - 12:30	Ozone Observations (ground based)
	13:30 - 16:30	Ozone Observations (ground based)
	16:30 - 18:30	Happy Hour Poster Sessions (ozone observations, developments of observational techniques)
	Wednesday, 10 Aug.	
	09:00 - 12:40	Ozone Observations (Satellite)
	Thursday, 11 Aug.	
	09:00 - 12:40	Other relevant trace constituents
	13:30 - 14:30	Other relevant trace constituents
	14:30 - 16:10	Ozone and Radiation, Solar effects and Mesosphere
	16:10 - 18:10	Happy Hour Poster Sessions (Relevant trace constituents, radiation and mesosphere)
	20:00 - 22:00	IOC second (open) meeting
	Friday, 12 Aug.	
	09:00 - 12:40	Tropospheric Ozone (including review of workshop)
	13:30 - 16:30	Chemical-radiative-dynamical model Studies: Composition and Climate
	16:30 - 18:30	Happy Hour Poster Sessions (Modelling, Reaction Kinetics and Laboratory Studies)
	Saturday, 13 Aug.	
	09:00 - 10:20	Modelling
	10:50 - 12:40	Developments in Observational Techniques

in collaboration with

Max–Planck–Institut für Aeronomie
Georg–August–Universität Göttingen
Stadt Göttingen – Fremdenverkehrsamt
Studentenwerk Göttingen

Substantial support provided by the following sponsors is gratefully acknowledged:

Bundesministerium für Forschung und Technologie, Bonn
Niedersächsisches Umweltministerium, Hannover
Commission of European Communities, Brussels
World Meteorological Organization
American Meteorological Society
United Nations Environmental Program
Max–Planck–Institut für Aeronomie, Katlenburg–Lindau
Alcan Deutschland GmbH, Werk Göttingen
Commerzbank, Göttingen
B.E.S.T. Ventil + Fitting GmbH, Hamburg
Deutsche Bank, Filiale Göttingen
Fischer & Porter, Göttingen
Lambda–Physik, Göttingen
Messer Griesheim, Kassel
Spindler & Hoyer, Göttingen
Städtische Sparkasse, Göttingen
Adolf Thies GmbH + CoKG, Göttingen

Workshop Tropospheric Ozone

University of Göttingen and Max-Planck-Institut für Aeronomie

August 4-6, 1988

Scientific Program

THURSDAY, AUGUST 4

9:30 – 13:00 Tropospheric Ozone Observations: global and regional aspects

Chairperson: P. Fabian

J. Fishman and H.G. Reichle, Jr.
Studies of tropospheric ozone using satellite observations

G. Vaughan and J.D. Price
Ozone transport into the troposphere in a cut-off low episode

H.G.J. Smit, D. Kley and S. Gilge
The latitudinal and vertical distribution of tropospheric ozone over the Atlantic Ocean in the southern and northern hemisphere

P. Winkler
Meridional distribution of surface ozone over the Atlantic Ocean (83°N - 76°S)

A. Marenco
Study of tropospheric ozone in the tropical belt from STRATOZ campaigns

V.W.J.H. Kirchhoff
A comparative study of tropospheric ozone in the Amazonian rainforest, the Cerrado, and a coastal site

E.V. Browell
Ozone variability over the Amazon rain forest of Brazil and in the Ozone Hole over Antartica determined from airborne LIDAR measurements

G. Helas, M.O. Andreae, J. Fontan, B. Cros
Comparison of ozone measurements in an equatorial region between the dry and rainy season

H. Tsuruta, K. Shinya, T. Mizoguchi, T. Ogawa
Seasonal behavior of the tropospheric ozone in rural Japan

T. Ogawa and Ninong-Komala
Diurnal and seasonal variations of the tropospheric ozone in tropical Asia

14:00 – 17:30 Solar radiation and ozone

Chairperson: R. Bojkov

M. Nicolet
The radiation field in the troposphere (invited)

U. Feister, K.H. Grasnick, W. Warmbt, A. Jaenicke
Surface ozone and solar radiation

A. Volz, H. Geiss, D. Kley
Correlation of O_3 and solar radiation at Montsouris and Hohenpeissenberg: Indications for photochemical production

M. Schmidt
Relationships between tropospheric ozone and global solar radiation in Germany

K.J. Weston, P.J.A. Kay, D. Fowler, A. Martin, J. Bower
On the extent of local photochemical ozone production in the U.K.

D. Paffrath, W. Peters, H. Stangl
Aircraft measurements of photochemical ozone in the atmospheric mixing layer

FRIDAY, AUGUST 5

9:00 – 13:00 Tropospheric ozone and precursors

Chairperson: D. Kley

R. Guicherit
Concentrations and patterns of ozone in Western Europe (invited)

F. Rohrer and D.H. Ehhalt
Measurements of NO and NO_y obtained during TROPOZ I

D.W. Fahey, D.M. Murphy, M.H. Proffitt, S.C. Liu, K.R. Chan
Measurements of total reactive odd nitrogen and ozone from the NASA Stratosphere-Troposphere Exchange Project

R.R. Dickerson, W.T. Luke, D.P. McNamara, L.J. Nunnermacker, K.E. Pickering, J.P. Greenberg, P.R. Zimmerman
Profiles of photochemically active trace gases in the troposphere

M.A. Carroll, B. Ridley, M. Trainer, D. Dunlap, G. Gregory, E. Condon
Tropospheric NO_x distributions and photochemistry

G. Hübler et al.
Airborne measurements of reactive odd nitrogen during NASA's GTE/CITE 2 Project

14:00 - 1600 Tropospheric ozone and precursors Contd.

Chairperson: U. Feister

M. Trainer, D.D. Parrish, E.J. Williams, M. Buhr, F.C. Fehsenfeld, S.C. Liu
Observations and models of the NO_y photochemistry at a rural site

Ph. Gerhard, D. Poppe, A. Marenco
Ozone and carbon monoxide distribution during STRATOZ III

M. Luria, C.C. van Valin, J.D. Ray, J.F. Boatman
The relationship between ozone and hydrogen peroxide: field observation and model evaluation

M. Paffrath, U. Meier, R. Dlugi, L. Kins, G. Roider
Measurements of peroxides in remote and urban areas in central Europe

16:00 - 18:00 Happy Hour Poster Session Tropospheric Ozone

Chairperson: R. Guicherit

W. Carnuth
Development of a mobile differential absorption LIDAR system for tropospheric ozone measurements from near ground level up to 13 km altitude

G. Ancellet, A. Papayannis, G. Megie, J. Pelon
Tropospheric ozone LIDAR measurements

W. Balzer, J. Ettl, H. Güsten, G. Heinrich, K. Reiniger
An autonomous station for remote measurement of tropospheric ozone in distant areas

C.J. Nappo, J.A. Herwehe, A.M. Thompson
Observations of ozone profiles in a developing boundary layer

D.F. Karchilava, A.G. Amiranashvili
Variations of ozone concentration in atmospheric boundary layer in Georgia

H. Hakola, P. Taalas, S. Joffre, H. Lättilä
Ozone concentration behaviour at a marine site and a forested site in Finland

I. Colbeck
Nocturnal ozone concentrations at a rural site in north-west England

I. Colbeck and A. Simmons
Comparison of methods to measure ozone dry deposition

D.R. Matt and J.D. Womack
A comparison study of measured to inferred ozone flux to a spruce conifer forest

N.F. Elansky
Investigation of mesoscale dynamics processes over mountain region by deformation of ozone field

S.J. Oltmans, R.C. Schnell, P.J. Sheridan, W.D. Komhyr
Seasonal surface ozone behavior in the high Arctic

G. Baumbach, K. Baumann, F. Dröscher
O_3 and NO_x measurements in forest and urban areas

W.A. Matthews, N.B. Jones, Y. Kondo, M. Takagi
In situ NO and NO_x measurements at Lauder, New Zealand

S. Semke and U. Platt
Observation of nitrate radical concentrations and lifetimes in free tropospheric air

F. Nolting and C. Zetzsch
The ozone forming potential of α-Pinene in comparison with Propene and cis-Butene-(2)

W. Behnke and C. Zetzsch
Smog chamber investigations of the influence of NaCl aerosol on the concentration of O_3 in a photo-smog system

D. Poppe, U. Platt, H.-P. Dorn, D.H. Ehhalt
Model calculations of tropospheric OH mixing ratio and comparison with field data

P. Zimmermann, H.J. Feichter, P.J. Crutzen
Modelling global tropospheric ozone trends in a 3-D model

Luis G.R. Suárez
Development of a model for the photochemical smog in Mexico City

H.A. Bravo, R.J. Torres, R.E. Sosa
Side effects of lead reduction in petrol on air quality in the metropolitan area of Mexico City

E. Jauregui
Local and heat-island induced circulation effects upon ozone dispersion patterns in the Mexico basin

N.M. Haleem
Pollutants interaction (ozone variation) on climate of Quetta

E. Schüpbach
Ozone measurements in the Jura mountains

W. Markiewicz and A. Sztyler
Ozone concentration measurements in selected places in the Katowice Province

M. Rahbar
Ozone variability at a semi rural site in France

SATURDAY, AUGUST 6

9:00 - 10:30 Tropospheric Ozone Trends

Chairperson: A. Volz

R.D. Bojkov
On the long-term changes of tropospheric ozone

H. Muramatsu
Trends in tropospheric ozone and ozone transport from the stratosphere

R. Sladkovic and W. Seiler
Trend of tropospheric ozone observed over Bavaria, Germany

S.J. Oltmans and W.D. Komhyr
Tropospheric ozone variations from surface and ECC ozonesonde observations

11:00 - 13:00 Tropospheric Modelling

Chairperson:

H. Levy II
Global reactive nitrogen: a numerical simulation of its distribution and deposition

I.S.A. Isaksen, J.E. Jonson, C. Reeves, R.B. Chatfield, S. Solberg
Model studies of free tropospheric ozone formation from pollution sources

S. Solberg, R.B. Chatfield, I.S.A. Isaksen
Design of a channel model to assess mid-latitude pollution effects including the increase of global tropospheric ozone

S. McKeen, D. Kley, A. Volz
The recent historical trend of tropospheric ozone over western Europe: a model perspective

O. Hov
Formation of ozone in the boundary layer and the exchange of ozone and precursors with the free troposphere over Europe

End of Workshop

Quadrennial Ozone Symposium

University of Göttingen, Fed. Rep. of Germany

August 8-13, 1988

Scientific program

(March 1988)

MONDAY, AUGUST 8

9:00-11:00	Formal opening (Aula der Universität) Invited speaker: M. Nicolet
11:00	Bus transfer to the symposium site (Zentrales Hörsaalgebäude)
11:30-12.30	Reviews on Ozone Trends and Polar Ozone Chairperson: R. Stolarski
	D.L. Albritton: The present status of understanding of ozone change as reflected in the NASA/WMO Ozone Trends Panel Report 1988 (invited) G. Brasseur: Review of the Ozone Hole Issue (invited)

13:30 - 16:30 Polar Ozone

Chairperson:

B.G. Gardiner
Antarctic ozone depletion measured by balloon sondes at Halley Bay

A.J. Krueger
Antarctic and Arctic total ozone extreme minima

P.A. Newman and M.R. Schoeberl
The breakup of the southern hemisphere spring polar ozone minimum from 1979-1987

Y. Sekigyuchi
Ozone and stratospheric temperature changes in the Antarctic correlated to the solar activity

W.G. Mankin and M.T. Coffey
Airborne observations of chemical constituents in the Antarctic winter stratosphere

J. Austin, A. Buckland, R.L. Jones, D.S. McKenna
The spring 1987 ozone decrease in Antarctica: Results from photochemical modelling along trajectories

M.J. Prather and M.M. Garcia
Simulation of the decay of the Antarctic ozone hole with a 3-D chemical transport model

R.L. de Zafra, M. Jeramillo, J. Barrett, L. Emmons, P. Solomon, A. Parrish
Observed diurnal variation of anomalous ClO concentrations in the Antarctic spring stratosphere over McMurdo station, 1986 and 1987

D.J. Hofmann, J.W. Harder, J.M. Rosen
A comparison of Antarctic ozone profiles during spring 1986 and 1987

16:30 - 18:00 Happy Hour Poster Session Polar Ozone

Chairperson: D. Rusch

K. Kawahira
Roles of the temperature decline on the ozone hole

P.-W. Wang, A. Ghazi, M.P. McCormick
Radiative relaxation in the polar stratosphere: The role of ozone

M.K.W. Ko, D.K. Weisenstein, N.-D. Sze, J.M. Rodriguez
A 2-D model study of the redistribution of ozone to compensate for the Antarctic ozone loss: Impact on global ozone

P.M. Solomon, R.L. de Zafra, M. Jaramillo, J. Barrnett, A. Parrish, L.K. Emmons
Secular trend of stratospheric ClO during the spring ozone hole formation, 1986 and 1987, at McMurdo station, Antarctica

A.Y. Hou, M.K.W. Ko
A gravity wave mechnism for exciting localized ozone minima in the Antarctic Spring

S. Chubachi
Summary of the ozone observation at Syowa, Antarctica, from February 1982 to January 1983

R.K.R. Vupputuri
2-D model simulation of anomalous spring variations of temperature, ozone and other minor trace constitutents over the south Polar regions

J.M. Rodriguez, N.D. Sze, M.K.W. Ko
Photochemistry of the Antarctic stratosphere in light of observations from the Airborne Antarctic Ozone Experiment

M. Jaramillo, R.L. de Zafra, J. Barrett, L.K. Emmons, P.M. Solomon, A. Parrish
A measured mixing ratio profile for HCN over McMurdo station during the Antarctic spring

R.P. Kane and Y. Sahai
Comparison of total ozone variations in the S. American continent and in the Antarctic

J.E. Rosenfield and M.R. Schoeberl
Ozone and nitrous oxide changes computed from 1987 springtime Antarctic data

Wang GuiQin
Computer modeling of the Antarctic ozone depletion

R.D. Bojkov
Polar ozone changes and the role of the stratospheric vortex

Y. Makino, H. Muramatsu, M. Tanaka, T. Ogawa, S. Kawaguchi
Antarctic ozone hole and column abundance of nitrous oxide observed at Syowa station in 1983

S. Chubachi, RF. Kajiwara, K. Kondoh
Total ozone by lunar Dobson at Syowa, Antarctica

D.J. Hofmann, J.M. Rosen, J.W. Harder
Antarctic stratospheric aerosol measurements during formation of the ozone hole in 1986 and 1987

M.T. Coffey and W.G. Mankin
Comparison of winter stratospheric chemistry in Arctic and Antarctic regions

N. Kundu and S.L. Jain
Depletion of ozone over Antarctica: Is it only in this decade?

W.S. Heaps, J.F. Burris, T.J. McGee
LIDAR studies in the Antarctic during the October 1987 ozone depletion

W.D. Komhyr, S.J. Oltmans, R.D. Grass, P.R. Franchois, R.K. Leonard
Changes in total ozone and ozone vertical distribution at South Pole, 1962-1987

G. Pitari, M. Verdecchia, G. Visconti
Ozone hole simulation using a prescribed dynamic field

C. Brühl and P.J. Crutzen
The modeling of the Antarctic "ozone hole"

M. Loewenstein, J.R. Podolske, S.E. Strahan, W.L. Starr, M.H. Proffitt, K. Kelly
N_2O as a dynamical tracer in the 1987 Airborne Antarctic Ozone Experiment (AAOE)

J.-P. Blanchet
The response of polar stratospheric clouds to increasing carbon dioxide

N.I. Brezgin, I.N. Ivanova, G.A. Kokin, V.N. Cuznetsova, E.M. Larin, V.I. Pavlov, S.P. Perov, V.V. Sazonov, A.F. Chizhov, O.V. Shtyrkov, V.A. Ikonnikov
Study of Antarctic ozone anomaly in spring by rocket and ground-based ozonometers

E.A. Jadin and E.V. Lisenko
Long-term variations in the dynamics of the stratosphere and the ozone deplection in the Antarctic

A.I. Voskresensky, V.A. Ikonnikov, I.L. Karol, V.V. Nesterov, K.I. Romashkina, V.V. Sazonov, A.M. Shalamyanskiy, L.N. Yurganov
Atmospheric ozone variations in spring-summer period according to the data of Soviet Antarctic stations

M.A. Tolbert
Heterogeneous reactions of $ClONO_2$ and N_2O_5 on ice and sulfuric acid surfaces at stratospheric temperatures

Le–Quan Li
The equilibrium properties of O_3, O dynamic system in stratosphere and Antarctic ozone hole

TUESDAY, AUGUST 9

9:00 - 10:20 Polar Ozone

Chairperson: Y. Sekigyuchi

M.J. Molina
The chemistry of some reactions believed to be important for ozone depletion in Antarctica

R.W. Sanders, S. Solomon, M.A. Carroll, A.L. Schmeltekopf
Ground-based measurements of O_3, NO_2, OClO and BrO during the 1987 Antarctic ozone depletion event

M.H. Proffitt, E.V. Browell, K.R. Chan, M. Loewenstein, J. Podolske
Ozone trends and latitude dependence near and within the polar vortex during the August-September 1987 Airborne Antarctic Ozone Experiment

K. Kawahira and T. Hirooka
Interannual variations of zonal mean temperature in the southern hemisphere stratosphere: With reference to the Antarctic ozone depletion

RESERVE PAPERS

R.F. Pueschel et al.
Ozone depletion over Antarctica: Evidence of heterogeneous phase chemistry

U. Feister, P. Plessing, H. Gernandt
Ozone soundings at the Antarctic station Georg Forster (70°46'S, 11°50'E)

10:50 - 12:30 Analysis of ozone observations (ground-based)

Chairperson: G. Megie

R.D. Bojkov
A critical assessment of published total ozone data and the need for re-evaluation

H.U. Dütsch and J. Staehelin
How much information on anthropogenic influences on the ozone layer can be obtained from a set of one station observations

R.E. Newell and H.B. Selkirk
Recent decreases of ozone over the globe

W.D. Komhyr, R.D. Grass, R.K. Leonard
Calibration history of World Standard Ozone Dobson Spectrophotometer 83 and total ozone trends at Mauna Loa Observatory, Hawaii, and at American Samoa, South Pacific

L.S. Hingane
Meridional reversal of total ozone gradient over India

13:30 - 16:30 Analysis of ozone observations (ground-based)

Chairperson: H.U. Dütsch

K. Yamazaki
El Nino and Ozone

G. Vaughan and D.A. Begum
Correlation between total ozone and potential vorticity

E.P. Röth and D.H. Ehhalt
Analysis of the temporal variance of stratospheric ozone

J.J. DeLuisi, D.U. Longenecker, C.L. Mateer
An objective method of correcting Umkehr measurements for volcanic aerosol error

J.B. Kerr et al.
Long term intercomparison between the Brewer and Dobson spectrophotometers at Canadian Network Stations

K. Wege, H. Claude, R. Hartmannsgruber
Some results from 20 years of ozone observations at Hohenpeißenberg

L. Froidevaux and M. Allen
The effect of waves on ozone-temperature correlations

A. Jayaraman et al.
The Indo-Soviet ozone experiment at Tumba to study diurnal variations in the vertical distribution over the tropics

RESERVE PAPER:

A.N. Gruzdev, I.I. Mokhov
Peculiarities of global dynamics of total ozone in annual course by surface measurements

16:30 - 18:30 Happy Hour Poster Session Observations of Ozone (ground based)
Chairperson: R.D. Bojkov

G.C. Reinsel and G.C. Tiao
Analysis of SBUV satellite and Dobson total ozone data for trends: An update through 1986

M. Degorska and B. Rajewska-Wiech
The interannual ozone variability at Belsk Station, Poland

S.H.H. Larsen
Total ozone amount in the Arctic, the Scandinavian sector.
Changes of ozone in connection with the 30 HPa temperature.
The ground based- and satellite measurements are used.

N. Kundu, N.K. Sethi, V.C. Jain
Consolidated study of ozone variability over tropical region

P.D. Angreji
Evaluation of various factors affecting homogenity of the total ozone time-series

C.M. Liu
Spatial variation of ozone trend

G.D. Nastrom and J.L. Green
Tropopause folding events observed using a VHF Radar

R.P. Kane
Long-term trends of total ozone

W.A. Matthews, W. Komhyr, P. Franchois
Vertical distribution and variation in atmospheric ozone over Lauder, New Zealand

R.D. Bojkov
Ozone distribution and changes from Central Europe to the Far East based on Filter Ozonometer data since 1972

R.D. Bojkov and W. Attmannspacher
Ozone variability and trends from ozone soundings at Hohenpeißenberg

R.D. Bojkov and A.C. Hansson
On the systematic differences between Mast and ECC ozonesondes

R.D. Bojkov and D.A. Simidchiev
On the ozone regime over south-east Europe

D. DeMuer
Trend analysis of the vertical distribution of ozone over middle latitudes, from 19 years of ozone soundings at Uccle (Belgium) •

D. Offermann and C.-P. DeBakker
Three dimensional stratospheric ozone distribution in Western Europe during MAP/GLOBUS 1983

B.H. Subbaraya, Y.B. Acharya, A. Jayaraman, R. Borchers, P. Fabian
Vertical distribution of ozone over Hyderabad using a balloon-borne suntracking photometer

K.S. Zalpuri, S.C. Garg, T. John
Measurement of ozone concentration in the equatorial middle atmosphere

G.K.Y. Hassan
Annual and semi-annual oscillations of the vertical ozone distribution over Cairo and Aswan, Egypt

D.R. Johnson and T.K. Schaack
On the exchange of mass, energy, water vapor and other trace constituents within isentropic coordinates

R.D. Grass and W.D. Komhyr
Traveling Standard Lamp calibration checks on Dobson ozone spectrophotometers during 1985-1987

W.D. Komhyr, S.J. Oltmans, P.R. Franchois, W.F.J. Evans, W.A. Matthews
The latitude distribution of ozone to 30 km altitude from ECC ozonesonde observations

R.D. Evans, G.L. Koenig, W.D. Komhyr, J. Easson, P. Marché
Umkehr observations at five automated Dobson spectrophotometer stations in 1986 and 1987

P.R. Franchois, R.D. Evans, G. Koenig, J.J. De Luisi, W.D. Komhyr
Comparison of ECC Ozonesonde and Umkehr ozone profiles at Boulder, Colorado, and Hilo-Mauna Loa, Hawaii, 1985-1987

B.C. Krueger, R. Neuber, S. Elnaggar, H. Walther
Measurements of O_3 profiles by a LIDAR method during the Arctic winter and spring 1988

W. Steinbrecht, K.W. Rothe, S. El Naggar, H. Walter, E. Augstein
Measurement of the stratospheric ozone concentration in the Arctic and at the Equator using a LIDAR system

Wang GuiQin and Xiao WenJun
An investigation on atmospheric ozone and atmospheric circulation

K. Indira and B.K. Mukherjee
El Nino and ozone variability

J.B. Kerr and D.I. Wardle
Long term stability of the Brewer reference triad

R. Atkinson et al.
Reevaluation of Australian total ozone record

D. Perner et al.
Spectroscopic observations in the springtime Arctic stratosphere

F.S. Rowland, N. Harris, R.D. Bojkov, P. Bloomfield
Statistical error analysis of ozone trends - winter depletion in the northern hemisphere

V.I. Bekorynkov
On the technique of compiling an empirical model of the vertical distribution of ozone in the northern hemisphere

I.L. Karol, L.P. Klyagina, A.D. Frolov, A.M. Shalamyansky
Seasonal features of total ozone and temperature fields in basic air masses of the northern hemisphere in the QBO phases

Ding-Wen Wei
The observations and studies of atmospheric ozone in Beijing and Kunming, China

E. Farkas
Long term variations and recent negative anomalies of New Zealand total ozone and Umkehr vertical ozone distribution

16:30-18.30 Happy Hour Poster Session Observations of Ozone (Satellites)

Chairperson: J. Miller

J. London and L.M. Perliski
Hemispheric differences in observed stratospheric ozone

E. Hilsenrath and S. Chandra
Satellite total ozone climatology covering 18 years

W.G. Planet, J.H. Lienesch, H.D. Bowman, A.J. Miller, R.M. Nagatani
Evaluation of operational SBUV/2 total ozone data

A.J. Fleig, D.S. Silberstein, C.G. Wellemeyer, P.K. Bhartia
An assessment of the SBUV/TOMS ozone data quality, based on comparison with external data

E.M. Weinstock and J.G. Anderson
On the continuing disagreement between in situ and satellite stratospheric ozone measurements

K.F. Klenk
Observational data for atmospheric ozone research at NASA's national Space Science Data Center

C.A. Reber
The Upper Atmosphere Research Satelalite (UARS)

R.M. Nagatani, A.J. Miller, K.W. Johnson, M.E. Gelman
An eight-year climatology of ozone and meteorology

D.F. Heath
Changes in the vertical distribution of stratospheric ozone and the associated global scale changes in total ozone from observations with the Nimbus 7 SBUV instrument: 1978-1986

W.G. Planet, J.H. Lienesch, M.L. Hill, A.J. Miller, J.K. Angell
Total ozone amounts from satellite and ground-based systems

W.F.J. Evans and J.B. Kerr
Ground truthing of TOMS and SBUV against Brewer ozone measurements in the Canadian network

Y. Yamazaki, Y. Nakamura, M.R. Cabral, C.H.S. Ning
Low latitude total ozone from TOVS data

R.T. Clancy and D.W. Rusch
Longitudinal and latitudinal variations of trends in SBUV ozone and NMC temperatures near the stratopause (1982-1986)

P.-H. Wang, M.P. McCormick, L.R. McMaster, S. Schaffner, G.E. Woodbury
Time-periodic variations in stratospheric ozone from satellite observations

M.J. Newchurch
A comparison of SAGE I, SAGE II, SBUV, and Umkehr ozone profiles including a search for Umkehr aerosol effects

R.P.; Cebula, D.S. Silberstein, C.G. Wellemeyer, A.J. Fleig, P.K. Bhartia
An assessment of the Chopper non-synchornized condition impact on the SBUV total ozone determination

L.A. Baranski and K. Rozenski
Determination of the ozone mixing ratio profiles in the atmosphere from TOVS/TIP satellite data

D.M. Cunnold, W.P. Chu, R.A. Barnes, M.P. McCormick, R.E. Vaiga, D.A. Chu
Error analysis and validation of SAGE II ozone measurements

G. Schuster, R. Rood, M. Schoeberl
Quasibiennial and interannual variability in high resolution total ozone data (TOMS)

McPeters, R.D. and W.D. Komhyr
Long-term changes in SBUV/TOMS relative to the World Primary Standard Dobson instrument

Happy Hour Poster Session Recent Developments of Observational Techniques
Chairperson: W.A. Matthews

D. Hanson, J. Morton, K. Mauersberger
Evaluation of an absolute ozone calibration system based on ozone vapor pressures

K. Mauersberger, J. Morton, B. Schueler, J. Barnes
The isotopes of ozone: Results from laboratory studies and stratospheric measurements

A. Bais, A. Kelessis, Z. Ziomas, G. Amanatidis, C. Zerefos
Six years of regular total ozone observations with a Brewer Spectrophotometer at Thessaloniki, Greece

A. Goldman et al.
Isotopic abundances of stratospheric ozone from balloon-borne observations of high resolution infrared solar spectra

T. Kitaoka
Increase of the temperature difference betweeen day and night (T-T) in upper level over northern area of Japan during spring, measured by the thermistor sensor of radiosonde

Y.V. Ramanamurty
A simple radio method of inferring the global stratospheric ozone

J.B. Kerr
The Brewer ozone data management system

J.B. Kerr, C.L. Mateer, C.T. McElroy, W.F.J. Evans
Umkehr measurements made with the Brewer' ozone spectrophotometer

D.I. Wardle, W.F.J. Evans, J.B. Kerr, G.M. Shah
Ultraviolet radiation monitoring with the Brewer spectrophotometer

J.B. Kerr
Scans of the solar ultraviolet spectrum using the Brewer spectrophotometer

J.B. Kerr
Observations of Arctic total ozone with the Brewer spectrophotometer during the polar winter using the moon as a light source

C.T. McElroy, A. Goldman, D.G. Murcray
Tunable Diode Laser Heterodyne Spectrophotometry of ozone

J.J. DeLuisi, E.G. Dutton, G. Koenig, D.U. Longenecker
Zenith sky cloud detection for automated Dobson Umkehr measurements

H. Claude and K. Wege
The Hohenpeißenberg LIDAR: First results and intercomparison with Brewer/Mast soundings

Y. Sasano, H. Nakane, S. Hayashida-Amano
Multiple-wavelength DIAL and a new analysis technique to deduce ozone profile without systematic errors due to aerosol effects

W. Speuser, S. Sahand, U. Schurath
A novel fast-response chemiluminescence sonde for routine soundings of stratospheric ozone up to 1.5 mbar

G. Rubin, K.F. Künzi
Ground-based microwave sensor to retrieve ozone profiles between 15 km and 80 km

S. Godin, G. Megie, J. Pelon
Systematic LIDAR measurements of the stratospheric ozone vertical distribution

C.A. Varotsos, C.G. Helmis, M.D. Loisidou, D.N. Asimakopoulos
A suggested correction of the background current of the ECC ozone sonde field measurements

O. Uchino, I. Tabata, K. Kai
A new mobile LIDAR for simultaneous measurements of ozone, temperature and aerosols in the stratosphere

Da-ren Lu, Jin-huan Qui
Observation of ozone column content with solar visible spectroscopy at several sites in China

G. Moreau and C. Robert
A new multipass ozonometer

B. Nikolaus, B. Basting, W. Mückenheim
100-W narrow-band excimer laser for lidar

M. Carlotti, A. Barbis, B. Carli, S. Piccioli
"Global FIT" analysis for the retrieval of O_3 distribution from for -IR Limb-scanning measurements

Wang Gengchen
Detection of ozone concentration in the atmosphere by using CO_2 Laser and error analysis

M. Degorska, B. Rajewska-Wiech
The effect of stray light on total ozone measurements at Belsk station, Poland

W. Lahoz
The 183 GHz O_3 channels on MLS

C. Muller and P.C. Simon
Backscattered U.V. ozone and aerosol measurements from the geostationary orbit: Feasability and interpretations

V.M. Dorokhov, T.N. Klimova, V.A. Torgovichev, V.U. Khattatov
An automated spectral complex for atmospheric ozone and UV-radiation measurements

P. Hartogh
High resolution Chirp Transform Spectrometer for Ozone Measurements

G.E. Brueckner, E.O. Hulbert, D.J. Strickland, R.P. Barnes
High Precision Ozone Densities from Solar UV Occultation Measurements

WEDNESDAY, AUG. 10

9:00 - 10:40 Ozone observations from satellites

Chairperson: C.L. Mateer

R.S. Stolarski, A.J. Krueger, M.R. Schoeberl
Total ozone trends from TOMS data

R.D. Hudson and J.R. Herman
On the determination of long-term trends from SBUV ozone data

P.K. Bhartia, S. Taylor, A.J. Fleig
Estimation of errors in the long-term change in ozone reported by the SBUV/TOMS instruments - a new way of looking at the old data

C.G. Wellemeyer, A.J. Fleig, P.K. Bhartia
Internal comparisons of SBUV and TOMS total ozone measurements

A.J. Miller, R. Nagatani, W.G. Planet, D.-W. Wei
Comparison of vertical ozone profiles as obtained from SBUV/2 and SBUV

11:00 - 12:40 Ozone observations from satellites

Chairperson: A. Ghazi

M.P. McCormick, R.C. Viega, J. Zawodny
A comparison of SAGE I and II stratospheric ozone measurements

J.J. DeLuisi, J.U. Longenecker, P.K. Bhartia, C.L. Mateer
Ozone profiles by Umkehr, SBUV and ozonesonde: A comparison including the inversion alogrithms for Umkehr and SBUV

T. Ogawa, M. Koike, K. Suzuki
Observation of vertical ozone profiles with the EXOS-C backscattered UV Spectrophotometer

D.W. Rusch and R.T. Clancy
A study of the time and spatial dependence of ozone near 1.0 mb with emphasis on the springtime

R.A. Barnes, M.P. McCormick, C.L. Parsons
Satellite/Rocket ozone comparisons at Natal, Brazil

RESERVE PAPER

S. Godin, G. Megie, J. Pelon, A.J. Miller, R.M. Nagatani, A.J. Fleig
Comparison of the ozone vertical distribution obtained by LIDAR and by the SBUV instrument

THURSDAY, AUG. 11

9:00 - 10:40 Other relevant trace constituents

Chairperson: R.L. McKenzie

J.P. Pommereau and H.K. Roscoe
Overview on the GLOBUS NO_x campaign

Y. Kondo, A. Iwata, P. Aimedieu, W.A. Matthews, W.R. Sheldon, J.R. Benbrook
Nitric oxide in the upper stratosphere

K. Shibasaki, N. Iwagami, K. Kita, T. Ogawa
Variability of NO_2 profiles at northern middle latitudes

J. Lenoble et al.
SAGE 2 nitrogen dioxide data comparison with infrared and visible balloon remote sensing measurements during GLOBUS NO_x

M.M. Abbas et al.
Nighttime measurements of stratospheric NO_x from balloon-borne limb thermal emissions observations

11:00 - 12:40 Other relevant trace constituents

Chairperson: J.P. Pommereau

J. Brasunas et al.
High resolution, balloon-borne emission spectroscopy of trace species in the lower stratosphere: N_2O_5

B.J. Kerridge and H.K. Roscoe
Balloon-borne IR emission measurements of stratospheric nitrogen oxides and aerosol

R. Borchers
The vertical distribution of source gases at tropical latitudes

P. Fabian
$CHClF_2$ (CFC-22): Distribution, budget and environmental impact

U. Schmidt, R. Bauer, A. Khedim, E. Klein, G. Kulessa, B. Schubert
In-situ observations of long-lived source gases in the arctic stratosphere during winter

13:30 - 14.30 Other relevant trace constituents

Chairperson: H. Roscoe

S.R. Wilson, G. Schuster, G. Helas
Measurements of COClF and $COCl_2$ near the tropopause

M. Helten, H. Fark, H.W. Pätz, D. Kley
HO_2 measurements in the lower stratosphere using matrix isolation and ESR spectroscopy

B.M. Dinelli, B. Carli, M. Carlotti
Stratospheric distribution of hydroxyl radiacal from far-IR emission spectra

RESERVE PAPERS

R.L. McKenzie, P.V. Johnston, M.P. McCormick
NO_2 column measurements: Comparison between results from 45°S and SAGE II

M. Helten, H. Fark, H.W. Pätz
In situ detection of stratospheric NO_3 by matrix isolation spectroscopy

14:30 - 16:10 Ozone and radiation, solar effects and mesosphere

Chairperson: P. Simon (Co-sponsored by IAGA)

G.M. Keating, M.C. Pitts, G. Brasseur
Recent detection of the response of the middle atmosphere to short-term solar UV variability

L.L. Hood and A.R. Douglass
Stratospheric responses to solar UV variations: Comparisons with radiative-photochemical models

R. Ramaroson, M. Pirre, P. Rigaud
Diurnal variations of the ozone concentration in the upper stratosphere

M.Yu. Danilin and G.I. Kouznetsov
On the influence of inertial gravity waves on radiation field and photodissociation rates of gases of the atmosphere

A.C. Aikin
Mesospheric ozone behaviour determined from satellite data

RESERVE PAPER

R.J. Thomas
Annual ozone variations in the upper mesosphere

16:10 - 18.10 Happy Hour Poster Session on Relevant Trace Constitutents

Chairperson: D.H. Ehhalt

H. Roscoe
Intercomparison aspects of balloon-borne measurements of NO_2 (BIC and GLOBUS)

M. Helten et al.
Comparison between NO_2 in situ measuremetns and remote NO_2 visible absorption measurements during MAP/GLOBUS 1985 NO_x campaign

R.L. McKenzie, W.A. Matthews, Y. Kondo, P. Fabian, R. Zander, F.J. Murcray, J.A. Pyle
NO column measurements: intercomparison during MAP/GLOBUS 1985

R. Zander, W.A. Matthews, J.P. Pommereau, N. Iwagami
Measurements of column abundances of NO_2 from the ground during the GLOBUS-NO_x campaign

J.P. Naudet and M. Pirre
Diurnal variation of stratospheric NO_2 from MAP/GLOBUS 1985

J.P. Pommereau, F. Goutail, Y. Kondo, W.A. Matthews
An NO_x source in the upper troposphere

F. Goutail and J.P. Pommereau
O_4 and H_2O atmospheric absorption measuremetns in the visible during the GLOBUS NO_x campaign

Y. Kondo, M. Takagi, A. Iwata, W.A. Matthews
Dirunal variation of nitric oxide at 26 km

G. Amanatidis, A. Bais, A. Kelessis, C. Zerefos, J. Ziomas
Two years of regular stratospheric NO_2 measurements in Greece using twilight photometry

J.B. Kerr
Ground based measurements of nitrogen dioxide using the Brewer spectrophotometer

C. Schiller, U. Platt, U. Schmidt
Measurements of column densities of NO_2 and other species by visible and UV spectroscopy at 68°N

D.K. Chakrabarty, G. Beig, J.S. Sidhu
Winter variabilities of O_3 and NO_2 at low latitude

F. Karcher, M. Amodei, G. Armand, C. Besson, B. Dufour, G. Froment, J.P. Meyer
Simultaneous measurements of stratospheric HNO_3, NO_2, HCl, O_3, N_2O, CH_4, H_2O and CO from the "STRATOZ 3" flights

J.P. Pommereau and F. Goutail
An ozone, nitrogen dioxide and chlorine dioxide permanent monitoring station at Dumont D'Urville, Antarctica (66°S, 134°E)

G. Helas, G. Schuster, S.R. Wilson
Measurements of ozone and other trace components near the tropopause over northern Europe

U. Schmidt, F.J. Johnen, A. Khedim, E. Klein, G. Kulessa, J. Rudolph, B. Schuster
The seasonal variability of long-lived source gases in the stratosphere at mid-latitudes

R. Borchers and P. Fabian
Halocarbon distribution: Data base for northern mid-latitudes

O.N. Singh, R. Borchers, P. Fabian, S. Lal, B.H. Subbaraya
Source gases for BrO_x in the atmosphere

H. Duscha
First results of RASMUS: Source gases in the mesosphere

W.A. Matthews, P.V. Johnston, D.G. Murcray, F.H. Murcray, R. Blatherwick
Column abundance of hydrogen chloride above Lauder, New Zealand

H. Fark, M. Helten, D. Kley, U. Schmidt
Measurement of formaldehyde in the lower stratosphere

H. Roscoe
Measurements of the cycles of pressure, temperature and transmission in an NO_2 pressure modulator

G.R. Davis
Laboratory measurements of the far infrared transmission of water vapor using a pressure modulation radiometer and the implications for the SAMS water vapor data

G.T. Burova, A.G. Katrich, E.V. Ivanov, E.A. Chayanova
The distribution of some halogenated hydrocarbons in the free atmosphere and atmospheric sea surface layer

J.R. Podolske, M. Loewenstein, S.E. Strahan, T.E. Blackburn
ATLAS - An Airborne Tunable Laser Absorption Spectrometer for rapid trace gas measurements from high altitude aircraft

S.E. Strahan, J.R. Podolske, T.E. Blackburn
CO mixing ratios over tropical Australia obtained with a new airborne Laser spectrometer

A.S. Elokhov, N.F. Elansky, O.V. Makarov, A.S. Smirnov
Automatic complex for measurements of minor species in the atmosphere

To'oru Sasaki
Improvements of Dew-Point Sonde in Japan

I. Nouchi, Y. Harazano, K. Yagi, M. Yokosawa, H. Seino, Naoto Owa, K. Minami, T. Uchijima
Emission of Methane from Rice Plants and Paddy Rice Fields

H. Fischer, T. Gulde, H. Oelhaf, D. Trieschmann, F. Fergg, D. Rabus, W. Völker
Simultaneous detection of atmospheric trace constituents with the MIPAS instrument

16:10-18:10 Happy Hour Poster Session Radiation and Mesosphere
Chairperson: K. Mauersberger (Co-sponsored by IAGA)

M.E. Summers et al.
Response of mesospheric ozone to short-term ultraviolet variations

M. Ilyas
Atmospheric ozone and solar UV radiation: The tropical scene

J. Barnes and K. Mauersberger
Ozone absorption cross-section at 253.7 nm and its temperature dependence

C. Schiller, U. Schmidt, U. Platt, M. Helten
Measurements of the solar flux at wavelength intervals responsible for photolysis of O_3 and NO_2 at stratospheric altitudes

G.P. Anderson, J.C. Gille, J.R. Herman, S.A. Clough
Presunset variation in the 9.6μ O_3 signature

T.V. Matveeva, M.Yu. Danilin, G.I. Kouznetsov
Short-term solar radiation variations and atmospheric ozone

S.P. Perov and S.V. Tishin
Diurnal variations of ozone in the mesosphere

D.B. Bromberg and S.P. Perov
Concentration of mesospheric water vapor inferred from ozone concentration

K. Henriksen and K. Stamnes
Solar irradiance and ozone column density at 78°N derived with an UV sensitive double monochromator

FRIDAY, AUG. 12

9:00 - 12:40 Tropospheric Ozone
Chairperson: I.S.A. Isaksen

H. Levy II: The Tropospheric Ozone Issue - reflected in the main results of the Workshop (Invited)

S. McKeen, H. Hsie, M. Trainer, R. Tallamraju, S.C. Liu
Ozone and odd nitrogen chemistry: Comparison of 3-D model results to observations

J. Rudolph, R. Koppmann, F.J. Johnen, A. Khedim
The distribution of light nonmethane hydrocarbons in the troposphere and their potential impact on photochemical ozone formation

D. Poppe, J. Drummond, D.H. Ehhalt
Tropospheric ozone budget during STRATOZ III: A model study

R.A. Brost, R.B. Chatfield, J.P. Greenberg, P.L. Haagenson, B.G. Heikes, S. Madronich, B.A. Ridley, P.R. Zimmermann
Three-dimensional modeling of the ozone budget over the north Atlantic Ocean

P.H. Zimmermann, J. Feichter, P.J. Crutzen
Global tropospheric ozone simulations with the MOGUNTIA-model

R.B. Chatfield
How lightning and human activity determine tropical ozone levels

I.S.A. Isaksen, T. Berntsen, S. Solberg
Estimates of past and future tropospheric ozone changes from changes in human released source gases

A.M. Thompson
Atmospheric chemical and climate change: Possible effects on tropospheric ozone

13:30 - 16:30 Chemical-radiative-dynamical model studies: Composition and Climate

Chairperson: D. Cariolle

M.P. Chipperfield, R.S. Eckmann, J.D. Haigh, L.J. Gray, J.A. Pyle
On the two-dimensional modelling of the stratosphere

A.Y. Hou, H.R. Schneider, M.K.W. Ko
The role of parameterized eddy mixing modelling the stratospheric circulation and tracer distribution

H.R. Schneider and M.K.W. Ko
The effect of parameterized friction and eddy diffusion on ozone perturbation calculations with 2-D models

K.P. Shine
Solar radiative heating by ozone

W.-C. Wang, G. Molnar, H.R. Schneider, M.K.W. Ko.
The effect of stratospheric changes due to increases of trace gases on the climate of the troposphere

D.J. Wuebbles and D.E. Kinnison
A two-dimensional model study of past trends in global ozone

B. Rognerud, I.S.A. Isaksen, F. Stordal
Model studies of stratospheric ozone depletion

K.K. Tung and H. Yang
Two-D model simulation of ozone climatology and year-to-year variations

16:30 - 18:30 Happy Hour Poster Session on Modelling

Chairperson:

D.E. Kinnison, D.J. Wuebbles, H.S. Johnston
A sensitivity study of global ozone to NO_x emissions from aircraft

M.K.W. Ko, D.K. Weisenstein, H.R. Schneider, N.D. Sze, C. Peterson
Tracer distributions in an interactive 2-D model of dynamics, chemistry and radiation

K.E. Grant and D.J. Wuebbles
A two-dimensional modeling study of the sensitivity of ozone to radiative flux uncertainties

P. Connell, K. Grant, D. Wuebbles
Aspects of CFC relative ozone destruction efficiencies determined in the LLNL 2-D model

C.H. Jackman, A.R. Douglass, P.A. Newman, P.D. Guthrie
Effect of computed horizontal diffusion coefficients on two-dimensional ozone assessment studies

A. Douglass, C.H. Jackman, R.S. Stolarski
Two-dimensional model comparisons of odd nitrogen family chemistry with separate calculations for the odd nitrogen species

P.D. Guthrie, T.L. Kucsera, C.H. Jackman
On the latitude dependence of ozone depletion predictions

M. Natarajan and L. Callis
Model study of ozone photochemistry using data from ATMOS and SAGE-II experiments

L.S. Hingane
Photochemical effects of increasing CO_2 on the level of O_3

J.D. Haigh
Radiative heating rates and the distribution of minor constituents in a two-dimensional model of the stratosphere

L.B. Callis, M. Natarajan, J.D. Lambeth, R.E. Boughner
Effects of precipitating relativistic electrons on stratospheric odd nitrogen and ozone

L.J. Gray and J.A. Pyle
A two dimensional model of the quasi biennial oscillation of ozone

R.S. Eckman, J.D. Haigh, J.A. Pyle
Ozone perturbations due to changing levels of atmospheric trace gases

J. Ballard, L.J. Gray, B.J. Kerridge, J.A. Pyle, R. Toumi
Test of photochemical theories using ATMOS observations

C. Gay
On the coupling of photochemistry, radiation and dynamics

G. Brasseur, M.H. Hitchman, A. DeRudder
Perturbations of the ozone layer studied with a coupled chemical-radiative-dynamical 2-D model

F. Stordal
Global stratospheric budgets of odd nitrogen and chlorine species

G.X. Cheng and K.K. Tschu
Effects of natural and man-made relevant trace constituents on production and depletion of ozone

F. Sassi and G. Visconti
Eddy diffusion parameterization from LIMS data

J.A. Kaye and R.B. Rood
Variability of chlorine containing species as revealed by three-dimensional stratospheric transport and chemistry models

R.B. Rood, M.A. Geller, W.C. Chao, M.R. Schoeberl, B. Boville, J.E. Nielsen, S. Steenrod
Stratospheric general circulation models at NASA/Goddard Space Flight Center

I.L. Karol, V.V. Babanova, A.A. Kiselev, A.P. Kondryavtzev, E.V. Rosanov
Model assessment of ozone and ozone active atmospheric gas content changes due to anthropogenic influence and solar activity changes

16:30-18:30 Happy Hour Poster Session on Reaction Kinetics and Laboratory Studies
Chairperson: R. Zellner

C. Zetsch
UV absorption cross sections of SF_6 and CH_3CN

A. Horowitz, W. Schneider, F. Simon, P. Crutzen, G.K. Moortgart
Ozone generation in the 214 nm photolysis of oxygen at 25°C

A. Schriver, L. Schriver, C. Barreau, J.-P. Perchard
Infrared matrix isolation studies of photochemical processes of atmospheric interest within water, SO_2 and ozone aggregates

S. Lal, B.H. Subbarya, S.V. Ramani, R. Borchers, P. Fabian, B.C. Krüger
Absorption cross-section of O_2 derived from the measured attenuation profiles of solar radiation at two wavelength bands in the Herzberg continuum

M.E. Jenkin and R.A. Cox
Kinetics and mechanisms of reactions of organic peroxy radicals and HO_2

G.D. Hayman and R.A. Cox
Characterisation of the reaction ClO + ClO — products and the products formed at low temperatures

O. Horie and G.K. Moortgart
The reaction of ozone with alkenes under atmospheric conditions

D. Hartmann, J. Karthäuser, G. Weibring, R. Zellner
Elementary processes of methane oxidation and their impact on atmospheric HO_x and O_x - a combined experimental and modelling study

M. Cacciani, A. di Sarra, G. Fiocco, A. Amoruso
Determination of ozone absorption coefficient at various temperatures in the 339-360 spectral region

A.M.Bass, J. Norris, R.J. Paur
A stable ozone generator for use as a transfer standard

S.J. Smith, P.T. Wassell, R.P. Wayne
Production and loss of the NO_3 radical in the $NO_2 + O_3$ system

D. Gillotay, P.C. Simon, L. Dierickx
Temperature dependence of ultraviolet absorption cross-sections of brominated methanes and ethanes

D. Daumont, A. Barbe, J. Brion, J. Malicet
New absolute cross-sections of O_3 in the 195-300 nm spectral range

Saturday, Aug. 13

9:00 - 10:20 Chemical-radiative-dynamical model studies: Composition and Climate
Chairperson: J.S. Chang

J. Austin
Two dimensional chemical modelling on an isentropic surface

D. Cariolle, M. Déqué, P. Simon
Simulation of the northern hemisphere ozone distribution with a general circulation model

G. Brasseur, K. Rose, C.M. Smythe
The behavior of ozone and other trace gases during dynamically perturbed situations: A three dimensional model

R.B. Rood, D.J. Allen, W. Baker, D. Lamich, J.A. Kaye
The use of winds and temperatures from a four-dimensional stratospheric assimilation model in constituent transport studies

RESERVE PAPERS

H.U. Dütsch and H. Oeschger
Another possible link between the ozone layers and the greenhouse effect

R.E. Turner, W.L. Grose, W.T. Blackshear, J.E. Nealy
A three-dimensional model simulation of the distribution of ozone in the stratosphere

M.R. Marinucci and G. Visconti
A self-consistent two dimensional photochemical model of the atmosphere

10:50 - 12:40 Developments in observational techniques

Chairperson: K.F. Künzi

J. Callies, H.-P. Dorn, U. Platt, D.H. Ehhalt
Measurement of tropospheric OH concentrations by Laser long-path absorption spectroscopy

W.D. Komhyr, R.D. Grass, R.D. Evans, R.K. Leonard
Results of international Dobson spectrophotometer and Brewer Spectrometer Intercomparisons, Arosa, Switzerland, 1986

W.F.J. Evans, J.B. Kerr, C.T. McElroy, D.I. Wardle
Modernization of the Canadian ozone network

J. de la Noe and M.B. Zekri
A study of temporal variations of stratospheric and mesospheric ozone profiles

W. Zommerfelds, K.F. Künzi
Operational ground based ozone sensor (OGOS)

13:00 Adjourn

AUTHOR LIST